MANUAL OF

Industrial Microbiology and Biotechnology

SECOND EDITION

MANUAL OF

Industrial Microbiology and Biotechnology

SECOND EDITION

EDITORS IN CHIEF

Arnold L. Demain
Department of Biology, Massachusetts Institute of Technology, Cambridge, Massachusetts 02139

Julian E. Davies
Department of Microbiology, University of British Columbia, Vancouver, B.C. V6T IZ3 Canada

EDITORS

Ronald M. Atlas
Department of Biology, University of Louisville, Louisville, Kentucky 40292

Gerald Cohen
Department of Molecular Microbiology and Biotechnology, Tel Aviv University, Ramat Aviv 69978, Israel

Charles L. Hershberger
Natural Products Research & Development, Eli Lilly & Company, Indianapolis, Indiana 46285

Wei-Shou Hu
Department of Chemical Engineering, University of Minnesota, Minneapolis, Minnesota 55455

David H. Sherman
Department of Microbiology, University of Minnesota, St. Paul, Minnesota 55108

Richard C. Willson
Department of Chemical Engineering, University of Houston, Houston, Texas 77204

J. H. David Wu
Department of Chemical Engineering, University of Rochester, Rochester, New York 14627

ASM
PRESS

WASHINGTON, D.C.

1325 Massachusetts Avenue, N.W.
Washington, DC 20005-4171

Library of Congress Cataloging-in-Publication Data

Manual of industrial microbiology and biotechnology / editors-in chief, Arnold L. Demain, Julian E. Davies; editors, Ronald M. Atlas . . . [et al.].—2nd ed.
p. cm.
Includes bibliographical references and indexes.
ISBN 1-55581-128-0
1. Industrial microbiology—Handbooks, manuals, etc.
2. Industrial microbiology—Handbooks, manuals, etc.
3. Biotechnology—Handbooks, manuals, etc. I. Demain, A. L. (Arnold L.), 1927– . II. Davies, Julian E. III. Atlas, Ronald M., 1946– .
QR53.M33 1999
660.6′2—dc21 98-44884
CIP

Printed in the United States of America

Contents

Editors in Chief

JULIAN E. DAVIES
TerraGen Diversity, Inc., and the Department of Microbiology and Immunology, University of British Columbia, Vancouver, BC V6T 1Z3, Canada

ARNOLD L. DEMAIN
Department of Biology, Room 68-223, Massachusetts Institute of Technology, Cambridge, MA 02139

Editors

RONALD M. ATLAS
Department of Biology, University of Louisville, Louisville, KY 40292-0002

GERALD COHEN
George S. Wise Faculty of Life Sciences, Department of Molecular Microbiology & Biotechnology, Tel Aviv University, Ramat Aviv 69978, Israel

CHARLES L. HERSHBERGER
Natural Products Research & Development, Lilly Corporate Center 3224, Eli Lilly & Company, Indianapolis, IN 46285

WEI-SHOU HU
Department of Chemical Engineering and Materials Science, University of Minnesota, 421 Washington Avenue SE, Minneapolis, MN 55455-0132

DAVID H. SHERMAN
Department of Microbiology, 240 Gortner Laboratories, University of Minnesota, St. Paul, MN 55108-1041

RICHARD C. WILLSON
Department of Chemical Engineering, 4800 Calhoun Avenue, University of Houston, Houston, TX 77204-4792

J.-H. DAVID WU
Department of Chemical Engineering, Room 206, Gavett Hall, University of Rochester, Rochester, NY 14627-0166

Contributors

MICHAEL W. W. ADAMS
Department of Biochemistry and Molecular Biology and Center of Metalloenzyme Studies, University of Georgia, Athens, GA 30602

THOMAS H. ADAMS
Department of Functional Genomics, Cereon Genomics, LLC, 45 Sidney Street, Cambridge, MA 02139

G. E. ALLISON
Department of Microbiology, Southeast Dairy Foods Research Center, North Carolina State University, Raleigh, NC 27695-7624

HIROYUKI ARAKI
Division of Microbial Genetics, Department of Cell Genetics, National Institute of Genetics, 1-111 Yata, Mishima-shi, Shizuoka 411-0801, Japan

FRANCES H. ARNOLD
California Institute of Technology, MC 210-41, Pasadena, CA 91125

ANNA ASTROMOFF
Department of Biochemistry, Beckman Center, B400, Stanford University School of Medicine, Stanford, CA 94305

RONALD M. ATLAS
Department of Biology, University of Louisville, Louisville, KY 40292-0002

M. J. BAILEY
Laboratory for Molecular Microbial Ecology, NERC, Institute of Virology and Environmental Microbiology, Mansfield Road, Oxford OX1 3SR, U.K.

FRANÇOIS BANEYX
Department of Chemical Engineering, Box 351750, University of Washington, Seattle, WA 98195

JOHN BARFORD
Department of Chemical Engineering, University of Sydney, Sydney 2006, NSW, Australia, and Hong Kong University of Science and Technology, Clear Water Bay, Kowloon, Hong Kong

ANGELA BELT
Blue Sky Research, 22049 Yerba Santa Road, Sonora, CA 95370

ALAN BERRY
Bio-Technical Resources Inc., 1035 South 7th Street, Manitowoc, WI 54220

ELAINE A. BEST
Somatogen Incorporated, 2545 Central Avenue, Boulder, CO 80301

DONALD B. BORDERS
BioSource Pharm, Inc., 135 Route 59 East, Spring Valley, NY 10977

DIRK FRANZ BROLLE
Mikrobiologie/Biotechnologie, Universität Tübingen, Auf der Morgenstelle 28, D-72074 Tübingen, Germany

SIERD BRON
Department of Genetics, Biomolecular Sciences and Biotechnology Institute, University of Groningen, Kerklaan 30, 9751 NN Haren, The Netherlands

RHYS BRYANT
15 Whitby Circle, Lincolnshire, IL 60069-3405

TAUSEEF R. BUTT
Research and Development, LifeSensors Inc., 271 Great Valley Parkway, Malvern, PA 19355, and Department of Biochemistry and Biophysics, University of Pennsylvania School of Medicine, Philadelphia, PA 19104-6059

GRAHAM BYNG
MDS Panlabs Inc., Bothell, WA 98011

J. CALANDRANIS
Intelligen, Inc., 2326 Morse Avenue, Scotch Plains, NJ 07076

WENDY CHAMPNESS
Department of Microbiology, Michigan State University, East Lansing, MI 48824-1101

J. DON CHEN
Department of Pharmacology and Molecular Toxicology, University of Massachusetts Medical Center, 55 Lake Avenue, N. Worcester, MA 01655-0126

JONG-IL CHOI
Department of Chemical Engineering, Korea Advanced Institute of Science and Technology (KAIST), 373-1 Kusong-dong, Yusong-gu, Taejon 305-701, Korea

GOPAL CHOTANI
Genencor International, 925 Page Mill Road, Palo Alto, CA 94304-1013

C. L. COONEY
Department of Chemical Engineering, Massachusetts Institute of Technology, Cambridge, MA 02139

SAMUN K. DAHOD
Chemical and Agricultural Products Division, Abbott Laboratories, 1401 Sheridan Road, North Chicago, IL 60064

TOM O. DAVIS
Department of Molecular Microbiology, Research Division, Centre for Applied Microbiology and Research, Porton Down, Salisbury, Wiltshire SP4 0JG, U.K.

FRANS J. DE BRUIJN
MSU-DOE Plant Research Laboratory, Department of Microbiology, and NSF Center for Microbial Ecology, 306 Plant Biology Building, Michigan State University, East Lansing, MI 48824

F. A. A. M. DE LEIJ
School of Biological Sciences, University of Surrey, Guildford, Surrey GU2 5XH, U.K.

VÍCTOR DE LORENZO
Centro Nacional de Biotecnología, Consejo Superior de Investigaciones Científicas, Campus de Cantoblanco, 28049 Madrid, Spain

BRADLEY S. DEHOFF
Lilly Research Laboratories, Eli Lilly and Company, Lilly Corporate Center, Indianapolis, IN 46285

L. DIJKHUIZEN
Department of Microbiology, Groningen Biomolecular Sciences and Biotechnology Institute, University of Groningen, Kerklaan 30, 9751-NN, Haren, The Netherlands

TIMOTHY C. DODGE
Genencor International, 925 Page Mill Road, Palo Alto, CA 94304-1013

MARK EGERTON
Incyte Pharmaceuticals, 3174 Porter Drive, Palo Alto, CA 94304

RICHARD FINK
Biohazard Assessment Office, Massachusetts Institute of Technology, Cambridge, MA 02139

MADILYN FLETCHER
Belle W. Baruch Institute for Marine Biology and Coastal Research, University of South Carolina, Columbia, SC 29208

CHRISTOPHER FRYE
Lilly Research Laboratories, A Division of Eli Lilly and Company, Lilly Corporate Center 3224, Indianapolis, IN 46285

PENG-CHENG FU
Department of Biochemical Engineering and Science, Faculty of Computer Science and Systems Engineering, Kyushu Institute of Technology, Iizuka, Fukuoka 820, Japan

M. M. GAGLIARDI
Natural Products Drug Discovery, Merck Research Laboratories, P.O. Box 2000, Rahway, NJ 07065-0900

JOSÉ A. GIL
Area of Microbiology, Faculty of Biology, University of León, 24071 León, Spain

JENNIFER GORDON
Pennie & Edmonds, 1155 Avenue of the Americas, New York, NY 10036

GUILLERMO GOSSET
Instituto de Biotecnología, Universidad Nacional Autónoma de México, Apdo. Postal 510-3, Cuernavaca, Morelos 62250, Mexico

BRUCE K. HAMILTON
BioDevelopment Associates, Inc., 15229 Watergate Road, Silver Spring, MD 20905

CHAE J. HAN
Laboratory of Cell Biology, National Heart, Lung and Blood Institute, National Institutes of Health, Bethesda, MD 20892

COLIN R. HARWOOD
School of Microbiological, Immunological and Virological Sciences, Medical School, University of Newcastle upon Tyne, Framlington Place, Newcastle upon Tyne NE2 4HH, U.K.

CHARLES L. HERSHBERGER
Natural Products Research & Development, Lilly Corporate Center 3224, Eli Lilly & Company, Indianapolis, IN 46285

MATTHEW D. HILTON
Natural Products Research and Development, Lilly Research Laboratories, Eli Lilly and Company, Lilly Corporate Center, Indianapolis, IN 46285

MASAHIKO HOSOBUCHI
Sankyo Co., 389-4 Ohtsurugi, Shimokawa, Izumi, Iwaki, Fukushima 971, Japan

WEI-SHOU HU
Department of Chemical Engineering and Materials Science, University of Minnesota, 421 Washington Avenue SE, Minneapolis, MN 55455-0132

LEEYUAN HUANG
Natural Product Discovery, Merck Research Laboratories, Merck & Co. Inc., P.O. Box 2000, R80Y-315, Rahway, NJ 07065

ANNALIESA S. HUDDLESTON-ANDERSON
Merck & Co., Inc., P.O. Box 2000, R80Y-300, Rahway, NJ 07065

J. C. HUNTER-CEVERA
Center for Environmental Biotechnology, E. O. Lawrence Berkeley National Laboratory, Building 70A/MS3317, One Cyclotron Road, Berkeley, CA 94720

C. RICHARD HUTCHINSON
School of Pharmacy, University of Wisconsin, 425 N. Charter Street, Madison, WI 53706

JANET K. JANSSON
Department of Biochemistry, Arrhenius Laboratories for Natural Sciences, Stockholm University, S-10691 Stockholm, Sweden

R. JOCKERS
Laboratoire d'Immuno-Pharmacologie Moléculaire, Institut Cochin de Génétique Moléculaire, CNRS UPR 0415, 22, rue Méchain, 75014 Paris, France

ERIC A. JOHNSON
Department of Food Microbiology and Toxicology, Food Research Institute, University of Wisconsin, Madison, WI 53706

VLADIMIR KABERDIN
Institute of Microbiology and Genetics, Vienna Biocenter, University of Vienna, Dr. Bohr-Gasse 9, A-1030 Vienna, Austria

TAKUO KAWAMOTO
Department of Synthetic Chemistry and Biological Chemistry, Graduate School of Engineering, Kyoto University, Yoshida, Sakyo-ku, Kyoto 606-8501, Japan

ROBERT M. KELLY
Department of Chemical Engineering, North Carolina State University, Raleigh, NC 27695-7905

ANURAG KHETAN
Department of Chemical Engineering and Materials Science, University of Minnesota, 421 Washington Avenue SE, Minneapolis, MN 55455-0132

T. R. KLAENHAMMER
Departments of Microbiology and Food Science, Southeast Dairy Foods Research Center, North Carolina State University, Raleigh, NC 27695-7624

RICHARD J. LADUCA
Genencor International, 925 Page Mill Road, Palo Alto, CA 94304-1013

SANG YUP LEE
Department of Chemical Engineering, Korea Advanced Institute of Science and Technology (KAIST), 373-1 Kusong-dong, Yusong-gu, Taejon 305-701, Korea

JAMES C. LIAO
Chemical Engineering Department, School of Engineering and Applied Science, Box 951592, University of California, Los Angeles, Los Angeles, CA 90095-1592

DANIEL F. LIBERMAN
Biology Department, Massachusetts Institute of Technology, Cambridge, MA 02139

A. K. LILLEY
Laboratory for Molecular Microbial Ecology, NERC, Institute of Virology and Environmental Microbiology, Mansfield Road, Oxford OX1 3SR, U.K.

STEVEN E. LINDOW
Department of Plant and Microbial Biology, 111 Koshland Hall, University of California, Berkeley, CA 94720

RUSSELL B. LINGHAM
Natural Product Discovery, Merck Research Laboratories, Merck & Co. Inc., P.O. Box 2000, R80Y-315, Rahway, NJ 07065

URBAN LUNDBERG
Institute of Microbiology and Genetics, Vienna Biocenter, University of Vienna, Dr. Bohr-Gasse 9, A-1030 Vienna, Austria

J. M. LYNCH
School of Biological Sciences, University of Surrey, Guildford, Surrey GU2 5XH, U.K.

JAN MAARTEN VAN DIJL
Department of Genetics, Biomolecular Sciences and Biotechnology Institute, University of Groningen, Kerklaan 30, 9751 NN Haren, The Netherlands

VOLKER MAI
Department of Microbiology and Center for Biological Resource Recovery, University of Georgia, Athens, GA 30602-2605

JUAN F. MARTÍN
Area of Microbiology, Faculty of Biology, University of León, 24071 León, and Institute of Biotechnology INBIOTEC, 24006 León, Spain

RICHARD I. MATELES
Candida Corporation, 175 W. Jackson Blvd., Suite A-1706, Chicago, IL 60604

MARGARET L. MAUCHLINE
Department of Molecular Microbiology, Research Division, Centre for Applied Microbiology and Research, Porton Down, Salisbury, Wiltshire SP4 0JG, U.K.

ROB MEIMA
Department of Genetics, Biomolecular Sciences and Biotechnology Institute, University of Groningen, Kerklaan 30, 9751 NN Haren, The Netherlands

MICHAEL MENKE
Lilly Research Laboratories, A Division of Eli Lilly and Company, Lilly Corporate Center 3224, Indianapolis, IN 46285

GUIDO MEURER
Terragen Diversity, Inc., Vancouver, BC V6T 1Z3, Canada

P. C. MICHELS
EnzyMed Inc., 2501 Crosspark Road, Oakdale Research Park, Iowa City, IA 52242-5000

NIGEL P. MINTON
Department of Molecular Microbiology, Research Division, Centre for Applied Microbiology and Research, Porton Down, Salisbury, Wiltshire SP4 0JG, U.K.

BRUNO MIROUX
The Medical Research Council Laboratory of Molecular Biology, Hills Road, Cambridge CB2 2QH, U.K.

R. L. MONAGHAN
Natural Products Drug Discovery, Merck Research Laboratories, P.O. Box 2000, RY80Y-220, Rahway, NJ 07065-0900

JEFFREY C. MOORE
Merck & Co., Inc., RY80Y-110, P.O. Box 2000, Rahway, NJ 07065

HARUHIKO MORI
Laboratory of Microorganism, Development Division, Takaki Bakery Co., Ltd., 3-7-1 Nakano-higashi, Aki-ku, Hiroshima-shi, Hiroshima 739-0323, Japan

JOHN P. MUELLER
Respiratory, Allergy, Inflammation, Immunology, and Infectious Diseases, Central Research Division, Pfizer, Inc., Eastern Point Road, Groton, CT 06340

GÜNTHER MUTH
Mikrobiologie/Biotechnologie, Universität Tübingen, Auf der Morgenstelle 28, D-72074 Tübingen, Germany

TERRY K. NG
Fort Dodge Laboratories, Division of American Home Products, 800 5th Street NW, Fort Dodge, IA 50501

R. NIR
SBH Sciences, Inc., 4 Strathmore Road, Natick, MA 01760

FRANKLIN H. NORRIS
Lilly Research Laboratories, Eli Lilly and Company, Lilly Corporate Center, Indianapolis, IN 46285

YASUJI OSHIMA
Department of Biotechnology, Faculty of Engineering, Kansai University, 3-3-35 Yamate-cho, Suita-shi, Osaka 564-8680, Japan

NANCY L. PAIVA
Plant Biology Division, The Samuel Roberts Noble Foundation, P.O. Box 2180, 2510 Sam Noble Parkway, Ardmore, OK 73401

MADHUSUDAN V. PESHWA
Dendreon Corporation, 291 N. Bernardo Avenue, Mountain View, CA 94043

D. P. PETRIDES
Intelligen, Inc., 2326 Morse Avenue, Scotch Plains, NJ 07076

SCOTT D. POWER
Genencor International, 925 Page Mill Road, Palo Alto, CA 94304-1013

HAROLD B. REISMAN
Biotechnology Results, 15 October Drive, Weston, CT 06883

KRISTINA D. RINKER
Department of Biomedical Engineering, Duke University, Durham, NC 27708

PAMELA K. ROCKEY
Lilly Research Laboratories, Eli Lilly and Company, Lilly Corporate Center, Indianapolis, IN 46285

J. P. N. ROSAZZA
Center for Biocatalysis and Bioprocessing and Division of Medicinal and Natural Products Chemistry, College of Pharmacy, University of Iowa, Iowa City, IA 52242-5000

JOHN T. ROSS
Consolidated Fermentation Service, National Cancer Institute-Frederick Cancer Research and Development Center, Frederick, MD 21702

PAUL R. ROSTECK, JR.
Lilly Research Laboratories, Eli Lilly and Company, Lilly Corporate Center, Indianapolis, IN 46285

JUAN M. SÁNCHEZ-ROMERO
Centro Nacional de Biotecnología, Consejo Superior de Investigaciones Científicas, Campus de Cantoblanco, 28049 Madrid, Spain

KAZUO SATO
National Research Institute of Brewing, 2-6-30 Takinogawa, Kita-ku, Tokyo 114, Japan

FREDERICK SCHAEFER
Consultant, Nashville, TN 37232

HILDGUND SCHREMPF
FB Biologie/Chemie, Universität Osnabrück, Barbarastrasse 11, 49069 Osnabrück, Germany

DAVID H. SHERMAN
Department of Microbiology, 240 Gortner Laboratories, University of Minnesota, St. Paul, MN 55108-1041

BERNHARD SONNLEITNER
Department for Chemistry and Biotechnology, University of Applied Sciences, Winterthur, P.O. Box 805, CH-8401 Winterthur, Switzerland

JEFFREY L. STEIN
Diversa Corporation, 10665 Sorrento Valley Road, San Diego, CA 92121-1623

JANE STERNER
Lilly Research Laboratories, A Division of Eli Lilly and Company, Lilly Corporate Center 3224, Indianapolis, IN 46285

SIOBHAN STEVENS-MILES
Natural Product Discovery, Merck Research Laboratories, Merck & Co. Inc., P.O. Box 2000, R80Y-315, Rahway, NJ 07065

S. L. STREICHER
Natural Products Drug Discovery, Merck Research Laboratories, P.O. Box 2000, Rahway, NJ 07065-0900

R. J. STROBEL
Lilly Research Laboratories, Eli Lilly and Company, Indianapolis, IN 46285

A. D. STROSBERG
Laboratoire d'Immuno-Pharmacologie Moléculaire, Institut Cochin de Génétique Moléculaire, CNRS UPR 0415, 22, rue Méchain, 75014 Paris, France, and University of Paris, Paris, France

SHIGETOSHI SUDO
National Research Institute of Brewing, 2-6-30 Takinogawa, Kita-ku, Tokyo 114, Japan

G. R. SULLIVAN
Lilly Research Laboratories, Eli Lilly and Company, Indianapolis, IN 46285

JOYCE A. SUTCLIFFE
Respiratory, Allergy, Inflammation, Immunology, and Infectious Diseases, Central Research Division, Pfizer, Inc., Eastern Point Road, Groton, CT 06340

EDWARD M. SYBERT
Engineering Research Center, University of Maryland, College Park, MD 20742

ATSUO TANAKA
Department of Synthetic Chemistry and Biological Chemistry, Graduate School of Engineering, Kyoto University, Yoshida, Sakyo-ku, Kyoto 606-8501, Japan

I. P. THOMPSON
Laboratory for Molecular Microbial Ecology, NERC, Institute of Virology and Environmental Microbiology, Mansfield Road, Oxford OX1 3SR, U.K.

JAMES M. TIEDJE
Center for Microbial Ecology, Plant & Soil Sciences Building, Michigan State University, East Lansing, MI 48824-1325

RONALD UNTERMAN
Envirogen, Inc., 4100 Quackenbridge Road, Lawrenceville, NJ 08648

KOHEI USHIO
Research Laboratory, Higashimaru Shoyu Co. Ltd., 28 Tatsuno-cho, Tatsuno-shi, Hyogo 679-4167, Japan

ERIC A. UTT
Biodiscovery, Animal Health Division, Pfizer, Inc., Eastern Point Road, Groton, CT 06340

FERNANDO VALLE
Genencor International, 925 Page Mill Road, Palo Alto, CA 94304-1013

ALBERT J. J. VAN OOYEN
Gist-brocades, P.O. Box 1, 2600 MA Delft, The Netherlands

EVIE L. VERDERBER
Somatogen Incorporated, 2545 Central Avenue, Boulder, CO 80301

JAN C. VERDOES
Department of Food Technology and Nutritional Sciences, Industrial Microbiology Group, Wageningen Agricultural University, P.O. Box 8129, 6700 EV Wageningen, The Netherlands

VICTOR A. VINCI
Lilly Research Laboratories, Eli Lilly and Company, Lilly Corporate Center, Indianapolis, IN 46285

ALEX A. VOLKOV
California Institute of Technology, MC 210-41, Pasadena, CA 91125

ALEXANDER VON GABAIN
Institute of Microbiology and Genetics, Vienna Biocenter, University of Vienna, Dr. Bohr-Gasse 9, A-1030 Vienna, Austria

JOHN E. WALKER
The Medical Research Council Laboratory of Molecular Biology, Hills Road, Cambridge CB2 2QH, U.K.

JOY E. M. WATTS
Department of Biological Sciences, University of Warwick, Coventry CV4 7AL, U.K.

JAMES C. WEAVER
Harvard-MIT Division of Health Sciences and Technology, Harvard-MIT Biomedical Engineering Center, 77 Massachusetts Avenue, 16-319, Cambridge, MA 02139

ELIZABETH M. H. WELLINGTON
Department of Biological Sciences, University of Warwick, Coventry CV4 7AL, U.K.

JAN WERY
Department of Food Technology and Nutritional Sciences, Industrial Microbiology Group, Wageningen Agricultural University, P.O. Box 8129, 6700 EV Wageningen, The Netherlands

J. M. WHIPPS
Plant Pathology and Microbiology, Horticulture Research Internations, Wellesbourne, Warwickshire CV35 9EF, U.K.

JUERGEN WIEGEL
Department of Microbiology and Center for Biological Resource Recovery, Biological Sciences Building 215, University of Georgia, Athens, GA 30602-2605

RICHARD C. WILLSON
Department of Chemical Engineering, 4800 Calhoun Avenue, University of Houston, Houston, TX 77204-4792

MARK WILSON
Biology Department, The Colorado College, 14 East Cache La Poudre, Colorado Springs, CO 80903

ANIL WIPAT
School of Microbiological, Immunological and Virological Sciences, Medical School, University of Newcastle upon Tyne, Framlington Place, Newcastle upon Tyne NE2 4HH, U.K.

WOLFGANG WOHLLEBEN
Mikrobiologie/Biotechnologie, Universität Tübingen, Auf der Morgenstelle 28, D-72074 Tübingen, Germany

JIMMY YONG-XIAO
Genencor International, 925 Page Mill Road, Palo Alto, CA 94304-1013

HIROJI YOSHIKAWA
Sankyo Co., 389-4 Ohtsurugi, Shimokawa, Izumi, Iwaki, Fukushima 971, Japan

JAE-HYUK YU
Department of Functional Genomics, Cereon Genomics, LLC, 45 Sidney Street, Cambridge, MA 02139

AN-PING ZENG
Biochemical Engineering Division, Gesellschaft für Biotechnologische Forschung mbH, Mascheroder Weg 1, D-38124 Braunschweig, Germany

HUIMIN ZHAO
California Institute of Technology, MC 210-41, Pasadena, CA 91125

Preface

Jackson W. Foster once said, "Never underestimate the power of the microbe." Demain and Solomon's *Manual of Industrial Microbiology and Biotechnology*, published by ASM Press in 1986, helped workers in the field harness the industrial "power of the microbe." That first edition has become a classic, and not just because it is now on the bookshelves of most microbiology laboratories with covers stained and pages well thumbed. "Demain and Solomon" was the forerunner of other manuals of microbiology outside of medical practice. This Second Edition has been expanded to include still more information on applications of molecular biology, not in an attempt to compete with Maniatis and others, but to provide more information on modern technical advances in the study and application of microorganisms to industrial situations. Areas covered in the First Edition are included, but the chapters have been written by new authors in an up-to-date manner. None of the original chapters remains.

Much has changed in the past decade; *Escherichia coli* and *Saccharomyces cerevisiae* have become industrial workhorses in a way no one could have predicted. New organisms have entered the biotechnology workplace to compete with these two stalwarts. Thus, covered in the Second Edition are *Bacillus*, actinomycetes, non-*Saccharomyces* yeasts, filamentous fungi, insect cells, mammalian cells, and plant cell culture. The genetics of thermophiles and other extremophiles, pseudomonads, and corynebacteria are also included. New techniques of combinatorial biosynthesis, metabolic engineering, directed evolution of enzymes, gel microdroplet technology, and informatics have been added.

Nothing can be like the original; it has been our goal, with the help of our section editors and authors and the encouragement (otherwise known as constant prodding) of ASM Press (Ellie Tupper, Greg Payne), to bring the *Manual of Industrial Microbiology and Biotechnology*, Second Edition, into a millennium format by broadening coverage. For example, recombinant DNA applications as applied to aromatic compounds and polyhydroxyalkanoates (PHAs) have been added as well as chapters on biodiversity, bioprospecting, bioremediation, biofilms, and release of recombinant microbes into the environment. Data analysis, contract fermentations, and quality assurance and quality control are now included. Greater emphasis has been given to aspects of secondary metabolism such as studies of resistance, regulation, and biosynthesis; bacteriocins; downstream processing of small and large molecules; and protein production and secretion. Perhaps this increased coverage has resulted in providing fewer recipes, but we believe that answers to all technical questions can be found in the references provided with each article.

This Second Edition should be of great use to all members of our field, including students, postdoctoral associates, faculty, technicians, senior researchers, factory personnel and managers, research and development managers, top management, patent agents, and technology transfer personnel. It is hoped that the manual finds its way into offices, laboratories, and university libraries, as well as pharmaceutical, chemical, energy, and food companies and the thousands of biotechnology companies and institutes throughout the world. If so, we will derive great satisfaction in that all who participated in this volume have helped to contribute to the spirit expressed by Louis Pasteur: "Non, mille fois non, il n'existe pas une catégorie de sciences auxquelles on puisse donner le nom de sciences appliquées. *Il y a la science et les applications de la science*, liées entre elles comme le fruit à l'arbre qui l'a porte." ("No, a thousand times no, there does not exist a category of science to which one can give the name applied science. There is science and the applications of science, bound together like the fruit carried on a tree.")

ARNOLD L. DEMAIN
JULIAN DAVIES
Editors in Chief

CULTURES

I

ALMOST ALL INDUSTRIAL MICROBIOLOGY PROCESSES REQUIRE THE initial isolation of cultures from nature, followed by small-scale cultivations and optimization, before large-scale production can become a reality. The chapters in this section deal with the most essential element of any fermentation process, the culture. They describe how to isolate cultures from nature, screen for biological activities, develop high-quality inoculum, carry out small-scale fermentations, and effectively improve and store the cultures.

Microorganisms, with their extremely diverse metabolic activities, represent an almost unlimited source of biological activities for industrial applications. The first chapter of this section, "Isolation of Cultures" by Hunter-Cevera and Belt, describes general procedures and media for collecting and isolating microorganisms from nature, including filamentous bacteria (actinomycetes), bacteria, and fungi. In the following chapter, "Screening for Activities," Huang et al. present methods for screening cultures for desired biological activities, using several industrial cases as examples.

After the desired culture is obtained, maintenance of the essential characteristics during long-term culture storage is crucial. The procedures and media for preserving various microorganisms are outlined in the chapter "Culture Preservation and Inoculum Development" by Monaghan et al., who also describe methods for developing inoculum from a preserved culture.

Once a culture producing the desired product is obtained, the feasibility of large-scale production is first tested in small-scale fermentors. Hilton describes fermentation equipment and methods in his chapter "Small-Scale Liquid Fermentations." Although liquid fermentation is used in most industrial fermentations, solid-state fermentation remains the method of choice for many traditional industrial processes, particularly for the food industry. The subsequent chapter, "Small-Scale Solid-State Fermentations" by Sato and Sudo, describes the methods for this type of fermentation, including fermentor design and process control strategies.

Small-scale fermentations provide an economical means for examining a large number of fermentation parameters for process optimization, largely through empirical approaches. One of the most important parameters is the medium, which contains various components regulating the production of the desired product. In their chapter "Experimental Design for Improvement of Fermentations," Strobel and Sullivan present experimental design and statistical analysis techniques for analyzing combinations of nutritional as well as nonnutritional parameters for rapidly improving a fermentation process. Several hypothetical models and authentic case studies are presented to illustrate the design techniques.

Besides growing cells, immobilized microbial cells or extracted enzymes can be used for production of the desired product. Tanaka and Kawamoto describe various methods for immobilizing cells and enzymes in their chapter "Cell and Enzyme Immobilization." The subsequent chapter deals with strain improvement. In addition to process optimization, successful industrialization of a fermentation process always depends on genetic improvement of strains. Despite the development of recombinant DNA techniques, the use of nonrecombinant methods for strain improvement remains important, particularly for organisms lacking well-developed genetic tools. Vinci and Byng present key methods for mutagenesis and subsequent screening for improving cultures in their chapter, "Strain Improvement by Nonrecombinant Methods." The recombinant approach for strain improvement is addressed in a different section of this manual.

Finally, this section includes two special chapters. Weaver describes a flow cytometric technique enabling culture of microbial cells and functional and compositional measurement at the individual cell level in his chapter, "Culture and Analysis Using Gel Microdrops." In the last chapter of the section, "Cultivation of Hyperthermophilic and Extremely Thermoacidophilic Microorganisms," Rinker et al. address culture methods for extremophiles, which are important sources of thermostable enzymes.

Isolation of Cultures

J. C. HUNTER-CEVERA AND ANGELA BELT

1

The use of microorganisms to produce natural products and processes that benefit and improve our socioeconomic lifestyles has been a part of human history since the days of early civilization. Isolating microorganisms from the environment is the microbiologist's first step in screening for natural products such as secondary metabolites and enzymes. Unfortunately for industry, no single isolation method will reveal the total number and variety of microorganisms present in a sample (43, 103). It is possible to isolate many different microorganisms by enrichment techniques (1, 3, 13, 37, 38, 52, 100, 114) or even to perform single-cell isolation by capillary methods (83). However, for industrial screening, such enrichment techniques usually require an inordinate amount of time, labor, and money, since only a few species of a particular genus arise from any one sample (6). Also, the industrial screen's assay procedures may require modification to "suit" the growth and metabolism of very different genera, and this is often time-consuming and costly. Thus, a more focused and less ad hoc approach to isolation is required (120). The literature of the past 10 years contains many articles that describe new species and new twists on classical isolation techniques that may even use molecular information to optimize the isolation procedure (16, 42, 49, 75, 93, 94, 101, 102, 117, 119).

There are quite a few publications in the literature that describe microbial diversity at the molecular level based on 16S rRNA or biosignatures such as FAME (fatty acid methyl esters). Some molecular biologists claim that current isolation procedures can recover only perhaps 1% of all microorganisms existing on the planet Earth. Others claim that there is no longer a need to isolate microorganisms since DNA and genes of interest can be isolated directly from the environment, cloned, and screened for specific activity. However, not all microbial products and processes can be cloned and expressed in *Escherichia coli* or *Pichia* spp. or, in some instances, be successfully scaled up for manufacturing.

One successful approach to the discovery of natural products such as novel antibiotics (76, 118) and enzymes (48, 62) involves (i) considering the desired product characteristics and process development, and (ii) using ecophysiological methods of isolation and screening (47). Consideration of the desired product characteristics, process development, or both may aid in answering the first major question, i.e., what do I isolate and from where? For example, if an enzyme were needed that functioned optimally at a moderately high pH and salinity yet at variable temperatures, the odds would certainly favor its discovery among microorganisms found in a desert such as Death Valley, rather than those from a Midwest or East Coast hardwood forest. If the optimal process required high temperatures, then composts, hot springs, and thermal vents would be the logical places to sample. The choice of the microorganisms, in turn, may be influenced by existing capacities for production, biomass and product yield, recovery costs, stability of the microorganism in large-scale fermentors, and ease of genetic manipulation. For example, the company undertaking the process may have capabilities and experience only with bacteria, not with fungi.

Using ecophysiological approaches to isolation can provide a screen with both a large number and a wide variety of microorganisms to examine for the product of interest (47). Even through microorganisms are highly adapted, specific microbial types are associated with different niches or samples within a variety of ecosystems (44). If different niches within a specific ecosystem are systematically sampled, a greater diversity of microbial types can be isolated. Medium selection, diluents, incubation conditions, and sample handling dictate the numbers and types of microorganisms isolated from plants, soils, and water (20, 47, 92, 109). Therefore, what is isolated from the defined ecosystem or habitat is a reflection of the isolation procedures and conditions set by the microbiologist, as well as the fluctuating conditions found in nature. This chapter illustrates different ways of treating and isolating single samples from specific ecosystems to isolate actinomycetes, bacteria, and fungi. The following isolation procedures essentially "milk" the sample for both autochthonous and zymogenous microorganisms that may be more representative of the biota associated with that particular sample. In the long run, such milking saves money and time and results in the isolation of a large, representative, and diverse microbial population. This chapter does not describe selective methods and techniques for specific groups of microorganisms but, rather, general procedures that will result in heterotrophic, aerobic microorganisms. For methods to isolate specific species of microorganisms, see references 5, 60, and 84.

Some general rules for applying ecological approaches to isolation are listed below and generally can be applied in the isolation of any one particular group of microorganisms (47).

1. List the groups of microorganisms that are to be isolated.
2. Describe the ecosystem or habitat from which the samples are to be collected.
3. Group samples into types, e.g., plants and plant parts, soils (types and horizons), rocks, water and insects.
4. List the environmental parameters to be considered and measured, such as pH, salinity, E_h, temperature, soil composition, and geochemical matrix.
5. Consider the constraints that large-scale conditions (e.g., the abiotic environment) can place on small-scale conditions (e.g., individuals, populations, and communities.) Species richness and numbers are often a seasonal response to the "windows of opportunity" which exist (125).
6. List the available natural substrates in the ecosystem, e.g., lignin and cellulose in forest soils, chitin in salt marshes.
7. Design isolation techniques around data obtained from steps 1 through 6, i.e., diluents, substrates, natural extracts, and incubation conditions.
8. Evaluate "ecophysiological isolation methods" by using standard methods as controls.
9. Modify known procedures as required by the ecological parameters of the material to be examined.
10. Use specific enrichment procedures for microbial groups that may be of screening interest.

1.1. ISOLATION OF ACTINOMYCETES

1.1.1. Sampling and Collection Methods

To isolate a representative actinomycete population from a particular ecosystem, especially species occurring in unique environmental niches, considerable attention should be paid to sampling. Samples should be collected as aseptically as possible with the aid of sterile spatulas, soil profile samplers, forceps, scalpels, gloves, Nasco Whirl-Pak bags, and plastic bottles. Samples should be representative of a site, e.g., a particular soil type and its horizons, leaf litter and detritus, rhizosphere plane and zone, marine sand, sediment and mud, plant surfaces and parts, or water column. Samples should be fully labeled with a description and date. Seasonal and temporal aspects of collecting should be considered, since true autochthonous population may occur transiently; e.g., actinomycete numbers decrease after a heavy rainfall. Once the samples are brought into the laboratory, they should be examined immediately or stored overnight at 4°C in an area separate from the actual plating, screening, and culture collection facilities to minimize the chances of mite infestations.

1.1.2. Ecophysiological Parameters and Media

Ecophysiological parameters to consider include high or low temperature and pH, ionic strength, E_h, and even substrate concentration. Most of these parameters are measured in the field or in the laboratory (13, 89). The actinomycete isolation media mentioned below are then modified to fit the ecosystem or habitat being examined. For example, seawater can be added to isolation agars at different concentrations to match the salt gradient in an estuary (45). The pH can be lowered to suit the different soil horizons present in a forest (36). The temperatures of incubation can be lowered for psychrophiles (4 to 15°C) or raised for thermophiles (55 to 70°C). "Natural extracts" (from plants, rocks, or compost) can be added as growth factors (50 to 250 ml/liter) for the initial isolation.

Actinomycetes are bacteria that are most efficient at utilizing substrates available at very low concentrations as well as complex substrates such as chitin. Therefore, most actinomycetes are isolated in lean or complex agars rather than in a rich growth medium. Except for thermophilic and psychrophilic forms, they are slow colonizers and usually appear on isolation plates within 4 to 20 days when incubated at 65 to 80°C. Isolation media have been developed which favor the growth of actinomycetes over other microbial groups. Media such as arginine-glycerol salts (31), colloidal chitin (41, 61), GAC (71), M_3 (98), starch-casein (55), and water agar (31) are well suited for the isolation of most actinomycetes. The antifungal agents nystatin (Squibb) and cycloheximide (Calbiochem) are usually incorporated into actinomycete isolation agars to retard fungal development. To increase the number of specific genera, other antibiotics may also be added (24). The isolation agar plates are air dried for 3 days to minimize bacterial development. See the descriptions of actinomycete isolation media below (section 1.1.8).

1.1.3. Nonselective Isolation of Actinomycetes from Soil

Standard dilution spread plate techniques may be used to isolate actinomycetes from soil. However, based on our experience, the numbers and types of organisms increase when dried soil is used. All sediments, muds, and soils are air dried in sterile glass petri dishes at room temperature for 3 to 10 days (depending upon the moisture content) to aid in reducing the bacterial population. The dried samples are then gently powdered with a sterile pestle and "stamped" (111) onto actinomycete isolation agars. A small circular sponge (Dispo culture plug; diameter, 16 mm [Scientific Products]) is directly pressed into the dried soil powder and then removed, and excess soil is shaken off. A stack (9 to 12 plates) of alternating different isolation agars are then inoculated by successive "stamping" of the agar surface 13 times (10 stamps round the perimeter and 3 in the middle) with the sponge to achieve a dilution effect. Discrete isolated colonies will result after incubation of the plates right side up at 4 to 15°C for psychrophiles, 22 to 35°C for mesophiles, and 55°C and higher for thermophiles. Saturated filter paper disks are used to prevent the drying of agar plates at higher temperatures.

1.1.4. Selective Isolation of Actinomycetes from Soil

There are many ways to pretreat soils to increase the overall number (count) of actinomycetes isolated or to favor the growth of specific genera. Table 1 summarizes some of the more practical pretreatments of soils to aid in isolating actinomycetes for industry. A combination of a pretreatment with a suitable selective medium is necessary for the efficient isolation of certain genera (31, 120).

1.1.5. Isolation of Actinomycetes from Plant Material

In some ecosystems, there appears to be a distinct seasonal fluctuation in the numbers of actinomycetes isolated from plants, as well as a specific association with plant parts (44, 46). For example, the numbers may be low in early spring

TABLE 1 Methods of pretreating soils to enhance isolation of some specific genera

Method of pretreatment	Recommended isolation agar(s)	Antibiotics incorporated (μg/ml)	Incubation time (days)	Genus	Reference(s)
Air dry soil, grind, heat for 1 h at 100 to 120°C. Plate out dilution of distilled water-soil suspensons.	AV, GAC, MGA-SE	Cycloheximide (50), nystatin (50), polymyxin (50), penicillin (1–8)	20–40 at 30, 32, or 40°C	*Actinomadura*, *Microbispora*, *Microtetraspora*, *Streptosporangium*	70–74
Mix soil with distilled water, heat for 1 h at 50 to 55°C.	CYC	Cycloheximide (50), novobiocin (25)	1 at 50–55°C	*Thermoactinomyces*	25
Air dry soil, grind, heat for 2 h at 60 to 65°C. Stamp.	Arginine glycerol salts, starch-casein-nitrate, thin Pablum agar	Cycloheximide (75), nystatin (75)	10–14 at 26–28°C	*Micromonospora*, *Nocardia*	44
Mix powdered chitin with soil (1:1). Incubate for 2 to 3 weeks at 26°C. Grind or stamp.	Arginine glycerol salts, starch-casein-nitrate, thin Pablum agar	Cycloheximide (75), nystatin (75)	10–14 at 28°C	*Micromonospora*	44
Mix $CaCO_3$ with soil (1:1). Incubate for 10 days in an inverted petri plate with saturated disk of filter paper. Dilute in water and plate.	Arginine glycerol salts, water agar	None	10 at 28°C	Overall numbers of actinomycetes increase	31, 112
Mix dead yeast cells with soil (1:3). Adjust pH to 5.0. Incubate for 10 days at 28°C in high humidity. Dilute and plate.	Czapek plus yeast extract	Cycloheximide (75), nystatin (50)	10 at 26–28°C	*Oerskovia*	56a
Store soil at 4°C; dilute in 0.25 × Ringer's solution plus 0.02% (wt/vol) gelatin (pH 7.0), and plate.	Diagnostic sensitivity test agar	Chlortetracycline (50), cycloheximide (45), mycostatin (59), methacycline (10)	7, 14, 21 at 25°C	*Nocardia*	77, 78

when plants are young but quite high in late summer, when plants begin to senesce and decay. Actinomycetes isolated from the flower usually differ from those isolated from stem or roots.

Plant parts (flowers, leaves, upper stem, middle base, and roots) are chopped aseptically with scissors or a scalpel, preferably in a laminar-flow hood, and dried (for 2 to 7 days) in sterile glass petri dishes. The plant pieces are then either implanted into the agar surface or gently rolled over the surface of the agar in a streaking manner. An alternative method is to shake the plant pieces for 20 min (at 150 rpm) in quarter strength plant extract or phosphate buffer, serially dilute samples in an appropriate diluent, and spread them onto plates of isolation agars into which 1 to 5 ml of plant extract has been incorporated (see Section 1.1.8 for a list of diluents). Decomposed leaf litter can be examined with an Andersen sampler that is connected to a sedimentation chamber (86) or can be exposed to bacteriophage suspensions before being inoculated onto isolation plates to isolate thermophilic actinomycetes (54). Differential centrifugation works well for separating the bacteroides of *Frankia* species from plant tissue and organelles in homogenates prepared from nodules taken from the host plant (4, 8, 59).

1.1.6. Isolation of Actinomycetes from Water Samples

Pretreatment of water samples often enhances the number and types of actinomycetes isolated. Centrifugation (6,000 × *g*) (76) or filtration (0.45-μm-pore-size filter) (2) of water samples, followed by serial dilution and plating, works well for the isolation of "aquatic actinomycetes." *Rhodococcus* and *Micromonospora* spp. can be selectively isolated by heating 2-ml water samples (previously stored at 4°C) in a glass tube (100 by 12 mm) sealed with silicone rubber bungs for 6 min at 55°C (98). The samples are diluted in quarter-strength Ringer's solution (Oxoid tablets) containing 0.02% (wt/vol) gelatin (pH 7.0) (109) and plated onto M_3 agar (98). The plates are incubated at 30°C. *Rhodococcus* colonies appear within 5 to 7 days, whereas *Micromonospora* colonies will take between 10 and 21 days to appear. *Actinoplanes* spp. can be isolated from flowing water by simply plating water samples on water agar plates made with filtered lake water or chitin agar and incorporating 0.1% (wt/vol) potassium tellurite (122).

1.1.7. Subculturing and Purification

Once colonies develop, they can be tentatively identified by differences in gross morphology such as colony form, aerial spore color, and diffusible and reverse pigments (28). With the aid of a long-working-distance objective (40×), aerial and substrate spore formation can be determined. The success of isolating diverse actinomycete genera, as well as species, depends somewhat upon the sample itself and the agars used; even more important to success, however, is the ability to recognize the different growth forms initially and identify them tentatively. Therefore, a "trained eye" is of great value in isolating actinomycetes. To confirm identification, many scientists employ chemical methods such as FAME (MIDI, Wilmington, Del.) or molecular techniques based on 16SrRNA.

Colonies are picked with a flamed, bent, L-shaped needle or a sterile wood stick and transferred to appropriate maintenance agars or replica plated for screening. All isolates should be properly maintained and preserved to ensure genetic stability and high titer. Many methods for preserving actinomycetes and other microorganisms have been published (45, 51).

1.1.8. Actinomycete Isolation Media and Diluents

PROTOCOLS

ARGININE GLYCEROL SALTS MEDIUM (31)

Arginine monohydrochloride	1.0	g
Glycerol (specific gravity not less than 1.249 at 25°C)	12.50	g
K_2HPO_4	1.0	g
NaCl	1.0	g
$MgSO_4 \cdot 7H_2O$	0.5	g
$Fe_2(SO_4)_3 \cdot 6H_2O$	0.010	g
$CuSO_4 \cdot 5H_2O$	0.001	g
$ZuSO_4 \cdot 7H_2O$	0.001	g
$MnSO_4 \cdot H_2O$	0.001	g
Agar	15.0	g
Distilled water	1,000	ml

Adjust the pH to 6.9 to 7.1. Autoclave for 15 min at 121°C. Add sterile additions of cycloheximide (Calbiochem) and nystatin (Squibb) so that the final concentrations are between 50 and 75 μg of each per ml. *Note:* Cycloheximide will go into solution in warm water (45°C). Nystatin is not totally soluble in water at pH 7.0. Increase the pH to 11 with 1 N NaOH, filter sterilize, and drop the pH to 7.0 immediately with HCl (nystatin is unstable at high pH).

AV AGAR (70)

L-Arginine	0.3	g
Glucose	1.0	g
Glycerol	1.0	g
K_2HPO_4	0.3	g
$MgSO_4 \cdot 7H_2O$	0.2	g
NaCl	0.3	g
Agar	15.0	g
Distilled water	1,000	ml

Autoclave for 15 min at 121°C. Add the solutions listed below. Quantities given are final concentrations per liter. Filter sterilize.

Vitamin solution

Thiamine hydrochloride	0.5	mg
Riboflavin	0.5	mg
Niacin	0.5	mg
Pyridoxine hydrochloride	0.5	mg
Inositol	0.5	mg
Calcium pantothenate	0.5	mg
p-Aminobenzoic acid	0.5	mg
Biotin	0.25	mg

Mineral solution plus antifungal antibiotics

$Fe_2(SO_4)_3$	10.0	mg
$CuSO_4 \cdot 5H_2O$	1.0	mg
$ZnSO_4 \cdot 7H_2O$	1.0	mg
$MnSO_4 \cdot 7H_2O$	1.0	mg
Nystatin	50.0	mg
Cycloheximide	50.0	mg

BENEDICT AGAR (7)

Glycerol	20.0 g
L-Arginine	2.5 g
NaCl	1.0 g
$CaCO_3$	0.1 g
$FeSO_4 \cdot 7H_2O$	0.1 g
$MgSO_4 \cdot 7H_2O$	0.1 g
Agar	20.0 g
Distilled water	1,000 ml

Autoclave for 15 min at 121°C, and adjust the pH to 7.0.

COLLOIDAL CHITIN AGAR (41, 61)

Colloidal chitin	2.0 g*
K_2HPO_4	0.7 g
KH_2PO_4	0.3 g
$MgSO_4 \cdot 5H_2O$	0.5 g
$FeSO_4 \cdot 7H_2O$	0.01 g
$ZnSO_4$	0.001 g
$MnCl_2$	0.001 g
Agar	20.0 g
Distilled water	to 1,000 ml

*Dry weight

Adjust to pH 6.8 to 7.0 before autoclaving.

In flask A, mix colloidal chitin with 500 ml of distilled water. In flask B, mix the remaining salts and agar. Autoclave separately for 15 min at 121°C. After autoclaving, swirl both flasks, slowly pour contents of flask A into flask B, and gently mix. Antifungal agents may be added once the agar has cooled to 45°C.

Preparation of colloidal chitin

1. Wash 100 g of unbleached chitin at 28°C, and shake overnight with 1 N HCl and 1 N NaOH alternately (24 h), four times with each solvent. Use a Büchner funnel with no. 1 or 4 Whatman paper to dry the chitin between washings.
2. Wash the chitin in cold ethanol four times at 4°C. This washing may be carried out overnight.
3. Moisten the brownish granular material thoroughly with cold acetone in a large beaker (e.g., 4,000 ml).
4. Add cold concentrated HCl to the beaker slowly and carefully in the cold (wear a mask) while stirring with a mechanical stirrer. Continue this process overnight. Within 2 h after adding the HCl, a viscous, dark brown solution should result. If not, add more acid.
5. Prepare several 4-liter or larger side arm flasks with about half the volume of ice-cold water (distilled).
6. Cover large Büchner funnels with three layers of glass wool, and carefully filter the chitin through. This process is slow, and the glass wool must be changed often. This is best accomplished by using several Büchner funnels, each with glass wool. The larger the funnel diameter, the better.
7. When the chitin enters the water, it flocculates into a fluffy white mass. Change the cold-water flasks when they are full with the flocculent chitin.
8. After the filtration, the chitin chunks in the glass wool may be dissolved in more concentrated HCl and refiltered into cold water.
9. Remove the flocculent chitin by centrifuging at 9,000 × *g* or more for 15 min. Carefully decant the bottles to minimize loss of some chitin.
10. Wash the chitin twice with water and centrifuge.
11. Place the chitin (still quite acidic) into large dialysis bags and dialyze, preferably with a continuous flow of tap water. Use a stirring bar to aid in solute-solvent exchange.
12. After continuous flow has been stopped for a few hours, the pH of the solvent is checked until it reaches a pH of 7.0. This may take several days.
13. Cut the bags and check the pH. If the pH is above 3.0, the chitin may be easily adjusted to pH 7.0 with a small volume of 1 N NaOH.
14. The dry weight of the chitin may be determined.
15. Sterilize the chitin with an ethylene oxide sterilizer, although autoclaving is adequate.

CYC (25)

Czapek-Dox liquid medium powder (Oxoid)	33.4 g
Yeast extract	2.0 g
Vitamin-free Casamino Acids	6.0 g
Agar	16.0 g
Distilled water	1,000 ml

The final pH is 7.2.

Autoclave for 20 min at 121°C. Filter sterilize novobiocin (Albamycin; Upjohn) and cycloheximide (Calbiochem) and add them at final concentrations of 25 and 50 μg/ml, respectively, after the agar has cooled to 45°C.

Czapek-Dox liquid medium powder ingredients (per liter)

Na NO_3	2.0 g
KCl	0.5 g
Magnesium glycerol phosphate	0.5 g
$FeSO_4 \cdot 7H_2O$	0.01 g
K_2SO_4	0.35 g
Sucrose	30.0 g

CZAPEK PLUS YEAST EXTRACT (56a)

Yeast extract	4.0 g
Sucrose	15.0 g
$NaNO_3$	2.0 g
$FeSO_4 \cdot 7H_2O$	0.01 g
K_2HPO_4	0.5 g
KCl	0.5 mg
$MgSO_5$	0.5 g
Agar	15.0 g
Distilled water	1,000 ml

Mix the ingredients and autoclave at 121°C for 20 min. Add filter-sterilized cycloheximide and nystatin (50 μg/ml [final concentration] each) once the agar has cooled to 45°C.

DIAGNOSTIC SENSITIVITY TEST AGAR (OXOID) (77, 78)

Proteose peptone	10.0 g
Veal infusion solids	10.0 g
Dextrose	2.0 g
NaCl	3.0 g
Na_2 HPO_4	2.0 g
Sodium acetate	1.0 g
Adenine sulfate	0.01 g
Guanine hydrochloride	0.01 g

Uracil	0.01	g
Xanthine	0.01	g
Thiamine	0.00002	g
Oxoid Agar no. 1	12.0	g
Distilled water	1,000	ml

Mix the ingredients, and autoclave for 15 min at 121°C. Add as sterile additions at the indicated final concentrations:

Cycloheximide	50 μg/ml
Nystatin	50 μg/ml
Chlortetracycline	45 μg/ml
Methacycline	10 μg/ml

GAC AGAR (70)

Bottom plate agar

Glucose	1.0 g
L-Asparagine	1.0 g
K_2HPO_4	0.3 g
$MgSO_4 \cdot 7H_2O$	0.3 g
NaCl	0.3 g
Trace salts solution	1.0 ml
Antibiotic solution	2.0 ml
Agar	20.0 g
Distilled water	1,000 ml

Mix and autoclave all the ingredients except the trace salts solution and antibiotic solution, which are filter sterilized and added once agar has cooled to 45°C. The final pH is 7.4.

Trace salts solution (per ml)	
$FeSO_4$	10.0 mg
$MgSO_4 \cdot 7H_2O$	1.0 mg
$CuSO_4 \cdot H_2O$	1.0 mg
$ZnSO_4 \cdot 7H_2O$	1.0 mg

Antibiotic solution (per ml)	
Nystatin	50.0 mg
Cycloheximide	50.0 mg
Polymyxin B	4.0 mg
Penicillin	0.8 mg

Top plate agar

Casamino Acids	50.0 mg
Antibiotic solution	4.0 ml
Agar	20.0 g
Distilled water	1,000 ml

Mix and autoclave all ingredients except the antibiotic solution. Pour bottom plate agar (15 ml), and let it solidify. Pour 4 ml of top plate agar.

M_3 AGAR (98)

Distilled water	1,000	ml
KH_2PO_4	0.466	g
Na_2HPO_4	0.732	g
KNO_3	0.10	g
NaCl	0.29	g
$MgSO_4 \cdot 7H_2O$	0.10	g
$CaCO_3$	0.02	g
Sodium propionate	0.02	g
$FeSO_4 \cdot 7H_2O$	200	μg
$ZnSO_4 \cdot 7H_2O$	180	μg
$MnSO_4 \cdot 4H_2O$	20	μg
Agar	18.0	g

Autoclave for 15 min at 121°C. The final pH after autoclaving is 7.0. Once the agar has cooled to 45°C, add a filter-sterilized solution of 4.0 mg of thiamine and 50.0 mg of cycloheximide.

MGA-SE AGAR (72)

Glucose	2.0 g
L-Asparagine	1.0 g
K_2HPO_4	0.5 g
$MgSO_4 \cdot 7H_2O$	0.5 g
Soil extract	200 ml
Agar	15.0 g
Distilled water	1,000 ml

pH 8.0 before autoclaving.

Soil extract

Autoclave 1,000 g of soil with 1 liter of tap water for 30 min. Decant and filter. Mix the ingredients and autoclave at 121°C for 20 min. Filter sterilize the following antibiotic solution, and add to agar once it has cooled to 45°C.

Antibiotic solution (final concentration per liter)	
Penicillin	1.0 mg
Polymyxin B	5.0 mg
Cycloheximide	50.0 mg
Nystatin	50.0 mg

STARCH-CASEIN-NITRATE AGAR (55)

Starch	10.0	g
Casein (Difco; vitamin free)	0.3	g
KNO_3	2.0	g
NaCl	2.0	g
$MgSO_4 \cdot 7H_2O$	0.05	g
$CaCO_3$	0.02	g
$FeSO_4 \cdot 7H_2O$	0.01	g
Agar (Difco)	18.0	g
Distilled water	1,000	ml

The pH is 7.0 to 7.2 before autoclaving at 121°C for 15 min. Add filter-sterilized nystatin and cycloheximide at 75 μg/ml each once the agar has cooled to 45°C.

THIN PABLUM AGAR (57, 58)

Pablum	7.5 g
Agar	15.0 g
Tap water	1,000 ml

Boil in cheesecloth (tea bag effect) for 20 to 30 min. If necessary, adjust the pH to 6.8 to 7.0. Autoclave for 20 min at 121°C. Add antifungal antibiotics.

WATER AGAR (31, 56a)

Tap water	1,000 ml
Crude agar flakes (SUP flake no. 1)	17.5 g

Autoclave for 15 min at 121°C. Add antifungal antibiotics.

1.2. ISOLATION OF UNICELLULAR BACTERIA FROM NATURE

1.2.1. Sampling and Collection Methods

Sampling procedures are essentially the same as those described for actinomycetes (see section 1.1.1). However, water samples for bacterial examination should be collected in clean, sterile, wide-mouth 100-ml plastic bottles. The bottles should be cleaned thoroughly before being used (91). Samples should be collected from still waters several inches deep because the surface water contains dust particles. The bottle should be grasped near the base and plunged open mouth downward, to a depth of 12 in. (ca. 30 cm), with the hand and bottle making a wide arc as they quickly pass into and out of the water (92). If a current exists, the mouth of the bottle is directed against it. Sample bottles should not be filled to the top. All water samples should be examined immediately after collection or stored at 5°C for no longer than 24 h. As soon as a water sample is collected, its condition of equilibrium is upset and a change in the bacterial population begins (91). For further information about sampling techniques and problems, see reference 20.

1.2.2. Ecological Parameters and Media

Incorporation of natural extracts and environmental biophysical parameters into media, as well as the diluent used in plating out the sample, can affect the numbers and variety of bacteria isolated in the laboratory (47).

Many different media are used to isolate bacteria from different samples within an ecosystem. Some agars should contain extracts (see section 1.2.8) at concentrations of 10 to 50% of the total liquid volume. Infusions and extracts are usually made from materials (soils, muds, leaves, roots, rocks, compost, detritus, barks, etc.) collected within the ecosystem being examined. Other agars contain multiple carbon and nitrogen sources or complex natural carbon-nitrogen sources, such as chitin, cellulose, or pectin. Agars selective for gram-negative organisms, such as nutrient agar plus crystal violet, red-violet bile agar, brilliant green bile agar, and others manufactured by BBL Microbiology Systems, Difco, or Oxoid, can have 5 to 15% natural extracts added. All isolation agars have the antifungal agents cycloheximide and nystatin incorporated at 50 to 100 μg/ml to retard fungal growth. Agar plates are also air dried for 1 to 2 days before use. Biophysical parameters such as medium pH and salinity are also adjusted to match the sample's ecosystem.

1.2.3. Direct Isolation of Soil Bacteria

A 5-g (wet weight) sample of soil is mixed with 99 ml of sterile quarter-strength soil infusion or extract contained in a 250-ml flask and shaken at 100 to 150 rpm for 25 min at 26°C. The soil-liquid suspension is serially diluted in the appropriate diluent (Table 2), and 0.1-ml volumes of three appropriate dilutions, based on turbidity, are spread onto plates of three to seven different agars to which natural extracts have been added. The soil composition will affect the turbidity observation; e.g., a high clay content may make the suspension more turbid, yet microbial numbers may be quite low. Written records of the soil type and dilutions used to plate out the soil sample should be kept so that database correlation can be made with respect to soil matrix, composition, and resulting microbial isolates. Plates are incubated upside down at 22 to 26°C for 4 to 14 days. The incubation temperature may vary depending upon the original soil temperature. The plates should be examined every 2 days for up to 3 weeks for new colony formation.

It is generally accepted that the microbial population associated with plant roots differs in numbers and types from that associated with the surrounding soil (17). The rhizosphere population is affected by the plant species; therefore, in isolating bacteria from the rhizosphere, it is important to consider the plant species involved. Plant, root, and soil extracts (see section 1.2.8) are incorporated into the isolation agars. Direct isolation or enrichment with carbohydrates, cellulose, and various nitrogen sources can be used to isolate many gram-negative rods and pleomorphic forms.

1.2.4. Simple Soil Enrichment for Bacteria

Some of the approximately 683 genera listed in *Bergey's Manual of Systematic Bacteriology* (53, 106, 107, 121) require specific enrichment or selection techniques for isolation in pure culture (1, 52). Most enrichments are feasible due to the induction of enzymes, i.e., activation of a specific genome. Thus, many precursors to antibiotics, complex substrates, and growth factors may be used. Enrichments may also promote antagonism. Therefore, in utilizing an enrichment technique that would result in two microbes possessing different metabolic pathways for utilizing the same substrate, one may dominate. The object here is to enrich for bacteria that are active in the ecosystem, i.e., those which can utilize the available nutrients present in the soil, and not dormant microorganisms that are capable of growing rapidly under appropriate artificial selection pressure.

A 1-g (wet weight) sample of soil is placed in a sterile 50-ml beaker and mixed with 2 to 10 ml of a sterile substrate suspended in an appropriate quarter-strength soil extract, half-strength Ringer's solution, or distilled water. Substrates can either be those present in the ecosystem or those that will favor the development of bacteria with specific or desired enzyme systems, such as chitin, cellulose, pectin, amino acids, metals, antibiotic precursors, inhibitors, and alcohols. The pH can be adjusted to the original soil pH or changed to act as a biophysical stress or enrichment.

1.2.5. Direct Isolation of Bacteria from Plants

Plant parts are cut with sterile scalpels or scissors, preferably in a laminar-flow hood or clean room. A 1-g sample of plant materials is placed in a 250-ml flask containing 99 ml of appropriate diluent and glass beads (optional). The flask is shaken for 15 to 20 min (150 rpm) at room temperature to loosen the attached surface microbiota. A 1-ml sample is withdrawn and serially diluted in the appropriate diluent (Table 2), and 0.1-μm volumes of three to four dilutions are spread onto plates containing plant isolation agars.

1.2.6. Simple Enrichment for Plant Bacteria

Chopped plant parts are mixed with 7.5 ml of a plant polysaccharide, sugar, or protein (5%) suspended in quarter-strength plant extract in a 50-ml beaker. The beaker is covered with a sterile paper cup and incubated in the manner described for soil enrichment. Samples should be periodically withdrawn, diluted in quarter-strength plant extract or half-strength phosphate buffer, and spread (0.1 ml) onto plates containing the appropriate plant isolation agars. For additional methods and a general discussion of phyllosphere bacteria, see references 14, 27, and 90.

1.2.7. Direct Isolation of Bacteria from Waters

A 50-ml water sample is filtered through a 0.22-μm disposable filter. Once the sediment has settled on the filter membrane, 1 ml of sterile diluent (usually half-strength mud or soil extract made from mud or soil surrounding the

TABLE 2 Isolation media for unicellular bacteria[a]

Sample type	Suggested media	Suggested diluents
Soil mud	Multiple extract agar, nutrient agar plus crystal violet, soil infusion agar, violet red bile agar, violet red bile agar plus extracts	Soil extract, phosphate buffer, Ringer's solution plus gelatin
Detritus, compost, leaf litter	Compost extract agar, cellulose agar, multiple carbon-nitrogen source agar, plant extract agar	Distilled water plus artificial seawater (pH adjusted), plant/soil extract, saline, plant extract
Fresh and dried plant parts and roots	Moss extract agar, nutrient agar plus crystal violet, plant extract agar, root extract agar	Plant extract, saline
Fresh and marine water	Cellulose agar, colloidal chitin agar, marine agar, violet-red bile agar plus extracts	Artificial seawater, filtered freshwater collected from the sampled site

[a]For diluent recipes and for media recipes, see section 1.2.9.

body of water) is added. The membrane is gently scraped with a flat, wide-mouth 2-ml sterile pipette. The suspension is transferred to 9 ml of corresponding diluent and vortexed for 5 min. Next, the sample is serially diluted and spread in the manner described for soils (section 1.2.3). For further techniques in aquatic microbiology, see reference 95.

1.2.7.1. Filter Membrane Imprints

A 50-ml water sample is filtered through a 0.22-μm disposable filter. The filter is gently cut out. With sterile forceps, the filter is successively laid (collected silt face down) on a stack of six agars to achieve a diminution effect. Agars can be alternately mixed to isolate a wider variety of types (see Table 2 for media).

1.2.7.2. Enrichment for Isolation of Aquatic Bacteria

The simplest enrichment is to incubate 50 ml of the water sample in a sterile, 500-ml wide-mouth flask, covered with a cotton gauze plug. The water is sampled periodically and plated out onto suitable isolation agars that have the sample water (filter sterilized) incorporated at 50 to 100%. It is also possible to bait for aquatic microorganisms by adding certain substrates such as sugars, polysaccharides, and proteins to different volumes of the sample. It is a good idea to duplicate the enrichment and incubate one flask in the dark and one flask in the light. Low concentrations of antifungal agents (15 to 35 μg/ml) may be added to retard fungal development. Another enrichment involves adding "attachment material," such as sterile plant material or ground rock material, to the flasks. Incubation periods and temperatures can be varied to identify psychrophiles, mesophiles, and thermophiles.

For additional enrichment procedures for the selective isolation of budding and prosthecate bacteria, see reference 39.

1.2.8. Subculturing and Purification

Once colonies develop, the agar plates are sorted according to types. Similar colonies or duplicates isolated from plating the different dilutions are sorted and then picked with sterile wooden sticks. The colonies are transferred either to master plates for screening or to appropriate maintenance agar slants that have a small amount (50 ml/liter) of natural extracts incorporated. Bacteria can be initially grouped according to morphological characteristics, such as pigment and colony forms, on plates. Pinpoint colonies are then picked, transferred to additional analogous isolation agars, and reincubated.

1.2.9. Bacteria Isolation Media, Extracts, and Diluents

All bacterial isolation agars should have nystatin and cycloheximide incorporated at 50 to 75 μg/ml each to retard fungal growth. See section 1.8.1 for instructions for preparing these antifungal agents.

PROTOCOLS

COMPOST EXTRACT AGAR

Yeast extract	5.0 g
Glucose	8.0 g
K_2HPO	4.5 g
KH_2PO_4	0.5 g
Compost extract	400 ml
Distilled water	600 ml

Mix and autoclave for 20 min at 121°C. Adjust the pH to match the detritus or soil sample. Add filter-sterilized antibiotic solutions when the agar has cooled to 45°C.

Compost extract

Tap water	3,000 ml
Compost	1,000 g

Bring to a boil, and steep for 30 min. Decant the supernatant, and let it cool. Filter it through cheesecloth. Centrifuge at 3,000 × g for 10 min.

CELLULOSE AGAR

Swollen cellulose	7.0 ml
Yeast extract	1.0 g

Agar	15.0	g
Water (distilled, sea, lake, river, etc., or combinations)	1,000	ml
K_2HPO_4	0.7	g
KH_2PO_4	0.3	g
$MgSO_4 \cdot 7H_2O$	0.5	g
$FeSO_4 \cdot 7H_2O$	0.01	g
$ZnSO_4$	0.001	g

In flask A, mix swollen cellulose and salts with 500 ml of distilled water. In flask B, mix agar, yeast extract, and the remaining water. Autoclave separately and then mix together slowly.

Method for preparing swollen cellulose

1. Slurry 30 g of air-dried cellulose powder in a little acetone, and slowly add to 800 ml of 85% phosphoric acid while stirring vigorously with a mechanical stirrer for 2 h to prevent clumping.
2. Add 2 liters of distilled ice water with rapid stirring.
3. Collect the suspended material by centrifugation at 4,000 × *g*.
4. Suspend the pellet in 5 liters of distilled ice water; centrifuge again.
5. Bring the product into solution with 1 liter of 2% $NaCO_3$.
6. Homogenize for 5 min at full speed in Waring blender and store for 12 h in a refrigerator. *Do not freeze.*
7. Wash the product in a suction filter with 5 liters of distilled ice water; suspend in 15 liters of distilled water.
8. Pellet by centrifugation at 10,000 × *g* for 5 min and homogenize for 5 min. This should yield a suspension with a pH within 0.1 unit of the pH of distilled water (pH 6.5). The total yield is calculated based on dry weights of three 1-ml samples of the suspension (44).

MARINE AGAR (DIFCO) 2216

Bacto-Peptone	5.0	g
Yeast extract	1.0	g
$FeCl_3 \cdot 6H_2O$	0.1	g
NaCl	19.45	g
$MgCl_2 \cdot 6H_2O$	8.8	g
Na_2SO_4	3.24	g
$CaCl_2 \cdot 6H_2O$	1.8	g
KCl	0.55	g
$NaHCO_3$	0.16	g
KBr	0.08	g
$SrCl_2 \cdot 6H_2O$	0.034	g
H_3BO_4	0.022	g
$Na_2SiO_3 \cdot 9H_2O$	0.004	g
NaF	0.0024	g
NH_4NO_3	0.0016	g
$NaHPO_4 \cdot 7H_2O$	0.008	g
Agar	15.0	g

Autoclave for 15 min at 121°C. The final pH is 7.6.

MOSS EXTRACT AGAR

Distilled water	700	ml
Moss extract	300	ml
Peptone	5.0	g
Yeast extract	5.0	g
Glucose	3.5	g
Trace mineral solution	1.0	ml
Agar	17.5	g

Moss extract

Distilled water	3,000	ml
Fresh moss	1,000	g

Trace mineral solution

$MgSO_4 \cdot 7H_2O$	0.25	g
$FeSO_4$	0.01	g
$CuSO_4 \cdot 5H_2O$	0.001	g
$ZnSO_4 \cdot 7H_2O$	0.001	g
$MnSO_4 \cdot 4H_2O$	0.001	g
Distilled water	1,000	ml

Bring water to a boil, and add the moss. Steep for 20 min, and decant supernatant. Cool, and filter through double-layered cheesecloth. Centrifuge at 3,000 × *g* for 10 min. Decant the supernatant.

MULTIPLE CARBON-NITROGEN SOURCE AGAR

Asparagine	0.1	g
NH_4Cl	0.1	g
Yeast extract	4.0	g
Glucose	5.0	g
Succinate	0.2	g
K_2HPO_4	0.1	g
$MgSO_4 \cdot 7H_2O$	0.1	g
Agar	18.0	g
Distilled water	1,000	ml

Mix the ingredients, and autoclave for 20 min at 121°C. The final pH should be approximately 6.5.

MULTIPLE EXTRACT AGAR

Soil extract	100	ml
Plant extract	100	ml
Root extract	100	ml
Artificial sea salts	0.25	g
Vitamin solution	1.0	ml
Peptone	4.0	g
Glucose	2.0	g
Agar	15.0	g
Distilled water	700	ml

Autoclave peptone, glucose, agar, and water for 15 min at 121°C. Add extracts. The final pH will depend on the pH of the extracts. Filter sterilize and add vitamin solution.

Soil extract

Collect garden, farm, forest, marsh, swamp, etc., soil or mud, preferably representative of the ecosystem. Spread the soil out to dry under a fume hood for at least 6 h.

Steeping tap water	960	ml
Dried soil	400	g

Autoclave the soil and water for 1 h at 121°C. Decant the liquid, and centrifuge to obtain a golden supernatant. Use immediately or filter sterilize and store in the cold until needed.

Plant extract

Collect plants from the ecosystem, and cut parts up with scissors.

Plant parts	500	g
Distilled water	1,000	ml

Bring water to a boil, and simmer the plant parts gently for 30 min. Decant the liquid, and filter it through double-layered cheesecloth. Filter sterilize, and store in the cold until needed.

Root extract
Collect roots from the plant. Gently shake soil particles off, and simmer the roots for 15 min in hot water. Filter sterilize and store in the cold until needed.

Artificial sea salts
Instant Ocean Aquarium Systems, Eastlake, Ohio

Vitamin solution
Quantities are final concentration per liter.

Thiamine hydrochloride	0.5 mg
Riboflavin	0.5 mg
Niacin	0.5 mg
Pyridoxine hydrochloride	0.5 mg
Inositol	0.5 mg
Calcium pantothenate	0.5 mg
para-Aminobenzoic acid	0.5 mg
Biotin	0.25 mg

NUTRIENT AGAR PLUS CRYSTAL VIOLET

Soil or other extract	100 ml
Tap water	900 ml
Beef extract	5.0 g
Yeast extract	5.0 g
Agar	15.0 g

Mix the ingredients, and autoclave for 3 min at 121°C. Add 1.0 ml of a filter-sterilized 0.1% solution of crystal violet. *Note*: Difco Nutrient Agar plus soil extract can be used.

PLANT EXTRACT AGAR

Plant extract	400 ml
Distilled water	600 ml
Glucose	8.5 g
Yeast extract	4.5 g
Crude agar flakes	17.5 g

See the description of multiple extract agar (above) for the method of preparing plant extracts. The final pH should be around 6.6 to 6.8, depending upon the pH of the plant extract used. Mix the ingredients, and autoclave for 15 min at 121°C.

ROOT EXTRACT AGAR

Root extract	375 ml
Plant extract	25.0 ml
Asparagine	0.5 g
Arginine	0.5 g
Tryptophan	0.5 g
Starch	2.0 g
Yeast extract	1.0 g
Artificial sea salts	0.5 g
Agar	16.0 g
Distilled water	1,000 ml

See the description of multiple extract agar (above) for the methods of preparing root and plant extracts and for the supplies of artificial sea salts.

SOIL INFUSION AGAR

Distilled water	600 ml
Soil infusion	400 ml
Yeast extract	3.0 g
Glucose	7.5 g
K_2HPO_4.	0.5 g
KH_2PO_4.	0.5 g
$MgSO_4 \cdot 7H_2O$	0.5 g
Agar	16.0 g

Mix the ingredients, and autoclave for 15 min at 121°C. The final pH will vary depending upon the ecosystem being examined.

Soil infusion

Fresh Super Soil (any good loam potting soil)	500 g
Distilled water	1,000 ml

Mix the soil and water. Gently heat without boiling for 15 min. Strain through double-layered cheesecloth, and filter sterilize. Store in the cold until needed.

VIOLET-RED BILE AGAR (DIFCO) PLUS EXTRACTS

Difco violet-red bile agar	28.0 g
Agar	7.0 g
Natural extract	250 ml
Distilled water	750 ml

The final pH will vary depending upon the extract and ecosystem being examined. Mix the ingredients, and autoclave for 15 min at 121°C.

DILUENTS
Filter sterilize all diluents, and dispense at 9 ml per tube.

Artificial seawater
Instant Ocean Aquarium Systems, Eastlake, Ohio. Prepare according to directions.

Distilled water plus artificial seawater

Distilled water	980.0 ml
Artificial seawater	20.0 ml

Adjust the pH to suit the sample pH.

Extract diluent

Natural extract	10 to 40%
Distilled water	60 to 90%

Adjust the pH to suit the sample pH.

Phosphate buffer diluent

K_2HPO_4	5.62 g
KH_2PO_4	2.13 g
Distilled water	to 1,000 ml

Mix and autoclave for 15 min at 121°C. This will yield a 0.05 M solution at pH 7.0. The molarity and concentration can be varied to obtain different pH values.

Saline

Distilled water	1,000 ml
NaCl	0.9 g

Ringer's solution plus gelatin

NaCl	8.6 g
KCl	0.3 g
CaCl	0.33 g

Gelatin	0.01%	(wt/vol)
Distilled water	1,000	ml

Mix until clear.

Mineral spring water
Filter sterilize and dispense. Many brands are available, and the composition varies in different areas. Make sure that the water has been collected from a natural spring and has not been distilled.

1.3. ISOLATION OF FUNGI

1.3.1. Sampling and Collecting Methods

Fungi are ubiquitous, heterotrophic organisms that occupy nearly every conceivable environmental niche. Before initiating an isolation program, it is important to define its goals (product of interest) and to understand the structure of the ecosystem being sampled. This will help determine the types of environments sampled and the isolation approach. Samples should be collected by using sterile gloves, tweezers, scalpels, or scoops, aseptically placed in sterile containers, e.g., Nasco Whirl-Pak bags or bottles (91), and either used immediately or stored overnight at 5°C. When collecting soil samples, all horizons of a particular soil, leaf litter and detritus, rhizosphere, sands, gravel, and rocks should be included.

Any part of a plant can be useful for isolating fungi (leaf, stem, bark, root, and flower), and the tissue sampled will influence the types and numbers of fungi isolated. For example, fungi that are predominant in leaf petioles may be found less frequently in the leaf tips and vice versa (29, 65). The physiological state and age of the plant sample influences the types of fungi isolated (85). Within a single ecosystem, different genetic strains can coexist depending upon which plants or parts of plants are sampled (65). The shape and size of the sample and the surface conditions (glabrous, pubescent, or glandular) can also influence the taxa isolated.

Water samples may be collected from estuaries, sewers, lakes, marine environments, streams, etc. At least 50 ml of water should be collected, and sterile bottles should be used.

1.3.2. Ecological Parameters and Media

Fungal forms and nutritional requirements are as diverse as their habitats. Newhouse and Hunter (69) have demonstrated that changing one constituent of a medium can favor the isolation of one fungus over another. Therefore, there is no "all-purpose" fungal isolation medium. The physical conditions (e.g., pH, salinity, nutrient sources, and temperature) of the ecosystem being sampled, as well as the "type" of fungi (lignicolous, halophilic, etc.) being isolated, should be considered when developing isolation agars. See section 1.3.11 for fungal isolation agars. Using media with low C/N ratios yield more discrete countable colonies, resulting in more effective isolation and identification (23). Such lean media may be effective because of the decrease in competition and restriction of spreading fungi (6).

Plates are incubated in stacks and should not be inverted. Low temperatures (5 to 15°C) are necessary for isolating psychrophiles, 20 to 35°C is typical for isolating mesophiles, and temperatures above 45 to 50°C select for thermophiles. Different species may be selected by incubation at different temperatures (19). Light is also an important factor, since it is sometimes needed to induce fructification (116). Usually a combination of selective enrichment and inhibition media is used (Table 3).

1.3.3. Selective Enrichment for Fungi and Inhibition of Other Organisms

Selective enrichments are designed to enhance the isolation of one or more species over others. Such enrichments may function on the species level (e.g., the separation of *Fusarium* species based on nutritional requirements) or on a more general scale (e.g., the separation of cellulolytic organisms, selective isolation of hymenomyctes [24], and isolation of melanolytic fungi [63]). Numerous other methods, such as the baiting technique (115), syringe sampling (32), soil sieving (35), adhesion flotation, sucrose centrifugation (66) and gelatin column centrifugation (104), are described elsewhere. These methods are often too laborious for effective large-scale isolations. However, they are effective for the isolation of some less ubiquitous species and for fungi in specialized niches, e.g., baiting with larvae to isolate fungi associated with insects (18, 126).

Incorporating sterile substrates (soil extracts, vegetable infusions) from the sampling environment into lean medium is a simple but effective way to enrich for fungi. For example, the number of varieties of yeasts isolated from grape leaves on a medium containing the same sugars as those found in vine sap (e.g., grape juice/yeast extract and grape juice/liver extract) was greater than the number isolated on standard yeast media (6). In addition, incorporating sterile plant parts into water agar (6) isolated osmotically sensitive yeasts. Likewise, for isolation of alkalophilic fungi with a high salt tolerance, it would be advantageous to include various levels of sea salts in the medium and adjust the pH to an alkaline value (47). There are as many ways to devise enrichment media as there are habitats.

Selective inhibition media function by discouraging "unwanted" organisms. Incorporating antibiotics such as chloramphenicol, kanamycin, penicillin, streptomycin, and tetracycline in the medium will inhibit bacterial colony development (113). Other methods of inhibiting bacteria include drying plates for 3 to 4 days before use, lowering the pH of the medium, or, when a low pH is undesirable, adding rose bengal (1:30,000) to the medium (21, 105).

Various antifungal antibiotics are available (nystatin, pimaricin, and cycloheximide) and can be excellent selective agents for specific groups of fungi. Some nonantibiotic inhibitory agents that have been used include high CO_2 levels, calcium proprionate salts, oxgall, and crystal violet (6). Commercial fungicides include Botran (2,6-dichloro-4-nitroaniline), benlate, pentachloronitrobenzene, and captan. These agents do not affect all fungi, and sometimes the effects that do occur are undesirable. Tsao (113) presents an excellent compilation of various selective inhibitors and their effectiveness. More recently, Dreyfuss (30) has pioneered effective measures against fast-growing fungi by adding cyclosporin A to low nutrient agar.

1.3.4. Isolation of Fungi from Soils

1.2.4.1. Implant Methods

With a sterile spatula or pick a small particle of soil is placed near the edge of the agar. The sample is gently pressed a little way into the agar, and the plate is incubated.

TABLE 3 Isolation media for lower fungi[a]

Sample type	Suggested media	Additions
Soil: sand, loam, humus, rocks, clay	Soil infusion agar, rose bengal medium, hay infusion agar, PCNB agar, tap water agar	As needed: salt, tetracycline, streptomycin, antifungal agents, etc.
Plant material: fresh or decaying plant matter, detritus	Hay infusion agar, leaf litter extract agar, PCNB agar, tap water agar	As needed: salts, chitin, pectin, cellulose, or sterile plant matter (e.g., carrot disks), inhibitors
Water: fresh, marine, sewage	Rose bengal medium, PCNB agar, tap water, agar, water (baited)	As needed: salts, antibiotics (e.g., chlortetracycline for sewage samples) (22), baits (e.g., seeds)

[a]See section 1.3.11 for medium formulas and instructions.

1.3.4.2. Dilution Plate Method

A series of dilution tubes containing 9 ml of sterile distilled water or other diluent is prepared. A 1-g sample of soil is placed in a dilution tube, and the tube is agitated for 15 min. Serial dilutions are performed, 0.1-ml samples of the final three dilutions are inoculated onto agar plates and spread with a sterile, bent glass rod. (Dilution should result in 40 to 60 colonies per plate.) The plates are incubated. The isolation plates may be used for replica plating by the same techniques described for bacteria, using damp velvet.

1.3.4.3. Warcup Method

A small amount of soil is placed in a sterile petri dish and dispersed with a small volume of sterile distilled water. Cooled molten agar is poured into the plate. This method is useful for the isolation of slow growers (115).

1.3.4.4. Stamping Method

Polyurethane foam cylinders (15 by 40 mm) are autoclaved. The end of a cylinder is moistened on the agar surface and pressed into the pulverized, dried soil sample. The plug is stamped 10 times in a circular pattern around the edge of the petri dish and 3 times in the center and is then used on six to eight more plates. This will create a diminution effect. Excessive moisture must be avoided. The use of this stamping technique (111) has resulted in the isolation of increased numbers and fungal types. The addition of NaCl (4 or 10%) to the medium further increases yields.

1.3.5. Isolation of Fungi from Dung

To eliminate yeasts and isolate more coprophilous species, the ethanol pasteurization technique of Bills and Polishook (11, 12) modified from that of Mahoney (64) can be used. Dried, pulverized dung is shaken in 60% ethanol for 30 s. After a 4-min settling period, the ethanol is decanted and the sample is transferred to molten agar (approximately 0.005 g [dry weight]/plate). The plates are incubated at 20°C.

1.3.6. Isolation of Fungi from Plant Material

1.3.6.1. Epibiotic Fungi

Many methods used for isolating fungi from the surfaces of plant parts can be found in the literature; some are more selective than others (9). Generally, a washing step, without surface sterilization, is followed by plating of samples.

Washing Method

Whole plant parts are placed in shake flasks containing 0.05% Tween 80 and water and shaken on a rotary shaker for 2 to 5 min. The inoculum is filtered and washed at this point or is immediately diluted and used in standard spread plate procedures. Mycelial forms tend to remain attached, while spores are removed more easily (26). The washed plant parts may also be aseptically teased or cut into fragments and used for the impression or implant methods below (68).

Implant Method

A sterile spatula or pick is used to place a small piece of plant matter near the edge of the agar plate. The sample is gently pressed a little way into the agar, and the plate incubated. Isolates will spread out from the point of inoculation across the plate. Growing fronts may be isolated.

Impression Method

A leaf surface is gently pressed onto agar by an operator wearing sterile gloves. The leaf is removed and pressed onto three or four more plates successively. This will produce a dilution effect. The procedure is repeated with the opposite leaf surface. Parberry et al. (81) found an increased frequency of nonsporulators with the increased number of imprints. Although this method was originally designed for leaf surfaces, it will work for many different types of plant material.

Maceration Method

In this method, plant material is macerated and then handled as for the dilution plate method or implant method. Note that some plants release inhibitory compounds during the maceration process. In addition, maceration may adversely affect the viability of some forms of fungi or, alternatively, may give unreliable counts by increasing the numbers of propagules. Beech and Davenport (6), working with grape leaves, found that washing the sample first resulted in greater numbers. The effect may have been one of "drawing organisms out" of the leaves.

1.3.6.2. Leaf Litter Fungi

The particle filtration method, modified from that of Möller (67), Bills and Polishook (12), and Polishook et al. (87), can be used to quantify the number of fungi in leaf litter and other organic materials. Samples (5 g) of leaf litter are air dried. They are then briefly homogenized with a small amount (20 ml) of water in a sterile blender. The samples are placed in the largest in a series of three stacked sieves of approximately 1,000, 250, and 100 μm mesh sizes. The sample is for several minutes under running water. The sample collected from the 100-μm sieve is scraped into a sterile 15-ml conical centrifuge tube containing 10 ml of sterile water and agitate for 1 min. It is then allowed to rest to settle the particulates. The supernatant is decanted, and the washing procedure is repeated. The sample is resuspended in 5 ml of sterile water and diluted five-fold. Sterile plastic pipette tips (it sometimes helps if the tips have about 1 mm cut off to widen the opening) are used plate 0.1 ml of the resuspended sample onto each of 5 to 10 plates of AMEA isolation agar (see section 1.3.11) or Oxoid DRBC agar plates. Incubate. The plates are checked daily. Redundant colonies are burned out with a pyropen or welding torch (30).

1.3.6.3. Endophytic Fungi

The methods for isolating endophytic fungi involve surface sterilization of plant samples followed by washing and plating onto appropriate agars (29, 33, 65, 68, 82, 85, 96, 97). The duration of the sterilization step, which depends on the sterilizing agent used and the nature of the sample, should be determined experimentally before embarking on a larger-scale isolation experiment. Typical sterilization agents include 70 to 90% ethanol, bleach, 1% peracetic acid, and 3 to 30% hydrogen peroxide. A combination of agents is often used. For example, a regimen of 1 min in 75% ethanol, 10 min in 65% sodium hypochlorite, and 30 s in 75% ethanol, followed by a sterile rinse in distilled water prior to plating, was used to isolate endophytes from palm leaves (96). More thorough discussions of sterilizing procedures can be found in reference 10.

1.3.7. Isolation of Fungi from Waters

Samples should be diluted to yield an optimum of fewer than 50 colonies per plate. Cooke (22) suggests as a general guideline the following dilutions: for liquid, 10^{-1}; rich sludge, 10^{-2}; sludge with 4 to 6% dry matter, 10^{-4}; and rich sludge with 30 to 60% dry matter, 10^{-4}. Dilution plates are prepared as described in section 1.3.4.2.

Other methods for sampling waters have been described (35, 50). Baiting entails placing sterilized "bait," e.g., hair or wool for keratinophilic fungi, hemp for phycomycetes, plant parts, seeds, termite wings, paper, or cellophane, into mesh baskets in the body of water and sampling periodically over several weeks. Other techniques include continuous-flow centrifugation and scrapings. These methods are generally small scale and may be impractical for some industrial screening purposes.

1.3.8. Isolation of Fungi from Rocks

Microcolony fungi can be picked from rock or statuaries with the aid of a microscope, using a dissecting needle to detach the colonies and to place them on agar (123). The method is useful when the integrity of the sample must remain intact, but it is somewhat laborious and impractical from an industrial point of view. When preserving the sample is not important, the method of Hirsch et al. (40) can be used. Rock samples are aseptically crushed (e.g., in a mortar and pestle or crusher), and suspended in 0.15 N NaCl with 0.5% Tween 80. They are homogenized, diluted in saline, and plated. The samples may also be plated onto agar directly. Alternatively, agar containing gauze squares can be pressed against the intact surface, removed, and incubated (79, 80).

1.3.9. Isolation of Macrofungi

Macrofungi may be isolated from fruiting or vegetative structures, resting stages, and substrates. Suitable media include potato dextrose agar, yeast malt agar, and basidiomycete selection medium (see section 1.3.11).

1.3.9.1. Tissue Isolation

When isolating basidiomycetes from sporocarps the freshest specimens should be selected; old or excessively wet sporocarps tend to harbor more contaminants. A sturdy knife is used to sever the fruiting body from woody substrates or to dig below the surface of the soil so that the entire structure can be collected. Isolations may be made in the field by using tubes of medium, alcohol, tweezers, scalpel, and a good cigarette lighter.

To inoculate media from the sporocarp tissue, debris is brushed from the surface of the fungus. The exterior surfaces may be wiped with alcohol or a weak bleach solution, but this is seldom necessary. The sporocarp is aseptically torn open to expose the interior. A small piece of tramal tissue is quickly removed with a sterile scalpel or tweezers, taking care not to touch the exterior surfaces that have been exposed to the environment. The tissue may originate in the pileus or stripe. The tissue is placed on solid agar; alternatively, the block of tissue is floated on a piece of sterile nylon mesh or filter paper in liquid medium (116).

1.3.9.2. Spore Isolation

Basidiospores and ascospores will yield monokaryotic mycelia. The pileus is removed from the stripe and carefully attached to a petri dish lid with a small dab of petroleum jelly. The lid is inverted, so that the gills or spore surface is facing downwards, and placed on a crystallizing dish or deep petri dish. After the spores have been deposited on the medium surface, the lid is replaced with a fresh, sterile one. The medium may be solid or liquid. If liquid medium is used, the spore suspension may be suitably diluted and spread onto plates or inoculated onto cooled molten agar and poured into plates. This method should yield more discrete units of germinating spores, with fewer instances of mycelial fusion (6a). The same basic inoculating technique can be used with slant tubes of media. The spore surface is attached to the tube side opposite the agar surface with petroleum jelly or a small piece of water agar. The tube is laid on its side. After the spores are deposited, the plug and spore surface are removed.

Many of the higher fungi, particularly ectomycorrhizal species, have special requirements for spore germination. Selective stimulatory or inhibitory substances influence germination differences among genera, and growth on simple nutrient agars is not always successful (34). Both basidiospores and ascospores may be self-inhibitory (65a, 108).

The methods discussed in this section may be applied to isolation of spores from soils.

1.3.9.3. Isolation of Macrofungi from Vegetative Structures

Vegetative structures may also be used as sources of inoculum. Rhizomorphs, sclerotia, or even mycorrhizal short roots are collected aseptically in small paper bags. The samples are surface sterilized with dilute bleach, hydrogen peroxide, or mercuric chloride (1:1) and thoroughly rinsed several times with sterile distilled water. They are aseptically chopped, sliced, or ground and plated onto medium. Mycelia will slowly grow out of the tissue and can be further isolated. When wood infected with basidiomycetes is collected, areas of profuse, actively growing infection should be chosen. The samples are surface sterilized with mercuric chloride (1:1) and implanted onto the agar surface (see section 1.3.4.1).

1.3.10. Subculturing and Purification

Subculturing and purification procedures are usually a necessary prelude to screening fungal isolates. Initial observations may be made with a dissecting microscope; the frequency of observation will depend on the fungi being isolated. Some basidiomycetes can take 4 to 6 weeks to emerge from wood samples, whereas yeasts may be apparent in 24 h. A dissecting needle (an insect pin attached to a long shank works quite well) is used to make single spore isolations or hyphal tip excision. The propagules can then be removed from the isolation plate and inoculated onto media that will promote growth and formation of fruiting structures for identification purposes. An effective method to eliminate fast-growing fungi which might overtake slower-growing isolates requires daily burning-out of undesirable colonies with a pyropen (30) or cutting out and removing colonies with a sterile scalpel or needle.

1.3.11. Fungal Isolation Media and Diluents

PROTOCOLS

Unless otherwise indicated, autoclave all media for 15 min at 121°C.

ANTIBIOTIC MALT EXTRACT AGAR (AMEA) (67)

Malt extract	5.0 g
Agar	10.0 g
Distilled water	990 ml

After autoclaving, add filter-sterilized solutions A and B (which may be prepared in advance and frozen).

Solution A

Bacitracin	0.02 g
Neomycin	0.02 g
Penicillin G	0.02 g
Polymixin B	0.02 g
Tetracycline	0.02 g
Streptomycin	0.02 g
Distilled water	10 ml

Solution B

Cyclosporin A	0.001 g
Methanol	1 ml

BASIDIOMYCETE SELECTION MEDIUM (99)

Malt extract	3.0 g
Peptone	0.5 g
Agar	2.5 g
o-Phenylphenol	0.006 g
Distilled water	100 ml

Lower the pH to 3.5 after autoclaving.

HAY INFUSION AGAR (110)

Agar

Infusion filtrate	1,000 ml
K_2HPO_4	2.0 g
Agar	15.0 g

Adjust the pH to 6.2, and autoclave.

Infusion filtrate

Distilled water	1,000 ml
Decomposing hay	50.0 g

Autoclave for 30 min at 121°C, and filter.

LEAF LITTER EXTRACT AGAR

Leaf litter extract	400 ml
Tap water	600 ml
Yeast extract	1.5 g
Agar	17.5 g
Benlate (optional)	15 μg/ml

After autoclaving, the addition of penicillin (0.5 g/liter) and streptomycin (0.025 g/liter) is optional.

PCNB (PENTACHLORONITROBENZENE) AGAR

Agar	20.0 g
Peptone	5.0 g
$MgSO_4 \cdot 7H_2O$	2.5 g
KH_2PO_4	0.50 g
Distilled water	1,000 ml

Autoclave. Cool. Add filter-sterilized:

Pentachloronitrobenzene	200 μg/ml
Penicillin G	50 μg/ml
Lactic acid (85% solution)	1.3 ml
Sodium deoxycholate	130 μg/ml

POTATO CARROT AGAR (116)

Potatoes (washed, peeled, grated)	20.0 g
Carrots (washed, peeled, grated)	20.0 g
Distilled water	1,000 ml
Agar	20.0 g

Bring the carrots and potatoes to a boil in water, and simmer for 1 h. Add agar and distilled water to 1,000 ml. Autoclave.

POTATO DEXTROSE AGAR

Variation 1 (56)

Dehydrated potatoes (without preservatives)	22 g
Distilled water	178 ml

Rehydrate potatoes by heating gently. Add:

Glucose	20.0 g
Agar	17.0 g
Distilled water	to 1,000 ml

Autoclave.

Variation 2 (15)

Scrubbed potatoes (peeled and diced)	200 g
Distilled water	1,000 ml

Bring to a boil, and simmer for 1 h. Strain through a sieve. Add:

Agar	20.0 g
Water	to 1,000 ml
Glucose	15.0 g

Autoclave.

ROSE BENGAL MEDIUM (21)

KH_2PO_4	1.0 g
$MgSO_4$	0.5 g
Soytone or Phytone	5.0 g
Glucose	10.0 g
Rose bengal	0.035 g
Agar	20.0 g
Distilled water	1,000 ml

After autoclaving, add 35 μg of filter-sterilized chlortetracycline per ml.

SOIL INFUSION AGAR (MODIFIED)

Solution 1

Soil solution	1.25 liters
Glucose	12.5 g

Solution 2

Distilled	1.25 liters
Agar	37.5 g

Mix solutions 1 and 2. Autoclave.

Mix equal volumes of Super Soil or sandy loam with distilled water. Let heavy grains settle. Decant the supernatant through four layers of cheesecloth.

TAP WATER AGAR

Agar	15.0 g
Tap water	1,000 ml

Modifications: After autoclaving, pieces of sterile plant material, salts, etc., can be added.

YEAST MALT AGAR

Yeast extract	3.0 g
Malt extract	3.0 g
Peptone	5.0 g
Glucose	10.0 g
Distilled water	1,000 ml
Agar	20.0 g

Autoclave.

REFERENCES

1. **Aaronson, S.** 1970. *Experimental Microbial Ecology*. Academic Press, Inc., New York, N.Y.
2. **Al-Diwany, I. J., G. A. Unsworth, and T. Cross.** 1978. A comparison of membrane filters for counting thermoactinomyes endospores in spore suspensions and river water. *J. Appl. Bacteriol.* **45:**249–258.
3. **Ashdown, L. R., and S. G. Clarke.** 1992. Evaluation of culture techniques for the isolation of *Pseudomonas pseudomallei* from soil. *Appl. Environ. Microbiol.* **58:**4011–4015.
4. **Baker, D., J. G. Torrey, and G. H. Kidd.** 1979. Isolation by sucrose density fractionation and cultivation in vitro of actinomycetes from nitrogen fixing root nodules. *Nature* **281:**76–78.
5. **Balows, A., H. G. Trüper, M. Dworkin, W. Harder, and K.-H. Schleifer (ed.).** 1991. *The Prokaryotes: a Handbook on the Biology of Bacteria: Ecophysiology, Isolation, Identification, Applications*, vol. I to IV, 2nd ed. Springer-Verlag, New York, N.Y.
6. **Beech, F. W., and R. R. Davenport.** 1971. A survey of methods for the quantitative examination of the yeast flora of apple and grape leaves, p. 139–157. *In* T. F. Preece and C. H. Dickinson (ed.), *Ecology of Leaf Surface Microorganisms*. Academic Press, Ltd., London, United Kingdom.

6a. **Belt, A.** Unpublished data.

7. **Benedict, R. G., T. G. Pridham, L. A. Lindenfelser, H. H. Hall, and R. W. Jackson.** 1955. Further studies in the evaluation of carbohydrate utilization tests as aids in the differentiation of species of streptomycetes. *Appl. Microbiol.* **3:**1–6.
8. **Benson, D. R. and W. B. Silvester.** 1993. Biology of *Frankia* strains, actinomycete symbionts of actinorhizal plants. *Microbiol. Rev.* **57:**293–319.
9. **Bills, G. F.** 1995. Analyses of microfungi diversity from a user's perspective. *Can. J. Bot.* **73:**S33–S41.
10. **Bills, G. F.** 1996. Isolation and analysis of endophytic fungal communities from moody plants, p. 31–66. *In* S. C. Redlin and L. M. Carris (ed.), *Endophytic Fungi in Grasses and Woody Plants: Systematics, Ecology, and Evolution*. The American Phytopathological Society, St. Paul, Minn.
11. **Bills, G. F., and J. D. Polishook.** 1993. Selective isolation of fungi from dung of *Odocoileus hemionus* (mule deer). *Nova Hedwigia* **57:**195–206.
12. **Bills, G. F., and J. D. Polishook.** 1994. Abundance and diversity of microfungi in leaf litter of a lowland rain forest in Costa Rica. *Mycologia* **86:**187–198.
13. **Black, C. A. (ed.).** 1965. *Methods of Soil Analysis*, Part 2. *Chemical and Microbiological Properties*. American Society of Agronomists, Inc., Madison, Wis.
14. **Blakeman, J. P.** 1981. *Microbial ecology of the phylloplane*. Academic Press, Ltd., London, United Kingdom.
15. **Booth, C.** 1971. Fungal culture media. *Methods Microbiol.* **4:**84.
16. **Bowman, J. P., S. A. McCammon, and J. H. Skerratt.** 1997. *Methylosphaera hansonii* gen. nov. sp. nov, a psychrophilic, group 1 methanotroph from Antarctic marine salinity, meromictic lakes. *Microbiology* **143:**1451–1459.
17. **Brown, M. E.** 1975. Rhizosphere microorganisms, opportunists, bandits or benefactors, p. 21–38. *In* N. Walker (ed.), *Soil Microbiology*. Butterworths, London, United Kingdom.
18. **Brownbridge, M., R. A. Humber, B. L. Parker, and M. Skinner.** 1993. Fungal entomopathogens recovered from Vermont forest soils. *Mycologia* **85:**358–361.
19. **Carreiro, M. M., and R. E. Koske.** 1992. Room temperature isolations can bias against selection of low temper-

ature microfungi in temperate forest soils. *Mycologia* **84:** 886–900.

20. **Collins, V. G., J. G. Jones, M. S. Hendrei, J. M. Shewan, D. D. Wynn-Williams, and M. E. Rhodes.** 1973. Sampling and estimation of bacterial populations in the aquatic environment. *Soc. Appl. Bacteriol. Tech. Ser.* **7:** 77–110.
21. **Cooke, W. B.** 1954. Fungi in polluted water and sewage. II. Isolation techniques. *Sewage Ind. Waste* **26:**661–674.
22. **Cooke, W. B.** 1963. *A Laboratory Guide to Fungi in Polluted Waters, Sewage, and Treatment Systems.* U.S. Department of Health, Education and Welfare, Cincinnati, Ohio.
23. **Cooke, W. B.** 1968. Carbon nitrogen relationships of fungus culture media. *Mycopathol. Mycol. Appl.* **34:**305–316.
24. **Cross, T.** 1982. Actinomycetes: a continuing source of new metabolites. *Dev. Ind. Microbiol.* **23:**1–18.
25. **Cross, T., and R. W. Atwell.** 1974. Recovery of viable thermoactinomycete endospores from deep mud cores, p. 11–20. *In* A. N. Barker and G. W. Gould (ed.), *Spore Research Academic Press, Ltd., London, United Kingdom.*
26. **Dickinson, C. H.** 1971. Cultural studies of leaf saprophytes, p. 129–137. *In* T. F. Preece and C. H. Dickinson (ed.), *Ecology of Leaf Surface Microorganisms.* Academic Press, Ltd., London, United Kingdom.
27. **Dickinson, C. H., and T. F. Preece.** 1976. *Microbiology of Aerial Plant Surfaces.* Academic Press, Ltd., London, United Kingdom.
28. **Dietz, A., and D. W. Thayer.** 1980. *Actinomycete Taxonomy. SIM Special Publication no.* 6. Society for Industrial Microbiology, Arlington, Va.
29. **Dobranic, J. K., J. A. Johnson, and Q. R. Alikhan.** 1995. Isolation of endophytic fungi from eastern larch (*Larix larcina*) leaves from New Brunswick, Canada. *Can. J. Microbiol.* **41:**194–198.
30. **Dreyfuss, M. M.** 1986. Neue Erkenntnisse aus einem pharmakologischen Pilzscreening. *Sydowia* **38:**22–36.
31. **el-Nakeeb, M. A., and H. A. Lechevalier.** 1963. Selective isolation of aerobic actinomycetes. *Appl Microbiol.* **11:**75–77.
32. **Favero, M. S., J. J. McDade, J. A. Robertson, R. K. Hoffman, and R. W. Edwards.** 1968. Microbiological sampling of surfaces. *J. Appl. Bacteriol.* **31:**336–343.
33. **Fisher, P. J., F. Graf, L. E. Petrini, B. C. Sutton, and P. A. Wookey.** 1995. Fungal endophytes of *Dryas octopetala* from a high arctic polar semi-desert and from the Swiss Alps. *Mycologia* **87:**319–323.
34. **Fries, N.** 1984. Spore germination in higher basidiomycetes. *Proc. Indian Acad. Sci. Plant Sci.* **93:**205–222.
35. **Gerdemann, J. W., and J. M. Trappe.** 1974. The endogonaceae of the Pacific Northwest. *Mycol. Mem.* **5:**1–76.
36. **Hagedorn, C.** 1976. Influence of soil acidity of *Streptomyces* populations inhabiting forest soils. *Appl. Environ. Microbiol.* **32:**368–375.
37. **Hanada, S., A. Hiraishi, K. Shimada, and K. Matsaura.** 1995. Isolation of *Chloroflexus aurantiacus* and related thermophilic phototrophic bacteria from Japanese hot springs using an improved isolation procedure. *J. Gen Appl. Microbiol.* **41:**119–130.
38. **Hiraishi, A., and Y. Udea.** 1995. Isolation and characterization of *Rhodovulum strictum* sp. nov. and some other purple nonsulfur bacteria from colored blooms in tidal and seawater pools. *Int. J. Syst. Bacteriol.* **45:**319–326.
39. **Hirsch, P., M. Muller, and H. Schlesner.** 1977. New aquatic budding and prosthecate bacteria and their taxonomic position, p. 107–133. *In* F. A. Skinner and J. M. Shewan (ed.), *Aquatic Microbiology.* Academic Press, Ltd., London, United Kingdom.
40. **Hirsch, P., F. E. W. Eckhardt, and R. J. Palmer Jr.** 1995. Fungi active in weathering of rock and stone monuments. *Can. J. Bot.* **73:**S1384–S1390.
41. **Hsu, S. C., and J. L. Lockwood.** 1975. Powdered chitin as a selective medium for enumeration of actinomycetes in water and soil. *Appl. Microbiol.* **29:**422–426.
42. **Huddleston, A. S., N. Cresswell, M. C. P. Neves, J. E. Beringer, S. Baumberg, D. I. Thomas, and E. M. H. Wellington.** 1997. Molecular detection of streptomycin-producing streptomycetes in Brazilian soils. *Appl. Environ. Microbiol.* **63:**1288–1297.
43. **Hungate, R. E.** 1962. Ecology of bacteria, p. 95–119. *In* I. C. Gunsalus and R. Y. Stanier (ed.), *The Bacteria,* vol. IV. *The Physiology of Growth.* Academic Press, Inc., New York, N.Y.
44. **Hunter, J. C.** 1978. *Actinomycetes of a Salt Marsh.* Ph.D. thesis. Rutgers, the State University, New Brunswick, N.J.
45. **Hunter, J. C., and A. Belt.** 1996. *Maintaining Cultures for Biotechnology and Industry.* Academic Press, Inc., San Diego, Calif.
46. **Hunter, J. C., D. E. Eveleigh, and G. Casella.** 1981. Actinomycetes of a saltmarsh. *Zentbl. Bakteriol. Parasitenkd. Infektionskr. Hyg. Suppl.* **11:**195–200.
47. **Hunter, J. C., M. Fonda, L. Sotos, B. Toso, and A. Belt.** 1984. Ecological approaches in isolation. *Dev. Ind. Microbiol.* **25:**247–266.
48. **Hunter, J. C., and L. Sotos.** 1986. Screening for a "new" enzyme in nature: haloperoxidase production by Death Valley dematiaceous hyphomycetes. *Microl. Ecol.* **12:**121–127.
49. **Jeanthow, C., A. L. Reysenbach, S. L. Haridon, A. Gambacorta, N. R. Pace, P. Lenat, and D. Prieur.** 1995. *Thermotoga subterranea* sp. nov., a new thermophilic bacterium isolated from a continental oil reservoir. *Arch Microbiol.* **164:**91–97.
50. **Jones, E., and B. Gareth.** 1971. Aquatic fungi. *Methods Microbiol.* **4:**355–363.
51. **Kirsop, B. E., and A. Doyle (ed.).** 1991. *Maintenance of Microrganisms and Cultured Cells,* 2nd ed. Academic Press, Ltd., Cambridge, United Kingdom.
52. **Krieg, N. R.** 1981. Enrichment and isolation, p. 112–142. *In* P. Gerhardt, R. G. E. Murray, R. N. Costilow, E. W. Nester, W. A. Wood, N. R. Krieg, and G. B. Phillips (ed.), *Manual of Methods for General Bacteriology.* American Society for Microbiology, Washington, D.C.
53. **Krieg, N. R., and J. G. Holt (ed).** 1984. *Bergey's Manual of Systematic Bacteriology,* Vol. 1. The Williams & Wilkins Co., Baltimore, Md.
54. **Kurtboke, N., E. Murphy, and K. Sivasithamparam.** 1992. Use of bacteriophage for the selective isolation of thermophilic actinomycetes from composed eucalyptus bark. *Can. J. Microbiol.* **39:**46–51.
55. **Kuster, E., and S. T. Williams.** 1964. Selection of media for isolation of streptomycetes. *Nature* **202:**928–929.

56. **Lacy, M. L., and G. H. Bridgmon.** 1962. Potato dextrose agar prepared from dehydrated mashed potatoes. *Phytopathology* **53:**173.

56a. **Lechevalier, M. P.** Personal communication.

57. **Lechevalier, M. P., and H. A. Lechevalier.** 1957. *Waksmania* gen. nov., a new genus of the actinomycetes. *J. Gen. Microbiol.* **17:**104–111.

58. **Lechevalier, M. P., H. A. Lechevalier, and P. E. Holbert.** 1968. *Sporicthya*, un nouveau genre de streptomyceteae. *Ann. Inst. Pasteur Paris* **114:**227–286.

59. **Lechevalier, M. P., and H. A. Lechevalier.** 1990. Systematics, isolation and culture of *Frankia*, p. 35–60. *In* C. R. Schwintzer and J. D. Tjepkema (ed.), *The Biology of Frankia and Actinorhizal Plants*. Academic Press, Inc., New York, N.Y.

60. **Lederberg, J. (ed.).** 1992. *Encyclopedia of Microbiology*, vol. 1 to 4. Academic Press, Inc., New York, N.Y.

61. **Lingappa, Y., and J. L. Lockwood.** 1975. Chitin media for selective isolation and culture of actinomycetes. *Phytopathology* **52:**317–323.

62. **Liu, T.-N., T. M'Timkulu, J. Geigert, B. Wolf, S. Neidleman, D. Silva, and J. C. Hunter-Cevera.** 1987. Isolation and characterization of a novel non-heme chloroperoxidase. *Biochem. Biophys. Res. Commun.* **142:** 329–333.

63. **Liu, Y.-T., S.-H. Lee, and Y.-Y. Liao.** 1995. Isolation of a melanolytic fungus and its hydrolytic activity of melanin. *Mycologia* **87:**651–654.

64. **Mahoney, D. P.** 1972. *Soil and Litter Microfungi of the Galapagos Islands*. Ph.D. thesis. University of Wisconsin, Madison.

65. **McCutcheon, T. L., G. C. Carroll, and S. Schwab.** 1993. Genotypic diversity in populations of a fungal endophyte from Douglas firs. *Mycologia* **85:**180–186.

65a. **Miller, S. L.** Unpublished data.

66. **Miller, S. L., P. Torres, and T. M. McClean.** 1994. Persistence of basidiospores and sclerotia of ectomycorrhizal fungi and *Morchella* in soil. *Mycologia* **86:**89–95.

67. **Möller, C.** Isolation of fungi from leaf litter and other kinds of organic material. *In Society of Industrial Microbiology Workshop Manual: Isolation of Fungi from Natural Substrates and Introduction to their Characterization*. Society for Industrial Microbiology, Arlington, Va.

68. **Möller, C., and M. M. Dreyfuss.** 1996. Microfungi from Antarctic lichens, mosses and vascular plants. *Mycologia* **88:**922–933.

69. **Newhouse, J. R., and B. B. Hunter.** 1983. Selective media for recovery of *Cylindrocladium* and *Fusarium* species from roots and stems of tree seedlings. *Mycologia* **75:**228–233.

70. **Nonomura, H., and Y. Ohara.** 1969. Distribution of actinomycetes in soil. VI. A culture method effective for both preferential isolation and enumeration of *Microbispora* and *Streptosporangium* strains in soil. 1. *J. Ferment. Technol.* **47:**463–469.

71. **Nonomura, H., and Y. Ohara.** 1971. Distribution of actinomycetes in soil. VIII. Greenspore group of *Microtetraspora*, its preferential isolation and taxonomic characteristics. *J. Ferment. Technol.* **49:**1–7.

72. **Nonomura, H., and Y. Ohara.** 1971. Distribution of actinomycetes in soil. IX. New species of the genera *Microbispora* and *Microtetraspora* and their isolation method. *J. Ferment. Technol.* **49:**887–894.

73. **Nonomura, H., and Y. Ohara.** 1971. Distribution of actinomycetes in soil. X. New genus and species of monosporic actinomycetes. *J. Ferment. Technol.* **49:**895–903.

74. **Nonomura, H., and Y. Ohara.** 1971. Distribution of actinomycetes in soil. XI. Some new species of the genus *Actinomadura*, Lechevalier et al. *J. Ferment Technol.* **49:** 904–912.

75. **Odintsova, E. V., H. Jannasch, A. Mamone, and T. A. Langworthy.** 1996. *Thermothrix azorensis* sp. nov., and obligately chemolithoautotrophic, sulfur-oxidizing thermophilic bacterium. *Int. J. Syst. Bacteriol.* **46:**422–428.

76. **Okami, Y., and T. Okazaki.** 1972. Studies in marine actinomycetes. I. Isolation from the Japan Sea. *J. Antibiot.* **25:**456–460.

77. **Orchard, V., and M. Goodfellow.** 1974. The selective isolation of *Nocardia* from soil using antibiotics. *J. Gen. Microbiol.* **85:**160–162.

78. **Orchard, V. A., M. Goodfellow, and S. T. Williams.** 1977. Selective isolation and occurrence of nocardiae in soil. *Soil Biol. Biochem.* **9:**233–238.

79. **Palmer, F. E., D. R. Emery, J. Stemmler, and J. T. Staley.** 1987. Survival end growth of microcolonial rock fungi as affected by temperature and humidity. *New Phytol.* **107:** 155–162.

80. **Palmer, F. E., J. T. Staley, and B. Ryan.** 1990. Ecophysiology of microcolonial fungi and lichens on rocks in northeastern Oregon. *New Phytol.* **116:**613–620.

81. **Parberry, I. H., J. F. Brown, and V. J. Bofinger.** 1981. Statistical methods in the analysis of phylloplane populations, p. 47–65. *In* J. P. Blakeman (ed.), *Ecology of the phylloplane*. Academic Press, Ltd., London, United Kingdom.

82. **Pereira, J. O., J. L. Azevedo, and O. Petrini.** 1993. Endophytic fungi of *Stylosanthes*: a first report. *Mycologia* **85:** 362–364.

83. **Perfil'ev, B. V., and D. R. Gabe.** 1969. *Capillary Methods of Investigating Microorganisms*. University of Toronto Press, Toronto, Canada.

84. **Perry, J. J., and J. T. Staley.** 1997. *Microbiology: dynamics and diversity*. Saunders College Publishing, Harcourt Brace College Publishers, Fort Worth, Tex.

85. **Petrini, O.** 1991. Fungal endophytes of tree leaves, p. 179–197. *In* J. H. Andrews and S. S. Hirano (ed.), *Microbial Ecology of Leaves*. Springer-Verlag, New York, N.Y.

86. **Petrolini, B., S. Quaroni, M. Saracchi and P. Sardi.** 1992. A sporangiate actinomycete with unusual morphological features: *Streptosporangium claviforme* sp. nov. *Actinomycetes* **3:**45–50.

87. **Polishook, J. D., G. F. Bills, and D. J. Lodge.** 1996. Microfungi from decaying leaves of two rain forest trees in Puerto Rico. *J. Ind. Microbiol.* **17:**284–294.

88. **Postgate, J. R.** 1984. *The Sulfate Reducing Bacteria*, 2nd ed. Cambridge University Press, Cambridge, United Kingdom.

89. **Pramer, D., and E. L. Schmidt.** 1964. *Experimental Soil Microbiology*. Burgess Publishing Co., Minneapolis, Minn.

90. **Preece, T. F., and C. H. Dickinson (ed.).** 1971. *Ecology of Leaf Surface Microorganisms*. Academic Press, Ltd., London, United Kingdom.

91. **Prescott, S. C.** 1946. *Water Bacteriology*. John Wiley & Sons, Inc., New York, N.Y.

92. **Rand, M. C., A. E. Greenberg, M. J. Taras, and M. A. Franson.** 1976. *Standard Methods for the Examination of Water and Wastewater*, 14th ed. American Public Health Association, Washington, D.C.

93. **Ravot, G., M. Magot, M. L. Fardeau, B. K. C. Patel, G. Prensier, A. Egan, J. L. Garcia, and B. Ollivier.** 1995. *Therotoga elfii* sp. nov., a novel thermophilic bacterium from an African oil-producing well. *Int. J. Syst. Bacteriol.* **45:**308–314.

94. **Rees, G. N., B. K. C. Patel, G. S. Grassia, and A. J. Sheehy.** 1997. *Anaerobaculum thermoterrenum* gen. nov. sp. nov., a novel, thermophilic bacterium which ferments citrate. *Int. J. Syst. Bacteriol.* **47:**150–159.

95. **Rodina, A. G.** 1972. *Methods in Aquatic Microbiology*. Translated, edited, and revised by R. R. Colwell and M. S. Zambruski. University Park Press, Baltimore.

96. **Rodrigues, K. F.** 1994. The foliar fungal endophytes of the Amazonian palm *Euterpe oleracea*. *Mycologia* **86:**376–385.

97. **Rollinger, J. L., and J. H. Langenheim.** 1993. Geographic survey of fungal endophytic community composition in leaves of coastal redwood. *Mycologia* **85:**149–156.

98. **Rowbotham, T. J., and T. Cross.** 1977. Ecology of *Rhodococcus coprophilus* and associated actinomycetes in fresh water and agricultural habitats. *J. Gen. Microbiol.* **100:** 231–240.

99. **Russell, P.** 1956. A selective medium for the isolation of basidiomycetes. *Nature* **177:**1038–1039.

100. **Sakaguchi, T., N. Tsujimura, and T. Matsunaga.** 1996. A novel method for isolation of magnetic bacteria without magnetic collection using magnetotaxis. *J. Microbiol. Methods* **26:**139–145.

101. **Schut, F., E. J. De Vries, J. C. Gottschal, B. R. Robertson, W. Harder, R. A. Prins, and D. K. Button.** 1993. Isolation of typical marine bacteria by dilution culture: growth, maintenance, and characteristics of isolates under laboratory conditions. *Appl. Environ. Microbiol.* **59:**2150–2160.

102. **Shooner, F., J. Bousquet, and R. D. Tyagi.** 1996. Isolation, phenotypic characterization, and phylogenic position of a novel, facultatively autotrophic, and phylogenetic position of a novel, facultatively autotrophic, moderately thermophilic bacterium, *Thiobacillus thermosulfatus* sp. nov. *Int. J. Syst. Bacteriol.* **46:**409–415.

103. **Slater, J. H., R. Whittenbury, and J. W. T. Wimpenny (ed.).** 1983. *Microbes in Their Natural Environment*. Cambridge University Press, Cambridge, United Kingdom.

104. **Smith, G. W., and H. D. Skiller.** 1979. Comparison of methods to extract spores of vesicular-abuscular mycorrhizal fungi. *Soil Sci. Soc. Am J.* **43:**722–725.

105. **Smith, N. B., and V. T. Dawson.** 1944. The bacteriostatic action of rose bengal in media used for plate count of soil fungi. *Soil Sci.* **58:**467–471.

106. **Sneath, P. H. A., N. S. Mair, M. E. Sharpe, and J. G. Holt (ed.).** 1986. *Bergey's Manual of Systematic Bacteriology*, Vol. 2. The Williams & Wilkins Co., Baltimore, Md.

107. **Staley, J. T., M. P. Bryant, N. Pfennig, and J. G. Holt (ed.).** 1989. *Bergey's Manual of Systematic Bacteriology*, vol. 3. The Williams & Wilkins Co., Baltimore, Md.

108. **Stone, J. K., J. N. Pinkerton, and K. B. Johnson.** 1994. Axenic culture of *Anisogramma anomala*: evidence for self-inhibition of ascospore germination and colony growth. *Mycologia* **86:**674–683.

109. **Straka, R. P., and J. L. Stokes.** 1957. Rapid destruction of bacteria in commonly used diluents and its elimination. *Appl. Microbiol.* **5:**21–25.

110. **Thom, C., and K. B. Raper.** 1945. *The Manual of the Aspergilli*. The Williams & Wilkins Co., Baltimore, Md.

111. **Tresner, H. D., and J. A. Hayes.** 1970. Improved methodology for isolating soil microorganisms. *Appl. Microbiol.* **19:**186–187.

112. **Tsao, P., C. Leben, and G. W. Keitt.** 1960. An enrichment method for isolating actinomycetes that produce diffusible antifungal antibiotics. *Phytopathology* **50:**88–89.

113. **Tsao, P. H.** 1970. Selective media for isolation of pathogenic fungi. *Annu. Rev. Phytopathol.* **6:**157–185.

114. **Veldkamp, H.** 1970. Enrichment cultures of prokaryotic organisms. *Methods Microbiol.* **6:**305–361.

115. **Warcup, J. H.** 1950. The soil-plate method for isolation of fungi from soil. *Nature* **166:**117–118.

116. **Watling, R.** 1982. *How To Identify Mushrooms to Genus. V. Cultural and Developmental Features*. Mad River Press, Eureka, Calif.

117. **Weigel, J.** 1992. The obligately anaerobic thermophilic bacterium, p. 105–184. *In* J. R. Kristjansson (ed.), *Thermophilic Bacteria*. CRC Press, Inc., Boca Raton, Fla.

118. **Wells, J. C., J. C. Hunter, G. L. Astle, J. C. Sherwood, C. M. Rica, W. H. Trejo, D. P. Donner, and R. B. Sykes.** 1982. Distribution of β-lactam and β-lactone producing bacteria in nature. *J. Antibiot.* **35:** 814–821.

119. **Williams, S. T., R. Locci, A. Beswick, D. I. Kurtboke, V. D. Kuznetsov, F. J. Le Monnier, P. F. Long, K. A. Maycroft, R. A. Palma, B. Petrolini, S. Quaroni, J. I. Todd, and M. West.** 1993. Detection and identification of novel actinomycetes. *Res. Microbiol.* **14:**653–656.

120. **Williams, S. T., and E. M. H. Wellington.** 1982. Principles and problems of selective isolation of microbes, p. 9–26. *In* J. D. Bu Lock, L. J. Nisbet, and D. J. Winstanley (ed.), *Bioactive Microbial Products: Search and Discovery*. Academic Press, Ltd., London, United Kingdom.

121. **Williams, S. T., M. E. Sharpe, and J. G. Holt (ed.).** 1989. *Bergey's Manual of Systematic Bacteriology*, Vol. 4. The Williams & Wilkins Co., Baltimore.

122. **Willoughby, L. G.** 1971. Observations on some aquatic actinomycetes of streams and rivers. *Freshwater Biol.* **1:** 23–27.

123. **Wollenstein, U., G. S. de Hoog, W. E. Krumbein, and C. Urzi.** 1995. On the isolation of microcolonial fungi occurring on and in marble and other calcareous rocks. *Sci. Total Environ.* 287–294.

124. **Worrall, J. J.** 1991. Media for selective isolation of hymenomycetes. *Mycologia* **83:**292–302.

125. **Zak, J. C., R. Sinsabaugh, and W. P. MacKay.** 1995. Windows of opportunity in desert ecosystems: their implications to fungal community development. *Can. J. bot.* **73:**S1407–S1414.

126. **Zoberi, M. H., and J. K. Grace.** 1990. Fungi associated with the subterranean termite *Reticulitermes flavipes* in Ontario. *Mycologia* **82:**289–294.

Screening for Activities

LEEYUAN HUANG, SIOBHAN STEVENS-MILES, AND RUSSELL B. LINGHAM

2

2.1. INTRODUCTION

2.1.1. Microbes as Sources of Unique Products

Microbes are exceptionally rich, diverse, and easily accessible sources of novel metabolites that can inhibit enzyme pathways related to disease targets (50). These metabolites vary enormously in structural complexity and biological activity. To discover therapeutically useful metabolites, it is critical not only to design suitable and sensitive assays for screening microbial extracts but also to test extracts that contain most or all of the metabolites from culture broths with a minimum of interference. In addition to sensitive assays, novel and diverse producing microorganisms are critical to the success of any natural products program; implicit in this statement is that biological diversity may lead to chemical diversity (53). The emphasis of this chapter is on various novel screening tests rather than on the isolation of organisms, which is described in chapter 1.

From 1940 to 1980, the primary emphasis of natural products screening was to search for novel antibiotics to treat infectious disease. Many novel antibiotics, such as the penicillins, cephalosporins, and tetracyclines, were discovered during this period. After 1980, the emphasis of natural products screening shifted from antibacterial to other therapeutic areas, including antiparasitic, anticancer, antifungal, and antiviral targets. Other examples of unique targets are hypercholesterolemia (2) and immunomodulation (34). In this chapter, we describe previously reported enzyme and receptor assays for screening of natural products.

2.1.2. Types of Samples To Be Screened

The testing of microbial products presents several interesting challenges. For example, the question of whether to test whole broth or extracts depends on a variety of factors. Antimicrobial assays using whole organisms can tolerate whole broth; other assays, such as enzyme and receptor targets or mammalian cell-based assays, are susceptible to contaminating metabolites that may be present in the broth. One way to circumvent some of these problems is to use either single-phase or two-phase extraction procedures that utilize methanol or ethyl acetate. These types of extracts are usually devoid of proteases and are suitable for testing in enzyme and receptor assays. In some cases, solid-phase extraction protocols that make use of Sep Pak C-18 cartridges (Waters, Division of Millipore Corp., Milford, Mass.) or ultrafiltration filters (Millipore, Bedford, Mass.) (with 10,000 molecular weight cut-off) result in samples that are suitable for screening in enzyme and receptor binding assays. Because of the highly colored nature of the extracts, the effect of color interference in some assays needs to be taken into account and corrected. Test samples can be dissolved in any solvent that is tolerated in the assay. We routinely use dimethyl sulfoxide (DMSO) or 50% aqueous DMSO added directly into the assay. The final concentration of DMSO ranges from 1 to 5% and varies from assay to assay.

2.1.3. Developing and Semiautomating Screening Tests

Prior to the development of automatic pipetting stations, most assays were performed manually in volumes exceeding 1 ml and in large test tubes. The pressure to screen more samples as rapidly as possible led to the use of automatic pipetting stations as well as accelerating miniaturization of assays to preserve assay reagents. In 1984, a shift in screening strategies occurred. This was precipitated by the advent of small-volume polypropylene tube strips and the development of the Skatron cell harvester (Sterling, Va.), which greatly increased the number of samples that could be tested in receptor binding assays. Since then, new technologies for high-throughput screening have rapidly advanced into a fully integrated robotics system that has developed within the last 5 years. New assay technologies such as scintillation proximity assays (8), homogeneous time resolved fluorometric assays (46), FlashPlate assays (NEN, Boston, Mass.) (25), and reporter gene assays using luciferase (39) and β-lactamase (66) are being adapted in many productive high-throughput screens. All of the screening assays that are described in this chapter were performed either with semiautomated machines such as pipetting stations and cell harvesters or with 96-well pipetters in a 96-well format unless specified otherwise.

2.2. ANTIHYPERCHOLESTEROLEMIA SCREEN

2.2.1. Background

Hypercholesterolemia is an important risk factor for atherosclerosis and coronary heart disease (31, 59). The microsomal enzyme 3-hydroxy-3-methylglutaryl-coenzyme A reductase (HMG-CoA reductase, EC 1.1.1.34) is the major rate-limiting step in the cholesterol biosynthetic pathway (54). Mevinolin (lovastatin), a potent inhibitor of HMG CoA reductase, was discovered in several strains of *Aspergillus terreus* in 1978 (2). Mevinolin and related compounds are now the drugs of choice for treating hypercholesterolemia. Other enzymes involved in cholesterol biosynthesis can also serve as therapeutic targets. One such enzyme, squalene synthase (farnesyl-diphosphate:farnesyl-diphosphate farnesyltransferase, EC 2.5.1.2.1), catalyzes the reductive dimerization of farnesyl pyrophosphate to squalene and is the first committed step in the biosynthesis of cholesterol (52, 55). Inhibitors of squalene synthase have the potential to be useful antihypercholesterolemia agents. The search for natural product inhibitors of this enzyme has resulted in the discovery of several potent inhibitors, namely, zaragozic acids (5, 13, 14), squalestatins and their minor components (7, 11), TAN-1607A (35), and viridiofungins A, B, and C (19). Inhibitors of squalene biosynthesis have been reviewed (1). Below we describe the screening assays used to detect natural product inhibitors of HMG-CoA reductase and squalene synthase.

2.2.2. HMG-CoA Reductase

Assays are performed utilizing a resin technique in glass test tubes. [^{3}H]HMG-CoA (NEN, Boston, Mass.) and rat liver homogenates are used as substrate and enzyme, respectively. Radioactive enzyme products are separated by absorption of [^{3}H]HMG-CoA with Bio-Rex-5 resin suspension and unabsorbed radioactive enzyme products are counted. The assay mixture (final volume, 300 μl) contains 50 mM potassium phosphate (pH 7.0), 3 mM dithiothreitol (DTT), 1.5 mM potassium chloride, 300 μM EDTA, 10 μM HMG-CoA, and 0.09 μCi of [^{3}H]HMG-CoA. Rat liver enzyme homogenate (5 μl), prepared as described (22), is added with 10 μl of test sample into glass tubes. Tubes are preincubated at 37°C for 5 min, and the reaction is initiated with 10 μl of 20 μM NADPH; tubes are incubated further at 37°C for 15 min. The reaction is stopped with 70 μl of 5 N HCl, and the tubes are vortexed and incubated at 37°C for 15 min. Resin suspension (3.2 ml), prepared by suspending 250 g of Bio-Rex-5 (100 to 200 mesh) resin (Bio-Rad Laboratories, Hercules, Calif.) in 1,600 ml of water with three washes and repeated decantations, is added to each tube. Tubes are vortexed for 5 s and centrifuged at 1,450 × *g* for 10 min. Supernatants (2 ml) are pipetted into plastic counting vials, the scintillation cocktail is added, and the radioactivity in the vials is counted. Assay controls include tubes with no enzyme, tubes with enzyme and solvent, and tubes with enzyme and a known concentration of mevinolin.

2.2.3. Screening Assay from Mevalonate to Squalene for Squalene Synthase Inhibitors

To look more globally for inhibitors of cholesterol biosynthesis, a multienzyme assay has been developed. Rat liver homogenates (S20 fractions) containing at least six enzymes, namely, mevalonate kinase, 5-phosphomevalonate kinase, 5-pyrophosphatemevalonate decarboxylase, isomerase, prenyl transferase, and squalene synthase, are prepared by standard procedures. The assay mixture (final volume, 100 μl) contains 10 mM KH_2PO_4, 3 mM glucose-6-phosphate, 11 mM potassium fluoride, 3 mM DTT, 150 mM HEPES buffer at pH 7.5, 7.5 mM ATP, 7.5 mM $MgCl_2 \cdot 6H_2O$, 1 mM NADPH, and 0.2 mM sodium mevalonate plus 0.2 μCi of 5-[^{3}H]mevalonic acid (27). Organic microbial extracts are dissolved in 100% DMSO and diluted 20-fold into the reaction mixture. Reaction products are extracted with heptane after heating with a mixture of ethanol and KOH and analyzed by using a scintillation counter. Assays are also performed with an S20 fraction prepared from HepG2 cells, as described below. The final protein concentrations of rat liver homogenate and HepG2 cell extracts in the assays are 1.56 and 0.32 mg per ml, respectively. To measure selectivity of putative inhibitors, extracts are tested against various squalene synthases.

2.2.4. Preparation of Cell Extracts of HepG2 Cells

HepG2 cells are grown in minimum essential medium (MEM) with nonessential amino acids, sodium pyruvate, penicillin, L-glutamate, and 10% fetal bovine serum (30). The medium is changed twice weekly, and a confluent monolayer is achieved in 1 week. Forty-eight hours prior to harvest, cells are switched from MEM with 10% fetal calf serum to MEM with 10% delipidated serum, as described by Cham and Knowles (9). Cells are harvested and washed with phosphate-buffered saline. Fresh trypsin (0.25%)-EDTA (0.02%) with Hanks' balanced salt solution (GIBCO BRL, 310-4060AJ; Life Technologies, Gaithersburg, Md.) is then added and removed. The flasks are incubated at 37°C until the cells detach. Detached cells are resuspended in MEM centrifuged at 1,000 × *g* for 5 min, and the cell pellets are washed once with phosphate-buffered saline. Cells are resuspended in 50 mM HEPES containing 5 mM $MgCl_2$, 2 mM $MnCl_2$, and 10 mM DTT (pH 7.5) (enzyme suspension buffer), sonicated twice (Branson Sonifier, Cell Disruptor 200; Branson Sonic Power Co., Danbury, Conn.) (sonicator setting #60, pulse) on ice for 1 min, and then centrifuged for 10 min at 1,000 × *g*. Supernatants are transferred to clean tubes and centrifuged at 20,000 × *g* for 20 min, and S20 fractions are used directly in the screening assay. Some of the enzyme preparations (S20 fractions) are also further centrifuged at 100,000 × *g* for 1 h to obtain heavy microsomes that contain squalene synthetase. This membrane preparation is resuspended in the above-described buffer and used as the source of squalene synthase.

2.2.5. Squalene Synthase Assays

Squalene synthase assays are performed as described by Lingham et al. (45). The enzyme sources are heavy microsomes prepared from HepG2 cells and rat liver. The final protein concentrations of both enzymes vary from 1.2 to 3 μg per ml in the assay mixtures. Compounds to be screened are dissolved in 100% DMSO and diluted 20-fold into the assay mixture. The assay mixture (final volume, 100 μl) contains 150 mM HEPES, pH 7.5, 11 mM potassium fluoride, 3 mM DTT, 5.5 mM $MgCl_2$, 1 mM NADPH, 0.1 μg/ml of a squalene epoxidase inhibitor (L-688,709), 0.06 μM [^{3}H]farnesyl diphosphate ([^{3}H]FPP; 740 CBq/mmol, NEN, Boston, Mass.), 2.94 μM unlabeled FPP, and 1 to 3 μg/ml of enzyme preparation. Reactions are initiated by the addition of substrate, incubated at 30°C for 20 min, and stopped with 100 μl of 100% ethanol. One hundred microliters of resin (AG1-X8, chloride, 200 to 400 mesh; Bio-Rad Laboratories, Hercules, Calif.) is added to each tube, and [^{3}H]squalene is extracted with 300 μl of heptane, and radioactivity is counted in a TopCount Scintillation Counter (Packard Instrument Co., Downers Grove, Ill.).

2.2.6. Interpretation of Results

In typical HMG-CoA reductase assays, the amount of radioactivity recovered without enzyme and in the enzyme reaction controls ranges between 150–200 cpm and 7,000–8,000 cpm, respectively. In the multi-enzyme pathway assays, the amount of radioactivity recovered without enzyme and in the enzyme reaction controls ranges between 150–300 cpm and 5,000–8,000 cpm, respectively. The percent inhibition of test samples is calculated as shown in the equation: {[cpm (control) − cpm (test sample)]/[cpm (control) − (blank)]} × 100. Samples are considered active when the degree of inhibition exceeds 65% inhibition and the activity can be titrated. The active samples in the multienzyme assays are confirmed with squalene synthase using enzymes prepared from both rat liver ad HepG2 cells.

2.3. ANTICANCER SCREEN

2.3.1. Background

Historically, most of the anticancer screens reported in the literature are cytotoxicity tests using various types of tumor cells. These tests, while usually informative, are time-consuming and difficult to perform. In addition, the capacity of these assays is limited and not amenable to high-throughput screening. In this section, we illustrate a novel mechanism-based assay that takes advantage of the prominent role that the *ras* oncogene product, Ras (p21), plays in signal transduction and cellular growth. Among the more prevalent and pervasive cancer-causing mutations are those involving a mutated form of the *ras* oncogene that associates with about 25% of all human cancers. The incidence of mutated *ras* is significantly higher in colon tumors (>50%) and pancreatic tumors (>90%) (4, 16, 18). Ras functions, in part, by transmitting signals involved in cellular proliferation and differentiation. Ras is a GTP-binding protein that couples the activation of growth factor receptor tyrosine kinases (e.g., epidermal growth factor and platelet-derived growth factor) to intracellular signal transduction pathways. An intrinsic GTPase activity regulates Ras in normal cells, resulting in the formation of a Ras-GDP complex that is functionally silent. Oncogenic forms of Ras lack the GTPase activity. This results in a constitutively active, predominant GTP-bound form of Ras that continually transmits signals inside the cell. Transformation, in part, results from the unregulated stimulation of mitogenic pathways. Continued expression of oncogenic *ras* is obligatory to maintain the transformed phenotype; as such, Ras is an attractive target for therapeutic intervention.

Functionally, normal and oncogenic forms of Ras undergo a series of complex posttranslational processing events prior to associating with the plasma membrane. These include farnesylation, proteolysis, methylation, and palmitoylation. The first and obligatory step in Ras processing is farnesylation by farnesyl-protein transferase (FPTase). Genetic experiments have shown that farnesylation is necessary for Ras cell-transforming activity. The later processing events (proteolysis, methylation, and palmitoylation) are not necessary for Ras function (17, 23, 61, 65). Farnesylation of Ras by FPTase directs and anchors Ras in the cell membrane. FPTase consists of two subunits (an $\alpha\beta$ heterodimer) with molecular sizes of 48,000 and 45,000 Da, respectively. FPTase transfers a farnesyl group from farnesyl pyrophosphate (FPP) to the cysteine residue at the carboxyl terminus CaaX box (C, Cys; a, usually an aliphatic amino acid; X, another amino acid) of Ras. Inhibition of Ras farnesylation prevents Ras membrane localization and blocks the cell-transforming activity of Ras (29, 33). Several FPTase inhibitors have been reported to selectively inhibit Ras processing in cell lines (6, 15, 29, 36, 43), prevent tumorigenesis in nude mice (38), and promote regression of mammary and salivary carcinomas in Ha-*ras* transgenic mice (37).

2.3.2. Farnesyl-Protein Transferase

The FPTase assay is performed as described by Lingham et al. (45). Partially purified bovine FPTase and Ras peptides are prepared as reported by Schaber et al. (56) and Gibbs et al. (16), respectively. Human FPTase is prepared as described by Omer et al. (49). Compounds are dissolved in 100% DMSO and diluted 20-fold directly into the assay mixtures. Bovine FPTase is assayed in a mixture (final volume, 100 μl) containing 100 mM HEPES (pH 7.4), 5 mM $MgCl_2$, 5 mM DTT, 100 mM [^{3}H]FPP (740 Cbq/mmol), 650 nM Ras-CVLS (cysteine, valine, leucine, and serine), and 10 mg/ml of FPTase at 31°C for 60 min. Reactions are initiated with FPTase and stopped with 1 ml of 1.0 M HCl in ethanol. Precipitates are collected onto filtermats using a TomTec Mach II cell harvester (TOMTEC, Hamden, Conn.), washed with 100% ethanol, and dried, and radioactivity is counted in an LKB β-plate 1205 Scintillation Counter (Wallac Inc., Gaithersburg, Md.). Human FPTase activity is assayed as described above with the exception that 0.1% (wt/vol) PEG 20,000, 10 mM $ZnCl_2$, and 100 nM Ras-CVIM (cysteine, valine, isoleucine, and methionine) are added to the reaction mixture. After 30 min, reactions are stopped with 100 μl of 30% (wt/vol) trichloroacetic acid in ethanol and processed as described above for the bovine enzyme.

2.3.3. Interpretation of Results

Typical experimental results obtained without enzyme and in uninhibited enzyme controls range between 1,300–1,600 and 7,300–7,600 cpm, respectively. The percent inhibition of the test sample is calculated as described above. Similarly, samples are considered active when the degree of inhibition exceeds 65% and can be titrated. Test samples are first screened against bovine FPTase, and active samples are tested against human FPTase. To determine specificity, active samples are also treated against rat liver or human squalene synthase and bovine brain geranylgeranyl protein transferase (45).

2.4. ANTIHYPERTENSIVE SCREEN

2.4.1. Background

Cardiovascular disease is one of the major causes of death in modern society. One contributing factor to this process is hypertension, which is arguably caused by overwork, stress, lack of exercise, modern-day lifestyles, and diet. The search for antihypertensive agents has focused on the role of the renin-angiotensin system. This system plays a central role in the regulation of normal blood pressure and appears to be involved in hypertension as well as in congestive heart failure, cirrhosis, and nephrosis. Angiotensin II, the active hormone of the renin-angiotensin system, is a powerful arterial vasoconstrictor that exerts its action by interacting with specific receptors located on the cell membranes of various target organs. Endothelin, a peptide secreted by endothelial cells, is a potent constrictor of vascular smooth muscle. Elevated levels of endothelin are found in myocardial infarction, systemic hypertension, cardiac ischemia, and coronary vasospasm. Inhibitors of angiotensin-converting enzyme (which converts angiotensin to angiotensin II) and antagonists of angiotensin II or endothelin receptors are potential drugs for treating hypertensive conditions. The search for natural product inhibitors has resulted in the discovery of several inhibitors, including cochinmicins (42), cytosporins (60), osteromycin (57), and namibione (41).

2.4.2. Angiotensin-Converting Enzyme Assay

The angiotensin-converting enzyme assay is performed essentially as described by Huang et al. (28). Hippuryl-L-histidyl-L-leucine and rat lung homogenate are used as substrate and enzyme, respectively. Routinely, assay volumes are 0.2 ml; the mixture consists of 1 mM substrate, 0.5 M NaCl, 0.06 M potassium phosphate buffer (pH 8.3), and test samples (10 μl of 50% aqueous methanol or a 50% methanolic extract). The reaction is initiated with 25 μl of diluted enzyme suspension and incubated at 37°C for 30 min. The enzyme reaction is stopped by boiling for 10 min. One milliliter of 0.2 M potassium phosphate buffer (pH 8.3) is added, followed by 0.5 ml of 3% 2,4,6-trichloro-S-triazine in dioxane, after which the tube is immediately vortexed. The product is measured at a wavelength of 382 nm. This assay can easily be miniaturized by decreasing the volume of the reagents and using microtubes or microliter plates.

2.4.3. Angiotensin II Receptor Binding

Bovine arterial membranes are prepared as described by Stevens-Miles et al. (60). Membranes (about 100 μg/ml) are incubated in the presence of [^{125}I]Tyr^4-angiotensin II (40 pM) with 100 mM Tris-HCl (pH 7.4), 5 mM $MgCl_2$, 0.2% bovine serum albumin, and 0.2 mg/ml bacitracin. Assays are incubated at 37°C for 90 min, after which the mixtures are filtered using a TomTec Mach II cell harvester with GF/B filtermats. The filtermats are dried, placed in bags with scintillation fluid, and sealed, and radioactivity is counted in an LKB β-plate 1205 Scintillation Counter.

2.4.4. Endothelin Receptor Binding

Endothelin binding assays can be performed by using membrane preparations (42) of CHO cells expressing cloned endothelin receptors. Furthermore, whole-cell binding assays with these same cells can also be developed. Briefly, cells are detached from flasks using a cell dissociation buffer (20 mM phosphate-buffered saline and 2 mM EDTA) and collected by centrifugation. Cells are washed once and resuspended in assay buffer (100 mM HEPES with 5 mM EDTA and 0.1% human serum albumin [pH 7.5]) at 5 × 10^5 cells/ml. Membranes or cells (in assay buffer) are incubated with [^{125}I]endothelin (50 pM) and test samples at 37°C for 60 min. Assays are terminated by filtration using a TomTec Mach II (6 by 16 format) 96-well cell harvester with GF/B filtermats. Filtermats are dried, placed in bags with scintillation fluid, and sealed, and radioactivity is counted in an LKB β-plate 1205 Scintillation Counter.

2.4.5. Interpretation of Results

In a typical experiment for angiotensin II binding, the number of counts recovered for the blank and uninhibited controls is 100 and 1,000 cpm, respectively. Similarly, for endothelin binding, the blank and control counts are 600 and 6,000 cpm, respectively. The percent inhibition of test sample is calculated by the equation: {[cpm (control) − cpm (test sample)]/[cpm (control) − cpm (blank)]} × 100. When the test sample shows more than 65% inhibition and is dose-related, the sample is considered active. Similarly, for angiotensin-converting enzyme, the optical density (OD) values for blank and uninhibited enzyme are 0.1 and 0.5, respectively. Furthermore, the percent inhibition of the test sample is calculated as shown in the equation: {[OD (control) − OD (test sample)]/[(OD (control) − OD (blank)]} × 100. As in all the examples presented, when the test sample shows more the 65% dose-related inhibition, the sample is considered active.

2.5. ANTIVIRAL SCREEN

2.5.1. Background

Historically, microbes have been the most dangerous pathogens known to humans. However, the advent of potent and specific antibiotics led to the premature and ill-founded claim that there is no need to discover and develop new antibiotic agents. Viruses are, in general, more insidious infectious agents than microbes. To date, there are very few therapies that can completely eradicate viruses. Traditionally, prevention of viral infection relies upon vaccination with inactive virus or viral antigens. However, some viruses such human immunodeficiency virus (HIV-1, HIV-2) and influenza mutate rapidly. There are currently no effective vaccines available to prevent the spread of these and other viruses. One novel strategy is to target enzyme pathways that viruses use to replicate. Enzymes that lend themselves to the development of screening assays for HIV-1 and HIV-2 include integrase, reverse transcriptase, and protease. For the purpose of this article, the protease and reverse transcriptase activities of HIV are illustrated as in vitro mechanism-based screens.

HIV-1 is a retrovirus that causes AIDS. The genome of HIV-1 encodes a 99-residue protease that processes the Gag, Pol, and Env polyproteins (12, 32). Genetic disruptions in the HIV protease result in noninfectious viral particles (33, 63). Potent protease inhibitors halt the spread of viruses in cell culture; inhibitors such as ritonavir (33) and MK-639 (63) have been approved for use in humans. These compounds are very efficacious and lower plasma levels of virus by about 2 logs (24, 64). On the basis of results, the inhibition of HIV protease is believed to be a vital and effective mechanism for reducing and managing the HIV infection.

Another essential step in the cycle of HIV-1 is reverse transcription of the viral RNA genome to produce a double-stranded DNA copy. This process is mediated by the virally encoded reverse transcriptase. Inhibition of reverse transcriptase by nucleoside analogs such as 3′-azido-3′-deoxythymidine and dideoxyinosine are clinically effective in treating HIV-1 infection (12, 32), and certain nonnucleoside blockers of reverse transcriptase, such as nevirapine, lower plasma viremia by about 1.0 to 1.5 logs (24). However, one of the main problems with the nonnucleoside inhibitors is the rapid development of resistance, which can be attributed to mutation of the virus. Nevertheless, reverse transcriptase is still an important therapeutic target for intervention in the progress of AIDS (12, 32).

With respect to influenza virus, primary transcription is a unique antiviral target. As an obligatory step in the life cycle, the eight RNA segments of negative polarity that make up the genome are transcribed into positive-sense mRNAs by an associated viral transcriptase. Transcription is initiated by a novel mechanism in which capped and methylated (capl) RNAs are used to prime mRNA synthesis. Capped RNAs are derived from mammalian RNA polymerase II transcripts in the nuclei of infected cells by a virally encoded endonuclease (40, 51). Highly selective inhibitors, such as 4-substituted 2,4-dioxobutanoic acids, target the endonuclease activity of influenza viral transcriptase and inhibit the replication of influenza A and B viruses in cell culture assays in vitro and in a mouse challenge model in vivo (21). These results indicate that the viral transcriptase is an effective therapeutic target against influenza viruses.

2.5.2. HIV-1 Protease

The cloning, expression, and purification of the HIV protease is performed as described by Darke et al. (10). The peptide substrate Val–Ser–Gln–Asn–β-naphthylAla–Pro–Ile–Val–Gln–Gly–Arg–Arg is synthesized according to Merrifield (47) and labeled with [^{3}H]acetic anhydride as described by Lingham et al. (44). Five-microliter extracts in 100% DMSO are mixed with 0.4 μM [^{3}H]acetyl–Val–Ser– Gln– Asn– β-naphthylAla– Pro–Ile–Val–Gln–Gly–Arg–Arg, 100 μM Val–Ser–Gln–Asn–β-naphthylAla–Pro–Ile–Val–Gln–Gly–Arg–Arg, and 2 nM HIV protease in a final volume of 100 μl containing 100 mM 2-(N-morpholino)ethanesulfonic acid (MES) buffer, pH 6.0, 10 mM DTT, 1 mM EDTA, and 0.1% bovine serum albumin. After 60 min at 37°C, the reaction is stopped by placing the assay tubes on ice and by the addition of 200 μl of resin (Dowex AG50W-X8, 200 to 400 mesh, H^+ form; Bio-Rad Laboratories, Hercules, Calif.) slurry, which is prepared by repeated decantation of fine particles and with 500 g of resin in 750 ml of 0.1 N HCl. The tubes are capped, mixed for 15 min, and centrifuged for 10 min in a Savant centrifuge using the swing-bucket rotor to sediment the resin. Aliquots of the supernatants are mixed with scintillation fluid, and radioactivity is counted in a TopCount Scintillation Counter (Packard Instrument Co., Downers Grove, Ill.).

2.5.3. HIV-1 Reverse Transcriptase

The recombinant HIV-1 reverse transcriptase is prepared as described by Azzolina et al. (3) and Stahlhut and Olsen (58). Poly(rA)·oligo(dT)$_{12\text{-}18}$ (Pharmacia 27-7878-03; Uppsala, Sweden) is used as template primer, and [^{3}H]TTP (NET 221-X; NEN, Boston, Mass.) is used as a tracer. All reagents are prepared with sterile double-glass distilled water. The assay is performed in a reaction mixture of 100 μl containing 100 mM Tris-HCl (pH 8.2), 80 mm KCl, 12 mM $MgCl_2$, 2 mM DTT, 0.5 mM EGTA, 1 mg/ml of bovine serum albumin, 1.5 μg/ml of poly(rA)·oligo(dT), 10 μM nonradioactive TTP, 0.5 μCi of [^{3}H]TTP, 0.01% (wt /vol) Triton X-100, 5 μl of extracts, and 0.1 nM HIV-1 reverse transcriptase enzyme. After 45 min at 37°C, the reaction is stopped with 100 μl of 300% aqueous trichloroacetic acid in 10 mM sodium pyrophosphate. The mixtures are kept on ice for 30 min, and the precipitate is collected on a filtermat using a TomTec Mach II cell harvester. The filtermats are washed with 0.1 M HCl in 10 mM sodium pyrophosphate using the same harvester. The filtermats are baked in a microwave oven for 7 min and placed in an LKB counting bag. Thirty milliliters of scintillant is added, and radioactivity is counted in an LKB β-plate 1205 Scintillation Counter.

2.5.4. Influenza A Virus Transcriptase

All reagents are prepared with sterile double-glass distilled water, and gloves are worn at all times when performing the assay. Influenza virus polymerase cores are prepared and purified as described by Tomassini et al. (62) and Honda et al. (26). The Alfalfa mosaic virus (ALMV) substrates are prepared as described by Tomassini et al. (62). Briefly, ALMV segment 4 RNA containing cap 0 structures is purchased from J. Biol, Leiden, The Netherlands. The 5′ cap of ALMV RNA is methylated with nucleoside-2′-O-methyl-transferase and used directly for the assay (62). The transcriptions are performed in a microtiter plate using a final volume of 75 μl containing 90 mM HEPES (pH 7.3), 0.05% Triton N-101, 80 mM KCl, 5 mM $MgCl_2$, 1 mM DTT, 2 μg of tRNA per ml, 20 μg of purified polymerase cores per ml, 100 μM ATP, 50 μM CTP, 50 μM GTP, 1 μM UTP, 0.3 μM [^{35}S]UTP, 5 μl of sample, and 7 nM substrate ALMV capped primer of 880 nucleotides. After 30 min of incubation at 30°C, the reaction is stopped with 75 μl of sterile saturated sodium pyrophosphate solution and 50 μl of ice-cold 40% trichloroacetic acid, and the plates are mixed and placed on the ice for 15 min. The precipitated RNA samples are collected using a TomTec MACH III cell harvester and Packard GF/C filter plates. Filters are dried in an oven at about 60°C for 5 min. The undersides of the plates are sealed with the white sealer sheet, and 25 μl of Microscint 20 cocktail is added. The top of the plate is then sealed with Topseal-S sealing film. Radioactivity is counted in a TopCount Scintillation Counter.

2.5.5. Interpretation of Results

In a typical experiment, the blank and uninhibited enzyme control counts recovered are as follows: HIV-1 protease, 100 and 900 cpm, respectively; HIV-1 reverse transcriptase, 800 and 10,000 cpm; and influenza virus transcription, 1,000 and 20,000 cpm. As described earlier, activity of samples is calculated using the equation: {[cpm (control) − cpm (test sample)]/[cpm (control) − cpm (blank)]} × 100. A sample is considered active when inhibition exceeds 65% and is dose-dependent.

2.6. CONCLUSIONS

Natural products lie at the heart of organic chemistry, and organic chemistry begins at the end of secondary metabolism (20). In addition to the isolation of unique and diverse cultures and the availability of robust assays, the isolation and characterization of the compounds eliciting the inhibition are essential to the success of natural products discovery programs. One of the major problems with testing natural product extracts for compounds of interest is that the screening for the activity is usually not trivial. A significant commitment on the part of the assayist is necessary to ensure that success can be achieved. Another major issue concerns the identification of active compounds relatively early in the isolation process. One possible way to accomplish this would be to couple traditional methods of classification to more innovative technologies such as liquid chromatography/mass spectrometry (48). It was our intention to illustrate a few of the assays that were performed to screen natural products in our group over the last 10 years. There are numerous other assays that, when coupled to high-throughput technologies, can make positive contributions to the future of natural products discovery. Some of the assays presented here can be adapted by using high-throughput formats.

REFERENCES

1. **Abe, I., J. C. Tomesch, S. Wattanasin, and G. D. Prestwich.** 1994. Inhibitors of squalene biosynthesis and metabolism. *Nat. Prod. Reports* **7:**279–302.
2. **Alberts, A. W., J. Chen, G. Kuron, V. Hunt, H. Huff, C. Hoffman, J. Rothrock, M. Lopez, H. Joshua, E. Harris, A. Patchett, R. Monaghan, S. Currie, E. Stapley, G. Albers-Schonberg, O. Hensens, J. Hirshfield, K. Hoogsteen, J. Liesch, and J. Springer.** 1980. Mevinolin: a highly potent competitive inhibitor of hydroxymethylglutaryl-coenzyme A reductase and a cholesterol-lowering agent. *Proc. Natl. Acad. Sci. USA* **77:**3957–3961.
3. **Azzolina, B., N. D. Behrens, B. Chang, M. E. Dahlgren, H. George, J. T. Menke, D. L. Linemeyer, and D. J. Hupe.** 1990. Cloning, expression in *Escherichia coli*, and purification of HIV-1 reverse transcriptase. *FASEB J.* **4:** A2253.
4. **Barbacid, M.** 1987. Ras genes. *Annu. Rev. Biochem.* **56:** 779–827.
5. **Bergstrom, J. D., M. M. Kurtz, D. J. Rew, A. M. Amend, J. D. Karkas, R. G. Bostedor, V. S. Bansal, C. Dufresne, F. L. VanMiddlesworth, O. D. Hensens, J. M. Liesch, D. L. Zink, K. E. Wilson, J. Onishi, J. A. Milligan, G. Bills, L. Kaplan, M. Nallin Omstead, R. G. Jenkins, L. Huang, M. S. Meinz, L. Quinn, R. W. Burg, Y. L. Kong, S. Mochales, M. Mojena, I. Martin, F. Pelaez, M. T. Diez, and A. W. Alberts.** 1993. Zaragozic acids: a family of fungal metabolites that are picomolar competitive inhibitors of squalene synthase. *Proc. Natl. Acad. Sci. USA* **90:**80–84.
6. **Bishop, W. R., R. Bond, J. Petrin, L. Wang, R. Patton, R. Doll, G. C. Njoroge, J. Catino, J. Schwartz, W. Windsor, R. Syto, D. Carr, L. James, and P. Kirschmeier.** 1995. Novel tricyclic inhibitors of farnesyl protein transferase. *J. Biol. Chem.* **270:**30611–30618.
7. **Blows, W. M., G. Foster, S. J. Lane, D. Noble, J. E. Piercey, P. J. Sidebottom, and G. Webb.** 1994. The squalestatins, novel inhibitors of squalene synthase, produced by a species of *Phoma*. V. Minor metabolites. *J. Antibiot.* **47:**740–754.
8. **Bosworth, N., and P. Towers.** 1989. Scintillation proximity assay. *Nature* **341:**167–168.
9. **Cham, B. E., and B. R. Knowles.** 1976. A solvent system for delipidation of plasma or serum without protein precipitation. *J. Lipid Res.* **17:**176–181.
10. **Darke, P. L., C. T. Leu, L. J. Davis, J. C. Heimbach, R. E. Diehl, W. S. Hill, R. A. F. Dixon, and I. S. Sigal.** 1989. Human immunodeficiency virus protease. *J. Biol. Chem.* **264:**2307–2312.
11. **Dawson, M. J., J. E. Farthing, P. S. Marshall, R. F. Middleton, M. J. O'Neal, A. Shuttleworth, C. Stylli, M. R. Tait, P. M. Taylor, H. G. Wildman, A. D. Buss, D. Langley, and M. V. Hayes.** 1992. The squalestatins, novel inhibitors of squalene synthase, produced by a species of *Phoma*. *J. Antibiot.* **45:**639–647.
12. **Debouck, C., and B. W. Metcalf.** 1990. Human immunodeficiency virus protease: a target for AIDS therapy. *Drug Dev. Res.* **21:**1–17.
13. **Dufresne, C., K. E. Wilson, S. S. Singh, D. L. Zink, J. D. Bergstrom, D. Rew, J. D. Polishook, M. Meinz, L. Huang, K. Silverman, and R. B. Lingham.** 1993. Zaragozic acids D and D_2: potent inhibitors of squalene synthase and of Ras farnesyl-protein transferase. *J. Nat. Prod.* **56:**1923–1929.
14. **Dufresne, C., K. E. Wilson, D. Zink, J. Smith, J. D. Bergstrom, M. Kurtz, D. Rew, M. Nallin, R. Jenkins, K. Bartizal, C. Trainor, G. Bills, M. Meinz, L. Huang, J. Onishi, J. Milligan, M. Mojena, and F. Pelaez.** 1992. The isolation and structure elucidation of zaragozic acid C, a novel potent squalene synthase inhibitor. *Tetrahedron Lett.* **48:**10221–10226.
15. **Garcia, A. M., C. Rowell, K. Ackermann, J. J. Kowalczyk, and M. D. Lewis.** 1993. Peptidomimetic inhibitors of Ras farnesylation and function in whole cells. *J. Biol. Chem.* **268:**18415–18418.
16. **Gibbs, J. B., A. Oliff, and N. E. Kohl.** 1994. Farnesyltransferase inhibitors: Ras research yields a potential cancer therapeutic. *Cell* **77:**175–178.
17. **Gibbs, J. B., D. L. Pompliano, S. D. Mosser, E. Rands, R. B. Lingham, S. B. Singh, E. M. Scolnick, N. E. Kohl, and A. Oliff.** 1993. Selective inhibition of farnesyl-protein transferase blocks Ras processing in vivo. *J. Biol. Chem.* **268:**7617–7620.
18. **Gibbs, J. B., M. D. Schaber, T. L. Schofield, E. M. Scolnick, and I. S. Sigal.** 1989. Xenopus oocyte germinal-vesicle breakdown induced by $[Val^{12}]$Ras is inhibited by acytosol-localized Ras mutant. *Proc. Natl. Acad. Sci. USA* **86:**6630–6634.
19. **Harris, G. H., E. T. T. Jones, M. Meinz, M. Nallin-Omstead, G. L. Helms, G. Bills, D. Zink, and K. E. Wilson.** 1993. Isolation and structure elucidation of viridiofungins A, B and C. *Tetrahedron Lett.* **34:**5235–5238.
20. **Haslam, E.** 1986. Secondary metabolism—fact and fiction. *Nat. Prod. Reports* **3:**217–249.
21. **Hastings, J., H. Selnick, B. Wolanski, and J. Tomassini.** 1996. Anti-influenza virus activities of 4 substituted 2,4-dioxybutanoic acids. *Antimicrob. Agents Chemother.* **40:** 1304–1307.
22. **Heller, R. A., and M. A. Shrewsbery.** 1976. 3-Hydroxy-3-methylglutaryl coenzyme A reductase from rat liver. *J. Biol. Chem.* **251:**3815–3822.
23. **Hiwasa, T.** 1996. Ras inhibitors (Review). *Oncol. Reports* **3:**7–14.
24. **Ho, D. D., A. U. Neumann, A. S. Perelson, W. Chen, J. M. Leonard, and M. Markowitz.** 1995. Rapid turnover of plasma virions and CD_4 lymphocytes in HIV1 infection. *Nature* **373:**123–126.
25. **Holland, J. D., P. Singh, J. G. Brennand, and A. J. Garman.** 1994. A nonseparation microplate receptor binding assay. *Anal. Biochem.* **222:**516–518.
26. **Honda, A. J., A. Mukaigawa, A. Yokoiyama, S. Kato, K. Ueda, K. Nagata, M. Krystal, D. P. Nayak, and A.**

Ishihama. 1990. Purification and molecular structure of RNA polymerase from influenza virus A/PR8. *J. Biochem.* **107:**624–628.

27. **Huang, L., R. B. Lingham, G. H. Harris, S. B. Singh, C. Dufresne, M. Nallin-Omstead, G. F. Bills, M. Mojena, M. Sanchez, J. D. Karkas, J. B. Gibbs, W. H. Clapp, M. S. Meinz, K. C. Silverman, and J. D. Bergstrom.** 1995. New fungal metabolites as potential antihypercholesterolemics and anti-cancer agents. *Can. J. Bot.* **73**(Suppl. 1)**:**S898–S906.

28. **Huang L., G. Rowin, J. Dunn, R. Sykes, R. Dobna, B. A. Mayles, D. M. Gross, and R. W. Burg.** 1984. Discovery, purification and characterization of the angiotensin converting enzyme inhibitor, L-681,176, produced by *Streptomyces* sp. MA 5143a. *J. Antibiot.* **37:**462–465.

29. **James, G. L., J. L. Goldstein, M. S. Brown, T. E. Rawson, T. C. Somers, R. S. McDowell, C. W. Crowley, B. K. Lucas, A. D. Levinson, and J. C. Marsters, Jr.** 1993. Benzodiapine peptidomimetics: potent inhibitors of Ras farnesylation in animal cells. *Science* **260:**1937–1941.

30. **Javitt, N. B.** 1990. Hep G2 cells as a resource for metabolic studies: lipoprotein, cholesterol, and bile acids. *FASEB J.* **4:**161–168.

31. **Kannel, W. B., W. P. Castelli, T. Gordon, and P. M. McNamara.** 1971. Serum cholesterol, lipoproteins, and the risk of coronary heart disease. *Ann. Intern. Med.* **74:** 1–12.

32. **Kay, J., and B. M. Dunn.** 1990. Viral proteinases in strength. *Biochim. Biophys. Acta* **1048:**1–18.

33. **Kempf, D. J., K. C. Marsh, J. F. Denissen, E. McDonald, S. Vasavanonda, C. Flentge, B. E. Green, L. Fino, C. H. Park, X. P. Kong, N. E. Wideburg, A. Saldivar, L. Ruiz, W. M. Kati, H. L. Sham, T. Robins, K. D. Stewart, A. Hsu, J. F. Plattner, J. M. Leonard, and D. W. Norbeck.** 1995. ABT-538 is a potent inhibitor of human immunodeficiency virus protease and has high oral bioavailability in humans. *Proc. Natl. Acad. Sci. USA* **92:**2484–2488.

34. **Kino, T., T. Hatanaka, M. Hashimoto, M. Nishiyama, M. Goto, M. Okuroda, H. Kohsaka, H. Aoki, and H. Imanaka.** 1987. FK-506, a novel immunosuppressant isolated from a Streptomyces. I. Fermentation, isolation, and physico-chemical and biological characteristics. *J. Antibiot.* **40:**1249–1255.

35. **Kittano, K., R. Tozawa, and S. Harada.** 1993. Euro. Patent Appl. and Publ.EP 0 568 946 A1.

36. **Kohl, N. E., S. D. Mosser, S. J. Desolms, E. A. Giuliani, D. L. Pompliano, S. L. Graham, R. L. Smith, E. M. Scolnick, A. Oliff, and J. B. Gibbs.** 1993. Selective inhibition of ras-dependent transformation by a farnesyltransferase inhibitor. *Science* **260:**1934–1937.

37. **Kohl, N. E., C. A. Omer, M. W. Conner, N. J. Anthony, J. P. Davide, S. J. Desolms, E. Giuliani, R. P. Gomez, S. L. Graham, K. Hamilton, L. K. Handt, G. D. Hartman, K. S. Koblan, A. M. Kral, P. J. Miller, S. D. Mosser, T. J. O'Neill, E. Rands, M. D. Schaber, J. B. Gibbs, and A. Oliff.** 1995. Inhibition of farnesyl-transferase induces regression of mammary and salivary carcinomas in ras transgenic mice. *Nature Med.* **1:**792–797.

38. **Kohl, N. E., F. R. Wilson, S. D. Mosser, E. Giuliani, S. J. DeSolms, M. W. Conner, N. J. Anthony, W. J. Holtz, R. P. Gomez, T. J. Lee, R. L. Smith, S. L. Graham, G. D. Hartman, J. B. Gibbs, and A. Oliff.** 1994. Farnesyltransferase inhibitors block the growth of *ras*-dependent tumors in nude mice. *Proc. Natl. Acad. Sci. USA* **91:**9141–9145.

39. **Kolb, A. J., and K. Neumann.** 1996. Luciferase measurements in high throughput screening. *J. Biomol. Screening* **1:**85–88.

40. **Krug, R. M., F. V. Alonso-Caplan, I. Julkunen, and M. G. Katze.** 1989. Expression and replication of the influenza virus genome, p. 89–152. *In* R. M. Krug (ed), *The Influenza Viruses*. Plenum Press, New York.

41. **Lam, T. Y. K., O Hensens, G. Helms, D. Williams, Jr., M. Nallin, J. Smith, S. Gartner, L. Herranz-Rodriguez, and S. Stevens-Miles.** 1995. L-755,805, A new polyketide endothelin binding inhibitor from an actinomycete. *Tetrahedron Lett.* **36:**2013–2016.

42. **Lam, T. Y. K., D. L. Williams, J. M. Sigmund, M. Sanchez, O. Genilloud, Y. L. Kong, and S. Stevens-Miles.** 1992. Cochinmicins, novel and potent cyclodepsipeptide endothelin antagonists from a *Microbispora* sp. I. Production, isolation and characterization. *J. Antibiot.* **45:**1709–1716.

43. **Lerner, E. C., Y. Qian, A. D. Hamilton, and S. M. Sebti.** 1995. Disruption of oncogenic K-Ras4B processing and signaling by a potent geranylgeranyl-transferase I inhibitor. *J. Biol. Chem.* **270:**26770–26773.

44. **Lingham, R. B., A. Hsu, K. C. Silverman, G. F. Bills, A. Dombrowski, M. E. Goldman, P. L. Darke, L. Huang, G. Koch, J. Ondeyka, and M. A. Goetz.** 1992. L-696,474, a novel cytochalasin as an inhibitor of HIV-1 protease. *J. Antibiot.* **45:**686–691.

45. **Lingham, R. B., K. C. Silverman, G. F. Bills, C. Cascales, M. Sanchez, R. G. Jenkins, S. E. Gartner, M. T. Diez, F. Pelaez, S. Mochales, Y. L. Kong, R. W. Burg, M. S. Meinz, L. Huang, M. Nallin-Omstead, S. D. Mosser, M. D. Schaber, C. A. Omer, D. L. Pompliano, J. B. Gibbs, and S. B. Singh.** 1993. *Chaetomella acutiseta* produces chaetomellic acids A and B which are reversible inhibitors of farnesyl-protein transferase. *Appl. Microbiol. Biotechnol.* **40:**370–374.

46. **Mathis, G., F. Socquet, M. Viguier, B. Darbouret, and J. Ejpj.** 1993. Amplified homogeneous time resolved immunofluorometric assay of prolactin. *Clin. Chem.* **39:** 1251. (Abstract.)

47. **Merrifield, R. B.** 1963. Solid phase peptide synthesis. I. The synthesis of tetrapeptide. *J. Am. Chem. Soc.* **85:** 2149–2154.

48. **Mynderse, J. S., L. W. Crandall, D. C. Duckworth, S. M. Lawrence, J. W. Martin, and L. E. Sachs.** 1996. Natural products novelty determination—the basics, p. 41. Proceedings of the 37th Ann. Meet. American Society of Pharmacognosy. University of California, Santa Cruz.

49. **Omer, C. A., A. M. Kral, R. E. Diehl, G. C. Prendergast, S. Power, C. M. Allen, J. B. Gibbs, and N. E. Kohl.** 1993. Characterization of recombinant human farnesyl-protein transferase: cloning, expression, farnesyl diphosphate binding and functional homology with yeast prenyl-protein transferases. *Biochemistry* **32:**8341–8347.

50. **Omura, S.** 1992. Trends in the search for bioactive microbial metabolites. *J. Ind. Microbiol.* **10:**135–136.

51. **Plotch, S. J., M. Bouloy, I. Ulmanen, and R. M. Krug.** 1981. A unique cap (m^7GpppXm)-dependent influenza virion endonuclease cleaves capped RNAs to generate the primers that initiate viral RNA transcription. *Cell* **23:** 847–858.

52. **Polulter, C. D., and H. C. Rilling.** 1981. Conversion of farnesyl pyrophosphate to squalene, p. 413–441. *In* J. W. Porter and S. L. Spurgeon (ed.), *Biosynthesis of Isoprenoid Compounds*, vol. 1. John Wiley & Sons, Inc., New York.

53. **Porter, N., and F. M. Fox.** 1993. Diversity of microbial products—discovery and application. *Pestic. Sci.* **39:**161–168.

54. **Rodwell, V. W., J. L. Nordstrom, and J. J. Mitschellen.** 1976. Regulation of HMG-CoA reductase. *Adv. Lipid. Res.* **14:**1–74.

55. **Sasiak, K., and H. C. Rilling.** 1988. Purification to homogeneity and some properties of squalene synthetase. *Arch. Biochem. Biophys.* **260:**622–627.

56. **Schaber, M. D., M. B. O'Hara, V. M. Garsky, S. D. Mosser, J. D. Bergstrom, S. L. Moore, M. S. Marshall, P. A. Frieman, R. Dixon, and J. B. Gibbs.** 1990. Polyisoprenylation of Ras *in vitro* by a farnesyl-protein transferase. *J. Biol. Chem.* **265:**14701–14704.

57. **Singh, S. B., M. A. Goetz, E. T. Jones, G. F. Bills, R. A. Giacobbe, L. Herranz, S. Stevens-Miles, and D. L. Williams, Jr.** 1995. Osteromycin: a novel antagonist of endothelin receptor. *J. Org. Chem.* **60:**7040–7042.

58. **Stahlhut, M. W., and D. B. Olsen.** 1996. Expression and purification of retroviral HIV-1 reverse transcriptase. *Methods Enzymol.* **275:**122–133.

59. **Stamler, J.** 1978. Dietary and serum lipids in the multifactorial etiology of atherosclerosis. *Arch. Surg.* **113:**21–25.

60. **Stevens-Miles, S., M. A. Goetz, G. F. Bills, R. A. Giacobbe, J. S. Tkacz, R. S. L. Chang, M. Mojena, I. Martin, M. T. Diez, F. Pelaez, O. Hensens, T. Jones, R. W. Burg, Y. L. Kong, and L. Huang.** 1996. Discovery of an angiotensin II binding inhibitor from a *Cytospora* sp. using a semi-automated screening procedures. *J. Antibiot.* **49:**119–123.

61. **Tamanoi, F.** 1993. Inhibitors of Ras farnesyl-transferases. *Trends Biochem. Sci.* **18:**349–353.

62. **Tomassini, J., H. Selnick, M. E. Davies, M. E. Armstrong, J. Baldwin, M. Bourgeois, J. Hastings, D. Hazuda, J. Lewis, W. McClements, G. Ponticello, E. Radzilowski, G. Smith, A. Tebben, and A. Wolfe.** 1994. Inhibition of cap (m^7GpppXm)-dependent endonuclease of influenza virus by 4-substituted 2,4-dioxybutanoic acid compounds. *Antimicrob. Agents Chemother.* **38:**2827–2837.

63. **Vacca, J. P., B. D. Dorsey, W. A. Schleif, R. B. Levin, S. L. McDaniel, P. L. Darke, J. Zugay, J. C. Quintero, O. M. Blahy, Roth, E., V. V. Sardana, A. J. Schlabach, P. I. Graham, J. H. Condra, L. Gotlib, M. K. Holloway, J. Lin, I. W. Chen, K. Vastag, D. Ostovic, P. S. Anderson, E. A. Emini, and J. R. Huff.** 1994. L-735,524: an orally bioavailable human immunodeficiency virus type 1 protease inhibitor. *Proc. Natl. Acad. Sci. USA* **91:**4096–4100.

64. **Wei, X., S. K. Ghosh, M. E. Taylor, V. A. Johnson, E. A. Emini, P. Deutsch, J. D. Lifson, S. Bonhoeffer, M. A. Nowak, B. H. Hahn, M. S. Sang, and G. M. Shaw.** 1995. Viral dynamics in human immunodeficiency virus type 1 infection. *Nature* **373:**117–122.

65. **Zhang, F. L., and P. J. Casey.** 1996. Protein prenylation: molecular mechanisms and functional consequences. *Annu. Rev. Biochem.* **65:**241–269.

66. **Zlokarnik, G., P. A. Negulesar, T. E. Knapp, L. Mese, N. Burnes, L. Feng, M. Whitney, K. Roemer, and R. Y. Tsien.** 1998. Quantitation of transcription and clonal selection of single living cells with β-lactamase as reported. *Science* **279:**84–88.

Culture Preservation and Inoculum Development

R. L. MONAGHAN, M. M. GAGLIARDI, AND S. L. STREICHER

3

3.1. CULTURE PRESERVATION

Streptomyces aureofaciens NRRL 2209 was a groundbreaking culture in the history of microbiology in the United States. It was the first microorganism deposited in a culture collection in support of a microbially based patent application. In accepting the deposit in 1949, the receiving collection agreed to maintain the deposit for at least the lifetime of the patent (113). Similarly, the depositor was required to maintain a viable, productive culture for the period of the patent. From that point, preservation of microbial cultures was critical for all individuals and firms engaged in the search for patentable products from and patentable processes by microorganisms. Preservation of cultures by freezing, drying, or a combination of the two processes is highly influenced by resistance of the culture to the damage caused by rapid freezing, the dehydrating effects of slow freezing, or damage caused during recovery. To minimize damage caused by these processes, agents have been used that protect against ice formation by causing the formation of glasses upon cooling. Methods to protect against the negative effects of dehydration include adaptation to lower effective water activity by preincubation in high-osmotic-pressure solutions. Damage caused by thawing after freezing can be minimized by rapid melting and by the composition of the medium used for growth after preservation.

There are various preservation methods which accommodate the fastidiousness of the specific organism and also the needs of the scientific investigator. We have outlined several procedures currently in use. To date, preservation in liquid nitrogen is still the most successful long-term method.

3.1.1. Serial Transfer

Based upon its ease of use, serial transfer is often the first "preservation" technique used by microbiologists. The disadvantages of relying upon this method for culture maintenance include contamination, loss of genetic and phenotypic characteristics, high labor costs, and loss of productivity.

Experience with "control" stocks of the cultivated mushroom *Agaricus bisporus* by the U.S. Department of Agriculture, Beltsville, Md., has led them to conclude that maintenance of a culture with good production is unlikely to be possible for more than 2 years when uninterrupted serial passage is used. Morphological variation or sectoring and the need to rejuvenate or reisolate the culture to maintain production have been commonly observed (123). Subculturing algae on a regular schedule is the most common method of maintaining algae in culture collections. For many algae, serial transfer is the only proven preservation technique (1).

3.1.2. Preservation in Distilled Water

Over 50 years ago, Castellani introduced the concept that fungi could be preserved in distilled water (27). This preservation method was extensively tested on 594 fungal strains, with 62% of the strains growing and maintaining their original morphology (58). In another study, 76% of yeasts, filamentous fungi, and actinomycetes survived storage in distilled water for 10 years (139). A recently completed test in which this procedure was used to preserve strains of the pathogen *Sporothrix schencki* concluded that even though long-term survival was good when this procedure was used, there was a noted loss in virulence (18). Since long-term loss of culture characteristics is not an unusual outcome, whatever the preservation method used, the Castellani technique certainly should be considered as one of the options for practical storage of fungal isolates.

PROTOCOL FOR PRESERVATION IN DISTILLED WATER (66)

From the margins of well-grown fungal cultures, plugs are cut with sterile polypropylene transfer tubes and transferred to cryovials (Nalgene Labware) containing approximately 2 ml of sterile distilled water. Alternatively, Wheaton 4-ml glass vials with rubber lined caps may be used. The vials are stored at room temperature.

3.1.3. Preservation under Oil

One of the earlier preservation methods was the use of mineral oil to prolong the utility of stock cultures. Mineral oil has been found to prevent evaporation from the culture and to decrease the metabolic rate of the culture by limiting the supply of oxygen. This method is more suitable than lyophilization for the preservation of nonsporulating strains.

PROTOCOL FOR PRESERVATION WITH MINERAL OIL (24)

The culture to be conserved is grown on a slant of suitable nutrient agar until a young, vigorous colony has developed.

Heavy mineral oil (specific gravity 0.8 to 0.9) is autoclaved for 45 min at 15 lb/in^2 pressure in half-filled 250-ml cotton-plugged Erlenmeyer flasks or at 150°C for 90 min in an oven. After sterilization and cooling, the mineral oil is poured over the colony to a depth of about 1 cm above the tip of the slant. The culture is then stored upright at about 10°C.

3.1.4. Lyophilization

One of the best methods for long-term culture preservation of many microorganisms is freeze-drying (lyophilization). The most commonly used cryoprotective agents are skim milk (15% [wt/vol] for cultures grown on agar slants and 20% for pelleted broth cultures) or sucrose (12% [wt/vol] final concentration). It should be noted that some plasmid-containing bacteria are successfully preserved by this method.

PROTOCOL FOR FREEZE-DRYING (LYOPHILIZATION) (50)

A VirTis Unitop 400 and Freezemobile 12 affords excellent freeze-drying results without the need for solvent freezing before beginning the drying process. Borosilicate ampoules (5 by 90 mm), prescored 40 and 55 mm from the bottom (facilitates sealing off), from Bellco Glass, Inc., are loosely plugged with absorbent cotton and oven sterilized. Sterile skim milk (15 to 20%; 1 ml) is added to a well-grown agar slant. When a number of slants are to be used, the contents are pooled and mixed before the ampoules are filled (approximately 0.2 ml). The cotton plugs are replaced, and the filled vials are kept refrigerated until all the cultures have been dispensed. The cotton plug tops are burned off, the ampoule lips are wiped clean of carbon residue, and the remainder of the plug is slightly recessed (approximately 2 mm) into the vial. The vials are loaded horizontally onto shelves at −43°C and allowed to freeze completely. Once the vacuum is down to less than 100 millitorrs, the temperature is raised to −1°C and the drying process continues until the samples have fully dried (usually overnight). The vials are removed and held in a lidded metal container with Drierite (Fisher Scientific) until sealed to prevent condensation. The vials are connected to a manifold hooked up to a vacuum pump. Sealing is done immediately with an oxygen/gas torch, and vials are stored at 5°C. Alternatively, vials may be backfilled with nitrogen (51a) and sealed with an oxygen/gas torch. The vials are then stored at −70°C.

Residual moisture after lyophilization has been reported as a problem for effective preservation (126). A moisture indicator consisting of filter paper strips impregnated with 3% $CoCl_2$ solution that had been dried in a stream of air was used by Sourek (134) during a 30-min heat test to measure if dehydration was complete. A color change from blue indicates unsatisfactory levels of moisture above 2%.

3.1.5. Storage over Silica Gel

Neurospora has successfully been preserved over silica gel.

PROTOCOL FOR DRYING ON SILICA GEL (109, 110)

Screw-cap tubes (13 by 100 mm), half-filled with desiccant-activated silica gel (6 to 12 mesh, grade 40. Davison Chemical Corp.) are oven sterilized. After the tubes have cooled, a skim milk (10% vol/vol) suspension of conidia or mycelium is dispensed (0.5 ml) into each tube. The tubes are quickly cooled to reduce heat generated as the liquid is absorbed and then vortexed to break up clumps. After being dried at 25°C, they are stored in closed containers with desiccants.

3.1.6. Preservation on Paper

Spore-forming fungi, actinomycetes, and unicellular bacteria can be preserved by drying the spores on some inert substrates. Fruiting bodies of the myxobacteria, containing myxospores, may be preserved on pieces of sterile filter paper and stored at room temperature or at 6°C for 5 to 15 years (116).

PROTOCOL FOR DRYING ON FILTER PAPER (116)

Pieces of agar containing fruiting bodies are placed on sterile filter paper in a petri dish, dried in a desiccator under vacuum, and stored at room temperature. Alternatively, vegetative cells are transferred from the growth medium to small pieces of sterile filter paper (or small filter discs) on water agar and incubated until fruiting bodies develop. After the fruiting bodies develop, they are allowed to mature for about 8 days. The filter papers are then placed into sterile containers, such as screw-cap tubes, and dried over silica gel in an evacuated desiccator. After a few days, the containers are tightly closed and stored.

3.1.7. Preservation on Beads

The method involving preservation on beads, developed by Lederberg, is successful for many bacteria (63).

PROTOCOL FOR DRYING ON PORCELAIN BEADS (63)

Porcelian beads (10 to 12 "Fishspine" beads no. 2 [Taylor, Tunnicliffe and Co.]) are autoclaved in screw-cap glass vials (10 ml). Cell suspensions are prepared from 24 to 48 h culture slants with a 20% (wt/vol) sucrose solution. The sterile beads are transferred to a sterile petri dish and inoculated (0.2 to 0.3 ml per bead) with the cell suspension. The beads are returned to the vial with sterile forceps, and the vial is loosely capped and dried in a vacuum desiccator for 72 to 96 h. A sterile spatula may be used to break the beads apart. Storage is at 25°C in a closed metal cabinet containing Drierite.

3.1.8. Preservation on Soil

For many soil-borne species, survival in soil is necessary. Survival for up to 20 years in soil has been reported for *Pythium*, *Fusarium*, and *Verticillium* spp. Survival for >1 year in soil is relatively common for other plant pathogens (103). It is not surprising, therefore, that storage in soil would be an effective preservation method for these and other soil-inhabiting microorganisms. Two soil preservation techniques are outlined below.

PROTOCOL FOR DRYING ON SOIL (53)

Fungal cultures are grown on agar medium. A spore suspension is prepared and pipetted into tubes of dry, sterile soil (1 ml of suspension per 5 g of soil) and allowed to dry at 25°C. The tubes are stored at room temperature.

PROTOCOL FOR PRESERVING ACTINOMYCETES ON SOIL (79)

To 100 g of potting soil measured into a 250-ml Ehrlenmeyer flask (twice sterilized for 1 h, and twice incubated at 37°C overnight) is added 0.25% dry blood made up of blood fibrin and hemoglobin (10:1) plus 1% $CaCO_3$. The mixture is autoclaved for 1 h, and 20 ml of distilled H_2O is added. The flasks are inoculated with a water suspension of the culture (2 ml) and incubated. When mycelial growth is evident, the flasks are covered with Parafilm and stored at 5°C. Alternatively, screw-cap tubes (20 by 125 mm) filled to a depth of approximately 4 cm may be used.

3.1.9. Liquid Drying

To avoid the damage that freezing can cause, a liquid-drying preservation process has been developed. Liquid drying has effectively preserved organisms such as anaerobes that are damaged by or fail to survive freezing (88). This procedure was preferred over lyophilization for the maintenance of the biodegradation capacity of six gram-negative bacteria capable of degrading toluene. Malik's liquid-drying method was also found to be markedly superior to lyophilization for the preservation of unicellular algae (89).

PROTOCOL FOR LIQUID-DRYING PRESERVATION (78)

Freeze-dried disc

Small glass vials are filled with 0.1 ml of 20% (wt/vol) skim milk (Bacto; Difco no. 0032) containing 1.0% (wt/vol) neutral activated charcoal (medical grade) and 5% (wt/vol) *myo*-inositol (87). The vials are sterilized at 115°C for 13 min, frozen at about −40°C for a few hours, and then freeze-dried for 8 to 24 h by a standard freeze-drying technique. This results in a disc of freeze-dried carrier material.

Liquid-drying procedure

A solution of *myo*-inositol (5% [wt/vol]) and neutral activated charcoal (1.0 [wt/vol]) is prepared in distilled water and autoclaved at 115°C for 13 min (87). About 30 μl of cell suspension in this *myo*-inositol charcoal solution is added to each vial with a freeze-dried disc of carrier material, and the vials are subjected to liquid drying (the cell suspension is dehydrated in liquid under vacuum in a metallic jar maintained at 20°C). First-step drying is continued for about 2 h at 1.5 to 4 kPa at 20°C and second-step drying is continued for about 1 h at 0.01 to 0.001 kPa while maintaining the temperature at about 20°C. Then vacuum is replaced by sterile nitrogen or argon gas. The ampoules are transferred to soft glass tubes and sealed under vacuum.

3.2. CRYOPRESERVATION

Microorganisms may be preserved at −5 to −20°C for 1 to 2 years by freezing broth cultures or cell suspensions in suitable vials. Deep freezing of microorganisms requires a cryoprotectant such as glycerol or dimethyl sulfoxide (DMSO) when stored at −70°C or in the liquid nitrogen at −156 to −196°C. Broth cultures taken in the mid-logarithmic to late logarithmic growth phase are mixed with an equal volume of 10 to 20% (vol/vol) glycerol or 5 to 10% (vol/vol) DMSO. Alternatively, a 10% glycerol–sterile broth suspension of growth from agar slants may be prepared. In either case, the suspensions are pipetted into cryogenic vials or ampoules, frozen and stored.

3.2.1. Frozen Agar Plugs

Actinomycetes and fungi may also be preserved as agar plugs under 10% glycerol for short-term to intermediate term preservation.

PROTOCOL FOR PRESERVATION AS A FROZEN AGAR PLUG (65)

Cultures are inoculated onto solid media suitable for producing confluent growth. A 10 or 15% (wt/wt) solution of glycerol is prepared, and 2 ml is dispensed into 4-ml-capacity Wheaton borosilicate glass vials (no. 224882) fitted with rubber-lined caps. The vials are autoclaved twice for 45 min each. Transfer tubes (Spectrum no. 190195) are autoclaved for 25 min. Several plugs are cut and deposited into each of the vials, which are then frozen and stored at −70°C.

3.2.2. Preservation in Liquid Nitrogen

Storage in liquid nitrogen is clearly the preferred method for preservation of culture viability. Challen and Elliott (28) have determined that storage in sealed polypropylene straw ampoules is a space-efficient method for preservation and storage in liquid nitrogen.

PROTOCOL FOR PRESERVATION AND STORAGE IN LIQUID NITROGEN (28)

Polypropylene drinking straws are used to make ampoules. Straws of 4 mm in diameter are routinely used and are prepared by being sealed along their length with a conventional polythene heat welder (e.g., HM-710P model supplied by Hulme-Martin Ltd., London, England). A ruler backed with double-sided Scotch tape is used to present several straws for sealing in one operation. Five 40-mm-long ampoules are produced from a single 200-mm-long straw. Cryoprotectant (a 10% [vol/vol] aqueous solution of glycerol or DMSO) is placed into the ampoules within 10 mm of the top. To avoid the formation of air locks, the straws are filled from the bottom via a long hypodermic needle. To store fungi, agar plugs are removed from the periphery of an actively growing colony via a tungsten wire and introduced into the straw. Alternatively, for the preservation of spore-bearing fungi or unicellular microorganisms, a loop full of spores or cells can be simply suspended in the cryoprotectant. The ampoules are sealed so that an air space remains at the top of each one. This prevents cracking of the straw due to the expansion of the contents during freezing. Ampoules are tested for leaks by simply pressing the straws against a hard surface. Ten replicate ampoules of each strain are routinely prepared, nine of which fit into space that would normally be occupied by one standard ampoule. The cultures are frozen (over liquid nitrogen) in the evolving nitrogen vapor. After being precooled, the samples are completely immersed in the nitrogen. Survival is assessed after storage. One straw of each isolate to be tested is removed and thawed by direct immersion in a water bath at 30°C. Washing of the straw with alcohol before plating out of the culture eliminates contamination on retrieval. The top of the straw is cut off with flamed scissors, and its contents are squeezed onto an agar plate by drawing flamed forceps along the length of the straw.

PROTOCOL FOR FREEZING MICROORGANISMS IN LIQUID NITROGEN (127)

Preparation of cultures grown on slants

A volume (3.0–6.0 ml) of suitable broth containing 10% glycerol is used to suspend growth from agar slants to achieve a dense suspension of spores and mycellium. Screw-cap glass vials with rubber caps. (Wheaton Scientific) are prepared, and the suspension is dispensed. The screw caps are fully tightened. Before being frozen in liquid nitrogen, all the vials are refrigerated for 30 min.

Preparation from submerged cultures
Sterile cryoprotectants are added to well-dispersed broth cultures. For pelleted cultures, sterile porcelain balls may be used to achieve a homogeneous broth. To the fully grown broth culture, five previously sterilized 13-mm-diameter porcelain balls are added, and the culture is shaken at 220 rpm for 30 to 60 min until macerated and pipettable. Tissue grinders are useful for some actinomycete strains. Suspensions are dispensed into plastic ampoules or glass vials, refrigerated and frozen.

Controlled-rate freezing
Ampoules are placed in containers within the freezing chamber of a controlled-rate freezer (such as Cryo-Med). A cooling rate of 1 to 2°C/min is maintained until the temperature is slightly above that for the phase change. The phase change should occur as rapidly as possible. The cooling rate is adjusted to 1°C/min until at least −50° is reached. The ampoules are then transferred to liquid nitrogen storage.

Thawing
It is best to thaw samples rapidly at 37 to 40°C. To prevent contamination, a sterile gauze pad soaked in 70% ethanol is used to wipe the vials before the culture is transferred to 2 ml of broth. Before inoculating, the broth mixture must be allowed to equilibrate.

3.2.3. Two-Stage Freezing Process

Cryopreservation with cryoprotectants, using a two-stage freezing process of slow cooling to −30°C during which dehydration occurs followed by rapid cooling to a storage temperature ideally below −120°C, results in the cell content solidifying without ice formation (vitrification-glass formation). Rapid thawing is used to minimize cell damage. This procedure has effectively preserved a bacterial inoculum, *Alcaligenes eutrophus*, for direct use in a microbial biosensor without the need for an intermediate growth stage to recover viability and activity (13).

PROTOCOL FOR CRYOPRESERVATION WITH CRYOPROTECTANTS BY A TWO-STAGE FREEZING PROCESS, AND REVIVAL OF CULTURES (13)

After centrifugation the supernatant is removed and the pellet, consisting of *Alcaligenes eutrophus* JMP 134 cells, is dissolved in an ice-cold solution containing polyvinyl ethanol (10% [wt/vol]) and glycerol (10% [wt/vol]) in a 1:1 ratio. Due to the presence of polyvinyl ethanol, a viscous thick cell suspension (10^8 to 10^{10} cells/ml) is obtained, which is kept for about 30 minutes in an ice bath for equilibration. During equilibration, an aliquot of 0.5 to 1.0 ml of the cell suspension is dispensed into each plastic cryovial or glass ampoule. Immediately thereafter, they are tightly closed, clamped onto labeled aluminum canes, and placed at −30°C for about 1 h or for a few minutes in the gas phase of liquid nitrogen to achieve a freezing rate of about 1°C/min. The canes are then placed into canisters, racks, or drawers and frozen rapidly at −80°C or in liquid nitrogen.

For revival of cultures, the frozen ampoules are removed from the liquid nitrogen. For thawing, they are immediately immersed to the neck in a water bath at 37°C for a few seconds. The thawed cell contents of the ampoule or vial are immediately transferred to membranes to form a thick layer. The resulting bacterial membranes with immobilized cells are used as a biological component of a biosensor for activity measurements.

3.2.4. Culture Preparation before Freezing

Sourek (134) considered the suspending medium to be the most important factor in preserving the viability and biological activity of lyophilized pathogenic bacteria. A universal suspending medium has not been found, but the following suspension medium has often been found to be effective: after sterilization of calf serum at 100°C (running steam, 20 min, three consecutive days), a 10% lactose solution in distilled water is diluted with an equal volume of the sterilized calf serum (134).

In an extensive study of suspending media before lyophilization of fungi, Berny and Hennebert (10) found optimal recovery of viable cells when the suspending medium contained combinations of supplements. A supplement of 10 or 20% skim milk plus 10% trehalose and 5% glutamate, combined with an optimum cooling rate of 3°C/min, resulted in >50% viability recovery for all four cultures tested. A skim milk supplement (10 or 20%) at the same cooling rate had one total failure and only one of four cultures with recovered viability of >50%.

PROTOCOL FOR HEAT SHOCK TREATMENT (70)

Heat shock treatment of a microbial culture will induce the production of heat shock proteins (HSP). *Saccharomyces cerevisiae* cells subjected to a heat shock or cold shock were markedly more able to survive a freezing-and-thawing cycle than were untreated cells (10, 70, 71). Exponentially growing cells are pelleted and resuspended at a concentration of 10^8 cells/ml in test tubes containing prewarmed (43°C) YM medium. The cells are incubated at 43°C for 2 h.

PROTOCOL FOR COLD SHOCK TREATMENT (71)

For cold shock treatment before freezing, the cells are incubated at 10°C for 2 h, washed three times with phosphate buffer (pH 7.4), and frozen immediately.

3.2.5. Cryopreservation Supplements

The value of cryoprotectants in general preservation of strains has been demonstrated for so many cultures that inclusion of a cryoprotectant as part of a standard cryopreservation protocol is a general rule. Preservation of 122 strains of 43 species of basidiomycetes by ultra-low-temperature freezing resulted in 100% recovery after slow freezing over DMSO, 97% recovery after fast freezing over DMSO, 97% recovery after slow freezing over glycerol, and 81% recovery after fast freezing over glycerol (31).

For *Agaricus bisporus*, data support the conclusion that 10% (vol/vol) glycerol, 10% (vol/vol) glycerol presoak, and 10% (vol/vol) DMSO are comparable in their ability to retain mushroom productivity (123). Other commonly used cryoprotectants include methanol, sucrose, mannitol, trehalose, polyvinyl ethanol, ethylene glycol, and hydroxyethyl starch (35, 117).

3.2.6. Preservation of Other Biological Material

Extensive research into the cryopreservation of erythrocytes and organs has led to the selection of supplements that form glasses upon cooling in aqueous solution. Studies on the optimal levels of supplements for glass-forming ability during cooling, stability during warming and reduced cell toxicity are under way (7, 19). It may be worthwhile investigating 26% (wt/wt) 2,3-butanediol–4% (wt/wt) trehalose or 30% (wt/wt) 1,2-propanediol–4% sucrose as supplements for cryopreservation of microbial cells.

Antifreeze glycoproteins from sea cod at 20 mg/ml, when used with DMSO, were two to four times better at preserving mouse embryos after cryopreservation than was DMSO alone. The glycoproteins alone were ineffective (69).

Oxidative damage to cells and membranes appears to be important in cryopreservation. *S*-Adenosylmethionine has been reported to be a useful additive for heptocyte cryopreservation solutions (141).

Preservation of blood has involved three techniques that may have application for the preservation of microbial cells. They are rapid freezing in liquid nitrogen in 17.5% glycerol, slow cooling to −80°C in 40 to 50% glycerol, and rapid freezing with hydroxyethyl starch at 11.5% (wt/wt) (120). Preincubation in mannitol or sucrose before freezing has been used to minimize damage caused by dehydration (91, 120).

3.2.7. Retention of Characteristics—Comparison of Methods

Retention of viability is necessary but not sufficient for maintenance of the characteristics for which a culture is preserved. Under some conditions, the preparation used before preservation aids in the retention of the desired characteristic. Pigment production by *Serratia* spp. was retained when only one of the prelyophilization suspending media was used, even though other suspending media were more effective at protecting viability (134).

Rennet-producing strains of *Rhizomucor miehei* preserved by lyophilization, by freezing at −17°C, by storage under mineral oil, and by sequential transfer on agar slants were compared, and storage under oil was found to be the preferred preservation method (15).

Cryogenic storage is the preferred method of storage for fungi that are not effectively preserved by lyophilization, such as members of the *Entomophthorales*. However, Humber (62) notes that even if vigorous and sporulating when initially frozen, some members of this group will grow poorly and sporulate little, if at all, after storage for a few years. Hajek et al. (57) found that three isolates of *Entomophaga maimaiga* stored under liquid nitrogen retained the ability to kill gypsy moth larvae but that their ability to produce conidia and azygospores degenerated.

Preservation of yeasts by lyophilization has been problematic. Preservation of viability and function of a cyanide (CN)-resistant strain of *Rhodotorula rubra* was tested. Storage in distilled water at room temperature and freezing at −20°C with 10% glycerol resulted in the loss of viability and CN resistance. Freezing at −70°C with 10% glycerol, preservation under liquid nitrogen with 10% glycerol, and drying on paper all resulted in good viability and CN resistance being retained for 12 months (82).

The volumetric production and yield of butanol by *Clostridium acetobutylicium* decreased after all short-term (9 month) culture storage methods used. Volumetric production decreased from a control level of 8.5 g/liter to 8.1 g/liter after freezing at −20°C in glycerol, to 7.0 g/liter after storage at 4°C in water; to 6.5 g/liter after freezing at −20°C in phosphate buffer; to 5.5 g/liter after storage at 4°C in phosphate buffer, to 5.2 g/liter after lyophilization, and to 2.0 g/liter after storage in soil. Storage in a cooked-meat medium at 4°C gave the poorest volumetric productivity at 1.3 g/liter. Volumetric productivity was independent of viability (55).

A study comparing antibiotic production from eight microorganisms found evidence of loss of production by all the preservation methods tested. The overall recovery of antibiotic production was best when the cultures were preserved as frozen slants followed in decreasing order by frozen vegetative cells, drying of soil, silica gel drying, lyophilization, and drying on paper (97).

Fusarium merismoides organisms preserved under liquid nitrogen and preserved by storage in soil were grown and compared for evidence of altered DNA. DNA prepared from the liquid nitrogen-stored culture was more susceptible to the restriction enzymes *BclI* and *AvaI* than was the culture stored in soil (47). It is not clear if these measured differences in DNA susceptibility to restriction enzyme digestion has any phenotypic consequence, but perhaps each method might serve as an alternate preservation protocol until it is determined which is more effective at preserving the phenotypic characteristic(s) of interest. Clearly, there is no universally effective preservation method.

3.3. INOCULUM DEVELOPMENT

The handling of a microbial culture, from the time it is transferred from its preserved state until it is inoculated into the final location where the microbial activity of interest is expressed, is referred to as inoculum development. The primary purpose of inoculum development is to provide microbial mass, of predictable phenotype, at a specific time, and at a reasonable cost for the productive stage of a microbial activity. Until now, inoculum development has been more art than science. There remains a need, especially at the shake flask or spore-generating stages of the process, for time and "it looks good" criteria to be replaced with biochemical, physiological, or morphological markers as both descriptors of an optimum inoculum and indicators for optimum timing of inoculum transfer.

3.3.1. Inoculum Source

Inoculum preparations for standardized antimicrobial tests have recently been reviewed (2, 16, 21). The necessity for standardization of conditions used for the preparation and harvest of inocula has been emphasized, since analysis of error pointed to variability of the test inoculum as a major factor in the lack of test reproducibility. After various inoculum regimens were tested, the original source of culture from culture collections was implicated as a significant source of variability (17).

When fungal spores are used as the inoculum source, it is common for conidia produced on an agar slant to be dispersed in sterile distilled water containing 0.01 to 0.1% Tween 80.

Spore formation of *Streptomyces coelicolor* on agar was dependent upon the type of agar used, the inclusion of trace elements, the nitrogen source, and a C/N ratio between 40 and 100 (68). Nabais and da Fonseca (102) have optimized a medium for sporulation by *Streptomyces clavuligerus* (Table 1). Spore storage, however, could be a problem, since the spores lost 72% of their viability after storage for 1 week in buffer at 4°C. Many strains isolated from

TABLE 1 Spore production medium for *Streptomyces clavuligerus*[a]

Component[b]	Amt (%)
Dextrin	1.0
Yeast extract	0.1
Meat extract	0.1
Bacteriological peptone	0.2
$CaCO_3$	0.2
Agar	2.0

[a]Data from reference 102.
[b]The pH of the medium is 7.1.

nature and often strains that have been subjected to a mutation program result in an "unstable" culture, whose productivity can be rapidly lost. For such strains, a single spore selection step or its equivalent is a necessity for maintenance of productivity.

PROTOCOL FOR SUBCULTURE INCORPORATING A SINGLE SPORE SELECTION STEP (107)

To prepare a dense suspension of spores, distilled water is used as the suspending fluid; distilled water and a wetting agent such as 0.1% Tween 80 or Triton X-100 is used if the spores are hydrophobic. The suspension is transferred to a 25-ml wide-mouth McCartney bottle containing about 16 sterile spherical glass beads (diameter, 5 mm), shaken or vortexed to break up clumps or spores, and filtered through a bed of sterile glass wool retained in a filter funnel to remove debris and mycelium. The filtrate is collected in a sterile test tube and filtered through a membrane filter. Selection of the appropriate filter depends on the dimensions of the spores; the filter should exclude all but single spores. A dilution series of the suspension in water is prepared, and 0.1-ml aliquots of the dilution series are spread onto solid agar. By plating a range of dilutions, a count of 20 to 50 colonies per plate on one dilution will be achieved. The plates are incubated, and a colony is transferred on a sterile wooden cocktail stick into a wide-mouth McCartney bottle containing glass beads and 2 ml of sterile water or 1/4-strength Ringer's solution. The bottles are shaken to give a suspension of culture, the surface of an agar slant in a tube (150 by 25 mm) is inoculated, and the tube is incubated. Several slants can be prepared from each suspension. One slant of each isolate is screened in the product assay, and the highest-yielding isolates are retained as master cultures. These masters are subcultured as required to provide batches of working cultures. All mature cultures are retained at 4°C to give a shelf life of at least 4 weeks. A "first-generation" slant of the highest-yielding isolate is retained as a parent strain for the next single-spore isolation procedure.

Large-scale fermentations which involve strains containing genetically engineered or altered plasmids present an additional problem in inoculum development, that of plasmid stability and plasmid loss. While naturally occurring plasmids are usually stable under most growth conditions and maintain a nearly constant copy number of plasmids in each cell, engineered plasmids are usually less stable. The added burden to the cell caused by the expression of the engineered gene or the synthesis of the desired protein or biosynthetic pathway may result in a decreased growth rate compared to that of a cell containing the vector or a non-plasmid-containing cell. Selective advantage is thus given to cells that maintain the vector but lack the gene(s) of importance or lack the plasmid entirely.

Small-scale growth and early inoculum development steps may include selections, usually the presence of an antibiotic whose resistance is encoded on the plasmid, to ensure the maintenance of the plasmid. Large-scale fermentations, as well as late seed train steps, must be performed in the absence of the antibiotic because of the high cost of the antibiotic supplementation and potential complications in product isolation and purification.

Plasmids that are maintained by complementation of host auxotrophic mutations are easily maintained but only in defined minimal growth media. Many yeast and filamentous fungal vectors are maintained through biosynthetic pathway complementation. However, defined media can have a very high cost, which may be prohibitive on the scale of commercial fermentations.

Plasmid stability may be addressed in several ways. Different vectors may be compared to identify one that is less prone to delete or rearrange the inserted gene of interest or to have the least deleterious effect on growth rate. Plasmid stability has been achieved in some instances by the introduction of the *parB* (*hok/sok*) locus of plasmid R1, which mediates plasmid stabilization via postsegregational killing of plasmid-free cells (49, 80, 102).

Regulated expression of the cloned gene may also be exploited to stabilize the plasmid and maintain a population of cells containing the plasmid. When such a culture reaches the appropriate cell density in a production-scale fermentor, an inducer is added. Only then will the gene of interest be expressed. While this does not address any inherent instability of the plasmid as a result of the inserted DNA itself, it usually minimizes the loss of plasmid-containing cells caused by expression-related lower growth rates.

3.3.2. Acclimatization

A number of commercial-level microbiological processes use as the inoculum, at least in part, culture growth that has been part of a previous "production phase."

For fermentation processes involved in the degradation of waste materials, a very important variable is the extent of acclimatizion of the inoculum source (25). The process lag period before initiation of biodegradation decreases with increased numbers of competent microorganisms. High degradation rates are obtained when acclimated sewage sludge operated in a plant with low retention times is used as the inoculum (140). Assessment of the ability of a waste water plant to degrade a material could greatly be underestimated if the inoculum has not been acclimatized to the tested waste material.

The use of an acclimatized inoculum has been reported to result in significant improvements in operational efficiencies for xylose conversion to xylitol by *Candida guilliermondii* grown on a sugar cane hemicellulosic hydrolysate (44).

In the brewery industry, the reuse or pitching of yeast is a common practice. The effect of serial pitching of the yeast inoculum on subsequent refermentation has not been well characterized. The condition of the yeast cell surface as measured by flocculation can be predictive of subsequent fermentation performance (145). Procedures used to measure yeast flocculation have been reviewed by Speers et al. (135).

3.3.3. Recovery from the Preserved State

The following steps have generally allowed good recovery of cultures from a preserved state. Frozen ampules, wrapped in a sterile gauze pad, are open after being submerged in disinfectant. The contents are shaken into 2 ml of a suitable recovery medium and incubated for at least 1 h before being transferred to agar slants or liquid media. Agar plugs that have been frozen are removed and streaked onto agar slants or plates. For organisms preserved in soil, a small aliquot is removed to either broth or agar plates. Cultures stored on paper are placed upside down on agar plates and shifted after 24 h to allow for better aeration of the growing culture. Rehydration of lyophilized cultures in broth rather than in saline or distilled water has been recommended (134). *Neisseria meningitidis* was able to be inoculated directly from the frozen preserved state into a defined seed medium when incubated in a 5% CO_2-in-air atmosphere (46). This minimized the growth lag and allowed the elimination of a solid-medium cell proliferation step, a step which can generally result in large variations in initial inoculum density (144).

3.3.4. Seed Media

Song et al. (133) reported on the details of their search for an optimum inoculum medium for the production of *Lentinus edodes* spawn. In a process fairly typical for seed medium selection, they initially screened a series of media for their effect on cell mass and morphology. Once the best medium was determined and optimized, it was tested for productivity with respect to the control solid substrate spawn (Table 2).

For the design of media used for the production of cell mass, the determination of an elemental material balance is a useful exercise. For defined media, the determination is a straightforward calculation from the components. For complex media, Traders' Co. (149) and other manufacturers of complex nutrients provide the basic data needed to estimate the contribution of various components to the sum of an element. Hunt and Stieber (64) have calculated a material balance for an experimental inoculum medium (Table 3). The molecular formulas for biomass produced by a number of microorganisms have been reported (26, 33, 38, 119, 142).

3.3.4.1. pH

Nutritionally balanced seed media often result in pH values not far from the optimum for culture growth. To prevent pH extremes in shake flasks, phosphate salts and $CaCO_3$ and/or buffers such as 2-(*N*-morpholino)ethanesulfonic acid (MES) or 3-(*N*-morpholino)propanesulfonic acid (MOPS) are often used. In fermentor inoculum development stages, buffers are usually replaced with the more economical online pH control.

3.3.5. Immobilization

The production of microbial inoculum for use in bioremediation, agricultural applications, and waste treatment is limited by the ability of the microorganism to compete in these environments and to be metabolically effective. One of the methods by which microbial inocula are being improved for these applications is the use of immobilization technology. The unique characteristics of immobilized inocula include (i) enhanced inoculum viability, (ii) protection from stress during manufacture, (iii) enhanced ecological competence, (iv) increased metabolite production, (v) UV resistance, (vi) the opportunity to use immobilized cells as a source of continuous inoculum, and (vii) the opportunity to introduce mixed culture inocula into a process (93). A protocol for alginate immobilization is shown in Fig. 1.

Storage of the immobilized inoculum is enhanced if cells in beads are incubated in nutrient or supplemented with nutrient when prepared (29). Immobilization of *Phanero-*

TABLE 2 Optimized synthetic medium for production of *Lentinuss edodes*[a]

Nutrient	Concn (g/liter)		
	Range tested	Max growth	Original medium
KH_2PO_4	0–5.0	2.0	1.5
$CaCl_2$	0–1.6	0.2	0.33
Glucose	0–40.0	20.0	10.0
NH_4Cl	—[c]	0.5[c]	0
$(NH_4)_2SO_4$	—[b]	0	1.0
L-Aspartic acid	—[c]	1.2[c]	0
L-Asparagine	—[b]	0	1.12
DL-Serine	—[b]	0	2.0
Thiamine HCl	$0–10^3$	10^4	10^{-3}
$MgSo_4 \cdot 7H_2O$	0–4.0	1.0	2.0
$FeSo_4 \cdot 7H_2O$	0–0.1	0.01	0.02
$ZnSo_4 \cdot 7H_2O$	0–0.1	0.02	0.02
$MnSO_4 \cdot 7H_2O$	0–0.1	0.02	0.02

[a]Adapted from reference 133.
[b]—, not tested due to relatively poor performance.
[c]Calculated on C/N ratio of 30:1 (d-day incubation at 25°C and pH 4.5).

TABLE 3 Cephamycin C fermentation elemental material balance for inoculum medium[a]

Source[b] of:	Material	Fraction of element in material	Concn of material (g/liter)	Amt of element charged (g/liter)	Potential cell mass (g/liter)
Energy	Glucose	NA[c]	10.0	NA	20
Carbon	Yeast extract	0.1660	10.0	1.66	3.3
	Primary yeast	0.1920	10.0	1.92	3.8
	Glucose	0.4000	10.0	4.00	8.0
Nitrogen	Yeast extract	0.0780	10.0	0.78	5.6
	Primary yeast	0.0770	10.0	0.77	5.5
Phosphorus	Yeast extract	0.0959	10.0	0.96	32.0
	Primary yeast	0.0127	10.0	0.127	4.0
	KH_2PO_4	0.2276	0.5	0.114	3.8
Iron	Yeast extract	0.0003	10.0	0.0028	1.4
	Primary yeast	0.0001	10.0	0.0012	0.6
	$FeSO_4 \cdot 7H_2O$	0.2009	0.1	0.0201	10.1
Potassium	Yeast extract	0.0004	10.0	0.0041	0.1
	Primary yeast	0.0224	10.0	0.215	7.2
	KH_2PO_4	0.2873	0.5	0.144	4.8
Magnesium	$MgSO_4 \cdot H_2O$	0.0990	0.25	0.025	8.3
	Yeast extract	0.0003	10.0	0.003	1.0
	Primary yeast	0.0018	10.0	0.017	5.7

[a]Adapted from reference 64.
[b]Oxygen source is calculated as follows: air, Y_{O_2} (moles of cell carbon/moles of O_2 absorbed) = 0.85; maximum oxygen uptake ($OUR_{max} = uX/Y_{O_2}$ for 8 g of cells per liter = 31 mmol/liter/h.
[c]NA, not applicable.

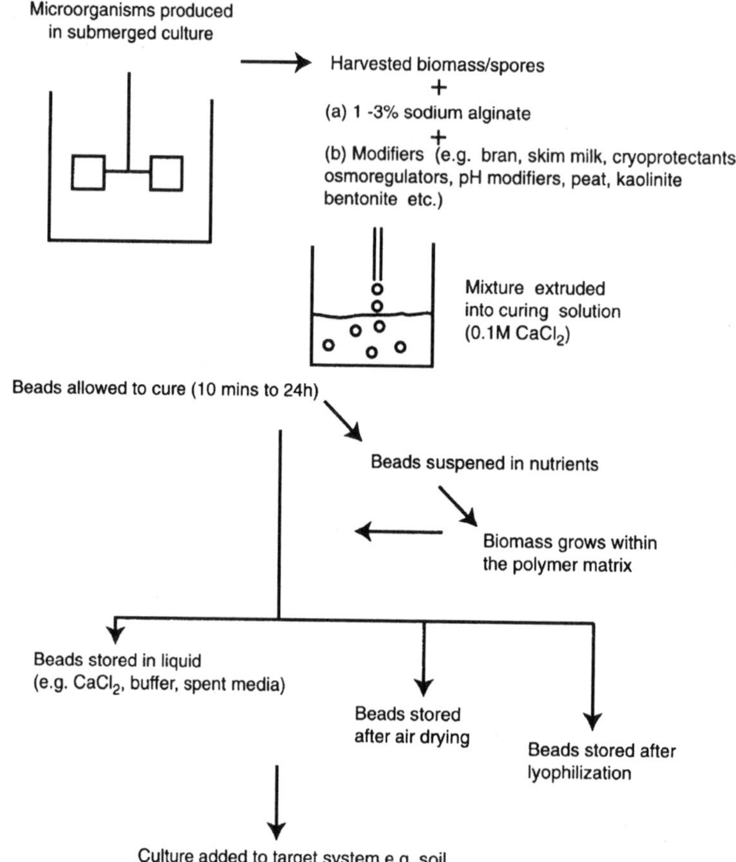

FIGURE 1 Immobilization in alginate. Reprinted from reference 93 with permission of the publisher.

chaete chrysoporium in calcium alginate for use in bioremediation applications was improved when immobilized *P. chrysosporium* in alginate was supplemented with either sawdust or corncob grits (9).

PROTOCOL FOR IMMOBILIZATION OF WHITE ROT FUNGI (9)

The entire liquid culture, both mycelial pellets and culture medium, is blended on the high setting in a Waring blender for 15 s to macerate the mycelium with a minimum of cell damage. Then 2 g of Kelgin high-viscosity sodium alginate (Kelco, San Diego, Calif.) 80 ml of fungal homogenate, 8 g of corncob grits (sieved through a 74-μm mesh), and 110 ml of deionized water are mixed with a magnetic stir bar in a beaker for 1 h. The resultant slurry is put into a separator funnel apparatus and dropped through four tubes with 3-mm orifices into 1 liter of sterile 0.25 M calcium chloride at a rate of about 90 beads/orifice/min. Calcium alginate gel beads form almost instantly upon contact with the calcium chloride solution. After 5 min in the solution, the beads are removed, rinsed with deionized water, spread on paper towels, and dried overnight in a fume hood.

Immobilization can affect cell morphology. For *Claviceps paspali*, mycelia immobilized in alginate beads upon incubation differentiated into swollen, arthrosporelike cells. Mycelial differentiation had previously been associated with alkaloid production and may in part be responsible for the multiple cycles of good alkaloid production by immobilized *C. paspali* (Fig. 2) (112).

Immobilization has effectively been used for the stabilization of genetically engineered cells. Stabilization of 22 plasmids in six genera with agarose, potassium carrageenan, calcium alginate, cotton, polyarcrylamide hydrazide, silicone foam, gelatin beads, biofilm, hollow-fiber membranes, and ultrafiltration (UF) membranes has been reported (6). One plausible explanation for the stabilization of plasmid-containing cultures is that immobilization provides a physical constraint to cell division, thus limiting growth-based competition in the immobilization matrix. To take advantage of plasmid stabilization by immobilization, a two-phase fermentation process in which the "growth" phase is performed in an immobilized matrix and cells released from the matrix are continuously transferred to a second reactor has been used (6, 11, 124). Vesicular-arbuscular mycorrhizal fungi have been proposed as plant inoculants to enhance growth, increase resistance to root pathogens, reduce the severity of disease, and minimize loss on transplanting. The key variables for successful production of inoculum are the use of expanded clay as the inoculum carrier, uncontaminated inoculum, optimum illumination of the plant host, optimum temperature for fungal growth, optimum plant nutrition, and moderate watering. The final step of the process is to stop watering the plant host to induce sporulation, followed by air drying of the substrate and storage in a cool dry place (43). Remarkably, *Glomus etunicatum* inoculum produced as described above retained 85 to 90% infectivity for 3 years when stored at 20 to 23°C and 30 to 50% relative humidity.

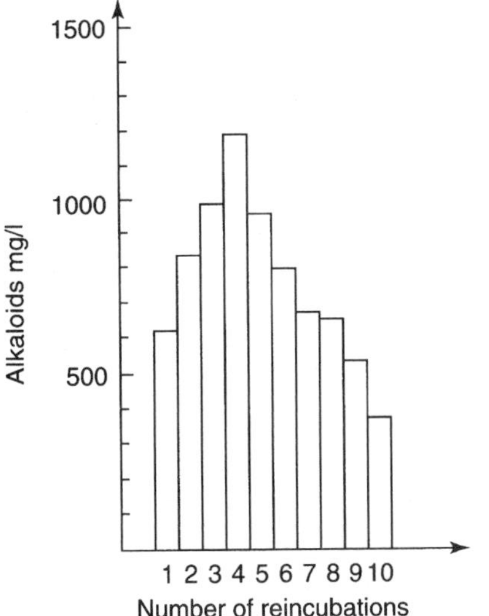

FIGURE 2 Alkaloid production by immobilized *Claviceps paspali*. Reprinted from reference 112 with permission of the publisher.

3.3.6. Sterilization

One of the more important variables in the scale-up of a fermentation process is the effect of sterilization on the productivity of the culture and on the production of foam.

To prevent the loss of important medium ingredients and to prevent the formation of caramelization products during sterilization, separate sterilization of sensitive or particularly reactive components or continuous sterilization protocols can be followed. Batch sterilization remains an important procedure for many industrial-scale fermentations. Singh et al. (129) found for batch sterilization that a series of operation parameters can be determined that will allow effective sterilization while minimizing the negative impact on productivity loss caused by caramelization.

The impact that sterilization procedures may have on an inoculum and ultimately on the productivity of a culture ideally should be determined at the shake flask level prior to pilot-scale fermentations. One problem, however, is how to correlate the data generated in a shake flask with what might occur on the fermentor scale beyond the generalization that the fermentation is or is not susceptible to damaging sterilization effects. Perhaps one can use the observation that the process of sterilization resulting in products of the Maillard reaction also results in a time- and temperature-related increase in foam formation by the resulting medium (77).

PROTOCOL FOR MEASUREMENT OF FOAMINESS (77)

Samples (200 ml) are taken to measure the foaminess (Σ) by the method of Bikerman (13a). Experiments are carried out in a glass tube 5.5 cm in diameter and with a porous glass plate (G4-fritt) 1 cm in diameter, using variable N_2 gas flow rates (0.25 to 10 ml s^{-1}). The foaminess is expressed as follows:

$$\Sigma = V_s/V_g$$

where V_s is the equilibrium volume (milliliters) of the foam above the liquid layer and V_g is the volumetric gas flow rate (milliliters per second).

3.3.7. Contamination

Microbial contaminant detection usually relies upon the use of differential media and conditions to encourage the growth of likely contaminants in the presence of the inoculated microbe. Detection of contamination in a mixed-culture fermentation provides the microbiologist with a different challenge. PCR has provided a rapid, effective technique for the detection of a contaminant present at low levels in a sample. This protocol can be applied to mixed-culture fermentations either for the detection of a particular contaminant of interest or for the detection of an indicator organism, such as the detection of *Escherichia coli* as an indicator of fecal contamination. The PCR detection of *Listeria monocytogenes* in silage by Torriani and Pallotta (138) is an example of the power of this technology, which was able to detect *L. monocytogenes* at a level of 15 CFU/g of silage.

3.3.7.1. Phages

Phage contamination is a constant threat to the productivity of any bacterial fermentation process. This problem is particularly prevalent in fermentations of dairy products. Selection of plasmids that confer phage resistance to lactic streptococci was proposed by Klaenhammer as a possible solution (73). Selection of phage-resistant strains for any bacterial process is an important piece of insurance, if multiple large-scale batches are contemplated. The report that alginate-immobilized streptococci were protected from attack by phages is potentially an interesting alternative approach (136).

3.3.7.2. Mites

Most microbiologists who work with filamentous fungi eventually are introduced to the problem of culture infestation by mites. These organisms can devastate a culture source or a series of culture sources either by eating the cultures and leaving no viable source or, more commonly, by causing marked levels of bacterial and fungal cross-contamination. Often the first indication of a problem are agar plates with bacterial or fungal tracks forming in a random-walk pattern across the plate. Smith (130) offers practical advice on how to prevent mite infestation. Treatment of incubators with acaricides on a preventative-maintenance schedule is also worth considering.

PROTOCOL FOR PREVENTION OF MITE INFESTATION (130)

Hygiene coupled with a quarantine procedure is perhaps the best protection against mite infestation. All work surfaces should be kept clean, and cultures should be protected from aerial infection by storage in cabinets. Benches should be washed down with acaricides. A "dirty" room should be available to which all incoming cultures are directed to ensure that that they are mite free. When mites are found, the infested cultures should, ideally, be sterilized and fresh isolates should be sought. If this is impossible, the culture should be covered with a quantity of liquid paraffin and subcultured at a later date. Surrounding cultures with oil, water, or Vaseline can prevent infection by crawling mites, although mites carried on clothing, by insects, or by air currents can still cause infestation. Sealing the necks of tubes or bottles with sterile cigarette paper attached with copper sulfate-gelatin glue (20 g of gelatin and 2 g of copper sulfate in 100 ml of distilled water) can prevent mite infection and still allow air to pass. Cold storage at 4 to 8°C reduces the spread of mites, but they soon become active on removal from the refrigerator. Deep-freezing below −18°C will kill most mite eggs.

3.3.8. Scale-Up

Fermentation processes expected to be scaled to a fermentor usually require the performance of experiments initially in shake flasks to improve the process (such as inoculum development). After positive results have been obtained, the somewhat disconcerting question is "will it work in a tank, will it scale up?" Kennedy et al. (72) studied the production of γ-linolenic acid by *Mucor hiemalis* and found that a 10-liter stirred fermentor (working volume of 8 liters equipped with two Rushton impellers rotating at 600 rpm with an air flow of 1.0 v.v.m. (volume of air per volume of broth per minute) and with pH control set for the average shake flask pH), compared with shake flasks (500-ml baffled flasks with 150 ml of medium, grown on a rotary shaker at 140 rpm), produced essentially the same cell dry weight, oil content of the cell, and fatty acid content of the oil in each of six media they tested. These results confirm the impression that performing shake flask studies followed by fermentor confirmation is an appropriate exercise.

3.3.9. Seed Train

Webb and Kamat (144) have shown that one highly variable transfer step in a typical inoculum seed train is the initial transfer, usually with a wire loop, from growth on agar medium to a liquid seed medium. To minimize the 11-fold range in inoculum level that occurred in their study, they recommend the following liquid transfer protocols for *Saccharomyces cerevisiae*. Using the first procedure decreased the range of inoculum variation to twofold. Using both procedures in sequence essentially eliminated the inoculum level variability. This procedure will work best with spores and cultures that do not have a hyphal morphology.

PROTOCOL FOR INOCULUM TRANSFER FROM AGAR (144)

Procedure 1

A 2-ml volume of sterilized medium is added to an agar slant, which is gently shaken in a vortex mixer to transfer the cells from the solid medium into the washing medium and to obtain a homogeneous suspension. Small (0.5-ml) aliquots of this suspension are then used to inoculate shaken flasks.

Procedure 2

Agar slants are first seeded with 0.01-ml aliquots of an exponentially growing shaken flask culture. After 72 h of incubation at 30°C, the slants are washed with medium as above.

Streptomycetes have often been reported to spontaneously lose the ability to sporulate and the ability to produce various secondary metabolites. McCann-McCormick et al. (92) and Novak et al. (104) have noted the rapid loss of sporulation, pigment production, and avermectin production by *Streptomyces avermitilis*. Clearly, rapid loss of the productivity of a strain during the serial transfer of the culture in liquid media has implications for the successful development of a liquid medium seed train. This problem was effectively handled for *S. avermitilis* by the selection of a subisolate that was stable under conditions of multiple seed passages (4).

Serial transfer of the culture in its seed medium in shake flasks for the same number of transfers expected in the fermentation plant seed train has been an effective predictor of large-scale performance in our hands for a number of secondary metabolite producers. If instability is evident during the shake flask seed train transfer experiments, substitution of seed media and the selection of a more stable variant should be undertaken. For *Streptomyces* strains including *S. avermitilis*, medium I (Table 4) has allowed for the expression and selection of multiple morphological variants.

3.3.10. Shear

The rheological behavior of a culture broth reflects the growth of the culture and its physiological state. In the course of determining the optimum seed transfer time for the production of nystatin by an industrial strain of *Streptomyces noursei*, Ettler (40) found that the optimum inoculum occurred in a very tight time window when the broth shear stress was between 80 and 120 mPa at a shear rate of 698 S^{-1}. It was also a period when the culture broth response shifted from a Newtonian to a non-Newtonian behavior. Perhaps shear measurements can become a valuable measure and indicator for the timing of optimum seed transfer. A procedure for measuring the shear rate of culture media in a shake flask fermentation has been reported and involves a rolling-sphere viscometer and movement of calibrated stainless steel spheres on inclined surfaces (22).

3.3.11. Antifoam

For commercial-scale production facilities, the final stages of inoculum development often take place in fermentors that require the addition of antifoam for normal operation. The effect of antifoam upon a culture and upon its productivity is an important study that can be undertaken at the shake flask level early in a culture development program. We have found that glycol-, oil-, and silicone-based antifoams often have different effects upon cell productivity.

The effect of antifoam addition on a cell or fermentation can be far reaching. *N*-Hexadecane, besides its antifoam properties, has been reported to cause changes in the morphology of *Penicillium chrysogenum* (108) and to be an oxygen transfer enhancer (59). Since the cell morphology of fungi including *P. chrysogenum* has been reported to be an important variable for productivity, the use of *N*-hexadecane in a seed fermentor could have an effect beyond its antifoam activity.

TABLE 4 Medium I—agar medium for the selection of *Streptomyces* morphological variants

Component[a]	Amt
Dextrose	10.1 g
Asparagine	1.0 g
K_2HPO_4	0.1 g
$MgSO_4 \cdot 7H_2O$	0.5 g
Yeast extract	0.5 g
Trace elements mix	10.0 ml
$FeSO_4 \cdot 7H_2O$	1.0 g
$MnSO_4 \cdot H_2O$	1.0 g
$CuCl_2 \cdot 2H_2O$	0.025 g
$CaCl_2$	0.1 g
H_3BO_3	0.056 g
$(NH_4)_6Mo_7O_{24} \cdot H_2O$	0.019 g
Distilled water	to 1 liter
$ZnSo_4 \cdot 7H_2O$	0.2 g
Distilled water	to 1 liter
Agar	18.0 g

[a]pH to 7.2 before autoclaving.

3.3.12. Seed Transfer Timing

The timing of transfer from seed fermentors into production can, if the fermentors have the appropriate instrumentation, take advantage of online measurements such as CO_2, fluorescence, dissolved oxygen, and the derived variables respiratory quotient (RQ) and specific growth rate to characterize the status of the inoculum. After optimization for productivity, these parameters can be used to direct the timing of the last inoculum transfer. For the earlier stages of a seed train and for fermentations where seed growth is performed only in shake flasks, other measures of the status of an inoculum are needed to standardize the optimum transfer time. Respiration rate, motility, shear, specific growth rate, product formation, dehydrogenase, and nucleic acid level have all been used for this purpose (Table 5).

One of the most common studies of inoculum optimization is the determination of the effect of the age of transfer of the culture on productivity (Table 6). The optimum titer of proteinase, for example, surprisingly clustered around a 6- to 7-day age of the slant used to prepare *Aspergillus oryzae* spores (106). Slants younger than 5 days and older than 9 days resulted in spore inocula with lower proteinase titers. The basis of this effect should be investigated because it points to an unexplained variation in spore inoculum competence. Unless all steps before that point are rigidly controlled, the age of the inoculum can be of disappointing predictive value for subsequent fermentations.

The protein/RNA ratio of *Streptomyces cattleya* grown in batch culture decreased to three intermittent minima when specific glucose uptake, specific ammonium uptake, and specific phosphate uptake approached zero (Fig. 3) (26). The intermittent peaks of the protein/RNA ratio just before specific uptake approached zero may be useful for inoculum development in that these peaks may provide a time-independent signal for when to transfer the seed growth into fresh media.

Yamamoto et al. (148) proposed elimination of lactic acid produced during inoculum development to prevent the lag in production during the final fermentation. Generally, product synthesis during inoculum development results in lower final titers.

The quality of the inoculum can be determined for some organisms by measuring a biochemical characteristic of the inoculum source. Nucleic acid content and enzyme levels have been used (Table 6).

3.3.13. Cell Morphology

Yeast in pellet form are the preferred morphological form for production in a continuous-process reactor. Induction of cell aggregation in a column reactor requires a high cell density, a low flow rate, and, in some cases, the presence of ethanol in the growth medium. When these aggregated cells are used as inoculum, they retain their aggregated form unless they are dispersed by blending before inoculation. Cells dispersed with a blender become nonflocculant and sediment at rates comparable to inoculum prepared from agar slants (128).

Using *Aspergillus terreus*, Gbewonyo et al. (48) were able to show that one of the very significant differences between inoculum in the form of pellets and inoculum in the form of dispersed mycelia was the dispersed mycelia under pro-

TABLE 5 Time-independent measures for optimum inoculum transfer

Measure	Transfer when measurement is:	Organism	Product and effect	Reference
Motility	Maximum	*Clostridium acetobutylicum*	2× > BuOH	55
RNA	Falling	*Claviceps paspali*	Maximum alkaloid	132
Declorization of methylene blue	Time ≤ 2.2 min	*Micromonspora* spp.	Peak gentamicin	85
Lactic acid level	Before inhibitory concn	*Lactococcus lactis*	Eliminate lag phase	148
Respiration	Maximum	*Acremoniuim chrysogenum*	Maximum cephalosporin	84
Specific growth rate	Maximum	*A. chrysogenum*	Maximum cephalosporin	84
Shear	80–120 mPa	*Streptomyces noursei*	Optimum nystatin	40

duction fermentor conditions are likely to experience dissolved-oxygen limitations.

Cell morphology has a dramatic impact on culture productivity by filamentous organisms. Production is usually optimal either when pellets are formed or when growth is in a dispersed mycelial form. Braun and Vecht-Lifshitz (20) have reviewed reports on the correlation between mycelial morphology and metabolite production. Their suggestions on how to induce pellet formation are summarized in Table 7. Since their review, other approaches to the induction of pellet formation have been presented; they include supplementation with $MgSO_4 \cdot 7H_2O$ or phosphate (30), nitrilotriacetic acid, or 8-hydroxyquinoline (39), and CO_2 supplementation of intake air (67). Carboxymethyl cellulose, on the other hand, enhanced the production of the dispersed mycelial morphological form (5).

For *Streptomyces akiyoshiensis*, an initial pH of 5.5 and a spore inoculum enhanced growth as dispersed mycelia (51). The production of dispersed mycelia in the inoculum stage of *Penicillium citrinum* was enhanced by the use of a concentrated medium which altered seed growth morphology from a pelleted to a dispersed hyphal form (60). Subsequent inoculation into the production medium resulted in the formation of small pellets and optimum production (61). Addition of 5% *n*-Hexadecane to a tank fermentation of *Penicillium chrysogenum* along with the inoculum causes the culture to develop in a dispersed mycelial form. Addition of *n*-hexadecane at the start of penicillin synthesis results in a change in morphology from tight pellets to disperse pellets and an increase in penicillin titer (108). Pellet formation by *Aspergillus anamori* was prevented by high agitation and pH control at 2.5 to prevent spore aggregation

TABLE 6 Key variables affecting inoculum quality

Variable	Organism	Optimum	Effect	Reference
Inoculum age	*S. hygroscopicus*	Juvenile (<25 h)	Decrease passage loss	143
Medium	*S. hygroscopicus*	Reduced carbohydrate	Decrease passage loss	143
Inoculum level	*S. hygroscopicus*	≥1%	Maximum productivity	143
No. of stages	*S. hygroscopicus*	≤2	Minimize loss on passage	143
Enzyme level	*S. antibioticus*	Higher than control	Earlier onset of oleandomycin	121
Krebs cycle				
Inoculum level	*A. niger*	10^2 spores/liter	Maximum polygalacturonase	45
Inoculum level	*A. niger*	10^8 spores/liter	Maximum pectin lyase	45
Inoculum age	*A. oryzae*	Spores from 6-day slant	Peak proteinase	106
Enoculum level	*A. oryzae*	≥6 × 10^7 spores/g of substrate	Proteinase peak	106
Enzyme ratio	*P. chrysogenum*	G6PDH/ICDH near 3	Optimum penicillin	131
Enzyme ratio	*P. patulum*	G6PDH/ICDH near 1	Optimum griseofulvin	131
Inoculum level	*A. mediterranei*	40 mg/flask	Highest specific production of rifamycin	41
RNA level	*C. purpurea*	≥ 6 μg/mg (dry weight)	Produced alkaloid >3 mg/liter	150
DNA level	*C. purpurea*	Maximum	Optimum alkaloid	37

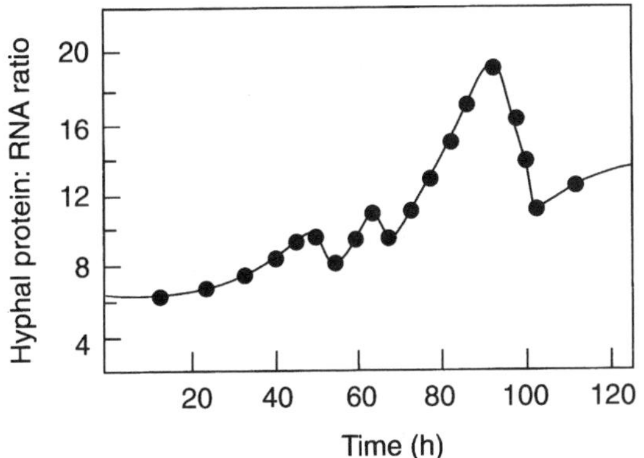

FIGURE 3 Protein/RNA ratio *Streptomyces cattleya*. Reprinted from reference 26 with permission of the publisher.

(122). The pellet diameter in a shake flask can be controlled by the presence of glass beads (125).

3.3.14. Water Activity

Water activity (A_w) is defined as the relative humidity of the gaseous atmosphere in equilibrium with the substrate (147). A_w is a critical variable for inoculum prepared on solid substrates. It is also a critical determinant for the survival of dried cultures (99). On-line sensors have been developed for A_w measurement in liquid and in solid substrate fermentations. Xavier and Karanth (147) have adapted the proximity equilibration cell method for A_w measurements of solid substrates.

PROTOCOL FOR A_w DETERMINATION OF SOLID SUBSTRATES (147)

Analytical reagent (AR) salts are used to prepare saturated slat slushes to obtain a standard graph: lead nitrate (A_w 0.98), potassium nitrate (A_w 0.94), zinc sulfate (A_w 0.90), potassium chromate (A_w 0.88), potassium bisulfate (A_w 0.86), potassium bromide (A_w 0.84), and ammonium sulfate (A_w 0.81). Plastic petri dishes and Whatman no. 44 circular filter papers (10.0 cm in diameter) are used. Preweighed filter papers are placed on the rims of the lower petri dishes containing the various saturated salt slushes in duplicate, on adhesive paper supports. The petri dishes are covered and sealed with adhesive tape to avoid moisture absorption. They are incubated at 20°C, and after exactly 24 h the filter papers are removed and weighted. The relative weight gain of the paper is calculated as a percentage of the initial weight. A standard graph is constructed by plotting the percent weight gain against log $(1 - A_w)$. For unknown samples, preweighted filter papers are placed in the lower petri dish containing 8 to 10 g of wheat bran moistened to predetermined levels and allowed to equilibrate for 24 h. The percent weight gain is calculated, and the A_w is read from the standard graph. The moisture content (percentage) of the wheat bran is plotted against log $(1 - A_w)$.

TABLE 7 Possible approaches for inducing mycelial pellet formation in fermentation culture[a]

Use nitrogen-limited medium.
Do not use readily assimilated carbon sources.
Add cationic polyelectrolytes.
Add mild nonionic detergents.
Add Ca^{2+} ions.
Use a small inoculum.
Reduce shear.
Reduce temperature.
Maintain medium at about pH 5–6.
Increase dissolved-oxygen concentration.

[a]Data from reference 20.

3.3.15. Inoculum Level

Inoculum level can have a dramatic effect upon productivity. *Aspergillus niger* produced maximum levels of polygalacturonase and maximum pectin lyase with inoculum levels of 10^2 and 10^8 spores/liter, respectively (45) (Table 6).

3.4. FERMENTATION PROCESS MEASUREMENTS

3.4.1. ATP

Measurement of the cellular ATP level has not been widely used as an indicator of cell growth because of the extensive variations in ATP concentrations in cells depending upon the carbon source, growth conditions, strain of the cell, and phase of growth (114). On the other hand, for inoculum development purpose, the ATP content may find utility as a marker of cell status and perhaps as a measure of its capacity to support energy-requiring biosynthetic processes. In cases when transfer at an early log phase of growth is optimal, the fact that the ATP content increased most rapidly during the early log phase of growth for *Candida utilis* may be germane.

PROTOCOL FOR ATP EXTRACTION DETERMINATION (114, 137)

Culture samples (1 ml) are rapidly withdrawn and extracted in 9 ml of boiling Tris-EDTA buffer for 5 min to release cellular ATP (137). The ATP is quantified by an internal standardization technique involving an ATP photometer (model 3000; SAI Technology, San Diego, Calif.) and luciferase-luciferin enzyme preparation (Picozyme P; Packard Instrument Co.).

3.4.2. Calorimetry

Heat produced by microbial activity can be measured by microcalorimetry with an LKB Bioactivity Monitor (105). Together with culture fluorescence, microcalorimetry can provide enough on-line information to monitor growth effectively (53) and perhaps be a valuable measure for inoculum quality and transfer timing (3, 8, 86).

3.4.3. Capacitance

Mishima et al. (95, 96) proposed that capacitance measured over a frequency range from 100 kHz to 1 mHz could be used to estimate the concentration of living cells in a bioreactor since this on-line measurement showed linearity between the capacitance value and living-cell concentrations of microorganisms as well as of animals and plant. For growth during the first 24 h (lag and log phases), capacitance and growth measurements of dry cell weight and viable-cell numbers showed a high correlation coefficient.

The general trend of capacitance measurements after log growth for *Saccharomyces cerevisiae* was similar to the curve generated by measuring viable-cell numbers (96). Capacitance measurements as a reflection of a combination of cell number and cell shape may become an interesting on-line measurement of monitoring inoculum development and timing of seed transfer.

3.4.4. Cell Products

Poly-3-hydroxybutyrate (PHB) is an intracellular storage product found in bacterial cells. With a procedure for the measurement of PHB containing cells using fluorescence and flow cytometry, population analysis for PHB content may become a useful measure of optimum seed preparation.

PROTOCOL FOR DETERMINATION OF PHB-CONTAINING CELLS (100)

For the staining of neutral lipids in microbial cells, 3×10^8 cells are preserved with 1 ml of 10% sodium azide in 0.9% sodium chloride for at least 30 min, resuspended in phosphate-buffered saline (pH 7.2), and stained with 20 to 100 μl of Nile red (1 mg/ml of acetone), with the amount being optimized for the bacterial strain used. The fluorescence of the Nile red-stained cells was detected through a 515 nm barrier filter and a BJP 495 excitation filter.

The measure of culture growth in a solid-substrate fermentation is a technical challenge that for filamentous fungi has often been estimated from the content of ergosterol. Matcham et al. (90) have found that this estimator fungal growth was directly proportional to the dry-weight increase for *Agaricus bisporus* in liquid media for 56 days of fermentation. Cultures grown on a solid substrate showed a correlation between linear extension of the mycelium and ergosterol content.

PROTOCOL FOR MEASUREMENT OF ERGOSTEROL CONTENT (90)

Extracts for ergosterol assay are prepared as follows. Total lipids are extracted from liquid cultured mycelium by the method of Garbus et al. (46a). The mycelium is homogenized in 2:1 (vol/vol) methanol-chloroform. The homogenate is allowed to stand for a minimum of 30 min, after which 1 volume each of water, chloroform, and 2 M KCl in 0.5 M phosphate buffer (pH 7.4) are added to establish a two-phase water/chloroform system from which the lipid-containing chloroform phase is separated and retained for further cleanup. The aqueous phase is discarded. Samples of dried, milled grain are extracted by mixing with ethanol in the proportion of 2 ml of ethanol to 1 g of grain. The mixture is steeped for 1 h, the supernatant is decanted, and the residue is washed with two further aliquots of ethanol. Following clarification by centrifugation for 10 min at $2{,}000 \times g$, the three washings are combined. Both sets of extracts are evaporated to dryness under oxygen-free nitrogen prior to saponification in a solution of 1 N KOH in 95% (vol/vol) ethanol for 1 h at 70°C. After being cooled, the mixtures are diluted with 2 volumes of water and the nonsaponifiable fraction containing ergosterol is extracted with three washings of diethyl ether or petroleum ether. The washings are combined and dried over sodium sulfate before being assayed. Extracts are assayed directly after saponification. In these later assays, ergosterol is differentiated from background UV absorbing material by means of electronically derived second-derivative spectra on a spectrophotometer.

Penicillium chrysogenum mycelial dry weight has been reported to correlate with the RNA content during log-phase growth. After log-phase growth there was no correlation (76). The RNA content has been used as an indicator for optimum timing of seed transfer (132). Koliander et al. (76) have developed a high-pressure liquid chromatography analysis protocol for RNA. An ultrasensitive RNA assay linked to light emission has also been described (98).

PROTOCOL FOR ASSAY OF RNA (98)

For the measurement of colony RNA, individual colonies are plucked, 5 μl of crystalline bovine serum albumin (5 mg/ml) is added as a carrier, and the macromolecules are precipitated with 100 μl of ice-cold 0.2 M perchloric acid. Extraction of ice for 20 min is followed by collection of the precipitated material by centrifugation. The pellet is dissolved in 10 μl of 0.1 M sodium hydroxide at room temperature for 5 min, 30 μl of a buffer consisting of 15 mM potassium acetate, 20 mM Tris acetate, 3 mM magnesium chloride (pH 7.7) is added, and then 10 μl of 0.1 M acetic acid is added. The final volumes of both standards and sample are therefore 50 μl. RNA standards (50 μl) or acid-precipitable material from cells is hydrolyzed by the addition of 50 μl of a solution containing 0.2 U of *Lysobacter enzymogenes* RNase per ml and 0.02 U of phosphodiesterase I per ml in a buffer consisting of 15 mM potassium acetate, 20 mM Tris, and 3 mM magnesium chloride (pH 7.7). The samples are incubated at 37°C for 1 h and then placed in a boiling-water bath for 5 min to inactivate the enzymes. AMP is converted to ATP by the addition of 100 μl of a solution consisting of 0.2 mM phosphoenolpyruvate, 0.05 mM ITP, 75 U of myokinase per ml, and 25 U of pyruvate kinase per ml in the buffer described above. The complete mixture (200 μl) is incubated for 30 min at 30°C. The ATP generated in this manner is then measured in duplicate by the luciferase-luciferin reaction. A 100-μl aliquot of sample is placed in a plastic cuvette, and 100 μl of luciferase-luciferin solution is injected. Light emission is measured and averaged over a 15-s interval by a photometer. A calibration curve is then constructed over the appropriate range by plotting light emission in arbitrary units against RNA in nanograms.

The total dehydrogenase level in the culture, as measured by methylene blue (85) resazurin reduction (83), has been used to characterize *Micromonospora* and activated sludge, respectively.

PROTOCOL FOR RESAZURIN REDUCTION ASSAY (83)

Resazurin solution (5 mg in 50 ml of distilled water) is stored in a brown bottle at 4°C and made up fresh weekly. Phthalate-HCl buffer (0.05 M) at pH 3.5 contains 1.02 g of potassium biphthalate in 90 ml of distilled water, adjusted to pH 3.5 with 6 N HCl. The volume is made up to 100 ml, and the solution is stored at room temperature. *n*-Amyl alcohol and sodium bicarbonate are laboratory reagent grade. A 2-ml sample of activated sludge is added to a test tube (2 by 15 cm) containing 2 ml of distilled water. Exactly 1 ml of resazurin solution is added, and the contents are gently mixed. After an exact incubation period (15 or 30 min) at room temperature (21°C) and in darkness, 10 ml of *n*-amyl alcohol and 0.1 ml of 0.05 M phthalate-HCl buffer are added. The contents are then vigorously mixed for 15 s on a vortex mixer; this is followed by centrifugation, in the same test tube, at $1{,}000 \times g$ for 5 min. Approximately 8 to 9 ml of the upper alcohol layer is transferred into a clean test tube containing approximately 2 g of sodium bicarbonate to adjust the pH and to clarify. The contents are gently mixed, and the absorbance of the supernatant is read on a spectrophotometer at 610 nm (the maximum absorbance of unreduced resazurin). The biological activity is quantified by comparing the absorbance with that of a reference prepared from the same resazurin solution. Biological activity is expressed in nanomoles of resazurin reduced per hour per milligram of mixed-liquor-suspended solids.

3.4.5. Fluorescence

The measurement of NADH-dependent culture fluorescence is an attractive on-line measurement for process development. It is based upon the observation that microbial cultures irradiated with UV at 366 nm fluoresce at 460 nm because of their NADH content (12). For some organisms and conditions fluorescence can be used to estimate cell growth, while in other cases it is ineffective as an estimator either because of the physiology of the organism or because of the production of fluorescing or quenching substances (94). Depending upon the culture, on-line fluorescence measurements may, under standardized conditions, provide information relevant to optimum inoculum and optimum transfer time.

Treating viable cells of *Pseudomonas fluorescens* with ethidium bromide results in little change. However, if the cells are damaged, ethidium bromide is able to enter the cell, and in combination with double-stranded DNA, it produces a strong fluorescence. Puchkov and Melkozernov (115) have validated the protocol of using ethidium bromide uptake as a rapid indicator of cell viability after freezing and thawing.

PROTOCOL FOR FLUORIMETRIC VIABILITY ASSESSMENT (115)

Samples (1 ml) of a bacterial suspension in cultivation medium are placed into polypropylene cryotubes. The samples are centrifuged at 4,000 × *g* for 2 min (Eppendorf Centrifuge 5415 C). The harvested cells are resuspended in a buffer containing 0.1 mol of Tris-HCl (pH 7.0) per liter; up to 10^{11} CFU per ml is used for fluorimetry. The assay mixture contains 10^{-5} mol of ethidium per liter and 10^{9} CFU of cells per ml. Fluorescence is measured on a Hitachi 850 spectrofluorophotometer with a thermostated cell holder in a quartz cuvette with a 1-cm light path. Ethidium fluorescence is excited at 500 nm and recorded at 600 nm. To induce the permeability of the bacterial envelopes for ethidium, 10^{-4} mol of cetyltrimethylammonium bromide per liter is added.

3.4.6. Optical Density

If not subject to interference by media or mycelial shape, turbidity is an excellent on-line and off-line measurement that correlates with cell growth. Turbidity can be a reproducible indicator for inoculum timing and quality. Optical density is most often measured at 600 nm. Continuous on-line measurement is available at the fermentor stage (34).

3.4.7. Respiratory Quotient

Respiratory quotient (RQ), the ratio of carbon dioxide concentration to the oxygen concentration, is a valuable calculated variable that effectively can be determined by mass spectrometric determination of exhaust air gas composition linked to a minicomputer. Changes in RQ can often be correlated with changes in the metabolic condition of a cell. An RQ of 1 indicates glucose-supported metabolism while a change to an RQ of 0.7 indicates oil-supported metabolism for *Nocardia lactamdurans* during the production of efrotomycin (23). RQ has proven to be an effective indicator for the control of nutrient feed in fed-batch processes. Changes in RQ have also been shown to be associated with the production of multiple products by *Streptomyces cattleya* (Fig. 4) (26). When used as a guide for inoculum development, RQ levels associated with a biomass accumulation phase should be maintained throughout the development of the inoculum. If any of the subsequent RQ phases are allowed to occur during seed growth, subsequent total productivity can be compromised.

3.4.8. Specific Growth Rate

The specific growth rate of a culture could be used as a key factor for the timing of inoculum transfer into the next stage of growth or into a production medium. Specific growth can be estimated from the oxygen uptake rate or the carbon dioxide evolution rate; however, measurement of these values requires expensive equipment or equipment that requires frequent calibrations. Lee et al. (81) suggest that if one assumes that cell maintenance is minimal and the cell yield coefficient is constant, then specific growth (μ) rate can be estimated from the time change in agitation speed when the dissolved oxygen concentration is maintained at a fixed percent saturation by altering the revolutions per minute (rpm) at a fixed aeration rate according to the equation

$$\mu = \text{RPM} / \int_0^t (\text{RPM})\, dt$$

3.5. RECOVERY OF ACTIVITY

Culture degeneration and instability is a constant risk to any microbiologist who has found a productive microbial culture. Over a period of 12 years, Bose and colleagues

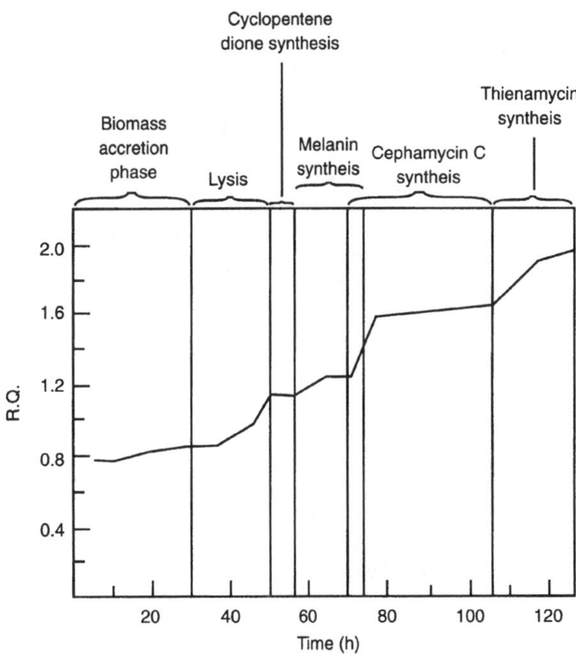

FIGURE 4 RQ and *Streptomyces cattleya* fermentation. Reprinted from reference 26 with permission of the publisher.

provided an important example of loss of productivity of a culture after repeated vegetative subculturing. They observed that during the process of loss, reisolation of productive subisolates was an effective method to recover the productive ability of an *Aspergillus veriscolor* strain if it was detected early enough (36). A decade later, one of the earlier stable "mutants" had yet again degenerated to a point where antibiotic production no longer could be detected (74). Although their procedure resulted in the apparent recovery of activity, the antifungal activity recovered did not match the original antifungal spectrum (75). Somehow the procedure had been selected for the production of a different antifungal agent. The lesson clearly appears to be preserve it or lose it.

Virulence of *Entomophaga maimaiga* was also found to decrease upon repeated subculturing but was recovered after the culture was passaged through insects (56). This recovery procedure is probably limited, however, to cases where the characteristics that are desired are linked to laboratory survival of the producing organism.

Since the recovery of activity has often been a futile effort, prevention of loss should be pursued more extensively. Early in the history of antibiotics, it was consistently noted that repeated transfer of *Streptomyces griseus* resulted in the loss of the ability of this culture to produce streptomycin (101, 111, 146). The loss of production upon serial transfer (up to 20 transfers) did not occur if the medium used was Fe limited (101). In medium that was not Fe limited, productivity was rapidly lost and could not be recovered. This observation may provide a guide to the prevention of loss upon serial transfer for other cultures.

Fazeli et al. (42) found that *S. griseus* ATCC 1247 was "highly unstable" when grown in a complex medium but could be maintained in continuous culture in a minimal medium at a dilution rate where growth was less than μ_{max}.

Reusser et al. (118) observed that repeated transfers on *Streptomyces niveus* resulted in the selection of a nonproducing strain that had, during log growth, an equivalent maximum growth rate comparable to that of the original high-titer producer. The low-titer producer, however, was able to use the carbohydrate supplied in the medium for growth much more completely than was the high-titer producer. Thus, the low-titer producer took over the population upon repeated transfer. If it is generally true that low-titer producers have similar maximum growth rates to their high-titer-producing relatives but are more efficient at using substrates provided for growth, an inoculum development scheme that transfers the culture before any nutrient becomes limiting to the high-titer producer would be preferred. Perhaps this is why transfer at early log phase is so often the optimum selected seed transfer condition.

REFERENCES

1. **Acreman, J.** 1994. Algae and cyanobacteria: isolation, culture and long-term maintenance. *J. Ind. Microbiol.* **13:** 193–194.
2. **Adams, G.** 1995. The preservation of inocula. Reproducibility and predictivity of disinfection and biocide tests, p. 89–120. *In* M. R. W. Brown and P. Gilbert (ed.), *Microbiological Quality Assurance Screening and Bioassay: a Guide towards Relevance and Reproducibility of Inocula*. CRC Press, Boca Raton, Fla.
3. **Andlid-Larsson, C., L. Gustafsson, I. Marison, and C. Liljenberg.** 1995. Enthalpy content as a function of lipid accumulation in *Rhodotorula glutinis*. *Appl. Microbiol. Biotechnol.* **42:**818–825.
4. **Baker, E.** Personal communication.
5. **Banerjee, U. C.** 1993. Effect of glucose and carboxymethyl cellulose on growth an rifamycin-oxidase production by *Curvularia lunata*. *Curr. Microbiol.* **26:**261–265.
6. **Barbotin, J. N.** 1994. Immobilization of recombinant bacteria. A strategy to improve plasmid stability. *Ann. N. Y. Acad. Sci.* **721:**303–309.
7. **Baudot, A., J. F. Peyridieu, P. Boutron, J. Mazuer, and J. Odin.** 1996. Effect of saccharides on the glass-forming tendency and stability of solutions of 2,3-butanediol, 1,2-propanediol, or 1,3-butanediol in water, phosphate-buffered saline, Euro-Collins Solution, or Saint Thomas Cardioplegic Solution. *Cryobiology* **33:**363–375.
8. **Beaubien, A., and C. Jolicoeur.** 1985. Application of flow microcalorimetry to process control in biological treatment of industrial waste-water. *J. Water Pollut. Control Fed.* **57:**95–100.
9. **Bennett, J. W., A. J. Turner, A. K. Loomis, and W. J. Connick.** 1996. Comparison of alginate and "pesta" for formulation of *Phanerochaete chrysosporium*. *Biotechnol. Tech.* **10:**7–12.
10. **Berny, J.-F., and G. L. Hennebert.** 1991. Viability and stability of yeast cells and filamentous fungus spores during freeze-drying: effects of protectants and cooling rates. *Mycologia* **83:**805–815.
11. **Berry, F., S. Sayadi, M. Nasri, D. Thomas, and J. N. Barbotin.** 1990. Immobilized and free cell continuous cultures of a recombinant *E. coli* producing catechol-2,3-dioxygenase in a two-stage chemostat improving plasmid stability. *J. Biotechnol.* **16:**199–210.
12. **Beyeler, W., A. Einsele, and A. Fiechter.** 1981. On line measurements of culture fluorescence: method and application. *Eur. J. Appl. Microbiol. Biotechnol.* **13:**10–14.
13. **Beyersdorf-Radeck, B., R. D. Schmid, and K. A. Malik.** 1993. Long-term storage of bacterial inoculum for direct use in a microbial biosensor. *J. Microbiol. Methods* **18:**33–39.

13a. **Bikerman, J. J.** 1972. Foam fractionation and drainage. *Separ. Sci.* **7:**647–651.

14. **Birgit, L., K. Baerbel, and L. Wuensche.** Jan. 1983. German patent DD-158481.
15. **Blatnik, J., N. Gunde-Cinerman, and A. Cinerman.** 1994. Preservation of rennet producing *Rhizomucor miehei* strain. *Biotechnol. Tech.* **8:**481–490.
16. **Bloomfield, S. F.** 1995. Reproducibility and predictivity of disinfection and biocide tests, p. 189–219. *In* M. R. W. Brown and P. Gilbert (ed.), *Microbiological Quality Assurance Screening and Bioassay: a Guide towards Relevance and Reproducibility of Inocula*. CRC Press, Boca Raton, Fla.
17. **Bloomfield, S. F., M. Arthur, H. Gibson, K. Morley, P. Gilbert, and M. R. W. Brown.** 1995. Development of reproducible test inocula for disinfectant testing. *Int. Biodeterior. Biodegrad.* **36:**311–331.
18. **Borba, C. M., A. M. Mendes da Silva, and P. C. de Oliveira.** 1992. Long-time survival and morphological stability of preserved *Sporothrix schenckii* strains. *Mycoses* **35:** 185–188.
19. **Boutron, P., and J. F. Peyridieu.** 1994. Reduction in toxicity for red blood cells in buffered solutions containing

high concentrations of 2,3-butanediol by trehalose, sucrose, sorbitol or mannitol. *Cryobiology* **31:**367–373.

20. **Praun, S., and S. E. Vecht-Lifshitz.** 1991. Mycelial morphology and metabolite production. *Trends Biotechnol.* **9:** 63–68.
21. **Brown, M. R. W., S. F. Bloomfield, and P. Gilbert.** 1995. Sources of biological variation and lack of inoculum reproducibility, p. 83–88. *In* M. R. W. Brown and P. Gilbert (ed.), *Microbiological Quality Assurance Screening and Bioassay: a Guide towards Relevance and Reproducibility of Inocula.* CRC Press, Boca Raton, Fla.
22. **Bryan, W. L., and R. W. Silman.** 1990. Rolling-sphere viscometer for *in situ* monitoring of shake-flask fermentations. *Enzyme Microb. Technol.* **12:**818–823.
23. **Buckland, B., T. Brix, H. Fastert, K. Gbewonyo, G. Hunt, and D. Jain.** 1985. Fermentation exhaust gas analysis using mass spectrometry. *Bio/Technology* **3:**982–988.
24. **Buell, C. B., and W. H. Weston.** 1947. Application of the mineral oil conservation method to maintaining collections of fungous cultures. *Am. J. Bot.* **34:**555–561.
25. **Buitrón, G., and B. Capdeville.** 1995. Enhancement of the biodegradation activity by the acclimation of the inoculum. *Environ. Technol.* **16:**1175–1184.
26. **Bushell, M. E., and A. Fryday.** 1983. The application of materials balancing to the characterization of sequential secondary metabolite formation in *Streptomyces cattleya* NRRL 8057. *J. Gen. Microbiol.* **129:**1733–1741.
27. **Castellani, A.** 1939. Viability of some pathogenic fungi in distilled water. *J. Trop. Med. Hyg.* **42:**225–226.
28. **Challen, M. P., and T. J. Elliott.** 1986. Polypropylene straw ampoules for the storage of microorganisms in liquid nitrogen. *J. Microbiol. Methods* **5:**11–23.
29. **Champagne, C. P., N. Gardner, and F. Dugal.** 1994. Increasing the stability of immobilized *lactococcus latis* cultures stored at 4°C. *J. Ind. Microbiol.* **13:**367–371.
30. **Chen, L. C. W.** 1996. Production of beta-fructofuranosidase by *Aspergillus japonicus*. *Enzyme Microbiol. Technol.* **18:**153–160.
31. **Chen, Y. Y.** 1987. The preservation of basidiomycetes cultures by ultra-low temperature freezing. *Acta Mycol. Sin.* **6:**110–117.
32. **Chirife, J., C. F. Fontan, and E. A. Benmergui.** 1980. The prediction of a_w of aqueous solutions in connection with intermediate moisture foods. 4. a_w predictions in aqueous non-electrolyte solutions. *J. Food Technol.* **15:**59–70.
33. **Cooney, C. L., H. Y. Wang, and D. I. C. Wang.** 1977. Computer-aided material balancing for prediction of fermentation parameters. *Biotechnol. Bioeng.* **19:**55–67.
34. **Cox, R. P., M. Miller, J. B. Nielson, M. Nielsen, and J. K. Thomsen.** 1989. Continuous turbidometric measurement of microbial cell density in bioreactors using a light-emitting diode and a photodiode. *J. Microbiol Methods.* **10:**25–31.
35. **Day, J. G., and M. R. McLellan.** 1995. Cryopreservation and freeze-drying protocols. Introduction. *Methods Mol. Biol.* **38:**1–5.
36. **Dhar, A. K., and S. K. Bose.** 1968, Mutagens for regeneration of an antibiotic-producing strain of *Aspergillus versicolor. Appl. Microbiol.* **16:**340–342.
37. **Didek-Brumec, M., V. Gaberc-Porekar, M. Alacevic, B. Druskovic, and H. Socic.** 1991. Characterization of sectored colonies of a high-yielding *Claviceps purpurea* strain. *J. Basic Microbiol.* **31:**27–35.
38. **Erickson, L. E., I. G. Minkewich, and V. K. Eroshin.** 1979. Utilization of mass-energy balance regularities in the analysis of continuous culture data. *Biotechnol. Bioeng.* **21:**575–591.
39. **Esuoso, K. O.** 1994. Influence of nitrilotriacetic acid and 8-hydroxyquinoline on the production of citric acid from molasses using *Aspergillus niger. J. Ferment. Bioeng.* **77:** 693–695.
40. **Ettler, P.** 1992. The determination of optimal inoculum quality for submerse fermentation process. *Collect. Czech. Chem. Commun.* **57:**303–308.
41. **Farid, M. A., M. R. Abu-Shady, A. I. El-Diwany, and H. A. El-Enshasy.** 1995. Production of rifamycin B and SV by free and immobilized cells of Amycolatopsis mediterranei. *Acta Biotechnol.* **15:**241–248.
42. **Fazeli, M. R., J. H. Cove, and S. Baumberg.** 1995. Physiological factors affecting streptomycin production by *Streptomyces griseus* ATCC 12475 in batch and continuous culture. *FEMS Microbiol. Lett.* **126:**55–62.
43. **Feldmann, F., and E. Idczak.** 1992. Inoculum production of Vesicular-arbuscular mycorrhizal fungi for use in tropical nurseries. *Methods Microbiol.* **24:**339–357.
44. **Felipe, M. G. A., M. Vitola, and I. M. Mancilha.** 1996. Xylitol formation by *Candida guilliermondii* grown in a sugarcane bagasse hemicellulosic hydrolyzate: effect of aeration and inoculum adaptation. *Acta Biotechnol.* **16:**73–79.
45. **Friedrich, J., A. Cimerman, and W. Steiner.** 1990. Production of pectolytic enzymes by *Aspergillus niger*: effect of inoculum size and potassium hexacyanoferrate II-trihydrate. *Appl. Microbiol. Biotechnol.* **33:**377–381.
46. **Fu, J., F. J. Bailey, J. J. King, C. B. Parker, R. S. R. Robinett, D. G. Kolodin, H. A. George, and W. I. Herber.** 1995. Recent advances in the large scale fermentation of *Neisseria meningitidis* group B for the production of an outer membrane protein complex. *Bio/Technology* **13:** 170–174.

46a. **Garbus, J., H. F. DeLuca, M. E. Loomans, and F. M. Strong.** 1963. The rapid incorporation of phosphate into mitochondrial lipids. *J. Biol. Chem.* **238:**59–63.

47. **Gaylarde, C., and J. Kelley.** 1995. Genetic variation in *Fusarium merismoides* preserved by two different methods. *World J. Microbiol. Biotechnol.* **11:**319–321.
48. **Gbewonyo, K., G. Hunt, and B. Buckland.** 1992. Interactions of cell morphology and transport processes in the lovastatin fermentation. *Bioprocess Eng.* **8:**1–7.
49. **Gerdes, K.** 1988. The *parB* (*hok/sok*) locus of plasmid R1: a general purpose plasmid stabilization system. *Bio/Technology* **6:**1402–1405.
50. **Gherna, R. L.** 1981. Preservation, p. 208–217. *In* P. Gerhardt, R. G. E. Murray, R. N. Gstilow, E. W. Nester, W. A. Wood, N. R. Krieg, and G. B. Phillips (ed.), *Manual of Methods for General Bacteriology*. American Society for Microbiology, Washington, D.C.
51. **Glazebrook, M. A., L. C. Vining, and R. L. White.** 1992. Growth morphology of *Streptomyces akiyoshiensis* in submerged culture: influence of pH, inoculum and nutrients. *Can. J. Microbiol.* **38:**98–103.
52. **Gordon, R. E.** Personal communication.

53. **Greene, H. C., and E. B. Fred.** 1934. Maintenance of vigorous mold stock cultures. *Ind. Eng. Chem.* **26:**1297–1299.

54. **Greer, C. W., D. Beaumier, and R. Samson.** 1989. Application of on-line sensors during growth of the dichloroethane degrading bacterium *Xanthobacter autotrophicus*. *J. Biotechnol.* **12:**261–274.

55. **Gutierrez, N. A., and I. S. Maddox.** 1987. The effect of some culture maintenance and inoculum development techniques on solvent production by *Clostridium acetobutylicum*. *Can. J. Microbiol.* **33:**82–84.

56. **Hajek, A. E., R. A. Humber, and M. H. Griggs.** 1990. Decline in virulence of *Entomophaga maimaiga* (Zygomycetes: Entomophthorales) with repeated *in vitro* subculture. *J. Invertebr. Pathol.* **56:**91–97.

57. **Hajek, A. E., M. Shimazu, and R. A. Humber.** 1995. Instability in pathogenicity of *Entomophaga maimaiga* after long-term cryopreservation. *Mycologia* **87:**483–489.

58. **Hartung de Capriles, C., S. Mata, and M. Middelveen.** 1989. Preservation of fungi in water (Castellani): 20 years. *Mycopathogia* **106:**73–79.

59. **Ho, C. S., L. K. Ju, and R. F. Baddour.** 1990. Enhancing penicillin fermentations by increased oxygen solubility through the addition of n-hexadecane. *Biotechnol. Bioeng.* **36:**1110–1118.

60. **Hosobuchi, M., F. Fukui, H. Matsukawa, T. Suzuki, and H. Yoshikawa.** 1993. Morphology control of preculture during production of ML-236B, a precursor of pravastatin sodium, by *Penicillium citrinum*. *J. Ferment. Bioeng.* **76:** 476–481.

61. **Hosobuchi, M., K. Ogawa, and H. Yoshikawa.** 1993. Morphology study in production of ML-236B, a precursor of pravastatin sodium, by *Pencillium citrinum*. *J. Ferment. Bioeng.* **76:**470–475.

62. **Humber, R. A.** 1994. Special considerations for operating a culture collection of fastidious fungal pathogens. *J. Ind. Microbiol.* **13:**195–196.

63. **Hunt, G. A., A. Gourevitch, and J. Lein.** 1958. Preservation of cultures by drying on procelain beads. *J. Bacteriol.* **76:**453–454.

64. **Hunt, G. R., and R. W. Stieber.** 1986. Inoculum development, p. 32–39. *In* A. L. Demain and N. A. Solomon (ed.), *Manual of Industrial Microbiology and Biotechnology*. American Society for Microbiology, Washington, D.C.

65. **Hwang, S.** 1968. Investigation of ultra-low temperature for fungal cultures. I. An evaluation of liquid-nitrogen storage for preservation of selected fungal cultures. *Mycologia* **60:**613–621.

66. **Jones, R. J., K. J. Sizmur, and H. G. Wildman.** 1991. A miniaturised system for storage of fungal cultures in water. *Mycologist* **5:**184–186.

67. **Ju, L. K., C. S. Ho, and J. F. Shanahan.** 1991. Effect of carbon dioxide on the rheological behavior and oxygen transfer in submerged penicillin fermentations. *Biotechnol. Bioeng.* **38:**1223–1232.

68. **Karandikar, A., G. P. Sharples, and G. Hobbs.** 1996. Influence of medium composition on sporulation by *Streptomyces coelicolor* A3(2) grown on defined solid media. *Biotechnol. Tech.* **10:**79–82.

69. **Karanova, M. V., L. M. Mezheivkina, and N. N. Petropavlov.** 1995. Study of cryoprotective properties of antifreeze glycoproteins from the white sea cod *Gadus morhua* on low temperature freezing of mouse embryos. *Biofizika* **40:**1341–1347.

70. **Kaul, S. C., K. Obuchi, H. Iwahashi, and Y. Komatsu.** 1992. Cryoprotection provided by heat shock treatment in *Saccharomyces cerevisiae*. *Cell. Mol. Biol.* **38:**135–143.

71. **Kaul, S. C., K. Obuchi, and Y. Komatsu.** 1992. Cold shock response of yeast cells: induction of a 33 kDa protein and protection against freezing injury. *Cell. Mol. Biol.* **38:**553–559.

72. **Kennedy, M. J., S. L. Reader, R. J. Davies, D. A. Rhoades, and H. W. Silby.** 1994. The scale-up of mycelial shake flask fermentations: a case study of gamma linolenic acid production *Mucor hiemalis* IRL 51. *J. Ind. Microbiol.* **13:**212–216.

73. **Klaenhammer, T. R.** 1987. Plasmid-directed mechanisms for bacteriophage defense in lactic streptococci. *FEMS Microbiol. Rev.* **46:**313–325.

74. **Kole, H. K., and S. K. Bose.** 1980. A note on the development of a technique for regeneration of a degenerated culture. *J. Appl. Bacteriol.* **48:**433–436.

75. **Kole, H. K., A. K. Samanta, S. K. Goswami, and S. K. Bose.** 1982. Attempt at gene reversion of a desired phenotype as a rescue against strain degeneration. *J. Appl. Bacteriol.* **53:**163–167.

76. **Koliander, B., W. Hampel, and M. Roehr.** 1984. Fermentation kinetics of *Penicillium chrysogenum* as determined by ribonucleic acid measurement. *Eur. Congr. Biotechnol.* **1:**693–697.

77. **Kotsaridu, M., B. Muller, V. Pfanz, and K. Schugerl.** 1983. Foam behavior of biological media. X. Influence of the sterilization conditions on the foaminess of PPL solution. *Eur. J. Appl. Microbiol. Biotechnol.* **17:**258–260.

78. **Lang, E., and K. A. Malik.** 1996. Maintenance of biodegradation capacities of aerobic bacteria during long-term preservation. *Biodegradation* **7:**65–71.

79. **Lechevalier, M. P.** Personal communication.

80. **Lee, S. Y., K. S. Yim, H. N. Chang, and Y. K. Chang.** 1994. Construction of plasmids, estimation of plasmid stability, and use of stable plasmids for the production of poly(3-hydroxybutyric acid) by recombinant *Escherichia coli*. *J. Biotechnol.* **32:**203–211.

81. **Lee, T. H., Y. K. Chang, and B. H. Chung.** 1996. Estimation of specific growth rate from agitation speed in DO-stat culture. *Biotechnol. Tech.* **10:**303–308.

82. **Linardi, V. R., M. C. Amancio, and N. C. M. Gomes.** 1995. Maintenance of *Rhodotorula rubra* isolated from liquid samples of gold mine effluents. *Folia Microbial.* **40:** 487–489.

83. **Liu, D.** 1983. Resazurin reduction method for activated sludge process control. *Environ. Sci. Technol.* **17:**407–411.

84. **Losev, V. A., B. A. Chagrin, T. N. Vadysheva, and L. V. Shurygina.** 1988. Influence of seed material quality on cephalosporin C biosynthesis. *Antibiot. Khimioter.* **33:**493–496.

85. **Losev, V. V., P. V. Kotlenska, D. P. Tsoneva, and T. N. Laznikova.** 1981. Oxidation-reduction activity of mycelium as criterion for estimation of inoculum quality in biosynthesis of gentamycin. *Antibiotiki* **26:**496–500.

86. **Luong, J. H. T., and B. Volesky.** 1983. Heat evolution during the microbial process: estimation measurement and applications. *Adv. Biochem. Eng. Biotechnol.* **28:**1–40.

87. **Malik, K. A.** 1990. A simplified liquid-drying method for the preservation of microorganisms sensitive to freezing and freeze-drying. *J. Microbiol. Methods* **12:**125–132.
88. **Malik, K. A.** 1991. Cryopreservation of bacteria with special reference to anaerobes. *World J. Microbiol. Biotechnol.* **7:**629–632.
89. **Malik, K. A.** 1993. Preservation of unicellular green algae by a liquid-drying method. *J. Microbiol. Methods* **18:** 41–49.
90. **Matcham, S. E., B. R. Jordan, and D. A. Wood.** 1985. Estimation of fungal biomass in a solid substrate by three independent methods. *Appl. Microbiol. Biotechnol.* **21:** 108–112.
91. **Matsumoto, T., and A. Sakai.** 1995. An approach to enhance dehydration tolerance of alginate-coated dried meristems cooled to −196°C. *Cryo Lett.* **16:**299–306.
92. **McCann-McCormick, P. A., R. L. Monaghan, E. E. Baker, R. T. Goegelman, and E. O. Stapley.** 1981. Studies on the avermectin fermentation, p. 69–75. *In* M. Moo-Young (ed.), *Advances in Biotechnology*, vol. 6. Pergamon Press, Oxford, United Kingdom.
93. **McLoughlin, A. J.** 1994. Controlled release of immobilized cells as a strategy to regulate ecological competence of inocula. *Adv. Biochem. Eng. Biotechnol.* **51:**1–45.
94. **Meyer, H. P., W. Beyeler, and A. Fiechter.** 1984. Experiences with the on-line measurement of culture fluorescence during cultivation of *Bacillus subtilis, Escherichia coli* and *Sporotrichum thermophile*. *J. Biotechnol.* **1:**341–349.
95. **Mishima, K., A. Mimura, Y. Takahara, K. Asami, and T. Hanai.** 1991. On-line monitoring of concentrations by dielectric-measurements. *J. Ferment. Bioeng.* **72:**291–295.
96. **Mishima, K., A. Mimura, and Y. Takahara.** 1991. On-line monitoring of cell concentrations during yeast cultivation by dielectric measurements. *J. Ferment. Bioeng.* **72:**296–299.
97. **Monaghan, R. L., and S. A. Currie.** 1985. Preservation of antibiotic production by representative bacteria and fungi. *Dev. Ind. Microbiol.* **26:**787–792.
98. **Moyer, J. D., J. F. Henderson, and R. Von Tigerstrom.** 1983. Ultrasensitive assay of RNA: application to 100–500 cells. *Anal. Biochem.* **131:**190–193.
99. **Mugnier, J., and G. Jung.** 1985. Survival of bacteria and fungi in relation to water activity and the solvent properties of water in biopolymer gels. *Appl. Environ. Microbiol.* **50:**108–114.
100. **Müller, S., A. Lösche, T. Bley, and T. Scheper.** 1995. A flow cytometric approach for characterization and differentiation of bacteria during microbial processes. *Appl. Microbiol. Biotechnol.* **43:**93–101.
101. **Musilek, V.** 1963. Prevention of loss of streptomycin production on repeated transfer of *Streptomyces griseus*. *Appl. Microbiol.* **11:**28–29.
102. **Nabais, A. M. A., and M. M. R. da Fonseca.** 1995. The effect of solid medium composition on growth and sporulation of *Streptomyces clavuligerus*: spore viability during storage at +4°C. *Biotechnol. Tech.* **9:**361–364.
103. **Naumann, K., and E. Griesbach.** 1993. The ability of plant pathogenic microorganisms to survive in soil a summarizing reflection. *Zentbl. Mikrobiol.* **148:**451–466.
104. **Novak, J., J. Kopecky, O. Kofronova, and Z. Vanek.** 1993. Instability of the production of avermectins, sporulation, and pigmentation in *Streptomyces avermitilis*. *Can. J. Microbiol.* **39:**265–267.
105. **Oriol, E., R. Contreras, and M. Raimbault.** 1987. Use of microcalorimetry for monitoring the solid state culture of *Aspergillus niger*. *Biotechnol. Tech.* **1:**79–84.
106. **Padmanabhan, S., M. V. R. Murthy, and B. K. Lonsane.** 1993. Potential of *Aspergillus oryzae* CFTRI 1480 for producing proteinase in high titres by solid state fermentation. *Appl. Microbiol. Biotechnol.* **40:**499–503.
107. **Parton, C., and P. Willis.** 1989. Strain preservation, inoculum preparation and development, p. 39–64. *In* B. McNeil and L. M. Harvey (ed.), Fermentation: a practical approach. IRL Press, Oxford, United Kingdom.
108. **Peng, C.-S., and T.-L. Chen.** 1994. Influence of n-hexadecane on penicillin fermentation with *Penicillium chrysogenum*: the morphological effect. *Biotechnol. Tech.* **8:**773–778.
109. **Perkins, D. D.** 1962. Preservation of Neurospora stock cultures with anhydrous silica gel. *Can. J. Microbiol.* **8:** 591–594.
110. **Perkins, D. D.** 1977. Details for preparing silica gel stocks. *Neurospora Newsl.* **24:**16–17.
111. **Perlman, D., R. B. Greenfield, Jr., and E. O'Brien.** 1954. Degeneration of a *Streptomyces griseus* mutant on repeated transfer. *Appl. Microbiol.* **2:**199–202.
112. **Pertot, E., D. Rozman, S. Milicic, and H. Socic.** 1988. Morphological differentiation of immobilized *Claviceps paspali* mycelium during semi-continuous cultivation. *Appl. Microbiol. Biotechnol.* **28:**209–213.
113. **Peterson, S. W.** 1994. The agricultural research service patent culture collection. *J. Ind. Microbiol.* **13:**199–200.
114. **Prior, B. A., N. H. M. Holder, S. G. Kilian, and J. C. Du Preez.** 1988. Measurement of *Candida utilis* growth using the adenosine triphosphate bioluminescent assay. *Syst. Appl. Microbiol.* **10:**191–194.
115. **Puchkov, E. O., and A. N. Melkozernov.** 1995. Fluorimetric assessment of *Pseudomonas fluorescens* viability after freeze-thawing using ethidium bromide. *Lett. Appl. Microbiol.* **21:**368–372.
116. **Reichenbach, H., and M. Dworkin.** 1991. The myxobacteria in the prokaryotes, p. 3416–3487. *In* A. Balows, H. G. Trüper, M. Dworkin, W. Harder, and K. H. Schleifer (ed.), *The Prokaryotes*, 2nd ed., vol. 4. Springer-Verlag, New York, N.Y.
117. **Reinhoud, P. J., E. W. M. Schrijnemakers, F. van Iren, and J. W. Kijne.** 1995. Vitrification and heat-shock treatment improve cryopreservation of tobacco cell suspensions compared to two-step freezing. *Plant Cell Tissue Organ Cult.* **42:**261–267.
118. **Reusser, F., H. J. Koepsell, and G. M. Savage.** 1961. Degeneration of *Streptomyces niveus* with repeated transfers. *Appl. Microbiol.* **9:**342–345.
119. **Roels, J. A.** 1980. Simple model for the energetics of growth on substrates with different degrees of reduction. *Biotechnol. Bioeng.* **22:**33–53.
120. **Rowe, A. W.** 1995. Cryopreservation of blood: an historical perspective. *Infusionsther. Transfusionmed.* **22**(Suppl. 1):36–40.
121. **Rudakova, A. V., and M. A. Malkov.** 1987. Characteristics of oleandomycin producing organism inoculum at

various levels of antibiotic biosynthesis. *Antibiot. Med. Biotekhnol.* **32:**434–437.

122. **Ruohang, W., and C. Webb.** 1995. Effect of cell concentration on the rheology of glucoamylase fermentation broth. *Biotechnol. Tech.* **9:**55–58.
123. **San Antonio, J. P.** 1978. Stability of spawn stocks of the cultivated mushroom stored for nine years in liquid nitrogen (−160 to −196°C). *Mushroom Sci.* **10:**103–113.
124. **Sayadi, S., M. Nasri, F. Berry, J. N. Barbotin, and D. Thomas.** 1987. Effect of temperature on the stability of plasmid, pTG201 and productivity of xylE gene product in recombinant *Escherichia coli* development of a two-stage chemostat with free and immobilized cells. *J. Gen. Microbiol.* **133:**1901–1908.
125. **Schuegerl, K., T. Bayer, J. Niehoff, J. Moeller, and W. Zhou.** 1988. Influence of cell environment on the morphology of molds and the biosynthesis of antibiotics in bioreactors. *Bioreactor Fluid Dyn.* **1988:**229–243.
126. **Scott, W. J.** 1958. The effect of residual water on the survival of dried bacteria during storage. *J. Gen. Microbiol.* **19:**624–633.
127. **Shearer, M. C.** 1979. Actinomycetes: permanent preservation. *Workshop on Preservation of Microorganisms by Freezing and Freeze-Drying*, p. 54–67. Society for Industrial Microbiology, Arlington, Va.
128. **Shieh, W.-J., and L.-F. Chen.** 1986. Effect of inoculum on yeast cell aggregation. *Appl. Microbiol. Biotechnol.* **25:** 232–237.
129. **Singh, V., W. Hensler, and R. Fuchs.** 1989. Optimization of batch fermentor sterilization. *Biotechnol. Bioeng.* **33:**584–591.
130. **Smith, D.** 1984. Maintenance of fungi, p. 83–107. *In* B. E. Kirsop and J. J. S. Snell (ed.), *Maintenance of Microorganisms*. Academic Press, Ltd., London, United Kingdom.
131. **Smith, G. M., and C. T. Calam.** 1980. Variations in inocula and their influence on the productivity of antibiotic fermentations. *Biotechnol. Lett.* **2:**261–266.
132. **Socic, H., V. Gaberc-Porekar, E. Pertot, A. Puc, and S. Milicic.** 1986. Developmental studies of *Claviceps paspali* seed cultures for the submerged production of lysergic acid derivatives. *J. Basic Microbiol.* **26:**533–539.
133. **Song, C. H., K. Y. Cho, and N. G. Nair.** 1987. A synthetic medium for the production of submerged cultures of *Lentinus edodes*. *Mycologia.* **79:**866–876.
134. **Sourek, J.** 1974. Long-term preservation by freeze-drying of pathogenic bacteria of the Czechoslovak National Collection of Type Cultures. *Int. J. Syst. Bacteriol.* **24:** 358–365.
135. **Speers, R. A., M. A. Tung, T. D. Durance, and G. G. Stewart.** 1992. Biochemical aspects of yeast flocculation and its measurement: a review. *J. Inst. Brew.* **98:** 293–300.
136. **Steenson, L. R., T. R. Klaenhammar, and H. E. Swaisgood.** 1987. Calcium alginate-immobilized cultures of lactic streptococci are protected from bacteriophages. *J. Diary Sci.* **70:**1121–1127.
137. **Theron, D. P., B. A. Prior, and P. M. Lategan.** 1983. Effect of temperature and media on adenosine triphosphate cell content in *Enterobacter aerogenes*. *J. Food Prot.* **46:**196–198.
138. **Torriani, S., and M. L. Pallotta.** 1994. Use of polymerase chain reaction to detect *Listeria monocytogenes* in silages. *Biotechnol. Tech.* **8:**157–160.
139. **van Gelderen de Komaid, A.** 1988. Viability of fungal cultures after ten years of storage in distilled water at room temperature. *Rev. Latinoam. Microbiol.* **30:**219–221.
140. **van Ginkel, C. G., A. Haan, M. L. Luitjen, and C. A. Stroo.** 1995. Influence of the size and source of the inoculum on biodegradation curves in closed-bottle tests. *Ecotoxicol. Environ. Saf.* **31:**218–223.
141. **Vara, E., J. Ariasdiaz, N. Villa, J. Hernandez, C. Garcia, P. Ortiz, and J. L. Balibrea.** 1995. Beneficial effect of S-adenosyl methionine during both cold-storage and cryopreservation of isolated hepatocytes. *Cryobiology.* **32:** 422–427.
142. **Wang, H. Y., C. I. Cooney, and D. I. C. Wang.** 1977. Computer aided bakers yeast fermentations. *Biotechnol. Bioeng.* **19:**69–86.
143. **Warr, S. R., C. J. Gershater, and S. J. Box.** 1996. Seed stage development for improved fermentation performance: increased milbemycin production by *Streptomyces hygroscopicus*. *J. Ind. Microbiol.* **16:**295–300.
144. **Webb, C., and S. P. Kamat.** 1993. Improving fermentation consistency through better inoculum preparation. *World J. Microbiol. Biotechnol.* **9:**308–312.
145. **Wilcocks, K. L., and K. A. Smart.** 1995. The importance of surface charge and hydrophobicity for the flocculation of chain-forming brewing yeast strains and resistance of these parameters to acid washing. *FEMS Microbiol. Lett.* **134:**293–297.
146. **Williams, A. M., and E. McCoy.** 1953. Degeneration and regeneration of *Streptomyces griseus*. *Appl. Microbiol.* **1:**307–313.
147. **Xavier, S., and N. G. Karanth.** 1992. A convenient method to measure water activity in solid state fermentation systems. *Lett. Appl. Microbiol.* **15:**53–55.
148. **Yamamoto, K., A. Ishizaki, and P. F. Stanbury.** 1993. Reduction in the length of the lag phase of L-lactate fermentation by the use of inocula from electrodialysis seed cultures. *J. Ferment. Bioeng.* **76:**151–152.
149. **Zabriskie, D. W., W. B. Armiger, D. H. Phillips, and P. A. Albano.** 1980. *Traders' Guide to Fermentation Media Formulation*. Traders Oil Mill Co., Fort Worth, Tex.
150. **Zamola, B., and S. Matosic.** 1987. Optimum lengths of seed development and fermentation in ergot alkaloid production by *Claviceps purpurea*. *Eur. Congr. Biotechnol.* **3:**309–310.

Small-Scale Liquid Fermentations

MATTHEW D. HILTON

4

4.1. INTRODUCTION AND SCOPE

Small-scale liquid fermentations are undoubtedly the most frequent fermentations. Why? Because they are operationally simple and inexpensive to conduct, allowing many to be run simultaneously. While small-scale fermentations can meet the requirements for many fermentation objectives, they have limitations that prevent them from solving all problems (Table 1). However, they can often save much expense by pointing the way for more effective use of innately more expensive larger equipment. The most frequent reason for disappointment in the results from small fermentations is in misalignment of expectations with this tool. The goal of this chapter is to describe the standard use of small-scale liquid fermentation, including reasonable expectations for this tool, and to put the use of this tool in the broader context of all fermentations, especially relative to alternative scales, vessels, and approaches. This chapter is intended to provide background on the basics of small-scale liquid fermentation as commonly practiced in industry today. I have tried to capture principles and literature as starting points for you to approach your particular objectives. Ultimately, however, fermentation science is an empirical and intelligent process, in which an intelligent and educated human works from general principles and specific empirical data to solve each unique problem.

"Fermentation" in this chapter is used in the common way, meaning simply the cultivation of microorganisms. This includes aerobic and anaerobic cultivation of bacteria and fungi but excludes cultivation of cells that would naturally be part of a plant or animal (i.e., cell culture). Discussion will be heavily weighted toward aerobic culture, because most of the higher value products are from aerobic industrial processes and because the scaling of aerobic processes is innately more complicated than the scaling of anaerobic processes. "Small scale" has been interpreted as individual fermentations ranging from microliter volumes, such as in microwell plates, up to the liter range in flasks or carboys. These scales are unified as much by a practical limitation to batch operations as by volume. In batch fermentations the sterilized media components are supplied at the beginning of the fermentation with no additional "feeds" after inoculation (see other chapters in this section for other modes of fermentation). This batch operation is often the greatest limitation of small-scale liquid fermentation. However, the cost of each fermentation vessel is very low compared with typical stirred and fed vessels.

There are many reasons to do small-scale liquid fermentations, and they can be conveniently divided into three major categories based on the "product." The three classes of products are (i) products associated with the growth of the cells, including whole cells (biomass) and subcellular components, (ii) microbial products produced after the growth phase (secondary metabolites or idiolites), and (iii) data.

Although batch fermentations are simple to set up and run, they are dynamic processes that are essentially never in a steady state; thus, they may be far more difficult to understand than more "controlled" processes such as those in stirred, monitored, and fed tank processes. The concept of control in fermentation has its roots in the operation of a chemostat, in which a continuous fermentation process is designed so that one nutrient will be "limiting" or "controlling." That is, the fermentation growth rate is proportional to the dilution rate of the medium. A corollary to the controlled growth rate achieved in a chemostat in true steady state is that the physiology is also controlled. Although small-batch fermentations cannot be in steady state, it is often productive to think of them as a succession of states in which some nutrient is limiting the process. If the goal is a reproducible process, then the immediate objective is to achieve the same sequence and duration of states, and to understand the nature of changes to this succession when experimental variables are introduced. Highly variable fermentation processes may constructively be viewed as the result of a lack of awareness of some parameter critical to the particular process, that is, a failure to search for, find, and control some parameter critical to that fermentation. Often in small liquid batch processes, that critical parameter is gas exchange or balance between respiration rate (oxygen demand) and oxygen transfer (see below).

A reasonably full mechanistic understanding of any fermentation process is costly. This is especially true when complex medium components are used. However, after appropriate background work it is possible to empirically define many aspects of the specific fermentation system. This requires beginning with a basic system that meets the basic needs of the microorganism and provides a solid, reproducible foundation for future research (Table 2). If im-

TABLE 1 Strengths and weaknesses of small-scale liquid fermentation

Strengths
- Potential for large numbers of parallel fermentations
  - Low-cost vessels
  - Low operating costs
  - Vessels take little space and require little infrastructure
- Easy to scale by increasing the number of identical vessels

Weaknesses
- Effectively limited to batch fermentation processes (or punctuated batch)
- Limited maximum oxygen transfer rate
- Discontinuous monitoring only
- No adaptive control of pH or substrate concentration

provements are sought, the scientist must learn, through experimental definition, what limits the cells of a fermentation at each stage of that process, e.g., what limits growth first in the process, and product yield later, if the product is not growth-associated. An empirical understanding of process may be obtained by systematically exploring the effect of various levels of each nutrient. Selection of the base fermentation system, including the vessel, closure, medium, associated equipment, and operating parameters, necessarily goes hand-in-hand with the objective of the fermentation and with the biological requirements of the microorganism to be cultured.

Shaken Erlenmeyer flasks with cotton plug closures are the classical small-scale liquid fermentation system (Fig. 1A), but variations on all aspects of this basic system have been developed. The basic shake flask system has been modified to increase mixing and mass transfer with indentations (baffles, Fig. 1B) and "corners." Alternative closures have been selected because they allow more gas transport in and out of the vessel, and simple sparged and lightly stirred carboys (Fig. 1C) are used for very large volumes when shaking is impractical or deleterious to the microbe.

4.2. FERMENTATION VESSELS

The standard Erlenmeyer flask has the virtue of being readily available in a variety of sizes at low cost. However, since it was not designed for fermentation, many other vessels have been used for small-scale batch fermentations to meet special objectives (Table 3). Alternatives to the basic shaken flask have included vessels with a square base (14), a tetrahedron shape (19), and a more barrel-like shape (49). Most often the vessel seems to be selected for oxygen transport, liquid mixing, convenience, and familiarity. However, it is important to periodically evaluate whether your standard fermentation vessel is appropriate to your current objectives.

There are no absolute limits on the range of volumes that could be considered small-scale liquid fermentations. The lower limit has historically been determined by requirements for ease of reproducible and aseptic operations and for reliable sampling and analysis methods. As liquid handling and analytical methods have been miniaturized, some fermentations have been similarly miniaturized. An entire industry has developed around automated liquid handling and biochemical analysis in microwell plates, especially the 96-well microwell plates. In addition, 48-, 24-, 12-, and 6-well plates are available with the same standard "footprint." This microwell format-based industry has increased the precision, accuracy, and ease of handling small volumes and measuring outcomes from reactions in these microwell plates. Those same plates and liquid handling technologies are now used for cultivation of microbes in some situations. They are especially useful when large numbers of strains need to be evaluated simultaneously, and growth (e.g., turbidity or pH change) is the measured end point. For example, enumeration of viable cells in a solution is sometimes more convenient by most probable number determination in microwell plates than by plating on agar (21). Furthermore, BiOLOG (Hayward, Calif.) has developed and markets a culture identification system based entirely on growth of unknown microbes in 96-well plates containing a fixed array of carbon sources. In addition to clinical culture identification, variations on the standard BiOLOG system have been used to survey microbial communities (22, 26) and to screen for microorganisms from nature that are capable of degrading toxic chemicals (27). Generic microwell plates have been used to evaluate the physiology of many strains in parallel (12) and to select improved strains with a variety of traits (17, 35, 48). Evaporation remains a significant limitation on miniaturization of lengthy fermentations. However, this barrier can be overcome by incubation in a high humidity chamber or ignored if the incubation time is short enough.

Test tubes of many sorts have been used effectively for volumes up to ca. 10 ml. They are cheap and easy to handle with common lab equipment. Traditionally, culture tubes were glass, but there are now many disposable plastic tubes, with caps, that are extremely convenient for use in the culture of microbes. Even the ubiquitous 1.5-ml microcentrifuge tubes have been used as shaken fermentation vessels to evaluate many strains in parallel fermentations (46). Gas transport was through an open hole with no barrier to the entrance of foreign microbes, but strain evaluation was successful because it was brief.

The upper limit on the volume of small-batch fermentations is typically set by oxygen uptake requirements of the culture to be fermented. Generally, the gas transfer rate at the gas-liquid interface is less as the volume gets larger. Shaken vessels with up to about 1 liter of working volume are fairly common and can be scaled for larger harvests by

TABLE 2 Components of a small fermentation system

Parameter selected	Parameters affected
Vessel	Culture volume, gas-liquid mass transfer, mixing
Closure	Gas transfer out-to-in, in-to-out; evaporation
Medium composition	Kinetics and extent of growth and metabolism
Temperature	Kinetics of growth and metabolism, evaporation
Agitation rate (cycles per min and throw)	Gas-liquid mass transfer, mixing time

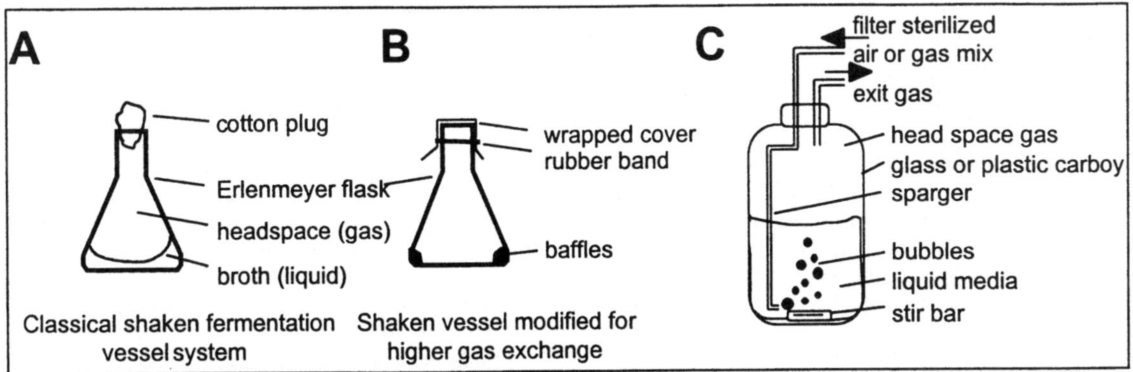

FIGURE 1 Schematic representations of common small-scale liquid fermentation systems.

simultaneous parallel batches. In addition, larger volume carboys with a bubble sparger are excellent fermentation vessels for microbes with low demand for gas exchange. These would include anaerobes, facultative organisms with anaerobic processes, aerobic organisms at low densities or low metabolic rates, and some photosynthetic processes. Carboys are not shaken, but rather are mixed with a magnetically coupled stir bar or simply by the lift of the bubble column (air lift mode). Maximum oxygen transfer rates would be much less than those in typical shaken flasks and stirred fermentors. Small-batch fermentations between these two extremes are traditional and convenient for most purposes. If the goal is biomass and the organism is aerobic, several well-balanced, shaken Erlenmeyer flask fermentations are likely to produce more cells than a harvest volume 10-fold greater from a bubbled carboy due to the substantially greater oxygen transfer rates possible in the shaken vessel.

4.2.1. Modification of Vessels

Common vessels, such as Erlenmeyer flasks, have a smooth inner surface. When the vessel is round-based and motion is orbital, mixing is very poor (40). However, Erlenmeyer flasks can be modified with baffles or stainless steel springs, or flasks with alternative shapes can be used (see above). Baffles have been used primarily to increase the turbulence of mixing to increase the liquid surface area and therefore gas transfer (24, 37, 38, 49). However, baffles tend to promote growth on the walls of the vessel and sometimes cause splash up to the vessel closure, which may alter gas transport in and out of the vessel (see below). The physiology of the cells growing on the walls is likely to be fundamen-

TABLE 3 Some vessels commonly used for small-scale liquid fermentations

Vessel	Throughput (~vessels/day)	Working volume range (ml)	Comments
96-Well microplates (200-μl nominal well volume)	1,000–10,000 or more	0.05–0.2	Vigorous mixing is not possible; automated liquid handling and readers are readily available
Test tubes or culture tubes	100–1000	1–10	
Erlenmeyer flasks	10–100	2.5–500 (5–25% of nominal)	
Fernbach flasks (2,800-ml nominal volume)	10–100	300–800	Similar to Erlenmeyer except lower height-to-width ratio
Tunair (Alpha Medco, Wayland, Mass.) flasks	10–100	100–800	Engineered shaken fermentation vessels, several sizes available (49)
Carboys (3,000–20,000-ml nominal volume)	1–10	500–10,000	Low O_2 transfer, can be bubbled; often used for algae with CO_2 bubbling

tally different from that in the liquid fraction of the same vessel, resulting in two simultaneous fermentations in the same vessel. Fermentations with much wall growth or splash are likely to be difficult to repeat or scale. Furthermore, some microorganisms may grow differently when subjected to conditions with higher shear (49).

Jensen and coworkers (33) used stainless springs, hooked head-to-tail for a toroidal insert into the bottom of the shaken vessel, for the same purpose as permanent baffles of increasing the gas transfer rate; Hopwood and coworkers (30) focused on the value of shear from the springs in encouraging dispersed growth by *Streptomyces* spp. Dispersion of growth by filamentous microorganisms may increase the cellular surface area exposed to the liquid medium. This may have benefit in relieving kinetic limitations of nutrient access to all of the cells in the culture. However, this dispersion may affect viscosity and result in greater peak oxygen demand, either of which could be deleterious to expression of many fermentation products if not balanced with reduced medium "concentration" and greater oxygen transfer (see below, under Liquid Media Composition).

4.2.2. Vessel Closures

The purpose of the closure is to serve as a selectively permeable barrier. Foremost, the closure must provide a sterile barrier to maintain an axenic culture and to prevent aerosols from the culture from escaping. The second set of requirements is to allow exchange of gases. A high rate of exchange of oxygen and carbon dioxide is most frequently critical, but the ideal closure would simultaneously limit evaporation. Limitation of gas exchange past the closure has two major consequences. First, it may limit oxygen availability to aerobic cells as their density and oxygen demand increase. Second, the gas mixture within the headspace of the flask will differ in CO_2 and N_2, as well as O_2, from the ambient gas mixture as the oxygen is consumed and converted to carbon dioxide (29, 41). Elevated CO_2 has the potential to alter the medium pH or have other specific biological effects. Few materials are ideal as closures, but many are in common use (Table 4).

Cotton fiber plugs are cheap and simple and therefore remain in common use. They are often wrapped in cotton cheesecloth to improve handling and to prevent fibers from sticking to the inside of the neck of the vessel. They are an excellent choice if high and reproducible rates of gas transport are not critical to your fermentation system. Cotton plugs are known to severely limit gas transport in and out of baffled fermentation vessels (9, 24, 37, 41), especially if wet from condensation following autoclaving or splash from vigorous mixing. Variability in the density of the hand-made cotton fiber plugs combines with variability in dryness to vary the maximum gas transport rate between ostensibly equivalent flasks. Foam plugs suffer from the same potential for wetting but may be more consistent with respect to density since they are commercially manufactured. They have the clear advantage over cotton plugs that they come ready to insert, with no preparation.

Solid caps for aerobic fermentation are typically friction-fit for ease of handling. They allow free air flow through a path between the cap and the neck of the vessel. Gas transport is limited by the annular area of that path and is often small enough to limit the maximum gas transport rate in the fermentation (i.e., gas transfer in and out of the vessel may be lower than that between the gas and liquid phases within the vessel). Since there is an open path, with no sterile barrier, there is a risk of airborne microbes or spores finding their way in or out of the vessel.

TABLE 4 Commonly used closures for small fermentation vessels

Closure description	Special source[a]	Comments
Cotton fiber plug		Highly variable in density; may be wrapped in cheesecloth
Open-cell foam plug	diSPO plugs	Very convenient plugs
Plastic or metal cap	KAP-UTS, Kim-Kap, S/P Morton stainless closures	
Cotton fiber-gauze sandwich	Reference 24	
Filter disks, e.g., for milk	Reference 51	Rubber band to hold in place
Sterile wraps	Bioshield II Sterile Wraps	Rubber band to hold in place
Silicone sponge closures	Silicone Sponge Closures 2004	Putatively limit evaporation by 50% versus cotton plugs
Membrane cover	CAP-SIL for Tunair flasks	Silicone is used by Tunair, but other membranes may be suitable

[a]KAP-UTS, Silicone Sponge Closures 2004, Bellco Glass, Inc., Vineland, N.J.; diSPO plugs, S/P Morton stainless closures, Bioshield II Sterile Wraps, Baxter Scientific Products, McGaw Park, Ill.; Kim-Kap, Kimble Glass, Inc., Vineland, N.J.; CAP-SIL, Tunair, Shelton Scientific, Shelton, Conn.

Solid caps are most often used for test tube vessels in short fermentations because of their ease in removal and reclosure. Screw-on solid caps and septum-type caps are commonly used for anaerobic fermentations.

A closure descended from the cotton plug but improved for maximum gas transport rates was made by constructing a filter "sandwich" from gauze and cotton fiber wrapped over the top of the vessel and held in place by a spring clip (24). Subsequently, milk filter disks (51) and specialized sterile-barrier "blue paper" (Bioshield; Baxter Scientific Products, McGaw Park, Ill.) have been substituted as economical alternatives to the sandwich with the advantage that they require less manual preparation. All can be held in place with thick rubber bands sold for office use. These covers all have the benefits of providing a cheap, convenient, and disposable sterile barrier with substantially improved maximum gas transport rates when compared with the plug-type closures. In addition, since the covers go over the top and around the outside of the vessel, they are less likely to be wetted and they protect the lip of the vessel fully from contamination, making aseptic handling easier. Evaporation is likely to increase with the improvements in gas transfer.

Variations providing potential improvements, with some trade-off in cost, are possible. Two noteworthy commercial variants are Bellco's silicone sponge closure (Bellco Glass, Inc., Vineland, N.J.) and the special membrane dome designed for Tunair flasks (Shelton Scientific Manufacturing, Inc., Shelton, Conn.) (49). These have many of the advantages of the elastic-banded covers described above and are easier to handle since they do not require a separate elastic band to hold them closed. Both can be reused, so clean-up may be a concern, especially for the foam, because mycelia or medium may clog the foam's cells, resulting in a loss in gas transport performance on repeated use. Tunair has designed its closure in two parts: a rigid plastic, dome-shaped frame (CAP-FLO) and a silicone membrane barrier (CAP-SIL). The dome shape of the closure provides more membrane surface area for gas transport than simple plug- and cover-type closure designs. In addition, there are vanes on the closure frame to increase air turbulence on the surface of the membrane. These features are logical design changes and may be of benefit to some fermentations.

4.3. SHAKERS

Machines for holding small liquid fermentation vessels and moving them to induce mixing are generically called shakers. There are many sizes, designs, and options available (reviewed in reference 23). Before purchasing shakers, it is vital to specify the functional requirements so you can carefully compare features, options, upgradability, price, and maintenance costs (Table 5).

Environmental control, minimally temperature and often humidity, is generally necessary for fermentation process control. Shakers can be purchased with or without environmental controls; if purchased without controls, the shaker would normally be placed in an environmentally controlled room or temperature-controlled box. Space, cost, and flexibility would determine which of these two alternatives is better. Maintenance of temperatures below 25 to 28°C generally requires chilling. This can be accomplished by built-in refrigeration or with a coil-type heat exchanger and externally supplied chilled water. Humidity control becomes important with long fermentation times, when much of the liquid volume will be lost to evaporation without elevated ambient humidity. However, high humidity may result in the unintended growth of mold on the walls of the environmental chamber due to condensation. Some humidified environmental chambers have heaters in their walls to prevent condensation.

TABLE 5 Items to consider in selecting a shaker

Shaker type
Floor
Chest
Cabinet
Stackable
Benchtop
Environmental control (see below under Facilities)
Mixing
Pattern of movement
Reciprocal
Orbital (rotary or gyrotory)
Speed range
Speed control (precision, accuracy, reliability)
Amplitude of motion (throw or diameter)
Slow acceleration (to limit splashing and wetting of closures)
Capacity
Clamps and racks
Vessel headspace
Number of vessels, volume per vessel
Facilities
Utility
Electrical
Chilled water
Lab space
Environmental control
Temperature
Humidity
Photosynthesis lights
Gas make-up control (typically for elevated CO_2)
Equivalence of all positions for temperature and air flow (for equal evaporation)
Convenience of use and maintenance
Ease of reconfiguring clamp arrangement
Ease of loading and unloading
Ease of cleaning after a spill
Service and maintenance
Durability
Resistance to damage from spills (e.g., electronics)
Corrosion resistance
Precise, durable friction parts
Routine maintenance
Frequency
Cost
Service availability
Service technician distance
Base cost per call
Time for service and parts locally available
Other considerations
Purchase price
Noise
Safety
Electronic data output (time, actual rpm, actual temperature, alarms)

Fermentation broth temperature can be controlled by bathing the vessels in constant-temperature water or air. Water bath shakers have the advantages of efficient heat transfer and the potential for sampling without removing the vessel from the bath; this may result in less effect on fermentation. But the shaker speed of water bath incubators is limited by splash outside of vessels, and substantial maintenance is required to keep the bath water clean. Culture vessels obviously come out wet, so they need to be dried prior to aseptic operations. Air incubators are more commonly used for shakers because of their overall greater ease of handling and because of their higher potential shaking speeds. With either type of temperature control, but especially with air incubators, there is a very real potential for unevenness in temperature within the chamber. This may be inherent in the design of the chamber (e.g., warmer near the motor in a multi-tier incubator-shaker) or may be due to the operation of the chamber (e.g., loading that blocks air flow or frequent opening). Also, the heat production in a rapidly growing vessel may exceed the heat removal capacity in some air incubator systems, so the temperature of the broth may be transiently elevated during the growth phase relative to the chamber temperature. Monitoring temperature within fermentations or mock fermentations at various positions within the chamber under normal operating conditions (culture, medium, vessel, and closure all interact) is the only sure way to know what temperature or temperature profile the cells "feel."

Shaker motion is in two common patterns: orbital (also called rotary or gyrotory) and reciprocal (also called linear). Shaking vigor (which determines gas-liquid mass transfer and shear) is a function of the diameter of the orbit or stroke of a reciprocal shaker (commonly called "throw" and measured in inches or centimeters) and the cycle time (or speed, measured in rpm or cpm). The throw for either motion is typically between 1 and 8 cm and is most often fixed for each shaker, or adjustable over a limited range. Many models with fixed throw have several choices available at the time of order. The shaker speed is generally variable, by highly accurate electronic speed controllers. Reciprocal shakers mix standard smooth flasks much better than orbital shakers do but run into more problems at higher agitation rates (e.g., severe splashing [40]).

Shakers come in a variety of physical sizes and configurations. Vessel capacity requirements, space availability, and cost are the major considerations to balance in selection of the basic shaker. The highest capacity per shaker and per unit of floor space comes from multi-tier shakers. A recent variation is front-loading modular incubator shakers. These can be used as bench-top units individually or can be stacked for multiple incubator shakers with independent shaker and temperature control yet still requiring a limited floor space. Modern shakers can be fitted with a wide variety of clamp sizes. Often the clamps are fitted to a board that can be quickly swapped for another board populated with a different set of clamp sizes or loaded with vessels outside the shaker.

4.4. LIQUID MEDIA COMPOSITION

At first look, selection of medium for small-scale liquid fermentations is the same as that for larger vessels (including animals). The ideal starting point is to supply a "diet" balanced for carbon, nitrogen, phosphorus, minerals, and essential vitamins. However, for many processes the selection of a fermentation medium is inextricably linked to selection of the other fermentation parameters. Therefore, if fermentation performance (efficiency, titer, etc.) is a goal, optimization will require empirical determination. Statistical experimental designs are most effective in optimization but depend on prior establishment of a reproducible base system (see, for example, references 10, 18, 25, 31, and 39). The principles of statistical experimental design are introduced succinctly by Deming (16), and their application to medium development is covered at length in chapter 6 of this volume. However, I will try to describe the principles special to small-scale batch liquid fermentation systems.

Medium composition can be divided into two extreme categories: (i) nutrient-balanced media typically used for biomass and growth-associated products (primary metabolism) and (ii) unbalanced media that support only low growth rates but ongoing metabolism (6, 7, 15, 45, 50). Hybrid media compositions between these two extremes are most common. In this instance, balanced nutrients support rapid growth to an initial cell mass (a vegetative stage); the remaining nutrients are unbalanced to trigger a secondary metabolic phase or sequence of secondary phases (6, 8).

Nutrient-balanced media for growth in small-scale batch processes are typically no different from those that would be used for stirred and fed fermentors, except that it is important to adjust the "strength" to be commensurate with the oxygen transfer capability of the fermentation system. If not adjusted, the fermentation process cannot be expected to be similar between scales, despite the use of the same medium composition. The whole medium can be adjusted by dilution or concentration of all ingredients, or the concentration of a key ingredient (e.g., glucose or ammonium) that is intended to limit rapid growth at all scales can be adjusted.

In larger, stirred tank fermentations, a secondary phase is often sustained and controlled by delivery of an exogenous feed, supplying limiting quantities of an essential nutrient, especially carbon, nitrogen, or phosphorus. With the restriction to batch operations of small liquid fermentations, a similar nutrient limitation may be achievable by introduction of a kinetic limit using a slowly metabolized nutrient source. The limitation may be directly on the rate of metabolism or it may be related to transport into the cells or availability for transport. Introduction of a limit to availability is commonly accomplished with ingredients such as starch, flour, oils, and proteins that must be degraded by extracellular enzymes before their subunits can be transported and further metabolized. Ingredients engineered into sustained-release formulations constitute a possible alternative approach to this kinetic limitation (36).

Sterilization of media can be a cause of variability. Although it is common to sterilize media by autoclaving, too often the operation is done with inadequate control, resulting in failure to sterilize (especially *Bacillus* spores, if present in high numbers) or in inconsistency due to differences in the chemical reactions driven by the heat of autoclaving (2, 3). Common mistakes include (i) failure to presteam long enough to bring the liquid volume to 100°C prior to pressurizing the chamber so the liquid never reaches 121°C; (ii) overcrowding the autoclave chamber or using closed-basin trays so steam flow is restricted, effectively insulating the material to be sterilized from the steam; and (iii) failure to deliver an equivalent amount of heat per volume to media in different sizes of vessels. Effective sterilization can be proved by including commercial spore test ampoules in a "dummy" load, equivalent to the largest liquid load in the cycle. Sterilization time can be varied experimentally to test whether your particular fermentation system is significantly affected by the heat of sterilization.

4.5. GAS EXCHANGE IN SMALL LIQUID FERMENTATIONS

The mass transfer rate between phases (e.g., oxygen transfer from gas to liquid) is very often one of the most critical operational parameters for any aerobic fermentation. Too often it is neglected as small fermentation systems are set up and is therefore undoubtedly a frequent cause of irreproducibility. Gas-liquid mass transfer, or an estimate of the maximum potential mass transfer (expressed as K_La [20]), is a standard parameter defining operating limits in sparged fermentors but has equal potential to limit small batch cultures. The importance of oxygen mass transfer in aerobic fermentations should be obvious since the maximum oxygen transfer rate will determine the cell mass that can be supported with a maximally aerobic physiology. The mass transfer that ultimately matters is that transferred to and from the cells per unit time. It is best to take a whole-system view toward this (28). For example, whether oxygen reaches the cells depends on, sequentially (Fig. 2): (i) diffusion of oxygen from the ambient air outside the vessel through the closure to the interior compartment (vessel headspace), (ii) oxygen transfer from the gas phase to the liquid phase within the vessel, and (iii) soluble oxygen transfer from the liquid to the cells. Any one of these steps can be limiting in an aerobic shaken fermentation, and different ones can limit at different times. The direction of diffusion of the product carbon dioxide will be opposite that of oxygen (i.e., from cells to liquid to gas and out), again depending on mass action as the driving force in this reverse pathway.

The rate of diffusion of gases through the vessel closure of a shaken fermentation depends on the diameter of the vessel opening, the depth of the closure, the nature of the closure, and the difference in concentrations of the specific gas in the inside and outside compartments (14, 24, 37, 41). Diffusion through the closure has been reported to limit oxygen availability in some fermentation systems based on inference from oxygen absorption rate experiments and on cell yield (14, 28, 37). Furthermore, even when the oxygen diffusion rate through the closure has not become rate limiting, that closure may cause very large increases in the concentration of CO_2 (to >15%) in the vessel headspace (29, 41). This situation is fundamentally different from that for sparged vessels, in which CO_2 is constantly stripped by the excess of N_2 in the sparge air and thus may be a barrier to scale-up. Generally, it is best to have greater mass transfer at the vessel closure than across the gas-liquid interface, although this does not eliminate this fundamental difference between shaken and sparged vessels.

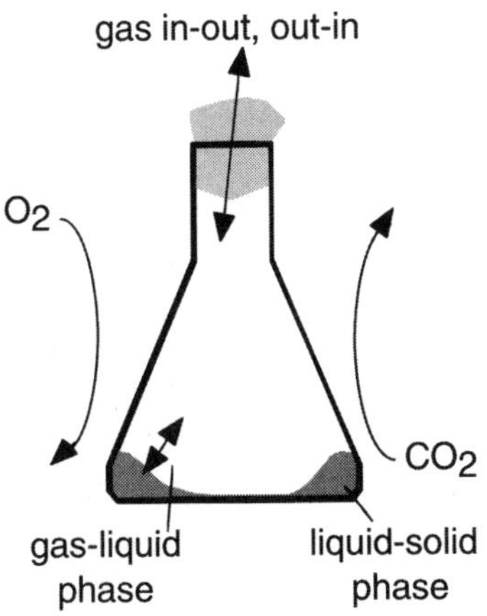

FIGURE 2 Schematic representation of gas exchange in a respiring shaken flask.

The transfer of oxygen across the gas-liquid barrier frequently limits aerobic stirred and sparged fermentations and has received much attention in the fermentation literature. If shaken fermentations are to be scaled to stirred vessels, it is probably best to attempt to match this rate between scales, or to do experiments that help extrapolate toward optimal for the destination. The gas-liquid mass transfer term K_La is the product of the gas-liquid mass transfer coefficient, K_L, and the gas-liquid interfacial area, a. There is normally no way to determine either of these separately, so they are commonly treated as a single value, with units in reciprocal time (typically h^{-1} or min^{-1}). More thorough treatment of gas-liquid mass transfer and derivation of K_La can be found elsewhere (e.g., references 1 and 4).

Although K_L and a cannot be determined separately, they should be thought of separately and can be affected independently. K_L is affected by medium composition, temperature, addition of antifoam agents, and growth of microorganisms in the medium. The interfacial area, a, is affected by the volume of liquid, the volume and shape of the fermentation vessel, the pattern of agitation, and the vigor of agitation. Power input per unit liquid volume is a major determinant of bubble size (and therefore interfacial area, a) and is very different between shaken vessels and standard baffled stirred tanks. Specifically, there is much greater power and shear in the vicinity of impellers of stirred fermentors as commonly configured and operated than can be achieved in shaken cultures. In addition, because of the mechanism for power delivery, the instantaneous power input in shaken broths is much lower and more homogeneous throughout the liquid than for stirred broths (40).

Liquid-solid mass transfer in fermentations generally will limit only when mixing is poor (40). This may occur with poor agitation, in highly viscous broths, or in cultures growing in large colonies (clumps, balls, or pellets). Mixing times in nonviscous solutions ranged from 52 s to less than 2 s in several standard fermentation configurations and were affected by liquid volume, vessel type (e.g., smooth versus baffled flask), agitation rate, and agitation pattern (40). However, if the broth is highly viscous, it may not be possible to improve mixing except by limiting cell mass or the nature of the growth (e.g., "pelleted" growth rather than diffuse). If viscosity becomes high enough, what began as a liquid fermentation may finish as something more akin to a solid-state fermentation where mixing does not occur.

Many estimates of the overall rate of oxygen absorption into shaken aqueous solutions have been done by one of two general methods. The first method is estimation of oxygen absorption potential by measurement of liquid-phase oxygen reaction with sodium sulfite in the presence of copper (13, 14). The reaction is believed to be essentially instantaneous and complete so when measured over time, the quantity of sulfite oxidized reflects the amount of oxygen absorbed into the liquid. This method allows ranking of fermentation operating configurations but generally overestimates the actual transfer that can be expected in fermentation medium with cells growing (53). The other method, with many variations, depends on measurement of oxygen concentration (typically using a polarographic probe) in a dynamic situation (4). This method is good because fermentation medium and growing cells can be present during the determination, but it depends on a homogeneous medium constantly covering the oxygen probe; this is easy in stirred tanks but difficult to achieve with low volumes or baffles in high-mass-transfer shaken vessels.

Although the variety of flasks, closures, shakers, volumes, media, and methods does not allow simple comparisons between the rates of mass transfer reported in the literature (see Table 6 for selected examples), some generalizations can be made. First, gas transfer in and out of the flask closure is better if the gas path has a larger area (e.g., wide-necked flasks) and if the thickness (path) of the closure is minimal (e.g., a filter rather than a plug). Anything that increases the gas-liquid interfacial area (a) will clearly increase K_La. If all else is held constant, this can generally be done by (i) decreasing the liquid volume relative to the vessel nominal volume, (ii) increasing shaker cycle rate, (iii) modifying the vessel, such as with baffles, to increase turbulent flow, and (iv) having a larger diameter or stroke on the shaker. Improving K_La through K_L is often constrained by other system requirements. However, it is worth being aware that generally K_L will be decreased when (i) medium components are added to water (51), (ii) cells grow in a culture medium (38), (iii) antifoam is added (see below under Foam Control), or (iv) viscosity increases (whether from medium components, mycelia, or polymer accumulation).

Some type of time-averaged rate of oxygen transfer to the cells is ultimately what determines cellular physiology and the sequence of stages of the fermentation. Thus, estimates of oxygen absorption in a shaken system serve as a guide to help balance the maximum transfer with the maximum cell mass that may be developed in that system. There is no generic rule as to whether oxygen transfer should exceed oxygen demand at all stages, since there are examples of different processes that depend on either oxygen limitation (11) or maintenance of oxygen transfer above demand sufficient to keep the dissolved oxygen concentration above some critical value (summarized in reference 4). However, in commercial processes the latter situation appears to outnumber the former, and respiration is so central to the energy generation systems of aerobes that it is prudent to estimate the likelihood of oxygen limitation on theoretical grounds and follow with empirical determinations. Brown (4) compiled a list of the specific oxygen demands for fermentations of a variety of cultures from literature reports; they fell in the range of 3 to 11 mmol per g of dry cell mass per h (0.05 to 0.18 mmol g^{-1} min^{-1}). Many common fermentation media have enough carbon, nitrogen, and phosphorus to support growth to densities of 10 g $liter^{-1}$ dry cell weight or more. When

TABLE 6 Range of reported oxygen absorbance rates determined by the measurement of the oxidation of sulfite

Vessel, closure	Liquid volume (% of nominal volume)	Range of oxygen absorbance rates (mmol/liter/min sulfite)	Reference
Test tube, cotton plug		7.9	41
Erlenmeyer flask, cotton plug	10–30	0.25–0.15	14
Erlenmeyer flask			
Cotton plug	18–60	0.25–0.05	24
Open	18–60	0.3–0.09	
Erlenmeyer flask, cotton plug	40	0.3–2.3	41
Erlenmeyer flask, cotton plug		2–0.27	44
Fernbach flask, filter disks			
Smooth	10	0.2	14
With indentations	10	3–3.5	
Fernbach flask, with indentations			
Cotton plug	10	1.6	14
Filter disks	10	3.5	
Open	10	5.1	
Stirred fermentor, fully baffled		5.3 at 500 rpm 1.8 at 300 rpm	14
Stirred fermentor			
760 liters		4.2	49
30 liters		2.6	

these values are multiplied together to estimate the potential range of oxygen demand of a small fermentation system (0.05 to 1.8 mmol liter^{-1} min^{-1}), it is clear that the demand is of the same magnitude as can be supplied by shaken vessels (Table 6) but would have to be determined empirically for any given system. If the microbe can grow rapidly (e.g., *Escherichia coli* or *Saccharomyces cerevisiae*) and the medium will support cell masses above 10 g liter^{-1} dry weight, it is likely the shaken system will become oxygen limited. The consequences of oxygen limitation in an uncharacterized fermentation system cannot be predicted a priori. The theoretical possibilities are (i) no effect or a positive effect if the product is produced under conditions of mixed aerobic and anaerobic metabolism, (ii) a negative effect if there is a critical oxygen concentration for the product or if cell yield is severely reduced, and (iii) a variable outcome if there is a critical oxygen concentration and the process kinetics are not tightly controlled. There are several possible approaches to empirically determine whether oxygen transfer limitation is having a major effect on a fermentation, but no simple approach is perfect. One standard method is to perform the same fermentation in several liquid volumes while all else, including the vessel nominal volume, is held constant. This is simple to perform and if there are no effects on fermentation outcome (e.g., on cell yield on carbon, product titer, or specific product yield), it can be concluded that oxygen transfer is not limiting the process. If effects are seen, they could be due to oxygen limitation, carbon dioxide accumulation, effects on mixing, or any combination of these (29, 41, 42). Deconvolution of these possibilities is not trivial but has been done by using shaken vessels at a constant liquid volume and flushing the headspaces with known mixtures of air enriched with each gas (29, 33).

4.6. FOAM CONTROL

Many fermentation media or fermentation broths will develop a foam head when mixed vigorously. A stable foam constitutes another potential barrier to gas exchange between the vessel's headspace and the liquid phase (47). Antifoams generally decrease gas-liquid mass transfer rates under simple circumstances, but may either increase or decrease overall fermentation mass transfer rates depending on where the limitation exists. Sparged and stirred fermentors do not depend primarily on gas diffusion from the headspace to the liquid, hence are more likely to lose mass transfer potential with antifoam addition (34). Although stirred fermentors have the option of either chemical or mechanical foam control, shaken vessels are limited to chemical control with antifoams (reviewed in reference 5). There are many antifoams in use. In perusing 220 recent abstracts containing the character strings "antifoam" and "ferment" from Derwent's Biotechnology Abstracts (Derwent Technology Ltd, London, U.K.), I found about 40 apparently different antifoams reported. This may be evidence that no single universal chemical antifoam has yet been found. Silicon-based antifoams were reported most often, followed by those based on polyglycols and oils. Antifoams are typically selected on the basis of past experience and empirical evaluation. Common criteria for selection of an antifoam are (i) effectiveness in controlling foam, (ii) absence of toxicity to the specific fermentation, (iii) compatibility with scale-up or harvest procedures, (iv) a minimal negative effect or a positive effect on gas-liquid oxygen transport, and (v) cost of use. Silicon-based emulsions seem to be most effective on the basis of final concentration of the active antifoam component (43), but natural oils are often used to control foam while serving as a carbon source to the microbe so the final concentration can be allowed to decline to very low levels to minimize impact on harvest (52).

4.7. FERMENTATION SAMPLING AND ANALYSES

The importance of analysis of fermentations cannot be overstated; clearly, without valid measurements of outcomes it is impossible to understand, improve, or scale a fermentation, and it may not be possible to reproduce it. There are two critical components to effective fermentation analysis. First is sample preparation. A representative sample of the culture must be taken and prepared appropriately for the specific analysis. Second, the specific analysis must be proved valid in the presence of the fermentation broth matrix; media ingredients often interfere with chemical analyses. Since analysis of small-scale fermentations is not fundamentally different from analysis of any other fermentation, the following is a limited discussion focusing on small-scale batch processes. Matrix effects depend primarily on the medium composition and are otherwise common to all scales. Sampling methods and sample preparation are often linked to the scale of fermentation.

It is best if small fermentations are sampled only once, at harvest. Time points can be obtained by harvesting replicate flasks over time. Miniaturized on-line analysis is possible but may not be sensible since it invariably adds complexity to a system whose strength is simplicity. "Grab" samples, if small relative to the total fermentation volume, can be taken throughout the fermentation but carry several risks: (i) sampling normally requires that the vessel be opened, exposing the fermentation to additional risk of contamination; (ii) each sample removed will lower the volume of the fermentation, thus changing the mass transfer characteristics of the fermentation; (iii) interrupting shaking while sampling may induce a stress response in the microbes (e.g., due to oxygen limitation or temperature change during handling) that is not reproducible.

Whole fermentation broth samples often need to be further processed prior to analysis. Most frequently, the cells (along with other particulates) and a clear liquid fraction are separated from one another by centrifugation or filtration for different analyses. Centrifugation is effective for essentially all fermentation processes. If only small volumes are needed, the method is very rapid and convenient with the use of disposable labware and a microcentrifuge or other small centrifuge. Filtration can be convenient with some microorganisms, but many fermentations will clog the filter quickly, making the method inconvenient. If your fermentation can be easily filtered, the method has the advantage that many broths can be filtered in parallel with the use of a manifold filter holder, and the cells can be washed easily.

Fermentations invariably have one primary end point to be measured that is aligned with the objective of the process. This may be biomass, a cellular product, or data, as outlined above. This is the obvious first analytical objective, and if there is no intention to scale or improve the process, there may be no need for any further analyses.

Conversely, if there is an intention to scale or transfer the process to a different fermentation system, it will likely be necessary to do additional analyses to characterize the process and the kinetics of events in the fermentation (Table 7).

Estimates of biomass are very useful when viewed temporally because they allow inferences about growth phase (when plotted versus fermentation time), cell yield on substrates (e.g., grams of cells per gram of glucose consumed) and specific product yields (grams of product per gram of cells). The best estimate of biomass would come from direct measurement of the dry weight after harvesting cells by centrifugation or filtration from a known volume of soluble broth, washing those cells to remove residual soluble materials, and weighing the heat-dried cell pellet. For many unicellular microorganisms a rapid turbidimetric determination can substitute effectively for dry weight determinations after establishing a standard curve (e.g., relating optical density around 600 nm to dry cell weight). The biomass of filamentous microbes grown on soluble media can sometimes be estimated from simple packed cell volume measurements (fraction of broth volume pelleted by centrifugation), again following establishment of a standard curve relating packed cell volume to dry cell weight. Biomass estimates of small fermentations may be inaccurate or impossible by any means if the broth contains large amounts of insoluble ingredients (e.g., flours, grits, or fish meal), as are common in many antibiotic fermentations. In practice, industrial scientists often forgo estimates of biomass until processes with insoluble media are scaled to stirred tanks and then use exit-gas measurements (especially CO_2 and O_2) for data to infer the growth stage; yield calculations that depend on biomass determination may still not be straightforward.

Determination of residual soluble components of the fermentation, such as glucose, organic acids, ammonium, or phosphate (present in a supernatant or filtrate fraction), constitutes the other broad class of common analyses. These may be for general characterization of the process (for example, to help determine whether carbon or nitrogen is limiting first) or for quantification of a specific component thought to be critical to the process.

4.8. BEGINNING WITH THE END IN MIND

It is always sound advice to be clear about the ultimate objective of a research line from the very start. However, it is not uncommon that small-scale liquid fermentations are done with an ultimate objective of scaling a process to much larger stirred equipment, yet the small-scale experiments are set up to optimize the small-scale fermentation, not the stirred equipment. If it is to be scaled up, the small-scale work should, if possible, be matched to the larger scale. This would include such things as obtaining media ingredients from the same sources and of the same quality as will be used in the larger scale. This may be as simple as obtaining the water and media feedstocks, such as glucose, peptone, or yeast extract, from the pilot plant or production area rather than from a laboratory supplier. Equivalent operating parameters may have to be determined through experimentation. When operating parameters cannot be matched, it becomes necessary to design experiments to allow extrapolation to the next larger scale or to test hypotheses about the best way to operate the larger equipment. This "scale-down" philosophy is expressed well and in detail by Jem (32) with regard to scaling from small stirred vessels to full production scale. The principles also apply to scaling of small-batch fermentations to larger fermentors. The primary product of process development at any scale destined for a larger scale should be data defining the biological responses to variations in parameters over the ranges that can be achieved in the larger scale. Ultimately, the goal of bridging between scales

TABLE 7 Common generic measurements of fermentations

Measurement (units)	Method
Dry cell weight (g/liter)	Gravimetric (packed, washed cells from a known harvest volume are dried fully and weighed) or indirect (optical density is related to dry weight in an empirical standard curve and measured as a surrogate)
Packed cell volume (%)	Centrifuge in calibrated tube, read directly
pH	pH dip strips or meter
Ammonium (μM or ppm)	Ion chromatography or blood analyzer[a]
Phosphate (μM or ppm)	Ion chromatography or blood analyzer[a]
Glucose (mM or g/liter)	Test strips (e.g., blood glucose) for semiquantitative estimate; HPLC or glucose oxidase-based analyzers for quantitative estimate
Amino and organic acids (mM or g/liter)	Ion chromatography or specialized assays

[a]Clinical blood analyzers and Kodak Biolyzer (Eastman Kodak Co.) can do enzyme-based determination of many chemicals (e.g., glucose, ammonia, phosphate) and enzymes (e.g., amylase, lipase) of potential interest to fermentations.

is knowledge that leads to a process with the equivalent optimized physiology, for an equivalent outcome, at the destination scale. It would be naive to expect equivalent outcomes from a simple duplication of those medium ingredients and operational set points that can be duplicated literally, when so many other parameters change as a consequence of the scale (such as hydrostatic pressure, shear, pH control, and mixing energy input). It is generally necessary to empirically establish the parameters that are most important to the process and their interactions and to use that data to guide initial operations and optimization at the destination scale.

4.9. SMALL-SCALE FERMENTATION EQUIPMENT AND SUPPLIES

There are many vendors for small-scale fermentation supplies and equipment. Given the constant changes in corporate relationships and the number of manufacturers, vendors, and affiliates around the world and their constant changes in models, it is impractical to try to index all the vendors here in any balanced way. Sources for specialized and potentially hard-to-find items have been identified in the text. Beyond that, the large scientific meetings, annual buyers' guides (e.g., from American Scientist, International Scientific Communications, Inc., Shelton, Conn; The Source Book, Cold Spring Harbor Laboratory Press, Plainview, N.Y.), and directories on the Internet (e.g., http://guide.nature.com, http://www.biosupplynet.com, and http://sciquest.com) are the most effective sources for current supplier and vendor information.

REFERENCES

1. **Bailey, J., and D. F. Ollis.** 1986. *Biochemical Engineering Fundamentals*, 2nd ed., McGraw-Hill Co., New York.
2. **Boeck, L. D., J. S. Alford, R. L. Pieper, and F. M. Huber.** 1989. Interaction of media components during bioreactor sterilization: definition and importance of R_0. *J. Ind. Microbiol.* **4:**247–252.
3. **Boeck, L. D., R. W. Wetzel, S. C. Burt, F. M. Huber, G. L. Fowler, and J. S. Alford, Jr.** 1988. Sterilization of bioreactor media on the basis of computer-calculated thermal input designated as F_0. *J. Ind. Microbiol.* **3:**305–310.
4. **Brown, D. E.** 1970. Aeration in the submerged culture of microorganisms. *Methods Microbiol.* **2:**125–174.
5. **Bryant, J.** 1970. Anti-foam agents. *Methods Microbiol.* **2:** 187–203.
6. **Bu'lock, J. D.** 1974. Secondary metabolism of microorganisms, p. 335–346. *In* B. Spencer (ed.), *Industrial Aspects of Microorganisms*, vol. 1. Elsevier, Amsterdam.
7. **Bushell, M. E.** 1989. The process physiology of secondary metabolite production, p. 95–120. *In* S. Baumberg (ed.), *Microbial Products: New Approaches*. Cambridge University Press, Cambridge, U.K.
8. **Bushell, M. E., and A. Fryday.** 1983. The application of materials balancing to the characterization of sequential secondary metabolite formation in *Streptomyces cattleya* NRRL 8057. *J. Gen. Microbiol.* **129:**1733–1741.
9. **Chain, E. B., and G. Gualandi.** 1954. Aeration studies. II. *Rend. 1st Super. Sanita* **17:**5–60.
10. **Chen, K. C., T. C. Lee, and J. Y. Houng.** 1992. Search method for the optimal medium for the production of lactase by *Kluyveromyces fragilis*. *Enzyme Microb. Technol.* **14:** 659–664.
11. **Clark, G. J., D. Langley, and M. E. Bushell.** 1995. Oxygen limitation can induce microbial secondary metabolite formation: investigations with miniature electrodes in shaker and bioreactor culture. *Microbiology* **141:**663–669.
12. **Collins, M. A., and S. Y. R. Pugh.** 1987. Rapid determination of bacterial pH growth optima using microtiter plates, p. 330. *In* O. M. Neijssel, R. R. Van der Meer, and K. C. A. M. Luyben (ed.), *Proceedings of the 4th European Congress of Biotechnology*, vol. 3. Elsevier, Amsterdam.
13. **Cooper, C. M., G. A. Fernstrom, and S. A. Miller.** 1944. Performance of agitated gas-liquid contactors. *Ind. Eng. Chem.* **36:**504–509.
14. **Corman, J., H. M. Tsuchiya, H. J. Koepsell, R. G. Benedict, S. E. Kelley, V. H. Feger R. G. Dworschack, and R. W. Jackson.** 1957. Oxygen absorption rates in laboratory and pilot plant equipment. *Appl. Microbiol.* **5:**313–318.
15. **Demain, A. L.** 1985. Control of secondary metabolism, *in* Biological, Biochemical and Biomedical Aspects of Actinomycetes. p. 215–225. *In* G. Szabo, S. Biro, and M. Goodfellow (ed.), Proceedings of the 6th International Symposium on Actinomycetes Biology. Akademiai Kiado, Budapest, Hungary.
16. **Deming, S. N.** 1990. Quality by design. *Chemtech.* Feb. 1990, p. 118–126.
17. **Dunn-Coleman, N. S., P. Bloebaum, R. M. Berka, E. Bodie, N. Robinson, and G. Armstrong.** 1991. Commercial levels of chymosin production by *Aspergillus*. *Bio/Technology* **9:**976–981.
18. **Esgalhado, M. E., J. C. Roseiro, and M. T. A. Collaco.** 1995. Interactive effects of pH and temperature on cell growth and polymer production by *Xanthomonas campestris*. *Process Biochem.* **30:**667–671.
19. **Falch, E. A., and C. G. Heden.** 1963. Disposable shaker flasks. *Biotechnol. Bioeng.* **5:**211–220.
20. **Finn, R. K.** 1954. Agitation-aeration in the laboratory and in industry. *Bacteriol. Rev.* **18:**254–274.
21. **Flemming, C. A., H. Lee, and J. T. Trevors.** 1994. Bioluminescent most-probable-number method to enumerate lux-marked *Pseudomonas aeruginosa* UG2L2r in soil. *Appl. Environ. Microbiol.* **60:**3458–3461.
22. **Fredrickson, J. K., D. L. Balkwill, J. M. Zachara, S.-M. W. Li, F. J. Brockman, and M. A. Simmons.** 1991. Physiological diversity and distributions of heterotrophic bacteria in deep cretaceous sediments of the Atlantic coastal plain. *Appl. Environ. Microbiol.* **57:**402–411.
23. **Freedman, D.** 1970. The shaker in bioengineering. *Methods Microbiol.* **2:**175–185.
24. **Gaden, E. L., Jr.** 1962. Improved shaken flask performance. *Biotechnol. Bioeng.* **4:**99–103.
25. **Garcia Sanchez, J. L., J. A. Sanchez Perez, F. Garcia Camacho, J. M. Fernandez Sevilla, and E. Molina Grima.** 1996. Optimization of light and temperature for growing *Chlorella* sp. using response surface methodology. *Biotechnol. Techniq.* **10:**329–334.
26. **Garland, J. L., and A. L. Mills.** 1991. Classification and characterization of heterotrophic microbial communities on the basis of patterns of community-level sole-carbon-source utilization. *Appl. Environ. Microbiol.* **57:**2351–2359.

27. **Gorden, R. W., T. C. Hazen, and C. B. Fliermans.** 1993. Rapid screening for bacteria capable of degrading toxic organic compound. *J. Microbiol. Methods* **18:**339–347.

28. **Henzler, H. J., and M. Schedel.** 1991. Suitability of the shaking flask for oxygen supply to microbiological cultures. *Bioprocess Eng.* **7:**123–131.

29. **Hirose, Y., H. Sonoda, K. Kinoshita, and H. Okada.** 1968. Studies on oxygen transfer in submerged fermentation. VIII. The effects of carbon dioxide and agitation on product formation in glutamic acid fermentation under controlled pressure of dissolved oxygen. *Agric. Biol. Chem.* **32:**851–854.

30. **Hopwood, D. A., M. J. Bibb, K. F. Chater, T. Kieser, C. J. Bruton, H. M. Kieser, D. J. Lydiate, C. P. Smith, J. M. Ward, and H. Schrempf.** 1985. Growth of *Streptomyces* mycelium, p. 7. *In Genetic Manipulation of Streptomyces. A Laboratory Manual.* John Innes Foundation, Norwich, England.

31. **Hujanen, M., and Y. Y. Linko.** 1994. Optimization of L(+)-lactic acid production employing statistical experimental design. *Biotechnol. Tech.* **8:**325–330.

32. **Jem, J.** 1989. Scale-down techniques for fermentation. *BioPharm.* March 30, p. 39.

33. **Jensen, A. L., M. A. Darken, J. S. Schultz, and A. J. Shay.** 1964. Relomycin: flask and tank fermentation studies, p. 49–53. *Antimicrob. Agents Chemother. 1963.*

34. **Kawase, Y., and M. Moo-Young.** 1990. The effect of antifoam agents on mass transfer in bioreactors. *Bioproc. Eng.* **5:**169–173.

35. **Legan, J. D., and J. D. Owens.** 1985. A note on the selection of a non-adhesive methylotrophic bacterium for use in continuous culture studies. *J. Appl. Bacteriol.* **58:** 163–165.

36. **Matelova, V., A. Brecka, and J. Matouskova.** 1972. New method of intermittent feeding in penicillin biosynthesis. *J. Appl. Microbiol.* **23:**669–670.

37. **McDaniel, L. E., and E. G. Bailey.** 1969. Effect of shaking speed and type of closure on shake flask cultures. *Appl. Microbiol.* **17:**286–290.

38. **McDaniel, L. E., E. G. Bailey, and A. Zimmerli.** 1965. Effect of oxygen supply rates on growth of *Escherichia coli*. *Appl. Microbiol.* **13:**109–119.

39. **Park, K., and K. F. Reardon.** 1996. Medium optimization for recombinant protein production by *Bacillus subtillis*. *Biotechnol. Lett.* **18:**737–740.

40. **Rhodes, R. P., and E. L. Gaden.** 1957. Characterization of agitation effects in shaken flasks. *Ind. Eng. Chem.* **49:** 1233–1236.

41. **Schultz, J. S.** 1964. Cotton closure as an aeration barrier in shaken flask fermentations. *Appl. Microbiol.* **12:**305–310.

42. **Shibai, H., A. Ishizaki, H. Mizuno, and Y. Hirose.** 1973. Effects of oxygen and carbon dioxide on inosine fermentation. *Agric. Biol. Chem.* **37:**91–97.

43. **Sie, T.-L., and K. Schugerl.** 1983. Foam behavior of biological media. XI. Efficiency of antifoam agents with regard to their suppression effect on BSA solutions. *Eur. J. Appl. Microbiol. Biotechnol.* **17:**221–226.

44. **Smith, C. G., and M. J. Johnson.** 1954. Aeration requirements for the growth of aerobic microorganisms. *J. Bacteriol.* **68:**346–350.

45. **Spizek, J., and P. Tichy.** 1995. Some aspects of overproduction of secondary metabolites. *Folia Microbiol.* **40:**43–50.

46. **Screenath, H. K., and T. W. Jeffries.** 1996. A variable-tilt fermentation rack for screening organisms in microfuge tubes. *Biotech. Tech.* **10:**239–242.

47. **Starks, O. B., and H. Koffler.** 1949. Aerating liquids by agitating on a mechanical shaker. *Science* **109:**495–496.

48. **Stieglitz, B., and P. J. Weimer.** 1985. Novel microbial screen for detection of 1,4-butanediol ethylene glycol and adipic acid. *Appl. Environ. Microbiol.* **49:**593–598.

49. **Tunac, J. B.** 1989. High-aeration capacity shake-flask system. *J. Ferment. Bioeng.* **68:**157–159.

50. **Uchida, M., H. Sawada, T. Asai, and M. Suzuki.** 1981. Effect of inorganic phosphate and dissolved oxygen on production of maridomycin. *J. Ferment. Technol.* **59:**399–401.

51. **van Suijdam, J. C., N. W. F. Kossen, and A. C. Joha.** 1978. Model for oxygen transfer in a shake-flask. *Biotechnol. Bioeng.* **20:**1695–1709.

52. **Vardar-Sukan, F.** 1987. A quantitative comparison of natural oils as antifoams in bioprocesses, p. 50–53. *In* O. M. Neijssel, R. R. van der Meer, and K. C. A. M. Luyben (ed.), *Proceedings of the 4th European Congress of Biotechnology*, vol. 1. Elsevier, Amsterdam.

53. **Wise, W. S.** 1950. The aeration of culture media: a comparison of the sulphite and polarographic methods. *J. Soc. Chem. Ind.* (Suppl. 1):S40–S41.

Small-Scale Solid-State Fermentations

KAZUO SATO AND SHIGETOSHI SUDO

5

Solid-state fermentation (SSF) is a microbial process in which a solid material is used as the substrate or the inert support of microorganisms growing on it. In SSF, microorganisms can sometimes grow well and produce larger amounts of extracellular enzymes and other metabolites than they do in submerged (liquid) fermentation. Although SSF was developed for the manufacturing of traditional foods and alcoholic beverages, its application has been extended to the pharmaceutical and biochemical industries. Despite the unknown factors contributing to the high production of enzymes or other metabolites, further applications of SSF in these industries are anticipated. It has evolved into a new type of biotechnology because of its unique production characteristics. Many reviews are available (3, 28, 48, 49, 95, 118, 125, 176, 179). In this chapter, we review the basic science and technology of SSF, with an emphasis on the cultivation of molds.

5.1. GENERAL CONSIDERATIONS

5.1.1. History

The origin of SSF can be traced back to bread making in ancient Egypt. Over the years, some major advances were established in the process engineering of SSF. In the 1960s, many turkeys died from ingestion of aflatoxin, which is produced by a mold, *Aspergillus flavus*, that grows on peanut meal (52). Hesseltine (62) found that a large amount of this toxin was produced in SSF, but not in submerged fermentation. These results attracted much attention (66, 157, 166). SSF had also been widely used in the production of traditional Oriental foods and alcoholic beverages: tempeh and ontjom in Indonesia, shao-hsing wine and kaoliang (sorghum) liquor in China, and miso, soy sauce, and sake in Japan. Hence, SSF has played an important role in the traditional food-making processes of Oriental countries.

In 1896, Takamine (176) produced a digestive enzyme, Takadiastase, by SSF employing *Aspergillus oryzae* on wheat bran. This led to the application of SSF to other food and beverage industries. After World War II, Underkofler et al. (191) and Terui et al. (181–185) used highly heaped bed cultures with forced aeration. Thus, the industrial-scale fermentor was developed for the production of enzymes and citric acid by SSF. Nowadays, the industrial utilization of SSF has expanded to include composting, mushroom cultivation, and the production of other foods such as bread and mold-ripened cheese. Table 1 lists examples of the applications of SSF.

5.1.2. Characteristics

As reviewed by Viniegra-Gonzàlez (194), SSF has unique characteristics and limitations. Extracellular hydrolytic enzymes and other metabolites are sometimes produced in greater quantities in SSF. Although the reasons are not clear, this characteristic is very important for the application of SSF. Martinez et al. (89) reported that the wheat straw-degrading fungus *Pleurotus* sp. produced isoenzymes differing in molecular mass and pI during SSF and submerged fermentation. These isoenzymes shared several characteristics, including two N-terminal amino acid consensus sequences under liquid or SSF conditions. Recently, Ishida et al. (70) reported that *A. oryzae* produced different types of glucoamylase depending on the cultivation method; i.e., the activity of raw starch digestion was found only in the submerged fermentation. This result was confirmed by DNA sequential analysis; the DNA sequences of the two genes were quite homologous, but a domain related to the activity of raw starch digestion was deleted in the gene of the glucoamylase produced in SSF. Minjares-Carranco et al. (90) found that *Aspergillus niger* produced isoenzymes with differences in pectinase properties depending on these cultivation means.

The basic differences between SSF and submerged fermentation are summarized as follows.

1. In SSF, the microbial distribution occurs on the solid surface, and microbial growth and product formation also occur mainly on the surface. The substrate is not uniform and not easily agitated. The culture environment is therefore heterogeneous. Figure 1 shows a scanning electron micrograph of wheat koji, in which the mycelia of *Aspergillus kawachii* are growing on the surface and penetrating into the endosperm of wheat grains.

2. The moisture content of a solid substrate is normally low, depending on the physical or chemical characteristics of the substrate. For media with a high moisture content, steady aeration throughout the substrate bed is difficult, and channeling of airflow often occurs.

3. Heat derived from the metabolism and growth of the microorganism raises the temperature of the solid substrate

TABLE 1 Examples of solid-state fermentations on natural substrates

Product	Microorganisms	Materials	References
Enzymes			
α-Amylase	*Aspergillus oryzae, Rhizopus* sp., *Bacillus licheniformis, Bacillus* sp.	Wheat bran, cassava	14, 116, 127, 128, 142, 164, 176
Glucoamylase	*Aspergillus niger, Aspergillus* sp., *Rhizopus* sp.	Cassava, wheat bran, corn	6, 7, 119, 164, 178
Cellulase	*Trichoderma reesei, A. niger* *Penicillium* sp., *Thermoascus aurantiacus*	Wheat bran, wheat straw Beet pulp, cellulosic biomass	31, 35, 76, 80, 88, 117, 131, 171, 172, 189
Xylanase	*Aspergillus fumigatus, Thermoascus lanuginos*	Wheat bran, jute fiber + wheat germ	6, 79
Pectinase	*Talaromyces flavus, A. niger, A. carbanerius*	Fruit pomace, wheat bran, coffee pulp	8, 47, 67, 104, 158
Glucose oxidase	*Penicillium notatum, Penicillium* sp.		122
β-Galactosidase	*Kluyveromyces lactis*	Whey + corn or wheat bran	26
Protease	*Penicillium caseicolum, Mortierella renispora, A. oryzae, A. niger*	Wheat bran, dried skim milk	98, 115, 122, 193
Rennin	*Mucor pusillus, Mucor miehei*	Wheat bran	8, 43, 65, 80, 188
Metabolites			
Ethanol	*Saccharomyces cerevisiae*	Fruit pomace, sweet sorghum, beet, corn, carob pods	48–51, 56, 58, 60, 74, 75, 78, 121, 135, 140, 146
Citric acid	*A. niger*	Sugarcane bagasse, fruit pomace, wheat bran	29, 59, 61, 81, 84, 121, 154, 201
Lactic acid	*Lactobacillus* sp., *Rhizopus oryae*	Sweet sorghum, sugarcane bagasse + glucose	129, 165, 199
Gibberellic acid	*Gibberella fujikuroi*	Wheat bran	82
Red pigment	*Monascus anka*	Rice, bread flake	57, 66
Antibiotics			
Penicillin	*Penicillium chrysogenum*	Sugarcane bagasse	16, 23, 25
Tetracycline	*Streptomyces viridifaciens*	Sweet potato residue	203
Cephalosporins	*Cepharlosporium acremonium*	Barley	73
Iturin, surfactin	*Bacillus subtilis*	Soybean curd residue	106, 107
Foods			
Natto	*Bacillus natto*	Soybean	3, 40, 64
Tempeh	*Rhizopus oligosporus*	Soybean	3, 64, 120, 164, 196, 197
Tape	*Amylomyces rouxii, Rhizobium chinensis*	Rice, cassava, maize	120, 139, 173
Ontjom	*Neurospora sitophila*	Peanut meal	3, 64, 120
Cheese	*Penicillium roqueforti*	Milk curd	36, 77
Bread dough	*Saccharomyces cerevisiae, Lactobacillus sanfrancisco*	Wheat powder	12, 21
Koji			
Sake, shochu	*A. oryzae, A. kawachii*	Rice, barley	3, 64, 71, 112, 167, 187, 204
Soy sauce	*Aspergillus sojae*	Soybean, wheat	3, 40, 64, 120, 122
Miso	*A. oryzae*	Soybean, rice	3, 40, 64, 120, 122
Shao-hsing wine	*Rhizopus* sp., *Mucor* sp. (*A. oryzae*)	Wheat (rice)	64
Kao-liang liquor	*Rhizopus* sp., *Mucor* sp.	Sorghum	64, 68, 195
Ragi	*Rhizopus* sp., *Saccharomycopsis* sp.	Rice	3, 64, 120, 139, 173
Single-cell proteins	Many yeasts and molds	Starchy or cellulosic biomass	14, 22, 30, 33, 85, 105, 108, 124, 161, 173, 174, 202, 205
Compost	White-rot fungi (mixed culture)	Cellulosic biomass	1, 12, 16, 17, 19, 32, 83, 198

bed and causes the loss of moisture. This phenomenon creates challenges for the control of the SSF process.

4. SSF substrates are usually natural materials, e.g., cereals, soybeans, agricultural biomass, and solid waste. Sometimes the product is the entire fermented substrate, as in the case of traditional foods, e.g., miso, natto, and tempeh.

5. The microorganisms generally used in SSF are molds that can produce amylases to degrade starch and penetrate into the solid substrate. In SSF, molds have morphological

FIGURE 1 Scanning electron micrograph of the mycelia of *A. kawachii* grown on barley koji. Cultivation time was 2 days. The mold mycelia penetrates deeply into the endosperm of the barley grain. The large aerial mycelia on the koji surface are sporangiophores which would bear spores. Before the sample was fixed for microscopy, the koji was treated with enzymes (amylase and pectinase) to dissolve the endosperm. (Courtesy of Dr. Kume, Institute of Bio-Science, Usuki, Oita, Japan.)

differences in the types of mycelia (aerial and submerged). These two types of mycelia have different physiological activities, complicating process control.

6. Since agitation of the substrate bed is very difficult and some activities are sensitive to shear stress, cultivation is normally stationary, except for the rotating drum and fluidized-bed fermentors.

These characteristics are unique to SSF and give rise to both advantages and disadvantages, as discussed below.

5.1.3. Advantages and Disadvantages of SSF

In the industrial production of extracellular enzymes such as amylase, cellulase, and pectinase, both SSF and submerged fermentation are used. The decision is likely based on the cost and efficiency of the process. It is therefore important to know the advantages and disadvantages of SSF as compared with submerged fermentation.

5.1.3.1. Advantages

1. SSF is relatively resistant to bacterial contamination. Since bacterial growth is restricted by low water activity, serious contamination on a solid medium rarely occurs.

2. Fermentors or the fermentative facilities are compact. The volumetric loading of the substrate is much higher in SSF than in submerged fermentation because the moisture content of the solid substrate is lower.

3. If extraction of the product from SSF is necessary, it requires much less solvent and lower recovery cost than from submerged fermentation.

4. Treatment of the fermented residue is very simple. Since the moisture content of the fermented residue is very low, it can be dried and used as animal feed or fertilizer.

5. Microbial utilization of gaseous oxygen reduces the energy cost of aeration. The air supply and temperature of the solid substrate bed can be controlled by forced aeration, in which the large surface area of the solid substrate promotes heat transfer and gas exchange of oxygen and carbon dioxide.

6. Conidiospores can be used as the fermentation starter in the conventional koji method. Since these spores can be preserved for a long time, they can be used repeatedly.

5.1.3.2. Disadvantages

1. Agitation of the substrate bed is difficult. Hence, uneven distribution of the cell mass, nutrients, temperature, moisture content, etc., occurs, resulting in a heterogeneously physiological, physical, and chemical environment in the substrate bed. This complexity makes process control very difficult.

2. Temperature control of the heat evolved by microbial respiration or metabolism is very difficult because the solid substrate bed has a low thermal conductivity. Usually, forced aeration is the only means of controlling cultivation temperature.

3. Rapid determination of microbial growth and other fermentative parameters is very difficult. There is no available sensor to measure them directly.

4. The microflora is limited on media with low moisture content. Molds or other filamentous fungi are suitable, but bacterial growth is rare, except for xerophiles.

5. Usually, temperature is the only means of controlling microbial growth or product formation in SSF. Since there is no feasible means of analyzing the fermentation conditions, continuous operation and automation are difficult.

6. Since the factors contributing to high production in SSF are not fully understood, cultivation strategy is based more or less on empirical results and the experience of the operator.

5.2. MICROBIAL GROWTH AND PRODUCTION OF ENZYMES

5.2.1. Microbial Growth

Since product yield is often strongly dependent on microbial growth, an understanding of growth characteristics is essential. Moreover, such an understanding is important for rational design, scale-up, and process control in SSF. The characteristics of microbial growth and the effects of environmental conditions are as follows.

5.2.1.1. Mycelial Growth of Mold

The growth of molds on solid substrates is unique since it is hyphal and heterogeneous in SSF. Varzakas et al. (192) observed a 2-mm depth of hyphal penetration in a 40-h cultivated soybean tempe. Sangsurasak et al. (138) and Guitiérrez-Rojas et al. (55) reviewed the growth kinetics of molds using mathematical growth models. As shown in Fig. 1, generally two types of mycelia are observed in the vegetative growth of the mold. Growth penetrating the surface of the solid substrate leads to the formation of submerged mycelia, and aerial growth leads to the formation of aerial mycelia. However, it is difficult to discriminate between these mycelia on the surface. After germination, the submerged mycelia grow first, and then the aerial mycelia develop above them. The sporangiophores that produce the conidiospores are formed later among the aerial mycelia. Sudo et al. (168) studied the difference between these two types of mycelia on the production of α-amylases in rice koji.

Usually, the solid substrate is porous, and the cluster of the substrate has void space inside. Oxygen is supplied through the void spaces. To avoid excessive microbial growth which might clog the airflow through the substrate bed, mixing or agitation of the bed should be carried out. Lindenfelser and Ciegler (86) studied the effect of agitation on the production of ochratoxin by *Aspergillus ochraceus* to find an optimum agitation rate. In a roller-type cultivation of *A. niger*, the production of citric acid was increased by agitation (201). However, in a case of anaerobic cultivation using a drum fermentor, the productivity of ethanol by *Saccharomyces cerevisiae* was decreased with increased rotation speed (74). Although the reason was not made clear, strong agitation of the substrate bed was thought to be occasionally harmful.

5.2.1.2. Temperature

Temperature control in the substrate bed is very important for SSF since growth and production of enzymes or metabolites are usually sensitive to temperature (175). It must be noted that the optimum temperature for the production of an enzyme or metabolite does not always coincide with that for growth. Thus, a strategy may be needed for changing the temperature during the cultivation. Smits et al. (162) studied the application of fuzzy logic for the prediction of glucosinolate during the cultivation of *Aspergillus clavatus* on rapeseed meal, in which temperature, relative humidity, and fermentation time were chosen as the input variables.

Another problem is that the metabolic heat evolved raises the temperature of the solid substrate bed. Without an efficient method for heat removal, the temperature may exceed the upper limit for growth. Since the thermal conductivity of the solid substrate bed is generally low, there is often a long delay in response to temperature change. Moreover, there is usually a temperature gradient in the bed, causing heterogeneous growth and production. The temperature gradient may occur vertically and horizontally, depending on both the thickness and the width of the bed, as will be discussed later. This problem illustrates the difficulty of measuring an average temperature.

5.2.1.3. Moisture Content

Generally, microbial growth is affected by the moisture content in the solid substrate. This effect is discussed using water activity (a_w) defined by Scott (152) as a cell growth parameter as follows.

$$a_w = P_w / P_0 \quad (1)$$

where P_w and P_0 are the equilibrated partial vapor pressures over the substrate and pure water at a given temperature, respectively.

Microbial growth is restricted strongly in a medium of low a_w. This low limit of a_w is dependent on the species of microorganism. Bacteria are usually less tolerant than yeasts and molds at a given a_w value (Table 2). This suggests that SSF with low a_w is more resistant to bacterial contamination, allowing a nonaseptic condition. However, contamination may still occur at the early stage of cultivation. For example, Ouchi et al. (114) reported growth of *S. Cerevisiae* at a cell density of 10^4 to 10^6 cells/g of koji, and Narahara (100) reported that bacterial growth (*Micrococcus* sp. and *Bacillus* sp.) at a density of 10^3 to 10^7 cells/g of koji reduced the growth extent of *A. oryzae* on rice koji.

Water evaporation depends on the amount of metabolic heat evolved. In the case of koji making for Japanese sake, in which the cultivation of *A. oryzae* is carried out on steamed rice grains for 2 days, the moisture content of the steamed rice decreases from approximately 40% to 20% (a_w, 0.99 to 0.92). The problem associated with the change of moisture content during SSF is discussed in section 5.3.2 below.

Gervais et al. (45) studied the effects of moisture content on the growth and metabolism of molds in SSF. Concerning the measuring of a_w in SSF, Xavier and Karanth (200) reported a convenient method, and Gervais and Bazelin (44) developed an a_w sensor.

5.2.1.4. Oxygen and Carbon Dioxide

Okazaki et al. (111) reported that oxygen, down to a concentration of 0.2%, did not affect the growth rate of *A. oryzae* on steamed rice grains; however, the lag time of germination was increased by the low oxygen concentration. They observed that an extremely high concentration (>10%) of carbon dioxide prevented growth.

Although the theoretical respiratory quotient (RQ) of aerobic microorganisms is 1.0, the cultivation data of molds sometimes show an RQ below 1.0 in SSF. This means that oxygen transfer to the mycelia may be insufficient in conventional SSF. Poor growth of the deeply submerged mycelia is observed in rice koji of high moisture content, possible because the pathways of oxygen transfer inside the rice grain are blocked.

The mycelial growth of mold is restricted by the diffusion of oxygen in the solid substrate bed. Mitchell et al. (91) proposed that oxygen diffusion limits the rate of growth of *Rhizopus oligosporus* into a model solid substrate. Gowthaman et al. (54) reported that the overall oxygen transfer coefficient (k_La) in a packed bed of wheat bran was affected by moisture content during the cultivation of *A. niger*. It was reduced at a high aeration rate, presumably due to the reduction of moisture content. Auria et al. (13) studied the effective diffusivities of CO_2 and O_2 in a packed bed of solid substrate in which *A. niger* was cultivated on umberlite absorbing a liquid medium. They observed that the diffusivities were reduced to below 5% of those in air

TABLE 2 Water activities (a_w) for the growth of microorganisms[a]

Microorganism	Optimal a_w	Minimal a_w
Aspergillus ochraceus		0.77
Aspregillus niger	0.98	0.86
Saccharomyces cerevisiae	0.98–1.0	0.94–0.95
Saccharomyces rouxii		0.65
Torulopsis candida		0.65
Hansenula anomala		0.75
Klebsiella aerogenes	1.0	0.95
Pseudomonas fluorescens	~0.99	0.94–0.98
Staphylococcus saprophyticus		0.90–0.98
Staphylococcus aureus	0.94–1.0	0.86
Lactobacillus sp.		0.90–0.93

[a]Data from reference 132.

when the mold growth reached 27 mg/g of substrate. Rajagopalan and Modak (126) simulated the growth of *A. niger* on a solid substrate which was limited by heat and oxygen transfer. Their model assumed that the increase of the thickness of the mold biofilm around the solid particles decreased the porosity of the substrate bed and the diffusivity of oxygen in the bed.

Sudo et al. (168) measured the dissolved oxygen (DO) concentration in steamed rice grains during the cultivation of *A. kawachii*. Figure 2 shows that the decrease of DO concentration was accelerated after 20 h from the beginning of the cultivation. The reduction of DO was associated with growth, and this meant that the mycelia could still grow at low DO concentrations in the steamed rice grains. This may be due to the microporosity of steamed rice, or to a characteristic of mycelial growth in which nutrients and oxygen are moved from the external to the internal mycelia by protoplasmic streaming through the protoplasmic connection.

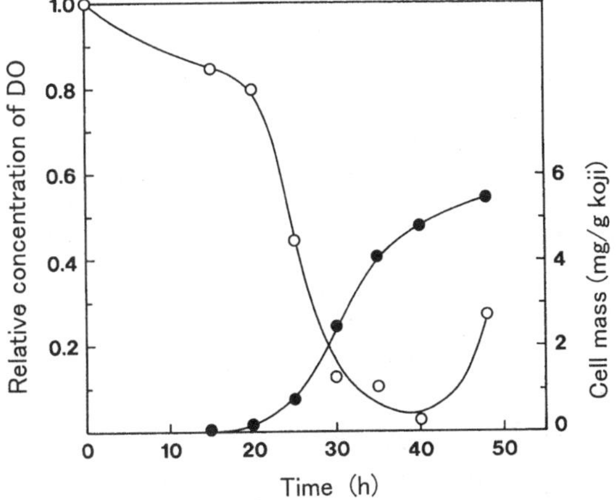

FIGURE 2 DO concentration in a moldy rice grain during solid-state cultivation. A microhole array DO sensor was inserted into a steamed rice grain on which *A. kawachii* was inoculated. The measurement point of the DO sensor was set at the center of the rice grain. Cultivation was carried out at 35 to 40°C (168). ○, relative value of DO against the initial concentration; ●, cell mass.

In koji, as the substrate is degraded by the mold amylase, the porosity of the substrate increases, allowing gaseous oxygen to be used by the submerged mycelia which penetrate deeply into the substrate. The mechanisms of mycelial penetration and oxygen transfer into the substrate are one of the significant problems in the analysis of SSF.

5.2.1.5. Particle Size and Physical Properties of Solid Substrate

Substrates with a smaller particle size have a larger specific surface area, which is advantageous for the surface growth of mold. It is also advantageous for heat transfer and exchange of oxygen and carbon dioxide between the air and the solid surface. Huang et al. (68) found that the optimal growth of *Aspergillus koppan* occurred on rice grains of smaller particle size. However, when the particle size is too small, it raises the packing density of the substrate and causes the particles to adhere to each other. This condition leads to compression and contraction of the bed, which blocks or creates channeling of the airflow. Sun et al. (170) found that there was an optimum particle size of the substrate (rice chaff) in the production of fibrinolytic enzyme by *Fusarium oxyporum*.

The compressibility of the solid substrate is variable depending on the physical properties of the solid substrate, i.e., the moisture content, particle size, and content of inert solids. The packing density of the solid substrate bed could be changed by compression or water evaporation during cultivation. These characteristics are serious since they reduce the effective surface area for growth. Terui et al. (182) studied these phenomena with wheat bran koji and developed the use of ledges attached to the fermentor wall to fill the gap created by the contraction of the substrate.

Nandakumar et al. (99) observed the reduction of particle size and weight of barley bran during the cultivation of *A. niger*, accompanied by the production of hydrolytic enzymes. Barrios-Gonzàlez et al. (24) studied the effect of particle size and packing density of sugarcane bagasse on penicillin production by *P. chrysogenum*.

5.2.2. Production of Enzymes and Other Metabolites

As seen in Table 1, the production of enzymes and other metabolites in SSF has been studied by many researchers. Some of these results are utilized in industrial processes.

5.2.2.1. Enzymes

Molds often produce large amounts of extracellular hydrolytic enzymes in SSF, such as α-amylase, glucoamylase, cellulase, pectinase, protease, and so on. Okazaki and Sugama (110) reported that α-amylase production by *A. oryzae* was strongly growth associated in rice koji. Thus the control of mold growth directly affects the yields of these enzymes. Sometimes carbon dioxide also affects the enzyme production. Villegas et al. (193) reported that the production of acid protease by *A. niger* was accelerated at 4% CO_2.

Enzyme production is regulated by physiological mechanisms. For example, the production of hydrolytic enzyme is often repressed by the catabolites of glucose (catabolite repression) in liquid culture. Reduced catabolite repression in SSF may explain the unusually high enzyme productivity in SSF, but the reason is not clear. Terui and coworkers (94, 187) suggested that the diffusivity of glucose is small in solid substrates of low moisture content, thus reducing the overall effective concentration of glucose and catabolite repression. On the other hand, Okazaki and Sugama (110) reported that *A. oryzae* grew well on rice koji at a low moisture content. Sudo et al (167) found that the production of an acid-stable α-amylase of *A. kawachii* was not repressed by the addition of glucose when it grew logarithmically. These results mean that the production of glucose is restricted at low moisture content and any glucose available is consumed rapidly in SSF. Some starvation may occur on solid substrates of low moisture content, which may reduce catabolite repression.

In the rice-koji method for Japanese sake, amylase production on a high-moisture-content substrate is usually reduced, suggesting that control of the moisture content is important for enzyme production. Sudo et al. (168) reported that the submerged mycelia of *A. kawachii* produced much more acid-stable α-amylase than did the aerial mycelia, correlating the production of this enzyme to the morphology of the mycelia.

The production of enzyme is affected by temperature. Figure 3 shows the time course of temperature and production of enzymes during cultivation of rice koji for Japanese sake. The conidiospores of *A. oryzae* were first scattered on steamed rice and the temperature was kept at 30 to 32°C, which is optimal for germination. After 1 day of cultivation, the temperature of the koji was raised gradually to approximately 35°C. After crushing and mixing, the koji was spread widely to allow uniform growth in the bed. The temperature was kept at approximately 35°C for several hours and then raised gradually to 37 to 38°C. After about 2 days of cultivation, the maximum temperature of the koji reached about 40°C. This control strategy is based on the observation that the optimum temperature for the production of amylase is rather high (37 to 38°C), while that for protease is lower (about 35°C). The temperature is regulated by controlling the room conditions (temperature and humidity), besides the thickness and mixing frequency of the koji bed.

Many authors have reviewed the production of enzymes by SSF. Selvakumar et al. (153) reviewed glucoamylase production, and Mitra et al. (92) reviewed the production of proteolytic enzymes. Babu and Satyanarayana (15) examined bacterial enzyme production. Nigam and Singh (103) reviewed the production of cellulolytic enzymes.

5.2.2.2. Citric Acid

Citric acid is an important organic acid used for foods, beverages, pharmaceuticals, chemicals, cosmetics, and other products. It can be produced by *A. niger* in SSF using a shallow pan (96) and also by submerged fermentation. In industrial-scale SSF, *A. niger* is usually cultured on potato starch residue for 6 to 7 days between 30 and 40°C. Sometimes fruit pomace or molasses with some solid support, e.g., sugarcane bagasse (69) or sawdust with rice hulls (181), is used as the substrate. The yield of citric acid is usually more than 60% of the theoretical yield on a sugar basis (185). Terui and coworkers (181, 182, 185) studied the various parameters of citric acid production in a highly heaped fermentor with forced aeration, inducing the time, heat evolution, and changes in physical properties of the solid substrates during cultivation. They further studied the effects of moisture content supplements. Calcium carbonate is often added to prevent the inactivation of enzymes involved in the production of citric acid. Lakshminarayana et al. (84) reported that adding 3% methanol to sucrose or molasses increased the yield of citric acid by 56–114% on a sugar basis after 6 days of cultivation at 30°C, as compared to 14–19% with no addition. Shankaranand and Lonsane (155) reported that the addition of both copper and zinc increased the production of citric acid to 49.21 g per kg of wheat bran, as compared to 22.49 g with no addition.

In the production of shochu, a Japanese distilled alcoholic beverage, *A. kawachii* or *Aspergillus awamori* is em-

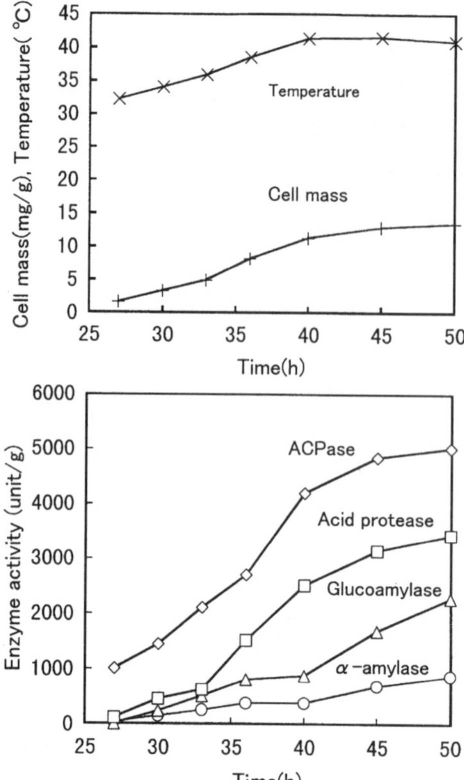

FIGURE 3 Time course of production of several hydrolytic enzymes during SSF cultivation of *A. oryzae* on rice koji. Cultivation was carried out by a traditional koji method, using a wooden tray. ×, temperature (°C); +, cell mass (milligrams per g of koji); ○, α-amylase (units per gram of koji); △, glucoamylase (units per 10 grams of koji); □, acid protease (units per gram of koji); ◇, acid carboxypeptidase (units per gram of koji).

ployed for koji using steamed rice or barley grains. The production of citric acid on the koji is important for lowering pH to prevent bacterial contamination. In this case, a proper temperature control strategy may facilitate citric acid production. At first, inoculated steamed rice or barley grains are kept at 38 to 40°C for 30 h. Then the substrate is cooled to 30 to 35°C until the end of cultivation (40 to 45 h). Citric acid production is enhanced by this operation.

5.2.2.3. Ethanol

Many applications of SSF to ethanol production have been reported, using natural materials including beets (30, 32, 65), fruit pomace (43, 46), sweet sorghum (31, 59, 62), corn (36, 112), and cassava or other tubers (29, 112, 140). As an alternative to submerged fermentation, the SSF ethanol process is attractive because of the cost advantage of the distillation process, high volumetric productivity, the ease of treatment of distillation residues, and the potential use of the solid wastes or biomass as the substrate for another SSF operation.

However, some difficulties are found in these applications besides the mechanical difficulties regarding fermentor design. Sato and Yoshizawa (146) reported that maximum ethanol production is restricted by moisture content. They developed a unique fermentor (Stripping Bio-Reactor) which stripped ethanol by gas circulation of evolved carbon dioxide for the simultaneous recovery of ethanol during fermentation (137, 140, 141, 144, 146, 148–150).

5.2.2.4. Mycotoxins

Hesseltine (63) studied the production of aflatoxin by *A. flavus*. Only a small amount of aflatoxin was produced in a submerged cultivation. In contrast, increased production was observed in SSF in shaken flasks containing wheat, oat hulls, groats, or whole oats as the substrate. Silman and Black (159) studied the large-scale production of aflatoxin using corn in a standard Butler-type corn storage bin 18 feet in diameter. They reported that the moisture content of the corn, the temperature, and the airflow rate affected the yield of aflatoxin.

Hesseltine (63) studied the production of ochratoxin by a continuous shaking culture of *Aspergillus parasiticus* and *Penicillium viridicatum*, using pearled wheat as the substrate. Lindenfelser and Ciegler (86) studied the production of ochratoxin by *A. ochraceus* in a four-section fermentor. The yield was affected by the agitation of the substrate, the highest ochratoxin yields being observed at 16 rpm although a long cultivation time (12 to 19 days) was required.

Barrios-Gonzàlez and Tomasini (24) reviewed the production of aflatoxins in SSF.

5.2.2.5. Antibiotics and Other Secondary Metabolites

As shown in Table 1, many antibiotics, e.g., penicillin, cephalosporins, tetracyclines, and cyclosporin, and other secondary metabolites, e.g., ergot alkaloids and gibberellic acid, can be produced by SSF. Ohno et al. (107) reported the successful production of surfactin, a lipopeptide antibiotic which acts as an inhibitor of fibrin clotting and erythrocyte lysing, by a recombinant strain of *Bacillus subtilis* grown on soybean curd residue. The product yield in SSF was eight times as high as that of the original strain and also four to five times more efficient than in submerged fermentation. These findings were reviewed by Balakrishnan and Pandey (20).

5.3. PROCESS CONTROL

Preparation of the solid substrate and subsequent inoculation are very important because the initial conditions strongly affect the entire SSF process. Usually, it is very difficult to correct the initial conditions, retrospectively. Among the process variables, the temperature and moisture content of the substrate are the most important for the control of SSF. However, there is always an uneven distribution of these physical and chemical conditions in the solid substrate bed, and the problem is that temperature is the only parameter that can be monitored rapidly. Control of the SSF process is therefore difficult and requires an empirical approach. Kinetic analyses of the thermal and mass (water) transfer processes are needed to improve process control of SSF.

5.3.1. Preparation of Materials and Inoculation

5.3.1.1. Preparation of Materials

Many kinds of natural solid materials are used as the substrate in SSF. These materials are usually pretreated before inoculation to facilitate mycelial penetration or to provide some chemical constituents for growth and product formation. In the case of rice koji, dehulled rice is polished by a mill to 40 to 70% of its original weight to remove the excess protein, fat, and minerals. Then it is washed and soaked for a few hours and drained overnight. Since the amount of water absorbed by the solid substrate depends on the soaking time and temperature, this operation determines the initial moisture content of the steamed rice. The moisture content of the solid substrate is increased by condensation of steam on the surface. Usually the amount of condensed water is equivalent to 10 to 15% by weight of the initial amount of solid substrate. In the case of finely cracked or powdered materials, the moisture content can be controlled by spraying and mixing an adequate amount of water directly. Then, the material is steamed for about 1 h at normal atmospheric pressure. Steaming or cooking of the solid substrate is carried out as a means of sterilization and thermal reconstruction of the starch, making it digestible by microbial amylase. In the koji process for soy sauce, soybeans and wheat are used. The wheat is roasted and crushed instead of being steamed or cooked, as roasting brings about the characteristic brown color and enriches the flavors of the soy sauce.

In the case of traditional kao-liang (sorghum) liquor brewing in China and Southeast Asia, soybeans, peas, wheat, and/or other powdered cereals are crushed and moistened without steaming or heating. They are mixed and formed into cakes in the shape of a brick (30 by 20 by 8 cm) or a doughnut with an outer diameter of approximately 25 cm.

5.3.1.2. Inoculation

The solid substrate is cooled to near the optimum temperature for germination of the mold spores. For inoculation, pure cultured spores of *Aspergillus* species are typically used in the industrial koji process. In contrast, *Rhizopus* or *Mucor* spp. and other *Mucorales* are isolated from the koji of Chinese kao-liang liquor. The traditional process does not involve any pure culture technique. The starter typically contains a mixture of molds, yeasts, and bacteria. *Aspergillus* spp. cannot grow well on the raw substrate used in the process because their amylase can barely hydrolyze raw starch. In contrast, *Rhizopus* sp., *Mucor* sp., and other *Mucorales* have a greater ability to hydrolyze raw starch.

In the rice-koji process for Japanese sake, 1×10^5 to 5×10^5 spores of *A. oryzae* per g of rice are usually inoculated. Okazaki et al. (109) reported that a large inoculum shortens the lag time of germination, and the specific growth rate and production of enzymes are affected only slightly. Conidiospores are produced in 5 to 7 days of cultivation on slightly polished rice to which 1% wood ash is added to supplement minerals. After drying, the moldy rice can be preserved for a long time and contains about 8×10^8 spores per g. The inoculating temperature is about 30°C, which is the optimum temperature for germination, and the temperature is controlled at below 38°C after inoculation.

Inoculation is done simply by scattering spores uniformly on the solid substrate. If an adhesive substrate is used, it is important to mix the solid substrate, preferably immediately after inoculation, to prevent lumping. The solid substrate should continue to be mixed at certain intervals to obtain uniform mold growth.

5.3.2. Temperature and Moisture Control

Table 3 shows the metabolic heat evolved in SSF (180, 182, 184). The evolved heat raises the temperature of the solid substrate and creates temperature gradients across the substrate bed vertically and horizontally, especially with forced aeration in which the circulated air usually flows upward from the bottom of the packed bed. The temperature gradients depend mainly on the thickness of the bed and the airflow rate. Nagatani et al. (97) simulated the vertical temperature profile during the cultivation of *A. oryzae* for rice koji as follows (Fig. 4):

$$[c_1 - (1/k_Ha)] \ln [1 + c_2(T - T_1)] = Z/G \qquad (2)$$

where Z is the thickness of the substrate bed (meters); T and T_1 are temperatures (°C) at thickness Z and at the bottom of the bed, respectively; G is the airflow rate (kilograms of dry air per meter^2 per hour); k_Ha is the volumetric coefficient of water evaporation (kilograms per meter^3 per hour); and c_1 and c_2 are constants. In this equation, metabolic heat evolution and the enthalpy of air are regarded as linear functions of temperature. The air condition on the surface of the koji particles is assumed to be saturated humidity, and heat is transferred according to the difference of enthalpy between the koji surface and the flowing air. Nagatani (96) simulated the transient state of the temperature profile of the koji bed before it reaches the steady state under intermittent aeration.

The evolved heat reduces the moisture content of the solid substrate by evaporation. In the case of aerobic microorganisms, the catabolic heat evolution rate is regarded to be proportional to the consumption rate of oxygen. If the transfer ratio of latent heat of vaporization is constant, the change of moisture content can be calculated by monitoring the oxygen consumption. Sato et al. (145) showed (Fig. 5) the relationship between moisture content and oxygen consumption during the cultivation of *A. oryzae* on wheat bran medium, assuming the value of the transfer ratio is 0.8. Figure 5 indicates that the more immediate decrease of moisture content occurs on media of lower initial moisture content. Since growth yield is regarded to be proportional to oxygen consumption, maximum cell growth should also be limited on media of low initial moisture content.

Narahara et al. (101) observed that, in the koji bed of a cylindrical fermentor of 20 cm thickness with forced aeration, in which *A. oryzae* was cultivated on rice, the activity of acid protease decreased toward the top of the bed. This gradient is related to the gradient of the moisture content created by the upward airflow.

Fernandez et al. (42) developed a control system for SSF in which the temperature and moisture content in the solid substrate bed were automatically controlled.

5.3.3. Monitoring of Microbial Growth

The cell mass is a very important variable in the control of any fermentation process. However, in the case of mold, the mycelium penetrates the solid substrate and is difficult to separate from it. Terui et al. (186) estimated the mycelial cell mass of *A. oryzae* in rice koji directly by subtracting the residual weight of the rice after a digestive reaction by takadiastase. However, this method requires a long reaction time, and error is unavoidable because the dry weight of the residue is changed by the filtration.

Instead of direct measuring, methods that assay cellular components, e.g., DNA (18), protein (69, 123, 151), chitin (9, 136), and *N*-acetyl glucosamine, a chitin component, are usually used. The *N*-acetyl glucosamine content in solid substrate is particularly often used as the mycelial growth parameter. Aidoo et al. (2) compared the efficiencies of various chemical assay methods for *N*-acetyl glucosamine. Arima and Uozumi (11) and Sakuerai et al. (136) measured *N*-acetyl glucosamine by a colorimetric assay, which degraded the mold chitin by acids or enzymatic reactions. Gomi et al. (53) proposed an enzymatic method using a chitinase derived from *Oerskovia* sp. Smits et al. (163) assayed *N*-acetyl glucosamine levels in moldy wheat bran by using high-pressure liquid chromatography–ion-exchange chromatography with pulse amperometric detection.

The internal or exit gas of the fermentor is often analyzed as an indirect method. Measurement of the oxygen uptake rate (OUR) or the carbon dioxide evolution rate (CER) is often used to estimate microbial growth indirectly (37, 86, 88, 104, 114, 117, 175). In the case of aerobic microorganisms, cell mass is estimated from OUR by the following equations, assuming the value of the respiration quotient to be constant:

$$dO_2/dt = X(\mu/Y_{x/o} + m) \qquad (3)$$

$$X = X_0 \exp(\mu t) \qquad (4)$$

where dO_2/dt is OUR (grams of O_2 per gram of solid per hour); X is cell mass (grams of cells per gram of solid); t is fermentation time (hours); μ is the specific growth rate (per hour); $Y_{x/o}$ is the yield based on oxygen (grams of cell per gram of O_2); m is the maintenance coefficient on OUR (grams of O_2 per gram of cell per hour); and X_0 is the initial cell mass.

In the case of CER:

$$dCO_2/dt = X(\mu/Y_{x/c} + m') \qquad (5)$$

where dCO_2/dt represents CER (grams of CO_2 per gram of solid per hour); $Y_{x/c}$ is the yield based on carbon dioxide (grams of cell per gram of CO_2); and m is the maintenance coefficient on CER (grams of CO_2 per gram of cell per hour). The values of $Y_{x/o}$ and m for *A. oryzae* are reported to be 0.47–2.17 and 0.057–0.096 on rice koji, respectively

TABLE 3 Metabolic heat evolved by some microorganisms in SSF[a]

Microorganism	Substrate	Initial moisture content (%)	Maximum heat evolution rate (kJ/kg of dry solid/h)	Peak time (h)	Total heat evolution (kJ/kg of dry solid)
Aspergillus oryzae	Wheat bran:rice hulls (1:1)	50~60	161	21~29	2.6×10^3
	Rice (sake koji)	35	29	30	4.2×10^2
	Soybean:wheat (1:1) (soy sauce koji)	40~45	131	22	1.8×10^3
Aspergillus niger	Sweet potato pulp:rice bran: rice hulls (1:0.1:0.5)	60~65	71	15~24	2.4×10^3
Bacillus amylosolvens	Wheat bran:rice hulls:soybean meal (3:1.25:1)	55	142	18	2.2×10^3

[a]Data from references 180, 182, and 185.

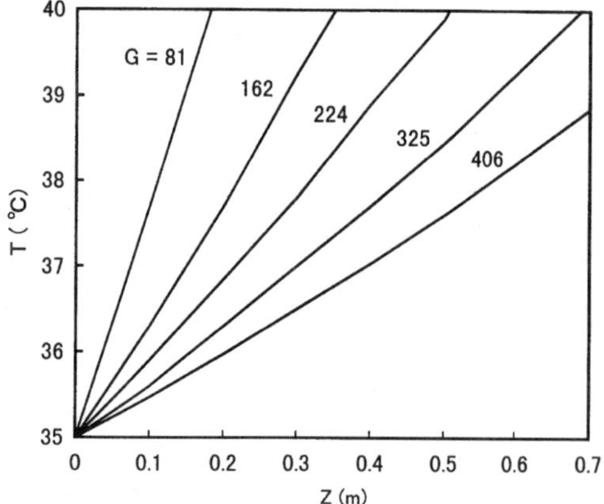

FIGURE 4 Simulation of the vertical temperature gradient in a packed substrate bed (79). Calculations were performed according to equation 2, in which the values of the constants were assumed as follows: $k_H a$ = 36,000 kg/m^3/h; c_1 = 0.018 m^3·h/kg; c_2 = 0.028/°C. Z, bed height.

(102, 112). The values of $Y_{x/c}$ and m' are 0.22 and 2.3, respectively (69).

Generally the value of μ varies throughout the cultivation. Okazaki et al. (112) reported that the value of *A. oryzae* growing on steamed rice can be expressed by the following equation:

$$\mu = \mu_{max}(1 - X/X_{max}) \tag{6}$$

where "max" means the maximum value. The observed values of μ_{max} were reported as 0.22 at 30°C and 0.295 at 35°C. The values of X_{max} were 0.0297 on steamed rice and 0.142 on steamed wheat bran (169, 180).

Okazaki and Sugama (110) developed a laboratory-scale device to measure OUR automatically and proposed a calculation method for the cell mass of *A. oryzae*. Sato et al. (143) proposed another calculation model for cell mass

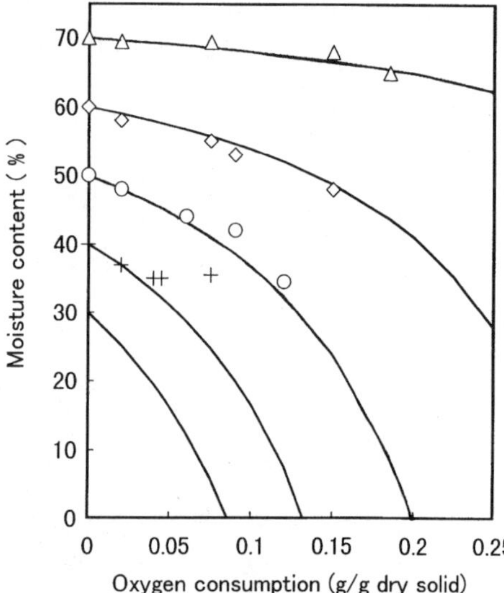

FIGURE 5 Relationship between moisture content and oxygen consumption by *A. oryzae* on wheat bran. Lines represent calculated data. Data points represent experimental results at different initial moisture content (147): △, 70%; ◇, 60%; ○, 50%; +, 40%.

based on OUR. This method was applied to the cultivation of an aerobic yeast, *Candida lipolytica*, using an oxygen meter to monitor OUR. Rodriguez Leon et al. (133) and Oriol et al. (113) applied this model to the growth estimation of *A. niger* on citrus peel and sugarcane bagasse, respectively. Similarly, Sato and Yoshizawa (148) calculated the anaerobic growth of *S. cerevisiae* according to equation 5, estimating growth and ethanol production from CER as monitored by a thermal mass flow meter. In this case, both growth and ethanol production were strongly restricted by the initial moisture content, which influences the ethanol concentration and hence product inhibition. Rinzema et al. (130) monitored the growth of *R. oligosporus* by analyzing of CO_2 in the exit gas from a packed-bed fermentor.

Aramaki et al. (9) applied a near-infrared analyzer to the indirect measurement of the mold cell mass in koji. The near-infrared spectra were analyzed statistically after differential treatment, and a correlation was found between the spectral patterns and *N*-acetyl glucosamine content. The advantages of the method are its ease of use and rapid response; however, it is sometimes difficult to discriminate among the spectral patterns of proteins (160).

Durand et al. (39) reviewed indirect means of cell mass estimation in SSF.

5.4. FERMENTOR DESIGN

Fermentors used in industrial SSF are classified into several groups (3, 97). Generally, the only means for heat removal from the substrate bed in these fermentors is forced aeration. However, for small-scale uses, especially for manufacturing foods and alcoholic beverages, simple trays with no forced aeration are sometimes used. The tray method requires much labor for setting up and mixing the substrate bed, and large-scale fermentors are normally automated and computer controlled (87). Fasidi et al. (41) reviewed fermentor types for the degrading of lignocellulosics. De Araujo et al. (34) reviewed several laboratory-scale SSF fermentors, including one monitored and controlled by computer.

5.4.1. Tray Fermentors

Tray fermentors, usually using wooden, plastic, or metal trays, are simple and widely used in traditional SSF. The cultivation is done in stationary trays with no mechanical agitation. Much labor is required for handling the trays. Since the efficiency of the heat exchange between substrate and air is not so high in this fermentor, the solid substrate cannot be highly heaped.

In the case of the rice-koji process for Japanese sake or soy sauce, the maximum thickness of the solid substrate bed is about 15 cm. Usually these trays are placed in a cultivating room where the temperature and the humidity are controlled for optimal mold growth. The steamed rice inoculated with the conidiospores of *A. oryzae* is placed in a lump and covered with pieces of cloth overnight. After about 20 h of cultivation, the rice is spread out widely on the tray. The thickness of the substrate bed can be varied to facilitate temperature control. Water evaporation from the bed is facilitated by reducing the bed thickness, which also leads to the lowering of temperature. The humidity of the cultivation room can be varied to facilitate the control of moisture content of the substrate; however, the response is quite slow. Mixing and adjusting the thickness of the substrate bed are performed at moderate intervals, based on knowledge obtained empirically.

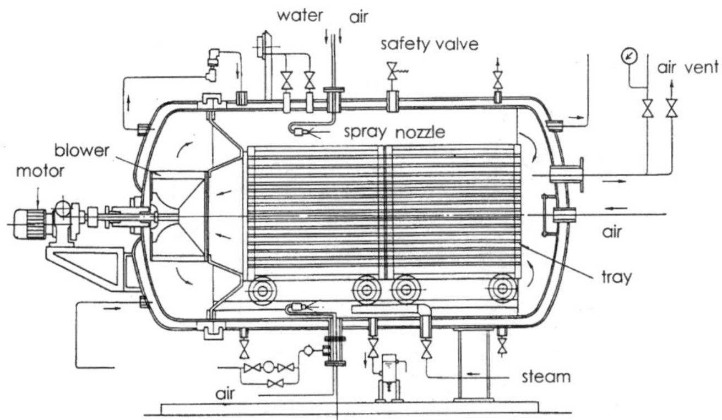

FIGURE 6 A tray fermentor for the production of rice koji starter. About 3 kg of rice is loaded on a 1-m-square tray. The maximum scale of cultivation is 200 kg, using 63 trays. (Courtesy Yamazaki Iron Works Co. Ltd., Kawaguchi, Japan.)

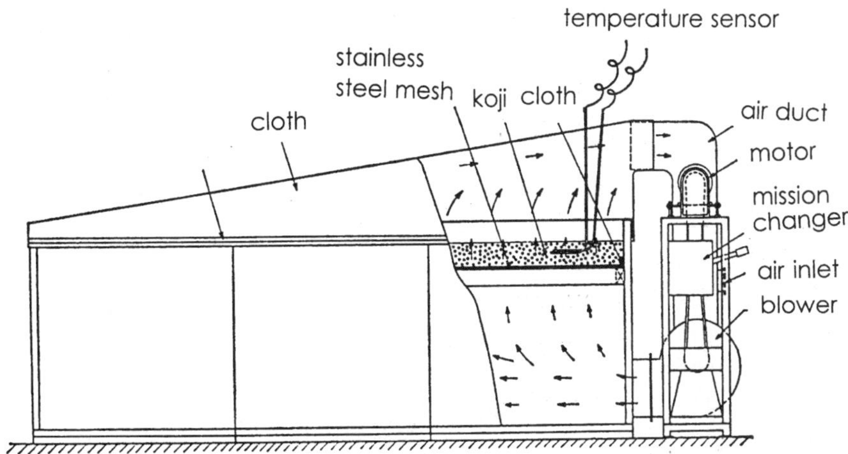

FIGURE 7 A packed-bed fermentor with forced aeration. This type of fermentor can be used only in an air-conditioned cultivation room. The height of the substrate bed should be less than 20 cm. The biggest fermentor of this type can be used for the cultivation of 1,000 kg of rice koji. (Courtesy Churitsu Industry Co. Ltd., Tokyo, Japan.)

Figure 6 shows a multiple-tray fermentor used for the cultivation of conidiophores. It is also used for koji making for shochu, which is a Japanese distilled alcoholic beverage. In this fermentor, all the operations are done in the cultivation vessel, including steam sterilization of the substrate, inoculation by air, moisture supply to the substrate by a double-flow nozzle with water and air, and forced aeration by a blower.

For reducing labor, a continuous-tray fermentor system has been developed. In this system all operations, including preparing the inoculum, sterilizing, cooling, inoculating, loading, drying, grinding, and bagging, are carried out continuously (72, 97). Byndoor et al. (27) reviewed industrial tray fermentors. Recently, computer-controlled "robots" have been used to carry out these operations in the factories of Japanese sake companies.

5.4.2. Packed-Bed Fermentors

Packed-bed fermentors are usually installed with a forced aeration device. As discussed in section 6.3.2, temperature adjustment between the top and bottom of the substrate depends on the thickness of the bed and the aeration rate. Terui et al. (181, 182, 185) studied a highly heaped fermentor (maximum thickness, 1.5 m) for the cultivation of *A. niger* or *A. oryzae* on sugarcane bagasse or wheat bran, in which rice hulls were used as the porous packing material. They calculated the heat evolution and estimated a 10°C temperature deviation between the top and bottom of the packed bed in this fermentor by intermittent and reversible aeration. Durand et al. (38) reviewed the application and control of packed-bed fermentors over several volume scales of SSF.

Commercial models of packed-bed fermentors have some differences in mechanical performance. Figure 7 shows a simple design with forced aeration for rice koji making. This fermentor is placed in a cultivation room, since it is not equipped with a heater. The forced air flows upward through the substrate bed from the bottom. The time course of the cultivation temperature is programmed, and aeration is done intermittently with an on-off regulation, based on the temperature in the substrate bed. However, some empirical knowledge is also needed for the operation, e.g., mixing and making up of the solid bed (usually the maximum thickness is 20 cm), determination of the temperature of the cultivation room, and so on.

Figure 8 shows a rotary double-disk fermentor for rice koji making. This fermentor is installed with screw mixers, and the wall is thermally insulated and supplied with heat-

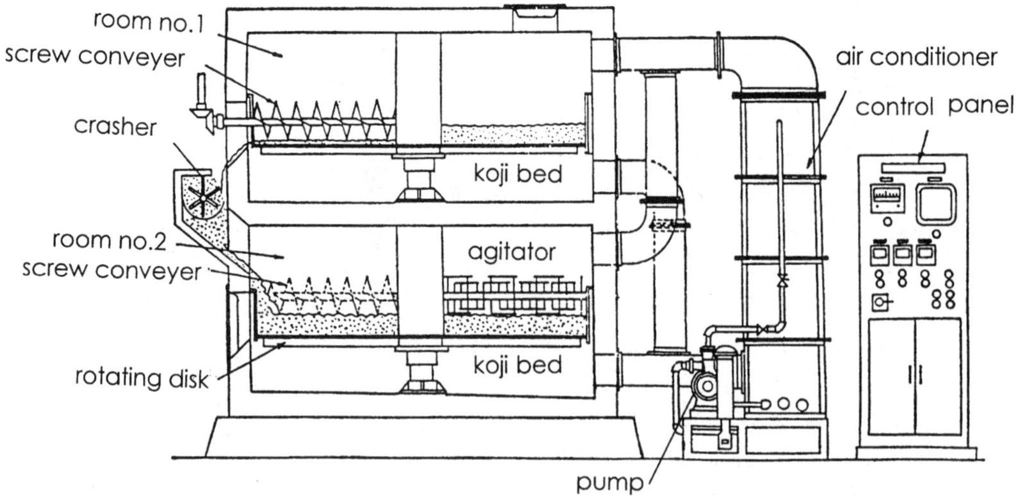

FIGURE 8 An automated packed-bed fermentor with forced aeration. The disk beds are continuously rotated during cultivation. The upper chamber is used for the first step of cultivation. Then, after about 1 day, the koji is delivered automatically to the lower chamber for the second step of cultivation. The biggest fermentor of this type has a cultivation chamber of 12-m disk size, in which 15,000 kg of rice koji or barley koji can be cultivated. (Courtesy Nagata Brewing Industry Co. Ltd., Takarazuka, Japan.)

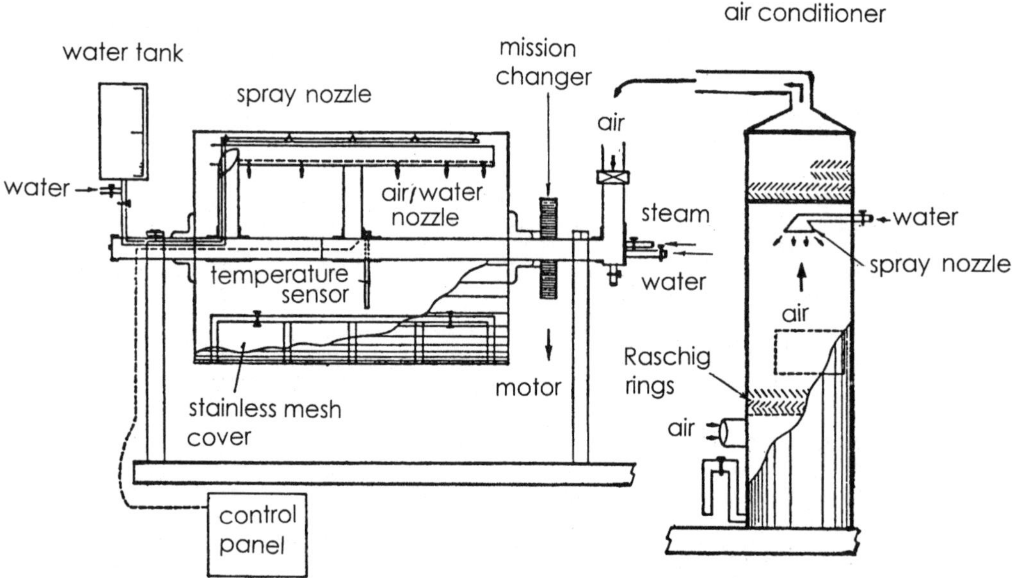

FIGURE 9 A rotary drum fermentor. The drum of this fermentor is rotated intermittently. Steeping, draining, steaming, and cultivation are all done in one drum. A maximum of 1,500 kg of rice koji or barley koji can be cultivated. (Courtesy Kawachi-Gen-Ichiro Co. Ltd., Kagoshima, Japan.)

ers. The conditioning of the circulated air is done by a cooling tower. The solid substrate is added and discharged by means of a conveyor and agitated by a screw at certain time intervals. The inoculated solid substrate is charged into the upper chamber of this fermentor and delivered to the lower chamber after about 1 day. The time course of the culture's temperature is programmed, and the operations are often controlled automatically by a computer. Recently knowledge technology, e.g., fuzzy inference and neural networks, has been used for the computer control system to simulate operation by an expert (156).

5.4.3. Rotary Drum Fermentors

Figure 9 shows a rotary drum fermentor, in which the drum rotates intermittently during cultivation to agitate and mix the substrate. Since this rotation is designed to work simultaneously with aeration, contact between the substrate and the fresh air supply is accelerated, facilitating rapid

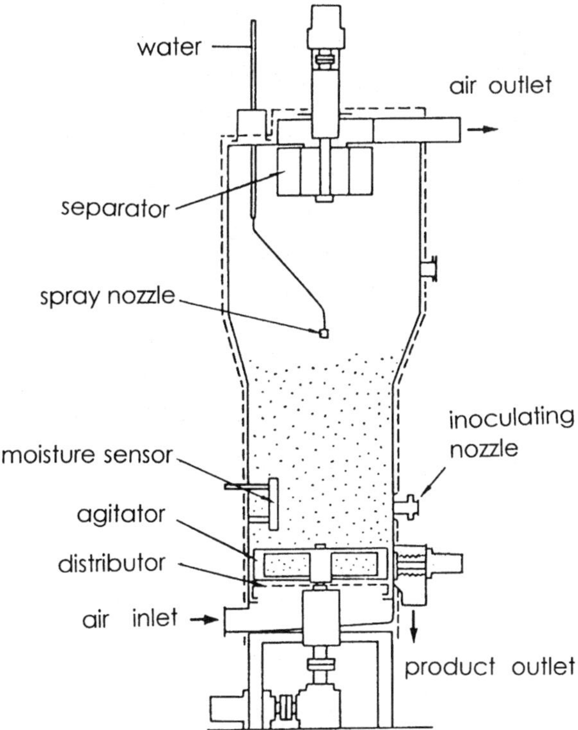

FIGURE 10 A fluidized-bed fermentor. The powdered substrate is fluidized by the airflow. The moisture content of the substrate is monitored by a electric capacity sensor. Water is supplied automatically by a spray nozzle. In a bench-scale fermentor with a 3.3-m^3 working volume, 500 kg of wheat bran could be fluidized for the cultivation of molds (5). (Courtesy Kikkoman Co. Ltd., Noda, Japan.)

heat removal and oxygen supply. Air is circulated, with the airflow varied by a damper in the air duct.

Accurate control of temperature is difficult in the rotary drum fermentor. In Japanese shochu making, therefore, this fermentor is usually used only at the first stage of the koji process, and the substrate is moved to a packed-bed fermentor with forced aeration in the second stage. All operations, i.e., washing, steeping, draining, steaming, and inoculation, are done in this fermentor. The raw substrate is fed into the drum with water and then steamed. Inoculation is done pneumatically, and the cultivation is started with forced aeration. The maximum substrate load is approximately 1,500 kg.

Recently Ueno et al. (190) developed a rotary drum fermentor with a temperature and moisture control system based on the control of latent heat transfer. They used this fermentor for the production of glucoamylase by *Rhizopus japonicus* on potato waste.

5.4.4. Fluidized-Bed Fermentor

In the fluidized-bed fermentor, the solid substrate is fluidized by upward airflow. The operative condition of the fermentor is estimated theoretically as follows:

$$u_{mf}/u_t = 0.02{\sim}0.10 \tag{5}$$

where u_{mf} and u_t are the minimum fluidization velocity and terminal velocity of the substrate pellet in air (meters per second), respectively. Actually the substrates tend to make lumps which adhere to the inner wall of the fermentor. This phenomenon causes an increase in the value of u_{mf}. Thus, u_{mf}/u_t is actually estimated to be 0.2~0.5 (4).

Fluidization of the substrate offers some advantages for SSF, as follows (5).

1. The effective surface area of the substrate powder is large for the microbial growth.
2. The entire bed of solid substrate is kept under uniform conditions.
3. Supplying water, nutrients, acid, and alkali and pH control are simple.
4. The removal of evolved heat and gas exchange of oxygen and carbon dioxide are easy.

These characteristics significantly enhance productivity in fluidized-bed fermentors (5, 93, 134, 177). Akao and Okamoto (5) studied the production of mold enzymes in a pilot-scale fluidized fermentor (Fig. 10). This fermentor could be used for the production of soy sauce koji using *Aspergillus sojae* or *A. oryzae* cultivated on wheat bran. It had a cylindrical shape and was installed with a wire net or a porous plate in the bottom, where the agitator rotated slowly to crush the lump of sedimented substrate. Fluidizing air was filtered and controlled at a moderate flow rate.

Table 4 compares the productivity of these three cultivation methods. Leucine aminopeptidase, neutral protease, acid protease, and α-amylase were remarkably higher in the fluidized-bed fermentor than with the other methods. How-

TABLE 4 Comparison of fermentation methods for production of enzymes by *A. oryzae*[a]

Enzyme	Submerged[b] (U/ml)	Solid state[c] (U/g of dry koji)	Fluidized bed[c] (U/g of dry koji)
Leucine aminopeptidase	1.6~4.3	3.2~17	38~100
Acid carboxypeptidase	0.09	16	7.6
Neutral protease	5.5×10^2	3.4×10^3	2.3×10^4
Acid protease	11	2.6×10^2	3.7×10^2
α-Amylase	6.4×10^2	1.9×10^4	5.8×10^4
Pectinase		28	
CMCase	1.4	87	72

[a]Data from references 4 and 5.
[b]Cultivation was done in a shaking flask with liquid medium containing 2% wheat bran and 0.5% defatted soybean.
[c]Cultivation was done with wheat bran.

ever, despite the advantages of fluidization of the solid substrate bed, the energy cost is high.

REFERENCES

1. **Abdullah, A. L., R. P. Tengerdy, and V. G. Murphy.** 1985. Optimization of solid substrate fermentation of wheat straw. *Biotechnol. Bioeng.* **27:**20–27.
2. **Aidoo, K. E., R. Hendry, and B. J. B. Wood.** 1981. Estimation of fungal growth in a solid state fermentation system. *Eur. J. Appl. Microbiol. Biotechnol.* **12:**6–9.
3. **Aidoo, K. E., R. Hendry, and B. J. B. Wood.** 1982. Solid substrate fermentations. *Adv. Appl. Microbiol.* **28:**201–237.
4. **Akao, T.** 1986. Solid-state fermentors, p. 235–258. *In* The Society of Fermentation Technology (ed.), *Bioengineering*. Suita City, Osaka, Japan. (In Japanese.)
5. **Akao, T., and Y. Okamoto.** 1983. Cultivation of microorganisms in an air-solid fluidized bed, p. 63–68. *In* Proceedings of 4th International Conference on Fluidization, Kashikojima, Japan. Engineering Foundation, New York.
6. **Alam, M., M. M. Hoq, I. Gomes, and G. Mohiuddin.** 1994. Production and characterization of thermostable xylanase by *Thermomyces lanuginosus* and *Thermoascus aurantiacus* grown on lignocelluloses. *Enzyme Microb. Technol.* **16:**298–302.
7. **Alazard, D., and M. Raimbault.** 1981. Comparative study of amylolytic enzymes production by *Aspergillus niger* in liquid and solid-state cultivation. *Eur. J. Appl. Microbiol. Biotechnol.* **8:**113–117.
8. **Antier, P., S. Roussos, M. Raimbault, A. Minjares, and G. Viniegra Gonzalez.** 1993. Pectinase-hyper producing mutants of *Aspergillus niger* C28B25 for solid-state fermentation of coffee pulp. *Enzyme Microb. Technol.* **15:**254–260.
9. **Aramaki, I., K. Hukuda, T. Hashimoto, T. Ishikawa, K. Kisaki, and N. Okazaki.** 1995. Near infrared diffuse reflectance spectrophotometric analysis of mycelial weight in rice-koji and search for characteristic wavelength for mycelia. *Seibutsu-kogaku Kaishi* **73:**3.
10. **Arima, K., S. Iwasaki, and G. Tamura.** 1967. Milk clotting enzyme from microorganisms. *Agric. Biol. Chem.* **31:** 540–545.
11. **Arima, K., and T. Uozumi.** 1967. A new method for estimation of the mycelial weight in koji. *Agric. Biol. Chem.* **41:**119–123.
12. **Auria, R., P. Christen, and J. P. Guyot.** 1996. Biotreatment of liquid, solid or gas residues: an integrated approach, p. 221–236. In M. Moo-Young et al. (ed.), *Environmental Biotechnology: Principles and Applications*. Kluwer Academic Publishers, Dordrecht, The Netherlands.
13. **Auria, R., J. Paracios, and S. Revah.** 1992. Determination of the interparticular effective diffusion coefficient for CO_2 and O_2 in solid state fermentation. *Biotechnol. Bioeng.* **39:**898–902.
14. **Babu, K. R., and T. Satyanarayana.** 1995. α-Amylase production by thermophilic *Bacillus coagulans* in solid state fermentation. *Process Biochem.* **30:**305–309.
15. **Babu, K. R., and T. Satyanarayana.** 1996. Production of bacterial enzyme by solid state fermentation. *J. Sci. Ind. Res.* **55:**464–467.
16. **Bach, P. D., M. Shoda, and H. Kubota.** 1984. Rate of composting of dewatered sewage sludge in continuously mixed isothermal reactor. *J. Ferment. Technol.* **62:**285–292.
17. **Bach, P. D., M. Shoda, and H. Kubota.** 1987. Thermal balance in composting operations. *J. Ferment. Technol.* **65:**199–209.
18. **Bajracharya, R., and R. E. Mudgett.** 1980. Effects of controlled gas environments in solid-substrate fermentations of rice. *Biotechnol. Bioeng.* **22:**2219–2235.
19. **Balagopalan, C.** 1996. Improving the nutritional value of cassava by solid state fermentation: CTCRI experience. *J. Sci. Ind. Res.* **55:**479–482.
20. **Balakrishnan, K., and A. Pandey.** 1996. Production of biologically active secondary metabolites in solid state fermentation. *J. Sci. Ind. Res.* **55:**365–372.
21. **Barber, S., M. J. Torner, M. A. Martnez-Anaya, and C. B. Barber.** 1989. Microflora of the sour dough of wheat flour bread. IX. Biochemical characteristics and baking performance of wheat doughs elaborated with mixtures of pure microorganisms. *Z. Lebensm. Unters. Forsch.* **189:**6–11.
22. **Barreto, T. J., J. G. Salva, V. L. Baldini, R. S. Papini, and A. M. Sales.** 1989. Protein enrichment of citrus wastes by solid substrate fermentation. *Process Biochem.* **24**(10)**:**167–171.
23. **Barrios-Gonzàlez, J., H. Gonzalez, and A. Mejia.** 1993. Effect of particle size, packing density and agitation on penicillin production in solid state fermentation. *Biotechnol. Adv.* **11:**539–547.
24. **Barrios-Gonzàlez, J., and A. Tomasini.** 1996. Production of aflatoxins in solid state fermentation. *J. Sci. Ind. Res.* **55:**424–430.
25. **Barrios-Gonzàlez, J., A. Tomasini, J. Viniegra-Gonzàlez, and J. Lopes.** 1988. Penicillin production by solid state fermentation. *Biotechnol. Lett.* **10:**793–798.
26. **Becerra, M., and M. I. Gonzalez-Siso.** 1996. Yeast β-galactosidase in solid state fermentations. *Enzyme Microb. Technol.* **19:**39–44.
27. **Byndoor, M. G., N. G. Karanth, and G. V. Rao.** 1997. Efficient and versatile design of a tray type solid state fermentation bioreactor, p. 113–119. *In* S. Roussos, B. K. Lonsane, M. Raimbault, and G. Viniegra-Gonzalez (ed.), *Advances in Solid State Fermentation*. Kluwer Academic Publishers, Dordrecht, The Netherlands.
28. **Cannel, E., and M. Moo-Young.** 1980. Solid-state fermentation systems. *Process Biochem.* **15**(5)**:**2–7, **15**(6)**:** 24–28.
29. **Chaudhary, K., S. Ethiraj, K. Lakshminarayama, and P. Tauro.** 1978. Citric acid production from Indian cane molasses by *Aspergillus niger* under solid state fermentation conditions. *J. Ferment. Technol.* **56:**554–557.
30. **Christen, P., R. Auria, R. Vega, E. Villegas, and S. Revah.** 1993. Growth of *Candida utilis* in solid state fermentation. *Biotechnol. Adv.* **11:**549–557.
31. **Considine, P. J., A. O'Rorke, T. J. Hackett, and M. P. Coughlan.** 1988. Hydrolysis of beet pulp polysaccharides by extracts of solid-state cultures of *Penicillium capsulatum*. *Biotechnol. Bioeng.* **31:**433–438.
32. **Crawford, J. H.** 1983. Composting of agricultural wastes, a review. *Proc. Biochem.* **18**(1)**:**15–18.
33. **Das, P., and M. N. Karim.** 1995. Mass balance and thermodynamic description of solid state fermentation of lignocellulosics by *Pleurotus ostreatus* for animal feed production. *J. Ind. Microbiol.* **15:**25–31.

34. **De Araujo, A. A., S. Roussos, C. Lepilleur, S. Delcourt, and P. Colavitti.** 1997. Laboratory scale bioreactors for study of fungal physiology and metabolism in solid state fermentation system, p. 93–111. *In* S. Roussos, B. K. Lonsane, M. Raimbault, and G. Viniegra-Gonzalez (ed.), *Advances in Solid State Fermentation*. Kluwer Academic Publishers, Dordrecht, The Netherlands.
35. **Deschamps, F., C. Giuliano, M. Asher, M. C. Huet, and S. Roussos.** 1985. Cellulase production by *Trichoderma harzianum* in static and mixed solid-state fermentation reactors under nonaseptic conditions. *Biotechnol. Bioeng.* **27:**1385–1388.
36. **Desfarges, C., C. Larroche, and J.-B. Gros.** 1987. Spore production of *Penicillium roqueforti* by solid state fermentation: stoichiometry, growth and sporulatin behavior. *Biotechnol. Bioeng.* **29:**1050–1058.
37. **Durand, A., R. Renaud, J. Maratray, and S. Almanza.** 1997. INRA-Dijon reactors for solid state fermentations: designs and applications, p. 71–92. *In* S. Roussos, B. K. Lonsane, M. Raimbault, and G. Viniegra-Gonzalez (ed.), *Advances in Solid State Fermentation*. Kluwer Academic Publishers, Dordrecht, The Netherlands.
38. **Durand, A., R. Renaud, J. Maratray, S. Almanza, and M. Diez.** 1996. Biomass estimation in solid state fermentation. *J. Sci. Ind. Res.* **55:**317–332.
39. **Durand, A., C. Vergoignan, and C. Desgranges.** 1997. The INRA-Dijon reactors: designs and applications, p. 23–37. *In* S. Roussos, B. K. Lonsane, M. Raimbault, and G. Viniegra-Gonzalez (ed.), *Advances in Solid State Fermentation*. Kluwer Academic Publishers, Dordrecht, The Netherlands.
40. **Ebine, H.** 1981. Fermented soybean foods in Japan, p. 41–52. *In* S. Saono, F. G. Winarno, and D. Karjadi (ed.), *Traditional Food Fermentation as Industrial Resources in ASCA Countries*. Proceedings of a Technical Seminar. The Indonesian Institute of Sciences, Jakarta, Indonesia.
41. **Fasidi, I. O., O. S. Isikhuemhen, and F. Zadrazil.** 1996. Bioreactors for solid state fermentation of lignocellulosics. *J. Sci. Ind. Res.* **55:**450–456.
42. **Fernandez, M., J. Ananias, I. Solar, R. Perez, L. Chang, and E. Agosin.** 1997. Advances in the development of a control system for a solid substrate pilot bioreactor, p. 155–168. *In* S. Roussos, B. K. Lonsane, M. Raimbault, and G. Viniegra-Gonzalez (ed.), *Advances in Solid State Fermentation*. Kluwer Academic Publishers, Dordrecht, The Netherlands.
43. **Fraile, E. R., J. O. Muse, and S. E. Bernardinelli.** 1981. Milk-clotting enzyme from *Mucor bacilliformis*. *Eur. J. Appl. Microbiol. Biotechnol.* **13:**191–193.
44. **Gervais, P., and C. Bazelin.** 1986. Procédé de mesure en continu et de régulation de l'activité de l'eau dans un milieu peu hydraté et dispositif por sa mise en ouevre. French patent 860572.
45. **Gervais, P., P. A. Marechal, and P. Molin.** 1996. Water relations of solid state fermentation. *J. Sci. Ind. Res.* **55:** 343–357.
46. **Ghildyal, N. P., M. K. Gowthaman, K. S. M. S. Raghava-Rao, and N. G. Karanth.** 1994. Interaction of transport resistances with biochemical reaction in packed-bed solid-state fermentors: effect of temperature gradients. *Enzyme Microb. Technol.* **16:**253–257.
47. **Ghildyal, N. P., S. V. Ramakrishna, P. N. Devi, B. K. Lonsane, and H. N. Asthana.** 1981. Large scale production of pectolytic enzyme by solid state fermentation. *J. Food Sci. Technol.* **18:**248–251.
48. **Gibbons, W. R.** 1989. Batch and continuous solid-phase fermentation of Jerusalem antichoke tubers. *J. Ferment. Bioeng.* **67:**258–265.
49. **Gibbons, W. R., and C. A. Westby.** 1986. Effect of pulp pH on solid phase fermentation of fodder beets for fuel ethanol production. *Biotechnol. Lett.* **8:**657–662.
50. **Gibbons, W. R., C. A. Westby, and T. L. Dobbs.** 1984. A continuous, farm-scale, solid-phase fermentation process for fuel ethanol and protein feed production from fodder beets. *Biotechnol. Bioeng.* **26:**1098–1107.
51. **Gibbons, W. R., C. A. Westby, and T. L. Dobbs.** 1986. Intermediate-scale, semicontinuous solid-phase fermentation process for production of fuel ethanol from sweet sorghum. *Appl. Environ. Microbiol.* **51:**115–122.
52. **Goldblatt, L. A.** 1969. Introduction, p. 1–11. *In* L. A. Goldblatt (ed.), *Aflatoxin*. Academic Press Inc., New York.
53. **Gomi, K., N. Okazaki, T. Tanaka, C. Kumagai, H. Inoue, Y. Iimura, and S. Hara.** 1987. Estimation of micelial weight in rice-koji with use of fungal cell wall lytic enzyme. *J. Soc. Brew.* (Japan) **82:**130–133.
54. **Gowthaman, M. K., K. S. M. S. Raghava-Rao, N. P. Ghildyal, and N. G. Karanth.** 1995. Estimation of kLa in solid-state fermentation using a packed-bed bioreactor. *Proc. Biochem.* **30:**9–15.
55. **Gutiérrez-Rojas, M., S. Auriar, S. Revah, and J.-C. Benet.** 1997. A phenomenological model for solid state fermentation of fungal mycelial growth, p. 131–142. *In* S. Roussos, B. K. Lonsane, M. Raimbault, and G. Viniegra-Gonzalez (ed.), *Advances in Solid State Fermentation*. Kluwer Academic Publishers, Dordrecht, The Netherlands.
56. **Han, I. Y., and M. P. Steinberg.** 1987. Solid-state yeast fermentation of raw corn with simultaneous koji hydrolysis. *Biotechnol. Bioeng. Symp.* **17:**449–462.
57. **Han, O., and R. E. Mudgett.** 1992. Effects of oxygen and carbon dioxide partial pressure on monascus growth and pigment production in solid-state fermentations. *Biotechnol. Prog.* **8:**5–10.
58. **Hang, Y. D., C. Y. Lee, and E. E. Woodams.** 1982. A solid state fermentation system for production of ethanol from apple pomace. *J. Food Sci.* **47:**1851–1852.
59. **Hang, Y. D., C. Y. Lee, and E. E. Woodams.** 1986. Solid-state fermentation of grape pomace for ethanol production. *Biotechnol. Lett.* **8:**53–56.
60. **Hang, Y. D., B. S. Luh, and E. E. Woodams.** 1987. Microbial production of citric acid by solid state fermentation of kiwifruit peel. *J. Food Sci.* **52:**226–227.
61. **Hang, Y. D., and E. E. Woodams.** 1987. Effect of substrate moisture content on fungal production of citric acid in a solid state fermentation system. *Biotechnol. Lett.* **9:** 183–186.
62. **Hesseltine, C. W.** 1972. Solid stat fermentations. *Biotechnol. Bioeng.* **14:**517–532.
63. **Hesseltine, C. W.** 1977. Solid state fermentation (part I–II). *Process Biochem.* **12(6):**24–32.
64. **Hesseltine, C. W.** 1983. Microbiology of oriental fermented foods. *Annu. Rev. Microbiol.* **37:**575–601.
65. **Higashio, K., and Y. Yasuoka.** 1982. Milk clotting enzyme production by NTG induced mutant. *Nippon Nogeikagaku Kaishi* **56:**777–785.
66. **Hiroi, T., T. Takahashi, T. Shima, T. Suzuki, M. Tsukioka, and N. Ogasawara.** 1981. Production of red-koji in solid culture. *Nippon Nogeikagaku Kaishi* **55:**1–6.

67. **Hours, R. A., C. E. Voget, and R. J. Ertola.** 1988. Apple pomace as raw material for pectinase production in solid state culture. *Biol. Wastes* **23:**221–228.

68. **Huang, S. Y., H. H. Wang, C-J. Wei, and G. W. Malaney.** 1985. Kinetic responses of koji solid state fermentation process. *Top. Enzyme Ferment. Biotechnol.* **10:**88–108.

69. **Huang, T.-L., Y. W. Han, and C. D. Callihan.** 1971. Application of the Lowry method for determination of cell concentration in fermentation of waste cellulosics. *J. Ferment. Technol.* **49:**574–576.

70. **Ishida, H., Y. Kojima, Y. Hata, E. Ichikawa, A. Kawado, K. Suginami, and S. Imayasu.** 1996. Cloning of the glucoamylase-encoding gene expressed in solid culture from *Aspergillus oryzae*, p. 137–138. *In* Proc. Ann. Meet. Soc. Ferment. Bioeng. Society of Fermentation and Bioengineering, Japan.

71. **Ito, K., A. Kimizuka, N. Okazaki, and S. Kobayashi.** 1989. Mycelial distribution in rice koji. *J. Ferment. Bioeng.* **68:**7–13.

72. **Jeffreys, G. A.** 1948. Mold enzymes produced by continuous tray method. *Food Ind.* **20:**688–690.

73. **Jermini, M. F. G., and A. L. Demain.** 1989. Solid state fermentation for cephalosporin production by *Streptomyces clavuligerus* and *Dephalosporium acremonium*. *Experientia* **45:**1061–1065.

74. **Kargi, F., and J. A. Curme.** 1985. Solid-state fermentation of sweet sorghum to ethanol in a rotary-drum fermentor. *Biotechnol. Bioeng.* **27:**1122–1125.

75. **Kargi, F., J. A. Curme, and J. J. Sheehan.** 1985. Solid-state fermentation of sweet sorghum to ethanol. *Biotechnol. Bioeng.* **27:**34–40.

76. **Kim, J. H., M. Hosobuchi, M. Kishimoto, T. Seki, T. Yoshida, H. Taguchi and D. D. Y. Ryu.** 1985. Cellulase production by a solid state culture system. *Biotechnol. Bioeng.* **27:**1445–1450.

77. **Kinsella, J. E., and D. Hwang.** 1976. Biosynthesis of flavors by *Penicillium roqueforti*. *Biotechnol. Bioeng.* **18:**927–938.

78. **Kirby, K. D., and C. J. Mardon.** 1980. Production of fuel ethanol by solid-phase fermentation. *Biotechnol. Bioeng.* **22:**2425–2427.

79. **Kitpreechavanich, V., M. Hayashi, and S. Nagai.** 1984. Production of xylan-degrading enzymes by themophilic fungi, *Aspergillus fumigatus* and *humicola langinosa*. *J. Ferment. Technol.* **62:**63–69.

80. **Kuhad, R. C., and A. Singh.** 1993. Enhanced production of cellulases by *Penicillium citrinum* in solid state fermentation of cellulosic residue. *World J. Microbiol. Biotechnol.* **9:**100–101.

81. **Kumagai, K., S. Usami, and T. Hattori.** 1981. Citric acid production from mandarin orange waste by solid culture of *Aspergillus niger*. *Hakkokogaku* **59:**461–464.

82. **Kumar, P. K. R., and B. K. Lonsane.** 1988. Batch and fed-batch solid-state fermentations: kinetics of cell growth, hydrolytic enzymes production, and gibberellic acid production. *Proc. Biochem.* **23**(April)**:**43–47.

83. **Kurunanandaa, K., S. L. Fales, G. A. Varga, and D. J. Royse.** 1992. Chemical composition and biodegradability of crop residues colonized by white-rot fungi. *J. Sci. Food Agric.* **60:**105–112.

84. **Lakshminarayana, K., K. Chaudhary, S. Ethiraj, and P. Tauro.** 1975. A solid state fermentation method for citric acid production using sugar cane bagasse. *Biotechnol. Bioeng.* **17:**291–293.

85. **Laukevics, J. J., A. F. Apsite, and U. E. Viesturs.** 1984. Solid substrate fermentation of wheat straw to fungal protein. *Biotechnol. Bioeng.* **26:**1465–1474.

86. **Lindenfelser, L. A., and A. Ciegler.** 1975. Solid-state fermentation for ochratoxin A production. *Appl. Microbiol.* **29:**323–327.

87. **Lonsane, B. K., N. P. Ghildyal, M. Ramakrishna, M. M. Krishnaiah, G. Saucedo-Castaneda, G. Viniegra-Gonzalez., M. Raimbault, and S. Roussos.** 1992. Scale-up strategies for solid state fermentation systems. *Proc. Biochem.* **27:**259–273.

88. **Madamwar, D., S. Patel, and H. Parikh.** 1987. Solid state fermentation for cellulases and β-glucosidase production by *Aspergillus niger*. *J. Ferment. Bioeng.* **67:**424–426.

89. **Martinez, M. J., B. Boeckle, S. Camarero, F. Guillen, and A. T. Martinez.** 1996. MnP isoenzymes produced by two *Pleurotus* species in liquid culture and during wheat-straw solid state fermentation. *ACS Symp. Ser.* **655:**183–196.

90. **Minjares-Carranco, A., G. Viniegra-Gonzalez, B. A. Trejo-Aguilar, and G. Aguilar.** 1997. Physiological comparison between pectinase-producing mutants of *Aspergillus niger* adapted either to solid-state fermentation or submerged fermentation. *Enzyme Microb. Technol.* **21:** 25–31.

91. **Mitchell, D. A., P. F. Greenfield, and H. W. Doelle.** 1990. Mode of growth of *Rhizopus oligosporus* on a model substrate in solid-state fermentation. *World J. Microbiol. Biotechnol.* **6:**201–208.

92. **Mitra, P., R. Chakraverty, and A. L. Chandra.** 1996. Production of proteolytic enzyme by solid state fermentation—an overview. *J. Sci. Ind. Res.* **55:**439–442.

93. **Moebus, O., and M. Teuber.** 1982. Production of ethanol by solid particles of *Saccharomyces cerevisiae* in a fluidized bed. *Eur. J. Appl. Microbiol. Biotechnol.* **15:**194–197.

94. **Morimoto, T., and G. Terui.** 1961. Time-course of metabolism in rice-koji. IV. Effects of moisture content of medium upon enzyme formation. *Hakkokogaku* **39:**200–203.

95. **Mudgett, R. E.** 1984. Solid-state fermentations, p. 66–83. *In* A. Demain and N. A. Solomon (ed.), *Manual of Industrial Microbiology and Biotechnology*. American Society for Microbiology, Washington, D.C.

96. **Nagatani, M.** 1978. Transient response of temperature of a solid culture of koji under intermittent aeration. *Hakkokogaku* **56:**217–221.

97. **Nagatani, M., Y. Hattori, and Y. Nunokawa.** 1977. Temperature distribution of a solid culture of koji under continuous aeration. *Hakkokogaku* **55:**175–180.

98. **Nakadai, T., and S. Nasuno.** 1988. Culture conditions of *Aspergillus oryzae* for production of enzyme preparation. *J. Ferment. Technol.* **66:**525–533.

99. **Nandakumar, M. P., M. S. Thakur, K. S. M. S. Raghava-Rao, and N. P. Ghildyal.** 1994. Mechanism of solid particle degradation by *Aspergillus niger* in solid state fermentation. *Proc. Biochem.* **29:**545–551.

100. **Narahara, H.** 1981. Growth characteristics of koji-infecting bacteria incubated on steamed rice. *Hakkokogaku* **59:**207–216.

101. **Narahara, H., Y. Koyama, T. Yoshida, S. Pichangkura, and H. Taguchi.** 1984. Control of water content in a

solid-state culture of *Aspergillus oryzae*. *J. Ferment. Technol.* **62:**453–459.

102. **Narahara, H., Y. Koyama, T. Yoshida, S. Pichangkura, R. Ueda, and H. Taguchi.** 1982. Growth and enzyme production in a solid-state culture of *Aspergillus oryzae*. *J. Ferment. Technol.* **60:**311–319.
103. **Nigam, P., and D. Singh.** 1996. Proceeding of agricultural wastes in solid state fermentation for cellulolytic enzyme production. *J. Sci. Ind. Res.* **55:**457–463.
104. **Nishio, N., and S. Nagai.** 1981. Single cell protein production from mandarin orange peel. *Eur. J. Appl. Microbiol. Biotechnol.* **11:**156–160.
105. **Nishio, N., K. Tai, and S. Nagai.** 1979. Hydrolase production by *Aspergillus niger* in solid-state cultivation. *Eur. J. Appl. Microbiol. Biotechnol.* **8:**263–270.
106. **Ohno, A., T. Ano, and M. Shoda.** 1993. Production of the antifungal peptide antibiotic, Iturin by *Bacillus subtilis* NB22 in solid state fermentation. *J. Ferment. Bioeng.* **75:** 23–27.
107. **Ohno, A., T. Ano, and M. Shoda.** 1995. Production of a lipopeptide antibiotic, surfactin, by recombinant *Bacillus subtilis* in solid state fermentation. *Biotechnol. Bioeng.* **47:**209–214.
108. **Okamoto, M., M. Yamakawa, and H. Abe.** 1992. Improvement of nutritive value of cereal straw by solid state fermentation using *Pleurotus ostreatus*. *Trop. Agric. Res. Ser.* **25:**178–185.
109. **Okazaki, N., Y. Fukuda, and S. Sugama.** 1979. Relationship among the appearance, cell mass and enzyme activities. *J. Brew. Soc.* (Japan) **74:**687–691.
110. **Okazaki, N., and S. Sugama.** 1979. A new apparatus for automatic growth estimation of mold cultured on solid media. *J. Ferment. Technol.* **57:**413–417.
111. **Okazaki, N., S. Sugama, and T. Tanaka.** 1980. Mathematical model for surface culture of koji mold. *J. Ferment. Technol.* **58:**471–476.
112. **Okazaki, N., K. Takeuchi, and S. Sugama.** 1979. A new apparatus for automatic growth estimation of mold cultured on solid media. *J. Ferment. Technol.* **57:**413–417.
113. **Oriol, E., B. Schettino, G. Viniegra-Gonzales, and M. Raimbault.** 1988. Solid-state culture of *Aspergillus niger* on support. *J. Ferment. Technol.* **66:**57–62.
114. **Ouchi, K., T. Ishido, S. Sugama, and K. Nojiro.** 1967. Growth characteristics of yeast in koji. *J. Brew. Soc.* (Japan) **62:**1029–1033.
115. **Padmanabhan, S., M. V. R. Murthy, and B. K. Lonsane.** 1993. Potential of *Aspergillus oryzae* CFTRI1480 for producing proteinase in high titres by solid state fermentation. *Appl. Microbiol. Biotechnol.* **40:**499–503.
116. **Padmanabhan, S., M. Ramakrishna, B. K. Lonsan, and M. M. Krishnaiah.** 1992. Enhanced leaching of product at elevated temperatures: alpha-amylase produced by *Bacillus licheniformis* M27 in solid state fermentation system. *Lett. Appl. Microbiol.* **15:**235–238.
117. **Panayotov, C. A., A. P. Atev, I. M. Stojanov, and P. A. Peev.** 1987. A cellulase enzymic product: industrial processing and possible applications. *Acta Biotechnol.* **7:**461–467.
118. **Pandey, A.** 1992. Recent process developments in solid-state fermentation. *Proc. Biochem.* **27:**109–117.
119. **Pandey, A., P. Selvakumar, and L. Ashakumary.** 1996. Performance of a column bioreactor for glucoamylase synthesis by *Aspergillus niger* in SSF. *Proc. Biochem.* **31:** 43–46.
120. **Paredes-Lopez, O., and G. I. Harry.** 1988. Food biotechnology review: traditional solid-state fermentations of plant raw materials—application, nutritional significance, and future prospects. *Crit. Rev. Food Sci. Nutr.* **27:**159–187.
121. **Prescott, S. C., and C. G. Dunn.** 1959. *Industrial Microbiology*, 3rd ed., p. 533–577. McGraw-Hill, New York.
122. **Prescott, S. C., and C. G. Dunn.** 1959. *Industrial Microbiology*, 3rd ed., p. 666–682. McGraw-Hill, New York.
123. **Raimbault, M., and D. Alazard.** 1980. Culture method to study fungal growth in solid fermentation. *Eur. J. Appl. Microbiol. Biotechnol.* **8:**263–270.
124. **Raimbault, M., S. Revah, and F. Pina.** 1985. Protein enrichment of cassava by solid substrate fermentation using molds isolated from traditional foods. *J. Ferment. Technol.* **63:**395–399.
125. **Raimbault, M., S. Roussos, and B. K. Lonsane.** 1997. Solid state fermentation at ORSTOM: evolution and perspectives, p. 577–612. *In* S. Roussos, B. K. Lonsane, M. Raimbault, and G. Viniegra-Gonzalez (ed.), *Advances in Solid State Fermentation*. Kluwer Academic Publishers, Dordrecht, The Netherlands.
126. **Rajagopalan, S., and J. M. Modak.** 1995. Modelling of heat and mass transfer for solid state fermentation process in tray bioreactor. *Bioprocess Eng.* **13:**161–169.
127. **Ramesh, M. V., and B. K. Lonsane.** 1987. Solid state fermentation for production of α-amylase by *Bacillus megaterium* 16M. *Biotechnol. Lett.* **9:**323–328.
128. **Ramesh, M. V., and B. K. Lonsane.** 1987. A novel bacterial thermostable alpha-amylase system produced under solid state fermentation. *Biotechnol. Lett.* **9:**501–504.
129. **Richter, K., and A. Traeger.** 1994. L(+)-Lactic acid from sweet sorghum by submerged and solid-state fermentations. *Acta Biotechnol.* **14:**367–378.
130. **Rinzema, A., J. C. De Reu, J. Oostra, F. J. I. Nagel, G. J. A. Nijhuis, A. A. Scheepers, J. R. Nout, and J. Tramper.** 1997. Models for solid-state cultivation of *Rhizopus oligosporus*, p. 143–154. *In* S. Roussos, B. K. Lonsane, M. Raimbault, and G. Viniegra-Gonzalez (ed.), *Advances in Solid State Fermentation*. Kluwer Academic Publishers, Dordrecht, The Netherlands.
131. **Roche, N., and A. Durand.** 1996. Kinetics of *Thermoascus aurantiacus* solid-state fermentation on sugarbeet pulp-polysaccharide alteration and production of related enzymatic activities. *Appl. Microbiol. Biotechnol.* **45:**584–588.
132. **Rodney, P. J., and P. F. Greenfield.** 1986. Role of water activity in ethanol fermentations. *Biotechnol. Bioeng.* **28:** 29–40.
133. **Rodriguez Leon, J. A., L. Sastre, J. Echevarria, G. Delgado, and W. Bechstedt.** 1988. A mathematical approach for the estimation of biomass production rate in solid state fermentation. *Acta Biotechnol.* **8:**307–310.
134. **Rottenbacher, L., M. Scossler, and W. Bauer.** 1987. Modelling a solid-state fluidized bed fermentor for ethanol production with *S. cerevisiae*. *Bioprocess Eng.* **2:**25–31.
135. **Roukas, T.** 1994. Solid-state fermentation of carob pods for ethanol production. *Appl. Microbiol. Biotechnol.* **41:** 296–301.
136. **Sakurai, Y., T. H. Lee, and H. Shiota.** 1977. On the convenient method for glucosamine estimation in koji. *Agric. Biol. Chem.* **41:**619–624.

137. **Samuta, T., M. Nakajima, T. Ota, H. Saeki, K. Sato, T. Oba, and K. Yoshizawa.** 1990. Production of spirits from raw corn grits by a solid-state fermentation. *J. Brew. Soc.* (Japan) **85:**269–276.
138. **Sangsurasak, P., M. Nopharatana, and D. A. Mitchell.** 1996. Mathematical modeling of the growth of filamentous fungi in solid state fermentation. *J. Sci. Ind. Res.* **55:**333–342.
139. **Saono, J. K. D.** 1981. Microflora of ragi: its composition and as a source of industrial yeasts, p. 189–200. *In* S. Saono, F. G. Winarno, and D. Karjadi (ed.), *Traditional Food Fermentation as Industrial Resources in ASCA Countries*. Proceedings of a Technical Seminar. The Indonesian Institute of Sciences, Jakarta, Indonesia.
140. **Sato, K., S. Miyazaki, N. Matsumoto, K. Yoshizawa, and K. Nakamura.** 1988. Pilot-scale solid-state ethanol fermentation by inert gas circulation using moderately thermophilic yeast. *J. Ferm. Technol.* **66:**173–180.
141. **Sato, K., S. Miyazaki, and K. Yoshizawa.** 1988. Application of a solid-state fermentation to the production of a spirits. *J. Brew. Soc.* (Japan) **83:**559–562.
142. **Sato, K., M. Nagatani, K. Nakamura, and S. Sato.** 1983. Growth estimation of *Candida lipolytica* from oxygen uptake in a solid-state culture with forced aeration. *J. Ferment. Technol.* **61:**623–629.
143. **Sato, K., M. Nagatani, and S. Sato.** 1982. A method supplying moisture to the medium in a solid-state culture with forced aeration. *J. Ferment. Technol.* **60:**607–610.
144. **Sato, K., K. Nakamura, and S. Sato.** 1985. Solid-state ethanol fermentation by means of inert gas circulation. *Biotechnol. Bioeng.* **27:**1312–1319.
145. **Sato, K., K. Osada, and M. Nagatani.** 1979. Oxygen consumption and water evaporation on a solid-state culture with forced aeration. *Hakkokogaku* **57:**360–365.
146. **Sato, K., and K. Yoshizawa.** 1988. Growth and growth estimation of *Saccharomyces cerevisiae* in solid-state ethanol fermentation. *J. Ferment. Technol.* **66:**667–673.
147. **Sato, K., and K. Yoshizawa.** 1989. Diffusivities of glucose and ethanol in corn grits. *J. Brew. Soc.* (Japan) **84:** 47–50.
148. **Sato, K., and K. Yoshizawa.** 1989. Dissolution rate of corn grits in a solid-state fermentation. *J. Brew. Soc.* (Japan) **84:**794–799.
149. **Sato, K., K. Yoshizawa, and T. Nishiya.** 1990. Recovery rate of ethanol in a stripping bioreactor. *J. Brew. Soc.* (Japan) **85:**645–650.
150. **Sato, K., K. Yoshizwa, and T. Nishiya.** 1990. Simulation study of the fermentation in a stripping bioreactor. *J. Brew. Soc.* (Japan) **85:**745–749.
151. **Schmidell, W., and M. V. Fernandes.** 1976. The measurement of cellular protein content as a method for determining mold concentration. *J. Ferment. Technol.* **54:** 225–226.
152. **Scott, W. J.** 1957. Water relations of food spoilage microorganisms. *Adv. Food. Res.* **7:**83–127.
153. **Selvakumar, P., L. Ashakumary, and A. Pandey.** 1996. Microbiol synthesis of starch saccharifying enzyme in solid cultures. *J. Sci. Ind. Res.* **55:**443–449.
154. **Shankaranand, V. S., and B. K. Lonsane.** 1992. Wheat bran as a substrate for production of citric acid by *Aspergillus niger* CFTRI 6 in solid state fermentation system: titre and yield improvements. *Chem. Mikrobiol. Technol. Lebensm.* **14:**33–39.
155. **Shankaranand, V. S., and B. K. Lonsane.** 1994. Ability of *Aspergillus niger* to tolerate metal ions and minerals in a solid-state fermentation system for the production of citric acid. *Proc. Biochem.* **29:**29–37.
156. **Shiba, H., K. Matsuura, M. Hirotsune, M. Hamachi, and C. Kumagai.** 1996. The model construction of mycelial growth in the koji making process, p. 366. *In* Proc. Annu. Meet. Soc. Ferment. Bioeng. Society for Fermentation and Bioengineering, Japan.
157. **Shotwell, O. L., C. W. Hesseltine, R. D. Stubblefield, and W. G. Sorensen.** 1966. Production of aflatoxin on rice. *Appl. Microbiol.* **4:**425–428.
158. **Siessere, V., and S. Said.** 1989. Pectic enzymes production in solid-state fermentation using citrus pulp pellets by *Talaromyces flavus*, *Tubercularia vulgaris* and *Penicillium charlessi*. *Biotechnol. Lett.* **11:**343–344.
159. **Silman, R. W., and L. T. Black.** 1983. Assay of solid-substrate fermentation by means of reflectance infrared analysis. *Biotechnol. Bioeng.* **25:**603–607.
160. **Silman, R. W., H. F. Conway, R. A. Anderson, and E. B. Bagley.** 1979. Production of aflatoxin in corn by a large-scale solid-substrate fermentation process. *Biotechnol. Bioeng.* **21:**1977–1808.
161. **Singh, K., A. K. Puniya, and S. Singh.** 1996. Biotransformation of crop residues into animal feed by solid state fermentation. *J. Sci. Ind. Res.* **55:**472–478.
162. **Smits, J. P., R. J. Janssens, P. Knol, and J. Bol.** 1994. Modelling of the glucosinolate content in solid-state fermentation of rapeseed meal with fuzzy logic. *J. Ferment. Technol.* **77:**579–581.
163. **Smits, J. P., H. M. Van Sonsbeek, W. Knol, A. Rinzema, and J. Tramper.** 1996. Solid-state fermentation of wheat bran by *Trichoderma reesei* QM9414: substrate composition changes, C balance, enzyme production, growth and kinetics. *Appl. Microbiol. Biotechnol.* **46:**489–496.
164. **Soccol, C. R., I. Iloki, B. Martin, and M. Raimbault.** 1994. Comparative production of alpha-amylase, glucoamylase and protein enrichment of raw and cooked cassava by *Rhizopus* strains in submerged and solid state fermentations. *J. Food Sci. Technol.* **31:**320–323.
165. **Soccol, C. R., B. Martin, and M. Raimbault.** 1994. Potential of solid state fermentation for production of L(+)-lactic acid by *Rhizopus oryzae*. *Appl. Microbiol. Biotechnol.* **41:**286–290.
166. **Stubblefield, R. D., D. L. Shotwell, C. W. Hesseltine, M. L. Smith, and H. Hall.** 1967. Production of afratoxin on wheat and oats: measurement with a recording densitometer. *Appl. Microbiol.* **15:**186–190.
167. **Sudo, S., T. Ishikawa, K. Sato, and T. Oba.** 1994. Comparison of acid-stable α-amylase production by *Aspergillus kawachii* in solid-state and submerged cultures. *J. Ferment. Bioeng.* **77:**483–489.
168. **Sudo, S., S. Kobayashi, A. Kaneko, K. Sato, and T. Oba.** 1995. Growth of submerged mycelia of *Aspergillus kawachii* in solid-state culture. *J. Ferment. Bioeng.* **79:** 252–256.
169. **Sugama, S., and N. Okazaki.** 1979. Growth estimation of *Aspergillus oryzae* cultured on solid media. *J. Ferment. Technol.* **57:**408–412.
170. **Sun, T., P. Li, B. Liu, D. Liu, and Z. Li.** 1997. A novel design of solid state fermentor and its evaluation for cellulase production by *Trichoderma viride* SL-1. *Biotechnol. Lett.* **19:**465–467.

171. **Sun, T., Z. Li, and D. Liu.** 1996. Effect of elevated temperature on *Trichoderma viride* SL-1 in solid state fermentations. *Biotechnol. Tech.* **10:**889–894.

172. **Sun, T., B. Liu, D. Liu, and Z. Li.** 1997. Solid state fermentation of rice chaff for fibrinolytic enzyme production by *Fusarium oxysporum*. *Biotechnol. Lett.* **19:**171–174.

173. **Suprianto, R. Ohba, T. Koga, and S. Ueda.** 1989. Liquefaction of glutinous rice and aroma formation in tape preparation by ragi. *J. Ferment. Bioeng.* **67:**249–252.

174. **Swaminathan, S., S. R. Joshi, and N. M. Parhad.** 1989. Bioconversion to single cell protein: a potential resource recovery from paper mill solid waste. *J. Ferment. Bioeng.* **67:**427–429.

175. **Szewczyk, K. W., and L. Myszka.** 1994. The effect of temperature on the growth of *A. niger* in solid state fermentation. *Bioprocess Eng.* **10:**123–126.

176. **Takamine, J.** 1914. Enzymes of *Aspergillus oryzae* and the application of its amyloclastic enzyme to the fermentation industry. *Ind. Eng. Chem.* **6:**824–828.

177. **Tanaka, M., A. Kawaide, and R. Matsuno.** 1985. Cultivation of microorganisms in an air-solid fluidized bed fermentor with agitators. *Biotechnol. Bioeng.* **28:**1294–1301.

178. **Tani, Y., V. Vongsuvanlert, and J. Kumnuanta.** 1986. Raw cassava starch-digestive glucoamylase of *Aspergillus* sp. N-2 isolated from cassava chips. *J. Ferment. Technol.* **64:**405–410.

179. **Tengerdy, R. P.** 1985. Solid substrate fermentation. *Trends Biotechnol.* **3:**96–99.

180. **Terui, G.** 1962. Automation of koji-making. *Kagaku to Seibutsu* **1:**32–40. (In Japanese.)

181. **Terui, G., and T. Morimoto.** 1956. Time-course of metabolism in rice-koji. *Hakkokogaku* **34:**575–578.

182. **Terui, G., and T. Morimoto.** 1961. Time-course of metabolism in rice-koji. III. A characteristic feature observed in the time-courses of specific growth rate and respiratory activity. *Hakkokogaku* **39:**196–200.

183. **Terui, G., and I. Shibazaki.** 1959. Industrial fermentation by a highly heaped culture with forced aeration. III. Citric acid fermentation in sweet potato pulp media. *Hakkokogaku* **37:**332–338.

184. **Terui, G., I. Shibazaki, and T. Mochizuki.** 1957. Industrial fermentation by a highly heaped culture with forced aeration. I. Citric acid fermentation. *Hakkokogaku* **35:**105–116.

185. **Terui, G., I. Shibazaki, and T. Mochizuki.** 1957. Industrial fermentation by a highly heaped culture with forced aeration. II. General description of improved process. *Hakkokogaku* **36:**109–116.

186. **Terui, G., I. Shibazaki, and T. Mochizuki.** 1959. Industrial fermentation by a highly heaped culture with forced aeration. IV. Some industrial heap cultures with wheat bran. *Hakkokogaku* **37:**479–494.

187. **Terui, G., I. Shibazaki, T. Mochizuki, and M. Takano.** 1960. Industrial fermentation by a highly heaped culture with forced aeration. V. Rice-koji for miso-making. *Hakkokogaku* **38:**29–40.

188. **Thakur, M. S., N. G. Karanth, and K. Nand.** 1993. Downstream processing of microbial rennet from solid state fermented moldy bran. *Biotechnol. Adv.* **11:**399–407.

189. **Toyama, N.** 1976. Feasibility of sugar production from agricultural and urban cellulosic wastes with *Trichoderma viride* cellulase. *Biotechnol. Bioeng. Symp.* **6:**207–219.

190. **Ueno, T., H. Tanaka, and T. Maekawa.** 1995. Development of a rotating drum reactor with a novel temperature control system for solid state fermentation. *Nogyo-Shisetsu* **25:**231–237.

191. **Underkofler, L. A., G. M. Severson, K. J. Goering, and L. M. Christensen.** 1947. Commercial production and use of mold bran. *Cereal Chem.* **24:**1–22.

192. **Varzakas, T. H., D. L. Pyle, and K. Niranjan.** 1997. Mycelial penetration and enzymic diffusion on soybean tempe, p. 59–70. *In* S. Roussos, B. K. Lonsane, M. Raimbault, and G. Viniegra-Gonzalez (ed.), *Advances in Solid State Fermentation*. Kluwer Academic Publishers, Dordrecht, The Netherlands.

193. **Villegas, E., S. Revah, S. Aubague, L. Alcantara, and R. Auria.** 1993. Solid state fermentation: acid protease production in controlled CO_2 and O_2 environments. *Biotechnol. Adv.* **11:**387–397.

194. **Viniegra-Gonzàlez, G.** 1997. Solid state fermentation: definition, characteristics, limitations and monitoring, p. 5–22. *In* S. Roussos, B. K. Lonsane, M. Raimbault, and G. Viniegra-Gonzalez (ed.), *Advances in Solid State Fermentation*. Kluwer Academic Publishers, Dordrecht, The Netherlands.

195. **Wang, H. H., and T. C. Hsieh.** 1972. Kao-liang brewing by pure cultures, p. 651–658. Proc. IV IFS. *Ferment. Technol. Today*

196. **Wang, H. L., and C. W. Hesseltine.** 1966. Wheat tempeh. *Cereal Chem.* **43:**563–570.

197. **Winarno, F. G.** 1981. The nutritional potential of fermented foods in Indonesia, p. 31–40. *In* S. Saono, F. G. Winarno, and D. Karjadi (ed.), *Traditional Food Fermentation as Industrial Resources in ASCA Countries*. Proceedings of a Technical Seminar. The Indonesian Institute of Sciences, Jakarta, Indonesia.

198. **Wujcik, W., and W. J. Jewell.** 1980. Dry anaerobic fermentations. *Biotechnol. Bioeng. Symp.* **10:**43–65.

199. **Xavier, S., and N. G. Karanth.** 1992. A convenient method to measure water activity in solid state fermentation systems. *Lett. Appl. Microbiol.* **15:**53–55.

200. **Xavier, S., and B. K. Lonsane.** 1994. Sugar-cane pressmud as a novel and inexpensive substrate for production of lactic acid in a solid-state fermentation system. *Appl. Microbiol. Biotechnol.* **41:**291–295.

201. **Xu, W. Q., and Y. D. Hang.** 1988. Roller culture technique for citric acid production by *Aspergillus niger*. *Proc. Biochem.* **23**(8):117–118.

202. **Yang, S. S.** 1988. Protein enrichment of sweet potato residue with amylolytic yeasts by solid-state fermentation. *Biotechnol. Bioeng.* **32:**886–890.

203. **Yang, S. S., and M. Y. Ling.** 1989. Tetracycline production with sweet potato residue by solid state fermentation. *Biotechnol. Bioeng.* **33:**1021–1028.

204. **Yoshizawa, K.** 1981. Traditional alcoholic beverage industry in Japan, p. 53–62. *In Industrial Resources in ASCA Countries*. Proceedings of a Technical Seminar. The Indonesian Institute of Sciences, Jakarta, Indonesia.

205. **Zvauya, R., and M. I. Muzondo.** 1994. Some factors affecting protein enrichment of cassava flour by solid state fermentation. *Lebensm. Wiss. Technol.* **27:**590–591.

Experimental Design for Improvement of Fermentations

R. J. STROBEL AND G. R. SULLIVAN

6

Even with the great gains made in our understanding of microbial physiology and molecular biology, improvement of fermentations remains largely an empirical process. The relevant literature and "prior art" serve only as a starting point for the development of new fermentations. When an investigator is working with microbial strains that are newly isolated and taxonomically undefined, empirical approaches to fermentation improvement are the only options available. Because these circumstances are so common in an industrial setting, this chapter will focus on experimental design and statistical analysis of data as the primary means of rapidly improving fermentations (17, 24). Several hypothetical examples as well as authentic case studies will be presented to illustrate the benefits of experimental design and to clarify how statistically designed experiments are conducted and analyzed.

In most instances, the microbiologist begins with some medium and set of conditions that allow for at least modest expression of the metabolite or activity of interest. The task then is to improve that expression to a level sufficient for isolation and characterization of the desired product(s). Preliminary experiments are usually done with shaken flasks or similar culture vessels to minimize time and cost involved in the study of nutritional and physical variables or factors (Tables 1 and 2). Many of these factors will need to be evaluated to identify a fermentation process yielding an optimal result (response). A list of response measurements common in commercial fermentations is shown in Table 3.

A fermentation improvement program may begin by measuring product yield as a response to factors like medium strength (e.g., increase the concentration of all nutrients by 25%), incubation temperature, and culture pH (buffers may be used to control pH). It is also common practice in the early stages of development to replace the original carbon and nitrogen sources with other widely used ingredients (38). Alternative sources of phosphorus and sulfur may need to be considered as well.

Fermentation studies are often done by using classical methods of experimentation, with one factor at a time changed while all others are held constant (5, 17, 34). Consider a hypothetical example in which a newly discovered antibiotic is produced by a microbial culture grown in a medium composed of 4% glucose and 2% soybean flour. The microbiologist begins to study the fermentation by testing alternative carbon sources (Table 4). In this initial experiment, replacing glucose with glycerol results in an increase in the antibiotic titer from 100 to 140 μg/ml. In the next experiment, glycerol is used as the carbon source and alternative nitrogen sources are considered (Table 5). The nitrogen source experiment shows that if soybean flour is replaced by cottonseed flour, the antibiotic titer increases further, to 160 μg/ml. Based on these two studies, further experimentation would build on the new medium composed of 4% glycerol and 2% cottonseed flour and the titer of 160 μg/ml as a benchmark.

Experimental design techniques present a more balanced alternative to the one-factor-at-a-time approach to fermentation improvement (2, 3, 18, 28). For example, instead of testing carbon and nitrogen sources in two separate experiments as described above, one may test all combinations of carbon and nitrogen simultaneously (26). Consider the results of the full factorial design experiment shown in Table 6 (30). Three carbon sources are tested with each of five nitrogen sources, resulting in 15 trials. Because all factor combinations are studied, another medium which produces even larger amounts of the new antibiotic is found. In the culture flask containing 4% potato dextrin and 2% cottonseed flour, the antibiotic has a titer of 190 μg/ml. Potato dextrin shows no improvement over other carbon sources except when used with cottonseed flour. The synergy between these two medium components would be missed if the classical approach to experimentation were used.

The hypothetical experiment presented above shows how more information can be obtained by considering all factor combinations. We used an example in which each factor was nonnumeric, or categorical (e.g., soybean flour versus cottonseed flour), rather than numerical, or continuous (e.g., 2% soybean flour versus 4% soybean flour). A factor is said to be categorical if its factor levels, or settings, represent disjoint entities which follow no numerical hierarchy. When categorical factors need to be tested, a full factorial is usually the experimental design of choice. However, as the number of factors increases, full factorial designs result in an excessively large number of experimental trials. Most experimental situations deal with continuous factors which are amenable to screening designs requiring fewer trials than full factorials. Although experiments containing both categorical and continuous factors can be de-

TABLE 1 Nutritional factors important in commercial fermentations

Factor(s)	Range of factor settings (concn, % [wt/vol])	Examples
Carbon source(s)	0.5–20	Glucose, sucrose, starch, molasses, dextrins, corn meal, whey, glycerol, alcohols, lipids
Nitrogen source(s)	0.1–10	Ammonia gas, ammonium salts, casein hydrolysates, corn gluten meal, corn steep liquor, cottonseed flour, fish meal, glutamic acid, nitrates, peptones, soybean flour and meal, urea, whole yeast, and yeast extract
Phosphorus	0.1–2	Corn steep liquor, phosphates
Sulfur	0.1–1	Methionine, proteins, sulfates
Other nutrients	<1	Iron salts, magnesium salts, oxygen, potassium, trace elements, vitamins

signed and statistically analyzed, our focus will be on designs containing only continuous factors (3, 18).

6.1. SEQUENTIAL NATURE OF STATISTICALLY DESIGNED EXPERIMENTS

Ideally, experimental design is a sequential process (3, 18). First, categorical factors are studied to determine which nutrients and physical conditions hold the most promise for optimizing the fermentation. Then, a large number of continuous factors (typically 5 to 12) are screened and insignificant ones are eliminated in order to obtain a smaller, more manageable set of factors. The remaining factors are optimized by response surface modeling. Finally, after model building and optimization, the predicted optimum is verified (34, 35). An example of the first step is given in the introduction of this chapter; screening of factors at two numerical levels will be discussed next.

6.2. SCREENING DESIGNS

Screening should be done when the investigator is faced with a large number of factors and is unsure which settings are likely to produce optimal or nearly optimal responses. Identifying the key response(s) and identifying all possible process factors are crucial steps in experimental design methodology. Information can be obtained by studying process flow charts, investigating relevant literature, and consulting with individuals who have knowledge of the process. Once a list of factors has been made, the settings for each factor must be determined. In a screening experiment, two settings are chosen for each factor studied. The screening design will contain only a subset of all possible factor-setting combinations, resulting in fewer trials and, ultimately, a large savings in time and cost.

6.2.1. Plackett-Burman Designs

Plackett-Burman designs comprise one type of two-level screening design (18, 29). Recent reports on the use of Plackett-Burman designs in fermentation include their application toward improved asperlicin yield (27), improved insecticide production (34), and an increased specific growth rate of recombinant *Escherichia coli* (16). These designs allow for the study of up to $n - 1$ factors with n trials. Typical Plackett-Burman designs consist of 8, 12, 16, 20, and 24 trials. It is common practice to test six to eight factors with a 12-trial design. For example, in the development of a new antibiotic fermentation at Eli Lilly, seven factors were investigated with a 12-trial Plackett-Burman design (Table 7). Choice of factors was based on prior experience growing the antibiotic-producing microorganism, and choice of settings reflected a wide but reasonable nu-

TABLE 2 Some physical and other nonnutritional factors important in commercial fermentations

Factor	Range of factor settings
Temperature	10–60°C
pH	3–9
Agitation	50–500 rpm
Aeration	0.1–2 vvm[a]
Pressure	1–15 lb/in^2 gauge
Inoculum size	1–15% (vol/vol)
Inoculum age	0.5–5 days
Fermentation time	0.8–14 days

[a]vvm, volume of air per volume of liquid broth per minute.

TABLE 3 Common response values and units of measurement

Response	Typical units
Titer or product yield	g/liter, mg/ml, mg/liter, μg/ml
Productivity	g/liter/day, g/liter/h
Cell dry weight	g/liter
Yield on cell dry weight	g/g
Yield on glucose used	g/g
Cost per unit weight	Dollars/g (of product)
Volumetric activity	U/ml, U/liter
Specific activity	U/mg (of protein), U/g (of cell dry weight)

TABLE 4 Carbon source study[a]

Carbon source in flask fermentation	Antibiotic titer (μg/ml)
4% Glucose (control)	100
4% Potato dextrin	110
4% Glycerol	140

[a]Soybean flour is held constant at 2%.

merical range. Also, some change in the response (antibiotic titer) was expected for each factor over the range of settings selected. To choose factor settings for any two-level screening design, consider the following criteria: (i) the factor range ideally should contain the optimum response for that factor, (ii) the range should be wide enough for any effect or trend to be exposed, and (iii) the range should avoid combinations of low and high factor settings which are likely to produce an outright process failure. Factor level selection can be a difficult part of the experimental process. Experience, prior experimentation, and the literature can be valuable resources for choosing factor settings.

In Plackett-Burman designs, low and high factor settings are coded as -1 and $+1$, respectively, so the factors are computationally equivalent. The coded arrangement for Plackett-Burman designs of any size may be found in statistical software packages or in the experimental design literature (3, 18, 29). The coding for a 12-trial design like the one used in our new antibiotic fermentation is shown in Table 8. The table displays the trials in a standardized format, but in practice, trials should always be randomized to remove any unknown bias which could hinder the identification of a critical effect (3, 33). Notice that each factor is tested an equal number of times at its low and high settings. Because of this equal allocation, the resulting data provide a fair and efficient estimate of the linear effect of each factor. In addition to equal allocation within each factor, balance exists between each pair of factors. For example, of the six trials tested at the low (-1) setting for factor 1, three are tested at each of the low (-1) and high ($+1$) settings of factor 2. This balance exists for every pair of factors in the design. Equal allocation and balance characterize all two-level designs and make statistically designed experiments complete and efficient from the standpoint of resources used and information gained.

The results for the new antibiotic fermentation are shown in Table 9. Once data are obtained for each trial (trials may be run simultaneously or sequentially if necessary), statistical analysis can be performed to evaluate and rank factors by their degree of impact on the fermentation process. This ranking begins with the calculation of the

TABLE 5 Nitrogen source study[a]

Nitrogen source in flask fermentation	Antibiotic titer (μg/ml)
2% Soybean flour (new control)	140
2% Cottonseed flour	160
2% Fish meal	70
2% Casein hydrolysate	110
2% Urea	80

[a]Glycerol (4%) is the carbon source; glucose is omitted from the medium.

TABLE 6 Full factorial study[a]

Carbon source	Nitrogen source	Antibiotic titer (μg/ml)
Glucose	Soybean flour	100
	Cottonseed flour	110
	Fish meal	60
	Casein hydrolysate	100
	Urea	70
Potato dextrin	Soybean flour	110
	Cottonseed flour	190
	Fish meal	80
	Casein hydrolysate	100
	Urea	60
Glycerol	Soybean flour	140
	Cottonseed flour	160
	Fish meal	70
	Casein hydrolysate	110
	Urea	80

[a]All combinations of carbon source (4%) and nitrogen source (2%) are tested simultaneously.

parameter estimate of each factor. For Plackett-Burman experiments, parameter estimates (or simply estimates) can be easily explained and depicted in graphical form. The estimate is the slope of a line that connects the mean of the responses at a low factor setting to the mean of the responses at a high factor setting. For example, the estimate or slope for cottonseed oil is 1.5565. Figure 1 shows the line drawn between the mean titers for cottonseed oil at its low (-1) and high ($+1$) coded settings. The equation for this line is $y = 1.5565x + 3.821$, where y is the antibiotic titer. The intercept, 3.821, is the predicted titer at the zero (coded) setting for cottonseed oil and, in this design, is simply the average of all titers. A similar plot and calculation can be generated for each factor.

Statistical software, such as JMP (SAS Institute, Inc., Cary, N.C.), can be used to quickly calculate parameter estimates and perform a statistical analysis of the data (Table 10). Like the interpretation of any slope, the estimate is the expected change in antibiotic titer for every unit increase in the coded factor level. Therefore, two times the factor estimate represents the change in titer over the en-

TABLE 7 Plackett-Burman design for new antibiotic case study[a]

Factor	Factor setting (g/liter)	
	Low (-1)	High ($+1$)
1. Beet molasses	5	40
2. Cottonseed oil	20	100
3. Corn steep liquor	5	40
4. Casein hydrolysate	5	20
5. Soybean meal	10	50
6. Cottonseed flour	10	50
7. $MgSO_4 \cdot 7H_2O$	6	24

[a]The coded levels are -1 for the low factor setting and $+1$ for the high factor settings and are shown in parentheses.

TABLE 8 Twelve-trial Plackett-Burman design used to study seven factors in new antibiotic fermentation[a]

Trial	Coded setting for factor:						
	1	2	3	4	5	6	7
1	−1	−1	−1	+1	−1	−1	+1
2	−1	−1	+1	−1	−1	+1	−1
3	−1	−1	+1	−1	+1	+1	+1
4	−1	+1	−1	−1	+1	−1	+1
5	−1	+1	−1	+1	+1	+1	−1
6	−1	+1	+1	+1	−1	−1	−1
7	+1	−1	−1	−1	+1	−1	−1
8	+1	−1	−1	+1	−1	+1	+1
9	+1	−1	+1	+1	+1	−1	−1
10	+1	+1	−1	−1	−1	+1	−1
11	+1	+1	+1	−1	−1	−1	+1
12	+1	+1	+1	+1	+1	+1	+1

[a]Factors 1 through 7 refer to those listed in Table 7.

tire factor range (−1 to +1). The change in response over the entire range is called the main effect of a given factor. The main effect of cottonseed oil is 3.113 (2 × 1.5565). The sign of a factor estimate indicates, on average, which factor setting results in higher titers. A large estimate, either positive or negative, indicates that a factor has a large impact on titer, while an estimate close to zero means that a factor has little or no effect.

A number of tools may be used to help assess the significance of each process factor. These include *P* values, normal plots, and Pareto charts. The most common means of assessing significance is the *P* value. The *P* value is the probability that the magnitude of a parameter estimate is due to random process variability (3). A low *P* value indicates a "real" or significant effect and provides a baseline for deeming some factors critical and others less important. Traditionally, a *P* value of <0.05 has been used as a cutoff point for significance. Because *P* values generally depend on the size and type of experiment, caution should be used in establishing cutoff values for significance (18). A second method for assessing significance is the normal probability plot (9). Parameter estimates not significantly different from zero will form a straight line on the normal probability plot. Estimates which deviate from the line are deemed significant. JMP and other statistical packages will construct normal plots and identify critical factors. Finally, a Pareto chart of estimates can be constructed to judge the relative importance of factors simply by the magnitude of the parameter estimates (18).

In the statistical analysis for the new antibiotic shown in Table 10, cottonseed oil has a large, positive estimate and a low *P* value (0.0008). This means that the titer increases significantly as the cottonseed oil setting is increased from 20 to 100 g/liter. Magnesium sulfate has a smaller estimate than cottonseed oil, and its estimate is negative, meaning that the titer decreases as the factor setting increases from 6 to 24 g/liter. The impact of magnesium sulfate, although smaller than that of cottonseed oil,

TABLE 9 Results of Plackett-Burman study for new antibiotic fermentation

Trial	Factor concn (g/liter)							Titer (g/liter)
	Beet molasses	Cottonseed oil	Corn steep liquor	Casein hydrolysate	Soybean meal	Cottonseed flour	$MgSO_4$	
1	5	20	5	20	10	10	24	1.21
2	5	20	40	5	10	50	6	2.95
3	5	20	40	5	50	50	24	2.89
4	5	100	5	5	50	10	24	5.10
5	5	100	5	20	50	50	6	6.46
6	5	100	40	20	10	10	6	5.61
7	40	20	5	5	50	10	6	2.92
8	40	20	5	20	10	50	24	0.69
9	40	20	40	20	50	10	6	2.92
10	40	100	5	5	10	50	6	5.20
11	40	100	40	5	10	10	24	3.35
12	40	100	40	20	50	50	24	6.55

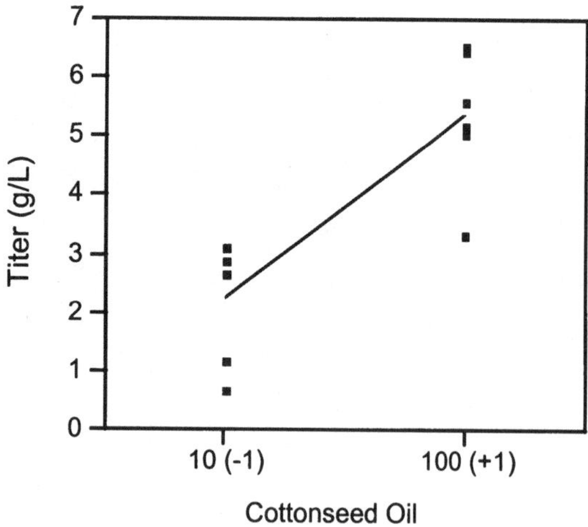

FIGURE 1 Plot of observed antibiotic titers versus cottonseed oil concentration (grams per liter) in a Plackett-Burman study. Coded settings are shown in parentheses.

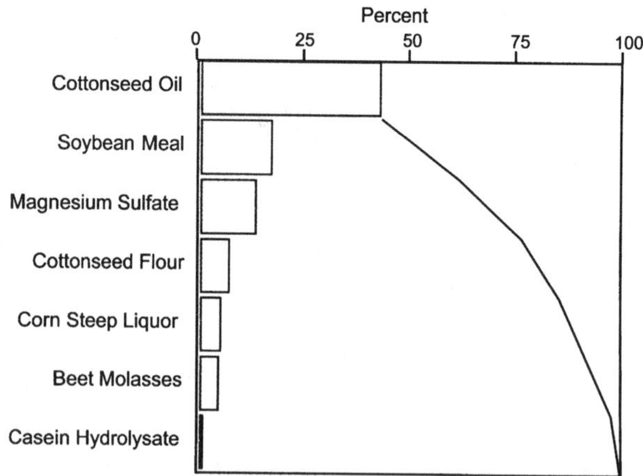

FIGURE 2 Pareto plot for Plackett-Burman parameter estimates. ——, cumulative percent.

is also statistically significant ($P = 0.0376$). A ranking of the factor estimates for this experiment is shown in a Pareto chart (Fig. 2). The Pareto chart displays the magnitude of each factor estimate and is a convenient way to view the results of a Plackett-Burman experiment. If we return to the parameter estimates in Table 10, we find that cottonseed oil, soybean meal, and magnesium sulfate have the lowest *P* values. We conclude that these three factors account for most of the variation in antibiotic titer and are the most critical factors in the fermentation.

Plackett-Burman designs are well suited for detecting the significant factors in a fermentation (cottonseed oil, soybean meal, and magnesium sulfate in our example). But unlike full factorial designs, Plackett-Burman designs cannot test for interaction between factors. An alternative to the Plackett-Burman and full factorial designs is the fractional factorial design, which estimates all main effects and some or all two-factor interactions (2, 3, 18, 28).

6.2.2. Fractional Factorial Designs

Fractional factorial designs are used in place of Plackett-Burman designs when estimates of two-factor interactions are desired (3, 18). A statistical interaction between two factors occurs when the main effect of one factor depends on the level or setting of the second factor. Although two-factor interactions are very common and play an important role in optimization, it is not always necessary to estimate them at the screening stage. However, under some circumstances the investigator may already suspect that an interaction exists between two factors. When this is the case, a fractional factorial design may be used. The application of a fractional factorial design to model the effects of temperature, pH and salts on growth of *Staphylococcus aureus* has been reported (4).

TABLE 10 Parameter estimates and *P* values for Plackett-Burman study of new antibiotic fermentation

Factor	Estimate	*P* value
Beet molasses	−0.2147	0.2767
Cottonseed oil	1.5565	0.0008
Corn steep liquor	0.2231	0.2612
Casein hydrolysate	0.0858	0.6413
Soybean meal	0.6507	0.0189
Cottonseed flour	0.3031	0.1504
Magnesium sulfate	−0.5227	0.0376

Fractional factorial designs are two-level designs which include only a fraction of the total trials that comprise a full factorial (3). Fractional factorial designs consist of 2^{k-p} trials, where "2" refers to the two levels or factor settings, k is the total number of factors, and $(1/2)^p$ is the fraction of the full factorial design (p is an integer which is less than k). For example, a 2^{6-1} design is a $(1/2)^1$ fraction of a 2^6 factorial which consists of 2^5, or 32, runs. The utility of fractional factorial designs is based on the notion that main effects and two-factor interactions are initially the most important effects for modeling and understanding a multifactor process. Since higher-order interactions (involving three or more factors) are relatively rare, their effects are assumed to be negligible, allowing the critical main effects and two-factor interactions to be estimated in a smaller number of trials. Therefore, instead of testing six factors with either a 12-run Plackett-Burman design or a 64-run full factorial design, the investigator can select a 1/2 fraction of the full factorial design and estimate all main effects and two-factor interactions with just 32 trials. The procedure for selecting the proper half of the 64 factorial trials is discussed in the experimental design literature (2, 3). Fractional factorial designs are very efficient and offer a valuable compromise between Plackett-Burman designs and full factorial designs.

An example of a fractional factorial design is given in Table 11. In this experiment, we are seeking to improve the growth of a seed-stage flask culture. The seed culture is to serve as an inoculum for bioreactors in the production of a newly discovered metabolite. In the seed culture study, growth is determined by centrifugation of the whole broth after 48 h of incubation with shaking. The response is cell solids (percent by volume) as measured in a graduated centrifuge tube. Six factors are to be tested, each at a low and a high numerical setting. A full factorial study of two levels of six factors requires 64 (2^6) flask trial runs. In this experiment, exactly one half of the complete set of combinations will be run. The chosen design will be sufficient to assess all main effects and two-factor interactions. The variability in the process is estimated by using the three-factor interactions which are not included in the model. Because

TABLE 11 Fractional factorial design for seed culture case study

Trial	Factor concn (g/liter)						Cell solids (%)
	Yeast	Magnesium sulfate	Glucose	NaCl	KH_2PO_4	Soy peptone	
1	0	0	30	0	0	10	7.5
2	0	0	30	0	5	30	15
3	0	0	30	10	0	30	15
4	0	0	30	10	5	10	10
5	0	0	80	0	0	30	13
6	0	0	80	0	5	10	5
7	0	0	80	10	0	10	5
8	0	0	80	10	5	30	10
9	0	4	30	0	0	30	19
10	0	4	30	0	5	10	7
11	0	4	30	10	0	10	9
12	0	4	30	10	5	30	18
13	0	4	80	0	0	10	10
14	0	4	80	0	5	30	15
15	0	4	80	10	0	30	15
16	0	4	80	10	5	10	5
17	6	0	30	0	0	30	20
18	6	0	30	0	5	10	10
19	6	0	30	10	0	10	16
20	6	0	30	10	5	30	21
21	6	0	80	0	0	10	10
22	6	0	80	0	5	30	15
23	6	0	80	10	0	30	20
24	6	0	80	10	5	10	7
25	6	4	30	0	0	10	15
26	6	4	30	0	5	30	22
27	6	4	30	10	0	30	15
28	6	4	30	10	5	10	16
29	6	4	80	0	0	30	20
30	6	4	80	0	5	10	9
31	6	4	80	10	0	10	10
32	6	4	80	10	5	30	13

three-factor interactions are rare, these effects are assumed to be negligible and close to zero. Any deviation from zero, therefore, represents random error and is used to estimate that random error (3). The JMP data analysis gives an estimate for each main effect and two-factor interaction (Table 12). Based on the *P* values obtained, the individual effects of yeast, glucose, and soy peptone are of primary importance. Of secondary importance are two interactions (glucose-KH_2PO_4 [$P = 0.08$] and magnesium-NaCl [$P = 0.11$]) as well as the individual effect of KH_2PO_4 ($P = 0.15$). The glucose-KH_2PO_4 interaction is depicted in Fig. 3. The plot shows the negative effect of glucose, since the higher level (80 g/liter) results in a lower level of cell solids. The plot also indicates the negative effect of KH_2PO_4, but only at the high glucose setting. This observation emphasizes the importance of interactions. Without knowledge of the interaction, the investigator may believe that higher levels of cell solids can be attained only at low levels of KH_2PO_4. However, at the "optimal" level of glucose (30 g/liter), KH_2PO_4 has no effect.

TABLE 12 Parameter estimates and *P* values for fractional factorial seed culture study

Term[a]	Estimate	*P* value
Yeast	1.890625	0.0013
Glucose	−1.671875	0.0030
Magnesium-NaCl	−0.765625	0.1054
KH_2PO_4	−0.671875	0.1493
Glucose-KH_2PO_4	−0.828125	0.0831
Soy peptone	3.578125	<0.0001

[a]All main effect and interaction terms where *P* is >0.20 are not shown.

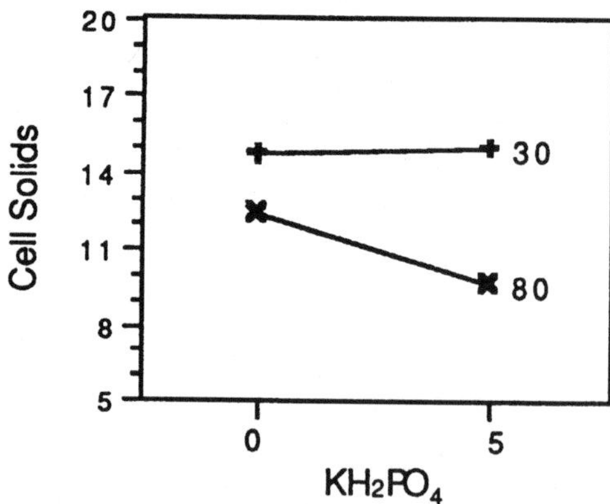

FIGURE 3 Interaction plot of cell solids (percent by volume) versus KH_2PO_4 (grams per liter) at glucose settings of 30 and 80 g/liter.

6.2.3. Curvature in Fractional Factorial Designs

The fractional factorial design, like the Plackett-Burman design, can determine factor estimates and main effects very efficiently, with only a few trial runs. In addition, the fractional factorial design can estimate two-factor interactions. Neither design, however, provides an estimate of curvature in the response between two factor settings. Curvature exists for a given factor when the response at the midpoint between two factor settings deviates from a straight line (Fig. 4). Since low and high factor settings are coded as -1 and $+1$, the midpoint is coded as 0.

The introduction of curvature raises several questions which can now be addressed.

(i) Why don't we assess curvature, linear effects, and interactions in just one carefully designed experiment? One reason is that such an experiment would be large and difficult to interpret. A process with 10 factors would require approximately 90 trials; we would need a large amount of data to accurately estimate 10 main effects, 45 two-factor interactions, 10 curvature effects, and an intercept. Also, because we may not have selected factor settings which yield the optimal response, we may not be in the correct design space, in which case the optimum will not be found. The design space is the multidimensional region whose boundaries are defined by the low and high factor settings. For example, three factors each at two settings define a cubic design space.

(ii) If only two settings of a factor are tested at the screening stage, how can we be sure we will not bypass a factor which may have no linear effect or interactions but has strong curvature? We cannot be sure; however, in most cases, a factor which is critical in the process will be critical in more than one way. If a factor has a large curvature effect, it also likely has a strong linear effect and appears in some important interactions. Also, if the screening design does not identify a factor as critical and the investigator strongly believes that it is critical, there is nothing wrong with taking that factor forward in subsequent experimentation.

(iii) Screening designs identify critical factors, but how do we know that our design space contains the optimal settings so that we can proceed to the optimization phase of experimentation? We can run center points to help assess whether we have identified the proper design space. A center point is an experimental trial at the middle level of all process factors and is in the center of the design space. Just as the response at the midpoint can be used to assess individual factor curvature, the response at the center point is used to assess overall, or gross, curvature. Therefore, if the response at the center point is significantly greater (or less) than the mean of the responses at all fractional factorial points, then gross curvature exists in the form of a maximum (or minimum). Several center point trials (typically three to six) are run to accurately estimate the response at the center of the design space. Replication of the center point also provides an estimate of pure process variability needed to determine statistical significance. With the aid of center points, gross curvature can be assessed in any two-level design. If curvature is not evident, the method of steepest ascent should be used to find the proper design space (19, 28). The application of center points to factorial designs in fermentation has been reported for the production of xylitol (20), lactic acid (21), and fumaric acid (22).

In the fractional factorial seed culture experiment discussed above, six center point replicates were actually included in the original study (Table 13). Note that all of the flask trials are identical and that each has factor settings midway between the two settings used in the fractional factorial. The center point data allow us to assess gross curvature in the cell solids response. A t test comparing the mean of the center points (15.2% cell solids) to the mean of the fractional factorial point (13.0% cell solids) yields a P value of 0.04, providing strong evidence of gross curvature (33). Since the mean of the center points exceeds the mean of the factorial points and our intention is to maximize cell solids, we now know that the optimum is near or within the experimental design space. If the mean of the center points had been less than the mean of the factorial points, the optimum would be outside the experimental design space and the method of steepest ascent should be applied (28).

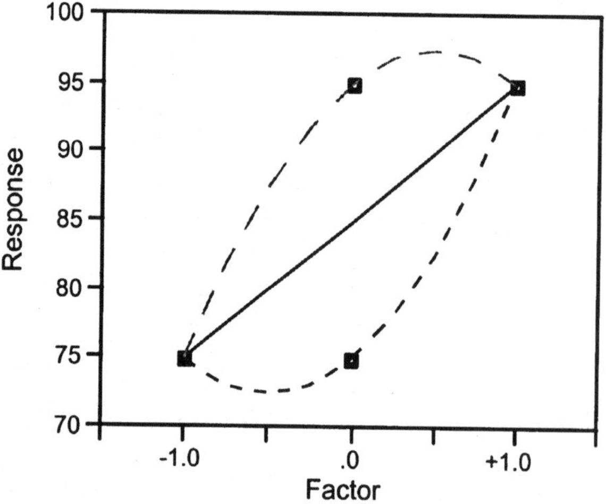

FIGURE 4 Response plot showing a linear effect and two examples of curvature for an individual factor.

6.3. OPTIMIZATION DESIGNS AND VERIFICATION OF EMPIRICAL MODELS

6.3.1. The Transition from Screening to Optimization

Once critical factors have been identified via screening and significant gross curvature has been detected in the design

TABLE 13 Center point data for fractional factorial seed culture study

Trial	Factor concn (g/liter)						Cell solids (%)
	Yeast	Magnesium sulfate	Glucose	NaCl	KH_2PO_4	Soy peptone	
33	3	2	55	5	2.5	20	14
34	3	2	55	5	2.5	20	15
35	3	2	55	5	2.5	20	17
36	3	2	55	5	2.5	20	14
37	3	2	55	5	2.5	20	17
38	3	2	55	5	2.5	20	14

space, the investigator can proceed to the optimization stage of experimental design. Information from screening experiments must first be used to fix certain factor settings and possibly revise the design space of the critical factors. All noncritical factors, i.e., those which will not be tested further, must be fixed at one setting. Typically, the investigator will either select the middle level of the factor setting range or select the level at which the factor provides the best response in light of individual effects and interactions seen earlier. In addition to fixing some factor levels, the investigator may choose to shift the design space for the critical factors (i.e., change the low and high settings for the optimization study). A shift should be considered if the most desirable responses are located near a "corner" of the design space rather than near the center.

A strategy can now be devised for optimizing cell solids based on our interpretation of the fractional factorial screening experiment. In the screening study, the less important factors (KH_2PO_4, magnesium, and NaCl) were primarily involved in marginally significant interactions. KH_2PO_4 can be excluded from optimization, since it has a negative individual effect. The positive magnesium effect, in conjunction with the negative magnesium-NaCL interaction, suggests fixing magnesium at its high setting and NaCl at its low setting. (When one is maximizing, a negative two-factor interaction suggests setting one factor at its high level and the other at its low level.) For the critical factors, a shift of the design space appears necessary, since all factors have very strong individual effects. For example, the high level of soy peptone clearly resulted in higher cell solids than the low level and the center points. Therefore, the investigator may consider changing the factor range for soy peptone from 10 to 30 g/liter to possibly 15 to 35 g/liter or even 20 to 40 g/liter, depending on any prior knowledge of the fermentation at those higher settings. The factor ranges for glucose and yeast may be similarly shifted in the direction of their optimal responses. Hence, the optimization study may investigate soy peptone at 20 to 40 g/liter, glucose at 20 to 50 g/liter, and yeast at 4 to 10 g/liter, with KH_2PO_4, magnesium, and NaCl fixed at 0, 4, and 0 g/liter, respectively. When this strategy was implemented in a subsequent optimization experiment, cell solids were increased to a level which resulted in a 37% increase in the production stage titer of the new metabolite (data not shown).

6.3.2. Response Surface Methodology

The objectives of a statistically designed optimization study are to (i) confirm previous effects and interactions, (ii) estimate specific curvature or quadratic effects, and (iii) determine optimal settings for the critical factors. Once the factors have been reduced to a reasonable number (between two and five), the first objective can often be accomplished with a factorial design. In the cell solids study, two levels of three factors can be tested in a 2^3- or eight-run factorial design. (For five factors, a 1/2 fraction results in a very efficient 16-run design which estimates main effects and all two-factor interactions.) We can add center points to estimate the pure process variability and reassess gross curvature, but what can be done to estimate the curvature specific for each factor? Consider the design space depicted in Fig. 5, which denotes factorial points on the corners and center points in the middle of a cube. Curvature can be assessed if trials are run along axes drawn from the middle of the cube through the centers of each face of the cube. These six axial runs, two for each factor, are typically placed on the middle of each face of the cube but also may be extended outside the cube. Data from these additional trials allow quadratic effects to be estimated. This type of optimization design is known as a Box-Wilson central composite design (2, 3, 18, 28). The use of central composite designs in fermentation has been widely reported and includes optimization for ethanol (1, 7), citric acid (6), cellulose (15), streptomycin (32), macrolides (25, 34), antifungal agents (36), proteins and enzymes (11, 23), and microbial growth (31, 37).

It is now appropriate to consider a polynomial rather than linear model to predict the behavior of the fermentation. The general form of the polynomial model for three factors is the following:

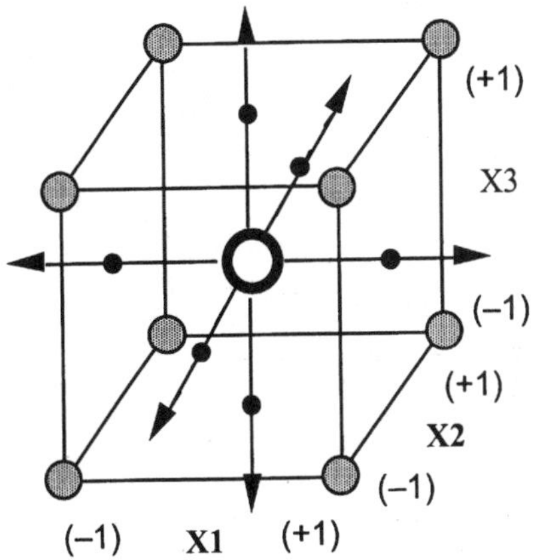

FIGURE 5 Design space for a three-factor optimization design. Axial runs are indicated by solid dots on the middle of each face of the cube but may be extended outside the cube (arrows).

$$y = \beta_0X_0 + \beta_1X_1 + \beta_2X_2 + \beta_3X_3 \quad \text{(intercept and main effects)}$$
$$+ \beta_{12}X_1X_2 + \beta_{13}X_1X_3 + \beta_{23}X_2X_3 \quad \text{(interactions)}$$
$$+ \beta_{11}X_1^2 + \beta_{22}X_2^2 + \beta_{33}X_3^2 \quad \text{(quadratic effects)}$$

where y is the response and X_1, X_2, and X_3 are the coded settings for the three factors. The β terms are the calculated parameter estimates for each main effect, interaction, and quadratic effect. Once the appropriate data are collected, this model can be estimated and optimal process conditions can be determined.

The model-building optimization approach described above is known as response surface methodology (2, 28). Typically, between two and five factors are examined at three (face-centered axial points) or five (extended axial points) factor settings. A full factorial design for five factors at five settings would entail 5^5, or 3,125, flask trials with no replicate trials for estimating random error. Response surface methodology utilizes a fraction of the full factorial and develops a polynomial model from only 32 trials, including six center point replicates.

At Eli Lilly, we were faced with the need to optimize the volumetric activity of a particular enzyme in a yeast fermentation. A Box-Wilson central composite design with five settings (i.e., extended axial trials) for each of four factors was run to optimize the process. Control conditions (5% glucose, 2.8% corn steep liquor, 1.2% DL-methionine, and 0.12% $MgSO_4 \cdot 7H_2O$) were used for the center point replicates. The factors and their settings are shown in Table 14, and the results of the experiment are shown in Table 15. The JMP output displays a fit value, termed R^2, for the estimated second-order model. The R^2 for the enzyme experiment was 97% (not shown). This means that 97% of the variability in volumetric enzyme activity seen in the flask trials can be accounted for by the second-order polynomial prediction equation given below (P values for each term are shown below each term in parentheses):

$$y = 0.887 + 0.209X_1 + 0.041X_2 - 0.026X_3 - 0.010X_4 \quad \text{(intercept and main effects)}$$
(0.001) (0.001) (0.001) (0.01) (0.31)

$$+ 0.003X_1X_2 + 0.041X_1X_3 - 0.004X_1X_4 - 0.005X_2X_3 \quad \text{(interactions)}$$
(0.81) (0.002) (0.74) (0.68)

$$+ 0.0002X_2X_4 + 0.004X_3X_4$$
(0.99) (0.71)

$$- 0.015X_1^2 - 0.0055X_2^2 + 0.012X_3^2 - 0.016X_4^2 \quad \text{(quadratic effects)}$$
(0.09) (0.53) (0.19) (0.08)

where y is volumetric enzyme activity and X_1, X_2, X_3, and X_4 are the coded settings for glucose, corn steep liquor, methionine, and magnesium sulfate. The high value of R^2, together with the many significant effects, provides strong evidence that the model accurately quantifies the process. Other diagnostic techniques like residual analysis and lack-of-fit testing should also be performed before a model is deemed appropriate (2, 14, 28).

We may now proceed to determine optimal process conditions. Many tools exist to examine and illustrate the factor-response relationship. Some of the most useful include contour plots, prediction profiles in JMP, and mathematical optimization of the prediction equation (28). Mathematical optimization is performed by most statistical software packages and involves finding the critical value (the factor settings at which the derivative of the fitted response surface model with respect to each coded factor is zero). Mathematical optimization then determines whether the predicted response at the critical value (and the corresponding response surface) is a global maximum, a global minimum, or a saddle surface (Fig. 6). If the scientist is maximizing (or minimizing) and the global maximum (or minimum) is within the experimental region, new trials are run at the critical value to verify the predicted response. When the maximum (or minimum) is outside the experimental region, additional trials are necessary. This situation will be addressed below.

The mathematical analysis of our model for enzyme activity performed with JMP shows the response surface to be a saddle. In this situation, we suggest examining the prediction equation and then using a profiler like the one in JMP. This interactive tool allows the user to see individual factor effects and factor interactions by actually changing the settings and observing the change in profiles as well as the predicted response. A profile enables the scientist to construct meaningful contour plots and ultimately determine optimal conditions or follow-up experiments. An example of such profiles is depicted in Fig. 7. Notice how the effect of methionine changes when glucose is moved from its middle level (top panel) to its high level (bottom panel). This shift is the result of a two-factor interaction between glucose and methionine (see X_1X_3 in the prediction equation above). A contour plot may now be constructed for these two factors with corn steep liquor fixed at its high setting and magnesium set at its middle setting. (These appear to be optimal based on the profiles.) The contour plot in Fig. 8 indicates that optimum titers are attainable at the highest settings of glucose, corn steep liquor, and methionine and the middle setting of magnesium. This combination of factors, though within the range of each factor's settings, is outside the experimental space (a hypersphere of radius two). Prediction outside the experimental region may be subject to large variability.

TABLE 14 Factor settings for Box-Wilson central composite design for yeast enzyme case study

Factor	Concn (%) for indicated coded setting				
	−2	−1	0	+1	+2
Glucose	2	3.5	5	6.5	8
Corn steep liquor	1.4	2.1	2.8	3.5	4.2
DL-Methionine	0	0.6	1.2	1.8	2.4
Magnesium sulfate	0	0.06	0.12	0.18	0.24

TABLE 15 Results for Box-Wilson central composite design for yeast enzyme optimization

Trial	Factor concn (%)				Volumetric enzyme activity (U/ml)
	Glucose	Corn steep liquor	Methionine	Magnesium sulfate	
1	3.5	2.1	0.6	0.06	0.711
2	3.5	2.1	0.6	0.18	0.646
3	3.5	2.1	1.8	0.06	0.543
4	3.5	2.1	1.8	0.18	0.543
5	3.5	3.5	0.6	0.06	0.742
6	3.5	3.5	0.6	0.18	0.735
7	3.5	3.5	1.8	0.06	0.591
8	3.5	3.5	1.8	0.18	0.596
9	6.5	2.1	0.6	0.06	1.03
10	6.5	2.1	0.6	0.18	0.979
11	6.5	2.1	1.8	0.06	1.034
12	6.5	2.1	1.8	0.18	1.051
13	6.5	3.5	0.6	0.06	1.09
14	6.5	3.5	0.6	0.18	1.081
15	6.5	3.5	1.8	0.06	1.137
16	6.5	3.5	1.8	0.18	1.052
17	2	2.8	1.2	0.12	0.437
18	8	2.8	1.2	0.12	1.274
19	5	1.4	1.2	0.12	0.767
20	5	4.2	1.2	0.12	1.02
21	5	2.8	0	0.12	1.001
22	5	2.8	2.4	0.12	0.922
23	5	2.8	1.2	0	0.861
24	5	2.8	1.2	0.24	0.842
25	5	2.8	1.2	0.12	0.848
26	5	2.8	1.2	0.12	0.865
27	5	2.8	1.2	0.12	0.893
28	5	2.8	1.2	0.12	0.889
29	5	2.8	1.2	0.12	0.852
30	5	2.8	1.2	0.12	0.879
31	5	2.8	1.2	0.12	0.982

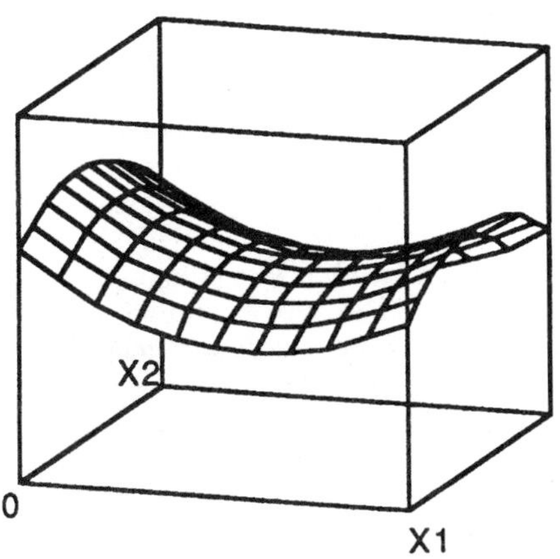

FIGURE 6 Three-dimensional saddle-shaped response surface.

Rather than taking one large "step" outside the experimental region and risking missing the optimal conditions (we only expect our model to be relevant within the experimental region), we will take smaller, incremental steps along the steepest path outside our region. This method is called ridge analysis (2, 6, 13, 28). We will not discuss the mathematical details here, but the proper "path" can be found via matrix algebra or, more directly, by using the PROC RSREG procedure in SAS. Trials are now run in triplicate along the steepest path. The experiment is summarized in Table 16. The data clearly show that the volumetric activity declines after the second step on the path. Although further experimentation could be performed under conditions near the second step, we decided to end the study and fix the medium at 2.2 coded units of glucose (8.3%), 0.45 coded units of corn steep liquor (3.1%), 1.05 coded units of methionine (1.2%), and −0.12 coded units of magnesium (0.1%). The optimization experiment resulted in an increase in volumetric activity from 0.887 U/ml (mean of the center point control trials) to 1.342 U/ml, an increase of over 50%.

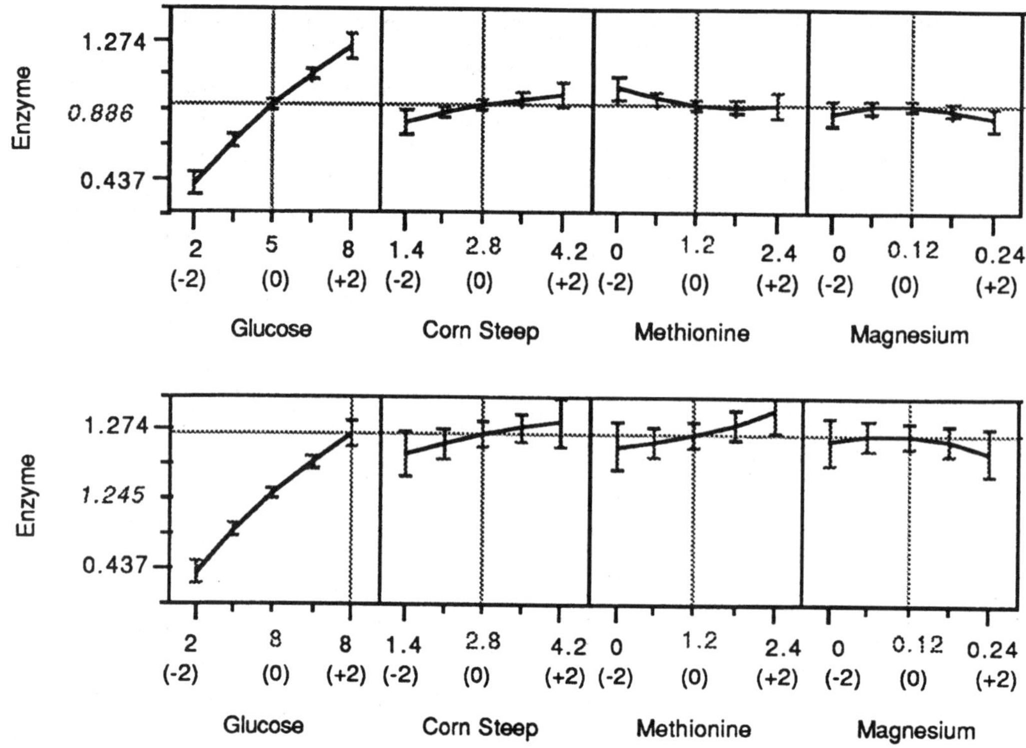

FIGURE 7 Example of JMP profiler showing the interaction between glucose and methionine.

6.4. IMPROVING FERMENTATION IN BIOREACTORS: SCREENING

Compared with shaken flasks, bioreactor vessels provide an opportunity to better control physical and nutritional factors such as pH, temperature, dissolved oxygen, nutrient mixing, and nutrient feeding. Also, mutant strains frequently need to be evaluated in bioreactors to manifest their full potential. Experimental design may be applied in laboratories or pilot plants that have five or preferably six identical bioreactors, regardless of the reactor size. For a bioreactor screening study, select two or three fermentation factors that are likely to affect the titer. Run a two-level design with center point (control) replicates. Possible designs are provided in Table 17. From a statistical standpoint, these designs provide important information in just a few trials. The factorial design will provide estimates of the linear effects of factors 1 and 2, as well as their interaction. The fractional factorial designs will provide estimates of only the linear effects of factors 1, 2, and 3. The inclusion of center points will provide a test for gross curvature.

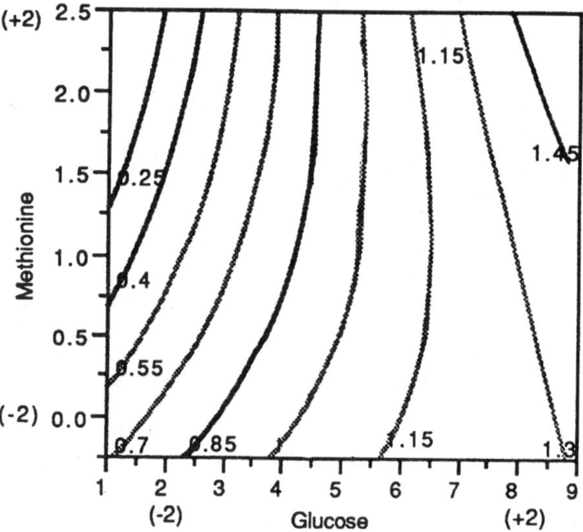

FIGURE 8 Contour plot for yeast enzyme optimization experiment.

Another approach to improving a fermentation is to use an eight-run, two-level design to screen from three to six factors, with an additional blocking factor to account for differences in time of inoculation (2, 3, 18, 33). If five or six reactors are available, center points can be included to assess gross curvature, or replicates can be run to quantify process variability. A randomized screening design for three to six factors with center points is shown in Table 18. (Note: For three factors, the block is confounded with the three-factor interaction, so all main effects and two factor interactions are estimable). Assuming five reactors can be run at a time, the block represents two sequential sets of bioreactor trials. Simply choose a number of columns in Table 18 that is equal to the number of factors and assign each factor to a different column. For three factors, assign columns labeled factor 1 through factor 3; for more factors, columns can be selected randomly. The resulting data from these designs will provide the same basic information that the four-run designs provide: critical main effects, a direction for process improvement, and a possible immediate increase in titer.

TABLE 16 Ridge analysis and model verification for enzyme optimization case study

Step[a]	Factor concn (coded units)				Volumetric enzyme activity (U/ml)[b]
	Glucose	Corn steep liquor	Methionine	Magnesium	
2.0	1.85	0.40	0.65	−0.10	1.246, 1.203, 1.269
2.5	2.2	0.45	1.05	−0.12	1.314, 1.352, 1.359
3.0	2.55	0.48	1.45	−0.12	1.282, 1.237, 1.276
3.5	2.9	0.5	1.9	−0.12	1.118, 1.065, 1.077

[a]Distance in coded units from the center point.
[b]Values are from trials run in triplicate.

TABLE 17 Three alternative screening designs recommended for study of two or three factors in bioreactors

Trial	Coded settings for factors		
	Factorial design (factor 1, factor 2)	Fractional factorial 1 (factor 1, factor 2, factor 3)	Fractional factorial 2 (factor 1, factor 2, factor 3)
1	(−1, −1)	(−1, +1, −1)	(+1, −1, +1)
2	(0, 0)	(0, 0, 0)	(0, 0, 0)
3	(+1, −1)	(−1, −1, +1)	(+1, +1, −1)
4	(+1, +1)	(+1, −1, −1)	(−1, −1, −1)
5	(0, 0)	(0, 0, 0)	(0, 0, 0)
6	(−1, +1)	(+1, +1, +1)	(−1, +1, +1)

6.5. OPTIMIZATION IN BIOREACTORS

Optimization of two process factors can be done with as few as five bioreactors. A 10-run central composite design (four factorial trials, four axial trials, and two center points) can be performed in two blocks of five fermentation vessels. First, the factorial points and a center point are run, followed by the axial points and another center point. The trials for this design (nonrandomized) are given in Table 19. The axial trials may be extended outside the cubic model (as in Table 19) or placed on the face of the cube (±1 coded units). This design is very efficient from the standpoint of sequential experimentation. Since the factorial trials and center point are run first, those data can be used to determine whether curvature exists and whether the optimum is within the chosen experimental region. If so, the axial points and an additional center point can be run in the next block of experiments. This provides the components of a second-order model (linear, interactive, and quadratic effects) which can identify optimal conditions for the two fermentation factors. If no curvature exists after the first set of five trials, steepest ascent can be used to determine the optimal region before additional experiments are conducted (2, 3, 18, 28).

6.6. SIMPLEX PROCEDURE

The simplex approach can be used to continuously improve an existing fermentation process (10, 28). The procedure

TABLE 18 Eight-run screening design with two center points[a]

Trial	Block	Coded setting for factor:					
		1	2	3	4	5	6
1	1	+1	+1	−1	−1	−1	+1
2	1	−1	+1	+1	+1	−1	−1
3	1	0	0	0	0	0	0
4	1	−1	−1	−1	+1	+1	+1
5	1	+1	−1	+1	−1	+1	−1
6	2	+1	+1	+1	+1	+1	+1
7	2	+1	−1	−1	+1	−1	−1
8	2	−1	−1	+1	−1	−1	+1
9	2	0	0	0	0	0	0
10	2	−1	+1	−1	−1	+1	−1

[a]Trials are blocked to run five bioreactors at a time.

TABLE 19 Trials for a central composite design with five bioreactors

Trial	Block	Coded setting for factor:		Description
		1	2	
1	1	−1	−1	Factorial
2	1	−1	+1	Factorial
3	1	+1	−1	Factorial
4	1	+1	+1	Factorial
5	1	0	0	Center point
6	2	−1.4	0	Axial
7	2	+1.4	0	Axial
8	2	0	−1.4	Axial
9	2	0	+1.4	Axial
10	2	0	0	Center point

begins by selecting two (or more) critical fermentation factors. Three bioreactors are needed to study two factors. One bioreactor is run under control conditions, while the other two are set up to form an equilateral triangle in the two-dimensional design space (Fig. 9). The results from these three fermentations establish a direction for improvement of the process response (titer). The triangle is "folded over" in the direction of the two highest titer values, and the procedure is repeated on the trials, forming the new triangle. The iteration continues until the procedure stalls (i.e., the new point of the triangle produces the lowest titer of the three), indicating that a local optimum has been identified. The simplex method is not amenable to modeling but is an efficient way of searching a given space with a minimum number of trials.

6.7. OTHER EXPERIMENTAL DESIGNS AND TECHNIQUES

Several other optimization designs are available to the investigator to meet certain needs and experimental conditions. Mixture designs are commonly used in pharmaceutical formulation and have been applied to medium composition in fermentation (8, 28). Mixture designs are useful when an investigator is faced with a fermentation requiring, for example, three nitrogen sources. An optimum response can be found by varying the ratio of the three nitrogen-containing components without changing the total amount of elemental nitrogen in the overall process.

D-optimal designs are computer-generated designs in which the scientist may be restricted by the number of trials that can be run (2, 28). D-optimal designs are useful in fermentation when certain trials cannot be run due to physical or operational constraints. The computer-generated design will substitute a different point (trial) for the constrained point and still allow the scientist to estimate desirable effects or determine optimal conditions.

An issue frequently faced by scientists using experimental design is the optimization of several responses. One method for handling the multiple response problem is desirability analysis (12, 28). This procedure, which standardizes and reduces multiple responses to a single response, is supported by various software packages, including JMP.

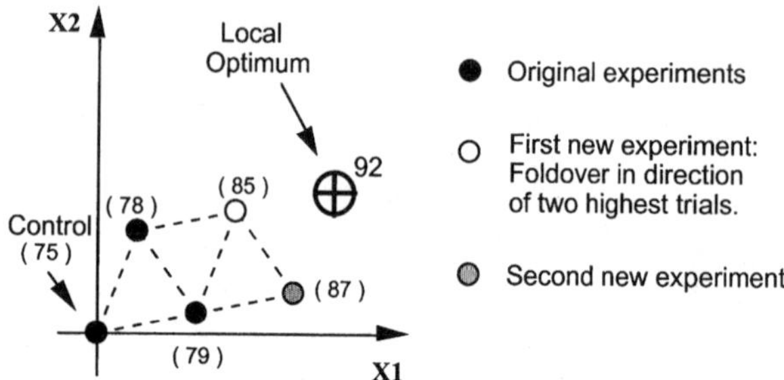

FIGURE 9 Simplex method for optimizing two factors in a fermentation.

We thank K. Cox and L. Nguyen for use of experimental data and T. Tietz and A. Cockshott for review of the manuscript.

REFERENCES

1. **Bowman, L., and E. Geiger.** 1984. Optimization of fermentation conditions for alcohol production. *Biotechnol. Bioeng.* **26:**1492–1497.
2. **Box, G. E., and N. R. Draper.** 1987. *Empirical Model-Building and Response Surfaces*. John Wiley & Sons, Inc., New York, N.Y.
3. **Box, G. E., W. G. Hunter, and J. S. Hunter.** 1978. *Statistics for Experimenters*. John Wiley & Sons, Inc., New York, N.Y.
4. **Buchanan, R. L., J. L. Smith, C. McColgan, B. S. Marmer, M. Golden, and B. Bell.** 1993. Response surface models for the effects of temperature, pH, sodium chloride, and sodium nitrate on the aerobic and anaerobic growth of *Staphylococcus aureus* 196E. *J. Food Saf.* **13:**159–175.
5. **Chen, H.-C.** 1994. Response-surface methodology for optimizing citric acid fermentation by *Aspergillus foetidus*. *Proc. Biochem.* **29:**399–405.
6. **Chen, H.-C.** 1996. Optimizing the concentrations of carbon, nitrogen and phosphorus in a citric acid fermentation with response surface method. *Food Biotechnol.* **10:**13–27.
7. **Chen, S. L.** 1981. Optimization of batch alcoholic fermentation of glucose syrup substrate. *Biotechnol. Bioeng.* **23:**1827–1836.
8. **Cornell, J. A.** 1990. *Experiments with Mixtures: Designs, Models, and the Analysis of Mixture Data*. John Wiley & Sons, Inc., New York, N.Y.

9. **Daniel, C.** 1959. Use of half-normal plots in interpreting factorial two-level experiments. *Technometrics* **1:**311–342.
10. **Deming, S. N., and S. L. Morgan.** 1973. Simplex optimization of variables in analytical chemistry. *Anal. Chem.* **45:**278–283.
11. **de Roubin, M.-R., M. D. Cailas, S.-H. Shen, and D. Groleau.** Some factors influencing the proportion of periplasmic hepatitis B virus pre-S2 antigen in the recombinant yeast *Hansenula polymorpha*. *J. Ind. Microbiol.* **9:**69–72.
12. **Derringer, G., and R. Suich.** 1980. Simultaneous optimization of several response variables. *J. Qual. Technol.* **12:** 214–219.
13. **Draper, N. R.** 1963. Ridge analysis of response surfaces. *Technometrics* **5:**469–479.
14. **Draper, N. R., and H. Smith.** 1981. *Applied Regression Analysis*. John Wiley & Sons, Inc., New York, N.Y.
15. **Embuscado, M. E., J. S. Marks, and J. N. BeMiller.** 1994. Bacterial cellulose. II. Optimization of cellulose production by *Acetobactr xylinum* through response surface methodology. *Food Hydrocolloids* **8:**419–430.
16. **Galindo, E., F. Bolivar, and R. Quintero.** 1990. Maximizing the expression of recombinant proteins in *Escherichia coli* by manipulation of culture conditions. *J. Ferment. Bioeng.* **69:**159–165.
17. **Greasham, R., and E. Inamine.** 1986. Nutritional improvement of processes, p. 41–48. *In* A. L. Demain and N. A. Solomon (ed.), *Manual of Industrial Microbiology and Biotechnology*. American Society for Microbiology, Washington, D.C.
18. **Haaland, P. D.** 1989. *Experimental Design in Biotechnology*. Marcel Dekker, Inc., New York, N.Y.
19. **Haltrich, D., M. Preiss, and W. Steiner.** 1993. Optimization of a culture medium for increased xylanase production by a wild strain of *Schizophyllum commune*. *Enzyme Microb. Technol.* **15:**854–860.
20. **Horitsu, H., Y. Yahashi, K. Takamizawa, K. Kawai, T. Suzuki, and N. Watanabe.** 1992. Production of xylitol from D-xylose by *Candida tropicalis*: optimization of production rate. *Biotechnol. Bioeng.* **40:**1085–1091.
21. **Hujanen, M., and Y.-Y. Linko.** 1994. Optimization of L(+)-lactic acid production employing statistical experimental design. *Biotechnol. Tech.* **8:**325–330.
22. **Kautola, H., and Y.-Y. Linko.** 1989. Fumaric acid production from xylose by immobilized *Rhizopus arrhizus* cells. *Appl. Microbiol. Biotechnol.* **31:**448–452.
23. **Lekha, P. K., N. Chand, and B. K. Lonsane.** 1994. Computerized study of interactions among factors and their optimization through response surface methodology for the production of tannin acyl hydrolase by *Aspergillus niger* PKL 104 under solid state fermentation. *Bioprocess Eng.* **11:**7–15.
24. **Maddox, I. S., and S. H. Reichert.** 1977. Use of response surface methodology for the rapid optimization of microbiological media. *J. Appl. Bacteriol.* **43:**197–204.
25. **McDaniel, L. E., E. C. Bailey, S. Ethiraj, and H. P. Andrews.** 1976. Application of response surface optimization techniques to polyene macrolide fermentation studies in shake flasks. *Dev. Ind. Microbiol.* **17:**91–98.
26. **Miller-Wideman, M., N. Makkar, M. Tran, B. Isaac, N. Biest, and R. Stonard.** 1992. Herboxidiene, a new herbicidal substance from *Streptomyces chromofuscus* A7847. *J. Antibiot.* **45:**914–921.
27. **Monaghan, R. L., E. Arcuri, E. E. Baker, B. C. Buckland, R. L. Greasham, D. R. Houck, E. D. Ihnen, E. S. Inamine, J. J. King, E. Lesniak, P. S. Masurekar, C. A. Schulman, B. Singleton, and M. A. Goetz.** 1989. History of yield improvements in the production of asperlicin by *Aspergillus alliaceus*. *J. Ind. Microbiol.* **4:**97–104.
28. **Myers, R. H., and D. C. Montgomery.** 1995. *Response Surface Methodology: Process and Product Optimization Using Designed Experiments*. John Wiley & Sons, Inc., New York, N.Y.
29. **Plackett, R. L., and J. P. Burman.** 1946. The design of optimum multifactorial experiments. *Biometrika* **33:**305–325.
30. **Prapulla, S. G., Z. Jacob, N. Chand, D. Rajalakshmi, and N. G. Karanth.** 1992. Maximization of lipid production by *Rhodotorula gracilis* CFR-1 using response surface methodology. *Biotechnol. Bioeng.* **40:**965–970.
31. **Sanchez, J. L. G., J. A. S. Perez, F. G. Camacho, J. M. F. Sevilla, and E. M. Grima.** 1996. Optimization of light and temperature for growing *Chlorella* sp. using response surface methodology. *Biotechnol. Tech.* **10:**329–334.
32. **Saval, S., L. Pablos, and S. Sanchez.** 1993. Optimization of a culture medium for streptomycin production using response-surface methodology. *Bioresour. Technol.* **43:**19–25.
33. **Snedecor, G. W., and W. G. Cochran.** 1980. *Statistical Methods*, 7th ed. Iowa State University Press, Ames.
34. **Strobel, R. J., and W. M. Nakatsukasa.** 1993. Response surface methods for optimizing *Saccharopolyspora spinosa*, a novel macrolide producer. *J. Ind. Microbiol.* **11:**121–127.
35. **Swanson, T. R., J. O. Carroll, R. A. Britto, and D. J. Duhart.** 1986. Development and field confirmation of a mathematical model for amyloglucosidase/pullulanase saccharification. *Starch* **38:**382–387.
36. **Tracz, J. S., R. A. Giacobbe, and R. L. Monaghan.** 1993. Improvement in the titer of echinocandin-type antibiotics: a magnesium-limited medium supporting the biphasic production of pneumocandins A_0 and B_0. *J. Ind. Microbiol.* **11:**95–103.
37. **Watier, D., H. C. Dubourguier, I. Leguerinel, and J. P. Hornez.** 1996. Response surface models to describe the effects of temperature, pH, and ethanol concentration on growth kinetics and fermentation end products of a *Pectinatus* sp. *Appl. Environ. Microbiol.* **62:**1233–1237.
38. **Zabriskie, D. W., W. B. Armiger, D. H. Phillips, and P. A. Albano.** 1980. *Traders' Guide to Fermentation Media Formulation*. Traders Protein, Memphis, Tenn.

Cell and Enzyme Immobilization

ATSUO TANAKA AND TAKUO KAWAMOTO

7

The immobilized enzyme is defined as "the enzyme physically confined or localized in a certain defined region of space with retention of its catalytic activity, which can be used repeatedly and continuously" (8). The term "immobilized" is applicable not only to enzymes but also to cellular organelles, microbial cells, plant cells, and animal cells, i.e., to all types of biocatalysts.

In 1916, Nelson and Griffin (34) found fortuitously that yeast invertase adsorbed on activated charcoal was able to catalyze the hydrolysis of sucrose. This was the first demonstration of the immobilized enzyme concept. After this finding, various reports were published on the immobilization of physiologically active proteins by covalent binding to various supports. However, immobilized enzymes were not used in practice until Grubhofer and Schleith (18) immobilized various enzymes, such as carboxypeptidase, diastase, pepsin, and ribonuclease, on diazotized polyaminostyrene resin by covalent binding. Thereafter, various principles of immobilization, e.g., ionic binding, physical adsorption, and entrapment, were developed.

The first industrial application of immobilized enzymes was performed by Chibata and coworkers of Tanabe Seiyaku Co., Japan, in 1969. In this process, a fungal aminoacylase was immobilized on DEAE-Sephadex through ionic binding and used for the enantioselective hydrolysis of *N*-acyl-D,L-amino acids to produce L-amino acids and *N*-acyl-D-amino acids (11). The first industrial application of immobilized microbial cells was also successfully carried out by Chibata and coworkers in 1973 to produce L-aspartate from ammonium fumarate by polyacrylamide gel-entrapped *Escherichia coli* cells containing a high level of aspartase activity.

Several industrial applications of immobilized biocatalysts have been established, and a large number of references on immobilization of biocatalysts are available (7, 10, 12, 26, 31, 32). Furthermore, a number of companies have commercialized immobilized enzymes, as well as the carriers for immobilization.

In this chapter, we describe promising methods and present some general guidelines for immobilization.

7.1. GENERAL CONSIDERATIONS

Various reactions that occur in biological systems, such as microbial cells, plant cells, and animal cells, are catalyzed by enzymes that are water-soluble globular proteins. Compared with chemical catalysts, enzymes as catalyst have the following superior properties. (i) Enzyme-mediated reactions take place under mild conditions, such as ordinary temperatures, atmospheric pressure, and pH values around neutrality. (ii) Enzymes have strict substrate specificity, stereospecificity, regiospecificity, and reaction specificity.

These facts suggest that energy-saving, resource-saving, and low-pollution processes could be designed by using biocatalysts that include not only enzymes but also cellular organelles, microbial cells, plant cells, and animal cells, benefiting from operation at relatively low temperatures and ambient pressures with little by-product formation. However, biocatalysts are, in general, easily inactivated under conditions such as high temperature, high or low pH, presence of organic solvents, and even conditions suitable for the catalytic reactions. The recovery of biocatalysts from spent reaction mixture is another problem when biocatalysts are used in a free form. Immobilization is one way of circumventing these problems. In general, immobilized biocatalysts are stable and can be used either repeatedly in a series of batchwise reactions or continuously in flow systems.

The immobilization of enzymes and cells is a practical method when bioprocesses employing biocatalysts are considered. At present, the applications of immobilized biocatalysts in bioprocessing include (i) the production of useful compounds by stereospecific and/or regiospecific reactions, (ii) the production of energy by biological processes, (iii) the selective treatments of pollutants to solve environmental problems, (iv) analyses of various compounds with high sensitivity and specificity, and (v) utilization in manufacturing new drugs, artificial organs, etc. These processes require the immobilization not only of single enzymes but also of multienzyme systems that mediate more complex reactions. In these systems, regeneration of ATP and/or coenzymes is often required. Cells and cellular organelles contain metabolic systems that mediate complicated reactions. Therefore, immobilization of cells or organelles serves to incorporate multienzyme systems. Furthermore, in the case of immobilization of cells, the need for extracting enzymes from the cells is eliminated. This avoids potential inactivation of the enzymes during the time-consuming and expensive purification procedures and makes it possible to use the enzymes under more stable conditions.

However, immobilized cells have some disadvantages that must be considered for practical application: (i) By-products may be synthesized, because the cells may contain enzymes catalyzing undesirable reactions. Strain selection, mutation, proper treatment of the cells, or genetic improvement of the cell line may help to overcome this problem. (ii) The cell wall and cell membrane of intact cells often prevent substrates, products, and other reaction components from permeating into and out of the cells. If this is the case, the barriers to permeability must be broken by appropriate treatment of the cells before or after immobilization.

The immobilized cells can be kept in a growing state by a continuous supply of suitable nutrients. The technique of immobilizing growing cells is advantageous, because the immobilized growing cells serve as self-proliferating and self-regenerating biocatalysts. However, immobilized growing cells also have some disadvantages: (i) Immobilized growing cells require nutrients and energy sources to maintain the cells in the living or growing state. The yield of the product may be lowered by the consumption of the substrate as growth nutrient or energy source. (ii) The product is likely to be contaminated by cells leaking from the carriers.

For the application of immobilized biocatalysts, screening of biocatalysts having the desired activity and characteristics is very important. In addition, selection of the appropriate carrier and immobilization procedure is essential. Both the carrier and immobilization procedure should be suitable for each biocatalyst to be immobilized. There is, however, no universally ideal immobilization method for all types of biocatalysts.

The immobilization methods can be classified into four categories: carrier-binding, cross-linking, entrapping, and combined. The combined method is the combination of the other three methods. Because each method has its own merits and demerits, as described below, the selection should be carried out based on the biocatalyst, reaction, reactor, etc.

An ideal carrier should have adequate functional groups for immobilization of biocatalyst. It should also have mechanical strength, physical, chemical, and biological stability, and a lack of toxicity. In addition, the ability to form different shapes is required for applying the immobilized biocatalyst to various types of reactors. Finally, the economic feasibility should be considered.

7.2. CARRIER-BINDING METHOD

The carrier-binding method is based on binding of the biocatalyst to a water-insoluble carrier through a covalent linkage, ionic bonds, physical adsorption, or biospecific binding. Various types of insoluble materials such as water-insoluble polysaccharides (e.g., cellulose, dextran, and agarose derivatives), proteins (e.g., gelatin and albumin), synthetic polymers (e.g., polystyrene derivatives, ion-exchange resins, and polyurethane), and inorganic materials (e.g., brick, sand, glass, ceramics, and magnetite) can be utilized directly or after proper modification or activation.

7.2.1. Covalent-Binding Method

The covalent-binding method is one of the most frequently used techniques for the immobilization of enzymes, but this method is not so widely used for cells because of the toxicity of the agents and the difficulty of finding the proper conditions of immobilization. Navarro and Durand (33) immobilized *Saccharomyces carlsbergensis* by covalent binding to porous silica beads that had been treated with aminopropyltriethoxysilane and activated with glutaraldehyde. *Serratia marcescens*, *Saccharomyces cerevisiae*, and *Saccharomyces amurcea* were also bound covalently to borosilicate glass or zirconia ceramics by using an isocyanate coupling agent (28). Organic carriers can also be used for immobilization of cells by the covalent-binding method. *Micrococcus luteus* cells were immobilized to the carboxyl group of agarose beads in a two-step process that avoided exposure of the cells to carbodiimide (21).

In the case of enzymes, the amino acid residues that are not involved in the active site or substrate-binding site can be used for covalent binding with carriers. The ϵ-amino group of lysine, the β-carboxyl group of aspartic acid, the γ-carboxyl group of glutamic acid, the hydroxyl groups of serine and threonine, the mercapto group of cysteine, the phenolic hydroxyl group of tyrosine, and the imidazole group of histidine can react with carriers having a reactive functional group, such as diazonium salt, acid azide, isocyanate, activated alkyl halide, or aldehyde.

The covalent-binding method has the following advantages. (i) The biocatalyst does not leak or detach from the carrier. (ii) The biocatalyst can easily interact with the substrate because it is on the surface of the carrier. (iii) The biocatalyst stability is often increased because of the strong interaction between it and the carrier.

On the other hand, the covalent-binding method has the following disadvantages. (i) The activity yield is likely to be low owing to exposure of the biocatalyst to toxic reagents or severe reaction conditions. (ii) The optimal conditions for the immobilization procedure are difficult to find. (iii) Renewal of the carrier and recovery of the biocatalyst from the carrier are, in general, impossible. Hence, this method is better suited to expensive enzymes whose stability is significantly improved by covalent binding to the carrier.

Several typical covalent-binding methods are described below.

7.2.1.1. Cyanogen Bromide Method

The cyanogen bromide method was first used by Axén and coworkers in 1967 (1). This method involves the activation of a carrier having vicinal hydroxyl groups, such as polysaccharides, glass beads, or ceramics, with cyanogen bromide, to yield reactive imidocarbonate derivatives. The subsequent reaction between the activated carrier and the amino group of the biocatalyst gives N-substituted isourea, N-substituted imidocarbonate, or N-substituted carbamate derivatives. This method has been widely used for the immobilization of various enzymes, and the CNBr-activated carriers are available commercially. For example, the CNBr-activated Sepharose 4B is a product of Pharmacia Fine Chemicals, Uppsala, Sweden. An aminated carrier can also be used. It is also possible to insert a spacer such as hexamethylenediamine to reduce the hindrance to movement of the enzyme molecule, caused by the interaction between the enzymes and the carriers.

Examples of the procedures for preparing the CNBr-activated Sepharose 4B and for immobilizing the enzyme on the activated carrier are as follows.

Preparation of CNBr-activated Sepharose 4B (31)

Sepharose 4B (40 ml) is washed with 2 M potassium phosphate buffer (pH 12.1) and recovered dry using slightly reduced pressure on a filter. Then, 10 ml of cold 5 M potassium phosphate buffer (pH 12.1) is added to 10 g of gel, followed by 20 ml of distilled water and an aqueous solution of CNBr (100 mg/ml). It is necessary to exercise extreme caution when using CNBr, because CNBr is a highly toxic reagent. The addition of CNBr should take about 2 min at 5 to 10°C. The reaction is allowed to continue for about 10 min with gentle stirring. The gel is transferred to a glass filter and washed with cold water. The activated gel can be preserved in acetone at −10 to −15°C for over 3 months but is better used immediately.

Immobilization of Ascorbate Oxidase with CNBr-Activated Sepharose 4B (3)

CNBr-activated Sepharose 4B (2 to 3 ml) is placed in a small culture tube (1 cm by 6 cm), and 2 to 3 ml of 0.1 M $NaHCO_3$ containing 2 to 3 mg ascorbate oxidase (lyophilized powder) is added. The tube is rotated slowly overnight at 4°C. The resin obtained is washed with 200 ml of water on a glass fritted filter over vacuum. Finally, the resin is treated with 2 ml of 1 M glycine at room temperature to decompose the residual active group. The final immobilized ascorbate oxidase preparation is obtained after rinsing the resin with 250 ml of water.

7.2.1.2. Acid Azide Derivative Method

The acid azide method, which is also known as a method for peptide synthesis, has been used for the immobilization of various enzymes (5, 30). The carboxyl group of the carrier is converted first to the methyl ester and then to the hydrazide with hydrazine. The hydrazide reacts with nitrous acid to form the azide derivative, which can then react with the amino groups of the enzyme molecules at low temperature to yield the immobilized enzyme.

As an example, immobilization of an enzyme by this method using carboxymethylcellulose (CMC) as the carrier is described (30). CMC is washed with 0.5 M HCl, distilled water, methanol, and ether, and dried. The dried CMC (10 g) is suspended in 500 ml of a 2.6 M solution of HCl in methanol and stirred vigorously for 24 h at 30°C to esterify CMC with methanol. The reaction mixture is filtered, and the filter cake precipitate is dried after washing with methanol and ether to yield the methyl ester of CMC (9 g). The methyl ester (5 g) is suspended in 500 ml of methanol. To the suspension, 25 ml of 80% aqueous solution of hydrazine is added, and the reaction mixture is refluxed at 66 to 67°C. The reaction mixture is filtered after refluxing for 3 h. The filter cake is dried after washing with methanol and ether to yield CMC-hydrazide (4.5 g). CMC-azide is then prepared by suspending 1 g of CMC-hydrazide in 150 ml of a 2% solution of HCl, and the mixture is stirred while adding 9 ml of a 3% solution of sodium nitrate. After stirring at 0°C for 20 min, CMC-azide is separated by centrifugation and washed with 150 ml of dioxane and then with distilled water to neutrality. The CMC-azide is stirred with 250 to 450 mg of enzyme in 100 ml of 0.05 M phosphate buffer at pH 8.7 for 2 h. The insoluble derivative is then adjusted to pH 4 with 0.1 M phosphoric acid, separated, washed twice with 150 ml of water (pH 3) and once with distilled water, and finally lyophilized.

7.2.1.3. Condensing Reagent Method

The carboxyl group or amino group of the carriers and the biocatalysts can be condensed through the formation of peptide linkages by the action of condensing reagents such as carbodiimides [dicyclohexylcarbodiimide, 1-ethyl-3-(3-dimethylaminopropyl)carbodiimide, 1-cyclohexyl-3-(2-morpholino-ethyl)carbodiimide metho-*p*-toluene-sulfonate, etc.] or Woodward's reagent K (*N*-ethyl-5-phenylisoxazolium-3′-sulfonate).

Immobilization of alkaline phosphatase by using carbodiimide is as follows (24). The carrier, synthesized by an emulsion polymerization process of styrene, acrylic acid, and divinyl benzene with the initiator, potassium peroxydisulfate, is suspended in 15 ml of water. 1-Ethyl-3-(3-dimethylaminopropyl)carbodiimide (40 mg) and 8 mg of alkaline phosphatase are added to the suspension. The pH of the mixture is kept at 4.0, and the mixture is incubated for 1 h at room temperature before being continuously stirred at 4°C overnight. The reaction mixture is centrifuged (36,500 × *g*) for 10 min at 4°C, and the pellet is washed with pH 9.2 glycine buffer containing 8 mM $MgCl_2$. The washing is carried out several times to obtain immobilized alkaline phosphatase (typically 10 mg enzyme/g carrier).

7.2.1.4. Diazo Coupling Method

Carriers having aromatic amino groups can be diazotized with nitrous acid to form the diazonium derivatives, which react with enzyme molecules to yield immobilized enzymes. The functional groups of an enzyme molecule that may participate in this reaction include free amino groups, imidazole groups, phenolic hydroxyl groups, and so on.

As an example, immobilization of an enzyme by this method using *p*-aminobenzylcellulose as carrier is described (6). *p*-Aminobenzylcellulose (5 g) is suspended in 10 ml of 2 M HCl, mixed with 20 ml of water, and chilled in an ice bath. With constant stirring, a 0.5 M $NaNO_2$ solution is slowly added until the test result with KI-starch test paper remains positive for 15 min after addition of the last portion of the $NaNO_2$ solution. Stirring is continued for another 15 min and the material is filtered and washed with weakly acidic ice water. The obtained filter cake of diazonium salt is added to an ice-cooled 2% solution of the enzyme in borate buffer, pH 8.75. The amount of dry enzyme used is generally one fourth to one fifth of the dry weight of *p*-aminobenzylcellulose. The mixture is stirred at low temperature for 2 h, the pH is adjusted to 7.3, and the mixture is stored at 4°C for at least 36 h. To eliminate unreacted diazonium groups, the mixture is incubated at 37°C for 1 h. The product is then filtered and washed with a large amount of buffer.

7.2.1.5. Alkylation Method

Carriers having an alkylation group can react with the phenolic hydroxyl groups or sulfhydryl groups of the enzyme molecules. Halogenated acetyl derivatives (36), triazinyl derivatives (22), and so on can be used as carriers for this method.

As an example, the use of halogenoacetylcellulose as a carrier with an alkylation group for enzyme immobilization is described below.

Preparation of Bromoacetylcellulose (36)

Cellulose (10 g, 100 to 200 mesh) is dispersed in a solution of bromoacetic acid (100 g) in 30 ml of dioxane, and the mixture is stirred at 30°C for 16 h. To the mixture, 75 ml of bromoacetylbromide is added and the stirring is continued at 30°C for 7 h. The mixture is poured into a large volume of water, and the white solid bromoacetylcellulose is collected and washed with 0.1 M sodium bicarbonate and water and dried.

Preparation of Chloroacetylcellulose

Chloroacetylcellulose is prepared in the same manner as bromoacetylcellulose except for the use of chloroacetic acid and chloroacetylchloride.

Preparation of Iodoacetylcellulose

Iodoacetylcellulose is prepared by the exchange reaction of bromoacetylcellulose or chloroacetylcellulose with iodine. To 6 g of bromoacetylcellulose or chloroacetylcellulose, 36 g of sodium iodine and 300 ml of 95% ethanol are added, and the mixture is stirred at 30°C for 20 h. The solid is collected, washed successively with ethanol, 0.1 M sodium bicarbonate, water, and ethanol, and dried.

Immobilization of Aminoacylase with Halogenoacetylcellulose

To a solution of aminoacylase (10 to 30 mg) in 5 ml of 0.2 M phosphate buffer (pH 8.5), 100 mg of halogenoacetylcellulose and 1 g of 1.5 M ammonium sulfate are added. The mixture is stirred for 24 h at 7°C and then centrifuged. The precipitate is washed with physiological saline (0.85% NaCl) and 0.2 M phosphate buffer (pH 8.5).

7.2.1.6. Carrier Cross-Linking Method

In the carrier cross-linking method, both the carriers and the biocatalyst having amino groups are cross-linked with a bi- or multifunctional reagent that reacts with the free amino groups. Glutaraldehyde is the most popular reagent that cross-links amino groups through a Schiff's base linkage. Diisocyanate is the other type of cross-link reagent.

As an example, aminopeptidase is immobilized on glass beads as described (13). Amino glass beads (550-μm average bead diameter, 80 Å pore size, 0.14 meq NH_2/g; Corning, New York, N.Y.) are suspended in a freshly prepared 2.5% glutaraldehyde solution in water (2 ml) and mixed in vacuo at room temperature. After 60 min, the beads are thoroughly washed with distilled water and 200 ml of 50 mM sodium phosphate buffer (pH 7.4)/1.0 mM EDTA. A solution (2 ml) of aminopeptidase (0.1 mg/ml) is mixed with the activated glass beads in an ice bath for 3 h. The beads are thoroughly washed with 50 mM sodium phosphate (pH 7.4) containing 1.0 mM EDTA and 0.1 M NaCl to remove any noncovalently bound protein and then with buffer without NaCl.

7.2.2. Ionic Binding Method

Since catalase was found to be able to bind to the ion-exchange cellulose (29), the ionic binding method has been applied for the immobilization of many biocatalysts. The procedure is very simple, renewal of the carrier and recovery of the biocatalyst from the carrier are easy, and the conditions of immobilization are mild. In fact, the first industrial application of immobilized enzymes employed this method for the immobilization of aminoacylase on DEAE-Sephadex to produce L-amino acids (11). In ionic binding, binding of the biocatalyst on the carrier is affected by the buffer used, pH, ionic strength, and temperature. Although renewal of the carrier and recovery of the biocatalyst from the carrier are easy, the biocatalyst is likely to detach from the carrier. The polysaccharide derivatives having ion-exchange groups, as well as various ion-exchange resins, can be utilized for this purpose.

Aminoacylase is immobilized by this method using DEAE-Sephadex as the carrier as follows (42). DEAE-Sephadex A-25 (hydroxy form, 18.5 g; Pharmacia), suspended in 100 ml of distilled water, is stirred with 1,670 ml of aminoacylase solution (560 U) at 5°C for 5 h before incubation at the same temperature overnight. After filtration, 1,000 ml of distilled water is added to the filter cake, and the mixture is stirred for 1 h and filtered again. This washing process is repeated one more time. For further washing, 1,000 ml of 0.2 M sodium acetate is added to the insoluble complex, and the mixture is stirred for 1 h and filtered. The resulting material is suspended in 500 ml of distilled water and lyophilized after washing with 1,000 ml of distilled water. Using this procedure, 19.6 g of the immobilized aminoacylase is typically obtained.

7.2.3. Physical Adsorption Method

The physical adsorption method is based on the physical interactions between the biocatalyst and the carrier, such as hydrogen bonding, hydrophobic interaction, van der Waals force, and their combinations. Although the biocatalyst may be immobilized without any modification, the physical interaction is, in general, weaker than ionic binding and more sensitive to environmental conditions, such as temperature and concentration of solutes. Cellular organelles and various types of cells can also be immobilized with ease by this method. Renewal of the carrier can be accomplished under appropriate conditions. Physical adsorption followed by cross-linking with glutaraldehyde sometimes helps to stabilize the immobilized biocatalysts.

Although many carriers that efficiently adsorb biocatalysts have been developed, the one with immobilized tannin (46) is among the most useful. Immobilization of tannin can be carried out in a reaction of aminohexyl cellulose with CNBr-activated tannin. Immobilization of aminoacylase using immobilized tannin is as follows (46). In 20 ml of the aminoacylase solution (3,200 U), 1 g (wet weight; corresponding to 0.4 g dry weight) of immobilized tannin (prepared as described above) is suspended. The suspension is shaken for 1 h at 25°C and then filtered. The immobilized aminoacylase is obtained after washing the residue on the filter with water.

7.2.4. Biospecific Binding Method

The biospecific binding method is based on the biospecific interaction between the enzymes and other substances, such as coenzymes, inhibitors, effectors, lectins, and antibodies, which are often utilized for affinity separation processes. If the interaction is strong, the enzyme can be immobilized on the carrier conjugated with one of these substances. However, antibodies and inhibitors are not good choices because the enzyme is usually inactivated by binding to them. Interaction between a lectin and the carbohydrate moiety of an enzyme is useful for this application (39). Lectins that are glycoproteins bind tightly to specific carbohydrate residues. One of the most useful lectins is concanavalin A, which is obtained from the jack bean. Since many enzymes are glycoproteins, lectins can be widely used.

Immobilization of ascorbate oxidase, an example of immobilization using a lectin, is carried out as follows (27). A column containing 0.5 ml of concanavalin A-Sepharose (Pharmacia) is washed with several volumes of 0.1 M sodium acetate buffer (pH 5.5) before ascorbate oxidase (4.5 U) is added to the buffer (flow rate, 0.85 ml/min). The column containing the immobilized ascorbate oxidase is washed with 1 M NaCl to remove unspecifically bound enzyme and then with 0.1 M sodium acetate buffer (pH 5.5).

7.3. CROSS-LINKING METHOD

The cross-linking method utilizes a bi- or multifunctional compound as in the carrier cross-linking method; however, a carrier is not used in this method. The bi- and multifunctional compound serves as the reagent for intermolecular cross-linking of the biocatalyst. The cross-linked biocatalyst becomes water insoluble.

In addition to glutaraldehyde (35, 37), which is the most popular cross-linking reagent, several other bi- or multifunctional compounds, such as toluene diisocyanate and hexamethylene diisocyanate, can be used. The activity of the biocatalyst immobilized by this method is, in general, reduced.

As an example, cross-linking of catalase is as follows (37). Crystalline catalase is dissolved in 10% NaCl and diluted with 0.05 M phosphate buffer (pH 7.2) until the concentration of catalase is 2 mg/ml. A solution of 4% glutaraldehyde (4 ml) in the same buffer is added to 4 ml of the enzyme solution and stirred for about 1 h at room temperature, until green lumpy precipitates appear. This cross-linking reaction can be performed in the cold room overnight with comparable results. The precipitates are separated by centrifugation (5 min, 4,000 rpm) and washed repeatedly with a 10% NaCl solution (6 to 8 times) until the supernatant fluid is free of catalase activity. The precipitates are homogenized in water by a Teflon pestle, to a fine suspension containing 1 mg/ml of the immobilized catalase.

7.4. ENTRAPMENT METHOD

The entrapment method is classified into five major types: lattice, microcapsule, liposome, membrane, and reversed micelle.

In the lattice type, the biocatalyst is entrapped in the matrix of one of the various polymers. The microcapsule type involves entrapment within microcapsules of a semipermeable synthetic polymer. The liposome type employs entrapment within an amphiphatic liquid-surfactant membrane prepared from lipid. In the membrane type, the biocatalyst is separated from the reaction solution by an ultrafiltration membrane, a microfiltration membrane, or a hollow fiber. In the reversed micelle type, the biocatalyst is entrapped within the reversed micelles, which are formed by mixing a surfactant with an organic solvent.

One advantage of the entrapment method is that not only single enzymes but also multiple enzymes, cellular organelles, and intact or treated cells can be immobilized. However, the disadvantages are that (i) the high-molecular-weight substrate may not be able to access the entrapped biocatalyst, and (ii) renewal of the carrier is difficult. Among the entrapment methods, the lattice type is the most widely used, and there are a large number of references available. Several representative techniques for the lattice-type method are described below.

7.4.1. Polyacrylamide Gel Method

Since Bernfeld and Wan (2) reported the entrapment of several enzymes in polyacrylamide gel in 1963, various types of biocatalysts, including cellular organelles, microbial cells, plant cells, and animal cells, have been immobilized by this method. This method has also been applied to the industrial production of useful compounds such as L-aspartate, L-malate, and acrylamide. The procedure for preparation of the gel is identical to that employed for electrophoresis. For the immobilization of the biocatalyst by this method, acrylamide and *N,N'*-methylenebisacrylamide (BIS) as the cross-linking reagent are mixed with the biocatalyst and polymerized in the presence of an initiator, such as potassium persulfate, and a stimulator, such as 3-dimethylaminopropionitrile (DMAPN) or *N,N,N',N'*-tetramethylethylenediamine (TEMED).

Although this technique is used for various purposes, a major disadvantage is the toxicity of the acrylamide monomer, the cross-linking reagent, the initiator, and the stimulator. In some cases, free-radical polymerization results in a decrease in the activity of the biocatalyst. Several analogs or derivatives of acrylamide are also useful in this method.

An example for entrapment of *E. coli* with a polyacrylamide gel is as follows (9). In 4 ml of physiological saline, 1 g of packed intact *E. coli* cells are suspended. To the suspension, acrylamide monomer (0.75 g), BIS (40 mg), 5% DMAPN (0.5 ml), and 2.5% potassium persulfate (0.5 ml) are added, and the mixture is incubated at 37°C for 30 min to yield the polyacrylamide gel containing the cells. Polymerization should be carried out under anaerobic conditions, because oxygen prevents the polymerization. The gel is washed with physiological saline after being made to the proper shape.

7.4.2. Alginate Gel Method

Several natural polysaccharides, such as alginate, κ-carrageenan, agar, and agarose, are excellent gel materials and are used widely for entrapment of biocatalysts.

Alginate is a linear copolymer of D-mannuronic and L-guluronic acids and can be gelled by multivalent ions. For immobilization of the biocatalyst, sodium alginate, which is soluble in water, is mixed with a solution or suspension of the biocatalyst and dropped into a calcium chloride solution to form the water-insoluble calcium alginate gel droplets (23). However, calcium alginate gels are gradually solubilized in the presence of a calcium ion-trapping reagent such as phosphate ion. In addition, leaking of the biocatalyst is likely to occur. The calcium ion-trapping agent frees the cells trapped in the gel, enabling the examination of the characteristics, e.g., the cell viability, of the trapped cells. Aluminum ion, strontium ion, or several other divalent metal ions can also be used instead of calcium ion. Treatment of the calcium alginate gel with a cationic polymer such as polyethyleneimine can improve the stability of the gel in the presence of phosphate (45). It is also known that addition of polyacryl acid to sodium alginate increases the physical strength of the gel and prevents the gel from being dissolved by bridging the polyacrylate chains and alginate chains with calcium ions (25).

The alginate gel method is widely applicable to the immobilization of various biocatalysts, including very fragile ones such as plant cells (4, 41, 44) and protoplasts (38), because of the simplicity of the procedure, availability of sodium alginate, and mild immobilization conditions.

The entrapment of *S. cerevisiae* in calcium alginate is described here as an example (45). *S. cerevisiae* (25 g wet weight) is suspended in sterile tap water and mixed with 50 ml of 4% sodium alginate. The resulting suspension is passed through a narrow tube of about 1 mm diameter and dropped into a calcium chloride solution (50 mM, pH 6 to 8). The beads (2.8 to 3.0 mm diameter) obtained are incubated in the calcium chloride solution at 20 to 22°C for 2 h to harden the gel.

7.4.3. κ-Carrageenan Gel Method

κ-Carrageenan is a readily available, nontoxic polysaccharide, which is obtained from seaweed. It is widely used in the food and cosmetic industries as a gelling, thickening, and stabilizing agent.

κ-Carrageenan forms a gel upon cooling or in the presence of a gel-inducing reagent such as potassium chloride (43). Various cations, such as ammonium, calcium, aluminum, and magnesium, also serve as good gel-inducing reagents. The conditions of immobilization by this method are mild. Another advantage of this method is that various shapes of the immobilized biocatalyst can be made (43).

κ-Carrageenan gel can be easily solubilized in saline or water to free the biocatalyst from the gel for the investigation of its properties. On the other hand, leaking of the biocatalyst can occur easily owing to the dissolution of the gel, when a gel-inducing reagent is not present in the reaction mixture. To increase the stability of the κ-carrageenan-entrapped biocatalyst, treatment with glutaraldehyde or hexamethylenediamine after entrapment is often effective.

Typical procedures for preparing various shapes of κ-carrageenan gel (43) are as follows.

7.4.3.1. Cube Type

The enzyme (100 mg) or microbial cells (16 g wet weight) are dissolved or suspended in 32 or 16 ml of physiological saline, respectively, at 25 to 50°C, and 3.4 g of carrageenan is dissolved in 68 ml of physiological saline at 37 to 60°C. The two are mixed, and the mixture is cooled at about 10°C for 30 min. To increase the gel strength, the gel is soaked in cold 0.3 M potassium chloride solution. After this treatment, the resulting stiffer gel is cut into cubes (3 by 3 by 3 mm).

7.4.3.2. Bead Type

A mixture (5 ml) of carrageenan and an enzyme or microbial cells is dropped into a 0.3 M potassium chloride solution through a nozzle having an orifice of 1 mm in diameter at a constant speed. Gel beads of 3 mm in diameter are obtained by this procedure.

7.4.3.3. Membrane Type

A mixture (5 ml) of carrageenan and an enzyme or microbial cells is spread on a plate to form a thin layer (1 by 250 by 200 mm) and soaked in cold 0.3 M potassium chloride solution to obtain a gel membrane.

7.4.4. Synthetic Resin Prepolymer Method

With the application of the immobilized biocatalyst in a variety of bioreactions, including synthesis, transformation, degradation, or analysis, each having a different desirable chemical environment, finding a suitable gel among the natural polymers may be difficult. Synthetic resin prepolymers such as photo-crosslinkable resin prepolymers and urethane prepolymers extend the list of polymers used for entrapment.

In the synthetic resin prepolymer method, the size of the gel matrix, which affects the substrate and product diffusion in the gel, the mechanical strength of the gel formed, the biocatalyst-holding capacity, and the growth capability of the cells inside the gel, can be regulated by the chain length of the prepolymer and the content of the reactive functional group. Furthermore, the hydrophilicity or hydrophobicity of the gel can be controlled by selecting a suitable prepolymer synthesized in advance in the absence of biocatalyst. The hydrophilicity-hydrophobicity balance of the gel is often critical in affecting the partition of the reactant to the gel in a bioconversion reaction. The ionic properties of the gel, which are an important factor for a bioconversion reaction, can be introduced to the prepolymers beforehand. Other advantages are that (i) the entrapment procedures are very simple and proceed under very mild conditions, and (ii) the prepolymers do not contain monomers that may have an unfavorable effect on the biocatalyst. Therefore, the synthetic resin prepolymer method may be one of the most widely applicable immobilization methods at present.

7.4.4.1. Photo-Crosslinkable Resin Prepolymer Method

The use of photo-crosslinkable resin prepolymer (Fig. 1) that is hydrophilic (compound 1) or hydrophobic (compound 2), cationic or anionic, and of different chain lengths has been developed by Fukui and coworkers (14,

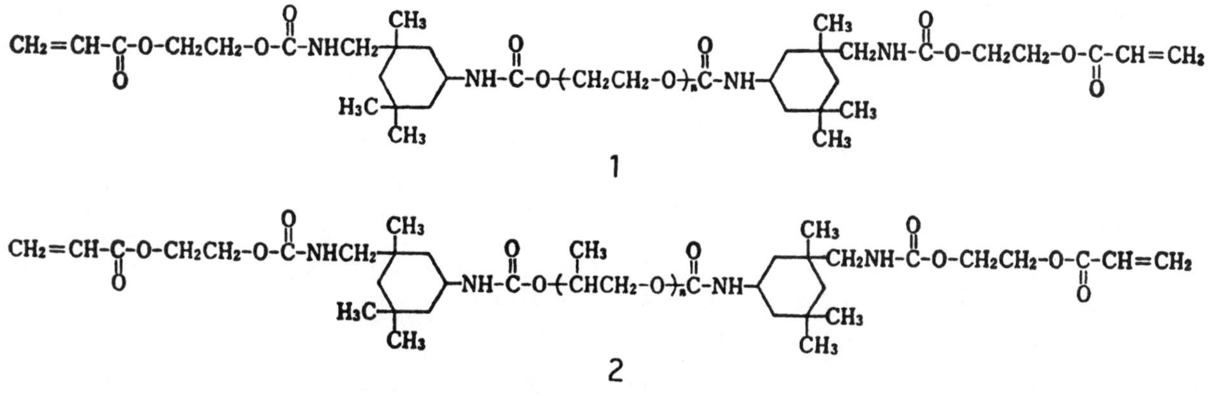

1

2

FIGURE 1 Structures of typical photo-crosslinkable resin prepolymers.

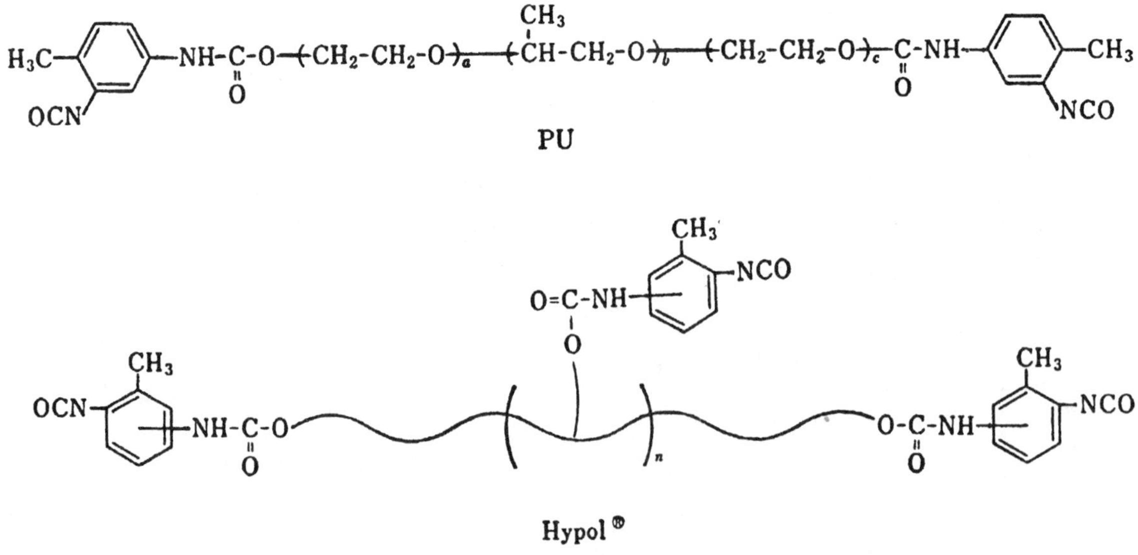

PU

Hypol®

FIGURE 2 Structures of typical urethane prepolymers.

15, 16, 40). A mixture of the prepolymer and the biocatalyst is gelled by irradiation with near-UV light for several minutes in the presence of a proper photosensitizer (initiator) such as benzoin ethyl ether. This method has been applied for the pilot-scale continuous production of ethanol by immobilized, growing yeast cells (19). Photo-crosslinkable resin prepolymers can be autoclaved at 120°C.

Examples of the procedure for use of photo-crosslinkable resin prepolymers are as follows.

Entrapment of Catalase with Photo-Crosslinkable Resin Prepolymer (16)

The photo-crosslinkable resin prepolymer (1 g; Kansai Paint Co., Tokyo, Japan) is mixed with 10 mg of an initiator, benzoin ethyl ether, and 0.5 ml of 50 mM potassium phosphate buffer (pH 7.2) and melted by heating at 60°C. To the molten mixture is added 2.2 ml of the chilled buffer to cool the mixture and then 0.3 ml of the solution of bacterial catalase (Nagase Industries Co., Osaka, Japan). The mixture is layered on a sheet of transparent polyester (7 by 10 cm) bounded by adhesive tape (thickness, 0.5 mm). The layer is covered with the same kind of sheet to eliminate exposure to the air and is illuminated by near-UV light (wavelength range, 300 to 400 nm; maximum intensity at 360 nm) for 3 min. The enzyme film thus formed (thickness, ca. 0.5 mm) is cut into small pieces (ca. 5 by 5 mm each) and used as the immobilized catalase.

Bead Type (20)

Ten parts (by weight) of photo-crosslinkable resin prepolymer are mixed with 0.08 parts of the initiator, 2 parts of 3% sodium alginate, 2 parts of distilled water, and 2 parts of centrifuged *Zymomonas mobilis* cells. This mixture is dropped into a 1.5% calcium chloride solution to form the beads. The initial cell concentration in this mixture is about 4×10^7 cells/ml. The beads thus formed are irradiated by near-UV light with a wavelength of 300 to 400 nm. After irradiating for 5 min, the immobilized cell beads obtained are washed with sterile water. For practical application, the bead type is often preferred to the cube type or the film type.

7.4.4.2. Urethane Prepolymer Method

The entrapment by urethane prepolymer (Fig. 2) is quite simple (14, 15, 17). There are many types of urethane prepolymers having different hydrophilicity or hydrophobicity and chain length. The urethane prepolymers are synthesized by the Toyo Tire and Rubber Industry Co., Osaka, Japan, but are not commercially available. Hypol can be obtained from W. R. Grace Co., Boca Raton, Fla. The urethane prepolymers react with each other in the presence of water to form urea bonds and liberate carbon dioxide. Therefore, the entrapment can be carried out simply by mixing the prepolymer with an aqueous solution or suspension of the biocatalyst as described below (17, 32). The urethane prepolymers can be sterilized at 120°C by avoiding moisture.

The urethane prepolymer (2.0 g), melted by incubation at 50°C (if necessary), is cooled to 4°C and mixed quickly and well with 2 ml of chilled 0.1 M phosphate buffer (pH 5.0) containing 20 mg of invertase (Grade V; Sigma, St. Louis, Mo.) in a small beaker. When gelation starts after a few minutes of mixing at room temperature, the mixture is kept at 4°C for 30 to 60 min to complete the polymerization. The resin gel thus formed is cut into small pieces (about 5 by 5 by 5 mm), washed thoroughly with the buffer, and used as immobilized invertase for the hydrolysis of sucrose (20). If the mixture, quickly and well stirred, is immediately layered on a glass plate framed with adhesive tape with the proper dimensions before the gelation starts, a film of the immobilized biocatalyst can be obtained (14).

REFERENCES

1. **Axén, R., J. Porath, and S. Ernback.** 1967. Chemical coupling of peptides and proteins to polysaccharides by means of cyanogen halides. *Nature* **214:**1302–1304.
2. **Bernfeld, P., and J. Wan.** 1963. Antigens and enzymes made insoluble by entrapping them into lattices of synthetic polymers. *Science* **142:**678–679.
3. **Bradberry, C. W., R. T. Borchardt, and C. J. Decedue.** 1982. Immobilization of ascorbic acid oxidase. *FEBS Lett.* **146:**348–352.

4. **Brodelius, P., B. Deus, K. Mosbach, and M. H. Zenk.** 1979. Immobilized plant cells for the production and transformation of natural products. *FEBS Lett.* **103:** 93–97.
5. **Brümmer, W., N. Hennrich, M. Klockow, H. Lang, and H. D. Orth.** 1972. Preparation and properties of carrier-bound enzymes. *Eur. J. Biochem.* **25:**129–135.
6. **Campbell, D. H., E. Luescher, and L. S. Lerman.** 1951. Immunologic adsorbents. I. Isolation of antibody by means of a cellulose-protein antigen. *Proc. Natl. Acad. Sci. USA* **37:**575–578.
7. **Chibata, I.** 1978. *Immobilized Enzymes—Research and Development.* John Wiley & Sons, Inc., New York.
8. **Chibata, I.** 1978. *Immobilized Enzymes, Research and Development.* Kodansha, Tokyo. (In Japanese.)
9. **Chibata, I., T. Tosa, and T. Sato.** 1974. Immobilized aspartase-containing microbial cells: preparation and enzymatic properties. *Appl. Microbiol.* **27:**878–885.
10. **Chibata, I., T. Tosa, and T. Sato.** 1983. Immobilized cells in the preparation of fine chemicals. *Adv. Biotechnol. Proc.* **10:**203–222.
11. **Chibata, I., T. Tosa, T. Sato, T. Mori, and Y. Matuo.** 1972. Preparation and industrial application of immobilized aminoacylases, p. 383–389. *In* Proc. IVth Int. Ferment. Symp. Fermentation Technology Today. Society for Fermentation Technology, Japan.
12. **Chibata, I., and L. B. Wingard, Jr. (ed.).** 1983. Immobilized microbial cells. *Appl. Biochem. Bioeng.*, vol. 4.
13. **Fleminger, G., and A. Yaron.** 1983. Sequential hydrolysis of proline-containing peptides with immobilized aminopeptidases. *Biochim. Biophys. Acta* **743:**437–446.
14. **Fukui, S., K. Sonomoto, and A. Tanaka.** 1987. Entrapment of biocatalysts with photo-cross-linkable resin prepolymers and urethane resin prepolymers. *Methods Enzymol.* **135:**230–252.
15. **Fukui, S., and A. Tanaka.** 1984. Application of biocatalysts immobilized by prepolymer methods. *Adv. Biochem. Eng. Biotechnol.* **29:**1–33.
16. **Fukui, S., A. Tanaka, T. Iida, and E. Hasegawa.** 1976. Application of photo-crosslinkable resin to immobilization of an enzyme. *FEBS Lett.* **66:**179–182.
17. **Fukushima, S., T. Nagai, K. Fujita, A. Tanaka, and S. Fukui.** 1978. Hydrophilic urethane prepolymers: convenient materials for enzyme entrapment. *Biotechnol. Bioeng.* **20:**1465–1469.
18. **Grubhofer, N., and L. Schleith.** 1953. Modifizierte Ionenaustausher als spezifishe Adsorbentien. *Naturwissenshaften* **40:**508.
19. **Iida, T.** 1993. Fuel ethanol production by immobilized yeasts and yeast immobilization, p. 163–182. *In* A. Tanaka, T. Tosa, and T. Kobayashi (ed.), *Industrial Application of Immobilized Biocatalysts.* Marcel Dekker, Inc., New York.
20. **Iida, T., M. Sakamoto, H. Izumida, and Y. Akagi.** 1993. Characteristics of *Zymomonas mobilis* immobilized by photo-crosslinkable resin in ethanol fermentation. *J. Ferm. Bioeng.* **75:**28–31.
21. **Jack, T. R., and J. E. Zajic.** 1977. The enzymatic conversion of L-histidine to urocanic acid by whole cells of *Micrococcus luteus* immobilized on carbodiimide activated carboxymethyl cellulose. *Biotechnol. Bioeng.* **19:**631–648.
22. **Kay, G., and E. M. Crook.** 1967. Coupling of enzymes to cellulose using chloro-*s*-triazines. *Nature* **216:**514–515.
23. **Kierstan, M., and C. Bucke.** 1977. The immobilization of microbial cells, subcellular organelles, and enzymes in calcium alginate gels. *Biotechnol. Bioeng.* **19:**387–397.
24. **Kitano, H., K. Nakamura, and N. Ise.** 1982. Kinetic studies of enzyme immobilized on anionic polymer latices: alkaline phosphatase, α-chymotrypsin, and β-galactosidase. *J. Appl. Biochem.* **4:**34–40.
25. **Mano, T., S. Mitsuda, E. Kumazawa, and Y. Takeshita.** 1992. New Immobilization method of mammalian cells using alginate and polyacrylate. *J. Ferm. Bioeng.* **73:** 486–489.
26. **Mattiasson, B.** 1983. *Immobilized Cells and Organelles.* Chemical Rubber Co., Cleveland.
27. **Mattiasson, B., and B. Danielsson.** 1982. Calorimetric analysis of sugars and sugar derivatives with aid of an enzyme thermistor. *Carbohydr. Res.* **102:**273–282.
28. **Messing, R., R. A. Oppermann, and F. B. Kolot.** 1979. Pore dimensions for accumulating biomass, in immobilized microbial cells. *Am. Chem. Soc. Symp. Ser.* **106:**13–28.
29. **Mitz, M. A.** 1956. New insoluble active derivative of an enzyme as a model for study of cellular metabolism. *Science* **123:**1076–1077.
30. **Mitz, M. A., and L. J. Summaria.** 1961. Synthesis of biologically active cellulose derivatives of enzymes. *Nature* **189:**576–577.
31. **Mosbach, K. (ed.).** 1976. Immobilized enzymes. *Methods Enzymol.*, vol. 44.
32. **Mosbach, K. (ed.).** 1987. Immobilized enzymes and cells. Part B. *Methods Enzymol.*, vol. 135.
33. **Navarro, J. M., and G. Durand.** 1977. Modification of yeast metabolism by immobilization onto porous glass. *Eur. J. Appl. Microbiol.* **4:**243–254.
34. **Nelson, J. M., and E. G. Griffin.** 1916. Adsorption of invertase. *J. Am. Chem. Soc.* **38:**1109–1115.
35. **Quiocho, F. A., and F. M. Richards.** 1964. Intermolecular cross linking of a protein in the crystalline state: carboxypeptidase-A. *Proc. Natl. Acad. Sci. USA* **52:**833–839.
36. **Sato, T., T. Mori, T. Tosa, and I. Chibata.** 1971. Studies on immobilized enzymes. IX. Preparation and properties of aminoacylase covalently attached to halogenoacetylcelluloses. *Arch. Biochem. Biophys.* **147:**788–796.
37. **Schejter, A., and A. Bar-Eli.** 1970. Preparation and properties of crosslinked water-insoluble catalase. *Arch. Biochem. Biophys.* **136:**325–330.
38. **Scheurich, P., H. Schnabel, U. Zimmerman, and J. Klein.** 1980. Immobilization and mechanical support of individual protoplasts. *Biochim. Biophys. Acta* **598:** 645–651.
39. **Sulkowski, E., and M. Laskowski.** 1974. Venom exonuclease (phosphodiesterase) immobilized on concanavalin-A-Sepharose. *Biochem. Biophys. Res. Commun.* **57:** 463–468.
40. **Tanaka, A., N. Hagi, S. Yasuhara, and S. Fukui.** 1978. Immobilization of catalase with photo-crosslinkable resin prepolymers. *J. Ferm. Technol.* **56:**511–515.
41. **Tanaka, A., and H. Nakajima.** 1990. Application of immobilized growing cells. *Adv. Biochem. Eng. Biotechnol.* **42:**97–131.
42. **Tosa, T., T. Mori, and I. Chibata.** 1969. Studies on continuous enzyme reactions. VI. Enzymatic properties of the

DEAE-Sephadex-aminoacylase complex. *Agric. Biol. Chem.* **33:**1053–1059.

43. **Tosa, T., T. Sato, T. Mori, K. Yamamoto, I. Takata, Y. Nishida, and I. Chibata.** 1979. Immobilization of enzymes and microbial cells using carrageenan as matrix. *Biotechnol. Bioeng.* **21:**1697–1709.
44. **Veliky, I. A., and A. Jones.** 1981. Bioconversion of gitoxigenin by immobilized plant cells in a column bioreactor. *Biotechnol. Lett.* **3:**551–554.
45. **Veliky, I. A., and R. E. Williams.** 1981. The production of ethanol by *Saccharomyces cerevisiae* immobilized in polycation-stabilized calcium alginate gels. *Biotechnol. Lett.* **3:**275–280.
46. **Watanabe, T., T. Mori, T. Tosa, and I. Chibata.** 1979. Immobilization of aminoacylase by adsorption to tannin immobilized on aminohexyl cellulose. *Biotechnol. Bioeng.* **21:**477–486.

Strain Improvement by Nonrecombinant Methods

VICTOR A. VINCI AND GRAHAM BYNG

8

Strain improvement can generally be described as the use of any scientific techniques that allow the isolation of cultures exhibiting a desired phenotype. The technology has been utilized for more than 50 years in conjunction with modern submerged culture fermentations and perhaps in a less systematic way for as long as fermented products have been made by humans. Most commonly, the ability of a strain to exhibit increased product accretion is the desired phenotype. However, the spectrum of improvements can include other traits, such as the elimination of toxic cometabolites (8, 17) or those problematic in downstream processing (37), the ability to degrade complex waste materials (38), or greater genetic stability of recombinant hosts (25).

The utility of strain improvement arises because of the existence of rate-limiting steps within all metabolic pathways. Most of these events are not readily measurable owing to the nature of metabolic transients or reaction fluxes during the course of an industrial fermentation. The metabolic flux involved in the biosynthesis of a secondary metabolite generally includes numerous specific reactions from the biosynthetic enzymes, primary metabolism for the supply of precursors to growth and secondary metabolites, and regulatory circuits involved in cell growth and differentiation. Likewise, heterologous protein expression in bacterial or fungal systems offers a significantly complex pathway. Because the rate-limiting enzymatic reactions or flux nodes are often unknown, an empirical process such as classical strain improvement is well suited to manipulation of the pathway. A screening program can be initiated with limited knowledge of the physiology or genetics associated with production of the molecule of interest. Classical mutagenesis and screening, also referred to as nonrecombinant strain improvement, can thus offer a significant advantage over genetic engineering approaches alone by yielding gains with minimal start-up time and sustaining such gains over years despite a lack of detailed knowledge concerning the physiology of the producing microorganism.

This empirical approach has a long history of success, as best exemplified by the improvements achieved for penicillin production in which reported penicillin titers are 50 g/liter, an improvement of at least 4,000-fold over the original parent (7). Examples of fungal or actinomycete cultures capable of overproducing metabolites in quantities as high as 80 g/liter can be found in the literature (7, 8, 26). Thus, application of strain improvement to new fermentation processes continues to be documented in the literature despite the age of the technology. One part of this continued interest in strain improvement is the marriage of classical techniques and molecular genetics to create a synergistic effect for process improvement. A growing interest in combining the approaches is evident from literature published since the first edition of this manual. Fermentation processes for products as diverse as antibiotics and human proteins have benefited from this combination of approaches (20, 34). A second area of interest has been the application of new approaches or technology to strain improvement. The greater availability of user-friendly equipment and enhanced detection limits for mass spectroscopy and high-pressure liquid chromatography (HPLC) have made their use more common (see below). In addition to using design to improve media during fermentation development, statistical analysis can also lead to enhancements in screening programs, as will be discussed. The availability of user-friendly software such as JMP (JMP Statistical Discovery Software; SAS Institute Inc., Cary, N.C.) has allowed the wider use of such analyses by the scientist. Finally, the growing field of metabolic flux analysis or quantitative physiology will likely become a tool in directing screening work or explaining the success of such work.

It is our intention to discuss key concepts in designing and running a strain improvement program. Particular emphasis will be placed on practical aspects of screens, including those programs relying predominantly on manual labor as well as high-throughput automated systems. Methods and concepts from the literature published since the last edition of this manual will predominate as a source of scientific literature. The reader is also directed to a number of excellent reviews that discuss basic concepts and the theoretical basis for mutagenesis and screening (2, 21, 22, 29). Regardless of the methods of strategy, strain improvement relies on the iteration of three operations: genetic alteration, fermentation, and assay.

8.1. GENETIC ALTERATIONS

8.1.1. Mutagenesis

The first key step is the generation of mutants. This can be accomplished by using either chemical or physical treatments to modify the genome of the target organism. It should be stressed that safety of the handler should be a consideration before starting such work. All mutagens should be considered potential carcinogens, and care should be taken to avoid exposure. Biosafety cabinets and protective equipment should be used, surfaces should be decontaminated, and used equipment should be decontaminated or disposed of by incineration.

There are many excellent reviews that list protocols for mutagenesis (2, 6) with a variety of different mutagens. Different mutagens are presumed to have different mechanisms of action, such as genetic alteration by base transitions or by frameshifts. During a long-term strain improvement program, it is advisable to change mutagens periodically to take advantage of these different mechanisms of action. The detailed procedure for the isolation of mutations and the sensitivity of an organism to a particular mutagen will vary considerably from organism to organism. For example, a highly pigmented organism will show increased resistance to the killing effects of UV light exposure. Alkalophilic organisms need to be harvested and resuspended in a neutral pH buffer before treatment with chemical mutagens because of the inactivation of the mutagen at high pH. The degree of killing versus the frequency of observed mutants will vary with different organisms. This can easily be verified by using antibiotic resistance as an indicator of mutation frequency. It is recommended that the multiplicity of mutation be determined in addition to monitoring survival rates as a more meaningful measure of efficiency (2).

In addition to vegetative cells and spore preparations, protoplasts can be used as starting material for mutagenesis. This is especially useful for basidiomycetes and other mycelial organisms. The nature of the protoplasting condition itself may prove to be mutagenic, and this can be enhanced by exposure of protoplasts to *N*-methyl-*N*′-nitro-*N*-nitrosoguanidine (MNNG) or UV light (18).

8.1.2. Protoplast Fusion

Protoplast fusion is another tool to be used to achieve genetic alterations in the lineage of industrial microorganisms. The technique can offer a means of combining favorable traits from two lineages or parental cultures. Fusion unfortunately does not allow the scientist to direct specific genes or DNA segments and thus, like mutagenesis, relies on empirical measurements to determine the success at combining two or more traits. The need for genetic markers thus becomes important in measuring efficiency of the approach. Phenotypic determinants such as auxotrophy, extracellular enzyme production, morphological differences, levels of antibiotic production, or antibiotic resistance can offer selectable traits. However, the use of auxotrophic markers may not be desirable for industrial microorganisms because of the cost of supplementing medium at the production scale. Protoplast fusion also offers the advantage of not requiring significant knowledge of the genetics of a particular culture. In addition, fusion is considered natural or homologous recombination and thus can avoid the regulatory constraints of fermenting the resulting strains at large-scale. A detailed description of protocols and considerations for the application of protoplast fusions to a variety of microorganisms is available from Matsushima and Baltz (23).

As with lignin degradation (see below), nodulation and nitrogen fixation are determined on multiple alleles. The requirement for agricultural application is ultimately to identify strains capable of enhanced fixation but also robust enough for field application. Despite the economic importance, the application of protoplast fusion to enhance nitrogen fixation capability in soybean rhizobia has been limited, perhaps because of the difficulty of establishing conditions as well as the slow growth of the cultures. Eisa et al. (13) reported the successful intraspecific fusion of *Bradyrhizobium japonicum* strains, resulting in isolates capable of forming 50% more nodules than parental strains. In addition, some intraspecific and interspecific fusion products (with *Sinorhizobium fredii*) exhibited increased nitrogenase activity. A procedure for preparing and fusing such protoplasts is described below.

PROTOCOL

Protoplast fusion

1. Grow cells in yeast extract mannitol broth for 48 h at 30°C (per liter: 5 g mannitol, 3 g yeast extract, 0.5 g $MgSO_4$, 0.7 g K_2HPO_4, 0.1 g KH_2PO_4, and 0.04 g $FeCl_3$).
2. Harvest cells in log phase by centrifugation.
3. Pre-wash the cells with 1% *N*-laurylsarcosine. Wash three times with Tris-HCl, pH 7.5, containing 0.6 M $MgSO_4$.
4. Incubate cells with 5 mg/ml lysozyme in the same buffer as above for 30 mins; spin out cells at 2,000 × *g* for 10 min and add fresh enzyme solution to the pellet.
5. Centrifuge cells again at 2,000 × *g* for 10 min and wash the precipitate with buffer. Resuspend the cells in the same buffer.
6. Add the suspension to a minimal medium (MM) with 7 g/liter agar and overlay onto the same medium containing 15 g/liter agar to determine protoplast efficiency. MM contains the following (per liter): 6.5 g HEPES, 5.5 g MES (morpholineethanesulfonic acid), 0.067 g $FeCl_3$, 1.8 g $MgSO_4 \cdot 7H_2O$, 0.13 g $CaCl_2$, 2.5 g Na_2SO_4, 3.2 g NH_4Cl, 1.25 g Na_2HPO_4, and 1 g L-(+) arabinose.

The exchange of the enzyme-buffer solution midway through the reaction in step 4 was found to be critical. Regeneration of the prepared protoplasts ranged in efficiency from 3×10^{-2} to 6.4×10^{-3}. Fusion of protoplasts was accomplished by using a polyethylene glycol treatment. Regeneration of protoplasts following fusion was achieved on the order of 10^{-7} (13).

8.1.3. Natural Recombination

Natural recombination mechanisms using naturally occurring conjugative plasmids have been used for strain improvement of *Lactococcus* starter cultures in the dairy industry. Phage infection can lead to slow acid production, which can have a significant economic effect in a large cheese factory. It has been established that many naturally occurring phage-resistant strains have a number of resistance mechanisms that in many cases are carried on conjugative plasmids (10, 11, 16). The phage resistance mechanisms can include abortive infection, restriction/modification, and adsorption inhibition. These approaches have been applied successfully to dairy starter cultures. As these bacteria are part of a finished food product, it is important to avoid antibiotic resistance markers or DNA from nonfood accepted organisms. Introduction of the phage resistance mechanism using conjugation is seen as exploiting a method of gene transfer used naturally by these organisms. The methods for conjugative transfer involve solid surface matings on milk agar; cells can then be harvested in 0.85% saline solutions and plated on selective medium (24). A more rapid method for conjugation uses a direct plate technique, whereby donor and recipient cells are mixed directly on selective medium. The advantages of this technique include improvement in transfer frequency, time savings, and the detection of low-frequency conjugal events because cells can be concentrated prior to mating. This method has been applied to the construction of nisin-producing *Lactococcus* strains (3).

8.2. OPERATIONAL CONSIDERATIONS

The ultimate success of a strain improvement program charged with developing and improving a fermentation process will be based largely on resource allocation. The key labor-intensive steps in strain improvement include the segregation and isolation of individual clones, preparation and dispensing of sterile media, transfer of the isolates in order to initiate the vegetative and fermentation stages, and assay of the fermentation broth from individual flasks or other containers. In general, the number of isolates screened will determine the success of detecting improved strains. If a manual operation is employed, the number of strains examined will be roughly proportional to the number of workers available. Alternative approaches toward minimizing the number of manipulations include bioassays or selective agents (see rational screening below). However, such methods have limited applications in a lineage and may not be effective over the longer lifetimes of some processes. Alternatively, automating the key steps in the process can drive throughput higher without adding labor.

8.2.1. Automated Screening

Toward the goal of increasing throughput without adding significant labor, automation of the key steps is a useful approach. It is generally desirable to miniaturize where possible to reduce the cost of equipment required as well as to reduce volumes of solvents and fermentation waste streams in the laboratory, which have come under more stringent control by environmental regulations. Efforts can range from automating single steps in a process (14) to construction of an integrated system. A schematic of an automated, high-throughput system is shown in Fig. 1 (27). In this industrial system, both agar and liquid media are charged into a vessel or fermentor, where they are sterilized under conditions that can simulate those at pilot-scale. Media are robotically dispensed in sterile laminar flow hoods into plates or fermentation bottles. The bottles, each of which will serve as a fermentation vessel, are arrayed in fixed positions in sterilizable, cleanable modules, each containing 100 bottles. Aeration is ensured by the use of covers capable of allowing the passage of air. The modules allow the easy handling and tracking of multiple bottles. Bottles generally contain 5 to 10 ml of medium, approximately 3- to 10-fold less than that typically used in 250-ml flasks. The inoculation of the vegetative stage for the primary screen is performed by a robot. Individual colonies are detected by an optical system and plugged from agar-based medium. Liquid transfer of the grown vegetative cultures is also accomplished robotically in a laminar flow hood. Extraction and HPLC analysis of the fermentation broth are also automated to match the throughput of the fermentation stage. A custom-designed unit that dispenses solvent and mixes the bottle contents is employed. Commercially available equipment transfers extract to microtiter dishes, and the extract is analyzed by HPLC units capable of performing isocratic separations. The use of isocratic separations of less than 5 mins is preferred to maintain throughput. One significant advantage of such a system is the high throughput, which can exceed that of a manual process by 10-fold based on comparable manpower (5, 35). In addition, the automation facilitates the capture and downloading of process data, which allows statistical process control approaches to be used. Perhaps most significantly, the replication of fermentation results crucial to making informed decisions (see below) becomes more feasible. The disadvantage of such high-throughput systems is the initial capital investment and continual maintenance of equipment and software. Compared to a smaller manual operation, an automated screen can also result in less flexibility in moving operations to new processes. However, the continued evolution of newer commercial equipment capable of handling microtiter plates or smaller formats could alleviate these concerns.

8.2.2. Manual Screening

A manual screen had advantages in that improved yield can be obtained readily with more limited capital investment. If a manual operation is employed, the number of strains examined will roughly be proportional to the number of workers available. Thus, labor tends to be the key driver of cost. The desired target of a program will dictate approaches. For a single protein product, it is often advisable to employ recombinant techniques as a first step. There may be multiple regulatory steps, precursor availability, export functions, and end product sensitivity that must be addressed, however, for complex antibiotic pathways. The flexibility of a manual screening system and the use of bioassays and selective agents (see below) allow for specific mutations to be built into the genealogy. When this approach is coupled with ongoing medium optimization, yields can be increased rapidly. In a typical strain improvement project, this usually equates to examination of around 500 mutants in shake flasks per week. The expected frequency of gains could be in the order of 1 in 10,000. Thus, new strains may only be found every 10 to 20 weeks. The use of prescreens and rational selection allows for a significant increase in efficiencies for the process. With a large pathway, the frequency of detecting an improved strain from random selection is around 1 in 1,000 to 2,000, depending on the organism and the product and the history of the production strain. This can be accomplished in 1 to 2 months per round, depending on the fermentation cycle time (5).

A manual screen starts in a similar fashion as any strain improvement program, with mutagenesis of conidia, spores, cells, or hyphal fragments (following maceration). If possible, an enrichment step is included from the rational screen, strains are selected, and working slants are produced that are then used to start the fermentation. After assay, the top few producing strains are selected for reisolation or natural selection. A number of reisolates are then reevaluated in shake flasks. The highest producer is then selected as the parent for the next round of mutagenesis. If the producing strain has already undergone considerable mu-

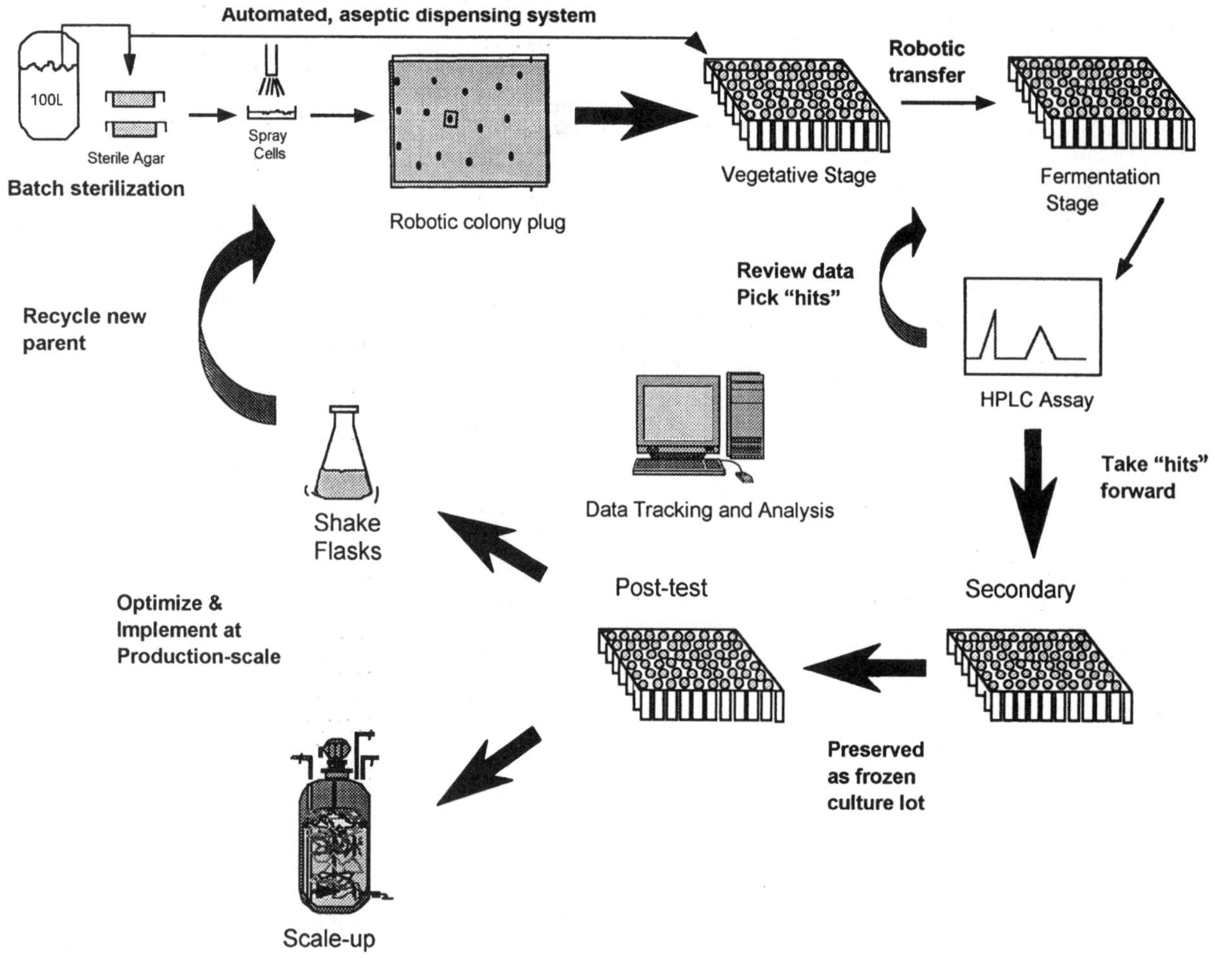

FIGURE 1 Schematic representation of an automated screening system.

tagenesis, then the level of increase may be close to the noise level of the system. It is still possible to continue strain improvement by using a recycling technique whereby a percentage of the highest producing strains are pooled and treated as a parent strain for mutagenesis in a succeeding round. This can continue for several iterations followed by single spore or cell reisolation to examine the end productivity (Fig. 2). One of the advantages of the manual screen is the trained eye of the operator. With a trained person, random selection is not quite as random as it appears. Differences in morphology, pigmentation, and growth rate can all be taken into account by the person doing the selection. In fact, selection for pelleted strains of filamentous organisms is quite commonly based on morphological criteria.

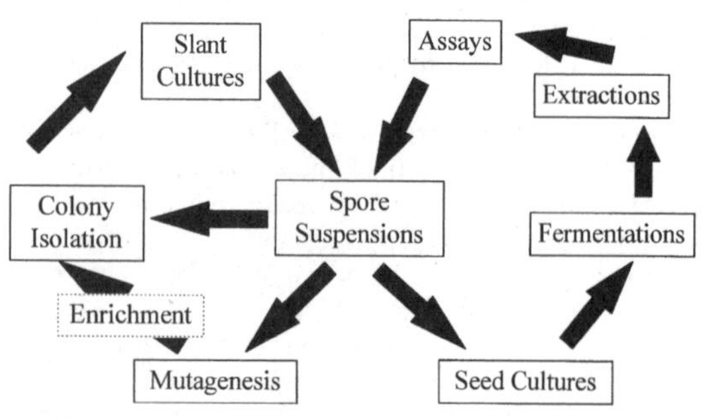

FIGURE 2 Key steps in the progression of cultures through a manual screen.

8.2.3. Selection and Rational Screening

Use of a selection strategy can greatly increase the efficiency of a strain improvement project. In random screening, a high percentage of putative mutants examined will be carried over as survivors from the mutagenesis and will exhibit the same yields or lower yields than the parent strain (2). However, direct selection on plates, for example, allows for a higher throughput, and only mutants are examined in shake flask experiments. An example of such an enrichment would be selection of amino acid overproducers using amino acid analogs. Only mutants that either overproduce the desired amino acid or are modified in transport or degradation are selected against a background of sensitive organisms that may be on the order of 10^8.

Rational screening requires some knowledge or inferred knowledge of the biosynthetic pathway to a product. A general selection scheme is outlined in Fig. 3. As an example, a polyketide-derived secondary metabolite offers several potential targets of attack for yield improvement (5). Mutants that survive treatment and overcome the toxic or inhibitory analog are often designated as resistant;

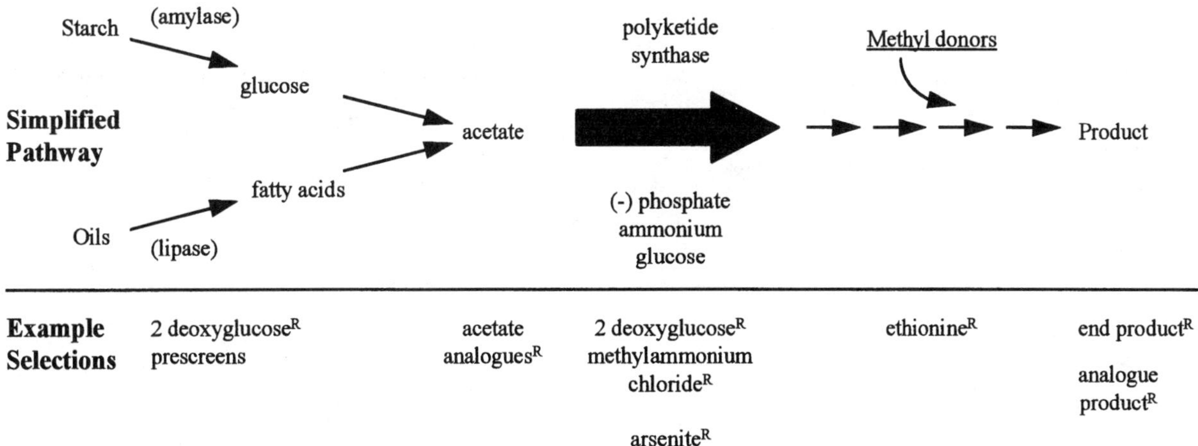

FIGURE 3 Strategies for screening of strains in rational selections

e.g., chloroacetate-resistant mutants can overproduce polyketide products. If the carbon source in the fermentation is starch or dextrin, then overproduction of amylase may lead to increased production of available acetate as a precursor. Similarly, if lipids are the carbon source, overproduction of lipase may lead to increased pools of acetyl-coenzyme A. Direct selection against an acetate analog may then be useful to increase precursor production. It should be noted that some organisms are incapable of taking up free acetate analogs such as fluoroacetate or chloroacetate. In cases such as these, amides of the analog (e.g., chloroacetamide) are sometimes useful alternatives. The polyketide synthase can also be a target, using inhibitors such as cerulenin to select overproducing mutants. The end product of a pathway itself may inhibit growth or overproduction. It is relatively easy to select mutants that are capable of growth in the presence of the end product. If stability of the compound is an issue, more stable analogs can be used, assuming a similar mechanism of action. Additional targets will depend on the physiology of the production strain. Other examples of phenotypes to overcome include repression by glucose (2-deoxyglucose resistance), ammonium ion, or phosphate. In these examples, selection strategies would encompass 2-deoxyglucose, methylammonium chloride, or arsenite resistance, respectively.

If there is a specific property of a desired product, this can be incorporated into a selection scheme. For example, carotenoids have been shown to protect *Phaffia rhodozyma* against singlet oxygen damage (31). Combinations of Rose Bengal and thymol in visible light select for carotenoid-overproducing strains, thus enriching the population for pigmented organisms. Direct enrichment with this singlet oxygen system led to an increase in certain carotenoids but a decrease in astaxanthin (32). As with any empirical selection, mutants with many phenotypes other than that of interest might be isolated.

8.3. OTHER APPLICATIONS

8.3.1. Applications to Genetically Engineered Strains

Strain improvement for therapeutic and biocatalytic protein products is approached initially by recombinant means, splicing the heterologous gene of interest into an appropriate expression vector and subsequent insertion into a bacterial or fungal host. However, the economics of a new process or mature process are likely to require higher production levels than those obtained by recombinant means alone. As discussed earlier, classical strain improvement derives its success from the empirical nature of the screening process and its application to complex pathways. Transcription, translation, protein secretion, activation, and folding offer one or more rate-limiting steps crucial to overproduction of a therapeutic protein of interest. Achieving overproduction of active therapeutic proteins in bacterial or fungal heterologous gene expression systems should thus be amenable to empirical methods such as classical strain improvement. In fact, fermentation processes for the manufacture of proteins used for food additives, biocatalysis, and human therapeutics have benefited from nonrecombinant strain improvement. Decreased development time and lower process costs are key incentives. There have been a few recent descriptions of the application of classical strain improvement to these products.

The complementary effects of performing nonrecombinant strain improvement methods on a genetically engineered fungal host have been demonstrated (9, 12, 20, 38). An agar plate screen was used to enhance the expression of recombinant chymosin by *Aspergillus awamori* (20). The prochymosin B construct designated pGRG3 utilized a glucoamylase promoter and signal peptide in conjunction with a glucoamylase terminator and was used to transform the host. The expression of chymosin was measured with a solid medium that utilizes the zone of clotting in overlays of milk. Unfortunately, *A awamori* produced a native acid protease that also clots milk. The authors were able to add the acid protease inhibitor diazoacetyl-norleucine methyl ester to the medium, thus allowing a reduction in the background. By this method, approximately 5,000 to 50,000 isolates per screen were examined and more than 500,000 in 8 months. As shown in Fig. 4, a shaken microtiter screen was employed for a secondary stage followed by further testing in shake flasks for top producers. Generally, 20 to 100 putative mutants were chosen to go forward to the secondary stage, and fewer than 5 were selected for the final, more labor-intensive confirmation stage. The power of the approach is evident in the fourfold increase in chymosin expression that was observed over five generations (20).

A second example of this synergy involved in *Aspergillus* system for production of human lactoferrin. This protein

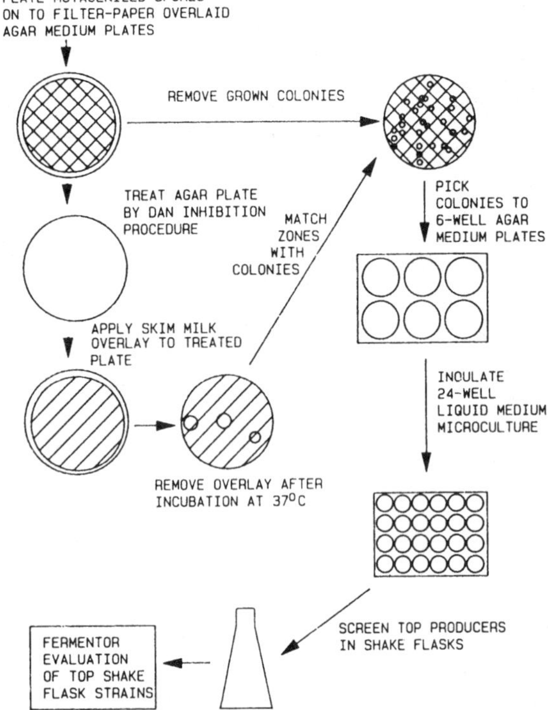

PLATE MUTAGENIZED SPORES ON TO FILTER-PAPER OVERLAID AGAR MEDIUM PLATES
REMOVE GROWN COLONIES
TREAT AGAR PLATE BY DAN INHIBITION PROCEDURE
MATCH ZONES WITH COLONIES
PICK COLONIES TO 6-WELL AGAR MEDIUM PLATES
APPLY SKIM MILK OVERLAY TO TREATED PLATE
INOULATE 24-WELL LIQUID MEDIUM MICROCULTURE
REMOVE OVERLAY AFTER INCUBATION AT 37°C
SCREEN TOP PRODUCERS IN SHAKE FLASKS
FERMENTOR EVALUATION OF TOP SHAKE FLASK STRAINS

FIGURE 4 Schematic representation of the major steps of the screening techniques. DAN, diazoacetyl-norleucine methyl ester. (From reference 20 with permission.)

was also expressed in *A. awamori* as a glucoamylase fusion polypeptide that was secreted into the growth medium and processed to the mature protein on export. Combining genetic engineering with classical strain improvement resulted in increases in lactoferrin production from approximately 20 mg/liter to >2 g/liter (39). Yields improved from a 70-fold increase over initial yields to a 200-fold increase following the classical work (Fig. 5).

Applications are not limited to processes for enzymes or traditional antibiotics. *Escherichia coli*-based expression systems are successfully used for a number of commercial pro-

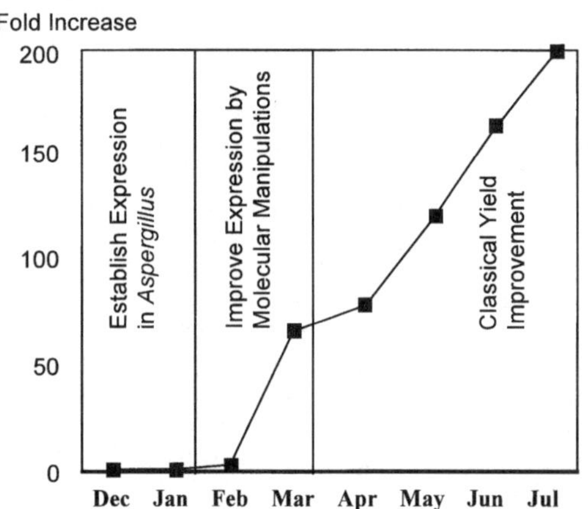

Fold Increase
200
150
100
50
0
Establish Expression in *Aspergillus*
Improve Expression by Molecular Manipulations
Classical Yield Improvement
Dec
Jan
Feb
Mar
Apr
May
Jun
Jul

FIGURE 5 Improvements in lactoferrin production using molecular and classical approaches.

cesses to make recombinant protein products. Such processes can also benefit from traditional screening efforts. It is common during the expression of high levels of heterologous proteins in *E. coli* that the host cells exhibit weakened growth or cell death. To examine a solution to this problem, Miroux and Walder (25) wished to select for strains that could support recombinant protein overexpression by presumably avoiding toxic effects within the cells. The authors examined the overproduction of membrane proteins using a T7 RNA polymerase expression vector, a common system for overexpression of cloned genes in *E. coli*. Overexpression of most of the membrane proteins as well as a number of globular proteins resulted in significant cell death during an initial experiment in a liquid fermentation medium. In a subsequent experiment, plating from a liquid culture onto agar-based medium was done to determine what culture phenotypes could be identified among the surviving cells. In addition to plating the culture on the control growth medium, agar medium containing either added isopropyl-2-D-thiogalactopyranoside (IPTG) or IPTG plus ampicillin was inoculated. IPTG was used as a selective agent because it acts as an inducer of the promoter contained on the cloned plasmids, whereas ampicillin serves to assist in maintenance of the plasmid, which contains an ampicillin resistance gene. It was thought that a strain that was resistant to the burden of overexpressing a cloned protein or maintaining a plasmid could be found. Following IPTG induction in a growing liquid culture, a culture was subsequently inoculated onto the plating medium above. Toxicity was demonstrated by significantly reduced viable cell counts on plates containing the selective agents. It was noticed that surviving cells resistant to ampicillin and IPTG occurred at a ratio of 3×10^{-3}. Two populations of cells appearing as large and small colonies were obvious in the small number of survivors, indicating again that a selective process was involved.

The authors next established a selection by using *E. coli* BL21(DE3) transformed with a plasmid containing the gene for oxoglutarate-malate carrier protein (OGCP). A 100-fold dilution of cells from liquid culture induced by IPTG 4 h earlier was plated onto TY agar medium containing IPTG. Three large colonies and one small colony were selected and subsequently tested in liquid medium for their ability to express OGCP. Interestingly, only the small colony-derived culture produced OGCP and continued to grow in the presence of inducer. Mutation in the genome of the strain designated C41(DE3) and not alteration of the plasmid was subsequently demonstrated. Significantly, strain C41(DE3) was able to express 10-fold more OGCP than the parental strain and could be grown in broth without ampicillin but with IPTG. Cell densities were six-fold higher for the mutant, thus indicating that product made per cell (productivity) was also higher. Thus, strain improvement using traditional selection methods can be a powerful and complementary tool in concert with molecular engineering of industrial strains and has yet to be fully exploited.

8.3.2. Plant Cell Culture

Interest in the use of plant cell cultures for the production of metabolites has grown dramatically and has led to increased efforts to develop techniques. Limited knowledge about the physiology of secondary metabolite production by plants and the culturing of plant cells at large-scale suggests that many techniques employed in microbial process development may have relevance. The empirical approach inherent in strain improvement should be applicable to plant cell culture and other tissue culture.

An example is the development of a unique aerobic fermentation process for the production of L-ascorbic acid (vitamin C) by Running et al. (30). The objective was to develop a fermentation-based process for ascorbic acid production to compete with the Reichstein process (one bacterial fermentation plus seven chemical steps). Screening efforts successfully identified *Chlorella pyrenodoxa*, a heterotrophic green microalga, as a producer of L-ascorbic acid. Mutagenesis and treatments common to microbial work were employed. Cells were incubated in growth medium for 4 to 8 h in the dark before plating. Initially, only 200 isolates were examined per week using shake flasks. However, the use of 2,6-dichlorophenol-indophenol to spray chloroform-lysed cells on plates led to an increase in throughput to greater than 25,000 isolates per week. A secondary screen of tubes was used to test putative improved strains followed by a shake flask screen. As with microbial fermentation, a trial in an agitated fermentor served as the final test and allowed additional optimization. Titers of 1.8 g/liter ascorbic acid were eventually achieved (Fig. 6). Thus, development of the process paralleled that of many microbial-based systems.

8.4. STATISTICAL ANALYSIS

8.4.1. Power to Detect Improvements

The ultimate function of any screening operation is to allow the user to detect a gain mutant among isolates performing at the control or parental titer. The factors that influence the ability to detect such improved strains are the variability of the process and the actual titer difference between the improved strain and the control. Assuming that mutagenesis conditions are optimal (single-hit kinetics), it is the variability inherent in a screen that determines whether improved strains are found. The challenge becomes greater in more mature processes in which the incremental titer difference between a mutant and a parent is generally similar to that increment early in the lineage but the percent improvement is considerably less. Empirical determinations have traditionally been made about the success or failure or screens as opposed to using statistical analysis. Unfortunately, considerable time and expense can be incurred before the user realizes that a screen has little chance of detecting improved mutants.

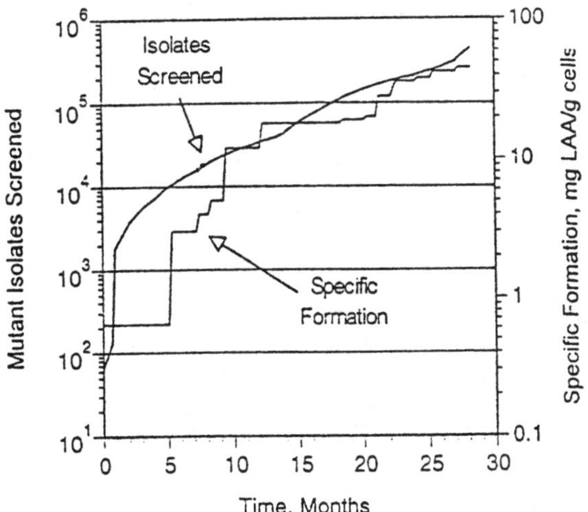

FIGURE 6 Progress as a function of the total number of *C. pyrenoidosa* mutant isolates screened during the first 2.5 years of the project. At a fermentor cell density of 40 g/liter, 45 mg of L-ascorbic acid (LAA) per g of cells results in a concentration of 1.8 g/liter of LAA. (From reference 30 with permission.)

Fortunately, the ability of a screen to detect improvements can be determined. Statistical design and analysis can be extended to strain improvement beyond the traditional use of statistical process control or medium optimization. A strain improvement screen or head-to-head run of a parental culture and putative improved strains is essentially a test of significance. As such, a successful test should show a statistical difference between control and improved means, i.e., a rejection of the null hypothesis. The probability that such a test accomplishes that is called the power of the test (33). Power calculations can be applied to any screening stage as a diagnostic tool. The analysis may be particularly useful in its application to a final or tertiary stage screen, as the data from this screen will likely be used to justify scale-up studies and some parameters can be more easily altered (number of replicates). For a given test, the key factors will include the chosen significance level of the test (confidence), the actual titer differences expected, the variability inherent in the system (from vegetative growth through to the assay), and the amount of data collected (replication).

An industrial screen offers an example of the success resulting from employing power calculations. The original final stage screen or tertiary test protocol examined nine putative improved strains and a control. Each strain was represented by two frozen vials of inocula that separately seeded two vegetative flasks, each of which became the inoculum for 10 fermentation bottles ($n = 20$ fermentations per strain). Following fermentation and HPLC analysis, decisions about the best new strain were made. However, problems in reproducing titer improvements in pilot-scale fermentors prompted a reevaluation of the screen. An experimental run under the original screening conditions employing three well-studied strains was used to gather additional data. Figure 7 graphically shows the results from a power determination for a final-stage fermentation test plotted as power versus sigma (standard deviation of the difference). The assumption for the calculations is that a statistical confidence level of 0.05 (95%) is desired. Note that sigma corresponds inversely to the number of vials per strain used and thus the number of vegetative flasks per strain (35). The curves represent targeted titer differences or gains ranging from 5 to 20% between mutant and control. As shown, the power to detect a 10% improvement in the original system was about 30%, which was clearly not acceptable (Fig. 7A). To improve the power of the system, efforts were directed in two key areas. Considerable work was directed at troubleshooting important handling steps to identify key contributors to error in the process. It was felt that if error or variation in the process could be reduced, the power of the system should improve. A key factor for improvement was found to be the temperature at the time of inoculation. The other major focus was to build more replication into the tests to allow for a better determination of error. It was evident

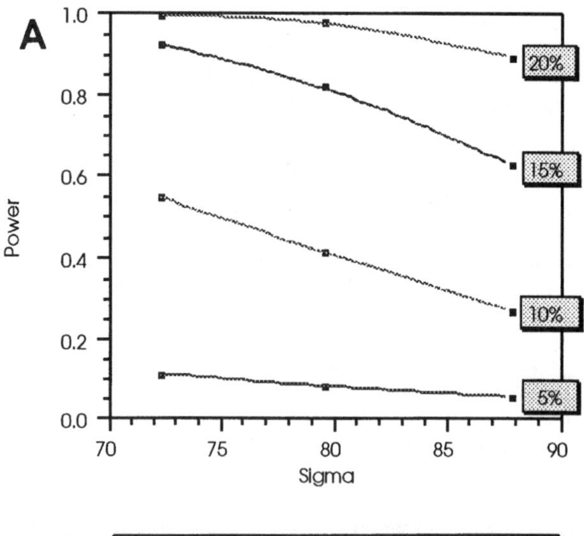

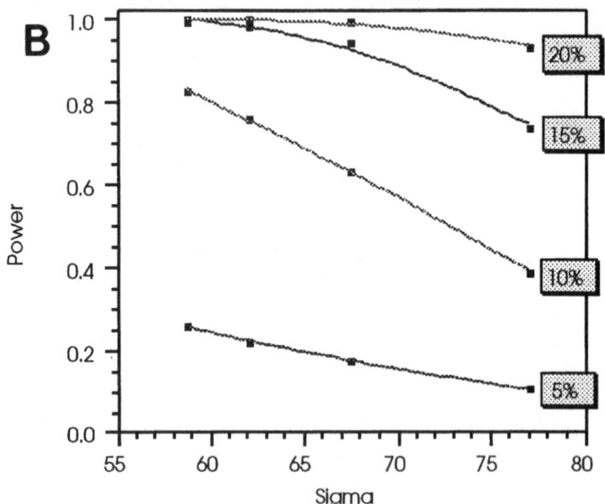

FIGURE 7 (A) Power curves for a 10-strain test design with a statistical significance level of 0.05 (95%). (B) Power curves for a five-strain test design with a statistical significance level of 0.05 (95%).

from the experimental run that a final test protocol would be enhanced by examining fewer strains, thus allowing greater replication at the fermentation stage ($n = 40$) and in the vegetative stage ($n = 4$ from four vials), which reduces sigma (Fig. 7A). Using these results, the final stage of the screening program was revised to reflect this five-strain/four-vegetative flask approach. Figure 7B shows the results of a subsequent study employing the new design and eliminating the key source of error. The power in this revised final screen gave a greater than 80% chance of detecting a new strain producing 10% more product. If titer increases of 15 or 20% are possible, the probability of detection is near theoretical. Enhancement of the screening system in this manner subsequently facilitated the scale-up of new cultures. As this study shows, screens built around too little replication are less likely to detect an improvement or will require considerably more throughput in the hope of overcoming the lack of power. Thus, the benefit of such statistically based work is not only to target optimal efficiencies in the number of replicates but also as a diagnostic tool to target poorly performing screens. Benefits will accrue as an efficient use of manpower and equipment, which ultimately determine the significant cost of running strain improvement projects.

8.4.2. Analytical Methods

As described earlier, analytical methods for strain improvement typically utilize methods that provide accurate measurements of the product of interest balanced by the need to limit time and manpower requirements. This balance of accuracy and speed is inherent in the need to screen large numbers of samples economically. One of the key issues in establishing an assay is to determine a robust means of extracting a metabolite for thin-layer chromatography or HPLC. The use of an extraction step is generally required for secondary metabolites as well as other aromatic substrates and products for biocatalysis. The time and manpower or automation needed for handling the extractions is costly. It is beyond the scope of this chapter to offer analytical separations; however, HPLC method development has become relatively straightforward as a result of enhanced detection and software applications.

The use of enzyme-linked immunosorbent assays (ELISAs) for sensitive and quantitative measurements of a number of antigens has become common. However, problems can be encountered with applications to strain improvement conducted in complex fermentation broths because of cross-reactivity. This issue is especially true if applied to secondary metabolites observed as multiple factors of the same pathway. One solution is to employ a rapid ELISA for preliminary plate or microtiter screening in a primary screen. ELISA techniques should be ideal for enzymes and proteins, especially those overexpressed in the available expressions systems. The power of an ELISA becomes especially relevant in attempts to miniaturize fermentation screening systems, as discussed above. The assays are done in microtiter format, which facilitates automation of handling and analysis. An example of an ELISA system for quantification of an enzyme in a complex milieu is a double-antibody sandwich assay developed for the endoglucanase I of *Trichoderma reesei* (4). The measuring range of the assay was 5 to 500 ng/ml, an improvement over earlier attempts, and as little as 20 pg of endoglucanase I in a microtiter well could be detected. Supernatants were used for the work described. The key to minimizing the potential for cross-reactivity with other cellulases in the culture broth was to employ a monoclonal antibody to endoglucanase I as the capture antibody.

8.5. NOVEL APPROACHES AND APPLICATIONS

8.5.1. Informed Strain Improvement

An interesting approach to the breeding of improved strains of *Phanerochaete chrysosporium* was proposed by Wyatt and Broda (40). The objective was to identify strains capable of improved lignin degradation while maintaining the robustness inherent in wild-type strains (as opposed to cultures crippled by extensive passage through mutagenesis and screening). Earlier mapping studies had demonstrated the abundance of restriction fragment length polymorphisms (RFLPs) in *P. chrysosporium* (8). The consideration in applying mutagenesis to lignin-degrading strains derives from the observation that multiple alleles contribute to the degradation of the natural substrate, lignocellulose. Enzymes involved in lignin degradation interact with gene products from alleles involved in growth and differentiation to give an optimal system or strain. Compounding the difficulty in detecting improved strains was the knowledge that simpler substrates to model degradation are not particularly accurate, because low-molecular-weight compounds do not model the large three-dimensional natural substrate.

Following an essential validation of the assay conditions, the authors employed a ^{14}C-labeled synthetic lignin called dehydrogenative polymerisate to assay mineralization in order to rank homokaryotic strains. Comparisons were made of 26 strains related to *P. chrysosporium* strain ME446 using mineralization activity and RFLP allele distributions of 38 genetic loci. Following a multifactorial analysis, no statistical correlation was found between loci and high mineralization. This finding most likely reinforces the empirical findings from other strain programs that suggest that an extensive array (more than 38) of loci are involved in complex interactions such as lignin degradation and secondary metabolite formation. It was noted that heterokaryotic strains gave a narrower range of levels of performance than expected and a slightly higher mean percent mineralization. Thus, the authors' assertion of multifactorial determinance for activity on lignin suggests that breeding within a parental lineage or more targeted genetic manipulation (such as recombinant means) would yield advantages as compared to chemical mutagenesis alone. They suggested gene knockouts or duplication of rate-limiting enzymes as useful strategies (40). Similar strategies have been developed and attempted with secondary metabolite pathways to varying degrees of economic success (27).

8.5.2. Metabolic Flux

Another attempt to look at multiple factors is embodied in metabolic flux analysis or balancing. The growth in recent years in readily accessible software and desktop computational power, as well as advances in mathematical modeling strategies, has led to growing interest in this area. The interest in the technology resides in its potential to provide leads for strain improvement, genetic manipulation, and process optimization. The approach is to determine stoichiometrically allowable flux distributions or a "metabolic genotype" for a strain (and process) of interest (36). The goal is to define the "metabolic phenotype" or optimal flux distribution by further analysis (36). The approach is largely unproved for natural product fermentations at this time.

The application of a multifactorial analysis such as informed strain improvement has perhaps only been examined empirically during classical lineage development. However, the approach could have important implications for strain improvement in the food and antibiotic industries. Traits of strains not directly measurable in screens, such as suitability in agitated fermentors, genetic stability, and interactions with complex raw material sources, could be compared with genetic mapping or parental histories. It is anticipated that metabolic flux analysis will also eventually become an invaluable tool in directing recombinant and nonrecombinant strain improvement. A synergy between metabolic flux analysis and expression monitoring using high-density oligonucleotide arrays has the potential to aid in directing strain improvement efforts.

8.5.3. Miniaturization and Automation

The need to automate the key steps in strain improvement to achieve high-throughput is discussed above. Essential for cost-effective automation is the miniaturization of the vessels used for growth of strains and production of product. As discussed above, plate handlers for microtiter plates are readily available, as is liquid-dispensing equipment capable of accurately delivering microliter volumes. Thus, microliter-scale fermentation followed by HPLC or possibly mass spectroscopy can be the basis of a smaller-scale program. More important, equipment and techniques that allow the examination of a small population of cells or even individual cells have become available. A combination of flow cytometry and cell sorting was demonstrated by An et al. (1) to enrich for and isolate mutants containing more carotenoid per cell. An argon ion laser (FACStar; Becton Dickinson, San Jose, Calif.) pulse emitting at 488 nm was used for excitation in conjunction with band-pass filter to minimize quenching of astaxanthin and β-carotene by an unwanted but related carotene. The authors first showed that cell agglomeration was not associated with increased fluorescence. A known parental culture and overproducing mutant were then used to empirically select the flow cytometry cell-sorting (FCCS) parameters. Mutants exhibiting higher carotenoid contents gave higher fluorescence than the parent, using the same forward scatter values within a limited range. Negative controls in the form of white mutants gave very faint fluorescent images, confirming that other metabolites were not confounding the observation. A yeast culture mutated with MNNG provided a test case for the system. A 10,000-fold enrichment of hyperproducing mutants was demonstrated by using FCCS compared to random isolation on plates.

A system developed by Huber et al. (19) resulted in the successful cloning of 16S rRNA and the isolation of a hyperthermophilic archaeum. The procedure employs visual recognition of single cells by staining them with fluorescently labeled oligonucleotide probes specific to 16S rRNA. Following staining, individual cells are isolated and cloned using a cell separation unit. The unit consists of a microscope equipped with a Nd-YAG laser at 1,064 nm wavelength (ADLAS, Lubeck, Germany). A single cell is optically trapped in the laser beam and separated from other cells by moving the microscope stage and later physically washing the cell into fresh medium. Importantly, the approach was the first to allow the determination of sequence identify without cultivation from the native habitat. The authors estimated that representatives from other cultures were present by at least four orders of magnitude higher than the isolate of interest. Interestingly, strain improvement schemes generally are designed to isolate a single improved strain from a background of about 5,000 to 50,000 parental or low-producing isolates, i.e., about three to four orders of magnitude enrichment. It is thus conceivable that overproducing mutants could be separated from control and low-producing mutants in a manner similar to that employed above. The approach also would lend itself to automation. Development would need to be done to identify proper fluorescent label or to develop the optics for the compounds of interest. Most significantly, such an approach would take strain improvement from macro-fermentation screens of several milliliters to the single-cell stage. Process economics and speed would shift dramatically.

8.5.4. Recent Analytical Approaches

Methods such as near-infrared spectroscopy and Fourier transform infrared spectroscopy have been increasingly used for analysis of fermentation broths and offer the ability to assay for detection in situ, thus avoiding sample preparation. However, these methods can be less desirable because of a lack of assay sensitivity for strain improvement, especially if conducted early in product development. In addition, commercial near-infrared spectroscopy probes are generally not made for the small volumes of broth inherent in higher throughput systems, which may be in the microliter to milliliter range. Mass spectroscopy has rarely been utilized for assay of nonvolatile compounds in a complex broth. The limitations have been with mass spectroscopy systems that require much routine assay work and the resulting sample preparation time and expense of equipment. On-line mass spectroscopic applications have tended to be limited to volatile compounds produced during fermentation. Goodacre et al. (15) made use of pyrolysis mass spectroscopy combined with multivariate calibration and artificial neural networks to demonstrate the quantitative analysis of antibiotics in fermentation broth. Pyrolysis mass spectra of complex mixtures are created not only by subpatterns of the pure components but also by spectral contributions from intermolecular reactions during pyrolysis. The authors compared the deconvolution of the mass spectra by using ampicillin mixed with either *E. coli* or *Staphylococcus aureus* fermentation broth as a test system. The three calibration methods gave similarly acceptable root mean squared errors on predictions of concentration versus actual. The range tested was 0 to 5 g/liter ampicillin. Moreover, the authors were able to demonstrate the ability of pyrolysis mass spectroscopy and artificial neural networks to predict increases in production of an unnamed fermentation-derived drug product. Artificial neural networks were trained using pyrolysis mass spectroscopy data compared with HPLC data for a range of concentrations of the drug up to double the parental level (0 to 87 mg/liter). A plot of the resulting estimate by the network versus HPLC concentrations gave a linear fit; the slope of the linear regression line was 0.82. The authors list the throughput for a single machine at 300 samples in one shift (less than 2 min per sample) or about 12,000 isolates per month screened using two shifts. The estimated costs of $75,000 for equipment and running costs of $1.50 per sample make the approach competitive with HPLC. The lack of solvent-handling equipment and the environmental problems inherent in disposal offer incentive for this methodology.

8.6. CONSIDERATIONS ON SCALE-UP

The intent of this chapter was to describe methods and strategies to develop a successful strain improvement program. It cannot be overstated that continuous improvement of media and growth conditions must be a sustained effort for any program. Less appreciated is the importance of establishing an open working relationship with scale-up teams. The ultimate destination of improved mutants is a fermentor or other bioreactor in which a product is made for clinical trials or marketed use. The importance of communication between scientists and engineers in each team cannot be overemphasized. To maximize the potential for success from the limited number of experiments to be devoted to a new strain, information about growth conditions, medium changes, and special properties of the new strain should be provided by the screening team. Feedback from the scale-up team should include data about growth and productivity in the fermentors in addition to the obvious information about success of initial trials. Other examples of parameters that strain changes can impact include viscosity, feeding patterns, foaming, and oxygen demand. In addition to the communication concerns, the performance of the fermentation broth in laboratory or pilot-scale downstream purification is essential for approval of the strain for a marketed product. These interactions can ultimately avoid the generation of many cycles of mutagenesis and creation of an extended culture lineage that leads to a dead-end because of scale-up problems that cannot be readily solved. The benefit is the efficient use of manpower and capital.

8.7. SUMMARY

Classical or nonrecombinant strain improvement strategies offer compelling scientific and economic advantages for product development and process improvements. Fermentation processes as varied as those employed for primary metabolites, antibiotics, and proteins can benefit from the technology. It is anticipated that unique applications of existing technology and the introduction of novel equipment will drive a continual renewal of this essential scientific methodology.

REFERENCES

1. **An, G.-H., J. Bielich, R. Auerbach, and E. A. Johnson.** 1991. Isolation and characterization of carotenoid hyperproducing mutants of yeast by flow cytometry and cell sorting. *Bio/Technology* **9:**70–73.
2. **Baltz, R. H.** 1986. Mutagenesis in *Streptomyces* spp., pp. 184–190. *In* A. L. Demain and N. A. Solomon (ed.), *Manual of Industrial Microbiology and Biotechnology*. American Society for Microbiology, Washington, D.C.
3. **Broadbent, J. F., and J. K. Kondo.** 1991. Genetic construction of nisin-producing *Lactococcus lactis* subsp. *cremoris* and analysis of a rapid method for conjugation. *Appl. Environ. Microbiol.* **57:**517–524.
4. **Buhler, R.** 1991. Double-antibody sandwich enzyme-linked immunosorbent assay for quantitation of endoglucanase I of *Trichoderma reesei*. *Appl. Environ. Microbiol.* **57:**3317–3321.
5. **Byng, G.** Unpublished results.
6. **Carlton, B. C., and B. J. Brown.** 1981. Gene mutation. pp. 222–242. *In* P. Gerhardt, R. G. E. Murray, R. N. Costilow, E. W. Nester, W. A. Wood, N. R. Krieg, and G. B. Phillips (ed.), *Manual of Methods for General Bacteriology*. American Society for Microbiology, Washington, D.C.
7. **Crueger, W., and A. Crueger.** 1984. Antibiotics, p. 203. *In* T. D. Brock (ed.), *Biotechnology: A Textbook of Industrial Microbiology*. Akademische Verlagsgesellschaft, Wiesbaden, Germany.
8. **Crueger, W., and A. Crueger.** 1984. Citric acid, pp. 115–116. *In* T. D. Brock (ed.), *Biotechnology: A Textbook of Industrial Microbiology*. Akademische Verlagsgesellschaft, Wiesbaden, Germany.
9. **Cullen, D., G. L. Gray, L. J. Wilson, K. L. Hayena, M. H. Lamsa, M. W. Rey, S. Norton, and R. M. Berka.** 1987. Controlled expression and secretion of bovine chymosin in *Aspergillus nidulans*. *Bio/Technology* **5:**369–376.
10. **Davidson, B. E., and A. J. Hillier.** 1995. Developing new starters for fermented milk products. *Aust. J. Dairy Technol.* **50:**6–9.

11. **Dinsmore, P. K., and T. R. Klaenhammer.** 1995. Bacteriophage resistance in *Lactococcus*. *Mol. Biotechnol.* **4:** 297–314.

12. **Dunn-Coleman, N. S., P. Bloebaum, R. M. Berka, E. A. Bodie, N. Robinson, G. Armstrong, M. Ward, M. Przetak, G. L. Carter, R. LaCost, L. J. Wilson, K. H. Kodama, E. F. Baliu, B. Bower, M. Lamsa, and H. Heinson.** 1991. Commercially viable levels of chymosin production by *Aspergillus*. *Bio/Technology* **9:**976–981.

13. **Eisa, E. G., K. Maeta, N. Mori, and Y. Kitamoto.** 1995. Enhanced nitrogen fixation capabilities of soybean rhizobia by inter and intra-specific cell fusion. *Jpn. J. Crop Sci.* **64:**272–280.

14. **Freysoldt, C.** 1986. A method for selection of antibiotic producing strains of *Penicillium chrysogenum* on solid media by means of a microcup cultivation and image analyzing of inhibition zones. *J. Microbiol. Methods* **5:**167–175.

15. **Goodacre, R., S. Trew, C. Wrigley-Jones, M. J. Neal, J. Maddock, and T. W. Ottley.** 1994. Rapid screening for metabolite overproduction in fermentor broths, using pyrolysis mass spectroscopy with multivariate calibration and artificial neural networks. *Biotechnol. Bioeng.* **44:** 1205–1216.

16. **Hill, C., P. Garvey, and G. F. Fitzgerald.** 1996. Bacteriophage- host interactions and resistance mechanisms, analysis of the conjugative bacteriophage resistance plasmid pNP40. *Lait* **76:**67–79.

17. **Hodges, R. L., D. W. Hodges, K. Goggans, X. Xuei, P. Skatrud, and D. McGilvray.** 1994. Genetic modification of an echinocandin B-producing strain of *Aspergillus nidulans* to produce mutants blocked in sterigmatocystin biosynthesis. *J. Indust. Microbiol.* **13:**372–381.

18. **Homolka, L., P. Vyskocil, and P. Pilat.** 1988. Use of protoplasts in the improvement of filamentous fungi. I. Mutagenesis of protoplasts of *Oudemansiella mucida*. *Appl. Microbiol. Biotechnol.* **28:**166–169.

19. **Huber, R., S. Burggraf, T. Mayer, S. M. Barns, P. Rossnagel, and K. O. Stetter.** 1995. Isolation of hyperthermophilic archaeum predicted by *in situ* RNA analysis. *Nature* **376:**57–58.

20. **Lamsa, M., and P. Bloebaum.** 1990. Mutation and screening to increase chymosin yield in a genetically-engineered strain of *Aspergillus awamori*. *J. Indust. Microbiol.* **5:** 229–238.

21. **Lein, J.** 1983. Strain development with non-recombinant DNA techniques. *ASM News* **49:**576–579.

22. **Lein, J.** 1986. Random thoughts on strain development. *SIM News* **36:**8–9.

23. **Matsushima, P., and R. H. Baltz.** Protoplast fusion, pp. 170–183. *In* A. L. Demain and N. A. Solomon (ed.), *Manual of Industrial Microbiology and Biotechnology*. American Society for Microbiology, Washington, D.C.

24. **McKay, L. L., K. A. Baldwin, and P. M. Walsh.** 1980. Conjugal transfer of genetic information in group N streptococci. *Appl. Environ. Microbiol.* **40:**84–91.

25. **Miroux, B., and J. E. Walder.** 1996. Over-production of proteins in *Escherichia coli*: mutant hosts that allow synthesis of some membrane proteins and globular proteins at high levels. *J. Mol. Biol.* **260:**289–298.

26. **Miyazaki, Y.** 1978. Method of producing polyether antibiotics as well as salinomycin antibiotics. European Patent Application 78100061.7.

27. **Queener, S. W., and D. H. Lively.** 1986. Screening and selection for strain improvement, pp. 155–169. *In* A. L. Demain and N. A. Solomon (ed.), *Manual of Industrial Microbiology and Biotechnology*. American Soceity for Microbiology, Washington, D.C.

28. **Raeder, U., W. Thompson, and P. Broda.** 1989. RFLP-based genetic map of *Phanerochaete chrysosporium* ME446. *Mol. Microbiol.* **3:**911–918.

29. **Rowlands, R. T.** 1984. Industrial strain improvement: mutagenesis and random screening procedures. *Enzyme Microbiol. Technol.* **6:**3–10.

30. **Running, J. A., R. J. Huss, and P. T. Olson.** 1994. Heterotrophic production of ascorbic acid by microalgae. *J. Appl. Phycol.* **6:**99–104.

31. **Schroeder, W. A., and E. A. Johnson.** 1995. Singlet oxygen and peroxyl radicals regulate carotenoid biosynthesis in Phaffia rhodozyma. *J. Biol. Chem.* **270:**18374–18379.

32. **Schroeder, W. A., and E. A. Johnson.** 1995. Carotenoids protect *Phaffia rhodozyma* against singlet oxygen damage. *J. Indust. Microbiol.* **14:**502–507.

33. **Snedecor, W., and W. G. Cochran.** 1980. Power of a test of significance, pp. 68–70. *Statistical Methods*, 7th ed. Iowa State University Press, Ames.

34. **Solenberg, P. J., C. A. Cantwell, A. J. Tietz, D. McGilvray, S. W. Queener, and R. H. Baltz.** 1996. Transposition mutagenesis in *Streptomyces fradie*: identification of a neutral site for the stable insertion of DNA by transposon exchange. *Gene* **168:**67–72.

35. **Sullivan, G., L. Nguyen, and V. A. Vinci.** Unpublished results.

36. **Varma, A., and B. O. Palsson.** 1994. Metabolic flux balancing: basic concepts, scientific and practical use. *Bio/Technology* **12:**994–998.

37. **Vinci, V. A., T. d. Hoerner, A. D. Coffman, T. G. Schimmel, R. L. Dabora, A. C. Kirpekar, C. L. Ruby, and R. W. Stieber.** 1991. Mutants of a lovastatin-hyperproducing *Aspergillus terreus* deficient in the production of sulochrin. *J. Indust. Microbiol.* **8:**113–120.

38. **Ward, M., L. J., Wilson, K. H. Kodama, M. J. Rey, and R. M. Berka.** 1990. Improved production of chymosin in *Aspergillus* by expression as a glucoamylase-chymosin fusion. *Bio/Technology* **8:**435–440.

39. **Ward, P. P., C. S. Piddington, G. A. Cunningham, X. Zhou, R. D. Wyatt, and O. M. Coneely.** 1995. A system for production of commercial quantities of human lactoferrin: a broad spectrum natural antibiotic. *Bio/Technology* **13:**498–503.

40. **Wyatt, A. M., and P. Broda.** 1995. Informed strain improvement for lignin degradation by *Phanerochaete chrysosporium*. *Microbiology* **141:**2811–2811.

Culture and Analysis Using Gel Microdrops

JAMES C. WEAVER

9

Rapid measurement of microbial growth and biochemical activity has long been important. Gel microdrop (GMD)-based microbial assays accomplish this by providing both functional and compositional measurements at the individual cell level. The basic idea is encapsulation of individual microorganisms (strictly, CFU) within microscopic gel particles, such that both the cell(s) and its extracellular environment can be individually measured.

Traditional microbial assays such as viable plating indeed involve individual cell growth, but they are generally viewed as slow because the optical methods used to measure colonies usually require relatively large colonies. However, this is not a fundamental limitation; microorganisms encapsulated within GMDs provide clonal growth assays within a few (often one to two) generation times. Similarly, traditional microbial assays based on the biochemical activity of a population of cells are often viewed as slow, because the cell density (number of cells per volume) is often too small to generate rapid changes in extracellular metabolite concentrations. This is also not a fundamental limitation, because confinement of individual microbial cells/CFU to the small volume of GMDs can provide an effective cell density that is much larger than that of the sample.

Optical instrumentation applicable to microbial measurements now extends beyond spectrometers for cell suspension turbidity measurements, and includes complex and powerful systems such as flow cytometry and image analysis (6). When used with conventional microbial suspensions, these systems can provide constitutive measurements of individual cells (e.g., amount of cell wall or membrane material, specific surface markers for viable microorganisms, amount of intracellular protein or nucleic acids for fixed microorganisms). GMDs extend this capability to assays involving the immediate extracellular environment, such that assays can provide measurements both for individual cells and for molecules and cells within their extracellular environments.

9.1. GMDs FOR CULTURE AND ASSAYS

For microbial culture and assays, GMDs are typically small (10 to 100 μm in diameter; corresponding volumes of 5×10^{-10} ml to 5×10^{-7} ml), nearly spherical agarose particles that are prepared such that most GMDs initially contain zero or one microorganism. GMDs can be created that are in a "closed" state, in which the GMDs are surrounded by a permeability barrier that confines most secreted molecules to the aqueous interior of the GMD. Thus, closed GMDs are essentially microminiaturized microtiter wells, such that the effective cell density due to one cell per GMD is large (e.g., $\rho_{eff} \approx 2 \times 10^6$ to 2×10^9 cells/ml). Alternatively, GMDs can be used in an "open" state, in which no significant permeability barrier exists, such that microorganisms encapsulated within GMDs readily exchange molecules with the medium external to the GMDs (9).

9.1.1. Protocols for Making GMDs

GMDs are formed by starting with a conventional microbial cell suspension, which is mixed with a gelable material (usually molten agarose at 37°C), and then gently dispersed within a nonaqueous fluid (certain mineral oils or silicone fluids) to form an emulsion. Although early studies succeeded in using a variety of laboratory devices to form GMDs, these methods were generally variable and tedious; currently, a specialized microdrop-making device (CellSys 100; One Cell Systems, Cambridge, Mass.) is used. The generic protocol involves dispersion of the microbial cell suspension to form a large number (e.g., 10^7) of liquid microdrops, which are the discontinuous phase of the emulsion. Gelation is forced by briefly cooling the emulsion, resulting in GMDs surrounded by the nonaqueous fluid. Such GMDs are closed GMDs, so designated because of the permeability barrier of the nonaqueous fluid. If open GMDs are desired, the GMDs are separated from the nonaqueous fluid by gentle centrifugation or filtration, resulting in GMDs that readily exchange water-soluble molecules with the external medium.

The following is an example of a protocol for encapsulating yeast cells in GMDs formed from biotinylated agarose. Suspended yeasts (*Candida albicans*; several different strains) from overnight cultures are diluted 1:10 in phosphate-buffered saline, pH 7.0, and 0.1% Tween 80 (PBS-Tween), with cell clumps separated by repeated up-and-down pipetting until predominantly two-cell clumps (presumably mother-daughter pairs) remain. Such clumps (CFU) are encapsulated into GMDs by the following procedure. Step 1: A 450-μl aliquot of CelBioGel (3% biotinylated agarose [One Cell Systems] in PBS) is melted at 100°C, supplemented with 100 μl of 10% pluroic acid, and equilibrated to 37°C. Step 2: The yeast cell suspension is added and mixed by vortexing. Step 3: The yeast/agarose mixture is added dropwise to 37°C emulsion oil (CelMix, One Cell Systems) with controlled stirring (1,800 rpm for 2 min) in a microdrop maker (CellSys 100). This step provides gentle emulsification for the creation of liquid microdrops containing cells with high viability. Step 4: With continued stirring (1,800 rpm for 1 min, then 1,100 rpm for 10 min), the emulsion temperature is lowered temporarily to induce gelation of the biotinylated agarose, thereby converting liquid microdrops into GMDs. At this point in the protocol, closed GMDs are present, because the nonaqueous fluid provides a permeability barrier for most biological molecules, particularly those with exposed charge groups. Step 5: The GMDs are removed from the nonaqueous fluid by centrifugation for 10 min at 600 × *g*. At this point, open GMDs are present, as only the highly porous gel limits molecular exchange between encapsulated cells and the external aqueous medium. Step 6: The GMDs are washed once in PBS and then resuspended in an appropriate medium. As is easily imagined, many variations of such protocols can be used, depending entirely on the microorganisms, media, etc., that are used. For sterility and safety, most or all of the above steps are carried out in a biosafety cabinet.

The above emulsification creates many GMDs at essentially the same time and can be used with cell suspensions containing some debris and cell clumps. A serial GMD-making process based on a vibrating orifice can also be used, but it requires well-filtered medium, removal of cell clumps, and scale-up leads to long run times. GMDs can be made from sodium alginate and other ionically cross-linked polymers, or from agarose if imposed temperature conditions are carefully controlled (3, 10). However, in the latter case, microbial viability can suffer.

9.1.2. Fluorescence-Labeled Gel Matrix

Assays based on open GMDs do not require uniform GMD size (volume). Growth assays only require that microcolonies be confined, and protein secretion assays initially capture most secreted molecules in the pericellular region and are then relatively insensitive to GMD size. However, prolonged incubation can saturate the capture sites near the cell, and GMD size then becomes important. Determination of individual GMD size can be accomplished by providing a volumetric fluorescence label (1). For closed GMDs, a soluble reference dye can be used.

9.1.3. Statistical Occupation of GMDs

Suspended microorganisms are distributed randomly in the suspending medium. In most cases, creation of GMDs therefore leads to random occupation of GMDs by the initial cells (Table 1). Usually, single-cell analysis is sought and can be reasonably achieved by seeking an average occupation, $\bar{n}$, which does not exceed $\bar{n} \approx 0.15$. This is based on Poisson statistics, which governs the inoculation of GMDs, microtiter wells, and petri dishes. The main difference is that a very large number of microscopic GMDs are used, so that initially most contain zero or one cell. Following a brief incubation, protein molecules are captured in the pericellular space within the originating GMD, progeny cells are retained within the GMD, or metabolites are retained within the small volume of a GMD which is immersed in a nonaqueous fluid such as an inert silicone oil. GMDs which initially contain no cells are "unoccupied,"

TABLE 1 Terminology relating to the statistical inoculation of GMDs[a]

Sketch[b]	Symbol	Terminology
○	$n = 0$	Unoccupied GMD
⊙	$n = 1$	Individually occupied GMD
⊙	$n \geq 1$	Multiply occupied GMD

[a]From reference 5.
[b]Dot, individual cell; circle, GMD.

those with one initial cell are "singly occupied," and those with two or more individual cells are "multiply occupied." Optical instruments such as flow cytometers and imaging systems can distinguish unoccupied from occupied GMDs.

9.1.4. Important Properties of GMDs

Although larger than microorganisms, GMDs are nevertheless microscopic and can be handled and manipulated much like cells. Thus, GMDs can be filtered, pipetted, centrifuged, and washed, provided it is recognized that because of their larger size, GMDs are subjected to greater hydrodynamic shear than cells. The gel matrix of GMDs is microporous, with a large molecular weight cutoff (about 5×10^5 g mol^{-1}). This results in high permeability, which allows microorganisms encapsulated within GMDs to be cultured under controlled conditions. Specifically, diffusion through the gel allows most nutrients to be readily supplied and most metabolic wastes to be removed. It also allows straightforward extension of fluorescence staining protocols, with only the need to use somewhat longer times.

9.2. OPEN GMDs FOR CLONAL GROWTH AND PROTEIN SECRETION ASSAYS

Culture of encapsulated microorganisms with externally imposed conditions can be accomplished by using open GMDs, and this includes an ability to expose microorganisms to candidate antibiotic levels for MIC and other traditional tests, but now at the individual cell level. Because individual cells/microcolonies are analyzed, this allows much more rapid assays, which can even be extended to mixed cultures (8).

Clonal growth assays can be constructed, because progeny are retained by the gel matrix, allowing microcolonies to form within GMDs. Protein secretion assays are also possible, by providing a large number of molecular capture sites within the gel matrix, such that secreted or released molecules have a high probability of capture within the originating GMD. The captured molecules are subsequently labeled with fluorescence by a reporter molecule, so that individual GMDs achieve reporter fluorescence in proportion to the number of secreted protein molecules.

9.2.1. Flexible Culture Conditions

GMDs are usually incubated under conditions that provide a homogeneous biochemical environment: there is no nearby air/medium interface, and the use of microcolonies limits the occurrence of microenvironments due to a locally high density of microorganisms. In this sense, GMD culture is similar to pour plate culture. Because GMDs are microscopic, diffusion times are relatively rapid (seconds to minutes), which allows exposure of entrapped cells and microcolonies to different molecules and stains for various times. Furthermore, the effective overall density of cell can be varied. For example, two preparations of GMDs can be used. The first preparation contains inert (coated) magnetic beads coentrapped in the gel with microorganisms, and either features a large $\bar{n}$ or contains preexisting microcolonies. A second (test) GMD preparation has a small value of $\bar{n}$ and is used for the subsequent assay. Because of the high permeability of GMDs, a mixture of the two GMD preparations results in medium conditioning that is dominated by the numbers and types of microorganisms in the first GMD preparation. Following incubation, the "influencing" GMDs can be removed by a magnet, and the test GMDs can be analyzed for the amount of growth per individual cell/CFU.

9.2.2. Clonal Growth Assays

GMD growth assays involve microcolony formation in singly occupied GMDs, followed by optical measurements that determine microcolony size (Fig. 1). By using fluorescence stains, the optical measurements are capable of resolving the difference in fluorescence due to as few as 1 and 2 cells. For viable cells, relatively nonspecific stains for cell membranes, cell walls, and vital stains provide generality. These can be supplemented with surface marker-specific stains (e.g., fluorescence-labeled antibodies) to provide specificity and a means of isolating viable cells. If cell viability is not required, fixation (e.g., exposure to 50% methanol) of cells within GMDs allows intracellular constituents (proteins, nucleic acids) to be fluorescence stained. GMDs thus provide the ability to determine clonal growth rapidly, within the first few cell divisions (5, 8).

9.2.3. Susceptibility Assays

Drugs that inhibit or kill cells readily penetrate the gel matrix of GMDs (5, 8). Because of the relatively short diffusional times (seconds to a few minutes), culture within GMDs allows various times of drug exposures to be used. Comparison of fluorescence signals from several thousand drug-exposed GMDs with zero-drug and methanol-fixation controls provides a rapid determination of the minimum drug concentration that inhibits or kills cells. Use of a

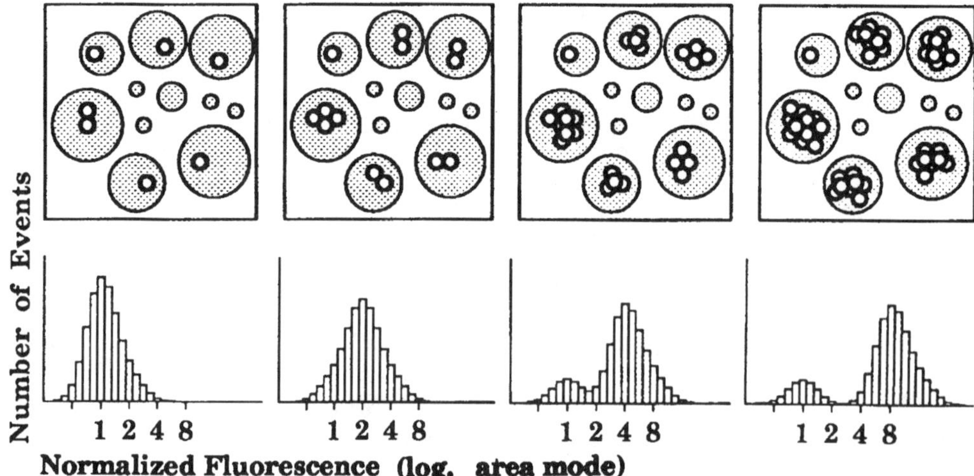

FIGURE 1 Illustration of the basic idea of microbial culture and associated growth measurement using GMDs for individual microcolony formation (8). (Top) Microcolony formation within GMDs, for times corresponding to zero, one, two, and three average cell divisions. The larger, gray circles represent GMDs, and the smaller, light circles represent individual cells. Most GMDs initially contain zero or one individual cells or individual CFU. Only a few of a large number of GMDs are shown, with most shown occupied in order to save space. One GMD (top left) has a two-cell CFU, while the other GMDs contain either zero or one cell. During incubation, viable cells grow to form microcolonies of two or more cells as shown. In experiments to date, the observed plating efficiency was high (generally greater than 90%) for bacteria, yeast cells, and mammalian cells. (Bottom) Clonal growth measurement based on microcolony formation followed by fluorescence staining and measurement by flow cytometry. Shown are expected (computer model) flow cytometric results, which consist of log fluorescence histograms based on measuring a large number of GMDs. At t = 0, the distribution of the biomass signal for one- and two-cell CFU is shown. There is variance in this signal due to cell-to-cell variations in biomass and instrumentation noise. Any stain (e.g., fluorescence-labeled antibodies to surface markers) can be used. In our initial work we emphasized staining of generic biomass, viz. (i) proteins (using fluorescein isothiocyanate), or (ii) double-stranded nucleic acids (using propidium iodide), even though stained individual cells exhibit a fairly broad fluorescence distribution. As the biomass increases in GMDs containing growing cells, the corresponding subpopulation of the log fluorescence distribution shifts to the right. In this case there is a one-unit (mean individual cell signal) increase for each doubling in biomass. Simple measures (e.g., mode or median of the distribution) can be used to quantify changes in microcolony size. DNA staining protocols can also be used but have significantly smaller fluorescence signals.

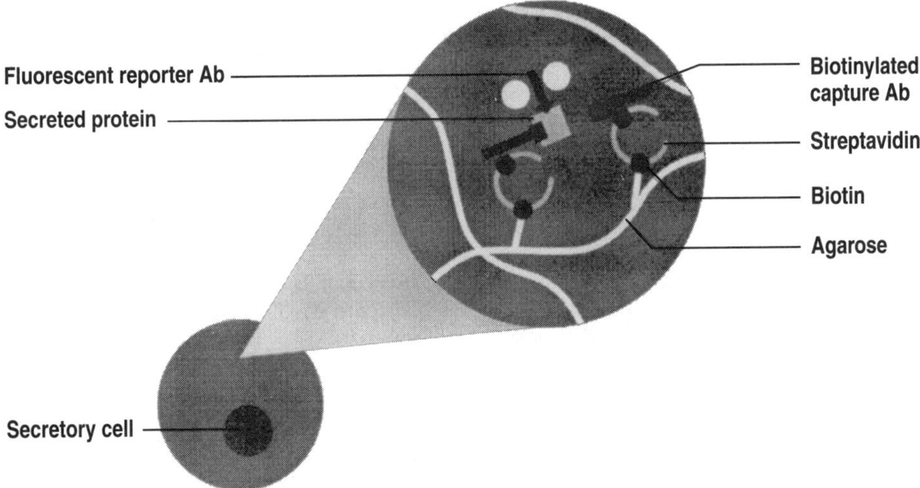

FIGURE 2 Illustration of the secreted antibody (Ab) capture matrix that is constructed within GMDs following cell encapsulation. The agarose has pre-attached biotin (One Cell Systems). After streptavidin is added to the medium in which the GMDs are suspended, it rapidly enters by diffusion and binds to the biotinylated sites. A capture Ab is similarly introduced and bound, resulting in a large number of capture sites distributed throughout the highly permeable agarose matrix. Ab molecules secreted from the individual cell are first captured near the cell, followed by a progressive involvement of sites further from the cell. Eventually some secreted Ab molecules escape from the originating GMD and migrate to other GMDs, particularly unoccupied GMDs. This "molecular cross-talk" has an important benefit: cell-free GMDs have negligible reporter Ab until secreted Ab escape begins, so the appearance of reporter Ab staining signals the beginning of saturation of the originating GMDs. This internal indication provides quality control for the overall assay and isolation process. Rapid physical isolation of the GMDs containing high producer cells can be carried out using fluorescence-activated cell sorting at the rate of 10^6 single cells per hour. After sorting the GMDs, the cells are either grown out of the GMDs or released by enzymatically digesting the GMDs with agarase. The overall result is a high-throughput isolation process based on quantitative assessment of Ab secretion from individual cells.

membrane exclusion stain (e.g., propidium iodide, ethidium bromide) allows microcolonies with inhibited but viable cells to be distinguished from microcolonies with dead cells (the latter have red fluorescence).

9.2.4. Protein Secretion Assays

Although originally developed for antibody secretion assays for mammalian cells, this general assay should be broadly applicable to microbial secretion. Secreted molecules are captured near the originating cell by providing a large number of molecular capture sites within the GMDs. Coentrapment of small beads with surface capture sites can be used (4), or, for less light scatter, GMDs formed from biotinylated agarose can be used to construct capture sites following cell encapsulation (2, 7). Following cell encapsulation, the capture matrix is constructed in situ by diffusing in a steptavidin bridge and then the capture molecule (usually an antibody). After a secretion incubation, a reporter fluorescence-labeled antibody is diffused into the GMDs to generate a fluorescence signal that is proportional to the number of secreted molecules that were captured (Fig. 2). This assay should be directly extensible to quantifying the number of molecules released by other stimuli, such as electroporation (7).

9.2.5. Double GMDs

The clonal growth assay can be used to quantitatively distinguish slowly growing cells from rapidly growing cells, but this is best accomplished if the initial cells are located near the centers of the GMDs. Otherwise, cells near the GMD surface grow out and can inoculate other GMDs, i.e., the rapidly growing cells can contaminate GMDs that contain slowly growing cells. Approximate centering of the inoculum is accomplished by forming a first preparation of relatively small GMDs, which is removed from the nonaqueous fluid. The first GMDs are then themselves encapsulated, forming double GMDs. If, for example, the first GMD preparation has $\bar{n} \leq 0.15$, then most first GMDs are empty ($n = 0$), and the second GMD preparation can consist of several first GMDs per second GMD. A detailed example of this approach has been described for the isolation of slowly growing mutant yeasts from a population of a more rapidly growing wild type (1).

9.3. CLOSED GMDS FOR BIOCHEMICAL ACTIVITY ASSAYS

Microorganisms can be rapidly detected and enumerated by measuring the extracellular accumulation of secreted small molecules within the small volumes of GMDs. If these molecules remain in solution, closed GMDs are appropriate, e.g., the GMDs are used while surrounded by a nonaqueous fluid (10, 11). This allows fluorescence-based measurements, or even simple counting the GMDs with different fluorescence colors, to score individual GMDs according to whether or not they contain biochemical activity above some threshold. Individual acid-producing bacteria and yeast cells can be detected in a few tens of minutes by suspending the microorganisms in a poorly buffered medium (to accentuate the pH change) and then encapsulating them in GMDs surrounded by mineral oil or silicone fluid.

The protocol for biochemical activity assays is as follows. One or more fluorescent dyes or indicators are added to the suspending medium prior to GMD formation. Essentially any enzymatic or indicator-based assay that uses changes in fluorescence can be used. In the case of activity determinations based on extracellular pH shifts, a pair of fluorescent dyes can be used: one responds to pH changes, the other provides a reference. For example, 300 μM sulforhodamine 640 (Exciton; red fluorescence) and 25 μM fluorescein isothiocyanate-dextran (Sigma; green fluorescence) provides a fluorescence indicator system based on one pH-sensitive fluorescent dye and one pH-insensitive (reference) fluorescent dye (12). Individual fluorescence indicators that change color have become available and are also suitable.

I am indebted to the many coworkers who have contributed to GMD-based assays and isolation procedures. The protocol for C. albicans *encapsulation is due to Anthony A. Ferrante of One Cell Systems.*

REFERENCES

1. **Gift, E. A., H. Park, G. A. Pardis, A. L. Demain, and J. C. Weaver.** 1996. FACS-based isolation of slow-growing microorganisms: a model system involving double encapsulation of yeast in gel microdrops. *Nature Biotechnol.* **14:**884–887.
2. **Gray, F. J. S. Kenney, and J. F. Dunne.** 1995. Secretion capture and report web: use of affinity derived agarose microdroplets for the selection of hybridoma cells. *J. Immunol. Methods* **182:**155–163.
3. **Nir, R., R. Lamed, L. Gueta, and E. Sahar.** 1990. Single-cell entrapment and microcolony development within uniform microspheres amenable to flow cytometry. *Appl. Environ. Microbiol.* **56:**2870–2875.
4. **Powell, K. T., and J. C. Weaver.** 1990. Gel microdroplets and flow cytometry: rapid determination of antibody secretion by individual cells within a cell population. *Bio/Technology* **8:**333–337.
5. **Ryan, C., B.-T. Nguyen, and S. J. Sullivan.** 1995. A rapid assay for mycobacterial growth and antibiotic sensitivity using gel microdrop encapsulation. *J. Clin. Microbiol.* **33:**1720–1726.
6. **Shapiro, H. M.** 1994. *Practice Flow Cytometry*, 3rd ed. Alan R. Liss, New York.
7. **Weaver, J. C.** 1995. Electroporation theory: concepts and mechanisms. *Methods Mol. Biol.* **47:**1–26.
8. **Weaver, J. C., J. G. Bliss, G. I. Harrison, K. T. Powell, and G. B. Williams.** 1991. Microdrop technology: a general method for separating cells by function and composition. *Methods* **2:**234–247.
9. **Weaver, J. C., P. McGrath, and S. Adams.** 1997. Gel microdrop technology for rapid isolation of rare and high producer cells. *Nat. Med.* **3:**583–585.
10. **Weaver, J. C., P. E. Seissler, S. A. Threefoot, J. W. Lorenz, T. Huie, R. Rodrigues, and A. M. Klibanov.** 1984. Microbiological measurements by immobilization of cells with small volume elements. *Ann. N.Y. Acad. Sci.* **434:**363–372.
11. **Weaver, J. C., G. B. Williams, A. M. Klibanov, and A. L. Demain.** 1988. Gel microdroplets: rapid detection and enumeration of individual microorganisms by their metabolic activity. *Bio/Technology* **6:**1084–1089.
12. **Williams, G. B., J. C. Weaver, and A. L. Demain.** 1990. Rapid microbial detection and enumeration using gel microdroplets: colorimetric versus fluorescent indicator systems. *J. Clin. Microbiol.* **28:**1002–1008.

Cultivation of Hyperthermophilic and Extremely Thermoacidophilic Microorganisms

KRISTINA D. RINKER, CHAE J. HAN, MICHAEL W. W. ADAMS, AND ROBERT M. KELLY

10

10.1. GENERAL CONSIDERATIONS

Over the past decade, there has been considerable interest in microorganisms growing at extremes of temperature, pressure, pH, and ionic strength (Table 1). The unusual conditions under which extremophiles grow have necessitated the development of cultivation methodologies that, while based on conventional approaches to growing microorganisms, require modifications related to the organism's growth environment. As a result, many interesting schemes for cultivating extremophiles have been reported. These have been the basis for both the discovery of new biocatalysts and investigations into physiological and metabolic issues associated with these microorganisms (1, 2, 4).

To date, most efforts on extremophilic microorganisms have centered on growth at high temperatures, for reasons stemming from the interest in thermostable biocatalysts (2, 4), as well as the proposed placement of thermophilic organisms in evolutionary biology (139). While there has been some confusion concerning nomenclature for this group of organisms, for the purposes of this discussion, extreme thermophiles are considered to be those growing optimally at 75°C and above; hyperthermophiles, a subset of extreme thermophiles, have maximal growth at or above 90°C (Table 2). Extreme thermophiles and, for that matter, many extremophilic microorganisms are predominantly in the domain of Archaea (139); two exceptions among the extreme thermophiles are the bacterial genera *Aquifex* (29, 60) and *Thermotoga* (14, 59), both of which are also hyperthermophilic. Although it is clear that the constituent biomolecules of extreme thermophiles are designed for function at high temperatures, it remains to be seen how the physiological and metabolic characteristics of these organisms are related to the growth environment. This, of course, becomes important in developing strategies for isolation and subsequent cultivation of specific microorganisms.

10.2. CULTIVATION STRATEGIES AND CHALLENGES

Planning and executing the cultivation of extremely thermophilic microorganisms under laboratory conditions brings into account a number of issues. These include the location from which the organism was isolated, media development, methodology, and equipment necessary to culture particular organisms on a specified scale. While many approaches for less thermophilic organisms apply, there are complications that arise because of the extremely high growth temperatures, e.g., the stability of media components, evaporation of fluid from high-temperature baths, logistics for culture maintenance and storage, lack of biological equipment designed for high temperatures, and measurement of cell growth. For the isolation of specific gene products, recombinant DNA technology has provided a means of circumventing the need to cultivate extreme thermophiles, i.e., by the direct cloning and expression of genes derived from DNA present in environmental samples (107). However, this approach is not suitable for gene products difficult to express in an active form in mesophilic systems or for which screening assays are not available. Hence, any comprehensive evaluation of the scientific or technological potential of an extremely thermophilic microorganism necessitates its cultivation under laboratory conditions, in many cases on a scale that provides sufficient cellular material for subsequent evaluation of a particular biomolecule.

10.2.1. Availability and Handling of Extremely Thermophilic Microorganisms

A wide variety of geothermal sites have been examined for the presence of culturable extreme thermophiles. These range from continental hot springs, such as those found in Yellowstone National Park (22), to deep sea hydrothermal vent systems, such as those associated with midocean ridges (12, 119). Initially, the general availability of these organisms was limited, forcing those interested in their study to visit geothermal sites and isolate their own. Fortunately, an ever-growing number of these microorganisms are now available from culture collections. In particular, DSMZ (Braunschweig, Germany) has an extensive stock of extreme thermophiles, including many hyperthermophiles, available for a modest fee with accompanying recommendations for growth media.

Methodology for isolation and handling of extreme thermophiles has been developed to some extent and is based on techniques used previously for less thermophilic microorganisms (12, 106). Storage and handling techniques can be straightforward. Some anaerobic hyperthermophiles, for example, can be maintained in liquid growth

TABLE 2 Extremely thermophilic genera (T_{opt} 75°C)[a]

Order/genus	T_{max}[b] (°C)	Metabolism[c]	Substrates[d]	Acceptors[e]
Archaea				
Thermoproteales				
Thermoproteus	92	Het (auto)	PEP, CBH (H_2)	S°
Thermofilum	100	Het	PEP	S°
Staphylothermus	98	Het	PEP	S°
Desulfurococcus	90	Het	PEP	S°
Pyrolobus	113	Het	PEP	?
Pyrobaculum	102	Het (auto)	PEP (H_2)	–, S°, mO_2
Sulfolobales				
Acidianus	96	Auto	S°, H_2	S°, O_2
Desulfurolobus	87	Auto	S°, H_2	O_2
Metallosphaera	80	Auto (het)	S°, PEP	O_2
Sulfolobus	90	Auto (het)	S °, PEP, CBH	O_2
Pyrodictiales				
Pyrodictium	110	Het (auto)	PEP, CBH (H_2)	S°
Thermodiscus	98	Het	PEP	S°
Thermococcales				
Pyrococcus	105	Ferm	PEP, CBH	–, S°
Thermococcus	97	Ferm	PEP, CBH	–, S°
Hyperthermus	110	Ferm	PEP (H_2)	–, S°
Archaeoglobales				
Archaeoglobus	95	Het (auto)	CBH (H_2)	SO_4, S_2O_3
Methanococcales				
Methanococcus	91	Auto	H_2	CO_2
Methanobacteriales				
Methanothermus	97	Auto	H_2	CO_2
Methanopyrales				
Methanopyrus	110	Auto	H_2	CO_2
Bacteria				
Thermotogales				
Thermotoga	90	Ferm	PEP, CBH	–, S°
Aquificales				
Aquifex	95	Auto	Sß (H_2)	mO_2, NO_3

[a]Modified from Kelly and Adams (64).
[b]T_{max}, maximum growth temperature within genus.
[c]Het, heterotrophic; auto, autotrophic; ferm, fermentative.
[d]PEP, peptides; CBH, carbohydrates.
[e]–, growth in the absence of an added electron acceptor; mO_2, microaerophilic.

media at room temperature for extended periods, although storage at 4°C limits the possibility of mesophilic contaminants arising from the handling procedures. While oxygen is usually toxic to anaerobic hyperthermophiles if exposed at or near growth temperatures (126), there are cases of such isolates being recovered from oxygenated, refrigerated samples stored for 5 years (52). Lyophilization has been used successfully for some hyperthermophilic archaea, although there are also reports of limited storage times for organisms stored in this way (39). *Pyrococcus furiosus*, a hyperthermophile that grows optimally at 100°C (44), can be stored in dimethyl sulfoxide for extended periods in glass capillary tubes over liquid nitrogen. Plastic cryotubes have been found to be too permeable to oxygen and, hence, ineffective for extended storage (32, 39).

Plating techniques have been reported for *Thermotoga* species (35, 51), *Aquifex pyrophilus* (60), *P. furiosus* (103), and *Sulfolobus* strains (49), presenting the possibility of genetic manipulation of these organisms. Methodology to carry out mutagenesis experiments for hyperthermophilic pyrococcii has been developed and used to produce a uracil auxotrophic mutant of *Pyrococcus abyssi* (136). It should be mentioned that, in many cases, extreme thermophiles can be very difficult to develop into a pure, laboratory culture, whether from isolation efforts or even from storage. This is especially true for some of the hyperthermophilic, sulfur-reducing chemolithotrophs, such as *Pyrodictium occultum* (93), which grow to very low cell densities and are highly sensitive both to shearing forces and to various metals (63a). The pioneering contributions of Karl Stetter (University of Regensburg, Germany), Holger Jannasch (Woods Hole Marine Biology Laboratory), Thomas Brock (University of Wisconsin), and others who have isolated a number of extreme thermophiles and developed media and protocols for their laboratory cultivation are best appreciated by those who endeavor to grow and study these often fastidious microorganisms.

10.2.2. Media Development

Considerable effort has been expended in the development of growth media for the isolation of extreme thermophiles, and subsequently for their cultivation in laboratory settings

TABLE 1 Biodiversity of extreme environments

Extreme condition	Archaea	Bacteria	Habitat	Extreme growth conditions	Metabolic characteristics
High temperature		*Thermatoga neapolitana*	Geothermal marine sediments	100°C	Anaerobic heterotroph; facultative S^0 reducer
	Pyrococcus furiosus		Geothermal marine sediments	85°C	Anaerobic heterotroph; facultative sulfur reducer
Low temperature		*Bacillus* TA41	Antarctic sea water	4°C	Aerobic heterotroph
High pressure + high temperature	*Methanococcus jannaschii*		Deep sea hydrothermal vent	250 atm 85°C	Growth/methane production stimulated by pressure
High pH		*Clostridium paradoxum*	Sewage sludge	pH 10.1 56°C	Anaerobic heterotroph
Low pH + high temperature	*Metallosphaera sedula*		Thermal, acid pools	pH 2.0 75°C	Facultative chemolithotroph
High salt	*Halobacterium halobium*		Hypersaline waters	4–5 M NaCl	Aerobic heterotroph

(12, 22, 106). For aerobic thermoacidophiles belonging to the order *Sulfolobales*, media containing complex carbon sources, such as yeast extract and tryptone, and chemolithotrophic substrates, such as elemental sulfur, ferrous iron, or iron pyrite, have been used in the isolation efforts (21, 23). Although there has been much debate concerning the direct or indirect dissolution of iron pyrite by thermoacidophiles (34, 133, 138), it is clear that these cells attach to such solid substrates (138). The importance of organic as compared to inorganic carbon and energy sources for these organisms is often unclear, with some isolates appearing to grow as mixotrophs (96, 97).

There are variations within the extreme thermoacidophiles in terms of nutritional requirements and growth physiology. Although most are aerobes, members of the genus *Acidianus* are facultative anaerobes that grow anaerobically by reducing elemental sulfur with H_2 (118). Several members of the *Sulfolobales* are facultative or obligate heterotrophs that grow on sugars, yeast extract, and peptone (48, 49, 124, 125) and are known to have a modified (nonphosphorylated) Entner-Doudoroff pathway (41). However, species of *Metallosphaera* (*M. sedula* and *M. prunae*) do not grow on sugars (46, 55), and *M. prunae* does not utilize ferrous sulfate, which is somewhat unusual among the extreme thermoacidophiles (46). Members of the *Sulfolobus*, *Acidianus*, and *Metallosphaera* genera also grow chemolithoautotrophically by aerobic oxidation of S^0 (and S^{2-}) and H_2 (49, 55, 118). In particular, *Sulfolobus metallicus* grows strictly by chemolithoautotrophy; no growth occurs on beef extract, casamino acids, peptone, tryptone, yeast extract, arabinose, fructose, or other sugars (56). *Desulfurolobus ambivalens*, an obligate chemolithoautotroph, has been found to contain several plasmids (143). In our laboratory, many of the extreme thermoacidophiles (*Acidianus brierleyi*, *Acidianus infernus*, *Sulfolobus acidocaldarius*, *Sulfolobus solfataricus*, *Sulfolobus shibatae*, *M. sedula*) grow on the same medium, with some minor differences: *Acidianus* strains and *S. acidocaldarius* grow better when trace elements are added. We have found the nutritional requirements of *M. sedula* (see section 10.4) to be the simplest, which may explain why it is one of the most effective leaching organisms (37).

As indicated in Table 2, virtually all of the currently known hyperthermophiles are obligate anaerobes, including methanogens, sulfur-respiring chemolithotrophs, and both sulfur-respiring and fermentative heterotrophs. In contrast to their mesophilic cousins, only hydrogenotrophic hyperthermophilic methanogens have been identified with no evidence for acetate utilization within this group. Sulfur-reducing chemolithotrophs, such as *Pyrodictium brockii*, grow by the reduction of sulfur with molecular hydrogen. Small amounts of complex substrates, e.g., yeast extract, may be required or stimulate growth of these organisms (98, 121, 127). For unknown reasons, efforts to eliminate organic carbon sources entirely from these chemolithotrophs often result in very poor growth (114a). Electron acceptors other than sulfur, such as sulfate (63, 128) and nitrate (134), can be used by some hyperthermophiles; however, such physiology has been encountered to only a limited extent.

By far, most attention among the extreme thermophiles has been directed toward hyperthermophilic, fermentative anaerobes. These organisms have been isolated from various marine hydrothermal sources using peptides and/or sugars (oligosaccharides) as carbon and energy sources. While there have been several successful efforts to develop defined media for members of this group (15, 52, 53, 62, 69, 75, 99, 103, 104, 135), the fact that most also require sulfur or polysulfides (16) to grow complicates interpretation of bioenergetic patterns.

Of the fermentative hyperthermophiles, the archaea *P. furiosus* (44) and *Thermococcus litoralis* (90) and the bacterium *Thermotoga maritima* (59) are the best studied, probably because they were among the first in this group to be isolated and they grow well in the absence of reducible sulfur compounds (64, 68, 117). *T. maritima* grows with glucose as a carbon and energy source, apparently carrying out glycolysis by a conventional Embden-Meyerhof pathway (117). *P. furiosus* (103) and *T. litoralis* (104) grow on sulfur-free defined media containing a number of single amino acids. However, *P. furiosus* requires an oligosaccharide in defined medium, while *T. litoralis* grows without the addition of any sugars. In contrast to *T. maritima*, *P. furiosus* apparently cannot utilize a monosaccharide as the carbon and energy source. Disaccharides such as maltose enhance growth (27) and appear necessary in the absence of reducible sulfur compounds (17, 115). *T. litoralis* possesses a high-affinity maltose transport system (140), and *P. furiosus* shows a regulatory response induced by maltose (108). Both organisms contain several glycosyl hydrolyses capable of breaking down starch to glucose (25, 27). The inability of these organisms to grow on simple sugars, such as glucose, is interesting in light of reports that monosaccharides are taken up by resting cells of *T. litoralis* and *P. furiosus* (132).

Media development must also take into account some additional features of hyperthermophilic growth physiology. For example, the growth of *P. furiosus* and several hyperthermophiles examined to date is stimulated by tungsten, an unusual requirement for biological systems (70). *P. furiosus* increases in cell density threefold in continuous culture by low concentrations of tungsten (116). Hyperthermophilic heterotrophs typically accumulate acetate as the predominant liquid-phase product (126), thereby reducing pH and, hence, growth yields. The adverse effects of acetate buildup may be partly due to uncoupling reactions, in addition to the resulting drop in pH, as has been seen in *Clostridium thermoaceticum* (11). High levels of acetate inhibit growth owing to the drop in the internal pH of the cell brought about by the influx of acid. However, this may not be a significant problem in all cases, since *P. furiosus* grows well in filtered spent media, suggesting a lack of product inhibition (15). Recent work has also revealed the presence of compatible solutes in hyperthermophiles, whose formation may be induced by environmental conditions (79).

In addition to acetate, alanine is a significant by-product of heterotrophic hyperthermophiles. This is especially true if molecular hydrogen is allowed to accumulate in the medium, i.e., if it is not removed by sparging with an inert gas. Significant alamine excretion in the presence of ammonia has been found for *Thermococcus profundus* when grown on pyruvate- or maltose-based media (71, 72). In *P. furiosus*, alanine production increases with the NH_4Cl concentration and decreases in the presence of sulfur or in coculture with *Methanococcus jannaschii*, a hyperthermophilic methanogen (67). Propionate, isobutyrate, isovalerate, and butyrate have also been detected by gas chromatography at low levels in *P. furiosus* cultures (115). The presence of a methanogen stimulates organic acid production in deep sea heterotrophic isolates growing at 85°C (31).

Central metabolic pathways have been well studied in *P. furiosus*, and the available evidence suggests that the same metabolic routes are also present in *T. litoralis* and related organisms (3). So far, these studies have revealed some surprising results. In contrast to the conventional glycolytic pathway in the bacterium *T. maritima*, the heterotrophic archaea metabolize sugars via a modified Embden-Meyerhof pathway. This route contains ADP-rather than ATP-dependent hexose kinases (66), and the novel enzyme glyceraldyde-3-phosphate ferredoxin oxidoreductase (GAPOR) replaces the expected NAD-dependent dehydrogenase (87). Interestingly, GAPOR contains at its catalytic site the metal tungsten, an element rarely used in biological systems, and this, at least in part, explains why the addition of tungsten stimulates growth (70). The main product of carbohydrate catabolism is acetate. This is generated by acyl coenzyme A synthetase (ACS), which converts acetyl coenzyme A, ADP, and phosphate into ATP and acetate in a single step. ACS is yet another unusual enzyme, which has so far been found only in archaea (112).

The heterotrophic archaea also contain unusual pathways for peptide metabolism. The first step involves various transaminases that convert peptide-derived amino acids to their corresponding 2-ketoacid derivatives. These are oxidized to the corresponding coenzyme A ester by four different types of 2-ketoacid oxidoreductase, which are specific for pyruvate, 2-ketoglutarate, branched chain, or aromatic 2-ketoacids (3). Interestingly, no known microorganism contains more than two enzymes of this type, with the notable exception of species of *Pyrococcus* and *Thermococcus*. The various coenzyme A esters that these enzymes generate are used to conserve energy in the form of ATP by two different isoenzymes of ACS that differ in their substrate specificities (77). Between them, they produce various branched chain and aromatic acids, as well as acetate, and these, like acetate, are excreted into the medium. This explains the presence of organic acids, such as isovalerate, isobutyrate, and phenyl acetate, in spent media after growth of various *Pyrococcus* and *Thermococcus* species.

As mentioned above, a unique aspect of metabolism by these hyperthermophilic archaea is their requirement for tungsten. Virtually all other life forms utilize the analogous element, molybdenum, with only a few microorganisms additionally able to use tungsten (70). Yet, *Pyrococcus* and *Thermococcus* species are obligatorily dependent upon tungsten and cannot utilize molybdenum (88). They uniquely contain three different types of tungstoenzyme: GAPOR (involved in glycolysis), aldehyde oxidoreductase (AOR) (85), and formaldehyde oxidoreductase (FOR) (86). The latter two enzymes are involved in oxidizing various aldehydes, which are generated as side-products by the four 2-ketoacid oxidoreductases, to the corresponding acid. Hence, two tungstoenzymes, AOR and FOR, produce acetate and various organic acids in spent growth media, in addition to those generated by the ACS isoenzymes (3).

10.2.3. Cultivation Equipment and Strategies

The growth of microorganisms at elevated temperatures has both advantages and disadvantages when compared to the cultivation of mesophiles. One potential advantage is the reduced risk of contaminant growth at high temperatures, although care must be taken to maintain media (sometimes containing high levels of yeast extract and oligosaccharides) under sterile conditions prior to utilization. Other favorable aspects of high-temperature cultivation include reduced viscosity of fermentation media and increased rates of gas-liquid mass transfer. The latter advantage may, however, be offset in the growth of aerobic extreme thermoacidophiles by the concomitant reduced solubility of oxygen at elevated temperatures. It well may be that the reduction in dissolved O_2 levels with increasing temperature is consistent with the growth requirements of extreme thermoacidophiles. For example, growth of the thermoacidophilic archaeon *S. acidocaldarius* is inhibited at enhanced dissolved oxygen levels applied by overpressurization with air (129).

For anaerobic hyperthermophiles, only autotrophic species utilize gaseous substrates (e.g., H_2 and CO_2), while heterotrophs in this group typically produce H_2S, H_2, and

CO_2 as by-products (126). Since many hyperthermophiles have relatively high growth rates, gas-liquid or liquid-gas mass transfer may be a factor under certain circumstances. The growth of *P. brockii*, a chemolithotrophic hyperthermophile growing by H_2-S^0 autotrophy, is limited by gas-liquid mass transfer under quiescent conditions (94). This cannot be overcome by increased agitation or high rates of gas sparging, since *P. brockii* is sensitive to even moderate shear rates (98). Other hyperthermophiles, such as members of the genus *Pyrococcus*, are also shear-sensitive (75, 111). The release of inhibitory by-product gases, especially H_2, is thought to be important in the growth of hyperthermophilic, heterotrophic anaerobes (126) and has been dealt with by using high inert gas sparging rates (102) and dialysis reactors (75). Alternatively, since these heterotrophs typically reduce polysulfide to H_2S during growth, the presence of sulfur may serve as a mechanism by which H_2 inhibition is reduced or averted. The consequence, of course, is the formation of large amounts of corrosive and toxic H_2S, creating a new set of problems involving the durability of cultivation equipment and safety of laboratory personnel.

Key considerations in the cultivation of aerobic and anaerobic extreme thermophiles are low cell density and biomass yields (65). While these may be related to, in general, a poor understanding of nutritional requirements, they may be characteristic of this group of microorganisms. There have been reports of cell densities in excess of 10^9 cells/ml being obtained (75, 102, 103, 104a), but this level of growth is not typical and usually comes after an extensive effort at optimizing media, cultivation strategy, and equipment. In fact, even the most experienced hands at cultivating extremely thermophilic organisms have witnessed problems in maintaining cultures over a prolonged period of time, in addition to observing erratic patterns in growth rates and cell density levels within a particular set of experiments.

The earliest efforts to cultivate extreme thermophiles attempted to mimic the characteristics of their peculiar habitats. In some cases, this involved sophisticated high-pressure, high-temperature bioreactors that were used to simulate and explore the effects of deep sea hydrostatic pressure on growth (for example, see references 36, 65, 82, and 100). At pressures of several hundred atmospheres, growth was stimulated for certain organisms; at high pressure, optimal growth temperatures were increased in some cases by several degrees (82). Ceramic-lined fermentors designed for operation with slight overpressures to prevent boiling at temperatures above 100°C were also utilized (58). Such systems were built to withstand the potentially corrosive attack by H_2S, a product of anaerobic growth, and by H_2SO_4, which is produced by thermoacidophiles.

The apparent need for bioreactors capable of high-pressure, high-temperature operation, as well as being resistant to corrosion, initially limited the number of laboratories with the capability to study extreme thermophiles. This was especially true when significant amounts of biomass were needed for protein purification. However, strategies to minimize equipment damage by corrosion have been developed, and high pressure has not been found to be essential for the growth of many high-temperature, deep sea isolates. Now, fermentations of up to 600 liters (working volume) have been performed for various hyperthermophilic heterotrophs isolated from both shallow and deep sea environments, both in the presence and absence of elemental sulfur. Yields of over 1 kg (wet weight) per 600-liter run have been reported, facilitating the purification of a growing list of thermostable proteins (1). Stainless steel fermentors have been used over long periods of time without significant corrosive damage, if proper care was taken with welds during equipment fabrication and if the system was meticulously cleaned after a particular run. Alternatively, biomass can be generated using continuous culture. This approach is also useful for determining growth yields and providing a steady-state environment to investigate the effect of changes in the cellular environment (nutrients, temperature, etc.) on cell physiology (102, 113–116). For example, a continuous culture of the hyperthermophile *P. furiosus* with a 1-liter liquid volume operating at a dilution rate of 0.4 h^{-1} produces almost 10 liters of biomass per day, from which approximately 5 g (wet weight) can be harvested. Under some conditions, *P. furiosus* can be cultivated at dilution rates up to 0.8 h^{-1} (26). In a gas-lift reactor, *P. furiosus* has been found to have an optimal biomass yield of 1.5 g of cells h/liter at a dilution rate of 0.4 h^{-1} (102). Dialysis reactors have also been utilized for biomass production with reported yields of 2.6 g/liter for *P. furiosus* and 114 g/liter for *S. shibatae* (75).

In some cases, increases in cell densities can be achieved by coculturing hyperthermophilic heterotrophs and methanogens (18). For a rapidly growing hyperthermophilic methanogen, *M. jannaschii*, and a hydrogen-inhibited heterotroph, *T. maritima*, very close association of the two species has been noted, with a five-fold enhancement of final cell densities for the heterotroph (89). Since the methanogen is often at a 50- to 100-fold lower concentration than the heterotroph, this may be a convenient route for obtaining cell mass for protein purification purposes. A possible problem is the significant amount of methane formed during cell growth, which could be hazardous.

While improving biomass yield is an ongoing effort with many extreme thermophiles, it is also important to be able to create physiological environments that lead to optimal production of an enzyme or other metabolites. This can only be achieved with some understanding of the connection between growth physiology and the regulation of genes encoding proteins of interest. Unfortunately, this type of cause-and-effect relationship is not yet well understood for many extreme thermophiles. As this information becomes available, better strategies for producing biomolecules of interest in particular extreme thermophiles will evolve.

10.3. GROWTH OF HYPERTHERMOPHILIC, HETEROTROPHIC ANAEROBES: *T. litoralis*

To illustrate cultivation procedures and protocols for hyperthermophilic, heterotrophic anaerobes, *T. litoralis* (13, 90), with an optimal growth temperature of 88°C, will be used as an example. *T. litoralis* was isolated from shallow geothermal waters near Vulcano Island, Italy. It is a coccus with an approximate diameter of 1 μm. Unlike many heterotrophic hyperthermophiles, *T. litoralis* grows well in the presence or absence of elemental sulfur (104). Several proteins have been purified and characterized from *T. litoralis*, including DNA polymerase (73), amylopullulanase (27), ferredoxin (30) and formaldehyde ferredoxin oxidoreductase (86).

10.3.1. Media Formulation

Several different growth media have been used for cultivation *T. litoralis*. These range from complex media, such

TABLE 3 Hyperthermophile media

Compound	Concentration (g/liter)		Adams[c] (*T. litoralis*)
	RDM[a]	ASW[b]	
Initial pH	6.0	6.5	5.5
NaCl	25	23.9	38
$MgCl_2·6H_2O$	1	1.8	2.75
$MgSO_4·7H_2O$			3.5
NH_4Cl			1.2
Na_2SO_4	1	4.0	
$NaHCO_3$		0.2	
$CaCl_2·2H_2O$	0.075	1.5	0.75
KCl	0.35	0.7	0.325
KBr		0.1	
NaBr	0.05		0.05
H_3BO_3	0.02	0.03	0.015
KI	0.02		0.05
$SrCl_2·6H_2O$	0.01	0.025	0.0075
Na_2SeO_4	0.001		
Rezasurin	0.001	0.001	0.0025
Trace elements[d]	10 ml		10 ml
Vitamin solution[e]	10 ml		
K_2HPO_4	0.14		
KH_2PO_4			0.5
Morpholine-ethanesulfonic acid	2		
Citric acid			0.005
$Na_2S·9H_2O$	0.2	0.25	0.25

[a]0.2 g/liter of each of 16 protein amino acids (minus Ala, Asn, Gln, Glu) and 0.01 g/liter adenine and uracil are added.
[b]1 g/liter yeast extract and 5 g/liter tryptone are typically added.
[c]5 g/liter yeast extract, 5 g/liter tryptone, and 1.25 g/liter maltose are typically added (106).
[d]See Table 4.
[e]See Table 5.

as that used in its isolation, to a sulfur-free, defined medium, RDM (104). Tables 3, 4, and 5 summarize the composition of media commonly used. In development of the RDM medium (104), alanine, asparagine, glutamine, and glutamate are the only naturally occurring amino acids that could be eliminated without adverse affects to either the

TABLE 4 Trace element solutions[a]

Component	Concentration (g/liter)	
	RDM	Adams
Nitrilotriacetate	1.5	1
$MnSO_4·H_2O$	0.5	0.5
$FeSO_4·7H_2O$	1.4	
FeCl3		1.1
$NiCl_2·6H_2O$	0.2	0.2
$CoSO_4·7H_2O$	0.362	0.1
$ZnSO_4·7H_2O$	0.1	0.1
$CuSO_4·5H_2O$	0.01	0.01
$Na_2MoO_4·2H_2O$	0.01	0.01
$NiCl_2·6H_2O$		0.2
$Na_2WO_4·2H_2O$	0.003	0.3

[a]Protect from light; store at 4°C.

TABLE 5 Vitamin solution[a]

Compound	Concentration (mg/liter)
Folic acid	2
Pyridoxine-HCl	10
Thiamine-HCl	5
Riboflavin	4
Nicotinic acid	5
Biotin	2
DL-Ca-pantothenate	5
Vitamin B_{12}	0.1
p-Aminobenzoic acid	5
DL-6,8-Thioctic acid	5

[a]Protect from light; store at 4°C.

cell density or growth rate of the organism. Glycine, isoleucine, threonine, and valine have the strongest adverse effect on growth rates and cell densities when omitted from a medium containing the other 19 common amino acids at 0.1 g/liter each, maltose at 2 g/liter, and yeast extract at 0.05 g/liter. However, only glycine increases cell densities when added to a defined medium containing 16 amino acids (minus Ala, Asn, Gln, and Glu) at 0.05 g/liter. The concentration of amino acids significantly affects cell densities. In media containing 5 g/liter maltose, an increase in the concentrations of the 16 amino acids from 0.05 to 0.1 g/liter each results in stimulation of cell densities by more than twofold, while a further increase to 0.2 g/liter increases cell densities threefold. The yeast extract requirement can be eliminated when 0.01 g/liter adenine and uracil are included, with cytosine able to replace adenine.

PROTOCOL

10.3.1.1. Media Containing Sulfur

When sulfur is needed in the medium, one of several forms of sulfur may be chosen. Most simply, elemental sulfur or flowers of sulfur can be added directly to the culture, after sterilizing at 100°C for 24 h. Since it has been established that several hyperthermophiles grow on soluble polysulfides (16, 105), this form can be used as well. Sulfur can also be added as colloidal suspensions (113–115); this is particularly useful if sulfur is being fed continuously to a fed-batch or continuous culture. Since colloidal sulfur preparations require the use of concentrated acid and generate toxic fumes, gloves and safety glasses should be worn, and work should be performed in a chemical fume hood. Briefly, the method for generating colloidal sulfur is as follows:

1. Prepare two solutions: solution A (500 ml) contains 64 g of $Na_2S \cdot 9H_2O$ in H_2O, and solution B (500 ml) contains 36 g of Na_2SO_3 in H_2O.
2. Add 15 ml of solution B to solution A, and add 30 ml of concentrated H_2SO_4 to the remaining solution B.
3. Slowly add 80 to 100 ml of dilute H_2SO_4 (add 15 ml of acid to 100 ml of H_2O) to solution A until the solution remains turbid.
4. Over a period of 0.5 to 3 min, pour solution A into solution B with vigorous stirring, and leave the mixture to stand for 1 h. Bring the volume to 2 liters with water,

and promote the settling of the sulfur by adding 20 g of NaCl.

5. After the suspension stands for at least 4 h, remove the solution above the colloid by aspiration and again add water and NaCl, thereby initiating a washing procedure that is repeated twice more.
6. After the washing, resuspend the colloid in 1 liter of H_2O. This suspension is stable for several months. Before use, dilute the colloid 10 to 20 times in H_2O to a turbidity of A_{850} = 1.25, which decreases its storage time to 1 month. This absorbance corresponds to approximately 85 mg of elemental sulfur per liter (113).

10.3.2. Culture Preservation

The most common storage procedure is to leave liquid cultures at room temperature or at 4°C, where they may be viable for several months or longer. Many hyperthermophiles have been found to tolerate oxygen better when maintained well below growth temperatures. For longer storage, lyophilization can be used; however, the equipment for doing this is not available to all laboratories. An alternative method is to grow 10 ml of the culture in a Hungate tube and add sterile, anaerobic glycerol to 10% (vol/vol). The cultures can then be stored at −20 or −70°C for years. This procedure has worked successfully in our laboratory for *T. litoralis*, *P. furiosus*, and *Thermotoga neapolitana*.

10.3.3. Batch Cultivation in Serum Bottles

For media development, culture maintenance, culture storage, and scale-up, *T. litoralis* (and other heterotrophic hyperthermophiles) can be grown in 125-ml serum bottles, with 50 to 100 ml of liquid volume. The procedure used has been adapted from the Hungate technique (61, 83) for cultivation of anaerobic microorganisms. Although ovens can be used to maintain culture temperatures, an alternative is a static or shaking bath filled with a heat transfer medium, such as silicon oil. This provides easy access to culture bottles for sampling at intermediate times, without affecting temperature control. The recommended procedure is as follows.

1. Prepare medium and sterilize by filtration or autoclaving. Some components that cannot be autoclaved, such as saccharides and vitamins, must be filter sterilized and added separately to the sterile medium.
2. Autoclave serum bottles and butyl rubber stoppers (Fisher Scientific, Pittsburgh, Pa.).
3. Prepare a reducing agent, such as $Na_2S \cdot 9H_2O$, filter sterilize, and store under N_2. This solution should remain stable as long as it is kept under an O_2-free atmosphere.
4. Add appropriate amount of medium to the serum bottle using a sterile technique. Add sulfur at this point if required.
5. Incubate the bottle at 98°C for 30 min.
6. Remove the bottle from the incubator and immediately sparge with high-purity N_2 to remove oxygen. If the stopper is slotted, it will rest on the top of the bottle with the pipet in the slot. If a plug stopper is used, place it to the side for the sparging procedure. Masterflex tubing size 15 (Cole-Parmer, Vernon Hills, Ill.) is used to connect the cylinder regulator to a sterile glass pipet that is inserted into the bottle.
7. Add $Na_2S \cdot 9H_2O$ and continue to sparge until the resazurin turns clear. The inoculum can be added at this point if desired, or added through a syringe at a later time. Place a rubber stopper on the bottle. Quickly remove the pipet, affix the stopper, and secure an aluminum seal using a hand crimper (Fisher Scientific). Seals can be removed with a hand decapper (Fisher Scientific).

It is strongly recommended to use butyl rubber stoppers owing to the oxygen permeability of other types of rubber. Sterile glass syringes, preflushed with anaerobic liquid, are commonly used for inoculation. The amount of reducing agent necessary is dependent on the medium formulation and the organism to be cultivated. Typically, 500 μl of 50 g/liter $Na_2S \cdot 9H_2O$ is added to 50 ml of the medium; however, 100 μl of 50 g/liter $Na_2S \cdot 9H_2O$ is added to the medium RDM, used for cultivation of *T. litoralis*. In some media, addition of reducing agent may cause significant precipitation of the medium components. If precipitation occurs, reduce the amount of reducing agent added.

10.3.3.1. Cell Enumeration by Epifluorescence Microscopy

Epifluorescence microscopy is typically used to quantify cell growth, especially in cultures containing precipitates (54, 142). The following procedure is recommended:

1. Fix a cell sample (1 ml) in 100 μl of 2.5% glutaraldehyde for at least 5 min.
2. Add an appropriate volume of sample or dilution to 200 μl of 1 g/liter acridine orange and sterile water to make 5 ml.
3. Soak a black 0.2-μm, 25-mm polycarbonate filter (Poretics, Livermore, Calif.) in water for 2 to 5 min. Place the filter on a pad on a 15-ml vacuum tower (Fisher Scientific) and vacuum filter well-mixed samples.
4. Remove the filter, place it on a glass slide, cover it with a drop of Type A immersion oil (Cargille Laboratories, Inc., Cedar Grove, N.J.), a no. 1 coverslip, and another drop of oil. This allows viewing under a microscope adapted for epifluorescence (HBO 100W, 440-490 nm excitation) with a 100× oil-immersion objective.
5. Count the cells in the 10 by 10 grid composed of 1 × 1-μm squares in the eyepiece of the microscope. Count 10 randomly chosen grids with approximately 30 cells per grid, which gives a cell density (cells/ml) of [(average number of cells/grid) × (dilution factor) × (1.24E4)]/volume of cells, with an approximate standard deviation of 10% for a 95% confidence interval (104).

10.3.4. Continuous Cultivation

An advantage of *T. litoralis* cultivation, and that of most extremophiles, is the lowered risk of contaminant growth at elevated temperatures. This allows utilization of continuous and semicontinuous methods of cultivation over extended periods of time (123). Details of this approach were first discussed for *P. furiosus* growth at 98°C (26), and methodology has been extended in recent years (75, 102). The design of a continuous culture is somewhat dependent on the organism, the type of samples needed for analysis, sterility requirements, and the level of operating complexity desired. One particular system that we have used for culturing hyperthermophiles continuously is shown in Fig. 1. The procedure for start-up of continuous culture is as follows.

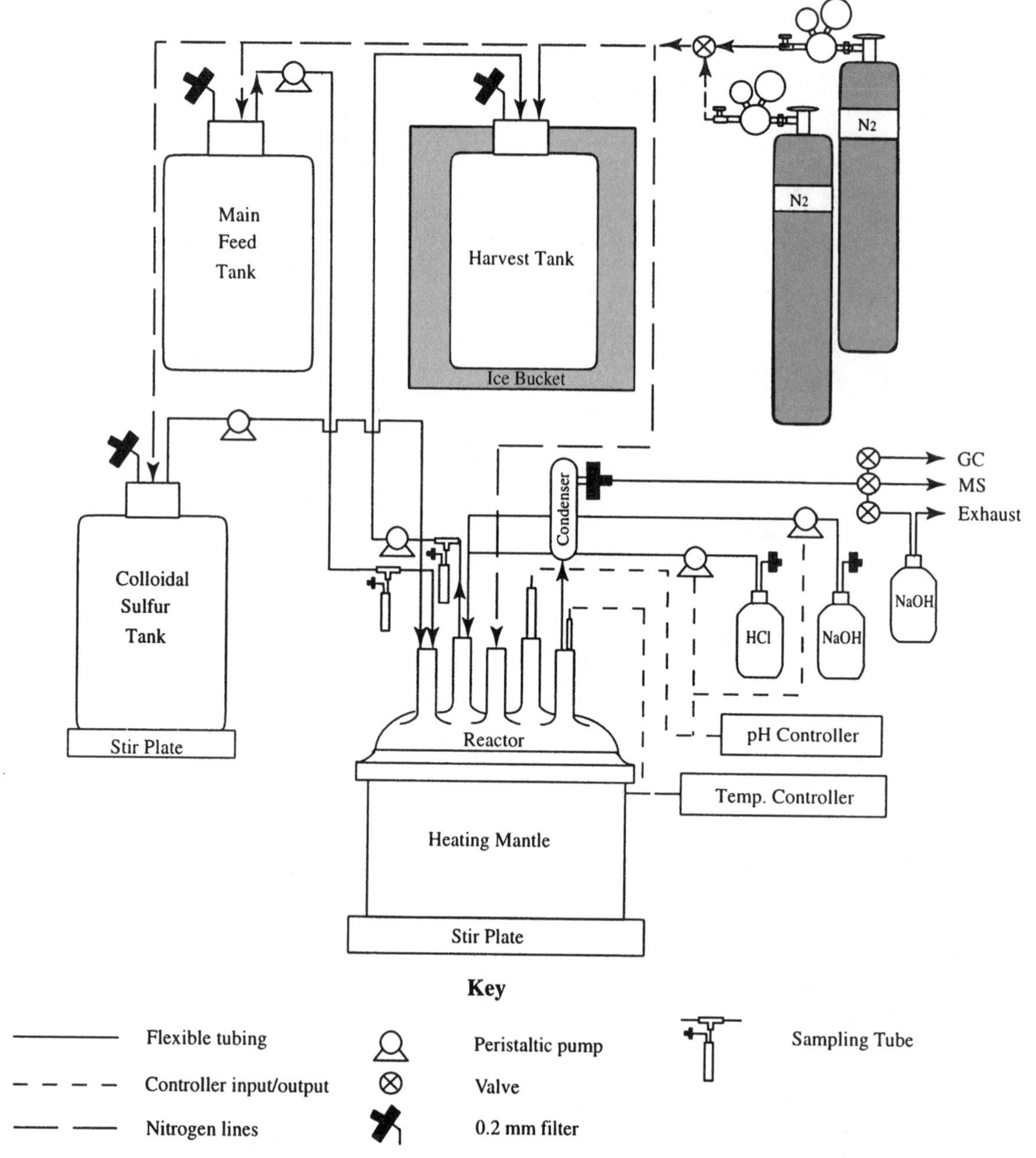

FIGURE 1 Chemostate for continuous cultivation of anaerobic extreme thermophiles. GC, gas chromatograph; MS, mass spectrometer.

PROTOCOL

1. Heat 1 liter of medium to 98°C in a 2-liter (5-neck) round bottom flask (Ace Glass, Vineland, N.J.) using a J thermocouple (Cole-Parmer), a Glas-Col heating omantle with splash guard (Glas-Col, Terre Haute, Ind.), and a temperature controller (Cole-Parmer).
2. Grow up a batch culture in a serum bottle to late exponential phase for use as the seed culture.
3. A stir plate and an egg-shaped stir-bar or a stirrer provide adequate mixing. Sparge the medium with high-purity N_2 through a gas dispersion tube. Add reducing agent and adjust temperature to the growth temperature of the organism. A Graham condenser (Ace Glass) is connected to the reactor to limit water loss. A gas manifold can be attached to the exit of the condenser to send gas to a mass spectrometer, gas chromatograph, or the exhaust. If sulfur is included in the medium, a solution of NaOH should be used to scrub out the H_2S from the exhaust.
4. Inoculate the continuous culture with the seed culture, grow the cells to late log phase, and start the feed and product pumps (peristaltic) to begin continuous operation. Liquid and gas lines are made of norprene Masterflex tubing, chosen because of its low oxygen permeability, autoclavability, and durability. Masterflex tubing in the pump head decreases in resilience with

time. This will alter flow rates, necessitating an in-line flowmeter for flow rate measurement and adjustment during long operation (123). Polypropylene luer-lock fittings are used for connectors. The product tank should have quick-connects to allow for rapid exchanging of tanks. Media for feed (10 liters) should be autoclaved for at least 105 min (74), if filter sterilization is not included in the continuous culture setup. The medium should be sparged and reduced before use. The feed and product tanks are continuously sparged with N_2. The product tank should also be kept on ice.

5. Control the pH by using a pH controller and autoclavable electrode. Teflon adapters with Viton "O" rings (Cole-Parmer) can be used for pH, temperature, and tubing adapters.
6. After the product is collected, filter the culture broth if elemental sulfur is used, centrifuge to harvest the cells at 10,000 × *g* for 30 min at 4°C, and wash the cells with sterile medium or in pH 7.0 buffer containing 10 mM phosphate and 100 mM NaCl. If processing samples for anaerobic assays (oxygen sensitive), more stringent anaerobic protocols are required (116).

10.3.5. Large-Scale Cultivation

T. litoralis has been successfully grown in conventional stainless steel fermentors up to the 600-liter scale (86). The procedure is based on scale-up from the batch cultivation method described above with the following modifications. The organism is grown at 85°C under all conditions.

PROTOCOL

1. The basic medium is summarized in Table 3. It is typically supplemented with maltose (1.25 g/liter), yeast extract (5 g/liter), and tryptone (5 g/liter), and the pH is adjusted to 5.5 prior to autoclaving. For the 100-ml to 16-liter cultures, add elemental sulfur (7 g/liter), presterilized by incubation at 100°C for 24 h, after autoclaving. After degassing and flushing with Ar, make the culture medium anaerobic by adding a titanium chloride/nitriloacetic acid mixture (1 ml/liter of medium), which contains $TiCl_3$ (10 g/liter) and nitriloacetic acid (32 g/liter) adjusted to pH 7.5.
2. Successively transfer cultures from the 100-ml to the 1-liter to the 16-liter scale by using 10% (vol/vol) inocula. The 16-liter cultures are grown in 20-liter glass carboys (#06-406J, Fisher Scientific) sealed with butyl rubber stoppers (#06-447H, Fisher Scientific). Two 16-liter cultures are used to inoculate a large-scale fermentor, such as the 600-liter (500-liter working volume) fermentor from W. B. Moore (Easton, Pa.). Stir the fermentor medium, lacking elemental sulfur, at 75 rpm and continuously sparge with Ar (5 liter/min).
3. Routinely monitor cell growth by optical density at 600 nm. After approximately 16 h, when the A_{600} reaches ~1.0 (late exponential phase), chill the fermentor to 20°C (this takes ~30 min) and harvest the cells using a continuous centrifuge (#AS-16 equipped with stainless steel cooling coils, Sharples-Strakes, Warminster, Pa.) at 4°C with a flow rate of 2 liter/min.
4. Quickly transfer the cell paste to a metal pan and rapidly freeze the cells by immersion in liquid N_2. Approximately 500 g (frozen net weight) is obtained from 500 liters of medium. Store the frozen cells at −80°C until needed.

10.3.6. Product Formation

The primary gaseous products formed by *T. litoralis* are CO_2 and H_2, along with H_2S formed from H_2 when sulfur is included in the medium. Acetate is the primary liquid-phase product, with smaller amounts of propionate, isobutyrate, and isovalerate also produced (17). Alanine production has been noted in other thermococci (72) and *P. furiosus* (67) and presumably is formed by *T. litoralis*. Growth inhibition by H_2 has been reported (17), although this may not be true in all cases (104). The inhibitory effects of H_2 can be reduced or eliminated by the presence of S^0 or a methanogen (17, 18, 31, 67, 78, 115) leading to the production of H_2S or methane, respectively.

10.3.7. Exopolysaccharide Production and Biofilm Formation

Exocellular polymeric substances (EPSs) (47) can be simple chains containing one sugar or complex polymers containing several different sugars linked together with acetate, pyruvate, formate, sulfate, phosphate, and/or other ester or N-linked side groups (137). Several different functions have been proposed for EPSs, including a role in cell protection, cell-to-cell recognition, excess reducing equivalent elimination, and biofilm formation (137). EPSs have been investigated for hydrothermal vent bacteria (50, 101), halobacteria (7), methanogens (122), autotrophic acidophiles (91), and heterotrophic hyperthermophiles (104). A common feature is the presence of sulfate groups in the exopolymer, which is a characteristic found in all domains of life. The EPS from the mesophilic archaeon *Haloferax mediterranei* contains 6 to 8% sulfate (7); that from *S. solfataricus*, 5 to 12% (91); and that from *T. litoralis*, 1 to 2% (104). A more uncommon feature is the predominance of mannose in the EPS from *T. litoralis* (104), a feature usually confined to the Eukarya (91).

Many techniques have been utilized for EPS extraction, including high-speed centrifugation, steaming, sodium hydroxide treatment, ultrasonication, and precipitation with ethanol or acetone (24, 109, 110). The EPS from *T. litoralis* can be precipitated from the culture supernatant with 2 vol of 95% ethanol at 4°C (104). After three successive H_2O washes and ethanol precipitations, the material can be dried in a Speed-Vac, lyophilizer, or vacuum oven. The stringency of acid hydrolysis of the polymer depends on its constituents. Typically, 2 to 3 N HCl is added to 2 to 5 mg of the polysaccharide, which is then incubated at 100°C for 2 to 5 h (95). A more suitable hydrolyzing agent is hydrofluoric acid, which produces fewer by-products but is more hazardous (141).

Characterization of EPSs ranges from simple, qualitative observations to in-depth analysis. Physical assessment includes gelling behavior, precipitation with certain salts, melting temperature, and color. The molecular weight of the polymer can be determined by gel permeation chromatography (45, 80). Many techniques are available to identify saccharidic components in the acid-hydrolyzed polymer, such as thin-layer chromatography (33) or high-performance liquid chromatography, which provides more accurate quantification (38, 43, 76, 131). Composition and linkage information can be obtained by using nuclear magnetic resonance with or without acetolysis of the polymer (19, 45, 80, 120) or by derivitization of the polymer followed by gas-liquid chromatography and mass spectrometry (95). Amino acid, acetyl, pyruvate, phosphorous, and sulfate groups may also be components of the polymer and should be quantified (80, 120, 137).

Biofilms, which typically contain large amounts of EPS, can be analyzed by a variety of microscopic techniques, each with their own pros and cons (130). Many procedures can be performed with few, if any, modifications for hyperthermophilic biofilms as compared to mesophilic samples. A representation of three techniques—acridine orange staining of cells, Congo red staining of EPS, and electron microscopy of cells in a biofilm—is shown in Fig. 2. The polymeric substances involved in the attachment of cells to each other and the support are typically not visible in electron microscopy techniques owing to the dehydration steps involved in the sample processing. One method, confocal scanning laser microscopy, allows a more quantitative analysis of biofilms in vivo with few artifacts (42, 130). This technique has allowed much progress to be made in the analysis of biofilms, no longer thought of as simply flat sheets of polymer and cells (40, 42). Two methods are presented here for visualizing the *T. litoralis* biofilm. They are recommended because of the common availability of light and fluorescent microscopes and the ease of use in determining the extent of cell adhesion.

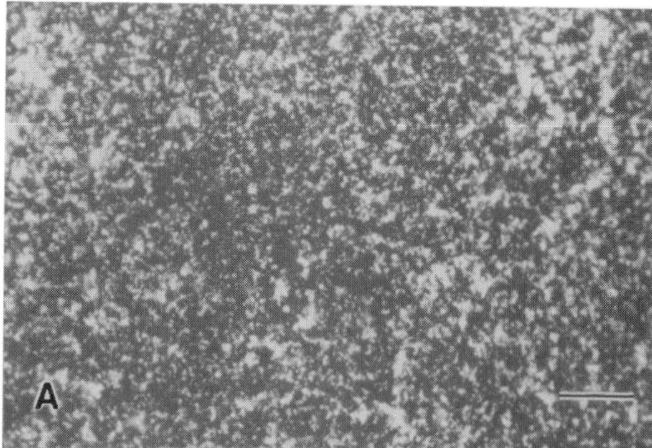

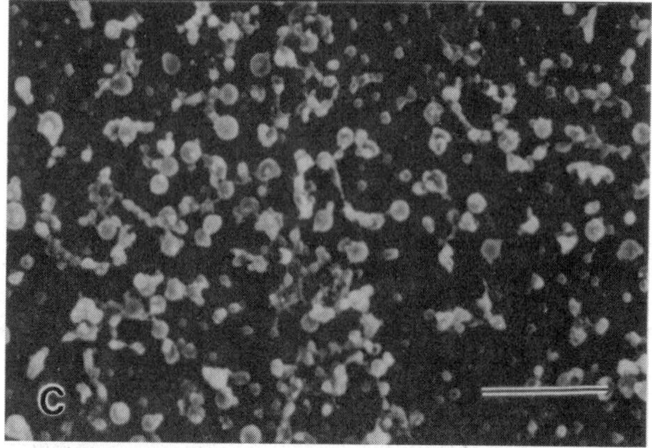

FIGURE 2 *T. litoralis* biofilm formation. (A) Epifluorescent microscopy of cells attached to polycarbonate filter. Bar = 60 μm. (B) Congo red-stained polysaccharide coating of biofilm formed on glass slide (torn with tweezers at top). Bar = 10 μm. (C) Scanning electron microscopy of cells in biofilm on a polycarbonate filter. Bar = 5 μm (104).

PROTOCOL

Acridine Orange Staining of *T. litoralis* Cells in Biofilm

1. Autoclave a 125-ml serum bottle containing a black polycarbonate filter (25-mm, Poretics) and a butyl rubber stopper.
2. Prepare the bottle and the medium as described in section 10.3.3.
3. Grow the culture well into stationary phase. The time point of optimal biofilm development will vary for different organisms and growth conditions.
4. Remove culture from the incubator and allow it to cool to room temperature.
5. Open the bottle and carefully remove the filter.
6. Rinse the filter twice with the sterile medium.
7. Incubate the filter in 2.5% glutaraldehyde for 30 min.
8. Stain nucleic acids in the cells by immersing the filter in 0.04 g/liter acridine orange for 5 min.
9. Dry the filter and place it on a glass slide with a coverslip. If an oil-immersion objective is used, add a drop of oil between the filter and the coverslip and on top of the coverslip.

Congo Red Staining of the Polysaccharide Component of the Biofilm (6)

1. Insert a chemically clean glass slide into a 100-ml large-mouth bottle and autoclave.
2. Follow steps 2 to 6 in the above procedure but replace the filter with a glass slide.
3. Cover the slide in 10 mM cetyl pyridinium chloride to precipitate the polysaccharide and allow to dry for 30 min.
4. Gently heat the slide over a Bunsen burner to fix slide and let it cool to room temperature.
5. Stain the polysaccharide in the biofilm for 15 min in a 2:1 solution of saturated Congo Red and 10% (vol/vol) Tween 80. Tween 80 intensifies the polysaccharide stain.
6. Rinse slide twice in sterile medium.
7. Stain cells with 10% (vol/vol) Ziehl carbol fuchsin.
8. Rinse the slide twice in the sterile medium and dry the slide at 37°C.

Quantification of polysaccharide in the biofilm can be obtained by the total sugar assay (6), and the biomass content can be estimated by the protein assay.

10.4. GROWTH OF EXTREMELY THERMOACIDOPHILIC ARCHAEA: *M. SEDULA*

All thermoacidophiles of the archaeal crenarchaeota branch belong to the *Sulfolobaceae* family (124, 139). They are members of the genera *Sulfolobus*, *Acidianum*, *Desulfurolobus*, and *Metallosphaera* (23, 55, 118, 143). *M. sedula*, the type species of the genus *Metallosphaera*, will be used as an example of general cultivation procedures for thermoacidophiles. *M. sedula* was isolated from a solfataric field in Italy (55) and can grow at temperatures between 50 and 79°C (optimum around 75°C) and at pH 1.0 to 4.0 (optimum around 2.0). This organism has a highly efficient ore-leaching (metal-extracting) capability and has a potential to be used for in situ leaching of geothermally heated ore deposits (37, 55, 96, 97). Thermoacidophiles can also be utilized for investigating unusual bioenergetic features (84), since these organisms must maintain a large transmembrane pH gradient under strongly acidic as well as thermal growth parameters.

10.4.1. Media Formulation

Most thermoacidophiles can be cultivated with the basal medium of Allen (5) and Brock et al. (23), which is similar in composition to the water (of hot springs or solfataric field) from which the organisms were isolated. Unlike other thermoacidophiles, the medium for *M. sedula* is somewhat less complicated, with no need for trace elements. The composition of this modified medium is (in g/liter): 0.4 K_2HPO_4, 0.4 $(NH_4)_2SO_4$, 0.4 $MgSO_4 \cdot 7H_2O$, and 0.2 acid-hydrolyzed casein or enzymatically hydrolyzed casein. The medium is adjusted to pH 2.0 with 10 N H_2SO_4 and autoclaved for at least 20 min. Since *M. sedula* is a chemolithotroph or a mixotroph, either sulfur or iron can be used as energy source. Flowers of elemental sulfur or colloidal sulfur can be directly added to the basal medium (see section 10.3.1 for details in preparation of sulfur). In the case of iron, 10 g of $FeSO_4 \cdot 7H_2O$ in 100 ml of deionized water is filter sterilized using a 0.2-μm-pore-size filter and acidified to pH 2.0 by adding 10 N H_2SO_4. The iron solution (10 ml) is mixed with 80 ml of the basal medium for batch cultures. Alternatively, iron pyrite (FeS_2) is used as an energy source instead of sulfur or $FeSO_4 \cdot 7H_2O$ to study the leaching capacity of *M. sedula*. Iron pyrite (2 g of NIST National Institute of Standards and Technology standard reference pyrite [92]) is added to 100 ml of the basal medium before inoculation. No extra autoclaving or filter sterilization is required when pyrite is added to the sterilized medium.

10.4.2. Culture Preservation

For long-term storage, lyophilization is the method most often used in our laboratory for extreme thermoacidophiles. Alternatively, cultures can be stored over liquid nitrogen. Both methods keep the culture viable for at least 1 year. For short-term storage, a late-log-phase sample of active culture can be placed at room temperature in a serum bottle or test tube with a foam stopper; such cultures remain viable for up to 2 months. After this period, the culture can be transferred and will remain viable for a similar period.

10.4.3. Batch Cultivation in Flasks

M. sedula can be grown in 250-ml flasks with 100 ml of total liquid volume, using either a silicon oil bath or an air shaker (with capability to operate at 75 to 80°C) for temperature control and agitation. Periodically, samples can be withdrawn, fixed with glutaraldehyde, diluted, and stained with acridine orange for enumeration of the cell population by epifluorescence microscopy ([142]; see also section 10.3.3.). One particular protocol that has been used in our laboratory is as follows.

PROTOCOL

1. Autoclave flasks with foam stoppers.
2. Prepare the medium by autoclaving the basal medium and by filter sterilizing the ferrous sulfate medium. The pH of the medium should be preadjusted to 2.0.
3. Add an appropriate amount of the sterile basal medium (e.g., 90 ml or 90% of the total culture volume) to the autoclaved flask, supplemented with the organic substrates (e.g., acid-hydrolyzed casein) and inorganic substrates (e.g., ferrous sulfate, sulfur, or pyrite), using the sterile technique.
4. Heat the flask to the culture temperature (~75°C) and then inoculate with an appropriate amount of the *M. sedula* cells (e.g., 10 ml or 10% of the total culture volume). The cell density of the inoculum used is typically between 8×10^5 cells per ml and 2×10^6 cells per ml.
5. Check the cell density of the culture periodically. As ferrous sulfide is oxidized to ferric hydroxide and other oxidation products as a consequence of growth, orange discoloration of the medium and walls of the culture vessel can be seen.

10.4.4. Continuous Cultivation

Figure 3 shows a schematic drawing of the 10-liter continuous culture apparatus developed for continuous cultivation of extreme thermoacidophiles; this system is somewhat simpler to operate than that used for anaerobic cultures. Basal medium is first autoclaved and then fed to a 12-liter (5-neck) round bottom flask (Ace Glass). The inorganic feed is filter sterilized through a 0.2-μm filter to avoid contamination of the culture, although contamination problems are rare owing to the highly thermal, and strongly acidic, growth conditions. However, because of the acidic environment, all internal parts of the system should be glass- or Teflon-coated to avoid corrosion problems. The culture is agitated at 150 rpm with an impeller and aerated by filtered (0.2-μm filter) house air at a rate of 100 ml/min. A condenser is used to decrease loss of water vapor in exit gas stream. A Digi-Sense temperature controller (Cole Parmer), connected to a Teflon-coated type-J probe, is used to monitor and adjust culture temperature. System operation is initiated by pumping the sterilized medium (pH 2.0) into the round-bottom vessel containing cells at mid-log phase (5×10^7 cells/ml). Dilution rate can be varied but should be chosen with regard to the demand for and, thus, preparation of the feed medium. In routine cases, we use a dilution rate of 0.04 h^{-1} (corresponding to a 17-

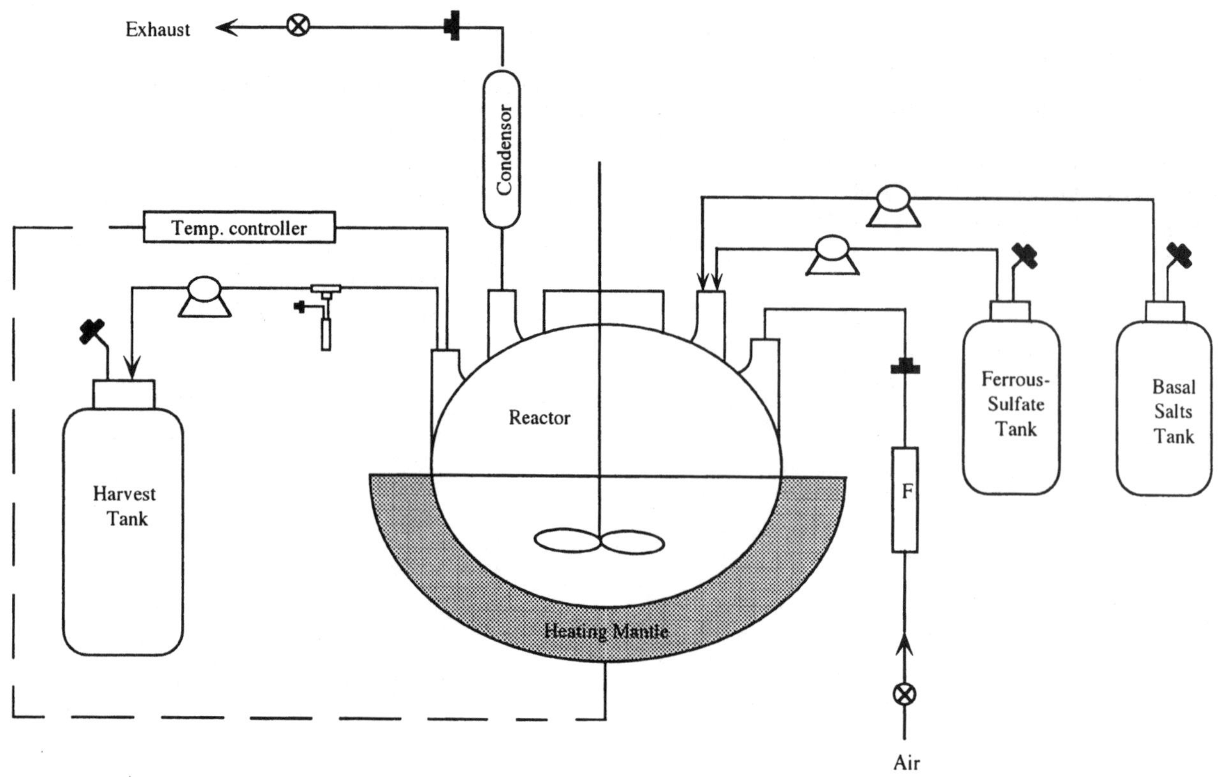

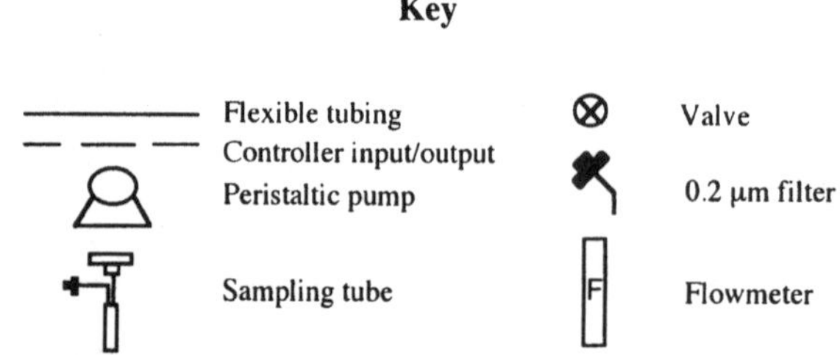

FIGURE 3 Chemostate for continuous cultivation of extreme thermoacidophiles.

h doubling time). Cells and the spent medium are collected in sterile polypropylene vessels, with the cells stored at 4°C for later analysis or processed immediately for enzyme purification. Samples are taken regularly to check cell density.

The suggested protocol for operation of the continuous culture setup shown in Fig. 3 is as follows.

PROTOCOL

1. Grow M. *sedula* (500 ml) in flasks as the seed culture.
2. Assemble the sterile connections (ports, aerator, etc.). Add 10 liters of water and boil the water to sterilize the reactor. Cooling water should be turned on to the condenser.
3. After the reactor sterilization, drain the water out carefully. Immediately add the autoclaved basal salts medium and the sterilized inorganic medium (e.g., ferrous-sulfate, pyrite) through the port into the reactor.
4. Heat the reactor to the M. *sedula* culture temperature (~75°C) and aerate before inoculation.
5. Using sterile technique, inoculate the vessel with the prepared active M. *sedula* batch culture. Start the continuous culture by pumping the sterilized medium (pH 2.0) at an appropriate dilution rate.
6. Maintain and monitor the culture periodically.

Caution: The culture vessel and other parts should be acid-cleaned to remove the scale that builds up inside the reactor, usually after a few months of continuous operation. For the vessel cleanup, carefully add 2 liters of concentrated HCl (through a funnel) into the top of the reactor and close the port immediately. Be extremely aware of toxic acid fumes and wear proper protection all the time.

10.5. NEW DEVELOPMENTS IN THE ISOLATION AND CHARACTERIZATION OF HIGH-TEMPERATURE MICROORGANISMS

There have been several interesting developments that affect the discovery and isolation of new extremely thermophilic microorganisms. Based on the work of Woese and coworkers (139) with 16S rRNA phylogeny, RNA from numerous extreme thermophiles has been sequenced and compared on this basis. While 16S rRNA analysis does not directly reveal physiological information, it is useful for placing a newly discovered microorganism within a framework, which provides a starting point for metabolic and physiological characterization. It has also been used to reveal the yet untapped biodiversity of geothermal environments.

Estimates of the fraction of total microorganisms of all types that have been isolated and characterized are usually on the order of 1%. This implies that the fraction of extreme thermophiles that have been identified to date is also likely to be small, a point that is supported by some recent investigations using 16S rRNA signatures from samples taken from hot springs (10).

Methods for isolation of extreme thermophiles using specific 16S rRNA sequence information have been described. Pure cultures of archaea containing 16S rRNA signatures previously observed from in situ analysis of samples from hot springs were obtained by the use of "optical tweezers" (8, 9). This technique combines the use of an inverse microscope and an infrared laser to pull a single cell with a particular 16S rRNA signature into a separate section of a capillary and then put it into cultivation medium, where it can reproduce and subsequently be characterized (57). This approach short-circuits the often laborious methods of serial dilution, from which the result might be an organism that has previously been isolated and characterized.

Recombinant DNA technology has provided another powerful avenue for recovering enzymes of potential commercial importance from geothermal environments, in some cases bypassing the need to culture organisms. The difficulty in culturing extreme thermophiles in the laboratory and the untapped potential represented by the many extreme thermophiles present in natural samples that are difficult to culture and, perhaps, uncultivatable have been the motivation for expression cloning and robotic screening (107). DNA is extracted from geothermal waters or sediments and subsequently placed in expression libraries from which enzyme activities of interest can be determined by automated screening techniques. Although the source of an interesting biocatalyst may never be known, this approach ensures its availability as a recombinant product (20).

The physiological and metabolic biodiversity of extremely thermophilic microorganisms that have yet to be isolated and characterized is difficult to assess. Not only is much more to be learned about the ecology of geothermal environments, but individual organisms must be studied more completely. Emerging tools to accomplish the latter include the availability of genome sequence information, recently reported for the hyperthermophilic methanogen *M. jannaschii* (28) and currently being sought for several other extreme thermophiles. The relationship between gene sequence and function will have to be established through a combination of approaches ranging from classical microbiology to molecular biology.

We thank the U.S. Department of Energy and the National Science Foundation for support of this research.

REFERENCES

1. **Adams, M. W. W. (ed.).** 1996. Enzymes and proteins from hyperthermophilic microorganisms. *Adv. Protein Chem.* **48:**1–509.
2. **Adams, M. W. W., and R. M. Kelly.** 1995. Extremozymes. *Chem. Eng. News* **73:**32–42.
3. **Adams, M. W. W., and A. Kletzin.** 1996. Oxidoreductase-type enzymes and redox proteins involved in the fermentative metabolisms of hyperthermophilic archaea. *Adv. Protein Chem.* **48:**101–180.
4. **Adams, M. W. W., F. B. Perler, and R. M. Kelly.** 1995. Extremozymes: expanding the limits of biocatalysis. *Bio/Technology* **13:**662–668.
5. **Allen, M. B.** 1959. Studies with *Cyanidium caldarium*, an anomalously pigmented chlorophyte. *Arch. Mikrobiol.* **32:** 270–277.
6. **Allison, D. G., and I. W. Sutherland.** 1984. A staining technique for attached bacteria and its correlation to extracellular carbohydrate production. *J. Microbiol. Methods* **2:**93–99.
7. **Antón, J., I. Meseguer, and F. Rodríguez-Valera.** 1988. Production of an extracellular polysaccharide by *Haloferax mediterranei*. *Appl. Environ. Microbiol.* **54:**2381–2386.
8. **Ashkin, A., and J. M. Dziedzic.** 1987. Optical trapping and manipulation of viruses and bacteria. *Science* **235:** 1517–1520.
9. **Ashkin, A., J. M. Dziedzic, and T. Yamane.** 1987. Optical trapping and manipulation of single cells using infrared laser beams. *Nature* **330:**769–771.
10. **Barns, S. M., R. E., Fundyga, M. W., Jeffries, and N. R. Pace.** 1994. Remarkable archaeal diversity detected in a Yellowstone National Park hot spring environment. *Proc. Natl. Acad. Sci. USA* **91:**1609–1613.
11. **Baronofsky, J. J., W. J. A. Schreurs, and E. R. Kashket.** 1984. Uncoupling by acetic acid limits growth of an acetogenesis by *Clostridium thermoaceticum*. *Appl. Environ. Microbiol.* **48:**1134–1139.
12. **Baross, J. A., and J. W. Deming.** 1995. Growth at high temperatures: isolation and taxonomy, physiology, and ecology, p. 169–217. *In* D. M. Karl (ed.), *The Microbiology of Deep-Sea Hydrothermal Vents*. CRC Press, Boca Raton, Fla.
13. **Belkin, S., and H. W. Jannasch.** 1985. A newly extremely thermophilic, sulfur-reducing heterotrophic, marine bacterium. *Arch. Microbiol.* **141:**181–186.
14. **Belkin, S., C. O. Wirsen, and H. W. Jannasch.** 1986. A new sulfur-reducing, extremely thermophilic eubacterium from a submarine thermal vent. *Appl. Environ. Microbiol.* **51:**1180–1185.
15. **Blumentals, I. I., S. H. Brown, R. N. Schicho, A. K. Skaja, H. R. Constantino, and R. M. Kelly.** 1990. The hyperthermophilic archaebacterium, *Pyrococcus furioasus*: development of culturing protocols, perspectives on scaleup, and potential applications. *Ann. NY Acad. Sci.* **589:** 301–314.
16. **Blumentals, I. I., M. Itoh, G. J. Olson, and R. M. Kelly.** 1990. Role of polysulfides in reduction of elemental sulfur by the hyperthermophilic archaebacterium *Pyrococcus furiosus*. *Appl. Environ. Microbiol.* **56:**1255–1262.
17. **Bonch-Osmolovskaya, E. A., and M. L. Miroshnichenko.** 1994. Effect of molecular hydrogen and elemental sulfur on metabolism of extremely thermophilic archaebacteria of the genus *Thermococcus*. *Microbiology* **63:**433–436.
18. **Bonch-Osmolovskaya, E. A., and K. O. Stetter.** 1991. Interspecies hydrogen transfer in cocultures of thermophilic *Archaea Syst. Appl. Microbiol.* **14:**205–208.

19. **Bradbury, J. H., and J. G. Collins.** 1979. An approach to the structural analysis of oligosaccharides by NMR spectroscopy. *Carbohydr. Res.* **71:**15–24.

20. **Brennan, M. B.** 1996. Enzyme discovery heats up. *Chem. Eng. News* **74:**31–33.

21. **Brierley, C. L., and J. A. Brierley.** 1973. A chemoautotrophic and thermophilic microorganism isolated from an acid hot spring. *Can. J. Microbiol.* **19:**183–188.

22. **Brock, T. D.** 1978. *Thermophilic Microorganisms and Life at High Temperatures*. Springer-Verlag, New York.

23. **Brock, T. D., K. M. Brock, R. T. Belly, and R. L. Weiss.** 1972. *Sulfolobus*: a new genus of sulfur-oxidizing bacteria living at low pH and high temperature. *Arch. Microbiol.* **84:**54–68.

24. **Brown, M. J., and J. N. Lester.** 1980. Comparison of bacterial extracellular polymer extraction methods. *Appl. Environ. Microbiol.* **40:**179–185.

25. **Brown, S. H., H. R. Costantino, and R. M. Kelly.** 1990. Characterization of amylolytic enzyme activities associated with the hyperthermophilic archaebacterium, *Pyrococcus furiosus*. *Appl. Environ. Microbiol.* **56:**1985–1991.

26. **Brown, S. H., and R. M. Kelly.** 1989. Cultivation techniques for hyperthermophilic archaebacteria: continuous culture of *Pyrococcus furiosus* at temperatures near 100°C. *Appl. Environ. Microbiol.* **55:**2086–2088.

27. **Brown, S. H., and R. M., Kelly.** 1993. Characterization of amylolytic enzymes, having both α-1,4 and α-1,6 hydrolytic activity, from the thermophilic archaea *Pyrococcus furiosus* and *Thermococcus litoralis*. *Appl. Environ. Microbiol.* **59:**2614–2621.

28. **Bult, C. J., O. White, G. J. Olsen, L. Zhou, R. D. Fleischmann, G. G. Sutton, J. A. Blake, L. M. FitzGerald, R. A. Clayton, J. D. Gocayne, A. R. Kerlavage, B. A. Gougherty, J.-F. Tomb, M. D. Adams, C. I. Reich, R. Overbeek, E. F. Kirkness, K. G. Weinstock, J. M. Merrick, A. Glodek, J. L. Scott, N. S. M. Geoghagen, J. F. Weidman, J. L. Fuhrmann, D. Nguyen, T. R. Utterback, J. M. Kelly, J. D. Peterson, P. W. Sadow, M. C. Hanna, M. D. Cotton, K. M. Roberts, M. A. Hurst, B. P. Kaine, M. Borodovsky, H.-P. Klenk, C. M. Fraser, H. O. Smith, C. R. Woese, and J. C. Venter.** 1996. Complete genome sequence of the methanogenic archaeon, *Methanococcus jannaschii*. *Science* **273:**1058–1073.

29. **Burggraf, S., G. J. Olsen, K. O. Stetter, and C. R. Woese.** 1992. A phylogenetic analysis of *Aquifex pyrophilus*. *Syst. Appl. Microbiol.* **15:**352–356.

30. **Busse, S. A., G. N. La Mar, L. P. Yu, J. B. Howard, E. T. Smith, Z. H. Zhou, and M. W. W. Adams.** 1992. Proton NMR investigation of the oxidized three-iron clusters in the ferredoxins from the hyperthermophilic archaea, *Pyrococcus furiosus* and *Thermococcus litoralis*. *Biochemistry* **31:**11952–11962.

31. **Canganella, F., and W. J. Jones.** 1994. Fermentation studies with thermophilic Archaea in pure culture and in syntrophy with a thermophilic methanogen. *Curr. Microbiol.* **28:**293–298.

32. **Carlsson, J., G. P. D. Granberg, G. K. Nyberg, and M. B. K. Edlund.** 1979. Bactericidal effect of cysteine exposed to atmospheric oxygen. *Appl. Environ. Microbiol.* **37:**382–390.

33. **Chaplin, M. F.** 1994. Monosaccharides, p. 1–42. *In* M. F. Chaplin and J. F. Kennedy (ed.), *Carbohydrate Analysis: A Practical Approach*, 2nd ed. IRL Press, Oxford.

34. **Chen, C. Y., and D. R. Skidmore.** 1988. Attachment of *Sulfolobus acidocaldarius* cells on coal particles. *Biotechnol. Prog.* **4:**25–30.

35. **Childers, S. E., M. Vargas, and K. M. Noll.** 1992. Improved methods for cultivation of the extremely thermophilic bacterium *Thermotoga neapolitana*. *App. Environ. Microbiol.* **58:**3949–3953.

36. **Clark, D. S., and R. M. Kelly.** 1990. Microorganisms at extreme temperatures and pressures: engineering insights. *Chemtech* **20:**641–648.

37. **Clark, T. R., F. Baldi, and G. J. Olson.** 1993. Coal depyritization by the thermophilic archaeon *Metallosphaera sedula*. *Appl. Environ. Microbiol.* **59:**2375–2379.

38. **Clarke, A. J., V. Sarabia, W. Keenleyside, P. R. MacLachlan, and C. Whitfield.** 1991. The compositional analysis of bacterial extracellular polysaccharides by high-performance anion-exchange chromatography. *Anal. Biochem.* **199:**68–74.

39. **Connaris, H., D. Cowan, M. Ruffett, and R. J. Sharp.** 1991. Preservation of the hyperthermophile *Pyrococcus furiosus*. *Lett. Appl. Microbiol.* **13:**25–27.

40. **Costerton, J. W.** 1995. Overview of microbial biofilms. *J. Ind. Microbiol.* **15:**137–140.

41. **Danson, M. J.** 1989. Central metabolism of the archaebacteria: an overview. *Can. J. Microbiol.* **35:**58–64.

42. **De Beer, D. P. Stoodley, F. Roe, and Z. Lewandowski.** 1994. Effects of biofilm structures on oxygen distribution and mass transport. *Biotech. Bioeng.* **43:**1131–1138.

43. **Farnsworth, V., and K. Steinberg.** 1993. The generation of phenlythicarbamyl or anilinothiazolinone amino acids from the postcleavage products of the Edman degradation. *Anal. Biochem.* **251:**200–210.

44. **Fiala, G., and K. O. Stetter.** 1986. *Pyrococcus furiosus* sp. nov. represents a novel genus of marine heterotrophic archaebacteria growing optimally at 100°C. *Arch. Microbiol.* **145:**56–61.

45. **Flatt, J. H., R. S. Hardin, J. M. Gonzalez, D. E. Dogger, E. N. Lightfoot, and D. C. Cameron.** 1992. An anionic galactomannan polysaccharide gum from a newly-isolated lactose-utilizing bacterium. I. Strain description and gum characterization. *Biotechnol. Prog.* **8:**327–334.

46. **Fuchs, T., H. Huber, K. Teiner, S. Burggraf, and K. O. Stetter.** 1995. *Metallosphaera prunae*, sp. nov., a novel metal-mobilizing, thermoacidophilic archaeum, isolated from a uranium mine in Germany. *Syst. Appl. Microbiol.* **18:**560–566.

47. **Geesey, G. G.** 1982. Microbial exopolymers: ecological and economic considerations. *ASM News* **48:**9–14.

48. **Grogan, D., P. Palm, and W. Zillig.** 1990. Isolate B-12, which harbors a virus-like element, represents a new species of the archaebacterial genus *Sulfolobus*, *Sulfolobus shibatae*, sp. nov. *Arch. Microbiol.* **154:**594–599.

49. **Grogan, D. W.** 1989. Phenotypic characterization of the archaebacterial genus *Sulfolobus*: comparison of five wild-type strains. *J. Bacteriol.* **171:**6710–6719.

50. **Guezennec, J. G., P. Pignet, G. Raguenes, E. Deslandes, Y. Lijour, and E. Gentric.** 1994. Preliminary chemical characterization of unusual eubacterial exopolysaccharides of deep-sea origin. *Carbohydr. Polym.* **24:**287–294.

51. **Harriott, O. T., R. Huber, K. O. Stetter, P. W. Betts, and K. M. Noll.** 1994. A cryptic miniplasmid from the

hyperthermophilic bacterium *Thermotoga* Sp Strain Rq7. *J. Bacteriol.* **176:**2759–2762.

52. **Hoaki, T., M. Nishijima, M. Kato, K. Adachi, S. Mizobuchi, N. Hanzawa, and T. Maruyama.** 1994. Growth requirements of hyperthermophilic sulfur-dependent heterotrophic archaea isolated from a shallow submarine geothermal system with reference to their essential amino acids. *Appl. Environ. Microbiol.* **60:**2898–2904.
53. **Hoaki, T., C. O. Wirsen, S. Hanzawa, T. Maruyama, and H. W. Jannasch.** 1993. Amino acid requirements of two hyperthermophilic archaeal isolates from deep-sea vents, *Desulfurococcus* Strain SY and *Pyrococcus* Strain GB-D. *Appl. Environ. Microbiol.* **59:**610–613.
54. **Hobbie, J. E., R. J. Daley, and S. Jasper.** 1977. Use of nucleopore filters for counting bacteria by fluorescence microscopy. *Appl. Environ. Microbiol.* **33:**1225–1228.
55. **Huber, G., C. Spinnler, A. Gambacorta, and K. O. Stetter.** 1989. *Metallosphaera sedula* gen. and sp. nov. represents a new genus of aerobic, metal-mobilizing, thermophilic archaebacteria. *Syst. Appl. Microbiol.* **12:**38–47.
56. **Huber, G., and K. O. Stetter.** 1986. *Sulfolobus metallicus*, sp. nov., a novel strictly chemolithoautotrophic thermophilic archaeal species of metal-mobilizers. *Syst. Appl. Microbiol.* **14:**372–378.
57. **Huber, R., S. Burggraf, T., Mayer, S. M. Barns, P. Rossnagel, and K. O. Stetter.** 1995. Isolation of a hyperthermophilic archaeum predicted by *in situ* RNA analysis. *Nature* **376:**57–58.
58. **Huber, R., J. K. Kristjannson, and K. O Stetter.** 1987. *Pyrobaculum* gen. nov., a new genus of neutrophilic, rod-shaped archaebacteria from continental solfataras growing optimally at 100°C. *Arch. Microbiol.* **149:**95–101.
59. **Huber, R., T. A. Langworthy, H. König, M. Thomm, C. R. Woese, U. B. Sleytr, and K. O. Stetter.** 1986. *Thermotoga maritima* sp. nov. represent a new genus of unique extremely thermophilic eubacteria growing up to 90°C. *Arch. Microbiol.* **144:**324–333.
60. **Huber, R., T. Wilharm, D. Huber, A. Trincone, S. Burggraf, H. König, R. Rachel, I. Rockinger, H. Fricke, and K. O. Stetter.** 1992. *Aquifex pyrophilus* gen. nov. sp. nov., represents a novel group of marine hyperthermophilic hydrogen oxidizing bacteria. *Arch. Microbiol.* **15:** 340–351.
61. **Hungate, R. E.** 1969. A roll tube method for cultivation of strict anaerobes. *Methods Microbiol.* **3B:**117–132.
62. **Jannasch, H. W., C. O. Wirsen, S. J. Molyneaux, and T. A. Langworthy.** 1992. Comparative physiological studies on hyperthermophilic archaea isolated from deep-sea hot vents with emphasis on *Pyrococcus* strain GB-D. *Appl. Environ. Microbiol.* **58:**3472–3481.
63. **Jørgensen, B. B., M. F. Isaksen, and H. W. Jannasch.** 1992. Bacterial sulfate reduction above 100°C in deep-sea hydrothermal vent sediments. *Science* **258:**1756–1757.

63a. **Kelly, R. M.** Unpublished observation.

64. **Kelly, R. M., and M. W. W. Adams.** 1994. Metabolism in hyperthermophilic microorganisms. *Antonie van Leeuwenhoek* **66:**247–270.
65. **Kelly, R. M., and J. W. Deming.** 1988. Extremely thermophilic archaebacteria: biological and engineering considerations. *Biotechnol. Prog.* **4:**47–62.
66. **Kengen, S. W. M., F. A. M. Debok, N. D. Vanloo, C. Dijkema, A. J. M. Stams, and W. M. de Vos.** 1994. Evidence for the operation of a novel Embden-Meyerhof pathway that involves ADP-dependent kinases during sugar fermentation by *Pyrococcus furiosus*. *J. Biol. Chem.* **269:**17537–17541.
67. **Kengen, S. W. M., and J. M. Stams.** 1994. Formation of L-alanine as a reduced end product in carbohydrate fermentation by the hyperthermophilic archaeon *Pyrococcus furiosus*. *Arch. Microbiol.* **161:**168–175.
68. **Kengen, S. W. M., A. J. M. Stams, and W. M. deVos.** 1996. Sugar metabolism of hyperthermophiles. *FEMS Microbiol. Rev.* **18:**119–137.
69. **Klages, K. U., and H. W. Morgan.** 1994. Characterization of an extremely thermophilic sulphur-metabolizing archaebacterium belonging to the Thermococcales. *Arch. Microbiol.* **162:**261–266.
70. **Kletzin, A., and M. W. W. Adams.** 1996. Tungsten in biological system. *FEMS Microbiol. Rev.* **18:**5–63.
71. **Kobayashi, T., S. Higuchi, K. Kimura, T. Kudo, and K. Horikoshi.** 1995. Properties of glutamate dehydrogenase and its involvement in alanine production in a hyperthermophilic archaeon, *Thermococcus profundus*. *J. Biochem.* **118:**587–592.
72. **Kobayashi, T., Y. S. Kwak, T. Akiba, T. Kudo, and K. Horikoshi.** 1994. *Themococcus profundus* sp. nov., a new hyperthermophilic archaeon isolated from a deep-sea hydrothermal vent. *Syst. Appl. Microbiol.* **17:**232–236.
73. **Kong, H., R. B. Kucera, and W. E. Jack.** 1993. Characterization of a DNA polymerase from the hyperthermophile archaea *Thermococcus litoralis*. *J. Bacteriol.* **268:** 1965–1975.
74. **Korczynski, M. S.** 1981. Sterilization, p. 476–486. *In* P. Gerhardt, R. G. E. Murray, R. N. Costilow, E. W. Nester, W. A. Wood, N. R. Krieg, and G. B. Phillips (ed.), *Manual of Methods for General Bacteriology*. American Society for Microbiology, Washington, D.C.
75. **Krahe, M., G. Antranikian, and H. Märkl.** 1996. Fermentation of extremophilic microorganisms. *FEMS Microbiol. Rev.* **18:**271–285.
76. **Lu, H.-S., and P.-H. Lai.** 1986. Use of narrow-bore high-performance liquid chromatography for microanalysis of protein structure. *J. Chromatogr.* **368:**215–231.
77. **Mai, X., and M. W. W. Adams.** 1996. Characterization of a fourth type of 2-keto acid oxidizing enzyme from hyperthermophilic archaea: 2-ketoglutarate ferredoxin oxidoreductase from *Thermococcus litoralis*. *J. Bacteriol.* **178:** 5890–5896.
78. **Malik, B., W.-W. Su, H. L. Wald, I. I. Blumentals, and R. M. Kelly.** 1989. Growth and gas production for hyperthermophilic archaebacterium, *Pyrococcus furiosus*. *Biotechnol. Bioeng.* **34:**1050–1057.
79. **Martins, L. O., and H. Santos.** 1995. Accumulation of mannosylglycerate and di-myo-inositol-phosphate by *Pyrococcus furiosus* in response to salinity and temperature. *Appl. Environ. Microbiol.* **61:**3299–3303.
80. **Mbawala, A., S. A. Mahmood, V. Loppinet, and R. Bonaly.** 1990. Acetolysis and ^{1}HNMR studies on mannans isolated from very flocculent and weakly flocculent cells of *Pichia pastoris* IFP 206. *J. Gen. Microbiol.* **136:**1279–1284.
81. **Millane, R. P., and T. L. Hendrixson.** 1994. Crystal structures of mannan and glucomannans. *Carbohydr. Polym.* **25:**245–251.

82. **Miller, J. F., E. L. Almond, N. N. Shah, J. M. Ludlow, J. A. Zollweg, W. B. Streett, S. H. Zinder, and D. S. Clark.** 1988. High-pressure-temperature bioreactor for studing pressure-temperature relationships in bacterial growth and productivity. *Biotech. Bioeng.* **31:**407–413.

83. **Miller, T. L., and M. J. Wolin.** 1974. A serum bottle modification of the Hungate technique for cultivating obligte anaerobes. *Appl. Microbiol.* **27:**985–987.

84. **Mitchell, P.** 1966. Chemiosmotic coupling in oxidative and photosynthetic phosphorylation. *Biol. Rev.* **41:**445–504.

85. **Mukund, S., and M. W. W. Adams.** 1991. The novel tungsten-iron-sulfur protein of the hyperthermophilic archaebacterium, *Pyrococcus furiosus*, is an aldehyde ferredoxin oxidoreductase: evidence for its participation in a unique glycolytic pathway. *J. Biol. Chem.* **266:**14208–14216.

86. **Mukund, S., and M. W. W. Adams.** 1993. Characterization of a novel tungsten-containing formaldehyde ferredoxin oxidoreductase from the extremely thermophilic archaeon, *Thermococcus litoralis*: a role for tungsten in peptide catabolism. *J. Biol. Chem.* **268:**13592–13600.

87. **Mukund, S., and M. W. W. Adams.** 1995. Glyceraldehde-3-phosphate ferredoxin oxidoreductase, a novel tungsten-containing enzyme with a potential glycolytic role in the hyperthermophilic archaeon, *Pyrococcus furiosus*. *J. Biol. Chem.* **270:**8389–8392.

88. **Mukund, S., and M. W. W. Adams.** 1996. Molybdenum and vanadium do not replace tungsten in the three tungstoenzymes of the hyperthermophilic archaeon *Pyrococcus furiosus*. *J. Bacterol.* **178:**163–167.

89. **Muralidharan, V., K. D. Rinker, I. S. Hirsh, E. J. Bouwer, and R. M. Kelly.** 1997. Hydrogen transfer between methanogens and fermentative heterotrophs in hyperthermophilic cocultures. *Biotechnol. Bioeng.* **56:**268–278.

90. **Neuner, A., H. W. Jannasch, S. Belkin, and K. O. Stetter.** 1990. *Thermococcus litoralis* sp. noc.: a new species of extremely thermophilic marine archaebacteria. *Arch. Microbiol.* **153:**205–207.

91. **Nicolaus, B., M. C. Manca, I. Romano, and L. Lama.** 1993. Production of an exopolysaccharide from two thermophilic archaea belonging to the genus *Sulfolobus*. *FEMS Microbiol. Lett.* **109:**203–206.

92. **Olson, G. J.** 1991. Rate of pyrite bioleaching by *Thiobacillus ferrooxidans*: results of an inter-laboratory comparison. *Appl. Environ. Microbiol.* **57:**642–644.

93. **Parameswaran, A. K., C. N. Provan, F. J. Sturm, and R. M. Kelly.** 1987. Sulfur reduction by the extremely thermophilic archaebacterium *Pyrodictium occultum*. *Appl. Environ. Microbiol.* **53:**1690–1693.

94. **Parameswaran, A. K., R. N. Schicho, J. P. Soisson, and R. M. Kelly.** 1988. Effect of hydrogen and carbon dioxide partial pressures on growth and sulfide production of the extremely thermophilic archaebacterium *Pyrodictium brockii*. *Biotechnol. Bioeng.* **32:**438–443.

95. **Pazur, J. H.** 1994. Neutral polysaccharides, p. 73–124. *In* M. F. Chaplin and J. F. Kennedy (ed.), *Carbohydrate Analysis: A Practical Approach*, 2nd ed. IRL Press, Oxford.

96. **Peeples, T. L., and R. M. Kelly.** 1993. Bioenergetics of the metal/sulfur oxidizing extreme thermoacidophile, *Metallosphaera sedula*. *Fuel* **72:**1619–1624.

97. **Peeples, T. L., and R. M. Kelly.** 1995. Bioenergetic response of the extreme thermoacidophile *Metallosphaera sedula* to thermal and nutritional stress. *Appl. Environ. Microbiol.* **61:**2314–2321.

98. **Pihl, T. D., R. N. Schicho, L. K. Black, B. A. Schulman, R. J. Maier, and R. M. Kelly.** 1990. Hydrogen sulfur autotrophy in the hyperthermophilic archaebacterium *Pyrodictium brockii*. *Biotechnol. Genet. Eng. Rev.* **8:**345–377.

99. **Pledger, R. J., and J. A. Baross.** 1991. Preliminary description and nutritional characterization of a chemoorganotrophic archaeobacterium growing at temperatures of up to 110°C isolated from a submarine hydrothermal vent environment. *J. Gen. Microbiol.* **137:**203–211.

100. **Pledger, R. J., B. C. Crump, and J. A. Baross.** 1994. A barophilic response by two hyperthermophilic hydrothermal vent Archaea: acceleration of growth rate at supra-optimal temperature by elevated pressure. *FEMS Microbiol. Ecol.* **14:**233–241.

101. **Raguenes, G., P. Pignet, G. Gauthier, A. Peres, R. Christen, H. Gougeaux, G. Barbier, and J. Guezennec.** 1996. Description of a new polymer-secreting bacterium from a deep-sea hydrothermal vent, *Alteromonas macleodii* subsp. *fijiensis*, and preliminary characterization of the polymer. *Appl. Environ. Microbiol.* **62:**67–73.

102. **Raven, N., N. Ladwa, D. Cossar, and R. Sharp.** 1992. Continuous culture of the hyperthermophilic archaeum *Pyrococcus furiosus*. *Appl. Microbiol. Biotechnol.* **38:**263–267.

103. **Raven, N., and R. Sharp.** 1997. Development of defined and minimal media for the growth of the hyperthermophilic archaeon *Pyrococcus furiosus* Vc1. *FEMS Microbiol. Lett.* **146:**135–141.

104. **Rinker, K. D., and R. M. Kelly.** 1996. Growth physiology of the hyperthermophilic archaeon *Thermococcus litoralis*: development of a sulfur-free defined medium, characterization of an exopolysaccharide, and evidence of biofilm formation. *Appl. Environ. Microbiol.* **62:**4478–4485.

104a. **Rinker, K. D., and R. M. Kelly.** Unpublished data.

105. **Ritzau, M., M. Keller, P. Wessels, K. O. Stetter, and Z. Zeeck.** 1993. Secondary metabolites by chemical screening, 25-new cyclic polysulfides from hyperthermophilic archaea of the genus *Thermococcus*. *Liebigs Ann. Chem.* **1993:**871–876.

106. **Robb, F. J. (ed.)** 1900. *Archaea: A Laboratory Manual.* Cold Spring Harbor Laboratory Press, Cold Spring Harbor, N.Y.

107. **Robertson, D. E., E. J. Mathur, R. V. Swanson, B. L. Marrs, and J. M. Short.** 1996. The discovery of new biocatalysts from microbial diversity. *SIM News* **46:** 3–8.

108. **Robinson, K. A., and H. J. Schreier.** 1994. Isolation, sequence and characterization of the maltose-regulated *mlrA* gene from the hyperthermophilic archaeum *Pyrococcus furiosus*. *Gene* **151:**1173–1176.

109. **Rudd, T., R. M. Sterritt, and J. N. Lester.** 1982. The use of extraction methods for the quantification of extracellular polymer production by *Klebsiella aerogenes* under varying cultural conditions. *Eur. J. Appl. Microbiol. Biotechnol.* **16:**23–27.

110. **Rudd, T. R. M. Sterritt, and J. N. Lester.** 1983. Extraction of extracellular polymers from activated sludge. *Biotechnol. Lett.* **5:**327–332.

111. **Rüdiger, A., J. C. Ogbonna, H. Märkl, and G. Antranikian.** 1992. Effect of gassing, agitation, substrate supplementation and dialysis on the growth of an extremely thermophilic archaeon *Pyrococcus woesei*. *Appl. Microbiol. Biotechnol.* **37:**501–504.

112. **Schäfer, T., M. Selig, and P. Schönheit.** 1993. Acetyl-CoA synthetase (ADP-forming) in archaea, a novel enzyme involved in acetate formation and ATP synthesis. *Arch. Microbiol.* **159:**72–83.

113. **Schicho, R. N.** 1992. Ph.D. thesis. Johns Hopkins University, Baltimore.

114. **Schicho, R. N., S. H. Brown, I. I. Blumentals, T. L. Peeples, G. D. Duffaud, and R. M. Kelly.** 1995. Continuous culture techniques for thermophilic and hyperthermophilic archaea, p. 31–36. *In* F. T. Robb (ed.), *Archaea: A laboratory Manual*. Cold Spring Harbor Laboratory Press, Cold Spring Harbor, N.Y.

114a. **Schicho, R. N., and R. M. Kelly.** Unpublished data.

115. **Schicho, R. N., K. Ma, M. W. W. Adams, and R. M. Kelly.** 1993. Bioenergetics of sulfur reduction in the hyperthermophilic archaeon *Pyrococcus furiosus*. *J. Bacteriol.* **175:**1823–1830.

116. **Schicho, R. N. Snowden, L. J., S. Mukund, J. B. Park, M. W. W. Adams, and R. M. Kelly.** 1993. Influence of tungsten on metabolic patterns in the hyperthermophile *Pyrococcus furiosus*. *Arch. Microbiol.* **159:**380–385.

117. **Schönheit, P., and T. Schäfer.** 1995. Metabolism of hyperthermophiles. *World J. Microbiol. Biotechnol.* **11:**26–57.

118. **Segerer, A., A. Neuner, J. K. Kristjansson, and K. O. Stetter.** 1986. *Acidianus infernus* gen. nov., sp. nov. and *Acidianus brierleyi* comb. nov.: facultatively aerobic, extremely acidophilic thermophilic sulfur-metabolizing archaebacteria. *Int. J. Syst. Bacteriol.* **36:**559–564.

119. **Seyfried, W. E., and M. J. Mottl.** 1995. Geologic setting and chemistry of deep-sea hydrothermal vents. p. 1–34. *In* D. M. Karl (ed.), *The Microbiology of Deep-Sea Hydrothermal Vents*. CRC Press, Boca Raton, Fla.

120. **Skjåk-Braek, G., H. Grasdalen, and B. Larsen.** 1986. Monomer sequence and acetylation pattern in some bacterial alginates. *Carbohydr. Res.* **154:**239–250.

121. **Smith, P. F., T. A. Langworthy, and M. R. Smith.** 1975. Polypeptide nature of growth requirement in yeast extract for *Thermoplasma acidophilum*. *J. Bacteriol.* **124:** 884–892.

122. **Sowers, K. R., and R. P. Gunsalus.** 1988. Adaptation for growth at various saline concentrations by the archaebacterium *Methanosarcina thermophila*. *J. Bacteriol.* **170:**998–1002.

123. **Stafford, K.** 1986. Continuous fermentation, p. 137–151. *In* A. L. Demain and N. A. Solomon (ed.), *Manual of Industrial Microbiology and Biotechnology*. American Society for Microbiology, Washington, D.C.

124. **Stetter, K. O.** 1989. Order III. *Sulfolobales* ord. nov., p. 2250–2253. *In* J. G. Hold (ed.), *Bergey's Manual of Systematic Bacteriology*, vol. 3. Williams & Wilkins, Baltimore.

125. **Stetter, K. O.** 1995. Microbial life in extreme environments. *ASM News* **61:**285–290.

126. **Stetter, K. O., G. Fiala, G. Huber, R. Huber, and G. Segerer.** 1990. Hyperthermophilic microorganisms. *FEMS Microbiol Rev.* **75:**117–124.

127. **Stetter, K. O., H. König, and E. Stackebrandt.** 1983. *Pyrodictium* gen. nov., a new genus of submarine disc-shaped sulfur reducing archaebacteria growing optimally at 105°C. *Syst. Appl. Microbiol.* **4:**535–551.

128. **Stetter, K. O., G. Lauerer, M. Thomm, and A.-M. Neuner.** 1987. Isolation of extremely thermophilic sulfate reducers: evidence for a novel branch of archaebacteria. *Science* **236:**822–824.

129. **Su, Wei-Wen, and R. M. Kelly.** 1988. Effect of hyperbaric oxygen and carbon dioxide on the heterotrophic growth of the extreme thermophile, *Sulfolobus acidocaldarius*. *Biotechnol. Bioeng.* **31:**750–754.

130. **Surman, S. B., J. T. Walker, D. T. Goddard, L. H. G. Morton, C. W. Keevil, W. Weaver, A. Skinner, K. Hanson, D. Caldwell, and J. Kurtz.** 1996. Comparison of microscope techniques for the examination of biofilms. *J. Microbiol. Methods* **25:**57–70.

131. **Tarr, G. E.** 1981. Rapid separation of amino acid phenylthiohydantoins by isocratic high-performance liquid chromatography. *Anal. Biochem.* **111:**27–32.

132. **Usenko, I. A., L. O. Severina, and V. K. Plakunov.** 1993. Uptake of sugars and amino acids by extremely thermophilic archae- and eubacteria. *Microbiology* **62:** 272–277.

133. **Vitaya, V. B., and K. Toda.** 1991. Physiological adsorption of *Sulfolobus acidocaldarius* on coal surfaces. *Appl. Microbiol. Biotechnol.* **35:**690–695.

134. **Völkl, P., R. Huber, E. Brobner, R. Rachel, S. Burggraf, A. Trincone, and K. O. Stetter.** 1993. *Pyrobaculum aerophilum* sp. nov., a novel nitrate-reducing hyperthermophilic archaeum. *Appl. Environ. Microbiol.* **59:**2918–2926.

135. **Watrin, L., V. Martin-Jezequel, and D. Prieur.** 1995. Minimal amino acid requirements of the hyperthermophilic archaeon *Pyrococcus abyssi*, isolated from deep-sea hydrothermal vents. *Appl. Environ. Microbiol.* **61:**1138–1140.

136. **Watrin, L., and D. Prieur.** 1996. UV and ethyl methanesulfonate effects in hyperthermophilic archaea and isolation of auxothrophic mutants of *Pyrococcus* strains. *Curr. Microbiol.* **33:**377–382.

137. **Weiner, R., S. Langille, and E. Quintero.** 1995. Structure, function, and immunochemistry of bacterial exopolysaccharides. *J. Ind. Microbiol.* **15:**339–346.

138. **Weiss, B. L.** 1973. Attachment of bacteria to sulphur in extreme environments. *J. Gen. Microbiol.* **77:**501–507.

139. **Woese, C. R., O. Kandler, and M. L. Wheelis.** 1990. Towards a natural system of organisms: proposal for the domains archae, bacteria and eucarya. *Proc. Natl. Acad. Sci. USA* **87:**4576–4579.

140. **Xavier, K. B., L. O. Martins, R. Peist, M. Kossmann, W. Boos, and H. Santos.** 1996. High-affinity maltose/trehalose transport system in the hyperthermophilic archaeon *Thermococcus litoralis*. *J. Bacteriol.* **178:**4773–4777.

141. **Yadav, M. P., J. N. BeMiller, and M. E. Embuscado.**

1994. Compositional analysis of polysaccharides via solvolysis with liquid hydrogen fluoride. *Carbohydr. Polym.* **25:**315–318.

142. **Yeh, T. Y., J. R. Godshalk, G. J. Olson, and R. M. Kelly.** 1987. Use of epifluorescence microscopy for characterizing the activity of *Thiobacillus ferrooxidans* on iron pyrite. *Biotechnol. Bioeng.* **30:**138–146.

143. **Zillig, W., S. Yeats, I. Holz, A. Böck, M. Rettenberger, F. Gropp, and G. Simon.** 1986. *Desulfurolobus ambivalens*, gen. nov., sp. nov., an autotrophic archaebacterium facultatively oxidizing or reducing sulfur. *Syst. Appl. Microbiol.* **8:**197–203.

PROCESSES

II

THIS SECTION INCLUDES CHAPTERS DESCRIBING METHODOLOGIES used for process development after initial discovery. In the process development stage, economic factors become a concern. If the product being developed eventually becomes successful, a manufacturing process must be put in place. Therefore, the availability of raw materials for producing the product, the robustness of the process, and the feasibility of employing the process for large-scale production all need to be taken into consideration. For sound decision making in process development, one needs not only knowledge of manufacturing process technology for the particular product, but also a basic understanding of potential alternative processes. In the past decade, a number of less conventional production systems have been increasingly employed. These process technologies, though less prevailing, are nonetheless important and play pivotal roles in the development of many products. The chapters included in this section are thus divided into two categories: the first group discusses unconventional culture processes, and the second group deals with the methodology generally used in process development.

Chapter 11 is a concise and yet comprehensive treatment of fermentation processes for anaerobic microbes. Microbes of this category are among the most classic industrial microorganisms used for the production of solvents. Recent exploration of protein products (toxins, enzymes) produced by anaerobes has reinvigorated research in this area.

The use of continuous culture in process development, the topic of chapter 12, has been rather limited. However, its power in probing microbe-environment interactions and in elucidating growth regulation mechanisms has long made it a tool of choice for many researchers. Recent advancements in equipment design and in better user interface for various process instruments have made traditionally mundane and even cumbersome tasks more enjoyable. Hence, the use of continuous culture in process research and development is increasing. Furthermore, the emergence of a number of products manufactured in continuous mammalian cell culture has had a very positive impact on the application of continuous culture for process development.

This chapter is followed by an overview of biotransformation, the catalytic conversion of organic molecules by microbial cells. The increasing use of biotransformations in the fermentation and chemical industries makes chapter 13 an extremely important contribution. This chapter discusses the means by which a biotransformation process is established, from obtaining and preparing the microorganisms to performing the reaction and detecting the products. It contains a number of examples to illustrate the applications of the principles outlined in the chapter.

The next three chapters deal with culture technologies that have matured only in the last decade: cultures of mammalian cells (chapter 14), plant cells (chapter 15), and insect cells (chapter 16). Both mammalian and insect cells are used in bioproduct manufacturing and in producing biological reagents for product development. The use of plant cell culture for manufacturing is less extensive. However, with the increasing use of transgenic plants, many of which incorporate genes from microbes, the role of plant cell culture in process development will only become more important in the future.

The final four chapters fall into the second category of this section. The raw material, i.e., the medium, for industrial fermentation plays a key role in the economics and consistency of the process. Even for process development on a laboratory or pilot scale, one cannot overlook the distinctions between raw materials for research and those for manufacturing. Chapter 17, on media for large-scale fermentation, provides essential information for formulating media for manufacturing processes.

Overall, the fermentation industry has improved its efficiency significantly in the last decade. Contributing to this improvement is the wide application of robust and more versatile environmental monitoring and control. The ability to better control environmental factors has also made the scale-up of processes less a purely experience-based black box: a rational strategy to deal with issues related to scale translation can be devised. Chapter 18 discusses instrumentation for small-scale processes, and chapter 19 is devoted to the challenges of scale-up.

Last, chapter 20 discusses a very important aspect of process development: data analysis, i.e., the transformation of raw data into information for process development. With the decreasing cost of computation and more and more user-friendly software, many data analysis tools that were not manageable by most process scientists and engineers are now desktop fixtures. This final chapter is an introduction to these techniques.

Anaerobic Fermentations

ERIC A. JOHNSON

11

11.1. THE ADVENT OF METHODS IN ANAEROBIC MICROBIOLOGY

Humans have used anaerobic fermentations since ancient times for many important industrial fermentations such as production of ethanol from yeasts, lactic acid preservation of foods, and anaerobic digestion of polysaccharides and proteins in ruminant cultivation and waste treatment. Anaerobic fermentations have contributed greatly to the development and success of industrial microbiology. Many ancient and contemporary industrial processes involve anaerobic fermentations. In recent years, the discovery of exotic and diverse anaerobic habitats such as deep-sea thermal vents has led to the isolation of anaerobic organisms of biotechnological potential. The isolation of specialty enzymes such as DNA polymerase from *Thermus aquaticus* (*Taq* polymerase) has further increased the contribution of anaerobes to biotechnology. One of the largest areas of industrial microbiology is anaerobic digestion of wastes, and anaerobes are having an increasing role in detoxification of hazardous wastes and pollutants.

Anaerobic organisms have a rich history in industrial and medical microbiology. Although Leeuwenhoek in 1680 reported that certain microorganisms could be envisioned in habitats lacking oxygen, the discovery of anaerobiosis is attributed to Pasteur (100, 137) who showed in the mid-1850s that butyric acid fermentation occurred in the absence of oxygen. Pasteur produced anaerobiosis by boiling the medium to drive out absorbed oxygen and by introducing inert gases for cultivation. He introduced the terms "anaerobies" and "aerobies." Pasteur showed that an organism was responsible for the butyric fermentation, and he called this organism *Vibrion butyrique*; it was probably *Clostridium butyricum*, the type species of the genus *Clostridium*. Pasteur also demonstrated the presence of strictly anaerobic life in his studies of quality defects in beer and wine (25). In 1877, Pasteur and Joubert also described the first pathogenic anaerobe, now known as *Clostridium septicum* (110).

The systematic evolution of techniques for the isolation and study of anaerobic bacteria occurred as a result of two major pursuits: during World War I and II by medical microbiologists who needed methods to isolate anaerobes causing infections in human wounds (136, 137), and by microbial ecologists. In the late 1800s and early 1900s, ecologists including Beijerinck, Kluyver, van Niel, and their collaborators in the Delft School (43, 102, 126, 130), Winogradsky (109), Omeliansky, and other pioneering microbiologists demonstrated that bacteria had diverse mechanisms of energy generation and that enrichment or elective cultures could yield strictly anaerobic bacteria. Hungate provided the crowning accomplishment in anaerobic methodology by isolating strict anaerobes from the rumens of cows by using strict anaerobic methods and specialized culture media (especially roll tubes) (57, 58). Many of the developments for studying anaerobic bacteria are based on Hungate's methods. Pressurized serum tubes containing 80% H_2 and 20% CO_2 were introduced for the study of methanogens (7). In the 1960s and 1970s, anaerobic chambers were invented that allowed the cultivation of numerous anaerobic cultures on agar media and their manipulation by techniques used routinely for aerobes and facultative anaerobes (4, 74).

Medical microbiology and the importance of anaerobes in human and animal diseases contributed significantly to the invention of important equipment and techniques for anaerobic culture (47, 54, 118, 136, 137). Reducing agents including alkaline pyrogallol and cysteine hydrochloride, as well as other sulfhydryl compounds, were used in the late 1800s for the study of pathogenic clostridia (28, 47). The meticulous techniques required for obtaining anaerobic isolates made it clear that there are few shortcuts in the study of anaerobic bacteria. The exclusive use of liquid media for anaerobic culture resulted in many incidences of mixed cultures, and contamination of anaerobic cultures remains a problem today. The invention of the anaerobic jar by McIntosh and Fildes in 1916 (79) enabled agar plate cultures isolation of pure colonies. Modifications of the anaerobic jar including the introduction of heated palladium or platinum catalysts were adapted by Brewer in the United States (118). The Brewer apparatus was used until about 1960, when much concern was raised about the possibility of explosions of hydrogen gas. It should be kept in mind that concentrations of pure hydrogen of >5% (vol/vol) can be explosive (22) and hydrogen levels in anaerobic vessels and chambers should be kept below 5%. Comparative studies have indicated that present-day methods (roll tubes, GasPak, anaerobic tubes, and anaerobic glove box) result in comparable recovery of anaerobes from clinical specimens (72). This chapter will emphasize current developments and applications in anaerobic fermentations, since methodologies have been thoroughly reviewed in previous treatises (1, 2, 5, 15, 18, 26, 32, 37, 38, 42, 43, 54, 58, 71, 72, 77, 81, 120, 121, 124, 136).

11.2. IMPORTANCE OF ANAEROBIC MICROBIOLOGY TO INDUSTRIAL AND MEDICAL MICROBIOLOGY

Anaerobic microbiology has had an important impact on medical and industrial microbiology. In the late 1800s, it was shown that certain strict anaerobic organisms, including the clostridia *C. botulinum*, *C. tetani*, and *C. welchii* (*C. perfringens*), produced some of the most toxic substances known to cause diseases in humans and animals (28, 49, 50, 62, 136, 137). The first demonstration of vaccination and development of immunity against protein toxins was accomplished by Kempner and Kitasato with tetanus and botulinus toxoids to vaccinate animals (69, 70).

The importance of anaerobic organisms in industrial fermentations was further demonstrated before and during World War I, when an industrial fermentation process was developed for butanol as a precursor of isoprene and of acetone used in the manufacture of munitions (67, 140). Perkins and Weizmann worked on the acetone-butanol fermentation with *Clostridium acetobutylicum* (31). Economically, the most important industrial fermentation worldwide is the anaerobic production of ethanol by *Saccharomyces* and other yeasts (29, 30). Thus, anaerobic fermentations have contributed and continue to contribute greatly to the science and industry of microbiology.

11.3. CHARACTERISTICS, ADVANTAGES, AND DISADVANTAGES OF ANAEROBIC FERMENTATIONS

A fundamental property of microorganisms of importance in biotechnology is their extremely small size, which gives a very large surface-to-volume ratio with consequent high metabolic rates. This property is of major importance in both aerobic and anaerobic industrial fermentations. Depending on their taxonomic group and physiology, microorganisms also contain specialized enzymes and enzyme systems that allow for specific degradation of substrates and synthesis of products. Anaerobic microorganisms do not utilize molecular oxygen in biosynthesis and are not capable of using oxygen as a terminal electron acceptor. Instead, they use a diverse array of organic and inorganic electron donors and acceptors in their energy metabolism (127, 142, 143). As expected, this wide diversity in metabolism involves an extremely diverse assemblage of microorganisms in the domains *Archaea*, *Bacteria*, and *Eucarya*. Because of the widely diverse metabolic patterns among the anaerobes, anaerobic fermentations have certain properties that are not found in aerobic processes; some of these properties appear to be advantageous and others are detrimental to the overall process.

11.3.1. Advantages of Anaerobic Fermentations

1. *The product yield may be higher.* Because of their characteristic energy metabolism and consequent formation of small quantities of ATP, anaerobes usually produce less biomass than aerobic organisms. With less biomass production, more carbon can be converted to end products and high specific product yields can be obtained. For high-density cultivation of anaerobes, it is necessary to limit the accumulation of end products such as acids, which can result in feedback inhibition of the process (76).

2. *Anaerobic fermentations require less mass and energy input than aerobic processes and may be more economical.* Most anaerobic fermentations require relatively little mass transfer (e.g., O_2) and energy input into the fermentors. The primary requirements for energy input into reactors are to keep cells in suspension and to make nutrients available to the suspended cells. Thus, fermentor operating costs can be considerably lower than those for oxygen-requiring aerobic fermentations. Anaerobes can use a variety of substrates including polysaccharides, sugars, molasses, and other complex substrates, which may be obtained from agricultural waste streams and reduce the overall cost of the fermentation process.

3. *Anaerobes catabolize complex substrates and produce unique products.* Certain anaerobic organisms efficiently ferment complex organic substrates such as polysaccharides or proteins, and many strict anaerobes are also capable of chemolithotrophic growth with simple compounds such as CO, CO_2, H_2, H_2S as electron donors or acceptors, with production of high yields of CH_4, organic acids, or other compounds that are not produced by aerobic organisms. The ecology and environments of anaerobes may be distinct from those of aerobes, and consequently anaerobes may possess unusual enzymes and catabolic pathways of potential value in biotechnological processes. The ability to grow under adverse conditions minimizes contamination during the fermentations. Many anaerobes grow at high temperatures at which oxygen is poorly soluble, and growth under these conditions can contribute to efficient product recovery.

4. *Anaerobes from consortia that enable the catabolism of complex substrates with formation of unique products.* Anaerobic life is well known to be dependent on the interaction of various groups of microorganisms (38). The interaction may be chemical, or the organisms may also physically associate, forming flocs or biofilms. This physical association promotes efficient metabolism and efficient degradative and synthesis reactions not achieved in aerobic processes. The involvement of mixed species of anaerobes may permit the complete utilization of complex substrates. For example, cellulose can be completely degraded to methane and hydrogen by anaerobic consortia. Mixed anaerobic cultures possess a wider range of enzyme activities than do individual organisms and are able to degrade a wide range of substrates and to achieve complete mineralization. Mixed anaerobic cultures are able to carry out multistep transformation of substrates that would not be possible for a single microorganism. Mixed-culture fermentations also enable the utilization of cheap and impure substrates including agricultural and industrial wastes.

5. *Anaerobes can thrive in extreme environments, providing an extremely rich reservoir of biodiversity and potential products and processes.* Anaerobes that grow under extreme environmental conditions such as thermophilic, acidic, and alkalophilic conditions have been found. These "extremophiles" utilize diverse genetic and physiological mechanisms and thus may provide highly stable enzymes and unique biosynthetic and catabolic capabilities.

11.3.2. Disadvantages of Anaerobic Fermentations

1. *Many anaerobic fermentations involving pure cultures are prone to contamination, bacteriophage infection, and spon-*

taneous degeneration. The associative nature of anaerobic ecological systems makes processes involving pure cultures particularly susceptible to contamination. Pure cultures of clostridia used in solvent fermentations and for production of certain toxins may lose their capacity for high solvent or toxin yields by mutation or other degeneration processes (68). Many clostridia also are subject to infection by bacteriophages that will decrease solvent or toxin yields (66, 90, 91, 108).

2. *Microbial communities (consortia) required for industrial processes may be unstable.* The component species of microbial consortia can change according to environmental changes and availability of nutrients. This can lead to disruption or inefficiency of the process. Mixed-culture microbial processes are also inherently difficult to study and model from a scientific perspective.

3. *Specialized media and apparatus are required for cultivation of strict anaerobes in the laboratory and in certain industrial processes.* By definition, obligate anaerobes are inactivated by exposure to oxygen. Considerable skill and meticulous methods are necessary for cultivation and manipulation of strictly anaerobic microorganisms.

4. *Many anaerobes are difficult to manipulate genetically.* Compared to certain groups of aerobic organisms such as actinomycetes, facultatively aerobic bacteria, and pseudomonads, relatively little is known about the methods for genetic manipulation and expression of desired genes and biosynthetic pathways.

5. *The inability of anaerobes to utilize oxygen in biosynthesis limits the production of certain primary metabolites and many secondary metabolites.* Obligate anaerobes do not possess oxygenases and oxidases (with certain exceptions) and are incapable of synthesizing many classes of compounds that utilize oxygen in biosynthesis. Included among these are many primary metabolites such as sterols (12) and secondary metabolites including specific classes of pigments (64), antibiotics (132), and other families of metabolites utilizing oxygen in their biosynthesis.

6. *Anaerobes can generate toxic and noxious products such as putrefactive amines and sulfide compounds.* The sulfate-reducing bacteria produce sulfides as principal end products; these compounds can cause physical damage to metals and concrete. Many putrefactive clostridia produce noxious products such as cadaverine, putrescine, and others that make them difficult to work with, especially on a large scale.

7. Yields of products can be low. Because of their characteristic low energy yields, low titers of products depending on energy for biosynthesis may be experienced. Consequently, it is sometimes useful to use specialized culture apparatus or techniques such as dialysis culture (42).

8. *Patentability and protection of intellectual property can be difficult.* Many anaerobic processes are extremely complex and difficult to define quantitatively. Consequently, it is difficult in many cases to obtain patents on these complex processes.

11.4. DEFINITION OF ANAEROBIOSIS AND OBLIGATELY ANAEROBIC MICROORGANISMS

It is difficult to provide an accurate definition of anoxia (38). The Earth's atmosphere contains 20.9% oxygen by volume. The solubility of oxygen in water is low and varies with temperature and solute concentration. At atmospheric pressure, the solubility in pure water is 291 μM at 20°C.

Obligately anaerobic microorganisms have an energy metabolism that is independent of oxygen, and they can complete their life cycle in the absence of oxygen. Facultative anaerobes are capable of growing by oxidative phosphorylation but can also grow in the absence of oxygen. Microaerophilic microorganisms require reduced oxygen concentrations for growth. Obligately anaerobic microorganisms lack the ability to grow in the presence of oxygen because of the absence of protective mechanisms to reactive oxygen species.

The sensitivity of anaerobic organisms to molecular oxygen varies considerably among different microbial groups. Some, such as methanogens, are extremely sensitive and are inhibited at levels far below the detectable limit of oxygen (58). Others, such as many clostridia, are relatively tolerant to oxygen and can withstand brief exposure. Obligate anaerobes also do not utilize oxygen in any energy-generating step or in any anabolic or catabolic reactions.

The inactivation of anaerobes by oxygen probably does not involve direct reactions of cellular components with oxygen. Molecular oxygen contains two unpaired electrons with unpaired spins, and its direct reaction with most organic compounds is inefficient. Oxygen gains oxidative capacity either by absorbing energy to the excited singlet state (singlet oxygen, $^1O_2$) (64) or by single-electron reductions to superoxide anion ($O_2^{\cdot}$), hydrogen peroxide (H_2O_2), and hydroxyl radical ($HO^{\cdot}$). These reactive oxygen species, particularly $^1O_2$ and the radicals ($O_2^{\cdot}$ and $HO^{\cdot}$), are highly reactive toward many organic molecules and can cause cellular damage and death. Aerobic organisms contain enzymatic and nonenzymatic means of protection from reactive oxygen species (113, 115).

Anaerobes are sometimes categorized in relation their ability to grow at different oxidation-reduction potentials, redox potential or E_h. The E_h is a measure of the tendency of substances in a solution to donate or receive electrons (i.e., to become oxidized or reduced) (15). Quantitatively, E_h is defined by the equation $E_h = E_0 + (RT/nF) \ln ([Ox]/[Red])$ where E_0 ($= -\Delta G_0/nF$) is the potential of the half-oxidized system (15, 38). In practice, E_h is expressed in units of electrical potential (millivolts) relative to the potential of the hydrogen electrode. E_0 is termed the standard redox potential of a 50% reduced compound, in reference to the standard hydrogen electrode. E_0' is the E_h at pH 7.0 at 25°C under 1 atm H_2 and has a value of −413 mV. The E_0' is useful in describing the standard redox potential of any 50% reduced substance at pH 7.0 in reference to the hydrogen electrode.

Positive E_h values will inhibit the growth of anaerobes, but the limiting E_h depends on the oxidant(s) involved in setting the E_h. Some anaerobes are capable of initiating growth at positive redox potential when a substance(s) other than O_2 has increased the E_h (84, 133). Depending on the microbial species, facultative anaerobes can grow between +300 and −420 mV and obligate anaerobes can grow between −150 and −420 mV. When the E_h is increased by the presence of dissolved O_2, most obligate anaerobes are inhibited at an E_h higher than −100 mV. Stringent anaerobes such as certain methanogens will not initiate growth at an E_h higher than −330 mV.

The pH of the medium affects the E_h and hence the growth and survival of anaerobes. Generally, the E_h increases as the pH is lowered in redox reactions in which protons are liberated. The E_h increases during culture when acids are produced. In industrial or laboratory fermentations, the measured E_h represents the sum of the E_h of a number of individual redox reactions (84). Methods involving electrodes can be used to determine the E_h of a medium, although in practice, redox-active dyes such as resazurin are often used and serve as good indicators of the degree of E_0' in a medium (see section 11.6).

As described by Hungate (58), the quantity of oxygen in 1 liter of water at 30°C in equilibrium with air at 1 atm is 1.48×10^{-56} molecule per liter. Hungate points out that this calculation illustrates that it is difficult to obtain low redox potentials for growth of strict anaerobes and that to obtain the needed low potential, some reduced system at a lower potential must be added. The use of reducing agents is described in section 11.6.

11.5. FUNDAMENTALS OF ANAEROBIC METABOLISM

11.5.1. Energy Production

There are several fundamental differences between aerobic and anaerobic energy metabolism. Energy production in anaerobic bacteria can be achieved by fermentation, by respiration using inorganic electron acceptors other than oxygen, and by anaerobic photosynthesis. Anaerobic energy metabolism is characterized by low production of ATP and low cell mass formation (10, 127, 142, 143). Anaerobic substrate metabolism in nearly all natural habitats and in many industrial processes is characterized by the participation of communities of organisms to break down complex substrates. For example, fermentation of cellulose to methane requires the participation of several microbial species. During this process, the organisms involved and the electron donors and acceptors used depend on the thermodynamic efficiencies of energy transformations (10, 127, 143). A wide variety of electron donors and acceptors are used by anaerobes, resulting in a multitude of end products (Table 1).

11.5.1.1. Fermentation

Fermentation utilizes organic substrates and organic electron acceptors, and reduced organic substances are produced as end products. ATP is obtained by substrate-level phosphorylation. If glucose is the primary substrate, the energy needed for biosynthesis is low compared to that for two- and three-carbon substrates (127). Therefore, in microbial consortia, the ability of organisms to compete will depend on their primary substrate for energy production. The end products will also influence the ATP acquired: fermentation of glucose to volatile fatty acids and H_2, as occurs in many clostridia, can yield more than two ATP molecules per hexose fermented, whereas fermentation to two lactate molecules (homolactic acid bacteria) or to ethanol (fermentative yeasts) molecules yields a maximum of two ATP molecules of hexose fermented. Other factors influencing the predominant species and yield of cells and end products is the substrate concentration and tolerance of an organism to end products such as ethanol. For most industrial anaerobic fermentations, the process rate is increased if end products are removed simultaneously. Energy-yielding degradative pathways unique to anaerobes have been described (44, 45, 127, 143).

Several groups of anaerobes utilize nitrogen-containing compounds as sources of energy (8, 44). Anaerobic fermentations of amino acids (8, 44) are particularly abundant in clostridia and in rumen bacteria. The availability of nitrogen is also critical and may be limiting for the growth in and consequent pollution of oceans and lakes by anaerobes (129).

11.5.1.2. Anaerobic Respiration, Phototrophy, and Methanogenesis

Anaerobic respiration yields energy by the oxidation of organic or inorganic substrates and the reduction of inorganic or organic electron acceptors by using cytochrome/FeS-membrane based systems (127, 143). Many organisms utilizing anaerobic respiration can generate more ATP

TABLE 1 Nutritional diversity of anaerobes[a]

Process	Electron donors utilized	Electron acceptors and reduced end product(s)
Fermentation	Organic molecules: carbohydrates, amino acids, purines, pyrimidines	Organic molecules, protons: alcohols, fatty acids, ketones, hydrogen gas
Anaerobic respiration	Organic molecules: as above Inorganic molecules or ions: carbon monoxide, hydrogen gas, metallic sulfides, ammonium, nitrite, ferrous and manganous salts, elemental sulfur	Nitrate, nitrite, ammonia, nitrogen gas, acetate, fumarate, succinate, dimethylsulfide, ferric salts, ferrous salts, trimethylamine, trimethylamine oxide
Methanogenesis	Hydrogen gas, formate, acetate	Carbonate, methane
Phototrophy	Organic molecules: alcohols, fatty acids, organic acids Inorganic compounds or ions: Hydrogen gas, ferric sulfide, elemental sulfur, thiosulfate	Carbonate, cellular components

[a]Adapted from reference 75.

molecules per substrate than can fermenters that acquire energy solely by substrate-level phosphorylations. The energetics of chemolithotrophy and phototrophy and its influence on anaerobic processes has been reviewed (127, 128, 142, 143).

11.5.2. Unique Role of Acetic Acid in Anaerobic Metabolism

In anaerobic environments in which levels sulfate or other inorganic electron acceptors are low, methane production from CO_2 and H_2 is the usual process (38). Many eubacteria are capable of synthesizing acetate from organic substrates or from carbon dioxide (125), and acetate is a key intermediate in anaerobic decomposition. Methanogens can ferment acetate to CH_4 and CO_2, whereas sulfate reducers utilize it for the production of H_2S. The final products of anaerobic decomposition are usually H_2S, CH_4, and CO_2, with acetate serving as a key intermediate.

11.5.3. Thermodynamics of Anaerobic Metabolic Processes and Modeling of the Processes

Zehnder and Stumm (143) have provided an excellent review of the thermodynamics of anaerobic growth as it is influenced by electron donors and acceptors. In particular, they have provided insights into the roles that electron acceptors and donors play in determining the bacterial groups that will predominate in specific environments and the physiology of substrate degradation and product formation. The oxidation states of the substrates and the final products determine the quantity of energy and biomass that can be obtained in a process. Thermodynamic models and chemical balance equations for biomass formation, growth kinetics, and catabolic-product and non-catabolic-product yields have also been described for simple anaerobic fermentations (37, 52, 80). Further research is needed to develop quantitative models for process design and optimization of industrial anaerobic processes.

11.5.4. Fermentation Balances

The stoichiometry of conversion of substrates to products can be derived by quantitative determination of the substrate(s) used and the product(s) formed (138). This analysis is valuable in evaluating the efficiency of an anaerobic process. The methods for calculation of carbon recovery and redox balance have been described previously (138). Physical and chemical techniques for determining product utilization, cell mass formation, and end product concentrations have also been described previously (27, 139).

11.6. METHODOLOGIES FOR WORKING WITH ANAEROBES

Special considerations are needed for the isolation, characterization, manipulation, and cultivation of anaerobic organisms. These considerations may differ depending on whether the process under study uses pure or mixed cultures and on the oxygen sensitivity of the species under study.

11.6.1. Pure Cultures versus Microbial Consortia

11.6.1.1. Pure Cultures

To the modern microbiologist, the isolation, purification, and identification of anaerobic bacteria and fungi present some problems not commonly encountered in other areas of microbiology. The necessity for special culture conditions and media is well known, but practical problems are still encountered, one being the maintenance of pure cultures. Several highly skilled investigators have inadvertently found that the "pure cultures" under study were actually mixed cultures (see e.g., reference 17). Careful attention must be given to the preparation of pure seed cultures for anaerobic fermentations. The presence of pure cultures can generally be confirmed by showing that a single colony type appears on streak cultures on several media. Phase microscopy is also useful for providing evidence of a pure culture.

11.6.1.2. Cultivation of Microbial Consortia and Communities

Mixed microbial consortia are required in many anaerobic industrial processes. Methods for the cultivation of microbial consortia and communities have been described previously (18). Various community culture methods have been used, including chemostats, nutristats, and microstats (18). Technically, these can be difficult to maintain and to study quantitatively.

11.6.2. Methodological Considerations in Cultivation of Anaerobes

Anaerobes vary considerably in their oxygen tolerance and in their nutritional requirements for growth. Therefore, their isolation and cultivation vary according to the species. The methods used for isolation, cultivation, and identification of anaerobes for industrial fermentations have been complemented by research in clinical (72, 136) and environmental (59) microbiology. Methods for growing and maintaining anaerobes have already been thoroughly described (15, 54, 77, 103), and only selected aspects are described here. Current anaerobic techniques derive from the principles of Hungate in his pioneering studies of rumen anaerobes (57). Subsequent work at the Virginia Polytechnic Institute Anaerobe Laboratory (54, 83) and the Centers for Disease Control and Prevention (32) and the advent of the anaerobe chamber (4, 35, 74) have further facilitated work with anaerobic cultures. The development of the Wolin-Miller tube system has enabled the culture of gas-producing organisms that create pressure in the vials fitted with septa (7, 81).

11.6.2.1. Use of Air-Impermeable Culture Materials

Anaerobic bacteria are routinely grown in "Hungate" tubes or in serum vials sealed with black butyl rubber stoppers. The tubes or vials are commonly sealed with screw caps or crimped aluminum seals (useful for gas utilizers such as methanogens or acetogens) (77). Nutrients can be added and cultures can be inoculated through the rubber stopper with syringes.

For storage of anaerobic cultures, the organisms should be maintained in complex media in anaerobic serum vials or tubes with black butyl rubber stoppers. Plastic tubes or grey stoppers are not suitable since they allow oxygen to diffuse into the cultures. Caution is needed when transferring cultures or adding reagents via hypodermic needles through rubber septa, particularly during the culture of pathogens such as C. *botulinum* and C. *tetani*. Researchers should receive proper immunization and reports of adequate antibody titers in serum prior to handling these pathogens (62). Precautions are also needed when working with organisms that utilize or produce pressurized gases. Vials or serum bottles should be placed behind a safety shield during pressurization with gases.

Anaerobic cultures should be stored frozen (−20 and/or −70°C) in glass or metal containers with butyl rubber stoppers that are impermeable to oxygen. Clostridia can be sporulated by being cultured in cooked meat medium before storage which will enhance their survival.

11.6.2.2. Reducing Agents and Redox Indicators

Although oxygen is routinely removed from anaerobic cultures by heating or evacuation, these physical methods are not usually sufficient to lower the redox potential to the degree required for the growth of certain strict anaerobes (58). Furthermore, oxygen tends to adhere to or dissolve in glassware and plastic used in manipulations. Therefore, reducing agents are commonly added to anaerobic media to lower the redox potential and to poise it at a desired value. The more commonly used agents are cysteine hydrochloride (0.1 to 1%; −210 mV), sodium thioglycolate (0.05%; −100 mV), sodium sulfide (0.01 to 0.025%; −270 mV), H_2 + $PdCl_2$ (variable concentrations; −420 mV), and titanium(III) citrate for strict anaerobes (0.5 to 2 mM; −480 mV). Recipes and practical aspects of working with these agents have been described previously (15, 77).

Dyes sensitive to redox potential are routinely used in laboratory culture to estimate the E_h (15, 77). Commonly used redox indicators are methylene blue (E_0' = 11 mV), resorufin from resazurin (E_0' = −51 mV), phenosafranine (E_0' = −252 mV), and benzyl viologen (E_0' = −359 mV). Redox-sensitive dyes should be tested for toxicity for the microorganism(s) under study before use.

11.6.2.3. Removal of Oxygen from Gases

O_2 is removed from gases with a column of heated copper. The design of columns and the safety precautions needed have been described previously (58, 77, 121).

11.6.2.4. Removal of Oxygen from Media and Other Solutions

The simplest means of removing oxygen from heat-stable solutions is to boil them vigorously for about 1 min (58). As the solution cools, oxygen-free gas can be added to the container. If the solution is heat labile or contains heat-labile constituents, it can be freed of oxygen by having oxygen-free gas bubbled through it, but removal by this method is slow, unless the bubbles are very abundant, and generally requires 30 min to 1 h depending on the surface area of the exiting gas. One of the most efficient methods to ensure a low oxygen content in a medium is to allow growth of an aerobic of facultatively anaerobic organism in the medium. In practice, this is carried out widely in food fermentations and in anaerobic digestion processes. The aerobic organisms must not produce metabolic products that are inhibitory to the anaerobic culture.

11.6.2.5. Safety of Palladium and Anaerobic Gas Mixtures

Palladium catalysts are used to convert oxygen to water, but they become inactivated by excess water, H_2S, and other metabolic end products. For routine use in jars and anaerobe chambers, catalyst materials should be reactivated daily by being heated at 160°C for 2 h (22). The catalyst should also be occasionally cleaned by heating at 200°C and then exposed, while still hot, to an anaerobic gas mixture. It has been suggested that a maximum of 5% hydrogen be used in anaerobic gas mixtures since higher concentrations can lead to explosions (22). A 5% hydrogen concentration in the gas mixture is sufficient to maintain anaerobic conditions below 5 to 10 parts per million of oxygen, which is sufficiently low for the growth of methanogens and other strict anaerobes.

11.6.2.6. Anaerobic Chambers

For genetic manipulations and other procedures with strict anaerobes, it is useful to work with isolated colonies on agar media. For this purpose, anaerobes can be grown on the surface of agar media in anaerobic chambers or anaerobic jars. Several commercial units are available, and guidelines for their use are available from the manufacturers (77). A walk-in anaerobic room used at the National Institutes of Health has been useful for certain manipulations of anaerobic organisms (96).

11.6.2.7. Specialized Apparatus and Fermentors for Large-Scale Production of Strict Anaerobes

Many anaerobes of industrial importance, including clostridia, methanogens, and anaerobic thermophiles, can be grown in glass carboys or stainless steel fermentors (26, 108, 120). Toxigenic clostridia grow optimally and produce toxins with minimal agitation. However, methanogens and other anaerobes that use gases as substrates or are nonmotile require agitators for mass transport of the substrate gases.

11.6.2.8. Media for Strict Anaerobes

It is not feasible in this short chapter to list media for the growth of various anaerobic microorganisms. Media appropriate for representative bacterial groups have been described (6, 21). Many anaerobic bacteria require rich media and lack biosynthetic pathways for many growth factors. Consequently, complex media are used in practice for cultivation of most anaerobes. Typically, media are prepared from three separate components, (i) the carbon source, (ii) growth factors such as vitamins, amino acids, and yeast extract, and (iii) buffer; these are combined after autoclaving or filter sterilization (for heat-labile nutrients). Defined media are useful for understanding the physiology and genetics of microorganisms at the molecular level (63, 86, 134).

11.7. ANAEROBIC ORGANISMS OF INDUSTRIAL IMPORTANCE

Anaerobic organisms comprise a wide range of genera in the domains *Bacteria* and *Archaea*. Obligately anaerobic microorganisms exhibit considerable physiological diversity and include autotrophs, heterotrophs, and chemolithotrophs. Anaerobic extremophiles are also prevalent and include acidophiles, alkalophiles, halophiles, thermophiles, and hyperthermophiles (34, 53). The major groups of current and potential importance in biotechnology and anaerobic fermentations are briefly described. The numerous species of medically important anaerobes are not described unless they are used in industrial anaerobic fermentations such as for vaccine production.

11.7.1. Bacteria

A diversity of organisms in the domain *Bacteria* are used or have potential uses in industrial processes.

11.7.1.1. Clostridia

The genus *Clostridium* is made up of an extremely diverse group of anaerobes characterized by a gram-positive cell wall and formation of endospores (49, 50, 122). The clostridia are important medically and for the production of metabolites including enzymes and solvents. The origins and relationships of clostridia used for solvent production have been described (65, 66, 98), with the interesting conclusion that the solventogenic clostridia are highly polyphyletic. Recent studies of the taxonomy of the clostridia suggest that the genus actually comprises numerous genera and species (20, 122), but a major effort will be needed to accurately redefine this group.

Toxigenic clostridia are industrially valuable for the production of human and animal vaccines (16, 28, 40, 87, 100, 141) and for human therapeutics (82, 108, 114). Selection of high-producing strains is of considerable importance in obtaining high titers and good-quality toxin (87, 108). Precautions must be taken, including immunization of workers, to avoid accidental intoxications.

11.7.1.2. Lactic Acid Bacteria

The lactic acid bacteria are important in several food fermentations as well as in the production of certain organic acids. The most important genera are *Lactococcus*, *Lactobacillus*, *Propionibacterium*, *Streptococcus*, and *Leuconosotoc*. Recent reviews have described the importance of the lactic acid bacteria in industrial microbiology (13, 41).

11.7.1.3. Bacterial Thermophiles

Anaerobic thermophiles are prominent in the domains *Bacteria* and *Archaea*. Most of the anaerobic thermophiles in the *Bacteria* have an optimum growth temperature below 75°C (135). Thermophiles offer advantages for biotechnological processes including high growth rates, processes resistant to the negative effects of contamination, and enhanced recovery of volatile end products. The main genera of interest in the *Bacteria* are *Clostridium*, *Thermoanaerobacter*, *Dictyoglomus*, and *Thermoanaerobium*. The sulfate-reducing thermophiles include species in the genera *Thermodesulfotobacterium*, *Desulfotomaculum*, *Desulfovibrio*, and *Desulfurella*. Thermophilic methanogens including *Methanobacterium*, *Methanothermus*, and *Methanopyrus* species have received attention for their potential in industrial processes (73, 135).

11.7.1.4. Sulfate-Reducing Bacteria

Sulfur-reducing bacteria have been demonstrated to be active in biotransformation of a variety of organic substrates including hazardous compounds and toxic wastes (9, 143). They are also industrially important in the oil and gas industry; in corrosion of iron, steel, and concrete; and in sewage treatment (89). Important genera include *Desulfovibrio*, *Desulfotomaculum*, and *Desulfobacter*, although the taxonomy of the bacterial group has been expanded and other genera are certainly also of importance (116).

11.7.1.5. Anoxygenic Phototrophic Bacteria

The concept of anoxygenic photolithotrophy was discovered by Buder and van Niel (126, 130). These bacteria lack photosystem II and thus differ from cyanobacteria. They depend on the oxidation of reduced substrates including simple organic compounds, reduced sulfur compounds, elemental sulfur, and molecular hydrogen. They have been considered for industrial production of single-cell protein, vitamins, ubiquinones, enzymes and biopolyesters and for use in biodegradations and waste treatment (106, 107), but their actual utilization has been largely unexplored.

11.7.1.6. Other Eubacteria of Consideration in Industrial Processes

Certain other obligately anaerobic bacteria are of importance in food fermentations, anaerobic digestion and detoxification of hazardous wastes, and anaerobic biotransformations. The most commonly encountered anaerobic bacteria isolated from human infections are *Bacteroides fragilis*, pigmented *Prevotella*, *Porphyromonas*, *Fusobacterium nucleatum*, *Peptostreptococcus*, and the pathogenic clostridia (39, 50). Precautions should be taken when these species are involved in anaerobic consortia.

11.7.2. Archaea

The major organisms of importance in anaerobic fermentations in the domain *Archaea* are the thermophiles and methanogens.

11.7.2.1. Methanogens

Methanogens require not only anoxic conditions but also highly reduced conditions (E_h below −330 mV) for growth (117, 121). Methanogens use carbon dioxide or methyl groups as electron acceptors and from methane as their end product. Substrate requirements include $H_2 + CO_2$, formate, acetate, methanol, and methylamines (119). The taxonomic groups of methanogens have been described previously (119). Hyperthermophilic methanogens include the genera *Methanococcus*, *Methanothermus*, and *Methanopyrus*, which have maximum growth temperatures ranging from 91 to 110°C (1).

11.7.2.2. Heterotrophic Anaerobic Thermophiles

The majority of thermophiles discovered in recent years belong within the *Archaea* (1, 119). Most of these organisms are dependent upon organic carbon and nitrogen as electron donors for growth, although some genera can oxidize H_2. Elemental sulfur is commonly used as an electron acceptor (1). The major orders include the *Thermoproteales*, *Thermococcales*, and several unclassified organisms (1). The maximum growth temperature of these organisms ranges from 67 to 110°C. Large-scale (600-liter) fermentation methods have been described for certain hyperthermophiles (1, 2). Certain practical aspects of isolating and growing these hyperthermophiles have recently been described (103). Advances have also been made in the genetics and molecular biology of thermophiles and hyperthermophiles (88, 103).

11.7.2.3. Sulfate Reducers

The unique sulfate-reducing genus *Archaeoglobus* has a maximum growth temperature of 95°C (1). It is capable of oxidizing organic carbon substrates or H_2 and utilizing SO_4^{2-} or $S_2O_3^{2-}$ as electron acceptors. It could be valuable in transformations of industrial sulfur residues and in petroleum microbiology. The mesophilic and thermophilic sulfate reducers are economically important in corrosion of metals and concrete.

11.7.3. Yeasts

Yeasts used industrially are not obligate anaerobes, but they are mentioned here since anaerobic fermentations by yeasts are economically the most important industrial microbiology processes. Yeasts are fungi that at some point in their life cycle grow as single cells and do not produce true hyphae (95). A new edition of *The Yeasts* is due to be published in 1998 and will contain invaluable information on the properties of various yeasts, including a summary of their industrial uses (29). Although yeasts have numerous industrial and agricultural applications, the main processes involve bioethanol production, baking, and other food fermentations. Nearly all yeasts used industrially for these processes belong to the species *Saccharomyces cerevisiae*.

11.7.4. Mixed Cultures

Mixed cultures are used in numerous anaerobic fermentations. Mixed-culture fermentations, including those used for the production of foods and beverages, have been used since antiquity (144). More recently, the utility of mixed cultures of anaerobes is being realized in the bioremediation of xenobiotics and pollutants, in the manufacture of fine chemicals (59, 85), and in the production of novel pharmaceuticals (108) and novel tumor treatments (82).

11.8. SELECTED ANAEROBIC INDUSTRIAL PROCESSES

Selected anaerobic processes are listed in Table 2, with emphasis on the potential of obligate anaerobes in industrial microbiology. The traditional and major processes have been thoroughly described in the earlier literature (24, 101), and only selected processes are illustrated to emphasize the unique aspects of obligately anaerobic fermentations. Emphasis is also given to largely uninvestigated uses of anaerobes.

TABLE 2 Selected anaerobic industrial processes

Product or process	Representative organism(s)	Reference(s)
Degradative and bioremediation processes		
Biopolymer degradation	*Clostridium* spp.	105
	Thermophiles	19, 66, 73, 135
Hydrolases from extremophiles	Various	56, 60, 61, 97
Salt-resistant enzymes	Various	92, 99
Bioremediation of toxic wastes	Consortia	11, 93, 94, 131
Aromatic metabolism	Various	51
Benzene, benzoate degradation	*Rhodopseudomonas* spp.	36
Degradation of chlorinated ethenes	Methanogens	55
Dehalogenation of wastes	Various, consortia	78
Degradation of nitroaromatics	Various	23
Bioremediation of metals	Sulfate reducers, others	3, 48
Biodegradation of agricultural chemicals	Various	14
Metal corrosion	Sulfate reducers, various	33
Wastewater treatment	Consortia	93, 94, 123, 130
Solvent, acid, and fuel production		
Ethanol	*Saccharomyces* spp.	29
	Clostridium spp.	135
	Thermoanaerobacter, others	135
Methane	*Methanobacterium spp*.	
	Methanococcus spp.	
	Thermophilic methanogens	1
Acetic acid	*Clostridium thermoaceticum*	19
Acetone and butanol	*Clostridium acetobutylicum*	67
Conversion of coal-derived synthesis gases to fuels	Various	46
Selective oxidation of crude oil	Various	104
Conversion of waste sulfides	Extremophiles	60, 61
Biotransformations and novel enzymes		
Redox enzymes	Various	85
DNA polymerases	*Thermus* spp., others	112
β-Lactam precursor	*Clostridium tetanomorphum*	112
Enzymes for increased rumen efficiency	Various	111
Therapeutics and vaccines		
Vaccine production	*Clostridium* spp.	16, 100, 114
Botulinum toxin as therapeutic	*Clostridium botulinum*	108
Tumor targeting	*Clostridium* spp.	82

11.9. CONCLUSION AND PROSPECTS

Anaerobic processes have contributed greatly to the advances and successes of industrial microbiology. The anaerobic world is an extremely diverse one in the organisms it contains and in the physiological processes it performs. Technical obstacles to the cultivation and maintenance of strict anaerobes have led to delays in realizing the industrial potential of certain groups of anaerobes. Because of their ability to utilize organic and inorganic substances as electron donors and acceptors, anaerobes exist successfully in several habitats (e.g., thermal vents) that have only begun to be explored. The application of modern technologies to ancient and newly discovered mircoorganisms and microbial processes should result in their sustained use in existing processes such as ethanol production and in new processes in industrial microbiology.

REFERENCES

1. **Adams, M. W. W.** 1995. Thermophilic archaea: an overview, pp. 3–7. *In* F. T. Robb and A. R. Place (ed.), *Archaea: a Laboratory Manual. Thermophiles.* Cold Spring Harbor Laboratory Press, Cold Spring Harbor, N.Y.
2. **Adams, M. W. W.** 1995. Large-scale growth of hyperthermophiles, pp. 47–50. *In* F. T. Robb and A. R. Place (ed.), *Archaea: a Laboratory Manual. Thermophiles.* Cold Spring Harbor Laboratory Press, Cold Spring Harbor, N.Y.
3. **Ahmann, D.** 1997. Bioremediation of metal-contaminated soil. *SIM News* **47:**218–233.
4. **Aranki, A., S. A. Syed, E. B. Kenney, and R. Freter.** 1969. Isolation of anaerobic bacteria from human gingiva and mouse cecum by means of a simplified glove box procedure. *Appl. Microbiol.* **17:**568–576.
5. **Atlas, R.** 1995. *Principles of Microbiology.* The C. V. Mosby Co., St. Louis, Mo.
6. **Atlas, R. M.** 1993. *Handbook of Microbiological Media.* CRC Press, Inc., Boca Raton, Fla.
7. **Balch, W. E., G. E. Fox, L. J. Magrum, C. R. Woese, and R. S. Wolfe.** 1979. Methanogens: revevaluation of a unique biological group. *Microbiol. Rev.* **43:**260–296.
8. **Barker, H. A.** 1981. Amino acid degradation by anaerobic bacteria. *Annu. Rev. Biochem.* **50:**23–40.
9. **Barton, L. L.** (ed.). 1995. *Sulfate-Reducing Bacteria.* Plenum Press, New York, N.Y.
10. **Bauchop, T., and S. R. Elsden.** 1960. The growth of microorganisms in relation to their energy supply. *J. Gen. Microbiol* **23:**457–469.
11. **Bhatnager, L., and B. Z. Fathepure.** 1991. Mixed cultures in detoxification of hazardous waste, pp. 293–340. *In* J. G. Zeikus and E. A. Johnson (ed.), *Mixed Cultures in Biotechnology.* McGraw-Hill Book Co., New York.
12. **Bloch, K.** 1994. Oxygen and evolution, pp. 37–64. *In* K. Bloch (ed.), *Blondes in Venetian Paintings, the Nine-Banded Armadillo, and Other Essays in Biochemistry.* Yale University Press, New Haven, Conn.
13. **Bozoglu, T. F., and B. Ray (ed.).** 1996. *Lactic Acid Bacteria: Current Advances in Metabolism, Genetics, and Applications.* Springer-Verlag KG, Berlin.
14. **Bradley, S. N., T. B. Hammill, and R. L. Crawford.** 1997. Biodegradation of agricultural chemicals, pp. 815–821. *In* C. J. Hurst, G. R. Knudsen, M. G. McInerney, L. D. Stetzenbach, M. V. Walter (ed.), *Manual of Environmental Microbiology.* ASM Press, Washington, D.C.
15. **Breznak, J. A., and R. N. Costilow.** 1994. Physiochemical factors in growth, p. 137–154. *In* P. Gerhardt, W. A. Wood, and N. R. Krieg (ed.), *Methods for General and Molecular Bacteriology.* American Society for Microbiology, Washington, D.C.
16. **Brown, J. E., and E. D. Williamson.** 1997. Molecular approaches to novel vaccines for the control of clostridial toxemias and infections, p. 505–524. *In* J. I. Rood, B. A. McClane, J. G. Songer, and R. W. Titball (ed.), *The Clostridia: Molecular Biology and Pathogenesis.* Academic Press, Inc., San Diego, Calif.
17. **Bryant, M. P., E. A. Wolin, M. J. Wolin, and R. S. Wolfe.** 1967. *Methanobacillus omelianski*, a symbiotic association of two species of bacteria. *Arch. Microbiol.* **59:** 20–31.
18. **Caldwell, D. E., G. Wolfaardt, D. R. Korber, and J. R. Lawrence.** 1997. Cultivation of microbial consortia and communities, p. 79–90. *In* C. J. Hurst, G. R. Knudsen, M. G. McInerney, L. D. Stetzenbach, and M. V. Walter (ed.), *Manual of Environmental Microbiology.* ASM Press, Washington, D.C.
19. **Cheryan, M., S. Parekh, M. Shah, and K. Witjitra.** 1997. Production of acetic acid by *Clostridium thermoaceticum. Adv. Appl. Microbiol.* **43:**1–33.
20. **Collins, M. D., P. A. Lawson, A. Willems, J. J. Cordoba, J. Fernandez-Garayzabal, P. Garcia, J. Cai, H. Hippe, and J. A. E. Farrow.** 1994. The phylogeny of the genus *Clostridium*: proposal of five new genera and eleven new species combinations. *Int. J. Syst. Bacteriol.* **44:**812–826.
21. **Cote, R. J., and R. L. Gherna.** 1994. Nutrition and media, p. 155–178. *In* P. Gerhardt, W. A. Wood, and N. R. Krieg (ed.), *Methods for General and Molecular Bacteriology.* American Society for Microbiology, Washington, D.C.
22. **Cox, M.** 1997. Testing, reactivating, and cleaning palladium catalysts. *Clin. Infect. Dis.* **25**(Suppl. 2):S139.
23. **Crawford, R. L.** 1995. The microbiology and treatment of nitroaromatic compounds. *Curr. Opin. Biotechnol.* **6:** 329–336.
24. **Crueger, W., and A. Crueger.** 1982. *Biotechnology. A textbook of Industrial Microbiology.* Sinauer Associates, Inc., Sunderland, Mass.
25. **Cuny, H.** 1966. *Louis Pasteur. The Man and His Theories.* Paul S. Eriksson, Inc., New York.
26. **Daniels, L.** 1995. Large-scale culture techniques for methanogenic archaea, p. 63–74. *In* F. T. Robb (ed.), *Archaea: a laboratory manual. Methanogens* Cold Spring Harbor Laboratory Press, Cold Spring Harbor, N.Y.
27. **Daniels, L., R. S. Hanson, and J. A. Phillips.** 1994. Chemical analysis, p. 512–554. *In* P. Gerhardt, W. A. Wood, and N. R. Krieg (ed.), *Methods for General and Molecular Bacteriology.* American Society for Microbiology, Washington, D.C.
28. **Davidson, I.** 1976. An international survey of clostridial sera and vaccines. *Dev. Biol. Stand.* **32:**3–14.
29. **Demain, A. L., H. J. Phaff, and C. P. Kurtzman.** 1998. The industrial and agricultural significance of yeasts. *In* C. P. Kurtzman and J. W. Fell (ed.), *The Yeasts*, p. 13–19. Elsevier Science B.V., Amsterdam.
30. **Demain, A. L., and N. A. Solomon.** 1981. Industrial microbiology. *Sci. Am.* **245:**66–75
31. **Dixon, B.** 1996. Chaim Weizmann: from fermentation chemist to President of Israel. *Anaerobe* **2:**195–196.

32. **Dowell, V. R., Jr., and T. M. Hawkins.** 1990. *Laboratory Methods in Anaerobic Bacteriology*. U.S. Department of Health and Human Services, Public Health Service, Centers for Disease Control, Atlanta, Ga.

33. **Dowling, N. J., and J. Guezennec.** 1997. Microbially induced corrosion, p. 842–855. *In* C. J. Hurst, G. R. Knudsen, M. G. McInerney, L. D. Stetzenbach, and M. V. Walter (ed.), *Manual of Environmental Microbiology*. ASM Press, Washington, D.C.

34. **Edwards, C. (ed.).** 1990. *Microbiology of Extreme Environments*. McGraw-Hill Book Co., New York.

35. **Edwards, T., and B. C. McBride.** 1975. New method for the isolation and identification of methanogenic bacteria. *Appl. Microbiol.* **29:**540–545.

36. **Egland, P. G., D. A. Pelletier, M. Dispensa, J. Gibson, and C. S. Harwood.** 1997. A cluster of bacterial genes of anaerobic benzene degradation. *Proc. Natl. Acad. Sci. USA* **94:**6468–6489.

37. **Erickson, L. E.** 1988. Kinetics of growth and product formation, p. 119–146. *In* L. E. Erickson and D. Y. C. Fund (ed.), *Handbook on Anaerobic Fermentations*. Marcel Dekker, Inc., New York.

38. **Fenchel, T., and B. J. Finlay.** 1995. *Ecology and Evolution in Anaerobic Worlds*. Oxford University Press, Oxford, United Kingdom.

39. **Finegold, S. M.** 1995. Anaerobic infections in humans: an overview. *Anaerobe* **1:**3–9.

40. **Galazka, A., and F. Gasse.** 1995. The present status of tetanus and tetanus vaccination, p. 31–53. *In* C. Montecucco (ed.), *Clostridial Neurotoxins*. Springer-Verlag KG, Berlin.

41. **Gasson, M. J., and W. M. Devos (ed.).** 1994. *Genetics and Biotechnology of Lactic Acid Bacteria*. Blackie Academic and Professional, London.

42. **Gerhardt, P., and S. W. Draw.** 1994. Liquid culture, p. 224–247. *In* P. Gerhardt, W. A. Wood, and N. R. Krieg (ed.), *Methods for General and Molecular Bacteriology*. American Society for Microbiology, Washington, D.C.

43. **Gottscahl, J. C., W. Harder, and R. A. Prins.** 1991. Principles of enrichment, isolation, cultivation, and preservation of bacteria, p. 149–196. *In* A. Balows, H. G. Trüper, M. Dworkin, W. Harder, and K.-L. Schleifer (ed.), *The Prokaryotes*, 2nd ed. Springer-Verlag, New York.

44. **Gottschalk, G.** 1986. *Bacterial Metabolism*. Springer-Verlag, New York.

45. **Gottschalk, G., and S. Peinemann.** 1992. The anaerobic way of life, p. 300–311. *In* A Balows, H. G. Trüper, M. Dworkin, W. Harder, and K.-L. Schleifer (ed.), *The Prokaryotes*, 2nd ed. Springer-Verlag, New York.

46. **Grethlein, A. J., and M. K. Jain.** 1992. Bioprocessing of coal-derived synthesis gases by anaerobic bacteria. *Trends Biotechnol.* **10:**418–423.

47. **Hall, I. C.** 1929. A review of the development and application of physical and chemical principles in the cultivation of obligately anaerobic bacteria. *J. Bacteriol.* **17:** 255–301.

48. **Harding, G. L.** 1997. Bioremediation and the dissimilatory reduction of metals, p. 806–810. *In* C. J. Hurst, G. R. Knudsen, M. G. McInerney, L. D. Stetzenbach, and M. V. Walter (ed.), *Manual of Environmental Microbiology*. ASM Press, Washington, D.C.

49. **Hatheway, C. L.** 1990. Toxigenic clostridia. *Clin. Microbiol. Rev.* **3:**67–98.

50. **Hatheway, C. L., and E. A. Johnson.** 1998. *Clostridium*: the spore-bearing anaerobes, p. 731–782. *In* L. Collier, A. Balows, and M. Sussman (ed.), *Topley and Wilson's Microbiology & Microbial Infections*, 9th ed. Volume 2, *Systematic Bacteriology*. Arnold, London.

51. **Heider, J., and G. Fuchs,** 1997. Microbial anaerobic aromatic metabolism. *Anaerobe* **3:**1–22.

52. **Heijnen, S. J.** 1994. Thermodynamics of microbial growth and its implications for process design. *Trends Biotechnol.* **12:**483–492.

53. **Herbert, R. A., and R. J. Sharp (ed.).** 1992. *Molecular Biology and Biotechnology of Extremophiles*. Blackie Academic and Professional, London.

54. **Holdeman, L. V., E. P. Cato, and W. E. C. Moore.** 1977. *Anaerobe Laboratory Manual*, 4th ed. Virginia Polytechnic Institute, Blacksburg, Va.

55. **Holliger, C.** 1995. The anaerobic microbiology and biotreatment of chlorinated ethenes. *Curr. Opin. Biotechnol.* **6:**347–351.

56. **Honda, H., H. Naito, M. Taya, S. Iijima, and T. Kobayashi.** 1987. Cloning and expression of a *Thermaoanaerobacter cellulolyticus* gene coding for heat-stable β-glucanase. *Appl. Microbiol. Biotechnol.* **25:**480–483.

57. **Hungate, R. E.** 1950. The anaerobic mesophilic cellulolytic bacteria. *Bacteriol. Rev.* **14:**1–49.

58. **Hungate, R. E.** 1969. A roll tube method for cultivation of strict anaerobes. *Methods Microbiol.* **3B:**117–132.

59. **Hurst, C. J., G. R. Knudsen, M. J. McInerney, L. D. Stetzenbach, and M. V. Walter (ed.).** 1997. *Manual of Environmental Microbiology*. ASM Press, Washington, D.C.

60. **Jannasch, H. W.** 1995. Deep-sea vents as sources of biotechnologically relevant microorganisms. *J. Mar. Biotechnol.* 3:5–8.

61. **Jannasch, H. W., C. O. Wirsen, and T. Hoaki.** 1995. Isolation and cultivation of heterotrophic hyperthermophiles from deep-sea hydrothermal vents, p. 9–14. *In* F. T. Robb and A. R. Place (ed.), *Archaea: a Laboratory Manual. Thermophiles*. Cold Spring Harbor Laboratory Press, Cold Spring Harbor, N.Y.

62. **Johnson, E. A., and M. C. Goodnough.** 1998. Botulism, p. 723–741. *In* L. Collier, A. Balows, and M. Sussman (ed.), *Topley and Wilson's Microbiology & Microbial Infections*, 9th ed. Volume 3, *Bacterial Infections*. Arnold, London.

63. **Johnson, E. A., A. Madia, and A. L. Demain.** 1981. Chemically defined minimal medium for growth of the anaerobic cellulolytic thermophile *Clostridium thermocellum*. *Appl. Environ. Microbiol.* **41:**1060–1062.

64. **Johnson, E. A., and W. S. Schroeder.** 1995. Microbial carotenoids. *Adv. Biochem. Eng.* **53:**119–178.

65. **Johnson, J. L., and J.-S. Chen.** 1995. Taxonomic relationships among strains of *Clostridium acetobutylicum* and other phenotypically similar organisms. *FEMS Microbiol. Rev.* **17:**233–240.

66. **Jones, D. J., and S. Keis.** 1995. Origins and relationships of industrial solvent-producing clostridial strains. *FEMS Microbiol. Rev.* **17:**223–232.

67. **Jones, D. T., and D. R. Woods.** 1986. Acetone-butanol fermentation revisited. *Microbiol. Rev.* **50:**484–524.

68. **Kashket, E. R., and Z.-Y. Cao.** 1995. Clostridial strain degeneration. *FEMS Microbiol. Rev.* **17:**307–315.

69. **Kempner, W.** 1897. Further contributions to the knowledge of meat poisoning. The antitoxin to botulism. *Z. Hyg. Infektionskr.* **26:**481–500.

70. **Kitasato, S.** 1989. Ueber den Tetanobacillus. *Z. Hyg. Infektionskr.* **7:**225–234.

71. **Koch, A. L.** 1994. Growth measurement, p. 248–277. *In* P. Gerhardt, W. A. Wood, and N. R. Krieg (ed.), *Methods for General and Molecular Bacteriology*. American Society for Microbiology, Washington, D.C.

72. **Koneman, E. W., S. D. Allen, W. M. Janda, P. C. Schreckenberger, and W. C. Winn, Jr. (ed.).** 1997. The anaerobic bacteria, p. 709–784. *In* E. W. Koneman, S. D. Allen, W. M. Janda, P. C. Schreckenberger, and W. C. Winn, Jr. (ed.), *Color Atlas and Textbook of Diagnostic Microbiology*, 5th ed. J. R. Lippincott, Philadelphia.

73. **Kristjansson, J. K., and K. O. Stetter.** 1992. Thermophilic bacteria, p. 1–18. *In* J. K. Kristjansson (ed.), *Thermophilic Bacteria*. CRC Press, Inc., Boca Raton, Fla.

74. **Leach, P. A., J. J. Bullen, and I. D. Grant.** 1971. Anaerobic CO_2 cabinet for the cultivation of strict anaerobes. *Appl. Microbiol.* **22:**824–827.

75. **Leadbetter, E. R.** 1997. Prokaryotic diversity, form, ecophysiology, and habitat, p. 14–24. *In* C. J. Hurst, G. R. Knudsen, M. G. McInerney, L. D. Stetzenbach, and M. V. Walter (ed.), *Manual of Environmental Microbiology*. ASM Press, Washington, D.C.

76. **Ljungdahl, L. G., J. Hugenholtz, and J. Wiegel.** 1989. Acetogenic and acid-producing clostridia, p. 145–191. *In* N. P. Minton and D. J. Clarke (ed.), *Clostridia. Biotechnology Handbook 3*. Plenum Press, New York.

77. **Ljungdahl, L. G., and J. Wiegel.** 1986. Working with anaerobic bacteria, p. 84–96. *In* A. L. Demain and N. A. Solomon (ed.), *Manual of Industrial Microbiology and Biotechnology*. American Society for Microbiology, Washington, D.C.

78. **McCarty, P. L.** 1993. In situ bioremediation of chlorinated solvents. *Curr. Opin. Biotechnol.* **4:**323–330.

79. **McIntosh, J., and P. Fildes.** 1916. A new apparatus for the isolation and cultivation of anaerobic microorganisms. *Lancet* **i:**768–770.

80. **Meyer, C. L., and E. T. Papoutsakis.** 1988. Detailed stoichiometry and process analysis, p. 83–118. *In* L. E. Erickson and D. Y. C. Fung (ed.), *Handbook on Anaerobic Fermentations*. Marcel Dekker, Inc., New York.

81. **Miller, T. L., and M. J. Wolin.** 1974. A serum bottle modification of the Hungate technique for cultivating obligate anaerobes. *Appl. Microbiol.* **27:**985–987.

82. **Minton, N. P., M. L. Mauchline, M. J. Lemmon, J. K. Brehm, M. Fox, N. P. Michael, A. Giaccia, and J. M. Brown.** 1995. Chemotherapeutic tumour targeting using clostridial spores. *FEMS Microbiol. Rev.* **17:**357–364.

83. **Moore, W. E. C.** 1966. Techniques for routine culture of fastidious anaerobes. *Int. J. Syst. Bacteriol.* **16:**173–190.

84. **Morris, J. G.** 1976. The physiology of obligate anaerobiosis. *Adv. Microb. Physiol.* **12:**169–246.

85. **Morris, J. G.** 1994. Obligately anaerobic bacteria in biotechnology. *Appl. Biochem. Biotechnol.* **48:**75–106.

86. **Mueller, J. H., and P. A. Miller.** 1942. Growth requirements of *Clostridium tetani*. *J. Bacteriol.* **43:**763–772.

87. **Mueller, J. H., and P. A. Miller.** 1945. Production of tetanal toxin. *J. Immunol.* **50:**377–384.

88. **Noll, K. M., and M. Vargas.** 1997. Recent advances in genetic analyses of hyperthermophilic Archaea and Bacteria. *Arch. Microbiol.* **168:**73–80.

89. **Odom, J. M.** 1993. Industrial and environmental activities of sulfate-reducing bacteria, p. 189–249. *In* J. M. Odom and R. Singleton, Jr. (ed.), *The Sulfate-Reducing Bacteria: Contemporary Perspectives*. Springer-Verlag, New York.

90. **Ogata, S., and M. Hongo.** 1979. Bacteriophages of the genus *Clostridium*. *Adv. Appl. Microbiol.* **25:**241–273.

91. **Oguma, K.** 1976. The stability of toxigenicity in *Clostridium botulinum* types C and D. *J. Gen. Microbiol.* **92:** 67–75.

92. **Ollivier, B., P. Caumette, J.-L. Garcia, and R. A. Mah.** 1994. Anaerobic bacteria from hypersaline environments. *Microbiol. Rev.* **58:**27–38.

93. **Pavlosthathis, S. G., G. Misra, M. Prytula, and D. Yeh.** 1996. Anaerobic processes. *Water Environ. Res.* **67:**459–470.

94. **Pavlosthathis, S. G., G. Misra, M. Prytula, and D. Yeh.** 1996. Anaerobic processes. *Water Environ. Res.* **68:**479–497.

95. **Phaff, H. J., M. W. Miller, and E. M. Mrak.** 1978. *The Life of Yeasts*, 2nd ed. Harvard University Press, Cambridge, Mass.

96. **Poston, J. M.** 1995. The anaerobic chamber at the National Institutes of Health. *Anaerobe* **1:**183–184.

97. **Prieur, D.** 1997. Microbiology of deep-sea hydrothermal vents. *Trends Biotechnol.* **15:**242–244.

98. **Rainey, F. A., and E. Stackebrandt.** 1993. 16S rDNA analysis reveals phylogenetic diversity among the polysaccharolytic clostridia. *FEMS Microbiol. Lett.* **113:**125–128.

99. **Rainey, F. A., T. N. Zhilina, E. S. Boulygina, E. Stackebrandt, T. P. Tourova, and G. A. Zavarzin.** 1995. The taxonomic status of the fermentative halophilic anaerobic bacteria: description of the Haloanaerobiales ord. nov., Halobacteroidaceae fam. nov., Orenia gen. nov., and further taxonomic rearrangements at the genus and species level. *Anaerobe* **1:**185–199.

100. **Regamey, R. H., E. C. Hulse, and M. Sebald (ed.).** 1976. Clostridial products in veterinary medicine. *Dev. Biol. Stand.*, vol. 32.

101. **Rehm, H., and G. Reed (ed.).** 1981–1988. *Biotechnology: A Comprehensive Treatise*, vol. 1 to 8. Verlag Chemie International, Inc., Deerfield Beach, Fla.

102. **Riviere, J. W. M.** 1997. The Delft School of Microbiology in historical perspective. *Antonie Leeuwenhoek* **71:** 3–13.

103. **Robb, F. T., and A. R. Place (ed.).** 1995. *Archaea: a Laboratory Manual. Thermophiles*. Cold Spring Harbor Laboratory Press, Cold Spring Harbor, N.Y.

104. **Rueter, P., R. Rabus, H. Wilkes, F. Aekersberg, F. Rainey, and H. W. Jannasch.** 1994. Anaerobic oxidation of hydrocarbons from crude oil by new types of sulfate-reducing bacteria. *Nature* **372:**455–458.

105. **Saha, B. C., R. Lamed, and J. G. Zeikus.** 1989. Clostridial enzymes, p. 227–263. *In* N. P. Minton, and D. J. Clarke (ed.), *Clostridia. Biotechnology Handbook 3*. Plenum Press, New York.

106. **Sasikala, C., and C. Ramana.** 1995. Biotechnological potentials of anoxygenic phototrophic bacteria. I. Pro-

duction of single-cell protein, vitamins, ubiquinones, hormones, and enzymes used in waste treatment. *Adv. Appl. Microbiol.* **41:**173–226.

107. **Sasikala, C., and C. Ramana.** 1995. Biotechnological potentials of anoxygenic phototrophic bacteria. II. Biopolyesters, biopesticide, biofuel, and biofertilizer. *Adv. Appl. Microbiol.* **41:**227–278.
108. **Schantz, E. J., and E. A. Johnson.** 1992. Properties and use of botulinum toxin and other microbial neurotoxins in medicine. *Microbiol. Rev.* **56:**80–99.
109. **Schlegel, H. G.** 1996. Winogradsky discovered a new Modus Vivendi. *Anaerobe* **2:**129–136.
110. **Sebald, M., and D. Hauser.** 1995. Pasteur, oxygen, and the anaerobes revisited. *Anaerobe* **1:**11–16.
111. **Selinger, L. B., C. W. Forsberg, and K.-J. Cheung.** 1996. The rumen: a unique source of enzymes for enhancing livestock production. *Anaerobe* **2:**263–284.
112. **Sharp, R., and R. Williams (ed.).** 1995. *Thermus Species*. Plenum Press, New York.
113. **Shigenaga, M. K., T. M. Hagen, and B. N. Ames.** 1994. Oxidative damage and mitochondrial decay in aging. *Proc. Natl. Acad. Sci. USA* **91:**10771–10778.
114. **Shone, C. C., and P. Hambleton.** 1989. Toxigenic clostridia, p. 265–292. *In* N. P. Minton and D. J. Clarke (ed.), *Clostridia*. Plenum Press, New York.
115. **Sies, H.** 1991. *Oxidative Stress: Oxidants and Antioxidants*. Academic Press, Inc., Orlando, Fla.
116. **Singleton, R., Jr.** 1993. The sulfate-reducing bacteria: an overview, p. 1–20. *In* J. M. Odom and R. Singleton, Jr. (ed.), *The Sulfate-Reducing Bacteria: Contemporary Perspectives*. Springer-Verlag, New York.
117. **Smith, P. H., and R. E. Hungate.** 1958. Isolation and characterization of *Methanobacterium ruminantium* n. sp. *J. Bacteriol.* **75:**713–718.
118. **Sonnenwirth, A. C.** 1972. Evolution of anaerobic methodology. *Am. J. Clin. Nutr.* **25:**1295–1298.
119. **Sowers, K. R.** 1995. Methanogenic archaea: an overview, p. 3–13. *In* F. T. Robb (ed.), *Arachaea. A Laboratory Manual. Methanogens*. Cold Spring Harbor Laboratory Press, Cold Spring Harbor, N.Y.
120. **Sowers, K. R.** 1995. Large-scale growth of methanogens that utilize acetate and formate in a pH auxostat, pp. 75–77. *In* F. T. Robb and A. R. Place (ed.), *Archaea: a Laboratory Manual. Methanogens*. Cold Spring Harbor Laboratory Press, Cold Spring Harbor, N.Y.
121. **Sowers, K. R., and K. M. Noll.** 1995. Techniques for anaerobic growth, p. 15–47. *In* F. T. Robb and A. R. Place (ed.), *Archaea: a Laboratory Manual. Methanogens*. Cold Spring Harbor Laboratory Press, Cold Spring Harbor, N.Y.
122. **Stackebrandt, E., and F. A. Rainey.** 1997. Phylogenetic relationships, p. 3–19. *In* J. I. Rood, B. A. McClane, J. G. Songer, and R. W. Titball (ed.), *The Clostridia: Molecular Biology and Pathogenesis*. Academic Press, Inc., San Diego, Calif.
123. **Stams, A. J. M., and S. J. W. H. Oude Elferink.** 1997. Understanding and advancing wastewater treatment. *Curr. Opin. Biotechnol.* **8:**328–334.
124. **Summanen, P., E. J. Baron, D. M. Citron, C. A. Strong, H. M. Wexler, and S. M. Finegold.** 1993. *Wadsworth Anaerobic Bacteriology Manual*, 5th ed. Star Publishing Co., Belmont, Calif.
125. **Tanner, R. S., E. Stackebrandt, G. E. Fox, R. Gupta, L. J. Magrum, and C. R. Woese.** 1982. A phylogenetic analysis of anaerobic eubacteria capable of synthesizing acetate from carbon dioxide. *Curr. Microbiol.* **7:**127–182.
126. **Thauer, R.** 1997. Biodiversity and unity in biochemistry. *Antonie Leeuwenhoek* **71:**21–32.
127. **Thauer, R. K., K. Jungermann, and K. Dekker.** 1977. Energy conservation in chemotrophic anaerobic bacteria. *Bacteriol. Rev.* **41:**100–180.
128. **Thiele, J. H.** 1991. Mixed-culture interactions in methanogenesis, p. 261–292. *In* J. G. Zeikus and E. A. Johnson (ed.), *Mixed Cultures in Biotechnology*. McGraw Hill Book Co., New York.
129. **Turner, R. E., and N. N. Rabalais.** 1994. Coastal eutrophication near the Mississippi river. *Nature* **368:**619–621.
130. **van Niel, C. B.** 1931. On the morphology, physiology, and classification of the green and purple sulfur bacteria. *Arch. Mikrobiol.* **3:**1–112.
131. **Verstraete, W., D. de Beer, M. Pena, G. Lettinga, and P. Lens.** 1996. Anaerobic processing of organic wastes. *World J. Microbiol. Biotechnol.* **12:**221–238.
132. **Waksman, S. A.** 1947. *Microbial Antagonisms and Antibiotic Substances*. The Commonwealth Fund, New York.
133. **Walden, W. C., and D. J. Hentges.** 1975. Differential effects of oxygen and oxidation-reduction potential on the multiplication of three species of anaerobic intestinal bacteria. *Appl. Microbiol.* **30:**781–785.
134. **Whitmer, M. E., and E. A. Johnson.** 1988. Development of improved defined media for *Clostridium botulinum* serotypes A, B, and E. *Appl. Environ. Microbiol.* **54:** 753–759.
135. **Wiegel, J.** 1992. The obligately anaerobic thermophilic bacteria, p. 105–184. *In* J. K. Kristjansson (ed.), *Thermophilic Bacteria* CRC Press, Inc., Boca Raton, Fla.
136. **Willis, A. T.** 1964. *Anaerobic Bacteriology in Clinical Medicine*. Butterworths, London.
137. **Willis, A. T.** 1969. *Clostridia of Wound Infection*. Butterworths, London.
138. **Wood, W. A.** 1961. Fermentation of carbohydrates and related compounds, p. 59–149. *In* I. C. Gunsalus and R. Y. Stanier (ed.), *The Bacteria. A Treatise on Structure and Function, vol. II. Metabolism*. Academic Press, Inc., New York.
139. **Wood, W. A., and J. R. Paterek.** 1994. Physical analysis, p. 465–511. *In* P. Gerhardt, W. A. Wood, and N. R. Krieg (ed.), *Methods for General and Molecular Bacteriology*. American Society for Microbiology, Washington, D.C.
140. **Woods, D. R.** 1995. The genetic engineering of microbial solvent production. *Trends Biotechnol.* **13:**259–264.
141. World Health Organization Expert Committee. 1977. *Manual for the Production and Control of Vaccines*. World Health Organization, Geneva, Switzerland.
142. **Zehnder, A. J. B. (ed.).** 1988. *Biology of Anaerobic Organisms*. John Wiley & sons, Inc., New York.
143. **Zehnder, A. J. B., and W. Stumm.** 1988. Geochemistry and biogeochemistry of anaerobic habitats, p. 1–38. *In* A. J. B. Zehnder (ed.), *Biology of Anaerobic Organisms*. John Wiley & Sons, Inc., New York.
144. **Zeikus, J. G., and E. A. Johnson (ed.).** 1991. *Mixed Cultures in Biotechnology*. McGraw-Hill Book Co., New York.

Continuous Culture

AN-PING ZENG

12

Since the fundamental work of Monod (38), Novick and Szilard (41), and Herbert et al. (25) on the theory of continuous culture, the advantages and potentials of this type of cultivation have been widely recognized. From an application point of view, these include high volumetric productivity, savings in labor and energy costs (e.g., inoculum preparation, reactor cleaning, and sterilization), uniform product quality, better automation and process control, and the use of more efficient and economical methods of medium preparation and downstream processing. As a research tool, continuous culture provides well-defined cultivation conditions for genetic, biochemical, and physiological characterization of cells (11, 20, 56, 59). It allows independent variation of growth parameters, enabling reliable kinetic studies of cell growth and metabolism for process optimization (21, 61). Competition and/or interaction of different species can be studied more easily in continuous mixed cultures than in batch cultures (20). Some phenomena, such as synchronized growth and oscillation, which are hardly observed in batch cultures, can be "trapped" and reproduced in continuous culture (6, 23, 34). The transition behavior of a continuous culture upon shift-up or shift-down of a variable is also a powerful tool to study the regulation of growth and metabolism (3, 5, 12, 27, 33). This kind of transient behavior can also be used for an improved medium design and optimization (16). In addition, continuous culture is a useful means of selecting strains and/or subclones with improved growth characteristics and productivity (15, 35). On the other hand, mutation and instability of some producing strains due to the selection pressure in continuous culture is a major obstacle for the industrial application of this cultivation technique, particularly in the case of genetically modified cells (7, 29, 49, 55). Another obstacle is the difficulty in maintaining monoseptic conditions over a long period of operation at a large scale. Cell adhesion on the reactor wall and sensors is sometimes a problem, preventing long-term operation of a culture.

The use of continuous culture received great interest in the 1960s for basic research in biology and cell physiology and led to a renaissance in the 1970s and earlier 1980s, mainly because of its potential industrial application for the production of single-cell protein, alcohols, solvents, and food-related products and for wastewater treatment. In today's biotechnological applications, continuous operation is fully established in wastewater treatment, production of some primary metabolites (e.g., ethanol and organic acids), and fermented foods and for some reactions catalyzed by enzymes. It is also used for the production of monoclonal antibodies and recombinant proteins by animal cell cultures. The major applications of continuous culture are, however, still found in fundamental studies and process optimization at laboratory scale. A number of new biological processes, such as animal cell cultures and microbial cultures with genetically modified cells, have been quantitatively studied with the help of continuous culture (4, 7, 17, 18, 24, 25, 32, 37, 40, 52, 54). Continuous culture has also been used for bioreactor characterization, control, and scale-up (10, 21, 61). Variations of the chemostat culture (e.g., the auxostat) have been introduced and exploited (19, 36, 46, 47). These variations overcome some of the disadvantages associated with chemostat culture and open new ways of continuous cultivation. With the development of more stable recombinant strains and cell lines and more reliable and accurate on-line measurement and control techniques, continuous culture may receive more acceptance in the future, both as a tool for basic research and for possible industrial application.

This chapter gives a brief introduction to the general concept and theory of different types of continuous culture. The design and operation of equipment and experiments are discussed from an application point of view. For more detailed treatment and special aspects of continuous culture, the reader is referred to several excellent textbooks and review articles (2, 11, 16, 20, 23, 30, 39, 44, 55, 56).

12.1. GENERAL CONCEPT AND THEORY OF CONTINUOUS CULTURE

Depending on the control parameter and the operation mode, continuous culture can be classified into four general types (Fig. 1). A common feature of these cultures is that they consist of one or more culture vessels into which fresh medium or culture from a preceding vessel is continuously introduced at a rate, F, expressed in liters per hour, and that the culture volume, V, expressed in liters, is kept constant by continuous removal of the culture. The general concept and theory of these four types of continuous culture are described in more detail as follows.

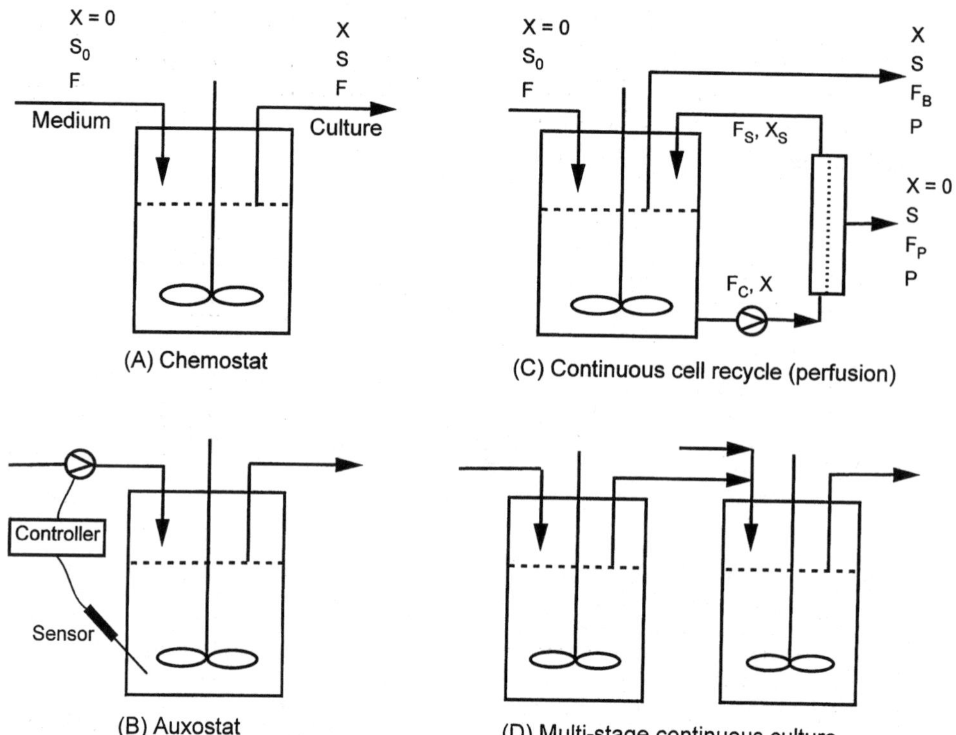

FIGURE 1 Schematic diagram of four types of continuous culture. (A) Chemostat. (B) Auxostat. (C) Continuous culture with cell recycle. (D) Multistage continuous culture.

12.1.1. Chemostat

The chemostat (Fig. 1A) is defined as a continuous culture system in which the feed rate is set externally and cell growth is limited by a selected nutrient. The second condition means that the specific growth rate, μ ($hour^{-1}$), of the organism is a function of a single growth-limiting nutrient. However, this definition may be relaxed to include continuous culture limited simultaneously by more than one nutrient component (20). Continuous cultures that are fed with an inhibitory nutrient or that form toxic products can be limited by growth inhibition under the condition of excess of all nutrients. Strictly speaking, they cannot be referred to as chemostat cultures.

A chemostat is usually started as a batch culture. Before a nutrient becomes limiting, the nutrient feed is started. Cells grow until the chosen nutrient becomes limiting. After this, cell growth is limited by the rate of addition of medium. The specific growth rate of a chemostat culture can be determined from a material balance for biomass:

Net increase in biomass = Biomass in incoming medium + Growth − Output − Death

In mathematical form this is:

$$\frac{dX}{dt} = \frac{F}{V}X_0 + \mu X - \frac{F}{V}X - \alpha X \quad (1)$$

where dX/dt is the rate of accumulation of biomass per unit of time and per volume (grams of dry cell weight [DCW] per hour per liter); X_0 is the biomass concentration (grams of DCW per liter) in the incoming medium; X is biomass concentration (grams of DCW per liter) in the bioreactor; and α is specific death rate of cells ($hour^{-1}$). This expression can be simplified if one assumes that the chemostat is at a steady state, that there are no cells in the incoming medium, and that cell death is negligible. The assumption of steady state implies that there is no net accumulation of cells in the bioreactor, so the left-hand side of the equation is equal to zero. Furthermore, if the dilution rate, D, is introduced for F/V, equation 1 becomes:

$$\mu X = DX \quad (2)$$

hence

$$\mu = D \quad (3)$$

Thus, by maintaining a constant volume and changing the nutrient feed rate, one can precisely control the specific growth rate of a culture over a range up to the maximum specific growth rate (μ_{max}) and let the system come to a steady state. This is one of the most important properties of a chemostat. Note that the following additional assumptions are made in the derivation of equations 1 to 3.

1. Cells are distributed randomly in the bioreactor, i.e., the cells do not adhere to each other or to the walls of the reactor, and the suspension is well mixed.
2. The population is nonsegregated and consists of physiologically identical cells.
3. The population density, X, is a continuous variable, i.e., the number of cells is sufficiently high and the size of each cell is sufficiently small for the discreteness of biomass to be ignored.
4. The volume occupied by the cells is negligible compared to the total volume of the culture.

Furthermore, assuming that as soon as medium enters the bioreactor it is instantly distributed uniformly throughout the culture, a material balance on the growth-limiting substrate can be formulated as:

Net increase = Input − Output
− Substrate consumed by cells

For an infinitely small time interval, dt, this balance for the whole culture is:

$$\frac{dS}{dt} = DS_0 - DS - q_s X \tag{4}$$

where S_0 is the concentration (grams per liter) of substrate in feed; S is the concentration (grams per liter) of substrate leaving the reactor; and q_s is the specific consumption rate of substrate (grams per gram of DCW per hour). Equation 4 can also be written in terms of growth yield:

$$\frac{dS}{dt} = DS_0 - DS - \frac{\mu X}{Y_{x/s}^{app}} \tag{5}$$

where $Y_{x/s}^{app}$ is the apparent yield of biomass on substrate (grams of DCW per gram of substrate), which is defined as:

$$Y_{x/s}^{app} = \frac{\mu}{q_s} \tag{6}$$

In general, both q_s and $Y_{s/s}^{app}$ are functions of μ and can also be affected by a number of other growth parameters, such as the nature of the limitation and the level of substrate concentration (62, 63). To illustrate the basic characteristics of a chemostat, $Y_{x/s}^{app}$ is assumed here to be constant. Thus, under steady-state conditions, equation 5 reduces to:

$$D S_0 - D\bar{S} - \frac{\mu \bar{X}}{Y_{x/s}^{app}} = 0 \tag{7}$$

where $\bar{S}$ and $\bar{X}$ denote steady-state values. For a chemostat in which growth inhibition by product formation and excess of other nutrients does not occur, the well-known Monod kinetic (38) may be used to describe the relationship between specific cell growth rate, μ, and the concentration of a limiting nutrient in culture:

$$\mu = \mu_{max} \frac{S}{S + K_s} \tag{8}$$

where μ_{max} is the maximal specific growth rate (hour^{-1}) and K_S is a saturation constant corresponding to the substrate concentration at which the maximum specific growth rate is reduced by one-half. Substituting μ by D in equation 8, we obtain for a steady state:

$$\bar{S} = K_S \frac{D}{\mu_{max} - D} \tag{9}$$

and from equations 3 and 7, the following expression for the steady-state biomass concentration is obtained:

$$\bar{X} = Y_{x/s}^{app} (S_0 - \bar{S}) = Y_{x/s}^{app} \left(S_0 - K_S \frac{D}{\mu_{max} - D} \right) \tag{10}$$

The characteristics of a chemostat can best be viewed by examining the behavior of the governing equations derived above. Figure 2 shows plots of $\bar{S}$ and $\bar{X}$ against dilution rate, D, with typical parameter values. The residual substrate concentration is only a function of the dilution rate. It increases slowly with D at low values but very rapidly as D approaches μ_{max}. The biomass concentration depends on the feed substrate concentration. At a given S_0, biomass is

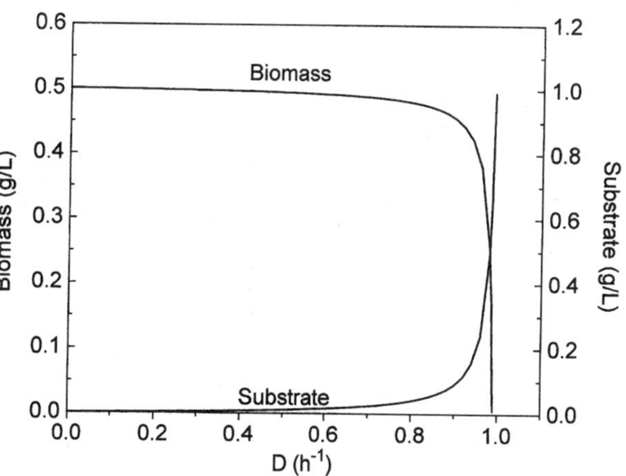

FIGURE 2 Steady-state values of biomass and growth-limiting substrate concentration in a chemostat according to equations 9 and 10. Parameters: $\mu_{max} = 1.0$ h^{-1}; $K_s = 0.01$ g/liter; $Y_{x/s}^{app} = 0.5$ g/g.

almost constant at low values of D but rapidly declines at dilution rates close to μ_{max}.

12.1.2. Deviations from the Ideal Chemostat

The suitability of the simple chemostat theory has been demonstrated for a number of organisms, including bacteria, yeasts, and plant and animal cell cultures. In most cases, these cultures were carried out at very low feed concentrations of nutrient, in small-scale bioreactors, and/or in a relatively narrow range of dilution rate. Under these conditions, the assumptions made in the derivation of the above equations may be approximately fulfilled. Nevertheless, deviations from chemostat theory have been frequently observed. Figure 3 shows typical biomass profiles as a function of dilution rate in continuous culture that depart from chemostat behavior. The deviations may originate from interactions of cells with the equipment and reactor hydrodynamics, such as wall growth and imperfect mixing. Wall growth is caused by the adhesion of cells to glass and metal surfaces. As shown in Fig. 3A, wall growth can increase the steady-state biomass concentration and enable an operation to be carried out at dilution rates higher than the maximum specific growth rate (53). Brown et al. (9) studied the effect of imperfect mixing (stagnant zone) on the performance of a continuous culture (Fig. 3B). Depending on the relative size of the stagnant zone and the velocity of liquid exchange between the well-mixed zone(s) and the stagnant zone, the biomass concentration can be markedly reduced at a certain range of the dilution rate. Similarly, as in the case of wall growth, the operational dilution rate can be higher than μ_{max}. Simple chemostat theory assumes perfect mixing of medium, i.e., a drop of medium entering the vessel should instantly be distributed uniformly throughout the culture. Quantitatively, this means that the time constant required for mixing should be smaller than the time constants for mass transfer and bioreactions (64). In laboratory-scale reactors with a culture volume of a few liters, nearly perfect mixing can be obtained at proper agitation rates unless the reactor contains zones that screen the culture from the agitation or the culture becomes very viscous. In large-scale reactors, however, imperfect mixing is inevitably encountered.

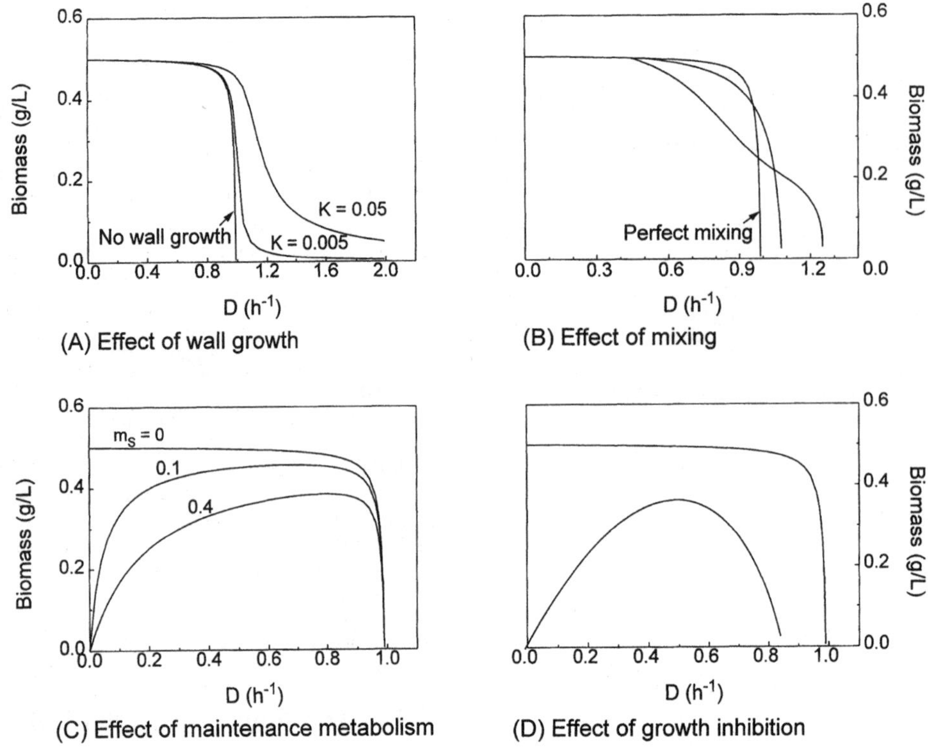

FIGURE 3 Typical profiles and reasons for continuous cultures deviating from chemostat behavior. The biomass curve of a corresponding chemostat culture (see Fig. 2) is also shown for comparison. (A) Effect of wall growth. K is the amount of growing biomass attached to the reactor surface per unit volume of culture. (B) Effect of imperfect mixing. (C) Effect of maintenance metabolism. m_s is the maintenance requirement of substrate (g/g · h). (D) Effect of growth inhibition by product(s) and/or substrate.

The deviations observed in laboratory-scale reactors are quite often caused by biological effects. Among others, these include maintenance requirement (43), variations of growth yield due to metabolic overflow (51, 62, 63), growth inhibition by products and/or substrate (22, 58), and regulatory and segregating effects at low and high growth rates (3, 33). As shown in Fig. 3C and D, both the effects of maintenance metabolism and growth inhibition tend to decrease the steady-state biomass concentration, particularly at a low dilution rate. Under conditions of strong growth inhibition, the operation of a continuous culture may be possible only at dilution rates significantly lower than μ_{max}. The factors mentioned above may be simultaneously involved in a culture system. In such a case, the relationship between the biomass concentration and the dilution rate will be more complicated.

12.1.3. Auxostat

An auxostat is a continuous culture in which a growth-dependent parameter is kept constant by adjusting the feeding rate of medium (Fig. 1B). The dilution rate and hence the specific growth rate of the culture adjust accordingly. The choice of the feedback parameter for an auxostat is quite broad. It includes cell density (turbidity), pH, dissolved oxygen concentration, CO_2 in effluent gas, and concentrations of nutrients and products (19). Sometimes, the term nutristat is also used to refer to auxostats using a nutrient concentration as the feedback growth parameter. Of the different kinds of auxostats, the turbidostat and the pH-auxostat have so far found the most applications (8, 19, 36, 46).

In a turbidostat, the biomass (cell density) is used as a control parameter. A sensor detecting the biomass density gives a signal to a pump to add more medium when the biomass density rises above a chosen level. By means of turbidostat control, therefore, the biomass density is set and the dilution rate adjusts itself to the steady-state value, in contrast to the chemostat in which the dilution rate is fixed and the biomass concentration adjusts itself to the steady-state level. The governing equations of a turbidostat can be derived from the general equations 1, 5, and 8. Under steady-state conditions, we have

$$\bar{S} = S_0 - \frac{\bar{X}}{Y^{app}_{x/s}} \tag{11}$$

$$D = \mu_{max} \frac{\bar{S}}{\bar{S} + K_S} \tag{12}$$

Here $\bar{S}$ is linearly dependent on both $\bar{X}$ and S_0, while D is a hyperbolic function of these two parameters. These relationships are illustrated in Fig. 4.

In principle, turbidostat culture is suitable for the cultivation range in which the biomass concentration varies significantly with change in dilution rate, such as near the critical dilution rate (Fig. 4). This is of particular interest because operation near the maximum growth rate can be very unstable in a chemostat. The turbidostat also provides

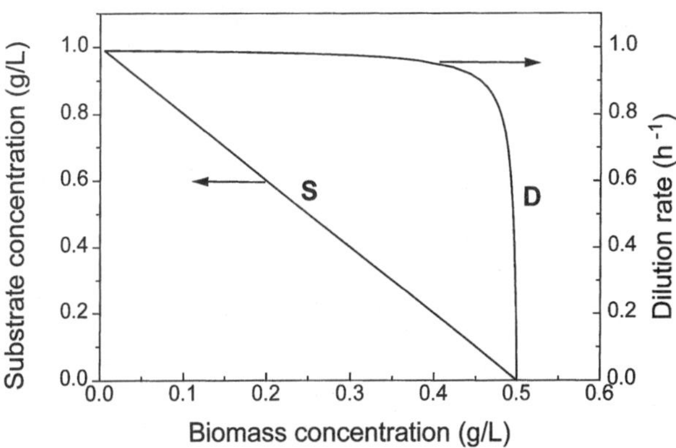

FIGURE 4 Steady-state limiting-nutrient concentration, $\bar{S}$, and dilution rate, D, functions of biomass concentration in a turbidostat culture. Growth parameters are as given in the legend to Fig. 2.

a means of maintaining cultures in a constant environment with excess of substrate. Since the growth rate is not fixed, the system will select for fast-growing organisms. Increase in the maximum growth rate will result from both selection of genetically different organisms (or mutants) and an adaptation of cellular metabolism. In the past, a photoelectric cell was the main sensor used to monitor cell density. This technique is limited to unicellular organisms and is difficult for long-term cultures, largely because of adhesion of organisms to the surface of the sensor. With the development of optical sensors based on laser or infrared light transmission/scattering, problems with long-term monitoring of biomass may be partly overcome.

In a pH-autostat, the addition of fresh medium is coupled to pH control. As the pH drifts from a given set point, fresh medium is added to bring the pH back to the set point. The rate of addition of medium is determined by the buffering capacity and the concentration of nutrients in the medium. Buffering capacity is defined as the equivalents of titrant required to change the medium pH to the culture pH. The governing equation is the mass balance on the H^+ ion concentration, expressed as milliequivalents (meq) per liter, in the bioreactor (19):

$$\frac{dH^+}{dt} = D(H_F^+ - H^+) + \mu Xh - D(BC) \quad (13)$$

where H_F^+ is the H^+ ion concentration in feed (meq per liter), h is the acid production yield of cells (meq per gram of biomass), and BC is the buffering capacity of medium (meq per liter).

By using fresh medium for pH control, the limiting nutrient level in the reactor can be manipulated by adjusting the buffering capacity and/or limiting nutrient concentration in the feed. For cultures producing high amounts of acid or alkali, the buffering capacity of the medium may not be sufficient. In this case, separate feeding lines for titrant and medium can be used (8). The concentration of nutrient determines the amount of ionic species produced and ions released during nutrient uptake. Assuming that the difference between feed and bioreactor H^+ concentrations is very small, the following relationships can be obtained from equation 13 and substrate balance under steady state:

$$\bar{X} = \frac{BC}{h} \quad (14)$$

$$\bar{S} = S_0 - \frac{BC}{hY_{x/s}^{app}} \quad (15)$$

The specific growth rate achieved in a pH-auxostat is inversely correlated with the buffering capacity and the concentration of nutrients in the medium. Roughly speaking, the specific growth rate is about 10 times greater with low buffering capacity than with a well-buffered system (46). The biomass concentration is also mainly determined by the buffering capacity of the medium.

The pH change of a culture is a good indication of cell growth and metabolic activity. However, the exact cause of pH change varies among organisms. It represents the summation of the production of different ionic species and ion release during nutrient uptake. The pH of a culture is normally reduced by the production of organic acids and by the uptake of ammonium. However, for microorganisms growing on protein- or amino acid-rich medium, the pH will rise with growth because of the release of excess ammonia.

Generally speaking, auxostats have the following advantages over conventional chemostats. First, auxostats permit stable operation in the "high gain" areas near the maximum growth rate. Second, they reach steady state more rapidly at high dilution rate than the open-loop chemostat. Third, population selection pressures in an auxostat lead to cultures that grow rapidly. Finally, it is possible to design a dual set-point auxostat that controls two growth parameters (e.g., concentrations of two nutrients) simultaneously (19). This kind of dual setpoint auxostat permits control of nutrient concentration ratios with ease, which tends to be very difficult with a chemostat. On the other hand, the experimental setup and operation of an auxostat are somewhat complicated.

12.1.4. Continuous Culture with Cell Recycle (Perfusion)

Cell recycle is a useful means for increasing the concentrations of biomass and product in a continuous culture (4, 26, 31, 60). This system can be operated at a dilution rate higher than the maximum specific growth rate, leading to a much higher output of the reactor. Another property is that the dilution rate is almost independent of the growth rate. Recycle of biomass can also protect against shock loading with an inhibitory substrate, because the critical dilution rate is raised. It is particularly advantageous in the following cases: (i) the growth-limiting substrate is unavoidably dilute, for example, in the treatment of effluents; (ii) the substrate has a low solubility, such as when a gaseous substrate is used; (iii) the concentration of growth-limiting substrate has to be limited because of the formation of inhibitory product(s); and (iv) product formation is not associated with growth.

Several methods can be used for the retention of biomass, such as filtration, sedimentation, centrifugation, and immobilization. Depending on the position of the separation device inside or outside of the reactor, the methods can be further divided into internal and external systems (44). The concentration of culture effluent outside of the reactor by means of membrane filtration has so far found the most frequent application, except for the biological treatment of wastewater, in which the recycling of sludge after sedimentation has been used for a long time. Figure

1C schematically shows a continuous culture with membrane cell recycle. The culture is circulated at an outflow rate, F_c (liters per hour), by a pump over a microfiltration membrane from which a cell-free filtrate is withdrawn at a rate of F_p (liters per hour). In addition, a portion of the liquid culture containing biomass is withdrawn from the reactor at a bleed rate of F_B (liters per hour).

Note that in a membrane bioreactor, the circulation rate of liquid culture, F_c, is normally much higher than the filtration rate, F_p, so that the concentration factor, $\alpha = X_s/X$, is very close to 1:

$$\alpha = \frac{X_s}{X} = \frac{F_c}{F_s} = \frac{F_c}{F_c - F_p} \approx 1 \tag{16}$$

where X_s and F_s are the biomass concentration and flow rate at the outlet of the filtration module, respectively. Thus, the position of biomass withdrawal (bleed) does not affect the governing equations for the membrane bioreactor, which can be written as follows:

$$\frac{dX}{dt} = (\mu - B)X \tag{17}$$

$$\frac{dS}{dt} = D(S_0 - S) - q_s X \tag{18}$$

$$= (B + f)(S_0 - S) - q_s X$$

where B is the bleed rate (hour^{-1}), defined as $B = F_b/V = 1/\tau$; τ is the mean cell residence time (hour); and f is the filtration rate, defined as $f = F_p/V$ (hour^{-1}). The dilution rate, D, is the sum of rates of bleed and filtration. Assuming a constant apparent biomass yield (equation 6) and Monod kinetics (equation 8), the following steady-state solution of equations 17 and 18 is obtained:

$$\mu = B \tag{19}$$

$$\bar{S} = K_s \frac{B}{\mu_{max} - B} \tag{20}$$

$$\bar{X} = \left(1 + \frac{f}{B}\right) Y_{x/s}^{app} \left(S_0 - K_s \frac{B}{\mu_{max} - B}\right) \tag{21}$$

Thus, cell growth rate in a chemostat with cell recycle is determined by the bleed rate, B. From equation 17, it is obvious that in a system of total cell recycle ($B = 0$), the biomass would build up to reach a point where the cultivation becomes inoperable owing to high viscosity and difficulties in mixing and nutrient supply. Thus, a controlled bleed of biomass is necessary to maintain an active biomass. In some cases, cell death and lysis at high biomass concentration may be considerable so that total cell retention is possible. A comparison of equation 21 with equation 10 reveals that at the same growth rate, the chemostat culture with cell recycle can increase the biomass concentration of a simple chemostat by a factor of f/B. In practice, a value of f/B in the range 10 to 100 is feasible. The increase of final product concentration is normally much lower because of the reduction of productivity of cells at high concentrations.

12.1.5. Multistage Continuous Culture

The two-stage continuous culture system shown in Fig. 1D can extend the range of application of continuous culture. For example, the second stage may be used to extend the growth rate downward to zero, and the first stage may be used to achieve stable conditions with maximum growth rate, both of which conditions may be desired in certain cases but are impossible in the simple chemostat (44). The latter property is particularly useful when the substrate is also a growth inhibitor. In the production of secondary metabolites and enzymes by continuous culture, the second stage may be used to provide a nongrowing situation in which product formation occurs. For products of foreign gene expression, the second stage can be used for induction of expression (52). Another useful application of the two-stage continuous culture is for reactor scale-down studies (10). The two-stage culture system can be extended to include more stages and more feeding streams with or without biomass recycle. For more information on these systems, the reader is referred to Pirt (44) and Moser (39).

12.2. EQUIPMENT SETUP AND DESIGN

12.2.1. General Setup of a Continuous Culture

Figure 5 shows a typical setup of a bench-scale reactor system for continuous cultivation. A continuous reactor system consists of at least three parts:

1. Reaction vessel, i.e., a bioreactor (at center) equipped with elements for aeration, mixing of liquid (medium) and solid (cells or insoluble substrate) phases, sampling, and control of temperature, pH, dissolved oxygen, foaming, and reaction volume. Equipment for on-line effluent gas measurement is also depicted in Fig. 5. Some of these control units may be omitted, depending on the type of fermentation.
2. Storage vessels for medium and solutions for control of pH and foam, suitable for sterilization. Normally, the entire homogeneous, mixed medium is stored therein, containing the substrate as carbon and energy source (e.g., glucose) and other nutrients at given concentrations. More vessels are necessary if heat-sensitive components are used, because they need separate sterilization. Components may be mixed together before being fed into the reactor by a metering device (e.g., a pump) or may have separate connecting lines to the reactor.
3. Broth withdrawal and collection vessel (at right). The collection vessel is not mandatory but is useful for measuring the effluent rate of the culture and avoiding the release of organisms into the environment.

For some applications, such as selection of strains (or mutants), much simpler configurations of continuous culture can be used (30).

12.2.2. Equipment Selection and Design

Stafford (50) provides an excellent overview of potential problems and solutions concerning selection, design, and operation of individual equipment used in continuous culture. The following discussion is mainly based on Stafford and includes some updated information and the author's own experience with continuous culture. In addition, necessary equipment and its operation for continuous culture with cell recycle are addressed. As pointed out by Stafford (50), the type of fermentation may dictate the choice of equipment. Medium composition (chemically defined or complex) and organism morphology (filamentous or unicellular) set restrictions on the type of equipment that can be used. Therefore, the suggestions below serve only as a

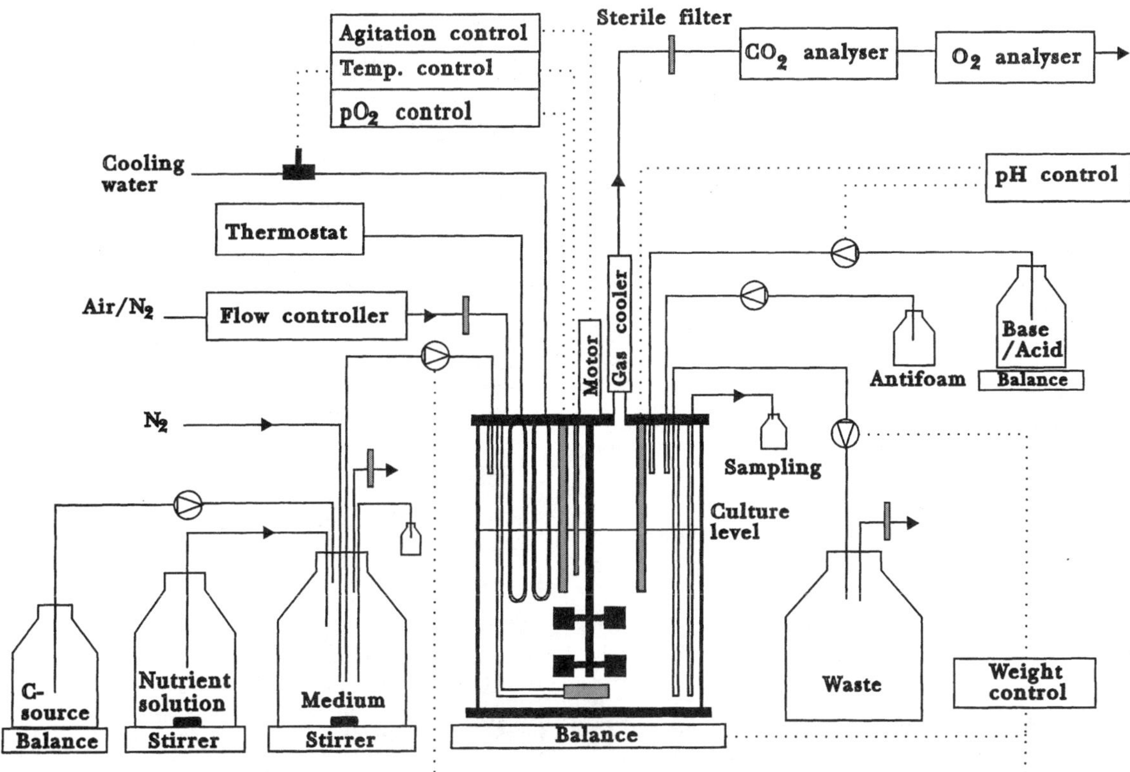

FIGURE 5 A typical setup of a bench-scale continuous-culture system. The feeding rate and the culture volume are controlled by a weight control unit.

guideline and may need to be modified or even replaced according to the actual requirements.

12.2.2.1. Reactor

The scale of reactor is an important factor in selection of equipment and setup of a continuous culture. The consumption of medium and the need for other peripheral equipment are largely determined by the working volume of the reactor. For research purposes, a benchtop-scale reactor is convenient to operate. In addition, investment and space requirements are modest, and peristaltic pumps of adequate reliability are available. A benchtop reactor is easily dismantled for cleaning.

Small-scale culture vessels have some unique problems, however. When operated at volumes of less than 500 ml, sampling can significantly reduce the volume, affecting the dilution rate and leading to fluctuation of the system. Flow rates for obtaining low dilution rates can be very difficult to control, and feeding complex medium at low flow rates can result in settling out of insoluble medium components in feed lines, causing plugging. It is often difficult to implement an on-line gas analysis owing to low back-pressure and low aeration rate. In addition, wall growth may represent a considerable fraction of the total population in a small vessel. For microbial cultures, a working volume of 1 to 5 liters (total volume of 2 to 7 liters) is in most cases appropriate. A lower working volume (e.g., 0.5 to 2 liters) may be desired for animal cell culture owing to the relatively high cost of medium. The alternative use of a reactor for batch or fed-batch cultures should also be considered.

Stirred tank reactors are usually used in benchtop scale, although other reactor types, such as airlift and fluidized bed reactors, are now available in small scale. The choice of a reactor type depends very much on the goal of the research. For small-scale stirred tank reactors, there are also several variations in design, including flat-bottom and round-bottom vessels. The flat bottom eases handling on the bench and in the autoclave. On the other hand, round- or dished-bottom vessels ensure better mixing and are mostly used for culturing shear-sensitive cells. Jacketed reactors are also available. The glass jacket gives better temperature control than the use of a heat exchanger inside the reactor. The jacketed reactor is more suitable for heat-sensitive cultures.

12.2.2.2. Nutrient Feed Reservoir and Auxiliaries

Nutrient reservoirs for continuous culture should have ports for feeding, addition and/or mixing of heat-labile nutrients and substrate, venting, and sparging of the medium (Fig. 5). The feed and sparging lines should be stainless-steel tubes going to the bottom of the vessel. The venting line should be attached with a sterile filter. A magnetic stirring bar must be included to suspend any particulates and mix nutrients that are added after sterilization of the bulk medium. A separate sampling line is also useful for taking samples to determine the exact substrate concentration after refilling. A carboy or glass flask with a medium volume large enough for at least one steady state may be used as an intermediate reservoir. This can be calibrated by volume or placed on a weighing device so that the nutrient feed rate can be checked over several hours or days. It is also flexible enough for adjusting the nutrient concentration if different growth limitations are to be used.

Silicon rubber tubing or special gas-impermeable tubing should be used when tubing is required. It can be sterilized repeatedly, retains its resilience, and has excellent chemical resistance. The method of connecting tubing should be aseptic, fast, and reliable. One should be aware that the flow rate through a peristaltic pump is dependent on the resilience of the tubing. Because silicone rubber tubing loses its resilience with use, a long piece of tubing should be used so that the pump can be moved every other day to a new section. For extended operation, replacement of the tubing may be necessary.

The medium inlet line may be connected with a drip-tube very close to the reactor. A drip-tube provides a barrier in the medium flow path and prevents miroorganisms from growing into the feed line. It can be purchased from fermentor manufacturers or constructed as shown in Fig. 6. The drip-tube may fail if pressure builds up in the reactor.

12.2.2.3. Broth Removal and Level Control

To achieve an accurate and constant control of the dilution rate, the working volume of the culture must be determined and controlled. The working volume of a culture is normally defined as the ungassed liquid volume. Two parameters can be used to control the ungassed liquid volume: gassed liquid level (assuming constant gas holdup) and weight of the reactor content. Because gassed volume is easier to control, it is widely used. However, constant gas holdup is difficult to achieve. Alternatively, weight control can be more accurate if properly installed, but it is much more expensive to implement on the bench scale of operation. The working volume of the reactor system shown in Fig. 5 is kept constant by a weight control unit. The signal from the balance is used to control the pump for culture removal.

There are basically two ways of controlling gassed liquid level. The simplest method is to make use of the overpressure in the reactor from aeration. The effluent gas outlet can be placed at the broth surface. The effluent gas is directed into a waste reservoir that is vented aseptically to the atmosphere. When the culture volume increases above the effluent gas outlet, broth is forced out into the reservoir with the gas. Sometimes, a pump is used to withdraw broth instead of using effluent gas. Both of these systems suffer from the problem that the cell concentration at the air/liquid interface can be quite different from that in the bulk liquid, especially if foam is formed. This can cause a nonrepresentative withdrawal of culture from the reactor.

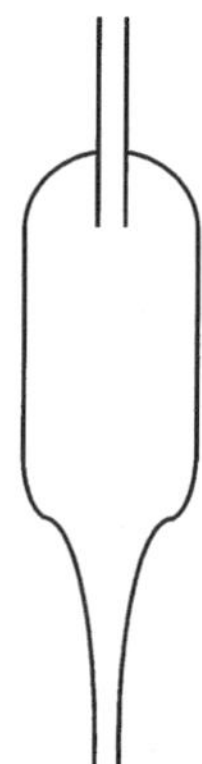

FIGURE 6 Diagram of a drip-tube.

Another method is to use a liquid level controller. Although this technique requires more equipment, it allows withdrawal of broth from below the liquid surface. A number of liquid level controllers are available. Most of them are based on a sensor within the vessel that makes physical contact with the culture fluid. Because the sensor is located within the reactor, failure of the probe will end the process. Depending on the type of sensor used, inaccuracies in volume can be caused by variations in agitation, aeration, foam generation, and wall growth. Ultrasonic level controllers are available that can be attached to the outside of a glass or steel vessel. They can be attached after autoclaving, and if a sensor malfunctions, it can be replaced without violating sterility.

Stafford (50) proposed the use of a redundant control scheme as shown in Fig. 7. In this system, a liquid level controller is backed up by the effluent gas system. The culture exit tube is located just below the liquid level, while the effluent gas exit is located just above the liquid level. If the level controller fails in the "on" position, the waste broth pump cannot reduce the liquid level below the culture take-off line. If the level controller fails in the "off" position, or the pump stops or the tubing becomes clogged, then the liquid level cannot rise above the effluent gas exit tubing. However, problems caused by the change of gas holdup and foam formation still cannot be fully eliminated. If an antifoam agent is added during operation, fluctuation in the liquid level may result. To avoid error in determining the dilution rate, the real ungassed liquid volume may need to be checked from time to time by a brief switch-off of aeration and agitation (in case of strong foaming), for example, after a steady state has been achieved.

A stable continuous culture requires accurate and reliable pumps. The pump heads and the pumping lines should be sterilizable. Most important, it should be capable of long-term aseptic operation. On the benchtop scale, peristaltic pumps meet these criteria. As fermentation scale increases, they become impractical. Silicone tubing should be replaced by piping. Generally, in situ steam sterilization is required for reactors greater than the 10-liter scale. Diaphragm pumps can be used that provide reliable, adjustable flow rates. These pumps have several advantages over centrifugal and gear pumps. First, they lack a mechanical seal. Over time, mechanical seals wear out and can be a source of contamination. Second, the pump head and the lines leading to and from the head can be sterilized by steam. Finally, the flow rate can be maintained over a wide range of pressure. This means that changes in the backpressure of the vessel can be accomplished without substantially changing the pump flow rate.

12.2.2.4. Gas Inlet and Exit

An example of a gas supply and effluent system is shown in Fig. 7. Humidification of the gas prevents evaporation of medium from the vessel and improves the estimation of dilution rate if the culture effluent rate is measured in the outlet lines. This may be significant when operating at a relatively high temperature (e.g., above 35°C), when using compressed air that is very dry, or when operating at dilution rates of less than 0.1 h^{-1}. Sterilization of the inlet gas can be done by a membrane filter with proper pore size or a packed glass-wool filter. Two filters connected by a tee are recommended as a precaution, especially when using

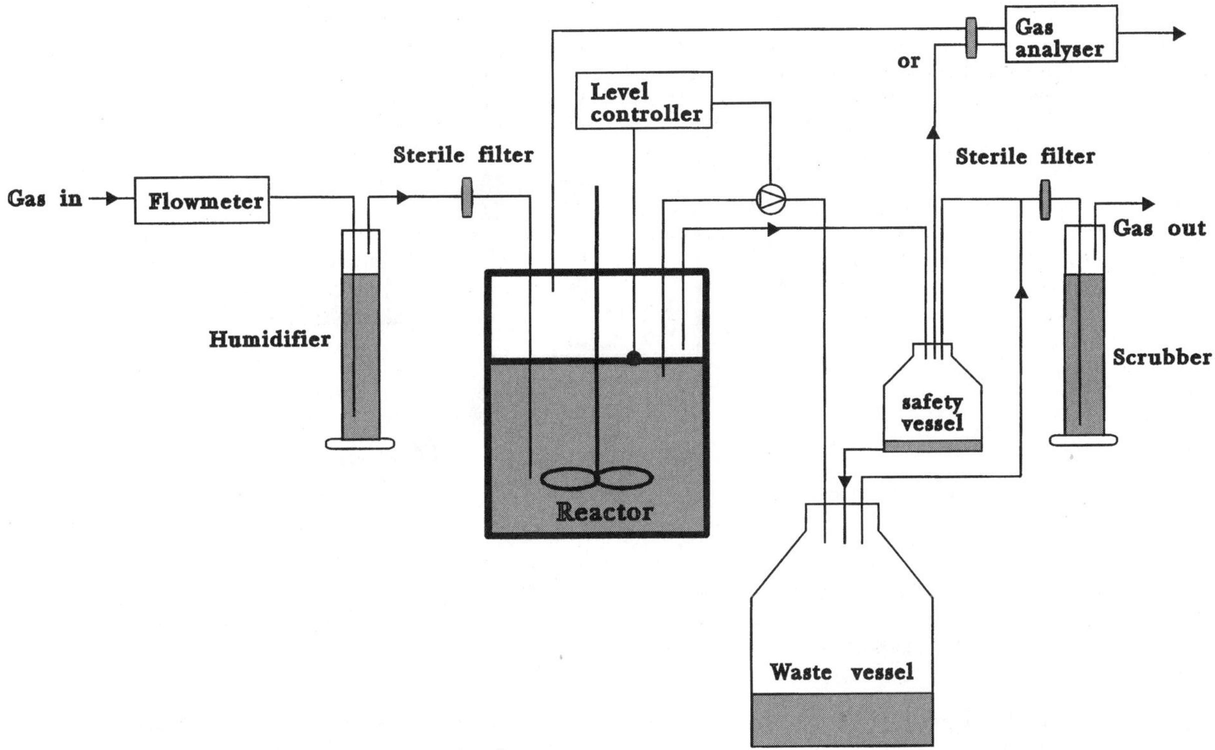

FIGURE 7 Control scheme for the reactor volume of a continuous culture that uses a liquid level controller backed up by an effluent gas system (modified from reference 50).

humidifiers. In the event that one filter becomes plugged, the other filter can be unclamped and used immediately. A regulated pressure of gas supply is important for safety. In case the gas exit becomes plugged, the gas pressure should be sufficiently low to prevent the vessel from exploding.

The effluent gas should be directed to an overflow reservoir (safety vessel). In this way, foam-out can be contained. The overflow reservoir is then connected to a graduated cylinder filled with 0.5% bleach. If an on-line gas measurement is desired, a bypass gas line from the reservoir can be used. The rest of the effluent gas is sparged through the liquid in the cylinder and vented through a sterilizing filter. The cylinder serves several functions. First, it gives a certain degree of back-pressure to the vessel, which is needed for effluent gas analyzers. It also facilitates sampling. Second, it provides a means to visually check whether the gas flow rate is high enough for the gas analyzer and if the reactor system is well sealed. Absence of bubbling may indicate loss of containment, which is of particular concern if pathogenic or recombinant organisms are being cultured. Finally, it acts as a scrubber to reduce unpleasant odors in the laboratory.

12.2.2.5. Cell Recycle

At laboratory scale, cell recycle is easily achieved by using a steam-sterilizable membrane microfiltration module. Fig. 8 shows a simplified structure of such a module. It is composed of microporous membrane capillaries. The membrane is usually made of polypropylene or similar materials and is originally hydrophobic. After wetting with ethanol, it becomes hydrophilic. The capillaries normally have a diameter of about 8 mm and a pore diameter of about 0.2 μm. Membrane modules with different diameters, pore sizes, and filtration areas are available.

An often observed problem of microfiltration is the decrease of filtration rate during operation. This is mainly caused by the formation of biofilms and adsorption of proteinaceous materials on the membrane surface (fouling) and/or loss of hydrophilicity. Factors affecting membrane fouling include concentrations and properties of cells, circulation rate, and properties of the broth and filtration rate. The use of antifoam may also significantly affect membrane performance. To reduce membrane fouling, a minimal broth circulation velocity of about 0.5-1 m/s should be maintained, normally by a gear-pump. This may impose problems for shear-sensitive cells. The filtration module should also be periodically back-flushed with permeate. Such a back-flushing configuration is shown in Fig. 8 (60). A small permeate reservoir is used for the back-flushing. It is controlled by a computer system or by a timer. When back-flushed, the gas (N_2) line is opened and the permeate line and pump are switched off. The permeate is forced back into the membrane module through a back-flushing line. The back-flushing pressure (e.g., 10^5 Pa) should not exceed the maximum operation pressure of the module given by the manufacturer (about 2×10^5 Pa to 3×10^5 Pa for most modules). For long-term operation, the module may be back-washed daily with sterilized distilled water. A previous cleaning with a NaOH solution or wetting with ethanol may substantially recover the filtration performance of the module. Care should be taken to prevent NaOH and/or ethanol from running into the reactor. It is advisable to use two parallel filtration modules; while one module is being cleaned, the other remains in operation.

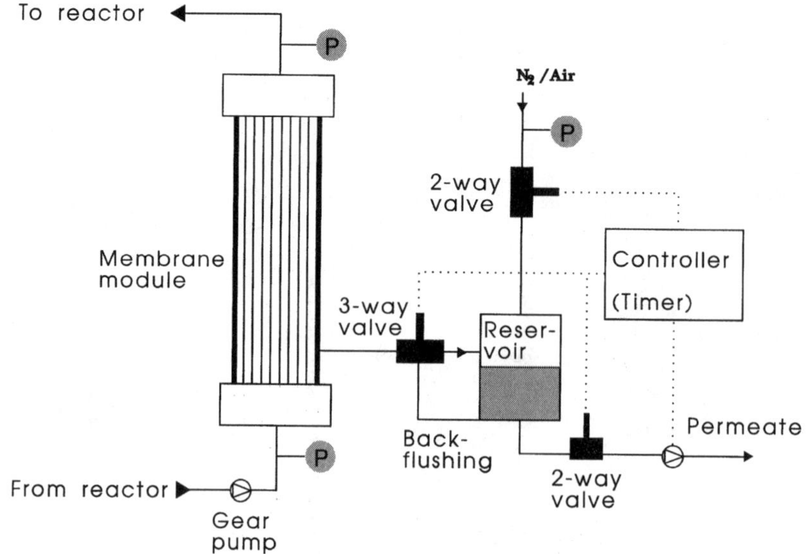

FIGURE 8 Schematic diagram of a back-flushing configuration for a cell recycle reactor with membrane microfiltration. The permeate in the reservoir is back-flushed through the membrane module by using pressurized nitrogen gas or air and controlled by a timer.

12.3. EXPERIMENTAL DESIGN AND OPERATION

12.3.1. Experimental Design

In designing a continuous culture for long-term operation, one must be aware of the possibility of spontaneous mutations because of selective pressure. The rate of spontaneous mutation is estimated at about 1 in 10^6 cell divisions (50). Sooner or later, cells with a competitive advantage will take over the original strain in the reactor. Thus, if one is interested in studying the physiology of a homogeneous cell population, the operation of a culture should not be too long in order to distinguish between metabolic responses of cells to changes in environmental conditions and genetic changes due to mutation. Some cultures form aggregates or exhibit strong wall growth after prolonged cultivation (e.g., after about 2 to 3 weeks for some *Enterobacter* strains) under metabolic stress (e.g., substrate excess, limitation by growth factors, product inhibition, etc.). For these kinds of culture, length of operation is also limited. Care should be taken when interpreting data obtained under conditions of aggregate formation and/or wall growth. If a long-term operation is planned, one should consider the necessity of replacing tubes (damage due to pumping and actions of acid and alkaline solutions) and filters for the incoming and effluent gases used during the process.

12.3.2. Medium Formulation

The basis for medium design is the nutritional requirement for cell growth and product synthesis. In addition to the identification of all components that are necessary for balanced growth or that may interfere with the kinetics of growth and product formation, one has also to establish the appropriate concentrations of the components in a continuous culture, especially if a certain kind of growth limitation is desired. As the use of new biotechnological techniques (e.g., recombinant DNA) is becoming important for production of high-value products, chemically defined media and tailored feeding are finding more applications. Considerations of stoichiometry and regulation of metabolism are important for these purposes (32).

The stability of medium formulation and interactions of nutrient components should be considered to avoid precipitation. Precipitation of medium components may be caused by high concentrations of metal ions (e.g., Fe, Ca, and Mg), high temperature during autoclaving, and high pH. This may result in the loss of metal ions and some other trace elements, impairing cell growth and product formation. To achieve a high cell density, the concentrations of metal ions and trace elements must be carefully controlled (or reduced). The use of chelating agents such as EDTA and citrate, particularly in combination, can effectively reduce precipitation. The proper selection of metal ions is also important, because the formation of insoluble salts, such as those associated with phosphate and magnesium, is often a problem.

12.3.3. Medium and Equipment Sterilization

The sterilization of a continuous-culture system is somewhat more complicated than that for a batch-culture system because of the necessity of having more and larger reservoirs for feed medium and acid and alkali solutions and the need to refill or replace them during the process (Fig. 5). Some components of the medium or the acid and alkali solutions may need to be sterilized separately. The lines connecting the various reservoirs and the waste broth vessel to the reactor can become sources of contamination and therefore should be sterilized and connected carefully. The connection should be done under aseptic conditions, e.g., on a clean bench if the reactor system is relatively small. For reactors larger than 5 liters, in situ sterilization may be necessary. This can be done with wet steam at 121°C, either indirectly through the double wall or by a heat exchanger inside the reactor, or directly by means of steam injection into the reactor. Steam injection is faster and easier but will cause an increase of medium volume by 10 to 15% owing to condensation. The piping lines for large reactors are normally sterilized in situ by using steam.

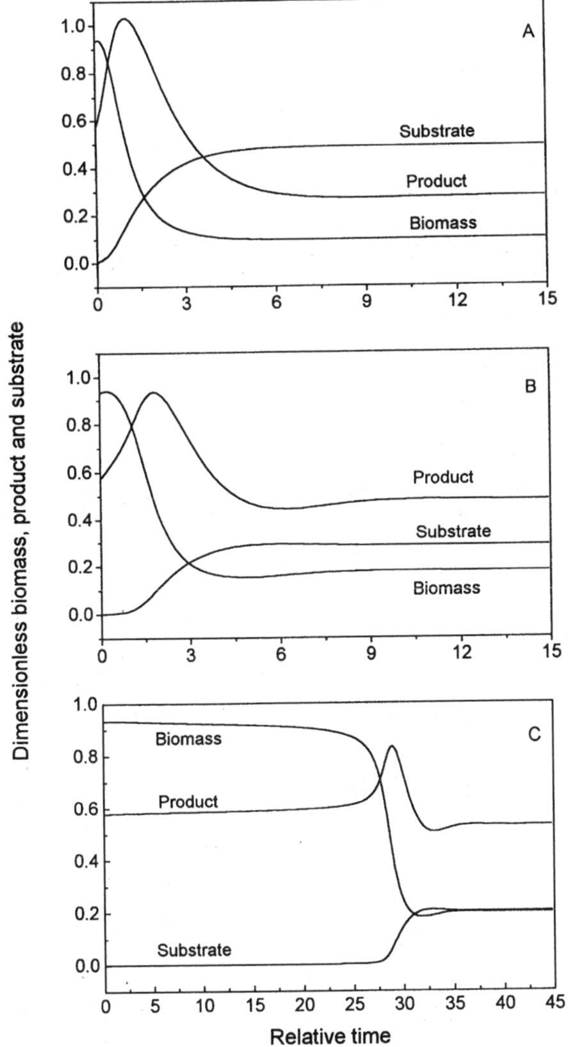

FIGURE 9 Transient behavior of a steady state after step changes in a continuous culture operated at a constant dilution rate. The substrate concentration in the feed (y_0) is increased from (A) 0.334 to 0.550 (in dimensionless units), (B) 0.334 to 0.400, and (C) 0.334 to 0.335. The relative time is numerically equal to the number of reactor volume exchanges (residence times) (57).

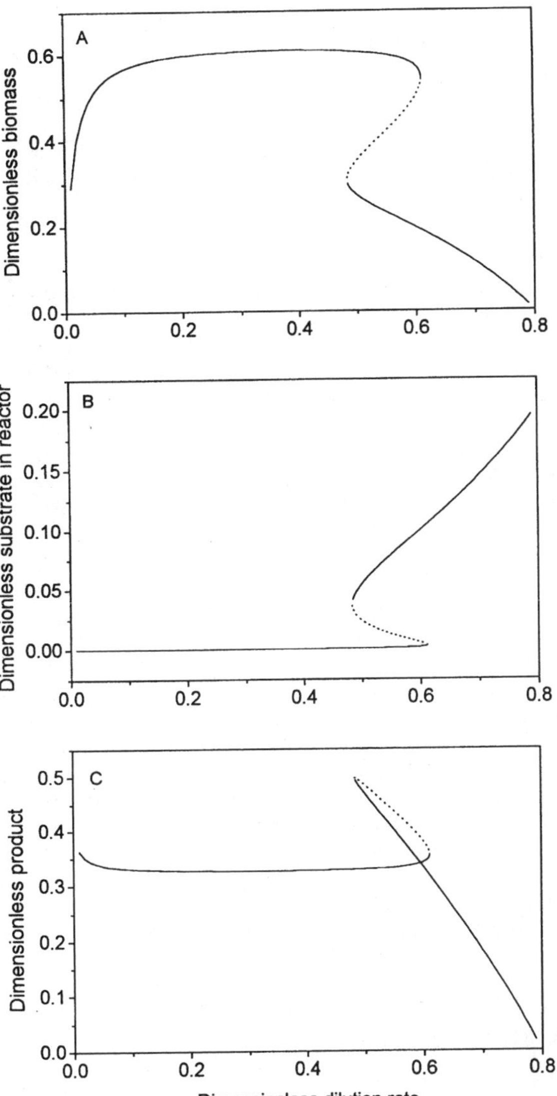

FIGURE 10 Steady-state concentrations of biomass (A), substrate in reactor (B), and product (C) in a continuous culture at a constant feed substrate concentration but varied dilution rates (dimensionless feed substrate concentration, y_0 = 0.2). This culture is subject to metabolic overflow and growth inhibition by product. Multiple steady states are found in a given range of dilution rates. Depending on the operation mode, different steady states can be obtained under identical operation conditions (57).

Piping lines should be as short as possible, straight, and laid with a downward slope to ensure the outflow by gravity. T connections should be avoided if possible. To keep the danger of contamination to a minimum, all piping in the sterile stage of a reactor may need to be welded, with no detachable connections. This is very cost-intensive and is therefore only used for special processes (e.g., cell culture processes). In the practical operation of bioreactors, detachable connections are desired to achieve a high degree of flexibility. In this case, only flanges designed in accordance with sterilization requirements should be employed for all connections. The cross sections of piping should be neither reduced nor increased at the flange connection points to avoid accumulation of dirt. The two most popular flange systems are the so-called quick-connection clamping ring system and the screw-on system (13). A common feature of these flanges is that a sterile connection is achieved by "O" rings. O rings are automatic sealing elements, the total sealing pressure of which increases with increasing pressure in the system.

12.3.4. Environmental Control

Temperature, pH, and dissolved oxygen (pO_2) measurement and control are similar in batch and continuous cultures. However, pH and pO_2 may present problems that are not usually encountered in short fermentations. Drifts in the pH and pO_2 probes can occur because of biofilm formation or adsorption of proteins and other materials on the tips of the probes. Whereas the pH probe could be roughly recalibrated by briefly changing the culture pH values and comparing with the off-line measurements, this cannot be done with the oxygen probe. Whenever possible, polarographic oxygen probes should be used that have much better stability than galvanic probes after autoclaving, and the electronic zero feature can be used to zero them. For more detailed information on the environmental control of bioreactors, the reader is referred to Schügerl (48).

12.3.5. Start-Up

Continuous cultivation of microorganisms is started as a batch culture. It is normally switched to continuous culture during the exponential growth phase before substrate is completely consumed. The transition of a culture after the switch-over can display very different dynamic behaviors, depending on the nature of the culture and the reactor conditions at the time of switch-over (6, 14). Oscillatory transitions may occur as a result of sudden changes of environmental conditions, particularly during transition from substrate limitation to substrate excess (14, 23, 34). If the culture is suddenly exposed to a very high concentration of toxic substrate, the oscillation may become so severe that cells are washed out. If a system can display multiplicity, the timing and operation mode of start-up of a continuous culture may lead to different transition behaviors and different steady states (28, 57).

In most cases, the use of a half-strength medium for the initial batch growth can give a smooth transition from batch to continuous cultivation. Nutrient feed is started when the cell concentration is about half of that expected for steady state. In this way, the culture initially approaches a nutrient-limited fed batch after the switch-over. The specific growth rate declines as the cell concentration increases until the steady-state concentration is achieved. At this time, the level controller is turned on and the continuous fermentation begins.

12.3.6. Dynamic Behavior and Determination of Steady State

Steady state is defined as the condition of a continuous culture in which changes in the process parameters (e.g., concentrations of substrate, biomass, and product) and the physiological state of cells are no longer detectable. Normally, only the process parameters that can be measured either on-line or off-line are used to judge if a culture has reached a steady state. In this case, as many process parameters as possible should be included. If product(s) is formed, particular attention should be given to the concentration of the product(s). The product usually reaches constant values later than the concentrations of biomass and substrate. The physiological state of cells reaches a steady state even more slowly, especially if long-term adaptations take place (42, 45).

In an ideal chemostat culture governed by Monod kinetics, the time required to achieve a steady state is about 3 to 4 times of culture replacement after a condition changes (44). The culture replacement time (residence time) is given by $1/D$. In reality, much longer time is often needed to achieve steady state, depending on the dynamic behavior of the culture and the magnitude of changes (25, 34, 45). Figure 9 gives some examples of transient behavior of a continuous culture after different step changes of substrate concentration in the feed medium at a constant dilution rate (57). In the case of the new feed concentration (y_0) being just above the previous steady-state value (y_{01}), the new steady state is reached after almost 40 residence times. The change is visible only after about 24 residence times. On the other hand, if y_0 is increased by a large step, the new steady state is quickly reached after about six residence times. Overshot and underwing of biomass and product are observed in this culture system. More profound oscillatory transient behaviors are observed under certain conditions (5, 6, 16, 23, 34, 45).

In some culture systems, multiplicity of steady states can occur (1, 28, 34). Depending on the operation mode, two markedly different stable steady states can be obtained under the same cultivation conditions. Xiu et al. (57) recently presented a theoretical and mathematical analysis of multiplicity and instability of microorganisms in continuous culture and compared them with recent experimental results. Figure 10 depicts results for a culture operated at a constant feed substrate concentration. For cells grown on one substrate without limitation, the combined effects of product inhibition and enhanced formation rate of product under substrate excess appear to be the main reasons for the multiplicity and hysteresis phenomena so far experimentally observed. These combined effects can also lead to unusual dynamic behaviors, such as a prolonged time to reach a steady state, oscillatory transition from one steady state to another, and sustained oscillations (Fig. 9).

REFERENCES

1. **Axelsson, J. P., T. Münch, and B. Sonnleitner.** 1992. Multiple steady states in continuous cultivation of yeast, p. 383–386. *In* M. N. Karim and G. Stephanopoulos (ed.), *Proceedings of ICCAFT5/IFAC Modeling and Control of Biotechnical Processes*, Keystone, Colo. Pergamon Press, Oxford.
2. **Bailey, J., and D. F. Ollis.** 1986. *Biochemical Engineering Fundamentals*. 2nd ed. McGraw-Hill, New York.
3. **Baloo, S., and D. Ramkrishna.** 1991. Metabolic regulation in bacterial continuous cultures. I. *Biotechnol. Bioeng.* **38:**1337–1352.
4. **Banik, G. C., and C. A. Heath.** 1995. Hybridoma growth and antibody production as a function of cell density and specific growth rate in perfusion culture. *Biotechnol. Bioeng.* **48:**289–300.
5. **Barford, J. P., J. H. Johnston, and P. K. Mwesigye.** 1995. Continuous culture study of transient behavior of *Saccharomyces cerevisiae* growing on sucrose and fructose. *J. Ferm. Bioeng.* **79:**158–162.
6. **Benthin, S., J. Nielsen, and J. Villadsen.** 1993. Two uptake systems for fructose in *Lactococcus lactis* subsp. *cremoris* FD1 produce glycolytic and gluconeogenic fructose phosphates and induce oscillations in growth and lactic acid formation. *Appl. Environ. Microbiol.* **59:**3206–3211.
7. **Bentley, W., and D. S. Kompala.** 1990. Stability in continuous cultures of recombinant bacteria: a metabolic approach. *Biotechnol. Lett.* **12:**329–334.
8. **Biebl, H.** 1991. Glycerol fermentation to 1,3-propanediol by *Clostridium butyricum*. Measurement of product inhibition by use of a pH-auxostat. *Appl. Microbiol. Biotechnol.* **35:**701–705.
9. **Brown, D. E., D. J. Halsted, and C. G. Sinclair.** 1979. Application of an aerated mixing model to a continuous culture system. *Biotechnol. Lett.* **1:**159–164.
10. **Byun, T.-G., A.-P. Zeng, and W.-D. Deckwer.** 1994. Reactor comparison and scale-up for the microaerobic pro-

duction of 2,3-butanediol by *Enterobacter aerogenes* at constant oxygen transfer rate. *Bioproc. Eng.* **11:**167–175.

11. **Calcott, P. H.** 1981. *Continuous Culture of Cells*. CRC Press, Boca Raton, Fla.
12. **Cooney, C. L., H. M. Koplove, and M. Häggstrom.** 1981. Transient phenomena in continuous culture, p. 143–168. *In* P. H. Calcott (ed.), *Continuous Culture of Cells*, vol. 2. CRC Press, Boca Raton, Fla.
13. **Deckwer, W.-D., R. Luttmann, H.-G. Reng, and S. Yonsel.** 1987. *Bioreaktoren: Ein Leitfaden für Anwender*. GBF Texte 4, Braunschweig-Stöckheim, Germany.
14. **Dunn, I. J., and J. R. Mor.** 1975. Variable volume continuous cultivation. *Biotechnol. Bioeng.* **17:**1805–1822.
15. **Dykhuizen, D. E., and D. L. Hartl.** 1983. Selection in chemostats. *Microbiol. Rev.* **47:**150–168.
16. **Fiechter, A.** 1981. Batch and continuous culture of microbial, plant and animal cells, p. 454–505. *In* H. J. Rehm and G. Reed (ed.), *Biotechnology: A Comprehensive Treatise*, vol. 1. Verlag Chemie, Weinheim, Germany.
17. **Frame, K. K., and W.-S. Hu.** 1991. Kinetic study of hybridoma cell growth in continuous culture. I. A model for non-producing cells. *Biotechnol. Bioeng.* **37:**55–64.
18. **Frame, K. K., and W.-S. Hu.** 1991. Kinetic study of hybridoma cell growth in continuous culture. II. Behavior of producers and comparison to non-producers. *Biotechnol. Bioeng.* **38:**1020–1028.
19. **Gostomski, P., M. Mühlemann, Y.-H. Lin, R. Mormino, and H. Bungay.** 1994. Auxostats for continuous culture research. *J. Biotechnol.* **37:**167–177.
20. **Gottschal, J. C.** 1992. Continuous culture, p. 559–572. *In* J. Lederberg (ed.), *Encyclopedia of Microbiology*, vol. 1. Academic Press, New York.
21. **Gregory, M. E., M. Bulmer, I. D. L. Bogle, and N. Titcherner-Hooker.** 1996. Optimising enzyme production by bakers yeast in continuous culture: physiological knowledge useful for process design and control. *Bioproc. Eng.* **15:**239–245.
22. **Han, K., and O. Levenspiel.** 1988. Extended Monod kinetics for substrate, product, and cell inhibition. *Biotechnol. Bioeng.* **32:**430–437.
23. **Harrison, D. E. F., and H. H. Topiwala.** 1974. Transient and oscillatory states of continuous culture, p. 168–219. *In* T. H. Ghose and A. Fiechter (ed.), *Advanced Biochemical Engineering*, Vol. 3. Springer Verlag, Berlin.
24. **Helmuth, K., D. J. Korz, E. A. Sanders, and W.-D. Deckwer.** 1994. Effect of growth rate on stability and gene expression of recombinant plasmids during continuous and high cell density cultivation of *Escherichia coli* TG1. *J. Biotechnol.* **32:**289–298.
25. **Herbert, D., R. Elsworth, and R. C. Telling.** 1956. The continuous culture of bacteria: a theoretical and experimental study. *J. Gen. Microbiol.* **14:**601–622.
26. **Hiller, G. W., D. S. Clark, and H. W. Blanch.** 1993. Cell retention-chemostat studies of hybridoma cells: analysis of hybridoma growth and metabolism in continuous suspension culture on serum-free medium. *Biotechnol. Bioeng.* **42:**185–195.
27. **Hiller, G. W., D. S. Clark, and H. W. Blanch.** 1994. Transient responses of hybridoma cells in continuous culture to step changes in amino acid and vitamin concentrations. *Biotechnol. Bioeng.* **44:**303–321.
28. **Imanaka, T., T. Kaieda, K. Sato, and H. Taguchi.** 1972. Optimization of α-galactosidase production by mold. I. α-Galactosidase production in batch and continuous culture and a kinetic model for enzyme production. *J. Ferm. Technol.* **50:**633–646.
29. **Kiss, R. D., and G. Stephanopoulos.** 1992. Culture instability of auxotrophic amino acid producers. *Biotechnol. Bioeng.* **40:**75–85.
30. **Kubitschek, H. E.** 1970. *Introduction to Research with Continuous Culture*. Prentice Hall, Englewood Cliffs, N.J.
31. **Lee, C. W., and H. N. Change.** 1987. Kinetics of ethanol fermentation in membrane cell recycle fermentors. *Biotechnol. Bioeng.* **29:**1105–1112.
32. **Linz, M., A.-P. Zeng, R. Wagner, and W.-D. Deckwer.** 1997. Stoichiometry, kinetics, and regulation of glucose and amino acid metabolism of a recombinant BHK cell line in batch and continuous cultures. *Biotechnol. Prog.* **13:**453–463.
33. **Marr, A. G.** 1991. Growth rate of *Escherichia coli*. *Microbiol. Rev.* **55:**316–333.
34. **Menzel, K., A.-P. Zeng, H. Biebl, and W.-D. Deckwer.** 1996. Kinetic, dynamic and pathway studies of glycerol metabolism by *Klebsiella pneumoniae* in anaerobic continuous culture. Part I. The phenomena and characterization of oscillation and hysteresis. *Biotechnol. Bioeng.* **52:**549–560.
35. **Merten, O.-W., D. Moeurs, H. Keller, M. Leno, G. E. Palfi, L. Cabanie, and E. Couve.** 1994. Modified monoclonal antibody production kinetics, kappa/gamma mRNA levels, and metabolic activities in a murine hybridoma selected by continuous culture. *Biotechnol. Bioeng.* **44:**753–764.
36. **Minkevich, I., G. Krynitskaya, and V. K. Eroshin.** 1989. Bistat: a novel method of continuous cultivation. *Biotechnol. Bioeng.* **33:**1157–1161.
37. **Mohan, S. B., S. R. Chohan, J. Eade, and A. Lyddiatt.** 1993. Molecular integrity of monoclonal antibodies produced by hybridoma cells in batch culture and in continuous-flow culture with integrated product recovery. *Biotechnol. Bioeng.* **42:**974–986.
38. **Monod, J.** 1950. La technique de culture continue: theorie et application. *Ann. Inst. Pasteur (Paris)* **79:**390.
39. **Moser, A.** 1985. Continuous cultivation, p. 285–309. *In* H. J. Rehm and G. Reed (ed.), *Biotechnology: A Comprehensive Treatise*, vol. 2. Verlag Chemie, Weinheim, Germany.
40. **Nancib, N., and J. Boudrant.** 1992. Effect of growth rate on stability and gene expression of a recombinant plasmid during continuous culture of *Escherichia coli* in a non-selective medium. *Biotechnol. Lett.* **14:**643–648.
41. **Novick, A., and Szilard, L.** 1950. Description of the chemostat. *Science* **112:**715.
42. **Petrik, M., O. Käppeli, and A. Fiechter.** 1983. An expanded concept for the glucose effect in the yeast *Saccharomyces uvarum*: involvement of short- and long-term regulation. *J. Gen. Microbiol.* **129:**43–49.
43. **Pirt, S. J.** 1965. The maintenance energy of bacteria in growing cultures. *Proc. R. Soc. Lond. Ser. B.* **163:**305–314.
44. **Pirt, S. J.** 1975. *Principles of Microbe and Cell Cultivation*. Blackwell Scientific Publications, Oxford.

45. **Postma, E., C. Verduyn, W. A. Scheffers, and J. P. van Dijken.** 1989. Enzymic analysis of the Crabtree effect in glucose-limited chemostat cultures of *Saccharomyces cerevisiae*. *Appl. Environ. Microbiol.* **55:**468–477.

46. **Rice, C. W., and W. P. Hempfling.** 1985. Nutrient limited growth in the pH-auxostat. *Biotechnol. Bioeng.* **27:**187–191.

47. **Sanchez, O., H. van Gemerden, and J. Mas.** 1996. Description of a redox-controlled sulfidostat for the growth of sulfide-oxidizing phototrophs. *Appl. Environ. Microbiol.* **62:**3640–3645.

48. **Shoham, Y., and A. L. Demain.** 1991. Kinetics of loss of a recombinant plasmid in *Bacillus subtilis*. *Biotechnol. Bioeng.* **37:**927–935.

49. **Schügerl, K.** 1991. Common instruments for process analysis and control, p. 5–25. *In* H.-J. Rehm and G. Reed (ed.), *Biotechnology*, 2nd ed., vol. 4. VCH, Weinheim, Germany.

50. **Stafford, K.** 1986. Continuous fermentation. In A. L. Demain and N. A. Solomon (ed.) *Manual Industrial Microbiology and Biotechnology*. American Society for Microbiology, Washington, D.C.

51. **Tempest, D. W., and O. M. Neijssel.** 1992. Physiological and energetic aspects of bacterial metabolite overproduction. *FEMS Microbiol. Lett.* **100:**169–176.

52. **Togna, A. P., J. Fu, and M. L. Shuler.** 1993. Use of a simple mathematical model to predict the behavior of *Escherichia coli* overproducing β-lactamase within continuous single-and two-stage reactor systems. *Biotechnol. Bioeng.* **42:**557–570.

53. **Topiwala, H. H., and G. Hamer.** 1971. Effect of wall growth in steady-state continuous cultures. *Biotechnol. Bioeng.* **13:**919–922.

54. **Vierheller, C., A. Goel, M. Peterson, M. M. Domach, and M. M. Ataai.** 1995. Sustained and constitutive high levels of protein production in continuous cultures of *Bacillus subtilis*. *Biotechnol. Bioeng.* **47:**520–524.

55. **Werner, R. G., F. Walz, W. Noé, and A. Konrad.** 1992. Safety and economic aspects of continuous mammalian cell culture. *J. Biotechnol.* **22:**51–68.

56. **Weusthuis, R. A., J. T. Pronk, P. J. A. van den Broek, and J. P. van Dijken.** 1994. Chemostat cultivation as a tool for studies on sugar transport in yeasts. *Microbiol. Rev.* **58:**616–630.

57. **Xiu Z.-L., A.-P. Zeng, and W.-D. Deckwer.** 1998. Multiplicity and stability analysis of microorganisms in continuous culture: effect of metabolic overflow and growth inhibition. *Biotechnol. Bioeng.* **57:**251–261.

58. **Yang, R. D., and Humphrey A. E.** 1975. Dynamic and steady states of phenol biodegradation in pure and mixed cultures. *Biotechnol. Bioeng.* **17:**1211–1235.

59. **Zeng, A.-P., H. Biebl, and W.-D. Deckwer.** 1990. 2,3-Butanediol production by *Enterobacter aerogenes* in continuous culture: role of oxygen supply. *Appl. Microbiol. Biotechnol.* **33:**264–268.

60. **Zeng, A.-P., H. Biebl, and W.-D. Deckwer.** 1991. Production of 2,3-butanediol in a membrane bioreactor. *Appl. Microbiol. Biotechnol.* **34:**463–468.

61. **Zeng, A.-P., and W.-D. Deckwer.** 1992. Utilization of the tricarboxylic acid cycle, a reactor design criterion for the microaerobic production of 2,3-butanediol. *Biotechnol. Bioeng.* **40:**1078–1084.

62. **Zeng, A.-P., and W.-D. Deckwer.** 1995. Mathematical modeling and analysis of glucose and glutamine utilization and regulation in cultures of continuous mammalian cells. *Biotechnol. Bioeng.* **47:**334–346.

63. **Zeng, A.-P., and W.-D. Deckwer.** 1995. A kinetic model for substrate and energy consumption of microbial growth under substrate-sufficient conditions. *Biotechnol. Prog.* **11:** 71–79.

64. **Zeng, A.-P., and W.-D. Deckwer.** 1996. Bioreaction techniques under low oxygen tension and oxygen limitation: from molecular level to pilot plant reactor. *Chem. Eng. Sci.* **51:**2305–2314.

Methods for Biocatalysis and Biotransformations

P. C. MICHELS AND J. P. N. ROSAZZA

13

13.1. BACKGROUND

Biotransformations are reactions of organic compounds by either enzyme or whole-cell biocatalysts. Biocatalysis is widely applied in industry for pharmaceutical, agrochemical, chemical, fragrance and flavor, nutritional, and bioremediation purposes. Since the first edition of this volume was published in 1986 (17), significant new advances have occurred, spurring broader applications of biotransformation. Many new enzymes have been characterized from microbial cells, and methods for their isolation, stabilization, and use have dramatically expanded. Importantly, biocatalysis has increasingly been extended to reactions in organic solvents, in which many compounds of interest are soluble and additional reactions are possible. At the same time, advances in biocatalyst improvement by recombinant technologies have provided the bases for unprecedented means of biocatalyst alteration and use. Continued advancements in the use and development of biocatalysis have drawn upon more sensitive, rapid, and informative analytical methods. Further advances in biocatalysis will be achieved from the diverse areas of organic chemistry, analytical chemistry, biochemistry, molecular biology, microbiology, and engineering. Indeed, as this chapter will make clear, the most successful practitioners of biotransformations have an appreciation and understanding of the highly interdisciplinary nature of biocatalysis development. Yet biotransformation techniques have evolved such that the synthetic chemist can utilize biocatalysts just as many other synthetic reagents are used.

Enzyme catalysts have several features that render them attractive as a class of "reagents" for organic synthesis. Enzymes are chiral catalysts. They are proteins that have evolved into specific biocatalytic structures. They often bind substrates very specifically and display high regio-, stereo-, and enantioselectivities. These desirable traits obviate the need to block undesirable reactions that commonly occur with other functional groups in traditional organic synthesis. Catalysis occurs under mild reaction conditions requiring no strong acids or bases, temperature extremes, rigorously controlled atmospheres, heavy metals, or other conditions commonly associated with chemical catalysts. Multistep biocatalytic processes can occur efficiently with a single microorganism. Enzyme reactions require little energy input because they occur very efficiently between 20 and 70°C. Most intriguingly, the maturation of genomics, molecular biology, and in vitro evolution techniques promise to provide highly efficient and tunable catalysts tailored for specific synthetic goals. While we realize this potential is still years off, biocatalysis is today a viable alternative for conducting many synthetic reactions.

This chapter is devoted to a consideration of biocatalysis methods applicable to the solution of problems in organic chemistry. We hope it will serve as a practical and concise guide to an enormous literature and provide a basis for simple and productive experimentation by scientists of many disciplines who may benefit from biotransformations.

13.2. CONCEPT AND GENERAL FEATURES OF BIOTRANSFORMATIONS

Organic reactions catalyzed by microorganisms are referred to as microbial transformations, biotransformations, or bioconversions. Biotransformation reactions are catalyzed by enzymes produced by microbial cells and by all living organisms. In their natural functions, enzymes catalyze, and indeed control, anabolic and catabolic reactions necessary to life processes such as bioenergetics, growth, and replication. Anabolic enzymes, involved in biosynthetic pathways, are usually substrate specific, while many catabolic enzymes, involved in digestive, defensive, and similar degradation roles in living organisms, seem to have evolved broader ranges of specificity. The substrate specificities of some enzymes are remarkably broad. A well-known example is the reactivity of cytochromes P-450 (35, 44a), with substrates of broadly different structures; some human hepatic forms catalyze hydroxylations or dealkylations on more than 50 diverse pharmaceutical structures. More general and convincing demonstrations are the many isolations of microorganisms capable of degradation of synthetic organic compounds only recently introduced to the environment (55). Many other examples are present in the literature, and, in fact, few enzymes are specific enough to catalyze reactions with only their natural substrates. Thus, most organic compounds (excluding unstable and highly reactive compounds) can serve as substrates for enzyme-catalyzed transformations.

From a synthetic perspective, as we shall summarize, it is possible to select experimental conditions that favor the production of desired enzyme catalysts that can be used to perform single and highly specific reactions. This is done by controlling the preinoculation, growth, and transformation environment of the culture and the physical form of the organic substrate, and by establishing reproducible experimental protocols. Knowledge of the natural substrates of enzymes used to catalyze organic reactions, or even the identity of the enzyme itself, is not necessary.

Biotransformation enzymes may be present within (endo) or outside (exo) of the cells that produce them. Bacteria often contain soluble enzymes within the cytosol and particulate enzymes bound to membrane structures. Yeasts and fungi are more complicated; their enzymes are often compartmentalized within various organelles, including mitochondria, nuclei, vacuoles, as well as in cell membranes. A priori, there is no way to know the location of useful enzymes within the biotransforming cell. Therefore, experimental methods are designed to allow the transport of reactants to catalytic sites, wherever they might be, favoring the highest possible solubility and dispersion in the reaction medium and enhancing permeability of the cells to the reactants.

Many microbial enzymes are constitutive in nature: they are always produced by the growing cell. If enzymes are not constitutive, their formation in microorganisms can sometimes be induced by the substrate of interest, or by related compounds. Inducible enzymes may catalyze a variety of reactions on the inducers, substrates, and even unrelated compounds. When multienzyme pathways are induced, environmental conditions can be controlled to favor desirable single or multistep reactions.

With this basic understanding of biocatalysis, the rest of this chapter summarizes methods used to conduct experiments in biocatalysis and biotransformations, particularly of organic compounds. Reviews of many types of enzyme-catalyzed reactions were cited earlier (17), and space permits us to list only a few selected additional literature references (9, 20, 24, 27, 40, 42).

13.3 PROCEDURES

Huge numbers of microbes coexist in almost all natural environments, particularly soils (estimated to have 10^9 cells/g of soil), waters, and sewage. The makeup of the flora in these ecosystems is determined by the availability of oxygen and water, light exposure, temperature, and the nutrients present. Widely different mixtures of bacteria, fungi, algae, and other microscopic life can be isolated from nature by using different natural ecosystems (e.g., soils from different locations) as sources of inocula (chapter 1, this volume). While it may be possible to predict good candidate biocatalysts for a given transformation on a given substrate, the best catalyst most often must be identified from small-scale test reactions. This catalyst identification stage can generally be performed in two ways. The first strategy involves "screening" large numbers of individual reactions with pure cultures for a specified transformation. The second approach involves "selection" of a strain from mixed culture, usually using the ability to grow on the test substrate as the selective pressure.

While the "selection" strategy is commonly used for bioremediation studies, synthetic biotransformations greatly benefit by work with pure cultures. Pure cultures are identifiable by their morphological, nutritional, and other characteristics that allow classification of organisms into taxonomic strata. Since pure cultures are definable reagents, they are easier to maintain, possible to control, and their use helps ensure experimental reproducibility. Moreover, multistep reactions can be more easily characterized with single biocatalyst strains than with microbial consortia. In essence, pure cultures are to biocatalysis as pure reagents are to chemistry! Simpler reactions are performed with straightforward and uncomplicated reagents, and the same is true with pure cultures vs. mixed cultures as specific biocatalysts.

13.3.1. Taxonomy

A detailed review of the science of taxonomy is beyond the scope of this chapter (5, 21). Nevertheless, it is useful to consider the basic organizational framework for classification of microorganisms commonly used as biocatalysts. Classifications are constructed from individual organisms that occur within populations of species. The species is the fundamental or most basic level of organization in taxonomy. Species are individuals sampled from populations closely resembling each other in many different characteristics. Species that share many common characteristics are placed in a group termed the genus. Families consist of still more inclusive groups, and so on to higher organizational levels. The taxonomic classification for a given organism, *Streptomyces cretosus*, for example, is as follows: kingdom *Procaryotae*, division *Bacteria*, class *Actinomycetes*, order *Actinomycetales*, family *Streptomycetaceae*, genus *Streptomyces*, species *cretosus*.

Taxonomy can be instructive when considering biocatalysts for synthetic applications. For example, when a culture such as *Mycobacterium fortuitum* performs an interesting biotransformation reaction, but in low yield, it is logical to examine taxonomically related cultures to find other candidates that might provide higher yields or that might perform related reactions on different starting materials. To do so, other *Mycobacterium* spp., other genera of the actinomycetes, and members of closely related families such as the *Nocardiaceae* would also be examined. This approach is reasonable, since there are often similarities in enzymatic makeup among members of related genera and families. Conversely, when building and screening a library of microbial catalysts, for synthetic applications, it makes sense to sample biocatalyst diversity by including as broad a range of genera as possible.

An understanding of cell structure and morphology can be helpful when designing biocatalytic processes and equipment. Bacterial, fungal, and yeast cells all contain a variety of organelles in the cytosol, such as storage granule ribosomes, mitochondria, spores, and membranous structures. Enzymes useful for bioconversions have been found to occur in all of these structures. Importantly, cell membranes are semipermeable barriers to nutrients and waste products entering and leaving the cell. This membrane must be penetrated by organic substrates for reactions to occur with intracellular enzymes. If enzymes are located in periplasmic spaces between the membrane and the rigid cell wall, organic substrate need only to penetrate this outer wall. In addition to rigid cell walls, the shells of some microorganisms may consist of an extra membrane (outer membrane of gram-negative bacteria), "slime" layers, and capsules that affect mass transport into the cell.

13.3.2. Biocatalyst Acquisition and Preservation (see Chapters 1 and 3)

Selection of the biocatalyst is the most critical of all the operations in a biotransformation experiment. Cultures with desirable properties are obtained from other investigators in the biotransformation field, from standard culture collections (12), or by isolation from natural habitats (17). Comprehensive lists of most of the significant culture collections around the world are available on the World Wide Web (www.sv-cict.fr/bacterio/collections.html). Some of the largest and most useful culture collections include:

- American Type Culture Collection (ATCC), 12301 Parklawn Dr., Rockville, Md. 20825; www.atcc.org
- Centrallbureau voor Schimmelcultures (CBS), Baarn, The Netherlands; www.cbs.knaw.nl
- German Collection of Microorganisms (DSM), Grisebachstr. 8, D-34 Goettingen; www.gdfbraunschweig.de/DSMZ
- National Institute of Bioscience and Human-Technology, Tsukuba-shi, Higashi 1-1, Ibarak; 305, Japan
- USDA Agriculture Research Service (NRRL), 1815 N. University Ave., Peoria Ill. 60604

Well-established culture collections catalog microorganisms by number so that it is possible to obtain the same organism (biocatalyst) each time. The purchase of a culture (they may be expensive) by catalog number is identical to the process whereby a specific catalyst or reagent is purchased from a chemical catalog. In many collections, strains are maintained in lyophilized form or on agar slants suitable for mailing. Investigators usually maintain their own culture collections of 100 to 400 strains. In addition, microbiology and mycology departments on many college campuses maintain culture collections. Investigators are usually willing to share their cultures with researchers in other laboratories. Alternatively, enrichment techniques are used to isolate cultures from nature (2, 53).

With access to a huge number of microbial strains, some intuition about narrowing the search for an acceptable biocatalyst can be valuable. The literature of the last 25 years provides excellent leads to available organisms with specific, desirable enzymatic capabilities. Electronic databases of this literature, allowing relational and structure-based searching, provide an excellent tool for selecting good candidate biocatalysts. Other catalogs of chemical reactions catalyzed by microorganisms have been assembled (12) with specific attention to groups of compounds such as the alkaloids (17, 42), steroids (17, 27), and nonsteroidal cyclic compounds, including various drugs (17, 24) and other xenobiotics.

Microbial strains must be handled like all complex reagents. When new cultures are procured, the following information should be recorded: culture name and number, source, acquisition and lyophilization dates, lyophilization medium, growth medium and temperature, literature source, reactions known to be catalyzed, and unusual properties and comments. Upon receipt, it is also important to establish the purity of new cultures and to transfer them to appropriate fresh media for propagation and storage. For short-term storage and routine culturing, slants in screw-top vials should be used. After significant growth, many strains can be stored in a refrigerator. For longer-term storage, lyophilization or ultra-low-temperature freezing is recommended (see chapter 3, this volume).

13.3.3. Growth Fundamentals

Details of culture growth are given elsewhere (chapter 4). Growth is the cumulative process resulting in the orderly increase of all chemical components of the living cell (5). Different groups of organisms behave quite differently when grown in liquid culture. As with enzyme-catalyzed reactions, an approximately linear relationship exists between the amount of microbial cells (biocatalyst) present and the rate of reaction in the incubation mixture. Thus, when biotransformations are conducted with growing cultures, conditions that favor enhanced growth usually result in greater yields of reaction products. Growth rates of unicellular bacteria and yeasts are usually regular (5). Growth of multicellular organisms such as fungi and filamentous procaryotes (actinomycetes) such as *Streptomyces* spp. is more difficult to define.

The complement of enzymes produced by microbial cells varies greatly at five times during the life cycle of the cell. The desired enzyme activity may be present from the start of the growth cycle, or it may not appear until the late exponential phase. The changes in enzyme activities during growth reflect the changes occurring within the cell and the culture medium as the organism grows and metabolizes nutrients. Thus, the optimum time for adding organic reactants or for harvesting cells must be established by experimentation. This is another motivation for using pure cultures for transformations.

13.3.4. Measurement of Cell Mass

For quantitative estimations of catalytic efficiency it may be necessary to estimate biomass, the amount of biocatalyst present. Wet weights of microbial cells typically reach levels of 20 to 30 g/liter of culture medium. These are estimated by filtering known volumes of fungi or actinomycetes, or by centrifuging known culture volumes of bacteria and yeasts. Culture dry weights are obtained by placing aliquots of filtered or pelleted cells in a 120°C oven overnight to drive off moisture. Alternatively, optical density readings at 600 or 660 nm can be correlated to dry weights. A calibration curve between 0.1 and 1.0 absorbance units is made from a series of dilutions of the original culture. A plot of cell densities (dry weight) versus optical density should yield a usable straight line falling off somewhat in the more turbid region. A separate calibration curve needs to be constructed for each cultured strain. This procedure works well for nonfilamentous bacteria and yeasts.

Viable cell counts can be determined by plating of serial culture dilutions. A handy rule of thumb for estimating cell numbers in suspensions is that a suspension with just barely visible turbidity contains about 10^6 cells per ml. This estimate is fairly accurate for bacteria of average size such as *Escherichia coli* (17).

13.3.5. Forms of the Biocatalyst

The use of live, growing microbial cells as biocatalysts for biotransformations of organic compounds has been extensively documented. Pure cultures are grown to a point where desired enzyme activities are maximal, at which time organic chemical substrates are added to the incubation mixture in which the transformations take place. However, much experimental latitude is now possible in the use and form of the biocatalyst.

13.3.6. Growing Cultures

Both batch and continuous cultures are used in bioconversion experiments. In the batch culture technique, a pure culture is grown in a suitable medium. At an experimentally determined time, the substrate is added, and reaction is continued until transformation of the substrate ceases or additional reactions seriously begin to affect yields. In batch processes, the biocatalyst is used only once and then discarded. Necessary equipment is inexpensive and simple, and the procedure is straightforward for screening purposes. However, it requires the repetitive production of cells for each experiment, and the isolation of reaction products from complex fermentation media can be complicated. The physiological state of cells in batch culture varies continuously throughout the growth cycle.

This is not true of cells in continuous culture, in which cells can be maintained in a steady physiological state for long periods of time by means of continuous addition of fresh nutrient medium and simultaneous withdrawal of equal amounts of spent medium. Details of the continuous culture method are discussed in chapter 12 of this volume. Continuous culture has not been used widely in biotransformation screening, owing to the relative complexity of the fermentor equipment, but it may be useful for scale-up of the biotransformation.

13.3.7. Resting Cells

Resting cells are nongrowing, live cells that retain most of the enzyme activities of growing cells. Resting cells are obtained from the culture medium at a time in the growth cycle when enzyme activities are highest, or at least present at useful levels. Mycelial growth can be removed by filtration, but yeasts and bacteria are best harvested by centrifugation. Concentrated cells are resuspended in buffers, modified culture media (usually without some required nutrient for growth), distilled water, or even nonaqueous solvent mixtures for use as biocatalysts.

As biotransformation catalysts, resting cells have several advantages versus growing cells or isolated enzymes. They are much cleaner than reactions with growing cells, resulting in easier product isolation. The cell concentration can be made higher, enhancing the sensitivity of the biocatalyst screening. Moreover, control of undesirable secondary or side reactions can be done more easily with nongrowing cells. Compared with isolated enzymes, whole resting cells can accomplish efficient multistep enzymatic reactions without the need for expensive coenzymes. For single-step reactions, the direct use of resting cells minimizes loss of activity, which is unavoidable during isolation and purification of enzymes. Enzymes in intact cells usually are more stable than their isolated counterparts.

Cometabolism and enzyme induction increase the usefulness of resting-cell biocatalysis. For some reactions to occur, it is necessary to use a cosubstrate such as glucose, glycerol, succinic acid, or another oxidizable metabolite along with the organic compound to be transformed (37). These cosubstrates drive reactions to completion by providing the necessary energy derived during their utilization. They also provide energy for recycling of coenzymes for the enzymatic reactions.

In some cells, desired enzyme activities are dramatically increased by cultivating the organism in the presence of the organic compound to be transformed (enzyme induction). Conversely, the use of resting cells in which enzyme activities have not been induced can result in poor yields. Many examples of the use of resting cells in biotransformations have been described (17, 36, 40, 58), and a typical application is summarized in section VII of this volume.

13.3.8. Dried Cells

In some cases microbial cells can be dehydrated and still maintain enzyme activities. The resulting powders are convenient biocatalysts. Esterases, amidases, oxidoreductases, and dehydrogenases, among other enzyme classes, can survive cell-drying procedures (36). The two most common methods of drying are lyophilization and acetone dehydration. For lyophilization, harvested cells are suspended in distilled water or dilute buffer. The suspension is then frozen over a dry ice-acetone (or ethanol) bath in a thin shell inside a round-bottomed flask (about 5 mm thick). Water sublimes from frozen cells to form a fluffy, dry powder that may be used immediately or frozen to preserve enzyme activities for many years. Many lyophilized yeasts and bacteria can be obtained from biochemical supply houses (e.g., Sigma Chemical Co.). When freeze-dried cells are used as biocatalysts, they must be evaluated for surviving enzyme activities before use.

Treatment of cell pastes or cakes with cold (−20°C) acetone is a simple method for preparing dried cells. Cells obtained by centrifugation or filtration are slurried in cold acetone and suction filtered. The drying process is repeated twice more, followed by a cold ether wash to remove residual acetone, which can be detrimental to enzyme activities. Removal of solvent from the dried cells under vacuum gives an "acetone powder," which is most stable when stored frozen (19).

Dried cells offer many of the same advantages of resting cells compared to growing cultures (36). Both lyophilized and acetone-treated cells are easy to prepare. An adequate supply of powdered cells helps to ensure experimental reproducibility, since many experiments can be performed with one batch and it is not necessary to establish rigorous fermentation protocols for each experiment. Incubations with dried cells do not require sterile manipulations.

13.3.9. Permeabilized Cells

Microbial cell membranes may be made permeable to organic chemical substrates (51). Permeabilizing agents include surfactants (30), solvents, and antibiotics. Permeabilizing agents are usually applied after growth stops in the stationary phase, or with resting cells. Alternatively, the addition of inhibitors of cell wall synthesis during growth enhances permeability. The solvents dimethylformamide (DMF) and dimethyl sulfoxide (DMSO), commonly used to disperse steroids and other lipophilic compounds, also increase the permeability of the cells. These solvents should be used with care since at higher concentrations they may adversely affect the viability of the cells.

13.3.10. Isolated Enzymes and Cell-Free Preparations

Isolated enzymes are already commonly used as reagents for organic synthesis. Isolated enzymes have advantages as synthetic reagents in that they are simple, well defined, and usually catalyze a single reaction step with little side reaction. Compared with whole cells, methodology and equipment are simplified with isolated enzymes, as concerns for sterility and cell viability are minimal. A wider range of reaction environments can be used, since the stability of cellular structures is irrelevant. Removal of an enzyme from the cell also allows cleaner reactions, with greatly reduced mass transfer concerns, and no undesirable reactions catalyzed by other enzymes present in the cell. Like whole-cell fermentations, isolated enzymes have been demonstrated for use on industrial scale for a number of processes (50). For these reasons, the use of enzymatic catalysis has expanded considerably. Several excellent general reviews on catalysis with pure enzymes are available (7, 9, 11, 15, 20, 40).

The vast majority of published enzyme-catalyzed reactions are acyl hydrolysis and reduction, for which the regio- and stereospecificity of enzymes allows chiral syntheses. However, a wide variety of enzymes catalyzing other synthetically useful reaction chemistries, such as epoxidation, halogenation, oxidations, phosphorylation, glycosylation, condensations, nitrile hydrolysis, decarboxylation, isomerization, and many other reactions, are available commercially. In fact, the full range of synthetic chemistries present in the structural diversity of natural products is enzyme-mediated.

However, some important enzymes, including the broad class of oxygenases that catalyze hydroxylation reactions, require multiple cofactors, multiple enzyme species, or are not sufficiently stable when purified. Other enzymes may be difficult or time-consuming to obtain in purified form. For these enzymes, catalysis with whole cells, as described above, is still very prevalent. In many cases, however, crude cell-free preparations may be a suitable alternative (19).

Crude enzyme preparations are obtained by breaking cells under mild conditions so that the contents are released in an active state to the buffer medium. Common techniques for the disruption of microbial cells are pressure shearing, enzymatic digestion of cell membranes, osmotic lysis, autolysis, and freezing-thawing (45). Ultrasonic disintegration and pressure shearing are the most widely used in exploratory bench research.

The French press (American Instrument Co.) is a reliable device for pressure shearing on a small scale. It consists of a solid steel cylinder containing a well that holds 5 to 40 ml of a thick cell suspension. The cylinder is fitted with a solid stainless-steel piston that can be forced into the well under high pressure (4,000 to 20,000 lb/in^2). Cells are broken by the high shear forces as they are squeezed through a small release orifice in the well. Ultrasonic oscillators also break cells by shearing. Rapidly moving bubbles in the sonic field at the probe tip cause high shear forces capable of breaking the toughest cell walls. Since high-frequency sonic oscillations generate much heat, the operation must be conducted in short bursts in an ice bath. Short bursts of 15 to 20 s break cell suspensions of 30 to 40 g (wet weight) per 100 ml.

Simple enzyme fractionation can be achieved by centrifugation to remove solid debris (cell membranes, unbroken cells) from crude enzyme mixtures. Centrifugation at 5,000 × *g* for 10 min will produce murky supernatant mixtures containing soluble and particulate enzymes and traces of cofactors from the microbial cell. More involved and time-consuming procedures (e.g., column chromatography) are necessary to obtain pure enzymes from these crude mixtures.

Crude broken-cell suspensions and supernatant fractions with minimal purification can catalyze useful organic reactions. Appropriate volumes of these suspensions are added to reaction vessels along with organic substrates in buffer or organic solvent. If coenzymes are required, these can be added in stoichiometric amounts for small-scale reactions; otherwise, coenzyme regeneration strategies are becoming well established (9, 13, 20, 40).

Naturally occurring, exocellular enzymes may be efficiently used with little isolation. Typical exocellular enzymes, including peroxidases (ligninases), laccases, and hydrolases are stable and can be useful biocatalysts. For convenience of use and storage, such enzymes may be concentrated by ultrafiltration, precipitation by solvents or salts, or adsorption to a carrier.

For example, *Polyporus anceps* grown in a defined medium secretes laccase (a copper oxidase). When enzyme titers are highest, cells are filtered from the fermentation broth and dry DEAE-cellulose (H^+ form, 3 g/liter) is added to the filtrate with stirring for 30 min at 4°C to bind most of the enzyme activity. The resin-bound enzyme is removed by simple filtration, washed twice with distilled water, and moist resin-enzyme is stored in small portions in a freezer. Elution of DEAE-cellulose with 0.2 M phosphate buffer (pH 5.0) gives quantitative recoveries of the active enzyme that catalyzes the oxidations of alkaloids.

13.3.11. Immobilized Systems

In developing biotransformations with an eye on practical applications, the issue of biocatalyst immobilization with often be important. Immobilization theoretically allows more convenient continuous processing and product isolation, improved biocatalyst stability, and reuse of the biocatalyst. The use of an immobilization support, however, does introduce additional development costs and mass transfer considerations. In practice, interestingly, relatively few industrial biocatalytic processes have employed immobilization (3, 50).

In general, the major approaches that have resulted in successful biocatalyst immobilization include noncovalent adsorption, covalent attachment, covalent crosslinking, physical entrapment within a porous support, and compartmentalization within a membrane. These methods for immobilizing cells and enzymes are discussed in greater detail elsewhere in this volume (see also reference 3).

Although each of the general approaches has proved useful for different applications, the most common immobilization format for whole-cell biocatalysts on small scales is gel entrapment. Empirically, this simple-to-apply procedure has generally resulted in improved biocatalyst stability and handling, while introducing moderate mass transfer barriers. In a typical laboratory-scale application, a 1 to 8% sodium alginate solution is mixed with an equal amount of a concentrated resting-cell suspension in a syringe. The mixture is injected through a small-gauge needle and dropped into a 0.1 M ca^{+2} (or other divalent cation) solution to promote gelling. After several minutes the gelled beads can be filtered off and used for transformation.

13.4. TECHNIQUES

Success in the application of microorganisms and microbial enzymes as catalysts for organic reactions requires a working knowledge of simple microbiological laboratory techniques. Sterile or aseptic techniques, medium preparation, and the use of the light microscope are the basics.

Aseptic technique is essential to the maintenance and use of pure cultures and can be simplified to two requirements: (i) the use of sterile equipment, vessels, substrates, media, etc., and (ii) the exclusion of airborne particles containing contaminating organisms when making additions or transfers to medium. For many synthetically useful microorganisms, practicing aseptic technique adds little to the complexity of a properly conducted reaction.

Sterilization is the complete removal or destruction of all living entities from materials by using heat, filtration, radiation, or chemicals. The best and most convenient sterilization method for heat-resistant materials likely to be used in bioconversion research is steam sterilization or autoclaving with steam under pressure (15 lb/in^2, 121°C). Flasks, culture media, and pipettes are all suitably sterilized by autoclaving. Common problems that can be encountered during sterilization are the precipitation of inorganic or organic salts and the destruction of sugars when heated in the presence of nitrogenous organic nutrients. Ovens heated to 180°C can be used to dry-sterilize glassware, pipettes, and other utensils in 1 h. Slow cooling for about 2 h minimizes breakage of glassware. Filtration through 0.1- to 0.2-μm membranes is a reliable method for obtaining sterile solutions of heat-labile materials or of materials dissolved in nonaqueous solvents. Flame sterilization or chemical sterilization with ethanol is usually the method of choice for small-scale sterilization of hands and utensils used during the course of an experiment.

The maintenance of aseptic conditions while working with sterilized equipment commonly requires the use of a laminar flow hood and common sense. The laminar flow hood provides a positive pressure of filtered air over the work area to prevent airborne particles from contaminating a culture. Common sense dictates that hands, tools, and other objects that are brought into the laminar flow hood, or contact cultures within the hood, must be sterilized before use with another culture. Proper use of aseptic techniques and equipment will ensure maximum reliability and reproducibility of cultures and the biotransformations they catalyze.

13.5. MEDIA

Catalytically active microbial cells are obtained by growth in balanced nutrient media, especially those containing inducers of the desired enzyme activities (sections I and II). Different microorganisms have special requirements for optimal growth. In addition to environmental factors (temperature and pH), the ratios and amounts of carbon, nitrogen, phosphorus, trace minerals, and special growth factors are important in proper nutrition (39). The elemental composition of a dried cell provides insight into the nutritional requirements of that cell. Water makes up 80 to 90% of the cell weight. For dried *E. coli* cells, the relative concentrations of the various elements are as follows: carbon (50%), nitrogen (15%), phosphorus (3.2%), sulfur (1.1%), sodium (1.3%), potassium (1.5%), magnesium (0.5%), calcium (1%), and iron (0.24%), with trace amounts of manganese and copper. Hydrogen and oxygen account for the balance. Conversely, cell composition reflects the growth medium used. For instance, the high amount of sodium in this particular analysis reflects the fact that the *E. coli* was grown in nutrient medium containing sodium chloride, which is customarily added even though there may be no requirements for it. Several different types of culture media have been used to accommodate these nutritional principles for the growth of microorganisms used in biotransformations.

13.5.1. Chemically Defined Media

Chemically defined media are made by the addition of specified ingredients to distilled water. Although more expensive, these media offer the very important advantages of reproducibility and greater simplicity in the analysis of biotransformation end products. A variety of carbohydrates, organic acids, alcohols, hydrocarbons, and lipids can serve as carbon sources. Nitrogen sources can be salts other than ammonium sulfate, such as sodium or ammonium nitrate. Urea is another good nitrogen source, as are certain amino acids (e.g., asparagine). Vitamins, purines and pyrmidines, and amino acids are added to the defined media to stimulate growth of more fastidious organisms.

13.5.2. Semidefined Media

Small amounts (from 0.05 to 0.5%) of single vegetable or meat extracts or preparations added to chemically defined media result in semidefined culture media. For growth and screening of a large number of microorganisms, semirefined media can serve as more generally applicable standard media. Yeast extract, meat peptones, soy peptones, malt extract, casein hydrolysates, and corn steep liquor are some of the most common and useful additives. Small amounts of these organic nutrients supply growth factors and vitamins that often enhance growth for a variety of microorganisms quite significantly compared to the completely defined medium, without greatly complicating reaction analysis.

13.5.3. Complex Media

Most of the nutrients in complex media are provided by extracts or enzyme digests of plant and animal products. The importance of reproducibility and ease of analysis and interpretation make complex media rarely the best choice for biotransformation. Of course, the use of pregrown resting-cell or dried-cell preparations in simple aqueous or nonaqueous reaction media largely alleviates these concerns.

A list of culture media used in biotransformations has been compiled (17). Representative recipes are given below for convenience. A longer list of media cross-referenced with cultures is available (1).

13.5.3.1. Defined Medium

Carbon source, 2 g; ammonium sulfate, 1 g; dipotassium phosphate, 1 g; salt solution A, 10 ml; distilled H_2O, 990 ml. Adjust medium to pH 7.0 before autoclaving.

Salts solution A: magnesium sulfate · $7H_2O$, 25 g; ferrous sulfate · $7H_2O$, 2.8 g; manganous sulfate · H_2O, 1.7 g; sodium chloride, 0.6 g; calcium chloride · $2H_2O$, 0.1 g; sodium molybdate · $2H_2O$, 0.1 g; zinc sulfate · $7H_2O$, 0.06 g; HCl (0.1 M), 1 liter.

13.5.3.2. Complex Biotransformation or Maintenance Media

Nutrient broth: beef extract, 3.0 g; peptone, 5.0 g; distilled H_2O, 1.0 liter; pH 6.8 after autoclaving.

Sabouraud dextrose (or maltose) broth/agar: neopeptone, 10 g; glucose (or maltose), 40 g; distilled H_2O, 1 liter; pH 5.7 before autoclaving.

Glucose, 20 g; soybean meal (or soy flour), 5 g; yeast extract, 5 g; sodium chloride, 5 g; potassium phosphate, dibasic, 5 g; distilled H_2O, 1 liter. Adjust to pH 7 with 6 N HCl.

Corn steep liquor (60% solids), 20 g; glucose, 10 g; tap H_2O, 1 liter. Adjust to pH 4.9.

13.6. REACTIONS IN SOLVENT MIXTURES

The use of organic solvents or aqueous/organic solvent mixtures as media for biocatalytic reactions using enzymes or suspended cells is a powerful and often necessary modification. From a practical, synthetic perspective, reactions in nonaqueous media provide three primary advantages: the ability to shift the thermodynamic equilibrium of hydrolytic reaction toward synthesis, the ability to solubilize a broad range of organic molecules at synthetically useful concentrations, and the ability to rapidly separate soluble reaction products from the insoluble biocatalyst. The inclusion of organic solvents may also minimize certain side reactions and permit continuous extraction and recovery of reaction products. Much has been presented in the literature about other advantages of nonaqueous solvents, including improved thermostability, altered specificity, or decreased chance of contamination, which are of more limited applicability.

The primary difficulty with conducting biotransformations in the presence of organic solvents is lower catalytic activity and catalyst stability. Substantial literature is devoted to determining reasons for loss of catalytic activity and methods for preventing it. A complete description is well beyond the scope of the present work but is available from many excellent reviews (13). In many cases, the practice of nonaqueous biocatalysis, widely regarded as untenable less than 20 years ago, is today a successful reality. Practical guidelines for conducting biocatalytic reactions in the presence of organic solvents will be the focus here.

The behavior of whole-cell biocatalysts in the presence of organic solvents is somewhat distinct from that of isolated enzyme catalysts and will be treated separately. Salter and Kell (44) offer an excellent review of activity preservation and solvent toxicity of whole-cell-catalyzed reactions in organic solvents. Much is still unknown about mechanisms for whole-cell solvent tolerance, owing to the complex nature of the living cell, and some individual strains exhibit large deviations from general trends. Empirically, however, several general recommendations can be made.

First, although no single solvent property has been definitively correlated to solvent tolerance, cell biocatalysts (both growing and resting) tend to maintain higher activity for a longer period with solvents of high hydrophobicity (normally expressed as solvent octanol/water partition coefficient, log P). Solvents with log $P > +4$–5 tend to make the most compatible media, while solvents of intermediate hydrophobicity (log $P = 0$ to 4) are often most toxic (6). Water-immiscible solvents are much better choices for bulk organic phase than are water-miscible ones. In general, cell immobilization, usually by entrapment or encapsulation, significantly improves organic solvent tolerance while permitting substrate access to the catalyst (44). In contrast to the behavior of isolated enzymes in predominantly nonaqueous environments, whole cells often exhibit decreased thermotolerance (52), and reactions are often more favorably run at lower temperatures than usual for a given strain.

While hydrophobic solvents are biocompatible media for reactions with hydrophobic substrates, such as steroids, many organic molecules of interest are of intermediate polarity and are not highly soluble in either nonpolar solvents or aqueous media. In many of these cases, the use of small quantities of a "good solvent" in either the organic or aqueous phase will provide successful reaction conditions. The incorporation of organic "carrier solvents," such as DMSO, DMF, or ethanol in final concentrations of <5% is one very important form of this strategy. Likewise, the addition of a toxic solvent with good solvating power to a biocompatible, hydrophobic bulk solvent can provide the positive attributes of both (43). The biocompatible solvent extracts the toxic one away from the catalyst, yielding an organic phase capable of holding a suitable concentration of reactant molecules.

Isolated enzymes tend to exhibit much better retention of catalytic activity in the presence of organic solvents than whole cells. This is likely due to the lack of a cell membrane and other cellular substructures that are likely targets for general solvent toxicity. Like whole cells, enzymes tend generally to prefer more hydrophobic solvents with log P values > 2–4 but tolerate a much broader range of solvents and solvent mixtures than do whole-cell catalysts. Polar solvents such as THF (tetrahydrofuran), acetonitrile, *tert*-amyl alcohol, methyl *tert*-butyl ether, and monglyme preserve adequate catalytic activity of many enzyme catalysts.

As mentioned previously, however, general rules for nonaqueous biocatalysis are rare; indeed, some highly tolerant whole-cell strains, such as pseudomonads may serve as excellent recombinant hosts for nonaqueous biocatalysts, while important enzyme catalysts, such as cytochromes P-450, exhibit a low tolerance even to minor levels of organic carrier solvents. Thus the most prudent strategy at present is to screen an abbreviated list of good candidate organic solvents with each chosen biocatalyst.

In summary, to be considered general catalysts on par with other, traditional chemical catalysts for organic synthesis, biocatalysts must be functional in a fair range of organic solvents. Practically, many organic molecules of interest for transformation have limited solubility in aqueous media, or in the highly lipophilic solvents most often described in the literature for use with biocatalysts. Moreover, thermodynamic control of normally hydrolysis-favoring equilibria, ease of product recovery, and minimization of certain side reactions are also important motivations for conducting biocatalytic reactions in organic media. Over the last 10 to 15 years, significant strides have been made toward making practical, synthetic biocatalysis feasible.

13.7. ADDITION OF ORGANIC COMPOUNDS TO REACTION MIXTURES

Most organic chemical reactions are run in nonaqueous solvents. Since microbial growth and biological reactions typically take place in aqueous environments, there is a natural tendency to restrict biotransformation reactions to aqueous media, and therefore to water-soluble organic substrates. In fact, biotransformations occur equally well with both lipophilic and hydrophilic substrates as long as an adequate concentration of reactants can be delivered to the biocatalyst. More directly, the key to success with biotransformations of lipophilic compounds is the enhancement of substrate availability to the active site of the appropriate enzyme.

It is generally assumed that access to the active site of microbial enzymes is possible only for compounds dissolved or dispersed in the reaction medium. An examination of growth rates of bacteria using poorly water-soluble hydrocarbons as their carbon source illustrates this point. Bacteria grown on naphthalene, phenanthrene, or anthracene have generation times of 1.5, 10.5, and 29 h, respectively. These growth rates are directly proportional to the water solubilities of the hydrocarbons but independent of the total amount of solid substrate present. For whole-cell reactions, once contact with the cell occurs, substrates can penetrate the cell wall and membrane by passive or active transport. Cell surfaces and membranes, as well as enzymes themselves, have hydrophobic domains that facilitate transport, binding, and reaction with lipophilic compounds. In addition, microorganisms produce a variety of endogenous emulsifiers that promote these reactions.

Thus, several methods have been developed to improve the solubility and dispersion of reactants in water. As we just discussed, one strategy is to conduct the biotransformation in bulk organic solvent media. Another successfully practiced approach is the delivery and dispersion of lipophilic substrates through chemical agents or physical methods that have a minimal impact on the bulk aqueous reaction medium.

13.7.1. Organic "Carrier Solvents"

The most common method for adding water-insoluble substrates to a bulk aqueous reaction medium is in water-miscible organic carrier solvents. Preferably, these solvents should have low toxicity to the biocatalyst and excellent solvation capacity. Common carriers include many of the same solvents used for organic compound transfer and storage, such as DMSO, DMF, ethanol, methanol, and acetone.

In a typical application of this strategy, the organic compound is dissolved in dry DMF (stored over molecular sieves) at a 20- to 100-fold higher concentration than desired in the reaction. Dissolution may be hastened by gentle treatment with a sonifier for a few seconds. If necessary, the substrate stock should be sterilized by filtration (with a solvent-resistant cartridge filter) before addition to the medium. Concentrated substrate solution is then added to incubation mixtures at the required level by sterile pipette. A milky precipitate forms instantly as the DMF mixes with the aqueous medium. To prevent reaggregation of the substrate and to ensure even distribution, each flask or vessel should be shaken immediately upon addition of substrate stock.

This technique works with most water-miscible solvents. Since many hygroscopic organic solvents lose solvent power as they take up water, they should be kept dry. As a case in point, a small amount of water in DMF greatly reduces the capacity of this excellent solvent. The use of these carrier agents is also exceptionally well tolerated by most microbial strains and isolated enzymes. However, there are some exceptions. For example, microsomal cytochrome P-450 enzymes tolerate only very low levels (<1%) of organic carrier solvents. Thus, additional strategies for compound delivery must sometimes also be considered.

A good empirical comparison of appropriate protocols for lipophilic substrate delivery to biocatalysts is offered by Lee and coworkers (28). They examined in detail the aggregation and solubilization phenomena of steroid substrates. A mixed culture of *Arthrobacter simplex* and *Curvularia lunata* catalyzed the simultaneous 1-dehydrogenation and 11β-hydroxylation of 16-α-hydroxycortexolone-16,17-acetonide. Substrate was prepared (i) as a suspension in cold solvent, (ii) in 0.1% (wt/vol) aqueous Tween 80 surfactant, or (iii) as solutions in hot and cold solvents. The substrates were added to the cultures immediately after preparation and again 25 h later. Best yields (60 to 90%) were obtained with hot solvents and cold DMF. Yields were related to the particle size of the substrate. Hot solvents gave dispersions with 0.5- to 2-μm particles; those from cold solvents ranged from 10 to 100 μm. Apparently, ultrafine, amorphous particles are more accessible to enzyme active sites than crystalline forms. This may be due to improved rates of compound dissolution and improved cell permeability.

Additional vehicles can be used to solubilize substrates and improve cell permeability. For instance, Chien and Rosazza (in reference 17) used polyvinylpyrrolidones (PVPs) to enhance the hydroxylation of ellipticine by *Aspergillus alliaceus*. This brilliant yellow alkaloid is barely soluble in water (<5 mg/liter), but solutions of 30% PVP (40,000 average molecular weight) gave the highest initial water solubilities of ellipticine, and the solubility increased proportionately with the concentration of PVP (60 g of PVP per liter solubilized 100 mg of ellipticine per liter). At that concentration, yields of hydroxylated ellipticines were doubled.

PVP disperses many types of aromatic compounds in aqueous media by formation of coprecipitates. The aromatic substrate and PVP 40,000 are dissolved in chloroform-methanol (9:1), and the mixture is evaporated to dryness in a rotary evaporator. Ratios of aromatic compound to PVP of 1:5 up to 1:120 should be used. To prepare concentrated solutions or dispersions, suspend compounds in 10 to 30% solutions of PVP 40,000 by grinding them together with a glass mortar and pestle. Cyclodextrins also enhance solubilities of water-insoluble substrates, and their uses in small-scale reactions have been documented (22).

Nakamatsu et al. (34) evaluated the influence of surfactants on the oxidation of sterols by *Nocardia corallina*. The bioconversions of substrates dispersed with several different surfactants were compared with the performance of substrates sonicated to reduce their particle size. Cationic, nonionic, and anionic surfactants were used at 0.01% concentration. Two cationic detergents (Sanisol C and Amiet 106) and one nonionic detergent (Nonal 106) significantly inhibited cell growth. Most nonionic surfactants did not inhibit growth and provided good emulsification. Emal 10C, Emulbon T-83, Sorbon T-40, and Tween 80 surfactants stimulated the oxidation of soy sterols.

Inert supports may be used to adsorb or dissolve a variety of compounds within the lattices of inert materials, such as zeolites, molecular sieves, diatomaceous earth, and polymers such as divinylbenzene-polystyrene (17, 54). The resulting ultrafine particle sizes and large surface areas promote a high degree of dispersion of lipophilic substrates. Adsorbed substrates are remarkably available to biocatalysts. Liquid paraffins are adsorbed to the supports from solvent solutions. After evaporation of the solvent, bound compounds are added directly to incubation mixtures.

A physical milling and wetting method has been used effectively to disperse steroids such as progesterone (57), as follows. USP-grade progesterone was ground with a Jet-O-Mizer model 202 grinder (Fluid Energy Processing and Equipment Co., Philadelphia, Pa.) to a fine particle size of about one-third the density of the starting material. The ground progesterone was added to 250-ml Erlenmeyer flasks wetted with suitable amounts of 0.01% aqueous Tween 80 surfactant, and sterilized by exposure to steam at atmospheric pressure for abut 30 min. By this technique, 20 to 50 g of steroid substrate was dispersed per liter in the aqueous fermentation medium, and yields of the hydroxylated product were 60 to 90%.

Gases, volatile solvents, and other compounds with high vapor pressure are relatively simple to handle in small fermentors and other types of bioreactors, except for the danger of explosions. Gases and other volatile substances can be carried into vessels along with sterile air. Air that is bubbled over the surface of a volatile solid or through a liquid, like toluene, will carry the compound into the reaction mixture. Enclosed incubator-shakers are the best way to handle volatile compounds and gases in shake-flask reactions. Extreme care must be taken to avoid the accumulation of explosive mixtures. For a small number of flasks, individual spargers or a gas manifold could be used.

13.7.2. Timing of Substrate Additions

When growing cells are used for biotransformations, the time of addition of the organic substrate profoundly influences the yield of product. Toxic substances, such as many antibiotics and antitumor compounds, often inhibit growth and enzyme production if they are added early in the growth cycle. In many cases, however, it is advantageous to add at least small amounts of substrate at the beginning of the growth phase to promote enzyme induction. The addition of substrate during the late logarithmic growth phase minimizes toxicity effects while promoting enzyme induction. At this point in their growth cycle, cells are still capable of enzyme synthesis, while the proliferation of the biomass is less inhibitable by toxic substrates. The same reasoning applies to substrates added in toxic solvents.

The timing of substrate addition must take into account the physiological state of the microorganism. The position of the cell in its growth cycle determines its enzyme capabilities, and, in general, enzyme levels will be determined by the competing rates of enzyme expression and degradation/inactivation. Enzymes of interest may be expressed only at specific times during the growth cycle, such as late log or stationary growth phases. Looking for reaction when the enzyme is not present would be futile. Enzyme concentration may be subject to the presence of inducers of expression, fluctuations in the pH or temperature of the medium, the amount and kinds of carbon and nitrogen nutrients in the medium, and the degree of oxygenation of the medium (37). The optimal time for substrate addition is difficult to predict and is best determined experimentally.

Toxic substrates or substrates in toxic solvents may be added incrementally by "dosing." The addition of progesterone semicontinuously to growing *Aspergillus ochraceus* cultures minimizes toxicity as well as the mechanical loss of starting material through aggregation. Unexpectedly, fewer side reactions occur. Lee et al. (29) used a dosing technique to obtain 38 g/liter of β-hydroxy-β-methylbutyric acid from the substrate β-methylbutyric acid, which was toxic to the *Galactomyces reesii*. Dosing techniques also sidestep the undesirable phenomenon of substrate inhibition, which almost invariably occurs when large amounts of substrate are added at a single time. The effect tends to be most prominent with water-soluble materials, but well-dispersed lipids display the same phenomenon.

13.8. EQUIPMENT

Details of small- and larger-scale fermentations and equipment used for them have been covered elsewhere (4). Erlenmeyer flasks on shakers have been the traditional reactors for aerobic culturing. For screening, the smaller sizes (50 to 250 ml) are recommended. Larger flasks (up to 2 liters) are convenient for scaled-up experiments. At this size, 2.8-liter Fernback flasks are used frequently. Normally, only 10 to 20% of the volume of the flask is filled with medium, since this allows for maximum agitation and aeration (see below) without splashing or excessive evaporation. Sterile closure of the flasks, while permitting gas transfer, is essential. Cotton and plastic foam plugs are widely used, but cotton-gauze filter disks are better; these are held in place by stainless-steel springs or rubber bands. Flasks with flush necks (DeLong flasks) are closed with special stainless-steel or plastic caps. Petri dishes about 100 mm in diameter and sterile, cotton-plugged pipettes are routinely used. From the standpoint of housekeeping in the laboratory, sterile, disposable labware is very convenient.

13.8.1. Incubators and Shakers

Bioconversion experiments require some form of temperature-controlled environment. Incubators that control temperatures from below ambient temperature to 50°C or higher are available in benchtop sizes up to full-room size.

The value of shakers in fermentation experimentation is well known. It is important to have the capability for shaking aerobic organisms, as this is an economical and practical way to screen a large number of cultures. The two common types of shakers are the rotary (or orbital) shaker and the reciprocal shaker. Shaking speeds are continuously adjustable from 0 to about 350 rotations or oscillations per min. Shakers range from desktop sizes to ones that can accommodate hundreds of 250-ml flasks. Reciprocal shakers are best for tube cultures, and orbital shakers are best for flasks. With either type, different platforms are available for holding various sizes of flasks. It is generally important, however, to mate the stroke length of the shaker with the diameter of the vessel.

13.8.2. Sterilizers (Autoclaves)

Steam sterilizers are essential for microbiological work. Culturing vessels and media must be sterilized. Automatic autoclaves with capacities for several hundred 250-ml flasks are generally used. For small-scale work, a desktop size or even a pressure cooker can be used.

13.8.3. Fermentors

Fermentors are useful for larger-scale biotransformations. Although the general sophistication of the devices has increased, the basic design of the vessel has not changed for many years. With a fermentor it is possible to control culture parameters in ways and degrees not possible in flasks and tubes. Stirring and air-sparging devices allow the maximum possible aeration. Many parameters (e.g., pH) can be measured and controlled continuously. Therefore, it is useful to have access to several benchtop fermentors (1 to 10 liters) for experiments that cannot be done conveniently in flasks and for scaling-up processes. Larger-scale fermentor studies (20 to 1,000 liters) often involving sophisticated downstream processing will require cooperation with more specialized laboratories. Fermentor monitoring and computer control of fermentations are discussed in section II.

13.8.4. Microscale Conversions

Often, one is faced with a small quantity of precious organic compound and a large number of biocatalysts to screen for transformations. In these cases, microscale equipment should be used. Essentially, all of the protocols and guidelines mentioned above can be scaled down for individual fermentation volumes of less than 1 ml. However, several issues require additional consideration when working with small volumes in nontraditional fermentor geometries. Several considerations relate to the supply of oxygen to microfermentors. As mentioned before, the shaker stroke should also be short, consistent with the reaction vessel diameter, to ensure adequate agitation. Because of the gross morphology of some microbial strains, the small fermentor dimensions may result in significant surface effects and poor mass transfer. As a result, resting cells and immobilized or soluble isolated enzymes may be more appropriate and reproducible catalysts for microscale applications. To reduce heterogeneity in resting-cell preparations, an efficient cell homogenizer, used in short bursts on ice, is an important tool.

Microscale experiments will also require micropipettes able to handle 1- to 10-μl volumes of test compound stock or analytical aliquots. With less reaction sample available, sensitive analytical techniques providing more information per analysis should be favored. Finally, care should be taken to prevent evaporation of the small reaction volumes by sealing the reactions or saturating the reaction headspace with reaction medium.

13.9. AUTOMATION

Biotransformations, especially involving whole cells, can be labor-intensive. The repetition involved with reproducibly preparing, inoculating, adding test compound to, monitoring, sampling, and working-up a large number of cultures and reactions encourages the consideration of automated equipment to speed the process and reduce tedium and human error. However, sterility requirements and the vast morphological, growth rate, and medium differences between strains greatly complicates the task of automation. For these reasons, automated culturing has been applied when only one or a few strains need to be accommodated in an automated process, such as for screening large numbers of recombinant mutants, testing for specific pathogens, or processing host cell lines for genomic studies. However, commercially available automated equipment can also greatly facilitate the processing of samples from biotransformations, since sterility becomes a minor issue after the biocatalytic reaction is complete.

13.10. STANDARDIZATION, QUALITY CONTROL, AND QUALITY ASSURANCE

As established earlier, screening biocatalysts typically requires the execution and evaluation of large numbers of individual reactions. Therefore, once reliable and efficient biocatalyst reaction protocols have been established, the best improvement in the frequency of identifying new biotransformations comes from improving the throughput, sensitivity, and interpretation of reaction analysis.

Prerequisite for most high-efficiency analytical methods is the development of rapid, parallel methods for the preparation of samples for analysis; catalyst removal, macromolecule (protein, polysaccharide, polynucleotide, etc.) removal, and/or solvent exchange can be important steps prior to reliable use of many analytical techniques described below. Useful in this regard are the wide variety of filtration, ultrafiltration, liquid-liquid extraction, and solid-phase adsorption products that are now commercially available. Multicartridge manifold and 96-well-plate-based techniques are particularly increasing in popularity owing to the large number of samples that can be processed simultaneously with automated liquid handlers or manually.

The best analytical method may be different for different lead molecules and different objectives. However, for almost all high-throughput screening studies, thin-layer chromatography (TLC), gas chromatography (GC), and high performance liquid chromatography (HPLC) have been the proven workhorses. Improvements in laboratory-scale mass spectrometry (MS) equipment, however, have made MS an important addition to the biotransformation practitioner's repertoire, either in direct flow injection mode or in tandem with other analytical techniques (e.g., LC/MS).

13.10.1. TLC

Traditionally, TLC has been the primary method for analysis of biotransformations. TLC is well suited for this work because it is an inexpensive method for parallel analysis and, thus, can be used to analyze a large number of reaction samples simultaneously. TLC is also simple to set up; the basic apparatus includes developing chambers, common laboratory solvents, and chromatography plates with various absorbents that can be prepared directly in the laboratory or are available commercially.

The theory and techniques of TLC are discussed in detail in several excellent texts (48). Silica gel, alumina, kieselguhr, and cellulose are the most commonly used absorbents and have a wide range of properties that can be altered to suit the particular need by pretreatment with acids, bases, buffers, or specific reagents (e.g., $AgNO_3$). A suitable solvent system will depend on the nature of the compounds to be separated. However, established procedures are available for a large number of organic compounds.

TLC can work very well as an initial screen if a sensitive, and preferably specific, indicator reagent is available for visualizing reaction components. It may also be an excellent technique if degradation or conversion of the test substrate is the primary endpoint. However, its application to highly polar compounds can be complicated if they cannot be simply extracted from an aqueous reaction into a volatile organic solvent. Moreover, improvements in equipment and methodology for other, higher-resolution and more informative techniques, such as HPLC and MS, have made them increasingly attractive for the rapid characterization of reaction products.

13.10.2. GC

GC is another commonly applied tool for analysis of biotransformation mixtures. GC typically permits rapid separation of any compounds that are volatile or that can be derivatized to a volatile substance. Although it is a serial analysis technique, separations are typically rapid (3 to 20 min). Sample introduction and data processing are typically automated on most modern equipment, further accelerating the analytical process. Furthermore, when these instruments are used in combination with MS, detailed structural analyses of many samples can be obtained.

Separation by GC is achieved by partitioning of the analytes between a mobile gas phase and a liquid or solid adsorbant stationed in the column. Retention times of compounds on particular columns at specified temperatures and gas flow allow characterization of the compounds. Many stationary-phase-column chemistries are commercially available for high-resolution separations of structurally similar derivatives, and in some cases even enantiomers.

GC permits rapid analysis of microliter quantities of sample with high resolution. The method can be quantitative or qualitative, it is highly sensitive (parts per billion), it is simple to operate, and it can be combined conveniently with MS. However, like TLC, GC usually requires extraction of the reaction mixture into a volatile organic solvent for application to the column. More important, many nonvolatile, functionalized organic compounds or thermally labile compounds are poorly analyzed by GC; for analysis of unanticipated products of biotransformation screens, these limitations can be undesirable. Nonetheless, for many volatile test substrates, GC is the method of choice.

13.10.3. HPLC

HPLC methods have the advantage of providing detailed analytical information for a very broad range of substrate molecules (and their derivatives). Analogous with GC, resolution in HPLC is achieved by partitioning of analytes between a mobile liquid and a solid adsorbant. The general versatility of HPLC methods makes them very attractive for analysis of biotransformation screens. HPLC analyses are not limited by the molecular weight, volatility, thermal stability, or organic extractability of test compounds and derivatives. Moreover, a wide variety of high-resolution separation columns with different solid-phase chemistries and analyte detection methods are commercially available. However, typical analysis times of 15 to 45 min per sample for a serial analytical technique have limited the use of HPLC for biotransformation screening from large biocatalyst collections.

Recently, very rapid, high-throughput HPLC methods have been described for the analysis of large libraries produced by combinatorial chemistry or natural product discovery (16). These approaches also apply "universal" gradients of acetonitrile (or methanol) with water to extract from a nonpolar (octyl or octyldecyl) solid phase to separate a broad diversity of compound classes. Automated instruments allow the convenient processing of many samples. Such approaches have also been adapted for general application to biotransformation screening (31). Several hundreds to thousands of injections a day, yielding resolution adequate for biocatalyst screening, can be performed using 1- to 10-min run times. Proper sample preparation, however, is more critical; sharp solvent gradients at high column pressures on high-efficiency, small-particle-size packed columns result in a higher susceptibility to plugging with microbial debris or precipitated proteins.

As described below, HPLC is highly compatible for both analytical and preparative work; analytical methods may be extended for the preparation of adequate quantities of products for structural determination by nuclear magnetic resonance (NMR) or for other characterization. Recent advances in HPLC with direct NMR detection of the eluant are notable for providing the promise of immediate structural characterization of biotransformation products, but this method is currently cost-effective for only a limited number of postscreening biotransformation analyses.

13.10.4. MS

Traditionally, MS has been a useful method for confirmation of a biotransformation product, most commonly used in tandem with GC separation during later stages of biotransformation product characterization. MS provides a spectrum of molecular weights of the parent molecule, and of submolecular fragments, that help identify the nature of a transformation and its position on the test compound. Improvements in "soft" methods for nondestructive compound ionization from liquid samples have recently transformed MS into a more powerful tool for high-throughput direct analysis, as well as general characterization of analytes eluting from GC or HPLC columns. In direct injection mode, MS can deliver specific molecular weight information that can identify products in less than 1 min per sample. Information can be gleaned even from relatively crude samples from a reaction mixture, although impurities may interfere by suppressing ionization of the desired analytes. Even more powerful is a connection of MS with GC or HPLC separations, as this allows two-dimensional resolution of complex samples (initially by chromatographic retention time, then by mass). Especially coupled with high-throughput separation methods described above, HPLC/MS yields a broadly applicable, rapid analysis (~5 min per sample) giving an unparalleled degree of information.

Depending upon the test compound, different ionization interfaces may be most appropriate for introduction of a sample or chromatography column stream to MS. Typically, chemical ionization represents a good interface for smaller, and less functionalized analytes, while electrospray offers a better ion source for more polar and functionalized organic test compounds and products. After ionization of the sample, there are a number of alternative designs for separation and detection of molecular species by weight. From the perspective of biotransformation analysis, commercially available quadrapole, triple quadrapole, ion trap, and time-of-flight instruments have relatively minor distinctions in analysis time, sensitivity, versatility, ruggedness, and cost that are beyond the scope of this review (see reference 56).

The advantages of MS analysis include broad applicability, high sensitivity, large information content, relative ease of interpretation, and very small volumes of sample required. By itself, however, MS does not provide accurate quantitative information on yields. And although equipment costs are decreasing rapidly, MS remains a very expensive and technically demanding tool, especially in comparison with techniques such as TLC.

13.10.5. Summary

Careful consideration of analytical strategies for biotransformation analysis is very important, especially in the common case when a large number of biocatalysts and several test compounds result in a considerable number of analyses to be performed. Analysis and interpretation can easily be the most time- and labor-consuming step of the process—and an important one, since undetected or unidentified products are lost, along with the work to produce them. For convenient initial screening for major transformations or degradation, TLC is a proven, cost-effective, efficient parallel technique. GC and HPLC can be more informative but are more expensive to run and will typically require more time per sample. MS and GC or HPLC/MS are very expensive but will likely give the most information per unit time and the greatest level and clarity of information. Selection of the best methods will ultimately depend on the type of test compounds and expected products, number of samples, time and resources available, stage of the biotransformation development, degree of information needed, and cost of missed information.

13.11. OPTIMIZATION PROCEDURES

To obtain enough product for identification and further testing, preliminary optimization studies may be necessary. Yields can be improved substantially by systematic studies of environmental and nutritional parameters (see section II of this volume). Such studies are of use in scaling up the process.

13.11.1. Environmental Parameters

Changes in temperature can drastically affect biocatalytic reactions. The temperature should be varied between 20 and 50°C for mesophilic organisms and enzymes. Large differences in yields can occur with a 1 or 2°C difference in temperature, as demonstrated with the bioconversion of isobutyric acid to L-(+)-3-hydroxybutyric acid by stationary-phase cells of *Pseudomonas putida* (17). The optimum temperature for growth of cells may well be different from the optimum for biocatalysis. Resting cells, stationary-phase cells, and even enzyme preparations frequently perform well at high temperatures.

It is difficult to study the effect of changes in pH in small-scale flask cultures. Systematic pH variations studies with growing cultures should be done in small fermentors where pH can be controlled automatically. With enzyme reactions or resting-cell reactions, the selection of appropriate, noninhibitory buffers is important. A pH stat may be useful to study reactions that produce pH changes.

Aerobic organisms require oxygen for growth. It may be difficult to achieve optimum growth in shaken flasks because almost always, as cultures proliferate, oxygen becomes the growth-limiting factor. The medium is easily saturated with oxygen at the beginning of growth. However, rapid cell growth during the logarithmic phase depletes oxygen faster than it can be dissolved in the medium. As growth slows toward late log phase, oxygen levels rise. Even in highly aerated vessels, rapidly growing cultures can reduce medium dissolved oxygen concentrations to zero. In culture flasks, the efficiency of aeration is determined by the shape of the flask, the volume of liquid it contains, the type of shaking (reciprocal or orbital), the gaseous environment, culture medium composition, the type of flask closure, and ambient conditions (17 and references therein).

Oxygen is also a substrate in many important biocatalytic reactions, such as aromatic hydroxylation, N dealkylation, O dealkylation, and sulfur oxidation. These reactions are catalyzed by monooxygenases, dioxygenases, and other enzymes that activate molecular oxygen. The dynamics of medium oxygenation are important to all these types of biotransformations, whether the reactions are performed with growing cells, dried cells, or enzyme preparations.

A few simple rules can be used to obtain the best aeration in shaken flasks. Maximum aeration is attained when the liquid volume is no more than 20% of total flask volume. Shaking rates should be high but adjusted so that splashing is not excessive. Rates of 100 to 250 rpm, depending on the flask size, are usually best. For larger flasks, baffles improve aeration efficiency.

Aeration rates achieved in culture media can be defined in terms of oxygen absorption rates (OAR). OAR is defined as millimoles of oxygen absorbed per liter of solution per minute. The OAR of any vessel incubated under any condition can be determined by iodometric titration, which measures the amount of sodium sulfite oxidized by molecular oxygen (8).

The OAR varies with the type of aeration used. By far the best aeration is achieved with stirred and sparged fermentors. Shaking is essential to achieve reasonable aeration with flasks or tubes. Typical OAR values are (i) 0.27 for 100 ml of medium and 0.60 for 50 ml of medium in 500-ml conical flasks shaken at 250 rpm and (ii) 2.0 in a 20-liter fermentor with a sparger operated at 250 rpm.

In modern fermentors, the measurement of dissolved oxygen is done with oxygen electrodes. When procedures are scaled up from shake-flask cultures to fermentors, it is important to increase OAR values as much as possible.

13.11.2. Nutritional Parameters

A discussion of the nutritional improvement of processes is given in sections I and II of this volume. Generally, the components of the culture medium should be tested for effect one at a time, if possible. Sampling times for monitoring the effect on biocatalytic reactions should be arranged to account for possible changes in growth kinetics. Variations in medium components should be checked in the order of their decreasing concentration. Concentrations and types of the carbon sources should be evaluated first by using different sources at the same concentration, then at different concentrations. Carbon sources to compare include glucose, other carbohydrates, and glycerol. Citric acid cycle intermediates are good candidates, as are pyruvate and acetate. Combinations of carbon sources can be effective for improvement of cell growth and enzyme induction and for diminishing catabolite repression of enzyme expression.

After carbon, the next most abundant nutrient is the nitrogen source. The first nitrogen compounds to compare are the simple organic salts, ammonium sulfate, ammonium nitrate, and potassium nitrate. Afterward, urea, glutamate, asparagine, and glutamine should be tested, then the various complex nitrogen sources such as yeast extracts, peptones, and tryptones. Combinations of inorganic salts may then be tried, followed by vitamins, purines and pyrimidines, amino acids, sulfur and phosphorus sources, and the various required inorganic salts and trace elements.

13.12. EXAMPLES OF TYPICAL BIOCONVERSION PROCEDURES: PULLING IT ALL TOGETHER

As should be evident, biotransformation is an interdisciplinary field involving microbiology, biochemistry, organic chemistry, analytical chemistry, and engineering. An appreciation for the contributions of all these fields is very important for the successful application of biotransformations. However, a complete understanding of all of these fields is not a prerequisite for success if the practical guidelines arising from these disciplines are recognized and followed.

To help illustrate the successful practice of biotransformation, the following examples serve as model solutions for some typical situations encountered when applying biocatalysis to organic compounds.

13.12.1. Aerobic Screening (17)

Of central importance for microbial transformations is a basic, general strain-screening method for checking the activity of large numbers of microorganisms on large numbers of compounds. For the first screening experiments, the protocol should be simple so that many samples can be processed. With a long history in the literature, the most reliable procedures involve a two- or three-stage incubation. During the first stage, the culture is grown to late logarithmic phase in a rich medium to provide a heavy inoculum for the second stage. The compounds to be screened are added to the growing second-stage culture when it has reached near-maximum growth. Alternatively, if nongrowing cells are desired, the second-stage culture is prepared by using small amounts of known inducers or substrate to raise desired enzyme activities. The fully grown second-stage cell mass is then processed (e.g., made resting, dried, permeabilized, or immobilized) for use as catalyst in a third stage, in which the compound to be transformed is added immediately. Reaction progress should be monitored occasionally from approximately 6 h up to 1 week using work-up and analytical techniques of suitable throughput and sensitivity to indicate which catalysts convert the substrate.

For improved efficiency, the first screen can be done with few controls: the similar reaction setup for multiple biocatalysts should act as internal controls. However, to confirm presumed transformation products, initial optimization of the most promising biocatalysts must be done with suitable controls. Controls should include cultures without substrate and substrates without microorganisms in sterile medium and in buffers at pH 3, 6, and 8 to account for the range of pH typically observed in cultures. Following screening, biocatalysts that produce even very low yields of desired products can be optimized by using the procedures described above.

By incorporating slight variations, this general procedure can be used to screen for biotransformation reactions using different forms of microbial biocatalysts and, when appropriate, different reaction environments. Some examples of typical modifications are described below.

13.12.2. Microscale Screening of Resting Cells

A good example of the screening of resting cells in a microscale format is given by Semba et al. (46). The objective was to identify microbial catalysts with efficient *para* hydroxylation activity on aromatic substrates. After an initial growth stage that isolated 23,400 strains from soil, colonies were regrown and induced on a solid medium. Each grown strain was transferred to separate wells of a microplate containing 50 μl of buffer, phenol as a probe substrate for the reaction, and glucose as an electron donor for the biocatalyst. A rapid dye indicator of the transformation of phenol to hydroquinone identified 1,263 biocatalysts with the desired activity.

13.12.3. Permeabilized Cells

Cell permeabilization procedures are a good solution if substrate transport through a cell membrane is likely to be poor and use of an isolated enzyme is impractical. As an illustration, D-malate is a rare isomer of malic acid in nature that may be produced by the conversion of the cheap bulk chemical maleic acid with maleate hydratase (33). Normal intact microbial cells do not catalyze the hydration reaction, and the pure enzyme is unstable. To overcome these difficulties, permeabilized cells were developed and analyzed versus pure maleate hydratase. *Pseudomonas pseudoalcaligenes* was cultivated (200 liters) using a mineral salts medium containing yeast extract and 3-hydroxybenzoate. Cells harvested by centrifugation (251 g) were suspended in 1.5 liters of 50 mM potassium phosphate buffer, pH 7, and incubated with 1% Triton X-100 for 0.5 min before being frozen at −80°C for bioconversion studies. Enantiomerically pure (ee [enantiomeric excess] 99.97%) D-malate could be produced using this permeabilized, nongrowing biocatalyst.

Permeabilized *Cephalosporium acremonium* cells were used to transform the antibiotic rifamycin S. A 10-ml reaction mixture containing 2.5 ml of cells, 0.5 mM rifamycin S, 0.5 mM NADH or NADPH, 1.5 mM $MgCl_2$ and 0.05 M phosphate buffer (pH 7.6) is incubated with shaking at 250 rpm and 288°C in an Erlenmeyer flask. The antibiotic is converted to the related rifamycins B and L by the permeabilized cells. The results compare favorably with those obtained with resting cells and cell-free preparations of *Nocardia mediterranei* (14).

13.12.4. Reductions with Yeast

Bioconversions catalyzed by dried baker's yeast are some of the most commonly applied processes because of their simplicity and their utility for generating chiral centers. The asymmetric reduction of carbonyl compounds frequently occurs in good yield. The yeast powder is often rehydrated by simply mixing with a tap water reaction medium. The only other required reagent is an ultimate electron donor, such as sugar, which serves to recycle yeast cell cofactors. Typically, a mixture containing substrate (10 g), baker's yeast (100 g), and sucrose (150 g) in 800 ml of water can be incubated at room temperature for 1 to 2 days, after which it may be extracted with solvent such as ethyl acetate. Evaporation of the solvent leaves a residue that may be subjected to chromatography to obtain stereoselectively reduced product in high yields. By utilizing a variety of available yeast catalysts, this general procedure can be used to reduce a broad range of carbonyl-containing compounds.

13.12.5. Catalysis with Dried Cells

Bioconversions with dried cells resemble the procedure for yeast cells. However, dried bacterial and fungal cells usually are not available commercially; they may be prepared as described earlier. 9α-Fluoro-16α-hydrocortisol (20 g), acetone-dried cells of *A. simplex* (250 g), phosphate buffer (pH 7.0, 0.1 M), and 2-methyl-1,4-naphthoquinone (3 g) in 2.5 liters of water are agitated and aerated for 2 to 3 h. Triamcinolone is obtained by $\Delta^{1,2}$-dehydrogenation 90% yield. In this reaction, naphthoquinone serves as a hydrogen acceptor cofactor. Acetone-dried cells remain active for relatively long periods when stored in a refrigerator. These cells also may be used for nonaqueous solvent procedures.

13.12.6. Use of Metabolic Inhibitors

During screening experiments, microorganisms frequently destroy compounds without the accumulation of recognizable products. Whole-cell biocatalysts contain many enzyme activities, which can result in undesirable side reactions or overmetabolism. The addition of metabolic inhibitors will stall utilization of certain intermediates to allow for the accumulation of products.

This strategy is illustrated in the synthesis of derivatives of monensin, a structurally complex polyether antibiotic (38). Biosynthetic studies revealed that the compound derives from a variety of precursors and that epoxidases and hydroxylases participate in antibiotic formation. Metyrapone, a potent cytochrome P-450 inhibitor, was added (9 mM) to cultures of *Streptomyces cinnamonensis* to cause partial inhibition of monensin biosynthesis. The use of metyrapone selectively inhibited the cytochrome P-450-mediated hydroxylation of a 26-methyl group, resulting in the synthesis of a rare monensin analog. As a result of inhibition, two new metabolites were obtained and designated 26-deoxymonensins A and B.

Another excellent example is the accumulation of a steroid intermediate in the presence of an iron chelator during the degradation of sterols by *A. simplex* and *N. cor-*

allina (17). The addition of α,α'-dipyridyl (0.001 M), *o*-phenanthroline (0.0001 M), or 8-hydroxyquinoline (0.001 M) to logarithmic-growth-phase cultures with cholesterol as substrate causes the accumulation of androstadienedione by blocking the 9α-hydroxylase reaction that leads to complete destruction of the sterol. This procedure was first used to elucidate the pathway of sterol degradation in microorganisms, but for economic reasons it is not the method of choice for large-scale androstadienedione production.

13.12.7. Blocked Mutants

Mutants blocked at various stages of a metabolic pathway accumulate isolable intermediates in their culture medium. Such mutants have been used for large-scale production of biotransformation products. The method can be laborious, since it involves the induction and selection of specifically blocked mutants. Coupled with enrichment culture techniques, the use of mutants is a powerful tool for producing biotransformation products of a wide variety of organic compounds. Thus, if it is possible to induce metabolic pathways for the degradation of a particular compound, it is reasonable to assume that mutants can be obtained that are blocked at points in this pathway, such that transformation products of the starting material will accumulate. Mutants have been used with great success (section I of this volume) for producing useful intermediates of sterol degradation, for example, with *Mycobacterium* species.

13.12.8. Solid Adsorbents

Solid adsorbents can be used when it is desired to maintain bulk substrate in a separate phase, owing to compound instability, toxicity, or for ease of handling. As an example, 3,4-methylene-dioxyphenyl acetone was stereoselectively reduced to S-3,4-methylenedioxyphenyl isopropanol in 95% yield and 99.9% enantiomeric excess by *Zygosaccharomyces rouxii* (54). Both substrate and product were toxic to the biocatalyst, so polymeric hydrophobic resins such as XAD-7 were used to supply substrate to and remove product from the reaction mixture as it formed. Using the solid adsorbent increased yields from 6 to 40 g/liter, allowing 75 g/day to be produced in a 300-liter-scale reaction.

13.12.9. Practical Application of Biodegradation in Aromatic Desulfuration

Microbial biocatalysts are useful for complete or controlled degradation reactions, in addition to syntheses. As a case in point, *Rhodococcus* sp. strain IGTS8 is able to release inorganic sulfur from dibenzothiophene and other sulfur heterocyclic compounds in petroleum using a multienzyme pathway consisting of two monooxygenases and a desulfinase (18). Expression of the gene cluster coding for these enzymes is regulated by sulfate- and sulfur-containing amino acids. This organism enzymatically oxidizes the sulfur atom of dibenzothiophene to the sulfoxide, then to the sulfone, which hydrolyzes to the sulfinate. Cleavage of the sulfur-carbon bond affords 2-hydroxybiophenyl and sulfate. Recombinant strains of this and other organisms are being used to desulfurize fuels.

13.12.10. Catalysis with Purified Enzymes (23)

Commercially available enzymes are useful in preparative biocatalytic reactions. Often, enzyme-catalyzed reactions are highly regioselective, stereoselective, or both. The use of an enzyme in a simple yet highly stereoselective synthetic reaction is illustrated by the use of pig liver esterase to produce a chiral monoester enantiomer by selective hydrolysis of the symmetrical diester, 3-methylglutaric acid, dimethyl ester. A 15-g sample of the diester is suspended in 100 ml of 0.01 M phosphate buffer (pH 8.0), and 1,000 U of pig liver esterase (Sigma) is added with vigorous stirring. The pH is kept constant by the addition of 1 N NaOH. After consumption of 1 mol equivalent of base (overnight), the mixture is homogeneous. The pH is adjusted to 9, and the reaction mixture is extracted with ether. Subsequent chemical treatment of the monoacid/monoester by reduction with borane-methyl sulfide complex of the free carboxyl group affords a hydroxyester that yields the R-six-membered lactone (90% ee).

13.12.11. B-12 Synthesis by a Multienzyme Packed Column (41)

The final example we offer illustrates the practical potential of biocatalysis for multistep fine organic synthesis. Roessner and coworkers (41) cloned and expressed 12 separate enzymes used by nature for the biosynthesis of the complex core structure of vitamin B-12. Each enzyme catalyst was immobilized on Sepharose and packed together in a column reactor. A continuous feedstream of a readily available 5-carbon aminolevulinic acid precursor was transformed in 17 consecutive catalytic steps within the column to the 45-carbon 10-chiral-center B-12 precursor hydrogenobyrnic acid. All reaction chemistries necessary for the synthesis were catalyzed under identical, mild conditions on the benchtop, without intermediate purification, in >90% yield per step and 50-mg scale. It must be pointed out that this synthesis was the culmination of about 25 years of development to discover and obtain every enzyme in catalytically active and pure form. However, it clearly illustrates the potential of biocatalysis for clean, efficient, mild, uniform, and selective synthetic processes, especially considering recent, vast improvements in molecular biology.

13.13. ENABLEMENT TECHNOLOGIES

Enablement technologies are already starting to make an impact on this promise for future development of biocatalysis. Through advances in molecular biology, new biocatalysts are being produced solely from their DNA and RNA blueprints. Using RNA extracted from environmental samples, many new enzymes have been made available by expressing PCR-amplified sequences in suitable generic microbial hosts (47). Shotgun cloning and the effort devoted to genomic sequencing are providing many more opportunities to make additional genes, and the encoded biocatalysts, accessible to the synthetic chemist (10). Building on this, in vitro evolution approaches, such as "directed evolution" (25) and "gene shuffling" (49), permit the tailoring of enzymes for broader ranges of operation, higher efficiency, and new synthetic applications.

Additional processes are under development to take greater advantage of the synthetic potential of biocatalysis. Combinatorial biology (26) attempts to engineer biosynthetic pathways within microorganisms to create modified versions of commercially important natural products. Combinatorial biocatalysis (32) purports to make a general synthetic platform for compound derivatization by combining enzymatic, microbial, and chemical synthetic techniques. With the increasing importance of chiral synthesis and environmental safety, it is likely that biocatalytic techniques will continue to increase in importance.

REFERENCES

1. **American Type Culture Collection.** *American Type Culture Collection Catalogue.* American Type Culture Collection, Rockville, Md.
2. **Beck, J. V.** 1971. Enrichment culture and isolation techniques particularly for anaerobic bacteria. *Methods Enzymol.* **22:**57–70.
3. **Bickerstaff, G. F.** 1997. *Methods in Biotechnology*, vol. 1. *Immobilization of Enzymes and Cells.* Humana Press, Totowa, N.J.
4. **Blanch, H. W., and D. S. Clark.** 1997. *Biochemical Engineering.* Marcel Dekker, Inc., New York.
5. **Brock, T. D., and M. T. Madigan.** 1991. *Biology of Microorganisms*, 6th ed. Prentice Hall, Englewood Cliffs, N.J.
6. **Bruce, L. J., and A. J. Daugulis.** 1991. Solvent selection strategies for extractive biocatalysis. *Biotechnol. Prog.* **7:** 116–124.
7. **Cabral, J. M. S., M. R. Airesbarros, H. Pinheiro, and D. M. E. Prazeres.** 1997. Biotransformations in organic media by enzymes and whole cells. *J. Biotechnol.* **59:**133–143.
8. **Corman, J., H. M. Tsuchiya, H. J. Koepsell, R. G. Benedict, S. E. Kelley, V. H. Feger, R. G. Dworschack, and R. W. Jackson.** 1957. Oxygen adsorption rates in laboratory and pilot plant equipment. *Appl. Microbiol.* **5:**313–318.
9. **Davies, H. G., R. H. Green, D. R. Kelley, and S. M. Roberts.** 1989. *Biotransformations in Preparative Organic Chemistry. The Use of Isolated Enzymes and Whole Cell Systems in Synthesis.* Academic Press, New York.
10. **Doolittle, R. F.** 1998. Microbial genomes opened up. *Nature* **392:**339–342.
11. **Dordick, J. S.** 1989. Enzymatic catalysis in monophasic organic solvents. *Enzyme Microb. Technol.* **11:**194–211.
12. **Edwards, M. J.** 1991. *ATCC Microbes and Cells at Work*, 2nd ed. American Type Culture Collection, Rockville, Md.
13. **Faber, K.** 1997. *Biotransformations in Organic Chemistry*, 3rd ed. Springer-Verlag, Berlin.
14. **Ghisalba, O., R. Roos, T. Schupp, and J. Nuesch.** 1982. Transformation of rifamycin S into rifamycins B and L. A revision of the current biosynthetic hypothesis. *J. Antibiot.* **35:**74–80.
15. **Godfrey, T., and S. West.** 1996. *Industrial Enzymology*, 2nd ed. Stockton Press, New York.
16. **Goetzinger, W. K., and J. N. Kyranos.** 1998. Fast gradient RP-HPLC for high throughput quality control analysis of spatially-addressable combinatorial libraries. *Am. Lab.* 27-37.
17. **Goodhue, C. T., J. P. Rosazza, and G. P. Peruzzotti.** 1986. Methods for transformation of organic compounds p. 97–121. A. L. Demain and N. S. Solomon (ed.), *Manual of Industrial Microbiology and Biotechnology.* American Society for Microbiology, Washington, D.C.
18. **Gray, K. A., O. S. Pogrebinsky, G. T. Mrachko, L. Xi, D. J. Monticello, and C. H. Squires.** 1996. Molecular mechanisms of biocatalytic desulfurization of fossil fuels. *Nat. Biotechnol.* **14:**1705–1709.
19. **Gunsalus, I. C.** 1955. Extraction of enzymes from microorganisms. *Methods Enzymol.* **1:**51–62.
20. **Holland, H. L.** 1992. *Organic Synthesis with Oxidative Enzymes.* VCH Publishers, New York.
21. **Holt, J. G., N. R. Krieg, P. H. A. Sneath, J. T. Staley, and S. T. Williams (ed.).** 1994. *Bergey's Manual of Determinative Bacteriology*, 9th ed. The Williams & Wilkins Co., Baltimore.
22. **Jadoun, J., and R. Bar.** 1993. Microbial transformations in a cyclodextrin medium. Part 4. Enzyme vs. microbial oxidation of cholesterol. *Appl. Microbiol. Biotechnol.* **40:** 477–482.
23. **Jones, J. B., C. J. Sih, and D. Perlman.** 1976. *Applications of Biochemical Systems in Organic Chemistry*, parts 1 and 2. *Techniques in Organic Chemistry.* John Wiley & Sons, New York.
24. **Kieslich, K.** 1976. *Microbial Transformation of Nonsteroid Cyclic Compounds.* John Wiley & Sons, New York.
25. **Kuchner, O., and F. H. Arnold.** 1997. Directed evolution of enzyme catalysts. *Trends Biotechnol.* **15:**523–530.
26. **Jacobsen, J. R., C. R. Hutchinson, D. E. Cane, and C. Khosla.** 1997. Precursor-directed biosynthesis of erythromycin analogs by an engineered polyketide synthase. *Science* **277:**367–369.
27. **Laskin, A., and H. Lechevalier (ed.).** 1984. *Chemical Rubber Handbook of Microbiology*, vol. VII. *Microbial Transformation.* CRC Press, Boca Raton, Fla.
28. **Lee, B. K., W. E. Brown, D. Y. Ryu, H. Jacobson, and R. W. Thoma.** 1970. Influence of mode of steroid substrate addition on conversion of steroid and growth characteristics in a mixed culture fermentation. *J. Gen. Microbiol.* **61:**97–105.
29. **Lee, I. Y., S. L. Nissen, and J. P. N. Rosazza.** 1997. Conversion of β-methylbutyric acid to β-hydroxy-β-methylbutyric acid by *Galactomyces reesii*. *Appl. Environ. Microbiol.* **63:**4191–4195.
30. **Martin, C. K. A., and D. Perlman.** 1972. Stimulation by organic solvents and detergents of conversion of L-sorbose to L-sorbosone by *Gluconobacter melanogenus* IFO 3293. *Biotechnol. Bioeng.* **17:**1473–1484.
31. **Michels, P. C.** Personal communication.
32. **Michels, P. C., Y. L. Khmelnitsky, J. S. Dordick, and D. Clark.** 1998. Combinatorial biocatalysis: a natural approach to drug discovery. *Trends Biotechnol.* **16:**210–215.
33. **Michielsen, M. J. F., E. A. Meijer, R. H. Wijffels, J. Tramper, and H. H. Beeftink.** 1998. Kinetics of D-malate production by permeabilized *Pseudomonas pseudoalcaligenes*. *Enzyme Microb. Technol.* **22:**621–628.
34. **Nakamatsu, T., T. Beppu, and K. Arima.** 1983. Microbial production of 3α-H-4α($3'$-propionic acid)5α-hydroxy-7β-methylhexahydro-1-indanone-δ-lactone from soybean sterol. *Agric. Biol. Chem.* **47:**1449–1454.
35. **Ortiz de Montellano, P. R.** 1986. *Cytochrome P-450. Structure, Mechanism and Biochemistry.* Plenum Press, New York.
36. **Perlman, D.** 1976. Procedures useful in studying microbial transformations of organic compounds, p. 47–68. *In* J. B. Jones, C. J. Sih, and D. Perlman (ed.), *Techniques of Chemistry*, vol. 10, part 1. John Wiley & Sons, New York.
37. **Perry, J. J.** 1979. Microbial cooxidations involving hydrocarbons. *Microbiol. Rev.* **43:**59–72.

38. **Pospisil, S., P. Sedmera, V. Havlicek, and J. Tax.** 1994. Production of 26-deoxymonensins A and B by *Streptomyces cinnamonensis* in the presence of metyrapone. *Appl. Environ. Microbiol.* **60:**1561–1564.

39. **Ribbons, D. W.** 1970. Quantitative relationships between growth media constituents and cellular yields and composition. *Methods Microbiol.* **3A:**297–304.

40. **Roberts, S. M., K. Wiggins, and G. Casy.** 1993. *Preparative Biotransformations. Whole Cells and Isolated Enzymes in Organic Synthesis*. John Wiley & Sons, Inc.

41. **Roessner, C. A., J. B. Spencer, S. Ozaki, C. Min, B. A. Atshaves, and A. I. Scott.** 1994. Genetically engineered synthesis of precorrin 6-x and the complete corrinoid hydrogenobyrinic acid, an advanced precursor of vitamin B-12. *Chem. Biol.* **1:**119–124.

42. **Rosazza, J. P., and M. W. Duffel.** 1986. Metabolic transformations of alkaloids, p. 323–405. *In* A. Brossi (ed.), *The Alkaloids*. Academic Press, New York.

43. **Salter, G. J., and D. B. Kell.** 1992. Rapid determination of the toxicity of organic solvents to intact cells, p. 291–297. *In* J. Tramper, M. H. Vermue, H. H. Beeftink, and V. von Stockar (ed.), *Biocatalysis in Non-Conventional Media*. Elsevier, Amsterdam.

44. **Salter, G. J., and D. B. Kell.** 1995. Solvent selection for whole cell biotransformations in organic media. *Crit. Rev. Biotechnol.* **15:**139–177.

44a. **Sariaslani, F. S.** 1991. Microbial cytochromes P-450 and xenobiotic metabolism. *Adv. Appl. Microbiol.* **36:**133–178.

45. **Schnaitman, C. A.** 1981. Cell fractionation, p. 52–61. *In* P. Gerhardt, R. G. E. Murray, R. N. Costilow, E. W. Nester, W. A. Wood, N. R. Krieg, and G. B. Phillips (ed.), *Manual of Methods for General Bacteriology*. American Society for Microbiology, Washington, D.C.

46. **Semba, H., M. Mukouyama, and K. Sakano.** 1996. A para-site specific hydroxylation of various aromatic compounds by *Mycobacterium* spp. strain 12523. *Appl. Microbiol. Biotechnol.* **46:**432–437.

47. **Short, J. M.** 1997. Recombinant approaches for accessing biodiversity. *Nat. Biotechnol.* **15:**1322–1323.

48. **Stahl, E.** 1969. *Thin-Layer Chromatography: a Laboratory Handbook*. Springer-Verlag, Berlin.

49. **Stemmer, W. P. C.** 1994. DNA shuffling by random fragmentation and reassembly: in vitro recombination for molecular evolution. *Proc. Natl. Acad. Sci. USA* **91:**10747–10751.

50. **Uhlig, H.** 1998. *Industrial Enzymes and Their Applications*. John Wiley & Sons, New York.

51. **Vaara, M.** 1992. Agents that increase the permeability of the outer membrane. *Microbiol. Rev.* **56:**395–411.

52. **van Uden, N.** 1984. Effects of ethanol on the temperature relations of viability and growth in yeast. *Crit. Rev. Biotechnol.* **1:**263–272.

53. **Veldkamp, H.** 1970. Enrichment cultures of prokaryotic organisms. *Methods Microbiol.* **3A:**305–361.

54. **Vicenzi, J. T., M. J. Zmijewski, M. R. Reinhard, B. E. Landen, W. L. Muth, and P. G. Marler.** 1997. Large-scale stereoselective enzymatic ketone reduction with *in situ* product removal via polymeric adsorbent resins. *Enzyme Microbiol Technol.* **20:**494–499.

55. **Wackett, L. P.** 1998. Natural evolution of improved and new enzymes for biodegradation. *In* R. L. Ornstein (ed.), *Enzyme Catalysis: Screening, Evolution and Rational Design*. Marcel Dekker, Inc. New York.

56. **Watson, J. T.** 1997. *Introduction to Mass Spectrometry*, 3rd ed. Lippincott-Raven, New York.

57. **Weaver, E. A., H. E. Kenney, and M. E. Wall.** 1960. Effect of concentration on microbiological hydroxylation of progesterone. *Appl. Microbiol.* **8:**345–348.

58. **Zhou, B., A. S. Gopalan, F. VanMiddlesworth, W. R. Shieh, and C. J. Sih.** 1983. Stereochemical control of yeast reductions. Asymmetric synthesis of L-carnitine. *J. Am. Chem. Soc.* **105:**5925–5926.

Mammalian Cell Culture

MADHUSUDAN V. PESHWA

14

Only a decade and a half ago, the use of mammalian cell cultures was limited to the production of human and animal vaccines and a few other biological substances (4, 50). Cell culture has since become an important aspect of biotechnology, both for therapeutic protein production (48), and for delivering living-cell therapies (72). Stimulating this expansion was the realization in the early '80s that some complex biological molecules produced in recombinant bacteria did not exhibit the desired function or activity. Many of the molecules of therapeutic value, such as tissue plasminogen activator (tPA) (59, 78), erythropoietin (EPO) (79), Factor VIII (58), and protein C (144), to name a few, are complex molecules with intricate tertiary structures anchored by disulfide bond formation and posttranslational modifications such as glycosylation and γ-carboxylation. In some cases, such as immunoglobulins, progress has been made to assemble these molecules in bacterial (122, 130) or plant (44) systems; however, these processes are still inefficient. In recent years, the advent of insect cell culture has allowed some of these molecules to be processed with posttranslational modifications similar to higher eukaryotic cells (89). Mammalian cell culture technology has significantly matured, and a number of products have reached the market, including a variety of vaccines, monoclonal antibodies, tPA, EPO, and Factor VIII, and a variety of others are currently undergoing human testing. While the production of such molecules of therapeutic value had been the driving force behind the development of mammalian cell culture technology throughout the last decade, the emergence of tissue engineering and cell/gene therapy in the early 1990s provided a new avenue for expansion of mammalian cell culture technology. The emphasis shifted from the cells being a preferred system for manufacturing a therapeutic product to include the cells themselves as the final product, such as skin, cartilage, stem cells, and cytotoxic T cells (72, 95). The coming years are likely to feature significantly more maturation in technologies leading to therapeutic protein production, with the focus being on improved automation, process monitoring and characterization, and process control to maintain the physiological state of the cells in culture. The immediate applications of these technologies in the field of living-cell therapies will lead to the development of customized application-specific systems which allow for regulation and control of differentiated cellular function in complex cell populations.

14.1. MAMMALIAN CELL CHARACTERISTICS

Mammalian cells are usually thought of as belonging to two classes: those that grow in suspension (suspension cells) and those that require a surface for attachment and growth (anchorage-dependent cells). Mammalian cells are typically much larger then bacterial and yeast cells, with an average diameter between 12 and 20 μm. Water accounts for 80 to 85% of the cell mass; the other major constituents are protein (10 and 20%) and carbohydrates (1 to 5%), along with lipids, DNA, and RNA. The typical dry cell weight of mammalian cells is between 3×10^{-10} and 6×10^{-10} g/cell, in contrast to bacteria and yeast cells, which have a dry cell weight of 10^{-12} and 10^{-11} g/cell, respectively. Mammalian cells are surrounded by a lipid bilayer-based cell membrane and are thus more shear sensitive than bacterial or yeast cells. Glucose and glutamine in the culture medium and oxygen are the key nutrients for mammalian cell growth. Amino acids, growth factors, cytokines, and trace elements may also contribute to the nutritional requirements of mammalian cells. Additionally, the presence of bicarbonate buffer in culture medium or 1 to 10% carbon dioxide (CO_2) in the gas phase is essential for cell growth. CO_2 participates in the de novo synthesis of purines and pyrimidines and appears to be necessary to prime energy metabolism reactions. However, excess CO_2 may suppress cell growth and productivity by inhibiting respiration, and it can also alter the intracellular pH by diffusing across the cell membrane. Other than the nutrients and metabolites, physical environmental factors such as temperature, pH, osmolarity, and shear may affect cell growth and product formation in mammalian cell cultures.

Mammalian cells are typically cultured at 37°C at near-neutral pH and physiological osmolarity. Both cell growth and product formation are temperature dependent, with the optimal temperature for each being potentially different. The cell growth rate is lower at lower temperatures, as the cell metabolism is downregulated. The optimal temperature for product formation may depend on a variety of factors, for example, the use of a heat shock promoter-driven host expression system or the presence of constituents in the culture medium that lead to temperature-dependent degradation of a secreted protein product. The pH in culture medium is usually, at least initially, controlled by the presence of bicarbonate buffer in the culture medium; base addition may be required for subsequent control of culture pH, especially in cases of programmed feeding in fed-batch cultures. The addition of base for control of pH leads to an increase in the salt concentration in the culture medium and hence to an increase in osmolarity. Osmolarity in the range of 280 to 360 mOsm/kg of H_2O usually is well tolerated by mammalian cells. Higher osmolarities may lead to cell shrinkage and death due to transport of water out of the cells into the culture medium, although higher osmotic pressures have been used to increase the specific productivities of certain secreted products, such as antibodies.

The main thrust in the development of mammalian cell culture was the production of therapeutic proteins requiring complex posttransational modifications to achieve the desired function or activity (13). Most mammalian cell culture processes are designed for secreted protein products. The production of proteins can be either cell growth associated or nonassociated. In the event of growth-associated product formation, optimization of protein production often requires maintenance of high concentrations of viable cells over a prolonged duration. With the advent of recombinant technology, cell lines can be selected for high specific protein productivity in the range of 3 to 6 pg/cell/h. Optimization of the host cell includes not just selection for a high rate of protein production in the desired form but also selection for the integrity, stability and ability for prolonged synthesis in an optimized defined medium, which translate into increased yield and reduced manufacturing costs (80). Established, permanently transfected cell lines rather than transient-expression systems are the preferred hosts in protein manufacturing because of the expense associated with cell line characterization (1).

14.2. CELL GROWTH KINETICS

Growth of mammalian cells, in the most simplistic manner, is often characterized by the Monod model,

$$\frac{\partial x}{\partial t} = \mu x \Rightarrow t_d = \frac{0.693}{\mu}$$

where μ is the specific growth rate and x is the cell concentration at time t. In contrast to bacterial and yeast fermentations, growth of mammalian cell cultures is often characterized in terms of cell numbers rather than dry cell weight. The main reason is that at typical cell concentrations achieved in mammalian cell cultures, 1×10^6 to 2×10^6 cells/ml, the dry cell weight is in the range of 0.5 to 1.0 g (dry weight)/liter and hence cannot be reliably measured throughout the culture process. The time required to increase the cell concentration by twofold is defined as the doubling time (t_d) and can be calculated by integration of the Monod equation. The doubling time for mammalian cells in exponential growth phase is typically between 12 and 36 h depending on the cell type. The specific growth rate, μ, is simplistically characterized in terms of the Michaelis-Menton kinetics for a limiting substrate (25). Additionally, as the viable cells in culture are multiplying, nonviable cells are dying. Cell death usually follows first-order kinetics. Thus, the overall growth of viable cells can be represented as

$$\frac{\partial x_v}{\partial t} = \mu x_v - k_d x_d = \left(\frac{\mu_{\max}\,[S]}{K_m + [S]}\right) x_v - k_d x_d$$

where x_v and x_d refer to the viable- and dead-cell concentrations, respectively, [S] represents the limiting substrate concentration, $\mu_{\max}$ is the maximal growth rate, K_m is the Michaelis-Menten constant, and k_d is the first-order rate constant for cell death. An extension of the simple Monod model, the multiplicative Monod model (82), allows for dependence on multiple rate-limiting substrates and has been used to describe the initial growth rate as a function of initial substrate and metabolite concentrations, including glutamine, serum, ammonium, and lactate (35). Thus, cell growth is comprised of three distinct phases, an initial lag phase, a rapid exponential growth phase, and a stationary phase; after this, the rate of cell death exceeds the rate of cell growth and hence the cell number drops.

For mammalian cell cultures, the growth rate following inoculation is low and is usually proportional to the inoculum concentration because of the requirement of conditioning factors and autocrine factors to build up in the medium (73). Cells are typically inoculated at a concentration of 1×10^5 to 3×10^5 cells/ml, and cultures usually achieve a maximum cell concentration of 1×10^6 to 3×10^6 cells/ml, a 10-fold increase in cell number, which corresponds to about three to five doublings. Given typical doubling times in the range of 12 to 36 h for mammalian cells, a batch culture usually lasts 4 to 7 days, with the exponential phase of rapid growth accounting for about 100 h. The same observations hold true for anchorage-dependent cells, where the typical inoculum density of 10^4 cells/m^2 leads to a maximum cell density of 1×10^5 to 2×10^5 cells/m^2 for monolayer formation, after which cell growth is affected by contact inhibition. For cells which grow in multilayers, nutrient or oxygen limitation and not contact inhibition determines the maximum cell density achieved. For microcarrier cultures, the optimum inoculum requirements are determined by cell-to-bead attachment kinetics. An attachment process following the Poisson distribution requires that the optimum inoculum be 6 to 8 cells per bead to ensure that a significant number (>99%) of beads have at least one cell attached to them (45).

Cell cultures can be operated either in the batch, fed-batch, or continuous-perfusion mode. In fed-batch mode, critical nutrients are added as the culture progresses. In perfusion mode, continuous medium replenishment and waste removal is performed. In most cases of continuous-perfusion systems, the bioreactors are linked to a cell retention device to retain the cells within the bioreactor (125, 126). The rate of nutrient feeding in fed-batch cultures and the dilution rate in continuous perfusion usually follow either a preset algorithm or are determined by the operator after sampling of cultures and off-line estimation of cell, nutrient, and metabolite concentrations. A more recent trend is to perform dynamic on-line measurements to characterize the physiological state of the cells and use this information to perform nutrient feeding (11, 48, 98, 147). A variety of such strategies have been reported to result in higher cell concentrations and product titers and are discussed in detail below (19, 64, 142).

14.3. CELL METABOLISM

Historical development of the growth media commonly used for the propagation of cultured mammalian cells may best be described as conservative and pragmatic. To date, the function of many components of the growth medium essential for cellular proliferation and biological production has not been precisely defined at the molecular level (54). Animal or human serum has been a major component of all culture media for provision of these "essential" growth factors. Significant progress has been made, leading in most cases to replacement of serum-supplemented media by defined media for most industrial mammalian cell culture applications (17, 81, 86).

Glucose and glutamine are the two main energy sources in the medium and are present at concentrations much higher than the other nutrients; typically 5.5 to 55 and 2.0 to 7.0 mmol/liter, respectively (91). They can be oxidized to generate CO_2 or can be anaerobically converted to lactic acid. Lactate and ammonium are the two metabolic products of glucose and glutamine metabolism, respectively. They are growth inhibitory if allowed to accumulate to high levels. In batch cultures, concentrations of lactate and ammonia as high as 30 and 5 mmol/liter, respectively, may accumulate at the end of the cultivation period (100). Accumulation of lactate leads to a reduction in the pH of the cultures, resulting in loss of cell viability. A characteristic feature of mammalian cell culture metabolism is the interchangeability of glucose and glutamine as the energy source. The fluxes of glucose and glutamine to the oxidative and anaerobic pathways are affected by the glucose concentration in the medium (148). At low glucose concentrations, below 0.25 mmol/liter, a large portion of glucose and glutamine is shunted via the oxidative pathway. Thus, it is desirable to maintain a low glucose concentration in the bioreactor to prevent the accumulation of lactate (49, 64). Accumulation of ammonia has been implicated in a reduction of growth rates and of maximal cell densities in batch cultures; changes in metabolic rates, perturbation of protein processing, and virus replication have also been reported (128). In addition to metabolism, glutamine in the medium can be degraded spontaneously to form ammonia (92). The cellular mechanisms of ammonia toxicity are still subjects of controversy. The physical and chemical characteristics of ammonia and ammonium are important, with the former capable of readily diffusing across cellular membranes and the latter competing with other cations for active transport by means of carrier proteins. Strategies to overcome toxic ammonia accumulation include substitution of glutamine by glutamate or other amino acids, nutrient control (i.e., controlled addition of glutamine at low concentrations), and removal of ammonia or ammonium from the culture medium by means of ion-exchange resins, ion-exchange membranes, gas-permeable membranes, or electrodialysis (120).

Oxygen is required for energy metabolism of glucose and glutamine and for synthesis of cellular components such as cholesterol. Specific oxygen consumption rates for mammalian cells are in the range of 0.05 to 5 pmol of O_2/cell/h. Oxygen transport can be a potential limitation in large-scale mammalian cell culture. The optimal dissolved-oxygen concentration (DO) is not generally the same for cell growth and product formation. Typically, the oxygen uptake rate (OUR) increases with an increase in DO up to about 10 to 30% saturation with air. Increase in DO beyond 30% does not lead to any further increase in OUR (30, 36). Higher DO concentrations are toxic and lead to oxidative damage, including DNA degradation, lipid peroxidation, polysaccharide depolymerization, and metabolism of nutrients in culture medium at rates greater than that required for consumption. The steady-state concentration of viable cells increases with decreasing DO until a critical DO is reached. At DO lower than this critical concentration, the viable-cell concentration declines because of incomplete glutamine oxidation. This is accompanied by an increase in specific lactate production from glucose to offset the reduced energy production from glutamine (83). The most simple and commonly used "large-scale" technique to provide oxygen is through the introduction of gas bubbles (52). However, almost since the beginning of in vitro cell culture, empirical observations have indicated that bubbles can be detrimental to the cells. Experimental correlations of cell damage with bubbles, cell attachment to bubbles, the hydrodynamics of bubble rupture, bioreactor studies, visualization studies, and computer simulations and qualification of cell death as a result of bubble rupture have been extensively reviewed (21). Additionally, these detrimental effects resulting from the shear sensitivity of mammalian cells can be abated by supplementing the cell culture medium with nonnutritional polymer additives, such as Pluronic F-68 or methylcellulose, without a significant detrimental impact on cell growth or protein production (51, 101, 145).

The range of optimal concentrations of amino acids in the medium for cell growth is wide and spans a concentration range of 1 log on either side of the amino acid concentration deemed optimal (42). Amino acid consumption is often cell line dependent and affects cell growth but not protein synthesis (99, 119). When the amount of glutamine directed toward energy metabolism is constant, the stoichiometric ratio of amino acids to glutamine is also constant. However, there is no single (or more) amino acid which may be limiting; rather, it is the balance of amino acids (or nutrients) which is critical for cell growth (71). Therefore, improved stoichiometric modeling can lead to improved medium design and feeding strategies to optimize cell growth in mammalian cell cultures (141). The first step in the synthesis of proteins is the binding of tRNA to the amino acids. The steady-state rate constant (k_M) for this binding reaction is of the order of 1 μM. The intracellular amino acid concentration is in the range of 0.1 to 1.0 mM and hence is present at a 100- to 1,000-fold excess over what would be required for protein synthesis. Therefore, the amino acid concentrations almost never limit protein synthesis (118).

Mammalian cell growth is controlled by "balanced growth" and not by rate-limiting conditions, as is often used as a model. Optimal medium design should thus strive to aim for reduced metabolite formation or amino acid production (53, 76). In addition to cell growth, the quantity and quality of protein production are influenced by the culture environment, which is subject to change over the course of cell cultivation. Therefore, in addition to cell growth, it is essential to integrate factors affecting the culture environment into the design of culture media for mammalian cell cultures (142).

14.4. BIOREACTOR DEVELOPMENT

Prior to the advent of recombinant DNA technology, the manufacturing of mammalian cell-derived products was carried out primarily in tissue culture flasks and roller bottles. Scaling up of cell culture processes in such a system presented some major drawbacks, e.g., the labor-intensive nature of the system and lack of process control. The simple suspension culture bioreactors, similar to the systems used in microbial processes, alleviated some of these scale-up issues and also provided improved pH and DO control. The roller-bottle system, however, provided manufacturing flexibility in terms of expansion or reduction in production capacity with minimum capital investment and short turnaround times for product changeovers. Such flexibility, coupled with the fact that such systems were in use during the infancy of mammalian cell culture technology, make roller bottles, even to this day, the system of choice for the manufacture of certain vaccines and a few biologicals. The process technology leading to the application of such roller bottles at the manufacturing scale has focused more on automating the process and reducing the risks of operator-induced contaminations (28, 68).

The use of stirred-tank bioreactors provided a homogeneous environment for suspension cells. Ease of operation and scale-up of cultures has led to an easier acceptance of stirred-tank bioreactors in cell culture processing. The introduction of microcarrier cultivation expanded the scope of stirred tanks to include culture of adherent cells (132). Original microcarriers were based on cross-linked dextran; subsequent development was focused on use of other materials including glass (133), polystyrene (69), cellulose (114), and gelatin (124). Modification of the microcarrier surface chemistry and surface charge density has been shown to have a significant effect on cell attachment and subsequent growth in these cultures (45, 74, 112, 113). The rate of initial attachment of inoculated cells to these microcarriers is, however, critical for successful progress of microcarrier cultures (31, 45). At present, microcarrier cultures are routinely used for manufacture of protein products and vaccines (7, 56, 85). Besides stirred-tank bioreactors, airlift or bubble column bioreactors have found limited application in suspension cultures of mammalian cells (14, 40, 57), with homogeneous cultures resulting from fluid flow rather than mechanical agitation. The main concern about microcarrier and airlift bioreactors was that the shear forces resulting from mixing would lead to loss of cell viability, especially when the bioreactor scale was large (5, 24, 101). Another concern was the inability to obtain high cell concentrations in suspension and microcarrier cultures. The next wave of development thus focused on obtaining high cell concentrations in bioreactors. This effort was channeled primarily along two paths: developing different designs for bioreactors (such as hollow-fiber and ceramic core bioreactors) and developing modifications to application and operation of stirred-tank bioreactors. The latter was based on two strategies: (i) modification to carrier technology, resulting in the use of macroporous microcarriers, aggregate cultures, or entrapment cultures; and (ii) operation of stirred-tank bioreactors in a perfusion mode with cell separation devices on the recycle loop to prevent washing out of cells from the bioreactor. The cell separation devices include external centrifuge (61, 123, 125), external filtration device (60), external membrane separator (18), external settling device (8, 26), rotating wire cage separator (134), internal dialysis membrane (23), and spin filter (6, 143).

Hollow-fiber bioreactors are more frequently used for suspension cells than for anchorage-dependent cells. The technology, introduced by Knazek et al. in 1972 (62), has found extensive application in hybridoma cultures for antibody production (46, 67). Cells are essentially inoculated into the extracapillary space of the hollow fibers, and medium is perfused along the intracapillary space (77). Nutrients and oxygen from the intracapillary stream diffuse across the membrane into the extracapillary space. Similar transmembrane permeation leads to the removal of low-molecular-weight metabolites via the intracapillary stream. Selection of fiber material, surface area, ultrafiltration rate, and pore size characteristics allows for retention of high-molecular-weight nutrients, such as serum components on the extracapillary side, thus significantly reducing the consumption of serum and growth factors (77). Additionally, product accumulates on the extracapillary side and can be harvested at considerably higher concentrations than in stirred-tank bioreactors (70). The ceramic core bioreactor operates on a similar principle, except that there is no separation between the cells and the medium stream, the cells being retained in the bioreactor by physical entrapment in the porous ceramic core. Both these reactor types operate on the plug-flow principle; hence, scale-up usually involves increasing the diameter of the bioreactor. However, as the diameter increases, the pressure drop across the bioreactor increases and thus impacts the flow distribution through the extracapillary side or through the bulk porous ceramic core, resulting in concentration gradients for viable cells, nutrients, and products (20, 102, 109). Various approaches, including alternating the direction of luminal flow and using cross-flow, have been suggested to overcome these limitations. Additionally, hollow-fiber designs have been modified, including the use of either spiral-wound hollow fibers or three-compartment hollow fibers, wherein a separate set of luminal fibers is allocated for gas transfer; however, these designs are mechanically complicated and have not found wide acceptance in mammalian cell culture.

Macroporous microcarriers are highly porous beads, made of collagen or gelatin, with convoluted surfaces and large interconnected pores extending throughout the bead (94, 111, 136). Cells attach to the outside pores of these beads and subsequently grow and migrate to occupy the internal pores (93). In addition to providing a larger surface area for cell attachment and growth, macroporous microcarriers protect cells from shear forces due to agitation or sparging in the stirred-tank bioreactors since the cells grow inside the pores, and not on the surface, of the beads. If the cell growth rate is higher than the migration rate, the cells which had initially attached to the external pores multiply and form clumps, which are no longer able to migrate through the beads to occupy the internal pore surfaces (75, 105). Hence, all the available surface area for cell attachment and growth in the macroporous beads is not utilized. Culture of cells as aggregates can alleviate some of these limitations (127). The ratio of settled volume of inert carrier to cells is lower than that in microcarrier cultures at the same cell concentrations; thus, the agitation conditions required for aggregate cultures are milder than those for microcarrier cultures and are more akin to those for suspension cultures (50). Also, the larger size of the aggregates than of single-cell suspensions facilitates the retention of the cell aggregates in the bioreactor during perfusion. Different methods have been used to induce cells to form aggregates in suspension, including medium modification (86), modulation of the Ca^{2+} concentration in the culture medium (106), lectin-induced aggregation (55), microsphere-induced aggregation (38, 103), and physical entrapment in gels such as calcium-alginate (32). In all such systems, for optimal performance it is critical to maintain and control the aggregate size within desired size specifications. The kinetics of aggregation and growth is, however, still unclear and needs further elaboration (135).

In summary, a large variety of reactor designs and configurations have been evaluated for cultivation of mammalian cells. Although these different bioreactors all have individual characteristics, the selection of a bioreactor depends on the specific process. In other words, all reactors work and can be adequately scaled up. The challenge lies in understanding the physiology, metabolism, and kinetics of mammalian cells and in developing optimized protocols for monitoring and controlling the cell cultures.

14.5. PROCESS MONITORING AND CONTROL

Modeling of cell growth kinetics is an area of active interest, since an incomplete understanding of mammalian cell culture kinetics hinders the ability of the biochemical engineer or biologist to design and control mammalian cell culture systems and to develop operating strategies (12, 48). A variety of unstructured/structured and unsegregated/segregated models have been proposed to simulate mammalian cell culture kinetics under a variety of bioreactor operating conditions (129). An important feature of in vitro mammalian cell metabolism in conventional cell culture media is the partial replaceability of the substrates glucose and glutamine for provisiȯn of energy to the cell. The utilization of glucose and glutamine by cells can therefore vary substantially, and these changes can profoundly affect the dynamics of substrate consumption and energy metabolism (27).

Another limitation to implementing robust process control schemes for mammalian cell cultures in the recent past had been the unavailability of robust on-line monitoring of culture parameters for assessment of physiological state of cells in culture. Temperature, pH, DO, and dilution rate in the reactor were the only accessible parameters for on-line monitoring and process control (30, 78). Temperature is typically controlled at 37 ± 1°C. pH control is initially not required, since the buffering capacity of the culture medium is sufficient to counteract small changes in pH. After exhaustion of the buffering capacity, pH is often controlled by either base addition at constant pCO_2 or lowering the pCO_2 to maintain pH prior to initiating base addition. DO control is usually combined with pH control to avoid accumulation or excessive stripping of CO_2 from the culture medium. Most DO controllers employ a proportional plus integrative and derivative (PID) algorithm with gas composition, gas flow rate, and agitation speed being the manipulated variables. The dilution rate, in perfusion cultures, is manipulated to essentially provide nutrients and remove metabolites from the cultures. In all these efforts, the fundamental idea in the control of mammalian cell culture processes is the usage of a standard personal computer, connected to pumps, valves, and sensors via signal transformation and coupled to free programming, allowing for the implementation of user-oriented software (19).

Subsequent efforts in the control of mammalian cell cultures were aimed at optimizing the metabolic and stoichiometric relationships between the key nutrients and metabolites, with these concentrations being measured off-line (33). The fluxes of glucose and glutamine to the oxidative and anaerobic pathways are affected by the glucose concentration in the medium (148). At low glucose concentrations, below 0.25 mmol/liter, a large portion of glucose and glutamine is shunted via the oxidative pathway. Control strategies aimed at maintaining a low glucose concentration in the bioreactor and preventing lactate accumulation, using a variety of algorithms, have been reported (49, 64). In one instance, which exemplifies the complexity of mammalian cell metabolism, lactate production rate was measured by the rate of base addition and OUR was measured on-line. These parameters were then used in a stoichiometric ratio to calculate "real-time" glucose feeding rate, the desired glucose feeding rate being calculated by the ATP flux assuming that a stoichiometric amount of ATP is produced from the fluxes of glucose and glutamine via the oxidative and anaerobic pathways (37). Another approach to nutrient feeding has been on-line assessment of cell concentration by using a turbidity probe and calculation of OUR on-line to determine the specific oxygen uptake rate for control of nutrient feeding (71, 146, 147). Thus, programmed nutrient feeding, although a common practice in microbial fermentations, has been only marginally realized in mammalian cell cultures.

Recent advances in sensor technologies have led to development of on-line monitoring devices for assessment of cell concentrations (88, 146, 147), nutrient (glucose, glutamine, and amino acids) and metabolite (lactate, ammonium, and amino acids) concentrations (9, 15, 87, 131), and product titers (41). These developments, along with the advent of on-line aseptic sampling devices, have resulted in the application of more advanced systems for bioprocess monitoring and control, including real-time feedback and expert system technology (64, 65). The experience with successful application of such technology to a variety of microbial processes on the laboratory and industrial scale presents an argument for the application of expert systems to high-complexity mammalian cell cultures processes. The organization and functionality of these intelligent control systems and their implementation are, however, still dependent on the elucidation of the physiological states of the cells in culture.

Our understanding of the physiological states of mammalian cells in culture and their switch between these states has been lacking, and much effort concerning medium optimization work in mammalian cell culture has focused on optimizing the initial conditions. Medium design and nutrient feeding strategies must be reevaluated from the perspective of maintaining optimal conditions during cell growth and product formation by better understanding the physiological state of the cells (11, 53, 142). From a more practical perspective, "programmed feeding" has found limited application because of the concern that manipulating cell metabolism in the bioreactor could affect product quality (3, 39) and hence product acceptability. This concern needs to be laid to rest in the light of ongoing experience with programmed feeding.

14.6. PROCESS ISSUES IN LARGE-SCALE MAMMALIAN CELL CULTURES

A variety of different fermentation processes have been successfully used to produce consistent lots of protein-based biopharmaceuticals from mammalian cell cultures. In one study, which evaluated the production of recombinant human soluble CD4, tPA, and EPO from genetically engineered Chinese hamster ovary (CHO) cells, the final product was shown to be consistent in terms of both biological traits at the cellular level and potency, purity, and structure of the product (79). These data exemplify how process design, process validation, and in-process and quality control assays can be used effectively to ensure the consistency of recombinant products derived from cell culture fermentations.

The approval of several recombinant DNA-derived biologicals addresses a number of issues concerning the safety of recombinant products (66). It is critical to identify the limits of acceptable process changes in scale-up of the fermentation and downstream processing of biopharmaceuticals (110) and to define the demand in production validation to prove product equivalency and identity of the isolated, purified therapeutic product (16). Assurance of the safety of such highly complex proteins requires that the following topics be investigated: characterization of the recombinant production organism (140); control of large-scale cell culture production conditions (78); design of the purification process and consistency of manufacture; the purity, safety, and stability testing of the final product (47, 80); and clinical studies (2). Each of the topics needs to be evaluated with respect to the key quality control issues that ensure safety of the final product (34).

Additionally, sterile technology becomes an important factor in long-term bioprocesses (137). Operator-induced biological contamination of cell cultures is a multifaceted problem involving the unexpected introduction of other animal cell, microbial, and viral contaminants (43). Awareness of the potential of this problem and the ability to detect and characterize the contaminations are key factors for its control. For example, fluorescent-antibody staining, isoenzyme analyses, cytogenetic evaluations, and DNA fingerprinting with molecular probes are characterizations used for quality assurance on master seed stocks. Detection of microbial contamination is relatively straightforward, but the prevalence of mycoplasma infections in cell cultures used in general research is still a significant problem (139). The utilization of prescreened reagents and antibiotic-free cultivation and the application of improved procedures, such as fluorescent dyes and molecular probes for detection, provide effective means of avoiding mycoplasma infection and facilitating control. For many viruses, the presence of mycoplasmas reduces immunoreactivity, suppresses transcriptase and other enzyme activities, and reverses viral neutralization. The introduction of viral contaminants into cell cultures is perhaps the most problematic, especially when no cytopathic effect is produced (84). Few cases are documented where technicians infected with specific viruses have introduced these unwittingly into cultures in their care (43). The potential exists, however, as reports have appeared documenting the considerable stability of rhinoviruses, respiratory syncytial virus, rotaviruses, and others in aerosols on workers' hands and safety hood surfaces. The use of multiple cell lines in a facility compounds these problems and necessitates the application of preventive methods both to avoid cross-infections and to document freedom from contamination (97). The facility itself needs to be designed and operated in conformance with current good manufacturing practices (cGMPs) for the production of biopharmaceuticals (29).

14.7. TISSUE ENGINEERING AND CELL THERAPY

The early 1990s witnessed the emergence of tissue engineering and cell and gene therapy and provided a new territory for the natural extension of mammalian cell culture technology (72). The emphasis shifted from the cells being a preferred system for manufacturing biotherapeutic products to include the cells themselves as the final product. A variety of diseases have been targeted for therapy, including stem cell transplantation (63, 95), treatment of cancer (90, 108, 116), viral infections (104, 107, 115), and other immune system disorders; artificial organ systems (10, 22, 96, 117, 138), and gene therapies (121). The requirements for process design and scale-up for such systems is akin to that for the production of biopharmaceuticals, with an added complexity being the need to maintain differentiated cellular function in in vitro culture systems. Current efforts in these fields have focused on the translation of cell culture technologies; the lessons learned in the use of mammalian cell cultures for biopharmaceutical production are equally applicable to tissue engineered and cell therapy products. A major portion of the current effort is, however, directed toward designing novel application-specific bioreactors for specific cell types (63, 95, 96, 117). The needs for such customized designs arise because of the importance of cell-cell, cell-matrix, and secreted-cytokine-induced interactions for maintenance of cell- and tissue-specific differentiated functions. The key elements are isolation, expansion, and manipulation for subsequent activation of effector cell subsets in mixed-cell populations in customized, cGMP, clinical-scale systems for implementation of therapy in a clinical or commercial setting.

14.8. SUMMARY

Process development for the production of biopharmaceuticals in mammalian cell culture systems is a very complex operation. It involves recombinant genetics, verification of a strong expression system, gene amplification, characterization of a stable host cell expression system, optimization and design of the mammalian cell culture fermentation system, and development of an efficient recovery process resulting in high yields and product quality. Among these, mammalian cell culture continues to draw major research efforts. A great deal of progress has recently been made in bioreactor design, understanding of cell culture kinetics, and assessment of cellular metabolism, especially with regard to factors adversely affecting cell growth or viability. Through molecular genetic manipulation, cells are more readily cultivated in a medium free of animal proteins. Achieving a high cell concentration and high viability continues to be a common theme in engineering research. Understanding the physiological states of mammalian cells in culture for the purposes of optimally controlling these cultures must be integrated with other realized advances in mammalian cell culture for ensuring product quality and process consistency.

REFERENCES

1. **Adamson, S. R., and T. S. Charlebois.** 1994. Genetic and phenotypic markers and their relationship to product quality and consistency. *Dev. Biol. Stand.* **83:**31–44.
2. **Akers, J., J. McEntire, and G. Sofer.** 1994. Biotechnology product validation. I. Identifying the pitfalls. *BioPharm* **7:**40–43.
3. **Andersen, D. C., and C. F. Goochee.** 1995. The effect of ammonia on the O-linked glycosylation of granulocyte colony-stimulating factor produced by chinese hamster ovary cells. *Biotechnol. Bioeng.* **47:**96–105.
4. **Arathoon, W. R., and J. R. Birch.** 1986. Large-scale cell culture in biotechnology. *Science* **232:**1390–1395.
5. **Aunins, J. G., M. S. Croughan, and D. I. C. Wang.** 1986. Engineering developments in homogeneous culture of animal cells: oxygenation of reactors and scaleup. *Biotechnol. Bioeng. Symp.* 17:699–723.
6. **Avgerinos, G. C., D. Drapeau, J. S. Socolow, J.-I. Mao, K. Hsiao, and R. J. Broeze.** 1990. Spin filter perfusion system for high density cell culture: production of recom-

binant urinary type plasminogen activator in CHO cells. *Bio/Technology* **8:**54–58.

7. **Baijot, B., M. Duchene, and J. Stephenne.** 1987. Production of aujeszky vaccine by the microcarrier technology. "From the ampoule to the 500 litre fermentor." *Dev. Biol. Stand.* **66:**523–530.
8. **Batt, B. C., R. H. Davis, and D. S. Kompala.** 1990. Inclined sedimentation for selective retention of viable hybridomas in a continuous suspension bioreactor. *Biotechnol. Prog.* **6:**458–464.
9. **Becker, T., W. Schuhmann, R. Betken, H. L. Schmidt, M. Leible, and A. Albrecht.** 1993. An automatic dehydrogenase-based flow-injection system: application for the continuous determination of glucose and lactate in mammalian cell-cultures. *J. Chem. Technol. Biotechnol.* **58:** 183–190.
10. **Bellamkonda, R., and P. Aebischer.** 1994. Tissue engineering in the nervous system. *Biotechnol. Bioeng.* **43:**543–554.
11. **Bibila, T. A., and D. K. Robinson.** 1995. In pursuit of the optimal fed-batch process for monoclonal antibody production. *Biotechnol. Prog.* **11:**1–13.
12. **Biener, R. K., W. Waldraff, W. Noe, J. Haas, M. Howaldt, and E. D. Gilles.** 1996. Model-based monitoring and control of a monoclonal antibody production process. *Ann. N. Y. Acad. Sci.* **782:**272–285.
13. **Birch, J. R., and S. J. Froud.** 1994. Mammalian cell culture systems for recombinant protein production. *Biologicals* **22:**127–133.
14. **Birch, J. R., K. Lambert, P. W. Thompson, A. C. Kenny, and L. A. Wood.** 1987. Antibody production with airlift fermentors, p. 1–20. *In* B. K. Lyderson (ed.), *Large Scale Cell Culture Technology*. Hanser Publications, New York, N.Y.
15. **Blankenstein, G., U. Spohn, F. Preuschoff, J. Thommes, and M. R. Kula.** 1994. Multichannel flow-injection-analysis biosensor system for on-line monitoring of glucose, lactate, glutamine, glutamate and ammonia in animal cell. *Biotechnol. Appl. Biochem.* **20:**291–307.
16. **Brass, J. M., K. Krummen, and C. Moll-Kaufmann.** 1996. Quality assurance after process changes of the production of a therapeutic antibody. *Pharm. Acta Helv.* **71:** 395–403.
17. **Broad, D., R. Boraston, and M. Rhodes.** 1991. Production of recombinant proteins in serum-free media. *Cytotechnology* **5:**47–55.
18. **Buntemeyer, H., C. Bohme, and J. Lehmann.** 1994. Evaluation of membranes for use in on-line cell separation during mammalian cell perfusion processes. *Cytotechnology* **15:**243–251.
19. **Buntemeyer, H., R. Marzahl, and J. Lehmann.** 1994. A direct computer control concept for mammalian cell fermentation processes. *Cytotechnology* **15:**271–279.
20. **Callies, R., M. E. Jackson, and K. M. Brindle.** 1994. Measurements of the growth and distribution of mammalian cells in a hollow-fiber bioreactor using nuclear magnetic resonance imaging. *Bio/Technology* **12:**75–78.
21. **Chalmers, J. J.** 1994. Cells and bubbles in sparged bioreactors. *Cytotechnology* **15:**311–320.
22. **Cieslinski, D. A., and H. D. Humes.** 1994. Tissue engineering of a bioartificial kidney. *Biotechnol. Bioeng.* **43:** 678–681.
23. **Comer, M. J., M. J. Kearns, J. Wahl, M. Munster, T. Lorenz, B. Szperalski, S. Koch, U. Behrendt, and H. Brunner.** 1990. Industrial production of monoclonal antibodies and therapeutic proteins by dialysis fermentation. *Cytotechnology* **3:**295–299.
24. **Croughan, M. S., and D. I. C. Wang.** 1989. Growth and death in overagitated microcarrier cell cultures. *Biotechnol. Bioeng.* **33:**731–744.
25. **Dalili, M., G. D. Sayles, and D. F. Ollis.** 1990. Glutamine-limited batch hybridoma growth and antibody production: experiment and model. *Biotechnol. Bioeng.* **36:** 74–82.
26. **Davis, R. H.** 1995. Cell aggregation and sedimentation. *Bioprocess Technol.* **20:**135–185.
27. **DiMasi, D., and R. W. Swartz.** 1995. An energetically structured model of mammalian cell metabolism. 1. Model development and application to steady-state hybridoma cell growth in continuous culture. *Biotechnol. Prog.* **11:** 664–676.
28. **Edwards, J.** 1996. Automated cell culture for the production of cellular therapy and tissue engineered products, abstr. 13. *In Cell & Tissue BioProcessing Conference*.
29. **Fitzpatrick, S. W., A. Maayan, and J. F. Waggett.** 1990. Implement good manufacturing practices for the production of biopharmaceuticals. *Chem. Eng. Prog.* **86:**26–31.
30. **Fleischaker, R. J., J. C. Weaver, and A. J. Sinskey.** 1981. Instrumentation for process control in cell culture. *Adv. Appl. Microbiol.* **27:**137–167.
31. **Forestell, S. P., N. Kalogerakis, and L. A. Behie.** 1992. Development of the optimal inoculation conditions for microcarrier cultures. *Biotechnol. Bioeng.* **39:**305–313.
32. **Fremond, B., C. Malandain, C. Guyomard, C. Chesne, A. Guillouzo, and J.-P. Campion.** 1993. Correction of bilirubin conjugation in the gunn rat using hepatocytes immobilized in alginate gel beads as an extracorporeal bioartificial liver. *Cell Transplant.* **2:**453–460.
33. **Fu, P., and J. P. Barford.** 1994. Methods and strategies available for the process control and optimization of monoclonal antibody production. *Cytotechnology* **14:**219–232.
34. **Garnick, R. L.** 1989. Safety aspects in the quality control of recombinant protein products from mammalian cell culture. *J. Pharm. Biomed. Anal.* **7:255–266.**
35. **Glacken, M. W., E. Adema, and T. J. Sinskey.** 1988. Mathematical description of hybridoma culture kinetics. I. Initial metabolic rates. *Biotechnol. Bioeng.* **32:**491–500.
36. **Glacken, M. W., R. J. Fleischaker, and A. J. Sinskey.** 1983. Large-scale production of mammalian cells and their products: engineering principles and barriers to scale-up. *Ann. N.Y. Acad. Sci.* **413:**355–372.
37. **Glacken, M. W., R. J. Fleischaker, and A. J. Sinskey.** 1986. Reduction of waste product excretion via nutrient control: possible strategies for maximizing product and cell yield on serum in culture of mammalian cells. *Biotechnol. Bioeng.* **28:**1376–1389.
38. **Goetghebeur, S., and W.-S. Hu.** 1991. Cultivation of anchorage-dependent animal cells in microsphere-induced aggregate culture. *Appl. Microbiol. Biotechnol.* **34:**735–741.
39. **Goochee, C. F., and T. Monica.** 1990. Environmental effects on protein glycosylation. *Bio/Technology* **8:**421–427.
40. **Gudermann, F., D. Lutkemeyer, and J. Lehmann.** 1994. Design of a bubble-swarm bioreactor for animal cell culture. *Cytotechnology* **15:**301–309.

41. **Handa-Corrigan, A., S. Nikolay, D. Jeffery, B. Heffernan, and A. Young.** 1992. Controlling and predicting monoclonal antibody production in hollow-fiber bioreactors. *Enzyme Microb. Technol.* **14:**58–63.

42. **Hansen, H.A., and C. Emborg.** 1994. Extra- and intracellular amino acid concentrations in continuous chinese hamster ovary cell culture. *Appl. Microbiol. Biotechnol.* **41:** 560–564.

43. **Hay, R. J.** 1991. Operator-induced contamination in cell culture systems. *Dev. Biol. Stand.* **75:**193–204.

44. **Hiatt, A., R. Cafferkey, and K. Bowdisk.** 1989. Production of antibodies in transgenic plants. *Nature* **342:**76–78.

45. **Himes, V. B., and W.-S. Hu.** 1987. Attachment and growth of mammalian cells on microcarriers with different ion-exchange capacities. *Biotechnol. Bioeng.* **29:**1155–1163.

46. **Hirschel, M. D., and M. L. Gruenberg.** 1987. An automated hollow-fiber system for large-scale manufacture of mammalian cell secreted product, p. 113–144. *In* B. K. Lyderson (ed.), *Large Scale Cell Culture Technology*. Hanser Publications, New York.

47. **Horaud, F.** 1995. Viral vaccines and residual cellular DNA. *Biologicals* **23:**225–228.

48. **Hu, W.-S., and J. G. Aunins.** 1997. Large-scale mammalian cell culture. *Curr. Opin. Biotechnol.* **8:**148–153.

49. **Hu, W.-S., T. C. Dodge, K. K. Frame, and V. B. Himes.** 1987. Effect of glucose and oxygen on the cultivation of mammalian cells. *Dev. Biol. Stand.* **66:**279–290.

50. **Hu, W.-S., and M. V. Peshwa.** 1991. Animal cell bioreactors—recent advances and challenges to scale-up. *Can. J. Chem. Eng.* **69:**409–420.

51. **Hua, J., L. E. Erickson, T. Y. Yiin, and L. A. Glasgow.** 1993. A review of the effects of shear and interfacial phenomena on cell viability. *Crit. Rev. Biotechnol.* **13:**305–328.

52. **Ishida,M., R. Haga, N. Nishimura, H. Matuzaki, and R. Nakano.** 1990. High cell density suspension culture of mammalian anchorage independent cells: oxygen transfer by gas sparging and defoaming with a hydrophobic net. *Cytology* **4:**215–225.

53. **Jayme, D. W.** 1991. Nutrients optimization for high density biological production applications. *Cytotechnology* **5:** 15–30.

54. **Jayme, D. W., and K. E. Blackman.** 1985. Culture media for propagation of mammalian cells, viruses, and other biologicals. *Adv. Biotechnol. Proc.* **5:**1–30.

55. **Jones, M. N., and R. Perry.** 1980. The application of particle size analysis to the aggregation of chinese hamster ovary cells by concanavalin A. *Exp. Cell Res.* **128:**41–46.

56. **Junker, B. H., F. Wu, S. Wang, J. Waterbury, G. Hunt, J. Hennessey, J. Aunins, J. Lewis, M. Silberklang, and B. C. Buckland.** 1992. Evaluation of a microcarrier process for large-scale cultivation of attenuated hepatitis A. *Cytotechnology* **9:**173–187.

57. **Katinger, J. W. D., W. Scheirer, and E. Kromer.** 1979. Bubble column reactor for mass propagation of animal cells in suspension culture. *Ger. Chem. Eng.* **2:**31–38.

58. **Kaufman, R. J.** 1989. Genetic engineering of Factor VIII. *Nature* **342:**207–208.

59. **Kaufman, R. J., L. C. Wasley, A. J. Spiliotes, S. D. Gossels, S. A. Latt, G. R. Larsen, and R. M. Kay.** 1985. Coamplification and coexpression of human tissue-type plasminogen activator and murine dihydrofolate reductase sequences in chinese hamster ovary cells. *Mol. Cell. Biol.* **5:**1750–1759.

60. **Kawahara, H., S. Mitsuda, E. Kumazawa, and Y. Takeshita.** 1994. High-density culture of FM-3A cells using a bioreactor with an external tangential-flow filtration device. *Cytotechnology* **14:**61–66.

61. **Kempken, R., A. Preissmann, and W. Berthold.** 1995. Clarification of animal cell cultures on a large scale by continuous centrifugation. *J. Ind. Microbiol.* **14:**52–57.

62. **Knazek, R. A., P. M. Gullino, P. O. Kohler, and R. L. Dedrick.** 1972. Cell culture on artificial capillaries: an approach to tissue in vitro. *Science* **178:**65–66.

63. **Koller, M. R., S. G. Emerson, and B. O. Palsson.** 1993. Large-scale expansion of human stem and progenitor cells from bone marrow mononuclear cells in continuous perfusion cultures. *Blood* **82:**378–384.

64. **Konstantinov, K. B., Y. Tsai, D. Moles, and R. Matanguihan.** 1996. Control of long-term perfusion Chinese hamster ovary cell culture by glucose auxostat. *Biotechnol. Prog.* **12:**100–109.

65. **Konstantinov, K. B., W. Zhou, F. Golini, and W.-S. Hu.** 1994. Expert systems in the control of animal cell culture processes: potentials, functions, and perspectives. *Cytotechnology* **14:**233–246.

66. **Kozak, R. W., C. N. Durfor, and C. L. Scribner.** 1992. Regulatory considerations when developing biological products. *Cytotechnology* **9:**203–210.

67. **Ku, K., M. J. Kuo, J. Delente, B. S. Wildi, and J. Feder.** 1981. Development of a hollow-fiber system for large-scale culture of mammalian cells. *Biotechnol. Bioeng.* **28:**79–95.

68. **Kunitake, R., A. Suzuki, H. Ichihashi, S. Matsuda, O. Hirai, and K. Morimoto.** 1997. Fully-automated roller bottle handling system for large scale culture of mammalian cells. *J. Biotechnol.* **52:**289–294.

69. **Kuo, M. J., C. Lewis, Jr., R. A. Martin, R. E. Miller, R. A. Schoenfeld, J. M. Scheck, and B. S. Wildi.** 1981. Growth of anchorage-dependent mammalian cells on glycine-derivatized polystyrene in suspension culture. *In Vitro* **17:**901–906.

70. **Kurkela, R., E. Fraune, and P. Vikho.** 1993. Pilot-scale production of murine monoclonal antibodies in agitated, ceramic matrix or hollow-fiber cell culture systems. *BioTechniques* **15:**674–683.

71. **Kyung, Y.-S., M. V. Peshwa, D. M. Gryte, and W.-S. Hu.** 1992. High density culture of mammalian cells with dynamic perfusion based on-line oxygen uptake rate measurements. *Cytotechnology* **14:**183–190.

72. **Langer, R., and J. P. Vacanti.** 1993. Tissue engineering. *Science* **260:**920–926.

73. **Lauffenburger, D., and C. Cozens.** 1989. Regulation of mammalian cell growth by autocrine growth factors: analysis of consequences for inoculum cell density effects. *Biotechnol. Bioeng.* **33:**1365–1378.

74. **Levine, D. W., D. I. C. Wang, and W. G. Thilly.** 1979. Optimization of growth surface parameters in microcarrier cell culture. *Biotechnol. Bioeng.* **21:**821–845.

75. **Lim, H.-S., B.-K. Han, J.-H. Kim, M. V. Peshwa, and W.-S. Hu.** 1992. Spatial distribution of mammalian cells

grown on macroporous microcarriers with improved attachment kinetics. *Biotechnol. Prog.* **8:**486–493.

76. **Ljunggren, J., and L. Haggstrom.** 1992. Glutamine limited fed-batch culture reduces the overflow metabolism of amino acids in myeloma cells. *Cytotechnology* **8:**45–56.

77. **Lowrey, D., S. Murphy, and R. A. Goffe.** 1994. A comparison of monoclonal antibody productivity in different hollow fiber bioreactors. *J. Biotechnol.* **36:**35–38.

78. **Lubiniecki, A., R. Arathoon, G. Polastri, J. Thomas, M. Wiebe, R. Garnick, A. Jones, R. van Reis, and S. Builde.** 1989. Selected strategies for manufacture and control of recombinant tissue plasminogen activator prepared from cell culture, p. 442–451. *In* R. Spier (ed.), *Animal Cell Biology and Technology for Bioprocesses*. Butterworths Publications, Oxford, United Kingdom.

79. **Lubiniecki, A. S., K. Anumula, J. Callaway, J. L'Italien, M. Oka, B. Okita, G. Wasserman, D. Zabriskie, R. Arathoon, and S. Builder.** 1992. Effects of fermentation on product consistency. *Dev. Biol. Stand.* **76:**105–115.

80. **Lubiniecki, A. S., and J. H. Lupker.** 1994. Purified protein products of rDNA technology expressed in animal cell culture. *Biologicals* **22:**161–169.

81. **McVicar, D. W., F. Li, C. W. McCrady, and R. E. Merchant.** 1991. A comparison of serum-free medias for the support of *in vitro* mitogen-induced blastogenic expansion of cytolytic lymphocytes. *Cytotechnology* **6:**105–113.

82. **Miller, W. M., H. W. Blanch, and C. R. Wilke.** 1988. A kinetic analysis of hybridoma growth and metabolism in batch and continuous suspension culture: effect of nutrient concentration, dilution rate and pH. *Biotechnol. Bioeng.* **32:**947–965.

83. **Miller, W. M., C. R. Wilke, and H. W. Blanch.** 1987. Effects of dissolved oxygen concentration on hybridoma growth and metabolism in continuous culture. *J. Cell. Physiol.* **132:**524–530.

84. **Minor, P. D.** 1994. Ensuring safety and consistency in cell culture production processes: Viral screening and inactivation. *Trends Biotechnol.* **12:**257–260.

85. **Montagnon, B. J., B. Fanget, and J. C. Vincent-Falquet.** 1984. Industrial-scale production of inactivated poliovirus vaccine prepared by culture of Vero cells on microcarriers. *Rev. Infect. Dis.* **6:**S341–S344.

86. **Moreira, J. L., P. M. Alves, A. S. Feliciano, J. G. Aunins, and M. J. T. Carrondo.** 1995. Serum-free and serum-containing media for growth of suspended BHK aggregates in stirred vessels. *Enzyme Microb. Technol.* **17:** 437–444.

87. **Moser, I., G. Jobst, E. Aschauer, P. Svasek, M. Varahram, G. Urban, V. A. Zanin, G. Y. Tjoutrina, A. V. Zharikova, and T. T. Berezov.** 1995. Miniaturized thin film glutamate and glutamine biosensors. *Biosens. Bioelectronics* **10:**527–532.

88. **Murashashi, F., S. Murakami, and K. Baba.** 1994. Automated monitoring of cell concentration and viability using an image analysis system. *Cytotechnology* **15:**281–289.

89. **Murhammer, D. W.** 1991. Review of patents and literature. The use of insect cell cultures for recombinant protein synthesis: engineering aspects. *Appl. Biochem. Biotechnol.* **31:**283–310.

90. **Murphy, G., B. Tjoa, H. Ragde, G. Kenny, and A. Boynton.** 1996. Phase I clinical trial: T-cell therapy for prostate cancer using autologous dendritic cells pulsed with HLA-A0201-specific peptides from prostate-specific membrane antigen. *Prostate* **29:**371–380.

91. **Neermann, J., and R. Wagner.** 1996. Comparative analysis of glucose and glutamine metabolism in transformed mammalian cell lines, insect and primary liver cells. *J. Cell. Physiol.* **166:**152–169.

92. **Newland, M., P. F. Greenfield, and S. Reid.** 1990. Hybridoma growth limitations: the roles of energy metabolism and ammonia production. *Cytotechnology* **3:**215–229.

93. **Nikolai, T. J., and W.-S. Hu.** 1992. Cultivation of mammalian cells on macroporous microcarriers. *Enzyme Microb. Technol.* **14:**203–208.

94. **Nilsson, K., F. Buzsaky, and K. Mosbach.** 1986. Growth of anchorage-dependent cells on macroporous microcarriers. *Bio/Technology* **4:**989–990.

95. **Nordon, R. E., and K. Schindhelm.** 1996. Ex vivo manipulation of cell subsets for cell therapies. *Artif. Organs* **20:**396–402.

96. **Nyberg, S. L., R. A. Shatford, M. V. Peshwa, J. G. White, F. B. Cerra, and W.-S. Hu.** 1993. Evaluation of hepatocyte function in a bioartificial liver: a potential device for the treatment of liver failure. *Biotechnol. Bioeng.* **41:**194–203.

97. **Odum, J. M.** 1994. Developing a program for biopharmaceutical facility design. *BioPharm* **7:**36–39.

98. **Omasa, T., K.-I. Higashiyama, S. Shioya, and K.-I. Suga.** 1992. Effects of lactate concentration on hybridoma culture in lactate-controlled fed-batch operation. *Biotechnol. Bioeng.* **39:**556–564.

99. **Ozturk, S. S., and B. O. Palsson.** 1991. Growth, metabolic, and antibody production kinetics of hybridoma cell culture. 1. Analysis of data from controlled batch reactors. *Biotechnol. Prog.* **7:**471–480.

100. **Ozturk, S. S., M. R. Riley, and B. O. Palsson.** 1992. Effects of ammonia and lactate on hybridoma growth, metabolism and antibody production. *Biotechnol. Bioeng.* **39:**418–431.

101. **Papoutsakis, E. T.** 1991. Media additives for protecting freely suspended animal cells against agitation and aeration damage. *Trends Biotechnol.* **9:**316–324.

102. **Patkar, A. Y., J. Koska, D. G. Taylor, B. D. Bowen, and J. M. Piret.** 1995. Protein transport in ultrafiltration hollow-fiber bioreactors. *AIChE J.* **41:**415–425.

103. **Perusich, C. M., S. Goetghebeur, and W.-S. Hu.** 1991. Virus production in microsphere-induced aggregate culture of animal cells. *Biotechnol. Tech.* **5:**145–148.

104. **Peshwa, M. V., C. Benike, M. Dupuis, S. K. Kundu, E. G. Engleman, T. C. Merigan, and W. C. A. van Schooten.** 1998. Generation of primary peptide-specific $CD8^+$ cytotoxic T-lymphocytes in vitro using allogeneic dendritic cells. *Cell Transplant.* **7:**1–9.

105. **Peshwa, M. V., W.-S. Hu, J.-H. Kim, H.-S. Lim, and B.-K. Han.** 1992. Characterization of cell growth and improvement of attachment kinetics on macroporous microcarriers, p. 77–80. *In* H. Murakami, S. Shirahata and H. Tachibana (ed.), *Animal Cell Technology: Basic and Applied Aspects*. Kluwer Academic Publishers, Dordrecht, The Netherlands.

106. **Peshwa, M. V., Y.-S. Kyung, D. B. McClure, and W.-S. Hu.** 1993. Cultivation of mammalian cells as ag-

gregates in bioreactors—effect of calcium concentration on spatial distribution of viability. *Biotechnol. Bioeng.* **41:** 179–187.

107. **Peshwa, M. V., L. A. Page, L. Qian, D. Yang, and W. C. A. van Schooten.** 1996. Generation and ex vivo expansion of HTLV-1 specific CD8$^+$ cytotoxic T-lymphocytes for adoptive immunotherapy. *Biotechnol. Bioeng.* **50:**529–540.

108. **Pierson, B. A., A. F. Europa, W.-S. Hu, and J. S. Miller.** 1996. Production of human natural killer cells for adoptive immunotherapy using a computer-controlled stirred-tank bioreactor. *J. Hematother.* **5:**475–483.

109. **Piret, J. M., D. A. Devens, and C. L. Cooney.** 1991. Nutrient and metabolite gradients in mammalian cell hollow fiber bioreactors. *Can. J. Chem. Eng.* **69:**421–428.

110. **Reisman, H. B.** 1993. Problems in scale-up of biotechnology production processes. *Crit. Rev. Biotechnol.* **13:** 195–253.

111. **Reiter, M., O. Hohenwater, T. Gaida, N. Zach, C. Schmatz, G. Bluml, F. Weigan, K. Nilsson, and H. Katinger.** 1990. The use of macroporous gelatin carriers for the cultivation of mammalian cells in fluidized bed reactors. *Cytotechnology* **3:**271–277.

112. **Reuveny, S., A. Mizrahi, M. Kotler, and A. Freeman.** 1983. Factors affecting cell attachment, spreading and growth on derivatized microcarriers. I. Establishment of working systems and effect of types of amino-acharged groups. *Biotechnol. Bioeng.* **25:**469–480.

113. **Reuveny, S., A. Mizrahi, M. Kotler, and A. Freeman.** 1983. Factors affecting cell attachment, spreading and growth on derivatized microcarriers. II. Introduction of hydrophobic elements. *Biotechnol. Bioeng.* **25:**481–496.

114. **Reuveny, S., L. Silberstein, A. Shahar, E. Freeman, and A. Mizrahi.** 1982. Cell and virus propagation on cylindrical cellulose based microcarriers. *Dev. Biol. Stand.* **50:** 115–123.

115. **Riddell, S. R., and P. D. Greenberg.** 1995. Principles for adoptive T cell therapy of human viral diseases. *Annu. Rev. Immunol.* **13:**545–586.

116. **Rooney, C. M., C. A. Smith, C. Y. C. Ng, S. Loftin, C. Li, R. A. Krance, M. K. Brenner, and H. E. Heslop.** 1995. Use of gene-modified virus-specific T lymphocytes to control epstein-barr-virus-related lymphoproliferation. *Lancet* **345:**9–13.

117. **Rozga, J., F. Williams, M. S. Ro, D. F. Neuzil, T. D. Giorgio, G. Backfisch, A. D. Moscioni, R. Hakim, and A. A. Demetriou.** 1993. Development of a bio-artificial liver: properties and function of a hollow-fiber module inoculated with liver cells. *Hepatology* **17:**258–265.

118. **Schmid, G., and H. W. Blanch.** 1992. Extra- and intracellular metabolite concentrations for murine hybridoma cells. *Appl. Microbiol. Biotechnol.* **36:**621–625.

119. **Schmid, G., H. Zilg, and R. Johannsen.** 1992. Repeated batch cultivation of rBHK cells on Cytodex # microcarriers: antithrombin III, amino acid, and fatty acid metabolic quotients. *Appl. Microbiol. Biotechnol.* **38:**328–333.

120. **Schneider, M., I. W. Marison, and U. von Stockar.** 1996. The importance of ammonia in mammalian cell culture. *J. Biotechnol.* **46:**161–185.

121. **Shankar, R., C. B. Whitley, D. Pan, S. Burger, J. McCullough, and D. Stroncek.** 1997. Retroviral transduction of peripheral blood leukocytes in a hollow-fiber bioreactor. *Transfusion* **37:**685–690.

122. **Skerra, A., and A. Plukthun.** 1988. Assembly of a functional immunoglobulin F_c fragment in *Escherichia coli*. *Science* **240:**1038–1041.

123. **Takamatsu, H., K. Hamamoto, K. Ishimaru, S. Yokoyama, and M. Tokashiki.** 1996. Large-scale perfusion culture process for suspended mammalian cells that uses a centrifuge with multiple settling zones. *Appl. Microbiol. Biotechnol.* **45:**454–457.

124. **Tao, T.-Y., G.-Y. J, and W.-S. Hu.** 1988. Serial propagation of mammalian cells on gelatin-coated microcarriers. *Biotechnol. Bioeng.* **27:**1466–1476.

125. **Tokashiki, M., T. Arai, K. Hamamoto, and K. Ishimaru.** 1990. High density culture of hybridoma cells using a perfusion culture vessel with an external centrifuge. *Cytotechnology* **3:**239–244.

126. **Tokashiki, M., and H. Takamatsu.** 1993. Perfusion culture apparatus for suspended mammalian cells. *Cytotechnology* **13:**149–159.

127. **Tolbert, W., M. M. Hitt, and J. Feber.** 1980. Cell aggregate suspension culture. *In Vitro* **16:**486–490.

128. **Trisch, G. L., and G. E. Moore.** 1962. Chemical decomposition of glutamine in cell culture media. *Exp. Cell Res.* **28:**360–364.

129. **Tziampazis, E., and A. Sambanis.** 1994. Modeling of cell culture processes. *Cytotechnology* **14:**191–204.

130. **van Brunt, J.** 1990. Monoclonal fine-tuning continues in *E. coli*. *Bio/Technology* **8:**276.

131. **van der Pol, J. J., U. Spohn, R. Eberhardt, J. Gaetgens, M. Biselli, C. Wandrey, and J. Tramper.** 1994. On-line monitoring of an animal cell culture with multi-channel flow injection analysis. *J. Biotechnol.* **37:**253–264.

132. **van Wezel, A. L.** 1967. Growth of cell strains and primary cells on microcarriers in homogeneous culture. *Nature* **216:**64–65.

133. **Varani, J., M. Dame, J. Fediske, T. F. Beals, and W. Hillegas.** 1983. Growth of three established cell lines on glass microcarriers. *Biotechnol. Bioeng.* **25:**1359–1372.

134. **Varecka, R., and W. Scheirer.** 1987. Use of a rotating wire-cage for the retention of animal cells in a perfusion fermentor. *Dev. Biol. Stand.* **66:**269–272.

135. **Vits, H., M. V. Peshwa, and W.-S. Hu.** 1992. Oscillatory behaviour of continuous aggregate cultures of mammalian cells, p. 121–127. *In* H. Murakami, S. Shirahata, and H. Tachibana (ed.), *Animal Cell Technology: Basic and Applied Aspects*. Kluwer Academic Publishers, Dordrecht, The Netherlands.

136. **Vournakis, J. N., and P. W. Runstadler, Jr.** 1991. Optimization of the microenvironment for mammalian cell culture in flexible collagen microspheres in a fluidized-bed bioreactor. *Bio/Technology* 17:305–326.

137. **Werner, R. G., F. Walz, W. Noe, and A. Konrad.** 1992. Safety and economic aspects of continuous mammalian cell culture. *J. Biotechnol.* **22:**51–68.

138. **Wilkins, L. M., S. R. Watson, S. J. Prosky, S. F. Meunier, and N. L. Parenteau.** 1994. Development of a bilayered living skin construct for clinical applications. *Biotechnol. Bioeng.* **43:**747–756.

139. **Wirth, M., E. Berthold, M. Grashoff, H. Pfutzner, U. Schubert, and H. Hauser.** 1994. Detection of myco-

plasma contaminations by the polymerase chain reaction. *Cytotechnology* **16:**67–77.

140. **Wurm, F., G. Polastri, J. Hilfenhaus, H. Harth, and H. Zankl.** 1985. Long term cultivation of Namalva cells for interferon production: stable cytogenetic markers for identification of cells in spite of drastic chromosomal variation. *Dev. Biol. Stand.* **60:**393–403.

141. **Xie, L., and D. I. C. Wang.** 1994. Applications of improved stoichiometric model in medium design and fed-batch cultivation of animal cells in bioreactor. *Cytotechnology* **15:**17–29.

142. **Xie, L., and D. I. C. Wang.** 1997. Integrated approaches to the design of media and feeding strategies for fed-batch cultures of animal cells. *Trends Biotechnol.* **15:**109–113.

143. **Yabannavar, V. M., V. Singh, and N. V. Connelly.** 1992. Mammalian cell retention in a spinfilter perfusion bioreactor. *Biotechnol. Bioeng.* **40:**925–933.

144. **Yan, S. C. B., P. Razzano, Y. B. Chao, J. D. Walls, D. T. Berg, D. B. McClure, and B. W. Grinnell.** 1990. Characterization and novel purification of recombinant human protein C from three mammalian cell lines. *Bio/Technology* **8:**655–660.

145. **Zhang, Z., M. Al-Rubeai, and C. R. Thomas.** 1992. Effect of pluronic F-68 on the mechanical properties of mammalian cells. *Enzyme Microb. Technol.* **14:**980–983.

146. **Zhou, W., and W.-S. Hu.** 1994. On-line characterization of a hybridoma cell culture process. *Biotechnol. Bioeng.* **44:**170–177.

147. **Zhou, W., J. Rehm, and W.-S. Hu.** 1995. High viable cell concentration fed-batch cultures of hybridoma cells through on-line nutrient feeding. *Biotechnol. Bioeng.* **46:** 579–587.

148. **Zielke, H. R., P. T. Oxand, and J. T. Tildon.** 1978. Reciprocal regulation of glucose and glutamine by cultured human diploid fibroblasts. *J. Cell. Physiol.* **95:**41–48.

Plant Cell Culture

NANCY L. PAIVA

15

Plant cell culture has found wide applications ranging from studies of basic plant biochemistry and molecular biology to mass propagation and genetic engineering of crop species. The media, conditions, culture vessels, and other critical parameters vary widely depending on the intended use of the culture. This chapter covers the basic requirements for common procedures in plant cell culture. For work with a particular species, it is recommended that one survey the literature to see if procedures and media for the chosen species or a close relative have already been developed, since optimum conditions can be species or genus dependent. Several books and reviews containing detailed protocols have been published (2, 10, 15, 18, 24, 29, 30, 32, 37, 39, 41, 43, 44, 46), and several journals are dedicated to reporting the latest breakthroughs in plant tissue culture.

In general, plant tissue culture begins with the initiation of callus cultures, i.e., dedifferentiated masses of rapidly dividing cell cultures, on solid, agar-based nutrient media. The callus is generated by exposing sterile pieces of a plant to plant growth regulators. Normally, cell division in an intact plant is restricted to various meristematic regions (for example, root and shoot tip meristems) by the plant regulating its natural hormone distribution. The plant growth regulators are either natural plant hormones or synthetic analogs which bind to receptors in the plants cells, causing all cells to resume division and growth.

Callus cultures can be maintained indefinitely by passing clumps of cells to fresh agar media at regular intervals. Alternatively, callus pieces can be placed in shake flasks containing liquid media to eventually form suspension cultures. These grow faster than callus cultures, probably due to better nutrient exchange with the medium. Plant cell suspensions consist of some single cells and many clumps of a few to hundreds of cells; the size of the clumps depends upon the species, the medium, and the length of time since the culture was initiated. Suspension cultures can be used as a uniform, year-round source of plant cells for biochemical and plant physiological studies or as a source of valuable plant secondary metabolites following appropriate treatments.

The callus cultures can also be allowed to differentiate (regenerate) back into plants; this process is often accelerated by the omission of or changes in the types of plant growth regulators. Each piece of callus can give rise to one to hundreds of plantlets, each with the same genome as the parent, giving rise to the industry of micropropagation. Either early or late in the callus stage, foreign DNA can be introduced into the dividing plant cells. If the cells are then allowed to regenerate, a portion of the plantlets will be transgenic and can be selected for in a number of ways. At one time, the main way of introducing foreign DNA and organelles into plant cells involved the production of protoplasts (wall-less plant cells released by the action of cellulases and pectinases). Currently, several methods of DNA delivery are used, including delivery by the "tamed" plant pathogen *Agrobacterium tumefaciens* or delivery by direct shooting of DNA-coated particles into the cells.

Since the last edition of this volume, breakthroughs have been made in the transformation of important crop species such as corn, wheat, and soybean, and transgenic crops are now being marketed. Given good sterile technique and some patience for the comparatively slow growth, much can be accomplished today with plant cell culture.

15.1. CULTURE MEDIA

The composition of a plant cell culture medium resembles that of an all-purpose plant fertilizer, with the addition of a carbon source, since most cultures cannot photosynthesize enough sugars for growth, and plant growth regulators to stimulate cell growth (11, 15, 17, 27, 36, 43, 44, 46). Most plant cell lines can be cultured in completely defined media, although for some applications complex ingredients such as coconut milk were used in the early formulations and are still used today. The compositions of two of the most commonly used media are shown in Table 1. Murashige and Skoog medium (MS) (27) was one of the first developed for the growth of plant cells, using tobacco as the test species; it is still used today for the growth of callus and suspension cultures, as well as the growth of axenic cultures of plants. Gamborg's B5 medium (B5) (17) was an adaptation of MS, optimized for the growth of soybean cells, and is used for many species. There is some confusion in the terminology used in the literature, but generally a medium is defined by its formulation of inorganic salts and organic compounds (the macro- and micronutrients), often excluding the major carbon sources (usually sucrose). Some authors include either the carbon source or growth regu-

TABLE 1 Composition of common plant cell culture media

Component	Concn in: MS mM	MS mg/liter	B5 mM	B5 mg/liter
Macronutrients				
NH_4NO_3	20.6	1,650	—	—
KNO_3	18.8	1,900	25	2,500
$CaCl_2 \cdot 2H_2O$	3.0	440	1.0	150
$MgSO_4 \cdot 7\ H_2O$	1.5	370	1.0	250
KH_2PO_4	1.25	170	—	—
$(NH_4)_2\ SO_4$	—	—	1.0	134
$NaH_2PO_4 \cdot H_2O$	—	—	1.1	150
Micronutrients				
KI	0.005	0.83	0.0045	0.75
H_3BO_3	0.100	6.3	0.050	3.0
$MnSO_4 \cdot 4\ H_2O$	0.100	22.3	—	—
$MnSO_4 \cdot H_2O$	—	—	0.060	10.0
$ZnSO_4 \cdot 7\ H_2O$	0.030	8.6	0.007	2.0
$Na_2MoO_4 \cdot 2\ H_2O$	0.001	0.25	0.001	0.25
$CuSO_4 \cdot 5\ H_2O$	0.0001	0.025	0.0001	0.025
$CoSO_4 \cdot 6\ H_2O$	0.0001	0.025	0.0001	0.025
Na_2EDTA	0.100	37.3	0.100	37.3
$FeSO_4 \cdot 7\ H_2O$	0.100	27.8	0.100	27.8
Vitamins/amino acid				
Inositol	0.49	100	0.49	100
Nicotinic acid	0.0047	0.5	0.0094	1.0
Pyridoxine·HCl	0.0024	0.5	0.0048	1.0
Thiamine·HCl	0.0003	0.1	0.0300	10.0
Glycine	0.0005	2.0	—	—
Carbon source				
Sucrose	30 g/liter		20 g/liter	
pH	5.8		5.5	

lators or both in their definitions, so it is important to check the formulations in the literature carefully.

15.1.1. Macronutrients

The macronutrients provide the sources of N, P, K, S, and Ca, all required for healthy plant growth. The Na and Cl included as counterions are not essential; MS contains no Na source other than a trace amount included with the Fe source. Na, K, and Cl ions can be tolerated at levels up to 50 to 60 mM and can be used as counterions for additives or for pH adjustment without concern. Nitrogen is provided mainly as nitrate, although some media contain both ammonium and nitrate. Utilization of ammonium causes the culture pH to drop, while utilization of nitrate causes the pH to rise. Some species have a requirement for ammonium or greatly benefit from the inclusion of ammonium or glutamine, but high concentrations of ammonium can be toxic either directly or by causing extremely low pH shifts. Phosphate salts provide both a source of P and the majority of the buffer capacity of the media.

15.1.2. Micronutrients

The micronutrients are sources of various trace elements required for growth. These are similar to those found microbial defined media, with the addition of iodide. Studies have indicated that iodide is not essential but greatly stimulates cell growth. Iron is always added in a chelated form, either as an EDTA salt or as agricultural chelates (e.g., Sequestrene 330 Fe [Novartis Crop Protection, Inc., Greensboro, N.C.]).

15.1.3. Carbon Sources

The most common carbon source is sucrose, generally in the range of 20 to 30 g/liter; the original formulation for MS called for 30 g/liter, while B5 called for 20 g/liter. Glucose is used less often, and fructose is used only occasionally. In many cultures, much of the sucrose is rapidly taken up by the cells in the first 1 or 2 days of culture and converted to starch granules inside the cells. If different carbon sources are to be tested for growth, one must take into account that the cells previously cultured on sucrose may have substantial starch reserves, allowing growth for several days. Cultures of other species often hydrolyze the sucrose to fructose and glucose and utilize the glucose first and then the fructose. Very high sugar concentrations (40 to 100 g/liter) have been used in specialized secondary-metabolite production media (18, 20, 21, 24, 29) and to adjust the osmotic potential of the media in short-term treatments for regeneration (2, 15).

15.1.4. Vitamins

Although intact, soil-grown plants normally synthesize all of the vitamins required for growth, vitamins are generally added to all plant culture media. Very early work indicated an absolute requirement for thiamine (vitamin B_1) in the growth of tomato cells (49), but this has not been determined for all species. Pyridoxine, nicotinic acid, and *myo*-inositol have each been shown to increase the growth of certain species and should also be included, especially when establishing new callus cultures. In many cases, however, these may not be required for growth, and some species have been reported to be unable to take up exogenous *myo*-inositol. All of the common vitamins listed here (thiamine, pyridoxine, nicotinic acid, and *myo*-inositol) are sufficiently stable during autoclaving at common medium pH values. Other vitamins have been added (1 mg/liter or less) in some instances, especially when cells or protoplasts are cultured at very low densities (18, 46); these include biotin, pantothenic acid, folic acid, choline chloride, *p*-aminobenzoic acid, riboflavin, vitamin B_{12}, and ascorbic acid. Ascorbic acid and cysteine•HCl are often added at very high concentrations (25 to 100 mg/liter) as antioxidants to prevent browning of cultures (18, 43).

15.1.5. pH

Most cell culture media are adjusted to pH 5.5 to 5.8 before autoclaving. Following inoculation, the cells may rapidly shift the pH of the media by the release of stored ions or by the uptake and utilization of nutrients such as ammonium. After a week or more of growth, suspension cell culture media may exceed pH 6.0. Rapid alkalinization of the medium is also common after elicitation (see below), with values reaching as high as pH 7.0.

15.1.6. Plant Growth Regulators

Plant growth regulators include both the naturally occurring plant hormones (phytohormones) and synthetic compounds, most of which are structural analogs of the natural hormones and bind to the same receptors. Commonly used regulators are listed in Table 2, along with their abbreviations. One class of regulators are called auxins, which cause plant cells to grow, the naturally occurring plant auxin being indoleacetic acid (IAA). IAA is unstable in solution and is easily oxidized and conjugated to inactive forms by plant cells. However, IAA and the slightly more stable indolebutyric acid (IBA) are still used when low doses or short pulses of hormone are needed, as in certain regeneration systems or in rooting micropropagated plants or cuttings. Naphthaleneacetic acid (NAA) and 2,4-dichlorophenoxyacetic acid (2,4-D) are the most frequently used auxins. 2,4-D is more potent than NAA, followed by IBA, then IAA. IAA and IBA must be added after autoclaving, and the media must be used immediately thereafter; stock solutions should be frozen or made weekly and refrigerated. 2,4-D and NAA can be added before autoclaving and will not break down in the media during storage; stock solutions can be stored indefinitely at 4°C. Many callus cultures can grow with auxins alone; the original formulation for B5 contained 1 ppm of 2,4-D (i.e., 1 mg/liter, or 4.5 μM).

TABLE 2 Growth regulators and additives commonly used in plant tissue cultures

Abbreviation	Full name
Auxins	
IAA	Indole-3-acetic acid
IBA	Indole-3-butyric acid
N AA	1-Naphthaleneacetic acid
2,4-D	(2,4-Dichlorophenoxy)acetic acid
2,4,5-T	(2,4,5-Dichlorophenoxy)acetic acid
PCPA or CPA	(*para*- or 4-Chlorophenoxy)acetic acid
PIC	Picloram (4-amino-3,5,6-trichloropicolinic acid)
NOA	2-Naphthoxyacetic acid
Cytokinins	
BA or BAP	6-Benzylaminopurine
ZEA	Zeatin
KIN	Kinetin
2iP	(2-Isopentenyl)adenine
ADE	Adenine
Other hormones	
GA or GA3	Gibberellic acid (gibberellin A3)
ABA	Abscisic acid
Other additives:	
CH or C	Casein hydrolysate
CW	Coconut water
EDTA	Ethylenediaminetetraacetic acid

A second class of regulators, the cytokinins, promote cell division and are sometimes required for culture initiation or long-term growth. Cytokinins are often added to induce shoot initiation during plant regeneration from callus (15), at which time the auxins may be decreased or omitted from the medium. The first naturally occurring cytokinin to be identified was zeatin, a purine derivative (19). Zeatin is occasionally used in culture media, but the most common cytokinins in use today are kinetin (6-furfurylaminopurine) and 6-benzylaminopurine (BAP; also known as N6-benzyladenine [BA]); adenine and adenosine can also have cytokinin activity, particularly at high levels. Most cytokinins, except for zeatin and its derivatives, can be sterilized by autoclaving.

Gibberellins (usually GA3, gibberellic acid) and abscisic acid are sometimes used to stimulate embryo or shoot development (2, 15). Other plant growth regulators such as the recently discovered brassinosteroids are still being evaluated in culture systems.

Auxins and cytokinins are used in the range of 1 to 50 μM. Most plant growth regulators can be dissolved in a small volume of reagent grade ethanol and then diluted in warm water to the final stock concentration; the ethanol will evaporate during autoclaving. Alternatively, NaOH is used to dissolve most auxins and HCl can be used to dissolve cytokinins, but these should be kept to a minimum to avoid altering the pH of the media.

15.1.7. Other Additives

Complex ingredients such as coconut milk, yeast extract, fruit juices, and casein hydrolysates (such as NZ Amine) were used in many early culture experiments and are still used today. They provide plant growth regulators, vitamins, and amino acids which can enhance plant cell growth rates and survival of some cultures. However, they are expensive, sometimes variable in quality, and sometimes detrimental, such as when yeast extract induces stress responses in dilute cell suspensions.

Mannitol, sorbitol, or combinations of these two sugar alcohols are routinely added as an osmoticum (i.e., a supplemental agent increasing osmolarity). A concentration of 100 g of sugar alcohol per liter (0.6 M) is used in most media for the isolation of plant protoplasts, to prevent the cells from bursting once the rigid cellulose cell walls are removed (46). The addition of 0.2 to 0.5 M total osmoticum (300 to 900 mOsm/kg of H_2O, added as an equimolar mixture of sorbitol and mannitol) to the culture medium increases the number of transformants produced by bombardment techniques (34).

Organic acids such as citric, fumaric, malic, or succinic acids, or synthetic buffers such as Tris, MES, or HEPES have been used to help maintain culture pH, especially when ammonium salts are used as the nitrogen source (18). Neutralized activated charcoal is occasionally added to young regenerating cultures to remove toxic phenolics released by the stressed plant cells, or to help remove plant growth regulators introduced at an earlier stage. Silver thiosulfate serves as an ethylene biosynthesis inhibitor, preventing this gaseous plant hormone from accumulating to detrimental concentrations.

"Nurse" medium or "conditioned" medium is the liquid medium removed from a suspension of fast-growing cells. It contains uncharacterized growth factors released by growing cells, or it may be depleted of growth-inhibiting substances, such as high levels of ammonium ions. Conditioned medium is usually removed aseptically from a culture and never autoclaved. A nurse culture or "feeder" layer is usually a thin layer of fine, fast-growing cells spread on the agar surface and covered with a moist sterile filter paper. Conditioned media and nurse cultures are sometimes used in the culture of regenerating protoplasts, single cells, or very dilute cell suspensions (25, 40). The nurse media or cultures are often not from the same species as the species of interest; e.g., a tobacco cell culture may be used as the nurse culture for carrot cells.

15.2. MEDIUM PREPARATION

Only high-purity (reagent grade) chemicals should be used for plant cell culture, especially when new cultures are being initiated or sensitive cell lines are being used. Water should be double-distilled (glass stills preferred) or deionized, such as 18 MΩ water produced by MilliQ devices (Millipore, Bedford, Mass.). Glassware must be free of residual detergents. If glassware is cleaned with chromic acid (or any solution containing heavy metal ions), it must be treated with EDTA or another chelator to remove all traces of chromium, since some cultures are very sensitive. In fermentors or bioreactors, all parts in contact with the medium must be made of high-quality stainless steel, glass, or plastic for work with sensitive cell lines. In studies with opium poppy cells (*Papaver somniferum*), the use of bioreactors containing even a small amount of silver solder or brass screws results in the browning and death of the cultures, while periwinkle (*Catharanthus roseus*) and tobacco (*Nicotiana tobacum*) cells grow well (personal observation).

The options for gelling agents for plate cultures include bacteriological grade agar (such as Bacto Agar [Difco, Detroit, Mich.]), Noble agar (agar reduced in ionic impurities [Difco]), tissue culture grade Phytagar (specially prepared agar [Gibco-BRL]), agarose (highly purified agar), and Phytagel (Gellam gum agar substitute [Sigma] or Gelrite [Merck/Kelco Division]). For routine work, our laboratory prefers Phytagar at 0.6 to 0.8% or agarose at 0.55%, although many cultures can tolerate the impurities in bacteriological agar. Carageenan (vegetable gelatin) and calcium alginate have been used for cell immobilization, particularly in bioreactor studies.

A number of strategies have been used to handle medium preparation. One traditional method is to prepare four liquid stock solutions, containing (i) the five or six macroelements (20× stock), (ii) the microelements omitting the iron source (200×), (iii) the iron chelate (200×), and (iv) the vitamins (200 to 1,000×). The first three are stored at 4°C, while the vitamins are stored frozen. The carbon source is added as a solid. If hormones are to be included (Table 2), heat-stable hormones are predissolved and added as a 200 to 1,000× stock; otherwise, they are filter-sterilized and added after autoclaving. Agar, if required, is added after pH adjustment. The use of multiple concentrated stocks of N and P sources make medium variations easier.

Once an acceptable medium is identified and large quantities are needed, a convenient practice is to prepare a 10× concentrated stock and freeze it in appropriately sized aliquots. An example of this strategy for the preparation of an SH stock (supplemented with three growth regulators and 30 g of sucrose per liter for use with alfalfa cell cultures [2, 11, 36] is provided in Table 3. A 4-liter volume of 10× concentrate, including three growth regulators but excluding sucrose, is dissolved, distributed into plastic Whirlpak bags (Nasco, Grand Prairie, Tex.) in 100-ml aliquots, and frozen at −20°C or lower. To prepare the medium, one bag of concentrate is thawed for every liter of liquid or agar needed, and the contents of the bag are transferred with rinsing to a large flask or beaker and dissolved by stirring for 30 to 60 min at room temperature. Sucrose (30 g/liter) is added, the pH is adjusted to 5.7, and the medium is distributed into flasks (with or without agar) for autoclaving. We find it advantageous to add the sugar at the time the concentrate is thawed for use, because otherwise the stock would contain 30% sucrose, which is difficult to dissolve, distribute, and freeze. The pH of the stock is slightly acidic and should not be adjusted prior to freezing. A slight precipitate may form in the bags during freezing, but it will redissolve with extended stirring.

Various companies now produce a variety of premixed plant cell culture media. Usually, the medium is divided into "basal salts" (the macro- and micronutrients), sold as a dried powder, and vitamins, sold separately as a solution. These components may also be sold as a combined dried powder ("basal salts with minimal organics"), with or without sugar included. The medium is usually dispensed into preweighed packets such that each makes 1 to 10 liters of medium; one can also purchase a jar of well-blended powder that makes 50 liters and weigh the powder as needed.

TABLE 3 Preparation of SH frozen stock[a]

Component	Concn (mM) in final medium	Amt (g/40 liters) per batch of 10 × stock
KNO_3	25	101
$MgSO_4 \cdot 7H_2O$	1.5	14.8
$NH_4H_2PO_4$	2.5	11.6
$CaCl_2 \cdot 2H_2O$	1.5	8.8
myo-Inositol	5.6	40
$MnSO_4$	0.059	0.358
H_3BO_3	0.13	0.200
$ZnSO_4 \cdot 7H_2O$	0.0035	0.040
KI	0.0060	0.040
$CuSO_4 \cdot 5H_2O$	0.0008	0.008
$FeSO_4 \cdot 7H_2O$	0.0054	0.600
Na_2EDTA	0.0054	0.800
Thiamine·HCl	0.0150	0.200
Nicotinic acid	0.0410	0.200
Pyridoxine·HCl	0.0024	0.020

[a]For 10× concentrate without sugar or pH adjustment. The method is as follows. Dissolve all of the above solids in 2.5 liters of distilled H_2O. Add 8 ml of Mo/Co microstock (50 mg of Na_2MoO_4 and 50 mg of $CoCl_2 \cdot 6H_2O$ dissolved together in 100 ml and stored at 4°C). Add 20 ml of kinetin stock (43 mg/200 ml). Add 80 ml of 2,4-D stock (225 mg/liter). Add 400 ml of PCPA stock (186.6 mg/liter). Adjust the volume to 4 liters. Mix well. Divide into 100-ml lots in Whirlpak bags. Store at −20 to −40°C. Note: Each 100-ml bag is equivalent to 1 liter of SH. To use, thaw 1 bag/liter needed, add 30 g of sucrose/liter, adjust the volume, and titrate to pH 5.7 with 1 N NaOH. For plates, add 8 g of agar/liter.

Plant growth regulators and agar are rarely included. Due to inconsistencies in formulations and terminology in the literature, manufacturers usually provide very detailed tables of their formulations; one manufacturer supplied 15 variations of MS salts. If one is careful to select the correct preparation and labor is limited, these commercial powders offer a slightly expensive but reliable option.

For many liquid media including MS, a small amount of white precipitate may be visible in shake flasks, especially after extended autoclaving or storage. For routine culture maintenance, this medium will work well, since the precipitate will gradually be used by the cells. Media should be autoclaved at 121°C for a minimum of 15 min for small volumes and for longer times for larger volumes or thick-walled vessels. The required times may need to be evaluated on-site; 30 min for 500 ml of agar or 1 h for a 2-liter bioreactor vessel is not unusual (5).

15.3. FACILITIES

Successful plant cell culture requires extremely close attention to proper sterile technique and temperature control. The growth of plant cell cultures is relatively slow compared to common microbial cultures, necessitating very long culture times for experiments; a low level of contamination which would be outgrown and perhaps unnoticed in a bacterial culture would have time to overwhelm plant cultures. It may take months to establish a high-producing cell line or to regenerate transgenic plants from the original explants.

15.3.1. Sterile Transfer Facilities

Laminar-flow hoods are strongly recommended. An even flow of HEPA (high-efficiency particulate air) filter-sterilized air comes from the back of the hood, protecting the cultures when the vessels are opened. Unlike primary animal cell cultures, no biohazard containment of the cultures is required. Some laboratories prefer to use containment hoods when introducing microbes into plant cell cultures, such as during genetic transformation or inoculation with plant pathogens. Each hood should be equipped with a natural gas outlet for a Bunsen burner and lighting (usually overhead). A UV germicidal lamp is strongly recommended for use in decontaminating the hood work surface after the HEPA filter is replaced, if the hood is switched off for more than 30 min, or if the hood is accidentally contaminated. A selection of stainless-steel forceps, spatulas, and scalpels with replaceable blades are required for culture initiation, callus subculture, and transformation and regeneration. For quantitative transfers, an electronic (digital) top-loading balance (400-g to 4-kg capacity) with auto-tare feature is essential.

15.3.2. Temperature

Most cell cultures grow well only in a narrow temperature range. The most common growth temperature is 25°C. Some plant species, especially heat-tolerant plants such as tobacco or *Catharanthus roseus*, can continue to grow at temperatures of 30°C and above, while heat-sensitive cultures like opium poppy will cease to grow above 27°C and will die at higher temperatures.

Cyclic temperature changes may occur during the day-night cycles used for some culture procedures. Localized heating of shelves occurs during the day, and they then cool during the night. Although not directly lethal to cultures, this contributes to the formation of condensation inside the lids of petri dishes and other culture vessels, which can dry out the cultures or contribute to contamination. Stacking an uninoculated petri dish containing water agar on top of the important cultures will greatly reduce condensation yet still allow sufficient light for regeneration.

15.3.3. Light

Callus and shake flask cultures generally do not require light for growth. Usually they are grown in dark rooms, equipped with lights which are turned on when inspecting or subculturing the cultures. Absolute darkness is not usually required. High light intensities or low levels of continuous illumination can cause cultures to turn green (form small numbers of chloroplasts) or brown (accumulate polymerized phenolics). Enough light to induce either color change can be sufficient to inhibit secondary-metabolite accumulation, and browning may greatly slow the growth of cultures. If callus and shake flasks must be incubated in lighted rooms (such as rooms shared with regenerating cultures or aseptically grown plants), they can be shielded with foil or boxes.

For regeneration, cool white fluorescent lights are recommended. Interspersing incandescent bulbs will greatly increase the quality of the light, providing a spectrum close to that of sunlight. Lights are generally suspended 12 to 14 in. above the culture vessels. Sufficient distance must be maintained between the lights and vessels to allow cooler air to circulate and avoid overheating the plates. Wire shelving, positioned a few inches from the culture room walls, are often preferred, providing good air circulation at a reasonable cost. The shelves can be partially covered with aluminum foil or Plexiglas to provide a more stable, easily cleaned surface. If several layers of closely lit shelves are used, heat from a lower shelf may warm the shelf above; plates can be partially insulated by placing them on foil-covered sheets of styrofoam, or lights should be suspended by chains to allow heat dissipation.

Too much light can actually be detrimental to culture growth and regeneration. One indication of a high light level is the accumulation of anthocyanins (red to purple pigments) in callus or differentiated tissues. Continuous light is sometimes detrimental to whole-plant growth and is rarely used in plant cell cultures. The use of household timers is adequate for generating day-night cycles (16 to 18 h of light, 6 to 8 h of dark). One exception is the culture of photoautotrophic cells, which require high-intensity and high-quality light to enable the cells to photosynthesize sufficient sugars from CO_2 (29).

15.3.4. Aeration

Under most conditions, shake flask cultures of plant suspension have a lower volumetric oxygen demand than do microbial cultures. The respiration rates (Q_{O_2}) reported for plant cell cultures range from 0.2 to 3.6 mmol of O_2/h/g [dry weight], compared with 2 to 16 mmol/h/g for *Escherichia coli*, 39 to 570 mmol/h/g for *Bacillus* spp., and 1.9 to 2.8 mmol/h/g for *Saccharomyces* spp. (39). Shaker speeds (100 to 150 rpm with a 1-in. throw) are generally lower than those used for microbial cultures, and the fill ratios in the flasks (volume of medium/volume of flask generally 0.2 to 0.4) are generally higher. Baffled flasks are rarely used to increase aeration but are occasionally used for breaking up clumps of cells, especially when a suspension is first being initiated.

Generally, shake flask cultures can be removed from a shaker for subculturing and observation for 1 h or more without killing the cells. However, this procedure will affect the growth rate and can affect the ability of the culture to produce secondary metabolites. High oxygen demands have been reported immediately following elicitation (see below), since a short-term burst of O_2 utilization occurs as a result of activation of many defense processes. If elicitation is used to induce secondary-metabolite accumulation, one should prevent even brief interruptions in aeration of shake flask cultures, and it may be to necessary to increase the aeration of production bioreactors.

A wide variety of closures have been used for shake flask cultures. Traditional cotton plugs in Erlenmeyer flasks can work well, particularly if the cotton plugs "mushroom" over the rim of the flask, keeping it free of microbial contamination. Since the cultures may be on the shakers for weeks, particulates in the culture room air will tend to collect around the rim if unprotected, making it impossible to open the flask for subculture without contamination. A preferred option is straight-walled Delong-style flasks with metal culture tube closures; a small piece of folded cheesecloth or gauze is placed in the top of the closure to act as a sterile filter, and the prongs on the closures hold it securely on the flask. These caps have the advantage of providing the same sterility barrier and good aeration as cotton plugs while covering the rim and upper edge of the flasks. Any dust from the shaker rooms or medium storage areas can be easily removed by a brief flaming before the flask is opened.

Petri dishes containing agar media, and callus cultures or regenerating plants are incubated agar side down, not inverted like microbial cultures. Plates are always sealed with Parafilm or porous medical tape (such as Urgopore tape [Karlan Research Products Co., Santa Rosa, Calif.]) or can be placed inside a sterilized box or new plastic self-sealing freezer bag for short-term incubation. Otherwise, contaminants will enter the plate during the long cultures or the agar media will dehydrate.

Closures on culture vessels must not only allow oxygen and CO_2 to pass but, in sensitive species, must also allow ethylene and other volatiles to escape. Ethylene, a gas, is a plant hormone, often released by plant cells in response to wounding and infection or during senescence. It can have a number of effects on axenically grown plants and cell cultures (15). In very sensitive species, such as *Arabidopsis* and potato, the addition of ethylene biosynthesis inhibitors such as silver ions (as silver thiosulfate) greatly increases the yield of regenerated transgenic plants. Without these, the wounded tissue produces enough ethylene to inhibit the development or lead to browning and death of nonwounded cells.

Dedicated growth rooms provide the most efficient and flexible use of space. Commercial growth chambers equipped with lights, temperature control, and (optional) humidity control are available; several growth chambers allow many different conditions to be maintained, which may be important if diverse species are to be cultivated. For important cultures, it is recommended that several separate cultures be maintained, preferably in separate culture rooms or incubators. Temperature recorders are strongly recommended to diagnose possible problems in temperature control. Motion sensors and temperature monitors with audible alarms can also be used to minimize the damage caused by heating and cooling and by shaker equipment breakdowns. Even brief overheating to 35°C or more can greatly reduce regeneration and secondary-metabolite production if it occurs at critical times, while regenerated plantlets are relatively resistant. Long-term exposures (hours) to high temperatures can stop the growth or rapidly kill suspension cultures.

15.4. CULTURE INITIATION

Rapidly growing plant tissues are generally best for rapidly producing callus cultures, although any living plant cell can in theory be induced to divide. While authors often indicate the plant organ from which a callus was derived (i.e., root, leaf, or hypocotyl), after several generations the cultures are fully dedifferentiated and there is little evidence that the original tissue source influences the behavior of the culture, such as the products it will accumulate. In contrast, for short-term cultures such as for transformation and quick regeneration, the explant source can be critical; for wheat and some recalcitrant legumes, immature seeds or cotyledonary nodes will produce easily regenerable callus while leaf or root callus cultures will not.

15.4.1. Sterile Explants

There are two basic sources for obtaining surface-sterilized plant tissues. One is to surface sterilize seeds, which are dormant and often dried and can therefore tolerate very harsh sterilization conditions. The seeds are then germinated under sterile conditions, and the various tissues are placed on callus-inducing media. The other is to surface sterilize undamaged tissue from a growing plant, which may be the only option for vegetatively propagated species or tree species which are slow to set seed. The same sterilization strategies can be used in each case. Most commonly, a brief ethanol wash serves to remove and kill spores and bacteria on the plant surface and to thoroughly wet the tissue; it is followed by a hypochlorite soak to kill any remaining microbes.

15.4.1.1. Seeds

The following protocol works well for soybeans and other medium-sized to large seeds. The seeds are placed in a clean plastic petri dish (25 by 100 preferred) or beaker and covered with 70% ethanol for 5 min. The dish is stirred or gently swirled occasionally to ensure that the ethanol contacts all the seed surfaces. The seeds are transferred with a sterile (flamed) spoon-shaped spatula to a sterile dish with enough bleach solution (20% Clorox in distilled water with 1 drop of Tween 20 or 2 drops of 10% sodium dodecyl sulfate solution per 100 ml; equivalent to 1% hypochlorite [final concentration]) to cover the seeds. They are soaked for 15 to 20 min with stirring. Any cracked, floating, or noticeably damaged seeds are removed. The seeds may begin to swell during this time. The seeds are transferred with a sterile spatula to a sterile dish containing distilled water and are rinsed two or three more times by transferring to new dishes of sterile water. They are then transferred to sugar-water agar plates (5 g of sucrose and 8 g of Phytagar/liter; 50 ml/25-mm-deep plate) and sealed with Parafilm. If possible, the seeds are spread out so that they do not touch each other, and several plates are used, so that unsterilized seeds will not contaminate clean seeds. The seeds are incubated at approximately 25°C.

Germinating the seeds on media containing sugar or other nutrients helps locate contaminants by stimulating their growth on the agar surface. Seeds may also be germinated on damp filter paper or cotton. Tiny seeds, such as those of tobacco, lettuce, and *Arabidopsis* spp., may be more easily sterilized by being treated in sterile 1.5-ml plastic snap-cap tubes, briefly centrifuging them to the bottom of the tubes to remove solutions. The seeds can be transferred to new tubes by being suspended and pipetted. To make tiny seeds easier to distribute, after sterilization they can be suspended in 0.1% agarose and dripped onto the agar surface. If a special treatment is recommended before planting in soil, the same treatment should be applied before sterilization. Some species require extended cold treatments, exposure to light, or scarification (scratching the seed surface with sandpaper or nicking the seed coat with a razor blade).

15.4.1.2. Leaves

Young leaves, free of insect and microbial damage, should be selected. They are rinsed with distilled water to remove dirt and kept in a moist container to prevent wilting. For leaves which are thick or have a thick waxy cuticle, the above protocol should work well. For very thin leaves or for delicate leaves such as those grown in a very humid environment, less harsh conditions may be required. For example, for tender alfalfa leaves, I reduce the ethanol concentration to 50% and the exposure time to 1 to 2 min to avoid killing the leaf cells.

15.4.1.3. Other Tissues

Depending upon the species and environment, other plant parts may be available. For species which make tubers or rhizomes, these organs should be first scrubbed free of soil and may be surface sterilized, peeled, then sterilized briefly again. Young stems and petioles are often more resistant to damage during sterilization than are leaves.

Other sterilizing agents have been used, such as $HgCl_2$ or antibiotics, especially for seeds which have natural bacterial population under their seed coats. A procedure involving chlorine gas is recommended for dry seeds such as soybeans. The seeds are placed in a dish in a larger, sealable glass container such as a desiccator, together with a beaker containing bleach. The bleach is acidified (3 to 5 ml of concentrated HCl per 100 ml of Clorox), which releases chlorine gas, and the container is quickly sealed. After 6 h to overnight, the container is opened and the seeds are rinsed with sterile water and transferred to plates for germination. Procedures have also recommended acidifying the bleach solution in the leaf and seed procedures described above to cause less damage to the tissues and to "activate" the chlorine. Any procedures in which bleach and acid are mixed should be carried out in a chemical fume hood, not a sterile transfer hood, due to human toxicity of the chlorine gas.

15.4.2. Callus Culture Initiation

Once surface-sterilized seedlings or explants are available, these are cut into small pieces (0.5- to 1-cm segments or squares) with a sterile scalpel and placed on callus-inducing media. Hypocotyl and root segments are most commonly used. The edges of leaves and any cut surface that was exposed to bleach should be trimmed away and discarded, since these will not yield callus. The tissues will increase in size and should show visible callus formation within 1 month. Some tissues, such as older leaf and stem segments, may show callus development only at the cut edges, while younger tissues, such as roots, hypocotyls, and very young leaves, may completely dedifferentiate. If tissues curl and lose contact with the medium, they should be cut and pushed gently into the agar surface.

Much time can be saved by consulting the literature and finding medium formulations which have been previously developed for the species of interest (2, 15, 18, 43). If this information is not available, one should try media which have been successful with closely related species or try a number of different basal salt and hormone combinations. MS and B5 salts with 1 mg of 2,4-D per liter and SH medium with the two auxins and one cytokinin described above (Table 3) provide a wide range of media for dicots. While callus may form on several media, the texture and growth rate may vary greatly. For example, peanut callus on B5 salts with 1 mg of 2,4-D per liter is very soft and friable, while the callus after several subcultures on SH medium (Table 3) is extremely dense and hard. Monocots often require very high concentrations of auxins (up to 20 mg of 2,4-D per liter) for callus initiation but require reduced levels for extended rapid callus growth (2). If a particular cultivar or variety of a species is not required, several should be tried, since callus formation may vary with the genotype. Several plates of each type should be initiated, due to variations in callus formation response between individual explants and to allow for possible losses to contamination.

The callus should be subcultured every 2 weeks to 1 month by simply moving clumps of rapidly dividing cells to fresh plates with a cooled sterilized spatula. During the first transfers, it may be beneficial to cut any remaining intact explants to expose fresh tissue to the medium, or to break up large callus pieces. If sufficient material is available, the softest, most friable calli should be selected for subculturing.

15.4.3. Suspension Culture Initiation

Suspension cultures are initiated by inoculating liquid media in shake flasks with callus clumps. Approximately 1 to 2 in^3 of callus (5 to 15 g [fresh weight]) should be suspended in 50 ml of medium. Generally, the same medium used for callus culture is used for suspension cultures. Callus from a single explant may be used, or calli from several plates of the same type can be combined. Large clumps can be gently broken or cut into smaller pieces. Softer, friable, rapidly growing callus works best, since it will more quickly disperse into a fine suspension, whereas hard callus clumps may simply enlarge without dispersing. If only a small amount of callus is available, less medium should be used, since high liquid-to-cell ratios may inhibit growth.

The growth rate of suspension cultures will gradually increase, as faster-growing cells take over the culture. Initially, growth may be slow, and it may take weeks for the culture to double or triple in biomass. At this time, the culture should be divided into two new medium flasks. This is often done by simply swirling the culture and pouring cells and old medium into the new flasks. If the major cell aggregates are white but the medium is brown and contains much cell debris, it may be beneficial to pour off the old medium and transfer the cells with a spatula. As the growth rate increases, the cultures will have to be divided more frequently. Also, the aggregate size should decrease, and the ratio of old culture to new medium can be reduced.

A portion of the original callus should be reserved and subcultured. If the initial suspension does not grow well, after additional months of subculturing, a callus which was too hard to form good suspensions may improve in texture and can be tried again. After a good suspension culture is established, a portion of the cells may be returned to agar to form a callus culture again; this callus will usually grow faster than the original. If the shaker stops and/or the shaker room overheats, suspension cultures can be killed or may lose desirable properties such as the ability to produce secondary metabolites; healthy callus cultures serve as a back-up in case such problems arise.

15.4.4. Bioreactors and Scale-Up

The same principles regarding scale-up that apply to microbial or mammalian cell cultures also apply to the scale-up of plant cell cultures (7, 10, 23, 24, 28, 29, 39). Because aeration is usually not limiting for growth in shake flasks, scale-up to larger shake flasks (up to 4 liters) is often easily accomplished for biomass production, keeping the ratio of medium volume to flask volume constant.

Plant cell suspension cultures have been grown successfully in a wide range of bioreactor designs, but there are two important design considerations. First, plant cells tend to be very shear sensitive, and impeller damage can greatly impair the growth of a culture. Strategies to decrease impeller damage include decreasing the angle of the impeller, changing to a marine-style impeller, or using air-lift fermentors. By manipulating the hormones or medium components, the culture may be induced to form larger aggregates, which can be less shear-sensitive. Second, the cultures can be very heterogeneous. Large cell aggregates tend to settle to the bottom of the reactor vessels, while small aggregates may be carried out of the medium to stick to the reactor walls or float on foam, forming a "meringue" layer above the useful area of the reactor.

To reduce the problems of shear and cell floating or settling, many immobilization matrices have been tested. Cells have been embedded in beads of calcium alginate and carrageenan or entrapped in porous matrices such as foam rubber or polyester fabric (10, 25, 39). For cells which release products into the medium or carry out biotransformation reactions, immobilization allows the cells to be recycled. Unfortunately, new problems arise, such as decreased mass transfer to and from the cells and decreased volumetric productivity of the bioreactor, plus the added cost associated with immobilization.

Plant cell suspension cultures have been grown in industrial-scale fermentors. An extremely fast-growing cell line of tobacco (TBY-2) was grown in a 369-liter, then a 1,500-liter, and finally a specially designed 20,000-liter culture tank (23, 28). For the commercial production of shikonin, cells are first cultured in a 250-liter fermentor in a growth medium and then transferred to a production medium in a 750-liter fermentor (7).

15.5. GROWTH QUANTITATION

Many methods have been used to estimate the growth of plant cell cultures both directly and indirectly. The method chosen will depend on the requirements for maintaining the sterility of the culture and the speed and accuracy of the measurement. One must keep in mind that the doubling time for most plant cell cultures exceeds 12 h and that a typical culture experiment will continue for 1 to 4 weeks.

15.5.1. Fresh Weight

The growth of callus cultures can be measured by simply moving the callus mass to a preweighed sterile container and measuring the weight of the callus at regular intervals. The new agar medium plates can be weighed before and after inoculation at each subculture.

Cells can be collected from suspension cultures by filtration by using Miracloth (Calbiochem, San Diego, Calif.), nylon mesh filters (40- to 200-μm mesh), or sintered glass filters. Filter paper may also be used, but it may clog quickly due to the abundant cell debris or polysaccharides present in most cultures. Due to their extremely high water content, the cells will quickly become too dry, so the length of time the vacuum is applied should be closely monitored, either by timing the filtration or by stopping it as soon as the liquid stops dripping from the filter. The weight of cells collected in this manner is often referred to as "filtered fresh weight."

An alternative measurement is "wet fresh weight" or "drained fresh wet." By placing a pipette loosely against the bottom of a flask, much of the liquid medium can be withdrawn from a suspension culture while the majority of the cells will remain in the flask. The cell mass will retain a high percentage of medium (often about 50%), but this percentage will be fairly constant if the cell aggregate size remains constant. The wet fresh weight is calculated by subtracting the weight of the culture flask. This procedure can be carried out aseptically with cotton-plugged pipettes, preweighed flasks, and a clean top-loading digital balance in a laminar-flow transfer hood. These cells can then be weighed into new culture flasks, giving fairly reproducible inoculum levels without risking damage to the cells by filtering them dry. This measurement has also been adapted to immobilized cells, for which the weight of the medium-soaked immobilization matrix is subtracted from the weight of the inoculated, drained matrix.

15.5.2. Dry Weight

Collected fresh cells can be dried on preweighed filters or dishes to obtain the dry weight. Due to their high water content, the dry weight can be as low as 1 to 2% of the fresh weight. Cells should be dried at low temperatures (55 to 65°C), preferably either under vacuum or in a convection oven, to avoid being caramelized. Filter-collected cells should be washed briefly with distilled water to remove medium solids before being dried. Cells which are drained but not filtered can also be dried for a relative comparison of biomass, but the solids in the medium retained by the cells will contribute greatly to the dry weight. If a more accurate estimate of the dry weight is needed, the percentage of the wet fresh weight contributed by the medium, multiplied by the solids content (largely the dissolved sugars), can be subtracted from the dry weight (25).

Increases in dry weight can be misleading, particularly soon after cells are subcultured. Cell cultures of many species will quickly absorb the sugars from the medium and convert them to starch granules inside the cells. This starch reserve will then be slowly utilized during the growth of the culture. If a growth curve of the culture is determined, starch formation will cause the dry weight to increase dramatically for 1 to 2 days and then to decrease slowly, while the fresh weight is still increasing. (The presence of starch granules can be confirmed by examining cells under a high-power microscope; starch granules will polarize light and will stain blue with 1% iodine solution.) Thus, for initial studies, researchers will often monitor more than one parameter and then determine which is the best indictor of growth for the particular culture. Some cell lines will also concentrate and store substantial amounts of phosphate. Each new cell line must be evaluated for starch and phosphate accumulation, since some cell lines will accumulate neither, both, or only one of these substances.

15.5.3. Packed-Cell Volume

Cell suspensions can be transferred to calibrated conical screw-cap tubes and gently centrifuged at 200 to 500 × *g* for 10 min. The volume of cells and medium can be read directly from the tube.

15.5.4. Indirect Measurements

In some instances, the entire cell mass cannot be harvested. For example, in long-term bioreactor experiments or immobilized-cell culture systems, only a small sample of the culture may be obtained. If the suspension is homogeneous and a representative sample can be obtained, any of the above methods can be used to estimate the entire contents of a bioreactor. At other times, the biomass can be estimated indirectly by monitoring the consumption of nutrients. A correlation coefficient will have to be established for each cell line, usually from small-scale shake flask experiments. If the culture does not accumulate starch, the growth of the culture can be correlated with the disappearance of sugars from the medium. If sucrose is the carbon source, many cultures will release invertase, which hydrolyzes sucrose to fructose and glucose, and utilize the glucose before the fructose. In such cases, the combination of all three carbohydrates must be monitored by a method such as ion-exchange high-pressure liquid chromatography (HPLC) with refractive index detection. If glucose is used as the sole carbon source, a simple reducing sugar assay may be sufficient. If the culture does not accumulate phosphate reserves (immediately depleting the medium of phosphate), phosphate uptake can be inversely correlated with growth by using either colorimetric or chromatographic detection.

Nitrate is the major nitrogen source in most cell culture media. Very few cultures accumulate and store nitrate, and therefore depletion of nitrate from the medium is often well correlated with the growth of the culture. Several methods of nitrate quantitation have been developed, including colorimetric methods, HPLC, and capillary electrophoresis. Since nitrate (NO_3^-) is also the highest-molarity ion (40 mM in MS and 25 mM in B5), it is the major source of conductivity in the medium. It is often possible to correlate a decrease in conductivity with cell growth (11). For some experiments, the exact biomass is not required. For example, if the same inoculation and culture regimen is used each time, a certain conductivity value can be simply correlated with the correct time for harvest or elicitation.

An example of a typical plant suspension cell culture growth curve and the corresponding pattern of sugar hydrolysis and consumption in shake flasks is given in Fig. 1. The poppy cell line used in this experiment does not accumulate starch, and the total sugar consumption closely parallels the increases of dry and fresh weight.

15.5.5. Viability Assays

At times, the percentage of live cells in a culture must be determined. This is often needed when cells have been subjected to harsh treatments as part of an experiment, such as heat shock, chemical or pathogen addition, or freezing during the development of cryopreservation conditions. It is also useful when cells have been unintentionally stressed such as during a culture room malfunction. Callus cultures can be dead but may not brown or collapse for months.

Several assays have been used, depending on the equipment available and the texture of the tissue. The best assay for viability is to subculture the cells and watch for a visible increase in cell mass. Unfortunately, this may take several days or weeks. Assays in which the stain (such as fluorescein diacetate or triphenyltetrazolium chloride) is meant to be taken up only by living cells are known to sometimes give false-positive results with cells which no longer divide.

15.5.5.1. Cytoplasmic Streaming

A quick estimate of viability can be made by examining individual cells under a high-power microscope. The cytoplasm and some organelles can be seen circulating inside live cells. This procedure works well for fine cell suspensions but not for large aggregates or regenerating embryos and plantlets. If the researcher is unfamiliar with this phenomenon, both positive and negative controls (a sample of a known well-growing culture and a heat- or freezing-killed sample, respectively) should be used. False-positive results can come from mistaking Brownian motion for cytoplasmic streaming.

15.5.5.2. FDA Staining

The diester fluorescein diacetate (FDA) will not fluoresce. However, it can pass through the plant cell membrane, where esterases will remove the acetate groups. The

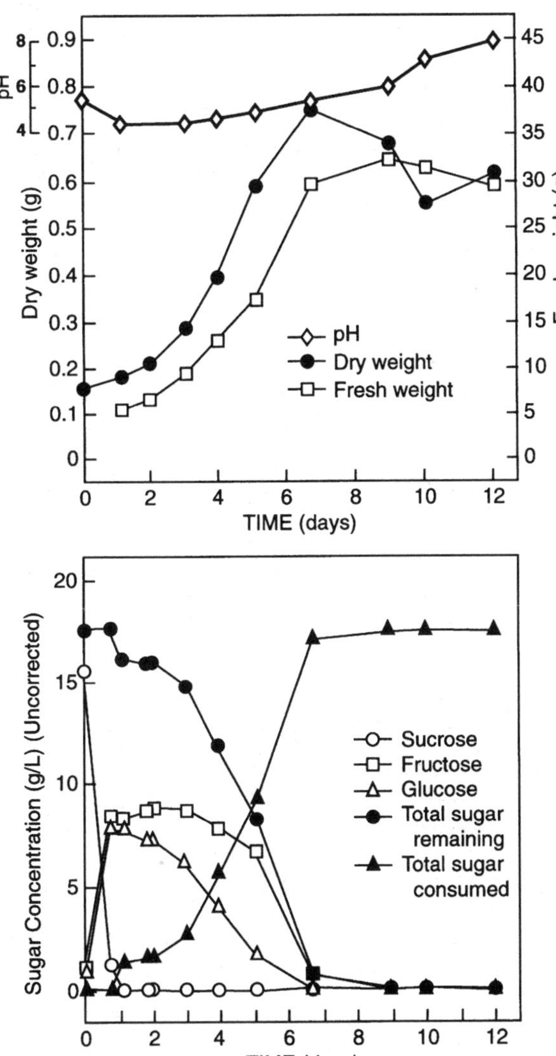

FIGURE 1 Growth of opium poppy cell suspension cultures in 250-ml shake flasks. The inoculum (*Papaver somniferum* L. cv. Marianne, PBI 2009) (25) was 10 g (wet fresh weight) (13%, wt/vol) per 75 ml of medium per 250-ml flask. Cells were cultured at 25°C and 150 rpm. The medium was 1B5C (B5 salts supplemented with 20 g of sucrose per liter, 1 mg of 2,4-D per ml, and 1 g of casein [NZ-Amine] per liter; pH 5.5). The initial sucrose level was less than 20 g/liter, due to dilution by sucrose-depleted medium in the inoculum.

charged product molecule is unable to leave the intact cell, due to a pH gradient, and fluoresces strongly. Only live cells should emit fluorescence, since dead cells will leak the stain or will not have active esterases. A 0.5% stock solution of FDA in anhydrous acetone can be stored at −20°C indefinitely. When needed, the stock is diluted 50-fold (final concentration, 0.01%) in ice-cold culture medium. One drop of FDA solution is mixed with one drop of culture on a microscope slide, covered with a coverslip, and monitored for 5 to 30 min. The slide is observed under a microscope with a UV/fluorescence attachment (excitation filter, 450 to 490 nm; barrier filter, 520 nm). If relevant, the total number of fluorescing cells can be counted and divided by the total number of cells (counted under visible light) to calculate the percent viability (46).

15.5.5.3. Evan's Blue Staining

The blue stain is able to enter dead cells but cannot penetrate intact cell membranes. A solution of Evan's blue (5 mg/ml) in culture medium is prepared. The cells are mixed with stain solution and incubated for 5 min, and then live (colorless) versus dead (blue) cells are counted under a microscope with bright-field illumination (47).

15.5.5.4. TTC Reaction

Viable cells will reduce triphenyltetrazolium chloride (TTC) to a pink formazan product, which can diffuse out of the cells and into the reaction medium. Viability can be assessed visually or quantitatively in a spectrophotometer. This procedure can be used for both fine and coarse cells and callus, as well as whole embryos or leaf pieces (3). A 0.6% solution of TTC in 0.5 M phosphate buffer (pH 7.0) is prepared. The tissue (100 mg of cells or plant tissue/3 ml of solution) is suspended and incubated overnight (15 h) at 30°C. The TTC solution is removed, the cells are washed with distilled water, and 7 ml of 95% ethanol is added. The mixture is incubated at 80°C for 5 min. The extract is cooled to room temperature, and the volume is adjusted to 10 ml with 95% ethanol. The absorbance at 530 nm is measured.

15.5.5.5. Phenosafranin Staining

Dead cells will be stained red with phenosafranin, but the dye cannot penetrate living cells, similar to Evan's blue. A 0.1% solution dissolved in culture medium is used, and the cells are observed under bright-field microscopy (11).

15.6. SECONDARY-METABOLITE PRODUCTION

One application of plant cell cultures is the study of plant secondary-metabolic biosynthesis and the commercial production of such valuable natural products. Plant secondary products have been used for centuries in human medicine, such as codeine and morphine (painkillers) from opium poppy (*Papaver somniferum*) latex, atropine and scopolamine (neuroregulatory), quinine (antimalarial) from *Chinchona* sp., and diosgenin and digoxin (cardioactive) from the foxglove plant (*Digitalis* sp.). Modern bioactivity-guided screens have identified a range of new important compounds, especially the anticancer compounds vincristine and vinblastine from periwinkle (*Catharanthus roseus*), taxol from the yew tree (*Taxus* sp.), and camptothecin (*Camptotheca acuminata*). Some medicinal plant parts are consumed whole or as a complex extract, such as ginseng root. There are also valuable food ingredients including coloring agents such as anthocyanins from many species and flavor extracts such as vanilla from the vanilla bean (a species of orchid). Fragrant oils such as mint, rose, and jasmine have always commanded high prices in the perfume industry. Some plant extracts are used as pesticides in organic gardening, including rotenone from tropical legumes, pyrethrins from the flowers of the pyrethrum daisy (*Chrysanthemum coccineum* and other species), and azadirachtin from the neem tree (*Azadirachta indica*).

Many studies have investigated the production of these and hundreds of other secondary metabolites in plant tissue culture (7, 10, 11, 18, 20, 24, 29, 30, 37, 39, 41, 42, 44). In some cases, cell cultures provide a source of homogeneous, highly active cells, ideal for studying the biosynthetic pathways by labeled-precursor feeding, enzyme purification, and cloning of the genes encoding pathway enzymes. Other studies have focused on increasing the productivity of cultures, with a goal of profitably producing these metabolites. Many of the above products are currently produced by growing the plants, often in tropical regions, and harvesting and processing the appropriate parts. This means that production is subject to seasonal and environmental variation, disease, competition with other crops and land uses, and political pressures such as environmental concerns or international embargoes. The growth of the source plant may be slow, such as yew trees (for taxol) or ginseng roots. Occasionally, cell cultures make high levels of one or a few compounds, which would not normally accumulate in the intact plant, decreasing downstream purification costs. For example, elicited *Papaver somniferum* cultures produce very high levels of sanguinarine, while the most competitive whole-plant system (*Macleaya cordata*) accumulates more sanguinarine but only as a complex mixture with other undesirable alkaloids, adding several expensive chromatographic steps to the purification.

In some cases, plant tissue cultures might appear to be an economic source of metabolites. Unfortunately, due to their slow growth and often low volumetric productivity, only extremely valuable compounds may be profitably produced; one analysis estimated that a product must be worth over $1,000 per kg, given average productivities and growth rates (41). Cell cultures are also often capable of degrading the desired products, and cell lines can suddenly lose their productivity, making industrial production difficult. Many strategies have been used to improve production, including selecting for higher-producing cell lines, developing production media, and inducing production by elicitation.

15.6.1. Constitutive Production and Strain Improvement

In certain species, media which allow good growth of the cultures will also allow good accumulation of secondary metabolites. In general, compounds which accumulate in root tissues or throughout all organs of a plant have been most easily produced at high levels in cell cultures. This may in part be due to the lack of light or the hormone regimes; it is often said that cell culture metabolism most closely resembles that of plant roots.

Production can vary greatly among cultures of the same species. When initiating cell cultures for secondary-metabolite production, many separate cultures should be initiated from individual seedlings, and plants or seedlings should be obtained from as many sources (companies, botanical gardens, and research laboratories) as possible. These should be evaluated separately to determine if there is a variation in production between individuals, cultivars, varieties, or other genotype designations.

In some cases, product accumulation decreases either rapidly or gradually as the culture is subcultured, possibly due to natural selection for faster-growing cells, especially in suspension (9, 12). This natural variation can also be used advantageously, in that occasionally a subpopulation of highly producing cells can be selected. Visual selection was successfully applied to the production of shikonin, a red alkaloid pigment with antibacterial activity. For initial rounds of selection, portions of callus which were darker red were picked by hand and subcultured separately; for later rounds of selection, quantitative chemical analysis was used (7). Fluorescence microscopy was used to select cell lines with high berberine production. The berberine content was increased from approximately 5% in unselected cells to 13.2% in a selected line (35).

15.6.2. Production Media

Many cell lines will accumulate only low levels of secondary metabolites in standard growth media, but production can be increased severalfold by altering one or more medium components. A wide range of production media have been developed, and a few trends have been established. In general, the growth of the culture may slow or stop completely. Production media often contain lower levels of growth regulators or none at all, or they contain weaker auxins such as NAA or IBA instead of 2,4-D. For anthocyanins (colored phenolic compounds), limiting the growth of the culture by restricting either the nitrate or phosphate supply while simultaneously providing very high concentrations of sugars has greatly improved production. Increasing the sucrose concentration from 20 to 80 g/liter and decreasing the nitrate concentration from 25 to 2.5 mM in B5 increases anthocyanin accumulation 10-fold (21), while phosphate limitation together with a high sucrose concentration is also successful (50). In the production of alkaloids (nitrogen-containing compounds), low phosphate levels are beneficial, but high nitrate levels increase accumulation in some cases and are inhibitory in others.

In many cases, fully dedifferentiated tissue cultures will not accumulate the same valuable product as the parent plant or will accumulate only small amounts, sometimes only when cells are allowed to differentiate back into shoots or roots (48). For example, thiophenes are light-activated insecticidal and antimicrobial compounds that are highly toxic to both plant pests and host plant cells; they are accumulated only in specialized veins running longitudinally inside the roots (24). For unknown reasons, opium alkaloids accumulate only in specialized lacticifers in the above-ground portions of plants, and production in tissue culture has been associated only with redifferentiation into shoots (26). Similar observations have been made for the famous anticancer compounds vincristine and vinblastine from *Cantharanthus roseus*. These molecules are heterodimers derived from the indole alkaloids catharanthine and vindoline. While high levels of catharanthine (5 to 20 times that in the plant) accumulate, no significant levels of vindoline, vincristine, or vinblastine have been found in homogeneous cell cultures, even though this has been one of the most thoroughly investigated systems (4). One of the enzymes in vindoline biosynthesis is light activated, while another may require the differentiation of functional chloroplasts, which are absent under standard cell culture conditions.

15.6.3. Elicitation

Many economically valuable plant secondary metabolites serve as defense compounds for the producing species, protecting them against attacks by insect herbivores and microbial pathogens. Some of these compounds are absent or present only at very low levels in healthy tissues, but they accumulate to very high levels in a narrow zone surrounding the site of microbial infections; these are referred to as phytoalexins (from the Greek for "plant warding-off agents"). In 1981, it was observed that contamination of a plant callus culture by a fungus resulted in the accumulation of a pigmented antifungal alkaloid (38). Later, it was determined that live fungus was not required to cause the response but that either autoclaved fungal culture filtrates or a number of purified compounds could "elicit" phytoalexin production and other defense responses, in both callus and suspension cultures. The process was referred to as "elicitation," because the regulatory mechanisms involved were not known at the time; it has subsequently been shown that in most cases, the increase in the production of secondary metabolites is the result of an increase in the transcription of the genes encoding the corresponding biosynthetic pathway enzymes.

Elicitors are classed as biotic or abiotic. Abiotic elicitors include heavy metals such as copper or mercury salts, and UV light. Biotic elicitors include many components of fungal cell walls (chitin, chitosan, β-glucan, peptidoglycans, and arachidonic acid), compounds excreted by plant pathogens (cellulase, ribonuclease, pectinase, antibiotics, various sugar polymers, and proteins), and compounds released from plant cell walls during degradation by plant pathogens (pectin fragments) (8). The type of elicitor that is most effective in inducing secondary metabolism in a particular plant cell culture varies widely. The elicitor does not have to be derived from a pathogen of the source plant. In general, members of the same plant genus will respond well to the same elicitors but members of the same family may not. For example, cultures of several legume species (alfalfa, soybean, and green bean) are highly elicited by a crude β-glucan derived from *Saccharomyces cerevisiae* (24, 38) but are unresponsive to chitin. Solanaceous species (tobacco, potato, and tomato) are all elicited by cellulases, arachidonic acid, and β-glucans, but while two members of the poppy family, *Papaver somniferum* and *Eschscholtzia californica*, are both elicited by β-glucans, only the former is responsive to chitin hydrolysates (25, 28).

In the *Eschscholtzia* system, in which elicitation was first studied, the control cultures accumulate a low level of one alkaloid. Following addition of the elicitor, the level of this alkaloid increases and five related alkaloids appear; the total benzophenanthridine alkaloid content of suspension cultures increases more than 20-fold compared to that in unelicited cultures (28). In some cultures, such as soybean and green bean, no pterocarpan phytoalexins are present before elicitor addition but reach maximum levels (13, 16) within 24 to 48 h after elicitation. In an attempt to use elicitation to stimulate the accumulation of opium (morphinan) alkaloids in *Papaver somniferum* cultures, the authors were surprised to induce instead the accumulation of sanguinarine, a member of a different (benzophenanthridine) class of alkaloids but one which is also inducible by pathogen attack on poppy plants (14).

The optimal time for addition of the elicitor is usually in the middle to late growth phase of the culture, often 3 to 7 days after subculture. Elicitor addition before or after this time results in little or no response from the plant cells. The optimal time may shift; for example, as a cell suspension is repeatedly subcultured and the net growth rate increases, the optimal time will become earlier by as much as 2 or 3 days. Elicitors are often effective when added to standard growth media, although combining a production medium with elicitation may result in even higher production levels (6). Many elicitors cause a browning of the cell cultures and may result in the death of a high percentage of the cells.

The first secondary metabolite commercially produced by plant cell cultures was shikonin, a pigmented antibacterial alkaloid which was normally extracted from the older (3- to 7-year old) roots of *Lithospermum erythrorhizon*. By using a combination of strain selection and production medium, the investigators found that the cultures could accumulate 12% dry weight of shikonin, versus 1 to 2% in older roots (7). While other products have come close to commercialization, in general, producers have defaulted to using field-grown plants as the source. The most recent promising candidate for tissue culture production is taxol, due to its high value and environmental concerns regarding the harvesting of wild trees. While plantation production and semisynthetic methods are currently the favored production routes, recent manipulations in explant sources and culture conditions have resulted in the accumulation of taxol in plant cell cultures at levels several times higher than that found in the tree bark (22).

15.7. REGENERATION

Even individual plant cells or protoplasts, if cultured properly, can regenerate into fully developed, fertile plants. This property is referred to as "totipotency" and was first demonstrated in plants in 1965 (15). In current applications, small explants (rather than individual cells) are first induced to form calli, which are then induced to regenerate into a number of plants. There are two modes of regeneration, organogenesis and somatic embryogenesis.

In organogenesis, leaf explants, callus cultures, or suspension cultures differentiate directly to form shoots, which can then be induced to form roots. Generally, the procedures involve induction of rapid cell division and/or callus formation on media containing high auxin levels, followed by a medium containing low auxin and high cytokinin levels to induce shoot formation. For example, for tobacco, young leaves are surface sterilized, cut into 1-cm squares, and placed onto MS agar containing 4.5 μM 2,4-D and 30 g of sucrose per liter for 2 to 4 weeks to induce callus formation. The explants are then transferred to MS agar supplemented with 5 μM BAP and placed under lights; shoots appear within a few weeks. Young shoots are cut off with a scalpel and rooted in MS with 20 g of sucrose per liter without hormones. Organogenesis is easily accomplished for most solanaceous species (tobacco, tomato, potato, petunia, etc.) and has been reported for many dicot species but is rarer among monocots.

In somatic embryogenesis, embryos or embryo-like structures are produced from somatic cells, as opposed to the maturation of zygotic cells in plant flowers. Somatic embryogenesis was first demonstrated for carrot cultures in 1964 but has since been demonstrated for many species, both dicots and monocots. For carrot (15, 46), callus cultures are simply initiated on MS agar supplemented with 4.5 μM 2,4-D and 30 g of sucrose per liter for 1 month; then a fine, rapidly growing suspension is initiated in the same medium without agar. After 2 to 4 weeks, the cells are washed with the same medium without 2,4-D and incubated in the dark to induce embryo formation. Embryos appear after 8 to 10 days and continue to increase in size for 2 weeks. The embryos are then transferred to MS agar medium (with sucrose and no hormones) for plantlet formation; this last step is referred to as conversion or germination of the embryos. In some systems, many embryos are formed but not all of them will germinate. The addition of abscisic acid may aid in the development of normal embryos, and supplementation of the medium with proline and glutamine can greatly increase embryogenesis.

15.8. MICROPROPAGATION

In both organogenesis and somatic embryogenesis systems, hundreds of plantlets can be generated from a single explant. This is one method by which plants can be multiplied for commercial purposes. Another method is simply establishing sterile shoot cultures of plants, from which axenic cuttings are made and rooted in culture; cuttings can be made at any time of the year and will usually grow more quickly under plant tissue conditions than in the field. Either method is referred to as micropropagation. Many houseplants and ornamentals are propagated in this manner, allowing producers to introduce new varieties in a short time. Crop species such as blueberries, strawberries, garlic, and potatoes are also propagated by micropropagation to produce virus-free plants. The regeneration step can also be used to induce secondary-metabolite accumulation (see the section on production media, above).

15.9. TRANSFORMATION

Genetic transformation of plant cells, and their subsequent regeneration into transgenic plants was first accomplished in the early 1980s for tobacco and is now routine for many species. Important crop species such as corn, wheat, rice, and soybean, which were deemed impossible to transform or regenerate only a few years ago, are now routinely manipulated. Usually, breakthroughs in transformation protocols of so-called recalcitrant species are first made with a highly regenerable cultivar. This transformed cultivar can then be crossed with elite high-producing cultivars to introduce the transgene into a number of plant lines.

In very early versions of plant transformation protocols, foreign DNA was introduced into plant protoplasts (plant cells treated with cellulase and other digestive enzymes to remove the cellulose cell walls) by using electroporation or polyethylene glycol (31). Since many plants are regenerated from protoplasts only with great difficulty, this method has not been widely applied. Most plant transformation has been carried out by *Agrobacterium tumefaciens*-mediated DNA delivery (32). Wild-type *A. tumefaciens* causes crown gall disease by transferring a segment of plasmid DNA (called the T-DNA, for transfer DNA) containing bacterial genes encoding auxin biosynthesis and other pathways into wounded plants, which in turn cause dedifferentiated callus tumors to form, releasing nutrients to the surrounding *A. tumefaciens* organisms in the soil. *A. tumefaciens* strains have been modified (disarmed) such that the auxin biosynthetic genes have been deleted so they can no longer cause disease but the strains still contain the *vir* genes required for T-DNA transfer. Foreign genes of interest, including a selectable marker, are inserted into binary vectors, such that the foreign genes are between the ends of the T-DNA region. When in contact with cultivated plant cells, usually at the beginning of the callus induction phase of either organogenesis or somatic embryogenesis, the disarmed *A. tumefaciens* strain transfers the foreign gene construct. Selection pressure is applied (usually an antibiotic or herbicide in the medium), so that the only cells which are able to grow are those which received the foreign T-DNA (including the selectable marker, usually a drug resistance or herbicide resistance gene). The *A. tumefaciens* strain is then killed by antibiotics (such as penicillin derivatives) which do not act against plant cells. The surviving transformed plant cells are then induced to regenerate into transgenic plants.

A. tumefaciens-mediated transformation has been used for many dicot species but has been less effective with monocots. A more recently developed DNA delivery system is called particle bombardment or the biolistic approach (33, 34). Early versions of biolistic equipment used the explosive force of gunpowder to propel the 1-μm tungsten or gold particles, coated with the desired foreign gene construct. The current version of the equipment (patented by Dupont but marketed through Bio-Rad) uses high-pressure helium gas and calibrated membranes (rupture disks) as a safer and more reproducible driving force. Transgenic plants have been produced by "shooting" leaf explants, embryogenic suspensions, or young embryos followed by applying drug or herbicide selection pressures, as described for the *A. tumefaciens* system, and then by regeneration. The biolistic approach was the first system to allow routine transformation of many monocot species such as corn, wheat (45), and rice (1), but it is also popular for dicot transformation (33).

Plant transformation and regeneration has resulted in the release and marketing of many genetically engineered plants. Current crops in the field include insect-resistant corn and cotton expressing the insecticidal *Bacillus thuringensis* toxin protein, herbicide-resistant soybeans that are resistant to glyphosate and bialophos, and ripening-delayed tomatoes. Many more products are under development, including plants with altered nutritional properties (e.g., high-lysine corn) and altered primary-metabolite (fatty acids, oil) and secondary-metabolite accumulation.

REFERENCES

1. **Aldemita, R. R., and T. K. Hodges.** 1996. *Agrobacterium tumefaciens*-mediated transformation of *japonica* and *indica* rice varieties. *Planta* **199:**612–617.
2. **Ammirato, P. V., D. A. Evans, W. R. Sharp, and Y. Yamada.** 1984. *Handbook of Plant Cell Culture, vol. 3. Crop Species.* Macmillan Publishing Co., New York.
3. **Bajaj, Y. P. S., and J. Reinert.** 1977. Cryobiology of plant cell cultures and establishment of gene banks, p. 757–777. *In* J. Reinert and Y. P. S. Bajaj (ed.), *Applied and Fundamental Aspects of Plant Cell, Tissue, and Organ Culture.* Springer-Verlag KG, Berlin.
4. **Balsevich, J., and G. Bishop.** 1989. Distribution of catharanthine, vindoline and 3′,4′-anhydrovinblastine in the aerial parts of some *Catharanthus roseus* plants and the significance thereof in relation to alkaloid production in cultured cells, p. 149–153. *In* W. G. W. Kurz (ed.), *Primary and Secondary Metabolism of Plant Cell Cultures II.* Springer-Verlag, New York.
5. **Burger, D. W.** 1988. Guidelines for autoclaving media used in plant tissue culture. *Hort. Sci.* **23:**1066–1068.
6. **Cline, S. D., and C. J. Coscia.** 1988. Stimulation of sanguinarine production by combined fungal elicitation and hormonal deprivation in cell suspension cultures of *Papaver bracteatum*. *Plant Physiol.* **86:**161–165.
7. **Curtin, M. E.** 1983. Harvesting profitable products from plant tissue culture. *Bio/Technology* **1:**649–657.

8. **Darvill, A. G., and P. Albersheim.** 1984. Phytoalexins and their elicitors—a defense against microbial infection in plants. *Annu. Rev. Plant Physiol.* **35:**243–275.
9. **Deus-Neumann B., and M. H. Zenk.** 1984. Instability of indole alkaloid production in *Catharanthus roseus* cell suspension cultures. *Planta Med.* **50:**427–431.
10. **DiCosmo, F., and M. Misawa.** 1996. *Plant Cell Culture Secondary Metabolism: Toward Industrial Application*. CRC Press, Inc., Boca Raton, Fla.
11. **Dixon, R. A., and R. A. Gonzales.** 1994. *Plant Cell Culture. A Practical Approach*. Oxford University Press, Oxford, United Kingdom.
12. **Dougall, D. K., and D. L. Vogelien.** 1987. The stability of accumulated anthocyanin in suspension cultures of the parental line and high and low accumulating subclones of wild carrot. *Plant Cell Tissue Org. Cult.* **8:**113–123.
13. **Ebel, J., A. R. Ayers, and P. Albersheim.** 1976. Host-pathogen interactions XII. Response of suspension-cultured soybean cells to the elicitor isolated from *Phytophthora megasperma* var. *sojae*, a fungal pathogen of soybeans. *Plant Physiol.* **57:**775–779.
14. **Eilert, U., W. G. W. Kurz, and F. Constabel.** 1984. Stimulation of sanguinarine accumulation in *Papaver somniferum* cell cultures by fungal elicitors. *J. Plant Physiol.* **119:** 65–76.
15. **Evans, D. A., W. R. Sharp, P. V. Ammirato, and Y. Yamada.** 1983. *Handbook of Plant Cell Culture*, vol. 1. *Techniques for Propagation and Breeding*. Macmillin Publishing Co., New York.
16. **Funk, C., K. Gügler, and P. Brodelius.** 1987. Increased secondary product formation in plant cell suspension cultures after treatment with a yeast carbohydrate preparation (elicitor). *Phytochemistry* **26:**401–405.
17. **Gamborg, O. L., R. A. Miller, and K. Ojima.** 1968. Nutrient requirements of suspension cultures of soybean root cells. *Exp. Cell Res.* **50:**151–158.
18. **George, E. F., D. J. M. Puttock, and H. J. George.** 1987. *Plant Culture Media*, vol. 1. *Formulations and Uses*. Exegetics Ltd., Edington, United Kingdom.
19. **Goodwin, T. W., and E. I. Mercer.** 1983. *Introduction to Plant Biochemistry*. Pergamon Press, Oxford, United Kingdom.
20. **Hay, C. A., L. A. Anderson, M. F. Roberts, and J. D. Phillipson.** 1988. Alkaloid production by plant cell cultures, p. 97–140. *In* A. Mizrahi (ed.), *Biotechnology in Agriculture*. Alan R. Liss, Inc., New York, N.Y.
21. **Hirasuna, T. J., M. L. Shuler, V. K. Lackney and R. M. Spanswick.** 1991. Enhanced anthocyanin production in grape cell cultures. *Plant Sci.* **78:**107–120.
22. **Hirasuna, T. J., L. J. Pestchanker, V. Srinivasan, and M. L. Shuler.** 1996. Taxol production in suspension cultures of *Taxus baccata*. *Plant Cell Tissue Org. Cult.* **44:**95–102.
23. **Kato, A., S. Kawazoe, M. Iijima, and Y. Shimizu.** 1976. Continuous culture of tobacco cells. *F. Ferment. Technol.* **54:**82–87.
24. **Kurz, W. G. W.** 1989. *Primary and Secondary Metabolism of Plant Cell Cultures II*. Springer-Verlag, New York.
25. **Kurz, W. G. W., N. L. Paiva, R. T. Tyler.** 1990. Biosynthesis of sanguinarine by elicitation of surface-immonilized cells of *Papaver somniferum* L., p. 682–688. *In* H. J. J. Nikamp, L. H. W. VanDerPlas, and J. Van Aartrijk (ed.), *Progress in Plant Cellular and Molecular Biology*. Kluwer, Boston.
26. **Kutchan, T. M., S. Ayabe, R. J. Krueger, E. M. Coscia, and C. J. Coscia.** 1983. Cytodifferentiation and alkaloid accumulation in cultured cells of *Papaver bracteatum*. *Plant Cell Rep.* **2:**281–284.
27. **Murashige, T., and F. Skoog.** 1962. A revised medium for rapid growth and bioassays with tobacco tissue cultures. *Physiol. Plant.* **15:**473–497.
28. **Nagata, T., Y. Nemoto, and S. Hesazawa.** 1992. Tobacco BY-2 cell line as the "HeLa" cell in the cell biology of higher plants. *Int. Rev. Cytol.* **132:**1–30.
29. **Neumann, K.-H.,W. Barz, and E. Reinhard.** 1985. *Primary and Secondary Metabolism of Plant Cell Cultures*. Springer-Verlag KG, Berlin.
30. **Parr, A. J.** 1988. Secondary products from plant cell culture, p. 1–34. *In* A. Mizrahi (ed.), *Biotechnology in Agriculture*. Alan R. Liss, Inc., New York.
31. **Paszkowski, J., and M. W. Saul.** 1986. Direct gene transfer to plants. *Methods Enzymol.* **118:**668–684.
32. **Rogers, S. G., R. B. Horsch, and R. T. Fraley.** 1986. Gene transfer in plants: production of transformed plants using Ti plasmid vectors. *Methods Enzymol.* **118:**627–640.
33. **Russell, D. R., K. M. Wallace, J. H. Bathe, B. J. Martinell, and D. E. McCabe.** 1993. Stable transformation of *Phaseolus vulgaris* via electric-discharge mediated particle acceleration. *Plant Cell Rep.* **12:**165–169.
34. **Sanford, J. C., F. D. Smith, and J. A. Russell.** 1993. Optimizing the biolistic process for different biological applications. *Methods Enzymol.* **217:**483–509.
35. **Sato, F., and Y. Yamada.** 1984. High berberine-producing cultures of *Coptis japonica* cells. *Phytochemistry* **23:**281–285.
36. **Schenk, R. U., and A. C. Hildebrandt.** 1972. Medium and techniques for induction and growth of monocotyledonous and dicotyledonous plant cell cultures. *Can. J. Bot.* **50:**199–204.
37. **Schripsema, J., and R. Verpoorte.** 1995. *Primary and Secondary Metabolism of Plants and Cell Cultures III*. Kluwer Academic Publishers, Dordrecht, The Netherlands.
38. **Schumacher, H.-M., H. Gundlach, F. Fiedler, and M. H. Zenk.** 1987. Elicitation of benzophenanthridine alkaloid synthesis in *Eschscholtzia* cell cultures. *Plant Cell Rep.* **6:** 410–413.
39. **Shargool, P. D., and T. T. Ngo (ed.).** 1994. *Biotechnological Applications of Plant Cultures*. CRC Press, Inc., Boca Raton, Fla.
40. **Shneyour Y., A. Zelcer, S. Izhar, and J. S. Beckmann.** 1984. A simple feeder-layer technique for the plating of plant cells and protoplasts at low densities. *Plant Sci. Lett.* **33:**293–302.
41. **Staba, E. J.** 1980. *Plant Tissue Culture as a Source of Biochemicals*. CRC Press, Boca Raton, Fla.
42. **Staba, E. J.** 1985. Milestones in plant tissue culture systems for the production of secondary products. *J. Nat. Prod.* **48:**203–209.
43. **Thorpe, T. A.** 1981. *Plant Tissue Culture. Methods and Applications in Agriculture*. Academic Press, Inc., New York.
44. **Trigiano, R. N., and D. J. Gray.** 1996. *Plant Tissue Culture Concepts and Laboratory Exercises*. CRC Press, Inc., Boca Raton, Fla.

45. **Weeks, J. T., O. D. Anderson, and A. E. Blechl.** 1993. Rapid production of multiple independent lines of fertile transgenic wheat (*Triticum aestivum*). *Plant Physiol.* **102:** 1077–1084.
46. **Wetter, L. R., and F. Constabel (ed.).** 1982. *Plant Tissue Culture Methods.*, 2nd ed. National Research Council of Canada, Prairie Regional Laboratories, Saskatoon, Saskatchewan, Canada.
47. **Widholm, J. M.** 1972. The use of fluorescein diacetate and phenosafranine for determining viability of cultured plant cells. *Stain Technol.* **47:**189–194.
48. **Wierman, R.** 1981. Secondary plant products and cell and tissue differentiation, p. 85–116. *In* E. E. Conn (ed.), *The Biochemistry of Plants*, vol. 7. Academic Press, Inc., New York.
49. **White, P. R.** 1937. Vitamin B1 in the nutrition of excised tomato roots. *Plant Physiol.* **12:**803–811.
50. **Yamakawa, T., S. Kato, K. Ishida, T. Kodama, and Y. Minoda.** 1983. Formation and identification of anthocyanins in cultures cells of *Vitis* sp. *Agric. Biol. Chem.* **47:** 997–1001.

Insect Cell Cultures

TERRY K. NG

16

16.1. CULTURE TECHNOLOGIES FOR INSECT CELLS

Insect cell and baculovirus expression vector systems have been widely used as powerful heterologous protein expression systems for vaccines and therapeutic applications. Potentially, insect cell and baculovirus expression vector systems can also be used to produce occluded viruses as biopesticides. A commonly used cell line, Sf-9, can be obtained from American Type Culture Collection, Rockville, Md. Sf-9 was cloned from the parent line, IPLB-Sf 21 AE, which originally was derived from pupal ovarian tissues of the fall armyworm *Spodoptera frugiperda*. This cell line is highly susceptible to infection by *Autographa californica* multinuclear polyhedrosis virus (AcMNPV) and other baculoviruses and can be used with baculovirus expression vector systems. Recently, BTI-Tn-5B1-4 cells derived from *Trichoplusia ni* egg cell homogenate have been introduced (Invitrogen Corp., Carlsbad, Calif.) and may be capable of expressing higher levels of secreted recombinant proteins than other insect cell lines. The need for large-scale productions has led to the commonplace adaptation of free suspension cultures with the use of shear protective agents such as Pluronic Polyol F-68 (BASF Corp., Parsippany, N.J.) at 0.1 to 0.3% (17, 23) and methylcellulose at 0.5% (wt/vol) and dextran at 4.5% (wt/vol) (8). Scaling up from frozen cell banks to large-scale bioreactors can be achieved easily by expanding passages in shakers, spinners, and small bioreactors to large bioreactors. It is advantageous to prepare cell banks already adapted in free suspension grown in serum-free cultures. Recently, advances have been made in serum-free media and fed-batch culture techniques that allow the scale-up of large quantities of recombinant therapeutic proteins. The economics of scale play a still larger part in the production of baculovirus as biopesticides in bioreactor volumes larger than 10,000 liters.

16.1.1. Flasks and Roller Bottles

Monolayer cultures of Sf-9, Sf-21, and BTI-Tn-5B1-4 cells may be cultivated in T flasks or roller bottles at a density of 2.0 to 5.0 $\times$ 10^4 cells/cm^2 in TNM-FH medium supplemented with 10% fetal bovine serum. Cultures grown to 80 to 100% confluence should be subcultured in 2 to 4 days. Logarithmically grown cells of greater than 98% viability are desirable for cell propagation and optimal protein expression. Cell viability can be determined by the trypan blue exclusion test (30) and cell counts by a hemocytometer with an inverted microscope or a pulse counter such as a Coulter counter (Hialeah, Fla.). The use of trypsin to dislodge attached cells from flask surfaces is not recommended. Attached cells may be removed by carefully scraping with pipettes or cell scrapers. Flask and bottle caps may be loosened to facilitate oxygen exchange, although sterility risks are greater. Gas-permeable membrane vented caps are also available if desired. The scale-up of monolayers is limited by the surface available. Pleated rollers and multiplate systems such as the Nunc Cell Factory (Naperville, Ill.) and Costar Cell Cubes (Cambridge, Mass.) may be used. However, these systems are more often used for the culture of anchorage-dependent animal cell lines without essentially changing the hydrodynamic aspects of the culture environments.

16.1.2. Shaker and Spinner Flasks

Shaker flasks have been widely used in scale-up to mass-produce insect cells in suspension. The volume of the medium and the volume of the flasks are important factors for the optimal transfer of oxygen in shakers and spinners. Shakers from 125 to 1,000 ml should be filled with 25 to 40% of the total volume. Rotational speeds of 120 to 150 rpm are commonly used for orbital shakers. Spinner flasks of similar volumes are agitated at 80 to 100 rpm. The use of larger spinners and shaker flasks is not recommended without sparging or oxygen supplementation, since oxygen limitation is quickly reached for culture volumes above 1 to 2 liters, depending on the cell density. The addition of air flow to the headspace of spinners has been shown to improve Sf-9 cell growth and recombinant β-galactosidase expression (16) by five-fold over nonaerated cultures. Cell suspensions may be diluted to 2.0 to 3.0 $\times$ 10^5 viable cells/ml in serum-free and serum-supplemented Sf-9 or Sf-21 cultures. Expansion or subculturing may be carried out in 3 to 4 days at 27°C without CO_2 when a density of 3.0 to 5.0 $\times$ 10^6 viable cells/ml is reached. Significant cell death is usually observed after 4 or 5 days, with deleterious effects on the culture.

It is advantageous to prepare frozen cell banks already adapted to suspension. Godwin et al. (7) found that cells frozen in serum-free medium are easily damaged. However, Vaughn and Fan (33) managed to recover Sf-21 cells frozen in serum-free medium by adding commercially prepared lipid supplements. Cell suspensions may also be frozen without serum in 5 to 10% dimethyl sulfoxide, 0.5 to 1.0% bovine serum albumin, 30 to 50% conditioned medium and the balance fresh medium. The use of bovine serum albumin is optional since it is a carrier of lipids. However, the issue of regulatory concerns regarding bovine spongiform encephalopathy may be a point of contention for human and veterinarian biopharmaceutical production. Cells grown in serum-containing medium may be adapted to serum-free medium by serially subculturing in a stepwise fashion by lowering the serum content. Multiple passages for serum-free adaptation and a higher inoculation density of 3.0 to 5.0 × 10^5 cells/ml may be required. Clumped cells are often observed in Sf-9 cultures (9) revived from the frozen state or when grown in serum-free medium in a low-shear environment. Similarly, BTI-Tn-5B1-4 cells grown in suspension in serum-free medium tend to clump as well. The effect of clumping on cell growth and recombinant protein expression is not entirely clear. Aggregation could complicate the infection process and protein production owing to nutrient mass transfer limitations. A majority of singlets and doublets may be selected for subculturing by removing the cell suspension from the top after allowing clumps to settle to the bottom. Shear may be applied by pipetting against the bottom of the recipient flask to break up clumped cells. No significant cell viability should be lost by using this simple maneuver. Recently, dextran sulfate and other sulfated polyanions have been used effectively by Dee et al. (6) to maintain single-cell suspension cultures of BTI-Tn-5B1-4 cells without loss of recombinant alkaline phosphatase and β-galactosidase productivity.

16.1.3. Stirred Tank Reactors

Insect cell scale-up and bioreactor design in relation to protein expression systems has been reviewed (1, 18). The issues of cell density, protein expression level, product quality, and downstream purification have been discussed (28) with reference to production cost. The choice of production system—stirred tank reactors, airlift fermentors, or packed-bed reactors—should be based on overall product yield and productivity. Although very high cell densities (>1 × 10^7 cells/ml) can be achieved in stirred tanks and special reactors such as perfusion (5) and packed-bed reactors, Shuler et al. (28) observed low protein productivity when cultures were infected at high cell densities. Suspension cultures grown in stirred tanks are not subjected to contact inhibition. Protein productivities per cell were observed to have doubled in Sf-21 and BTI-Tn-5B1-4 cultures supplemented with oxygen and glutamine in serum-free medium such as ExCell 400 (JRH Biosciences, Lenexa, Kans.).

Stirred tanks, already widely used in the biologics and pharmaceutical industries, can be used for insect cell cultures with minimal adaptation (27). Rushton impellers, commonly used in fermentors, can be replaced with low-shear marine ABEC (Bethlehem, Pa.) LS, or Lightning (Kansas City, Mo.) A315 impellers. The aeration system can be modified with flow-metering mass controllers and pure oxygen enrichment using dissolved-oxygen proportional-integral-derivative control. The culture may be sparged with air at a rate of 5 to 10 ml/min/liter of culture and maintained at 30 to 70% of air saturation. This minimizes cell damage caused by high-volumetric gas sparging at the sparger and the bursting of bubbles at the gas-liquid interfaces (13). High cell density will require supplementation with pure oxygen. Caution should be exercised to avoid the accumulation of CO_2 when stripping by air flow is limited. Flushing the bioreactor headspace was demonstrated to alleviate the adverse effect of CO_2 accumulation in SF-9 and BTI-Tn-5B1-4 cells in a 3-liter bioreactor (19). Culture pH should be controlled to between 6.2 and 6.5 with 0.5 to 1.0 M sodium hydroxide or hydrochloric acid, although insect cell culture media such as Ex-Cell 401 (JRH Biosciences), Express Five, and Sf-900 II SFM (Life Technologies, Grand Island, N.Y.) are fairly well buffered. The optimal temperature for cell propagation in shakers or bioreactors is 27°C. During sterile operations, care should be taken when steaming connections to industrial bioreactors, especially those connections located below the vessel. Substantial deviation from the optimal temperature may result in loss of cell and protein yields.

16.1.4. Airlift Fermentors

The effect of air flow and sparging on Sf-21 cells grown in TNM-FH medium supplemental with 10% fetal bovine serum was investigated by Tramper et al. (31) using a bubble column reactor with a height of 0.18 m and inside diameter of 0.035 m. It was found that the death kinetics were proportional to the air flow rate. The shear stress of cells adhered to the bubble-liquid interface and bursting bubbles was estimated to be about 2 orders of magnitude higher than the critical value for cell viability. This may explain the unsuccessful use of airlift fermentors to grow insect cells early on. However, with the use of Pluronic Polyol F-68 at 0.1% as a shear protectant, Maiorella et al. (17) achieved a cell density of greater than 5 × 10^6 cells/ml in a 2-liter airlift fermentor. A gas sparging rate of 0.02 to 0.06 VVM (volume of gas per volume of culture per minute) was used. The sparger orifice produced a bubble diameter of 0.5 to 1.0 cm. It was suggested that gas bubbles in this range rise without changing the size and therefore the velocity. Cell damage due to shear stress with bubble interactions is minimized. The difference between an airlift fermentor and a bubble column is that an airlift fermentor contains an inner draft tube where the bubbles rise and disengage at the surface. The difference in the liquid densities of the outside (riser) and the inner core (downcommer) causes gentle recirculation of the culture fluid. The constant sparging rate, which controls the recirculation rate, can be controlled with either oxygen or nitrogen supplementation, depending on the cell density and culture oxygen demand (14). The gentle and largely homogeneous bubbling environment offered in an airlift fermentor may be an advantage over a sparged bioreactor, in which shear stress is created at the sparger and the tips of the stirrer.

16.2. FED-BATCH CULTURES

With the growing prospects of using insect cell–baculovirus systems for human therapeutic proteins and the commodity-like economic requirements for the production of occlusion bodies as biopesticides (22), the need for increasing scale is imminent. The development of high cell density and high-level protein expression systems is an alternative to shear increase in the size of bioreactors. It was initially observed by Caron et al. (4) that the expression level of recombinant bovine rotavirus capsid protein, VP6, was lowered when Sf-9 cultures were infected by Bac-BRV6L baculovirus at high cell densities in batch cultures. This process was reversible by culture medium renewal after infection, indicating possible nutrient deficiency or toxic metabolite accumulation in the spent medium at high cell densities. More recently, advances have been made in high-density fed-batch cultures by using sophisticated media designs and feeding schemes beyond the addition of carbon and energy substrates such as glucose and glutamine (35). The formulation of nutrient supplements will have to be optimized for both cell growth kinetics and protein expression levels, since the multiplicity of infection (MOI), infection time, and cell density difference may lead to completely different responses. Cell growth and infection kinetics should be optimized to give the highest protein expression or occluded virus production for pesticide manufacturing. For the production of recombinant human nerve growth factor (rhNGF) using Sf-9 cells infected with recombinant AcMNPV, Nguyen et al. (24) tested the multiple feeding of glucose, glutamine, yeastolate, and lipid concentrates. Sf-9 cell growth data showed that glucose and glutamine were required at a certain level to achieve high cell densities. However, the addition of glucose, glutamine, and lipids to growing cells had no effect on cell growth or rhNGF yields. The feeding of yeastolate, however, provided significant increases in both cell density and protein expression. It was suggested that amino acids and vitamin B complex may have been limiting. Feeding of lipids with yeastolate, however, did not increase the cell density further but yielded significantly more rhNGF. A more systematic factorial approach was examined by Bedard et al. (3) on the effect of a one-time addition to late-log-phase Sf-9 cells growing in Sf-900 II serum-free medium of four solutions of amino acids, vitamin-salts, yeastolate, or lipids (see Tables 1 and 2 and protocol, below) on the production of recombinant AcMNPV-expressing *Escherichia coli* β-galactosidase. The low specific protein yields in late-log-phase infected cultures were reversed by the addition of ultrafiltered yeastolate and a mixture of amino acids. The results suggest that it is the limitation of one or more of the nutrients rather than the accumulation of toxic metabolites that impedes high recombinant protein production in high-density Sf-9 cultures.

TABLE 1 Composition of amino acid solution

Amino acid	Concentration (mM)
L-Arginine	40.2
L-Asparagine	26.5
L-Aspartic acid	26.3
L-Glutamic acid	40.8
L-Glutamine	68.4
L-Glycine	86.6
L-Histidine	12.9
L-Isoleucine	57.2
L-Leucine	19.1
L-Lysine	34.2
L-Methionine	3.4
L-Phenylalanine	9.1
L-Proline	30.4
DL-Serine	104.7
L-Threonine	14.7
L-Tryptophan	4.9
L-Valine	8.5
L-Cysteine·HCl·H_2O	4.3

TABLE 2 Composition of vitamin-salts solution[a]

Component	Concentration (mM)
Solution 1	
Thiamine·HCl·1/2H_2O	0.237
Riboflavin	0.213
D-Calcium pantothenate	0.181
Pyridoxine-HCl	1.945
para-Aminobenzoic acid	2.334
Nicotinic acid	1.300
i-Inositol	2.220
Biotin	0.655
Choline chloride	143.000
Vitamin B_{12}	0.177
Solution 2	
Folic acid	0.181
Solution 3	
Molybdic acid, ammonium salt	0.035
Cobalt chloride·6H_2O	0.210
Cupric chloride·2H_2O	0.117
Manganese chloride·4H_2O	0.104
Zinc chloride	0.294
Solution 4	
Ferrous sulfate·7H_2O	1.980
Aspartic acid	2.675

[a]The vitamin-salt solution is prepared immediately before addition to cultures by mixing together equal amounts of solutions 1 through 4. Reprinted from reference 3 with permission.

PROTOCOL

PREPARATION OF LIPID EMULSION

Lipid mix

Cod liver oil	10 mg
Tween 80	25 mg
Cholesterol	4.5 mg
α-Tocopherol	2 mg
Ethanol	1 ml

Pluronic Polyol solution

10% Pluronic Polyol F-68 in water 10 ml

Vortex to emulsify. Add to 1 liter of basal medium. Lipid mix and aqueous Pluronic Polyol solution are individually sterile-filtered (0.2 μm). Emulsified lipid supplement is then prepared by slow dropwise addition of aqueous Pluronic Polyol solution to the organic lipid mixture at 37°C while vortexing. The mix will initially turn cloudy but will clear as the addition of Pluronic Polyol is completed. (Protocol from reference 17 is used with permission.)

In a robust manufacturing environment, it is advantageous to simply add nutrient supplements to bioreactors and reduce the chance of contamination. This negates the need for other complex and costly procedures for medium changes and cold room storage that are commonly associated with perfusion cultures (17) and packed-bed bioreactors. The supplements may be added using sterile filtration at the point of use by spiking or adding continuously or semicontinuously in multiple steps. The best feeding schemes may be devised by supplementing various limiting components to maintain certain optimal levels at various times of the culture. The dynamic interactions of cell and virus metabolism should be taken into account. The accumulation of alanine and ammonia can reach high levels that may be inhibitory in fed-batch cultures of BTI-Tn-5B1-4 cells. The levels of nutrients can be monitored with the aid of a flow injection system or near-infrared spectroscopy. The concentrations of a number of sugars and amino acids should be controlled to minimize substrate inhibition and maximize cell production or protein expression.

16.3. MOI AND INFECTIVITY

High MOI, up to 100, has been used for infection of insect cell cultures. A substantial amount of seed virus is needed to infect high-cell-density cultures, especially when fed-batch feeding techniques are used. A typical seed virus titered at 10^8 PFU/ml would require a 30% volumetric inoculation with an MOI of 10 for a culture reaching a density of 3×10^6 cells/ml. High carryover of spent medium would not be beneficial to an infected culture. With the use of the BTI-Tn-5B1-4 cell line as a high recombinant protein expression system, the ability to produce high-titer seed virus preparations may pose a problem. Anderson et al. (2) found that recombinant bMON14272 DNA containing the *uidA* gene encoding glucuronidase transfected into Sf-9 and Sf-21 cells yielded virus titers of 2.0×10^7 to 3.0×10^7 PFU/ml, whereas the titers of BTI-Tn-5B1-4 cells were 10- to 50-fold lower. Therefore, infecting high-density cultures at high MOI may pose problems in scaling up the virus seed train, since a large amount of seed will have to be used. Concentrating the seed virus by ultrafiltration may be required. Alternatively, the seed virus can be produced in either Sf-9 or Sf-21 cells at high titers, although the cell debris and cellular DNA carryover may have to be removed by purification or validated downstream.

Many studies have been carried out to examine the effect of MOI on protein expression. Licari and Bailey (15) showed that the expression of recombinant β-galactosidase in Sf-9 cells infected with baculovirus (AcNPV) during mid- and late-exponential phases has a logarithmic relationship to MOI from 0.1 to 100 (Fig. 1). However, the protein yields are relatively constant if the cells are infected at early exponential phase, with a decreasing trend toward higher MOI. Therefore, it is logical to infect young populations with a low MOI from the scale-up point of view. This method allows the secretion of nonoccluded viruses for secondary infection. It is also critical that nutrient depletion does not occur. Nutrients may be supplemented by using fed-batch feeding techniques as described in section 16.2.

To distinguish between virus infectivity and cell death due to cytotoxicity or other adverse environmental conditions, indirect immunofluorescence assays can be used. Infected cells can be fixed onto a slide and stained with fluorescein isothiocyanate, conjugated antiserum (29), or monoclonal antibody against baculovirus Gp64 envelope fusion protein, Gp64 EFP. Fluorescence can be viewed with a fluorescence microscope equipped with a mercury vapor or halogen lamp and an exciting filter system. The hydrophobic cluster, i.e., region I of the Gp64 EFP gene, plays a major role in fusion activities of baculovirus (34). Gp64 EFP has also been shown to be responsible for virus entry into the cell via receptor-mediated endocytosis (10). A neutralizing antibody generated against a synthetic 21-amino-acid peptide of region I was shown to reduce virus infectivity. Quantitatively, cell infectivity can be measured by enzyme-linked immunoabsorbent assay, utilizing the antibody against Gp64 EFP (20). Cell surface Gp64EFP can be quantified by monoclonal antibody MabAcV5 conjugated to goat anti-mouse immunoglobulin G and a chromogenic substrate. The assay can be developed in 96-well plates for infected cell exfoliation. Infected and uninfected insect cells in a population can also be separated and further analyzed using flow cytometry by tagging infected cells with a murine monoclonal antibody against the Gp64 protein (21). These analyses can reduce the amount of time for viral quantification usually associated with baculovirus plaque assays (36).

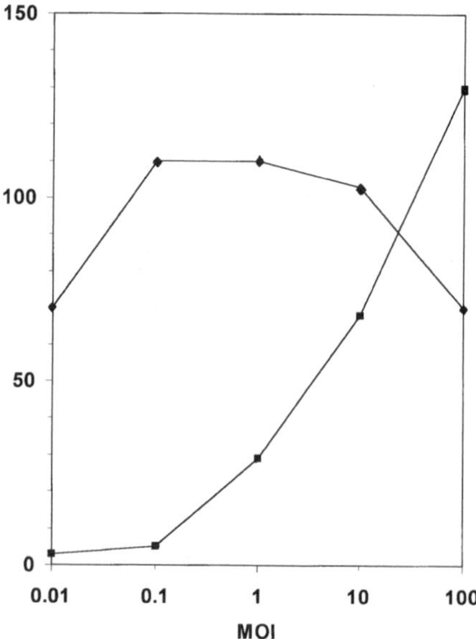

FIGURE 1 The production of β-galactosidase (activity units per ml of culture) as a function of the MOI on a logarithmic scale. Symbols: ♦, cells infected in the early-exponential-growth phase (4.3×10^5 cells/ml); ■, cells infected in the mid- to late-exponential-growth phase (1.1×10^6 cells/ml). (From reference 15 with permission.)

16.4. RECOVERY OF INSECT CELLS

The recovery of recombinant proteins by downstream purification will not be discussed in detail here since the recoveries are highly variable depending on the proteins in question. However, the recovery of mammalian or insect cells may be required for scale-up or as the first step in downstream purification in the high-level production of in-

TABLE 3. Medium replacement by concentrating BTI-Tn-5B1-4 cells using a Membrex axially rotating filter

Rotor speed (rpm)	Recirculation rate (liter/min)	Starting cell concentration (cells/ml)	Starting cell viability (%)	Ending cell viability (%)
1,500	4	1.0×10^6	98.0	82.0
1,000	4	1.0×10^6	98.2	95.8
750	4	1.0×10^6	98.0	97.0

tracellular or membrane-associated recombinant proteins. The recovery of insect cells using tangential flow membrane microfiltration is a viable alternative to centrifugation at low *g* forces if sterility is desired for cell scale-up. Reuveny et al. (25) showed that the specific protein production is more efficient if infection is performed after medium replacement. Sf-9 cells grown in serum-free medium were collected by polysulfone hollow-fiber cross-flow microfiltration (AG Technology, Needham, Mass.) using 0.75-mm-diameter lumens, 0.45-μm pore size, and a cross-flow shear rate of 14,000 s^{-1}. A high permeate flux rate of 500 liters/m^2/h was obtained at a cell concentration of 2.5×10^6 cells/ml without loss of cell viability (32). The operating parameters for permeate flux optimization are temperature, transmembrane pressures, and tangential velocities. A significant drop in fluxes may be encountered if cells are grown in medium supplemented with serum. Polysulfone hollow-fiber cartridges can be autoclaved or steam-sterilized for sterile medium replacement operations, although the pore size may be altered slightly after steaming or autoclaving. The slight change in pore size for microfiltration is immaterial since insect cells are sized at 15 to 20 μm.

Similar concentration operations for trypsinized animal cells or insect cell suspensions can be performed by using an axially rotating filtration device (12), e.g., a Membrex Pacesetter (Fairfield, N.J.), equipped with a steam-sterilizable stainless steel screen of 3- to 5-μm pore size. BTI-Tn-5B1-4 cells were grown in a 40-liter spinner using serum-free Ex-Cell 401 medium. Cells at 1.0×10^6 cells/ml were concentrated to 15-fold using a 400-cm^2 Membrex filtration device equipped with a 3.0-μm stainless steel filter. Results in Table 3 show that no significant loss in viability is observed when the rotor speed is lowered from 1,500 rpm to 750 or 1,000 rpm. Permeate flux rates, however, may decrease if the rotating speed is reduced further. The recovery of cells as a scale-up step may not be necessary if sufficiently high cell densities are achieved so that a minimum amount of spent medium is carried over.

After cell removal, ultrafiltration membranes such as Viresolve (Millipore Corp., Bedford, Mass.) and Planova (Asahi Chemical Industries, Tokyo, Japan) may be used as validatable membranes specifically designed for the removal of viruses from the process fluids. This may negate the need for chemical or heat inactivation of the virus during which the protein activities may be destroyed.

16.5. PROTEIN EXPRESSION USING STABLE CELL LINES

The study of insect cell cultures largely focuses on the manipulation of the transcriptional signals of the viral promoters to increase heterologous protein expression. However, it has been pointed out (11) that heterologous gene expression using baculovirus expression vector systems for membrane-targeted and secretory proteins is not as high as that for proteins transported to the cytoplasm. The integrity of plasma membranes at the end of the infection cycle is very much compromised. Recently, some studies have been conducted on the use of Sf-9 as a stable insect cell line for the expression of recombinant neurotransmitter receptors such as chick nicotinic acetylcholine receptor, bovine GABA receptor, and hamster β_2-adrenergic receptor. A number of inducible promoters, heat shock promoters, and eukaryotic virus promoters, including AcNPV *ie-1* promoter, are being examined. Others have used *Drosophila* S2 cells to express recombinant human immunodeficiency virus (HIV) envelope protein and HIV gp 120 glycoprotein, which are secreted and are highly glycosylated (26). A high density of 1×10^7 cells/ml can be obtained readily by using serum-free medium and growth at room temperature in stirred vessels. These cells can be transfected in short time frames with selectable markers, carrying multiple high copy numbers of transfected DNAs. Thus, the use of transformed lepidopteran and dipteran insect cell lines to achieve high protein levels in the absence of viruses is possible.

I thank David Murhammer of the Department of Chemical Engineering, University of Iowa, for reviewing this chapter.

REFERENCES

1. **Agathos, S. N., Y. H. Jeong, and K. Venkat.** 1990. Growth kinetics of free and immobilized insect cell cultures. *Ann. N.Y. Acad. Sci.* **589:**372–396.
2. **Anderson, D., R. Harris, D. Polayes, V. Ciccarone, R. Donahue, G. Gernard, J. Jessee, and V. Luckow.** 1995. Rapid generation of recombinant baculovirus and expression of foreign genes using the Bac-to-Bac baculovirus expression system. *Focus* **17:**53–58.
3. **Bedard, C., A. Kamen, R. Tom, and B. Massie.** 1994. Maximization of recombinant protein yield in the insect cell/baculovirus system by one-time addition of nutrients to high density batch cultures. *Cytotechnology* **15:**129–138.
4. **Caron, A. W., J. Archambault, and B. Massie.** 1990. High level recombinant protein production in bioreactors using the baculovirus-insect cell expression system. *Biotechnol. Bioeng.* **36:**1133–1140.
5. **Caron, A. W., R. L. Tom, and A. A. Kamen.** 1994. Baculovirus expression system scale-up by perfusion of high density Sf-9 cell cultures. *Biotechnol. Bioeng.* **43:**881–891.
6. **Dee, K. U., M. L. Shuler, and H. A. Wood.** 1997. Inducing single-cell suspension of BTI-Tn-5B1-4 insect cells. I. The use of sulfated polyanions to prevent cell

aggregation and enhance recombinant protein production. *Biotechnol. Bioeng.* **54:**191–205.

7. **Godwin, G., B. Belisle, A. DeGiovanni, K. Kohler, T. Gong, and D. Wojchowski.** 1989. Excell 400, for serum-free growth of insect cells and expression of recombinant proteins. *In Vitro* **25:**178–183.
8. **Goldblum, S., Y. K. Bae, W. F. Hink, and J. Chalmers.** 1990. Protective effect of methylcellulose and other polymers on insect cells subjected to laminar shear stress. *Biotechnol. Prog.* **6:**383–390.
9. **Hensler, W. T., and S. Agathos.** 1994. Effect of insect cell aggregation on growth and β-galactosidase production. Presented at the American Chemical Society Meeting, San Diego, Calif., March 13–17, 1994.
10. **Keddie, B. A., and L. E. Volkman.** 1985. Infectivity difference between the two phenotypes of *Autographa californica* nuclear polyhedrosis virus: importance of the 64K envelope glycoprotein. *J. Gen. Virol.* **66:**1195–1200.
11. **King, L.** 1995. Expression technology: stable insect cell lines. Presented at the Baculovirus and Insect Cell Gene Expression Conference, Pinehurst, N.C., March 26–30, 1995.
12. **Kroner, K. H., and V. Nissinen.** 1988. Dynamic filtration of microbial suspensions using an axially rotating filter. *J. Membr. Sci.* **36:**85–100.
13. **Kunas, K. T., and E. T. Papoutsakis.** 1990. Damage mechanisms of suspended animal cells in agitated bioreactors with and without bubble entrainment. *Biotechnol. Bioeng.* **36:**476–483.
14. **Lazarte, J. E., P.-F. Tosi, and C. Nicolau.** 1992. Optimization of the production of full length rCD4 in baculovirus-infected Sf-9 cells. *Biotechnol. Bioeng.* **40:**214–217.
15. **Licari, P., and J. E. Bailey.** 1990. Factors influencing recombinant protein yields in an insect cell-baculovirus expression system: multiplicity of infection and intracellular protein degradation. *Biotechnol. Bioeng.* **37:**238–246.
16. **Linsay, D. A., and M. J. Batenbaugh.** 1992. Quantification of cell culture factors affecting recombinant protein yields in baculovirus infected insect cells. *Biotechnol. Bioeng.* **39:**614–618.
17. **Maiorella, B., D. Inlow, A. Shauger, and D. Harano.** 1988. Large-scale insect cell-culture for recombinant protein production. *Bio/Technology* **6:**1406–1410.
18. **Malinowski, J. J., and A. J. Daugulis.** 1993. Bioreactor design for insect cell cultivation, p. 51–68. *In* M. A. Goosen (ed.), *Insect Cell Culture Engineering*. Marcel Dekker, Inc., New York.
19. **Mitchell-Logean, C., and D. W. Murhammer.** 1997. Bioreactor headspace purging reduces dissolved carbon dioxide accumulation in insect cell cultures and enhances cell growth. *Biotechnol. Prog.* **13:**875–877.
20. **Monosama, S. A., and G. W. Blissard.** 1995. Identification of a membrane fusion domain and an oligomerization domain in the baculovirus Gp64 envelope fusion protein. *J. Virol.* **69:**2583–2595.
21. **Murhammer, D. W.** 1995. Personal communication.
22. **Murhammer, D. W.** 1995. Use of viral insecticides for pest control and production in cell culture. *Appl. Biochem. Biotechnol.* **59:**199–220.
23. **Murhammer, D. W., and C. F. Goochee.** 1988. Scale-up of insect cell cultures: protective effects of Pluronic F-68. *Bio/Technology* **6:**1411–1418.
24. **Nguyen, B., K. Jarnagin, S. Williams, H. Chan, and J. Barnett.** 1993. Fed-batch culture of insect cells; a method to increase the yield of recombinant human nerve growth factor (rhNGF) in the baculovirus expression system. *J. Biotechnol.* **31:**205–217.
25. **Reuveny, S., Y. J. Kim, C. W. Kemp, and J. Shiloach.** 1993. Production of recombinant proteins in high density insect cell culture. *Biotechnol. Bioeng.* **42:**235–239.
26. **Rosenberg, M., and A. R. Shatzman.** 1995. Using Drosophila cells for stable continuous efficient expression of foreign gene products. Presented at the Baculovirus and Insect Cell Gene Expression Conference, Pinehurst, N.C., March 26–30, 1995.
27. **Schwartz, J. I., J. Terracciano, J. Troyanovich, I. Gunnrsson, A. D. Kwong, E. Ferrari, J. Wright-Minogue, and J. W. Chan.** 1995. Production of recombinant herpes simplex virus protease in 10 liter stirred vessels using an insect cell–baculovirus expression system. Presented at the Baculovirus and Insect Cell Gene Expression Conference, Pinehurst, N.C., March 26–30, 1995.
28. **Shuler, M. L., H. A. Wood, R. R. Granados, and D. A. Hammer.** 1995. *Baculovirus Expression Systems and Biopesticides*. Wiley-Liss, New York.
29. **Specter, S., and G. Lancz.** 1992. *Clinical Virology Manual*, 2nd ed., p. 117–128. Elsevier, New York.
30. **Summers, M. L., and G. E. Smith.** 1987. A manual method for baculovirus vectors and insect cell culture procedures. Texas Agricultural Experiment Station Bulletin No. 1555. Texas A&M University, College Station.
31. **Tramper, J., J. B. Williams, D. Joustra, and J. M. Vlak.** 1986. Shear sensitivity of insect cells in suspension. *Enzyme Microbiol. Technol.* **8:**33–36.
32. **Trinh L., and J. Shiloach.** 1995. Recovery of insect cells using hollow fiber microfiltration. *Biotechnol. Bioeng.* **48:** 401–405.
33. **Vaughn, J. L., and F. Fan.** 1989. Use of commercial serum replacements for the culture of insect cells. *In Vitro Cell Dev. Biol.* **25:**143–145.
34. **Volman, L. E.** 1986. The 64K envelope protein of budded *Autographa californica* nuclear polyhedrosis virus. *Curr. Top. Microbiol. Immunol.* **131:**103–118.
35. **Wang, M.-Y., S. Kwong, and W. E. Bently.** 1993. Effect of oxygen/glucose/glutamine feeding on insect cell baculovirus protein expression: a study on epoxide hydrolase production. *Biotechnol. Prog.* **9:**355–361.
36. **Webb, N. R., and M. D. Summers.** 1990. Expression of proteins using recombinant baculoviruses. *Technique* **40**(2)**:**173–188.

Raw Materials Selection and Medium Development for Industrial Fermentation Processes

SAMUN K. DAHOD

17

The fermentation medium forms the environment in which the fermentation microorganisms live, reproduce, and carry out their specific metabolic reactions to produce useful products. The importance of this environment cannot be overemphasized when it comes to the development of a productive fermentation process. Over the years, substantial progress has been made in developing fermentation medium design as a systematic science. However, experienced industrial microbiologists and biochemical engineers will be the first to point out that this field is as much an art as it is a science. In most industrial fermentations, where the product is something other than the cell mass itself, there are two distinct biological requirements for medium design. First, nutrients have to be supplied to establish the growth of the organism. Second, after growth is established, proper nutritional conditions have to be provided to maximize product formation. Besides these obvious biological requirements, one needs to worry about selection of nutrient components that are cost-effective, readily available, and consistent from lot to lot. In recent years, as integrated approaches to fermentation and downstream processing have been developed, it has also been recognized that the fermentation medium should not unduly hinder the downstream processing and, if possible, should even facilitate downstream processing. For new fermentation processes brought up from microbiology laboratories, considerable flexibility and latitude in medium design are possible. The process is not locked into a fixed set of raw materials (for example, due to a Food and Drug Administration [FDA] filing), and the medium components can be freely selected for the sole purpose of maximizing the product yield and minimizing the cost. For an established fermentation process, the choice of medium components may be limited by such factors as FDA filing, the cost structure for the product, and the requirements of downstream processing. In spite of these limitations, continued medium development remains a necessity so that an established product retains its competitive edge in the marketplace.

While literature reports for medium development in specific fermentation processes are plentiful, a general treatment of broad principles involved in fermentation medium development is rare. Readers may find the review by Corbett (2) informative. Another review, by Kennedy and Reader (3), although describing only raw materials available in New Zealand, may have general applicability. Miller and Churchill (4) list many fermentation raw materials by their trade names along with their applications in various types of fermentation processes.

This review focuses primarily on raw materials and medium development for microbial fermentation processes. Although general principles also apply to it, mammalian cell culture will not be emphasized. The chapter is not intended to provide a literature search or a review of specific medium types used in specific fermentation processes. Rather, it is designed to provide practicing microbiologists and biochemical engineers with a rational basis for medium development and improvement. At the start of the chapter, a chemically defined fermentation medium is considered with its pros and cons. Then various commercially available ingredients for key nutrient components of traditional complex fermentation media are described in generic terms. This is followed by a discussion of general considerations and a set of guidelines for medium development and improvement. The information provided is derived from experiences in the fermentation industry, and no effort has been made to cite references for specific examples and dictums mentioned in the chapter, even though similar information may also be present in the literature.

17.1. CHEMICALLY DEFINED FERMENTATION MEDIA

In industrial fermentation, it is very rare that a chemically defined medium is used. Media derived from complex ingredients such as flours and by-products of the brewery, meat, and corn-milling industries are common. Complex media usually give higher fermentation yields at a lower cost. There are rare cases, however, when downstream processing considerations may prohibit the use of complex raw materials. For example, when peptide products are made by cell culture processes, the use of serum in the medium may make the downstream processing more difficult. In recombinant *Escherichia coli* fermentations for production of biological proteins, the use of complex ingredients can introduce a myriad of proteins and polypeptides that can interfere with the recovery process. In many such cases, for reasons of growth and productivity, a small amount of complex nutrients such as yeast extract or bovine serum is still used. While not considered the ideal medium formulation, the completely defined medium can be designed to support most organisms to at least some extent. A completely defined medium is indeed the medium of choice for studying the metabolic pathways of a fermentation process, for example, when determining substrate uptake kinetics or when studying nutritional control and nutrient requirements for product formation. The metabolic information obtained with a defined medium can then be used to understand and further improve the industrial process involving complex ingredients. Typically, the design of a chemically defined medium is based on the elemental composition of the microorganism being cultivated. Since the elemental compositions of organisms differ according to the type of organism and the growth conditions, excess nutrients are generally used. The cell density is then controlled by using a limiting nutrient such as glucose or ammonia. When glucose is used as the limiting nutrient, it can be assumed that the cell yield on a dry weight basis will be about 50% of the glucose consumed. As a rule of thumb, to calculate the nitrogen requirement of a nitrogen-limited culture, it can be assumed that 10% of the dry weight of the organism will be nitrogen. Table 1 illustrates the typical components and their proportions used in a glucose-limited defined fermentation medium when a cell density of 10 g (dry cell weight) per liter is desired. It should be noted that a medium formulation based on the information in Table 1 will not give the desired results with all cell types. In addition to these basic medium components, individual cell types may require for growth other unique medium ingredients such as amino acids, vitamins, purines, pyrimidines, additional metal ions, and chelating agents. Furthermore, the medium given in Table 1 does not take into account the additional nutrients and other unique components required for successful synthesis of the desired product. If the metabolic processes involved in product formation are reasonably known, the nutrients needed for product synthesis can be calculated from the stoichiometry and yield factors and included in the medium formulation. One of the most important aspects of medium design that is often overlooked is the pH balance of the medium. In complex fermentation media, the natural amino acids, peptides, and other organic compounds provide pH buffering. In a chemically defined medium, inorganic buffering agents such as phosphates, organic acids, and carbonates have to be used or provision has to be made for external addition of acid or base to control the pH during the fermentation process.

TABLE 1 Components of a chemically defined fermentation medium needed to obtain about 10 g of dry cell weight per liter

Source of:	Typical ingredient[a]	Concn (g/liter)
Carbon	Glucose	20
	Sucrose	20
	Glycerol	20
Nitrogen	$(NH_4)_2SO_4$	5
	$NaNO_3$	7
	NH_4NO_3	3
	Alanine or other amino acids	7
Phosphorus	KH_2PO_4	1
	K_2HPO_4	1
Sulfur	K_2SO_4	0.4
	$MgSO_4 \cdot 7H_2O$	0.5
	Methionine	0.3
Metals		
Mg	$MgSO_4 \cdot 7H_2O$	0.1
K	K_2SO_4	0.1
Ca	$CaCl_2$	0.05
Fe	$FeSO_4 \cdot 7H_2O$	0.001
Zn	$ZnSO_4 \cdot 7H_2O$	0.001
Cu	$CuSO_4 \cdot 5H_2O$	0.0004
Mn	$MnSO_4 \cdot H_2O$	0.0004

[a]Typically, one component from each grouping is used based on the substrate preference of the organism being cultivated.

17.2. COMPONENTS OF INDUSTRIAL FERMENTATION MEDIA

As noted above, most industrial fermentation media are complex formulations containing poorly defined ingredients. Often these ingredients contain multiple nutrients for the growth of fermentation microorganisms. However, for the purposes of medium development, a given ingredient is thought to provide primarily a single nutrient. For example, soy flour is used primarily to supply complex nitrogen or protein for the growth of an organism. However, soy flour also contains substantial amounts of metabolizable carbohydrates and minerals. In the discussion below, the medium ingredients are classified according to their primary role in the fermentation process. On this basis, we can classify the fermentation raw materials in four broad nutrient categories: materials used primarily as sources of carbon, nitrogen, or minerals, and materials used for special purposes.

17.2.1. Carbon Sources

17.2.1.1. Carbohydrates

Glucose is the most frequently used carbohydrate in the fermentation industry. In the United States, it is derived from the corn-processing industry. Two types of products are in use, dextrose monohydrate and hydrolyzed corn syrups containing glucose at a level greater than 95% (called DE95 or dextrose equivalent of 95%). While dextrose monohydrate comes in the form of easy-to-handle crystalline material, it is more expensive. This material is used primarily in small-scale applications as in seed fermentors and when consistency is of the utmost importance. For the bulk of the glucose needs, such as for large-scale fermentations and for in-process feeding, the hydrolysate is the more economical material. If the fermentation organism is able to hydrolyze low-molecular-weight saccharides, less expensive corn syrups of various lower degrees of hydrolysis can be used. Industrial fermentation processes such as those for the production of penicillin can readily utilize hydrolysates with a dextrose equivalent as low as 20 (DE20). In fact, some processes give higher yields with these higher-molecular-weight saccharides than they do with pure glucose. The next level of complexity in these glucose-based carbohydrates comes in the form of various dextrins. These are primarily cornstarch products with just enough hydrolysis carried out to make them soluble in the fermentation medium. The dextrins, corn starch, and other starches (such as potato starch) are never used for in-process feeding. They are generally used as batched-in carbon sources for initial growth of the organism or as carbon sources that are gradually assimilated by the microorganism during the product synthesis phase. In the United States, the crudest and the cheapest source of complex carbohydrate is corn flour. This product is primarily starch but also contains about 5% protein. An important cost reduction strategy used by many fermentation companies is to use crude starch or corn flour in the batch along with the commercially available enzyme amylase. The amylase breaks down starch molecules to generate more readily utilizable carbohydrates.

Sucrose is often used in fermentation processes. In its crystalline form, sucrose is available as table sugar of various degrees of refinement. The white crystalline sucrose is generally used in small-scale applications and in seed fermentors. However, it can also be used as a gradually utilized carbon source in some fermentations in which the organism has a limited ability for metabolizing sucrose. The crudest form of sucrose comes as molasses, which contains anywhere from 3 to 10% protein. In some fermentations (for example, glutamic acid fermentation), this product gives excellent results as a combined carbon-nitrogen feed.

In the early days of penicillin fermentations, the carbon source of choice was lactose. This sugar is gradually metabolized by the penicillin-producing organism and hence can be batched into the medium from the beginning of the process. However, since the advent of controlled feeding of glucose, the importance of lactose in the fermentation industry has decreased. Lactose is available in granular form for small-scale applications, and it is still used in some fermentations, especially in Europe, where it is more readily available than dextrose and corn syrups. The most economical source of lactose is derived from the cheese industry by-product cheese whey. This product is available in a spray-dried form and is an excellent source of protein and minerals besides being a source of lactose.

Other sugars that are used less frequently in the fermentation industry include maltose, mannitol, sorbitol, and xylose. All of these are generally used in their purified forms. A related carbon source for the fermentation industry is glycerol. It is useful in many processes as a gradually metabolized carbon source. Additionally, organic acids, such as acetic acid, may be used on rare occasions as combination pH control agents and carbon nutrients. Minoda (5) has reported on the potential uses of other unusual carbon sources for amino acid fermentations.

17.2.1.2. Oils

Various oils are widely used as carbon sources in the fermentation industry, especially in antibiotic fermentations. Oils can supply both the energy and the growth carbon needs of the organism. In many antibiotic fermentations, where the antibiotic backbone is synthesized from low-molecular-weight fatty acids, the oils make ideal carbon sources since they gradually supply these fatty acids during the fermentation process. The oils are used both as batched-in ingredients and as continuous feeds. In some fermentations, oils play an important auxiliary role even when they are not actively metabolized by the fermentation organism. The yield-enhancing effect of the oil when it is not metabolized is not well understood. It is possible that it provides protection to cells from excessive shear forces or that it makes a key micronutrient from the complex medium more available to the organism in the form of micelle. In fermentations in which oils can be utilized as carbon feeds, they offer important benefits. First, the caloric content and the corresponding energy availability per unit volume of feed is appreciably higher for oils than for carbohydrates. One liter of vegetable oil has more than twice the utilizable energy as 1 liter of a 55% solution of glucose. This high energy density allows for lower feed rates and smaller feed vessels. Consequently, the fermentor volume management for long-cycle fermentations is easier with oil-fed fermentations than with sugar-fed fermentations. This is true not only because less feed is introduced into the fermentor but also because the metabolism of oil does not produce as much water as the metabolism of sugars. The antifoaming property of the oils is also beneficial for most fermentation processes. Before the advent of synthetic defoamers, oils were used for foam control in many fermentation processes even when the carbon source of choice was a sugar. However, the oil added for foam control is metabolized by the organism, and continuous addition is required to control foam. The synthetic defoamers are more effective because they are not readily degraded by the fermenting organism and they are cost-effective. In special cases, when the presence of synthetic defoamer interferes with the downstream processing, oils are still used as defoamers. The antifoaming properties of several natural oils are reviewed by Vardar-Sukan (6).

The most important oil in the United States fermentation industry is soybean oil. It is abundant and relatively inexpensive. Other oils that are often used are lard oil and fish oil and oils of other plants such as corn, cottonseed, peanut, sunflower, and safflower. One specialty oil product that is synthetically made and has found application in the fermentation industry is methyl oleate. Methyl oleate is often used as a supplemental feed in conjunction with another feed such as soybean oil. The fatty acid contents of various oils vary according to their source, and there may be a theoretical basis for one type of oil to perform better than another type. However, the choice of oil in a given fermentation is generally determined empirically. The oil that is used in the shake flask fermentations during screening of the producing strains very often also gives better results in large-scale fermentations.

17.2.2. Sources of Organic Nitrogen or Protein

There are principally three classes of raw materials available to supply the organic nitrogen or protein requirement of a fermentation process: (i) those derived from agricultural products, (ii) those derived from brewery industry by-products, and (iii) those derived from meat and fish by-products. All of these products supply other important fermentation nutrients in addition to organic nitrogen.

17.2.2.1. Nitrogen Sources Derived from Agricultural Products

The sources derived from agricultural products are the workhorse ingredients of fermentation industry. They include the products of commodities such as various grains and soybean. The soybean flours, meals, and grits head the list of applications in antibiotic fermentations. The popularity of the soy products is based on the fact that after the soy oil is extracted from the soybeans, the residue is about 50% protein, which is readily available for cell growth. In addition, soy flour, meals, and grits contain up to 30% utilizable carbohydrates. Most minerals required for microbial growth are also present in soy-based products. In many seed medium applications, where growth is the primary consideration, all that is required in the medium is soy flour along with salts such as magnesium sulfate and potassium phosphate. A product that is processed very similarly to soy flour is cottonseed flour. In recent years, cottonseed flour has become the nitrogen source of choice in penicillin fermentations. The protein in the cottonseed flour is less readily available and thus makes a good slow-releasing nitrogen source. Corn gluten meal is another readily available product that is suitable as a slow-releasing nitrogen source. Corn steep liquor, a by-product of the corn milling industry, was very extensively used in the early years of the antibiotic fermentation industry. In recent years, though, due to the variability in the product quality, the liquid form of corn steep liquor has fallen out of favor. Spray-dried corn steep liquor is now available and is used in many antibiotic fermentations because it is less variable. Other agricultural commodities used as nitrogen sources in fermentation industry include peanut meal, linseed meal, wheat flour, barley meal, and rice meal.

17.2.2.2. Nitrogen Sources Derived from Brewery Industry By-Products

The brewing industry is an important source of fermentation raw materials. The principal product is the yeast left over after beer fermentation. The suitability of the yeast by-product for a given fermentation depends upon the method of drying. The yeast may be drum dried or spray dried. It is also sold as a paste produced by water evaporation in an industrial evaporator. All of these products have found applications in the fermentation industry as sources of nitrogen. However, the yeast is never used as the primary source of nitrogen. Instead, it is thought of as a nitrogen supplement with additional beneficial nutrients that are not available from grain-based nitrogen sources. Generally, these additional nutrients are organic phosphorus and unknown micronutrients. Brewery yeast is also refined into yeast extracts of different water solubilities, which are more expensive and used in smaller quantities. Yeast extract is often the single undefined component used in so-called semidefined fermentation media to provide micronutrients. The brewing and distilling industries supply two other by-products that are sometimes used in the fermentation industry: distillers' solubles, in the form of a concentrate or a spray-dried powder, and leftover grains from the brewing process.

17.2.2.3. Nitrogen Sources Derived from Meat and Fish By-Products

Meat and fish products are very rich in protein. So are the by-products of these industries. The primary meat-based product is generically known as spray-dried lard water. This is a by-product of lard processing. The animal bones and tissues are boiled in water, sometimes in the presence of proteases, to free the fat. The resulting liquor is separated into fat and water layers. The water part is rich in proteins and peptides. This lard water, when spray dried, gives a product with a protein content of 80% or greater. The lard water can be obtained with different degrees of chemical or enzymatic hydrolysis. Hydrolyzed lard water products are sold as meat peptones under various brand names. A parallel line of products labeled fish meals and fish hydrolysates is derived from heat and enzymatic treatment of fish wastes. These products are generally about 70% protein.

17.2.3. Minerals

Minerals are used in fermentation media to serve many purposes, e.g., as major nutrients, as trace metal suppliers, as ionic strength-balancing agents, as precursors for secondary metabolite syntheses, as buffering agents, as pH control agents, and as reactants to remove specific inhibiting nutrients from the medium. The nitrogen-containing salts (e.g., ammonium sulfate, ammonium nitrate, sodium nitrate, and potassium nitrate) can provide a substantial portion of the nitrogen requirement for cell growth when combined with organic nitrogen. When these salts are used as nitrogen nutrients, their metabolism invariably results in pH changes in the medium. For example, when ammonium sulfate is utilized by the organism, the pH tends to fall, and when sodium nitrate is utilized, the pH tends to rise. Therefore, it is very important that adequate buffering or pH control be provided to counterbalance these pH effects. Ammonia used for pH control has the advantage of regulating pH while replenishing ammonium nitrogen used up from ammonium sulfate in the medium. Another major nutrient supplied as inorganic salt is phosphorus in the form of phosphate salts. Phosphorus from soluble phosphate salts is more readily available to the organism than the phosphorus derived from organic nutrients such as yeast. As a result, it is possible to control the rate of growth by balancing organic phosphorus against inorganic phosphorus salts.

Although most organic nitrogen sources such as grain meals and yeast extracts contain many of the minerals required for growth, the fermentation medium is often supplemented with salts that provide elements that are required in greater than trace quantities. For example, magnesium and potassium salts and the salts containing sulfate are generally included in the medium if they have not already been included for other purposes. Trace elements such as iron, zinc, manganese, copper, cobalt, and molybdenum are generally not included in fermentation medium containing high concentrations of complex ingredients unless they serve specific purposes in metabolism. For example, if product synthesis is known to be carried out by an enzyme complex containing cobalt, this element will be included in the medium at a few parts per million concentration to ensure that it is not scarce. When a medium contains low concentrations of complex ingredients, it is important to include a trace element mixture in the fermentation medium.

In fermentations where the ionic strength has to be relatively high, sodium chloride or sodium sulfate is included in the medium. The insoluble salt calcium carbonate is added to prevent the fermentation pH from falling below 6.0. As the pH drops below 6, calcium carbonate dissolves in the medium raising its pH. Phosphate salts are rarely used for buffering in fermentation medium because the phosphorus balance has to be based on the metabolism rather than on the buffering needs. The soluble calcium salts such as calcium chloride and calcium acetate are often used to precipitate out soluble phosphate (in the form of calcium phosphate) from the media of fermentations in which the product synthesis is strongly inhibited by phosphate. Minerals also serve as precursors in antibiotic fermentations. In penicillin and cephalosporin fermentations, sufficient sulfate salts have to be included in the medium to supply the sulfur required for the syntheses of these sulfur-containing antibiotics. Similarly chloride salts must be included in the medium for vancomycin fermentation since the vancomycin molecule contains several chlorine atoms.

17.2.4. Specialty Chemicals

Several types of specialty chemicals are added to large-scale fermentation media. The most important of these chemicals are the defoamers. The defoamers reduce the interfacial surface tension between air and water to facilitate bubble coalescence. In the fermentation industry, silicone and polyol-based defoamers have largely replaced vegetable oils as defoamers. The advantages of the synthetic defoamers are that they are cost-effective and very slowly metabolized and do not have appreciable metabolic side effects. The two most popular defoamers in use in the fermentation industry are polypropylene glycol and silicone emulsion. The defoamers are generally batched with the starting medium. In many fermentations, however, it is necessary to supply defoamer throughout the fermentation cycle to control foam and to control air holdup. Emulsifiers used in fermentations (such as Tween and Span) play a role opposite to that of defoamers. They are added to stabilize small droplets of oily nutrients by increasing the surface tension between oil and water. The small droplets have a dramatically increased surface area and thus allow oily substrates to be more readily utilized by the fermentation organism. Metal-chelating agents such as EDTA are often included in fermentation media. The chelating agents have two diametrically opposed effects. On the one hand, they can tie up metal ions that are toxic to the organism. On the other hand, they can prevent the precipitation of a required trace metal by forming a soluble complex. The availability of the metal to the fermenting organism depends upon whether the organism can effectively compete with the complexing agent for the required metal.

An important class of specialty products used in the fermentation industry is made up of various enzyme preparations. Crude preparations of enzymes such as amylase, protease, and cellulase are used to precondition the medium. Invariably, these enzymes are used at the mixing stage before medium sterilization. A partial breakdown of the starch of medium components such as corn flour can be achieved by the addition of amylase enzyme. The cellulase complex can be used to reduce the viscosity of a medium containing a high concentration of ingredients such as soy or cottonseed flour. The protease enzymes can predigest the medium proteins before sterilization. Enzymatic pretreatment of a fermentation medium thus allows a crude and cheaper raw material to be substituted for a more refined and expensive raw material.

17.2.5. Sources of Information on Fermentation Raw Materials

The best source of information on a given class of fermentation raw material is the industry in which it is generated. Information about such things as the protein, fat, carbohydrate, and mineral contents of various raw materials is readily available from the supplier of the raw materials. However, this information is not necessarily generated for the use of the fermentation industry. It is generated for the benefit of the primary users, which in most cases are the animal feed and food industries. As a result, interpretation of the information for fermentation use is up to the fermentation scientist. For example, while the total nitrogen value of a grain-based product may be meaningful from the point of view of a weight gain calculation when the product is fed to a farm animal, it may not necessarily have the same meaning as the nitrogen available for the fermentation organism to grow on. For the same reason, the carbohydrate value provided by the manufacturer of one product may be higher than the value provided for a second product, and yet the second product could have more available carbon for a particular fermentation organism. The information provided by the manufacturer is a good approximation for the initial evaluation and for preliminary cost calculations. Actual fermentation experiments are necessary in all cases to justify a change of raw material. In recent years, some of the raw-material suppliers have taken it upon themselves to evaluate their products for various fermentation processes and publish the results in their own manuals or in the scientific journals. One such publication with a wealth of information on cottonseed flour and other fermentation raw materials is *Trader's Guide to Fermentation Media Formulation* (7). A similar information booklet by Cargill, Inc., titled *Soy Protein Products in Fermentation* (1), attempts to provide information on soy products that is relevant to the fermentation process. Table 2 lists some of the major suppliers of fermentation raw materials and their products. Many of the companies sell their products under brand names. Only the generic designations of the products are listed. Most products listed in Table 2 are available in large quantities at competitive prices. It should be noted, however, that the list in Table 2 is by no means an exhaustive list of the companies offering raw materials for fermentation application. In addition, the companies listed may have other products of interest to fermentation scientists.

17.3. GENERAL CONSIDERATIONS FOR INDUSTRIAL (COMPLEX) MEDIUM DEVELOPMENT OR IMPROVEMENT

17.3.1. Product Yield and Its Market Value in Relation to Cost, Availability, and Usage Rate of Raw Materials

In a typical industrial antibiotic fermentation, the medium development work undertaken to improve product yield always has a larger impact on the overall process cost than does simple medium cost reduction. Very often, a yield improvement not only improves the economy of the fermentation process itself but also has beneficial effects in the downstream processing. The product-to-impurity ratio increases as the fermentation yield increases, making the recovery process more efficient. In some mature fermentations, the productivity improvement beyond a certain level is difficult to attain due to genetic limitations or to the inability of the organism to tolerate increasingly higher concentration of the product. In such cases, fermentation raw-material cost reduction alone can have a substantial impact on overall cost reduction efforts. The value of the final product and the volume of the product produced are other important considerations. First, consider the final product value relative to the cost of the raw materials used. In the fermentation industry, the contribution of fermentation raw materials to the overall production cost may vary from as little as 5% (for example, the production of high-value biological agents such as interferon or the production of steroids) to as much as 50% (for example, the production of commodities such as ethanol). The scientist working on the former type of product has a much greater flexibility in selecting raw materials, since the overall production cost is not appreciably increased by introduction of a relatively costly raw material. The goal here is to reduce the overall cost by increasing the fermentation yield. In the latter case, however, the incremental cost increase

TABLE 2 Major suppliers of fermentation raw materials and their products

Company and address	Products
American Protein Corp. 2325 North Loop Dr., Ames, IA 50010	Hydrolyzed meat proteins, blood protein, spray-dried lard water
Archer-Daniels-Midland or ADM P.O. Box 1470, Decatur, IL 62525	Soybean grits, soybean flour, soybean meal, corn flour, cornstarch, dextrin, corn syrups, dextrose, sorbitol, soy oil, corn oil
Cargill, Inc. P.O. Box 9300, Minneapolis, MN 55440	Soybean flour, corn flour, corn syrups, corn gluten meal, cornmeal, soy oil
Champlain Industries P.O. Box 3055, Clifton, N.J. 07012	Yeast extracts
CPC International P.O. Box 8000, Englewood Cliffs, NJ 07632	Cornstarch, dextrins, corn syrups, dextrose, corn oil
Grain Processing Corp. 1600 Oregon St., Muscatine, IA 52761	Soy flour, soy proteins, corn gluten meal, corn flour, corn syrups, dextrins
Lauhoff Grain Co. P.O. Box 571, Danville, IL 61834	Soy grits, soy flour, corn flour, cornmeal, corn gluten meal
Roquette Freres F.62136 Lestrem, France	Corn steep liquor, corn steep powder, corn gluten, wheat gluten, potato protein, starch, dextrin, dextrose, starch hydrolysates, modified sugars, corn oil
Sheffield Products P.O. Box 630, Norwich, NY 13815	Hydrolyzed proteins derived from casein, meat, soy, and cottonseed
A. E. Staley Manufacturing Co. P.O. Box 151, Decatur, IL 62525	Cornstarch, dextrins, corn syrups
Trader's Protein P.O. Box 80367, Memphis, TN 38108	Cottonseed flour, cottonseed meal
Universal Foods Corp. 433 E. Michigan St., Milwaukee, WI 53202	Baker's yeast, brewer's yeast, yeast extracts, casein hydrolysates

due to the introduction of a new raw material has to be more than compensated by the increase in yield and product quality. The agricultural commodity products and by-products from the brewery and corn wet-milling industries are the typical raw materials used in fermentation processes for low- and medium-value products such as organic acids and well-established antibiotics. On the other hand, exotic raw materials such as refined yeast extracts, specialty meat peptones, and exotic growth factors can be cost-effective in fermentation processes of high-value products such as biological peptides. The usage rate of a given raw material and the overall volume of the fermentation broth processed also have to be taken into account for medium development decisions. If an ingredient is used at a few parts per million, its unit cost does not significantly affect the overall process cost. If the volume of the fermentation broth is very large, however, the overall cost may still be significantly affected.

The availability of a given raw material in a given geographical location is another consideration. Should a specific material be shipped long distance, or should the medium formulation be changed so that a readily available material can be used in its place? This depends largely upon how sensitive the fermentation yield is to the type of material used. While a readily available raw material may give a somewhat reduced yield, in the long run it may be more cost-effective to standardize the medium with that material than to depend upon a material that gives higher yield but may be subject to supply disruption. Other factors to take into account are whether the quality of the material will remain high during long-distance shipping and/or prolonged storage. Raw materials such as yeast paste and corn steep liquor are not stable enough for prolonged storage. On the other hand, raw materials with a low moisture content such as cottonseed meal, soy flour, and spray-dried yeast are reasonably stable over long periods of storage.

17.3.2. Nature of Fermentation Raw Materials

Most raw materials used in the fermentation industries are not designed for that use. They are generally designed to supply commodities for the food and feed industry. Thus, soy meals, cottonseed meals, and corn gluten meals are designed primarily as animal feed protein sources. Various yeast products are designed for both human food and animal feed applications. Corn syrups of different levels of hydrolysis are made for application in the food-processing industry. Since the fermentation industry is not the primary user of these raw materials, the industry does not have much control over their processing and the resulting quality from the point of view of their use in fermentation processes. Also, agricultural products are subject to variation due to growing seasons, soil conditions, and storage conditions. In short, raw-material variability is the rule rather than the exception. In medium design then, it is necessary to use multiple sources of the same class of nutrient to reduce process variability. Thus, including two complex nitrogen sources in the medium formulation is more desirable than depending upon a single ingredient. It is also recommended that several lots of the same raw material be tested before settling on a given medium formulation. If the product yield varies excessively due to lot-to-lot variability, it is better to avoid that raw material in the medium formulation altogether.

At this point, it should be noted that water used to prepare fermentation medium is the major component of the medium. In large fermentation plants, this water is never distilled or deionized water, as may be the case in the laboratory. As a result, certain metal ions and organic components that come dissolved in the water as impurities become part of the fermentation medium. These impurities and their concentrations may vary on a seasonal basis. In addition, the profile of inorganic and organic components that come with the water may vary when the municipal water treatment plant experiences upsets in its operations. Many fermentation plants use readily available water from adjacent water sources such as lakes, rivers, or deep wells with a minimal pretreatment. These water sources are also subject to seasonal variability. Water quality is an important variable when fermentation processes are scaled up from the laboratory, where deionized or distilled water may be used. The water quality is also an important consideration when fermentations involving identical raw materials perform differently at differing physical plant locations. Most fermentation plants monitor the water quality only superficially, and it is seldom known which water quality parameters are important for a given fermentation process.

17.4. GENERAL GUIDELINES FOR FERMENTATION MEDIUM DEVELOPMENT

17.4.1. Seed Medium and Product Synthesis Medium

Generally, the purpose of the seed culture is to grow cells as fast as possible and limit the growth on the basis of predefined criteria such as dissolved-oxygen level, oxygen uptake rate, or centrifuged cell volume. This can be readily achieved by supplying the required nutrients for growth without regard to the product formation needs. Cell growth can be limited by restricting either the carbon source or the nitrogen source so that the cells stop growing when the predefined criteria are met. When developing a medium for the fermentation stage of a process, the selection of medium components and optimization of their concentrations in the medium are more involved. The objective is not only to develop cell mass but also to synthesize the product at the highest rate possible. The cell density attained, the growth rate during the cell growth, and the subsequent maintenance metabolism are all important factors in maximizing product formation. Typically, in an antibiotic fermentation, the medium nitrogen concentration determines the maximum cell density that can be achieved. The rate of cell growth often can be controlled by controlling the level of readily available nutrients such as glucose, amino acids, and soluble phosphate and by controlling the growth temperature. The slow-growth and maintenance metabolism during the product synthesis phase of the fermentation process is generally controlled by supplying additional nutrients slowly. This controlling nutrient feed is usually glucose or vegetable oil. In some cases, ammonia or complex nitrogen sources are also supplied during this phase. Another way of controlling the slow-growth and maintenance phase of the fermentation process is to include in the medium a carbon (or a nitrogen) source that is only gradually utilized by the organism. Often, carbohydrates such as lactose or starch are used for this purpose. The organism being cultivated must produce specific enzymes such as β-galactosidase or amylase to be able to utilize these carbon sources. Various oils are frequently added as the source of carbon that is gradually consumed. Coarse raw materials such as soybean grits and corn gluten meal are used to supply slow-releasing nitrogen. Enzymes such as cellulase and protease must be induced for the organism to utilize these coarse nitrogen sources. In addition to maintenance nutrients, some secondary metabolite fermentations require the addition of precursor compounds. For example, the precursors phenylacetic acid and uracil are added to fermentations of the antibiotics penicillin and nikkomycin, respectively.

17.4.2. Using Laboratory Fermentation Medium as the Starting Point

Industrial fermentation organisms are generally highly mutated organisms that are developed in strain development laboratories over many years. The fermentation conditions under which these organisms have been selected must be taken into consideration during medium development work. If a strain has been developed with a laboratory fermentation medium that is based on cotton seed meal as the primary nitrogen source, it may not perform well in a medium based on corn steep liquor as the primary nitrogen source. This is not to say that more closely related medium ingredients such as soy flour and peanut meal may not give yield improvements. The relationship between the laboratory carbon source and the carbon source used in the large-scale fermentation is often not straightforward. Because in shake flask fermentations external pH control is not possible, a readily utilized carbon source such as glucose is very seldom used unless the medium is heavily buffered. Typically, a carbon source such as sucrose, lactose, dextrin, or starch is used to maintain the pH in a reasonable range. When these processes are scaled up, similar complex carbohydrates are initially used in a batch mode. As the process is developed further, however, they are often replaced with an external feed of a readily utilizable carbon source such as glucose, accompanied by pH control. A similar situation can also arise with regard to simple nitrogen sources. While nitrates or amino acids have to be used in a laboratory fermentation medium for the purpose of pH balancing and slow nitrogen release, they can be replaced in large-scale fermentations with more readily available and cheaper materials such as ammonium sulfate or ammonia, with appropriate control mechanisms. The overall efficiency of nutrient utilization may also change when the fermentation process is scaled up to large fermentors, in which the agitation and aeration conditions are more intense than those in shake flasks. More often than not, the nutrient requirement increases when going from shake flasks to large fermentors.

17.4.3. Considerations of the Fermentation Medium as a Whole

A fermentation medium is typically prepared by dissolving or suspending various raw materials in water. Before the medium is inoculated with the desired organism, it is heat sterilized. The batch sterilization involves heating the medium to over 121°C for a period ranging from 30 to 60 min. Continuous sterilization is carried out by rapidly bringing up the temperature to 145 to 155°C and holding it at that temperature for 5 to 10 min. This heat sterilization of a mixture of ingredients in water has a profound effect on the resulting fermentation medium. A number of chemical and physical changes occur during sterilization. Insoluble ingredients such as grain flours and meals are partially solubilized. Macromolecules such as proteins and starch are partly degraded to more soluble and readily metabolizable lower-molecular-weight peptides and oligosaccharides. The inorganic components of the medium react among themselves and with organic components to give new compounds. For example, various metal ions complex with protein molecules to alter protein solubility, organic phosphorus compounds release phosphate into the medium, dissolved phosphorus is precipitated as insoluble metal phosphates, etc. In some cases, the heat sterilization generates toxic chemicals from relatively benign medium ingredients. A well-known example of this is the Maillard reaction between reducing sugars and amino compounds to give growth-inhibiting amino sugars. To prevent this reaction, reducing sugars such as glucose are sterilized separately from the medium containing amino acids and ammonia. The two components of the medium are mixed after they are cooled to about 40°C. Since various medium components interact during sterilization, it is important to examine the effect of an ingredient being added or removed on the overall chemistry of the medium. The organism may not require calcium salt for growth or for product formation. However, calcium may play a critical role by precipitating out excess phosphate from the medium in the form of insoluble calcium phosphate and allowing a phosphate-regulated product to be synthesized. The elimination of soluble phosphate will also change the medium's buffering capacity. The pH of the medium during sterilization is important because the chemical reactions occurring in an aqueous medium are affected by pH. The pH can have an effect on both the rates of reactions and the equilibrium composition. For this reason, it is generally necessary to experiment with sterilization pH to optimize the performance of the medium under development. It is well known that by manipulating sterilization pH, one can increase or decrease protein solubilization from a medium containing insoluble protein sources such as grain flours and meals.

Some components of the medium may have an indirect effect even in the absence of heat sterilization. For example, seemingly inert oils and defoamers may create micelles in the broth that solubilize proteinaceous components and fats that may otherwise be unavailable for metabolism. Some surface-active agents have no metabolic effect but may have substantial effects on the oxygen transfer characteristics of the fermentation broth by changing the surface tension at the air-liquid interface. Many fermentation media containing complex proteins tend to foam heavily during sterilization, and defoamer addition may be necessary even though the fermentation process itself does not require foam control chemicals. On rare occasions, the order of addition of various ingredients when the medium is prepared, the temperature at which the presterilized medium is prepared and the length of time the medium is held before sterilization will affect the performance of the fermentation process.

Because the fermentation medium after sterilization (and hence after the chemical and physical changes have taken place) is the real medium in which the organism of interest is to be grown, it is important to characterize the sterilized medium. Certain overall indices such as soluble nitrogen, reducing-sugar equivalent, and soluble phosphate are often used to characterize the sterilized fermentation medium. However, these indices give only a gross measure of the properties of the medium. Only by understanding the chemical and physical phenomena taking place in the medium during batching, during sterilization, and during the fermentation process itself can one truly master the art of fermentation medium development.

I thank R. Dale Cooper, Bioprocess Development Department, Abbott Laboratories, for proofreading the manuscript and for making important suggestions to improve its content and its readability.

REFERENCES

1. **Cargill, Inc.** *Soy Protein Products in Fermentation.* Cargill, Inc., Cedar Rapids, Iowa.
2. **Corbett, K.** 1985. Design, preparation and sterilization of fermentation media, p 127–139. *In* A. T. Bull and H. Dalton (ed.), *Comprehensive Biotechnology*, Vol. 1. *The Principles of Biotechnology: Scientific Fundamentals.* Pergamon Press, Inc., New York.
3. **Kennedy, M. J., and S. L. Reader.** 1991. Industrial fermentation substrates available in New Zealand and a strategy for industrial fermentation medium formulation. *Aust. Biotechnol.* **1:**116–120.
4. **Miller, T. L., and B. W. Churchill.** 1986. Substrates for large scale fermentations, p. 122–136. *In* A. L. Demain and N. A. Solomon (ed.), *Manual of Industrial Microbiology and Biotechnology.* American Society for Microbiology, Washington, D.C.
5. **Minoda, Y.** 1986. Raw materials for amino acid fermentation—culture medium C—source development. *Prog. Ind. Microbiol.* **24:**51–66.
6. **Vardar-Sukan, F.** 1988. Efficiency of natural oils as antifoaming agents in bioprocesses. *J. Chem. Technol. Biotechnol.* **43:**39–47.
7. **Zabriskie, D. W., W. B. Armiger, D. H. Phillips, and P. A. Albano.** 1994. *Traders' Guide to Fermentation Media Formulation.* Traders' Protein, Memphis, Tenn.

Instrumentation of Small-Scale Bioreactors

BERNHARD SONNLEITNER

18

Modern bioprocesses are monitored by on-line sensors or devices and by manual analytical methods. This chapter will deal with the on-line methods only.

The number of monitored state variables is usually greater in small-scale reactors, used for process research and development, whereas in large-scale production reactors, only a minimal set of the most relevant variables is monitored. The major reasons for the reduction of monitored variables are minimizing the contamination risk and avoiding wrong controller actions caused by potentially malfunctioning sensors. In fact, industrial production remains conservatively on the safe side and prefers open-loop control, while research and development exploit monitoring and closed-loop techniques.

Besides sensor probes, more and more analytical subsystems appropriately interfaced to bioreactors are being used to monitor the state of a bioprocess on-line and in real time. Some of these subsystems deliver signals that are useful for closed-loop process control.

18.1. PROCESS MONITORING BACKGROUND AND STANDARDS

Cellular activities such as those of enzymes, DNA, RNA, and other components are the primary variables determining the performance of microbial or cellular populations. The development of specific analytical tools for measurement of these activities in vivo is therefore essential in order to get direct analytical access to these primary variables.

The number of sensors useful for in situ monitoring of biotechnological processes is comparatively low compared to other disciplines, such as physics or engineering. The established sensors measure physical and chemical variables rather than biological ones (90). The reasons are manifold but, generally, biologically relevant variables are much more difficult and complex than others (e.g., temperature, pressure). A further important reason derives from typical constraints, namely:

- sterilization procedures
- stability and reliability over extended periods
- application over an extended dynamic range
- no interference with the sterile barrier
- insensitivity toward protein adsorption and surface growth
- resistance toward degradation or enzymatic breakdown

There are undoubtedly a few variables that are generally regarded as a "must" in bioprocess engineering. Among these are several physical, fewer chemical, and even fewer biological variables. Figure 1 gives a summary of what is currently believed to be a minimum set of required measurements in a bioprocess. Such equipment is typical for standard production of material; see, for example, reference 127. However, monitoring of only these variables is insufficient to characterize the microenvironment and activity of cells.

Besides some environmental and operational variables, the most important state variables, namely, the concentrations of active biocatalysts (e.g., cell density), of starting materials (e.g., limiting substrate), and of products must be known.

18.2. TERMINOLOGY AND FOCUS

"On-line" is synonymous for fully automatic. There is no manual interaction involved in signal generation. This chapter is focused on on-line methods, sensors, and devices only.

Depending on the site of installation, one discriminates further between in situ, which means built-in, and ex situ, which can mean in a bypass or in an exit line; in the latter case, the withdrawn volumes are lost for the process. Depending on the mode of operation of the sensing device, one can discriminate between continuous and discontinuous or discrete signal generation; in the latter case, a signal is repeatedly generated once in a while, but in between there is no signal available. The signal is called real time when the delay of signal generation is negligible with respect to the relaxation time of the (bio)process.

18.3. IN SITU INSTRUMENTS

18.3.1. Temperature

Temperature is an environmental variable that is not only monitored but usually maintained at a desired set point by closed-loop control; this is standard. For the interesting

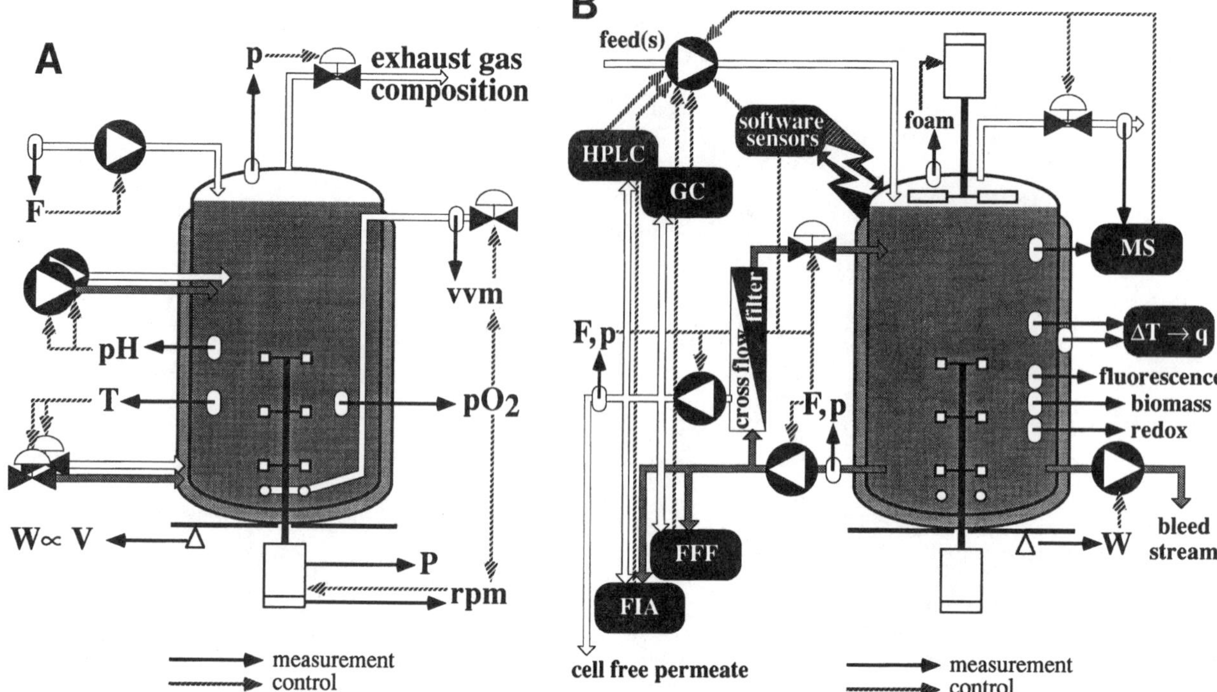

FIGURE 1 Survey of generally accepted standard monitoring and control equipment (A) and possible extensions that have been shown to be desirable and realistic in many instances (B). Measurement and control loops: F: flow rate of a liquid component; vvm: flow rate of gas; p: pressure; pH: pH measurement and control by addition of either alkali or acid (split range controller); T: temperature (either cooling or heating; split range), rpm: stirrer speed; pO_2: oxygen partial pressure measurement and probably control via a cascaded set-point control of the slave-loops gas flow rate and/or stirrer speed. Measurement only: W: weight which is proportional to volume (V); P: power consumption. The extensions are nonstandard in situ sensors such as those for monitoring of fluorescence, biomass, or redox potential, an additional temperature sensor for the coolant circuit (in order to calculate the heat transferred, q, from the temperature difference, ΔT) and a foam sensor. As indicated here, this sensor can activate a mechanical foam separator, which is more desirable than the addition of a chemical antifoam. Further extensions are probably multiple feed lines, a controlled harvest line (bleed stream), and a cell recycle system consisting of a circulation pump and a cross flow filter; this recycle system can be operated in a preparative or an analytical mode. In the first instance, cell-free permeate containing the product can be obtained in large quantities, and flow rate as well as pressure drops should be controlled. In the second case, both pump and filter are a much smaller design, and the amounts of cell-free permeate removed are typically in the order of 1 ml min^{-1} or less. This stream can also be created by a built-in filter and can then be fed to one or more analytical subsystems as indicated. It is reasonable for flow injection analysis and field flow fractionation to supply only those subsystems with complete biosuspensions, provided the cells or cell-associated components are to be analyzed. A mass spectrometer can be connected either to the biosuspension directly via a membrane inlet interface or to the headspace or exhaust gas (as shown) via a capillary interface. Software sensors can exploit any available monitoring data and affect the bioprocess via appropriate control strategies; the software sensors are calculated either on the front end or on the supervisory controller, depending on the sophistication and resource demand of the algorithms.

range between 0 and 130°C, the preferred sensors are thermoelements or thermometers based on resistance changes, e.g., of a platinum wire (a so-called Pt-100 or Pt-1000 sensor; the resistance is either 100 or 1,000 Ω at 0°C). A sensor is typically embedded in thermal conductivity paste and mounted in a stainless-steel housing. Temperature is one of the most reliable measures in bioprocesses but not necessarily the most accurate, because calibration of the absolute value is very cumbersome. With a sound control system it is possible to obtain a precision of ±10 mK in laboratory-scale bioreactors, whereas ±500 mK is usually acceptable in large-scale reactors.

18.3.2. Pressure

Pressure is of paramount interest during sterilization for safety reasons. A variety of sterilizable sensors exist, e.g., piezoresistive, capacitive, or resistance-strain gauge sensors, but not all of them are sufficiently temperature compensated. The direct dependence of microorganisms on pressure changes is negligible provided they do not exceed many bars (1 bar = 10^5 Pa) (5, 61, 72, 170). Both partial pressure of dissolved gases and their solubility are affected by the total pressure (p): an increase of pressure increases partial pressure and solubility:

$$p_{gas} = y_{gas}\, p$$

and

$$c_{gas} = k\, p_{gas}$$

where y_{gas} is the molar fraction of a component in the gas mixture, p_{gas} is its partial pressure, c_{gas} is the dissolved concentration, and k is the reciprocal of the Henry coefficient, for instance: 36.5 mg liter^{-1} bar^{-1} for oxygen at 30°C in pure water. A reduction of infection risks by a controlled overpressure is desirable.

18.3.3. Fluxes of Gases and Liquids

Previously, gas fluxes were determined by using so-called rotameters or turbine flowmeters. The signals of these instruments depend greatly on the pressure conditions, and the useful dynamic range is low (usually less than 1:10). Today, the use of thermal mass flowmeters is routine. Their advantage is that they directly determine a mass flow rather than a volume flow. However, the latter can be converted to a mass flow value provided that temperature and pressure during measurement are known.

Thermoanemometers are also available for liquids. However, the liquids must be "clean" enough to avoid fouling, and the instruments currently available have a restricted dynamic range and are currently limited by a low maximum. Magnetic-inductive instruments can be used for liquids containing particles, but they need a certain minimal flow; the fluid must contain ions. Coriolis-type measuring cells determine a mass flow and need a certain minimal flow as well. Sensors determining both the frequency and mechanical forces of circular eddies behind a solid obstacle in the flow are independent of density, temperature, and pressure. The most reliable and independent method is a gravimetric one. The derivative of weight loss of a reservoir or weight gain of a harvest vessel with respect to time is the mass flow. Most of the good devices are expensive but much more suitable for sterile operation than the use of metering pumps or peristaltic pumps.

18.3.4. Weight and Volume

For bioreactors, in which the biosuspension is vigorously agitated and aerated, the gravimetric principle is the only acceptable one to determine the working mass or working "volume." Reactors and vessels are usually mounted on balances or load cells. Larger reactors require more load cells (3 to 4) in order to improve accuracy and precision. Connections to peripheral equipment must be force-neutral.

18.3.5. Stirrer and Pump Speed

Tacho-generators on a rotating shaft product a direct current with a voltage linearly proportional to the shaft speed. Magnetic-inductive or optical impulse counters are very reliable noncontact transducers, sufficiently accurate, and inexpensive; the same holds true for the additionally necessary electronic circuits.

18.3.6. Electrical Power

It is the electrical power consumption of the motor that has to be paid for, not the shaft power of a stirrer. It is therefore reasonable to monitor the effective power consumption, including the fraction necessary to compress air. Standard watt-meters are sufficient for this purpose. If, however, the power input of the stirrer shaft should be of importance (very unlikely in small-scale reactors), a resistance strain gauge can be glued or welded to the rotating shaft, and the resistance change of the strain due to tension and/or compression caused by the torque on the shaft can be measured. This signal needs to be transmitted; nowadays, small battery-powered wireless transmitters are used.

18.3.7. pH

There is only one reasonable possibility to keep pH within a narrow range in shake-flask cultures, namely, the use of a very strong buffer, usually phosphate buffer. This is why many culture media contain a tremendous excess of phosphate.

The pH of biosuspensions is measured potentiometrically by using sterilizable glass electrodes filled with liquid or gel electrolytes. A brief comparison of properties is given in reference 32. The glass electrodes develop a gel layer when dipped into an aqueous solution. Protons can easily diffuse through this layer, and pH changes cause ion diffusion processes, shifting the electrode potential. This is measured against a reference electrode, usually Ag/AgCl. The electric circuit is closed via a porous diaphragm that physically separates the reference electrolyte from the measuring solution. The measured potential, U, depends on the construction (U_{st}), temperature (T), ion charge (z) (1 for a proton), Faraday (F), and gas (R) constant and on the target, the proton activity (a_P):

$$U = U_{st} + \frac{RT}{zF}\ln(a_P)$$

The pH is the negative decadic logarithm of proton activity, $-\log(a_P)$.

Spoiling of the reference electrolyte or electrode is one of the major problems during long-term cultivations. Monzambe et al. (108) reported discrepancies of one pH unit caused by clogging of the porous diaphragm. Pressurization of the electrolyte or use of a second "bridge"-electrolyte are useful solutions to this serious problem. Alternatives to the glass electrode are optical measurements of pH (1, 23) or use of pH-sensitive field effect transistors (136). These alternatives are, however, not yet mature enough for routine use.

pH is one of the variables often controlled in bioprocesses operated in bioreactors because enzymatic activities and, therefore, metabolism are very sensitive to pH changes. pH can be maintained within a few hundredths of a pH unit, provided mixing time is sufficiently small.

If pH is well controlled, it is useful to monitor the added quantities of alkali and/or acid as well, because they reflect the changing activities of the culture, e.g., the transition from one physiological state to another.

18.3.8. Redox Potential

Redox potential is measured potentiometrically with electrodes made of noble metals (Pt, Au). The mechanical construction is similar to that of pH electrodes. Accordingly, the reference electrode must meet the same requirements. The redox potential, as measured by a redox electrode, is related to an "overall availability of electrons" rather than to a specific compound. The extracellular redox measurement is very instructive, specifically, under microaerobic conditions where the pO_2 sensor signal becomes inaccurate (164). The signal generation is faster than that of pO_2 because the diffusion through the membrane is omitted (29). It should be remembered that pH is one of the determinants of the redox potential and, therefore, variations in pH do cause changes of the redox signal. The use and control of redox potential is reviewed in reference 74. Reports on the successful application of redox sensors (see, for example, references 8 and 62) are confined to detailed description of observed phenomena rather than their interpretation.

The application of a redox sensor in a control loop has been reported (100) for controlled xylanase production of *Bacillus amyloliquefaciens* by defined oxygen limitation: redox electrodes refer essentially to dissolved oxygen concentration below 10 μmol liter^{-1} O_2. This property was also used to determine the quality of anaerobic processes (145). Berovic and Cimerman (9) used closed-loop control of redox potential for improving citric acid production.

18.3.9. Oxygen Partial Pressure (pO_2)

Oxygen solubility is low in aqueous solutions, namely, 36 mg liter^{-1} bar^{-1} at 30°C in pure water. The concentration of dissolved oxygen is difficult to measure, but in equilibrium, the partial pressure is proportional to the concentration: $p_{gas} = H\,c_{gas}$; H is known as the Henry coefficient and depends on temperature as well as on the concentrations of other dissolved components (137). This coefficient is therefore very likely to change during (batch or fed-batch) cultivation. However, the partial pressure is much easier to measure. The signal is usually expressed in engineering units such as [mbar], [torr], or [% of air saturation]. Generally, oxygen is reduced via a cathode operated at a polarizing potential of 600 to 750 mV, which is generated either externally (polarographic method) or internally (galvanic method) (16, 27, 52, 118).

The cathode reaction is:

$$O_2 + 2\,H_2O + 4\,e^- \rightarrow 4\,OH^-$$

and the corresponding anode reaction is:

$$4\,Ag + 4\,Cl^- \rightarrow 4\,AgCl + 4\,e^-$$

A membrane separates the electrolyte from the medium to create some selectivity for diffusible gases. The membrane made of silicon or Teflon or both (sandwiched) is responsible for the dynamic characteristics of the sensor, which are diffusion controlled. The measured current depends on the geometry of the electrode, the diffusion coefficient and solubility of O_2 in the membrane, the surface area of the cathode, and the target, the pO_2; it depends inversely on the thickness of the membrane.

Merchuk et al. (101) investigated the dynamics of oxygen electrodes when analyzing mass transfer. *Bacillus subtilis* cultures respond rapidly to changes of pO_2 around 80 to 90 ppb by changing the ratio of acetoin and butanediol produced (29, 104). This fact could be used for the characterization of a bioreactor's oxygen transport capabilities. A control loop for low pO_2 (<100 ppb) based on a fast but nonsterilizable sensor (Marubishi DY-2) has been devised (50).

18.3.10. Oxygen in the Gas Phase

Analyzers of oxygen in the gas phase are often based on its paramagnetic properties. Any change of the mass concentration of O_2 in the measuring cell affects the density of a magnetic field and thus the forces on any (dia- or para-) magnetic material in this field. These forces on, for instance, an electro-balance can be compensated electrically and the current can be converted into mass concentrations. Further conversion into a molar ratio, e.g., % O_2, requires the knowledge of total pressure in the measuring cell or, at least, frequent (re)calibration. The same holds for magneto-acoustic monitoring principles. Analysis of O_2 as well as CO_2 in the exhaust gas is becoming generally accepted and is likely to be applied as a standard measuring technique in bioprocessing because the analysis does not interfere with the sterile barrier. It is possible to select the exhaust gas lines from several reactors via a multiplexer in order to reduce costs. However, it should be taken into account that the time delay of measurements can be in the order of several minutes, depending on how the gas is transported (active, passive) and pretreated (drying, filtering of the gas aliquot).

18.3.11. Carbon Dioxide Partial Pressure (pCO_2)

pCO_2 can be measured indirectly. The pH value of a bicarbonate buffer, separated from the medium by a gas-permeable membrane, drops whenever CO_2 diffuses into this compartment and vice versa; the pH in the electrolyte chamber depends on the logarithm of pCO_2 in the measuring solution and on electrode properties (lumped together in K^*) (121). Either a glass electrode or optical principles (157) can be used for pH determination:

$$pH_{ElectrolyteChamber} = pK^* - \log(pCO_{2,MeasuringSolution})$$

This direct signal is qualitatively very informative since it represents the logarithm of a growth-associated variable. If growth is exponential, the time trajectory of this signal is most probably linear, and this can be resolved easily by the human eye. This signal is therefore extremely valuable for rapid checking of the growth stage (in a batch culture). If the numerical value of pCO_2 is of interest, rearrangement gives:

$$pCO_{2,MeasuringSolution} = 10^{(pK^* - pH_{ElectrolyteChamber})}$$

CO_2 affects microbial growth in various ways according to its appearance in catabolism as well as in anabolism. Morphological changes, (see, for example, reference 26) and variations of growth and metabolic rates (65, 116) in response to pCO_2 have been demonstrated.

A step up in external pH results in a pCO_2 downward spike and vice versa because of the equilibration of $CO_2 \cdot H_2O$ and HCO_3^- in the biosuspension. Pressure shifts in the range of 1 to 2 cause pCO_2 fluctuations to an extent of >10%. Mass transfer is assumed to control the dynamics of CO_2 equilibration. The biocarbonate buffer solution in the electrode must be replaced regularly owing to the limited capacity. Otherwise, baseline drifts occur and the dynamic behavior of the electrode deteriorates.

18.3.12. Carbon Dioxide in the Gas Phase

CO_2 in the gas phase can be determined by means of its significant infrared absorbance at wavelengths $\lambda < 15$ μm, particularly at 4.3 μm (105), or with photoacoustic devices. CO_2 flux measurements at very large scale are among the simplest that can be done and probably are also important for economic reasons, for instance, in the brewing industry; this has, indeed, been selected (138–140) to get important on-line information for automatic control of such processes. Knowledge of the mass of CO_2 produced is mandatory for calculation of the carbon recovery.

Supercritical CO_2—an interesting extraction fluid—was found to be moderately tolerated by yeasts (61, 85). It is most likely that an optimum CO_2 level exists that is generally accepted for mammalian cells but also reported for bacteria, e.g., the growth rate of *Escherichia coli* (79, 126) or the biomass yield, glucose uptake, and ethanol production of *Zymomonas mobilis* (116).

Park et al. (191) assumed a linear correlation between biomass growth rate and carbon dioxide evolution rate and exploited this model for the estimation of cell concentration, an elegant tool for processes using technical media, such as highly colored molasses media with large amounts of particles. We have designed a simple algorithm to estimate the specific growth rate from exhaust-CO_2 data on-line, which also requires the assumption to hold true that q_{CO_2} is proportional to μ [see also the use of log(pCO_2) in section 18.3.11].

18.3.13. Gas Balance

The respiratory quotient (RQ) can be calculated directly when the composition of the fresh and exhaust gas is known. The oxygen consumption rate and carbon dioxide evolution rate can be calculated when the gas flow rate is additionally known, and the respective specific rates can be calculated provided the biomass in the reactor is known. All these values are interesting because they characterize physiology, specific performance of a reactor, or total performance of plant.

The gas phase balance for gaseous components is:

$$\frac{d(V_G y_G)}{dt} = \text{MAFR}^{\text{in}} y_G^{\text{in}} - \text{MAFR}^{\text{out}} y_G - \text{GasTransfer}_{G \leftrightarrow L}$$

and the liquid balance for gaseous compounds is:

$$\frac{d(V_L c_L)}{dt} = \text{GasTransfer}_{G \leftrightarrow L} + \text{Consumption/Production} + F_L^{in} c_L^{in} - F_L^{out} c_L$$

V_G is the active gas volume in the reactor, V_L is the volume of the continuous liquid phase, y_G is the molar fraction of a gas component in the gaseous phase, and c_L is the dissolved concentration in the liquid phase. MAFR is the mass air flow rate, and F_L is the flow rate of the liquid phase. $y^{\text{out}} = y$ provided the gas phase is well mixed. In the notation used here, $\text{GasTransfer}_{G \leftrightarrow L}$ has a positive value for transport from gas to liquid and a negative value for liquid to gas direction, i.e., opposite for O_2 and CO_2. Likewise, Consumption has a negative sign and Production a positive sign. Be aware of identical engineering units when using the formulas shown above; conversion factors, e.g., from mole to milligram, are not included here.

Although the differentials are zero in steady state only, their numerical values are very small compared to the right-hand side terms in the balance equations. Therefore, it is reasonable to assume pseudo-steady state, i.e., let the differentials vanish. It is further plausible to assume that the transport terms in the liquid phase are negligibly small because of the low gas solubility and to let the contributions of the $F_L c_L$ terms vanish, too. However, some care is recommended for CO_2, which equilibrates with bicarbonate under pH-neutral conditions or even with carbonate under alkalophilic conditions; then, this assumption is no longer justified. However, with these simplifying assumptions, one can combine the two balances into one by substituting the term for $\text{GasTransfer}_{G \leftrightarrow L}$:

$$0 = \text{MAFR}^{\text{in}} y^{\text{in}} - \text{MAFR}^{\text{out}} y + \text{Consumption/Production}$$

The values of MAFR, in, and out, and the values of y, in, and out, are either known a priori or measured. Consumption and Production can now be calculated from the gas flow and composition data. However, it is important to note that the values of MAFR need not be identical for in and out. This is always the case when Consumption (of O_2) and Production (of CO_2) are not equimolar. However, this does not mean that both gas flows, in and out, need to be measured. One measurement is sufficient if the assumption holds true that all components of the gas mixture used are inert with the exception of O_2 and CO_2. Then the inert gas entering the reactor must also quantitatively leave the reactor, provided the reactor pressure, volume, and temperature do not change. The respective balance of inert gases is simply:

$$\text{MAFR}_{\text{inert}}^{\text{in}} y_{\text{inert}}^{\text{in}} = \text{MAFR}_{\text{inert}}^{\text{out}} y_{\text{inert}}$$

with $y_{\text{inert}} = 1 - y_{O_2} - y_{CO_2}$.

Usually, the mass flow rate of the fresh gas is measured directly with a thermoanemometer; hence, the exhaust gas flow rate can be calculated by solving the inert gas balance and knowing the gas compositions. If a volumetric method is used to determine the gas flow, the temperature, pressure, and water vapor pressure during the measurement must be known to calculate the mass flow from the measured volumetric value.

The dimensions of the terms Consumption and Production are [mass-of-gas time^{-1}]. They characterize the turnover of the entire reactor and, of course, depend on reactor volume and biomass density. If one divides these values by the reactor volume, V_L, one gets characteristic values for the reactor performance, namely, the transfer rates in the dimensions [mass-of-gas volume^{-1} time^{-1}]. If one further divides by the cell density, x, one gets characteristic values for the physiological performance of the cells in [mass-of-gas cell-mass^{-1} time^{-1}]. The respective variables are called specific oxygen consumption rate, q_{O_2}, and specific carbon dioxide evolution rate, q_{CO_2}. The latter diverges from the true specific carbon dioxide production rate the higher the pH of the culture, because the fraction of CO_2 trapped as bicarbonate or even carbonate is not included in this measure (of exhaust gas); under acidic conditions, both variables cannot be distinguished experimentally. The ratio of produced CO_2 and consumed O_2 is called the respiratory quotient, RQ. It is another state variable characterizing the respiratory state of a population. It always has a positive value and can be simply calculated from either $|q_{CO_2}/q_{O_2}|$ or from $|x\ q_{CO_2}/x\ q_{O_2}|$ or from $|V_L\ x\ q_{CO_2}/V_L\ xq_{O_2}|$ which is the same as |Production/Consumption|. The correct engineering units for RQ are [mol mol^{-1}]. If the pseudostationary assumption holds true, the RQ can be derived from the gas composition data alone. Since those can be made available on-line, the RQ can be calculated on-line simply as:

$$\text{RQ} = \frac{y_{CO_2} - y_{CO_2}^{\text{in}} - y_{O_2}^{\text{in}} y_{CO_2} + y_{O_2} y_{CO}^{\text{in}\,2}}{y_{O_2}^{\text{in}} - y_{O_2} - y_{O_2}^{\text{in}} y_{CO_2} + y_{O_2} y_{CO_2}^{\text{in}}}$$

where the product terms in this formula account for the inert gas balance. Since the RQ is not measured directly but is calculated from other measures, it is called a "software sensor."

The GasTransfer$_{G \leftrightarrow L}$ terms used above can also be evaluated directly assuming a pseudo-steady-state condition. For oxygen, this term is usually expressed as $|OTR| = k_La$ $(c_{O_2}{}^* - c_{O_2})$. If the values for the dissolved concentration, $c_{O_2}{}^*$ (saturation) and c_{O_2} (current), can be approximated by calculation from pO_2 assuming a reasonable Henry coefficient, the k_La value can also be estimated on-line from the gas balance.

However, the limits of assuming a pseudo-steady state condition become obvious at this point. Applying the same for carbon dioxide must result in an identical k_La for CO_2. This is not likely because the diffusion coefficients for O_2 and CO_2 differ: 2.3 10^{-9} for O_2 and 1.6 10^{-9} m^2 s^{-1} for CO_2. Hence, the k_La values should differ—depending on the model applied—by the ratio or by the square root of these coefficients, which is 1.45, or 1.2, respectively (114). Nevertheless, assuming a pseudo-steady state is reasonable and, most useful and important, permits calculation of many important data.

18.3.14. Culture Fluorescence

Fluorescence measurements have been used both for characterization of technical properties of bioreactors (42, 82, 134) and for basic scientific investigations of cellular physiology. Technically, either intra- or extracellular fluorophores are excited by visible or UV light. The light generated by an appropriate source is filtered according to the fluorophor of interest prior to immission into the reactor. Fluorescent light is emitted by the excited fluorophores at a longer characteristic wavelength. Only the backward fluorescence can be collected with appropriate (fiber) optics. It is filtered and the residual light is detected by a sensitive photodetector. Descriptions of typical sensors are given in references 10, 132, and 133.

Intensity measurement is prone to many interferences and disturbances from the background. These drawbacks can be avoided by measuring the fluorescence life time, but this is more demanding (4, 99, 147).

Most investigators measure NAD(P)H-dependent culture fluorescence, but other fluorophores are also interesting. Humphrey (58) gave a (nonexhaustive) survey of the historical evolution of fluorescence measurements for bioprocess monitoring. All fluorescence data have to be interpreted very carefully. Turbidity of the culture suspension should be low, and the bubble distribution should remain constant (10).

NAD(P)H-dependent culture fluorescence has mainly been exploited for metabolic investigations (66, 78, 122–124, 148). The signal is sensitive to variables such as substrate concentration or oxygen supply. Thus, all attempts to exploit this signal as a biomass sensor (172) are limited to conditions in which no metabolic alterations occur (92, 141, 142). It is well known that a mechanistic or causal-analytical interpretation of the signal trajectory in secondary metabolite cultivations may be very difficult (112). The outstandingly rapid principle of fluorescence measurements served excellently for the controlled suppression of ethanol formation during continuous baker's yeast production (102).

18.3.15. Biomass

Many commercially available biomass sensors rely on optical measuring principles; others exploit filtration characteristics, density changes of the suspension as a consequence of cells, or (di)electrical properties of suspended intact cells. Direct comparisons of some representative sensors to estimate biomass in bacterial and yeast cultures have been made (28, 75, 115, 165). These studies are of importance because the sensors were mounted in situ and used in parallel. Most of the sensors measured optical density (OD); one measured autofluorescence of the cultures (fluorosensor) and another was a capacitance sensor ("Bugmeter").

18.3.15.1. Optical Density

Commercially available OD sensors are based on the determination of either transmission, reflection, or scatter of light, or a combination thereof. A direct, a priori calculation of dry weight concentration from any OD measurement is not realistic, but the systems can be calibrated from case to case. The wavelength for bacteria should be chosen in the visible range; for larger organisms, infrared. Even large plant cell cultures (155) or insect cell cultures (7) can be estimated with turbidimetric methods. Many media absorb increasingly with decreasing wavelength; therefore, green filters, infrared diodes, laser diodes, or lasers between 780 and 900 nm are often used. Inexpensive variants can be made by using stabilized light-emitting diodes (emitting at around 850 nm) or arrays thereof (47); modulation with a few 100 Hz ("light chopping") should be used in order to minimize the influence of ambient light (173). Fiber sensors with high-quality spectrophotometers outside the reactor in a protected room are a valuable but probably expensive alternative (22). Interferences with gas bubbles or particulate matter other than cells are common to almost all sensors.

The FundaLux system, for instance, aspirates a liquid aliquot with a Teflon piston into an external glass cell, allows for some (selectable) time (typically 2 min) to degas, measures transmission versus an air blank, and releases the aliquot back to the reactor. Problems with contamination have been communicated, since the measuring cell is external to the bioreactor and the sensor is probably insufficiently sterilized in situ.

A sensor based on the same principle of sample degassing in a void volume but mounted completely inside the reactor (Foxboro/Cerex) has been described (57). The minimal time interval between individual measurements is 30 s. Both 90° scatter and transmission measurements could be made simultaneously. A linear correlation between OD and *Saccharomyces cerevisiae* density from 0.1 to 150 g liter^{-1} is claimed for this instrument (46). Geppert and coworkers (33, 34) have used a similar method but a different instrument to measure a suspension aliquot outside the bioreactor and reported a fairly good linear correlation between OD and biomass concentration.

The LT 201 (ASR/Komatsugawa/Biolafitte) instrument protects the optical path from gas bubbles by a cylindrical stainless-steel screen arranged around this region. The mounting position of the sensor is critical. There are several reports of good experience (28, 54, 168, 169), and the signal was found so reliable that it was used for automatic control of fed-batch processes (67). Iijima et al. (59) described a sensor that measures both transmission (1 fiber) and 90° scatter (2 fibers), which may allow compensation for interferences mathematically. The MEX-3 sensor (BTG, Bonnier Technology Group, Lausanne) compensates internally for errors due to deposition on the optical windows, temperature, or aging of optical components; this is made possible by evaluating quotients of intensities from four different light beams (straight and cross beams from two emitters to two detectors; multiplexed). The Monitek sensor has a special optical construction (prior to the receiver, called a spatial filtering system) to eliminate scattered light not originating from particles or bubbles in the light path. Aquasant minimizes interference by depositions on the optical window by the special design of the precision receiver optics of the AF 44 S sensor. Other sensors, used in different industrial areas, are equipped with mechanical wipers.

18.3.1.15.2. Electrical Properties

The capacitance of a suspension at low radio frequencies is correlated with the concentration of the suspended phase of fluid elements that are enclosed by a polarizable membrane, i.e., intact cells (3, 45, 73, 96, 97). The capacitance covered by the biomonitor (Aber Instruments, Aberystwyth, Wales) ranges from 0.1 to 200 pF, the radio frequency from some 200 kHz to 10 MHz. A severe limit to this principle is the maximally acceptable conductivity (of the continuous phase) of approximately 24 mS cm^{-1}. However, this conductivity is easily reached in more concentrated media necessary for high-cell-density cultures. Noise is also created by gas bubbles.

18.3.15.3. Thermodynamics

A noninvasive method is the determination of the heat generated during growth, which is proportional to the amount of active cells (14). Under well defined conditions, calorimetry can be an excellent tool for the estimation of total (active) biomass (11, 12, 93, 160), even for slowly growing organisms such as hybridoma cells or for anaerobic bacteria growing with a low biomass yield (135, 152, 159).

The net heat released during growth depends on both the biomass concentration and the metabolic state of the cells. Its general use in biotechnology has been reviewed (151). A theoretical thermodynamic derivation for aerobic growth gives a heat yield coefficient $Y_{Q/O}$ of 460 kJ mol^{-1} O_2. This prediction was experimentally confirmed as an excellent estimate: the average value found in many different experiments was 440 ± 33 kJ mol^{-1} O_2 (11). Among three different approaches applied (micro-, flow- and heat flux calorimetry), heat flux calorimetry is certainly the best choice for bioprocess monitoring (95). In a dynamic calorimeter, the temperatures of reactor and jacket are measured. Various heat fluxes (e.g., heat dissipated by stirrer or lost due to vaporization of water) need to be known in order to calculate the heat flux from the bioreaction. The global heat transfer coefficient k_w (W K^{-1} m^{-2}) can be simply determined by electrical calibration; the heat exchange area A (m^2) is usually constant; both parameters can be lumped together ($k_w A$):

$$q = k_w\, A\, (T_R - T_J)$$

where q (W) is the heat exchanged. The temperature controller needs appropriate tuning; vanKleeff et al. (159) gave practical hints for tuning a very simple system. Heat flux calorimeters are bioreactors equipped with special temperature control tools. They provide a sensitivity that is approximately two orders better than that of micro calorimeters (13, 93). The larger the scale, the simpler is heat flux calorimetry.

18.4. EX SITU METHODS IN BYPASS OR EXIT LINE

18.4.1. Sampling

"Sampling" does not necessarily mean manual operation; fully automatic sampling via various interfaces is routine today. Samples must be removed from the reactor in such a way that they can be analyzed with devices that are not (yet) suitable or available to be mounted in situ. Depending on the analyte of interest, i.e., whether it is soluble or in the dispersed phase, one needs to sample either the entire biosuspension or just the filtrate or supernatant.

No separation of phases or contamination can be allowed during sampling when the cells are of interest. This requires the use of no-return valves and probably some repetitive sterilization of the sampling line(s). Time is critical for the representativity of the analyses, because cells continue to consume substrates during sampling (144).

Whenever the analyte of interest is not in or on the cells, sample removal via a filtering device is the most reasonable solution. In my experience, bypass filters should be operated in cross flow mode with ≥2 ms^{-1} of superficial liquid velocity. Then, a useful life time of a few weeks can be achieved even in cultures of filamentous organisms.

18.4.2. Interfaces

The interface between the reaction site, the monoseptic space, and the site of analysis is of decisive importance: the monoseptic space must be protected from contamination, and the sample specimen must be transported to the analytical device without significant change in composition. This goal can be achieved by various strategies with varying effort/effect ratio. Rapid sampling is, in any case, advantageous. A couple of methods separate catalysts from reactants by filtration, and other approaches are poisoning or inactivating the catalyst by either addition of strongly inhibitory material (e.g., heparin or KCN, both of which may interfere with the analytical method) or temperature variations (heating or cooling) (98). These aspects are not restricted to samples from monoseptic bioprocesses; they are equally important for environmental analyses (35).

18.4.3. Flow Injection Analysis (FIA)

Ruzicka and Hansen (130) characterized FIA as: "information gathering from a concentration gradient formed from an injected, well-defined zone of a fluid, dispersed into a continuous unsegmented stream of a carrier." Accordingly, basic components of FIA equipment are a transport system consisting of tubing, pumps, valves, and a carrier stream into which a technical system injects a sample or reagents (44, 129). A (bio)chemical reaction, which is typical for the substance to be measured, usually occurs during the flow, and products or residual (co)substrates are measured by the sensing system. However, physical sample treatment can also be easily implemented, such as extraction (109), separation (161), or diffusion (86). The detector choice is highly flexible. Optical or electrical devices have been widely used, in addition to thermistors, mass spectrometers, biosensors, microbial electrodes, and others.

FIA does not generate continuous signals due to the injections. But there are several important advantages: a high sampling frequency (up to >100 h^{-1}), small sample volumes, low reagent consumption, high reproducibility, and total versatility with respect to sensing methods. Even separation of compounds by high-pressure liquid chromatography prior to FIA analysis has been reported (171). Kroner (77) reported on good automation properties of FIA used for enzyme analysis. There is no interference with the sterile barrier since the entire apparatus works outside the bioreactor. Special emphasis must therefore be given to the sampling device interfacing the sterile barrier. FIA easily allows one to meet validation requirements. For instance, two different FIA methods for penicillin V monitoring were compared (18).

Comprehensive surveys of various applications have been reported (24, 111). FIA has been used for on-line determination of glucose, to estimate biomass directly or indirectly by means of an extended Kalman filter, to determine chemical oxygen demand (COD), for water monitoring, measuring amino acids, enzymes, or peptides, antibiotics, DNA and RNA, simple metabolites such as lactic acid or ethanol, and many more. A novel development is the combination with cytometry (83, 84, 131). More and more are biosensors used as detectors in FIA systems.

A high degree of automation is, however, necessary and desirable (55, 56). FIA is expected to become one of the most powerful tools for quantitative bioprocess monitoring in the near future provided that nonlinear calibration models are also used and the data evaluation techniques improve (15, 31, 53, 60, 94). Wu and Bellgardt (166) were able to detect faults of the analytical system automatically. The present tendency is toward using multichannel FIA systems that work either in parallel or with sequential injection (6, 103, 158, 167), miniaturization of FIA devices (17, 43, 153, 154), and automation (30, 55, 56, 120). A FIA can also be operated without injection and gives valuable results and even a continuous signal. We have stained DNA within yeast cells removed from the reactor and could quantify the amount of DNA on-line, thus giving evidence for the cell-cycle dependence of oscillations (143).

18.4.4. Field Flow Fractionation (FFF)

FFF is an elution technique suitable for molecules with a molecular weight of >1,000, up to particles with a size of some 100 μm. The separating, driving, external field forces are applied perpendicular to a liquid carrier flow, causing different species to be placed in different stream lines. Useful fields are gravity, temperature, cross flow, electrical charge, and others (36–39). The range of the (molecular) size of the analytes usually exceeds what can be determined with classical laboratory analytical methods. Reports are found on characterization of proteins and enzymes, viruses, separation of human and animal cells, isolation of plasmid DNA, or particle size distribution of polymers. The approach is relatively new in biotechnology, but it is as simple to use as FIA. Langwost et al. (80) provide a comprehensive survey on various applications in biomonitoring.

18.4.5. Chromatography

Both liquid chromatography and gas chromatography have been applied in numerous cases to off-line analyses of biotechnological samples, but the on-line application is recently developing. Progress is a function of improving interfacing, sample pretreatment, automation, and data evaluation. A nonexhaustive list of some examples includes monitoring of ethanol, butanol-acetone, acetaldehyde, fusel alcohols, CO_2, components of highly polluted waste water, cephalosporin, penicillin and precursors, 3-chlorobenzoate, naphthalenesulfonic acid, amino acids, alkanes, alkanoic acids, sugars, and many more.

18.4.6. Mass Spectrometry (MS)

MS has been applied mainly for the on-line detection and quantification of gases such as O_2, CO_2, N_2, H_2, CH_4, or H_2S and volatiles (alcohols, acetoin, butanediol, carboxylic acids). The detection principle allows instantaneous and simultaneous monitoring, and consequently control, of important volatile metabolites. The principles, sampling systems, control of the measuring device, and application of MS for bioprocesses have been summarized (48, 49, 51). Samples are introduced into vacuum ($<10^{-5}$ bar) via a capillary (heated, stainless-steel or fused silica, 0.3 × 1,000 mm or longer) or a direct membrane inlet, for instance, silicon or Teflon (21, 149). Electron impact ionization is commonly applied. Mass separation can be obtained either by quadruple or magnetic field, and the detection is performed by (fast and sensitive) secondary electron multipliers or (slower and less sensitive) Faraday cups. Almost all volatile substances can be analyzed from the gas phase using a capillary inlet MS, provided their partial pressure in the exhaust gas is ≥ 1 μbar (117). Software for data evaluation is necessary and available (107). Closed-control loops based on H_2 or CH_4 measurements have been set up (87, 163).

An important problem with membrane inlet systems is quantification, because the membrane behavior is quite unpredictable (41, 128), but the membrane inlet can be even more rapid than the capillary (19, 20). This problem can be solved by combining two techniques: the discrete concentration signal of an on-line gas chromatography can be used to recalibrate the continuous partial pressure signal of the MS (29).

MS requires expensive equipment. However, it should be taken into account that automatic multiplexing of different sample streams is possible, and a great variety of different substances can be determined simultaneously.

18.4.7. Biosensors

The rationale for using biosensors is to combine the high specificity of biological components with the capabilities of electronic tools (i.e., the usual sensors). Biosensors consist of a sensing biological module of either catalytic (e.g., enzymes, organism) or affinity reaction type (e.g., antibodies, cell receptors) in intimate contact with a physical transducer. The former does not withstand sterilization, and the latter converts the chemical finally into an electric signal (64). Principles and typical examples have appeared (25, 63, 68–71). Some principles are sketched in the following:

Electrochemical transducers are based on an amperometric, potentiometric, or conductometric principle. Further, chemically sensitive semiconductors are under development. Sensors are commercially available today for carbohydrates, such as glucose, sucrose, lactose, maltose, galactose, the artificial sweetener Nutra Sweet, urea, creatinine, uric acid, lactate, ascorbate, aspirin, alcohol, amino acids, and aspartate.

Optical biosensors typically consist of an optical fiber, which is coated with the indicator chemistry for the material of interest at the distil tip. The quantity or concentration is derived from the intensity of absorbed, reflected, scattered, or re-emitted electromagnetic radiation. These sensors are ideal for miniaturization, are of low cost, and the fiber optics are sterilizable (even if the analyte is not!).

Calorimetric biosensors exploit the fact that enzymatic reactions are exothermic (5 to 100 kJ mol^{-1}). The biogenic heat can be detected by thermistors or temperature-sensitive semiconductor devices. Applications to measure biotechnologically relevant substances are ATP, glucose, lactate, triglycerides, cellobiose, ethanol, galactose, lactose, sucrose, penicillin, and others.

The piezo-electrical effect of deformations of quartz under alternating current (at a frequency in the order of 10 MHz) is used by coating a crystal of an acoustic/mechanical sensor with a selectively binding substance, e.g., an antibody. When exposed to the antigen, the antibody-antigen complex will be formed on the surface and shift the resonance frequency of the crystal proportional to the mass increment, which is, in turn, proportional to the antigen concentration. A similar approach is used with surface acoustic wave detectors or with the surface plasmon resonance technology.

18.4.8. Biomass

High-density cultures present a problem to OD measurements because of the inner filter effect; i.e., light intensity is lost due to absorbance and scatter by cells over the length of the light path. A very elegant method to obtain a great dynamic range for OD has been published (113). The workers exploited the dilution equipment of a flow injection analysis and measured a steady-state absorbance. The group of Reuss (81, 125) has further developed an automatic filtration device (110, 156) that allows one to estimate the biomass concentration in a relatively large sample (approximately 100 ml) according to its filtration properties. Each sample is filtered through a fresh filter, and the flux of filtrate is monitored as well as the buildup of the filter cake. The latter restricts this method to filamentous organisms.

18.5. SOFTWARE SENSORS

Software sensors are virtual sensors that calculate the target variable from related physical measurements. There must always be a model available that relates the measured variable(s) reliably with the target variable. The most prominent software sensor is the respiratory quotient (RQ value), which can characterize the physiological state of a culture.

A complete characterization of the physiological state of a culture involves more than one (measurable) variable at a time. Physiological state estimation requires recognition of complex patterns. With various algorithms used for this purpose (2, 40, 76, 88, 106), it is not the present values alone that are evaluated, but also the recent history of signal trajectories. Data describing the current state and the recent history can be compared with so-called reference patterns. An expert associates data from historical experiments with a typical physiological state. A distinct physiological state can then be automatically recognized either if the current constellation matches any one of the reference sets best or if the match exceeds a predefined degree of certainty, e.g., 60%.

Data trajectories can be translated into trend-qualities via shape descriptors, such as oxygen uptake rate is decreasing (concave down) while redox potential is increasing (linear up) and nitrogen source remains constant. These combinations of trends of the trajectories of various state variables or derived variables define a certain physiological state; the advantage of this definition is that the association is no longer dependent on time and on the numerical values of variables and rates (150).

Calculation of carbon recovery is the only objective check for the correctness of assumptions and results. Bioreactors with a higher degree of (automatic) process analyses are mandatory for this purpose and are, therefore, the only acceptable standard for quantitative metabolic research and process development. Villadsen (162) brought this to the point when he argued that, since it is possible today to determine many state variables such as the biomass concentration or the RNA content with a precision of better than $\pm 1\%$, it is a matter of scientific credibility and responsibility to use the appropriate techniques and exploit the available possibilities; anything else would be ignorance. The carbon recovery is determined as the ratio of the sum of carbon found analytically (at the outlet of a reactor) to the sum of carbon fed initially (to the inlet of the reactor). Theoretically, the recovery must be 1, because bioreactions are chemical and not nuclear reactions; hence, carbon cannot be converted.

18.6. DATA HANDLING AND CONTROL

18.6.1. The Importance of Selecting Data To Keep

Automated measurement and control of bioprocesses, presently an art but certainly to be routine in the near future, generates a tremendous amount of data (91). This requires us to judge the importance of these data for documentation and to reduce data effectively without loss of valuable information. The reduction must not be based on an intention to make one or the other variable a parameter; rather, it must keep a true image of the real data. We adopted a simple algorithm to achieve this goal (89): all data, independent of whether measured or calculated, are treated as variables and kept in a circular buffer (holding some 2 days) in the frequency with which they were generated. This volatile data base serves as a rapid graphic review of the short history. However, only values (data points) of variables that change significantly in time are written to the archive. The significance of a change is judged by a reasonable window individually defined for each variable, including all intended culture parameters, the width of which is usually determined by the noise on the respective signal. In any instance, every "first" data point (of an experiment) is archived, together with a time stamp. A next entry to the archive is made only if the variable moves outside the respective window that had been centered around the last archived value; concomitantly, the window is recentered around the new entry, and so on. This technique ensures that all relevant changes, including those not intended, of any considered variable are trapped and that the dynamics of all signal trajectories are fully documented. The only inconveniences with this data treatment are that one needs to time-stamp every data entry individually and one ends up with nonequidistant data vectors. The benefit is a data-to-archive reduction of usually between 10^{-2} and 10^{-4} and the assurance that no important data are lost.

18.6.2. Exploitation of Available Data

Modern database management systems can provide a professional solution to handle a large amount of data. However, such systems cannot decide which data are valuable and decisive, and at what frequency they are required for either control or documentation. They are just a technical tool to improve the data accessibility and organization. The bottleneck is no longer the availability of computer memory or storage capacity; it is primarily the necessity to concentrate on important and relevant data and to avoid unintentional data losses due to whatever cause.

So-called culture parameters are process variables that are made constant by means of closed-loop controllers. Theoretically, there is not much difference between classical analog and digital controllers. Practically, there are significant differences: plants operated with discrete control units very seldom have free (spare) units to expand the equipment for a new controlled variable. Electronic units in classical technique cannot be changed very much after installation as there are always certain constraints due to the physical elements used in construction. While such a configuration is quite inflexible, a direct digital control system normally provides for system extensions with a minimum of hardware installations by just adding the necessary sensors, wiring, and actuators. Software such as additional configuration, calibration factors, and controller parameters can be updated simply; even the type of controller can be changed easily since it is just another software algorithm that has to be used. As a consequence, controller performance is no longer a compromise; it can be made excellent with respect to performance.

Process documentation is not just a good practice, it is often necessary for regulatory and forensic reasons. Automated systems are governed by rules that can easily comply with the documentation needs. All available data including the temporal state matrix (process variables) provide a sufficiently complete documentation, even a sound basis for tracing back and troubleshooting. Automation of (bio)processes requires the realization of several different functional or hierarchical levels (146). A primary task of a bioreactor is to provide strictly aseptic conditions. This imposes extreme constraints on the use of on-line sensors and analyzers and actuators, still drastically reducing the number of suitable available instruments. Most of the historical problems with electronic components have been solved satisfactorily, and a series of alternatives is commercially available today.

The front end or "slave" part of a hierarchical automation system (Fig. 2) deals with relatively unintelligent tasks that were formerly performed by discrete electrical, electromechanical, or electropneumatical units. A major demand is the real-time behavior of this component. Some of the simple control loops such as those for pressure control, flux control, speed control, etc., require cycle times of $<<1$ s for laboratory and pilot-scale equipment. An operator's interface is mandatory.

The next higher level component, the supervisor or "master," deals with intelligent and more demanding tasks. It should have a comfortable man/machine interface (operator's console and visualization) with extra power to store, retrieve, and organize different sets of data, calculate resource-consuming algorithms (filters, models, complex controllers, patterns, state estimators, etc.), provide a library of recipes, and prepare the desired documentation. It must check the users' authorization and manage the communication with other systems (e.g., coordinating computers or intelligent analytical subsystems). It runs typically with a professional multiuser, multitasking operating system. Further levels, if implemented, must provide a unique,

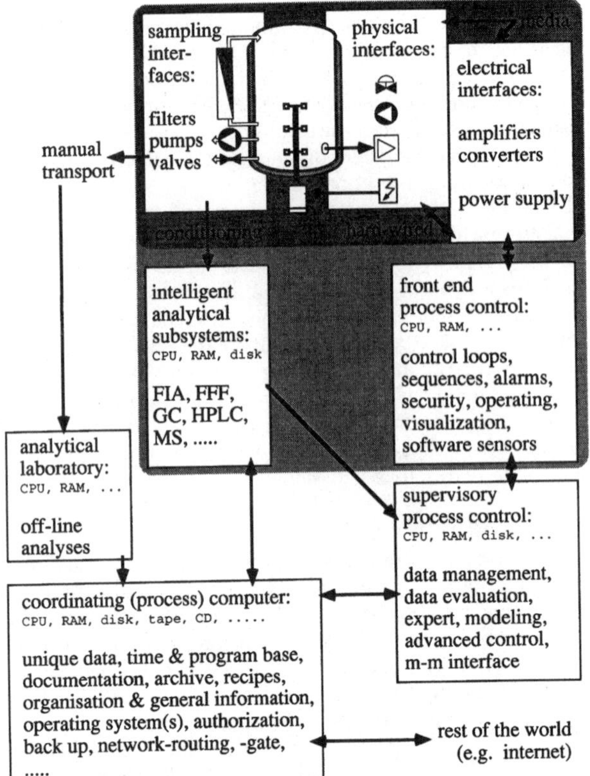

FIGURE 2 Concept of a hierarchical bioprocess automation system. The physical and sampling interfaces are in intimate contact with the bioreactor and provide the sterile barrier; some of them may, in fact, be built in. Wires and tubes or pipes connect these with either analytical subsystems or amplifiers, converters, and power and air supplies; among the converters are also electric pneumatic versions. The data exchange between this hardware level and the front-end process controller is advantageously realized with a professional bus system to reduce the degree of wiring and faults. Up to this level, all components are in close proximity. The front-end component does not necessarily have a storage device; then, the visualization of trends must be outsourced to the next (supervisory) level, which is typically realized on a graphic workstation with plenty of memory and storage space accessible. The major task is sound management of data, including acquisition from lower hierarchical levels, plausibility checking, reduction, evaluation, and storage in an appropriate database. Along with the pure data management, the supervisory controller should provide sufficient resources to allow efficient on-line modeling, to calculate advanced control algorithms, and to hold expert knowledge (rules besides data). It usually serves as a comfortable man-machine (m-m) interface. This level may supervise more than one front-end component, for instance, one for upstream, one for downstream processing, and another one for the bioconversion. In larger installations, e.g., pilot plants or production facilities, it is reasonable to use a plant manager or coordinating computer. This level must give the only and unique time base for all lower-level components as well as the database organization. An important task is the correct rerouting of off-line data delivered from analytical laboratories to the relevant database; the same holds true for data from analytical subsystems that are multiplexed i.e., switched between several reactors or processes. This top-level computer will further provide programs, general documents such as recipes, libraries, backup, and archive facilities for the entire network. Depending on the owner's restrictions, this component may be the end node or act as a fire wall toward the outside world connected via networks.

common time and database and may increase flexibility, power, and comfort deliberately.

Support of this work through the Swiss Priority Program in Biotechnology is greatly acknowledged.

REFERENCES

1. **Agayn, V., and D. Walt.** 1993. Fiber-optic sensor for continuous monitoring of fermentation pH. *Bio/Technology* **11**:726–729.
2. **Albiol, J., C. Campmajo, C. Casas, and M. Poch.** 1995. Biomass estimation in plant cell cultures: a neural network approach. *Biotechnol. Prog.* **11**:88–92.
3. **Asami, K., and T. Yonezawa.** 1995. Dielectric analysis of yeast cell growth. *Biochim. Biophys. Acta* **1245**:99–105.
4. **Barnbot, S., J. Lakowicz, and G. Rao.** 1995. Potential applications of lifetime-based, phase-modulation fluorimetry in bioprocess and clinical monitoring. *Trends Biotechnol.* **13**:106–115.
5. **Bavouzet, J., C. Lafforguedelorme, C. Fonade, and G. Goma.** 1995. The effect of an abrupt stepwise reduction in pressure on the integrity of the eukaryotic and prokaryotic cell envelope. *Enzyme Microb. Technol.* **17**:712–718.
6. **Beck, H. P., and C. Wiegand.** 1995. Development and optimization of a multichannel FIA-cell allowing the simultaneous determination with a multiwavelength photometric device based on light emitting diodes. *Fresenius J. Anal. Chem.* **351**:701–707.
7. **Bedard, C., M. Jolicoeur, B. Jardin, R. Tom, S. Perret, and A. Kamen.** 1994. Insect cell density in bioreactor cultures can be estimated from on-line measurements of optical density. *Biotechnol. Tech.* **8**:605–610.
8. **Berovic, M.** 1987. Redox potential, p. 327–345. *In* Y. U. Otocec and B. Kidric (ed.), *Bioreactor Engineering Course.* Slov. Chem. Soc., Ljubljana.
9. **Berovic, M., and A. Cimerman.** 1993. Presented at the BHRA 3rd International Congress on Bioreactor Bioprocess Fluid Dynamics, Cambridge.
10. **Beyeler, W., A. Einsele, and A. Fiechter.** 1981. On-line measurements of culture fluorescence: method and application. *Eur. J. Appl. Microbiol. Biotechnol.* **13**:10–14.
11. **Birou, B., I. Marison, and U. V. Stockar.** 1987. Calorimetric investigation of aerobic fermentations. *Biotechnol. Bioeng.* **30**:650–660.
12. **Birou, B., I. Marison, and U. V. Stockar.** 1987. Observing microbial growth and product formation by monitoring the heat generated by the culture. *Eur. Congr. Biotechnol.* **3**:105–108.
13. **Birou, B., and U. V. Stockar.** 1989. Application of bench-scale calorimetry to chemostat cultures. *Enzyme Microb. Technol.* **11**:12–16.
14. **Boe, I., and R. Lovrien.** 1990. Cell counting and carbon utilization velocities via microbial calorimetry. *Biotechnol. Bioeng.* **35**:1–7.
15. **Brandt, J., and B. Hitzmann.** 1993. Computer-aided detection of failures in flow injection analysis systems. *Am. Biotechnol. Lab.* **11**(5):78.
16. **Brookman, J.** 1969. The design, construction, and characteristics of a new long-lived steam sterilizable oxygen electrode. *Biotechnol. Bioeng.* **6**:323–335.
17. **Busch, M., J. Schmidt, S. Rothen, C. Leist, B. Sonnleitner, and E. Verpoorte.** 1996. Presented at the 2nd International Symposium on Micro Total Analysis Systems at ILMAC '96, Basel.
18. **Carlsen, M., C. Johansen, R. Min, J. Nielsen, H. Meier, and F. Lantreibecq.** 1993. On-line monitoring of penicillin V during penicillin fermentations: a comparison of two different methods based on flow-injection analysis. *Anal. Chim. Acta* **279**:51–58.
19. **Chauvatcharin, S., K. Konstantinov, K. Fujiyama, T. Seki, and T. Yoshida.** 1995. A mass spectrometry membrane probe and practical problems associated with its application in fermentation processes. *J. Ferm. Bioeng.* **79**:465–472.
20. **Chauvatcharin, S., T. Seki, K. Fujiyama, and T. Yoshida.** 1995. On-line monitoring and control of acetone-butanol fermentation by membrane-sensor mass spectrometry. *J. Ferm. Bioeng.* **79**:264–269.
21. **Cox, R.** 1987. Membrane inlets for on-line liquid-phase mass spectrometric measurements in bioreactors, p. 63–74. *In* E. Heinzle and E. M. Reuss (ed.), *Mass Spectroscopy in Biotechnological Analysis and Control.* Graz, Austria.
22. **Danigel, H.** 1995. Fiber optical process measurement technique. *Optical Eng.* **34**:2665–2669.
23. **Deboux, B., E. Lewis, P. Scully, and R. Edwards.** 1994. A novel technique for optical fibre pH sensing based on methylene blue adsorption. Presented at Optical Fibre Sensors Conference, Glasgow.
24. **Decastro, M. D. L., and M. Valcarcel.** 1995. Flow injection analysis, p. 35–89. *In* W. J. Hurst (ed.), *Automation in the Laboratory.* VCH Publ., New York.
25. **Delaguardia, M.** 1995. Biochemical sensors: the state of the art. *Mikrochim. Acta* **120**:243–255.
26. **Edwards, A., and C. Ho.** 1988. Effects of carbon dioxide on *Penicillium chrysogenum*: an autoradiographic study. *Biotechnol. Bioeng.* **32**:1–7.
27. **Fatt, I.** 1976. *Polarographic Oxygen Sensors.* CRC Press Inc., Cleveland.
28. **Fehrenbach, R., M. Comberbach, and J. Pêtre.** 1992. On-line biomass monitoring by capacitance measurement. *J. Biotechnol.* **23**:303–314.
29. **Filippini, C., J. Moser, B. Sonnleitner, and A. Fiechter.** 1991. On-line capillary gas chromatography with automated liquid sampling, a powerful tool in biotechnology. *Anal. Chim. Acta* **255**:91–96.
30. **Fillippini, C., B. Sonnleitner, and A. Fiechter.** 1992. "Intelligent" analytical subsystems for on-line control and monitoring of bioprocesses. *Anal. Chim. Acta* **265**:63–69.
31. **Forster, R., and D. Diamond.** 1992. Nonlinear calibration of Ion-Selective electrode arrays for flow injection analysis. *Anal. Chem.* **64**:1721–1728.
32. **Gary, K., P. Meier, and K. Ludwig.** 1988. General aspects of the use of sensors in biotechnology with special emphasis on cell cultivation. Presented at Canbiocon 1988. *Biotechnol. Res. Appl.*, p. 155–164.
33. **Geppert, G., and H. Thielemann.** 1984. Streulichtphotometer zur kontinuierlichen Bestimmung der Biomassekonzentration in Fermentationsmedien. *Acta Biotechnol.* **4**:361–367.
34. **Geppert, G., H. Thielemann, and G. Langkopf.** 1989. Industrial meters for measuring the turbidity of liquids. *Acta Biotechnol.* **9**:541–545.
35. **Gere, D., C. Knipe, P. Castelli, J. Hedrick, L. Frank, H. Schulenbergschell, R. Schuster, L. Doherty, J. Orolin,**

and **H. Lee.** 1993. Bridging the automation gap between sample preparation and analysis—an overview of SFE, GC, GC-MS, and HPLC applied to environmental samples. *J. Chromatogr. Sci.* **31:**246–258.

36. **Giddings, J.** 1989. Field-flow fractionation of macromolecules. *J. Chromatogr.* **470:**327–335.

37. **Giddings, J.** 1993. Field-flow fractionation—analysis of macromolecular, colloidal, and particulate materials. *Science* **260:**1456–1465.

38. **Giddings, J. C.** 1995. Measuring colloidal and macromolecular properties by FFF. *Anal. Chem.* **67:**A592–A598.

39. **Giddings, J., and M. Moon.** 1991. Measurement of particle density, porosity, and size distributions by sedimentation stearic field-flow fractionation—application to chromatographic supports. *Anal. Chem.* **63:**2869–2877.

40. **Gollmer, K., and C. Posten.** 1995. Pattern recognition for phase detection in bioprocesses, p. 41–46. *In* K. Schügerl and A. Munack (ed.), *CAB6 Preprints*. IFAC & EFB, Garmisch-Partenkirchen, Germany.

41. **Gariot, M., E. Heinzle, I. Dunn, and J. Bourne.** 1987. Optimization of a MS-membrane probe for the measurement of acetoin and butanediol. *Mass Spectrom. Biotechnol. Proc. Anal. Contr*, p. 75–90.

42. **Gschwend, K., W. Beyeler, and A. Fiechter.** 1983. Detection of reactor nonhomogenities by measuring culture fluorescence. *Biotechnol. Bioeng.* **25:**2789–2793.

43. **Haemmerli, S., A. Schaeffler, A. Manz, and H. Widmer.** 1992. An improved micro enzyme sensor for bioprocess monitoring by flow injection analysis. *Sensors Actuators B(Chemical)* **7:**404–407.

44. **Hansen, E.** 1995. Flow-injection analysis: leaving its teenyears and maturing. A personal reminiscence of its conception and early development. *Anal. Chim. Acta* **308:**3–13.

45. **Harris, C., R. Todd, S. Bungard, R. Lovitt, J. Morris, and D. Kell.** 1987. Dielectric permittivity of microbial suspensions at radio frequencies: a novel method for the real-time estimation of microbial biomass. *Enzyme Microb. Technol.* **9:**181–186.

46. **Hatch, R., and B. Veilleux.** 1995. Monitoring of *Saccharomyces cerevisiae* in commercial bakers' yeast fermentation. *Biotechnol. Bioeng.* **46:**371–374.

47. **Hauser, P., T. Rupasinghe, and N. Cates.** 1995. A multiwavelength photometer based on light-emitting diodes. *Talanta* **42:**605–612.

48. **Heinzle, E.** 1987. Mass spectrometry for on-line monitoring of biotechnological processes. *Adv. Biochem. Eng Biotechnol.* **35:**1–45.

49. **Heinzle, E.** 1992. Present and potential applications of mass spectrometry for bioprocess research and control. *J. Biotechnol.* **25:**81–114.

50. **Heinzle, E., J. Moes, M. Griot, E. Sandmeier, I. Dunn, and R. Bucher.** 1986. Measurement and control of dissolved oxygen below 100 ppb. *Ann. N.Y. Acad. Sci.* **469:** 178–189.

51. **Heinzle, E., and E. M. Reuss (ed.)** 1987. *Mass Spectroscopy in Biotechnological Analysis and Control*. Graz, Austria.

52. **Hemert, P. V., D. Kilburn, R. Righelato, and A. V. Wezel.** 1969. A steam sterilizable electrode of the galvanic type for the measurement of dissolved oxygen. *Biotechnol. Bioeng.* **6:**549–560.

53. **Hernandez, O., A. I. Jimenez, F. Jimenez, and J. J. Arias.** 1995. Evaluation of multicomponent flow-injection analysis data by use of a partial least squares calibration method. *Anal. Chim. Acta* **310:**53–61.

54. **Hibino, W., Y. Kadotani, M. Kominami, and T. Yamane.** 1993. 3 automated feeding strategies of natural complex nutrients utilizing on-line turbidity values in fed-batch culture—a case study on the cultivation of a marine microorganism. *J. Ferm. Bioeng.* **75:**443–450.

55. **Hitzmann, B., F. Lammers, B. Weigel, and A. V. Putten.** 1993. Die Automatisierung von Fliessinjektionsanalyse-Systemen. *BIOforum* **16:**450–454.

56. **Hitzmann, B., A. Lohn, M. Reinecke, B. Schulze, and T. Scheper.** 1995. The automation of immun-FIA systems. *Anal. Chim. Acta* **313:**55–62.

57. **Hopkins, D., and R. Hatch.** 1990. Monitoring and control of fermentations with an in-situ steam sterilizable optical density probe. Abstr. Amer. Chem. Soc., 199th Mtg., Pt. 1, p. BIOT115.

58. **Humphrey, A.** 1988. The potential of on-line fluorometric measurements for the monitoring and control of fermentation systems. *Aust. J. Biotechnol.* **2:**141–147.

59. **Iijima, S., S. Yamashita, K. Matsunaga, H. Miura, M. Morikawa, K. Shimizu, M. Matsubara, and T. Kobayashi.** 1987. Use of a novel turbidimeter to monitor microbial growth and control glucose concentration. *J. Chem. Technol. Biotechnol.* **40:**203–213.

60. **Isaacs, S., H. Soeberg, L. Christensen, and J. Villadsen.** 1992. A computational technique for simulating the dynamic response of a flow injection analysis system. *Chem. Eng. Sci.* **47:**1591–1600.

61. **Isenschmid, A., I. Marison, and U. V. Stockar.** 1995. The influence of pressure and temperature of compressed CO_2 on the survival of yeast cells. *J. Biotechnol.* **39:**229–237.

62. **Jee, H., N. Nishio, and S. Nagai.** 1987. Influence of redox potential on biomethanation of H_2 and CO_2 by *Methanobacterium thermoautotrophicum* in Eh-stat batch cultures. *J. Gen. Appl. Microbiol.* **33:**401–408.

63. **Jobst, G., G. Urban, A. Jachimowicz, F. Kohl O. Tilado, I. Lettenbichler, and G. Nauer.** 1993. Thin-film Clark-type oxygen sensor based on novel polymer membrane systems for in vivo and biosensor applications. *Biosens. Bioelectron.* **8:**123–128.

64. **Jones, J., and D. Zhou.** 1994. A 1st look at biosensors. *Biotechnol. Adv.* **12:**693–701.

65. **Jones, R., and P. Greenfield.** 1982. Effect of carbon dioxide on yeast growth and fermentation. *Enzyme Microb. Technol.* **4:**210–223.

66. **Ju, L. K., X. Yang, J. F. Lee, and W. B. Armiger.** 1995. Monitoring of the biological nutrient removal process by an on-line NAD(P)H fluorometer. *Biotechnol. Prog.* **11:** 545–551.

67. **Kadotani, Y., K.k Miyamoto, N. Mishima, M. Kominami, and T. Yamane.** 1995. Acquisition of data from on-line laser turidimeter and calculation of some kinetic variables in computer-coupled automated fed-batch culture. *J. Ferm. Bioeng.* **80:**63–70.

68. **Karube, I., Y. Nomura, and Y. Arikawa.** 1995. Biosensors for environmental control. *Trends Anal. Chem.* **14:**295–299.

69. **Karube, I., K. Sode, and E. Tamiya.** 1989. Current trends in microbiosensor development. *Swiss Biotechnol* **7:**25–32.

70. **Karube, I., K. Sode, E. Tamiya, M. Gotoh, Y. Kitagawa, and H. Suzuki.** 1988. New microbiosensors for estimation

of fermentation parameters. 8th Int. Biotechnol. Symp. Pt. 1, p. 537–546.

71. **Karube, I., K. Yokoyama, K. Sode, and E. Tamiya.** 1989. Microbial bod sensor utilizing thermophilic bacteria. *Anal. Lett.* **22:**791–801.

72. **Kato, C., T. Sato, and K. Horikoshi.** 1995. Isolation and properties of barophilic and barotolerant bacteria from deep-sea mud samples. *Biodiv. Conserv.* **4:**1–9.

73. **Kell, D., G. Markx, C. Davey, and R. Todd.** 1990. Real-time monitoring of cellular biomass: methods and applications. *TRAC* **9:**190–194.

74. **Kjaergaard, L.** 1977. The redox potential: its use and control in biotechnology. *Adv. Biochem. Eng.* **7:**131–150.

75. **Konstantinov, K., S. Chuppa, E. Sajan, Y. Tsai, S. Yoon, and F. Golini.** 1994. Real-time biomass-concentration monitoring in animal-cell cultures. *Trends Biotechnol.* **12:** 324–333.

76. **Konstantinov, K., and T. Yoshida.** 1992. Knowledge-based control of fermentation processes. *Biotechnol. Bioeng.* **39:**479–486.

77. **Kroner, K.** 1988. On-line determination of enzymes in bioprocessing with the emphasis on flow injection analysis and continuous sampling. *Fresenius Z. Anal. Chem.* **329:** 718–725.

78. **Kwong, S., L. Randers, and G. Rao.** 1993. On-line detection of substrate exhaustion by using NAD(P)H fluorescence. *Appl. Environ. Microbiol.* **59:**604–606.

79. **Lacoursiere, A., B. Thompson, M. Kole, D. Ward, and D. Gerson.** 1986. Effects of carbon dioxide concentration on anaerobic fermentations of *Escherichia coli*. *Appl. Microbiol. Biotechnol.* **23:**404–406.

80. **Langwost, B., G. Kresbach, T. Scheper, M. Ehrat, and H. Widmer.** 1995. Field-flow fractionation—an analytical tool for many applications. European Congress on Biotechnology 7, European Federation of Biotechnology, Nice, France.

81. **Lenz, R., C. Boelcke, U. Peckmann, and M. Reuss.** 1986. A new automatic sampling device for the determination of filtration characteristics and the coupling of an HPLC to fermentors. Presented at 1st Modelling Contr. Biotechnol. Process, p. 85–90.

82. **Li, J., P. Gomez, and A. Humphrey.** 1990. The use of fluorimetry for on-line measurement of mixing time and hold-up in fermentations. *Biotechnl. Technol.* **4:**293–298.

83. **Lindahl, B., and B. Gullberg.** 1991. Flow cytometrical DNA and clinical parameters in the prediction of prognosis in stage-I-II endometrial carcinoma. *Anticancer Res.* **11:**397–401.

84. **Lingberg, W., J. Ruzicka, and G. Christian.** 1993. Flow injection cytometry—a new approach for sample and solution handling in flow cytometry. *Cytometry* **14:**230–236.

85. **L'Italien, Y., J. Thibault, and A. LeDuy.** 1989. Improvement of ethanol fermentation under hyperbaric conditions. *Biotechnol. Bioeng.* **33:**471–476.

86. **Ljunggren, E., and B. Karlberg.** 1995. Determination of total carbon dioxide in beer and softdrinks by gas diffusion and flow injection analysis. *J. Autom. Chem.* **17:**105–108.

87. **Lloyd, D., and T. Whitmore.** 1988. Hydrogen-dependent control of the continuous thermophilic anaerobic digestion process using membrane inlet mass spectrometry. *Lett. Appl. Microbiol.* **6:**5–10.

88. **Locher, G., B. Sonnleitner, and A. Fiechter.** 1990. Pattern recognition: a useful tool in technological processes. *Bioproc. Eng.* **5:**181–187.

89. **Locher, G., B. Sonnleitner, and A. Fiechter.** 1991. Automatic bioprocess control. 2. Implementations and practical experiences. *J. Biotechnol.* **19:**127–144.

90. **Locher, G., B. Sonnleitner, and A. Fiechter.** 1992. On-line measurement in biotechnology: techniques. *J. Biotechnol.* **25:**23–53.

91. **Locher, G., B. Sonnleitner, and A. Fiechter.** 1992. Software and implementation for automated decision making in bioprocess control. *Proc. Contr. Qual.* **2:**257–274.

92. **Luong, J., and D. Carrier.** 1986. On-line measurement of culture fluorescence during cultivation of *Methylomonas mucosa*. *Appl. Microbiol. Biotechnol.* **24:**65–70.

93. **Luong, J., and B. Volesky.** 1982. A new technique for continuous measurement of the heat of fermentation. *Eur. J. Appl. Microbiol. Biotechnol.* **16:**28–34.

94. **Maclaurin, P., P. Worsfold, P. Norman, and M. Crane.** 1993. Partial least squares resolution of multianalyte flow injection data. *Analyst* **118:**617–622.

95. **Marison, I., and U. V. Stockar.** 1989. The use of calorimetry in biotechnology. *Adv. Biochem. Eng. Biotechnol.* **40:**93–136.

96. **Markx, G., C. Davey, and D. Kell.** 1991. The permittistat: a novel type of turbidostat. *J. Gen. Microbiol.* **137:** 735–743.

97. **Markx, G., and D. Kell.** 1995. The use of dielectric permittivity for the control of the biomass level during biotransformations of toxic substrates in continuous culture. *Biotechnol. Prog.* **11:**64–70.

98. **Mattiasson, B., and H. Hakanson.** 1993. Sampling and sample handling—crucial steps in process monitoring and control. *Trends Biotechnol.* **11:**136–142.

99. **McGown, L., S. Hemmingsen, J. Shaver, and L. Geng.** 1995. Total lifetime distribution analysis for fluorescence fingerprinting and characterization. *Appl. Spectrosc.* **49:** 60–66.

100. **Memmert, K., and C. Wandrey.** 1987. Continuous production of Bacillus exoenzymes through redox-regulation. *Ann. N.Y. Acad. Sci.* **506:**631–636.

101. **Merchuk, J., S. Yona, M. Siegel, and A. Zvi.** 1990. On the first-order approximation to the response of dissolved oxygen electrodes for dynamic k_La estimation. *Biotechnol. Bioeng.* **35:**1161–1163.

102. **Meyer, C., and W. Beyeler.** 1984. Control strategies for continuous bioprocesses based on biological activities. *Biotechnol. Bioeng.* **26:**916–925.

103. **Min, R. W., J. Nielsen, and J. Villadsen.** 1995. Simultaneous monitoring of glucose, lactic acid and penicillin by sequential injection analysis. *Anal. Chim. Acta* **312:** 149–156.

104. **Moes, J., M. Griot, J. Keller, E. Heinzle, I. Dunn, and J. Bourne.** 1985. A microbial culture with oxygen-sensitive product distribution as a potential tool for characterizing bioreactor oxygen transport. *Biotechnol. Bioeng.* **27:**482–489.

105. **Molt, K.** 1992. Grundlagen und Anwendungen der modernen NIR-Spektroskopie. *GIT* **36:**107–113.

106. **Montague, G., and A. Morris.** 1994. Neural-network contributions in biotechnology. *Trends Biotechnol.* **12:** 312–324.

107. **Montesinos, J., C. Campmajo, J. Iza, F. Velero, J. Lafuente, and C. Sola.** 1993. Software development to fermentation gas analysis using mass spectrometry. *Biotechnol. Tech.* **7:**429–434.

108. **Monzambe, K., H. Naveau, E. Nyns, N. Bogaert, and H. Bühler.** 1988. Problematics and stability of on-line pH measurements in anaerobic environments: the jellied combined electrode. *Biotechnol. Bioeng.* **31:**659–665.

109. **Moskvin, L., and J. Simon.** 1994. Flow injection analysis with the chromatomembrane—a new device for gaseous/liquid and liquid/liquid extraction. *Talanta* **41:** 1765–1769.

110. **Nestaas, E., and D. Wang.** 1983. A new sensor—the "filtration probe"—for quantitative characterization of penicillin fermentation. III. An automatically operating probe. *Biotechnol. Bioeng.* **25:**1981–1987.

111. **Nielsen, J.** 1992. On-line monitoring of microbial processes by flow injection analysis. *Proc. Contr. Qual.* **2:** 371–384.

112. **Nielsen, J., C. Johansen, and J. Villadsen.** 1994. Culture fluorescence measurements during batch and fed-batch cultivations with penicillium chrysogenum. *J. Biotechnol.* **38:**61–62.

113. **Nielsen, J., K. Nikolajsen, S. Benthin, and J. Villadsen.** 1990. Application of flow-injection monitoring of sugars, lactic acid, protein and biomass during lactic acid fermentations. *Anal. Chim. Acta* **237:**165–176.

114. **Nielsen, J., and J. Villadsen.** 1994. *Bioreaction Engineering Principles.* Plenum Press, New York.

115. **Nipkow, A., C. Andretta, and O. Käppeli.** 1990. Biomasseschätzung in Hefezüchtungen mittels Trübungsmesssonden und einer Kapazitanzmesssonde. *Chem. Ing. Tech.* **62:**1052–1053.

116. **Nipkow, A., B. Sonnleitner, and A. Fiechter.** 1985. Effect of carbon dioxide on growth of *Zymomonas mobilis* in continuous culture. *Appl. Microbiol. Biotechnol.* **21:** 287–291.

117. **Oeggerli, A., and E. Heinzle.** 1994. On-line exhaust gas analysis of volatiles in fermentation using mass spectrometry. *Biotechnol. Prog.* **10:**284–290.

118. **Ohashi, M., T. Watabe, T. Ishikawa, Y. Watanabe, K. Miwa, M. Shode, Y. Ishikawa, T. Ando, T. Shibata, T. Kitsunai, N. Kamiyama, and Y. Oikawa.** 1979. Sensors and instrumentation: steam-sterilizable dissolved oxygen sensor and cell mass sensor for on-line fermentation system control. *Biotechnol. Bioeng. Symp.* **9:**103–116.

119. **Park, S., K. Hong, J. Lee, and J. Bae.** 1983. On-line estimation of cell growth for glutamic acid fermentation system. *Eur. J. Appl. Microbiol.* **17:**168–172.

120. **Pasquini, C., and L. Defaria.** 1991. Operator-free flow injection analyser. *J. Autom. Chem.* **13:**143–146.

121. **Puhar, E., A. Einsele, H. Bühler, and W. Ingold.** 1980. Steam-sterilizable pCO_2 electrode. *Biotechnol. Bioeng.* **22:**2411–2416.

122. **Rao, G., and R. Mutharasan.** 1989. NADH levels and solventogenesis in *Clostridium acetobutylicum*: new insights through culture fluorescence. *Appl. Microbiol. Biotechnol.* **30:**59–66.

123. **Reardon, K., T. Scheper, and J. Bailey.** 1987. Metabolic pathway rates and culture fluorescence in batch fermentations of *Clostridium acetobutylicum*. *Biotechnol. Prog.* **3:** 153–167.

124. **Reardon, K., T. Scheper, and J. Bailey.** 1987. Use of a fluorescence detector for the determination of the NAD(P)H-dependent, immobilized cell culture fluorescence system. *Chem. Ing. Tech.* **59:**600–601.

125. **Reuss, M., C. Boelcke, R. Lenz, and U. Peckmann.** 1987. Entwicklung und Einsatz einer Apparatur zur automatischen Erfassung der Filtrationseigenschaften von Biosuspensionen und zur Ankoppelung von Analysenautomaten bei industriellen Fermentationsprozessen. *BTF-Biotech Forum* **4:**3–12.

126. **Riesenberg, D.** 1991. High-cell density cultivation of *Escherichia coli*. *Curr. Opin. Biotechnol.* **2:**380–384.

127. **Riesenberg, D., V. Schulz, W. Knorre, H. Pohl, D. Korz, E. Sanders, A. Ross, and W. Deckwer.** 1991. High cell density cultivation of *Escherichia coli* at controlled specific growth rate. *J. Biotechnol.* **20:**17–28.

128. **Rohner, M., G. Locher, B. Sonnleitner, and A. Fiechter.** 1988. Kinetics and modeling of the stereoselective reduction of acetoacetic acid esters by continuously growing cultures of *Saccharomyces cerevisiae*. *J. Biotechnol.* **9:**11–28.

129. **Ruzicka, J.** 1994. Discovering flow injection: journey from sample to a live cell and from solution to suspension—tutorial review. *Analyst* **119:**1925–1934.

130. **Ruzicka, J., and E. Hansen.** 1981. *Flow Injection Analysis.* Wiley, New York.

131. **Ruzicka, J., and W. Lindberg.** 1992. Flow injection cytoanalysis. *Anal. Chem.* **64:**A537-A545.

132. **Scheper, T.** 1991. *Bioanalytik.* Vieweg, Braunschweig, Germany.

133. **Scheper, T., T. Lorenz, W. Schmidt, and K. Schügerl.** 1987. On-line measurement of culture fluorescence for process monitoring and control of biotechnological processes. *Ann. N.Y. Acad. Sci.* **506:**431–445.

134. **Scheper, T., and K. Schügerl.** 1986. Characterization of bioreactors by in-situ fluorometry. *J. Biotechnol.* **3:**221–229.

135. **Schill, N., and U. V. Stockar.** 1995. Thermodynamic analysis of *Methanobaterium thermoautotrophicum*. *Thermochim. Acta* **251:**71–77.

136. **Schügerl, K.** 1991. *Analytische Methoden in der Biotechnologie.* Vieweg, Braunschwedig, Germany.

137. **Schumpe, A., G. Quicker, and W. Deckwer.** 1982. Gas solubilities in microbial culture media. *Adv. Biochem. Eng.* **24:**1–38.

138. **Simutis, R., I. Havlik, M. Dors, and A. Lübbert.** 1993. Training of artificial neural networks extended by linear dynamic subsystems. *Proc. Contr. Qual.* **4:**211–220.

139. **Simutis, R., I. Havlik, and A. Lübbert.** 1993. Fuzzy-aided neural network for real-time state estimation and process prediction in the alcohol formation step of production-scale beer brewing. *J. Biotechnol.* **27:**203–215.

140. **Simutis, R., I. Havlik, F. Schneider, M. Dors, and A. Lübbert.** 1995. Artificial neural networks of improved reliability for industrial process supervision, p. 59–65. *In CAB6 Preprints.* IFAC & EFB, Garmisch-Partenkirchen, Germany.

141. **Sonnleitner, B.** 1991. Dynamics of yeast metabolism and regulation. *Bioproc. Eng.* **6:**187–193.

142. **Sonnleitner, B.** 1991. Quantitation of microbial metabolism. *Antonie van Leeuwenhoek Int. J. Gen. Mol. Microbiol.* **60:**133–143.

143. **Sonnleitner, B.** 1993. Experimental verification of rapid dynamics in biotechnological processes, p. 143–153. *In* V. Mortensen and H. J. Noorman, (ed.), *Bioreactor Per-*

formance. Biotechnology Research Foundation, Lund, Sweden.

144. **Sonnleitner, B.** 1996. New concepts for quantitative physiological research. *Adv. Biochem. Eng. Biotechnol.* **54:**155–188.
145. **Sonnleitner, B., A. Fiechter, and F. Giovannini.** 1984. Growth of *Thermoanaerobium brockii* in batch and continuous culture at supraoptimal temperatures. *Appl. Microbiol. Biotechnol.* **19:**326–334.
146. **Sonnleitner, B., G. Locher, and A. Fiechter.** 1991. Automatic bioprocess control. 1. A general concept. *J. Biotechnol.* **19:**1–18.
147. **Soper, S. A., B. L. Legendre, and D. C. Williams.** 1995. On-line fluorescence lifetime determinations in capillary electrophoresis. *Anal. Chem.* **67:**4358–4365.
148. **Srinivas, S., and R. Mutharasan.** 1987. Culture fluorescence characteristics and its metabolic significance in batch cultures of Clostridium acetobutylicum. *Biotechnol. Lett.* **9:**139–142.
149. **Srinivasan, N., N. Kasthurikrishnan, R. G. Cooks, M. S. Krishnan, and G. T. Tsao.** 1995. On-line monitoring with feedback control of bioreactors using a high ethanol tolerance yeast by membrane introduction mass spectrometry. *Anal. Chim. Acta* **316:**269–276.
150. **Stephanopoulos, G., G. Locher, and M. Duff.** 1995. Pattern recognition methods for fermentation database mining, p. 195–198. *In* A. Munack and K. Schügerl (ed.), *CAB6 Preprints*. IFAC & EFB, Garmisch-Partenkirchen, Germany.
151. **Stockar, U. V., and I. Marison.** 1989. The use of calorimetry in biotechnology. *Adv. Biochem. Eng. Biotechnol.* **40:**93–136.
152. **Stockar, U. V., I. Marison, and B. Birou.** 1988. On-line calorimetry for process control. Presented at 1st Swiss-Japanese Joint Meeting on Bioprocess Development, Interlaken.
153. **Suda, M., T. Sakuhara, and I. Karube.** 1993. Miniaturized detectors for a chemical analysis system. *Appl. Biochem. Biotechnol.* **41:**3–10.
154. **Suda, M., T. Sakuhara, Y. Murakami, and I. Karube.** 1993. Micromachined detectors for an enzyme-based FIA. *Appl. Biochem. Biotechnol.* **41:**11–15.
155. **Tanaka, H., H. Aoyagi, and T. Jitsufuchi.** 1992. Turbidimetric measurement of cell biomass of plant cell suspensions. *J. Ferm. Bioeng.* **73:**130–134.
156. **Thomas, D., V. Chittur, J. Cagney, and H. Lim.** 1985. On-line estimation of mycelial cell mass concentrations with a computer-interfaced filtration probe. *Biotechnol. Bioeng.* **27:**729–742.
157. **Uttamlal, M., and D. Walt.** 1995. A fiber-optic carbon dioxide sensor for fermentation monitoring. *Bio/Technology* **13:**597–601.
158. **Vanderpol, J. J., U. Spohn, R. Eberhardt, J. Gaetgens, M. Biselli, C. Wandrey, and J. Tramper.** 1994. On-line monitoring of an animal cell culture with multi-channel flow injection analysis. *J. Biotechnol.* **37:**253–264.
159. **vanKleeff, B.** 1995. On-line heat flow measurement in laboratory fermenters. *Thermochim. Acta* **251:**111–118.
160. **vanKleef, B., J. Kuenen, and J. Heijnen.** 1993. Continuous measurement of microbial heat production in laboratory fermenters. *Biotechnol. Bioeng.* **41:**541–549.
161. **Vanstaden, J.** 1995. Membrane separation in flow injection systems. 1. dialysis. *Fesenius J. Anal. Chem.* **352:** 271–302.
162. **Villadsen, J.** 1995. Fundamental aspects of the physiology-engineering approach, abstr. MAC 161. Abstr. ECB7, Nice, France.
163. **Whitmore, T., G. Jones, M. Lazzari, and D. Lloyd.** 1987. Methanogenesis in mesophilic and thermophilic anaerobic digestors: monitoring and control based on dissolved hydrogen, p. 143–162. *In* E. Heinzle and E. M. Reuss (ed.), *Mass Spectroscopy in Biotechnological Analysis and Control*. Graz, Austria.
164. **Winter, E., G. Rao, and T. Cadman.** 1988. Relationship between culture redox potential and culture fluorescence in *Corynebacterium glutamicum*. *Biotechnol. Tech.* **2:**233–236.
165. **Wu, P., S. Ozturk, J. Blackie, J. Thrift, C. Figueroa, and D. Naveh.** 1995. Evaluation and applications of optical cell density probes in mammalian cell bioreactors. *Biotechnol. Bioeng.* **45:**495–502.
166. **Wu, X. A., and K. H. Bellgardt.** 1995. On-line fault detection of flow-injection analysis systems based on recursive parameter estimation. *Anal. Chim. Acta* **313:** 161–176.
167. **Xie, B., M. Mecklenburg, B. Danielson, O. Ohman, P. Norlin, and F. Winquist.** 1995. Development of an integrated thermal biosensor for the simultaneous determination of multiple analytes. *Analyst* **120:**155–160.
168. **Yamane, T.** 1993. Application of an on-line turbidimeter for the automation of fed-batch cultures. *Biotechnol. Prog.* **9:**81–85.
169. **Yamane, T., W. Hibino, K. Ishihara, Y. Kadotani, and M. Kominami.** 1992. Fed-batch culture automated by uses of continuously measured cell concentration and culture volume. *Biotechnol. Bioeng.* **39:**550–555.
170. **Yano, Y., A. Nakayama, and K. Yoshida.** 1995. Population sizes and growth pressure responses of intestinal microfloras of deep-sea fish retrieved from the abyssal zone. *Appl. Environ. Microbiol.* **61:**4480–4483.
171. **Yao, T., Y. Matsumoto, and T. Wasa.** 1990. Development of a FIA system with immobilized enzymes for specific post-column detection of purine bases and their nucleosides separated by HPLC column. *J. Biotechnol.* **14:**89–97.
172. **Zabriskie, D., and A. Humphrey.** 1978. Estimation of fermentation biomass concentration by measuring culture fluorescence. *Enzyme Microb. Technol.* **35:**337–343.
173. **Zhang, F., P. Scully, and E. Lewis.** 1994. An optical fibre sensor for on line yeast measurement, abstr. AS13.1, p. 173–174. Presented at the Institute of Physics Applied Optics and Optoelectronics Conference, York.

Scale-Up of Microbial Processes

MASAHIKO HOSOBUCHI AND HIROJI YOSHIKAWA

19

Development of bioprocesses is usually carried out in three steps: (i) bench scale or flask scale where basic screening is carried out, (ii) pilot scale where optimal operating conditions are established, (iii) plant scale where the process is brought to an economically favorable level.

The fundamental idea of scale-up is that the optimal physiological conditions that are obtained in small-scale studies should be maintained on the large scale by controlling environmental conditions. These include not only physical factors (mass transfer ability, mixing ability, shear field, power consumption, etc.) and chemical factors (medium composition and concentration) but also process factors (number of precultures, sterilization conditions, etc.). Many physical and chemical parameters that influence the behavior of cultured organisms change as the scale of process becomes larger. For conditions that are chosen in small-scale fermentors to be reproduced in large fermentors, the factors significant to the cultivation should be determined and investigated in small-scale cultivation.

In this chapter, we discuss the factors to be considered in the scale-up procedure, such as construction of a new large-scale facility, translation of a new bioprocess with a new microorganism, and bioprocess improvement from the pilot plant into an existing fermentation facility.

19.1. CRITERIA USED FOR SCALE-UP

Historically, many investigations concerning physical factors that have an influence on scale-up, especially oxygen transfer (1), have been done by many researchers. Several criteria used for scale-up that focus on mass transfer and on the characteristics of a microorganism have been proposed. These are as follows.

1. Constant agitation power per unit volume
2. Constant oxygen transfer coefficient (k_La)
3. Constant mixing time
4. Constant impeller tip speed
5. Feedback controlling the values of key environmental factors as nearly as possible between two scales of cultivation

When a new fermentor is to be designed, criteria 1 and 2 are useful for comparing oxygen transfer ability between two different sizes of fermentation equipment. In these, the agitation power per unit volume is the most frequently used parameter for scale-up. As sensor technologies developed for dissolved oxygen (DO) concentration, scale-up methods to maintain DO (13) at a constant level became popular. The scale-up method related to oxygen transfer is established by controlling the DO concentration at a suitable level. This is not difficult to control by changing agitation speed and the shape of impellers.

On the other hand, adjusting factors that are important to the physiological condition of the microbial cells is very difficult. Many biotechnology companies already possess large fermentors for multipurpose use. When a new fermentative process is ready for scale-up, the new process should be adjusted to the existing fermentor by changing a limited number of parameters.

19.2. PHYSICAL FACTORS

19.2.1. Mass Transfer

For almost all microorganisms, oxygen is a very important factor for growth and production. Many useful microorganisms that produce important secondary metabolites require a high level of oxygen for production. A lack of oxygen often causes a reduction of the yield or death of the microorganism.

The DO is the most commonly used indicator for monitoring conditions in a fermentor. To control DO in a fermentor, one changes agitation speed, aeration level, or pressure. Oxygen mass transfer coefficient and agitation power per unit volume are very important parameters for scaling up. However, as it is rather dangerous to rely solely on the calculation of these parameters, these values are often used as reference values for experiments to be carried out before designing a new large fermentor.

19.2.1.1. Oxygen Mass Transfer Coefficient (k_La)

k_La (1, 3, 15) is an excellent parameter for comparing oxygen transfer ability in different size fermentors. k_La is an index of oxygen transfer ability from the gas phase to the liquid phase. It is defined as follows:

$$dc/dt = k_La(C^* - CL).$$

where C is oxygen concentration, t is time, and k_La is an oxygen transfer coefficient. The value of k_La is affected by the size of fermentor, the shape and number of turbine impellers and baffles, agitation speed, aeration rate, pressure, and so on. One major problem is that the k_La value should be measured using cultured broth for accurate estimation, since the rheological characteristics of cultured broth significantly affect the k_La value. In cultivation of filamentous microorganisms, such as fungi or actinomycetes, the viscosity of culture broth changes with culture time and sometimes increases to a very high level. In such cases, the characteristics of the fermentation broth are those of a non-Newtonian fluid. Addition of water to the culture broth (14) occasionally is effective in decreasing the viscosity.

19.2.1.2. Agitation Power per Unit Volume (Pg/V)

A scale-up method using Pg/V (1, 16) has often been applied to industrial fermentations. The benefit of this method is that the cultivation data obtained from small-scale fermentations are often available. Factors that have great influence, such as viscosity or non-Newtonian fluid characteristics, are already included in the experimental value of Pg/V. When the motor power of the existing fermentor is not strong enough to increase DO for cultivation of a new microorganism, a decrease in volume of fermentation broth is sometimes effective as a means to maintain DO.

19.2.2. Shear Stress

When filamentous microorganisms, such as fungi or actinomycetes, are cultured in a large fermentor, cells are damaged by shear stress (23), and yield of product is reduced. Shear stress is considered to be proportional to the tip speed of the agitation impeller. Oldshue (12) showed the relationship between the tip speed of the impeller and the amount of a nucleic acid-related compound leaked from the cells. A microorganism that is sensitive to shear stress should be cultured with large-diameter impellers or multiple impellers (2), to maintain a suitable DO level with a low agitation speed.

19.2.3. Effect of Time Delay in a Large Fermentor

Mixing time is often a key factor in the bioprocesses in which pH control or substrate feeding is needed. When concentrated alkali or acid are fed and mixing time is long, a pH gradient is formed in the fermentor. Especially when a fed substrate inhibits cell growth or when substrate concentration has to be controlled at a very low level, long mixing time in a large fermentor results in low production yield. To shorten mixing time, one can use large-diameter or sloped impellers or feeding from many points in the fermentor.

As the size of the fermentor is increased, fermentor depth also increases, and DO at the bottom of a fermentor is higher than that at the surface. Microorganisms in a large fermentor repeatedly travel up and down. When the fermentor depth is 10 m, the pressure and the DO difference (10, 17) are about 1 atm and about 8 ppm, respectively. For microorganisms that show low productivity at high DO levels, production yield sometimes decreases in a large fermentor because of a high DO level at the bottom. In bialaphos production (19, 20) by *Streptomyces hygroscopicus*, production yield under a high DO condition is low. When the process was carried out and the DO was measured in the middle of a fermentor (which was considered to represent the DO in the fermentor) the yield of bialaphos was about 85% of that in small-scale fermentation. To improve the fermentation yield, the DO was measured at the bottom of the fermentor and controlled at 0.5 ppm. As a result, bialaphos production yield increased to 96% of that in small-scale culture.

19.3. CHEMICAL FACTORS

For economic reasons, low-grade or cheap natural ingredients are often used for medium preparation in large-scale fermentations instead of special-grade ingredients. The products obtained with these low-grade ingredients may be of lower purity than those obtained with high-grade media or may contain harmful by-products. When medium composition or grades are changed, several small-scale studies are necessary to ensure the purity of the final product.

When the nutrient requirements of a large-scale fermentation are different from those at the small scale, nutrient feeding is effective. Amounts of nutrients in a culture broth can be measured by the combination system of a continuous filtration module and high-performance liquid chromatography (4, 5, 7, 9). When a main medium component that affects the culture conditions is known, controlling the nutrient concentration to a suitable level is an effective way to scale up the chemical condition of culture broth.

19.4. PROCESS FACTORS

As the size of the fermentor increases, scale-up processes have to change. To increase cell mass, the number of precultures (i.e., seed stages) has to increase as the size of the production fermentor becomes larger. When a large amount of medium, especially natural medium, is used, lot-to-lot differences sometimes cause a decrease of production yield. For sterilization it has to be understood that the heat transfer speed in a large fermentor is much lower than that in a small-scale fermentor. Thus, excess sterilization is often needed for complete sterilization, and some medium components may be degraded. These process factors are very important for scale-up.

19.4.1. Number of Precultures

To increase a cell mass that is prepared for a production culture in a large-scale fermentor, several repeated precultures are required. The condition of the cells that are inoculated into the production culture is extremely important, and the following guidelines should be used. (i) Growing cells can shorten growth lag time in the production culture. (ii) There should be a high enough concentration of cells. (iii) There should be suitable growth morphology (22) of the cells.

Many experiments are needed to obtain knowledge about points i and iii. Establishment of the scale-up method for filamentous microorganisms is often very difficult because there are many factors that affect growth morphology during cultivation. In this section, control of growth morphology for scale-up is described.

In cultivation of fungi or actinomycetes, there are two main growth morphologies to be considered: pellet and filamentous (11). For such types of microorganisms, growth morphology is very important for secondary metabolite production.

There are some reports about scale-up from the point of view of preculture process. Hosobuchi et al. (6, 8) reported on growth morphology control for ML-236B (compactin) production by *Penicillium citrinum*. In those studies, a quantitative approach was used to control growth morphology in the production fermentor for high yield. The suitable morphological form for the production culture was found to be that of small pellets. However, *P. citrinum* formed large pellets (21) in the production culture when preculture was carried out more than three times. To form small pellets in the production culture, a relationship between growth morphology in the production culture and the number of mycelia in the last preculture was determined by using flasks and small fermentors. It was found that the morphology in the last preculture had to be of filamentous form in order to increase the number of mycelia. Next, the factors that influenced growth morphology were tested. Metz and Kossen (11) had reported on the factors that affect growth morphology in filamentous microorganisms. Medium concentration was found to be an effective factor in changing growth morphology from the pellet form to the filamentous form. More work is required to develop preculture processes to yield the appropriate number and suitable quality of mycelia in the last preculture.

To determine scale-up criteria that are optimal for a particular microorganism used for metabolite production is a difficult task. For this reason, characteristics of a microorganism should be well known before attempting scale-up.

19.4.2. Sterilization Factors

Sterilization time in a large-scale fermentor is longer than that in a small fermentor. The long sterilization often causes caramelization of media and degradation of important components in the medium. These chemically changed media often inhibit the growth rate of cells and negatively affect production yield. Since mixed sterilization of carbon sources, nitrogen sources, and inorganic chemicals often causes caramelization, separate sterilization is sometimes necessary.

The growth rate with medium that is sterilized separately is usually higher than that with mixed sterilized medium. Several experiments using separately sterilized medium are necessary at the small scale for ensuring the suitability of the sterilization process.

19.5. CLOSING COMMENTS

Scale-up studies are inevitable for the industrialization of bioprocesses. When we compare two different scales, several points of difference have to be taken into consideration. For example, perfect mixing is impossible in a large-scale vessel. Heterogeneity leads to a distribution of DO values, nutrients, temperature, pH, etc. The increase in the number of preculture steps influences not only cell morphology but also stability of the strain. To move the culture from small to large scale, it is necessary to simulate important factors for culture conditions as nearly as possible. An appropriate scale-down process should be constructed for studying these factors. This is the most troublesome part of constructing a scale-up process. When these factors are examined, several methods are to be considered to solve the problems, and strain breeding (18) will be one way to overcome the effect of specific factors for a large-scale bioprocess.

There is still no standard method for fermentation scale-up. Since scale-up is very much related to the characteristics of the microorganism, more studies and examples about large-scale fermentation are needed to construct a general scale-up method.

REFERENCES

1. **Aiba, S., A. E. Humphrey, and N. F. Millis.** 1973. *Biochemical Engineering.* Academic Press, New York.
2. **Bader, F. G.** 1986. Modeling mass transfer and agitator performance in multiturbine fermentor. *Biotechnol. Bioeng.* **30:**37–51.
3. **Constance, M. T., and F. Pinho.** 1970. Determination of oxygen-transfer coefficients in viscous Streptomycete fermentations. *Biotechnol. Bioeng.* **12:**849–871.
4. **Dincer, A. K., M. Kalyanpur, W. Skea, M. Ryan, and T. Kierstead.** 1984. Continuous on-line monitoring of fermentation proesses. *Dev. Ind. Microbiol.* **25:**603–609.
5. **Gbewonyo, G., B. C. Buckland, and M. D. Lilly.** 1991. Development of a large-scale continuous substrate feed process for the biotransformation of simvastatin by *Nocardia* sp. *Biotechnol. Bioeng.* **37:**1101–1107.
6. **Hosobuchi, M., F. Fukui, H. Matsukawa, T. Suzuki, and H. Yoshikawa.** 1993. Morphology control of preculture during production of ML-236B, a precursor of pravastatin sodium, by *Penicillium citrinum*. *J. Ferm. Bioeng.* **76:** 476–481.
7. **Hosobuchi, M., K. Kurosawa, and H. Yoshikawa.** 1993. Application of computer to monitoring and control of fermentation process: microbial conversion of ML-236B Na to pravastatin. *Biotechnol. Bioeng.* **42:**815–820.
8. **Hosobuchi, M., K. Ogawa, and H. Yoshikawa.** 1993. Morphology study in production of ML-236B, a precursor of pravastatin sodium, by Penicillium citrinum. *J. Ferm. Bioeng.* **76:**470–475.
9. **Kleman, G. L., J. J. Chalmers, G. Luli, and W. R. Strohl.** 1991. A predictive feedback control algorithm maintains a constant glucose concentration in fed-batch fermentation. *Appl. Environ. Microbiol.* **57:**910–917.
10. **Manfredini, R., and V. Cavellera.** 1983. Mixing and oxygen transfer in conventional stirred fermentors. *Biotechnol. Bioeng.* **25:**3115–3131.
11. **Mezt, B., and N. W. F. Kossen.** 1977. The growth of mold in the form of pellets. *Biotechnol. Bioeng.* **19:** 781–799.
12. **Oldshue, J. Y.** 1966. Fermentation mixing scale-up techniques. *Biotechnol. Bioeng.* **8:**3–24.
13. **Oosterhuis, N. M. G., and N. W. F. Kossen.** 1983. Dissolved oxygen concentration profiles in a production-scale bioreactor. *Biotechnol. Bioeng.* **26:**546–550.
14. **Taguchi, H.** 1971. The nature of fermentation fluids. *Adv. Biochem. Eng.,* **1:**1–30.
15. **Taguchi, H., and A. E. Humphrey.** 1966. Dynamic measurement of volumetric oxygen transfer coefficient in fermentation systems. *J. Ferm. Technol.* **44:**881–889.
16. **Taguchi, H., and S. Miyamoto.** 1966. Power requirement in non-newtonian fermentation broth. *Biotechnol. Bioeng.* **8:**43–54.
17. **Takamatsu, T., S. Shioya, and H. Nakatani.** 1981. Analysis of a deep U-tube aeration system from the viewpoint of oxygen transfer. *J. Ferm. Technol.* **59:**287–294.
18. **Takebe, H., S. Imai, H. Ogawa, A. Satoh, and H. Tanaka.** 1989. Breeding of bialaphos producing strain from

a biochemical engineering viewpoint. *J. Ferm. Technol.* **67:**226–232.

19. **Takebe, H., M. Matsunahga, O. Hiruta, A. Satoh, and H. Tanaka.** 1993. Effect of oxygen partial pressure on bialaphos synthesis, sugar metabolism and the activity of tricarboxylic acid cycle enzymes in submerged culture of *Streptomyces hygroscopicus*. *J. Ferm. Bioeng.* **75:**283–287.
20. **Takebe, H., and H. Tanaka.** 1994. Scale-up of bialaphos production. *J. Ferm. Bioeng.* **78:**93–99.
21. **Whitaker, A., and P. A. Long.** 1973. Fungal pelleting. *Proc. Biochem.* **11:**27–31.
22. **Wittler, R., H. Baumgartl, W. Lubbers, and K. Schugerl.** 1986. Investigation of oxygen transfer into *Penicillium chrysogenum* pellets by microprobe measurement. *Biotechnol. Bioeng.* **28:**1024–1036.
23. **Ziegler, H., I. J. Dunn, and J. R. Bourine.** 1980. Oxygen transfer and mycelial growth in a tubular loop fermentor. *Biotechnol. Bioeng.* **22:**1613–1635.

Data Analysis

JOHN BARFORD AND PENG-CHENG FU

20

Biochemical engineers and biologists are usually faced with an overwhelming mass of data from both the literature and their own laboratories/plants. Quite often they may find that there is a general lack of sources to extract information from their collection of data. This situation calls for effective and hands-on means of transformation of raw data into a format useful for the acquisition of the quantitative information about biological systems under investigation. The gap between data generation and data comprehension is widening. With the advent of powerful computers and cheap storage devices for mass data, efficient computational methods for analyzing data can now narrow the existing gap and allow people to work with data for a variety of purposes, such as plant design, process control, fault diagnosis, economic evaluation, model validation, and theoretical prediction, as well as employing a variety of computationally intelligent techniques (19).

This chapter does not intend to introduce the abovementioned new developments in detail. Instead, it will focus on the practical issues of data presentation and analysis for biochemical engineers and biologists. The main objectives of data analysis for biotechnological practitioners are to search for patterns and variations of a bioprocess, draw meaningful conclusions from events represented as numbers, and interpret and report the findings based on their observations. The means of presentation of experimental data to visualize the general trends and relationships of process variables will be mentioned first. Correlation of process variables and mathematical models to test the applicability of a particular model to a process, which is a useful way to verify our understanding of the underlying mechanism of bioprocesses, will then be illustrated. Specific rates are important concepts and useful parameters for bioprocesses. The calculation of specific bioprocess rates from experimental data will be presented third. Data-based modeling approaches, which are useful tools to generate a model from input-output data, will follow. Finally, a spreadsheet application to fermentation data presentation will be used as an example of computer-aided interpretation and analysis of bioprocess data. At the end of this chapter, readers may be able to gain sufficient insight into how to obtain the hidden information from their collection of data and how to maximize the use of such information.

20.1. DATA PRESENTATION

There are three general forms for data summary, as follows: tables, graphs, and equations.

20.1.1. Tables

The function of tables is to enable the reader to compare statistical information of biological systems more easily. Although tables usually consist of numerical data, they can also be used to present other kinds of information. For example, a table was used to present the observation from petri dish cultivation and rotating biological reactors mixed liquor (20). The authors used a nonnumerical data table to show the presence of 22 species of ciliates.

When the construction of a table seems extremely complicated, then it is time to divide such a table into two or more simpler tables. For example, a two-table set was used to accommodate the experimental results of lipase hydrolysis in both water and a 50% water and 50% dioxane mixture (12).

20.1.2. Figures

One of the simplest and most useful types of figure is the bar chart (5, 33). The bars are sometimes vertical, sometimes horizontal. An example in reference 33 shows that the acidic and most polar amino acids predominantly come from yeast extract, whereas the basic and more hydrophobic ones mainly come from Bacto Tryptone (33). The type of information presented in a table can also be presented in a bar chart. The latter makes it easier for the reader to grasp the information. By data combination and visualization, the reader can be given a clear and comparative impression. There are variations of the bar chart. For example, it is good practice to indicate the precision of data on graphs by using error bars (5). On the other hand, the bars on a bar chart may be "stacked" for a pseudo-3D visualization (16).

Like the bar chart, the graph or curve is a means of presenting data that might also be presented as a table. In industry or research, data are often collected to examine relationships between process variables. The roles of these variables in the experiments or plant measurements are clearly defined. Dependent variables are uncontrolled during the process operation. They are measured as responses to changes in one or more independent controlled variables. A graph or curve can be used to show how the dependent variable(s) varies as the independent variable(s) varies. Most likely, more than one curve will appear on the same graph. An example is the cell distribution in the G_0, G_1, S, G_2, and M phases during the cell cycle (21).

Useful information on the relationship between the dependent variables, and the independent-dependent variables as well, may be obtained from raw experimental data. One such example is the plotting of oxygen uptake rate (OUR) and carbon dioxide evolution rate (CER) with respect to time (17). The figure reveals an interesting finding of OUR-CER phase shift during the authors' study on the dynamic oscillatory behavior of *Saccharomyces cerevisiae* that is caused by the difference between the intracellular and extracellular pH during continuous culture.

There are many other diagrams that are used to illustrate scientific data, e.g., the recovered activity of lysozyme obtained at various urea and LiCl concentrations by the contour lines (25). With the gradual change of the colors, it may be viewed as a 3D presentation where lysozyme percentage is the dependent variable, and urea and LiCl concentrations are the two independent variables. Another 3D diagram illustrates the relationships of kojic acid and glucose/polypeptone concentrations with pH and temperature (30). One part was designed to focus attention on the effect of initial glucose and polypeptone concentrations on the final production rate of kojic acid and the other part showed the response surface of kojic acid production rate.

20.1.3. Equations

The use of equations is another way to present our understanding of the real world. The equations can be either mechanistic or empirical. Mechanistic models are founded on theoretical assessment of the phenomenon being measured. For example, the time interval (t_d) required to double the cell population during exponential batch growth in a continuous stirred tank reactor is given by:

$$\bar{t_d} = \ln 2/\mu \tag{1}$$

where μ is the specific growth rate. Equation 1 is a solution of the mass balance equation for the cell growth in the exponential phase:

$$\frac{dX}{dt} = \mu X \text{ or } \frac{1}{X}\frac{dX}{dt} = \mu \tag{2}$$

with

$$x = x_0 \qquad \text{at } t = t_{\text{lag}} \tag{3}$$

where x_0 is the cell concentration at the end of the lag phase t_{lag}. Equation 2, together with other mass balance equations, is widely used for the mathematical description of batch cultures of microorganisms.

On the other hand, the empirical formulas, in which no theoretical hypothesis can be postulated, are also used widely in the engineering fields. In many cases, they are the only feasible option for correlating data from actual processes. For instance, the following relationship has been developed to relate the volumetric mass transfer coefficient k_La to the superficial gas velocity γ_{gs} for the bubble column, water and coalescing operation condition:

$$k_La = 0.32(\gamma_{gs})^{0.7} \tag{4}$$

where γ_{gs} is equal to the gas feed volumetric flow rate divided by vessel cross-section area times the gas hold-up. The relationship (equation 4) is based on examination of mass transfer data from many different bubble columns. Experimental results are often reported in the form of k_La because of the lack of knowledge of a directly. Also, in some cases, this combined parameter is used to account for other effects, such as long residence times of small bubbles in highly viscous fermentations (1). There is no exact theoretical explanation for the relationships summarized in empirical formulas. Therefore, to use them for the interpretation of experimental observations requires not only a large set of data, but a great deal of judgment. The latter sometimes involves difficult decisions.

20.1.4. Combination for Data Presentation

Tables listing data have the highest accuracy but can easily become too long, and the overall result or trend of the data cannot be readily visualized. Graphs can convey, in an understandable manner, more information than numbers alone. This is because the graphs or plots of data can create immediate visual impact, since relationships between variables are represented directly. Graphs, with their accompanying legends, should provide an explanation of the research that runs parallel to the text. In presenting graphics, the legend is almost as important as the graph. It should not be necessary for a reader to refer to the text to understand what information is being provided by the graph. The graph and its legend should form a self-contained unit. Graphs also allow easy interpolation of data, which can be difficult with tables. By convention, independent variables are plotted along the abscissa (the x axis), while one or more dependent variables are plotted along the ordinate (y axis). Plots show at a glance the general pattern of data and can help identify whether there are anomalous points; it is good practice to plot raw experimental data as they are being measured. In addition, graphs can be directly used for quantitative data analysis.

Equations are generally more convenient and accurate in applications than graphs. Graphical correlations have the great advantage over equations of demonstrating visually the nature and magnitude of deviations of individual data points. A combination of the two is the best solution—a graph for demonstration and an equation for use.

20.2. DEVELOPMENT OF DATA CORRELATION

The basic purpose behind correlation is to find out if two or more variables are related to one another. Regression then allows the use of the relationship in the prediction of one variable given a score on the other variables. A correlation is usually developed with the objectives of summarizing and generalizing data for prediction and synthesis. Being a form of data presentation, correlation may consist of a table of numbers representing the relationship between variables; a graph providing a more convenient and revealing representation of this set of numbers; or an equation illustrating a still more concise and convenient representation, in which the set of numbers in the table is reduced to a value or set of values for the constants in the equation.

20.2.1. Graphical Analysis

Graphics can be used to visualize the relationships between variables, establish correlations of variables, and indicate which variables are more important. They also show sample patterns in the data, similarities, dissimilarities, groups, and unusual samples. Therefore, prior visual examination of the data is suggested for the mechanical development of a correlation. This is because the trial scatterplots may outline the form for an empirical equation to be considered, may provide a quick test of the applicability of a theoretical model or indicate the best choice between different models, or may identify anomalous data points (outliers).

Consider the data representing the relationship of antibody production and viable cells for animal cell line BTSN6 in batch culture (21). It is a perfect positive correlation. From the figure, it is reasonable to decide that a straight line can be used to approximate the experimental result. We can have a further impression that the relationship can be presented by a linear equation. When the plot for the cell-specific DNA contents with respect to time during the same experiment is examined (21), it may be concluded that a curve, instead of a straight line, needs to be drawn for the representation of the trends of DNA contents during the culture.

When the scatterplots have been previewed, one may wish to smooth the data by drawing a curve through the points. Different smoothing techniques are available for curve-drawing. A simple method is called hand smoothing. As the name suggests, it is a manual smoothing that relies heavily on human judgment. The advantage of hand smoothing is that judgments about the significance of individual data points can be taken into account. The risk of this type of curve sketched freehand through the data is that, when we try to moderate the effects of experimental error, our subjectivity may drive the expected response into the data. Another method is called machine smoothing, which involves a computer software package. By this method, the subjective element may be eliminated with the aid of preprogrammed mathematical or statistical algorithms. However, bias may still be introduced into the results. For example, abrupt changes in the trend of data are generally not recognized by statistical analysis.

20.2.2. Curvefitting and Least-Squares Methods

Curvefitting is the determination of coefficients for an equation so that the resulting data model meets some desired criteria, such as minimum absolute error. Standard curvefitting techniques fit data by the use of what is called the least-squares algorithm. This means that the sum of the squares of the absolute errors is minimized. Other curvefit criteria, such as minimum percent error, are often highly desirable but until now have not been practically available.

20.2.2.1. Least-Squares Methods

Consider a data set consisting of n pairs of points $[(x_i, y_i);\ i = 1, 2, \ldots, n]$ where $y = (y_1, y_2, \ldots, y_n)$ are the observations referenced to a basic set of indices x_i. Least-squares theory assumes that a set of model functions $M(x_i, b)$, $i = 1, 2, \ldots, n$, are given as a function of m parameters $b = (b_1, b_2, \ldots, b_m)$. Ideally, one wants to find parameters $b^0 = (b_1^0, b_2^0, \ldots, b_m^0)$ such that $y_i = M(x_i, b^0) + e_i$, where e_i is an experimental error drawn at random from a population with zero mean.

It can be shown that the best estimator of b^0 is given by the minimum $\bar{b}^0$ of the residual $L(b)$, or

$$\bar{b}^0 \text{ minimizes } L(b) = \sum_{i=1}^{n} [y_i - M(x_i, b]^2 \tag{5}$$

The estimator $\bar{b}^0$ is best in the sense that it represents the maximum-likelihood estimation of b^0.

Linear least-squares analysis is a special case of least-squares theory where the model functions are linear functions of the parameters b,

$$M(x_i, b) = \sum_{j=1}^{m} b_j F_{ji} = (F^* b)_i \tag{6}$$

The star denotes the transposition operation. The solution to this linear least-squares problem is the well-known normal equation:

$$b = (FF^*)^{-1} Fy \tag{7}$$

If the model functions M are not linear, the problem can be solved approximately by linearizing M,

$$M(x_i, b') \approx M(x_i, b) + \sum_{j=1}^{m} \frac{\partial M(x_j, b)}{\partial b_j} (b'_j - b_j) \tag{8}$$

and iterative application of the normal equation (equation 7).

Sometimes there are not enough independent observations, i.e., the number of data points n is less than the number, m, of adjustable parameters. In this case, the least-squares problem cannot be solved unless additional "constraints" on the adjustable parameters can be introduced. Under other circumstances, we may want to introduce constraints on the adjustable parameters to add additional information, e.g., the sum of angles of a triangle. In more complex cases, the forms of the constraints are unknown. Here we confine ourselves to least-squares fit problems in which the forms of constraints are known and thus the least-squares problem can be solved. (In the case where not enough independent observations are available, a minimum number of sufficient constraint equations have to be provided.) Details of the solution of the least-squares methods can be found in statistics textbooks (8, 14, 34).

Linear Least-Squares Fit

Given a set of observations, (x_j, y_i), we seek a pair of numbers (a, b) such that the sum of the squares

$$\sum_{i=1}^{n} [y_i - (a\,x_i + b)]^2 \tag{9}$$

attains a minimum value. Differentiating with respect to a, dividing by -2, and setting the result equal to zero, we get

$$\sum_{i=1}^{n} [y_i - (a\,x_i + b)](x_i) = 0 \tag{10}$$

and doing the same with respect to b, we get

$$\sum_{i=1}^{n} [y_i - (a\,x_i + b)] = 0 \tag{11}$$

Equations 8 and 9 can be solved, giving

$$a = \frac{\left(\sum_{i=1}^{n} x_i y_i\right)\left(\sum_{i=1}^{n} x_i\right) - \left(\sum_{i=1}^{n} x_i^2\right)\left(\sum_{i=1}^{n} y_i\right)}{\left(\sum_{i=1}^{n} x_i\right)^2 - N \sum_{i=1}^{n} x_i^2} \tag{12}$$

and

$$b = \frac{N \sum_{i=1}^{n} x_i y_i - \left(\sum_{i=1}^{n} x_i\right)\left(\sum_{i=1}^{n} y_i\right)}{N \sum_{i=1}^{n} x_i^2 - \left(\sum_{i=1}^{n} x_i\right)^2} \tag{13}$$

These are the values of a and b for which the sum of the squares of the deviations is minimum. The resulting equation is as follows:

$$y = 10.463095 - 1.709524\,x \tag{14}$$

The linear least-squares problem is usually solved by explicitly forming and inverting the normal equations or by supplying a standard linear algebra matrix equation solver (such as LU or QR decomposition). If the normal equations are close to singular, the singular value decomposition method may be used.

Nonlinear Fitting

Nonlinear least-squares algorithms are discussed in the literature (4, 6, 10, 29). Interested readers should refer to these books.

In practice, sometimes we can reexpress data in such a fashion that we can find a least-squares line formula that fits the reexpressed data, and by working backwards, we can find a formula that fits the original data. To decide how to reexpress the data, often it's a good idea to see a scatterplot of the data to see what curve it resembles. Then, by experimenting with reexpressions and graphing the scatterplot of the reexpressions, we can determine how to reexpress the data so the scatterplot looks very much like a line.

Example 1. Observed data:

t (time in days)	M (Mathium in kg)
0	100
5	37
10	13.5
15	5
20	1.8
25	0.67
30	0.25

The problem is to determine a formula relating Mathium (M) to time (t). If the data are graphed, it can be seen that t and M are not related by a linear relationship. In fact, it looks as if M is an exponential function of t, maybe something of the form:

$$M = M_0 e^{Kt}$$

The data can be reexpressed as:

t	ln (M)
0	4.605
5	3.610
10	2.603
15	1.609
20	0.588
25	−0.401
30	−1.386

ln (M) and t can then be shown to be linearly related. We can use the least-squares line program to get an equation:

$$\ln(M) = -0.2t + 4.609$$

We can convert to an exponential equation to solve for M.

$$M = e^{(-0.2t+4.609)} = e^{(-0.2t)}\,e^{4.609}$$

Therefore

$$M = 100.38e^{(-0.2t)}$$

Now we have a formula relating t to M.

20.2.3. Linearization

When one collects experimental data, one does not expect the data to fit exactly to some theoretical function. In fact, any data that do fit exactly to the theory should be considered suspect. When you are analyzing experimental data, you do not want your curve to go through all the points, as splining and other interpolations do, since that would imply an exact relation between the data and the theory. In practice, one has to experiment with many different orders and other functions (such as exponential) before an adequate fit can be found.

When developing a correlation, it is desirable to plot the data in such a form as to yield a linear relationship over part or all of the range of the variables. A model may provide guidance to the choice of the appropriate coordinates for this purpose. A combination of equations and graphs is one of the effective ways for the linearized presentation of the correlations. Equations can be rearranged for plotting data as straight lines (7) (Table 1).

Proper graphing of tabulated data can be divided into several steps:

1. Select appropriate graph type.
2. Select appropriate axes locations.
3. Select range and scale of axes.
4. Graduate and calibrate the axes. Scale graduations should be selected so that the smallest division of the axis is an integer power of 10 times 1, 2, or 5.
5. Completely label the axes.
6. Plot data points.
7. Plot or fit curves.
8. Add titles and notes.
9. Darken appropriate lines.

A more complex example in which a nonlinear relationship can be linearized involves mixing theory (27). A number of researchers have tried to estimate the mean circulation time and/or the terminal mixing time from simple models for the fluid flow in the tank by predicting a characteristic length of the circulation path and using an appropriate expression for an averaged circulation velocity. Mean circulation time is then calculated from:

$$\theta = \frac{L}{\gamma} \tag{15}$$

where θ is the mean circulation time, L is the characteristic length of the circulation path, and γ is fluid velocity. From systematic experimental observations, it has been summarized that the relationship of the ratios of liquid height/reactor diameter (H/D_T) and the reactor diameter/impeller diameter (D_T/d_i), and the mean circulation time, can be correlated as follows:

$$n\theta = C_1' \left(\frac{H}{D_T}\right)^{0.6} \left(\frac{D_T}{d_i}\right)^{2.7} \tag{16}$$

It is assumed that the fluid flow through the impeller region

TABLE 1 Methods for plotting data as straight lines (21)

Equation	Method
$y = Ax^n$	Plot y versus x on logarithmic coordinates
$y = A + Bx^2$	Plot y versus x^2
$y = A + Bx^n$	First obtain A as the intercept on a plot of y versus x, then plot $(y - A)$ versus x on logarithmic coordinates
$y = B^x$	Plot y versus x on semilogarithmic coordinates
$y = A + \frac{B}{x}$	Plot y versus $\frac{1}{x}$
$y = \frac{1}{Ax + B}$	Plot $\frac{1}{y}$ versus x
$y = \frac{1}{A + Bx}$	Plot $\frac{x}{y}$ versus x or $\frac{1}{y}$ versus $\frac{1}{x}$
$y = 1 + (Ax^2 + B)^{1/2}$	Plot $(y - 1)^2$ versus x^2
$y = A + Bx + Cx^2$	Plot $\frac{v - v_n}{x - x_n}$ versus x, where (y_n, x_n) are the coordinates of any point on a smooth curve through the experimental points
$y = \frac{x}{A + Bx} + C$	Plot $\frac{x - x_n}{y - y_n}$ versus x where (y_n, x_n) are the coordinates of any point on a smooth curve through the experimental points

is equal to the corresponding circulation through the cross-section of the reactor, characterized by:

$$\gamma = C_1' \, n \, d_i \left(\frac{d_i}{D_T}\right)^2 \tag{17}$$

The lengths of the two circulation paths L_1 and L_2 are given by:

$$L_1 = D_T + 2\,H - 2.5\,d_i \tag{18}$$

and

$$L_2 = D_T + 1.5\,d_i \tag{19}$$

respectively. This results in an averaged value for the path length

$$L = \frac{L_1 - L_2}{2} = D_T + H - 0.5\,d_i \tag{20}$$

Making use of equations 17 and 20, the mean circulation time (equation 15) can be estimated to be:

$$\theta = \left(1 + \frac{H}{D_T} - 0.5\,\frac{d_i}{D_T}\right)\left(\frac{D_T^3}{nd_i}\right) \tag{21}$$

The term in brackets in equation 21 can be approximated by a simple power function, resulting in the following correlation:

$$\left(1 + \frac{H}{D_T} - 0.5\,\frac{d_i}{D_T}\right) = 1.64\left(\frac{H}{D_T}\right)^{0.6}\left(\frac{D_T}{d_i}\right)^{0.1} \tag{22}$$

This correlation holds in a region $0.2d_i/D_T$ 0.5 and 1 H/D_T2. The final equation for the mean circulation time is, therefore, given by:

$$n\theta = C_1'\left(\frac{H}{D_T}\right)^{0.6}\left(\frac{D_T}{d_i}\right)^{3.1} \tag{23}$$

A comparison between equations 23 and 16 shows an excellent agreement in the exponent for (H/D_T). The theoretical value of the exponent for (D_T/d_i) is, however, higher than that observed in the experiments. This deviation is caused by the uncertainties in the estimation of the characteristic circulation velocity.

20.2.4. Model Validation—An Advanced Fitting Technique

The curvefitting techniques that we have discussed so far are based on the correlations between independent and dependent variables of a process. They can be used to visualize your experimental data or plant measurements. However, they usually do not give out the physical significance describing internal physical states of the system automatically. By contrast, when we combine the data analysis method with mathematical modeling for biological systems, it will result in an advanced fitting technique for the improvement of the model behavior. The validated models are capable of predicting the physiological states of microorganisms or other cells during a bioprocess.

In the biotechnological field, the application of mathematical models is widespread. The establishment of mathematical models in biological systems may be considered to consist of:

1. Model construction—the initial choice of particular mathematical forms to describe and analyze phenomena of interest
2. Model analysis—a study of the models to determine their characteristics and suitability

3. Parameter analysis—using the models to determine physiologically meaningful parameters

The advanced fitting techniques can be used in steps 2 and 3: first, visualizing both the model predictions and the experimental data; second, implementing sensitivity analysis; and third, repeatedly adjusting the model structure and parameters to reduce the errors between model predictions and the actual observations.

As an example, consider a detailed, structured model of animal cells growing in batch culture (28). When the model was employed for the simulation of cell growth by the hybridoma cell line AFP27, it could be seen that the model prediction for antibody production fit the experimental observation quite well.

Quite often, models, constructed by the initial choice of particular mathematical forms, are imprecise when they are directly used for describing a real bioprocess; then the data-based fitting approach can be employed for iterative performance improvement of the models. We have developed such an approach for modeling complex and partly known systems, in which both knowledge-based systems and neural networks have been incorporated. The resulting artificial intelligence-based framework is called a hybrid model, to distinguish it from the conventional deterministic models. Data analysis has played a key role in the proposed scheme, which can be defined as a two-step task:

1. Compare the actual behavior of a process, as manifested by the values of the operating variables, against the behavior predicted by a model, and generate the residuals that reflect model discrepancies.
2. Evaluate the residuals and through a model-based inversion process identify the inputs that cause the observed behavior.

State predictions of both the conventional mathematical model (parameter untuned) and the proposed artificial intelligence-based learning (parameter adjusted with the aid of artificial intelligence tools and the experimental database) may be compared to the actual experimental observation [11]. In order to validate mathematical models by the use of advanced fitting techniques, simple or detailed, a large database is required to give accurate values for the model parameters. In addition, an independent and equally large set of experimental data are required to confirm that the model has predictive strength when operating at conditions that were not used when the parameters were fitted.

20.3. CALCULATION OF SPECIFIC RATES AND OTHER METABOLIC PARAMETERS OF INTEREST

The most important metabolic parameters are discussed in this section.

20.3.1. Yields and Specific Rates

Specific rates and yields (see Table 2) may be related. For example,

$$Q_S = \mu / y_{XS} \tag{24}$$

$$Q_{ATP} = \mu / Y_{ATP} \tag{25}$$

In addition, there are other "derived" parameters that are commonly used in descriptions of cell growth. For example,

$$\mu = \mu_{max}\,(S/K_S + S) \tag{26}$$

and

$$Q_S = \alpha\mu + \beta \tag{27}$$

The first expression is the Michaelis-Menten relationship commonly used to describe microbial growth. Values of K_S and μ_{max} are used to estimate (importantly) K_S and μ_{max}. The second expression is the commonly used Leudeking-Piret relationship. This expression assumes that there is a growth-related and a nongrowth-related component for substrate use. Values of Q_S and μ are used to estimate the relative sizes of α and β.

The calculation of each of these parameters is now discussed and illustrated by using a combination of hand calculations, spreadsheets, and linear and nonlinear regression methods.

20.3.2. Calculation of Yields and Specific Rates

Typical batch data from our laboratories for the growth of *S. cerevisiae* are used to illustrate the calculation of overall yields and overall specific rates and "point to point" yields and specific rates (see Table 3A). The most convenient way of analyzing these data is by using a spreadsheet. The spreadsheet batch.xls (25) gives the following analysis:

Overall Specific Rates and Yields:

$\mu = 0.3855\ h^{-1}$

$Q_S = 2.2941$ g substrate/g biomass/h

$y_{XS} = 0.1680$ g biomass/g biomass

"Point To Point" Specific Rates and Yields:
(a) From Point to Next Point

$Q_S = 4.2049 \pm 5.022$ g substrate/g biomass/h

$y_{XS} = 1.11049 \pm 1.7485$ g biomass/g substrate

(b) From Point to Second Next Point

$Q_S = 3.1644 \pm 1.8144$ g substrate/g biomass/h

$y_{XS} = 0.2840 \pm 0.1016$ g biomass/g substrate

It can be seen from this analysis that:

(a) the overall yields/specific rates and "point to point" yields/specific rates vary considerably;

(b) averaging over a wider range of data results in a rapid convergence to the overall yields/specific rates from the "point to point" yields/specific rates.

Clearly, the accuracy of the data from point to point must be of the highest quality to expect a close correlation between "point to point" and overall rates. Nevertheless, real variations in the values from "point to point" are of importance, if they can be obtained accurately.

20.3.3. Mass Balances, "% Unaccounted for" Carbon, and Other Inferential Calculations

Spreadsheets also allow us to conveniently undertake mass balances around the cell. This can serve a number of purposes:

(a) To calculate major metabolic parameters, e.g., Q_{CO_2}, Q_{O_2}, and Q_P

TABLE 2 Yields and specific rates

Parameters	Definition	Units	Equation
Yield			
y_{XS}	Biomass from substrate yield	gram of biomass/gram of substrate	
y_{PS}	Product from substrate yield	gram of product/gram of substrate	
Y_{ATP}	Biomass from ATP yield	gram of biomass/mol of ATP	
Specific rates			
Q_S	Specific substrate uptake rate	gram of substrate/gram of biomass/hour	$(1/X)\ dS/dt$
Q_P	Specific product production rate	gram of product/gram of biomass/hour	$(1/X)\ dP/dt$
Q_{ATP}	Specific ATP production rate	mol of ATP/gram of biomass/hour	$(1/X)\ d\ ATP/dt$

(b) To calculate "% unaccounted for" carbon, "% unaccounted for" oxygen, etc.

(c) To make inferential calculations including Y_{ATP} from known μ, product distribution and cell stoichiometry or μ from assumed values of Y_{ATP}.

These are illustrated in the following manual calculations. The spreadsheets associated with these calculations (metaerob.xls; metanaer.xls) are available (2).

Sample Problem:
A microorganism has the following composition (grams per gram of biomass) during batch growth:

Carbon	0.48
Hydrogen	0.06
Nitrogen	0.08
Oxygen	0.29
Ash	0.09

The microorganism is capable of growing on glucose both aerobically (producing carbon dioxide and water) and anaerobically (producing ethanol and carbon dioxide). Under aerobic conditions, the biomass from substrate yield is 0.5 g biomass/g glucose, and the specific growth rate is 0.2 h^{-1}. Under anaerobic conditions, the biomass from substrate yield is 0.17 g biomass/g glucose and the specific growth rate is 0.45 h^{-1}.

(a) Calculate the theoretical specific oxygen uptake rate (mmol O_2/g biomass/h) required for catabolism (aerobic) and the theoretical specific ethanol production rate (mmol ethanol/g biomass/h) by catabolism (anaerobic).

(b) Under experimental conditions (anaerobic), an ethanol from substrate yield of 0.31 g ethanol/g glucose and a specific carbon dioxide production rate of 76.26 mmol CO_2/g biomass/h is obtained. Under experimental conditions (aerobic), a specific oxygen uptake rate of 10.95 mmol O_2/g biomass/h is obtained. Calculate the "unaccounted for" carbon (anaerobic) and "unaccounted for" oxygen (aerobic).

(c) For the experimental conditions in (b), calculate the experimental Y_{ATP} (assuming a PO ratio of 2.2 mol ATP/0.5 mol O_2 for aerobic metabolism).

(d) Assuming a Y_{ATP} (g biomass/mol ATP) of 10.5, estimate the anaerobic and aerobic specific growth rate from the experimentally measured specific oxygen uptake rate (aerobic) and the end product distribution (anaerobic).

Solution:

Figures 1 and 2 represent the carbon flow occurring in the two cases (anaerobic and aerobic metabolism). The are obtained by mass balances around the cell. The spreadsheet calculations for these figures are metaerob.xls and metanaer.xls (25).

(a) Anaerobic:

$Q_{CO_2} = Q_E$ = (0.008844 mol/0.17 g biomass) * 0.45 h^{-1} * 1,000 mmol/mol
= 23.4 mmol/g biomass/h

Aerobic:

$Q_{CO_2} = Q_{O_2}$ = (0.0066 mol/0.5 g biomass) * (0.2 h^{-1})/(1,000 mmol/mol)
= 5.328 mmol/g biomass/h

(b) Anaerobic:

"unaccounted for" carbon
= [(0.38824 − 0.31) g ethanol * (24 g carbon/46 g ethanol)] + [(0.0844 − 0.07626) mol CO_2/g biomass * (12 g carbon/mol CO_2) * (0.17 g biomass/0.45 h^{-1})]
= (0.04082 + 0.0369) g carbon = 0.07772 g carbon

"unaccounted for" carbon
= (0.07772 g carbon/0.4 g carbon) * 100
= 19.43%

Aerobic:

"% unaccounted for" oxygen
= [(13.22 − 10.95) mmol O_2/13.22 mmol O_2] * 100
= 12.39%

(c) Anaerobic:

glucose → 2 ethanol + 2 CO_2 + 2 ATP

Hence $Q_{ATP} = Q_E$
= 2*(0.31 g ethanol)/(46 g ethanol/mol)*(0.45 h^{-1})/(0.17 g biomass)
= 0.03568 mol ATP/g biomass/h

Hence $Y_{ATP} = \mu/Q_{ATP}$
= 0.45 h^{-1}/(0.03568 mol ATP/g biomass/h)
= 12.56 g biomass/mol ATP

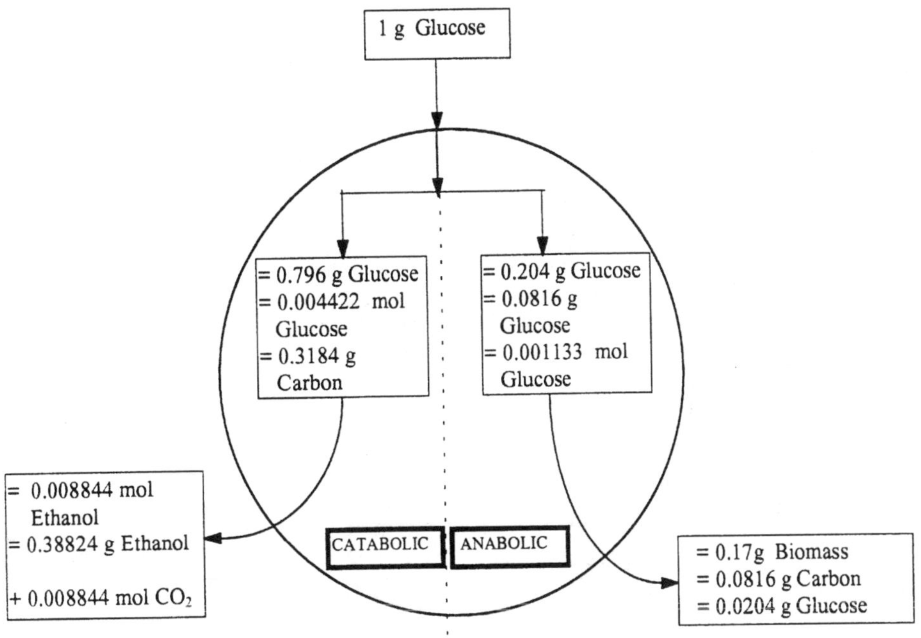

FIGURE 1 Anaerobic metabolism of glucose-carbon flow chart.

Aerobic:

glucose + 6 $O_2 \rightarrow$ 6 CO_2 + 6H_2O

hence $Q_{ATP} = Q_{O_2} * PO$

$= (4.95 \text{ mmol } O_2/\text{g biomass/h}) * (\text{mol } O_2/1{,}000 \text{ mmol } O_2) * (2.2 \text{ mol ATP}/0.5 \text{ mol } O_2)$

$= 0.02178 \text{ mol ATP/g biomass/h}$

hence $Y_{ATP} = \mu/Q_{ATP}$

$= 0.20 \text{ h}^{-1}/(0.02178 \text{ mol ATP/g biomass/h})$

$= 9.18 \text{ g biomass/mol ATP}$

(d) Anaerobic:

$\mu = Y_{ATP} * Q_{ATP}$

$= (10.5 \text{ g biomass/mol ATP}) * (0.03568 \text{ ATP/g biomass/h})$

$= 0.375 \text{ h}^{-1}$

Aerobic:

$= Y_{ATP} * Q_{ATP}$

$= (10.5 \text{ g biomass/mol ATP}) * (0.02178 \text{ ATP/g biomass/h})$

$= 0.223 \text{ h}^{-1}$

20.3.4. K_S and V_{max} Calculations

The estimation of these parameters is usually undertaken by graphical analysis, most typically the Lineweaver-Burk, Haynes, or Eadie-Hofstie plot. The methods may be accessed through a number of computer packages using nonlinear regression analysis. A typical one is Eukinetic (9), and the analysis is performed as follows.

The program determines K_m and V_{max} values for enzyme-kinetic data by nonlinear regression analysis. Starting values of K_m and V_{max} are obtained by linear regression fitting of a Haynes plot. These values are used to seed the nonlinear regression, which fits the data to the best hyperbola. A Taylor series of partial derivatives of K_m and V_{max} is constructed, and their errors are estimated by multivariate, linear regression. Weighting may also be applied to the data. This weighting is a simple weighting in which the variance is assumed proportional to the square of the velocity. This has the effect of providing some correction for "outfliers" at high velocities. The program also "flags" if the data are considered inaccurate.

Examples of the application of this package, with data obtained in our laboratories for the batch and continuous growth of *S. cerevisiae* and the continuous growth of *Candida utilis*, are shown in Table 3. The results indicate a broad agreement with V_{max} values but a wide variation in possible values of K_m. This demonstrates the need for accurate experimental data as well as the difficulty in measuring K_S experimentally. It reinforces the point that models of a process always need to be capable of experimental verification. In this case, such verification of the model basis is extremely difficult due to experimental limitations.

20.3.5. Growth- and Nongrowth-Associated Metabolism

The Leudeking-Piret relationship has been used for many years to describe the use of substrates by cells for growth- and nongrowth-related activities. In a review (3), this concept was examined for its ability to fit data from a range of biochemical processes, including yeast metabolism, amino acid production, and monoclonal antibody production. It was shown to be an inadequate description for anything but the most basic metabolism. It was concluded that more sophisticated modeling techniques were likely needed if a unifying relationship between specific substrate uptake rate and specific growth rate was sought.

The fitting of this relationship to the continuous data for *C. utilis* and *S. cerevisiae* is now used to illustrate this point. It also illustrates the caution that is necessary when

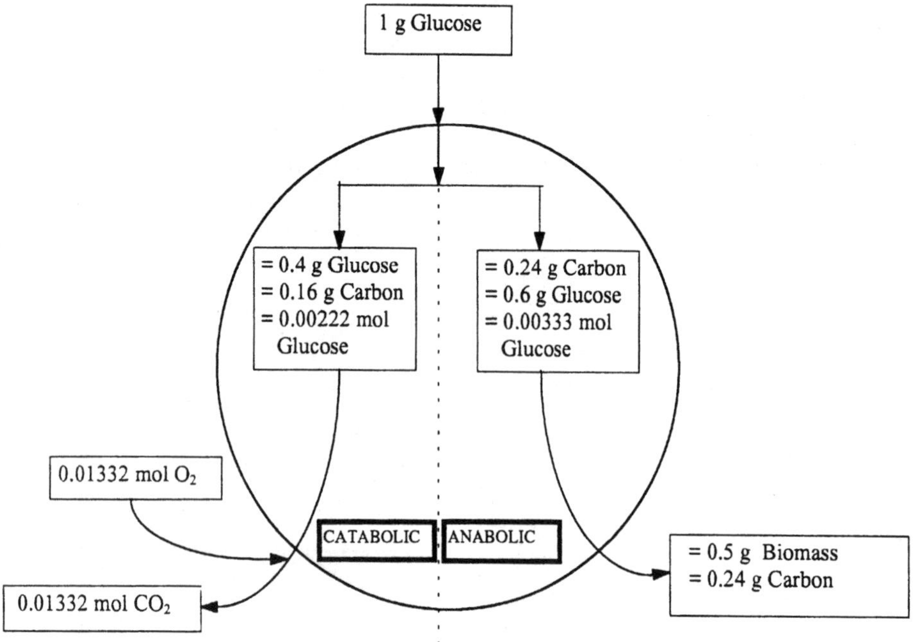

FIGURE 2 Aerobic metabolism of glucose-carbon flow chart.

attempting to verify relationships by fitting experimental data to them. The spreadsheets for these analyses (ccutilis.xls and ceserev.xls) are available (2).

The following results were obtained:

C. utilis:	α, 0.1033;	β, 0.0323;	R^2, 0.9383
S. cerevisiae:	α, 0.1854;	β, −0.3789;	R^2, 0.7719

It can be seen from the regression analysis that *C. utilis* may broadly be described by the relationship, but *S. cerevisiae* cannot be. This is to be expected, because one cell has fixed stoichiometry and one major catabolic pathway and the other has variable stoichiometry and more than one major catabolic pathway (2). Hence, each growth situation should be considered a separate metabolism and not made to fit prescribed patterns or relationships.

20.3.6. Least-Squares Analysis of Biochemical Oxygen Demand (BOD) for Wastewater Treatment

The BOD relationship is expressed as:

$$L_t = L_o \left(e^{-kt}\right) \tag{28}$$

where L_t is BOD remaining at time t, L_o is ultimate BOD demand (= total amount of oxygen consumed in the reaction), and k is average reaction rate constant.

The BOD relationship is used frequently in wastewater to evaluate the relative potential pollutant strength of wastes and to monitor the rate of remediation for polluted sites. Normally this is rearranged to:

$$y = L_o - L_t = L_o \left(1 - e^{-kt}\right) \tag{29}$$

where y = amount of oxygen consumed after time t.

It is then necessary to estimate L_o and k. To do this, experimental measurements of y at various times are used to estimate L_o and k. This normally will involve some least-squares analysis. There are also graphical methods. The most popular methods are the least-squares, log-mean, and Fujimoto methods (graphical) (30). These methods are best suited to spreadsheet analysis, and they are available at leastsq.xls, logmean.xls, and fujimoto.xls (2). An example is illustrated in Table 4A. When used in the spreadsheets, the results shown in Table 4B are obtained. It can be seen that consistent results are obtained.

20.4. DATA-BASED MODELING

The fundamental goal of data-based models is to build a model from the input-output data of a process that can be used for generating future predictions for some variables of interest whose value is unknown. In this context, it is critical to select suitable methods to turn the data into useful patterns and models that allow computer calculations or simulations of physical quantities. Many approaches may be available for the formulation of the data-based modeling. Following are tools from probabilistic modeling, neural networks, and genetic programming.

20.4.1. Probabilistic Modeling

Modern statistical modeling tools allow an analyst to think about the problem at a higher level, try numerous approaches, view data from different perspectives, estimate the uncertainty of conclusions arising out of different processes, iterate through several stages before coming up with a representation scheme, and so on. Among the statistical models, probabilistic models provide a good language for problem decomposition and knowledge refinement. Because of their flexibility, they are suitable for representing a wide variety of learning models. Also, probabilistic mod-

TABLE 3 Examples of calculation of K_S and μ_{max} by graphical analysis

A. Batch data: *S. cerevisiae*

Time, t (h)	Biomass concentration, X (g/liter)	Residual sugar concentration, S (g/liter)
14.2	0.12	9.7
14.7	0.175	8.5
15.2	0.36	8.1
15.7	0.45	7.3
16.4	0.6	6.1
17.2	0.89	5.0
18.0	1.25	3.1
18.8	1.5	0.3
19.7	1.75	0

B. Continuous data

	Dilution rate, D (h^{-1})	Biomass concentration, X (g/liter)	Residual sugar concentration, S (g/liter)
S. cerevisiae	0.08	5.3	0.01
	0.17	5.4	0.01
	0.20	5.5	0.01
	0.22	5.55	0.01
	0.24	5.6	0.01
	0.30	4.4	1.8
	0.35	1.7	3.25
	0.42	1.75	3.7
C. utilis	0.05	4.8	0
	0.10	5.2	0.2
	0.20	5.3	0.4
	0.30	5.2	0.8
	0.35	5.0	1.4
	0.40	4.8	2.2
	0.45	4.1	3.8

C. Results (using the last three points)

Data	V_{max}	K_m
Batch, *S. cerevisiae*		
Hanes plot	0.12556	0.01551
Lineweaver-Burk	0.11923	0.00089
Eadie-Hofstee	0.11950	0.00089
Continuous, *C. utilis*		
Hanes plot	0.49719	0.49218
Lineweaver-Burk	0.23323	0.00367
Eadie-Hofstee	0.30197	0.00508
Continuous, *S. cerevisiae*		
Hanes plot	0.44097	0.02419
Lineweaver-Burk	0.36094	0.00588
Eadie-Hofstee	0.37315	0.00645

TABLE 4 Spreadsheet analysis of BOD for wastewater treatment

A. Data

Time (h)	Oxygen consumed (mg/liter)
0	0
1	7.3
2	12.8
3	16
4	20.1
5	22.5
6	23.8
7	25.3

B. Results

	L_o (mg/liter)	k (h^{-1})
Least-squares	29.106	0.2897
Log-difference	33.452	0.2828
Fujimoto	29.361	0.2918

els allow a computational approach to developing learning and discovery algorithms.

We take a dynamic linear model for tank-level monitoring (18) as the illustrative statistic computing/graphing example. For 177 underground wastetanks, in-tank instrumentation has been taken to monitor liquid level and determine leaks (a decrease) or hydrogen buildup (an increase). This instrumentation generates large quantities of data that in turn require the development of efficient and timely methods of analysis. The purpose of this work was to incorporate a model of the steady state along with models of various alternative states (features) that are of interest for some reason. Subsets of actual tank-level data that show features of interest are illustrated (18). The features shown are:

- Steady state (SS), which refers to behavior that is constant over time or is linear with time. The decrease in level over time (linear with time) is attributed to evaporation.
- Level change (LC), which refers to an abrupt jump in a series of tank-level measurements.
- Slope change (SC), which indicates that the rate of level change is changing. Events that have produced SC include leaks and changes in the rate of evaporation, the latter being primarily due to a change in the tank's venting system.
- Transient (T), or outlier, which refers to a single measurement that has a value far from other measurements taken at nearly the same time. Transcription errors are one source of transients.

For each new observation, the strategy is to determine whether a state change has occurred and, if so, what is the likely (i.e., most probable) new state of the system. The sets of probabilities for the four states of the dynamic linear model are used to evaluate the state of the system after each new observation. An observation is "flagged" whenever the probability that the system is in steady state falls below some threshold. The observation y_t might be flagged, for example, whenever

$$\text{Prob}\{\text{SS at time } t\} < 0.8$$

The following diagnostic strategy is outlined (28). In summary, consider the decision to be made after observing the value y_t at time t. In this case, y_t does not fit the steady-state model. However, at time t, it is not possible to identify the new state of the system. The next observation, y_{t+1}, is needed to make this call. Accordingly, the probability that the system is in each of the four states must be updated in light of the new observation y_{t+1}. A graphical summary of the one-step-back probabilities is shown for a typical series of liquid-level observations (28). The identification of level and slope changes by the dynamic linear model analysis also turned out to be an important preprocessing step in analysis of tank-level data to identify trapped gas, a process not previously directly accessible (18).

20.4.2. Neural Networks

Artificial neural networks (ANNs) are another type of black box methods, which have become very popular lately. They are powerful at modeling even strongly nonlinear relationships. An ANN is an information-processing paradigm that is inspired by the way biological nervous systems, such as the brain, process information. The key element of this paradigm is the novel structure of the information processing system. It is composed of a large number of highly interconnected processing elements (neurons) working in unison to solve specific problems. ANNs, like people, learn by example. An ANN is configured for a specific application, such as pattern recognition or data classification, through a learning process.

ANNs can work when all other modeling efforts fall short. In addition, they are quite useful when a mix of floating-point numbers and digital switches must be combined. It has been shown mathematically that an ANN is capable of learning any continuous nonlinear input-output mapping. ANNs are extremely useful data-modeling tools, but like all models, they have their pros and cons. They are best used when a data set can be described as incomplete or poorly characterized. Additionally, if it is suspected that some relationships exist in a given data set but these relationships are unknown, ANNs can usually identify them.

Tholdur and Ramirez (32) have reported their work on the optimization of fed-batch bioreactors using ANN parameter function models. They have proved that neural networks can be trained to learn any arbitrary function using prior examples of input-output data. Therefore, ANN can be used as general purpose function approximators to identify and mimic unknown functions. This eliminates the need for trial-and-error techniques of assuming a functional form followed by some form of nonlinear regression to identify parameters in the function. The neural network modeling coupled with a dynamic programming optimization technique was illustrated using the Park-Ramirez bioreactor system (26) for the secretion of foreign protein from baker's yeast. The data set consisted of 250 data points, of which 200 were used for training and 50 for model validation. In this work, neural networks with 5, 7, 7, and 5 hidden neurons, respectively, were chosen for the following parameter functions: the specific growth rate, the biomass-to-glucose yield, the secretion rate constant, and the protein expression rate. The approximation capabilities of the neural network parameter functions have been summarized (26). The result clearly demonstrated that the neural network parameter functions are excellent representations of the actual parameter functions. After training, these neural networks were used in place of the parameter functions in the differential equations, and dynamic programming was used to obtain the optimal control policy (32).

20.4.3. Genetic Programming

The performance of an individual organism in its environment determines the likelihood of its passing on its genetic material to future generations. This basic biological principle is known as Darwinian "survival of the fittest" and has inspired a class of algorithms known as genetic algorithms (GAs). Like evolution, GA operates on a population of individuals that represent potential solutions to a given problem. It then seeks to produce better (more fit) individuals (solutions) by combining the better of the existing ones (breeding). Using a survival-of-the-fittest tactic, it weeds out the bad and tends to produce more of the good individuals. Not only does it produce more of the good solutions but better and better solutions. This is because it combines the best traits of parent individuals to produce superior children. This combination operator is called crossover. The term genetic algorithm comes from the fact that individuals are represented as strings of bits analogous to chromosomes and genes. In addition to recombination by crossover, we also throw in random mutation of these bit-strings every so often. This keeps the GA from getting stuck at good but nonoptimal solutions.

GAs operate on populations of strings, with the string coded to represent some underlying parameter set. Reproduction, crossover, and mutation are applied to successive string populations to create new string populations. These operators are very simple, involving nothing more complex than random number generation, string copying, and partial string exchanging; yet the resulting search performance is wide-ranging and impressive. GAs realize an innovative notion exchange among strings and thus connect to our own ideas of human search or discovery. The strings of an artificial genetic system are analogous to chromosomes in biological systems. The total package of strings is called a structure. These structures decode to form a particular parameter set, solution alternative, or point (in the solution space). Strings are composed of features, or detectors, that take on different values. Features may be located on different positions on the strings.

GA has been applied for on-line determination of the culture temperature of an ethanol fermentation process (24). In this work, the fermentation was optimized by the control of culture temperature for a maximum ethanol concentration at the terminal. Two computers were employed, one for the monitoring, control, and data processing and the other for the GA calculation. After receiving the on-line data of cell and ethanol concentrations from computer 1, computer 2 uses the GA to calculate the optimal temperature profile. The calculation procedure of GA is schematically illustrated as follows:

1. Coding of culture temperature:
   Let T_i ($i = 1, 2, \ldots, m$) be the culture temperature of ith period. The algorithm starts with coding each culture temperature into binary bit string of N bits as follows:
   N bits

   $$\underset{n_1}{11\ldots10}\quad \underset{n_2}{011\ldots11}\quad \ldots\quad \underset{n_m}{100\ldots00}$$

where $N = n_1 + n_2 + \ldots + n_m$. Note that the number of possible temperature profiles is 2^N.

2. Initializing population:
   To generate an initial population of L strings randomly, and set
   G (generation) = 1.
3. Calculation of fitness value:
   Each binary bit string is decoded into the sequence of culture temperature, which can be expressed by
   $T_i = [g_{r(i+k)} * 2^{k-1} (T^* - T_*)/2^{n_i}] + T_*$
   where $S_i = n_1 + n_2 + \ldots + n_{i-1}$, g_r is the rth gene of a string, and T^* and T_* are the upper and lower limits of the culture temperature, respectively. Once the culture temperature is determined for each string, the state equations that describe the dynamic behavior of the bioprocess are integrated to evaluate the final product concentration $P(t)$, where t is the specific culture time.
4. Selection and reproduction:
   Several strings (20% of the population) whose fitness values are high are selected, and reproduction is implemented through roulette wheel slots weighted in proportion to the fitness value of the individual string in the old generation. A common way to implement roulette wheel selection is to:
   a. Sum up all the fitness values in the current population, call this value SumFitness. SumFitness is in effect the total area of the roulette wheel.
   b. Generate a random number between 0 and 1, called Rand.
   c. Multiply SumFitness by Rand to get a number between 0 and SumFitness which we will call Roulette Value (Roulette Value = SumFitness × Rand). Think of this value as the distance the imaginary roulette ball travels before falling into a slot.
   d. Finally, we sum up the fitness values (slot sizes) of the individuals in the population until we reach an individual that makes this partial sum greater or equal to the RouletteValue. This will then be the individual that is selected.
5. Crossover:
   Two strings are picked up randomly and the decision is made whether or not to cross them over according to the specified crossover probability.
6. Mutation:
   A bit of the string is altered according to the specific mutation probability. First select one portion of the string randomly which corresponds to one temperature and then apply mutation with 20% of probability to each bit contained in the portion.
7. New individuals are randomly generated and are added to the population, which becomes the specified value.

Steps 2 to 7 are repeated from generation to generation such that the average fitness value of the population increases. The effect of population size on the convergence of the fitness value in the simulation may be generated, and the experimental result using the GA-aided optimal temperature profile for the ethanol fermentation may be illustrated (26). It can be concluded that the optimization using on-line data with the aid of GA is promising (24).

20.5. SPREADSHEET APPLICATION IN DATA ANALYSIS

One of the problems of developing custom laboratory data-processing programs is that even relatively simple jobs often take a substantial amount of time to write, debug, and document. In recent years, the problem has been exacerbated by the growing popularity of "point and click" graphical user interface environments, which can make software easier to use but which have the reputation of being difficult to program. There are a number of possible solutions to these problems, including improved conventional languages (e.g., Microsoft's Visual Basic), graphical programming systems (e.g., National Instrument's LabView), and spreadsheet macro languages.

20.5.1. Spreadsheet Basics

There are currently many different high-powered spreadsheets that have good graphical and presentation capabilities. These include programs such as Lotus 1-2-3, Quattro Pro, Excel, Wingz, and integrated packages such as Microsoft Works and Claris Works. The spreadsheet skills learned in one program are easily transferred to other programs.

Coincidentally, spreadsheet systems have many analogs with the structure and capabilities of a computer. For example, the spreadsheet display is very similar to the memory of a computer in that it is composed of a series of "cells," or value locations, each capable of storing a single value. For ease of manipulation and display, cells are grouped into blocks or columns. The size of each block (column) is system-dependent, but 8K or 16K block sizes are quite common. The total capacity (i.e., number of cells) of the sheet is also system-dependent, with 2 million and 4 million available cells being quite common. Cells have various properties that can be manipulated to allow the current value, behavior, and display characteristics of the cell to be modified. The current value of a cell may be changed by overwriting it with a new value. The ability to change a cell value can be disabled by "protecting" the cell. Protected cells are essentially read-only locations. The size (width) of a cell can be changed to enable it to display more data so that a cell can be made to look like an 8-bit, 16-bit, 32-bit, or 64-bit computer word.

Cells are uniquely identifiable by using cell references, or addresses, which are assigned sequentially beginning with the first cell. This provides the capability for cells to refer to each other using equivalents of the various addressing styles found in most computer architectures, for example, direct (absolute) and indirect (relative) addressing.

20.5.2. Spreadsheets and Computerized Data Analysis

The large amount of data generated in an experiment or plant operation makes thorough analysis of historical data a difficult and cumbersome task. For example, in order to account for all carbon flows going in and out of the bioreactor, a tedious but necessary bookkeeping of all measurements of relevant extracellular components and biomass assays must be performed. In addition, biotic phase volume variations, experienced in fed-batch processes, must be taken into account. Spreadsheets make this bookkeeping easier. They include a number of advanced features, namely, graphical analysis and multiple regression techniques, which allow the direct computation of fermentation metabolic quotients (e.g., specific growth rate) and other culture parameters (e.g., substrate to biomass yield). Quick analysis of various metabolic quotient profiles with predefined models can provide a quantitative characterization of the fermentation. It further constitutes a diagnostic tool for applying timely corrective actions, avoiding suboptimal runs.

Using spreadsheets allows the developer to focus on key isues associated with vizualizing an algorithm and its associated data structure. Liberated from the mundane detail of providing code to realize the pictorial events, manipulate and control various object instances, and establish storage structures to support object persistence on disk, the developer is able to devote attention to the pedagogic issues of how the data structure will be depicted on the screen, what cues (if any) will be required to aid comprehension, as well as decisions regarding the interaction scenarios permitted.

In an integrated software environment, such as Microsoft Windows, where built-in data exchange capabilities, multitasking, and a consistent graphical interface are available, spreadsheets are ideal tools for the type of analysis described. Personal computers are the appropriate hardware, since their current computational power and affordability makes them acceptable in an industrial environment, and applications for them are more likely to succeed in terms of compatibility and portability than for workstations or mainframes.

20.5.3. PROMETHEUS—A Spreadsheet Application of Data Analysis

Menezes et al. (22) have developed a computerized data analysis and modeling software called PROMETHEUS. The program was written for the penicillin-G fermentation with industrial media in stirred tank pilot-bioreactors and can be run under the Excel environment of Windows for direct experimental data analysis and process learning. It is a successful application example that covers a broad spectrum from data analysis through process modeling to process optimization.

The main menu, PROMETHEUS.xls, is the functional heart of PROMETHEUS, and this menu provides links to the submenus—the entries to satellite spreadsheets that perform different data analysis, model, and parameter evaluation tasks through direct inspection of fermentation data. A session with PROMETHUS would include:

1. Filling the satellite spreadsheets with data
2. Activating any of the spreadsheet options for "FIRST ESTIMATES" or the option for fermentation "OVERVIEW"
3. Examining "KINETICS," with options for numerical or graphical analysis for each of the fermentation state variables

In all cases, hard copies and general printouts of results are available to the users.

The main spreadsheet window is split into two main areas: the information overview and the individual table/graph displays. Excel has a variety of tools to assist in the analysis of data. These tools include Pivot Tables, Charts, Filters, Subtotals, etc., which allow a user to quickly analyze data. By exploiting Excel's object set, over 100 macros are hidden behind PROMETHEUS to perform the tasks shown in the buttons. The application enables:

- bookkeeping of all relevant fermentation data
- graphical displays of fermentation data along the cultivation period (state variable and metabolic quotient profiles)
- several rapid approximate ways for obtaining first estimates for model parameters
- nonlinear regression of process kinetics
- several scenarios to be examined for substrate uptakes; for this, substrates can be lumped as one or two substrates
- analysis of different biomass growth and penicillin production models

PROMETHEUS uses spreadsheets to model the penicillin-G fermentation described by the mathematical equations (23). The user can plot the equations and study the model performance under different conditions. It is also possible to change the values of the constants for the equations to see the effect of the individual parameters. Such a modeling approach enables the user to watch graphically the kinetic analysis of raw data, follow the parameter estimation on-line, and be able to see from the squared relative errors which data points should be discarded.

A strategy for parameter evaluation was developed to overcome practical problems:

- Model sensitivity to parameters is examined; key parameters are chosen to be estimated by nonlinear least-squares analysis.
- The fermentation is divided into phases (growth and production), and information is extracted directly from experimental data for each phase.
- Literature information is used for the less sensitive parameters.
- Nonlinear analysis is performed on a limited number of parameters.

Fitting the kinetic expressions to the time series of the experimentally measured values of the model parameters produces first estimates for μ_{max}, τ_{max}, k_x, and m_s. Also, Y_{XS} can be computed from the yield definition, $Y_{XS} = X/S$, with data from the growth phase (X and S are absolute mass variations). This enables the separate study of kinetics of three basic phenomena: biomass growth, penicillin production, and substrate consumption.

From the modeling viewpoint, the most important contribution made by the application is the nonlinear regression of process kinetics, which can be followed graphically. This uses Excel solver—a quasi-Newton iterative method—which is called by pressing the corresponding "SOLVE" button, available in all subscreens under the "KINETICS/DATA" options.

Another feature is the capacity to test any combination of the pre-defined substrates (four in this example) by lumping them together in one or two pseudo-substrates. During the study of growth kinetics, the graphical analysis allowed us to find that glucose consumption alone could not be responsible for all biomass formed. With the spreadsheet interactive capacity to test combinations of substrates, it could be readily concluded that both glucose and organic nitrogen had to be accounted for as substrates, or otherwise a mass balance violation would result. Furthermore, several kinetic expressions are available for test. The availability of graphical outputs typifies the capabilities of PROMETHEUS in a first assessment of some fermentation mechanisms, in this case showing that no repression of penicillin production could be practically identified (23).

20.5.4. Conclusion

Spreadsheets are computational tools for doing calculations, plotting data, and presenting numerical information. Spreadsheets originated in a business and accounting environment, but their usefulness has extended to many different disciplines, including biochemistry and engineering sciences. A review of the full set of features provided by the current generation of spreadsheet systems suggests that the spreadsheet environment is essentially a dedicated operating system supporting the development of visualization software. Spreadsheets have been phenomenally successful systems because they provide a simple and convenient interface for those interested in developing computational models and subsequently examining the model's response to various scenarios and input variable changes. The spreadsheet systems are also programmable. This enables inventive users to extend the system capabilities, disable features unnecessary for particular applications, and expand the range of applications.

REFERENCES

1. **Bailey, J. E., and D. F. Ollis.** 1986. Transport phenomena in bioprocess systems, p. 457–532. *Biochemical Engineering Fundamentals*, 2nd ed. McGraw-Hill, New York.
2. **Barford, J. P.** 1997. htpp://www.chem.eng.usyd.edu.au Spreadsheets available from: anonymous/dropbox/staff/johnb/asm:
   batch.xls
   ccserev.xls
   ccutilis.xls
   fujimoto.xls
   leastsq.xls
   logmean.xls
   metaerob.xls
   metanaer.xls
3. **Barford, J. P., P. J. Phillips, C. P. Marquis, and C. Harbour.** 1996. Biosynthesis of protein products by animal cells. Are growth and non-growth associated concepts valid or useful? *Cytotechnology* **21:**133–148.
4. **Bates, D. M., and D. G. Watts.** 1988. *Nonlinear Regression Analysis and Its Applications*. John Wiley & Sons, Inc., New York.
5. **Blackburn, M., U. Fauth, D. Renno, and S. Trew.** 1996. Optimization of fermentation conditions for the production of a novel GABA-benzodiazepine receptor agonist by Acremonium strictum. *J. Ind. Microbiol.* **17:**36–40.
6. **Dennis, J. E., and R. B. Schnabel.** 1983. *Numerical Methods for Unconstrained Optimization and Nonlinear Equations*. Prentice-Hall, Englewood Cliffs, N.J.
7. **Doran, P. M.** 1995. Presentation and analysis of data, p. 27–48. *In Bioprocess Engineering Principles*. Academic Press, New York.
8. **Draper, N. R., and H. Smith.** 1981. *Applied Regression Analysis*, 2nd ed. John Wiley & Sons, Inc., New York.
9. **Easterby, J. S.** 1992. Hyperbolic regression analysis of enzyme kinetic data. jse@liverpool.ac.uk
10. **Fletcher, R.** 1987. *Practical Methods of Optimization*, 2nd ed. John Wiley & Sons, Inc., New York.
11. **Fu, P.-C.** 1996. Hybrid modelling and simulation of animal cell cultures by the synergism of mathematical models and AI techniques. Ph.D. thesis, University of Sydney.
12. **Furutani, T., R.-H. Su, H. Okamoto, H. Oshima, and J. Kato.** 1995. Effect of organic solvents on substrate specificity for lipases, p. 11–16. *In* Special Study Group for Biochemical Engineering, (ed.), *Biochemical Engineering as a Key Technology for Bioindustry*. Society of Chemical Engineers, Japan.
13. **Gill, P. E., W. Murray, and M. H. Wright.** 1981. *Practical Optimization*. Academic Press, New York.
14. **Himmelblau, D. M.** 1970. *Process Analysis by Statistical Methods*, John Wiley, New York.
15. **Hoshino, K., M. Taniguchi, H. Ueoka, M. Ohkuwa, C. Chida, S. Morohashi, and T. Sasakura.** 1996. Repeated utilization of α-glucosidase immobilized on a reversibly soluble-insoluble polymer for hydrolysis of phloridzin as a model reaction producing a water-insoluble product. *J. Ferm. Bioeng.* **82:**253–258.
16. **Kawakami, K., and S. Yoshida.** 1996. Thermal stabilization of lipase by sol-gel entrapment in organically modified silicates formed on Kieselguhr. *J. Ferm. Bioeng* **82:** 239–245.
17. **Keulers, M., T. Asaka, and H. Kuriyama.** 1995. Clarifying yeast physiology using a graphical front end simulation tool, p. 43–48. *In* Special Study Group for Biochemical Engineering, (ed.), *Biochemical Engineering as a Key Technology for Bioindustry*. Society of Chemical Engineers, Japan.
18. **Liebetrau, A. M.** 1995. Dynamic linear models for tank level monitoring. *Statistical Computing and Graphics* **6:**22–25.
19. **Liu, X.-H.** 1996. Intelligent data analysis: issues and challenges. *Knowledge Engineering Review* **11**(4), or http://web.des.bbk.ac.uk/ida97.html.
20. **Luna-Pabello, V. M., M. A. Aladro-Lubel, and C. Duran-de-Bazua.** 1996. Biomonitoring of wastewaters in treatment plants using ciliates. *J. Ind. Microbiol.* **17:**62–68.
21. **Marquis, C. P.** 1995. Batch and fed-batch culture of murine hybridomas and human lymphocyte cell lines. Ph.D. thesis, University of Sydney.
22. **Menezes, J. C., S. S. Alves, J. M. Lemos, and S. F. Azevedo.** (1994). Computer-aided fermentation modelling and data analysis. PROMETHEUS and Excel application, p. 649–653. Proc. PSE '94, Korea.
23. **Menezes, J. C., S. S. Alves, J. M. Lemos, and S. F. Azevedo.** (1994). Mathematical modelling of industrial pilot-plant penicillin-G fed-batch fermentations. *J. Chem. Technol. Biotechnol.* **21:**123–138.
24. **Moriyama, H., and K. Shimizu.** 1995. Application of genetic algorithm for the on-line optimization of ethanol fermentation, p. 90–94. *In* Special Study Group for Biochemical Engineering, (ed), *Biochemical Engineering as a Key Technology for Bioindustry*. Society of Chemical Engineers, Japan.
25. **Nohara, D., M. Matsubara, K. Tano, and T. Sakai.** 1996. Design of optimal refolding solution by combination of reagents classified by specific function. *J. Ferm. Bioeng.* **82:**401–403.
26. **Park, S., and W. F. Ramirez.** 1988. Optimal production of secreted protein in fed-batch reactors. *Am. Inst. Chem. Eng. J.* **3:**1550–1558.
27. **Reuss, M., and R. Bajpai.** 1991. Stirred tank models. *In* K. Schügerl (ed.), *Biotechnology*, 2nd ed. vol. 4. *Measuring, Modelling, and Control*. VCH, Weinheim, Germany.
28. **Sanderson, C.** 1996. The development and application of a structured model for animal cell metabolism. Ph.D. Thesis, University of Sydney.
29. **Seber, G. A. F., and C. J. Wild.** 1989. *Nonlinear Regression*. John Wiley & Sons, Inc., New York.

30. **Takamizawa, K., S. Nakashima, Y. Yahashi, K. B. Kubara, T. Suzuki, K. Kawai, and H. Horitsu.** 1996. Optimization of kojic acid production rate using the Box-Wilson method. *J. Ferm. Bioeng.* **82:**414–416.

31. **Tchobanooglous, G., and F. L. Burton.** 1991. *In* Metcalf and Eddy, Inc., *Wastewater Engineering: Treatment, Disposal and Reuse*, 3rd ed. McGraw-Hill, New York.

32. **Tholdur, A., and W. F. Ramirez.** 1996. Optimization of fed-batch bireactors using neural network parameter function models. *Biotechnol. Prog.* **12:**302–309.

33. **van de Merbel, N. C., P. Zuur, M. Frijlink, J. J. M. Holthuis, and H. Longerman.** 1995. Automated monitoring of amino acids during fermentation processes using on-line ultrafiltration and column liquid chromatography: application to fermentation medium improvement. *Anal. Chim. Acta* **303:**175–185.

34. **Walpole, R. E., and R. H. Myers.** 1989. *Probability and Statistics for Engineers and Scientists*, 4th ed., MacMillan, New York.

DOWNSTREAM PROCESSES

"DOWNSTREAM PROCESSING," WHICH OFTEN DESCRIBES THE purification of fermentation products, here encompasses a whole range of activities involved in the development of an investigation or product beyond solely microbiological concerns. These include purification, economic analysis, legal and regulatory affairs, and production on larger scales.

Microbes are often used as sources of valuable products, which must be isolated from them. Recovery and identification of small-molecule products, especially secondary metabolites, is discussed by Borders, while Willson addresses the recovery and characterization of proteins.

Economic and regulatory considerations are of dominant importance in industrial practice. The economics of fermentation processes are discussed by Reisman, while recent progress in computer-based design, analysis, and optimization of manufacturing processes is summarized by Petrides et al. Safety- and regulation-driven aspects of biosafety and containment are discussed by Liberman, Fink, and Schaefer.

The protection of intellectual property is of increasing interest to academic researchers, as well as being a critical part of industrial efforts. The essentials of intellectual property protection are discussed by Gordon, including record-keeping, applying for patent protection, and the special features of patent law in the biotechnology area.

The establishment of larger-scale pilot plant facilities is a milestone in the development of many processes, individuals, and companies. Hamilton, Sybert, and Ross provide a detailed, step-by-step consideration of the required steps, from scoping and contractor selection through construction, start-up, validation, and operations.

Quality assurance and quality control (QA/QC) are important to many manufacturing operations and are particularly emphasized in the pharmaceutical industry. Bryant reviews the functions used to control manufacturing elements including labels and packaging, raw materials, in-process items, and the finished product, as well as the documentation and regulations controlling QA/QC activities.

In some cases there is need for fermentation more promptly or on a larger scale than can be performed by in-house facilities. This situation arises often in commercial development, but outsourced fermentation services also are increasingly used in support of basic research, e.g., to provide starting material for isolation of a microbial product of interest. The selection and use of outside contract fermentation services are discussed by Mateles.

Isolation and Identification of Small Molecules

DONALD B. BORDERS

21

21.1. BACKGROUND

There are now over 17,000 antibiotics and other biologically active microbial metabolites that have been reported in the literature. These compounds are relatively small molecules with great diversity in chemical structures. A majority have molecular weights (MWs) in the 300 to 800 range. They frequently have a number of asymmetric centers, yet they are usually synthesized by the microorganism as a single isomer. One of the smallest biologically active microbial metabolites that have been used in medicine is the antibiotic phosphonomycin, which has only three carbon atoms and a MW of 138. This compound illustrates the advantages of screening microbial metabolites, since only three carbon atoms are in the structure and no chemist had previously synthesized a molecule like this. It is very doubtful that any combinatorial synthesis would result in this type of molecule unless it was patterned after the natural product. Most of the biologically active microbial metabolites discovered up to this time are antibacterial or antifungal antibiotics. This tendency is a consequence of available screening methods to target molecules with these types of activities. Some of the commercially important antibiotics are listed in Table 1. All of the most recent antibiotics in this listing are based on fermentation products but are semisynthetic or totally synthetic compounds with improved properties over the original compounds (1). A number of metabolites have other types of biological activity such as anti-inflammatory, antihypertensive, immunomodulating, or analgesic, which are referred to as "pharmacological activities." Within the last several years, fundamentally new screening approaches have been developed, resulting in reliable assays for compounds with pharmacological activity. A number of the microbial metabolites discovered in this area have been reviewed (20). This change in screening emphasis is illustrated by the 1996 index of the *Journal of Antibiotics*, which has the following biological activities for novel microbial metabolites reported in the year: pharmacological, 42%; anticancer, 29%; antiviral, 4%; antifungal, 14%; and antibacterial, 10%. Table 2 lists some of the nonantibiotic microbial metabolites that are being used commercially for medical and agricultural applications.

21.2. FACTORS INFLUENCING ISOLATION AND IDENTIFICATION

Two of the most important factors in determining the optimal downstream processing steps for microbial metabolites are the physical properties of the molecules (MW, polarity, and ionic charge) and their concentrations in the fermentation broths. In general, the physical properties of a molecule determine the general approaches to separations such as types of chromatography: ion-exchange, gel filtration, partition, or adsorption. The concentration frequently determines the number of steps required for adequate purification because of the ratio of the active component to the impurities. Processing requirements dependent on physical properties of a molecule are more obvious than the requirements to deal with the concentration effects. A compound occurring in a fermentation at very low concentrations may require several extraction and chromatography steps for adequate purification, whereas the same compound at much higher fermentation yields may require only a couple of steps. In some cases where concentration of a metabolite is high enough, only that portion of a chromatography peak giving the highest purity of the desired compound can be used for spectral data to determine identity. This type of approach can greatly expedite the identification procedure since fewer chromatography steps are required, compared to a purification sequence where the entire biologically active peak is used, frequently including overlapping impurity peaks.

Most microbial metabolites have been observed in fermentations of wild-type cultures at concentrations of 0.1 to 10 μg/ml. Occasionally the fermentation titers are near 100 μg/ml. For most microbial metabolites with titers in the 1- to 10-μg/ml range, a 2- to 5-liter fermentation is adequate to isolate several milligrams of relatively pure metabolite and to identify the compound by UV, nuclear magnetic resonance (NMR), mass spectrometry (MS), and database methods. The original fermentation titers of calicheamicin represent the low extreme near 1 ng/ml (9, 15). Even after titer improvement, a 1,500-liter fermentation was required to obtain 18 mg of the major component, calicheamicin β_1^{Br}, for definitive characterization to establish novelty. Throughout these studies, the detection and processing of calicheamicin was directed by an assay highly sensitive to this compound, the biochemical prophage in-

TABLE 1 Year of discovery or market introduction of some of the more important antibiotics

Antibiotic	Year	
	Discovery	Introduction
Penicillin	1929	
Tyrothricin	1939	
Griseofulvin	1939	
Streptomycin	1944	
Bacitracin	1945	
Chloramphenicol		1947
Polymyxin	1948	
Chlortetracycline		1948
Cephalosporin C,N,P	1948	
Neomycin	1949	
Oxytetracycline		1950
Nystatin	1950	
Erythromycin		1952
Amphotericin B	1955	
Novobiocin	1955	
Vancomycin	1957	
Kanamycin		1957
Ampicillin[a]		1962
Fusidic acid		1961
Cephalothin[a]	1962	
Lincomycin		1963
Gentamicin		1963
Carbenicillin[a]	1964	
Cephalexin[a]	1966	
Clindamycin[a]		1967
Cephalothin[a,b]		1969
Cephaloxidine[a,b]		1969
Minocycline[a]		1971
Amoxicillin[a]		1972
Cefoxitin[a,c]		1978
Tricarcillin[a]		1979
Mezlocillin[a]		1980
Piperacillin[a]		1980
Cefotaxime[a]		1980
Moxalactam[a]		1981
Augmentin[d]		1984
Aztreonam[e]		1984
Imipenem[a,f]		1985

[a]Semisynthetic product.
[b]First oral cephalosporins.
[c]First commercial cephamycin.
[d]First β-lactamase inhibitor combination.
[e]First monobactam, a synthetic product designed after a natural product.
[f]First carbapenem.

duction assay, which detects DNA-damaging agents. The lower concentration limits for detection and purification of metabolites with unknown structures are for the most part dependent on assay sensitivity and reliability. Economic factors may become very important for very low-titer fermentations, since the purifications may be very difficult and time consuming. For compounds detected by affinity-binding assays, it may be possible to simplify the purification of compounds present in very low concentrations by use of affinity-binding columns. When a microbial metabolite generates special interest and deserves a large commitment of effort, its fermentation titers can be improved significantly by fermentation media studies, strain selection, and mutagenesis of the producing organism. The titers of some antibiotics have been increased to as much as 30 g/liter, but this type of increase requires intensive long-term efforts (6).

The antibiotics and other biologically active microbial metabolites are produced in complex mixtures containing other metabolites and residual media ingredients. This usually means that hundreds of other compounds are found in the fermentations along with the desired compounds and must be separated away. Since the desired products and impurities are usually quite different from the fermentation of one organism to the next, the purification is usually an involved process that cannot be accomplished by highly routine procedures. However, when affinity-binding procedures are used in the screen for compounds with special structural features, as in the screens for glycopeptide antibiotics by the Smith Kline and French group, affinity-binding columns can be used for purification in a relatively routine manner (22).

TABLE 2 Examples of commercial nonantibiotic applications of microbial metabolites and their derivatives in medicine and agriculture

Therapeutic area	Compound	Source
Anticancer	Andriamycin	Adria
Anticancer	Mitomycin	Bristol-Myers Squibb
Anticancer	Doxorubicin	Astra
Anticancer (leukemia)	Pentostatin	SuperGen
Cholesterol lowering	Lovastatin	Merck
Cholesterol lowering	Pravastatin	Bristol-Myers Squibb
Antidiabetic	Acarbose	Bayer
Immunosuppressant	Cyclosporin A	Novartis
Immunosuppressant	Mycophenolate mofetil	Syntex
Immunosuppressant	Tacrolimus (FK506)	Fujisawa
Antistrongyloidiasis	Ivermectin	Merck
Growth promoter in animals	Avoparcin	Cyanamid
Growth promoter in animals	Monensin	Eli Lilly
Endectocide in animals	Ivermectin	Merck
Endectocide in animals	Moxidectin	Cyanamid

21.3. IDENTIFICATION APPROACHES

If screening is being conducted for novel microbial metabolites with biological activity or for known microbial metabolites with new types of biological activity, rapid identification of these fermentation products is essential. The process of identifying known compounds is frequently referred to as dereplication, and the sequence of steps to accomplish this during the discovery process is shown in Table 3. Probably one of the most useful instrumental techniques for any natural product identification study at this time is high-pressure liquid chromatography (HPLC), coupled with a photodiode array detector that allows the chemist to identify compounds not only by retention times but also by UV/visible chromophores. More recently, it has been possible to couple HPLC with thermospray (2, 11, 19) or electrospray MS to obtain retention times and molecular weights. A critical aspect of these approaches is to associate the biological activity with the correct HPLC peak, especially when peaks are overlapping or close together.

A detailed analysis of the dereplication process accompanying the discovery of the calicheamicin and LL-D49194 antitumor antibiotics is outlined in Fig. 1 (13). During the year of discovery of each of these antitumor agents, approximately 10,000 microorganisms were screened by a very selective and sensitive assay, the biochemical induction assay (BIA). The BIA detects DNA-damaging agents with the aid of a genetically engineered strain of *Escherichia coli*. The selective nature of the assay simplified the dereplication process by greatly narrowing the field of known metabolites that would be positive; however, the extreme sensitivity of the assay for calicheamicin and its extremely low fermentation titers greatly complicated the identification process. The initial dereplication scheme involved HPLC and was mainly designed to eliminate anthracycline antibiotics, the most prominent group of known compounds positive in the BIA. However, owing to the cultures selected, most of the antibiotics eliminated were related to antibiotic M92, the macromolecular antibiotics like the neocarzinostatin complex, and other miscellaneous known antibiotics.

21.3.1. Identification Methods

There are two general methods used for metabolite identification: (i) direct comparison to reference compounds by techniques like thin-layer chromatography or HPLC and (ii) isolation of a few milligrams of relatively pure metabolites and identification/structure determination by MS, UV, and NMR spectral methods. The first method can be applied to relatively crude samples at early stages of the work but requires an adequate library of reference compounds. In some cases, as little as 50 μl of broth filtrate was adequate for identification of a metabolite utilizing HPLC photodiode array screening (18) through direct comparison to reference compounds. Microbial metabolites that occur at low concentrations in crude fermentations can be adsorbed onto C_{18} reverse-phase Sep-Pak cartridges, which are then washed with water and eluted with a solvent such as methanol. The eluate can be concentrated and used for HPLC studies to compare to reference compounds. To obtain a few milligrams of purified metabolite for identification by the second procedure involving spectroscopy and databases may require two or more liters of fermentation broth. The second approach requires access to specialized databases on fermentation products, since there are over 17,000 of these compounds reported in the literature. Computer searches with STN, the scientific and technical information network provided by the Chemical Abstracts Service, do not readily resolve this problem. Searches with these databases are complicated significantly by inclusion of large numbers of synthetic compounds and the inability to search the databases for UV absorptions and other spectral features obtainable at early stages of the metabolite identification process.

TABLE 3 Typical screening sequence for the discovery of novel microbial metabolites

Selection of unusual microorganisms
Fermentation of microorganisms
Bioassays (antimicrobial, competitive binding, enzyme inhibitor, etc.)
Chemical dereplication of metabolites
HPLC with photodiode array detector
HPLC with thermospray or electrospray mass spectrometry
Isolation of novel microbial metabolites
In vivo testing
Structure determination
Extensive biological evaluation

21.3.2. Databases for Microbial Metabolites

A few of the large pharmaceutical firms conducting research on fermentation products have developed and maintained their own excellent databases in this field. Some of the commercially available databases specializing in fermentation products are listed in Table 4. Some of these databases have been expanded into other areas of natural products, such as botanical and marine products. As a result, the numbers of compounds entered into the first two databases, Berdy and the Chapman & Hall, are much larger than the number of fermentation products. In contrast, the DEREP database has been organized to have representative compounds from known structural families and, as a result, the database has a lower number of entries than the total number of reported compounds. The DEREP database is currently undergoing major modifications and may have over 15,000 compounds soon. Searches of the Berdy, DEREP, Antibase, and Kitasato databases are usually based on UV chromophores, MWs, and general biological properties. The Chapman & Hall database contains information only on compounds that are well characterized and, in almost all cases, that have known chemical structures. Searches of this database utilize MW, molecular formula (MF), structure, and substructure approaches. The Antibase system is one of the more recent commercial databases for antibiotics and other microbial metabolites. This database provides for searches by biological properties, UV, MW, MF, and ^{13}C NMR signals. The NAPRALERT database is not included in Table 4, although it contains information on microbial products. This database focuses more on biological properties and cannot be searched by MW, MF, or UV chromophore. An excellent review of the various natural product databases and a method of utilizing a subset of the STN files for natural products dereplication is given by Corley and Durley (5).

Early identification of microbial metabolites is crucial to avoid wasting time with known compounds. At the early phases of the work, the characterization data most readily available for the metabolite are usually the UV chromophore, MW, general type of producing organism, and type of biological activity. Some substructure information may also be available from ^{1}H and ^{13}C NMR data. If a search

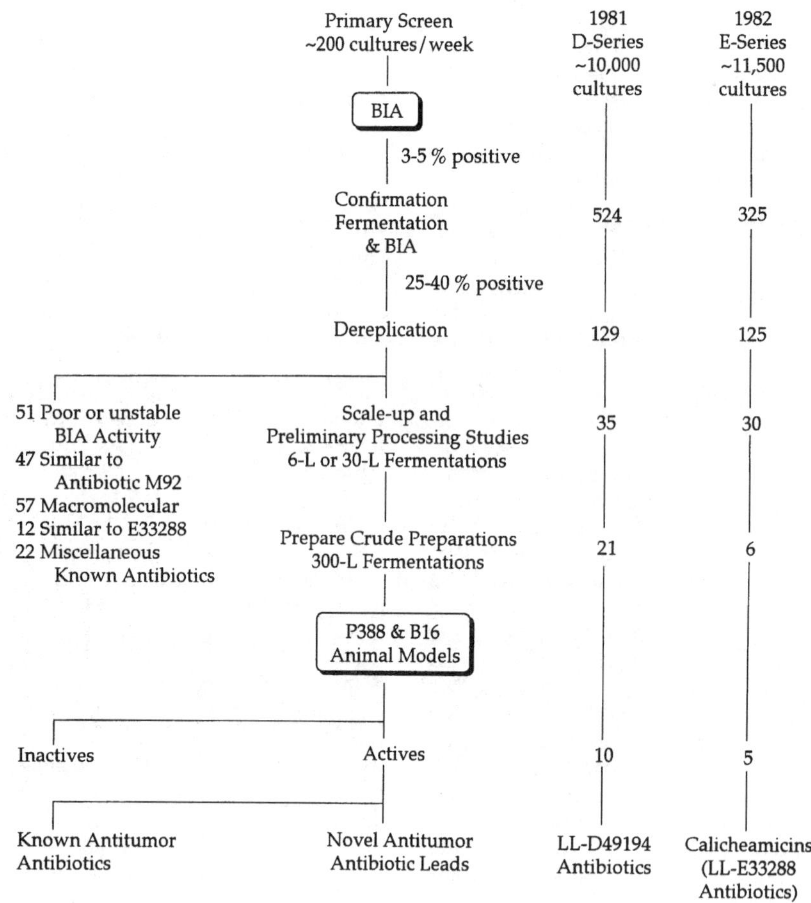

FIGURE 1 Screening approach used for the discovery of the calicheamicins and the D49194 antibiotics. BIA, biochemical induction assay.

with the exact spectral information from a particular metabolite does not seem to find a match in the databases, searches should incorporate ranges for the MW and UV peaks. If this is not considered, a known compound can be mistakenly classified as novel. For example, if a compound has λ_{max} 295 nm, it would be best to search the databases with a range of 290 to 300 nm. This allows for differences from one laboratory to another that may occur from different solvents or instruments or small errors in reading the peaks. In most cases, matching data to a known compound is relatively easy. The most difficult aspect of a search is to clearly define a compound as novel, since there is always concern that a known compound with the defined properties has been overlooked because of the search method or errors in the database. If a compound appears novel, it is best to verify its assignment by searching more than one database.

In some searches involving novel compounds, it is useful to find compounds that have an identical UV chromophore but not the same overall structure, since the chromophore can define a significant structural unit within a molecule. The most effective way of searching a database for this type of information is to search for chromophores that appear to give the same shifts under acidic, basic, and neutral conditions, since pH shifts are more definitive in identifying a chromophore. Not all databases allow this type of search.

There are many ways that microbial metabolites can be grouped or classified, such as by mechanism of action, biosynthetic pathway, type of biological activity, or chemical structure class. The most useful grouping is by chemical structures, which allows categorization into general structure families such as aminoglycosides, tetracyclines, macrolides, or β-lactams, and this is the approach for all of the commercially available databases.

21.4. ISOLATION METHODS

Antibiotics and other microbial metabolites are usually observed in the fermentation filtrates when fermentation titers of the metabolites are relatively low. As the fermentation titers become higher through strain development or media manipulation, the product frequently becomes more associated with the cells. The initial step for a solvent-extractable metabolite may involve solvent extraction of the whole broth, the fermentation filtrate, or the separated cells. If the cells contain most of the desired metabolite, separation of the cells from the whole broth prior to solvent extraction of the cells will most likely result in an extract that is much cleaner than one obtained from extraction of the whole broth.

Frequently, the isolation of microbial metabolites involves fractionation of the compounds in a fermentation

TABLE 4 Commercially available databases for microbial metabolites

Database and source	No. of compounds
Bioactive Natural Products Database (Berdy) MEDIMPEX H-1808 Budapest, Hungary Phone 36-1-142-2796 Fax 36-1-117-7179	>23,000
Chapman & Hall Dictionary of Natural Products Chapman & Hall Limited Scientific Data Division 2-6 Boundary Row London SE1 8HN, United Kingdom U.S. CD-ROM Rep. Phone 301-699-7777 Fax 301-699-1110	>103,000
DEREP Database Dr. Chris Beecher University of Illinois 833 South Wood Street, M-C781 Chicago, Ill. 60612 Phone 312-996-9035 Fax 312-996-7107	>7,000[a]
Kitasato Microbial Chemistry Database USACO Corporation Marketing Department Tsutsumi Bldg. 1-13-12, Simbashi, Minato-ku Tokyo 105, Japan Phone 81-3-3502-6472 Fax 81-3-3593-2709	>17,000
Antibase Chemical Concepts GmbH Boschstrasse 12 D-69469 Weinheim Germany Phone 49(0)6201 606 433 Fax 49(0)6201 606 430	>17,000

[a]Undergoing major revisions and may be expanded to >15,000 compounds soon.

broth by solvent extraction procedures, followed by a sequence of chromatography steps that are monitored by some in vitro bioassay that targets the type of biological activity under investigation. For antibiotics, the assay usually involves inhibition of certain target bacteria or fungi. For pharmacologically active compounds, the assays may involve enzyme inhibition, receptor binding, or other biochemical processes.

21.4.1. Bioassay Directed Separations

Microbial fermentations invariably result in complex mixtures of hundreds of compounds. When compound identification studies are conducted beyond direct comparisons to reference compounds, the necessary purification steps are guided by bioassays. Efficient bioassays for this process can selectively detect compounds with the desired biological activity in complex mixtures. Bioassays can be accomplished by a number of techniques. Fractions can be adsorbed onto paper discs, dried to remove interfering solvents, and then placed on assay plates such as Nunc plates or petri dishes containing agar seeded with a specific bacterium or fungus to detect antimicrobial substances. Fractions can also be collected in microtiter plates and subjected to bioassays after interfering solvents are removed by evaporation. Most bioassays utilizing live cells can tolerate only 1 to 2% of solvents such as methanol or dimethyl sulfoxide, and these two solvents are perhaps the least toxic for most assay systems.

Assay sensitivity is a critical factor, since all fractionation procedures usually result in dilution with solvents. The active compound in a sample placed on a chromatography column may be diluted 10-fold when it emerges from the column. In addition, it may be spread out over several fractions. Solvent extraction of a fermentation may result in distribution of the active component between two phases, giving a net dilution effect. Therefore, if the bioassay used in the screen is relatively insensitive, the dilution effects during fractionation can result in an apparent loss of activity, which is sometimes erroneously blamed on compound instability. For example, if a column charge cannot be diluted 10-fold and retain bioactivity, it is likely that activity will not be detected in the fractions unless all of the fractions are first concentrated, which is a labor-intensive step. A simple method to avoid these types of problems is to check dilution effects on the bioassay of an extract or fermentation before it is processed. It is highly recommended to check actual dilution effects rather than checking a response at a single concentration and extrapolating possible dilution values from a point on a curve. A highly sensitive assay that can tolerate large dilutions of compounds encountered at usual fermentation concentrations avoids these types of problems.

For the most efficient fractionation of an active compound, it is best to follow the separation not only with bioassays but also with physical methods such as UV absorption or with total solids using instruments like evaporative light-scattering detectors. Following preparative column fractions with analytical HPLC and a photodiode array detector can be very informative. By comparing the bioassay results with data from physical methods, it is possible to combine fractions or select fractions in a very rational way to avoid impurities and save significant amounts of time. If the entire biologically active peak is pooled as one fraction, significant amounts of impurities may be included from overlapping peaks that may require several additional chromatography steps to remove. If division of an active peak is based on physical methods along with the bioassay, it may be possible to select only that portion of the peak that is relatively pure and to proceed directly with identification or other characterization studies on that portion without further fractionation.

In the mid-1960s, it was common to evaluate the stability of antibiotics in crude extracts or fermentation broths prior to any isolation studies. This usually involved checking the stability of the active compound with a bioassay by exposing crude material to different pH and temperature conditions. Current separation techniques involving reverse-phase HPLC or centrifugal partition chromatography (CPC) provide rapid and relatively mild procedures with minimum tendencies to degrade compounds. As a result, stability studies are usually conducted on metabolites only after a problem has been observed.

The whole broths from fermentations of various organisms invariably contain enzymes that can, in many cases, degrade the desired metabolites if the fermentations are allowed to stand too long before processing. Many of these enzymes are proteolytic and can also have an effect on proteins of the receptor binding assays that are used to screen fermentations and track the active components during isolation and purification studies. Screens and tracking procedures with receptor binding assays should be designed to avoid these problems.

With some microbial metabolites, certain impurities can have a pronounced effect on the response of the metabolite to bioassays. Inorganic salts, for example, can greatly influence the response of polymyxin in an agar diffusion assay. If the purification procedure concentrates salts along with the antibiotic, the quantitative aspects of the assay results can be misleading. Some compounds, like the polyene antifungal antibiotics, that are relatively insoluble can be solubilized or "formulated" by lipids or other impurities found in crude extracts, making them bioavailable for certain types of assays. I observed a good example of this effect when a crude antifungal antibiotic preparation was found to be highly active in an animal model, but the purified crystalline compound was inactive. It could be mixed with dimethyl sulfoxide, however, to given an active material upon oral administration.

21.4.2. Methods of Purification

The purification of microbial metabolites requires a wide range of techniques since these compounds have very diverse chemical structures. They range from extremely water-soluble compounds to very solvent-extractable substances that have all types of ionic character and a wide range of molecular weights. The techniques and instrumental methods for the solvent-extractable compounds are in general more refined than for the very water-soluble compounds. The most difficult compounds to purify are the very water-soluble, neutral, or amphoteric substances. This is especially true for compounds of this type that are also very unstable. Ion-exchange methods are very effective for acidic and basic water-soluble metabolites, but the methods for these compounds are not as straightforward and general as those for the solvent-extractable materials. Reverse-phase chromatography has expanded the range of solvent-extractable compounds that can be readily purified by chromatography, since it allows processing of some of the more polar compounds, like netropsin, that are difficult or almost impossible to extract from fermentation broths with solvents. This technique is especially useful when applied to HPLC, which, particularly with gradient elution, is perhaps one of the most widely used methods of analyzing and purifying microbial metabolites. The most effective general method for purification of microbial metabolites and other natural products is countercurrent chromatography, which is a liquid-liquid partition chromatography technique employing two immiscible liquid phases. Separation of the components of a mixture is achieved by partitioning the components between the two phases as the mobile phase is passed over the stationary phase. This method prevents loss of compounds from irreversible adsorption to a solid support that can occur with any column chromatography method, and recoveries are usually near 100%. Solvent requirements are usually low. The disadvantage of countercurrent chromatography is the manual effort required to perform separations, and over the years this disadvantage has greatly reduced its application. Frequently, this separation method is held in reserve for problems not readily resolved by HPLC or other, simpler approaches. Newer versions of countercurrent chromatography, such as CPC, significantly expedite the separations (19), but HPLC approaches still are more generally used.

The isolation of a microbial metabolite from a fermentation usually involves separation of the cells from the fermentation filtrate, an initial extraction of the cells or filtrate, concentration of the extract, and one or more steps of chromatography to obtain the pure compound. Frequently, the activity is not a single component but instead a complex of several related compounds with one or more major components. The detailed fractionation procedure for calicheamicin is given in Fig. 2, which shows the separation and purification of the various components by reverse-phase chromatography and final purification of these components by chromatography on Sephadex LH-20 or Silica Woelm (15). For purification of any microbial metabolite, the chromatography is tailored for the particular compound under study and the impurities with which it is associated. The next two sections describe initial extraction and final chromatography steps in more detail.

21.4.2.1. Initial Extraction Steps

Secondary metabolites are found in the fermentation filtrates and cells of the microorganisms. While there are many exceptions, with high-titer fermentations the metabolites frequently are found mostly in the cells or in both cells and filtrates, whereas low-titer fermentations often have the metabolites mainly in the filtrates. To recover solvent-extractable metabolites, a typical extraction can be accomplished by mixing two volumes of the fermentation filtrate with one volume of ethyl acetate, separating the organic layer, and repeating the process. The combined extracts are concentrated under vacuum, and the concentrate is purified further by chromatography. If the metabolite is an acid, the extraction will in general be more efficient if carried out under acidic conditions, and the opposite effect is observed for basic metabolites. This type of pH adjustment to the whole broth or fermentation filtrate (within stability limits for the metabolite) minimizes ionization of the ionic groups and improves extraction efficiency. In situations where the metabolites are found mainly in the cells, the cells can be separated by filtration, washed with water, and extracted with a solvent such as methanol. This type of procedure eliminates many of the more polar impurities that are separated off in the filtrate, but usually incorporates more of the lipid-type impurities that can occur in the cells. To expedite the filtration step, filter aids such as Celite are frequently used. Usually 1% (wt/vol) filter aid is stirred into the whole broth, followed immediately by filtration. Extraction of the filtrate or the cells in the filter cake can then be accomplished. In some situations, extraction of the whole broth with one-half volume of a solvent such as ethyl acetate may prove to be the most effective procedure. The resulting extract can be separated from the mycelium and aqueous phase, concentrated, and further processed by chromatography.

Another approach to solvent-extractable and some near-solvent-extractable metabolites is to adsorb the metabolite from the fermentation filtrate with a reverse-phase resin such as Amberlite XAD-2, Diaion HP-20, or Diaion HP-21. As a typical example in the isolation of pterulinic acid and pterulone, 16 liters of a fermentation filtrate was passed through a column (30 by 6.5 cm) of Diaion HP-21, and the resin was washed with water and eluted with two liters of acetone to obtain an initial crude extract of the metabolites (8).

With very water-soluble ionic metabolites, the initial concentration steps may be accomplished by treating the

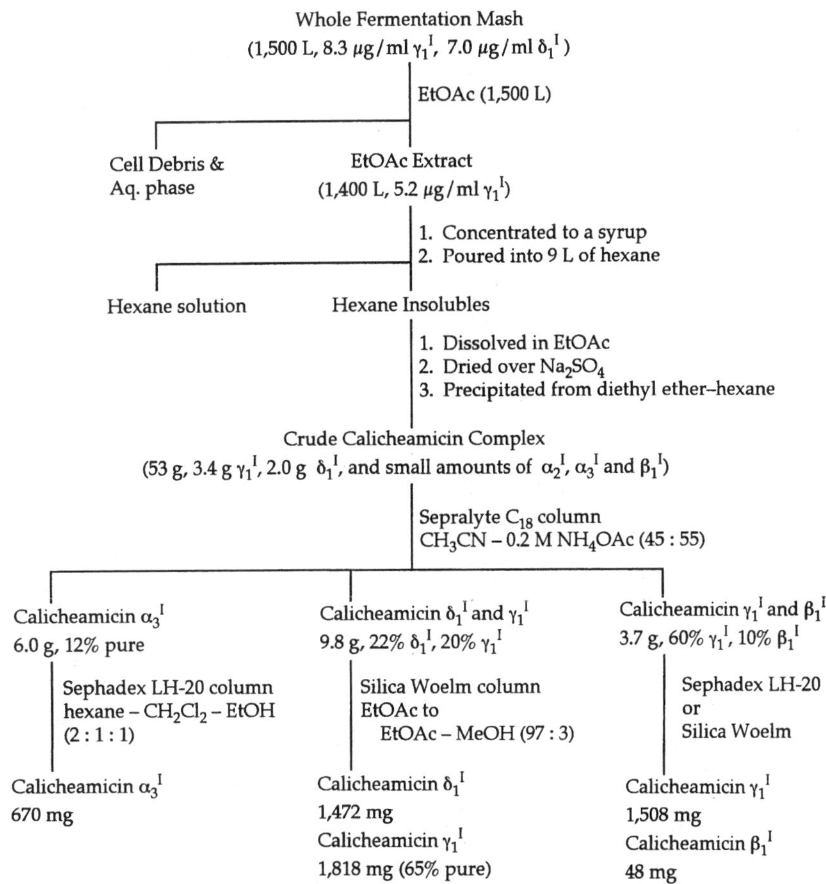

FIGURE 2 Process for the isolation of the iodinated calicheamicins from the fermentation of strain UV785.

fermentation filtrates with ion-exchange resins. Before adsorption of a metabolite, commercial ion-exchange resins are usually pretreated by wash cycles with 1 N hydrochloric acid and sodium hydroxide to remove impurities and then converted to the desired ionic form by treatment with 1 N solutions of the appropriate salts, acids, or bases. The resins are then washed with water prior to application of the fermentation filtrate. Basic metabolites that are alkali stable, such as neomycin or kanamycin, can be adsorbed onto the NH_4^+ form of a weak cationic resin such as Amberlite IRC-50, washed with water, and eluted off the resin with 1 N NH_4OH or a gradient of 0.1 N to 1.0 N NH_4OH. The bioactive peak can then be concentrated by evaporation to remove ammonia and water.

In the case of highly water-soluble neutral or amphoteric compounds, the initial concentration step can be very difficult, since solvent extraction, reverse-phase resins, and ion-exchange resins can be ineffective. In these cases, carbon or activated charcoal can sometimes be used, but these methods are not very desirable since they frequently involve low recoveries and require tailor-made processes. For the unstable water-soluble antitumor antibiotic LL-DO5139β, 8 liters of fermentation filtrate was adjusted to pH 9.0 to stabilize the antibiotic and cooled in an ice bath. The antibiotic was then adsorbed from the filtrate onto a 3.2 by 85 cm column of granular carbon, which was washed with 1 bed volume of water and eluted with a gradient of 0 to 60% aqueous methanol. The active fraction was concentrated, neutralized, and lyophilized to obtain a material for further purification (14).

21.4.2.2. Chromatography Methods

After an initial step to obtain a concentrate or extract that is frequently only 1 to 5% pure with respect to the active metabolite, more refined purification methods are applied. Usually, column chromatography provides the most effective method, and in recent years commercial preparative HPLC systems have greatly improved these purification methods. The preparative HPLC utilizes chromatography supports with very small uniform beads or particles similar to those in the analytical systems, thus resulting in separations with high resolution. An additional advantage to using these preparative HPLC systems is the close carryover from methods developed for analytical HPLC at the initial characterization and identification of the biologically active metabolites. The net effect is to reduce the number of chromatography steps required and the effort needed to develop additional chromatography methods. Previously, analytical HPLC systems were developed to analyze fermentations, and standard chromatography systems with totally different supports and solvents were designed for preparative isolation of the metabolites. In the case of the ganefromycins, growth-promoting agents produced by *Streptomyces lydicus* subsp. *tanzanius* and also referred to as LL-E19020α and β (3, 4), the analytical and preparative chromatography utilized reverse-phase C_{18} columns employing solvent mixtures of acetonitrile and 0.1 M NH_4OAc. The preparative system provided multigram quantities of material for structure determinations and extensive biological evaluation. When kilogram quantities of ganefromycin α were required for field trials, the same basic purification method was used in the form of process-scale reverse-phase HPLC on a Millipore Kiloprep unit with a 12-liter cartridge packed with 55 to 105 μm of μBondapak C_{18}, which allowed processing of 100-g batches (23). This approach greatly reduced the time required to obtain kilogram quantities of the ganefromycins, since the development of a new isolation process was not required.

For most screening situations, scale-up will involve preparative purification systems capable of handling metabolites produced in fermentation broths only up to 5 or 10 liters in volume. This usually provides enough metabolite to determine if the compound is novel and to determine in a preliminary way if it has interesting in vitro and perhaps in vivo activity. Reviews describing analytical HPLC columns and solvent systems for a large number of antibacterial and antifungal antibiotics have been given (10, 18). In most cases, these systems can be readily scaled up for preparative chromatography of similar or identical compounds. Systems involving reverse-phase HPLC on C_{18} columns with mobile phases such as methanol or acetonitrile with buffers like aqueous ammonium acetate have a wide range of application. For two FK506 analogs produced by targeted gene disruption in *Streptomyces* sp. strain MA6548, a crude concentrate of the metabolites from a 1.5-liter fermentation was purified by HPLC using a semipreparative Phenomenex C_{18} reverse-phase column, 9.2 mm by 250 mm, eluted with an isocratic solvent system containing 0.1% aqueous phosphoric acid and acetonitrile (30:70) at a flow rate of 2 ml/min (21). Final purification of the two metabolites was obtained by preparative thin-layer chromatography on silica (E. M. Merck) using a dichloromethane-methanol (9:1) solvent system. The purified compounds were then analyzed by liquid chromatography/MS/MS and NMR.

For illudinic acid, a novel sesquiterpene antibiotic, the antibiotic was characterized as anionic by electrophoresis. As a result, 500 ml of the whole broth from a fungal fermentation was extracted with MeOH and the antibiotic in the extract was adsorbed onto an anion ion-exchange resin for an initial concentration step. The antibiotic was eluted from the ion-exchange resin with NH_4Cl, desalted with Diaion HP-20 resin, and then purified by chromatography on reverse-phase HPLC (Zorbax RX-C8, 5 mm, 25 by 250 mm) eluting with 30% acetonitrile-water at a rate of 10 ml/min. The purified illudinic acid (1.6 mg) obtained from the active fractions was characterized by ^{1}H NMR, HMBC, HMQC, and NOE experiments (7).

Another frequently used approach for purification of microbial metabolites is low-pressure chromatography utilizing a sequence of steps, such as reverse-phase chromatography on Sephralyte C_{18} followed by chromatography on Sephadex LH-20 or silica gel, as shown in Fig. 2 for the calicheamicin complex. In this separation scheme, the extraction and chromatography procedures are given in detail and the weights of the various calicheamicin components are presented for each step. This also gives general information on losses that might be expected in any multistep isolation process (14).

In many situations, the metabolites are concentrated by solvent extraction and purified by a combination of steps using both low-pressure chromatography and preparative HPLC. The enzyme inhibitor SNA-8073-B was isolated from 5 liters of culture broth by extraction with ethyl acetate to obtain an initial concentrate, which was partially purified by chromatography on a silica gel column (2.5 by 33 cm) eluted with $CHCl_3$-CH_3OH (20:1, 10:1). The active fraction provided a crude material that was purified further by preparative HPLC on a column of Nucleosil $5C_{18}$ (20 by 250 mm) eluted with 55% methanol to obtain 27 mg of pure SNA-8073-B (12).

The isolation of fusaricide represents a very desirable extreme for a purification process in which the metabolite is crystallized from the initial fermentation extract. A 4-liter hexane extract of an agar plate surface fermentation (20 by 20 by 20 cm Nunc plates) was concentrated to a small volume that caused crystallization of 58 mg of fusaricide (17). Occasionally, metabolites crystallize directly in solid or liquid fermentations, but these events are relatively rare in screening wild-type cultures.

Forms of counter current chromatography such as CPC are very effective for purifying microbial metabolities (16, 19); however, these methods are usually considered more labor intensive. In these two-phase liquid-liquid separations, solvent systems such as ethylacetate-water and *n*-butanol-acetic acid-water are used. A microbial metabolite screening protocol utilizing CPC with solvent systems containing hexane, ethyl acetate, methanol, and water has been reported (18).

21.5. NOVEL MOLECULES

As previously noted, novel microbial metabolites can be identified by evaluation of preliminary MS, UV, and NMR data in conjunction with searches of the databases. These identifications are based on UV chromophores, MWs, and some information on other fragments or functional groups. However, it is not until the structure of a novel molecule is known that the novelty and relationship to other metabolites is certain. When a new microbial metabolite falls within a known family of compounds, its structure in many cases can be quickly determined by comparing NMR, MS, and UV data with data for other members in the family. In other situations, especially with compounds in novel structure classes with MWs greater than 1,000, the structure determination may require significantly more time.

An accurate determination of MW and MF can be readily obtained from high-resolution FABMS (fast atom bombardment mass spectrometry). When the MW is greater than approximately 700, it is necessary to have additional supporting evidence for which elements are present in the molecule. Carbon and proton counts by NMR are a first step, but other analyses are necessary and can involve techniques like electron spectroscopy for chemical analysis to detect and quantitate elements beyond the expected carbon, hydrogen, nitrogen, and oxygen (15). Electrospray MS provides a method of readily obtaining MWs of compounds, small and large. MWs in the range of 20,000 for proteins are routine for this technique.

NMR techniques, especially two-dimensional techniques such as COSY, HMBC, and HMQC, are extremely powerful methods for structure determination. These methods establish molecular connectivity. NOE, HMQC, HMBC, and ^{1}H NMR techniques were used to readily determine the structure of illudinic acid, $C_{15}H_{18}O_4$, a compound closely related to illudin. These studies were performed with the compound isolated from 500 ml of fermentation broth (7).

REFERENCES

1. **Borders, D. B.** 1992. Survey of antibiotics, p. 893–904. *Kirk-Othmer Encyclopedia of Chemical Technology*, 4th ed., vol. 2. John Wiley & Sons, New York.
2. **Carter, G. T.** 1997. LC/MS and MS/MS procedures to facilitate dereplication and structure determination of natural products, p. 3–20. *In* D. M. Sapienza (ed.), *Natural Products Drug Discovery. II. New Technologies to Increase Efficiency and Speed*. Biomedical Library Series, International Business Communications, Southborough, Mass.
3. **Carter, G. T., D. W. Phillipson, J. J. Goodman, T. S. Dunne, and D. B. Borders.** 1988. LL-E19020α and β, novel growth promoting agents: isolation, characterization and structures. *J. Antibiot.* **41:**1511–1514.
4. **Carter, G. T., D. W. Phillipson, R. R. West, and D. B. Borders.** 1993. Chemistry and structure of ganefromycin. *J. Org. Chem.* **58:**6588–6595.
5. **Corley, D. G., and R. C. Durley.** 1994. Strategies for database dereplication of natural products. *J. Natl. Prod.* **57:**1484–1490.
6. **Creuger, W., and A. Creuger.** 1989. *Biotechnology: A Textbook of Industrial Microbiology*, 2nd ed., p. 238. Sinauer Associates, Sunderland, Mass.
7. **Dufresne, C., K. Young, F. Pelaez, A. G. Del Val, D. Valentino, A. Graham, G. Platas, A. Bernard, and D. Zink.** 1997. Illudinic acid, a novel illudane sesquiterpene antibiotic. *J. Natl. Prod.* **60:**188–190.
8. **Engler, M., T. Anke, O. Sterner, and U. Brandt.** 1997. Pterulinic acid and pterulone, two novel inhibitors of NADH: ubiquinone oxidoreductase (complex I) produced by a *Pterula* species. I. Production, isolation and biological activities. *J. Antibiot.* **50:**325–329.
9. **Fantini, A. A., and R. T. Testa.** 1995. Taxonomy, fermentation, and yield improvement. *In* D. B. Borders and T. W. Doyle (ed.), *Enediyne Antibiotics as Antitumor Agents*. Marcel Dekker, New York.
10. **Isaacson, D. M., and J. Kirschbaum.** 1986. Assays of antimicrobial substances. p. 410–435. *In* A. L. Demain and N. A. Solomon (ed.), *Manual of Industrial Microbiology and Biotechnology*. American Society for Microbiology, Washington, D.C.
11. **Kenion, G. B., G. T. Carter, J. Ashraf, M. M. Siegel, and D. B. Borders.** 1990. Qualitative analysis of pharmaceuticals by thermospray liquid chromatography/mass spectrometry. p. 381–390. *In* M. A. Brown (ed.), *Liquid Chromatography/Mass Spectrometry: Applications in Agriculture, Pharmaceutical, and Environmental Chemistry*. ACS Symposium Series No. 420. American Chemical Society, Washington, D.C.
12. **Kimura, K.-I., F. Kanou, H. Koshino, M. Uramoto, and M. Yoshihama.** 1997. SNA-8073-B, a new isotetracenone antibiotic inhibits prolyl endopeptidase. I. Fermentation, isolation and biological properties. *J. Antibiot.* **50:**291–296.
13. **Lee, M. D.** 1995. Identification, isolation, and structure determination. p. 49–73. *In* D. B. Borders and T. W. Doyle (ed.), *Enediyne Antibiotics as Antitumor Agents*. Marcel Dekker, New York.
14. **Lee, M. D., A. A. Fantini, N. A. Kuck, M. Greenstein, R. T. Testa, and D. B. Borders.** 1987. New antitumor antibiotic, LL-D05139β. Fermentation, isolation, structure determination and biological activities. *J. Antibiot.* **40:**1657–1663.
15. **Lee, M. D., J. K. Manning, D. R. Williams, N. A. Kuck, R. T. Testa, and D. B. Borders.** 1989. Calicheamicins, a novel family of antitumor antibiotics. 3. Isolation, purification and characterization of calichemicins β_1^{Br}, γ_1^{Br}, α_2^{I}, α_3^{I}, β_1^{I}, γ_1^{I} and δ_1^{I}. *J. Antibiot.* **42:**1070–1087.
16. **McAlpine, J. B., and J. E. Hochlowski.** 1989. Countercurrent chromatography. p. 1–53. *In* G. H. Wagman and R. Cooper (ed.), *Natural Products Isolation. Journal of Chromatography Library*, vol. 43. Elsevier, New York.
17. **McBrien, K., Q. Gao, S. Huang, S. E. Klohr, R. R. Wang, D. M. Pirnik, K. M. Neddermann, I. Bursuker, K. F. Kadow, and J. E. Leet.** 1996. Fusaricide, a new cytotoxic N-hydroxypyridone from *Fusarium* sp. *J. Natl. Prod.* **59:**1151–1153.
18. **Mierzwa, R., J. A. Marquez, M. Patel, and R. Cooper.** 1989. HPLC detection methods for microbial products in fermentation broth, p. 55–110. *In* G. H. Wagman and R. Cooper (ed.), *Natural Products Isolation. Journal of Chromatography Library*, vol. 43. Elsevier, New York.
19. **Mocek, U.** 1995. Chemical purification and structure elucidation of lead compounds in a high-throughput natural products screening program, p. 2.5–2.26. *In* N. Mulford (ed.) *Natural Products, Rapid Utilization of Sources for Drug Discovery and Development*. Biomedical Library Series, International Business Communications, Southborough, Mass.
20. **Nash, C., J. Hunter-Cevera, R. Cooper, D. E. Eveleigh, and R. Hamill (ed.).** 1992. *Microbial Metabolites*. Developments in Industrial Microbiology Series, vol. 32 Wm. C. Brown Publishers, Dubuque, Iowa.
21. **Shafiee, A., H. Motamedi, F. J. Dumont, B. H. Arison, and R. R. Miller.** 1997. Chemical and biological characterization of two FK506 analogs produced by targeted gene disruption in *Streptomyces* sp. MA6548. *J. Antibiot.* **50:**418–423.
22. **Sitrin, R. D., and G. F. Wasserman.** 1989. Affinity and HPLC purification of glycopeptide antibiotics, p. 111–152. *In* G. H. Wagman and R. Cooper (ed.), *Natural Products Isolation. Journal of Chromatography Library*, vol. 43. Elsevier, New York.
23. **Williams, D. R., G. T. Carter, F. Pinho, and D. B. Borders.** 1989. Process-scale reversed-phase high-performance liquid chromatography purification of LL-E19020α, a growth promoting antibiotic produced by *Streptomyces lydicus* ssp. *tanzanius*. *J. Chromatogr.* **489:**381–390.

Purification and Characterization of Proteins

RICHARD C. WILLSON

22

22.1. BACKGROUND

This chapter provides an introduction to the fascinating, useful, and sometimes dreaded world of protein purification and characterization. It is intended to provide an introduction to the most commonly used laboratory-scale methods, along with pointers to detailed references for further reading.

The goals of purification vary with the intended use of the purified protein. The amount of protein needed can range from micrograms for microsequencing or the raising of antiserum, through milligrams for activity characterization or structural studies, to grams or more for pharmaceutical products. Purity is defined by the general level of other protein contaminants, but also by the absence of contaminants of special interest, which can include endotoxins, highly antigenic proteins, high levels of salts, viruses, and competing enzymatic or biological activities.

Protein purification can generally be divided into five broad stages, which need not all occur sequentially: preparation of the source, gathering of all available information about the protein's properties, development of an assay, primary isolation, and final purification. Each of these steps can often be made easier, or even avoided, by effective searching of the literature. Only rarely is the protein of interest unrelated to any that has previously been purified. Hints and tricks from previous work can greatly simplify development of a novel purification; it may even be possible to "purify by phone" if a previously isolated protein will serve the intended purpose.

22.2. PREPARATION OF THE SOURCE

Preparation begins with selection of the raw material from which the protein will be isolated. The source is here assumed to be a microbial or cultured metazoan cell line, although transgenic animals and plants will be of increasing importance in the foreseeable future.

It is important to start with ample quantities of target protein in the raw material. Overall purification yields for optimized pharmaceutical protein processes often exceed 80%, but the first few developmental purifications of a poorly characterized target may achieve levels closer to 5%, or even less. Make sure there is at least 10 to 20 times as much protein present in the raw material as the amount required for the intended use; there have been attempts in which even a 100% yield would not have sufficed! An abundance of raw material permits greater exploration of alternatives and allows each step to achieve a greater degree of purification, at the expense of yield.

Protein supplies can be increased by increasing the cultivation volume, by growing more cells per unit volume (e.g., by computer-controlled fermentation with pH control, enhanced oxygen transfer, and regulated feeding of substrate), or by producing more of the desired protein per cell (2). Expression levels can be increased by screening of high-producing cell lines; cloning and overexpression; use of antibiotic selection pressures other than ampicillin, which is usually ineffective at high cell densities; use of tightly regulated, strong promoters such as T7 or alkaline phosphatase; and often by expression at lower temperatures, e.g., as low as 10°C for *Escherichia coli* (22).

In addition to its quantity, the quality of the raw material is subject to improvement. These improvements can include elimination of key contaminants (e.g., proteases, competing enzymatic activities), optimization of posttranslational modifications, maximization of extracellular secretion or correct folding, and the genetic addition of purification "handles" such as His_6 or protein fusions, as discussed below.

22.3. KNOWLEDGE OF THE PROTEIN'S PROPERTIES

It is essential to collect as much information on the protein as possible. Knowledge of almost any characteristic of the protein may find application in the development of a purification process; properties of interest are summarized as follows.

1. Source: cell type, expression level, intra/extracellular location, folding state, presence of proteases/glycosidases
2. Stability: To endogenous proteases, temperature range, pH range, ionic strength, hydrophobic surfaces, aggregation tendency, cofactor or metal ion loss/requirement, freeze-thaw stability (for storage during and after purification)
3. Size: multimeric state, molecular weight, peptide chain(s), hydrodynamic radius

4. Charge: isoelectric point, titration curve, electrophoretic mobility

5. Binding partners: substrates and cofactors (and analogs), screening-derived binding agents (antibodies, peptides, aptamers), metal affinity. Binding to any of these in immobilized forms, for affinity chromatography. Stabilizing effects of binding partners.

22.4. ASSAY

An assay for the desired activity or protein is required. Assays for use in purification must be convenient and rapid; extreme precision is less important. Typical methods include UV/visible absorbance, dot-blot and gel-shift binding assays, chromogenic enzyme activity assays (e.g., in microtiter plates), and biosensor (BIAcore) testing of fractions.

22.5. INITIAL ISOLATION

Initial isolation steps separate the product from the majority of the water in the cultivation medium and from the majority of the host cell components. Recovery of both extracellular and intracellular proteins usually begins with cell separation by centrifugation or filtration.

22.5.1. Concentration

Extracellular (secreted) proteins are usually concentrated from the cell-free broth by ultrafiltration or adsorption on chromatographic media. Ultrafiltration can be carried out at small scale in centrifugally driven cells, or at larger scale in stirred cells or tangential-flow units (17, 31). Adsorption can employ any of the chemistries noted below, although expensive and/or delicate adsorbents (e.g., immobilized antibodies) are normally used for initial capture only on the small scale, or for unstable proteins. An increasingly popular approach at larger scale is direct protein capture in fluidized beds of dense adsorbent particles from culture media and cell lysates, thus reducing the number of early process steps.

Sometimes a secreted protein is adsorbed to the outside of the cells and can be concentrated along with them, then liberated by washing, often with a high-salt buffer.

22.5.2. Cell Lysis

Intracellular proteins are liberated by cell lysis. Total lysis of cells is required for liberation of cytoplasmic proteins. Microbial lysis often employs chemical agents such as chloroform, toluene, sodium dodecyl sulfate (SDS), EDTA, lysozyme (especially for gram-positive organisms), freeze-thaw cycling, and mechanical stress. Chemical, enzymatic, and freeze-thaw methods are often less than completely effective and may best be used on smaller scales. Mechanical methods include sonication (for small scale), grinding, glass bead mills, the French press, and the Manton-Gaulin homogenizer. Each of these methods destroys proteins as well as cells and requires optimization of the release of active protein, but each can give reliable, satisfactory results. The French press in particular is a standard tool for lysis of some tens of grams of wet cell paste by forcing cells through a small orifice at high pressure. Primary isolation yields may benefit from cooling, or from addition of protease or glycosidase inhibitors. These measures can be particularly valuable with cell lysis using mechanical methods, which tend to heat samples.

Periplasmic proteins can be selectively liberated by osmotic shock, which ruptures only the outer membrane by sudden transfer from hyper- to hypotonic medium. Successful performance and scale-up of this operation can be tricky and may benefit from measurement of release of enzyme activities known to be cytoplasmic (e.g., β-galactosidase) or periplasmic (e.g., alkaline phosphatase).

Cell lysates are unattractive substances, rich in degradative activities, sticky lipids, and viscous nucleic acids along with thousands of protein contaminants. Soluble proteins are often recovered from cell lysates using precipitation or (less commonly) liquid partitioning, as these methods are tolerant of viscosity, particulates, and fouling. Ribosomes and nucleic acids can first be precipitated with streptomycin, protamine, or polyethyleneimine. The desired protein is often concentrated and partially purified by precipitation with ammonium sulfate or polyethylene glycol (25, 30). Stepwise addition of optimized levels of precipitant can first precipitate some contaminants, and then (at a higher concentration) the desired protein, leaving other contaminants in the final supernatant. Exact results depend on the rate of precipitant addition, mixing, and aging for precipitate formation, but these are generally robust methods from which proteins can be recovered in high yields.

22.5.3. Refolding

Recombinant proteins often misfold to form dense, insoluble aggregates of inactive protein. (Such aggregation can be detected by SDS-polyacrylamide gel electrophoresis [SDS-PAGE] analysis of the soluble and insoluble fractions of a cell lysate, or inclusion bodies can be directly observed by microscopy.) Proteins in inclusion bodies can be harvested by centrifugation at purities exceeding 70%, but they are valueless unless the native tertiary structure can be restored. The first step in the renaturation process is dissolution of the inclusion bodies (sometimes even whole cells) in a strong chaotrope solution such as 6 M urea or guanidine hydrochloride. Dissolution in denaturant is rapid and reliable. The dissolved, denatured protein is then (hopefully) allowed to renature to its native conformation by removing the denaturant through dialysis, dilution, or chromatographic separation. In the renaturation step, unimolecular refolding processes must compete with multimolecular reaggregation reactions, so operation with immobilized or highly dilute protein molecules is required. Development of a process that recovers the protein in native form while losing as little as possible to reaggregation can be difficult and is always unpredictable. In many cases, a low refolding yield is acceptable because the protein is abundantly available from inclusion bodies.

Several rules of thumb for refolding have emerged.

1. Small, disulfide-free, single-peptide-chain proteins are easiest to refold, but proteins of all types have been refolded.
2. Refold at low protein concentrations.
3. If possible, immobilize the refolding protein by adsorption to prevent multimolecular reactions, which promote reaggregation.
4. Control redox potential (through dithiothreitol and β-mercaptoethanol addition) and oxygen supply.
5. Add arginine, polyethylene glycol, or cyclodextrins.
6. Reduce denaturant concentration in two or more steps, rather than attempting a drastic reduction in a single step.

7. Allow up to 7 to 10 days for refolding.
8. Recycle the reaggregated protein back to the refolding process.

22.6. HIGH-RESOLUTION PURIFICATION

Chromatography is the usual method of preparing highly purified, active proteins (3, 5, 7, 9–12, 25, 28, 32). Most forms of chromatography involve selective adsorption of proteins on the surface of porous particles, through interactions that can be broadly classified as ion-exchange, hydrophobic, and affinity. Chromatographic operations are also classified as low-pressure, medium-pressure (including Pharmacia's popular FPLC) and high-pressure (HPLC), depending on the pressure used to force liquid through the packed bed of adsorbent particles (Table 1). In general, high pressures imply higher costs but also finer adsorbent particles and better resolving power. Samples loaded onto high- and medium-pressure columns must be rigorously free of particulates.

In addition to the adsorbent matrix, apparatus for chromatography includes a column into which the particles are packed (glass, plastic, or steel, depending on operating pressure), a pump or height differential to drive liquid flow, some method for introducing the sample into the flow before the column (a switching valve, or manual pipetting onto the top of the packed bed), and a fraction collector that deposits the emerging, separated proteins into different vessels. Possible enhancements include UV/visible absorbance detectors for monitoring the emergence of the separated proteins from the column, a conductivity monitor for tracking changes in salt concentration used to elute proteins from the column, and a computer for control and for storage of the results.

Not all adsorptive separations involve particles in a column. Particles can be used for batch adsorption by gently mixing them into a solution containing proteins to be captured. The resolution obtained in this manner is low, but the method is tolerant of fouling and particulates and can easily be applied to large volumes of solution. Nonparticulate supports, notably membranes, are available in many of the surface chemistries discussed below. These give rapid adsorption and desorption kinetics, although with somewhat reduced capacity.

There are a variety of excellent references for the methodological details of chromatographic separation of proteins (see below); the present discussion will address general characteristics, along with some guidelines for method development.

22.6.1. Ion-Exchange Chromatography

Ion-exchange chromatography (20) is the most common high-resolution method for preparative separation of proteins and is used in most protocols (Table 2). It will be discussed in particular detail because of its importance, and also to illustrate the general issues involved in development of chromatographic separations. Adsorbents with fixed positive charges (from immobilized amines, e.g., DEAE, QMA) are called anion exchangers and are employed at pHs above the isoelectric point of the protein to be adsorbed so that the protein's net charge will be of sign opposite to that of the adsorbent. Matrices bearing negative charges (carboxylates and sulfonates, e.g., CM, SP) are called cation exchangers and are used at lower pHs. In practice, the protein of interest is usually adsorbed rather than passed through the column, as higher resolution and a degree of concentration can be achieved. Ion-exchange adsorption depends strongly on the ionic strength (I), equal to half the sum of the concentrations of all ions present, each multiplied by the square of the ion's charge. For 1 M NaCl, I = 1 M; for 1 M $(NH_4)_2SO_4$, I = 3 M. Initial trials might employ ionic strengths of 10 to 50 mM for loading and 1 M for elution.

The combinations of pH and matrix type (anion or cation exchange) at which the protein will be adsorbed can be predicted with some confidence from its pI, although the detailed distribution of charges on the protein's surface plays a role, as do nonspecific interactions with the adsorbent backbone. If the pI is not known, it can sometimes be measured through isoelectric focusing or can be estimated from the sequence. Otherwise, a sample containing the protein is loaded (at low ionic strength) on an anion-exchange column at the lowest pH at which it is known to be stable, or on a cation-exchange column at its highest stable pH. The column is then washed with 2 to 3 volumes of the starting buffer, or until the initial peak of unretained protein has fallen to baseline. Most proteins will be retained on the column and can be eluted with a gradient of increasing salt concentration up to 1 to 2 M applied over a volume 5 times that of the column packing. Failure to bind can be due to overloading of the column (5 to 10 mg of total protein per ml of column packing is a conservative starting point); high loading ionic strength; failure to regenerate the column sufficiently at low ionic strength after the previous elution (this can require more than 5 column volumes of low-salt buffer and is prudently monitored with a conductivity probe); instability of the protein while adsorbed (this can improve at lower temperatures, or with slightly higher loading ionic strengths); or instability at too extreme a loading pH.

TABLE 1 Types of chromatography

Feature	Low-pressure	Medium-pressure	High-pressure (HPLC)
Particle size	40–150 μm	10–75 μm	2–15 μm
Flow driver	Gravity, peristaltic	Piston or syringe	Positive displacement
Run time	40–1,000 min	15–60 min	0.5–30 min
Apparatus cost	Low	Medium high	High
Resolving power	Lowest	Intermediate	Highest
Particulate tolerance	Low	Very low	Lowest

TABLE 2 Modes of adsorptive chromatography

Feature	Ion-exchange	Hydrophobic	Metal chelate	Biospecific
Adsorbent	Carboxyl, amine, sulfonate	Propyl, butyl, phenyl	Chelator, loaded with Ni^{2+}, Zn^{2+}	Antibody, cofactor, receptor
Selectivity	Moderate-high	Moderate-high	Moderate-high	High-very high
Capacity	High	High	Moderate-high	Moderate-high
Matrix cost	Low	Low	Moderate	High
Elution	High salt, pH	Low salt	pH, imidazole	pH, chaotrope
Initial salt	Low	High	Indifferent	Often indifferent

Optimization of the method involves increasing the fraction of the protein recovered in active form, the purity of the recovered material, and the rate at which purified protein is isolated (through loading of larger samples or shorter separation times). Steps in process development often include loading under less permissive conditions (higher salt, pH closer to neutrality) to cause more contaminants to wash through unadsorbed, higher operating temperatures, faster liquid flow rates (to the limits specified by the adsorbent manufacturer), and testing of shorter gradients, or gradients tailored to be shallow only near the ionic strength at which the protein of interest is eluted.

22.6.2. Hydrophobic Interaction Chromatography

Hydrophobic interaction chromatography (HIC) (21) exploits the presence of exposed hydrophobic groups on the surfaces of proteins, which can interact with immobilized nonpolar moieties such as short alkyl chains and phenyl rings (i.e., immobilized propane and benzene). Adsorption is promoted by high ionic strength, and elution often employs a salt gradient of decreasing concentration. As burial of nonpolar groups is a major contributor to protein stability, there is reason for concern that exposure to hydrophobic surfaces may be denaturing. Many very hydrophobic polymer surfaces are in fact rapidly coated with denatured protein, and this is a major constraint on the selection of support materials for chromatographic matrices. Isolated hydrophobic groups sparsely immobilized on polar supports such as polysaccharides and acrylates, however, are not nearly as denaturing as bulk materials, and recoveries of many proteins from optimized hydrophobic separations can approach 100%. The cost and capacity of HIC matrices are attractive, as is the method's tolerance of high salt concentrations, which can allow immediate processing of proteins eluted in high salt from a previous ion-exchange chromatography or ammonium sulfate precipitation step. HIC is widely used in industrial processes and is probably underutilized in academic laboratories.

22.6.3. Affinity Chromatography

The remaining adsorptive techniques are affinity methods, which exploit selective interactions characteristic of biological systems. The normal biological functions of many proteins depend on their ability to associate with substrates, cofactors, etc., and proteins often will bind selectively to immobilized forms of these partners. Alternatively, an affinity agent such as an antibody can be used to selectively capture the protein of interest. Finally, it has become increasingly common to append DNA encoding a "purification handle" protein with affinity for an available ligand to the gene encoding a recombinant protein of interest. The resulting fusion protein can be purified by its affinity for this ligand. The purification handle sequence can be removed by proteolysis, or left in place for many applications. Typical purification handles include cationic polyarginine, the immunoglobulin-binding domain of protein A of *Staphylococcus aureus*, cellulose- and maltose-binding proteins, and the increasingly popular immobilized-metal-binding hexahistidine (His_6) described in the next section.

Immobilized-metal affinity chromatography (IMAC) exploits the affinity of certain chelated metal ions (e.g., Ni^{2+} and Zn^{2+}) for the amino acid side chains of cysteine and especially histidine. Elution is achieved by pH gradient or the addition of competitive metal-chelating compounds such as imidazole. IMAC adsorbent matrices are relatively inexpensive and robust, and selectivity can be quite high. The method is also tolerant of variations in ionic strength. The popularity of IMAC has increased dramatically with the widespread use of genetically added chelating-peptide purification handles on recombinant proteins, especially His_6.

Biospecific affinity chromatography describes separations based on molecular recognition interactions normally found in nature, such as antibody/antigen, enzyme/cofactor, and hormone/receptor affinity (29). Many widely applicable premade affinity matrices (at least 50 types) are commercially available. Among the most popular affinity methods are Cibacron Blue dye affinity for purification of enzymes recognizing ATP or NAD(P)(H), to which this textile dye bears a coincidental structural resemblance, and chromatography on protein A of *S. aureus* for purification of antibodies (4, 18). There are also available many "activated" matrices that have been chemically derivitized for facile coupling of affinity ligands through surface amines, carbohydrate groups, etc.

The selectivity of affinity separations can be enormously high. This advantage is balanced in some cases by the cost and delay of obtaining and coupling suitable ligands. Also, the harsh conditions that may be required for elution of the bound protein can damage both target protein and affinity column. An active area of development in purification technology is the screening of large numbers of potential ligands, both phage-displayed peptides, aptamers, and small organic molecules, for use as improved affinity separation ligands.

22.6.4. Size Exclusion Chromatography

Size exclusion chromatography (SEC, also known as gel filtration), differs from the adsorptive methods in that protein interactions with the chromatographic matrix are not exploited, but minimized (19). SEC media display pores of varying sizes, and differential access to these pores is the basis of separation. Large molecules run faster, as they enter fewer pores and flow through an effectively smaller volume. All molecules above a certain size run together, "excluded," as they enter very few pores. Small molecules below the "included" size limit run together, exploring nearly all pores. Useful resolving power is confined to a range of hydrodynamic radii (*not* molecular weights) between the effective sizes of the largest and smallest pores; matrices of differing average pore size are used to purify proteins of different sizes. The inclusion and exclusion limits of a given matrix are given by the manufacturer. It is important to remember that larger molecules run faster (the opposite of a sieve or filter), and that separation is on the basis of molecular radius, not molecular weight. This implies that an elongated protein runs more rapidly than a compact globular protein of the same molecular weight, and both run much more slowly than a nucleic acid or random coil carbohydrate of the same molecular weight as the proteins.

SEC has relatively low sample capacity, and the packing and sample loading of long, narrow SEC columns is more demanding than preparation of relatively squat columns of adsorptive media. It is well worth testing the quality of a newly packed SEC column using a tracer such as acetone or blue dextran. SEC is a gentle, nondenaturing method, and in contrast to most methods is capable of removing aggregated forms from nearly pure protein samples. An SEC column can be "calibrated" by running a set of proteins differing in molecular weight, but all of similar shape (in practice, all globular). The retention times will ideally give a linear dependence on the log of the molecular weight of the protein (deviations often indicate interactions with the matrix, which can be suppressed by adding salts or traces of organic solvents to the running buffer). The retention time of the unknown protein indicates its largest dimension. This information is especially valuable if it can be compared with the peptide molecular weight from SDS-PAGE, as large differences in the results of the two methods can often be the first sign of aggregation or multimerization.

A few general rules for protein purification are given here.

1. Tailor the process to the intended purpose.
2. Obtain a convenient assay.
3. Select the right source of raw material.
4. Work quickly, or in the cold, or both.
5. Eliminate proteases early, through host selection, separation, or inhibition.
6. Identify stable intermediate storage forms.
7. Avoid freeze-thaw cycles.
8. Use columns of appropriate capacity.
9. Build on previous work.

22.7. CHARACTERIZATION

Complete characterization of the purified product involves study of the idiosyncratic activities of each protein, which are beyond the scope of this chapter. There are also, however, general structural properties that are subject to determination by generic methods (13, 16, 26).

22.7.1. Electrophoresis

Probably the most common method of protein characterization is SDS-PAGE, which gives the molecular weight(s) of the peptide chain(s) of the protein, and also a good indication of the purity of the sample (6). SDS converts nearly all proteins into regular rodlike forms of constant charge density per unit mass, so that electrophoretic mobility is independent of amino acid composition. The gel serves to suppress mixing induced by temperature gradients, which would reduce resolution. SDS-PAGE can use commercial pre-cast gels (e.g., Pharmacia's pHAST system), or gels can be cast at the time of use. Related electrophoretic methods include isoelectric focusing (IEF; for measuring isoelectric point), native gel electrophoresis (in which proteins are not denatured with SDS), and two-dimensional gel electrophoresis, in which IEF separation by charge is crossed with orthogonal separation by mass using SDS-PAGE. Native electrophoresis and IEF also show promise as preparative methods (6).

Capillary electrophoresis (CE) is conducted in tubing of very small diameter (<100 μm) using high voltages (e.g., 30 kV). CE can achieve very high resolution in separations requiring only a few minutes and is widely regarded as an extremely promising method (33). The methods required for CE analysis of protein samples are still relatively sophisticated, however, and this technique is not yet recommended for the beginner.

22.7.2. Peptide Sequencing

Amino-terminal sequencing is used to identify the first few amino acids of the protein (13, 26). Most workers will choose to have this procedure performed by a specialist core facility. This sequence information can be used to confirm the identity of the protein, to establish the degree of removal of N-terminal methionines, purification handles, or secretion leader peptides, and to detect exoproteolytic degradation. Protein sequencing depends on sequential, stepwise removal of N-terminal amino acids in the form of labeled derivatives that are separated by HPLC and identified by their characteristic retention times. Blocked N termini are not uncommon and require removal of the blockage, isolation and sequencing of internal peptides, or use of mass spectrometric sequencing as discussed below. Carboxy-terminal sequencing, while successfully demonstrated, is not yet readily available in most core facilities.

22.7.3. Tryptic Mapping

Tryptic mapping, in which small peptides derived from the protein by endoprotease action are separated by high-resolution reverse-phase HPLC, allows sensitive detection of variations in protein (especially pharmaceutical) preparations (13, 26). Individual peptides can be subjected to sequencing as described above (e.g., avoiding problems with blocked N termini) to confirm the identity of a given peptide, to develop information for DNA probe development, or even to establish the entire sequence of the protein.

22.7.4. Analytical Ultracentrifugation

Analytical ultracentrifugation has undergone a renaissance with the recent introduction of the first commercial instrument in many years (Beckman's XL-A). This technique, which requires considerable investment in equipment and training, allows measurement of a variety of properties of a protein sample, including solution molecular weight, self-association, interaction with other molecules, and sample homogeneity (24). Sedimentation field-flow fractionation is useful for characterizing similar properties of larger particles (23).

22.7.5. Spectroscopy

Circular dichroism gives an indication of the fraction of the polypeptide that is composed of specific secondary structural features, such as α-helix and β-sheet. Raman spectroscopy has been used (less widely) for similar purposes, and especially for characterizing metal-containing protein cofactors (8, 15).

22.7.6. Biosensors

Biosensor devices such as Pharmacia's BIAcore, Fisons' IASys, and ASI's Integrated Optics biosensor are now widely used for detecting and characterizing intermolecular interactions. These devices share the property of continuous detection of proteins, cells, or nucleic acids present in a narrow region adjacent to the detector surface. If, for example, an antibody to a particular protein is immobilized on the sensor surface, addition of a sample containing that protein will produce an immediate signal. Because of their sensitivity, low material consumption, and convenience, these devices are very useful for tracking the desired protein in the course of purification. They also have a role to play in characterizing the purified protein, as they can readily answer questions such as: Does this protein bind to this immobilized ligand/antibody/oligonucleotide? Does this compound inhibit binding of this pair of molecules? What is the (approximate) concentration of the molecule of interest in this preparation? Do the binding/desorption kinetics of this batch of this material match those of the last batch? It is necessary, however, to guard against the (often severe) effects of mass-transfer limitation, rebinding, flow, multivalency, etc., in attempting to derive quantitative biophysical information from these devices; errors have been published as users have gradually become familiar with these devices.

Use of a biosensor device requires covalent coupling of one member of an interacting pair to the surface of the detector, in active form. The chemistries involved are derived from those of affinity chromatography and are now generally well characterized by the instrument manufacturers. For a signal to be produced, a meaningful amount of mass must be captured by the interaction. For example, in studying the interaction of an antibody with a small hapten, it would be better to immobilize the hapten and capture the much larger antibody, as capturing the hapten might not produce a useful signal.

22.7.7. Analysis of Glycosylation

Analysis of glycosylation is an actively developing field, in part because the characterization of glycoproteins remains quite challenging (14). Many eukaryotic proteins contain sugars, which are sometimes essential to the protein's function. Available methods include stains and blots for detection of (particular classes of) glycosylation, deglycosylating enzymes, sugar-specific lectins for blotting and separations, and nuclear magnetic resonance and mass spectrometric methods of characterizing individual side chains. An emerging area is the controlled creation and remodeling of glycosylation to create what are called "neoglycoproteins."

22.7.8. Mass Spectrometry

Mass spectrometry (MS) has blossomed in recent years with continuing advances in instrumentation and methods of sample introduction (27). MS services are now commonly available in core facilities and are probably somewhat underutilized by potential users, considering the amount of information MS can provide for a small investment of sample and time. While there exists a family of related MS techniques, the present discussion will introduce the uses of MS as a whole, all of which are driven by the ability to determine the mass of proteins and peptides with remarkable precision, of the order of a few hydrogen atoms in a typical protein. The technique, therefore, can detect and characterize almost any important modification or variation in a protein's structure. These include post-translational modifications, N-terminal methionine addition or removal, methionine oxidation, proteolytic trimming, and many sequence mutations and deamidation events (1). Because these modifications are detected without being specifically looked for, MS is an excellent way to characterize a newly purified protein.

Mass spectrometry is advancing into new applications. These include protein sequencing, on-line coupling to HPLC and CE, and identification of "hits" from combinatorial libraries. A new application driven by the availability of genome sequences and cDNA sequence banks is the identification of unknown proteins separated by electrophoresis or HPLC from a diseased tissue, or a cell line exposed to an agent thought to alter expression of one or more proteins. MS determination of the molecular weight of a protein showing interesting behavior (or, if digestion can be done, of one or more peptides derived from the protein) gives mass values that can then be searched against the (large) database of all proteins/peptides encoded by that genome or cDNA bank. Remarkably, this type of search can often be successful and in many cases will avoid the need to purify the protein in question at all.

22.8. EMERGING TRENDS

1. Automation. Computer control of chromatographic processes is increasingly common. Several manufacturers have introduced software capable of testing a large number of gradient designs, pHs, etc., to minimize the human effort required to develop optimal methods.
2. Purification handles. Genetic addition of biotinylation signals, or proteins or domains having affinity for particular ligands, continues to be popular. Probably the newest development is the use of self-removing intein fusions designed to recover the native protein with no added sequences.
3. Combinatorially derived separation ligands. Phage display, screening of libraries of small molecules, and nucleic-acid-based ligand (aptamer) libraries are all promising methods of identifying optimized ligands for affinity separations.
4. CE and MS. Each of these methods is advancing rapidly. MS should and will be more widely used, largely through core facilities, in the near term. CE will likely replace SDS-PAGE in a few more years.

SUGGESTED READING

1. **Asenjo, J. A., and J. Hong (ed.).** 1986. *ACS Symposium Series*, vol. 314. *Separation, Recovery, and Purification in Biotechnology: Recent Advances and Mathematical Modeling.* American Chemical Society, Washington, D.C.
2. **Ataai, M. M., and S. K. Sikdar (ed.).** 1992. *AIChE Symposium Series*, vol. 88. *New Developments in Bioseparation.* American Institute of Chemical Engineers, New York.
3. **Belter, P., E. L. Cussler, and W.-S. Hu.** 1988. *Bioseparations: Downstream Processing for Biotechnology.* John Wiley & Sons, New York.
4. ***Bollag, D. M., and S. J. Edelstein.** 1991. *Protein Methods.* Wiley-Liss Inc., New York.
5. **Burgess, R. (ed.).** 1987. *UCLA Symposia on Molecular and Cellular Biology New Series*, vol. 68. *Protein Purification: Micro to Macro.* Alan R. Liss Inc., New York.
6. **Dechow, F. J.** 1989. *Separation and Purification Techniques in Biotechnology.* Noyes Publications, Park Ridge, N.J.

7. ***Deutscher, M. P. (ed.).** 1990. *Methods in Enzymology*, vol. 182. *Guide to Protein Purification*. Academic Press, Inc., New York.
8. **Fiechter, A. (ed.).** 1992. *Advances in Biochemical Engineering/Biotechnology*, vol. 47. *Bioseparation*. Springer-Verlag Publishers, New York.
9. **Hamel, J.-F. P., J. B. Hunter, and S. K. Sikdar (ed.).** 1990. *ACS Symposium Series*, vol. 419. *Downstream Processing and Bioseparations: Recovery and Purification of Biological Products*. American Chemical Society, Washington, D.C.
10. **Harris, E. L. V., and S. Angal (ed.).** 1990. *Protein Purification Applications: A Practical Approach*. Oxford University Press, Oxford, U.K.
11. **Kaplan, N. O.** 1983. Historical perspectives of purification of biomolecules, p. 407–420. *In Affinity Chromatography and Biological Recognition*. Academic Press, Inc., New York.
12. **Kula, M.-R.** 1985. Recovery operations, p. 727–760. *In Biotechnology: A Comprehensive Treatise*. VCH Publishers, Deerfield Beach, Fla.
13. **Ladisch, M. R., R. C. Willson, C.-D. C. Painton, and S. E. Builder (ed.).** 1990. *ACS Symposium Series*, vol. 427. *Protein Purification: From Molecular Mechanisms to Large-Scale Processes*. American Chemical Society, Washington, D.C.
14. **Moo-Young, M. (ed.).** 1985. *Comprehensive Biotechnology: The Principles, Applications and Regulations of Biotechnology in Industry, Agriculture and Medicine*. Volume 2, C. L. Cooney and A. E. Humphrey (ed.), *The Principles of Biotechnology: Engineering Considerations*. Pergamon Press, New York.
15. **Shirley, B. A. (ed.).** 1995. *Methods in Molecular Biology*, vol. 40. *Protein Stability and Folding: Theory and Practice*. Humana Press, Totowa, N.J.
16. **Verrall, M. S., and M. J. Hudson (ed.).** 1987. *Separations for Biotechnology*. Society of Chemical Industry. Ellis Horwood Limited Publishers, Chichester, U.K.
17. **Walker, J. M. (ed.).** 1984. *Methods in Molecular Biology*, vol. 1, *Proteins*. Humana Press, Clifton, N.J.
18. ***Wheelwright, S. M.** 1991. *Protein Purification: Design and Scale Up of Downstream Processing*. Hanser Publishers, New York.

REFERENCES

An asterisk indicates a useful starting point for the new practitioner.

1. **Aswad, D. W. (ed.).** 1995. *Deamidation and Isoaspartate Formation in Peptides and Proteins*, CRC Series in Analytical Biotechnology. CRC Press, An Arbor, Mich.
2. **Blanch, H. W., and D. Clark.** 1997. *Biochemical Engineering*. Marcel Dekker, New York.
3. **Fiechter, A. (ed.).** 1993. *Advances in Biochemical Engineering/Biotechnology*, vol. 49. *Chromatography*. Springer-Verlag, New York.
4. ***Gagnon, P.** 1996. *Purification Tools for Monoclonal Antibodies*. Validated Biosystems, Tucson, Ariz.
5. **Gooding, K. M., and F. E. Regnier (ed.).** 1990. *Chromatographic Science Series*, vol. 51. *HPLC of Biological Macromolecules: Methods and Applications*. Marcel Dekker Inc., New York.
6. **Hames, B. D., and D. Rickwood (ed.).** 1990. *Gel Electrophoresis of Proteins: a Practical Approach*, 2nd ed. Oxford University Press, Oxford.
7. **Hancock, W. S. (ed.).** 1990. *High Performance Liquid Chromatography in Biotechnology*. John Wiley & Sons, New York.
8. **Havel, H. A.** 1996. *Spectroscopic Methods for Determining Protein Structure in Solution*. VCH Publishers, New York.
9. **Horváth, C., and L. S. Ettre.** 1993. *ACS Symposium Series*, vol. 529. *Chromatography in Biotechnology*. American Chemical Society, Washington, D.C.
10. ***Janson, J.-C., and L. Ryden (ed.).** 1998. *Protein Purification: High Resolution Methods and Applications*, 2nd ed. VCH Publishers, New York.
11. **Karger, B. L., and W. S. Hancock (ed.).** 1996. *Methods in Enzymology*, vol. 270. *High Resolution Separation and Analysis of Biological Macromolecules, Part A: Fundamentals*. Academic Press, New York.
12. **Karger, B. L., and W. S. Hancock (ed.).** 1996. *Methods in Enzymology*, vol. 271. *High Resolution Separation and Analysis of Biological Macromolecules, Part B: Applications*. Academic Press, New York.
13. **Kellner, R., F. Lottspeich, and H. E. Meyer (ed.).** 1994. *Microcharacterization of Proteins*. VCH Publishers, New York.
14. **Lennarz, W. J., and G. W. Hart (ed.).** 1994. *Methods in Enzymology*, vol. 230. *Guide to Techniques in Glycobiology*. Academic Press, New York.
15. **Mantsch, H. H., and D. Chapman (ed.).** 1996. *Infrared Spectroscopy of Biomolecules*. Wiley-Liss Inc., New York.
16. **Marshak, D. R. (ed.).** 1997. *Techniques in Protein Chemistry VIII*, Academic Press, New York.
17. **McGregor, W. C. (ed.).** 1986. *Membrane Separations in Biotechnology*. Marcel Dekker Inc., New York.
18. **Östlund, C.** 1986. Large-scale purification of monoclonal antibodies. *Trends. Biotechnol.* November, p. 288–293.
19. ***Pharmacia, Inc.** 1990. *Gel Filtration, Principles and Methods*, 6th ed. Pharmacia, San Francisco.
20. ***Pharmacia, Inc.** 1990. *Ion Exchange Chromatography, Principles and Methods*. Pharmacia, San Francisco.
21. ***Pharmacia, Inc.** 1993. *Hydrophobic Interaction Chromatography, Principles and Methods*. Pharmacia, San Francisco.
22. **Reznikoff, W., and L. Gold (ed.).** 1986. *Maximizing Gene Expression*. Butterworths, Boston.
23. **Schallinger, L. E., and L. A. Kaminski.** 1985. Sedimentation field-flow fractionation: a promising new bioseparations technique. *BioTechniques* March/April, p. 124–135.
24. **Schuster, T., and T. Laue (ed.).** 1994. *Modern Analytical Ultracentrifugation*. Birkhauser, Boston.
25. ***Scopes, R. K.** 1994. *Protein Purification: Principles and Practice*. 3rd ed., Springer Verlag, New York.
26. **Shively, J. E. (ed.).** 1986. *Methods of Protein Microcharacterization: A Practical Handbook*. Humana Press, Clifton, N.J.
27. **Snyder, A. P. (ed.).** 1996. *ACS Symposium Series*, vol. 619. *Biochemical and Biotechnological Applications of Electrospray Ionization Mass Spectrometry*. American Chemical Society, Washington, D.C.
28. **Sofer, G. K., and L.-E. Nyström.** 1991. *Process Chromatography: A Guide to Validation*, Academic Press, New York.
29. **Street, G. (ed.).** 1994. *Highly Selective Separations in Biotechnology*. Blackie Academic & Professional, New York.
30. **Sumner, J. B.** 1926. The isolation and crystallization of the enzyme urease. *J. Biol. Chem.* **69:**435–441.
31. **Vieth, W. R.** 1988. *Membrane Systems: Analysis and Design. Applications in Biotechnology, Biomedicine and Polymer Science*. Hanser Publishers, New York.
32. ***Vydac, Inc.** *The Handbook of Analysis and Purification of Peptides and Proteins by Reverse Phase HPLC*. Vydac, Hesperia, Calif.
33. **Weinberger, R.** 1993. *Practical Capillary Electrophoresis*, Academic Press, New York.

Economics

HAROLD B. REISMAN

23

Recently, scientists analyzed a yellowish residue on a potsherd found some 20 years ago at Hajji Firuz Tepe in modern Iran (3). Analyses indicated that the vessel held wine; dating of the remains was 5400 to 5000 BC. This means that evidence of wine making (and storage) has been found dating to the late Stone Age, some 2,000 years earlier than previous findings. While biotechnology can be traced to the time of the earliest permanent human settlements, no evidence relating to cost of production has been unearthed. As biotechnology has advanced, economics has become a more important factor. It seems that the more advanced the bioactive entity, the greater the problem in paying for discovery, scale-up, facilities, regulatory compliance, and overall investment. In recent years, reimbursement planning has evolved as a novel, highly complex, yet necessary adjunct to biotechnology economics.

In recent years, cost pressures have been exerted earlier and earlier in the development cycle. No matter how great or functional the drug or device discovery, someone must pay for its use. That "someone" has evolved to a few third parties (some say only one). The complexity of the endeavor is heightened by the immediate need to efficiently produce the drug or device. Cost of goods has always been of importance, but only in recent years has it become another key factor in the earliest stages of the project cycle. Pressures on selling price were exerted by competitors; that pressure remains, but perhaps greater pressure is exerted now by third-party payers. As pressure on selling price mounts, pressure on all costs mounts as well. Understanding and controlling the cost picture will very often spell success or failure independently of the value, efficacy, and quality of the bioproduct.

23.1. SCOPE

Economics in biotechnology would cover these areas:

1. Research and development (R&D) costs (including initial funding for a start-up company)
2. Process development costs (including pilot operations)
3. Plant and total investment and cost of goods
4. Clinical and regulatory costs
5. Reimbursement issues (third-party payers)
6. Return on investment

While there will be some mention of all these areas, each one could be the subject of a chapter (or book). The main emphasis in this chapter will be on the second and third headings listed above.

23.2. DISCOVERY AND PROCESS DEVELOPMENT

Just about a decade ago (mid-1980s), a cost estimate for new drug development, including regulatory clearance, gave a range of \$50 to \$75 million. Just recently, the cost range for discovery and development has been estimated at \$350 to \$500 million per drug (37). This is a compounded growth rate of approximately 19%. This is a rather remarkable value and may lead to a suspension of belief. However, there are other, related values that tend to support such a steep "growth curve."

Data have been presented on R&D spending for a number of pharmaceutical firms from 1985 to 1995 (2). For three of these firms, the compounded rate of growth of research has been calculated: Pfizer, 17.5%; Merck, 12%; and Schering-Plough, 14.1%. Since it is likely that not all increased clinical and regulatory spending is included in these figures, the rapid growth in drug discovery and development cost is far more reasonable. "Where will it end?" is an interesting philosophic question; of more immediate interest is how money can be invested more wisely today. That is, how can discovery and process development dollars be expended more efficiently? Furthermore, how can overall investment be controlled while concurrently shortening time to market?

One must begin with a research finding of some entity (whether drug, biologic, or device) coupled with some indication that the entity is reasonably functional and reasonably safe. At a very early stage, little or no human experimentation has been performed. Process development can begin at any time, but in the usual case there must be some incentive for a planned expenditure of significant funds. Process development begins with a laboratory (or, at best, a small-scale) process for preparation of the material. In what follows, the word drug or compound will be used; this in no way means that the discussion is limited to that strict term. The entity of interest could be a natural product mixture, a monoclonal antibody, an isotope, a blood derivative, a medical device, and so forth. A single word is used for simplicity to define some active entity or apparatus.

Once larger quantities of the drug are required, questions arise as to (i) timely production, (ii) method(s) for production, (iii) cost of equipment and auxiliaries needed, and (iv) total cost of production. All these questions must be addressed within a fairly well defined spectrum of regulatory constraints. The method of production or manufacture must be validatable, must give stable and predictable performance (especially that relating to rate, yield, efficiency, and purity), and must be cost-effective. The general subject headings for process development are

- raw materials/reagents
- strain isolation
- strain modification
- strain maintenance
- media optimization
- catalyst selection
- process optimization
- stage-wise stability
- materials of construction
- unit operation selection/sequence
- purification methods
- waste streams
- waste treatment
- degree of process isolation
- dosage (or device) form
- packaging/storage conditions
- product stability
- analytical procedures (test methods)

Obviously, the needs of a cell culture process are different from those of an organic synthesis; appropriate subject headings depend upon the specific process or product under study. Throughout the development stage (which, in rare cases, may last for months but more often lasts for years), there is a set of guidelines that govern the effort. Perfection is not always achieved, but the final selected process should fulfill all regulatory requirements, give consistent product, be safe, result in no environmental problems, and have product costs falling within a predetermined range. As part of the process development package, a process flow sheet should be developed with inputs and outputs (step and overall yield), and an economic model should be given (with capital and operating cost estimates).

This is a formidable picture made even more complex by regulatory constraints. Once a process is selected to prepare clinical material, significant changes to the process become costly and cause unwanted delays. A significant change in operation that might improve yield and save reagent costs would require extensive checks (and revalidation) before regulatory bodies would be convinced that the final product had not been altered. Therefore, any procedure that saves time and money in process selection and optimization early on would be welcomed. Such techniques have been in use in the chemical industry for many years; they have been applied to biotechnology processes in recent years. The generic title for the procedure is computer-aided process design. A chapter by Petrides et al. elsewhere in this volume (chapter 24) covers this important area.

23.3. GENERAL REFERENCES

There are a number of general references that should be consulted for basic economic concepts and for information on costing, including pertinent factors to use. (These references will be reviewed in a later section.) A chemical engineering handbook has a section on process economics (26). Investment and profitability are reviewed, and alternative methods and examples are given. While not specific for biotechnology, the detail for each cost determinant is very useful in avoiding errors due to an incomplete picture of costs. The classic work on plant design and costing, by Peters and Timmerhaus, is now in its fourth edition (40). While the coverage involves all chemical plants, the chapters on cost estimation and on profitability are clear and inclusive. Another handbook has an entire section on biotechnology economics (4). A book on the project cycle describes costs at various stages of development and during plant construction (45). The entire book relates to biotechnology; the example employed throughout is that for the production of citric acid.

The first edition of this handbook contains a chapter on cost estimation for biotechnology projects (27). Other economic aspects of biotechnology (such as evaluation of product development and health technology assessment) are reviewed in another handbook (7). In a review article on biotechnology economics, I discuss contract manufacturing, clinical, regulatory, and validation costs, as well as return on investment and reimbursement issues (47). Yet another review article covers problems encountered in the scale-up of biotechnology processes (46). Project planning and potential difficulties are emphasized, and a project planning checklist is given. The checklist covers items to consider from inception of a scale-up program through selection and monitoring of a contractor, up to receipt of as-built drawings. A detailed question list is also given; this can be used as a guide for in-house production or for use with a contract manufacturer.

The entire subject of process development is extremely well covered in a concise manner by Vogel (56). The same encyclopedia contains an excellent summary of plant design and construction issues (34). The detail given should be understood by any owner contemplating plant construction.

One of the best and most up-to-date references for commercial biotechnology (product, process, marketing) is a lengthy two-part review by Prokop (41). It covers discovery, moves through development and scale-up, and combines aspects of the business with the technology of biotechnology. This article is very valuable for its insights, descriptions of alternatives, explanations of needs at each step of the project cycle, and extensive coverage of the literature. A chapter on fermentation economics is included in a handbook (24); 14 process flow sheets are given in an appendix. Another textbook on fermentation technology includes a chapter on economics (52). A recent book is useful for modern approaches (and interesting case studies) to project management (5). An important section covers manufacturing economic analysis, reviews sources of capital, and lists issues involved in contract management; another section covers commercial software packages for project management and criteria for selection. Two books are noteworthy for covering many aspects of both business and science that are specifically related to biotechnology (35, 36). The spectrum of economic matters (including fund-raising for a start-up) is covered.

Finally, there are a number of publications (usually published annually) that list names and locations of biotechnology companies and organizations that supply specific information, assistance, or even contract facilities. One such listing (6) gives consultants, contract manufacturing organizations, contract research companies, engineering and construction companies, and others who might assist in pilot plant design, scale-up, plant design, validation, regulatory assistance, and technology transfer. Another such directory (10) gives international companies, universities, government departments, institutes, and societies, as well as a buyer's guide to products and services; the focus is biotechnology.

23.4. CAPITAL COST

Once the process flow diagram exists and some throughput is assumed, an estimate of capital cost can be made. This is only one aspect of investment required. Total capital investment consists of fixed capital and working capital. Fixed capital includes not only equipment cost but also design cost, construction/installation cost, and start-up cost. None of these individual elements is insignificant; furthermore, regulatory and validation costs must be included as significant factors under start-up cost. Depreciation is normally determined from fixed capital (excluding land). There is some discussion as to what part of start-up and regulatory costs can be included to calculate depreciation; these determinations are probably best left to legal/financial personnel with appropriate input from operations. Once the corporate "logic" is established, depreciation can be set. It should be remembered that the *total* dollar expenditures *prior* to any product being made must be estimated and included in required funds. Working capital refers to inventory, goods, supplies, intermediates, and any other dollar expenditure needed to keep the facility in operation. For biotechnology plants, metrology and ongoing calibration and testing (including quality assurance and quality control) must be included, as salable product may not be made for some time.

The determination of capital cost is important, because many other key cost elements are factored from this value. The software programs mentioned earlier can be used, or a more laborious item-by-item costing can be done manually. Calling individual vendors probably results in the best and most up-to-date equipment estimates, but this is very time-consuming. It would likely be more efficient to consult general reviews or publications that list equipment costs. It is possible to use capital cost indices to correct for inflation (from the date of publication of the original cost). Even after the cost of a capital item has been established, one must add transportation, erection, and installation (and taxes where applicable). Factors or percentages are used to establish "installed" cost. One example of the use of factors to determine fixed capital cost is given in Table 1. Note that total capital investment is some 7.5 times higher than delivered equipment cost. Even if the low end of each range is used, total capital investment is about 5.7 times higher than delivered equipment cost. Such factored cost estimates depend almost totally on accurate values for equipment cost and installation cost for each class of equipment. The value of computer-aided design (with routinely updated costing information in the appropriate module) should now be readily apparent.

TABLE 1 Fixed capital cost estimate[a]

Direct costs	
Delivered equipment (DE)	100
Equipment installation (% of DE)	35 to 45
Process piping (% of DE)	40 to 60
Installed instrumentation (% of DE)	20 to 40
Installed insulation (% of DE)	3 to 5
Buildings (% of DE)	20 to 40
Installed electrical (% of DE)	10 to 12
Yard improvements (% of DE)	10 to 14
Auxiliaries (% of DE)	40 to 60
Land, if purchased (% of DE)	3 to 5
Subtotal: direct plant cost (DPC)	331
Indirect costs (IPC)	
Engineering/supervision (20–30% of DPC)	83
Construction (30–40% of DPC)	116
Subtotal: DPC + IPC	530
Contractor's fee (negotiable; use 7.5% of DPC + IPC)	40
Contingency (use 10% of DPC + IPC)	53
Total fixed capital investment (TFC)	623
Working capital (15 to 25% of TFC)	125
Total capital investment	748

[a]All subtotals and totals assume use of the midpoint of each range shown.

Scaling factors may be used if cost information is available for a piece of equipment and if a cost is desired for a similar device having another size or capacity (a rule of thumb is not to exceed a 10-fold change in size up or down). If a scaling factor is unknown, then the six-tenths rule is applied. This is represented by: cost of X = cost of A × (capacity of X/capacity of A)$^{0.6}$, where information for A is known. This procedure is prone to error, as scaling factors vary widely around 0.6. Some examples of actual exponents for equipment cost versus capacity are available (40), and exponents for biotechnology-oriented equipment are also available (48). It is very important that the similarity between two pieces of equipment goes beyond mere functionality. Material of construction must be the same (the cost difference between 304 and 316 stainless steel [SS] plate may be as high as 40%), type of construction (including finish) must be evaluated, and pressure and temperature ratings must be considered. Other factors (such as ease of installation and types of required alarms and controls) are involved when factored estimates are used.

As noted, historical cost data must be corrected for inflation. Cost data from dated estimates (or actual past purchases) or from publications may be converted to estimated current cost by means of a cost index (Table 2). This historical cost is multiplied by the ratio of the present cost index divided by the index applicable at the earlier date. One index cannot cover all costs, since prices or rates change in differing manners. Major elements (such as labor cost, construction cost, equipment cost) should be indexed individually. Using an index for utilities' cost and raw materials' cost is possible; a far better alternative would be to use current published information or to call appropriate sellers. Geographic differences (even within the United States) should be considered. Obtaining costs overseas

TABLE 2 Selected cost indices

Year	Chemical engineering plant cost index (1957–1959 = 100)	Marshall & Swift equipment cost index (1926 = 100, all industry)
1990	357.6	915.1
1991	361.3	930.6
1992	358.2	943.1
1993	359.2	964.2
1994	368.1	993.4
1995	381.1	1,027.5
1996	381.7	1,039.2
1997	386.5	1,056.8
1998	386.8 (May)	1,061.8 (second quarter)

requires additional effort, but indices exist for these calculations also. One of the major causes of underestimation is not including all equipment and associated utilities. There are specific requirements for ductwork and piping in biotechnology plants; in addition, the need for cleaning-in-place (CIP) or sterilization-in-place (SIP) is often neglected or minimized. The complexities added to the design of a facility (aside from the cleaning apparatus and piping itself) are noteworthy (8). Special handling, special surface treatments, and special construction protocols are only some of the factors that add to cost. Even if one remembers to include a necessary water-for-injection (WFI) generator (a costly piece of equipment), every use point (or "drop") is expensive. The need for special controls, special valves, costly heat exchangers, and special installation and testing methods is often neglected or minimized. The danger of underestimation is one reason to use employees, consultants, or engineering firms that have experience in these areas. The need for back-up equipment (redundancy) must be recognized and provisions made; a single equipment failure should not cause a complete plant shutdown.

A complete picture for fixed capital cost would include a host of factors (26, 34). While not every item can be included in a preliminary estimate, existence of all the inputs must be kept in mind and quantified as the estimation process continues. Even if a quantity is carried as an unknown, it will not be forgotten or neglected and result in a later underestimation.

Whether or not an in-house estimate has been prepared, it is usually necessary to use a contractor or construction manager to prepare final estimates. Even large companies with extensive in-house engineering expertise will not be fully staffed to handle all new construction tasks. Construction managers usually have their own staff of estimators and purchasing personnel and appropriate contacts with proper vendors. The information needed for appropriate final costing is listed in Table 3. Indirect costs will be included by the construction manager (note that these costs are incurred even if all work is performed by in-house staff). These costs may include those for detailed engineering, supervision, contractor's fees, construction-related costs, insurance, and permits.

One of the major factors in achieving success in implementation of a bioprocess is selection of the construction manager or general contractor. This is not a straightforward decision, and there is no quantitative method to choose the best contractor. Lowest cost is not the key criterion

TABLE 3 Information required for final plant cost estimate

Plant site	
Process buildings	Site plan
Nonprocess buildings	Prior-use review
Lighting	Soil samples
Erosion control	Transportation (roads, railways)
Equipment and process definition	
Process flow diagram	Piping and instrumentation diagram
Equipment list (including finishes)	Interconnections
Utilities (including standby)	Waste handling/treatment/connections
Material and energy balances	HVAC system (air-handling units)
Underground piping or conduit	Electrical system
Lighting	Motor control centers
Type of controls	Materials transport
Floor plan	Critical room layouts
Special finishes (rooms, floors)	Welding procedures
Allocated facilities	
Administration	Process management information system
Laboratories	R&D and technical services
Maintenance	Hazardous storage
Sanitary plumbing	Lockers/change areas/personnel flow
Other	
Expandability	Containment strategy
Environmental planning	Permits
cGMP review	Hazard analysis review
Code review	Long delivery items
Plan for validation and start-up	

(but neither is highest cost). What qualifications should the contractor have? What questions should be posed during interviews? Table 4 gives a short list in an attempt to aid any owner in selecting a contractor. Of some importance is the discussion of "value engineering." It is rather common to have the final cost estimate fall well outside the owner's earlier estimate of overall cost. It is never too early to broach the subject to value engineering, i.e., examples of more economic alternatives without losing overall functionality. All key personnel from the owner's side should be involved in selection of the contractor. An optimum choice at this stage will be of inestimable aid in achieving timely and economic goals. The detailed estimate (meaning the estimate of the contractor) is based upon complete specifications and drawings and is the "tightest" estimate; its accuracy is normally within 5% of the final cost.

Working capital is normally given as 15 to 25% of total fixed capital investment. As noted earlier, working capital

TABLE 4 Qualifications of construction manager/engineering firm

Prior experience	
Dollar value installed	Specific area (bulk chemical, medical product, etc.)
Firms where jobs completed	Type of installation (including aseptic processing)
Comparable experience	Radioactive compounds
Degree of isolation (containment)	Controlled substances
Résumés of those involved	Computer-aided design (specific programs used)
GMP compliance	
Validation expertise	Résumés of persons who will direct effort
Check with FDA (under Freedom of Information)	Examples of validation master plans
Inspection history	Calibration and metrology capabilities
Software and programmable logic controller validation	
Construction management	
Personnel (including turnover)	Change-order control (and history of comparable tasks)
Construction manager background	Photography and videotaping capability
Process engineering staff	Reporting (how often? where? how? who?)
Economic/technical evaluation	Value engineering and examples
Laboratories, quality control facilities	Pilot and full-scale plants
Scheduling (how?)	Publications and awards (comparable tasks)
Organization (especially of the group responsible for the task to be performed)	
Utilities	
CIP/SIP design and installation	Sterilization methods and validation
Instrumentation/degree of automation	Decontamination
Gases (systems installed)	Installed WFI and purified water systems (plus histories and recommendations)
Specific needs (lyophilization, room pressurization, barrier technology)	
Finished forms (packaging)	
Installed aseptic systems	Transdermals, time release, implantable drugs
Fill and seal	Tablets, capsules, powders
Historical records	
Cost control	
Detail of design/cost estimation	Retainage
Billing cycle and job cost tracking	Bid analysis
Recommended vendors	Degree of involvement by owner
Start-up	
Schedules and practices	Tests by contractor
Punch-list development	Coordination with owner
Examples	
Construction schedules	Drawings (including isometrics and as-builts)
FDA findings on inspections	Sterility records
Disqualified vendors (reasons?)	"Best" practices, especially for automation and control
Savings due to value engineering	Adherence to schedule (comparable-dollar-value task)
Validation protocols	Contracts

costs cover a host of individual items. Money must be available for the following items: raw materials for plant start-up, supplies for start-up, all in-process and finished goods inventories, inventory control, accounts receivable (this is essentially credit to customers), accounts payable, and unforeseen emergencies. Not every plant start-up results in acceptable product; delays and problems may be more severe for biotechnology plants. If a novel bioprocess is involved, delays should probably be programmed into the plan. Provision for needed funds must be made. It is also true that working capital needs vary with production rate. Raw material inventory may increase, as will in-process and finished goods inventory if there is a step increase in output. It is likely that accounts receivable will rise. These issues must be taken into account during start-up and during the planning for significant throughput changes.

Often forgotten or underestimated are start-up costs. Costs of entering or expanding a business will not be included here. Start-up costs include:

1. Expenses due to process equipment changes after construction is complete. Changes in project scope are not included here. Examples of expenses that should be included: additional electrical or control lines or connections to improve operability, addition of water or air drops not planned, alarms not included originally, modifications needed to improve regulatory compliance.
2. Labor costs incurred after construction, especially those related to equipment functionality. Validation costs should be included; unless installation qualification, operational qualification, and performance qualification are complete, the process will not be approved. (A separate budget for validation is a suggested alternative. Both equipment and process validation must be completed.) In many cases, validation tests point to the need for equipment modification. A common example is that, very often, seals or diaphragms are found to be torn or malfunctioning during start-up; costs are incurred (both labor and materials) to correct the problems.
3. Costs incurred during the start-up period other than "normal" operating expenses. Examples include special cleaning agents to remove film or debris related to construction and costs incurred for water batching.
4. Training costs. There are requirements for qualification and training in a biotechnology facility. Obvious examples are gowning, sterile sampling, and data recording (requiring multiple signatures in correct sequences).
5. All R&D and other departmental costs incurred during the start-up. Management information systems and laboratory and quality assurance personnel not normally involved in day-to-day operations will probably be heavily involved during the start-up period. Computer systems and software rarely operate "normally" from the onset of the process.

It is important that normal operations be defined well before the start-up begins. If an outside contractor is involved, that firm will strive for completion as soon as the last bolt is tightened. While an infinite period is not suitable for determination of completion, a mutually agreed upon definition is essential to protect both owner and contractor. Such a definition will also aid in setting the start-up period. The definition might include a certain rate of production, a certain yield of acceptable product, preset sterility conditions, days of continuous operation at defined levels, or any combination of these factors.

Formulas exist for estimating start-up cost as a function of direct capital cost (battery-limits capital). "Inside battery-limits" refers to all costs related to manufacturing equipment but usually excludes costs related to administrative offices, warehousing, utilities, and other auxiliaries. Those correlations are more useful for large-volume chemical plants or plants similar to those built at an earlier time. Start-up costs may be roughly estimated at between 2 and 20% of battery-limits fixed capital cost; for biotechnology plants, the value near the high end of the range should be used. For novel processes, a range of 15 to 20% has been given (34). Much of the variation is due to a qualitative determination of the novelty of the process and the technology. For example, if a totally new process involving a new biosynthesis coupled with a chemical step using a new catalyst is followed by a purification in a novel chromatography operation, one can expect that both start-up cost and time will be higher than similar values for a start-up of a penicillin plant using conventional technology (as a percentage of fixed capital). Very often, start-up difficulties are neglected and the pressure for usable product often results in more problems and more delay. It should be recognized that start-up is a difficult and psychologically trying period even under the best of circumstances. Careful and reasoned planning (and budgeting for both money and personnel) is essential. Allowances for downtime should be built into the plan. Lines of authority and approval levels must be established early on, and alternatives should be identified for both needed work on-the-floor and immediate management decisions when circumstances warrant. There are many instances where a biotechnology plant or process did not work or took years to come on-line; the circumstances are normally not published, and the stories are relayed by word-of-mouth with its concurrent exaggeration. In general, difficulties can be traced to one or more of the following: overoptimism, use of unproved technology, poor relations between owner and contractor, and neglect of key process steps or in selection of process equipment. Start-up planning may be a very late stage for recognition of such problems, but it is not too late. Failure and high economic loss are not foreordained at this point, and there is still time for problem solving and the needed additional work. This may be costly and even embarrassing, but the alternative, i.e., continuing into a very problematic and uncertain start-up, would probably be even more costly.

There is insufficient space here to detail all the factors involved in cost control for the construction project. However, the subject is important and should be noted. There are specific and detailed concerns that must be addressed in design and construction of biotechnology manufacturing facilities (39). Certain considerations applied in design of a gene therapy manufacturing facility can be extended to other biotechnology projects (54). The design approach taken and team members included are important (51). Last, but not least, is the understanding that change orders will occur. Change in design after drawings are complete is costly, but that is dwarfed by the cost of changes made in the course of construction. The issue is not to strive for elimination of change orders, but to work toward managing and controlling change and its concurrent cost impact (11).

Problems in time and cost to completion are significant in their own right and may be indicative of problems to come. Once funds are authorized and a decision is made on how the construction is to be managed (in-house or outside contractor), the schedule should be set. Presumably, tentative schedules and approximate delivery times have already been reviewed; a definitive schedule must now be set. For biotechnology projects, much of the equipment has long lead times. For example, it is highly unlikely that an SIP continuous centrifuge or a production-scale lyophilizer will be available from vendor inventory. Delivery of a double-door, stainless-steel sterilizer may take 4 to 6 months. If it is feasible, early orders (with defined cancellation clauses) should be placed. A complete schedule will indicate the critical node(s), and appropriate alternatives may be selected so that construction can proceed smoothly and logically. With the schedule and involvement of purchasing personnel, a standard code of accounts should be established. An expenditure plan can be set and approval methods for authorizing payments established. For biotechnology projects, methods of shipping and storage of components must assume a heightened awareness. Piping and ductwork may require caps or protection, special inspection on receipt, and special storage prior to installation. This is only one simple example.

On a preset schedule (semimonthly or monthly), degree of commitment and degree of completion must be known for each of the standard accounts. Craft, nonmanual, and supervisory labor must be known. The owner's representative(s) would be present for some, or all, of the time during construction but must be present at routine meetings with the construction manager and all involved trades. The reporting system (including distribution of meeting minutes) should have been established. Problems of delivery, field conditions, and overruns should become obvious during the weekly or biweekly meetings. One should not await written minutes to initiate remedial actions as needed. Examples are authorization of overtime, cancellation of one item and replacement with an acceptable alternative, and correction of a safety hazard. Attention to detail and the owner's intimate participation will ensure, at the least, that cost consciousness, efficiency, and adherence to regulatory and safety requirements pervade the site. Any overrun must be recognized and flagged. It is essential that proper documentation be prepared and as-built drawings be created in a timely fashion. Manuals and drawings should be controlled and filed. Documentation for installation qualification, operational qualification, and performance qualification must be completed and filed. Training documentation must be available. At all stages, actual completion and expenditure should be compared with the plan. Validation planning and execution are key elements of the plan and have the potential for disruption if responsibility and timing are not clearly delineated.

An estimate is available that gives percentage ranges for various cost elements for a biopharmaceutical facility (51). These values, given in Table 5, are only estimates derived from historical data. They should be used with caution in comparing various options under study for a new plant. The categories are useful for estimation purposes and to be certain nothing is left out. The same reference has interesting data on four typical cell culture facilities completed in recent years (1990 to 1995); the term "typical" refers to both pilot plants and small production plants.

Table 6 gives cost data for four different facilities, and the costs are fairly inclusive; the range of \$500 to \$800/ft^2 can be used for comparative purposes or to quickly prepare a very early cost projection. The International Society for Pharmaceutical Engineering (ISPE, 3816 W. Linebaugh Ave., Suite 412, Tampa, FL 33624), in conjunction with the Food and Drug Administration (FDA), has developed a series of facilities engineering guides. The goal is to enable manufacturers to reduce cost while maintaining regulatory standards and product quality.

TABLE 5 Initial cost ranges for a biopharmaceutical facility

Category	Percentage range	Major items
Pre-investment	0 to 10	Design for the future
Equipment costs	25 to 40	Production equipment
		Laboratory equipment
		Office furniture/fixtures
		Consumables
		Automation
		Information
		Communications
		Materials handling
Land costs	2 to 5	Acquisition
		Survey/geotechnology
		Environmental
Hard costs	35 to 60	Site development
		Utilities
		Building envelope
		Building systems
		Process services
Soft costs	10 to 20	Design
		Consultants
		Legal/finance
		Compliance
		In-house resources
		Construction

23.5. OPERATING COST

Operating, or often manufacturing, costs are divided into fixed and variable costs. Fixed costs are incurred regardless of volume of product output. (Another term sometimes used for fixed cost is indirect cost.) The clearest case of a fixed cost is depreciation. Even if a plant were shut down for a month, depreciation would be unaffected. Variable costs are those that are directly related to the volume of production. The clearest case of a variable cost would be

TABLE 6 Pilot plant cost for a typical cell culture facility[a]

Facility	Cost (\$)	Area (ft^2)	Cost/ft^2 (\$)
Cell culture 1	10,000,000	12,500	800
Cell culture 2	5,972,000	10,000	597
Cell culture 3	12,200,000	21,824	559
Cell culture 4	7,000,000	13,400	522

[a]Assumptions: GMP pilot or production launch facilities; build out of existing shell (no site or shell costs included); ranges as given above; mix of production and support spaces; costs include both hard and soft costs; costs reflect both east and west coasts of the United States; costs include validation, external engineering, facilities, and equipment; time period: 1990 to 1995.

that for raw materials. There would be no use of raw materials during a plant shutdown. Unfortunately, not all costs are simply fixed or variable. Some element of choice or interpretation is needed for appropriate costing. "Appropriate," in the sense meant here, refers to an estimated cost that is closest to the real cost of goods.

Many operating costs have a fixed and a variable element; i.e., such costs can best be expressed by a formula with a constant and a variable component that is related to output. A few examples will suffice. Even during a plant shutdown, there will be a need for electricity (or almost any utility). This load may be a small percentage of full-output load but is usually significant, in any case. During a shutdown (even without a major maintenance overhaul), there is also usually an expenditure for maintenance labor and maintenance materials. Even if a cleanroom is used for production purposes for 3 h per day and fan speed is reduced for the other 21 h, there is significant power requirement for no product output. In these and other cases, some judgment is needed to construct a cost estimate that reflects both fixed and variable elements. In some of the literature, terms such as semi-variable or semi-fixed are used to reflect this complexity. A meaningful operating cost estimate must include these considerations. Furthermore, cost estimates are prepared for one output (or a limited range of outputs). Serious error results if the output is doubled or tripled and costs are factored up or down. A major change in throughput might mean that an additional shift or many more operators would be required. That is, once some output threshold is reached, there is a step change in some cost element (it is often labor and/or supervision). The first additional unit of output over this threshold is rather costly. It is good practice to completely redo the cost of goods estimate for significant changes in output. Certain warnings should be discussed. For biotechnology processes, one comes across one or more of these items: membranes, ion-exchange resins, recoverable solvents, chromatography gels, catalysts, carbon beds. It must be recognized that while each of these (and related) items has a relatively long useful life, the life is not infinite. Solvent recovery is never 100% efficient, and membranes, while cleanable, do not last forever. A reasonable estimate of useful life must be established, and cost of replacement must be included in a cost determination. This is in addition to ongoing cost of cleaning or regeneration. Vendors can usually give a useful life estimate based on their experiences. Assuming too short a life cycle is preferable to assuming too long a life cycle. The reason is obvious, even beyond the fact that catastrophic losses have been known to occur.

There are also certain indirect costs that should be added. These are often costs incurred by a manufacturing site from other departments within the corporation. Examples are quality assurance, routine analyses (central laboratory, such as nuclear magnetic resonance, mass spectrometry), computer services, sales and marketing, corporate technical services, and R&D. It is fair to say that many of these internal transfer costs are not very well accepted by the recipient, and discussion concerning quality, quantity, and cost of the specific effort is ongoing. Cost allocation within a corporation is beyond the scope of this chapter, but the problem (and the cost impact) should be recognized. To be fair, even in a multiproduct plant, allocation of overhead costs within a strictly limited geographic space and within one department causes much heated discussion. Such costs can be allocated by head count, by floor area, or by strict cost detailing and projection. There is a very good discussion of transfer pricing from an accounting perspective (17); the matter is neither trivial nor straightforward and deserves appropriate consideration. The point is that, for all allocations noted above, all costs must be included to develop a reasonable and useful cost of goods. If the cost of goods is inaccurate, then the profit margin is inaccurate. All return on investment (or return on assets or equity) values are similarly inaccurate. The unfortunate circumstance is that, in a multiproduct plant, certain products might be underburdened while others are overburdened. Future investment decisions will be made in error and compound the existing problems. A list of factors used in determination of operating costs is given in Table 7.

In any manufacturing operation, certain general expenses are incurred. These may be allocated (see above), added directly to cost of goods, or charged elsewhere. The point here is that these costs cannot be neglected or forgotten. Certain examples have been noted; other such costs are administration, marketing, sales and distribution, and financing expenses. For a novel biotechnology application (a new drug moiety), legal fees for intellectual property protection may be quite high initially. General reviews on the importance of patent protection to the pharmaceutical industry are available (36, 37). As if the rapidly changing

TABLE 7 Operating cost determinants[a]

Fixed cost (major element)	
Salaries and benefits	Maintenance
Clerical help	Operating supplies
Labor and benefits	Outside services (including contracts)
Engineering	Depreciation
Fixed utilities	Taxes and insurance
Uniforms	Purchasing
Office supplies	Revalidation
Postage/courier	Administration
Rentals and leases	Computer services
Safety/fire protection	Permits and fees
Cafeteria/locker rooms	Training
Plant overhead	Communications
Variable cost (major element)	
Raw materials	Operating supplies
Maintenance materials	Cleaning materials
Utilities (including waste treatment)	Packaging
Water	Royalties
Demurrage	Specialty gases
Quality control	Catalysts/reagents
Solvents	
Relevant utility systems[b]	
Cooling water	Steam
Process water	Condensate return
Well water	Refrigeration
Sanitary piping	Electricity
Air (other gases)	Waste treatment
Vacuum	Inert gas
Fuel	

[a]Include all chemical and treatment requirements.
[b]Utilities' costs usually have both fixed and variable components.

field of patent coverage in biotechnology in the United States were not difficult enough, most biotechnology companies must be concerned with worldwide patent law that is anything but static (49). Isolation and structural determination of parts of the genome have brought forth a novel series of issues relating to intellectual property (16). The point here is that the cost of obtaining appropriate worldwide coverage is high, ongoing maintenance is costly, and litigation to protect one's rights is probably the highest of all such related expenses. Quality assurance-related costs, including the cost of preparing required reports and documentation (such as a pre-market-approval application), may be high. For a company with a single product, no allocation is possible; these costs must be recovered through revenues derived from the single product.

Raw materials for biotechnology processes often involve complex materials that are not fully defined, natural products of varying composition, materials available from a few (or even a single) suppliers, materials of very high purity, substances that involve use-testing, labile substances, and so forth. Sometimes one or more of the raw materials in use have a number of these attributes. Provision must be made to counter any untoward result in testing. Cost of special shipping, receipt, and appropriate storage must be included. Provision must be made for complex use-testing. Not only are extra costs incurred, but problems in ordering lead time and storage of acceptable raw material must be overcome. Generally, for the novel biotechnology products, raw material cost (including reagents) is one of the most, if not the most, significant factors in cost of goods. Since extraction and purification costs make up the greatest percentage of cost of goods for the newer products, cost of reagents for those steps must be carefully considered and minimized. Needed steps for qualification of raw materials for clinical product manufacturing are given by Del Tito et al. (14). An example of raw material evaluation with estimates of product cost (for a microbial polymer) is given by Lee (29).

Labor cost is derived from (i) the process flow diagram and scheduling; (ii) analysis of unit processes and unit operations selected; (iii) number of shifts required and head count per shift; (iv) expected labor rate (plus fringe benefits) at the site; and (v) overall personnel count and necessary supervision per shift. Plant and equipment layout is often planned with a view toward ease of installation and need for maintenance. The same consideration should be given to operating labor efficiencies (span of coverage).

Consumption of utilities should be calculated from the process flow diagram, flow balances, and scheduling (for peak and average loads). Even with recovery systems and insulation, losses must be recognized. For biotechnology processes (where a single operation may continue for days or weeks or even longer), it is essential that redundancy be included. A power failure or fuel supply stoppage should not be catastrophic. Many facilities are equipped with an emergency generator and with steam boilers that can be fired with alternative fuels. This will affect not only equipment cost but also personnel, training, routine testing of standby equipment, and validation costs. Every geographic locale has specific costs for various fuels, for water, and for electricity. Engineering firms and contractors normally have these cost data available.

Maintenance and metrology costs are usually significant in a biotechnology facility. Process controls are complex, and contact surfaces (e.g., seals, diaphragms, control elements, gaskets) are made of expensive materials. Routine replacement based on time in use is common, as opposed to replacement on failure. Use of preventative maintenance procedures (such as vibration analysis, changes in pressure drop, specialized software programs) is common. Therefore, maintenance cost (labor, materials, supervision) is high relative to that of many other chemical processes. There is yet another area of facility planning and economics that receives too little attention. Very often, it is neglected altogether. The cost factor is related to, but is not the same as, maintenance. Any new building requires funds for facilities management. This cost segment includes the many elements that go toward keeping the facility in optimum running condition (9). Such costs for a plant producing a regulated biotechnology product are significant. The design of a plant and process will go a long way toward minimizing facilities management cost while allowing the responsible personnel to perform necessary tasks in a reasonable and timely fashion. The simple aspects of facilities management involve utilities, maintenance of surfaces, security, moving equipment, cleaning, materials flow, and warehousing, as well as changes to wiring and communications. With ever-changing regulations, which almost always add requirements, it is important to use historical data in an attempt to predict cost for these elements for maintenance of the facility.

Cost of goods may be estimated by using various factors and percentages that are derived from historical data and experience. Table 8 gives an overview of such estimation using various sources, including experience. These formulations are obviously not used to set final costing but are

TABLE 8 Cost of goods via factored estimate[a]

Direct costs	
Raw materials[b]	40–60% of product cost
Labor	10–20% of product cost
Supervision	10–25% of labor cost
Utilities[b]	10–20% of product cost
Payroll charges	20–30% of labor and supervision cost
Maintenance	2–10% of fixed capital
Laboratory	10–20% of labor cost
Operating supplies	0.5–1% of fixed capital
Waste treatment[b]	Actual calculations
Royalty, patents	1–5% of product cost
Contingencies	1–5% of product cost
Indirect costs	
Depreciation (11-year straight-line)	9% of fixed capital (2–3% of buildings)
Property taxes	1–3% of fixed capital
Insurance	0.5–1% of fixed capital
Interest	0–10% of fixed capital
Plant overhead	50–100% of labor and supervision
Other	
Packaging	5–30% of selling price (pharmaceuticals)
Distribution and selling	2–20% of product cost
Shipping	1–3% of product cost
R&D	5–15% of product cost

[a]It is assumed that there is no by-product or scrap value resulting from the process.

[b]Direct calculation is the preferred method (from unit usage, yield data, direct quantities, price lists, and current costs).

extremely useful in preliminary calculations and in determining which individual costs are reasonable, whether estimated fixed capital is reasonable, and whether a range for cost of production needs revision or modification. To give an example: a hypothetical fixed capital investment of $40 million will result in a facility that has an output of 1 million units (or devices or grams or doses) having an estimated or desired cost of $60/unit. Using the low end of the range, total direct cost is $40.60/unit, total indirect cost is $11.50/unit, and total "other" cost is $9.80/unit. Waste treatment cost is assumed at $1/unit. The total is $61.90/unit, or very close to the original estimate of $60/unit. If needed capital were $80 million, or if R&D cost were 3 times the value (15% of product cost versus 5%), further review and analysis would be needed. The factored estimate presents a goal, and sensitivity to necessary cost inputs becomes clearer. Furthermore, if the selling price were $500/unit, a greater elasticity in the cost picture would be allowed, compared with a selling price of $100/unit. A projected gross margin of 88% allows more freedom in outlay than a projected gross margin of 40%. It should be recognized that biotechnology companies strive for, and are more comfortable with, the 80 to 90% gross margin range.

23.6. CASE STUDIES

A number of general references related to capital and operating cost estimation have been given earlier. There are also reports and articles that give case studies that are useful for the biotechnologist involved in economic analyses. The case study by Kalk and Langlykke (27), which involves an intracellular protein made by a recombinant microorganism, was noted above. Unit cost came to $1,510/kg, and the capital investment was estimated at $13.6 million. While the conditions that resulted did not give an economically attractive process, the analysis was used to direct further research efforts that might drive the costs down. A book on the economics of bioprocesses (45), already cited, discusses citric acid as the process example. Unit cost (year 1) was $0.74/lb with total capital investment of $61.2 million; output was 40 million pounds per year. Individual equipment costs and profitability and sensitivity analyses are given. Datar (12) presented information on primary separation for an *Escherichia coli* synthesis of an intracellular enzyme. Capital and operating costs are given, as is a listing of off-sites and all assumptions made. Approaches for cost reduction are given. Datar is senior author on another excellent paper that gives a cost breakdown for two alternative processes for production of tissue plasminogen activator (13). The basis for design for each case is clearly given. Careful process descriptions are given with flow sheets. Only summaries of capital estimates are given; the total capital investment for the mammalian process was $61.4 million, whereas the bacterial process investment was far higher, at $389 million. The start-up cost estimate was set at 11% of fixed capital investment for each process. Unit production cost was $10,660/g for the mammalian process and $22,030/g for the bacterial process. Profitability studies and sensitivity analyses are given.

Recovery of an enzyme (polygalacturonase) from broth has been described with cost information (23). If 100 batches per year are run, a payout time of less than 1 year is claimed. Lonza workers present a cost example for a fine chemical, L-carnitine (25). Costs of raw materials, personnel, overhead, and depreciation are given. The purified product is a commercial material. A review is available for production of recombinant human insulin (28). Work by Petrides et al. utilizing computer-aided process design in insulin production is reviewed elsewhere in this volume (chapter 24).

A chapter by Haritatos in a book noted earlier (35) is entitled "Process Economics." While some biotechnology applications are noted, the prime example given, rather surprisingly, is that for a petrochemical. However, the stepwise detail and methodology are both clear and useful for anyone involved in plant design of any sort. A mathematically oriented article using protein configuration and refolding strategy is available (32). The importance of *global* optimization is emphasized; an example of maximum biosynthetic yield that leads to a suboptimal global solution is given. The formulation includes direct fixed capital, consumables, waste treatment, labor, utilities, and R&D. Sensitivity analyses and control charts are given. Swartz (53) presents an analysis of investment and costs for production of bulk penicillin. The review is based on a McKee and Co. report. Use of an agricultural product (erucic acid) as a starting material for production of nylon 1313 is studied (55). Material balance, cost of production, and sensitivity analyses are given.

There are limited published reports that cover economics of isolation and purification. Dwyer (15) reviews high-value protein and peptide purification. Equipment and scale-up are covered, and equipment and operating costs for a chromatographic purification process are given. Gerstner (22) compares effectiveness and economics of displacement and linear gradient chromatography. An oligonucleotide costing about $200/g to synthesize is to be purified; displacement chromatography is more cost-effective. Detailed assumptions and a sensitivity analysis are given. Costs are divided into reagents and buffers, production labor, quality control/analytical, and crude feed. Process water is costed at $1/liter. Anyone considering any type of process chromatography should review this article for its coverage of various cost inputs and assumptions for each step. For an up-to-date review of various chromatography purification methods in biotechnology applications, see Freitag and Horvath (19). There are 383 references listed. Limited economic data (but a good deal of useful laboratory and process data) are given in a review of aqueous two-phase extraction for downstream processing of proteins (42).

Production of purified water (PW) and water-for-injection (WFI) is of great importance in many biotechnology facilities. A few articles are now available on alternatives and costs. One such article covers six systems, each at three different capacities, and gives resulting capital and operating costs (18). The 5-year present-value system cost is given for each configuration. Another paper discusses ceramic membrane ultrafiltration that has been validated for production of water meeting current requirements for pyrogen-free WFI (44). Schematics and cost comparisons are given for current and suggested methods. Challenges to the ceramic membrane system (during validation) are described. Outputs of 300 gallons per hour (GPH) (pilot or small facility) and 3,000 GPH (a parenteral facility) are analyzed. It is stated that Merck & Co. (where the work was performed) produces over 80 million gallons of WFI annually on a worldwide basis. Savings related to substitution with ceramic membrane systems are calculated to be in the range of $10 to $15 million over the next 17 years.

A specific reference to biopharmaceutical process equipment is included here, even though a single process or utility is not described (48). In this important paper, exponential scaling factors are given for 58 different types and sizes of bioprocess equipment. The dangers of using a single scaling factor are emphasized. The data presented are valuable for more accurate capital cost estimation.

Table 9 is a summary of selected references discussed. Some articles list cost of individual items, some review overall costs, and others include profitability and sensitivity analyses. Some have combinations of information. There is sufficient published economic data to allow reasonable cost analysis and economic model building for modern biotechnology projects.

23.7. CONTRACT MANUFACTURING

One estimate for construction of a fully equipped biopharmaceutical facility is $20 to $60 million (31). The time estimate for the facility to become operational is 1 to 5 years. Another author gives the cost range for a biotechnology production plant at $40 to $75 million (50). Furthermore, the time frame for design/construction/validation of a licensed current Good Manufacturing Practices (cGMP) facility is estimated to be 30 to 36 months, with an additional 12 months for completion of the first conformance lot. Whatever the dollar values selected, the amount is high, and time to production of salable quantities is long. With the potential for either product failure during clinical trials or the appearance of a superior product in the interim, there is abundant stimulus for use of a contract manufacturer (which may also include contract process development). A listing of needed documentation in technology transfer as well as details to cover (issues and concerns) in selection of a contract manufacturer is available (47). Use of contract organizations is growing, and this should be a viable choice for both speeding time to market and saving corporate capital. This subject is merely noted here and is more fully developed elsewhere in this book.

23.8. RETURN ON INVESTMENT

A number of references will be given for readers interested in pursuing the subject in greater depth. Earlier, a number of specific processes were highlighted in which profitability and return were discussed in some depth. Table 9 lists some references that review these subjects. One book is remarkably free of jargon and presents definitions and concepts in a user-friendly manner (20). The book by Peters and Timmerhaus (40) has a chapter (somewhat more complex) on selection of alternative investments, project profitability, and replacement. One of the symposium series published by the American Institute of Chemical Engineers is devoted to investment appraisal (57). Biotechnology is not specifically highlighted. All the papers are written for the nonfinancial specialist, but some technical knowledge is presumed. One paper is really a review of many technical and financial texts; serious and persistent problems (actually, deficiencies in overall analysis) are noted. This is worth reviewing if for nothing else than to be sure the specific analysis being performed by the reader does not suffer from the same difficulties. Another paper in the book reviews the strengths and weaknesses of internal rate of return and net present value and suggests where each might best be applied. Yet another paper discusses variation in Lang factor for different sorts of plant modifications. (The Lang factor is one method of estimating total plant investment from equipment cost.) Unsurprisingly, pharmaceutical projects exhibit some of the highest factors for a solid-liquids plant. Once again it is shown that use of a constant factor for all projects can lead to underestimation for a biotechnology facility.

TABLE 9 Equipment costing and profitability/sensitivity analyses: selected references

Authors	Reference	Item or process	Individual equipment costs	Sensitivity/ profitability studies
Akers	1	Isolators	X	
Datar	12	Bacterial enzyme		X
Datar and Cartwright	13	Recombinant tissue plasminogen activator		X
Dwyer	15	Protein purification	X	
Fournier et al.	18	Purified water	X	
Gerstner	22	Displacement chromatography	X	
Haritatos	35	Petrochemical and enzyme		X
Harsa et al.	23	Bacterial enzyme		X
Hepner and Male	24	Various (review)		
Holland et al.	26	All chemical processes	X	X
Kalk and Langlykke	27	Recombinant protein	X	X
Reinholtz	44	WFI	X	
Reisman	45	Citric acid	X	X
Remer and Idrovo	48	Equipment (biotechnology)	X	
Swartz	53	Penicillin		X
Van Dyne and Blase	55	Nylon 1313		X
Weaver and Thorne	57	All chemical processes		X

Another paper presents a modification of return evaluation to screen different projects under consideration (58). In most cases, a single determinant (such as net present value) cannot give a complete picture of project profitability. Additional calculated quantities (such as discounted cash flow rate of return, payback period) are used to differentiate projects. For a novel bioproduct, who will pay and how much will be allowed as payment are probably major confounding factors. The situation is far simpler for a product produced to compete in an existing market.

Sensitivity analyses have already been mentioned. Capital and operating costs can be estimated with some degree of clarity; changes in one or more inputs (such as utilities cost or cost of labor or overall yield) are readily introduced, and financial impacts are noted. Sensitivity analysis on the sales end is far more difficult to perform. A change in sales price, sales volume, clinical and clearance costs, impact of changes in government and private insurance payment policies, and cost of maintaining an intellectual property position are all factors that are (i) difficult to quantify and (ii) apt to have a significant impact on rate of return. Still, it is an exercise that should be performed. The factors that impact profitability will be shown and, furthermore, those factors having the greatest impact will be highlighted. Examples of factors to be studied (meaning the factors that would lower profitability) are revenue reduction (price and/or volume), increased investment (in the first year and in subsequent years), increased operating expense (reduced throughput will increase unit overhead cost), and ongoing legal and regulatory costs. Combinations of factors should be studied as well, since more than one negative circumstance can occur.

23.9. COSTS OF SELECTED EQUIPMENT

Pricing of selected items will be given with specifications, where applicable. Since only certain vendors are mentioned, the information is not meant to be comprehensive, but the values are useful for estimation purposes. These items will be reviewed: fermentors; modular systems (including fermentors, biowaste decontamination [batch and continuous], and CIP systems [one-tank and three-tank]); chromatography columns; cleanrooms; isolators; and robotics.

23.9.1. Fermentors (New Brunswick Scientific Co., Inc.)

Pricing of fermentors is given in Table 10. The design basis for the units listed there is as follows:

TABLE 10 Pricing of fermentors

Max. working volume (liters)	Cost ($)
50	118,000
112.5	135,000
187.5	149,500
375	187,000
750	212,000
1,125	230,000
1,500	295,000
5,600	490,000

- aerobic fermentation, non-GMP
- 316 SS jacketed American Society of Mechanical Engineers (ASME) pressure vessel, mechanically polished (interior) to less than 20 microinch Ra
- double mechanical seal with top-entry agitator
- monitor and control of airflow, pressure, dissolved oxygen, rpm, pH, high foam level
- semi-automatic sterilization
- skid-mounted, fully piped and wired (larger units have free-standing vessel)
- *NOT* included: addition vessels, utility supplies, shipping, installation
- base period for costs: 1994

For satisfaction of regulatory requirements for validation, add 20 to 40% to the costs shown.

23.9.2. Modular Systems (W.H.E. Bio-Systems, Inc.)

All prices given are for the third quarter of 1996.

1. Fermentors
   a) Design basis:
      - all 316 SS including frame, ASME rated to 45 psig, full vacuum, 240 grit, electropolished
      - double mechanical seal with bottom-entry drive
      - log orbital butt welding (sanitary lines) with boroscope inspection
      - 316 SS sanitary diaphragm valves, EPDM diaphragm
      - controls: temperature, pH, airflow, nitrogen, pressure, rpm, antifoam, unit is equipped for other gases
      - monitor: temperature, pH, airflow, pressure, vessel weight, dissolved oxygen, rpm, clean steam, plant steam
      - SIP: automatic and programmed
      - CIP: manual set-up, controlled by CIP station

   b) Seed fermentor: 150 liters, $135,000
   c) Production fermentor: 1,500 liters, $325,000
   d) Footprint for the seed unit: 4′ × 6′ × 9′
   e) Footprint for the production unit: 4′ × 10′ × 12′
2. Batch waste sterilization system
   a) Design basis:
      - all 316 SS including frame, ASME rated to 45 psig, full vacuum, 240 grit, electropolished
      - log orbital butt welding (sanitary lines) with boroscope inspection
      - 316 SS sanitary diaphragm valves, EPDM diaphragms
      - control via programmable logic controller, proportional-integral-derivative control loops
      - controls: temperature, pH, pressure, transfer, level
      - monitor: temperature, pH, pressure
      - SIP: automatic
      - CIP: manual

   b) 600-liter heat-inactivation tank
   c) 250-liter sump, BL-2 containment
   d) Cost: $154,000
   e) Footprint for the 600-liter tank: 5′ × 6′ × 10′
   f) Footprint for the 250-liter tank: 6′ × 4′ × 5′

- continuous system: 240 GPH
- hold time: 30 s at 240 GPH (215 to 285°F in heaters and retention section)
- similar design basis as above
- SIP and CIP: automatic

g) Cost: $198,400
h) Footprint: 12′ × 5′ × 10′

3. CIP systems
   a) Design basis: Similar to above (single tank)
      - controls: temperature, pressure, CIP flow
      - monitor: temperature, pressure, process flow
      - 150-liter tank
      - cost: $50,000
      - footprint: 4′ × 4′ × 5′
   b) Three-tank system: 300-liter wash tank, 300-liter pre-rinse tank, 300-liter WFI rinse tank
      - controls: temperature, pressure, level, CIP flow
      - monitor: temperature, pressure, total organic carbon, conductivity, CIP flow
      - cost: $180,000
      - footprint: 10′ × 4′ × 10′

Many other modular systems for biotechnology applications are manufactured by the same company. Advantages are shortened time line to delivery, reduced costs for labor and installation, validation prior to shipment (including operational testing prior to shipment); in some cases, a guaranteed capital cost execution program is offered.

23.9.3. Chromatography Columns

Pharmacia Biotech is one manufacturer of columns for use in processing of biologic materials. One series (BTG) includes glass columns designed for sanitary use and recommended for use in clinical manufacturing and small-scale production. Inside diameters (i.d.) are 10, 14, 20, and 30 cm; each has a single screw adapter. Columns with an i.d. of 20 cm giving varying bed heights to 60 cm would range in cost from $6,200 to $7,800 (height dependent). A 45-cm i.d. column is also available; price range is $32,000 to $38,000. All prices given in this section are as of the fourth quarter of 1996. Sanitary fixed-bed-height columns made of electropolished stainless steel are also available. Standard diameters range from 40 to 120 cm; standard bed heights are 15, 30, 60, and 100 cm. Examples of approximate prices are given in Table 11.

Yet another series of more advanced columns is available for process-scale chromatography. These columns are often selected when the diameter is 40 cm or greater. These columns are packed, operated, unpacked, and cleaned with the adapter and lid in place. A special nozzle is installed in both the top and bottom of the column. An optional packing station is available for separate purchase. Column material is either cast acrylic or stainless steel. Costs must be quoted by the manufacturer, since it is important to note that in large-scale chromatography, almost all columns are made to order. Standard diameters are 18, 28, 40, 60, 80, and 100 cm. Columns with a diameter greater than 100 cm can be custom ordered. All diameters can be made with any fixed height. The 28- to 100-cm columns are available with adjustable bed heights (10- to 70-cm range). In this series of production columns, the 100-cm by 15-cm-high column would cost approximately $80,000; for a 60-cm fixed-bed column (same diameter), the price would be about $92,000. If a variable height range (10 to 70 cm) is desired for these columns, there is a cost increment of approximately 10 to 15% compared with the 60-cm fixed-bed column. It should also be noted that in the larger scale (60- to 80-cm diameter), high-grade acrylic and stainless steel are about the same price.

23.9.4. Cleanrooms

Cleanroom construction costs include a number of design variables. Each manufacturer and many installers have different systems, methods of interlocking panels, sealing systems, drivers, and types of protective surfaces. There is no intent here to give a fixed cost for a given space, only a working estimate for preliminary design purposes. For cleanroom estimates, it is suggested that at least three companies be contacted and prepared questions be presented to each. If a contractor (internal or external) has been selected, the owner should be involved in interviews with original equipment vendors. The lowest bid is not always the most acceptable one. It must be remembered that maintenance and ongoing validation must be performed by the owner. An installer of cleanrooms (Hodess Building Co.) has listed costs. It is assumed that an existing structure exists and that the cleanroom is to be built and installed within an acceptable building shell. The room is basic; i.e., there are no pass-throughs and no connections included, there is one door, no windows, and standard lighting (dependent upon class) is installed. Approximation costs, based on values in the fourth quarter of 1996, are given in Table 12. The costs shown are at the low end of values normally seen; however, these are also very basic designs. The rooms are stick-built, floor is seamless vinyl, walls are one-quarter-inch laminated panel with metal studs, and ceiling is a flush grid system. The smaller Class 100 room has a unit cost of $276/ft^2. If a seamless epoxy floor is added and enamel steel walls are used, the unit cost would rise by about $30/ft^2. Validation is not included. Controlled entrance and exit spaces are not included. Certain additional options and costs are listed (add per ft^2 of room area):

- Dycem (antimicrobial mat) on polyvinyl chloride flooring, $6.50

TABLE 11 Pricing of chromatography columns

Diameter (cm)	Height (cm)	Approximate cost ($)
40	15	40,000
80	15	59,000
100	15	67,000
40	60	43,000
80	60	61,000
100	60	70,000

TABLE 12 Pricing of cleanrooms

Class	8′ × 12′ (96 ft^2)	12′ × 24′ (288 ft^2)
10	29,200	58,500
100	26,500	53,000
1,000	22,500	41,100
10,000	19,900	33,100
100,000	17,900	27,200

- Seamless epoxy flooring, $5.85
- Enamel steel walls, $5.20/ft^2 of panel
- 209E certification (*not* validation), $1,000 for an 8′ × 12′ room, $3,000 for a 12′ × 24′ room

Modular air showers would add $9,000 to $10,000 each to the overall cost. There are many other options that should be discussed with the vendor. Some such items are variable speed or two-speed fan drives, type of room sealing, allowable cleaning agents, amount of validation effort provided by vendor, extent of monitoring and control, pressurization levels, ease of filter changes, and appropriate interlocks.

23.9.5. Isolators

There is not a great deal of published cost information on barrier technology. A useful example is an article by Akers (1). A detailed operating cost comparison is given for an isolator system and a conventional cleanroom. Items included are direct operating expenses, gown-changing time, maintenance/decontamination, utilities cost, DOP testing, and start-up. Total annual expenses for the isolator were $280,815 and for the cleanroom $629,815, or more than double the amount for the isolator. There may be some smaller percentage savings in capital investment, as well. It appears to me that a significant shift to isolator technology (most likely within a less stringent, controlled environment) is occurring and that growth will be exponential for the installation of isolators. Aseptic cleanroom space (either Class 10 or Class 100) can cost 4 to 5 times more per square foot than Class 100,000 space. The advantage (aside from the potential for improved sterility assurance) of barrier technology is apparent.

Cleanroom use in the biotechnology and chemical industries is discussed by Moore et al. (33). There are a number of references that are useful for introducing isolator technology and its advantages (21) and also for giving design and construction details for actual operating systems (30, 38). In the article by McDonald and Walker (30), there is the claim that the line described was the first to receive FDA approval for aseptic filling using isolator technology where the unit was *not* installed in a Class 100 or 10,000 room. With the advent of chemical processes employing extremely toxic materials and bioprocesses involving ultra-high-potency compounds (which might impact reproductive function or induce allergic responses), there is a need to consider elimination of outflow of materials just as there has been a drive to eliminate the influx of contaminants to an aseptic process. Isolator/barrier technology must be considered for both operator protection and product or process protection (43).

One manufacturer of isolators is M. Braun, Inc. A single-station mini-environment/isolator/glove box has these characteristics:

1. Construction:
   - 304 SS, brushed interior or Teflon coat
   - radius corners, one-piece welded or modular flange
   - epoxy-coated steel stand
   - 49″ L × 31″ W × 35″ H
   - closed-loop HEPA filtered air circulation
   - oxygen or moisture control (optional)
   - Class 100
2. Standard equipment:
   - antechamber or pass-through
   - programmable logic control for pressure, environmental control, safety
   - vacuum pump
   - feed-throughs for power, gas, water, etc.
   - lighting and shelving
   - Lexan front panel (10 mm) with 9″ glove ports
3. Range of cost:
   - $30,000 to $40,000 (based on fourth quarter of 1995)
4. Options:
   - particle counter
   - data logging
   - heat treatment/packaging chamber
   - sterile fill/packaging system integration
   - rapid transfer ports or mini-antechamber
   - mobile docking antechambers with isolation doors
5. Alternatives:
   - double-sided and/or multiple-length workstations
   - sloping floor, common drain
   - other heights and widths
   - conveyor material transfer systems

   Special equipment (such as a freeze dryer) can also be incorporated into the design.
6. Range of cost:
   - $50,000 to $150,000, depending on options (based on fourth quarter of 1995)

23.9.6. Robotics

One of the more interesting developments in automated aseptic processing is a robotic system called the Cellmate (The Automation Partnership). This is a programmable robot operating in a Class 100 environment (laminar flow cabinet or room). The robot can perform all the standard manipulations needed for mammalian cell culture in roller bottles or T-flasks. Vessels are fed into the cabinet on a conveyor, and the robot performs one or more of these operations:

- removal/holding of cap
- liquid dispensing
- gassing via nozzles
- collection of liquids
- precise volume control
- cell seeding
- media changes
- cell sheet rinsing
- aspiration
- conveyance out

The basic unit and laminar flow hood, including components for all basic manipulations, is priced at $434,200 (at $1.67 per pound sterling, 1996 list). Various options are available. Add the following costs:

- In-process incubator for removal of cells from culture vessels by trypsinization, $116,900
- Automated scraping of cells from roller bottles, $10,000
- Additional HEPA filters and revised flow scheme to create a negative-pressure environment, $23,400

Since the equipment is designed and built to satisfy cGMP requirements for biologics or pharmaceuticals, full documentation (including installation qualification and operational qualification validation) is supplied. Installation, training, and commissioning are also included in the price. The Cellmate system is in production use worldwide; products include vaccines (human and animal), modified cell lines for cancer treatment, and cells for use in medical products. There are multiple installations in the United States at a number of diverse biotechnology companies.

REFERENCES

1. **Akers, J.** 1994. Biotechnology product validation. Part 6: Isolation technology applications. *BioPharm* **7**(6)**:**43–47. (Erratum, *BioPharm* **7**(8)**:**10.)
2. **Anonymous.** 1996. Industrial R&D barely paced inflation in '90s. *Chem. Eng. News* **August 26, 1996:**57–60.
3. **Anonymous.** 1996. New evidence of wine 7,000 years old from northern Iran. *Biblical Archaeol. Rev.* **22**(6)**:**24.
4. **Atkinson, B., and F. Mavituna (ed.).** 1991. Principles of costing and economic evaluation for bioprocesses, p. 1059–1109. *In* B. Atkinson and F. Mavituna (ed.), *Biochemical Engineering and Biotechnology Handbook*. Stockton Press, New York.
5. **Badiru, A. B.** 1996. *Project Management in Manufacturing and High Technology Operations*, 2nd ed. John Wiley & Sons, Inc., New York. See especially chapter 7, Manufacturing economic analysis, p. 301–364, and sections 10.11 to 10.14, Project management software, p. 478–496.
6. **Biotechnology Industry Services.** 1996. Special International Buyers' Guide Issue for 1997. *Nat. Biotechnol.* **14**(12). See especially p. 1362–1366.
7. **Bronzino, J. D. (ed.).** 1995. *The Biomedical Engineering Handbook*. CRC Press, Inc., Boca Raton, Fla.
8. **Chisti, Y., and M. Moo-Young.** 1993. Clean-in-place systems for bioreactors; design, validation and operation, p. 5–12. *In* B. Henon (ed.), *Bioprocess Engineering, 1993*, BED-vol. 27. American Society of Mechanical Engineers, New York.
9. **Christian, J., and A. Pandeya.** 1997. Cost predictions of facilities. *J. Mgmt. Eng.* **13:**52–61.
10. **Coombs, J., and Y. R. Alston.** 1995. *The Biotechnology Directory, 1996*, Stockton Press, New York.
11. **Cox, R. K.** 1997. Managing change orders and claims. *J. Mgmt. Eng.* **13:**24–29.
12. **Datar, R. V.** 1986. Economics of primary separation steps in relation to fermentation and genetic engineering. *Proc. Biochem.* **21:**19–26.
13. **Datar, R. V., and T. Cartwright.** 1993. Process economics of animal cell and bacterial fermentations: a case study analysis of tissue plasminogen activator. *Bio/Technology* **11:**349–357.
14. **Del Tito, B. J., Jr., M. A. Tremblay, and P. J. Shadle.** 1996. Qualification of raw materials for clinical biopharmaceutical manufacturing. *BioPharm* **9**(10)**:**45–49.
15. **Dwyer, J. L.** 1993. Process purification, p. 533–572. *In* F. Franks (ed.), *Protein Biotechnology*. Human Press, Totowa, N.J.
16. **Eisenberg, R. S.** 1996. Intellectual property issues in genomics. *Trends Biotechnol.* **14**(8)**:**302–307.
17. **Engler, C.** 1993. Decentralized operations and transfer pricing, p. 827–850. *In Managerial Accounting*, 3rd ed., R. D. Irwin, Inc., Homewood, Ill.
18. **Fournier, C., B. Rothenberg, and G. V. Zoccolante.** 1995. The impact of proposed pharmaceutical water quality test revisions on system design, performance and costs. *Pharm. Eng.* **15**(6)**:**48–62.
19. **Freitag, R., and C. Horvath.** 1996. Chromatography in the downstream processing of biotechnological products, p. 17–59. *In* A. Fiechter (ed.), *Advances in Biochemical Engineering/Biotechnology*, no. 53. Springer-Verlag, Heidelberg, Germany.
20. **Friedlob, G., and F. J. Plewa, Jr.** 1996. *Understanding Return on Investment*. John Wiley and Sons, Inc., New York.
21. **Galatowitsch, S.** 1996. Isolators: the future of aseptic processing. *Cleanrooms* **10**(11)**:**14–21.
22. **Gerstner, J. A.** 1996. Economics of displacement chromatography—a case study: purification of oligonucleotides. *BioPharm* **9**(1)**:**30–35.
23. **Harsa, S., C. A. Zaror, and D. L. Pyle.** 1993. Production of polygalacturonases from *Kluyveromyces marxianus* fermentation: preliminary process design and economics. *Proc. Biochem.* **28:**187–195.
24. **Hepner, L., and C. Male.** 1987. Economic aspects of fermentation processes, p. 685–697. *In* P. Prave, U. Faust, W. Sittig, and D. A. Sukatsch (ed.), *Fundamentals of Biotechnology*. Weinheim, Deerfield Beach, Fla.
25. **Hoeks, F. W. J., J. Muhle, L. Bohlen, and I. Psenicka.** 1996. Process integration aspects for the production of fine chemicals illustrated with the biotransformation of gamma-butyrobetaine into L-carnitine. *Chem. Eng. J.* **61:** 53–61.
26. **Holland, F. A., and J. K. Wilkinson.** 1997. Process economics, p. 9.1–9.79. *In* R. H. Perry and D. W. Green (ed.), *Perry's Chemical Engineers' Handbook*, 7th ed. McGraw-Hill, Inc., New York.
27. **Kalk, J. P., and A. F. Langlykke.** 1986. Cost estimation for biotechnology projects, p. 363–385. *In* A. L. Demain and N. A. Solomon (ed.), *Manual of Industrial Microbiology and Biotechnology*. American Society for Microbiology, Washington, D.C.
28. **Ladish, M. R., and K. L. Kohlmann.** 1992. Recombinant human insulin. *Biotechnol. Prog.* **8:**469–478.
29. **Lee, S. Y.** 1996. Plastic bacteria? Progress and prospects for polyhydroxyalkanoate production in bacteria. *Trends Biotechnol.* **14**(11)**:**302–307.
30. **McDonald, A., and N. Walker.** 1996. Design, implementation and operation of an isolation network for aseptic processing. *Pharm. Eng.* **16**(4)**:**18–30.
31. **McKown, R. L.** 1996. Contract manufacturing operations come of age in the biotechnology industry. *Genet. Eng. News* **16**(17)**:**6, 25, 28.
32. **Middleberg, A. P. J.** 1996. The influence of protein refolding strategy on cost for competing reactions. *Chem. Eng. J.* **61:**41–52.
33. **Moore, S., G. Parkinson, and G. Ondrey.** 1995. Cleanrooms go mainstream. *Chem. Eng.* **102**(5)**:**33–37.
34. **Mosberger, E., et al.** 1992. Chemical plant design and construction, p. 477–558. *In* B. Elvers, S. Hawkins, and G. Schulz (ed.), *Ullmann's Encyclopedia of Industrial Chemistry*, 5th ed., vol. B4, *Principles of Chemical Reaction Engineering and Plant Design*. VCH Verlagsgesellschaft mbH, Weinheim, Germany.
35. **Moses, V., and R. E. Cape.** (eds.). 1991. *Biotechnology, the Science and the Business*. Harwood Academic Publishers, Chur, Switzerland.
(See especially chapter 8, R. B. Nicholas and B. Ager,

The regulation of biotechnology in the U.S. and Europe, p. 103–115, and chapter 15, N.J. Haritatos, Process economics, p. 225–246.)

36. **Moses, V., and S. Moses.** 1995. *Exploiting Biotechnology.* Harwood Academic Publishers, Chur, Switzerland.

37. **Mossinghoff, G. J.** 1996. The road to pharmaceutical progress lies in patent protection for inventions and innovations. *Genet. Eng. News* **16**(17)**:**4, 35, 36, 43.

38. **Noble, N., D. Rouse, and G. Wyrick.** 1996. The Glaxo Wellcome barrier isolation filling line. *Pharm. Eng.* **16**(4)**:** 8–16.

39. **Odum, J. N.** 1992. Construction concerns for biotech manufacturing facilities. *Pharm. Eng.* **12**(1):8–12.

40. **Peters, M. S., and K. D. Timmerhaus.** 1991. *Plant Design and Economics for Chemical Engineers,* 4th ed., McGraw-Hill, Inc., New York.
(See especially chapter 6, Cost estimation, p. 150–215, and chapter 10, profitability, alternative investments, and replacements, p. 295–340.)

41. **Prokop, A.** 1995. Challenges in commercial biotechnology. Part I: Product, process, and market discovery, p. 95–154, and Part II: Product, process, and market development, p. 155–236. *In* S. Neidleman and A. I. Laskin (ed.), *Advances in Applied Microbiology,* vol. 40, Academic Press, San Diego, Calif.

42. **Raghavarao, K. S. M. S., N. K. Rastogi, M. K. Gowthaman, and N. G. Karanth.** 1995. Aqueous two-phase extraction for downstream processing of enzymes/proteins. *Adv. Appl. Microbiol.* **41:**97–171.

43. **Rahe, H.** 1996. Consider barrier/isolation technology for batches. *Chem. Eng. Prog.* **92**(12)**:**72–75.

44. **Reinholtz, W.** 1995. Making water for injection with ceramic membrane ultrafiltration. *Pharm. Technol.* **19**(9)**:** 84–96.

45. **Reisman, H. B.** 1988. *Economic Analysis of Fermentation Process.* CRC Press, Inc., Boca Raton, Fla.

46. **Reisman, H. B.** 1993. Problems in scale-up of biotechnology production processes. *Crit. Rev. Biotechnol.* **13:** 195–253.

47. **Reisman, H. B.** Economics. *In* M. C. Flickinger and S. W. Drew (eds.), *The Encyclopedia of Bioprocess Technology: Fermentation, Biocatalysis and Bioseparation,* in press. John Wiley & Sons, Inc., New York.

48. **Remer, D. S., and J. H. Idrovo.** 1990. Cost-estimating factors for biopharmaceutical process equipment. *BioPharm* **3:**36–42.

49. **Rosendal, G. K.** 1995. The politics of patent legislation in biotechnology: an international view, p. 453–476. *In* M. R. El-Gewely (ed.), *Biotechnology Annual Review,* vol. 1. Elsevier Science, Amsterdam.

50. **Scarlett, J. A.** 1996. Outsourcing process-development and manufacturing of rDNA-derived products. *Trends Biotechnol.* **14**(7)**:**239–244.

51. **Schroeder, R., M. J. Hanchar, J. Magyar, and P. J. Wald.** 1995. The team construct approach to biopharm facilities. Presented at BioPharm West '95. April 25–27, 1995, Fairfield, Calif.

52. **Stanbury, P. F., A. Whitaker, and S. J. Hall.** 1995. Fermentation economics, p. 331–349. *In Principles of Fermentation Technology,* 2nd ed. Pergamon, Elsevier Science, Oxford.

53. **Swartz, R. W.** 1985. Penicillins, p. 7–47, section 1.2 of Comprehensive biotechnology. *In* M. Moo-Young (ed.), *The Practice of Biotechnology, Current Commodity Products,* vol. 3. Pergamon Press, Elmsford, New York.

54. **Tolbert, W. R., B. Merchant, J. A. Taylor, and R. G. Pergolizzi.** 1996. Designing an initial gene therapy manufacturing facility. *BioPharm* **9**(10)**:**32–40.

55. **Van Dyne, D. L., and M. G. Blase.** 1990. Process design, economic feasibility, and market potential for nylon 1313 produced from erucic acid. *Biotechnol. Prog.* **6:**273–276.

56. **Vogel, H.** 1992. Process development, p. 437–475. *In* B. Elvers, S. Hawkins, and G. Schulz (ed.), *Ullman's Encyclopedia of Industrial Chemistry,* 5th ed., vol. B4, *Principles of Chemical Reaction Engineering and Plant Design.* VCH Verlagsgesellschaft mbH, Weinheim, Germany.

57. **Weaver, J. B., and H. C. Thorne (eds.).** 1991. *Investment Appraisal for Chemical Engineers,* Symp. Ser. 285, vol. 87. American Institute of Chemical Engineers, New York.
(See especially J. B. Weaver, Persistent problems overlooked by most authors, p. 10–19, and D. M. Stephenson and V. W. Hill, Factor approach to investment estimates for plant modifications, p. 102–111.)

58. **Woinsky, S. G.** 1996. Use simple payout period to screen projects. *Chem. Eng. Prog.* **92**(6)**:**33–37.

Introduction to Bioprocess Simulation

D. P. PETRIDES, R. NIR, J. CALANDRANIS, AND C. L. COONEY

24

Bioprocess simulation is a powerful analysis tool for professionals involved in the development, design, and operation of integrated biochemical processes. It enables process engineers and scientists to describe complex and integrated biochemical processes in the form of computer models. Once the computer model representation of a process is completed, engineers can conduct "experiments" to better understand the behavior of the real system under various equipment configurations and/or operating conditions. If these experiments were to be performed in the laboratory, they would require a significant investment of time and money.

Given a product and a desired annual production rate (plant throughput), bioprocess simulation generally endeavors to answer the following questions: What are the required amounts or raw materials and utilities? What is the required size of process equipment and supporting utilities? What is the total capital investment? What is the manufacturing cost? How long does a single batch take? How many batches can we carry out in a year? During the course of a batch, what is the demand for various resources (e.g., raw materials, labor, utilities)? What is the total amount of resources consumed? Which process steps or resources constitute bottlenecks? What can we change to increase throughput? What is the environmental impact of the process (i.e., amount and type of waste materials)? Which design is the "best" among several plausible alternatives?

24.1. BENEFITS FROM THE USE OF BIOPROCESS SIMULATION

The primary motivation for performing bioprocess simulation depends on the type of product, the stage of development, and the size of the investment. For commodity-like, low-priced biochemicals, simulation typically aims at minimizing the manufacturing cost. For high-priced biopharmaceuticals, engineers use simulation to improve process characteristics (environmental impact, safety, flexibility, operability) and accelerate commercialization (reduce the time-to-market). Figure 1 shows a pictorial representation of the objectives and benefits from the use of simulation tools at the various stages of the commercialization process (3).

1. *Idea generation*. At the early stages, when product and process ideas are conceived, bioprocess simulation is used for project screening/selection and strategic planning.
2. *Process development*. The goal at this stage is to improve upon the process characteristics that are often neglected yet are difficult to modify in a later stage: reduce environmental impact, improve resource utilization, increase process flexibility. Simulation tools enable engineering teams tackling the above issues to function cooperatively with minimal duplication or loss of data. Being able to experiment on the computer with alternative process setups and operating conditions reduces the costly and time-consuming laboratory and pilot plant effort. Furthermore, since such tools pinpoint the most cost-sensitive areas—the economic "hot spots"—of a complex process, they can be used to judiciously focus further laboratory and pilot plant studies. The result is an accelerated and cost-effective process development.
3. *Facility design*. With process development near completion at the pilot plant level, simulation tools are used to systematically design and optimize the manufacturing facility. Issues of operational flexibility, safety, and process scheduling must be considered at this stage; simulation tools greatly facilitate and improve the outcome of these tasks.
4. *Manufacturing*. During product manufacturing, simulation tools are primarily used for plant debottlenecking, process scheduling, and overall plant optimization. For existing batch manufacturing facilities, when throughput needs to be increased, either the equipment capacity of a processing step or a utility supply becomes a bottleneck. Simulation tools that are capable of tracking equipment utilization for overlapping batches can identify bottleneck candidates and guide the user through the debottlenecking effort.

The next two sections elaborate on the role of process simulation in debottlenecking batch operations and minimizing environmental impact.

24.1.1. Debottlenecking of Batch Operations

Most biochemical plants operate in batch (cyclical) mode. Several upstream and downstream process steps are required to convert the raw materials into the purified final product. Each step is usually initiated after the previous step is completed. However, a new batch is usually initiated before the previous one is fully completed. The maximum possible overlap between consecutive batches is determined by the process step that has the longest cycle time (usually the bioreactors). For a given plant, to increase annual production we reduce the time between consecutive batches until we reach the limit set by the longest process step. At that point, the only way to increase throughput is by adding extra capacity to the bottlenecking equipment or by introducing multiple processing lines that operate in a staggered mode.

Let us define Batch Equipment Capacity Utilization (BECU) as the product of Batch Equipment Uptime (BEU) and Cycle Equipment Capacity Utilization (CECU):

$$\text{BECU} = \text{BEU} \times \text{CECU}$$

where

BEU = (Total time equipment is utilized per batch)/(Effective plant batch time)

and

CECU = (Fraction of equipment capacity utilized during a cycle)

Effective plant batch time is the time between consecutive plant batches. The value of BEU is always in the [0, 1] range. A BEU value of 0.7 indicates that a unit is occupied 70% of the time and idle 30% of the time. CECU is unity for units that fully utilize their capacity during a cycle (e.g., membrane filters, disk-stack centrifuges). CECU is less than 1 for units with storage capacity (e.g., reactors and storage tanks) that do not operate at full capacity. For instance, if a vessel is 60% full during its operation, then its CECU is 0.6. If we plot the value of BECU for each piece of equipment involved during a batch production (Fig. 2), the resulting bar graph directly shows all equipment bottleneck candidates.

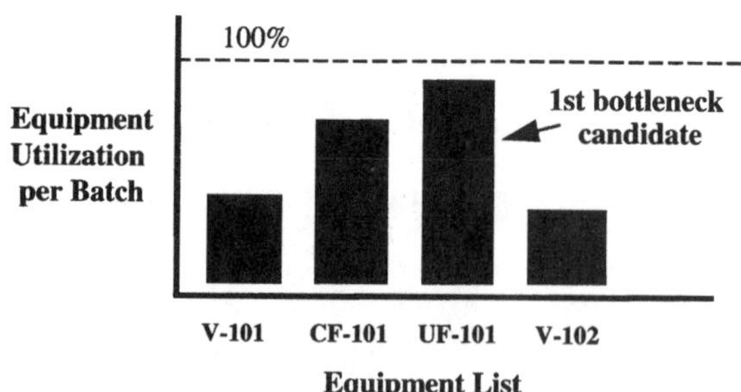

FIGURE 2 Equipment utilization.

In addition to allocating appropriate equipment, process steps require various utilities (e.g., water-for-injection [WFI], various types of steam, cleaning materials, power) and labor. Utilities and labor (called resources in the rest of the chapter) are usually required at the same time by several steps of multiple overlapping batches at various rates. Figure 3 shows typical instantaneous and cumulative demands for a resource. As we increase throughput in a batch plant, the demand for resources increases. If a resource demand exceeds its maximum availability rate, we reach a resource-related bottleneck. Very often, adjustments in scheduling can be made that eliminate this type of resource bottleneck. Another type of resource-associated bottleneck can be reached if the total consumption (cumulatively) of a given resource rises over its available quantity. In this case, the only way to eliminate the bottleneck is by installing extra capacity for that resource.

Simulation tools that can calculate equipment utilization and resource demand as a function of time can greatly facilitate the debottlenecking effort in a batch manufacturing plant. During the design stage, the same tools can be used to judiciously size process equipment and utilities.

FIGURE 1 Benefits from the use of bioprocess simulation.

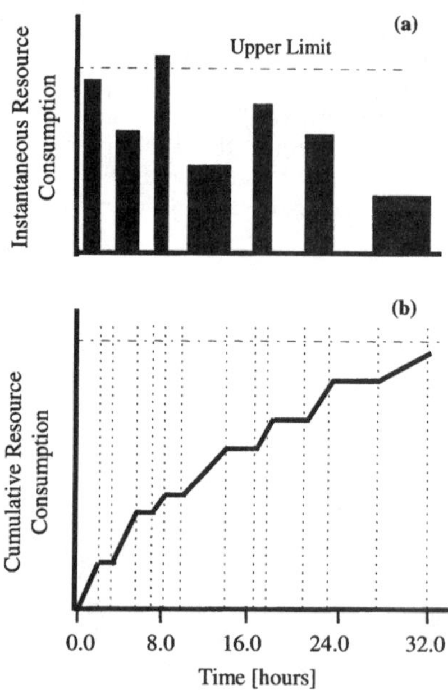

FIGURE 3 Instantaneous and cumulative demand of a resource.

24.1.2. Environmental Impact Assessment

Biochemical plants generate a wide range of liquid, solid, and gaseous waste streams that require treatment prior to discharge. The cost associated with waste treatment and disposal has skyrocketed in recent years owing to increasingly stricter environmental regulations. This cost can be reduced through minimization of waste generation at the source. However, generation of waste from a chemical or biochemical process is a function of the process design and the manner in which the process is operated. Thus, reducing waste in an industrial process requires intimate knowledge of the process technology, in contrast to waste treatment, which essentially is an add-on at the end of the process (8). Minimization of waste generation must be considered by process engineers at the early stages of process development. Once a process has undergone significant development, it is difficult and costly to make major changes. Furthermore, regulatory constraints that are unique to the pharmaceutical industry restrict process modifications once clinical efficacy of the drug is established.

Process simulators enable the user to evaluate the impact of alternative technologies on the type and amount of waste generated from a process. For instance, the impact of alternative extraction solvents or the selection of alternative chromatography elution buffers can readily be assessed. Certain simulators also facilitate waste stream characterization by calculating common environmental stream properties, such as biochemical oxygen demand (BOD), chemical oxygen demand (COD), total Kjeldahl nitrogen (TKN), total phosphorus (TP), and total suspended solids (TSS). Other simulators allow users to simulate end-of-pipe treatment processes and predict the fate of any waste component during treatment (4).

24.2. COMMERCIALLY AVAILABLE TOOLS

Currently, there are three process simulation tools that specifically address the needs of the biotechnology industry. BioProcess Simulator (BPS) from Aspen Technology (Cambridge, Mass.) was the first tool of this type. BPS is an extension of Aspen Plus, an established simulator for chemical and petrochemical processes. BPS, for a given flow sheet, carries out material and energy balances, estimates the size and cost of equipment and carries out economic evaluation.

BioPro Designer, the second product of this category, was initially developed at the Biotechnology Process Engineering Center (BPEC) of the Massachusetts Institute of Technology. INTELLIGEN, Inc. (Scotch Plains, N.J.) completed the development and commercialized first the Apple Macintosh version and later the Microsoft Windows version of BioPro Designer. This tool handles material and energy balances, equipment sizing and costing, economic evaluation, process scheduling, and debottlenecking of batch operations in a facile manner. Extensions of BioPro have been created to focus on design and evaluation of synthetic pharmaceutical and environmental processes.

Biotechnology Design Simulator (BDS), the third tool of this family, was developed by Life Sciences International (Philadelphia, Pa.). BDS runs on top of Gensym's G2 system and focuses on scheduling of batch operations and resource utilization as a function of time.

A number of other simulation and economic evaluation software tools have found applications in the biotechnology field despite the fact that they are not specifically designed for bioprocesses. For instance, Icarus Process Evaluator (IPE) from Icarus Corporation (Rockville, Md.) is used for detailed capital cost estimation and investment analysis. BATCHES from Batch Process Technologies (West Lafayette, Ind.) is a generic batch process simulator that has found applications in synthetic pharmaceuticals, biochemicals, and food processing. It is especially useful in fitting a new process into an existing facility and analyzing resource demand as a function of time. Rapid Access Cost Estimating (RACE) from Richardson Engineering Services (Mesa, Ariz.) is an advanced spreadsheet that performs detailed capital cost estimation based on the Richardson Engineering database.

24.3. ILLUSTRATIVE EXAMPLES

The use of BioPro Designer is illustrated to analyze and evaluate the production of two biological products. The first example focuses on the production of monoclonal antibodies (Mabs) using mammalian cells cultured in stirred tank bioreactors. The second example deals with the production of β-galactosidase using recombinant *Escherichia coli*. The generation of the flow sheets for the production of Mabs and β-galactosidase was based on information available in the patent and technical literature, combined with our engineering judgment and experience with other biological products. We use these flow sheets to draw general conclusions on the manufacturing cost of high-value biological products.

24.3.1. Monoclonal Antibody Production

Monoclonal antibodies are used in diagnostic tests as well as for therapeutic purposes. World demand for currently approved Mabs is in the order of a few kilograms per year. However, new therapeutic Mabs are under development that require doses of several hundred milligrams to a gram over the course of therapy (6). The world demand for such products will exceed 100 kg per year.

Current production choices for Mabs are limited to three well-established systems: ascites, stirred tank bioreactors, and hollow-fiber bioreactors. Alternative technologies under development include transgenic animals and genetically altered crop plants (1, 2, 7). Currently, stirred tank bioreactors tend to be favored for production of Mabs in kilogram quantities. They are operated under batch, fed-batch, or perfusion mode.

This example analyzes the production of a typical therapeutic Mab. In the base case, approximately 6 kg of purified product is produced per year in 44 batches. The manufacturing cost for producing larger quantities is estimated as part of the sensitivity analysis.

24.3.1.1. Process Description

Upstream Section

The entire flow sheet is shown in Fig. 4. The serum-free and low-protein-content media powder is dissolved in WFI in a stainless-steel tank (V-101), and the solution is sterilized using a 0.1-μm dead-end polishing filter (DE-101). The concentration of media powder in the feed solution is 10 g/liter. A stirred tank bioreactor (R-101) is used to grow the cells that express the therapeutic imunoglobulin G. The bioreactor operates in fed-batch mode. A cycle time of 156 h (132 h for fermentation and 24 h for turnaround) was assumed for the bioreactor. The volume of broth generated per bioreactor batch is approxi-

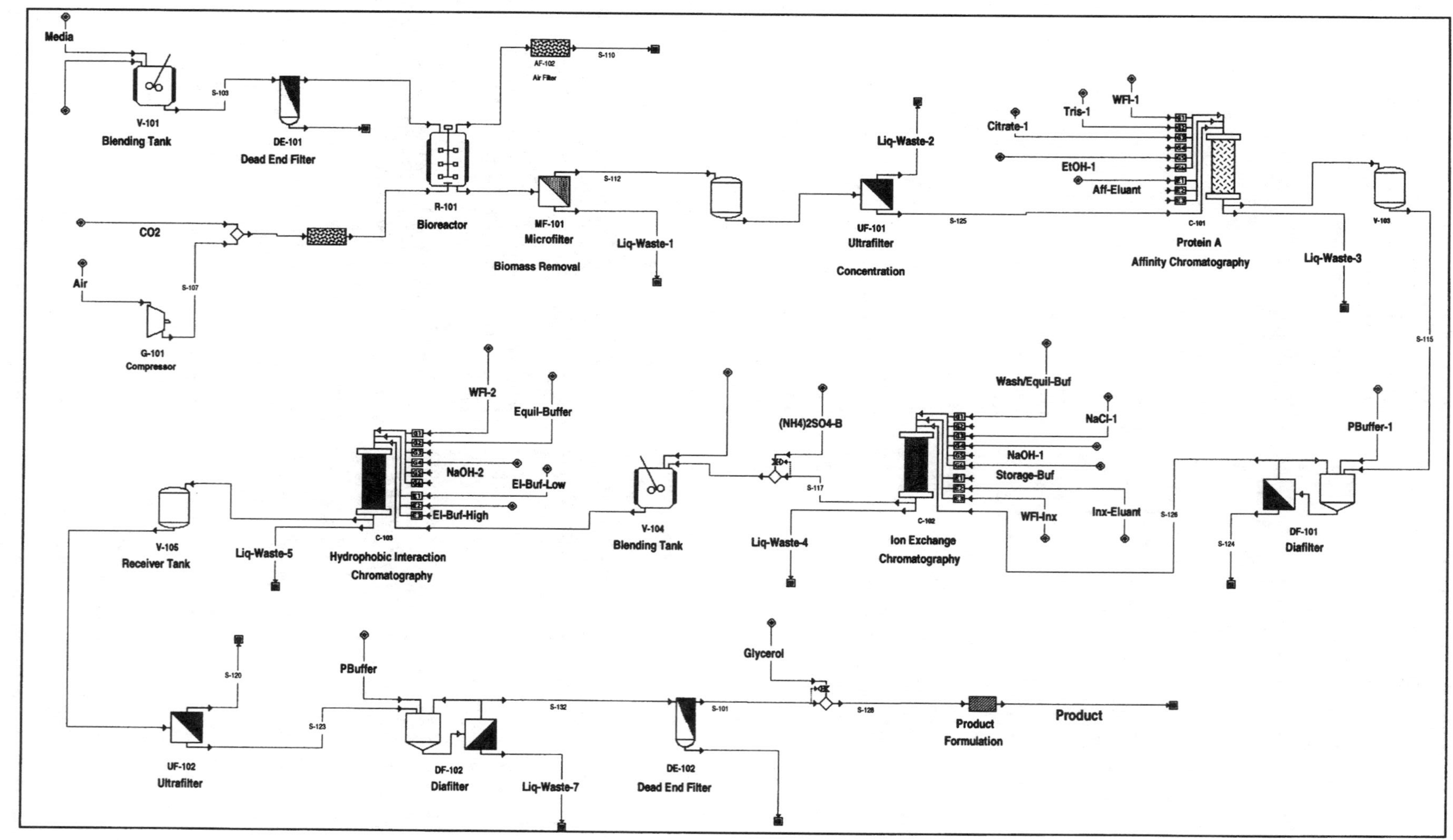

FIGURE 4 Monoclonal antibody production flow sheet.

mately 2,200 liters containing 220 g of product (the product titer is 100 mg/liter). The total volume of the bioreactor vessel is 3,000 liters.

Downstream Section

The generated biomass and other suspended compounds are removed using a 0.65-μm membrane microfilter (MF-101). The product recovery yield of this step is 95%. This filtration step takes 4 h and requires a membrane area of around 30 m^2. The clarified solution is concentrated 20-fold using a 50,000 MW cutoff ultrafilter (UF101). The recovery yield of this step is 95%. This step takes 4 h and requires a membrane area of 27 m^2. The bulk of the contaminant proteins are removed using a protein A affinity chromatography column (C-101). The following operating assumptions were made: (i) Resin-binding capacity is 15 mg of product per ml of resin. (ii) The eluant volume is equal to 6 column volumes (CVs) and is a 0.1 M solution of sodium citrate. (iii) The product is recovered in 3 CVs of eluant buffer with a recovery yield of 95%, and pH is immediately neutralized to ensure product stability. (iv) The total volume of the solutions for column equilibration, wash, and regeneration is 13 CVs. This step takes around 10 h and requires a resin volume of 20.2 liters (assuming an overdesign factor of 50%). The protein A elution buffer is exchanged with phosphate buffer using a 50,000 MW cutoff diafilter (DF-101). The product recovery yield of this step is 95%. The purification proceeds using a cation-exchange chromatography column (C-102). The following operating assumptions were made. (i) Resin-binding capacity is 20 mg of product per ml of resin. (ii) A gradient elution step is employed with a sodium chloride concentration ranging from 0.0 to 1.0 M and a volume of 6 CVs. (iii) The product is recovered in 3 CVs of eluant buffer with a recovery yield of 90%. (iv) The total volume of the solutions for column equilibration, wash, and regeneration is 17 CVs. This step takes around 8 h and requires a resin volume of 14 liters (assuming an overdesign factor of 50%). Ammonium sulfate is added to a concentration of 2.0 M to increase the ionic strength of the solution and prepare it for the hydrophobic interaction chromatography (HIC) step (C-103) that follows. The following operating assumptions were made for the HIC step. (i) Resin-binding capacity is 20 mg of product per ml of resin. (ii) A gradient elution step is used in which the concentration of ammonium sulfate changes linearly from 2.0 M to 0.0 M. (iii) The product is recovered in 2 CVs of eluant buffer with a recovery yield of 95%. (iv) The total volume of the solutions for column equilibration, wash, and regeneration is 22 CVs. This step takes around 9.5 h and requires a resin volume of 12.3 liters (assuming an overdesign factor of 50%). The purified product solution is concentrated two-fold using a 50,000 MW cutoff ultrafilter (UF-102), and the HIC elution buffer is exchanged with phosphate buffer using a diafilter (DF-102). Glycerol is added to a concentration of 100 g/liter, and the solution is sent to product formulation. The product concentration in the final solution is around 10 g/liter. The following additional assumptions were made. (i) WFI is used for the preparation of water solutions and buffers. (ii) To calculate the cycle time of chromatography steps, it was assumed that loading and elution operate at a linear velocity of 100 cm/h, while equilibration, washing, and regeneration operate at a linear velocity of 200 cm/h.

24.3.1.2. Material Balances

Table 1 provides a summary of the overall material balances per batch. The quantities are in kilograms per batch. The duration of a single batch is 162-h. The overall recovery yield of IgG (the product) is 62% (140 g of IgG is recovered out of the 220 g that is present in the fermentation broth). Note the large amount of WFI utilized per batch. The materials for equipment cleaning (primarily solutions of water) and maintenance were not considered in the material balances.

24.3.1.3. Process Scheduling

Figure 5 shows the results of process scheduling in the form of a Gantt chart generated by BioPro Designer. The Gantt chart shows the execution of the various process steps as a function of time for an isolated batch. The black portion of a bar represents process time while the white portion represents turnaround time. The bioreactor, which is the limiting step, has a cycle time of 156 h (132 h for fermentation and 24 h for turnaround). While the bioreactor is preparing a new batch, the downstream equipment is being utilized to purify the product of the previous batch. The downstream section requires two shifts per day for 5 days a week. The actual time between consecutive batches (effective batch time) is 180 h. On an annual basis, the plant processes 44 batches and produces 6 kg of purified IgG.

24.3.1.4. Economic Evaluation

Table 2 shows the key economic evaluation results generated by using the built-in cost functions of BioPro Designer. For the base case (6 kg/year of IgG), the total capital investment is around $36 million. The floor area of the production facility is around 2,000 m^2. The unit production cost is around $1,620/g of purified IgG. Assuming a selling price of $7,000/g, the project yields an after-tax internal rate of return of 52% and a net present value of $132 million (assuming a discount interest of 7%). If amortization of up-front research and development (R&D) cost is considered in the economic evaluation, the numbers change drastically. For instance, a modest amount of $25 million for up-front R&D cost amortized over a period of

TABLE 1 Overall material balances

Component	Material (kg/batch)[a]		
	Total inlet	Total outlet	Product
Ammonium sulfate	45.56	45.56	
Biomass	0.00	1.79	
Glycerol	1.34	2.34	
IgG	0.00	0.22	0.14
Growth medium	22.00	8.41	
Na_3 citrate	0.33	0.33	
Sodium hydrophosphate	5.51	5.51	
Sodium chloride	24.74	24.74	
Sodium hydroxide	1.55	1.55	
Tris-HCl	0.61	0.61	
WFI	3,329.16	3,340.73	
Total	3,430.80	3,430.80	0.14

[a]1 batch = 2,200 liters of feed material.

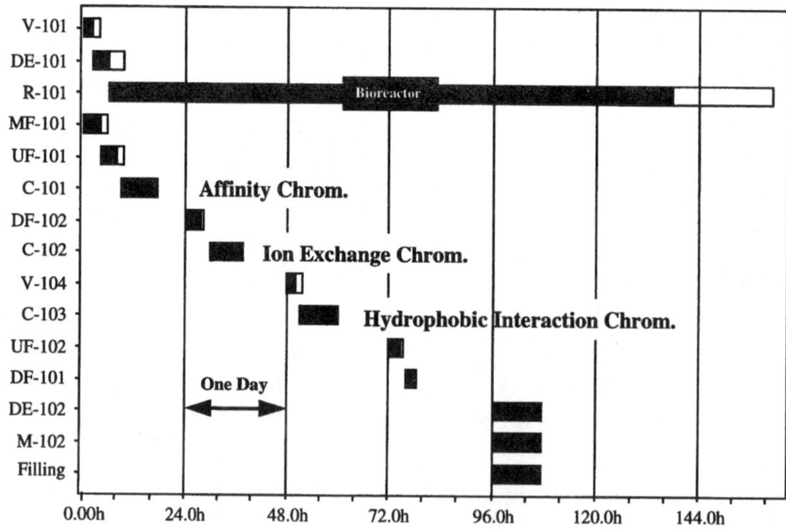

FIGURE 5 IgG production scheduling Gantt chart.

10 years reduces the internal rate of return to 28.5% and the net present value to $107 million.

Figure 6 breaks down the operating cost. The direct fixed capital (DFC)-dependent cost is the most important item, accounting for 61% of the manufacturing cost. This is common for high-value products that are produced in small quantities. Labor lies in the second position, accounting for 18% of the total cost. A total of 14 operators are required to run the plant around the clock, supported by 10 scientists for quality control/quality assurance work. Administration and overhead account for 8% of the total manufacturing cost. Raw materials and consumables account for 6% each. Consumables include the cost of chromatography resins and membrane filters that need to be replaced on a regular basis. In terms of cost distribution per section, 64% of the cost is associated with the upstream section and 36% with the downstream.

Key assumptions for the economic evaluation include the following. (i) A new manufacturing facility will be built and dedicated to production of 6 kg/year of IgG. (ii) The entire DFC is depreciated linearly over a period of 10 years. (iii) The project lifetime is 15 years. (iv) The unit cost of WFI is $0.12/liter. (v) The cost of media is $5/liter (based on volume of solution fed to bioreactors). (vi) All of the chemicals used are of high purity grade. (vii) The unit cost of membranes is $500/m^2; the unit cost of chromatography resins is (per liter) $10,000, and $3,200 for columns C-101, C-102, and C-103, respectively. (viii) The chromatography resins are replaced every 20 cycles. (ix) The waste disposal cost is $0.5/kg. (x) The cost of cleaning chemicals is not considered.

TABLE 2 Key economic evaluation results

Direct fixed capital	$32.2 million
Total capital investment	$36.1 million
Plant throughput	6 kg of IgG/year
Manufacturing cost	$9.7 million/year
Unit production cost	$1,620/g of IgG
Selling price	$7,000/g of IgG
Revenues	$42.0 million/year
Gross profit	$32.3 million/year
Taxes (40%)	$12.9 million/year
Net profit	$19.4 million/year
Internal rate of return (after taxes)	52.4%
Net present value (for 7% discount interest)	$132 million

24.3.1.5. Sensitivity Analysis

After a model for the entire process is developed on the computer, tools like BioPro Designer can be used to ask and readily answer "what if" questions and carry out sensitivity analysis with respect to key design variables. In this example, we looked at the impact of product titer (in the bioreactor) and production rate on unit production cost. For a product titer of 100 mg/liter, the cost drops considerably for production rates of up to 50 kg/year of purified IgG. For higher production rates, the cost levels off, approaching a value of $500/g. Increasing the titer from 100 to 250 mg/liter reduces the production cost by $200 to 350/g, depending on production rate. The reduction in cost is smaller (in the range of $100 to $180/g) when the product titer is increased from 250 to 500 mg/liter. As can be seen from Fig. 7, the production cost reaches a minimum

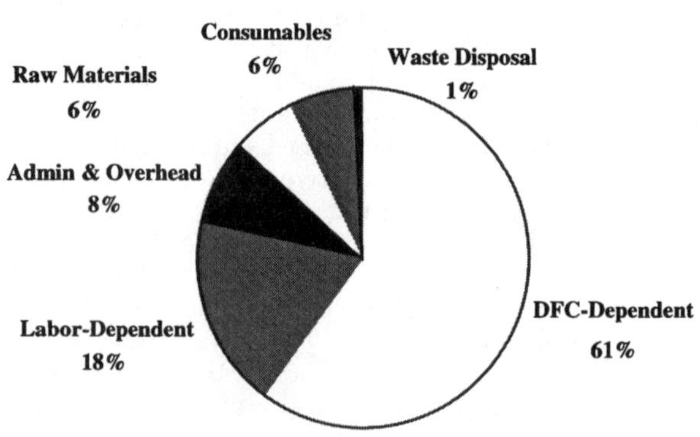

FIGURE 6 Breakdown of manufacturing cost.

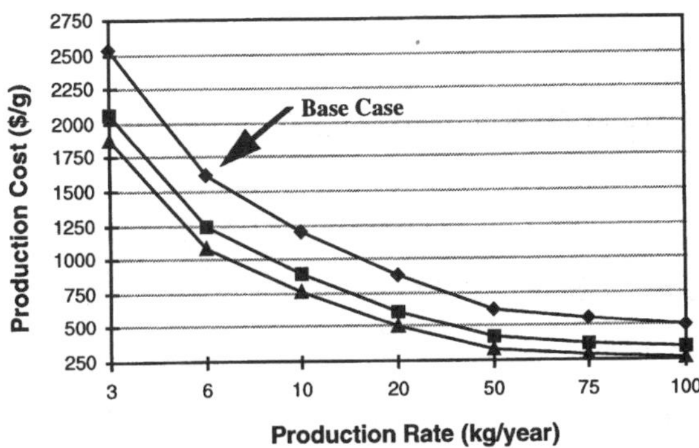

FIGURE 7 Production cost as a function of product titer (♦, 100 mg/liter; ■, 250 mg/liter; ▲, 500 mg/liter) and production rate.

of $250/g as we increase throughput and product titer. For throughputs in the range of 100 kg/year and titers of 500 mg/liter, almost 80% of the manufacturing cost is associated with the downstream section. Furthermore, under such conditions the cost of purification scales pretty much linearly with production rate, because most of the cost is associated with purification of raw materials and consumables. Therefore, less expensive product formation options, such as transgenic animals and genetically altered crop plants, can have an impact on only the 20% of the total cost. In other words, the cost of Mabs will not drop below $200/g (80% of $250/g), no matter what upstream technology is used. The only way to go below the $200/g barrier is by developing less expensive product purification technologies and deploying them in combination with inexpensive upstream technologies (such as transgenic animals).

Key assumptions for the sensitivity analysis are that (i) the amount and cost of fermentation media is independent of product titer, and (ii) the scheduling is independent of plant throughout (as we increase throughput, we continue to process 44 batches per year by utilizing larger and multiple pieces of the same type of equipment).

24.3.2. Production of an Intracellular Protein Using Recombinant *Escherichia coli*

This example analyzes the manufacturing cost of β-galactosidase, an intracellular enzyme produced by recombinant *E. coli*. The purpose of the analysis is to determine whether 5,000 kg of this enzyme can be manufactured for less than $3/g. This enzyme is normally produced by *E. coli* up to 1 to 2% of total cell protein under conditions of induction of the *lac* operon. Using genetic engineering, however, the level can go up to 20 to 25% of total protein. In the base case of this analysis, an easily attainable level of 10% was assumed. There are several types of β-galactosidase formed by different microorganisms. The one formed by *E. coli* has a molecular weight of 540,000 and an isoelectric point of 4.8.

β-Galactosidase is formed intracellularly and allows growth of *E. coli* on lactose (a disaccharide present in milk). It has found limited industrial applications until now. It is mainly used in the utilization of cheese whey. More specifically, immobilized reactors with β-galactosidase have been developed to convert lactose found in cheese whey to glucose and galactose, yielding a sweetened product that can be used as an additive to ice cream, eggnog, yogurt, and other dairy products. Another application of β-galactosidase is in the treatment of milk products. A large number of people are lactose intolerant and cannot digest milk or milk products. Production of lactose-free milk products, using immobilized β-galactosidase reactors, allows those people to consume dairy products. The increasing demand for lactose-free dairy products is expected to create a world demand for β-galactosidase in excess of 20,000 kg/year by the year 2005.

24.3.2.1. Process Description

Upstream Section

The entire flow sheet is shown in Fig. 8. Fermentation media are prepared in a stainless-steel tank, V-101, and are continuously sterilized in ST-101. The filling of the production fermentor, R-101, with sterilized media takes 4 h. Two seed fermentors (not shown on the flow sheet), one of 0.2 m^3 and the other of 3.5 m^3 total volume, respectively, are used to provide inoculum to the production fermentor of 68.7 m^3. The fermentation broth processed per batch is 52,000 liters. An axial compressor in combination with an absolute filter provides sterile air to the fermentors at an average rate of 0.5 volumes of air per volume of fermentation broth per minute. The fermentation time in the production fermentor is about 18 h, and the turnaround time is 12 h. The final concentration of *E. coli* in the production fermentor is about 30 g/liter dry cell weight. The fermentation temperature is 37°C. After the end of fermentation, the broth is cooled to 5°C using a heat exchanger while it is being transferred to a holding tank, V-102.

Downstream Section

The first step of the downstream section is cell harvesting. It is carried out in three hollow-fiber membrane microfilters (MF-101), each having a membrane area of 55 m^2 and a membrane pore diameter of 0.45 μm. A twofold volumetric reduction is achieved in two cycles of 6 h each (4 h for filtration and 2 h for cleaning). The concentrated broth is mixed (M-103) with a buffer solution that stabilizes the product after its release. A high-pressure homogenizer, HG-101, breaks the cells and releases the intracellular product. The broth undergoes two passes through the homogenizer at a pressure drop of 800 bar (1 bar = 10^5 Pa). Cell debris particles, generated by homogenization, are removed using a disk-stack centrifuge, CF-1-1, over a period of 36 h (two cycles of 18 h each). A dead-end polishing filter, DE-101, removes any remaining particulate material, which can block the chromatography units and impair the final purification. The dilute protein solution is concentrated using a hollow-fiber membrane ultrafilter (UF-101) of 100,000 molecular weight cutoff to a total protein concentration of 3 to 4% (wt/wt). The membrane area of the ultrafilter is 80 m^2. The concentration is accomplished in two 8-h cycles (6 h process time and 2 h turnaround time). The primary purpose of the ultrafiltration step is to reduce the amount of liquid that needs to be processed by the expensive chromatography steps. It also facilitates product purification, because low-molecular-weight proteins and other solutes pass through the membrane while the product remains in the concentrate stream. Ammonium sulfate is added (M-102) to a concentration of 2 M to increase the ionic strength of the solution and prepare it for the HIC step (C-101). The HIC column is loaded with Alkyl Superose resin (5). Four columns are required, each having a resin volume of 800 liters.

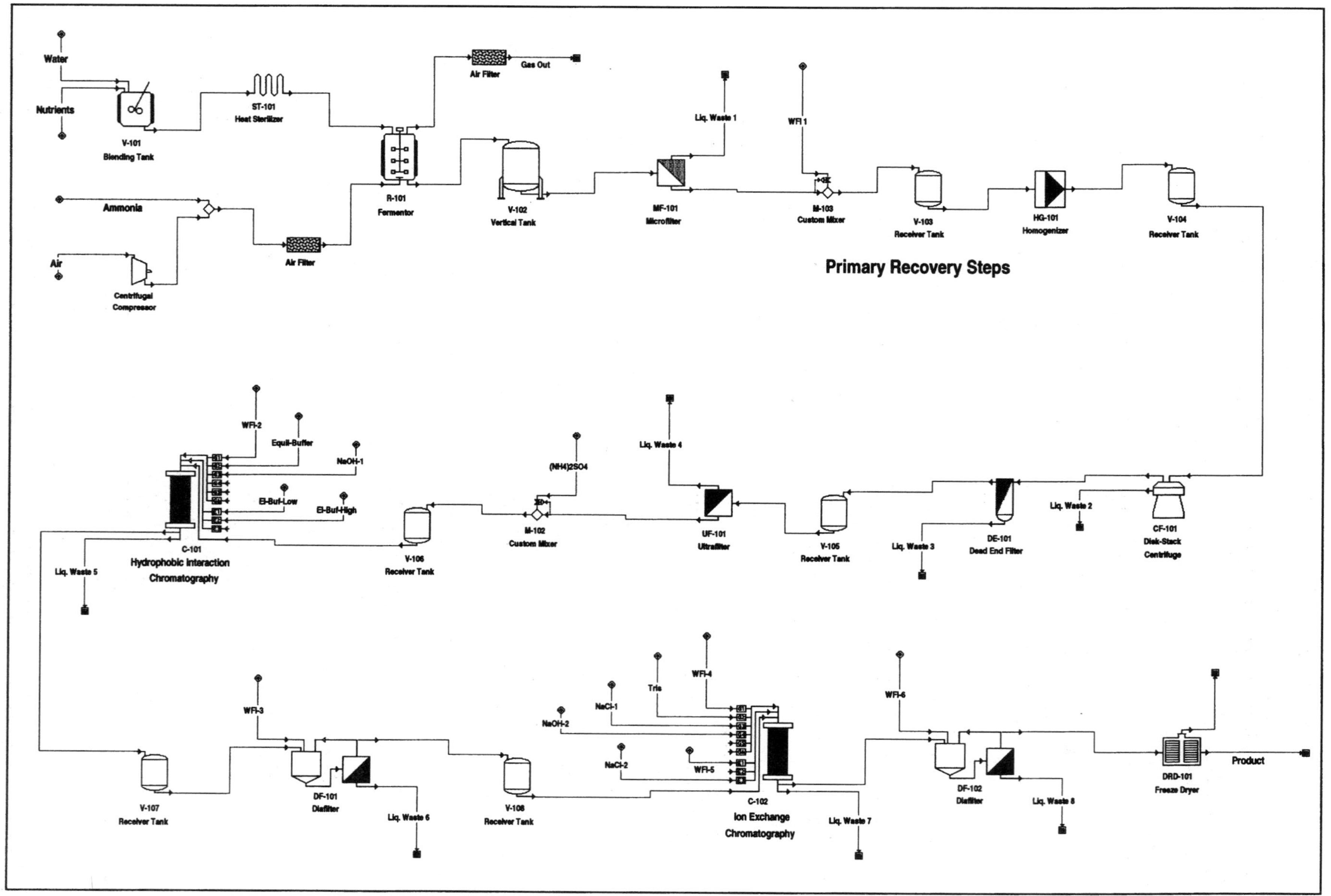

FIGURE 8 Process flow sheet of β-galactosidase production.

Each column operates four cycles per batch. The duration of a cycle is 4.9 h. The following operating assumptions were made for the HIC step. (i) Resin-binding capacity is 20 mg/ml (based on all proteins that bind to the resin). (ii) A gradient elution step of 2 CVs is used, during which the ammonium sulfate concentration varies linearly between 0.0 and 2.0 M. (iii) The product is recovered in 0.5 CV of eluant buffer with a recovery yield of 95%. (iv) The total volume of the solutions for column equilibration, wash, and regeneration is 25 CVs. (v) The linear velocity for loading is 100 cm/h and for elution, equilibration, wash, and regeneration is 200 cm/h. The HIC elution solution is exchanged with phosphate buffer using a diafilter (DF-101). Diafiltration is accomplished in one 8-h cycle (6 h process time and 2 h turnaround time) with a product recovery yield of 95%. The total membrane area of the diafilter is 90 m^2. The final purification to 99.5% purity is accomplished using an ion-exchange chromatography step, C-102. An anion exchange is used since β-galactosidase is negatively charged in the range of pH where it is stable. Three columns are required, each having a resin volume of 620 liters. Each column operates two cycles per batch. The duration of a cycle is 6.9 h. The following operating assumptions were made for the ion-exchange step. (i) Resin-binding capacity is 20 mg/ml (based on all proteins that bind to the resin). (ii) A gradient elution step of 2 CVs is used, during which the sodium chloride concentration varies linearly between 0.05 and 0.3 M. (iii) The product is recovered in 0.5 CV of eluant buffer with a recovery yield of 90%. (iv) The total volume of the solutions for column equilibration, wash, and regeneration is 25 CVs. (v) The linear velocity for loading is 150 cm/h and for elution, wash, and regeneration is 200 cm/h. The elution buffer is exchanged with phosphate buffer using a second diafilter (DF-102). Diafiltration is accomplished in one 8-h cycle (6 h process time and 2 h turnaround time) with a product recovery yield of 95%. The enzyme is recovered in solid form using freeze-drying (DRD-101).

24.3.2.2. Process Scheduling

Figure 9 displays the Gantt chart for the production scheduling of β-galactosidase. The dark portion of a bar represents process time while the white portion represents turnaround time. Multiple rectangles (e.g., MF-101, UF-101, C-101, and C-102) represent multiple cycles of that process step within the same plant batch time. The fermentor (R-101, which is the limiting step, has a cycle time of 30 h (18 h fermentation and 12 h turnaround time). The entire plant batch time is almost 6 days (close to 140 h). However, a new batch is initiated every 3 days (72 h). The entire plant operates around the clock for 7,400 h per year, processing 100 batches and producing 5,000 kg of purified product. Eighteen operators are required to run the plant around the clock, supported by 10 scientists that handle quality control/quality assurance tasks.

24.3.2.3. Material Balances

Table 3 provides a summary of the overall material balances. The quantities are in kilogram per batch. The overall product recovery yield is 66% (50.56 kg of product is recovered out of the 76.6 kg that is present in the fermentation broth). Note the large amounts of ammonium sulfate, Tris buffer, and WFI utilized per batch. The materials for equipment cleaning (primarily solutions of water) and maintenance were not considered in the material balances.

24.3.2.4. Economic Evaluation

Table 4 shows the key cost estimation results generated by using BioPro Designer. For the base case (5,000 kg/year of β-galactosidase), the total capital investment is around \$35.6 million and the unit production cost is around \$4.4/g of purified β-galactosidase. In other words, the unit cost is above the threshold value of \$3/g of purified enzyme.

Figure 10 breaks down the manufacturing cost. The DFC-dependent and raw materials costs each account for 32% of the manufacturing cost. The cost of raw materials includes the fermentation media as well as the product purification chemicals (e.g., various chromatography solutions). Consumables, which include the cost of chromatography resins and filter membranes, constitute 18% of the manufacturing cost. Labor and administration and overhead account for 15% of total cost. In terms of cost distribution per section, 13% of the cost is associated with the upstream section and 87% with the downstream. In terms of contribution of specific process steps, 45% is associated with the HIC step and 14.7% with the ion-exchange chromatography step. In other words, almost 60% of the total manufacturing cost is associated with the chromatography steps. This is typical for high-value proteins produced by using bacterial fermentation.

Key assumptions for the economic evaluation include the following. (i) A new manufacturing facility will be built and dedicated to production of 5,000 kg of β-galactosidase per year. (ii) The entire DFC is depreciated linearly over a period of 10 years. (iii) The project lifetime is 15 years. (iv) The unit cost of WFI is \$0.05/liters. (v) The chemicals used are of high purity grade (the following prices are assumed: \$2.0/kg of ammonium sulfate, \$3.75/kg of NaCl, \$3.5/kg of NaOH, and \$0.6/kg of glucose). (vi) The unit cost of membranes is \$200/m^2. (vii) The unit cost of chromatography resins is \$700/liter and \$500/liter for columns C-101 and C-102, respectively. (viii) The chromatography resins are replaced every 300 and 200 cycles for columns C-101 and C-102, respectively. (ix) The waste disposal cost is \$0.01/kg for mixed liquid waste. (x) The cost of cleaning chemicals is not considered.

24.3.2.5. Sensitivity Analysis

Cost analysis is frequently used in industry as a decision-making tool for planning future R&D efforts. The previous analysis showed that the target of \$3/g of purified product could not be met with the base case assumptions. However, reduction in cost from the estimated value of \$4.4/g to the target of \$3/g is not unrealistic. This can be achieved through increased product expression levels, through higher product recovery, through changes in the recovery process steps (by using less expensive alternatives), through increased production rates (economy of scale), and others. Figure 11 shows the impact of expression level and production rate on unit cost. As can be seen, for a production rate of 5,000 kg/year a product expression level of around 22% of total cell protein is required to reach the target cost. For a production rate of 10,000 kg/year, an expression level of around 17% is adequate for reaching the goal. Since these levels of product expression are readily achievable using genetic engineering techniques, the R&D efforts should proceed with emphasis on maximizing expression level and optimizing downstream processing.

Efforts to minimize the cost of downstream processing should focus on the two chromatography steps, HIC and

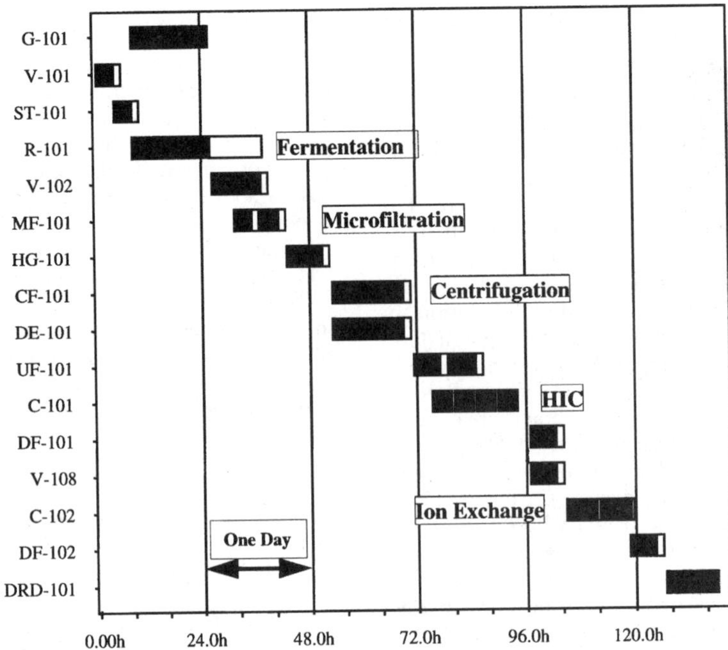

FIGURE 9 β-Galactosidase production scheduling Gantt chart.

ion-exchange, which account for 45% and 14.7% of the total cost, respectively. Alternatives that can reduce the cost of HIC include (i) use of less expensive resins, (ii) replacement of ammonium sulfate with a less expensive salt for the elution buffer, and (iii) partial removal of impurity proteins and nucleic acids via precipitation prior to HIC.

Since a large fraction (32%) of the manufacturing cost is dependent on DFC, savings also can be realized through better utilization of equipment. The effective plant batch time (the time between consecutive batches) of the base case is around 72 h (3 days), and the product is manufactured in 100 batches per year. However, the cycle time of the limiting process step, which is the fermentor in the base case, is only 30 h (18 h fermentation time and 12 h turnaround time). Theoretically, this means that a new batch can start every 30 h, resulting in up to 250 batches per year. For a fixed product throughput of 5,000 kg/year, a higher number of batches per year leads to higher labor cost (more operators are needed around the clock) and to lower capital cost (equipment of smaller size is required to produce the same annual amount of product). The results of the analysis are shown in Fig. 12 for product expression levels (based on total cell protein) of 10% (base case) and 20%. A cost reduction of around 16% can be realized by increasing the annual number of batches from 100 to 250.

TABLE 3 Overall material balances

Component	Material (kg/batch)[a]		
	Total inlet	Total outlet	Product
Ammonia	293.32	79.9	
β-Galactosidase	0.00	76.60	50.56
Biomass	0.00	8.11	
Carbon dioxide	0.00	2,698.51	
Debris	0.00	460.22	
Glucose	4,584.30	967.00	
Nucleic acids	0.00	306.81	
Contaminant proteins	0.00	700.09	
Salts	766.24	259.81	
Sodium chloride	119.09	119.09	
Sodium hydroxide	214.87	214.87	
Tris buffer	36,795.92	36,795.92	
Process water	47,807.69	47,894.49	
WFI	276,536.26	276,536.26	
Total	389,690.49	389,690.49	50.56

[a]1 batch = 52,130 liters of feed material.

24.4. SUMMARY

In the first part of this chapter we discussed the role that process simulation can play during the various stages of process development and product commercialization. The benefits from the use of such tools vary depending on the type of product, the stage of development, and the size of the investment. For commodity-like, low-priced biochemicals, minimization of manufacturing cost is the primary benefit. For high-value, regulated products, such as biopharmaceuticals, efficient evaluation of alternatives during process development is the primary benefit. That is the case because experience shows that poor choices made early in

TABLE 4 Key cost estimation results

Direct fixed capital	\$38.7 million
Total capital investment	\$39.2 million
Plant throughput	5,000 kg of β-galactosidase/year
Manufacturing cost	\$22.0 million/year
Unit production cost	\$4.4/g of β-galactosidase

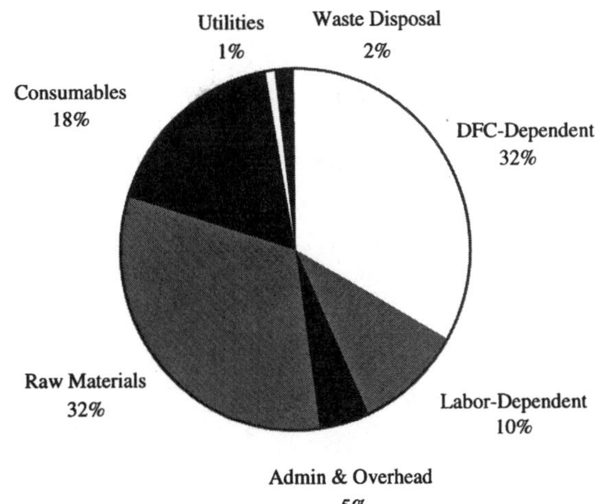

FIGURE 10 Breakdown of manufacturing cost for β-galactosidase production.

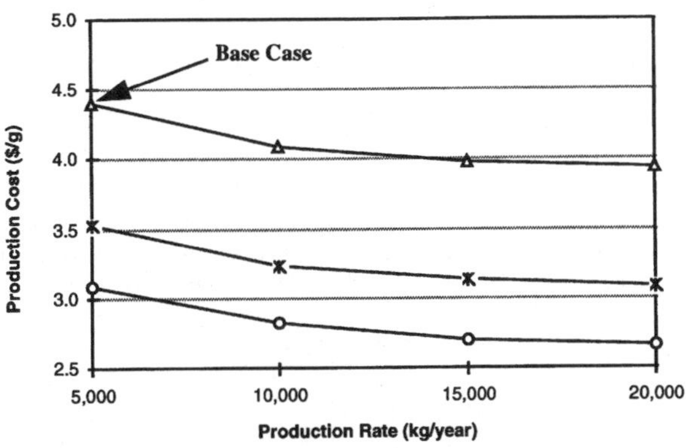

FIGURE 11 Unit cost as a function of product expression level (△, 10%; *, 15%; ○, 20%) and production rate.

process development are difficult to change later owing to regulatory constraints. This often leads to manufacturing processes that are far from optimum from the viewpoints of process economics, environmental impact, and safety.

In the second part of this chapter we analyzed the production of two biological products from a systems point of view. The analysis has clearly shown that depreciation of capital tends to be the most significant item of the manufacturing cost, especially for low production rates. As production rate goes up, raw materials and consumables tend to dominate the cost. The sensitivity analysis of Mab production has shown that the manufacturing cost of Mabs is unlikely to drop below $250/g even if transgenic animals or genetically altered crop plants are used for product formation. New, less expensive recovery technologies are required in combination with inexpensive upstream options to reduce production cost below $250/g. For bacterial systems, the second example has shown that a combination

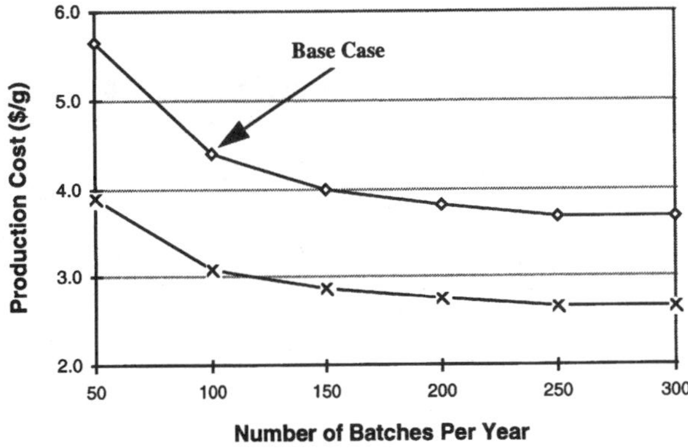

FIGURE 12 Effect of process scheduling on project economics. Symbols indicate product expression levels (◇, 10%; ×, 20%)

of high product expression levels and high production rates is required to achieve unit production cost of $3/g of β-galactosidase.

The biochemical industry has just begun making significant use of process simulation to support process development and optimize manufacturing. Increasingly, universities are incorporating the use of these tools in design courses. Thus, in the future, we can expect to see increased use of this technology and integration with other enabling technologies, such as advanced process control, computerized batch recipe generation, and on-line optimization. The result will be better processes, developed faster and at lower cost, making higher-quality products.

REFERENCES

1. **Atkinson, B., and F. Mavituna.** 1991. *Biochemical Engineering and Biotechnology Handbook*, 2nd ed., p. 447–519. Stockton Press, New York.
2. **DeYoung, H. G.** 1996. Multiple choices for monoclonal production. *Genet. Eng. News* **16(14):**12–33.
3. **Petrides, D. P., J. Calandranis, and C. L. Cooney.** 1996. Bioprocess optimization via CAPD and simulation for product commercialization. *Genet. Eng. News* **16(16):**24–40.
4. **Petrides, D. P., J. Calandranis, and J. Flora.** 1997. A comprehensive computer simulator can simplify the task of designing a cost-effective wastewater treatment plant. *Ind. Wastewater* (Water Environment Federation), May/June.
5. **Pharmacia Biotech.** 1997. *BioDirectory*, p. 235. Pharmacia Biotech, Inc., Piscataway, N.J.
6. **Seaver, S. S.** 1997. Monoclonal antibodies: using new techniques to reduce development time. *Genet. Eng. News* **17(1):**13–28.
7. **Smith, G. L.** 1994. Large scale animal cell culture, p. 69–84. *In* B. K. Lydersen, N. A. D'Elia, and K. L. Nelson (ed.), *Bioprocess Engineering*. Wiley Interscience, New York.
8. **Valentino, F. W., and G. E. Walmet.** 1986. Industrial waste reduction: the process problem. *Environment* **28:**16–34.

Biosafety and Biotechnology

DANIEL F. LIBERMAN, RICHARD FINK, AND FREDERICK SCHAEFER

25

Biotechnology has been defined as the application of biological systems to technical and industrial processes. This has involved the integration of biology, including molecular biology, genetics, microbiology, cell biology, and biochemistry, with chemical and process engineering in a way that develops the full potential of each of these systems (1, 12). From a practical standpoint, academic and industrial applications of biotechnology can be sorted into two general categories: research/development and production. Traditionally, the differences between research and production have involved both the goal and the scale of operation for these activities.

Production is the manufacture of specific materials (organisms or metabolites) by carefully developed and controlled procedures. Research, on the other hand, represents an investigative effort that may result in the development of a procedure, in new information concerning the properties of an existing procedure, or in process improvement. Production is restricted to industrial activity, whereas research is within the domain of both academia and industry.

Biosafety considerations of biotechnology are associated with three properties of microorganisms: (i) the potential of a few strains to cause disease; (ii) the potential for undetected genotypic or phenotypic changes to alter a tested and approved process; and (iii) the ubiquity of organisms that can contaminate the system.

While it is unlikely that dangerous human and animal pathogens will be used in very large-scale fermentation, the industrial use of plant pathogens is increasing (6, 16, 40). The literature concerning plant diseases and the agents that cause them and the potential impact of their release to the environment (see references 16 and 31 for a review) is somewhat limited. The aerosol dissemination of fungal spores (3), bacteria (19), and viruses that are pathogenic for plants (4, 31, 32) has been described. The economic consequences of the inadvertent release of a pathogen in an area where there are susceptible hosts is of concern, and efforts to prevent the escape of these organisms to the environment are in order.

The opportunity for new genotypes to become established in batch production is limited. This is due to the relatively few generations of the organisms involved. However, in continuous fermentations such as in single-cell protein production, the content of the bioreactor may be maintained for hundreds of days. This could involve many thousands of generations. With natural mutation rates as high as 1 in 10^5 (39, 40) and with additional chance of contamination from outside the system, the potential for genetic change is significant. However, it is important to keep in mind that the majority of these mutational events would be selected against by the fermentation conditions, and, unless some specific advantage was conferred, the mutant or contaminant probably would not become established (40). In contrast, if the contaminant that entered was able to maintain itself, it could disrupt the system in several ways. The contaminant could directly inhibit or interfere with the biocatalyst (enzyme, cell, or microorganism) or could destroy the catalyst or destroy the product by using it as an energy source. In addition, the contaminant could introduce substances that are difficult to separate from the product, thereby rendering the product unusable.

Since phenotypic change is a response to a changed environment and is maintained only while the new conditions persist, such change should be prevented by the close process control used to maintain maximum efficiency during the fermentation.

In practice, an industrial fermentation process is extremely unlikely to become contaminated with a highly pathogenic microorganism, because the environment inside the fermentor is so different from the human body that pathogenicity confers no advantage upon the organism. On the other hand, unless steps are taken to preclude the introduction of environmental microorganisms, they can and will get in and disrupt the system.

While the orientation of this chapter is toward worker safety, it must be kept in mind that the steps taken to protect the worker are equivalent to the steps one takes to protect the integrity of the system. Whenever there is a significant potential for introducing undesirable organisms into the bioreactor, there is an equivalent opportunity for organisms to escape into the environment. It is therefore appropriate to examine more closely the various stages involved in manipulating organisms in a process and to describe the practices that historically have been shown to minimize or eliminate contamination of personnel, product, and the environment. We first consider overall containment strategy and subsequently will examine various stages of manipulation to see how these containment strategies apply.

25.1. BIOSAFETY CONSIDERATIONS

The actual causes for most laboratory-acquired illnesses are not known (18, 25, 27, 30, 38). Fewer than 20% of documented infections have been attributed to accidental contact, ingestion, or injection with infectious material. The remaining 80% have been attributed to unknown or unrecognized causes (18, 25, 27, 38, 41, 42). These analyses suggest that personnel engaged in research are at a higher risk than personnel associated with diagnostic, educational, or industrial activities (15). The risk to personnel who are in direct contact with the agent is higher than that for personnel who are only remotely involved. Eighty percent of the 300 illnesses at Fort Detrick reviewed by Wedum et al. (43) involved trained laboratory personnel. Support personnel such as janitors, dishwashers, and maintenance and clerical workers were at lower risk. These results serve to confirm previous estimates reported by Sulkin and Pike (41).

Laboratory studies on the potential sources of infection have focused on the hazard potential of routine laboratory techniques. A number of studies suggest that laboratory techniques create aerosols and that inhalation exposure to undetected infectious aerosols may contribute significantly to occupational illness among laboratory workers who handle infectious material (5, 20, 25, 35–37, 42). Based upon these limited data, preventive measures have been proposed that will provide safeguards for the protection of scientific and support personnel, the experiment, and the environment. These safeguards are collectively referred to as containment practices (5, 18, 27).

25.2. TYPES OF CONTAINMENT

Practical planning for safety is hampered by the fact that safety cannot be measured directly. The words "safe" and "safety" represent ideal concepts that, while desirable, are unattainable in absolute terms. Practical planning for safety is therefore performed by evaluating its opposite, risk. Safety in laboratory research activity then becomes an exercise in recognizing what the risks are and then introducing procedures, practices, etc., to control the identified hazards or reduce them to acceptable levels.

The purpose of physical containment is to reduce the potential for exposing workers (inside as well as outside the laboratory environment) and the public to infectious agents that are under study within the laboratory. The elements of physical containment include the scrupulous use of safe laboratory techniques in well-designed facilities and the use of equipment that is appropriate for the given activity (24, 33, 36, 37).

The procedures and equipment used by the workers to control potentially hazardous agents are as important a barrier as the laboratory facility itself to protect the workers, the public, and the product from exposure to potentially detrimental material (hazardous organisms and their metabolites as well as saprophytic contaminants).

It would be difficult to describe in detail all the specific rules and procedures designed to prevent or control laboratory-acquired illness or product contamination, because these procedures vary and depend to a great extent on the agent, the type of experiment, the equipment, the facilities, and the proficiency of the personnel. However, there are several basic practices that should be mentioned. They constitute the basic recommended practices that have been shown to be effective for the protection of personnel, product, and the environment. There are a number of excellent reviews on this subject, and the reader is encouraged to consult them (5, 21, 24, 25, 27, 36–38). It is incumbent on each laboratory that is involved with or contemplates the use of hazardous or potentially hazardous agents to examine these references and to develop protocols that meet the associated safety concerns.

25.3. PERSONNEL PRACTICES

Laboratory supervisors should prohibit eating, smoking, drinking, food storage, and the application of cosmetics in the laboratory. They should not permit pipetting by mouth; the use of mechanical devices must be required. These devices are more accurate than standard pipettes and clearly eliminate aspiration as a source of laboratory-acquired illness. These simple personnel protective practices are designed to eliminate ingestion as a mode of exposure.

A second group of protective practices includes the use of appropriate gloves when working with hazardous material. Gloves function to prevent the worker's hands or fingers from being contaminated, which further reduces the hazards associated with ingestion (hand-to-mouth transfer) or the penetration of material through broken or unbroken skin. Although there are relatively few organisms that can penetrate unbroken skin, one should not lose sight of the fact that experiments performed today also involve chemicals (some of which are toxic) and radionuclides. Wearing proper gloves should be regarded as a *minimum* requirement.

Laboratory clothing, which is designed to protect one's personal clothing, should be worn. The purpose of laboratory clothing is to keep street clothing and the worker's forearms free of contamination. It is important to realize that wearing potentially contaminated clothing (laboratory coats) to the cafeteria, library, meetings, or other buildings provides a mechanism for spreading contamination to others as well as to oneself. Laboratory costs should never be worn outside the laboratory area.

Work surfaces must be decontaminated both daily and immediately after spills. This will reduce the spread of contamination to the worker and at the same time reduce the potential for contaminating the experiment or process at hand or elsewhere in the laboratory.

As indicated previously, considerable information has been accumulated that clearly indicates that nearly *all* routine laboratory procedures are capable of producing aerosols, including particles of respirable size (2, 8, 14, 23).

Any operation that involves an animal or plant pathogen or an agent that could disrupt the environment if inadvertently released should be contained within safety equipment or facilities. These containment systems must be subjected to periodic inspection and certification to ensure proper function (36). All biological waste should be decontaminated or inactivated before disposal. This is especially true if known pathogens are involved. Contaminated materials such as bioreactors, glassware, laboratory equipment, etc., should be decontaminated before washing, reuse, or disposal. This again will help protect personnel who are not directly associated with the laboratory activity (e.g., glassware workers, janitors, repair personnel, technical support).

Finally, all employees working with or potentially exposed to hazardous organisms or substances must receive sufficient information and training to enable them to work safely and to understand and appreciate the relative importance of potential hazards. This instruction should include a thorough review of operations and procedures, with emphasis on material transfer and other possible sources of exposure; adverse health effects and early warning symptoms; specific acceptable and unacceptable work practices; engineering controls (such as hazard control ventilation, contained centrifuges, and safety cabinets) in use or being considered for use to limit personnel or environmental exposure; proper disposal of contaminated waste; decontamination of surfaces; and specific emergency procedures to be followed in the event of an accident or spill. Each member of the work team must be familiar with the biology of the system or process under way.

These basic practices represent a commonsense approach to safety that is based on over four decades of experience with laboratory management of microorganism and associated illnesses. While there are risks associated with activity involving infectious agents, these risks can be minimized if appropriate attention is paid to the safety practices described above as well as to primary and secondary containment barriers.

25.4. PRIMARY CONTAINMENT BARRIERS

The selection of specific items of equipment is dictated by the hazard potential of the organism (10, 18, 30). Safety equipment used as primary barriers include biological safety cabinets, safety blenders, safety centrifuge cups, and a variety of enclosed containers (e.g., fermentors). These items are designed to prevent the escape of aerosols into the laboratory environment.

Biological safety cabinets are used as a primary barrier to prevent the escape of aerosols from the point of origin into the laboratory environment. Three types of cabinets (class I, II, and III) are used in laboratories (36). Open-fronted class I and II biosafety cabinets are partial containment cabinets that offer protection to laboratory personnel and to the environment when used in conjunction with good microbiological technique. Since the inward air velocity (face velocity) is similar for both classes, they provide an equivalent level of personnel protection. Class II cabinets offer the additional capability of protecting materials contained within them from extraneous airborne contaminants. This is provided by a laminar stream of HEPA (high-efficiency particulate air)-filtered air that passes over (vertically) within the workspace.

Neither of these cabinet classes is appropriate for the containment of the highest-risk infectious agents because of the chance for inadvertent escape of aerosols across the open front (5, 36, 37). The class III cabinet, commonly referred to as a glove box, provides this level of personnel and material protection. Protection is provided by the physical isolation of the space in which the infectious agent is manipulated. When class III cabinets are required, all procedures must be conducted within them. These cabinets are frequently designed as a system of interconnected units that contain all the equipment required by the laboratory program, e.g., incubators, refrigerators, centrifuges, and even animal storage cages (36).

25.5. SECONDARY CONTAINMENT BARRIERS

Secondary barriers are the features of the facility that surround the primary barriers. These barriers provide a separation of the laboratory from the outside environment as well as from other facilities in the same building. Examples of secondary barriers are floors, walls, ceilings, airlocks, doors, differential pressures, exhaust air filters, and provisions for treating contaminated wastes. These and other design features provide barriers that serve to prevent the escape of hazardous microorganisms in the event of a failure in a primary barrier and can be designed to prevent external organisms from contaminating the workspace (34). Actually, the more effective the primary barriers are, the less need there is for emphasis on secondary barriers. In the design of a research laboratory, it is an important and economic necessity to determine and select the primary containment to be used, thereby reducing the complexity and cost of the secondary barriers (11).

The primary function of the facility is to provide a physical environment in which work can be performed efficiently and safely. A well-designed facility will facilitate good laboratory practice, contain equipment designed to protect the worker from the potential hazards associated with the activity or the system, and ensure the protection of people and the environment outside of the laboratory or building.

Three categories of facility safeguards have been established for hazardous or potentially hazardous research projects. These categories are classified as the basic facility, the containment facility, and the maximum containment facility. The levels of protection increase with the risks associated with the agents to be used and, in general, correspond with levels of containment required for low-, moderate-, and high-risk agents or procedures. The basic facility is suitable for research, development, or production activity that requires biosafety levels 1 or 2 containment and procedures and practices. Organisms such as *Escherichia coli*, *Bacillus subtilis*, and *Saccharomyces cerevisiae*, along with the GRAS (generally recognized as safe list of important industrial organisms that pose little or no risk), can be safely managed at level 1. Clinically significant organisms such as enteropathic *E. coli*, *Salmonella*, and their relatives among the *Enterobacteriaceae* require level 2 containment and work practices. Biosafety level 3 is used for organisms that are hazardous and can be transmitted by an aerosol route. Maximum containment is required for work that involves extremely dangerous organisms for which there is no known therapy and that are associated with high mortality. Industrial interest in these latter organisms is limited to vaccine development. The design criteria for these facilities are published elsewhere (24, 34, 36). It should be emphasized that, while the facility can provide an environment in which work can be conducted safely, the facility itself does not ensure that the work environment is safe. Well-designed facilities primarily protect the environment and cannot be considered a replacement for good laboratory procedures and practices.

These comments on containment are consistent with those put forth by the National Institutes of Health for research with hazardous materials such as carcinogens (35), oncogenic viruses (21), and recombinant DNA molecules (37). The Centers for Disease Control and Prevention (CDC), in conjunction with the National Institutes of Health (NIH), has developed a similar approach in its recommendations for microbiological and biomedical laboratories (42). The bottom line in all these documents is that hazardous organisms or materials must be contained in a fashion that is commensurate with the level of hazard associated with the organisms or materials (42).

The principles that underlie the containment practices for large-scale activity are based upon the recognition that there may be risks associated with the activity. The best way to control potential risk is through the use of techniques and equipment designed to contain aerosols and prevent their release to the work and general environment. The requirements to treat effluent air and liquid to render them biologically inactive before release are as important as the requirements to use aseptic technique (36, 37, 40).

The CDC/NIH publication describes four levels of biosafety (BL-1 through BL-4), classifies a variety of infectious agents into one or more of these levels, provides basic recommendations on practical safeguards that can reduce the risks of laboratory associated diseases (see below), and presents work practices, containment equipment, and facility design requirements for practitioners to consider (42). It also describes a process to evaluate the biosafety requirements as the type of activity changes. In other words, it describes how one can evaluate containment requirements for activity that takes place in a hospital or research laboratory or animal research laboratory or even within a pilot plant. In addition, it describes various conditions (e.g., volume concentration, equipment, facilities, hazards) that could alter biosafety requirements (42).

This approach is appropriate because the hazard(s) may depend on the nature of the activity. For example, the diagnosis of viral illness via cytopathic effects involves lower concentrations than does the isolation and cultivation of the same virus for purposes of research. Challenging animals to determine antigenicity or pathogenicity of a virus has different containment implications than does the production of large amounts of the same virus for vaccine development purposes. For a single virus, the biological containment may vary from BL1 to BL3 for these various activities.

The facility design requirements for each biosafety level are outlined below. In addition, the types of work that are generally regarded as acceptable and the recommended practices for that level of activity are also presented.

25.5.1. Biosafety Level 1 Containment

BL-1 is the typical facility for work with low-hazard agents. Viable microorganisms not known to cause disease in healthy adult human beings are used at this level. Work activity is conducted on an open bench. It is believed that any hazard present can be controlled by standard laboratory practices. Biological safety cabinets are not required, and it is more common to find a chemical fume hood for the containment of acutely hazardous chemical substances. Standard facility features consist of easily cleaned, impervious bench surfaces, sturdy furnishings, handwashing sinks, and window screens where windows can be opened. The laboratory space should be separated from general offices, food service, patient, animal, or production areas.

A BL-1 facility is suitable for experiments involving:

1. Recombinant DNA activity requiring BL-1 containment (37)
2. Microorganisms of minimal or no biohazard potential under ordinary handling conditions. Such agents are designated class 1 (10) or BL-1 (37, 42)
3. Nonrecombinant cell or tissue culture studies that do not involve infectious plant or animal virus (24, 33)
4. Management of animal populations that are free of zoonotic organisms or are not part of a study that involves organisms or chemical substances that may require higher containment considerations
5. Production activity with organisms classified as requiring BL-1 containment (37, 42)

The control of potential biohazards at the BL-1 level is provided by use of standard microbiological practices.

25.5.2. Biosafety Level 2 Containment

Since the BL-2 facility is identical in construction to the BL-1 facility, the term basic facility is frequently used as an identifier. Work that involves organisms that pose some risk can be conducted here. While work that does not produce significant aerosols can be conducted on an open bench, biological safety cabinets (for biological exposure control) and chemical fume hoods (for chemical exposure control) are frequently present.

This laboratory is suitable for experiments involving:

1. Recombinant DNA activity requiring BL-2 physical containment, including animal studies that involve the construction of transgenic animals (37, 42)
2. Microorganisms of low biohazard potential, such as those in class 2 (10) or BL-2 (37, 42)
3. Nonrecombinant cell and/or tissue culture systems that require this level of containment (33)
4. Oncogenic viral systems classified as low risk (24, 33)
5. Certain suspected carcinogens and other toxic chemicals
6. Introduction of BL-1 materials into experimental animals (37)
7. Production activities with BL-2 organisms (37, 42)

It is important to understand that at these levels we are dealing with research, diagnostic or production activities thought to pose few or minimal hazards to workers and that standard procedures and practices are considered sufficient to protect the work force.

25.5.3. Biosafety Level 3 Containment

At BL-3, facility design plays a significant role in safety (27, 42). BL-3 activity involves organisms or systems that pose a significant risk or represent a potentially serious threat to health and safety of workers. Such facilities include special engineering design features and containment equipment. These facilities are usually separated from the general traffic flow by controlled access corridors, air locks, locker rooms, or other double-door entries. Biosafety cabinets are required for all technical manipulations that involve viable cultures (no work is allowed on an open bench). The surfaces of all walls, floors, and ceilings are sealed and therefore impervious to liquids that may spill onto them. This means that all penetrations (telephone, lights, plumbed lines for gas, vacuum, electrical lines, electrical switches, etc.) are caulked, collared, or sealed to prevent leaks. The collars and seals are made of material that can be cleaned (34, 36).

The ventilation system in the BL-3 facility is designed to exhaust more air than is supplied, resulting in a directional airflow from the outer corridors, which are regarded as clean, into the laboratory, which is regarded as contaminated. The air is usually discharged to the outdoors and not recirculated to other parts of the building without appropriate filtration treatment (34, 36).

This laboratory design is suitable for experiments involving:

1. Recombinant DNA molecules requiring physical containment at the BL-3 level, including animal studies with BL-3 and some BL-2 agents (37, 42)
2. Microorganisms of moderate biohazard potential, such as those in class 3 (10) or BL-3 (37, 42)
3. Oncogenic viruses that have human cells in their host range (24, 33)
4. The production of large volumes or high concentrations of certain BL-2 and all BL-3 microorganisms or virus-infected cells (where the virus is infectious for humans and requires BL-3 containment) (24, 33)
5. Certain carcinogens and other toxic chemicals
6. Production activity that involves BL-3 and some BL-2 recombinant DNA organisms (37, 42)

BL-3 organisms differ from those at BL-2 by being infectious through an aerosol route of exposure and may cause serious illness. Organisms such as *Mycobacterium tuberculosis*, St. Louis encephalitis virus, and *Coxiella burnetii* belong in this category (1).

The procedures and practices should be reviewed carefully, and only workers who are trained properly should be permitted to work at this level.

25.5.4. Biosafety Level 4 Containment

BL-4 facilities are extremely sophisticated in terms of design and provide a very high level of containment for research involving biological agents that present a life-threatening potential to the worker or may initiate a serious epidemic disease (34, 36, 42). The distinguishing characteristic is the use of barriers to prevent the escape of hazardous material to the environment, as well as additional barriers to protect laboratory personnel.

Barriers that serve to isolate the laboratory area from the immediate area include:

1. Monolithic walls, floors, and ceilings in which all penetrations such as air ducts, electrical conduits, and utility pipes are sealed to ensure the physical isolation of the laboratory area
2. Air locks through which supplies and materials can be brought safely into the facility
3. Contiguous clothing change rooms and showers through which personnel enter and exit the facility
4. Double-door autoclaves to sterilize wastes and other materials prior to removal from the facility
5. Biowaste treatment systems to sterilize liquid waste
6. Separate ventilation systems that control air pressures and airflow directions within the facility
7. Treatment systems to decontaminate exhaust air before discharge into the atmosphere
8. Back-up systems for electrical and equipment failure

A description of specific operational procedures and laboratory practices recommended for use in maximum containment facilities can be found in references 36 and 42. These facilities are usually operated by very well trained workers who work under rigorous supervision.

This laboratory is suitable for experiments involving:

1. Recombinant DNA molecules requiring physical containment at BL-4 (42)
2. Microorganisms of high biohazard potential such as those classified as class 4 or 5 (10) or of BL-4 (42)
3. High-risk oncogenic viruses (24, 34, 42)
4. Carcinogens on the Occupational Health and Safety Administration restricted list
5. Introducing any of the above into experimental animals

As indicated, these are extremely sophisticated facilities that can provide the highest level of containment for working with agents that have life-threatening potential. More detailed descriptions of these facilities and associated research activities are beyond the scope of this chapter.

25.5.5. Containment for Production Activities

The previous discussion of laboratory containment requirements was initially developed by the NIH for research activity that involved recombinant DNA molecules (37). Subsequently, the containment descriptions (P1, P2, P3, and P4) were used to describe the containment requirements for the research activity that involved infectious agents as well (24, 42).

Industrial applications that involve recombinant organisms usually involve the propagation of the organism in large scale. The NIH guidelines define large scale as a volume equal to or greater than 10 liters and describe the containment requirements for production activity that exceeds this volume.

The specific recommendations are similar to the research recommendations and are found in Appendix K of the Guidelines (37). Originally, three levels of containment were defined, e.g., BL-1LS, BL-2LS, and BL-3LS, where LS stands for large scale (see sections 25.5.1 to 25.5.3). Basically, if an organism requires BL-2 containment under research-scale conditions, the cultivation of 10 or more liters would require BL-2LS containment. Most large-scale recombinant DNA applications have been conducted at BL-1 containment. Subsequently, the NIH and the Food and Drug Administration came to the realization that certain organisms are inherently safe and that a variety of organisms can be managed under containment conditions that are lower than BL-1LS. The NIH Guidelines were modified to include a fourth containment level called GLSP, which defines a set of Good Large-Scale Practices that can be used with organisms which are generally regarded as safe and therefore pose little or no threat to the worker or the environment (37).

25.5.6. Practical Considerations

In the absence of information and/or the lack of experience with new microorganisms, laboratory personnel should select work practices that offer increased protection. Laboratory manipulations of the human immunodeficiency virus (HIV) have prompted biosafety committees and safety personnel to take a hard look at containment practices (18). Based upon observations that HIV does not have a documented aerosol route of exposure, the need for the additional containment of a BL-3 facility does not seem critical at this time. BL-2 facilities used in conjunction with BL-3 procedures and work practices with the appropriate safety equipment (safety centrifuge cups, biosafety cabinets, etc.) afford a greater margin of safety for personnel than merely increasing unneeded containment. This hybrid containment (which is referred to as BL-2+) reduces the risk to laboratory personnel who are working with infectious agents that are not environmental hazards. This containment level would also be suitable for activity with agents for which there is insufficient worker safety information available.

25.6. BIOTECHNOLOGY

Biotechnology was defined above as the application of biological systems to basic and applied technical and industrial processes. These processes are based on the use of some form of organic catalyst in the conversion of a substrate to a desired product. These catalysts may be simple enzymes or highly complex microorganisms or cells (23).

The selection of the appropriate catalyst depends on the process and product of interest. In the production of beer or wine, yeasts are selected for the conversion of sugars into ethanol and carbon dioxide. In the production of a biologically active protein like interferon, the catalyst can be a bacterium, a yeast, or a cultured cell that has been engineered to express proteins.

A common concern in all applications is the need to maintain aseptic conditions (11). The reason for this is that most products of such processes are made by a pure culture that was derived from a single strain or clone. Introduction of another type of organism into this culture could affect the efficiency of the process or even destroy the culture altogether. To avoid contamination, all phases of the process must be carefully evaluated and monitored.

Biotechnology encompasses a wide range of activities. While each activity is associated with its own starting material (bacterial, fungal, yeast, plant, or animal cells or enzymes), technical manipulation (cell fusion, cell or enzyme immobilization, genetic engineering, etc.), processes (continuous or batch fermentation), and scale (small to very large) that are unique to that activity, there are common features or stages that lend themselves to a generic treatment. Regardless of the goal, biotechnology will involve (i) isolation and preservation of the organism or cell that is the biocatalyst or the source of the specific catalyst of interest, (ii) preparation of biocatalyst, (iii) process scale-up or large-scale growth, (iv) separation of the desired product from the biocatalyst, (v) purification of the product, and (vi) waste treatment (29).

25.6.1. Culture Preservation

Organisms to be preserved are usually cultivated in small volumes (less than 100 ml), and the resultant mixture is transferred in quantities of up to 1 ml into small tubes with a cryopreservative. These tubes are sealed and stored in or above liquid nitrogen (−180°C or −196°C), in a low-temperature freezer (−80°C), or in a standard freezer (−20°C) (17). Alternatively, samples can be lyophilized and stored between room temperature and 0°C (17, 22). These samples serve as the starting material for all future cultivations. Each sealed tube may contain a few micrograms (dry weight) of microorganism. The number of organisms per tube may vary from 10^8 for yeast and molds to 10^{10} for bacteria (40).

Even though the volume of medium or the number of organisms in the starting sample is comparatively small, the biology of the organism (pathogen or saprophyte, mechanism of pathogenicity, identification of suitable hosts at risk, type of potential illness, etc.) must be understood before work begins. Appropriate work practices, in an appropriate biological safety cabinet, should protect both the worker and the biological system from gross contamination. When a lyophilizer is used, care should be taken to prevent the vacuum lines from being contaminated. The insertion of a suitable filter or trap is usually sufficient. All lines, tubing, or glass connectors that come in contact with the primary vessel or tube should be decontaminated by chemical disinfection, followed by extensive washing to remove residual disinfectant, or by steam sterilization before reuse.

25.6.2. Preparation of the Product

The next stage is to provide the required number and quality of organisms or enzymes to initiate a larger-scale cultivation or process.

All plastic or glassware that comes in contact with the organism should be decontaminated or disposed of promptly after use and not allowed to accumulate. Unattended bottles, flasks, plates, etc., containing medium (or even a residual film of medium) can support the growth of saprophytes, which may result in unnecessary contamination of the system. Keep in mind that many of these environmental organisms can produce spores and therefore are capable of extensive contamination.

If pathogens are to be used, as in the production of live/attenuated vaccines, precautions to prevent accidental release (breakage of flasks during amplification) must be prevented. All glassware should be examined before use; any item that is chipped or cracked should not be used.

When a dry air shaker or a shaker platform is used with pathogens (especially for organisms known to be respiratory pathogens), the unit should be enclosed in a chamber that operates under a negative pressure, and the exhaust air should be cleaned by filtration. If a vessel should break or a cotton plug or stopper be dislodged, then any aerosol that might be generated would be trapped in the filter.

If such an accident should occur, then personnel who are suitably trained, clothed, and provided with appropriate respiratory protection should remove the undamaged vessels, clean up the debris, and decontaminate the shaker and chamber surfaces appropriately. A listing of appropriate disinfectants/decontaminants and procedures for their use can be found in reference 35.

25.6.3. Scale-Up of Process

To maximize the yield of a desired product, microorganisms are grown in bioreactors under rigorously controlled conditions. Often, the final process is preceded by growth in successively larger vessels, e.g., 20, 200, and 2,000 liters (7), usually constructed of stainless steel.

There are three basic fermentor-bioreactor designs:

1. Small, portable units that are filled with medium and sterilized in an autoclave
2. Portable units that are sterilized in an autoclave and filled aseptically with sterile medium
3. Fixed units that are sterilized in place (7, 9, 11). In a batch process, most of the constituents of the medium are combined with the biological catalyst at the start.

The process vessel is sterilized before the medium is introduced. Depending on the scale of operation, the starting materials are added by means of tubes or pipes (7, 11).

In the vessel, the catalyst and constituents of the medium are mixed by a rotating control shaft that carries several impeller rotor blades. As the biological conversion proceeds, any additional nutrients needed can be added via additional tubes and lines. This requires that conditions in the reaction vessel be monitored during the conversion and that sensors be inserted through the vessel wall at various locations. A vessel for a continuous process is similar, except that nutrient is continuously added and the products of the reaction are continuously removed (7).

In either the batch or continuous mode of operation, the design of fermentation equipment and facilities is central to the containment strategy for biological control. The term "biological control" is used, because both the worker and the integrity of the culture must be protected. To avoid contamination of the culture, all materials that come in contact with it, as well as the medium constituents, must be sterilized. Air is frequently filtered through a deep bed of glass wool or by special HEPA filters. Reactor vessels, pipelines, and other surfaces with which the medium comes in contact are steam sterilized. The mechanical system is designed and operated so that the opportunities for contaminants to enter, or for organisms to escape, are eliminated or minimized. Maintaining the integrity of various entry and exit points in the system is both crucial and often difficult. Since the potential for human error and mechanical failure exists, culture contamination, unfortunately, does occur.

While massive contamination is usually caused by mistakes, slight contamination that may develop gradually is frequently due to inadequate sterilization. If the contamination is chronic, the whole system should be given a detailed, critical inspection.

In the initial installation and during contamination review, care must be taken to ensure that all piping and fittings are welded properly and that the use of connecting fittings is minimized (7, 9). Dead spaces, crevices, and nondraining portions of lines should be avoided. Sterile and nonsterile segments of the system should be separated by steam blocks. Valves should be examined for dead spots or crevices before being installed.

When the activity involves infectious organisms, exhaust gases must be either filter sterilized or incinerated to remove or destroy any organisms suspended in them. An effluent stream with a high moisture content, i.e., in the event of a "foam-out," would reduce HEPA filtration efficiency and containment efficacy. In such a situation, hydrophobic prefilters, in conjunction with a catch basin, demister, or tank to collect condensate or foam, are to be included in the design.

Agitator seals must be examined for leaks. The possibility that leaks can occur is not trivial. The reason for the recommendation for top-mounted agitation for large-scale processes is that if the impeller shaft is located at the bottom, a seal leak could result in the vessel's contents' contaminating the motor, the floor, etc. If the system is located at the top, then the extent of contamination is reduced substantially.

Three types of seals (packed seal, lip or oil seal, and mechanical seal) are commonly used (9, 11). The current trend in design of contained fermentors is toward double mechanical seals.

Sampling ports must also be considered possible routes of contamination. Samples should be taken via a closed system to avoid the generation of aerosols (28, 36). Flush-mounted valves that eliminate dead spaces should be used where possible. Piston and ball valves can be designed to include a steam block, in which the sample flows from the vessel to the sample port when the valve is open, and steam flows back through the valve toward the sample port and the vessel when the valve is closed. Butterfly valves are also very popular.

Ports must also be provided for probes to monitor the progress of the fermentation. Probes that can be sterilized along with the reaction vessel should be used. It is important to avoid crevices on the ports and to incorporate both internal and external seals in the design of probe ports.

25.6.4. Recovering the Product

The specific method of product recovery and purification depends on the properties of the product, such as its location (intracellular, dissolved in medium, or both) and its stability to heat and chemical disinfectants (reagents used in the extraction and purification may inactivate the product). Yeasts and fungi are usually harvested by continuous filtration, whereas bacteria, viruses, and cells are routinely harvested by centrifugation. It is possible to recover on the order of 100 kg of wet cell material from 1,000 liters of culture (9, 40).

Separation and subsequent processing always involve the disruption of air/liquid interfaces and, therefore, may result in the release of aerosols. If the biocatalyst is a pathogen, it is desirable to inactivate the organism first. This can be accomplished by "in place" sterilization within the reactor vessel (batch sterilization). Alternatively, the reactor contents can be sterilized by being passed through a heat exchanger (continuous-flow sterilization) (28, 37). Either way is acceptable if the product of interest is stable under the conditions of sterilization.

If the product is heat labile, then it is necessary to explore the possibility of chemical inactivation of the organism. This requires that the product be unaffected by the disinfectant selected.

In the event that the product is not stable, then it must be separated from the cells before further processing. Such bulk processing must be performed under "closed conditions" to minimize personnel and environmental exposures (28, 37).

The vessel contents are transferred to a separation or treatment tank, a filtration column or centrifuge, by a series of plumbed lines. The integrity of these lines should be examined in the same fashion as that of the vessel itself. The columns and centrifuge should be closed systems (primary containment) or placed in a chamber or room which is specifically designed to control the spread of aerosols (secondary containment). Such rooms or chambers are maintained under negative pressure to prevent dissemination of organisms to neighboring areas or to the environment (28, 34).

25.6.5. Processing the Product

Subsequent processing methods vary widely and depend on the product to be recovered. In some cases, the whole culture may be used without separation of cells. Separated microorganisms may be used without further processing beyond drying, such as the yeasts, fungi, and bacteria used as single-cell protein (31). In other instances, the cells may be subjected to some form of chemical or physical disruption to release the desired product (9).

Cells may be disrupted by nonmechanical methods (osmotic shock or enzymatic or chemical lysis) or by mechanical methods. Because of difficulties in scaling up the former procedures, mechanical methods are more popular on the industrial scale. Two types of mechanical disrupters are widely used: high-pressure homogenizers and high-speed agitator mills. Since both have the potential for generating aerosols, the system design must ensure either that any aerosol generated be retained within the unit itself or, as mentioned above for filtration systems and centrifuges, that the unit be placed within a suitable chamber or room that will provide the necessary aerosol control. It is also often essential to protect the product from contamination. In this case, efforts must be directed toward keeping external organisms out rather than keeping process organisms in. This requires facilities and equipment that are operated under positive pressure conditions. Containment facilities then are transformed into barrier facilities (34). In such circumstances it is essential that any pathogenic organism used as the biocatalyst be inactivated before processing to avoid personnel exposures.

25.7. WASTE TREATMENT

Although some of the activities associated with biotechnology are comparatively new (e.g., genetic engineering, enzyme and cell immobilization, cell fusion), applied microbiology (fermentation), which is the cornerstone of this technology, is not new (11–13). The practice of industrial fermentation since World War II has provided us with sufficient background to anticipate the kinds of waste that will be generated and, of equal import, the methods available to treat them.

The important point to remember is that if pathogens are used, they must be rendered inactive before being introduced into the environment. This also applies to genetically engineered organisms, which are regulated by the NIH Guidelines (37). Such cells can be inactivated by steam sterilization or chemical treatment. Clearly, the scale of operation and the stability of the product are essential parameters in the method(s) used.

More often than not, the product is heat labile, and the use of a disinfectant is not feasible or possible owing to product instability. In such cases, the biocatalyst is separated from the product, and the waste that is generated is processed by an appropriate method. Frequently, the product is intracellular and the cells are disrupted to release the product into the milieu. While disruption should render the organism nonviable, it is not certain that all the cells will be disrupted. Care should be exerted whenever pathogens are used to avoid the introduction of viable organisms into the environment. This will necessitate the treatment of all biowaste generated in any step that has not been subjected to previous treatment.

In principle, the disposal of waste is fairly straightforward. Whether the waste is processed on-site or off-site, standard sewage treatment should be effective with microbial waste streams to reduce the amount of oxidizable organic material. The components of a biological waste stream are organic chemicals containing nitrogen, phosphorus, and carbon and inorganic chemicals, including buffers, acids, and alkalis, that, once inactivated, pose limited risk to the environment.

Several monographs deal with the disposal or processing of biological waste (see Liberman [26]).

25.8. POSSIBLE HAZARDS FROM INDUSTRIAL PRODUCTION AND USE OF PATHOGENS

A pathogen that is still virulent after growth in a fermentor cannot be handled safely on a very large scale. Nevertheless, if the kinds of precautions described above are employed, there is little reason to expect that industrial processing of pathogens on a scale not exceeding a few thousand liters will lead to their release, either as bulk liquid or, much more important, as an aerosol. Therefore, at this scale, we need only consider the possible consequences of a system failure.

The possibility of such a failure can be reduced to a very low level but never eliminated. Accidental release might follow a mechanical failure, filter failure, or foam-out. The negative consequence of such a failure would depend on a number of factors: (i) the strain of microorganism used (10, 42); (ii) the quantity of microorganism released (24, 35, 40, 42); (iii) the physiological state of the microorganism, including its pathogenicity, infectivity, spore production, and aerostability as determined by growth conditions (16, 43); (iv) the local environmental conditions that will determine the organism's dispersal, dilution, survival, and ability to enter the ecosystem; (v) the location of the facility with respect to susceptible hosts (18, 30, 37); and (vi) the susceptibility of possible hosts to infection, as determined by a variety of factors (18).

All of these factors are interrelated and subject to variation. At present, it is quite impossible to predict their relative importance. The best way to eliminate the hazard of a pathogenic organism is not to use that organism at all, but instead to employ a substitute.

Less hazardous organisms that would fulfill the same purpose as the undesirable pathogen can be discovered, selected, or created. Industry for many years has employed selection techniques to screen large numbers of related organisms for nonpathogens, as well as more complex methods involving mutation and selection of desirable strains. Genetic engineering techniques have been widely used to remove the information (genes) of interest from potentially harmful pathogens and introduce this information into nonpathogenic organisms such as *E. coli*, *B. subtilis*, or *S. cerevisiae* (37).

25.9. CONCLUSIONS

Potential applications for biological systems are so vast and diverse that it is impossible to predict all the areas of academic or industrial activity that will benefit from their use. All that is certain is that biotechnology will play an increasingly important role in future industrial processes.

We must try to ensure that each individual who works with microorganisms is properly trained, that each program of activity is adequately reviewed, and that equipment necessary to protect personnel, the product, and the environment is provided and used correctly, to avoid personal and environmental contamination.

The widely divergent activities associated with biotechnology make it difficult, if not impossible, for safety specialists to anticipate the use of potentially hazardous materials and to monitor appropriately each and every operation that involves these materials. It is obvious, then, that the user must have sufficient knowledge of the hazard potential presented by these materials so that the precautions necessary to conduct the activity safely can be instituted.

REFERENCES

1. **Abelson, P. H.** 1983. Biotechnology: an overview. *Science* **219:**609.
2. **Anderson, R. E., L. Stein, M. L. Moss, and N. H. Gross.** 1952. Potential infectious hazards of common bacteriological techniques. *J. Bacteriol.* **64:**473–481.
3. **Aylor, D. E., and P. E. Waggoner.** 1980. Aerial dissemination of fungal spores. *Ann. N.Y. Acad. Sci.* **353:**116–122.

4. **Banttaii, E. E., and J. R. Venette.** 1980. Aerosol spread of plant viruses: potential role in disease outbreaks. *Ann. N.Y. Acad. Sci.* **353:**167–173.

5. **Barkley, W. E.** 1981. Containment and disinfection, p. 487–503. *In* P. Gerhardt (ed.), *Manual of Methods for General Bacteriology*. American Society for Microbiology, Washington, D.C.

6. **Barton, K. A., and W. J. Brill.** 1983. Prospects in plant genetic engineering. *Science* **219:**671–676.

7. **Blakebrough, M.** 1969. Design of laboratory fermenters. *Methods Microbiol.* **1:**473–504.

8. **Brown, J. H., K. M. Cook, F. G. Ney, and T. Hatch.** 1950. Influence of particle size upon the retention of particulate matter in the human being. *Am. J. Publ. Health* **40:**450–458.

9 **Calam, C. T.** 1969. Culture of microorganisms in liquid medium. *Methods Microbiol.* **1:**255–326.

10. **Center for Disease Control.** 1974. *Classification of Etiologic Agents on the Basis of Hazard*. Center for Disease Control, Atlanta.

11. **Cooney, C. L.** 1983. Bioreactors: design and operation. *Science* **219:**728–733.

12. **Demain, A. L.** 1981. Industrial microbiology. *Science* **214:** 987–995.

13. **Demain, A. L.** 1983. New applications of microbial products. *Science* **219:**709–714.

14. **Druett, H. A., D. W. Henderson, L. Packman, and S. Peacock.** 1953. Studies on respiratory infection. 1. The influence of particle size on respiratory infection with anthrax spores. *J. Hyg.* **51:**359–371.

15. **Ducatman, A. M., and D. F. Liberman.** 1992. Toxic hazards of industries and sites, p. 560–562. *In Worker Hazards in the Biotechnology Industry*. Williams & Wilkins Co., Baltimore.

16. **Evans, C. G. T., T. F. Preece, and K. Sargeant.** 1981. *Microbial Plant Pathogens: Natural Spread and Possible Risks in Their Industrial Use*. Commission of the European Communities, London, United Kingdom.

17. **Ghema, R. L.** 1981. Preservation, p. 208–217. *In* P. Gerhardt (ed.), *Manual of Methods for General Bacteriology*. American Society for Microbiology, Washington, D.C.

18. **Harding, L., and D. F. Liberman.** 1995. Epidemiology of laboratory-associated infections, p. 7–15. *In* D. O. Fleming, J. H. Richardson, J. J. Tulis, and D. Vesley (ed.), *Laboratory Safety: Principles and Practices*, 2nd ed. American Society for Microbiology, Washington, D.C.

19. **Harrison, M. D.** 1980. Aerosol dissemination of bacterial plant pathogens. *Ann. N.Y. Acad. Sci.* **353:**94–104.

20. **Hatch, T. F.** 1961. Distribution and deposition of inhale articles in respiratory tract. *Bacteriol. Rev.* **25:**237–240.

21. **Hellman, A. (ed.).** 1969. *Biohazard Control and Containment in Oncogenic Virus Research*. National Cancer Institute, Bethesda, Md.

22. **Kennett, R. H.** 1980. Freezing of hybridoma cells, p. 375. *In* R. H. Kennett, T. J. McKearn, and K. B. Bechtol (ed.), *Monoclonal Antibodies*. Plenum Press, New York.

23. **Klibanov, A. L.** 1983. Immobilized enzymes and cells as practical catalysts. *Science* **219:**722–727.

24. **Liberman, D. F.** 1980. Facility description and personnel practices for research activity of comparable hazard. *Public Health Lab.* **37:**118–129.

25. **Liberman, D. F.** 1987. Identification and control of human health hazards associated with current and emerging technology, p. 193–219. *In* S. S. Draggen, J. J. Cohrssen, and R. E. Morrison (ed.), *Environmental Impacts on Human Health: the Agenda for Long-Term Research and Development*. Praeger, New York.

26. **Liberman, D. F.** 1993. Biowaste management in bioprocessing, p. 769–787. *In* G. Stephanopolos (ed.), *Biotechnology*, 2nd ed., Vol. 3. VCH Verlagsgesellschaft mbH, Wienheim, Germany.

27. **Liberman, D. F., and R. Fink.** 1991. Containment considerations for the biotechnology industry. *Occup. Med. State of the Art Rev.* **6:**271–283.

28. **Liberman, D. F., R. C. Fink, and F. Schaefer.** 1984. Biosafety and bioreactors. *Biotechnol. Rep.* **1:**103–114.

29. **Liberman, D.F., R. Fink, and F. Schaefer.** 1986. Biosafety and biotechnology, p. 402–409. *In* A. L. Demain and N. A. Solomon (ed.), *Manual of Industrial Microbiology and Biotechnology*. American Society for Microbiology, Washington, D.C.

30. **Liberman, D. F., E. Israeli, and R. Fink.** 1991. Risk assessment of biological hazards in the biotechnology industry. *Occup. Med. State of the Art Rev.* **6:**285–299.

31. **Liberman, D. F., L. Wolfe, R. S. Fink, and E. Gilman.** 1996. Biological safety considerations for environmental release of transgenic organisms and plants. *In* M. Levin and E. Israeli (ed.), *Engineered Organisms in Environmental Settings: Biotech and Agricultural Applications*. CRC Press, Boca Raton, Fla.

32. **Maramorosch, K.** 1980. Spread of plant viruses and spiroplasmas through airborne vectors. *Ann. N.Y. Acad. Sci.* **353:**179–185.

33. **Medical Research Council of Canada.** 1977. *Guidelines for the Handling of Recombinant DNA Molecules and Animal Viruses and Cells*. Medical Research Council of Canada, Ottawa.

34. **National Cancer Institute.** 1979. *Design of Biomedical Research Facilities*. National Institutes of Health, Bethesda, Md.

35. **National Institutes of Health.** 1974. *Biohazards Safety Guidelines*. National Institutes of Health, Bethesda, Md.

36. **National Institutes of Health.** 1979. *Laboratory Safety Monograph*. National Institutes of Health, Bethesda, Md.

37. **National Institutes of Health.** 1983. *Guidelines for Research Involving Recombinant DNA Activity*. *Fed. Reg.* **48:** 24555–24581.

38. **Pike, R. M.** 1978. Past and present hazards of working with infectious agents. *Arch. Pathol. Lab. Med.* **102:**333–336.

39. **Powell, E. O.** 1958. Criteria for growth of contaminant and mutants in continuous culture. *J. Gen. Microbiol.* **18:** 255–268.

40. **Sargeant, K., and C. G. T. Evans.** 1979. *Hazards Involved in the Industrial Use of Micro-Organisms*. Commission of the European Communities, Brussels.

41. **Sulkin, S. E., and R. M. Pike.** 1951. Survey of laboratory acquired infections. *Am. J. Publ. Health* **41:**769.

42. **U.S. Department of Health and Human Services.** 1984. *Biosafety in Microbiological and Biomedical Laboratories*. HHS Publication no. (CDC) 84-8395. U.S. Government Printing Office, Washington, D.C.

43. **Wedum, A. G., W. E. Barkley, and A. Hellman.** 1972. Handling of infectious agents. *JAMA* **161:**1557–1567.

Intellectual Property

JENNIFER GORDON

26

Scientists and engineers, like the general population, would undoubtedly prefer to keep their dealings with lawyers to an absolute minimum. Justifiably, they perceive their time is more productively (and enjoyably) spent in their laboratories or production plants, conducting research, developing new products and processes, and implementing their discoveries for the benefit of humankind. However, given the importance of patents to the biotechnology and pharmaceutical industries, innovators in all areas of the biological sciences—from molecular biology, genetics, protein biochemistry, and pharmaceutical chemistry to industrial microbiology, biochemical engineering, agriculture, and medicine—are wise to consider patent attorneys as friends, not foes. Interactions with patent practitioners are essential, lest important advances end up as the intellectual property of a competitor.

Patents are one form of intellectual property. Intellectual property is a subset of personal property that also includes trademarks and copyrights (not addressed in this chapter) and trade secrets (treated briefly). Patents have some of the attributes of real property insofar as they define the boundaries of an invention. Patents are valuable assets that can be exploited commercially. They can be bought and sold, licensed for royalties, and used as bargaining chips when engaging in business deals (e.g., seeking licenses under others' patents). Perhaps most significant for start-up biotechnology companies, patents can increase the value (real and perceived) of a business to investors. Thus, a patent portfolio may enable a start-up company to obtain the capital required to bring a nascent product to market.

Patents provide an exclusive right, preventing others from practicing the invention for the life of the patent. Competitors who cross the lines of demarcation circumscribed by the patent claims do so at the risk of incurring infringement liability (injunctions can shut down business operations and/or awarded damages can be trebled if the infringement is considered willful). A patent provides protection for almost two decades. Thus, with even one patent protecting a commercially important invention, a small business can potentially catapult into a highly profitable company of international significance.

Nevertheless, even brilliant ideas and developments will not translate into patentable inventions, nor will issued patents be maintained when litigated in courts, unless certain precautions are taken and specific requirements met. Because of the importance of patents, it is useful for scientists and engineers to have some general knowledge of patent laws and procedures, as well as some familiarity with the unique requirements, and pitfalls, involved in obtaining patent protection on biotechnology-related inventions. This chapter provides an overview of the U.S. patent system. It must be emphasized, however, that obtaining competent patent counseling is always advisable to help you successfully negotiate the big hurdles and the minutiae of the patent system to obtain valid patent protection on your inventions.

Over the past 20 years, great strides have been made in the biological sciences, transforming an historically observational discipline into one in which biological matter can be manipulated by human intervention and put to commercial use. We have witnessed breakthroughs such as (i) recombinant DNA technology, including the mass production by fermentation of therapeutically useful human polypeptides and the creation of transgenic plants and animals; (ii) hybridoma technology and humanized antibodies; (iii) automated DNA and protein sequencing; (iv) powerful nucleic acid amplification methodologies, such as PCR; (v) advances in the diagnosis and treatment of genetic diseases and cancers; (vi) gene therapy; (vii) combinatorial molecular libraries and ligand/receptor screening techniques; and (viii) computer-aided drug design, to name a few. These developments have transformed the pharmaceutical industry and have literally changed society.

During this same time period, our legislature, courts, and the United States Patent and Trademark Office (PTO), while adhering to fundamental principles, have nevertheless written, interpreted, and promulgated, respectively, our patent laws in ways to accommodate the unique characteristics of inventions based on life forms and other biological materials. Since the first edition of this manual was published in 1986, there have been significant developments in biotechnology patent law. Furthermore, there have been several profound changes in the patent system as the United States attempts to "harmonize" its laws with those of other nations in the interest of global trade.

In the discussion that follows, aspects of U.S. patent law (and to a minor extent, the patent laws of other countries), particularly as they apply to biotechnology, are presented. Although some of the legal concepts may seem dense or arcane, it is hoped that the practical tips interspersed throughout will facilitate your interactions with your patent attorneys and help you obtain and successfully utilize patents on your inventions.

26.1. GENERAL PATENT INFORMATION

26.1.1. U.S. Patent Laws

The rights afforded by a patent are derived from the Constitution (art. 1, sec. 8), which entitles an inventor to a limited monopoly on his or her invention in exchange for full disclosure of the discovery to the public. The U.S. patent system, therefore, rewards the inventor for his or her discovery by securing the exclusive rights to the invention for a limited time and, most importantly, benefits the public by promoting the full disclosure of the invention, which can be used by anyone after the expiration of the limited monopoly. The full disclosure of the invention also benefits the public by enabling those so disposed to make improvements on the basic invention. U.S. patent laws are enacted by Congress in accordance with the power conferred by the Constitution and are codified in Title 35 of the *U.S. Code* (35 USC). The statutory laws set forth in 35 USC are interpreted and applied by the Patent Office Board of Patent Appeals and Interferences, federal district courts, the U.S. Court of Appeals for the Federal Circuit, and the U.S. Supreme Court. Thus, in addition to the patent statutes themselves, there has developed a body of case law that governs resolution of disputes over issues such as patentability, inventorship, priority of invention, claim interpretation, patent validity, patent infringement, enforceability, and patent misuse. Furthermore, Title 37 of the *Code of Federal Regulations* (37 CFR) sets forth the rules, based on the patent statutes and various international treaties and conventions, which govern practice before the PTO.

26.1.2. Patentable Subject Matter

Under U.S. law (35 USC § 101), whoever invents or discovers any *new and useful* (i) process, (ii) machine, (iii) manufacture, or (iv) composition of matter, or any new and useful *improvement* of the above items ("statutory subject matter"), may obtain a patent, provided that the invention meets further requirements for patentability, which are discussed in more detail below. The term "process" is defined by statute to mean process, art, or method and includes a *new use* of a known composition of matter or material, process, machine, or manufacture [35 USC § 100(b)].

This chapter addresses the requirements for what are known as utility patents, the most common and generally sought type of patent. Another type of patent that may be of some interest to biotechnologists is the plant patent, which protects specific varieties of asexually reproduced plants (35 USC § 161). Plant Variety Protection Act (PVPA) certificates can be obtained from the U.S. Department of Agriculture for specific varieties of sexually reproduced plants. Plant patents and PVPA certificates are very limited in scope. If there is a generic attribute of a plant (e.g., it overproduces an amino acid), it is advisable to claim the generic plant invention by way of utility patent and obtain, as appropriate, either a plant patent or PVPA certificate on the specific variety. Plant patents/PVPA certificates and utility patents are not mutually exclusive.

While the scope of potentially patentable subject matter is broad, it is not unlimited. In particular, laws of nature, physical phenomena, and abstract ideas remain outside the realm of patent protection. Thus, new minerals found in the earth, plants discovered in the wild, and Einstein's theory of relativity are not within the ambit of patent protection.

This is not to say, however, that all products of nature are unpatentable. On the contrary, useful microorganisms (e.g., antibiotic producers) isolated in substantially pure culture from natural sources have been patentable for many years. Indeed, the landmark 1980 Supreme Court case of *Diamond v. Chakrabarty* affirmed the patentability of microorganisms with useful characteristics (in that case, the capacity to metabolize oil) as compositions of matter. The Supreme Court's holding on the patentability of unicellular organisms has since been extended by the Patent Office Board of Patent Appeals and Interferences to multicellular plants (*Ex parte Hibberd*; 1985) and to multicellular animals (*Ex parte Allen*; 1988). Furthermore, natural compounds, such as vitamins and proteins, if purified into a form that makes them new, useful, and nonobvious, i.e., patentably distinct from the way they exist in nature, are considered patentable inventions under U.S. law. Finally, recombinant microorganisms, transgenic plants, and transgenic animals are patentable subject matter; although they are "natural" in the sense that they are living, they are human creations.

26.1.3. Who May Apply For A Patent

In the United States, a patent application must be filed in the name of the actual inventor or inventors. This requirement stems from the Constitutional grant as well as from the patent statute (35 USC § 115). Thus, it is important to determine who actually invented the subject matter sought to be patented. While the law in this area is quite complicated, and each case depends on its own set of facts, the rules may be summarized as follows.

Sole invention occurs when one person conceives the solution to a problem and the means to the desired end that constitutes the subject matter of the invention. Conception is the mental formulation and disclosure by the inventor of a complete and operative idea for a product or process. The idea must be of a specific means, not just a desirable end or result, and must be sufficiently complete so as to enable anyone of ordinary skill in the art to reduce the concept to practice.

Determining inventorship is more difficult when two or more persons are involved in the development of a project. The patent statute does not define joint inventorship (35 USC § 116). Case law provides this definition. Generally, a joint invention occurs when two or more persons, *collaborating together, working toward the same end, contribute to the inventive thought and final result which constitutes the invention*. Coinventors need not conduct their work in the same physical place; they can collaborate from separate locations. It is frequently difficult to ascertain who has, in fact, made an original contribution to the solution to a problem; however, the contribution must consist of more than suggesting a desired result or following the instructions of another. Generally, a coinventor must have contributed an essential feature recited in at least one claim. Joint inventorship is not, as is frequently the practice in determining authorship in academia, a collection of all those who worked on the project. Inventorship, although difficult to define, has a precise meaning, and it is important to determine inventorship at the outset of the patenting process. Because inventorship is defined by what is claimed, if claims are deleted or added during prosecution of the patent application, the inventorship may change.

26.1.4. The Requirements for Patentability

To be patentable, an invention must be more concrete than an idea, requiring both *conception and reduction to practice*. Conception must be mental formulation and disclosure of the complete ideas for the invention. Reduction to practice can be either (i) *actual*: for example, a method or process was practiced, a compound was synthesized, or a prototype of a mechanical invention was made, or (ii) *constructive*: a patent application, describing the invention in sufficient detail that one with ordinary skill in the art can practice the invention without undue experimentation, was filed in the PTO.

An invention that constitutes statutory subject matter must then satisfy several other requirements for patentability: (i) the invention must be useful, (ii) the invention must be novel, (iii) the invention must have been nonobvious at the time it was made, and (iv) the patent application must be sufficiently detailed so as to (a) provide a written description of the invention, (b) enable those skilled in the art to practice the invention, and (c) set forth the best mode contemplated by the inventor for practicing the invention as of the time of filing the application. Each requirement is discussed briefly below.

26.1.4.1. Utility

U.S. patent law requires that the subject matter claimed in patents be useful (35 USC § 101). There must be practical, "real-world" utility, eliminating claims to things like perpetual motion machines (if such actually exist). At first blush, utility would not seem to be a particularly controversial requirement. The very notion of invention suggests that a problem has been solved and that whatever brought about the solution must, therefore, be useful. Almost by definition, an invention should meet the utility requirement. However, as discussed below, the utility requirement has recently been somewhat controversial in biotechnology-related inventions.

26.1.4.2. Novelty

To be patentable, an invention must be novel. Novelty is defined by statute (35 USC § 102), which provides that a patent shall not be granted if certain events occur. The events that obviate patent protection for your invention are listed below:

1. The invention was known or used by others in the United States before you invented it.
2. The invention was patented or published (in a printed publication) by others anywhere in the world before you invented it.
3. The invention was patented or published (in a printed publication) by you or others anywhere in the world more than 1 year before you filed an application for a patent in the United States.
4. The invention was in public use or on sale, or offered for sale in the United States more than 1 year before you filed an application for a patent in the United States.
5. You abandoned the invention.
6. You patented the invention in a foreign country based upon an application in that foreign country that you filed more than 1 year before filing an application in the United States.
7. The invention was described in a U.S. patent or an international patent under the Patent Cooperation Treaty granted to another inventor who filed in the United States or internationally before you invented it.
8. You, yourself, did not invent the subject matter you wish to patent (e.g., you derived it from another).
9. Before you invented it, the invention was made in the United States by another who has not abandoned, suppressed, or concealed it.

Events 1 through 4, 6, 7, and 9 constitute "prior art" with respect to your invention.

In general, for purposes of determining novelty, the date of invention is either the date you actually reduced the invention to practice in the United States, or the date of your constructive reduction to practice, e.g., the date you filed an application for patent in the United States. Under recent legislation (see section 26.2.4. below), actual reductions to practice are now recognized when they take place in certain countries outside the United States. In certain specific instances, the date you first conceived of your invention can be factored into the determination of novelty. Novelty of an invention is assessed by determining whether a single prior art reference anticipates the claims of the invention. That is, the prior art reference must disclose each and every element of the claimed invention. If the prior art does *not* identically disclose your invention, then the novelty requirement for patentability is met.

26.1.4.3. Nonobviousness

In addition to the requirement that the invention be novel, the invention must also be nonobvious; that is, the differences between the prior art and the invention as a whole must have been unobvious to a person of ordinary skill in the art at the time the invention was made (35 USC § 103). To render a claimed invention obvious, the prior art must both provide a suggestion to make it and imbue the ordinary artisan with a reasonable expectation of successfully making and using it. Furthermore, the prior art must suggest the particular form of the invention and how to make it; general guidance is insufficient.

The question of obviousness is often difficult to resolve since it requires measuring the invention against the prior art through the eyes of a hypothetical person with ordinary skill in the art at the time the invention was made; reliance upon hindsight is strictly forbidden. To assess nonobviousness, the courts make the following factual determinations:

1. The scope and content of the prior art
2. The level or ordinary skill in the art
3. The differences between the claimed invention and the prior art
4. So-called secondary considerations that indicate nonobviousness, such as whether the invention fills a long-felt need, whether others have tried and failed, whether others expressed disbelief or skepticism that the invention would work, whether the invention yields unexpected results, or whether the invention is commercially successful

26.1.4.4. Sufficiency of Disclosure

Full disclosure of the invention in sufficient detail to permit persons skilled in the art to make and use the invention is the quid pro quo for obtaining a patent. The requirements of 35 USC § 112 ensure that the public will be able to practice the invention once the patent expires. In this way, the public fully benefits from the patentee's invention and the advancement of the useful arts and science is promoted. In this context, "the invention" is what is claimed in the numbered paragraphs at the end of the patent. Accordingly, an application for patent in the United States must contain not only a set of numbered "claims," but also a "specification," the portion of the application preceding the claims. The specification is a written description of the invention and the manner and process of making and using it. The written description is required to be in such full, clear, concise, and exact terms as to enable any person skilled in the art to make and use the invention. The specification must also describe the *best mode* of practicing the invention contemplated by the inventor at the time of filing the application.

"Written description," "enablement," and "best mode" are three separate and distinct requirements, all of which, as discussed below, can affect patent applications on biotechnology-related inventions. The written description requirement is fairly self-explanatory. The specification of the patent application must convey through words, drawings, and/or other figures what is considered to be the invention and must provide support for the claims that appear at the end of a patent. Be aware that there can be pitfalls with respect to the written description requirement. For example, if you want to claim an invention both generically and specifically, it must be described both ways. Thus, if your invention is a process that can be performed over a temperature range of 10 to 65°C, and if you want to have a specific claim to your process run at its optimum temperature range, 45 to 55°C (in addition to a generic claim reciting the 10 to 65°C temperature range), your specification should set forth in writing both the entire 10 to 65°C range and the optimum temperature range of 45 to 55°C. The generic claim setting forth the entire range, of course, will provide protection against the infringer who operates the process at 45 to 55°C. However, if prior art turns up in which someone performed the process at 37°C, the patentability/validity of the generic claim will be destroyed. If your specification described the optimum temperature range, 45 to 55°C, you potentially have a fallback position, i.e., support for a claim to the process run at the particular optimum temperature range. If performing the process at 45 to 55°C provides a nonobvious advantage over the 37°C process of the prior art, your written description is sufficient to support a patentable "species" claim that will still protect your optimum range.

The specification of a patent application must be "enabling." There have to be sufficient technical details such that the invention will be reproducible in the hands of persons skilled in the particular field of technology. This does not mean that every last detail must be included if such would already be known to skilled artisans. However, patents will not be granted (or granted patents can be invalidated) if it appears that it would take "undue experimentation" to figure out how to make and use the invention. What is "undue" is, of course, considered on a case-by-case basis. Undue experimentation should be determined by applying factors that include (i) the quantity of experimentation necessary (time and expense), (ii) the amount of direction or guidance presented, (iii) the presence or absence of working examples of the invention, (iv) the nature of the invention, (v) the state of the prior art, (vi) the relative skill of those in the art, (vii) the predictability or unpredictability of the art, and (viii) the breadth of the claims. No one factor is dispositive of the question of enablement. Suffice it to say, the public should not have to engage in research to determine how to practice a patented invention, although routine experimentation or optimization is permissible.

The "best mode" requirement ensures that an inventor cannot conceal what he or she considers to be the preferred way of practicing the invention at the time the application is filed. In other words, the inventor should not be allowed the limited monopoly afforded by a patent if the public has been deprived of information that the inventor could then use to his or her advantage. If the inventor considers there to be a best mode with respect to the claimed invention (even if others might disagree it was the best), it must be disclosed in a manner understandable to one of skill in the art. If an inventor subsequently files what are known as continuation-in-part applications—where the specification is expanded to contain "new matter" (e.g., new data, descriptions of new compounds)—then the inventor must disclose the best mode for practicing the newly disclosed inventions as of the filing date of the continuation-in-part application.

26.1.5. Rights Obtained Upon Issuance of a Patent

Do not think that just because you obtain a patent on your invention that you are free to practice it. In the United States, the granting of a patent gives the patentee, and his or her heirs or assignees, the right to exclude others from making, using, offering to sell, selling, or importing the patented invention within the United States for a limited time period (35 USC § 154). The patent statute does not give the patentee the right to manufacture, use, or sell anything. Someone else may have a patent with a claim broad enough to encompass your specific invention. *That* patent owner can prevent you from practicing your invention. You, however, may also prevent the other patent owner from practicing your invention (often, such impasses are alleviated by cross-licensing agreements). To repeat, the patent right is a right to *exclude others* from practicing your invention. Practicing your own patented invention does not necessarily absolve you from infringement of another's patent.

The time period of U.S. patent protection has changed significantly as a result of recent legislation implementing the General Agreement on Tariffs and Trade (GATT). Prior to GATT, the term of a U.S. patent was 17 years from the date of grant (i.e., the day on which the PTO issued the patent), provided all maintenance fees required to keep the patent in force were paid. For patent applications filed after June 8, 1995, the term of protection provided by a U.S. patent starts on the date of grant (the issue date) and terminates 20 years from the filing date in the United States of the application for the patent. Furthermore, if the patent issues from an application claiming priority of an earlier filed application, the 20-year patent expiration date is measured from the filing date of the *earliest* U.S. priority application relied on by the applicant. Patents in force as of June 8, 1995, and patents issuing on applications filed before that date, are granted a term that is the greater of the new 20-years-from-filing term or the old 17-years-from-grant term. Limited extensions are also available with respect to certain inventions that have undergone lengthy periods of regulatory (e.g., Food and Drug

Administration [FDA]) approval that decreased the effective patent term.

Cognizant that delays which can erode the 20-year time period occur in the PTO during prosecution of a patent application, the patent statute (35 USC § 154) has been amended to provide extensions in certain instances (e.g., delays due to interference proceedings or appeals to the Board of Patent Appeals and Interferences [see section 26.2.4., below]). However, many people, particularly those within the biotechnology industry, believe the present provisions for term extensions are insufficient. This is because biotechnology-related patent applications frequently take many years (sometimes more than a decade) to percolate through the PTO. Measured from the filing date, biotechnology patents can therefore end up robbed of valuable patent life, to the point of making the massive research investments that go into biotechnology-related inventions impractical.

26.1.6. Infringement of a Patent

While the owner of a patent can grant licenses authorizing others to practice his invention, whoever without authority makes, uses, offers to sell, sells, or imports the patented invention within the United States during the life of the patent infringes the patent [35 USC § 271(a)]. When the patented invention is a process, whoever uses, offers to sell, or sells in the United States, or imports into the United States, the product of the patented process is liable as an infringer. The statute 35 USC § 271(g), which governs infringement of process patents, was enacted, in part, because of biotechnology-related inventions. For example, in the 1980s, inventors obtained U.S. patents on processes for making known (hence, unpatentable) proteins by recombinant means. The recombinant process was considered novel and nonobvious, but the product was not. Before the enactment of 35 USC § 271(g), others could employ the patented process outside the territorial reach of the U.S. patent (e.g., in a foreign country where there was no counterpart patent protection) and import the unpatented protein with impunity. Recognizing this loophole, Congress enacted a statute that would make overseas manufacturers liable for infringement if the product of the patented process was shipped into the United States.

Legal remedies for infringement include equitable relief (such as an injunction to stop the infringement), monetary damages, and, in exceptional cases, attorneys' fees (35 USC §§ 281, 283, 284, and 285). Important exemptions from patent infringement liability exist, such as when one is using a patented invention for purposes solely and reasonably related to obtaining regulatory (e.g., FDA) approval in the United States. [35 USC § 271(e)(1)].

26.2. OBTAINING PATENT PROTECTION

26.2.1. It Starts with Record Keeping

Laboratory record keeping can be as important to obtaining and maintaining patent protection as the invention itself. When you have had that great idea and are anxious to get the work under way, or when you are in the midst of experiments and generating exciting results, probably the last thing on your mind is record keeping. However, because U.S. patent law is based on the proposition that the "first to invent" is entitled to a patent on the invention, it is often critical to have legally sufficient contemporaneous documentation to prove conception and actual reduction to practice of the invention. Furthermore, a detailed record of activities between conception and reduction to practice is required to prove you were diligent in your efforts to reduce the invention to practice.

For example, when separate inventors file patent applications claiming the same invention, the PTO may declare an interference, which is a proceeding to determine which inventor was first. The PTO considers the respective dates of conception and reduction to practice and whether the inventor was reasonably diligent during the period from conception to reduction to practice. Usually, the first inventor to reduce the invention to practice (either by actually performing or making the invention, or, constructively, by filing the patent application) prevails. However, the second inventor to reduce the invention to practice will prevail if he or she can show (i) an earlier date of conception and (ii) reasonable diligence from a time just before the first inventor to reduce the invention to practice entered the field until the second reduction to practice. Clearly, without solid evidence of early conception, diligence, and reduction to practice, one cannot hope to win an interference (or a litigation) where the major issue is, Who did it first?

Another example in which good records come in handy is during patent prosecution. Patent prosecution is a term used to describe the process by which a patent is obtained. Patent applications are scrutinized by PTO examiners to determine whether the requirements for patentability are met. Often, applications are rejected for various reasons delineated by the examiner in an "office action." The inventor may then respond to this office action by amending patent claims or making arguments to persuade the examiner. The interplay between examiner and inventor may go on for some time.

Sometimes an examiner rejects your claims by citing against them a reference about which you were unaware and which was published less than a year before your filing date. If you have kept good records, you can antedate or "swear behind" the reference (i.e., overcome the examiner's rejection) by showing you had an actual reduction to practice of your invention before the publication date of the reference, or by showing that you had conceived of your invention before the publication and diligently reduced it to practice.

Thus, it is highly advisable to make a written record of the conception of your invention. Conception is more than just a bright idea, a wish, or a vague goal. You must express your invention clearly and completely and have an operative game plan for putting it into practice (e.g., a protocol or a sketch of a device). If there are several aspects of the invention, note them all. You should also describe an intended utility for the invention. A good record of conception reflects a complete idea of the invention, including the means for making or performing it in a way that a person skilled in the art could follow.

In addition to documenting your conception, it is also advisable to keep a record of all the experimental work that went into putting the invention into practice, at least up to the time when a patent application was filed. Hopefully, there will not be gaps (i.e., large blocks of time unaccounted for) in your record keeping or evidence of your being sidetracked by unrelated experiments or projects. Both can be fatal to subsequently proving diligence.

The mechanics of record keeping are fairly straightforward but include an important wrinkle: the need for corroboration or witnessing. You should record your ideas and experiments chronologically in ink in a bound notebook with numbered pages. Bound notebooks are preferable to loose-leaf binders as there can be no question of insertion or removal of pages. Numbering notebooks and using tables of contents can be helpful. You should sign and date the pages on which you make your entries, on the same day as you undertook the various tasks reflected on the page (e.g., setting up experiments, performing experiments, recording results, and summarizing conclusions).

Although you may be very busy, do not skimp on details in your notebook. These details will help you prove what was done and when it was done. Include your ideas, not just work performed. For work actually done, provide full experimental protocols, including reagents and conditions used. Give full recipes for, e.g., media, or if you use reagents or equipment of a particular manufacturer, it is wise to record the product information or even tape product inserts into your notebook. If you are following a method described in the literature, cite the reference and/or transcribe or tape a copy of the relevant materials and methods into the notebook.

Record results completely. Where appropriate, using tape, glue, or some other reasonably permanent means, affix original data or copies (e.g., printouts from analytical machines, photographs of gels, reduced copies of strip charts) into your laboratory notebook. Often, a technician will carry out some work and obtain data. Cross-referencing notebooks is advisable. It is also a good idea to obtain a copy of the data, or a copy of the page from the technician's notebook, and tape it into yours. Discuss your results, conclusions, and plans for future work legibly and understandably in your notebook. If you make mistakes, note them when realized rather than crossing out, erasing, or otherwise altering the original writings in the notebook. Such markings and erasures only raise questions of tampering. If you must alter an entry, initial and date the alteration.

Contemporaneously with your own notebook entries (i.e., on the same day, or within a few days), have a coworker (but not a coinventor), who is knowledgeable in your scientific discipline and who witnessed you perform or knows you conducted the experiments, read your notebook and witness it. That is, the witness should indicate that each page was read and understood by him or her and then should sign and date it. This will provide corroboration that the recorded work was actually done on or by the dates indicated and that one skilled in the art understood the details well enough to be able to perform the same experiment, comprehend the results, etc. Under U.S. patent law, a coinventor cannot corroborate another coinventor's conception or reduction to practice of aspects of the invention. Thus, if you are working on a project with one or several people who are *all* contributing inventive ideas to the invention, have a noninventor witness your notebook. Finally, periodically have your notebooks microfiched and store the original notebooks and microfiche in separate, safe repositories.

26.2.2. Applying for Patent Protection

26.2.2.1. Preliminary Steps

Besides the record keeping discussed in the previous section, several other preliminary steps should be undertaken prior to filing a patent application on an invention. First, it is a good idea to determine if the invention will be patentable. What seems like a great idea may have already been done by someone else, or made obvious by something someone else has done. Therefore, it is advisable to perform a patentability search. A key word or author search utilizing computer databases covering both patents and the literature can aid in determining whether there are prior art references that anticipate the invention or render it obvious. Searching can be done without the help of a patent attorney, and independent searching can save money. Nevertheless, it is a good idea for a patent attorney to be involved in the search design and the evaluation of the search results. Also, a patent attorney can arrange for more comprehensive searches, such as manual searches in the PTO or chemical structure searches, or even searches performed by the European Patent Office, which has a good reputation for accurate and thorough searching.

If it appears from the search that the invention is novel and nonobvious, one must decide whether to apply for a patent. Another option (besides, of course, publishing and dedicating it to the public) is to maintain the invention as a trade secret. Products that are impossible to reverse-engineer, or processes for which infringement is not easily discovered, are candidates for trade secret protection. If the decision is made to keep the invention a trade secret, care must be taken to maintain it as such. Employees should be bound by employment contracts to keep the product or process a secret during and after their tenure with the company. Nonemployees should not have access to the trade secret. All documents related to the trade secret should be clearly marked "CONFIDENTIAL." The gist of maintaining the trade secret is to define what the secret is and do everything possible to protect the confidentiality of the secret.

Pros and cons to trade secret protection must be carefully considered. A trade secret can provide an indefinite competitive advantage. There is no time limit on the term of protection, if the secret remains truly secret. No monies are spent on patenting the invention. However, the effort to keep an invention a secret can be considerable. Also, if someone else independently secures a patent on the development, prior trade secret use does not absolve the user of infringement liability under our present law: the trade secret holder will have to get a license from the patent holder to continue to use the invention.

26.2.2.2. Going Forward with Patenting the Invention

If the decision is made to patent the invention, find a competent patent attorney or patent agent. Look for a patent attorney with education and experience in the area of technology of your invention. Your attorney must understand the technical details of your invention and be able to distinguish them from the prior art. Patent attorneys registered to practice before the PTO had to take a specialized exam on patent claim drafting and PTO procedures and rules of practice. To qualify to take the exam, the attorney (or nonattorney agent) had to have had formal education in a technical area, such as chemistry, biology, or engineering. A registered patent attorney or patent agent is qualified to prepare and prosecute patent applications, whereas general attorneys with no science background or who are not registered before the PTO are not qualified.

An initial face-to-face meeting with the patent attorney is highly recommended. There are several things you can do prior to the meeting that will make the application process more expeditious and less costly. First, gather all the documents necessary to explain the invention to the attorney. Manuscripts or preprints of unpublished papers describing the work are very helpful. If such are not available, you can facilitate the discussion, and the subsequent drafting of the application, if you provide a written disclosure detailing your materials and methods and results. An introduction briefly outlining the current state of the art along with references is also helpful. Have a good idea of what you want to claim (e.g., a compound, a method, or both) and how this invention is different from previous discoveries. Second, think through substitutes for chemicals actually used or alternative process parameters in advance, even if these substitutions will not work as well. The substitutions and alternatives will help the patent attorney draft claims to give you the broadest possible coverage. Broader patent protection encompasses not just what you have actually done but realistic, workable variations on the theme. Third, remember to provide the attorney with information about what you consider to be your preferred or most preferred way of practicing the invention. This will ensure compliance with the best-mode requirement. Finally, if practical, economically or otherwise, let your patent attorney review the relevant underlying notebooks. This may permit him or her to spot aspects of your work you may have forgotten or to verify what you consider to be your best mode. Remember, the more organized you are and the more legwork you do in advance, the less your attorney (who is probably billing by the hour) will have to do and the smaller your bills will be.

Prior to the initial meeting, serious thought should be given to who should be named as the inventor or coinventors. Your patent attorney should carefully go over with you the reasons for inclusion of each inventor. If a coinventor is left off the patent in bad faith, patent rights can be jeopardized. Although seemingly harmless, naming as inventors people who did not make an inventive contribution (e.g., technicians who worked at your direction) should not be done. Inventorship is often one of the first issues discussed when a patent is involved in litigation. Thus, inventorship should be reviewed thoroughly.

Tell your patent attorney about events that could potentially prevent you from obtaining patent protection in the United States or abroad. It is always advisable to keep your invention in confidence prior to filing for a patent. If you must disclose the invention to others prior to filing the application, do so under a confidentiality agreement. Inform your patent attorney of the date and nature of the disclosure so that its confidentiality can be confirmed. Otherwise, a "time bar" as described in section 26.1.4.2 may have started running.

Publication, sales, offers for sale, disclosure to others, and public use of an invention, prior to filing a patent application, may jeopardize your patent rights. For example, if you publish your invention in the literature, or sell it, more than 1 year before your patent application's filing date, you are barred from obtaining a patent. By statute, your invention is no longer novel. Even an abstract can be sufficiently detailed to enable the invention and hence potentially bar its patentability. Note, also, that even a single copy of a thesis placed in a university library may potentially bar patentability of an invention described in it, if it is sufficiently publicly accessible (e.g., cataloged).

The United States provides for a 1-year grace period. (Most foreign countries have no grace period. Hence, to preserve foreign patent rights, it is important to have a patent application on file before publishing an invention.) Thus, if you disclose your invention within the year prior to filing your patent application, you will not automatically be barred from obtaining a U.S. patent. Provided certain procedures are followed, your pre-filing-date disclosure will not count against you during patent prosecution. Similarly, if someone else publishes a paper on the subject matter of your invention and if you file a patent application within 1 year of that paper's publication date, you will be allowed to "swear behind" this reference during prosecution of your U.S. application, *if* you can prove (through your records) your prior conception and/or reduction to practice.

It cannot be overemphasized that you must be forthcoming with your attorney concerning prior and planned publications or commercialization of your invention. That way, your attorney can establish critical dates as to when a potential 1-year time bar began or will begin running and can prepare and file an application without any loss of U.S. (and foreign) patent rights.

All patent applicants and attorneys prosecuting their applications owe a duty of candor to the PTO (37 CFR § 1.56). All publications and information material to the patentability of an invention, known to the applicant or attorney, should be brought to the attention of the PTO. If relevant references and information are intentionally withheld, any patent issuing on the application (and, sometimes, related patents) can be found to be unenforceable for inequitable conduct. This is a very serious duty, and defendants in patent litigation frequently raise inequitable conduct or "fraud on the Patent Office" as a defense to patent infringement. You do not want to be accused someday of procuring your patent through fraud. Therefore, it is highly advisable to bring all prior art and other relevant information to the attention of your patent attorney so that it can be disclosed in the application and/or submitted to the PTO during prosecution. The duty of candor continues throughout the pendency of the patent application. If new, material, non-cumulative prior art or information is discovered after filing the application, it should be disclosed in time for the examiner to consider it before a patent is issued.

26.2.2.3. Types of Patent Applications: Provisional and Regular

Because of GATT, there is a new choice with regard to the type of patent application to file. The GATT legislation established a new type of patent application called a "provisional application." This application is not examined, except for certain formal requirements, and may be used by the applicant to establish an effective filing date that does not form the basis for measuring the 20-year expiration date of the patent. A provisional application (which does not require claims or nucleic acid or protein sequence listings) cannot claim the benefit of the filing date of an earlier filed application. The provisional application automatically becomes abandoned 12 months from its filing date, at which point it cannot be revived. A U.S. application that seeks to obtain the benefit of the filing date of the provisional application must be filed while the provisional application is pending; foreign convention applications (see, section 26.2.5, below) must be filed within this 12-month period as well.

Although the provisional application is characterized as a simple and inexpensive means for establishing an early filing date, do not get your hopes up too high. To establish the early priority date, the provisional application must still meet the written description, enablement, and best mode requirements of 35 USC § 112 (i.e., it must adequately describe, enable, and set forth the best mode of practicing the invention ultimately claimed). If preserving a filing date without starting the 20-year time clock and delaying prosecution provides an advantage to you, then provisional applications can be useful. Provisional applications can buy a year of time to determine commercial viability, procure investment money, or permit further research and development. Furthermore, the fees associated with the filing are considerably reduced. However, in rapidly changing technologies, where inventions become obsolete before the end of the patent term, the extra year a provisional application affords is probably commercially meaningless. Also, because filing a provisional application starts the 12-month period for filing foreign patent applications, you can be faced with the difficult decision of whether to file expensive foreign applications without the benefit of having received an office action from the PTO on the merits of your case (see section 26.2.3, below). In most cases, it is still recommended that a regular patent application be filed.

26.2.2.4. Parts of the Patent Application

A regular patent application must contain a written specification, drawings if necessary to understand the invention, a set of claims, an abstract, and an oath or declaration signed by the inventor. This last requirement is very important. The oath or declaration is a sworn statement by you, acknowledging that you have read the application, *including the claims*, and that you believe yourself to be the original and first inventor (or joint inventor) of the process, machine, manufacture, or composition of matter or improvement thereof for which you are applying for a patent.

A regular patent application should contain the following (and, as the PTO suggests, should be arranged as follows):

1. Title of the Invention.
2. Reserved Section. This section usually consists of a short "Introduction" or a description of the "Field of the Invention" and a section called "Background of the Invention" that describes prior work done in the field of the invention. It is helpful to describe the shortcomings of these works and to point out that your invention overcomes these shortcomings.
3. Cross-references to related applications on file, if any.
4. Brief Summary of the Invention. This section should set forth your invention in a manner that tracks your broadest (independent) claims and summarizes the specific embodiments claimed in your narrower (dependent) claims. This section should also describe the advantages obtained by using your invention.
5. Brief Description of Drawings or Figures (if there are drawings or figures). This section essentially consists of abbreviated figure legends.
6. Detailed Description of the Invention. This section should describe your invention as completely as possible. It is not limited to what has actually been reduced to practice, but rather should describe as many permutations of your invention as are within the scope of the invention. The description must be enabling to one with skill in the art of the invention. This section, in addition to disclosing a more generic description of the invention, usually sets forth specific experimental examples. If the experiments have actually been performed, the examples are preferably written in the past tense. "Prophetic examples," which describe experiments that will or could be done, may also be explained in this section of the specification. Prophetic examples are preferably worded in the present tense.
7. Claim or Claims. The claims define the metes and bounds of your invention. They define the actual property you are trying to protect. They are generally drafted like an inverse pyramid (broad, narrower, narrowest). The patent attorney will draft the claims, but you should pay very close attention to them to ensure they define all aspects of your invention.
8. Abstract of the Disclosure. This section provides a concise summary of the invention. It should be useful for others performing key word searches.
9. Signed oath or declaration (usually a form prepared by your attorney). As described above, when you sign this form you are swearing that you have read and understood the application, including the claims, and that you believe yourself to be the first inventor.

Finally, applications that involve proteins, peptides, nucleic acids and fragments thereof, identified by amino acid sequences and nucleic acid sequences, respectively, have certain special requirements. The patent application must include "sequence listings." Sequences referred to in the application are given a numerical sequence listing (e.g., SEQ ID NO.: 1, SEQ ID NO.: 2, etc.). The sequences themselves are presented on separate pages immediately preceding the claims. The sequences also have to be provided to the PTO in computer-readable form. There are very specialized rules that govern the presentation of sequence listing and computer discs containing them (see 37 CFR §§ 1.821 to 1.825). Patent offices worldwide have been following suit, requiring that protein and nucleic acid sequences be presented in computer-readable form.

26.2.3. Patent Prosecution

Patent prosecution is the "back and forth" that takes place with the PTO after the application is filed and before it issues as a patent. An examiner, experienced in the technological area of the invention, examines the application. The PTO, through its examiners, issues office actions that state the results of the examination. More often than not, the initial claims will be rejected. The examiner is likely to raise numerous grounds for rejection, including novelty and obviousness rejections based on prior art. Also, indefiniteness, overbreadth, or nonenablement rejections based on lack of claim precision or clarity or insufficiency of disclosure will often be included.

Have your patent attorney send you copies of all office actions. Work with your attorney to prepare an appropriate response, whether by amending claims, setting forth arguments as to why the examiner is incorrect, or providing additional evidence. Such evidence may be in the form of publications or sworn declarations presenting specific facts or expert opinion that support the patentability of your invention. It can also be very helpful to have an interview (in person or by telephone) with the examiner to discuss the bases for rejection.

For biotechnology inventions, there are usually several rounds of office actions and responses. The group in the PTO that examines biotechnology patent applications is notorious for long, arduous, and therefore expensive prosecution. This situation has been changing recently, a reflection of the fact that the new 20-year clock begins as of the filing date.

During the prosecution of a patent application, a point will be reached (usually after the PTO has issued at least two office actions) when your claims will either be allowed or "finally" rejected. When the former occurs, the application will subsequently issue as a U.S. patent, provided the issue fee is paid in a timely manner. When an office action is made "final," there are several recourses. Under appropriate circumstances, the finality of the last round of rejections can be removed and patent prosecution can continue. Alternatively, the application can be refiled as a continuation, and prosecution can begin anew, but the 20-year time period continues from the filing date of the earlier application, i.e., the clock is not restarted. If there appears to be no hope of ever persuading the examiner to allow the claims of the application, an appeal can be taken.

26.2.4. Appeals and Interference Proceedings

The examiner is not the final arbiter of whether a patent application issues as a patent. Within the PTO itself, appellate review by the Board of Patent Appeals and Interferences can be sought if the examiner finally rejects the claims of the application. If unsuccessful before the board, the case can be appealed to the court of Appeals for the Federal Circuit, the specialized court that hears appeals in patent cases. The U.S. Supreme Court can hear appeals from the Federal Circuit, but this happens only rarely. If the invention is important enough, and you have the funds to support an appeal, it can be well worth your while, since the PTO has been known to misapply the law, and the Federal Circuit does an exceptionally good job of straightening things out.

When two or more applications are filed in the PTO claiming the same invention, the United States, unlike all other countries, does not necessarily grant the patent to the party who is first to file but rather to the party who is the first to invent. The PTO conducts what are known as interference proceedings (interferences) to determine who was the first to invent. As previously described, conception, diligence, and reduction to practice all play a role in this determination [35 USC § 102(g)]. Interferences are very complicated proceedings and frequently involve multiple parties, all trying to prove they are the first inventor and therefore the one entitled to the U.S. patent. The evidentiary standards that must be met to prove priority of invention are extremely strict. Thus, the unwitting inventor, who may truly be the first to invent, can end up losing a priority battle because of poor record keeping or lack of corroborating witnesses.

As a result of the North American Free Trade Agreement (NAFTA) and GATT, significant changes have been made to U.S. patent law concerning evidence that the PTO will permit a foreign patent applicant to present to establish the date of invention. Prior to NAFTA and GATT, only acts within the United States, the filing of a foreign patent application, or proof of introduction of the invention into the United States could be used to establish the date of invention. Now, reliance on inventive acts occurring in a NAFTA or World Trade Organization (WTO, created under GATT) country can be used to prove a date of invention for the purpose of obtaining a patent in the United States (35 USC § 104). As is the case for U.S. inventors, those who have made inventions in NAFTA and WTO countries can present laboratory notebooks and other contemporaneously created documentation as evidence of dates of conception, reduction to practice, and diligence. This change in the patent laws will seriously affect the outcome of interference contests and undoubtedly will further complicate interference proceedings.

26.2.5. Applying for Foreign Patent Protection

Most industrialized nations have patent laws providing protection for worthy inventions. The patent laws vary from country to country and are beyond the scope of this chapter, but a few highlights of foreign filings should be noted. Many countries are signatories to international treaties, such as the Patent Cooperation Treaty (PCT), that facilitate obtaining patent protection in countries outside the United States. A single international application that designates the countries where protection is sought is filed in an authorized receiving office. (The PTO is such a receiving office.) There is a period of delay before the applicant has to make the decision whether to file the application in the patent offices of all the designated countries. A preliminary examination, much like the examination that occurs in the PTO, takes place during this time period and provides information that is helpful in deciding whether to go forward to the "national stage" in the designated countries. This decision is important, since very substantial fees are incurred in prosecuting patents in some foreign countries. During the national stage, copies of the patent application are provided to the patent offices of each of the designated countries, where they are examined according to patent laws of each particular country.

One of the benefits of the PCT is that the international application can claim priority to an earlier filed U.S. application, as long as the international application is filed within 1 year of the U.S. filing date. (Note that even a U.S. provisional application starts this year running [see section 26.2.2.3.].) This is known as the convention year. The term has its origins in what is known as the Paris Convention, essentially an international treaty to which many countries are signatories. According to the convention, a patent application filed in a signatory or convention country can form the basis for a claim to priority of invention as of its filing date, if foreign counterpart applications claiming priority are filed within 1 year. These priority dates can be extremely important in determining patentability. Some countries are nonconvention countries, i.e., they are not parties to the Paris Convention or to the PCT and, hence, the Paris convention and PCT rules do not apply. It is advisable to obtain a list of nonconvention countries and determine whether a patent application should be filed in any of those countries. If filings are desired in nonconvention countries, it may be advisable to file applications in those national patent offices at the same time the U.S. application is filed. Remember that many nonconvention countries are "strict novelty" countries where any publication of the invention, even if after filing a U.S. application, can bar a patent if an application is not on file in that country. In other words, in a nonconvention country, you may not be able to rely on your U.S. filing to antedate the publication.

26.3. UNIQUE ASPECTS OF BIOTECHNOLOGY PATENT LAW AND PROCEDURES

26.3.1. Fulfilling the Utility Requirement

Until recently, many patent applications on biotechnology-related inventions were plagued by utility rejections. A strong pronouncement by the Federal Circuit in *In re Brana* (1995) on what constitutes utility, followed by the issuance of Utility Examination Guidelines by the PTO consistent with the court decision, has reduced the number of utility rejections in biotechnology-related cases. Prior to 1995, the examiners in the biotechnology groups within the PTO were requiring the equivalent of FDA phase III clinical trials before accepting claims drawn to compositions that had human therapeutic utility. These rejections were completely at odds with general patent policy, which encourages early disclosure of inventions to the public. Undoubtedly, many worthy inventions were abandoned as a result of these onerous rejections. Now there is clear court precedent that animal testing can be sufficient to support a claim to therapeutic utility in humans, i.e., that potentially human therapeutic compounds can meet the utility requirement without proof of human efficacy. Indeed, under the new Utility Examination Guidelines, if a patent applicant provides a sworn affidavit or declaration by a competent expert stating the disclosed human therapeutic utility of the invention would be considered credible by a person of ordinary skill in the art (e.g., in view of animal data or even in vitro data that are reasonably predictive of human efficacy), that ends the inquiry. The invention will be deemed to meet the utility requirement. Despite this progress, issues about utility continue to arise for biotechnology-related inventions, e.g., new DNA sequences, resulting from the Human Genome Project, that are not yet known to encode a useful protein or to have a useful control function. (Several expression sequence tags have been patented by commercial ventures.) Utility questions may also arise about compounds that pass early combinatorial screening assays but are otherwise of unknown therapeutic utility. The patent applicant is advised not to delay filing patent applications until all the therapeutic data are in, but to disclose additional known utilities besides potential therapeutic ones (e.g., use as standards, use in tissue typing, use as probes, use as PCR primers, or some other diagnostic use) in the patent application.

26.3.2. Fulfilling the Novelty Requirement

Biotechnology-related inventions are judged by the same standards as other inventions, including the requirement of novelty. Note, however, that U.S. patent law permits the patenting of naturally occurring microorganisms and other naturally occurring compounds such as proteins, cofactors, nucleic acids, vitamins, and antibiotics. In one sense, these things are not "new" because they exist in nature. However, their natural surroundings are such that they are unknown and waiting to be discovered through, e.g., screening and isolation or purification techniques. When a natural product is made to exist in a form other than the one in which it exists in nature, it is considered novel for patentability purposes.

26.3.3. Fulfilling the Nonobviousness Requirement

Many worthy biotechnology-related inventions have been barraged with obviousness rejections by patent examiners. This is due in large part to the standardization of cloning methodologies and the routine nature of screening cDNA or genomic libraries. Also, the ability to determine the protein sequence from the gene sequence makes certain aspects of biotechnology predictable. For example, examiners were rejecting claims to DNA sequences encoding particular proteins for which partial or complete amino acid sequences of the encoded proteins were known. Examiners reasoned that certain techniques for making hybridization probes and screening cDNA libraries or for making antibodies to proteins and screening, e.g., expression libraries, had become sufficiently routine that the gene encoding the protein became obvious.

Rejections of claims to nucleic acids (e.g., genes) identified by their specific nucleotide sequences based on the obviousness of the methodologies used to clone genes have largely been vanquished by recent decisions by the Federal Circuit. Two important cases are *In re Bell* (1993) and *In re Deuel* (1995). In *Deuel*, the Court looked to whether claimed compounds (nucleic acids), defined by chemical structure (nucleotide sequence), were obvious, not to whether obvious methodologies were employed to obtain the sequences. In *Bell*, the applicants claimed the cDNA sequence encoding a protein, the entire amino acid sequence of which was known. Given the particular amino acid sequence and the degeneracy of the genetic code, the Federal Circuit reasoned that the likelihood of predicting the exact sequence of the cDNA was exceedingly small. Hence, the structurally defined claimed compound was not obvious.

In *Deuel*, the applicants were claiming specific DNA sequences, structurally identified, encoding a protein, the partial amino acid sequence of which was known in the prior art. The examiner rejected the claims for obviousness, reasoning that the known partial amino acid sequence provided general motivation to search for the gene and that the gene could be found by well-known general cloning strategies. The Federal Circuit overturned the rejection, again focusing on the nonobviousness of the claimed specific chemical structures, i.e., the nucleic acid molecules defined by nucleotide sequence.

The logic the Federal Circuit applied in the preceding cases is consistent with earlier precedent regarding conception of structurally defined gene sequences. In *Amgen, Inc. v. Chugai Pharmaceutical Co. Ltd.* (1991), the Federal Circuit held that knowledge of a protein sequence and a general plan for locating the gene encoding this sequence is not a conception of a chemically identifiable nucleic acid molecule. Similarly, in *Fiers v. Sugano* (1993), which arose out of an interference, the Federal Circuit held that conception of the claimed structurally identified DNA did not occur when a method for obtaining it was conceived. The Federal Circuit held that for a structurally identified gene, conception and reduction to practice take place simultaneously. The structure of the gene cannot be known until the compound is isolated and analyzed (sequenced). The notion that a specifically identified gene cannot be obvious in view of the methodologies used to obtain it is consistent with this view of conception: what cannot be contemplated or conceived, until you actually have it, cannot be obvious. The take-home lesson is: do not delay sequencing your gene. The date you obtain the complete sequence may be the earliest date you will be able to rely on for proof of conception of a structurally identified nucleic acid sequence.

For many years, biotechnology-related inventions were also plagued by what were known as "*Durden*-type" obviousness rejections. Based on a case called *In re Durden*, certain processes (e.g., expressing a gene and making the encoded protein) were considered obvious even though the reactants and resulting products were novel and nonobvious. Relief from such rejections has come in the form of a new statute, 35 U.S.C. § 103(b) and two recent Federal Circuit cases, *In re Ochiai* and *In re Brower*. The statute provides that, if certain PTO procedural requirements are met, a biotechnological process using or resulting in a composition of matter that is novel and nonobvious shall be considered nonobvious. "Biotechnological process" is specifically defined as meaning:

A. A process of genetically altering or otherwise inducing a single- or multicelled organism to—
   1. express an exogenous nucleotide sequence,
   2. inhibit, eliminate, augment, or alter expression of an endogenous nucleotide sequence, or
   3. express a specific physiological characteristic not naturally associated with said organism.
B. Cell fusion procedures yielding a cell line that expresses a specific protein, such as a monoclonal antibody.
C. A method of using a product produced by a process defined by subparagraph A or B, or a combination of subparagraphs A and B.

26.3.4. Fulfilling the Sufficiency of Disclosure Requirements

It is difficult to provide specific recommendations as to how to ensure that your biotechnology-related invention is adequately described, because descriptions vary on a case-by-case basis. Even though a written description is called for, sometimes words are inadequate to convey completely the intended scope of the invention. Always consider using figures and photographs. For example, suppose you are claiming a plant that is identifiable by some special physical trait, e.g., flower, leaf or seed shape, or color. Including a photograph can often easily solve the inherent ambiguities and vagaries of the English language. Other pictorial examples include (i) photomicrographs or electron microscope images revealing particular characteristics of the thing claimed, (ii) photographs of protein or nucleic acid gel analyses (e.g., sodium dodecyl sulfate-polyacrylamide gel electrophoresis or Southern blot results) to demonstrate size distributions, or (iii) photographs of treated and untreated conditions to show the effect of a claimed cosmetic or therapeutic compound.

For many biotechnology-related inventions, nucleotide sequences and amino acid sequences are very frequently presented in a patent application. The presentation of a complete sequence does not necessarily provide adequate written description for gene or protein fragments that fulfill the same function as the complete gene or protein. Therefore, if you wish to claim less than the entire gene or protein sequence, part of your written description should provide some explicit guidance as to what the specific fragments are. Do not rely on your entire sequence to adequately describe these smaller nucleic acid sequences or polypeptides. In the same fashion, make sure any mutations, deletions, or insertions are adequately described. If you wish to claim a genus of nucleic acids, do not rely solely on functional language (e.g., the type of protein encoded). Be sure to describe the essential chemical structural features, or, better yet, give examples of as many complete sequences within the genus as possible.

A truly unique aspect of certain biotechnology-related inventions is the necessity of making a "microorganism deposit" to fulfill the enablement requirement. Recognizing that words may not be adequate to enable someone else to make and use a new life form, procedures have evolved since *In re Argoudelis* (a 1970 case approving the deposit of a claimed microorganism with a recognized depository as a means of fulfilling the enablement requirement of 35 U.S.C. § 112) whereby deposits of biological materials (e.g., bacteria, viruses, hybridomas, plasmids, or plant seeds) are made in connection with filing U.S. and foreign patent applications. Specimens needed to practice the biological invention are deposited with an internationally recognized depository, e.g., the American Type Culture Collection (ATCC) in Rockville, Md. There are fees associated with making the deposit and additional fees, e.g., if the depositor wishes to deposit under the Budapest Treaty (for foreign filing purposes) or if he or she wishes to know who requests the deposit after a patent issues. The depositor must agree to replace the stock if it becomes contaminated, depleted, or loses viability. The depositor is expected to notify the ATCC when a U.S. patent issues, at which time the ATCC is authorized to release the culture to requestors (who, hopefully, use it for noninfringing purposes if requested during the life of the patent). At the time the deposit is made, accession numbers are assigned to the deposits, which are then referenced in the U.S. patent application. This way, when the U.S. patent issues, the public can access the biological material from the ATCC through the accession number.

It is recommended that biological materials be deposited before the patent application is filed, although this is not always required under U.S. law (*In re Lundak*). The patent rules in many foreign countries, however, strictly require that the deposit of biological material be made before filing the application to which the foreign patent application claims priority and that, within a specified time period, the application be updated to contain the accession number. For example, if you wish to file for a patent in the European Patent Office (which examines for many European countries), claiming priority of your U.S. application, the deposit of biological material under the Budapest Treaty must be made before filing the U.S. case. It is preferable to include the accession number in the specification and claims as of the time of filing, although the accession number can be added to the application for a limited time after filing. Also, it is a good idea to make the deposit sufficiently far ahead of filing the application so that the viability of the culture can be determined by the depository. Certain foreign countries require that the deposit be viable as of the filing date of the application. If the culture is not viable, and the application was already filed, the priority date may be lost. However, if the application has not yet been filed, there will be time to redeposit a viable culture with the depository.

New procedures recently went into effect in the European Patent Office for depositing biological materials in connection with patent filings. A competent European patent practitioner should be consulted when applying for European patent protection on a biotechnology-related invention. European patent applications do not have to be filed in the name of the inventor (as in the United States) and are usually filed in the name of the company that owns the application. To avoid various procedural requirements, it is recommended that the applicant for the European patent and the depositor of the biological material be one and the same. Thus, if the European patent claims priority to a U.S. application, the owner of the U.S. application should be the depositor. Also, you should be aware that in the European Patent Office, deposited biological materials become publicly accessible to authorized individuals when the patent application is published. (Most foreign jurisdictions publish patent applications within a certain number of months of the claimed priority date.) There are procedures that the applicant can adopt for limiting the accessibility of the cultures to specified experts. Obviously, the goal is to provide public disclosure of the complete invention, yet prevent misappropriations.

Since deposits of biological material are usually necessary to fulfill the enablement requirement for biotechnology-related inventions, the question often arises, Must I deposit the biological material that I consider to be my best mode? The answer is, not necessarily. Enablement and best mode are distinct requirements. However, an inventor must not only disclose the best mode, but must disclose it in a way that skilled artisans can understand it and practice it. That is, the best mode has to be enabled. Thus, it may not be necessary to deposit your best genetic construct if the materials you made it from are publicly available or if you deposit the necessary starting biological materials and you provide enabling guidance or instruction in your specification for constructing what you consider to be the biological material that is or best performs the claimed invention. If you cannot adequately describe how to make the best mode, then you have not enabled it and you should deposit the biological material you consider to be your best mode.

26.4. CONCLUSION

The above overview of general aspects of U.S. patent law and unique aspects of patent protection for biotechnology-related inventions is by no means exhaustive. Furthermore, because of global trade concerns and other influences (e.g., health care concerns about the patenting of medical procedures), U.S. patent laws are probably in a greater state of flux than at any other point in history. Consequently, you should work closely with your patent attorney to plan your company's or university's patent strategy with respect to obtaining patent protection on biotechnology-related inventions as well as licensing and litigating your issued patents. Your company's or university's financial well-being could depend on it.

Pilot Plant

BRUCE K. HAMILTON, EDWARD M. SYBERT, AND JOHN T. ROSS

27

Microbiology and biotechnology pilot plants are charged with a wide variety of assignments. Some examples are presented in Table 1. Over the years, significant changes have occurred in the types of host organisms used in pilot plants, in the complexity of the processes developed, and in the level of sophistication of the equipment and instrumentation employed. In addition, with the introduction of validation requirements by the Food and Drug Administration (FDA) in the late 1970s and with the associated upgrading of current Good Manufacturing Practices (GMPs), a pilot plant that is expected to produce clinical supplies or commercial product must be prepared to come under the scrutiny of the FDA.

When the first edition of the *Manual of Industrial Microbiology and Biotechnology* (7) was published only a little more than a decade ago, the first assignment listed in this chapter's Table 1—preparation of retroviral vector lots for human gene therapy clinical trials—had never been made to any pilot plant on the planet. This example illustrates that it is not always possible to predict exactly what microbiology and biotechnology pilot plants will be assigned to do a few years down the road. Flexibility and versatility may therefore be among the principal design guidelines for a new pilot plant, if the new facility is not to become rapidly obsolete.

Given that change is to be expected, it is nonetheless best to make a diligent effort to "scope" or analyze the mission of a new pilot plant before starting to design it. In scoping the mission, it is useful to consider seven factors that turn out to be highly interrelated. These factors are discussed in the next section.

27.1. PILOT PLANT MISSION

Perhaps the most critical aspect of conceiving and commissioning a pilot plant is developing a clear and concise mission for the pilot plant facility. Careful definition of mission is of obvious fundamental importance, so it is imperative to make sure that the pilot plant mission statement is fully developed and complete. The mission statement document should become the basis for design, construction, and operation of the facility. Initial generation of such a document mandates the involvement of research and development (R&D), manufacturing, engineering, validation, regulatory affairs, and management. Support groups, such as information systems and finance, may also participate in formulation of the mission statement.

Seven highly interrelated factors to consider when formulating the mission statement of a new pilot plant are (i) scope of tasks, (ii) scope of regulatory compliance, (iii) process scope, (iv) scope of organisms, (v) scope of containment, (vi) budget and schedule constraints, and (vii) long-term vision.

27.1.1. Scope of Tasks

Obviously, an organization would not consider building a pilot plant without a need or directive. Before the introduction of genetic engineering a score of years ago, many biological pilot plants were employed as dedicated models for scaling up new processes from R&D into a full-scale manufacturing plant. Currently, the biotechnology pilot plant is more likely to be utilized for multiple campaigned projects, probably including production of clinical supplies (e.g., Investigational New Drug-stage biopharmaceuticals), or, after licensing approval, even early commercial lots. The FDA is responding to this trend of the industry and is in the process of revising guidelines and regulations to accommodate these needs.

Begin as early as possible to generate the conceptual plan (sometimes referred to as a "white sheet"), which explicitly defines the requirements of the pilot plant: what tasks it will be expected to perform, how much of how many products it will be required to make in what time periods, and under what regulations (e.g., GMPs suitable for phase I/II, or, even more rigorous, phase III or licensed product) and policies it will operate. Defining this scope of tasks well requires the development of a project team, which will be discussed in greater detail later in this chapter. For now, suffice it to say that at this early point in scoping out the pilot plant, a project manager should be selected and the project team coalition should be formed.

Some words of caution are in order regarding the project team and the general development of a conceptual design. The rule of thumb is that the generation of quality conceptual design documents, and the number of work sessions required to do so, will consume more time than is often anticipated. Many organizations underestimate the investment of time that will be required by their staff and initially envision that there will be little impact on regular work

TABLE 1 Some examples of pilot plant assignments[a]

Assignment	Product	Organism(s)	Culture vessel	Product isolation
Prepare 3,000-vial GMP lots of a retroviral vector for human gene therapy clinical trials	Retrovirus	Mammalian cells, retrovirus	Roller bottles	Aqueous based
Produce gram quantities of purified GMP recombinant glycoprotein, with CHO cells, for human clinical trials	Glycoprotein	CHO cells	Stirred tank bioreactor	Aqueous based
Prepare gram quantities of purified GMP recombinant protein, with *E coli*, for human clinical trials	Protein	*E. coli*	Fermentor	Aqueous based
Prepare gram quantities of purified GMP MAb, with hybridoma cells, for human clinical trials	MAb	Hybridoma cells	Hollow fiber bioreactor	Aqueous based
Prepare gram quantities of GMP conjugated toxoid vaccine for human clinical trials	Conjugated toxoid vaccine	Bacteria	Fermentor	Aqueous based
Prepare kilogram quantities of purified GMP low-molecular-weight antibiotic for human clinical trials	Antibiotic	Actinomycete	Fermentor	Organic solvent based
Improve any of the above processes by raising product titer, increasing yield on substrate, increasing throughput (productivity), substituting with lower-cost raw materials				

[a]Abbreviations: GMP, good manufacturing practices; MAb, monoclonal antibody.

scopes. This is a false and dangerous assumption. Something will have to give, and what can happen is that development of the pilot plant scope is shortchanged. If you must do everything, be prepared to extend deadlines and scale back routine work so that adequate attention can be devoted to conceptual design of the pilot plant. Skimping on pilot plant conceptual development will cost more in time, money, and lost productivity later. Skibo (22) suggests that for every dollar's worth of design input missed at the conceptual planning stage, the cost will be $15 to accommodate it later at the detailed design stage, or $150 to correct for it during construction.

The pilot plant conceptual plan must address whether operations for different products will be done in multiple, dedicated, separate facility lines, or else in a single line so that equipment, with proper design and use of suitable protocols, will have to be shared in common. How much processing will be done concurrently versus how much will be sequentially campaigned must be well defined. Regulatory aspects of a facility are directly affected by these decisions and dictate a multitude of requirements integral to the design of the facility. Therefore, a project team composed of management, engineering, development, validation, quality control (QC), quality assurance (QA), and manufacturing personnel is crucial for decision making at this stage of conceptual development. The organization must determine whether the facility will be geared toward development and scale-up, clinical phase I/II or phase III production, early commercial production, or combinations of these. The design must be oriented such that the most rigorous operation envisioned for the pilot plant is considered. Major equipment systems must be appropriate for all tasks within contemplated work scopes. Obviously, minimizing revalidation operations, simplifying cleaning operations by judicious design selection, and having common equipment standard operating procedures (SOPs) will reduce time, effort, and money required for switching from one process to another. However, these aspects must be taken into consideration during conceptual design and can be appraised accurately only if the facility's intended use is clearly defined initially.

Pilot facilities dedicated to a single process, while the least expensive for accomplishing their dedicated objective, lack versatility and are likely to require renovation and revalidation as the organization's requirements change. Separate/concurrent operations, while the most versatile, are the most expensive to build and the most difficult to validate. Extensive process segregation, including piping, waste disposal, and heating–ventilation–air conditioning (HVAC) as a short list, will be required. Concomitant increases in validation costs will accompany the choice of a separate/concurrent facility. Currently, a common compromise is the construction of a multiuse facility targeted for campaigned operations (2, 11). The use of a facility for a single product initiative, followed by cleaning for product changeover, is a practical solution to the requirement of many organizations for simplicity, versatility, and cost-effectiveness.

27.1.2. Scope of Regulatory Compliance

Unless the facility will be dedicated to a single use, one must evaluate several key regulatory issues with regard to multiuse facilities of any kind and plan accordingly to accommodate these concerns. These issues include the potential for preparation of poorly characterized products (resulting in difficulty in cleaning/decontamination evaluations), cross-contamination between product campaigns (equipment and cleaning validation issues), backflow of utility systems or waste systems, separate HVAC systems, area pressurization issues, and appropriate personnel, raw material, and product flows. Contained systems should be utilized wherever practical.

While the FDA is currently gravitating toward minimizing intensive oversight of facilities for production of well-characterized biological products, this does not imply regulatory lenience. On the contrary, manufacturers will be required to assume major responsibility for their facility, including assurance of validation and documentation. Inspections may be fewer but more detailed. The FDA, however, should be involved early on with review of the owner's plan. They will offer constructive criticism at a point where changes can be accommodated on paper rather than through physical change orders. Do not, however, contact the FDA for a meeting without a complete and detailed version of your conceptual design, or else you will waste their time and your own. The FDA provides excellent guidance documents that should be obtained, reviewed, and understood at the outset of a project. One such document is entitled "Use of Pilot Manufacturing Facilities for the Development and Manufacture of Biological Products" (10). Another is an FDA field inspector's guide entitled "Guide to Inspection of Bulk Pharmaceutical Chemicals" (revised 1991), available from the FDA Center for Drug Evaluation and Research (CDER, Rockville, Md.) or on the Internet at http://www.fda.gov/cber/cberftp.htm. Other documents, available from such pharmaceutical organizations as the Parenteral Drug Association (PDA, Bethesda, Md; http://www.pda.org) and the International Society for Pharmaceutical Engineers (ISPE, Tampa, Fla.; http://www.ispe.org), also provide valuable guidance and insight and should be on the mandatory reading list. Examples include:

- Baseline Pharmaceutical Engineering Guide, vol. 1: Bulk Pharmaceutical Chemicals, November 1995, ISPE/FDA
- Baseline Pharmaceutical Engineering Guide, vol. 4: Pharmaceutical Water and Steam Guide, November 1996, ISPE/FDA
- Baseline Pharmaceutical Engineering Guide, vol. 3: Sterile Manufacturing Facilities, November 1996, ISPE/FDA

Numerous other publications relating to regulatory compliance are cited throughout this chapter.

27.1.3. Process Scope

The conceptual design document should include a section that describes the intended process(es), including, in as much detail as possible, flow of raw materials, products, waste streams, and personnel. Also included should be raw materials acceptance, testing, and storage facilities; general warehousing, administrative areas, QC laboratories, raw materials staging, and the various production trains involved, through final product quarantine, release, and shipment. Key equipment, and all utility drops, needed for all of the intended processes should be identified, keeping in mind that equipment not immediately required can be stored outside the process area in a controlled environment.

Although time-consuming to perform, the analysis just described has multiple benefits. First, performing it forces the project team to scrutinize the layout of the pilot plant with regard to the intended process(es) and, in so doing, recognize possible problems that otherwise might not have been anticipated. Where will raw materials be received and stored? How will final bulk product(s) be transferred to the quarantine area? Are environmental storage areas adequate? If a clean-in-place (CIP) skid is to be employed, how will it be moved from point A to point B? Questions of this type are frequently neglected by overworked project personnel concentrating on a multitude of other imminent decisions and daily distractions. Second, a process scope will be required for discussions with the architectural and engineering (A&E) firm, consultants, and the FDA. Finally, a process scope will be a crucial requirement of the facility description section of a validation master plan and any registration documents that may be required, including product license applications.

27.1.4. Scope of Organisms

Biological pilot plants handle a range of organisms of differing characteristics. Included are animal cells (mammalian, insect), bacteria (e.g., *Escherichia coli*, bacilli, streptomycetes), fungi (yeasts, molds), algae, and plant cells. The organism type influences the design of culturing vessels (fermentors, bioreactors), batching equipment, support laboratories, etc. Therefore, one should attempt to define early on the types of organisms that will be cultivated in the pilot plant. Note that a biologics facility (as defined by the FDA) that cultivates gram-positive sporulating bacteria must be dedicated solely to such use.

27.1.5. Scope of Containment

Containment refers to the protection of the environment, operations personnel, and the product(s) being produced. The use of closed systems (primary containment engineering controls) should always be the design goal.

In many instances, the organism utilized for production of a biopharmaceutical product will dictate the level of biocontainment required. For guidance in this area, the primary document to review is the *NIH Guidelines for Research Involving Recombinant DNA Molecules* (10a), which indicates in great detail the level of containment assigned to larger-scale production (usually defined as greater than 10 liters of cultured organism) and the containment requirements associated with particular levels of containment. One must also be cognizant of the fact that some products, even after removal of the microorganism, may be inherently hazardous, and containment, decontamination, and cleaning for these products must be addressed accordingly.

Many organizations establish a facility design criterion at least one level above that stipulated by the NIH Guidelines (10a) for their target organism. This element of insurance may be wise when one considers that the costs involved to install modest extras during design and construction may be minimal compared with the costs after the facility is built or after a new product is developed that requires additional containment.

Issues that will require significant consideration and warrant the assistance of personnel or agencies well versed in these specific topics include biowaste handling and decontamination systems, HVAC filtration, area pressurization and balancing, containment drain systems, material transport systems, and prevention of process stream backflow. Equipment and process vessel vent and pressure relief filtration are also important design considerations. Validation and maintenance of these systems will be heavily scrutinized by regulatory agencies.

In addition to biohazard and product-associated toxicities, the possible use of hazardous chemicals and solvents must be addressed in the conceptual design. What may have been seemingly insignificant waste disposal problems and costs at the R&D stage can quickly become a major cost factor at the pilot scale. If in-house expertise is not available to address these issues, use of an A&E firm or consultant well versed in these specific design considerations is advisable.

Regardless of what types of hazards are inherent in the pilot plant process(es), spill control is an issue that must be adequately addressed. Drain covers, diked thresholds, pressurized positive barrier mechanical seals on fermentor agitator shafts, emergency spill control systems, SOPs, and intensive personnel training are all integral parts of containment and spill control procedures.

27.1.6. Budget and Schedule Constraints

It is the rare organization that can afford to lavish unlimited time and money on a construction project. As a rule, a project is on a capped budget with a 10 to 20% cost contingency and is also on a "fast-track" schedule. Real-world project execution is fraught with unforeseen obstacles (e.g., unanticipated deterioration of existing interconnecting drain piping, foundation bedrock, third-party vendor delays) which can voraciously consume resources. Change orders can burn time and money at an alarming rate. While controls must be in place to contain spiraling costs, the project team must resist the temptations to take resources from one area to accommodate an unexpected problem in another. If the conceptual plan was originally developed with due diligence and approved by all appropriate parties, then the overall design was, and likely remains, adequate for the facility's intended function. Impromptu borrowing of resources usually results in complications later and can lead to a domino effect.

Although not all problems can be anticipated during the conceptual design phase of a project (which is a reason for the contingency budget), most construction-phase delays and cost overruns are the result of skimping on up-front planning. Although it is at the preliminary design and drawing review stage that problems are easiest and least costly to rectify, this stage is unfortunately often the most undeveloped step of pilot plant construction. The pressure to "get on with the job" can be exceedingly strong. There is sometimes the pervasive mind-set that the design development and review process is taking too long and there is concurrent pressure to see tangible construction progress. Quite frequently, the design development and review process is undertaken by engineers and managers who are expected to continue with their usual workloads in addition to handling facility design issues, all without the benefit of the intended senior pilot plant staff on board. (Such staff are unfortunately often added only after construction has been completed.)

Avoid these pitfalls if at all possible. Review all design documents thoroughly, and compare these against the pilot plant process design, facility layout, personnel and materials flows, and other GMP design requirements. Explain to management the requirement for the investment of adequate time to prepare the design carefully, and offer decreased construction time and faster start-up as returns on such investments. Try to negotiate around the allure of cutting storage space or downgrading equipment to decrease costs. Cause-and-effect relationships apply to pilot plant construction, and the "pay now or pay more later" rule applies. For a given conceptual scope, if the pilot plant costs too much on paper, the thing to do is to take the hard choice and reduce the conceptual scope of the facility rather than promise to somehow magically construct a facility of the same scope at unrealistically low cost (17).

27.1.7. Long-Term Vision

Today's competitive markets demand pilot plant strategies with built-in versatility. Starting at the conceptual design stage, the organization must consider long-term goals and future product requirements. By way of example, consider the case where results with a pilot lot of product exceed the organization's expectations in clinical trials (we all want this problem!), precipitating the requirement for immediate production. Because such a degree of clinical trial success of the product was not anticipated, preparations for a full-scale production plant are not far along, and so start-up of the full-scale plant is at least 2 years in the future, resulting in lost potential revenues of many millions of dollars. Although it may be possible to make production lots in the pilot plant, the economy of scale using existing pilot equipment may preclude this avenue. If adequate utilities (e.g., wiring, air compressors, water systems, space for expansion) were installed initially, simple modifications and minimal revalidation would be necessary to shift the pilot plant to a small production facility quickly and with minimal investment. On the other hand, if this eventuality was not foreseen during pilot plant design, no amount of resources will enable conversion of the pilot plant to a production facility. Similar scenarios can be envisioned for transitions from dedicated to multiuse facilities.

27.2. DESIGN AND CONSTRUCTION DEVELOPMENT

27.2.1. Development of a Project Team

As already discussed, an organization's management should establish a pilot plant project team early on when pilot plant construction is being considered. The team is charged with the review of facility and pilot production requirements and the subsequent development of equipment and service requirements listings and a conceptual plan (white sheet). The team will also be responsible for review of requests for proposals (RFPs) and selection of consultants and architectural, engineering, and construction firms. As the project progresses, the team may also be responsible for review of various cost estimates, change orders, vendor selections, time-line plans, and construction progress reports. Although the general composition of the project team may evolve throughout the project, a basic core must maintain oversight, starting from conceptual design through start-up, and possibly through the validation phases of the facility. If an outside A&E firm and contractor are selected to design and build, appropriate representatives of these firms must also become members of the project team.

It is important that a project team leader or project manager be selected at the outset. This individual should be well versed in operating policies of the owner's organization, have excellent organizational and communication skills, have a good basic understanding of engineering, process, and production requirements, and have sufficient time available to devote to the project, ideally 100%. The individual must have the authority (directly or indirectly) to resolve conflict, determine project team composition at any given point in time, and ensure the smooth transition from one project phase to another.

The project team is also responsible for identification of project-related problems, formulation of problem resolutions, and reports to upper management. A properly run project team will generate the appropriate documents necessary to avoid owner/contractor litigation, avert engineering or regulatory conflicts, and minimize surprises. The composition of the team can facilitate a forum for idea exchange, and, when properly led, the team membership provides a mechanism for formulating mutually beneficial solutions to problems, with all relevant organizational sectors having an opportunity to participate in problem resolution.

27.2.2. Selection of an Architectural and Engineering Firm

Selection of an A&E firm is one of the most crucial decisions that the project team will make having a considerable impact on cost, scheduling, and the success of the overall project effort. Although a detailed review of the various approaches for A&E selection are beyond the scope of this chapter, excellent articles are available that cover this topic in detail (14, 22–24).

The team members should begin to develop lists of potential A&E contractors as soon as the conceptual design is nearing completion. A list of selection criteria must also be developed and used to narrow the field of potential firms from an initial four to eight companies down to about three or four. The narrowed-down field of candidate companies should provide formal presentations of their capabilities to the prospective client that outline their experience, capabilities, and expertise. Many organization- and process-specific requirements will be involved in development of the selection criteria, but several universally applied requirements are worth mentioning. These include the following:

- Demonstrated A&E track record of building *fully operational* facilities similar to the one planned, including the provision of past customer references that may be contacted.
- A complete review of what in-house expertise the owner organization will utilize, what aspects of the project will be farmed out, and what input the owner will have over selection of subcontracted services. Some A&E firms offer full services and have total capabilities from process development through process start-up and validation. Be sure of what you are buying and what options are available at what cost.
- A guarantee from the A&E firm that their key project staff will not change for the duration of your project unless exceptional circumstances arise. If a change is unavoidable, the owner should reserve the right to approve the change after review of the proposed substitute's credentials. The A&E firm should identify in advance, in writing, which of their key staff will perform the project tasks. The percentage of key A&E staff devoted to the owner's project should also be stipulated in writing.
- Ideally, the A&E firm should be geographically close to the pilot plant construction site, or a field office may be established if the project warrants such an approach.
- Look for breadth and depth in an A&E firm. Individual résumés of the A&E team should be requested and reviewed. The ability of the A&E project manager to devote dedicated time and effort to the owner's project is of extremely high importance. This is the key individual who makes sure that the architect talks to the HVAC person who talks to the piping specialist. Without the assurance of this communication and concerted effort, you are most assuredly heading for a host of problems, delays, and cost overruns.

Finally, while perhaps rudimentary, there are several additional important aspects of A&E selection that warrant emphasis. The generation of the RFP is a critical task and is one area in which finance and contracts personnel earn their keep. It is important that all companies selected for RFPs receive the same information and that the scope is as complete as possible. Particularly if the owner organization is in the public sector, it is wise to avoid phone conversations between the bidder and operations/engineering personnel at this point, and to rely instead on the organization's contracts personnel to shuttle the inevitable questions and answers back and forth between the A&E firm and the owner's project team. Although cumbersome, and sometimes a bit frustrating, this is the only way to ensure fair treatment and avoid potential litigation. When a contract is awarded, it is hoped that the determination has been made on the basis of written, weighted, and tangible criteria such as cost, expertise, presentation quality, demonstrated track record, geographical location, and similar factors.

27.2.3. New Construction versus Renovation

The consideration of whether to renovate or build new is a highly individual decision based upon many unique factors. What is almost certain, however, is that a complete renovation will be a more complex endeavor, will result in more design and review time, will encounter more obstacles, and will result in some compromise. This is especially true of older facilities that are being renovated and must meet new codes and regulatory requirements. A complete and careful paper exercise that evaluates the merits and drawbacks of each scenario is called for. One must resist the urge to renovate the process rather than to adapt the facility to accommodate the process. As Robins and DelCiello (20) perceptively state, "Experience suggests . . . that more often than not all those involved end up looking for ways to adapt—and sometimes compromise—the *process* to fit the facility."

By way of example, older facilities may have inadequate surface finishes, drain systems, and space for upgraded HVAC or enclosed piping closets. Design and retrofit of these systems are always more expensive than new construction. An older facility may contain hazardous materials such as asbestos or lead paint. Abatement or containment of these materials during construction will add significantly to job costs, compared with building from the ground up. Drawings of existing structure and piping may not be available or may be hopelessly dated. If drawings are more than 6 to 10 years old, they are probably not available on electromagnetic media format (computer-aided design [CAD]). Generation of as-built CAD drawings, important for both renovation and facility validation, is very costly and time consuming.

Ideally, the decision to renovate a facility rather than build new should be driven by a careful analysis of all relevant factors, including total costs involved and a thoroughly thought-out time line. Depending upon the extent of the renovation involved, there may be a considerable period of time required for the demolition phase, including demolition planning, demolition phase drawings, physical plant marking, execution, and cleanup. Usually, demolition work is subcontracted through the prime contractor. For obvious reasons, it is therefore imperative that there is clear communication between the owner and the demolition subcontractor such that equipment and structures scheduled for demolition are clearly identified, both physically and via demolition drawings.

If the decision is made to renovate an existing facility, detailed attention must be given to removal and reinstallation (R&R) of process equipment to be retained (or adequate protection if the equipment is to remain in place during construction). The equipment must be properly decommissioned, stored indoors with adequate security, and eventually reinstalled to the owner's design specifications. Clear definition of the R&R process, the division of responsibilities of all parties involved, and a clear understanding of the equipment conditions before and after reinstallation are necessary.

27.2.4. Project Pre-Scope and Scheduling

Project pre-scope may be simply stated as an initial collection of the intended facility design and operating requirements, approach, and preliminary schedules that must be prepared by the owner's project team before soliciting bids for design services. As previously indicated, a clear definition of the facility requirements is vital for a successful project. This aspect cannot be overemphasized, as failure to ensure that everyone involved in the project is in agreement and acting in concert is a recipe for disaster. At the very least, a poorly defined project precludes accurate scheduling and thus makes it impossible to predict accurate milestones and completion dates. This predicament translates to lost revenue, re-work, change orders, and disgruntled management and ultimately results in a project that is fraught with cost overruns and delays.

At a minimum, the project pre-scope (which will eventually evolve into a formal project scope of work) should include preliminary design requirements, process flowcharts (or surrogate process flowcharts), conceptual facility layout, tentative construction schedule, and preliminary cost estimates. Lists of personnel, safety concerns, security needs, space requirements, equipment, and SOPs must be developed, as should preliminary materials and personnel flow diagrams. If the facility is being renovated or retrofitted, complete and accurate as-built architectural, engineering, and piping and instrumentation diagrams will be required as well for the existing facility and equipment.

It is important at this stage to determine several design specifications that significantly affect cost. The facility shell configuration should be defined, because multiple floors (versus single-story construction) may enhance product flow via gravity and reduce overall costs. Process and ancillary space requirements must be estimated, because utility areas, laboratory space, office space, warehouse space, and classified GMP processing areas all have different associated costs. There is usually a "bottom-line" dollar figure budgeted for the total project, and one must be cognizant that internal preliminary cost estimates are invariably crude, usually low by a significant factor. Contingency factors must be included with the project cost estimate (usually 20 to 30%), and the required target date for completion should be determined. One must bear in mind that the time from conceptual design to final plant turnover can consume 36 to 48 months on a conventional construction track, and fluctuating labor and materials costs and inflation must be factored into the cost estimate. The requirement for fast-track, or what's becoming known as "hyper-fast-track," design and build will increase associated costs significantly, so the benefits of decreased build time (particularly faster time to market) must be weighed against what will probably be the lower costs associated with a conventional design-and-build scenario.

27.2.5. Conceptual Design Phase

The conceptual, or preliminary, design phase involves preparatory design and engineering development work that is usually executed by outside design firms to completely define the project scope. This phase usually includes the preparation of a conceptual estimate of project costs (a reality check against the owner's initial estimates), establishment of target dates, verification and location of required space and area specifications (air classification, biocontainment levels, emissions control, etc.), process flow diagrams (PFDs) and equipment block diagrams, and evaluation of existing equipment resources (both utilities and process equipment that may be relocated). Federal, state, and local building codes must also be considered at this time. Interaction with the owner's corporate managers, area managers, QA, QC, and safety personnel, facilities engineers, finance and contracts personnel, and project manager is mandated at this time, both in project team format and through individual interviews. A clear mutual understanding regarding the extent of funding required and available, projected completion dates, and intermediate milestones should be established, and the facility's expansion plans and life expectancy should be articulated.

27.2.6. Detailed Design Phase

The detailed design phase is the preamble for construction and represents the final opportunity for making changes at low cost. As such, this phase is extremely important in terms of scrupulous review. It should come as no surprise that extensive reviews will take considerable time and thought, both by individuals versed in particular aspects of the scope and through brainstorming sessions by the project team. All lists generated during the preliminary design phase should be finalized at this time. Engineering and construction drawings, including all equipment specifications, personnel flow diagrams, piping and instrumentation diagrams (P&IDs), material flow diagrams with material balance grids, HVAC air and heat balance calculations, and utility load assessments, must be completed, reviewed, and approved for build. Updated estimates of cost and scheduling must be generated.

Note that it is extremely important that long-lead-time items be given high priority and that the appropriate project personnel check the status of construction and equipment deliveries on a regular basis. Lack of attention to these items, which are sometimes easily forgotten, can have a cascade effect resulting in a devastating impact on construction schedules.

An extremely valuable and effective method for ensuring that all aspects of the design and layout are adequately considered is to utilize the project team to conduct a conceptual start-up exercise using the final P&IDs, PFDs, and block diagrams to "walk through" a virtual process. This serves to test the concepts from start to finish, including how raw materials and personnel will flow, how final product will be sampled, removed, and stored, and how equipment will be changed out and serviced. Frequently overlooked issues such as the proximity of lighting or parking to air-intake ducts (resulting in the attraction of insects and filter clogging or aspiration of vehicle exhaust, respectively) will become evident through this exercise if done with diligence.

27.3. VALIDATION PLANNING

27.3.1. Importance of Early Validation Planning

If the pilot plant will be used to prepare drugs or biologics for human or animal use, even if only at the clinical trial stage, the facility will be governed by FDA regulations, which include validation requirements (see Title 21 of the *Code of Federal Regulations*, Parts 210 and 211, available from the U.S. Government Printing Office). According to the FDA "Guideline on General Principles of Process Validation," validation is "a documented program which provides a high degree of assurance that a specific process will consistently produce a product meeting its predetermined specifications and quality attributes." To meet validation requirements, it is best to design a facility that is consistent with good practices for flow of personnel and materials, and it is mandatory to provide utilities (e.g., process water such as water for injection, clean steam, HVAC, waste treatment) that can be validated, as well as surface finishes that will support cleaning validation by area monitoring. A consequence of all these points is that planning for validation should begin as soon as a facility that will require validation is conceived. Waiting until the facility is built to begin validation planning is asking for a huge problem; even beginning construction without a validation plan in place is asking for trouble. Therefore, if validation will be required, put a validation plan in place early, and continuously act on it as the pilot plant project proceeds, starting with the conceptual stage of the pilot plant. An ideal time line for validation activities in relation to engineering, construction, and start-up work is presented by Baird and De Santis (3).

27.3.2. Validation Master Plan

If the pilot plant is to be validated, an excellent practice is to prepare a written validation master plan (VMP). It is best to complete the VMP during the early stage of detailed engineering of the pilot plant, since detailed engineering should be impacted by validation plans. A description of the contents of a VMP is presented by Baird and De Santis (3).

27.3.3. Document Control

During the planning, design, procurement, construction, validation, start-up, and operation phases of the pilot plant, a large number of documents will be generated, reviewed, revised, and reissued, with input coming from many people (for a sample list of documents through the construction phase, see Table 2 in reference 6). To maintain the usefulness and accessibility of these documents, a document control system must be implemented.

27.3.4. Enhanced Turnover Packages

To facilitate rapid and efficient validation, arrange for enhanced turnover packages to be provided whenever possible. In the case of certain pieces of process equipment, such as pilot-scale chromatography skids, vendor turnover packages may be enhanced to the extent that the vendor supplies, executes, and documents installation qualification (IQ) and operational qualification (OQ) packages. For utility systems, such as area HVAC systems, arrange to have component equipment operating and maintenance manuals turned over early and in a complete and orderly fashion to facilitate the writing of IQs and OQs. Chin (6) discusses turnover packages in detail.

27.3.5. Validation Control

The VMP should set out a scope and timetable for the validation of the pilot plant. What documents will be generated, when, and by whom, along with what procedures will be performed, when, and by whom, should all be specified in the VMP, preferably in tabular form. Progress toward completion of validation can then be tracked and measured against the plan specified in the VMP, with elements that fall behind schedule being highlighted. This tracking activity will then provide a basis for managing the validation effort, assessing whether the right resources are being utilized in adequate amounts, and evaluating the status of the validation program throughout the time course of establishing a new or renovated pilot plant.

27.4. BUDGETING

27.4.1. Capital Budget Phases

Typically, as soon as the notion of a new pilot plant comes up, a "ballpark" budget figure is required for it. It is usually not easy to come up with a very precise figure early in the pilot plant planning cycle, but rarely can the requirement for such a figure be escaped.

One approach to generating an early estimate is first to rough out the mission and scope of the facility, then to block out a basic process and utility equipment list for purposes of sketching out a crude facility layout, drawn roughly to scale. Based on the crude layout, the rough square footage of the facility can be estimated. The estimate of capital required is then typically the facility area in square feet multiplied by a figure of \$500 to \$1,000 or more per square foot, depending on equipment sophistication and density. For example, a 50,000-ft^2 pilot plant for manufacturing advanced clinical lots of several products simultaneously can easily cost \$25 million or more. Similarly, a 10,000-ft^2 facility expansion to increase cell culture, microbial fermentation, and purification capacity can easily cost \$5 million or more. Of course, if half of the new facility is left as open warehouse space, the overall cost per square foot will be significantly decreased, as it will if much of the new facility is given over to offices or the like.

A caveat is that the rough budget figure delivered early on may become "chiseled in stone" as soon as it is uttered. Therefore, proceed with caution. An estimate that is either grossly high (rare) or low (common) will be worse than useless, because it will give rise to false expectations at higher management levels. Also, cost will be very sensitive to the scope of the actual facility constructed: for example, if the ballpark estimate is based on a single train of culture vessels, and later it is decided to include a second train in the construction, the increase in facility cost will be substantial. In fact, the final cost for a facility may well be 5 to 10 times the cost of process equipment purchased for it.

For example, HVAC for a biologics facility is very costly and is typically not considered as part of the process equipment cost. Take care not to neglect or slight costs for engineering, design coordination and review (including travel), and permits.

More precise capital estimates will become available as engineering work progresses (see Fig. 1 and section 26.3.2 in reference 13). Once again, however, the actual capital required can be greatly affected by changes in facility scope, so be alerted to this cost sensitivity.

27.4.2. Capital Budget Control

Set up a cost-tracking system early on, with variances calculated against a working budget. Keep a unified and coherent record of purchase orders, invoices, payments, etc., and keep the record current. Review updated cost figures in detail monthly, and do not allow "cost data lag" to grow. All this is conceptually elementary, but make sure it is executed well so that financial control does not start to drift.

27.4.3. Validation Budget and Control

A substantial effort is required to validate a pilot plant. Many protocols must be written, revised, approved, and executed, with results reported and documented in both detailed and summary form. One approach is to use the personnel that will operate and maintain the pilot plant to perform the required validation activities. In this case, adequate budgeting of personnel in terms of number and time must be provided. If a contractor is used to perform the validation work, adequate funds must be budgeted. One rule of thumb is that the cost of validation will be in the neighborhood of 10 to 20% of the capital required to design, construct, and equip the facility.

Just as capital expenditures should be tracked and controlled (see section 27.4.2), so should validation costs. The VMP is a useful guide to track against (see sections 27.3.2 and 27.3.5).

27.4.4. Operating Budget and Control

A good estimate of the annual cost to operate and maintain a completed pilot plant obviously should be made and accepted before capital is paid out to build it. Pilot plants are not inexpensive to run. It is painful when a pilot plant stands idle for any reason, including the absence of a realistic operating budget to run it.

The principal component of a pilot plant operating budget is personnel cost. Often, an overall operating budget is estimated by factoring against "head count," e.g., perhaps $150,000 to $200,000 per person per year multiplied by the number of personnel staffing the plant. The dollar multiplier in this calculation is heavily affected by organization-specific accounting practices, such as how much overhead is allocated against the pilot plant and which personnel are, or are not, counted in the pilot plant head count. Obviously, the dollar multiplier is also determined by the nature and scale of pilot plant operations, e.g., recombinant cell culture at the 10-liter scale versus simple microbial fermentation at the same or perhaps a much larger scale, or GMP versus non-GMP operations. Make every effort to perform a careful analysis to arrive at a good estimate.

Means to monitor and control expenditures during operations are also critical and can be straightforwardly handled by standard accounting procedures.

27.5. EQUIPMENT LIST AND SIZING

27.5.1. Equipment Identification Lists

The project team should begin generating master equipment lists as early as possible, preferably during the conceptual design phase. While the pilot plant operating department should have primary responsibility for equipment determinations, the engineering department is concerned with engineering aspects of the installation, utility support requirements, and structural considerations. The QA department will also be interested in these and other aspects of the equipment, such as clean-in-place (CIP) capabilities, assignment of equipment numbering systems, and the generation of equipment master file documents, including calibration certificates. Input from the technical operations staff regarding their experience with specific equipment from various manufacturers should be sought. Obviously, management will be concerned with the cost of such equipment, and the contracts department must consider approved vendor lists and sole source and other procurement regulations.

The question of who should order the equipment—the general contractor or the owner—must be answered as early as possible. There are advantages and disadvantages to both scenarios. Having the general contractor responsible for procurement and installation (including engineering and start-up) may be the least troublesome avenue, provided clear specifications from the owner have been generated. Of course, equipment purchased or supplied by the owner will cost less, since there will be no contractor overhead or profit added onto the base price. Coordination of installation, however, will be more critical.

27.5.2. Matching Utility Load Requirements

After equipment has been specified, detailed material and energy balances for critical systems must be calculated and documented. Major utility systems such as electrical, oil-free air compressors, circulating liquid chiller systems, plant and clean steam generators, purified water systems, emergency relief systems, HVAC systems (air volumes and heat loads), and even drain pipe flows should be evaluated for adequate capacity, and a specified contingency should be added for future expansion and efficiency loss over the life of the systems. These calculations can be daunting and are best left to experienced design engineers using specialized computer software. The calculations, however, should be carefully reviewed by a second party (e.g., the owner's engineering department if the A&E firm was responsible for their generation) for accuracy. Documentation of these calculations is crucial to the project's master file and equipment IQs as well as for future equipment upgrades, modifications, additions, or construction of a large-scale production plant from pilot plant specifications.

27.5.3. Maintaining Scale-Up Geometry

When designing a pilot plant it is important to consider equipment configuration and proportion from the outset. Although detailed discussion of scale-up geometry is beyond the scope of this chapter (a myriad of excellent articles exist on this issue), the project team must always remain focused on the possible, if not eventual, scale-up of the equipment and processes developed in the pilot plant to 10×, 50×, or even 100×. The issues of fermentor geometry, vessel mixing, and heat and oxygen transfer are particularly important to keep in mind. If technology and engineering transfer is a possible outcome of pilot plant trials, then every piece of equipment procured should be specified and installed with those aspects duly considered. These choices range from the selection of either top or bottom fermentor agitator drives to valve types, liquid transport piping size, and gravity or pump-fed systems, to name just a few.

27.5.4. Requirements of Versatility

To say that a multiuse pilot plant must be versatile and flexible probably qualifies as tautological. Today's pilot plant might be required to produce multiple campaigned lots of the same product, followed by a decontamination and switch to a different product. Some pilot plants might be required to produce multiple product lots within the same facility concurrently, while still others, with adequate segregation and containment, might produce different compounds in the same facility concurrently. Design for each of these scenarios must be adequately considered and addressed at the outset of conceptual design for a multiuse facility. Increased versatility, however, is not realized without a concomitant (and sometimes exponential) increase in complexity and cost, and a compromise must be struck.

One approach to design of a versatile pilot plant is the use of dedicated (contained) process suites or equipment trains. The use of individual header systems for product transport with common raw material transport headers is sometimes employed, as are sanitary spool pieces, lockable flow-path transfer stations, and flex lines for construction of semipermanent material transport trains. Mobile equipment in adequate variety and duplication is also used, with appropriate cleaning validation and proper storage when the equipment is not in use. Pilot plants with concurrently run multiple product trains typically employ separate HVAC systems for each distinct product area. Particular attention must be paid to the avoidance of cross-contamination between products, including raw materials, personnel, equipment, and even housekeeping equipment and storage. Critical items for consideration in multiple product facilities include the following: What happens when one HVAC system goes down? What is the fail-safe area pressurization? Can air exchange between one process suite and another occur via a common drain system due to excessive pressure differentials? Are all waste treatment discharges and condensate drains equipped with air breaks to prevent backflow? How are pipe chases serviced and segregated? How do maintenance service personnel access the systems, and how are pieces of equipment to be serviced uniquely identified? Is separate interim storage available for each process area? How is the incidental or inadvertent introduction of product into purified water, gas, or steam systems avoided? How are intermediate product samples segregated and stored such that cross-contamination or mixups are avoided?

Finally, in addition to these considerations, the more mundane aspects of multiple processing must be addressed. These include the chore of more frequent waste decontamination and discharge, the requirement for additional office space, additional training for personnel, more detailed SOPs, a larger number of SOPs, and increased storage requirements (for spare parts, process equipment, and housekeeping supplies).

Whenever possible, equipment used for one product processing area should be identical to that found in another. This reduces training expenditures, parts in inventories, and the number of SOPs required and enables the economy of bulk buying for parts and supplies.

27.5.5. Computers and Controllers

The degree of computerization appropriate for a given facility is always a matter for consideration and is, to a large degree, a matter of personal preference. Pilot plant computer systems can be as simple as a personal computer that monitors and logs alarms or acquires process data or as complex as a full-blown distributed data system that monitors and controls entire building and process systems. Computer control systems have advanced very rapidly over the past 10 to 15 years. It is this exponential advancement in technology, however, that should give one pause before deciding to "install the best and computerize everything." Although computerized mega-systems are an attractive concept initially, the costs and ramifications of a major computer system must not be underestimated. Consider that the time from conceptual design to plant start-up may easily span a period of 3 years. What was specified over 3 years ago as a state-of-the-art distributed control system (DCS) may now be on the verge of obsolescence. Although one may hedge against such situations by carefully wording RFPs as well as specifying only manufacturers with a long track record of supplying systems for comparable use in validated GMP facilities, hardware and software will sooner or later (usually sooner) require upgrading. A DCS system will require extensive configuration, engineering design, and programming. IQs, OQs, and performance qualifications (PQs) of DCS systems can be an onerous task, gobbling financial and time resources at an insatiable rate. Field wiring requirements and the number of remote sensors/switches increase proportionally to the amount of monitoring and control via a DCS. While today's computer systems are usually equipped with redundant microprocessors and power supplies and are in themselves extremely reliable, the field instruments reporting to the computer as well as the field elements controlled by the system are still subject to failure. All of these sensors are potentially subject to preventive maintenance and calibration requirements. Maintenance of the additional sensors as well as the computer systems management may ultimately mandate the increase of plant staffing, which obviously increases overall costs. Perpetual training must also be factored in for those individuals responsible for oversight of the computers and associated systems. Troubleshooting problems, making repairs, and making software changes usually require specialized skills and training. Although rare, the complete failure of a computer system that monitors and controls an entire facility would obviously have disastrous results.

The discussion above should not be misconstrued as negativism against computer control. On the contrary, computer systems are today an essential and integral part of any facility and aid in the assurance that systems are reliable and provide reproducible processing results. Computer control also facilitates and enhances reliability, documentation, and alarm-reporting functions, while the requirement for SOPs and extensive operator training decreases. It is, however, extremely important that a careful initial assessment regarding the degree of required computerization be made and that the hidden costs of extensive automation are realized. Computer control of sterilization cycles, CIP systems, and HVAC systems is quite common and, in many instances, highly desirable. The enhanced monitoring and reporting capabilities (real-time data, archiving, alarm reporting, etc.) obtained for these systems is frequently well worth the added investment. A larger operation may further realize the benefits of automated batch records and reporting, resulting in a significant reduction in time and personnel requirements.

The degree and extent of computerization must be scrutinized carefully at the earliest opportunity. Extensive consideration during the planning stages must not be overlooked. While a DCS system may be appropriate (or even mandated) for some facilities, a smaller operation having the requirement for multiple products may do better by substituting multiple programmable logic controllers (PLCs) and personal computer monitoring/data acquisition for extensive dedicated control systems. PLCs and personal computers are inexpensive, durable, reliable, and easily maintained, can be validated and calibrated economically, are available everywhere, and provide adequate control for a smaller operation when flexibility is mandated.

27.6. SITE LOCATION

Factors important in choosing a site for a new pilot plant include proximity to R&D laboratories, from which new processes typically are "transferred in," and proximity to manufacturing, to which new processes typically are "transferred out"; labor availability, skill, and cost; labor climate; local weather; surrounding community receptivity; expansion options; supply and cost of utilities (water, natural gas, electricity); waste (sewage, solids, vapors) treatment availability, costs, and regulations; compatibility with trucking needs; time lags and costs for obtaining required permits issued locally; and any relevant ordinances, such as any that involve work with recombinant DNA. Factors for site location are discussed further by Hamilton et al. (12).

27.7. DESIGN

27.7.1. Use of an Engineering Company

While some larger companies will have their own A&E department, most will be faced with selecting an outside contractor to fulfill the role. The work product expected from the A&E provider is a detailed design and specifications for the new pilot plant building or renovations to an existing structure. The client, or owner, will provide the A&E firm with a description of needs in terms that can be converted to process flow plans, equipment selection, room and building sizes, and utility selection and sizing. The final design documents will include sections for civil engineering, landscaping, structural engineering, architecture, plumbing, fire protection, HVAC, electrical engineering, and specialty piping. Each of these major categories will contain 5 to 20 or more individual construction drawings. Specifications will be submitted in bound books with chapters keyed to standardized construction "divisions." For example, division 1 covers general requirements; division 2 contains specifications for site work; each trade (concrete, masonry, metals, etc.) is covered in its own division. This structure allows potential subcontractors to quickly review the owner's requirements in order to prepare a bid.

27.7.2. Communications

A project team with the composition already discussed should be the mechanism used to establish and maintain rapid and efficient communications among research, development, quality assurance, internal engineering, and contracted engineering groups. The project team will have one key representative from each of the groups listed above, and it is the responsibility of each such representative to communicate input from his or her group. Normally, the A&E firm will assign one individual to record the minutes of each meeting and distribute them for comment or correction. Bi-weekly meetings are the norm during the planning and design phase, and weekly or more frequent meetings may be needed during construction.

27.7.3. Drawing Approvals

Most contracts will call for drawing and specifications reviews at key milestones during the design period. Packages are typically reviewed at the 50%, 95%, and 100% completion points. The final drawings that will be issued for construction bids should be signed off by the owner's authorized representative. Any changes made after this time are subject to change orders and attendant additional charges.

27.7.4. Personnel and Material Flow

Current practice calls for a one-way flow of personnel, materials, and waste through the controlled process space. The use of an entry or "clean" corridor leading to the production suites and a separate exit or "dirty" corridor leading away precludes the possibility of tracking contaminants back into the clean process areas. Air locks or doors with electronically controlled interlocks are often used to ensure that once a process area is entered by one door, only the other door will allow exit. Emergency override provisions are incorporated to avoid trapping personnel, and input from fire and safety authorities should be obtained for decisions involving these engineering details.

27.7.5. HVAC

Individual air zones may be used to reduce the possibility of airborne cross-contamination. Each zone is typically designed with its own air handlers, and if any portion of air is recirculated, it is not shared with another zone. When a higher risk warrants, such as would be the case with BL3-LS biohazard cell lines or aerosols or vapors of a toxic nature, the HVAC system should be designed for once-through flow, with no recirculation. Many details of HVAC design for biotechnology facilities are given by Dobie (9).

27.7.6. Water Systems

Water meeting the USP quality standards for purified water and water for injection (WFI) is used in many current processes. Make-up of microbial fermentation media and routine washing typically use purified water. Cell culture media, purification buffers, and final equipment rinses typically call for WFI.

City water (potable water) is often the starting water source. From the starting water source, purified water may be produced, perhaps with a reverse osmosis, deionized system. Purified water may then be used to feed a clean steam generator. Clean steam, in turn, may be used directly for pharmaceutical steam applications and may also be condensed to yield WFI. Pharmaceutical water systems are discussed in detail by Slabicky (25).

The purification systems selected to produce the quality of water meeting USP standards must be selected to address local water contaminants. When possible, raw water analyses over the course of a full year should be studied to be sure that treatment systems have the capacity to address seasonal variations in organics, sediment, chlorine, etc.

27.7.7. Clean Steam

Clean steam differs from plant steam in that it contains no chemical additives and meets USP quality standards. It is used in sterilizing interior surfaces of process equipment, interconnecting piping, and any additional sampling and harvesting valves that might come in contact with the process stream. The clean steam generator may be fed water from the USP purified water system to reduce scaling and consequent maintenance. When WFI volume demands are modest or easily scheduled, a single unit may be employed to produce clean steam and WFI. Clean steam systems are discussed by Adams (1).

27.7.8. Process Air

Compressed air that will be used to provide oxygen to the cultures in bioreactors, drive liquids through filters, or otherwise come in contact with the process stream must be free of entrained oil mist. This requirement necessitates the use of either an efficient oil-removal system or, more commonly, an oil-free air compressor. Both rotary screw and reciprocating piston compressors are available with lubrication points well isolated from the air stream. In any case, the compressed process air must be dried and filtered to remove particles down to at least 0.2 μm. Drying may be accomplished with either mechanical refrigeration systems or adsorbents. Particulate filtration employs a series of elements with decreasing pore size. Final filtration, with presterilized or resterilizable point-of-use filters rated for 0.2 μm, protects the process stream from contamination.

27.7.9. Process Gases

When needed, oxygen, carbon dioxide, and nitrogen may be supplied from banks of cylinders or from large liquid storage tanks. In either case, gases are routed to point of use at medium pressure and fed into the process via local pressure-reducing stations.

27.7.10. CIP/SIP

At the pilot plant scale, much of the process equipment is fixed in place and cannot be brought to a glassware washer or autoclave for cleaning and sterilization. It must, therefore, be designed to be cleaned in place (CIP) and sterilized in place (SIP).

CIP is accomplished by using a series of liquid solutions sprayed or flushed through the equipment for a predetermined time and at established flow rates and temperatures. For CIP to be effective, the equipment must have crevice-free, well-polished interior surfaces, be free of dead-legs where solutions could accumulate, and be fully drainable. The CIP systems themselves may be centrally located, with distribution and return piping from a minimum of a single tank used to prepare a series of various wash solutions in sequence, or else a series of dedicated solution tanks. Alternatively, portable CIP units may be used; these are brought to each piece of equipment in turn. CIP systems are covered extensively by Brunkow et al. (5).

SIP is carried out by using clean steam on interior (product contact) surfaces. Plant steam is usually used on non-process-contacting surfaces, such as vessel jackets, because clean steam is expensive and corrosive. Fully saturated steam, at 15 to 18 psig (pounds per square inch gauge), is used to raise the temperature of the equipment and any contents to achieve sterilization. Provisions must be made for removal of air and steam condensate, which would otherwise prevent complete steam contact. SIP is discussed in detail by Oakley (18).

27.7.11. Transfer Pipes and Transfer Panels

Process fluids, buffers, cleaning solutions, and waste liquids must be transferred from one tank or process system to another throughout the plant. Most pilot plants use a system of pipes, valves, and routing junctions, or transfer panels, to move the fluids from point to point. The pipes must be sized and installed to convey the liquids efficiently and generally under sanitary (and sometimes sterile or aseptic) conditions. The latter requires that the piping system be self-draining, with a slope, one-eighth inch to one-fourth inch per foot, between drain points, and that all branch valves observe the "6D" rule with the sealing surface of the valve no farther than six times the smaller (branch) pipe's diameter from the center line of the main pipe. The purpose of these two design principles is to preclude the accumulation of liquids that could nurture the growth of microorganisms or adulterate subsequent transfers. For maximum process flexibility, it is desirable to be able to connect any tank or process unit with any other. The routing of fluids may be controlled by manually actuated or automated valves. Alternatively, transfer panels may be used at branching points to allow one pipe to be connected to any of several others. In one commonly used format, a rigid "U"-shaped tube is used to connect a central pipe to any one of a group of pipes arrayed in a circle around the central one. In another format, a flexible hose is used to connect any two pipes that may be arranged in a square or rectangular pattern. To facilitate CIP and sanitization, the interior surface of all pipes, valves, and fittings should be suitably polished. All fixed pipe-to-pipe and pipe-to-fitting joints should be welded, preferably with automated equipment. Where pipes or fittings may need to be dismantled, the most widely used connections employ a clamped, sanitary flange design using a unitized gasket and "O" ring to seal the joint. A discussion of multipurpose plant connectivity is found in reference 4.

27.7.12. Finishes

The interior surfaces of process equipment, tanks, and process piping must be polished to enhance cleanability. Previously, finishes were specified in "grit" sizes such as 180 or 240; the current practice is to refer to the surface roughness in terms of a profilometer reading. The appropriate interior finishes for fermentors, mix tanks, filters, etc., might be specified as "25 to 30 Ra," where Ra is a peak-to-valley measurement in microinches. Buffer tanks, product holding tanks, final sterilizing filter housings, etc., that come in contact with the highly purified process stream might call for a 12 to 20 Ra. To achieve these surface finishes, many manufacturers rely on electropolishing after the mechanical polishing steps. Electropolishing is a form of reverse electroplating, in which the peaks of the surface roughness are dissolved by direct electric current.

Room interior finishes are a critical element in maintaining airborne particulate counts and in controlling surface microbial growth. Wall, ceiling, and floor finishes should be hard-surfaced, nonporous, nonshedding, nonreactive, and easily cleanable. They should also be free of crevices and dust-collecting horizontal surfaces or ledges. Walls may be epoxy painted, covered with a smooth laminate, or metal skinned. Ceilings may be constructed similar to the walls, avoiding if possible, T-grid and drop-in panels. (Although these can be sealed in place, they offer little advantage over other forms of construction.) Floors may be sealed concrete, heat-welded (seamless) vinyl, preferably self-coved at the floor to wall joint, troweled epoxy, or other crevice-free material suited to the chemical resistance and traffic requirements.

27.7.13. Waste Disposal

The layout of the facility should provide for orderly removal of solid waste from the process areas, using the "dirty" corridor and avoiding crossing over or into clean areas. If the waste is biologically contaminated, it should be autoclaved prior to being removed from the controlled area. Separate autoclaves, designated for decontamination, are frequently used as pass-through points for removal of solid waste. Liquid waste streams should be piped to a "kill system" designed to inactivate any cultures that may be present. Either heat or chemical decontamination may be employed. The system should be thoroughly tested with an innocuous surrogate prior to actual use with biohazardous materials. For more information on biowaste decontamination systems, see Nelson (16).

27.8. CONSTRUCTION MANAGEMENT AND EXECUTION

27.8.1. Construction Control

During the construction phase of the project, weekly progress meetings should be held between the owner's representative and representatives of the construction team. The construction team will be made up of workers from numerous trades, e.g., plumbing and mechanical, electrical, framing and finishing, etc. The owner most likely will employ a general contractor that directly schedules and supervises the work done by the trades. At the weekly meeting, the general contractor should report progress to the owner's representative, identify any problems that may have become apparent during the past week (e.g., late deliveries, field installation problems), and propose approaches for handling these problems. The owner's representative can then assess the impacts of the problems and the proposed solutions on the project and gauge progress against the project construction schedule. In addition, the owner's representative should frequently inspect the actual construction to verify progress and attempt to anticipate field problems that may be developing. Throughout all these activities, the project construction schedule is a key tool to use to measure progress, and so the quality of the schedule is important. Project construction scheduling for biotechnology and pharmaceutical facilities is discussed in detail by Odum (19).

27.8.2. Early Procurement and Construction Activities

It is important to realize that certain items have a long lead time and must be handled up front if the project is not to be greatly delayed. Long-lead-time items usually include process and utility equipment such as fermentors, boilers, clean steam generators, WFI systems, air compressors, air handlers, chillers, cleanrooms, etc., unless these items can be found on the used-equipment market. Certain permits may also be long-lead-time items.

For both utility and process equipment, make sure that all long-lead-time items get sized and specified rapidly, that potential vendors are identified quickly, that quotations are obtained and evaluated without delay, and that the items are then placed on order. Following order placement, diligently track and expedite each order until the equipment is delivered into the hands of the owner. Do not assume that delivery dates promised up front will be met without continual checking along the way. Even then, be prepared to coax, cajole, demand, and, if necessary, threaten in order to get critical path equipment delivered as required.

27.8.3. Procurement

For major pieces of equipment, written specifications should be developed, reviewed by the project team, and then approved for requesting quotations. Next, requests for quotations (RFQs) should be provided to qualified vendors. Typically, for each major piece of equipment, RFQs are provided to at least three vendors for purposes of competitive bidding. To encourage quick responses from the vendors, a closing date should be specified in the RFQ. After responses to the RFQ are obtained, they should be evaluated comparatively against decision factors (e.g., compliance with RFQ specifications [carefully note any exceptions taken], price, delivery time), a vendor choice should be made, and the order should be placed. Once again, after the order is placed, expediting as already described is prudent.

27.8.4. Demolition for Renovation

If an area previously used for other purposes is to undergo demolition for renovation in order to install a new pilot plant, a number of special challenges may arise. For example, if the area was previously classified for biological containment (e.g., BL-3), it will obviously be important to verify and certify that appropriate decontamination procedures have been executed, validated, and documented before demolition begins. If demolition is to occur in an area that is next to an ongoing work area, then precautions will have to be taken to protect that work area from demolition dust. Furthermore, both demolition and subsequent construction may require the temporary shutdown of certain critical existing services (e.g., electricity, clean steam, WFI) to remove existing lines or install new tie-ins. This can be very challenging if ongoing operations should not be interrupted. In any case, careful and extensive communication and scheduling involving all parties affected is crucial. If service shutdowns cannot be avoided, they should be planned in advance and scheduled together to keep lost time to a minimum.

27.8.5. Change Orders

Construction proceeds based upon detailed design drawings. Typically, even after all the detailed drawings have been completed, and perhaps as construction is under way, change requests come forward from various interested parties. These change requests should be systematically reviewed, evaluated for importance and impact on budget, schedule, and overall engineering design intent, and then, if appropriate, approved. An approved change should be documented in a change order and incorporated into the drawings used for construction, either by issuing revisions of those drawings that are involved or through a formal sketch that becomes part of the construction documentation. Requested changes may, however, be too expensive or noncritical to implement.

27.8.6. "As-Built" Drawings

Upon completion of construction, a set of drawings called as-builts should be issued that document what was actually built in the field, which may differ, at least in fine detail, from what was originally planned. As-builts are important for keeping track of the physical facility and serve as a key tool for validation. Additionally, every effort should be made to keep the as-builts up to date as changes are made to the field over the months and years, because accurate field drawings are an invaluable diagnostic aid when breakdowns occur or when new equipment is to be installed.

27.8.7. Reporting to Management

New pilot plant design, construction, staffing, start-up, validation, operation, and maintenance are costly and also often on the critical path for introduction of important new products into clinical use and the marketplace. Therefore, it is appropriate to keep management up to date on status, progress, plans, problems, and problem solutions relating to the pilot plant. Communication with management may occur through meetings and presentations, written reports, telephone, electronic mail, or any combination of these, depending upon circumstances and the preferences of those involved. In any case, be well prepared to provide accurate key information, preferably in a proactive mode, and to answer questions that may come up at any moment and demand immediate response. It may well be possible that the future of the entire organization heavily depends on results involving the pilot plant.

27.9 STAFFING AND TRAINING

27.9.1. Staffing

Staffing in the pilot plant must address both time and skill coverage of the plant's projects. By their nature, biological processes do not readily fit into standard 8-h days and 5-day work weeks. Depending on the specifics of the process, the actual operations may require one, two, or three shifts (of 8 h) per operation day to prepare media, seed culture, buffers, etc.; to set up and operate the process equipment; to monitor and sample the process; and to harvest and purify the product. There are certain natural break points in most processes, but these may not correspond to the standard 8-h work periods. Likewise, a process may need to be started, run through, or ended on a weekend. Therefore, most pilot plants will provide 24-h-a-day, 7-days-a-week coverage. Many plant managers will find it expedient to schedule the most critical or labor-intensive work for the day shift when additional resources in the form of laboratory, scientific, or engineering staff and service groups are present. When the process must continue beyond the 8-h day shift, the evening shift will normally be able to complete it. Then, the third, or "graveyard," shift can be assigned the tasks of decontamination, cleaning, disassembly, and storage of the equipment just used, and set-up and preparation for the process steps to follow. Alternatively, the third shift may be assigned to monitor the process and prepare solutions and equipment for the next day's operations. Depending on company policy and/or plant "culture," a separate cleaning crew may be employed for general room housekeeping and sanitation. However, since much of the pilot plant equipment is CIP and SIP, the regular production staff will normally carry out these operations.

The personnel and disciplines needed to support the plant will include electrical, mechanical, and instrument repair, preventive maintenance, calibration, QA/QC, purchasing, inventory control, clerical (document control), process equipment operation, critical cleaning, process development (chemical engineering), microbiology, biochemistry, and two or three levels of management. The division of management responsibility may take several forms. In one style, there may be a maintenance manager, a production (fermentation) manager, and a purification manager, all reporting to a plant manager. Operators may be permanently assigned to each of the managers, or they may be assigned to the project and given direction by different managers as the process proceeds. In another style, each project in the plant may be assigned a project manager, who can direct the operators throughout the course of the process. Here, the several project managers would report to the plant manager. In either scenario, the plant manager is responsible for setting priorities and allocating resources to avoid conflicts.

27.9.2. Training

All pilot plant personnel should be given training before they are assigned duties and periodically thereafter. Ideally, all training should be conducted with the aid of approved SOPs, and in any case must be documented. General training should cover biological, chemical, electrical, and fire safety, plant personnel policy, and, if appropriate, GMPs. If the plant is operating under GMP, a formally established and fully documented training program is required.

Training specific to the person's duties should include both theory and practice. It is especially important that no one be assigned to operate process equipment or handle materials until they have been trained in that operation and have demonstrated their understanding. After initial training sessions, new personnel may be assigned to work with experienced staff who have demonstrated their ability to communicate the proper (officially approved) operations. This on-the-job training can be very effective but must be monitored by supervisors to ensure that the new personnel are gaining the correct knowledge. Not every experienced plant operator has the temperament or ability to be a trainer.

27.10. START-UP

27.10.1. Contractor Turnover

Acceptance of the physical plant from the contractor should never be relegated to a single group of owner representatives walking through the new plant with a "punch list" (a list of items to be completed) in hand. In fact, preparation for contractor turnover should be performed on a continuous basis in concert with the construction validation process. The start-up team, a subset of the project team, should have most of the punch-list issues at least conceptually resolved prior to the plant inspection and turnover. Turnover must also include a complete and thorough review of all deliverables, with special attention to review and verification of as-built drawings, warranty certificates, calibration stickers and certificates, weld coupons and certificates, materials certificates, American Society of

Mechanical Engineers code documents, security systems (keys, computer passwords), software codes, and labeling of piping, utilities, and equipment, to mention a few. The turnover process, inclusive of the above delineated tasks, may involve several weeks of review by the appropriate shops involved with the team. If turnover planning and preparations are left until the last day before the construction trailers pull out, both the owner and the contractor will end up disgruntled. The turnover documents will become an integral part of the facility master files, and each critical section should be reviewed and signed by the appropriate groups. To avoid the arduous tasks of retrieving information or rectifying design problems after close-out, the construction time-line must include a sequenced turnover plan. Every effort must be made to maintain this turnover schedule without sacrificing quality of review, or else the inevitable result will be delay in plant start-up and eventually loss of product revenue.

27.10.2. Plant Acceptance, Warranties, and Service Contracts

Avoid having the owner and the contractor end up trying to resolve discrepancies involving acceptance criteria, warranties, and service contracts at the last minute. This pitfall is invariably the result of poorly or nebulously worded construction contracts. If the A&E firm and the construction contractor are different organizations, the potential for problems is compounded. It is therefore imperative that the project team, including the contractor's authorized representatives, spend time reviewing and scrutinizing the contract documents to ensure that each party agrees upon what the deliverables are, the effective warranty (or service contract) dates, what is warranted, and who will be the after-turnover contact(s).

27.10.3. Installation Qualifications

Taken together, the IQs document that the facility and major equipment systems have been installed according to design specifications. The IQ phase is usually the last phase of the validation life cycle that involves the construction contractor (unless a design/build/validate firm has been retained). If proper planning has been applied, certificates and documents obtained, and proper turnover procedures implemented, each IQ should be the relatively simplistic exercise of formalizing the information that the owner will already have in hand. Sawyer and Stotz (21) suggest that a modular approach to individual IQ protocols and checklists be developed for general types of equipment that share common attributes, such as pumps, heat exchangers, centrifuges, and reactor vessels. In this way, generic protocols can be generated and executed throughout the construction process, with specific information being filled in as required. In addition to the physical check of the equipment (i.e., proper power supplied to a pump, with correct direction of pump rotation), the IQ documents should generally include the following information:

- Definition of qualification
- Pre-approval signatures
- Scope of qualification
- Equipment or system description
- Post-installation review
- Change control procedures
- List of change orders and review
- Review/verification of final as-built drawings
- List of expendables and consumables
- List of spare parts and qualified replacements
- Review of materials of construction
- Instrumentation/calibration
- Preventive maintenance schedule/procedures
- Certificates, stamps, and materials/weld coupons
- Cleaning and passivation procedures and verification
- Shop/operations manuals
- Establishment of master equipment files
- Standard operating procedures index
- Equipment training
- References

27.10.4. Operational Qualifications

The OQs constitute documented confirmation that each system or subsystem performs as intended throughout all anticipated operating ranges in a consistent and reliable manner. Obviously, the operating ranges and system parameters must be predefined by the appropriate subset of the project team. The intended operation of a system or subsystem incorporates the design specifications, the manufacturer's recommendations, minimum and maximum operating ranges, and the appropriate SOPs (usually formalized draft procedures at this stage for new equipment).

The persons responsible for providing an OQ protocol for each equipment system must be designated. The OQ protocol should specify who will review the protocol prior to the commencement of actual qualification operations and who will review and approve the results of the study after execution. The test protocol may rely heavily upon the equipment or system manufacturer's original equipment manuals and minimum/maximum operating range specifications. Where applicable, key utility services and requirements (e.g., water, steam, air) and operating parameters must be identified and recorded during testing.

OQs may be performed using the owner's (expendable) product or a suitable placebo having similar characteristics to the product or products to be produced. Although water is frequently used for this purpose, the rationale for selection of process simulants must be provided. The data generated during the qualification runs must be recorded and reviewed. Excursions, deviations, failures, and remedial actions taken must also be documented. Appropriate change control procedures must be followed.

The following are generally included in an OQ protocol:

- Definition of qualification
- Pre-approval signatures
- Scope of qualification
- Equipment or system description
- Description of the qualification to be performed
- Rationale for the qualification, including objectives
- Rationale for selection of product/simulant used for testing
- Verification of successfully completed IQs for all systems involved
- List of affected equipment
- List of critical equipment and support systems
- Verification of calibration status
- Identification of all test instruments or equipment used

- Target, minimum, and maximum operating ranges and parameters for equipment
- List of all SOPs, including support equipment used for OQ
- List and identification of all qualified personnel involved
- Test conclusion report and comparison of data with acceptance criterion
- Reviews and approvals
- References

27.10.5. Performance Qualifications

A PQ documents that a piece of equipment or an equipment system is able to consistently meet the design specifications and function as intended throughout its entire anticipated operating ranges, including the minimum and maximum ranges for a particular process. A PQ study generally tests that the equipment actually performs the unit operation required to provide an acceptable product. The minimum and maximum operating ranges are those at the outer boundaries of the intended production process or the system's design criteria. For example, an autoclave should be qualified fully loaded with product or product simulants designed to mimic actual operating configurations of minimum and maximum loads. A filtration system should be tested with the highest percentage of solids that the filtration system is designed to clarify. HVAC systems and room pressurizations (balances) must be tested with all heat loads and room activities present (dynamic testing). In the above examples, the constituents of the autoclave loads (test product and load configuration) must be documented, the test media used for the filtration system must be identified and quantified, and the activities within a room being tested for temperatures or air flows must be described, including the number of personnel, a description of their activities, equipment in operation, and similar reference information.

Since the PQs are executed in the final step in qualification of the physical plant systems prior to process validation, all IQs and OQs should have been completed and approved previously, and all repairs or corrective actions should have been taken. Control setpoints and parameters within microprocessor-based controllers must have been set and documented (e.g., temperature setpoints; alarm limits; proportional, integral, and derivative values in programmable logic controllers). In addition, operations personnel must be appropriate for the task and adequately trained. Personnel training should also be documented. Effective execution of PQs requires that all process support utilities, such as purified water systems, CIP/SIP systems, and decontamination equipment, are utilized as intended. These systems, therefore, should also have been tested and qualified with formal documentation. Although not formally mandated by federal regulations, the completion of three consecutive successful tests is generally accepted as the industry standard for satisfactory qualification (8, 15).

The product produced during these qualifications is generally tested extensively after each key process unit operation, and, of course, the final product is subjected to full QC analysis. Draft instructional worksheets (batch records) are maintained and reviewed as in a routine production operation. Although the product produced is usually not manufactured for release and sale, some products or raw materials are so expensive that qualifications and production must proceed concurrently. Although this is a somewhat risky approach, careful planning in conjunction with the QA and QC departments will minimize the probability of lot rejection. Conversely, some products may be extremely hazardous, infectious, or toxic. In this situation, the use of a suitable simulant may be more appropriate for trial runs prior to the actual introduction of product into the facility. Media and a simulant organism are sometimes employed. This is an acceptable practice provided that the rationale for simulant selection is valid. Placebo trials must be handled with the same attention to detail, including all the documentation, batch records, testing, and reviews, that a product trial would receive, and possibly more. Of course, the first several lots of actual product produced should be subjected to extensive testing as well.

27.10.6. Process Validation

As stated earlier, the objective of process validation is the assurance that the specific production process being employed is under predefined control and that process variables and acceptable limits have been duly considered, identified, and tested. In the context of a pilot plant, the form and degree of process validation must be consistent with the intended use of the product. All products that will be introduced into clinical studies or marketed must be made under GMP conditions and therefore must be made with a validated process. However, it is recognized by the FDA that much of the detailed information that will be included in the validation package for a product that is ready for market will not be available at the time that same product is submitted for phase I clinical trials. It is also recognized that, in some instances, perhaps between phase II and phase III trials, production operations may be moved from a pilot plant, or a pilot area of a larger plant, to a full-scale facility to achieve the needed capacity. Also, many products fail to perform as intended during phase I and II trials. Thus, developing a complete process validation package before the product advances to phase III trials would not be warranted.

Those plant systems that have a direct influence on the product's safe use with humans must be validated prior to clinical usage. These include, but are not limited to, clean steam, purified water, autoclaves, bioreactor sterilization, and cleaning (CIP/SIP). Process steps critical to product safety include final product (terminal) sterilization, viral clearance, and removal of host DNA and extraneous protein. The focus should be on validation of steps that affect the safety, potency, and stability of the product. Special emphasis should be given to those properties that cannot be tested in the final product. In addition, those systems that are critical to the protection of plant personnel and the environment must be validated prior to their use in the decontamination of autoclaves, HEPA filters, vent systems, etc.

27.11. OPERATIONS

27.11.1. Standard Operating Procedures

SOPs are the basic protocols by which all personnel perform a given procedure in the same manner. Without SOPs, there is no assurance that the equipment is operated, or a procedure is performed, consistently. SOPs also provide a means for updating operating methods and are an invaluable training tool. The FDA is very cognizant of these facts, and during an inspection, a review of current procedures is almost guaranteed.

Each major piece of equipment and each critical unit operation should have an SOP. Format should be standardized but is not as important as SOP content, adequate tracking, review, and change control programs. Draft form procedures should ideally be developed during the IQ phase of construction if possible, and certainly before or during operational testing. PQs are best conducted with approved and completed procedures in hand.

27.11.2. Safety Procedures

Personnel and facility safety would be placed at the top of all but the most myopic of individuals' list of important items for consideration. Unfortunately, while usually not completely overlooked, safety is frequently, if unintentionally, shortchanged; production targets, schedules, and physical plant engineering are the usual and predominant driving forces.

Certainly, procedures driven by regulatory and other governmental agencies (e.g., Occupational Safety and Health Administration, Environmental Protection Agency, National Institutes of Health, National Fire Protection Association) help to ensure that the basics of safety issues are addressed. Many larger companies are fortunate enough to have departments that review and address safety concerns. This is a luxury that many smaller start-up companies cannot afford, however, and safety often ends up combined with the responsibility of QA.

One method that helps to avoid overlooking safety issues is to form an internal safety committee. The safety committee should ideally be formed no later than at the conceptual design stage of the new facility. Regular safety committee meetings should be held; a monthly or bimonthly frequency is usually adequate. A chairperson should be selected, and representatives from the process operations group, engineering, maintenance, QA/QC, management, and the A&E/contractor should constitute the quorum. Both contractors and owners have their own sets of safety rules. These rules do not always agree, and a safety committee is the perfect forum for resolving differences in procedure and developing a common set of guidelines during construction. The impromptu safety concerns that arise during engineering, start-up, or production can be addressed and resolved by a group of persons who represent all interests and who are focused on the single topic of safety. Although "one more meeting" is apt to be scowled at, the infrequency of the safety meetings and the focused nature of the group should foster the diligence and attention that safety matters require.

27.11.3. Spill Emergency Procedures

Another safety-related item that may receive scant attention is spill control and procedures to adequately address appropriate personnel actions and clean-up protocols in the event of a spill. The detailed nature of spill control procedures is obviously a plant- and project-specific item. There are, however, some spill control-related procedures, including both primary and secondary engineering controls, that are common to all plants:

- Double mechanical seals with pressurized interstitial spaces on rotating shafts that require containment, including regular leak-check and preventive maintenance procedures
- Normally closed and sealed floor drains, opened only during use and closed immediately afterwards
- Diked door thresholds (adequately marked to avoid tripping accidents and sloped to avoid equipment movement hazards) of sufficient height to contain a catastrophic failure of the largest vessel installed within the area. Calculations supporting the diked containment volume are required.
- Appropriate HVAC filtration, pressurization, and fail-safe (fail-static interlock) mechanisms
- Alarms for advanced warning of carbon monoxide, oxygen deficiency, solvent vapors, or HVAC failure
- Fail-safe valving, system over-pressure sensors, and rupture disk/relief valve discharge containment vessels
- Validated decontamination equipment and approved SOPs
- Spill control procedures, evacuation plan, and regularly scheduled spill-control training for all involved personnel
- Approved personal gowning and respiratory protection program with regular training and certification
- Environmental monitoring program and employee health monitoring program (if applicable)
- Approved waste handling and disposal methods and procedures
- Regularly checked and restocked spill control cabinets

27.11.4. Process Waste Handling

Planning for and handling process waste must be given the same priority and consideration as any other process system. Strict federal, state, and local codes mandate that written policy and procedure be established. Waste-handling equipment should be qualified in terms of installation, operation, and performance. Change control and preventive maintenance programs are also an important aspect of waste holding, processing, and disposal systems, as is maintenance of processing and disposal records.

The choice of waste holding, processing, neutralization, and disposal systems is a matter contingent upon company philosophy and is frequently process driven. For treatment of purely biological waste, steam sterilization is the usual method of choice, as this process is well established, readily available, easy to validate, and usually appropriate for killing all microorganisms as well as inactivating bioactive molecules. Although chemical sterilants may be appropriate for specific compounds and waste formulations, the users risk dumping inadequately treated waste if even moderate changes to waste formulations (e.g., concentration, protein load, resistant organisms) are made. Chemical treatment also requires an increased level of initial validation of the inactivation processes, it is generally more difficult, and subtle changes in processes may require revalidation. An additional drawback to the use of chemical agents is the requirement for poststerilization chemical neutralization. The sole application of chemical inactivation agents in multiuse facilities should be discouraged unless the streams to be decontaminated are very similar in composition. Of course, the process/methods development groups should always be aware of the need to minimize the generation of waste, particularly difficult waste.

27.11.5. Documentation

A saying goes, "If it is not documented, it never happened." In addition to the obvious pharmaceutical documentation requirements, some additional consideration may be in order. The FDA considers all documents related to the production of a regulated product within the scope of an audit. This includes all documents referenced in master files and even operations staff shift logs. Multiuse pilot plants should be exceptionally strong in the documentation of change control, cleaning procedures, containment/barrier systems, personnel training, waste treatment, and flow of personnel, raw materials, and product.

It is advisable to maintain documents that are strong on factual content pertinent to the process and facility. Consequently, instruction on documentation should be incorporated into the regularly scheduled GMP training program for the operations staff.

27.11.6. Maintenance and Calibration

Once a pilot plant is established, the owner must ensure continued service of the validated utilities and process equipment systems. A written procedure that defines the maintenance and calibration strategy should be instituted, and specific SOPs for these activities must be written. Levels of equipment criticality should be indicated, because instruments that control or otherwise directly affect the safety, potency, or efficacy of the product must receive the highest level of control and attention. If maintenance and calibration are outsourced, the owner is still held responsible for the contractors' actions, and therefore maintenance and calibration contractors must be audited and qualified by the owner. Certification of reference instruments, weights, and measures used for calibration must be documented, and recertification to controlled standards (e.g., National Institute of Standards and Technology [NIST]) must follow an established frequency. All reference standards used for calibration, as well as critical process instruments, should have a sticker physically affixed to the unit that includes a unique identification number, the date when the device was calibrated, the date when calibration is next due, and who performed the calibration. A database associated with these instruments should include historical "as-found/as-left" calibration data as well.

It is the owner's responsibility to establish appropriate calibration frequency and ranges for the equipment. Although little guidance is provided by the Code of Federal Regulations on this issue, manufacturers' suggestions, historical information, and common sense should dictate the reasonable (and justifiable) selection of criteria. Setting unreasonably frequent calibration and programmed maintenance schedules or parameters will result in constant interruption of processing, process deviation reports, and excessive expenditure of resources. Conversely, the established of extended servicing schedules places product and personnel at risk. It is therefore important to maintain the history of the equipment, review this data carefully, and adjust maintenance schedules according to the requirements dictated by statistical findings.

Although preventive maintenance and calibration draw substantial resources, careful planning during the construction of a pilot plant will minimize capital and labor expenditures over the life of the facility.

A very significant savings in all categories involving calibration and maintenance can be realized by judicious selection of equipment during the conceptual design and engineering phase of the construction project. Selection of standardized process controllers, valves, vessels, pumps, etc., results in less training (in both operations and service), a reduced number of procedures, reduced spare-parts inventory, decreased validation costs, and the opportunity for volume-based procurement opportunities.

27.11.7. Inventory and Warehousing

During the conceptual planning stages of pilot design, due consideration must be given to adequate handling of raw materials flow, intermediate and bulk product storage, and final product storage.

The design of materials storage areas must consider the requirements for raw materials receipt, quarantine, release, and weigh/dispersement in a sequential, logical, and methodical manner. Controlled environments and restricted access appropriate to the operation must be observed. Intermediate and final product storage areas frequently require tempered storage (i.e., refrigeration or freezing). If tempered storage is required, the area must be equipped with instrumentation enabling the continuous trending of temperature and the capability of alarm notification.

Estimates of storage space (with room for expansion) in a controlled area for spare parts and general materials and supplies must be developed at the outset of the project, and, once established, the space allocated must be maintained.

As with other critical unit operations, a written plan for raw materials flow, inventory control, and materials balance must be generated and approved.

27.11.8. Housekeeping

It is an unfortunate fact that housekeeping considerations are given less attention during construction planning than the subject is due. Although this neglect is somewhat understandable, the omission usually becomes obvious during commissioning of the facility. Biotechnology plant real estate is expensive, and there is a reluctance to designate highly coveted space to the storage of mops, cleaning supplies, HEPA vacuum cleaners, and the like. Although housekeeping and support pact is the least expensive space per square foot relative to classified biotechnology processing spaces (by a factor of 10 to 20), it is usually the first area to be cut when reduction of cost overrun is necessary. Nowhere, however, is adequate housekeeping more important than in a multiuse pharmaceutical pilot plant facility. Sufficient and appropriate space must be provided to accommodate housekeeping requirements.

Careful consideration must be given to location of the housekeeping storage areas, access to the controlled areas by custodial personnel, and avoidance of cross-contamination by common use equipment. Regarding cross-contamination, the custodial staff should be included in basic GMP training programs and specifically trained in gowning and other SOPs germane to facility policy and procedure.

27.12. MAINTAINING A FLOW OF NEW ASSIGNMENTS

Designing, constructing, starting up, and validating a pilot plant takes (to put it mildly) a lot of time, money, and effort. Once the pilot plant is ready to run, the ideal situation is that it has plenty of work to do for a long period of time. Circumstances, however, can arise that result in a less than optimal stream of new projects coming into the pilot plant: perhaps a clinical trial or two do not go well, perhaps discovery hits a streak of dry holes, or whatever. The expected sources of workload just may not be sufficient. In this case, a focused and high-priority effort toward getting new business becomes mandatory. New internal sources of projects could be sought. Projects might be brought in from the outside. At any rate, do not assume that a flow of new projects will continue without attention being paid to prospecting. Be on the lookout for attractive new projects to keep that pilot plant humming.

Portions of this chapter draw from the corresponding chapter published in the first edition of the Manual of Industrial Microbiology and Biotechnology *(12).*

REFERENCES

1. **Adams, D. G.** 1994. Utilities for biotechnology production plants, p. 575–610. *In* B. K. Lydersen, N. A. D'Elia, and K. L. Nelson (ed.), *Bioprocess Engineering: Systems, Equipment and Facilities.* John Wiley & Sons, Inc., New York.
2. **Bader, F. G., A. Blum, B. D. Garfinkle, D. MacFarlane, T. Massa, and T. L. Copmann.** 1992. Multiuse manufacturing facilities for biologicals. *BioPharm.* **(Sept.):**32–40.
3. **Baird, B., and P. De Santis.** 1994. Validation of biopharmaceutical facilities, p. 745–781. *In* B. K. Lydersen, N. A. D'Elia, and K. L. Nelson (ed.), *Bioprocess Engineering: Systems, Equipment and Facilities.* John Wiley & Sons, Inc., New York.
4. **Brocklebank, M. P.** 1992. Multi-purpose pharmaceutical plants—solutions to key design problems. *Pharm. Eng.* **(Nov./Dec.):**17–31.
5. **Brunkow, R., D. Delucia, G. Green, S. Haft, J. Hyde, J. Lindsay, J. Meyers, R. Murphy, J. McEntire, K. Nichols, R. Prasad, B. Terranova, J. Voss, C. Weil, and E. White.** 1996. *Cleaning and Cleaning Validation: A Biotechnology Perspective.* Parenteral Drug Association, Bethesda, Md.
6. **Chin, M.** 1988. TOP: a rational approach for ensuring proper biopharmaceutical plant construction, p. 73–85. *In* S. Schuber (ed.), *PharmTech Conference '88 Proceedings.* Aster Publishing Corp., Eugene, Oreg.
7. **Demain, A. L., and N. A. Solomon (ed.).** 1986. *Manual of Industrial Microbiology and Biotechnology.* American Society for Microbiology, Washington, D.C.
8. **DeSain, C.** 1993. *Documentation Basics That Support Good Manufacturing Practices.* Aster Publishing Corp., Eugene, Oreg.
9. **Dobie, D.** 1994. Heating, ventilation, and air conditioning (HVAC), p. 641–668. *In* B. K. Lydersen, N. A. D'Elia, and K. L. Nelson (ed.), *Bioprocessing Engineering: Systems, Equipment and Facilities.* John Wiley & Sons, Inc., New York.
10. **Food and Drug Administration, Code of Federal Regulations.** 1995. FDA guidance document concerning use of pilot manufacturing facilities for the development and manufacture of biological products. *Fed. Regist.* **60:**35750.

10a. **Food and Drug Administration and U.S. Department of Health and Human Services.** 1996. NIH guidelines for research involving recombinant DNA molecules. *Fed. Regist.* **61:**10004.

11. **Hamers, M. N.** 1993. Multiuse biopharmaceutical manufacturing. *Bio/Technology* **11:**561–570.
12. **Hamilton, B. K., J. J. Schruben, and J. P. Montgomery.** 1986. Establishment of pilot plant, p. 321–344. *In* A. L. Demain and N. A. Solomon (ed.), *Manual of Industrial Microbiology and Biotechnology,* American Society for Microbiology, Washington, D.C.
13. **Kalk, J. P., and A. F. Langlykke.** 1986. Cost estimation for biotechnology projects, p. 363–385. *In* A. L. Demain and N. A. Solomon (ed.), *Manual of Industrial Microbiology and Biotechnology,* American Society for Microbiology, Washington, D.C.
14. **Kladko, M.** 1992. Purchasing engineering contract services in the pharmaceutical industry. *Pharm. Eng.* **(May/June):**26–30.
15. **Lovejoy, C. K.** 1986. Validation and facility design, p. 29–46. *In* F. J. Carleton and J. P. Agalloco (ed.), *Validation of Aseptic Pharmaceutical Processes.* Marcel Dekker, Inc., New York.
16. **Nelson, K. L.** 1994. Bowaste decontamination systems, p. 611–639. *In* B. K. Lydersen, N. A. D'Elia, and K. L. Nelson (ed.), *Bioprocess Engineering: Systems, Equipment and Facilities.* John Wiley & Sons, Inc., New York.
17. **Nelson, K. L., and C. A. Perkowski.** 1987. Important points in the design and start-up of a biopharmaceutical plant. *Gene. Eng. News* **7**(10)**:**58, 59, 68.
18. **Oakley, T.** 1994. Sterilization of process equipment, p. 499–521. *In* B. K. Lydersen, N. A. D'Elia, and K. L. Nelson (ed.), *Bioprocess Engineering: Systems, Equipment and Facilities.* John Wiley & Sons, Inc., New York.
19. **Odum, J. N.** 1992. Project schedule evaluation for biotech/pharmaceutical manufacturing facilities. *Pharm. Eng.* **(May/June):**16–24.
20. **Robins, D., and R. DelCiello.** 1988. To renovate or build new: guidelines for evaluating construction alternatives. *BioPharm* **(Jan.):**35–38.
21. **Sawyer, C. J., and Stotz, R. W.** 1992. Validation requirements for bulk pharmaceutical chemical factories. *Pharm. Eng.* **(Sept./Oct.):**44–52.
22. **Skibo, A. D.** 1989. Project management and contracting issues, part I: project team development. *BioPharm.* **(Apr.):**46–52.
23. **Skibo, A. D.** 1989. Project management jand contracting issues, part II: project scope development. *BioPharm.* **(May):**24–28.
24. **Skibo, A. D.** 1989. Project management and contracting issues, part III: contracting format and contractor selection. *BioPharm.* **(July/Aug.):**34–40.
25. **Slabicky, R.** 1994. Pharmaceutical water systems: design and validation, p. 523–573. *In* B. K. Lydersen, N. A. D'Elia, and K. L. Nelson (ed.), *Bioprocess Engineering: Systems, Equipment and Facilities.* John Wiley & Sons, Inc., New York.

Quality Assurance and Quality Control

RHYS BRYANT

28

28.1. DEFINITIONS

28.1.1. Quality

"Quality is meeting the requirements." This definition, attributed to Philip B. Crosby, is the essence of quality assurance and quality control. In the pharmaceutical industry, the requirements involve meeting specifications and developing a system to make certain that materials are sampled, tested, and evaluated against those specifications.

28.1.2. Quality Control

Quality control is defined as follows: "The operational techniques and activities that are used to fulfill requirements for quality." This is an ISO 9000 definition. (The International Organization for Standardization, Geneva, Switzerland, is an organization that sets international standards. It deals with all fields except electrical and electronics, which is governed by the International Electrotechnical Commission (IEC). The organization is commonly known as the International Standards Organization [ISO]. Its standards for quality have been issued as policy ISO 9000.) These techniques and activities are usually performed by the Quality Control Laboratory.

28.1.3. Quality Assurance

Quality assurance is defined as follows: "All those planned and systematic actions necessary to provide adequate confidence that a product or service will satisfy given requirements for quality." This is an ISO 9000 definition (see above). Quality assurance is performed by the Quality Assurance Department, which oversees the activities of research and development, manufacturing, and quality control. This function is defined in the current Good Manufacturing Practice Regulations (GMPs) as the Quality Control Unit. The basic principles of quality assurance have as their goal the production of articles that are fit for their intended use. These principles may be stated as follows.

1. Quality, safety, and effectiveness must be designed and built into the product.
2. Quality cannot be inspected or tested into the finished product.
3. Each step of the manufacturing process must be controlled to maximize the probability that the finished product meets all quality and design specifications.

28.2. CHEMICALS

Organic chemicals for commercial use are made by chemical synthesis, recombinant DNA technology, fermentation, enzymatic reactions, recovery from natural materials, or combinations of these processes.

When these chemicals are designated for use as components of drug products, they are typically known as bulk pharmaceutical chemicals (BPCs) or active pharmaceutical ingredients (APIs).

28.3. PHARMACEUTICALS

A "pharmaceutical" is a fancy word for what the Food Drug & Cosmetic (FD&C) Act calls a drug. In section 201(g)(1) of the Act, drugs are defined as

1. articles recognized in the official U.S. Pharmacopeia, official Homeopathic Pharmacopeia of the United States, or official National Formulary, or any supplement to any of them; and
2. articles intended for use in the diagnosis, cure, mitigation, treatment, or prevention of disease in man or other animals; and
3. articles (other than food) intended to affect the structure or any function of the body of man or other animals; and
4. articles intended for use as a component of any articles specified in clause (1), (2), or (3).

From this definition, under U.S. law, the component of a drug is also a drug. Thus, any chemical used as an ingredient of a drug product, whether pharmacologically active or not, is also a drug.

28.4. CHEMICAL AND PHARMACEUTICAL PRODUCTION

Regardless of whether the product is a chemical or a drug, production typically takes place by using a batch system.

In fact, producing a product is very similar to making cookies. To make one batch of bookies, all of the needed raw materials are weighed out individually in the correct proportions. They are then subjected to various operations, such as mixing, homogenizing, reacting, granulating, milling, compressing into the correct shape, etc. The specified number of cookies are counted and placed into a container, the cap is screwed on, and the label is affixed. The container may then be placed in an outer carton, and the flap is closed or sealed.

If a second batch is desired, one would weigh out a second set of raw materials, making more in-process items, filling another container, and affixing a new label. The whole process would give us a new batch of finished product.

As with cookies, so it is with drug tablets, capsules, vials, etc.

A chemical, although produced by a series of chemical or biological transformations, is more often than not produced by a similar batch process. For example, in the preparation of a typical antibiotic by fermentation, an inoculum preparation is made by transferring a carefully preserved bacterium into a vessel containing a mixture of nutrients (chemicals) in solution or suspension (the growth medium) and subjecting it to initial growth. The inoculum preparation is then transferred to fermentation tanks containing greater quantities of the growth medium. The by-product of the bacterial growth is the desired chemical, which is then extracted and purified, and the final product is packaged and labeled.

Because of the limited size of vessels, the whole process is done as a batch operation.

28.5. THE FIVE VARIABLES

The batch process is built on five types of materials, which I call the five variables:

- labels
- packaging materials
- raw materials
- in-process items
- finished products

They are termed variables because minor variations in any one of these areas affects the other areas and in turn can affect the quality control function dramatically.

Quality assurance and quality control are the procedures used to document and control these five variables.

(Labels, packaging materials, and raw materials are usually purchased by a manufacturer and are then used to manufacture and package in-process items and finished products. However, because labels and packaging materials are less likely to be tested for chemical or microbiological properties, the five variables are treated in a different order below from that in which they are listed above.

28.6. RAW MATERIALS

Raw materials are the solvents and chemicals, whether active ingredients or inactive excipients, used in the manufacture of the dosage form of a drug product; or the solvents, chemicals, biological materials, and nutrients used to produce chemicals.

28.6.1. Types of Raw Materials

Raw materials may be categorized in several ways: they may be active or inactive, depending upon their physiological action; they may be organic or inorganic, depending upon their chemical composition; they may be solid, liquid, or gas; or they may be classified according to their origin as synthetic, animal, vegetable, or mineral. These classifications are important when assessing microbiological properties.

28.6.2. Method of Manufacture

Raw materials, whether solids, liquids, or gases, all must be purified for use in drug products. Although the bench chemist can ensure purity at each step because of the small scale of such production, the full-scale version requires a great deal of quality control.

For the most part, raw materials are purchased from outside the company. Because of differences in the method of manufacture, handling, and packaging, quality will vary from company to company or even from lot to lot. It is essential to recognize that water used in the process is a raw material and that because it is a hospitable environment for many bacteria, microbiological contamination is a serious concern. For most organic solvents, microbiological contamination is less of a problem. For parenterally used products, liquid purity is crucial; for other products, such a high level of purity might not be necessary. Finally, gases are sometimes used in drug production. Some products are processed under nitrogen to avoid oxidation.

These quality concerns involve testing and inspection of incoming solids, of holding and distribution systems for liquids, and of distillation systems for gases. The goal in each case is to make full-scale batches as similar as possible to the bench chemist's reference standard but at a reasonable cost.

28.6.3. Controls

Raw materials of any form are sampled, tested, and released to the production department by the quality control department. Many suppliers routinely provide raw materials of consistently high quality, in which case the amount of sampling and testing may be reduced or even replaced by a certificate of analysis supplied by the vendor. Just the same, the quality of the raw material is always defined by certain specifications, with tests required to determine its identity, assay, quality, and purity by using physical, chemical, and microbiological procedures.

Many reference compendia supply information on raw materials. The compendia are generally written in "monograph" format; that is, for each item there is a collection of data under one heading, giving the name of the material, a chemical formula, a description, and specifications and test methods for such parameters as identity, strength, quality, and purity. These items usually include chemical, physical, and microbiological limits for the raw materials. Official compendia include American Chemical Society (ACS) Reagent Chemicals, British Pharmacopeia (BP), British Pharmaceutical Codex (BC), Deutsches Arzneibuch (DAB), European Pharmacopeia (EP), Food Chemicals Codex (FCC), Japanese Pharmacopeia (JP), Official Methods of Analysis (AOAC), and U.S. Pharmacopeia (USP) including the National Formulary (NF).

28.7. IN-PROCESS ITEMS

An in-process item is any item that is produced and held by the manufacturing department for further processing or packaging.

28.7.1. Types of In-Process Items

In-process items include intermediate chemicals, premixes, granulation, bulk tablets or capsules, bulk liquids, ointments, and creams. Also included in this classification are packaged products that require additional processing. For example, sealed vials must undergo terminal sterilization. Excluding medicinal gases, there are only four types of in-process items: liquids (and solutions), liquid-liquid emulsions, solid-liquid suspensions, and pure solids.

28.7.2. Method of Manufacture

For pharmaceutical liquids, the raw materials are weighed out. The solvent is accurately measured, and the raw materials are added in the order prescribed by the manufacturing instructions. A solution is produced by stirring; the temperature, pH, and other variables are controlled to maintain solubility and to avoid decomposition. Mixing and transfer of liquids are relatively straightforward steps.

Liquid-liquid emulsions, however, require special instructions and care to avoid separation of the phases. The same holds true for solid-liquid suspensions, such as ointments and creams.

Powders and tablets are made by a more complicated procedure than liquids. They are generally mixed, wetted, granulated, milled, dried, and compressed, although advances in the field of direct compression have obviated some of these steps.

Sterile products of all types require special handling techniques: they must be produced in microbiologically clean environments by validated sterilization procedures under tight microbiological controls.

Before packaging, in-process items are stored in bulk. Solutions and liquids are often held in storage tanks, while bulk powders and tablets are generally held in sealed drums.

28.7.3. Controls

In-process items are subjected to in-process controls and are sampled, tested, and released by the Quality Control Laboratory before being used in the next step. A time limit is set so that in-process items do not remain unused for long periods. In-process items are tested for identity, strength, quality, purity, and other physical properties.

28.8. FINISHED PRODUCTS

A finished pharmaceutical product is the article received by the retailer for sale to a consumer. It consists of a dosage form inside a container/closure system. A finished chemical product is usually in the form of a powder or liquid packaged in a suitable container.

28.8.1. Types of Finished Products

Finished pharmaceutical products include tablets, capsules, powders, liquids, ointments, creams, and suppositories. These are packed in bottles, jars, vials, ampules, syringes, tubes, canisters, or packets, which are often packed inside a cardboard carton.

A finished chemical product is packaged into plastic or metal drums, large tote bins, or, for bulk shipments, tank cars.

28.8.2. Method of Packaging

If desired, the packaging process for tablets, powders, and liquids can be managed entirely by hand. The procedure, whether done manually or automatically, is as follows. The material is weighed or counted, containers are filled, closures are applied, labels are affixed, and containers may be placed in cartons. It is usual in modern processing for packaging lines to be partially or fully automated. The only manual operations in automated systems are the filling of hoppers with packaging materials (for example, bottles and caps) and with bulk material (tablets or liquids). High-speed packaging lines are sophisticated and efficient. Indeed, the industry's earliest approaches to robotics came in this area.

28.8.3. Controls

Line clearance checks are performed on the packaging lines before any new product or packaging material is brought into the area. The bill of materials or packaging ticket details the components that are to be assembled. The line is set up, the batch code and expiration date are assigned and checked, and packaging begins.

In-process samples are checked for fill volume or quantity at regular intervals, and samples are taken by the Quality Control Laboratory to check against the specifications for the finished product. The identity, strength, quality, and purity of the finished, packaged product are reconfirmed by a complete records check, and this may be confirmed by additional laboratory testing. The master control documents for the finished product dictate the specifications, methods, and sampling procedures, and the data are recorded in the batch record.

28.9. LABELS AND LABELING

In regulatory usage (as defined in the FD&C Act), a label is a display of written, printed, or graphic matter on the immediate container, that is, on the container received by the consumer. The term "labeling" encompasses all written, printed, and graphic matter accompanying the product. Thus, labeling includes labels, but labels do not include labeling.

The necessity for complete agreement between the label and the product contained within the container is both obvious and essential. The label is therefore closely controlled to prevent a mix-up of products that could harm a patient. While it is just as important to control other kinds of labeling, such as the outside of a shipping case, the potential to harm a patient certainly would not be as great as an error in printing the bottle's label.

28.9.1. Types of Labels and Labeling

The most common type of label is made of paper. This is sometimes preglued like a postage stamp and then attached to the container during the packaging operation. The label, however, may be printed on the bottle, tube, or jar before receipt, in which case the packaging component must be examined for both labeling and packaging defects.

Labeling appears on the following: instructions to physicians, patient package inserts, cartons, outer wrapping, cardboard stock for shipping containers, shelf display cartons, and advertising or promotional materials accompanying the product.

28.9.2. Method of Printing

First, the copy (i.e., the text of the labeling) is typed on a sheet of paper. A mechanical art board is then prepared with the required layout. The layout and format are agreed upon by individuals or a group of individuals with knowledge of the regulations, medical aspects of the product, marketing needs, and packaging and printing technology. The printer prepares printing plates and cutting dies to produce labels of the correct size. The printer then prints a few copies, often in black ink. These are called the proofs or stats. After proofreading, the press is set up to produce the correct color scheme and the material is printed as the final labeling.

Labels are printed in any of three ways. Gang-printing is the term used when many different types of labels are printed on one sheet, which is then cut and sorted into bundles of cut labels. The use of gang-printed labeling for different drug products, or different strengths or net contents of the same drug product, is prohibited unless the labeling from gang-printed sheets is adequately differentiated by size, shape, or color. A second, preferable method, is to print only one type of label on each sheet, thereby avoiding errors in sorting. A third printing method, which is the best for drug labels, uses roll labels, where the labels are printed on a large roll and not cut into individual labels. Many companies also now use their own computers to generate the labels they need. The cut labels or roll labels are received, sampled, tested, and then released by the Quality Control Laboratory.

28.9.3. Control of Labeling Quality

Sampling a cut label involves fanning the bundle to ensure uniformity in the basic color and style of the labels. To assist with this, bar codes are usually printed on the label's edge. A few labels are then removed for comparison with the proof and for necessary checks of paper weight, dimensional measurement, and so on.

If acceptable, the labels are signed, dated, and placed in the quality control file for use as references for comparison with future labels, as well as for checking against the final packaged product. The labels are counted by hand, by machine, or by weight and are then released to the Packaging Department, where they are held under lock and key to prevent mix-ups. Roll labels are similarly evaluated, although they are generally counted electronically. For computer-generated labels, the computer may be programmed to produce the exact number of labels required.

The control of labeling quality varies from extremely detailed examinations of labels, such as would be the case with patient package inserts, to less stringent examinations of packaging materials, such as cardboard for shipping cases, where the printing is limited to such information as name and identification number.

28.10. PACKAGING MATERIALS

A packaging material is anything used in packaging a product, including the immediate container and closure; desiccants, cotton, or rayon coil inside the container; and cartons, cases, and shipping containers.

28.10.1 Types of Packaging Materials

Packaging materials are generally categorized by their compositions and by the testing methods used to ensure their quality. The following is a list of the basic classifications: bottles (glass or plastic), caps (metal or plastic), paper stock, carboard stock, plastic parts, plastic film, metal foil, metal or fiber drums, metal cans, tubes, vials, jars, aerosol cans, and tote bins. (In the cosmetic industry, a greater variety of packaging components is used, including brushes, wipers, plugs, pencils, and compacts.)

28.10.2. Methods of Manufacture

Packaging materials are produced in large quantities by highly sophisticated industries. Glass bottles, for instance, are produced in vast quantities by glass manufacturers. The manufacturers begin with sand to produce the glass and then mold and blow a seemingly endless variety of shapes and sizes of bottles to exceedingly precise tolerances.

The manufacturers of plastic containers normally buy the various types of polymers, such as polyethylene, polystyrene, and polypropylene, and mix them with the necessary pigments. They are extruded and pelletized and then molded into various shapes. Metal cans and drums, by contrast, are produced from sheet steel or aluminum (with or without soldered side seams), and special equipment may be needed for sealing the lid. Paper and cardboard are produced from various grades of pulp. In essence, each packaging material has its own distinctive source and manufacturing method. The variability possible from lot to lot or manufacturer to manufacturer must therefore be considered in any quality control program.

28.10.3. Controls of Packaging Quality

Packaging materials are produced in large numbers of units, so it is possible to apply known statistical methods in their evaluation. They are accordingly sampled, inspected, and tested for the number of variables by quantitative measurements and for different general defects or attributes. The Quality Control Laboratory must consider the final configuration of the package and assess the quality of the packaging components when they are combined in that configuration. In other words, a bottle and a cap may each pass individual specifications, but the filled, capped bottle could still leak. The types of defects possible will depend on which category of packaging material they are in and how much information is available from the vendors as an aid to setting the acceptable quality levels (AQLs). All packaging materials are ultimately tested against the master control documents and then released to the Production Department or rejected as unsuitable. Documentation of the packaging process is discussed below.

28.11. DOCUMENTATION

Quality assurance and quality control require the use of many different types of documents, including general procedures, master documents, and batch records. Many years ago, I combined all these documents into one list which I called my "30-Document List." It is now a 31-document list, but it still contains basically all the documents needed to manufacture and control a chemical or pharmaceutical product. The documents are listed, in alphabetical order, in Table 1.

It is not possible in this short chapter to explain each document in detail, but in most cases the titles are self-explanatory.

28.12. REGULATIONS

The FD&C Act, as amended, is the law in the United States covering processed foods, drugs, cosmetics, and medical devices.

The act is administered by the Food and Drug Administration (FDA), and its regulations are published in the Code of Federal Regulations (21 CFR). Proposed changes to the regulations and other notices are published in the *Federal Register* (FR).

Of particular interest to those in quality assurance and quality control are the GMPs found in 21 CFR (Table 2) and the many Guidelines issued by the FDA (Table 3) to

TABLE 1 Documents needed for manufacture and control of a chemical or pharmaceutical product

No.	Document title	Brief explanation
1.	Alternative Names	List of the various names used with the product (U.S. approved name; generic name; trade name; code number, etc.)
2.	Batch Manufacturing Record	Complete record of the manufacturing process
3.	Batch Packaging Record	Complete record of the packaging process
4.	Bill of Materials	List of all materials used in the finished packaged product
5.	Certificate of Analysis	List of the tests done, specifications, and results for the batch
6.	Certificate of Disposal	Certification that a rejected product has been properly disposed of
7.	Disposition	Authorized document identifying the batch as accepted, rejected, quarantined, for re-work, etc.
8.	Flow Diagram	Schematic showing in simplified form how the product is made
9.	General Instructions/Hazards	Special instructions (e.g. "wear gloves," flammable material") for the safety of the operators
11.	Master Manufacturing Instructions	Preapproved recipe for manufacturing the product
12.	Master Packaging Instructions	Preapproved method for packaging the product
13.	Quarantine Report	Report of all materials held in quarantine
14.	Resample Request	Request to resample from the batch
15.	Receiving Report	Record that a material has been received in the company or in the department
16.	Sampling Frequency	Preapproved document specifying the frequency of sampling
17.	Sample Procedure	Request to resample from the batch
18.	Sample Taken	Record that the sample was taken
19.	Standard Operating Procedures (SOPs)	Compilation of procedures that cover such things as equipment, facilities, personnel, etc., and the way they are used to manufacture a product according to the approved documentation
20.	Specifications: Labels	Preapproved specifications for the labels and labeling
21.	Specifications: Packaging	Preapproved specifications for the packaging materials
22.	Specifications: Raw Materials	Preapproved specifications for the raw materials
23.	Specifications: In-Process Items	Preapproved specifications for the in-process items
24.	Specifications: Finished Products	Preapproved specifications for the finished products
25.	Stickers	Stickers physically attached to materials to display their status ("Accepted," "Rejected," "Quarantine," etc.)
26.	Stability	Documentation confirming the stability of the product
27.	Test Frequency	Preapproved document indicating the frequency of testing (e.g., every batch; every fifth batch)
28.	Test Methods	Analytical test methods
29.	Validation: Cleaning	Validation of the cleaning procedures
30.	Validation: Methods	Validation of the analytical methods
31.	Validation Process	Validation of the manufacturing process

TABLE 2 GMPs in 21 CFR

Part	Title
Part 110	Current good manufacturing practice in manufacturing, packing, or holding human food
Part 210	Current good manufacturing practice in manufacturing, processing, packing, or holding of drugs; general
Part 211	Current good manufacturing practice for finished pharmaceuticals
Part 225	Current good manufacturing practice for medicated feeds
Part 226	Current good manufacturing practice for type A medicated articles
Part 606	Current good manufacturing practice for blood and blood components
Part 820	Quality systems regulation

TABLE 3 FDA guidelines

Title	Date
Glossary of computerized system and software development terminology	August 1995
Guideline for submitting documentation for packaging for human drugs and biologics	February 1987
Guideline for submitting documentation for the stability of human drugs and biologics	February 1987
Guideline for submitting documentation for the manufacture of and controls for drug products	February 1987
Guideline for submitting samples and analytical data for methods validation	February 1987
Guideline for submitting supporting documentation in drug applications for the manufacture of drug substances	February 1987
Guideline for submitting supporting documentation in drug applications for the manufacture of drug substances	March 1991
Guideline for the format and content of the microbiology section of an application	February 1987
Guideline for the format and content of the nonclinical pharmcology/toxicology section of an application	February 1987
Guideline for the format and content of the chemistry, manufacturing, and controls section of an application	February 1987
Guideline for the format and content of the summary for new drug and antibiotic applications	February 1987
Guideline for the study of drugs likely to be used in the elderly	November 1989
Guideline on sterile drug products produced by aseptic processing	June 1987
Guideline on the preparation of investigational new drug products (human and animal)	March 1991
Guideline on validation of the *Limulus* amebocyte lysate test as an end-product endotoxin test for human and animal parenteral drugs, biological products, and medical devices	March 1989
Guidance: Immediate-release solid oral dosage forms; scale-up and post-approval changes: chemistry, manufacturing, and controls; in vitro dissolution testing; in vivo bioequivalence documentation	November 1995
Guide to inspections of bulk pharmaceutical chemicals	May 1994
Guide to inspections of cosmetic product manufacturers	February 1995
Guide to inspections of dosage form drug manufacturers—CGMPRs[a]	October 1993
Guide to inspections of high-purity water systems	July 1993
Guide to inspections of lyophilization of parenterals	July 1993
Guide to inspections of microbiological pharmaceutical quality control laboratories	July 1993
Guide to inspections of oral solid dosage forms pre/post-approval issues for development and validation	January 1994
Guide to inspections of oral solutions and suspensions	August 1994
Guide to inspections of pharmaceutical quality control laboratories	July 1993
Guide to inspections of sterile drug substance manufacturers	July 1994
Guide to inspections of topical drug products	July 1994
Guide to inspections of validation of cleaning processes	July 1993

[a]CGMPRs, Current Good Manufacturing Practice Regulations.

TABLE 4 Sources of publications mentioned in this chapter

ISO 9000 Standards
International Organization for Standardization (ISO), 1, Rue de Verembé, Casse Postale 56, CH-12 Geneva 20, Switzerland/Suisse Telephone: 011-41-22-749-01-11; Fax: 011-41-22-733-34-30
or
American National Standards Institute (ANSI), 11 W. 42nd St., 13th Floor, New York, NY 10036 Telephone: 1-212-642-4900; Fax: 1-212-398-0023
Federal FD&C Act, as amended:
DHSS Publication no. (FDA) 93-1051. For sale by the U.S. Government Printing Office, Superintendent of Documents, Mail Stop SSOP, Washington, DC 20402-9328
GMPs
Code of Federal Regulations, 21 CFR parts 200 to 299, published by the Office of the Federal Register, National Archives and Records Administration.
For sale by the U.S. Government Printing Office, Superintendent of Documents, Mail Stop SSOP, Washington, DC 20402-9328
FDA Guidelines
An index of documents available may be obtained on request from CDER Executive Secretariat Staff (HFD-8), Center for Drug Evaluation and Research, U.S. Food and Drug Administration, 7500 Standish Place, Rockville, MD 20855
FDA Guides
Available from the Division of Field Investigations, Office of Regional Operations (HFC-100), Office of Regulatory Affairs, U.S. Food and Drug Administration, 5600 Fishers Lane, Rockville, MD 20857-0001

further explain some of the regulations. Sources for these publications are listed in Table 4.

SELECTED READINGS

1. **Bryant, R.** 1984. *The Pharmaceutical Quality Control Handbook.* Aster Publishing Corp., Springfield, Oreg.
2. **Denyer, S., and R. Baird (ed.).** 1990. *Guide to Microbiological Control in Pharmaceuticals.* Ellis Horwood, New York, N.Y.
3. **Huxsoll, J. F.** 1994. *Quality Assurance of Biopharmaceuticals.* John Wiley & Sons, Inc., New York, N.Y.
4. **Juran, J. M., et al.** 1998. *Quality Control Handbook.* McGraw-Hill Book Co., New York, N.Y.

Contract Fermentations

RICHARD I. MATELES

29

A contract manufacturer manufactures a product for a customer which, prior to the manufacture, has agreed to purchase the product. Thus, the vendor knows that if the product meets specifications, the purchaser will buy it at the agreed-upon price. The vendor benefits through utilizing excess plant capacity and through not having to produce a product for which a market and price are not guaranteed. The purchaser benefits through obtaining a material which may otherwise not be available on the market or obtaining the product with less financial risk than if it were to build a plant for its production.

Other similar terms for this vendor-purchaser relationship are custom manufacture and toll manufacture. Although distinctions can be made in the various designations (toll, contract, or custom manufacture), for the purposes of this review they are not important and are considered synonyms. Because the use of the word "toller" can be ambiguous and could be understood to refer to either party in a tolling relationship, the terms "client" and "vendor" are used in this chapter to denote the two parties to the toll arrangement. Thus, the vendor supplies toll services to the client.

Toll manufacture is common in the chemical industry, where many companies specialize in efficiently carrying out a single step in a synthetic pathway, e.g., hydrogenation, fluorination, or phosgenation, with the client furnishing the initial reactant and receiving the product. Such reactions may involve unusual process hazards or demand highly specialized equipment. They may also involve the custom or contract manufacturer's proprietary technology, which is made available to clients through the use of the technology to carry out the reaction step(s).

A more comprehensive manufacturing relationship is sometimes sought, in which the client wishes to contract out the entire manufacturing process. In this case, the vendor may arrange for the supply of raw materials, may carry out several reactions and/or purifications, and may deliver an end product that meets specifications. If it is destined for pharmaceutical use, the product may be of U.S. Pharmacopeia standard for bulk products.

Thus, the spectrum of toll, contract, or custom manufacture can run from contracting out (tolling out) a single step of a synthetic process to contracting for the production of a product, whose manufacture involves fermentation followed by recovery, purification, and formulation of the product.

Although tolling relationships have been common in the chemical industry, they have been relatively unusual in the fermentation industry. This is indicated by the lack of any mention of toll or contract fermentation in a comprehensive book, published in 1988, devoted to fermentation economics (6). The changing economics of the pharmaceutical industry, toward increased efficiency of manufacturing, has forced serious consideration of toll or contract manufacturing as evidenced by the publication in recent years of several articles dealing with this subject (3, 5, 7).

29.1. WHY ENTER INTO A TOLLING RELATIONSHIP?

There can be a number of situations in which a company might want to seek a vendor to manufacture a product via fermentation.

A company already in the fermentation business, e.g., producing added-value food ingredients by fermentation of whey or producing antibiotics, may find itself with capacity limitations. Perhaps it believes that although capacity is strained at the moment, process improvements that are on the way will relieve the production bottlenecks within a year or two, and therefore the company wishes to avoid a capital investment which may eventually not be justified. The manufacturer may also be concerned that the demand for the product is transient, so that by the time the added production capacity is onstream, the additional demand will have disappeared.

A different situation would be that of a company which is entering a new business with a new product, manufactured by fermentation and requiring the construction of a fermentation facility. Because of uncertainty about how the market for the new product will develop and in order to furnish large samples for market studies and initial commercial activities, the company may wish to toll out the manufacture for a year or two while it is designing the new facility, finalizing decisions about the size of facility that will be required, and generally demonstrating the market demand for a new product before making what may be a substantial capital investment in a facility.

Still another situation would be that of a research-and-development-oriented biotechnology company with an exciting product that it believes is ready for commercial production. The use of contract manufacture may not be an option but, rather, a necessity: the company may not be able to make the very large investment required for a product produced by microbial fermentation or mammalian cell culture. Contracting out the production is a way of maintaining some control over the manufacture, compared to an alternative strategy of seeking a partner or joint venturer which would take responsibility for the manufacturing.

In these instances, the client may seek a contract manufacturer to produce some quantities of the products or possibly to produce all of the needed production of one of the products being marketed by the established client.

From the perspective of the vendor, there are also some advantages to entering into a toll relationship. In fact, there are two main types of vendors: (i) those which have a captive business based on fermentation and which expect to have excess production capacity for some period, and (ii) those whose business is composed primarily of undertaking toll production for other companies.

Historically, some companies that have produced pharmaceuticals such as antibiotics, enzymes for the food or detergent industries, or similar products have found that over the course of time, as a consequence of various factors (e.g., process improvements, product obsolescence, decreased demand, and insufficient generation of new products), some of their capacity became excess to their own needs. Given the high capital costs of a fermentation plant, it makes a great deal of sense to seek to recover these fixed costs by finding a client that wishes to toll out some of its production. Furthermore, the staff of the facility may also be underutilized. Although labor is not exactly a fixed cost, skilled or professional staff have many of the characteristics of a fixed cost, at least for the short or medium term, i.e., up to a year or so. Thus, there are a number of veteran fermentation companies which have spare capacity and have been willing to carry out toll production for clients on a more or less regular basis (Table 1).

As of late 1996, there are no *veteran* fermentation companies which unequivocally fit into the category of companies whose business is composed exclusively or mainly of providing toll services to clients. One company, FermPro Manufacturing L.P., of Kingstree, S.C., has been engaged in providing large-scale toll fermentation services to clients as its sole business for several years. The facility was acquired from Gist-Brocades, which found it to be in excess of its own requirements and sold it to a management group, which runs it today. For the client, the important advantage of dealing with such a provider of services is that there is no competition for scheduling or other resources between the manufacturer's own products and those of its clients.

TABLE 1 Some companies which provide toll microbial fermentation services on a regular basis[a]

Company	Location
Abbott Laboratories	United States
Biochemie GmbH	Austria
Gist-Brocades	The Netherlands
Lonza/Celltech Biologics	Switzerland/Czech Republic/ United States
MultiFerm	Sweden
Pharmacia & Upjohn	United States
Recordati	Italy

[a]These companies all have their own established product lines and provide toll services for a variety of products by using spare capacity.

The increased demand for production, by either mammalian cell or microbial culture, of recombinant proteins for pharmaceutical use has resulted in the formation of a number of companies which intend to devote themselves to the provision of toll production for clients. In this area of technology, facilities with the capability of producing recombinant proteins for pharmaceutical use and of meeting the strict regulatory guidelines for their production were not commonly available; therefore, there was essentially no spare capacity which established producers could offer to the numerous biotechnology companies that had recombinant-DNA-based products in development. Table 2 provides some information on such purpose-built vendors. It remains to be seen whether these companies will be successful in maintaining themselves for a full cycle of plant investment and will generate sufficient profits to replace and maintain the plant and provide adequate returns on the investment.

In addition to the purpose-built facilities listed in Table 2, there are several biotechnology start-up companies which built cell culture facilities, found the facilities to be larger than their actual needs, and in some cases (e.g., Synergen and Centocor) suffered severe financial damage when the biopharmaceutical destined for the plant failed prior to commercialization. Some of these biotechnology companies also serve as vendors of cell culture services based on the excess capacity they have.

The large multinational pharmaceutical companies are building new facilities aimed at microbial or mammalian cell culture systems for production of recombinant proteins for pharmaceutical use. Occasional excess capacity is likely to appear eventually with cell culture facilities as it did years ago in the traditional large-scale microbial fermentation facilities.

29.2. POSSIBLE DRAWBACKS TO A TOLLING RELATIONSHIP

While there are reasons, discussed above, why companies seek to toll out fermentation production, there are also problems with such a relationship.

One of the strongest objections, which has both emotional and realistic components, is the desire of a manufacturer to maintain control of the proprietary technology involved. This is particularly the case when the manufacturer is a technologically oriented company, often highly dependent on patents to protect its proprietary position and to obtain a correspondingly large gross margin. This objection is sometimes phrased as "we're not gonna let them touch our family jewels," or, in more academic terms, "we look upon manufacturing as a means of securing a strategic advantage, which we do not want to dissipate."

Indeed, since the manufacturer will usually have to disclose to the vendor all of the process details, it is clear that there is a risk of the technology becoming known to a larger body of people than if it were maintained as a trade secret by the client. Despite secrecy agreements, there is always a greater risk of technology "leaking" if people in different companies are privy to the secrets. Furthermore, not only will the vendor know all the process details, but also its personnel will learn key manufacturing cost parameters, e.g., fermentation yields, concentrations, and productivities; suitable and unsuitable raw materials and

TABLE 2 Examples of new companies[a] which specialize in the provision of toll services to clients and which do not have their own products

Company	Location	Area (ft^2)	Investment (millions)	Start-up date
Gist Brocades/Bio-Intermediair	Groningen, The Netherlands	20,000	$20	1994
Contract Bioscience	Baltimore, Md.	54,000	$26	1996
Gist Brocades/Bio-Intermediair	Montreal, Canada	56,000	$26	1997
Covance Biotechnology Services	Research Triangle Park, N.C.	109,000	$55	1997

[a]These companies are oriented to recombinant proteins for pharmaceutical use and similar high-value products.

suppliers; raw material, in-process, and end product specifications; recovery methods; and all the panoply of technical and economic data that comprise process technology.

On the other hand, even if the process is not tolled out, the client's own staff will be privy to all this information, and if they are inept, casual, or disloyal regarding the importance of maintaining trade secrets, the information can leak out anyway. In addition, owing to increased mobility of personnel engendered by the current fashion for corporate "reengineering," it is certainly harder today to maintain trade secrecy than it was when employment of capable employees was assumed to be for very long periods, if not for life.

Another important and common problem is that of conflicting priorities. If a company is producing in its own plant and the demand for the product increases, the company has more or less unilateral control over the scheduling: it can determine which product gets priority and which can wait. If the production is tolled out, a change in production volume or schedule from that agreed upon requires negotiation with the vendor. The vendor's priorities will probably be very different from those of the client. This is often a source of conflict between vendor and client, particularly when the vendor is using excess capacity which over the course of several years becomes needed for its own production, in which case the client may be confronted with the choice of paying a much higher price or looking for another facility.

A recent article (8) addresses many of the issues for clients interested in securing toll production of pharmaceutical products, and there is a directory (4) that provides considerable information on various toll fermentation or cell culture facilities located in the United States and other countries.

29.3. ECONOMICS

29.3.1. Vendor

The typical vendor of toll fermentation services confronts a situation where its plant is utilized at significantly less than full capacity. If it can partly or completely fill the excess capacity with toll production for a client, and receive a price which covers all of the incremental costs associated with the new production and at least some of the fixed costs, the plant's profitability will be enhanced by the relationship. These fixed costs include the depreciation attributed to the facility and also much of the labor which typically cannot be laid off for a short time. Therefore, there may be excess personnel available if the plant is not running at full capacity, and the ability to recover some or all of the wages for these personnel is also a motivation for providing toll services.

It is important to recognize that a vendor in this position does not have to recover all of the overhead and fixed costs associated with the plant to justify the decision to provide toll services. Even a partial recovery of such costs can be very attractive. On the other hand, if the management has to invest in additional facilities to enlarge the plant's capacity, a normal rate of return on this investment will be desired and toll production will have to be priced accordingly.

However, a vendor with a purpose-built facility, which has as its sole or primary business the provision of toll services to clients, must recover all of the overhead, fixed costs, and variable costs from the clients, along with some margin of profit. While the facility has the freedom to charge some clients more and some less, the toll business has to pay for itself without having captive products to absorb most or all of the fixed costs. In this situation, selling toll services based on incremental cost is a quick route to bankruptcy.

The economic benefits that the vendor derives from providing toll fermentation services are accompanied by some problems. The vendor has to consider the client's priorities in scheduling production. Some clients and some processes take a great deal more attention than might have been anticipated. There is frequently an attitude of looking upon the company's "own" production as important and that of the client as a "nuisance" which distracts from the main mission of the plant.

It is important for the vendor, or potential vendor, of toll fermentation services to consider these issues before committing to toll fermentation. The staff involved in selling the services and in fulfilling the production need to feel comfortable with a mission which differs from the normal mission of efficient production of the company's own products.

29.3.2. Client

In many instances, the client seeking toll fermentation services is prepared to pay a relatively high price for this production. The reason is that the production may be destined for sampling to potential customers, for market development, for regulatory purposes, or for other activities where a high production cost of relatively small quantities of product is not very important compared with the time saved by using a toll producer. In the long run, the cost of production may be very important, but for these trial runs, other factors may outweigh cost.

Sophisticated clients recognize that aseptic fermentation and the associated downstream purification operations are capital intensive and require relatively skilled staff. Thus, the provision of these services will not be inexpensive, except possibly for a short time based on excess capacity or because a vendor is in a bad business situation and is pressed for revenue. In the latter case, the vendor is likely to disappear more or less suddenly, with embarrassing consequences for clients which depend on the vendor for production.

If a client needs some production of small or moderate volume and for a short (a few months) or intermediate (up to 1 to 2 years) period, toll production can usually be found at costs which are acceptably low. However, if the client seeks a large-scale production commitment for a period of years, the client should be prepared to pay a price reflecting the commitment by the vendor of substantial resources and the expectation that the vendor will legitimately expect a normal return on investment and will charge the toll production with all of the associated overhead and distribution of fixed costs.

In the long run, if a client expects to have increasing need for toll services in the fermentation area, it will eventually make sense to build a plant. However, this decision often can be delayed several years because of the availability of toll production services, and that time can be used to diversify a product line and/or secure greater confidence in the sales potential of the product or products being produced.

29.3.3. Pricing

For instances of toll production where the vendor is asked to produce a final product, the payment can be based on the units of final product that meet specifications. The contract ought to provide for what happens when process problems occur and the product fails to meet specifications. The contract must also consider the effects of changing prices of raw materials, utilities, and labor, particularly if the arrangement is anticipated to extend for years. The vendor and the client must discuss what happens if the vendor is unable to deliver the required quantities at the specified times. They also should discuss what happens if the client finds that the demand for product is not as large as had been planned or is much greater than anticipated.

In many cases, the pricing of the fermentation part of the process is based on production volumes and times, and a common unit is the liter-day (or cubic meter-day, gallon-day, etc.) An example will aid in understanding this unit and its use. Suppose a plant has a fermentor with a total capacity of 75,000 liters. The actual working volume depends on the tendency of the medium to foam when aerated, so assume that the working volume is 50,000 liters. Assume further that the fermentation duration, from inoculation until harvest, is 40 h and that the time it takes to clean and sterilize the fermentor and fill it in preparation for a new fermentation is 8 h. Thus, a fermentation cycle is 2 days. Assume also that the product yield is 40 g/liter, that the recovery and purification process has an overall yield of 85%, and that 25,000 kg of final product is wanted. Therefore, the total volume of fermentation broth which must be generated is (25,000 kg)(1,000 g/kg)/(40 g/liter)(0.85 g/g) = 735,294 liters, which rounds off to 750,000 liters. It will take (750,000 liters)(2 days) = 1,500,000 liter-days of fermentor capacity to provide this. Since the fermentor has an output of 50,000 liter-days per day, it will take 30 days of operation to satisfy this client's needs.

On an annual basis, a fermentor of this size has the capability of providing (50,000 liters)(365 days per years) (0.94 running factor) = 17,155,000 liter-days per year. The advantage of using liter-days as the units for calculating capacity is that the units are additive. Another fermentation, which has a 9-day cycle and needs to be run six times during the year, will use 6 × (450,000 liter-days) = 2,700,000 liter-days of capacity, so together these two fermentations require 4,200,000 liter-days from the plant, which is about 24% (4,200,000/17,155,000) of the plant's capacity.

It is possible to obtain estimates or quotes from various vendors for the price of a liter-day, and Table 3 shows such estimates as of early 1996 for several different vendors and fermentations. It is crucial, however, to realize that these estimates are not sufficient to enable a client to choose which vendor will be the least expensive.

For commodity fermentations, (e.g., bulk antibiotics, detergent enzymes, and amino acids), the fermentation operation, with raw material costs, will represent the major share of the final manufacturing costs. However, recovery and purification costs will be far more important than fermentation or growth stage costs for products (e.g., recombinant pharmacological proteins) with a complex recovery or purification scheme. Furthermore, it is necessary to clarify whether the "liter-day" referred to is based on the nominal capacity of the fermentor or the actual working volume. These two numbers can easily differ by 65%. Also, the price per liter-day generally does not include (i) the costs of the nutrient medium ingredients; (ii) the costs of materials for downstream processing and purification; or (iii) the costs of nonroutine downstream processing operations, where each facility has its own definition of "routine." The equipment demands of a particular fermentation product, and its downstream processing, may not fit each vendor's facility. Parameters such as oxygen transfer capacity, cooling requirements, batch feeding, and instrumentation requirements vary from fermentation to fermentation. Costs for filling, formulating, and packaging have to be factored in according to the specific needs. Overall, it is often true that downstream processing is a limiting factor in deciding which facility can handle a particular job.

The rather large differences in prices shown in Table 3 are influenced by factors such as whether the plant complies with the pharmaceutical (1) or food (2) cGMP (current Good Manufacturing Practice) regulations of the U.S. Food and Drug Administration or equivalent bodies in other countries; the volume of the contract and the likelihood of it being repeated several times; the higher labor and utility costs in western Europe than in the United States; and, of course, how hungry the vendor is for additional business. Not reflected in these costs are the variations in raw material prices, which tend to be significantly higher in Europe than the United States, particularly for carbohydrates and raw material derived from agriculture.

29.4. PROCESS DEVELOPMENT

One of the more serious problems that arises during discussion between a would-be client and a would-be vendor is the failure of the client to understand the manufacturing

TABLE 3 Estimated prices (1995 to 1996) provided by some toll fermentation vendors for microbial production

Company	Price per liter-day ($)	Annual volume of job (10^6 liter-days)	Comments
A (United States)	0.10	1	Doesn't include medium. Does include routine downstream processing. Pharmaceutical cGMP.
B (United States)	0.35	4	Doesn't include medium or downstream processing. Food cGMP.
C (Europe)	0.50	2	Doesn't include medium or downstream processing. Pharmaceutical cGMP.
C (Europe)	0.35	5	Doesn't include medium or downstream processing. Pharmaceutical cGMP.
D (United States)	0.60	4	Doesn't include medium or downstream processing. Pharmaceutical cGMP.
E (Europe)	0.16	1	Doesn't include medium or downstream processing.
F (Europe)	0.14	1	Doesn't include medium or downstream processing.

operations involved in production of products by fermentation and subsequent purification. This has been particularly common since the advent of biotechnology companies populated by scientists with limited industrial experience in production. The belief that manufacturing can quickly and readily proceed ("the process is ready to give to the plumbers") after the biologists have finished constructing appropriate recombinant organisms and the biochemists have worked out purification schemes in the laboratory has led to general grief and even the disappearance of some biotechnology companies. Today, there are many more people in the biotechnology industry who have had experience in process development and commercialization, but basic principles of process implementation are still neglected too often.

It has been my experience that even when both vendor and client have a good understanding of the general and specific process technology, the implementation of such technology by the vendor is rarely smooth and free of trouble. It is advisable for the client to take into consideration the time and cost of trial runs and to attempt to resolve technical issues at the pilot plant level before moving into large-scale plant production. This relates in particular to operations that cannot be conducted as quickly and easily on a large scale as on a laboratory or pilot-plant scale, e.g., heating and cooling of liquids and mixing of liquids or solids. When these operations are critical, it is better to simulate the plant conditions in the laboratory or pilot plant to determine the important parameters than to expect the production facility to be able to reproduce the working methods of the laboratory or pilot plant.

29.5. CONCLUSIONS

Obtaining toll production services can be a highly desirable step for a company interested in commercializing a product. It can save time and money in the commercialization process, and it may provide a financially attractive long-term solution to manufacturing a product for which the company does not have an adequate facility. From the standpoint of the vendor, increasing the intensity of plant utilization by providing toll services to clients can be a highly desirable means of increasing the return on its investment.

Both sides need to understand the possible difficulties in a toll relationship and attempt to provide for their resolution during the negotiation of the contract. While there may be some successful companies whose philosophy is that in every deal there is a winner and a loser, a satisfactory toll relationship is more likely to be based on a business philosophy that expects that both sides in a business relationship should look upon themselves as winners.

SELECTED READINGS

1. Current good manufacturing practice in manufacturing, processing, packing, or holding of drugs; general. *Code of Federal Regulations*, Title 21, Part 210. Current good manufacturing practice for finished pharmaceuticals. *Code of Federal Regulations*, Title 21, Part 211.
2. Current good manufacturing practice in manufacturing, packing, or holding human food. *Code of Federal Regulations*, Title 21, Part 110.
3. **Hamers, M. N.** 1993. Multiuse biopharmaceutical manufacturing. *Bio/Technology* **11**:561–570.
4. **Mateles, R. I. (ed.).** 1997. *Directory of Toll Fermentation and Cell Culture Facilities*, 3rd ed. Candida Corp., Chicago, Ill.
5. **Nicholson, I. J., and P. Latham.** 1994. When "make or buy" means "make or break." *Bio/Technology* **12**:473–477.
6. **Reisman, H. B.** 1988. *Economic Analysis of Fermentation Processes*. CRC Press, Inc., Boca Raton, Fla.
7. **Seaver, S. S.** 1995. Lessons learned from working with contract manufacturers. *BioPharm* **8**(2):20–25.
8. **Tedesco, J. L., T. Olson, S. Tse, and J. M. Patrick.** 1994. Evaluation of biopharmaceutical contract manufacturing facilities: a client perspective. *Pharm. Technol.* **18**:174–187.

GENETICS

IN RETROSPECT IT SEEMS SURPRISING THAT THE FIRST EDITION OF THE *Manual of Industrial Microbiology and Biotechnology* did not contain a section devoted solely to the genetics of industrial microorganisms. This situation is now amply remedied by the inclusion in this Second Edition of nine chapters covering the genetics of *Bacillus*, clostridia, corynebacteria, *Pseudomonas*, *Streptomyces* and non-*Streptomyces* actinomycetes, *Saccharomyces* and non-*Saccharomyces* yeasts, and filamentous fungi, thereby encompassing most of the major groups of microorganisms of industrial importance. *Escherichia coli*, now firmly entrenched in this category, is treated elsewhere in this Manual.

These chapters, written by leading authorities, present an accurate and comprehensive account of the field with special emphasis on recent developments and technologies in genomics and post-genomics. Even a cursory perusal of the contents of these chapters will make it evident that remarkable advances have been made during the past decade in our understanding of the genetics of microorganisms. Powerful molecular genetic tools, aptly termed by David Hopwood "enabling technologies," have been instrumental in gene cloning, in DNA transfer, in gene mapping, sequencing, and expression, and most recently in genome sequencing. These are discussed in detail here. Vast amounts of information are becoming available on the genetic makeup of microbial cells, and new disciplines are being forged to handle the complexity of the data. At this writing, several microbial genomes of industrial importance have been completely sequenced, including that of *Saccharomyces cerevisiae*—which determines over 6,000 genes—and that of *Bacillus subtilis*, and the sequencing of several other microbial genomes treated in this section is also well under way.

Undoubtedly, the functional assignment of gene sequences in open reading frames presents the immediate challenge to analyzing genome complexity, and it is to this end that gene inactivation techniques have acquired paramount importance. The various approaches to carrying out gene knockouts are detailed in these chapters. In parallel with these developments, a battery of powerful new technologies have been created that permit monitoring of the expression, in defined conditions, of the entire set of a cell's mRNAs and proteins—transcriptome and proteome analysis—while other technologies aim at a systematic exploration of protein-protein interactions. One recent example will suffice to indicate the power of these approaches, in which the transcriptional pattern of expression of some 97% of the known or predicted *Saccharomyces cerevisiae* genes was analyzed during spore morphogenesis, providing clues into the potential functions of hundreds of previously uncharacterized genes (1).

The relevance of these studies for understanding cellular processes should be obvious, for it is the coordinated action and interaction of the products of multiple genes that determine and control such diverse activities as DNA synthesis, cellular differentiation, and secondary metabolite production. It is no mere guess to suppose that in the next few years we can expect the application of the above technologies, together with bioinformatics, control theory, systems analysis, and other tools, to revolutionize our ability to manipulate microbial metabolism for the benefit of humankind.

REFERENCE

Chu, S., J. DeRisi, M. Eisen, J. Mulholland, D. Botstein, P. O. Brown, and I. Herskowitz. 1998. The transcriptional program of sporulation in budding yeast. *Science* **282:**699–705.

Genetics of *Streptomyces*

GÜNTHER MUTH, DIRK FRANZ BROLLE, AND WOLFGANG WOHLLEBEN

30

Streptomycetes have become industrial microorganisms of growing importance since their discovery as antibiotic producers. *Streptomyces* is a genus within the order *Actinomycetales*, belonging to the high-G + C-content, gram-positive eubacteria (26). *Streptomyces* comprises a large number of species that are often closely related and can taxonomically be grouped by physiological tests (48) and 16S rRNA comparisons (26). Most *Streptomyces* strains synthesize antibacterial, antifungal, antitumor, antiparasitic, and herbicidal compounds as well as many other agents and enzymes used in medicine, agriculture, and various technical processes. Because of their ability to synthesize numerous compounds that exhibit extreme chemical diversity, *Streptomyces* strains are major parts of industrial strain collections used in screenings for new bioactive molecules.

Most of these compounds are synthesized as secondary metabolites. Their production initiates with the onset of the idiophase. The switch from primary to secondary metabolism not only constitutes a complex biochemical differentiation, but coincides with a process of morphological differentiation (18). On solid media, colonies develop substrate mycelium, then form aerial hyphae that later give rise to the formation of spores (Fig. 1).

Complex physiological and morphological differentiation is highly regulated, probably by internal and external signals (including nutrient limitations). At least some steps of the regulation cascade govern both morphological and physiological differentiation (18).

Yield improvement in industrial processes has mainly been achieved in combination with medium optimization and process development. Strain development programs made use of only the classical techniques such as mutation, selection, and recombination. But during the last few years, an increasing number of molecular genetic tools have been developed that are summarized in an excellent laboratory manual (41) of which an updated version is in preparation. These techniques have been used to get a better understanding of the biology of streptomycetes in general and of the processes of secondary metabolite synthesis in particular. This will enable the development of rational approaches to intervene in primary metabolism for a better supply of precursors and to influence the regulation and synthesis of secondary metabolites. Furthermore, these techniques are the tools to perform such targeted manipulations.

The aim of this chapter is to summarize the methodologies that are available for analyzing and manipulating *Streptomyces* strains, highlighting the recent developments in gene-inactivating techniques.

30.1. DNA TRANSFER TECHNIQUES

In general, conjugation and transformation are the major mechanisms for the genetic exchange in *Streptomyces*. Although transducing phages have also been characterized (33, 104), transduction is of minor importance for the introduction of foreign DNA into streptomycetes. However, recently successful generalized transduction was reported for some *Streptomyces* strains (16a).

30.1.1. Conjugation

As in other bacteria, self-transmissible plasmids are responsible for the conjugal gene transfer in *Streptomyces*. Following the discovery of plasmids SCP1 and SCP2 involved in the genetic exchange in *Streptomyces coelicolor* (91, 105), a great variety of plasmids have been detected in other *Streptomyces* strains. These plasmids include giant linear plasmids, integrative plasmids that arise by the excision of chromosomal fragments, large low-copy-number plasmids, and small multicopy plasmids replicating via the rolling-circle mechanism (reviewed in references 42 and 111). With the exception of deletion derivatives, all naturally isolated *Streptomyces* plasmids not only encode their own transfer but have also been shown to mobilize chromosomal markers (Cma, chromosome mobilizing ability). For the low-copy plasmid SCP2*, a high-fertility mutant of SCP2, a stimulation of homologous recombination during the conjugative transfer has been suggested to be involved in the Cma (113).

Conjugative plasmid transfer takes place only on solid medium and does not occur in liquid culture. Conjugation in *Streptomyces* is very efficient. Up to 100% (titer of transconjugants in relation to the titer of the recipient) of the progenies of a mating receive the conjugative plasmid, and up to 0.1 to 1% are recombinants (15, 52). Although the mechanism of conjugative gene transfer in *Streptomyces* remains unclear, it seems to be different from transfer systems of other gram-positive and gram-negative bacteria. The transfer event is macroscopically visible on agar plates by the formation of "pocks." These pocks resemble character-

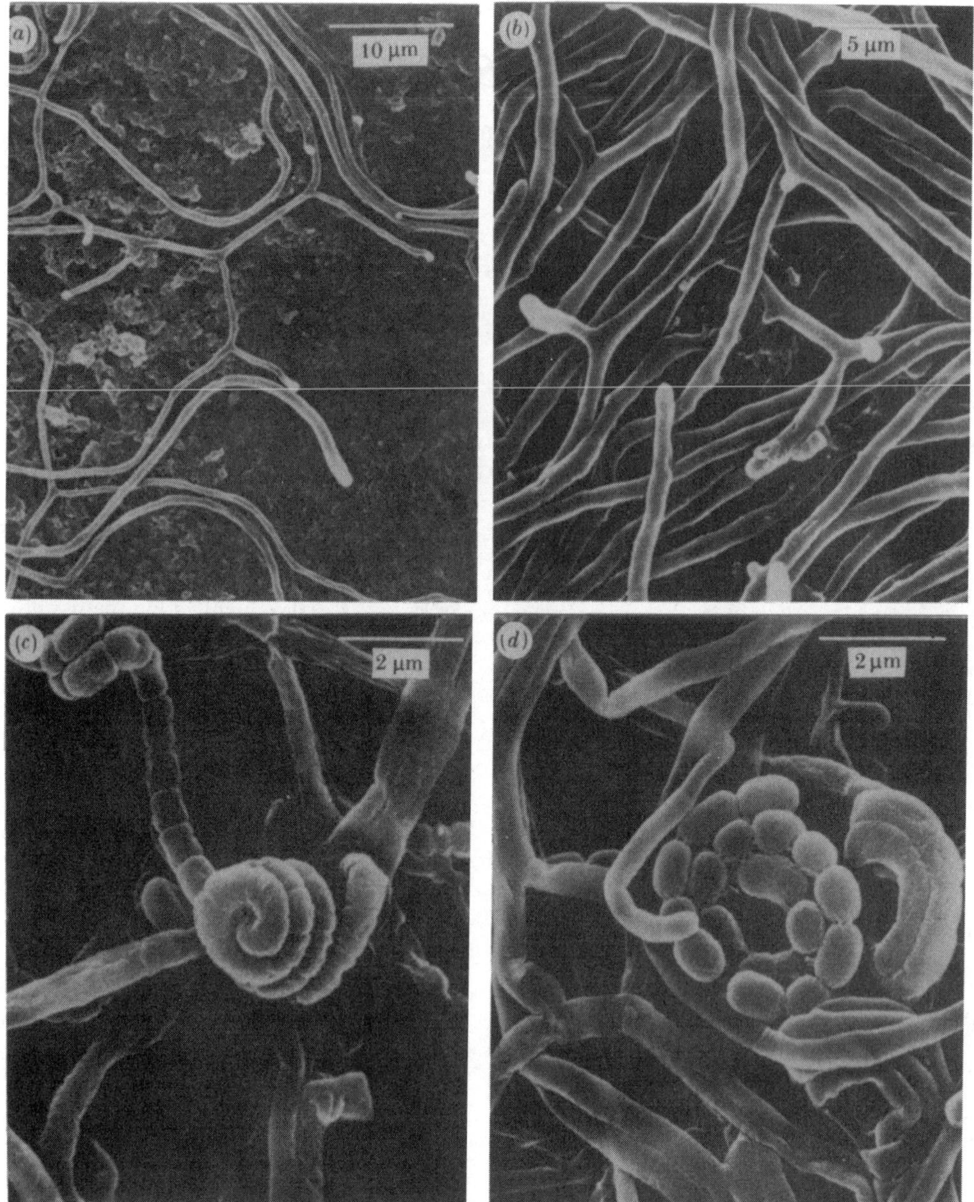

FIGURE 1 Scanning electron micrographs of *Streptomyces lividans* at different stages in the cycle of development. (a) Young substrate mycelium at the margin of a colony. (b) Young aerial mycelium. (c,d) Two stages in the metamorphosis of aerial hyphae into spore chains. Photographs courtesy of M. Bibb and J. Burgess.

istic inhibition zones, surrounding the donor colonies where the growth of the aerial mycelium is retarded, and indicate the spreading of a plasmid within the recipient mycelium (6, 52).

Although conjugation is not a widespread technique for the introduction of foreign DNA into streptomycetes, it is the only route to transfer very large DNA fragments that have the capacity to encode even complex pathways between *Streptomyces* strains.

30.1.2. Interspecific (*Escherichia coli–Streptomyces*) Transfer

Interspecific conjugation proved to be a convenient way to introduce DNA into *Streptomyces* strains (10, 66). Since the protocol for transfer of DNA from *E. coli* to *Streptomyces* does not require any strain-specific optimizations of protoplasting conditions, etc., intergeneric conjugation is of particular interest for industrial *Streptomyces* strains that normally are genetically poorly characterized. Intergeneric conjugation used the broad-host-range transfer system of the IncP plasmid RK2 (=RP4=RP1). The mobilizable vector system carries the *oriT* region of RK2. The mostly used 780-bp *Hae*II fragment (34) contains the relaxation nicking site (*oriT*) and encodes TraJ, which, in addition to the nicking enzyme TraI, is essential to form the relaxosome (109), thus providing the single-stranded molecule to be transferred to the recipient.

The mechanisms of intergeneric conjugation were investigated by Giebelhaus et al. (31). By electron microscopy, they showed that even in the absence of RP4 in the *E. coli* donor cell, so-called nonconjugative junctions between *E. coli* cells and the *Streptomyces lividans* mycelium were formed. However, for the mobilization of *oriT*-carrying plasmids, not only the TraI core region encoding the genes required for the conjugal DNA metabolism was essential, but also the Tra2 core region, which is responsible for the mating pair formation (31).

The most widely used host strain S17-1 carries a RP4 derivative integrated into the chromosome of a *res/mod E. coli* strain (92). An *E. coli* with particularly efficient mobilization properties has been developed by Flett et al. (30a) and Smith (93). This strain is based on the methylation (*dam, dcm, hsdM*)-defective *E. coli* ET12567 strain (62) and carries the RP1 (RP4) derivative pUB307 as the mobilization plasmid.

Intergeneric conjugation is performed by mixing late-log-phase *Streptomyces* mycelium with the plasmid-carrying *E. coli* donor and plating onto agar plates. After 12 to 18 h the plates are overlaid with antibiotic-containing soft agar to select the transconjugants. Nalidixic acid (50 μg/ml) was shown to be very efficient in preventing outgrowth of the *E. coli* donor (10).

30.1.3. Polyethylene Glycol (PEG)-Induced Protoplast Transformation

PEG-mediated introduction of DNA into protoplasts is the standard procedure for transformation. *Streptomyces* mycelium grown in the presence of 0.5 to 1% glycine can be converted into protoplasts by the action of hen eggwhite lysozyme (79). These protoplasts are stable in an isotonic buffer and can be stored at −20 or −70°C for years. The membrane-active substance PEG mediates the efficient introduction of DNA in these protoplasts (9). Mostly PEG 1000 is used, but other molecular weights of PEG have also been reported to be effective. There seems to be great variation in the transformation efficiency of PEG from the different suppliers. Following regeneration of the protoplasts on isotonic media (79), transformants are selected by overlaying with the desired antibiotic. The transformation efficiency dramatically depends on the host strain. Using *S. lividans*, transformation rates of up to 10^7 to 10^8 per μg of CCC DNA can be obtained. The main barrier for introducing foreign DNA into *Streptomyces* protoplasts seems to be the potent restriction systems of streptomycetes. To overcome restriction, several modifications of the standard protocol, such as heat attenuation of the restriction system or the transformation of single-stranded DNA, have been worked out. Particularly effective is alkaline denaturation/renaturation of plasmid DNA prior to transformation, as shown by Oh and Chater (78).

30.1.4. Electroporation

Electroporation protocols for the efficient introduction of DNA have also been developed for *Streptomyces* (61, 80). Mycelium can be electroporated directly and does not necessarily require protoplasting and protoplast regeneration steps, which otherwise have to be carefully worked out for each *Streptomyces* strain. Critical parameters for further strain-specific optimizations are cultivation conditions, pretreatment of mycelia, buffer composition, and electrical field strength. Application of an electrical field strength of 10 kV/cm at 400 W/25 mF to a 24-h culture pretreated with lysozyme (100 μg/ml) and resuspended in a sucrose-glycerol-PEG buffer yielded transformation rates of up to 10^5 to 10^6 per μg of DNA for *Streptomyces rimosus* R6. These frequencies are 10^2 to 10^3 times higher than the efficiency of PEG-mediated protoplast transformation for that strain (80).

30.1.5. Electroduction

Electroduction is a simple and fast method that allows the direct transfer of bifunctional plasmids from streptomycetes to *E. coli* (107). However, the transfer from *E. coli* to streptomycetes by this technique has not been shown so far. This procedure depends on the formation of transient pores in cell membranes under conditions of electroporation (see above). Plasmid DNA is released into the medium through these pores but can also be taken up by the pores. When *Streptomyces* mycelium carrying a shuttle plasmid is mixed with electroporation-competent *E. coli* cells, the plasmids are transferred to *E. coli* after an electrical pulse is applied. This method is of particular interest for the rapid analysis of recombinant *Streptomyces* plasmids in *E. coli* to overcome the difficulties associated with plasmid preparations obtained directly from *Streptomyces*.

30.2. HOST-VECTOR SYSTEMS

30.2.1. Vector Systems

30.2.1.1. Replicative Vectors

pIJ101 Derivatives

The most commonly used vectors for *Streptomyces* are based on the 8.83 kb high-copy-number *S. lividans* broad-host-range plasmid pIJ101 (52). pIJ101-based vectors have been used to construct genomic libraries (41), to analyze promoter activity (108), to create a positive selection vector (54), and as inducible high-copy-number expression vectors (101). In addition, Birch and Cullum (11) isolated a temperature-sensitive (*ts*) replicon from a pIJ101 derivative after in vitro mutagenesis with hydroxylamine.

pSG5 Derivatives

pSG5 is a naturally temperature-sensitive, high-copy-number rolling-circle-replicating plasmid of *Streptomyces ghanaensis* (73, 76). It is distinguished from pIJ101 by its moderate copy number (approximately 50 copies per chromosome [56] instead of up to 300 copies in the case of pIJ101 [52]). Many useful vectors have been constructed on the basis of the *ts* minimal replicon, such as pGM9 and pGM160 (75), and are used particularly in gene disruption and gene replacement experiments. Derivatives of pSG5 have also been used as cloning vectors to identify genes of interest in *Streptomyces* (e.g., reference 4).

SCP2* Derivatives

The 31.4-kb conjugative plasmid SCP2* is an intensively studied low-copy-number plasmid from *S. coelicolor* A3(2) (6, 15, and references therein). The advantage of SCP2*-derived vectors is their high stability and their ability to carry large fragments of DNA. They have been used to clone the entire biosynthetic gene cluster encoding the antibiotic actinorhodin (65) and to produce the first hybrid antibiotics by genetic engineering (43). In addition, a SCP2* derivative has been used to express novel hybrid polyketide antibiotics in *S. coelicolor* CH999 (60).

30.2.1.2. Integrative Plasmid and Phage Vectors

Several integrating vectors have been constructed to allow stable integration into the host chromosome. Mostly, they are based on naturally integrative genetic elements of *Streptomyces*, some of which have a broad host range. Examples are IS*117*-derived vectors (70) and pSAM2-derived vectors (94).

ϕC31 is a temperate phage with a relatively wide host range (17). The wild-type ϕC31 phage lysogenizes nearly all its hosts by integration into the chromosomal *att* site. Derivatives lacking the *att* site integrate via homologous recombination; cloning of a host DNA segment enables usage of ϕC31 as a vector system for mutagenic gene disruption experiments, e.g., with antibiotic gene clusters (19). Furthermore, Bruton et al. (16) constructed ϕC31::*xylE* vectors, allowing the monitoring of transcription of genes in *Streptomyces*.

30.2.1.3. Bifunctional Cosmids

The most frequently used bifunctional cosmid vector in *Streptomyces* genetics is pKC505 (84). This cosmid was constructed by fusing the SCP2* transfer and replication region with a modified *E. coli* cosmid vector carrying the *aac(3)IV* gene for conferring apramycin resistance (Am^r) both in *Streptomyces* and in *E. coli*. However, despite its advantages, pKC505 appears to be quite unstable, at least in *S. lividans* (only 27% of the progeny retained the Am^r resistance in *Streptomyces* after one round of sporulation on nonselective media, possibly owing to the lack of the stability region of SCP2* [84]).

30.2.2. Marker Genes

Many resistance genes originating from antibiotic biosynthetic pathways have been cloned and used as marker genes for construction of cloning vectors. Furthermore, several marker genes from gram-negative bacteria have been tested in streptomycetes. The neomycin resistance gene *aphII* of Tn*5* (47) and the gentamicin resistance gene *aacC1* of Tn*1696* (110) are expressed in streptomycetes by their own promoters. Both the *xylE* gene from *Pseudomonas putida* (46) and *luxAB* from *Vibrio fischeri* (88) have been used for construction of promoter probe plasmids. Expression of the green fluorescent protein from *Aequorea victoria* (23) was shown in *S. lividans* when cloned on plasmids based on pSG5 (72).

Detailed lists of marker genes and their use in *Streptomyces* are summarized by Hopwood et al. (41) and Wohlleben and Muth (111).

30.2.3. Host Strains

In contrast to other eubacterial species, such as *E. coli* K-12 or *Bacillus subtilis*, in which research has been focused on one particular strain, many species of the genus *Streptomyces* have been analyzed because of their industrial importance. However, two strains have gained importance for molecular biologists: *S. coelicolor* A3(2) and the closely related *S. lividans* 66. *S. coelicolor* A3(2) is the best-studied strain and has therefore developed into a model strain. However, *S. coelicolor* shows a strong restriction barrier against DNA from *E. coli*. Therefore, the largely nonrestricting and plasmid-free *S. lividans* (no plasmid-free *S. coelicolor* strain is available) is used in general as an intermediate host for cloning experiments. DNA manipulated in *E. coli* and destinated for *S. coelicolor* is passaged through either *S. lividans* or *E. coli* ET12567 (63), a completely nonmethylating strain. (Note: the passage through *E. coli* ET12567 is less time-consuming than that via *S. lividans* but has the disadvantage of an incomplete DNA repair system due to the *dam* and *dcm* mutations.)

MacDaniel et al. (60) constructed *S. coelicolor* CH999 by deleting the entire chromosomal *act* gene cluster using pGM160. This strain will be particularly useful for rational design of aromatic polyketides by genetic engineering.

Until now no completely recombination-deficient *streptomyces* strain has been isolated. Despite several attempts, Muth et al. (74) were unable to construct a knockout *recA* mutant of *S lividans*: only the *S. lividans recA* mutant FrecD3 putatively expressing a truncated RecA protein proved to be viable, leading to the hypothesis that the *recA* gene is indispensible. *S. lividans* FrecD3 is UV-sensitive and impaired in homologous recombination but still has residual recombination activity. Furthermore, this mutant enhances deletions of chromosomal ends and shows a deficient amplification process (106). Therefore, this partial *recA* mutant would *not* be useful as primary cloning host owing to increased genetic instability.

30.3. GENE ISOLATION

30.3.1. Direct Cloning

In principle, for the isolation of *Streptomyces* genes, the standard techniques for isolating bacterial genes can be used, such as complementation of null mutants. If the expression of the desired gene (e.g., resistance genes) can directly be detected, *S. lividans* is used routinely as *Streptomyces* host. If *E. coli* is the cloning host, the use of expression vectors is recommended since *Streptomyces* promoters are normally not recognized in *E. coli*.

30.3.2. Reverse Genetics

An effective approach for cloning genes that cannot easily be selected for is to use reverse genetics. The probes for isolating genes are often fragments derived from heterologous organisms owing to the high G + C content, preferably from other *Streptomyces* strains. The high similarity of secondary metabolite genes enables the identification of antibiotic biosynthetic genes from other producers: for example, 16 strains containing polyketide biosynthetic genes could be isolated out of 25 strains tested using *actI* and *actII* probes from the actinorhodin gene cluster (64), and 6-desoxysugar biosynthetic genes were isolated with *strDELM* probes originating from the streptomycin producer *Streptomyces griseus* (99).

Instead of heterologous gene probes, oligonucleotide probes can be used either in hybridization or in PCR approaches. To design the probes, known protein sequences are aligned, and highly conserved regions are identified. Based on consensus sequences, the correct DNA sequence can be deduced with rather high probability because of the biased codon usage. Codon usage tables for *Streptomyces* have been assembled by Wright and Bibb (112) by compiling all sequenced *Streptomyces* genes. They are also available on the Internet (http://www.dna.affrc.go.jp). With this approach, housekeeping genes (such as *recA* [77]) as well as secondary metabolite genes (such as dNDP-glucose dehydratase genes [25]) have been cloned. Recent progress in the *Mycobacterium* genome project offers a further alternative, particularly for the isolation of primary metabolite genes, since *Streptomyces* and *Mycobacterium* genes are very similar in most cases. Many of those genes are already included in the database (http://kiev.physchem.kth.se/MycDB.html) and can serve for the design of probes to amplify the corresponding gene from streptomycetes.

The PCR protocol for *Streptomyces* can be simplified in comparison to the standard protocols, since the annealing reaction does not need a particularly low temperature. A reliable procedure that may require some primer-specific optimization for *Streptomyces* thus includes the following (modified according to reference 25):

1. First denaturation: 94°C, 2 min
2. Mixing of reaction components: 72°C, 4 min
3. 30 cycles of denaturation (94°C, 1.5 min) and annealing-extension (72°C, 1.5 min)
4. Final extension: 72°C, 10 min
5. Concentrations:
   a. chromosomal DNA: 0.2 μg/100 μl incubation volume
   b. primers: 1 μM
   c. deoxyribonucleoside 5′-triphosphates (dNTPs): 200 μM
   d. $MgCl_2$: 2 mM
   e. 1% gelatine (or dimethyl sulfoxide)

Use *Taq* polymerase in standard reaction buffer and a GeneAmp 2400 thermocycler from Perkin Elmer.

An elegant way to identify and isolate genes is to use two-dimensional gel electrophoresis. This technique allowed the monitoring of changes in gene expression related to bialaphos synthesis in *Streptomyces hygroscopicus*. By identifying and determining the N-terminal sequence of bialaphos-specific proteins, they could be related to the corresponding genes (39). In the case of the nikkomycin producer *Streptomyces tendae* Tü901, the protein pattern of nonproducing mutants that carried deletions of the nikkomycin cluster were compared with patterns of wild-type *S. tendae*, leading to the identification of 10 gene products specific for nikkomycin production (69). N-terminal sequences could be obtained by microsequencing of protein spots excised from preparative two-dimensional gels, and the derived oligonucleotides were successfully used for the first isolation of nikkomycin biosynthetic genes.

The gene inactivation techniques (see section 30.4) enable the construction of mutants harboring genes disrupted by vector sequences. This can serve as an alternative approach for isolating *Streptomyces* genes. By restricting the total DNA of the mutant with suitable restriction enzymes, large DNA fragments can be generated, including the intact replication genes of the vector. These fragments can be circularized, transferred in a *Streptomyces* host (or *E. coli* if shuttle vectors were used for the disruption experiment), and selected. Thus, genes flanking a mutated gene were isolated either with ϕC31 vectors (19) or with a pGM vector (75). If genes were inactivated either by a marker cassette or by a transposon, the isolation of the mutated gene and adjacent regions is possible after restricting the DNA of the mutant with suitable endonucleases, cloning, and selecting for the resistance encoded by the cassette or the transposon.

30.4. GENE MANIPULATION AND GENE INACTIVATION METHODOLOGIES

30.4.1. Classical Mutagenesis

Streptomycetes mycelia contain several chromosomes per compartment (18). Spores are the only stage of the *Streptomyces* life cycle where a haploid genome status with a single chromosome is present. Therefore, the selection of mutants always requires a sporulation cycle on solid medium. The most widely used mutagenic agent for *Streptomyces* is UV light. Since streptomycetes are in general pigmented, the UV irradiation of mycelium is not very efficient, and a high UV irradiation is required to achieve a killing rate of more than 99%. UV mutagenesis of *Streptomyces* is done by short-wavelength (254 nm) UV irradiation of spores. A spore suspension is placed in a petri dish on a magnetic stirrer in a darkroom. After UV irradiation, aliquots of the mutagenized spore suspension are taken, and appropriate dilutions are plated to determine the killing ratio. Irradiation times that result in killing rates of 99 to 99.9% are used for the mutagenesis (41).

30.4.2. Directed Gene Inactivation (Knockout Experiments)

Targeted gene inactivation is of particular importance for the analysis of complex pathways, e.g., antibiotic production to identify the different genes and to elucidate their function.

30.4.2.1. Gene Inactivation Strategies

To inactivate genes by gene disruption, internal gene fragments that lack translational start and stop sites are cloned into plasmids or phages. Recombination between the defective, plasmid-encoded copy and the intact chromosomal copy integrates the whole plasmid into the chromosome (Fig. 2). This results in the presence of two defective copies of the gene, separated by vector sequences within the chromosome. One copy is truncated at the 5′ end. It lacks the promoter region, the ribosome binding site, the start codon, and a region encoding the N-terminal amino acids. Therefore, it is very unlikely that any gene product is synthesized from this defective copy. The second copy possesses the original promoter and lacks only a fragment at the 3′ end. From this copy a C-terminally truncated protein is synthesized. One has to keep in mind that this truncated copy might still possess some residual activity, at least in the case of large multifunctional proteins. Since gene disruption mutants carry a duplication of homologous fragments in the chromosome, they are not completely stable. Recombination between the two copies results in the excision of the gene disruption plasmid and in the restoration of the chromosomal gene. Therefore, it is important to maintain selection for the integrated plasmid.

Completely stable mutants can be generated only via gene replacement, i.e., the exchange of a part or the complete chromosomal gene by an antibiotic resistance cassette or a defective copy. Since streptomycetes possess particularly efficient recombination systems, knockout experiments requiring two crossover events are very efficient. A counter-selection system for the direct selection of vector loss and gene replacement in *Streptomyces* has been described (3, 44). This system used the *rpsL* gene, encoding the ribosomal protein that can determine resistance against streptomycin. The streptomycin-sensitive wild-type allele is dominant over the resistant one. If the streptomycin-sensitive *rpsL* variant is introduced on a plasmid into a host carrying the resistance-determining allele, the resulting strain cannot survive streptomycin treatment. Therefore, growth on streptomycin-containing medium is a strong selection for the loss of the plasmid (and for the plasmid-encoded streptomycin sensitivity determinant). Using this counter-selection system, even rare double-crossover events

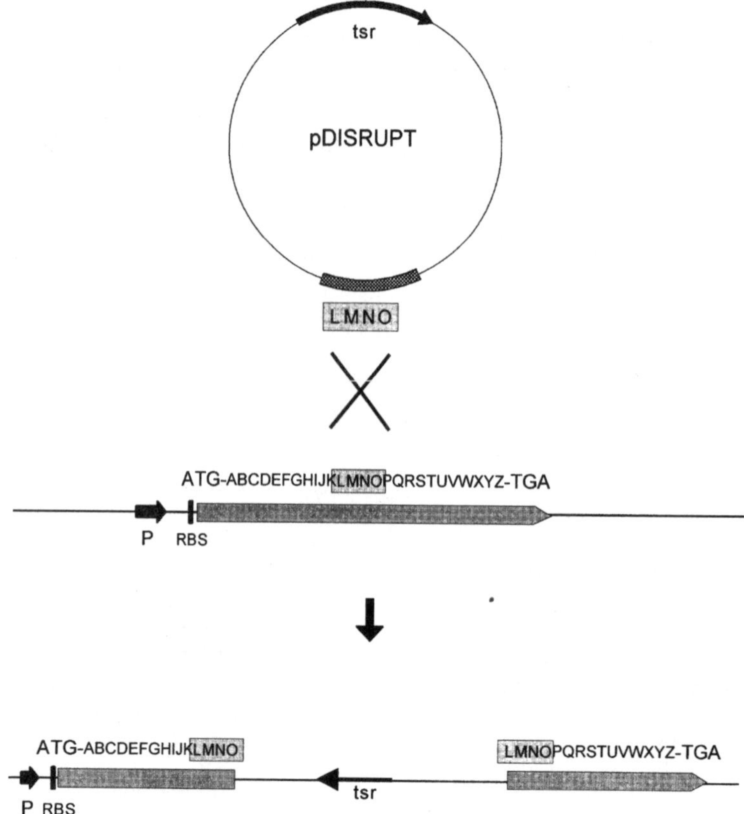

FIGURE 2 Generation of mutants expressing C-terminal truncated proteins by gene disruption. Recombination between the intact chromosomal gene and the defective plasmid-borne copy as described in the text.

resulting in the replacement of DNA fragments can easily be selected.

30.4.2.2 Suicide Vectors for Gene Inactivation

For convenient DNA transfer and the subsequent selection of recombination events resulting in the disruption or replacement of chromosomal genes, several suicide vector systems have been established for streptomycetes (Table 1).

Nonreplicative Plasmids

Each *E. coli* vector carrying a marker gene that can be selected in *Streptomyces* can be used as a suicide vector. However, the success of such experiments is often restricted by the poor transformation efficiencies caused by the potent restriction systems of *Streptomyces* strains. To overcome this restriction barrier, several procedures have been developed:

TABLE 1 Suicide delivery systems for mutagenesis

Vector	*Streptomyces* replicon	Suicide system	Comments	Reference
pDH5		Nonreplicative	f1 origin, *tsr*, *bla*, *lacZ*	37
pWHM3	pIJ101	Unstable, minus origin	Shuttle vector, *tsr*, *bla*	103
pIJ4680	SCP2*	Unstable, *par*$^-$	Shuttle vector, *tsr*, *bla*	51
pMT660	pIJ101	Rep^{ts}	*ts* replication mutant, *tsr*, *mel*	11
pGM11	pSG5	Rep^{ts}	fd terminator to stop transcription, *aphII*	111
pGM160	pSG5	Rep^{ts}	Shuttle vector, *tsr*, *aacC1*	75
pRHB538	pSG5	Rep^{ts}, *rpsL* selection	Shuttle vector, positive selection, *acc(3)IV*	3
KC860	ϕC31	Phage, Δ*att*	Shuttle vector, *tsr*, *vph*, *xylE* fusion	16
pKC1139	pSG5	Rep^{ts}	Shuttle vector, *aac(3)IV*, *lacZ*	10
pOJ446	SCP2*	Unstable, *par*	Shuttle vector, cosmid, *aac(3)IV*	10
KC857	ϕC31	Phage, Δ*att*	Shuttle vector, *tsr*, *vph*, *bla* fd terminator to stop transcription	16

1. DNA preparation from methylation-deficient *E. coli* strains, such as *E. coli* ET12567 (63).
2. Denaturation of the plasmid DNA to generate single-stranded DNA, which is a poor substrate for restriction enzymes.
3. Introduction of the intergenic region of the *E. coli* phage f1 into a commonly used *E. coli* cloning vector. Using a helper phage, the plasmid DNA can be isolated as a single strand. Single-stranded (ss) plasmid DNA for the transformation and subsequent integration into the chromosome is up to 100 times more effective than double-stranded DNA (37).
4. Introduction of a suicide vector by intergeneric conjugation from *E. coli* (36).

An additional advantage of the introduction of ss DNA is the fact that it is highly recombinogenic and may induce an SOS response that enhances *recA* expression.

Phage Derivatives

Integration vectors have also been developed from the actinophage ϕC31. Derivatives that lack the *att* site can integrate via homologous recombination only if they share homologous DNA fragments with the chromosome (19). A variety of ϕC31 derivatives have been engineered for gene disruption and replacement experiments (Table 1).

Unstable Replicative Plasmids

Two different plasmids have successfully been used to select for recombination events:

1. Derivatives of the low-copy plasmid SCP2* that lack the stability region. Plasmid pIJ4680 lacking the SCP2* stability region (*par*) is not stably inherited to the progeny and therefore disappears from the host cell at high frequency (51). The function of the *par* region in stabilizing SCP2* is unknown so far.
2. Plasmid pWHM3 (103) is a pIJ101 derivative lacking the minus origin and fused to the *E. coli* vector pUC19. Very often, shuttle plasmids are unstable in *Streptomyces* and are maintained only under selection. In particular, plasmids replicating via the rolling-circle mode are lost at high frequency if the minus origin for the initiation of the lagging strand synthesis was deleted.

Temperature-Sensitive Plasmids

ts plasmids represent the most successful suicide vector systems for *Streptomyces*. They are very efficiently eliminated by shifting the incubation temperature. Since *ts* plasmids can be maintained stably at permissive temperature, it is possible to provide sufficient time for recombination to occur. Therefore, even rare events can be selected. There are two different *ts* plasmid systems available for *Streptomyces*:

1. pMT660 (11) is a *ts* mutant of pIJ702 (50). pMT660 replicates stably at temperatures below 39°C.
2. Based on the naturally temperature-sensitive plasmid pSG5 from *S. ghanaensis* DSM2932 (76), several widely used vectors with *ts* replication were developed (75, 111). Plasmid pSG5 replicates stably at temperatures below 35°C. At elevated temperatures (>37°C) replication of pSG5 is inhibited, and the remaining plasmid copies are diluted out during growth (86).

The best way to eliminate autonomously replicating plasmids is to incubate a nonselective liquid culture on a rotary shaker at 37 to 39°C for 3 days. Following homogenizing, appropriate dilutions of the mycelial fragments are plated to selective media and incubated at 37 to 39°C to select integration events into the chromosome. In general, integration rates of 29 to 90% (titer on selective medium in relation to the titer on nonselective medium; insert size, 450 bp) are readily achieved.

Most of the small pGM plasmids lack the minus origin and therefore accumulate ss plasmid molecules. This accumulation of ss DNA might be one of the reasons that the pGM vectors are so effective in the generation of integration mutants. In particular, for the generation of gene replacement mutants, which require two crossover events, the pGM plasmids proved to be more efficient than other suicide vector systems (111). Following the integration of the plasmid into the chromosome, the frequency of the second crossover resulting in the replacement was analyzed by Khosla et al. (51). Compared to a SCP2* derivative, the second recombination event occurred about 10 times more often when using a pSG5-based delivery vector (51).

30.4.3. Transposon Mutagenesis

30.4.3.1 *Streptomyces* Transposons

Only recently have useful transposon mutagenesis systems for *Streptomyces* emerged. Among several known transposable elements from streptomycetes, the transposon Tn*4556* (21) and the insertion element IS*493* of *S. lividans* (97) and their derivatives have been successfully used as genetic tools for transposon mutagenesis in *Streptomyces*.

The Tn*3*-type transposon Tn*4556* of *Streptomyces fradiae* was discovered by Chung (21). Tn*4556* was modified by inserting a viomycin resistance gene (*vph*, viomycin phosphotransferase) as a selectable marker to create Tn*4560* and Tn*4563* (22, 89). Tn*4560* has been used in several experiments to mutagenize DNA clones on the plasmid SCP2* (24), to localize loci involved in plasmid transfer of the SCP2* derivative pIJ903 (15), and to identify biosynthetic genes for the antibiotic avermectin (45).

The insertion element IS*493* of *S. lividans* was engineered by inserting the apramycin resistance gene (95) or the hygromycin resistance gene (*hph*, hygromycin phosphotransferase [35]), resulting in Tn*5096* or Tn*5099*, respectively. Tn*5099* has already proved its use in identifying genes involved in the biosynthesis of the antibiotic daptomycin in *Streptomyces roseosporus* (68). Hypertransposing Tn*5099* derivatives, such as Tn*5099-10*, were discovered by Solenberg and Baltz (96). Tn*5099-10* carries a spontaneous deletion close to its left inverted repeat, which results somehow in a 1,000-fold higher transposition frequency in *Streptomyces griseofuscus* and a 10-fold higher frequency in *S. fradiae*, compared to the parental transposon.

Both Tn*4556* and IS*493* derivatives transpose in different *Streptomyces* strains at sufficient frequencies to identify desired genes by insertional mutagenesis. Furthermore, their insertion loci were shown to be random by several researchers (15, 95).

Volff and Altenbuchner (106a) established a transposon mutagenesis system for *S. lividans* by using a mini-Tn*5* transposon derivative on a *ts* replicon. By exchanging the native transposase promoter against the *S. lividans mer* promoter, transposition of the mini-Tn*5* derivative Tn*5493* was achieved in *S. lividans* with a frequency as high as 3% (106a).

TABLE 2. Available transposable elements for mutagenizing streptomycetes

Transposon	Features	Reference
IS*493* derivatives		
Tn*5096*	IS*493*::Tn*acc(3)IV* (Am^r)	95
Tn*5099*	IS*493*::*xylE*::*hyg*; promoter probe transposon, promoterless *xylE* gene, Hm^r	35
Tn*5099-10*	IS*493*::*xylE*::*hyg*; hypertransposing TN5099 derivative, lacking the promoterless *xylE* gene, Hm^r	96
Tn*4556* derivatives		
Tn*4560*	Tn*4556*::*vph*; Vm^r	22
Tn*5351*	Tn*4556*::*luxAB*::*vph*; promoter probe transposon, promoterless *luxAB* gene, Vm^r	94a
Tn*5353*	Tn*4556*::*luxAB*::*aph*; promoter probe transposon, promoterless *luxAB* gene, Neo^r	94a
Tn*5* derivative		
Tn*5493*	*tsr* gene flanked by the IS50 repeats, *tnpA* under control of the *S. lividans mer* promoter	106a

Table 2 summarizes available transposable elements. A general scheme for mutagenizing chromosomal DNA using *Streptomyces* transposons is shown in Fig. 3.

30.4.3.2. Transposon Delivery Systems

Both of the two known types of *ts* plasmids in *Streptomyces* mentioned above have been used as delivery systems for Tn*4556* and IS*493* derivatives. McHenney and Baltz (67) used the pIJ101 derivative pMT660 to mutagenize several *Streptomyces* strains with the IS*493* derivative Tn*5096*. pCZA213 (96), an example of a mobilizable delivery vector based on pGM160 (75), is shown in Fig. 4.

Another procedure to introduce transposons in *Streptomyces* was shown by Solenberg and Baltz (95) with Tn*5097* (another IS*493* derivative), which transposed from pUC19 into the chromosome of *S. griseofuscus*. The hypertransposing Tn*5099-10* was shown to transpose from the pUC118 derivative pWWB11 into the *S. coelicolor* M124 chromosome (14), thus enabling the isolation of Tn*5099-10* mutants of this strain.

Brolle et al. (15) designed an *S. lividans* 1326 strain (designated *S. lividans* TK503) carrying a copy of Tn*4560* introduced into its chromosome. This strain might be particularly useful to mutagenize DNA cloned on SCP2* de-

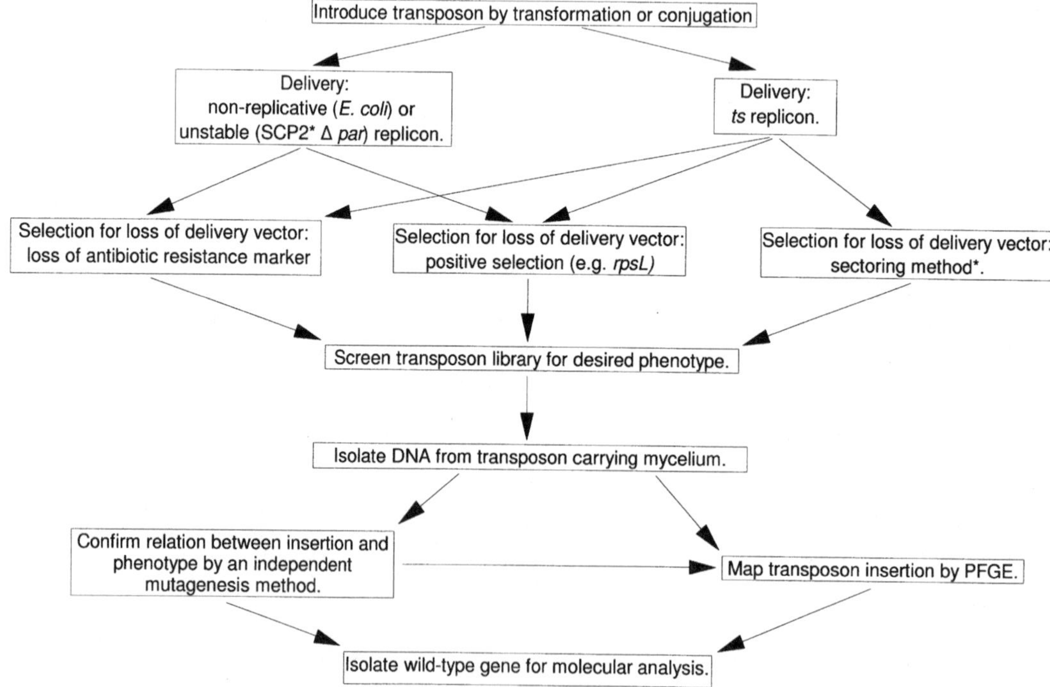

FIGURE 3 Transposon mutagenesis of streptomycetes. Sectoring method (95): colonies carrying a *ts*-delivery vector as pCZA213 (see Fig. 4) were plated on agar containing hygromycin and grown at 29°C, followed by a temperature shift to 39°C to eliminate the delivery plasmid. Sectors developing at the edges of the small colonies were hygromycin resistant. Each sector, even from the same colony, should have the transposon inserted into a different chromosomal sequence.

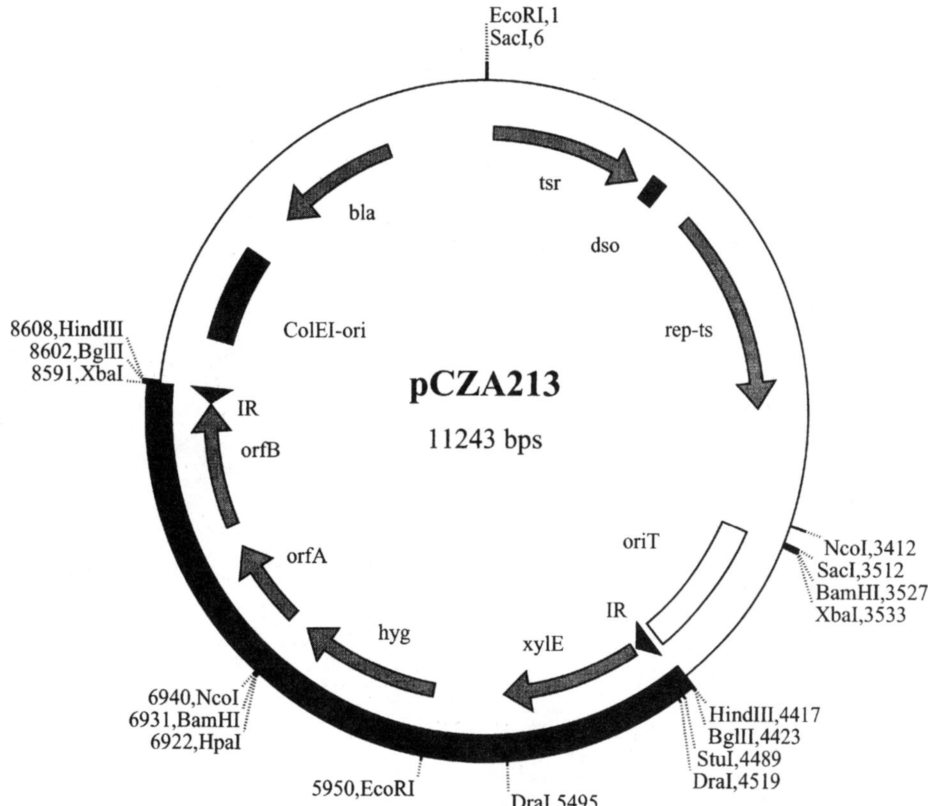

FIGURE 4 Transposon delivery vector pCZA213. pCZA213 (96) carries transposon Tn*5099* on a mobilizable pGM160 derivative. pCZA213 can be transferred to *Streptomyces* by intergeneric conjugation. Transposition into the chromosome results in the fusion of the promoterless *xylE* gene to chromosomal sequences, allowing not only the inactivation of a gene but also the detection of promoter activity. tsr, thiostrepton resistance gene; dso, double stranded origin; rep-ts, *rep* gene of pSG5; oriT, origin of transfer and *traJ* gene from RK2; IR, inverted repeats; xylE, promoterless catechol 2,3-dioxygenase encoding gene; hyg, hygromycin phosphotransferase gene; orfA, orfB, open reading frames of IS*493*; bla, β-lactamase gene. Tn*5099* is indicated by a solid bar.

rivatives, because high transposition frequencies were detected from the chromosome into the SCP2* derivative pIJ903.

30.5. GENOME SEQUENCING AND MAPPING

30.5.1. Genome/Gene Mapping

S. coelicolor A3(2) is the best-characterized *Streptomyces* species and is a typical strain of this genus. By classical genetic means, a genetic map comprising about 70 loci was constructed in 1967 by Hopwood (40) and subsequently refined. With the development of pulsed-field gel electrophoresis (PFGE), a combined physical and genetic map (Fig. 5) could be established (53). It was the basis for constructing an ordered cosmid bank consisting of 319 overlapping cosmids with only three short gaps (83). By hybridization, more than 170 genes were mapped to specific cosmids. These data are currently updated and available on the Internet (http://www.uni-kl.de/FB-Biologie/Streptomyces.genome/map.html). The cosmids containing wild-type DNA can also be used to complement newly isolated mutants and to rapidly isolate the complementing genes.

The *S. coelicolor* genome project began at the Sanger Centre with the sequencing of overlapping cosmids. The preliminary sequence data are available via ftp (ftp://ftp.sanger.ac.uk/pub/S_coelicolor/sequences/). Furthermore, the *S. coelicolor* Blast server (http://www.sanger.ac.uk/Projects/S_coelicolor/blast_server.shtml) can be searched by submitting a query sequence.

The use of PFGE for other *Streptomyces* strains may be hindered by the occurrence of Tris-dependent strand sissions during electrophoresis, as observed for *S. lividans* and *Streptomyces avermitilis* (82). However, this problem can be solved by using a HEPES buffer (16 mM HEPES-NaOH [pH 7.5], 16 mM sodium acetate, 0.8 mM EDTA) as shown for *S. lividans* (29). All chromosomes of *Streptomyces* species analyzed by PFGE contain a linear chromosome (58, 59) whose ends consist of long inverted repeats with proteins attached to the 5′ ends (57). The well-known genetic instability of many *Streptomyces* species, resulting in deletions, sometimes amplifications, and chromosomal circularization (20), can be attributed to this structure.

For physical mapping, restriction endonucleases, such as *AseI*, *AsnI*, *DraI*, and *SspI* (57, 68, 83), are routinely used

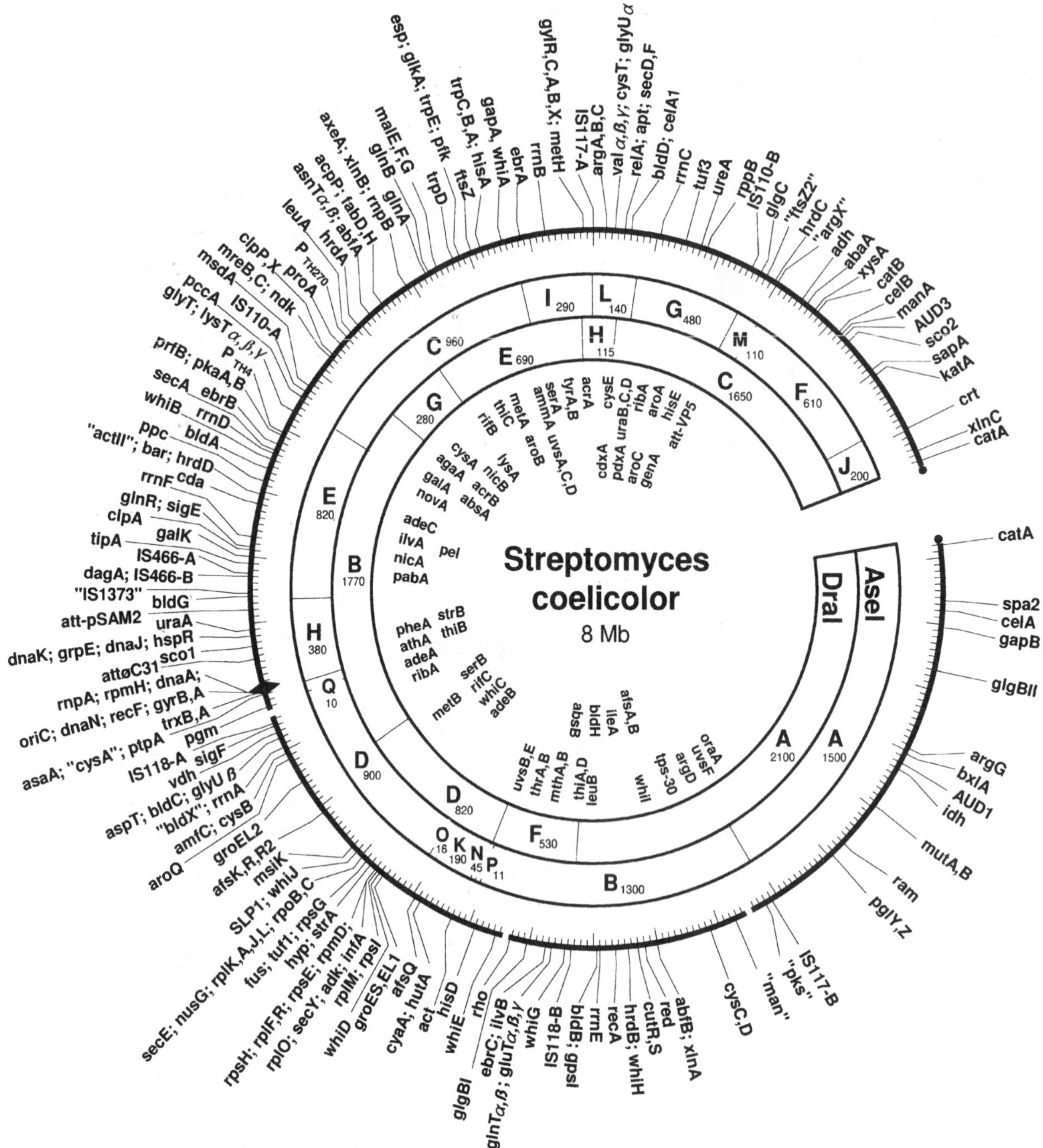

FIGURE 5 Combined genetic and physical map of the *S. coelicolor* M145 chromosome. The positions of markers on the outside of the circle come from their locations on the cosmid contig. Markers inside the circle have been mapped only genetically. Sizes of the *Ase*I and *Dra*I fragments are in kb. ●, chromosomal telomeres. ▲, *ori*C. (The map is reprinted from reference 83.)

to cut the *Streptomyces* genome into 10 to 30 pieces. The transposon Tn*5099* (35) contains such restriction sites (*Asn*I, *Dra*I). Therefore, the location of transposon-mutated genes can easily be mapped by PFGE.

30.5.2. DNA Sequencing

Enzymatic sequencing (87) has been used to determine many DNA sequences in both gram-negative and gram-positive bacteria. However, sequencing projects of *Streptomyces* DNA had to overcome a major experimental problem caused by the high G + C content of up to 77 mol%. Autoradiograms of sequencing gels very frequently show bands in all four lanes caused by unspecific termination of the sequencing reaction or compressions within sequencing ladders due to secondary structures developing during gel electrophoresis. 7-Deaza GTP and highly processive enzymes can be used in sequencing protocols, helping to overcome some unspecific terminations and compressions. Additionally, use of ss DNA rather than denatured CCC DNA seems to be helpful in several cases. However, to determine the correct sequence, three main procedures have been used by several researchers:

1. DNA sequencing at higher temperature using a *Taq* polymerase instead of a T7 DNA polymerase
2. Cycle sequencing at higher temperatures
3. Use of specific oligonucleotides priming sequencing reactions approximately 50 to 100 bases upstream of areas of the gel compression

Not all compressions or unspecific terminations can be dissolved by either method, and often combinations of the three procedures have to be used.

In recent years many laboratories have started using automated sequencing techniques rather than manual sequencing. Automatic sequencing has already proved useful for sequencing of *Streptomyces* DNA in many cases, with average readings of ~400 bases per reaction.

The identification of open reading frames actually encoding a gene product is again greatly facilitated in *Streptomyces* by means of the high G + C content. Algorithms that consider the *Streptomyces* codon usage (e.g., reference 98) and/or the typical distribution of bases (72 mol% G + C in the first, 49 mol% G + C in the second, and 94 mol% G + C in the third position of each codon) (5) can clearly indicate those open reading frames that are transcribed and translated. Furthermore, sequencing errors resulting in frameshift mutations can easily be detected. Even if only parts of a gene have been sequenced, an unequivocal prediction is possible in most cases. This can be exploited in the analysis of large gene clusters. If one gene of a cluster has been isolated, other genes of interest can be identified by random sequencing of a large piece of DNA (e.g., cosmids) that includes the isolated gene. Sequence data of short pieces (shotgun-cloned in suitable vectors) will then allow a rapid detection of genes with specific functions without time-consuming, detailed, and expensive DNA sequencing.

30.6. EXPRESSION/SECRETION SYSTEMS

30.6.1. Expression Systems for *Streptomyces* Genes

Streptomycetes exhibit a high degree of promoter sequence heterogeneity (12, 100). This is in agreement with the occurrence of at least eight different factors identified in *S. coelicolor* (81 and citations therein). Transcription of many *Streptomyces* genes is initiated at different promoters. It is therefore not surprising that only a few *Streptomyces* genes could be expressed from their own promoters in foreign hosts such as *E. coli*. Expression in *E. coli* is therefore normally obtained by transcriptional fusions to *E. coli* promoters subsequently using the *E. coli* expression systems. However, there is one major problem caused by the fact that *Streptomyces* codons contain almost exclusively G or C in the third position (112). Since some of these codons are rather rare in *E. coli*, translation may be obstructed, thus leading to a low, sometimes undetectable expression level. One possibility to overcome this is to exchange the codons of the N-terminal amino acids with typical *E. coli* codons, as shown by Gramajo et al. (32), who thus could increase the level of proteins using the T7 RNA polymerase-dependent pT7-7 expression system.

For expression in *Streptomyces*, either strong constitutive promoters or inducible promoters are available. The most frequently used promoters are those from the aminoglycoside phosphotransferase gene (*aph* = *aphI*) from *S. fradiae* (102), or that (*aphII* = *neo*) from the *E. coli* transposon Tn*5* (47), the erythromycin resistance gene (*ermE*) from *Saccharopolyspora erythraea* (8) or an *ermE* up-mutant (7). In addition, strong promoters have been isolated from actinophages (e.g., reference 55) and characterized as strong and suitable expression signals.

In recent years, adjustable expression systems for *Streptomyces* have been described. They mainly make use of the thiostrepton-inducible *tipA* promoter originating from the chromosomal *tipA* gene of *S. lividans*, which is apparently induced at least 200-fold by low levels of the antibiotic thiostrepton (71). This promoter is available on integrative (94) and autonomously replicating (101) vectors. For successive protein purification, His-tag fusion signals were combined with the *tipA* promoter sequences (28), multiple cloning sites, and nucleotide sequences encoding a protease cleavage site (101).

Gene expression using the *tipA* vectors is not completely independent of chromosomal genes (38), and their application is limited to certain *Streptomyces* strains (e.g., *S. lividans*, *S. coelicolor*). Therefore, a temperature-inducible expression system based on a plasmid gene has been developed (49). It consists of a promoter regulating the transfer operon of plasmid pSN22 and the corresponding repressor gene (*traR*), which was mutated to temperature sensitivity. Most of the expression vectors include termination signals in front of the (regulatable) promoter and/or behind the multiple cloning site in which the gene to be expressed is inserted. As termination signals, synthetic sequences (90) or terminators from phages such as λ or fd (94, 101) were used. Whether the presence of the terminator signals in the vector can really improve the expression level in streptomycetes has not been shown (90).

30.6.2. *Streptomyces* as a Host for Protein Secretion

To achieve secretion of proteins, N-terminal excretion signals of *Streptomyces* exoenzymes have been ligated to promoter sequences (reviewed in reference 27). Different secretion signals that all exhibit a similar structure (27) have been used in these experiments. By site-directed mutagenesis, the net charge (naturally +3) of the N-terminus of the leader of the α-amylase inhibitor tendamistat from *S. tendae* has been modified, resulting in a variation of excretion. Introduction of additional positive charges in the first seven amino acids significantly decreased the amount of secreted tendamistat, whereas a charge reduction to +2 resulted in the doubling of α-amylase production (27, 30). However, no generally useful secretion system has been developed, and the suitability for the expression of genes strongly depends on the individual gene to be expressed, particularly if the expressed gene does not originate from *Streptomyces* (1, 27).

The yield of protein is also influenced by the expression of proteases, of which only some have been analyzed. Even *S. lividans*, known as a relatively protease-poor strain, contains at least seven different protease activities (2, 3), some of which are partially characterized (13). Until now, the construction of protease-free expression hosts, such as those described for *B. subtilis*, has not been reported for *Streptomyces*, although inactivation of individual protease genes has been shown to increase the stability of excreted proteins (85).

Research work in the department of the authors was supported by BMFT (0310814) and DFG (SFB323) to W.W. by DFG (Mu1219/1-2) to G.M., and by DFG (Br1558/2-1 and Br1558/2-2) to D.F.B. We are grateful to D. A. Hopwood and M. J. Bibb for permission to include their figures in this chapter.

REFERENCES

1. **Anné, J., and L. VanMellaert.** 1993. *Streptomyces lividans* as host for heterologous protein production. *FEMS Microbiol. Lett.* **114:**121–128.
2. **Aretz, W., K. P. Koller, and G. Riess.** 1989. Proteolytic enzymes from recombinant *Streptomyces lividans* TK24. *FEMS Microbiol. Lett.* **65:**31–36.
3. **Baltz, R. H., and T. J. Hosted.** 1996. Molecular genetic methods for improving secondary-metabolite production in actinomycetes. *Trends Biotechnol.* **14:**245–249.
4. **Behrmann, I., D. Hillemann, A. Pühler, E. Strauch, and W. Wohlleben.** 1990. Overexpression of a *Streptomyces viridochromogenes* gene (*glnII*) encoding a glutamine synthetase similar to those of eucaryotes confers resistance against the antibiotic phosphinothricyl-alanyl-alanine. *J. Bacteriol.* **172:**5236–5334.
5. **Bibb, M. J., P. R. Findlay, and M. W. Johnson.** 1984. The relationship between base composition and codon usage in bacterial genes and its use in the simple and reliable identification of protein coding sequences. *Gene* **30:**157–166.
6. **Bibb, M. J., R. F. Freeman, and D. A. Hopwood.** 1977. Physical and genetical characterisation of a second sex factor, SCP2, for *Streptomyces coelicolor* A3(2). *Mol. Gen. Genet.* **154:**155–166.
7. **Bibb, M. J., and G. R. Janssen.** 1986. Unusual features of transcription and translation of antibiotic resistance genes in antibiotic-producing *Streptomyces*, p. 309–318. *In* M. Alacevic, D. Hranueli, and Z. Toman (ed.), *Fifth International Symposium on the Genetics of Industrial Microorganisms*. Ognjen Prica Printing Works, Karlovac.
8. **Bibb, M. J., G. R. Janssen, and J. M. Ward.** 1985. Cloning and analysis of the promoter region of the erythromycin-resistance gene (*ermE*) of *Streptomyces erythraeus*. *Gene* **38:**E375–E368.
9. **Bibb, M. J., M. J. Ward, and D. A. Hopwood.** 1978. Transformation of plasmid DNA into *Streptomyces* at high frequency. *Nature* **284:**526–531.
10. **Bierman, M., R. Logan, K. O'Brien, E. T. Seno, R. Nagaraja-Rao, and B. E. Schoner.** 1992. Plasmid cloning vectors for the conjugal transfer of DNA from *Escherichia coli* to *Streptomyces* spp. *Gene* **116:**43–49.
11. **Birch, A. W., and J. Cullum.** 1985. Temperature-sensitive mutants of the *Streptomyces* plasmid pIJ702. *J. Gen. Microbiol.* **131:**1299–1303.
12. **Bourn, W. R., and B. Babb.** 1995. Computer assisted identification and classification of streptomycete promoters. *Nucleic Acids Res.* **23:**3696–3703.
13. **Brawner, M. E.** 1994. Advances in heterologous gene expression by *Streptomyces*. *Curr. Opin. Biotechnol.* **5:**474–481.
14. **Brolle, D. F., and T. Henzler.** 1996. Unpublished data.
15. **Brolle, D. F., H. Pape, D. A. Hopwood, and T. Kieser.** 1993. Analysis of the transfer region of the *Streptomyces* plasmid SCP2*. *Mol. Microbiol.* **10:**157–170.
16. **Bruton, C. J., E. P. Guthrie, and K. F. Chater.** 1991. Phage vectors that allow monitoring of transcription of secondary metabolism genes in *Streptomyces*. *Bio/Technology* **9:**652–656.

16a.**Burke, J., D. Schneider, and J. Westpheling.** 1998. Generalized transduction in *Streptomyces* species. Poster Abstract. Society for General Microbiology, 141st Ordinary Meeting, Norwich, U.K.

17. **Chater, K. F.** 1986. *Streptomyces* phages and their applications to *Streptomyces* genetics, p. 119–158. *In* S. W. Queener and L. E. Day (ed.), *The Bacteria. A Treatise on Structure and Function*, vol. IX. Academic Press, Orlando, Fla.
18. **Chater, K. F.** 1993. Genetics of differentiation in *Streptomyces*. *Annu. Rev. Mircrobiol.* **47:**685–713.
19. **Chater, K. F., and C. J. Bruton.** 1983. Mutational cloning in *Streptomyces* and the isolation of antibiotic production genes. *Gene* **26:**67–78.
20. **Chen, C. W., Y.-S. Lin, Y.-L. Yang, M.-F. Tsou, H.-M. Chang, H. M. Kieser, and D. A. Hopwood.** 1994. The linear chromosomes of *Steptomyces*: structure and dynamics. *Actinomycetologica* **8:**103–112.
21. **Chung, S.-T.** 1987. Tn*4556*, a 6.8-kilobase-pair transposable element from *Streptomyces fradiae*. *J. Bacteriol.* **169:**4436–4441.
22. **Chung, S.-T., and L. L. Crose.** 1989. *Streptomyces* transposon Tn*4556* and its applications, p. 168–175. *In* C. L. Hershberger, S. W. Queener, and G. Hegeman (ed.), *Genetics and Molecular Biology of Industrial Microorganisms*. American Society for Microbiology Press, Washington, D. C.
23. **Crameri, A., E. A. Whitehorn, and W. P. C. Stemmer.** 1996. Improved Green Fluorescent Protein by molecular evolution using DNA shuffling. *Nature Biotechnol.* **14:**315–319.
24. **Davis, N. K., and K. F. Chater.** 1990. Spore colour in *Streptomyces coelicolor* A3(2) involves the developmentally regulated synthesis of a compound biosynthetically related to polyketide antibiotics. *Mol. Microbiol.* **4:**1679–1691.
25. **Decker, H., S. Gaisser, S. Pelzer, P. Schneider, L. Westrich, W. Wohlleben, and A. Bechthold.** 1996. A general approach for cloning and characterization dNDP-glucose dehydratase genes from actinomycetes. *FEMS Microbiol. Lett.* **141:**195–201.
26. **Embley, T. M., and E. Stackebrandt.** 1994. The molecular phylogeny and systematics of the actinomycetes. *Annu. Rev. Microbiol.* **48:**257–289.
27. **Engels, J. W., and K.-P. Koller.** 1992. Gene expression and secretion of eucaryotic foreign proteins in *Streptomyces*, p. 31–53. *In* J. A. H. Murray (ed.), *Transgenesis, Application of Gene Transfer*. John Wiley & Sons, Inc., New York.
28. **Enguita, F. J., J. L. de la Fuente, J. F. Martín, and P. Liras.** 1996. An inducible expression system of histidine-tagged proteins in *Streptomyces lividans* for one-step purification by Ni^{2+} affinity chromotography. *FEMS Microbiol. Lett.* **137:**135–140.
29. **Evans, M., and P. Dyson.** 1993. Pulsed-field electrophoresis of *Streptomyces lividans* DNA. *Trends Genet.* **9:**72.

30. **Fass, S. H., and J. W. Engels.** 1996. Influence of specific signal peptide mutations on the expression and secretion of the α-amylase inhibitor tendamistat in *Streptomyces lividans*. *J. Biol. Chem.* **271:**15244–15252.

30a. **Flett, F., V. Mersinias, and C. P. Smith.** 1997. High efficiency intergeneric conjugal transfer of plasmid DNA from *Escherichia coli* to methyl DNA-restricting streptomycetes. *FEMS Microbiol. Lett.* **155:**223–229.

31. **Giebelhaus, L. A., L. Frost, E. Lanka, E. P. Gormley, J. E. Davies, and B. Leskiw.** 1996. The Tra2 core of the incP plasmid RP4 is required for the intergeneric mating between *Escherichia coli* and *Streptomyces lividans*. *J. Bacteriol.* **178:**6378–6381.

32. **Gramajo, H. C., J. White, C. R. Hutchinson, and M. J. Bibb.** 1991. Overproduction and localization of components of the polyketide synthase of *Streptomyces glaucescens* involved in the production of the antibiotic tetracenomycin C. *J. Bacteriol.* **173:**6475–6483.

33. **Green, B., and J. Westpheling.** 1996. Generalized transduction of inserted chromosomal markers by the *Streptomyces venezuelae* phage SV1. Abstr. 6th GMBIM Conference, Bloomington, Ind.

34. **Guiney, D. G., and E. Yacobson.** 1983. Location and nucleotide sequence of the transfer origin of the broad host range plasmid RK2. *Proc. Natl. Acad. Sci. USA* **80:**3595–3598.

35. **Hahn, D. R., P. J. Solenberg, and R. H. Baltz.** 1991. Tn*5099*, a *xylE* promoter probe transposon for *Streptomyces* spp. *J. Bacteriol.* **173:**5573–5577.

36. **Hillemann, D., T. Dammann, A. Hillemann, and W. Wohlleben.** 1993. Genetic and biochemical characterization of the two glutamine synthetases GS1 and GSII of the phosphinothricyl-alanyl-alanine producer *Streptomyces viridochromogenes* Tü494. *J. Gen. Microbiol.* **139:**1773–1783.

37. **Hillemann, D., A. Pühler, and W. Wohlleben.** 1991. Gene disruption and gene replacement in *Streptomyces* via single stranded DNA transformation of integration vectors. *Nucleic Acids Res.* **194:**727–731.

38. **Holmes, D. J., J. L. Caso, and C. T. Thompson.** 1993. Autogenous transcriptional activation of a thiostrepton induced gene in *Streptomyces lividans*. *EMBO J.* **12:**3183–3191.

39. **Holt, T. G., C. Change, C. Laurent-Winter, T. Murakami, J. I. Garrels, J. E. Davies, and C. T. Thompson.** 1992. Global changes in gene expression related to antibiotic synthesis in *Streptomyces hygroscopicus*. *Mol. Microbiol.* **6:**969–980.

40. **Hopwood, D. A.** 1967. Genetic analysis and genome structure in *Streptomyces coelicolor*. *Bacteriol. Rev.* **31:**373–403.

41. **Hopwood, D. A., M. J. Bibb, K. F. Chater, T. Kieser, C. J. Bruton, H. M. Kieser, D. J. Lydiate, C. P. Smith, J. M. Ward, and H. Schrempf.** 1985. *Genetic Manipulation of Streptomyces: A Laboratory Manual*. John Innes Foundation, Norwich, England.

42. **Hopwood, D. A., and T. Kieser.** 1993. Conjugative plasmids in *Streptomyces*, p. 293–312. *In* D. B. Clewell (ed.), *Bacterial Conjugation*. Plenum Press, New York.

43. **Hopwood, D. A., F. Malpartida, H. M. Kieser, H. Ikeda, J. Duncan, I. Fujii, B. A. M. Rudd, H. G. Floss, S. Ōmura.** 1985. Production of "hybrid" antibiotics by genetic engineering. *Nature* **314:**642–644.

44. **Hosted, T. J., and R. H. Baltz.** 1996. Mutants of *Streptomyces roseosporus* that express enhanced recombination within partially homologous genes. *Microbiology* **142:**2803–2813.

45. **Ikeda, H., Y. Takada, C.-H. Pang, H. Tanaka, and S. Ōmura.** 1993. Transposon mutagenesis by Tn*4560* and applications with avermectin-producing *Streptomyces avermitilis*. *J. Bacteriol.* **175:**2077–2082.

46. **Ingram, C., M. Brawner, P. Youngman, and J. Westpheling.** 1989. *xylE* functions as an efficient reporter gene in *Streptomyces* spp.: use for the study of *galP1*, a catabolite-controlled promoter. *J. Bacteriol.* **171:**6617–6624.

47. **Jorgensen, R. A., S. J. Rothstein, and W. S. Reznikoff.** 1979. A restriction enzyme cleavage map of Tn*5* and location of a region encoding neomycin resistance. *Mol. Gen. Genet.* **185:**223–238.

48. **Kämpfer, P., R. M. Kroppenstedt, and W. Dott.** 1991. A numerical classification of the genera *Streptomyces* and *Streptoverticillium* using miniaturized physiological tests. *J. Gen. Microbiol.* **137:**1831–1891.

49. **Kataoka, M., T. Tatsuta, I. Suzuki, S. Ksono, T. Seki, and T. Yoshida.** 1996. Development of a temperature-inducible expression system for *Streptomyces* spp. *J. Bacteriol.* **178:**5540–5542.

50. **Katz, E., C. J. Thompson, and D. A. Hopwood.** 1983. Cloning and expression of the tyrosinase gene from *Streptomyces antibioticus* in *Streptomyces lividans*. *J. Gen. Microbiol.* **129:**2703–2714.

51. **Khosla, C., S. Ebert-Khosla, and D. A. Hopwood.** 1992. Targeted gene replacements in a *Streptomyces* polyketide synthase gene cluster: role for the acyl carrier protein. *Mol. Microbiol.* **6:**3237–3249.

52. **Kieser, T., D. A. Hopwood, H. M. Wright, and C. J. Thompson.** 1982. pIJ101, a multi-copy broad host-range *Streptomyces* plasmid: functional analysis and development of DNA cloning vectors. *Mol. Gen. Genet.* **185:**223–238.

53. **Kieser, H. M., T. Kieser, and D. A. Hopwood.** 1992. A combined genetic and physical map of the *Streptomyces coelicolor* A3(2) chromosome. *J. Bacteriol.* **174:**5496–5507.

54. **Kieser, T., and R. E. Melton.** 1988. Plasmid pIJ699, a multi-copy positive-selection vector for *Streptomyces*. *Gene* **65:**83–91.

55. **Labes, G., M. J. Bibb, and W. Wohlleben.** 1997. Isolation and characterisation of a strong promoter element from the *Streptomyces ghanaensis* phage I19 using the gentamicin resistance gene (*aacC1*) of Tn*1696* as reporter. *Microbiology* **143:**1503–1512.

56. **Labes, G., R. Simon, and W. Wohlleben.** 1990. A rapid method for the analysis of plasmid content and copy number in various streptomycetes grown on agar plates. *Nucleic Acids Res.* **18:**2197.

57. **Leblond, P., G. Fischer, F.-X. Francou, F. Berger, M. Guérineau, and B. Decaris.** 1996. The unstable region of *Streptomyces ambofaciens* includes 210 kb terminal inverted repeats flanking the extremities of the linear chromosomal DNA. *Mol. Microbiol.* **19:**261–271.

58. **Leblond, P., M. Redenbach, and J. Cullum.** 1993. Physical map of *Streptomyces lividans* 66 chromosome and comparison with that of the related strain *Streptomyces coelicolor* A3(2). *J. Bacteriol.* **175:**3422–3429.

59. **Lin, Y.-S., H. M. Kieser, D. A. Hopwood, and C. W. Chen.** 1993. The chromosomal DNA of *Streptomyces lividans* 66 is linear. *Mol. Microbiol.* **10:**923–933.

60. **MacDaniel, R., S. Ebert-Khosla, D. A. Hopwood, and C. Khosla.** 1993. Engineered biosynthesis of novel polyketides. *Science* **262:**1546–1550.

61. **MacNeil, D. J.** 1987. Introduction of plasmid DNA into *Streptomyces lividans* by electroporation. *FEMS Microbiol. Lett.* **42:**239–244.

62. **MacNeil, D. J.** 1988. Characterization of a unique methyl-specific restriction system in *Streptomyces avermitilis*. *J. Bacteriol.* **170:**5607–5612.

63. **MacNeil, D. J., K. M. Gewain, C. L. Ruby, G. Dezeny, P. H. Gibbons, and T. MacNeil.** 1992. Analysis of *Streptomyces avermitilis* genes required for avermectin biosynthesis utilizing a novel integration vector. *Gene* **111:**61–68.

64. **Malpartida, F., S. E. Hallam, H. M. Kieser, H. Motamedi, C. R. Hutchinson, M. J. Butler, D. A. Sugden, M. Warren, C. McKillop, C. R. Bailey, G. O. Humphreys, and D. A. Hopwood.** 1987. Homology between *Streptomyces* genes coding for synthesis of different polyketides used to clone antibiotic biosynthetic genes. *Nature* **325:**818–821.

65. **Malpartida, F., and D. A. Hopwood.** 1984. Molecular cloning of the whole biosynthetic pathway of a *Streptomyces* antibiotic and its expression in a heterologous host. *Nature* **309:**462–464.

66. **Mazodier, P., R. Petter, and C. Thompson.** 1989. Intergeneric conjugation between *Escherichia coli* and *Streptomyces* species. *J. Bacteriol.* **171:**3583–3585.

67. **McHenney, M. A., and R. H. Baltz.** 1991. Transposition of Tn*5096* from a temperature-sensitive transducible plasmid in *Streptomyces* spp. *J. Bacteriol.* **173:**5578–5581.

68. **McHenney, M. A., and R. H. Baltz.** 1996. Gene transfer and transposition mutagenesis in *Streptomyces roseosporus*: mapping of insertions that influence daptomycin or pigment production. *Microbiology* **142:**2363–2373.

69. **Möhrle, V., U. Roos, and C. Bormann.** 1995. Identification of cellular proteins involved in nikkomycin production in *Streptomyces tendae* Tü901. *Mol. Microbiol.* **15:** 561–571.

70. **Motamedi, H., A. Shafiee, and S.-J. Cai.** 1995. Integrative vectors for heterologous gene expression in *Streptomyces* spp. *Gene* **160:**25–31.

71. **Murakami, T., T. G. Holt, and C. J. Thompson.** 1989. Thiostrepton-induced gene expression in *Streptomyces lividans*. *J. Bacteriol.* **171:**1459–1466.

72. **Muth, G.** 1996. Unpublished data.

73. **Muth, G., M. Farr, V. Hartmann, and W. Wohlleben.** 1995. *Streptomyces ghanaensis* plasmid pSG5: nucleotide sequence analysis of the self-transmissible minimal replicon and characterization of the replication mode. *Plasmid* **33:**113–126.

74. **Muth, G., D. Frese, A. Kleber, and W. Wohlleben.** 1997. Mutational analysis of the *Streptomyces lividans recA* gene suggests that only mutants with residual activity are viable. *Mol. Gen. Genet.* **255:**420–428.

75. **Muth, G., B. Nussbaumer, W. Wohlleben, and A. Pühler.** 1989. A vector system with temperature-sensitive replication for gene disruption and mutational cloning in streptomycetes. *Mol. Gen. Genet.* **219:**341–348.

76. **Muth, G., W. Wohlleben, and A. Pühler.** 1988. The minimal replicon of the *Streptomyces ghanaensis* plasmid pSG5 identified by subcloning and Tn*5* mutagenesis. *Mol. Gen. Genet.* **211:**424–429.

77. **Nussbaumer, B., and W. Wohlleben.** 1994. Identification, isolation and sequencing of the *recA* gene of *Streptomyces lividans*. *FEMS. Microbiol. Lett.* **118:**57–64.

78. **Oh, S. H., and K. F. Chater.** 1997. Denaturation of linear DNA facilitates targeted integrative transformation of *Streptomyces coelicolor* A3(2): possible relevance to other organisms. *J. Bacteriol.* **179:**122–127.

79. **Okanishi, M., K. Suzuki, and U. Umezawa.** 1974. Formation and reversion of streptomycete protoplasts: cultural conditions and morphological study. *J. Gen. Microbiol.* **80:**389–400.

80. **Pigac, J., and H. Schrempf.** 1995. A simple and rapid method of transformation of *Streptomyces rimosus* R6 and other streptomycetes by electroporation. *Appl. Environ. Microbiol.* **61:**352–356.

81. **Potúcková, L., G. H. Kelemen, K. C. Findlay, M. A. Lonetto, M. J. Buttner, and J. Kormanec.** 1995. A new RNA polymerase sigma factor, σ^F, is required for the late stages of morphological differentiation in *Streptomyces* spp. *Mol. Microbiol.* **17:**37–48.

82. **Ray, T., A. Mills, and P. Dyson.** 1995. Tris-dependent oxidative DNA strand sission during electrophoresis. *Electrophoresis* **16:**888–894.

83. **Redenbach, M., H. M. Kieser, D. Denapaite, A. Eichner, J. Cullum, H. Kinashi, and D. A. Hopwood.** 1996. A set of ordered cosmids and a detailed genetic and physical map for the 8 Mb *Streptomyces coelicolor* A3(2) chromosome. *Mol. Microbiol.* **21:**77–96.

84. **Richardson, M. A., S. Kuhstoss, P. Solenberg, N. A. Schaus, and R. Nagaraja Rao.** 1987. A new shuttle cosmid vector, pKC505, for streptomycetes: its use in the cloning of three different spiramycin-resistance genes from a *Streptomyces ambofaciens* library. *Gene* **61:**231–241.

85. **Rohling, A., and G. Muth.** Unpublished data.

86. **Roth, M., C. Hoffmeier, R. Geuther, G. Muth, and W. Wohlleben.** 1994. Segregational stability of pSG5-derived vector plasmids in continuous cultures of *Streptomyces lividans* 66. *Biotechnol. Lett.* **16:**1225–1230.

87. **Sanger, F., S. Miklen, and A. R. Coulson.** 1977. DNA sequencing with chain-termination inhibitors. *Proc. Natl. Acad. Sci. USA* **74:**5463–5467.

88. **Schauer, A., M. Ranes, R. Santamaria, J. Guijarro, E. Lawlor, C. Mendez, K. Chater, and R. Losick.** 1988. Visualizing gene expression in time and space in the filamentous bacterium *Streptomyces coelicolor*. *Science* **240:** 768–772.

89. **Schauer, A. T., A. D. Nelson, and J. B. Daniel.** 1991. Tn*4563* transposition and its application to isolation of new morphological mutants. *J. Bacteriol.* **173:**5060–5067.

90. **Schmitt-John, T., and J. W. Engels.** 1992. Promoter constructions for efficient secretion expression in *Streptomyces lividans*. *Appl. Microbiol. Biotechnol.* **36:**493–498.

91. **Schrempf, H., H. Bujard, D. A. Hopwood, and W. Goebell.** 1975. Isolation of covalently closed circular deoxyribonucleic acid from *Streptomyces coelicolor* A3(2). *J. Bacteriol.* **121:**416–421.

92. **Simon, R., U. Priefer, and A. Pühler.** 1983. A broad host range mobilization system for in vivo genetic engineering: transposon mutagenesis in gram negative bacteria. *Bio/Technology* **1:**784–791.

93. **Smith, C. P.** Personal communication.

94. **Smokvina, T., P. Mazodier, F. Boccard, C. J. Thompson, and M. Guérineau.** 1990. Construction of a series of pSAM2-based integrative vectors for use in actinomycetes. *Gene* **94:**53–59.

94a. **Sohaskey, C. D., H. Im, and A. T. Schauer.** 1992. Construction and application of plasmid and transposon-based promoter-probe vectors for *Streptomyces* spp. that employ a *Vibrio harveyi* luciferase reporter cassette. *J. Bacteriol.* **174:**367–376.

95. **Solenberg, P. J., and R. H. Baltz.** 1991. Transposition of Tn*5096* and other IS*493* derivatives in *Streptomyces griseofuscus*. *J. Bacteriol.* **173:**1096–1104.

96. **Solenberg, P. J., and R. H. Baltz.** 1994. Hypertransposing derivatives of the streptomycete insertion sequence IS*4943*. *Gene* **147:**47–54.

97. **Solenberg, P. J., and S. G. Burgett.** 1989. Method for selection of transposable DNA and characterization of a new insertion sequence, IS*493*, from *Streptomyces lividans*. *J. Bacteriol.* **171:**4807–4813.

98. **Staden, R., and A. D. McLachlan.** 1982. Codon preference and its use in identifying protein coding regions in large DNA sequences. *Nucleic Acids Res.* **10:**141–156.

99. **Stockmann, M., and W. Piepersberg.** 1992. Gene probes for the detection of 6-deoxyhexose metabolism in secondary metabolite-producing streptomycetes. *FEMS Microbiol. Lett.* **90:**185–190.

100. **Strohl, W. R.** 1992. Compilation and analysis of DNA sequences associated with apparent streptomycete promoters. *Nucleic Acids. Res.* **20:**961–974.

101. **Takano, E., J. White, C. T. Thompson, and M. J. Bibb.** 1995. Construction of thiostrepton-inducible, high-copy-number expression vectors for use in *Streptomyces* spp. *Gene* **166:**133–137.

102. **Thompson, C. T., and G. S. Gray.** 1983. Nucleotide sequence of a *Streptomyces* aminoglycoside phosphotransferase gene and its relationship to phosphotransferases encoded by resistance plasmids. *Proc. Natl. Acad. Sci. USA* **80:**5190–5194.

103. **Vara, J., M. Lewowska-Skarbek, Y. G. Wang, S. Donadio, and C. R. Hutchinson.** 1989. Cloning of genes governing the deoxysugar portion of the erythromycin biosynthesis pathway in *Saccharopolyspora erythrea* (*Streptomyces erythreus*). *J. Bacteriol.* **171:**5872–5881.

104. **Vats, S., C. Stuttard, and L. C. Vining.** 1987. Transductional analysis of chloramphenicol biosynthesis genes in *Streptomyces venezuelae*. *J. Bacteriol.* **169:**3809–3813.

105. **Vivian, A.** 1971. Genetic control of fertility in *Streptomyces coelicolor* A3 (2): plasmid involvement in the interconversion of UF and IF strains. *J. Gen. Microbiol.* **69:** 353–364.

106. **Volff, J. N., and J. Altenbuchner.** 1997. Influence of the disruption of the *recA* gene on genetic stability and genome rearrangements in *Streptomyces lividans*. *J. Bacteriol.* **179:**2440–2445.

106a. **Volff, J. N., and J. Altenbuchner.** 1997. High frequency transposition of the Tn*5* derivative Tn*5493* in *Streptomyces lividans*. *Gene* **194:**81–86.

107. **Vujaklija, D., and J. Davies.** 1995. Direct transfer of plasmid DNA between *Streptomyces* spp. and *E. coli* by electroduction. *J. Antibiot.* **48:**635–638.

108. **Ward, J. M., G. R. Jannsen, T. Kieser, and M. J. Bibb.** 1986. Construction and characterization of a series of multicopy promoter-probe plasmid vectors for *Streptomyces* using the aminoglycoside phosphotransferase gene from Tn*5* as indicator. *Mol. Gen. Genet.* **203:**468–478.

109. **Wilkins, B., and E. Lanka.** 1993. DNA processing and replication during plasmid transfer between Gram-negative bacteria, p. 105–136. *In* D. B. Clewell (ed.), *Bacterial Conjugation*. Plenum Press, New York.

110. **Wohlleben, W., W. Arnold, L. Bisonette, A. Pelletier, A. Tanguay, P. H. Roy, G. C. Gamboa, G. F. Barry, E. Aubert, J. Davies, and S. A. Kagan.** 1989. On the evolution of Tn*21*-like multiresistance transposons: sequence analysis of the gene (*aacC1*) for gentamicin acetyltransferase-3-I(AAC(3)-I), another member of the Tn*21*-based expression cassette. *Mol. Gen. Genet.* **217:** 202–208.

111. **Wohlleben, W., and G. Muth.** 1993. *Streptomyces* plasmid vectors, p. 147–175. *In* K. G. Hardy (ed.), *Plasmids: A Practical Approach*. Oxford University Press, Oxford.

112. **Wright, F., and M. J. Bibb.** 1992. Codon usage in the G + C-rich *Streptomyces* genome. *Gene* **113:**55–65.

113. **Xiao, J., R. E. Melton, and T. Kieser.** 1994. High-frequency homologous plasmid-plasmid recombination coupled with conjugation of plasmid SCP2* in *Streptomyces*. *Mol. Microbiol.* **14:**547–555.

Genetics of Non-*Streptomyces* Actinomycetes

L. DIJKHUIZEN

31

Actinomycetes are gram-positive bacteria with a high G + C content of their DNA (60 to 70 mol%) belonging to the order *Actinomycetales*. A characteristic of many actinomycetes is that they form branching filaments, giving them a fungal appearance. Actinomycetes can be separated into different genera on the basis of morphological, chemical, molecular systematic, and microbiological criteria (15, 22). Examples of chemotaxonomic properties used for classification of these organisms are cell wall, whole cell, and lipid compositions. Some actinomycetes lack diaminopimelic acid (DAP) in their cell walls (e.g., the genus *Actinomyces*). Most actinomycetes, however, possess so-called cell wall type I to IV peptidoglycans containing L-DAP and glycine (type I; e.g., the genus *Streptomyces*), *meso*-DAP and glycine (type II; e.g., the genus *Micromonospora*), *meso*-DAP (type III; e.g., the genus *Thermoactinomyces*) or *meso*-DAP, arabinose, and galactose (type IV; e.g., the genera *Rhodococcus*, *Nocardia*, *Mycobacterium*, and *Amycolatopsis*) (22).

Compared to the wealth of genetic techniques and experience available for *Streptomyces*, work on the non-*Streptomyces* actinomycetes is clearly only in a fairly early stage. Suitable genetic tools for many of these nonstreptomycetes remain to be established. Often these organisms are genetically inaccessible because of the presence of very efficient restriction barriers (2, 61, 66). Early work on the biology and genetics of the nonstreptomycete actinomycetes has been reviewed previously (5, 40). The statement of Lechevalier and Lechevalier (40), that most of our knowledge of these actinomycetes comes from studies in the fields of microbial taxonomy and industrial microbiology, to a large extent still appears to be true today.

Very limited information is available about the genetics of *Actinomyces* (70) and *Micromonospora* (25). Only methods developed for the genetic characterization and manipulation of the non-*Streptyomyces* actinomycetes with cell wall type IV peptidoglycans warrant review here. A high priority should be given to the genetic analysis of these non-*Streptomyces* actinomycetes. Many genera include strains of medical importance (e.g., *Mycobacterium tuberculosis* and *Mycobacterium leprae*, causative agents of tuberculosis and leprosy), or industrially valuable strains (e.g., *Saccharopolyspora erythraea* and *Amycolatopsis orientalis*, producers of the antibiotics erythromycin and vancomycin, and *Rhodococcus erythropolis*, used in biotransformations). Furthermore, species of the genera *Rhodococcus* and *Mycobacterium* are frequently isolated from polluted soils, displaying interesting abilities to degrade xenobiotic compounds. These organisms are drawing increasing attention for application in bioremediation processes. Several attempts to use ready-made *Streptomyces* vectors in these genera were reported to be unsuccessful. Often an extreme instability of *Streptomyces* vectors is observed in these strains, resulting in rapid segregation of the vector under nonselective conditions. It therefore appears desirable to develop cloning systems based on plasmid vectors from the specific nonstreptomycetes replicons.

In this chapter some of the plasmid vectors in use for the above-mentioned genera are discussed, with emphasis on shuttle vectors for *Escherichia coli*. The electrotransformation protocols developed for DNA transfer into whole cells are also outlined. Finally, the first examples of gene replacement methods reported for the genus *Mycobacterium* are described.

31.1. THE GENUS *RHODOCOCCUS*

Members of the genus *Rhodococcus* are aerobic, nonsporulating bacteria, exhibiting nocardioform morphology. The phylogenetic position of *Rhodococcus* species has been reexamined (54). There is increasing evidence that *Rhodococcus* spp. play a key role in the biodegradation of a wide range of organic pollutants in the environment (69). One of the major bottlenecks in characterizing *Rhodococcus* genes encoding catabolic enzymes is the lack of gene cloning systems for these species. Systematic study of *Rhodococcus* plasmids is the more interesting, because it is likely that some of these plasmids will encode catabolic pathways.

31.1.1. *Rhodococcus* Vectors

Information on genetic systems for rhodococci is rather limited, although a number of plasmids, cloning vectors, and DNA transfer systems have been described (18). Various *Rhodococcus*–*E. coli* shuttle cloning vectors have also become available (4, 10, 26, 53, 61). Unfortunately, there is increasing evidence that these vectors are rather restricted in their host range and stability, even within the genus *Rhodococcus*. The molecular basis for this is unknown.

Plasmid pMVS301 was constructed by cloning a 3.8-kb *Hin*dIII fragment (*rep*) from the cryptic *Rhodococcus* sp. H13-A plasmid pMVS300 into pIJ30 (6.3 kb), a pBR322 derivative containing an *E. coli* origin of replication (*ori*) and an ampicillin resistance determinant (*amp*), together with the *Streptomyces* thiostrepton resistance gene (*tsr*) (61). This recombinant plasmid (pMVS301, 10.1 kb; Fig. 1) replicates in *R. erythropolis*, *Rhodococcus equi*, and *Rhodococcus globerulus*, but not in several other *Rhodococcus* species tested. Using a protocol for polyethylene glycol (PEG 8000)-assisted transformation of protoplasts, frequencies of 10^5 thiostrepton (and ampicillin) resistant transformants per microgram of pMVS301 DNA (isolated from a *Rhodococcus* host) are obtained (61). Approximately 500-fold lower transformation frequencies are observed with *E. coli*-derived pMVS301 DNA, indicating the presence of a restriction-modification system in the *Rhodococcus* recipient.

Quan and Dabbs (53) combined a *Rhodococcus* arsenic resistance plasmid (*rep*) with an *E. coli* ampicillin resistance (*amp*) plasmid (*ori*) carrying the *Eco*RI endonuclease gene (placed under the control of λ repressor, λ*pr*). Further size reduction to the minimal replicon of the *Rhodococcus* plasmid, and replacement of the arsenic marker by a chloramphenicol resistance marker (*cat*), resulted in the development of plasmid pDA71 (53), which needs only cloramphenicol for selection and maintenance in *Rhodococcus* and *E. coli* (8.8 kb; Fig. 2). Plasmid pDA71 contains a unique *Bgl*II restriction site in the *Eco*RI gene, allowing positive selection for inserts in *E. coli*. This plasmid is stably maintained in several *R. erythropolis* strains and in *R. equi*, but not in other *Rhodococcus* species.

De Mot et al. (10) determined the complete nucleotide sequence of the 5,936-bp cryptic plasmid pFAJ2600 from *R. erythropolis* NI86/21. Chimeras composed of a pUC18-Cm^r derivative inserted in the *parA-repA* integenic region of vector pFAJ2600 produced vectors that could be shuttled between *E. coli* and several *Rhodococcus* species (*R. erythropolis*, *R. fascians*, *R. rhodochrous*, *R. ruber*). This shuttle vector pFAJ2574 was stably maintained in *R. erythropolis* and *R. fascians* growing under nonselective conditions (10).

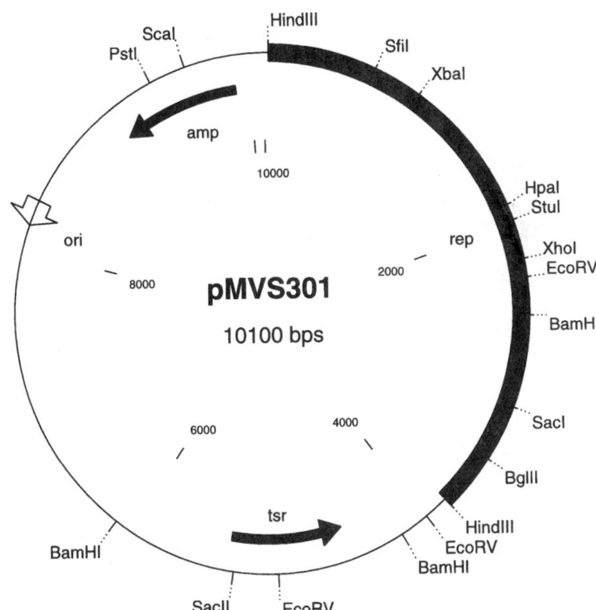

FIGURE 1 Restriction map of the *E. coli–Rhodococcus* shuttle plasmid pMVS301. Reprinted from reference 61 with permission.

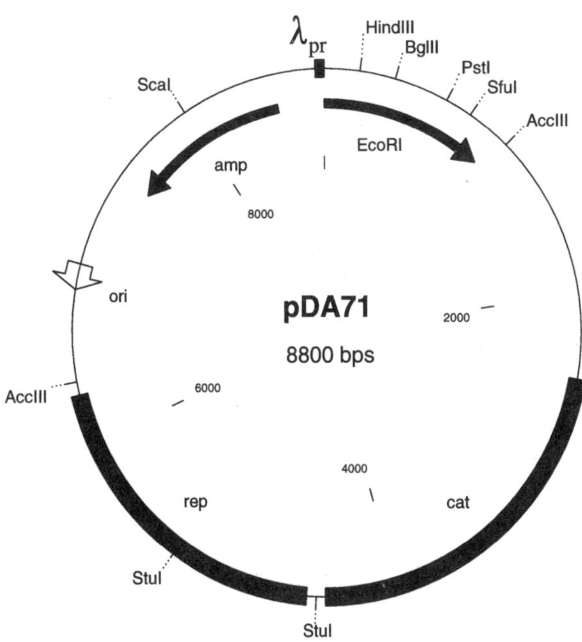

FIGURE 2 Restriction map of the *E. coli–Rhodococcus* shuttle plasmid pDA71. Reprinted from reference 53 with permission.

31.1.2. Electrotransformation of *Rhodococcus*

Various protocols for transformation of *Rhodococcus* protoplasts have been described. However, in general these procedures are rather laborious. Using plasmids pMVS301 (Fig. 1) and pDA71 (Fig. 2), isolated from *E. coli* strain DH5α, an efficient electrotransformation protocol has been developed for *R. erythropolis* SQ1 (63a). *R. erythropolis* SQ1 is grown in liquid medium (LBP) containing 1% (wt/vol) Bacto Peptone (Difco), 0.5% (wt/vol) yeast extract (BBL), and 1% (wt/vol) NaCl. For growth on solid medium, LBP is supplemented with 1.5% (wt/vol) Bacto Agar (Difco). LBP medium (5 ml) is inoculated with *R. erythropolis* SQ1 cells from a freshly prepared agar plate and incubated at 30°C with shaking for 24 h. The 5-ml preculture is used to inoculate 300 ml of LBP medium supplemented with 3% (wt/vol) glycine and incubated overnight. The cells are harvested in the late exponential phase, by centrifugation in a Sorvall GSA rotor for 10 min at 5,000 rpm, and washed twice with cold distilled water. The pellet is resuspended in 2 ml of cold 30% (vol/vol) PEG 1450 (Sigma). Competent cells are divided in 100-μl portions (10^{10} CFU ml^{-1}) and frozen until use at −80°C. For transformation, cells are thawed on ice. Plasmid DNA (1 μg) is added, and cells are kept on ice for 1 min. Electrotransformation is performed in 2-mm gapped cuvettes, using a BTX600 electroporation apparatus set at 1750 V, 186 Ω, and 50 μF. These settings result in observed field strength and pulse time constants of 6.7 kV cm^{-1} and 8 ms, respectively. Immediately after the electropulse, the cell suspension is diluted with 1 ml of LBP medium and incubated at 30°C with shaking for 4.5 h. Appropriate dilutions of the cell suspension are then spread on LBP agar plates supplemented with 40 μg ml^{-1} chloramphenicol (for pDA71) or 10 μg ml^{-1} thiostrepton (for pMVS301), or without antibiotics (to determine total viable counts). Transformants that have appeared after 3 days are counted. Using the procedure outlined above, frequencies up to 1 $\times$ 10^6 transformants per μg of plasmid DNA are obtained reproducibly (63a).

31.1.3. Mutagenesis of *Rhodococcus*

No specific protocol for random mutagenesis of *Rhodococcus* spp. has yet been described. There is also a general lack of information about transposons in rhodococci. Insertion elements from *Rhodococcus* strains have been sequenced, but in most cases their transposition activity has not been demonstrated (11, 36). Nagy et al. (50) reported identification of the IS*21*-related element IS*1415* in *R. erythropolis* NI86/21. A transposon constructed from IS*1415*, by inserting a chloramphenicol resistance marker, was capable of faithful transposition following delivery from a suicide vector to *R. erythropolis* SQ1 cells (50).

An alternative procedure has been described by Desomer et al. (12) using electrotransformation of *R. fascians* by nonreplicating plasmids carrying a ColE1 replicon and an ampicillin resistance gene. Chromosomal (illegitimate) integration of these plasmids at different sites results in insertional mutagenesis. Various mutants of *R. fascians* blocked in amino acid biosynthetic pathways or in pigment biosynthesis were identified. Tagged genes in these transformants are subsequently cloned in *E. coli* (under selection for ampicillin resistance) and used for the isolation of wild-type genes from a cosmid library of *R. fascians*. No further reports about the use of this system in other *Rhodococcus* spp. have appeared.

31.2. THE GENUS *MYCOBACTERIUM*

The genus *Mycobacterium* is the subject of extensive molecular biological research in many laboratories worldwide, mainly because of its clinical importance, with the aim to gain insight into mechanisms of pathogenicity, but also for various biotechnological applications (29, 35). After a relatively slow start, rapid progress has been made in recent years in the development of methods and techniques, making many species of *Mycobacterium* more accessible for molecular research (27, 29, 35). Recently the complete genome sequence of *Mycobacterium tuberculosis* has become available (7a).

31.2.1. *Mycobacterium* Vectors and Transformation

Both replicative and integrative plasmid vectors have been developed. The *Mycobacterium fortuitum* plasmid pAL5000 (4.8 kb) is one of the best-characterized mycobacterial plasmids to date (63). The minimal replicon of pAL5000 provides a basis for the construction of *Mycobacterium* shuttle vectors for both *E. coli* (e.g., pAL8 and pRR3) (56) and *Streptomyces lividans* (e.g., pYUB12) (62). Using a temperature-sensitive derivative of pAL5000, pCG79 has been constructed carrying the *E. coli* transposon Tn*611* with a kanamycin resistance gene. By raising the temperature to the nonpermissive level (39°C), a large number of random insertional mutants of *Mycobacterium smegmatis* have been isolated using this vector (19, 24). Plasmid pAL8 is relatively large (9.2 kb) and lacks unique restriction sites, however.

The development of techniques for electroporation of whole cells has rendered a simple, fast, and efficient tool for the transformation of *Mycobacterium* species (7, 30). The cell wall appears to be the major barrier for the introduction of DNA into intact mycobacterial cells (30, 57). Cells grown with glycine show an increased susceptibility to the action of wall-degrading chemicals like isoniazid; such treatments significantly enhance electroporation efficiency of *Mycobacterium aurum* cells (2.3 $\times$ 10^4 transformants per μg of plasmid pAL8 DNA) (30).

Frequently, plasmid vectors are found to be unstable in mycobacterial hosts. To avoid this problem, several integrative plasmids were developed, based on mycobacteriophage L5 (42) and *Streptomyces ambofaciens* chromosomally integrating plasmid pSAM2 (60). The integrative plasmid pMH94, constructed by cloning the attachment site (*attP*) and the integrase gene (*int*) of the mycobacteriophage L5 into an *E. coli* plasmid, transformed *M. smegmatis* very efficiently and integrated site-specifically in the host genome (42). Electroporation of *Mycobacterium intracellulare* 1403 with pMH94 yields 10^4 to 10^6 transformants per μg of plasmid DNA; no transformants are obtained with the plasmid derivative pMH94Δ*int*, which lacks *int* as well as *attP* (46).

Hermans et al. (31, 32) reported electrotransformation (at low frequencies) of *M. aurum*, *M. smegmatis*, and *Mycobacterium parafortuitum* (but not *M. fortuitum* or *Mycobacterium phlei*) with the gram-negative cosmid vector pJRD215. This plasmid was stably maintained in the *Mycobacterium* species, and both the kanamycin and streptomycin antibiotic resistance determinants were expressed. Attempts to transform *Nocardia* sp., *R. erythropolis*, and *R. rhodochrous* with pRJD215 were unsuccessful (32). Using pJRD215 and electroporation, the successful transformation of *Actinomyces* spp. has also been reported (70). Cosmid pJRD215 is based on RSF1010, a broad-host-range IncQ plasmid that replicates in most gram-negative bacteria. Conjugative transfer of RSF1010 from *E. coli* to *M. smegmatis* (at high frequencies) subsequently was demonstrated as well (23). This opens up the possibility to shuttle recombinant cosmid DNA with large inserts directly between *E. coli* and *Mycobacterium* species, provided no further plasmid rearrangements occur in the *Mycobacterium* background.

31.2.2. Gene Replacement in *Mycobacterium*

Several systems allowing positive selection of mutants resulting from gene replacement (double-crossover events) have now been developed for *M. smegmatis* (52, 58), making use of a two-step selection strategy. One example is *rpsL*, a dominant marker conferring susceptibility to streptomycin to a streptomycin-resistant strain with a mutation in the endogenous *rpsL* allele. When used as a counter-selectable marker, *rpsL* leads to allelic-exchange mutants resistant to streptomycin (58). Pelicic et al. (52) used the vector pPR34 for gene replacement (Fig. 3). This vector carries an antibiotic marker (gentamicin resistance) for the primary selection of transformants and a second marker with a conditionally dominant lethal effect in mycobacteria (the *Bacillus subtilis* levansucrase *sacB* gene conferring sucrose sensitivity) to counterselect clones that have lost the vector DNA, eliminating the need for extensive screening. Thus, unmarked uracil auxotrophic mutants of *M. smegmatis* mc^{2}155 are selected in two steps (Fig. 3), using an unmarked and mutated (frameshift) copy of the *pyrF* gene

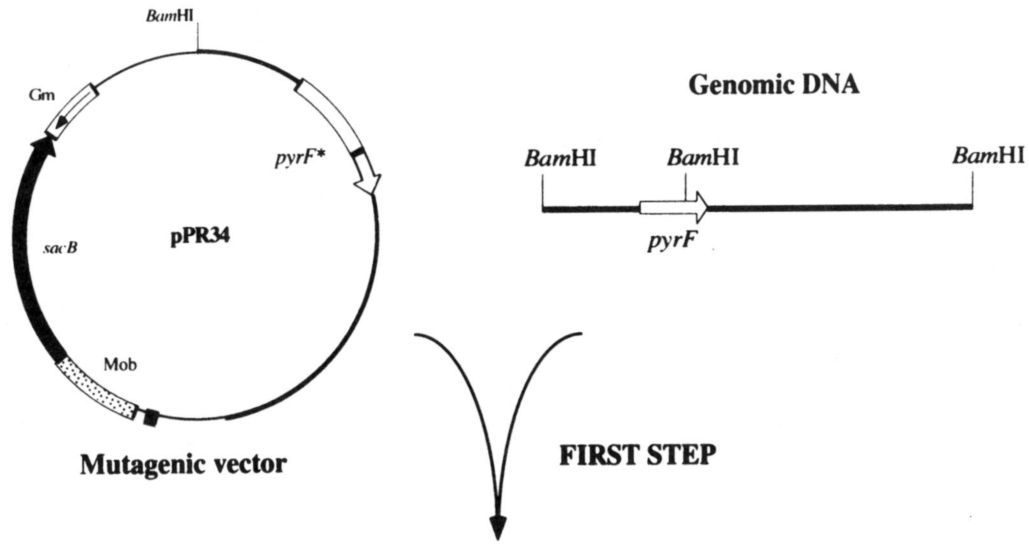

Single recombination (Ura$^+$, Gm^r, Suc^S)

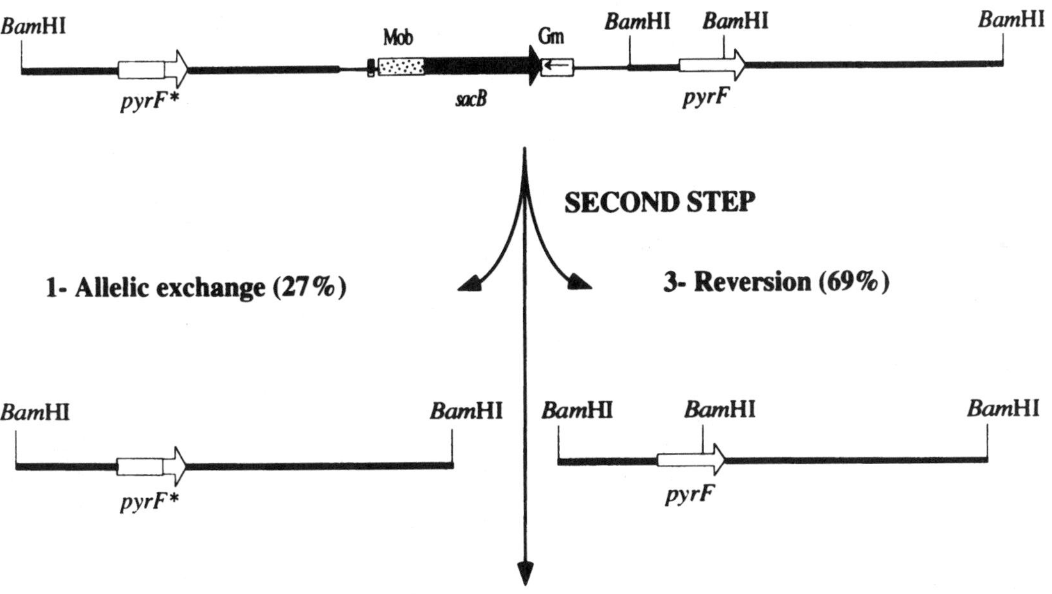

2- Mutation of *sacB* (4%)

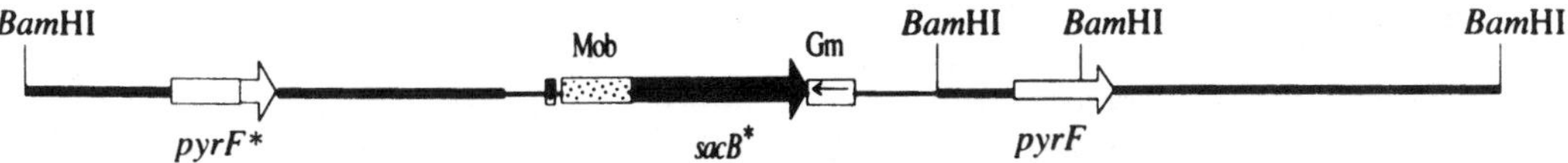

FIGURE 3 General strategy for generating unmarked mutations in mycobacteria using a two-step selection method. The first selection is performed on plates with gentamycin. The second selection step is performed on plates with 10% sucrose, allowing positive selection for clones that have lost the *sacB* gene. Mob, RP4 origin of transfer; *pyrF** and *sacB** indicate mutant alleles of those genes. Reprinted from reference 52 with permission.

(*pyrF**). Following electrotransformation, clones resulting from a single homologous recombination event are selected on Luria-Bertani medium supplemented with uracil and gentamicin (LUG). Phenotypical and Southern blot analyses confirmed that the selected clones result from a single crossover in the *pyrF* gene. Clones from the LUG plates subsequently are grown overnight on liquid Middlebrook 7H9 medium (Difco) supplemented with uracil but without antibiotic selection, to allow a deletion-recombination event to occur at the *pyrF* locus. Any such event may eliminate either the wild-type or the mutant *pyrF* allele. Samples of such cultures are spread on LUS medium (with 10% sucrose) to select clones that have lost the *sacB* gene. Colonies obtained are replicated onto mineral medium minus uracil; approximately one third of the clones obtained are unable to grow, being uracil auxotrophs (Fig. 3). The remaining clones are either revertants or presumably carry a mutation in the *sacB* gene (52).

Homologous recombination in mycobacteria is a process that is little understood. While homologous recombination has been easily achieved in the rapidly growing nonpathogen M. *smegmatis*, it has been a difficult problem in the slowly growing pathogenic species (48). Its use for genetic analysis is hampered by the high frequencies of illegitimate recombination. Very recently, promising systems for gene replacement (and transposon mutagenesis) in the human pathogens *Mycobacterium marinum* (55) and M. *tuberculosis* (51) were reported, both using the *sacB* gene for counterselection. Using vectors with a thermosensitive origin of replication, powerful methods for transposon mutagenesis of M. *tuberculosis* were developed as well (3, 51).

31.2.3. Reporter Systems for *Mycobacterium*

The green fluorescent protein of the jellyfish *Aequorea victoria* has been used successfully as a reporter molecule in mycobacteria, allowing development of promoter-probe vectors, in vivo assessment of drug susceptibility, tracing of mycobacterial cells within infected macrophages, or as free-living cells (13, 37).

31.3. THE GENUS *AMYCOLATOPSIS*

The genus *Amycolatopis* was established relatively recently when it was recognized that many nocardioform bacteria in fact lack mycolic acids in their cell walls, clearly differing from the *Rhodococcus* and *Mycobacterium* genera in this property (9, 14, 41). The species *Amycolatopsis mediterranei* and A. *orientalis* are of special interest as producers of the commercially and medically important rifamycin and vancomycin antibiotics.

31.3.1. Isolation of *Amycolatopsis* Mutants

The isolation of mutants is an important procedure allowing characterization of metabolic pathways, e.g., identification of separate enzyme steps, the possible presence of isoenzymes, and the overall organization of these enzymes, such as in enzyme complexes. This approach is also widely used for the construction of industrial strains overproducing interesting compounds. Positive selection procedures, e.g., for the isolation of mutants resistant to inhibitory compounds such as antibiotics, or mutants able to develop a specific color reaction, are the most preferred methods but are often not available or are impossible to use. This is the case when attempting the isolation of mutants in biosynthetic and catabolic pathways.

Selection of auxotrophic mutants of filamentous bacteria such as actinomycetes necessitates isolation of spores containing single genomes. General procedures to efficiently induce sporulation of actinomycete species are not available. Attempts to isolate auxotrophic mutants of, for instance, A. *mediterranei*, to identify the branch point between aromatic amino acid biosynthesis and rifamycin biosynthesis, have met with limited success (21). Procedures for the isolation of mutants from single cells of *Amycolatopsis methanolica*, combined with a simple protocol for the identification of metabolic lesions, have been reported (16). This has allowed the isolation and characterization of a large number of recessive and auxotrophic mutants of A. *methanolica*.

In liquid media, A. *methanolica* grows in long chains of individual cells containing clearly visible cell walls. Instead of isolating spores or generating protoplasts, *Amycolatopsis* cells may be separated from each other by using vortexing, pipetting at high shear, or ultrasonic vibration treatments in a sonication bath or with a sonication probe. Any one of these methods may work for specific *Amycolatopsis* species. Individual cells of A. *methanolica* are rather tightly linked together, and only sonication with a probe directly inserted into the cell suspension, a rather harsh and potentially lethal treatment, is successful in generating single cells, according to the following procedure. Late-exponential-phase cells (5 ml, 10^9 CFU) grown in glucose mineral medium are sonicated for 15 s at an amplitude of 6 μm, using an MSE sonicator and an ethanol-sterilized probe (10 mm). No cell lysis is observed under optimal conditions (15-s sonication), with the number of CFU even increasing 10-fold. (In the late-exponential phase, 1 CFU of A. *methanolica* thus is the result of 10 cells on average.) Similar procedures can also be used for other nocardioform bacteria, but optimization of cultivation and sonication procedures may be required. During sonication the progress in cell separation can be followed by microscopical observations, looking for the presence of single cells, and determining viable counts.

Cell samples are spread on glucose mineral agar with limiting amounts of the aromatic amino acids and quinate (final concentration 1 mg liter^{-1} of each) as growth supplements. Mutations are introduced by UV irradiation for 10 to 30 s with a UV lamp (Philips TAW 15 W) at a distance of 20 cm. The UV irradiation procedure should result in 70 to 90% killing. After 7 days of incubation, survivors forming pinpoint colonies are purified on glucose agar with an excess of the above supplements (25 mg liter^{-1} each) and further characterized for growth requirements and enzyme lesions. A typical experiment yields about 10% pinpoint colonies among the survivors, using 1 mg liter^{-1} of the supplements. Pinpoint colonies may arise by leaky mutations in glucose metabolism or in various biosynthetic pathways, or by knockout mutations in the aromatic amino acids pathways; the latter type of mutants can only form pinpoint colonies because of the low supplement concentrations used in the medium. Growth tests show that 0.1 to 0.5% of these pinpoint colony-forming mutants strictly depend on the presence of aromatic amino acids. Using these procedures, about 150 mutants of A. *methanolica* have been isolated, all displaying stable genotypes. Enzyme analysis showed that mutants blocked in virtually all single steps in aromatic amino acid biosynthesis had been obtained. No mutants were obtained in three separate steps; biochemical evidence subsequently was obtained that there are isoenzymes for these steps in *A. methanolica* (16).

Using similar procedures, mutants of *A. methanolica* blocked in enzymes involved in utilization of primary alcohols as growth substrates have also been isolated (28). Methanol is abundantly produced from lignin degradation in soils. Thus far, very few methanol-utilizing actinomycetes have been described, but this ability may be more widespread. Growth on the one-carbon substrate methanol is relatively poor and may easily be overlooked. Spreading cells of *A. methanolica* on mineral medium agar plates already results in development of pinpoint colonies; addition of methanol to these agar plates does not further stimulate colony growth. This is surprising, because in liquid medium the organism grows readily on methanol mineral medium, reaching relatively high cell densities. The use of washed agars, Noble agar, and various types of agarose also failed to improve methanol utilization. Further studies have revealed that the problem is caused by the presence of impurities in the agar and the limited capacity of *A. methanolica* to degrade agar and agarose; the metabolism of these alternative substrates apparently causes repression of methanol utilization. No such problems are caused by the agar alternative Gelrite (Kelco), allowing development of larger colonies with methanol mineral medium. After UV mutagenesis on glucose agar plates as described above, mutants blocked in methanol utilization can be identified by their inability to grow on methanol Gelrite plates (28).

31.3.2. Transformation of *Amycolatopsis*

Several methods have been reported for transformation of *Amycolatopsis* strains with plasmid DNA. Transformation protocols and plasmids initially developed for *Streptomyces* spp. (33) allow efficient transformation (10^6 transformants per μg of DNA) of *A. orientalis* (47). This transformation method uses protoplast formation and regeneration, which is a rather laborious process. Electrotransformation of whole cells of *A. mediterranei* (2.2×10^3 transformants per μg of DNA) and *A. orientalis* (1.1×10^5 transformants per μg of DNA) has also been reported (39). A relatively simple and rapid transformation procedure, using PEG and cesium chloride, has been described for whole cells of *A. mediterranei* (44). This method gives approximately 10^6 transformants per μg of DNA and could also be applied for transformation of *Amycolatopsis* (*Nocardia*) *lactamdurans* (38). The plasmid used for transformation of *A. mediterranei* was derived from its integrative plasmid pMEA100 (49) and contained the erythromycin resistance gene as a selective marker. Plasmid pMEA100 has not been developed into a suitable cloning vector, because it has a low copy number, integrates into the chromosome (45), and thus is difficult to isolate from liquid cultures.

The transformation protocol and plasmid vectors used for *A. methanolica* probably have been studied in the most detail and will be discussed here. None of the existing actinomycete plasmid vectors tested are stably maintained in *A. methanolica*, differing from observations made with *A. orientalis* (47) but similar to the situation reported for *A. mediterranei* (44, 59). Cloning vectors for *A. methanolica* based on the indigenous plasmid pMEA300 (Fig. 4A) therefore were developed. This 13.3-kb plasmid is present both in the free state and integrated at a unique chromosomal location (65). A pMEA300-free derivative, strain WV1, has been isolated (see section 31.3.3). A full sequence and functional analysis of pMEA300 has been carried out. The data obtained allow construction of stable pMEA300-derived *A. methanolica–E. coli* shuttle vectors (64–68).

The transformation technique used for wild-type *A. methanolica* and strain WV1 (66) is an adaptation of the method described by Madoń and Hütte (44) for *A. mediterranei*. *A. lactamdurans* has also been transformed successfully with this method (38). Cells of *A. methanolica* grown in TS broth (44) are harvested in the stationary phase at an OD_{433} of 5.0 to 8.0 and used immediately. Cells are centrifuged (10 min, 3,600 $\times$ *g*), washed once in one volume of TE buffer (10 mM Tris-HCl [pH 8.0], 1 mM EDTA), and resuspended in TE to an OD_{433} of 160. The standard transformation mixture (400 μl), containing 100 μl of cell suspension (final OD_{433} of 40), 5 mM $MgCl_2$, 0.625 M CsCl, 7.5 μg of sonicated calf thymus DNA, 0.1 μg of plasmid pMEA300 DNA, and 32.5% (wt/vol) PEG 1000 (Koch Light) (44), is incubated for 40 min at 37°C. Samples of the transformation mixture (2 to 50 μl) are plated in 3 ml of R2L overlay medium (with low-melting-point agarose) on S27M agar plates. R2L contains (per liter): 73.2 g of D-mannitol (Fluka), 0.25 g of K_2SO_4, 10.1 g of $MgCl_2 \cdot 6H_2O$, 10 g of glucose, 0.1 g of Casamino Acids (Difco), 5 g of yeast extract (Difco), 100 ml of 0.25 M *N*-tris(hydroxymethyl)methyl-2-aminoethanesulfonic acid (TES) (pH 7.2), 80 ml of 3.68% (wt/vol) $CaCl_2 \cdot 2H_2O$, 10 ml of 0.5% (wt/vol) KH_2PO_4, 2 ml of trace elements solution (33), 20 ml of 10% (wt/vol) asparagine, 5 ml of 1 N NaOH, and 7 g of low-melting-point agarose (type VII; Sigma). Mannitol, K_2SO_4, $MgCl_2$, glucose, Casamino Acids, yeast extract, and low-melting-point agarose can be autoclaved together and stored at room temperature. The other ingredients must be autoclaved separately and added (in the following sequence: TES, $CaCl_2$, KH_2PO_4, trace elements, asparagine, NaOH) to the main mixture shortly before use. NaOH must be added dropwise with vigorous mixing. NaOH added too fast causes the immediate appearance of a precipitate, which then dramatically decreases the transformation efficiency (44). After the addition of NaOH, the medium has to be mixed intensively for about 30 s. S27M agar contains (per liter) 73.2 g of D-mannitol (Fluka), 5 g of peptone (Difco), 3 g of yeast extract (Difco), and 17 g of Bacto Agar (Difco) (45). First, S27M medium (1 ml, minus agar) is added to the suspension, which is centrifuged after gentle mixing. The cell pellet is gently resuspended in S27M (1 ml), centrifuged, and resuspended in S27M medium (400 μl). This wash step is necessary to avoid toxic effects of CsCl. Plating of cells on S27M agar medium in the R2L low-melting-point agarose overlay and selection for pock formation or antibiotic (thiostrepton) resistance are performed as described (44). After 5 h of incubation at 37°C, a thiostrepton solution (final concentration 10 μg ml^{-1}) is added to the plates. The plates are allowed to dry for 1.5 h in a laminar flow cabinet and incubated at 37°C.

Transformation of *A. methanolica* is successful (66) in selecting for pock formation using the cell density, PEG, and CsCl concentrations previously described for *A. mediterranei* (44). The number of transformants of *A. methanolica*, however, further increases with the cell density up to an OD_{433} of 40 (Fig. 5A) with stationary-phase cells. The fact that more transformants are obtained when the transformation mixture is further diluted for plating led to the identification of CsCl as a toxic compound, causing a five-fold reduction in plating efficiency and in number of transformants at a final concentration of 30 mM. This problem can be overcome by introducing a wash step before plating

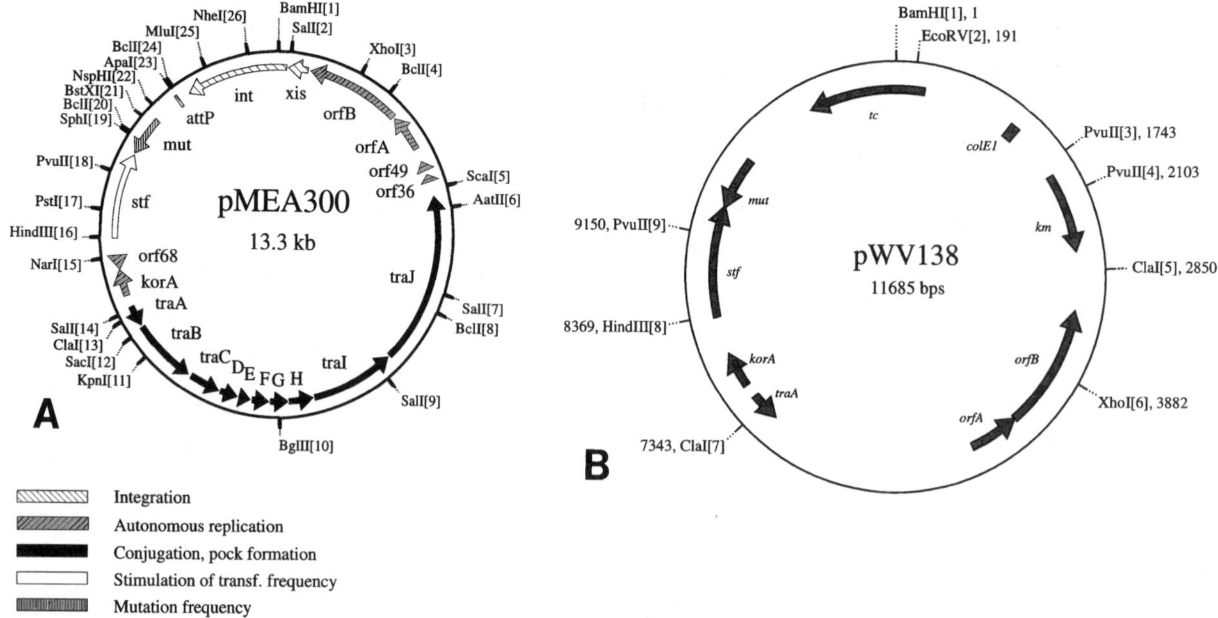

FIGURE 4 Restriction maps of pMEA300 (A) and pWV138 (B). Numbers between brackets indicate restriction sites. Other numbers (B) indicate distances from the unique *Bam*HI site. Bars and arrows indicate functional regions and genes identified. Reprinted from reference 68 with permission.

cells in the R2L overlay. CsCl and PEG are essential for transformation (Fig. 5B and C). The optimal CsCl concentration is 0.625 M, as for *A. mediterranei* (44) but approximately 60% lower than that for *A. lactamdurans* (38). The transformation protocol for strain WV1 with pMEA300 can be further optimized by changing the PEG concentration (Fig. 5C). PEG exhibits its stimulatory effect in a narrow concentration range of 30 to 37.5%, the optimal concentration being 32.5% (Fig. 5C). In addition to the normal nutrients, mannitol (0.4 M) in the R2L and S27M media is essential for efficient transformation. With this optimized transformation protocol (see above), pMEA300 routinely yields 9×10^5 tranformants per μg of DNA in the plasmid pMEA300-free strain WV1 and 1×10^4 transformants per μg of DNA in the wild-type strain (66).

31.3.3. Curing of Integrative Plasmids

Protoplast formation and regeneration, ethidium bromide, and heat treatments are effective for curing of the free plasmid pA387 from *Amycolatopsis* sp. strain DSM 43387 and of the integrative plasmid pMEA100 from *A. mediterranei* (49). Such treatments, however, do not result in loss of pMEA300 from *A. methanolica* wild type. The procedure outlined in the following, however, can be successfully used for the isolation of a derivative of the NCIB 11946 strain that has lost both the integrated and free pMEA300 sequences. This procedure may be more widely applicable for curing of integrative plasmids, after optimization for specific hosts.

Plasmid pMEA301 contains the thiostrepton resistance gene as a 1.1-kb *Bcl*I fragment of pIJ702 in the unique *Bgl*II[10] site of pMEA300 (Fig. 4A and 6A). Plasmid pMEA302 is a spontaneously occurring 20.8-kb hybrid plasmid of pMEA300 and pMEA301, except for a 6.4-kb deletion extending over the *attP* site, with one endpoint located between the unique *Pst*I[17] and *Pvu*II[18] sites, and the other between the *Bcl*I[8] and the *Sal*I[9] sites of pMEA300 (Fig. 4A, 6B). Transformation of wild-type *A. methanolica* with pMEA302 yields the thiostrepton-resistant strain AM302.

Strain AM302 is grown for 30 generations in complete medium with thiostrepton, protoplasted, and regenerated to obtain colonies arising from individual cells. Southern hybridization shows that pMEA302 replaces both the integrated and the free forms of pMEA300 in all AM302-derived strains. Subsequent loss of plasmid can now be detected by loss of thiostrepton resistance. Strain AM302/1 is therefore cultivated for 30 generations in CM medium without thiostrepton, protoplasted, and regenerated. When testing approximately 1,000 colonies, about 20 thiostrepton-sensitive colonies are obtained. In our own studies only one of these, strain WV1, was found to be completely devoid of both free and integrated forms of pMEA300. No differences are noticed between strain WV1 and wild type with respect to antibiotic sensitivity, ability to use methanol, or growth rates on various media (66).

31.3.4. *Amycolatopsis–E. coli* Shuttle Vectors

The construction of pRL1, an *Amycolatopsis–E. coli* shuttle vector carrying a 5.1-kb DNA fragment with the origin of replication of plasmid pA387 from *Amycolatopsis* sp. strain DSM 43387 and the 5.1-kb *E. coli* plasmid pDM10, has been described by Lal et al. (39). The latter plasmid has its origin from pBR322 and carries the Tn*5* kanamycin resistance gene and the gentamycin resistance of Tn*1696*. Plasmid pRL1 has not yet been characterized in more detail. The vectors used for transformation of *A. lactamdurans* were based on pA387 (see above) and comprise useful cloning vectors and promoter probe vectors (38).

A. methanolica–E. coli shuttle vectors were constructed using pMEA300 and the ColE1-derived plasmid pHSS6

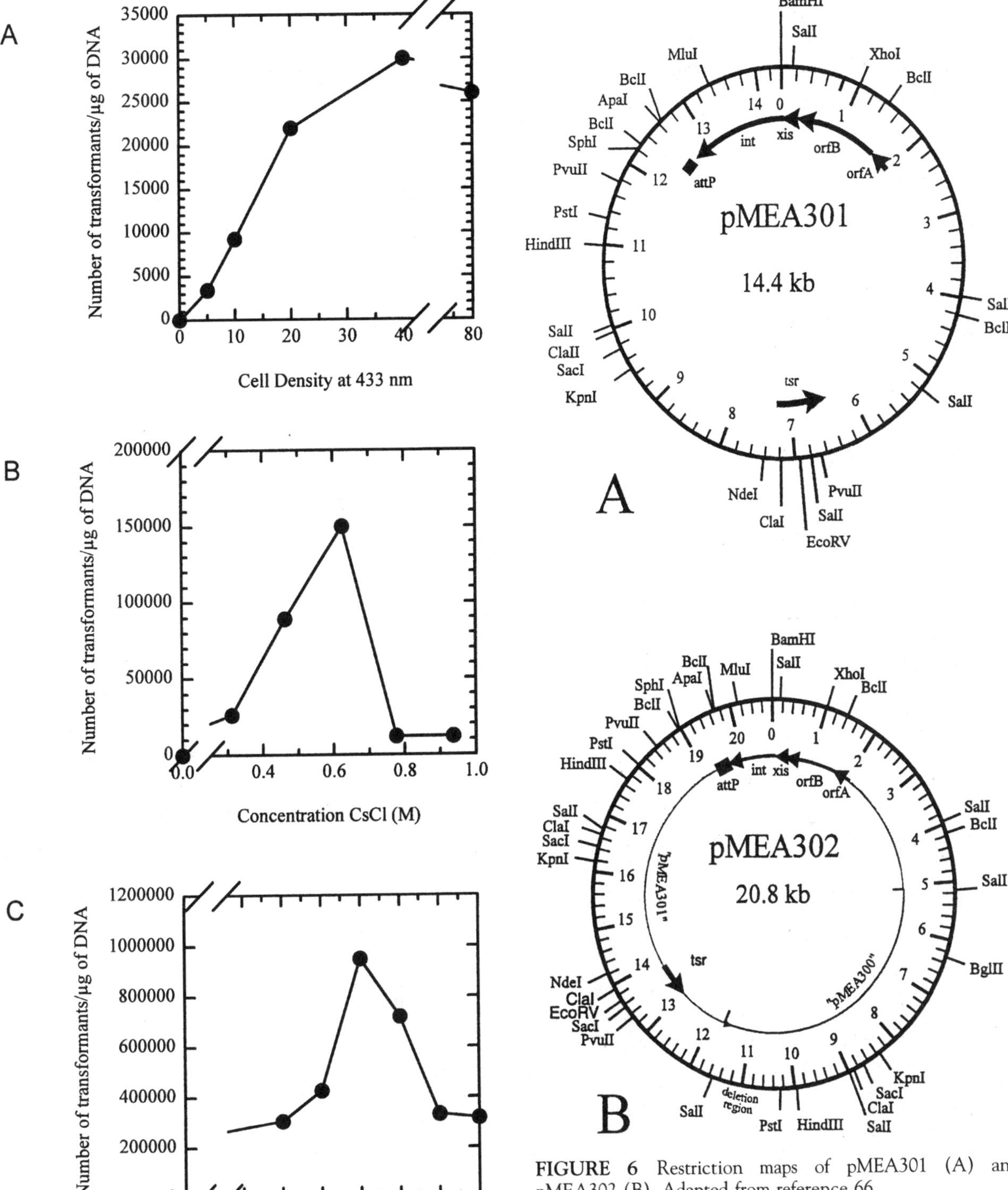

FIGURE 5 Effects of varying cell density (A), concentrations of CsCl (B), and concentrations of PEG 1000 (C) in the transformation mixture on the transformation frequency of strain WV1 with pMEA300. (A) CsCl and PEG concentrations according to Madoń and Hütter (45). (B) Optimal cell and PEG concentrations as in A. (C) Optimal cell and CsCl concentrations as in A and B. Reprinted from reference 66 with permission.

FIGURE 6 Restriction maps of pMEA301 (A) and pMEA302 (B). Adapted from reference 66.

carrying the kanamycin resistance gene (*km*) of Tn*5* for selection in *A. methanolica*. First, the minimal regions of pMEA300 involved in replication as a free plasmid were identified. Subsequent deletion of the *int* gene and most of the Tra region yielded the nonintegrative, nonconjugative derivative pWV136 (64). Cloning of DNA fragments may be greatly facilitated by the use of transcriptional terminators (6). Therefore, the tetracycline (*tc*) resistance marker flanked by omega (Ω) fragments from pHP45Ω-Tc

(17) was cloned into the *Bam*HI site of pWV136, yielding pWV138 (Fig. 4B) (68). This plasmid contains a unique *Bam*HI site that can be used for cloning, resulting in insertional inactivation of the Tc resistance gene in *E. coli*. Attempts to construct a gene bank of *A. methanolica* DNA in *E. coli* were successful only when using the pWV138 vector with these transcriptional terminators (68). The reason for this instability is not known; possibly the inserts generate a high transcriptional activity that is lethal for *E. coli*.

To construct gene banks, the ligation mixture of pWV138 DNA digested with *Bam*HI and a partial *Sau*3A digest (>3 kb) of total DNA of *A. methanolica* was used to transform *E. coli* strain MC1061, yielding approximately 15,000 independent transformants in six different batches. The average insert frequency was 84%; plasmid DNA of 34 Tc-sensitive colonies was analyzed by restriction analysis, showing an average insert size of 4 kb. This gene bank has been used successfully for the cloning of a number of *A. methanolica* genes (1, 68).

Attempts to transform *A. methanolica* strain WV1 with plasmid pMEA300-derived shuttle vectors isolated from methylating *E. coli* strains generally fail. This indicates the presence of a restriction barrier between *E. coli* and *A. methanolica*, as reported for several non-*Streptomyces* species (8, 43, 61). Such restriction barriers can be avoided by performing DNA isolations from nonmethylating strains, e.g., *E. coli* JM110 (*dam, dcm*) and *S. lividans* (2, 43). Using this procedure, the restriction system active in *A. methanolica* can also be avoided (68).

Neither pRL1 (39) nor pWV138 (63b) could be introduced into *S. lividans*, indicating the presence of some maintenance function specific to organisms closely related to the genus *Amycolatopsis*.

31.4. CONCLUDING REMARKS

Relatively simple methods for electrotransformation of whole cells, reaching frequencies of 10^6 transformants per μg of plasmid DNA, are now available for a number of *Rhodococcus*, *Mycobacterium*, and *Amycolatopis* species. Various plasmid replicons have been developed for the introduction of (recombinant) DNA into these species, generally as parts of *E. coli* shuttle vectors. Ideally, these shuttle vectors should be developed on the basis of a detailed nucleotide sequence and functional analysis of the indigenous plasmids in these organisms. Most of these shuttle vectors, however, are constructed by trial and error, since limited information is available about replication, maintenance, and compatibility functions of plasmids from these organisms. There is clearly a strong need for further studies of these plasmid functions. This is the more urgent task, in view of the emerging evidence that most of the plasmids studied have a rather restricted host range, most likely dictated by specific maintenance functions. This would require development of specific gene cloning vectors for specific hosts, unless this problem can be overcome at a fairly early stage.

There is also a general lack of plasmid delivery systems for transposon mutagenesis and gene replacement studies in these organisms; much work clearly remains to be done for the further development of such essential genetic tools.

REFERENCES

1. **Alves, A. M. C. R., W. G. Meijer, J. W. Vrijbloed, and L. Dijkhuizen.** 1996. Characterization and phylogeny of the *pfp* gene of *Amycolatopsis methanolica* encoding pyrophosphate-dependent phosphofructokinase. *J. Bacteriol.* **178:**149–155.
2. **Ankri, S., O. Reyes, and G. Leblon.** 1996. Improved electro-transformation of highly DNA-restrictive coryne-bacteria with DNA extracted from starved *Escherichia coli*. *FEMS Microbiol. Lett.* **140:**247–251.
3. **Bardarov, S., J. Kriakov, C. Carriere, S. Yu, C. Vaamonde, R. A. McAdam, B. R. Bloom, G. F. Hatfull, and W. R. Jacobs, Jr.** 1997. Conditionally replicating mycobacteriophages: a system for transposon delivery to *Mycobacterium tuberculosis*. *Proc. Natl. Acad. Sci. USA* **94:** 10961–10966.
4. **Bigey, F., B. Grossiord, C. K. N. Chan Kwo Chion, A. Arnaud, and P. Galzy.** 1995. *Brevibacterium linens* pBL33 and *Rhodococcus rhodochrous* pRC1 cryptic plasmids replicate in *Rhodococcus* sp. R312 (formerly *Brevibacterium* sp. R312). 1995. *Gene* **154:**77–79.
5. **Brownell, G. H., and K. Denniston.** 1984. Genetics of the nocardioform bacteria, p. 201–228. *In* M. Goodfellow, M. Mordarski, and S. T. Williams (ed.), *The Biology of the Actinomycetes*. Academic Press, London.
6. **Chen, J., and D. A. Morrison.** 1988. Construction and properties of a new insertion vector, pJDC9, that is protected by transcriptional terminators and useful for cloning DNA from *Streptococcus peneumoniae*. *Gene* **64:**155–164.
7. **Cirillo, J. D., T. R. Weisbrod, and W. R. Jacobs, Jr.** 1993. Efficient electrotransformation of *Mycobacterium smegmatis*. *Bio-Rad Bull.* **1360:**1–4.

7a. **Cole, S. T., R. Brosch, L. Parkhill, T. Garnier, C. Churcher, D. Harris, S. V. Gordon, K. Eiglmeier, S. Gas, C. E. Barry III, F. Tekaia, R. Badcock, D. Basham, D. Brown, T. Chillingworth, R. Connor, R. Davies, K. Devlin, T. Feltwell, S. Gentles, N. Hamlin, S. Holroyd, T. Hornsby, K. Lagels, and B. G. Barrell.** 1998. Deciphering the biology of *Mycobacterium tuberculosis* from the complete genome sequence. *Nature* **393:**537–544.

8. **Cox, K. L., and R. H. Baltz.** 1984. Restriction and bacteriophage plague formation in *Streptomyces* spp. *J. Bacteriol.* **159:**499–504.
9. **de Boer, L., L. Dijkhuizen, G. Grobben, M. Goodfellow, E. Stackebrandt, J. H. Parlett, D. Whitehead, and D. Witt.** 1990. *Amycolatopsis methanolica* sp. nov., a facultatively methylotrophic actinomycete. *Int. J. Syst. Bacteriol.* **40:**194–204.
10. **De Mot, R., I. Nagy, A. De Schrijver, P. Pattanapipitpaisal, G. Schoofs, and J. Vanderleyden.** 1997. Structural analysis of the 6 kb cryptic plasmid pFAJ2600 from *Rhodococcus erythropolis* NI86/21 and construction of *Escherichia coli-Rhodococcus* shuttle vectors. *Microbiology* **143:** 3137–3147.
11. **Denome, S. A., and K. D. Young.** 1995. Identification and activity of two insertion sequence elements in *Rhodococcus* sp. strain IGTS8. *Gene* **161:**33–38.
12. **Desomer, J., M. Crespi, and M. Van Montagu.** 1991. Illegitimate integration of non-replicative vectors in the genome of *Rhodococcus fascians* upon electrotransformation as an insertional mutagenesis system. *Mol. Microbiol.* **5:**2115–2124.
13. **Dhandayuthapani, S., L. E. Via, C. A. Thomas, P. M. Horowitz, D. Deretic, and V. Deretic.** 1995. Green fluorescent protein as a marker for gene expression and cell

biology of mycobacterial interactions with macrophages. *Mol. Microbiol.* **17**:901–912.

14. **Embley, T. M.** 1991. The family *Pseudonocardiaceae*, p. 996–1027. *In* A. Balows, H. G. Trüper, M. Dworkin, W. Harder, and K.-H. Schleifer (ed.), *The Prokaryotes*. Springer-Verlag, New York.

15. **Embley, T. M., and E. Stackebrandt.** 1994. The molecular phylogeny and systematics of the actinomycetes. *Annu. Rev. Microbiol.* **48**:257–289.

16. **Euverink, G. J. W., G. I. Hessels, J. W. Vrijbloed, and L. Dijkhuizen.** 1996. Isolation and analysis of mutants of the actinomycete *Amycolatopsis methanolica* blocked in aromatic amino acid biosynthesis. *FEMS Microbiol. Lett.* **136**:275–281.

17. **Fellay, R., J. Frey, and H. Krisch.** 1987. Interposon mutagenesis of soil and water bacteria: a family of DNA fragments designed for in vitro insertional mutagenesis of Gram-negative bacteria. *Gene* **52**:147–154.

18. **Finnerty, W. R.** 1992. The biology and genetics of the genus *Rhodococcus*. *Annu. Rev. Microbiol.* **46**:193–218.

19. **Gavigan, J. A., C. Guilhot, B. Gicquel, and C. Martin.** 1995. Use of conjugative and thermosensitive cloning vectors for transposon delivery to *Mycobacterium smegmatis*. *FEMS Microbiol. Lett.* **127**:35–39.

21. **Ghisalba, O., J. A. L. Auden, T. Schupp, and J. Nuësch.** 1984. The rifamycins: properties, biosynthesis, and fermentation, p. 281–327. *In* E. J. Vandamme (ed.), *Biotechnology of Industrial Antibiotics*. Marcel Dekker, Inc., New York.

22. **Goodfellow, M.** 1989. The actinomycetes. I. Suprageneric classification of actinomycetes, p. 2333–2339. *In* S. T. Williams, M. E. Sharpe, and J. G. Holt (ed.), *Bergey's Manual of Systematic Bacteriology*, vol. 4. Williams & Wilkins, London.

23. **Gormley, E. P., and J. Davies.** 1991. Transfer of plasmid RSF1010 by conjugation from *Escherichia coli* to *Streptomyces lividans* and *Mycobacterium smegmatis*. *J. Bacteriol.* **173**:6705–6708.

24. **Guilhot, C., I. Otal, I. van Rompaey, C. Martin, and B. Gicquel.** 1994. Efficient transposition in mycobacteria: construction of *Mycobacterium smegmatis* insertional mutant libraries. *J. Bacteriol.* **176**:535–539.

25. **Hasegawa, M., T. Dairi, T. Ohta, and E. Hashimoto.** 1991. A novel, highly efficient gene-cloning system for *Micromonospora* strains. *J. Bacteriol.* **173**:7004–7011.

26. **Hashimoto, Y., M. Nishiyama, F. Yu, I. Watanabe, S. Horinouchi, and T. Beppu.** 1992. Development of a host-vector system in a *Rhodococcus* strain and its use for expression of the cloned nitrile hydratase gene cluster. *J. Gen. Microbiol.* **138**:1003–1010.

27. **Hatfull, G. F.** 1993. Genetic transformation of mycobacteria. *Trends Microbiol.* **1**:310–314.

28. **Hektor, H. J., and L. Dijkhuizen.** 1996. Mutational analysis of primary alcohol metabolism in the methylotrophic actinomycete *Amycolatopsis methanolica*. *FEMS Microbiol. Lett.* **144**:73–80.

29. **Hermans, J., and J. A. M. de Bont.** 1996. Techniques for genetic engineering in mycobacteria. *Antonie van Leeuwenhoek* **69**:243–256.

30. **Hermans, J., J. G. Boschloo, and J. A. M. de Bont.** 1990. Transformation of *Mycobacterium aurum* by electroporation: the use of glycine, lysozyme and isonicotinic acid hydrazide in enhancing transformation efficiency. *FEMS Microbiol. Lett.* **72**:221–224.

31. **Hermans, J., C. Martin, G. N. M. Huyberts, T. Goosen, and J. A. M. de Bont.** 1991. Transformation of *Mycobacterium aurum* and *Mycobacterium smegmatis* with the broad-host-range Gram-negative cosmid vector pJRD215. *Mol. Microbiol.* **5**:1561–1566.

32. **Hermans, J., I. M. L. Suy, and J. A. M. de Bont.** 1993. Transformation of Gram-positive microorganisms with the Gram-negative broad-host-range cosmid vector pJRD215. *FEMS Microbiol. Lett.* **108**:210–204.

33. **Hopwood, D. A., M. J. Bibb, K. F. Chater, T. Kieser, C. J. Bruton, H. M. Kieser, D. J. Lydiate, C. P. Smith, J. M. Ward, and H. Schrempf.** 1985. *Genetic Manipulation in Streptomyces: A Laboratory Manual*. John Innes Foundation, Norwich, United Kingdom.

35. **Jacobs, W. R., Jr., G. V. Kalpana, J. D. Cirillo, L. Pascopella, S. B. Snapper, R. A. Udani, W. Jones, R. G. Barletta, and B. R. Bloom.** 1991. Genetic systems for mycobacteria. *Methods Enzymol.* **204**:537–555.

36. **Jäger, W., A. Schäfer, J. Kalinowski, and A. Pühler.** 1995. Isolation of insertion elements from Gram-positive *Brevibacterium*, *Corynebacterium*, and *Rhodococcus* strains using the *Bacillus subtilis sacB* gene as a positive selection marker. *FEMS Microbiol. Lett.* **126**:1–6.

37. **Kremer, L., A. Baulard, J. Estaquier, O. Poulain-Godefroy, and C. Locht.** 1995. Green fluorescent protein as a new expression marker in mycobacteria. *Mol. Microbiol.* **17**:913–922.

38. **Kumar, C. V., J. J. R. Coque, and J. F. Martin.** 1994. Efficient transformation of the cephamycin C producer *Nocardia lactamdurans* and development of shuttle and promoter-probe cloning vectors. *Appl. Environ. Microbiol.* **60**:4086–4093.

39. **Lal, R., S. Lal, E. Grund, and R. Eichenlaub.** 1991. Construction of a hybrid plasmid capable of replication in *Amycolatopsis mediterranei*. *Appl. Environ. Microbiol.* **57**:665–671.

40. **Lechevalier, M. P., and H. Lechevalier.** 1985. Biology of actinomycetes not belonging to the genus *Streptomyces*, p. 315–358. *In* A. L. Demain and N. A. Solomon (ed.), *Biology of Industrial Microorganisms*. Benjamin/Cummings Publishing Company, Inc., London.

41. **Lechevalier, M. P., H. Prauser, D. P. Labeda, and J.-S. Ruan.** 1986. Two new genera of nocardioform actinomycetes: *Amycolata* gen. nov. and *Amycolatopsis* gen. nov. *Int. J. Syst. Bacteriol.* **36**:29–37.

42. **Lee, M. H., L. Pascopella, W. R. Jacobs, Jr., and G. F. Hatfull.** 1991. Site-specific integration of mycobacteriophage L5: integration-proficient vectors for *Mycobacterium smegmatis*, *Mycobacterium tuberculosis*, and bacille Calmette-Guérin. *Proc. Natl. Acad. Sci. USA* **88**:3111–3115.

43. **MacNeil, D. J.** 1988. Characterization of a unique methyl-specific restriction system in *Streptomyces avermitilis*. *J. Bacteriol.* **170**:5607–5612.

44. **Madoń, J., and R. Hütter.** 1991. Transformation system for *Amycolatopsis mediterranei*: direct transformation of mycelium with plasmid DNA. *J. Bacteriol.* **173**:6325–6331.

45. **Madoń, J., P. Moretti, and R. Hütter.** 1987. Site-specific integration of pMEA100 in *Nocardia mediterranei*. *Mol. Gen. Genet.* **209**:257–264.

46. **Marklund, B.-I., D. P. Speert, and R. W. Stokes.** 1995. Gene replacement through homologous recombination in *Mycobacterium intracellulare*. *J. Bacteriol.* **177**:6100–6105.

47. **Matsushima, P., M. A. McHenney, and R. H. Baltz.** 1987. Efficient transformation of *Amycolatopsis orientalis* (*Nocardia orientalis*) protoplasts by *Streptomyces* plasmids. *J. Bacteriol.* **169:**2298–2300.

48. **McFadden, J.** 1996. Recombination in mycobacteria. *Mol. Microbiol.* **21:**205–211.

49. **Moretti, P., G. Hintermann, and R. Hütter.** 1985. Isolation and characterization of an extrachromosomal element from *Nocardia mediterranei*. *Plasmid* **14:**126–133.

50. **Nagy, I., G. Schoofs, J. Vanderleyden, and R. De Mot.** 1997. Transposition of the IS*21*-related element IS*1415* in *Rhodococcus erythropolis*. *J. Bacteriol.* **179:**4635–4638.

51. **Pelicic, V., M. Jackson, J.-M. Reyrat, W. R. Jacobs, B. Gicquel, and C. Guilhot.** 1997. Efficient allelic exchange and transposon mutagenesis in *Mycobacterium tuberculosis*. *Proc. Natl. Acad. Sci. USA* **94:**10955–10960.

52. **Pelicic, V., J.-M. Reyrat, and B. Gicquel.** 1996. Generation of unmarked directed mutations in mycobacteria, using sucrose counter-selectable suicide vectors. *Mol. Microbiol.* **20:**919–925.

53. **Quan, S., and E. R. Dabbs.** 1993. Nocardioform arsenic resistance plasmid characterization and improved *Rhodococcus* cloning vectors. *Plasmid* **29:**74–79.

54. **Rainey, F. A., J. Burghardt, R. M. Kroppenstedt, S. Klatte, and E. Stackebrandt.** 1995. Phylogenetic analysis of the genera *Rhodococcus* and *Nocardia* and evidence for the evolutionary origin of the genus *Nocardia* from within the radiation of *Rhodococcus* species. *Microbiology* **141:** 523–528.

55. **Ramakrishnan, L., H. T. Tran, N. A. Federspiel, and S. Flakow.** 1997. A *crtB* homolog essential for photochromogenicity in *Mycobacterium marinum*: isolation, characterization, and gene disruption via homologous recombination. *J. Bacteriol.* **179:** 5862–5868.

56. **Ranes, M. G., J. Rauzier, M. Lagranderie, M. Gheorghiu, and B. Gicquel.** 1990. Functional analysis of pAL5000, a plasmid from *Mycobacterium fortuitum*: construction of a "mini" *Mycobacterium-Escherichia coli* shuttle vector. *J. Bacteriol.* **172:**2793–2797.

57. **Sadashiva Karnik, S., and K. P. Gopinathan.** 1983. Transfection of *Mycobacterium smegmatis* SN2 with mycobacteriophage I3 DNA. *Arch. Microbiol.* **136:**275–280.

58. **Sander, P., A. Meier, and E. C. Böttger.** 1995. rpsL$^+$: a dominant selectable marker for gene replacement in mycobacteria. *Mol. Microbiol.* **16:**991–1000.

59. **Schupp, T., and M. Divers.** 1986. Protoplast preparation and regeneration in *Nocardia mediterranei*. *FEMS Microbiol. Lett.* **36:**159–162.

60. **Seoane, A., J. Navas, and J. M. Garcia Lobo.** 1997. Targets for pSAM2 integrase-mediated site-specific integration in the *Mycobacterium smegmatis* chromosome. *Microbiology* **143:**3375–3380.

61. **Singer, M. E. V., and W. R. Finnerty.** 1988. Construction of an *Escherichia coli-Rhodococcus* shuttle vector and plasmid transformation in *Rhodococcus* spp. *J. Bacteriol.* **170:** 638–645.

62. **Snapper, S. B., L. Lugosi, A. Jekkel, R. E. Melton, T. Kieser, B. R. Bloom, and W. R. Jacobs, Jr.** 1988. Lysogeny and transformation in mycobacteria: stable expression of foreign genes. *Proc. Natl. Acad. Sci. USA* **85:**6987–6991.

63. **Stolt, P., and N. G. Stoker.** 1996. Functional definition of regions necessary for replication and incompatibility in the *Mycobacterium fortuitum* plasmid pAL5000. *Microbiology* **142:**2795–2802.

63a. **van der Geize, R., J. W. Vrijbloed, and L. Dijkhuizen.** 1998. Development of an efficient electrotransformation protocol for *Rhodococcus erythropolis* SQ1 and its use in gene cloning. Unpublished data.

63b. **Vrijbloed, J. W.** Unpublished data.

64. **Vrijbloed, J. W., M. Jelínková, G. I. Hessels, and L. Dijkhuizen.** 1995. Identification of the minimal replicon of plasmid pMEA300 of the methylotrophic actinomycete *Amycolatopsis methanolica*. *Mol. Microbiol.* **18:**21–31.

65. **Vrijbloed, J. W., J. Madoń, and L. Dijkhuizen.** 1994. A plasmid from the methylotrophic actinomycete *Amycolatopsis methanolica* capable of site-specific integration. *J. Bacteriol.* **176:**7087–7090.

66. **Vrijbloed, J. W., J. Madoń, and L. Dijkhuizen.** 1995. Transformation of the methylotrophic actinomycete *Amycolatopsis methanolica* with plasmid DNA: stimulatory effect of a pMEA300-encoded gene. *Plasmid* **34:**96–104.

67. **Vrijbloed, J. W., N. M. J. van der Put, and L. Dijkhuizen.** 1995. Identification and functional analysis of the transfer region of plasmid pMEA300 of the methylotrophic actinomycete *Amycolatopsis methanolica*. *J. Bacteriol.* **177:**6499–6505.

68. **Vrijbloed, J. W., J. van Hylckama Vlieg, N. M. J. van der Put, G. I. Hessels, and L. Dijkhuizen.** 1995. Molecular cloning with a pMEA300-derived shuttle vector and characterization of the *Amycolatopsis methanolica* prephenate dehydratase gene. *J. Bacteriol.* **177:**6666–6669.

69. **Warhurst, A. M., and C. A. Fewson.** 1994. Biotransformations catalyzed by the genus *Rhodococcus*. *Crit. Rev. Biotechnol.* **14:**29–73.

70. **Yeung, M. K., and C. S. Kozelsky.** 1994. Transformation of *Actinomyces* spp. by a Gram-negative broad-host-range plasmid. *J. Bacteriol.* **176:**4173–4176.

Corynebacteria

JUAN F. MARTÍN AND JOSÉ A. GIL

32

Corynebacteria are pleomorphic, asporogenous gram-positive bacteria widely distributed in nature (6, 38). In the last edition of the *Bergey's Manual of Systematic Bacteriology*, corynebacteria are included in Section 15 as irregular, nonsporing, gram-positive rods together with a diverse collection of 22 genera (29).

Some nonpathogenic soil corynebacteria are widely used for the production of amino acids (1), for steroid conversions, for degradation of hydrocarbons, and for terpenoid oxidation (41). Amino-acid-producing strains include members of the genera *Corynebacterium* and *Brevibacterium* and of the related genera *Arthrobacter* and *Microbacterium*. Other corynebacteria are animal (e.g., *Corynebacterium diphtheriae, Corynebacterium xeroxis*) or plant pathogens (e.g., *Rhodococcus* [formerly *Corynebacterium*] *fascians*). Many of the plant pathogenic corynebacteria are only distantly related to the amino-acid-producing corynebacteria, and they are being reclassified into *Rhodococcus* and other genera. The results of 16S rRNA studies confirmed that corynebacteria is a heterogeneous group of different taxa (39, 67).

Corynebacterium glutamicum has a genome mass of 1.7 $\times$ 10^9 (approximately 2,250 kb) as determined by DNA renaturation studies (13). Genome mapping by pulsed-field gel electrophoresis (PFGE) has confirmed that the genome size ranges from 2,987 kb in *C. glutamicum* to 3,105 kb in *Brevibacterium linens* (4, 11). A physical and genetic map of *C. glutamicum* has been published (4). A large effort has been made in the last decade in several European, Japanese, and American laboratories for the development of advanced molecular genetics in this group of bacteria.

32.1. PLASMIDS AND CLONING VECTORS

During the 1980s a relatively large number of endogenous plasmids (most of them cryptic) were found in soil coryneform bacteria and pathogenic corynebacteria (Fig. 1) (43). Plasmids from C *diphtheriae* and *C. xeroxis* contained genes encoding resistance to chloramphenicol, erythromycin, kanamycin, streptomycin, and tetracycline (34, 61). In addition, plasmids isolated from other human pathogenic corynebacteria showed resistance markers to penicillin, tetracycline, gentamicin, neomycin, and fusidic acid (32). Since these plasmids were large and had low copy number, their resistance markers were subcloned into small, multi-copy cryptic plasmids; alternatively, resistance markers were derived from other gram-positive bacteria. The replication origins of pBL1 of *Brevibacterium lactofermentum* (57) and pCG1 of *C. glutamicum* (51) (also known as pHM1519 [46], pSR1 [77], and pCG100 [63]) have been used to construct vectors. The initial vectors containing a single marker were later converted into vectors with two or more selective markers (56). The best markers for corynebacteria (and therefore most widely used in different vectors) are the kanamycin resistance gene of transposon Tn*5* (which confers resistance to more than 200 μg of kanamycin per ml), the streptomycin and spectinomycin resistance genes from the native plasmid pCG4 of *C. glutamicum*, the erytromycin resistance gene from pTP10 of *C. xeroxis*, the chloramphenicol resistance gene of *Streptomyces acrimycini*, the hygromycin resistance gene of *Streptomyces hygroscopicus* (56), and the phleomycin resistance gene of transposon Tn*5* (19).

The host range of most replicons used for constructing vectors for corynebacteria is narrow. Some vectors are able to replicate in other corynebacterial hosts; e.g., pCG4-derived plasmids of *C. glutamicum* replicate in *Corynebacterium herculis, Brevibacterium flavum*, and *Microbacterium ammoniaphilum* (31). Similarly, pBL1-derived vectors transform *C. glutamicum, Corynebacterium callunae, B. linens* (56), and *R. fascians*, although the efficiency of transformation in the latter organism is very low (15). pGA1, a plasmid isolated from *C. glutamicum* LP-6, was shown to be compatible in this species with plasmids pBL1, pCC1, and pHM1519, which are the replicons routinely used in corynebacteria (66). Chion and coworkers (10) tested replicons from several cryptic plasmids of coryneform bacteria to construct a host-vector system for *Brevibacterium* sp. R312, a strain of industrial interest for its ability to produce nitrile hydratase and amidase. Only the *C. glutamicum* pSR1 replicon was found to be suitable for transforming *Brevibacterium* sp. R312.

An additional problem related to vector development is the instability of large hybrid plasmids in corynebacteria owing to the fact that all available *Corynebacterium* and *Brevibacterium* strains are likely to be recombination-proficient. However, plasmid rearrangement might be advantageous under some conditions. The large plasmid pUL61 containing the *kan, bla*, and *tsr* genes was spontaneously trimmed down in *B. lactofermentum*, giving rise to

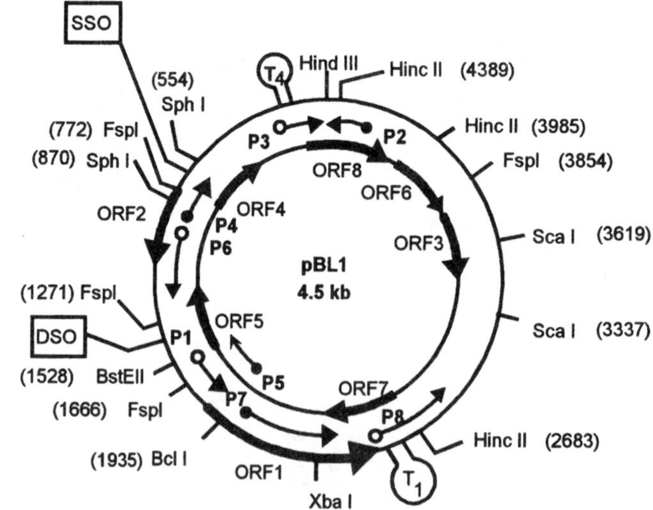

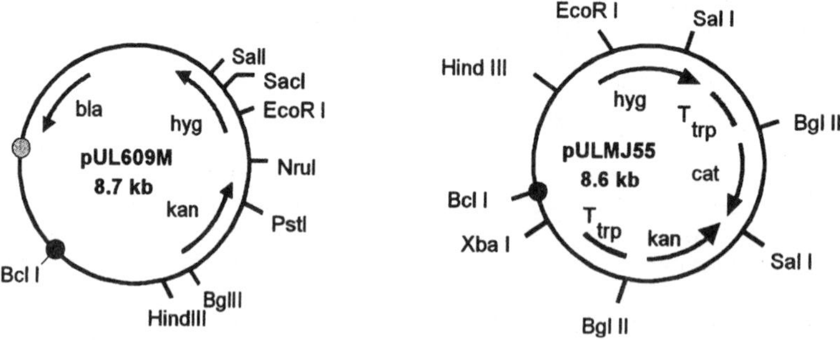

FIGURE 1 (Upper) Detailed map of pBL1, an endogenous plasmid of *B. lactofermentum*, showing the encoded open reading frames (ORF) and the promoters P_1 to P_8 found by using promoter-probe vectors. Open circles with arrows indicate promoters that are expressed both in *Escherichia coli* and in corynebacteria. Full circles with arrows indicate promoters that are expressed only in corynebacteria. T_1 and T_4 correspond to transcriptional terminators, SSO to the single-strand replication origin, and DSO to the double-strand origin. (Lower) Restriction maps of the bifunctional (shuttle) plasmid pUL609M (left) and the positive selection vector pULMJ55 (right). T*trp* indicates the tryptophan operon terminators. When the stuffer region of pULMJ55 is removed by *Bgl*II digestion, the resulting plasmid is lethal unless a foreign DNA is inserted between the two terminators (see the text for details). *bla*, β-lactamase gene; *hyg*, hygromycin phosphotransferase gene; *kan*, kanamycin phosphotransferase gene.

two new small vectors, pUL330 and pUL340, that were, thereafter, completely stable and retained the kanamycin resistance marker (56). These plasmids have lost all of the pBR322 moiety of pUL61 and part of pBL1, retaining, however, the pBL1 replication origin. The frequency of transformation of *B. lactofermentum* by these small vectors was in the range of 10^5 to 10^6 transformants per μg of DNA, whereas for pUL61 the transformation frequency was 10^3 to 10^4 transformants per μg of DNA (56). Similarly, a high-copy-number plasmid, pAJ43, was isolated after deletion of part of the pAM330::pBR325 cointegrate (45). Little is known about the control of the number of plasmid copies in corynebacteria. They range from 1 to 2 for the very large plasmids to about 140 copies for the smaller ones (see review by Martín et al. [43]). We obtained plasmid variants in some transformants of *B. lactofermentum* with more than 500 copies per chromosome. Some of these plasmids were lethal when reintroduced into *B. lactofermentum*, but others were stable (9a). Moreover, no detailed studies have been carried out on plasmid stability in corynebacteria, specially under long-term batch fermentation conditions. Plasmids pGX1415, pGX1416, and pGX1418 were apparently lost from *B. lactofermentum* in the absence of selection for kanamycin resistance (65).

The replication functions of some plasmids of corynebacteria has been studied. Kurusu et al. (37) reported a 673-bp fragment of plasmid pBY503 of *Brevibacterium stationis* involved in plasmid partition. This sequence is also functional in *C. glutamicum* and *B. flavum*. The minimal replication region of plasmid pSR1 was identified by Archer and Sinskey (3). Trautwetter and Blanco (71) identified five proteins encoded by plasmid pCG100 (=pCG1) of 12.4, 18, 32, 14.2, and 16 kDa and proposed that the 32-kDa protein was involved in plasmid replication since inactivation of this protein prevented plasmid replication.

The minimal region for autonomous replication of pBL1, a 4.5-kb cryptic plasmid of *B. lactofermentum* ATCC 13869 that has been used to construct a variety of corynebacterial vectors, was shown to be contained on a 1.8-kb *Hin*dII-*Sph*I DNA fragment. This region contains two open reading frames (ORFs) (ORF1 and ORF5), which are essential for pBL1 replication. Accumulation of single-strand intermediates in some of the constructions indicates that plasmid pBL1 replicates via the rolling-circle replication model; its plus and minus strands were identified by hybridization with two synthetic oligonucleotide probes complementary to each pBL1 strand (15). ORF1 seems to encode the Rep protein and showed homology with sequences for Rep proteins from *Streptomyces* plasmids that replicate via rolling-circle replication, such as pIJ101, pSB24, and pJV1.

32.2. SHUTTLE VECTORS WITH SEVERAL SINGLE RESTRICTION SITES

A series of bifunctional (shuttle) plasmids that may be used in *Escherichia coli* and corynebacteria have been constructed using the origin of replication and stability sequences of the corynebacterial plasmid pULRS6 (a pBL1 derivative) and the ColE1 replication origin of *E. coli* plasmids.

The pUL600M series of shuttle plasmids contain kanamycin and hygromycin resistance markers that are expressed in corynebacteria as well as in *E. coli*. They also carry the ampicillin resistance marker subcloned from pUC13 for the selection of transformants in *E. coli*. The best plasmids of this series are pUL610M and pUL609M, of 8.8 kb and 8.7 kb, respectively (42). pUL609M (Fig. 1) contains seven unique sites for cloning with different restriction endonucleases. Insertions at the *Bgl*II or *Pst*I sites inactivate the kanamycin resistance gene, whereas insertions in *Eco*RI, *Sal*I, and *Sac*I inactivate the hygromycin resistance marker.

Plasmids of the pUL600M series have been routinely used for subcloning and expressing genes of corynebacteria. Transformation of protoplasts and electroporation of *B. lactofermentum* using these plasmids is obtained with high efficiency (about 10^6 transformants per μg of DNA), similar to that obtained with previously developed plasmids pULRS6 and pULRS8 (58). A similar shuttle plasmid was constructed by Yeh et al. (75), and a different vector, pCEM500, able to replicate in *E. coli* and *B. flavum* was constructed by Nesvera and coworkers (49).

Restriction analysis after amplification of the recombinant plasmids in corynebacteria revealed that fragments of DNA subcloned in these vectors are usually stable and do not suffer deletions or rearrangements of the heterologous DNA.

32.3. POSITIVE SELECTION VECTORS AND MOBILIZABLE VECTORS

Positive selection vectors, e.g., plasmid pULMJ55 (Fig. 1), have been developed in our laboratory (7). pULMJ55 is a pBL1 derivative containing the hygromycin resistance gene, and a stuffer region flanked by two copies of the terminator of the tryptophan operon (20) inserted in opposite orientations (Fig. 1). When the stuffer region is removed by *Bgl*II digestion and recircularization, the resulting plasmid, which contains a 376-bp perfect palindrome, is lethal in *B. lactofermentum* unless a DNA fragment is inserted between the two terminators. Using this positive selection vector, 90 to 95% of the transformants obtained contain recombinant plasmids.

Schwarzer and Pühler (62) reported the construction of mobilizable shuttle vectors that can be transferred by conjugation between *E. coli* and *C. glutamicum*. The conjugation efficiency could be increased by a heat treatment step of *C. glutamicum* cells prior to mating with *E. coli*. Mobilizable shuttle vectors have been used for conjugative transfer of genes from *E. coli* to corynebacteria (see below).

32.4. PROMOTER-PROBE VECTORS

Promoter-probe vectors are useful to clone (or subclone from the genes already available) sequences active in transcription initiation. Construction of promoter-probe vectors has been reported by Morinaga et al. (47) and Cadenas et al. (9). We have constructed several monofunctional (for corynebacteria) and bifunctional (for *E. coli* and corynebacteria) promoter-probe vectors using the promoterless kanamycin phosphotransferase gene of Tn*5* (*kan*) or the promoterless α-amylase gene (*amy*) of *Streptomyces griseus* (7). Three of these vectors, pULMJ77, pULMJ88, and pULMJ51 (Fig. 2), contain (i) replication origins of plasmids of *B. lactofermentum* (pBL1) and/or *E. coli* (pBR322), (ii) a promoterless kanamycin resistance marker that is expressed both in *E. coli* and in corynebacteria as an indicator gene, and (iii) a selectable gene for plasmid amplification (*bla*) in *E. coli*. The fourth vector (pULMJ95) contains the promoterless *amy* gene as reporter.

To prevent readthrough from upstream promoters, the terminator of the tryptophan operon of *B. lactofermentum* (as a 180-bp *Sau*3AI fragment) was inserted upstream of the kanamycin resistance gene. The terminator of the *B. lactofermentum* tryptophan operon terminates transcription very efficiently in both corynebacteria (pULMJ51, pULMJ77, pULMJ88) and *E. coli* (pULMJ51), whereas the phage *fd* terminator does not terminate transcription from upstream promoters incorynebacteria.

The promoterless *kan* gene in plasmids pULMJ51, pULMJ77, or pULMJ88 carried the ribosome binding site needed for kanamycin phosphotransferase synthesis. Expression of the reporter gene will be regulated by those control sequences located in the promoter being cloned, but it does not require a separate ribosome binding sequence in the cloned DNA fragment.

Promoter-probe vector pULMJ95 contains the promoterless *amy* gene downstream from the tryptophan terminator and is particularly suitable for the detection of promoters that are activated late during the growth phase. α-Amylase is very efficiently secreted in corynebacteria (8) without any apparent bottleneck, and its activity can be detected using simple plate tests.

32.5. PHAGE-DERIVED VECTORS

Several phages, other than corynephage β of *C. diphtheriae*, have been identified in soil corynebacteria. Ozaki and coworkers (51, 52) first described three phages, CG1, CG3, and CG5, that transfect protoplasts of species of *Corynebacterium*, *Brevibacterium*, and *Mycobacterium*. Later Yeh et

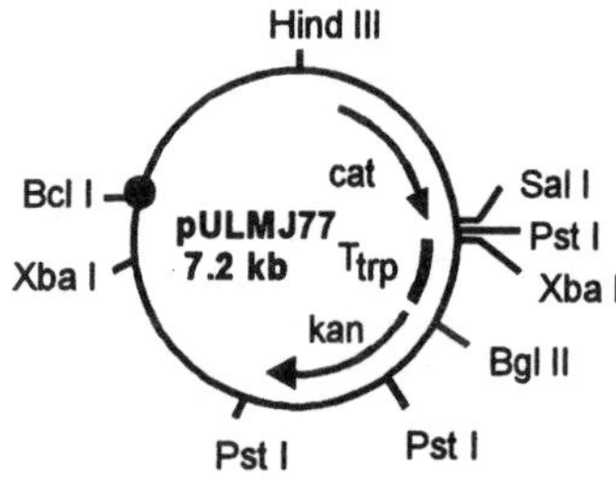

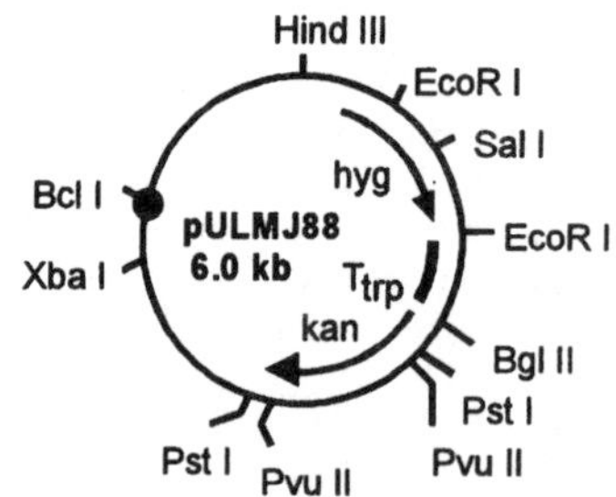

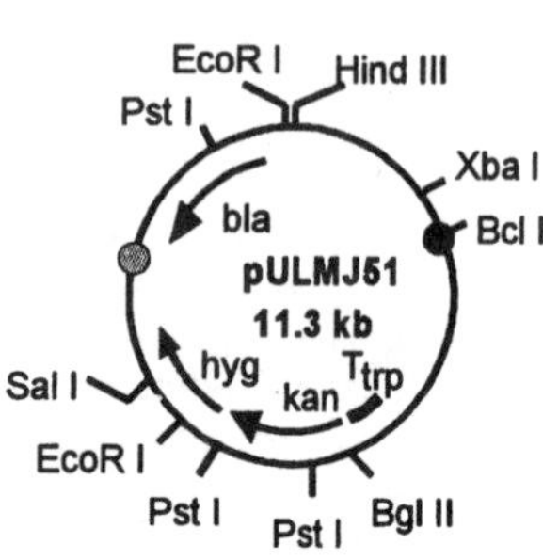

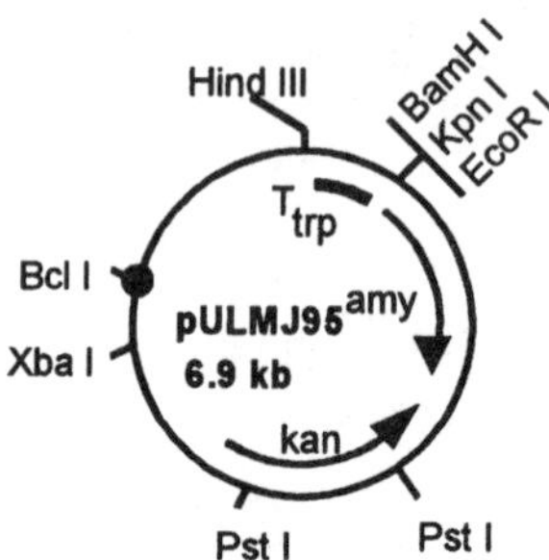

FIGURE 2 Restriction maps of the promoter-probe vectors pULMJ77, pULMJ88, pULMJ51, and pULMJ95. Note that pULMJ77, pULMJ88, and pULMJ95 are monofunctional vectors containing the origin of replication of pBL1 (●), whereas pULMJ51 is a bifunctional vector containing the origins of replication of pBR322 (○) and pBL1 (●). T*trp*, tryptophan operon terminator; *cat*, chloramphenicol acetyltransferase gene; *hyg*, hygromycin phosphotransferase gene; *bla*, β-lactamase gene; *amy*, α-amylase gene.

al. (75) reported a lytic phage, CS1, which infects *Corynebacterium lillium* but not other corynebacteria, and Trautwetter and Blanco (70) described 20 different phages that infect *Brevibacterium* and *Arthrobacter*. More recently, Koptides et al. (35) described a new phage BFK20 that infects *B. flavum*.

However, development of corynebacteria phage derivatives has advanced very little. The genome of phage BFK20 was used as a source of promoters and transcriptional terminators (35). Miwa et al. (45) constructed a cosmid by ligating the cohesive ends of the *B. lactofermentum* phage f1A to plasmid pAJ43, but no further developments of this vector have been published and the use of this cosmid is limited because of the lack of an in vitro packaging system.

32.6. TRANSPOSONS AND INSERTION ELEMENTS

Several insertion sequences and transposons have been identified in corynebacteria in the last few years. Two different insertion elements, IS*1206* and IS*31831*, were isolated by different groups (5, 74) from *C. glutamicum*. IS*1206* is related to the IS3 group of insertion elements, whereas IS*31831* contains a single ORF that encodes a transposase similar to that of IS*1096* of *Mycobacterium smegmatis*.

A different element, IS*13869* of 1.4 kb, isolated from *B. lactofermentum* (12), contained a long ORF (ORF1) inside a perfect 26-bp terminal inverted repeat that was flanked by an 8-bp direct repeat. The protein encoded by ORF1 (Mr 49380) showed extensive similarity to the putative transposases of the insertion elements IS*1096* of *M. smegmatis*, *tpn*A of *Pseudomonas* sp., and IS*31831* of *C. glutamicum*. Four copies of IS*13869* were observed in the wild-type *B. lactofermentum*, but only three occurred in a *rec*A mutant. The IS*13869* element is not related to IS*1206* of *C. glutamicum* (12).

Vertès and coworkers (73) described the construction of transposons and minitransposons derived from IS*31831* that are useful to perform transposon mutagenesis in corynebacteria.

32.7. INTRODUCTION OF EXOGENOUS DNA INTO CORYNEBACTERIA

Amino-acid-producing corynebacteria have been found to be highly restrictive and, therefore, difficult to transform with exogenous DNA. Three different strategies have been followed to introduce exogenous DNA: transformation of protoplasts, electroporation, and conjugation between *E. coli* and corynebacteria, as discussed below.

32.7.1. Transformation of Protoplasts

Obtaining and regenerating protoplasts from corynebacteria is a tricky procedure (30) because of the mycolic acid content in the cell walls of these bacteria. Weakening of the cell walls by growing the cells in subinhibitory concentrations of penicillin (31, 57) or in the presence of glycine (77) improved the transformation efficiencies. Indeed, we observed that a short lysozyme treatment without removal of the cell wall (e.g., formation of spheroplasts) was sufficient to obtain a good transformation. Under these conditions, regeneration of protoplasts was facilitated (69). Efficiencies of up to 10^6 transformants per μg of DNA were obtained by optimizing the age of the cultures used to prepare protoplasts and the DNA and polyethylene glycol (PEG) concentrations (31, 56, 77; see review by Martín et al. [43]). To bypass the restriction systems, Thierbach and coworkers (69) used a heat shock treatment (48–49°C for 9 min) of the protoplasts prior to transformation.

32.7.2. Electroporation

Introduction of DNA by electrotransformation (usually known as electroporation) has the advantage that it does not require obtaining and regenerating protoplasts, which is a limiting step for PEG-mediated transformation. Although the reported efficiencies of electroporation with plasmid DNA are relatively high (up to 10^7 transformants per μg of DNA) (14), the routine results in the laboratory using ligated DNA are less satisfactory (10^4 to 10^5 transformants per μg of DNA). Haynes and Britz (23, 24) reported improved electroporation efficiencies by growing the cells in the presence of glycine or Tween 80 (23, 24), suggesting that weakening of the cell wall or alteration of the mycolic acid layer by these treatments makes the cell more sensitive to electroporation. A simple selection system after electroporation based on the complementation of auxotrophic markers and selection of complemented clones has been reported by Kurusu et al. (36).

Electroporation of highly restrictive corynebacteria has also been achieved by loading the cells with high amounts of synthetic DNA obtained by PCR (2).

32.7.3. Conjugal Plasmid Transfer from *E. coli* to Corynebacteria

Conjugal transfer by broad-host-range IncP-type resistance plasmids within gram-negative bacteria is well known. Schäfer and coworkers (60) applied this system to corynebacteria. For mating experiments, plasmids were introduced into the mobilizing strain *E. coli* S17-1 carrying an RP4 derivative integrated into the chromosome, which provides the transfer functions necessary for mobilization. Introduction of plasmids into corynebacteria by conjugal transfer was markedly increased by heat treatment of the recipient cells or when the restriction system of the recipient strain was eliminated by mutation. A similar stimulation of conjugal transfer was observed by treatment of the recipient cells with organic solvents or detergents or by pH changes (59).

32.8. GENE DISRUPTION AND REPLACEMENT

Disruption of genes in *B. lactofermentum* and *C. glutamicum* by single crossing-over with a homologous DNA fragment in the plasmids has been obtained after transformation (50) using nonreplicative plasmids. Similar results were reported in *Brevibacterium divaricatum* (68).

Gene disruption and replacement in corynebacteria was facilitated by introducing via conjugation *E. coli* plasmids carrying manipulated *C. glutamicum* DNA fragments (62). A 650-bp internal fragment of the *lysA* gene was deleted to generate a *lysA* mutation in the plasmid. This was used to replace by double recombination the endogenous gene previously interrupted by a chloramphenicol insertion.

Small mobilizable vectors based on the *E. coli* plasmids pK18mob and pK19mob with the broad-host-range transfer machinery of plasmid RP4 and a modified *sacB* gene, which confers sucrose sensitivity to gram-negative and some gram-positive bacteria (27), were constructed by Schäfer and coworkers (59). These plasmids (named pK18mob SacB and pK19mob SacB) facilitate gene disruption and allelic exchange by homologous recombination. They have been used to generate targeted deletions in *C. glutamicum*.

Transformation with a repliconless construction called an integron that contains a selectable marker and the gene (encoding citrate synthase) has been reported to be a suitable method for stable integration in highly restrictive corynebacteria (55).

32.9. PROMOTERS

Bacterial promoters recognized by the major RNA polymerase σ-factors, e.g., σ^{70} of *E. coli* or σ^A of *Bacillus subtilis*, are similar in their −10 regions and, to some extent, also in their −35 regions (21, 22, 25, 40). The consensus sequences conserved in all σ^{70} and σ^A promoters contain two hexameres, TTGaca and TAtaaT, located at −35 and −10, respectively, separated by 17 ± 1 bp. Graves and Rabinowitz (18) found a conserved TG motif located 1 bp upstream of the TATAAT hexamer in more than 50% of all promoters of gram-positive bacteria in addition to the −35 and −10 consensus sequences. Promoters of different gram-positive bacteria also contain an A+T-rich track upstream of the −35 region, including species of *Bacillus* (25), *Lactobacillus* (44, 54), and *Streptococcus* (48, 72). Similar features were observed in the promoters of a few genes cloned from corynebacteria (reviewed in references 28 and 41).

Several promoters of corynebacterial genes were not expressed in *E. coli*, whereas they were efficiently transcribed in *C. glutamicum* or *B. lactofermentum*. Therefore, at least two different types of promoters occur in *E. coli*, as described also for *Streptomyces* species. Type I or corynebacterium specific promoters are expressed only in corynebacteria, and type II or *E. coli*-like promoters are expressed efficiently in *E. coli* in addition to corynebacteria (9, 42).

Pátek and coworkers (53) cloned random *Sau*3A fragments from the chromosomal DNA of *C. glutamicum* and the corynebacterial phage pGA1 in a promoter-probe vector and identified their transcriptional start sites by primer extension analysis. Results revealed consensus sequences centered in the −35 (ttGcca) and −10 (TAnaaT) comparable to the −35 and −10 regions of *E. coli*-type. It remains unclear which is the structure of other corynebacterial promoters that are not expressed in *E. coli*. Graves and Rabinowitz (18) found in a set of 29 different promoters, obtained mainly from the genera *Bacillus* and *Staphylococcus*, a conserved TG motif located 1 bp upstream of the −10 hexamer and an A-track at position −43 relative to transcriptional start site (in more than 50% of the characterized promoters). Similar sequence motifs have been described in an analysis of 236 *B. subtilis* promoters (25) and in promoters from *Lactobacillus* and *Streptococcus* (44, 48, 54, 72). Corynebacterial specific promoters may have similar characteristics.

PROTOCOLS

Based on the previously described information, a series of protocols for molecular genetics of corynebacteia have been developed over the last two decades in several laboratories. We describe here the optimized procedures as used in our laboratory.

Corynebacteria cultures on agar plates are used for isolation of a single colony or a single transformant by standard microbiological methods. Because corynebacteria do not sporulate, only those strains that are needed for short-term use can be kept as cultures on agar at 4°C and should be subcultured after 2 to 3 months. The best method for keeping strains for longer periods is to prepare cell suspensions in 20% glycerol and freeze them at −20°C or by lyophilization in skim milk.

CORYNEBACTERIA CULTURES ON AGAR

The most frequently used solid media for corynebacteria are the following.

TSA medium

Casein peptone	17	g
Soy peptone	3	g
NaCl	5	g
Glucose	2.5	g
Agar	20	g
Distilled water	1,000	ml

TSA is a commercially available medium and is prepared by dissolving 30 g of Oxoid Tryptone Soy Broth (TSB) powder in 1,000 ml of distilled water. The medium is distributed in 100-ml quantities in 200-ml bottles containing 2 g of agar and is sterilized by autoclaving at 121°C for 20 min.

Minimal medium for corynebacteria (MMC) (30)

MMC is used for the isolation and characterization of auxotrophic mutants. Some corynebacteria may require additional growth factors (38).

Agar	20	g
$(NH_4)_2SO_4$	10	g
Urea	3	g
KH_2PO_4	1	g
$MgSO_4 \cdot 7H_2O$	0.41	g
$FeSO_4 \cdot 4H_2O$	2	mg
$MnSO_4 \cdot 4H_2O$	2	mg
NaCl	50	mg
Biotin	50	μg
Thiamine	200	μg
Glucose	20	g

Biotin and thiamine are sterilized by filtration and added to the medium. Glucose is sterilized by autoclaving separatedly.

Place 2 g of agar powder in each 200-ml bottle; dissolve the other ingredients (except glucose, biotin, and thiamine) in distilled water, adjust to pH 7.3, and pour 100 ml per bottle; close the bottles and autoclave. Remelt the medium and add glucose, biotin, and thiamine before use.

RMMC (56)

RMMC, used for the regeneration of *B. lactofermentum* protoplasts, is MMC supplemented with sucrose (0.3 M) and 5.73 g/liter TES buffer (pH 7.2).

R2YE (26)

Sucrose	103	g
K_2SO_4	0.25	g
$MgCl_2 \cdot H_2O$	10.12	g
Glucose	10	g
Difco Yeast Extract (10%)	5	ml
Difco Casamino Acids	0.1	g
Distilled water	800	ml

Distribute 2.2 g of Difco Bacto Agar in 200-ml bottles and pour in 80 ml of the above solution. Sterilize by autoclaving.

Remelt the medium and add to each bottle before use:

KH_2PO_4 (0.5%)	1 ml
$CaCl_2 \cdot 2H_2O$ (3.68%)	8 ml
L-Proline (20%)	1.5 ml
TES buffer (5.73%, adjusted to pH 7.2)	10 ml
Trace element solution (see below)	0.2 ml
NaOH (1 N)	0.5 ml (does not need sterilization)

Trace element solution (per liter)

$ZNCl_2$	40 mg
$FeCl_3 \cdot 6H_2O$	200 mg
$CuCl_2 \cdot 2H_2O$	10 mg
$MnCl_2 \cdot 4H_2O$	10 mg
$Na_2B_4O_7 \cdot 10H_2O$	10 mg
$(NH_4)_6Mo_7O_{24} \cdot 4H_2O)$	10 mg

Soft nutrient agar (SNA)

Difco nutrient broth	8 g
Agar	3 g
Distilled water	1,000 ml

Melt in a microwave oven to dissolve the agar. Pour 50 ml in 200-ml bottles and sterilize by autoclaving.

GROWTH OF CORYNEBACTERIAL CELLS IN LIQUID MEDIA

For many purposes (e.g., preparation of protoplasts, DNA and RNA isolation, gene expression), corynebacteria are grown in different liquid media from isolated colonies or from cell suspensions maintained in 20% glycerol. Cultures are aerobically incubated at 30°C in an orbital shaker at 250 rpm in Erlenmeyer flasks (with or without baffles or with stainless-steel springs).

The most frequently used media are the following.

Tryptone soy broth (TSB)

General-purpose medium for corynebacteria. TSB is TSA without agar (see above).

MMCYC medium (56)

Medium used for growth of *B. lactofermentum* to prepare protoplasts. MMCYC is MMC supplemented with 1 g of Bacto Yeast Extract and 0.5 g of Casamino Acids per liter.

Antibiotic concentrations for selecting resistant strains

(Stock solutions are in parentheses.)
Ampicillin (100 mg/ml): 50–100 μg/ml
Kanamycin (100 mg/ml): 5–80 μg/ml
Chloramphenicol (100 mg/ml, in 50% ethanol): 50–100 μg/ml
Hygromycin (200 mg/ml): 50–100 μg/ml

Growth factor supplements

Dissolve amino acids in water at 10 mg/ml, sterilize by autoclaving, and use at a final concentration of 50 μg/ml.

Prepare vitamins in water at 0.1 mg/ml, sterilize by filtration, and use at a final concentration of 0.5 μg/ml.

PREPARATION OF PROTOPLASTS FROM CORYNEBACTERIA

Transformation of protoplasts by plasmid DNA was the first method used for introducing DNA in corynebacteria. However, this method has limitations and cannot be used in all corynebacteria.

The protocol described is used in our laboratory for *C. glutamicum* and *B. lactofermentum*; transformation efficiencies of 10^5 to 10^6 transformants per μg of DNA are obtained. Protoplasts are prepared by a modification of the method of Kaneko and Sakaguchi (30) as described by Santamaría et al. (57) as follows:

1. Prepare 50 ml of MMCYC medium in a 250-ml flask. Inoculate with a single colony or with 0.2 ml of corynebacterial cells frozen in 20% glycerol. Incubate overnight at 30°C in an orbital incubator.
2. Inoculate a 500-ml flask containing 100 ml of MMCYC with 2 ml of the overnight inoculum and incubate the culture until the optical density at 600 nm (OD_{600}) is approximately 0.6 (10^8 cells/ml).
3. Add 0.3 units/ml of penicillin G and continue the incubation for 3 h. The amount of penicillin G should be established for each strain. Some investigators use glycine or Tween 80 prior to the lysozyme treatment.
4. Harvest cells by centrifugation (in a bench centrifuge) at 3,000 rpm for 10 min. Discard supernatant.
5. Resuspend the pellet in 20 ml of FL solution. (FL is MMCYC diluted twice and supplemented with 0.41 M sucrose, 0.01 M $MgSO_4$, and 300 μg/ml lysozyme.)
6. Incubate 4 to 6 at 30°C. Monitor protoplast formation by microscopy or by plating in hypertonic (R2YE) or hypotonic (TSA) media.
7. Sediment protoplasts gently by spinning in a bench centrifuge (3,500 rpm for 5 min).
8. Discard supernatant and resuspend pellet in 10 ml of TSM buffer (Tris-HCl, 50 mM [pH 7.5]; sodium succinate, 500 mM; $CaCl_2 \cdot 2H_2O$, 30 mM).
9. Repeat step 7 and resuspend protoplasts in 2 ml of TSM buffer. Aliquots (0.2 ml) of the protoplast suspension can be frozen at −70°C or used directly for transformation.

Transformation of corynebacteria protoplasts and selection of transformants

This protocol for transformation was developed by Santamaría et al. (57).

1. Dispense 0.1 ml of protoplasts (10^9 to 10^{10} protoplasts per ml) into tubes.
2. Add DNA (CCC plasmid or ligation mixture) in up to 20 μl of TSM buffer per tube.
3. Immediately add 0.7 ml of 25% PEG 6000 dissolved in TSM buffer and mix by pipetting up and down two or three times.
4. Spread protoplast suspension (100 to 200 μl) on R2YE plates. Use TSM buffer to make dilutions if required.
5. Incubate plates at 30°C. After 24 h of incubation, overlay with 3 ml of nutrient broth (Difco) containing 0.4% agarose and the antibiotic used for selection.
6. Score for resistant colonies after 2 days.

ELECTROPORATION

Preparation of corynebacterial cells for electroporation

The protoplast transformation technique has been replaced in some cases by the electroporation technique for several reasons: (i) sometimes a particular strain is insensitive to lysozyme, (ii) it is difficult to generate protoplasts suitable for DNA transformation, (iii) the regeneration into viable cell is not always possible, or (iv) the transformation efficiencies are too low.

This protocol was taken from the work of Dunican and Shivnan (14), who showed that electroporation can be an efficient method for introducing plasmid DNA into coryneform bacteria. The method uses very dense cell suspensions (10^9 to 10^{10} cells/ml) with high electrical fields. The DNA and cells used for electroporation have to be free of ions; therefore, extensive washing of the cells and purification of the DNA is required. Efficiencies of 10^6 to 10^7 transformants per μg of DNA are routinely achieved in our hands with *C. glutamicum* and *B. lactogermentum*.

1. Add 50 ml of Luria-Bertani (LB) medium to a 250-ml flask and inoculate with a single colony or with 0.1 ml of corynebacterial cells (from frozen stock in 20% glycerol). Incubate overnight at 30°C in an orbital incubator.
2. Inoculate a 2-liter flask containing 1 liter of TSB with 100 ml of the overnight culture and incubate the cultures until the OD_{600} reaches 0.6 to 0.8.
3. Cool cells in an ice water bath (30 min) and keep them at 0 to 4°C during the following steps.
4. Centrifuge for 15 min at 4,000 rpm and 4°C. Discard the supernatant.
5. Resuspend the pellet in 100 ml of cold sterile water and repeat step 4.
6. Resuspend cells again in 50 ml of cold sterile water and repeat step 4.
7. Resuspend cells in 2 ml of 10% glycerol, freeze in liquid nitrogen, and thaw slowly in ice water. Then add 8 ml of 10% glycerol. Repeat step 4.
8. Resuspend in 0.2 ml of 10% glycerol.

The transformation capability of these cells is retained for long periods of time when frozen and can be used for electroporation as soon as they are thawed. To freeze the cells for storage, place aliquots (40 μl) in sterile Eppendorf tubes in liquid nitrogen, and then store at −70°C.

Transformation of corynebacterial cells by electroporation and selection of transformants

1. Thaw frozen cells (50 μl) in an ice water bath and pour them into as many cold electroporation cuvettes as transformation experiments are required.
2. Add DNA (CCC plasmid or ligation mixture) in up to 5 μl in TE buffer.
3. Insert the electroporation cuvette into the Gene Pulser (Bio-Rad Laboratories). Electroporate with an electric current of 2.5 kV using a 25 μF capacitor and a resistance of 200 Ω. These conditions give an electric field of 12.5 kV/cm with a pulse time constant of 4 to 5 ms.

4. Add 960 μl of TSB to the cuvette, transfer the electroporated cells to a 10-ml sterile tube, and incubate at 30°C for 1 h in an orbital shaker at 220 rpm.
5. Spread cell suspension (100 to 200 μl) on TSA plates supplemented with the appropriate antibiotic. Use TSB medium to make dilutions if required.
6. Incubate plates at 30°C and score for antibiotic-resistant colonies after 24 to 48 h of incubation.

PLASMID TRANSFER OF *E. coli* TO CORYNEBACTERIA BY MATING

Plasmids to be transferred by conjugation from *E. coli* to coryneform bacteria are introduced by transformation into the donor strain *E. coli* S17-1 (64). *E. coli* S17-1 is a mobilizing donor strain that has a derivative of plasmid RP4 integrated into the chromosome and provides the transfer functions (*tra*) for mobilization of plasmids carrying the mobilization fragment (*mob*) and the *oriT* of plasmid RP4.

The method is useful for a variety of coryneform bacteria (60); in our hands the best recipient strain is *B. lactofermentum* R31 (56), probably because of the lack of a DNA restriction system. The fertility of this strain does not increase after a heat treatment prior to mating (16).

Transfer frequencies are expressed as the number of transconjugants per final donor colony, and values ranging from 10^{-2} to 10^{-4} are achieved with *B. lactofermentum* R31.

This procedure is similar to the protocol developed by Schäfer et al. (60), except that in our procedure the donor-recipient ratio is 1:1.

Preparation of donor cells

1. Introduce the desired mobilizable plasmid construction into *E. coli* S17-1 using any of the standard *E. coli* transformation protocols.
2. Inoculate 100 ml of LB medium supplemented with the desired antibiotic to select for the mobilizable plasmid (in a 250-ml flask) with a single colony of transformed *E. coli* S17-1. Incubate at 37°C in a rotary shaker (200 rpm) until OD_{600} = 1 to 1.5.
3. Harvest cells by centrifugation (in a bench centrifuge) at 3,000 rpm for 10 min. Discard supernatant.
4. Resuspend in pellet in 10 ml of LB medium and repeat step 3.
5. Resuspend cells in 5 ml of LB medium and repeat step 3.
6. Resuspend cells in LB medium to obtain about 3 × 10^8 cells/ml.

Preparation of recipient cells

1. Inoculate 100 ml of TSB medium with 1 ml of an overnight culture of *B. lactofermentum* R31. Incubate at 30°C in a rotary shaker (250 rpm) until OD_{600} = 3 to 4.
2. Harvest cells by centrifugation (in a bench centrifuge) at 3,000 rpm for 10 min. Discard supernatant.
3. Resuspend the pellet in 10 ml of TSB medium and repeat step 2.
4. Resuspend cells in 5 ml of TSB medium and repeat step 2.
5. Resuspend cells in TSB medium to obtain about 3 × 10^8 cells/ml.
6. Treat with heat shock for 9 min at 49°C (not required for *B. lactofermentum* R31) and then place at 37°C before mating.

Mating

1. Mix 3 × 10^8 donor and recipient cells and centrifuge at room temperature (3,000 rpm) for 5 min.
2. Carefully resuspend cells in 5 ml of TSB medium and centrifuge as in step 1.
3. Carefully resuspend cells in 0.5 ml of TSB medium and spread the mating mixture onto a 0.45-μm-pore-size cellulose acetate filter (Millipore) placed on a prewarmed (too room temperature) TSA plate.
4. Incubate at 30°C for 20 h (mating).
5. Wash the cells from the filter with 1 ml of TSB and spread cell suspension (100 to 200 μl) on TSA plates with the appropriate antibiotic and nalidixic acid (30 μg/ml) (see below). Use TSB medium to make dilutions if required.
6. Incubate plates at 30°C and score for resistant colonies after 24 to 48 h of incubation.

Note that most of the corynebacteria tested showed natural resistance to 30 μg/ml of nalidixic acid; whereas *E. coli* was sensitive.

METHODS FOR ISOLATING CORYNEBACTERIA "TOTAL" DNA

There are several methods for DNA isolation reported in the literature; we will describe methods used in our laboratory with *B. lactofermentum* and *C. glutamicum*. Two of them are similar to methods 3 and 4 described by Hopwood et al. (26), and the other two were developed for the isolation of high-molecular-weight chromosomal DNA to prepare gene libraries in cosmids or to separate large DNA fragments for chromosome physical mapping.

Isolation of corynebacterial total DNA

This method is a modification of procedure 3 (developed by C. P. Smith) for the isolation of *Streptomyces* total DNA (Hopwood et al. [26]). The average DNA fragment size obtained using this method is around 30 kb. The procedure is rapid (6 h) and allows 6 to 12 samples to be processed in parallel.

1. Inoculate 100 ml of TSB medium with 1 ml of an overnight culture of coryneform bacteria. Incubate at 30°C in a rotary shaker (220 rpm) until the end of the exponential phase (24 to 30 h).
2. Harvest cells by centrifugation at 10,000 rpm for 10 min at room temperature. Discard supernatant.
3. Resuspend the pellet in 10% sucrose and repeat step 2.
4. Resuspend cells in 3 ml of TES buffer (see below) and add lysozyme to a final concentration of 5 mg/ml. Incubate for 2 to 3 h at 30°C.
5. Add 3 ml of 2× Kirby mixture (see below) and agitate for 1 min on a vortex mixer.
6. Add 6 ml phenol/chloroform/isoamyl alcohol and agitate for 30 s by vortexing.
7. Centrifuge for 10 min at 8,000 rpm to separate phases.
8. Transfer upper (aqueous) phase to a new tube containing 3 ml of phenol/chloroform/isoamyl alcohol and agitate as in step 6. Centrifuge as in step 7.
9. Transfer the upper phase to another tube containing 3 ml of chloroform/isoamyl alcohol and agitate as in step 6. Centrifuge as in step 7.

10. Transfer the upper phase to a fresh tube; add 1/10 volume of 3 M sodium acetate (pH 6.0) and 0.6 volume of isopropanol (see note below); mix gently but thoroughly until DNA precipitates.
11. Recover DNA on a sealed Pasteur pipette and transfer to a battery of 10-ml tubes containing 5 ml of 70% ethanol, and finally to a 10-ml tube containing 5 ml of absolute ethanol.
12. Pour off the ethanol, dry the DNA in the laminar flow cabinet, and resuspend in 1 ml of TE buffer (see below).

Note that addition of 0.6 volume of isopropanol is used for the selective precipitation of DNA, leaving most of the RNA in solution.

Solutions

TES: Tris-HCl, 25 mM (pH 8.0); EDTA, 25 mM (pH 8.0); sucrose, 10.3%.

Kirby mixture (1×): sodium tri-isopropylnaphthalene sulfonate, 1%; sodium 4-amino-salycilate, 6%; phenol mixture, 6% (vol/vol); Tris-HCl, 50 mM (pH 8.0).

Phenol mixture: 500 g of phenol; 0.5 g of 8-hydroxyquinoline saturated with 50 mM Tris-HCl (pH 8.0).

Phenol/chloroform/isoamyl alcohol: 50 ml of phenol mixture; 50 ml of chloroform; 1 ml of isoamyl alcohol.

Chloroform/isoamyl alcohol: 24 ml of chloroform; 1 ml of isoamyl alcohol.

TE: Tris-HCl, 10 mM (pH 8.0); EDTA, 1 mM (pH 8.0).

Rapid small-scale isolation of corynebacterial total DNA

This method is similar to that described by Hopwood et al. (26) as procedure 4 (contributed by S. H. Fisher).

1. Inoculate 10 ml of TSB medium with a single colony or with 0.2 ml of a cell suspension in 20% glycerol. Incubate overnight at 30°C in a rotary shaker at 220 rpm.
2. Harvest cells by centrifugation at 10,000 rpm for 10 min at room temperature. Discard supernatant.
3. Resuspend the pellet in 5 ml of 10% sucrose and repeat step 2.
4. Resuspend cells in 0.5 ml of TES buffer (see above) containing lysozyme and RNase at a final concentration of 5 mg/ml and 50 μg/ml, respectively. Incubate for 1 to 2 h at 37°C.
5. Add 0.25 ml of 2% sodium dodecyl sulfate (SDS) and mix for about 1 min by vortexing until the viscosity of the solution has clearly decreased.
6. Add 0.25 ml of phenol/chloroform/isoamyl alcohol and mix by vortexing for 30 s.
7. Centrifuge at 14,000 rpm for 5 min.
8. Remove the upper phase and repeat steps 6 and 7 until no white interface is seen.
9. Precipitate the DNA from the upper phase by adding 0.1 volume of sodium acetate (pH 4.8) and 1 volume of isopropanol.
10. Incubate for 10 min at room temperature. Centrifuge at 14,000 rpm for 5 min and resuspend the pellet in 0.5 ml of TE buffer.

Isolation of high-molecular-weight chromosomal DNA from corynebacteria

This method was developed by Correia et al. (11) to prepare gene libraries of corynebacteria in cosmids. Because DNA fragments to be cloned in cosmids should have a size of about 40 kb, the isolated and undigested DNA should be at least 80 to 100 kb. This method is time-consuming, but the quality of the DNA and the average size of DNA fragments are acceptable for cloning. The size of the DNA obtained is determined by PFGE of the isolated DNA.

1. Inoculate 200 ml of TSB medium (supplemented with 1% glycine) with 2 ml of an overnight culture of coryneform bacteria. Incubate at 30°C in a rotary shaker (220 rpm) until the OD_{600} = 0.5.
2. Harvest cells by centrifugation at 8,000 rpm for 10 min at room temperature. Discard supernatant.
3. Resuspend pellet in 10% sucrose and repeat step 2.
4. Resuspend cells in 5 ml of TES buffer and add lysozyme to a final concentration of 5 mg/ml. Incubate for 2 to 3 h at 30°C.
5. Add 6 ml of phenol/chloroform and shake thoroughly for 10 min at room temperature.
6. Centrifuge for 10 min at 8,000 rpm to separate phases.
7. Transfer upper (aqueous) phase to a new tube containing 3 ml of phenol/chloroform/isoamyl alcohol and agitate as in step 5. Centrifuge as in step 6.
8. Again transfer upper phase to a new tube containing 3 ml of chloroform/isoamyl alcohol and agitate as in step 5. Centrifuge as in step 6.
9. Transfer upper phase to a dialysis tube (see note below) and dialyze for 16 h against 4 liters of TE containing 0.1 M NaCl and then against 4 liters of TE for 16 h.
10. Reduce the volume of the DNA solution by placing the dialysis tube on pure sucrose. Dialyse against 4 liters of TE and keep the DNA solution at 4°C until use.

The dialysis tube is prepared by boiling in water containing 2% sodium bicarbonate and 1 mM EDTA according to the method described by Hopwood et al. (26).

Preparation of high-molecular-weight DNA for the physical mapping of the chromosome of corynebacteria

This method was used by Correia et al. (11) to study the size and organization of chromosomes of several *B. lactofermentum*, *C. glutamicum*, and *B. linens* strains. The DNA obtained was digested with endonucleases that recognize octanucleotide or AT-rich hexanucleotide sequences and was electrophoresed by PFGE. Digestion of DNA produced a discrete pattern of bands useful for fingerprinting analysis and for establishing a physical map of the chromosome.

Preparation of high-molecular-weight DNA

1. Inoculate 100 ml of TSB medium (supplemented with 1% glycine and 0.5% glucose) with 1 ml of an overnight culture of corynebacteria. Incubate at 30°C in a rotary shaker (220 rpm) until the OD_{600} = 0.8 to 1.0.
2. Harvest cells by centrifugation at 8,000 rpm for 10 min at room temperature. Discard supernatant.
3. Resuspend the pellet in TE buffer and repeat step 2.

4. Resuspend the pellet in 2 ml of TE buffer, mix with 2 ml of low-melting-point agarose gel (1%) at 42°C, and allow to solidify in 100-μl plastic molds to form blocks.
5. Incubate agarose blocks at 37°C for 24 h in lysis solution (see below) containing lysozyme to a final concentration of 1 mg/ml.
6. Remove the lysis solution and incubate blocks by a 24-h treatment at 50°C with 1 μg/ml proteinase K in ES solution (see below) supplemented with 0.25 M EDTA and 1% Sarkosyl.
7. Remove proteinase K by rinsing twice every 2 h in ES solution. Keep agarose blocks at 4°C in ES solution.

Lysis solution: 1 M NaCl; 10 mM Tris-HCl (pH 7.0); 0.5% Sarkosyl; 0.2% sodium deoxycholate; 0.25 M EDTA.
ES solution: 0.25 M EDTA; 1% Sarkosyl.

Restriction enzyme digestion

1. Cut agarose blocks into pieces containing about 1 μg of DNA.
2. Rinse agarose blocks for 1 h at room temperature in 1 ml of TE buffer containing 1 mM phenylmethylsulfonyl fluoride.
3. Rinse agarose blocks for 1 h at room temperature in 1 ml of Tris-HCl (pH 8.0) and then equilibrate with the appropriate restriction buffer for 1 h.
4. Incubate each agarose block in 100 μl of the restriction buffer with 10 to 20 U of restriction enzyme.

METHOD FOR THE ISOLATION OF PLASMID DNA FROM CORYNEBACTERIA

There are several methods for the isolation of plasmid DNA from coryneform bacteria; we will describe the method used in our laboratory, which is a slight modification of the alkaline lysis method of Kieser (33). It is used for the isolation of low- and high-molecular-weight plasmid DNA from *B. lactofermentum*, *C. glutamicum*, and *R. fascians*. The protocol takes advantage of the differences between plasmid DNA and chromosomal DNA. Because each strand of plasmid DNA is a covalently closed circle, the strands cannot be separated by treatment with alkali (up to pH 12.5), which breaks most of the hydrogen bonds in DNA. Closed circular molecules regain their native configuration when returned to neutral pH, whereas chromosomal DNA remains in the denatured state.

1. Inoculate 100 ml (or 10 ml) of TSB medium with 1 ml (or 0.1 ml) of an overnight culture of coryneform bacteria. Incubate for 24 to 30 h at 30°C in a rotary shaker (220 rpm).
2. Harvest cells by centrifugation at 8,000 rpm for 10 min at room temperature. Discard supernatant.
3. Resuspend cells in a total volume of 5 ml (or 500 μl) with TES buffer containing 5 mg/ml of lysozyme. Incubate for 2 to 3 h at 37°C.
4. Add 2.5 ml (or 250 μl) of freshly made alkaline solution (see below). Mix thoroughly by using a syringe or by vortexing.
5. Incubate for 15 min at 70°C (for high-copy-number plasmids) or for 30 min at 55°C (for low-copy-number plasmids).
6. Add 1 ml (or 100 μl) of acid phenol/chloroform (see below) and mix by vortexing.
7. Centrifuge at 8,000 rpm for 10 min and transfer the supernatant to a clean tube, measuring the volume.
8. Add 0.1 volume of unbuffered 3 M sodium acetate and 0.6 volume of isopropanol.
9. Let stand at room temperature for 15 min. Recover the DNA by centrifugation at 8,000 rpm for 10 min.
10. Dissolve DNA in 0.5 ml (or 50 μl) of TE buffer.

TES buffer: Tris-HCl, 25 mM (pH 8.0); EDTA, 25 mM (pH 8.0); sucrose, 10.3%.
Acid phenol/chloroform: 500 g of phenol; 0.5 g of γ-hydroxyquinoline, 500 ml of chloroform, and 100 ml of water.
Alkaline solution: NaOH, 0.2 N; SDS, 1%.

ISOLATION OF RNA FROM CORYNEBACTERIA

This method was developed by M. Malumbres and is based on the method of Glisin et al. (17). The procedure involves a quick cellular lysis using a French press and high-speed cesium chloride centrifugation that specifically pellets RNA in the bottom of the tube since RNA is the most dense macromolecule.

1. Inoculate 100 ml of TSB medium with 1 ml of an overnight culture of coryneform bacteria. Incubate at 30°C in a rotary shaker (220 rpm) until the OD_{600} = 1.5 to 2.0.
2. Harvest cells by centrifugation in 8,000 rpm for 10 min at 4°C. Discard supernatant.
3. Resuspend cells in TENS buffer (see below) to get a dense suspension, avoiding the formation of bubbles.
4. Pass the suspension through a French press at 15,000 lb/in^2 and freeze in liquid nitrogen.
5. The cell extract can be kept at −70°C until use. To utilize, thaw it slowly on ice.

TENS buffer: 10 mM EDTA; 20 mM Tris-HCl (pH 8.0); 100 mM NaCl; 1% SDS.

Gradient preparation and centrifugation

1. Pour 2 ml of G2 solution (see below) in the bottom of an Ultraclear tube (Beckman) and 1 ml of G1 solution (see below) to form two phases.
2. Mix 0.5 ml of G1 solution with 0.5 ml of the lysate and add as a layer over the gradient formed in step 1.
3. Centrifuge in a swing bucket rotor (Kontron TST 55.5) for 16 to 19 h at 35,000 rpm at 20°C.
4. Remove DNA and proteins from the interface and invert the tube.
5. Cut off the bottom of the tube, collect the RNA pellet, and dissolve it in 200 ml of TE buffer containing 0.2% SDS; precipitate RNA with 1/10 volume of 3 M sodium acetate (pH 7.0) and 2 volumes of cold absolute ethanol.

G1 solution: 50 mM Tris-HCl; 25 mM EDTA; 0.5% sarcosyl sulfate; 5.94 M CsCl.
G2 solution: 5.7M CsCl; 0.1 M EDTA.
Both solutions are sterilized by filtration.

This work was supported in part by grants of the European Union (BIO2-CT92-0483, BIOT-CT91-0264 RZJE and BIO4-CT96-0145). We thank R. Santamaría, H. Sandoval, G. del Real, L. M. Mateos, R. F. Cadenas, M. Malumbres, C. Fernández-González, A. Pisabarro, and A. Correia for their contributions to the development of these procedures.

REFERENCES

1. **Aida, K., K. Chibata, K. Nakayama, K. Takinami, and H. Yamada.** 1986. *Biotechnology of Amino Acid Production*. Elsevier, Amsterdam.
2. **Ankri, S., O. Reyes, and G. Leblon.** 1996. Electrotransformation of highly DNA-restrictive corynebacteria with synthetic DNA. *Plasmid* **35:**62–66.
3. **Archer, J. A. C., and A. J. Sinshey.** 1993. The DNA sequence and minimal replicon of the *Corynebacterium glutamicum* plasmid pSR1: evidence of a common ancestry with plasmid from *C. diphtheriae*. *J. Bacteriol.* **139:**1753–1759.
4. **Bathe, B., J. Kalinowski, and A. Pühler.** 1996. A physical and genetic map of the *Corynebacterium glutamicum* ATCC 13032 chromosome. *Mol. Gen. Genet.* **252:**255–265.
5. **Bonamy, C., J. Labarre, O. Reyes, and G. Leblon.** 1994. Identification of IS1206, a *Corynebacterium glutamicum* IS3-related insertion sequence and phylogenetic analysis. *Mol. Microbiol.* **14:**571–581.
6. **Bousfield, I. J., and A. G. Callely.** 1978. *Coryneform Bacteria*. Academic Press, London.
7. **Cadenas, R. F., C. Fernández-González, J. F. Martín, and J. A. Gil.** 1996. Construction of new cloning vectors for *Brevibacterium lactofermentum*. *FEMS Microbiol. Lett.* **137:**63–68.
8. **Cadenas, R. F., J. A. Gil, and J. F. Martín.** 1992. Expression of *Streptomyces* genes encoding extracellular enzymes in *Brevibacterium lactofermentum*: secretion proceeds by removal of the same leader peptide as in *Streptomyces lividans*. *Appl. Microbiol. Biotechnol.* **38:**362–369.
9. **Cadenas, R. F., J. F. Martín, and J. A. Gil.** 1991. Construction and characterization of promoter-probe vectors for corynebacteria using the kanamycin-resistance reporter gene.

9a. **Cadenas, R. F., R. Santamaría, J. A. Gil, and J. F. Martín.** Unpublished results.

10. **Chion, C. K., R. Duran, A. Arnaud, and P. Galzy.** 1991. Cloning vectors and antibiotic-resistance markers for *Brevibacterium* sp. R312. *Gene* **105:**119–124.
11. **Correia, A., J. F. Martín, and J. M. Castro.** 1994. Pulsed-field gel electrophoresis analysis of the genome of amino-acid-producing corynebacteria: chromosome sizes and diversity of restriction patterns. *Microbiology* **140:**1–7.
12. **Correia, A., A. Pisabarro, J. M. Castro, and J. F. Martín.** 1996. Cloning and characterization of an IS-like element present in the genome of *Brevibacterium lactofermentum* ATCC 13869. *Gene* **170:**91–94.
13. **Crombach, W. H. J.** 1978. DNA base ratios and DNA hybridization studies of coryneform bacteria, mycobacteria and nocardiae, p. 161–179. *In* I. J. Bousfield and A. G. Callely (ed.), *Coryneform Bacteria*. Academic Press, London.
14. **Dunican L. K., and E. Shivnan.** 1989. High-frequency transformation of whole cells of amino acid producing coryneform bacteria using high voltage electroporation. *Bio/Technology* **7:**1067–1070.
15. **Fernández-González, C., R. F. Cadenas, M. F. Noitot-Gros, J. F. Martín, and J. A. Gil.** 1994. Characterization of a region of plasmid pBL1 of *Brevibacterium lactofermentum* involved in replication via rolling circle model. *J. Bacteriol.* **176:**3154–3161.
16. **Fernández-González, C., J. A. Gil, L. M. Mateos, A. Schwarzer, A. Schäfer, J. Kalinowski, A. Pühler, and J. F. Martín.** 1996. Construction of L-lysin-overproducing strains of *Brevibacterium lactofermentum* by targeted disruption of the *hom* and *thr*B genes. *Appl. Microbiol. Biotechnol.* **46:**554–558.
17. **Glisin, V., R. Crkvenjakov, and C. Byus.** 1974. Ribonucleic acid isolated by cesium chloride centrifugation. *Biochemistry* **13:**2633–2637.
18. **Graves, M. C., and J. C. Rabinowitz.** 1986. *In vivo* and *in vitro* transcription of the *Clostridium pasteuriannum* ferredoxin gene. Evidence for extended promoter elements in Gram-positive organisms. *J. Biol. Chem.* **261:**11409–11415.
19. **Guerrero, C., L. M. Mateos, M. Malumbres, and J. F. Martín.** 1992. The bleomycin resistance gene of transposon Tn5 is an excellent marker for transformation of corynebacteria. *Appl. Microbiol. Biotechnol.* **36:**759–762.
20. **Guerrero, C., L. M. Mateos, and J. F. Martín.** 1988. Cloning and expression of the complete tryptophan operon of *Brevibacterium lactofermentum*, p. 305. Abstr. 8th International Biotechnology Symposium, Paris.
21. **Haldenwang, W. G.** 1995. The sigma factors of *Bacillus subtilis*. *Microbiol. Rev.* **59:**1–30.
22. **Hawley, D. K., and W. R. McClure.** 1983. Compilation and analysis of *Escherichia coli* promoter sequences. *Nucleic Acids Res.* **11:**2237–3355.
23. **Haynes, J. A., and M. L. Britz.** 1989. Electrotransformation of *Brevibacterium lactofermentum* and *Corynebacterium glutamicum*: growth in Tween 80 increases transformation frequencies. *FEMS Microbiol. Lett.* **61:**329–334.
24. **Haynes, J. A., and M. L. Britz.** 1990. The effect of growth conditions of *Corynebacterium glutamicum* on the transformation frequency obtained by electroporation. *J. Gen. Microbiol.* **136:**255–263.
25. **Helmann, J. D.** 1995. Compilation and analysis of *Bacillus subtilis* σ^A-dependent promoter sequences: evidence for extended contact between RNA polymerase and upstream promoter DNA. *Nucleic Acids Res.* **23:**2351–2360.
26. **Hopwood, D. A., M. J. Bibb, K. F. Chater, T. Kieser, C. J. Bruton, H. M. Kieser, D. J. Lydiate, C. P. Smith, J. M. Ward, and H. Schrempf.** 1985. *Genetic Manipulation of Streptomyces: A Laboratory Manual.* The John Innes Foundation, Norwich, United Kingdom.
27. **Jager, W., S. Schafer, A. Pühler, G. Labes, and W. Wohlleben.** 1992. Expression of the *Bacillus subtilis sac*B gene leads to sucrose sensitivity in the gram-positive bacterium *Corynebacterium glutamicum* but not in *Streptomyces lividans*. *J Bacteriol.* **174:**5462–5465.
28. **Jetten, M. S. M., and A. J. Sinskey.** 1995. Recent advances in the physiology and genetics of amino acid-producing bacteria. *Crit. Rev. Biotechnol.* **15:**73–103.
29. **Jones, D., and M. D. Collins.** 1986. Irregular non-sporing Gram-positive rods, p. 1261–1434. *In* P. H. A. Sneath,

N. S. Mair, M. E. Sharpe, and J. G. Holt (ed.), *Bergey's Manual of Systematic Bacteriology*, vol. 2. Williams and Wilkins, Baltimore, pp. 1261–1434.

30. **Kaneko, H., and K. Sakaguchi.** 1979. Fusion of protoplasts and genetic recombination of *Brevibacterium flavum*. *Agric. Biol. Chem.* **43:**867–868.
31. **Katsumata, R., A. Ozaki, T. Oka, and A. Furuya.** 1984. Protoplast transformation of glutamate-producing bacteria with plasmid DNA. *J. Bacteriol.* **159:**306–311.
32. **Kerry-Williams, S. M., and W. C. Noble.** 1984. Plasmid associated bacteriocin production in a JK-type coryneform bacterium. *FEMS Microbiol. Lett.* **25:**179–182.
33. **Kieser, T.** 1984. Factors affecting the isolation of CCC DNA from *Streptomyces lividans* and *Escherichia coli*. *Plasmid* **12:**19–36.
34. **Kono, M., M. Sasatsu, and T. Aoiki.** 1983. R plasmids in *Corynebacterium xerosis* strains. *Antimicrob. Agents Chemother.* **23:**506–508.
35. **Koptides, M., Y. Barak, M. Sisova, E. Balaghova, J. Ugorcakova, and J. Timko.** 1992. Characterization of bacteriophage BFK20 from *Brevibacterium flavum*. *J. Gen. Microbiol.* **138:**1387–1891.
36. **Kurusu, Y., M. Kainuma, M. Inui, Y. Satoh, and H. Yukawa.** 1990. Electroporation-transformation system for coryneform bacteria by auxotrophic complementation. *Agric. Biol. Chem.* **54:**443–447.
37. **Kurusu, Y., Y. Satoh, M. Inui, K. Kohama, M. Kobayashi, M. Terasawa, and H. Yukawa.** 1991. Identification of plasmid partition function in coryneform bacteria. *Appl. Environ. Microbiol.* **57:**759–764.
38. **Liebl, W.** 1992. The genus *Corynebacterium*—nonmedical, p. 1157–1171. *In* A. Balows, H. G. Trüger, M. Dworkin, W. Harder, and K. H. Schleifer. (ed.), *The Prokaryotes*, 2nd ed., vol. II. Springer-Verlag, New York.
39. **Liebl, W., M. Ehrmann, W. Ludwig, and K. H. Schleifer.** 1991. Transfer of *Brevibacterium divaricatum* DSM 20411, *Brevibacterium lactofermentum* DSM 20412 and DSM 1412, and *Corynebacterium lilium* DSM 20137 to *Corynebacterium glutamicum* and their distinction by rRNA gene restriction patterns. *Int. J. Syst. Bacteriol.* **41:**255–260.
40. **Lisser, S., and H. Margalit.** 1993. Compilation of *E. coli* mRNA promoter sequences. *Nucleic Acids Res.* **21:**1507–1516.
41. **Martín, J. F.** 1989. Molecular genetics of amino-acid producing corynebacteria, p. 25–59. *In* S. Baumberg, I. Hunter, M. Rhodes (ed.), *Microbial Products: New Approaches*. Cambridge University Press, Cambridge.
42. **Martín, J. F., R. F. Cadenas, M. Malumbres, L. M. Mateos, C. Guerrero, and J. A. Gil.** 1990. Construction of promoter-probe and expression vectors in corynebacteria. Characterization of corynebacterial promoters, p. 283–292. *In* H. Heslot, J. Davies, J. Florent, L. Bobichon, G. Durand, and L. Penasse (ed.), Proceedings of the 6th International Symposium on Genetics of Industrial Microorganisms. Sociétè Française de Microbiologie, Strasbourg, France.
43. **Martín, J. F., R. I. Santamaría, H. Sandoval, G. Del Real, L. M. Mateos, J. A. Gil, and A. Aguilar.** 1987. Cloning systems in amino acid-producing corynebacteria. *Bio/Technology* **5:**137–146.
44. **Matern, H. T., J. R. Klein, B. Henrich, and R. Plapp.** 1994. Determination and comparison of *Lactobacillus delbrueckii* subsp. *lactis* DSM7290 promoter sequences. *FEMS Microbiol. Lett.* **122:**121–128.
45. **Miwa, K., K. Matsui, M. Terabe, K. Ito, M. Ishida, H. Takagi, S. Nakamori, and K. Sano.** 1985. Construction of novel shuttle vectors and a cosmid vector for the glutamic acid-producing bacteria *Brevibacterium lactofermentum* and *Corynebacterium glutamicum*. *Gene* **39:**281–286.
46. **Miwa, K., H. Matsui, M. Terabe, S. Nakamori, K. Sano, and H. Momose.** 1984. Cryptic plasmids in glutamic acid-producing bacteria. *Agric. Biol. Chem.* **48:**2901–2903.
47. **Morinaga, Y., M. Tsuchiya, K. Miwa, and K. Sano.** 1987. Expression of *Escherichia coli* promoters in *Brevibacterium lactofermentum* using the shuttle vector pEB003. *J. Bacteriol.* **5:**305–312.
48. **Morrison, D. A., and B. Jaurin.** 1990. *Streptococcus pneumoniae* possesses canonical *Escherichia coli* (sigma 70) promoters. *Mol. Microbiol.* **4:**1143–1152.
49. **Nesvera, J., M. Patek, J. Hochmannova, and P. Pinkas.** 1990. Plasmid shuttle vector with two insertionally inactivable markers for coryneform bacteria. *Folia Microbiol.* (Praha) **35:**273–277.
50. **Oguiza, J. A., M. Malumbres, G. Eriani, A. Pisabarro, L. M. Mateos, F. Martin, and J. F. Martín.** 1993. A gene encoding arginyl-tRNA synthetase is located in the upstream region of the *lysA* gene in *Brevibacterium lactofermentum*: regulation of *argS-lysA* cluster expression by arginine. *J. Bacteriol.* **175:**7356–7362.
51. **Ozaki, A., R. Katsumata, T. Oka, and A. Furuya.** 1984. Functional expression of the genes of *Escherichia coli* in Gram-positive *Corynebacterium glutamicum*. *Mol. Gen. Genet.* **196:**175–178.
52. **Ozaki, A., R. Katsumata, T. Oka, and A. Furuya.** 1984. Transfection of *Corynebacterium glutamicum* with temperate phage CG1. *Agric. Biol. Chem.* **48:**2597–2601.
53. **Pátek, M., B. J. Eikmanns, J. Pátek, and H. Sahm.** 1996. Promoters from *Corynebacterium glutamicum*: cloning, molecular analysis and search for a consensus motif. *Microbiology* **142:**1297–1309.
54. **Pouwels, P. H., and R. J. Leer.** 1993. Genetics of lactobacilli: plasmids and gene expression. *Antonie van Leeuwenhoek* **64:**85–107.
55. **Reyes, O., A. Guyonvarch, C. Bonamy, V. Salti, F. David, and G. Leblon.** 1991. "Integron"-bearing vectors: a method suitable for stable chromosomal integration in highly restrictive Corynebacteria. *Gene* **107:**61–68.
56. **Santamaría, R., J. A. Gil, and J. F. Martín.** 1985. High-frequency transformation of *Brevibacterium lactofermentum* protoplasts by plasmid DNA. *J. Bacteriol.* **162:**463–467.
57. **Santamaría, R., J. A. Gil, J. M. Mesas, and J. F. Martín.** 1984. Characterization of an endogenous plasmid and development of cloning vectors and a transformation system in *Brevibacterium lactofermentum*. *J. Gen. Microbiol.* **130:** 2237–2246.
58. **Santamaría, R., J. F. Martín, and J. A. Gil.** 1987. Identification of a promoter sequence in plasmid pUL340 of *Brevibacterium lactofermentum* and construction of new cloning vectors for corynebacteria containing two selectable markers. *Gene* **56:**199–208.
59. **Schäfer, A., J. Kalinowski, and A. Pühler.** 1994. Increased fertility of *Corynebacterium glutamicum* recipients in intergeneric matings with *Escherichia coli* after stress exposure. *Appl. Environ. Microbiol.* **60:**756–759.
60. **Schäfer, A., J. Kalinowski, R. Simon, A. H. Seep-Feldhaus, and A. Pühler.** 1990. High-frequency conjugal transfer from gram-negative *Escherichia coli* to various

gram-positive coryneform bacteria. *J. Bacteriol.* **172:**1663–1666.

61. **Schiller, T., N. Groman, and M. Coyle.** 1980. Plasmids in *Corynebacterium diphtheriae Antimicrob. Agents Chemother.* **18:**814–821.
62. **Schwarzer, A., and A. Pühler.** 1991. Manipulation of *Corynebacterium glutamicum* by gene disruption and replacement. *Bio/Technology* **9:**84–87.
63. **Shaw, P. C., and B. S. Hartley.** 1985. A host-vector system for an *Arthobacter* species *J. Gen. Microbiol.* **134:**903–911.
64. **Simon, R., U. Priefer, A. Pühler.** 1983. A broad host range mobilization system for in vivo genetic engineering: transposon mutagenesis in Gram-negative bacteria. *Bio/Technology* **1:**784–791.
65. **Smith, M. D., J. L. Flickinger, D. W. Lineberger, B. Schmidt.** 1986. Protoplast transformation in coryneform bacteria and introduction of an α-amylase gene from *Bacillus amyloliquefaciens* into *Brevibacterium lactofermentum*. *Appl. Environ. Microbiol.* **51:**634–639.
66. **Sonnen, H., G. Thierbach, S. Kautz, J. Kalinowski, J. Schneider, A. Pühler, and J. Kutzner.** 1991. Characterization of pGA1, a new plasmid from *Corynebacterium glutamicum* LP-6. *Gene* **107:**69–74.
67. **Stackebrandt, S. E., and C. R. Woese.** 1981. Towards a phylogeny of the actinomycetes and related organisms. *Curr. Microbiol.* **5:**197–202.
68. **Su, Y. C, and S. T. Jane.** 1995. Construction of lysine-producing strains by gene disruption and replacement in *Brevibacterium divaricatum*. *Proc. Natl. Sci. Counc. Repub. China* (*B*)**19:**113–122.
69. **Thierbach, G., A. Schwarzer, and A. Pühler.** 1988. Transformation of spheroplasts and protoplasts of *Corynebacterium glutamicum*. *Appl. Microbiol. Biotechnol.* **29:** 356–362.
70. **Trautwetter, A., and C. Blanco.** 1988. Isolation and preliminary characterization of twenty bacteriophages infecting either *Brevibacterium* or *Arthrobacter* strains. *Appl. Environ. Microbiol.* **54:**1466–1471.
71. **Trautwetter, A., and C. Blanco.** 1991. Structural organization of the *Corynebacterium glutamicum* plasmid pCG100. *J. Gen. Microbiol.* **137:**2093–2101.
72. **van der Vossen, J. M., D. van der Lelie, and R. Venema.** 1987. Isolation and characterization of *Streptococcus cremoris* Wg2-specific promoters. *App. Environ. Microbiol.* **53:**2452–2457.
73. **Vertès, A. A., Y. Asai, M. Inui, M. Kobayashi, Y. Kurusu, and H. Yukawa.** 1994. Transposon mutagenesis of coryneform bacteria. *Mol. Gen. Genet.* **245:**397–405.
74. **Vertès, A. A., M. Inui, M. Kobayashi, Y. Kurusu, and H. Yukawa.** 1994. Isolation and characterization of IS31831, a transposable element from *Corynebacterium glutamicum*. *Mol. Microbiol.* **11:**739–746.
75. **Yeh, P., J. Oreglia, F. Prévots, and A. M. Sicard.** 1986. A shuttle vector system for *Brevibacterium lactofermentum*. *Gene* **47:**301–306.
76. **Yeh, P., J. Oreglia, and A. M. Sicard.** 1985. Transfection of *Corynebacterium lilium* protoplasts. *J. Gen. Microbiol.* **131:**3179–3183.
77. **Yoshihama, M., K. Higashiro, E. A. Rao, M. Akedo, W. G. Shanabruch, M. T. Follettie, G. C. Walker, and A. J. Sinskey.** 1985. Cloning vector system for *Corynebacterium glutamicum*. *J. Bacteriol.* **162:**591–597.

Molecular Biology and Genetics of *Bacillus* spp.

SIERD BRON, ROB MEIMA, JAN MAARTEN VAN DIJL,
ANIL WIPAT, AND COLIN R. HARWOOD

33

33.1. INTRODUCTION

33.1.1. History, Natural History, and Taxonomy

33.1.1.1. Distribution in the Environment

Bacteria of the gram-positive genus *Bacillus* (type strain *Bacillus subtilis* Marburg ATCC 6051) are among the most widely distributed microorganisms in nature, with representatives commonly isolated from soil and water environments (90). The genus includes a variety of commercially important species, responsible for the production of a range of products including enzymes, fine biochemicals, antibiotics, and insecticides (38, 94). Most species are harmless to humans and animals, and only a few pathogens are known. The latter include *Bacillus anthracis*, the causative agent of anthrax, *Bacillus cereus*, which causes food poisoning, and several insect pathogens. Bacilli have also been used in several traditional food fermentations, including the production in Japan of natto from soybeans by *B. subtilis* var. *natto*. The low level of reported incidence of pathogenicity and the widespread use of its products and those of its close relatives (*Bacillus amyloliquefaciens*, *Bacillus licheniformis*) in the food, beverage, and detergent industries, has resulted in the granting of GRAS (generally regarded as safe) status to *B. subtilis* by the U.S. Food and Drug Administration.

The primary reservoir for *Bacillus* species is the soil, where they secrete a variety of biopolymer-degrading enzymes that allow them to grow at the expense of plant material and other nutrients. Most species sporulate well, and environmental samples are usually prepared by heating at about 80°C for 10 min, followed by germination and outgrowth of spores on suitable media. The major soil types are members of the *B. subtilis* and *Bacillus sphaericus* groups, while representatives of the more fastidious *Bacillus polymyxa* group tend to accumulate in soils with a high organic content. Some species of this group form a close relationship with plant roots. Several species (e.g., *Bacillus azotofixans*, *Bacillus macerans*, and *B. polymyxa*) fix nitrogen under anaerobic conditions. Certain strains of *B. subtilis* are categorized as plant growth-promoting rhizobacteria, receiving in return nutrients in the form of plant exudates (55).

33.1.1.2. Taxonomy

Representatives of the genus *Bacillus* are aerobic or facultatively anaerobic, rod-shaped bacteria that can differentiate into endospores. The genus was created in 1872 by Cohn, with *B. subtilis* as the type species (17). Members of the genus show extraordinary metabolic diversity and include thermophiles, psychrophiles, alkalophiles, and acidophiles (91). *B. subtilis* is a prototroph capable of growing at mesophilic temperatures on chemically defined salts media with glucose or other simple sugars as carbon sources. In contrast, some insect pathogens are nutritionally fastidious and require highly specialized growth media.

The metabolic diversity of the genus is matched by its genetic diversity. The G+C content of the genomic DNA varies among species from about 33 to 67 mol% (94). This indicates that the 60 to 70 currently recognized species of *Bacillus* should be reassigned to an increased number of more clearly defined genera. *Bacillus acidocaldarius* and some other acidophilic thermophiles have already been reassigned to the genus *Alicyclobacillus* (126).

The classification of bacteria within the genus *Bacillus* was originally based on their ability to sporulate and their biochemical, morphological, and physiological characteristics. However, these criteria do not provide information on the cladistic relationships between species. More recent data from numerical taxonomic (phenetic) and 16S rRNA (cladistic) analyses (93) show good congruence and have resulted in *Bacillus* species' being assigned to at least six groups (94). This approach may eventually form the basis for the reassignment of species to new genera.

Group I, the so-called *B. polymyxa* group, is only loosely related to other bacilli and includes organisms with ellipsoidal spores that distend the sporangium. They are facultative anaerobes that exhibit either a mixed or butanediol type of fermentation, growing at the expense of sugars and polysaccharides.

Group II, based on *B. subtilis*, includes many of the better known bacilli. They produce ellipsoidal spores that do not distend the sporangium. They include facultative anaerobes such as *B. licheniformis* that grow fermentatively in the absence of exogenous electron acceptors, and aerobes such as *B. subtilis* that grow weakly in the absence of oxygen except in the presence of nitrate, which they can use as an alternative electron acceptor.

Group III, based on *Bacillus brevis*, are strict aerobes that produce oval endospores that distend the sporangium. Group IV, including *B. sphaericus* and other species that produce spherical endospores, are virtually unique among bacilli in having the *meso*-diaminopimelic acid usually present in cell walls replaced by lysine or ornithine. Group V includes thermophiles with various types of energy metabolism, including chemolithotrophic autotrophs. Group VI includes *B. acidocaldarius* and the other acidophilic thermophiles that have been reassigned to the genus *Alicyclobacillus* (126).

The development of robust methods for the identification of new isolates of *Bacillus*, whether of commercial or environmental interest, has not proved easy. Traditional methods, based on morphological features (particularly of spores) and dichotomous keys, have largely been abandoned in favor of computerized schemes based on biochemical tests. One such system (6) is the API 50 CHB test strip system (API, Plainview, N.Y.). An alternative scheme (92) uses 30 classical phenotypic tests to identify representatives of 44 species.

33.1.2. Culture Conditions and Preservation of Strains

33.1.2.1. Culture Conditions

The majority of *Bacillus* species will grow at mesophilic temperatures on commercially prepared nutrient media, although in some cases it is necessary to modify the pH or salt concentration. Obligate thermophilic species, such as *Bacillus stearothermophilus*, are usually grown at 60°C. Moderate thermophiles, such as *Bacillus coagulans*, are grown between 45 and 50°C. The more fastidious insect pathogens, *Bacillus larvae* and *Bacillus popilliae*, require the addition of thiamine to the growth medium and are usually grown between 25 and 30°C. *B. stearothermophilus* requires additional calcium and iron, while *Bacillus pasteurii* requires the addition of 0.5 to 1% urea.

B. subtilis and many other species are able to grow in simple salts media containing ammonium or amino acids as sources of nitrogen and glucose or other simple sugars as sources of carbon. Commonly used is Spizizen's minimal medium (110). *B. subtilis* is able to use a number of amino acids as nitrogen sources (e.g., arginine, glutamine, glutamate, asparagine, and aspartate), the catabolic pathways of which are induced by these compounds. Consequently, many amino acids recommended to overcome auxotrophy are actually growth limiting. Many studies are carried out on *B. subtilis* 168, which requires tryptophan for growth, even in media with acid-hydrolyzed casein as the main source of nitrogen.

Although many *Bacillus* species sporulate readily, special media and growth protocols are required for efficient sporulation. Sporulation is induced in response to nutrient deprivation, normally carbon, nitrogen, or phosphate, and occurs after exponential growth. Widely used is Schaeffer's sporulation medium (101).

33.1.2.2. Preservation of Strains

Most strains of *Bacillus* survive well on agar plates, either at room temperature or at 4°C, although it is recommended to subculture on a weekly basis. Viable cells may even be recovered from severely dehydrated plates, particularly from minimal agar plates, which encourage sporulation. Long-term stocks of *Bacillus* may be preserved as glycerol or lyophilized cultures, in the form of spores or vegetative cells (39). Glycerol cultures are prepared by scraping cells from the surface of an agar plate (pellets from liquid cultures may also be used) and resuspending in nutrient broth containing 15% (vol/vol) glycerol. Suspensions are frozen rapidly in 1-ml Nunc tubes and stored at −70°C or in liquid nitrogen. It is not necessary to thaw the stock prior to use, and small amounts of iced culture can be scraped from the frozen stock and streaked onto nutrient medium containing any required nutrient supplements. It is not advisable to apply selection pressure at the resuscitation stage.

Lyophilized cultures are prepared by suspending spores or vegetative cells in double-strength skim milk (Difco), distributing samples (0.2 ml) into freeze-drying ampoules, and freezing at −70°C. The contents are then freeze-dried and sealed under vacuum. Ampoules are stored at 4°C in the dark. Strains that sporulate well may also be preserved as a spore suspension. Spores need to be washed extensively to remove any nutrients and are stored at 4°C in sterile water. Spore suspensions are generally stable for many years. Most *Bacillus* strains can be transported on freshly inoculated nutrient agar slopes (not stabs) or as spore suspensions spotted on sterilized filter paper disks (25 mm) encased in sterilized aluminum foil.

33.1.3. Culture Collections

Strains of *Bacillus* are available from a variety of international culture collections; a comprehensive list has been published (18). The American Type Culture Collection (ATCC, 12301 Parklawn Drive, Rockville, MD 20852) has a World Wide Web site at http://www.atcc.org/, and the Japan Collection of Microorganisms (JCM: Riken, Wakoshi, Saitama 351, Japan) is available at http://www.wdcm.riken.go.jp/wdcm/JCM. General links to culture collections are available at the World Data Collection for Microorganisms at http://www.wdcm.riken.go.jp/.

In addition, the *Bacillus* Genetic Stock Center (BGSC) (Department of Biochemistry, Ohio State University, 484 West 12th Avenue, Columbus, Ohio 43210-1292; Fax 614-292-1538) has an extensive collection of mutant *B. subtilis* strains, bacteriophages, and plasmids. The collection also includes strains of *B. cereus*, *B. licheniformis*, *Bacillus megaterium*, *Bacillus pumilus*, *B. stearothermophilus*, and *Bacillus thuringiensis*. BGSC produces a catalog that can be requested by e-mail from dzeigler@magnus.acs.ohio-state.edu.

33.1.4. Industrial Uses

Bacillus species are an important source of industrial enzymes, fine biochemicals, antibiotics, and insecticides (38), and the ease with which they can be grown and their well-proven safety has also made them prime candidates for the production of heterologous proteins.

33.1.4.1. Enzymes

The world annual sales of industrial enzymes was recently valued at $1 billion, with strong growth in the paper, textile, and waste treatment markets. Three quarters of the market is for enzymes involved in the hydrolysis of natural polymers, of which about two-thirds are proteolytic enzymes used in the detergent, dairy, and leather industries, and one third are carbohydrases used in the baking, brewing, distilling, starch, and textile industries.

Fermentation from *Bacillus* accounts for about half of the world's production of industrial enzymes; the main classes of enzymes and their producer strains are listed in Table 1. Two of these enzymes dominate the industrial enzymes market: alkaline (serine) proteinase (protease) and

TABLE 1 Industrial enzymes produced by *Bacillus* species[a]

Enzyme	Producer strains
α-Amylase	*B. amyloliquefaciens, B. circulans, B. licheniformis, B. stearothermophilus, B. subtilis*
β-Amylase	*B. polymyxa, B. cereus, B. megaterium*
Alkaline phosphatase	*B. licheniformis*
Cyclodextran glucanotransferase	*B. macerans, B. megaterium, Bacillus* sp.
β-Galactosidase	*B. stearothermophilus*
β-Glucanase	*B. subtilis, B. circulans*
β-Glucosidase	*Bacillus* sp.
Glucose isomerase	*B. coagulans*
Glucosyl transferase	*B. megaterium*
Glutaminase	*B. subtilis*
Galactomannase	*B. subtilis*
β-Lactamase	*B. licheniformis*
Lipase	*Bacillus* sp.
Metalloprotease	*B. lentus, B. polymyxa, B. subtilis, B. thermoproteolyticus*
Metalloprotease	*B. amyloliquefaciens*
Penicillin acylase	*Bacillus* sp.
Pullulanase	*Bacillus* sp., *B. acidopullulans*
Serine protease	*B. amyloliquefaciens, B. amylosaccharicus, B. Licheniformis, B. subtilis*
Urease	*Bacillus* sp.
Uricase	*Bacillus* sp.

[a]Data from reference 94.

α-amylase. The catalytic properties of these secreted enzymes vary from one producer strain to another. Alkaline proteinases are the single largest enzyme market and are used extensively as detergent additives. α-Amylases are used extensively in the starch industry, where they need to be used at high temperatures. Industry has sought to obtain thermostable amylases by screening for new sources and by improving the stability of existing enzymes. Related enzymes from *B. amyloliquefaciens*, *B. stearothermophilus*, and *B. licheniformis* have very different thermostabilities at 90°C, and salt bridges are responsible for the stability (113). The homology between the genes encoding these amylases has been used for in vivo recombination and for subsequent screening of hybrid enzymes combining beneficial characteristics (53).

B. coagulans is an important source of glucose isomerase, an intracellular enzyme required for the conversion of glucose (0.75 times as sweet as sucrose) to fructose (twice as sweet as sucrose) in the production of high-fructose corn syrup.

33.1.4.2. Metabolites

Bacillus species are used for the production of a number of primary metabolites for the food and health care industries. *B. subtilis* has been used for the production of the nucleotides xanthanylic acid, inosinic acid, and guanylic acid, which are of commercial importance as flavor enhancers (94). Attempts to develop strains of *Bacillus* for the production of amino acids such as tryptophan, histidine, and phenylalanine, and vitamins such as biotin, folic acid, and riboflavin, have given promising results.

33.1.4.3. Peptide Antibiotics

B. subtilis 168 and other *Bacillus* species produce a variety of peptide antibiotics that enhance their survival under conditions of nutritional stress (132). In most cases these are short (up to about 20 amino acid residues) peptides that are synthesized by a nonribosomal mechanism within multienzyme complexes (peptide synthetases). Individual amino acid residues are often modified. Peptide synthetases range in size from 100 to 600 kDa and are among the largest known natural polypeptides. Gramicidin-S is a cyclic decapeptide from *B. brevis* with antibacterial and surfactant properties. Bacitracin is a branched cyclic dodecapeptide produced by *B. licheniformis* and is used as a topical antibiotic directed against bacterial cell well synthesis. Surfactin, produced by most strains of *B. subtilis*, has both antibacterial and powerful surfactant properties.

A minority of the peptide antibiotics are synthesized on ribosomes and modified extensively posttranslationally. The products are usually somewhat larger than those produced by the peptide synthetases. They include subtilin, a 32-residue lantibiotic produced by *B. subtilis* that shows antibacterial and antitumor activity.

33.1.4.4. Heterologous Proteins

Despite the high-level secretion of certain native enzymes, attempts to use *B. subtilis* for the production of heterologous proteins have met with only limited success. While extracellular proteins from close relatives can be produced at high concentrations, the yield of proteins from unrelated species, including eukaryotes, remains disappointingly low. This is, at least in part, due to the production of at least seven extracellular (83) and cell wall-associated proteinases and to incompatibilities with the *Bacillus* protein secretion pathway (see section 33.5.5). The isolation of strains defective in the identified proteinases has helped in some cases (127). *B. brevis*, naturally producing low levels of extracellular proteinases, has a proteinaceous crystalline surface (S) layer that can accumulate in the culture medium to concentration up to 35 g/liter. Attempts have been made to incorporate the expression and signal sequences of their genes into secretion vectors (114).

B. subtilis has also proved useful for the intracellular production of outer membrane proteins of gram-negative pathogens that have potential for use as vaccines and immunodiagnostics. These proteins are produced in *B. subtilis* to avoid contamination with endotoxins from the native host (96).

33.1.4.5. Insecticides

The use of chemical insecticides, with a world market worth $5 billion, is increasingly seen as suffering from significant disadvantages, including the development of resistance in target insect populations, lack of specificity, and toxicity to humans and other animals. An alternative is the use of insect pathogens, most notably *B. thuringiensis* (4). However, even now, their use represents less than 1% of the insecticide market. *B. thuringiensis* strains have been identified against each of the main groups of insect pests and, more recently, against nematodes, mites, and protozoa.

The toxicity of *B. thuringiensis* is due to proteinaceous δ-endotoxins produced during sporulation. The toxins form a crystal within the mother cell and are encoded by *cry* genes found on large plasmids. They are approximately 300 times more potent than pyrethroid-based and 8,000 times more potent than organophosphate-based insecticides. *B. thuringiensis* toxins therefore combine high toxicity and specificity for their target pests with little or no toxicity for nontarget insects and other animals. A large number of *cry* genes have been cloned, and chimeric toxins are currently being developed that combine the specificity and toxicity regions of different natural toxins. As an alternative to the use of whole cells, a variety of plant crops, including tomatoes, tobacco, potatoes, and cotton, have been transformed with genes encoding δ-endotoxins, and field trials have confirmed their activity against target pests.

33.2. GENOMICS

33.2.1. Mutagenesis

33.2.1.1. In Vivo and In Vitro Mutagenesis

Excision repair, inducible repair, and error-prone SOS-like translesion DNA synthesis have been detected in *B. subtilis* (129), and many of the classical mechanisms for inducing mutations can be used with *Bacillus*. Protocols for whole-cell mutagenesis by UV light and *N*-methyl-*N*′-nitro-*N*-nitrosoguanidine are given in Cutting and Vander Horn (21). An alternative approach is in vitro mutagenesis on cloned copies of a gene(s). Use of a *Bacillus* replicon allows the mutant gene to be maintained autonomously in the host (see section 33.4.1.4), or integration vectors such as pDY6 (Fig. 1) (78) can be used to reintroduce the gene into the chromosome (see section 33.2.2.3). pDY6 has an *Escherichia coli* origin of replication and antibiotic resistance genes for selection in *E. coli* (ampicillin, Ap) and *B. subtilis* (chloramphenicol, Cm). It also has the *lacI* gene encoding the *E. coli* Lac repressor and an isopropyl-β-D-thiogalactopyranoside (IPTG)-inducible P_{spac} promoter (128) (see section 33.3.1.2). Mutations in the target gene can be generated with a mutagen such as hydroxylamine and the mutagenized plasmid DNA amplified in *E. coli*. It is then used to transform *B. subtilis* 168 using natural competence (see section 33.2.2.1). Mutations can be in the upstream (wild-type controlled) or downstream (IPTG-inducible) copy of the gene, depending on the locations of the lesion and crossover event. Selection therefore needs to be made in the presence or absence of IPTG.

Site-specific changes are nowadays introduced by PCR techniques, extensive changes being engineered by splicing PCR methodologies (e.g., splicing by overlap extension; 47).

33.2.1.2. Integrational Mutagenesis

Insertion Vectors

The ability to generate chromosomal insertion mutations is one of the great technical strengths of *B. subtilis*, and similar techniques may be used for other transformable bacilli. A fuller discussion of the technique is given in section 33.2.2.3.

Transposons

Transposons native to *B. subtilis* have not been discovered; instead, transposons for other genera have been adapted to function in this bacterium: Tn*971* from *Enterococcus faecalis* and Tn*10* from *E. coli*. Although the available genome sequence (58) has diminished the value of transposons as a tool for the analysis of *B. subtilis*, transposons remain important for other *Bacillus* species, e.g., for mutagenesis, the cloning of DNA adjacent to the site of integration, the generation of transcriptional fusions to reporter genes, controlling the expression of adjacent genes, and as phenotypic tags for mapping.

Tn*917*, a 5.3-kb Tn*3*-like transposon (130, 131), has been used to generate mutants in *B. subtilis*, *B. amyloliquefaciens*, *B. licheniformis*, and *B. megaterium* (94). Tn*917* has a number of relevant properties, including (i) the ability to insert relatively randomly, although a limited preference for specific loci has been observed; (ii) a relatively high transposition frequency; (iii) the ability to accept DNA inserts (at least 8 kb) without influencing the frequency of transposition; and (iv) a host range that includes gram-positive and gram-negative bacteria. DNA adjacent to the site of integration can be recovered in *E. coli* by integrating an *E. coli* replicon within Tn*917* and using methods analogous to those described for integrational vectors (see section 33.2.2.3). Vectors have also been developed for generating transcriptional fusions to chromosomal genes, using reporter genes such as *lacZ* and/or *cat-86* (see sections 33.3.3.1 and 33.3.3.2).

An alternative to Tn*917* is a miniderivative of the *E. coli* transposon Tn*10*, which consists of a Cm resistance gene flanked by 307-bp fragments derived from IS*10* (85). The transposase gene is incorporated into the delivery vector rather than the minitransposon itself. Tn*10* derivatives transpose at a significantly higher frequency and insert more randomly than does Tn*917*. In conjunction with the temperature-sensitive vector pE194Ts, a series of special-purpose delivery systems has been developed for gene inactivation, the recovery of adjacent chromosomal sequences, and transcription fusions (85).

33.2.2. Transformation Systems

The most frequently used method for introducing DNA into *B. subtilis* is transformation of competent cells, although protoplasts of *B. subtilis* and several other *Bacillus* species can be efficiently transformed by naked DNA. Electrotransformation usually results in low efficiencies and is not discussed here. DNA can also be introduced in *B. subtilis* by transducing phages (see sections 33.2.3 and 33.4.2).

33.2.2.1. Transformation of Competent Cells

Transformation of competent *B. subtilis* cells was first described in 1958 by Spizizen (110), and several reviews exist (22, 23). Natural competence is one of several postexponential phase phenomena that are a characteristic of this bacterium (for review, see reference 77) and that also include peptide antibiotic production (see section 33.1.4.3), secretion of proteins (see sections 33.1.4.1 and 33.5), and sporulation.

Maximal competence develops shortly after the transition from exponential to stationary phase (2), and high cell densities promote the initiation of competence via a quorum-sensing mechanism in which secreted oligopeptides are involved (32). Maximally, only 10 to 20% of the cells in the population are able to take up DNA. Natural competence is best documented for *B. subtilis* 168 and is known for only a limited number of other *Bacillus* species. The size of DNA fragments taken up is about 20 to 30 kb (23). Transformation frequencies with homologous chromosomal DNA are maximally a few percent of the cells

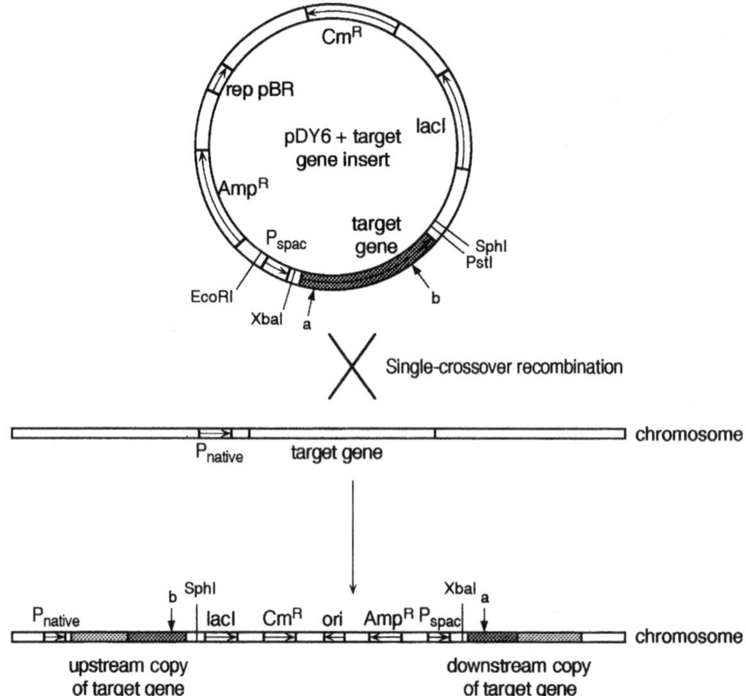

FIGURE 1 Structure and mode of integration of plasmid pDY6. The target gene for mutagenesis is cloned downstream of the P_{spac} promoter. The consequences of a single-crossover recombination into the bacterial chromosome for mutations located upstream (region a) or downstream (region b) of the crossover site are shown. In the former case (a), the mutation will reside in the P_{spac}-controlled downstream copy of the target gene; in the latter case (b) it will be in the upstream copy of that gene under the control of its native promoter. Amp, ampicillin resistance gene active in *E. coli*; Cm, chloramphenicol resistance gene active in *B. subtilis*; *lacI*, gene encoding the *E. coli* LacI repressor; rep pBR, pBR322 replication functions active in *E. coli*; P_{native}, native promoter for the target gene; P_{spac}, IPTG-inducible promoter.

with saturating amounts of DNA (>1 μg/ml of culture). Under these conditions, the cotransfer of unlinked genetic markers is possible. This phenomenon, called congression, can be used for the introduction of nonselectable genes.

Transformation with plasmid DNA is also possible, although the frequency with which free replicons are established is usually low: between 0.001 and 0.01% for covalently closed circular (CCC) DNA and about 10-fold lower for ligation mixtures. A major reason for these low efficiencies is that donor DNA becomes single-stranded and randomly fragmented during entry into the competent cell. As a consequence, only plasmid multimers or monomers containing internal repeat sequences, required for recircularization, are effective in plasmid-mediated transformation. Such molecules are naturally present in plasmid preparations and are also formed during in vitro ligations. Another consequence of the processing to single-stranded DNA is that removal of 5′-phosphate groups from linearized vectors, to prevent self-closure, cannot be used, because the single-stranded gaps that remain after ligation to target DNA will prevent recircularization following DNA uptake.

Competent cell transformation has several advantages. First, the method is simple, cheap, and efficient enough for most applications, in particular, single- and double-crossover recombinations with the chromosome (see section 33.2.2.3). Second, competent cells can be stored at −80°C, and aliquots from the same batch will have known, reproducible levels of competence. Third, a wide variety of mutants, including restriction-deficient mutants and mapping strains, are available from the *Bacillus* Genetic Stock Center (BGSC) (see section 33.1.3).

PROTOCOL

Strains and Genes

Media and detailed procedures can be found in Bron (9). Strains should be derived from *B. subtilis* 168 (BGSC 1A1). We prefer to use highly transformable derivatives of the G-type, such as 8G-5 (BGSC 1A437) (12) or 6GM (BSGC 1A685). The latter lacks the *BsuM* restriction/modification system, which affects plasmid-mediated transformation but not transformations with homologous DNA (11, 35).

Several antibiotic resistance genes are available for selection in *B. subtilis*: chloramphenicol (Cm, 5 μg/ml); erythromycin (Em, 1 μg/ml); clindamycin (Cli, 1 μg/ml); lincomycin (Lin, 25 μg/ml); kanamycin (Km, 10 or 50 μg/ml, depending on the origin of the resistance gene); spectinomycin (Spc, 100 μg/ml); blasticidin S (Bls, 400 μg/ml); tetracycline (Tet, 10 μg/ml); and phleomycin (Plm, 1 μg/ml). An efficient procedure for the preparation of competent cells (59) is described below.

Media and Solutions

1. Phosphate-citrate buffer stock solution (10× PC) (per liter: 107 g of K_2HPO_4 [anhydrous]; 60 g of KH_2PO_4

[anhydrous]; 10 g of trisodium citrate·$7H_2O$). Dilute stock solution, check pH of 1× PC buffer, and adjust (if necessary) to pH 7.0. (1× PC corresponds to Spizizen's salts, commonly used in other procedures, without ammonium sulfate.)

2. MD medium (per 10 ml: 9.2 ml of 1× PC buffer; 0.4 ml of glucose [50% wt/vol]; 0.1 ml of L-tryptophan [5 mg/ml]; 0.05 ml of ferric ammonium citrate [2.2 mg/ml]; 0.25 ml of potassium aspartate [100 mg/ml]; 0.03 ml of 1 M $MgSO_4$). Potassium glutamate can be used instead of potassium aspartate but is slightly less efficient.
3. Luria-Bertani (LB) broth (per liter: 10 g of Bacto Tryptone; 5 g of Bacto yeast extract; 10 g of NaCl; 1.0 ml of NaOH [1 M]).
4. LM broth: LB broth with any required growth factors and 3 mM $MgSO_4$.
5. MDCH medium: 10 ml of MD medium and 0.2 ml of casein hydrolysate (5%).

Procedure

1. Grow a 2-ml preculture overnight at room temperature in LM broth.
2. Use the preculture to inoculate MDCH medium at an OD_{600} of about 0.05, and culture at 37°C with shaking to T_0 (transition from exponential to stationary phase).
3. Add 1 volume of fresh MD medium to 1 volume of culture and continue shaking at 37°C for 1 h.
4. Add DNA and continue incubation with shaking for 1 h for the expression of antibiotic resistance.
5. Plate 100 ml of the culture, and 10- and 100-fold dilutions in LB broth, on LB agar containing selective antibiotics, and incubate the plates at 37°C.

33.2.2.2. Transformation of Protoplasts

In the presence of polyethylene glycol, protoplasts of bacilli can be stabilized and incorporate DNA from the medium. Cell walls can subsequently be regenerated and transformed cells selected. A protoplast transformation system was developed for *B. subtilis* by Chang and Cohen (15). The plasmid DNA internalized into the protoplasts is double-stranded and usually unfragmented. Transformation frequencies up to 10% can be obtained with plasmid DNA. In this system, plasmid monomers are active and the method is applicable to several *Bacillus* species, such as *B. subtilis*, *B. amyloliquefaciens*, *B. licheniformis*, *B. stearothermophilus*, *B. anthracis*, and *B. firmus*. Removal of 5′-phosphate groups from linearized vector molecules, to increase the frequency of recombinant DNA molecules, is possible. The method is, however, laborious, and results are difficult to reproduce. Moreover, selection for prototrophic markers is not possible, and the regeneration of cell walls may occur at low efficiencies (<1% of the protoplasts).

The G-type strains developed for efficient competent cell transformation are not suitable for protoplast transformation, because they are susceptible to lysis. The restrictionless strain *B. subtilis* 1012 (BGSC 1A447) (11) is preferred. Detailed protocols and recommendations for protoplast transformation of *B. subtilis*, *B. licheniformis*, and *B. stearothermophilus* can be found in reference 9.

33.2.2.3. Chromosomal Integration Systems

Chromosomal integration systems provide powerful tools for gene technology in *B. subtilis*. The versatility of these systems, together with the availability of the entire DNA sequence of the *B. subtilis* genome (58) (see section 33.2.4.1), render this bacterium currently one of the best known and most amenable prokaryotes for research and commercial exploitation. Transformation of competent cells with homologous DNA fragments cloned on plasmids that do not replicate in *B. subtilis* is the preferred method. Protoplast transformation is much less efficient. Integration can occur by either single-crossover (SCO), or double-crossover (DCO) recombination; we briefly describe some of the major applications below. For details, specific literature is recommended, e.g. reference 82.

Single-Crossover Recombination

In SCO events, also known as Campbell-type integrations, vectors such as the *E. coli* pUC plasmids are routinely used, although low-copy-number *E. coli* plasmids (e.g., based on pSC101) can be used for genes that are toxic if expressed at high levels in *E. coli*. In the latter case, an alternative is to use pUC plasmids in *pcnB* mutants of *E. coli* that maintain such plasmids at a copy number of about 10 per cell (64). The integration vectors contain antibiotic resistance genes (e.g., for Cm, Km, Em, Tet, Plm, or Bls) that can be selected in *B. subtilis* in the single-copy state. A fragment of homologous *B. subtilis* DNA (no smaller than about 0.15 kb; preferably larger to obtain higher efficiencies) is cloned in the vector. After propagation in *E. coli*, competent *B. subtilis* cells are transformed with the constructs, and integrants are selected using the appropriate antibiotic. The chromosomes of the integrants contain the vector and a duplication of the cloned fragment (Fig. 2A). The integrated structures are usually stable; reversal of the process occurs at a frequency of about 10^{-4} to 10^{-5} per cell generation. The integrated plasmid plus insert represent an amplifiable unit (Fig. 2A), and the selection of cells carrying amplifications (up to 50- to 70-fold; see, e.g., reference 71) is favored if the selection pressure is increased.

SCO has been used for the following applications.

Directed Gene Inactivation and Mapping of Transcription Units. The outcome of an SCO event with respect to the functionality of target genes is dependent on the structure of the cloned fragment. If the fragment carries an intact gene (or one of its ends), two (or one) functional copies will be present in the chromosome (Fig. 2A). However, when regions internal to a gene are used, no functional copies are formed after integration (Fig. 2B). This is the basis of directed gene inactivation, providing an important tool for gene function analyses (see section 33.2.4.2). Using nested sets of increasingly smaller fragments, transcription units can be delineated.

Cloning, Plasmid Walking, and Map Extension. Integrated vectors and cloned fragments, together with adjacent regions, can be excised from the chromosome with restriction enzymes and, after ligation, recovered in *E. coli*. This cycle of integration and excision can be repeated to extend the mapping of a particular region, in a process known as plasmid walking (see section 33.2.3). Extensions of various lengths can be obtained if several different restriction enzymes are used. If the inserted DNA results in plasmid instability in *E. coli* (see section 33.4.1.2), the use of low-copy-number plasmids, as discussed above, is recommended. Alternatively, an antibiotic resistance gene can be introduced at the target site by DCO recombination (see below). The marker gene with adjacent sequences is

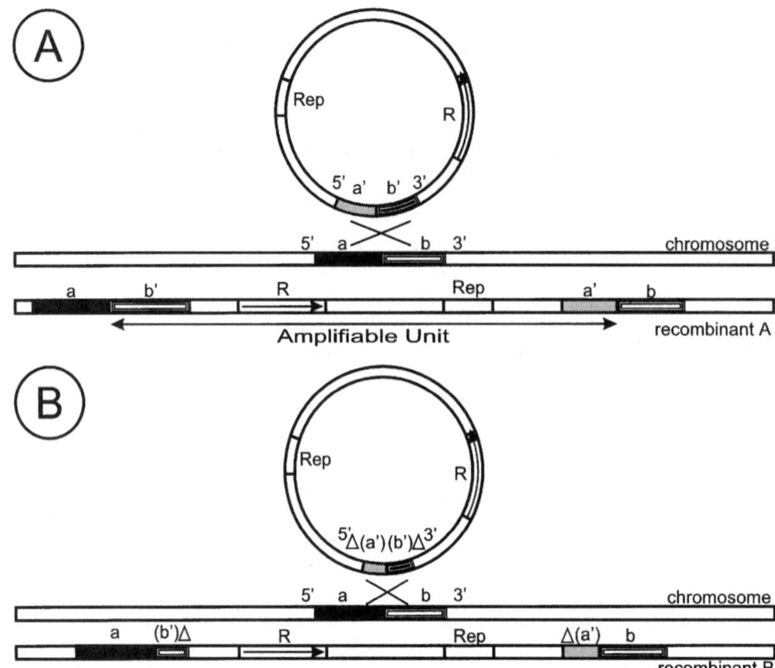

FIGURE 2 Single-crossover integration and delineation of transcription units. (A) SCO with intact transcription unit. (B) SCO with internal gene fragment. The principle of the method is described in the text. For the delineation of transcription units, the integrational plasmid should contain progressively smaller parts of the cloned fragment, e.g., nested sets of deletions. The letters a and a′ indicate intact 5′ ends of the unit on the chromosome and plasmid, respectively. Similarly, b and b′ indicate intact 3′ ends on the chromosome and plasmid. Deletions of the 5′ ends (or 3′ ends) are indicated as Δ(a′) and (b′)Δ. Rep, replication functions of *E. coli* pUC-type plasmid; R, antibiotic resistance gene. The amplifiable unit is indicated.

then excised from the chromosome and ligated to a plasmid that replicates and can be selected for in *B. subtilis*, thereby avoiding the need to propagate in *E. coli*. Suitable vectors for this purpose are pHV1431 and pHB201, described in section 33.4.1.4.

Gene Expression/Gene Fusion. Special purpose vectors carrying a promoterless reporter gene (see section 33.3.3) are suitable for assaying gene expression. A typical example is plasmid pMUTIN2, developed for the *B. subtilis* gene functional analysis project (see section 33.2.4.2 for a description of this vector). The reporter gene (*lacZ*) in this vector is preceded by a ribosome binding site appropriate for *B. subtilis* and a multiple cloning site (MCS) in which part of the target gene is cloned.

In one application, SCO integration of internal fragments of the target gene results in the transcriptional fusion of the reporter gene to the target gene's promoter, the activity of which can then be monitored using a suitable assay. In a second application, the gene, deprived of its own promoter but carrying intact 5′-sequences including the ribosome binding site, is cloned in the MCS. Integration will place one chromosomal copy of the gene under the control of the P_{spac} promoter, which can be induced with IPTG (see sections 33.2.1.1 and 33.3.1.2). With essential genes, cell survival will become dependent on the addition of IPTG (conditional mutants).

Gene Amplification. Integrations can be used for the amplification of native and foreign DNA. This is often preferred to the use of plasmids, which may be unstable (see section 33.4.1.2).

Random and Site-Specific Mutations. The gene of interest is subjected to in vitro random or site-directed mutagenesis (see section 33.2.1.1). After SCO integration, one copy of the duplicated fragment will carry the mutation. Reversal of the process, which occurs at a low frequency, may either restore the intact gene or leave the mutant copy behind. The principle of this procedure is discussed in reference 60.

Double-Crossover Recombination

In contrast to SCO integration, DCO integration (also known as replacement recombination) results in only one copy of the target DNA fragment. Typically, a region of chromosomal DNA is replaced by another region, either foreign DNA or mutationally altered homologous DNA. In the integration vector the target gene sequence is flanked on both sides by chromosomal DNA sequences that, normally, are in close proximity on the chromosome. Before being transferred to the host strain, the integration vector is linearized at a site outside the flanking homologous regions. This forces double-crossover rather than single-crossover events, since the latter are lethal to the host.

A special application of DCO recombination is shown in Fig. 3, where the homologous fragments illustrated are the front (5′) and back (3′) ends of the *B. subtilis amyE* gene, specifying α-amylase. Integration places the cloned fragment within the *amyE* gene. The latter is inactivated, providing a selectable phenotype on starch plates (see section 33.5.7.1). The integrants are normally checked by Southern hybridization, in which chromosomal DNA is digested with the same restriction enzyme used for the line-

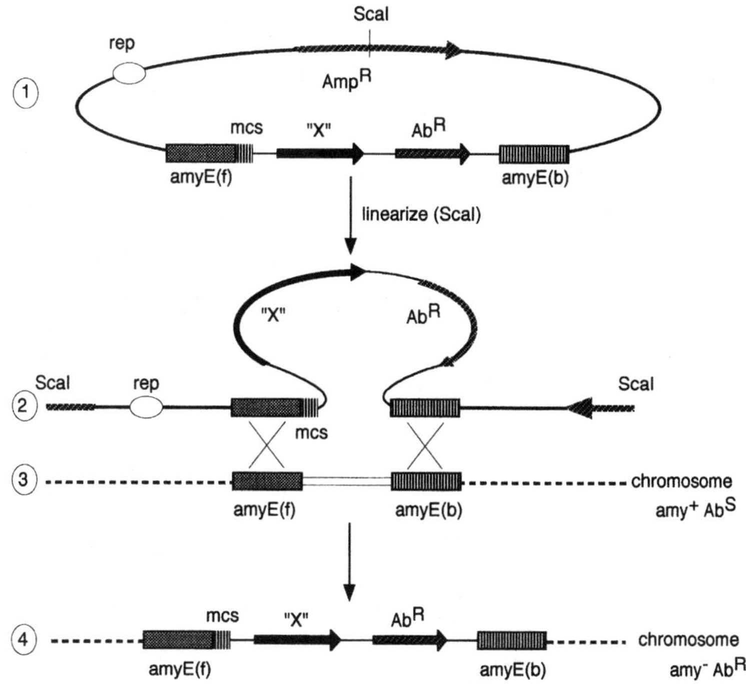

FIGURE 3 Double-crossover recombination. The principle of the method is described in the text. The vector is usually an *E. coli* plasmid containing a region of homology with the *B. subtilis* chromosome, which is interrupted by other DNA sequences. The integration in the *amyE* locus is shown here. The two regions of homology with the chromosome are provided by the 5′ end [*amyE*(f)] and the 3′ end [*amyE*(b)] of the α-amylase gene. Ab indicates a selectable antibiotic resistance marker, and "X" is the fragment of cloned DNA between the homologous regions. MCS, multiple cloning site.

arization of the plasmid (*Sca*I in the example in Fig. 3). With the insert-containing vector as probe, a single hybridization signal should be observed following a DCO event if genomic DNA is cleaved with *Sca*I. In the case of SCO, two signals will normally be observed.

DCO has been used for the following applications.

Cloning. The method is used for the stable introduction in single-copy of native and foreign DNA, e.g., for gene expression studies in the *amyE* locus.

Site-Specific Mutation/Gene Function Analysis. (i) Antibiotic resistance genes, or other DNA sequences, can be used to disrupt the chromosomal copy of the gene. (ii) Site-specific mutations can be introduced in the chromosome to study, e.g., gene function and structure/function relationships.

Genome Engineering. DCO recombination can be used to delete genes or larger regions from the chromosome.

33.2.3. Mapping

33.2.3.1. Genetic Maps

Detailed genetic maps of the *B. subtilis* chromosome have been available since 1980, and the last one to be based primarily on genetic mapping studies was published in 1993 (1). The *B. subtilis* genome sequence (see section 33.2.4.1) now provides an accurate structure/function map (7, 58) that, not unexpectedly, has revealed several errors in previously published genetic maps. Two methods have mainly been used for genetic mapping in *B. subtilis*: bacteriophage PBS1-mediated transduction and transformation of competent cells. With the available genome sequence the need for these procedures in *B. subtilis* will be limited, but similar procedures may be valuable for other *Bacillus* species. Protocols for the large-scale mapping by PBS1-mediated transduction can be found in Hoch et al. (44). Transformation of competent cells (see section 33.2.2.1) provides a good method for fine-structure analyses of closely linked genes.

33.2.3.2. Physical Maps

Several procedures have been used to construct physical maps of the *B. subtilis* chromosome. Although the available DNA sequence (58) obviates the need for further mapping in this bacterium, methods similar to those described below are likely to be valuable for other *Bacillus* species.

Plasmid Walking and Inverse PCR

Plasmid walking, introduced in section 33.2.2.3, can be used to extend existing maps. An attractive alternative is to use inverse PCR, which can be applied if part of the sequence of a cloned fragment is known. Since inverse PCR products can be sequenced directly, this technique can be used to overcome problems of plasmid instability and/or gene product toxicity (see section 33.4.1.2). The method has been applied successfully in the *B. subtilis* genome sequencing project (see section 33.2.4.1), particularly when combined with long-range PCR procedures. For many purposes, this is the recommended walking technique for *B. subtilis*. Commercial kits, such as Expand (Boehringer, Mannheim, Germany), can yield amplified products of up to about 25 kb.

Lambda Libraries

Suitable lambda vectors for constructing DNA libraries of *B. subtilis* and probably other *Bacillus* species are the λGEM11 replacement vector (Promega, Madison, Wis.) and λFixII (Stratagene, La Jolla, Calif.), which accommodate inserts of about 9 to 20 kb. Plaque hybridization can be used to identify linking clones for map extension. Since lambda libraries are propagated in *E. coli*, inserts should be checked for integrity by means of Southern hybridization and/or PCR.

Ordered pYAC Library

A *B. subtilis* genome library in yeast artificial chromosomes (YACs) has been constructed (103). Most of the *B. subtilis* DNA-carrying minichromosomes are stably maintained in *Saccharomyces cerevisiae*; the inserts have an average size of 115 kb, and an ordered set of 59 pYAC clones has been assembled that covers 98% of the chromosome. Individual YAC clones can be purified by pulsed-field gel electrophoresis. The YAC library has been used for DNA sequencing (109), for global mapping of DNA fragments, and as a hybridization probe (125).

Long-Range Restriction Maps

A long-range map that was based on restriction sites that occur infrequently (49) has been of great value in the *B. subtilis* genome sequencing project, and no major deviations from this map were observed.

33.2.4. Genome Analysis

A large number of European and Japanese groups (coordinated by F. Kunst [Institut Pasteur, Paris] and N. Ogasawara [Nara Institute of Science and Technology, Nara, Japan]) have recently jointly published the complete *B. subtilis* genome sequence (4, 124, 807 bp) (58). The project was sponsored by the Commission of the European Union and the Japanese government. The available sequence will greatly increase our fundamental knowledge of bacteria in general and this bacterium in particular. It will also facilitate the directed manipulation of related bacilli and other gram-positive bacteria of industrial and medical importance.

At the end of 1997, the genomes of several other eubacteria (*Haemophilus influenzae*, *Mycoplasma genitalium*, *Synechocystis* sp. strain PCC 6803, *Mycoplasma pneumoniae*, *Helicobacter pylori*, *E. coli* K-12, *Borrelia burgdorferi*), the archaea *Methanococcus jannaschii* and *Methanobacterium thermoautotrophicum* (references can be found in reference 58), and the eukaryotic yeast *S. cerevisiae* (31) had been sequenced. These sequences, together with the nearly completed sequences of over 30 other bacterial genomes, including those of *Mycobacterium tuberculosis* and *Vibrio cholerae*, will provide invaluable knowledge about, for example, gene function, physiology, biochemistry, molecular adaptation, and genome evolution. Moreover, genome analysis of widespread pathogens will be of paramount importance for the development of new therapeutic drugs and diagnostics. A listing of microbial genomes that have been published or that are known to be in the process of being sequenced can be found at the web site of the Institute for Genomic Research's Microbial Database: http://www.tigr.org./tdb/mdb.html.

33.2.4.1. Genome Sequencing and Databases

Sequencing Methodology

High-copy-number pUC-, or M13- and phage λ-derived vectors were used for obtaining clones, although instability problems (see section 33.4.1.2) were frequently observed with particular fragments of *B. subtilis* DNA. Clones derived from the YAC collection were also used (109), but difficulties in obtaining sufficient quantities of DNA limited the use of this application. Inverse and long-range PCR techniques were ultimately responsible for increasing the rate of sequencing. The amplified DNA was either sequenced directly or used to generate shotgun libraries of fragments ranging from 1 to 1.5 kb. The latter were obtained by mild DNase I treatment and cloning in linearized, dephosphorylated pUC-based vectors (125). Randomly selected clones were sequenced by using automated techniques, yielding about 90 to 95% of the data. The sequences of the remaining 5 to 10% were obtained by using PCR-based gap-filling approaches.

Management of Sequence Data

The genome sequence of *B. subtilis* 168 is available as a dedicated relational database called SubtiList (76) (http://www.pasteur.fr/Bio/SubtiList.html), which is managed by A. Danchin and I. Moszer (Institut Pasteur, Paris). It provides a dataset of nonredundant sequences, associated to relevant annotations and protein sequences. The database can be interrogated by using various criteria (gene names, keywords, location, etc.). DNA and protein sequences can be viewed as HTML files or downloaded as text files, and the annotated features can be shown as graphical files. Finally, the sequences in SubtiList can be analyzed by using BLAST, FASTA, and pattern-searching algorithms. A similar site, NRsub, is provided in Japan (84) (http://ddbjs4h.genes.nig.ac.jp from Japan; http://acnuc.univ-lyon1.fr/nrsub/nrsub.html from Europe).

General Conclusions

The genome of *B. subtilis* has a very high coding capacity; at least 87% of the DNA codes for putative open reading frames, 4,100 of which have currently been annotated (58). High levels of sequence redundancy, such as that found in the telomeres of some yeast chromosomes (31), were not observed. Nevertheless, a large proportion of the genome (47%) comprised paralogous genes, some of which were highly expanded; the largest class comprised 77 genes, the proteins of which are likely to belong to the family of ABC-type transporters. Many of these genes have diverged significantly from their progenitor and are likely either to be expressed at different times in the cell cycle or to encode proteins with distinct functions.

Another striking feature is the presence of at least 18 genes putatively encoding sigma factors (see section 33.3.1.1) and the identification of 10 prophages or the remnants thereof. Several components of the protein secretion apparatus were identified (see section 33.5.3).

33.2.4.2. Functional Analysis of *B. subtilis* Genes

In line with the observation of other genome sequences of similar size, 40% of the identified open reading frames of *B. subtilis* could not have a function ascribed to them. Because of the extent of the prior biochemical and physiological knowledge about it and its extreme amenability to genetic manipulation, the *B. subtilis* sequence has formed the basis for a systematic functional analysis project in which the open reading frames of unknown function are investigated. The project parallels a similar program established for the yeast sequence (31).

Structure and Management

The *B. subtilis* Function Analysis project, in which about 25 European and Japanese groups participate, started early in 1996. The project, coordinated by S.D. Ehrlich (Jouy en Josas, France) and N. Ogasawara (Nara, Japan), is divided into two consortia. The Resource consortium has nodes for (i) the construction and initial characterization of mutants (about 1,500 in total) by standardized procedures; (ii) the construction of transcription maps; (iii) the analyses of cellular proteins and cell composition; and (iv) the development of databases.

The function consortium has nodes for (i) the metabolism of small molecules and inorganic compounds (carbon, nitrogen and sulfate); (ii) macromolecule metabolism (DNA, RNA, and proteins); (iii) cell structures and mobility (cell envelope, motility); (iv) stress and stationary phase; and (v) cell processes (cell cycle, competence, sporulation, and germination). Mutants are characterized at three levels. At the primary level, mutants are screened by using relatively simple high-throughput tests. Mutants showing relevant characteristics are then subjected to secondary- and tertiary-level tests of increasing complexity and specificity by groups with the required specific expertise. Mutants that cannot be classified in the primary-level tests will be released to the scientific community after 9 months, through a public domain in the Micado database (see below).

Methodology for the Generation of Mutants

The mutants are constructed by using the same basic methodology. Target genes are inactivated via SCO integration using plasmid pMUTIN2 (Fig. 4A: see also section 33.2.2.3). Internal gene fragments, generated by PCR and cloned into pMUTIN2, are propagated in *E. coli*, and the constructs are integrated into the chromosome of *B. subtilis* 168 (Fig. 4B). The *lacZ* gene of pMUTIN2 is located within the transcriptional unit of the target gene, facilitating its activity to be reported via the synthesis of β-galactosidase. In the case of polycistrons, any downstream gene is placed under the control of P_{spac} and can be expressed by the addition of IPTG to the growth medium to avoid potential polar effects of the upstream target gene. Once a mutant has been verified by PCR and/or Southern hybridization, growth in nutrient and minimal medium (with or without IPTG) is monitored in parallel with the formation of β-galactosidase. If the target gene appears to be essential, its expression can be made conditional, and the phenotype rescued, by fusing its 5′-end to the P_{spac} promoter (see section 33.2.1.1 and 33.3.1.2).

Progress

By the end of 1997 more than 500 mutants had been constructed in the European consortium. Of these, 11 appeared to be essential and, on the basis of primary-level tests, phenotypes were tentatively assigned to a number of the mutants. In addition, several hundred kilobases of sequence had been transcriptionally mapped, and many proteins had been identified by two-dimensional gel electrophoresis (see section 33.2.4.3).

A number of problems that will limit the assignment of gene functions have already been identified. For example, the existence of paralogs and the limited range of conditions that can be applied in high-throughput analyses are both likely to limit phenotypic characterization.

Functional Analysis Database

The Micado (Microbial Advanced Database Organization) database (http://138.102.88.140/cgi-bin/genmic/madbase_home.pl) is dedicated to the European *B. subtilis* Function Analysis project (7). It contains a nonpublic domain for the participants in the project and includes (i) contiguous sequences (contigs) and gene names (hyperlinked to Swiss-Prot for accessing DNA or amino acid sequences); (ii) the coordinates of the fragment used for insertional mutagenesis; (iii) hybridization patterns of the integrants; and (iv) growth and expression data. The Micado server also contains data on a variety of other gram-positive and gram-negative bacteria. The corresponding public database of the Japanese consortium is the BSORF-DB database (http://bacillus.genome.ad.jp/BSORF-DB.html). The general structure is similar to that of Micado, but it contains some additional features, such as gene category classification.

33.2.4.3. Protein Cataloging by Two-Dimensional Gel Electrophoresis

The *B. subtilis* genome encodes a little over 4,000 polypeptides and, as part of the *B. subtilis* Function Analysis project, two-dimensional gel electrophoresis is used to catalog the polypeptides synthesized under various growth and stress regimes. In vivo radiolabeled polypeptides are separated by gel electrophoresis on the basis of net charge and size. The data is stored as coordinates, which allows polypeptide spots from independent gels to be compared and expression profiles to be constructed. Individual polypeptides can be excised from the gel and identified, after reference to the *B. subtilis* protein database, by N-terminal sequencing or mass spectrometry of peptides (e.g., by using MALDI [matrix-assisted laser desorption ionization]). This method has been of particular value for the identification of components of regulons induced in response to various stresses (28, 42).

33.3. GENE EXPRESSION

Bacillus species live in heterogeneous environments in which the supply of nutrients is discontinuous and its range variable. They are also subjected to conditions that are potentially life-threatening, e.g., high osmolarity, heat, antibiotics, irradiation, and oxidation. Exposure to starvation and other stresses in their natural environment is likely to be the norm rather than the exception, and soil-living bacteria need the capacity to respond in an appropriate manner. *Bacillus* species respond either by switching up/on or down/off the synthesis of specific proteins that improve their potential to survive, or by differentiating into resistant endospores.

The control of gene expression is achieved through (i) specific regulation, in which metabolites or catabolites regulate genes encoding enzymes involved in their own metabolism; (ii) global regulation, in which the cells respond to general stimuli such as limitations in nutrient sources, or physical or chemical insult (e.g., DNA damage, heat shock, osmotic shock); and (iii) temporal regulation, in which regulation is coupled to other events, such as the cell cycle or differentiation (e.g., sporulation). All three types of regulation have been recognized and studied in *B. subtilis*.

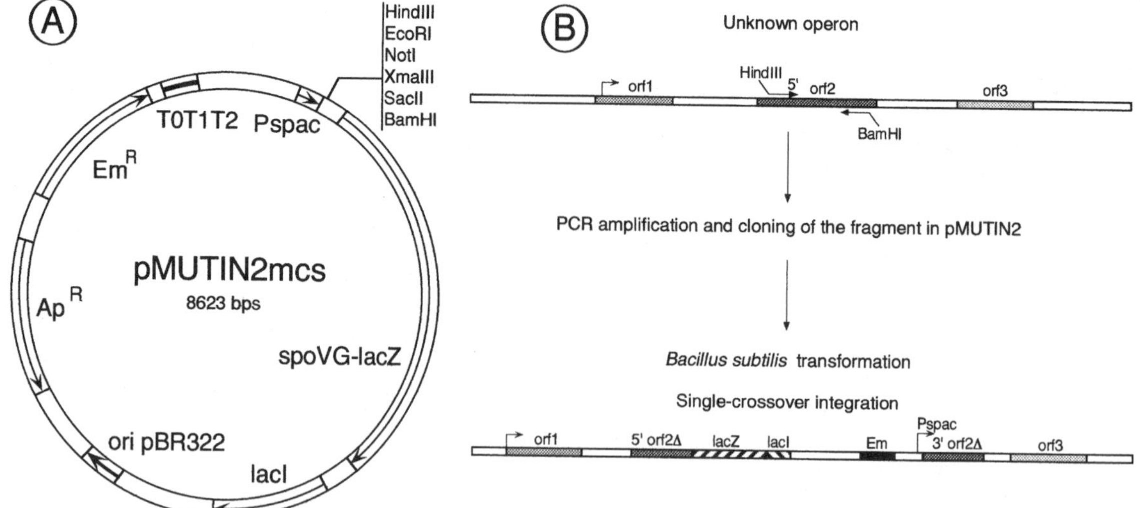

FIGURE 4 Plasmid pMUTIN2 and its use in insertional mutagenesis. (A) pMUTIN2 is an integrational vector used by the European *B. subtilis* Function Analysis (BSFA) group to introduce loss-of-function mutations in genes of unknown function. It was constructed by V. Vagner and S. D. Ehrlich (INRA, Jouy en Josas, France) and redrawn with their permission. This work was recently published (114a). The plasmid is based on the replication functions (ori) of pBR322 and carries two selectable markers (Ap for use in *E. coli*, and Em for *B. subtilis* and *E. coli*). T_0 and T_1T_2 indicate transcription terminators from phage lambda and the *E. coli* rRNA genes, respectively. P_{spac} is the LacI-controlled and IPTG-inducible promoter (see section 33.3.1.2). The *lacZ* reporter gene (see section 33.3.3.1) lacks its own promoter but is preceded by an efficient ribosome binding site (from the *spoVG* gene) for *B. subtilis*. (B) Principle of insertional mutagenesis. ORF2 represents the gene to be inactivated; it forms part of an operon that also includes the promoter-proximal ORF1 and distal ORF3. The internal fragment produced by PCR from ORF2 is provided with *Hin*dIII and *Bam*HI sites at its 5′ and 3′ ends, respectively, enabling insertion into the corresponding sites of pMUTIN2. Single-crossover integration places the *lacZ* reporter gene under the control of the native promoter of the operon, while expression of the downstream ORF3 is controlled by the IPTG-inducible P_{spac} 5′orf2Δ and 3′orf2Δ represent the mutant copies of ORF2, missing the 5′ or 3′ end.

33.3.1. Transcription and Transcriptional Control

Transcription, being the primary means of regulating gene expression, has been extensively characterized in *B. subtilis*. The proteins involved in this process are generally well conserved between *B. subtilis* and *E. coli*, although details of the structure of the RNA polymerase, and the activities of transcription regulators, may differ between the two organisms.

33.3.1.1. RNA Polymerase Core Enzyme and Sigma Factors

The core enzyme of the *B. subtilis* RNA polymerase is structurally similar to that of *E. coli*, being composed of four subunits, denoted α (×2), β, and β' (75). In addition, a *B. subtilis* gene encodes a 24.4-kDa polyanionic protein, the δ subunit (RpoE), which is reported to displace RNA from RNA polymerase and may be involved in enhancing promoter specificity (63).

The σ subunit, which associates with the core enzyme to generate the holoenzyme of RNA polymerase, determines promoter specificity; each σ subunit directs expression from a unique set of promoters. At least 10 distinct σ factors (Table 2) have been characterized in *B. subtilis* (37), and another 8 sigma-like factors have been identified by homology (58).

σ^A is the main factor during growth. It shows extensive homology with *E. coli* σ^{70} and recognizes the same promoter consensus sequences: TTGACA (−35) and TATAAT (−10) with an optimal spacing of 17 bp. This means that most σ^A-controlled promoters are expressed well in *E. coli*, which may account for the toxicity of some *B. subtilis* genes in this organism. Other σ factors in *B. subtilis* are responsible for the expression of genes required for sporulation, the production of flagella, the response to certain stresses, and the utilization of levan.

33.3.1.2. Promoters for Controlled Gene Expression

Relatively few systems have been developed for controlled, high-level expression in *B. subtilis*. Industry has developed expression systems that can direct the synthesis, in some cases over several days, of extracellular proteins to about 20 g/liter, although for commercial reasons the details of these systems are not generally available. Here we review systems that are widely used in research laboratories; those based on the P_{spac} promoter and *XylR*-controlled promoters seem to be preferred in most cases.

P_{spac} Promoter

A widely used system for controlling gene expression is based on the P_{spac} promoter. It was constructed by fusing the 5′-sequences of a promoter from the *B. subtilis* phage SPO1 and the 3′-sequences of the *E. coli lac* promoter,

TABLE 2 Sigma factors characterized in *B. subtilis*[a,b]

Factor	Function
Sigma factor	
σ^A	Housekeeping/early sporulation
σ^B	General stress responses
σ^C	Postexponential gene expression
σ^D	Chemotaxis/autolysin/flagellar synthesis
σ^H	Postexponential gene expression; competence and early sporulation genes
σ^L	Degradative enzyme gene expression
Sporulation-specific factor	
σ^E	Early mother cell gene expression
σ^F	Early forespore gene expression
σ^G	Late forespore specific
σ^K	Late mother cell specific

[a]In addition to the sigma factors shown in this table, eight new sigma-like factors were identified from the genome sequence (58).

[b]Data from references 37 and 58.

including the operator (128). The controllability of P_{spac} is dependent on the repressor encoded by the *lacI* gene, which has been adapted to achieve constitutive expression in *B. subtilis*. When present in the same cell as *lacI*, genes located downstream of P_{spac} are inducible with 1 to 10 mM IPTG; a 50-fold induction can be obtained (128). The promoter has been used for the expression of numerous genes; examples are described in sections 33.2.1.1 and 33.2.4.2.

Advantages/Disadvantages. The P_{spac} promoter functions in plasmid and chromosomal locations and, when present in multicopy situations, can direct the synthesis of a protein to significant proportions of total cellular protein. Moreover, P_{spac} is functional in *E. coli*, so constructs can be tested in this bacterium before transfer to *B. subtilis*. Potential disadvantages are that this promoter is not sufficiently strong and its inducer is too expensive for large-scale fermentations. Also, IPTG is hazardous and cannot be used in food-grade applications. Moreover, this promoter directs the synthesis of small amounts of protein in the absence of IPTG, even when located downstream of a strong transcription terminator to prevent transcriptional read-through from adjacent genes. This problem can be reduced by increasing the number of copies of *lacI*, for example by providing it on a multicopy plasmid.

XylR-Controlled Promoters

The *B. subtilis* xylose-inducible promoter/operator elements have been used without modification to control gene expression (30). A copy of the *xylR* gene, encoding the repressor for this system, is usually included on high-copy-number expression vectors to maintain a balance between the number of repressor molecules and operator sites. As with P_{spac}, *XylR*-controlled promoters are active in *E. coli*, where they also respond to the presence of xylose. Although genes in the xylose regulon are subject to catabolite repression, the catabolite responsive element (CRE) is not included in the vectors.

Advantages/Disadvantages. *xylR*-controlled promoters direct moderately high levels of expression. Xylose is a cheap and readily available substrate and can be used in large-scale fermentations. Potential disadvantages are that *xylR*-controlled promoters direct the synthesis of small amounts of protein even in the absence of xylose.

sacB Promoters

The inducible expression of *sacB*, encoding extracellular levansucrase, by sucrose involves a number of regulatory mechanisms, not all of which are fully understood. This gene is controlled positively by sucrose, the SacY antiterminator, and the products of *degQ* and *sacU*, and negatively by SacX, a putative phosphotransferase system enzyme $II^{sucrose}$ (20). Various expression cassettes have been based on the *sacB* promoter (see reference 127), which is particularly effective in *sacU*^{h} backgrounds (134).

Advantage/Disadvantages. The *sacB* promoter can be induced during exponential growth when extracellular proteinase concentrations are generally low; moreover, catabolite repression does not occur on this promoter. The level of induction can be modulated by using different concentrations of sucrose, from 1 to 30 mM. Sucrose is a readily available, cheap, and nontoxic substrate. Potential disadvantages are that vector/regulatory elements need to be developed further, a process that is hampered by the limited current knowledge of the molecular biology of the regulatory pathways involved. Another disadvantage is that the *sacB* promoter is not strong enough for very large scale protein production.

Phage Vector Expression Systems

Bacteriophages ϕ105 and PBSX (a defective phage of *B. subtilis* 168) have both been developed for the production of proteins in *B. subtilis* (26; see also section 33.4.2). The phages have been modified to make them temperature inducible by mutating the genes for the immunity repressors. Target genes are introduced downstream of a strong prophage promoter via SCO or DCO recombination (see section 33.2.2.3). Temperature induction can lead to the production of the target gene product to 0.5 mg/ml culture supernatant (111).

Advantages/Disadvantages. The ϕ105 and PBSX constructs are relatively stable and maintained as a single copy during growth, while the copy number increases at the time of induction. The promoter is tightly controlled by the immunity repressor, and induction by increasing temperature is cheap and favored in industrial fermentations. The systems can be arranged so as to lead to cell lysis if desired. The main disadvantage is that further vector development is needed to maximize the potential of this system.

33.3.1.3. Analysis of Transcription

Many of the techniques commonly used to measure transcription and translation in bacteria are applicable to *B. subtilis* with only a few modifications. Several methods have been reported for the isolation of mRNA from *B. subtilis* (see reference 120), and the standard methods to avoid contamination with RNases need to be employed (e.g., use of rubber gloves, treatment of glassware with diethylpyrocarbonate [99]). The RNeasy kit (Qiagen, Hilden, Germany) has successfully been used in our laboratories for the isolation of *B. subtilis* mRNA. Reverse transcription (RT)-PCR methods, such as the one based on the Access RT-PCR system kit provided by Promega, can be used for the qualitative and quantitative analysis of *B. subtilis* mRNA. Methods for S1 endonuclease mapping and primer extension have been reported previously (75, 99).

33.3.2. Translation and Translational Control

The ribosomes of *Bacillus* species show strong structural and functional similarities to those of *E. coli* (119). *B. subtilis* has 10 rRNA operons with the same organizational structure (16S, 23S, 5S) as the rRNA operons of *E. coli*. 16S rRNA sequencing formed the basis for the division of representatives of the genus into subgroups (91; see section 33.1.1). Eighty-eight tRNA genes have been identified (58).

The large 50S subunit of *B. subtilis* ribosomes, like that in *E. coli*, has two species of rRNA (5S and 23S) and about 35 ribosomal proteins. The 30S subunit has a single species of rRNA (16S) and 20 ribosomal proteins but lacks a homolog of protein S1 (34). In *E. coli*, protein S1 has been implicated in translation initiation. Its absence in *Bacillus* spp. may account for the high stringency between the ribosome binding site and 3′ end of the 16S rRNA (average ΔG ca. −18 kcal/mol, cf. −11 kcal/mol for *E. coli*) (98) and, consequently, for the inability of *Bacillus* spp. to translate mRNAs from most *E. coli* genes.

RNA- and DNA-directed in vitro translation and coupled transcription/translation systems have been developed for *Bacillus* spp. (14, 118), although the presence of many proteinases requires additional precautions to avoid ribosomal protein degradation. It should also be remembered that *Bacillus* mRNA can usually be translated in commercially available *E. coli*-based systems.

In *Bacillus* spp., control of protein synthesis at the level of translation is known in a number of cases. Translational control permits immediate synthesis of the product in the presence of inducer, which, in the case of the Cm resistance gene *cat-86*, is chloramphenicol itself. This reduces the potentially fatal (in competitive terms, at least) delay between challenge and response. The translational attenuation mechanism controlling *cat-86* has been discussed by Lovett and Rogers (65).

In vivo synthesized proteins can be analyzed with a variety of techniques, including Western blotting, pulse-chase labeling with or without immunoprecipitation (see section 33.5.7.2), and the use of minicells (see section 33.3.4).

33.3.3. Reporter Genes

Gene fusion has proved to be an effective means of studying gene expression in *B. subtilis*. Several reporter genes from other organisms are applicable for use in this bacterium.

33.3.3.1. Chromogenic Reporters

lacZ/bgaB

E. coli lacZ, when fused to the 5′ end of plasmid-borne or chromosomal genes, is a widely used reporter gene for monitoring gene expression in *Bacillus* spp. This gene has also been used to detect gene expression at the single-cell level by using combined cytochemical and video microscopy techniques (62).

lacZ expression can be detected on solid media by using the chromogenic substrate X-Gal (5-bromo-4-chloro-3-indolyl-β-D-galactopyranoside) or the fluorogenic substrate MUG (4-methylumbelliferyl-β-D-galactopyranoside) (130). In liquid cultures the assay of Miller (74) is most commonly used to assay β-galactosidase activity. In this system the hydrolysis of the colorless substrate *o*-nitrophenyl-β-D-galactopyranoside (ONPG) gives rise to a yellow compound that can be assayed colorimetrically. Since *B. subtilis* is impermeable to ONPG, cells must be permeabilized with toluene or lysozyme. The use of MUG enables *lacZ* expression to be assayed without cell lysis and is more sensitive.

The *lacZ* reporter gene is unsuitable for studying heat shock gene expression, since *E. coli* β-galactosidase is degraded rapidly under these conditions (W. Schumann, personal communication). More stable reporter genes, such as the *bgaB* gene (β-galactosidase from *B. stearothermophilus*) and chloramphenicol acetyltransferase (see section 33.3.3.2), have thus been proposed for this application.

xylE

The *xylE* gene from *Pseudomonas putida*, specifying catechol-2,3-dioxygenase, is a useful reporter gene in *Bacillus* spp. for analyzing expression from strong promoters (133). Expression is measured spectrophotometrically as the production of the yellow compound 2-hydroxymuconic-semialdehyde upon the addition of the substrate catechol. Expression of *xylE* in colonies may also be visualized by spraying plates with a 1% aqueous solution of catechol, which leads to the development of yellow colonies after about 5 min.

33.3.3.2. Antibiotic Resistance Genes

Since chloramphenicol acetyltransferase (CAT) is relatively easily assayed, and the amount of enzyme generally shows a good correlation with the level of resistance, *cat* genes have been used in *Bacillus* species to report expression in a number of contexts. Two *cat* genes are mainly used, *cat-86* from *B. pumilus* (40) and that of plasmid pC194 (46). Both genes are induced in the presence of chloramphenicol by a translational attenuation mechanism (65). CAT is assayed by monitoring the change in absorbance at 412 nm when free coenzyme A-sulfhydryl groups, generated by the action of CAT on chloramphenicol, react with 5,5′-dithio-*bis*-nitrobenzoic acid, releasing a molar equivalent of 5-thio-2-nitrobenzoate (104).

33.3.3.3. Fluorescent and Luminescent Reporters

Over the past few years reporter systems have been developed that are based on light-producing enzymes or fluorescent proteins (41). An advantage of fluorescent proteins is that the uptake of substrates by the host cell is not required for activity. Although both types of reporter have been used for cytological studies, in practice, the fluorescence output from a single cell is too low for quantitative analyses.

Luciferase

The *luxAB* genes of *Vibrio harvei* code for a luciferase that emits light when exposed to a suitable substrate (e.g., decanal). A chromosomally located *luxAB* gene fusion has been used as a reporter for tracking *B. subtilis* in soil (19).

Green Fluorescent Protein

One of the most versatile reporters is green fluorescent protein, specified by a gene that was isolated from the jellyfish *Aequorea victoria* (89). This small protein (27 kDa) fluoresces owing to an autocatalytic cyclization between amino acids 65 (Ser) and 67 (Tyr) and subsequent oxidation (43). The wild-type protein is excited at 395 nm and emits green light at 590 nm, and blue- and red-shifted fluorescent derivatives have been isolated that facilitate dual-labeling experiments. Green fluorescent protein has been expressed in *B. subtilis* to locate sporulation proteins (123) and to demonstrate compartment-specific gene expression by fluorescence microscopy (61).

33.3.4. Minicells

In vivo synthesized translation products of genes cloned onto high-copy-number *Bacillus* plasmids can be visualized by using minicells. Such cells result from aberrant cell divisions and, although they lack chromosomal DNA, they often contain plasmid DNA. A number of minicell-producing strains have been isolated for *B. subtilis* (97), and these small cells can be separated from normal cells by rate-zonal centrifugation. Minicell suspensions retain the ability to synthesize proteins and, since bacterial mRNA is relatively unstable, de novo protein synthesis is directed from plasmid-encoded genes. Methods for studying proteins synthesized by minicells have been described by Moran (75).

33.4. HOST/VECTOR SYSTEMS

33.4.1. Plasmid-Based Systems

B. subtilis 168, the naturally transformable strain (see section 33.2.2.1), does not contain endogenous plasmids, and most plasmids present in other *Bacillus* strains are cryptic. This is why plasmid vectors for *B. subtilis* were initially taken from other gram-positive bacteria, such as *Staphylococcus aureus* and *Lactococcus lactis*. Several of these plasmids, such as pUB110, pC194, pE194, and pWVO1, are still in common use (for reviews, see references 9 and 51). More recently, vectors based on endogenous *Bacillus* plasmids have been developed (see section 33.4.1.4).

33.4.1.1. Replication of Plasmids from Gram-Positive Bacteria

Most plasmids from gram-positive bacteria use the rolling-circle (RC) mode of replication, which is characterized by the uncoupled synthesis of leading and lagging strands. Whereas RC plasmids are small (usually <10 kb), theta replicating plasmids are generally considerably larger, although exceptions to this rule exist.

Rolling-Circle Plasmids

RC plasmids are highly interrelated and are organized in a modular way (for reviews, see references 9, 33, and 51). The primary replication functions consist of the *rep* gene, encoding the replication initiation protein (Rep), and the origin of plus-strand synthesis (ori^+; also called double-strand origin). Rep initiates replication through the introduction of a site-specific single-strand (ss) nick in the double-strand origin. Characteristic for RC replication is the formation of ssDNA intermediates. In the conversion of this DNA to double-stranded (ds) DNA, the secondary replication function, SSO (single-strand origin), serves as the major initiation site for synthesis. SSOs are usually active in a limited number of hosts and, although these functions are dispensable, they affect the efficiency of replication and plasmid stability (see section 33.4.1.2). A *mob* gene, enabling the conjugative transfer of the plasmid to other gram-positive bacteria, is present on several RC plasmids. The copy number of RC plasmids in *B. subtilis* can vary from about 5 to 200 per chromosome.

Theta Plasmids

A number of theta plasmids that replicate in *B. subtilis* are known. One is the enterococcal plasmid pAMβ1 (13, 51). pLS20 is an endogenous *B. subtilis* theta plasmid with a host range that is probably limited to *Bacillus* spp. (66).

33.4.1.2. Plasmid Instability

Frequently, recombinant plasmids are unstable in *B. subtilis*. Both segregational instability (loss of the plasmid population from a cell) and structural instability (usually deletions) occur.

Segregational Instability

An important cause of segregational instability of RC plasmids is the accumulation of ssDNA and linear high-molecular-weight plasmid DNA (9, 33). Since SSOs are usually host specific, substantial amounts of ssDNA can accumulate with nonnative RC plasmids. A possible explanation for the reduced stability of, in particular, nonnative RC plasmids is that the accumulation of ss and high-molecular-weight DNA reduces the cell's growth rate, which, because of the growth advantage of plasmid-free cells, can drastically increase the rate of plasmid loss. With native RC plasmids from *B. subtilis*, such as pTA1060 and its derivative pHB201 (see section 33.4.1.4), these problems can largely be avoided.

Structural Instability

Deletions in plasmids can occur between short direct repeat sequences (3 to 20 bp), or between nonrepeated sequences. For details of possible mechanisms, see references 24, 25, 33, and 70. One mechanism involves copy-choice replication errors due to slippage of the replication machinery at short direct repeat sequences. The generation of ssDNA, in particular with RC plasmids, stimulates this event. Copy-choice errors can largely be avoided with theta plasmids or with RC plasmids such as pTA1060 (see section 33.4.1.4) that have an efficient SSO.

Errors resulting from aberrant initiation and/or termination by the Rep protein of RC plasmids have also frequently been observed (33). Furthermore, deletions may arise from cleavage of inappropriately modified DNA by the *Bsu*R restriction system (35). When this is likely to be a problem, it is recommended that restriction-deficient strains, such as *B. subtilis* 1012 (BGSC 1A447), are used. In yet another class of deletions, resulting from breakage and reunion between nonrepeated sequences, topoisomerases (72, 81) have been implicated. A positive selection system for the analysis of deletion formation has been developed by R. Meima and S. Bron (72).

33.4.1.3. Methodologies

We globally describe a number of methods for the use and isolation of plasmids in *B. subtilis*. For details of procedures, see reference 9 and general textbooks, e.g., reference 99.

Isolation of Plasmid DNA

Standard procedures for the isolation of plasmid DNA frequently give unsatisfactory results when applied to *B. subtilis*, in particular with stationary phase cultures: the yields may be low and the purity of the DNA insufficient. Details of a suitable miniscale (1 ml to about 5 ml of culture) alkaline-lysis procedure have been described previously (9). For plasmids with copy numbers of >20 per chromosome, yields are about 0.5 to 5 μg of plasmid DNA from 1 ml of culture; for lower-copy-number plasmids, yields are about 0.1 to 2 μg.

For plasmid DNA isolations at a midiscale (5 ml to about 100 ml of culture), Qiagen-tip-100 columns (Qiagen) can efficiently be used, following the protocols provided by the manufacturer. The plasmid DNA obtained is suitable for restriction analysis, cloning, sequencing, and PCR.

Distinguishing Rolling-Circle from Theta Plasmids

The modular organization described in section 33.4.1.1, and sequence homology with known RC plasmids, are characteristics for RC plasmids. Further evidence for RC replication is the demonstration, by Southern hybridization, of ss plasmid DNA replication products. S1 nuclease is used to reveal the presence of ssDNA bands in the gel. Details of this procedure are given in reference 9.

Conclusive evidence for theta replication involves the demonstration of replication intermediates containing bubbles. These can be visualized by electron microscopy and two-dimensional gel electrophoresis (13).

Copy-Number Determinations

Plasmid copy numbers are usually given per chromosome equivalent. Since they reflect population means, deviations can occur in individual cells. This is important for plasmid maintenance, since cells with a lower than average number of plasmid copies have a higher probability of generating plasmid-free daughter cells. Two methods for copy-number determination have mainly been used for *B. subtilis*. In the first, plasmid and chromosomal DNA are radiolabeled during cell growth. Total DNA is fractionated by agarose gel electrophoresis, plasmid and chromosomal DNA bands are cut from the gel, and radioactivities in the fractions are measured. After normalization for the difference in molecular mass of the plasmid and the chromosome, plasmid copy numbers per chromosome are obtained. The second method involves the comparison of total DNA from cells carrying the plasmid of interest with a reference DNA mixture (9).

Assays for Plasmid Instability

Segregational Instability. Cells containing the plasmid are grown overnight in batch culture under selective conditions so that no plasmid-free cells can grow. The culture is then diluted into fresh medium without antibiotics and grown for about 100 generations. Samples taken as a function of time are plated on nonselective agar, and the colonies are tested for antibiotic resistance by replica-plating or transfer to selective plates by toothpicking. The kinetics of appearance of plasmid-free cells is plotted as the fraction of antibiotic-resistant colonies against the number of cell doublings (9).

Precise Excision. Precise excision of transposons is a model system for measuring deletion frequencies between short direct repeat sequences. The transposon is inserted in an antibiotic resistance gene, and recombination between the duplicated target site sequences of the transposon will result in precise excision of the transposon and restoration of gene activity. This can be scored positively by the appearance of antibiotic-resistant cells.

A more general system for the selection of deletions is based on plasmid pGP100 (72), an *E. coli*/*B. subtilis* shuttle derived from plasmid pGK12 (see section 33.4.1.4). It contains a promoterless *cat-86* reporter gene, preceded by the highly efficient *E. coli rrnB* T_1T_2 transcriptional terminator. Deletions removing this terminator result in the transcription of the *cat-86* gene from a promoter upstream of the terminator sequence. The deletion events can be recognized by the Cm resistance acquired by the host cell. This system allows the identification of several types of deletions, also between nonrepeated sequences (72).

A potential problem with these systems is that, eventually, the growth advantage of cells carrying deleted plasmids will lead to overestimation of deletion frequencies. This can be avoided by using a modification of the fluctuation test, such as that described by Chédin et al. (16).

33.4.1.4. Cloning Vectors for *B. subtilis*

Several regularly used *S. aureus*-derived plasmids (e.g., pUB110, pC194, and pE194) are frequently unstable in *B. subtilis*, particularly when carrying large inserts (9). Here we describe a number of versatile and more stable vectors.

Rolling-Circle Plasmids for *L. lactis*

pWVO1 is a small (2178-bp) cryptic RC plasmid from *L. lactis* (60), replicating in many gram-positive bacteria (including *B. subtilis*) and even in the gram-negative *E. coli*. pGK12 (Fig. 5A) is the prototype of several special-purpose vectors based on pWVO1 (60). The copy numbers of pWVO1 derivatives are low (about 5 per chromosome) in *B. subtilis* and *L. lactis* but high (50 to 100 per chromosome) in *E. coli*. Recombinant pGK12 plasmids are relatively stable in *B. subtilis*.

Plasmids from *Bacillus* spp.

RC plasmids from *Bacillus* spp. which in their native form are normally devoid of selectable markers, were recently reviewed (69). pTA1060 is a 8.6-kb RC plasmid from *B. subtilis* with a copy number of 5 per chromosome. Its SSO is very efficient in *B. subtilis* (68), and derivatives of pTA1060 can stably carry inserts up to at least 30 kb. pTA1060 is superior to nonnative RC plasmids with respect to both segregational and structural plasmid stability and has been used to develop a series of versatile cloning vectors (9, 10, 35). A recent variant is pHB201 (Fig. 5B), a *B. subtilis*/*E. coli* shuttle carrying a modified *lacZα* gene that is expressed in *B. subtilis*. An extended MCS is present in the *lacZα* gene. Read-through transcription from inserts in the *lacZα* gene is prevented by the *E. coli* T_1T_2 transcription terminators. If used in combination with the restriction-deficient host *B. subtilis* 1012M15 (BGSC 1A748), which carries a modified *lacZΔM15* gene, selection of colonies carrying recombinant plasmids can be carried out by the convenient blue-white assay on plates containing X-Gal. Strains containing pHB201 can be obtained from the BGSC (see section 33.1.3) (*E. coli*: ECE59; *B. subtilis*: 1E59). Since the copy number of this plasmid is high in *E. coli*, this is the preferred host for the amplification of plasmid DNA.

Theta Plasmids

Theta plasmids are generally more stable than RC plasmids (51). Derivatives of the enterococcal plasmid pAMβ1 (see section 33.4.1.1) have been used as the basis of a series of vectors for *B. subtilis*. pHV1431 (Fig. 6A) (50) is a *B. subtilis*/*E. coli* shuttle plasmid carrying the pBR322 replication functions for *E. coli* and the pAMβ1 replication functions for *B. subtilis*. The plasmid has a very high copy number in *B. subtilis* (about 200), but it is slightly unstable. pIL252, another pAMβ1-based vector (105), is unable to replicate in *E. coli* and lacks the stability-promoting resolvase function of the parental plasmid. A variant of pIL252, pAMS100 (Fig. 6B), has been constructed in which the

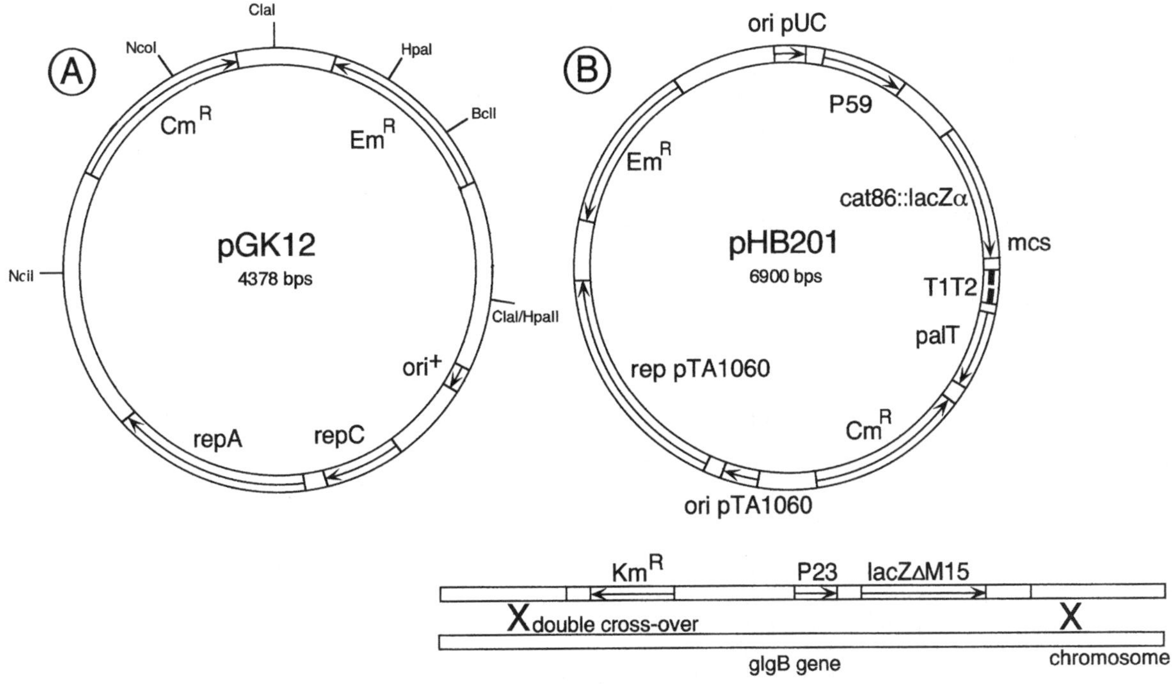

FIGURE 5 Plasmids pGK12 and pHB201. (A) pGK12 carries the replication functions (ori$^+$, *repA*, and *repC*) of the lactococcal plasmid pWVO1, which is a natural broad-host-range plasmid, replicating in many gram-positive bacteria, including *B. subtilis*. Several unique restriction sites are indicated, as are two selectable markers (Cm and Em). pWV01 and other special-purpose derivatives of this plasmid family are described in reference 60. (B) pHB201 is a shuttle plasmid carrying the ori-pUC (for *E. coli*) and ori-pTA1060 (for *B. subtilis*) replication functions. The plasmid contains two selectable markers (Cm and Em) and a *lacZα* gene, the expression signals of which have been modified (36) to enable the synthesis of the LacZα peptide in *B. subtilis*. An extended MCS is present in the *lacZα* gene. The plasmid contains the highly active *palT* SSO of pTA1060 (9, 68) and the T_1T_2 transcription terminator. In conjunction with host strain 1012M15 (BGSC 1A748), which is restriction-deficient and carries a *lacZΔM15* gene in a nonessential part (*glgB* gene) of the chromosome (lower part of figure; reference 36), the blue-white assay for recombinant plasmids on X-Gal plates by *lacZα* complementation can be used.

resolvase stability function has been reintroduced, together with the strong T_1T_2 transcription terminator (54). Cloning in pHV1431 and pAMS100 is highly efficient, and long inserts are generally stably maintained.

Integrative Plasmids

Advantages and applications of integrative plasmids have been described in section 33.2.2.3.

33.4.2. Bacteriophage-Based Vector Systems

In addition to plasmid vectors, phage vectors have been developed for *B. subtilis*. ssDNA phages, such as M13, are not known for *B. subtilis*. Most frequently used in this bacterium are the temperate phages ϕ105 and SPβ. Extensive reviews on this subject, including protocols, are available (26, 27, 87). We briefly summarize these systems and their potential uses.

B. subtilis phage vectors have mainly been used for the cloning of homologous genes and the construction of *B. subtilis* genomic libraries. In the lysogenic state, cloned genes are very stable and facilitate complementation assays. Two basic approaches can be used with *B. subtilis* phage vectors: direct transfection and prophage transformation.

33.4.2.1. Direct Transfection

In most ϕ105 vectors the functions for site-specific integration are intact, enabling the recovery of recombinants as lysogens (26, 27). Close to the phage attachment site is a dispensable region that, when removed, provides space for inserts up to about 6 kb. About 10^6 PFU/μg of DNA can be obtained when genome libraries are constructed. Ligation mixtures are used to transfect protoplasts (see section 33.2.2.2), and phage lysates are subsequently obtained by plaquing on sensitive (nonlysogenic) host cells.

33.4.2.2. Prophage Transformation

In this procedure, competent ϕ105-lysogenic cells are transformed with DNA ligated in ϕ105 vectors, which results in a double-crossover replacement in the prophage sequences. Standard vectors do not replicate in *B. subtilis*, and they contain only the fragments of ϕ105 DNA that flank the dispensable region in the genome. The principle of this method is shown in Fig. 7. Cloned inserts become inserted in the prophage, the lytic cycle of which can still be induced (26, 27).

Unlike ϕ105, SPβ is present as a prophage in nearly all derivatives of *B. subtilis* 168. This precludes the use of di-

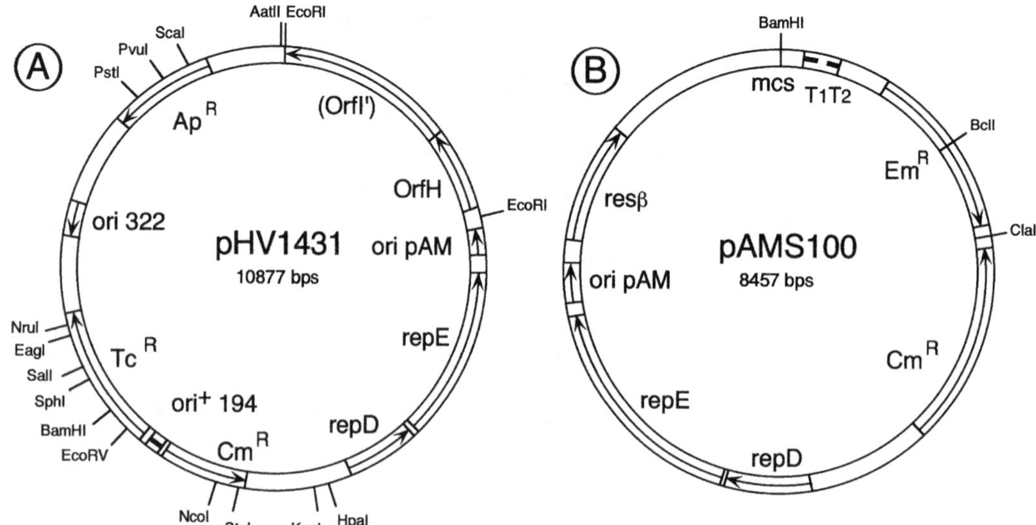

FIGURE 6 Theta plasmids pHV1431 and pAMS100. These plasmids carry the replication functions of the enterococcal plasmid pAMβ1 (ori pAM, *repD*, and *repE*), enabling replication in several gram-positive bacteria, including *B. subtilis*. (A) pHV1431 is a shuttle vector that can also replicate in *E. coli* (using the pBR322 replication functions). Selectable markers are Ap (for *E. coli*) and Tc or Cm (for *B. subtilis*). Several suitable restriction sites are indicated. (B) pAMS100 lacks the pBR322 replication functions, and its stability is improved over that of pHV1431 by the introduction of the resolvase gene (*resβ*) and the T_1T_2 transcription terminator.

rect transfection methods. Little is known about the genomic organization of this phage, and the size of its genome (≈120 kb) causes difficulties in the handling of its DNA. Despite these limitations, a versatile prophage transformation system has been developed for SPβ, which facilitates the cloning of fragments of more than 10 kb. An advantage of this system is that DNA inserts can be recovered on plasmid vectors, in both *B. subtilis* and *E. coli*, without the need for additional in vitro cloning steps (87).

33.5. PROTEIN SECRETION

33.5.1. Protein Secretion in *B. subtilis*

B. subtilis and related bacilli secrete specific proteins to high concentrations into the growth medium. Several secreted proteins are enzymes of commercial interest (see section 33.1.4), and this is a major motivation for the extensive industrial use of bacilli.

The protein secretion process in *B. subtilis* can be divided into three stages: early stages, involving the targeting of secretory proteins to the plasma membrane; middle stages, involving the translocation of these proteins across the membrane; and late stages, involving signal peptide processing, the release of the secretory proteins at the *trans* side of the membrane, their folding into a native conformation, their passage through the cell wall, and release into the growth medium.

33.5.2. Properties of Secreted Proteins

33.5.2.1. Signal Peptides

Most *Bacillus* secretory proteins are synthesized as precursors with N-terminal signal peptides. These are required during the early stages of the secretion process to target precursors to the membrane, to keep them in an unfolded conformation, and to initiate their interaction with the secretion machinery. During or shortly after translocation of the precursor across the membrane, the signal peptide is removed by signal peptidases (see section 33.5.3), which is a prerequisite for the release of the mature protein from the membrane.

Signal peptides from *Bacillus* species vary in length between 18 and 35 amino acid residues (106) and are, on average, ≈5 to 7 residues longer than those from gram-negative bacteria and eukaryotes. As with these other organisms, three regions can be distinguished in the signal peptides of bacilli (Fig. 8): (i) an N-terminal (n-) region of 2 to 8 amino acids, containing at least two positively charged residues; (ii) a hydrophobic core (h-region) with, on average, 17 residues; and (iii) a polar C-terminal (c-) region of ≈8 residues. The n- and h-regions are required for the initiation of precursor transport across the membrane through interactions with phospholipids and the protein transport machinery in the membrane (see section 33.5.3 and references 95 and 121). The c-region contains a signal peptidase I recognition sequence with the consensus A-X-B (residues −3 to −1), where A is Ala, Gly, Leu, Ser, Thr, or Val; B is usually Ala, Gly, Ser, or Thr; and X can be almost any residue, although Met and Pro are very rare. Cleavage by signal peptidase I occurs at the C-terminal side of residue B. Signal peptidase cleavage site predictions can be performed on the web site of the Center for Biological Sequence Analysis (http://www.cbs.dtu.dk/services/SignalP/).

Lipoproteins remain attached to the *trans* side of the membrane and have different c-regions with different structures to facilitate recognition by the lipoprotein-specific signal peptidase II (88). The consensus signal peptidase II cleavage site consists of four residues: the N-terminal residue is Leu, Val or Ile; the second is Ala, Gly, Ser or Thr; the third is Gly or Ala; and the fourth is

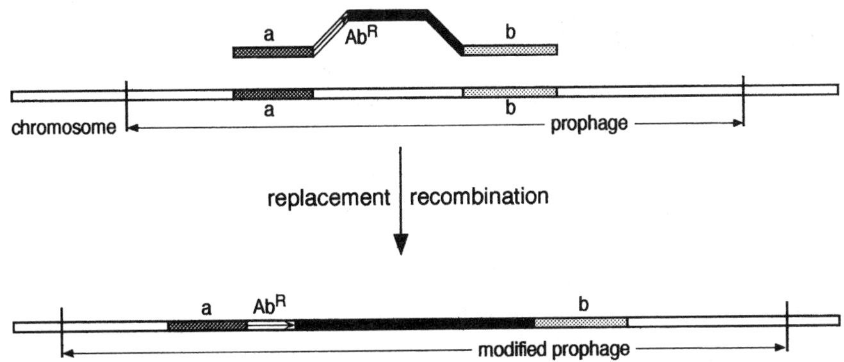

FIGURE 7 Prophage transformation. The letters a and b indicate fragments of prophage DNA in a plasmid vector, and Ab, which is present between a and b, is a selectable marker for *B. subtilis*. The filled line represents a cloned chromosomal DNA fragment that, by double-crossover recombination with a chrosomal copy of the prophage, can be integrated into the host chromosome. Recombinant phage can be obtained by induction of the lytic cycle, e.g., by thermoinduction of phage mutants carrying a thermosensitive immunity repressor (26, 27, 111).

invariably Cys. Cleavage by signal peptidase II takes place at the N-terminal side of the Cys residue. Diacylglyceryl modification of the Cys residue is a prerequisite for signal peptidase II cleavage (100).

33.5.2.2. Propeptides

Many exported proteins from gram-positive bacteria are synthesized as preproproteins. The propeptide is located between the signal peptide (pre-) and the mature protein (Fig. 8). Propeptides vary in length from ≈8 (*B. subtilis* α-amylase) to ≈200 amino acids (various *Bacillus* neutral proteinases). Propeptides are probably not involved in translocation of secretory proteins across the membrane, but rather in the folding of the mature protein into its native conformation on the *trans* side of the membrane (5, 106). Cleavage of the propeptide occurs after membrane translocation and is catalyzed by extracellular proteinases; cleavage of the propeptide of exported *Bacillus* proteinases occurs autocatalytically (106).

33.5.2.3. Mature Protein

In addition to signal peptides, certain properties of the mature part of exported proteins are important for secretion. First, the signal peptide, mature region, and export machinery must interact to prevent the formation of tertiary structures in the secretory protein that would render it secretion incompetent. Second, secreted proteins do not usually contain long (8 to 20-residue) hydrophobic domains flanked by positively charged residues, since such domains act as "stop transfer" or "membrane anchor" sequences (122). Finally, relatively few *Bacillus* exoproteins form disulfide bridges.

33.5.3. Secretory Pathway

33.5.3.1. Early Stages

In the early stages of protein export, molecular chaperones and/or elements of the signal recognition particle (SRP)-like pathway are usually involved. These assist in maintaining precursors of secretory proteins in an unfolded conformation, preventing their aggregation, and targeting them to the secretory machinery. Several chaperones have been identified in *B. subtilis* (DnaJ, DnaK, GrpE, GroEL, and GroES) (102, 124), but their relevance for protein secretion has not yet been demonstrated. Components of the SRP-like pathway have been identified in *B. subtilis*, including (i) scRNA, a small RNA molecule with structural similarity to the 7S RNA of the mammalian SRP; (ii) Ffh, a homolog of the 54-kDa subunit of the mammalian SRP (79); and (iii) Srb, a homolog of the α subunit of the mammalian SRP receptor (80).

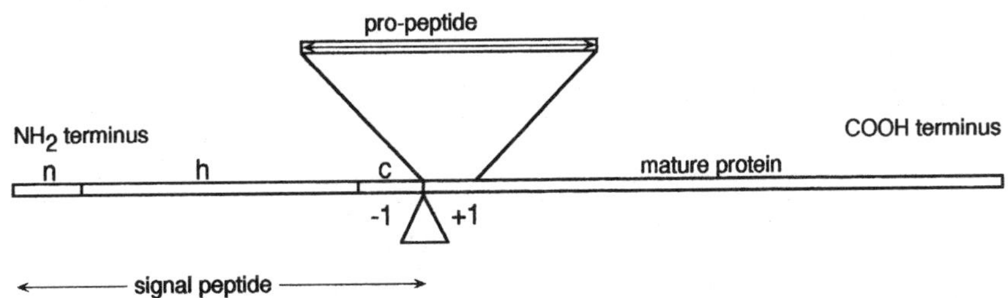

FIGURE 8 Schematic representation of the precursor of a secretory protein. The three regions characteristic of the signal peptide (positively charged n-region [n], hydrophobic core [h], and c-region [c]), the propeptide, and the mature protein are indicated. The signal peptidase cleavage site is indicated by −1/+1.

33.5.3.2. Middle Stages

The middle stages of secretion involve the translocase complex in the membrane, which consists of at least four subunits: SecA (the translocation ATPase), SecE, SecDF, and SecY. This complex is responsible for the ATP-dependent transport of proteins across the membrane (8a, 52, 106). In addition, a fifth component, SecG, is likely to be present, as in *E. coli*.

33.5.3.3. Late Stages

Gram-positive bacteria lack an outer membrane and generally have thicker (10- to 50-nm) cell walls than do their gram-negative counterparts. The anionic polymers (teichoic and teichuronic acids) that make up approximately half of the cell wall by weight (the other polymer is peptidoglycan) confer a high concentration of immobilized negative charge (3). These properties profoundly influence the late stages of secretion, in which signal peptidases, cell-associated proteases, and folding factors play an important part. Five distinct chromosomally encoded type I signal peptidases, denoted SipS (117), SipT, SipU, SipV, and SipW (112), and one type II signal peptidase, LspA (88), have been identified in *B. subtilis*. In addition, some strains of *B. subtilis* contain plasmids specifying a type I signal peptidase, denoted SipP (67).

Metal ions and proteins can act as folding factors for secreted proteins. For instance, Fe^{3+} and Ca^{2+} act as folding catalysts for levansucrase (86). Similarly, the lipoprotein PrsA is crucial for the folding of certain secreted *B. subtilis* proteins, such as α-amylase (57). Efficient folding of the mature protein is also required to prevent the degradation of the translocated protein by the proteinases that are secreted by *B. subtilis* in high amounts. A model that combines the known properties of the cellular components involved in protein export in *B. subtilis* is presented in Fig. 9.

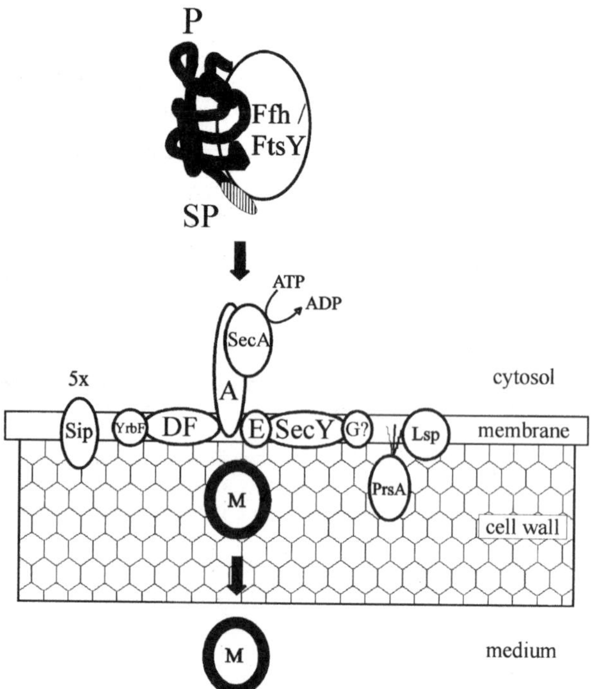

FIGURE 9 Schematic representation of the secretory pathway of *B. subtilis*. The secretory precursor protein (P) is kept in an unfolded, secretion-competent conformation and targeted to the membrane by chaperones and targeting factors such as Ffh and FtsY. The precursor-chaperone complex interacts with the translocation ATPase, SecA (A), which is probably functional as a dimer that directs the precursor into the translocation channel. This channel consists of at least two components: SecE (E) and SecY (Y). In addition, SecDF (DF), YrbF, and, probably, SecG (G?) are also associated with the translocase complex. Upon translocation, the signal peptide (SP) is cleaved by one of the five type I signal peptidases encoded by *B. subtilis* genes (SipS, SipT, SipU, SipV, or SipW) or, in the case of a lipoprotein, by the type II signal peptidase (Lsp). The mature protein folds into its native conformation, which may be catalyzed by folding factors such as PrsA and divalent cations. Finally, the mature, folded protein (M) is released from the membrane and transported through the cell wall into the growth medium.

33.5.4. Rate-Limiting Steps in the Secretion of Native Proteins

Frequently, the level of transcription constitutes the limiting step in the production of native secretory proteins from *Bacillus* strains. This is a complicated phenomenon, since the transcription of most genes encoding secreted proteins is tightly controlled in this organism. The system usually involves regulatory networks (see section 33.3.1) in which several transcriptional activators (e.g., DegQ, DegR, DegU, and SenS) and repressors (e.g., AbrB, Hpr, and Sin; for review, see reference 29) participate. Although there is little evidence that components of the secretion apparatus can limit protein secretion under natural conditions, the situation is different when native proteins are overproduced. For example, *B. amyloliquefaciens* pre-α-amylase accumulates in *B. subtilis* upon overproduction of the enzyme (56). Unexpectedly, this accumulation appeared to be prevented in strains from which the gene for the type I signal peptidase SipS had been deleted, suggesting that the interaction of pre-α-amylase with SipS leads to reduced efficiencies of processing of this particular precursor (8). Furthermore, it was shown that the folding factor PrsA sets a limit for high-level secretion of various enzymes, such as α-amylase and subtilisin (57).

33.5.5. Secretion of Heterologous Proteins

In contrast to native proteins, the extracellular production of heterologous proteins, especially from eukaryotic sources, is frequently inefficient in bacilli. This may, at least in part, be due to the inherent properties of the *Bacillus* protein secretion apparatus.

33.5.5.1. Strategies for Improving the Secretion of Heterologous Proteins

To achieve efficient extracellular production of heterologous proteins, secretion vectors have been constructed that contain a strong promoter, followed by an efficient signal sequence and an MCS (38, 106). An alternative strategy involves the random selection of signal sequences from the *B. subtilis* chromosome with the help of suitable reporter proteins, the secretion of which could be monitored by using plate assays (107, 108). The latter strategy has the advantage that efficient signal peptides can be selected for the secretion of a particular heterologous protein, provided that an appropriate assay for the protein in question is available.

33.5.5.2. Secretion Bottlenecks

The secretion of heterologous proteins can be blocked at various stages. One problem is that precursors may accumulate in the cytoplasm as insoluble aggregates, owing to a failure to maintain them in an export-competent state or to inefficient interactions with components of the host secretory machinery (see section 33.5.3). A second problem could be the failure of signal peptidases to process the precursor correctly, which is likely to be of particular importance if the product is biotherapeutic. A third problem is the folding of the processed proteins into their native conformation at the *trans* side of the membrane, an environment that is biochemically and physiologically very different from that encountered in other organisms. Aberrant or slow folding may be due to nonproductive interactions with components of the cell wall, with the hosts' folding factors, such as PrsA, or with disulfide bond oxidoreductases. The absence of appropriate folding factors is likely to result in inefficient release of the protein from the membrane, increased susceptibility to degradation, and reduced biological activity. Finally, because of its high density of immobilized negative charge and limited porosity, the cell wall may form a barrier to heterologous protein secretion in gram-positive bacteria (3).

33.5.6. Optimization of Heterologous Protein Secretion

33.5.6.1. Optimization of the Secreted Protein

Three types of modification of the target protein may be considered. First, the inclusion of optimal signal peptides, because the efficiency of export of proteins fused to different signal peptides can vary significantly (107). Second, the removal of positively charged amino acid residues from the N-terminus of the mature protein may be of value (see section 33.5.2). Third, degradation of proteins may be avoided by the removal of proteinase target sites.

33.5.6.2. Optimization of the Host

The overproduction of chaperones or other components of the secretion machinery should be considered, as these may become limiting when exported proteins are produced in large quantities. Overproduction of chaperones, such as GroEL, GroES, and DnaK, can prevent the formation of insoluble aggregates in *E. coli*, and a combination of a mutation in SecY with overproduction of SecE resulted in improved export of interleukin-6 to the periplasm (45). Furthermore, overproduction of signal peptidase can result in improved efficiencies of processing of certain hybrid precursor proteins in *E. coli* and *B. subtilis* (115, 117). Alternatively, the yield of some heterologous proteins may be increased by the coproduction and secretion of heterologous folding factors, such as the human protein disulfide bond isomerase (48). Human protein disulfide bond isomerase may facilitate the folding of heterologous secreted proteins into an active and proteinase-resistant conformation. An alternative approach to solve the problem of degradation is the use of proteinase-negative strains (114, 127). Finally, the use of strains with an altered cell wall composition may improve the release of exported proteins from the cells.

33.5.7. Methods of Analysis

Various assays are available to monitor general protein secretion or the secretion of specific reporter proteins. The simplest and most convenient are plate assays in which the activity of secreted enzymes can be visualized as zones of hydrolysis around colonies growing on agar plates. More refined assays require the subcellular fractionation of cells or pulse-chase labeling of secretory precursor proteins. The latter assays facilitate the identification of certain secretion bottlenecks, such as the accumulation of secretory proteins in the cytosol, membrane, or cell wall, inefficient precursor processing, or slow release into the culture medium.

33.5.7.1. Plate Assays

α-Amylase

Standard agars are supplemented with 1% (wt/vol) potato starch (Sigma). Degradation of starch by secreted α-amylase can be visualized after exposure to iodine vapor by inverting the agar plate over a few crystals of iodine (in a fume hood). The iodine stains the starch in the plate blue, and zones of starch hydrolysis around positive colonies are colorless.

Penicillinase

Penicillinases catalyze the hydrolysis of penicillin to penicilloic acid, which can decolorize a blue starch-iodine complex. Cells are plated on LB agar (see section 33.2.2.1) containing 0.2% (wt/vol) starch (Sigma) and 50 mM K-phosphate buffer (pH 6.5). After overnight incubation, 6.5 ml of a deep-blue soft agar medium is poured on the plates. After solidification of the overlay agar, the plates are incubated for about 15 min at 30°C, after which decolorization haloes appear around colonies secreting penicillinase. The soft agar assay medium is prepared by mixing (at 60°C) 1 volume of starch-agar medium with 1 volume of a reaction mixture (770 mg of I_2; 3 g of KI; 300 mg of ampicillin; 2 ml of K-phosphate buffer [1 M, pH 6.5], in a total volume of 100 ml of H_2O).

Proteinase

To assay the activity of secreted proteinases on plates, 1 (wt/vol) skim milk (Oxoid, Basingstoke, U.K.) is added to agar media. The agar medium and a 10% stock solution of skim milk are autoclaved separately (15 min, 20 lb/in^2). After overnight incubation at 37°C, haloes are visible around proteinase-secreting colonies.

33.5.7.2. Subcellular Fractionation and Pulse-Chase Labeling

A protocol for the subcellular fractionation of *B. subtilis* cells is given in reference 73. Three subcellular fractions are obtained, representing the cytosol, plasma membrane, and cell wall.

Pulse-chase labeling experiments are performed to determine the kinetics of precursor processing. Useful protocols for the growth of cells in synthetic media, the depletion of intracellular methionine pools, the labeling of cells with [^{35}S] methionine, cell lysis, immunoprecipitation of the target protein, sodium dodecyl sulfate-polyacrylamide gel electrophoresis, and fluorography have been published previously by van Dijl et al. (116).

We thank Siger Holsappel for preparing the figures and V. Vagner and S.D. Ehrlich (INRA, Jouy en Josas, France) for permitting us to include Fig. 4A, which was adapted from their unpublished work. Parts of the work described in this chapter were financed by Gist-Brocades (Delft, The Netherlands), Genencor International (Rijswijk, The Netherlands), NOVO/Nordisk (Bagsvaerd, Denmark), the Dutch Foundation for Technical Research (STW 349-1622), the

Biotechnology and Biological Sciences Research Council (U.K.), and the Commission of the European Union (BIOT-CT91-0268; BIO2-CT93-0254; BIO2-CT93-0272; BIO2-PL96-0655; BIO2-CT95-0278).

References

1. **Anagnostopoulos, C., P. J. Piggot, and J. A. Hoch.** 1993. The genetic map of *Bacillus subtilis*, p. 425–462. *In* A. L. Sonenshein, J. A. Hoch, and R. Losick (ed.), *Bacillus subtilis and Other Gram-Positive Bacteria: Biochemistry, Physiology, and Molecular Genetics*. American Society for Microbiology, Washington, D.C.
2. **Anagnostopoulos, C., and J. Spizizen.** 1961. Requirements for transformation in *Bacillus subtilis*. *J. Bacteriol.* **81:**741–746.
3. **Archibald, A. R., I. C. Hancock, and C. R. Harwood.** 1993. Cell wall structure, synthesis, and turnover, p. 381–410. *In* A. L. Sonenshein, J. A. Hoch, and R. Losick (ed.), *Bacillus subtilis and Other Gram-Positive Bacteria: Biochemistry, Physiology, and Molecular Genetics*. American Society for Microbiology, Washington, D.C.
4. **Aronson, A. L.** 1993. Insecticidal toxins, p. 953–963. *In* A. L. Sonenshein, J. A. Hoch, and R. Losick (ed.), *Bacillus subtilis and Other Gram-Positive Bacteria: Biochemistry, Physiology, and Molecular Genetics*. American Society for Microbiology, Washington, D.C.
5. **Baker, D., A. K. Shiau, and D. A. Agard.** 1993. The role of pro-regions in protein folding. *Curr. Opin. Cell Biol.* **5:**966–970.
6. **Berkeley, R. C. W., N. A. Logan, L. A. Shute, and A. G. Capey.** 1984. Identification of *Bacillus* species. *Methods Microbiol.* **16:**291–328.
7. **Biaudet, V., F. Samson, C. Anagnostopoulos, S. D. Ehrlich, and Ph. Bessières.** 1996. Computerized genetic map of *Bacillus subtilis*. *Microbiology* **142:**605–618.
8. **Bolhuis, A., A. Sorokin, V. Azevedo, S. D. Ehrlich, P. G. Braun, A. de Jong, G. Venema, S. Bron, and J. M. van Dijl.** 1996. **Bacillus subtilis** can modulate its capacity and specificity for protein secretion through temporally controlled expression of the *sipS* gene for signal peptidase I. *Mol. Microbiol.* **22:**605–618.

8a. **Bolhuis, A., J. M. van Dijl, and S. Bron.** Unpublished results.

9. **Bron, S.** 1990. Plasmids, p. 75–175. *In* C. R. Harwood and S. M. Cutting (ed.), *Molecular Biological Methods for Bacillus*. John Wiley & Sons Ltd., Chichester, U.K.
10. **Bron, S., S. Holsappel, G. Venema, and B. P. H. Peeters.** 1991. Plasmid deletion formation between short direct repeats in *Bacillus subtilis* is stimulated by single-stranded rolling-circle replication intermediates. *Mol. Gen. Genet.* **226:**88–96.
11. **Bron, S., L. Jannière, and S. D. Ehrlich.** 1988. Restriction and modification in *Bacillus subtilis* Marburg 168: target sites and effects on plasmid transformation. *Mol. Gen. Genet.* **211:**186–189.
12. **Bron, S., and G. Venema.** 1972. Ultraviolet inactivation and excision repair in *B. subtilis*. I. Construction of a transformable eightfold auxotrophic strain and two ultraviolet-sensitive derivatives. *Mutat. Res.* **15:**1–10.
13. **Bruand, C., S. D. Ehrlich, and L. Jannière.** 1991. Unidirectional theta replication of the stable *Enterococcus faecalis* plasmid pAMβ1. *EMBO J.* **10:**2171–2177.
14. **Chambliss, G. H., T. M. Henkin, and J. M. Leventhal.** 1983. Bacterial in vitro protein synthesizing systems. *Methods Enzymol.* **101:**598–605.
15. **Chang, S., and S. N. Cohen.** 1979. High frequency transformation of *Bacillus subtilis* protoplasts by plasmid DNA. *Mol. Gen. Genet.* **168:**111–115.
16. **Chédin, F., E. Dervyn, R. Dervyn, S. D. Ehrlich, and P. Noirot.** 1994. Frequency of deletion formation decreases exponentially with distance between short direct repeats. *Mol. Microbiol.* **12:**561–569.
17. **Claus, D., and R. C. W. Berkeley.** 1986. Genus *Bacillus* Cohn 1872, p. 1105–1141. *In* P. H. A. Sneath, (ed.), *Bergey's Manual of Systematic Bacteriology*. Williams & Wilkins, Baltimore.
18. **Claus, D., and D. Fritze.** 1989. Taxonomy of *Bacillus*, p. 5–26. *In* C. R. Harwood (ed.), *Biotechnology Handbooks 2: Bacillus*. Plenum Publishing Corp, New York.
19. **Cook, N., D. J. Silcock, R. N. Waterhouse, J. I. Prosser, L. A. Glover, and K. C. Killham.** 1993. Construction and detection of bioluminescent strains of *Bacillus subtilis*. *J. Appl. Bacteriol.* **75:**350–359.
20. **Crutz, A.-M., M. Steinmetz, S. Aymerich, R. Richter, and D. LeCoq.** 1990. Induction of levansucrase in *Bacillus subtilis*: an antitermination mechanism negatively controlled by the phosphotransferase system. *J. Bacteriol.* **172:** 1043–1050.
21. **Cutting, S. M., and P. B. Vander Horn.** 1990. Genetic analysis, p. 27–74. *In* C. R. Harwood and S. M. Cutting (ed.), *Molecular Biological Methods for Bacillus*. John Wiley & Sons Ltd., Chichester, U.K.
22. **Dubnau, D.** 1991. Genetic competence in *Bacillus subtilis*. *Microbiol. Rev.* **55:**395–424.
23. **Dubnau, D.** 1993. Genetic exchange and homologous recombination, p. 555–584. *In* A. L. Sonenshein, J. A. Hoch, and R. Losick (ed.), *Bacillus subtilis and Other Gram-Positive Bacteria: Biochemistry, Physiology, and Molecular Genetics*. American Society for Microbiology, Washington, D.C.
24. **Ehrlich, S. D.** 1989. Illegitimate recombination in bacteria, p. 797–829. *In* D. Berg and M. Howe (ed.), *Mobile DNA*. American Society for Microbiology, Washington, D.C.
25. **Ehrlich, S. D., H. Bierne, E. d'Alençon, D. Vilette, M. Petranovic, P. Noirot, and B. Michel.** 1993. Mechanisms of illegitimate recombination. *Gene* **135:**161–166.
26. **Errington, J.** 1990. Gene cloning techniques, p. 175–220. *In* C. R. Harwood and S. M. Cutting (ed.), *Molecular Biological Methods for Bacillus*. John Wiley & Sons Ltd., Chichester, U.K.
27. **Errington, J.** 1993. Temperate phage vectors, p. 645–650. *In* A. L. Sonenshein, J. A. Hoch, and R. Losick (ed.), *Bacillus subtilis and Other Gram-Positive Bacteria: Biochemistry, Physiology, and Molecular Genetics*. American Society for Microbiology, Washington, D.C.
28. **Eymann, C., H. Mach, C. R. Harwood, and M. Hecker.** 1996. Phosphate-starvation-inducible proteins in *Bacillus subtilis*: a two-dimensional gel electrophoresis study. *Microbiology* **142:**3163–3170.
29. **Ferrari, E., A. S. Jarnagin, and B. F. Schmidt.** 1993. Commercial production of extracellular enzymes, p. 917–937. *In* A. L. Sonenshein, J. A. Hoch, and R. Losick (ed.), *Bacillus subtilis and Other Gram-Positive Bacteria: Biochem-*

istry, Physiology, and Molecular Genetics. American Society for Microbiology, Washington, D.C.

30. **Gartner, D., J. Degenkolb, J. A. E. Ripperger, R. Allmansberger, and W. Hillen.** 1992. Regulation of the *Bacillus subtilis* W23 xylose utilization operon: interaction of the Xyl repressor with the *xyl* operator and the inducer xylose. *Mol. Gen. Genet.* **232:**415–422.
31. **Goffeau, A.** 1997. The yeast genome directory. *Nature* **387:**5–6.
32. **Grossman, A. D.** 1995. Genetic networks controlling the initiation of sporulation and the development of genetic competence in *Bacillus subtilis*. *Annu. Rev. Genet.* **29:** 477–508.
33. **Gruss, A., and S. D. Ehrlich.** 1989. The family of highly interrelated single-stranded deoxyribonucleic acid plasmids. *Microbiol. Rev.* **53:**231–241.
34. **Hahn, V., and P. Stiegler.** 1986. An *Escherichia coli* S1-like ribosomal protein is immunologically conserved in Gram-negative bacteria, but not in Gram-positive bacteria. *FEMS Microbiol. Lett.* **36:**293–297.
35. **Haima, P., S. Bron, and G. Venema.** 1987. The effect of restriction on shotgun cloning and plasmid stability in *Bacillus subtilis* Marburg. *Mol. Gen. Genet.* **209:**2335–342.
36. **Haima, P., D. van Sinderen, S. Bron, and G. Venema.** 1990. An improved β-galactosidase α-complementation system for molecular cloning in *Bacillus subtilis*. *Gene* **93:** 41–47.
37. **Haldenwang, W. G.** 1995. The sigma factors of *Bacillus subtilis*, *Microbiol. Rev.* **59:**1–30.
38. **Harwood, C. R.** 1992. *Bacillus subtilis* and its relatives: molecular biological and industrial workhorses. *Trends Biotechnol.* **10:**247–256.
39. **Harwood, C. R., and A. R. Archibald.** 1990. Growth, maintenance and general techniques, p. 1–26. *In* C. R. Harwood and S. M. Cutting (ed.), *Molecular Biological Methods for Bacillus*. John Wiley & Sons Ltd., Chichester, U.K.
40. **Harwood, C. R., D. M. Williams, and P. S. Lovett.** 1983. Nucleotide sequence of a *Bacillus pumilus* gene specifying chloramphenicol acetyltransferase. *Gene* **24:**163–169.
41. **Hastings, J. W.** 1996. Chemistries and colors of bioluminescent reactions: a review. *Gene* **173:**5–11.
42. **Hecker, M., and U. Völker.** 1990. General stress protein in *Bacillus subtilis*. *FEMS Microbiol. Ecol.* **74:**197–213.
43. **Heim, R., D. C. Prasher, and R. Y. Tsien.** 1994. Wavelength mutations and posttranslational autooxidation of green fluorescent protein. *Proc. Natl. Acad. Sci. USA* **91:** 12501–12504.
44. **Hoch, J. A., M. Barat, and C. Anagnostopoulos.** 1967. Transformation and transduction in recombination-defective mutants of *Bacillus subtilis*. *J. Bacteriol.* **93:**1925–1937.
45. **Hockney, R. C.** 1994. Recent developments in heterologous protein production in *Escherichia coli*. *Trends Biotechnol.* **12:**456–463.
46. **Horinouchi, S., and B. Weisblum.** 1982. Nucleotide and function map of pC194, a plasmid that specifies inducible chloramphenicol resistance. *J. Bacteriol.* **150:**804–814.
47. **Horton, R. M., D. H. Hunt, S. N. Ho, J. K. Pullen, and L. R. Pease.** 1989. Engineering hybrid genes without the use of restriction enzymes: gene splicing by overlap extension. *Gene* **77:**61–68.
48. **Humphreys, D. P., N. Weir, A. Mountain, and P. A. Lund.** 1995. Human protein disulphide isomerase functionally complements a *dsbA* mutation and enhances the yield of pectate lyase C in *Escherichia coli*. *J. Biol. Chem.* **270:**28210–28215.
49. **Itaya, M.** 1993. Physical map of the *Bacillus subtilis* chromosome, p. 463–471. *In* A. L. Sonenshein, J. A. Hoch, and R. Losick (ed.), *Bacillus subtilis and Other Gram-Positive Bacteria: Biochemistry, Physiology, and Molecular Genetics*. American Society for Microbiology, Washington, D.C.
50. **Jannière, L., C. Bruand, and S. D. Ehrlich.** 1990. Structurally stable *Bacillus subtilis* cloning vectors. *Gene* **87:**53–59.
51. **Jannière, L., A. Gruss, and S. D. Ehrlich.** 1993. Plasmids, p. 625–644. *In* A. L. Sonenshein, J. A. Hoch, and R. Losick (ed.), *Bacillus subtilis and Other Gram-Positive Bacteria: Biochemistry, Physiology, and Molecular Genetics*. American Society for Microbiology, Washington, D.C.
52. **Jeong, S. M., H. Yoshikawa, and H. Takahashi.** 1993. Isolation and characterization of the *secE* homologue gene of *Bacillus subtilis*. *Mol. Microbiol.* **10:**133–142.
53. **Jørgensen, P. L., C. K. Hansen, G. B. Poulsen, and B. Diderichsen.** 1990. *In vivo* genetic engineering: homologous recombination as a tool for plasmid construction. *Gene* **96:**37–41.
54. **Kiewiet, R., J. Kok, J. F. M. L. Seegers, G. Venema, and S. Bron.** 1993. The mode of replication is a major factor in segregational plasmid instability in *Lactococcus lactis*. *Appl. Environ. Microbiol.* **59:**358–364.
55. **Kleopper, J. W., R. Lifshitz, and R. M. Zablotowicz.** 1989. Free-living bacterial inocula for enhancing crop productivity. *Trends Biotechnol.* **7:**39–44.
56. **Kontinen, V. P., and M. Sarvas.** 1988. Mutants of *Bacillus subtilis* defective in protein export. *J. Gen. Microbiol.* **134:** 2333–2344.
57. **Kontinen, V. P., and M. Sarvas.** 1993. The PrsA lipoprotein is essential for protein secretion in *Bacillus subtilis* and sets a limit for high-level secretion. *Mol. Microbiol.* **8:**727–737.
58. **Kunst, F., et al.** 1997. The complete genome sequence of the Gram-positive bacterium *Bacillus subtilis*. *Nature* **390:** 249–256.
59. **Kunst F., and G. Rapoport.** 1995. Salt stress is an environmental signal affecting degradative enzyme synthesis in *Bacillus subtilis*. *J. Bacteriol.* **177:**2403–2407.
60. **Leenhouts, K. J., and G. Venema.** 1993. Lactococcal plasmid vectors, p. 65–94. *In* K. G. Hardy (ed.), *Plasmids, A Practical Approach*. Oxford University Press, New York.
61. **Lewis, P. J., and J. Errington.** 1996. Use of green fluorescent protein for the detection of cell-specific gene expression and subcellular protein location during sporulation in *Bacillus subtilis*. *Microbiology* **142:**733–740.
62. **Lewis, P. J., C. E. Nwoguh, M. R. Barer, C. R. Harwood, and J. Errington.** 1994. Use of digitized video microscopy with a fluorogenic enzyme substrate to demonstrate cell and compartment-specific gene expression in *Salmonella enteritidis* and *Bacillus subtilis*. *Mol. Microbiol.* **13:**655–662.
63. **Lopez de Saro, F. J., W. A. Y. Moon, and J. D. Helmann.** 1995. Structural analysis of the *Bacillus subtilis* delta factor: a protein polyanion which displaces RNA from RNA polymerase. *J. Mol. Biol.* **252:**189–202.

64. **Lopilato, J., S. Bortner, and J. Beckwith.** 1986. Mutations in a new chromosomal gene of *Escherichia coli* K-12, *pcnB*, reduce plasmid copy number of pBR322 and its derivatives. *Mol. Gen. Genet.* **205:**285–290.

65. **Lovett, P. S., and E. J. Rogers.** 1996. Ribosome regulation by the nascent peptide. *Microbiol. Rev.* **60:**366–385.

66. **Meijer, W. J. J., A. de Boer, S. van Tongeren, G. Venema, and S. Bron.** 1995. Characterization of the replication region of the *Bacillus subtilis* plasmid pLS20: a novel type of replicon. *Nucleic Acids Res.* **23:**3214–3223.

67. **Meijer, W. J. J., A. de Jong, G. B. A. Wisman, H. Tjalsma, G. Venema, S. Bron, and J. M. van Dijl.** 1995. The endogenous *Bacillus subtilis* (*natto*) plasmids pTA1015 and pTA1040 contain signal peptidase-encoding genes: identification of a new structural module on cryptic plasmids. *Mol. Microbiol.* **17:**621–631.

68. **Meijer, W. J. J., G. Venema, and S. Bron.** 1995. Characterization of single strand origins of cryptic rolling-circle plasmids from *Bacillus subtilis*. *Nucleic Acids Res.* **23:**612–619.

69. **Meijer, W. J. J., G. B. A. Wisman, P. Terpstra, P. B. Thorsted, C. M. Thomas, S. Holsappel, G. Venema, and S. Bron.** 1998. Rolling-circle plasmids from *Bacillus subtilis*: complete nucleotide sequences and analyses of genes of pTA1015, pTA1040, pTA1050 and pTA1060, and comparisons with related plasmids from Gram-positive bacteria. *FEMS Microbiol. Rev.* **21:**337–368.

70. **Meima, R. B., B. J. Haijema, H. Dijkstra, G.-J. Haan, G. Venema, and S. Bron.** 1997. Roles of enzymes of homologous recombination in illegitimate plasmid recombination in *Bacillus subtilis*. *J. Bacteriol.* **179:**1219–1229.

71. **Meima, R., B. J. Haijema, G. Venema, and S. Bron.** 1995. Overproduction of the ATP-dependent nuclease AddAB improves the structural stability of a model plasmid system in *Bacillus subtilis*. *Mol. Gen. Genet.* **248:**391–398.

72. **Meima, R., G. Venema, and S. Bron.** 1996. A positive selection vector for the analysis of structural plasmid instability in *Bacillus subtilis*. *Plasmid* **35:**14–30.

73. **Merchante, R., H. M. Pooley, and D. Karamata.** 1995. A periplasm in *Bacillus subtilis*. *J. Bacteriol.* **177:**6176–6183.

74. **Miller, J.** 1972. *Experiments in Molecular Genetics*. Cold Spring Harbor Laboratory, Cold Spring Harbor, Cold Spring Harbor, N.Y.

75. **Moran, C. P., Jr.** 1990. Measuring gene expression in *Bacillus*, p. 267–293. *In* C. R. Harwood and S. M. Cutting (ed.), *Molecular Biological Methods for Bacillus*. John Wiley & Sons Ltd., Chichester, U.K.

76. **Moszer, I., P. Glaser, and A. Danchin.** 1995. *SubtiList*: a relational database for the *Bacillus subtilis* genome. *Microbiology* **141:**261–268.

77. **Msadek, T., F. Kunst, and G. Rapoport.** 1993. Two-component regulatory systems, p. 729–745. *In* A. L. Sonenschein, J. A. Hoch, and R. Losick (ed.), *Bacillus subtilis and Other Gram-Positive Bacteria: Biochemistry, Physiology, and Molecular Genetics*. American Society for Microbiology, Washington, D.C.

78. **Müller, J. P., A. Zhidong, T. Merad, I. C. Hancock, and C. R. Harwood.** 1997. The influence of *Bacillus subtilis phoR* on cell wall anionic polymers. *Microbiology* **143:**947–956.

79. **Nakamura, K., M. Nishiguchi, K. Honda, and K. Yamane.** 1994. The *Bacillus subtilis* SRP54 homologue, Ffh, has an intrinsic GTPase activity and forms a ribonucleoprotein complex with small cytoplasmic RNA. *Biochem. Biophys. Res. Commun.* **199:**1394–1399.

80. **Oguro, A., H. Kakeshita, K. Honda, H. Takamatsu, K. Nakamura, and K. Yamane.** 1995. Srb: a *Bacillus subtilis* gene encoding a homologue of the alpha-subunit of the mammalian signal recognition particle receptor. *DNA Res.* **2:**95–100.

81. **Peijnenburg, A. A. C. M., S. Bron, and G. Venema.** 1988. Plasmid deletion formation in *Bacillus subtilis*. *Plasmid* **20:**23–32.

82. **Perego, M.** 1993. Integrational vectors for genetic manipulation in *Bacillus subtilis*, p. 615–624. *In* A. L. Sonenshein, J. A. Hoch, and R. Losick (ed.), *Bacillus subtilis and Other Gram-Positive Bacteria: Biochemistry, Physiology, and Molecular Genetics*. American Society for Microbiology, Washington, D.C.

83. **Pero, J., and A. Sloma.** 1993. Proteases, p. 939–952. *In* A. L. Sonenshein, J. A. Hoch, and R. Losick (ed.), *Bacillus subtilis and Other Gram-Positive Bacteria: Biochemistry, Physiology, and Molecular Genetics*. American Society for Microbiology, Washington, D.C.

84. **Perrière, G., I. Moszer, and J. Gojobori.** 1996. NRsub—A nonredundant database for *Bacillus subtilis*. *Nucleic Acids Res.* **24:**41–45.

85. **Petit, M.-A., C. Bruand, L. Jannière, and S. D. Ehrlich.** 1990. Tn*10*-derived transposons active in *Bacillus subtilis*. *J. Bacteriol.* **172:**6736–6740.

86. **Petit-Glatron, M. F., L. Grajcar, A. Munz, and R. Chambert.** 1993. The contribution of the cell wall to a transmembrane calcium gradient could play a key role in *Bacillus subtilis* protein secretion. *Mol. Microbiol.* **9:**1097–1106.

87. **Poth, H., and P. Youngman.** 1988. A new cloning system for *Bacillus subtilis* comprising elements of phage, plasmid and transposon vectors. *Gene* **73:**215–226.

88. **Prágai, Z., H. Tjalsma, A. Bolhuis, J. M. van Dijl, G. Venema, and S. Bron.** 1996. The signal peptidase II (*lsp*) gene of *Bacillus subtilis*. *Microbiology* **143:**1327–1333.

89. **Prasher, D. C., V. E. Eckenrode, W. W. Ward, P. G. Prendergast, and M. J. Cormier.** 1992. Primary structure of the *Aequorea victoria* green-fluorescent protein. *Gene* **111:**229–233.

90. **Priest, F. G.** 1989. Isolation and identification of aerobic endospore-forming bacteria, p. 27–56. *In* C. R. Harwood (ed.), *Biotechnology Handbooks 2: Bacillus*. Plenum Publishing Corp., New York.

91. **Priest, F. G.** 1993. Systematics and ecology of *Bacillus*, p. 1–16. *In* A. L. Sonenshein, J. A. Hoch, and R. Losick (ed.), *Bacillus subtilis and Other Gram-Positive Bacteria: Biochemistry, Physiology, and Molecular Genetics*. American Society for Microbiology, Washington, D.C.

92. **Priest, F. G., and B. Alexander.** 1988. A frequency matrix for the probabilistic identification of some bacilli. *J. Gen. Microbiol.* **134:**3011–3018.

93. **Priest, F. G., M. Goodfellow, and C. Todd.** 1988. A numerical analysis of the genus *Bacillus*. *J. Gen. Microbiol.* **134:**1847–1882.

94. **Priest, F. G., and C. R. Harwood.** 1994. *Bacillus* species, p. 377–421. *In* Y. H. Hui and G. G. Khachatourians

(ed.), *Food Biotechnology*. VCH Publishers Inc., New York.

95. **Pugsley, A. P.** 1993. The complete general secretory pathway in Gram-negative bacteria. *Microbiol. Rev.* **57:** 50–108.
96. **Puohiniemi, R., S. Butcher, E. Tarkka, and M. Sarvas.** 1991. High level production of *Escherichia coli* outer membrane proteins OmpA and OmpF intracellularly in *Bacillus subtilis*. *FEMS Microbiol. Lett.* **83:**29–34.
97. **Reeve, J. N., N. H. Mendelson, S. I. Coyne, L. L. Hallock, and R. M. Cole.** 1973. Minicells of *Bacillus subtilis*. *J. Bacteriol.* **114:**860–873.
98. **Roberts, M. W., and J. C. Rabinowitz.** 1989. The effect of *Escherichia coli* ribosomal protein S1 on the translation specificity of bacterial ribosomes. *J. Biol. Chem.* **264:** 2228–2235.
99. **Sambrook, J., E. F. Fritsch, and T. Maniatis.** 1989. *Molecular Cloning: A Laboratory Manual*. Cold Spring Harbor Laboratory Press, Cold Spring Harbor, N.Y.
100. **Sankaran, K., and H. C. Wu.** 1994. Signal peptidase II, p. 17–29. *In* G. von Heijne (ed.), *Signal Peptidases*. R. G. Landes Company, Austin, Tex.
101. **Schaeffer, P., J. Millet, and P.-J. Aubert.** 1965. Catabolite repression of bacterial sporulation. *Proc. Natl. Acad. Sci. USA* **54:**704–711.
102. **Schmidt, A., M. Schieswohl, U. Völker, M. Hecker, and W. Schumann.** 1992. Cloning, sequencing, mapping, and transcriptional analysis of the *groESL* operon from *Bacillus subtilis*. *J. Bacteriol.* **174:**3993–3999.
103. **Serror, P., V. Azevedo, and S. D. Ehrlich.** 1993. An ordered collection of *Bacillus subtilis* DNA in yeast artificial chromosomes, p. 473–474. *In* A. L. Sonenshein, J. A. Hoch, and R. Losick (ed.), *Bacillus subtilis and Other Gram-Positive Bacteria: Biochemistry, Physiology, and Molecular Genetics*. American Society for Microbiology, Washington, D.C.
104. **Shaw, W. V.** 1975. Chloramphenicol acetyltransferase from chloramphenicol resistant bacteria. *Methods Enzymol.* **43:**737–755.
105. **Simon, D., and A. Chopin.** 1988. Construction of a vector plasmid family and its use for molecular cloning in *Streptococcus lactis*. *Biochimie* **70:**559–566.
106. **Simonen, M., and I. Palva.** 1993. Protein secretion in *Bacillus* species. *Microbiol. Rev.* **57:**109–137.
107. **Smith, H., S. Bron, J. van Ee, and G. Venema.** 1987. Construction and use of signal sequence selection vectors in *Escherichia coli* and *Bacillus subtilis*. *J. Bacteriol.* **169:** 3321–3328.
108. **Smith, H., A. de Jong, S. Bron, and G. Venema.** 1988. Characterization of signal-sequence-encoding regions selected from the *Bacillus subtilis* chromosome. *Gene* **70:** 351–361.
109. **Sorokin, A., V. Azevedo, E. Zumstein, N. Galleron, S. D. Ehrlich, and P. Serror.** 1996. Sequence analysis of the *Bacillus subtilis* chromosome region between the *serA* and *kdg* loci cloned in a yeast artifical chromosome. *Microbiology* **142:**2005–2016.
110. **Spizizen, J.** 1958. Transformation of biochemically deficient strains of *Bacillus subtilis* by deoxyribonucleate. *Proc. Natl. Acad. Sci. USA* **44:**1072–1078.
111. **Thornewell, S. J., A. K. East, and J. Errington.** 1993. An efficient expression and secretion system based on *Bacillus subtilis* phage ϕ105 and its use for the production of *B. cereus* β-lactamase. *Gene* **133:**47–53.
112. **Tjalsma, H., M. A. Noback, S. Bron, G. Venema, K. Yamane, and J. M. van Dijl.** 1997. *Bacillus subtilis* contains four closely related type I signal peptidases with overlapping substrate specificities; constitutive and temporally controlled expression of different *sip* genes. *J. Biol. Chem.* **272:**25983–25992.
113. **Tomazic, S. J., and A. M. Klibanov.** 1988. Why is one *Bacillus* α-amylase more resistant against irreversible thermoinactivation than another? *J. Biol. Chem.* **263:** 3092–3096.
114. **Udaka, S., and H. Yamagata.** 1993. High-level secretion of heterologous proteins by *Bacillus brevis*. *Methods Enzymol.* **217:**23–34.

114a. **Vagner, V., E. Dervyn, and S. D. Ehrlich.** 1998. A vector for systematic gene inactivation in *Bacillus subtilis*. *Microbiology* **114:**3097–3104.

115. **van Dijl, J. M., A. de Jong, H. Smith, S. Bron, and G. Venema.** 1991. Signal peptidase I overproduction results in increased efficiencies of export and maturation of hybrid secretory proteins in *Escherichia coli*. *Mol. Gen. Genet.* **227:**40–48.
116. **van Dijl, J. M., A. de Jong, H. Smith, S. Bron, and G. Venema.** 1991. Non-functional expression of *Escherichia coli* signal peptidase I in *Bacillus subtilis*. *J. Gen. Microbiol.* **137:**2073–2083.
117. **van Dijl, J. M., A. de Jong, G. Venema, and S. Bron.** 1992. Signal peptidase I of *Bacillus subtilis*: patterns of conserved amino acids in prokaryotic and eukaryotic type I signal peptidases. *EMBO J.* **11:**2819–2828.
118. **Vehmaanperä, J., A. Görner, G. Venema, S. Bron, and J. M. van Dijl.** 1993. *In vitro* assay for the *Bacillus subtilis* signal peptidase SipS: systems for efficient *in vitro* transcription-translation and processing of precursors of secreted proteins. *FEMS Microbiol. Lett.* **114:**207–214.
119. **Vellanoweth, R. L.** 1993. Translation and its regulation, p. 699–711. *In* A. L. Sonenshein, J. A. Hoch, and R. Losick (ed.), *Bacillus subtilis and Other Gram-Positive Bacteria: Biochemistry, Physiology, and Molecular Genetics*. American Society for Microbiology, Washington, D.C.
120. **Völker, U., S. Engelmann, B. Maul, S. Riethdorf, A. Völker, R. Schmid, H. Mach, and M. Hecker.** 1994. Analysis of the induction of general stress proteins of *Bacillus subtilis*. *Microbiology* **140:**741–752.
121. **von Heijne, G.** 1990. The signal peptide. *J. Membr. Biol.* **115:**195–201.
122. **von Heijne, G.** 1992. Membrane protein structure prediction: hydrophobicity analysis and the positive-inside rule. *J. Mol. Biol.* **225:**487–494.
123. **Webb, C. D., A. Decatur, A. Teleman, and R. Losick.** 1995. Use of green fluorescent protein for visualization of cell-specific gene expression and subcellular protein localization during sporulation in *Bacillus subtilis*. *J. Bacteriol.* **177:**5906–5911.
124. **Wetzstein, M., U. Völker, J. Dedio, S. Löbau, U. Zuber, M. Schiesswohl, C. Herget, M. Hecker, and W. Schumann.** 1992. Cloning, sequencing, and molecular analysis of the *dnaK* locus from *Bacillus subtilis*. *J. Bacteriol.* **174:**3300–3310.
125. **Wipat, A., N. Carter, S. B. Brignell, B. J. Guy, K. Piper, J. Saunders, P. T. Emmerson, and C. R. Harwood.** 1996. The region *dnaB-pheA* (256°–240°) of the

Bacillus subtilis chromosome encoding genes responsible for stress responses, the utilisation of plant cell walls and primary metabolism. *Microbiology* **142:**3067–3078

126. **Wisotzkey, J. D., P. J. Jurtshuck, G. E. Fox, G. Reinhard, and K. Poralla.** 1992. Comparative sequence analysis of the 16S rRNA (rDNA) of *Bacillus acidocaldarius* and *Bacillus cycloheptanicus* and proposal for creation of a new genus *Alicyclobacillus gen. nov. Int. J. Sys. Bacteriol.* **42:**236–269.

127. **Wu, X., W. Lee, L. Tran., S.-L. Wong.** 1991. Engineering a *Bacillus subtilis* expression-secretion system with a strain deficient in six extracellular proteases. *J. Bacteriol.* **173:**4952–4958.

128. **Yansura, D. G., and D. J. Henner.** 1984. Development of an inducible promoter for controlled expression in *Bacillus subtilis*, p. 249–263. *In* A. T. Ganesan and J. A. Hoch (ed.), *Genetics and Biochemistry of Bacilli*. Academic Press, Orlando, Fla.

129. **Yasbin, R. E., W. Firshein, J. Laffan, and R. G. Wake.** 1990. DNA repair and DNA replication in *Bacillus subtilis*, p. 296–320. *In* C. R. Harwood and S. M. Cutting (ed.), *Molecular Biological Methods for Bacillus*. John Wiley & Sons Ltd., Chichester, U.K.

130. **Youngman, P.** 1990. Use of transposons and integrational vectors for mutagenesis and construction of gene fusions in *Bacillus* species, p. 221–266. *In* C. R. Harwood and S. M. Cutting (ed.), *Molecular Biological Methods for Bacillus*. John Wiley & Sons Ltd., Chichester, U.K.

131. **Youngman, P.** 1993. Transposons and their applications, p. 585–596. *In* A. L. Sonenshein, J. A. Hoch, and R. Losick, (ed.), *Bacillus subtilis and Other Gram-Positive Bacteria: Biochemistry, Physiology, and Molecular Genetics*. American Society for Microbiology, Washington, D.C.

132. **Zuber, P., M. K. Nakano, and M. A. Marahiel.** 1993. Peptide antibiotics, p. 897–916. *In* A. L. Sonenshein, J. A. Hoch, and R. Losick (ed.), *Bacillus subtilis and Other Gram-Positive Bacteria: Biochemistry, Physiology, and Molecular Genetics*. American Society for Microbiology, Washington, D.C.

133. **Zukowski, M. M., D. F. Gaffney, D. Speck, M. Kauffmann, A. Findeli, A. Wisecup, and J. P. Lecocq.** 1983. Chromogenic identification of genetic regulatory signals in *Bacillus subtilis* based on expression of a cloned *Pseudomonas* gene. *Proc. Natl. Acad. Sci. USA* **80:**1101–1105.

134. **Zukowski, M. M., and L. Miller.** 1986. Hyperproduction of an intracellular heterologous protein in a *sacU*^{h} mutant of *Bacillus subtilis*. *Gene* **46:**247–255.

Filamentous Fungi

JAE-HYUK YU AND THOMAS H. ADAMS

34

The most important step in applying molecular genetic approaches to filamentous fungi is the development of gene transfer or DNA-mediated transformation systems. The first report of DNA-mediated transformation in a filamentous fungus was in 1973 when Mishra and Tatum (96) described transformation of *Neurospora crassa* inositol-requiring mutants to inositol prototrophy using DNA from a wild-type strain. During the early 1980s, such transformation systems were improved upon in *N. crassa* (2, 25, 26) and were developed in other fungi, like *Aspergillus nidulans*, that had also been studied for a long time in academic research laboratories with the aim of understanding basic biological problems (45, 70, 137, 141). As knowledge accumulated, transformation systems were reported for additional fungal species with medical, agricultural, or industrial relevance (for review see reference 145). This work has progressed to the point that transformation systems have been reported for representatives of all major fungal classes, and it seems likely that with the proper care, almost any fungus of interest can be used as a recipient in transformation.

Many reviews have been published that describe the development and use of DNA-mediated transformation systems in filamentous fungi (45, 141, 145, 153). This chapter will attempt to summarize some important features of these fungal transformation systems and how they can be applied in specific strategies for manipulating fungal genomes. These strategies are generally applicable in many fungal hosts, including the industrially important *Aspergillus* spp., *Trichoderma* spp., and *Fusarium* spp. (35, 37, 41, 50, 59, 65, 72, 77, 99, 107, 108, 138).

34.1. VECTORS AND SELECTABLE MARKERS

34.1.1. Plasmid Vectors Used in Fungal Transformation

Most transformation vectors used with filamentous fungi are hybrids containing *Escherichia coli* plasmid DNA with the appropriate selectable markers for the fungal recipient. In almost all cases, transformation of filamentous fungi results in integration of vector DNA into the fungal genome. Although autonomously replicating fungal transformation vectors have been reported, their use is limited (51, 52, 150). Some transformation-mediated cloning strategies based on the use of such vectors have been described, but generally they have not proved to be useful as shuttle vectors, like the replicating plasmids that have been used for transformation of the yeast *Saccharomyces cerevisiae* (17, 28, 49). This lack of autonomously replicating plasmid vectors can be a shortcoming in certain types of genetic manipulations, but for the most part, approaches have been developed to overcome these limitations.

While transformation vectors based on selectable markers that have no homology to the host genome always result in nonhomologous integration at presumably random sites, when vectors containing selectable markers that have homology to the host genome are used, they can integrate by any one of three different ways (139, 145). Type I integration involves homologous recombination with the host genome and results in duplication of the homologous sequences coupled with integration of vector sequences into the genome (see Fig. 3A). Type II integration presumably involves nonhomologous recombination (or recombination due to a region of limited similarity between the host genome and vector sequences) and results in ectopic integration of transforming DNA. Type III integration involves gene replacement resulting from homologous recombination between transformed DNA sequences and the genome (see Fig. 3B). The relative frequencies of each type of transformation seem to vary according to both the transformation host species and the homologous selectable marker used (10, 43, 58, 95, 156). In some cases, there is evidence that linearizing the transforming DNA by digesting with restriction enzymes that cut within the homologous sequences helps to stimulate homologous recombination, but this has not been carefully tested.

34.1.2. Selectable Markers

34.1.2.1. Auxotrophic Markers for Complementation of Recessive Mutations

When auxotrophic mutant strains are readily available, the use of cloned genes to complement the auxotrophies in selecting transformants has met with great success and has several advantages. First, because the transforming DNA simply replaces the function of a mutant gene, a single copy of the transforming plasmid is usually sufficient to complement the mutant phenotype. This contrasts with the use of dominant drug resistance markers (see section 34.1.2.2) where multiple copies of the resistance gene may be needed to overcome the drug selection. Similarly, complementation of auxotrophies usually provides a very clean selection with little or no background. Finally, complementation of point mutants using the appropriate portion of the wild-type gene can provide a means to select strains in which the transforming plasmid DNA has integrated at a homologous site (e.g., references 57 and 63; see section 34.3.2).

Many of the fungal genes that were originally isolated during the early development of fungal transformation have subsequently been employed as selectable markers in vectors designed for complementation of readily available auxotrophic mutant strains. Among the most frequently used genes for transformation of auxotrophic mutant strains in *N. crassa* are *qa-2* encoding catabolic dehydroquinase (26), *pyr-4* encoding orotidine-5′-phosphate (OMP) decarboxylase (9), *trp-1* encoding a trifunctional enzyme (glutamine aminotransferase, indoleglycerolphosphate synthetase, and phosphoribosylanthranilate isomerase) for tryptophan biosynthesis from chorismate (78), and *am* encoding NADP-specific glutamate dehydrogenase (81). Common markers used for selecting *A. nidulans* transformants include *amdS* encoding acetamidase (75), *argB* encoding ornithine carbamoyltransferase (69, 84), *niaD* encoding nitrate reductase (71, 89), *trpC*, which is equivalent to *Neurospora* sp. *trp-1* (160), and *pyrG*, which is equivalent to *Neurospora* sp. *pyr-4* (102). The frequent use of these markers relates largely to the fact that the wild-type genes were cloned early in the development of transformation systems and because corresponding mutant strains are already available. In addition, each of these markers provides a relatively simple selection for transformants and, because the corresponding genes have all been sequenced and engineered into a variety of vectors (plasmid and cosmid), they are relatively easy to use.

Another useful feature of many of the genes described above is that they are able to complement auxotrophies in fungal species other than the one they were derived from and are thus referred to as broad-range markers. For instance, the acetamidase gene of *A. nidulans* (*amdS* [75]) has been used as a dominant nutritional marker in several fungi, allowing selection of transformants that have gained the ability to use acetamide or acrylamide as a sole nitrogen and carbon source. Other examples include the use of *argB* and *pyrG* from *A. nidulans* to complement the corresponding auxotrophic mutations in a number of fungi, including *Trichoderma* spp. (60, 107). While this approach has been extremely valuable, it has also been noted that transformation frequencies often increase when a homologous selectable marker is used (59, 131, 144). For example, when the *Trichoderma* sp. *pyr4* gene was used in homologous transformation experiments, the transformation frequency increased at least 10-fold over the frequency observed using the heterologous *A. nidulans pyrG* gene (59). Similar results have been observed in several other fungal transformation systems (56, 131, 144).

The *pyrG* and *niaD* genes (or their equivalents) have been particularly useful in developing transformation systems for less studied fungal species because there are simple schemes for selecting loss-of-function mutations in these genes. Fungal strains that have a wild-type *pyrG* gene (encoding OMP decarboxylase) are sensitive to 5-fluoro-orotic acid (5FOA), presumably owing to toxicity of 5-fluoro-UMP produced from the combined activities of OMP pyrophosphorylase and OMP decarboxylase on 5FOA (13). Thus, *pyrG* loss-of-function mutants can be selected by plating the mutagenized spores onto minimal medium containing uracil and 5FOA (13, 56). Survivors resulting from loss of OMP decarboxylase should be uridine auxotrophs and can be used as recipients in transformation experiments with the wild-type *pyrG* gene, selecting strains that grow without uracil supplementation of media. Similarly, fungal strains that have a wild-type *niaD* gene (encoding nitrate reductase) are sensitive to chlorate, presumably because nitrate reductase can convert chlorate to chlorite, which is toxic, although other explanations have been offered (33, 34). In any case, selection for chlorate resistance on medium containing a nitrogen source other than nitrate allows isolation of *niaD* loss-of-function mutants. Because there are several different mutations that can cause chlorate resistance (e.g., *niiA*$^-$, *crnA*$^-$, *areA*$^-$), it is necessary to test survivors for the ability to grow using several different nitrogen sources to distinguish *niaD*$^-$ mutants. Detailed protocols for selection and scoring of chlorate resistant mutants have been described (e.g., reference 143). The ability to select loss-of-function mutations in the *niaD* and *pyrG* genes also provides the opportunity to use these genes as mutagenic markers. In this case, homologous integration of a circular plasmid carrying the internal DNA fragment (see section 34.3.2, Fig. 3A) from either the *niaD* or *pyrG* gene leads to disruption of the wild-type gene. Thus, transformants can be selected based on their resistance to chlorate or 5FOA, respectively.

Another advantage in using auxotrophic markers for which loss-of-function mutants can easily be selected is that successive rounds of transformation could, in principle, use the same selectable marker. This is accomplished by first transforming an auxotrophic mutant strain with the corresponding wild-type gene and selecting complemented transformants. Mutants that have lost the complementing marker can then be selected using the appropriate medium, and the transformation can be repeated. This approach is the basis for the so-called *ura*-blaster cassette that has been put to extensive use in *Candida albicans* (46) as well as the more recently constructed *pyrG*-blaster described by d'Enfert (40) for use in *Aspergillus* spp. In each case, the OMP decarboxylase-encoding gene is flanked by two identical elements that can form a direct repeat. As shown in Fig. 1, a crossover between two direct repeats (for example, the *neo* gene, which confers neomycin resistance in *E. coli*) will lead to deletion of the *pyrG* (or *URA3*) gene, resulting in a uracil auxotroph with resistance to 5FOA. In theory, any auxotrophic markers for which loss-of-function mutants can be selected can be used for a similar approach.

34.1.2.2. Drug Resistance Markers

While complementation of auxotrophic mutations has been very successful, a major disadvantage of this approach is the reliance on generating appropriate auxotrophic recipient strains. This limitation can be overcome by using dominant selectable markers. Several fungal transformation schemes based on the use of dominant or semidominant genes that provide resistance to various drugs have been described (6, 115, 145). The only requirement for using these systems is that the fungal strain of interest needs to be sensitive to available drugs. The most commonly used drug resistance markers in fungal transformations rely on bacterial genes like the *Streptoalloteichus hindustanus* or *E. coli* phleomycin (or bleomycin) resistance gene *ble*, encoding a phleomycin-binding protein (82, 92), or the *E. coli* hygromycin resistance gene *hph*, encoding hygromycin B phosphotransferase (116). Alternatively, mutant fungal genes that confer resistance to common antifungal compounds, like benomyl, have been employed (13, 93). In the

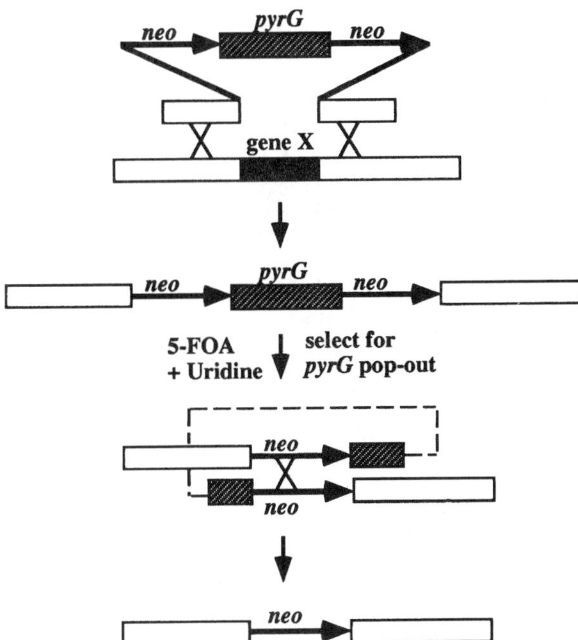

FIGURE 1 The *pyrG*-blaster cassette. The plasmid contains a DNA fragment in which the internal region of gene X (the black bar) was deleted and replaced by the *pyrG* gene flanked by direct repeats (*neo*). Replacement of the wild-type gene results in a ΔX, *pyrG*$^{+}$ transformant. Conidia from the disruption mutant can be plated on medium containing 5FOA and uridine to select resistant colonies that have resulted from popping-out of the *pyrG* gene via crossing over between the direct repeats. The resultant strain is ΔX, *pyrG*$^{-}$, and the *pyrG* gene can again be used as a selectable marker for further gene manipulation (40).

case of bacterial resistance genes (*hph* and *ble*), it has been necessary to make hybrid genes by fusing the bacterial genes with fungal transcriptional promoters and terminators. For example, the *S. hindustanus ble* gene was introduced in a fungal expression vector containing the promoter region of the highly expressed *A. nidulans gpdA* gene and the terminator region of the *A. nidulans tryC* gene (92). Similarly, vectors containing the *E. coli hph* gene have been developed and used for the transformation of various fungal species (88, 113, 134). These vectors are generally useful in a number of systems, but, as with complementation of auxotrophic mutations using heterologous genes, it has been reported that the use of homologous expression signals can increase transformation efficiency severalfold (88).

34.2. TRANSFORMATION SYSTEMS

34.2.1. Polyethylene Glycol-Mediated Transformation

The most commonly used transformation system in filamentous fungi involves polyethylene glycol (PEG)-enhanced spheroplast fusion. The various protocols for this procedure can generally be divided into three steps: (i) isolation of spheroplasts, (ii) introduction of DNA by protoplast fusion caused by PEG treatment, and (iii) regeneration of transformants. The fusion of protoplasts in the presence of DNA assists in uptake of the DNA into the cytoplasm of the fused protoplasts. It is not understood how this transforming DNA is transferred into the nucleus, but this can occur and is frequently followed by integration of the DNA into the fungal genome. As fused protoplasts generally have multiple nuclei and a single nucleus may be transformed, this typically results in generation of a heterokaryon. In some fungal species, like *A. nidulans*, the heterokaryon is easily resolved through the formation of uninucleate conidiospores. However, in many other fungi, several rounds of streaking for isolation may be necessary to purify homokaryotic strains.

Because these protocols rely on the use of spheroplasts, it is important to remember that osmotic stability needs to be maintained throughout the procedure. Commonly used osmotic stabilizers are sorbitol (1.0 to 1.2 M), and KCl (0.6 M). Occasionally, sucrose (1.2 M) is also used, but it typically gives a lower transformation efficiency. We describe the details of one of the protocols for PEG-mediated transformation in *Aspergillus* and *Trichoderma* spp.

PROTOCOL

TRANSFORMATION PROTOCOL

There are two common protocols in general use for generation of spheroplasts. One relies on metabolic activity in the living cytoplasm contributing to vesicle formation, which allows less-dense spheroplasts to be banded away from wall debris. The other protocol is somewhat simpler in that spheroplasts are separated from debris by filtering through sterile Miracloth (Calbiochem). Both protocols work well, but banding of spheroplasts gives somewhat better purification away from cellular debris and in our hands has usually given more consistent results. We have found that spheroplasts generated from some mutant strains are more fragile than others and may rupture easily in the solution used for generating banded spheroplasts (osmotic medium A, OM-A). In these cases, the use of the modified spheroplasting solution (osmotic medium B, OM-B) and filtering of protoplasts has sometimes been useful. While a search of the literature provides almost as many modified protocols as there are fungi that have been transformed, we describe the most commonly used protocols and attempt to provide useful suggestions for modifying them to different fungal systems.

Materials and supplies

Media

1. Defined solid minimal medium containing all appropriate supplements and 1.2 M sorbitol.
2. Defined solid minimal medium containing supplements to select for transformants and 1.2 M sorbitol.
3. 400 ml of defined liquid minimal medium containing all appropriate supplements in a 1-liter flask.

Note: Media 1 and 2 may be prepoured for spread plating protoplasts or kept molten at 45°C for pour plating.

Solutions

1. Mycelium wash: 0.6 M $MgSO_4$ (autoclaved and stored at 4°C)
2. a) OM-A:
   1.2 M $MgSO_4$
   10 mM NaPB (from 2 M NaPB stock, which has 90.88 g Na_2HPO_4 and 163.38 g NaH_2PO_4 per liter, pH 6.5)
   Adjust the pH to 5.8 with 1 M Na_2HPO_4.

Filter sterilize and store at 4°C.
b) OM-B:
1.2 M sorbitol
100 mM KH_2PO_4
pH 5.6 with 10 N KOH
Autoclave and store at 4°C.

3. Trapping buffer:
0.6 M sorbitol
0.1 M Tris-HCl, pH 7.0
Autoclave and store at 4°C.
4. STC buffer:
1.2 M sorbitol
10 mM $CaCl_2$
10 mM Tris-HCl, pH 7.5
Autoclave and store at 4°C.
5. PEG solution:
60% PEG 4000 (BDH)
10 mM $CaCl_2$
10 mM Tris-HCl, pH 7.5
Autoclave and store at room temperature.

Supplies

1. Sterile 250-ml flask
2. Washed sterile Miracloth
3. Sterile Büchner funnel
4. Two sterile 30-ml Corex tubes
5. Sterile Pasteur pipettes
6. Sterile spatulas

Preparation of inoculum

In most cases, fungal cultures for transformation can be initiated from spores. For *A. nidulans*, a spore inoculum can be prepared by first streaking spores from the intended recipient strain evenly over the surface of an agar plate containing minimal medium and appropriate supplements and then incubating for 2 to 3 days at 37°C. Spores are collected by suspending in ~10 ml of sterile 0.1% Tween-80 followed by vortexing to break up spore clumps. In cases where sufficient spores to inoculate the culture cannot be recovered, a hyphal macerate can be prepared as inoculum (16).

Growth of culture

As with production of inoculum, the way in which cultures are grown is somewhat dependent on the fungal recipient. In most cases with *A. nidulans*, we inoculate 400 ml of sterile defined minimal medium containing the appropriate supplements with about 10^9 fresh spores and shake at 28°C and 300 rpm for approximately 14 h. The amount of growth can be critical to the success of the transformation, and a proper culture should have an abundance of young germlings in small aggregates. If spores are slow to germinate and sufficient growth has not occurred by 14 h, the culture can be placed at 37°C and carefully monitored to assess growth. If the culture grows to the point that it is primarily mature hyphae, it is probably best to start over, as protoplasts from old cultures are rarely competent for transformation.

With other fungi, like *Trichoderma* spp., cultures are grown on agar plates overlaid with cellophane discs. These are inoculated with 5×10^6 conidia per plate and incubated at 30°C for 20 to 26 h. Five plates typically yield sufficient material for an experiment.

Making protoplasts using OM-A for banding

1. Harvest mycelia by filtration through sterile Miracloth and then wash the mycelia with 100 ml of 0.6 M $MgSO_4$ (mycelial wash solution). Place the mycelia in a preweighed sterile 15-ml disposable tube and determine the weight (usually treated as 1 g). Place mycelia on ice. With *Trichoderma* spp., suspend the mycelium from five cellophane discs in 15 ml of protoplasting solution OM-A (or OM-B).

2. Make enzyme solutions just before use. (It is not absolutely necessary to filter sterilize, as most Novozyme preps do not contribute to contamination.)
   a. Novozyme 234: 20 mg/ml in OM
   b. Bovine serum albumin (Sigma A-6003): 12 mg/ml in OM

3. Suspend the mycelia in OM-A (5 ml/g wet weight but use at least 5 ml) and vortex vigorously to break up fungal clumps. Transfer to a sterile 250-ml flask and add Novozyme 234 (1 ml per 5 ml of cells). Mix and incubate on ice for 5 min and add 0.5 ml bovine serum albumin per 5 ml of cells.

4. Shake at 80 rpm for 90 min at 28°C. Check every 30 min under the microscope to see that protoplasts are forming. The rate of spheroplast formation can vary greatly between Novozyme preps, so it is important to follow this stage carefully. Treating with Novozyme can lead to reduced transformation frequency (see section 34.2.1.2).

5. Transfer the material to a sterile 25-ml Corex tube and overlay gently with 10 ml of trapping buffer (Fig. 2). Centrifuge at 4,000 × *g* for 15 min and 4°C using a swinging bucket rotor (Sorvall HB-4 or equivalent).

6. During centrifugation, protoplasts float to the interface between the OM-A and the trapping buffer. Remove protoplasts using a sterile Pasteur pipette and transfer them to a sterile 15-ml tube (Fig. 2). (*Optional:* Remove the remaining trapping buffer and resuspend the pellet in the remaining OM-A using a Pasteur pipette. Overlay the cells with 10 ml of trapping buffer and centrifuge as before. Pool the protoplasts with those from the first spin.)

7. Dilute the protoplast suspension with at least 1 vol of STC buffer (any less and protoplasts will not pellet). Centrifuge at 6,000 × *g* for 5 to 8 min (a fixed angle or swinging bucket rotor can be used).

8. Decant the supernatant and resuspend the pellet using a Pasteur pipette in 10 ml of STC. Spin as above, then repeat wash once more.

9. Using a Pasteur pipette, resuspend the protoplasts in STC at a concentration of 10^8 to 10^9 protoplasts per ml (1 ml is usually added). Store on ice.

10. Dilute a sample of the protoplasts 1:100 in STC and count by using a hemocytometer. Save this dilution to plate and determine total viable protoplasts.

Making protoplasts using OM-B for filtering

1. Harvest cultures as described in step 1 above.

2. Resuspend mycelia (5 ml of OM-B per g of tissue) in OM-B containing 5 mg/ml Novozyme 234 in a 125-ml flask.

3. Incubate at 28°C, 80 rpm for 90 min. Check protoplast formation under microscope every 30 min, being careful not to overprotoplast.

4. Put two or three layers of sterile Miracloth in a sterile funnel on top of a 30-ml Corning centrifuge tube

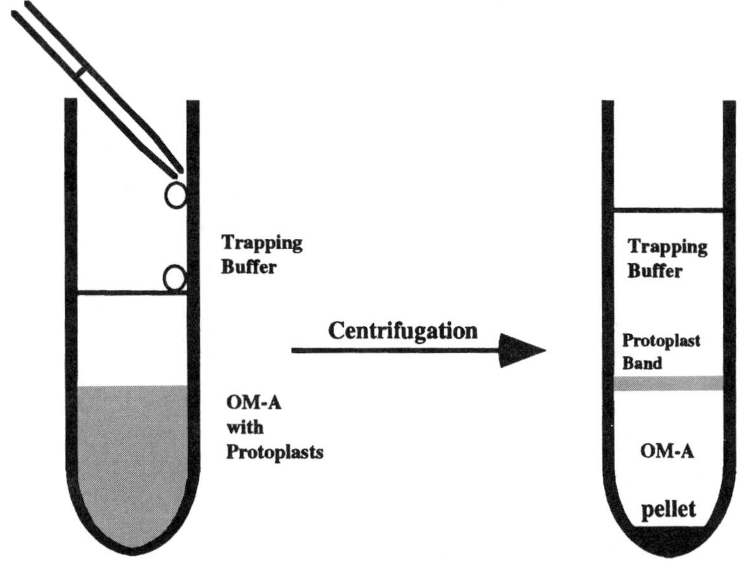

FIGURE 2 Protoplast banding technique. About 10 ml of trapping buffer is *gently* loaded on top of the protoplasting solution. After centrifugation at 4,000 × *g* for 15 min, turbid protoplast band should be visible between the solutions. The protoplast band is collected for further processes.

and pour the protoplast solution through to get rid of cell debris.

5. Add 1 volume of STC to the protoplast solution and centrifuge at 6,000 × *g* for 10 min at 4°C.

6. Decant the supernatant and resuspend the protoplast pellet with 10 ml of STC and transfer solution into a 15-ml disposable tube. Centrifuge as in step 5, then repeat step 6.

7. Resuspend the protoplast pellet in desired volume of STC (ca. 1 ml). Store on ice. Count the number of protoplasts.

Transformation

1. Dilute the DNA (use around 5 to 10 μg) to be transformed 1:1 with STC in a 15-ml disposable sterile tube (e.g., Falcon catalog number 2059). Keep the volume under 100 μl.
2. Add 100 μl of protoplasts and incubate 15 min at room temperature.
3. Add 1.25 ml of PEG solution to the tubes; mix gently by turning the tube on its side and rotating it. Incubate at room temperature for 10 min.
4. Add 5 ml of STC to the tubes and mix well.
5. Plate the transformed protoplasts by either:
   a. spreading on defined minimal medium containing 1.2 M sorbitol and the appropriate supplements to select for transformants, or
   b. placing protoplast mix in plates and adding molten agar medium (45°C) and swirling.

With *A. nidulans*, we typically divide the transformation mix among 10 plates, but this decision needs to be gauged by the expected number of transformants. This can vary from one to 10,000 transformants per μg of DNA, depending upon the fungus and the selectable marker used. For some fungi, protocols have been described for storing protoplasts at −80°C for later use (e.g., references 107 and 151).

6. Controls:
   a. Plate untransformed protoplasts on the same medium used to select for transformants, using at least as many protoplasts as for transformations to see revertants or contaminants.
   b. Plate untransformed protoplasts (100 μl of 10^{-3} to 10^{-5} dilutions) on medium containing all supplements with 1.2 M sorbitol to determine protoplast viability.

34.2.1.1. Troubleshooting

Because PEG-mediated transformation is the most commonly used protocol, and the success of transformation depends on the general quality of spheroplasts formed, we will discuss a few of the possible problems that may arise in PEG-mediated transformation. Occasionally, even if the protocol was followed exactly as suggested, transformation frequencies are unacceptably low. An important variable that is difficult to control relates to the quality of the Novozyme 234 used in generating spheroplasts. This is a crude enzyme preparation, and the transformation efficiency can vary greatly depending upon the Novozyme 234 lot number. Other possible problems we have met in *Aspergillus* transformation are generally discussed below, but there may be some other problems for different filamentous fungal species.

Poor Spheroplast Formation

Occasionally, few or no spheroplasts are observed even after prolonged incubation with Novozyme 234. Assuming that the enzyme was added to the solution, the most likely cause is that a particular lot of Novozyme 234 has poor activity. This situation may be corrected by simply adding more Novozyme 234 or by trying another lot. We have also found that some mutant strains repeatedly give low spheroplast yields. In these cases, we have found that omitting bovine serum albumin during the protoplasting step gave higher spheroplast yields. We have also had success in dealing with this problem by using mycelia from younger cultures or from cultures that have been grown in supplemented minimal medium, rather than complete medium, prior to spheroplast formation.

Loss of Spheroplasts during the Washing Step

We have frequently found that there is a significant loss of protoplasts during the washing steps. We do not know the cause of this, but again there seems to be some correlation with particular batches of Novozyme 234. We have had some success in eliminating the problem by shortening the time of Novozyme 234 digestion or by reducing the amount of Novozyme 234 used. More trivial explanations, such as failure to sufficiently dilute the protoplasts before pelleting, can also cause reduced yield.

Poor Transformation Frequency

One of the most frustrating problems is to go through the trouble of preparing protoplasts only to find that few or no stable transformants are recovered. This problem can result from poor competence of the protoplasts, low quality of the DNA, or some combination of the two. A frequent problem in preparation of protoplasts is that the fungi were grown too long before harvesting the mycelia. At least in *Aspergillus* spp., mature hyphae produce large numbers of spheroplasts that are apparently nontransformable. It is not clear what causes this reduced competence, but one possibility is that many of the spheroplasts produced lack nuclei and, therefore, although they can effectively take up transforming DNA, they are inviable. Another common problem is that prolonged treatment with Novozyme 234 can reduce spheroplast viability. Again, the reason for reduced viability is not known but may be associated with degradation of essential membrane proteins. We recommend careful microscopic observation during spheroplast formation to develop a sense of what is the right amount of Novozyme 234 treatment. If in doubt, we have generally found that it is always best not to be too greedy in going for huge protoplast yields.

While DNA prepared in any number of ways seems to work equally well for transformation, there are occasional problems in recovering transformants that can be linked to DNA quality. It is usually possible to recognize this problem if you have performed the proper positive control and, in many cases, a new prep of DNA inexplicably makes the difficulty disappear. One problem with fungi like *A. nidulans* where transformation frequencies are acceptable but not phenomenal is that only a small change in transformation frequency can cause a drop below acceptable levels. As with most DNA-mediated transformation protocols, there is some poorly defined optimum ratio of DNA to competent cells that must be met. If too much DNA is used, it precipitates after addition of PEG or too many multicopy transformants are recovered. Alternatively, too little DNA might fall below a threshold needed for efficient uptake into nuclei. We generally set up at least three different transformations with any given DNA sample using stepwise increases in DNA from 5 to 15 μg. This empirical approach is usually sufficient to overcome experimental variation.

34.2.2. Electroporation

One critical disadvantage in using the PEG-mediated transformation procedure described above is that there can be dramatic variation depending on batches of Novozyme 234. To minimize these problems, electroporation protocols for *Neurospora* spp. (27, 83), *Trichoderma* spp. (55), and *Aspergillus* spp. (104, 121) have been developed. In principle, a brief introduction of high voltage will rupture the plasma membrane, making the protoplast remains permeable to macromolecules such as DNA for a few minutes. Generally, DNA is introduced into either weakened cell wall germinated conidia or spheroplasts by overcoming any electrostatic repulsions, particularly at the moment the pulse is applied. In *Neurospora* spp., Kothe and Free (83) could generate 10,000 to 50,000 transformants per 10^9 spheroplasts. Transformation efficiency may vary according to amount, type, and topology of DNA used and field strength. In recent protocols, 3.5 kV/cm for *Neurospora* spp. (83) and 5.0 kV/cm for *Aspergillus* spp. (121) have been reported. Most protocols involve direct transformations of young germlings and thus bypass the need to generate spheroplasts (27, 104, 121). Detailed protocols for *Neurospora* and *Aspergillus* electroporation can be found in the *Fungal Genetics Newsletter* (83, 121).

34.2.3. Biolistic Bombardment

Potentially, the simplest method for fungal transformation involves the use of a particle gun. The basic principle of action is that a macroprojectile, the leading face of which is loaded with DNA-coated microprojectiles, is accelerated and then the macroprojectile is halted by a stopper plate. The DNA-coated microprojectiles pass through holes in the plate and strike the samples. Such DNA-coated microprojectiles have been shown to penetrate cell walls and carry DNA into the cell nucleus or organelle. Biolistic bombardment methods of DNA transfer have been successfully used for transfection of single cells of higher plants (including pollen), mitochondria in fungi, and chloroplasts in algae (for review see reference 153). For transformation of filamentous fungi, Lorito et al. (87) first reported successful introduction of a bacterial hygromycin resistance gene containing plasmid DNA into conidia of *Trichoderma harzianum* and *Gliocladium virens*. They showed that, compared to PEG-mediated transformation, biolistic protocols produced higher levels of transformation and increased heterokaryon stability. Thereafter, several biolistic transformation experiments for other fungal species have been reported (48, 66). Details of biolistic transformation procedure are described by Lorito et al. (87).

34.3. GENE ISOLATION AND MANIPULATION

So far we have discussed several ways of introducing DNA into a fungal recipient strain, ultimately leading to alteration of the strain's genotype. Depending upon the experimental objectives and the way in which DNA is integrated into the recipient genome, this can result in disruption, duplication, activation, or complementation of a particular gene or mutation. The ability to manipulate fungal genomes using the tools of molecular biology coupled with DNA-mediated transformation systems provides the means to address important questions relating to fungal biology, even in fungi with poorly developed genetic systems. In this section we will briefly discuss the various strategies currently available to isolate and manipulate interesting fungal genes.

34.3.1. Gene Cloning

Before beginning to discuss approaches for manipulating the activity of a given gene or for analyzing features of the gene's structure, function, or control, it is important to discuss strategies for isolating a cloned copy of the gene. Rather than describe the many different approaches available for isolating fungal genes that are generally applicable to all biological systems, we will limit our discussion to techniques that rely on the use of fungal molecular genetics. Detailed information regarding more general approaches can be found in a variety of molecular protocols books (7, 120).

34.3.1.1. Cloning Genes by Mutant Complementation

The ability to identify fungal genes based on the interesting properties of mutant strains was obviously a major consideration in the development of filamentous fungi as experimental systems. This functional approach to examining fungal genes is also the basis of the most common means of gene cloning. In most fungal species it is possible to use DNA-mediated transformation for genetic complementation of a recessive mutation using a library of recombinant DNA clones containing wild-type DNA. In taking this approach there are several things that must be considered, including the size of the fungal genome, the transformation frequency, the means for recovering transforming DNA, and the nature of the mutant phenotype.

The first step in isolating a gene is construction of a reliable genomic DNA library. The number of clones required for a complete library varies according to the type of cloning vector used (and how this affects average insert size) and the size of the genome, which is typically on the order of 3×10^7 bp for fungi. The probability of having a given DNA sequence in the library can be calculated from the equation $N = \ln(1 - P)/\ln(1 - f)$ where P is the desired probability, f is the fractional proportion of the genome in a single recombinant, and N is the necessary number of recombinants (30). Thus, with a genome size of 3×10^7 bp, to achieve a 99% probability ($P = 0.99$) of having a given DNA sequence represented in a plasmid library with 10 kb genomic DNA inserts, approximately 13,820 clones are needed, i.e., $N = \ln(1 - 0.99)/\ln(1 - [1 \times 10^4]/[3 \times 10^7]) = 13{,}814$. Similarly, the number of clones required for 99% genomic representation using a cosmid vector (average insert size 40 kb) or BAC vector (bacterial artificial chromosome, generally average insert size 100 kb [see reference 129]) is 3,452 or 1,380, respectively.

The next step after constructing a genomic library is transformation, leading to identification of a complementing DNA clone. The nature of the original mutation influences the approaches available here. Complementation of an auxotrophic mutation (or any other mutation that results in an inability to grow under certain conditions) can be accomplished by plating transformants directly on selective media. Alternatively, transformants can first be selected based on a marker included in the vector, and individual colonies can then be screened for complementation of the mutation of interest. In either case, the number of viable transformants that may need to be screened could in theory approach as many as three times the size of the library. However, in practice we usually find that using a cosmid library for *A. nidulans* transformation, we recover a wild-type transformant within the first 1,000 transformants.

After isolating an apparently complemented transformant, some strategy to recover the transforming DNA is required. Because the transforming sequence usually integrates into the fungal host genome with the complementing DNA insert adjacent to the bacterial vector sequences, it is generally possible to recover the responsible cosmid or plasmid directly from the transformant. Plasmid DNA carrying at least a portion of the insert can be isolated following restriction digestion of genomic DNA from this transformant using an enzyme that does not cut within the plasmid vector. Ligation of the restricted DNA and transformation of *E. coli* with selection for the antibiotic resistance marker carried on the bacterial vector results in recovery of the plasmid and sequences that were adjacent to it in the fungal genome. For cosmid transformants, it is sometimes possible to recover the entire cosmid by simply subjecting genomic DNA from the transformant to bacteriophage lambda packaging extract and using the resultant defective phage particles to transduce *E. coli* cells to antibiotic resistance (161). The mechanisms that lead to successful recovery of cosmid DNA by this method are not understood but can lead to isolation of cosmids containing either the wild-type or mutant allele of the desired gene. The status of the allele in the recovered cosmid can be distinguished by retransformation of the original mutant strains.

An important alternative to plasmid rescue strategies described above is the technique known as "sib-selection," which has been the preferred method for isolationg genes in *N. crassa* (2, 151). With this approach, individual cosmid or plasmid clones are pooled in several groups, and cosmid DNA prepared from each group is used to transform the mutant strain, screening for complemented transformants. This allows identification of a cosmid pool containing the complementing sequence, which can then be subdivided and the process repeated until a single clone responsible for the complementation is identified.

As described above, the isolation of genes by complementing recessive mutations involves use of a wild-type DNA library for transformation. However, not all mutations are recessive to the wild-type alleles. Sometimes mutagenesis results in gain-of-function mutations, and the mutant allele may be dominant to the wild-type allele. Thus, before transforming the mutant strain with a genomic DNA library prepared from wild type, it is important to determine the dominance relationships between wild-type and mutant alleles. In many fungi, like *A. nidulans*, this can be accomplished by constructing a heterozygous diploid strain (109, 110). Alternatively, it is sometimes possible to assess dominance by using heterokaryons (109). If the mutation turns out to be dominant, this means that a new genomic library needs to be constructed using genomic DNA from the mutant strain. This library can then be used to transform a wild-type strain, screening for the mutant phenotype (e.g., reference 163).

After a cosmid or plasmid clone has been identified that appears to contain the gene of interest, it is usually necessary to identify a smaller DNA fragment that includes the gene, which may represent just a few kilobases within a large insert (ca. 40 kb for a cosmid). A rapid method to localize the complementing region involves use of purified DNA fragments from the uncharacterized clone in cotransformation experiments with a plasmid that has a selectable marker (140). The fragment resulting in rescue of the wild-type phenotype can then be subcloned and further characterized. It is important to recognize at this point that in some cases the region identified does not truly complement the mutation but instead functions as a suppressor (e.g., references 21 and 90). Thus, it is critical to devise tests to learn if the cloned DNA sequence actually corresponds to the target gene. This can be accomplished by using the cloned DNA to construct a site-directed mutation (see section 34.3.2) and testing to see that the new mutation is linked to the original mutation and that the two mutations are not able to complement in *trans* (e.g., references 90, 94, 136, and 161). Alternatively, sequencing of the wild-type and putative mutant alleles can be used to show that sequence changes are associated with the mutation.

34.3.1.2. Cloning Genes by Insertional Mutagenesis

While the ability to isolate genes by complementation of a mutant phenotype has proved to be a powerful tool for genetic studies, it remains a significant undertaking to clone a gene by this approach; it can easily take several months to over a year to go from mutation to gene. An alternative approach that has a great deal of popular appeal is the use of mutagenesis methods that end up by creating a mutant allele marked by a molecular tag. This technique is especially useful for screening for loss-of-function mutations in nonessential genes.

Restriction Enzyme-Mediated Integration (REMI)

One recent innovation in isolating tagged mutations is termed restriction enzyme-mediated integration (REMI) (124, 125). This approach was first described in yeast, where it was reported that addition of restriction enzyme in transformation experiments with enzyme-cut vector resulted in increased transformation frequency (124, 125). Molecular analysis of these transformants showed that the transforming DNA fragment had predominantly integrated into the host genome such that it could be directly removed using the restriction enzyme included in the transformation (124, 125). This finding has led to the hypothesis that the introduced restriction enzyme cuts the host genome at one or more of its recognition sites and that the plasmid is then ligated to this breakpoint in a manner that preserves the compatible ends.

Since the original reports of REMI in yeast (124), it has been shown that the same approach can be used in a number of microorganisms, including *Dictyostelium* (85) and various filamentous fungi (14, 118, 128). Because REMI facilitates the integration of the plasmid DNA into broadly distributed sites throughout the genome, insertion can tag many different genes and allow their rapid isolation. For example, *Xho*I, like any restriction enzyme with a 6-bp recognition sequence (CTCGAG), will typically cut the genome about once every 4,096 bp (4^6) if the AT/GC ratio is near 1. Thus, in *A. nidulans*, whose genome size is approximately 2.7×10^7, there should be around 6,591 *Xho*I sites ($2.7 \times 10^7/4{,}096$). This means that there are approximately 6,591 places that the transforming DNA could integrate in the genome (Fig. 3C). By using additional enzymes having different recognition sequences, it should in theory be possible to recover transformants with the transforming DNA inserted into almost any gene of interest. After the desired mutants have been isolated, the plasmid along with flanking DNA can be recovered by digestion of genomic DNA with a restriction enzyme that does not cut within the plasmid sequences, followed by ligation and *E. coli* transformation.

As a prerequisite to REMI experiments, it is important to optimize conditions for a particular host and a given restriction enzyme to yield the highest number of transformants in which the vector has integrated at random restriction sites. A brief look at published reports of REMI in various filamentous fungi shows that the best conditions may vary considerably, using enzyme concentrations anywhere between 5 and 200 U per transformation (14, 118, 128). We recommend setting up a series of transformation experiments that vary both enzyme and DNA concentrations to arrive at the most effective conditions for a given organism. It is important to recognize that a high frequency of transformation does not necessarily indicate that the plasmid is integrating in the desired manner. For instance, in some cases when auxotrophic markers are used for selection, many of the transformants could represent gene conversion events or cases where the plasmid integrated into the homologous site by recombination. Such problems may be eliminated by using marker deletion strains (e.g., Δ*argB*), heterologous but functional markers (e.g., *pyr4* of *N. crassa* for *Aspergillus* transformation), or bacterial drug resistance genes. In any case, it is important to evaluate transformants to determine if plasmid integration is random and whether or not the restriction site used to direct REMI has been preserved. This can be accomplished by using Southern blot analysis in which restriction digests of genomic DNA from a random set of transformants is hybridized to the plasmid used for transformation. If REMI has taken place, digestion with the same enzyme used in the transformation will release a plasmid-sized band (14, 85, 124). Digestion with an enzyme that does not cut in transforming DNA should yield a single hybridizing band of varying size, indicating randomness.

Finally, it is important to recognize that transformation can be a metagenic process; it is not necessarily the case that the phenotype of an interesting mutant strain arose from integration of the transforming sequences. In organisms with well-developed genetic systems, this can be examined by following the segregation of the mutant phenotype and the integrated sequence through a sexual cross. A second approach that is also possible, even in imperfect fungi, is to examine the ability to regenerate the mutant strain by transforming with recovered plasmid sequences. In theory, if the plasmid was recovered from the mutant following restriction digestion with an enzyme that does not cut within the integrating sequences, the recovered DNA should already represent a mutagenesis vector (85). Thus, digestion of the recovered plasmid with the same enzyme that was used to cut the host DNA, followed by transformation of a wild-type strain, should regenerate the mutant phenotype.

Transposon Tagging

Another potential way to tag fungal genes is with the use of transposons. Such approaches have been widely used in other organisms, like plants and *Drosophila*, but have not been exploited in fungi (for general references see references 98 and 148). This presumably is primarily due to the great success of other gene cloning approaches in fungi. However, transposable elements that fit into several different mechanistic groups have been recovered from numerous fungi, including *A. niger* (53), *A. fumigatus* (100), *N. crassa* (24, 80, 158), *Fusarium oxysporum* (38, 86), and *Magnaporthe grisea* (73, 74), and in principle these could be adapted for mutagenic approaches. An important advantage of transposable elements over tagging techniques like REMI is that they are autonomous and frequently insert randomly around the host genome rather than relying on

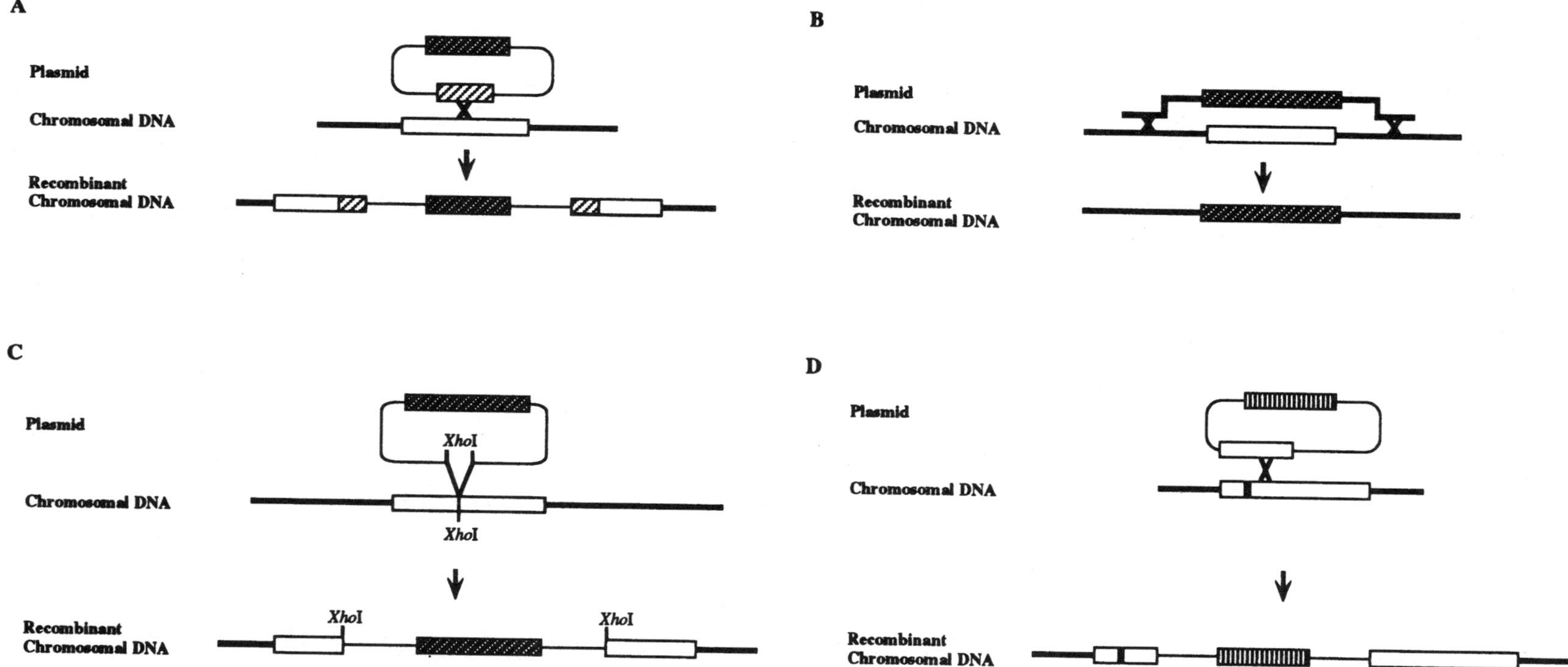

FIGURE 3 Targeted gene manipulation. (A) Gene inactivation by homologous integration of a circular plasmid. A plasmid containing an internal fragment from the target gene and a selectable marker can integrate into a chromosome by a single homologous recombination event (X) to result in formation of two partially deleted copies of the target gene separated by plasmid DNA sequences. (B) Gene inactivation by replacement. A linearized plasmid containing a mutant form (here, deleted and replaced by the selectable marker) of the target gene can recombine with the genome, leading to replacement of the wild-type target gene by the modified version. (C) Restriction enzyme-mediated integration. By transformation with restriction enzyme-digested (e.g., *XhoI*) plasmid DNA including the same enzyme in the transformation mix, it is possible to identify transformants in which the plasmid has integrated at the same restriction site. If this site is within a gene, this results in gene disruption. (D) Directed integration of the plasmid. A plasmid containing the one complete half of a host target gene can be used to direct integration to a specific genomic site by selecting transformants in which a target gene mutation (the vertical bar) has been rescued. This occurs by a single homologous recombination event (X) resulting in duplication of the transforming region, leaving a half copy of the target gene that includes the mutation separated by vector sequences from a second complete wild-type copy of the gene. This approach can be useful for experiments like promoter studies when it is important to know the genomic integration site. The shaded bar represents a selectable marker. The white bar represents the target gene. The bold line represents flanking DNA sequences of the target gene. The thin line represents a bacterial plasmid sequence.

the presence of restriction sites. Thus, developing transposon mutagenesis techniques in filamentous fungi could provide a useful additional tool for manipulating the genome.

Random Gene Activation

The general goal of insertional mutagenesis is usually to identify genes based on loss-of-function mutations. However, it should also be possible to use the same approaches to randomly activate genes and identify interesting mutants based on gain-of-function phenotypes. One approach that can be used to activate genes involves a modification of REMI mutagenesis in which the transforming DNA includes a strong controllable promoter (see section 34.4.4) along with a selectable marker. By using an enzyme that restricts the vector just downstream of the promoter, integration into the host genome can result in a gene fusion in which random host sequences are transcribed under the direction of the integrated promoter element. We have called this approach REMIGA, for restriction enzyme-mediated integrational gene activation, and have found some success in using this method to activate *A. nidulans* genes (162).

Another approach to randomly activating gene expression is to construct a plasmid library in which random sequences from the fungal host are placed downstream of an activatable promoter (91). In this case, fungal transformation results in the plasmid being integrated into the host genome, and transformants can be screened for the phenotype of interest. This has an advantage over REMIGA in that integration of plasmid sequences will not necessarily result in disruption of the gene to be activated. This should in theory make it possible to identify essential genes that cause abnormal growth phenotypes upon overexpression (91).

34.3.1.3. Cloning Genes by Position

In some cases, even though the gene of interest was identified based on the mutant phenotype, the screen for complementation of this phenotype is too expensive or too time consuming to make it practical to examine the large number of transformants required to identify the wild-type gene. In these cases, it may be possible to identify the clones covering a particular chromosome region and test individually for the ability to complement the mutations. For fungal species like *A. nidulans* and *N. crassa*, for which a relatively dense genetic map is available, it is usually possible to map the new mutation of interest between two known genetic markers. In many cases the genes corresponding to these markers have already been cloned, but even if they have not been isolated, the phenotypes associated with mutations in these genes may provide a simpler cloning strategy. In these cases, a useful way to reduce the number of clones that need to be screened to identify the gene of interest is to use chromosome walking to isolate all the clones spanning the targeted chromosomal region. Chromosome walking is carried out by the sequential isolation of overlapping clones, using the first clone to probe the gene library to isolate further clones, carrying parts of the same insert together with adjacent sequences, until contiguous clones representing the region bounded by the markers are obtained. Given the small size of fungal genomes and the close spacing of fungal genes, cloning a gene by chromosomal position is a good alternative strategy. For fungi with poorly developed genetic maps, production of restriction fragment length polymorphism (RFLP) maps covering the entire genome may be required (44, 132). Defined RFLP markers can then be used as a starting point for chromosome walking.

One valuable genetic resource that is very useful for identifying genes by chromosome walking is ordered chromosomal clone banks. Ordered chromosomal clone banks and information about the position (clones) of the identified genes for *N. crassa* and *A. nidulans* are available from the Fungal Genetics Stock Center (FGSC). For *A. nidulans*, these clones have been placed in chromosome specific sets and the order of clones within chromosomes has been established (18, 110a). Thus, in this case much of the most tedious work required for position-based cloning is already complete.

34.3.1.4. Cloning Genes by Heterologous Expression

Many of the earliest efforts to isolate fungal genes based on their function involved attempts to complement mutations in *E. coli* or *S. cerevisiae* by using genomic DNA libraries or cDNA expression libraries derived from filamentous fungi. Examples of fungal genes cloned by this approach include the *N. crassa qa-2* gene (79, 149), several fungal *trpC* genes (29, 123, 159), the *pyr-4* gene of *N. crassa* (22) and the *argB* gene of *A. nidulans* (12). While the development of functional transformation systems in filamentous fungi has greatly reduced the importance of this approach, the possibility of identifying genes by function from less genetically tractable fungi by heterologous complementation of mutations in more experimentally amenable systems remains a valuable approach. A good example of how heterologous expression between filamentous fungi can be effective is the cloning of the *Nectria haematococca PDA* gene. This gene encodes a pisatin-demethylating activity and was recovered based on expression in *A. nidulans* (155). Pisatin is a phytoalexin that functions as an antifungal toxin and is produced by the pea plant at the site of infection by *N. haematococca*. Demethlyation of pisatin by the fungus inactivates the toxin, thereby allowing fungal growth. To isolate the *PDA* gene, an *N. haematococca* cosmid library, constructed in a vector including the *A. nidulans trpC* gene, was used to transform a *trpC801 A. nidulans* strain (which does not produce PDA) selecting for tryptophan prototrophy. Transformants were then screened for the ability to produce pisatin demethylase activity, and one such transformant was found, allowing recovery of the *N. haematococca PDA* gene (155). Similarly, an *Aspergillus niger* invertase-encoding gene was cloned by expression in *Trichoderma reesei*, which lacks invertase and therefore cannot use sucrose as a sole carbon source (11). A different approach that relies on heterologous expression between fungal strains involves complementation of well-characterized mutants. For instance, *Gaeumannomyces graminis* genes that complement *A. nidulans argB* and *pyrG* mutants were identified by direct selection for prototrophic transformants (15).

34.3.2. Gene Inactivation by Integration/Replacement

One important feature of most fungal transformation systems is that it is usually possible to direct integration of plasmid sequences at a homologous genomic site. This genetic tool is particularly valuable because it allows assessment of gene function by construction of knockout or altered function alleles of a given gene. There are two basic approaches to gene disruption in filamentous fungi: integrative disruption and gene replacement. Both of these approaches require homologous integration of transforming DNA into a specific site in the recipient genome.

For integrative disruption, a circular plasmid carrying a selectable marker and an internal fragment from the target gene is used to transform a wild-type strain. Integration into the recipient genome at the homologous target site by a single recombination event leads to duplication of the transforming fragment and results in two partially deleted copies of the target gene separated by plasmid DNA and selectable marker sequences (Fig. 3A). This strategy is relatively simple and, in most cases, leads to gene inactivation. It is important to recognize, however, that the remaining partial gene fragments may maintain some activity, so the resultant mutant phenotype will not always be equivalent to the null mutant phenotype. Generally such duplicative gene disruptions are stable during mitotic growth, and revertants are only rarely observed (16, 94). However, during meiotic crosses, some duplications are particularly unstable (76, 137, 161), presumably owing to either unequal crossing over or intrachromosomal recombination. Whatever the mechanism, this can result in a high reversion frequency following sexual crosses so that the mutant is underrepresented in the progeny.

This approach of integrative duplication of a circular plasmid has also been used for directing plasmids to a particular site in the genome by rescuing a mutation. By transforming a recipient strain containing a mutation in the gene to be used for selecting transformants with a circular plasmid containing a second mutant allele of the same gene, it is possible to select transformants in which an homologous integration event gave rise to a wild-type copy of the gene (57, 63). For example, as shown in Fig. 3D, when the 5′ half of the *A. nidulans trpC* gene is used to transform a *trpC* mutant strain in which the mutation is known to be within this part of the gene, integration of the vector results in duplication of the transforming *trpC* gene sequence (63, 64). Thus, one copy containing the original mutation and a second complete wild-type *trpC* gene results. Typically, while we find that 20 to 40% of the *trpC*$^+$ transformants have resulted from such a recombination event, the remaining transformants apparently represent simple gene conversions. Finally, although this approach can be used to direct a single-copy integration of the transforming vector into a specific site, it is also important to realize that multiple copies of the transforming plasmid may be found in a trandem array. Thus, it is critical to select the appropriate restriction enzymes for Southern blot analysis to determine the plasmid copy number.

The second approach to disrupt a target gene involves gene replacement. This strategy can be used to create mutations ranging from a complete deletion of the coding region to subtle modifications expected to alter the gene activity. Here, the target gene is modified in vitro, and linearized DNA is used for transformation of the host. A double recombination event like that shown in Fig. 3B will lead to replacement of the functional gene with the modified target DNA fragment (95, 105, 139). Typically, the target gene modification involves replacement of gene sequences with a selectable marker so that colonies carrying the modified DNA can be identified directly. However, it is also possible to use cotransformation, selecting for acquisition of a plasmid containing a selectable marker and then screening to identify transformants that have also acquired the modified target sequence. When the modified target gene includes a selectable marker, it is important to include enough flanking DNA on each side of the marker to direct efficient recombination. This can usually be accomplished by including at least 1.0 kb of target DNA on each side of the marker (95). The frequency with which such gene replacement events occur varies between fungi and even between genes within a given fungus, but in most cases it is possible to design strategies to effectively modify target gene sequences.

Finally, it is important to consider the strategies necessary for demonstrating that a particular gene may be essential for the life of an organism. In *S. cerevisiae*, it is possible to make a directed mutation in one copy of an essential gene using a diploid strain and follow segregation of the mutant phenotype (death) by tetrad analysis (130, 135), but this approach is not feasible in filamentous fungi. In fungi like *N. crassa*, in which tetrad analysis is possible, there is no stable vegetative diploid stage in which to make the initial deletion (39). In other species, like *A. nidulans*, it is possible to construct stable diploids; however, tetrad analysis is not practical. Two alternative approaches involving the use of diploids or heterokaryons have been adopted for *A. nidulans*. In the first strategy, a diploid strain is transformed to disrupt one gene copy followed by regeneration of haploid strains by treatment of the diploid with the tubulin-destabilizing drug benomyl (31). Haploid segregants can then be tested for the presence of the disrupted gene and the transformation marker. If no haploid segregants contain the disrupted gene, but segregation of other markers is normal, this can be taken as indirect evidence that the gene is essential. However, it is important to note that it can be very difficult to recover slow-growing mutants as haploid segregants. Specht et al. (133) described an approach that avoided this problem by isolating ascospores from sexual crosses that took place during benomyl treatment. This allowed recovery of slow-growing mutants from disruption of a gene that had previously been described as essential (133, 157).

The second strategy that has been employed to demonstrate essentiality relies on the fact that transformation results in a high frequency of heterokaryon formation. In cases where gene disruption results in lethality, it is possible to recover heterokaryons in which one nucleus carries the selectable marker but lacks the essential gene while the second nucleus carries the essential gene (e.g., references 101 and 103). If the essential gene is not necessary for asexual spore formation, it is possible to recover uninucleate asexual spores containing both nuclear types. This can be demonstrated by plating spores on both selective and nonselective medium. Neither spore type should be able to grow on selective medium, because one nucleus lacks the transformation marker while the other lacks the essential gene. In contrast, the spores containing the untransformed nucleus will be able to grow on nonselective medium. In cases where this approach has been used, it has been possible to show by Southern blot analysis that colonies arising from transformation had both wild-type and mutant copies of the target gene, while colonies regenerated on nonselective medium had only the wild-type gene copy. In addition, in some cases it has been possible to get information about how death occurs (implicating gene function) by following germination of mutant spores on selective medium (101, 103).

34.3.3. In Vivo Site-Directed Mutagenesis

As described above, an important tool for analysis of gene function is the ability to construct specific alterations in a given gene and observe the effects on activity. Alternatively, when the effect of a given mutation is already known from studies in a model organism, it may be desirable for practical purposes to introduce this same mutation into an industrially important fungus to change its properties such that a desirable gene product is more easily secreted, becomes more stable, loses toxicity, or gains activity. Usually, such site-directed mutagenesis requires in vitro manipulations of a gene (84) and can require several time-consuming steps before a new strain has been constructed. Recently, Calissano and Macino (23) described a protocol for in vivo site-directed mutagenesis in the *N. crassa tub-2* gene encoding beta-tubulin. Their approach was to transform wild-type *N. crassa* spheroplasts using sense or antisense oligonucleotides complementary to a known mutant allele of the *tub-2* gene, followed by selection for benomyl-resistant transformants. In this experiment, they found that a 30-nucleotide-long antisense oligonucleotide could mutate the wild-type beta-tubulin gene to the benomyl-resistant allele in vivo. We have tried to employ this protocol in *A. nidulans* to cause a mutation that could not be directly selected by using cotransformation with a selectable marker but have not been able to get the desired mutant. Although the use of this protocol could be limited for cases involving strong selection of introduced mutations, it is certainly worth considering in other fungal species.

34.3.4. Gene Overexpression or Activation

The ability to provide controlled activation of certain genes under conditions where they may not normally be expressed can be useful both for analysis of gene function and for increasing production of beneficial fungal products. This can be accomplished by constructing gene fusions between a given structural gene and the regulatory regions for another gene that can be activated or inactivated at will. For basic science, production of unique phenotypes resulting from forced expression of a gene may provide useful insights into the gene's normal function. For example, Adams et al. (1) used the promoter from the *A. nidulans* gene encoding catabolic alcohol dehydrogenase (*alcA*) to force high levels of expression of a gene proposed to regulate conidiation (*brlA*). This experiment showed that activation of *brlA* in vegetative cells activated inappropriate sporulation, thus demonstrating a pivotal role for *brlA* in activating development.

There are two general classes of promoters that can potentially be used for forced gene expression experiments: constitutive and inducible. Related approaches can be used in other fungal species (e.g., references 3, 32, 47 and 126), but we will limit our discussion to the most commonly used *A. nidulans* expression systems. The best-characterized constitutive promoters in *A. nidulans* are those for *trpC* (63) and *gpdA* (the gene encoding glyceraldehyde-3-phosphate dehydrogenase A) (112). These promoters drive expression of the genes at relatively constant levels throughout the life cycle and under most growth conditions that have been examined. Thus, activation of these promoters does not require special growth medium.

Regulated promoters allow conditional assessment of gene function through controlled activation of expression at inappropriate times. If overexpression of a gene is detrimental to growth, the strain containing the gene fusion of interest can be maintained under noninducing conditions indefinitely, and expression can be induced at will by changing the growth medium. The best-characterized regulated promoters for directed gene expression in *A. nidulans* are those for the carbon source-regulated *alcA* gene (8) and the nitrogen source-regulated *niaD* and *niiA* genes (4). Decisions of which promoter to use for a given experiment should be based on expression levels required (*alcA* is both more tightly repressed and more strongly activated) and medium needs.

The promoter used most often for examining forced gene expression in *A. nidulans* has been *alcA*(*p*) (61, 62). Like the promoter for many genes in carbon catabolism, the *alcA*(*p*) is glucose repressed by a mechanism that requires the *creA* product (8, 68, 127). Repression of the *alcA*(*p*) is relieved when glucose is removed from the medium, but high levels of activation require the presence of an inducer (e.g., ethanol, threonine, or other inducer compounds such as methyl-ethyl ketone) and the product of the *alcR* gene, which encodes a positive activator (106, 127). All of the sequences essential for *alcA*(*p*) control have been localized within about 300 bp upstream of the translation start site (67, 152), and this region can easily be fused to any structural gene to be activated.

The *niaD* and *niiA* genes are divergently transcribed and are separated by a 1.3-kb intergenic region (71, 114). Expression of these genes is repressed in the presence of ammonia or glutamine, but such repression is relieved during growth on other nitrogen sources (4, 122). Activation requires the presence of nitrate and the absence of a repressing nitrogen source as well as the products of the *nirA* and *areA* genes (4, 19, 20, 117, 142). The sequences essential for *niiA*(*p*) and *niaD*(*p*) control have been localized within about 400 and 380 bp upstream of the respective translation start sites (114), and these regions can easily be fused to any structural gene to be activated.

34.3.5. Analysis of Fungal Promoters Using Reporter Genes

Understanding the mechanisms regulating expression of specific genes can be important for both basic and applied studies in fungal molecular biology. The use of transformation systems and gene manipulation strategies can make it possible to study the detailed mechanisms controlling fungal gene expression. Standard reporter genes, including *lacZ* (encoding β-galactosidase) and *uidA* (encoding β-glucuronidase [GUS]), have been used extensively in fungal promoter studies (63, 119, 146, 147). These reporter genes can be fused to the proposed regulatory segments from a specific gene, and its promoter activity is analyzed by qualitative or quantitative studies following activity of the reporter gene products (e.g., references 119, 146, and 147). More recently, strategies have been adapted to fungi by using the jellyfish *Aequorea victoria* green fluorescent protein (for review see references 36 and 111) as a reporter in gene fusions in which activity can be monitored in living tissue (5, 97).

An important requirement for quantitative analysis of expression signals is that a single copy of the expression unit is integrated into the chromosome of the recipient at a specific site. This can be achieved by using a partial homologous selection marker as described in section 34.3.2. Using this strategy, it is possible to make direct comparisons between transformants having modified gene fusions without concern for chromosome position effects.

Promoter analysis strategies can also be adapted to more random approaches for identifying strong transcription signals that may be of value in industry, where a large amount of a particular gene product may be needed. One method for easy isolation of promoter fragments from filamentous fungi using promoter-probe libraries was recently reported (154). Two libraries were constructed in which random 0.5- to 2.0-kb fragments of *Gibberella pulicaris* genomic DNA were inserted 5′ of a promoterless *hph* gene. These libraries were then used to transform *G. pulicaris* selecting for hygromycin resistance, and two separate genomic fragments capable of driving high-level expression of *hph* were identified. Similar approaches could be used to isolate similar promoter fragments from any number of filamentous fungi and, combined with reporter gene analysis, could be used to identify appropriate promoters for any purpose.

34.4. PERSPECTIVES

The development of fungal DNA-mediated transformation systems has provided an important tool for addressing both basic and applied problems in filamentous fungi. It is now relatively straightforward to identify an interesting gene through genetic analysis, isolate the gene by transformation to complement a mutation, modify the gene in vitro, and replace the wild-type copy in the fungus to assess the effect of the modification. Similarly, genes identified through biochemical studies or based on DNA similarity to well-studied genes from other organisms can be isolated and characterized to learn how they affect processes of interest in filamentous fungi. As this technology has moved from fungi that have been used traditionally for genetic studies to less tractable organisms that have importance in industry or medicine, it is providing new approaches to harvesting fungal diversity that will certainly bring new value to fungi in the biotechnology industry.

All of this leaves the question of what the next decade of fungal research will bring. To answer this question, perhaps it is worthwhile to consider the impact of the recent completion of the entire genomic sequence for the yeast *S. cerevisiae* (see references 42 and 54). This accomplishment provides the first look at all of the genes present in the genome of a simple eukaryote and of a simple fungus. This has changed the way research is approached in yeast similarly to how transformation technology changed research in the 1980s. Now it is possible to look at all 6,000 yeast genes to ask directed questions about their possible functions (42, 54).

The potential value of similar sequencing efforts in more complex fungi than yeast should also be considered. Based on the enormous genetic diversity of fungi and the relative simplicity of yeast, it seems certain that valuable information will be missing from the 6,000 yeast genes. It is significant that initial efforts to examine so-called expressed sequence tags (ESTs) from filamentous fungi like *N. crassa*, *A. nidulans*, and *M. grisea* have found that nearly half of the clones characterized do not have clear homologs in yeast. With regard to biotechnology, the sequences not present in yeast but found in other filamentous fungi are likely to be the most interesting. These novel genes presumably include those that give filamentous fungi their diverse properties, with all the good and bad consequences they can bring to humans. By moving from the era of gene-oriented questions to genome-oriented questions, it is likely that the vast information that becomes available will have an important impact on how we now manipulate fungal genomes.

REFERENCES

1. **Adams, T. H., M. T. Boylan, and W. E. Timberlake.** 1988. *brlA* is necessary and sufficient to direct conidiophore development in *Aspergillus nidulans*. *Cell* **54:**353–362.
2. **Akins, R. A., and A. M. Lambowitz.** 1985. General method for cloning *Neurospora crassa* nuclear genes by complementation of mutants. *Mol. Cell. Biol.* **5:**2272–2278.
3. **Aronson, B. D., K. A. Johnson, J. J. Loros, and J. C. Dunlap.** 1994. Negative feedback defining a circadian clock: autoregulation of the clock gene frequency. *Science* **263:**1578–1584.
4. **Arst, H. N., Jr., and D. J. Cove.** 1973. Nitrogen metabolite repression in *Aspergillus nidulans*. *Mol. Gen. Genet.* **176:**111–141.
5. **Atkins, D., and J. G. Izant.** 1995. Expression and analysis of the green fluorescent protein gene in the fission yeast *Schizosaccharomyces pombe*. *Curr. Genet.* **28:**585–588.
6. **Austin, B., R. M. Hall, and B. M. Tyler.** 1990. Optimized vectors and selection for transformation of *Neurospora crassa* and *Aspergillus nidulans* to bleomycin and phleomycin resistance. *Gene* **93:**157–162.
7. **Ausubel, F. M., R. Brent, R. E. Kingston, D. D. Moore, J. A. Smith, J. G. Seidman, and K. Struhl.** 1987. *Current Protocols in Molecular Biology*. John Wiley & Sons, New York.
8. **Bailey, C., and H. Arst.** 1975. Carbon catabolite repression in *Aspergillus nidulans*. *J. Eur. Biochem.* **51:**573–577.
9. **Ballance, D. J., F. P. Buxton, and G. Turner.** 1983. Transformation of *Aspergillus nidulans* by the orotidine-5′-phosphate decarboxylase gene of *Neurospora crassa*. *Biochem. Biophys. Res. Comm.* **112:**284–289.
10. **Benninger, O. M., C. Skrzynia, P. J. Pukkila, and L. A. Casselton.** 1987. DNA-mediated transformation of the basidiomycete *Coprinus cinereus*. *EMBO J.* **6:**835–840.
11. **Berges, T., C. Barreau, J. F. Peberdy, and L. M. Boddy.** 1993. Cloning of an *Aspergillus niger* invertase gene by expression in *Trichoderma reesei*. *Curr. Genet.* **24:**53–59.
12. **Berse, B., A. Dmochowska, M. Skrzypek, P. Weglenski, M. A. Bates, and R. L. Weiss.** 1983. Cloning and characterization of the ornithine carbamoyl-transferase gene from *Aspergillus nidulans*. *Genetics* **25:**109–117.
13. **Boeke, J. D., F. LaCroute, and G. Fink.** 1984. A positive selection for mutants lacking orotidine-5′-phosphate decarboxylase activity in yeast: 5-flouro-orotic acid resistance. *Mol. Gen. Genet.* **197:**345–346.
14. **Bölker, M., H. U. Böhnert, K. H. Braun, J. Görl, and R. Kahmann.** 1995. Tagging pathogenicity genes in *Ustilago maydis* by restriction enzyme-mediated integration (REMI). *Mol. Gen. Genet.* **248:**547–552.
15. **Bowyer, P., A. E. Osbourn, and M. J. Daniels.** 1994. An "instant gene bank" method for heterologous gene cloning: complementation of two *Aspergillus nidulans* mutants with *Gaeumannomyces graminis* DNA. *Mol. Gen. Genet.* **242:**448–454.
16. **Boylan, M. T., M. J. Holland, and W. E. Timberlake.** 1986. *Saccharomyces cerevisiae* centromere CEN11 does not induce chromosome instability when integrated into the *Aspergillus nidulans* genome. *Mol. Cell. Biol.* **6:**3621–3625.

17. **Broach, J. R.** 1983. Construction of high copy yeast vectors using 2 μm circle sequences. *Methods Enzymol.* **101:** 307–325.

18. **Brody, H., J. Griffith, A. J. Cuticchia, J. Arnold, and W. E. Timberlake.** 1991. Chromosome-specific recombinant DNA libraries from the fungus *Aspergillus nidulans*. *Nucleic Acids Res.* **19:**3105–3109.

19. **Burger, G., J. Strauss, C. Scazzocchio, and B. F. Lang.** 1991. *nirA*, the pathway-specific regulatory gene of nitrate assimilation in *Aspergillus nidulans*, encodes a putative GAL4-type zinc finger protein and contains four introns in highly conserved regions. *Mol. Cell. Biol.* **11:**5746–5755.

20. **Burger, G., J. Tilburn, and C. Scazzocchio.** 1991. Molecular cloning and functional characterization of the pathway-specific regulatory gene *nirA*, which controls nitrate assimilation in *Aspergillus nidulans*. *Mol. Cell. Biol.* **11:**795–802.

21. **Busby, T. M., K. Y. Miller, and B. L. Miller.** 1996. Suppression and enhancement of the *Aspergillus nidulans medusa* mutation by altered dosage of the *bristle* and *stunted* genes. *Genetics* **143:**155–163.

22. **Buxton, F. P., and A. Radford.** 1983. Cloning of the structural gene for orotidine-5′-phosphate carboxylase of *Neurospora crassa* by expression in *Escherichia coli*. *Mol. Gen. Genet.* **190:**403–405.

23. **Calissano, M., and G. Macino.** 1996. *In vivo* site-directed mutagenesis of *Neurospora crassa beta-tubulin* gene by spheroplasts transformation with oligonucleotides. *Fungal Genet. Newsl.* **43:**15–16.

24. **Cambareri, E. B., J. Helber, and J. A. Kinsey.** 1994. *Tad1-1*, an active LINE-like element of *Neurospora crassa*. *Mol. Gen. Genet.* **242:**658–665.

25. **Case, M. E.** 1982. Transformation of *Neurospora crassa* utilizing recombinant plasmid DNA. *Basic Life Sci.* **19:** 87–100.

26. **Case, M. E., M. Schweizer, S. R. Kushner, and N. H. Giles.** 1979. Efficient transformation of *Neurospora crassa* by utilizing hybrid plasmid DNA. *Proc. Natl. Acad. Sci. USA* **76:**5259–5263.

27. **Charkraborty, B. N., N. A. Patterson, and M. Kapoor.** 1991. An electroporation-based system for high-efficiency transformation of germinated conidia of filamentous fungi. *Can. J. Microbiol.* **37:**858–863.

28. **Chinery, S. A., and E. Hinchliffe.** 1989. A novel class of vector for yeast transformation. *Curr. Genet.* **16:**21–25.

29. **Choi, H. T., J. L. Revuelta, C. Sadhu, and M. Jayaram.** 1988. Structural organization of the *TRP1* gene of *Phycomyces blakesleeanus*: implications for evolutionary gene fusion in fungi. *Genetics* **71:**85–95.

30. **Clarke, L., and J. Carbon.** 1976. A colony bank containing synthetic Col E1 hybrid plasmids representative of the entire *E. coli* genome. *Cell* **9:**91–99.

31. **Clutterbuck, A. J.** 1974. *Aspergillus nidulans*, p. 447–510. *In* R. C. King (ed.), *Handbook of Genetics*. Plenum Press, New York.

32. **Collins, M. E., G. Briggs, C. Sawyer, P. Sheffield, and I. F. Connerton.** 1991. An inducible gene expression system for *Neurospora crassa*. *Enzyme Microbial Technol.* **13:** 400–403.

33. **Cove, D. J.** 1976. Chlorate toxicity in *Aspergillus nidulans*. Studies of mutants altered in nitrate assimilation. *Mol. Gen. Genet.* **146:**147–159.

34. **Cove, D. J.** 1976. Chlorate toxicity in *Aspergillus nidulans*: the selection and characterization of chlorate resistant mutants. *Heredity* **36:**191–203.

35. **Crowhurst, R. N., J. Rees-George, E. H. Rikkerink, and M. D. Templeton.** 1992. High efficiency transformation of *Fusarium solani* f. sp. *cucurbitae* race 2 (mating population V). *Curr. Genet.* **21:**463–469.

36. **Cubitt, A. B., R. Heim, S. R. Adams, A. E. Boyd, L. A. Gross, and R. Y. Tsien.** 1995. Understanding, improving and using green fluorescent proteins. *Trends Biochem. Sci.* **20:**448–455.

37. **Daboussi, M. J., A. Djeballi, C. Gerlinger, P. L. Blaiseau, I. Bouvier, M. Cassan, M. H. Lebrun, D. Parisot, and Y. Brygoo.** 1989. Transformation of seven species of filamentous fungi using the nitrate reductase gene of *Aspergillus nidulans*. *Curr. Genet.* **15:**453–456.

38. **Daboussi, M. J., T. Langin, and Y. Brygoo.** 1992. *Fot1*, a new family of fungal transposable elements. *Mol. Gen. Genet.* **232:**12–16.

39. **Davis, R. H., and F. J. de Serres.** 1970. Genetic and microbiological research techniques for *Neurospora crassa*. *Methods Enzymol.* **27A:**79–143.

40. **d'Enfert, C.** 1996. Selection of multiple disruption events in *Aspergillus fumigatus* using the orotidine-5′-decarboxylase gene, *pyrG*, as a unique transformation marker. *Curr. Genet.* **30:**76–82.

41. **Diolez, A., T. Langin, C. Gerlinger, Y. Brygoo, and M. J. Daboussi.** 1993. The *nia* gene of *Fusarium oxysporum*: isolation, sequence and development of a homologous transformation system. *Gene* **131:**61–67.

42. **Dujon, B.** 1996. The yeast genome project: what did we learn? *Trends Genet.* **12:**263–270.

43. **Durrens, P., P. M. Green, H. N. Arst, and C. Scazzocchio.** 1986. Heterologous insertion of transforming DNA and generation of new deletions associated with transformation in *Aspergillus nidulans*. *Mol. Gen. Genet.* **203:**544–459.

44. **Farman, M. L., and S. A. Leong.** 1995. Genetic and physical mapping of telomeres in the rice blast fungus, *Magnaporthe grisea*. *Genetics* **140:**479–492.

45. **Fincham, J. R. S.** 1989. Transformation in fungi. *Microbiol. Rev.* **53:**148–170.

46. **Fonzi, W. A., and M. Y. Irwin.** 1993. Isogenic construction and gene mapping in *Candida albicans*. *Genetics* **134:** 717–728.

47. **Fowler, T., and R. M. Berka.** 1991. Gene expression systems for filamentous fungi. *Curr. Opin. Biotechnol.* **2:**691–697.

48. **Fungaro, M. H., E. Rech, G. S. Muhlen, M. H. Vainstein, R. C. Pascon, M. V. de Queiroz, A. A. Pizzirani-Kleiner, and J. L. de Azevedo.** 1995. Transformation of *Aspergillus nidulans* by microprojectile bombardment on intact conidia. *FEMS Microbiol. Lett.* **125:**293–297.

49. **Futcher, A. B.** 1988. The 2 μm circle plasmid of *Saccharomyces cerevisiae*. *Yeast* **4:**27–40.

50. **Garciapedrjas, M. D., and M. I. G. Roncero.** 1996. A homologous and self replicating system for efficient transformation of *Fusarium oxysporum*. *Curr. Genet.* **29:**191–198.

51. **Gems, D., I. L. Johnstone, and A. J. Clutterbuck.** 1991. An autonomously replicating plasmid transforms *Aspergillus nidulans* at high frequency. *Gene* **98:**61–67.

52. **Gems, D. H., and A. J. Clutterbuck.** 1993. Co-transformation with autonomously-replicating helper plasmids facilitates gene cloning from an *Aspergillus nidulans* gene library. *Curr. Genet.* **24:**520–524.

53. **Glayzer, D. C., I. N. Roberts, D. B. Archer, and R. P. Oliver.** 1995. The isolation of *Ant1*, a transposable element from *Aspergillus niger*. *Mol. Gen. Genet.* **249:**432–438.

54. **Goffeau, A., B. G. Barrell, H. Bussey, R. W. Davis, B. Dujon, H. Feldman, F. Galibert, J. D. Hoheisel, C. Jacq, M. Johnston, E. J. Louis, H. W. Mewes, Y. Murakami, P. Philippsen, H. Tettelin, and S. G. Oliver.** 1996. Life with 6000 genes. *Science* **274:**546–567.

55. **Goldman, G. H., M. Van Montagu, and A. Herrere-Estrella.** 1990. Transformation of *Trichoderma harzianum* by high-voltage electric pulse. *Curr. Genet.* **17:**169–174.

56. **Goosen, T., G. Bloemheuvel, C. Gysler, D. A. De Bie, H. W. J. Van den Broek, and K. Swart.** 1987. Transformation of *Aspergillus niger* using the homologous orotidine-5′-triphosphate-decarboxylase gene. *Curr. Genet.* **11:**499–503.

57. **Gouka, R. J., J. G. Hessing, H. Stam, W. Musters, and C. A. M. J. J. Van den Hondel.** 1995. A novel strategy for the isolation of defined *pyrG* mutants and the development of a site-specific integration system for *Aspergillus awamori*. *Curr. Genet.* **27:**536–540.

58. **Goyon, C., and G. Faugeron.** 1989. Targeted transformation of *Ascobolus immersus* and *de novo* methylation of the resulting duplicated DNA sequences. *Mol. Cell. Biol.* **9:**2818–2827.

59. **Gruber, F., J. Visser, C. P. Kubicek, and L. H. De Graaff.** 1990. Cloning of the *Trichoderma reesei pyrG* gene and its use as a homologous marker for a high-frequency transformation system. *Curr. Genet.* **18:**447–451.

60. **Gruber, F., J. Visser, C. P. Kubicek, and L. H. de Graaff.** 1990. The development of a heterologous transformation system for the cellulolytic fungus *Trichoderma reesei* based on a *pyrG*-negative mutant strain. *Curr. Genet.* **18:**71–76.

61. **Gwynne, D. I., F. P. Buxton, S. Sibley, R. W. Davies, R. A. Lockington, C. Scazzocchio, and H. M. Sealey-Lewis.** 1987. Comparison of *cis*-acting control regions of two coordinately controlled genes involved in ethanol utilization in *Aspergillus nidulans*. *Gene* **51:**205–216.

62. **Gwynne, D. I., F. P. Buxton, S. A. Williams, M. Sills, J. A. Johnstone, J. K. Buch, Z. Guo, D. Drake, M. Westphal, and W. Davies.** 1989. Development of an expression system in *Aspergillus nidulans*. *Biochem. Soc. Trans.* **17:338–340.**

63. **Hamer, J. E., and W. E. Timberlake.** 1987. Functional organization of the *Aspergillus nidulans trpC* promoter. *Mol. Cell. Biol.* **7:**2352–2359.

64. **Han, S., J. Navarro, R. A. Greve, and T. H. Adams.** 1993. Translational repression of *brlA* expression prevents premature development in *Aspergillus*. *EMBO J.* **12:**2449–2457.

65. **Herrera-Estrella, A., G. H. Goldman, and M. Van Montagu.** 1990. High-efficiency transformation system for the biocontrol agents, *Trichoderma* spp. *Mol. Microbiol.* **4:**839–843.

66. **Hilber, U. W., M. Bodmer, F. D. Smith, and W. Koller.** 1994. Biolistic transformation of conidia of *Botryotinia fuckeliana*. *Curr. Genet.* **25:**124–127.

67. **Hintz, W. E., and P. A. Logosky.** 1993. A glucose-derepressed promoter for expression of heterologous products in the filamentous fungus *Aspergillus nidulans*. Bio/Technology **11:**815–818.

68. **Hynes, M., and J. M. Kelly.** 1977. Pleiotrophic mutants of *Aspergillus nidulans* altered in carbon metabolism. *Mol. Gen. Genet.* **150:**193–204.

69. **John, M. A., and J. F. Peberdy.** 1984. Transformation of *Aspergillus nidulans* using the *argB* gene. *Enzyme Microbiol Technol.* **6:**386–389.

70. **Johnstone, I. L.** 1985. Transformation of *Aspergillus nidulans*. *Microbiol. Sci.* **2:**307–311.

71. **Johnstone, I. L., P. C. McCabe, P. Greaves, S. J. Gurr, G. E. Cole, M. A. Brow, S. E. Unkles, A. J. Clutterbuck, J. R. Kinghorn, and M. A. Innis.** 1990. Isolation and characterisation of the *crnA-niiA-niaD* gene cluster for nitrate assimilation in *Aspergillus nidulans*. *Gene* **90:**181–192.

72. **Joutsjoki, V. V.** 1994. Construction by one-step gene replacement of *Trichoderma reesei* strains that produce the glucoamylase P of *Hormoconis resinae*. *Curr. Genet.* **26:** 422–429.

73. **Kachroo, P., S. A. Leong, and B. B. Chattoo.** 1994. Pot2, an inverted repeat transposon from the rice blast fungus *Magnaporthe grisea*. *Mol. Gen. Genet.* **245:**339–348.

74. **Kachroo, P., S. A. Leong, and B. B. Chattoo.** 1995. Mg-SINE: a short interspersed nuclear element from the rice blast fungus, *Magnaporthe grisea*. *Proc. Natl. Acad. Sci. USA* **92:**11125–11129.

75. **Kelly, J. M., and M. J. Hynes.** 1985. Transformation of *Aspergillus niger* by the *amdS* gene of *Aspergillus nidulans*. *EMBO J.* **4:**475–479.

76. **Kelly, J. M., and M. J. Hynes.** 1987. Multiple copies of the *amdS* gene of *Aspergillus nidulans* cause titration of *trans*-acting regulatory proteins. *Curr. Genet.* **12:**21–31.

77. **Keranen, S., and M. Pentilla.** 1995. Production of recombinant proteins in the filamentous fungus *Trichoderma reesei*. *Curr. Opin. Biotechnol.* **6:**534–537.

78. **Kim, S. Y., and G. A. Marzluf.** 1988. Transformation of *Neurospora crassa* with the *trp-1* gene and the effect of host strain upon the fate of the transforming DNA. *Curr. Genet.* **13:**65–70.

79. **Kinghorn, J. R., and A. R. Hawkins.** 1982. Cloning and expression in *Escherichia coli* K12 of the biosynthetic dehydroquinase function of the *arom* cluster gene from the eukaryote *Aspergillus nidulans*. *Mol. Gen. Genet.* **186:**145–152.

80. **Kinsey, J. A., P. W. Garrett-Engele, E. B. Cambareri, and E. U. Selker.** 1994. The *Neurospora* transposon *Tad* is sensitive to repeat-induced point mutation (RIP). *Genetics* **138:**657–664.

81. **Kinsey, J. A., and J. A. Rambosek.** 1984. Transformation of *Neurospora crassa* with the cloned *am* (glutamate dehydrogenase) gene. *Mol. Cell. Biol.* **4:**117–122.

82. **Kolar, M., P. J. Punt, C. A. M. J. J. Van den Hondel, and H. Schwab.** 1988. Transformation of *Penicillium chrysogenum* using dominant selection markers and expression of an *Escherichia coli lacZ* fusion gene. *Gene* **62:**127–134.

83. **Kothe, G. O., and S. J. Free.** 1996. Protocol for the electroporation of Neurospora spheroplasts. *Fungal Genet. Newsl.* **43:**31–32.

84. **Kunkel, T. A.** 1985. Rapid and efficient site-specific mutagenesis without phenotypic selection. *Proc. Natl. Acad. Sci. USA* **82:**488–492.

85. **Kuspa, A., and W. F. Loomis.** 1992. Tagging developmental genes in *Dictyostelium* by restriction enzyme-mediated integration of plasmid DNA. *Proc. Natl. Acad. Sci. USA* **89:**8803–8807.

86. **Langin, T., P. Capy, and M. J. Daboussi.** 1995. The transposable element impala, a fungal member of the Tc1-mariner superfamily. *Mol. Gen. Genet.* **246:**19–28.

87. **Lorito, M., C. K. Hayes, A. Di Oietro, and G. E. Harman.** 1993. Biolistic transformation of *Trichoderma harzianum* and *Gliocladium virens* using plasmid and genomic DNA. *Curr. Genet.* **24:**349–356.

88. **Mach, R. L., M. Schindler, and C. P. Kubicek.** 1994. Transformation of *Trichoderma reesei* based on hygromycin B resistance using homologous expression signals. *Curr. Genet.* **25:**567–570.

89. **Malardier, L., M. J. Daboussi, J. Julien, F. Roussel, C. Scazzocchio, and Y. Brygoo.** 1989. Cloning of the nitrate reductase gene (*niaD*) of *Aspergillus nidulans* and its use for transformation of *Fusarium oxysporum*. *Gene* **78:**147–156.

90. **Mann, B. J., R. A. Akins, A. M. Lambowitz, and R. L. Metzenberg.** 1988. The structural gene for a phosphorus-repressible phosphate permease in *Neurospora crassa* can complement a mutation in positive regulatory gene *nuc-1*. *Mol. Cell. Biol.* **8:**1376–1379.

91. **Marhoul, J. F., and T. H. Adams.** 1995. Identification of developmental regulatory genes in *Aspergillus nidulans* by overexpression. *Genetics* **139:**537–547.

92. **Mattern, I. E., P. J. Punt, and C. A. M. J. J. Van den Hondel.** 1988. A vector of *Aspergillus* transformation conferring phleomycin resistance. *Fungal Genet. Newsl.* **35:** 25.

93. **May, G. S., J. Gambino, J. A. Weatherbee, and N. R. Morris.** 1985. Identification and functional analysis of β-tubulin genes by site-specific integrative transformation in *Aspergillus nidulans*. *J. Cell Biol.* **100:**712–718.

94. **Mayorga, M. E., and W. E. Timberlake.** 1990. Isolation and molecular characterization of the *Aspergillus nidulans wA* gene. *Genetics* **126:**73–79.

95. **Miller, B. L., M. Y. Miller, and W. E. Timberlake.** 1985. Direct and indirect gene replacements in *Aspergillus nidulans*. *Mol. Cell. Biol.* **5:**1714–1721.

96. **Mishra, N. C., and E. L. Tatum.** 1973. Non-Mendelian inheritance of DNA-induced inositol independence in *Neurospora*. *Proc. Natl. Acad. Sci. USA* **70:**3875–3879.

97. **Monosov, E. Z., T. J. Wenzel, G. H. Luers, J. A. Heyman, and S. Subramani.** 1996. Labeling of peroxisomes with green fluorescent protein in living *P. pastoris* cells. *J. Histochem. Cytochem.* **44:**581–589.

98. **Morgan, B. A., F. L. Conlon, M. Manzanares, J. B. Millar, N. Kanuga, J. Sharpe, R. Krumlauf, J. C. Smith, and S. G. Sedgwick.** 1996. Transposon tools for recombinant DNA manipulation: characterization of transcriptional regulators from yeast, *Xenopus*, and mouse. *Proc. Natl. Acad. Sci. USA* **93:**2801–2806.

99. **Nakarisetala, T., N. Aro, N. Kalkkinen, E. Alatalo, and M. Penttila.** 1996. Genetic and biochemical characterization of the *Trichoderma reesei* hydrophobin HFBI. *Eur. J. Biochem.* **235:**248–255.

100. **Neuveglise, C., J. Sarfati, J. P. Latge, and S. Paris.** 1996. *Aftu1*, a retrotransposon-like element from *Aspergillus fumigatus*. *Nucleic Acids Res.* **24:**1428–1434.

101. **Oakley, B. R., C. E. Oakley, Y. Yoon, and M. K. Jung.** 1990. Gamma-tubulin is a component of the spindle pole body that is essential for microtubule function in *Aspergillus nidulans*. *Cell* **61:**1289–1301.

102. **Oakley, B. R., J. E. Rinehart, B. L. Mitchell, C. E. Oakley, C. Carmona, G. L. Gray, and G. S. May.** 1987. Cloning, mapping and molecular analysis of the *pyrG* orotidine-5′-phosphate decarboxylase gene of *Aspergillus nidulans*. *Gene* **61:**385–399.

103. **Osmani, S. A., R. T. Pu, and N. R. Morris.** 1988. Mitotic induction and maintenance by overexpression of a G2-specific gene that encodes a potential protein kinase. *Cell* **53:**237–244.

104. **Ozeki, K., F. Kyoya, K. Hizume, A. Kanda, and M. Hamachi.** 1994. Transformation of intact *Aspergillus niger* by electroporation. *Biosci. Biotechnol. Biochem.* **58:**2224–2227.

105. **Paietta, J. V., and G. A. Marzluf.** 1985. Gene disruption by transformation in *Neurospora crassa*. *Mol. Cell Biol.* **5:** 1554–1559.

106. **Pateman, J. H., C. H. Doy, J. E. Olson, U. Norris, E. H. Creaser, and M. Hynes.** 1983. Regulation of alcohol dehydrogenase (ADH) and aldehyde dehydrogenase (AldDH) in *Aspergillus nidulans*. *Proc. R. Soc. Lond. (Biol.)* **217:**243–264.

107. **Penttilä, M., H. Nevalainen, M. Rätto, E. Salminen, and J. Knowles.** 1987. A versatile transformation system for the cellulolytic filamentous fungus *Trichoderma reesei*. *Gene* **61:**155–164.

108. **Penttilä, M., T. T. Tuula, H. Nevalainen, and J. K. C. Knowles.** 1991. The molecular biology of *Trichoderma reesei* and its application to biotechnology, p. 85–102. *In* C. E. C. J. F. Peberdy, J. E. Ogden, and J. W. Bennett. (ed.), *Applied Molecular Genetics of Fungi*. Cambridge University Press, Cambridge.

109. **Pontecorvo, G.** 1963. Microbiol genetics: retrospect and prospect. *Proc. R. Soc. Lond. (Biol.)* **158:**1–23.

110. **Pontecorvo, G., J. A. Roper, L. M. Hemmons, K. D. Macdonald, and A. W. J. Bufton.** 1953. The genetics of *Aspergillus nidulans*. *Adv. Genet.* **5:**141–238.

110a. **Prade, R., J. Arnold, and W. Timberlake.** Personal communication.

111. **Prasher, D. C.** 1995. Using GFP to see the light. *Trends Genet.* **11:**320–323.

112. **Punt, P. J., M. A. Dingemance, A. Kuyvenhoven, R. D. M. Soede, P. H. Pouwels, and C. A. M. J. J. Van den Hondel.** 1990. Functional elements in the promoter region of the *Aspergillus nidulans gpdA* gene, encoding glyceraldehyde-3-phosphate dehydrogenase. *Gene* **93:** 101–109.

113. **Punt, P. J., R. P. Oliver, M. A. Dingemanse, P. H. Pouwels, and C. A. M. J. J. Van den Hondel.** 1987. Transformation of *Aspergillus* based on the hygromycin B resistance marker from *Escherichia coli*. *Gene* **56:**117–124.

114. **Punt, P. J., J. Strauss, R. Smit, J. R. Kinghorn, C. A. M. J. J. van den Hondel, and C. Scazzocchio.** 1995. The intergenic region between the divergently transcribed *niiA* and *niaD* genes of *Aspergillus nidulans*

contains multiple NirA binding sites which act bidirectionally. *Mol. Cell. Biol.* **15:**5688–5699.

115. **Punt, P. J., and C. A. M. J. J. van den Hondel.** 1992. Transformation of filamentous fungi based on hygromycin B and phleomycin resistance markers. *Methods Enzymol.* **216:**447–457.

116. **Queener, S. W., T. D. Ingolia, P. L. Skatrud, J. L. Chapman, and K. R. Kaster.** 1985. A system for genetic transformation of *Cephalosporium acremonium*, p. 468–472. *In* L. Leive (ed.), *Microbiology—1985*. American Society for Microbiology, Washington, D.C.

117. **Rand, K. N., and H. N. Arst, Jr.** 1978. Mutations in the *nirA* gene of *Aspergillus nidulans*. *Nature* **272:**732–734.

118. **Redman, R. S., and R. J. Rodriguez.** 1994. Factors affecting the efficient transformation of Colletotrichum species. *Exp. Mycol.* **18:**230–246.

119. **Roberts, I. N., R. P. Oliver, P. J. Punt, and C. A. M. J. J. Van den Hondel.** 1989. Expression of the *Escherichia coli* β-glucuronidase gene in industrial and phytopathogenic filamentous fungi. *Curr. Genet.* **15:**177–180.

120. **Sambrook, J., E. F. Fritsch, and T. Maniatis.** 1989. *Molecular Cloning*. Cold Spring Harbor Laboratory Press, Plainview, N.Y.

121. **Sánchez, O., and J. Aguirre.** 1996. Efficient transformation of *Aspergillus nidulans* by electroporation of germinated conidia. *Fungal Genet. Newsl.* **43:**48–51.

122. **Scazzocchio, C., and H. N. Arst, Jr.** 1989. Regulation of nitrate assimilation in *Aspergillus nidulans*, p. 299–313. *In* W. R. Wray and J. R. Kinghorn (ed.), *Molecular and Genetic Aspects of Nitrate Assimilation*. Oxford Science Publications, Oxford.

123. **Schechtman, M. G., and C. Yanofsky.** 1983. Structure of the trifunctional *trp-1* gene from *Neurosporo crassa* and its aberrant expression in *E. coli*. *J. Mol. Appl. Genet.* **2:** 83–99.

124. **Schiestl, R. H., and T. D. Petes.** 1991. Integration of DNA fragments by illegitimate recombination in *Saccharomyces cerevisiae*. *Proc. Natl. Acad. Sci. USA* **88:** 7585–7589.

125. **Schiestl, R. H., J. Zhu, and T. D. Petes.** 1994. Effects of mutations in genes affecting homologous recombination on restriction enzyme-mediated and illegitimate recombination in *Saccharomyces cerevisiae*. *Mol. Cell. Biol.* **14:**4493–4500.

126. **Schilling, B., R. M. Linden, U. Kupper, and K. Lerch.** 1992. Expression of *Neurospora crassa* laccase under the control of the copper-inducible metallothionein-promoter. *Curr. Genet.* **22:**197–203.

127. **Sealy-Lewis, H. M., and R. A. Lockington.** 1984. Regulation of two alcohol dehydrogenase genes in *Aspergillus nidulans*. *Curr. Genet.* **8:**253–259.

128. **Shi, Z., D. Christian, and H. Leung.** 1995. Enhanced transformation in *Magnaporthe grisea* by restriction enzyme mediated integration of plasmid DNA. *Phytopathology* **85:**329–333.

129. **Shizuya, H., B. Birren, U.-J. Kim, V. Mancino, T. Slepak, Y. Tachiiri, and M. Simon.** 1992. Cloning and stable maintenance of 300-kilobase-pair fragments of human DNA in *Escherichia coli* using an F-factor-based vector. *Proc. Natl. Acad. Sci. USA* **89:**8794–8797.

130. **Shortle, D., J. E. Haber, and D. Bobstein.** 1982. Lethal disruption of the yeast actin gene by integrative DNA transformation. *Science* **217:**371–373.

131. **Skory, C. D., J. S. Horng, J. J. Pestka, and J. E. Linz.** 1990. Transformation of *Aspergillus parasiticus* with the homologous gene (*pyrG*) involved in pyrimidine biosynthesis. *Appl. Environ. Microbiol.* **56:**3315–3320.

132. **Sone, T., M. Suto, and F. Tomita.** 1993. Host species-specific repetitive DNA sequence in the genome of *Magnaporthe grisea*, the rice blast fungus. *Biosci. Biotechnol. Biochem.* **57:**1228–1230.

133. **Specht, C. A., Y. Liu, P. W. Robbins, C. E. Bulawa, N. Iartchouk, K. R. Winter, P. J. Riggle, J. C. Rhodes, C. L. Dodge, D. W. Culp, and P. T. Borgia.** 1996. The *chsD* and *chsE* genes of *Aspergillus nidulans* and their roles in chitin synthesis. *Fungal Genet. Biol.* **20:**153–167.

134. **Staben, C., B. Jensen, M. Singen, J. Pollock, M. Schechtman, and J. Kinsey.** 1989. Use of a bacterial hygromycin B resistance gene as a dominant selectable marker in *Neurospora crassa* transformation. *Fungal Genet. Newsl.* **36:**79–81.

135. **Struhl, K.** 1983. The new yeast genetics. *Nature* **305:** 391–397.

136. **Tilburn, J., F. Roussel, and C. Scazzocchio.** 1990. Insertional inactivation and cloning of the *wA* gene of *Aspergillus nidulans*. *Genetics* **126:**81–90.

137. **Tilburn, J., C. Scazzocchio, G. G. Taylor, J. H. Zabicky-Zissman, R. A. Lockington, and R. W. Davies.** 1983. Transformation by integration in *Aspergillus nidulans*. *Gene* **26:**205–221.

138. **Timberlake, W. E.** 1990. Molecular genetics of *Aspergillus nidulans*. *Annu. Rev. Genet.* **24:**5–36.

139. **Timberlake, W. E.** 1991. Cloning and analysis of fungal genes, p. 51–83. *In* J. W. B. M. Alic and L. L. Lasure (ed.), *More Gene Manipulations in Fungi*. Academic Press, San Diego, Calif.

140. **Timberlake, W. E., M. T. Boylan, M. B. Cooley, P. M. Mirabito, E. B. O'Hara, and C. E. Willett.** 1985. Rapid identification of mutation-complementing restriction fragments from *Aspergillus nidulans* cosmids. *Exp. Mycol.* **9:**351–355.

141. **Timberlake, W. E., and M. A. Marshall.** 1989. Genetic engineering of filamentous fungi. *Science* **244:**1313–1317.

142. **Tollervey, D. W., and H. N. Arst, Jr.** 1981. Mutations to constitutivity and derepression are separate and separable in a regulatory gene of *Aspergillus nidulans*. *Curr. Genet.* **6:**79–85.

143. **Unkles, S. E., E. I. Campbell, D. Carrez, C. Grieve, R. Contreras, W. Fiers, C. A. M. J. J. Van den Hondel, and J. R. Kinghorn.** 1989. Transformation of *Aspergillus niger* with the homologous nitrate reductase gene. *Gene* **78:**157–166.

144. **Unkles, S. E., E. I. Campbell, Y. M. J. T. De Ruiter-Jacobs, M. Broekhuijsen, J. A. Macro, D. Carrez, R. Contreras, C. A. M. J. J. Van den Hondel, and J. R. Kinghorn.** 1989. The development of a homologous transformation system for *Aspergillus oryzae* based on the nitrate assimilation pathway: a convenient and general selection system for filamentous fungal transformation. *Mol. Gen. Genet.* **218:**99–104.

145. **Van den Hondel, C. A. M. J. J., and P. J. Punt.** 1991. Gene transfer systems and vector development for filamentous fungi, p. 1–28. *In* C. E. C. J. F. Peberdy, J. E. Ogden, and J. W. Bennett (ed.), *Applied Molecular Genetics of Fungi*. Cambridge University Press, Cambridge.

146. **Van Gorcom, R. F. M., P. H. Pouwels, T. Goosen, J. Visser, H. W. J. van den Broek, J. E. Hamer, W. E. Timberlake, and C. A. M. J. J. Van den Hondel.** 1985. Expression of an *Escherichia coli* β-galactosidase fusion gene in *Aspergillus nidulans*. *Gene* **40:**99–106.

147. **Van Gorcom, R. F. M., P. J. Punt, P. H. Pouwels, and C. A. M. J. J. Van den Hondel.** 1986. A system for the analysis of expression signals in *Aspergillus*. *Gene* **48:** 211–217.

148. **Van Haaren, M. J., and D. W. Ow.** 1993. Prospects of applying a combination of DNA transposition and site-specific recombination in plants: a strategy for gene identification and cloning. *Plant Mol. Biol.* **23:**525–533.

149. **Vapnek, D., J. A. Hautala, J. W. Jacobson, N. H. Giles, and S. R. Kushner.** 1977. Expression in *Escherichia coli* K12 of the structural gene for catabolic dehydroquinase of *Neurospora crassa*. *Proc. Natl. Acad. Sci. USA* **74:** 3508–3512.

150. **Verdoes, J. C., P. J. Punt, P. van der Berg, F. Debets, A. H. Stouthamer, and C. A. M. J. J. van den Hondel.** 1994. Characterization of an efficient gene cloning strategy for *Aspergillus niger* based on an autonomously replicating plasmid: cloning of the *nicB* gene of *A. niger*. *Gene* **146:**159–165.

151. **Vollmer, S. J., and C. Yanofsky.** 1986. Efficient cloning of genes of *Neurospora crassa*. *Proc. Natl. Acad. Sci. USA* **83:**4869–4873.

152. **Ward, P. P., G. S. May, D. R. Headon, and C. M. Conneely.** 1992. An inducible expression system for the production of human lactoferrin in *Aspergillus nidulans*. *Gene* **122:**219–223.

153. **Watts, J. W., and N. J. Stacey.** 1991. Novel methods of DNA transfer, p. 44–65. *In* C. E. C. J. F. Peberdy, J. E. Ogden, and J. W. Bennett. (ed.), *Applied Molecular Genetics of Fungi*. Cambridge University Press, Cambridge.

154. **Weltring, K. M.** 1995. A method for easy isolation of promoter fragments from promoter-probe libraries of filamentous fungi. *Curr. Genet.* **28:**190–196.

155. **Weltring, K.-M., B. G. Turgeon, O. C. Yoder, and H. D. VanEtten.** 1988. Isolation of a phytoalexin-detoxification gene from the plant pathogenic fungus *Nectria haematococca* by detecting its expression in *Aspergillus nidulans*. *Gene* **68:**335–344.

156. **Wernars, N., T. Goosen, L. M. J. Wennekes, J. Visser, C. J. Bos, H. W. J. Van den Broek, R. F. M. Van Gorcom, C. A. M. J. J. Van den Hondel, and P. H. Pouwels.** 1985. Gene amplification in *Aspergillus nidulans* by transformation with vectors containing the *amdS* gene. *Curr. Genet.* **9:**361–368.

157. **Yanai, K., N. Kojima, N. Takaya, H. Horiuchi, A. Ohta, and M. Takaki.** 1994. Isolation and characterization of two chitin synthase genes from *Aspergillus nidulans*. *Biosci. Biotechnol. Biochem.* **58:**1828–1835.

158. **Yeadon, P. J., and D. E. Catcheside.** 1995. Guest: a 98 bp inverted repeat transposable element in *Neurospora crassa*. *Mol. Gen. Genet.* **247:**105–109.

159. **Yelton, M. M., J. E. Hamer, E. R. De Souza, E. J. Mullaney, and W. E. Timberlake.** 1983. Developmental regulation of the *Aspergillus nidulans trpC* gene. *Proc. Natl. Acad. Sci. USA* **80:**7576–7580.

160. **Yelton, M. M., J. E. Hamer, and W. E. Timberlake.** 1984. Transformation of *Aspergillus nidulans* by using a *trpC* plasmid. *Proc. Natl. Acad. Sci. USA* **81:**1470–1474.

161. **Yelton, M. M., W. E. Timberlake, and C. A. M. J. J. van den Hondel.** 1985. A cosmid for selecting genes by complementation in *Aspergillus nidulans*: selection of the developmentally regulated *yA* locus. *Proc. Natl. Acad. Sci. USA* **82:**834–838.

162. **Yu, J.-H., and T. H. Adams.** Unpublished results.

163. **Yu, J.-H., J. Wieser, and T. H. Adams.** 1996. The *Aspergillus* FlbA RGS domain protein antagonizes G protein signaling to block proliferation and allow development. *EMBO J.* **15:**5184–5190.

Saccharomyces cerevisiae: Genetics and Genomics

ANNA ASTROMOFF AND MARK EGERTON

35

The budding yeast *Saccharomyces cerevisiae* has gained widespread acceptance as an experimental system. Moreover, its use has resulted in landmark discoveries in the investigation of numerous fundamental biological processes (e.g., transcription, translation, cell cycle, membrane transport), and in some of these fields it has emerged as the organism of choice. The key attributes that have propelled it to this status include the ease with which the organism can be grown, its highly developed molecular genetic systems, and the availability of the complete DNA sequence of its genome. The simplicity with which a yeast cell can now be manipulated is unparalleled for any other eukaryote, illustrated by the fact that mammalian genes are routinely introduced into yeast cells for systematic investigation of gene function. The yeast research community is well developed, with several thousand researchers across the world actively using the organism. Conveniently for these workers, and for individuals wishing to enter the field, much of the progress being made with *S. cerevisiae* is captured and collated at two web sites, one maintained by Stanford University (http://genome-www.stanford.edu) and the other by the Munich Information Center for Protein Sequences (http://speedy.mips.biochem.mpg.de/).

The objective of this chapter is to provide the reader with an overview of *S. cerevisiae* as an experimental system, with particular emphasis on the most recent developments that have been stimulated by completion of the genome sequence. It is beyond the scope of this chapter to provide a comprehensive review of basic experimental protocols or results obtained from particular studies. Where necessary, the reader is referred to original research articles or review papers that are germane to that topic. Several reviews have been written on yeast genetics and molecular biology (e.g., see references 1 and 30), and the reader can also gather further information at the web sites mentioned above.

35.1. THE MOLECULAR GENETICS OF *S. CEREVISIAE*

S. cerevisiae has a highly developed molecular genetic system that has been used for decades to identify the molecular components that participate in key biological pathways. Fundamental to this approach has been the ability to isolate mutants, classify mutants into complementation groups, characterize their phenotype, and clone the corresponding gene from plasmid libraries.

Strains of *S. cerevisiae* that are commonly used in the laboratory can exist in either a haploid or a diploid state. Two haploid cell types, referred to as mating type **a** (*MAT***a**) or mating type α (*MAT*α), can mate together to form a *MAT***a**/α diploid strain that is stable under most conditions. When exposed to the appropriate conditions, diploid cells can be induced to enter meiosis to yield four haploid progeny. The formation of a diploid strain from two haploids is regularly used to establish complementation groups and to assign a recessive or dominant phenotype to particular mutant alleles. This information uncovers the number of genes identified in a mutant screen and can be used to devise an experimental strategy to clone the corresponding gene.

S. cerevisiae plasmids are generally shuttle vectors that can be propagated in either *Escherichia coli* or *S. cerevisiae* (1, 30). Plasmids will, therefore, carry origins of replication and selectable markers for maintenance in bacteria and in yeasts. Families of plasmids have been developed that integrate directly into the yeast genome or exist autonomously, either in single copy or in multicopy (e.g., see reference 62). The ease with which plasmids can be introduced into yeast cells means that almost any gene corresponding to a mutant phenotype can be isolated via complementation with a plasmid library. Furthermore, using these plasmids, either by themselves or in combination with an inducible promoter, provides an opportunity to overexpress homologous or heterologous genes (55).

Integration of transformed DNA into the genome occurs via homologous recombination (57). Gene knockout or replacement techniques have, therefore, been developed that allow the wild-type version of a gene to be replaced with an altered or disrupted copy of the gene (see below). The phenotype that is associated with a particular mutant allele, or complete loss of the function, can thus be studied in either haploid or diploid cells. The efficiency with which these techniques can be carried out has increased to the extent that gene knockout programs can be considered on a genome-wide scale (see below).

35.2. OVERVIEW OF THE *S. CEREVISIAE* GENOME

The complete genome sequence of *S. cerevisiae* was deposited in public databases in April 1996 (28, 29). This event was significant not only because it represented the first eukaryote for which the genome sequence had been determined but also because of the manner in which it had been obtained. The genome of *S. cerevisiae* comprises 16 chromosomes, which were sequenced by a consortium of laboratories from three continents. This effort emphasizes the collaborative nature of *S. cerevisiae* research.

The 16 chromosomes range in size from 200 to 2,200 kb, and the total genome sequence deposited in the database is 12.1 Mb. Analysis of the complete genome sequence confirmed many previous studies but also revealed many surprises and challenged a number of preconceived views with respect to the content of the genome. The genome is apparently rich in information, with over 70% devoted to coding sequences (genes) (20). In total, 6,215 potential protein coding sequences have been identified, although this figure will evolve as our ability to predict genes improves. The average size of an open reading frame (ORF) is 1,450 bp, with an average integenic distance of 472 bp. Introns are rare, and those that exist are short (the longest is 1 kb) and primarily found at the 5′ ends of the gene (20).

Perhaps the most striking revelation from the genome sequence was the number of genes whose existence had gone undetected prior to the sequencing project. At the time that the sequence was released, only 30% of the predicted coding sequences had been discovered by conventional approaches and therefore could have some biological function assigned to them. For part of the remaining 70%, some functional information could be inferred from homology to other genes of known function. However, one-third of the genes either had no significant homology to any entry in the sequence databases or homology to uncharacterized or hypothetical ORFs (20).

Why had so many ORFs previously gone undetected? It is possible that these ORFs are not expressed and have no biological function. However, transcripts have been detected for 87% of the genes in the genome, indicating that this is probably not the case (73). The likeliest explanation is based on the way that previous genetic screens were conducted. Most screens were configured to identify genes that play a major role in the pathway under investigation and were therefore biased toward mutants with strong phenotypes, e.g., conditional lethal mutants. In these screens, mutations that resulted in a weak phenotype would have been excluded or undetected. It is also possible that many of the novel genes are required for growth under conditions not yet reproduced in the laboratory. A future challenge to geneticists is to identify more refined or completely different screening conditions that allow them to uncover the function of these novel genes.

An additional obstacle to genetic analysis is the high level of sequence redundancy and therefore the potential functional redundancy in the yeast genome. Sequence similarity analysis indicates that approximately 25% of yeast genes exist as homologous gene pairs (74). Of these, 376 gene pairs reside in regions of the genome referred to as cluster homology regions (CHRs). These are defined as regions having three or more homologous genes in the same order and relative orientation and no more than 50 kb apart. CHRs vary in length from 7.5 to 170 kb and occupy approximately 50% of the genome. Many CHRs have gained or lost components (genes, introns, transposable elements, and pseudogenes) with respect to one another so that within the CHRs only 25% of genes are duplicated. It has been proposed that the CHRs were generated simultaneously in a genome-wide duplication event approximately 100 million years ago (74). There are also a large number of homologous gene pairs located outside the CHRs as well as a smaller number of gene families. Some of these probably originated at the time of the ancient duplication while the origins of others are more recent (29). The relationship between sequence and functional redundancy is not known, but it is almost certain that functional redundancy has contributed to the failure of genetic screens to identify a significant number of genes in the yeast genome. Consistent with this hypothesis, a survey of essential genes has shown that only 2.5% are found in CHRs (71). Presumably, the homologs of these essential genes are either not functional or have evolved a different function.

The completion of the genome sequence has provided a genetic blueprint for how a yeast cell functions. With this in hand, systematic approaches can now be initiated to investigate the function of each gene within the genome. Large-scale approaches of this nature demand the development of new technologies that could not have been considered prior to obtaining the complete genome sequence. We are now in the initial phase of functional analysis studies on a genome-wide scale. We anticipate that these studies will provide a wealth of information about the yeast cell and will serve as models for experimental strategies in other organisms when their genome sequences become available.

35.3. POST-GENOME ANALYSIS OF *S. CEREVISIAE*

The first steps toward understanding the function of a gene involve obtaining a number of key pieces of information about that gene. For each gene, it is necessary to characterize gene expression, protein expression, protein modification, protein-protein interactions, and the consequences to the cell of losing that gene. Unlike the genome sequence, these data sets are dynamic and change in response to external and internal factors such as culture conditions and the presence of mutations. To perform these sorts of analyses on a genome-wide scale, new technologies have been developed that allow these parameters to be monitored for all 6,000 genes simultaneously. The trends in all of these technologies are toward increased throughput, miniaturization, and parallelization, all of which will result in the rapid generation of vast amounts of data. Collating and integrating this information represents a significant challenge, but the outcome will be a much more complete picture of how a yeast cell functions.

35.3.1. Characterization of the Yeast Transcriptome

The transcriptome is defined as the set of mRNAs that are present within a given cell. Characterization of the yeast transcriptome in a variety of conditions and genetic backgrounds will be essential for understanding regulatory pathways and networks in *S. cerevisiae*. Traditionally, gene expression has been studied by Northern hybridization (3), which is limited in the number of genes that can be easily analyzed in any one experiment. Expression technologies have been developed that allow the transcriptome to be characterized simultaneously either in parallel or serial fashion.

35.3.1.1. Parallel Analysis of Gene Expression Using Array Technologies

The ability to monitor expression of the complete genome in parallel was realized by the development of DNA array technologies (for review see reference 58). Gene expression arrays consist of probes immobilized at specified locations on a glass surface. Each probe represents a single gene, and every gene in the genome is represented on the array. The two common formats of expression arrays use either a DNA fragment (59, 60) or an oligonucleotide (44) as the probe. These are commonly referred to as DNA microarrays and oligonucleotide arrays, respectively (Fig. 1). In both methods the level of mRNA in a target sample is measured by converting the population of mRNAs in the sample to fluorescently labeled cDNA, cRNA, or RNA and hybridizing the entire population to the array. Hybridization is detected using a scanning confocal fluorescence microscope.

The two types of arrays differ significantly in the way they are constructed. DNA microarrays consist of probes that are spotted on a glass surface. Oligonucleotide arrays consist of 20- to 25-mer oligonucleotide probes, which can be synthesized directly on a silicon wafer using light-directed combinatorial chemistry (23, 24, 51). Other methods of generating oligonucleotide arrays have been developed (reviewed in reference 58). Access to the complete genome sequence allows primers to be designed in an automated fashion for every ORF that has been identified (43). Yeast expression microarrays contain double-stranded DNA probes generated by PCR using gene-specific primers. On the oligonucleotide arrays each ORF is represented by up to 20 different 25-mer probes. Both array formats require specialized equipment for their construction, but prefabricated oligonucleotide arrays are available from Affymetrix (Santa Clara, Calif.).

The DNA microarray and oligonucleotide microarray technologies also differ in the detection system used. In oligonucleotide array technology, a single-color detection system is used, and the abundance of a transcript within a sample is calculated by averaging the signal from all of the probes on the array that represent that particular gene. The data are quantitative and reproducible, so different samples can be hybridized to different arrays and compared directly. In DNA microarray technology, a two-color detection system is used, allowing simultaneous hybridization and analysis of two comparative samples. The arrays are therefore used to determine the relative abundance of transcripts in the two samples. The two-color detection system compensates for variation in the amount of DNA deposited on the array, variations in hybridization characteristics of the different DNA sequences, and variations in hybridization and wash conditions between experiments (60). In both technologies absolute RNA concentrations can be determined by comparison with internal standards such as the *TBP1* gene (TATA binding protein) (37), whose absolute expression level in rich medium has been determined.

These technologies have been used to characterize gene expression in cells grown on rich or minimal medium (73), during different phases of the growth cycle (19), in the cell division cycle (12), and during sporulation cycles (13). Both technologies are capable of detecting mRNAs present at a copy number as low as 1 per 10 cells, corresponding to a detection specificity of approximately 1:150,000 to 1:300,000 (18, 73). In the first study above (73), transcripts were detected for 87% of predicted genes when cells were grown on rich medium. Many of the transcripts corresponded to hypothetical ORFs identified by the genome sequencing project, confirming that these ORFs do indeed represent authentic genes. The 13% of the ORFs that were not detected were either not expressed in these conditions or were expressed at less than 1 copy per 10 cells. In addition, some of these ORFs may not represent authentic genes. In all of the studies, expression information was obtained both for genes that were well characterized and for genes that were novel. In many cases, new patterns of regulation were observed.

35.3.1.2. Serial Analysis of Gene Expression

Analysis of gene expression can also be performed by the sequential sequencing of cDNAs derived from a population of mRNA. This approach measures genome-wide expression patterns without the need to generate hybridization probes for each gene and can be used by any laboratory with the capability to perform high-throughput DNA sequencing. The first embodiment of this approach was the generation of expressed sequence tags, which are generated by sequencing approximately 250 bases from the 3′ ends of cDNA molecules (69). The frequency with which a particular 3′ end is sequenced in a given population is an indication of its relative abundance within the population.

The approach was enhanced by the development of SAGE technology (serial analysis of gene expression) (67). In this technique, 11 to 13 base sequence tags are isolated from defined locations at the 3′ ends of individual mRNAs, concatenated, and sequenced. This allows tags from multiple RNAs to be sequenced in the same sequencing reaction. The frequency with which a particular tag appears in the concatenated sequence indicates the abundance of the corresponding transcript. In the first study of yeast gene expression that used this technique, approximately 60,000 tags were analyzed from cells in log phase, S phase, and the G_2/M transition (68). Expression information was obtained for a total of 4,665 genes, representing 76% of the total number of predicted genes. Comparison of the log phase, S phase arrested, and G_2/M arrested cells indicated that there were only 29 transcripts whose abundance varied more than 10-fold among the three different states. Analysis of the data indicated that expression levels ranging from 0.3 to over 200 transcripts per cell were detected and that the number of unique transcripts detected reached a plateau at around 60,000 tags. This is consistent with 60,000 tags representing fourfold coverage for those genes expressed at one transcript per cell or more.

The study described above (68) indicates that SAGE cannot identify messages below 0.3 copy per cell and thus does not appear to be as sensitive as array technologies. A clear advantage of SAGE, however, is that it does not rely on previous identification of ORFs within the genome sequence and can therefore detect transcripts from nonannotated ORFs. The study identified 160 expressed ORFs that had not been annotated in the genome project because of their small size. SAGE technology can therefore be used in systems where no genome sequence is available, but it is also a valuable tool for the identification of genes missed by in silico analysis of a genome sequence.

35.3.2. Characterization of the Yeast Proteome

The proteome is defined as the complete set of proteins within a cell. Characterization of the proteome complements the information obtained in mRNA-based expres-

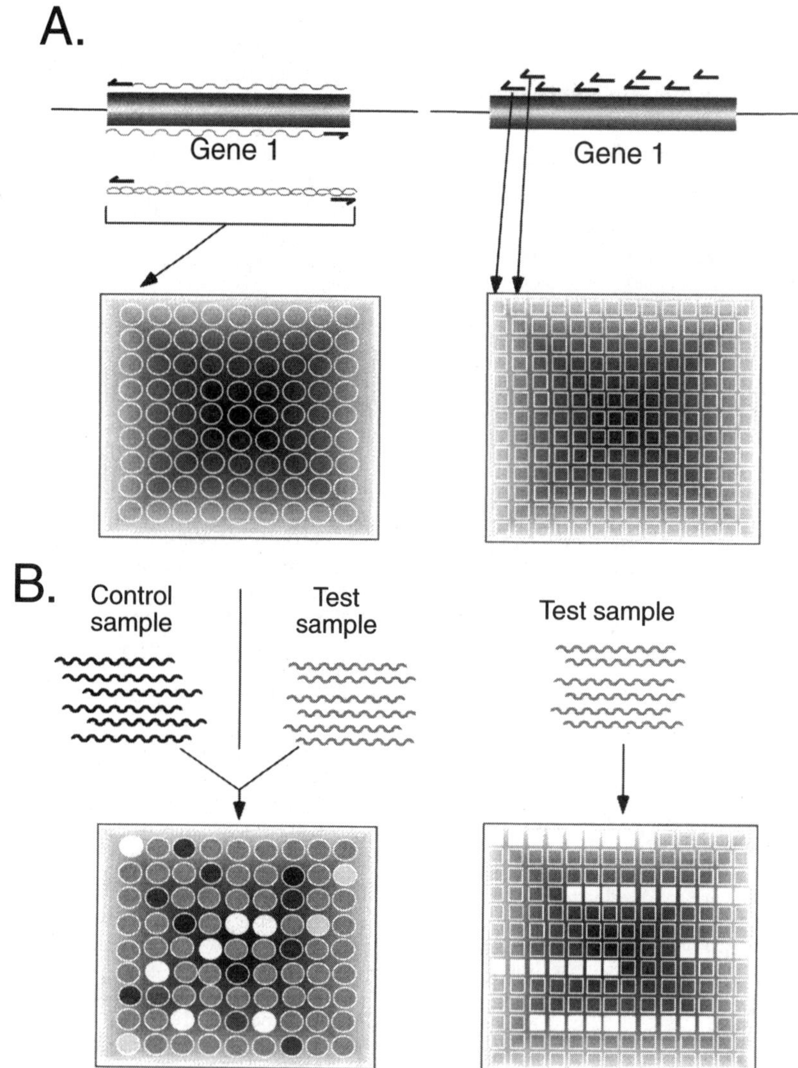

FIGURE 1 Comparison of DNA microarrays (left) and oligonucleotide arrays (right). (A) Probes for the yeast expression microarrays are generated by PCR amplification of genes from genomic DNA using gene-specific primers. Probes are then deposited at known locations on the array. Yeast expression oligonucleotide arrays contain up to 20 different oligonucleotide probes for each gene. Probes are synthesized directly on the solid support. (B) In microarray analysis, control (e.g., t = 0) and test samples (target) are labeled with different fluors and hybridized competitively to the array. Probes that hybridize predominantly to the control template (black) represent genes expressed at higher levels in the control sample than in the test sample. Probes that hybridize predominantly to the test template (white) represent genes that are induced under the test condition. Probes shown in gray represent genes that are not differentially expressed under the two conditions. In oligonucleotide array analysis, one sample is hybridized to the array. Probes complementary to expressed genes produce a fluorescent signal whose intensity is proportional to the amount of mRNA in the test sample

sion analysis. Like the transcriptome, the proteome is dynamic and will change in response to different stimuli. Therefore, one would like to monitor the proteome in a variety of conditions and genetic backgrounds, in terms of the level of protein expression, the posttranslational modifications, and the protein-protein interactions that occur within a cell.

35.3.2.1. A Biochemical Approach

The traditional approach used to investigate the proteome consisted of two-dimensional gel separation of proteins (49) followed by identification of the proteins via sequencing of proteolytic fragments. Mass spectrometry (MS) has now become the method of choice for protein identification in the proteomics field. The development of new ionization methods such as matrix-assisted laser desorption ionization and nanoelectrospray ionization has increased the sensitivity and accessible mass range of MS. In combination with numerous types of mass analyzers, these technologies allow peptide mass fingerprints or sequence tags to be rapidly generated for each spot within a two-dimensional protein gel (reviewed in references 22, 38, and

54). The identities of the proteins can then be determined by comparison of their experimentally determined fingerprints or tags with those predicted from the genome sequence. MS does not have to be linked to two-dimensional gel electrophoresis separations, and future studies may involve alternative separation or fractionation procedures. MS can also be used for other proteomic analyses, including the identification and mapping of secondary modifications and protein-protein interactions. Proteomics is now poised to provide a biochemical correlate of protein-protein interactions identified via the genetic approach described below.

35.3.2.2. A Genetic Approach

The development of the two-hybrid system, a genetic approach for identifying protein-protein interactions, was pioneered in *S. cerevisiae* (21) (Fig. 2). This system uses the transcription of a reporter gene to assay protein-protein interactions and is based on two key observations about transcriptional activators. These observations are (i) that the DNA-binding and transcriptional activation functions are separable and function independently (9, 35, 39) and (ii) that a functional transcriptional activator can be assembled via noncovalent interaction of the DNA-binding and activation domains (45).

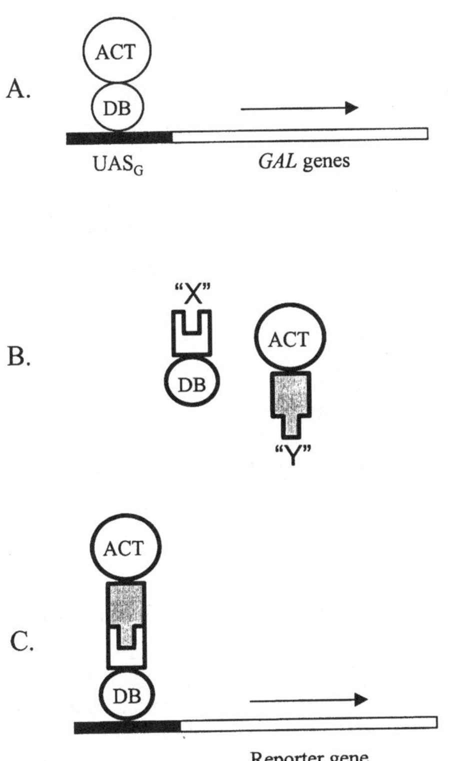

FIGURE 2 The yeast two-hybrid system. (A) Transcription factors are modular. DNA-binding and transcription activation functions can reside in different domains. Functional transcription factors can be generated from heterologous DNA-binding (DB) and activation (ACT) domains via covalent or noncovalent association. A common version of the two-hybrid strategy uses the transcription activator Gal4, which binds to the upstream activating sequence (UAS) upstream of the *GAL* genes. (B) To identify protein-protein interactions, the protein of interest (protein X) is fused to the DNA-binding domain and serves as the bait. The bait is then used to screen a library of preys in which proteins (protien Y) are fused to the activating domain. (C) Reconstitution of a functional transcription activator, indicating interaction between X and Y, is detected using a yeast strain in which a reporter gene is placed under the control of a UAS_G.

The system provides an opportunity to rapidly identify and characterize gene products that interact with a protein(s) of interest. This protein (protein X, the bait) is fused to the DNA-binding domain of the transcription factor Gal4. Potentially interacting proteins (protein Y, the prey) are fused to the domain of Gal4 that activates transcription. If proteins X and Y interact, a functional transcription factor is reconstituted and can direct expression of a reporter gene. Versions of the two-hybrid system that use alternative DNA-binding and activation domains have not been developed (31, 47). The initial two-hybrid system used the *E. coli* reporter gene *lacZ*, whose expression can be easily monitored. Selectable markers such as *HIS3*, *LEU2*, and *URA3* have subsequently been developed as reporter genes, and most two-hybrid systems now incorporate one of these in addition to the *lacZ* gene (reviewed in reference 26). This makes it possible to perform a selection step followed by a screening step, which increases the efficiency of the procedure and decreases the frequency of false positives. When a protein-protein interaction has been identified, the same system can be used to identify the amino acid residues or domains that participate in the interaction or to identify molecules that inhibit the interaction.

The availability of a genome sequence makes it possible to use the two-hybrid system to systematically identify protein-protein interactions. This type of approach was first exemplified for *E. coli* bacteriophage T7 (4). For *S. cerevisiae*, baits and preys for each gene can now be constructed in a directed manner, and every combination of bait and prey can be tested systematically (see reference 26). In an alternative iterative approach, large prey libraries can be screened with small number of bait fusions (27). Ultimately, this will allow us to construct a map of protein-protein interactions in the yeast cell.

35.3.3. Mutational Analysis of the Yeast Genome

The phenotype of a yeast cell in which a gene has been mutated is an important indicator of the function of the encoded protein. In a genetic screen the ultimate goal is to evaluate the phenotype of a complete set of mutants. Having a complete genome sequence provides a route toward the construction of a mutant allele for every gene. Three large-scale projects are currently in progress with the objective of generating and characterizing genome-wide sets of mutants. The first of these projects utilizes gene knockout technology to create a complete set of deletion strains that can be analyzed in parallel; the two other projects use transposon insertions to mutate and/or tag genes. It is anticipated that the three projects will provide not only useful overlapping and complementary data sets but also reagent sets for future functional analysis studies.

35.3.3.1. Parallel Analysis of Gene Deletion Strains

Deletion of genes in *S. cerevisiae* is routinely accomplished by transformation with a linear DNA fragment that replaces a target sequence in the chromosome by homologous recombination (6, 57). The DNA fragment generally consists of a selectable marker, flanked by targeting sequences that are homologous to the regions immediately upstream and downstream of the gene to be deleted (Fig. 3A). A consortium of European and North American laboratories has been organized with the purpose of generating a complete and uniform collection of yeast deletion strains. This has become a realizable goal because of the ability to rapidly generate deletion cassettes by PCR (70). The strains are released to the scientific community as they are being made, and the complete collection will be available in the next 1 to 2 years. For each gene, four different deletion strains will be generated: two haploids (*MAT***a** and *MAT*α), a heterozygous diploid, and a homozygous diploid.

The goal of the consortium is to delete each individual ORF from start to stop codon. The 6,000 deletion cassettes (one for each gene) are engineered to contain a heterologous selectable marker (Kan^R), flanked by regions of homology to direct integration into the genome. Each cassette also contains two unique 20-nucleotide tags, which are physically linked to specific deletions by virtue of being incorporated into the PCR primers containing the yeast homology. The presence of the tags in the deletion strains allows large numbers of strains, potentially all 6,000, to be pooled and analyzed in parallel (Fig. 3B). Individual strains in the resulting mixed population can be detected via amplification and hybridization of the identifier tags to an oligonucleotide array. This array contains, at defined positions, oligonucleotides that are complementary to the set of tag sequences used in the deletion cassettes. To test the phenotype of strains under a particular selection, strains are pooled together and grown for multiple generations. Mutants that have decreased levels of fitness, due to even subtle effects on growth, will over time become underrepresented in the population. The relative abundance of different strains over the course of the selection can be determined from the hybridization signal generated by each tag on the oligonucleotide array.

In the pilot study validating this technique, the growth of 11 auxotrophic mutants was carried out in a competitive assay in selective growth conditions (61). More recently, the approach has been extended to the analysis of a pool of the first 500 deletion strains generated by the deletion project (71). The relative fitness of each strain in both rich and minimal medium was determined. Many of the deletion strains had a growth defect in both media, and a smaller number had a growth defect specific to minimal medium. Each class included both known and unknown genes.

The construction of a complete set of deletion strains is labor-intensive, but once this set is available it will be an invaluable experimental resource. The tags allow the phenotypes of the set of 6,000 mutants to be tested quickly under a large number of selective conditions. In addition, experimental strategies can now be devised to characterize the transcriptome and proteome in every mutant strain. These profiles serve as molecular signatures for loss of the corresponding gene function. Comparing the complete set of expression profiles will uncover groups of genes that have similar signature profiles and can therefore be hypothesized to have similar functions.

35.3.3.2. Transposon-Based Strategies

Transposons are another tool for making genome-wide sets of mutants. In these procedures a large number of insertion mutants are generated by random integration of transposons into the genome. Insertion of a transposon into a coding sequence frequently disrupts the function of that gene and generates a mutant phenotype. These approaches differ from the deletion strategy described above in that both complete and partial loss-of-function alleles can be obtained. Transposons can also be engineered to contain a reporter gene that will allow transcription and/or translation of a tagged ORF to be monitored.

Genetic footprinting is a protocol that was developed to generate and analyze a large number of insertion mutants using the yeast transposon Ty1 (63). In this protocol, a wild-type yeast strain is transformed with a plasmid carrying the transposon. Transposition is induced transiently using conditions that lead to one or a few transposition events per cell, resulting in a population in which insertion events into each gene are represented. This population can then be subjected to growth in different selective conditions. As described for the deletion strategy above, mutant strains with reduced fitness will disappear from the population over the course of the experiment. The presence and relative abundance of individual transposon insertions in the population can be determined using a PCR protocol. Transposon insertions that affect the ability of a cell to grow under the selective condition will disappear at a rate determined by the magnitude of their effect on fitness.

In the first application of genetic footprinting, the effect of transposon insertions into all of the predicted ORFs on chromosome V was studied (64). In the seven different selection conditions employed, growth defects were detected for 157 of the 255 genes analyzed. For the ORFs that were included in both this and the deletion study (71), results were generally in good agreement.

Multipurpose transposons have been developed that allow gene expression, protein localization, and disruption phenotypes to be investigated (56). These transposons contain a reporter gene for assaying gene expression and an epitope tag that can be used in immunolocalization experiments. To date, more than 1,000 strains containing transposon insertions have been generated and are available to the public either individually or as ordered sets. Protein localization studies have been performed with many of these strains and have provided subcellular localization data for many novel gene products identified through the sequencing project (http://ycmi.med.yale.edu/YGAC/home.html).

35.3.4. Detection of Allelic Variation

The *S. cerevisiae* genome sequence was derived from the commonly used laboratory strain S288C. This strain is predicted to differ in its genome sequence from other laboratory strains by up to 1% (48) and has significant phenotypic differences from nonlaboratory strains. Analysis of genetic variation between strains will provide insight into the origins of their phenotypic differences and will help us understand evolutionary relationships between strains.

Oligonucleotide arrays can be used to detect and map polymorphisms between the sequenced strain of *S. cerevisiae* and any other strain (72). As described above, these arrays contain as many as 20 oligonucleotide probes for each gene. Polymorphisms in genomic DNA are detected by differences in the array hybridization pattern of genomic DNA from the two strains. Probes that consistently give a high signal in the laboratory strain and little or no signal in the test strain correspond to sites of allelic variation. Hybridization of DNA from segregants of a cross between the two strains can then be used to map phenotypic traits. This approach has been used to map the allelic variation

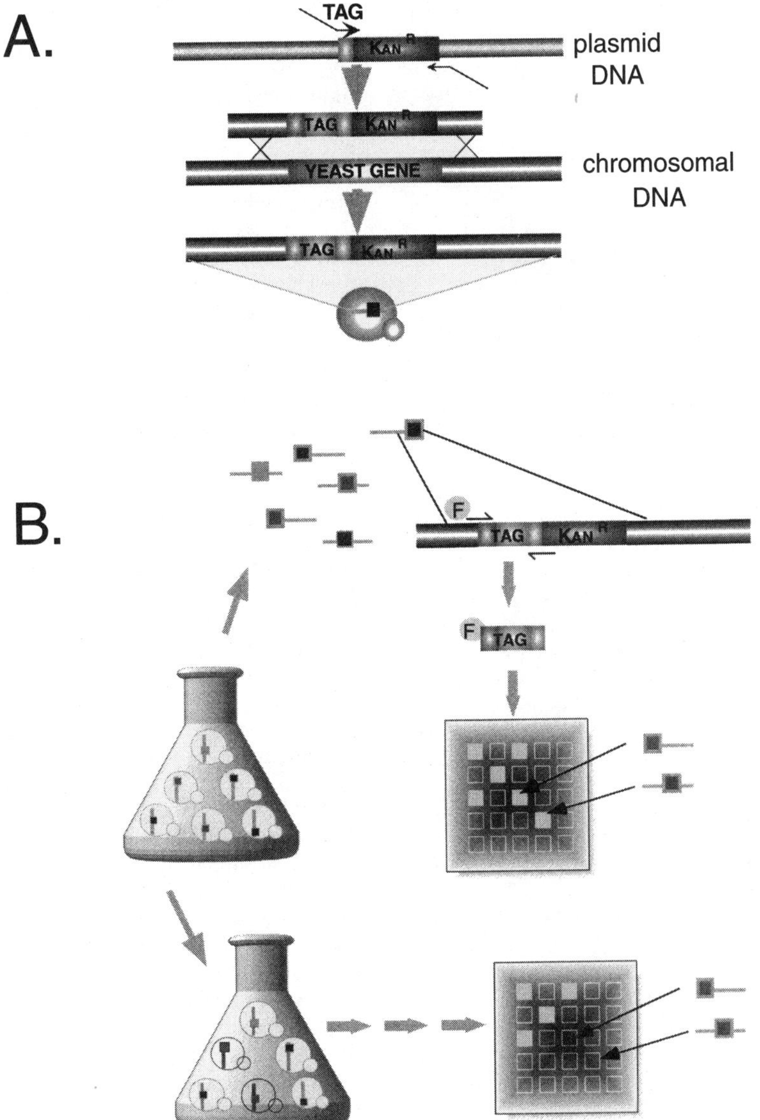

FIGURE 3 Parallel analysis of tagged deletion strains. (A) A deletion construct is generated for each gene by PCR amplification of the selectable marker. The PCR primers include (5′ to 3′) targeting sequences, unique oligonucleotide tags flanked by common amplification sequences, and homology to the marker. The yeast gene is replaced from start to stop via a double recombination event between the deletion construct and the yeast chromosome. Each yeast strain therefore contains a unique tag at the site of the deletion. (B) At the beginning of the experiment (t = 0), strains are present in the population at equal concentrations. After generations of growth under selective conditions, strains with reduced fitness become underrepresented in the population. The tags are PCR amplified from genomic DNA isolated from different samples taken over the course of the experiment. The PCR primers are fluorescently labeled, and the PCR amplicons are hybridized to the array. The array is composed of a set of oligonucleotides that corresponds to the set of tags used in the deletion project. In the t = 0 hybridization, all of the tagged strains produce equal intensity signals on the array. After selective growth, strains that have persisted produce a signal, and strains that have been depleted generate either a reduced signal or no signal.

between two strains that resulted in different sensitivities to cycloheximide. The high density of markers and the ease with which they can be scored also make this technique attractive for the mapping of quantitative or multigenic traits. This technology can now be used to identify loci that determine characteristics such as virulence or the generation of secondary metabolites. It will, therefore, facilitate the optimization of strains for use in industrial processes such as brewing.

35.4. USING *S. CEREVISIAE* AS A MODEL SYSTEM

35.4.1. Understanding Gene Function in Higher Eukaryotes

The relatively low number of genes in its genome, the extent of similarity between its genes and those of higher eukaryotes, and the tractability of its experimental system make *S. cerevisiae* an excellent model for other eukaryotic organisms. Comparative genomic analysis can be used to identify yeast genes that are related in sequence to genes in other organisms and are therefore potential functional homologs. Alternatively, heterologous genes and potentially complete pathways can be introduced into *S. cerevisiae* for functional analysis studies (55).

The utility of *S. cerevisiae* as a system to understand gene function in higher eukaryotes is exemplified by two studies that compared the sequence of genes within *S. cerevisiae* to those of other organisms. Comparison of all the predicted yeast protein sequences with mammalian sequences in the Genbank database revealed that approximately 31% of yeast genes have a mammalian homolog (using the BLAST algorithm and a P value of $<e^{-10}$ as a significant hit) (8). Although this analysis is encouraging, it almost certainly underestimates the true degree of sequence conservation, because mammalian expressed sequence tags were excluded from the analysis. An alternative approach, focused on genes that have been associated with human disease, produced results consistent with this study (25). Approximately 2,000 genetic diseases have been recorded in the OMIM database (www.ncbi.nlm.nih.gov/Omim), from which 250 disease-associated genes have been identified. From this subset, 105 (42%) have a homolog in *S. cerevisiae* with a P value of $<e^{-10}$, and for 74 of these the P value is $<e^{-40}$. In this analysis, the majority of yeast genes with a human homolog encode proteins involved in major metabolic pathways. Those human genes with no homolog in *S. cerevisiae* encode receptors, components of the blood or immune systems (which would not be anticipated to be found in *S. cerevisiae*), or bear no similarity to any gene of known function. A public database (XREFdb; www.ncbi.nlm.nih.gov/XREFdb/) has now been developed to provide comparative genomic analyses of key model organisms, including *S. cerevisiae* (5). This database will aid in the identification of gene homologs across a number of species and will cross-reference genes to mammalian phenotypes and human disease. The database provides researchers with the ability to query gene sequences of their choice.

Some caution must be exercised in the interpretation of these analyses, however, since it is not always the case that sequence homology can be extended to functional conservation. The cystic fibrosis transmembrane conductance regulator (CFTR) protein, for example, is a member of the ABC transporter family and transports chloride ions (53). The closest yeast gene to CFTR is *YCF1* (65), which also encodes an ABC transporter. Reciprocal sequence comparison analysis, however, indicates that the closest human neighbor to *YCF1* is MRP (75) and not CFTR. This is further reinforced by the fact that the *YCF1* and MRP transporters both function to transport glutathione S-conjugates (15, 75). Functional conservation, therefore, is indicated by sequence homology but cannot be assumed without additional studies. A common approach is to establish whether a homologous gene from another organism can complement a yeast mutant defective in the corresponding function. A preliminary list of human cDNAs that can complement *S. cerevisiae* mutants has been presented (66), and it is likely to become more extensive as the Human Genome Project progresses. The possession of a full-length cDNA for each human gene and a complete collection of yeast deletion mutants will provide the foundation for a systematic study to understand functional conservation between these two systems.

35.4.2. Applications of *S. cerevisiae* in Drug Discovery

The broad range of systematic functional analysis studies that are in progress will greatly increase our understanding of eukaryotic biology. Integrating these data sets with comparative genomic analyses, e.g., comparing *S. cerevisiae* and humans, will facilitate the identification and validation of targets for therapeutic intervention across a range of disease areas. In some therapeutic areas, e.g., fungal infection, *S. cerevisiae* is likely to play a very direct role in the drug discovery process.

Human fungal infections range from superficial diseases of the skin or nails to life-threatening systemic infections (16). There has been a dramatic increase in systemic infections in recent years, owing to an increase in the number of immunosuppressed patients in the population, coupled with a lack of suitable therapeutic agents. Thus, an urgent need to identify new antifungal drugs has been recognized, and antifungal discovery programs have been reemphasized in many pharmaceutical companies. Infections are caused by a wide range of fungal pathogens, with the principal offending pathogens being *Candida* spp. (41), *Aspergillus* spp. (17), and *Cryptococcus neoformans* (14). The experimental tractability of many of these species is limited, and in some cases a molecular genetic system is completely lacking. Data from *S. cerevisiae* functional analysis studies are, therefore, being used as the starting point for new drug discovery projects.

The current goal of most antifungal drug discovery projects is the identification of an agent that is active against a broad spectrum of fungal pathogens, via a fungicidal mechanism. A molecular target must, therefore, be present in a spectrum of fungal pathogens, it must be absent from (or significantly different) in humans, and it must encode a function that is essential for cell viability. The *S. cerevisiae* gene knockout program will deliver a definitive list of genes that are essential for cell viability. Presumably, many of the functions essential for the growth of *S. cerevisiae* will also be essential in other fungal species. This list, therefore, represents the starting point for comparative genomic analyses to identify essential genes in *S. cerevisiae* that are conserved across fungal species and that are absent from humans. Large-scale sequencing of human expressed sequence tags and genomic DNA and genomes of fungal pathogens have emerged in the public domain, and a number of commercial vendors such as Incyte Pharmaceuticals Inc. (www.incyte.com), Genome Therapeutics Corporations (www.genomecorp.com), and Human Genome Sciences (www.hgsi.com) have also reported ongoing sequencing projects. As these projects progress toward completion, we can begin to envisage a time when the process of antifungal target selection is performed completely in silico.

The impact of *S. cerevisiae* as a research tool is not limited to infectious disease. The ability to express heterologous coding sequences in this organism provides an opportunity to develop fast and effective screens for new therapeutic agents. This is especially true if expression of the heterologous coding sequence is associated with a phenotype (output) that can be easily measured. Thus, engineered yeast assays have been developed for numerous molecules, including human immunodeficiency virus integrase (11), phosphodiesterases (2), ion channels (33), and topoisomerase II (34). An excellent example of the approach is provided in the field of G protein-coupled receptors (GPCRs). In *S. cerevisiae*, GPCRs and heterotrimeric G proteins are used to control the mating process (for review see reference 42). Haploid yeast cells express specific receptors that bind mating pheromones secreted from cells of the opposite mating type (7, 32). Activation of receptors by ligand binding stimulates a signal transduction pathway, resulting in changes in gene expression and morphological shape that are necessary for the mating process. Several reports have now demonstrated the feasibility of replacing yeast mating receptors with mammalian GPCRs, including the $\beta 2$ receptor (40), muscarinic acetylcholine receptors (36, 50), and the somatostatin receptor (52). In each case, interaction of agonists and antagonists with the appropriate receptor could be monitored using assays based on the mating phenotype (e.g., growth arrest) (Fig. 4). These observations have now led to the development of systems to identify novel peptide ligands for GPCRs. Manfredi et al. (46) described an autocrine assay in which a random peptide expression library is introduced into a yeast strain. This strain is engineered so that receptor activation allows growth and the formation of a colony (Fig. 4). Cells expressing a peptide that binds and activates endogenous GPCRs can, therefore, be selected from the population of transformants. Potentially, this provides a generic approach for the identification of GPCR ligands, including agonists of orphan GPCRs discovered via genome sequencing projects. Such agonists can then be used to investigate the function of orphan receptors in their native environment. Moreover, since these assays are simple, rapid, and inexpensive, they can be employed in drug discovery projects as high-throughput screens to identify novel agonists and antagonists for therapeutic use (10).

35.5. FUTURE DIRECTIONS

When the proposal to sequence the entire genome of *S. cerevisiae* was first made, thoughts were no doubt triggered in many researchers' minds about both the feasibility and the value of the project. A few years after the sequence was completed, there is now ample evidence to demonstrate that the genome project was invaluable. Looking forward to the next 5 years, a number of other significant landmarks are already in sight. The technologies are in place to characterize the yeast transcriptome and proteome, and the initial data sets have already been reported. An international project to knock out every yeast gene is under way and will be completed within the next 2 years. This is significant for two reasons: (i) the collection represents the initial step toward defining the minimal gene set required for a eukaryotic cell to function; (ii) the complete collection of mutant strains can be exploited in further functional analysis studies. We can now begin to anticipate that phenotypic characterization of a mutant strain will typically be extended to include its effects on the transcrip-

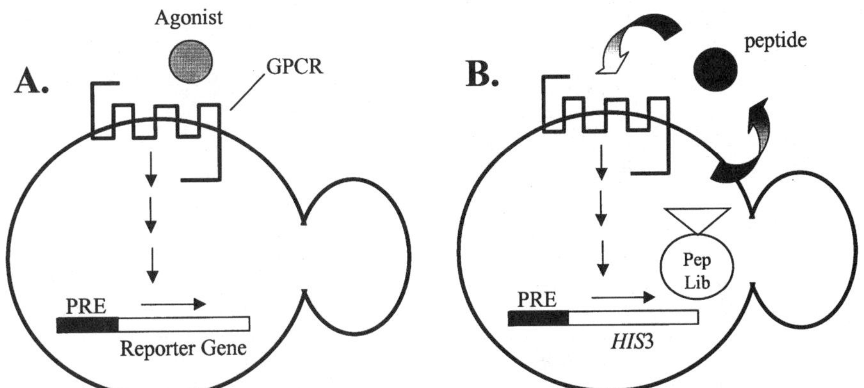

FIGURE 4 Yeast-based assays to identify GPCR ligands. (A) Mating in *S. cerevisiae* is controlled by GPCRs. Binding of pheromones (agonists) stimulates a signal transduciton pathway that culminates in changes in gene expression. Expression of a number of genes that have pheromone responsive elements (PRE) in their upstream regions is induced. Yeast GPCRs can be replaced with mammalian receptors, and the activity of corresponding agonists and antagoists is monitored by a reporter gene(s) placed downstream of a PRE. (B) A modification of the system can be used to identify agonists of orphan GPCRs. A yeast strain containing an orphan receptor of interest is transformed with a peptide library. Peptide agonists that induce expresison of a reporter gene (in this case, the nutritional marker *HIS3*) can be selected from a population, and the peptide can be identified.

tome and proteome. Quantitative and sensitive methods for detecting phenotypes will be essential to uncover the function of many of the novel genes identified in the genome, especially those that are members of redundant gene families. These studies will result in the identification of regulatory networks and will help to place genes into functional groups. The vast amounts of data that will be generated in these studies will require development of powerful computational tools both for processing and analyzing the data. The full value of the data will be recognized only if the scientific community can organize and integrate information from many data sets in an efficient manner. The collaborative nature of the *S. cerevisiae* community provides an excellent opportunity for this to become a reality.

REFERENCES

1. **Adams, A., D. E. Gottschling, C. A. Kaiser, and T. Sterns.** 1997. *Methods in Yeast Genetics*. Cold Spring Harbor Laboratory Press, Cold Spring Harbor, N.Y.
2. **Atienza, J. M., and J. Colicelli.** 1998. Yeast model system for studying mammalian phosphodiestrases. *Methods* **14:** 35–42.
3. **Ausubel, F. A., R. Brent, R. E. Kingston, D. D. Moore, J. G. Seidman, J. A. Smith, and K. Struhl (ed.).** 1998. *Current Protocols in Molecular Biology*. John Wiley & Sons, New York.
4. **Bartel, P. L., J. A. Roecklein, D. SenGupta, and S. Fields.** 1996. A protein linkage map of *Escherichia coli* bacteriophage T7. *Nat. Genet.* **12:**72–77.
5. **Basset, D. E., M. S. Boguski, F. Spencer, R. Reeves, M. Goebl, and P. Hieter.** 1995. Comparative genomics, genome cross referencing and XREFdb. *Trends Genet.* **11:** 372–373.
6. **Baudin, A., O. Ozier-Kalogeropoulos, A. Denouel, F. Lacroute, and C. Cullin.** 1993. A simple and efficient method for direct gene deletion in *Saccharomyces cerevisiae*. *Nucleic Acids Res.* **21:**3329–3330.
7. **Blumer, K. J., J. E. Reneke, and J. Thorner.** 1988. The *STE2* gene product is the ligand-binding component of the a-factor receptor of *Saccharomyces cerevisiae*. *J. Biol. Chem.* **263:**10836–10842.
8. **Botstein, D., S. A. Chervitz, and J. M. Cherry.** 1997. Yeast as a model organism. *Science* **277:**1259–1260.
9. **Brent, R., and M. Ptashne.** 1985. A eukaryotic transcriptional activator bearing the DNA specificity of a prokaryotic repressor. *Cell* **43:**729–736.
10. **Broach, J. R., and J. Thorner.** 1996. High-throughput screening for drug discovery. *Nature* **384(Suppl.):**14–16.
11. **Caumont, A. B., G. A. Jamieson, S. Pichuantes, A. T. Nguyen, A. Litvak, and C.-H. Dupont.** 1996. Expression of functional HIV-1 integrase in the yeast *Saccharomyces cerevisiae* leads to the emergence of a lethal phenotype: potential use for inhibitor screening. *Curr. Genet.* **29:** 503–510.
12. **Cho, R. J., M. J. Campbell, E. A. Winzeler, L. Steinmetz, A. Conway, L. Wodicka, T. G. Wolfsberg, A. E. Gabrielian, D. Landsman, D. J. Lockhart, and R. W. Davis.** 1998. A genome-wide transcriptional analysis of the mitotic cell cycle. *Mol. Cell* **2:**65–73.
13. **Chu, S., J. DeRisi, M. Eisen, J. Mullholland, D. Botstein, P. O. Brown, and I. Herskowitz.** 1998. The transcriptional program of sporulation in budding yeast. *Science* **282:**699–705.
14. **Chuck, S. L., and M. A. Sande.** 1989. Infections with *Cryptococcus neoformans* in the acquired immunodeficiency syndrome. *N. Engl. J. Med.* **321:**794–800.
15. **Cole, S. P., O. Bhardwaj, J. H. Gerlach, J. E. Mackie, C. E. Grant, K. C. Almquist, A. J. Stewart, E. U. Stewart, E. U. Kurz, A. M. Duncan, and R. G. Deeley.** 1992. Overexpression of a transporter gene in a multidrug-resistant human lung cancer cell line. *Science* **258:**1650–1654.
16. **Denning, D. W.** 1994. Evolving etiology of fungal infection in the 1990s. Infect. Dis. Clin. Pract. **3**(Suppl. 2)**:** S50–S55.
17. **Denning, D. W., S. Follansbee, M. Scolaro, S. Norris, S. Edelstein, and D. A. Stevens.** 1991. Pulmonary aspergillosis in AIDS. *N. Engl. J. Med.* **324:**654–662.
18. **DeRisi, J.** Personal communication.
19. **DeRisi, J. L., V. R. Iyer, and P. O. Brown.** 1997. Exploring the metabolic and genetic control of gene expression on a genomic scale. *Science* **278:**680–686.
20. **Dujon, B.** 1996. The yeast genome project: what did we learn? *Trends Genet.* **12:**263–270.
21. **Fields, S., and O. Song.** 1989. A novel genetic system to detect protein-protein interactions. *Nature* **340:**245–246.
22. **Figeys, D., S. P. Gygi, Y. Zhang, J. Watts, M. Gu, and R. Aebersold.** 1998. Electrophoresis combined with novel mass spectrometry techniques: powerful tools for the analysis of proteins and proteomes. *Electrophoresis* **19:**1811–1818.
23. **Fodor, S. P., R. P. Rava, X. C. Huang, A. C. Pease, C. P. Holmes, and C. L. Adams.** 1993. Multiplexed biochemical assays with biological chips. *Nature* **364:**555–556.
24. **Fodor, S. P. A., J. L. Read, M. C. Pirrung, L. Stryer, A. T. Lu, and D. Solas.** 1991. Light-directed, spatially addressable parallel chemical synthesis. *Science* **251:**767–773.
25. **Foury, F.** 1997. Human genetic diseases: a cross-talk between man and yeast. *Gene* **195:**1–10.
26. **Frederickson, R. M.** 1998. Macromolecular matchmaking: advances in two-hybrid and related technologies. *Curr. Opin. Biotechnol.* **9:**90–96.
27. **Fromont-Racine, M., J. C. Rain, and P. Legrain.** 1997. Toward a functional analysis of the yeast genome through exhaustive two-hybrid screens. *Nat. Genet.* **16:**277–282.
28. **Goffeau, A., et al.** 1997. The yeast genome directory. *Nature* **385:**5.
29. **Goffeau, A. B. G. Barrell, H. Bussey, R. W. Davis, B. Dujon, H. Feldmann, F. Galibert, J. D. Hoheisel, C. Jacq, M. Johnston, E. J. Louis, H. W. Mewes, Y. Murakami, P. Philippsen, H. Tettelin, and S. G. Oliver.** 1996. Life with 6000 genes. *Science* **274:**546, 563–567.
30. **Guthrie, C., and G. R. Fink (ed.).** 1991. *Methods in Enzymology*, vol. 194. *Guide to Yeast Genetics and Molecular Biology*. Academic Press, Inc., San Diego, Calif.
31. **Gyuris, J., E. Golemis, H. Chertkov, and R. Brent.** 1993. Cdi1, a human G1 and S phase protein phosphatase that associates with Cdk2. *Cell* **75:**791–803.
32. **Hagen, D. C., G. McCaffrey, and G. F. Sprague.** 1986. Evidence the yeast *STE3* gene encodes a receptor for the peptide pheromone **a** factor: gene sequence and implications for the structure of the presumed receptor. *Proc. Natl. Acad. Sci. USA* **83:**1418–1422.

33. **Hahnenberger, K. M., and S. E. Kurtz.** 1997. A drug screening program for ion channels in yeast. *Trends Biotechnol.* **15:**1–4.
34. **Hammonds, T. R., A. Maxwell, and J. R. Jenkins.** 1998. Use of a rapid throughput in-vivo screen to investigate inhibitors of eukaryotic topoisomerase II enzymes. *Antimicrob. Agents Chemother.* **42:**889–894.
35. **Hope, I. A., S. Mahadevan, and K. Struhl.** 1988. Structural and functional characterization of the short acidic transcriptional activation region of yeast GCN4 protein. *Nature* **333:**635–640.
36. **Huang, H. J., C. F. Liao, B. C. Yang, and T. T. Kuo.** 1992. Functional expression of the rat M5 muscarinic acetylcholine receptor in yeast. *Biochem. Biophys. Res. Commun.* **182:**1180–1186.
37. **Iyer, V., and K. Struhl.** 1996. Absolute mRNA levels and transcriptional initiation rates in *Saccharomyces cerevisiae*. *Proc. Natl. Acad. Sci. USA* **93:**5208–5212.
38. **James, P.** 1997. Of genomes and proteomes. *Biochem. Biophys. Res. Commun.* **231:**1–6.
39. **Keegan, L., G. Gill, and M. Ptashne.** 1986. Separation of DNA binding from the transcription-activating function of a eukaryotic regulatory protein. *Science* **231:**699–704.
40. **King, K., H. G. Dohlman, J. Thorner, M. G. Caron, and R. L. Lefkowitz.** 1990. Control of yeast mating signal transduction by a mammalian $\beta 2$ adrenergic receptor and Gs α subunit. *Science* **250:**121–123.
41. **Komshian, S. V., A. K. Uwaydah, J. D. Sobel, and L. R. Crane.** 1989. Fungemia caused by Candida species and Torulopsis glabrata in the hospitalized patient: frequency, characteristics, and evaluation of factors influencing outcome. *Rev. Infect. Dis.* **11:**379–390.
42. **Kurjan, J.** 1992. Pheromone response in yeast. *Annu. Rev. Biochem.* **61:**1097–1129.
43. **Lashkari, D. A., J. H. McCusker, and R. W. Davis.** 1997. Whole genome analysis: experimental access to all genome sequenced segments through larger-scale efficient oligonucleotide synthesis and PCR. *Proc. Natl. Acad. Sci. USA* **94:**8945–8947.
44. **Lockhart, D. J., H. Dong, M. C. Byrne, K. T. Follettie, M. V. Gallo, M. S. Chee, M. Mittmann, C. Wang, M. Kobayashi, H. Horton, and E. L. Brown.** 1996. Expression monitoring by hybridization to high-density oligonucleotide arrays. *Nat. Biotechnol.* **14:**1675–1680.
45. **Ma, J., and M. Ptashne.** 1988. converting a eukaryotic transcriptional inhibitor into an activator. *Cell* **55:**443–446.
46. **Manfredi, J. P., C. Klein, J. J. Herrero, D. R. Byrd, J. Trueheart, W. T. Wiesler, D. M. Fowlkes, and J. R. Broach.** 1996. Yeast **a** mating factor structure-activity relationship derived from genetically selected peptide agonists and antagonists of Ste2p. *Mol. Cell. Biol.* **16:**4700–4709.
47. **Mendelsohn, A. R., and R. Brent.** 1994. Applications of interaction traps/two-hybrid systems to biotechnology research. *Curr. Opin. Biotechnol.* **5:**482–486.
48. **Nelson, S. F., J. H. McCusker, M. A. Sander, Y. Kee, P. Modrich, and P. O. Brown.** 1993. Genomic mismatch scanning: a new approach to genetic linkage mapping. *Nat. Genet.* **4:**11–18.
49. **O'Farrell, P. H.** 1975. High resolution two-dimensional electrophoresis of proteins. *J. Biol. Chem.* **250:**4007–4021.
50. **Payette, P., F. Gossard, M. W. Whiteway, and M. Dennis.** 1990. Expression and pharmacological characterization of the human M1 muscarinic receptor in *Saccharomyces cerevisiae*. *FEBS Lett.* **266:**21–25.
51. **Pease, A. C., D. Solas, E. J. Sullivan, M. T. Cronin, C. P. Holmes, and S. P. Fodor.** 1994. Light-generated oligonucleotide arrays for rapid DNA sequence analysis. *Proc. Natl. Acad. Sci. USA* **91:**5022–5026.
52. **Price, L. A., E. M. Kajkowski, J. R. Hadcock, B. A. Ozenberger, and M. H. Pausch.** 1995. Functional coupling of a mammalian somatostatin receptor to the yeast pheromone response pathway. *Mol. Cell. Biol.* **15:**6188–6195.
53. **Riordan, J. R., J. M. Rommens, B. Kerem, N. Alon, R. Rozmahel, Z. Grzelczak, et al.** 1989. Identification of the cystic fibrosis gene: cloning and characterization of complementary DNA. *Science* **245:**1066–1073.
54. **Roepstorff, P.** 1997. Mass spectrometry in protein studies from genome to function. *Curr. Opin. Biotechnol.* **8:**6–13.
55. **Romanos, M. A., C. A. Scorer, and J. J. Clare.** 1992. Foreign gene expression in yeast. *Yeast* **8:**423–488.
56. **Ross-Macdonald, P., A. Sheehan, G. S. Roeder, and M. Snyder.** 1997. A multipurpose transposon system for analyzing protein production, localization, and function in *Saccharomyces cerevisiae*. *Proc. Natl. Acad. Sci. USA* **94:**190–195.
57. **Rothstein, R.** 1991. Targeting, disruption, replacement, and allele rescue: integrative DNA transformation in yeast. *Methods Enzymol.* **194:**281–301.
58. **Schena, M., R. A. Heller, T. P. Theriault, K. Konrad, E. Lachenmeier, and R. W. Davis.** 1998. Microarrays: biotechnology's discovery platform for functional genomics. *Trends Biotechnol.* **16:**301–306.
59. **Schena, M., D. Shalon, R. W. Davis, and P. O. Brown.** 1995. Quantitative monitoring of gene expression patterns with a complementary DNA microarray. *Science* **270:**467–470.
60. **Shalon, D., S. J. Smith, and P. O. Brown.** 1996. A DNA microarray system for analyzing complex DNA samples using two-color fluorescent probe hybridization. *Genome Res.* **6:**639–645.
61. **Shoemaker, D. D., D. A. Lashkari, D. Morris, M. Mittmann, and R. W. Davis.** 1996. Quantitative phenotypic analysis of yeast deletion mutants using a highly parallel molecular bar-coding strategy. *Nat. Genet.* **14:**450–456.
62. **Sikorski, R. S., and P. Hieter.** 1989. A system of shuttle vectors and yeast host strains designed for efficient manipulation of DNA in *Saccharomyces cerevisiae*. *Genetics* **122:**19–27.
63. **Smith, V., D. Botstein, and P. O. Brown.** 1995. Genetic footprinting: a genomic strategy for determining a gene's function given its sequence. *Proc. Natl. Acad. Sci. USA* **92:**6479–6483.
64. **Smith, V., K. N. Chou, D. Lashkari, D. Botstein, and P. O. Brown.** 1996. Functional analysis of the genes of yeast chromosome V by genetic footprinting. *Science* **274:**2069–2074.
65. **Szczypka, M. S., J. A. Wemmie, W. S. Moyle-Rolley, and D. J. Thiele.** 1994. A yeast metal resistance protein similar to human cystic fibrosis transmembrane conductance regulator (CFTR) and multidrug resistance-associated protein. *J. Biol. Chem.* **269:**22853–22857.

66. **Tugendreich, S., D. E. Bassett, V. A. McKusick, M. S. Boguski, and P. Hieter.** 1994. Genes conserved in yeast and humans. *Hum. Mol. Genet.* **3:**1509–1517.

67. **Velculescu, V. E., L. Zhang, B. Vogelstein, and K. W. Kinzler.** 1995. Serial analysis of gene expression. *Science* **270:**484–487.

68. **Velculescu, V. E., L. Zhang, W. Zhou, J. Vogelstein, M. A. Basrai, D. E. Bassett, Jr., P. Hieter, B. Vogelstein, and K. W. Kinzler.** 1997. Characterization of the yeast transcripotome. Cell **88:**243–251.

69. **Venter, J. C., H. O. Smith, and L. Hood.** 1996. A new strategy for genome sequencing. *Nature* **381:**364–366.

70. **Wach, A., A. Brachat, R. Pohlmann, and P. Philippsen.** 1994. New heterologous modules for classical or PCR-based gene disruptions in *Saccharomyces cerevisiae*. *Yeast* **10:**1793–1808.

71. **Winzeler, E., D. Shoemaker, A. Astromoff, H. Liang, K. Anderson, T. Jones, C. Connelly, C. Pai, C. Roberts, D. Lockhart, G. Yen, H. Liao, H. Bussey, J. Rine, J. Boeke, K. Yu, L. Riles, M. Snyder, M. Mittman, P. Ross-Macdonald, R. Bangham, C. Rebischung, P. Philippsen, S. Whelen, S. Véronneau, S. Friend, T. Ward, M. Johnston, and R. Davis.** Construction of a comprehensive set of yeast deletion strains. Unpublished data.

72. **Winzeler, E. A., D. R. Richards, A. R. Conway, A. L. Goldstein, S. Kalman, M. J. McCullough, J. H. McCusker, D. A. Stevens, L. Wodicka, D. J. Lockhart, and R. W. Davis.** 1998. Direct allelic variation scanning of the yeast genome. *Science* **281:**1194–1197.

73. **Wodicka, L., H. Dong, M. Mittmann, M.-H. Ho, and D. J. Lockhart.** 1997. Genome-wide expression monitoring in *Saccharomyces cerevisiae*. *Nat. Biotechnol.* **15:**1359–1367.

74. **Wolfe, K. H., and D. C. Shields.** 1997. Molecular evidence for an ancient duplication of the entire yeast genome. *Nature* **387:**708–713.

75. **Ze-Sheng, L., M. Szczypka, Y. P. Lu, D. J. Thiele, and P. A. Rea.** 1996. The yeast cadmium factor protein (YCF1) is a vacuolar glutathione S-conjugate pump. *J. Biol. Chem.* **271:**6509–6517.

Genetics of Non-*Saccharomyces* Industrial Yeasts

JAN WERY, JAN C. VERDOES, AND ALBERT J. J. VAN OOYEN

36

The yeast *Saccharomyces cerevisiae* has been extensively studied on the molecular level and has been accessible for recombinant DNA techniques since its development as a host for transformations 20 years ago (5, 31).

The vast genetic knowledge about *S. cerevisiae* and the many genes isolated have been useful in the development of transformation systems for an increasing number of industrially important yeasts that share ascomycetous features with *S. cerevisiae*.

In this chapter the development of a transformation system for the basidiomycetous yeast *Phaffia rhodozyma* is described in detail. The relatively unknown yeast *P. rhodozyma* has gained increasing interest, both academically and commercially, since its isolation and detailed description by Phaff et al. in 1972 (43). Like a number of other yeasts and fungi of basidiomycetous origin, *Phaffia* accumulates carotenoids that are responsible for the orange to red color of this yeast. Miller et al. (40) discovered in 1976 that the main carotenoid (or, preferably, xanthophyll) produced was astaxanthin. Besides *Phaffia*, no other yeasts or fungi have been reported to produce this particular xanthophyll.

Astaxanthin is a derivative of the yellow carotenoid β-carotene and is responsible for the pink to orange pigmentation of a variety of higher organisms, such as flamingos and other birds, crustaceans, and salmon, through their natural diet. In the past two decades salmon is increasingly produced in pens, where the use of astaxanthin as a food additive is the cause of its pink pigmentation.

The growing demand for astaxanthin has encouraged industries to find profitable ways of astaxanthin production.

We have developed a transformation system for this unconventional yeast in order to make it accessible for recombinant DNA technology, providing the means for the isolation and overexpression of genes. This is of key interest for the understanding of the genes and the conversions they carry out in the biosynthesis pathway of astaxanthin, which could ultimately lead to well-defined, stable astaxanthin-overproducing strains.

In this chapter the development of the *Phaffia* transformation system has been chosen as a case study because of the unconventional basidiomycetous nature of this yeast, which differs both phylogenetically and physiologically from the ascomycetous yeasts. Specific methods and problems in the various stages will be discussed, and a comparison will be made with the development of basic transformation systems for the industrially important yeasts *Kluyveromyces lactis*, *Pichia pastoris*, *Hansenula polymorpha*, and *Yarrowia lipolytica* to provide insight into the considerations for choosing a specific strategy. It is hoped that this will be useful for making other (unconventional) yeasts accessible for recombinant DNA technology.

36.1. DEVELOPMENT OF A TRANSFORMATION SYSTEM (IN GENERAL)

36.1.1. Introduction of Foreign DNA

A transformation system depends, first, on the introduction of foreign DNA into the host organism; second, on the maintenance of foreign DNA in the host organism; and third, on selection for this event.

For the introduction of foreign plasmid DNA into cells, transformation methods are described for other organisms. In general, the introduction of foreign DNA can be achieved by chemical treatment of the host cells. For *Escherichia coli*, treatment with divalent cations like Ca^{+2} ($CaCl_2$), alone or in combination with dimethyl sulfoxide, hexaminecobalt, and dithiothreitol is the most common method used for obtaining cells that are highly competent for taking up foreign DNA (12, 29). For yeast, sole treatment with these chemicals has proved not to be sufficient; moreover, the yeast cell wall appears to be a major obstacle for the passage of DNA.

It was shown by Hinnen et al. (31) that enzymatic degradation of the cell wall in a hypertonic sorbitol solution, generating protoplasts, followed by treatment with polyethylene glycol (PEG) and $CaCl_2$, is a prerequisite for obtaining highly competent cells. Although this method is highly efficient, it has proved to be somewhat tedious and laborious, especially the preparation and regeneration of protoplasts.

More recently, a simplified procedure for the preparation of intact competent yeast cells was achieved by treatment with the monovalent cation Li^+ (LiCl and LiAc) in combination with PEG (33). This method was optimized by using high concentrations of denatured carrier DNA (49).

The most recently developed method for yeast transformation (4, 16, 38) is based on exposure of yeast cells, in the presence of the transforming DNA, to high-voltage electric discharges, creating pores in the cell membrane that allow passage of the DNA. This method, called electroporation, was already well established for introducing macromolecules into mammalian cells (45) and *E. coli* (18).

36.1.2. Plasmid Maintenance

Examples of basic plasmids for *K. lactis*, *P. pastoris*, *H. polymorpha*, and *Y. lipolytica* and their properties are summarized in Table 1. Several of these plasmids have been further engineered for optimal heterologous gene expression by the introduction of secretion signals, inducible promoters, and other elements. For a detailed description of the latter, the following reviews are recommended: *K. lactis* (55); *P. pastoris* (47); *H. polymorpha* (21); *Y. lipolytica* (30).

For maintenance of plasmid DNA in yeasts, two possibilities exist: episomal replication or integration into the chromosomal DNA. Episomally maintained plasmids replicate autonomously and independently from the chromosomal DNA, whereas integrating plasmids replicate along with the chromosomal DNA. In general, episomal plasmids transform cells with significantly higher efficiency, because successful transformation is solely dependent on the transfer into the cell. In case of an integrative event at a single locus, transformation efficiencies drop 10^3- to 10^4-fold, because successful transformation is further dependent on homologous or heterologous integration.

For yeasts, several elements are known to promote episomal replication. Autonomous replication sequences (ARS) have been cloned from *S. cerevisiae* and from industrial yeasts like *P. pastoris* (13), *H. polymorpha* (46), *K. lactis* (14), and *Y. lipolytica* (25) (Table 1). ARS elements naturally reside in the chromosomal DNA at relatively high frequency (approximately one ARS per 50 kb), where they initiate chromosomal replication during the S phase of the cell. ARS-based vectors, which are present in multiple copies from 1 to 20, are inefficiently transmitted over the next generations under nonselective growth. This instability is a drawback for using these plasmids for industrial applications. In general, ARS elements can be easily isolated by cloning random genomic DNA fragments and screening for increasing transformation efficiency and instable inheritance (53). In *Y. lipolytica*, however, this procedure was not successful in several instances. In 1991 Fournier et al. (25) isolated two *Y. lipolytica* ARS elements (ARS18 and ARS68) with deviating characteristics, such as exceptionally high stability (Table 1).

Another type of episomal maintenance is provided by the 2μm plasmid that naturally occurs in *S. cerevisiae* strains at high copy numbers (100 copies) with high stability. Using 2μm sequences, involved in replication and stabilization, in *E. coli*–yeast shuttle expression vectors, transformants with stable high copy number are obtained (2, 9, 42). For *K. lactis*, similar episomal expression vectors have been constructed using Klori, the origin of replication isolated from the 2μm-like plasmid pKD1 which occurs naturally in *K. lactis* (Table 1). This system has been successfully used for heterologous gene expression (8, 23, 24).

Plasmid maintenance by integration usually involves the targeting of plasmid DNA to a specific locus in the chromosomal DNA. Although the transformation of circular plasmid may yield integration by a double crossover event, digestion of the plasmid in the homologous sequence prior to transformation will enhance targeted integration (single crossover). It is well established in a variety of yeasts that integrated plasmids are even more stably maintained without selective pressure than are 2μm- or pKD1-derived plasmids.

In *K. lactis*, systems based on integration in the *LAC4* locus are described for stable heterologous gene expression (58). In *P. pastoris*, stable heterologous gene expression was obtained by homologous integration into the alcohol oxidase gene (*AOX1*) and the *HIS4* gene (57). Furthermore, in *H. polymorpha*, plasmids that were initially episomally maintained by the presence of either a homologous or a fortuitous ARS element were shown to integrate after prolonged cultivation (stabilization). This was accompanied by a transition from rather instable to stable transformants (46). In addition, both heterologous integration and homologous integration in the alcohol oxidase gene (*AO*) were obtained by transformation of linearized plasmid (22). In *Y. lipolytica*, linearized plasmids targeted to the *LEU2* gene and the alkaline extracellular protease-encoding *XPR2* gene are used for efficient homologous integration (3, 15, 26, 41).

In general, integrative transformation yields one or a few integrated plasmid copies, which may impose limits on the expression of the foreign gene. Multicopy integration occasionally occurs, especially if high concentrations of transforming plasmid are used.

For *P. pastoris*, a procedure was described that was based on the screening for multiple copy transformants. The transposon 903 encoded the kanamycin gene (Km^r), which confers resistance to the antibiotic G418 in a variety of eukaryotes, was cloned in a *P. pastoris* expression vector. Since the resistance to G418 is highly correlated with the copy number of the Km^r gene, transformants could be screened by exposure to increasing concentrations of G418, carrying 3 to 18 plasmid copies (Table 1).

A different type of integrative transformation is based on the targeting of plasmids to the ribosomal DNA (rDNA). It was found by Szostak and Wu (56) that rDNA-containing plasmids could transform *S. cerevisiae*, upon linearization within the rDNA portion, with at least 100-fold higher efficiency than could linearized plasmids that were targeted to the *LEU* locus. This was attributed to the fact that yeast rDNA consists of approximately 140 copies, providing a much higher amount of homologous DNA for recombination. Moreover, a part of the rDNA gene cloned promoted high-frequency nonintegrative transformation, suggesting the presence of an ARS. More recently, Lopes et al. (37) and Bergkamp et al. (8) found for *S. cerevisiae* and *K. lactis*, respectively, that rDNA-containing plasmids could integrate at high copy numbers (up to 200) with even higher stability than other integrating systems.

Since these findings were made, rDNA has been used as a stabilizing sequence in plasmids for non-*Saccharomyces* yeasts like *Candida utilis* (35) and *Y. lipolytica* (36).

36.1.3. Selection Markers

The selection for cells that have acquired the foreign DNA is in many cases based on the presence of a marker gene that complements an auxotrophy of the host (auxotrophic selection marker) or renders resistance to an antibiotic (dominant selection marker).

In the first case, a host strain deficient in a particular conversion in the synthesis of, e.g., an amino acid (*TRP*, *HIS*, *LEU*, etc.) or a nucleoside (*URA*) will lose this phenotype if complemented by the marker gene coding for this conversion. This event can be easily selected for in medium lacking the relevant amino acid or nucleoside.

TABLE 1 Basic transformation systems for *K. lactis*, *H. polymorpha*, *P. pastoris*, and *Y. lipolytica*[a]

Yeast	Plasmid	Marker gene	Source of marker gene	Mode of plasmid maintenance	Max. trans eff.[b]	Copy no.[c]	Stability (%)[d]	Ref.
K. lactis	pKARS	*TRP1*	Het.(Sc)	KARS (ep.)	3×10^4	ND	10–15	14
	pKS105	Km[r] + ADHI prom.	Het.(Tn5)+het.(Sc)	*LAC4* (Hom. int.)	ND	Multiple	100 (50 gen.)	58
	YIprDI-LYS	*HIS3*	Het.(Sc.)	rDNA (repl. int.)	ND	4–40	100	48
	pMIRK1	*TRP1*	Het.(Sc.)	rDNA (add. int.)	ND	60	100	8
	pCJX10,11,12	*URA3, TRP1, LEU2*	Het.(Sc)	Klori (ep.)	1.2×10^4	19	12	10
	pCJX3,5,6	Km[r]	Het. (TN903)	Klori (ep.)	1.1×10^4	16	18	
	pCJX18,19,20	*URA3, TRP1, LEU2*	Het.(Sc)	KICEN2 (ep.)	8.2×10^3	1	37	
H. polymorpha	YEp13	*LEU2*	Het.(Sc)	*LEU2* ARS act. (ep.)	60	ND	8.5–86	28
	YIp5	*URA3*	Het.(Sc)	Fortuitous ARS/het. int.	40	1	100	46
	YRP17	*URA3*	Het.(Sc)	ScARS1 (ep.)	300	5	1	
	pHARS1	*URA3*	Het.(Sc)	HARS1 (ep.)	1.5×10^3	40	2	
	pHARS2	*URA3*	Het.(Sc)	HARS2 (ep.)/int	460	25 (ep.) 75 (int.)	1.5 (ep.) 100 (int.)	
	pHIP1	*LEU2*	Het.(Sc)	*LEU2* ARS act. (ep.)	5×10^3	6	15–45 (40 gen.)	22
	(linear)			het. int.	10^5	1–3	100	
	pHRP2	*LEU2*	Het.(Sc)	HARS1 (ep.)	1.5×10^3	6	2	
	pHIP11	*LEU2*	Het.(Sc)	*LEU2* ARS act. (ep.)	3×10^3	ND	15–45 (40 gen.)	
	(linear)			hom. int. AO	4.6×10^4	1–3	100	
	(linear)			het. int.	5.6×10^4	1–3	100	
	pCE36	*LEU2*	Het.(Sc)	HARS36 (ep.)/int.	3×10^3	ND	3 (ep.) 100 (int.)	51
P. pastoris	pYA2	*HIS4*	Het.(Sc)	Fortuitous ARS	1.6×10^4	6	10	13
	pYA4	*HIS4*	Hom.	*LEU2* ARS act. (ep.)	9.7×10^4	6	29	
	pYJ30	*HIS4*	Hom.	PARS1 (ep.)	1.8×10^5	13	50	
	pYJ32	*HIS4*	Hom.	PARS2 (ep.)	1.7×10^5	13	51	
	pPIC3K	Km[r]	Het. (Tn903)	*AOX1* (hom. add. int.)	400	3–8	ND	50
				AOX1 (hom. repl. int.)	20	7–18	ND	

TABLE 1 Basic transformation systems for *K. lactis*, *H. polymorpha*, *P. pastoris*, and *Y. lipolytica*[a] (*Continued*)

Yeast	Plasmid	Marker gene	Source of marker gene	Mode of plasmid maintenance	Max. trans eff.[b]	Copy no.[c]	Stability (%)[d]	Ref.
Y lipolytica	pINA46S (linear)	*LYS2* + fortuitous prom.	Het.(Sc) Hom.	Hom./het. int.	10	Multiple, tandem	ND	27
	pLD25 (linear)	*LEU2*	Hom.	Hom. int.	100 10^4	ND	100	15
	pINA65 (linear)	*LEU2*	Hom.	Hom. int.	10^4	1	ND	26
	pINA95 (linear)	*LEU2* + Tn*5ble* + *LEU2* prom.	Hom. Het. (Tn5) + hom.	Hom. int.	6.4×10^3 80	1	ND	
	pINA98 (linear)	*LEU2*, *lacZ* + *LEU2* prom.	Hom. Het. (Tn5) + hom.	Hom. int.	100	1	ND	
	pINA 169 (linear)	*SUC2* + *XPR2* prom., *LEU2*	Het.(Sc) +hom. Hom.	Hom. int. *XPR2*	5.4×10^4 2.4×10^4 (suc$^+$) 3.3×10^4 (leu$^+$)	1	ND	41
	pINA 119	*LEU2*	Hom.	ARS18	10^4	3	25 (Fil$^-$), 69–96	25
	pREB53 (linear)	*β*-GUS + *LEU2* prom., *LEU2*	Het.(Ec) + hom., hom.	Hom. int.	ND	ND	ND	3
	pIN 767, 772, 773 (linear)	*URA3* + trunc. prom.	Hom.	Hom. int. rDNA	$10–10^3$	5–60	80–100	36

[a]Abbreviations: act., activity; add., additive; ADH1, alcohol dehydrogenase; AO, *H. polymorpha* alcohol oxidase gene; *AOX1*, *P. pastoris* alcohol oxidase gene; ARS, autonomous replication sequence; *β*-GUS, *β*-glucuronidase gene; Ec, *E. coli*; ep., episomal; Fil$^-$, in a mutant unable to produce hyphae; gen., generations; HARS, *H. polymorpha* ARS; het., heterologous; hom., homologous; int., integration; KARS, *K. lactis* ARS; KICEN2, *K. lactis* centromeric sequence; Klori, *K. lactis* replication origin from pKD1; ND, not determined; PARS, *P. pastoris* ARS; prom., protmoter; repl., replacement; Sc, *S. cerevisiae*; trunc., truncated; *XPR2*, alkaline extracellular protease gene.

[b]Transformation efficiency in colonies per microgram of transforming DNA.

[c]Number of plasmid copies in transformations.

[d]Stability in percentage of cells that maintained the plasmid after 10 generations of nonselective cultivation, unless stated otherwise.

In case no well-defined auxotrophic strains are available, dominant selection markers, such as the Km^r and *ble* genes, which confer resistance to antibiotics, can be used if the host for transformation is sensitive.

The Km^r gene, which originates from transposons Tn*5* and Tn*903* and codes for a bacterial aminoglycoside 3′-phosphotransferase (7), is used in *S. cerevisiae*, *K. lactis*, and *P. pastoris* (Table 1). The selection procedure is based on the resistance to the aminoglycoside antibiotic G418, which is inactivated by this gene.

Y. lipolitica was shown to be insensitive to G418 (26). For this yeast the Tn*5*-encoding *ble* gene was used as a dominant marker. This gene confers resistance to phleomycin and is used as a marker in many filamentous fungi as well as in *S. cerevisiae* and *schizosaccharomyces pombe* (44, 60).

36.2. DEVELOPMENT OF A TRANSFORMATION SYSTEM FOR *PHAFFIA*

36.2.1. The Opportunistic Approach: Using Existing Heterologous Plasmids and Transformation Procedures

The main question that arises when developing a transformation system for any "new" organism is what strategy to employ: development of a heterologous system using existing techniques and plasmids or of a homologous system, which requires the laborious and time-consuming isolation and sequencing of marker and maintenance sequences.

Since a transformation system is usually only a step in reaching the final goal, which is isolating genes and producing foreign proteins, it is tempting to choose the fastest option and use plasmids and transformation procedures that already exist. The use of plasmids consisting entirely or partly of heterologous DNA has been shown to be successful (Table 1).

Since most industrial yeasts are more or less related to *S. cerevisiae* in that they share ascomycetous characteristics, the use of *S. cerevisiae* markers, promoters, and maintenance sequences has been of help in the development of transformation systems for several industrially important yeast species, like *K. lactis*, *H. polymorpha*, and *P. pastoris*.

Table 1 shows that the majority of the auxotrophic markers used in the basic transformation systems for these yeasts in fact originate from *S. cerevisiae*. *P. pastoris HIS4* strains could be efficiently transformed with a plasmid carrying the *S. cerevisiae HIS4* gene, whereas the the *S. cerevisiae TRP1*, *HIS3*, and *URA3* genes could be used for efficient transformation of a *K. lactis* strain with corresponding auxotrophies. Furthermore, the *S. cerevisiae LEU2* and *URA3* genes have been used to complement leu and ura auxotrophic strains of the methylotrophic yeast *H. polymorpha*.

In a few cases, the heterologous use of plasmid maintenance sequences has also been described. DNA sequences from very divergent eukaryotic species were shown to be capable of promoting high-frequency transformation with episomal maintenance in *S. cerevisiae* (52). In addition, the *S. cerevisiae LEU2* gene, for example, has been shown to promote autonomous replication in *P. pastoris* and *H. polymorpha* (6, 13, 22). Fortuitous ARS activity may also be present on heterologous plasmid sequences.

Heterologous sequences are also used to drive expression of antibiotic resistance markers. Expression of the Km^r gene was observed in *S. cerevisiae*, *K. lactis*, and *P. pastoris* under control of its native bacterial promoter (14, 34, 50). In addition, Chen and Fukuhara (11) showed that the Km^r gene driven by the *S. cerevisiae* ADH1 promoter conferred high resistance to G418 in both *S. cerevisiae* and *K. lactis*.

Y. lipolytica appears to be quite different in that most of the auxotrophic markers are homologous. In addition, the expression of the *S. cerevisiae LYS2* and *SUC2* and the bacterial Tn*5ble*, *lacZ*, and *β*-GUS genes is driven by homologous promoter sequences (Table 1).

The examples given here show that, in many instances, DNA-encoded properties can be universally applied. Therefore, we decided to first try the heterologous approach to develop a transformation system for *Phaffia*.

Since no well-defined stable auxotrophic *Phaffia* strains were available, we decided to use the Km^r gene as a dominant selection marker. First, the sensitivity to G418 was determined, since the selection procedure for the transformation was based on the resistance to that antibiotic (Table 2).

We found that growth inhibition of *Phaffia* cells on G418-containing YPD agar plates was influenced by the cell density. The G418 resistance was determined by plating *Phaffia* cells on YPD agar plates with increasing concentrations of G418. We used 10^7 cells per plate, anticipating the maximum amount of cells to be spread in our transformation experiments. Since background growth was negligible at 40 μg/ml G418, we decided to select for transformants on YPD plates containing 40 μg/ml G418 in various transformations.

Existing plasmids containing the Km^r gene downstream of the native Tn*903*, the *S. cerevisiae* ADHI, or the mammalian simian virus 40 promoter were used for the transformations. Although no *Phaffia* sequences were present, we anticipated expression in *Phaffia* by using these various plasmids, since the gene was previously shown to confer aminoglycoside resistance to a wide variety of prokaryotes and eukaryotes.

The transforming plasmid DNAs were prepared differently. Covalently closed circle forms, purified on CsCl gradients, were used in anticipation of the presence of fortuitous replicating sequences. In addition, linearization was performed prior to transformation to enhance integration by single cross-over.

Since we did not know what transformation procedure would work best, we used both the spheroplasting method of Hinnen et al. (31) and the LiCl method of Ito et al. (33). However, in both cases no transformants were obtained.

TABLE 2 Growth inhibition of *Phaffia* by G418[a]

G418 (μg/ml)	Colony count
0	Overgrown
20	Overgrown
30	30
40	2
50	0
100	0

[a]Approximately 10^7 exponentially growing *Phaffia* cells were plated on YPD (1% yeast extract, 2% Bacto peptone, 2% glucose) agar slants containing increasing amounts of G418. Colonies were counted after 3 days of incubation at 21°C.

At this stage of research a number of questions arose concerning the cause of the failure to obtain transformants. Are the marker genes on the plasmids used not sufficiently expressed? Do the plasmids lack sequences for maintenance in *Phaffia*? Are the transformation procedures used not suited for *Phaffia*?

It was decided to abandon the heterologous approach, because of the many uncertainties to be resolved and, more important, because we were committed to work with *Phaffia* on a long-term basis.

36.2.2. The Secure Approach: Development of Homologous Plasmids

36.2.2.1. Isolation of Total *Phaffia* DNA

Before isolating specific *Phaffia* DNA sequences we first developed a method for the isolation of total DNA. We found that procedures optimized for many ascomycetous yeasts, including *K. lactis*, *P. pastoris*, *H. polymorpha*, and *Y. lipolytica*, that were based on protoplasting methods with Zymolyase 20T or 100T were not suited for *Phaffia*. This was probably due to the double-layered *Phaffia* cell wall and its specific basidiomycetous physicochemical properties, such as the presence of xylose (54). A method using Novozym 234 (Novo Industri A/S, Denmark), which is used as a cell wall-degrading enzyme for several basidiomycetous yeasts and fungi, was applied for *Phaffia* (see protocol).

PROTOCOL

Isolation of total *Phaffia* DNA

Inoculate 20 ml of YPD medium (1% yeast extract, 2% Bacto Peptone, 2% glucose) with a single *Phaffia* colony. Cultivate at 21°C with vigorous shaking until the OD_{600} reaches 10 to 20. Harvest cells by centrifuging for 5 min at 5,000 × *g* and wash twice by resuspending in 20 ml of ESC (60 mM EDTA, 1.2 M sorbitol, 0.1 M trisodium citrate [pH 7.0]). Cells were resuspended in 3 ml of ESC containing 2.5 mg of Novozym 234 per ml to form protoplasts. Incubate for 3 h at 30°C and then centrifuge the protoplasts for 5 min at 2,000 × *g*. Resuspend the pellet in 10 ml of $T_{50}E_{20}$ (50 mM Tris, 20 mM EDTA [pH 7.5]) containing 0.5% sodium dodecyl sulfate. Extract the mixture twice with an equal volume of Tris-equilibrated phenol (pH 8.0)–chloroform–isoamylalcohol (25:24:1) by vortexing and then centrifuge for 5 min at 20,000 × *g*.

Transfer the upper aqueous phase to a fresh tube and add 1/10 volume of 3 M sodium acetate (pH 5.2) and 2 volumes of absolute ethanol. After vortexing, centrifuge the mixture for 10 min at 20,000 × *g*. Dissolve the pellet in 0.5 ml of $T_{10}E_1$ (pH 8.0) containing 0.1 mg of RNase A per ml. Finally, purify the DNA by phenol/chloroform extraction and ethanol precipitation and dissolve in 0.5 ml of $T_{10}E_1$.

36.2.2.2. Isolation of the *Phaffia* Glyceraldehyde Phosphate Dehydrogenase Gene

For the development of a homologous transforming plasmid, three issues that were raised previously—expression of the marker, plasmid maintenance, and the transformation procedure—must be addressed.

To optimize the chance of expression, the Km^r gene would best be driven by a strong constitutive promoter. For this purpose we decided to isolate a glycolytic gene. Besides having a high level of constitutive expression, glycolytic genes are well conserved, which is important for the isolation procedure.

For the isolation of these types of genes, two strategies can be considered: (i) heterologous hybridization and (ii) PCR using degenerate primers. The first method usually involves the use of a gene probe from a different species for hybridization with genomic or library DNA. If performed under nonstringent conditions, a specific hybridization signal may be obtained. However, if the genes differ too much, no hybridization will be obtained. In addition, there is a real risk of picking up aspecific products, as a result of hybridization conditions with too-low stringency.

This was illustrated by the fact that, in our efforts to pick up the triose phosphate isomerase (TPI) and phosphoglycerate kinase (PGK) genes with the use of heterologous probes from *K. lactis* and *S. cerevisiae*, respectively, we could not obtain a hybridization signal under nonstringent hybridization conditions with chromosomal *Phaffia* DNA. Even the use of a *K. lactis* actin probe resulted in only a very weak specific signal (62), despite the fact that actin is one of the best-conserved genes known.

We decided to isolate the *Phaffia* glyceraldehyde phosphate dehydrogenase (GAPDH) gene by an alternative approach, a method that involves synthesis of a homologous GAPDH probe by PCR on total *Phaffia* DNA, followed by screening of a *Phaffia* genomic DNA cosmid library with this probe (59).

The key step in this procedure is the choice of the sequence of the oligonucleotides (primers) used in the PCR. The sequence of the primers must be optimized for specific annealing to ensure optimal amplification of the target DNA. Since the nucleotide sequence of the target DNA (in this case, part of the *Phaffia* GAPDH gene) is unknown, the design of the primers must be derived from conserved regions, as determined by comparison of the sequences of several species of this gene from other organisms. In case none or only a few species of the gene are sequenced, it is better not to use this method, because a reliable estimate of conserved regions cannot be made.

The conserved regions of the *Phaffia* GAPDH gene were determined by comparing 11 GAPDH protein sequences from various organisms (39). Approximately 10 highly conserved peptide motifs were found throughout the protein. For the design of the PCR primers, the DNA sequence of these motifs must be inferred from the amino acid sequence. This procedure is complicated by the fact that the third and sometimes the first nucleotide in a codon may be variable (degenerate codon). This necessitates the use of a mixture of each primer variant in these nucleotides (degenerate primers) in the PCR assay.

The extent to which the primers are degenerate depends on the differential use of codons by the organism (codon usage). Codon usage is species specific (see reference 1 for a review) and can be determined if more genes are isolated.

For *Phaffia* the codon usage was determined in the previously isolated actin gene (62) and in eight ribosomal protein genes (unpublished data) that are, like the GAPDH gene, constitutively expressed. From Table 3 it can be deduced that *Phaffia* has a strong codon preference for various amino acids that differs from that in *S. cerevisiae*.

In general, a C is preferred in the third codon position, whereas an A in this position is strongly underrepresented.

TABLE 3 Codon usage in *Phaffia* compared with *S. cerevisiae*

Amino acid	Codon	In *Phaffia* genes[a] (%)	In *S. cerevisiae* genes[b] (%)	Amino acid	Codon	In *Phaffia* genes[a]	In *S. cerevisiae* genes[b] (%)
Phe	TTT	4	54	Tyr	TAT	0	50
	TTC	96	46		TAC	100	50
Leu	TTA	2	27	His	CAT	2	60
	TTG	15	36		CAC	98	40
	CTT	33	11	Gln	CAA	4	74
	CTC	47	5		CAG	96	26
	CTA	2	13	Asn	AAT	8	55
	CTG	2	9		AAC	92	45
Ile	ATT	20	50	Lys	AAA	4	52
	ATC	80	30		AAG	96	48
	ATA	0	20	Asp	GAT	36	62
Met	ATG	100	100		GAC	64	38
Val	GTT	25	44	Glu	GAA	2	74
	GTC	65	25		GAG	98	26
	GTA	1	16	Cys	TGT	26	67
	GTG	9	15		TGC	74	33
Ser	TCT	25	31	Trp	TGG	100	100
	TCC	58	18	Arg	CGT	1	17
	TCA	2	19		CGC	1	4
	TCG	12	8		CGA	86	5
	AGT	2	15		CGG	2	2
	CGC	1	9		AGA	3	54
Pro	CCT	35	29		AGG	8	17
	CCC	60	13	Gly	GGT	41	60
	CCA	2	49		GGC	7	15
	CCG	3	9		GGA	50	15
Thr	ACT	19	38		GGG	2	9
	ACC	75	24	Ala	GCT	35	44
	ACA	2	26		GCC	65	24
	ACG	3	11		GCA	0	24
					GCG	0	8

[a]A total of 10 *Phaffia* genes, including eight ribosomal protein genes (unpublished data), the GAPDH gene (59), and the actin gene (62), were analyzed.
[b]A total of 484 *S. cerevisiae* genes were analyzed. Data from Zhang et al. (63).

In the case of Gly and Arg, the opposite is true. Especially for the nine amino acids that are represented by two codons and for Arg, Thr, and Ile, the preference for one codon is evident (>75%).

This knowledge significantly facilitates the design of degenerate primers; e.g., for Arg, which is in theory represented by six different codons, the use of one codon (CGA, 86%) would be appropriate in a degenerate primer for *Phaffia*.

The presence of introns, which generally reside at the 5′ part of a gene, may interfere with specific priming or may cause amplification of a product with an unexpected size. The first will occur if the introns are located within the target sequence of one or both primers, whereas introns residing in the DNA fragment to be amplified will yield an amplified product that is larger than expected.

Since introns may occur frequently in the *Phaffia* genome, as indicated previously by the presence of four introns in the 5′ part of the actin gene (61), it was decided to design the upstream and downstream primers based on two amino acid motifs occurring at the carboxy-terminal end of the GAPDH protein. The distance between the motifs was 90 amino acids, which represents approximately 0.3 kb on the GAPDH gene.

A PCR assay with degenerate primers must be optimized for specific product generation. The annealing temperatures used are of especial importance. Too-low temperatures will give rise to a specific annealing, resulting in many nonsense products, whereas no product will be obtained if too-high temperatures are used (Fig. 1).

In our trials PCR was performed at annealing temperatures ranging from 40 to 50°C. At 45°C a DNA fragment was generated with an expected size of 0.3 kb. This fragment encodes the carboxy-terminal end of the *Phaffia* GAPDH gene by comparison with the *S. cerevisiae* GAPDH sequence (59).

From this critical stage on, the remaining part of the procedure involved straightforward homologous hybridizations using the PCR-generated *Phaffia* GAPDH fragment as a probe for screening a *Phaffia* genomic cosmid library. This resulted in the isolation of the complete *Phaffia* GAPDH gene, including promoter and terminator.

36.2.2.3. Construction of a Dominant Selection Marker for *Phaffia*

For construction of a dominant selection marker, the GAPDH promoter and terminator were used to mediate

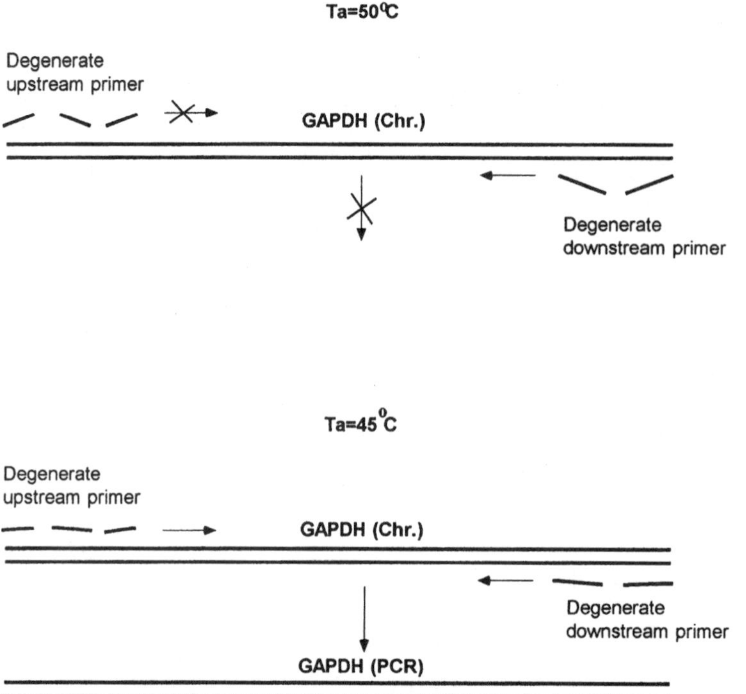

FIGURE 1 Schematic representation of the influence of the annealing temperature in a PCR assay using degenerate primers for the amplification of a fragment of the *Phaffia* GAPDH gene. At 50°C the degenerate primers do not anneal properly to the target sequence on the GAPDH gene, inhibiting first-strand synthesis by the *Taq* polymerase. At 45°C the primers anneal sufficiently for proper amplification. Ta, annealing temperature; Chr., chromosomal DNA.

expression of the Km^r gene. In this construction procedure we estimated it of great importance to maintain the GAPDH promoter unaffected in subsequent cloning steps, to avoid any chance of losing regions that are important for expression. Besides an intact promoter, optimal spacing between the TATA/CT boxes within the GAPDH promoter and the ATG start codon of the Km^r gene is important for efficient expression.

Since the use of restriction sites, naturally occurring or introduced, always causes mutations that may interfere with the optimal configuration, an alternative strategy was used.

A 0.4-kb 3′ part of the *Phaffia* GAPDH promoter containing TATA and CT boxes was precisely fused to the 241-bp 5′ end of the coding Km^r gene by fusion PCR (17, 32). This was done in a two-step PCR reaction as illustrated in Fig. 2.

In the first step, both the GAPDH and Km^r DNA fragments were generated. The GAPDH promoter fragment was synthesized by using an upstream primer (I) designed on a region at position −385 with respect to the ATG start codon. This primer also contained a *Kpn*I restriction sequence. The downstream primer (II) was designed on the utmost 3′ end of the GAPDH promoter and contained a 12-nucleotide overlap with the 5′ end of the Km^r gene. Using these primers in a PCR with the GAPDH gene as a template, a 385-bp GAPDH promoter fragment was obtained, including 12 bp of the Km^r coding sequence.

The 5′ part of the Km^r coding sequence was generated similarly, using an upstream primer (III) with several nucleotides overlapping the 3′ part of the GAPDH promoter and a downstream primer designed on an *Msc*I site-containing region around position +241 with respect to the ATG start codon.

In the second step, equimolar amounts (up to 10 ng) of both DNAs were used in a PCR with only the outer primers (I and IV) in the reaction mixture, giving rise to a fusion PCR DNA fragment of 0.6 kb flanked by a *Kpn*I site and an *Msc*I site. This fragment could be easily used in subsequent straightforward cloning steps to obtain a plasmid carrying the complete Km^r gene driven by the GAPDH promoter.

A PCR-generated *Bam*HI-*Hin*dIII DNA fragment containing the 0.3-kb sequence 3′ of the GAPDH coding sequence (terminator) was cloned in the corresponding sites present in the multiple cloning site downstream of the Km^r gene. We expected that in this case the spacing between the Km^r stop codon and the GAPDH terminator was less critical.

36.2.2.4. Isolation of a *Phaffia* rDNA Fragment as a Stabilizing Sequence

We decided to use the *Phaffia* rDNA as a stabilizing sequence in our transforming plasmids. This was primarily based on the high transformation efficiency and stability-promoting properties of these sequences in *S. cerevisiae* and *K. lactis*. In addition, we anticipated that the highly conserved nature of rDNA may be advantageous for isolation.

Indeed, we successfully isolated a 3-kb *Phaffia* rDNA fragment after heterologous hybridization with a *K. lactis* rDNA fragment as a probe. This sequence was cloned in the vectors containing the GAPDH promoter/terminator-driven Km^r genes as a marker.

Three different *Phaffia* plasmids were obtained and were used for transformation (Fig. 3). Plasmid pPR1T contains

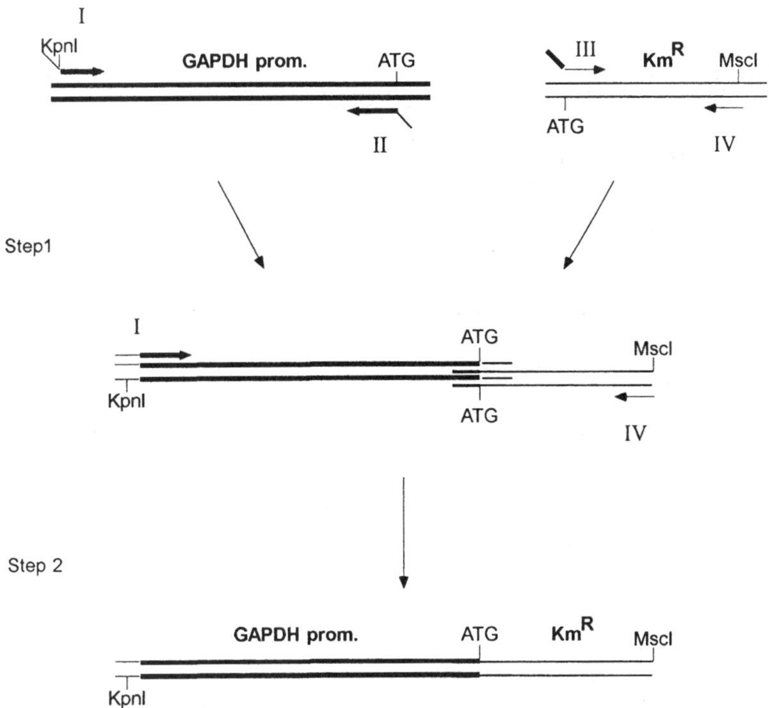

FIGURE 2 Schematic representation of the precise fusion of the *Phaffia* GAPDH promoter with the coding part of the Km^r gene by a two-step PCR procedure. The primers are depicted by Roman numerals. In step 1, the 5′ part of the GAPDH promoter (prom.) and the Km^r gene are amplified with primers (II, III) that contain overlapping sequences of both genes. In step 2, the partly overlapping products are used in equimolar amounts in a PCR assay with the outer primers I and IV, which contain a *Kpn*I restriction site and an *Msc*I restriction site, respectively. A product is obtained in which the GAPDH promoter is precisely fused to the coding part of the Km^r gene, flanked by a *Kpn*I site and an *Msc*I site.

the Km^r gene driven by the *Phaffia* GAPDH promoter and terminated by the GAPDH terminator. In pPR2T the 3-kb rDNA sequence is also present. Plasmid pPR2 lacks the GAPDH terminator.

36.2.2.5. Transformation of *Phaffia*

Two transformation procedures, based on chemical treatment and electroporation, were employed for introduction of the plasmids to *Phaffia*.

The chemical transformation procedure, based on a method by Elble et al. (19), was initially designed for *S. cerevisiae* and involves the prolonged incubation of *Phaffia* cells in a mixture of PEG 4000, LiAc, carrier DNA, and plasmid DNA, followed by spreading on G418-containing agar slants. Since there was no information as to what conditions were best to be used for *Phaffia*, the original procedure conditions were applied.

Indeed, tranformants were obtained at low frequency. The procedure was optimized (see protocol below) by varying parameters such as the growth stage of the cells (early/late log phase), incubation time of the cells in the transformation mix, and recovery in YPD medium prior to spreading on the selective plates.

PROTOCOL

Transformation of *Phaffia* by chemical treatment

Dilute overnight cultures of *Phaffia* 100 to 200-fold in YPD medium and grow at 21°C to an OD_{600} of 0.8. For each transformation, centrifuge 1.2 ml of culture at 12,000 × *g* for 10 s. Remove the supernatant and add both 10 μg of transforming DNA in a volume no larger than 10 μl and 10 μl (10 mg/ml) of heat-denatured herring sperm DNA. After mixing, add 0.5 ml of PLATE solution consisting of 40% PEG 4000 (Brocacef), 0.1 M LiAc·$2H_2O$ (sigma-Aldrich Chemie), 10 mM Tris-HCL (pH 7.5), and 1 mM EDTA and mix by vortexing. Incubate the mixtures without shaking at 21°C for 40 h and subsequently spread on selective YPD slants containing 40 μg of G418 (GIBCO-BRL) per ml. No more than 100 μl of transformation mixture is spread per slant, since in this procedure the sensitivity to G418 is lower in case of too-high cell densities, promoting background growth. For determination of cell survival, spread 100 μl of a 10^{-4} dilution of the suspension on nonselective YPD agar slants at 10-h intervals.

However, it was very difficult to reproduce successful transformation using this method, mainly because of great differences in the survival rates, ranging from 1 to 100% in the various transformation experiments, caused by events we could not trace. Apparently, the chemical treatment optimized for the ascomycetous yeasts was less suitable for *Phaffia*, probably due to the different cell wall structure.

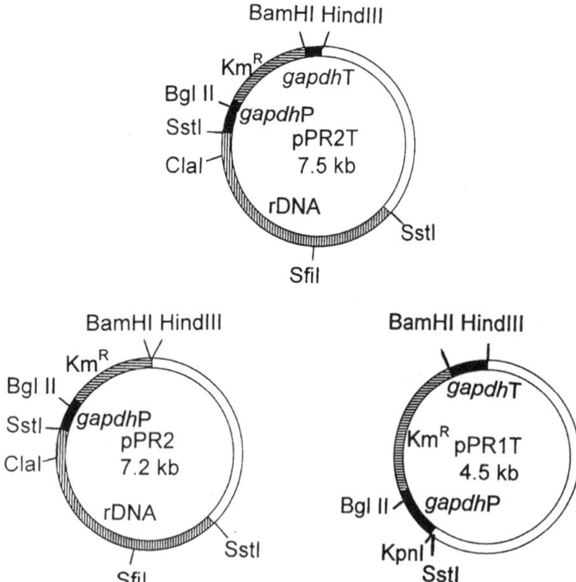

FIGURE 3 *Phaffia* plasmids. Plasmid pPR2T contains the following elements: pUC 18 (2.7 kb, white), a 3-kb *Phaffia* rDNA fragment with unique *ClaI* and *SfiI* restriction sites (vertical stripes), and the Km^r gene (1.1 kb, horizontal stripes) driven by the *Phaffia* GAPDH promoter (*gapdhP*, 0.4 kb, black) and terminated by the GAPDH terminator (*gapdhT*, 0.3 kb, black). Plasmids pPR2 and pPR1T differ from pPR2T in that the GAPDH terminator and the rDNA fragment, respectively, are missing.

Since electroporation has been proved to be widely applicable, being effective for the introduction of DNA into mammalian, plant, fungal, yeast, and bacterial cells, an electroporation procedure was optimized for *Phaffia*.

Yeast transformation by electroporation was first described in 1989 by Delorme (16). *S. cerevisiae* cells concentrated in 200 μl of YPD medium were electropermeabilized in the presence of 2μ-based plasmid, giving rise to maximally 4,500 transformants per microgram of plasmid DNA. Although 2μ-based plasmids transform yeast with much higher frequencies by spheroplasting or even chemical-based transformation, electroporation proved to be easier and quicker. The electroporation procedure was improved by providing osmotic support to the cells during transformation (4), ensuring stabilization of the cell membrane. Furthermore, it was found by Meilhoc et al. (38) and Faber et al. (20) that pretreatment of cells of different yeast species, such as *Y. lipolytica* and *H. polymorpha*, with the reducing agent dithiothreitol was in some cases an absolute requirement for obtaining transformants. Presumably, the dithiothreitol-mediated reduction of cell wall proteins provides the transforming DNA with better access to the cell membrane.

For the transformation of *Phaffia*, the protocol based on these procedures was optimized by focusing on growth stage, recovery, and electroporation parameters (see protocol below).

Phaffia cells were best used for electroporation in the mid-log phase (OD_{600} 1.0 to 1.4). Since the selection procedure was based on G418 resistance, a recovery period in YPD after electropulsing of 2.5 h was necessary to allow for expression of the marker gene. Prolonging this period did not improve efficiency.

Electropulsing was carried out by using a Bio-Rad Gene Pulser with Pulse Controller that rapidly generates a set voltage, holding that voltage for a specific time, and rapidly reducing it to zero (square wave-type electrical pulse).

In particular, the voltage and pulse duration applied were parameters to be varied in the optimization studies. The number of cells that actually become competent for DNA uptake is very dependent on the voltage applied, in that there is a minimal amplitude to generate membrane pores that allow DNA to pass and a maximum amplitude at which cell decay becomes limiting. Between these voltage limits, the duration of the electric pulse is important for DNA uptake. The duration of the electric pulse can be prolonged by increasing the internal resistance of the Gene Pulser apparatus.

Table 4 shows that *Phaffia* transformants can be obtained by using electric pulses ranging from 2 to 8 kv/cm. In addition, the dependence of transformation efficiency on pulse duration is illustrated. It is shown that for optimal efficiencies the cells are best exposed to relatively low electric field strength with longer exposure times.

PROTOCOL

Transformation of *Phaffia* by electroporation (61)

Transformation of *P. rhodozyma* CBS 6938 was performed as derived from a procedure previously described (20, 38). The procedure described here was designed for 15 transformations. Dilute overnight cultures of *Phaffia* 100 to 200-fold in 200 ml of YPD medium and grow at 21°C to an OD_{600} of 1.2, which takes approximately 16 h. Harvest cells by centrifugation (5,000 × *g*) at 21°C, resuspend in 25 ml of 50 mM potassium phosphate buffer (pH 7.0) containing 25 mM dithiothreitol, and incubate at 21°C for 15 min. Perform the following steps at 2°C, using precooled materials and solutions. Wash the cells twice in 25 ml of STM (270 mM sucrose, 10 mM Tris-HCl [pH 7.5], 1 mM $MgCl_2$) and resuspend in 0.5 ml of STM (3 × 10^9 cells/ml). Add transforming DNA to 60 μl of cell suspension. Transfer cell/DNA mixture to an electroporation cuvette. Electropulse with the Bio-Rad Gene Pulser with Pulse Controller and with Bio-Rad 0.2-cm cuvettes. Following the electric pulse (4 kV/cm, 1,000 Ω, 25 μF), add 0.5 ml of YPD medium. Incubate mixture for 2.5 h at 21°C and subsequently spread on selective YPD agar plates containing 40 μg/ml G418. Count the transformants after 5 days of growth.

36.2.2.6. Influence of the Plasmid Sequences of Transformation

The Ribosomal DNA Portion

The various *Phaffia* plasmids (Fig. 3) could be transformed only after linearization in either the rDNA portion (*ClaI*) or the GAPDH promoter (*BglII*), indicating the absence of ARS activity. In addition, the plasmids showed differences in efficiency of transformation and copy number (Table 5). From these differences the function of the three *Phaffia* elements—rDNA, GAPDH promoter, and GAPDH terminator—can be seen.

The rDNA portion clearly promoted high transformation efficiencies and high copy number integration. This is

TABLE 4 Influence of increasing electric field strength at different internal resistance values on tranformation efficiency of *Phaffia*

Field strength	No. of transformants at the following internal resistance values[a]:				
	200 Ω (5 ms)	400 Ω (8 ms)	600 Ω (11 ms)	800 Ω (14 ms)	1,000 Ω (16 ms)
0 kV/cm	0	0	0	0	0
2 kV/cm	0	0	2	8	13
3 kV/cm	0	9	63	330	688
4 kV/cm	1	36	233	433	**948**[b]
5 kV/cm	5	88	**430**[b]	**570**[b]	521
6 kV/cm	17	**210**[b]	181	196	69
7 kV/cm	41	75	46	50	33
8 kV/cm	**82**[b]	38	44	35	14

[a]Number of transformants per microgram of *Cla*I-linearized plasmid pPR2T at the indicated internal resistance values of the Bio-Rad Gene Pulser, which correlate with the duration of the electric pulse (shown in parentheses).

[b]Highest transformation efficiences are shown in boldface type.

seen by comparing plasmid pPR2T (+rDNA) and pPR1T (−rDNA).

Plasmid pPR2T transformed *Phaffia* with 100-fold higher efficiency, indicating that, as in *S. cerevisiae* and *K. lactis*, the presence of multiple rDNA copies in the host (61 for *Phaffia* [reference 61]) increases the chance of homologous integration.

Moreover, all *Phaffia* strains investigated that were transformed with the rDNA-containing plasmids pPR2 and pPR2T carried 20 to 70 plasmid copies. The −rDNA plasmid pPR1T yielded transformants by heterologous integration that contained only 2 to 10 copies. This observation clearly indicates that the rDNA portion has properties promoting high copy number integration.

The GAPDH Promoter/Terminator

The presence of the GAPDH promoter to drive the Km^r gene was shown to be important in obtaining transformants. The cloning of the GAPDH terminator downstream of the Km^r gene proved to be valuable for optimization of the transformation efficiency. Furthermore, by comparison of plasmid pPR2 (−terminator) and pPR2T (+terminator), it can be deduced that transformation efficiencies increase 10-fold if the GAPDH terminator is present.

The presence of the GAPDH terminator also affects the plasmid copy number. *Phaffia* strains transformed with pPR2T DNA contain on the average 20 plasmid copies, which is less than one-third of pPR2-transformed strains.

36.3. CONCLUSION

The choice of developing a homologous transformation system for *Phaffia* has yielded a maximum of possibilities for recombinant DNA applications for this yeast. First, the system is highly efficient, which is useful for the isolation of genes by complementation of defined *Phaffia* mutants. Second, the copy number of any homologous or heterologous gene introduced in *Phaffia* can be easily tuned by the vector used, from a few to up to 65. Tunable copy numbers may be important for a maximum yield, which is established by the delicate balance between expression levels and growth. Third, the transformation system is stable, which is a prerequisite for long-term and large-scale production. Finally, *Phaffia* is the only basidiomycetous yeast for which a transformation system with these properties has been developed. Given the fact that *Phaffia* differs, both physiologically and genetically, from ascomycetous yeasts that are traditionally used as a host for recombinant protein production, this may offer new possibilities for the overexpression of either new genes or genes that are poorly expressed in the organisms used so far.

TABLE 5 Influence of various transforming DNAs on transformation efficiency, copy number, G418 resistance, and stability

Transforming plasmid[a]	Enzyme[b]	Efficiency of transformation[c]	Copy no.[d]	G418 resistance[e] (mg/ml)	Stability[f] (60 generations)
pPR2T	*Cla*I	950	30	2	+
pPRIT (−rDNA)	*Bgl*II	8	2–10	0.4–1	ND
pPR2 (−term.)	*Cla*I	100	65	2	+

[a]−rDNA, plasmid lacks rDNA portion; −term., plasmid lacks the GAPDH terminator downstream of the Km^r gene.

[b]Enzyme used for linearizing the plasmid before transformation.

[c]Number of transformants per microgram of transforming plasmid.

[d]Approximate number of copies of the transformed plasmid.

[e]Maximum G418 concentration at which growth occurred.

[f]Resistance to 200 μg/ml G418 after 60 generations of nonselective growth. ND, not determined.

REFERENCES

1. **Andersson, S. G. E., and C. G. Kurland.** 1990. Codon preferences in free-living microorganisms. *Microbiol. Rev.* **54:**198–210.
2. **Armstrong, K. A., T. Som, F. C. Volkert, F. C. Rose, and J. R. Broach.** 1989. Propagation and expression of genes in yeast using 2-micron circle vectors, p. 165–192. *In* P. H. Barr, A. J. Brake, and P. Valenzuela, (ed), *Yeast Genetic Engineering*. Butterworths, London.
3. **Bauer, R., F. Paltauf, and S. D. Kohlwein.** 1993. Functional expression of bacterial β-glucuronidase and its use as a reporter system in the yeast *Yarrowia lipolytica*. *Yeast* **9:**71–75.
4. **Becker, D. M., and L. Guarente.** 1991. High-efficiency transformation of yeast by electroporation. *Methods Enzymol.* **194:**182–187.
5. **Beggs, J. D.** 1978. Transformation of yeast by a replicating hybrid plasmid. *Nature* **275:**104–109.
6. **Berardi, E., and D. Y. Thomas.** 1990. An effective transformation method for *Hansenula polymorpha*. *Curr. Genet.* **18:**169–170.
7. **Berg, D. E., J. Davies, B. Allet, and J. D. Rochaix.** 1975. Transposition of R-factor genes to bacteriophage λ. *Proc. Natl. Acad. Sci. USA* **72:**3628–3632.
8. **Bergkamp, R. J. M., I. M. Kool, R. H. Geerse, and R. J. Planta.** 1992. Multiple-copy integration of the α-galactosidase gene from *Cyamopsis tetraglonoloba* into the ribosomal DNA of *Kluyveromyces lactis*. *Curr. Genet.* **21:** 365–370.
9. **Broach, J. R.** 1983. Construction of high copy number yeast vectors using 2 μm circle sequences. *Methods Enzymol.* **101:**307–325.
10. **Chen, X. J.** 1996. Low- and high-copy-number shuttle vectors for replication in the budding yeast *Kluyveromyces lactis*. *Gene* **172:**131–136.
11. **Chen, X. J., and H. Fukuhara.** 1988. A gene fusion system using the aminoglycoside 3′-phosphotransferase gene of the kanamycin resistance transposon Tn 903: use in the yeasts *Kluyveromyces lactis* and *Saccharomyces cerevisiae*. *Gene* **69:**181–192.
12. **Cohen, S. N., A. C. Y. Chang, and L. Hsu.** 1972. Nonchromosomal antibiotic resistance in bacteria: genetic transformation of *Escherichia coli* by R-factor DNA. *Proc. Natl. Acad. Sci. USA* **69:**2110–2114.
13. **Cregg, J. M., K. J. Barringer, A. Y. Hessler, and K. R. Madden.** 1985. *Pichia pastoris* as a host system for transformations. *Mol. Cell. Biol.* **5:**3376–3385.
14. **Das, S., and C. P. Hollenberg.** 1982. A high frequency transformation system for the yeast *Kluyveromyces lactis*. *Curr. Genet.* **6:**123–128.
15. **Davidow, L. S., D. Apostolakos, M. M. O'Donnel, A. R. Proctor, D. M. Ogrydziak, R. A. Wing, I. Stasko, and J. R. DeZeeuw.** 1985. Integrative transformation of the yeast *Yarrowia lipolytica*. *Curr. Genet.* **10:**39–48.
16. **Delorme, E.** 1989. Transformation of *Saccharomyces cerevisiae* by electroporation. *Appl. Environ. Microbiol.* **55:** 2242–2246.
17. **Dillon, P. J., and C. A. Rosen.** 1990. A rapid method for the construction of synthetic genes using the polymerase chain reaction. *BioTechniques* **9:**298–299.
18. **Dower, W. J., J. F. Miller, and C. W. Ragsdale.** 1988. High efficiency transformation of *E. coli* by high voltage electroporation. *Nucleic Acids Res.* **16:**6127–6145.
19. **Elble, R.** 1992. A simple and efficient procedure for transformation of yeasts. *Biofeedback* **13:**18–20.
20. **Faber, K. N., P. Haima, W. Harder, M. Veenhuis, and G. Ab.** 1994. Highly-efficient electrotransformation of *Hansenula polymorpha*. *Curr. Genet.* **25:**305–310.
21. **Faber, K. N., W. Harder, G. Ab, and M. Veenhuis.** 1995. Review: methylotrophic yeasts as factories for the production of foreign proteins. *Yeast* **11:**1331–1344.
22. **Faber, K. N., G. J. Swaving, F. Faber, G. Ab, W. Harder, M. Veenhuis, and P. Haima.** 1992. Chromosomal targeting of replicating plasmids in the yeast *Hansenula polymorpha*. *J. Gen. Microbiol.* **138:**2405–2416.
23. **Fleer, R., A. Fournier, and P. Yeh.** 1993. Procédé de production de protéine recombinantes et cellules hôtes utilisées. European Patent Application EP-521767A.
24. **Fleer, R., A. Fournier, and P. Yeh.** 1993. Highly stable recombinant yeasts for the production of recombinant proteins. PCT Patent Application WO 9303159.
25. **Fournier, Ph., L. Guyaneux, M. Chasles, and C. Gaillardin.** 1991. Scarcity of ars sequences isolated in a morphogenesis mutant of the yeast *Yarrowia lipolytica*. *Yeast* **7:** 25–36.
26. **Gaillardin, C., and A.-M. Ribet.** 1987. *LEU2* directed expression of β-galactosidase activity and Phleomycin resistance in *Yarrowia lipolytica*. *Curr. Genet.* **11:**369–375.
27. **Gaillardin, C., A. M. Ribet, and H. Heslot.** 1985. Integrative transformation of the yeast *Yarrowia lipolytica*. *Curr. Genet.* **10:**49–58.
28. **Gleeson, M. A., G. S. Ortori, and P. E. Sudbery.** 1986. Transformation of the methylotrophic yeast *Hansenula polymorpha*. *J. Gen. Microbiol.* **132:**3459–3465.
29. **Hanahan, D.** 1983. Studies on transformation of Escherichia coli with plasmids. *J. Mol. Biol.* **166:**557–580.
30. **Heslot, H.** 1990. Genetics and genetic engineering of the industrial yeast *Yarrowia lipolytica*. *Adv. Biochem. Eng. Biotechnol.* **43:**43–73.
31. **Hinnen, A., J. B. Hicks, and G. R. Fink.** 1978. Transformation of yeast. *Proc. Natl. Acad. Sci. USA* **75:**1929–1933.
32. **Ho, S. N., H. D. Hunt, R. M. Horton, J. K. Pullen, and L. R. Pease.** 1989. Site-directed mutagenesis by overlap extension using the polymerase chain reaction. *Gene* **77:**51–59.
33. **Ito, H. I., Fukuda, Y., Murata, K., and A. Kimura.** 1983. Transformation of intact yeast cells treated with alkali cations. *J. Bacteriol.* **153:**163–168.
34. **Jiminez, A., and J. Davies.** 1980. Expression of a transposable antibiotic resistance element in *Saccharomyces*. *Nature* **287:**869–871.
35. **Kondo, K., T. Saito, S. Kajiwara, M. Takagi, and N. Misawa.** 1995. A transformation system for the yeast *Candida utilis*: use of a modified endogenous ribosomal protein gene as a drug-resistant marker and ribosomal DNA as an integration target for vector DNA. *J. Bacteriol.* **177:**7171–7177.
36. **Le Dall, M.-T., J.-M. Nicaud, and C. Gaillardin.** 1994. Multiple-copy integration in the yeast *Yarrowia lipolytica*. *Curr. Genet.* **26:**38–44.
37. **Lopes, T. S., J. Klootwijk, A. E. Veenstra, C. Van der Aar, H. Van Heerikhuizen, H. E. Raué, and R. J. Planta.** 1989. High-copy-number integration into the ri-

bosomal DNA of *Saccharomyces cerevisiae*: a new vector for high level expression. *Gene* **79:**199–206.

38. **Meilhoc, E., Masson, J.-M., and J. Teissié.** 1990. High efficiency transformation of intact yeast cells by electric field pulses. *Bio/Technology* **8:**223–227.

39. **Michels, P. A. M., A. Poliszczak, K. A. Osinga, O. Misset, J. Van Beeumen, R. K. Wierenga, P. Borst, and F. R. Opperdoes.** 1986. Two tandemly linked identical genes code for the glycosomal glyceraldehyde-phosphate dehydrogenase in *Trypanosoma brucei*. *EMBO J.* **5:**1049–1056.

40. **Miller, M. W., M. Yonoyama, and M. Soneda.** 1976. *Phaffia*, a new yeast genus in the Deuteromycotina (Blastomycetes). *Int. J. Syst. Bacteriol.* **26:**286–291.

41. **Nicaud, J.-M., E. Fabre, and C. Gaillardin.** 1989. Expression of invertase activity in *Yarrowia lipolytica* and its use as a selective marker. *Curr. Genet.* **16:**253–260.

42. **Parent, S. A., C. M. Fenimore, and K. A. Bostian.** 1985. Vector systems for the expression, analysis and cloning of DNA sequences in *Saccharomyces cerevisiae*. *Yeast* **1:**83–138.

43. **Phaff, H. J., M. W. Miller, M. Yoneyama, and M. Soneda.** 1972. A comparative study of the yeast florae associated with trees on the Japanese Islands and on the west coast of North America, p. 759–774. *In* G. Terui (ed.), Proceedings of the 4th IFS: Fermentation Technology Today. Kyoto Society of Fermentation Technology, Osaka, Japan.

44. **Prentice, H. L., and R. E. Kingston.** 1992. Mammalian promoter element function in the fission yeast Schizosaccharomyces pombe. *Nucleic Acids Res.* **20:**3383–3390.

45. **Rieman, F., U. Zimmermann, and G. Pilwat.** 1975. Release and uptake of haemoglobin and ions in red blood cells induced by dielectric breakdown. *Biochim. Biophys. Acta* **394:**449–462.

46. **Roggenkamp, R., H. Hansen, M. Eckart, Z. Janowicz, and C. P. Hollenberg.** 1986. Transformation of the methylotrophic yeast *Hansenula polymorpha* by autonomous replication and integration vectors. *Mol. Gen. Genet.* **202:**302–308.

47. **Romanos, M.** 1995. Advances in the use of *Pichia pastoris* for high-level gene expression. *Curr. Opin. Biotechnol.* **6:**527–533.

48. **Rossolini, G. M., M. L. Riccio, E. Gallo, and C. L. Galeotti.** 1992. *Kluyveromyces lactis* rDNA as a target for multiple integration by homologous recombination. *Gene* **119:**75–81.

49. **Schiestl, R. H., and R. D. Gietz.** 1989. High efficiency transformation of intact yeast cells using single stranded nucleic acid as a carrier. *Curr. Genet.* **16:**339–346.

50. **Scorer, C. A., J. J. Clare, W. R. McCombie, M. A. Romanos, and K. Sreekrishna.** 1993. Rapid selection using G418 of high copy number transformants of *Pichia pastoris* for high-level foreign gene expression. *Bio/Technology* **12:**181–184.

51. **Sohn, J.-H., E.-S. Choi, C.-H. Kim, M. O. Agaponov, M. D. Ter-Avanesyan, J.-S. Rhee, and S.-K. Rhee.** 1996. A novel autonomously replicating sequence (ARS) for multiple copy integration in the yeast *Hansenula polymorpha* DL-1. *J. Bacteriol.* **178:**4420–4428.

52. **Stinchcomb, D. T., M. Thomas, J. Kelly, E. Selker, and R. W. Davis.** 1980. Eukaryotic DNA segments capable of autonomous replication in yeast. *Proc. Natl. Acad. Sci. USA* **77:**4559–4563.

53. **Struhl, K., D. T. Stinchcomb, D. T. Scherer, and R. W. Davis.** 1979. High frequency transformation of yeast: autonomous replication of hybrid molecules. *Proc. Natl. Acad. Sci. USA* **76:**1035–1039.

54. **Sugiyama, J., M. Fukagawa, S.-W. Chiu, and K. Komagata.** 1985. Cellular carbohydrate composition, DNA base composition, ubiquinone systems, and Diazonium Blue B color test in the genera *Rhodosporidium*, *Leucosporidium*, *Rhodotorula* and related basidiomycetous yeasts. *J. Gen. Appl. Microbiol.* **31:**519–550.

55. **Swinkels, B. W., A. J. J. van Ooyen, and F. J. Bonekamp.** 1993. The yeast *Kluyveromyces lactis* as an efficient host for heterologous gene expression. *Antonie van Leeuwenhoek* **64:**187–201.

56. **Szostak, J. W., and R. Wu.** 1979. Insertion of a genetic marker into the ribosomal DNA of yeast. *Plasmid* **2:**536–554.

57. **Thill, G. P., G. R. Davis, C. Stillman, G. Holtz, R. Brierley, M. Engel, R. Buckholz, J. Kenney, S. Provow, T. Vedvick, and R. S. Siegel.** 1990. Positive and negative effects of multicopy-number integrated expression vectors on protein expression in *Pichia pastoris*, p. 477–490. *In* H. Heslot, J. Davies, J. Florent, L. Bobichon, G. Durand, and L. Penasse (ed.), Proceedings of the 6th International Symposium on Genetics of Microorganisms. Société Française de Microbiologie, Paris.

58. **Van den Berg, J. A., K. J. van der Laken, A. J. J. van Ooyen, T. C. H. M. Renniers, K. Rietveld, A. Schaap, A. J. Brake, R. J. Bishop, K. Schultz, D. Moyer, M. Richman, and J. R. Shuster.** 1990. Kluyveromyces as a host for heterologous gene expression: expression and secretion of prochymosin. *Bio/Technology* **8:**135–139.

59. **Verdoes, J. C., J. Wery, T. Boekhout, and A. J. J. van Ooyen.** 1996. Molecular characterization of the glyceraldehyde-3-phosphate dehydrogenase gene of *Phaffia* rhodozyma. Submitted for publication.

60. **Wenzel, T. J., A. Migliazza, H. Yde Steensma, and J. A. van den Berg.** 1992. Efficient selection of Phleomycin-resistant *Saccharomyces cerevisiae* transformants. *Yeast* **8:**667–668.

61. **Wery, J., D. Gutker, A. C. H. M. Renniers, J. C. Verdoes, and A. J. J. van Ooyen.** 1996. High-copy-number integration into the ribosomal DNA of the yeast Phaffia rhodozyma. *Gene*, in press.

62. **Wery, J., J. Ter Linde, M. J. M. Dalderup, T. Boekhout, and A. J. J. van Ooyen.** 1996. Structural and phylogenetic analysis of the actin gene from the yeast *Phaffia rhodozyma*. *Yeast* **12:**641–651.

63. **Zhang, S., G. Zubay, and E. Goldman.** 1991. Low-usage codons in *Escherichia coli*, yeast, fruit fly and primates. *Gene* **105:**61–72.

Genetic Engineering of Nonpathogenic *Pseudomonas* Strains as Biocatalysts for Industrial and Environmental Processes

JUAN M. SÁNCHEZ-ROMERO AND VÍCTOR DE LORENZO

37

37.1. BACKGROUND

The term *Pseudomonas* has been traditionally used to name a quite heterogeneous collection of nonenteric gram-negative strains, generally aerobes, nonfermenting and motile. In recent years, however, the development of classification criteria based on 16S RNA sequences has allowed us to put some rationale in the otherwise chaotic taxonomy of this important class of bacteria (48, 71). There are at least three types of industrial applications that have brought about a considerable interest in the genetic analysis of *Pseudomonas* and the construction of recombinant strains with engineered phenotypes. Many *Pseudomonas* strains, generally isolated from soils polluted with chemical wastes, are able to metabolize or co-oxidize recalcitrant chemicals that are perilous environmental pollutants, such as chloro- or nitro-aromatic compounds. Their genetic engineering may therefore result in strains with enhanced biodegradative activities with a potential for bioremediation applications (98, 119). Not infrequently, the partial metabolism of some organic compounds involves enzymes that are instrumental in giving rise to intermediates of high added value. Construction of strains with defined enzymatic activities has also, therefore, a remarkable interest for the fine-chemical industry (121). A second area of economic interest involving *Pseudomonas* is that of plant growth promotion and biological control (16, 32). Some *Pseudomonas* strains are plant pathogens, and understanding the biological and genetic basis of their virulence is required in order to develop effective protection strategies (47). Many other *Pseudomonas* strains that colonize the rhizosphere do just the contrary, exerting a protective effect on the roots through the production in situ of antibiotic compounds that prevent, among others, fungal infections (16, 32) or even have the potential to degrade xenobiotics (4). Finally, various *Pseudomonas* strains are opportunistic human and animal pathogens that play a determining role in the outcome of serious illnesses such as cystic fibrosis (76), not to mention infections of burns or surgical wounds. Similarly to the plant pathogens, the investigation of the virulence factors is of the essence to overcome the problems derived from the multi-resistance phenotypes frequently observed in these strains (41, 123).

Regardless of the origin or application of the specific *Pseudomonas* isolates, their genetic manipulation requires a number of tools based, with little variation, on the same basic functional modules. In general, virtually all the methods employed nowadays for *Pseudomonas*, described below, are not specific to the genus but have a broad application. In this chapter, we have focused on these genetic tools for manipulating nonpathogenic *Pseudomonas* species which are the subject of some industrial, environmental, or agronomical interest, typically, *Pseudomonas putida* or *Pseudomonas fluorescens*. Many of the techniques described below are also applicable to other interesting strains formerly classified as *Pseudomonas* (e.g., *Burkholderia*) and, in fact, to an entire range of gram-negative eubacteria, provided they fill a number of requisites specified later in this chapter. Readers are directed to reference 100 for detailed methods for handling *Pseudomonas aeruginosa* and other species more relevant to human and animal health.

37.2. GENE CLONING

Once extracted, *Pseudomonas* DNA (like any other DNA, prokaryotic or eukaryotic) can be cloned in one or more of the many plasmid-based or lambda phage-based cloning vectors available for propagation in *Escherichia coli* (97). Given the taxonomical proximity between the two genera and the likely expression of *Pseudomonas* genes in *E. coli*, this may suffice in some cases for cloning of genes or short gene clusters encoding scorable phenotypes (83). In this section, however, we will concentrate on tools that have been designed specifically for the cloning and manipulation of *Pseudomonas* genes. Any cloning scheme requires a procedure to extract DNA from the required source, its digestion with restriction enzymes, insertion of the DNA in a suitable vector (that should be able to replicate), and the transfer of the cloned genes back to the original strain. This final critical step requires that the recipient be amenable to some type of gene transfer (transformation, conjugation) and that the vector bearing the cloned gene(s) be endowed with a selection marker that is functional in the *Pseudomonas* host of choice.

37.2.1. Preparation of *Pseudomonas* DNA

Nearly all gram-negative eubacteria yield DNA of good quality for cloning when a protocol is used that includes sodium dodecyl sulfate and proteinase K, followed by selective precipitation with the cationic detergent cetyltrimethylammonium bromide (2). In other instances, only high-molecular-weight plasmid DNA might be desired, in which case the classical protocol of Kado and Liu (55), or some recent improvements (13, 84, 118), might be employed to yield supercoiled DNA for subsequent cloning procedures. Depending on the method (cloning in plasmid or cosmid vectors; see below), the DNA is then partially or totally digested with frequent or infrequent cutters that match the restriction sites available in the vector. Size fractionation of the DNA is best done through centrifugation in sucrose gradients (2). Simpler alternatives (although frequently yielding lower-quality DNA) involve the extraction of DNA from low-melting-point agarose gels (2) or the use of glass beads such as Geneclean (Bio101, Vista, Calif.).

37.2.2. Broad-Host-Range Cloning Plasmids

It is well known that some plasmids (27) are able to propagate only in very specific bacterial hosts, while others can thrive not only between intergeneric barriers (broad-host-range replicons) but even between gram-positive and gram-negative species (promiscuous replicons). This phenomenon involves two aspects (broad-host-range replication and broad-host-range transfer), both of which have been used as the source of functional elements for vector development (11). The host range of replicons generally relies on their relative dependence on host factors for the replication mechanism. The best-known narrow-range replicon is ColE1 plasmid, generally limited to *E. coli* and very closely related enteric bacteria; the plasmid encodes only an origin of replication (*oriV*) and depends entirely on the host for the other proteins required for its propagation. A second class of replicons (best exemplified by the RK2 plasmid) bears not only an *oriV*, but also one cognate replication protein (TrfA, in the case of RK2), which in some cases provides a degree of host tolerance (33). Finally, very promiscuous plasmids such as RSF1010 seem to encode by themselves most, if not all, the gene products required for autonomous replication, thus displaying a remarkable permissiveness with respect to the host (103).

With very few exceptions (89), plasmids have not been found that are able to replicate in a *Pseudomonas* host bearing only an *oriV*, although several replicons with a narrow host range for the genus have been described (3, 49, 85). Conversely, there are a large number of vectors derived from archetypical broad-host-range plasmids such as RK2 (also named RP4), pSa, RSF1010 (11), and R388 (29). In other cases, less well characterized DNA segments endowing autonomous replication in *Pseudomonas* hosts, such as the ori_{1600} of pRO1600 (52, 125) or replicons from *Bordetella* spp. (67, 68), have been employed for vector development, either alone or in combination with a second origin of replication for propagation in *E. coli* (34, 124). A survey of families of vectors based in these and other broad-host-range replicons is shown in Table 1. It should be noted that the degree of effectiveness of each vector for specific projects depends not only on plasmid organization itself, but also on its stability in a particular strain, which is difficult to predict ahead of time. In spite of the many vectors developed for *Pseudomonas* (Table 1), those derived from RSF1010 remain unmatched in their stability and in their wide host range (11). This makes up for the inconvenience of their large size and the presence of undesirable restriction sites in essential locations of the plasmid map (Fig. 1).

37.2.3. The pUCP Vector System

Schweizer and collaborators have produced over the years an entire, and increasingly improved, series of vectors (59, 104, 107) and cassettes (105–109) for genetic analysis of *P. aeruginosa* (the pUCP collection), although the authors claim them to be equally effective for a variety of other *Pseudomonas* species, including *P. putida* and *P. fluorescens*. These relatively small and convenient plasmid vectors (average size 4 to 5 kb) generally bear a segment of DNA (originally named the stabilizing fragment) from pRO1600 spanning the broad-host-range origin of replication (ori_{1600}) and the *rep* gene for its cognate replication protein (125), along with various antibiotic resistance markers, as well as many of the features of the pUC-type plasmids, including color identification of inserts and built-in expression devices (such as *Ptac*/*lacI*^{q} cassettes). More recent vectors of the same series have a trimmed-off stabilizing fragment joined to an RP4 origin of transfer (*oriT*) engineered as an adapter DNA segment, so that users may easily construct plasmid vectors a la carte for specific purposes. The organization and applications of the pUCP vectors have been the subject of a recent review (110). An alternative series of cloning and promoter-probing vectors based on the pRO1600 (Table 1, see below) has been developed by Farinha and Kropinski (34, 35).

37.2.4. Selection Markers

An essential trait of any cloning plasmid is the selection marker associated with the replicon. The sensitivity of the recipient strain under study to the corresponding selective agent (antibiotic or other) should be determined prior to the use of a particular vector. With a few exceptions, such markers involve genes of various origins conferring resistance to ampicillin/piperacillin (Ap^{r}), kanamycin (Km^{r}), streptomycin/spectinomycin (Sm^{r}), chloramphenicol (Cm^{r}), gentamicin (Gm^{r}), or tetracycline (Tc^{r}). Most nonpathogenic *Pseudomonas* strains are sensitive to at least one of these antibiotics, whereas the phenomenon of multiple antibiotic resistance (41) appears to be confined to clinical isolates. In addition to these standard markers, one broad-host-range plasmid free of antibiotic resistance and carrying a *thyA*$^{+}$/thymidylate synthase-autoselective marker has been constructed (99) and employs a *thyA*$^{-}$ mutant strain as the host. This may be useful in cases where resistance to antibiotics of clinical interest is undesirable. A different approach involves the use of nonantibiotic selection markers, such as those engineered in several mini-transposon vectors (see below), but to our knowledge only resistance to iodoacetate endowed by the haloacetate dehalogenase gene (*dehH2*) of a *Moraxella* strain has been engineered as a selective trait in plasmids (60).

37.2.5. Vector Transfer Systems

Genes cloned in broad-host-range vectors are normally targeted back to the *Pseudomonas* strain where they originated. In many cases, electroporation has made it possible to routinely transform *Pseudomonas* cells with a ligation mixture (50, 51, 62), although the outcome of the procedure may vary greatly from strain to strain. If the bacteria under study can be transformed or electroporated at high efficiencies, then it is possible to employ vectors devoid of conjugal transfer elements (such as various members of the pUCP or pCN plasmid series, Table 1), which may decrease the plasmid size and facilitate cloning procedures. An entire cloning system with *P. aeruginosa* as the host has been developed on these principles (59) and which faithfully reproduces the α-*lac* complementation scheme of *E. coli* pUC-type plasmids. Unfortunately, this approach cannot be generalized to other species. In most cases, cloning of genes into *Pseudomonas* requires the use of *E. coli* as the

TABLE 1 Families of plasmid/cosmid vectors for nonpathogenic *Pseudomonas* strains

Vector series	Relevant characteristics	References
pKT/pDSK	Mobilizable RSF1010 derivatives with various selection markers (Cm, Sm, Km), general-purpose cloning vectors	11, 61
pMMB/pNM/pVLT	Mobilizable RSF1010-derived expression vectors based on the *lacI*^{q}/*Ptac* or the regulatory elements of the TOL plasmid; Ap, Cm, Km, Tc, Sm versions available	2, 22, 77, 81, 96
pRK	Mobilizable RK2-based, α-*lacZ*-containing system	11, 31, 61, 75
pCN	*Pseudomonas*-specific cloning vectors based on pPS10 replicon of *P. sevastanoi*; double *oriV* in *E. coli*	85
pML	RSF1010-based expression and *lacZ* fusion vectors with built-in constitutive and *lacI*^{q}/*lac* inducible promoters	70
pLAFR	Mobilizable RK2-based cosmid vectors tailored for cloning DNA from environmental gram-negative strains	40, 61, 115
pJRD	RSF1010-based plasmid and cosmid vectors with restriction site banks	18, 19
pUCP/pQF	Vector collection based on the pRO1600 replicon; includes plasmids for general cloning purposes, *lac*-based insertion detection and expression systems, *lacZ*, *phoA*, *lux* reporter genes, mobilization cassettes, and allelic replacement	34, 35, 104–110, 125

immediate recipient of ligation mixtures, followed by transfer of the cloned gene to the final host by (electro)transformation or conjugation. In fact, conjugal transfer is the only procedure of gene transfer that works for the majority of environmental isolates of *Pseudomonas*. The most efficient system of conjugal transfer present in broad-host-range vectors is that based on the *tra/mob* system of the promiscuous plasmid RK2 (43). The only element of the conjugation machinery that is present in the cloning vectors developed for this purpose is an *oriT* that spans no more than 350 bp of the RK2 sequence. The conjugal functions are provided in *trans* by an RK2 integrated into the chromosome of the donor *E. coli* strain (such as *E. coli* S17-1 or *E. coli* SM10 [111]). In this scheme, the plasmid bearing the cloned genes of interest is transformed into any one of these *E. coli* strains, and the transformants are mated with the *Pseudomonas* recipient. Subsequently, the cells are plated on a medium selecting for the vector marker and the *Pseudomonas* recipient and inhibiting growth of the *E. coli* donor strain. Alternatively, the plasmid of interest can be passed on to *Pseudomonas* through triparental matings between the *E. coli* strain harboring the construct, the *Pseudomonas* recipient, and a specialized *E. coli* helper strain (Fig. 2). The latter harbors a plasmid with a ColE1 replicon (i.e., with a host range limited to *E. coli*) carrying the cloned transfer region of RK2. The most popular *mob*$^+$/*tra*$^+$ helper plasmids of this type are RK2013 (Km^r) (39) or its Cm^r derivative RK600 (26). The mixture of the three strains originates mating trios (Fig. 2) in which the transient transfer of RK2013 or RK600 to the *E. coli* strain harboring the plasmid results in its mobilization towards the *Pseudomonas* host. A detailed protocol to set up biparental or triparental matings can be found in reference 26. These must be assembled on a solid surface or a filter to minimize breakage of the very fragile RK2-encoded pillae. RSF1010 derivatives can also be mobilized by the RK2 transfer system by virtue of an *oriT* sequence within the native plasmid (Fig. 1) that is efficiently recognized by the heterologous conjugal machinery.

A simpler alternative approach to mobilizing cloned genes into *Pseudomonas* hosts consists in forming cointegrates in vivo between *E. coli* plasmid vectors and self-conjugative broad-host-range replicons mediated by a transposon. This has the advantage that the initial cloning steps are made in *E. coli* vectors, but it has the disadvantage that the final construct has a large size and sometimes the organization of the cointegrate is not well defined. Several procedures are available to pursue this approach. In one case (44), pUC-type *E. coli* vectors are used for cloning purposes and the resulting plasmids are then fused in vivo with an IncW broad-host-range plasmid (R388) carrying Tn*10*, thus giving rise to self-conjugative cointegrates. A second procedure exploits the remnants of Tn*3* present in many pBR322 derivatives to insert in vivo genes cloned in *E. coli* vectors in the broad-host-range shuttle and conjugative plasmid R751, catalyzed by the Tn*3* transposase gene provided in *trans* (64). Finally, formation in vivo of cointegrates between *E. coli* plasmids and the R751 replicon can be easily achieved through the transposase$^+$/resolvase$^-$ activity of a defective Tn*813* transposon carried by the broad-host-range plasmid (88).

37.2.6. Cosmids

Typically, genetic analysis of *Pseudomonas* strains starts with the preparation of a cosmid bank. Cosmids have the advantage (increased by the availability of commercially premade lambda phage packaging mixtures) of enabling the entire DNA content of a given strain to be assembled in a collection of some 700 to 1000 overlapping but separate clones spanning the whole genome, each containing approximately 15 to 30-kb segments. By far, the most valuable cosmid for this purpose is pLAFR3 (115), although it has been modified with some useful improvements (61). pLAFR-type cosmid vectors (Table 1) contain a large portion of the broad-host-range plasmid RK2, including the origin of replication, genes for replication proteins, the origin of transfer, and some stability determinants, as well as a λ*cos* site and a polylinker with multiple restriction sites, of which *Bam*HI is the most useful for cloning purposes. A critical step in the preparation of a cosmid bank (a detailed protocol for which can be found in reference 80) is the purity of the partially digested and size-fractionated *Pseudomonas* DNA (see above). For long-term maintenance, it is generally prudent to store the entire cosmid

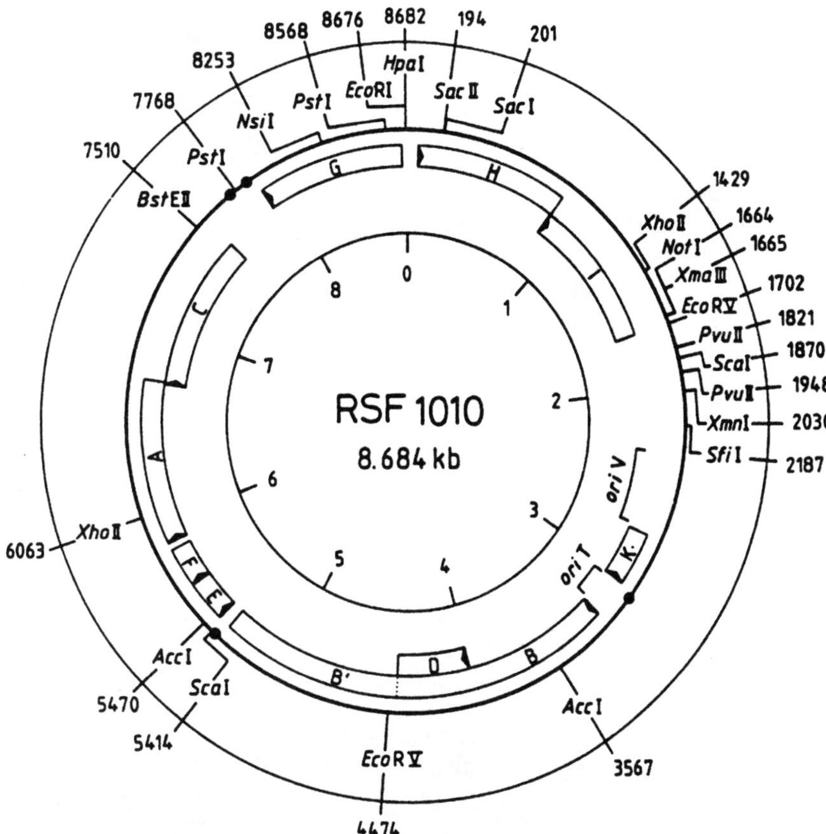

FIGURE 1 Organization of broad-host-range plasmid RSF1010. From the beginning of the molecular studies on *Pseudomonas*, this plasmid has been the most versatile broad-host-range replicon employed for vector development. Many of its very first derivatives, such as the pKT series (11), are still in widespread use owing to their superior stability. The map shows the positions of restriction sites that appear once or twice within the plasmid sequence. The numbers refer to the coordinates of the first nucleotide in the 5′ position. The boxes indicate the location of the genes determining 11 known proteins; two of the genes endow host strains with resistance to streptomycin (genes H and I), and one endows host strains with resistance to sulfonamide (gene G). Letters A, B, B′, and C indicate the replication genes *repA*, *repB*, *repB′*, and *repC*, respectively. Black dots (•) pinpoint the positions of the four major *E. coli* RNA polymerase binding sites. (Reproduced with permission from C. Thomas [ed.], *Promiscuous Plasmids in Gram-Negative Bacteria*. Academic Press Limited, London, 1989.)

collection in 10 to 15-microwell titer plates (Falcon). To each well (Fig. 3) is added approximately 100 μl of LB-Tc medium (for pLAFR3), and each well is inoculated individually with colonies from the selection plates for growth of cosmid-containing clones. Colonies bearing insertions in the cosmid can be easily identified in *lac* indicator media, since the pLAFR3 carries a built-in α-*lac* system (115). After incubation to allow growth of the clones, the wells are supplemented with glycerol up to 20% and then stored at −80°C. Each time the cosmid library is surveyed, a small aliquot of the individual clones is simply retrieved with a multichannel pipette or a square metal comb (Fig. 3). When required, the cosmids can be mobilized from the *E. coli* host into the *Pseudomonas* recipient through triparental conjugation with a helper strain, as described above.

37.2.7. In Vivo Cloning with Defective Phages

The technology developed for gene cloning in vivo in *E. coli* by Casadaban's group with mini-Mu phage elements in the early 1980s (42) has inspired the development of a conceptually similar system for *P. aeruginosa* based on phage D3112 (15). This approach is described in detail in reference 100. Although this system is very useful for a variety of genetic analyses in some *P. aeruginosa* strains, most other variants are not sensitive to the phage and cannot, therefore, be generally used. A more general system of in vivo cloning has been developed that is based on retrotransfer of *Pseudomonas* genes into *E. coli* mediated by RK2/mini-Mu hybrid plasmids (122). While this system has been instrumental for the cloning in vivo of some genes determining catabolic pathways for aromatic compounds, its use is not widespread for *Pseudomonas* because of the very low frequency of the retrotransfer events.

37.3. GENE MAPPING

Until very recently, the only procedure generally available to map genes on the *Pseudomonas* chromosome was that of

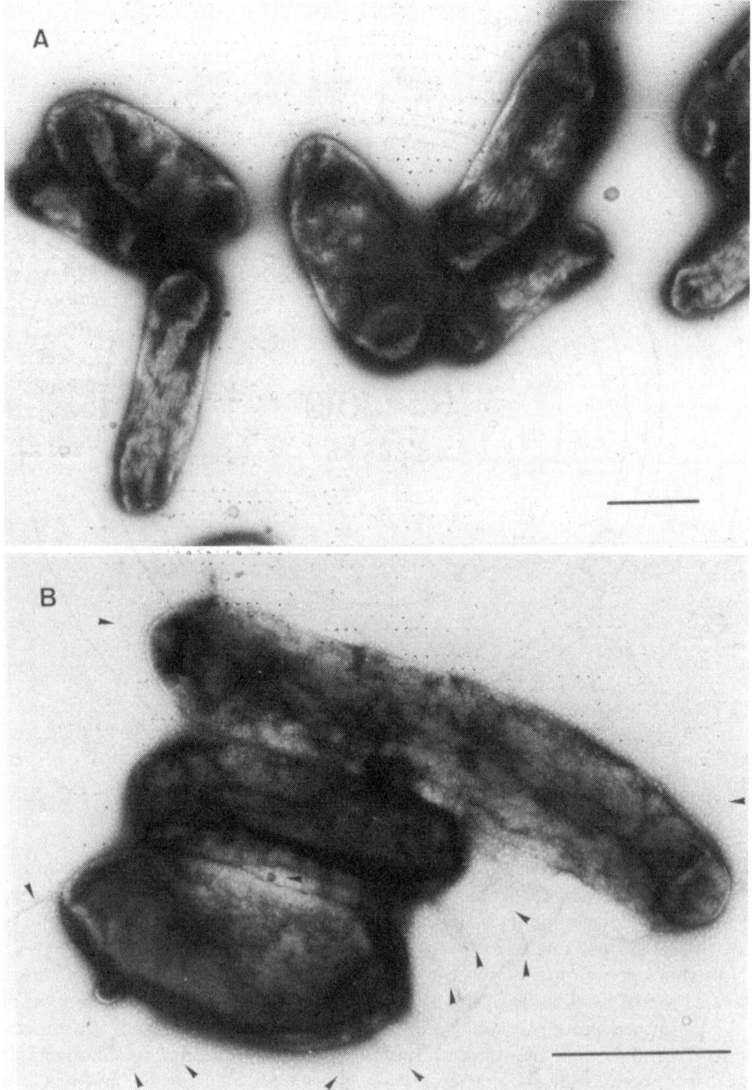

FIGURE 2 Formation of mating trios during triparental conjugation. (A) The electron micrograph shows discrete groupings of 3 cells observed during triparental matings between a donor *E. coli* CC118λ*pir* (pUT/mini-Tn*5* Km, bearing an *oriT*), a recipient *P. putida* KT2442, and a helper strain, *E. coli* HB101 (RK600 tra^+/mob^+). (B) The *Pseudomonas* recipient is the longer cell, while the two cells attached are *E. coli*. Some overproduced surface structures in the *E. coli* cells are indicated with arrows.

polarized chromosomal transfer mediated by conjugative and R-prime plasmids (48, 100). In the case of *P. aeruginosa*, the availability of transducing phages such as F116L and G101 was used to determine the order of some closely linked genes (48). These earlier (and extremely limited) procedures have now given way to far simpler methods of physical and genetic mapping of entire chromosomes based on the use of transposon mutagenesis together with analysis of the whole genome with pulsed-field gel electrophoresis (PFGE), a technology pioneered in *Salmonella* (126). The standard method involves the random mutagenesis of the target strain with mini-transposons (see below) bearing one or more very unusual restriction sites. Transposon insertion results in the introduction of a novel restriction site that can be quickly pinpointed on a physical restriction map of the chromosome made by PFGE. The phenotype created by the insertion can thus be associated with a specific chromosomal location (126). This approach, when combined with Southern blotting of the PFGE gels (2) and hybridization probing with DNA of cloned genes, enables localization of specific genes within particular restriction fragments of the physical map.

37.4. GENE INACTIVATION

Any genetic analysis starts with the generation of mutants. A major technical breakthrough for the manipulation of nonenteric bacteria was published in 1983 by Simon et al. (111). A general transposon mutagenesis procedure was described that is based on the delivery of the Tn*5* transposon into the host by a "suicide" plasmid via RK2-mediated con-

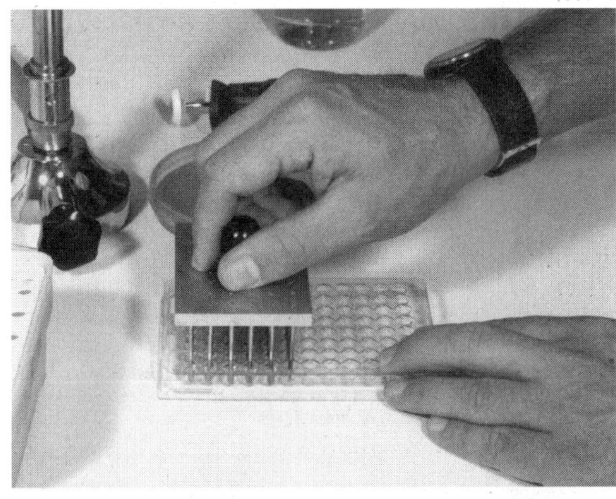

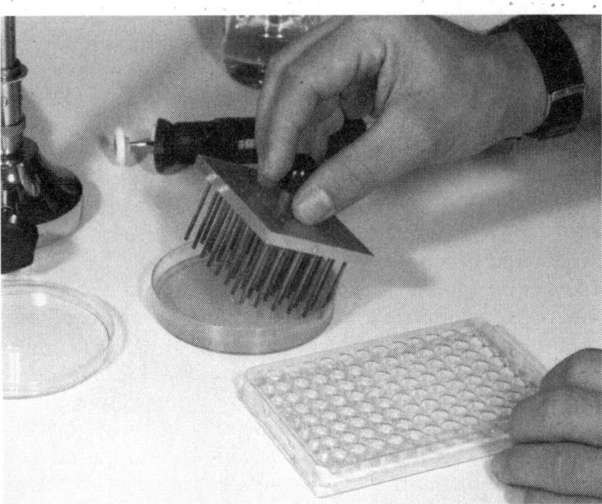

FIGURE 3 Manipulation of cosmid libraries of *Pseudomonas* strains kept in microtiter plates. The photograph shows an easy procedure for simultaneously retrieving individual clones from a cosmid library of *P. putida* KT2442 maintained in frozen *E. coli* cultures in a microtiter plate. By using a matching sterile comb, each clone can be transferred sequentially to a second microwell plate, along with an *E. coli* helper strain and a *Pseudomonas* recipient strain to allow conjugal transfer. Finally, each of the mating mixtures can again be transferred with the same comb to a selective medium or to a membrane for immunological or radioactive detection of the desired genes. This procedure avoids underrepresentation of certain clones and allows long-term storage of the same library for use over the years.

jugal transfer. The two key mobilizing strains of *E. coli* developed for this purpose (*E. coli* S17-1 and *E. coli* SM10) carry an RP4 derivative integrated into their chromosome which provides conjugal transfer functions to a Tn*5*-bearing pBR322 derivative carrying an RP4*oriT*. This arrangement allows the mating of the donor strains bearing mobilizable pBR322::Tn*5* derivatives with the recipient *Pseudomonas* strain, followed by plating on selective media for insertions of the Tn*5* in the chromosome of the recipient. Since the Tn*5*-bearing plasmid is a suicide vector, i.e., can pass into but cannot replicate in the recipient, selection for transposon markers (and loss of the other plasmid markers) can arise only as a consequence of the transposition of the mobile element to the chromosome of *Pseudomonas*. Although the specific tools available for transposition mutagenesis have been considerably refined over the years (see below), the concept developed by Simon et al. (111) remains unchallenged as the best way to insert transposons into the chromosomes of many otherwise intractable gram-negative strains. However, alternative, less used systems for mutagenesis of *Pseudomonas* strains with Tn*5* do exist and are based on broad-range plasmids such as P4 (92) or involve other transposons such as Tn*903* (28) and Tn*1721*/Tn*1722* (56, 120) or IS*1* (37).

37.4.1. Tn*5*-Derived Mini-Transposons

Tn*5* is a composite transposon, i.e., its mobility is determined by two insertion sequences (IS50L and IS50R) flanking the DNA region encoding the kanamycin/neomycin resistance genes. Interestingly, the transposase determined by IS50R (the product of the *tnp* gene) still functions when the gene is artificially placed outside the mobile unit, though preferably placed in *cis* to the cognate terminal sequences. This has allowed the construction of recombinant transposons (mini-transposons) in which only those elements essential for transposition (i.e., insertion element terminal sequences and transposase gene) have been retained and arranged such that the transposase gene is adjacent to but outside the mobile DNA segment (46). Since the elimination of nonessential sequences leads to a major reduction in transposon size, the resulting recombinant mini-Tn*5* transposons are much more convenient to handle than natural transposons. Because of the loss of the *tnp* gene after insertion, mini-transposons are stably inherited and are not prone to cause DNA rearrangements or other forms of genetic instability. A related feature of mini-transposons is that cells containing transposon insertions do not become "immune" to further rounds of transposition, thus allowing the organism to be remutagenized, providing, of course, that subsequent transposons contain distinct selection markers (24, 25, 46). A critical feature of the mini-transposons is the system used to deliver them into the target strain, based on the narrow-host-range plasmid R6K. Plasmids having the R6K origin of replication require the R6K-specified replication protein π and can be maintained only in host strains producing this protein. Mini-transposon delivery plasmids have the R6K origin of replication as well as the origin of transfer (*oriT*) of RK2. Thus, delivery plasmids can be maintained stably in λ*pir* lysogens (26) or in *E. coli* strains with the *pir* gene recombined in their chromosome (78) and can be mobilized into target *Pseudomonas* cells through RP4 transfer functions. A very detailed description of the practical use of mini-Tn*5* transposons can be found in reference 26. The use of mini-transposons as vectors for chromosomal insertion of heterologous DNA is discussed below. An alternative series of mini-transposons based on IS*1* have been constructed and successfully used for genetic analysis of environmental *Pseudomonas* isolates (37, 54).

37.4.2. Directed Mutations: Gene Targeting and Allelic Replacement

Various strategies of gene targeting have become available that are based on replacement of the wild-type gene in the chromosome, by homologous recombination, employing a variant of that sequence created on the cloned gene. Alternatively, a specific gene might be deleted or inserted with a heterologous DNA sequence encoding a selectable or scorable marker (105, 108, 109). Since only a few *Pseudomonas* strains (for example, *P. stutzeri*) are naturally transformable (73, 117), in most cases the task involves (i) the production of the desired mutation in the gene of interest (insertion, deletion, variation) carried on a narrow-host-range plasmid, (ii) the transfer (via conjugation or transformation) of the resulting construct into the target *Pseudomonas* cells for formation of a cointegrate through a single crossover between plasmid and chromosomal sequences, and (iii) the resolution of the cointegrate. This last step used to be the most time-consuming one, because duplications and inversions of identical sequences during chromosomal replication may give rise to DNA structures that do not match the expected resolution products. A simple vector for gene replacement in *Pseudomonas* is pJP5603 (90). Eventually, it is a Km^r derivative of the Ap^r precursor pGP704 (26) and, like this vector, has the R6K origin of replication and the origin of transfer (*oriT*) of RK2. To target a certain chromosomal site, the gene of interest is cloned in the polylinker region of the plasmid and inactivated through internal deletions, insertion of an antibiotic-resistance cassette, or site-directed mutagenesis. A substantial region of DNA (preferably >1 kb) should be present on both sides of the mutated sequence to allow homologous recombination to occur at a reasonable frequency. The resulting construct is then mobilized into the target strain by mating with a λpir^+ RK2*mob*$^+$ strain of *E. coli* transformed with the plasmid (or through triparental mating; see above) and transformants selected for on medium containing kanamycin. In the main, this results in cointegrates between the plasmid and the chromosome at the corresponding location. A second homologous recombination event deletes the plasmid markers. This is achieved by regrowth of the exconjugants in a medium devoid of antibiotics and the patching of individual colonies on agar plates with or without the antibiotic. Once the desired phenotype has been obtained, the physical structure of the chromosomal mutation (insertion, deletion) is analyzed by Southern blotting (2) or by the PCR of the DNA with primers targeted towards the region of interest. Recently, the use of the *sacB* gene of *Bacillus subtilis* as a conditionally toxic marker has considerably simplified the selection of the second crossover event leading to cointegrate resolution. A number of vectors, exemplified by pKGN101 (58), pEX100T (109), and pUCD4121 (57), are available that bear a *sacB* gene. As before, the constructs based on these or other plasmids (94, 102) are transferred into *Pseudomonas* strains by conjugation or transformation, and the exconjugants or transformants are selected for cointegrate formation. Following growth in the nonselective medium, clones bearing a second crossover can be selected for on a medium containing sucrose which is converted to toxic levanes upon exposure to the *sacB*-encoded enzyme levansucrase. This kills the cells containing nonresolved *sacB*$^+$ cointegrates and selects those that have lost it. A second strategy of gene inactivation involves selection for double recombination events leading to allelic replacement and is based on the presence in the plasmid vector of the *rpsL* gene (encoding ribosomal protein S12) which confers sensitivity to streptomycin when present together with the mutated Sm^r allele of the gene. In this case, cointegrates have an Sm^s phenotype, whereas those that have lost the marker return to being Sm^r (94, 113, 114). Clearly, an Sm^r mutant of the target strain should be available for this procedure.

37.5. CHROMOSOMAL INTEGRATION OF HETEROLOGOUS DNA SEGMENTS

Although plasmids have been the favorite workhorses for gene cloning and expression projects, they do present a number of difficulties when it is necessary to generate a recombinant phenotype in industrial or environmental conditions. This is due to the inherent instability of recombinant plasmids and the physiological stress that may be associated with their multicopy nature (25). For industrial strains, or for those destined for environmental or agronomical release, it is therefore essential that the desired traits be chromosomally encoded, thus ensuring the stability and predictability of the engineered phenotype (20, 21).

37.5.1. Transposon Vectors

The notion of using transposons as vectors for stable insertions of foreign DNA into a target chromosome was first developed in the late 1980s (reviewed in reference 25) but was not fully developed until the advent of the Tn*5*-derived mini-transposons described above (26). In addition to being useful for mutagenesis, mini-transposons provide a straightforward tool for cloning and stably inserting foreign genes into the chromosomes of various gram-negative bacteria, including, of course, *Pseudomonas*. Their original design included the presence of single *Not*I or *Sfi*I restriction sites internal to the Tn*5* inverted repeats. Thus, hybrid transposons bearing the DNA segment of interest (up to 20 kb) can be easily assembled with the help of the specialized cloning vector, pUC18Not, that flanks with *Not*I sites any restriction fragment cloned at the polylinker. Some modifications of the original design have been made that provide additional restriction sites for cloning purposes (1). The modular nature of the mini-transposons facilitates the construction of mobile elements a la carte, in which not only new selective markers can be introduced (Sm, Km, Cm, Tc, Gm), but also DNA fragments of various origins and a variety of expression devices can be engineered (see below). As mentioned above, multiple insertions in the same strain are limited only by the availability of distinct selection markers. On this basis, mini-transposons are, to this day, the most useful genetic tool for any metabolic engineering project that involves the combination in one strain of otherwise separate traits encoded by cloned DNA segments (119).

37.5.2. Excision of Selection Markers in the Chromosome

Previous work on the multimer resolution system (*mrs*) of the broad-host-range plasmid RP4 has shown that DNA segments flanked by tandem *res* sites and cloned in a multicopy vector can be precisely deleted in vitro and in vivo by the product of the *parA* gene. The *mrs* system has been exploited to develop a general method that permits the precise excision of antibiotic resistance markers present in mini-transposon vectors after the transposition event has taken place (Fig. 4). This is based on a site-specific recombination between two directly repeated 140-bp resolution (*res*) sequences of RP4 effected by a resolvase encoded by the *parA* gene and borne by a conditional replication plasmid. This strategy permits the stable inheritance of heterologous DNA segments virtually devoid of the sequences used initially to select their insertion (69). The mechanism of multimer resolution (which works over 60 kb in RP4) makes this strategy generally applicable to DNA fragments of almost any size. Although various applications of the

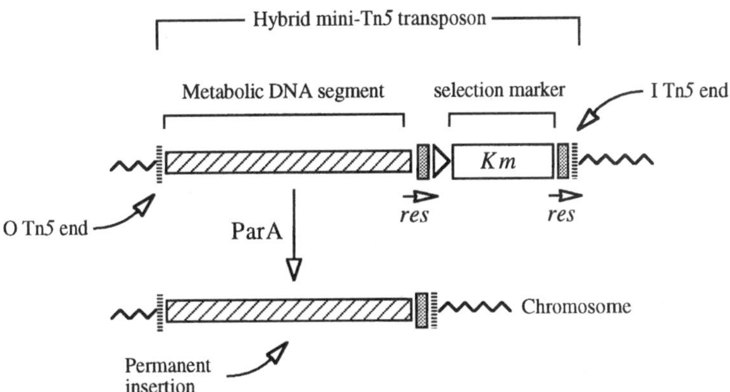

FIGURE 4 Metabolic engineering of *Pseudomonas* with mini-transposon vectors bearing excisable antibiotic resistance markers. The antibiotic resistance markers compatible with the collection of transposon vectors described in reference 26 can be placed within the mobile element flanked by tandem *res* sites of the multimer resolution system of plasmid RP4 (69). In this way, chromosomal insertion of a DNA segment (encoding a metabolic function of interest) cloned in the mini-transposon can be selected through the marker present in the vector (Km in the figure). Subsequent transient expression of the cognate resolvase *parA* gene, from an unstable plasmid, causes the selection marker gene to be specifically deleted, thus leaving behind only the DNA segment of interest stably inserted in the chromosome (69).

ParA-*res* system can be easily envisioned, it seems to be especially suitable for generating chromosomal insertions of heterologous DNA segments that are devoid of any selection marker, as is generally required for metabolic engineering of strains destined for environmental release (119).

37.6. ENGINEERING REGULATED GENE EXPRESSION

Expression cassettes bearing conditional promoters responsive to chemical or environmental signals have been engineered both in plasmids and in transposon vectors. A number of useful, mostly plasmid-based, expression systems are reviewed in reference 11. In this section, we will concentrate on a limited number of procedures which, in our hands, have been instrumental for various heterologous expression projects in *Pseudomonas*.

37.6.1. Plasmid-Based Expression Systems

For the reasons discussed above, we have systematically used RSF1010 replicons as the basis for further development of the constructs made by Bagadasarian's group (81), in which *lacI*^{q}-*Ptac* expression cassettes were assembled in front of a polylinker and were accompanied by genes endowing resistances to β-lactams or choramphenicol. Since many, if not most, *Pseudomonas* strains are sensitive to these antibiotics, a number of additional plasmids have become available (22, 70) with different markers—Sm, Tc, or Km—with the same basic design (Fig. 5). Although regulated promoters of widespread use in *E. coli*, such as *Ptac*, *Ptrp*, and $\lambda P_{L/R}$, are useful for manipulations of *Pseudomonas* strains in the laboratory, they may not be suitable for large-scale operations because of the high price of the inducer or complications involved in activating the system. An interesting contribution to the solution of this problem employs plasmid pNM185 (77), an RSF1010-derived expression plasmid that utilizes the *Pm* promoter of the TOL plasmid, together with its native regulator *xylS*, to achieve transcription of heterologous genes in response to the addition of very cheap inducers such as benzoate. This vector has been further improved (96) by creating mutant variants of the *xylS* gene that are responsive to different alkyl- and chloro-benzoates, some of which are environmental pollutants. Along the same lines, an expression cassette based on the *nahR*-regulated P_G (called also *Psal*) promoter of the NAH7 catabolic plasmid has also become available (129) and directs salicylate-inducible transcription of downstream heterologous genes. These types of expression devices open up the possibility of designing strains in which the expression of a recombinant gene is effected by a signal present in a contaminated location, in some cases even by the very compound that the strain has been designed to target (21). Other gene-inducible systems in *Pseudomonas* are based on the regulation of the tryptophan operon (86) and can be controlled through addition of indoleglycerol phosphate.

37.6.2. Transposon-Based Expression Systems

An easy procedure for having a gene or operon transcribed in response to a predetermined signal is to construct a specialized mini-Tn*5* transposon in which the promoterless gene or gene cluster is placed next to the I or O termini of the mobile element. The strain of interest is then mutagenized with such a transposon and insertions that happen to occur next to a promoter regulated by that signal are screened, e.g., by immunological procedures (21). In many cases, however, the construction of specialized transposons of this sort is impractical, and some information on promoter strength and regulation may be required prior to engineering expression of a heterologous gene placed artificially downstream. The range of promoter strengths available in a given gram-negative strain can be assessed with Tn*5* or mini-Tn*5* derivatives carrying a variety of convenient reporter genes, such as *lacZ*, *xylE*, or *luxAB* (24, 26, 112). A much better solution is to start with a promoter for which some information is available. As with their plasmid-based counterparts, various systems based on *lacI*^{q}

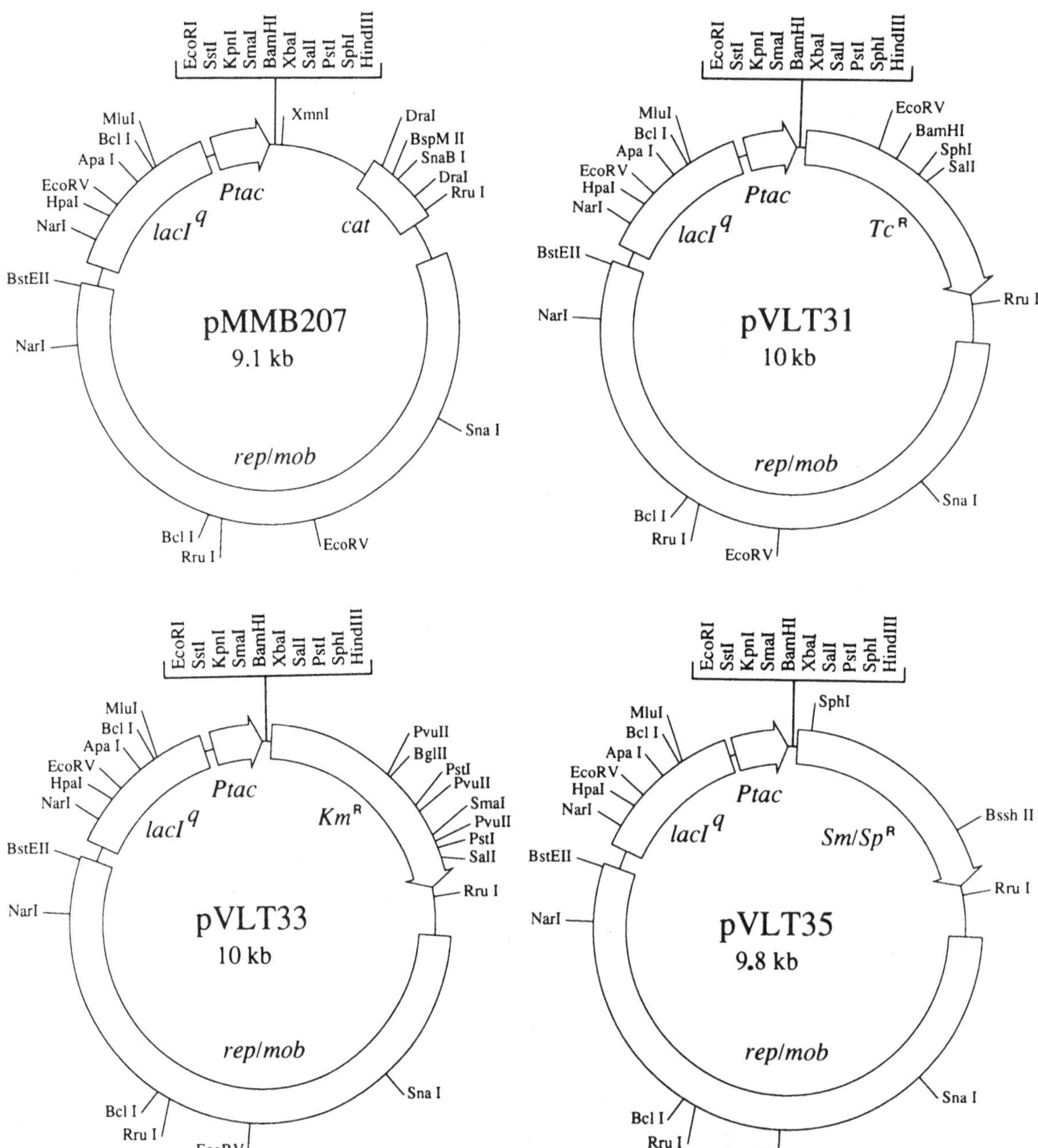

FIGURE 5 RSF1010-derived, *lacI^q/Ptac*-based expression vectors for *Pseudomonas*. pMMB207 is the basic plasmid from which the other three expression vectors, pVLT31, pVLT33, and pVLT35, are derived. For their construction, the *cat* gene of pMMB207 was substituted by either the Tc^r Km^r- or Sm^r/Sp^r-encoding genes of the Ω interposons (22). Functional elements of the plasmid, relevant restriction sites, and the location of the replication and mobilization functions (*rep/mob*) are indicated. Antibiotic selection of these plasmids can be applied in either rich or minimal medium. Suggested antibiotic concentrations for selection: 12 μg of tetracycline per ml, 50 to 75 μg of kanamycin per ml, or 50 to 100 μg of streptomycin per ml, 50 to 300 μg of chloramphenicol per ml. (Reproduced with permission from reference 22).

and *Ptac* or *Ptrc* have been assembled in mini-transposons that permit conditional expressions of the desired cloned genes in a fashion dependent on *lac* inducers (22, 91). However, the best-studied promoters in *Pseudomonas* are those that control expression of catabolic operons for biodegradation of aromatic compounds (74), and a number of recombinant mini-Tn*5* transposons have been constructed that contain outward-facing *Pm*, *Pu*, or *Psal* promoters from the catabolic plasmids TOL and NAH of *P. putida*, along with their cognate wild-type regulatory genes (*xylS*, *xylR*, *nahR*) or mutant varieties (*xylS2*). Transcription from such promoters is activated when the host bacteria is exposed to certain aromatic compunds, such as alkyl- and halo-benzoates (XylS, XylS2), alkyl- and halo-toluenes (XylR), or salicylates (NahR). These transposons permit the generation of conditional mutants dependent on the presence of specific effectors, as well as the engineering of strains expressing heterologous genes that are regulated by aromatic inducers (23). The broad host range of the Tn*5* transposition system and the stability of the inserted genes, due to the loss of the transposase gene during delivery of the mobile element, make these transposons well suited for the construction of stable strains exhibiting halo/alkyl aromatic-regulated conditional phenotypes in the absence of antibiotic selection, as is required for some uncontained bioremediation and biomonitoring applications (20, 21).

In our hands, the most practical expression system in *P. putida* is that based on *xylS/Pm*, since it possesses a relatively low basal level of expression and is quickly triggered (in monocopy gene dosage) 500- to 1,000-fold upon addition of the inducer *m*-toluate (23). For other purposes, the induction pattern of mini-Tn*5 xylR/Pu* is particularly interesting because *Pu* activity is virtually indistinguishable from background in the absence of inducer, whereas it reaches maximum activity after overnight exposure to vapors of the effector *m*-xylene (74). The distinct regulation of *Pu* results in the absence of significant transcription in uninduced cells and little or no promoter activity during the exponential growth of induced cells. Since the expression system can be activated upon exposure to vapors of inducer rather than direct addition to the medium, activation of a desired gene may be arranged in a nondisruptive manner.

37.6.3. Designing Transcriptional Cascades

The functionality of the polymerase of T7 phage in *Pseudomonas* (5, 17) has been employed not only to devise expression systems similar to their *E. coli* counterparts, but also to design regulatory cascades in which two or more promoters placed at different places on the chromosome are under the control of a single environmental signal (45). In particular, a specialized transposon, mini-Tn*5 xylS/Pm::T7pol*, is available that contains the structural T7 gene *1* (T7pol, encoding the phage RNA polymerase), downstream of the XylS protein/benzoate-regulated *Pm* promoter of the *meta*-operon of the TOL catabolic plasmid. When this transposon is stably inserted into the chromosome of a *P. putida* target strain, the T7 gene *1* is transcribed upon exposure of the bacteria to benzoate effectors of the XylS regulator. Genes whose expression is to be mediated by T7pol are cloned in mini-Tn*5* transposons containing T7 promoter sequences upstream of the cloning site, and then the hybrid transposons are inserted into different positions in the same chromosome. In this way, expression of the genes cloned within the mini-Tn*5* vectors is dependent on the T7pol/XylS/*Pm* system. This expression device is particularly well suited for applications in which the expression of two or more genes is to take place in response to a single chemical signal, such as exposure to certain aromatic compounds.

37.7. MONITORING GENE EXPRESSION

Following the pioneering use of the *E. coli* β-galactosidase system for monitoring the activity of the *lacZ* gene, the use of reporter genes has been at the basis of nearly every study on gene regulation in bacteria. A variety of indicator genes—*lacZ* (65, 70, 87), *phoA* (24, 35), *GUS* (128), *xylE* (12, 66, 101, 109, 116), *lux* (24, 35, 38), *luc* (8, 14), and *ice* (72)—encoding enzymatic, optical, and physical markers are currently available for *Pseudomonas*. Many of these genes have been used to study specific promoters in response to various environmental signals or, simply, to tag the cells in order to distinguish them from other members of a microbial community (36, 93). More recently, the use of fluorescent substrates of β-galactosidase along with digitized video microscopy and the popularization of the jellyfish green fluorescent protein as a universal reporter (6, 9, 10) have permitted the visualization of specific gene expression at the level of single cells.

37.7.1. Plasmid-Based and Transposon-Based Promoter Probe Vectors

A large number of plasmids are available that bear promoterless reporter genes (34, 35, 38, 66, 70). In other cases, specific reporter genes such as *lacZ* (65, 108) or *xylE* (101, 108, 109, 116) are available as DNA cassettes that can be conveniently excised and placed downstream of promoter clones in different plasmids. An alternative approach to promoter probing involves the use of transposons with built-in promoterless reporter genes such as *lacZ*, *ice* (or *ina* [ice nucleation]), *phoA*, *luxAB*, or *xylE* (24, 26, 54, 72, 112). A number of mini-Tn*5* derivatives have been adapted to this end (24). These constructs lack transcriptional terminators within the I and O terminal sequences of Tn*5* and, therefore, insertion of the mini-transposons in the appropriate orientation downstream of chromosomal promoters may activate the expression of a promoterless gene placed within the mobile unit. Procedures to measure the activity of the standard reporter genes in the mini-Tn*5* transposons have been described in detail (26). Besides Tn*5* derivatives, promoter-probe transposons are available that are based on Tn*903* (28), Tn*1721*/Tn*1722* (56, 120), and IS*1* (54) and have been successfully used in various *Pseudomonas* strains.

37.7.2. Surface Reporters: Immunological Detection of Promoter Activity

In addition to these general procedures, an immunological method to detect the switch on and off of *Pseudomonas* promoters at the level of single cells has been developed employing immunodetection of a reporter epitope expressed on the surface of bacterial cells (7). For this, the antigenic sequence Asp-Leu-Pro-Pro-Asn-Ser-Asp-Val-Val-Asp, from a coronavirus, was inserted genetically in the permissive site around amino acid position 153 of the LamB protein (receptor for lambda phage and specific pore for maltodextrins) of *E. coli*, giving rise to a hybrid protein named LamB-A6. When the hybrid *lamB* gene is transcribed, the epitope is presented on the surface of the bacterial cells in a configuration available to specific antibodies. The genetic coupling of promoter activity to the display of a strongly antigenic sequence on the surface of *Pseudomonas* cells in a conformation amenable to specific recognition by monoclonal antibodies permits visualization of events such as the activity of the *m*-toluate-responsive *Pm* promoter of the TOL plasmid pWW0 in individual cells of *P. putida* inoculated in the activated sludge of an oil refinery. In contrast to other procedures used so far to follow the activity of bacterial promoters, this indicator system relies on a physical property of the reporter product (its recognition by a cognate antibody) instead of an enzymatic or optical trait. The extremely rare amino acid sequence of the viral epitope used in this system guarantees its specific recognition by the matching monoclonal antibodies, even when the cells under study are a very minor component of a complex bacterial population. Although the LamB-A6 protein has been used so far only to examine the activity of TOL plasmid promoters for biodegradation of toluene, its use can be considered in cases where transcription is to be assessed in environments where the cells under scrutiny must be studied in situ (7). A similar approach with a hybrid PhoE-ColA as a reporter was successfully used, with some limitations, to detect activity in situ of promoters of the plant root-colonizing strain *P. putida* WCS358 with an anti-colicin A monoclonal antibody (130). Also, the OprF porin of *P. aeruginosa* has been found to bear permissive sites (127) suitable for genetic insertion of heterologous peptides, thus adding one more asset to those outer membrane proteins amenable for being developed as surface reporters.

37.8. ENGINEERING BIOLOGICAL AND GENETIC CONTAINMENT

An increasingly important application of *Pseudomonas* strains in biotechnology is their use as biocatalysts for environmental bioremediation (119), or as biocontrol agents for plant pests (16, 32). This implies the deliberate release of certain constructs into the environment in large amounts, thus creating a degree of impact on native bacterial populations. The uncertainty of such an impact is exacerbated when the strains in question have been modified genetically, thus entering the legal category of genetically modified organisms. Although the availability of specialized vectors for the construction of genetically modified organisms, in particular, transposon vectors, has notably increased the predictability of genetically improved strains (20, 119), there is general uncertainty about possible horizontal transfer of recombinant DNA into other members of the microbial community (95).

37.8.1. Biological Containment: Programming Cell Death

The pioneering conditional suicide systems developed in Molin's group (79) were based on the stochastic induction with time of a lethal function (the *hok* gene of the plasmid R1 and the *gef* gene of *E. coli*) engineered in a plasmid vector. Unfortunately, the same system does not work efficiently when placed in the chromosome, thus preventing its application to *Pseudomonas* strains destined for environmental release. However, the same killing functions work very efficiently when rationally assembled within regulatory elements of catabolic pathways for degradation of aromatic compounds (53), with the purpose of causing cell death once the compound to be degraded has disappeared from the medium. In one successful scheme, a *P. putida* strain able to grow on *m*-toluate was engineered such that the aromatic-responsive *Pm* promoter drove expression of a repressor which, in turn, controlled transcription of the toxic *gef* gene. In the presence of the substrate to be degraded (*m*-toluate), activity of the *Pm* promoter caused repression of the toxic gene, thus allowing growth. However, once the aromatic compound was depleted, the same circuit allowed expression of the *gef* gene, giving rise to cell death (53).

This scheme and others employing the same concept (79) have proved useful in restricting the viability of recombinant cells by factors of $>10^{-3}$ to 10^{-4}, and circuits with additional killer genes may prove to be even more effective.

37.8.2. Inhibiting Gene Transfer

A second level of containment deals with directly restricting horizontal gene transfer between the carrier of the recombinant DNA and its potential partners in a bacterial community. The most efficient design to this end is that developed in Timmis's group (30) for killing occasional recipients that may acquire the DNA by any of the gene transfer mechanisms operating in nature. This scheme is based on the properties of the *E. coli* colicin E3. The synthesis of this broad-spectrum bacterial toxin in naturally producing cells is nonlethal for the host because of the simultaneous production of an immunity function (imm^+) that blocks colicin activity among a wide range of microbial ribosomes. When the imm^+ gene is placed with a transposon vector in the chromosome of a genetically engineered organism whose containment is desired, the cells are able to tolerate the presence of plasmids bearing the colicin gene ($colE3^+$) or the insertion of the gene somewhere in the genome. Any transfer event that leads to the acquisition of the *colE3* gene without the immunity gene results in the death of the recipient. Depending on the particular scheme, such a circuit decreases the chances of fortuitous transfer to ecologically insignificant levels. *Pseudomonas* strains designed for degradation of polychloro biphenyls and endowed with such a gene containment circuit (82) decreased their transfer frequencies to $<10^{-7}$.

37.9. CONCLUSIONS AND OUTLOOK

As described in this chapter, *Pseudomonas* is perhaps second only to *E. coli* in the number of genetic tools available for its analysis and for the construction of phenotypes of interest for diverse industrial, agronomic, and environmental applications. In particular, the widespread use of transposon vectors for chromosomal insertion of an almost unlimited number of heterologous DNA segments affords the development of engineered strains with a high degree of predictability in their performance and behavior. In fact, current genetic technology permits construction of recombinant strains that are virtually indistinguishable from "natural" isolates in stability and predictability. Furthermore, the emerging tools and circuits for biological and genetic containment permit the ability not only to program a certain phenotype, but also to anticipate when and where it is to be displayed and even to program its physical disappearance. In this respect, it is worth mentioning a recent report on the possibility of lysing *Pseudomonas* cells upon exposure to an aromatic inducer such as *m*-toluate (63). This is based on the conditional expression of a lysis gene under the control of the *Pm* promoter of the TOL plasmid, which, as mentioned above, strongly responds to inexpensive inducers such as benzoate and its derivatives. This opens the way, for instance, to construction of industrial strains with a built-in ability to facilitate downstream processing operations, so that biocatalyst design can proceed in parallel with the requirements of process development.

This chapter was written under the auspices of the BIOTECH Program of the European Commission in Biosafety. Work was financed by the Comisión Interministerial de Ciencia y Tecnología Grant BIO95-0788 and Contract ENV4-CT95-0141 from the European

Union. JMSR was the recipient of a predoctoral fellowship from the Spanish Ministry for Education and Science (1992–1995). We are indebted to K. Timmis for his inspiration and support.

REFERENCES

1. **Alexeyev, M., I. Shokolenko, and T. Croughan.** 1995. New mini-Tn5 derivatives for insertion mutagenesis and genetic engineering in Gram-negative bacteria. *Can. J. Microbiol.* **41:**1053–1055.
2. **Ausubel, F., R. Brent, R. Kingston, D. Moore, J. Seidman, J. Smith, and K. Struhl (ed.).** 1989. *Current Protocols in Molecular Biology*. John Wiley & Sons, New York.
3. **Boivin, R., G. Bellemare, and P. Dion.** 1994. Novel narrow host range vectors for direct cloning of foreign DNA in *Pseudomonas*. *Curr. Microbiol.* **28:**41–47.
4. **Brazil, D., L. Kenefick, M. Callahan, A. Haro, V. de Lorenzo, D. Dowling, and F. O'Gara.** 1995. Construction of a rhizosphere *Pseudomonad* with potential to degrade polychorinated biphenyls and detection of *bph* gene expression in the rhizosphere. *Appl. Environ. Microbiol.* **61:** 1946–1952.
5. **Brunschwig, E., and A. Darzins.** 1992. A two component T7 system for expression of genes in *Pseudomonas aeruginosa*. *Gene* **111:**35–41.
6. **Burlage, R., Z. Yang, and T. Mehlhorn.** 1996. A transposon for green fluorescent protein transcriptional fusions: application for bacterial transport experiments. *Gene* **173:** 53–58.
7. **Cebolla, A., C. Guzmán, and V. de Lorenzo.** 1996. Nondisruptive detection of activity of catabolic promoters of *Pseudomonas* with an antigenic surface reporter system. *Appl. Environ. Microbiol.* **62:**214–220.
8. **Cebolla, A., M. Vázquez, and A. Palomares.** 1995. Expression vectors for the use of eukaryotic luciferases as bacterial markers with different colors of luminiscence. *Appl. Environ. Microbiol.* **61:**660–668.
9. **Chalfie, M.** 1995. Green fluorescent protein. *Photochem. Photobiol.* **62:**651–656.
10. **Christensen, B., C. Sternberg, and S. Molin.** 1996. Bacterial plasmid conjugation on semisolid surfaces monitored with the green fluorescent protein (GFP) from *Aequorea victoria* as a marker. *Gene* **173:**59–65.
11. **Christopher, F., H. Franklin, and H. Spooner.** 1989. Broad host range cloning vectors, p. 247–267. *In* C. Thomas (ed.), *Promiscuous Plasmids in Gram-Negative Bacteria*. Academic Press Limited, London.
12. **Clark, E., and G. Cirvilleri.** 1994. Cloning cassettes containing the reporter gene *xylE*. *Gene* **151:**329–330.
13. **Cork, D., and A. Khalil.** 1995. Detection, isolation and stability of megaplasmid encoded chloroaromatic herbicide degrading genes within *Pseudomonas* species. *Adv. Appl. Microbiol.* **40:**289–321.
14. **Coronado, C., M. Vázquez, A. Cebolla, and A. Palomares.** 1994. Use of firefly luciferase gene for plasmid copy number determination. *Plasmid* **32:**336–341.
15. **Darzins, A., and M. Casadaban.** 1989. *In vivo* cloning of *Pseudomonas aeruginosa* genes with mini-D3112 transposable bacteriophage. *J. Bacteriol.* **171:**3917–3925.
16. **Davison, J.** 1988. Plant beneficial bacteria. *Bio/Technology* **6:**282–286.
17. **Davison, J., N. Chevalier, and F. Brunel.** 1989. Bacteriophage T7 RNA polymerase controlled specific gene expression in *Pseudomonas*. *Gene* **83:**371–375.
18. **Davison, J., M. Heusterspreute, and F. Brunel.** 1987. Restriction site bank vectors for cloning in Gram-negative bacteria and yeast. *Methods Enzymol.* **153:**34–55.
19. **Davison, J., M. Heusterspreute, N. Chevalier, V. Ha-Thi, and F. Brunel.** 1987. Vectors with restriction site banks. pJRD215, a wide host range cosmid vector with multiple cloning sites. *Gene* **51:**275–280.
20. **de Lorenzo, V.** 1992. Genetic engineering strategies for environmental applications. *Curr. Opin. Biotechnol.* **3:** 227–231.
21. **de Lorenzo, V.** 1994. Designing microbial systems for gene expression in the field. *Trends Biotechnol.* **12:**365–371.
22. **de Lorenzo, V., L. Eltis, B. Kessler, and K. Timmis.** 1993. Analysis of gene products of *Pseudomonas* with *lacI^q/Ptrp-lac* plasmids and transposons conferring conditional phenotypes. *Gene* **123:**17–24.
23. **de Lorenzo, V., S. Fernández, M. Herrero, and K. Timmis.** 1993. Engineering of alkyl- and haloaromatic-responsive gene expression with mini-transposons containing regulated promoters of biodegradative pathways of *Pseudomonas*. *Gene* **130:**41–46.
24. **de Lorenzo, V., M. Herrero, and K. Timmis.** 1990. Mini-Tn5 transposon derivatives for insertion mutagenesis, promoter probing and chromosomal insertion of cloned DNA in gram-negative bacteria. *J. Bacteriol.* **172:**6568–6572.
25. **de Lorenzo, V., and K. Timmis.** 1992. Specialized host-vector system for the engineering of *Pseudomonas* strains destined for environmental release, p. 415–428. *In* E. Galli, S. Silver, and B. Witholt (ed.), *Pseudomonas: Molecular Biology and Biotechnology*. American Society for Microbiology, Washington, D.C.
26. **de Lorenzo, V., and K. Timmis.** 1994. Analysis and construction of stable phenotypes in Gram-negative bacteria with Tn5 and Tn*10*-derived mini-transposons. *Methods Enzymol.* **235:**386–405.
27. **del Solar, G., J. C. Alonso, M. Espinosa, and R. Díaz-Orejas.** 1996. Broad host range plasmid replication: an open question. *Mol. Microbiol.* **21:**661–666.
28. **Derbyshire, K.** 1995. An IS903-based vector for transposon mutagenesis and the isolation of gene fusions. *Gene* **160:**59–62.
29. **DeShazer, D., and E. Woods.** 1996. Broad host range cloning cassette vectors based on the R388 trimethoprim resistance gene. *BioTechniques* **20:**762–764.
30. **Díaz, E., M. Munthali, V. de Lorenzo, and K. Timmis.** 1994. Universal barrier to lateral spread of specific genes among microorganisms. *Mol. Microbiol.* **13:**855–861.
31. **Ditta, G., T. Schmidhauser, E. Yakobson, P. Lu, X. W. Liang, D. Finlay, D. Guiney, and D. Helinski.** 1985. Plasmids related to the broad host range vector, pRK290, useful for gene cloning and for monitoring gene expression. *Plasmid* **13:**149–153.
32. **Dowling, D., and F. O'Gara.** 1994. Metabolites of *Pseudomonas* involved in the biocontrol of plant disease. *Trends Biotechnol.* **12:**133–141.
33. **Fang, F., R. Durland, and D. Helinski.** 1993. Mutations in the gene encoding the replication initiation protein of plasmid RK2 produced elevated copy numbers of RK2 derivatives in *E. coli* and distantly related bacteria. *Gene* **133:**1–8.
34. **Farinha, M., and A. Kropinski.** 1989. Construction of broad host range vectors for general cloning and promoter selection in *Pseudomonas* and *E. coli*. *Gene* **77:**205–210.

35. **Farinha, M., and A. Kropinski.** 1990. Construction of broad host range plasmid vectors for easy visible selection and analysis of promoters. *J. Bacteriol.* **172:**3496–3499.

36. **Fedi, S., D. Brazil, D. Dowling, and F. O'Gara.** 1996. Construction of a modified mini-Tn*5 lacZY* non-antibiotic marker cassette: ecological evaluation of a *lacZY* marked *Pseudomonas* strain in the sugarbeet rhizosphere. *FEMS Microbiol. Lett.* **135:**251–257.

37. **Fellay, R., H. Krisch, P. Prentki, and J. Frey.** 1989. Omegon-Km: a transposable element designed for in vivo insertional mutagenesis and cloning of genes in Gram-negative bacteria. *Gene* **76:**215–226.

38. **Fernández-Pinas, F., and C. Wolk.** 1994. Expression in *luxCDE* in *Anabaena* sp. can replace the use of exogenous aldehyde for *in vivo* localization of transcription by *luxAB*. *Gene* **150:**169–174.

39. **Figurski, D., and D. Helinski.** 1979. Replication of an origin containing derivative of plasmid RK2 dependent on a plasmid function provided in *trans*. *Proc. Natl. Acad. Sci. USA* **76:**1648–1652.

40. **Freidman, A. M., S. Long, S. Brown, W. Buikema, and F. Ausubel.** 1982. Construction of a broad host range cosmid cloning vector and its use in the genetic analysis of *Rhizobium meliloti*. *Gene* **18:**289–296.

41. **George, A.** 1996. Multidrug resistance in enteric and other Gram-negative bacteria. *FEMS Microbiol. Lett.* **139:** 1–10.

42. **Groisman, E., B. Castilho, and M. Casadaban.** 1984. *In vivo* DNA cloning and adjacent gene fusing with a mini-Mu-*lac* bacteriophage containing a plasmid replicon. *Proc. Natl. Acad. Sci. USA* **81:**1480–1483.

43. **Guiney, D., and E. Lanka.** 1989. Conjugative transfer of IncP plasmids, p. 27–56. *In* C. Thomas (ed.), *Promiscuous Plasmids in Gram-Negative Bacteria.* Academic Press Limited, London.

44. **Harayama, S., and M. Rekik.** 1989. A simple procedure for transferring genes cloned in *E. coli* vectors into other Gram-negative bacteria: phenotypic analysis mapping of TOL plasmid gene *xylK*. *Gene* **78:**19–27.

45. **Herrero, M., V. de Lorenzo, B. Ensley, and K. Timmis.** 1993. A T7 RNA polymerase-based system for the construction of *Pseudomonas* strains with phenotypes dependent on TOL-meta pathway effectors. *Gene* **134:**103–106.

46. **Herrero, M., V. de Lorenzo, and K. Timmis.** 1990. Transposon vectors containing non-antibiotic resistance selection markers for cloning and stable chromosomal insertion of foreign genes in gram-negative bacteria. *J. Bacteriol.* **172:**6557–6567.

47. **Hirano, S., and C. Upper.** 1983. Ecology and epidemiology of foliar bacterial plant pathogens. *Annu. Rev. Phytopathol.* **21:**243–269.

48. **Holloway, B.** 1996. *Pseudomonas* genetics and taxonomy, p. 22–32. *In* T. Nakazawa, K. Furukawa, D. Haas, and S. Silver (ed.), *Molecular Biology of Pseudomonads.* American Society for Microbiology, Washington, D.C.

49. **Itoh, N., Y. Koide, H. Fukuzawa, S. Hirose, and T. Inukai.** 1991. Novel plasmid vectors for gene cloning in *Pseudomonas*. *J. Biochem.* **110:**614–621.

50. **Itoh, N., T. Kouzai, and Y. Koide.** 1994. Efficient transformation of *Pseudomonas* strains with pNI vectors by electroporation. *Biosci. Biotechnol. Biochem.* **58:**1306–1308.

51. **Iwasaki, K., O. Yagi, T. Kurabayashi, K. Ishizuka, and Y. Takamura.** 1994. Transformation of *Pseudomonas putida* by electroporation. *Biosci. Biotechnol. Biochem.* **58:** 851–854.

52. **Jansons, I., G. Touchie, R. Sharp, K. Almquist, M. Farinha, J. Lam, and A. Kropinski.** 1994. Deletion and transposon mutagenesis and sequence analysis of the pRO1600 OriR region found in the broad host range plasmids of the pQF series. *Plasmid* **31:**265–274.

53. **Jensen, L., J. L. Ramos, Z. Kaneva, and S. Molin.** 1993. A substrate dependent biological containment system for *Pseudomonas putida* based on the *E. coli gef* gene. *Appl. Environ. Microbiol.* **59:**3713–3717.

54. **Joseph-Liauzum, E., R. Fellay, and M. Chandler.** 1989. Transposable elements for efficient manipulation of a wide range of Gram-negative bacteria: promoter probes and vectors for foreign genes. *Gene* **85:**83–89.

55. **Kado, C., and S. Liu.** 1981. Rapid procedure for detection and isolation of large and small plasmids. *J. Bacteriol.* **145:** 1365–1373.

56. **Kahrs, A., S. Odenbreit, W. Schmitt, D. Heuermann, T. Meyer, and R. Haas.** 1995. An improved Tn*Max* mini-transposon system suitable for sequencing, shuttle mutagenesis and gene fusions. *Gene* **167:**53–57.

57. **Kamoun, S., E. Tola, H. Kamdar, and C. Kado.** 1992. Rapid generation of directed and unmarked deletions in *Xanthomonas*. *Mol. Microbiol.* **6:**809–816.

58. **Kaniga, K., I. Delor, and G. Cornelis.** 1991. A wide host range suicide vector for improving reverse genetics in Gram-negative bacteria: inactivation of the *blaA* gene of *Yersinia enterocolitica*. *Gene* **109:**137–141.

59. **Karkhoff-Schweizer, R., and H. Schweizer.** 1994. Utilization of a mini-Dlac transposable element to create an alpha-complementation and regulated expression system for cloning in *Pseudomonas aeruginosa*. *Gene* **140:**7–15.

60. **Kawasaki, H., H. Kuriyama, and K. Tonomura.** 1995. Use of haloacetate dehalogenase genes as selection markers for *E. coli* and *Pseudomonas* vectors. *Biodegradation* **6:** 181–182.

61. **Keen, N., S. Tamaki, D. Kobayashi, and D. Trollinger.** 1988. Improved broad host range plasmids for DNA cloning in Gram-negative bacteria. *Gene* **70:**191–197.

62. **Kilbane, J., and B. Bielaga.** 1991. Instantaneous gene transfer from donor to recipient microorganisms via electroporation. *BioTechniques* **10:**354–365.

63. **Kloos, D., M. Stratz, A. Guttler, R. Steffan, and K. Timmis.** 1994. Inducible cell lysis system for the study of natural transformation and environmental fate of DNA released by cell death. *J. Bacteriol.* **176:**7352–7361.

64. **Kok, M.** 1995. A new horizon for pBR322: *in vivo* insertion of plasmid fragments into wide host range shuttle vectors. *Nucleic Acids Res.* **23:**5085–5086.

65. **Kokotek, W., and W. Lotz.** 1989. Construction of a *lacZ*-kanamycin resistance cassette, useful for site directed mutagenesis and as a promoter probe. *Gene* **84:**467–471.

66. **Konyecsni, W., and V. Deretic.** 1988. Broad host range plasmid and M13 bacteriophage-derived vectors for promoter analysis in *Escherichia coli* and *Pseudomonas aeruginosa*. *Gene* **74:**357–386.

67. **Kovach, M., P. Elzer, D. Hill, G. Robertson, M. Farris, R. Roop, and K. Peterson.** 1995. Four new derivatives of the broad-host-range cloning vector pBR1MCS, carrying different antibiotic-resistance cassettes. *Gene* **166:** 175–176.

68. **Kovach, M., R. Phillips, P. Elzer, R. Roop II, and K. Peterson.** 1994. pBBR1MCS: a broad host range cloning vector. *BioTechniques* **16:**800–802.

69. **Kristensen, C., L. Eberl, J. M. Sánchez-Romero, M. Giskov, S. Molin, and V. de Lorenzo.** 1995. Site-specific deletions of chromosomally located DNA segments with the multimer resolution system of broad-host-range plasmid RP4. *J. Bacteriol.* **177:**52–58.

70. **Labes, M., A. Pühler, and R. Simon.** 1990. A new family of RSF1010 derived expression and *lac*-fusion broad host range vectors from Gram-negative bacteria. *Gene* **89:** 37–46.

71. **Li, X., M. Dorsch, T. DelDot, L. Sly, E. Stackebrandt, and A. Hayward.** 1993. Phylogenetics studies of the rRNA group II pseudomonads based on 16S rRNA gene sequences. *J. Appl. Bacteriol.* **74:**324–329.

72. **Lindgren, P., R. Frederick, A. Govindarajan, N. Panopoulos, B. Staskawicz, and S. Lindow.** 1989. An ice nucleation reporter system: identification of inducible pathogenicity genes in *Pseudomonas syringae* pv. *phaseolicola. EMBO J.* **8:**1291–1301.

73. **Lorenz, M., and W. Wackernagel.** 1990. Natural genetic transformation of *Pseudomonas stutzeri* by sand adsorbed DNA. *Arch. Microbiol.* **154:**380–385.

74. **Marqués, S., and J. L. Ramos.** 1993. Transcriptional control of the *Pseudomonas putida* TOL plasmid catabolic pathways. *Mol. Microbiol.* **9:**923–929.

75. **Mather, M., L. McReynolds, and C. Yu.** 1995. An enhanced broad host range vector for Gram-negative bacteria: avoiding tetracycline phototoxicity during the growth of photosynthetic bacteria. *Gene* **156:**85–88.

76. **May, T., D. Shinabarger, R. Maharaj, J. Kato, L. Chu, J. deVault, S. Roychoudhury, N. Zielinski, A. Berry, R. Rothmel, T. Misra, and A. Chakrabarty.** 1991. Alginate synthesis by *Pseudomonas aeruginosa*: a key pathogenic factor in chronic pulmonary infections of cystic fibrosis patients. *Clin. Microbiol. Rev.* **4:**191–206.

77. **Mermod, N., J. L. Ramos, P. Lehrbach, and K. Timmis.** 1986. Vector for regulated expression of cloned genes in a wide range of gram-negative bacteria. *J. Bacteriol.* **167:** 447–454.

78. **Metcalf, W., W. Jiang, and B. Wanner.** 1994. Use of the rep technique for allele replacement to construct new *E. coli* hosts for maintenance of R6K gamma origin plasmids at different copy numbers. *Gene* **138:**1–7.

79. **Molin, S., L. Boe, L. Jensen, C. Kristensen, M. Givskov, J. L. Ramos, and A. Bej.** 1993. Suicidal genetic elements and their use in biological containment of bacteria. *Annu. Rev. Microbiol.* **47:**139–166.

80. **Mondello, F.** 1989. Cloning and expression in *E. coli* of *Pseudomonas* strain LB400 genes encoding polychlorinated biphenyl degradation. *J. Bacteriol.* **171:**1725–1732.

81. **Monthali, M., K. Timmis, and E. Diaz.** 1996. Restricting the dispersal of recombinant DNA: design of a contained biological catalyst. *Bio/Technology* **14:**189–192.

82. **Morales, V., A. Backman, and M. Bagdasarian.** 1991. A series of wide host range low copy number vectors that allow direct screening for recombinants. *Gene* **97:**39–47.

83. **Nakazawa, T., and S. Inouye.** 1986. Cloning of *Pseudomonas* genes in *E. coli*, p. 357–382. *In* J. Sokatch (ed.), *The Bacteria. A Treatise on Structure and Function*, vol. 10. *The Biology of Pseudomonas*, Academic Press, Inc., San Diego, Calif.

84. **Nies, D., M. Mergeay, B. Friedrich, and H. Schlegel.** 1987. Cloning of genes encoding resistance to cadmium, zinc and cobalt in *Alcaligenes eutrophus* CH34. *J. Bacteriol.* **169:**4865–4868.

85. **Nieto, C., E. Fernandez-Tresguerres, M. Sánchez, M. Vicente, and R. Díaz.** 1990. Cloning vectors from a naturally occurring plasmid of *Pseudomonas sevastanoi*, specifically tailored for genetic manipulations in *Pseudomonas. Gene* **87:**145–149.

86. **Olekhnovich, I., and Y. Fomichev.** 1994. Controlled expression shuttle vector for pseudomonads based on the *trpIBA* genes of *Pseudomonas putida. Gene* **140:**63–65.

87. **Parales, R., and C. Harwood.** 1993. Construction and use of a new broad host range *lacZ* transcriptional fusion vector, pHRP309, for Gram-negative bacteria. *Gene* **133:** 23–30.

88. **Pemberton, J., and C. Harding.** 1986. Cloning of carotenoid biosynthesis genes from *Rhodopseudomonas spaeroides. Curr. Microbiol.* **14:**25–29.

89. **Pemberton, J., and R. Penfold.** 1992. High frequency electroporation and maintenance of pUC and pBR-based cloning vectors in *Pseudomonas stutzeri. Curr. Microbiol.* **25:**25–29.

90. **Penfold, R., and J. Pemberton.** 1992. An improved suicide vector for construction of chromosomal insertion mutations in bacteria. *Gene* **118:**145–146.

91. **Pérez-Martín, J., and V. de Lorenzo.** 1996. VTR expression cassettes for engineering conditional phenotypes in *Pseudomonas*: activity of the *Pu* promoter of the TOL plasmid under limiting concentrations of the XylR activator protein. *Gene* **172:**81–86.

92. **Polissi, A., G. Bertoni, F. Acquati, and G. Dehò.** 1992. Cloning and transposon vectors derived from satellite bacteriophage P4 for genetic manipulation of *Pseudomonas* and other Gram-negative bacteria. *Plasmid* **28:**101–114.

93. **Prosser, J.** 1994. Molecular marker systems for detection of genetically engineered microorganisms in the environment. *Microbiology* **140:**5–17.

94. **Quandt, J., and M. Hynes.** 1993. Versatile suicide vectors which allow direct selection for gene replacement in Gram-negative bacteria. *Gene* **127:**15–21.

95. **Ramos, J. L., E. Díaz, D. Dowling, V. de Lorenzo, S. Molin, F. O'Gara, C. Ramos, and K. Timmis.** 1994. Behavior of bacteria designed for biodegradation. *Bio/Technology* **12:**1349–1356.

96. **Ramos, J. L., M. González-Carreró, and K. Timmis.** 1988. Broad host range expression vectors containing manipulated meta-cleavage pathway regulatory elements of the TOL plasmid. *FEBS Lett.* **226:**241–246.

97. **Rodriguez, R., and D. Denhardt (ed.).** 1988. Vectors. A survey of molecular cloning vectors and their uses. *Bio/Technology* **10.**

98. **Ronchel, M. C., C. Ramos, L. Jensen, S. Molin, and J. L. Ramos.** 1995. Construction and behaviour of biologically contained bacteria for environmental applications in bioremediation. *Appl. Environ. Microbiol.* **61:** 2990–2994.

99. **Ross, P., F. O'Gara, and S. Condon.** 1990. Thymidylate synthase gene from *Lactococcus lactis* as a genetic marker: an alternative to antibiotic resistance genes. *Appl. Environ. Microbiol.* **56:**2164–2169.

100. **Rothmel, R., A. Chakrabarty, A. Berry, and A. Darzins.** 1991. Genetic systems in *Pseudomonas*. *Methods Enzymol.* **204:**485–514.

101. **Saint, C., S. Alexander, and N. McClure.** 1995. pTIM3, a plasmid delivery vector for transposon based inducible marker gene system in Gram-negative bacteria. *Plasmid* **34:**165–174.

102. **Schafer, A., A. Tauch, W. Jager, J. Kalinowski, G. Thierbach, and A. Puhler.** 1994. Small mobilizable multi-purpose cloning vectors derived from *E. coli* plasmid pK18 and pK19: detection of defined deletions in the chromosome of *Corynebacterium glutamicum*. *Gene* **145:**69–73.

103. **Scholz, P., V. Haring, B. Wittmann-Liebold, K. Ashman, M. Bagdasarian, and E. Scherzinger.** 1989. Complete sequence and gene organization of the broad host range plasmid RSF1010. *Gene* **75:**271–288.

104. **Schweizer, H.** 1991. *Escherichia-Pseudomonas* shuttle vectors derived from pUC18/19. *Gene* **97:**109–112.

105. **Schweizer, H.** 1991. Improved broad host range lac based plasmid vectors for the isolation and characterization of protein fusions in *Pseudomonas aeruginosa*. *Gene* **103:**87–92.

106. **Schweizer, H.** 1992. Allelic exchange in *Pseudomonas aeruginosa* using novel ColE1-type vectors and a family of cassettes containing a portable *oriT* and the counter-selectable *Bacillus subtilis sacB* marker. *Mol. Microbiol.* **6:** 1195–1204.

107. **Schweizer, H.** 1993. Small broad host range gentamycin resistance cassettes for site-specific insertion and deletion mutagenesis. *BioTechniques* **15:**831–833.

108. **Schweizer, H.** 1993. Two plasmids, X1918 and Z1918, for easy recovery of the *xylE* and *lacZ* reporter genes. *Gene* **134:**89–91.

109. **Schweizer, H., and T. Hoang.** 1995. An improved system for gene replacement and *xylE* fusion analysis in *Pseudomonas aeruginosa*. *Gene* **158:**15–22.

110. **Schweizer, H., T. Klassen, and T. Hoang.** 1996. Improved methods for gene analysis and expression in *Pseudomonas* spp., p. 229–237. *In* T. Nakazawa, K. Furukawa, D. Haas, and S. Silver (ed.), *Molecular Biology of Pseudomonads*. American Society for Microbiology, Washington, D.C.

111. **Simon, R., U. Priefer, and A. Pühler.** 1983. A broad host range mobilization system for in vivo genetic engineering: transposon mutagenesis in Gram-negative bacteria. *Bio/Technology* **1:**784–791.

112. **Simon, R., J. Quandt, and W. Klipp.** 1989. New derivatives of transposon Tn*5* suitable for mobilization of replication, generation of operon fusions and induction of genes in Gram-negative bacteria. *Gene* **80:**161–169.

113. **Skorupski, K., and R. Taylor.** 1996. Positive selection vectors for allelic exchange. *Gene* **169:**47–52.

114. **Skrzypek, E., P. Hassiz, G. Plano, and S. Straley.** 1993. New suicide vector for gene replacement in yersiniae and other Gram-negative bacteria. *Plasmid* **29:**160–163.

115. **Staskawicz, B., D. Dahlbeck, N. Keen, and C. Napoli.** 1987. Molecular characterization of cloned avirulence genes from Race 0 and Race 1 of *Pseudomonas syringae* pv. *glycinea*. *J. Bacteriol.* **169:**5789–5794.

116. **Stein, D.** 1992. Plasmids with easily excisable *xylE* cassettes. *Gene* **117:**157–158.

117. **Stewart, G., and C. Sinigalliano.** 1991. Exchange of chromosomal markers by natural transformation between the soil isolate, *Pseudomonas stutzeri* JM300, and the marine isolate, *Pseudomonas stutzeri* strain ZoBell. *Antonie van Leeuwenhoek* **59:**19–25.

118. **Takahashi, S., and Y. Nagano.** 1984. Rapid procedure for isolation of plasmid DNA and application to epidemiological analysis. *J. Clin. Microbiol.* **20:**608–613.

119. **Timmis, K., R. Steffan, and R. Unterman.** 1994. Designing microorganisms for the treatment of toxic wastes. *Annu. Rev. Microbiol.* **48:**525–557.

120. **Tsuda, M., and T. Nakazawa.** 1993. A mutagenesis system utilizing a Tn*1722* derivative containing an *Escherichia coli* specific vector plasmid: application to *Pseudomonas* species. *Gene* **136:**257–262.

121. **Van Beilen, J., M. Wubbolts, Q. Chen, M. Nieboer, and B. Witholt.** 1996. Effects of two liquid phase systems and expression of *alk* genes on the physiology of alkane oxidizing strains, p. 35–47. *In* T. Nakazawa, K. Furukawa, D. Haas, and S. Silver (ed.), *Molecular Biology of Pseudomonads*. American Society for Microbiology, Washington, D.C.

122. **Van-Gijsegmen, F., and A. Toussaint.** 1982. Chromosome transfer and R-prime formation by an RP4::mini-Mu derivative in *E. coli, Salmonella typhimurium, Klebsiella pneumoniae* and *Proteus mirabilis*. *Plasmid* **7:**30–44.

123. **Vezina, G., and R. Levesque.** 1991. Molecular characterization of the class II multiresistance transposable element Tn*1403* from *Pseudomonas aeruginosa*. *Antimicrob. Agents Chemother.* **35:**313–321.

124. **Watson, A., R. Alm, and J. Mattick.** 1996. Construction of improved vectors for protein production in *Pseudomonas aeruginosa*. *Gene* **172:**163–164.

125. **West, S., H. Schweizer, C. Dall, A. Sample, and L. Runyen-Janecky.** 1994. Construction of improved *E. coli-Pseudomonas* shuttle vectors derived from pUC18/19 and sequence of the region required for the replication in *Pseudomonas aeruginosa*. *Gene* **148:**81–86.

126. **Wong, K., and M. McClelland.** 1992. Dissection of the *Salmonella typhimurium* genome by use of a Tn*5* derivative carrying rare restriction sites. *J. Bacteriol.* **174:** 3807–3811.

127. **Wong, R., R. Wirtz, and R. Hancock.** 1995. *Pseudomonas aeruginosa* outer membrane protein OprF as an expression vector for foreign epitopes: the effects of positioning and length on the antigenicity of the epitope. *Gene* **158:**55–60.

128. **Yang, C., H. Azad, and D. Cooksey.** 1996. A chromosomal locus required for copper resistance, competitive fitness, and cytochrome c biogenesis in *Pseudomonas fluorescens*. *Proc. Natl. Acad. Sci. USA* **93:**7315–7320.

129. **Yen, K.** 1991. Construction of cloning cartridges for development of expression vectors in gram-negative bacteria. *J. Bacteriol.* **173:**5328–5335.

130. **Zaat, A., K. Slegtenhorst-Eegdeman, J. Tommansen, V. Geli, W. Wijffelman, and B. Lugtenberg.** 1994. Construction of *phoA-caa*, a novel PCR and immunogenically detectable marker gene for *Pseudomonas putida*. *Appl. Environ. Microbiol.* **60:**3965–3973.

Clostridia

MARGARET L. MAUCHLINE, TOM O. DAVIS, AND NIGEL P. MINTON

38

As we approach the end of the second millennium, the quest to convert our reliance on extractive technologies to those based on renewable resources gathers pace. One of the largest genera among prokaryotes, the genus *Clostridium*, exhibits extreme biocatalytic diversity. Its members are, therefore, likely to figure prominently in these emerging technologies. Indeed, the commercial development earlier in this century of the acetone-butanol-ethanol (ABE) fermentation, undertaken by *Clostridium acetobutylicum* and related solvent-producing bacteria, is considered by many to be the forerunner of the modern biotechnology industry. The successful exploitation of individual clostridial species will be largely dependent on the extent to which rational alterations can be made to their biocatalytic capabilities. A prerequisite is to more fully understand the molecular basis of fermentative metabolism/substrate utilization, and to have available the genetic tools to bring about defined changes. In the following pages we review the current status of this technology.

38.1. TRANSFORMATION PROCEDURES

There are no examples of natural competence in clostridia. All described transformation procedures are, therefore, reliant on the physical alteration of the cell wall/membrane to promote DNA uptake. As with other gram-positive bacteria, early progress in transformation technology revolved around polyethylene glycol (PEG)-induced DNA uptake by "naked" cells (134). Much effort was devoted to optimizing protoplast conversion and regeneration rates. The complexity and time-consuming nature of the steps involved in a protoplast transformation procedure, however, make them unsuitable for routine use.

Protoplast transformation has largely been superseded by electroporation technology. In this case, DNA gains access to the cell following the transient formation of pores in the cell membrane as a result of exposure to high-intensity electric field pulses. Instrumentation is readily available, and improved efficiencies can be achieved by a rational investigation of the parameters, conditions, and media used (106).

38.1.1. Transformation of Solvent-Producing Species

The solvent-producing clostridia, typified by *Clostridium beijerinckii* NCIMB 8052 (formerly *C. acetobutylicum* NCIMB 8052) and *C. acetobutylicum* ATCC 824, represent the species of greatest biotechnological potential. Not unsurprisingly, therefore, numerous methods have been published describing their successful transformation. In the following procedures, all stages are, wherever possible, performed under an anaerobic atmosphere (N_2-CO_2-H_2, 80:10:10). When cells are removed from an anaerobic cabinet for centrifugation steps, the bottles used remain sealed until they are returned to the cabinet, where supernatants are decanted. Similarly, electroporation cuvettes remain sealed throughout the time they may be outside the cabinet and, ideally, the electroporation apparatus should be sited within the cabinet.

38.1.1.1. Electroporation of *C. beijerinckii* NCIMB 8052

C. beijerinckii NCIMB 8052 was the first clostridial species to be successfully electroporated (85). Typically, an inoculated broth of 2× YTG (per liter: tryptone, 16 g; yeast extract, 10 g; NaCl, 5 g; glucose, 5 g) is serially diluted and grown overnight (16 h); from this broth, 10 ml of an exponential culture is used to inoculate 100 ml of the same broth. At an A_{600} of 0.6, the culture is cooled on ice, and cells are harvested by centrifugation, washed once in 10 ml of cold (4°C) electroporation buffer (270 mM sucrose, 1 mM $MgCl_2$, 7 mM $NaHPO_4$ [pH 7.4]), and resuspended in 5 ml of cold electroporation buffer for a further 10 min of incubation on ice. A 0.8-ml aliquot of the cell resuspension is then added to 0.5 μg of plasmid DNA and held on ice for a further 8 min prior to electroporation in a Bio-Rad Gene Pulsar. The pulse applied is 1.25 kV in a 0.4-cm inter-electrode distance cuvette, capacitance of 25 μF and resistance of 200 Ω. The cells are again held on ice for 10 min before being diluted in 10 volumes of 2× YTG and incubated at 37°C for 3 h to allow gene expression. Prior to plating, the culture is concentrated by centrifugation in Eppendorf tubes, resuspended in 100 μl of fresh medium, and plated onto 2× YTG plates supplemented with a suitable antibiotic, in most cases, 20 μg of erythromycin per ml. Colonies appear within 2 to 3 days.

Higher frequencies can result from the use of nutrient glucose medium (per liter: beef extract, 1 g; yeast extract, 2 g; peptone, 5 g; NaCl, 5 g; glucose, 10 g); alternative electroporation buffer (270 mM sucrose, 5 mM NaH_2PO_4 [pH 7.4]); or an increased pulse of 2 kV and the immediate transfer of electroporated cells to recovery medium (58). A transformation procedure for a different strain of *C. beijerinckii* has also been published (11).

38.1.1.2. Electroporation of *C. acetobutylicum* ATCC 824

Electroporation of *C. acetobutylicum* ATCC 824 was first described by Mermelstein and coworkers (70). In their procedure cells are grown at 37°C to late exponential phase in 60 ml of Reinforced Clostridial Medium (RCM, Difco) (pH 5.2), harvested by centrifugation, washed, and resuspended in 2.1 ml of chilled electroporation buffer (5 mM NaH_2PO_4, 272 mM sucrose [pH 7.4]). A 0.7-ml aliquot is placed in a 0.4-cm inter-electrode distance cuvette and cooled on ice for 5 min prior to the addition of plasmid DNA. After a further 2 min on ice, the sample is electroporated at 2.0 kV and 25 μF. Thereafter, the cell suspension is added to 10 ml of RCM and incubated for 4 h at 37°C before plating onto RCM agar (pH 5.8) with a suitable phenotypic selection for transformants.

In contrast to *C. beijerinckii* NCIMB 8052, *C. acetobutylicum* ATCC 824 possesses a formidable restriction barrier to inappropriately methylated DNA (70). The endonuclease responsible recognizes a G+C-rich palindrome which is particularly prevalent in *Escherichia coli*-derived vector sequences, resulting in discrimination against many of the shuttle vectors constructed for use in clostridia. Such vectors can, however, be protected by their growth in an *E. coli* host carrying a cloned modification system which methylates the vectors in a way equivalent to that of the ATCC 824 system (68). Restriction systems appear to be generally widespread in clostridia (134) and therefore may account for transformation difficulties in other clostridial species.

38.1.1.3. Protoplast Transformation of *C. acetobutylicum* NI-4081

Although superseded, a protoplast transformation procedure developed for *C. acetobutylicum* NI-4081 is worthy of note because of the high transformation frequencies attained (94). These were largely attributed to deficient production of autolysin by strain NI-4081 and the use of autolysin inhibitors, such as choline.

Protoplast Formation

Bacteria are grown in T69 medium (per liter: KH_2PO_4, 0.5 g; ammonium acetate, 2 g; $MgSO_4.7H_2O$, 0.3 g; $FeSO_4.7H_2O$, 0.01 g; yeast extract, 1 g; casamino acids, 0.5 g; bactotryptone, 0.5 g; cysteine-HCl, 0.5 g; glucose, 10 g; pH adjusted to 6.5 with NaOH) to mid-exponential phase (approximately 1×10^8 cells/ml). Solid sterile sucrose is added to a final concentration of 0.6 M, and the cells are converted to protoplasts by incubation for 1 h at 34°C with 100 μg of lysozyme per ml and 20 μg of penicillin G per ml. Protoplasts are recovered by centrifugation at 3,000 × g for 15 min at room temperature, washed twice with T69 medium containing 0.6 M sucrose, 0.5% (wt/vol) bovine serum albumin (BSA) and 1 mM $CaCl_2$, and then resuspended in protoplast buffer (T69 supplemented with 0.5 M xylose, 0.5% (wt/vol) BSA, 25 mM $MgCl_2$, and 25 mM $CaCl_2$). At this point the number of potential regenerant and L-form colonies can be estimated by diluting a quantity of cells in regeneration medium T69C (T69 containing 0.3 M sucrose or 0.25 M xylose, 0.5% [wt/vol] BSA, and 1 mM $CaCl_2$) and plated onto solid (2.5% [wt/vol] agar) T69C; protoplast numbers are estimated by means of a phase-contrast light microscope.

Plasmid Transformation Procedure

Plasmid DNA (50 to 800 ng) and PEG 4000 to a final concentration of 35% (wt/vol) are added to 10^9 protoplasts and incubated for 2 min at room temperature, then diluted with 10 volumes of T69 medium containing 0.5 M xylose, 1 mM $CaCl_2$, 0.5% (wt/vol) BSA, and 4 mg of choline per ml. Protoplasts are centrifuged, washed, and resuspended in the same medium. Dilutions are added to T69 soft agar (T69 supplemented with 1 mM $CaCl_2$, 0.5% [wt/vol] BSA, 25 M xylose, 4 mg of choline per ml, and 0.8% [wt/vol] agar) before being poured onto T69 agar (0.25 mM xylose, 2.5% [wt/vol] agar) and incubated at 34°C for 20 h. For selection of plasmids with erythromycin resistance genes, a further 3 ml of T69 soft agar overlay is added (T69 supplemented with 0.25 M xylose, 1 mg of erythromycin per ml, and 0.8% [wt/vol] agar). Colonies appear after 4 to 6 days.

38.1.2. Transformation Procedures for Thermophilic Clostridia

Transformation procedures for thermophilic clostridia remain poorly developed. Protoplasts of *C. thermocellum*, arguably the most commercially important clostridia, have reportedly been transformed with plasmids based on pUB110 and pMK4 (113), but beyond phenotypic conversion no direct evidence that transformation had occurred was presented. Restriction systems have been characterized in *Clostridium thermocellum* ATCC 27405, which may improve the prospects of bringing about gene transfer in this strain (134). Transformation of spheroplasts of another clostridial thermophile has also been reported, with the use of a plasmid (pCS1) specifying chloramphenicol resistance (51). The best-documented transformation system, however, is that developed in Staudenbauer's laboratory for *Clostridium thermohydrosulfuricum* (108).

C. thermohydrosulfuricum is of biotechnological interest because of its ability to produce ethanol from biomass, especially in coculture with *C. thermocellum*. The developed method relies on the induction of competence in a strain with low DNase activity (*C. thermohydrosulfuricum* DSM 568), by disruption or removal of the surface S-layer of the cell wall followed by PEG-induced membrane permeabilization. The same approach was subsequently used to transform *C. acetobutylicum* (133). In the case of *C. thermohydrosulfuricum* DSM 568, the plasmid transferred was the *Bacillus subtilis* plasmid pUB110. The detailed procedure employed is as follows.

A 50-ml volume of modified RCM broth (per liter: tryptone, 10 g; beef extract, 4 g; yeast extract, 3 g; glucose, 2.5 g; NaCl, 5 g; $Na_2S{\cdot}9H_2O$, 0.5 g; cysteine HCl, 0.5 g; resazurin, 0.5 g; K_2HPO_4, 2 g [pH 7.2]) is inoculated with an overnight culture of *C. thermohydrosulfuricum* DSM 568. Cells are grown stationary, in a strictly anaerobic environment at 68°C to an A_{578} of 0.45 (1.5×10^9 cells/ml) and harvested by centrifugation (6000 × g, for 8 min at 21°C). Cells are washed once in buffer (50 mM Tris-HCl [pH 8.3], 0.05% [wt/vol] $Na_2S{\cdot}9H_2O$, 0.05% [wt/vol] cysteine HCl), centrifuged as before, and resuspended in 5 ml of electroporation buffer (previous buffer with the addition of 0.35 M sucrose). Centrifugation is repeated, and the cell pellet is resuspended in 0.2 ml of electroporation buffer. Between 2 and 5 μg of plasmid DNA dissolved in TE buffer (25 mM Tris-CHl [pH 8.0], 10 mM EDTA) is added and incubated at 60°C for 5 min prior to the addition of 1.5 ml of 40% PEG 6000. Incubation is continued for a further 60 min, at which point 2 ml of buffer (10 mM Tris-HCl [pH 8.3], 0.05% [wt/vol] $Na_2S•9H_2O$, 0.15 M NaCL, 0.05% [wt/vol] cysteine HCl) is added, centrifugation is repeated as before, and cells are resuspended in 1 ml of RCM broth. Aliquots are then plated on RCM agar (2% wt/vol), supplemented with an appropriate antibiotic, and incubated at 55°C for 4 to 6 days until colonies appear. Manipulations are performed under an anaerobic environment (N_2-H_2, 92:8).

38.1.3. Transformation of Pathogenic Clostridia

The most notorious clostridial species are those associated with human disease, such as *Clostridium perfringens*, *C. difficile*, *C. tetani*, and *C. botulinum*. Alone among these pathogens, the genetic tools available for the manipulation of *C. perfringens* (see reference 97) rival in complexity and numeracy those developed for solvent-producing clostridia. However, despite the plethora of protoplast and electroporation transformation procedures described for *C. perfringens*, the species is of insufficient biotechnological importance to consider them here. Paradoxically, *C. botulinum* has far greater commercial value. This is due to the increasing therapeutic uses being found for botulinum toxin in the treatment of human disorders which arise through aberrant muscular dysfunctions (77). A typical method for transforming *C. botulinum* is as follows.

An overnight 20-ml culture cultivated in TPGY medium (per liter: trypticase, 20 g; peptone, 5 g; glucose, 1 g; yeast extract, 5 g; cysteine-HCl, 1 g) is used to inoculate a 300-ml volume of TPGY and allowed to grow to an A_{660} of 0.8. The culture is divided in two, cooled on ice for 10 min, and then centrifuged at 6000 × *g* outside of the anaerobic cabinet for 10 min at 4°C. The centrifuge pots are returned to the cabinet, and the supernatant is carefully removed. Cells are gently resuspended in 50 ml of chilled electroporation buffer (10% PEG 6000 or 270 mM sucrose, 7 mM sodium phosphate, 1 mM $MgCl_2$), and the centrifugation procedure is repeated. The cell pellet is eventually resuspended in 3 ml of fresh chilled electroporation buffer. DNA (between 0.1 and 2.0 mg/ml) is added to a chilled electroporation cuvette with an inter-electrode distance of 0.4 cm, followed by 0.8 ml of cells, and the two are mixed by 2 or 3 gentle inversions of the cuvette. The cuvette is wiped dry and subjected to a pulse of magnitude 2.5 kV, 25 μF, and 100 Ω. The cuvette is immediately incubated on ice for 5 min. The contents of the cuvette are resuspended in 10 ml of warm TPGY broth supplemented with 25 mM MgCl, and the cells are left to recover for 5 h. Thereafter, the 10 ml of broth is separated into 6 Eppendorf tubes and centrifuged for 3 min; the supernatant is decanted, and the cells are gently resuspended in 150 μl of TPGY broth and spread on individual TPGY agar (2.5% wt/vol) plates containing 5 μg/ml erythromycin. Plates are incubated for 24 to 48 h.

38.2. CONJUGATIVE PROCEDURES

Conjugative transfer, although technically more difficult than electrotransformation, has the advantage of not being limited by extracellular nucleases (e.g., see reference 70) because of the requirement for cell-to-cell (donor-to-recipient) contact. In addition, the frequency of transfer for conjugation can be higher than that for electrotransformation and, in some cases, may represent the only available option (e.g., see reference 5). From a regulatory perspective, however, the strategy has the disadvantage that the deployment of mobilizable genetic systems brings an additional element of risk.

38.2.1. Conjugative Plasmids

The only clostridial species known to possess conjugative plasmids is *C. perfringens*. Most of these plasmids encode similar, if not identical, antibiotic resistance genes, and the range of plasmids may well have arisen from a common progenitor (97). To date, there has been only one report of transfer of a *C. perfringens* plasmid to another clostridial species (42). In contrast, several broad-host-range enterococcal plasmids can be transferred into clostridia, e.g., pAMβ1 (87) and pIP501 (95). In certain instances, the mobilization of nonconjugative plasmids can occur (84).

While transfer of pAMβ1 and pIP501 is generally most efficient from gram-positive donors (e.g., *Lactococcus lactis*, *Enterococcus faecalis*), the conjugative strategy with most potential involves the use of gram-negative donors. This strategy is reliant on broad-host-range IncP plasmids and has been most successfully utilized for the transfer of genetic material to *C. beijerinckii* NCIMB 8052 from *E. coli* donor strains (126). Several vectors have been created (125, 126) which carry the origin of conjugative transfer, *oriT*, from an IncP plasmid. All other conjugation functions are provided in *trans* from another IncP plasmid contained in the *E. coli* (Tra^+) donor strain. A range of gram-positive replicons have been employed in vector construction, e.g., pAMβ1, pWV01, and pCB101 (see reference 126). A typical protocol is as follows.

Set up a fresh culture of *E. coli* HB101 R702 containing the conjugative plasmid of interest in 10 ml of Brain Heart Infusion broth (BHIB, Difco), plus necessary selective antibiotic(s), and incubate aerobically overnight at 37°C. Grow serial dilutions of the recipient clostridial strain anaerobically overnight in 2× YT. On the morning of the experiment, dilute back a culture which has not entered stationary phase to an A_{600} of 0.05–0.1 and grow until the A_{600} is 0.6. Put donor culture into an anaerobic cabinet. Harvest 2 ml of donor culture, wash with 2 ml (anaerobic) holding buffer (25 mM potassium phosphate, 1 mM $MgSO_4$), then resuspend in 2 ml of holding buffer. Add 0.2 ml of recipient culture to donor culture. Harvest the donor/recipient mixture onto a Whatman 0.2-μm filter by using a syringe. Place filter, bacteria, uppermost, onto RCM agar which has previously been spread with 0.2 ml of 10-mg/ml catalase, and incubate anaerobically overnight. Resuspend bacterial growth from the filter in 0.5 ml of holding buffer by vortexing. Spread 0.1-ml samples onto selective agar to select for transconjugants. Make serial dilutions from one 0.1-ml sample, and spread onto necessary agar for separate donor and recipient counts. Transconjugants are usually visible after 24 to 48 h of incubation.

38.2.2. Conjugative Transposons

Conjugative transposons (21) are generally large elements which encode all the functions necessary for their own transfer, confer fertility potential to each recipient cell, and are generally confined to gram-positive bacteria. If insertion occurs at random into the chromosome, they may be employed to generate phenotypic mutants. The mechanism of transposition differs from classical transposons in that transposition involves a covalently closed circular DNA intermediate produced by the excision of the transposon from donor DNA, and no duplication of the target sequence occurs (104). Transfer of conjugative transposons can occur between different species, and indeed between genera, and hence these transposons are implicated in the spread of antibiotic resistance genes among medically important bacteria, including clostridia (105).

Mutagenesis studies using conjugative transposons have resulted in the isolation of *C. acetobutylicum* mutants defective in protease activity (99), solvent production (5, 10, 67), the identification of possible solvent production regulatory areas (10), and mutants deficient in degeneration (i.e., the loss of the ability to produce both solvents and spores) (46). Further evaluation of such mutants could be valuable in the selection, or possibly even the construction, of strains of *C. acetobutylicum* which are more efficient in solvent production. Transposon mutagenesis has also been used to identify toxin production regulatory genes in *C. perfringens* (64) and to aid the physical mapping of the chromosome of *C. beijerinckii* NCIMB 8052 (123).

Certain *C. perfringens* conjugative plasmids carry transposons (1). Tn*4451* has been shown to carry six genes, including a site-specific recombinase (6) which mediates its excision by catalyzing the formation of a circular form of the transposon in a manner similar to that of conjugative transposons. *C. difficile* also carries elements with the characteristics of conjugative transposons which are able to transfer to other hosts (79).

In terms of classical mutagenesis studies in clostridia, the most widely used conjugative transposons have been those originally isolated from the streptococci, particularly Tn*916* (31). Tn*916* has been transferred into *C. tetani* (115), *C. botulinum* (61), *C. difficile* (78), *C. perfringens* (64), and a range of *C. acetobutylicum* strains, e.g., DSM 792 and DSM 1732 (9), NCIMB 8052 (132), P262 (5), and ATCC 824 (67). In all of these *C. acetobutylicum* strains, except NCIMB 8052, insertion of Tn*916* into the chromosome was found to occur at different sites, although insertion was not truly random, because Tn*916* has recognized target sequences (20). Since these target sites are A+T-rich sequences, Tn*916* is of particular use in clostridia because of the low G+C content of clostridial DNA. The identification of a "hot spot" for Tn*916* insertion in NCIMB 8052 (132) was one of many factors which highlighted the differences between this strain and others also identified as *C. acetobutylicum*, which subsequently led to the reclassification of NCIMB 8052 as *C. beijerinckii* (124).

38.3. HOST VECTOR SYSTEMS

Generally speaking, the efficiency with which DNA may be transformed into clostridia is relatively poor. This precludes direct cloning strategies. Rather, vectors must be bifunctional in their replicative abilities. This allows DNA first to be manipulated in a more genetically amenable bacterial host prior to the transformation (or, indeed, conjugative transfer) of the resultant recombinant plasmid into the intended clostridial host. The alternative host is more generally *E. coli*, and the vectors, therefore, are *Clostridium*/*E. coli* shuttle vectors.

38.3.1. Cloning Vectors

The vectors utilized to date (Table 1) are reliant on replication regions derived either from plasmids of a nonclostridial origin or from the replicons of indigenous clostridial plasmids.

38.3.1.1. Vectors Based on Heterogeneous Replicons

Many clostridial vectors are based on replicons derived from members of the single-stranded (ss) DNA family of plasmids. Such plasmids are ubiquitous in gram-positive bacteria (39) and replicate by a rolling circle mechanism, via an ssDNA intermediate. They include the staphylococcal plasmids pT181, pC221, pE194, pC194, and pUB110; the streptococcal plasmid pLS1; the bacillus plasmid pIM13; and the lactococcal plasmid pWV01. Their promiscuous nature and native resistance genes make them ideal candidates for the construction of vectors. Both *E. coli*/*C. acetobutylicum* and *B. subtilis*/*C. acetobutylicum* shuttle vectors have been successfully based on these plasmids (57, 58). Plasmid pWV01 is of special utility, because, through the unique properties of its replicon (45), it is able to function in both gram-negative (50) and gram-positive (62) bacterial hosts.

Despite their prevalence in vector construction, the replication strategy adopted by ssDNA plasmids is considered to be an inherent source of instability, because of the recombinogenic nature of ssDNA (72). This may be circumvented by the use of the broad-host-range *E. faecalis* plasmid pAMβ1 (53). pAMβ1 is well characterized (110) and replicates via a unidirectional theta mechanism (15). Theta replicons form the basis of structurally and segregationally more stable plasmids, and hence a variety of useful shuttle vectors have been constructed with pAMβ1 (58, 72).

38.3.1.2. Clostridial Replicons

Plasmids which naturally occur in various saccharolytic clostridia (54, 63, 74, 91, 112, 114) have been a logical, but not always successful, choice for shuttle vector construction. Thus, plasmid pCS86 was isolated from a strain of *C. acetobutylicum* (133) and employed to make the chloramphenicol and ampicillin resistance shuttle vectors pTY10 and pTY20. However, the deletion of *E. coli*-derived sequences was shown to occur in the clostridial host used, suggesting that either the construct was unstable or that restriction systems exist in this strain which discriminate against GC-rich DNA.

Strains of *Clostridium butyricum* have proven a more fruitful source of plasmid elements (63, 74, 114). The three plasmids isolated from *C. butyricum* IFO 3847 proved to be relatively large in size (13.5, 46, and 73.5 kb). Those isolated from *C. butyricum* NCIB 7423 (6.4 kb) and NCTC 7423 (6.3 and 8.4 kb) are smaller and have been both characterized (63, 74) and sequenced (14, 72). Both plasmids have been used as a basis of cloning vectors, particularly pCB101 (58, 72). While plasmid pCB101 clearly belongs to the ssDNA plasmid family (14), a 1.6-kb fragment encompassing the minimum replicon of pCB102 exhibits no features, either at the protein or DNA level, common to any known plasmid element. Nevertheless, a recombinant plasmid, pMTL540E (Fig. 1), incorporating this subfragment of pCB102 has been found to exhibit enhanced segregational stability in *C. beijerinckii* NCIMB 8052 and to be able to transform a variety of other clostridial hosts (24).

38.3.2. Bacteriophage Replicons

A number of bacteriophages have been isolated from clostridia of biotechnological interest, including CA1 (93) and

TABLE 1 Cloning vectors used in clostridia

Plasmid	Size (kb)	Source of replicon[a]	Host range[b]	Characteristics	Markers[c]	Reference
pCTC1	7.18	pAMβ1 (*E. faecalis*)	Ca Bs Ec	IncP mobilizable	Em^r[Ap^r]	125
pCTC511	7.85	pCB101	Ca Bs Ec	IncP mobilizable	Em^r[Ap^r]	125
pMTL30/31	4.36		Ec	IncP mobilizable	Em^r[Ap^r]	125
pMTL20/201E	3.6		Ec	Replicon probe vector	Em^r[Ap^r]	85
pIP501	35.0	pIP501 (*E. faecalis*)	Ca	Conjugative	Em^r Cm^r	95
pAMβ1	25.5	pAMβ1 (*E. faecalis*)	Ca Bs	Conjugative theta replicating	Em^r	87
pMTL500E	6.43	pAMβ1 (*E. faecalis*)	Ca Bs Ec	Broad-host-range cloning vector	Em^r[Ap^r]	85
pMTL502E	7.52	pAMβ1 (*E. faecalis*)	Ca Bs Ec	Low-copy version of pMTL500E	Em^r[Ap^r]	72
pMTL500F	6.69	pAMβ1 (*E. faecalis*)	Ca Bs Ec	Expression vector	Em^r[Ap^r]	72
pMTL513	7.29	pAMβ1 (*E. faecalis*)	Ca Bs Ec	Stability cloning vector	Em^r[Ap^r]	72
MTL710	7.38	pAMβ1 (*E. faecalis*)	Ca Bs Ec	Promoter probe vector	Em^r[Ap^r]	72
pMU1328	7.5	pAMβ1 (*E. faecalis*)	Ca Bs	Deletion variant in *C. acetobutylicum*	Em^r	72
pSYL9	8.9	pAMβ1 (*E. faecalis*)	Ca Bs Ec	Broad-host-range cloning vector	Em^r[Ap^r] Tc^r	111
pVA1	11.0	pAMβ1 (*E. faecalis*)	Ca Bs	Broad-host-range cloning vector	Em^r	94
pVA677	7.6	pAMβ1 (*E. faecalis*)	Ca Bs	Broad-host-range cloning vector	Em^r	94
PIM13	2.3	pIM13 (*B. subtilis*)	Ca Bs	ssDNA replication	Em^r	111
pKNT11	6.5	pIM13 (*B. subtilis*)	Ca Bs Ec	General-purpose cloning vector, pBR322	Em^r[Ap^r]	111
pKNT14	4.3	pIM13 (*B. subtilis*)	Ca Bs	No *E. coli* replicon	EM^r	111
pKNT15	6.8	pIM13 (*B. subtilis*)	Ca Bs Ec	Based on pBR322Δ	Em^r[Ap^r] Tc^r	4
pKNT19	4.9	pIM13 (*B. subtilis*)	Ca Bs Ec	pUC19 cloning sites	Em^r[Ap^r]	4
pSYL14	4.4	pIM13 (*B. subtilis*)	Ca Bs Ec		Em^r[Ap^r]	111
pFNK1	2.4	pIM13 (*B. subtilis*)	Ca Bs		Em^r	58
pIA		pIM13 (*B. subtilis*)	Ca Bs Ec	pACYC-based shuttle vector	EM^r	128

(*Table continued on next page*)

TABLE 1 Cloning vectors used in clostridia (*Continued*)

Plasmid	Size (kb)	Source of replicon[a]	Host range[b]	Characteristics	Markers[c]	Reference
pCB3	7.03	pCB101 (*C. butyricum*)	Ca Bs Ec	ssDNA replication	Em^r[Ap^r]	72
pSYL2	8.7	pCB101 (*C. butyricum*)	Ca Bs Ec		Em^r Tc^r	58
pMTL540E	5.23	pCB102 (*C. butyricum*)	Ca Ec	Good segregational stability	Em^r[Ap^r]	30
pCB5	9.5	pCB103 (*C. butyricum*)	Ca Ec	Uncharacterized clostridial replicon	Em^r[Ap^r]	72
pSYL7	9.2	pJU122 (*C. perfringens*)	Ca Ec		Em^r Tc^r	111
pRZL3	10.8	pJU122 (*C. perfringens*)	Ca Ec	Replicates in *C. beijerinckii*	[Ap^r] Tc^r	11
PRZE4	10.0	pJU122 (*C. perfringens*)	Ca Ec	Replicates in *C. beijerinckii*	Em^r[Ap^r] Cm^r	11
pTYD101	4.0	pSC86 (*C. acetobutyricum*)	Ca	Deletion variant	Cm^r	133
pTYD104	7.6	pSC86 (*C. acetobutyricum*)	Ca Ec	Deletion variant	[Ap^r] Cm^r	133
pT127	4.4	pT127 (*Staphylococcus aureus*)	Ca Bs	Unstable in *C. acetobutyricum*	Tc^r	111
pBC16Δ1	2.8	pBC161 (*Bacillus cereus*)	Ca Bs	Unstable in *C. acetobutyricum*	Tc^r	111
pGK12	4.4	pWVO1 (*L. lactis*)	Ca Bs Ec	Very-broad-host-range vector	Em^r Cm^r	50
pCAK1	11.6	CAK1 (*C. acetobutyricum*)	Ca Ec	Phagemid, ssDNA intermediates	Em^r[Ap^r]	48
pUB110	4.5	pUB110 (*S. aureus*)	(Ct) Bs Ca	ssDNA replication, thermostable	Km^r	108
pMK419	5.6	pUB110 (*S. aureus*)	(Ct) Bs Ca Ec		[Ap^r] Cm^r	113
pCL1	11.9	pTA688L (*Clostridium* sp.)	(Ct) Ec	Thermophilic clostridial source	Tc^r Cm^r	51
PCS1	7.2	pTA688S (*Clostridium* sp.)	(Ct) Ec	Thermophilic clostridial source	Tc^r Cm^r	51

[a]Progenitor plasmid and host from which it was originally isolated.
[b]Ca, *C. acetobutylicum*; Bs, *B. subtilis*; Ec, *E. coli*; Ct, clostridial thermophiles.
[c]Em^r, erythromycin resistance; Ap^r, ampicillin resistance; Tc^r, tetracycline resistance; Cm^r, chloramphenicol resistance; [Ap^r], gram-negative marker.

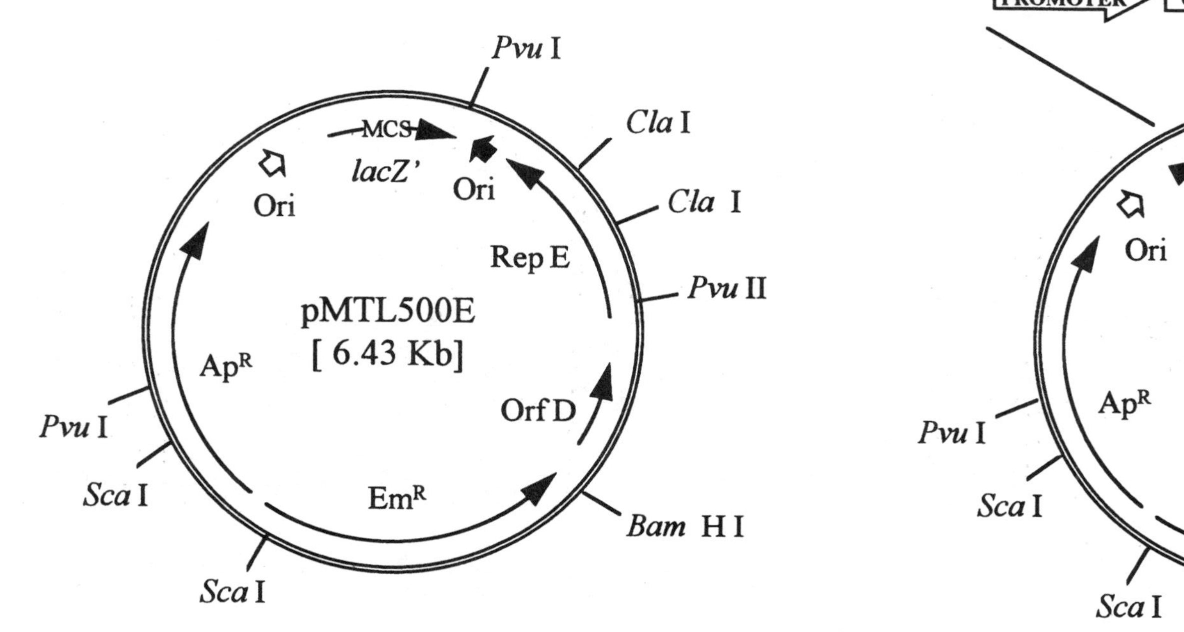

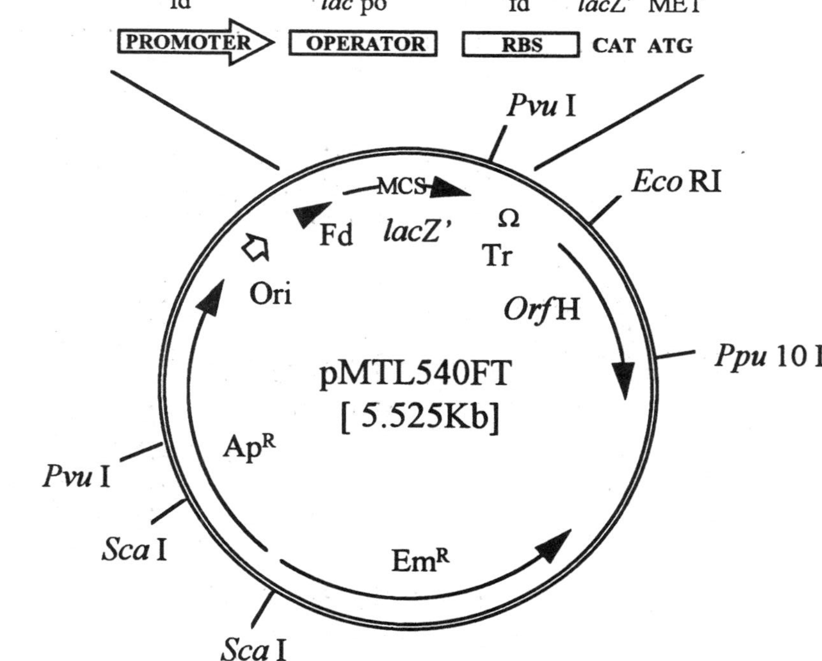

EcoR I *Sst* I *Kpn* I *Sma* I *Bam* H I *Xba* I *Sal* I *Aat* II *Mlu* I *Nco* I *Bgl* II *Xho* I *Stu* I *Pst* I *Sph* I *Hind* III

ATGACCATGATTACGAATTCGAGCTCGGTACCCGGGGATCCTCTAGAGTCGACGTCACGCGTCCATGGAGATCTCGAGGCCTGCAGGCATGCAAGCTTGGC

M T M I T L A

LacZ

FIGURE 1 The *C. acetobutylicum/E. coli* shuttle vectors pMTL500E and pMTL540FT. Plasmid pMTL500E was constructed by using the Col E1 replicon (open arrow) and erythromycin resistance gene from the pUC-derivative plasmid pMTL20E (85) and the gram-positive broad-host-range replicon (filled arrow) of pAMβ1. The expression vector pMTL540FT was constructed by combining the gram-positive Fd expression cartridge of pMTL500F (83) with the replicon (OrfH) of pCB4 (83). A transcriptional terminator (Tr) has been inserted 3′ to the expression cartridge to prevent transcripts from interfering with the replication mechanism. Note: *Bam*HI and *Eco*RI sites are not unique in pMTL500E and pMTL540FT, respectively.

CAK1 (49). A DNA fragment encoding novel viral replication sequences from CAK1 has been the basis of a phagemid construct, pCAK1, for use in *C. acetobutylicum* ATCC 824, *C. beijerinkii* NCIMB 8052, and *C. perfringens* 13 (48).

38.3.3. Expression Vectors

The generation of vectors designed specifically to bring about the overexpression of heterologous genes has received scant attention. One exception has been the construction of the vector pMTL500F (72), which carries the transcriptional signals of the *Clostridium pasteurianum* ferredoxin (Fd) gene (35). This promoter is capable of directing the production of *C. pasteurianum* Fd to levels equivalent to 2% of the cells' soluble protein (92). Plasmid pMTL500F has an expression cassette containing, in tandem, the Fd promoter and ribosome binding site optimally spaced before an *Nde*I site at which the initial formylmethionine codon of a gene can be fused (Fig. 1). Alternatively, heterologous gene fusions can be created with the 5′ codons of the *lacZ′*, or the gene can simply be inserted such that it relies on its own ribosome binding site. The inclusion of a multiple cloning site and the *lacZ′* gene allows "blue to white" selection of recombinants in *E. coli*. The operator sequence of *E. coli lac* promoter has been incorporated to allow regulatory control of transcription, at least in *E. coli*, through provision of LacI.

The replication origin of pMTL500F originates from pAMβ1. Construction of an alternative vector, pMTL540F, was based on the highly stable pCB102 replicon (Fig. 1). Initial stability problems encountered with pMTL540F, due to transcriptional read-through from the Fd promoter into the plasmid's replication region, were overcome by the introduction of the transcriptional terminator from the Fd gene.

38.3.4. Integrative Vectors

Vectors designed for targeted insertion into bacterial genomes are defective or deficient in their ability to replicate in the recipient organism. There are few examples of vectors which have been specifically constructed with this purpose in mind. Most notable are the suicide vectors, pMTL30/31, created by Williams and co-workers. Here, the gram-positive replicon-deficient vector pMTL20E was endowed with the RK2 *oriT* region (125) to facilitate their subsequent conjugative transfer from a Tra$^+$ *E. coli* donor into *C. beijerinckii* NCIMB 8052 (122).

In *C. acetobutylicum* ATCC 824, the generally higher transformation frequencies that can be achieved have meant that it has merely been necessary, subject to the provision of a gram-positive resistance gene, to use standard *E. coli* cloning vectors (e.g., see references 36, 37, and 128). Problems with "transformation efficiency" may be overcome through the use of plasmids with a thermosensitive replicon. The utility of such plasmids (66) has yet to be demonstrated in clostridia.

38.3.5. Plasmid Isolation Procedure

Various methods have been employed to isolate the parental plasmids used to construct vectors and to prepare vectors and their derivatives. A protocol commonly used in this laboratory is an adaptation of one such method.

For screening purposes, colonies are inoculated into a small volume (10 ml) of TPGY (*C. botulinum*, *C. perfringens*) or 2× YT (*C. beijerinckii*, *C. acetobutylicum*) broth supplemented with an appropriate antibiotic (e.g., 5 to 20 μg of erythromycin or chloramphenicol) and incubated anaerobically at 37°C. When in mid- to late exponential phase, a 1.8-ml aliquot is transferred to an Eppendorf tube and centrifuged at 7,000 × *g* for 5 min at 4°C. The supernatant is carefully aspirated, the cells are resuspended in 100 μl of GTE buffer (50 mM glucose, 25 mM Tris-HCl [pH 8.0], 10 mM EDTA) containing lysozyme (10 mg/ml) and incubated at 37°C for 15 min. A 200-μl aliquot of freshly made lysis solution (1% sodium dodecyl sulfate, 0.2 M NaOH, in sterile H_2O) is added, the solution is briefly vortexed to mix, and it is then held on ice for 5 min. The lysate is neutralized through the addition of 150 μl of ice-cold 3 M potassium acetate (pH 4.8) and returned to ice for a further 30 min. At this point the lysate is centrifuged at 6,000 × *g* for 30 min at 4°C, and the supernatant is promptly removed and extracted with an equal volume of phenol–chloroform–isoamyl alcohol (25:24:1). The resultant aqueous layer is removed and precipitated through the addition of 2.5 volumes of absolute ethanol and 0.1 volume of 3 M sodium acetate (pH 4.8) and then stored at −20°C for 30 min. The DNA is recovered by centrifugation at 7,000 × *g* for 20 min at 4°C. The supernatant is decanted, and the pellet is washed with 70% ethanol and then air dried and resuspended in 50 to 20 μl of sterile H_2O. For removal of RNA, 15 μl of 10-mg/ml RNase A is added, and the DNA is incubated for 30 min at 37°C. A 5-μl sample is analyzed on an agarose gel to check complete removal of RNA and, if complete, the phenol–chloroform–isoamyl alcohol step is repeated and the DNA is resuspended in 10 to 20 μl of 1× TE buffer.

This method can be scaled up to prepare larger quantities of DNA. Typically, a 1-liter culture would use 12 ml of GTE buffer, 24 ml of lysis solution, and 18 ml of potassium acetate solution. Larger preparations may contain significant contaminating chromosomal DNA and RNA. RNA may be removed by incubating at 4°C for 16 h. To obtain a good-quality DNA, free of chromosomal contamination, an isopycnic centrifugation step is usually necessary.

38.4. GENE ANALYSIS

38.4.1. Fermentative Metabolism

Clostridia have been important in industrial processes since the early part of this century. Fermentations carried out include homoacetate fermentation by *Clostridium thermoaceticum*; butyrate-acetate fermentation by *C. butyricum*; and ethanol fermentation by *C. thermocellum*, *C. thermohydrosulfuricum*, and *C. thermosaccharolyticum* (96). The ABE fermentation undertaken with *C. acetobutylicum* is, however, the most commercially important clostridia-based process. Although largely replaced in modern times, there is currently a revival of interest in ABE fermentation, driven by economic, political, and environmental concerns (131).

During exponential growth, *C. acetobutylicum* exhibits a typical butyric acid fermentation in which the major products are acetate, butyrate, carbon dioxide, and hydrogen (27). The switch to solventogenesis, in which the major products are acetone and butanol, occurs when the cells enter stationary phase (44). The trigger(s) for this shift remains unknown, although in vitro experiments have shown an association with high concentrations of carboxylic acids (71) and low pH due to the production of acetate and butyrate (27). Solventogenesis is linked to endospore formation, although each does not appear to be an essential requirement of the other, since mutants defective in only one of these functions can be isolated (e.g., see reference 5). The hypothesis that a complex response to cell stress occurs at this point in the growth cycle is enhanced by the observation that some heat shock proteins are transiently induced (100).

While many of the individual enzymes involved in acidogenesis and solventogenesis have been purified, not all have been cloned (88), and even less information on the organization and possible regulation of these genes is available. Some enzymes are predicted to have a role in both acidogenesis and solventogenesis (18). The enzyme thiolase, which catalyzes the conversion of two molecules of acetyl-coenzyme A (CoA) to one of acetoacetyl-CoA, is required during both acidogenesis and solventogenesis and, indeed, evidence for two different thiolase genes in one strain of *C. acetobutylicum* has been presented, suggesting that each is activated under different conditions (127). A putative operon encoding the genes for β-hydroxybutyryl-CoA dehydrogenase, crotonase, and butyryl-CoA dehydrogenase (all of which are required in both acidogenesis and solventogenesis), plus two possible electron transfer flavoprotein genes, has recently been reported (13). Butyryl-CoA dehydrogenase is analogous to an acyl-CoA dehydrogenase from mammalian tissue which requires electron transfer flavoprotein as an electron donor; hence, this operon is evidence that a similar requirement exists in *C. acetobutylicum* ATCC 824 (13).

Acetone production is catalyzed by the sequential action of two enzymes: acetoacetyl-CoA: acetate/butyrate: coenzyme A transferase (CoA transferase) and acetoacetate decarboxylase. The gene encoding the latter, *adc*, occurs in a monocistronic operon which is adjacent to a convergent operon (the *sol* operon) coding for the two subunits of the CoA transferase (*ctfA* and *ctfB*) and an aldehyde/alcohol dehydrogenase (*adhE* or *aad*) gene (29). The transcriptional terminator which separates these operons appears to function bidirectionally (89). The *sol* operon also contains another small putative gene which has been suggested to have a regulatory function (29). This discovery of more than one operon for the enzymes responsible for catalyzing acetone formation suggests that production of these enzymes may have different physiological triggers (27). Recently, evidence has been presented which suggests that degeneration (loss of the ability to produce solvents and form spores) is a result of the loss of the region of the chromosome carrying the *adc* and *sol* operons (109).

Butyryl-CoA is an intermediate in the formation of both butyrate and butanol. Butyrate formation is catalyzed by phosphotransbutyrylase and butyrate kinase in two consecutive steps, and this relationship is also reflected in their genetic structure; the genes encoding butyrate kinase (*buk*) and phosphotransbutyrylase (*ptb*) have been found in an operon with a single transcriptional start point in *C. acetobutylicum* ATCC 824 (119) and in *C. beijerinckii* NCIMB 8052 (83). Two other enzymes involved in butanol formation, butanol dehydrogenase I and II, have both been mapped to monocistronic operons, *bdhA* and *bdhB* (116).

Since solventogenesis and sporulation are generally induced by the same conditions, it is of interest at this point to consider recent research into the regulation of sporulation. Much work has been performed on sporulation in another gram-positive endospore-forming bacterium, *B. subtilis*, leading to the identification of sporulation-specific sigma (σ) factors which confer new promoter specificities on RNA polymerases (28). While it would appear that sporulation in clostridia is not induced by the same trigger (starvation) as in bacilli, there are similarities at the genetic level between the two (102). The genes encoding the sporulation-specific sigma factors, σ^E and σ^G, from *C. acetobutylicum* DSM 792 (103) and ATCC 824 (129) have been found to be organized in a way similar to those in bacilli, and all three sets of gene products have extensive homology (102). The similarity between sigma factors (and, therefore, possibly also in bacterial transcription control mechanisms) has been found to extend beyond those which are sporulation-specific; the principal sigma factor, σ^A, from *C. acetobutylicum* occurs in an operon analogous to that found in both *B. subtilis* and *E. coli* and exhibits significant sequence homology to other σ^A genes (103). In *B. subtilis*, the *spo0A* gene encodes a regulatory protein which controls many aspects of stationary phase metabolism (41). A similar role in *C. beijerinckii* NCIMB 8052 is suggested by the finding that inactivation of the *spo0A* gene results in no sporulation or solvent production, although it remains unclear if such control is direct or indirect (120).

38.4.2. mRNA Analysis

Examination of mRNA is useful in terms of mapping the exact start points of individual genes with consequent deductions about promoters, operons, etc., and also in terms of determining at which point during the cell growth cycle particular genes are turned on. Our laboratory has recently published a method for the gel electrophoresis and Northern blotting of RNA which avoids the use of volatile and toxic denaturants and has been particularly successful with a range of clostridial species (34).

38.4.2.1. Procedure for Extraction and Electrophoresis of Clostridial RNA

Grow 10 ml of culture to late exponential phase, harvest by centrifugation, and resuspend in 500 μl of chilled AE buffer (50 mM sodium acetate [pH 5.3], 10 mM EDTA). Lyse cells by adding 50 μl of 10% (wt/vol) sodium dodecyl sulfate, vortex briefly but thoroughly, then add an equal volume of AE-equilibrated phenol and incubate at 65°C for 5 min. Place in a dry ice/ethanol bath for 2 min, then centrifuge, extract upper aqueous layer with an equal volume of 1:1 phenol-chloroform. Centrifuge at 10,000 × *g* for 5 min, then precipitate RNA in the upper aqueous phase at −70°C by the addition of 0.1 volume of 3 M sodium acetate (pH 5.3) and 2.5 volumes of absolute ethanol. The RNA is recovered by centrifugation at 7,000 × *g* for 20 min at 4°C. The supernatant is decanted, and the pellet is washed with 70% ethanol and then air dried and resuspended in 50 μl of RNase-free distilled water.

Pre-soak gel tank and casting tray in 1% (wt/vol) sodium dodecyl sulfate overnight. Add 0.5 ml of freshly prepared 1 M guanidine thiocyanate and 2 μl of 10-mg/ml ethidium bromide to 100 ml of 1.2% (wt/vol) molten agarose. Pour gel and run at 8 V/cm in TBE buffer (90 mM Tris-HCl [pH 8], 90 mM sodium borate, 2 mM EDTA). After electrophoresis, soak gel in 0.05 M NaOH for 30 min and transfer RNA to a nylon membrane according to the manufacturer's instructions. Northern blotting can then be performed under standard conditions.

38.4.2.2. mRNA Analysis of Solventogenic Genes

mRNA analysis has proven to be particularly useful in the study of solventogenesis genes. In continuous culture of *C.*

acetobutylicum, differential induction of the *sol* operon genes and the two butanol dehydrogenase isozymes (*bdhA* and *bdhB*) was observed in the order *bdhA–sol* operon–*bdhB* during the pH-induced shift to solventogenesis (101). Transcription of the *sol* operon has previously been shown to occur just before the onset of solventogenesis (29). The difference between levels of mRNA of the *bdh* genes and the *sol* operon at given times is an indication that different regulatory mechanisms are involved (101).

In *C. acetobutylicum* P262, the genes encoding for 3-hydroxybutyryl-CoA dehydrogenase (*β-hbd*) and an NADPH-dependent alcohol dehydrogenase (*adh1*), which are adjacent on the chromosome and are both expected to be required during acidogenesis and solventogenesis, were not found to be transcribed at the same time (135). While the *β-hbd* gene was transcribed during both acidogenesis and solventogenesis, the *adh1* gene was expressed only during solventogenesis. As well as indicating that these two genes do not form an operon, this result suggests that a different NADPH-dependent alcohol dehydrogenase may be active during acidogenesis (135).

38.4.3. Substrate Utilization

There are numerous clostridia which are of industrial interest with respect to their substrate utilization, in terms of both end products and particular enzymatic activities. This area has been reviewed relatively recently (130), and it is the purpose of this chapter to simply highlight the more important areas.

One particularly useful feature of *C. acetobutylicum* is its ability to utilize a variety of industrial and agricultural byproducts as substrates in the ABE fermentation (38, 44). Carbohydrates such as starch, cellulose, xylans, and lactose can all be fermented, and a variety of enzyme activities associated with these substrates have been identified, e.g., α-amylase and glucoamylase (19), α-glucosidase (2), carboxymethylcellulase (3), cellulase (55), endo-β-1,4-glucanase (136), endoxylanse (56), β-galactosidase (40), and lactate dehydrogenase (26). Although cellulase enzymes have been identified in *C. acetobutylicum*, this species does not appear to be truly cellulolytic or to possess the complex organelle structure of *C. thermocellum* (see below).

Some of the above enzymes have been characterized at a genetic level, generally by cloning and expression in *E. coli*, or by complementation studies with *E. coli* mutants. There are, however, reports which suggest that crucial differences in genetic regulation between *E. coli* and clostridia exist; e.g., the expression in *E. coli* of the β-galactosidase gene from *C. acetobutylicum* requires a second putative regulatory gene product (40). Recently, a putative repressor gene concerned with starch degradation by *C. acetobutylicum* has also been described (25).

Within the clostridia, the best-described cellulase system is that of *C. thermocellum*. Organisms able to utilize cellulose generally produce a variety of hydrolytic enzymes, e.g., endoglucanases, cellobiohydrolases, and β-glucosidases, as well as hemicellulases, xylanases, and ligninases. The cellulases of *C. thermocellum* occur on the cell surface as high-molecular-weight (>1-2 MDa) organelles termed cellulosomes (52) and are composed of many enzymes; e.g., 15 endoglucanases, 2 xylanases, 2 β-glucosidases, and 2 lichenases are produced by a single *C. thermocellum* strain (8).

Some cellulosome components do not possess enzymic activity but play a structural role. Thus, the scaffold protein CipA acts as a cellulose-binding factor with which the other catalytic subunits associate in order to remain attached to cellulose (8). The majority of *C. thermocellum* endoglucanases contain a highly conserved, noncatalytic domain near their carboxyl terminus. This domain is called the dockerin domain (7) and is responsible for anchoring the catalytic subunits to the CipA scaffolding protein. The structure of CipA suggests that it is multifunctional, containing a cellulose binding domain and nine very similar modules (cohesion domains) which are responsible for binding the complementary dockerin domains carried by the catalytic subunits (33).

Although a large number of *C. thermocellum* cellulase enzymes have been characterized, little is known about their genetic regulation, and the lack of a means of genetic transfer within this organism does little to aid this problem. Cellulase biosynthesis is thought to be regulated by a mechanism analogous to catabolite repression (82). Most of the cellulase genes are scattered throughout the *C. thermocellum* chromosome, and their mRNA is monocistronic, although CipA occurs in a gene cluster with other cell surface proteins (32). These other cell surface proteins have been proposed to be cell envelope components involved in binding the cellulosome to the cell surface (60). Transcripts of some cellulase genes appear most abundant as the cells enter stationary phase (76).

Other clostridia also produce cellulosome-like complexes, e.g., *Clostridium cellulovorans* (107) and *Clostridium cellulolyticum* (65), although in both cases the structures themselves may not be as complex as that in *C. thermocellum* (8). In contrast to *C. thermocellum*, the cellulase system of *Clostridium papyrosolvens* C7 has been proposed to consist of at least seven distinct multiprotein complexes (90).

38.4.4. Toxins

With notable exceptions (e.g., production of botulinum toxin by certain strains of *C. butyricum*), toxins are not generally produced by industrially important clostridia. However, they are of great medical interest, being the major virulence determinants of *Clostridium* species associated with disease. Moreover, the increasing therapeutic deployment of botulinum toxin in the treatment of dystonias as an alternative to surgical intervention underlines that even the most toxicogenic species can be of commercial value (43). A review of the genetics of toxin production has recently been published (98).

38.5. GENE MANIPULATION

38.5.1. Recombinant Expression of Homologous Genes

The majority of examples of recombinant expression of homologous genes in clostridia have involved metabolic engineering of the key stages in the transition of fermentative metabolism from acid to solvent generation in *C. acetobutylicum*, e.g., acidogenesis genes (12), solventogenesis genes (13, 59, 69–71, 80, 81, 118), and sporulation-specific genes (128). All of these reports utilized genes from *C. acetobutylicum* ATCC 824, employed published electroporation methodologies (68), and either used vectors based on the *B. subtilis* pIM13 or *C. butyricum* pCB101 replicons or employed plasmids designed for recombination into the clostridial chromosome because of the absence of a clostridial replicon.

The majority of these reports resulted in the successful expression of the homologous genes. Increases in the activity of the targeted enzymes were generally recorded, but this was not always reflected in amounts of solvents formed. The use of plasmid-borne operons has a number of inherent problems, including increased metabolic load due to plasmid replication, the use of foreign DNA with a G+C DNA bias, the topological environment of the plasmid DNA, the removal of distant regulatory elements, copy-number duplication of potential operator sequences, and the use of selective agents to maintain plasmid segregational stability. Some of these factors may explain the conflicting results obtained (70, 81, 118).

It is of note that an "any plasmid" effect was also reported, in that *C. acetobutylicum* ATCC 824 containing any plasmid, with or without homologous DNA, exhibited increased solvent production (117). This finding is indicative of the role played by stress factors in the transition from acid to solvent production and underlines the complexity of the regulation of expression of solvent genes (131).

38.5.2. Recombinant Expression of Heterologous Genes

There are few examples of the expression of heterologous genes in clostridia, other than plasmid replication functions and antibiotic resistance genes. Certain clostridial genes have been expressed in other species; e.g., a *C. pasteurianum* gene for leucine biosynthesis has been transferred into an auxotrophic mutant of *C. beijerinckii* NCIMB 8052 by conjugation and shown to restore prototrophy (86). The *engB* gene from *C. cellulovorans*, encoding an endo-β-1,4-D-glucanase, has been expressed in *C. acetobutylicum* ATCC 824 (47) with the vector pMTL500E (85), and activity was confirmed by zones of hydrolysis on carboxymethyl cellulose. The same vector was used to express the *C. thermocellum celA* and *celC* genes in *C. beijerinckii* NCIMB 8052 (75). In all of these cases the heterologous genes were transferred via shuttle vectors with their native control elements intact.

Expression vectors specifically designed for clostridia have been developed (72). An example of their use is in the development of clostridia as a potential delivery system for cancer therapy; the bacterial genes coding for nitroreductase (73) and cytosine deaminase (30) have been successfully expressed in *C. beijerinckii* NCIMB 8052. In both these cases, the genes were under the control of the *C. pasteurianum* Fd promoter.

38.5.3. Gene Inactivation

While transposon mutagenesis can lead to gene inactivation by random insertion, it is possible to direct insertion by developing vectors designed to integrate into the chromosome (see section 38.3.4). Such gene inactivation studies using suicide systems have been performed with *C. beijerinckii* NCIMB 8052 and *C. acetobutylicum* ATCC 824. With *C. beijerinckii* NCIMB 8052 (122) a conjugative vector, pMTL30 (125), was used, integration was targeted into *gutD* or *spo0A*, and in both cases insertion was mutagenic, resulting in loss of the ability to grow on sorbitol as sole carbon source (*gutD*) or loss of the ability to form endospores (*spo0A*). For *C. acetobutylicum* ATCC 824 (36, 37) a derivative of pBR322 was used as vector, with integration targeted into the phosphotransacetylase (*pta*), butyrate kinase (*buk*), or aldehyde/alcohol dehydrogenase (*aad*) genes. The limitations caused by the inactivation of these genes were then assessed in fermentation studies. In all of these studies, the integrants were found to be relatively stable.

In another study with *C. acetobutylicum* ATCC 824 (128), vectors were constructed to bring about three different types of integration: gene duplication by single crossover, gene replacement by double crossover, and gene inactivation. Integration was directed at three sporulation-specific genes (*orfA*, *sigE*, and *sigG*), and although transformants were achieved with all three vector types, no gene inactivation was evident and Southern hybridization indicated that integration had not occurred at the target sites.

38.6. GENOME ANALYSIS

The characterization of entire bacterial genomes is now a firmly established science, in which pulsed-field gel electrophoresis (PFGE) and automated DNA sequencing technology are playing a central role. Their application to clostridia, however, is still at an early stage.

38.6.1. Pulsed-Field Gel Electrophoresis

The advent of PFGE allows whole genomes to be physically characterized through restriction enzyme mapping, and it has now been applied to many different prokaryotes. Among the clostridia, the clinically significant pathogen *C. perfringens* (16) and the biotechnologically important *C. acetobutylicum* ATCC 824 and *C. beijerinckii* NCIMB 8052 (123) have been investigated.

Good-quality genomic DNA is essential if defined restricted fragments of the chromosome are to be visualized and manipulated. Adaptations may be necessary between genera and between individual strains, and the medium and growth conditions may also be an influential factor. Essentially, the conditions required for the in situ cell lysis and subsequent removal of cell wall and cellular debris must be found with the concomitant preservation of intact DNA.

For *C. beijerinckii* NCIMB 8052, the following procedure was devised (121). Bacteria (250 ml) are grown until early to mid-exponential phase and pelleted by centrifugation. The pellet is washed in STE buffer (50% sucrose, 50 mM Tris-HCl, 1 mM EDTA [pH 7.6]) to a final density of 6 × 10^8 cells/ml. To this an equal volume of molten agarose is added (low-temperature gel agarose 1.5% [wt/vol], Sigma type VII in STE buffer), mixed gently with the cell suspension, and poured into an insert former (plug former). Cells in the solidified plugs are lysed in 4 volumes of lysis solution (10 mM Tris-HCl, 1 M NaCl, 100 mM EDTA, 50% [wt/vol] sucrose, 0.5% [wt/vol] sodium lauroylsarcosine, 10 mg of lysozyme per ml, 20 μg of DNase-free RNase [pH 7.6]) and incubated overnight at 37°C. All manipulations are conducted in an anaerobic environment. Following lysis, the plugs are deproteinized for 60–72 h at 48°C in 3 volumes of solution (0.5 M EDTA, 1.0% [wt/vol] sodium lauroylsarcosine, 2 mg of proteinase K per ml [pH 9.0]). To clean the DNA, the plug is washed by gentle agitation at room temperature for 1 h in 4 volumes of 1× TE (pH 8.0), containing 1.5 mM phenylmethlysulfonyl fluoride, and subsequently washed 6 times in 4 volumes of 1× TE (pH 8.0) for 1 h each. Restriction enzyme digests and PFGE are performed by standard methods.

38.6.2. Genome Mapping

The generation of a genetic map of a genome formed by the systematic elucidation of the contiguous restriction fragments found in the chromosome is accomplished in a number of ways. Type II restriction endonucleases are commonly used, typically with 8 bp of palindromic recognition sites. In most mesophilic clostridia, the bases A+T predominate. Restriction endonucleases recognizing G+C-rich sequences are, therefore, most useful, e.g., *Sma*I, *Not*I, *Sfi*I, *Sgr*AI, *Sse*83871, *Srf*I, *Sgf*I, *Fse*I, and *Asc*I. An alternative source of enzymes is intron-encoded "meganucleases" with 18 to 30 bp of recognition sequences available, of which *Ceu*I is particularly useful because its 26 bp site is conserved among many bacteria in the *rrl* gene of 23S rRNA. Restriction digestion with *Ceu*I is likely to reveal the distribution of *rrn* operons.

Linkage of restriction fragments can involve a number of approaches, such as the use of additional restriction enzymes, excision of fragments from one gel and subsequent restriction with another enzyme before running again (double digests), and the use or generation of linking fragments which can be used in Southern hybridization experiments to detect two contiguous fragments (22). An alternative approach is to generate an overlapping library of cloned fragments, using cosmid or plasmid clones. However, DNA clones from AT-rich organisms may not be stable in *E. coli*, or the encoded proteins may prove lethal, especially at high copy number (128). Indeed, many genes involved in a common process or metabolic pathway have been found to be located in close proximity to one another (123) and may reside on transmissible extrachromosomal elements (17). Genes associated with solventogenesis have been detected on a large plasmid-like element (210 kb) in one clostridial strain (23).

38.7. CONCLUSION

In the last decade there have been significant advances in the genetic tools available for manipulating clostridial species. Many of the genes involved in fermentative metabolism, substrate utilization, and toxin production have been cloned and physically characterized. The major preoccupation now is to understand how such genes are regulated, as a necessary prelude to the generation of strains with improved fermentation characteristics. One can anticipate that over the next decade molecular biologists should be able to create strains better able to meet the manufacturing demands of the modern era. It is, however, intriguing to note that certain members of the genus are now perceived to be of therapeutic value. Indeed, it may well transpire that saccharolytic species have far greater value in the area of cancer chemotherapy than in their more traditional role as a producer of chemical fuels.

We thank Michael Young and Sayed Goda for kindly supplying up-to-date protocols, and we acknowledge the financial support of the U.K. Department of Health, the U.K. Medical Research Council (G9207831CB), the U.K. Ministry of Agriculture Food and Fisheries (FS1529), the U.K. Biotechnology and Biological Sciences Research Council (T04089), the Anglo-German Research Council (667/800), and the U.S. National Cancer Institute (CA 64697-01A1).

REFERENCES

1. **Abraham, L. J., and J. I. Rood.** 1987. Identification of Tn*4451* and Tn*4452*, chloramphenicol resistance transposons from *Clostridium perfringens*. *J. Bacteriol.* **169:**1579–1584.
2. **Albasheri, K. A., and W. J. Mitchell.** 1995. Identification of two α-glucosidase activities in *Clostridium acetobutylicum* NCIMB 8052. *J. Appl. Bacteriol.* **78:**149–156.
3. **Allcock, E. R., and D. R. Woods.** 1981. Carboxylmethyl cellulase and cellobiase production by *Clostridium acetobutylicum* in an industrial fermentation medium. *Appl. Environ. Microbiol.* **41:**539–541.
4. **Azeddoug, H., J. Hubert, and G. Reysset.** 1992. Stable inheritance of shuttle vectors based on plasmid pIM13 in a mutant strain of *Clostridium acetobutylicum*. *J. Gen. Microbiol.* **138:**1371–1378.
5. **Babb, B. L., H. J. Collett, S. J. Reid, and D. R. Woods.** 1993. Transposon mutagenesis of *Clostridium acetobutylicum* P262: isolation and characterization of solvent deficient and metronidazole resistant mutants. *FEMS Microbiol. Lett.* **114:**343–348.
6. **Bannen, T. L., P. K. Crellin, and J. I. Rood.** 1995. Molecular genetics of the chloramphenicol-resistance transposon Tn*4451* from *Clostridium perfringens*: the TnpX site-specific recombinase excises a circular transposon molecule. *Mol. Microbiol.* **16:**535–551.
7. **Bayer, E. A., E. Morag, and R. Lamed.** 1994. The cellulosome—a treasure-trove for biotechnology. *Trends Biotechnol.* **12:**379–386.
8. **Béguin, P., and M. Lemaire.** 1996. The cellulosome: an exocellular multiprotein complex specialized in cellulose degradation. *Crit. Rev. Biochem. Mol. Biol.* **31:**201–236.
9. **Bertram, J., and P. Dürre.** 1989. Conjugal transfer and expression of streptococcal transposons in *Clostridium acetobutylicum*. *Arch. Microbiol.* **151:**551–557.
10. **Bertram, J., A. Kuhn, and P. Dürre.** 1990. Tn*916*-induced mutants of *Clostridium acetobutylicum* defective in regulation of solvent formation. *Arch. Microbiol.* **153:** 373–377.
11. **Birrer, G. A., W. R. Chesbro, and R. M. Zsigray.** 1994. Electro-transformation of *Clostridium beijerinckii* NRRL B-592 with shuttle plasmid pHR106 and recombinant derivatives. *Appl. Microbiol. Biotechnol.* **41:**32–38.
12. **Boynton, Z. L., G. N. Bennet, and F. B. Rudolph.** 1996. Cloning, sequencing and expression of genes encoding phosphotransacetylase and acetate kinase from *Clostridium acetobutylicum* ATTC 824. *Appl. Environ. Microbiol.* **62:** 2758–2766.
13. **Boynton, Z. L., G. N. Bennett, and F. B. Rudolph.** 1996. Cloning, sequencing and expression of clustered genes encoding β-hydroxybutyryl-coenzyme A (CoA) dehydrogenase, crotonase, and butyryl-CoA dehydrogenase from *Clostridium acetobutylicum* ATCC 824. *J. Bacteriol.* **178:** 3015–3024.
14. **Brehm, J. K., A. Pennock, H. M. S. Bullman, M. Young, J. D. Oultram, and N. P. Minton.** 1992. Physical characterisation of the replication origin of the cryptic plasmid pCB101 isolated from *Clostridium butyricum* NCIB 7423. *Plasmid* **28:**1–13.
15. **Bruand, C., S. D. Ehrlich, and L. Janniere.** 1991. Unidirectional theta replication of the structurally stable *Enterococcus faecalis* plasmid pAMβ1. *EMBO J.* **10:**2171–2177.
16. **Canard, B., and S. T. Cole.** 1989. Genome organization of the anaerobic pathogen *Clostridium perfringens*. *Proc. Natl. Acad. Sci. USA* **86:**6676–6680.

17. **Canard, B., B. Saint-Joanis, and S. T. Cole.** 1992. Genomic diversity and organisation of virulence genes in the pathogenic anaerobe *Clostridium perfringens*. *Mol. Microbiol.* **6:**1421–1429.

18. **Chen, J.-S.** 1993. Properties of acid and solvent forming enzymes of clostridia, p. 51–76. *In* D. R. Woods (ed.), *The Clostridia and Biotechnology*. Butterworth-Heinemann, Boston.

19. **Chojecki, A., and H. P. Blaschek.** 1986. Effect of carbohydrate source on alpha-amylase and glucoamylase formation by *Clostridium acetobutylicum* SA-1. *J. Ind. Microbiol.* **1:**63–67.

20. **Clewell, D. B., S. E. Flannagan, Y. Ike, J. M. Jones, and C. Gawron-Burke.** 1988. Sequence analysis of the termini of conjugative transposon Tn*916*. *J. Bacteriol.* **170:** 3046–3052.

21. **Clewell, D. B., and C. Gawron-Burke.** 1986. Conjugative transposons and the dissemination of antibiotic resistance in Streptococci. *Annu. Rev. Microbiol.* **40:**635–659.

22. **Cole, S. T., and I. S. Girons.** 1994. Bacterial genomics. *FEMS Microbiol. Rev.* **14:**139–160.

23. **Cornillot, E., and P. Soucaille.** 1996. Solvent-forming genes in clostridia. *Nature* (London) **380:**489.

24. **Davis, T. O., M. L. Mauchline, and N. P. Minton.** 1996. Unpublished results.

25. **Davison, S. P., J. D. Santangelo, S. J. Reid, and D. R. Woods.** 1995. A *Clostridum acetobutylicum* regulator gene (*regA*) affecting amylase production in *Bacillus subtilis*. *Microbiology* **141:**989–996.

26. **Diez-Gonzalez, F., J. B. Russell, and J. B. Hunter.** 1995. The role of an NAD-independent lactate dehydrogenase and acetate in the utilisation of lactate by *Clostridium acetobutylicum* P262. *Arch. Microbiol.* **164:**36–42.

27. **Durre, P., R. J. Fischer, A. Kuhn, K. Lorenz, W. Schreiber, B. Sturzenhofecker, S. Ullmann, K. Winzer, and U. Sauer.** 1995. Solventogenic enzymes of *Clostridium acetobutylicum*: catalytic properties, genetic organization, and transcriptional regulation. *FEMS Microbiol. Rev.* **17:**251–262.

28. **Errington, J.** 1993. *Bacllus subtilis* sporulation: regulation of gene expression and control of morphogenesis. *Microbiol. Rev.* **57:**1–33.

29. **Fischer, R. J., J. Helms, and P. Durre.** 1993. Cloning, sequencing and molecular analysis of the *sol* operon of *Clostridium acetobutylicum*, a chromosomal locus involved in solventogenesis. *J. Bacteriol.* **175:**6959–6969.

30. **Fox, M. E., M. J. Lemmon, M. L. Mauchline, T. O. Davis, A. J. Giaccia, N. P. Minton, and J. M. Brown.** 1996. Anaerobic bacteria as a delivery system for cancer gene therapy: *in vitro* activation of 5-fluorocytosine by genetically engineered clostridia. *Gene Ther.* **3:**173–178.

31. **Franke, A. E., and D. B. Clewell.** 1981. Evidence for a chromosome-borne resistance transposon (Tn*916*) in *Streptococcus faecalis* that is capable of conjugal transfer in the absence of a conjugative plasmid. *J. Bacteriol.* **145:** 494–502.

32. **Fujino, T., P. Béguin, and J.-P. Aubert.** 1993. Organization of a *Clostridium thermocellum* gene cluster encoding the cellulosomal scaffolding protein CipA and a protein possibly involved in the attachment of the cellulosome to the cell surface. *J. Bacteriol.* **175:**1891–1899.

33. **Gerngross, U. T., M. P. M. Romaniec, N. S. Huskisson, and A. L. Demain.** 1993. Sequence of *Clostridium thermocellum* gene (*cipA*) encoding the cellulosome SL-protein reveals an unusual degree of internal homology. *Mol. Microbiol.* **8:**325–334.

34. **Goda, S. K., and N. P. Minton.** 1995. A simple procedure for gel electrophoresis and Northern blotting of RNA. *Nucleic Acid Res.* **23:**3357–3358.

35. **Graves, M. C., G. T. Mullenbach, and J.-C. Rabinowitz.** 1985. Cloning and nucleotide sequence determination of the *Clostridium pasteurianum* ferrodoxin gene. *Proc. Natl. Acad. Sci. USA* **82:**1653–1657.

36. **Green, E. M., and G. N. Bennett.** 1996. Inactivation of an aldehyde/alcohol dehydrogenase gene from *Clostridium acetobutylicum* ATCC 824. *Appl. Biochem. Biotechnol.* **57–58:**213–221.

37. **Green, E. M., Z. L. Boynton, L. M. Harris, F. B. Rudolph, E. T. Papoutsakis, and G. N. Bennett.** 1996. Genetic manipulation of acid formation pathways be gene inactivation in *Clostridium acetobutylicum* ATCC 824. *Microbiology* **142:**2079–2086.

38. **Grobben, N. G., G. Eggink, F. P. Cuperus, and H. J. Huizing.** 1993. Production of acetone, butanol and ethanol (ABE) from potato wastes: fermentation with integrated membrane extraction. *Appl. Microbiol. Biotechnol.* **39:**494–498.

39. **Gruss, A. D., and S. D. Ehrlich.** 1989. The family of highly interrelated single-stranded deoxyribonucleic acid plasmids. *Microbiol. Rev.* **53:**231–241.

40. **Hancock, K. R., E. Rockman, C. A. Young, L. Pearce, I. S. Maddox, and D. B. Scott.** 1991. Expression and nucleotide sequence of the *Clostridium acetobutylicum* β-galactosidase gene cloned in *Escherichia coli*. *J. Bacteriol.* **173:**3084–3095.

41. **Hoch, J. A.** 1993. Regulation of the phosphorelay and the initiation of sporulation in *Bacillus subtilis*. *Annu. Rev. Microbiol.* **47:**441–465.

42. **Ionesco, H.** 1980. Transfért de la résistance à la tétracycline chez *Clostridium difficile*. *Ann. Microbiol. Inst. Pasteur* **131A:**171–179.

43. **Jankovic, J., and M. Hallett.** 1994. *Therapy with Botulinum Toxin*, vol. 25. *Neurological Disease and Therapy*. Marcel Dekker, New York.

44. **Jones, D. T., and D. R. Woods.** 1986. Acetone-butanol fermentation revisited. *Microbiol. Rev.* **50:**484–524.

45. **Jos, F., M. L. Seegers, A. C. Zhao, W. J. J. Meijer, S. A. Khan, G. Venema, and S. Bron.** 1995. Structural and functional analysis of the single-stranded origin of replication from the lactococcal plasmid pWV01. *Mol. Gen. Genet.* **249:**43–50.

46. **Kashket, E. R., and Z. Y. Cao.** 1993. Isolation of a degeneration-resistant mutant of *Clostridium acetobutylicum* NCIMB 8052. *Appl. Environ. Microbiol.* **59:**4198–4202.

47. **Kim, A. Y., G. T. Attwood, S. M. Holt, B. A. White, and H. P. Blaschek.** 1994. Heterologous expression of endo-beta-1,4-D-glucanase from *Clostridium cellulovorans* in *Clostridium acetobutylicum* ATCC 824 following transformation of the engB gene. *Appl. Environ. Microbiol.* **60:** 337–340.

48. **Kim, A. Y., and H. P. Blaschek.** 1993. Construction and characterization of a phage-plasmid hybrid (phagemid), pCAK1, containing the replicative form of virus-like particle CAK1 isolated from *Clostridium acetobutylicum* NCIB 6444. *J. Bacteriol.* **175:**3838–3843.

49. **Kim, A. Y., A. A. Vertes, and H. P. Blaschek.** 1990. Isolation of a single-stranded plasmid from *Clostridium acetobutylicum* NCIB 6444. *Appl. Environ. Microbiol.* **56:** 1725–1728.

50. **Kok, J., J. M. van der Vossen, and G. Venema.** 1984. Construction of plasmid cloning vectors for lactic streptococci which also replicate in *Bacillus subtilis* and *Escherichia coli*. *Appl. Environ. Microbiol.* **48:**726–731.

51. **Kurose, N., T. Miyazaki, T. Kakimoto, J. Yagyu, M. Uchida, A. Obayashi, and Y. Murooka.** 1989. Isolation of plasmids from thermophilic clostridia and construction of shuttle vectors in *Escherichia coli* and cellulolytic clostridia. *J. Fermentation Bioeng.* **68:**371–374.

52. **Lamed, R., E. Setter, R. Kenig, and E. A. Bayer.** 1983. The cellulosome: a discrete cell surface organelle of *Clostridium thermocellum* which exhibits separate antigenic, cellulose-binding and various cellulolytic activities. *Biotechnol. Bioeng. Symp.* **13:**163–181.

53. **Leblanc, D. J., and L. N. Lee.** 1984. Physical and genetic characterization of streptococcal plasmid pAMb1 and cloning of its replication region. *J. Bacteriol.* **157:**445–453.

54. **Lee, C. K., P. Durre, H. Hippe, and G. Gottschalk.** 1987. Screening for plasmids in the genus *Clostridium*. *Arch. Microbiol.* **148:**107–114.

55. **Lee, S. F., C. W. Forsberg, and L. N. Gibbins.** 1985. Cellulytic activity of *Clostridium acetobutylicum*. *Appl. Environ. Microbiol.* **50:**220–228.

56. **Lee, S. F., C. W. Forsberg, and J. B. Rattray.** 1987. Purification and characterisation of two endoxylanases from *Clostridium acetobutylicum* ATTCC 824. *Appl. Environ. Microbiol.* **53:**644–650.

57. **Lee, S. Y., G. N. Bennett, and E. T. Papoutsakis.** 1992. Construction of *Escherichia coli–Clostridium acetobutylicum* shuttle vectors and transformation of *Clostridium acetobutylicum* strains. *Biotechnol. Lett.* **14:**427–432.

58. **Lee, S. Y., L. D. Mermelstein, G. N. Bennett, and E. T. Papoutsakis.** 1992. Vector construction, transformation, and gene amplification in *Clostridium acetobutylicum* ATCC 824. *Ann. N. Y. Acad. Sci.* **665:**39–51.

59. **Lee, S. Y., L. D. Mermelstein, and E. T. Papoutsakis.** 1993. Determination of plasmid copy number and stability in *Clostridium acetobutylicum* ATCC 824. *FEMS Microbiol. Lett.* **108:**391–323.

60. **Leibovitz, E., and P. Beguin.** 1996. A new type of cohesion domain that specifically binds the dockerin domain of the *Clostridium thermocellum* cellulosome-integrating protein CipA. *J. Bacteriol.* **178:**3077–3084.

61. **Lin, W. J., and E. A. Johnson.** 1991. Transposon Tn*916* mutagenesis in *Clostridium botulinum*. *Appl. Environ. Microbiol.* **57:**2946–2950.

62. **Luchansky, J. B., P. M. Muirana, and T. R. Klaenhammer.** 1988. Application of electroporation for transfer of plasmid DNA to *Lactobacillus, Lactococcus, Leuconostoc, Listeria, Pediococcus, Bacllus, Staphylococcus, Enterococcus* and *Propionibacterium*. *Mol. Microbiol.* **2:**637–646.

63. **Luczak, H., H. Schwarzmoser, and W. L. Staudenbauer.** 1985. Construction of *Clostridium butyricum* hybrid plasmids and transfer to *Bacillus subtilis*. *Appl. Microbiol. Biotechnol.* **23:**114–122.

64. **Lyristis, M., A. E. Byrant, J. Sloan, M. M. Awad, I. T. Nisbet, D. L. Stevens, and J. I. Rood.** 1994. Identification and molecular analysis of a locus that regulates extracellular toxin production in *Clostridium perfringens*. *Mol. Microbiol.* **12:**761–777.

65. **Madarro, A., J. L. Pena, J. L. Lequerica, S. Valles, R. Gay, and A. Flors.** 1991. Partial purification and characterisation of the cellulases from *Clostridium cellulolyticum* H10. *J. Chem. Technol. Biotechnol.* **52:**393–406.

66. **Maguin, E., P. Duwat, T. Hege, S. D. Ehlrich, and A. Gruss.** 1992. New thermosensitive plasmid for gram-positive bacteria. *J. Bacteriol.* **174:**5633–5638.

67. **Mattsson, D. M., and P. Rogers.** 1994. Analysis of Tn*916*-induced mutants of *Clostridium acetobutylicum* altered in solventogenesis and sporulation. *J. Ind. Microbiol.* **13:**258–268.

68. **Mermelstein, L. D., and E. T. Papoutsakis.** 1993. In vivo methylation in *Escherichia coli* by the *Bacillus subtilis* phage phi 3-T I methyltransferase to protect plasmids from restriction upon transformation of *Clostridium acetobutylicum* ATCC 824. *Appl. Environ. Microbiol.* **59:**1077–1081.

69. **Mermelstein, L. D., E. T. Papoutsakis, D. J. Petersen, and G. N. Bennett.** 1993. Metabolic engineering of *Clostridium acetobutylicum* ATCC 824 for increased solvent production by enhancement of acetone formation enzymes activities using a synthetic operon. *Biotechnol. Bioeng.* **42:**1053–1060.

70. **Mermelstein, L. D., N. E. Welker, G. N. Bennett, and E. T. Papoutsakis.** 1992. Expression of cloned homologous fermentative genes in *Clostridium acetobutylicum* ATCC 824. *Bio/Technology* **10:**190–195.

71. **Mermelstein, L. D., N. E. Welker, D. J. Petersen, G. N. Bennett, and E. T. Papoutsakis.** 1994. Genetic and metabolic engineering of *Clostridium acetobutylicum* ATCC 824. *Ann. N. Y. Acad. Sci.* **721:**54–68.

72. **Minton, N. P., J. K. Brehm, T.-J. Swinfield, S. M. Whelan, M. L. Mauchline, N. Bodsworth, and J. D. Oultram.** 1993. Clostridial cloning vectors, p. 119–150. *In* D. R. Woods (ed.), *The Clostridia and Biotechnology*. Butterworth-Heinemann, Boston.

73. **Minton, N. P., M. L. Mauchline, M. J. Lemmon, J. K. Brehm, M. Fox, N. P. Micheal, A. Giaccia, and J. M. Brown.** 1995. Chemotherapeutic tumour targeting using clostridial spores. *FEMS Microbiol. Rev.* **17:**357–364.

74. **Minton, N. P., and J. G. Morris.** 1981. Isolation and partial characterisation of three cryptic plasmids from strains of *Clostridium butyricum*. *J. Gen. Microbiol.* **127:** 325–331.

75. **Minton, N. P., T. J. Swinfield, J. K. Brehm, S. M. Whelan, and J. D. Oultram.** 1992. Vectors for use in *Clostridium acetobutylicum*, p. 120–139. *In* M. Seblad (ed.), *Genetics and Molecular Biology of Anaerobic Bacteria*. Springer-Verlag, New York.

76. **Mishra, S., P. Béguin, and J.-P. Aubert.** 1991. Transcription of *Clostridium thermocellum* endoglucanse genes *celF* and *celD*. *J. Bacteriol.* **173:**80–85.

77. **Moore, A. P.** 1995. *Handbook of Botulinum Toxin Treatment*. Blackwell Scientific Ltd, Oxford.

78. **Mullany, P., M. Wilks, and S. Tabaqchali.** 1991. Transfer of Tn*916* and Tn*916* delta E into *Clostridium difficile*: demonstration of a hot-spot for these elements in the *C. difficile* genome. *FEMS Microbiol. Lett.* **63:**191–194.

79. **Mullany, P., M. Wilks, and S. Tabaqchali.** 1995. Transfer of macrolide-lincosamide-streptogramin B (MLS) resistance in *Clostridium difficile* is linked to a gene homologous

with toxin A and is mediated by a conjugative transposon, Tn*5398*. *J. Antimicrob. Chemother.* **35:**305–315.

80. **Nair, R. V., G. N. Bennett, and E. T. Papoutsakis.** 1994. Molecular characterization of an aldehyde/alcohol dehydrogenase gene from *Clostridium acetobutylicum* ATCC 824. *J. Bacteriol.* **176:**871–885.
81. **Nair, R. V., and E. T. Papoutsakis.** 1994. Expression of plasmid-encoded aad in *Clostridium acetobutylicum* M5 restores vigorous butanol production. *J. Bacteriol.* **176:** 5843–5846.
82. **Nochur, S. V., M. F. Roberts, and A. L. Demain.** 1993. True cellulase production by *Clostridium thermocellum* grown on different carbon sources. *Biotechnol. Lett.* **15:** 641–646.
83. **Oultram, J. D., I. D. Burr, M. J. Elmore, and N. P. Minton.** 1993. Cloning and sequence analysis of the genes encoding phosphotransbutyrylase and butyrate kinase from *Clostridium acetobutylicum* NCIMB 8052. *Gene* **131:** 107–112.
84. **Oultram, J. D., A. Davies, and M. Young.** 1987. Conjugal transfer of a small plasmid from *Bacillus subtilis* to *Clostridium acetobutylicum* by cointegrate formation with plasmid pAMβ1. *FEMS Microbiol. Lett.* **42:**113–119.
85. **Oultram, J. D., M. Loughlin, T.-J. Swinfield, J. K. Brehm, D. E. Thompson, and N. P. Minton.** 1988. Introduction of plasmids into whole cells of *Clostridium acetobutylicum* by electroporation. *FEMS Microbiol. Lett.* **56:** 83–88.
86. **Oultram, J. D., H. Peck, J. K. Brehm, D. E. Thompson, T. J. Swinfield, and N. P. Minton.** 1988. Introduction of genes for leucine biosynthesis from *Clostridium pasteurianum* into *C. acetobutylicum* by cointegrate conjugal transfer. *Mol. Gen. Genet.* **214:**177–179.
87. **Oultram, J. D., and M. Young.** 1985. Conjugal transfer of plasmid pAMβ1 from *Streptococcus lactis* and *Bacillus subtilis* to *Clostridium acetobutylicum*. *FEMS Microbiol. Lett.* **27:**129–134.
88. **Papoutsakis, E. T., and G. N. Bennett.** 1993. Cloning, structure and expression of acid and solvent pathway genes of *Clostridium acetobutylicum*, p. 157–200. *In* D. R. Woods (ed.), *The Clostridia and Biotechnology*. Butterworth-Heinemann, Boston.
89. **Petersen, D. J., J. W. Cary, J. Vanderleyden, and G. N. Bennett.** 1993. Sequence and arrangement of genes encoding enzymes of the acetone production pathway of *Clostridium acetobutylicum* ATCC 824. *Gene* **123:**93–97.
90. **Pohlschroder, M., E, Canale-Parola, and S. B. Leschine.** 1995. Ultrastructural diversity of the cellulase complexes of *Clostridium papyrosolvens* C7. *J. Bacteriol.* **177:**6625–6629.
91. **Popoff, M. R., and N. Truffaut.** 1985. Survey of plasmids in *Clostridium butyricum* and *Clostridium beijerinckii* strains from different origins and different phenotypes. *Curr. Microbiol.* **12:**151–156.
92. **Rabinowitz, J.** 1972. Preparation and properties of clostridial ferredoxins. *Methods Enzymol.* **24:**431–446.
93. **Reid, S., E. R. Allcock, D. T. Jones, and D. R. Woods.** 1983. Transformation of *Clostridium acetobutylicum* protoplasts with bacteriophage DNA. *Appl. Environ. Microbiol.* **45:**305–307.
94. **Reysset, G., J. Hubert, L., Podvin, and M. Sebald.** 1988. Transfection and transformation of *Clostridium acetobutylicum* strain N1-4081. *Biotechnol. Techniques* **2:**199–204.
95. **Reysset, G., and M. Sebald.** 1985. Conjugal transfer of plasmid-mediated antibiotic resistance from streptococci to *Clostridium acetobutylicum*. *Ann. Inst. Pasteur Microbiol.* **136:**275–282.
96. **Rogers, P., and G. Gottschalk.** 1993. Biochemistry and regulation of acid and solvent production in clostridia, p. 25–50. *In* D. R. Woods (ed.), *The Clostridia and Biotechnology*. Butterworth-Heinemann, Boston.
97. **Rood, J. I., and S. T. Cole.** 1991. Molecular genetics and pathogenesis of *Clostridium perfringens*. *Microbiol. Rev.* **55:**621–648.
98. **Rood, J. I., B. A. McClane, J. G. Sanger, and R. W. Titball. (eds.)** 1997. *The Clostridia: Molecular Biology and Pathogenesis*. Academic Press, San Diego, Calif.
99. **Sass, C., J. Walter, and G. N. Bennett.** 1993. Isolation of mutants of *Clostridium acetobutylicum* ATCC 824 deficient in protease activity. *Curr. Microbiol.* **26:**151–154.
100. **Sauer, U., and P. Durre.** 1993. Sequence and molecular characterization of a DNA region encoding a small heat shock protein of *Clostridium acetobutylicum*. *J. Bacteriol.* **175:**3394–3400.
101. **Sauer, U., and P. Durre.** 1995. Differential induction of genes related to solvent formation during the shift from acidogenesis to solventogenesis in continuous culture of *Clostridium acetobutylicum*. *FEMS Microbiol. Lett.* **125:** 115–120.
102. **Sauer, U., J. D. Santangelo, A. Treuner, M. Buchholz, and P. Durre.** 1995. Sigma factor and sporulation genes in *Clostridium*. *FEMS Microbiol. Rev.* **17:**331–340.
103. **Sauer, U., A. Treuner, M. Bucholz, J. D. Santangelo, and P. Durre.** 1994. Sporulation and primary sigma factor homologous genes in *Clostridium acetobutylicum*. *J. Bacteriol.* **176:**6572–6582.
104. **Scott, J. R.** 1992. Sex and the single circle: conjugative transposition. *J. Bacteriol.* **174:**6005–6010.
105. **Sebald, M.** 1994. Genetic basis for antibiotic resistance in anaerobes. *Clin. Infect. Dis.* **18**(Suppl 4)**:**S297–304.
106. **Shigekawa, K., and W. J. Dower.** 1988. Electroporation of eukaryotes and prokaryotes: a general approach to the introduction of macromolecules into cells. *BioTechniques* **6:**742–752.
107. **Shoseyov, O., and R. H. Doi.** 1990. Essential 170 kDa subunit for degradation of crystalline cellulose by *Clostridium cellulovorans* cellulase. *Proc. Natl. Acad. Sci. USA* **87:**2192–2195.
108. **Soutschek-Bauer, E., L. Hartl, and W. L. Staudenbauer.** 1985. Transformation of *Clostridium thermohydrosulfuricum* DSM 568 with plasmid DNA. *Biotechnol. Lett.* **7:**705–710.
109. **Stim-Herndon, K. P., R. Nair, E. T. Papoutsakis, and G. N. Bennett.** 1996. Analysis of degenerate variants of *Clostridium acetobutylicum* ATCC 824. *Anaerobe* **2:**11–18.
110. **Swinfield, T. J., J. D. Oultram, D. E. Thompson, J. K. Brehm, and N. P. Minton.** 1990. Physical characterisation of the replication region of the *Streptococcus faecalis* plasmid pAMβ1. *Gene* **87:**79–90.
111. **Truffaut, N., J. Hubert, and G. Reysset.** 1989. Construction of shuttle vectors useful for transforming *Clostridium acetobutylicum*. *FEMS Microbiol. Lett.* **49:**15–20.
112. **Truffaut, N., and M. Sebald.** 1983. Plasmid detection and isolation in strains of *Clostridium acetobutylicum* and related species. *Mol. Gen. Genet.* **189:**178–180.

113. **Tsoi, T. V., N. A. Chuvil'skaia, Atakishieva-lalu, T. Dzhavakhishvili, and V. K. Akimenko.** 1987. *Clostridium thermocellum*—a new object of genetic studies. *Mol. Genet. Mikrobiol. Virusol.* **11:**18–23.

114. **Urano, N., I. Karube, S. Suzuki, T. Yamada, H. Hirochika, and K. Sakaguchi.** 1983. Isolation and partial characterisation of large plasmids in hydrogen-evolving bacterium *Clostridium butyricum*. *Eur. J. Appl. Microbiol. Biotechnol.* **17:**349–354.

115. **Volk, W. A., B. Bizzini, K. R. Jones, and F. L. Macrina.** 1988. Inter- and intrageneric transfer of Tn*916* between *Streptococcus faecalis* and *Clostridium tetani*. *Plasmid* **19:** 255–259.

116. **Walter, K. A., G. N. Bennett, and E. T. Papoutsakis.** 1992. Molecular characterization of two *Clostridium acetobutylicum* ATCC 824 butanol dehydrogenase isozyme genes. *J. Bacteriol.* **174:**7149–7158.

117. **Walter, K. A., L. D. Mermelstein, and E. T. Papoutsakis.** 1994. Host-plasmid interactions in recombinant strains of *Clostridium acetobutylicum* ATCC 824. *FEMS Microbiol. Lett.* **123:**335–342.

118. **Walter, K. A., L. D. Mermelstein, and E. T. Papoutsakis.** 1994. Studies of recombinant *Clostridium acetobutylicum* with increased dosages of butyrate formation genes. *Ann. N. Y. Acad. Sci.* **721:**69–72.

119. **Walter, K. A., R. V. Nair, J. W. Cary, G. N. Bennett, and E. T. Papoutsakis.** 1993. Sequence and arrangement of two genes of the butyrate-synthesis pathway of *Clostridium acetobutylicum* ATCC 824. *Gene* **134:**107–111.

120. **Wilkinson, S. R., D. I. Young, J. G. Morris, and M. Young.** 1995. Molecular genetics and the initiation of solventogenesis in *Clostridium beijerinckii* (formerly *Clostridium acetobutylicum*) NCIMB 8052. *FEMS Microbiol. Rev.* **17:**275–285.

121. **Wilkinson, S. R., and M. Young.** 1993. Wide diversity of genome size among different strains of *Clostridium acetobutylicum*. *J. Gen. Microbiol.* **139:**1069–1076.

122. **Wilkinson, S. R., and M. Young.** 1994. Targeted integration of genes into the *Clostridium acetobutylicum* chromosome. *Microbiology* **140:**89–95.

123. **Wilkinson, S. R., and M. Young.** 1995. Physical map of the *Clostridium beijerinckii* (formerly *Clostridium acetobutylicum*) NCIMB 8052 chromosome. *J. Bacteriol.* **177:** 439–448.

124. **Wilkinson, S. R., M. Young, R. Goodacre, J. G. Morris, J. A. E. Farrow, and M. D. Collins.** 1995. Phenotypic and genotypic differences between certain strains of *Clostridium acetobutylicum*. *FEMS Microbiol. Lett.* **125:** 199–204.

125. **Williams, D. R., D. I. Young, J. D. Oultram, N. P. Minton, and M. Young.** 1990. Development and optimisation of conjugative plasmid transfer from *Escherichia coli* to *Clostridium acetobutylicum* NCIMB 8052, p. 239–246. *In* S. P. Boriello and J. M. Hardie (ed.), *Anaerobes in Medicine and Industry*. Wrightson Biomedical, Petersfield, U.K.

126. **Williams, D. R., D. I. Young, and M. Young.** 1990. Conjugative plasmid transfer from *Escherichia coli* to *Clostridium acetobutylicum*. *J. Gen. Microbiol.* **136:**819–826.

127. **Winzer, K., and P. Durre.** 1994. Cloning and sequencing of two thiolase genes from *Clostridium acetobutylicum* DSM 792. *Bioengineering* **48**(Special Issue 2)**:**191.

128. **Wong, J., and G. N. Bennett.** 1996. Recombination-induced variants of *Clostridium acetobutylicum* ATCC 824 with increased solvent production. *Curr. Microbiol.* **32:** 349–356.

129. **Wong, J., C. Sass, and G. N. Bennett.** 1995. Sequence and arrangement of genes encoding sigma factors in *Clostridium acetobutylicum*. *Gene* **153:**89–92.

130. **Woods, D. R. (ed.)** 1993. *The Clostridia and Biotechnology*. Butterworth-Heinemann, Boston.

131. **Woods, D. R.** 1995. The genetic engineering of microbial solvent production. *Trends Biotechnol.* **13:**259–264.

132. **Woolley, R. C., A. Pennock, R. J. Ashton, A. Davies, and M. Young.** 1989. Transfer of Tn*1545* and Tn*916* to *Clostridium acetobutylicum*. *Plasmid* **22:**169–174.

133. **Yoshino, S., T. Yoshino, S. Hara, S. Ogata, and S. Hayashida.** 1990. Construction of shuttle vector plasmid between *Clostridium acetobutylicum* and *Escherichia coli*. *Agric. Biol. Chem.* **54:**437–441.

134. **Young, M., N. P. Minton, and W. L. Staudenbauer.** 1989. Recent advances in the genetics of the clostridia. *FEMS Microbiol. Rev.* **5:**301–325.

135. **Youngleson, J. S., F.-P. Lin, S. J. Reid, and D. R. Woods.** 1995. Structure and transcription of genes within the *β-hbd-adh1* region of *Clostridium acetobutylicum* P262. *FEMS Microbiol. Lett.* **125:**185–192.

136. **Zappe, H., W. A. Jones, D. T. Jones, and D. R. Woods.** 1988. Structure of an endo-*β*-1,4-glucanase gene from *Clostridium acetobutylicum* P262 showing homology with endoglucanase genes from *Bacillus* spp. *Appl. Environ. Microbiol.* **54:**1289–1292.

RECOMBINANT DNA APPLICATIONS

RECOMBINANT DNA APPLICATIONS CONTAINS 13 CHAPTERS focused on three broad topics: organisms, expression, and engineering proteins and pathways. The three areas of focus represent three of the major approaches to develop recombinant DNA applications. The reader may readily identify additional sections and even additional chapters within the focus areas; however, the particular chapters were selected to represent important concepts rather than provide comprehensive coverage of all recombinant DNA applications.

The papers on organisms include chapters on bacterial genomics by Rosteck et al., streptomycetes by Schrempf, thermophiles by Mai and Wiegel, and *Zygosaccharomyces rouxii* by Oshima et al. The genomics chapter provides a snapshot of an exploding field that has assumed major importance because of the massive international campaign to discover new effective treatments for infectious diseases. Schrempf highlights the use of recombinant DNA technology to explore the cell physiology of actinomycetes, especially *Streptomyces*. Mai and Wiegel's chapter on thermophiles frames an enormously interesting group of organisms that have entered the industrial limelight as an important source of enzymes for biocatalysis. Oshima et al. describe the use of the unique *Z. rouxii*, which is a valuable yeast because it is used in the production of soy sauce.

The fascinating chapter by Butt and Chen provides a transition, with emphasis on both an organism and expression of a particular class of genes. The chapter focuses on the popular yeast *Saccharomyces cerevisiae*, used to express nuclear receptors for high-throughput screening to identify new drug candidates. Hershberger et al. contribute a chapter describing how to design polycistronic operons to express high levels of the proteins encoded by all the genes in the operon. The exciting chapter by Baneyx highlights the complicated subject of protein folding in vivo. Jockers and Strosberg highlight heterologous expression of the fascinating G protein-coupled receptors in a variety of microorganisms. Miroux and Walker describe novel methods to solve the vexing problem of expressing toxic proteins in *Escherichia coli*. Lundberg et al. provide methods to accumulate high levels of proteins when mRNA stability is the major barrier to high levels of expression.

The final topic, with chapters by Zhao et al., LaDuca et al., and Lee and Choi, deals with protein and pathway engineering. Zhao et al. provide an exceptionally clear treatment of directed evolution, which begins where nature stopped, to optimize the properties of industrial and commercial enzymes. LaDuca et al. provide an outstanding review of pathway engineering to produce aromatic compounds. The aromatic amino acids and products from the branches and extensions of aromatic amino acid biosynthetic pathways are valuable commercial products. Lee and Choi focus attention on polyhydroxyalkanoates to produce biodegradable plastics that may decrease the environmental pressure generated by petroleum-based plastics.

Chapters in this section illustrate 13 different methods to implement recombinant DNA applications for important industrial processes that use microorganisms. Such a sampling amply demonstrates the continued health of a diverse fermentation industry which flourishes worldwide.

Bacterial Genomics and Genome Informatics

PAUL R. ROSTECK, JR., BRADLEY S. DEHOFF, FRANKLIN H. NORRIS, AND PAMELA K. ROCKEY

39

The study of microbial genomes and, in particular, whole bacterial chromosomes historically included construction of genetic maps, construction of physical maps, and generation of limited DNA sequence information from individual genes and chromosomal segments. The use of classical genetic techniques allowed construction of maps of the positions and order of individual genes along the chromosome and estimates of the relative distances separating these genes by measuring the frequency of recombination between two or more markers of interest. This approach was practical only for organisms for which appropriate genetic techniques were available to construct linkage maps, however. Physical maps of many bacterial genomes can be constructed by using current molecular biology methods such as PCR (57) and analysis of large restriction enzyme digest fragments by gel-based methods (59). Recently, advances in large-scale DNA sequencing techniques and more extensive availability of this technology have enabled the determination of whole bacterial chromosome sequences. The large amount of bacterial genome mapping and sequencing information being generated requires an increased emphasis on genome informatics, the evolving discipline of genome data acquisition, management, analysis, and interpretation.

The emerging field of bacterial genomics includes the study of the structure and organization of entire bacterial chromosomes or multichromosome genomes and also the structure and function of individual genes. The complete encyclopedias of chromosome maps and whole genome DNA sequences becoming available will increase knowledge of bacterial cell biology and enable unprecedented comparative studies of the similarities and differences between organisms and genes (52). Comparisons of entire genome sequences will ultimately permit identification of the minimal set of genes required to support a free-living cell (11, 29, 33, 53) and allow studies of the physiological diversity provided by additional genes (15, 27, 34). In addition to the fundamental scientific value of such studies, bacterial genomics will advance our understanding of medically important organisms and permit increased industrial exploitation of microbes. The identification of all the genes of pathogenic organisms will aid development of new therapeutic treatments for bacterial infections or could result in new vaccines to prevent these infections.

In this chapter we briefly review applications of genetic and physical mapping techniques used for bacterial genome studies, describe current methods for whole genome sequencing, and introduce some computational concepts and tools for management and analysis of genome map and sequence data. Computerized database systems for managing and analyzing increasingly large amounts of molecular genetic data of diverse types are necessary for rapid and efficient access to the information and knowledge derived from these data (16). Such database tools for managing the data being generated are evolving rapidly, but significant additional needs still exist. Currently, a primary challenge in biology is to predict the true biochemical function of a protein from gene and amino acid sequence data (12). Simple comparisons of available sequences are often used to infer related function in similar sequences. A continuing challenge will be to understand the relationships between protein sequence or biochemical function and the physiology of the cell under study.

39.1. GENETIC AND PHYSICAL MAPPING

The distance between two or more genetic markers in a bacterium is estimated by measuring the frequency of coinheritance of the phenotypic traits they impart when bacterial cells are conjugally mated and DNA is transferred between them, or when DNA is transferred by transduction. It is assumed that the more often two traits are inherited together, the closer they are physically on a single chromosome or DNA element and the less likely they are to be separated by recombination during chromosomal replication or alteration. Extensive mapping of single genes and individual transcription units in bacteria by using genetic linkage analysis is limited to organisms for which these genetic transfer techniques have been developed. Consequently, only a small number of bacterial genomes, such as those of *Escherichia coli* (4), *Salmonella typhimurium* (58), *Bacillus subtilis* (2), *Pseudomonas aeruginosa* (35), and *Streptomyces coelicolor* (36), have been well characterized by this method. Genetic methods are imprecise and provide only estimates of physical distances between genes and markers. Physical techniques are required to accurately determine gene order and distance.

Physical mapping of individual chromosomes and bacterial genomes is needed to understand issues of genome evolution and physical relationships of individual genes (19, 28). Physical mapping techniques also ensure colinearity of the information provided by genetic mapping techniques and can provide purified reagents for subsequent analysis and use. For example, pulsed-field gel electrophoresis methods (59) have been used to produce physical maps of the genomes of *Mycoplasma genitalium* (20) and *Pyrococcus furiosus* (10). Manipulation of megabase (10^6 bp)-size DNA fragments by molecular cloning techniques advanced physical mapping of whole bacterial chromosomes, such as the genome of *B. subtilis* (3, 67). Further improvements on these initial strategies, such as ordering of physical subclones by combinatorial PCR, continue to accelerate construction of higher-resolution physical maps (63).

39.2. DNA SEQUENCING

Traditional strategies for generating genomic DNA sequence relied on construction of detailed physical maps of the target region or selection of specific subclones containing genes of interest. Such strategies require extensive genetic or physical mapping efforts before DNA sequence acquisition can begin (68). For example, sequencing of physically mapped bacteriophage lambda inserts has been used to study the genome of *E. coli* (62), and physically ordered sets of both lambda and yeast artificial chromosome clones have been used to support systematic sequencing of the *B. subtilis* genome (3, 50). Advances in DNA sequencing technology and strategies (24, 37, 65) and computational tools to assemble large, contiguous sequences from many random, but overlapping, short DNA sequences (66) have enabled the first total bacterial genome sequences to be fully determined (15, 27, 29). These projects made use of a whole-genome random sequencing strategy to obtain highly redundant sequence data for each genome studied, then identified all overlapping fragments computationally and assembled these into the most likely final order. After completion of the random sequencing phase, sequenced regions of lower coverage or higher ambiguity are targeted for additional study. This random, or shotgun, sequencing strategy still requires construction of molecular subclone libraries of relatively small, unordered fragments of the starting DNA but usually eliminates the need for all physical or genetic mapping studies at the beginning of the project. The genomes of a number of bacteria are currently being analyzed by complete sequencing (Table 1).

TABLE 1 Selected microbial genome projects completed or under way

Organism	Genome size (Mb)
Bacillus subtilis	4.2
Clostridum perfringens	3.6
Enterococcus faecalis	2.2
Escherichia coli	4.7
Haemophilus influenzae Rd	1.8
Helicobacter pylori	1.7
Methanococcus jannaschii	1.8
Mycobacterium leprae	1.2
Mycobacterium tuberculosis	4.2
Mycoplasma capricolum	1.2
Mycoplasma genitalium	0.6
Neisseria gonorrhea	2.2
Neisseria meningitis	1.9
Pyrococcus furiosus	2.1
Rhodobacter sphaeroides	3.8
Staphylococcus aureus	2.9
Streptococcus pneumoniae	2.2
Streptococcus pyogenes	1.8
Sulfolobus sulfataricus	3.0
Thermotoga maritima	2.0

39.3. CONSTRUCTION OF RANDOM FRAGMENT LIBRARIES

For construction of shotgun libraries, total chromosomal DNA (or large fragments of previously subcloned genomic DNA) is randomly fragmented, ligated into a vector suitable for subsequent use to purify each subclone for sequence analysis, and cloned in *E. coli*. Fragmentation of the target DNA is done enzymatically or by physical shearing. Partial digestion of chromosomal DNA with a restriction enzyme produces a set of random, overlapping fragments that can be cloned in either plasmid or bacteriophage vectors that are also digested with a compatible restriction enzyme. Physical shearing of chromosomal DNA for subcloning is accomplished by sonication (25, 27) or by nebulization (29). The success of shotgun DNA sequencing strategies is highly dependent on the initial completeness and quality of recombinant libraries of bacterial DNA prepared and used as sequencing templates. Care should be taken to ensure a high frequency of recombinant subclones in the final library and to maximize the proportion of subclones containing single inserts of the target DNA. In the method described below, fully purified chromosomal DNA is partially digested by the restriction enzyme of choice in a series of reactions containing various ratios of enzyme to DNA and incubated for various times. These considerations increase the randomness of cleavage of the genomic DNA and ensure the maximum number of overlapping fragments in the final recombinant library (60). The resulting DNA fragments are sorted by size by agarose gel electrophoresis, purified, and ligated to the bacteriophage vector M13mp19, which is linearized and enzymatically treated to prevent recircularization. The final vector treatment ensures that transformed clones contain foreign DNA inserts. The choice of restriction enzyme used for chromosomal DNA preparation should take into consideration the extent of cleavage observed in a limit, or complete, digestion of the starting DNA. It is desirable that the limit digest produce final digestion products of a few hundred base pairs in length or shorter. If the enzyme used does not digest the target DNA into sufficiently small fragments to allow subsequent partial digestion products to be included in the size range purified for cloning, a large proportion of the genome will not be present in the final clone library.

PROTOCOLS

DNA fragmentation by partial digestion with a restriction enzyme

1. On ice, mix 3 μg of chromosomal DNA and 4 U of *Sau*3AI (New England BioLabs) in the buffer recom-

mended by the supplier at the suggested concentrations. Place the reaction tube at 37°C and immediately remove and incubate at 70°C to inactivate the restriction enzyme. Similarly, prepare a reaction mixture containing 3 μg of DNA and 4 U of *Sau*3AI. Incubate at 37°C for 1 min, then at 70°C for 15 min. Prepare a reaction mixture containing 3 μg of DNA and 8 U of restriction enzyme. Place at 37°C, then immediately inactivate at 70°C for 15 min. Reaction mixtures should be immediately placed on ice at the conclusion of the 70°C incubation.

2. Small aliquots of each reaction mixture should be evaluated for extent of digestion by electrophoresis on an agarose gel. Correct digestion is indicated by a range of fragment sizes from nearly full length to less than a few hundred base pairs in length. If additional digestion is required to achieve the desired size distribution, reaction mixtures containing underdigested DNA should be pooled, redistributed into new reaction mixtures with additional fresh restriction enzyme, and incubated as above.

3. Final reaction mixtures are pooled and electrophoresed on a 1% low-melting-temperature agarose gel to size fractionate. After staining briefly in ethidium bromide, view the gel under long-wavelength UV, excise gel slices containing DNA fragments approximately 0.8–1.8 kb in length, and place in a 1.5-ml microcentrifuge tube. Melt the agarose slices by heating at 65°C for 10 min while vortexing 2–3 times during the incubation period. Extract the agarose solution with an equal volume of water-saturated phenol by vortexing for 20–30 s. Centrifuge the tube in a microcentrifuge at 14,000 rpm (12,000 × *g*) for 5 min and transfer the upper phase to a clean tube. Extract the aqueous phase with an equal volume of ether by vortexing for 20–30 s and remove the ether phase. Add 0.1 volume of 3 M sodium acetate (pH 8.0) to the final solution, mix, and add 2 volumes of absolute ethanol. Incubate solution at −20°C for several hours or at −80° for 20–30 min. Centrifuge tube at 14,000 rpm (12,000 × *g*) for 15 min to recover precipitated DNA, rinse pellet with 70% ethanol, and dry under vacuum for a few minutes. Dissolve the purified DNA in Tris-EDTA (TE) buffer (10 mM Tris-Cl, 1 mM EDTA [pH 7.5]).

Preparation of M13 vector and ligation to inserts

1. Linearize 5 μg of bacteriophage M13mp19 replicative-form DNA by digestion with *Bam*HI (New England BioLabs) according to the manufacturer's recommendations. Confirm complete linearization of the vector by analyzing an aliquot of the reaction mixture containing approximately 0.15 μg of DNA on a 1% agarose gel. When digestion is complete, inactivate the restriction enzyme by incubation at 70°C for 15 min. Purify the digested DNA by adding 0.1 volumes of 3 M sodium acetate and 2 volumes of absolute ethanol. Chill solution at −80°C for 10–15 min and centrifuge to recover precipitated DNA. Wash pellet with 70% ethanol and dry.

2. To prevent recircularization of the linearized vector during ligation to the insert fragments, dephosphorylate the exposed 5′ ends of the vector DNA by treatment with calf intestinal alkaline phosphatase (Boehringer) under the conditions and with the buffer provided with the enzyme. At the conclusion of the reaction, extract the sample with an equal volume of phenol by vortexing for 20–30 s. Remove the aqueous phase to a clean tube and extract a second time with an equal volume of phenol. Centrifuge briefly and transfer the aqueous phase to a clean microcentrifuge tube. Extract with an equal volume of water-saturated ether and again recover the DNA by ethanol precipitation. Wash the final pellet with 70% ethanol, and dry briefly under vacuum. Dissolve the resulting DNA in 50 μl of TE buffer and store at 4°C until ready to use.

3. The purified bacterial DNA insert fragments are ligated to the M13mp19 vector DNA in 50-μl reaction mixtures containing approximately 2 μg of insert DNA, 0.2 μg of linearized vector DNA, and 10 U of T4 DNA ligase (Boerhinger) in the buffer recommended by the manufacturer. Incubate the reaction mixture at 15°C for 12–16 h. Heat for 10 min at 70°C and store at 4°C until ready for transformation.

Bacterial transformation and preparation of sequencing templates

1. On ice, thaw 100-μl protions of transformation-competent *E. coli* XL1Blue MRF′ or JM101 cells (Stratagene), add 1-μl portions of the final ligation solution or a dilution of the solution in TE buffer, and mix gently. Incubate on ice for 30 min, heat-treat at 42°C for 2 min, and incubate on ice an additional 10 min. Alternatively, electroporation can be used to transform *E. coli* host cells with the ligation mixture (25). Immediately add 1-, 10-, and 100-μl aliquots of the transformation mixture to prepared 3-ml portions of molten YT soft agar containing isopropyl-β-D-thiogalactopyranoside (IPTG), 5-bromo-4-chloro-3-indolyl-β-D-galactopyranoside (X-Gal), and plating cells (see Media and Reagents, below). Mix by vortexing gently and pour onto YT agar plate. Cover and allow top agar overlay to solidify. Incubate plates at 37°C for 16–24 h.

2. White plaques containing foreign DNA inserts are selected for subsequent growth and purification for DNA sequencing (45). Prepare an overnight culture of *E. coli* host cells (JM101 or XL1Blue MRF′) grown in 2× YT medium at 37°C. Subculture host cells 1:100 into fresh 2× YT medium and incubate at 37°C for 1 h. Transfer 800 μl of culture into each 1.2-ml microtube in a 96-tube rack (VWR #20901-029 or equivalent). Wearing gloves, pick individual M13 plaques using sterile toothpicks and drop each toothpick into an individual tube. Let toothpicks sit for 10 min and remove. Cover the tube rack with the cover provided and incubate at 37°C for 6–15 h with vigorous shaking. Centrifuge tubes at 3,500 rpm (3,000 × *g*) for 20 min to pellet cells. Transfer 600 μl of supernatant containing bacteriophage particles to clean 1.2-ml tubes containing 120 μl of 20% polyethylene glycol (PEG) 2.5 M NaCl solution. Cover tubes and invert several times to mix well. Allow racks to sit at room temperature for at least 15 min. Centrifuge racks at 3,500 rpm (3,000 × *g*) for 15 min to pellet the bacteriophage. Discard supernatant, place tubes over a paper towel, and centrifuge inverted at 400 rpm to remove residual PEG solution. Add 20 μl of Triton-TE solution to bacteriophage pellets and seal boxes. Vortex tube racks 1–3 min to resuspend pellets. Briefly centrifuge tubes to collect the solution at the bottom of tubes. Incubate tube racks at 80°C for 10 min to lyse bacteriophage, cool, and centrifuge briefly. Add 20 μl of water and store template solution at −20°C until ready to sequence. Sequence templates using DYE-labeled M13 forward primers (Applied Biosystems Division, Perkin-Elmer) and *Taq*-FS DNA polymerase. Analyze sequencing reactions on model 373A or 377 DNA Sequencers (ABD).

The amount of initial sequence data required to effectively apply the random sequencing strategy is dependent on the size of the genome target and can be predicted statistically (27, 44). Shotgun sequencing projects typically require determination of 5 to 10 genome equivalents of primary sequence data to provide adequate coverage of the target DNA (68). Additional targeted sequencing is then also frequently required to cover regions not completely represented in the random sequence generated and to close gaps in the final sequence.

39.4. GENOME INFORMATICS

The final determination of the nucleotide sequence of a single gene or an entire genome depends on assembly of the complete sequence from the many short sequences of the fragments analysed initially. Relatively small sequencing projects, such as individual genomic fragments cloned in bacteriophage lambda or cosmid vectors, can be managed, assembled, and edited with software developed for use on desktop computers (e.g., AUTOASSEMBLER from ABD or Sequencher from GeneCodes, Madison, Wis.). The sequencing of complete bacterial genomes requires improved computational techniques for assembly of very large numbers of independent sequences (18, 64). For example, approximately 8,771 individual random sequencing reactions were required to complete the sequence of the smallest bacterial genome determined to date, that of *M. genitalium* (29). Such projects were made possible by development of more powerful computational tools to manage and assemble the primary sequence data required to effectively use the shotgun strategy (27, 66).

Computerized analysis of a final assembled bacterial genome sequence is used to predict rRNA and tRNA genes; protein coding regions, which are often identified as open reading frames (ORFs) greater than 100 amino acids in length (27, 29, 30); and potential regulatory or structural features in the sequence. The FRAMES program from the Genetics Computer Group (Madison, Wis.) is one example of such a program (23). FRAMES identifies ORFs by searching for translation initiation and termination codons in each reading frame in a nucleotide sequence and graphically representing these landmarks for the investigator. Statistical methods based on nonrandom base composition in protein coding sequences or codon usage biases also can be used to further increase the sensitivity and reliability of such predictions. Fickett's TestCode program (26) and the FRAME algorithm developed by Bibb et al. (8) are useful methods for analysis of bacterial sequences. These methods are especially effective for genomes that are biased in G+C base composition. For example, genomic DNA in the actinomycetes is approximately 75% G+C in base composition. ORFs in these genomes typically contain over 90% G or C in the first and third positions in codon triplets in legitimate coding regions, while the second nucleotide in codons is usually less than 50% G or C (8, 55). We determined more than 43 kb of contiguous sequence from the genome of the tylosin-producing organism *Streptomyces fradiae* to analyze the organization of the tylactone synthase gene *tylG* (GenBank accession no. U78289). The *tylG* gene encodes five ORFs ranging from 4,836 to 13,302 bp long. Figure 1 illustrates an analysis of 40 kb of this chromosomal region with the FRAME algorithm. The limits of each ORF are clearly distinguishable in this plot. Insertion or deletion errors in the sequence resulting in frame shifts in the coding sequence are detected by changes in the regular pattern observed in such an analysis.

Prediction of the biological function of newly identified genes is an initial goal of genome sequencing projects (7, 12, 17, 21, 22, 46, 51). Sequence comparisons of predicted genes or encoded gene products to sequences in molecular databases such as GenBank and SwissProt provide insight into the possible biological function of newly determined sequences. Extensive similarity between two or more sequences is generally assumed to suggest similar biological function for the genes or proteins. Commonly used programs for similarity searches include BLAST (1) and the Smith-Waterman algorithm (32). Better integrated systems to characterize and analyze gene sequences are being developed and made available to scientists studying newly determined genome sequences (61). Tools, such as GenMark, that are useful for gene identification in prokaryotes are now widely available (13). Relational databases are often used to organize the diverse data generated by genetic and physical mapping or sequencing of microbial genomes and analysis of individual genes and gene clusters (14, 42, 48, 56). Correlating functions of individual genes and their encoded proteins to the biochemical and metabolic pathways they comprise requires additional software and database products (39). Sequence similarity comparisons are also used extensively to assess ancestral origins of genes and evolutionary relatedness between genes (43). Multiple sequence comparisons and sequence alignments are useful for identifying highly conserved short sequences (32). These motifs frequently indicate conservation of functional domains within proteins and can provide further clues into physiological relevance. Analysis of extended sequence alignments is used to evaluate gene duplication and divergence during evolution.

39.5. SUMMARY

Complete bacterial genome sequencing is still impractical for many laboratories, because appropriate high-throughput sequencing technology and informatics tools for sequence assembly, editing, and analysis are not available. Statistical predictions of the amount of coverage obtained by using random requencing methods indicate that most genes are partially sequenced early in a random sequencing project (27, 38, 44). Therefore, sparse sequencing of whole genomes is a reasonable alternative for some applications, especially if access to partial gene sequences is sufficient for the project goals. For example, once a reference genome sequence is known, comparative sequence information can be obtained rapidly from the genomes of related species by a sequence sampling strategy in which less than onefold genome coverage is typically achieved and individual sequences are analyzed computationally and compared to existing database sequences (38). These genome sequence tags are useful for quickly comparing genomes of multiple organisms to identify potential antigenic determinants for vaccine development against bacterial pathogens (54). Regulatory elements such as transcriptional promoters can be tentatively identified by this method, avoiding the biological selection typically imposed when heterologous promoters are introduced into bacterial hosts (47). The physical clones representing genome sequence tags also provide reagents for use as probes to map genes onto the genome or to isolate complete genes from appropriate recombinant DNA libraries. Such a strategy provides rapid access to information and reagents to study additional bacterial chromosomes. This method, also called genome scanning (40), has been applied recently to the genome of the

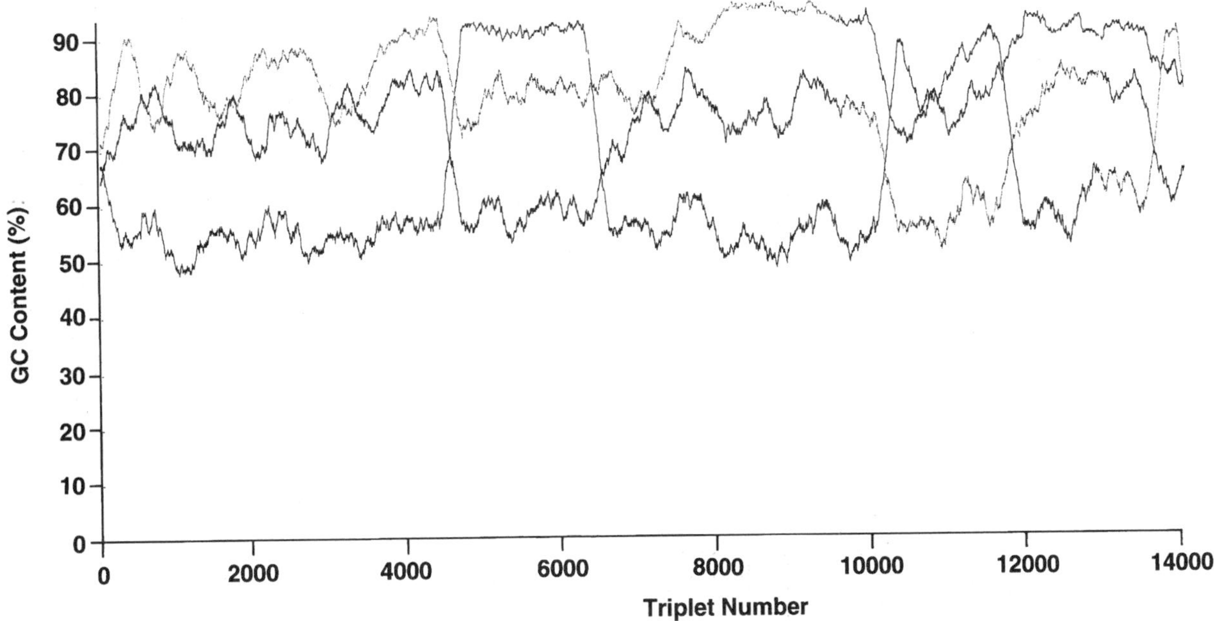

FIGURE 1 FRAMES analysis of the *S. fradiae tylG* gene. The gene is transcribed from left to right in this diagram and encodes five ORFs as described in the text.

archaeon *P. furiosus* (10, 49). Random sequencing of bacterial cDNA libraries is also useful for developing short sequences for comparison to known sequences (9, 31, 41). As many gene sequences become available, strategies for simultaneously surveying expression levels of many genes will emerge in importance (5). These approaches will incorporate combinations of automated biological methods and intelligent, computerized informatics support to monitor and interpret multiple simultaneous biological inputs from relevant experimental systems.

The techniques and strategies for whole genome mapping and sequencing described here are rapidly accelerating microbial research. The complete sequences being determined provide information about the organization of entire chromosomes and genomes. The central role microorganisms have played in evolution will become clearer as complete genomes of representatives of the *Bacteria* and the *Archaea* are sequenced. For example, many individual genes of the archaeon *Methanococcus jannaschii* (15) are more similar to those in the eukaryotes than in the bacteria. The increased understanding of genetic and molecular diversity in microbes provided by this work will stimulate research in energy production and bioremediation of the environment. Further insight into transfer of genes between bacteria, leading to the spread of antibiotic resistance and virulence factors (6) among natural populations of pathogenic bacteria, will be provided by comparison of gene sequences in these organisms. This understanding of emerging molecular targets and physiological mechanisms will lead to improved therapies to treat disease.

39.6. MEDIA AND REAGENTS

2× YT medium

Bacto Tryptone (Difco)	16 g
Yeast extract (Difco)	10 g
Sodium chloride	5 g
Water	900 ml

Stir until all ingredients are dissolved. Adjust pH of solution to 7.5 with 5 N sodium hydroxide solution and bring final volume to 1 liter. Sterilize by autoclaving for 20 min at 121°C and 15 lb/in^2 pressure.

YT agar

Bacto Tryptone (Difco)	10 g
Yeast extract (Difco)	5 g
Sodium chloride	5 g
Bacto Agar (Difco)	14 g
Water	900 ml

Stir until all ingredients are dissolved. Adjust pH of solution to 7.5 with 5 N sodium hyroxide solution and bring final volume to 1 liter. Sterilize by autoclaving for 20 min at 121°C and 15 lb/in^2 pressure. To prepare plates, cool agar to 50°C and pour 30–35 ml into sterile plastic petri plates (100 by 15 mm). Allow to solidify and store at 4°C until ready to use. Bring plates to room temperature for use. Dry plates at 37°C for a few hours if necessary to remove visible condensation before using, to prevent cross-contamination of plaques during growth.

YT soft agar
Same as YT agar but solidified with 7 g of Bacto Agar (Difco) per liter. Sterilize by autoclaving at 121°C for 15 min. Store at room temperature. For use, melt agar, distribute 3-ml portions to sterile tubes, and keep at 50°C. Add 12 μl of 10 mM IPTG and 7.5 μl of 2% X-Gal to the tube and mix. Add 50 μl of an overnight culture of lawn cells (XL1Blue MRF′ or JM101), mix, and keep at 50°C.

Triton-TE solution
(0.5% TritonX-100, 10 mM Tris-C1, 1 mM EDTA [pH 8.0])

Triton X-100	250 μl
1.0 M Tris-C1, pH 8.0	500 μl
0.5 M EDTA, pH 8.0	10 ml

Water.. to 50 ml

20% PEG/2.5 M NaCl
PEG 6000-8000.................................. 50 g
Sodium chloride............................... 36.56 g
Water.. to 250 ml

REFERENCES

1. **Altschl, S. F., W. Gish, W. Miller, E. W. Myers, and D. J. Lipman.** 1990. Basic local alignment search tool. *J. Mol. Biol.* **215:**403–410.
2. **Anagnostopoulos, C., P. J. Piggot, and J. A. Hoch.** 1993. The genetic map of *Bacillus subtilis*, p. 425–461. *In* A. L. Sonenshein, J. A. Hoch, and R. Losick (ed.), *Bacillus subtilis and Other Gram-Positive Bacteria*. American Society for Microbiology, Washington, D.C.
3. **Azevedo, V., E. Alvarez, E. Zumstein, G. Damiani, V. Sgaramella, S. D. Ehrlich, and P. Serror.** 1993. An ordered collection of *Bacillus subtilis* DNA segments cloned in yeast artificial chromosomes. *Proc. Natl. Acad. Sci. USA* **90:**6047–6051.
4. **Bachmann, B.** 1990. Linkage map of *Escherichia coli* K-12, edition 8. *Microbiol. Rev.* **54:**130–197.
5. **Bains, W.** 1996. Virtually sequenced: the next genomic generation. *Nat. Biotechnol.* **14:**711–713.
6. **Barinaga, M.** 1996. A shared strategy for virulence. *Science* **272:**1261–1263.
7. **Bassett, D. E., Jr., M. S. Boguski, and P. Hieter.** 1996. Yeast genes and human disease. *Nature* (London) **379:** 589–590.
8. **Bibb, M. J., P. R. Findlay, and M. W. Johnson.** 1984. The relationship between base composition and codon usage in bacterial genes and its use for the simple and reliable identification of protein-coding sequences. *Gene* **30:** 157–166.
9. **Boguski, M. S., and G. D. Schuler.** 1995. ESTablishing a human transcript map. *Nat. Genet.* **10:**369–371.
10. **Borges, K. M., S. R. Brummet, A. Bogert, M. C. Davis, K. M. Hujer, S. T. Domke, J. Szasz, J. Ravel, J. DiRuggiero, C. Fuller, J. W. Chase, and F. T. Robb.** 1992. A survey of the genome of the hyperthermophilic archaeon, *Pyrococcus furiosus*. *Genome Sci. Technol.* **1:**37–46.
11. **Bork, P., C. Ouzounis, G. Casari, R. Schneider, C. Sander, M. Dolan, W. Gilbert, and P. M. Gillevet.** 1995. Exploring the *Mycoplasma capricolum* genome: a minimal cell reveals its physiology. *Mol. Microbiol.* **16:**955–967.
12. **Bork, P., C. Ouzounis, and C. Sander.** 1994. From genome sequences to protein function. *Curr. Opin. Struct. Biol.* **4:**393–403.
13. **Borodovsky, M., and J. D. McIninch.** 1993. GENMARK: parallel gene recognition for both DNA strands. *Computers Chem.* **17:**123–133.
14. **Bouffard, G., J. Ostell, and K. E. Rudd.** 1992. GeneScape: a relational database of *Escherichia coli* genomic map data for Macintosh computers. *Comput. Appl. Biosci.* **8:**563–567.
15. **Bult, C. J., O. White, G. J. Olsen, L. Zhou, R. D. Fleischmann, G. G. Sutton, J. A. Blake, L. M. FitzGerald, R. A. Clanton, J. D. Gocayne, A. R. Kerlavage, B. A. Dougherty, J.-F. Tomb, M. D. Adams, C. I. Reich, R. Overbeek, E. F. Kirkness, K. G. Weinstock, J. M. Merrick, A. Glodek, J. L. Scott, N. S. M. Geoghagen, J. F. Weidman, J. L. Fuhrmann, D. Nguyen, T. R. Utterback, J. M. Kelley, J. D. Peterson, P. W. Sadow, M. C. Hanna, M. D. Cotton, K. M. Roberts, M. A. Hurst, B. P. Kaine, M. Borodovsky, H.-P. Klenk, C. M. Fraser, H. O. Smith, C. R. Woese, and J. C. Venter.** 1996. Complete genome sequence of the methanogenic Archaeon, *Methanococcus jannaschii*. *Science* **273:**1058–1073.
16. **Cannon, G.** 1990. Nucleic acid sequence analysis software for microcomputers. *Anal. Biochem.* **190:** 147–153.
17. **Cardon, L. R., C. Burge, G. A. Schachtel, B. E. Blaisdell, and S. Karlin.** 1993. Comparative DNA sequence features in two long *Escherichia coli* contigs. *Nucleic Acids Res.* **21:**3875–3884.
18. **Charnock-Jones, D. S., and S. A. J. R. Aparicio.** 1994. DNA sequence assembly on UNIX systems, p. 191–214. *In* M. J. Bishop (ed.), *Guide to Human Genome Computing*. Academic Press, London.
19. **Cole, S. T., and I. Saint Girons.** 1994. Bacterial genomics. *Microbiol. Rev.* **14:**139–160.
20. **Colman, S. D., P. C. Hu, W. Litaker, and K. F. Bott.** 1990. A physical map of the *Mycoplasma genitalium* genome. *Mol. Microbiol.* **4:**683–687.
21. **Daniels, D. L.** 1990. Constructing encyclopedias of genomes, p. 43–51. *In* K. Drlica and M. Riley (ed.), *The Bacterial Chromosome*. American Society for Microbiology, Washington, D.C.
22. **Dear, S., and R. Staden.** 1992. A standard file format for data from DNA sequencing instruments. *DNA Sequence* **3:**107–110.
23. **Devereux, J., P. Haeberli, and O. Smithies.** 1984. A comprehensive set of sequence analysis programs for the VAX. *Nucleic Acids Res.* **12:**387–395.
24. **Edgington, S. M.** 1993. Breaking open the bottlenecks in genomic DNA sequencing. *Bio/Technology* **11:**39–42.
25. **Favello, A., L. Hillier, and R. K. Wilson.** 1995. Genomic DNA sequencing methods. *Methods Cell Biol.* **48:**551–569.
26. **Fickett, J. W.** 1982. Recognition of protein coding regions in DNA sequences. *Nucleic Acids Res.* **10:**5303–5318.
27. **Fleischmann, R. D., M. D. Adams, O. White, T. A. Clayton, E. F. Kirkness, A. R. Kerlavage, C. J. Bult, J. Tomb, B. A. Dougherty, J. M. Merrick, K. McKenney, G. Sutton, W. FitzHugh, C. Fields, J. D. Gocayne, J. Scott, R. Shirley, L. Liu, A. Glodek, J. M. Kelley, J. F. Wiedman, C. A. Phillips, T. Springs, E. Hedblom, M. D. Cotton, T. R. Utterback, M. C. Hanna, D. T. Nguyen, D. M. Saudek, R. C. Brandon, L. D. Fine, J. L. Fritchman, J. L. Fuhrmann, N. S. M. Geoghagen, C. L. Gnehm, L. A. McDonald, K. V. Small, C. M. Fraser, H. O. Smith, and J. C. Venter.** 1995. Whole-genome random sequencing and assembly of *Haemophilus influenzae* Rd. *Science* **269:**496–512.
28. **Fonstein, M., and R. Haselkorn.** 1995. Physical mapping of bacterial genomes. *J. Bacteriol.* **177:**3361–3369.
29. **Fraser, C. M., J. D. Gocayne, O. White, M. D. Adams, R. A. Clayton, R. D. Fleischmann, C. J. Bult, A. R. Kerlavage, G. Sutton, J. M. Kelley, J. L. Fritchman, J. F. Weidman, K. V. Small, M. Sandusky, J. Fuhrmann, D. Nyuyen, T. R. Utterback, D. M. Saudek, C. A. Philips, J. M. Merrick, J.-F. Tomb, B. A. Daugherty, K. F. Bott, P.-C, Hu, T. S. Luccier, S. N. Peterson, H. O. Smith, C. A. Hutchison III, and J. C. Venter.** 1995. The

minimal gene complement of *Mycoplasma genitalum*. *Science* **270:**397–403.

30. **Fsihi, H., and S. T. Cole.** 1994. The *Mycobacterium leprae* genome: systematic sequence analysis identifies key catabolic enzymes, ATP-dependent transport systems and a novel *polA* locus associated with genomic variability. *Mol. Microbiol.* **16:**909–919.

31. **Fulton, L. L., L. Hillier, and R. K. Wilson.** 1995. Large-scale complementary DNA sequencing methods. *Methods Cell Biol.* **48:**571–582.

32. **Ginsburg, M.** 1994. Sequence comparison, p. 215–248. *In* M. J. Bishop (ed.), *Guide to Human Genome Computing*. Academic Press, London.

33. **Goffeau, A.** 1995. Life with 482 genes. *Science* **270:**445–446.

34. **Goffeau, A., B. G. Barrell, H. Bussey, R. W. Davis, B. Dujon, H. Feldmann, F. Galibert, J. D. Hoheisel, C. Jacq, M. Johnston, E. J. Louis, H. W. Mewes, Y. Murakami, P. Philippsen, H. Tettelin, and S. G. Oliver.** 1996. Life with 6000 genes. *Science* **274:**546–567.

35. **Holloway, B. W., S. V. Dharmsthiti, V. Kirshnapillai, A. Morgan, V. Obeyesekere, E. Ratnaningsih, M. Sinclair, D. Strom, and C. Zhang.** 1990. Patterns of gene linkage in *Pseudomonas* species, p. 97–105. *In* K. Drlica and M. Riley (ed.), *The Bacterial Chromosome*. American Society for Microbiology, Washington, D.C.

36. **Hopwood, D. A., and T. Kieser.** 1993. The chromosome map of *Streptomyces coelicolor* A3(2), p. 497–504. *In* A. L. Sonenshein, J. A. Hoch, and R. Losick (ed.), *Bacillus subtilis and Other Gram-Positive Bacteria*. American Society for Microbiology, Washington, D.C.

37. **Hunkapiller, T., R. J. Kaiser, B. F. Koop, and L. Hood.** 1991. Large-scale and automated DNA sequence determination. *Science* **254:**59–67.

38. **Kamb, A., C. Wang, A. Thomas, B. S. DeHoff, F. H. Norris, K. Richardson, J. Rine, M. H. Skolnick, and P. R. Rosteck, Jr.** 1995. Software trapping: a strategy for finding genes in large genomic regions. *Comput. Biomed. Res.* **28:**140–153.

39. **Karp, P. D., M. Riley, S. Paley, and A. Pellegrini-Toole.** 1996. EcoCyc: electronic encyclopedia of *E. coli* genes and metabolism. *Nucleic Acids Res.* **24:**32–40.

40. **Katayama, S.-I., B. Dupuy, T. Garnier, and S. T. Cole.** 1995. Rapid expansion of the physical and genetic map of the chromosome of *Clostridium perfringens* CPN50. *J. Bacteriol.* **177:**5680–5685.

41. **Kim, C. W., P. Markiewicz, J. J. Lee, C. F. Schierle, and J. H. Miller.** 1993. Studies of the hyperthermophile *Thermotoga maritima* by random sequencing of cDNA and genomic libraries. *J. Mol. Biol.* **231:**960–981.

42. **Kunst, F., A. Vassaroti, and A. Danchin.** 1995. Organization of the European *Bacillus subtilis* genome sequencing project. *Microbiology* **141:**249–255.

43. **Labedan, B., and M. Riley.** 1995. Widespread protein sequence similarities: origins of *Escherichia coli* genes. *J. Bacteriol.* **177:**1585–1588.

44. **Lander, E. S., and M. S. Waterman.** 1988. Genomic mapping by fingerprinting random clones: a mathematical analysis. *Genomics* **2:**231–239.

45. **Mardis, E. R.** 1994. High-throughput detergent extraction of M13 subclones for fluorescent DNA sequencing. *Nucleic Acids Res.* **22:**2173–2175.

46. **Miklos, G. L. G., and G. M. Rubin.** 1996. The role of the genome project in determining gene function: insights from model organisms. *Cell* **86:**521–529.

47. **Morrison, D. A., and B. Jaurin.** 1990. *Streptococcus pneumoniae* possesses canonical *Escherichia coli* (sigma 70) promoters. *Mol. Microbiol.* **4:**1143–1152.

48. **Moszer, I., P. Glaser, and A. Danchin.** 1995. *SubtiList*: a relational database for the *Bacillus subtilis* genome. *Microbiology* **141:**261–268.

49. **Nurminsky, D., and D. L. Hartl.** 1995. Sequence scanning: a method for rapid sequence acquisition from large-fragment DNA clones. *Proc. Natl. Acad. Sci. USA* **93:**1694–1698.

50. **Ogasawara, N., Y. Fujita, Y. Kobayashi, Y. Sadaie, T. Tanaka, H. Takahashi, K. Yamane, and H. Yoshikawa.** 1995. Systematic sequencing of the *Bacillus subtilis* genome: progress report of the Japanese group. *Microbiology* **141:**257–259.

51. **Oliver, S. G.** 1996. From DNA sequence to biological function. *Nature* (London) **379:**597–600.

52. **Olsen, G. J., C. R., Woese, and R. Overbeek.** 1994. The winds of (evolutionary) change: breathing new life into microbiology. *J. Bacteriol.* **176:**1–6.

53. **Pennisi, E.** 1996. Seeking life's bare (genetic) necessities. *Science* **272:**1098–1099.

54. **Rahme, L. G., E. J. Stevens, S. F. Wolfort, J. Shao, R. G. Tompkins, and F. M. Ausubel.** 1995. Common virulence factors for bacterial pathogenicity in plants and animals. *Science* **268:**1899–1902.

55. **Rosteck, P. R., Jr., P. A. Reynolds, and C. L. Hershberger.** 1991. Homology between proteins controlling *Streptomyces fradiae* tylosin resistance and ATP-binding transport. *Gene* **102:**27–32.

56. **Rudd, K. E., W. Miller, C. Werner, J. Ostell, C. Tolstoshev, and S. G. Satterfield.** 1991. Mapping sequenced *E. coli* genes by computer: software, strategies and examples. *Nucleic Acids Res.* **19:**637–647.

57. **Saiki, R. K., D. H. Gelfand, S. Stoffel, S. J. Scharf, R. Higuchi, G. T. Horn, K. B Mullis, and H. A. Erlich.** 1988. Primer-directed enzymatic amplification of DNA with a thermostable DNA polymerase. *Science* **239:**487–491.

58. **Sanderson, K. E., and J. R. Roth.** 1988. Linkage map of *Salmonella typhimurium*. *Microbiol. Rev.* **52:**485–532.

59. **Schwartz, D. C., and C. R. Cantor.** 1984. Separation of yeast chromosome-sized DNAs by pulsed-field gel electrophoresis. *Cell* **37:**67–75.

60. **Seed, B., R. C. Parker, and N. Davidson.** 1982. Representation of DNA sequences in recombinant DNA libraries prepared by restriction enzyme partial digestion. *Gene* **19:**201–209.

61. **Smith, D. R.** 1996. Microbial pathogen genomes—new strategies for identifying therapeutics and vaccine targets. *Trends Biotechnol.* **14:**290–293.

62. **Sofia, H. J., V. Burland, D. L. Daniels, G. Plunkett, III, and F. R. Blattner.** 1994. Analysis of the *Escherichia coli* genome. V. DNA sequence of the region from 76.0 to 81.5 minutes. *Nucleic Acids Res.* **22:**2576–2586.

63. **Sorokin, A., A. Lapidus, V. Capuano, N. Galleron, P. Pujic, and S. D. Ehrlich.** 1996. A new approach using multiplex long accurate PCR and yeast artificial chro-

mosomes for bacterial chromosome mapping and sequencing. *Genome Res.* **6:**448–453.

64. **Staden, R., and S. Dear.** 1991. A sequence assembly and editing program for efficient management of large projects. *Nucleic Acids Res.* **19:**3907–3911.
65. **Studier, F. W.** 1989. A strategy for high-volume sequencing of cosmid DNAs: random and directed priming with a library of oligonucleotides. *Proc. Natl. Acad. Sci. USA* **86:**6917–6921.
66. **Sutton, G., O. White, M. D. Adams, and A. R. Kerlavage.** 1995. TIGR Assembler: a new tool for assembling large shotgun sequencing projects. *Genome Sci. Technol.* **1:**9–19.
67. **Weinstock, G. M.** 1994. Bacterial genomes: mapping and stability. *ASM News* **60:**73–78.
68. **Wilson, R., R. Ainscough, K. Anderson, C. Baynes, M. Berks, J. Bonfield, J. Burton, M. Connell, T. Copsey, J. Cooper, A. Coulson, M. Craxton, S. Dear, Z. Du, R. Durbin, A. Favello, A. Fraser, L. Fulton, A. Gardner, P. Green, T. Hawkins, L. Hillier, M. Jier, L. Johnston, M. Jones, J. Kershaw, J. Kirsten, M. Laister, P. Latreille, J. Lightning, C. Lloyd, B. Mortimore, M. O'Callaghan, J. Parsons, C. Percy, L. Rifken, A. Roopra, D. Saunders, R. Shownkeen, M. Sims, N. Smaldon, A. Smith, N. Smith, E. Sonnhammer, R. Staden, J. Sulston, J. Thierry-Mieg, K. Thomas, M. Vaudin, K. Vaughan, R. Waterston, A. Watson, L. Weinstock, J. Wilkerson-Sproat, and P. Wohldman.** 1994. 2.2 Mb of contiguous nucleotide sequence from chromosome III of *C. elegans*. *Nature* (London) **368:**32–38.

Investigations of Streptomycetes Using Tools of Recombinant DNA Technology

HILDGUND SCHREMPF

40

Streptomycetes are gram-positive bacteria whose natural habitat is the soil. The first identified antibiotic was streptomycin, detected about 60 years ago; since then, thousands of low-molecular-weight, chemically different compounds with antibacterial, antifungal, antiparasitic, agroactive, cytostatic, and other activities have been found within many *Streptomyces* species and some other actinomycetes. The discovery of natural *Streptomyces* plasmids and their tailoring as vectors allowed gene cloning in streptomycetes; consequently, numerous genes could be cloned. Studies using molecular biological tools have focused on the cloning of genes required for the differentiation process and for the synthesis of antibiotics, some antifungal cytostatics, and other biologically active compounds. However, the role of streptomycetes in their natural habitat, their interaction with other procaryotic or eucaryotic microorganisms, plants, or animals, and their ability to degrade various macromolecules and to modify xenobiotics have scarcely been investigated.

This chapter is intended to demonstrate that the application of recombinant DNA technology in combination with physiological, biochemical, and genetic studies extends and deepens the knowledge of the biology of streptomycetes.

40.1. PHYLOGENY

Some members of the order of *Actinomycetales* were already identified about a century ago. Initially, the phylogeny of actinomycetes was primarily based on the great diversity of their morphological traits. More recently, the patterns of chemotaxonomic characters (i.e., type of peptidoglycan, menaquinone, phospholipids, cell wall sugars, fatty acids) showed that most actinomycete genera and nearly all families are composed of widely differing taxa (60). As a consequence, chemotaxonomy has been employed as an efficient tool for the rapid differentiation of genera. Recent comparisons of 16S rRNA sequences or corresponding rDNA have provided a basis for classifications. Additional studies revealed that all investigated actinomycetes contain a homologous insertion of about 100 nucleotides at the same position in the 23S rDNA gene (96). The relationships between different lineages within the actinomycetes are still largely uncertain, and interrelationships are only known for a few taxa. The current phylogenetic tree contains *Bifidobacteriaceae*, *Actinomycetaceae*, *Arthrobacteriaceae*, *Cellulomonadaceae*, *Microbacteriaceae*, *Dermatophilaceae*, *Propionibacteriaceae*, *Nocardioidaceae*, *Frankiaceae*, *Corynebacteriaceae*, *Mycobacteriaceae*, *Nocardiaceae*, *Actinoplanaceae*, *Pseudonocardiaceae*, *Streptomycetaceae*, and *Streptosporangiaceae*.

The family *Streptomycetaceae* (114) includes the genus *Streptomyces*, which contains strains formerly classified as *Chania*, *Elytrosporangium*, *Kitasatoa*, *Kitasosporia*, *Actinosporangium*, and *Streptoverticillium*. All members of the *Streptomycetaceae* family contain partially saturated (H2, H4, H6) menaquinones with nine isoprene units, and their peptidoglycan has LL-diaminopimelic acid as diagnostic amino acid, but not characteristic sugar (27).

An enormous number (>800) of *Streptomyces* species have been described, and their classification is frequently based on numerical taxonomy, applying more than 350 physiological traits (30). However, only about 70 species have been investigated by phylogenetic methods. Clearly, more molecular studies are required to determine the phylogenetic relationships. In addition, rDNA sequence similarities will have to be correlated with DNA:DNA reassociation data. Restriction fragment length polymorphism (RFLP) studies have been initiated. Preliminary studies suggest that analysis of DNA by pulsed-field gel electrophoresis (PFGE) in combination with hybridization is a valuable tool that has not been exploited sufficiently.

Repetitive DNA (mostly insertion sequences or directly repeated GC elements) was used for the efficient classification of mycobacterial strains (50). By current comparative investigations of genomic sequences, additional diagnostic elements for mycobacteria are to be deduced (86). It should be possible to use similar strategies after having sequenced the *Streptomyces coelicolor* A3(2) genome. Comparative analyses of the genomic sequences of *S. coelicolor* A3(2), of the G+C-rich mycobacteria, and of other organisms are expected to provide important taxonomic insights.

40.2. GROWTH CHARACTERISTICS

Streptomycetes grow as vegetative, long, branching hyphae that rarely contain septae and are thus multinucleoid. The formation of aerial hyphae is initiated under the control of several *bld* (bald) genes. Subsequent action of different *whi* (white) genes induces a curling of the aerial hyphae, their septation, and finally spore formation. The early regulatory *whi* genes (*A*, *B*, *G*, *H*, *I*, and *J*) from *S. coelicolor* A3(2) are needed for sporulation septation, the production of sigma F, and the polyketide spore pigment specified by the *whiE* cluster (17, 18). The recently identified five additional *whi* loci are being characterized in detail, to better understand the sporulation process (8). Partitioning of the chromosomes occurs within the aerial hyphae, in which a comparatively synchronous septation entails the formation of compartments where spores are formed after maturation. Recent sequencing results revealed that genes homologous to chromosomal partitioning genes from other bacteria are located in the vicinity of the chromosomal *S. coelicolor* A3(2) *oriC* region (29); their detailed role in the course of the developmental cycle needs to be explored.

In the unicellular bacterium *Escherichia coli*, in contrast, DNA replication, growth, and septum formation are coordinated. The earliest defined step of cell division is the formation of a Z-ring due to GTP-dependent polymerization of monomers of the protein FtsZ. The *ftsZ* genes of *S. coelicolor* A3(2) and *Streptomyces griseus* were recently cloned. A knockout *S. coelicolor* A3(2) mutant of the gene is viable but lacks the septae that are present occasionally in the vegetative hyphae and are found frequently in the aerial hyphae. Immunological studies have recently shown that the regular arrangement of FtsZ rings dissolves when septae are formed during sporulation (101). Transcription studies in *S. griseus* have suggested that *ftsZ* is expressed during both vegetative growth and sporulation (23). Either the production of FtsZ or its assembly in rings appears to be defective in some of the *S. coelicolor* A3(2) *whi* mutants (18). The *mre* and *pbp* genes required for the synthesis of murein- and penicillin-binding proteins, respectively, have been identified in *S. coelicolor* A3(2) (14). It will be interesting to analyze their role during development.

The differentiation cycle generally occurs on solid media, although some strains (e.g., *S. griseus*) (68) synchronously undergo sporulation in liquid culture. Several *Streptomyces* strains show a transient slowdown during growth in liquid culture, before entering the stationary phase. This transition phase is characterized by an increase in ppGpp, a decrease in GTP, and the activation of genes required for secondary metabolism. Quantification of proteins showed that the synthesis of two ribosomal proteins is drastically reduced when the culture approaches the stationary phase (10).

The transcription of some genes for morphogenesis is dependent on the interaction of DNA binding proteins with autoregulators (36) (see below). It has been proposed that by monitoring the intracellular GTP pool, the Obg GTP binding protein senses changes of nutrients in the environment, and that an unknown signal cascade induces differentiation (79).

40.3. THE CHROMOSOMAL DNA

Most *Streptomyces* strains contain about 8,000 bp of chromosomal DNA; however, some contain chromosomes that are smaller by 1,000 to 1,500 kbp. Genetic investigations had suggested that the numerous loci are distributed on a circular map in *S. coelicolor* A3(2), *S. lividans*, *S. rimosus*, *S. venezuelae*, and some other species. Recently, an ordered cosmid library was constructed from *S. coelicolor* A3(2). Further hybridizations, combined with PFGE studies, allowed the localization of many cloned genes. Comparisons of physical maps from the wild-type and mutant strains of *S. coelicolor* A3(2) and *S. lividans* gave rise to the assumption that the chromosome exists in a linear form and in a circular form (65, 92). Recent PFGE studies revealed the existence of a linear chromosome in other streptomycetes, including *S. ambofaciens* (63), *S. antibioticus*, *S. moderatus*, *S. lipmanii*, *S. parvulus*, *S. rochei* (65), and *S. griseus* (64). Linear topology of the chromosomal DNA has been demonstrated for a *Streptoverticillium* sp. belonging to the family of *Streptomycetaceae* and for representatives of other actinomycetes, including *Actinoplanes philippinensis*, *Nocardia asteroides*, *Saccharopolyspora erythraea* (93), and *Rhodococcus fascians* (22). In this context it is interesting that several genomes of various *Mycobacterium* spp. were found to be circular (86).

The structure of the chromosomal replication origin (*oriC*) region is highly conserved among the investigated *Streptomyces* strains *S. coelicolor* A3(2) (15), *S. lividans*, (116), *S. ambofaciens*, *S. antibioticus*, *S. chrysomallus*, and *S. reticuli* (46). Contrary to the high overall G+C content (69 to 73%) of *Streptomyces* DNA, the *oriC* region is comparatively rich in AT (64% G+C), and it contains 19 DnaA boxes arranged in two clusters. It is more complex than the *E. coli oriC* region (5 DnaA boxes). The consensus sequence for the *Streptomyces* DnaA box is TTGTCCACA, similar to other G+C-rich organisms such as mycobacteria. The initiator DnaA protein (interacting with DnaA boxes) differs from that in *E. coli* in that it harbors an additional, highly variable domain comprising preferentially acidic amino acids. The *oriC* of *S. lividans* and *S. coelicolor* A3(2) is located centrally, opposite to the supposed terminal ends located in the formerly genetically designated "silent region" of the chromosome. Further investigations have suggested that the characterized *oriC* is active in vivo and proceeds bidirectionally (72).

The chromosomal ends of a few *Steptomyces* species contain terminal inverted repeats covalently bound to protein, presumably at their 5′ ends. Depending on the strains, the length of the terminal inverted repeats varies between 24 and 550 kb, and they are quite divergent in sequence. Within the terminal ends of chromosomal and linear plasmid *Streptomyces* DNAs (see below), seven homologous palindromes varying only by a few nucleotide substitutions are distributed in a stretch of about 166 to 167 bp. Modeling predicts that the 3′ ends of these sequences can form extensive hairpin structures resembling those characterized within the 3′ ends of the single-stranded parvoviral DNA. Most of the putative hairpins contain a GCGCAGC sequence able to form a single C-residue loop closed by a sheared G:A base pair. As the *Streptomyces* chromosome is replicated from the centrally located replication origin, 3′-single-stranded gaps arise that are assumed to be patched by an unknown mechanism, involving the covalently bound terminal proteins, which might act as a nickase (40).

The chromosome of *S. coelicolor* A3(2) is being sequenced from an ordered encyclopedia of cosmids (92). Studies of the *Streptomyces* genome, in comparison with those of other bacteria (including the already known genomes of several mycobacteria), are expected to add to the rapidly growing knowledge about the architecture of bacterial genomes.

40.4. PLASTICITY OF THE CHROMOSOME

Among colonies—or sectors of colonies—of *Streptomyces* species, there is often a high variability of pigmentation, sporulation, antibiotic biosynthesis, or A-factor production. These variations result from large chromosomal deletions, which occur preferentially at the chromosomal ends and may include up to 1 Mbp of DNA. Analysis by PFGE demonstrated that macrodeletions are not commonly observed within other regions of the chromosomes (19, 44, 62, 100).

Several unstable genes encode resistance to various antibiotics, A-factor formation, synthesis of tyrosinase, or arginosuccinate (100). However, most genes within deletable regions have not yet been characterized. Deletions may occur in high frequencies (up to 10^{-2}), in several cases sequentially either at one or both ends of the chromosome. In addition, amplifications of DNA comprising the amplifiable units (AUDs) type 1 and type 2, which are flanked by small (up to 25 bp) or longer (0.7 to 2.2 kb) direct repeats, have been found (44, 62).

Contrary to the presence of one copy, two copies of an AUD lead to a predictable amplification; thus, the formation of a tandem duplication has been presumed to be the rate-limiting step in amplification (100). The repeats of the *S. lividans* AUD1 (88) are substrates for recombination, which promotes amplification. The *S. lividans* AUD2 carries mercury resistance genes and is flanked by two direct repeats that are similar to IS*112* (*Streptomyces albus*). These repeats represent a functional insertion sequence (111). The characterized AUDs are located relatively close to the chromosomal ends. To explain a sudden increase in the copy numbers of AUDs, the rolling circle mechanism has been proposed to occur in the *Streptomyces* genome and to differ from the operation of unequal crossover or circle excision and reinsertion mechanism.

In some cases, amplifications of certain genes correlate with enhanced levels of gene products. Accordingly, the 4.3-kb AUD (*S. lividans*) encodes a surface-located protein that is overproduced in large quantities after amplification (7). Upon selection for highly resistant spectinomycin (37) or chloramphenicol (25), variants comprising amplifications of the corresponding gene have been encountered. Recombinations and mutations are also expected to occur in amplified regions, leading to modified gene products.

The variability of the chromosomal DNA is enhanced by its interaction with linear and circular plasmids, phages, and transposable elements (see below). As streptomycetes inhabit quickly changing environments (see below), variants that are best adapted are expected to rapidly multiply. The high plasticity of the genome is assumed to be an effective prerequisite for quick adaptation.

40.5. PLASMIDS AND PHAGES

In addition to the chromosomal DNA, *Streptomyces* strains contain various types of circular and small or giant linear plasmids of different sizes. The low-copy-number circular plasmid SCP2 was described first; despite its relatively large size, it has been used as a cloning vector (34).

The natural plasmid pIJ101 (about 8.8 kb) is a high-copy-number conjugative *Streptomyces* plasmid. Regions required for its replication, stability, transfer, and distribution have been identified. A variety of nonconjugative derivatives carrying a selectable marker have been constructed (34). The most commonly used vector is pIJ702 (5.8 kb) containing a tyrosinase and a thiostrepton resistance gene. After in vitro mutagenesis, a derivative of pIJ702 that exhibits a temperature-sensitive replication phenotype was selected. Several *E. coli–Streptomyces* bifunctional vectors have been constructed using pIJ101 derivatives; however, some of them are not particularly stable. pWHM3 and pWHM4 (109) proved to be quite stable and have a number of suitable cloning sites.

The plasmid pSG5 (initially found in *Streptomyces ghanaensis*) is naturally temperature sensitive for replication, has a size of 12.3 kb, and has a copy number of about 50 per chromosome. A number of vectors have been constructed using pSG5, and some of them are also bifunctional. The replicon of pSG5 is naturally temperature sensitive; therefore, all vectors based on this replicon can be eliminated easily at nonpermissive temperatures and can be used as suicide delivery systems (see below) for transposons (6), gene disruption, and gene replacement experiments (73).

The multicopy plasmids pSN22 (51), pJV1 (4, 102), and pSMA2 (85) replicate via a rolling circle mechanism (32, 102). All Rep proteins encoded by these plasmids carry at their NH_2 terminus a motif resembling zinc fingers that is also found in DNA topoisomerase I and other unwinding proteins. The deduced Rep protein (73) of pSG5 differs from the above-mentioned ones; it is most closely related to the pC194 family in other gram-positive bacteria.

The 11-kb pSAM2 (initially identified within *S. ambofaciens*) occurs in an autonomous form and replicates via a rolling circle mechanism. The plasmid, pSAM2, is self-transmissible and mobilizes chromosomal markers. It can be integrated specifically into a chromosomal attachment site, and it can be maintained stably or reexcised. Replicase, excisionase, and integrase genes constitute an operon that is positively regulated by the regulator gene *pra* (104). Replicative and integrative vectors have been constructed on the basis of pSAM2.

Although linear plasmids are abundant among streptomycetes, only a few functions encoded in linear plasmids are known. These include antibiotic production (31, 56) and mercury resistance (90). Some linear plasmids were efficiently transferred during conjugation. The *tra* genes of the linear plasmid pBL1 have been characterized by insertional inactivation (117). Common to all linear plasmids are terminal inverted repeats of various lengths and a protein that has been linked in several cases to the 5′ ends of the plasmids SCP1 [*S. coelicolor* A3(2)] (56) or pSLA2 (*S. rochei*) (33).

The replication mechanism has best been investigated for pSLA2. In contrast to the linear ϕ29 *Bacillus* phage DNA, its replication is initiated bidirectionally near the center and proceeds toward its telomeric ends, generating 3′ leading-strand overhangs. A two-step telomere replication mechanism involving the fold-back pairing of distant palindromes and endonucleolytic processing has been proposed, showing features common to, but also divergent from, the modified rolling circle hairpin model suggested for the replication of the autonomous parvovirus (89). The existence of an internal origin of replication and of a mechanism for patching the ends allows the replication of pSLA2 either as circular or linear DNA. Thus, the linear *Streptomyces* plasmids have been proposed to occupy an evolutionarily intermediate position between circular plasmids and linear phage replicons (16). As shown for *S. lividans* (19) and *S. rimosus* (31), large linear plasmids can interact with the ends of the chromosome.

It is interesting that several other members of the order of *Actinomycetales* harbor on linear plasmids genes required for isopropylbenzene and trichlorethylene catabolism (*Rhodococcus erythropolis*) (52), for biphenyl degradation (*R. erythropolis* and *Rhodococcus globerulus*) (59), for hydrogen autotrophy (*Rhodococcus opacus*, formerly *Nocardia opaca*)

(49), and for fasciation in plants (*Rhodococcus fascians*) (22). In *Mycobacterium* species (*M. xenopi, M. branderi, M. celaturum, M. avrium*), linear plasmids of various sizes have been discovered. Their termini are related to those of linear plasmids from *Streptomyces* and *Rhodococcus* species (87). It should be of interest to explore whether the linear and circular plasmids are exchanged during conjugation among various actinomycetes.

The *Streptomyces* temperature phage ϕ31 has a relatively broad host spectrum and has thus been used to develop cloning and integration vectors, as well as cosmids (34). ϕ31 has a linear genome (41.5 kb) with 3′ overhanging cohesive ends (66). The transcription of the various phage genes is at present being intensively studied (38).

40.6. TRANSPOSONS

Compared to other bacteria, relatively little is known about *Streptomyces* transposons. The transposon Tn*4556* was discovered in *Streptomyces fradiae* and used for transposition studies (20, 45). On the basis of the *S. lividans* insertion element IS*493*, transposons containing an antibiotic resistance gene have been developed that are suitable for transposon mutagenesis of various *Streptomyces* strains (6). Recently, the IS*6100* element (originally identified in *Mycobacterium fortuitum*) (67) was transferred on plasmids to *S. lividans* and *Streptomyces avermitilis* and shown to transpose at random in their chromosomal DNA (113). An engineered derivative of the *E. coli* transposon Tn*5* also transposed randomly into the *S. lividans* genome (111).

40.7. AUTOREGULATORS

The formation of autoregulating factors and derivatives of γ-butyrolactone is widely distributed among streptomycetes (36, 54). These compounds have a 2,3-disubstituted γ-butyrolactone skeleton in common. They have been classified into three groups: the A-factor type (*S. griseus*), the virginiae butanolide (VB) type (*Streptomyces virginiae*), and the IM-2 type (*Streptomyces* sp.). Although the structural differences among the butyrolactones are slight, each of them is effective at very low concentration (10^{-9} M). They act like a highly specific hormone or morphogen for a given strain, controlling the formation of various secondary metabolities (such as antibiotics and pigments), as well as the formation of aerial mycelium.

The receptor proteins ArpA, BarA, and IM2 deduced from the corresponding genes (*arpA* in *S. griseus*, *barA* in *S. virginiae*, and *farA* in *Streptomyces* sp.) have sizes of about 24 to 29 kDa and share 38.5 to 48.5% amino acid identity with each other. All of them share an NH_2-terminally located helix-turn-helix motif, and they have a binding domain with a high specificity for the respective ligand (i.e., A-factors or virginiae butanolide). Recent studies suggested that ArpA and BarA represent transcriptional regulators recognizing a specific DNA motif, and it has been assumed that they prevent the expression of (a) certain key gene(s) during the early growth phase when a critical level of the corresponding ligand has not yet been reached. Upon binding to the ligands, the key gene(s) for secondary metabolites and morphogenesis is (are) expressed. Recent gene replacement studies confirmed that BarA acts as a repressor, controlling not only virgamycin production, but also autoregulator biosynthesis (75). CprB from *S. coelicolor* A3(2) acts, like ArpA, as a negative regulator and affects actinorhodin formation and sporulation. In contrast, the second *S. coelicolor* A3(2) homolog CprA is a positive regulator accelerating secondary metabolite synthesis and sporulation (80).

Autoregulating factors are produced in the late stage of growth, presumably under the control of global regulatory systems. A-factor-defective *S. griseus* strains produce neither streptomycin, aerial mycelium, nor spores. However, exogenously supplemented A-factor at very low concentrations (10^{-9} M) restores the wild-type phenotype. It has been presumed that a given γ-butyrolactone is also an effective chemical signal for communication between different parts of the mycelia within one strain (36). The role of autoregulatory compounds may resemble that of *N*-acetyl-homoserine lactones, which control a number of cell density-dependent processes and are widespread among gram-negative bacteria (42).

40.8. PHARMACOLOGICALLY ACTIVE SUBSTANCES

Streptomycetes synthesize an amazing variety of chemically different substances, many of them acting as antibiotics, fungicides, cytostatics, or modulators of immune response. Numerous efforts have been made in the past to elucidate their structure and mode of action. New natural compounds are continuously being identified from newly isolated streptomycetes.

Genes encoding these pharmacologically active substances have been found clustered within DNA stretches of 20 kb to more than 100 kb. Cloning has mostly been achieved by complementing mutants and by screening total genomic DNA or gene libraries with homologous or heterologous gene probes. The biosynthetic genes for the antibiotic are in general linked to one or more genes mediating resistance to the corresponding antibiotic. Accordingly, the initial identification of such resistance genes frequently has facilitated the identification of a biosynthetic cluster. Some enzymes catalyzing an important biosynthetic step could be purified to homogeneity and characterized. During the production of secondary metabolites, protein patterns change drastically. The NH_2-terminal sequences of characteristic proteins can be determined to design oligonucleotides, which are used to screen gene libraries in order to identify the desired gene cluster. Using various approaches, the biosynthetic genes for actinomycin (39, 106), aminoglycosides (24), β-lactams (1, 83), cyclophilins (82), macrolides (35), nikkomycin (12), rifamycin (3), tetracycline (69), and many other substances have been identified.

Polyketides constitute a diverse class of substances, including antibiotics, anticancer and antiparasitic pigments, and immunosuppressants. The synthesis of these compounds is catalyzed by polyketide synthases (PKSs). Type I PKSs are multifunctional enzymes consisting of modules that control the sequential addition of acylthioester units to a growing polyketide chain and the subsequent modification of the condensation products. Type II PKSs, in contrast, are composed of three to seven separate mono- or bifunctional proteins that are used repeatedly for the formation of the polyketide chain and for some of its modifications. Both PKS types are biochemically similar to fatty acid synthases and catalyze decarboxylative condensations between a thiester-linked nascent carbon chain and short-chain fatty acid extender units. However, PKSs can use different starter molecules (acetate, propionate, butyrate) and chain extenders (malonate, methylmalonate, ethylmalonate) at various steps. Whereas condensations of fatty acids are typically followed by trioreactions (β-ketoreduction, dehydration, enoylreduction), a PKS may employ all, some, or none of these reactions after each condensation step. Thus, the structural diversity of polyketides is due to the variable numbers and types of acyl units, reduction, dehydration, cyclization, and aromatization reactions of the initial β-ketoacyl condensation products (2, 26, 35, 43).

Using *pks* genes as probes, a continuously increasing number of biosynthetic gene clusters have been identified. Analysis by PCR has led to the discovery of many *pks* genes in various microorganisms, including unculturable ones. The natural diversity of these genes can be further extended by rational mutations. Having established the sequence of enzymatic steps, rules for polyketide design have emerged, enabling the rational designing of a wide range of modified polyketides.

Similar to polyketides, the pathways of nonribosomally synthesized peptide antibiotics, aminoglycosides, and other metabolites have been elucidated. A thorough analysis of the information gathered should allow the design of a broader diversity of biosynthetic pathways.

Current studies explore the regulation of genes encoding pharmacologically active compounds (5, 18, 53). The data gathered up to now suggest the presence of transport systems (25, 70) for antibiotics, and it will be interesting to gain deeper insights into their characteristics. Little is known about the primary metabolism of streptomycetes. It will be necessary to investigate the pathways and genes required for the biosynthesis of primary compounds, to examine the regulation (95) and to explore metabolic fluxes (77), because the precursors of pharmacologically active compounds are derived from primary metabolites.

The modular arrangement of biosynthetic gene clusters has facilitated the construction of hybrid gene clusters. Consequently, a designed tailoring of antibiotics will complement the "screening" for new antibiotics. A rational design strategy may help to overcome the problems linked with the broad repertoire of antibiotic resistances developed by pathogenic microorganisms.

40.9. DEGRADATION OF MACROMOLECULES

Streptomycetes play an important role in the turnover of chitin, the second most abundant polysaccharide in nature. Contrary to other chitinolytic bacteria, almost every *Streptomyces* species uses chitin not only as a carbon source but also as a nitrogen source (60). The few analyzed *Streptomyces* strains produce several chitinases (71, 94). Some of the chitinase genes were cloned, and the corresponding proteins have been shown to be successfully secreted. The efficient degradation of crystalline forms of chitin is dependent on the presence of a chitin-binding domain in a chitinase (9). During growth on chitin, many streptomycetes secrete, in addition to chitinases, small proteins (58) that act like glue to mediate contact between chitin-containing substrates or organisms.

Enzymes catalyzing the hydrolysis of soluble forms of cellulose have been investigated for a few *Streptomyces* strains (including *S. lividans*), and some of the genes have been cloned (21, 57, 74). *S. reticuli* specifically adheres to crystalline cellulose via a membrane-anchored protein (112) and efficiently degrades crystalline forms of cellulose to cellobiose (98). Cellobiose is taken up via a specific ABC transporter system (99).

Xylan consists of a β-1-4-linked xylose polymer and commonly contains side branches of arabinosyl, glycoronosyl, acetyl, uronyl, and mannosyl residues. Enzymes degrading this complex polymer are used for different purposes, including biopulping and bioleaching in the paper industry and bioconversion. Xylanases and their genes were identified from *S. lividans* (81), *Streptomyces halstedii* (97), and the thermophilic *Streptomyces thermoviolaceus* (108).

Extracellular proteases are abundant among streptomycetes, and several corresponding genes (55) have been characterized. Inhibitors of the proteases leupeptin, subtilisin, and others have been studied. Streptomycetes also contain a multitude of genes for protease inhibitors (107). Keratinases are frequently encountered among streptomycetes (60), and for the disintegration of chicken feathers by *Streptomyces pactum*, keratinolytic proteinases and extracellular reduction of disulfide bonds (11) were identified. The synthesis of amylases and their inhibitors is common among streptomycetes. Some of the genes (110) have been cloned, and studies to understand the regulation of the glucose-repressed genes have been initiated (76). *Streptomyces* strains are predominant among isolates of actinomycetes degrading latex rubber (47).

The interest in bacterial lipases has increased considerably. However, only a few extracellular lipases and their genes have been characterized from *Streptomyces* strains (103, 105).

In addition to enzymes degrading macromolecules, streptomycetes produce a large repertoire of enzymes, including those for the modification of pharmacologically relevant compounds and xenobiotics (84). Many enzymes are efficiently secreted; investigations of the secretion of homologous (81) and heterologous proteins (61) have been initiated.

40.10. ECOLOGY

Soil is the most important habitat of streptomycetes. Most soils contain 10^4 to 10^7 colony-forming units of streptomycetes representing 1 to 20% or even more of the total viable counts. Grass vegetation or soil rich in organic matter contains the highest numbers of streptomycetes. Most *Streptomyces* species prefer a neutral to mildly alkaline pH. However, several reports prove that some strains also grow under acidic (~pH 3.5) or alkalophilic (pH 8 to 11.5) conditions. As expected, streptomycetes settle in soil as microcolonies on particles of organic matter.

In terrestrial habitats, streptomycetes are the most abundant actinomycete (90% or more). Streptomycetes from marine habitats have also been investigated (78, 90). Thorough studies of sediment from lakes showed a predominance of *Micromonospora* species (between 40 and 90%); streptomycetes were the second most abundant microorganisms in sediments (48). Aquatic streptomycetes hydrolyze a wide range of macromolecules, including cellulose, chitin, and proteins, and produce compounds with antibacterial and antifungal activities. They are assumed to play a major part in the turnover of toxic compounds.

Actinomycetes have been detected and localized by the dilution plate technique, either without enrichment or under special selective isolation conditions. Recently, however, hybridization techniques using fluorescent, genus-specific 16S rRNA probes have been developed for a broad range of bacteria. Such oligonucleotides can be used to detect streptomycetes in situ.

Streptomycetes occur at higher concentrations in the gut of earthworms and arthropods. Some highly cellulolytic strains can be characterized from the gut of termites. The rhizosphere of plants is populated by various amounts of actinomycetes. It is assumed that secretion of antibacterial and antifungal compounds plays an important role in natural habitats (60). Using molecular detection methods, streptomycin-producing *Streptomyces* strains have been found to colonize the rhizosphere of soybeans (41). Previous investigations suggested that some streptomycetes serve as potential biocontrol agents against fungi within the rhizosphere (60), and recent studies support this interpretation (115).

Plant-pathogenic streptomycetes produce various types of potato scab (60). DNA-DNA hybridization and rRNA studies revealed that the pathogenic strains belong to different species. The production of thaxtomins (4-nitroindol-3yl-containing 2,5-dioxopiperazines) has been correlated with plant pathogenicity. The genes encoding biosynthesis of the toxins have not yet been identified, except for the new *nec1* gene leading to necrosis (13). A gene encoding an extracellular esterase of a *Streptomyces scabies* strain was cloned (91). It is expected that future molecular studies will yield more insights into the interaction of pathogenic streptomycetes and potatoes.

At the initial decomposition stages of organic material in compost, mesophilic streptomycetes play an important part. Later, thermophilic streptomycetes as well as other thermophilic actinomycetes (e.g., *Thermomonospora, Saccharopolyspora, Microbispora*) become dominant (60). The increasing application of molecular tools should considerably deepen our knowledge of the ecological role of actinomycetes.

The recent identification of *Streptomyces thermoautrophicus*, a thermophilic, CO- and H_2-oxidizing, obligate chemolithoautotrophic organism and nitrogen fixer, suggests that it will be interesting to investigate a wider range of biotopes to identify more strains with unknown properties (28).

40.11. CONCLUSIONS

In the course of the last two decades, many molecular tools, including vector systems, transformation systems, mutagenesis, and expression and efficient secretion systems, have been developed (for details, see chapter 30 in this volume). The successful application of molecular techniques for the cloning of genes, which are required for a number of antibiotics, is elucidated in detail in chapters 60 and 61. It is expected that available as well as future data can be successfully used to tailor hybrid antibiotics, cytostatics, antifungal and antiparasitic compounds, as well as other substances. The capability of actinomycetes to degrade a variety of biopolymers, to modify xenobiotics, and to detoxify harmful compounds is of great biotechnological and ecological relevance. Efficient secretion systems will be of practical use. Comparisons of streptomycete gene pools with those of mycobacteria and other bacteria will provide insights into the evolution of actinomycetes. Using available and newly developed tools, it will also be exciting to elucidate the molecular interaction of streptomycetes with other microbes, plants, and animals and to understand their ecological importance.

I am very grateful to M. Lemme for supporting the writing of the manuscript.

REFERENCES

1. **Aharonowitz, Y., and G. Cohen.** 1992. Penicillin and cephalosporin biosynthesis genes: Structure, organization, regulation, and evolution. *Annu. Rev. Microbiol.* **46:**461–495.
2. **Aparicio, J. F., P. Caffrey, A. F. A. Marsden, J. Stauton, and P. F. Leadlay.** 1994. Limited proteolysis and active-site studies of the first multienzyme component of the erythromycin-producing polyketide synthase. *J. Biol. Chem.* **269:**8524–8528.
3. **August, P. R., L. Tang, Y. J. Yoon, S. Ning, R. Muller, T. W. Yu, M. Taylor, D. Hoffmann, C. G. Kim, X. H. Zhang, C. R. Hutchinson, and H. G. Floss.** 1998. Biosynthesis of the ansamycin antibiotic rifamycin—deductions from the molecular analysis of the *rif* biosynthetic gene cluster of *Amycolatopsis mediterranei* S699. *Chem. Biol.* **5:**69–79.
4. **Bailey, C. R., C. J. Bruton, M. J. Butler, K. F. Chater, J. E. Harris, and D. A. Hopwood.** 1986. Properties of *in vitro* recombinant derivatives of pJV1, a multi-copy plasmid from *Streptomyces phaeochromogenes*. *J. Gen. Microbiol.* **132:**2071–2078.
5. **Baltz, R. H., and T. J. Hosted.** 1996. Molecular genetic methods for improving secondary-metabolite production in actinomycetes. *Trends Biotechnol.* **14:**245–250.
6. **Baltz, R. H., M. A. McHenney, C. A. Cantwell, S. W. Queener, and P. J. Solenberg.** 1997. Applications of transposition mutagenesis in antibiotic producing streptomycetes. *Antonie Leeuwenhoek* **71:**179–187.
7. **Betzler, M., I. Tlolka, and H. Schrempf.** 1997. A novel protein of *Streptomyces lividans* is overproduced and associated with vesicles upon amplification of a 4.3 kb DNA element. *Microbiology* **143:**1243–1252.
8. **Bibb, M. J., N. J. Ryding, V. Molle, K. F. Chater, and M. J. Buttner.** 1998. *whiK, whiL, whiM, whiN* and *whiO*—five new sporulation genes in *Streptomyces coelicolor* A3(2), abstr., p. 82. *In* G. Cohen and Y. Aharonowitz (ed.), *8th International Symposium of the Genetics of Industrial Microorganisms, June 28–July 2, 1998. Jerusalem, Israel.*
9. **Blaak, H., and H. Schrempf.** 1995. Binding and substrate specificities of a *Streptomyces olivaceoviridis* chitinase in comparison with its proteolytically processed form. *Eur. J. Biochem.* **229:**132–139.
10. **Blanco, G., M. R. Rodicio, A. M. Puglia, C. Méndez, C. J. Thompson, and J. A. Salas.** 1994. Synthesis of ribosomal proteins during growth of *Streptomyces coelicolor*. *Mol. Microbiol.* **12:**375–385.
11. **Böckle, B., and R. Müller.** 1997. Reduction of disulfide bonds by *Streptomyces pactum* during growth on chicken feathers. *Appl. Environ. Microbiol.* **63:**790–792.
12. **Bormann, C., V. Möhrle, and C. Bruntner.** 1996. Cloning and heterologous expression of the entire set of structural genes for nikkomycin synthesis from *Streptomyces tendae* Tü901 in *Streptomyces lividans*. *J. Bacteriol.* **178:** 1216–1218.

13. **Bukhalid, R. A., and R. Loria.** 1997. Cloning and expression of a gene from *Streptomyces scabies* encoding a putative pathogenicity factor. *J. Bacteriol.* **179:**7776–7783.

14. **Burger, A., K. Sichler, W. Wohlleben, G. Kelemen, and M. Buttner.** 1998. Identification and characterization of new cell division and differentiation genes of *Streptomyces coelicolor* A3(2), (abstr.), p. 83–84. *In* G. Cohen and Y. Aharonowitz (ed.), *8th International Symposium on the Genetics of Industrial Microorganisms, June 28–July 2, 1998. Jerusalem, Israel.*

15. **Calcutt, M. J., and F. J. Schmidt.** 1992. Conserved gene arrangement in the origin region of the *Streptomyces coelicolor* chromosome. *J. Bacteriol.* **174:**3220–3226.

16. **Chang, P.-C., E.-S. Kim, and S. N. Cohen.** 1996. *Streptomyces* linear plasmids that contain a phage-like, centrally located, replication origin. *Mol. Microbiol.* **22:**789–800.

17. **Chater, K. F.** 1993. Genetics of differentiation in *Streptomyces. Annu. Rev. Microbiol.* **47:**685–713.

18. **Chater, K. F.** 1998. Taking a genetic scalpel to the *Streptomyces* colony. *Microbiology* **144:**1465–1478.

19. **Chen, C. W.** 1995. The unstable ends of the *Streptomyces* linear chromosomes: a nuisance without cures? *Trends Biotechnol.* **13:**157–160.

20. **Chung, S.-T., and L. L. Crose.** 1989. *Streptomyces* transposon Tn*4556* and its applications, p. 168–175. *In* C. L. Hershberger, S. W. Queener, and G. Hegemen (ed.), *Genetics and Molecular Biology of Industrial Microorganisms.* American Society for Microbiology, Washington, D.C.

21. **Crawford, D. L., and E. McCoy.** 1972. Cellulases of *Thermomonospora fusca* and *Streptomyces thermodiastaticus. Appl. Microbiol.* **24:**150–152.

22. **Crespi, M., E. Messens, A. B. Caplan, M. Vanmontagu, and J. Desomer.** 1992. Fasciation induction by the phytopathogen *Rhodococcus fascians* depends upon a linear plasmid encoding a cytokinin synthase gene. *EMBO J.* **11:**795–804.

23. **Dharmatilake, A. J., and K. E. Kendrick.** 1994. Expression of the division-controlling gene *ftsZ* during growth and sporulation of the filamentous bacterium *Streptomyces griseus. Gene* **147:**21–28.

24. **Distler, J., A. Ebert, K. Mansouri, K. Pissowotzki, M. Stockmann, and W. Piepersberg.** 1987. Gene cluster for streptomycin biosynthesis in *Streptomyces griseus*: nucleotide sequence of three genes and analysis of transcriptional activity. *Nucleic Acids Res.* **15:**8041–8056.

25. **Dittrich, W., M. Betzler, and H. Schrempf.** 1991. An amplifiable and deletable chloramphenicol-resistance determinant of *Streptomyces lividans* 1326 encodes a putative transmembrane protein. *Mol. Microbiol.* **5:**2789–2797.

26. **Donadio, S., M. Staver, J. B. McAlpine, S. J. Swanson, and L. Katz.** 1991. Modular organization of genes required for complex polyketide biosynthesis. *Science* **252:**675–679.

27. **Embley, T. M., and E. Stackebrandt.** 1994. The molecular phylogeny and systematics of the actinomycetes. *Annu. Rev. Microbiol.* **48:**257–289.

28. **Gadkari, D., G. Mörsdorf, and O. Meyer.** 1992. Chemolithoautotrophic assimilation of dinitrogen by *Streptomyces thermoautotrophicus* UTB1: identification of an unusual N_2-fixing system. *J. Bacteriol.* **174:**6840–6843.

29. **Gal-Mor, O., I. Borovok, Y. Av-Gay, G. Cohen, and Y. Aharonowitz.** 1998. Gene organization in the *trxAB oriC* region of the *Streptomyces* chromosome and comparison with other bacteria, abstr., p. 52–53. *In* G. Cohen and Y. Aharonowitz (ed.), *8th International Symposium on the Genetics of Industrial Microorganisms, June 28–July 2, 1998. Jerusalem, Israel.*

30. **Goodfellow, M., L. J. Stanton, K. E. Simpson, and D. E. Minnikin.** 1990. Numerical and chemical classification of *Actinoplanes* and related actinomycetes. *J. Gen. Microbiol.* **136:**19–36.

31. **Gravius, B., D. Glocker, J. Pigac, K. Pandza, D. Hranueli, and J. Cullum.** 1994. The 387 kb linear plasmid pPZG101 of *Streptomyces rimosus* and its interactions with the chromosome. *Microbiology* **140:**2271–2277.

32. **Hagège, J., J.-L. Pernodet, A. Friedmann, and M. Guérineau.** 1993. Mode and origin of replication of pSAM2, a conjugative integrating element of *Streptomyces ambofaciens. Mol. Microbiol.* **10:**799–812.

33. **Hirochika, H., K. Nakamura, and K. Sakaguchi.** 1984. A linear DNA plasmid from *Streptomyces rochei* with an inverted terminal repetition of 614 base pairs. *EMBO J.* **3:**761–766.

34. **Hopwood, D. A., M. J. Bibb, K. F. Chater, T. Kieser, C. J. Bruton, H. M. Kieser, D. J. Lydiate, C. P. Smith, J. M. Ward, and H. Schrempf.** 1985. *Genetic Manipulation of Streptomyces: a Laboratory Manual.* John Innes Foundation, Norwich, United Kingdom.

35. **Hopwood, D. A., and D. H. Sherman.** 1990. Molecular genetics of polyketides and its comparison to fatty acid biosynthesis. *Annu. Rev. Genet.* **24:**37–66.

36. **Horinouchi, S., and T. Beppu.** 1992. Autoregulatory factors and communication in actinomycetes. *Annu. Rev. Microbiol.* **46:**377–398.

37. **Hornemann, U., D. J. Otto, and G. Hoffmann.** 1986. DNA amplification and spectinomycin resistance in *Streptomyces achromogenes* subsp. *rubradiris*, p. 29. *In* D. Alacevic, D. Hranueli, and Z. Toman (ed.), *Abstracts of the Fifth International Symposium on the Genetics of Industrial Microorganisms.* Split, Yugoslavia.

38. **Howe, C. W., and M. C. M. Smith.** 1996. Gene expression in the *cos* region of the *Streptomyces* temperate actinophage *ϕ*C31. *Microbiology* **142:**1357–1367.

39. **Hsieh, C.-J., and G. H. Jones.** 1995. Nucleotide sequence, transcriptional analysis, and glucose regulation of the phenoxazinone synthase gene (*phsA*) from *Streptomyces antibioticus. J. Bacteriol.* **177:**5740–5747.

40. **Huang, C.-H., Y.-S. Lin, Y.-L Yang, S. Huang, and C. W. Chen.** 1998. The telomeres of *Streptomyces* chromosomes contain conserved palindromic sequences with potential to form complex secondary structures. *Mol. Microbiol.* **28:**905–916.

41. **Huddleston, A. S., N. Cresswell, M. C. P. Neves, J. E. Beringer, S. Baumberg, D. I. Thomas, and E. M. H. Wellington.** 1997. Molecular detection of streptomycin-producing streptomycetes in Brazilian soils. *Appl. Environ. Microbiol.* **63:**1288–1297.

42. **Huisman, G. W., and R. Kolter.** 1994. Sensing starvation: a homoserine lactone-dependent signaling pathway in *Escherichia coli. Science* **265:**537–539.

43. **Hutchinson, C. R., and I. Fujii.** 1995. Polyketide synthase gene manipulation: a structure-function approach in

engineering novel antibiotics. *Annu. Rev. Microbiol.* **49:** 201–238.

44. **Hütter, R., and T. Eckhardt.** 1988. Genetic manipulation, p. 89–184. *In* M. Goodfellow, S. T. Williams, and M. Mordarski (ed.), *Actinomycetes in Biotechnology*. Academic Press, London.
45. **Ikeda, H., Y. Takada, C.-H. Pang, H. Tanaka, and S. Omura.** 1993. Transposon mutagenesis by Tn4560 and applications with avermectin-producing *Streptomyces avermitilis*. *J. Bacteriol.* **175:**2077–2082.
46. **Jakimowicz, D., J. Majka, W. Messer, C. Speck, M. Fernandez, M. Cruz Martin, J. Sanchez, F. Schauwecker, U. Keller, H. Schrempf, and J. Zakrzewska-Czerwinska.** 1998. Structural elements of the *Streptomyces oriC* region and their interactions with the DnaA protein. *Microbiology* **144:**1281–1290.
47. **Jendrossek, D., G. Tomasi, and R. M. Kroppenstedt.** 1997. Bacterial degradation of natural rubber: a privilege of actinomycetes? *FEMS Microbiol. Lett.* **150:**179–188.
48. **Jiang, C.-L., and L.-H. Xu.** 1996. Diversity of aquatic actinomycetes in lakes of the Middle Plateau, Yunnan, China. *Appl. Environ. Microbiol.* **62:**249–253.
49. **Kalkus, J., C. Dörrie, D. Fischer, M. Reh, and H. G. Schlegel.** 1993. The giant linear plasmid pHG207 from *Rhodococcus* sp. encoding hydrogen autotrophy: characterization of the plasmid and its termini. *J. Gen. Microbiol.* **139:**2055–2065.
50. **Kamerbeek, J., L. Schouls, A. Kolk, M. van Agterveld, D. van Soolingen, S. Kuijper, A. Bunschoten, H. Molhuizen, R. Shaw, M. Goyal, and J. D. A. van Embden.** 1997. Simultaneous detection and strain differentiation of *Mycobacterium tuberculosis* for diagnosis and epidemiology. *J. Clin. Microbiol.* **35:**907–914.
51. **Kataoka, M., T. Seki, and T. Yoshida.** 1991. Five genes involved in self-transmission of pSN22, a *Streptomyces* plasmid. *J. Bacteriol.* **173:**4220–4228.
52. **Kebeler, M., E. R. Dabbs, B. Averhoff, and G. Gottschalk.** Studies on the isopropylbenzene 2,3-dioxygenase and the 3′-isopropylcatechol 2,3-dioxygenase genes encoded by the linear plasmid of *Rhodococcus erythropolis* BD2. *Microbiology* **142:**3241–3251.
53. **Kelemen, G. H., M. Zalacain, E. Culebras, E. T. Seno, and E. Cundliffe.** 1994. Transcriptional attenuation control of the tylosin-resistance gene *tlrA* in *Streptomyces fradiae*. *Mol. Microbiol.* **14:**833–842.
54. **Khokhlov, A. S., I. I. Tovarova, L. N. Borisova, S. A. Pliner, L. A. Schevchenko, E. Y. Kornitskaya, et al.** 1967. A-factor responsible for the biosynthesis of streptomycin by a mutant strain of *Actinomyces streptomycini*. *Dokl. Akad. Nauk. SSSR* **177:**232–235.
55. **Kim, I. S., and K. J. Lee.** 1995. Physiological roles of leupeptin and extracellular proteases in mycelium development of *Streptomyces exfoliatus* SMF13. *Microbiology* **141:**1017–1025.
56. **Kinashi, H., M. Shimaji-Murayama, and T. Hanafusa.** 1991. Nucleotide sequence analysis of the unusually long terminal inverted repeats of a giant linear plasmid, SCP1. *Plasmid* **26:**123–130.
57. **Kluepfel, D., F. Shareck, F. Mondou, and R. Morosoli.** 1986. Characterisation of cellulase and xylanase activities of *Streptomyces lividans*. *Appl. Microbiol. Biotechnol.* **24:** 230–234.
58. **Kolbe, S., S. Fischer, A. Becirevic, P. Hinz, and H. Schrempf.** 1998. The *Streptomyces reticuli* α-chitin-binding protein CHB2 and its gene. *Microbiology* **144:** 1291–1297.
59. **Kosono, S., M. Maeda, F. Fuji, H. Arai, and T. Kudo.** 1997. Three of the seven *bphC* genes of *Rhodococcus erythropolis* TA421, isolated from a termite ecosystem, are located on an indigenous plasmid associated with biphenyl degradation. *Appl. Environ. Microbiol.* **63:**3282–3285.
60. **Kutzner, H. J.** 1981. The family Streptomycetaceae, p. 2028–2090. *In* M. P. Starr, H. Stolp, H. G. Trüper, A. Balows, and H. Schlegel (ed.), *The Prokaryotes: a Handbook on Habitats, Isolation and Identification of Bacteria*. Springer-Verlag, Berlin.
61. **Lammertyn, E., and J. Anné.** 1998. Modification of *Streptomyces* signal peptides and their effects on protein production and secretion. *FEMS Microbiol. Lett.* **160:**1–10.
62. **Leblond, P., and B. Decaris.** 1994. New insights into the genetic instability of *Streptomyces*. *FEMS Microbiol Lett.* **123:**225–232.
63. **Leblond, P., G. Fischer, F. Francou, F. Berger, M. Guérineau, and B. Decaris.** 1996. The unstable region of *Streptomyces ambofaciens* includes 210 kb terminal inverted repeats flanking the extremities of the linear chromosomal DNA. *Mol. Microbiol.* **19:**261–271.
64. **Lezhava, A., T. Mizukami, T. Kajitani, D. Kameoka, M. Redenbach, H. Shinkawa, O. Nimi, and H. Kinashi.** 1995. Physical map of the linear chromosome of *Streptomyces griseus*. *J. Bacteriol.* **177:**6492–6498.
65. **Lin, Y. S., H. M. Kieser, D. A. Hopwood, and C. W. Chen.** 1993. The chromosomal DNA of *Streptomyces lividans* 66 is linear. *Mol. Microbiol.* **10:**923–933.
66. **Lomovaskaya, N. D., K. F. Chater, and N. M. Mkrtumian.** 1980. Genetics and molecular biology of *Streptomyces* bacteriophages. *Microbiol. Rev.* **44:**206–229.
67. **Martin, C., J. Timm, J. Rauzier, R. Gomez-Lus, J. Davies, and B. Gicquel.** 1990. Transposition of an antibiotic resistance element in mycobacteria. *Nature* **345:**739–743.
68. **McCue, L. A., J. Kwak, J. Wang, and K. E. Kendrick.** 1996. Analysis of a gene that suppresses the morphological defect of bald mutants of *Streptomyces griseus*. *J. Bacteriol.* **178:**2867–2875.
69. **McDowall, K. J., D. Doyle, M. J. Butler, C. Binnie, M. Warren, and I. S. Hunter.** 1991. Molecular genetics of oxytetracyline production by *Streptomyces rimosus*, p. 105–116. *In* S. Baumberg, H. Krügel, and D. Noack (ed.), *Genetics and Product Formation in Streptomyces*. Plenum Press, New York.
70. **Méndez, C., and J. A. Salas.** 1998. ABC transporters in antibiotic-producing actinomycetes. *FEMS Microbiol. Lett.* **158:**1–8.
71. **Miyashita, K., T. Fujii, and Y. Sawada.** 1991. Molecular cloning and characterization of chitinase genes from *Streptomyces lividans* 66. *J. Gen. Microbiol.* **137:**2065–2072.
72. **Musialowski, M. S., F. Flett, G. B. Scott, G. Hobbs, C. P. Smith, and S. G. Oliver.** 1994. Functional evidence that the principal DNA replication origin of the *Streptomyces coelicolor* chromosome is close to the *dnaA-gyrB* region. *J. Bacteriol.* **176:**5123–5125.
73. **Muth, G., M. Farr, V. Hartmann, and W. Wohlleben.** 1995. *Streptomyces ghanaensis* plasmid pSG5: nucleotide sequence analysis of the self-transmissible minimal repli-

con and characterization of the replication mode. *Plasmid* **33:**113–126.

74. **Nakai, R., S. Horinouchi, and T. Beppu.** 1988. Cloning and nucleotide sequence of a cellulase gene, *casA*, from an alkalophilic *Streptomyces* strain. *Gene* **65:**229–238.

75. **Nakano, H., E. Takehara, T. Nihira, and Y. Yamada.** 1998. Gene replacement analysis of the *Streptomyces virginiae barA* gene encoding the butyrolactone autoregulator receptor reveals that BarA acts as a repressor in virginiamycin biosynthesis. *J. Bacteriol.* **180:**3317–3322.

76. **Nguyen, J., F. Francou, M.-J. Virolle, and M. Guérineau.** 1997. Amylase and chitinase genes in *Streptomyces lividans* are regulated by *reg1*, a pleiotropic regulatory gene. *J. Bacteriol.* **179:**6383–6390.

77. **Obanye, A. I. C., G. Hobbs, D. C. J. Gardner, and S. G. Oliver.** 1996. Correlation between carbon flux through the pentose phosphate pathway and production of the antibiotic methylenomycin in *Streptomyces coelicolor* A3(2). *Microbiology* **142:**133–137.

78. **Okami, Y., T. Okazaki, T. Kitahara, and H. Umezawa.** 1976. Studies on marine microorganisms. V. A new antibiotic, aplasmomycin, produced by a streptomycete isolated from shallow sea mud. *J. Antibiot. Ser. A* **29:**1019–1025.

79. **Okamoto, S., M. Itoh, and K. Ochi.** 1997. Molecular cloning and characterization of the *obg* gene of *Streptomyces griseus* in relation to the onset of morphological differentiation. *J. Bacteriol.* **179:**170–179.

80. **Onaka, H., T. Nakagawa, and S. Horinouchi.** 1998. Involvement of two A-factor receptor homologues in *Streptomyces coelicolor* A3(2) in the regulation of secondary metabolism and morphogenesis. *Mol. Microbiol.* **28:**743–753.

81. **Pagé, N., D. Kluepfel, F. Shareck, and R. Morosoli.** 1996. Effect of signal peptide alterations and replacement on export of xylanase A in *Streptomyces lividans*. *Appl. Environ. Microbiol.* **62:**109–114.

82. **Pahl, A., A. Gewies, and U. Keller.** 1997. ScCypB is a novel second cytosolic cyclophilin from *Streptomyces chrysomallus* which is phylogenetically distant from ScCypA. *Microbiology* **143:**117–126.

83. **Paradkar, A. S., K. A. Aidoo, A. Wong, and S. E. Jensen.** 1996. Molecular analysis of a β-lactam resistance gene encoded within the cephamycin gene cluster of *Streptomyces clavuligerus*. *J. Bacteriol.* **178:**6266–6274.

84. **Peczynska-Czoch, W., and M. Mordarski.** 1988. Actinomycete enzymes, p. 219–283. *In* M. Goodfellow, S. T. Williams, and M. Mordarski (ed.), *Actinomycetes in Biotechnology*. Academic Press, London.

85. **Pernodet, J.-L., J.-M. Simonet, and M. Guérineau.** 1984. Plasmids in different strains of *Streptomyces ambofaciens*: free and integrated form of plasmid pSAM2. *Mol. Gen. Genet.* **198:**35–41.

86. **Philipp, W. J., S. Poulet, K. Eiglmeier, L. Pascopella, V. Balasubramanian, B. Heym, S. Bergh, B. R. Bloom, W. R. Jacobs, Jr., and S. T. Cole.** 1996. An integrated map of the genome of the tubercle bacillus, *Mycobacterium tuberculosis* H37Rv, and comparison with *Mycobacterium leprae*. *Proc. Natl. Acad. Sci. USA* **93:**3132–3137.

87. **Picardeau, M., and V. Vincent.** 1998. Mycobacterial linear plasmids have an invertron-like structure related to other linear replicons in actinomycetes. *Microbiology* **144:**1981–1988.

88. **Piendl, W., C. Eichenseer, P. Viel, J. Altenbuchner, and J. Cullum.** 1994. Analysis of putative DNA amplification genes in the element AUD1 of *Streptomyces lividans* 66. *Mol. Gen. Genet.* **244:**439–443.

89. **Qin, Z., and S. N. Cohen.** 1998. Replication at the telomeres of the *Streptomyces* linear plasmid pSLA2. *Mol. Microbiol.* **28:**893–903.

90. **Ravel, J., H. Schrempf, and R. T. Hill.** 1998. Mercury resistance is encoded by transferable giant linear plasmids in two Chesapeake Bay *Streptomyces* strains. *Appl. Environ. Microbiol.* **64:**3383–3388.

91. **Raymer, G., J. M. A. Willard, and J. L. Schottel.** 1990. Cloning, sequencing, and regulation of expression of an extracellular esterase gene from the plant pathogen *Streptomyces scabies*. *J. Bacteriol.* **172:**7020–7026.

92. **Redenbach, M., H. M. Kieser, D. Denapaite, A. Eichner, J. Cullum, H. Kinashi, and D. A. Hopwood.** 1996. A set of ordered cosmids and a detailed genetic and physical map for the 8 Mb *Streptomyces coelicolor* A3(2) chromosome. *Mol. Microbiol.* **21:**77–96.

93. **Redenbach, M., J. Scheel, J. Cullum, and U. Schmidt.** 1998. The chromosome of various actinomycetes strains is linear, abstr., p. 69–70. *In* G. Cohen and Y. Aharonowitz (ed.), *8th International Symposium of the Genetics of Industrial Microorganisms, June 28–July 2, 1998. Jerusalem, Israel.*

94. **Robbins, P. W., C. Albright, and B. Benfield.** 1988. Cloning and expression of a *Streptomyces plicatus* chitinase (chitinase-63) in *Escherichia coli*. *J. Biol. Chem.* **263:**443–447.

95. **Rodríguez-García, A., M. Ludovice, J. F. Martín, and P. Liras.** 1997. Arginine boxes and the *argR* gene in *Streptomyces clavuligerus*: evidence for a clear regulation of the arginine pathway. *Mol. Microbiol.* **25:**219–228.

96. **Roller, C., W. Ludwig, and K. H. Schleifer.** 1992. Gram-positive bacteria with a high DNA G+C content are characterized by a common insertion within their 23S rRNA genes. *J. Gen. Microbiol.* **138:**167–175.

97. **Ruiz-Arribas, A., G. G. Zhadan, V. P. Kutyshenko, R. I. Santamariá, M. Cortijo, E. Villar, J. M. Fernandez-Abalos, J. J. Calvete, and V. L. Shnyrov.** 1998. Thermodynamic stability of two variants of xylanase (Xys1) from *Streptomyces halstedii* JM8. *Eur. J. Biochem.* **253:**462–468.

98. **Schlochtermeier, A., S. Walter, J. Schröder, M. Moormann, and H. Schrempf.** 1992. The gene encoding the cellulase (Avicelase) Cel1 from *Streptomyces reticuli* and analysis of protein domains. *Mol. Microbiol.* **6:**3611–3621.

99. **Schlösser, A., and H. Schrempf.** 1996. A lipid-anchored binding protein is a component of an ATP-dependent cellobiose-triose transport system from the cellulose degrader *Streptomyces reticuli*. *Eur. J. Biochem.* **242:**332–338.

100. **Schrempf, H., P. Dyson, W. Dittrich, M. Betzler, C. Habiger, B. Mahro, V. Brönneke, A. Kessler, and H. Düvel.** 1989. Genetic instability in *Streptomyces*, p. 145–150. *In* Y. Okami, T. Beppu, and H. Ogawara (ed.), *Biology of Actinomycetes '88*. Scientific Press, Tokyo.

101. **Schwedock, J., J. R. McCormick, E. R. Angert, J. R. Nodwell, and R. Losick.** 1997. Assembly of the cell division protein FtsZ into ladder-like structures in the aer-

ial hyphae of *Streptomyces coelicolor*. *Mol. Microbiol.* **25:** 858.

102. **Servin-Gonzalez, L.** 1993. Relationship between the replication functions of *Streptomyces* plasmids pJV1 and pIJ101. *Plasmid* **30:**131–140.

103. **Servín-González, L., C. Castro, C. Pérez, M. Rubio, and F. Valdez.** 1997. *bldA*-dependent expression of the *Streptomyces exfoliatus* M11 lipase gene (*lipA*) is mediated by the product of a contiguous gene, *lipR*, encoding a putative transcriptional activator. *J. Bacteriol.* **179:**7816–7826.

104. **Sezonov, G., A.-M. Duchêne, A. Friedmann, M. Guérineau, and J.-L. Pernodet.** 1998. Replicase, excisionase, and integrase genes of the *Streptomyces* element pSAM2 constitute an operon positively regulated by the *pra* gene. *J. Bacteriol.* **180:**3056–3061.

105. **Sommer, P., C. Bormann, and F. Götz.** 1997. Genetic and biochemical characterization of a new extracellular lipase from *Streptomyces cinnamomeus*. *Appl. Environ. Microbiol.* **63:**3553–3560.

106. **Stindl, A., and U. Keller.** 1994. Epimerization of the D-valine portion in the biosynthesis of actinomycin D. *Biochemistry* **33:**9358–9364.

107. **Taguchi, S., T. Endo, Y. Naoi, and H. Momose.** 1995. Molecular cloning and sequence analysis of a gene encoding an extracellular serine protease from *Streptomyces lividans* 66. *Biosci. Biotechnol. Biochem.* **59:**1386–1388.

108. **Tsujibo, H., T. Ohtsuki, T. Iio, I. Yamazaki, K. Miyamoto, M. Sugiyama, and Y. Inamori.** 1997. Cloning and sequence analysis of genes encoding xylanases and acetyl xylan esterase from *Streptomyces thermoviolaceus* OPC-520. *Appl. Environ. Microbiol.* **63:**661–664.

109. **Vara, J., M. Lewandowska-Skarbek, Y.-G. Wang, S. Donadio, and C. R. Hutchinson.** 1989. Cloning of genes governing the deoxysugar portion of the erythromycin biosynthesis pathway in *Saccharopolyspora erythraea* (*Streptomyces erythreus*). *J. Bacteriol.* **171:**5872–5881.

110. **Virolle, M. J., and M. J. Bibb.** 1988. Cloning, characterization and regulation of an alpha-amylase gene from *Streptomyces limosus*. *Mol. Microbiol.* **2:**197–208.

111. **Volff, J. N., and J. Altenbuchner.** 1997. High-frequency transposition of IS*1373*, the insertion sequence delimiting the amplifiable element *AUD2* of *Streptomyces lividans*. *J. Bacteriol.* **179:**5639–5642.

112. **Walter, S., E. Wellmann, and H. Schrempf.** 1998. The cell wall-anchored *Streptomyces reticuli* Avicel-binding protein (AbpS) and its gene. *J. Bacteriol.* **180:**1647–1654.

113. **Weaden, J., and P. Dyson.** 1998. Transposon mutagenesis with IS*6100* in the avermectin-producer *Streptomyces avermitilis*. *Microbiology* **144:**1963–1970.

114. **Wellington, E. M. H., E. Stackebrandt, D. Sanders, J. Wolstrup, and N. O. G. Jorgensen.** 1992. Taxonomic status of *Kitasatosporia*, and proposed unification with *Streptomyces* on the basis of phenotypic and 16S rRNA analysis and emendation of *Streptomyces* Waksman and Henrici 1943^{AL}. *Int. J. Syst. Bacteriol.* **42:**156–160.

115. **Yuan, W. M., and D. L. Crawford.** 1995. Characterization of *Streptomyces lydicus* WYEC108 as a potential biocontrol agent against fungal root and seed rots. *Appl. Environ. Microbiol.* **61:**3119–3128.

116. **Zakrzewska-Czerwinska, J., and H. Schrempf.** 1992. Characterization of an autonomously replicating region from the *Streptomyces lividans* chromosome. *J. Bacteriol.* **174:**2688–2693.

117. **Zotchev, S. B., and H. Schrempf.** 1994. The linear *Streptomyces* plasmid BL1: analyses of transfer functions. *Mol. Gen. Genet.* **242:**374–382.

Recombinant DNA Applications in Thermophiles

VOLKER MAI AND JUERGEN WIEGEL

41

Thermophilic microorganisms are characterized by their ability to grow at temperatures above 60°C with temperature optima above 55°C. Extreme thermophiles have a temperature optimum above 65°C, and hyperthermophiles can grow optimally at temperatures above 80°C. According to the above definition, true thermophiles have been described only in the prokaryotic world, and they include aerobic and anaerobic archaea as well as bacteria.

Over the past 15 years, the number of isolated and validly published thermophiles, especially extreme thermophiles and hyperthermophiles, has increased from less than five to over 500 species. Nevertheless, many characteristics involved in determining thermophilic growth of microorganisms or thermostability of proteins remain elusive (24). With respect to genetic manipulation in thermophiles, it may be important that some membrane characteristics, such as lower length of fatty acid chains and higher degree of substitution, seem to be conserved. These membrane characteristics may interfere with traditional protocols of molecular procedures such as transformation and DNA isolation. The previously hypothesized high mol% G+C content as a means of stabilizing DNA does not hold true, as thermophiles with low mol% G+C content have also been described, such as *Clostridium thermohydrosulfuricum* (32 mol% G+C [26]), *Thermosipho africanus* (30 mol% G+C [20]), and *Methanococcus jannaschii* (32 mol% G+C [59]).

Thermophiles are ubiquitous and can be isolated from naturally hot habitats such as hot springs, volcanic areas, and deep sea vents, as well as from various mesophilic environments (30, 63). Isolation methods and optimized growth conditions have been developed for many organisms. The reader interested in the principles for growth of thermophiles and hyperthermophiles is referred to reference 61 and chapter 10 in this volume.

Besides basic interest in their sometimes unique physiology and their place in evolution, thermophiles are investigated as a likely source for various industrial and biotechnological applications. The potential of thermophiles in biotechnology has been reviewed extensively (3, 34, 65) and is beyond the scope of this chapter. The unique temperature stability of enzymes from thermophilic microorganisms is often accompanied by other desired enzyme characteristics, such as wide pH range and solvent resistance (3). Genetic engineering of potential industrial microorganisms is an important tool in manipulating these microorganisms toward higher profitability. Industrial thermophilic fermentations might have advantages over mesophilic processes, such as energy-efficient product recovery, avoidance of cooling costs, wide substrate range, and possibly lower likelihood of contamination. Recombinant DNA polymerases from various thermophiles are being developed for the improvement of specific PCR applications. Thermotolerant polymer-degrading enzymes such as cellulases, hemicellulases, amylases, proteases, lipases, etc., are used in the food, chemical (detergent), and pulp and paper industries. These are just a few examples illustrating the need for the improvement of genetic systems presently available for thermophiles.

Many genes from thermophiles have been cloned and expressed successfully in mesophilic hosts. However, the development of complex genetic systems allowing for the efficient genetic manipulation of thermophiles is still in its infancy. This is not surprising if one considers the short amount of time many thermophiles have been available (e.g., *Pyrococcus furiosus*, the first bacterium able to grow above 100°C, was described in 1986 [15]) and how this compares to the huge efforts by many research groups in the development of the genetics in *Escherichia coli* or *Bacillus subtilis* over the last 50 years. Very recently, progress has been made in the development of shuttle vectors for various thermophiles as well as in the expression of heterologous genes in thermophilic hosts.

Advances in genetics of *Thermus thermophilus*, *Bacillus stearothermophilus*, Gram-type (60, 66) positive anaerobic thermophiles, and thermophilic archaea will be elaborated on in some detail below. (The term Gram-type is used to differentiate between the use of the term to describe the phylogenetic/taxonomic position and to report the Gram-staining reaction.) Transformation efficiencies are generally lower in thermophiles as compared to mesophiles with well-established genetic systems. Additionally, difficulties are often encountered with the level of heterologous gene expression and the isolation of plasmid DNA from thermophiles. Selectable markers are often nonfunctional at elevated temperatures, and many antibiotics are not stable at more extreme growth conditions (46). To develop new genetic systems and expand existing ones, these and other difficulties must be circumvented by careful and innovative experimental design.

41.1. CLONING OF THERMOPHILIC GENES

A large variety of genes with thermophilic origin have been cloned and expressed in *E. coli*. In most instances molecular techniques established for *E. coli* (6) can be adapted for the use in thermophiles. Thorough knowledge of the growth requirements of the DNA donor organism is essential for obtaining sufficient amounts of DNA/RNA for cloning. After complete lysis, genetic material can be purified by conventional methods and used for further restriction and ligation at ambient temperatures. Expression problems due to codon usage, misfolding, or intracellular accumulation have rarely been encountered, and these problems do not appear to be any different from problems with the expression of heterologous mesophilic genes in *E. coli*. One advantage of expressing thermophilic genes in *E. coli* is the ease of purification of the recombinant product. Single heat-treatment steps of 30 min at 60 to 70°C and above have been used to obtain highly enriched enzyme. However, when using organic buffers such as Tris-HCl, one must consider temperature-induced pH changes. Expression cloning of thermophilic genes is hampered by the high temperature optima of their gene products. Incubation of replica plates at higher temperatures for detecting enzyme activities is one option for circumventing this problem. Alternative thermophilic cloning hosts could potentially be useful but require efficient transformation systems in addition to well-defined genetic backgrounds. Complementation of auxotrophic mutants allows for the selection of specific gene products. Again, mutagenesis methods in thermophiles do not vary in principle from established techniques, and quite a few auxotrophic mutants are available for various thermophiles.

41.2. TRANSFER AND EXPRESSION OF DNA IN THERMOPHILES

41.2.1. General Considerations

41.2.1.1. Stability of Commonly Used Antibiotics at Elevated Temperatures

The development of genetic systems for thermophiles depends on the use of shuttle vectors carrying thermostable selectable markers. Besides complementation of auxotrophic mutants, antibiotic resistance markers frequently have been used for selection of transformants. However, because of their instability at elevated temperatures, certain antibiotics must be treated with caution if used at temperatures above 50°C. The thermostability of antibiotics varies in different media and at different pH values (46). High concentrations of calcium and lactose have been reported to destabilize certain antibiotics at elevated temperatures (68). Table 1 gives examples of the stability of antibiotics at 50 and 72°C in prereduced clostridial media (pH 7.3) as determined by Peteranderl et al. (46).

The above data indicate that inactivation of an antibiotic has to be considered in the development of a selective system in each specific background. In general, significant deviation from neutral pH seems to negatively affect the stability of an antibiotic. In some instances, elevated temperature not only failed to cause inactivation of antibiotics but, to the contrary, resulted in more potent products, e.g., chloramphenicol at 50°C and neomycin at 72°C.

41.2.1.2. Stability of Selective Markers at Elevated Temperatures

Expression of selective markers developed in mesophiles is often ineffective at temperatures above 50°C. The easiest way to circumvent this problem is to select for transformants at lower temperatures. Most thermophiles, except the hyperthermophiles, can be cultured at acceptable growth rates at temperatures around and below 45°C. Establishing temperature curves, especially for thermophiles with extended temperature spans and biphasic temperature curves (62), might thus prove an important prerequisite for the development of a genetic system for thermophiles. For some of these thermophiles, antibiotic susceptibility has been reported to exhibit temperature dependence (62), a property that can be used to advantage in isolating mutants, including temperature-sensitive mutants at both ends of the temperature spans.

Various antibiotic resistance markers conferring resistance to kanamycin, tetracycline, and chloramphenicol have been shown to function in various backgrounds at temperatures of 60°C and above (21, 68). In many instances, antibiotics with bacteriocidal activity can be successfully used at the elevated temperature even if they are not stable for a long time at those temperatures. Mutagenesis screens for thermostable variants of mesophilic markers can be used to develop new selective markers.

Examples of molecular techniques used in various groups of thermophilic microorganisms are given below. We have purposely excluded "thermophilic" eucaryotes.

41.2.1.3. Available Vectors

Useful vectors that have been described for the representative microorganisms discussed below are listed in Table 2.

41.2.2. Gram-Type-Negative Aerobic Thermophiles

Gram-type-negative thermophiles are best represented by *Thermus* species. *T. thermophilus*, a widespread bacterium in thermobiotic environments, is capable of growing at temperatures up to 85°C. The optimal temperature (T_{opt}) at pH_{opt} 7.5 is 72°C (44). A variety of genes have been cloned by conventional methods and successfully expressed in *E. coli*. Expression problems in *E. coli*, thought to be caused by the high G+C content of *T. thermophilus*, can be circumvented in various ways. For instance, the expression of a 3-isopropylmalate dehydrogenase can be achieved by elimination of a palindromic structure around the ribosome binding site or the introduction of a leader open reading frame into the upstream flanking region of the gene (22). For the *leuB* gene, the addition of a leader open reading fame appears more crucial for the efficient expression than the use of a potent promoter (23). The protease aqualysine from *Thermus aquaticus* is expressed in its soluble form only by an *E. coli* strain lacking an F′ episome. This soluble protein can be processed into the active protease by heat treatment (48). However, the aqualysine from *T. aquaticus* is correctly processed and secreted into the culture medium by *T. thermophilus* harboring an expression plasmid for the aqualysine gene (53).

Various auxotrophic *T. thermophilus* HB27 strains (Pro^-, Leu^-, Met^-, Trp^-) are available as genetic backgrounds for complementation (28). Natural transformation of *T. thermophilus* was first accomplished by complementation of auxotrophic strains with genomic DNA from the prototro-

TABLE 1 Stability of antibiotics in prereduced clostridial media at pH 7.3[a]

Antibiotic	$j_{1/2}$ (h) at 50°C	$t_{1/2}$ (h) at 72°C
Neomycin, kanamycin, monensin, trimethoprim	>150	>150
Polymyxin, bacitracin, chloramphenicol	>72	>26
Erythromycin	>26	>72
Streptomycin, vancomycin	>48	<24
Penicillin G, tetracycline[b]	>26	<24
Ampicillin	<24	<24

[a]See Peteranderl et al. (46) for values at pH 5 and detailed conditions.
[b]Tetracycline has been shown to be stable at 70°C in LB medium by Wu et al. (68).

phic parental strain (29). Competence was shown throughout the growth phase, and efficiency is improved by the addition of basal salts, 0.35 mM Ca^{2+}, or 0.4 mM Mg^{2+} to the growth medium. Transfer of a cryptic plasmid (pTT8) without selection has been shown by colony hybridization. A shuttle vector, pYK109, was constructed by ligation of the *trpB* gene into pTT8 (28). A selectable cloning vector was constructed by random insertion of a heterologous gene encoding a mutagenized thermostable nucleotidyltransferase from pUB110 into the cryptic multicopy plasmid pTT8 (38). Thermophilic origins of replication from cryptic *T. thermophilus* plasmids can be selected for by cloning randomly digested plasmid DNA into a pUC19-based vector in *E. coli*, followed by kanamycin selection of transformed *T. thermophilus* (57). Three novel thermophilic origins were obtained, and the smallest *ori*-containing sequence (4.2 kb) was refined further (2.3 kb). An open reading frame of 341 amino acids was found within the *ori* and postulated to encode a replication protein necessary for thermophilic plasmid replication. A new *T. thermophilus–E. coli* shuttle integration vector, pINV, was developed recently (52). The vector consists of *E. coli* plasmid pBluescript and a 2.1-kb segment of the *T. thermophilus leu* operon. The bactericidal compound 5-fluoroorotic acid can be used to isolate mutants in the orotate phosphoribosyltransferase gene. The vector can be integrated into the expected sites, and kanamycin resistance can be expressed in the thermophile. Promoter activities in *T. thermophilus* can be tested with the use of the promoterless kanamycine resistance gene on vector pPP11. Maseda and Hoshino (37) classified promoters into three groups and suggested a possible consensus promoter sequence in *T. thermophilus*.

41.2.2.1. Transformation

The following protocol from *T. thermophilus* can be used for transformation with chromosomal or plasmid DNA and should generally work with other *Thermus* spp. and related bacteria. Transformation efficiencies of up to 10^7 transformants per μg of plasmid DNA can be expected with this method. Total competence of *T. thermophilus* in any growth phase has been postulated (18), and efficiency was improved by the addition of basal salts, 0.35 mM Ca^{2+}, or 0.4 mM Mg^{2+} to the growth medium.

PROTOCOL

Transformation of *Thermus* spp.

1. Grow *T. thermophilus* cells overnight at 70°C in TM broth medium (0.4% polypeptone, 0.2% yeast extract, 0.1% NaCl, and basal salts, pH 7.5).

2. Dilute 1:100 into fresh TM medium and incubate with shaking for 2 h at 70°C.

3. Add up to 5 μg of plasmid DNA and incubate with shaking for 1 h at 70°C.

TABLE 2 Available vectors

Organism	Vector
Gram-type-negative thermophiles	pYK109 (28)
	Promoter probe vector pPP11 (37)
	Integration vector pINV (52)
	Thermophilic *ori* selecting vector (57)
Gram-type-positive thermophilic facultative anaerobes	Shuttle vector pRP9 (11)
	pTHT15; Tc (19)
	pLW05; Cm (68)
	pSTE33 and pSTK3 (40, 42)
Gram-type-positive thermophilic anaerobes	pIKM1; Km (35)
	pRPK; Cm, Km (36a)
	pCTC1; Em (67)
	pUB110; Km (51)
	pDP9-16 cryptic (58)
Hyperthermophiles	Cryptic plasmid pGT5 (14)
	Shuttle vectors pAG1/pAG2 (5)
	Intron-based pDMI1 (1)
	Phage SSV1 (45)

4. Cool on ice for 5 min, dilute with 0.9% NaCl, and plate on the appropriate selective medium.

5. DNA uptake can be stopped by incubating the mixture at 37°C for 15 min after the addition of DNase I (50 μg/ml).

41.2.2.2. Plasmid DNA Isolation

Plasmid DNA from *T. thermophilus* can be isolated as previously described for mesophilic bacteria (6) without any special treatment except for the growth temperature (generally 70°C). Commercially available plasmid kits such as Qiagen and Promega can be used if the cells can be lysed efficiently.

41.2.3. Gram-Type-Positive Thermophilic Facultative Anaerobes

Gram-type-positive thermophilic facultative anaerobes are represented by *B. stearothermophilus*, a well-studied spore-forming bacterium capable of growing at temperatures up to 75°C (above 50°C anaerobically). At pH_{opt} 7.0, the T_{opt} is around 60°C (10).

Cloning and overexpression of *B. stearothermophilus* genes, sometimes followed by single-step purification after heat treatment, has been achieved in many different hosts but mainly in *E. coli* and *B. subtilis*. Large-scale production of a cloned alanine dehydrogenase from *B. stearothermophilus* in *E. coli* can be achieved in a fermentor containing a HEPA filter as a biological barrier (47). Treatment of the cells with Triton X-100 (0.1%), lysozyme (100 mg/liter), and EDTA (10 mM) for 30 min followed by heat treatment at 60°C for 30 min leads to an approximate 98% recovery of thermostable enzyme. Enzyme yields of recombinant *B. stearothermophilus* alpha-amylase can be increased in *E. coli* by the co-overexpression of *prlF* (39). This enzyme has also been expressed and excreted in recombinant *Lactobacillus sanfrancisco* CB1 (16). Rapid enrichment of recombinant proteins in *B. subtilis* via His sub(6)-tagging vectors and metal chelate affinity chromatography has been shown for the GroES protein from *B. stearothermophilus* (50). The addition of GroES, GroEL, and ATP in vitro has been shown to improve the remaining activity of *Saccharomyces cerevisiae* alcohol dehydrogenase after heat treatment at 50°C for 6 min from 55 to 90% (27). Furthermore, effective extracellular production of *B. stearothermophilus* esterase in *Bacillus brevis* can be achieved in pH-stat modal fed-batch culture (55). This recombinant esterase has been shown to be stabilized by various sulfhydryl compounds (54).

An efficient polyethylene glycol (PEG) protoplast transformation system is available, and selective markers encoding tetracycline (pTHT15) and chloramphenicol (pLW05) resistance have been successfully expressed in *B. stearothermophilus* NUB36 (68). Vector pRP9 (2.9 kb) is based on the cryptic *Bacillus coagulans* plasmid pBC1 (11). This shuttle vector has been shown to possess all the properties required for efficient gene cloning in a variety of mesophilic as well as thermophilic hosts (12). Studies of the stability of recombinant pRP9-based plasmids in nonselective medium during continuous culture have shown that heterologous inserts did indeed decrease the structural plasmid stability (7). Furthermore, this study showed the influence of several fermentative conditions on the rate of plasmid loss and the specific growth rate in strains carrying recombinant plasmids. Plasmid stability decreases in minimal media, at higher temperature, and at lower dilution rates. The amount of dissolved oxygen does not affect plasmid maintenance. Plasmids pSTE33 and pSTK3 express a kanamycin resistance marker, and these plasmids are stably maintained at 67°C without selective pressure (42). These two potential shuttle vectors are based on a fusion between *E. coli* pUC19 and the cryptic plasmid pSTK1 from *B. stearothermophilus* and have been patented (40).

Expression of *E. coli* aspartate transcarbamylase in *B. stearothermophilus* has been demonstrated with the use of a novel oligonucleotide cassette linked to the upstream region of the *E. coli* gene (41). The oligonucleotide cassette required for efficient expression of the linked *E. coli* gene contained promoter sequences and a ribosome binding site from the host organism. Electrophoration was used to transform *B. stearothermophilus* with the recombinant plasmid. Detectable expression of the *Clostridium thermocellum celE* gene on the recombinant plasmid pHE9102 in *B. stearothermophilus* is prevented by significant structural plasmid rearrangement (4). Transposon mutagenesis has been used to generate a *B. stearothermophilus* BR219 strain producing catechol from phenol. The transposon Tn*916* can be electrophorated into *B. stearothermophilus* on a suicide vector and integrated into the resident plasmid pGG01 (43). Further genetic engineering of *B. stearothermophilus* could utilize two intrinsic transposable elements, IS*5376* and IS*5377*, described for *B. stearothermophilus* CU21 (69).

41.2.3.1. Transformation

PROTOCOL

PEG-Mediated Protoplast Transformation (68)

1. Harvest *B. stearothermophilus* cells grown in Luria-Bertani (LB) medium at 60°C until late exponential phase by centrifugation at 1,900 × *g* for 5 min at room temperature. Resuspend cells to a density of 3×10^9 cells/ml in modified P-medium (LB medium containing 10% [wt/vol] lactose, 20 mM $CaCl_2$, and 10 mm $MgCl_2$).

2. Convert cells to protoplast by incubation with lysozyme on a gyratory shaker at 130 rpm for 10 min at 50°C. Dilute protoplasts in 5 l of modified P-medium and centrifuge for 7 min at 800 × *g*. Gently resuspend protoplasts in original suspension volume (3×10^9 cells/ml).

3. Mix 5 to 20 μl (up to 5 μg) of plasmid DNA with 0.1 ml of protoplast suspension and add 0.9 ml of freshly prepared 40% (wt/vol) PEG 6000 in modified P-medium. Gently shake the mixture for 2 min at 50°C on a gyratory shaker (130 rpm). Dilute with 2.5 ml of modified P-medium and centrifuge for 7 min at 800 × *g* (room temperature). Gently resuspend protoplasts in 0.1 ml of modified P-medium.

4. Allow for phenotypic expression by incubating on a gyratory shaker at 130 rpm for 60 min at 50°C.

5. Regenerate the protoplasts by plating diluted protoplast suspension (in modified P-medium) on plates of R medium (P-medium containing 0.8% agar) containing the appropriate antibiotic. Incubate for 12 h at 50°C and transfer to 60°C for an additional 24 to 48 h. Transfer recombinant colonies immediately to LB medium containing the appropriate antibiotic.

Transformation efficiencies of up to 4×10^8 transformants per μg of plasmid DNA can be expected with this method.

Electroporation (56)

Successful electrotransformation has been shown for a variety of Gram-type positive bacteria. However, electroporation efficiencies are generally 3 to 4 orders lower than those reported for Gram-type negative bacteria. Important parameters such as the stabilization of the cell wall with isotonic buffers, growth phase of the cells, and ionic strength of the electrocompetent cell have to be optimized for each strain. The following protocol was initially developed for *Bacillus amyloliquefaciens* (56) but can be used as a general guide for adapting the procedure for similar thermophilic Gram-type positive aerobic bacteria.

1. Harvest cells grown in rich medium containing 270 mM sucrose and 250 mM potassium phosphate at an OD_{600} of < 1.0 via centrifugation at either room temperature or 4°C. Keep cells and buffers at the chosen temperature for the remainder of the procedure.
2. Wash cells three times with one-fourth starting volume of washing buffer: 270 mM sucrose, 1 mM HEPES, 1 mM magnesium chloride, 10% (vol/vol) glycerol.
3. Resuspend cells in 1/200 starting volume in washing buffer.
4. Use immediately or freeze at −80°C. (Cells to be frozen should be kept at 4°C throughout the procedure.)
5. Electroporate in a prechilled cuvette, and after the electroshock immediately dilute cells 1:10 in rich medium containing 10% (vol/vol) glycerol.
6. Allow for phenotypic expression by incubating on a gyratory shaker at 130 rpm for 60 min at 50°C.

41.2.3.2. Plasmid DNA Isolation

B. stearothermophilus should be grown for plasmid isolation at 50°C in LB medium containing the appropriate antibiotic. Plasmids can be isolated by the alkaline lysis method (6). Purification of the plasmid DNA can be achieved on cesium chloride density gradients or via various commercial kits.

41.2.4. Gram-Type Positive Thermophilic Anaerobes

Gram-type positive thermophilic anaerobes are represented by several thermophiles from various genera belonging to the *Bacillus–Clostridium* eubacterial branch. Several thermophilic anaerobic bacteria, including *Thermoanaerobacter*, *Thermoanaerobacterium*, and *Moorella* species are regarded as microorganisms with a high potential for biotechnological applications. These microorganisms have an optimal growth temperature between 60 and 70°C, and some can grow at nearly 80°C (63). They are certainly interesting organisms as a source of thermostable enzymes, such as hemicellulases and xylose isomerases, as well as being considered a prospective host organism for the industrial production of ethanol or the alternative road deicer Ca-Mg-acetate from renewable resources (65). The development of economically feasible industrial processes, however, is hampered by the lack of any developed genetic system for these organisms. Below, the beginning of a genetic system for *Thermoanaerobacterium* sp. and related species is described.

41.2.4.1. Transformation

Members of the genus *Thermoanaerobacterium* are widely distributed Gram-type positive anaerobic thermophiles (64). Several enzymes from this organism have been isolated and characterized, and various genes have been cloned and sequenced (31, 32, 33). Interesting features of this organism include a complex hemicellulolytic system as well as a proposed attachment mechanism of extracellular enzymes to the S-layer via conserved S-layer specific domains (32). A genetic system for *Thermoanaerobacterium* sp. strain JW/SL-YS485 is currently being developed. Various vectors (pIKM1-pIKM5, pRKM1) have been developed to express heterologous genes in this background (35, 36, 36a).

PROTOCOL

Electrotransformation of *Thermoanaerobacterium*

1. Grow *Thermoanaerobacterium* sp. strain JW/SL-YS485 cells in four Hungate tubes with 10 ml of prereduced mineral medium at 60°C. Add isonicotinic acid hydrazide (isoniacin) to cultures in early exponential phase at a final concentration of 4 $\mu g\ ml^{-1}$ to weaken the cell walls. After the addition of isoniacin, cultures are allowed to grow for an additional two to four doubling times until they reach an OD of 0.6 to 0.8.
2. Harvest cells via centrifugation in closed Hungate tubes (4,000 × *g*; IEC centra-8 centrifuge) at room temperature and wash in one-half original volume of sterile prereduced (nitrogen-flushed, 1 $\mu mol\ ml^{-1}$ Na_2S added) water. Repeat once.
3. After the second centrifugation step, resuspend into 270 mM sucrose and incubate at 48°C until the beginning of autoplast (spheroplast) formation can be observed by using light microscopy. Centrifuge in closed Hungate tubes and resuspend in 0.2 ml of either N_2-flushed water or 270 mM sucrose.
4. Transform via electroporation using a Bio-Rad gene pulser. Transfer 0.1 ml of resuspended cells into prechilled (4°C) 0.1-cm electroporation cuvettes containing up to 5 μg of plasmid DNA. Mix by shaking the cuvette. After a single electroporation pulse (1.25 kV, 400 Ω, 25 mF) with a time constant of 4 to 8 ms, transfer immediately into Hungate tubes with 5 ml of prereduced mineral medium. The cells are allowed to recover at 48°C for 4 h.
5. Suspend dilutions of the electroporation mixture into selective medium containing 50 $\mu g\ ml^{-1}$ kanamycin, and spread aliquots onto plates containing agar (1% [wt/vol])-solidified selective medium.
6. Incubate in anaerobic jars containing oxygen-free nitrogen as gas atmosphere.

Electrotransformation of *Thermoanaerobacterium* (formerly *Clostridium*) *thermosaccharolyticum* (25)

1. Harvest 100 ml of cells at an OD_{600} of ~1 via centrifugation at 10,000 × *g* for 10 min (4°C).
2. Wash three times in ice-cold sterile deionized water, and resuspend in 0.5 ml of water containing 20% glycerol.
3. Use immediately for electroporation or store at −80°C.
4. Add up to 5 μl of plasmid DNA and electroporate in a prechilled 0.2-cm cuvette at 2.0 kV, 800 Ω, 25 mF resulting in a time constant of 16 ms.

5. Allow for phenotypic expression by incubating for 1 h at 45°C, and spread dilutions onto selective plates.

41.2.4.2. Plasmid DNA Isolation

PROTOCOL

Plasmid DNA Isolation from *Thermoanaerobacterium* sp. (35)

1. Transfer single colonies of transformed *Thermoanaerobacterium* sp. strain JW/SL-YS485 into 10 ml of selective medium (100 μg ml^{-1} kanamycin). Transfer 2 ml into serum bottles containing 100 ml of the selective medium (75 μg ml^{-1} kanamycin) and incubate at 50°C for 16 to 24 h.

2. Harvest 6 to 8 ml of cells in 8–10 1.5-ml Eppendorf tubes by centrifuging repeatedly at maximum speed for 30–60 s.

3. Resuspend pellets in 0.2 ml of 25% sucrose containing 20 mg/ml lysozyme and incubate for 30 min at 37°C.

4. Add 0.4 ml of freshly made 3% sodium dodecyl sulfate, 0.2 N NaOH and mix by inverting the tubes. Incubate at room temperature for 5 to 7 min.

5. Add 0.3 ml of ice-cold 3 M sodium acetate (pH 4.8), mix, and spin for 15 min at 4°C.

6. Transfer supernatants to new tubes and add 0.65 ml of isopropanol, mix, and spin for 15 min at 4°C.

7. Remove liquid and resuspend pellet in 0.32 ml of sterile deionized water. Add 0.2 ml of 7.5 mM ammonium acetate containing 0.5 mg/ml ethidium bromide. Add 0.35 ml of phenol/chloroform/isoamylalchol (24:24:1). Mix well and spin for 5 min at room temperature.

8. Transfer upper phases to new tubes and add 1 ml of ethanol (−20°C), mix, and spin for 15 min at 4°C.

9. Wash pellet in 70% ethanol, allow to dry for 10 to 15 min, and resuspend pellets in 40 μl of TE containing 0.1 mg/ml RNase.

It is crucial for this procedure to resuspend the pellet properly, especially in step 7. Let the pellet dry for 5 to 10 min to allow for easier resuspension. Ethidium bromide is added in the phenol extraction step to break up DNA-protein interactions.

41.2.5. Hyperthermophiles

Hyperthermophiles are microorganisms capable of growth at or above 90°C. Members of this unique group include mostly archaea, but also a few bacterial species such as the anaerobic *Thermotoga* and *Aquifex*. Hyperthermophiles are at the center of interest for biotechnological applications for the extreme stability of their enzymes, e.g., in extreme cases with half-lives up to several days at above 120°C. A variety of genes from this group of microorganisms have been cloned and expressed in *E. coli* and other organisms (9). In most cases, enzymes can be readily expressed in the mesophile, and the recombinant enzymes exhibit equivalent thermostability compared to the native protein. Because of their thermostable characteristics, enzymes from extreme thermophiles and hyperthermophiles are easily purified by heat treatment. Multiple attempts have been made to develop systems for the genetic manipulation of extremophiles. Plasmids have been isolated and described for *Pyrococcales* (14), *Thermotogales* (17), and *Sulfolobales* (13) species. Recent advances in the genetic systems have been reviewed (1, 70). Shuttle vectors for hyperthermophilic archaea have been developed (5). The alcohol dehydrogenase from *Sulfolobus acidocaldaricus* is available as a selectable marker conferring resistance to butanol and benzyl alcohol. Future antibiotic resistance markers might include mutated 23S rRNA genes that have been shown to mediate chloramphenicol, carbomycin, and celesticetin (2).

41.2.5.1. Transformation/Transfection

PROTOCOL

Transformation of *S. acidocaldarius* (5)

1. Grow culture to an OD_{600} of 0.4 and wash three times in one-tenth original volume of 20 mM sucrose.

2. Add up to 1 μg of methylated plasmid DNA to cells resuspended in 100 μl of sucrose (20 mM) and leave on ice for 20 min.

3. Transform via electroporation using a gene pulser. Transfer 0.1 ml of resuspended cells into prechilled (4°C) 0.1-cm electroporation cuvettes. After a single electroporation pulse (1.25 kV, 400 Ω, 25 mF), leave on ice for 20 min.

4. Transfer to selective medium and incubate at 70°C.

Transformation of *P. furiosus* (1)

1. Harvest exponentially growing cultures by centrifugation.

2. Resuspend in growth medium and add 80 mM $CaCl_2$.

3. Flush with nitrogen and store anaerobically at 4°C for 30 min.

4. Add plasmid DNA and leave at 4°C for 1 h.

5. Heat-pulse cells for 3 min at 80°C and leave on ice for 10 min.

6. Transfer to growth medium and incubate at 95°C.

This procedure can also be used for other extremophiles. The heat-pulse procedure should be optimized for each microorganism.

Transfection of *Sulfolobus* sp. (49)

Phage SSV1 has been shown to have potential in the development of a genetic system for members of the *Sulfolobales*. A plaque assay has been established for *Sulfolobus solfataricus*, and transformation conditions have been optimized (49).

1. Grow two overnight cultures, reserving one for use as an indicator lawn.

2. Harvest the cells after cooling by centrifugation (perform all steps at 4°C).

3. Resuspend in equal volume of 20 mM sucrose, centrifuge, and wash twice in one-half the original volume and then in one-fiftieth the original volume. Resuspend in 100 to 500 μl of 20 mM sucrose (10^{10} cells/ml).

4. Add 1 μl of SSV1 DNA to 50 μl of the cell suspension and transfer to a prechilled electroporation cuvette.

5. Electroporate the cells with the gene pulser set at 15 kV/cm, 400 Ω, and 25 μF. Immediately add 1 ml of growth medium containing sucrose and incubate for 30 to 60 min.

6. Mix aliquots with 500 μl of 10-fold-concentrated mid-log cells (indicator lawn), pour in overlay, and incubate the plates in a moist atmosphere. Transfection efficiencies of 10^6 transfectants per μg of DNA and transfection frequencies of up to 10^{-4} transfectants per surviving cell have been reported for this method in *S. solfataricus*.

41.2.5.2. Plasmid DNA Isolation

DNA-protein interactions are thought to cause insufficient plasmid yields by conventional plasmid preparation methods. However, if efficient lysis of the cells can be followed by an effective disruption of the DNA-protein interactions, conventional commercial kits for the isolation of plasmid DNA can be employed.

PROTOCOL

Plasmid DNA Isolation from Archaea (8)

1. Harvest the cells by centrifugation 10 min at 500 × g (4°C).

2. Resuspend in buffer (pH 7 to 7.5). If required for efficient lysis, add up to 10 mg/ml lysozyme and incubate for 30 min at 37°C.

3. Add one-eighth volume of 10% *N*-lauroyl sarcosine and invert slowly several times.

4. Add one-eight volume of 10% sodium dodecyl sulfate and invert slowly several times.

5. Add one-sixteenth volume of 20 mg/ml proteinase K and invert gently; incubate up to 3 h at 50°C.

After this step the lysate should be viscous but translucent. Alternatively, cells can often be lysed by addition of NaOH or KOH (6). Proceed by separating the protein from DNA and purifying the plasmid DNA.

41.3.4. CONCLUSION

Throughout the last decade, interest in the unique properties of thermophilic and hyperthermophilic bacteria and archaea has led to a wealth of knowledge about these special microorganisms. Molecular tools have been modified to meet specific needs, new methods have been developed, and detailed genetic studies can now be performed in several thermophilic microorganisms. Numerous genetic systems for these organisms have been developed or are under development. Various thermophiles have been included in genomic sequencing projects, and available genome sequences will have an immense effect on the way we look at and work with thermophiles. The potential of biotechnological applications appears to increase with the amount of information accumulated from these unique microorganisms. However, for constantly described novel and other prospective microorganisms, modifications and optimizations of previously established methods are required for success. To do this effectively, comprehensive knowledge of the physiology of each microorganism is a prerequisite. Such information includes, but is not limited to, the determination of growth characteristics, elucidation of cell wall properties, and investigations into metabolic capabilities. The methods described here have shown that it is possible to apply genetic tools to the thermophilic extremophiles and that it should be possible in principle to further develop these systems to the level of today's genetics in *E. coli* and *B. subtilis*. Since some anaerobic thermophiles have even shorter doubling times than *E. coli* (e.g., *Thermobrachium celere*, t_d 10 min), it is conceivable that from the large number of novel microorganisms a (hyper)thermophilic pendant to the *E. coli* system will emerge in the future.

REFERENCES

1. **Aagaard, C., I. Leviev, N. A. Rajagopal, P. Forterre, D. Prieur, and R. A. Garrett.** 1996. General vectors for archaeal hyperthermophiles: strategies based on a mobile intron and a plasmid. *FEMS Microbiol. Rev.* **18:**93–104.
2. **Aagaard, C., H. P. Phan, S. Trevisanto, and R. A. Garrett.** 1994. A spontaneous point mutation in the single 23S rRNA gene of the thermophilic archaeon *Sulfolobus acidocaldarius* confers multiple drug resistance. *J. Bacteriol.* **176:**7744–7747.
3. **Aguilar, A.** 1996. Extremeophile research in the European Union: from fundamental aspects to industrial expectations. *FEMS Microbiol. Rev.* **18:**89–92.
4. **Aminov, R. I., N. P. Golovchenko, and K. Ohmiya.** 1995. Expression of a *celE* gene from *Clostridium thermocellum* in *Bacillus*. *J. Ferment. Bioeng.* **79:**530–537.
5. **Aravalli, R. N., and R. G. Garrett.** 1997. Shuttle vectors for hyperthermophilic archaea. *Extremophiles* **1:**183–191.
6. **Ausubel, F. M., R. Brent, R. E. Kingston, D. D. Moore, J. G. Seidman, J. A. Smith, and K. Struhl.** 1995. *Current Protocols in Molecular Biology*. John Wiley & Sons, Inc., New York.
7. **Brigidi, P., A. Gonzales-Vara, M. Rossi, and D. Matteuzzi.** 1997. Study of stability of predominant plasmids during continuous culture of *Bacillus stearothermophilus* NUB3621 in nonselective medium. *Biotechnol. Bioeng.* **53:**507–514.
8. **Charbonnier, F., P. Forterre, G. Erauso, and D. Prieur.** 1995. Purification of plasmids from thermophilic and hyperthermophilic archaea. *In* F. T. Robb and A. R. Place (ed.), *Archaea: A Laboratory Manual*. Cold Spring Harbor Laboratory, Cold Spring Harbor, New York.
9. **Ciamarella, M., R. Cannio, M. Moracci, F. M. Pisani, and M. Rossi.** 1995. Molecular biology of extemophiles. *World J. Microbiol. Biotechnol.* **11:**71–84.
10. **Claus, D., and R. C. W. Berkeley.** 1986. Genus *Bacillus* Cohn 1872, 174[AL], p. 1105. *In* P. H. A. Sneath, N. S. Mair, M. E. Sharpe, and J. G. Holt (ed.), *Bergey's Manual of Systematic Bacteriology*, vol. 2. Williams & Wilkins, Baltimore.
11. **De Rossi, E., P. Brigidi, M. Rossi, D. Matteuzzi, and G. Riccardi.** 1991. Characterization of Gram-positive broad host range plasmids carrying a thermophilic origin. *Res. Microbiol.* **142:**389–396.
12. **De Rossi, E., P. Brigidi, N. E. Welker, G. Riccardi, and D. Matteuzzi.** 1994. New shuttle vector for cloning in *Bacillus stearothermophilu*. *Res. Microbiol.* **145:**579–583.
13. **Elferink, M. G. L., C. Schleper, and W. S. O. Zillig.** 1996. Transformation of the extremely thermoacidophilic archaeon *Sulfolobus solfataricus* via a self-spreading vector. *FEMS Microbiol. Lett.* **137:**31–35.
14. **Erauso, G., S. Marsin, N. Benbouzid-Rollet, M. F. Baucher, T. Barbeyron, Y. Zivanovic, D. Prieur, and P. Forterre.** 1996. Sequence of plasmid pGT5 from the archaeon *Pyrococcus abyssi*: evidence for rolling-circle replication in a hyperthermophile. *J. Bacteriol.* **178:**3232–3237.
15. **Fiala, G., and K. O. Stetter.** 1986. *Pyrococcus furiosus* sp. nov. representing a novel genus of marine heterotrophic

archaebacteria growing optimally at 100°C. *Arch. Microbiol.* **145:**56.

16. **Gobbetti, M., A. Corsetti, L. Morelli, and M. Elli.** 1996. Expression of alpha-amylase from *Bacillus stearothermophilus* in *Lactobacillus sanfrancisco*. *Biotechnol. Lett.* **18:**969–974.
17. **Harriott, O. T., R. Huber, K. O. Stetter, P. W. Betts, and K. M. Noll.** 1994. A cryptic miniplasmid from the hyperthermophilic bacterium *Thermotoga* sp. strain RQ7. *J. Bacteriol.* **176:**2759–2762.
18. **Hidaka, Y., M. Hasegawa, T. Nkahara, and T. Hoshino.** 1994. The entire population of *Thermus thermophilus* cells is always competent at any growth phase. *Biosci. Biotechnol. Biochem.* **58:**1338–1339.
19. **Hoshino, I., T. Ikeda, H. Narushima, and N. Tomizuka.** 1985. Isolation and characterization of antibiotic resistance plasmids in thermophilic bacilli. *Can. J. Microbiol.* **31:**339–345.
20. **Huber, R., C. R. Woese, T. A. Langworthy, H. Fricke, and K. O. Stetter.** 1989. *Thermosipho africanus* gen. nov., represents a new genus of thermophilic eubacteria within the *Thermotogales*. *Syst. Appl. Microbiol.* **12:**32.
21. **Imanaka, T., M. Fuji, J. Aramori, and S. Aiba.** 1983. Transformation of *Bacillus stearothermophilus* with plasmid DNA and characterization of shuttle vector plasmids between *Bacillus stearothermophilus* and *Bacillus subtilis*. *J. Bacteriol.* **149:**824–830.
22. **Ishida, M., and T. Oshima.** 1994. Overexpression of genes of an extreme thermophile, *Thermus thermophilus*, in *E. coli* cells. *J. Bacteriol.* **176:**2767–2770.
23. **Ishida, M., and T. Oshima.** 1996. A leader open reading frame is essential for the expression in *E. coli* of GC-rich *leuB* gene of an extreme thermophile, *Thermus thermophilus*. *FEMS Microbiol. Lett.* **135:**137–142.
24. **Jaenicke, R.** 1996. Protein-folding and association. *Curr. Top. Cell Regul.* **34:**209–314.
25. **Klapatch, T. R., M. L. Guerinot, and L. R. Lynd.** 1996. Electrotransformation of *Clostridium thermosaccharolyticum*. *J. Ind. Microbiol.* **16:**342–347.
26. **Klaushofer, H., and E. Parkkinen.** 1965. Zur Frage der Bedeutung aerober und anaerober therm. Sporebildner als Infectionsurache in Ruebenzuckerfabriken. I. *Clostridium thermohydrosulfuricum*, eine neue Art eines saccharoseabbauenden, thermophilen, schwefelwasserstoffbildenden *Clostridiums*. *Z. Zuckerind. Boehm.* **15:**445.
27. **Kohda, K., Y. Tsuji, M. Takagi, and T. Imanaka.** 1996. Cloning and functional expression of molecular chaperone genes from *Bacillus stearothermophilus* SIC1. *Biotechnol. Lett.* **18:**1061–1066.
28. **Koyama, Y., Y. Arikawa, and K. Furukawa.** 1990. A plasmid vector for an extreme thermophile, *Thermus thermophilus*. *FEMS Microbiol. Lett.* **72:**97–102.
29. **Koyama, Y., T. Hoshino, N. Tomizuka, and K. Furukawa.** 1986. Genetic transformation of the extreme thermophile *Thermus thermophilus* and of other *Thermus* spp. *J. Bacteriol.* **166:**338–340.
30. **Kristjansson, J. K., and K. O. Stetter.** 1992. Thermophilic bacteria, p. 1–18. *In* J. K. Kristjansson (ed.), *Thermophilic Bacteria*. CRC Press, Inc., Boca Raton, Fla.
31. **Lee, Y. E., M. V. Ramesh, and J. G. Zeikus.** 1993. Cloning, sequencing and biochemical characterization of xylose isomerase from *Thermoanaerobacterium saccharolyticum* strain B6A-RI. *J. Gen. Microbiol.* **139:**1227–1234.
32. **Liu, Y.-S., F. C. Gherardini, M. Matuschek, H. Bahl, and J. Wiegel.** 1996. Cloning, sequencing, and expression of the gene encoding a large S-layer-associated endoxylanase from *Thermoanaerobacterium* sp. strain JW/SL-YS 485 in *Escherichia coli*. *J. Bacteriol.* **178:**1539–1547.
33. **Lorenz, W. W., and J. Wiegel.** 1997. Isolation, analysis and expression of two genes from *Thermoanaerobacterium* sp. strain JW/SL-YS485: a beta-xylosidase and a novel xylan esterase with cephalosporin C deacetylase activity. *J. Bacteriol.* **179:**5436–5441.
34. **Lowe, S. E., M. K. Jain, and J. G. Zeikus.** 1993. Biology, ecology, and biotechnological applications of anaerobic bacteria adapted to environmental stresses in temperature, pH, salinity, or substrates. *Microbiol. Rev.* **57:**451–509.
35. **Mai, V., W. W. Lorenz, and J. Wiegel.** 1997. Transformation of *Thermoanaerobacterium* sp. strain JS/SL-YS485 with plasmid pIKM1 conferring kanamycin resistance. *FEMS Microbiol. Lett.* **148:**163–167.
36. **Mai, V., and J. Wiegel.** 1998. Expression of mannanases and cellulases in *Thermoanaerobacterium*, abstr. O-63. Abstr. 98th Annu. Meet. Am. Soc. Microbiol. 1998. American Society for Microbiology, Washington, D.C.

36a. **Mai, V., and J. Wiegel.** Unpublished results.

37. **Maseda, H., and T. Hoshino.** 1995. Screening and analysis of DNA fragments that show promoter activities in *Thermus thermophilus*. *FEMS Microbiol. Lett.* **128:**127–134.
38. **Mather, M. W., and J. A. Fee.** 1992. Development of plasmid cloning vectors for *Thermus thermophilus* HB8: expression of a heterologous, plasmid borne kanamycin nucleotidyltransferase gene. *Appl. Environ. Microbiol.* **58:** 421–425.
39. **Minas, W., and J. E. Bailey.** 1995. Co-overexpression of *prlF* increases cell viability and enzyme yields in recombinant *E. coli* expressing *B. stearothermophilus* alpha-amylase. *Biotechnol. Prog.* **11:**403–411.
40. **Nagayama, N., and S. Nakamoto.** 1995. *Bacillus stearothermophilus* and *E. coli* plasmids. U.S. patent 5384258.
41. **Narumi, I., K. Sawakami, T. Kimura, S. Nakamoto, N. Nakayama, T. Yanagisawa, N. Takahashi, and H. Kihara.** 1992. A novel oligonucleotide cassette for the overproduction of *E. coli* aspartate transcarbamylase *Bacillus stearothermophilus* and *E. coli* plasmids. *Biotechnol. Lett.* **14:** 759–764.
42. **Narumi, I., N. Nakayama, S. Nakamoto, T. Kimura, T. Yanagisawa, and H. Kihara.** 1993. Construction of a new shuttle vector PSTE33 and its stabilities in *Bacillus stearothermophilus*, *B. subtilis* and *E. coli*. *Biotechnol. Lett.* **15:** 815–820.
43. **Natarajan, M. R., and P. Orie.** 1992. Production of catechol by a *Bacillus stearothermophilus* transpositional mutant. *Biotechnol. Prog.* **8:**78–80.
44. **Oshima, T., and K. Imahori.** 1974. Description of *Thermus thermophilus* (Yoshida and Oshima) comb. nov., a nonsporulating thermophilic bacterium from a Japanese thermal spa. *Int. J. Syst. Bacteriol.* **24:**102.
45. **Palm, P., C. Schleper, B. Grampp, S. Yeats, P. McWilliam, W. D. Reiter, and W. Zillig.** 1991. Complete nucleotide sequence of the virus SSV1 of the archaebacterium *Sulfolobus shihatae*. *Virology* **185:**242–250.

46. **Peteranderl, P., E. B. Shotts, Jr., and J. Wiegel.** 1990. Stability of antibiotics at growth conditions of thermophilic anaerobes. *Appl. Environ. Microbiol.* **56:**1981–1983.

47. **Sakamoto, Y., H. Nakajima, K. Nagata, N. Esaki, and K. Soda.** 1994. Large-scale production of thermostable alanine dehydrogenase from recombinant cells. *J. Ferment. Bioeng.* **78:**84–87.

48. **Sakamoto, S., I. Terada, M. Iijima, T. Ohta, and H. Matsuzawa.** 1995. Expression of aqualysin I (a thermophilic protease) in soluble form in *E. coli* under bacteriophage T7 promoter. *Biosci. Biotechnol. Biochem.* **59:**1438–1443.

49. **Schleper, C., and W. Zillig.** 1995. Transfection of *Sulfolobus solfataricus*. *In* F. T. Robb and A. R. Place (ed.), *Archaea: A Laboratory Manual*. Cold Spring Harbor Laboratory, Cold Spring Harbor, N.Y.

50. **Schoen, U., and W. Schumann.** 1994. Construction of His sub(6)-tagging vectors allowing single step purification of GroES and other polypeptides produced in *Bacillus subtilis*. *Gene* **147:**91–94.

51. **Soutschek-Bauer, E., L. Hartl, and W. L. Staudenbauer.** 1985. Transformation of *Clostridium thermohydrosulfuricum* DSM 568 with plasmid DNA. *Biotechnol. Lett.* **7:**705–710.

52. **Tamakoshi, M., M. Uchida, K. Tanabe, S. Fukuyama, A. Yamakashi, and T. Oshima.** 1997. A new *Thermus–E. coli* shuttle integration vector system. *J. Bacteriol.* **179:** 4811–4814.

53. **Touhara, N., H. Taguchi, Y. Koyama, T. Ohta, and H. Matsuzawa.** 1991. Production and extracellular secretion of aqualysin I (a thermophilic subtilisin-type protease) in a host-vector system for *Thermus thermophilus*. *Appl. Environ. Microbiol.* **57:**3385–3387.

54. **Tulin, E. E., Y. Amaki, T. Nagasawa, and T. Yamane.** 1993. A *Bacillus stearothermophilus* esterase produced by a recombinant *Bacillus brevis* stabilized by sulfhydryl compounds. *Biosci. Biotechnol. Biochem.* **57:**856–857.

55. **Tulin, E. E., S. Ueda, H. Yamagata, S. Ugaka, and T. Yamane.** 1992. Effective extracellular production of *Bacillus stearothermophilus* esterase by pH-stst modal fed-batch culture of recombinant *Bacillus brevis*. *Biotechnol. Bioeng.* **40:**844–850.

56. **Vehmaanperä, J.** 1989. Transformation of *B. amyloliquefaciens* by electroporation. *FEMS Microbiol. Lett.* **61:**165–170.

57. **Wayne, J., and S. Y. Xu.** 1997. Identification of a thermophilic plasmid origin and its cloning within a new *Thermus–E. coli* shuttle vector. *Gene* **195:**321–328.

58. **Weimer, P. J., L. W. Wagner, S. Knowlton, and T. G. Ng.** 1984. Thermophilic bacteria which ferment hemicellulose: characterization of organisms and identification of plasmids. *Arch. Microbiol.* **138:**31–36.

59. **Whitman, W. B.** 1989. Order II *Methanococcales* Balch and Wolfe 1981, p. 2185. *In* J. T. Staley, M. P. Bryant, N. Pfennig, and J. G. Holt (ed.), *Bergey's Manual of Systematic Bacteriology*, vol. 3. Williams & Wilkins, Baltimore.

60. **Wiegel, J.** 1981. Distinction between the Gram resection and the Gram type of bacteria. *Int. J. Syst. Bacteriol.* **31:** 88.

61. **Wiegel, J.** 1986. Methods for isolation and study of thermophiles, p. 17–37. *In* T. D. Brock (ed.), *Thermophiles: General, Molecular and Applied Microbiology*. John Wiley & Sons, New York.

62. **Wiegel, J.** 1990. Temperature spans for growth: a hypothesis and discussion. *FEMS Microbiol. Rev.* **75:**155–170.

63. **Wiegel, J.** 1992. The anaerobic thermophilic bacteria, p. 105–184. *In* J. K. Kristjansson (ed.), *Thermophilic Bacteria*. CRC Press, Inc., Boca Raton, Fla.

64. **Wiegel, J.** Unpublished results.

65. **Wiegel, J., and L. G. Ljungdahl.** 1986. The importance of thermophilic bacteria in biotechnology. *CRS Rev. Biotechnol.* **3:**39–107.

66. **Wiegel, J., and L. Quandt.** 1982. Determination of the Gram type using the reaction between polymyxin B and lipopolysaccharides of the outer cell wall of whole bacteria. *J. Gen. Microbiol.* **128:**2261–2270.

67. **Williams, D. R., D. I. Young, and M. Young.** 1990. Conjugative plasmid transfer from *E. coli* to *C. acetobutylicum*. *J. Gen. Microbiol.* **136:**819–826.

68. **Wu, L., and N. E. Welker.** 1989. Protoplast transformation of *Bacillus stearothermophilus* NUB36 by plasmid DNA. *J. Gen. Microbiol.* **135:**1315–1324.

69. **Xu, K., Z.-Q. He, Y.-M. Mao, R.-Q. Sheng, and Z.-J. Sheng.** 1993. On two transposable elements from *Bacillus stearothermophilus*. *Plasmid* **29:**1–9.

70. **Zillig, W., D. Prankishvilli, C. Schleper, M. Elferink, I. Holz, S. Albers, D. Janekovic, and D. Gotz.** 1996. Viruses, plasmids and other genetic elements of thermophilic and hyperthermophilic archaea. *FEMS Microbiol. Rev.* **18:** 225–236.

Zygosaccharomyces rouxii

YASUJI OSHIMA, HIROYUKI ARAKI, HARUHIKO MORI, AND KOHEI USHIO

42

An osmotolerant or halotolerant yeast, *Zygosaccharomyces rouxii*, is important in the soy sauce (shoyu) and miso (fermented soy bean paste) industries (10, 26). The genus *Zygosaccharomyces* also includes various yeasts that spoil foods of high sugar content, such as honey (17, 24), syrup (28), and marzipan (5). These yeasts have a haplotic life cycle, which means that they preferentially spend their vegetative life in a haploid phase and sporulate after going through meiosis immediately following zygote formation between two cells of opposite mating types. Because of the industrial importance of these yeasts, a number of studies on their physiology, in particular on their salt and sugar tolerance, have been performed, but few genetic studies of *Z. rouxii* have been carried out. In 1960, however, Wickerham and Burton (41) reported on the occurrence of heterothallism in *Z. rouxii*. (At that time, this yeast was erroneously classified in the genus *Saccharomyces*.) The occurrence of heterothallism in *Z. rouxii* was confirmed with strains isolated independently from various samples of soy sauce mash and miso paste collected from several factories in Japan (22). Most of the isolates (97 of 105) of this yeast species showed heterothallism and were classified into two complementary mating types. The remaining 8 clones failed to mate with any heterothallic strains. In contrast, some of the sugar-tolerant clones were found to be homothallic (23). With these heterothallic strains, general methodologies for cultivation, mating, sporulation (21, 22), and mutagenesis (20, 34, 38, 42, 43) of *Z. rouxii* were developed.

With an increased interest in recombinant DNA technology, considerable attention has been paid to detection of plasmids in various yeasts other than *Saccharomyces cerevisiae*. It was found that several species of circular cryptic DNA plasmids are distributed in some *Zygosaccharomyces* strains (31, 32). During a series of comparative studies of a plasmid in *Z. rouxii*, pSR1, and the 2μm DNA plasmid of *S. cerevisiae* (1–4, 15, 16), a host-vector system of *Z. rouxii* was developed (38). Here we describe these methods as applied to *Z. rouxii*.

42.1. LIFE CYCLE OF *Z. ROUXII*

The morphological characteristics of vegetative cells of haploid, diploid, triploid, and tetraploid phases, zygotes, and zygotic and azygotic asci of *Z. rouxii* are shown in Fig. 1. The vegetative cells isolated from natural habitats are preferentially haploid (Fig. 1A). The most characteristic feature in the life cycle of *Z. rouxii* is that meiosis immediately follows conjugation, resulting in direct conversion of the zygotes (Fig. 1B) to zygotic asci (Fig. 1D). Since most, if not all, halotolerant *Z. rouxii* strains are heterothallic, zygotic asci are formed by mixing cells of two haploid strains of opposite mating types, arbitrarily named **a** and α (21), or by inoculating cells of a single homothallic strain (23) onto a sporulation plate. Zygotes are also obtained by interrupting meiosis at the zygote stage on a specially designed mating plate. When zygotes are transferred into nutrient medium supplemented with 5% NaCl, or into fresh liquid mating medium, large diploid cells are occasionally formed at one or both ends of a zygote (Fig. 1E), but never at the part connecting the two cells (21). More often, haploid buds, showing the same size as the haploid cells, are produced from the zygotes in a manner similar to that described above, suggesting that there is a long heterokaryotic phase in a zygote of *Z. rouxii*. These observations are in contrast to the extremely short period of the heterokaryotic phase and the formation of diploid buds at the conjugation site in the typical diploid yeast *S. cerevisiae*. The large diploid cells formed on zygotes begin vegetative growth by budding in a standing liquid culture at 30°C (Fig. 1C). Thus, it is possible to maintain the vegetative diploid phase of *Z. rouxii* cells by preventing them from undergoing meiosis. When the diploid cells are placed on a sporulation plate, numerous azygotic asci (Fig. 1F) are produced within 3 days of incubation at 25°C. The ascospores germinate in nutrient medium and give rise to haploid vegetative cells.

Diploids showing **a** or α mating type, possibly due to homozygosity of a mating-type allele generated by mitotic recombination, occur spontaneously at substantially high frequencies in cell populations of the above **a**/α diploid cultures (21). Putative triploid vegetative cells are obtained by mating the above putative **a**/**a** or α/α diploid clones with a haploid clone having the opposite mating type, similar to the manner in which diploids are obtained, and the azygotic asci are formed under the same conditions as for meiosis of the diploid cells (Fig. 1G). Putative tetraploid vegetative cells (Fig. 1H) are derived from a cross between putative **a**/**a** and α/α diploids, and they form azygotic asci.

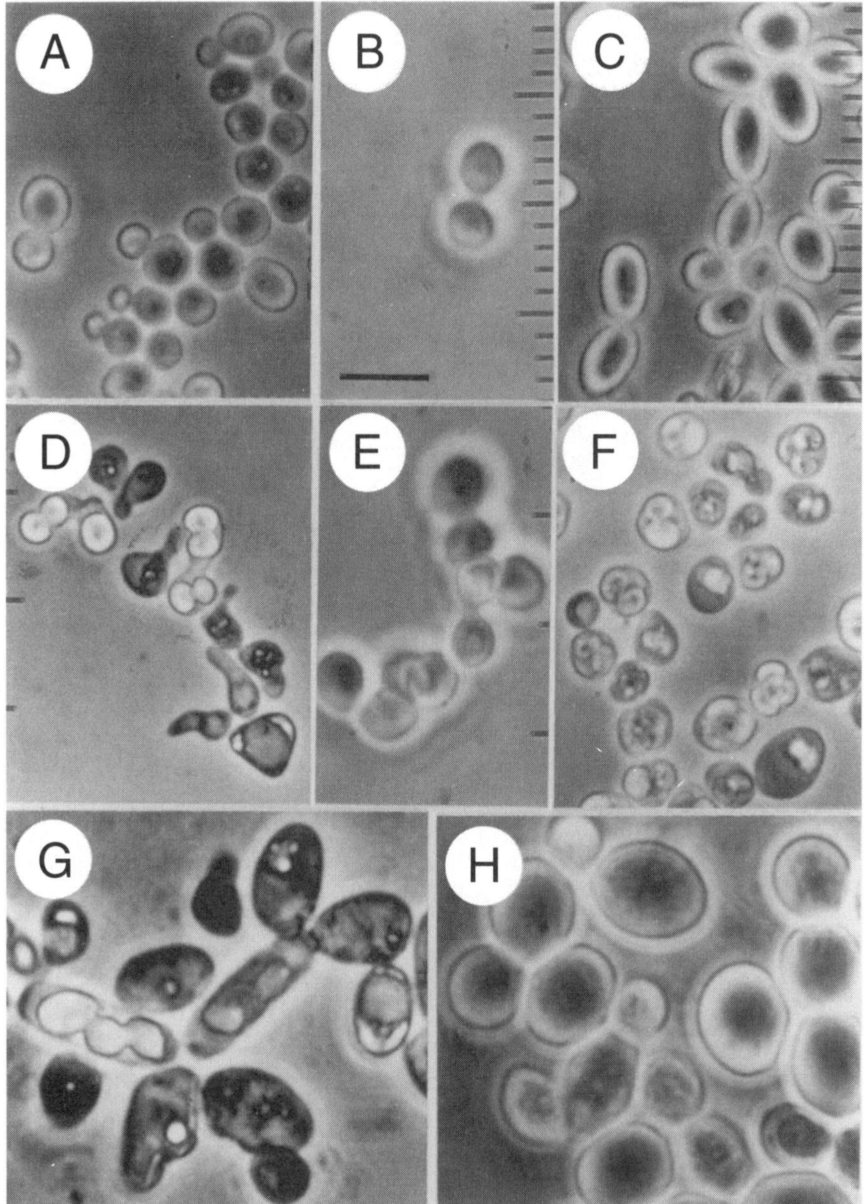

FIGURE 1 Morphological characteristics of vegetative cells of haploid to tetraploid phases, zygotes, zygotic and azygotic asci of *Z. rouxii*. (A) Haploid vegetative cells. (B) A zygote formed by mating haploid **a** and α cells. (C) Diploid vegetative cells. (D) Zygotic asci formed by mixing haploid cells of **a** and α mating types. (E) A zygote sprouting diploid buds. (F) Azygotic asci of a diploid clone. (G) Triploid cells and azygotic asci. (H) Tetraploid vegetative cells. Photomicrographs were taken at the same magnification. Bar (panel B), 10 μm.

42.2. STRAINS

The majority of our genetic work with *Z. rouxii* was performed using a pair of the heterothallic strains, NRRL 2547 (= ATCC 14679; **a**-mating type, prototrophic and [cir°], i.e., harboring no plasmid) and NRRL 2548 (= ATCC 14680; α-mating type, prototrophic and [cir°]), which were isolated from miso paste (41), and their mutants and pedigrees from various crosses between them (20, 34, 38). These two complementary mating types, **a** and α, were arbitrarily assigned by Mori (21) independently of those for *S. cerevisiae*. Several other *Z. rouxii* strains commonly used in genetic studies and as cloning hosts are IFO 1130 (**a** mating type, prototrophic and [pSR1], i.e., harboring plasmid pSR1 [32, 38]), ME3 (α-mating type marked with *leu*$^-$, which is complemented by the *LEU2* gene of *S. cerevisiae*, and [pSR1] [4, 11, 12, 15, 16, 38]), LST11 (α *leu*$^-$ [pSR1]) with the same genotype as ME3 but exhibiting a higher frequency of transformation (38), and MA11 (**a** *leu*$^-$ [cir°]) (1, 13, 42, 43).

42.3. MEDIA AND CULTIVATION METHODS

Nutrient-rich YPD medium containing 20 g of glucose, 20 g of polypeptone, and 10 g of yeast extract per liter and minimal medium consisting of 6.7 g of Yeast Nitrogen Base without amino acids (Difco) and 20 g of glucose per liter are used for general cultivation of *Z. rouxii*. These are the same media (pH 5.0 to 5.2) used for *S. cerevisiae*, but appropriate amounts of NaCl are added for optimal growth of *Z. rouxii* (see below). Auxotrophic markers can be scored using the minimal medium with addition of appropriate nutrients. Solid media are prepared by adding agar to a final concentration of 2%. Haploid strains are preserved on YPD agar at 4°C or in suspension in sterilized glycerol solution (15%) at −80°C.

For sporulation of *Z. rouxii*, use of YM agar (Difco) plates with or without addition of 2% NaCl, Gorodkowa agar, or cornmeal agar has been recommended (6). It is also said that use of potato-dextrose agar supplemented with 5 to 10% NaCl is effective for sporulation of *Zygosaccharomyces* sp. It is, however, impossible to interrupt the sporulation at the zygote stage on these plates. The sporulation media recommended for *S. cerevisiae*, such as sodium acetate agar, are unsuitable for *Z. rouxii*. The most effective conditions determined to date for sporulation by formation of zygotic asci are accomplished by mixing of cells of two haploid strains of opposite mating types on a plate containing diluted soy-koji extract agar added with 5% NaCl (sporulation plate) and incubation of the plates at 25°C for 4 to 10 days (21). Interruption of the meiotic process at the zygote stage is also possible by mixing cells of two haploid heterothallic strains of opposite mating types on plates containing diluted soy-koji extract agar added with 5% each NaCl and glucose (mating plate) and incubation of the plates at 25°C for 2 days. The zygotes of *Z. rouxii* formed on the mating plates are transformed to zygotic asci within a few days of their being smeared on sporulation plates and incubation of the plates at 25°C. It should be noted that the incubation temperature for general cultivation of *Z. rouxii* is 30°C, while mating and sporulation of this yeast are achieved preferably at 25°C.

Hybridization and diploid construction between two haploid **a** and α strains with complementary auxotrophic markers are performed more easily by mixing the cells on a mating plate than that in the prototrophic strains (38). To increase the rate of zygote formation, it is recommended to mix the cells again, thoroughly with sterile loop, on the mating plate after 2 days of incubation at 25°C, and continuation of the incubation for an additional 2 to 3 days. A loopful of cells should then be transferred from the mixed culture into 2 ml of sterilized water containing 2% glucose and the culture left to stand overnight at 25°C. The cells should then be washed and spotted on a minimal plate with a sterile loop for selection of prototrophs. Colonies appearing on the plates are isolated as diploids after incubation of the plates for several days at 30°C. The isolates are confirmed as hybrid diploids by their larger cell size than the parental haploid cells, their ability to sporulate, and segregation of the mutant traits in the spore-clones. Diploids are cultivated and preserved on YPD agar supplemented with 18% NaCl to prevent them from sporulating.

For tetrad analysis, both the zygotic and the azygotic asci are dissected under a microscope with a micromanipulator as for *S. cerevisiae*. Before dissection, asci are, in general, treated with Zymolyase (type 100T, Seikagaku-Kogyo, Tokyo, Japan; 30μg per ml) in 0.1 M KH_2PO_4-K_2HPO_4 buffer (pH 7.5) containing 25 mM EDTA, 50 mM dithiothreitol, and 0.8 M sorbitol at 30°C for at least 10 min (34). Nutrient YPD plates used for spore germination, but not for the dissection, should be supplemented with 0.8 M sorbitol.

PROTOCOL

PREPARATION OF SOY-KOJI EXTRACT (22)

For laboratory-scale preparation of soy-koji, mix 55 g of defatted soybeans and 45 g of parched and cracked wheat thoroughly with 50 ml of tap water. After the mixture is moistened at room temperature for about 0.5 h, steam it in an autoclave at 120°C for 30 min. Then cool the material to room temperature and inoculate with soy-koji seeds (conidia of *Aspergillus oryzae* or *Aspergillus sojae*; commercially available in Japan), and incubate the inoculated material at 30°C for 3 days. The material should be mixed thoroughly once per day during the incubation. Place a cotton pouch containing the resultant mold culture, i.e., soy-koji, in warm (45°C) water, equivalent to four times the weight of soy-koji used, for about 5 h. Next, boil the extract for 20 min after adding egg white (ca. 1 egg equivalent amount per liter of the extract) and filter the extract through a filter paper (Whatman no. 1). The clear soy-koji extract generally contains 2.0 to 2.5% reducing sugar as glucose and 0.6 to 0.7% total nitrogen.

For preparation of the mating and sporulation plates, the above soy-koji extract should be diluted approximately 10-fold to reduce the concentrations of reducing sugar and total nitrogen to less than 0.25% and 0.06%, respectively, and supplemented with 5% NaCl and 2 to 2.5% agar with 5% glucose (for mating plates) or without glucose (for sporulation plates). Adjust the pH of the mixture to 5.0 to 5.2 before sterilization.

42.4. MUTAGENESIS AND MUTANT ISOLATION

In the initial attempts to isolate *Z. rouxii* mutants, it was found that it was difficult to isolate auxotrophic mutants from vegetative haploid cells by mutagenesis with ethyl methanesulfonate or *N*-methyl-*N′*-nitro-*N*-nitrosoguanidine (MNNG) (20). To overcome this difficulty, a rather laborious method involving MNNG mutagenesis using ascospores was developed (20). It is, however, now possible to isolate various types of mutants from haploid vegetative cells using MNNG as the mutagen (34, 38, 42, 43). Cells grown in 5 ml of YPD medium with shaking at 30°C for 24 h are washed once with sterilized water. Then the cells are suspended in 1 ml of 0.1 M K-phosphate buffer (pH 7.5) containing 0.5 to 1 mg of MNNG per ml and the suspension is shaken at 30°C for 15 to 120 min. After being washed three times with sterilized water, the treated cells are suspended in sterilized water and a few drops of the suspension are inoculated into several tubes, each containing 2 ml of fresh YPD medium, and the subcultures are shaken at 30°C for 2 days. This step is important for fixing mutations in haploid vegetative cells of *Z. rouxii*. The cells of each subculture are then spread on YPD plates after appropriate dilution, and the plates are incubated at 30°C for 3 to 4 days. The colonies on the plates are replicated onto various test plates, e.g., to select auxotrophic mutants, in a way similar to the methods for *S. cerevisiae* (29) or selecting other mutants (e.g., see references 34, 35, and 37). To avoid isolation of identical mutants, only one mutant for a certain phenotype should be isolated from each of the subcultures.

42.5. TRANSFORMATION

42.5.1. Plasmids

Most *S. cerevisiae* strains of independent origin harbor a 2μm DNA plasmid, while DNA plasmids are rarely detected in other yeasts. Only six species of circular double-stranded DNA plasmids are known to be present in some strains of the genera *Zygosaccharomyces* (2, 31–33, 39, 40) and *Kluyveromyces* (8). Among them, four species of circular DNA plasmids, pSR1 and pSB3 in *Z. rouxii* (or *Zygosaccharomyces bisporus*) and pSB1 and pSB2 in *Zygosaccharomyces bailii*, have been isolated. All four plasmids resemble 2μm DNA of *S. cerevisiae* in that (i) they are about 6 kb in size, (ii) each molecule possesses a pair of inverted repeats, (iii) they exist as a mixture of two isomers, and (iv) their copy numbers in the native host are similar. None of them, however, show nucleotide sequence similarity to 2μm DNA or to each other as determined by Southern hybridization under moderately stringent conditions (31), or by comparison of their nucleotide sequences. (Note that the nucleotide sequence of pSB1 has not yet been determined.)

The plasmid pSR1 has been studied most extensively in terms of architecture (2), functions of the proteins encoded by the open reading frames (15) and of a *cis*-acting site (16), mechanisms of stable maintenance of the plasmid in the host cell (1), structure of the autonomously replicating sequences (ARSs) (4), gene conversion at a site in the inverted repeats (19), and the mechanism of site-specific recombination (3). These studies revealed striking similarities of pSR1 with 2μm DNA in terms of architecture and mechanisms of the functional regions. It is worth noting that the pSR1 ARS is functional in *S. cerevisiae* hosts (4), while that of 2μm DNA is not functional in *Z. rouxii* hosts (2, 38).

42.5.2. Host-Vector System

For convenience in recombinant DNA experiments with *Z. rouxii* hosts, strains having improved efficiency of transformation were constructed from the wild-type strains NRRL 2547 (**a**; ATCC 14679) and NRRL 2548 (α; ATCC 14680), by mutagenesis and selection of meiotic segregants from crosses among them (34, 38). The only genetic markers currently available for vectors for effective selection of *Z. rouxii* transformants are the *leu*$^-$ marker (which is complemented with the *LEU2* DNA of *S. cerevisiae*) and sensitivity to G418 (38). Vectors were ligated with a *S. cerevisiae* DNA fragment bearing the *LEU2* gene, prepared from YEp13 (27) of the *S. cerevisiae* vector as a 2.2-kb *Sal*I-*Xho*I fragment. A representative of this type of cloning vector is pKU24 (Fig. 2) (36), which is replicated with the pSR1 ARS. Wild-type strains of *Z. rouxii* are, in general, sensitive to 100 μg of G418 per ml of YPD medium (38). In contrast, transformants harboring a YRp or YEp type plasmid with an ARS of a *S. cerevisiae* chromosome or of pSR1, ligated with an 8.9-kb *Eco*RI fragment bearing the Tn*601* sequence prepared from pAJ50 (18), are resistant to 1,000 μg of G418 per ml of YPD. pAT143 (Fig. 2) (38), which is replicated with the pSR1 ARS, and pKU71 (Fig. 2) (1, 38), which is replicated with the *S. cerevisiae* ARS1, are examples of this type of vector. pAK43 (Fig. 2) (36) can be used as a YIp vector with the G418-resistance marker in *Z. rouxii* by ligating with an appropriate *Z. rouxii* DNA fragment. The plasmid pSRT303D (Fig. 2) (16), constructed by ligation of a 4.4-kb *Sal*I fragment of YIp105 (15) bearing *amp*^{r} and Tn*601* for G418 resistance, can be used as a vector because it is very stable in *Z. rouxii* hosts. For construction of an expression vector in *Z. rouxii*, a gene, *GAP-Zr*, encoding one of two isozymes of glyceraldehyde-3-phosphate dehydrogenase in *Z. rouxii* has been cloned (11). The promoter of *GAP-Zr* was found to be effective for expressing *Escherichia coli lacZ* as a reporter gene, which is expressed at a level comparable to that obtained by using the promoters of analogous genes, *GAP-Sc* and *PHO5* of *S. cerevisiae*, in both *Z. rouxii* and *S. cerevisiae*. The *PHO5* promoter shows constitutive expression, as efficient as that of the *GAP-Zr* promoter, in the *Z. rouxii* host. With the *GAP-Zr* promoter, an alkaline protease gene of *A. oryzae* ligated on pSR1 marked with Tn*601* (G418^r), has been expressed efficiently in *Z. rouxii*, and the enzyme is secreted into the culture medium (25).

42.5.3. Practical Methods for Transformation

Methods for preparation of genomic (9, 29, 30) and plasmid (7) DNA from *S. cerevisiae* are applicable to *Z. rouxii*, with minor modifications as described below. Transformation of *Z. rouxii* is performed with protoplasts as follows (38). Cultivate host cells in 5 ml of YPD supplemented with 5 to 10% NaCl, depending on the strain used, for 3 days at 30°C. Add to the culture an equal volume of fresh YPD medium containing the same amount of NaCl, then shake the culture at 30°C until the optical density of the culture at 600 nm is 0.5. Harvest the cells by centrifugation and resuspend them in 1.5 ml of 0.8 M sorbitol solution (pH 7.5) containing 50 mM dithiothreitol and 25 mM EDTA, and incubate the suspension at 30°C for 30 min. Wash the cells twice with 2 ml each of ice-cold 0.8 M sorbitol solution and suspend them in 1.5 ml of 0.1 M potassium phosphate buffer (pH 6.5) containing 10 mM EDTA, 0.8 M sorbitol, and 0.1 mg of Zymolyase-100T. Then incubate the reaction mixture at 30°C for 45 to 60 min. Wash the resultant protoplasts twice with 2 ml each of ice-cold 0.8 M sorbitol solution and once with 1 ml of ice-cold 0.8 M sorbitol solution containing 200 to 400 mM $CaCl_2$. Then suspend the protoplasts in 0.1 ml of the same ice-cold sorbitol-$CaCl_2$ solution and keep the suspension on ice for 30 to 60 min. Add 10 to 20 μl of DNA solution containing 1 μg of plasmid DNA per μl to the protoplast suspension and keep the entire mixture on ice for 30 min. Then add 10 volumes (1 ml) of 50% PEG 1000 solution in 10 mM Tris-HCl buffer (pH 7.4) containing 200 to 400 mM $CaCl_2$. Mix the reaction mixture once and keep on ice for 30 to 60 min. Harvest the protoplasts by centrifugation at 2,000 rpm in a microfuge for 5 min and resuspend them in 0.1 to 0.3 ml of YPD containing 0.8 M sorbitol and incubate the suspension at 30°C for 30 min. Mix a portion of the protoplast suspension with 8 ml of molten (40 to 45°C) soft agar (1.5% agar) of minimal or Leu-test medium containing 0.8 M sorbitol and pour the mixture immediately onto a plate containing Leu-test medium or G418-test medium (containing 100 μg of G418 per ml) supplemented with 0.8 M sorbitol and 2% agar. Incubate the plates at 30°C for several days. With the above pro-

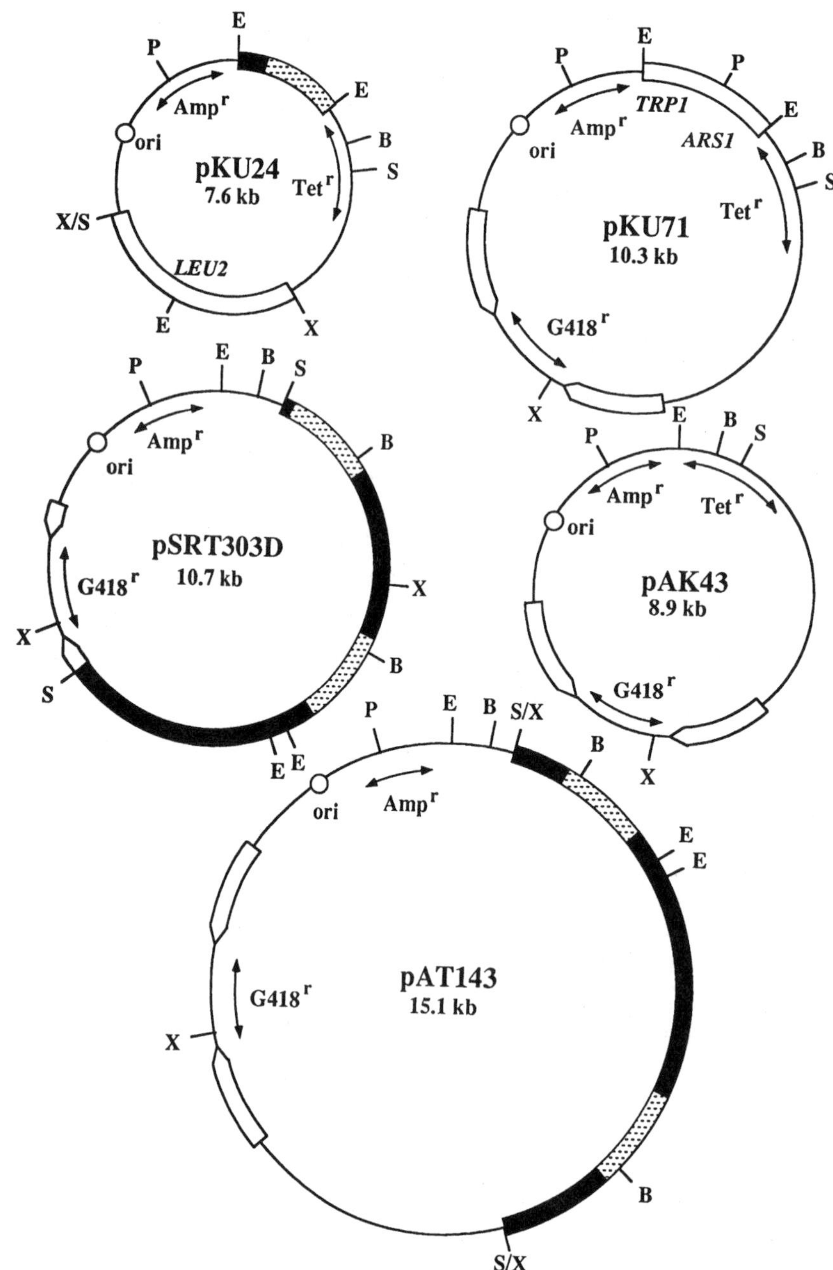

FIGURE 2 Structures of the *Z. rouxii* vectors. Thick lines indicate the pSR1 DNA and gray boxes are the regions of its inverted repeats. Open boxes represent the chromosomal DNA of *S. cerevisiae* bearing the *LEU2* or *ARS1-TRP1* fragment. Thin lines represent pBR322 DNA. Pairs of open arrows with a thin line between them indicate the Tn*601* DNA bearing the G418-resistance gene. Bidirectional arrows identify genes for antibiotic resistance. Abbreviations for restriction sites: B, *Bam*HI; E, *Eco*RI; P, *Pst*I; S, *Sal*I, X, *Xho*I. S/X indicates a ligation site between *Sal*I and *Xho*I cohesive ends.

tocol, we have achieved frequencies of 1,600 to 2,500 transformants per μg of plasmid DNA with the pSR1 ARS (38).

In contrast to the protocol for *S. cerevisiae* transformation, that for *Z. rouxii* transformation includes steps in which the cells are treated with 50 mM dithiothreitol and the protoplasts are washed and suspended in a solution of high-concentration $CaCl_2$ before and during contact with the transforming DNA. A feature of the *Z. rouxii* transformation is that an appropriate amount of NaCl in the YPD medium for host cell cultivation somewhat increases the transformation frequency, although the optimum concentration of NaCl differs for each strain. Although we do not have extensive experience in using the method of alkali cation treatment (14), which is effective for *S. cerevisiae* transformation, it has also been used for transformation in *Z. rouxii* (25).

We thank Mitsuo Okazaki and Shigetoshi Yoshikawa for their comments. Work from Y.O.'s laboratory was supported by a special fund under contract with the Science and Technology Agency of the Japanese government and by grants from the Ministry of Education, Science, Sports and Culture of Japan.

REFERENCES

1. **Araki, H., K. Awane, K. Irie, Y. Kaisho, A. Naito, and Y. Oshima.** 1993. A specific host factor binds at a *cis*-acting transcriptionally silent locus required for stability control of yeast plasmid pSR1. *Mol. Gen. Genet.* **238:** 120–128.
2. **Araki, H., A. Jearnpipatkul, H. Tatsumi, T. Sakurai, K. Ushio, T. Muta, and Y. Oshima.** 1985. Molecular and functional organization of yeast plasmid pSR1. *J. Mol. Biol.* **182:**191–203.
3. **Araki, H., N. Nakanishi, B. R. Evans, H. Matsuzaki, M. Jayaram, and Y. Oshima.** 1992. Site-specific recombinase, R, encoded by yeast plasmid pSR1. *J. Mol. Biol.* **225:**25–37.
4. **Araki, H., and Y. Oshima.** 1989. An autonomously replicating sequence of pSR1 plasmid is effective in two yeast species, *Zygosaccharomyces rouxii* and *Saccharomyces cerevisiae*. *J. Mol. Biol.* **208:**757–769.
5. **Ayres, R., K. H. Steinkraus, A. Olek, and D. Farr.** 1987. Characterization of the semi-purified membrane bound ATPase of *Zygosaccharomyces rouxii* adapted to 18% NaCl. *Int. J. Food Microbiol.* **4:**331–339.
6. **Barnett, J. A., R. W. Payne, and D. Yarrow.** 1990. *Yeasts: Characteristics and Identification*, 2nd ed. Cambridge University Press, Cambridge.
7. **Cameron, J. R., P. Philippsen, and R. W. Davis.** 1977. Analysis of chromosomal integration and deletions of yeast plasmids. *Nucleic Acids Res.* **4:**1429–1448.
8. **Chen, X. J., M. Saliola, C. Falcone, M. M. Bianchi, and H. Fukuhara.** 1986. Sequence organization of the circular plasmid pKD1 from the yeast *Kluyveromyces drosophilarum*. *Nucleic Acids Res.* **14:**4471–4481.
9. **Hereford, L., K. Fahrner, J. Woolford, Jr., M. Rosbash, and D. B. Kaback.** 1979. Isolation of yeast histone genes H2A and H2B. *Cell* **18:**1261–1271.
10. **Hesseltine, C. W.** 1983. Microbiology of oriental fermented foods. *Annu. Rev. Microbiol.* **37:**575–601.
11. **Imura, T., I. Utatsu, and A. Toh-e.** 1987. Glyceraldehyde-3-phosphate dehydrogenase genes of *Zygosaccharomyces rouxii*: the source of a promoter for a host-vector system for *Z. rouxii*. *Agric. Biol. Chem.* **51:**1641–1647.
12. **Imura, T., I. Utatsu, and A. Toh-e.** 1989. High expression vectors for a *Zygsaccharomyces rouxii* host. *Agric. Biol. Chem.* **53:**813–819.
13. **Irie, K., H. Araki, and Y. Oshima.** 1991. Mutations in a *Saccharomyces cerevisiae* host showing increased holding stability of the heterologous plasmid pSR1. Mol. Gen. Genet. **225:**257–265.
14. **Ito, H., Y. Fukuda, K. Murata, and A. Kimura.** 1983. Transformation of intact yeast cells treated with alkali cations. *J. Bacteriol.* **153:**163–168.
15. **Jearnpipatkul, A., H. Araki, and Y. Oshima.** 1987. Factors encoded by and affecting the holding stability of yeast plasmid pSR1. *Mol. Gen. Genet.* **206:**88–94.
16. **Jearnpipatkul, A., R. Hutacharoen, H. Araki, and Y. Oshima.** 1987. A *cis*-acting locus for the stable propagation of yeast plasmid pSR1. *Mol. Gen. Genet.* **207:**355–360.
17. **Jermini, M. F., O. Geiges, and W. Schmidt-Lorenz.** 1987. Detection, isolation and identification of osmotolerant yeasts from high-sugar products. *J. Food Prot.* **50:**468–472.
18. **Jimenez, A., and J. Davies.** 1980. Expression of a transposable antibiotic resistance element in *Saccharomyces*. *Nature* **287:**869–871.
19. **Matsuzaki, H., H. Araki, and Y. Oshima.** 1988. Gene conversion associated with site-specific recombination in yeast plasmid pSR1. *Mol. Cell. Biol.* **8:**955–962.
20. **Mori, H.** 1972. Induction of auxotrophic mutants in *Saccharomyces rouxii* by *N*-methyl-*N'*-nitro-*N*-nitrosoguanidine. *J. Ferment. Technol.* **50:**218–221.
21. **Mori, H.** 1973. Life cycle in a heterothallic haploid yeast, *Saccharomyces rouxii*. *J. Ferment. Technol.* **51:**379–392.
22. **Mori, H., and H. Onishi.** 1967. Diploid hybridization in a heterothallic haploid yeast, *Saccharomyces rouxii*. *Appl. Microbiol.* **15:**928–934.
23. **Mori, H., and S. Windisch.** 1982. Homothallism in sugar-tolerant *Saccharomyces rouxii*. *J. Ferment. Technol.* **60:** 157–161.
24. **Munitis, M. T., E. Cabrera, and A. Rodriguez-Navarro.** 1976. An obligate osmophilic yeast from honey. *Appl. Environ. Microbiol.* **32:**320–323.
25. **Ogawa, Y., H. Tatsumi, S. Murakami, Y. Ishida, K. Murakami, A. Masaki, H. Kawabe, H. Arimura, E. Nakano, H. Motai, and A. Toh-e.** 1990. Secretion of *Aspergillus oryzae* alkaline protease in an osmophilic yeast, *Zygosaccharomyces rouxii*. *Agric. Biol. Chem.* **54:**2521–2529.
26. **Onishi, H.** 1963. Osmophilic yeasts. *Adv. Food Res.* **12:** 53–94.
27. **Parent, S., C. M. Fenimore, and K. A. Bostian.** 1985. Vector systems for the expression, analysis and cloning of DNA sequences in *S. cerevisiae*. *Yeast* **1:**83–138.
28. **Restaino, L., S. Bills, K. Tscherneff, and L. M. Lenovich.** 1983. Growth characteristics of *Saccharomyces rouxii* isolated from chocolate syrup. *Appl. Environ. Microbiol.* **45:**1614–1621.
29. **Rose, M. D., F. Winston, and P. Hieter.** 1990. *Methods in Yeast Genetics: A Laboratory Course Manual.* Cold Spring Harbor Laboratory Press, Cold Spring Harbor, N.Y.
30. **Rothstein, R. J.** 1985. Cloning in yeast, p. 45–66. *In* D. M. Glover (ed.), *DNA Cloning: A Practical Approach*, vol. II. IRL Press, London.
31. **Toh-e, A., H. Araki, I. Utatsu, and Y. Oshima.** 1984. Plasmids resembling 2-μm DNA in the osmotolerant yeasts *Saccharomyces bailii* and *Saccharomyces bisporus*. *J. Gen. Microbiol.* **130:**2527–2534.
32. **Toh-e, A., S. Tada, and Y. Oshima.** 1982. 2-μm DNA-like plasmids in the osmophilic haploid yeast *Saccharomyces rouxii*. *J. Bacteriol.* **151:**1380–1390.
33. **Toh-e, A., and I. Utatsu.** 1985. Physical and functional structure of a yeast plasmid, pSB3, isolated from *Zygosaccharomyces bisporus*. *Nucleic Acids Res.* **13:**4267–4283.
34. **Ushio, K., and Y. Nakata.** 1989. Isolation and characterization of mutants defective in salt tolerance in *Zygosaccharomyces rouxii*. *J. Ferment. Bioeng.* **68:**165–169.
35. **Ushio, K., H. Ohtsuka, and Y. Nakata.** 1991. Lipid composition of an obligate osmophilic mutant in *Zygosaccharomyces rouxii*. *J. Ferment. Bioeng.* **72:**210–213.
36. **Ushio, K., H. Otsuka, S. Yoshikawa, G. Taguchi, M. Shimosaka, N. Mitsui, and M. Okazaki.** 1996. Cloning of the *SAT1* gene concerned with salt-tolerance of the

yeast *Zygosaccharomyces rouxii*. *J. Ferment. Bioeng.* **82:**16–21.

37. **Ushio, K., M. Sawatani, and Y. Nakata.** 1991. Effect of altered sterol composition on the salt tolerance of *Zygosaccharomyces rouxii*. *J. Ferment. Bioeng.* **71:**390–396
38. **Ushio, K., H. Tatsumi, H. Araki, A. Toh-e, and Y. Oshima.** 1988. Construction of a host-vector system in the osmophilic haploid yeast *Zygosaccharomyces rouxii*. *J. Ferment. Technol.* **66:**481–488.
39. **Utatsu, I., S. Sakamoto, T. Imura, and A. Toh-e.** 1987. Yeast plasmids resembling 2μm DNA: regional similarities and diversities at the molecular level. *J. Bacteriol.* **169:** 5537–5545.
40. **Utatsu, I., A. Utsunomiya, and A. Toh-e.** 1986. Functions encoded by the yeast plasmid pSB3 isolated from *Zygosaccharomyces rouxii* IFO 1730 (formerly *Saccharomyces bisporus* var. *mellis*). *J. Gen. Microbiol.* **132:**1359–1366.
41. **Wickerham, L. J., and K. A. Burton.** 1960. Heterothallism in *Saccharomyces rouxii*. *J. Bacteriol.* **80:**492–495.
42. **Yoshikawa, S., K. Chikara, H. Hashimoto, N. Mitsui, M. Shimosaka, and M. Okazaki.** 1995. Isolation and characterization of *Zygosaccharomyces rouxii* mutants defective in proton pumpout activity and salt tolerance. *J. Ferment. Bioeng.* **79:**6–10.
43. **Yoshikawa, S., N. Mitsui, K. Chikara, H. Hashimoto, M. Shimosaka, and M. Okazaki.** 1995. Effect of salt stress on plasma membrane permeability and lipid saturation in the salt-tolerant yeast *Zygosaccharomyces rouxii*. *J. Ferment. Bioeng.* **80:**131–135.

Reconstruction of Mammalian Nuclear Receptor Function in *Saccharomyces cerevisiae*

TAUSEEF R. BUTT AND J. DON CHEN

43

The emergence of the yeast *Saccharomyces cerevisiae* as a model eukaryote is due to its numerous advantages as an experimental organism. The doubling time of *S. cerevisiae* is 1.5 h and it grows in synthetic defined media, offering the virtues that *Escherichia coli* gave to geneticists in the middle of the 20th century when the era of modern molecular biology was launched. The classic and molecular genetic techniques make it possible to mutate virtually all of the 6,000 genes of *S. cerevisiae* and to analyze its phenotypes both before and after its complementation with homologous or heterologous genes. The molecular biology of *S. cerevisiae* has entered a new historic phase, as all of its 6,000 genes (5,885 verified open reading frames) have been sequenced (29). Developments in high-throughput sequencing technology and bioinformatics have allowed comparative genomics to emerge as a powerful tool in functional genomics (1, 50). The future challenge is to identify the function of newly discovered human genes. We believe that *S. cerevisiae* will play a pivotal role in the elucidation of human gene functions. Only a short time has passed since the *S. cerevisiae* genome database was released, and about 40% of the yeast's gene functions have been recognized (87). Thirty-one percent of the *S. cerevisiae* genes are homologous to the mammalian expression sequence tags database (87). Almost all of the mammalian genes have found functional or structural homologs in *S. cerevisiae*.

The remarkable finding that human estrogen receptor can function in a ligand-dependent fashion in *S. cerevisiae* has opened the possibility that mammalian nuclear receptor function can be studied in this yeast (55, 61). Currently, about 150 nuclear receptors have been discovered (24, 57). Ligands for most of these nuclear receptors have not been identified; hence, they are known as "orphan receptors." We predict that most (if not all) of the human cDNAs will be made public within a year, via expression sequence tag approaches (1, 50). It is reasonable to assume that with the completion of human genome sequences, the number of nuclear receptors is likely to increase to 500. With such a dramatic increase, the elucidation of their biological role will clearly lag behind. In this chapter we describe methods used to reconstruct higher eukaryotic nuclear receptor functions in *S. cerevisiae*. We describe various plasmid vector systems used to express nuclear receptors and the reporter and selection systems most widely used to analyze the receptor function. It has also become apparent that tissue- and ligand-specific responses of nuclear receptors are the function of coregulatory molecules that do not bind to ligands but do govern nuclear receptor function. We describe how to examine the receptor function in *S. cerevisiae* that is regulated by corepressors and activators.

43.1. NUCLEAR RECEPTORS AND THEIR MODE OF ACTION

The steroids, retinoids, thyroid hormones, and vitamin D_3 are small fat-soluble molecules that regulate cell differentiation, embryonic development, homeostasis, and adult organ physiology. Unlike the water-soluble ligands for cell surface receptors, these lipophilic hormones are able to penetrate through the lipid bilayer of the cell membrane and bind directly to their intracellular receptors. The nuclear hormone receptors constitute a large superfamily that regulates diverse genetic programs (4, 11, 46, 58, 75). These receptors share a common domain structure, including an N-terminal DNA-binding domain (DBD) that enables the receptors to bind to specific DNA sequences and a C-terminal ligand-binding domain (LBD) that enables the receptors to bind to specific hormones. The DBD consists of two evolutionary conserved zinc fingers that distinguish the receptors from other DNA binding proteins (6). The LBD possesses the essential capability for hormone recognition, which ensures the specificity of hormone response.

Over the past years, it has been shown that the steroid receptors form homodimers upon ligand binding, while the retinoid/thyroid hormone receptors and others form heterodimeric complexes with the retinoid X receptors (RXRs) (58). In these heterodimeric complexes, RXR was shown to help produce the appropriate DNA-binding and ligand-binding properties of the individual complexes (27, 49).

Several known human diseases have now been shown to arise as a result of genetic mutations of the nuclear hormone receptors. For example, the *DAX1* gene encodes an unusual member of the nuclear receptor superfamily that contains a nuclear receptor-like LBD but an unrelated DBD. Mutations in *DAX1* are responsible for the genetic basis of X-linked adrenal hypoplasia congenita and X-linked hypogonadotropic hypogonadism, as characterized by structural abnormalities of the adrenal glands and gonads (65, 89). Mutations in the thyroid hormone receptor b (TRβ) gene are found in patients with the dominantly inherited general resistance to thyroid hormone syndrome, characterized by high circulating levels of thyroid hormones and thyroid-stimulating hormone and by symptoms similar to those arising from thyroid hormone deficiency. A viral counterpart of the cellular thyroid hormone receptor (TR), v-*erbA*, contains several point mutations throughout the TR LBD (17, 19, 30, 71), and the oncogenic activity of v-*erbA* was shown to correlate with its abnormal transcriptional repression activity. The vitamin D_3 receptor (VDR) is associated with hypocalcemic vitamin D-resistant rickets in humans, a recessive disease characterized by high levels of 1,25-dihydroxyvitamin D_3, severe rickets, hypocalcemia, secondary hyperparathyroidism, and total absence of hair in severe cases. Homozygous point mutations in the VDR gene have been identified in several kindreds (42, 48, 85). A strong association has also been found between apparently functional VDR allelic variants, a decreased bone mineral density, and the rate of bone mineral density change in relation to calcium intake (63). The retinoic acid receptor (RARα) has long been known to be involved in the t(15;17) chromosomal translocation that causes human acute promyelocytic leukemia (20, 44). Two rare cases of translocation involving the RARα gene with PLZF and NPM genes have also been reported to cause acute promyelocytic leukemia (15, 21, 84).

Nuclear receptors are ligand-dependent transcriptional regulators. Once bound to a hormone response element, the receptor responds to hormone through the C-terminal LBD, which mediates not only ligand binding but also homo- and heterodimerization, as well as transcriptional activation and repression. Mutations in nuclear receptors that affect either transcriptional activation or repression activities have been found to be involved in the development of human diseases and oncogenic transformation, suggesting that both activation and repression are essential properties for the biological function of nuclear receptors. Hormone is thought to function as a key that may switch the function of the receptors from repressor to activator or vice versa.

43.2. ROLE OF ADAPTERS AND ACCESSORY PROTEINS IN RECEPTOR FUNCTION

Several nuclear receptors possess both transcriptional activation and repression activities. Previous studies have demonstrated that the TR and RAR can be a transcriptional repressor in the absence of ligand and an activator in the presence of hormone (2, 3, 72). Utilizing combinations of DNA binding and functional assays, the repressor activity of the nonliganded TR and RAR was demonstrated to depend on appropriate hormone response elements as well as the intact LBDs (39, 69). v-*erbA* represents an oncogenic form of TR that is one of the two oncogenes of the avian thyroblastosis virus. This virus is capable of inducing acute erythroleukemia and sarcoma in chickens as well as transformation of hematopoietic cells into fibroblasts in vitro. Previous studies have demonstrated that v-*erbA* acts as a constitutive repressor of the basal promoter activity and that a transformation-defective form of v-*erbA* (Td359) fails to suppress basal transcription (18, 64). Thus, the oncogenic activity of v-*erbA* is directly linked to its constitutive repressor activity. The repressor activity resides in the LBD of TR and is functionally separable from the C-terminal activation function and ligand-binding properties. Like TR, deletion of the activation domain of RAR converts it into a potent transcriptional repressor. This repression is as potent as that of the v-*erbA* oncogene, and when expressed in vivo it was shown to have dramatic functional consequences leading to defects in cellular differentiation in vitro and lethal developmental effects in vivo (43, 76, 77). These experiments suggest that both transactivation and repression are important components for the biological function of nuclear receptors. It is thus important to understand how these two distinct activities are performed and what are the molecular mechanisms underlying these effects. The adapters or accessory proteins connecting the nuclear receptors with the basal transcriptional machinery are believed to play important roles in mediating both transactivation and repression by nuclear receptors (see Table 2).

43.2.1. Yeast Strains and Growth Media

Yeast strains should be carefully selected according to the promoter system used to express the proteins and the method of selection or assay for transactivation response. We recommend the following yeast *S. cerevisiae* strains for biochemical studies involving overexpression of the receptors and transactivation studies using β-galactosidase (β-Gal) assays: strain BJ3505 (*MAT*α, *pep4*; *his3PR6-1'*Δ*1.6R his3 lys2-208 trp1-*Δ*1 ura3-52 gal2* CUP1^r) and strain BJ1991 (*MAT*α, *prb1-1122 pep4-3 leu2-1 trp1 ura3-52 gal2* CUP1^r). Not all the test compounds are permeable to *S. cerevisiae*, and it is advisable to check the receptor function in more than one strain. A hyperpermeable strain, RS 188N (*MAT*α, *ade2-1 his3-1 leu2-1-112 trp1-1 ura3-1*), was constructed by selecting the yeast on increasing concentrations of the antibiotic nystatin. This strain is permeable to a variety of drugs that do not normally enter wild-type strains.

Yeast strains are grown in rich YEPD medium or minimal synthetic medium as described previously (33). Examples illustrated in this chapter utilize a copper metallothionein (CUP1) promoter of yeast to induce synthesis of a gene (7). To ensure that efficient copper-inducible expression is observed in yeast, CUP1^r strains should be used. Addition of copper sulfate up to 100 μM does not affect the growth rate or nuclear receptor integrity. Synthetic minimal medium contains traces of copper sulfate; hence, CUP1 is partially constitutive, and synthesis of the receptor is adequate to promote transactivation. However, 100 μM copper is added to the synthetic medium culture to overexpress the receptor or adapter gene for band-shift analysis or other biochemical studies. Yeast expression vectors with glycolytic promoters constitutively express large amounts of protein in minimal medium with 2% glucose. Because expression of certain proteins in yeast may retard their growth rate, it is advisable to monitor the growth rate of the strain before comparative transactivation studies are carried out.

43.2.2. Strategies for Expressing Nuclear Receptors

Human nuclear receptors can be expressed under regulated or constitutive promoter systems. Yeast copper-regulated metallothionein gene or Gal1 promoters are examples of regulated promoters (5, 7). Phosphoglycerate kinase, glyceraldehyde-3-phosphate dehydrogenase, and alcohol dehydrogenase are some of the glycolytic enzyme promoters that are transcribed constitutively at high efficiency in *S. cerevisiae* (12, 78). Constitutive promoters are useful for producing large amounts of protein; however, for transactivation analysis, few molecules of the receptor per cell are sufficient to detect a functional response. As mentioned above, overexpression of some of the nuclear receptors can be toxic to the cells. Hence, it is advisable to use regulated promoter systems in which the level of the protein can be controlled by an inducer. The Gal1 promoter system has been used very extensively. However, a change of carbon source is required to fully derepress the Gal1 promoter system. On the other hand, the CUP1 promoter is inducible in a variety of media and simply requires addition of copper sulfate. Another advantage of the CUP1 promoter is that the level of protein can be controlled by the amount of copper sulfate in the medium (7). Previous studies have also demonstrated that when fully induced, the CUP1 promoter is the strongest and most efficient yeast promoter (7). Most of the nuclear receptor genes have been expressed using the CUP1 promoter.

43.2.3. Role of Ubiquitin-Fusion Technology in Expression of Active Nuclear Receptors

A biologically active receptor molecule is capable of binding its ligand with wild-type affinity, interact with its DNA response elements in a sequence-specific fashion, and transactivate a target gene(s) in a heterologous system. Despite the availability of strong promoter systems and efficient translation signals, it is not always possible to express biologically active nuclear receptors. Fusion of the human ubiquitin gene to the N-terminus of nuclear receptors dramatically increases the quantity as well as restores the biological properties of the nuclear receptors (31). Ubiquitin is the most conserved 76-amino-acid protein found in all eukaryotes. Eukaryotic cells contain ubiquitin C-terminal hydrolases that rapidly cleave ubiquitin-fusion proteins to generate an intact ubiquitin and the protein with novel N-termini. The exact mechanism of ubiquitin-mediated increase in expression of heterologous proteins in *S. cerevisiae* is not clear. However, it is thought that rapid folding of the nascent ubiquitin into a highly compact structure promotes efficient and proper folding of the fused protein, increased stabilization, and production of receptor molecules that are stoichiometrically active (7a). In this respect C-terminal attachment of ubiquitin acts as a chaperonin. We and others have expressed a variety of nuclear receptors as ubiquitin-fusion proteins under the control of the CUP1 promoter in *S. cerevisiae* (Table 1) and have demonstrated that the quality and quantity of proteins were increased severalfold compared to the proteins that were expressed as nonfusion (31, 47, 55, 60, 82).

43.3. REPORTER VECTORS AND SELECTION SYSTEMS

Several reporter and selection systems are possible in *S. cerevisiae*. Owing to the ease of assay and a convenient mode of selection, the *E. coli* β-galactosidase (β-Gal) gene has been most extensively employed as a reporter. The *S. cerevisiae* β-Gal expression vector contains CYC TATAA and initiation of transcription sequences (32). This vector employs *URA3* as a selection marker for transformation in *S. cerevisiae*. Regulatory sequences with *Xho*I ends can be inserted at a unique *Xho*I site at the 5′ end of the TATAA sequences. The reporter vector described in these studies contains two copies of palindromic estrogen response elements from *Xenopus vitellogenin* gene A2. A single or a double copy of the synthetic response elements can be inserted at the unique *Xho*I site, and the reporter vectors are named YRpE1 and YRpE2, respectively (55). For other modular DBDs that recognize a variety of DNA response element configurations, e.g., direct, inverted, or everted, repeats can be conveniently designed. For two-hybrid screening, in which the Gal4 DBD is fused to any one of the LBDs of a nuclear receptor or a cDNA library (25), a Gal4 response element is present upstream of the β-Gal reporter gene.

The yeast two-hybrid system originally developed by Fields and coworkers (16, 26) has been adapted for screening of nuclear receptor partners. This system is based on in vivo screening in *S. cerevisiae* by the fact that many transcriptional activators have discrete DNA binding and transcriptional activation domains. Provided the DBD of the yeast GL4 on the bacterial LexA protein is fused with a known nuclear receptor, and the transcriptional activation domain of Gal4 or VP16 is fused to its interacting partner, the physical interaction between these two proteins will then reconstitute a functional transcriptional activator and stimulate the reporter gene expression. Most of the nuclear receptors are weakly active in the presence of hormone in *S. cerevisiae* cells. These properties permit the use of the yeast two-hybrid system to detect their interacting proteins either in the absence or in the presence of hormone. One major drawback of the yeast two-hybrid system is the isolation of false-positive clones owing to the high degree of sensitivity and alternative mechanisms in the activation of target gene expression. This utilization of the double selection procedure (see below) eliminates many of the false-positive clones and significantly enriches the positive interacting clones.

The *HIS3* gene of *S. cerevisiae* has also been a useful selectable marker, especially from two-hybrid screens to identify nuclear receptor coactivators or repressors. In this construct, the hormone or other response element is inserted upstream of the imidazole-glycerol phosphate dehydratase gene (*HIS3*). After the initial selection, the transactivation potential of the selected clones can be titrated by adding increasing concentrations of 3-amino-1,2,4-triazole, a competitive inhibitor of HIS3 enzyme. This selection/reporter system is very powerful in two respects. First, the use of aminotriazole eliminates false-positive results to a great extent if the response element is partly constitutive. Second, the strength of the transactivator can be titrated by screening the transformants in increasing concentrations of aminotriazole. After initial selection with the *HIS3* gene, the transformant can be assayed for the second reporter gene, e.g., β-Gal, which will have a different promoter element but the same regulatory elements. This step can eliminate transformants that become His^+ owing to mutation, insertion, or rearrangement at the HIS locus or to cloning of a *HIS3* homolog from the cDNA library. The requirement that both genes, *HIS3* and β-Gal, be active eliminates artifactual transactivation as well as false nuclear receptor partners whose activity depends on sequences present only in the first reporter gene. The *URA3* gene has also been used as a reporter to investigate estrogenic compounds in *S. cerevisiae* that harbors human estrogen receptor (74). In this case, hormone response elements (HREs) are inserted upstream of the *URA3* gene, and after transformation the strains are grown on plates or in liquid media without uracil. *S. cerevisiae*

TABLE 1 Orphan and steroid receptor function in *S. cerevisiae*[a]

Receptor expressed	Transcriptional regulation in *S. cerevisiae*	Reference
Human estrogen receptor	Estradiol-dependent regulation of ERE transcription	55, 61
Human progesterone receptor	Progesterone-dependent regulation of PRE transcription. Mutant PR receptor identified that uses RU486 as an agonist.	79
Androgen receptor	Ligand-dependent transactivation demonstrated	56
Rat glucocorticoid receptor	Glucocorticoid and other corticosteroids stimulate GRE-dependent transactivation	68, 73
Human vitamin D receptor	Human vitamin D receptor regulates vitamin D-responsive transcription	47, 60
Thyroid hormone receptors	Homo- and heterodimerization to affect transcription of target genes in yeast. Dual ligand (9-*cis*-RA+T3)-dependent synergy observed in transcription.	35, 53, 66, 82
Retinoic acid receptor and retinoid X receptor	Homo- and heterdimerization to activate transcriptional target genes in response to retinoid ligands. Cloning of novel orphans that interact with RXR.	35, 36, 47, 53, 82
Orphan nuclear receptors		
Peroxosome proliferator-activated receptor	PPAR-RXR-dependent transactivation demonstrated. Petroselenic acid stimulated PPAR-dependent transactivation	37, 59
Dioxin receptor	Agonist and antagonist properties studied	8, 70
Hepatic nuclear factor 4	Constitutively active	28

[a]ERE, estrogen response element; GRE, glucocorticoid response elements; PPAR, peroxosome proliferator-activated receptor; PR, progesterone receptor; PRE, progesterone response element; 9-*cis*-RA, 9-*cis*-retinoic acid; RXR, retinoid X receptor; T3, L-triidothyronine.

grows very slowly in the absence of estrogen, but addition of estrogen restores full growth on uracil-minus plates (74).

43.3.1. Yeast Hormone Response Assay System

The yeast two-hybrid system has been extremely useful in identifying nuclear receptor coactivators and repressors from higher eukaryotic cells. It is not the ideal system to study interaction between two nuclear receptors that heterodimerize on a response element, and it is this configuration that is responsible for interaction with a coactivator or repressor. One limitation of the two-hybrid system is that the DBDs are swapped with Gal4 or LexA DBDs. It is becoming increasingly clear that the DBD and the hinge regions of the nuclear receptors play an important role in selection of the response element, especially in a heterodimeric situation, which ultimately governs its interaction with cofactors. To study the interaction between native nuclear receptors, their HREs, and the coactivators and repressors, respective receptor expression vectors can be transferred with selection markers TRP1 and LEU2. The reporter selection can be based on the *HIS3* system described above or on β-Gal as a reporter that can be selected with *URA3*. The cofactor expression vector or cDNA library can also be transferred to the same cells with an alternative selection marker. This four-plasmid system has been successfully used to study the effect of cognate ligands on heterodimeric receptors and their interaction with cofactors (83).

43.3.2. Estimating the Level of Functional Receptor in *S. cerevisiae*

We have used human estrogen receptor (hER) as a model system to assess the nuclear receptor function in *S. cerevisiae*. A shown in Table 1, any number of receptors can be substituted for ER. The *S. cerevisiae* strain BJ3505 is transformed with an hER expression plasmid (YEpE12) and with a corresponding reporter plasmid (YRpE2) containing two copies of the vitellogenin estrogen response element (ERE) and the iso-1-cytochrome *c* (CYC1) promoter in a *lacZ*-fusion vector (31, 55). The auxotrophic markers for the two plasmid systems are tryptophan (*trp*) for the expression plasmid and uracil (*ura*) for the reporter plasmid. Both plasmids are transferred to *S. cerevisiae* to the modified yeast transformation procedure as described earlier (38). The transformed clones are selected on synthetic media plates. Selected colonies are tested for β-Gal activity and the amount of receptor protein, and stock cultures are stored at −70°C. Growth of liquid cultures in synthetic medium is monitored spectrophotometrically at a wavelength of 600 nm. For all experiments, overnight cultures are diluted to an OD_{600} of 0.5 prior to the induction of hER expression and addition of hormones. Tests for ligand-dependent transactivation are performed in 5-ml cultures. hER expression is induced by addition of 10 μM $CuSO_4$. For all titration experiments, equal amounts of ethanol (96%, wt/vol) or dimethyl sulfoxide, depending on the solvent used for a compound, are added to the *S. cerevisiae* cultures. The incubation period (4 h) is terminated by collecting and washing the cells.

43.3.2.1. Preparation of Yeast Extracts

For transactivation assays, yeast pellets are extracted with LacZ buffer (100 mM sodium-phosphate buffer [pH 7.0] containing 10 mM KCl, 1 mM $MgSO_4$, and 50 mM *β*-mercaptoethanol). Yeast cells are disintegrated in test tubes by vortexing with glass beads in three intervals of 30 s (with 15-s rest on ice between intervals). The debris is removed by centrifugation (1,500 × g for 10 min at 4°C). The supernatant is recovered and assayed immediately. For radioligand assays, the yeast cells are resuspended in extraction buffer (20 mM Tris [pH 8.0], 150 mM NaCl, 10 mM Na_2MoO_4, 2 mM dithiothreitol, 1 mM phenylmethylsulfonyl fluoride, 0.1 mM *N*-tosyl-L-phenylalanine chloromethyl ketone, and 0.1 mM *N*-*p*-tosyl-L-lysine chloromethyl ketone) and lysed in a bead-mill. Extracts are used freshly for *β*-Gal assays or stored at −70°C.

43.3.2.2. β-Gal Assay and Protein Determination

β-Gal activity is determined essentially as described earlier (32, 55). In short, 2 to 5 *μ*l of extract are pipetted into microtiter plate wells and combined with 250 *μ*l of the chromogenic substrate *o*-nitrophenyl-*β*-D-galactopyranoside. The plates are incubated at 37°C until color is developed. The resulting absorption is measured at 405 nm with a microtiter plate reader. Total protein is estimated using the Bio-Rad protein assay reagent (Bio-Rad Laboratories, Richmond, Calif.). A bovine serum albumin (BSA) standard curve is created in parallel with each protein assay. The specific enzyme activity is expressed in Miller units (62), which take the amount of total protein into account. They are defined as follows:

$$1 \text{ Miller unit} = \frac{OD_{405}}{\mu g \text{ protein/ml}} \cdot \frac{1}{\Delta t} \cdot \frac{\text{sample volume protein assay}}{\text{sample volume } \beta\text{-Gal assay}} \cdot 1{,}000$$

where OD_{405} is the optical density at a wavelength of 405 nm, and Δt is the incubation time at 37°C in minutes. Each determination is done in duplicate. We have noticed that freeze-thaw of the yeast extract destroys the *β*-Gal activity. Hence, it is advisable that if the *β*-Gal activity is to be estimated the next day, the samples should be stored in the refrigerator.

43.3.2.3. Radioligand Assay

Yeast extract is combined with any radiolabeled ligand or, in this case, [2,4,6,7-^{3}H]17*β*-estradiol (NEN-DuPont, Boston) and incubated for 16 to 18 h. Receptor content is estimated by addition of increasing concentrations of radiolabeled 17*β*-estradiol. Excess (approximately 100- to 10,000-fold) unlabeled 17*β*-estradiol or any other ligands are added to a fixed radioligand concentration level above ligand saturation in competition assays. After 60-min incubation at 4°C, unbound radiolabeled and nonradiolabeled ligands are removed by dextran-coated charcoal treatment. Aliquots (100 *μ*l) are used for scintillation counting. Each determination is done in triplicate.

To estimate the amount of receptor protein made in *S. cerevisiae*, [2,4,6,7-^{3}H]17*β*-estradiol is used as a ligand. Saturation is achieved at 1 nM. These data are used to estimate the amount of ER from yeast cells, assuming a relative molecular mass of 66,260 Da (PC/GENE). Approximately 40 ng of receptor per ml of extract is expressed when cultures are induced with 10 *μ*M Cu^{2+} and 162.8 ng/ml extract with 100 *μ*M Cu^{2+} induction. These values correspond to 3,600 and 14,000 ER molecules per cell at 10 *μ*M Cu^{2+} and 100 *μ*M Cu^{2+}, respectively. Fungi contain low-affinity sex steroid-binding proteins, e.g., oxidoreductases, and dehydrogenases that can interfere with the ligand binding of the 17*β*-estradiol. This residual background binding is eliminated by using yeast strains that do not contain ER expression plasmid. In addition, the yeast estrogen-binding protein is unable to bind at the concentration of radiolabeled ligand used in these studies.

43.3.2.4. Filter Assays for Detection of Colonies Expressing β-Gal

If large numbers of colonies need to be screened for the *β*-Gal assay, it is convenient to perform a semiquantitative assay on the filters using 5-bromo-4-chloro-3-indolyl-*β*-D-galactopyranoside (X-Gal). The filter assay protocol is as follows. Prepare X-Gal/LacZ buffer solution containing 100 ml of Z buffer, 0.27 ml of *β*-mercaptoethanol, and 1.67 ml of X-Gal stock (20 mg/ml in dimethylformamide). For a standard 100-mm-diameter petri dish, pipette 1.8 ml of this solution onto a clean petri dish and soak a 75-mm Whatman no. 1 or VWR grade 413 filter paper. For a 150-mm petri dish, use 5 ml of solution and a 12.5-cm filter. Place the sterile filter onto a plate with transformants and orient it with the agar medium by using a needle to make holes in three or more asymmetric locations. Allow the filter to wet uniformly such that the colonies transfer. Lift the filter with forceps from one edge and submerge it, flipped so that the colonies are facing up, into liquid nitrogen. Allow the filter to freeze for 5 s, remove, and allow it to thaw a few minutes at room temperature. With the colonies side up, layer the filter on another filter presoaked with X-Gal/LacZ buffer solution. Avoid trapping air bubbles between the filters. Colonies producing *β*-Gal will turn blue in a few minutes to several hours. It is feasible to incubate overnight. Positive colonies can be identified by aligning the filter using the orienting marks.

43.3.2.5. DNA Band-Shift Assay

Synthesis and labeling of DNA response elements are similar to the method described by Wittliff et al. (86), with modifications. Synthetic double-stranded HRE is labeled by filling the ends with Klenow enzyme and [α-^{32}P]dCTP. For determining DNA band-shift activity, about 4 *μ*g of total protein of each extract are combined with 12 *μ*l of low-salt buffer (Tris buffer as described earlier with 100 mM instead of 150 mM NaCl) and 50 ng of nonspecific DNA [poly(dI-dC), Pharmacia Biotech, Uppsala, Sweden]. These reactions are incubated on ice for 15 min. Radiolabeled HREs are added to each reaction and incubated for another 25 min at room temperature to allow association. The samples are loaded on a 5% polyacrylamide gel (acrylamide to bis-acrylamide, 30:1% in 0.5 × TBE buffer, 14 by 15 by 0.15 cm) and separated in a vertical slab-gel electrophoresis apparatus (Hoefer Scientific, San Francisco) at 10 V/cm in 0.5 × TBE electrophoresis buffer (90 mM Tris base, 90 mM boric acid, 2 mM EDTA). Subsequently, the gels are dried (Hoefer Scientific Geldryer) and autoradiographed (Kodak Scientific Imaging Film X-OMAT AR).

43.3.2.6. Western Blot Detection

Small fractions of each yeast extract are boiled with 2% sodium dodecyl sulfate and separated on 4 to 20% polyacrylamide gels using a NOVEX electrophoresis system

(San Diego, Calif.). The proteins are transferred to a nitrocellulose membrane (Bio-Rad), and the membrane is blocked with 3% BSA in phosphate-buffered saline containing 1% Triton X-100. The blots are incubated with mouse anti-hER antibodies (AER 320, NeoMarkers, Fremont, Calif.). The anti hER-antibody is targeted by an anti-mouse antibody conjugated with alkaline phosphatase (Sigma, St. Louis, Mo). Color is developed with NitroBlue Tetrazolium and 5-bromo-4-chloro-3-indolylphosphate (Sigma). Each gel contains prestained low-molecular-weight marker proteins (Bio-Rad).

43.3.3. Effects of Various Estrogens on Transactivation in hER-Transformed *S. cerevisiae*

The receptor protein expressed as ubiquitin-fusion in *S. cerevisiae*, under the control of the CUP1 promoter, is capable of binding ligands with affinity similar to that observed in mammalian cells (31, 47, 60, 82). Under fully induced conditions (100 μM copper sulfate), up to 20 pmol of hER can by synthesized per milligram of yeast protein (68a). The yeast-expressed protein is also capable of binding ERE in a sequence-specific fashion. The exposure of 5 nM estradiol to *S. cerevisiae* containing hER and an ERE β-Gal reporter vector results in the stimulation of β-Gal activity that is highly specific for estrogens in that other steroids are ineffective in stimulating this activity. Figure 1 shows that metabolites of estrogens, estrone and estriol, are able to promote ER-mediated transactivation in a dose-dependent manner. However, compared to estrogen, 100-fold-higher concentrations of estrone and estriol are required to obtain full transactivation. Interestingly, estrogen, estrone, and estriol demonstrate levels of potencies in *S. cerevisiae* that are very much analogous to their biological activity observed in uterotrophic responses and human tissue culture cells (Fig. 1) (54, 55). In this respect, the yeast system faithfully replicates efficiency and potency of ligands that target hER in mammalian systems. These results demonstrate the usefulness of *S. cerevisiae* as a model organism to test and screen for ligands that target mammalian ligand binding or orphan nuclear receptors. Table 1 lists all the mammalian nuclear receptors whose function has been reconstructed in *S. cerevisiae*. In consideration of reconstructing the nuclear receptor function in *S. cerevisiae*, the answers to at least three questions are important. (i) Is it possible to achieve the same or greater sensitivity of ligand-dependent transactivation that is observed in mammalian or homologous cells? (ii) Is yeast cell wall or membrane permeable to a variety of chemicals for screening a wide range of ligands? (iii) Do the compounds that behave as antagonists for a receptor in mammalian cells do so in *S. cerevisiae* as well? Answers to these questions are discussed below.

43.3.3.1. Sensitivity of the Ligand-Dependent Nuclear Receptor System in *S. cerevisiae*

As demonstrated for the hER in *S. cerevisiae*, transactivation is extremely responsive to nanomolar concentrations of the hormone. Responsiveness of a ligand in *S. cerevisiae* is a function of permeability as well as the nature of inter-

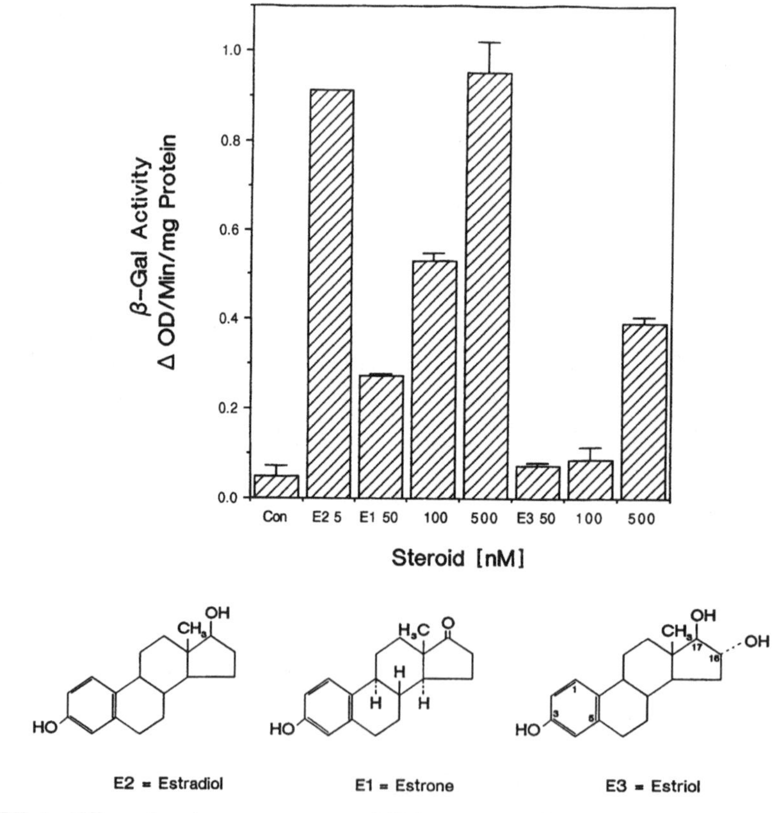

FIGURE 1 Effect of various estrogens on β-Gal activity in *S. cerevisiae* harboring hER. Yeast cultures were exposed to 5 nM estradiol (E2) and 50, 100, or 500 nM estrone (E1) and 50, 100, or 500 nM estriol (E3). For experimental details, see the text. Adapted from reference 55.

action between the receptor and the yeast RNA polymerase II machinery. We have tested a wide variety of potential estrogenic compounds, including phytoestrogens, in yeast and human hepatoma HepG2 cells that harbor hER. Comparative studies showed that some of the compounds were 1,000-fold more potent in *S. cerevisiae* than in tissue culture cells (5a). In testing the lower limit for receptor function, it was found that 1 to 10 pmol of estrogen is sufficient to evoke a transactivation response in *S. cerevisiae*. Although the precise reason for this difference is not known, it is likely that a portion of the hormone is quenched by serum in tissue culture media. *S. cerevisiae* is grown in minimal synthetic media, and presumably quenching of the ligand is minimal. It is also likely that a simple RNA polymerase II context of the yeast interacts with hER very efficiently to transduce the transcriptional response. It is important to note that hER and progesterone receptor-dependent yeasts are highly responsive to their ligands; we do not know whether all the ligands for the nuclear receptors described in Table 1 show high degree of potencies as demonstrated for ER and progesterone receptor. Whatever is the mechanism of greater sensitivity for human nuclear receptors in *S. cerevisiae*, it is clear that this property can be well exploited for identifying ligands for human nuclear receptors.

43.3.3.2. Permeability of *S. cerevisiae* to a Variety of Compounds

The impermeable nature of *S. cerevisiae* is well recognized. However, we have learned that even a standard laboratory yeast strain is very responsive to hER ligands. Several other compounds do not enter yeast cells. This lack of permeability or pumping of small molecules from *S. cerevisiae* makes it less desirable as a screening system. This problem can be partly circumvented by selecting yeast strains in an antibiotic, nystatin. Yeast strains become resistant to nystatin because of a mutation in the ergasterol biosynthetic pathway that renders yeast cell walls "leaky" to a variety of compounds. A hyperpermeable strain, RS 188N, has been used to test the biological activity of those estrogens that are highly insoluble in aqueous solvents. In these "leaky" cells (Fig. 2B), unlike in the wild-type cells (Fig. 2A), estradiol valerate and estradiol benzoate demonstrated β-Gal activity; however, estradiol stearate did not demonstrate any activity (6). It is possible that estradiol stearate did not enter the cells. These results show that it may be possible to select yeast strains that are permeable to most if not all of the compounds. We recommend that ligand-dependent transactivation properties of any receptor be tested in more than one strain with different genetic backgrounds.

43.3.3.3. Behavior of a Mammalian Receptor Agonist and Antagonist in *S. cerevisiae*

As outlined in Table 1, it is important to recognize that, although *S. cerevisiae* is quite faithful in profiling compounds that have shown agonist responses in mammalian cells, it does not faithfully profile molecules that act as antagonists. For example, hER and progesterone receptor antagonists behave as partial agonists in this system (31, 31a, 55, 79). We therefore suggest that the *S. cerevisiae* system not be used for detail profiling of antagonist activities. However, it is relevant to note that the agonist activities of various nuclear receptors in *S. cerevisiae* have provided a very sensitive cell-based system to monitor most ligand-dependent responses (31, 35, 36, 47, 55, 79, 82). As for the cell-specific agonist/antagonist function, a picture is emerging that suggests that the ligand-dependent function of nuclear receptors is governed by a new class of receptor-interacting factors (see below). These adapter proteins may very well be responsible for determining the agonist and antagonist behavior of a ligand. Therefore, it may be feasible to transfer as many mammalian genes in *S. cerevisiae* as necessary to resurrect a mammalian cell-specific antagonist or agonist response. Such an attempt to study the interaction of coactivator protein GRIP1 with ligand-dependent responses of TRβ:RXR heterodimers suggested that *S. cerevisiae* is very useful for studying mechanisms of adapter protein interaction with nuclear receptors (83). In mammalian cells, ligand-specific responses can vary not only from cell to cell but from promoter to promoter, even in a single cell. Hence, a cautionary note is that it may not be possible to resurrect precise ligand-dependent function while targeting a single receptor in *S. cerevisiae*.

43.3.4. Orphan and Steroid Receptor Function in *S. cerevisiae*

Table 1 summarizes all of the orphan and nuclear receptors that have been expressed in *S. cerevisiae*, and their ligand-dependent or -independent function has been reconstructed. Before reviewing the table, it is useful to mention the merits of the *S. cerevisiae* system. One of the advantages is that all of its genome has been sequenced. Thus far, to our knowledge, no endogenous RAR, RXR, TR, or VDR or orphan receptor-like sequences have been found in *S. cerevisiae*. While it is remarkable that most human nuclear receptor ligands are very responsive in *S. cerevisiae*, the main advantage of this system lies in its primitive environment that is devoid of nuclear receptors. Any transcriptional responses, constitutive or ligand dependent, are purely the result of the receptor genes that have been transformed into *S. cerevisiae*. In this respect, one is studying pure heterodimeric or homodimeric receptor interaction that is not possible in mammalian cells.

Another important feature of the *S. cerevisiae* system is that cells are grown in synthetic minimal medium; unlike tissue culture cells, which need fetal calf serum, the transcriptional responses observed in *S. cerevisiae* are not subject to endogenous ligands present in tissue culture medium. Some of the ligands that have been tested in our laboratory, i.e., 9-*cis*-retinoic acid (9-*cis*-RA), vitamin D_3, L-triidothyronine (T3), estrogen, and progesterone, do not undergo any significant chemical modification in *S. cerevisiae*, as has been the case with some of the ligands in tissue culture cells. Since the ligand-dependent responses observed in *S. cerevisiae* are pure and entirely driven by the added ligands, the *S. cerevisiae* system is therefore quite amenable to high-throughput screening to discover novel ligands.

As shown in Table 1, almost all of the classic steroid receptor functions and some of the nuclear receptor functions have been reconstructed in *S. cerevisiae*. Most of the receptors described in Table 1 are expressed as ubiquitin-fusion under the control of the *S. cerevisiae* CUP1 promoter. The transactivation of homodimeric receptors is tightly regulated by respective ligands in yeast. The heterodimeric receptors demonstrate various levels of ligand-responsive activities in *S. cerevisiae*. For example, VDR: RXR heterodimers show limited ligand-dependent activity

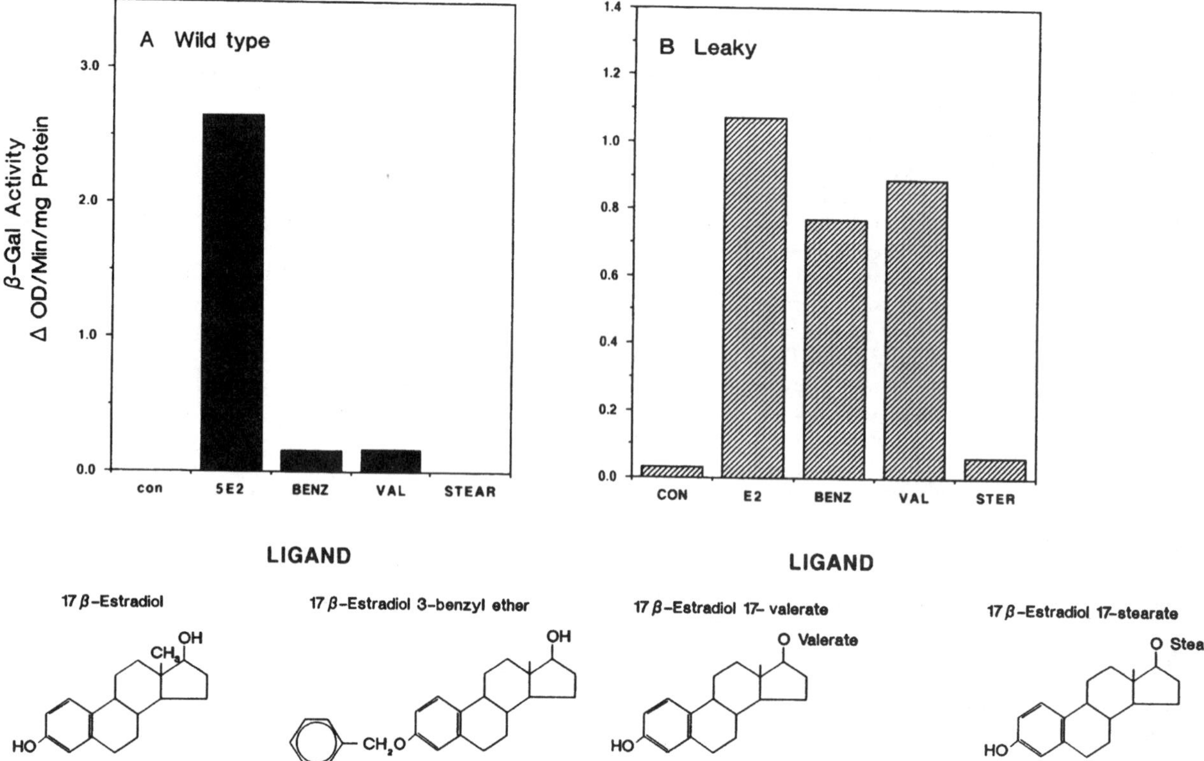

FIGURE 2 Effect of various conjugated estrogens on hER function measured by β-Gal activity. *S. cerevisiae* cultures were exposed to ethanol (con), estradiol (E2), estradiol benzoate (BENZ), or estradiol valerate (VAL); all compounds were tested at 5 nM. Experiments were performed using both the wild-type yeast (A) and the hyperpermeable or "leaky" yeast (B). Adapted from reference 55.

using rat 24-hydroxylase promoter (47), whereas TRβ:RXR heterodimers are quite responsive to dual ligands in the background of a variety of response elements. In contrast to mammalian cells, in which it may be difficult to assign a role for a particular RXR subtype in TRβ function, it is feasible to do so in *S. cerevisiae*, because pure heterodimeric strains can be conveniently constructed (82). In addition, the potential of single or dual ligands can be tested in a variety of heterodimeric configurations. While TRβ homodimer is not very responsive to T3, TRβ:RXR heterodimer is responsive to T3 as well as to 9-*cis*-RA (82).

43.3.5. Isolation of Nuclear Receptor Cofactors by the Yeast Two-Hybrid System

By using the yeast two-hybrid screening system, many nuclear receptor partners have been isolated, including several nuclear receptor coactivators that mediate transcriptional activation upon hormone binding and the nuclear receptor corepressors that mediate transcriptional repression in the absence of hormone (Table 2) (28a). Although the exact mechanism of transcriptional activation and repression by nuclear receptors is unclear, it is believed that activation of transcription by liganded receptors requires coactivators and that the repression activity of the liganded receptors requires corepressors. In an exciting series of studies, it was discovered that many of the coactivators, including SRC1 and CBP (43a), also possess intrinsic histone acetyltransferase activity. In contrast, association of histone deacetyl transferase inhibits the steroid receptor-mediated transcription. Recruitment of histone acetyltransferase by liganded receptors promotes remodeling of chromatin structure so as to allow transcriptionally repressed chromatin to assemble RNA polymerase II initiation complex. Several laboratories have focused on the isolation of such cofactors for nuclear receptors by using the yeast two-hybrid system (23, 26). Several proteins that interact with liganded receptors and potentially function as transcriptional coactivators and corepressors have been isolated (9, 28a, 32, 34, 43a, 51, 52, 67) (Table 2). It is becoming clear that cell and promoter-selective, ligand-dependent nuclear receptor responses will require multimeric interaction between ligand:receptors:adapter proteins. As the number of novel gene sequences increases in genomic databases, it will become increasingly important to reconstruct multimeric complexes of human genes in *S. cerevisiae* to elucidate the gene function in this yeast. *S. cerevisiae* is proving to be a good model organism in which these complex functions can be deciphered (83).

We are grateful to Paul Walfish of Mount Sinai Hospital, Toronto, Canada, for allowing us to quote unpublished data from coactivator GRIP1 function in yeast. The encouragement of Bill Bergman (Allegheny University of Health Sciences), James Wittliff (University of Louisville), Charles Hershberger (Eli Lilly), and Annette Tobia (QED

TABLE 2 Putative nuclear receptor coactivators and corepressors

Coactivators and corepressors	Targeted receptors[a]	Reference
Coactivator		
Trip1/Sug1	TR, RAR, RXR, ER, VDR	53, 81
TIF1	TR, RAR, RXR, ER, VDR	22
RIP140	ER	9
ERAP160	ER	34
SRC1	PR, ER, GR, RAR, RXR, TR	45, 67
GRIP1/TIF2	GR, ER, RAR, RXR, TR	10, 40, 80, 83
CBP/P300	ER, RAR, RXR	
ARA 70	AR	88
Corepressors		
SMRT	TR, RAR	13, 14
N-CoR	TR, RAR	41

[a]AR, androgen receptor; ER, estrogen receptor; GR, glucocorticoid receptor; PR, progesterone receptor; RAR, retinoic acid receptor; RXR, retinoid X receptor; TR, thyroid hormone receptor; VDR, vitamin D_3 receptor.

Technologies) is also acknowledged. Thanks to Janet Yacovelli for assistance in preparing the manuscript.

REFERENCES

1. **Adams, M. D., A. R. Kerlavage, R. D. Fleichmann, and R. A. Fuldner, et al.** 1995. Initial assessment of the human gene diversity and expression patterns based upon 83 million nucleotides of cDNA sequences. *Nature* **377**(Suppl. 28):3–20.
2. **Baniahmad, A., X. Leng, T. P. Burris, S. Y. Tsai, M. J. Tsai, and B. W. O'Malley.** 1995. The tau 4 activation domain of the thyroid hormone receptor is required for release of a putative corepressor(s) necessary for transcriptional silencing. *Mol. Cell. Biol.* **15:**76–86.
3. **Baniahmad, A., C. Steiner, A. C. Kohne, and R. Renkawitz.** 1990. Modular structure of a chicken lysozyme silencer: involvement of an unusual thyroid hormone receptor binding site. *Cell* **61:**505–514.
4. **Beato, M., P. Herrlich, and G. Schültz.** 1995. Steroid hormone receptors: many actors in search of a plot. *Cell* **83:**851–857.
5. **Boeke, J. D., D. J. Garfinkel, C. A. Styles, and G. R. Fink.** 1985. Ty elements transpose through an RNA intermediate. *Cell* **40:**491–500.

5a. **Breithofer, A., K. Graumann, M. S. Schicchitano, S. K. Karathanasis, T. R. Butt, and A. Jungbauer.** Regulation of estrogen receptor by phytoestrogens in yeast and human cells. *J. Steroid Biochem. Mol. Biol.*, in press.

6. **Burg, J. M.** 1989. DNA binding specificity of steroid receptors. *Cell* **57:**1065–1068.
7. **Butt, T. R., and D. J. Ecker.** 1987. Yeast metallothionein and applications in biotechnology. *Microbiol. Rev.* **51:** 351–369.

7a. **Butt, T. R., and J. L. Stadel.** Increasing expression of genes in heterologous system by fusion with ubiquitin: the role of ubiquitin structure and folding in yeast. Submitted for publication.

8. **Carver, L. A., V. Jackiw, and C. A. Bradfield.** 1995. The 90 kDa heat shock protein is essential for AH receptor (AHR) signaling in a yeast expression system. *J. Biol. Chem.* **269:**30109–30112.
9. **Cavaillès, V., S. Dauvois, F. L'Horset, G. Lopez, S. Hoare, P. J. Kushner, and M. G. Parker.** 1995. Nuclear factor RIP140 modulates transcriptional activation by the estrogen receptor. *EMBO J.* **14:**3741–3751.
10. **Chakravarti, D., V. J. LaMorte, M. C. Nelson, T. Nakajima, I. G. Schulman, H. Juguilon, M. Montminy, and R. M. Evans.** 1996. Role of CBP/P300 in nuclear receptor signaling. *Nature* **383:**99–103.
11. **Chang, C., A. Saltzman, S. Yeh, W. Young, E. Keller, H. J. Lee, C. Wang, and A. Mizokami.** 1995. Androgen receptor: an overview. *Crit. Rev. Eukaryot. Gene Express.* **5:**97–125.
12. **Chen, C. Y., H. Oppermann, and R. A. Hitzman.** 1984. Homologous versus heterologous gene expression in the yeast *Saccharomyces cerevisiae*. *Nucleic Acids Res.* **12:**8951–8970.
13. **Chen, J. D., and R. M. Evans.** 1995. A transcriptional co-repressor that interacts with nuclear hormone receptors. *Nature* **377:**454–457.
14. **Chen, J. D., K. Umesono, and R. M. Evans.** 1996. SMRT isoforms mediate repression and anti-repression of nuclear receptor heterodimers. *Proc. Natl. Acad. Sci. USA* **93:**7567–7571.
15. **Chen, Z., F. Guidez, P. Rousselot, A. Agadir, S. J. Chen, Z. Y. Wang, L. Degos, A. Zelent, S. Waxman, and C. Chomienne.** 1994. PLZF-RAR alpha fusion proteins generated from the variant t(11;17)(q23;q21) translocation in acute promyelocytic leukemia inhibit ligand-dependent transactivation of wild-type retinoic acid receptors. *Proc. Natl. Acad. Sci. USA* **91:**1178–1182.
16. **Chien, C. T., P. L. Bartel, R. Sternglanz, and S. Fields.** 1991. The two-hybrid system: a method to identify and clone genes for proteins that interact with a protein of interest. *Proc. Natl. Acad. Sci. USA* **88:**9578–9582.
17. **Damm, K., H. Beug, T. Graf, and B. Vennstrom.** 1987. A single point mutation in erbA restores the erythroid transforming potential of a mutant avian erythroblastosis virus (AEV) defective in both erbA and erbB oncogenes. *EMBO* **6:**375–382.
18. **Damm, K., and R. M. Evans.** 1993. Identification of a domain required for oncogenic activity and transcriptional

suppression by v-erbA and thyroid-hormone receptor alpha. *Proc. Natl. Acad. Sci. USA* **90:**10668–10672.

19. **Damm, K., C. C. Thompson, and R. M. Evans.** 1989. Protein encoded by v-erbA functions as a thyroid hormone receptor antagonist. *Nature* **339:**593–597.

20. **de The, H., C. Lavau, A. Marchio, C. Chomienne, L. Degos, and A. Dejean.** 1991. The PML-RAR alpha fusion mRNA generated by the t(15;17) translocation in acute promyelocytic leukemia encodes a functionally altered RAR. *Cell* **66:**675–684.

21. **Dirks, W. G., M. Zaborski, K. Jager, C. Challier, M. Shiota, H. Quentmeier, and H. G. Drexler.** 1996. The (2;5)(p23;q35) translocation in cell lines derived from malignant lymphomas: absence of t(2;5) in Hodgkin-analogous cell lines. *Leukemia* **10:**142–149.

22. **Douarin, B. L., C. Zechel, J.-M. Garnier, Y. Lutz, L. Tora, B. Pierrat, D. Heery, H. Gronemeyer, P. Chambon, and R. Losson.** 1995. The N-terminal part of TIF1, a putative mediator of the ligand-dependent activation function (AF-2) of nuclear receptors, is fused to B-raf in the oncogenic protein T18. *EMBO J* **14:**2020–2033.

23. **Durfee, T., K. Becherer, P. L. Chen, S. H. Yeh, Y. Yang, A. E. Kilburn, W. H. Lee, and S. J. Elledge.** 1993. The retinoblastoma protein associates with the protein phosphatase type 1 catalytic subunit. *Genes Dev.* **7:**555–569.

24. **Enmark, E., and J. Gustafsson.** 1996. Orphan nuclear receptors—the first eight years. *Mol. Endocrinol.* **10:**1293–1303.

25. **Fields, S.** The two-hybrid system to detect protein-protein interaction. *Methods Ezymol.*, in press.

26. **Fields, S., and I.-K. Song.** 1989. A novel genetic system to detect protein-protein interactions. *Nature* **340:**245–246.

27. **Forman, B. M., K. Umesono, J. Chen, and R. M. Evans.** 1995. Unique response pathways are established by allosteric interactions among nuclear hormone receptors. *Cell* **81:**541–550.

28. **Fuernkranz, H. A., Y. Wang, S. K. Karathanasis, and P. Mak.** 1994. Transcriptional regulation of the apo A1 gene by hepatic nuclear factor 4 in yeast. *Nucleic Acids. Res.* **22:**5665–5671.

28a. **Glass, C. K., D. W. Rose, and M. G. Rosenfeld.** 1997. Nuclear receptor coactivators. *Curr. Opin. Cell Biol.* **9:** 222–232.

29. **Goffeau, A., B. G. Barrell, H. Bussey, R. W. Davis, B. Dujon, H. Feldmann, F. Galibert, J. D. Hoheisel, C. Jacq, M. Johnston, E. J. Louis, H. W. Mewes, Y. Murakami, P. Philippsen, H. Tettelin, and S. G. Oliver.** 1996. Life with 6000 genes. *Science* **274:**546–567.

30. **Graf, T., and H. Beug.** 1983. Role of the v-erbA and v-erbB oncogenes of avian erythroblastosis virus in erythroid cell transformation. *Cell* **34:**7–9.

31. **Graumann, K., J. L. Wittliff, W. Raffelsberg, L. Miles, and T. R. Butt.** 1996. Structural and functional analysis of N-terminal point mutants of the human estrogen receptor. *J. Steroid Biochem. Mol. Biol.* **57:**293–300.

31a. **Graumann, K., et al.** Unpublished data.

32. **Guarente, L.** 1983. Yeast promoters and lac Z fusion designed to study expression of cloned genes in yeast. *Methods Enzymol.* **101:**181–191.

33. **Guthrie, C., and G. R. Fink.** 1991. Guide to yeast genetics and molecular biology. *Methods Enzymol.* **194:**3–21.

34. **Halachmi, S., E. Marden, G. Martin, H. MacKay, C. Abbondanza, and M. Brown.** 1994. Estrogen receptor-associated proteins: possible mediators of hormone-induced transcription. *Science* **264:**1455–1458.

35. **Hall, B. L., Z. Smit-McBride, and M. L. Privalsky.** 1993. Reconstitution of retinoid X receptor and combinatorial regulation of other nuclear receptors in the yeast *Saccharomyces cerevisiae*. *Proc. Natl. Acad. Sci. USA* **90:**6929–6933.

36. **Heery, D. M., T. Zacharewski, B. Pierrat, H. Gronemeyer, P. Chambon, and R. Losson.** 1993. Efficient transactivation by retinoic acid receptors in yeast require retinoid X receptors. *Proc. Natl. Acad. Sci. USA* **90:**4281–4285.

37. **Henry, K., M. L. O'Brien, W. Clevenger, L. Jow, and D. J. Noonan.** 1995. Peroxisome proliferator-activated receptor specificities as defined in yeast and mammalian cell transcription assays. *Tox. Appl. Pharmacol.* **132:**317–324.

38. **Hill, J., K. A. Ian, G. Donald, and D. E. Griffiths.** 1991. DMSO-enhanced whole cell yeast transformation. *Nucleic Acids Res.* **19:**5791.

39. **Holloway, J. M., C. K. Glass, S. Adler, C. A. Nelson, and M. G. Rosenfeld.** 1990. The C′-terminal interaction domain of the thyroid hormone receptor confers the ability of the DNA site to dictate positive or negative transcriptional activity. *Proc. Natl. Acad. Sci. USA* **87:**8160–8164.

40. **Hong, H., K. Kohli, A. Trivedi, D. L. Johnson, and M. R. Stallcup.** 1996. GRIP1, a novel mouse protein that serves as a transcriptional coactivator in yeast for the hormone binding domains of steroid receptors. *Proc. Natl. Acad. Sci. USA* **93:**4948–4952.

41. **Horlein, A. J., A. M. Naar, T. Heinzel, J. Torchia, B. Gloss, R, Kurokawa, A. Ryan, Y. Kamei, M. Soderstrom, C. K. Glass, and M. G. Rosenfeld.** 1995. Ligand-independent repression by the thyroid hormone receptor mediated by a nuclear receptor co-repressor. *Nature* **377:** 397–404.

42. **Hughs, M. R., P. J. Malloy, D. G. Kieback, R. A. Kesterson, J. W. Pike, D. Feldman, and B. W. O'Malley.** 1988. Point mutations in the human vitamin D receptor gene associated with hypocalcemic rickets. *Science* **242:** 1702–1705.

43. **Imakado, S., J. R. Bickenbach, D. S. Bundman, J. A. Rothnagel, P. S. Attar, X. J. Wang, V. R. Salczak, S. Wisniewski, J. Pote, J. S. Gordon, R. A. Heyman, R. M. Evans, and D. R. Roop.** 1995. Targeting expression of a dominant-negative retinoic acid receptor mutant in the epidermis of transgenic mice results in loss of barrier function. *Genes Dev.* **9:**317–329.

43a. **Jenster, G., T. E. Spencer, M. M. Burcin, S. Y. Tsai, M. J. Tsai, and B. W. O'Malley.** 1997. Steroid receptor induction of gene transcription: a two-step model. *Proc. Natl. Acad. Sci. USA* **94:**7879–7884.

44. **Kakizuka, A., W. Miller, Jr., K. Umesono, R. Warrell, Jr., S. R. Frankel, V. V. Murty, E. Dmitrovsky, and R. M. Evans.** 1991. Chromosomal translocation t(15;17) in human acute promyeloctic leukemia fuses RAR alpha with a novel putative transcription factor, PML. *Cell* **66:** 663–674.

45. **Kamei, Y., L. Xu, T. Heinzel, J. Torchia, R. Kurokawa, B. Gloss, S. C. Lin, R. A. Heyman, D. W. Rose, C. K. Glass, and M. G. Rosenfeld.** 1996. A CBP integrator

complex mediates transcriptional activation and AP-1 inhibition by nuclear receptors. *Cell* **85:**403–414.

46. **Kastner, P., M. Mark, and P. Chambon.** 1995. Nonsteroid nuclear receptors: what are genetic studies telling us about their role in real life? *Cell* **83:**859–869.
47. **Kephart, D. D., P. G. Walfish, H. DeLuca, and T. R. Butt.** 1996. RXR isotype identity directs human vitamin D receptor heterodimer transactivation from the 24-hydroxylase vitamin D response elements in yeast. *Mol. Endocrinol.* **10:**408–419.
48. **Kristjansson, K., A. R. Rut, M. Hewison, J. L. O'Riordan, and M. R. Hughs.** 1993. Two mutations in the hormone binding domain of the vitamin D receptor cause tissue resistance to 1,25-dihydroxyvitamin D3. *J. Clin. Invest.* **92:**12–16.
49. **Kurokawa, R., J. DiRenzo, M. Boehm, J. Sugarman, B. Gloss, M. G. Rosenfeld, R. A. Heyman, and C. K. Glass.** 1994. Regulation of retinoid signalling by receptor polarity and allosteric control of ligand binding. *Nature* **371:**528–531.
50. **Lander, E. S.** 1996. The new genomics: global views of biology. *Science* **274:**536–539.
51. **Le Douarin, B., C. Zechel, J. M. Garnier, Y. Lutz, L. Tora, B. Pierrat, D. Heery, H. Gronemeyer, P. Chambon, and R. Losson.** 1995. The N-terminal part of TIF-1, a putative mediator of the ligand-dependent activation function (AF-2) of nuclear receptors, is fused to B-raf in the oncogenic protein T18. *EMBO J.* **14:**2020–2033.
52. **Lee, J. W., H.-S. Choi, J. Gyurist, R. Brent, and D. D. Moore.** 1995. Two classes of proteins dependent on either the presence or absence of thyroid hormone for interaction with the thyroid hormone receptor. *Mol. Endocrinol.* **9:**243–254.
53. **Lee, J. W., F. Ryan, J. C. Swaffield, S. A. Johnston, and D. D. Moore.** 1995. Interaction of thyroid-hormone receptor with a conserved transcriptional mediator. *Nature* **374:**91–94.
54. **Littlefield, B. A., E. Gurpide, L. Markiewicz, B. McKinley, and R. B. Hochberg.** 1990. A simple and sensitive microtiter plate estrogen bioassay based on stimulation of alkaline phosphatase in Ishikawa cells: estrogenic action of delta 5 adrenal steroids. *Endocrinology* **127:**2757–2762.
55. **Lyttle, C. R., P. Damian-Matsumura, H. Juul, and T. R. Butt.** 1992. Human estrogen receptor regulation in a yeast model system and studies on receptor agonists and antagonists. *J. Steroid Biochem. Mol. Biol.* **42:**677–685.
56. **Mak, P., C. Y.-F. Young, and D. J. Tindall.** 1994. A novel yeast expression system to study androgen action. *Recent Prog. Horm. Res.* **49:**347–352.
57. **Mangelsdorf, D. J., and R. M. Evans.** 1995. The RXR heterodimers and orphan receptors. *Cell* **83:**841–850.
58. **Mangelsdorf, D. J., C. Thummel, M. Beato, P. Herrlich, G. Schütz, K., Umesono, B. Blumberg, P. Kastner, M. Mark, P. Chambon, and R. M. Evans.** 1995. The nuclear receptor superfamily: the second decade. *Cell* **83:**835–839.
59. **Marcus, S. L., K. Mityata, R. A. Rachubinski, and J. P. Capone.** 1995. Transactivation by PPAR/RXR heterodimers in yeast is potentiated by exogenous fatty acid via a pathway requiring intact peroxisomes. *Gene Express.* **4:**227–239.
60. **McDonnell, D. P., W. Pike, T. R. Butt, and B. O'Malley.** 1989. Reconstruction of the vitamin D-responsive osteocalcin transcription unit in yeast. *Mol. Cell. Biol.* **9:**3517–3523.
61. **Metzger, D., J. H. White, and P. Chambon.** 1988. The human oestrogen receptor functions in yeast. *Nature* **334:**31–36.
62. **Miller, J. H.** 1972. Assay for β-galactosidase. *In* J. H. Miller (ed.), *Experiments in Molecular Genetics*, Cold Spring Harbor Laboratory, Cold Spring Harbor, N.Y.
63. **Morrison, N. A., J. C. Qi, A. Tokita, P. J. Kelly, L. Crofts, T. V. Nguyen, P. N. Sambrook, and J. A. Eisman.** 1994. Prediction of bone density from vitamin D receptor alleles. *Nature* **367:**284–287.
64. **Munoz, A., M. Zenkee, U. Gehring, J. Sap, H. Beug, and B. Vennstom.** 1988. Characterization of the hormone binding domain of the chicken c-erbA/Thyroid hormone receptor protein. *EMBO J.* **7:**155–159.
65. **Muscatelli, F., T. M. Strom, A. P. Walker, E. Zanaria, D. Recan, A. Meindl, B. Bardoni, S. Guioli, G. Zehetner, W. Rabl, H. P. Schwarz, J.-C. Kaplan, G. Camerino, T. Meltinger, and A. P. Monaco.** 1994. Mutations in the Dax-1 gene give rise to both X-linked adrenal hypoplasia congenita and hypogonadotropic hypogonadism. *Nature* **372:**672–676.
66. **Ohashi, H., Y.-F. Yang, and P. G. Walfish.** 1991. Rat liver c-erb A β1 thyroid hormone receptor is a constitutive activator in yeast (*Saccharomyces cerevisiae*): essential role of domains D, E and F hormone dependent transcription. *Biochem. Biophys. Res. Commun.* **178:**1167–1175.
67. **Oñate, S. A., S. Y. Tsai, M. J. Tsai, and B. W. O'Malley.** 1995. Sequence and characterization of a coactivator for the steroid hormone receptor superfamily. *Science* **270:**1354–1357.
68. **Picard, D., D. B. Khursheed, M. J. Garabedian, M. G. Fortin, S. Lindquist, and Yamamoto.** 1990. Reduced levels of hsp90 compromise steroid receptor in vivo. *Nature* **348:**166–168.

68a. **Raffelsberger, W., and J. L. Wittliff.** Personal communication.

69. **Renkawitz, R.** 1993. Repression mechanisms of v-ERBA and other members of the steroid receptor superfamily. *Ann. N.Y. Acad. Sci.* **684:**1–10.
70. **Rowland, J. C., and J. A. Gustafsson.** 1995. Human dioxin receptor chimera transcription in yeast model system and studies on receptor agonists and antagonists. *Pharm. Toxicol.* **76:**328–333.
71. **Sap, J., A. Munoz, K. Damm, Y. Goldberg, J. Ghysdael, A. Leutz, H. Beug, and B. Vennstrom.** 1986. The erbA protein is a high affinity receptor for thyroid hormone. *Nature* **324:**635–640.
72. **Sap, J., A. Munoz, J. Schmitt, H. Stunnenberg, and B. Vennstrom.** 1989. Repression of transcription mediated at a thyroid hormone response element by the v-erb-A oncogene product. *Nature* **340:**242–244.
73. **Schena, M., and K. R. Yamamoto.** 1988. Mammalian glucocorticoid derivative enhance transcription in yeast. *Science* **241:**965–967.
74. **Shiau, S.-P., A. Glasebrook, S. D. Hardikar, N. Young, and C. Hershberger.** 1997 Activation of the human estrogen receptor by estrogenic and antiestrogenic compounds in *S. cerevisiae*: a positive selection system. *Gene* **179:**205–210.
75. **Thummel, C. S.** 1995. From embryogenesis to metamorphosis: the regulation and function of Drosophila nuclear receptor superfamily members. *Cell* **83:**871–877.

76. **Tsai, S., S. Bartelmez, R. Heyman, K. Damm, R. Evans, and S. J. Collins.** 1992. A mutated retinoic acid receptor-alpha exhibiting dominant-negative activity alters the lineage development of a multipotent hematopoietic cell line. *Genes Dev.* **6:**2258–2269.

77. **Tsai, S., and S. J. Collins.** 1993. A dominant negative retinoic acid receptor blocks neutrophil differentiation at the promyelocyte stage. *Proc. Natl. Acad. Sci. USA* **90:** 7153–7157.

78. **Valenzuela, P., D. Coit, and C. H. Kuo.** Synthesis and assembly in yeast of hepatitis B surface antigen particles containing the polyalbumin receptor. *Bio/Technology* **3:** 317–320.

79. **Vegelo, E., G. F. Allen, W. T. Schrader, M.-J. Tsai, D. P. McDonnell, and B. R. O'Malley.** 1992. The mechanism of RU486 antagonism is dependent on the human progesterone receptor. *Cell* **69:**703–713.

80. **Voegel, J. J., M. J. S. Heine, C. Zechel, P. Chambon, and H. Gronemeyer.** 1996. TIF2, a 160 kDa transcriptional mediator for the ligand-dependent activation function AF-2 of nuclear receptors. *EMBO J.* **15:**3667–3675.

81. **vom Baur, E., C. Zechel, D. Heery, M. J. Heine, J. M. Garnier, V. Vivat, B. Le Douarin, H. Gronemeyer, P. Chambon, and R. Losson.** 1996. Differential ligand-dependent interactions between the AF-2 activating domain of nuclear receptors and the putative transcriptional intermediary factors mSUG1 and TIF1. *EMBO J.* **15:** 110–124.

82. **Walfish, P. G., Y.-F. Yang, X. Zhu, L. Xia, and T. R. Butt.** 1996. Cross-talk between thyroid hormone and specific retinoid X receptor subtypes in yeast selectively regulates cognate ligand actions. *Gene Express.* **6:**169–184.

83. **Walfish, P. G., T. Yoganathan, Y.-F. Young, H. Hong, T. R. Butt, and M. R. Stallcup.** 1997. Yeast hormone response element-assay detect and characterize mouse GRIP1 coactivator dependent activation of thyroid and retinoid nuclear receptors. *Proc. Natl. Acad. Sci. USA* **94:** 3697–3702.

84. **Wellmann, A., T. Otsuki, M. Vogelbruch, H. M. Clark, E. S. Jaffe, and M. Raffeld.** 1995. Analysis of the t(2; 5)(p23;35) translocation by reverse transcription-polymerase chain reaction in CD30+ anaplastic large-cell lymphomas, in other non-Hodgkin's lymphomas of T-cell phenotype, and in Hodgkin's disease. *Blood* **86:**2321–2328.

85. **Wiese, R. J., J. Goto, J. M. Prahl, S. J. Marx, M. Thomas, A. Al-Ageel, and H. F. De Luca.** 1993. Vitamin D-dependency rickets type II: truncated vitamin D receptor in three kindreds. *Mol. Cell. Endocrinol.* **90:**197–201.

86. **Wittliff, J. L., L. L. Wenz, J. Dong, Z. Nawaz, and T. R. Butt.** 1990. Expression and characterization of an active human estrogen receptor as a ubiquitin fusion protein from *Escherichia coli*. *J. Biol. Chem.* **265:**22016–22022.

87. **Yeast Genetics and Human Disease.** 1996. Conference of the American Society for Microbiology in Collaboration with The Johns Hopkins University, November 14–17, Baltimore.

88. **Yeh, S., and C. Chang.** 1996. Cloning and characterization of a specific coactivator, ARA70, for the androgen receptor in human prostate cells. *Proc. Natl. Acad. Sci. USA* **93:**5517–5521.

89. **Zanaria, E., F. Muscatelli, B. Bardoni, T. M. Strom, S. Guioli, W. Guo, E. Lalli, C. Moser, A. P. Walker, E. R. B. McCabe, T. Meltinger, A. P. Monaco, P. Sassone-Corsi, and G. Camerino.** 1994. An unusual member of the nuclear hormone receptor superfamily responsible for X-linked adrenal hypoplasia congenita. *Nature* **372:** 635–641.

Design and Assembly of Polycistronic Operons in *Escherichia coli*

CHARLES HERSHBERGER, ELAINE A. BEST, JANE STERNER, CHRISTOPHER FRYE, MICHAEL MENKE, AND EVIE L. VERDERBER

44

Many recombinant proteins that are expressed in *Escherichia coli* require a cofactor for proper function. Although these essential cofactors are frequently produced by *E. coli*, their levels are often insufficient for high-level production of the recombinant protein. The literature contains several examples in which yields of soluble proteins were dramatically improved by supplementation with cofactors that associate with the proteins. For example, addition of exogenous hemin to *E. coli* cultures expressing recombinant human hemoglobin allowed accumulation of greater amounts of recombinant human hemoglobin (16). Accumulation of two other recombinant hemoproteins, human cytochrome P450 2D6 (18) and rat neuronal nitric oxide synthase (27), was significantly improved by addition of a heme precursor, δ-aminolevulinic acid (ALA), and ALA plus riboflavin, respectively. The yield of human cystathionine β-synthase in *E. coli* was increased by supplementation with ALA (19). Metabolic engineering of the host bacterium for enhanced production of the required cofactor is an attractive alternative to cofactor supplementation.

Many metabolic engineering problems such as overexpression of an amino acid or a secondary metabolite can be solved by redirecting carbon flow to produce a single metabolic product that is excreted into the medium and subsequently recovered. Product degradation can be a limitation in some metabolic engineering projects such as overproduction of amino acids, but it is not anticipated to be a limitation for protoheme synthesis because *E. coli* possesses only a very weak heme degradation pathway (30, 31). Coproduction of a recombinant protein and a cofactor requires methods for enhancing synthesis of the cofactor that do not diminish the availability of ribosomes to synthesize the protein of interest (20). Likewise, the methods should not drain metabolic capacity for production of the protein or inhibit growth by allowing accumulation of cofactor pathway intermediates that are toxic in the free form. The sections that follow provide suggestions, examples, and theoretical considerations to illustrate the design and assembly of polycistronic operons for redirecting carbon flow in *E. coli*, with protoheme biosynthesis used as a model system.

44.1. EXPRESSION CASSETTE

Figure 1 shows a schematic representation of a generic expression cassette. An expression cassette contains a regulatable promoter, an efficient Shine-Dalgarno sequence otherwise known as a ribosome binding site (RBS), sites for insertion of coding sequences that may be monocistronic or polycistronic, and a sequence specifying transcription termination. The expression cassette is bounded by unique restriction sites to allow subcloning into different plasmid vectors to vary copy number or to facilitate integration into the chromosome. Moreover, the junctions between the subcassettes also contain unique restriction sites, as do the junctions between the coding sequences for proteins in the polycistronic operon.

44.2. PROMOTER CASSETTE

Several promoters are available for expressing genes in *E. coli* (11). The *E. coli lac* promoter is an excellent choice for metabolic engineering projects because the *lac* operon has been studied extensively and because an extensive collection of mutants is available to manipulate a wide range of expression levels. Moreover, tools are available for genetically modifying an *E. coli* host or expression cassette to provide appropriate levels of repressor protein, LacI, for effective control of expression.

The wild-type *lac* promoter is subject to repression by the repressor protein (LacI), but it is also subject to catabolite repression by glucose because glucose metabolism regulates the cellular levels of cyclic AMP. Synthetic *lac* promoters for regulated gene expression are insensitive to catabolite repression because the binding site for catabolite repressor protein is not included in the promoter cassette. Insensitivity to catabolite repression makes these promoters much more amenable to large-scale fermentation in defined medium with glucose as the sole carbon source. Wild-type *lac* promoter exhibits a fairly weak transcriptional strength; however, it is well documented that modifications to the −35 and −10 regions of the *lac* promoter may modify the transcriptional activity significantly.

A series of five promoter cassettes derived from the *E. coli lac* promoter are shown in Fig. 2. All five cassettes are portable as *Bam*HI-*Nde*I fragments. All five promoters are inducible to varying levels by isopropyl thiogalactoside. These five promoters direct expression at varying efficiencies, with P*tac*16 as the most efficient and P*lac*(down) as the least efficient. Expression of the series of promoters shown in Fig. 2 can be rendered constitutive by deletion

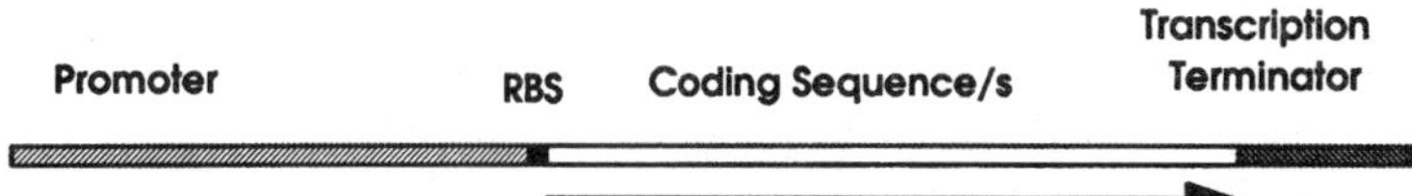

FIGURE 1 Expression cassette. Different fill patterns identify the different portions of the expression cassette. Diagonal hatches identify the promoter region. Solid fill identifies the RBS. The nonfilled region identifies a coding sequence that may represent more than one coding sequence in the case of a polycistronic operon. The arrow identifies the direction for transcription and translation of the coding sequence. The cross-hatched region identifies the transcription terminator.

of the operator sequence (10) or by mutagenesis of specific residues located in the operator sequence. Figure 3 shows a few examples of promoter cassettes containing operator constitutive (O^c) mutations (26). Deletion or mutation of the operator region should not affect the efficiency of transcription of these five promoters.

The reader should note that the sequences in Fig. 2 contain the promoter and RBSs in the same cassette that is designated the promoter cassette. The RBS contains the sequence TAAGGAGG separated from the ATG translation initiation codon by 7 nucleotides.

44.3. TRANSCRIPTION REPRESSOR

The wild-type *lac* repressor gene *lacI*, contained in many commonly used *E. coli* expression hosts, does not provide a sufficient level of repressor protein to control expression from multicopy plasmids containing *lac* promoters. Complete repression of expression is important when the gene products are detrimental to cell growth. Moreover, improper control of gene expression can sometimes lead to plasmid instability. The *lacI* mutation, *lacI*^{Q1}, provides a high level of the Lac repressor and adequately controls *lac* promoters contained on multicopy plasmids (1). The *lacI*^{Q1} allele is carried by the *E. coli* strain RS6177(NK7049-{F′*lacI*^{Q1}*lacZ*::Tn*5*}), which is available from Simons (29). The *lacI*^{Q1} allele can be introduced into the chromosome of almost any *E. coli* host using a two-step procedure that involves conjugation and allele exchange. In the first step, F′*lacI*^{Q1}*lacZ*::Tn*5* is conjugally transferred from RS6177 into a wild-type *lac* recipient using standard methods (23). Transconjugants are plated on Luria-Bertani agar (23) supplemented with 250 μg/ml acridine orange and 75 μg/ml kanamycin to select for allele exchange and loss of the F′ plasmid. The *lacI*^{Q1} allele mutation is modified by a 15-bp deletion in the *lacI* promoter region (6); therefore, replacement of the wild-type *lacI* with *lacI*^{Q1} can be verified by PCR amplification of the chromosomal DNA. The *lacI*^{Q1} allele may be transduced to other *E. coli* strains by selecting for cotransduction with the kanamycin resistance marker in the Tn*5* transposon (23). Mutants with the Tn*5* element in *lacZ* eliminated from the chromosome can be selected by growth on M9 + magnesium agar plates (23) with lactose as a sole carbon and energy source.

Rapid assays to detect F′ factors, Tn*5* transposons, *lacI* coding sequence, and *lacI* or *lacI*^{Q1} promoters can be performed by PCR. Table 1 summarizes the primers, template, purpose, and size of the PCR product for each assay. Pairs of primers with sequential numbers are used together in the assay. For example, MM236 and MM237 are used together, MM238 and MM239 are used together, MM261 and MM262 are used together, MM263 and MM264 are used together, and MM267 and MM268 are used together. However, MM256 is used with both MM257 and MM258.

44.4. DESIGN FEATURES FOR POLYCISTRONIC OPERONS

Many factors can influence expression of a synthetic operon. Some of these are the order in which the genes are assembled, their codon usage, the efficiency and placement of the ribosome binding sequence, and the spacer sequences that separate the genes. Naturally occurring RBSs and spacer sequences such as those listed in Fig. 4 may be used for an initial assessment of gene expression in operons. A theoretical RBS based on the consensus and the consensus spacer distance is also provided in Fig. 4. Indeed, our experience indicates that the highest expression levels are obtained when the theoretical RBS is used in the spacer sequence in polycistronic operons.

The DNA sequence immediately downstream of the ATG start codon can have a profound effect on translation. When cloning a coding sequence by PCR, it may be useful to design primers and synthetic oligonucleotides with AT-rich regions near the start of each coding sequence to enhance expression. Placing A and T bases at the "wobble" positions in the first 6 to 8 codons following the initiation codon can sometimes increase protein accumulation by as much as 100-fold (14a).

It is useful to add unique restriction enzyme cleavage sites or remove restriction enzyme cleavage sites from the coding sequences and intercistronic regions of a polycistronic operon. Sequence alterations can be done by using PCR site-directed mutagenesis by overlap extension (15), substituting synthetic oligonucleotides, or other site-directed mutagenic techniques. In the examples described below, restriction enzyme cleavage sites are placed near the beginning and end of each gene in the operon to allow rapid confirmation of the cloning intermediates, replacement of linkers, and removal or shuffling of coding sequences.

The order of the coding sequences in the assembled operon can affect their levels of expression and the pool size of intermediates. The coding sequences early in the operon may be more highly expressed because of polarity in expression of polycistronic operons (37). It may be necessary to shuffle the order of the coding sequences in an

Ptac16

BamHI **-35** **-10** **+1**

GGA TC CGAGCTG**TTGACA**ATTAATCATCGGCTCG**TATAAT**GTGTGG**A**ATTGTGAGCGGATAACAATTTCAC

CCT AG GCTCGAC**AACTGT**TAATTAGTAGCCGAGC**ATATTA**CACACC**T**TAACACTCGCCTATTGTTAAAGTG

NdeI

TAAGGAGGTTAATCATATG

ATTCCTCCAATTAGTATAC

Ptac17

SacII

BamHI **-35** **-10** **+1**

GGA TC CGCGGAGCTG**TTGACA**ATTAATCATCGAACTAG**TTTAAT**GTGTGG**A**ATTGTGAGCGGATAACAATTTCAC

CCT AG GCGCCTCGAC**AACTGT**TAATTAGTAGCTTGATC**AAATTA**CACACC**T**TAACACTCGCCTATTGTTAAAGTG

NdeI

TAAGGAGGTTAATCATATG

ATTCCTCCAATTAGTATAC

PlacUV5

SmaI

BamHI **-35** **-10** **+1**

GGATCCCGGGCCAGGC**TTTACA**CTTTATGCTTCCGGCTCG**TATAAT**GTGTGG**A**ATTGTGAGCGGATAACAATTTCAC

CCTAGGGCCCGGTCCG**AAATGT**GAAATACGAAGGCCGAGC**ATATTA**CACACC**T**TAACACTCGCCTATTGTTAAAGTG

NdeI

TAAGGAGGTTAATCATATG

ATTCCTCCAATTAGTATAC

Plac

BamHI XbaI **-35** **-10** **+1**

GGATCCTCTAGACCAGGC**TTTACA**CTTTATGCTTCCGGCTCG**TATGTT**GTGTGG**A**ATTGTGAGCGGATAACAATTTCAC

CCTAGGAGATCTGGTCCG**AAATGT**GAAATACGAAGGCCGAGC**ATACAA**CACACC**T**TAACACTCGCCTATTGTTAAAGTG

NdeI

TAAGGAGGTTAATCATATG

ATTCCTCCAATTAGTATAC

Plac(down)

BamHI SalI **-35** **-10** **+1**

GGATCCGTCGACCAGGC**TTTACA**CTTTATGCTTCCGGCTCG**TTTGTT**GTGTGG**A**ATTGTGAGCGGATAACAATTTCAC

CCTAGGCAGCTGGTCCG**AAATGT**GAAATACGAAGGCCGAGC**AAACAA**CACACC**T**TAACACTCGCCTATTGTTAAAGTG

NdeI

TAAGGAGGTTAATCATATG

ATTCCTCCAATTAGTATAC

FIGURE 2 *lac*-derived promoters for expression of synthetic operons. Construction was accomplished by synthesis of *Bam*HI-*Nde*I oligonucleotide cassettes containing the promoter sequences (−35 and −10 regions) and the RBS. Note that each cassette is "tagged" with a signature restriction site for diagnostic purposes. The −35 and −10 sequences are underlined and in boldface type. The RBSs are indicated in boldface italic type. The transcription start site is indicated in boldface type as +1.

PlacOc-1

```
        SmaI
BamHI               -35                        -10          +1
GGA TC CCGGGAGCTGTTGACAATTAATCATCGGCTCGTATAATGTGTGGAATGGTGAGCGGATAACAA
CCT AG GGCCCTCGACAACTGTTAATTAGTAGCCGAGCATATTACACACCTTACCACTCGCCTATTGTT
                    NdeI
TTTCACTAAGGAGGTTAATCATATG
AAAGTGATTCCTCCAATTAGTATAC
```

PlacOc-2

```
BamHI  XhoI          -35                        -10          +1
GGA TC CTCGAGAGCTGTTGACAATTAATCATCGGCTCGTATAATGTGTGGAATGGTGGGCGGATAACAATT
CCT AG GAGCTCTCGACAACTGTTAATTAGTAGCCGAGCATATTACACACCTTACCACCCGCCTATTGTTAA
                  NdeI
TCACTAAGGAGGTTAATCATATG
AGTGATTCCTCCAATTAGTATAC
```

Ptach

```
BamHI  PvuI          -35                        -10          +1
GGA TC CGATCGAGCTGGTGACAATTAATCATCGGCTCGTATAATGTGTGGAATTGAAT
CCT AG GCTAGCTCGACCACTGTTAATTAGTAGCCGAGCATATTACACACCTTAACTTA
                    NdeI
CGATATAAGGAGGTTAATCATATG
GCTATATTCCTCCAATTAGTATAC
```

FIGURE 3 Operator constitutive versions of *lac*-derived promoters. Oligonucleotide "cassettes" are mutant in the *lac*O binding region such that repression of the promoter sequences is disrupted. Repression may be disrupted by modifying a single nucleotide, by modifying multiple nucleotides, or by deletion of the entire *lac*O binding region.

operon or to alter the RBSs of specific genes to ensure appropriate balanced expression levels of the enzymes.

44.5. DESIGN OF OPERONS ENCODING PROTOHEME BIOSYNTHETIC ENZYMES

Two common pathways are used for the synthesis of ALA, which is the committed precursor in either pathway for synthesis of protoheme and other tetrapyrroles (Fig. 5). Many bacterial species, including *E. coli*, use the C5 pathway, which is a multistep pathway to synthesize ALA from the 5-carbon skeleton of glutamate. The bacteria *Rhodobacter* and *Bradyrhizobium* spp. use the C4 pathway to condense glycine and succinylcoenzyme A to form ALA in a single enzymatic reaction catalyzed by ALA synthase.

The *E. coli* genes for heme synthesis are widely scattered in the chromosome (Fig. 6). Enhancing expression of the entire complement of heme biosynthetic enzymes is a challenging problem, because to date no single protein or factor is known to control expression of the *E. coli hem* genes and because the *hem* genes are transcribed from many different promoters. One potential solution for enhancing *E. coli* heme production involves overexpression of the coding sequence for the entire complement of pathway enzymes using an inducible promoter that is insensitive to feedback regulation by heme.

The genes for the heme biosynthetic pathway enzymes are arranged in two operons in *Bacillus subtilis*. Genes encoding the conversion of glutamate to uroporphyrinogen III are clustered in an operon at 244° on the *B. subtilis* genetic map (14). Other *B. subtilis hem* pathway genes form an operon at 94° on the genetic map and encode enzymes that convert uroporphyrinogen III to heme (13). By using operons of *B. subtilis* as models, two synthetic operons can be constructed, one that specifies enzymes for synthesis of uroporphyrinogen III and another that specifies conversion of uroporphyrinogen III to heme.

We elected to substitute the C4 pathway *Rhodobacter* ALA synthase, encoded by *hemA*^{RC}, in place of the *E. coli gltX*, *hemA*(M), and *hemL* genes that provide enzymes for ALA formation by the *E. coli* C5 pathway. Substituting the C4 pathway for synthesis of ALA minimizes the number of genes required and avoids the possibility of depleting

TABLE 1 Diagnostic PCR

Oligo	Primer sequence	Gene amplified	Purpose	Size of PCR product (bp)
MM236	ATG AAT GCT GTT TTA AGT GTT CAG	*traA*	Assay for F plasmid	365
MM237	TCA GAG GCC AAC GAC GGC CAT AAC CAC AG	*traA*	Assay for F plasmid	365
MM238	GGC TCA ACA GGT TGG TGG TTC	*oriT*	Assay for F plasmid	289
MM239	CGC ACC GCT AGC AGC GCC CCT AGC	*oriT*	Assay for F plasmid	289
MM256	GAC ACC ATC GAA TGG CGC AAA ACC	*PlacI-wt*	Assay for *lacI* promoter	250
MM257	CGC CAG TTG TTG TGC CAC GC	*PlacI-wt* and *Iq1*	Assay for *lacI* promoter	250
MM258	GCA TGC ATT TAC GTT GAC ACC ACC TTT CGC	*PlacIq1*	Assay for *lacI* promoter	250
MM261	ATG ATT GAA CAA GAT GGA TTG CAC	Tn*5* Kan^r	Assay for Tn*5* insertion	795
MM262	TCA GAA GAA CTC GTC AAG AAG GC	Tn*5* Kan^r	Assay for Tn*5* insertion	795
MM263	TCA GAT CTT GAT CCC CTG CGC CAT C	Tn*5* transposase	Assay for Tn*5* insertion	1431
MM264	ATG ATA ACT TCT GCT CTT CAT CGT GC	Tn*5* transposase	Assay for Tn*5* insertion	1431
MM267	GTG AAA CCA GTA ACG TTA TAC GAT G	*lacI* coding sequence	Assay for entire *lacI* gene	1083
MM268	TCA CTG CCC GCT TTC CAG TCG GG	*lacI* coding sequence	Assay for entire *lacI* gene	1083

glutamyl tRNA, which is needed for protein synthesis. To further simplify the work, we have restricted our focus to aerobic conditions and thus have excluded *E. coli hemN* from the operons because *hemN* encodes an anaerobically active coproporphyrinogen oxidase.

44.6. OPERON FOR BIOSYNTHESIS OF UROPORPHYRINOGEN III

An operon for synthesis of uroporphyrinogen III was constructed using the *Rhodobacter capsulatus hemA*^{RC} gene, which specifies δ-aminolevulinic acid synthase, and the *E. coli hemB*, *hemC*, and *hemD* genes (Fig. 7). The DNA sequences encoding these heme synthetic enzymes were amplified by PCR with chromosomal DNA as the templates. The primers contain gene-specific sequences and extensions in the 5′ nucleotides containing unique restriction enzyme cleavage sequences, a ribosome binding sequence with a high degree of complementarity to the 3′ end of the 16S RNA, and a spacer sequence when appropriate. In *E. coli*, the TGA stop codon of *hemC* overlaps with the ATG initiation codon of *hemD* (28), suggesting that coordinate expression of *hemC* and *hemD* may be necessary. Accordingly, we preserved this organization of the *hemC* and *hemD* overlap in our synthetic operon. No changes were made in

Gene Junction	Sequence
lacZ-lacY	TA**AGGA**AATCCATTATGTACTA
lacY-lacA	**GGAG**TGATCGCATTGAACATG
galE-galT	TA**AGGA**ACGACCATGACG
galT-galK	TCC**GGAG**TGTAAGAAATGAGTCT
araG-araH	TTGCCTGAGTA**AGGAG**AGTATGATGTCTTCT
araB-araA	CTATA**AGGAC**ACGATAATGACGAT
trpE-trpD	**AGGAG**ACTTTCTGATG
trpC-trpB	TA**AGGAAAGGAA**CAATG
trpB-trpA	**CGAGGGGAA**ATCTGATG
Theoretical Design	TA**AGGAG**NNNNNNNNNATG

FIGURE 4 Representative naturally occurring intercistron sequences. Nonsense and start codons are underlined, and RBSs are in boldface type. The last sequence is the consensus sequence used to design optimized linkers.

the coding sequences of *hemA*^{RC}, *hemB*, *hemC*, or *hemD* (Fig. 7). The nucleic acid sequences of all of these genes are available in GenBank.

The spacer sequences separating *hemA*^{RC} and *hemB*, and *hemB* and *hemCD*, are shown in Fig. 8. The spacer sequences were designed to include 14 to 16 nucleotides between the stop codon of one cistron and the initiation codon of the downstream cistron (28a). The *Bam*HI-*Pst*I cassette depicted in Fig. 7 was inserted into pALTER-EX2 (Promega, Madison, Wis.), a medium-copy-number expression vector containing an efficient RBS and the strong *tac* promoter P*tac*16. Transcription of the operon is controlled by an opposing T7 promoter rather than by a transcription termination sequence.

44.7. OPERON TO CONVERT UROPORPHYRINOGEN III TO PROTOHEME

Initially, two operons were constructed to encode the enzymes to convert uroporphyrinogen III to protoheme (Fig. 9). One operon contains the *hemE*, *hemF*, *hemG*, and *hemH* genes. The other operon contains the *hemE*, *hemF*, *hemG*, *hemK*, and *hemH* genes. We constructed two operons because Nakayashiki et al. (25) suggested that the protein coded by the *hemK* gene may function as a subunit with the enzyme coded by the *hemG* gene; however, the conclusion was not firm. Each gene was cloned by PCR amplification using *E. coli* genomic DNA as template. The primers use degenerate codons to incorporate unique restriction sites close to the ends of each gene so that the encoded amino acid sequence is not altered. The restriction sites facilitate the use of synthetic oligonucleotides to assemble the genes into operons with a variety of intercistronic linkers. Several intercistronic linkers have been tested by substituting synthetic oligonucleotides between the unique restriction sites at the 3′ and 5′ ends of adjacent coding sequences (Fig. 10).

44.8. TRANSCRIPTION TERMINATOR SUBCASSETTE

It is useful to position a strong transcription terminator sequence after the last gene in a polycistronic operon to enhance mRNA stability and to prevent transcriptional readthrough into plasmid sequences. A large number of naturally occurring transcription terminator sequences from *E. coli*, and phages that infect *E coli* (e.g., λ and T4), are available for this purpose. Some of the more commonly used terminators are listed in Table 2. An extensive list of rho-independent transcription terminator sequences can be found in a review by D'Aubenton and coworkers (8).

44.9. COPY NUMBER AND CHROMOSOME INTEGRATION

Increasing the copy number of a gene, by providing additional chromosomal copies of the gene or a multicopy plasmid with the gene, often can increase enzyme production. Genes and operons should be expressed from relatively "weak" promoters on medium- or high-copy-number plasmids, because the goal of this work is a modest increase in levels of pathway enzymes rather than purification of large quantities of enzymes. Table 3 lists several low- to medium-copy-number plasmids that are compatible with pBR322 and other plasmids containing the ColE1 replicon (4) and that may be useful for metabolic engineering projects. Two plasmids with different genes or operons can be expressed in *E. coli* cells provided the two plasmids have compatible origins of replication and different selectable markers.

Integration of one or a few copies of a gene or operon into the chromosome can provide a significant increase in the level of enzyme produced, especially if the gene is expressed from a strong promoter such as P*tac*16. The sites for integration should be carefully selected. The integrated genes should not exert a negative effect on either recombinant protein production or the ability to grow to high cell density. Four "silent" integration sites are shown in Fig. 6: three sites, *galE*, *galT*, and *galK*, are located in the galactose operon, and one site is the integration site for transposon Tn*7*, *att*Tn*7*.

A functional *gal* operon, *galETK*, is not required for the expression of many recombinant proteins. Therefore, the *gal* genes may be targeted as the chromosomal integration sites for genes and assembled operons. The temperature-sensitive plasmid pMAK705 (12) was modified to contain polylinker cloning sites inserted into the middle of either *galE*, *galT*, or *galK*. The *gal* gene sequences flank genes or operons that are clones into the polylinker. Table 4 summarizes the sizes of the *gal* gene sequences that flank the inserted genes or operons. Homologous recombination between the flanking *gal* segments and the chromosomal *gal* genes integrates the inserted genes or operons into the *E. coli* chromosome. A restriction fragment or PCR amplicon containing an operon or a single gene can be inserted into one of the pMAK705-derived plasmids and integrated at the *gal* locus as described (12). A preliminary screen for isolates that do not metabolize galactose can be used to detect plasmid integration because integration inactivates one of the *gal* genes. Plasmid restriction analysis can be employed to verify recombinants. Putative recombinants are grown at 44°C without antibiotic selection to cure the plasmid from the strain. PCR is used to map the new *gal* allele in the chromosome, using primer sets that occur

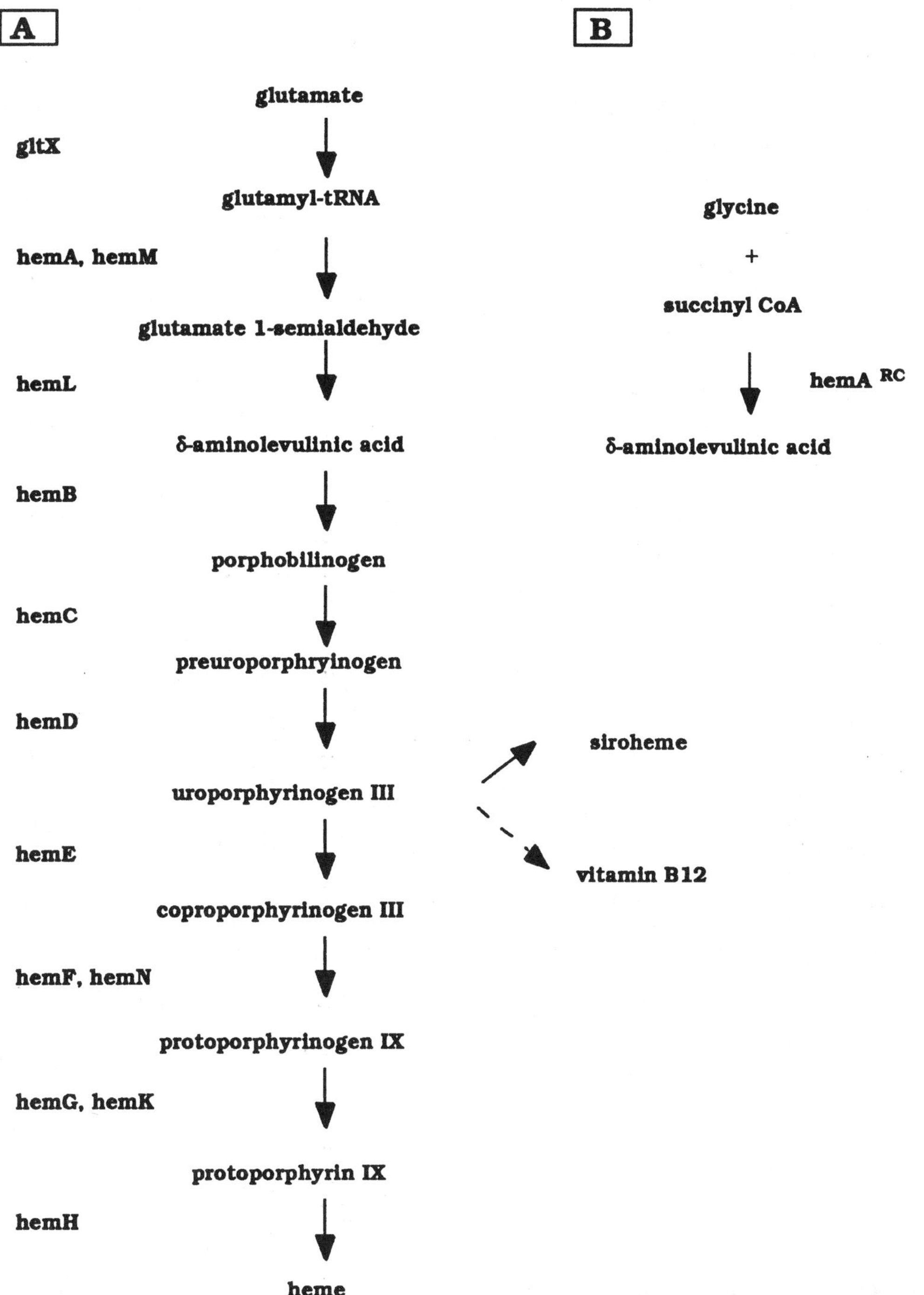

FIGURE 5 Pathways for biosynthesis of heme and intermediates. (A) Pathway and genes for conversion of glutamate to heme in *E. coli*. Only one of the two branches from uroporphyrinogen III (siroheme) is functional, because *E. coli* lacks most of the enzymes required for vitamin B_{12} synthesis. (B) Alternative pathway for synthesis of ALA.

within both the *gal* gene flanking sequences and the integrated operon.

A gene or operon can be integrated at a unique site in the bacterial genome using site-specific recombination with either transposon Tn*7* (3) or Tn*21* (9). Although it was designed for another application, we have found that the two-component transposition system developed by Leusch and coworkers (21) is an efficient method for gene integration at the *E. coli attTn7* locus. The components of the system include (i) thermosensitive donor plasmid, pMON18137, which contains a spectinomycin resistance gene and several unique cloning sites flanked by minimal

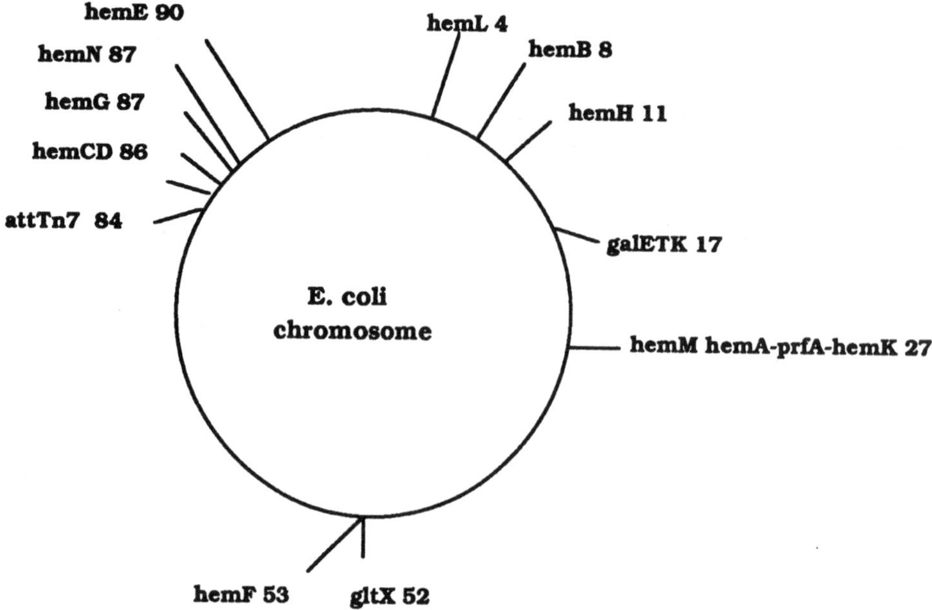

FIGURE 6 Map location of heme pathway genes and sites for integration of polycistronic operons. Map locations are given in minutes following the gene designation.

segments of Tn*7*, Tn*7L*, and Tn*7R*, required for transposition, and (ii) a tetracycline-resistant helper plasmid, pMON7124 (3), that provides Tn*7* transposase proteins.

An expression cassette containing a gene or operon is cloned into a site within Tn*7L* and Tn*7R* of pMON18137. A host strain is cotransformed with the resulting donor plasmid and the pMON7124 helper plasmid, and colonies resistant to both tetracycline and spectinomycin are selected at 30°C, the permissive temperature for pMON18137. Plasmid DNA is isolated from putative transformants to verify the presence of both plasmids. The Tn*7* element and integrated gene or operon are transposed from the donor plasmid into the chromosome with high frequency following overnight growth in Luria-Bertani broth at 44°C with selection for spectinomycin resistance. Integration at the *attTn7* site is verified by isolating chromosomal DNA and amplifying the integrated sequences by PCR using oligonucleotide primers specific for the two *attTn7* sequences flanking the integrated gene, or an *attTn7*-specific primer coupled with a primer specific for the integrated gene. The helper plasmid is lost readily from cells in the absence of selective pressure (2).

44.10. EVALUATING GENE EXPRESSION

Methods for evaluating the effects of enzyme overproduction are an essential component of any metabolic engineering program. Ideally, it is desirable to have methods for evaluating changes in pool sizes of intermediates as well as the end product of the pathway. Having these tools available allows an investigator to assess whether overexpression of particular enzymes results in a "bottleneck," generating increased pools of intermediates, which prevents carbon from being converted completely to end product.

If no changes in levels of either the end product or intermediates are observed, one must verify production of functional enzyme and demonstrate that the level of enzyme is greater than basal level of expression. This may be accomplished by in vitro measurement of enzyme activity. If a bacterial mutant is available, one can verify production, but not levels, of active enzyme by complementation of the defective activity in vivo. If specific antisera are available, Western blots can be used to demonstrate production of the protein of interest, but not to verify functionality or to quantify the amount of protein produced. If production of increased amounts of active enzyme is demonstrated and no changes in intermediates or end product are observed, one must consider the possibility that the enzyme of interest is subject to feedback inhibition, and that deregulated mutants must be identified and overproduced to provide increased levels of enzyme.

A variety of methods can be used to evaluate the operons shown in Fig. 7 and 9. Expression of the *hemA*^{RC}*BCD*

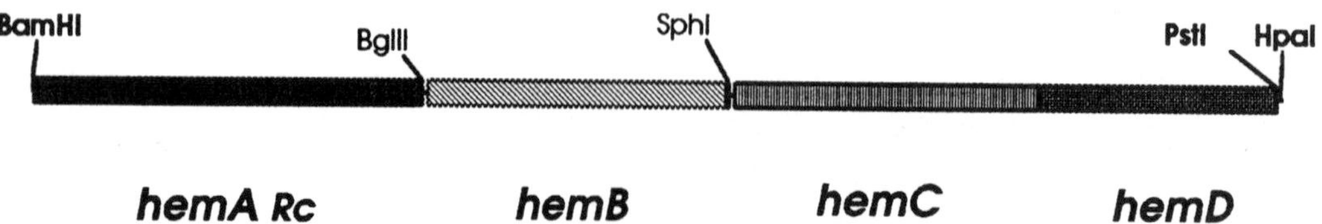

FIGURE 7 Synthetic *hemA*^{RC}*BCD* operon. Each coding sequence is indicated by a different fill pattern and labeled below the coding sequence.

HemA-B Junction

```
      BglII
TGAAGATCTCAGGAGGGTATTTCATATG
ACTTCTAGAGTCCTCCCATAAAGTATAC
```

Hem B-CD Junction

```
      SphI
TAAGCATGCGGAGGTATAAACATATG
ATTCGTACGCCTCCATATTTGTATAC
```

FIGURE 8 Intercistronic regions in a *hemA*^{RC}*BCD* operon. Each sequence begins with a nonsense codon and ends with an ATG codon. Sythetic RBS sequences are italicized. Restriction enzyme cleavage sequences are labeled above the sequence.

operon was evaluated in a quantitative fashion. Induction of the *hemA*^{RC}*BCD* operon in *E. coli* JM109 cells resulted in excretion of copious amounts of a red-pigmented compound, which co-chromatographs with a uroporphyrinogen III standard (34a). The enzyme ferrochelatase, encoded by *hemH*, was assayed as an indicator for expression of the *hemEFGH* and *hemEFGKH* operons. Ferrochelatase expression was highest when the junction sequences between the coding sequences included the intercistronic linkers designated JSopt (Fig. 10).

If no protein is expressed, one can assess whether the corresponding mRNA is produced by using any of several well-established methods, including Northern hybridization, ribonuclease protection, or reverse transcription of the mRNA coupled with PCR amplification of the target gene. The pros and cons of these methods and detailed protocols for their use can be found in any of several primers on molecular biology. Similarly, the *lacZ* gene can be fused in frame as the last coding sequence in an operon to provide a tool, β-galactosidase expression, for assessing transcription and translation of the polycistronic operon.

In our example, the protoheme pathway, many of the tetrapyrrole intermediates, uroporphyrinogen III, coproporphyrinogen III, protoporphyrin IX, and heme, can be extracted from *E. coli* cells, esterified in 5% H_2SO_4 in methanol, separated and quantified by reverse-phase high-pressure liquid chromatography, with commercially produced tetrapyrroles available from Porphyrin Products (Logan, Utah) as standards (33, 36). Some of the heme pathway intermediates, however, are highly reactive or are unstable in the presence of oxygen or light. For example, in the absence of the *hemD*-encoded uroporphyrinogen III synthase, preuroporphyrinogen cyclizes without rearrangement to form uroporphyrinogen I, an isomer that cannot be converted to heme by *E. coli*. Protoporphyrinogen IX is oxidized rapidly to protoporphyrin IX following extraction, and thus the oxidized and reduced forms of that tetrapyrrole are difficult to quantify separately. Two early intermediates, ALA and porphobilinogen, form a colored complex with Ehrlich's reagent that can be quantified spectrophotometrically. Because an enzymatic assay for the HemH protein ferrochelatase has been described (24), one can evaluate transcription and translation of the *hemEFGH* and *hemEFGKH* operons by monitoring ferrochelatase activity in cell extracts expressing the operons.

44.11. CONCLUSION

In conclusion, operons of biosynthetic pathway genes to overexpress a prosthetic group or cofactor can be used to circumvent a cell's regulation of that pathway. However, many factors must be considered carefully to ensure a final operon that is flexible enough to produce the optimum expression. Some of these factors are the selection of promoters, the efficiency of the RBSs, the strategy of construction, codon utilization, and base composition in the coding sequences, the linkers between the coding sequences, the order of the genes, the restriction sites deleted or added, suitable transcription terminators, and the copy number of the operons.

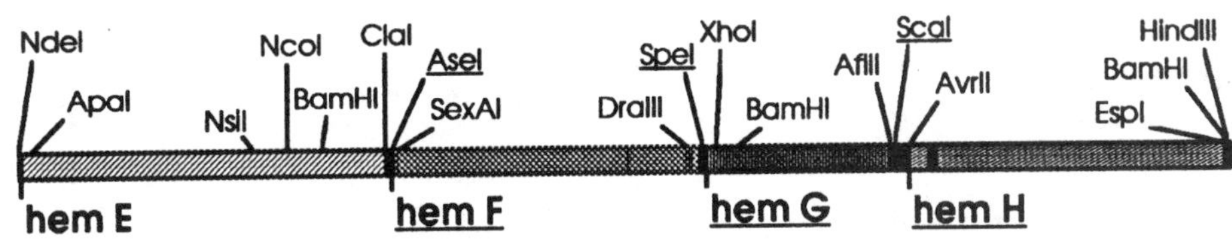

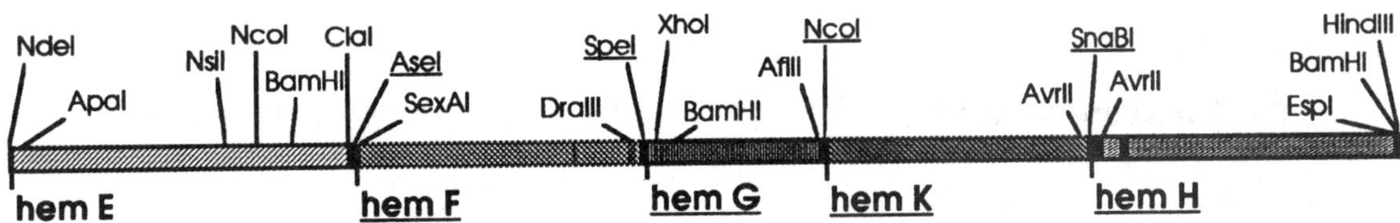

FIGURE 9 Synthetic *hemEFGH* and *hemEFGKH* operons. Each coding sequence is indicated by a different fill pattern and labeled below the coding sequence. Spacer sequences between the coding sequences contain solid fill.

HemE-F Junction

JSHemL1 Gal ET Linker

ClaI hemF Degen. Ts SexAI

ATCGATAA**GGAA**CGACATGAAACC**T**GA**T**GCACACCAGGT
TAGCTATT**CCTT**GCTGTACTTTGG**A**CT**A**CGTGTGGTCCA

JSoptL1

ClaI XbaI hemF Degen. Ts SexAI

ATCGATAA**GGAG**ATTCTAGAAATGAAACC**T**GA**T**GCACACCAGGT
TAGCTATT**CCTC**TAAGATCTTTACTTTGG**A**CT**A**CGTGTGGTCCA

HemF-G Junction

JShemL2 Ara AD Linker

DraIII hemG Degenerate T XhoI

CACTGAGTGAGTTTATTAAGGTCAGGGATTGGGTGTAA**GGAC**ACGATAATGAAAACATTAATTCTTTT**T**TCAACTCGAG
GTGACTCACTCAAATAATTCCAGTCCCTAACCCACATT**CCTG**TGCTATTACTTTTGTAATTAAGAAAA**A**AGTTGAGCTC

JSOpt L2

DraIII SpeI hemG Degenerate T XhoI

CACTGAGTGAGTTTATTAAGGTCAGGGATTGGGTGTAA**GGAG**ATACTAGTAATGAAAACATTAATTCTTTT**T**TCAACTCGAG
GTGACTCACTCAAATAATTCCAGTCCCTAACCCACATT**CCTC**TATGATCATTACTTTTGTAATTAAGAAAA**A**AGTTGAGCTC

HemG-H Junction

JSHemL3 Trp E-D Linker

AflII hemH Degenerate A & T AvrII

CTTAAGTAA**GGAG**ACTTTCTGAATGCGTCA**A**ACTAAAAC**T**GGTATCCTGCTGGCAAACCTAGG
GAATTCATT**CCTC**TGAAAGACTTACGCAGT**T**TGATTTTG**A**CCATAGGACGACCGTTTGGATCC

JSOpt L3

AflII ScaI hemH Degenerate A & T AvrII

CTTAAGTAA**GGAG**TAAGTACTTATGCGTCA**A**ACTAAAAC**T**GGTATCCTGCTGGCAAACCTAGG
GAATTCATT**CCTC**ATTCATGAATACGCAGT**T**TGATTTTG**A**CCATAGGACGACCGTTTGGATCC

HemK insertion

JSHemL4 Linker

Intermediate linker with opt RBS for hem K

AflII NcoI SnaBI hemH Degenerate A & T AvrII

CTTAAGTAA**GGAG**ATCGTATCCATGGATCTACGTATAATGCGTCA**A**ACTAAAAC**T**GGTATCCTGCTGGCAAACCTAGG
GAATTCATT**CCTC**TAGCATAGGTACCTAGATGCATATTACGCAGT**T**TGATTTTG**A**CCATAGGACGACCGTTTGGATCC

Sequence of the end of Hem K and RBS of Hem H opt added by PCR during construction of Hem K

AvrII SnaBI

CCTAGGCCGCTATTATCAATGA**GGAG**TACGTATC
GGATCCGGCGATAATAGTTACT**CCTC**ATGCATAG

FIGURE 10 Linker sequences joining coding sequences in *hemEFGH* and *hemEFGKH* operons. Two linkers were designed for each junction. The first linker used a naturally occurring intercistron sequence. The second linker used the consensus sequence for highly expressed genes. A linker with special sites was used for the introduction of *hemK* between *hemG* and *hemH*. The RBS for *hemH* downstream from *hemK* was added to the end of *hemK* during its cloning. The nonsense and start codons are underlined and the RBS is in boldface type. Degenerate positions in the first six codons of each coding sequence were changed to be A or T to increase expression. The changed degenerate nucleotides are underlined and in boldface type.

TABLE 2 Naturally occurring transcription termination sequences

Terminator	Sequence	Reference
λtL3	cagtggatttcggataacagaaaggccgggaaata cccagcctcgctttgtaacggagta	22
trp	taatcccacagccgccagttccgctggcggcattt taactttctttaatg	35
$rrnBT_1T_2$	ggaactgccaggcatcaaataaaacgaaaggctca gtcgaaagactgggcctttcgttttatctgttgtt ggtcggtgaacgctctcctgagtaggacaaatccg ccgggagcggatttgaacgttgcgaagcaacggcc cggagggtggcgggcaggacgcccgccataaactg ccaggcatcaaattaagcagaaggccatcctgacg gatggcctttttgcgtttctacaaactctt	5

TABLE 3 Plasmids containing replicons compatible with pBR322

Replication origin	Example	Copy number	Reference
F	pDF41	low; 1–2	17
RK2	pRK2501	low; 2–4	17
pSC101	pLG338, pMAK705	low; 2–6	32
R6K	pRK353	low; < 15	17
R1 (R1drd-17)	pBEU50	low at 30°C; high above 35°C	34
p15A	pACYC184 pALTER-EX2	moderate; 10–30	7

TABLE 4 Size in base pairs of *galETK* operon flanking sequences for pMAK705-based integration

Gene	5′	3′
galE	777	782
galT	753	794
galK	740	887

REFERENCES

1. **Amann, E., J. Brosius, and M. Ptashne.** 1983. Vectors bearing a hybrid *trp-lac* promoter useful for regulated expression of cloned genes in *Escherichia coli*. *Gene* **25:**167–178.
2. **Barry, G.** 1986. Permanent insertion of foreign genes into the chromosomes of soil bacteria. *Bio/Technology* **4:**446–449.
3. **Barry, G.** 1988. A broad-host range shuttle system for gene insertion into the chromosomes of Gram-negative bacteria. *Gene* **71:**75–84.
4. **Bolivar, F., R. L. Rodriguez, P. J. Greene, M. C. Betlach, H. L. Heynecker, and H. W. Boyer.** 1977. Construction of useful cloning vectors. *Gene* **2:**95–113.
5. **Brosius, J., T. J. Dull, D. D. Sleeter, and H. Noller.** 1981. Gene organization and primary structure of a ribosomal RNA operon from *Escherichia coli*. *J. Mol. Biol.* **148:**107–127.
6. **Calos, M., and J. H. Miller.** 1981. The DNA sequence change resulting from the *lacI*^{Q1} mutation, which greatly increases promoter strength. *Mol. Gen. Genet.* **183:**559–560.
7. **Chang, A. C. Y., and S. N. Cohen.** 1978. Construction and characterization of amplifiable multicopy DNA cloning vectors derived from P15A cryptic miniplasmid. *J. Bacteriol.* **134:**1141–1156.
8. **D'Aubenton, C., E. Brody, and C. Thermes.** 1990. Prediction of rho-independent *Escherichia coli* transcription terminators. A statistical analysis of their RNA stem-loop structures. *J. Mol. Biol.* **216:**835–858.
9. **Francia, M. V., and J. M. Garcia Lobo.** 1996. Gene integration in the *Escherichia coli* chromosome mediated by Tn*21* integrase (Int21). *J. Bacteriol.* **178:**894–898.
10. **Gilbert, W., and A. Maxam.** 1973. The nucleotide sequence of the lac operator. *Proc. Natl. Acad. Sci. USA* **70:**3581–3584.
11. **Goeddel, D. V. (ed.).** 1990. *Gene Expression Technology*, vol. 185. Academic Press, San Diego.
12. **Hamilton, C. M., M. Aldea, B. K. Washburn, P. Babitzke, and S. R. Kushner.** 1989. New method for generating deletions and gene replacements in *Escherichia coli*. *J. Bacteriol.* **171:**4617–4622.
13. **Hansson, M., and L. Hederstedt.** 1992. Cloning and characterization of the *Bacillus subtilis hemEHY* gene cluster, which encodes protoheme IX biosynthetic enzymes. *J. Bacteriol.* **174:**8081–8093.
14. **Hansson, M., L. Rutberg, I. Shroder, and L. Hederstedt.** 1991. The *Bacillus subtilis hemAXCDBL* gene cluster, which encodes enzymes of the biosynthetic pathway from glutamate to uroporphyrinogen III. *J. Bacteriol.* **173:**2590–2599.

14a. **Hershberger, C., et al.** Unpublished observation.

15. **Ho, S. N., H. D. Hunt, R. M. Horton, J. K. Pullen, and L. R. Pease.** 1989. Site-directed mutagenesis by over-

lap extension using polymerase chain reaction. *Gene* **77:** 51–59.

16. **Hoffman, S. J., D. L. Looker, J. M. Roehrich, P. E. Cozart, S. L. Durfee, J. L. Tedesco, and G. L. Stetler.** 1990. Expression of fully functional tetrameric human hemoglobin in *Escherichia coli*. *Proc. Natl. Acad. Sci. USA* **87:**8521–8525.
17. **Kahn, M., R. Kolter, C. Thomas, D. Figurski, R. Meyer, E. Remaut, and D. R. Helinski.** 1979. Plasmid cloning vehicles derived from plasmids ColE1, F, R6K, and RK2. *Meth. Enzymol.* **68:**268–280.
18. **Kempf, A. C., U. M. Zanger, and U. A. Meyer.** 1995. Truncated human P450 2D6: expression in *Escherichia coli*, Ni2+-chelate affinity purification, and characterization of solubility and aggregation. *Arch. Biochem. Biophys.* **321:** 277–288.
19. **Kery, V., D. Elleder, and J. P. Kraus.** 1995. δ-Aminolevulinate increases heme saturation and yield of human cystathionine β-synthase expressed in *Escherichia coli*. *Arch. Biochem. Biophys.* **316:**24–29.
20. **Kurland, C. G., and H. Dong.** 1996. Bacterial growth inhibition by overproduction of protein. *Mol. Microbiol.* **21:**1–4.
21. **Leusch, M. S., S. C. Lee, and P. O. Olins.** 1995. A novel host-vector system for direct selection of recombinant baculoviruses (bacmids) in *Escherichia coli*. *Gene* **160:**191–194.
22. **Luk, K.-C., and W. Szybalski.** 1982. Transcription termination: sequence and function of rho-independent tL3 terminator in the major leftward operon of bacteriophage lambda. *Gene* **17:**247–258.
23. **Miller, J. H.** 1992. *A Short Course in Bacterial Genetics: A Laboratory Manual and Handbook for Escherichia coli and Related Bacteria.* Cold Spring Harbor Laboratory Press, Plainview, N.Y.
24. **Miyamoto, K., S. Kanaya, K. Morikawa, and H. Inokuchi.** 1994. Overproduction, purification, and characterization of ferrochelatase from *Escherichia coli*. *J. Biochem.* **115:**545–551.
25. **Nakayashiki, T., K. Nishimura, and H. Inokochi.** 1995. Cloning and sequencing of a previously unidentified gene that is involved in the biosynthesis of heme in *Escherichia coli*. *Gene* **153:**67–70.
26. **Reznikoff, W. S.** 1984. Gene expression in microbes: the lactose operon model system. *In* D. P. Kelly and N. G. Carr (ed.), *The Microbes 1984*, vol. II. *Eukaryotes and Prokaryotes*. Society for General Microbiology. Cambridge Press, New York.
27. **Roman, L. J., E. A. Sheta, P. Martasek, S. S. Gross, Q. Liu, and B. S. S. Masters.** 1995. High-level expression of functional rat neuronal nitric oxide synthase in *Escherichia coli*. *Proc. Natl. Acad. Sci. USA* **92:**8428–8432.
28. **Sassarman, A., A. Nepevue, Y. Echelard, J. Dymetryszyn, M. Drolet, and C. Goyer.** 1987. Molecular cloning and sequencing of the hemD gene of Escherichia coli K-12 and preliminary data on the uro operon. *J. Bacteriol.* **169:**4257–4262.

28a. **Shoner, B., R. M. Belagaje, and R. G. Schoner.** 1990. Enhanced translational efficiency with two-cistron expression system. *Methods Enzymol.* **185:**94–103.

29. **Simons, R. W., F. Houman, and N. Kleckner.** 1987. Improved single and multicopy *lac*-based cloning vectors for protein and operon fusions. *Gene* **53:**85–96.
30. **Stojilkovic, I., and K. Hantke.** 1992. Hemin uptake system of *Yersinia enterocolitica*: similarities with other TonB-dependent systems in Gram-negative bacteria. *EMBO J.* **11:**4359–4367.
31. **Stojiljkovic, I., and K. Hantke.** 1994. Transport of haemin across the cytoplasmic membrane through a haemin-specific periplasmic binding-protein-dependent transport system in *Yersinia enterocolitica*. *Mol. Microbiol.* **13:**719–732.
32. **Stoker, N. G., N. F. Fairweather, and B. G. Spratt.** 1982. Versatile low-copy-number plasmid vectors for cloning in *Escherichia coli*. *Gene* **18:**335–341.
33. **Straka, J. G.** 1986. High-performance liquid chromatography of porphyrin methyl esters. *Methods Enzymol.* **123:** 352–363.
34. **Uhlin, B. E., V. Schweickart, and A. J. Clark.** 1983. New runaway-replication-plasmid cloning vectors and suppression of runaway replication by novobiocin. *Gene* **22:**255–266.

34a. **Van Dehy, J. A., and E. A. Best.** Unpublished data.

35. **Wu, A. M., G. E. Christie, and T. Platt.** 1981. Tandem termination sites in the tryptophan operon of *Escherichia coli*. *Proc. Natl. Acad. Sci. USA* **78:**2913–2917.
36. **Xu, K., J. Delling, and T. Elliott.** 1992. The genes required for heme synthesis in *Salmonella typhimurium* include those encoding alternative functions for aerobic and anaerobic coproporphyrinogen oxidation. *J. Bacteriol.* **174:**3953–3963.
37. **Zabin, I., and A. V. Fowler.** 1980. β-Galactosidase, the lactose permease protein, and thiogalactoside transacetylase, p. 89–121. *In* J. H. Miller and W. S. Reznikoff (ed.), *The Operon*, 2nd ed. Cold Spring Harbor Laboratory, Cold Spring Harbor, N.Y.

In Vivo Folding of Recombinant Proteins in *Escherichia coli*

FRANÇOIS BANEYX

45

In recent years, the gram-negative bacterium *Escherichia coli* has seen a virtual rebirth as a host for the production of recombinant polypeptides of commercial or therapeutic interest. Traditional advantages of *E. coli* include well-known genetics, an ability to grow at high density on relatively inexpensive substrates, and straightforward fermentation scale-up. More recently, a number of cloning vectors optimized for the high-level production of recombinant polypeptides in either the cytoplasm or the periplasmic space of the cell, and designed to facilitate the purification of the overexpressed gene products, have been developed. Despite these advantages, the production of heterologous proteins in *E. coli* remains hampered by the misfolding of overproduced polypeptides and their deposition into insoluble and biologically inactive aggregates known as inclusion bodies. This chapter reviews recent advances in the understanding of the in vivo protein folding process in *E. coli* and focuses on the use of pragmatic and mechanistic approaches to improve the recovery yield of recombinant gene products in a biologically active form.

45.1. PROPER FOLDING VERSUS AGGREGATION

45.1.1. Protein Folding In Vitro

In now-classic experiments, Anfinsen (4) proved that all the information necessary for the correct folding of RNase A is contained within its amino acid sequence. This result was soon generalized to other proteins by a number of laboratories. As a consequence, most of the current understanding of protein folding has been inferred from the study of the biochemical and biophysical behavior of purified polypeptides. When a full-length protein is denatured by incubation with a strong denaturant, or by exposure to acidic pH or high temperatures, it is often able to refold in a native conformation when physiological conditions are restored. The folding of a polypeptide chain is a reversible and highly cooperative event that follows first-order kinetics. Folding proceeds through a series of equilibrium steps involving folding intermediates that are very sparsely populated at any point in time. On a submillisecond time scale following the initiation of refolding, interactions between neighboring amino acids are believed to give rise to local hydrophobic clusters and elements of secondary structure that are in rapid dynamic exchange with the unfolded state (155). After 10 to 50 ms, the ensemble of partially folded conformations collapses into compact intermediates and molten globules that contain most of the secondary structure present in the native polypeptide but display a fluctuating and improperly packed hydrophobic core (48). To reach a native structure, compact intermediates must minimize the exposure of hydrophobic domains to the solvent, freeze internal size chains into a unique conformation, and pair all hydrogen-bonding groups with other groups or with the solvent. Such rate-limiting steps can take from hundreds of milliseconds to hours, depending on the protein studied. Thus, the progress of an unfolded protein to its native state is accompanied by a large entopic penalty, and compact intermediates must cross the highest free-energy barrier. Slow rearrangement reactions on the way to the native state include association of protein subdomains, *cis*/*trans* isomerization of proline bonds, and formation of native disulfide bonds. Since all these processes are accompanied by a reduction in free energy, they are, however, thermodynamically favorable. In the case of oligomeric proteins, assembly of subunits must further take place through a collision process. Since protein oligomerization involves the formation of intersubunit contacts requiring a precise match between hydrophobic side chains, ionic interactions, and backbone hydrogen bonds, the association of monomers occurs productively only after individual recognition sites have been formed (75).

The recovery yield of active proteins from their denatured state is highly dependent upon the protein concentration at which the refolding is performed. At high protein concentrations, second- or higher-order aggregation reactions compete kinetically with proper folding, thereby reducing the recovery yields of native proteins. The consensus view is that such off-pathway aggregation reactions result from the unproductive intermolecular association of compact intermediates through interactions between hydrophobic regions, or complementary domains of secondary structure (56, 105, 127).

45.1.2. Differences between In Vitro and In Vivo Protein Folding

While the principles of protein folding derived from in vitro experiments are also likely to be valid in vivo (at least for small single-domain proteins), there are pronounced differences between protein folding in a test tube and in the cellular environment. First, newly synthesized proteins emerge from the ribosome in a vectorial fashion and, if not impaired from doing so, should fold cotranslationally since the rate of spontaneous folding is typically faster than the rate of protein synthesis (108). Evidence that folding of nascent structural domains can occur while a protein is in the process of being translated has indeed been reported (46, 53, 84, 100, 167). However, it is not clear whether the majority of polypeptide chains synthesized by *E. coli* obey this mechanism, based on the following observations. (i) Certain small polypeptides such as RNase and staphylococcal nuclease, as well as multidomain proteins such as *Pecten jacobaeus* octopine dehydrogenase, require an intact C terminus to reach a native conformation following chemical denaturation (75, 135). (ii) The activity regain of two ribosome-bound proteins was dependent on the presence of a ca. 25-amino-acid (aa)-long extension at their C termini (84, 100). (iii) Folding modulators have been detected in association with nascent protein chains (54, 66, 85, 140) and are thus likely to hinder co-translational folding. (iv) Finally, since translocation through the *E. coli* inner membrane requires that preproteins be in a loosely folded or unfolded conformation to be efficiently engaged by the secretion machinery (117), co-translational folding would be expected to interfere with secretion.

A second important difference between in vivo and in vitro folding is that, while refolding experiments are typically performed at low protein concentrations in order to kinetically favor proper folding over aggregation, the cytoplasm of *E. coli* is a highly viscous environment where the concentration of folding chains has been argued to be as high as 30 to 50 μM (42), and a complete polypeptide chain is released from a ribosome at least every 35 s (98). Such conditions should be extremely deleterious to correct folding and heavily favor aggregation side reactions. Nevertheless, aggregation of host proteins is not observed in wild-type *E. coli* cells growing under physiological and balanced conditions owing to the presence of folding modulators that mediate intracellular protein folding. In contrast, the overproduction of foreign or host proteins in *E. coli* often results in their misfolding and subsequent deposition into refractile aggregates known as inclusion bodies.

45.1.3. Inclusion Bodies: Structure, Composition, and Formation

The aggregation problem has come into strong focus with the advent of recombinant DNA technology and the development of expression vectors, leading to the production of recombinant proteins at levels as high as 30 to 50% of the total cell protein. While overexpressed proteins can accumulate almost quantitatively in the soluble or insoluble fraction of the cell, they are typically partitioned between these two locations. Inclusion bodies can be found either in the cytoplasm or in the periplasm of *E. coli*, depending on whether or not the target polypeptide was designed to be translocated across the inner membrane. In the cytosol, aggregation-prone proteins form spherical or cylindrical inclusions with a maximum characteristic length of about 1 μm (21, 134). In contrast, periplasmic inclusion bodies of the *E. coli* enzyme TEM β-lactamase were found to be hemispherical and to have an approximate diameter of 0.5 μm (21).

Overexpressed proteins typically represent 40 to 95% of the inclusion body material. Contaminants, which include outer membrane proteins, ribosomal components, and a small amount of phospholipids and nucleic acids, appear to adsorb onto the aggregated material following cell lysis (142). Since the outer membrane protease OmpT can interact tightly with inclusion body material and degrade the target polypeptide during in vitro refolding (65, 150), *ompT* mutants (9) should be used for the production of polypeptides that must be renatured from their aggregated state.

The intrinsic characteristics of a polypeptide chain appear to play only an indirect role in the inclusion body formation process. Statistical analysis of proteins that are produced in either a soluble or an insoluble form in *E. coli* failed to establish a strong correlation between the propensity of a protein to aggregate and its molecular weight, hydrophobicity, cysteine or proline content, or solubility (79, 153). Interestingly, charge average and fraction of turn-forming residues—two variables that are likely to have a significant effect on protein folding—correlated more strongly with inclusion body formation (153).

By analyzing the behavior of *tsf* (temperature sensitive for folding) mutants of the phage P22 tailspike protein, King and coworkers (105, 127) first suggested that inclusion body formation resulted from the unproductive association of folding intermediates. Mutational analyses performed on interleukin-1β and other polypeptides (reviewed in reference 149), together with the demonstration that interleukin-1β (109) and β-lactamase (116) inclusion bodies contain elements of secondary structure, confirm this view. It is therefore generally well accepted that both in vitro and in vivo aggregation result from off-pathway reactions.

45.1.4. Influence of Pragmatic Approaches on Protein Aggregation

Despite considerable efforts aimed at optimizing protocols for the renaturation of heterologous proteins that accumulate in an insoluble form in *E. coli* (reviewed in reference 120), the refolding of aggregated polypeptides remains a delicate endeavor that can only be accomplished on a trial-and-error basis and is not readily amenable to scale-up. An alternative solution which applies to those polypeptides that are not toxic to the cell or are not very susceptible to proteolysis is to limit or prevent inclusion body formation and to purify the target protein from soluble extracts using traditional methods. Since inclusion body formation is likely to result from the interaction of folding intermediates that are kinetically trapped in a dead-end branch of their folding pathway, and because aggregation reactions become prominent at high protein concentrations, experimental conditions that favor on-pathway folding should lead to reduced aggregation. A search for easy-to-manipulate variables influencing inclusion body formation revealed that the growth temperature and the level of promoter induction were two powerful parameters for improving the production of aggregation-prone proteins in a soluble form. Table 1 lists a number of examples for which growth in the 15 to 23°C temperature range led to significant reduction in aggregation and a concomitant increase in the recovery of biologically active material. Figure 1A further shows that, in the case of preS2-S′-β-galactosidase (a tripartite fusion protein between the hepatitis B surface antigen preS2 and S′ sequences and *E. coli* β-galactosidase [92]), there is a direct

TABLE 1 Effect of growth temperature on inclusion body formation

Conditions	Protein (promoter)	Effect	Reference
42 or 37°C	Ricin A chain (P_L)	Soluble and fully active at 37°C; aggregated at 42°C	114
37 or 30°C	Human interferon-α2 (T7, colE1, *amp*)	2.2–3.5-Fold increase in soluble material at 30°C	123
37 or 30°C	Human interferon-γ (*trp*)	16.5-Fold less insoluble at 30°C	123
37 or 30°C	Murine protein Mx (*trp*)	Insoluble at 37°C; 50% soluble at 30°C	123
37 or 23°C	Subtilisin E (*lpp-lac*)	Periplasmic aggregation at 37°C with 2 mM IPTG; 16-fold higher activity at 23°C with 5 μM IPTG	133
37, 30, or 21°C	Fab fragments (*trp*)	10 times more soluble at 30 or 21°C	29
37 or 23°C	Luciferase $\alpha\beta$ fusion (T7)	50,000-Fold higher luminescence at 23°C	43
37—22°C	Yeast α-glucosidase (*tac*)	5-Fold increase in activity at 22°C	83
42, 37, or 28°C	PreS2-S′-β-galactosidase (*tac*)	7-Fold increase in activity at 28°C	92
37–20°C	TEM β-lactamase (*tac*)	90% decrease in periplasmic inclusion bodies at 20°C	31
37, 20, or 15°C	Soybean lipoxygenase L-1 (T7)	3.5-Fold increase in activity at 15°C	129
37 and 23°C	Kanamycin nucleotidyltransferase (T7)	10-Fold increase in specific activity at 23°C	94
37 and 22°C	Rabbit muscle glycogen phosphorylase (T7, *tac tac*)	Complete aggregation at 37°C; 50% soluble in T7 constructs with 25 μM IPTG at 20°C; 50–90% soluble in *tac-tac* constructs depending on host	25

correlation between the growth temperature and the amount of aggregated material present in the cell. It is, however, important to bear in mind that this simple solution to the inclusion body dilemma is not without penalty, since the total accumulation of the target protein decreases significantly when cells are cultivated at suboptimal growth temperatures (Fig, 1C, open circles). Furthermore, a reduction in the growth temperature does not always increase the solubility of aggregation-prone proteins.

For those genes placed under control of easy-to-regulate promoters (e.g., *lac*, *tac*, and *trc*), the level of aggregation can be reduced by using suboptimal concentrations of inducer. For instance, maximum yields of active yeast α-glucosidase PI were obtained when *tac*-driven transcription was induced with 10 μM isopropyl-β-D-thiogalactopyranoside (IPTG) (83). Similarly, the formation of periplasmic TEM β-lactamase inclusion bodies was observed only when the IPTG concentration used to induce the *tac* promoter exceeded 50 μM (20). Figure 1B shows the relationship between the IPTG concentration used to induce the *tac* promoter and the level of preS2-S′-β-galactosidase aggregation in JM109 cells growing at 30°C. From these data it is clear that above an IPTG concentration threshold of 50 μM, the *tac* promoter is fully induced, and the fraction of preS2-S′-β-galactosidase committed to aggregation (approximately 70 to 80% at 30°C)

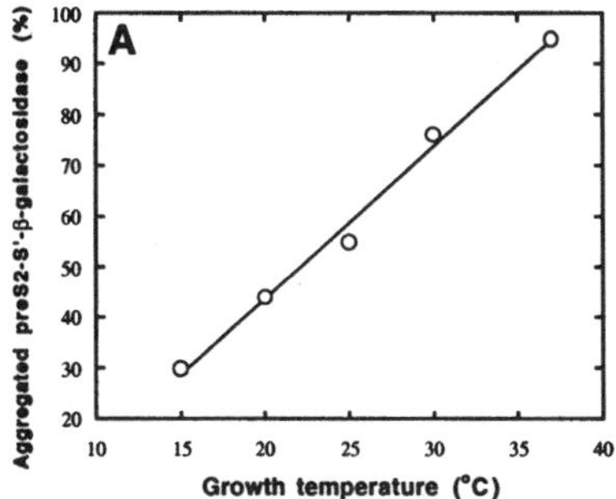

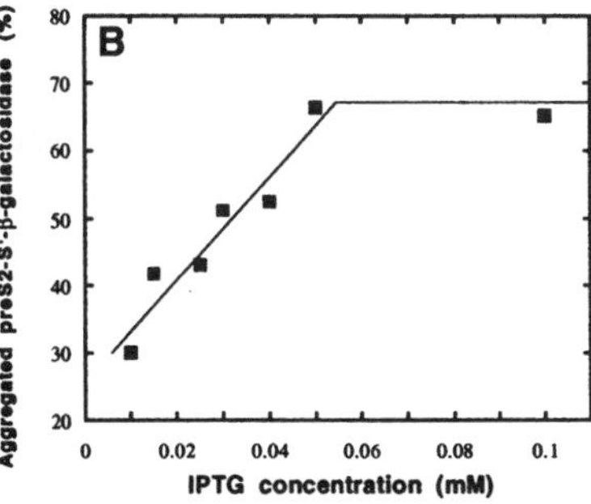

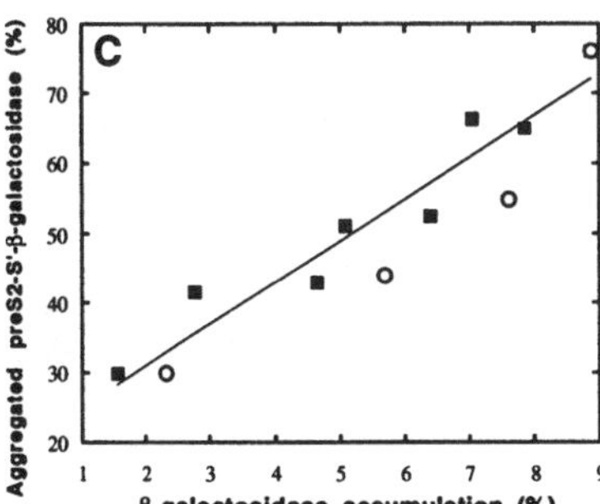

FIGURE 1 Effect of temperature and inducer concentration on the production of soluble preS2-S′-β-galactosidase. JM109 cells harboring plasmid pTBG(H+) (92) were grown in supplemented LB medium at various temperatures and induced with 1 mM IPTG (A) or at 30°C with various IPTG concentrations (B). The ratios of aggregated to total preS2-S′-β-galactosidase were evaluated by videodensitometric scanning of gels loaded with samples 2 h after induction. (C) Relationship between aggregation ratio and intracellular preS2-S′-β-galactosidase concentration. (○) Growth at 15, 20, 25, or 30°C and induction with 1 mM IPTG, from low to high accumulation levels. (■) Growth at 30°C and induction with 10, 15, 25, 30, 40, 50, or 100 μM IPTG, from low to high accumulation levels. Adapted from reference 145.

remains constant. A decrease in IPTG concentration between 50 and 10 μM directly correlates with increased production in the soluble form, at the cost of a reduction in yields (Fig. 1C, solid squares).

The use of low temperatures and suboptimal levels of gratuitous inducer may also synergistically improve the yield of active proteins. In the case of subtilisin E, growth at 23°C and induction of the *lpp-lac* promoter with 5 μM IPTG led to a 16-fold increase in activity compared with more traditional conditions (37°C and 2 mM IPTG). Since the transcription of genes placed under the control of the popular T7 bacteriophage promoter is indirectly regulated by a *lac*UV5-controlled T7 RNA polymerase gene integrated in the host chromosome, a similar strategy can be employed for these constructs, although less sensitivity should be expected. For instance, Browner et al. (25) observed that, whereas rabbit glycogen phosphorylase expressed from the T7 promoter aggregated quantitatively in BL21 (DE3) cells induced with 0.4 mM IPTG at either 22 or 37°C, about 50% of the protein was active at 22°C when the inducer concentration was decreased to 25 μM.

Inspection of Fig. 1C suggests that the mechanisms responsible for the increased partitioning of normally insoluble recombinant proteins in the soluble fraction of the cell at low temperatures and at suboptimal inducer concentrations are similar. An obvious consequence of both approaches is a reduction in the synthesis rate of overexpressed polypeptides, either through a decrease in transcription in the case of suboptimal promoter induction, or a decrease in transcription, translation initiation, and elongation in the case of growth at reduced temperature (23). It is likely that, when produced at low levels in the cytoplasm, aggregation-prone proteins are more efficiently engaged by the host molecular chaperone machinery with the net result of facilitated folding (see following section and references 136–138). Nevertheless, since suboptimal growth temperatures significantly affect *E. coli* physiology (23) and because in vitro folding is often facilitated at low temperatures (presumably owing to the suppression of hydrophobic interactions), other factors may contribute to the reduction in inclusion body formation at reduced growth temperatures. Cold shock inducible promoter systems may therefore be particularly valuable to take advantage of these additional benefits (143, 144, 146).

In addition to the growth temperature and the inducer concentration, a variety of additives and fermentation parameters have been shown to influence inclusion body formation in *E. coli* (Table 2). As a result of their differential effects on protein aggregation, these variables should be tested and optimized on a case-by-case basis, since the end result appears to be highly dependent on the substrate protein. The nature of the host strain has also been reported to affect inclusion body formation (25). However, no general rule can as yet be derived, and the choice of an optimal production host remains an empirical process.

45.2. MOLECULAR CHAPERONES

Although numerous studies have demonstrated that neither input of energy nor external factors are necessary for a protein to reach a native conformation in vitro, it has recently become obvious that protein folding in the cellular environment is an energy-dependent process that is mediated by folding catalysts and molecular chaperones. While the former group of folding modulators increases the recovery yields of active proteins by accelerating rate-limiting steps along the folding pathway, molecular chaperones appear to favor on-pathway folding by transiently interacting with folding intermediates and suppressing their aggregation. A brief summary of the characteristics of these proteins and a discussion of their usefulness for the production of recombinant proteins in a soluble form are presented below.

45.2.1. Stress Proteins and the Heat Shock Response

Early studies on the effect of temperature upshifts on protein synthesis in *E. coli* revealed that, while the expression of most cellular polypeptides was only marginally affected by growth at high temperatures, the synthesis rate of about 20 heat shock proteins (hsps) was transiently increased 10- to 20-fold under the same conditions. A similar response was obtained when cells cultured under physiological conditions were subjected to viral infection, treated with ethanol, or grown in the presence of agents leading to the accumulation of misfolded proteins (reviewed in reference 107). Hsps are therefore often referred to as stress proteins to reflect the fact that a variety of signals result in their induction. This class of protein has been highly conserved, and homologs of *E. coli* hsps have been identified in all organisms examined to date.

Although hsps are abundant proteins in unstressed cells, their transcription is transiently upregulated under stress conditions. Inspection of DNA sequences located upstream of *hsp* structural genes has revealed the presence of heat shock promoters whose consensus sequence is very different from that recognized by the vegetative RNA polymerase · σ^{70} ($E\sigma^{70}$) holoenzyme (63). The RNA polymerase core enzyme acquires specificity for heat shock promoters upon binding of the alternative sigma factor σ^{32}. The latter protein, which is encoded by the *rpoH* gene, is subject to a complex multilevel regulation mechanism that includes a feedback loop involving the DnaK-DnaJ-GrpE hsps themselves (for reviews, see references 63 and 162). Cells bearing the *rpoH165* mutation synthesize lower levels of hsps at all temperatures relative to their wild type, do not upregulate the synthesis of hsps upon temperature upshift, and are unable to form colonies at 42°C. A further consequence of *rpoH* lesions is the "wholesale" aggregation of host proteins in mutant cells grown at 42°C, a result indicating that hsps play an essential role in cellular protein folding (60). Indeed, overproduction of either the *groE* or *dnaKJ* operons, or expression of both sets of chaperones at physiological levels, can suppress this phenomenon (61). The usefulness of *rpoH* and other *hsp* mutants for producing normally soluble overexpressed proteins in an inclusion body form was recently reported (137).

Well-characterized hsps can be divided into two groups based on their function: molecular chaperones (e.g., DnaK, DnaJ, GrpE, GroEL, and GroES) and ATP-dependent proteases (e.g., Lon, ClpAP, ClpXP, and HflB). The obvious implication of this categorization is that hsps are part of a dedicated cellular machinery in which chaperones help those proteins that have misfolded as a result of cellular stress to regain an active conformation, while heat shock proteases degrade those polypeptides that have been irreversibly damaged. The most extensively studied molecular chaperone machines are the DnaK-DnaJ-GrpE and GroEL-GroES systems.

TABLE 2 Effect of environmental parameters on inclusion body formation

Parameter	Protein	Effect	Reference
pH	Protein A-β-galactosidase	Increased aggregation at low pH	130
	Salmon growth hormone	Increased aggregation at high pH	131
	Yeast α-glucosidase	Higher specific activity at low pH	83
Medium	Yeast α-glucosidase	Higher activity in minimal medium	83
	Phage T4 deoxycytidylate deaminase	20% soluble in rich medium; full aggregation in minimal medium	106
Betaine	Dimethylallyl pyrophosphatase 5′ AMP transferase	2.5 mM glycyl betaine and 660 mM sorbitol suppress aggregation	16
Sugars	TEM-β-lactamase	Periplasmic aggregation suppressed with 0.6 M sucrose or 0.3 M raffinose	19
	CC49 ScFv	Periplasmic aggregation decreased with 400 mM sucrose	122
Ethanol	Soybean lipoxygenase L-1	40% higher activity at 15°C with 3% (vol/vol) ethanol	129
	PreS2-S′-β-galactosidase	2–3-Fold higher activity with 3% (vol/vol) ethanol at 30 and 42°C	138
Aeration	*Arabidopsis* manganese-stabilizing protein	Precursor aggregation at high aeration; mature aggregation with weak aeration	15

45.2.2. The DnaK-DnaJ-GrpE System

A wealth of genetic and biochemical data indicate that DnaK, DnaJ, and GrpE interact with one another to form a molecular chaperone machine that (i) mediates the folding of nascent and newly synthesized proteins in the cell cytoplasm, (ii) maintains secreted proteins in a translocation-competent conformation, (iii) regulates the intracellular concentration of σ^{32} via a feedback loop involving the heat shock protease HflB, and (iv) is involved in general proteolysis in *E. coli* (for reviews, see references 63 and 162). In addition, DnaK, DnaJ, and GrpE have been implicated in bacteriophage and cellular DNA replication, flagellar synthesis, ribosome assembly, cell division, and mutagenesis (63).

DnaK is an abundant 69-kDa protein active as a monomer (125). It consists of a 44-kDa N-terminal ATP binding region connected to a 26-kDa C-terminal substrate binding domain through a protease-sensitive linker. The crystal structure of the ATPase domain of hsc70, the constitutively expressed mammalian homolog of DnaK, is reminiscent of that of actin and hexokinase, two proteins that undergo large conformational changes upon adenine nucleotide binding and hydrolysis (50, 51). The substrate binding region of DnaK consists of a β sandwich subdomain followed by five capping α helices (165). Polypeptides in an extended conformation bind within a channel that is defined by loops projecting from the β sandwich. Peptide binding studies (59) and analysis of the crystal structure (165) indicate that DnaK binds heptameric stretches of hydrophobic amino acids that do not contain negatively charged residues and are positively charged at their ends.

DnaJ is a 41-kDa dimeric protein that binds DnaK as well as nascent and denatured polypeptide chains and is therefore considered to be a bona fide molecular chaperone. The *dnaJ* gene is located downstream of *dnaK*, and these two proteins form an Eσ^{32}-transcribed operon. DnaJ and its homologs consist, from their amino to carboxyl termini, of a highly conserved 70- to 80-residue-long J domain; a 30-aa glycine-phenylalanine-rich G/F linker region; a cysteine-rich domain lying between residues 127 and 209 that is present in some, but not all, DnaJ homologs; and a conserved C-terminal stretch of 165 aa. While the first two domains are essential for the association of DnaJ to DnaK, the 90-aa cysteine-rich region forms a zinc finger-like structure that is required for the binding of DnaJ to denatured substrates (8, 132). The binding of DnaJ to DnaK accelerates the rate of ATP hydrolysis by DnaK (95) and promotes the stable binding of DnaK to some of its substrates. It is therefore possible that DnaJ targets DnaK to the exposed hydrophobic patches of unfolded or partially folded substrates. Alternatively, it may catalytically activate DnaK for substrate binding (96).

GrpE, the third member of the team, is a 21-kDa protein active as a dimer (125) that is transcribed by Eσ^{32} from a single, essential gene in *E. coli* (5). GrpE functions as a nucleotide exchange factor, since its binding to an exposed N-terminal loop of DnaK (26) accelerates the rate of release of both ATP and ADP (77, 95). Crystallographic data indicate that GrpE binds asymmetrically to DnaK monomers and contains two long α helices that may be involved in substrate release (64).

Figure 2 shows a possible model for the mode of action of the DnaK-DnaJ-GrpE molecular chaperone machine derived from current biochemical and structural data. Proteins damaged by heat shock or emerging from the ribosome are captured by DnaJ and presented to the substrate-accepting, ATP-loaded form of DnaK. Binding of a hydrophobic segment of the nonnative polypeptide to the channel present on the floor of the C-terminal β sandwich of DnaK, together with DnaJ interaction, stimulates the ATPase activity of DnaK. ATP hydrolysis leads to a major conformational change in DnaK which induces the capping of the DnaK substrate binding site by its α-helical C-terminal domain (step 1). This ADP-loaded, "closed" form of DnaK is stabilized by DnaJ and has a high affinity for the polypeptide substrate. Whether DnaJ is released at this stage or remains associated with the substrate remains unclear. Partial folding of the unbound portions of the substrate may then take place without the risk of aggregation reactions antagonized by the interaction of hydrophobic

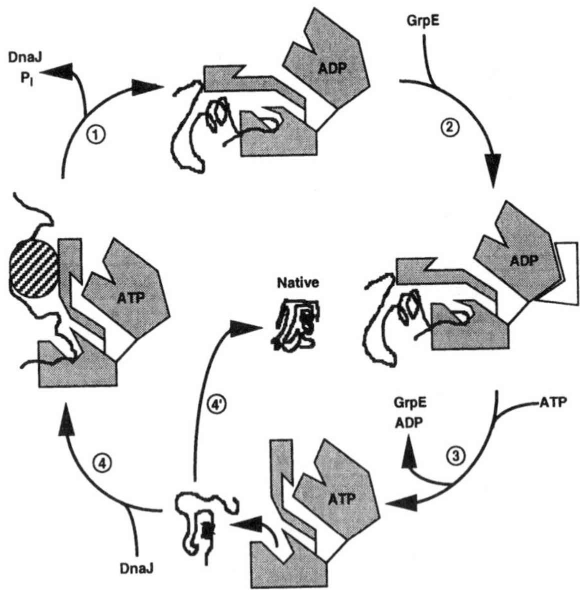

FIGURE 2 Possible mechanism of action of the DnaK-DnaJ-GrpE molecular chaperone system. The three structural domains of DnaK (grey) are shown. DnaJ is hatched and GrpE is white. See text for details.

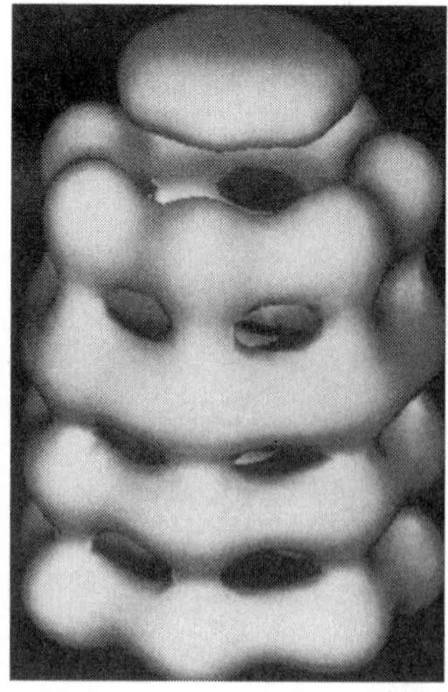

FIGURE 3 Structure of the GroEL-GroES complex determined by cryoelectron microscopy. Binding of the dome-shaped GroES (top) to GroEL triggers the outward opening of the apical domains. Reprinted from reference 32 with permission.

domains. Binding of GrpE to the complexes (step 2) promotes the release of ADP from DnaK and allows ATP binding to take place (step 3). This operation shifts DnaK from a "closed" to an "open" conformation and leads to the release of both GrpE and the bound polypeptide. The substrate protein may then fold in a native conformation (step 4′) or be rebound by DnaJ and DnaK to undergo additional cycles of folding (step 4). It has also been shown that certain small substrates such as rhodanese may be transferred to the GroEL-GroES chaperonins for further folding (87). However, recent data suggest that the two main *E. coli* molecular chaperone machines may function as parallel networks rather than in a cascade arrangement (27).

45.2.3. The GroEL-GroES System

E. coli contains a single *groESL* operon transcribed primarily by $E\sigma^{32}$, although a minor $E\sigma^{70}$-controlled promoter is also present (162). The GroE proteins are abundant in unstressed cells, and both the *groEL* and *groES* genes are essential for growth at all temperatures (45). GroEL is a weak ATPase organized as a double stack of seven identical 57-kDa subunits arranged around a central, solvent-exposed cavity that is 4 to 6 nm in diameter but can significantly expand upon binding of GroES (Fig. 3). Crystal structures of GroEL (22, 160) indicate that each monomer is composed of an equatorial domain containing the ATP binding sites and providing most of the subunit-subunit and all ring-ring interactions, a central "hinge" domain, and an apical domain defining the opening of the central channel. A ring of hydrophobic residues located on the internal face of the apical domain is important for both substrate and GroES binding (47). GroES, which is often referred to as a cochaperonin, consists of a single ring of seven identical subunits. *E. coli* GroES and its *Mycobacterium leprae* homolog are dome-shaped molecules pierced by an oculus in their roofs (72, 101). In both cases, the inner surface of the dome is very hydrophilic, and a ring of negatively charged residues surrounds the oculus. Furthermore, extended loops involved in GroEL binding (86, 160) project outward from the floor of the dome.

In the presence of adenine nucleotide, GroEL and GroES form a complex with a stoichiometry of one GroEL tetradecamer per GroES heptamer. Binding of GroES to one ring of GroEL triggers an outward movement of the apical domains of the bound ring, defining a dome-shaped cavity that can accommodate a peptide 50 to 60 kDa in size (Fig. 3). There is considerable controversy about the precise mechanism of action and the relevance of the chaperonin system in vivo (42, 98). Strong lines of evidence support the idea that nonnative proteins bind to the solvent-exposed GroEL ring of asymmetric chaperonin complexes of stoichiometry $GroEL_{14}$-$GroES_7$-ADP_7. Polypeptides in a molten globule-like conformation appear to be the preferred chaperonin substrate. Substrate binding to the GroES-free ring leads to the release of both ADP and GroES from the distal ring. The subsequent binding of ATP induces the binding of GroES to either GroEL ring, but not both. If the substrate is small enough, GroES binding may occur on the polypeptide-loaded GroEL ring. Late conformational changes would then favor the release of the bound polypeptide that begins to fold in a sequestered environment conducive to proper folding, since interactions with other folding intermediates are precluded (102, 148). It is tempting to hypothesize that GroES plays an active role in the process by binding to the charged and hydrophilic (and thus likely to be surface-exposed) regions of the substrate. Productive folding would be driven by conformational changes induced by ATP hydrolysis in the distal ring. Once all seven ATP molecules have been converted to ADP, release of the capping GroES molecule would occur (139), and the polypeptide would either be recaptured by the same or a different GroEL molecule if still partially folded (28), diffuse away from the chaperonin if in a nativelike conformation, or be transferred to a proteolytic system if successive folding cycles are unsuccessful (78). However, it has also been argued that net unfolding of the substrate may occur in the cavity (163), thus giving the substrate another chance to reach a proper conformation, perhaps in solution, or through recapture by the DnaK-DnaJ-GrpE system (27).

An interesting consequence of the above model is that only proteins smaller than 60 kDa should be efficiently chaperoned by GroEL in a GroES-dependent manner. This view is supported by the in vitro observation that, although large proteins can interact with GroEL, they do not exhibit efficient GroES-dependent refolding (7, 27, 62). Thus, one would expect that overexpression of the *groE* operon may be of limited value in mediating the folding of large recombinant proteins. However, it was also reported that the recovery of rat neuronal nitric oxide synthase, a 160-kDa protein, can be dramatically improved in cells overproducing GroEL and GroES (119). Therefore, GroEL-mediated chaperoning of large proteins may still occur in vivo, although a mechanism different from the one highlighted above would be at play.

45.2.4. Other Molecular Chaperones

45.2.4.1. SecB

Translocation across the *E. coli* plasma membrane requires that secreted polypeptides be maintained in an extended conformation to be efficiently engaged by the Sec secretion machinery (for a review, see reference 117). Although the GroEL-GroES and DnaK-DnaJ-GrpE systems can contribute to this process for a few host proteins, including β-lactamase and alkaline phosphatase, the major secretory molecular chaperone appears to be SecB, since it recognizes a group of about 10 periplasmic and outer membrane proteins (reviewed in references 34 and 118). SecB, which is *not* an hsp, represents less than 0.1% of the cytosolic protein and is organized as a homotetramer of 16-kDA subunits. Peptide binding studies have shown that SecB preferentially binds positively charged and flexible sequences at multiple sites. Simultaneous binding triggers the exposure of hydrophobic regions that allow the chaperone to form a strong complex with one precursor protein. To date, it is not clear how SecB specifically binds proteins that are destined for export. An obvious hypothesis would be that SecB recognizes the signal sequence of precursor proteins, since these N-terminal extensions typically contain positively charged amino acids at their N terminus, followed by a stretch of hydrophobic residues. Nevertheless, in addition to binding signal sequences, SecB recognizes many sites within the mature region of preproteins. It has therefore been argued that the signal sequence plays an indirect role in SecB binding by retarding protein folding and allowing interaction with the chaperone (118). Polypeptides complexed with SecB are delivered to the peripheral membrane protein SecA, which contains a SecB binding site. ATP-dependent translocation to the periplasmic space completes the process.

45.2.4.2. HtpG

HtpG is a cytoplasmic heat shock protein that belongs to the hsp90 family of stress proteins. In eukaryotic cells, Hsp90 is an abundant polypeptide that associates with a variety of steroid receptors, Hsp70 (the DnaK homolog), and peptidyl prolyl isomerases (73). Taken together with the fact that Hsp90 dimers exhibit an ATP-independent chaperone activity in vitro (152), it is likely that HtpG participates in protein folding in vivo. Since *E. coli* HtpG is highly homologous to eukaryotic Hsp90, it is likely to function in the same fashion, although this has not yet been established. The fact that multicopy clones containing *htpG* have been isolated as suppressors of *secY* mutations (141) also suggests that this protein may have a role in secretion.

45.2.4.3. IbpA and IbpB

The two 16-kDa hsps IbpA and IbpB were first identified as contaminant proteins tightly associated with human prorenin, renin, and bovine insulin-like growth factor-2 inclusion bodies (2) and subsequently by global mapping of mRNA transcripts (33). IbpA and IbpB (inclusion body-associated proteins) are homologous to the ubiquitous group of small hsps and probably result from a duplication event since they are 48% identical to each other. The *ibpA* and *ibpB* genes form an operon that exhibits the highest known heat shock induction (33). Although very little is known about these proteins, they are likely to be involved in protein folding since (i) several small hsps function as molecular chaperones in vitro (76) and (ii) murine Hsp25 (41) and pea Hsp18.1 (91) bind thermally denatured proteins on their surface and maintain them in a folding-competent form.

45.2.4.4. ClpB

The ca. 90-kDa ClpB hsp was identified by sequence homology to ClpA, the ATPase subunit of the energy-dependent ClpAP protease (80, 128). ClpB has been reported to be a weak ATPase organized as a tetramer; however, it did not substitute for ClpA in supporting the casein-degrading activity of ClpP (154). On the basis of the observation that ClpB is highly homologous to yeast Hsp104, a protein that resolubilizes heat-inactivated luciferase from insoluble aggregates (111), and the fact that ClpA functions as a chaperone in the absence of ClpP in vitro (151), ClpB may also be a molecular chaperone.

45.2.4.5. Periplasmic Chaperones

Despite extensive searches for periplasmic molecular chaperones with wide substrate specificity, no such protein has been discovered to date. Specific chaperones displaying an immunoglobin-like fold and responsible for the assembly of pili and fimbriae have, however, been identified (71). Since the periplasmic space of *E. coli* does not appear to contain ATP, putative periplasmic molecular chaperones may function in an energy-independent manner (as Hsp90 and small hsps do in vitro) or be associated with the plasma membrane to derive the energy necessary for folding from the cytoplasm, either directly through transmembrane segments or indirectly through transduced effects. In this respect, it is interesting to note that the SecD and SecF transmembrane proteins have been proposed to help prevent the aggregation of newly translocated proteins during the early stages of folding in the periplasm (157).

45.2.5. Influence of Chaperone Overproduction on Protein Aggregation

The observation that molecular chaperones play a key role in facilitating the folding and translocation of *E. coli* host proteins raised the possibility that an increase in their intracellular concentration could alleviate the misfolding of recombinant proteins. There is indeed ample evidence that co-overexpression of the *groE* operon can increase the soluble recovery yields of recombinant proteins produced in the cytoplasm of *E. coli* and facilitate the assembly of oligomeric structures (3, 6, 11, 24, 30, 36, 37, 40, 44, 57, 58,

89, 93, 119, 121, 138, 159, 161). Since folding intermediates are prone to aggregation and proteolytic degradation, a net reduction in their concentration due to their interaction with overproduced chaperonins and subsequent release in a form that is committed to proper folding may reduce either inclusion body formation (3, 6, 24, 30, 36, 58, 89, 138, 161) or proteolysis (11, 93, 119, 121). Despite these success stories, however, overproduction of the *groE* operon may have only limited effect or no effect on the solubility of other recombinant proteins in the *E. coli* cytoplasm (18, 30, 58, 89, 124, 138, 161). By maintaining secreted proteins in a translocation-competent (i.e., loosely folded) form, chaperonin overproduction may also facilitate the translocation of recombinant proteins to the periplasmic space (12, 113), although this effect is by no means general (14, 40, 52, 93, 112).

Despite the importance of the DnaK-DnaJ-GrpE molecular chaperone machine in early folding events, relatively little attention has been paid to the effect of its overproduction on heterologous protein production in *E. coli*. Co-overexpression of DnaK alone increases the amount of soluble human growth hormone (hgH) (18) and human procollagenase (93) produced in the cytoplasm of *E. coli*. Furthermore, cells overproducing DnaK and hgH contain an average of five hgH inclusion bodies 76 nm in diameter, while only one inclusion body of 270-nm average diameter is present in a control strain that does not overexpress the chaperone (18). Since overexpression of DnaK in the absence of DnaJ can lead to plasmid instability and defective filamentation and is ultimately bacteriocidal in *E. coli* (17), a complete *dnaKJ* operon should be used to study the role of the DnaK-DnaJ-GrpE system on heterologous protein aggregation. Because of its catalytic mechanism of action, however, overproduction of GrpE may not be necessary to achieve an increase in soluble protein accumulation. An increase in the *dnaKJ* operon copy number has been shown to enhance the cytoplasmic production of soluble protein tyrosine kinases (30), preS2-S′-β-galactosidase (138), and human SPARC (124). However, in two cases, the increase in solubility was accompanied by a decrease in overall yields (30, 124). Interestingly, although the DnaK-DnaJ-GrpE system has been shown to disaggregate heat-inactivated RNA polymerase and DnaA-phospholipid aggregates in vitro (74, 166), overproduction of the *dnaKJ* operon increased the recovery yields of active preS2-S′-β-galactosidase without solubilizing preformed inclusion bodies (138). It is therefore likely that higher intracellular levels of cytoplasmic chaperones improve the yields of bona fide aggregation-prone recombinant proteins by facilitating folding rather than by dissolving inclusion body material.

Overproduction of DnaK or of the *dnaKJ* operon may also be beneficial for protein secretion, as reported in the case of a membrane-jamming LamB-LacZ fusion protein (113), human granulocyte colony-stimulating factor (14, 112), and interleukin-13 (14). However, these effects appear to depend on the nature of the signal sequence (14) and that of the substrate, as a LamB-procollagenase protein destined for export was in fact retained in the cytoplasm of cells overexpressing DnaK (93), and an increase in the *dnaK* gene dosage had no effect on the secretion of periplasmic versions of penicillin-binding protein 3 (52).

Very little information is available on the effect of other molecular chaperones on recombinant protein expression in *E. coli*. Shotgun approaches consisting of co-overproduction σ^{32} to increase the intracellular concentration of all hsps often result in intermediate solubilization yields compared with co-overexpression of a specific chaperone operon (124, 138); such approaches should be attempted with care, since an increase in proteolysis may follow. Only a few reports have analyzed the effect of SecB overproduction on heterologous protein production. Higher SecB levels improved the secretion of truncated forms of penicillin-binding protein 3 (52) and enhanced the recovery yields of periplasmic granulocyte-macrophage colony-stimulating factor and interleukin-13 following osmotic shock (14). However, no effect was observed in the case of another secreted granulocyte colony-stimulating factor (112).

From the above observations, it can be concluded that, while chaperone co-overproduction can greatly increase the recovery yields of certain recombinant proteins in a soluble and active form, this approach remains highly dependent on the nature of the target polypeptide and the choice of the overexpressed chaperone. Therefore, the effect of several plasmid-encoded molecular chaperones on the folding of the desired polypeptide should be tested in a systematic fashion.

45.3. FOLDASES

45.3.1. Isomerization of Proline Residues

The *cis*/*trans* isomerization of X-Pro prolyl peptide bonds (where X is any amino acid) has long been recognized as a rate-limiting step during in vitro refolding of purified proteins, owing to the fact that nonnative prolyl bond conformations are present in denatured proteins. Based on the observations that (i) the *trans* conformation of peptide bonds is energetically favored in nascent protein chains (97), (ii) about 5% of all prolyl bonds are estimated to be in a *cis* conformation in native polypeptides (99), and (iii) *trans* to *cis* isomerization reactions of prolyl peptide bonds are slow in vitro, it is likely that the process is accelerated catalytically in vivo. This function is performed by peptidyl-prolyl *cis*/*trans* isomerases (PPIases; for a review, see reference 49). On the basis of homology, *E. coli* contains at least eight such enzymes, two of which are located in the periplasm and six in the cytoplasm (147). While the effect of many of these PPIases on in vivo protein folding is mostly unknown, the identification of trigger factor as a PPIase that associates with nascent polypeptides and interacts with GroEL suggests the existence of a ribosome-associated folding system that accelerates prolyl bond isomerization and cooperates with molecular chaperones to guarantee efficient in vivo folding (67). This hypothesis, and the effect of trigger-factor overproduction on the fate of aggregation-prone recombinant proteins, remains to be tested.

PPIase A, the *rotA* gene product, is a dispensable periplasmic PPIase that does not appear to play an essential role in protein folding, as judged by the lack of effect of a *rotA* deletion on the total yield or the kinetics of folding of periplasmic and outer membrane proteins (81). In agreement with these results, *rotA* overproduction did not affect the yields of active soluble Fab fragments or placental alkaline phosphatase expressed in the periplasm and had only a small effect on the recovery of ScFv fragments (13, 82). Interestingly, inactivation of *surA*, a gene originally identified as required for cell survival, compromised the folding of three outer membrane proteins and increased the levels of the periplasmic protease DegP, as would be expected if misfolded proteins accumulated in the periplasm (90). Since SurA is homologous to the cytoplasmic PPIase parvulin, it is possible that this protein is a PPIase involved in general periplasmic folding. However, the folding efficiency of four secreted host proteins remained unchanged in *surA* mutants (90), suggesting that SurA could exhibit specificity towards membrane proteins. The most potent periplasmic PPIase appears to be FkpA (103), a protein that holds great promise in facilitating the production of secreted polypeptides whose misfolding results from the improper isomerization of peptidyl-prolyl bonds.

45.3.2. Disulfide Bond Formation

The presence of a correct pattern of disulfide bonds is crucial for the function and stability of many secreted eukaryotic and prokaryotic proteins. In vitro, the formation of disulfide bonds between aligned cysteine residues is thought to arise during the folding process. However, since the number of incorrect disulfides increases with the polypeptide cysteine content, the isomerization (reshuffling) of disulfides is often rate-limiting (for review, see reference 35). Disulfide bond formation is an intermolecular process involving thiol-disulfide exchange reaction with an oxidant. While small molecules such as glutathione disulfide (GSSG) can perform this role (and probably do in the endoplasmic reticulum of eukaryotic cells), they react very slowly with protein thiols at neutral pH. As a result, in vivo disulfide bond formation is predominantly accomplished through the enzymatic action of thiol/disulfide oxidoreductases whose Cys-X-Y-Cys active sites (where X and Y are different amino acids) cycle between dithiol, and disulfide forms. Thiol-disulfide exchange between an oxidized oxidoreductase (O) and a reduced protein substrate (*P*) involves the formation of mixed disulfide species according to the scheme:

$$O^{S}_{S} + {}^{HS}_{HS}P \leftrightarrow O^{S-S}_{S\ \ HS}P \leftrightarrow O^{SH}_{SH} + {}^{S}_{S}P$$

The periplasm of *E. coli* contains at least four Dsb proteins that are involved in the formation of disulfide bonds in secreted and membrane proteins (reviewed in reference 10). DsbA is a 21-kDa protein that acts as a strong oxidant, owing to the presence of an unstable disulfide bond at its active site, and possesses the ability to bind peptides in a deep groove running along its accessible cysteine. Reduced DsbA is reoxidized by the inner membrane protein DsbB, which exposes four cys residues to the periplasm. DsbC is a 24-kDa homodimer that also contains a very reactive and unstable disulfide bond. However, while DsbA exhibits little disulfide isomerase activity, it has been suggested that the main function of DsbC is to catalyze disulfide bond rearrangement (164). The recently identified DsbD protein is an essential inner membrane protein that generates a reducing source in the periplasm and appears to be required at physiological levels to maintain a proper redox environment in this cellular compartment (104).

While DsbA overproduction had no effect on the yields of correctly folded antibody fragments (82) and human alkaline phosphatase (13), higher yields of recombinant products have been observed in the presence of additional factors. For instance, the degradation of secreted T-cell receptor fragments was reduced when both DsbA and σ^{32} were overproduced but not when the concentration of DsbA alone was raised (156). Similarly, the amount of a correctly folded secreted trypsin inhibitor could be increased 14-fold upon overproduction of DsbA and supplementation of the medium with reduced glutathione, while little effect was detected in the absence of this additive (158). Interestingly, inactivation of *dsb* genes may also improve yields, as was shown in the case of a metallo-β-lactamase from *Bacteroides fragilis* which is rapidly degraded in wild-type strains owing to the formation of aberrant disulfide bonds (1). Co-overexpression of DsbC has been found to improve the yields of secreted tissue plasminogen activator (55) and a mutant version of alkaline phosphatase (126), suggesting that the isomerization of disulfide bonds in the *E. coli* periplasm may be limited. In agreement with this hypothesis, plasmid-encoded eukaryotic protein disulfide isomerase can increase the recovery yields of bovine pancreatic trypsin inhibitor (110), *Erwinia carotovora* pectate lyase C (70), and antibody fragments (69) in *E. coli*.

In contrast to the periplasm, the cytoplasm of *E. coli* is considered to be a reducing environment that is not conducive to the formation of disulfide bonds. TrxB, the product of the thioredoxin reductase gene, appears to play an essential role in this process, although it probably does not directly reduce disulfide bonds in proteins (39). The production of disulfide-bonded recombinant proteins is possible in surprisingly healthy *E. coli* strains bearing a *trxB* mutation, as was shown for mutant alkaline phosphatases (39), mouse urokinase (39), a single-chain Fv (115), and human SPARC (124). It has also been reported that incubation of wild-type cells at low temperatures favors cytoplasmic disulfide bond formation (38). Indeed, in the case of urokinase and SPARC, incubation of *trxB* mutant cells at low temperatures was essential for the accumulation of disulfide-bridged species. Interestingly, while the ratios of aggregated to soluble recombinant proteins are not affected by inactivation of *trxB* (115, 124), the solubility of several eukaryotic gene products can be increased upon overproduction of TrxA, the thioredoxin reductase substrate (161), or when their coding sequence is fused to the *trxA* gene (83). The precise mechanisms by which TrxB prevents cytoplasmic disulfide bond formation and TrxA increases solubility remain unclear.

45.4. CONCLUSIONS

The past decade has revolutionized our perception of in vivo protein folding from an energy-independent, self-arising process to one that is precisely controlled by the complex interplay of molecular chaperones, folding catalysts, and proteases. It is thus not surprising that the high-level expression of recombinant protein interferes with the fine-tuning of this mechanism and results in the aggregation and degradation of target polypeptides. This review shows that, while there is no guarantee that co-overexpression of a given folding modulator or optimization of the culture conditions will increase the recovery yield of a particular protein, enough success stories have been reported to encourage one to investigate these approaches. An improved understanding of the mechanism of action of folding modulators will undoubtedly open the way to a more rational choice of options.

Due to space limitations, many relevant references could not be cited. These can be found in a number of excellent reviews cited in the text. I am grateful to Jeff Thomas for reading the manuscript. Protein folding work in my laboratory is supported by grants from NSF and the Whitaker Foundation.

REFERENCES

1. **Alksne, L. E., D. Keeney, and B. A. Rasmunssen.** 1995. A mutation in either *dsbA* or *dsbB*, a gene encoding a component of a periplasmic disulfide bond-catalyzing sys-

tem, is required for high-level expression of the *Bacteroides fragilis* metallo-β-lactamase, CcrA, in *Escherichia coli*. *J. Bacteriol.* **177:**462–464.

2. **Allen, S. P., J. O. Polazzi, J. K. Gierse, and A. M. Easton.** 1992. Two novel heat shock genes encoding proteins produced in response to heterologous protein expression in *Escherichia coli*. *J. Bacteriol.* **174:**6938–6947.
3. **Amrein, K. E., B. Takacs, M. Stieger, J. Molnos, N. A. Flint, and P. Burn.** 1995. Purification and characterization of recombinant human p50csk protein-tyrosine kinase from an *Escherichia coli* expression system overproducing the bacterial chaperones GroES and GroEL. *Proc. Natl. Acad. Sci. USA.* **92:**1048–1052.
4. **Anfinsen, C. B.** 1973. Principles that govern the folding of protein chains. *Science* **181:**223–230.
5. **Ang, D., and C. Georgopoulos.** 1989. The heat-shock regulated *grpE* gene of *Escherichia coli* is required for bacterial growth at all temperatures but is dispensible in certain mutant backgrounds. *J. Bacteriol.* **171:**2748–2755.
6. **Ashiuchi, M., T. Yoshimura, T. Kitamura, Y. Kawata, J. Nagai, S. Gorlatov, N. Esaki, and K. Soda.** 1995. In vivo effect of GroESL on the folding of glutamate racemase of *Escherichia coli*. *J. Biochem.* **117:**495–498.
7. **Ayling, A., and F. Baneyx.** 1996. Influence of the GroE molecular chaperone machine on the *in vitro* folding of *Escherichia coli* β-galactosidase. *Protein Sci.* **5:**478–487.
8. **Banecki, B., K. Liberek, D. Wall, A. Wawrzynów, C. Georgopoulos, E. Bertoli, F. Tanfani, and M. Zylicz.** 1996. Structure-function analysis of the zinc-finger region of the DnaJ molecular chaperone. *J. Biol. Chem.* **271:** 14840–14848.
9. **Baneyx, F., and G. Georgiou.** 1991. Construction and characterization of *Escherichia coli* strains deficient in multiple secreted proteases: protease III degrades high molecular weight substrates in vivo. *J. Bacteriol.* **173:**2696–2703.
10. **Bardwell, J. C.** 1994. Building bridges: disulphide bond formation in the cell. *Mol. Microbiol.* **14:**199–205.
11. **Battistoni, A., M. T. Carri, C. Steinkuhler, and G. Rotilio.** 1993. Chaperonins dependent increase of Cu, Zn superoxide dismutase production in *Escherichia coli*. *FEBS Lett.* **322:**6–9.
12. **Battistoni, A., A. P. Mazzetti, R. Petruzzelli, M. Muramatsu, G. Federici, G. Ricci, and M. Lo-Bello.** 1995. Cytoplasmic and periplasmic production of human placental glutathione transferase in *Escherichia coli*. *Protein Expression Purif.* **6:**579–587.
13. **Beck, R., H. Crooke, M. Jarsch, J. Cole, and H. Burtscher.** 1994. Mutation in *dipZ* leads to reduced production of active placental alkaline phosphatase in *Escherichia coli*. *FEMS Microbiol. Lett.* **124:**209–214.
14. **Bergès, H., E. Joseph-Liauzun, and O. Fayet.** 1996. Combined effects of the signal sequence and the major chaperone proteins on the export of human cytokines in *Escherichia coli*. *Appl. Environ. Microbiol.* **62:**55–60.
15. **Betts, S. D., T. M. Hachigian, E. Pichersky, and C. F. Yocum.** 1994. Reconstitution of the spinach oxygen-evolving complex with recombinant Arabidopsis manganese-stabilizing protein. *Plant Mol. Biol.* **26:**117–130.
16. **Blackwell, J. R., and R. Horgan.** 1991. A novel strategy for production of a highly expressed recombinant protein in an active form. *FEBS Lett.* **295:**10–12.
17. **Blum, P., J. Ory, J. Bauernfeind, and J. Krska.** 1992. Physiological consequences of DnaK and DnaJ overproduction in *Escherichia coli*. *J. Bacteriol.* **174:**7436–7444.
18. **Blum, P., M. Velligan, N. Lin, and A. Matin.** 1992. DnaK-mediated alterations in human growth hormone protein inclusion bodies. *Bio/Technology* **10:**301–304.
19. **Bowden, G. A., and G. Georgiou.** 1988. The effect of sugars on β-lactamase aggregation in *Escherichia coli*. *Biotechnol. Prog.* **4:**97–101.
20. **Bowden, G. A., and G. Georgiou.** 1990. Folding and aggregation of β-lactamase in the periplasmic space of *Escherichia coli*. *J. Biol. Chem.* **265:**16760–16766.
21. **Bowden, G. A., A. M. Paredes, and G. Georgiou.** 1991. Structure and morphology of inclusion bodies in *Escherichia coli*. *Bio/Technology* **9:**725–730.
22. **Braig, K., Z. Otwinowski, R. Hedge, D. C. Boisvert, A. Joachimiak, A. L. Horwich, and P. B. Sigler.** 1994. The crystal structure of the bacterial chaperonin GroEL at 2.8 Å. *Nature* **371:**578–586.
23. **Broeze, R. J., C. J. Solomon, and D. H. Pope.** 1978. Effect of low temperature on in vivo and in vitro protein synthesis in *Escherichia coli* and *Pseudomonas fluorescens*. *J. Bacteriol.* **134:**861–874.
24. **Bross, P., B. S. Andresen, V. Winter, F. Kräulte, T. G. Jensen, A. Nandy, S. Kølvraa, S. Ghisla, L. Bolund, and N. Gregersen.** 1993. Co-overexpression of bacterial GroESL chaperonins partly overcomes non-productive folding and tetramer assembly of *E. coli*-expressed human medium-chain acyl-CoA dehydrogenase (MCAD) carrying the prevalent disease-causing K304E mutation. *Biochim. Biophys. Acta* **1192:**264–274.
25. **Browner, M. F., P. Rasor, S. Tugendreich, and R. J. Fletterick.** 1991. Temperature-sensitive production of rabbit muscle glycogen phosphorylase in *Escherichia coli*. *Protein Eng.* **4:**351–357.
26. **Buchberger, A., H. Schröder, M. Büttner, A. Valencia, and B. Bukau.** 1994. A conserved loop in the ATPase domain of the DnaK chaperone is essential for stable binding of GrpE. *Struct. Biol.* **1:**95–101.
27. **Buchberger, A., H. Schröder, T. Hesterkamp, H.-J. Schönfeld, and B. Bukau.** 1996. Substrate shuttling between the DnaK and GroEL systems indicates a chaperone network promoting folding. *J. Mol. Biol.* **261:**328–333.
28. **Burston, S. G., J. S. Weissman, G. W. Farr, W. A. Fenton, and A. L. Horwich.** 1996. Release of both native and non-native proteins from a *cis*-only GroEL ternary complex. *Nature* **383:**96–99.
29. **Cabilly, S.** 1989. Growth at sub-optimal temperatures allows the production of functional, antigen-binding Fab fragments in *Escherichia coli*. *Gene* **85:**553–557.
30. **Caspers, P., M. Stieger, and P. Burn.** 1994. Overproduction of bacterial chaperones improves the solubility of recombinant protein tyrosine kinases in *Escherichia coli*. *Cell. Mol. Biol.* **40:**635–644.
31. **Chalmers, J. J., E. Kim, J. N. Telford, E. Y. Wong, W. C. Tacon, M. L. Shuler, and D. B. Wilson.** 1990. Effects of temperature on *Escherichia coli* overproducing β-lactamase or epidermal growth factor. *Appl. Environ. Microbiol.* **56:**104–111.
32. **Chen, S., A. M. Roseman, A. S. Hunter, S. P. Wood, S. G. Burston, N. A. Ranson, A. R. Clarke, and H. R. Saibil.** 1994. Location of a folding protein and shape

changes in GroEL-GroES complexes imaged by cryo-electron microscopy. *Nature* **371:**261–264.

33. **Chuang, S.-E., V. Burland, G. Plunkett III, D. L. Daniels, and F. R. Blattner.** 1993. Sequence analysis of four new heat-shock genes constituting the *hslTS/ibpAB* and *hslVU* operons in *Escherichia coli*. *Gene* **134:**1–6.
34. **Collier, D. N.** 1993. SecB: a molecular chaperone of *Escherichia coli* protein secretion pathway. *Adv. Protein Chem.* **44:**151–193.
35. **Creighton, T. E., A. Zapun, and N. J. Darby.** 1995. Mechanisms and catalysis of disulphide bond formation in proteins. *Trends Biotechnol.* **13:**18–23.
36. **Dale, G. E., H. J. Schönfeld, H. Langen, and M. Stieger.** 1994. Increased solubility of trimethoprim-resistant type S1 DHFR from *Staphylococcus aureus* in *Escherichia coli* cells overproducing the chaperonins GroEL and GroES. *Protein Eng.* **7:**925–931.
37. **Davie, J. R., R. M. Wynn, M. Meng, Y. S. Huang, G. Aalund, D. T. Chuang, and K. S. Lau.** 1995. Expression and characterization of branched-chain α-ketoacid dehydrogenase kinase from the rat. Is it a histidine-protein kinase? *J. Biol. Chem.* **270:**19861–19867.
38. **Derman, A. I., and J. Beckwith.** 1995. *Escherichia coli* alkaline phosphatase localized to the cytoplasm acquires enzymatic activity in cells whose growth has been suspended: a caution for gene fusion studies. *J. Bacteriol.* **177:** 3764–3770.
39. **Derman, A. I., W. A. Prinz, D. Belin, and J. Beckwith.** 1993. Mutations that allow disulfide bond formation in the cytoplasm of *Escherichia coli*. *Science* **262:**1744–1747.
40. **Dueñas, M., J. Vázquez, M. Ayala, E. Söderlind, M. Ohlin, L. Pérez, C. A. K. Borrebaeck, and J. V. Gavilondo.** 1994. Intra- and extracellular expression of an scFv antibody fragment in *E. coli*: effect of bacterial strains and pathway engineering using GroES/L chaperonins. *BioTechniques* **16:**476–483.
41. **Ehrnsperger, M., S. Gräber, M. Gaestel, and J. Buchner.** 1997. Binding of non-native protein to Hsp25 during heat shock creates a reservoir of folding intermediates for reactivation. *EMBO J.* **16:**221–229.
42. **Ellis, R. J., and F.-U. Hartl.** 1996. Protein folding in the cell: competing models of chaperonin function. *FASEB J.* **10:**20–26.
43. **Escher, A., D. J. O'Kane, J. Lee, and A. A. Szalay.** 1989. Bacterial luciferase $\alpha\beta$ fusion protein is fully active as a monomer and highly sensitive in vivo to elevated temperatures. *Proc. Natl. Acad. Sci. USA* **86:**6528–6532.
44. **Escher, A., and A. A. Szalay.** 1993. GroE-mediated folding of bacterial luciferases *in vivo*. *Mol. Gen. Genet.* **238:** 65–73.
45. **Fayet, O., T. Ziegelhoffer, and C. Georgopoulos.** 1989. The *groES* and *groEL* heat shock gene products of *Escherichia coli* are essential for bacterial growth at all temperatures. *J. Bacteriol.* **171:**1379–1385.
46. **Fedorov, A. N., and T. O. Baldwin.** 1995. Contribution of cotranslational folding to the rate of formation of native protein structure. *Proc. Natl. Acad. Sci USA* **92:** 1227–1231.
47. **Fenton, A., Y. Kashi, K. Furtak, and A. L. Horwich.** 1994. Residues in chaperonin GroEL required for polypeptide binding and release. *Nature* **371:**614–619.
48. **Fink, A. L.** 1995. Compact intermediate states in protein folding. *Annu. Rev. Biophys. Biomol. Struct.* **24:**495–522.
49. **Fischer, G.** 1994. Peptidyl-prolyl *cis/trans* isosmerases and their effectors. *Angew. Chem. Int. Ed. Engl.* **33:**1415–1436.
50. **Flaherty, K., D. B. McKay, W. Kabsch, and K. C. Holmes.** 1991. Similarity of the three-dimensional structure of actin and the ATPase fragment of a 70-kDa heat shock cognate protein. *Proc. Natl. Acad. Sci. USA* **88:** 5041–5045.
51. **Flaherty, K. M., C. DeLuca-Flaherty, and D. B. McKay.** 1990. Three-dimensional structure of the ATPase fragment of a 70K heat-shock cognate protein. *Nature* **346:** 623–628.
52. **Fraipont, C., M. Adam, M. Nguyen-Disteche, W. Keck, J. Van-Beeumen, J. A. Ayala, B. Granier, H. Hara, and J. M. Ghuysen.** 1994. Engineering and overexpression of periplasmic forms of the penicillin-binding protein 3 of *Escherichia coli*. *Biochem. J.* **298:**189–195.
53. **Friguet, B., L. Djavadi-Ohaniance, J. King, and M. E. Goldberg.** 1994. In vitro and ribosome-bound folding intermediates of P22 tailspike protein detected with monoclonal antibodies. *J. Biol. Chem.* **269:**15945–15949.
54. **Gaitanaris, G. A., A. Vysokanov, S. C. Hung, M. E. Gottesman, and A. Gragerov.** 1994. Successive action of *Escherichia coli* chaperones in vivo. *Mol. Microbiol.* **14:** 861–869.
55. **Georgiou, G.** Personal communication.
56. **Goldberg, M. E., R. Rudolph, and R. Jaenicke.** 1991. A kinetic study of the competition between renaturation and aggregation during the refolding of denatured-reduced egg white lysozyme. *Biochemistry* **30:**2790–2797.
57. **Goloubinoff, P., A. A. Gatenby, and G. H. Lorimer.** 1989. GroE heat-shock proteins promote assembly of foreign prokaryotic ribulose bisphosphate carboxylase oligomers in *Escherichia coli*. *Nature* **337:**44–47.
58. **Gordon, C. L., S. K. Sather, S. Casjens, and J. King.** 1994. Selective in vivo rescue by GroEL/GroES of thermolabile folding intermediates to phage P22 structural proteins. *J. Biol. Chem.* **269:**27941–27951.
59. **Gragerov, A., X. Zeng, W. Zhao, W. Brukholder, and M. E. Gottesman.** 1994. Specificity of DnaK-peptide binding. *J. Mol. Biol.* **235:**848–854.
60. **Gragerov, A. L., E. S. Martin, M. A. Krupenko, M. V. Kashlev, and V. G. Nikiforov.** 1991. Protein aggregation and inclusion body formation in *Escherichia coli rpoH* mutant defective in heat shock protein induction. *FEBS Lett.* **291:**222–224.
61. **Gragerov, A. L., E. Nudler, E. Komissarova, G. A. Gaitanaris, M. E. Gottesman, and V. G. Nikiforov.** 1992. Cooperation of GroEL/GroES and DnaK/DnaJ heat shock proteins in preventing protein misfolding in *Escherichia coli*. *Proc. Natl. Acad. Sci. USA* **89:**10341–10344.
62. **Grimm, R., G. K. Donaldson, S. M. van der Vies, E. Schafer, and A. A. Gatenby.** 1993. Chaperonin-mediated reconstitution of the phytochrome photoreceptor. *J. Biol. Chem.* **268:**5220–5226.
63. **Gross, C. A.** 1996. Function and regulation of the heat shock proteins, p. 1382–1399. *In* F. C. Neidhardt, R. Curtiss III, J. L. Ingraham, E. C. C. Lin, K. B. Low, B. Magasanik, W. S. Reznikoff, M. Riley, M. Schaechter, and H. E. Umbarger (ed.), *Escherichia coli and Salmonella: Cellular and Molecular Biology*. ASM Press, Press, Washington, D.C.

64. **Harrison, C. J., M. Hayer Hartl, M. DiLiberto, F.-U. Hartl, and J. Kuriyan.** 1997. Crystal structure of the nucleotide exchange factor GrpE bound to the ATPase domain of the molecular chaperone DnaK. *Science* **276:**431–435.

65. **Hellebust, H., M. Murby, L. Abrahmsén, M. Uhlén, and S.-O. Enfors.** 1989. Different approaches to stabilize a recombinant fusion protein. *Bio/Technology* **7:**165–168.

66. **Hendrick, J. P., T. Langer, T. A. Davis, F.-U. Hartl, and M. Wiedmann.** 1993. Control of folding and membrane translocation by binding of the chaperone DnaJ to nascent polypeptides. *Proc. Natl. Acad. Sci. USA* **90:**10216–10220.

67. **Hesterkamp, T., and B. Bukau.** 1996. The *Escherichia coli* trigger factor. *FEBS Lett.* **389:**32–34.

68. **Hesterkamp, T., S. Hauser, H. Lütcke, and B. Bukau.** 1996. *Escherichia coli* trigger factor is a prolyl isomerase that associates with nascent polypeptide chain. *Proc. Natl. Acad. Sci. USA* **93:**4437–4441.

69. **Humphreys, D. P., N. Weir, A. Lawson, A. Mountain, and P. A. Lund.** 1996. Co-expression of human protein disulphide isomerase (PDI) increase the yield of an antibody Fab′ fragment expressed in *Escherichia coli. FEBS Lett.* **380:**194–197.

70. **Humphreys, D. P., N. Weir, A. Mountain, and P. A. Lund.** 1995. Human protein disulfide isomerase functionally complements a *dsbA* mutation and enhances the yield of pectate lyase C in *Escherichia coli. J. Biol. Chem.* **270:** 28210–28215.

71. **Hung, D. L., S. D. Knight, R. M. Woods, J. S. Pinkner, and S. J. Hultgren.** 1996. Molecular basis of two subfamilies of immunoglobin-like chaperones. *EMBO J.* **15:** 3792–3805.

72. **Hunt, J. F., A. J. Weaver, S. J. Landry, L. Gierasch, and J. Deisenhofer.** 1996. The crystal structure of the GroES co-chaperonin at 2.8 Å resolution. *Nature* **379:**37–45.

73. **Hutchinson, K. A., K. D. Dittmar, and W. B. Pratt.** 1994. All of the factors required for assembly of the glucocorticoid receptor into a functional complex with heat shock protein 90 are preassociated in a self-sufficient protein folding structure, a foldosome. *J. Biol. Chem.* **269:** 27894–27889.

74. **Hwang, D. S., E. Crooke, and A. Kornberg.** 1990. Aggregated DnaA protein is dissociated and activated for DNA replication by phospholipase or DnaK protein. *J. Biol. Chem.* **265:**19244–19248.

75. **Jaenicke, R.** 1987. Folding and association of proteins. *Prog. Biophys. Mol. Biol.* **49:**117–237.

76. **Jakob, U., M. Gaestel, K. Engel, and J. Buchner.** 1993. Small heat shock proteins are molecular chaperones. *J. Biol. Chem.* **268:**1517–1520.

77. **Jordan, R., and R. McMacken.** 1995. Modulation of the ATPase activity of the molecular chaperone DnaK by peptides and the DnaJ and GrpE heat shock proteins. *J. Biol. Chem.* **270:**4563–4569.

78. **Kandror, O., L. Busconi, M. Sherman, and A. L. Goldberg.** 1994. Rapid degradation of an abnormal protein in *Escherichia coli* involves the chaperones GroEL and GroES. *J. Biol. Chem.* **269:**23575–23582.

79. **Kane, J. F., and D. L. Hartley.** 1988. Formation of recombinant protein inclusion bodies in *Escherichia coli. Trends Biotechnol.* **6:**95–101.

80. **Kitagawa, M., C. Wada, S. Yoshioka, and T. Yura.** 1991. Expression of ClpB, an analog of the ATP-dependent protease regulatory subunit in *Escherichia coli*, is controlled by a heat-shock sigma factor. *J. Bacteriol.* **173:**4247–4253.

81. **Kleerebezem, M., M. Heutink, and J. Tommassen.** 1995. Characterization of an *Escherichia coli rotA* mutant, affected in periplasmic peptidyl-prolyl *cis/trans* isomerase. *Mol. Microbiol.* **18:**313–320.

82. **Knappik, A., C. Krebber, and A. Plückthun.** 1993. The effect of folding catalysts on the in vivo folding of different antibody fragments expressed in *Escherichia coli. Bio/Technology* **11:**77–83.

83. **Kopetzki, E., G. Schumacher, and P. Buckel.** 1989. Control of formation of active soluble or inactive insoluble baker's yeast α-glucosidase PI in *Escherichia coli* by induction and growth conditions. *Mol. Gen. Genet.* **216:**149–155.

84. **Kudlicki, W., J. Chirgwin, G. Kramer, and B. Hardesty.** 1995. Folding of an enzyme into an active conformation while bound as peptidyl-tRNA to the ribosome. *Biochemistry* **34:**14284–14287.

85. **Kudlicki, W., O. W. Odom, G. Kramer, and B. Hardesty.** 1994. Activation and release of enzymatically inactive, full-length rhodanese that is bound to ribosomes as peptidyl-tRNA. *J. Biol. Chem.* **269:**16549–16553.

86. **Landry, S. J., J. Zeilstra-Ryalls, O. Fayet, C. Georgopoulos, and L. M. Gierasch.** 1993. Characterization of a functionally important mobile domain of GroES. *Nature* **364:**255–258.

87. **Langer, T., C. Lu, H. Echols, J. Flanagan, M. K. Hayer, and F.-U. Hartl.** 1992. Successive action of DnaK, DnaJ and GroEL along the pathway of chaperone-mediated protein folding. *Nature* **356:**683–689.

88. **LaVallie, E. R., E. A. DiBlasio, S. Kovacic, K. L. Grant, P. F. Schendel, and J. M. McCoy.** 1993. A thioredoxin gene fusion expression system that circumvents inclusion body formation in the *E. coli* cytoplasm. *Bio/Technology* **11:**187–193.

89. **Lawson, J. E., X. D. Niu, K. S. Browning, H. L. Trong, J. Yan, and L. J. Reed.** 1993. Molecular cloning and expression of the catalytic subunit of bovine pyruvate dehydrogenase phosphatase and sequence similarity with protein phosphatase 2C. *Biochemistry* **32:**8987–8993.

90. **Lazar, S. W., and R. Kolter.** 1996. SurA assists the folding of *Escherichia coli* outer membrane proteins. *J. Bacteriol.* **178:**1770–1773.

91. **Lee, G. J., A. M. Roseman, H. R. Saibil, and E. Vierling.** 1997. A small heat shock protein stably binds heat-denatured model substrates and can maintain a substrate in a folding-competent state. *EMBO J* **16:**659–671.

92. **Lee, S. C., Y. C. Choi, and M.-Y. Yu.** 1990. Effect of the N-terminal hydrophobic sequence of hepatitis B virus surface antigen on the folding and assembly of hybrid β-galactosidase in *Escherichia coli. Eur. J. Biochem.* **187:**417–424.

93. **Lee, S. C., and P. O. Olins.** 1992. Effect of overproduction of heat shock chaperones GroESL and DnaK on human procollagenase production in *Escherichia coli. J. Biol. Chem.* **267:**2849–2852.

94. **Liao, H. H.** 1991. Effect of temperature on the expression of wild-type and thermostable mutants of kanamycin nucleotidyltransferase in *Escherichia coli. Protein Expression Purif.* **2:**43–50.

95. **Liberek, K., J. Marzlalek, D. Ang, and C. Georgopoulos.** 1991. *Escherichia coli* DnaJ and GrpE heat shock proteins

jointly stimulate the ATPase activity of DnaK. *Proc. Natl. Acad. Sci. USA* **88:**2874–2878.

96. **Liberek, K., D. Wall, and C. Georgopoulos.** 1995. The DnaJ chaperone catalytically activates the DnaK chaperone to preferentially bind the sigma 32 heat shock transcriptional factor. *Proc. Natl. Acad. Sci. USA* **92:** 6224–6228.

97. **Lim, V. I., and A. S. Spirin.** 1986. Stereochemical analysis of ribosomal transpeptidation: conformation of nascent peptide. *J. Mol. Biol.* **188:**565–574.

98. **Lorimer, G. H.** 1996. A quantitative assessment of the role of the chaperonin proteins in protein folding in vivo. *FASEB J.* **10:**5–9.

99. **MacArthur, M. W., and J. M. Thornton,** 1991. Influence of proline residues on protein conformation. *J. Mol. Biol.* **218:**397–412.

100. **Makeyev, E. V., V. A. Kolb, and A. S. Spirin.** 1996. Enzymatic activity of the ribosome-bound nascent polypeptide. *FEBS Lett.* **378:**166–170.

101. **Mande, S. C., V. Mehra, B. R. Bloom, and W. G. J. Hol.** 1996. Structure of the heat shock protein chaperonin-10 of *Mycobacterium leprae*. *Science* **271:**203–207.

102. **Mayhew, M., A. C. da Silva, J. Martin, H. Erdjument-Bromage, P. Tempst, and F. U. Hartl.** 1996. Protein folding in the central cavity of the GroEL-GroES chaperonin complex. *Nature* **379:**420–426.

103. **Missiakas, D., J.-M. Betton, and S. Raina.** 1996. New components of protein folding in extracytoplasmic compartments of *Escherichia coli* SurA, FkpA and Skp/OmpH. *Mol. Microbiol.* **21:**871–884.

104. **Missiakas, D., F. Schwager, and S. Raina.** 1995. Identification and characterization of a new disulfide isomerase-like protein (DsbD) in *Escherichia coli*. *EMBO J* **14:** 3415–3424.

105. **Mitraki, A., and J. King.** 1989. Protein folding intermediates and inclusion body formation. *Bio/Technology* **7:**690–697.

106. **Moore, J. T., A. Uppal, F. Maley, and G. F. Maley.** 1993. Overcoming inclusion body formation in a high-level expression system. *Protein Expression Purif.* **4:**160–163.

107. **Neidhardt, F. C., and R. A. VanBogelen.** 1987. Heat shock response, p. 1334–1345. *In* F. C. Neidhardt, J. L. Ingraham, K. B. Low, B. Magasanik, M. Schaechter, and H. E. Umbarger (ed.), *Escherichia coli and Salmonella: Cellular and Molecular Biology*. ASM Press, Washington, D.C.

108. **Nilsson, B., and S. Anderson.** 1991. Proper and improper folding of proteins in the cellular environment. *Annu. Rev. Microbiol.* **45:**607–635.

109. **Oberg, K., B. A. Chrunyk, R. Wetzel, and A. L. Fink.** 1994. Nativelike secondary structure in interleukin 1-β inclusion bodies by attenuated total reflectance FTIR. *Biochemistry* **33:**2628–2634.

110. **Ostermeier, M., K. De Sutter, and G. Georgiou.** 1996. Eukaryotic protein disulfide isomerase complements *Escherichia coli dsbA* mutants and increases the yield of heterologous secreted protein with disulfide bonds. *J. Biol. Chem.* **271:**10616–10622.

111. **Parsell, D. A., A. S. Kowal, M. A. Singer, and S. Lindquist.** 1994. Protein disaggregation mediated by heat-shock protein Hsp 104. *Nature* **372:**475–478.

112. **Pérez-Pérez, J., C. Martínez-Caja, J. L. Barbero, and J. Gutiérrez.** 1995. DnaK/DnaJ supplementation improves the periplasmic production of human granulocyte-colony stimulating factor in *Escherichia coli*. *Biochem Biophys. Res. Commun.* **210:**524–529.

113. **Phillips, G. J., and T. J. Silhavy.** 1990. Heat-shock proteins DnaK and GroEL facilitate export of LacZ hybrid proteins in *E. coli*. *Nature* **344:**882–884.

114. **Piatak, M., J. A. Lane, W. Laird, M. J. Bjorn, A. Wang, and M. Williams.** 1988. Expression of soluble and fully functional ricin A chain in *Escherichia coli* is temperature sensitive. *J. Biol. Chem.* **263:**4837–4843.

115. **Proba, K., L. Ge, and A. Plückthun.** 1995. Functional antibody single-chain fragments from the cytoplasm of *Escherichia coli*: influence of thioredoxin reductase (TrxB). *Gene* **159:**203–207.

116. **Przybycien, T. M., J. P. Dunn, P. Valax, and G. Georgiou.** 1994. Secondary structure characterization of β-lactamase inclusion bodies. *Protein Eng.* **7:**131–136.

117. **Pugsley, A. P.** 1993. The complete general secretory pathway in gram-negative bacteria. *Microbiol. Rev.* **57:** 50–108.

118. **Randall, L. L., and S. J. S. Hardy.** 1995. High selectivity with low specificity: how SecB has solved the paradox of chaperone binding. *Trends Biochem. Sci.* **20:**65–70.

119. **Roman, L. J., E. A. Sheta, P. Martasek, S. S. Gross, Q. Liu, and B. S. S. Masters.** 1995. High-level expression of functional rat neuronal nitric oxide synthase in *Escherichia coli*. *Proc. Natl. Acad. Sci. USA* **92:**8428–8432.

120. **Rudolph, R., and H. Lilie.** 1996. In vitro folding of inclusion body proteins. *FASEB J.* **10:**49–56.

121. **Sato, K., M. H. Sato, A. Yamaguchi, and M. Yoshida.** 1994. Tetracycline/H+ antiporter was degraded rapidly in *Escherichia coli* cells when truncated at last transmembrane helix and this degradation was protected by overproduced GroEL/ES. *Biochem. Biophys. Res. Commun.* **202:**258–264.

122. **Sawyer, J. R., J. Schlom, and S. V. S. Kashmiri.** 1994. The effects of induction conditions on production of a soluble anti-tumor SFv in *Escherichia coli*. *Protein Eng.* **7:** 1401–1406.

123. **Schein, C. H., and M. H. M. Noteborn.** 1988. Formation of soluble recombinant proteins in *Escherichia coli* is favored by lower growth temperatures. *Bio/Technology* **6:** 291–294.

124. **Schneider, E. L., J. G. Thomas, J. A. Bassuk, E. H. Sage, and F. Baneyx.** 1997. Manipulating the aggregation and oxidation of human SPARC in the cytoplasm of *Escherichia coli*. *Nat. Biotechnol.* **15:**581–585.

125. **Schönfeld, H.-J., D. Schmidt, H. Schröder, and B. Bukau.** 1995. The DnaK chaperone system of *Escherichia coli*: quaternary structure and interactions of the DnaK and GrpE components. *J. Biol. Chem.* **270:**2183–2189.

126. **Sone, M., Y. Akiyama, and K. Ito.** 1997. Differential in vivo roles played by DsbA and DsbC in the formation of protein disulfide bonds. *J. Biol. Chem.* **272:**10349–10352.

127. **Speed, M. A., D. I. C. Wang, and J. King.** 1996. Specific aggregation of partially folded polypeptide chains: the molecular basis of inclusion body composition. *Nat. Biotechnol.* **14:**1283–1287.

128. **Squires, C. L., S. Pedersen, B. M. Ross, and C. Squires.** 1991. ClpB is the *Escherichia coli* heat shock protein F84.1. *J. Bacteriol.* **173:**4254–4262.

129. **Steczko, J., G. A. Donoho, J. E. Dixon, T. Sugimoto, and B. Axelrod.** 1991. Effect of ethanol and low-temperature culture on expression of soybean lipoxygenase L-1 in *Escherichia coli*. *Protein Expression Purif.* **2:**221–227.

130. **Strandberg, L., and S. Enfors.** 1991. Factors influencing inclusion body formation in the production of a fused protein in *Escherichia coli*. *Appl. Environ. Microbiol.* **57:** 1669–1674.

131. **Sugimoto, S., Y. Yokoo, N. Hatakeyama, A. Yotsuji, S. Teshiba, and H. Hagino.** 1991. Higher culture pH is preferable for inclusion body formation of recombinant salmon growth hormone in *Escherichia coli*. *Biotechnol. Lett.* **13:**385–388.

132. **Szabo, A., R. Korszun, F.-U. Hartl, and J. Flanagan.** 1996. A zinc finger-like domain of the molecular chaperone DnaJ is involved in binding to denatured substrates. *EMBO J.* **15:**408–417.

133. **Takagi, H., Y. Morinaga, M. Tsuchiya, H. Ikemura, and M. Inouye.** 1988. Control of folding of proteins secreted by a high expression secretion vector, pINIIIompA: 16-fold increase in production of active subtilisin E in *Escherichia coli*. *Bio/Technology* **6:**948–950.

134. **Taylor, G., M. Hoare, D. R. Gray, and F. A. O. Martson.** 1986. Size and density of inclusion bodies. *Bio/Technology* **4:**553–557.

135. **Teschner, W., R. Rudolph, and J. R. Garel.** 1987. Intermediates on the folding pathway of ODH from *Pecten jacobaeus*. *Biochemistry* **26:**2791–2796.

136. **Thomas, J. G., A. Ayling, and F. Baneyx.** 1997. Molecular chaperones, folding catalysts and the recovery of biologically active recombinant proteins from *E. coli*: to fold or to refold? *Appl. Biochem. Biotechnol.* **66:**197–238.

137. **Thomas, J. G., and F. Baneyx.** 1996. Protein folding in the cytoplasm of *Escherichia coli*: requirements for the DnaK-DnaJ-GrpE and GroEL-GrosES molecular chaperone machines. *Mol. Microbiol.* **21:**1185–1196.

138. **Thomas, J. G., and F. Baneyx.** 1996. Protein misfolding and inclusion body formation in recombinant *Escherichia coli* cells overproducing heat-shock proteins. *J. Biol. Chem.* **271:**11141–11147.

139. **Todd, M. J., P. V. Viitanen, and G. H. Lorimer.** 1994. Dynamics of the chaperonin ATPase cycle: implications for facilitated protein folding. *Science* **265:**659–666.

140. **Tokatlidis, K., B. Friguet, D. Deville-Bonne, F. Baleux, A. N. Fedorov, N. A., L. Djavadi-Ohaniance, and M. E. Goldberg.** 1995. Nascent chains: folding and chaperone interaction during elongation on ribosomes. *Philos. Trans. R. Soc. Lond. B* **348:**89–95.

141. **Ueguchi, C., and K. Ito.** 1992. Multicopy suppression: an approach to understanding intracellular functioning of the protein export system. *J. Bacteriol.* **174:**1454–1461.

142. **Valax, P., and G. Georgiou.** 1993. Molecular characterization of β-lactamase inclusion bodies produced in *Escherichia coli*. 1. Composition. *Biotechnol. Prog.* **9:**539–547.

143. **Vasina, J. A., and F. Baneyx.** 1996. Recombinant protein expression at low temperatures under the transcriptional control of the major *Escherichia coli* cold shock promoter *cspA*. *Appl. Environ. Microbiol.* **62:**1444–1447.

144. **Vasina, J. A., and F. Baneyx.** 1997. Expression of aggregation-prone proteins at low temperatures: a comparative study of the *E. coli cspA* and *tac* promoter systems. *Protein Expression Purif.* **9:**211–218.

145. **Vasina, J. A., and F. Baneyx.** Unpublished data.

146. **Vasina, J. A., M. S. Peterson, and F. Baneyx.** 1998. Scale-up and optimization of the low-temperature inducible *cspA* promoter system. *Biotechnol. Prog.* **14:**714–721.

147. **Wall, J. G., and A. Plückthun.** 1995. Effects of overexpressing folding modulators on the *in vivo* folding of heterologous proteins in *Escherichia coli*. *Curr. Opin. Biotechnol.* **6:**507–516.

148. **Weissman, J. S., C. M. Hohl, O. Kovalenko, Y. Kashi, S. Chen, K. Braig, H. R. Saibil, W. A. Fenton, and A. L. Horwich.** 1995. Mechanisms of GroEL action: Productive release of polypeptide from a sequestered position under GroES. *Cell* **83:**577–587.

149. **Wetzel, R.** 1994. Mutations and off-pathway aggregation of proteins. *Trends Biotechnol.* **12:**193–198.

150. **White, C. B., Q. Chen, G. L. Kenyon, and P. C. Babbitt.** 1995. A novel activity of OmpT proteolysis under extreme denaturing conditions. *J. Biol. Chem.* **270:** 12990–12994.

151. **Wickner, S., S. Gottesman, D. Skowyra, J. Hoskins, K. McKenney, and R. Maurizi.** 1994. A molecular chaperone, ClpA, functions like DnaK and DnaJ. *Proc. Natl. Acad. Sci. USA* **91:**12218–12222.

152. **Wiech, H., J. Buchner, R. Zimmermann, and U. Jakob.** 1992. Hsp90 chaperones protein folding in vitro. *Nature* **358:**169–170.

153. **Wilkinson, D. L., and R. G. Harrison.** 1991. Predicting the solubility of recombinant proteins in *Escherichia coli*. *Bio/Technology* **9:**443–448.

154. **Woo, K. M., K. I. Kim, A. L. Goldberg, D. B. Ha, and C. H. Chung.** 1992. The heat shock protein ClpB in *Escherichia coli* is a protein-activated ATPase. *J. Biol. Chem.* **267:**20429–20434.

155. **Wright, P. E., H. J. Dyson, and R. A. Lerner.** 1988. Conformation of peptide fragments of proteins in aqueous solution: implication for initiation of protein folding. *Biochemistry* **27:**7167–7175.

156. **Wülfing, C., and A. Plückthun.** 1994. Correctly folded T-cell receptor fragments in the periplasm of *Escherichia coli*. Influence of folding catalysts. *J. Mol. Biol.* **242:**655–669.

157. **Wülfing, C., and A. Plückthun.** 1994. Protein folding in the periplasm of *Escherichia coli*. *Mol. Microbiol.* **12:** 685–692.

158. **Wunderlich, M., and R. Glockshuber.** 1993. In vivo control of redox potential during protein folding catalyzed by bacterial protein disulfide-isomerase (DsbA). *J. Biol. Chem.* **268:**24547–24550.

159. **Wynn, R. M., J. R. Davie, R. P. Cox, and D. T. Chuang.** 1992. Chaperonins GroEL and GroES promote assembly of heterotetramers ($\alpha_2\beta_2$) of mammalian mitochondrial branched-chain α-keto decarboxylase in *Escherichia coli*. *J. Biol. Chem.* **267:**12400–12403.

160. **Xu, Z., A. L. Horwich, and P. B. Sigler.** 1997. The crystal structure of the asymmetric GroEL-GrosES-(ADP)7 chaperonin complex. *Nature* **388:**741–750.

161. **Yasukawa, T., C. Kanei-Ishii, T. Maekawa, J. Fujimoto, T. Yamamoto, and S. Ishii.** 1995. Increase of solubility

of foreign proteins in *Escherichia coli* by coproduction of the bacterial thioredoxin. *J. Biol. Chem.* **270:**25328–25331.

162. **Yura, T., H. Nagai, and H. Mori.** 1993. Regulation of the heat-shock response in bacteria. *Annu. Rev. Microbiol.* **47:**321–350.
163. **Zahn, R., S. Perrett, G. Stenberg, and A. R. Fersht.** 1996. Catalysis of amide proton exchange by the molecular chaperones GroEL and SecB. *Science* **271:**642–645.
164. **Zapun, A., D. Missiakas, S. Raina, and T. E. Creighton.** 1995. Structural and functional characterization of DsbC, a protein involved in disulfide bond formation in *Escherichia coli*. *Biochemistry* **34:**5075–5089.
165. **Zhu, X., X. Zhao, W. F. Burkholder, A. Gragerov, C. M. Ogata, M. E. Gottesman, and W. A. Hendrickson.** 1996. Structural analysis of substrate binding by the molecular chaperone DnaK. *Nature* **272:**1606–1614.
166. **Ziemienowicz, A., D. Skowyra, J. Zeilstra-Ryalls, O. Fayet, and C. Georgopoulos.** 1993. Both the *Escherichia coli* chaperone systems, GroEL/GroES and DnaK/DnaJ/GrpE, can reactivate heat-treated RNA polymerase. Different mechanisms for the same activity. *J. Biol. Chem.* **268:**25425–25431.
167. **Zipser, D., and D. Perrin.** 1963. Complementation on ribosomes. *Cold Spring Harbor Symp. Quant. Biol.* **28:** 533–537.

Expression of G Protein-Coupled Receptors in Microorganisms

R. JOCKERS AND A. D. STROSBERG

46

Recent years have witnessed a considerable expansion of studies involving G protein-coupled receptors (GPCRs), a major family of membrane-bound receptors (33). This increase is directly linked to progress in cloning and sequencing of the corresponding genes or cDNAs. The first GPCR proteins to be isolated and affinity-purified turned out not only to correspond to single polypeptide chains, but also to be encoded by intronless homologous genes. This was the case for the β- and α-adrenergic receptors (58). Cloning by homologous hybridization thus rapidly yielded a number of additional genes actually encoding additional receptor subtypes. PCR-based procedures contributed to the cloning of an array of cDNAs or genes, sometimes only identified by the fact that they displayed the telltale hallmark of GPCRs: nucleotide sequences corresponding to seven stretches of 22 to 28 hydrophobic residues, likely to correspond to as many transmembrane domains (37). Several of these genes still remain unidentified, since no natural (or even synthetic) ligand was identified, but many of these so-called "orphan" receptors gradually became reconciled with known pharmacological activities.

The fact that all GPCRs share very similar structural features led early on to the idea that all could be expressed in similar cellular systems to allow systematic ligand binding and possibly effector activation studies. Such "generic" expression systems were deemed necessary in view of the considerable importance of GPCRs, which intervene in all important physiopathological interactions in the human body.

Among the available expression systems, microorganisms were initially discarded in view of the lack of success in functionally expressing bacteriorhodopsin in *Escherichia coli*, an early structural model for GPCR. This protein, while not coupled to a G protein, displays seven transmembrane domains in its single polypeptide chain, but expression in *E. coli* of the protein originally found in *Halobacterium* sp. unfortunately results in an inactive protein. Partial activity could be recovered only after renaturation in solution.

For this reason, mammalian (CHO cells, 3T3 fibroblasts, etc.), amphibian (*Xenopus* oocytes, etc.) or baculovirus-infected insect cells became the generally preferred expression systems. Nevertheless, the demonstration that β-adrenergic receptors could actually be functionally expressed in *E. coli*, and also in *Saccharomyces cerevisiae*, has now also established microorganisms as valuable systems (60). A number of obvious advantages were revealed by the first examples: ease of use, low cost, low background binding, absence of endogenous GPCRs, G proteins, or effectors. Since these first studies, a number of additional GPCRs have been functionally expressed in *E. coli* or in yeasts, justifying a review of the currently available data.

We will discuss here the various properties and applications of the microbial expression systems for GPCR.

46.1. EXPRESSION IN MICROORGANISMS FOR FUNCTIONAL STUDIES

Several features have contributed to the success of microorganisms for functional studies of GPCRs. Most notable are well-developed genetics, easy access to site-directed mutagenesis, and the absence of endogenous signaling molecules in the case of *E. coli*.

46.1.1. Analysis of Receptor Topology and Ligand Binding Pocket

46.1.1.1. Bacteriorhodopsin

Bacteriorhodopsin (bR), naturally made in *Halobacterium salinarium*, was the first seven-transmembrane-spanning protein to be successfully expressed in *E. coli* (8, 27, 47). Although not coupled to heterotrimeric G proteins, bR serves as a valuable model for the topology of GPCR (20). Purification of bR expressed in *E. coli*, and subsequent functional reconstitution with retinal, has been reported by Braiman et al. (5). Site-directed point mutations were generated in 1987 by Hackett et al. (15) to analyze protein-retinal interactions. During the following years, a large number of mutant bR proteins were expressed in *E. coli*, then purified and reconstituted in liposomes for functional studies (28, 44, 56; see also reference 29 for review and references within). bR is today one of the best-characterized membrane proteins: it provides a detailed picture of amino acid residues important for the overall topological organization in membrane-spanning regions and connecting loops and of residues involved in retinal binding and proton pumping (29).

Successful expression of bR in *Schizosacharomyces pombe* and *H. salinarium* has been reported more recently (22, 23, 31, 32, 48) and has also contributed to the identification of functionally crucial amino acid residues. Homologous expression of bR in halobacteria revealed the limitations of expression of denatured bR in *E. coli* combined with in vitro renaturation, since the properties of some mutants turned out to be dependent on whether these proteins were expressed in bacteria or halobacteria (35). These differences might be due to the fact that bR expressed in halobacteria assembles into specialized membrane structures, purple membrane patches, forming two-dimensional crystalline arrays.

In terms of analyzing mutants, Teufel et al. (62) have gone beyond the replacement of single amino acids by inserting a 13-amino-acid-long peptide into each of the six helix-connecting loops. Characterization of *E. coli*- and *S. pombe*-expressed mutant bR not only revealed the structural importance of each loop but also opened the door to the construction of multifunctional membrane proteins via loop replacement.

46.1.1.2. Other Microbial GPCRs

Functional overexpression of two other microbial GPCRs, the cyclic AMP receptor 1 from *Dictyostelium discoideum* and the α-mating-factor receptor from *S. cerevisiae*, has been reported. Overexpression of wild-type and truncated α-mating-factor receptor contributed to its biochemical characterization and identification as the product of the STE2 gene (26). The cyclic AMP receptor is only expressed at early developmental growth states of *D. discoideum*. Constitutive and functional overexpression of this receptor revealed multiple insights into receptor regulation and its importance during development (3).

46.1.1.3. Expression of Mammalian β-Adrenergic Receptors in *E. coli*

The successful expression of two mammalian GPCRs, the β_1- and β_2-adrenergic receptors in *E. coli* (41, 42), triggered several interesting functional studies. The molecular basis of ligand binding selectivity was investigated by expressing chimeric β_1/β_2-adrenergic receptors in *E. coli* and studying a number of selective ligands. Multiple subsites were shown to contribute to the selectivity of ligands toward either one of the subtypes (43). An in situ screening procedure (Fig. 1) was developed to characterize mutations in the β_2-adrenergic receptor (6, 11). Receptor-expressing bacteria were plated out on agar, and binding of a radiolabeled ligand was monitored directly by autoradiography of nitrocellulose replica filters. After saturation mutagenesis in the second transmembrane helix of the β_2-adrenergic receptor, Breyer et al. (6) used this screening approach to identify essential amino acid residues involved in ligand binding of this transmembrane region.

Lacatena et al. (34) analyzed the topology of the β_2-adrenergic receptor expressed in *E. coli* by using β_2-adrenergic receptor bacterial alkaline phosphatase fusion proteins. Alkaline phosphatase has to be exported in the periplasma to become active. Fusion to receptor domains located in the periplasma leads to higher phosphatase activity than fusion to domains located in the cytoplasma. Using this approach, the overall topology of the β_2-adrenergic receptor was confirmed and the correct positioning of the receptor N terminus was shown to depend strongly on the presence of its C-terminal portion.

46.1.1.4. Expression of Other Mammalian GPCRs in *E. coli*

The list of GPCRs expressed in *E. coli* (see Table 1) has now been extended to several other members: the human

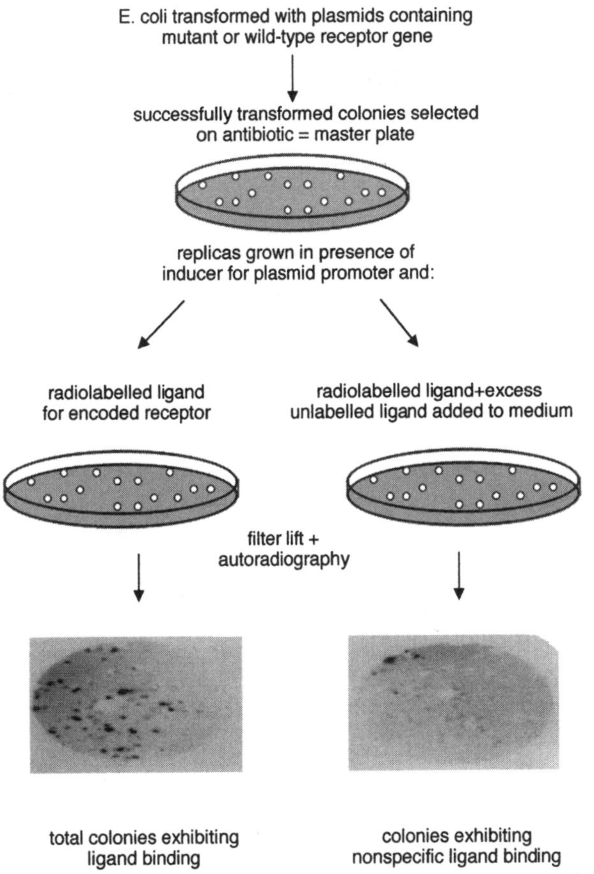

FIGURE 1 Replica filter assay for *E. coli* bacteria expressing functional receptors. *E. coli* transformed with genes coding for receptors may be directly screened for binding radiolabeled ligand. For example, colonies expressing β-adrenoceptors were identified after autoradiography. On the left, colonies were grown in the presence of the radioligand [^{125}I]iodocyanopindolol alone, while on the right, unlabeled propranolol, another β-adrenoceptor antagonist, was added to the medium.

serotonin 5HT$_{1A}$ (2), the human ET$_B$ endothelin (16, 17), the rat neurokinin A (13), the turkey β_1-adrenergic (7), the octopus rhodopsin (19), the human adenosine A1 (25), the human dopamine D$_3$ (64), the human neuropeptide Y1 (21), and the rat neurotensin receptor (12).

In several neuropeptide Y1 receptors mutated in the supposed ligand binding site, agonist binding was readily detectable in *E. coli* but poorly detectable when the receptors were expressed in mammalian cell lines (46). This finding confirms that GPCRs expressed in *E. coli* are functional and interact with bacterial phospholipids at least as well as do receptors expressed in mammalian cells.

46.1.2. Analysis of G Protein Coupling and Signaling

46.1.2.1. G Protein Coupling

Recombinant receptors expressed in *E. coli* retain not only their ligand binding properties and topology but also their ability to interact with various G proteins (10). Recombinant receptors interact in a manner indistinguishable from that of their counterparts in native membranes, with recombinant G protein α-subunits in the presence of purified

TABLE 1 G protein-coupled receptors expressed in microorganisms

Receptor	Type	Species	Host	Expression Level[a]	Reference[b]
Bacteriorhodopsin	bR	Halobacterial	*E. coli*	0.05% of total proteins;	8
				17 mg/liter	56
			S. pombe	10% of membrane protein	23
			H. salinarium	1.3 mg/liter;	48
				20–40% of wt	31, 32
Adrenergic receptor	β_1	Human	*E. coli*	50/cell (1 pmol/mg)[c]	42
	β_1	Turkey	*E. coli*	30/cell (0.6 pmol/mg)[c]	7
	β_2	Human	*E. coli*	220/cell (4 pmol/mg)[c]	41
			S. cerevisiae	115 pmol/mg	30
Muscarinic acetylcholine receptor	M1	Human	*S. cerevisiae*	0.02 pmol/mg	49
	M5	Rat	*S. cerevisiae*	0.12 pmol/mg	24
Serotonin receptor	$5HT_{1A}$	Human	*E. coli*	120/cell (2 pmol/mg)[c]	2
	$5HT_{5A}$	Mouse	*S. cerevisiae*	16 pmol/mg	1
			P. pastoris	22 pmol/mg	67
Endothelin receptor	ET_B	Human	*E. coli*	41/cell (0.8 pmol/mg)[c]	17
Neurotensin receptor		Rat	*E. coli*	15 pmol/mg	63
				800/cell (26 pmol/mg)[c]	
Opsin	oRh	Octopus	*E. coli*	1–10 mg/liter	19
Neurokinin A receptor	NK2	Rat	*E. coli*	2.5–7 pmol/mg	13
Neuropeptide Y1 receptor	NPY Y1	Human	*E. coli*	9 pmol/mg	21
Adenosine receptor	A_1	Human	*E. coli*	0.2–0.4 pmol/mg	25
	A_{2a}	Rat	*S. cerevisiae*	0.4 pmol/mg	51
Dopamine receptor	D_{2S}	Human	*S. cerevisiae*	1–2 pmol/mg	52
			S. pombe	15 pmol/mg	53
	D_3	Human	*E. coli*	0.5–1 pmol/10^9 cells	64
Somatostatin receptor	SST2	Rat	*S. cerevisiae*	0.1–0.2 pmol/mg	50
Opioid receptor	μ	Human	*P. pastoris*	0.4 pmol/mg	61

[a]Expression level of receptors; wt, wild type; 1 liter, liter of culture medium.
[b]For clarity, only the first reported reference has been listed (for more complete references, see the text). More than one reference is included if considerable improvement of expression levels has been obtained.
[c]For uniformity, this value has been estimated from the original value as sites per cell, using the approximation that 15 pmol/mg of crude membrane proteins corresponds to approximately 800 sites per *E. coli* cell.

$G_{\beta\gamma}$-dimers (Fig. 2). The idea to use bacterial membranes expressing GPCRs for reconstitution experiments relies on the absence of resident endogenous receptors or G proteins. This approach has been employed for the human β_1- and β_2-adrenergic receptors, the human $5HT_{1A}$ serotonin receptor, and the human A_1-adenosine receptor, to examine the G protein specificity of these recombinant receptors (2, 10, 25). Reconstitution of the *E. coli*-expressed $5HT_{1A}$ serotonin receptor with G proteins showed that this receptor interacts only with G_i and G_o proteins and not with G_s proteins, suggesting that the observed 5HT-mediated adenylyl cyclase activation observed in some tissues may be due to indirect effects but not to a direct interaction with G_s proteins. Reconstitution experiments with the A_1-adenosine receptor expressed in *E. coli* revealed species differences in the G protein selectivity between the human and the bovine receptor. The human receptor interacts equally with the three isoforms of the $G_{i\alpha}$-subunit whereas the bovine receptor couples preferentially to $G_{i\alpha\text{-}3}$-subunits, suggesting that species homologs of receptors may use different signaling mechanisms.

Vanhauwe et al. (64) recently expressed the human dopamine D_3 receptor in *E. coli*, with the aim of finding out to which G protein it couples best. Saturation binding experiments with two radiolabeled agonists yielded the expected binding properties. Interestingly, K_i values determined for various compounds detected no difference between the D_3 receptor expressed in CHO cells or in *E. coli*, suggesting either that the CHO cells were just as devoid of the endogenous appropriate G protein as the *E. coli* bacteria or, alternatively, that G protein coupling to D_3 had no effect on ligand binding. Reconstitution of D_3 receptors expressed in *E. coli* with various G proteins, as was previously done for the serotonin $5HT_{1A}$ and adenosine A_1 (2, 25), should help resolve this question.

46.1.2.2. Signaling

Transfection of mammalian GPCRs in yeasts such as *Saccharomyces cerevisiae* also results in expression of specific

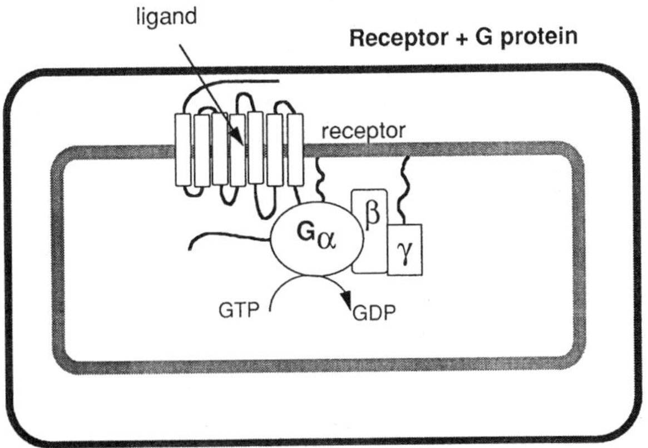

FIGURE 2 Topology of a reconstituted receptor-G protein complex in *E. coli*. Mammalian recombinant receptors expressed alone in microorganisms may serve as a self-replicating reconstitution system, allowing the study of coupling preferences between these two families of signaling proteins. Discrimination between agonistic or antagonistic properties of new ligands may also be determined in such systems because agonist activation may be monitored by the hydrolysis by G protein of radiolabeled GTP. In yeasts expressing mammalian recombinant receptors and G proteins, agonist-induced signaling may result in the activation of endogenous effectors that are normally linked to the response pheromones. Various examples now exist in which GPCRs have been expressed in *E. coli* (see Table 1), either alone or, more frequently, as initial fusion proteins. To make the GPCR-Gα complex fully functional, addition of purified Gβ- and Gγ-subunits, of mammalian origin, was found necessary (2, 10, 25).

binding sites. In addition the receptors may be made to couple to the intrinsic mating factor signal transduction pathway. In yeasts, pheromone receptors such as the α- or **a**-mating-factor receptors are coupled to heterotrimeric G proteins to result in growth arrest. King et al. (30) thus achieved the functional expression of the β_2-adrenergic receptor in *S. cerevisiae*. The 5′ untranslated and N-terminal coding sequence of the β_2-adrenergic receptor gene was replaced by the yeast α-mating-factor receptor gene, and the construct was placed under the control of the galactose-inducible GAL1 promoter.

As in *E. coli*, the receptor expressed in *S. cerevisiae* displayed the characteristic pharmacological profile. Agonist stimulation of β_2-adrenergic receptor-expressing cells was not sufficient to induce a yeast pheromone response. Only coexpression of the mammalian G_s protein in mutant yeast cells lacking endogenous G proteins activated the pheromone response pathway partially by β-adrenergic receptor agonists, demonstrating the in vivo reconstitution of functional signaling in *S. cerevisiae* for the first time.

Agonist-stimulation of somatostatin receptor-expressing yeast cells partially activated the pheromone response pathway even without the coexpression of mammalian G proteins (50). Full activation of this pathway was obtained by the coexpression of a chimeric G protein construct, formed from DNA sequences encoding an N-terminal domain from the yeast G protein Gpa1 and a C-terminal receptor interaction domain from rat $G_{\alpha i2}$, a G protein known to couple to somatostatin receptors.

Recently, functional expression of the G_s protein-coupled rat adenosine A_{2a} receptor in *S. cerevisiae* has been reported by the same group (51). The receptor retained typical pharmacological characteristics, and high-affinity agonist binding sites were detected. Coupling of the adenosine A_{2a} receptor to the endogenous G protein Gpa1 was confirmed by agonist-dependent activation of the yeast pheromone signaling pathway. Deletion of the endogenous α-mating-factor receptor gene significantly increased sensitivity of the pheromone pathway for adenosine agonists.

A number of other GPCR family members have now been expressed in yeasts (Table 1). These include the human M1 (49) and rat M5 muscarinic (24), the human dopamine D_{2s} (9, 54), the mouse serotonin $5HT_{1A}$ (1), and the human μ-opioid receptor (61). Expression levels were uniformly much lower than those reported by King et al. (30). Some of the GPCRs functionally coupled to endogenous or other G proteins. At present, no general prediction can be made as to whether a receptor will couple to G proteins in yeasts.

46.2. EXPRESSION IN MICROORGANISMS FOR STRUCTURAL ANALYSIS

Two recent reviews deal with the overexpression of membrane proteins; one of these focuses on structural studies (14, 55). The elucidation of the three-dimensional structure of GPCRs remains one of the most challenging tasks of the future for our understanding of signal transduction processes. Despite considerable efforts to obtain direct high-resolution structural information, most conclusions about the potential structure of GPCR still rely on indirect methods (14). The first step toward a structural analysis is the availability of sufficient amounts of functional protein. Among others, microorganisms are attractive and competitive expression systems increasingly used, according to the recent literature (see below). The major advantages of microbial expression systems are low-cost production, high-level expression, and the availability of protease-deficient strains. The lack of the enzymatic machinery for posttranslational modifications in most microorganisms, is, however, considered a disadvantage. Indeed, GPCRs are often modified after synthesis (i.e., N-glycosylated and palmitoylated). Although these modifications may not be essential for expression of functional receptors, they could play an important role in folding of the polypeptide chain, and thus in the elaboration of the conformation of the protein.

In 1990, Henderson et al. (20) determined by high-resolution electron cryomicroscopy the three-dimensional structure of bR from purple membranes. These specialized membrane structures have the unique feature of expressing the receptor in high quantities and high purity. bR, which is not coupled to G proteins but is structurally related to GPCRs, still serves as a valuable model for the topology of GPCRs.

Recent research efforts concentrate on very high level expression of various GPCRs in *E. coli* or different yeast species. The development of covalently attached suitable affinity tags (i.e., the Bio, Strep, or His tags) (Table 2), which do not interfere with high expression levels and which facilitate rapid purification by affinity chromatography, constitute a second major direction.

46.2.1. Expression in *E. coli* for Structural Purposes

Continuous efforts to increase expression levels of bR in *E. coli* permitted Mitra et al. (45) to produce sufficient

TABLE 2 Structure and specificity of currently used affinity tags

Tag	Structure[a]	Affinity partner
Bio	LGGIFEAMKMEWR	Avidin
His	HHHHH(H)	Ni^{2+}-NTA[b]
Strep	AWRHPQFGG	Streptavidin
c-*myc*	AAEQKLISEEDLN	c-*myc* antibody
FLAG	DYKDDDDK	FLAG antibody

[a]Structure of tags is expressed in the one-letter amino-acid-residue code.
[b]Ni^{2+}-NTA, nitrilotriacetic acid complexed with Ni^{2+}.

amounts of bR suitable for two-dimensional crystallization of wild-type and mutant bR.

Grisshammer et al. (12, 13) reported the functional expression of the rat NK-2 (neurokinin A) receptor and the rat neurotensin receptor in *E. coli*. Highest expression levels were obtained for both receptors when expressed as a fusion protein with the maltose-binding protein preceded by a signal peptide sequence. B_{max} values for NK-2 receptors and neurotensin receptors were 2.5 and 15 pmol/mg of protein, respectively. Both receptors proved to be functionally expressed in *E. coli*, although agonist binding studies on NK-2 receptors revealed the presence of two affinity states of the receptor despite the absence of endogenous G proteins. An agonist affinity chromatography step was suggested to be necessary by the authors to select for high-affinity receptors during purification (13). For immunodetection and purification of both receptors, a double C-terminal pentahistidine/c-*myc* tail was introduced (Table 2).

High expression levels of the maltose-binding protein neurotensin receptor fusion protein (15 pmol/mg protein) encouraged Tucker and Grisshammer (63) to purify this receptor from *E. coli* membranes. The C terminus of this fusion protein was modified by the addition of various affinity tags (Bio, Strep, or His) to study the influence on receptor expression level and to evaluate different purification procedures. The His tag contains six histidine residues and allows affinity purification by Ni^{2+}-NTA (nitrilotriacetic acid) chromatography. The Strep tag and Bio tag are 9 and 13 amino acid residues long, respectively. The Strep tag binds specifically to streptavidin, but not to avidin. The Bio tag is biotinylated in vivo in *E. coli* and will therefore bind to both avidin and streptavidin. Suitable affinity chromatography matrices are commercially available for all three tags.

The maltose-binding protein neurotensin receptor fusion protein was linked to 16 different combinations of these C-terminal tags, and surprising differences in expression levels were observed. The best results were obtained with a C-terminal fusion construct composed of *E. coli* thioredoxin and the Bio or Strep tag. Purification of histidine-tagged receptors by Ni^{2+}-NTA chromatography and receptors linked to the Strep tag by streptavidine chromatography was unsatisfactory, whereas purification of receptors linked to the Bio tag by avidin chromatography was satisfactory. Coexpression of *E. coli* biotin ligase further improved purification yields. Finally, the purification of functional fusion protein to apparent homogeneity was obtained by using a two-step procedure consisting of Bio tag binding to monomeric avidin followed by a neurotensin chromatography column.

New *E. coli*-based expression systems for GPCR have been explored using the rat neurotensin receptor as a model receptor. The receptor coding region was fused to six different membrane proteins from the membrane-containing bacteriophage PRD1 (18). The expression of heterologous membrane proteins in phages has several advantages: the phage is composed of only some twenty different proteins, which facilitates purification of the target protein and results in incorporated target proteins being less exposed to intracellular proteases. Two of six constructs expressed functional receptors in *E. coli*, demonstrating the usefulness of bacteriophage PRD1 has a heterologous expression system for GPCR. However, expression levels were low and optimization is needed to compete with current high-level expression systems.

46.2.2. Expression in Yeasts for Structural Purposes

A series of three articles described the functional expression of the human D_{2s} dopamine receptor in different types of yeast (14, 15, 52). Several different receptor constructs under the control of inducible or constitutive promoters have been transfected into *S. cerevisiae* by Sander et al. (52, 53). Once again, transfection into a protease-deficient *S. cerevisiae* strain improved expression levels, reaching approximately 2 pmol/mg of protein. Further improvement of receptor density was obtained by transformation of a plasmid harboring the unmodified receptor sequence in the yeast *S. pombe* (54); expression levels reached 15 pmol/mg of protein.

Recently, the functional expression of the mouse $5HT_{1A}$ serotonin receptor was achieved in the methylotrophic yeast *Pichia pastoris* (66). Expression of a c-*myc*-tagged receptor construct in a protease-deficient strain significantly improved the expression level, reaching a receptor concentration of 22 pmol/mg of protein. A comparable approach was used by Bach et al. (1) expressing the same receptor in a protease-deficient strain from *S. cerevisiae*. Several constructs were expressed at expression levels of up to 15 pmol/mg of protein. Both expression systems provide sufficient amounts of protein to start purification for structural analysis. *P. pastoris* has also been used to express the human μ-opioid receptor (61). The cDNA from which the first 32 N-terminal codons were substituted by the α-mating-factor signal sequence was placed in front of the host promoter of the alcohol oxidase-1 gene. This construct was expressed at 0.4 pmol/mg of protein while retaining the typical μ-opioid pharmacological binding profile. Functional coupling to endogenous G proteins was not detected. Large-scale production and structural analysis are ongoing.

46.3. EXPRESSION IN MICROORGANISMS FOR SCREENING PURPOSES

GPCRs are of particular interest for the pharmaceutical industry since they are the targets of many therapeutic compounds (38). Membrane preparations from animal tissues have been used for decades as sources of receptor for new-drug screening. However, the biological complexity of membrane preparations results in two undesired features: (i) high variability of the material and (ii) interferences due to the presence of other receptors.

The availability of cloned human receptors expressed in microorganisms has changed the way of drug screening and allowed the replacement of tissues as a receptor source, thus avoiding the sacrifice of animals. Microorganisms expressing GPCR provide an inexpensive, pure, and reproducible source of human receptors (for review see reference 60). Pharmacological characterization of more than 10 different GPCRs successfully expressed in different microorganisms (Table 1) revealed that the pharmacological profiles of receptors expressed in microorganisms remained unchanged from those determined in native tissues.

Reports of recombinant receptors used for drug screening are rare in the literature, since these experiments are mainly conducted in pharmaceutical industry laboratories and results are kept in-house. The first and best-documented example of receptor expression for screening purposes is the expression of β_1- and β_2-adrenergic receptors in *E. coli*. Marullo et al. (41, 42) demonstrated that both receptors expressed in *E. coli* are well adapted for screening purposes of β-adrenergic receptor-specific drugs. Luyten et al. (39), using the same bacteria, extended the pharmacological characterization of both β-adrenergic receptors further, confirming the complete pharmacological integrity of the heterologously expressed receptors.

Herzog et al. (21) have reported the expression of the human neuropeptide Y1 receptor in *E. coli*. This receptor retained its pharmacological properties, and its use in drug screening was discussed by the authors.

46.4. EXPRESSION OF GPCRs IN E. COLI FOR ANTIBODY PRODUCTION

Another valuable application of GPCRs expressed in microorganisms is their use for generating specific antibodies. Low cost and ease of use are indeed especially of interest when relatively large amounts are necessary, not only to induce the immune response but also to examine its characteristics and to develop the immunodetection procedures.

46.4.1. Antireceptor Antibody Production

GPCR are intrinsic proteins: nearly 160 to 170 of the 350 to 450 residues belong to the hydrophobic transmembrane domain. As such, these parts are poorly immunogenic and explain why antireceptor antibodies for this GPCR family of proteins remain rare.

To circumvent this difficulty, several groups have expressed the receptors or parts thereof as fusion proteins in *E. coli*. The first examples were the β_2AR (41, 42) and the β_1AR (7, 43), which were expressed either alone or as fusion partners with β-galactosidase, λ phage receptor, maltose E protein, or the promoter region of maltose E protein (6). Most of these partners were actually cleaved off the receptor.

The βAR-expressing *E. coli* were used to characterize autoantibody present in sera of patients with idiopathic cardiomyopathy (40). In these individuals, anti-β_1AR but not anti-β_2AR antibodies were found.

Autoantibodies were also characterized by using *E. coli* expressing the $5HT_{1A}$ serotonin receptor (2). The presence of such antibodies had been predicted in sera of autistic children, and this hypothesis was verified by Verdot et al. (65). An alternative strategy was pursued by Levey and his collaborators (36), who expressed fragments of the five muscarinic acetylcholine receptors in *E. coli* to induce the production of antisubtype receptor antibodies. The fragments selected for expression in *E. coli* corresponded to the intracellular loop i3 of each receptor subtype. The resulting antibodies were specific for every one of the immunogens.

46.4.2. Use of Bacterial Expression Systems for Detection of Antireceptor Antibodies

A number of situations have now been described in which antireceptor antibodies have been detected in sera of patients afflicted with a variety of autoimmune diseases. While the best-known example concerns antibodies against the channel protein, the nicotinic acetylcholine receptor in the disease myashenia gravis, several other instances concern GPCR: Graves' disease (stimulatory or inhibitory antibodies against the phrotropin-stimulating hormone receptor), Chagas disease, and idiopathic cardiomyopathy (antibodies against β-adrenergic receptors). In that last case, screening assays for serum autoantibodies were developed using receptors expressed in *E. coli* bacteria as the control antigen preparations. Magnusson et al. (40) thus used bacteria expressing either the β_1 or the β_2AR in immunoblot or enzyme-linked immunosorbent assay formats, to detect semiquantitatively subtype-specific antibodies raised against synthetic peptides derived from the known sequences. The resulting standardized assays were then used to screen sera of patients with Chagas disease (4, 57) or cardiopathy (66). Receptors expressed in bacteria provided a convenient and reliable source of control antigen. Results showed that 4% of Chagas patients, and 30% of cardiomyopathic patients indeed produced anti-β_1AR antibodies in their serum. It was subsequently proposed that in Chagas disease, these antibodies could have arisen by cross-reaction between the β_1AR antibodies and *Trypanosoma cruzei*, the parasite that causes the disease.

46.5. PERSPECTIVES

The increased use of GPCR expressed in microorganisms is the direct consequence of improvements in generic biochemical procedures, together with the development of automated assays suitable for high-throughput screening on large libraries of synthetic or natural compounds. These assays were mostly perfected because of the need to study GPCRs in environments devoid of endogenous receptors or G proteins. The fact that the levels of GPCR in *E. coli*, for instance, are apparently narrowly controlled constitutes a major advantage compared to the wide diversity in levels obtained serendipitously in mammalian cells. This diversity has led to widely divergent binding results, now well recognized and not seen in *E. coli*. Low cost, excellent reproducibility, low background, wide applicability, and now automation all combine to make expression in microorganisms a very valuable way to create tools for basic studies as well as for high-throughput search of new ligands.

Support for our work comes mostly from the Centre National de la Recherche Scientifique, the Institut National de la Santé et de la Recherche Médicale, the University of Paris, and the Ministry for Research (MENESRIP: Ministère de l'Education Nationale, de l'Enseignement Supérieur, de la Recherche et de l'Insertion Professionnelle). We are also grateful for help from the Ligue Nationale contre le Cancer, the Fondation pour la Recherche Médicale Française, and, last but not least, the Association pour la Recherche contre le Cancer. We thank D. Granger for excellent secretarial work on the manuscript. R. Jockers holds a fellowship from the Société de Secours des Amis de la Science.

REFERENCES

1. **Bach, M., P. Sander, W. Haase, and H. Reiländer.** 1996. Pharmacological and biochemical characterization of the mouse 5HT(5A) serotonin receptor heterologously produced in the yeast *Saccharomyces cerevisiae*. *Receptors & Channels* **4:**129–139.
2. **Bertin, B., M. Freissmuth, R. M. Breyer, W. Schütz, A. D. Strosberg, and S. Marullo.** 1992. Functional expression of the human serotonin 5HT1A receptor in *Escherichia coli*. *J. Biol. Chem.* **267:**8200–8206.
3. **Blumer, K. J., J. E. Reneke, and J. Thorner.** 1988. The STE2 gene product is the ligand-binding component of the alpha-factor receptor of *Saccharomyces cerevisiae*. *J. Biol. Chem.* **263:**10836–10842.
4. **Borda, E., J. Pascual, P. Cossio, M. De la Vega, R. Arana, and L. A. Sterin-Borda.** 1984. A circulating IgG in Chagas' disease which binds to beta-adrenoceptors of myocardium and modulates their activity. *Clin. Exp. Immunol.* **57:**679–686.
5. **Braiman, M. S., L. J. Stern, B. H. Chao, and H. G. Khorana.** 1987. Structure-function studies on bacteriorhodopsin. IV. Purification and renaturation of bacterio-opsin polypeptide expressed in *Escherichia coli*. *J. Biol. Chem.* **262:**9271–9276.
6. **Breyer, R. M., A. D. Strosberg, and J. G. Guillet.** 1990. Mutational analysis of ligand binding activity of beta 2 adrenergic receptor expressed in *Escherichia coli*. *EMBO J.* **9:**2679–2684.
7. **Chapot, M. P., Y. Eshdat, S. Marullo, J. G. Guillet, A. Charbit, A. D. Strosberg, and K. C. Delavier.** 1990. Localization and characterization of three different beta-adrenergic receptors expressed in *Escherichia coli*. *Eur. J. Biochem.* **187:**137–144.
8. **Dunn, R. J., N. R. Hackett, J. M. McCoy, B. H. Chao, K. Kimura, and H. G. Khorana.** 1987. Structure-function studies on bacteriorhodopsin. I. Expression of the bacterio-opsin gene in *Escherichia coli*. *J. Biol. Chem.* **262:** 9246–9254.
9. **Francken, B. J. B., W. H. M. L. Luyten, K. Josson, and J. E. Leysen.** 1996. Functional expression of human G protein-coupled receptors in the yeast *Saccharomyces cerevisiae*. Abstr. 8th International Catecholamine Symposium, Asilomar, Calif., 13–18 October 1996.
10. **Freissmuth, M., E. Selzer, S. Marullo, W. Schütz, and A. D. Strosberg.** 1991. Expression of two human β-adrenergic receptors in *Escherichia coli*: functional interaction with two forms of the stimulatory G protein. *Proc. Natl. Acad. Sci. USA* **88:**8548–8552.
11. **Galli, G., R. Matteoni, E. Bianchi, L. Testa, D. Marazziti, N. Rossi, and V. G. Tocchini.** 1990. Replica filter assay of human beta-adrenergic receptors expressed in *E. coli*. *Biochem. Biophys. Res. Commun.* **173:**680–688.
12. **Grisshammer, R., R. Duckworth, and R. Henderson.** 1993. Expression of a rat neurotensin receptor in *Escherichia coli*. *Biochem. J.* **295:**571–576.
13. **Grisshammer, R., J. Little, and D. Aharony.** 1994. Expression of rat NK-2 (neurokinin A) receptor in *E. coli*. *Receptors & Channels* **2:**295–302.
14. **Grisshammer, R., and C. G. Tate.** 1995. Overexpression of integral membrane proteins for structural studies. *O. Rev. Biophys.* **28:**315–422.
15. **Hackett, N. R., L. J. Stern, B. H. Chao, K. A. Kronis, and H. G. Khorana.** 1987. Structure-function studies on bacteriorhodopsin. V. Effects of amino acid substitutions in the putative helix F. *J. Biol. Chem.* **262:**9277–9284.
16. **Haendler, B., U. Hechler, A. Becker, and W. D. Schleuning.** 1993. Expression of human endothelin receptor ETB by *Escherichia coli* transformants. *Biochem. Biophys. Res. Commun.* **191:**633–638.
17. **Haendler, B., U. Hechler, A. Becker, and W. D. Schleuning.** 1993. Extracellular cysteine residues 174 and 255 are essential for active expression of human endothelin receptor ETB in *Escherichia coli*. *J. Cardiovasc. Pharmacol.* **22** Suppl. 8:4–6.
18. **Hänninen, A. L., D. H. Bamford, and R. Grisshammer.** 1994. Expression in *Escherichia coli* of rat neurotensin receptor fused to membrane proteins from the membrane-containing bacteriophage PRD1. *Biol. Chem. Hoppe-Seyler* **375:**833–836.
19. **Harada, Y., T. Senda, T. Sakamoto, K. Takamoto, and T. Ishibashi.** 1994. Expression of octopus rhodopsin in *Escherichia coli*. *J. Biochem.* (Tokyo) **115:**66–75.
20. **Henderson, R., J. M. Baldwin, T. A. Ceska, F. Zemlin, E. Beckmann, and K. H. Downing.** 1990. Model for the structure of bacteriorhodopsin based on high-resolution electron cryo-microscopy. *J. Mol. Biol.* **213:**899–929.
21. **Herzog, H., G. Münch, and J. Shine.** 1994. Human neuropeptide Y1 receptor expressed in *Escherichia coli* retains its pharmacological properties. *DNA Cell Biol.* **13:**1221–1225.
22. **Hildebrandt, V., F. Polakowski, and G. Büldt.** 1991. Purple fission yeast: overexpression and processing of the pigment bacteriorhodopsin in *Schizosaccharomyces pombe*. *Photochem. Photobiol.* **54:**1009–1016.
23. **Hildebrandt, V., R. M. Ramezani, U. Swida, P. Wrede, S. Grzesiek, M. Primke, and G. Büldt.** 1989. Genetic transfer of the pigment bacteriorhodopsin into the eukaryote *Schizosaccharomyces pombe*. *FEBS Lett.* **243:**137–40.
24. **Huang, H. J., C. F. Liao, B. C. Yang, and T. T. Kuo.** 1992. Functional expression of rat M5 muscrinic acetylcholine receptor in yeast. *Biochem. Biophys. Res. Commun.* **182:**1180–1186.
25. **Jockers, R., M. D. Linder, M. Hohenegger, C. Nanoff, B. Bertin, A. D. Strosberg, S. Marullo, and M. Freissmuth.** 1994. Species difference in the G protein selectivity of the human and bovine A(1)-adenosine receptor. *J. Biol. Chem.* **269:**32077–32084.
26. **Johnson, R. L., R. A. Vaughan, M. J. Caterina, H. P. Van, and P. N. Devreotes.** 1991. Overexpression of the cAMP receptor 1 in growing Dictyostelium cells. *Biochemistry* **30:**6982–6986.
27. **Karnik, S. S., M. Nassal, T. Doi, E. Jay, V. Sgaramella, and H. G. Khorana.** 1987. Structure-function studies on bacteriorhodopsin. II. Improved expression of the bacterio-opsin gene in *Escherichia coli*. *J. Biol. chem.* **262:** 9255–9263.
28. **Khorana, H. G.** 1988. Bacteriorhodopsin, a membrane protein that uses light to translocate protons. *J. Biol. Chem.* **263:**7439–7442.
29. **Khorana, H. G.** 1993. Two light-transducing membrane proteins: bacterio-rhodopsin and the mammalian rhodopsin. *Proc. Natl. Acad. Sci. USA* **90:**1166–1171.
30. **King, K., H. G. Dohlman, J. Thorner, M. G. Caron, and R. J. Lefkowitz.** 1990. Control of yeast mating signal transduction by a mammalian beta 2-adrenergic receptor and Gs alpha subunit. *Science* **250:**121–123.

31. **Krebs, M. P., T. Hauss, M. P. Heyn, B. U. Raj and H. G. Khorana.** 1991. Expression of the bacteriorhodopsin gene in *Halobacterium halobium* using a multicopy plasmid. *Proc. Natl. Acad. Sci. USA* **88:**859–863.

32. **Krebs, M. P., R. Mollaaghababa, and H. G. Khorana.** 1993. Gene replacement in *Halobacterium halobium* and expression of bacteriorhodopsin mutants. *Proc. Natl. Acad. Sci. USA* **90:**1987–1991.

33. **Labbé, O.** 1996. *Contribution à la Caractérisation de 3 Récepteurs Orphelins Couplés aux Protéines* G. Doctoral thesis. University of Brussels, Brussels, Belgium.

34. **Lacatena, R. M., A. Cellini, F. Scavizzi, and G. P. Tocchini-Valentini.** 1994. Topological analysis of the human beta (2)-adrenergic receptor expressed in *Escherichia coli*. *Proc. Natl. Acad. Sci. USA* **91:**10521–10525.

35. **Lanyi, J. K.** 1992. Proton transfer and energy coupling in the bacteriorhodopsin photocycle. *J. Bioenerg. Biomembr.* **24:**169–179.

36. **Levey, A. I., T. M. Stormann, and M. R. Brann.** 1990. Bacterial expression of human muscarinic receptor fusion proteins and generation of subtype-specific antisera. *FEBS Lett.* **275:**65–69.

37. **Libert, F., M. Parmentier, A. Lefort, C. Dinsart, J. Van Sande, C. Maenhaut, M. Simons, J. Dumont, and G. Vassart.** 1989. Selective amplification and cloning of four new members of the G protein-coupled receptor family. *Science* **244:**569–572.

38. **Luyten, W. H., and J. E. Leysen.** 1993. Receptor cloning and heterologous expression—towards a new tool for drug discovery. *Trends Biotechnol.* **11:**247–254.

39. **Luyten, W. H., P. J. Pauwels, H. Moereels, S. Marullo, A. D. Strosberg, and J. E. Leysen.** 1991. Comparative study of the binding properties of cloned human β1-and β2-adrenergic receptors expressed in *Escherichia coli*. *Drug Invest.* **3:**3–12.

40. **Magnusson, Y., S. Marullo, S. Höyer, F. Waagstein, B. A. V. Andersson, J. G. Guillet, A. D. Strosberg, A. Hjalmarsson, and J. Hoebeke.** 1990. Mapping of a functional autoimmune epitope on the beta1-adrenergic receptor in patients with idiopathic dilated cardiomyopathy. *J. Clin.Invest.* **86:**1658–1663.

41. **Marullo, S., C. Delavier-Klutcho, Y. Eshdat, A. D. Strosberg, and L. J. Emorine.** 1988. Human β2-adrenergic receptors expressed in *Escherichia coli* membranes retain their pharmacological properties. *Proc. Natl. Acad. Sci. USA* **85:**7551–7555.

42. **Marullo, S., C. Delavier-Klutchko, J.-G. Guillet, A. Charbit, A. D. Strosberg, and L. J. Emorine.** 1989. Expression of human beta1 and beta2 adrenergic receptors in *E. coli* as a new tool for ligand screening. *Bio/Technology* **7:**923–927.

43. **Marullo, S., L. Emorine, A. Strosberg, and C. Delavier-Klutchko.** 1990. Selective binding of ligands to β1, β2 or chimeric β1/β2-adrenergic receptors involves multiple subsites. *EMBO J.* **9:**1471–1476.

44. **Miercke, L. J., M. C. Betlach, A. K. Mitra, R. F. Shand, S. K. Fong, and R. M. Stroud.** 1991. Wild-type and mutant bacteriorhodopsins D85N, D96N, and R82Q: purification to homogeneity, pH dependence of pumping, and electron diffraction. *Biochemistry* **30:**3088–3098.

45. **Mitra, A. K., L. J. Miercke, G. J. Turner, R. F. Shand, M. C. Betlach, and R. M. Stroud.** 1993. Two-dimensional crystallization of *Escherichia coli*-expressed bacteriorhodopsin and its D96N variant: high resolution structural studies in projection. *Biophys. J.* **65:**1295–1306.

46. **Münch, G., P. Walker, J. Shine, and H. Herzog.** 1995. Ligand binding analysis of human neuropeptide Y1 receptor mutants expressed in *E-coli*. *Receptors & Channels* **3:** 291–297.

47. **Nassal, M., T. Mogi, S. S. Karnik, and H. G. Khorana.** 1987. Structure-function studies on bacteriorhodopsin. III. Total synthesis of a gene for bacterio-opsin and its expression in *Escherichia coli*. *J. Biol. Chem.* **262:**9264–9270.

48. **Ni, B. F., M. Chang, A. Duschl, J. Lanyi, and R. Needleman.** 1990. An efficient system for the synthesis of bacteriorhodopsin in *Halobacterium halobium*. *Gene* **90:**169–172.

49. **Payette, P., F. Gossard, M. Whiteway, and M. Dennis.** 1990. Expression and pharmacological characterization of the human M1 muscarinic receptor in *Saccharomyces cerevisiae*. *FEBS Lett.* **266:**21–25.

50. **Price, L. A., E. M. Kajkowski, J. R. Hadcock, B. A. Ozenberger, and M. H. Pausch.** 1995. Functional coupling of a mammalian somatostatin receptor to the yeast pheromone response pathway. *Mol. Cell. Biol.* **15:**6188–95.

51. **Price, L. A., J. Strnad, M. H. Pausch, and J. R. Hadcock.** 1996. Pharmacological characterization of the rat A2a adenosine receptor functionally coupled to the yeast pheromone response pathway. *Mol. Pharmacol.* **50:**829–837.

52. **Sander, P., S. Grünewald, M. Bach, W. Haase, H. Reiländer and H. Michel.** 1994. Heterologous expression of the human D2S dopamine receptor in protease-deficient *Saccharomyces cerevisiae* strains. *Eur. J. Biochem.* **226:**697–705.

53. **Sander, P., S. Grünewald, G. Maul, H. Reiländer, and H. Michel.** 1994. Constitutive expression of the human D2S-dopamine receptor in the unicellular yeast *Saccharomyces cerevisiae*. *Biochim. Biophys. Acta* **1193:**255–262.

54. **Sander, P., S. Grüewald, H. Reiländer, and H. Michel.** 1994. Expression of the human D2S dopamine receptor in the yeasts *Saccharomyces cerevisiae* and *Schizosaccharomyces pombe*: a comparative study. *FEBS Lett.* **344:**41–46.

55. **Schertler, G. F. X.** 1992. Overproduction of membrane proteins. *Curr. Opin. Struct. Biol.* **2:**534–544.

56. **Shand, R. F., L. J. Miercke, A. K. Mitra, S. K. Fong, R. M. Stroud, and M. C. Betlach.** 1991. Wild-type and mutant bacterioopsins D85N, D96N, and R82Q: high-level expression in *Escherichia coli*. *Biochemistry* **30:**3082–3088.

57. **Sterin-Borda, L., C. P. Leiros, M. Wald, G. Cremaschi, and E. Borda.** 1988. Antibodies to beta-1 et beta-2 adrenoreceptors in Chagas' disease. *Clin. Exp. Immunol.* **74:** 349–354.

58. **Strosberg, A. D.** 1993. Structure, function and regulation of adrenergic receptors. *Protein Sci.* **12:**1198–1209.

60. **Strosberg, A. D., and S. Marullo.** 1992. Functional expression of receptors in microorganisms. *Trends Pharmacol. Sci.* **13:**95–98.

61. **Talmont, F., S. Sidobre, P. Demange, A. Milon, and L. J. Emorine.** 1996. Expression and pharmacological characterization of the human μ-opioid receptor in the methylotrophic yeast *Pichia pastoris*. *FEBS Lett.* **394:**268–272.

62. **Teufel, M., M. Pompejus, B. Humbel, K. Friedrich, and H. J. Fritz.** 1993. Properties of bacteriorhodopsin deriv-

atives constructed by insertion of an exogenous epitope into extra-membrane loops. *EMBO J.* **12:**3399–3408.

63. **Tucker, J., and R. Grisshammer.** 1996. Purification of a rat neurotensin receptor expressed in *Escherichia coli*. *Biochem. J.* **317:**891–899.
64. **Vanhauwe, J., W. H. M. L. Luyten, K. Josson, N. Fraeyman, and J. E. Leysen.** 1996. Expression of the human dopamine D3 receptor in *Escherichia coli*: investigation of the receptor properties in the absence and presence of G proteins. Abstr. 8th International Catecholamine Symposium, Asilomar, Calif., 13–18 October 1996.
65. **Verdot, L., M. Ferrer di Martino, B. Bertin, A. D. Strosberg, and J. Hoebeke.** 1994. Production of anti-peptide antibodies directed against the first and second extracellular loop of the human serotonin 5HT1A receptor. *Biochimie* **76:**165–170.
66. **Wallukat, G., and A. Wollenberger.** 1987. Effects of the serum gamma globulin fraction of patients with allergic asthma and dilated cardiomyopathy on chronotropic beta adrenoceptor function in cultured neonatal rat heart myocytes. *Biomed. Biochim. Acta* **46:**634–639.
67. **Weiss, H. M., W. Haase, H. Michel, and H. Reiländer.** 1995. Expression of functional mouse 5-HT5A serotonin receptor in the methylotrophic yeast *Pichia pastoris*: pharmacological characterization and localization. *FEBS Lett.* **377:**451–456.

Selection of *Escherichia coli* Hosts That Are Optimized for the Overexpression of Proteins

JOHN E. WALKER AND BRUNO MIROUX

47

Over the past 20 years, an ever-increasing number of novel genes have been discovered in a wide variety of species. The functions of many of them are unknown, and so the demand for expression systems for producing large amounts of the encoded proteins is very high. Other proteins of medical interest, such as human insulin, many enzymes, receptors, and monoclonal antibodies also need to be made on an industrial scale, preferably by producing the highest yield of the target protein at the lowest cost. Laboratories working on protein structure determination are also faced with the same problem of optimization of overexpression of proteins. For instance, the formation of two- or three-dimensional crystals of a protein often demands a large amount of concentrated pure protein, and structure determination of proteins by nuclear magnetic resonance usually requires the proteins to be isotopically labeled, a very expensive process if the target protein can only be expressed at low levels.

For all of these applications, *Escherichia coli* has proved to be one of the most powerful vehicles for the overproduction of heterologous proteins, but its utility is limited because attempts to overexpress many proteins either have failed completely or have given only low levels of expression of the target protein. Foremost in the catalog of failures are most membrane proteins, which are particularly difficult to overexpress both in *E. coli* and in eukaryotic expression systems (16). The discovery that about one-third of more than 6,000 proteins encoded in the genome of *Saccharomyces cerevisiae* are predicated to be integral membrane proteins containing 1 to 14 membrane spans, many of them of unknown function (11), serves to emphasize the magnitude of the problem. Some membrane proteins have been overproduced in amounts sufficient to permit functional studies (41), but so far no membrane protein structure has been solved to atomic resolution with recombinant material. In contrast, the use of recombinant proteins in globular protein structural analysis is commonplace and has contributed significantly to the rapid growth in the number of structures determined in recent years. An additional factor is that most membrane proteins are not naturally abundant, and since they cannot be overexpressed, their structures cannot be analyzed (47).

As emphasized in this review, the overproduction of a protein in the extant *E. coli* hosts is not a passive process, and it always produces effects that are toxic in varying degrees to the host bacterium. These toxic effects impair the growth of the bacterial host, and frequently they kill it. Therefore, the strains in current use in laboratories are not adapted or "trained" for protein overproduction (27). We have addressed this problem by devising a simple method for selection of new, improved bacterial hosts that overcomes the toxic effects of the overproduction of some heterologous proteins (30). Two such mutant strains have been isolated, and they have proved to be superior to the parental strain for the overexpression of both globular and membrane proteins. Therefore, the selection procedure offers the possibility to obtain a collection of host strains that are optimized for the production of various classes of proteins. Analysis of these mutants is likely to provide a better understanding of the basis of the toxicity associated with overproduction of proteins in *E. coli*.

47.1. TOXIC EFFECTS ASSOCIATED WITH OVEREXPRESSION OF PROTEINS IN *E. COLI*

47.1.1. Examples of Toxic Effects Induced by Expression Plasmids

Toxicity associated with protein overexpression has been described in several independent studies. An early study by Brosius (2) showed that the target polypeptide (rat insulin) was toxic when overproduced under the control of a strong promoter such as *tac* or *rrnB*. The toxicity was attributed to the transport of insulin through the bacterial inner membrane into the periplasm. The toxic phenotype of this expression system was used to select and study strong transcriptional terminators. Many other proteins, including DNase (9), DNA helicase (15), and cell division proteins (8), have been described as having an endogenous toxicity toward the host cell. In these cases, the target gene was cloned with difficulty, transformation efficiency was low, the bacteria stopped growing immediately after induction of expression, and the cells formed very long filaments (15). Dong et al. (10) showed that overexpression of even nonfunctional proteins can lead to growth inhibition, accompanied by a cumulative breakdown of ribosomal RNAs and consequent loss of ribosomes and protein synthetic capacity. If this were a general effect, it would provide an explanation of the basis of the toxicity of overexpression.

Valenzuela (42) showed that insertion into a pBR322 vector of a 100-bp sequence containing poly(dA)-poly(dT) increased the copy number of the plasmid and caused growth inhibition of *E. coli* host cells containing the plasmid. The inhibitory effect was found by overexpressing the tetracycline resistance gene encoding the membrane-bound protein TetA. A mutant strain, selected for its resistance to the overproduction of TetA, had a constitutive slow-growth phenotype.

From these and other experiments, it is evident that the toxicity associated with overexpression of proteins in *E. coli* is a complex combination of various factors, including the plasmid copy number, the strength of the transcriptional promoter in the expression vector, the presence of cryptic promoters, the effect of antibiotic resistance genes in the expression plasmid, the efficiency of the coupling between transcription and translation, and the toxicity of the target gene and its product. The interaction of these and other biological processes with protein overexpression is poorly understood.

Nonetheless, many protein overexpression systems have been designed in which transcriptional promoters and terminators have been manipulated in a variety of plasmids and bacterial hosts. Two general ideas behind their design have been that the overproduction of a protein is likely to be favored by increasing the gene dosage of the target protein by using multicopy plasmids, and by maximizing the production of mRNAs by the use of strong inducible promoters. In consequence, the present-day experimenter who wants to overexpress a protein is faced with a multitude of bacterial expression systems that can be tried, usually in a nonsystematic manner. If the target protein cannot be overexpressed in one of these systems, the more elaborate and more costly eukaryotic expression systems may then be investigated. One of our aims is to develop a more rational approach for optimizing for overexpression of proteins, one that will eventually bring some understanding of the basis of toxicity of target proteins to the host cells.

47.1.2. The F_1F_0-ATPase and the Transport Protein Super-Family from Mitochondria

Our own work is centered on the multi-subunit enzyme F_1F_0-ATPase from the inner membranes of mitochondria and a super-family of transport proteins from the same source. The bovine mitochondrial F_1F_0-ATPase is a membrane-bound complex of 16 different proteins (3, 5, 46). It uses a proton electrochemical gradient generated by respiration as energy source to synthesize ATP. The enzyme has three distinct parts; the membrane domain, F_0; the globular domain, F_1; and the junction between F_1 and F_0, referred to as the stalk. The F_1 domain is an assembly of five different polypeptides (stoichiometry $\alpha_3\beta_3\gamma_1\delta_1\varepsilon_1$) with a combined molecular mass of 371 kDa. The ATPase inhibitor protein (known as IF_1), a basic protein of 84 amino acids, binds to a β-subunit in F_1 and is involved in its physiological regulation (44). The stalk contains three globular polypeptides, oligomycin sensitivity conferral protein (OSCP), F_6, and d, and the globular domain of a membrane-bound subunit b. Together these four proteins interact with the F_1 sector and join to F_0 through the b-subunit (6, 7). The bovine F_0 domain contains six other subunits, among which subunit a and c play central roles in the proton transport process. The other components are subunits A6L, e, f, and g (5). In the simpler F_1F_0-ATPase from *E. coli*, the F_0 domain contains three subunits, a, b, and c (13).

The mitochondrial carrier proteins are located in the inner membrane, where they transport essential substrates in and out of the organelle. They all have apparent molecular sizes of about 30 kDa, and their sequences and structures are related (26, 33, 36, 45). Each appears to contain six trans-membrane spans (29, 37). In all, about 12 distinct transport activities have been identified in mammalian mitochondria, of which the 4 most extensively characterized are the ADP/ATP translocator, the phosphate carrier, the oxoglutarate-malate carrier, and the uncoupling protein, a proton-anion transporter expressed only in brown adipocytes. The sequences of the citrate (24), dicarboxylate (34), and carnitine (20) carriers are also known. From its genomic sequence, it is now evident that *S. cerevisiae* contains about 35 members of the family, many of them of unknown function (34).

Although large amounts of the intact F_1F_0-ATPase and of the F_1-ATPase domain can be isolated readily from mitochondria, the isolation of substantial quantities of individual subunits of the enzyme is more difficult. Likewise, only the ADP/ATP carrier and the uncoupling protein can be isolated readily from mitochondria in the quantities approaching those demanded by structural analysis. Therefore, we wished to overexpress many members of this relatively large and diverse group of membrane and globular proteins in *E. coli*. This offered us the opportunity to investigate systematically the toxic effects of their overexpression in *E. coli*, as summarized below. Detailed accounts of some of these experiments have been published (7, 12, 32, 43). In many cases, we have been able to overcome the toxicities, and the means of doing so has a more general applicability.

47.1.3. Analysis of the Toxicity in the T7 RNA Polymerase-Based Expression System

In one of the most widely used expression systems, and the one investigated in our work, the target gene is transcribed from the plasmid vector by the bacteriophage T7 RNA polymerase (40). The T7 RNA polymerase itself is encoded in the λ-DE3 phage integrated into the genome of the *E. coli* host BL21(DE3). Its expression is under the control of the isopropyl-β-D-thiogalactopyranoside (IPTG)-inducible *lac* UV5 promoter. This host is used in conjunction with a wide variety of plasmids, such as the pET vectors, which have various features to aid cloning and isolation of the target protein. A high-copy-number derivative of pET vectors, called pMW7, is also useful (48).

The toxic effect of each expression plasmid listed in Tables 1 and 2 was investigated systematically by attempting to grow the cells containing the plasmid on two sets of agar plates, one containing IPTG and the other lacking the inducer. Introduction of expression plasmid lacking a target gene into the *E. coli* BL21(DE3) host prevented the cells from growing in the presence of IPTG, irrespective of the copy number of the plasmid (Table 1). Introduction of a target gene into the expression plasmid reinforced the toxic effects on the host (Table 2). Similar toxic phenotypes on plates containing IPTG were associated both with well-expressed target proteins (such as the OSCP or F_6-subunits of the bovine ATPase) and with poorly expressed proteins (exemplified by the δ- and γ-subunits of the enzyme [Table 3]). Plasmids encoding membrane proteins

TABLE 1 Phenotypes of T7-dependent expression plasmids *E. coli* BL21(DE3) host cells

Expression plasmid		Colony formation[a]	
Vector	Tag	−IPTG	+IPTG
pMW7	None	Normal	None
pET17b	N-t T7 tag	Normal	Very small
pET23a	N-t T7 tag, C-t His-tag	Normal	None
pET29a[b]	S-tag	Normal	None
pGEMX-1	Gene 10a fragment	Small	None

[a]On 2×TY agar plates.
[b]Contains a *lac* operator sequence between the T7 promoter and the ribosomal binding site sequence.

were all extremely toxic to *E. coli* BL21(DE3). For example, the b-subunit of the F_1F_0-ATPase, the bovine mitochondrial ADP/ATP carrier, and a fusion protein between the bacteriophage T7 10a protein and the alanine-H^+ carrier from *Bacillus* sp. PS3 (22) were unable to form colonies of viable cells on agar plates, even in the absence of the inducer (Table 2). Therefore, in summary, at least three phenotypes with increasing levels of toxicity were observed: first, and least toxic, an intrinsic level associated with the "empty" plasmid and the bacterial host *E. coli* BL21(DE3); second, a more severe level, typical of poorly expressed-globular proteins; and third, a very severe level of toxicity typical of membrane proteins, but also observed with some globular proteins.

47.2. SELECTION OF MUTANTS THAT OVERCOME THE TOXIC EFFECTS OF OVEREXPRESSION

47.2.1. The Selection Procedure

The procedure for the selection of mutant bacterial hosts that were resistant to the toxic effects of overexpression is summarized in Fig. 1 (30). The first round of selection was conducted with *E. coli* BL21(DE3) transformed with pMW7(OGCP), and expression plasmid encoding the oxoglutarate carried protein (OGCP) from bovine mitochondria. Cells were grown at 37°C until they had reached an optical density of 0.6 at 600 nm. Then expression of the OGCP was induced by addition of IPTG (final concentration 0.7 mM). Four hours after induction, serially diluted samples of the bacterial cells from the culture were plated on solid medium containing ampicillin and IPTG. Two populations of colonies, one large and the other small, survived on the plates. Both were examined for their ability to produce the OGCP in liquid media, but none of the target protein was produced by cells grown from large colonies. In contrast, a culture grown from a small colony produced large amounts of the OGCP, partly because the culture continued to grow and to divide in the presence of IPTG, eventually attaining a saturation optical density comparable with that of wild-type *E. coli* cells (Fig. 2). The mutant host strain was named C41(DE3). Its phenotype was stable, and the mutation that overcame the toxicity of overexpression of the OGCP was in the bacterial genome (30).

TABLE 2 Phenotypes in *E. coli* BL21(DE3), C41(DE3), and C43(DE3) arising from the presence of expression plasmids for various proteins

Expression plasmid		Colony formation[a]			
Vector	Encoded protein (m or g)[b]	BL21 (−IPTG)	BL21 (+IPTG)	C41 (+IPTG)	C43 (+IPTG)
pMW7(OGCP)	Bovine OGCP (m)	Normal	None	Small	Normal
pMW7(PC)	Bovine phosphate carrier (m)	Normal	None	Very small	Normal
pMW7(ATP/ADP)	Bovine ADP/ATP translocase (m)	Very small	None	None	Normal
pCGT180[c]	*Bacillus* PS3 alanine carrier (m)	Very small	None	None	Normal
pMW7(Ecb)	*E. coli* F-ATPase subunit b (m)	Very small	None	None	Small
pMW7(Ecc)	*E. coli* F-ATPase subunit c (m)	Small	None	None	Normal
pMW7(b)	Bovine F-ATPase subunit b (m)	Normal	None	None	Normal
pMW7(α)	Bovine F-ATPase α-subunit (g)	Normal	None	Normal	Normal
pMW7(β)	Bovine F-ATPase β-subunit (g)	Normal	None	Normal	Normal
pmW7(γ)	Bovine F-ATPase γ-subunit (g)	Normal	None	Normal	Normal
pMW7(δ)	Bovine F-ATPase δ-subunit (g)	Normal	None	Small	Small
pMW7(d)	Bovine F-ATPase d-subunit (g)	Normal	None	Small	Normal
pMW7(OSCP)	Bovine F-ATPase OSCP subunit (g)	Normal	None	Normal	Normal
pMW7(F_6)	Bovine F-ATPase F_6-subunit (g)	Normal	None	Small	Normal
pMW7(I)	Bovine F-ATPase inhibitor protein (g)	Normal	None	Normal	Normal
pET7 (*Staufen*)[d]	*Drosophila melanogaster* staufen protein (g)	Normal	None	Small	Normal
pMW7(GFP)	*Aequorea victoria* GFP (g)	Normal	None	Small	Normal

[a]On 2×TY agar plates.
[b]m, membrane protein; g, globular protein.
[c]Derived from pGEMX and encoding a fusion protein between the major capsid protein 10a of phage T7 and the *Bacillus* alanine carrier protein; see reference 22.
[d]See reference 39.

TABLE 3 Expression of various proteins in BL21(DE3), C41(DE3) and C43(DE3) hosts

Protein[a]	Location[b]	Expression level[c]		
		BL21	C41	C43
Bovine OGCP (m)	IB	10*	100*	84*
Bovine phosphate carrier (m)	IB	5*	35*	52*
Bovine ADP/ATP translocase (m)	IB	–	9*	18*
Bacillus PS3 alanine/H^+ carrier (m)	IB	–	19†	79†
E. coli F-ATPase subunit b (m)	IB/M	–	8†	25†
E. coli F-ATPase subunit c (m)	M	2†	10†	15†
Bovine F-ATPase subunit b[d] (m)	IB	ND	30†	ND
Bovine F-ATPase subunit a (g)	IB	35†	135†	ND
Bovine F-ATPase subunit b (g)	IB	50†	240†	ND
Bovine F-ATPase subunit g (g)	IB	11†	74†	ND
Bovine F-ATPase subunit d (g)	IB	4†	18†	ND
Bovine F-ATPase subunit d (g)	IB	10†	20†	3†
Bovine F-ATPase subunit OSCP (g)	IB	50*	300*	ND
Bovine F-ATPase subunit F_6 (g)	C	65†	130†	ND
Bovine F-ATPase inhibitor protein (g)	C	8†	70†	ND
D. melanogaster staufen protein[e] (g)	C	–	ND[d]	ND
Aequorea victoria GFP (g)	IB/C	37†	140†	ND

[a]m, membrane protein, g, globular protein.

[b]IB, inclusion bodies; C, soluble in cytosol; M, membrane. For *E. coli* F-ATPase subunit b, IB/M indicates that in strain C41(DE3), the protein was in a form that was difficult to solubilize in detergent, but in C43 it was in the membrane and was readily detergent extractable (see the text); for the GFP, IB/C indicates that the protein was partially soluble and partially in inclusion bodies in both strains BL21(DE3) and C41(DE3).

[c]The expression level is given as milligrams of protein per liter of bacterial cells, quantified by (*) bicinchoninic acid assay or (†) N-terminal sequencing. –, because of toxicity of the expression plasmid, no expression was obtained; ND, not determined.

[d]The bovine F-ATPase b-subunit probably has two trans-membrane-spanning α-helices and is not related in sequence to the *E. coli* b-subunit, which has one trans-membrane span.

[e]The staufen protein was detected in the soluble fraction of the cells by Western blotting by D. St. Johnston.

Subsequently, it has proved to be possible to overproduce many other proteins without toxic effect in *E. coli* C41(DE3); many of these examples are globular proteins (Tables 2 and 3). Therefore, although the OGCP is an intrinsic membrane protein, the mutation has corrected a general defect in overexpression in *E. coli* BL21(DE3), associated with both membrane and globular proteins.

However, the toxicity of overexpression of other proteins, including the membrane b-subunit of *E. coli* ATPase, persisted in strain C41(DE3) (Table 2). Therefore, a second round of selection was conducted on *E. coli* C41(DE3) transformed with pMW7(Ecb), a plasmid containing the gene for the bacterial b-subunit. In this experiment, a single small colony grew in the presence of IPTG and overproduced subunit b in liquid media. As described above for the expression of the OGCP in *E. coli* C41(DE3), after addition of the inducer, this strain also continued to grow to high cell density and produced three times more of the b-subunit of the *E. coli* ATPase (Table 3). As with *E. coli* C41(DE3), its phenotype was also stable and was associated with the bacterial genome. The main features of the two new bacterial hosts are described in sections 47.3.1 and 47.3.2.

47.2.2. Further Improvements of the Method

The frequency of the small colonies and the proportion of those small colonies that are competent for the expression of the target protein differed widely from one expression plasmid to another. With relatively nontoxic plasmids, such as pMW7(OGCP), small colonies competent for overexpression were common and easily identifiable, whereas with more toxic plasmids, such as pMW7(Ecb), mutants of C41(DE3) expressing subunit b were rare. The mechanism of mutation has not been established, but it could arise either from an SOS response (14, 31) caused by the introduction of the expression plasmid into the host (40) or alternatively, by an SOS-independent mechanism found in *E. coli* cells that are not actively dividing (14). At present, it is unclear how far it will be possible to improve bacteria for the purpose of overproduction of proteins. Therefore, we prefer to follow a stepwise approach by making sequential selection with plasmids of increasing toxicity. For example, mutant host *E. coli* C43(DE3) had to be selected from strain C41(DE3), since the expression plasmid for the *E. coli* b-subunit of the ATPase was so toxic that it could not be transformed into *E. coli* BL21(DE3) (see Table 2).

In principle, the selection procedure could be improved, for example, by increasing the frequency of mutants and by developing a more efficient screening procedure. Preliminary attempts to use chemical mutagenesis and mutagenesis with UV light on the host bacteria before the transformation with the plasmid provoked a considerable increase in frequency of mutation. However, most of the mutants had arisen from major and destructive recombination of the expression plasmid (30a). SOS-independent mutagenic compounds could also be tested. However, at present, the induction of the expression of the target gene

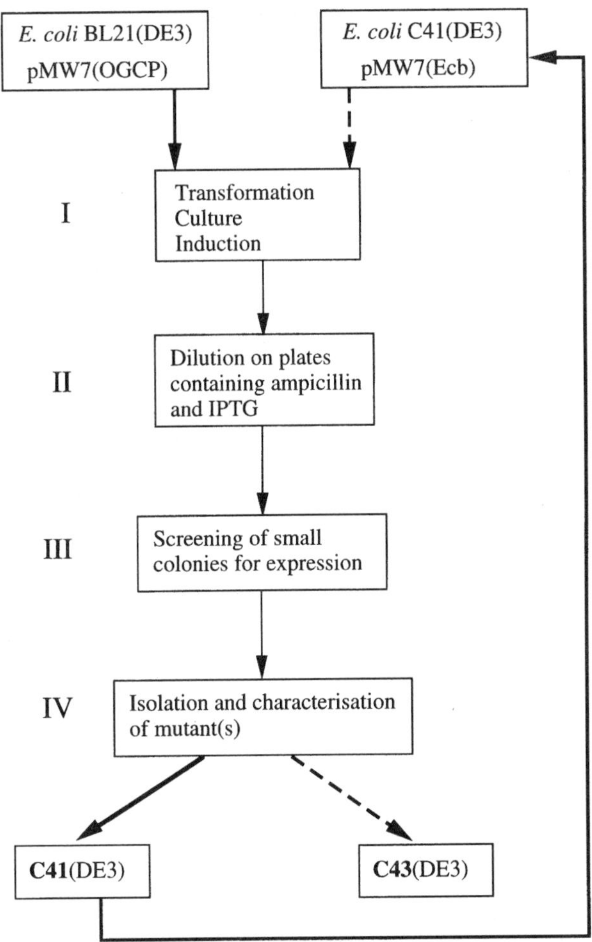

FIGURE 1 Selection of mutants of *E. coli* BL21(DE3) for overexpression of proteins. The OGCP was expressed in BL21(DE3) (step I). Three hours after induction of its expression, cells were diluted and plated onto agar medium containing ampicillin and IPTG (step II). The following morning, small colonies were screened for their ability to express the OGCP (step III). Clone C41(DE3) was selected for its ability both to continue growth in the presence of IPTG and to express the OGCP. In a second round of selection, plasmid pMW7(Ecb) encoding the b-subunit of F-ATPase [toxic in C41(DE3)] was used to select C43(DE3) from strain C41(DE3) by a similar procedure.

remains the best approach for obtaining mutant host cells and for reducing plasmid recombination events.

In the cases of the expression of the OGCP and of the b subunit of the *E. coli* ATPase, overexpressing mutants were identified simply by sodium dodecyl sulfate-polyacrylamide gel electrophoresis (SDS-PAGE) analysis of a cell extract of each clone. Alternative and more efficient approaches that have not yet been attempted would be either to use antibodies against the target protein or to grow the bacterial mutants on positive selective media.

The green fluorescent protein (GFP) from the jellyfish *Aequorea victoria* has a highly fluorescent chromophore and has been used extensively as a reporter of gene expression, particularly in eukaryotes (4, 35). It was used to analyze the two populations of mutants from *E. coli* BL21(DE3) that form the large and small colonies on plates in the presence of IPTG. A dispersed colony of strain BL21(DE3) containing pMW7(GFP) was plated out in the presence of both ampicillin and IPTG. The number of large and small colonies that were observed corresponded to the spontaneous level of mutation of *E. coli* (Fig. 3A). However, the characteristic fluorescence of the GFP was associated almost exclusively with small colonies (Fig. 3B). Only 2 small colonies out of 66 were not fluorescent, and no bright fluorescence was associated with any of the large colonies (four faintly fluorescent large colonies were observed [Fig. 3B]). Therefore, this experiment confirmed that only small colonies contain a population of mutants that can express target proteins. To screen mutants that are able to overexpress a toxic gene, the GFP could be fused to the toxic gene or, alternatively, it could be coexpressed with the target protein from a dicistronic operon.

47.3. OVERPRODUCTION OF PROTEINS IN C41(DE3) AND C43(DE3) BACTERIAL HOSTS

47.3.1. Mutant Strains C41(DE3) and C43(DE3) Are Superior to Their Parental Strains

The expression of a variety of proteins (7 membrane proteins, 10 globular proteins; [Table 3]) were compared in *E. coli* BL21(DE3), C41(DE3), and C43(DE3) hosts. For all seven of the membrane proteins, and particularly for the alanine-H^+ transporter and the *E. coli*-F-ATPase subunits b and c, expression in the mutant hosts was significantly better than that in *E. coli* BL21(DE3). In all three of these latter examples, the transformed *E. coli* BL21(DE3) host cells produced tiny nonviable colonies on plates in the absence of the inducer, and the induction of the expression both on plates and in liquid media was toxic for *E. coli* C41(DE3) but not for *E. coli* C43(DE3). The fusion protein between the gene 10a protein and the alanine-H^+ carrier was very well expressed in *E. coli* C43(DE3), and 200 mg of inclusion bodies (about 50% pure) was obtained per liter of culture (Table 3). Significant improvements in the expression of the ADP/ATP and phosphate carriers were also obtained in *E. coli* C41(DE3).

A general improvement in the expression of globular proteins was also found in mutant host *E. coli* C41(DE3), even where the protein was already well expressed in *E. coli* BL21(DE3). For example, the GFP was expressed at 37 and 140 mg per liter of culture in *E. coli* BL21(DE3) and C41(DE3) strains, respectively (Table 3). As usual in *E. coli* BL21(DE3), cell division stopped soon after induction of expression of GFP, and the cells did not grow on agar plates in the presence of IPTG, whereas in *E. coli* C41(DE3) the cells continued to grow, thereby producing more of the desired product per liter of culture.

In other examples of overexpression of globular proteins typified by the γ-subunit of bovine F_1F_0-ATPase, *E. coli* BL21(DE3) cells stopped growing at low cell density. Consequently, the γ-subunit could not be detected by SDS-PAGE analysis of the cells (although its presence was detected by sequencing), whereas in *E. coli* C41(DE3) the cells continued to divide, grew to high density, and produced a large amount of the γ-subunit (Table 3). Host strains C41(DE3) and C43(DE3) are superior to their parental strain for overexpression of all proteins that have been examined so far. It is possible that in special cases

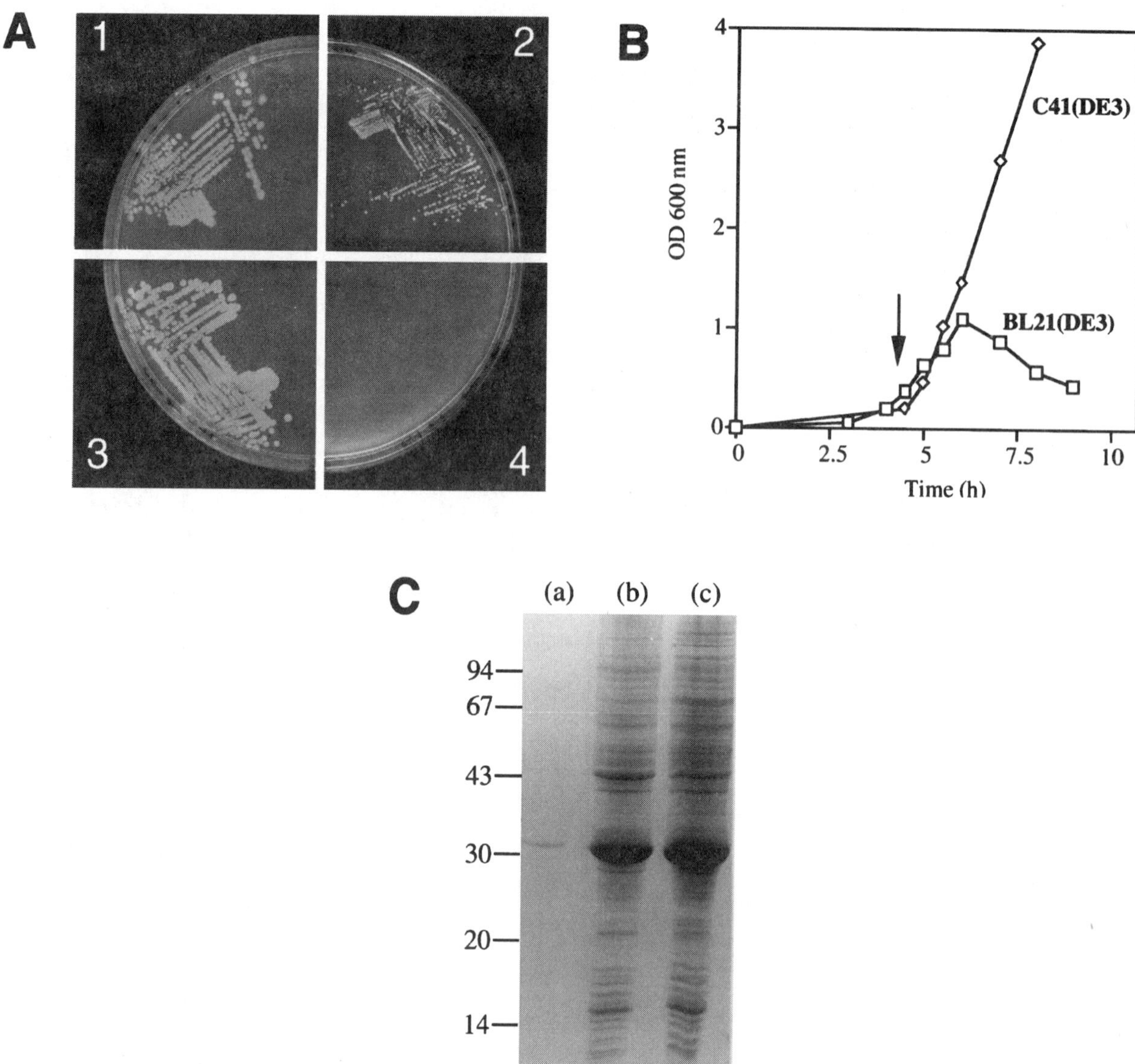

FIGURE 2 Comparison of the expression of the OGCP in *E. coli* BL21(DE3) and C41(DE3) hosts. (A) Comparison of phenotypes of *E. coli* BL21(DE3) and mutant C41(DE3), both containing pMW7(OGCP). Quadrants 1 and 2, *E. coli* C41(DE3) in the absence and presence of IPTG, respectively; quadrants 3 and 4, *E. coli* BL21(DE3) in the absence and presence of IPTG, respectively. (B) Growth of the two strains containing pMW7(OGCP). The arrow indicates the addition of the inducer IPTG (final concentration 0.7 mM) to the liquid culture. (C) Expression of the OGCP analyzed by SDS-PAGE. The cultures were grown in 250 ml of broth. In both cases, the protein formed inclusion bodies in bacterial cytoplasm. They were each resuspended in 4 ml of buffer, and 1 μl was analyzed on the gel, which was stained with PAGE 83 dye. At the left, the positions of molecular mass markers are indicated (in kDa). Lane (a), OGCP expressed in *E. coli* BL21(DE3) 3 h after induction; lane (b), OGCP expressed in C41(DE3) 3 h after induction in medium lacking ampicillin; lane (c), OGCP expressed in C41(DE3) 18 h after induction by IPTG added at the starting point of the culture.

BL21(DE3) could be superior, but there is no known example of each superiority to date.

47.3.2. Modulation of the Expression System Optimizes the Yield of the Target Protein

Studies of the transcription of the target gene in the bacterial hosts *E. coli* C41(DE3) and C43(DE3) revealed that the accumulation of the target mRNA encoding the OGCP was reduced 10-fold in comparison with the *E. coli* BL21(DE3) parental strain (30). In *E. coli* C43(DE3), the time course of expression of the target gene was delayed and more prolonged in comparison to that in *E. coli* C41(DE3), allowing more time for the cell to overproduce the target protein. In these examples, a slight modulation

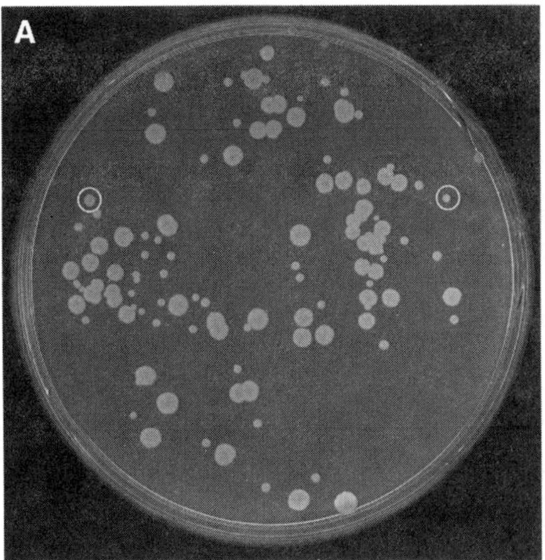

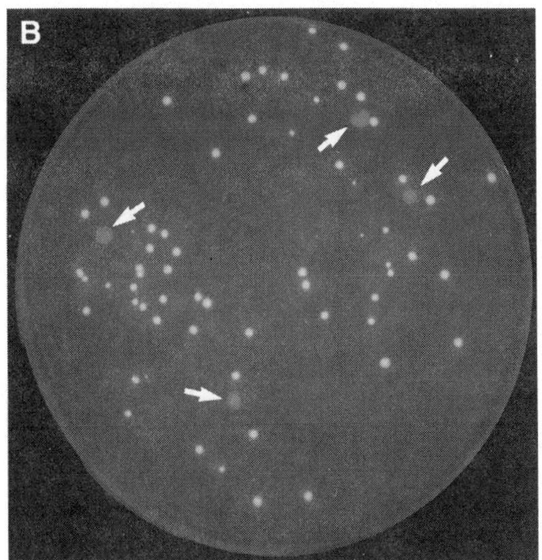

FIGURE 3 Analysis of the expression of the GFP in *E. coli* BL21(DE3). Colonies of strain BL21(DE3) containing plasmid pMW7(GFP) were grown on a plate in the presence of ampicillin and in the absence of IPTG. A single colony was dispersed in water and plated out again, this time in the presence of both ampicillin and IPTG. The resultant colonies are shown photographed in white light (A) and in UV light (B). In panel A, two small colonies that did not emit fluorescence are encircled. In panel B, arrows indicate four large colonies that emitted a diffuse fluorescence.

of transcription allowed the bacterial hosts to grow and to express the target protein successfully (Tables 2 and 3).

At present, the locations of the mutations in *E. coli* C41(DE3) and C43(DE3) strains are not known, but plausible hypotheses are that either they affect the activity of the T7 RNA polymerase or they reduce the level of production of the polymerase. Probably, both effects would help to prevent the uncoupling of transcription and translation (see reference 21). It is noteworthy that a mutant of T7 RNA polymerase able to transcribe three times more slowly than the wild-type enzyme yielded about four times more β-galactosidase from an appropriate expression vector (28). Whatever the molecular explanation for these mutants, their isolation has shown that it is possible to improve the T7 RNA polymerase-based expression system and to overproduce more of the target protein with less-toxic effects.

The optimization of this expression system is summarized in Fig. 4. According to this summary, if the transcription of the target gene in the host *E. coli* BL21(DE3) is too high, the capacity of the translation machinery is exceeded, transcription and translation become uncoupled from each other, and effects that are lethal for the host bacteria ensue. This toxic phenotype provided the basis for the isolation of new mutant bacterial hosts. In the *E. coli* C41(DE3) and C43(DE3) hosts, the transcription of the target gene was modulated in such a way that transcription and translation remained coupled. Although *E. coli* C41(DE3) had a general advantage for the expression of many globular proteins, it appeared that expression of many membrane proteins required further improvement of the expression system, as was obtained, for example, with the expression of the b-subunit of the *E coli* ATPase leading to the selection of the *E coli* C43(DE3) mutant host. It is likely that many other proteins will require specific bac-

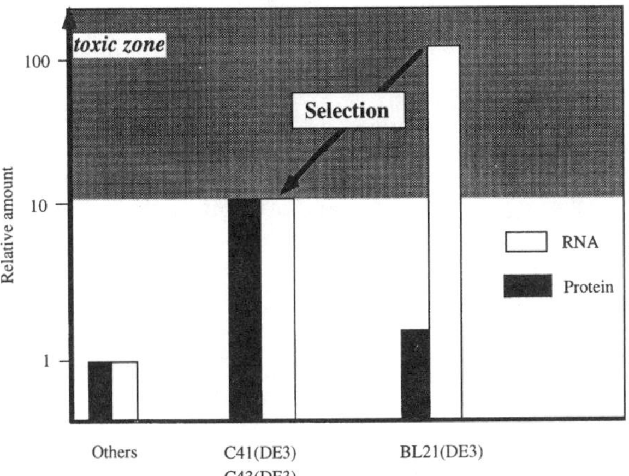

FIGURE 4 Modulation of the T7 expression system by selection of bacterial mutants. The bacterial host BL21(DE3) produces a toxic amount of the target RNA (□) after induction of the expression, leading to the loss of protein synthesis capacity and cell growth inhibition. Mutant hosts C41(DE3) and C43(DE3) produce 10 times less of the target RNA after induction of the expression. Since the expression of the target gene is no longer toxic, the translation machinery is more efficient and produces more of the target protein (■) relative to the parental host. In other expression systems using a weak promoter, the expression of the target gene is not toxic but the production of the target protein is not optimized. Therefore, the selection procedure applied to the T7 expression system allows optimization of the production of the target protein by modulation of the expression.

terial hosts optimized by selection for their overproduction. Because the T7 RNA polymerase-based expression system is one of the strongest systems for expression of proteins in *E. coli*, it is highly suitable for modulation of the expression by selection. The use of an expression system with a weak promoter may sometimes alleviate the problem of toxicity, but such a step is unlikely to be a useful general approach for overexpressing toxic proteins. First, the choice of promoter is arbitrary, and only in very rare, random cases will it provide an optimal combination of promoter and target protein. In other words, the power of selection to optimize the expression system in a rational way is not used. Second, by reducing the level of transcription to a low level by use of a weak promoter, the expression level will be suboptimal (Fig. 4).

For the production of folded and active membrane proteins, it is likely that not only must transcription and translation remain coupled to each other, but they must also remain coupled to the subsequent steps of folding and insertion into the bacterial membrane. *E. coli* C43(DE3) has an additional striking feature that is linked to the overexpression of the b-subunit of the *E. coli* ATPase, namely, the cells produce a network of internal membranes in which the overexpressed membrane protein accumulates (1). It may be possible to link this membrane proliferation to the overproduction of other membrane proteins.

Another example of optimization of expression by modulation of translation has been described (38). A library of expression vectors with various translational strengths was constructed from a randomly mutated translational initiation region associated with the membrane translocation signal sequence of the heat stable enterotoxin II. From this library, variants with translational strengths differing by an order of magnitude were selected and fused with a heterologous protein. The efficiency of translocation of the heterologous protein into the periplasm depended on the associated translational initiation region and was effective only within a narrow range of translational potency. These experiments add weight to the view that modulation of translation is critical for optimization of the subsequent processes of folding and targeting of the overexpressed protein.

47.4. PRACTICAL GUIDELINES FOR USING *E. COLI* C41(DE3) and C43(DE3) HOSTS

47.4.1. Choice of the Bacterial Host

To decide which of the available host strains is more appropriate for a particular expression plasmid, the plasmid is introduced into both *E. coli* C41(DE3) and C43(DE3). Then the transformed cells are grown on 2×TY agar plates containing ampicillin and on 2×TY plates containing both ampicillin and 0.7 mM IPTG. If the *E. coli* C41(DE3) containing the expression plasmid forms colonies in the presence of IPTG, then it will be preferred to the *E. coli* C43(DE3) strain. In all other cases, the expression level of the target protein and its physical state and location (soluble in cytosol, inclusion bodies in cytosol, inserted into the inner membrane, translocated into the periplasm) should be compared in both strains. It should be noted that even when toxicity remains in C41(DE3), the level of expression of the target protein can be high, and the overgrowth of the culture by nonexpressing cells can be delayed by adapting the growth conditions (see below). Only when an outcome of the expression trials in the existing mutant hosts is unfavorable should new selection experiments be undertaken.

Sometimes the transformation of plasmids containing toxic genes leads to the formation of small colonies on plates, even in the absence of IPTG. This is because the expression system is not completely repressed. Additional repression can be obtained either by inclusion of glucose in the agar medium or by incubation of the plate at room temperature, but selection of a bacterial host with a very low basal level of expression is the best option.

47.4.2. Parameters That Influence the Production of the Target Protein

In addition to the combination of appropriate host and expression plasmid, the time of induction, the temperature, and the composition of the medium are important parameters that influence the expression of the target gene.

In contrast to *E. coli* BL21(DE3), in many cases the *E. coli* C41(DE3) and C43(DE3) strains will continue to grow after addition of IPTG (0.7 mM final concentration; Fig. 2 and Table 2). Therefore, the duration of the induction period can be extended to last for 18 h (overnight). In practice, a freshly transformed colony is inoculated in 500 ml of prewarmed 2×TY medium in 2.5-liter flasks, and IPTG is added when the cells reach an optical density at 600 nm (OD_{600}) of 0.6. About 1, 3, and 16 h after induction, a sample (100 μl) of the cells is taken and mixed immediately with an equal volume of the loading buffer for SDS-PAGE analysis. It is very important to follow the growth of the culture after induction of the expression of the target gene. If the expression of the target gene is not toxic at all, the cells should reach an OD_{600} of 3 to 4 by 3 h after induction, and in the 2×TY-rich medium the final OD_{600} at 18 h after induction should be between 5 and 6. If the cells have reached an OD_{600} of only 2 by 3 h after introduction of the inducer, it is often advantageous to reduce the growth temperature from 37° to 25°C and to leave the culture to grow overnight. The lower growth temperature delays the overgrowth of the culture by cells that have lost the ability to express the target gene, and after overnight induction the culture usually contains two or three times more of the target protein. If the cells are unable to reach an OD_{600} of 2 by 3 h after introduction of the inducer, then the culture should be harvested.

In some instances, both the growth temperature and the choice of mutant host strain can influence whether the overexpressed protein is soluble in the bacterial cytoplasm or whether it accumulates in the form of inclusion bodies. The host strain C41(DE3) and a growth temperature of 25°C rather than 37°C both favor the production of a soluble protein. In experiments to investigate the influence of temperature, the culture is grown initially at 37°C and then reduced to 25°C immediately after the addition of IPTG. If inclusion bodies form in C41(DE3), then sometimes the same protein will be obtained in a soluble form in C43(DE3), under similar growth conditions. The effects of growth temperature and mode of induction on protein solubility have been described before (1a, 19). The behavior of C43(DE3) at 25°C is somewhat unpredictable. Sometimes reduction in the growth temperature improves the level of expression greatly (especially with membrane proteins), but in other cases the protein is overexpressed weakly. However, in general, host strain C43(DE3) has proved to be better than C41(DE3) for membrane protein overexpression (Table 3), especially when the culture is induced over an 18-h period.

47.5. CONCLUSIONS AND PERSPECTIVES: BACTERIAL OVEREXPRESSION OF POLYTOPIC MEMBRANE PROTEINS

The b- and c-subunits of *E. coli* F-ATPase are rather simple membrane proteins with one and two trans-membrane spans, respectively. It remains to be demonstrated that the T7 RNA polymerase-dependent expression system can be adapted further to permit the overexpression, folding, and membrane insertion of bacterial membrane proteins with more complex topologies. Overexpression of a family of related bacterial sugar transporters with 12 trans-membrane spans has been achieved by trying a range of different expression systems in each case, until an appropriate combination of vector and host gave the desired result (17). This random approach supports our proposal that, for successful overexpression of each particular membrane protein, it may be necessary to search for conditions in which overexpression is tolerated by the host. The selection procedure provides a more powerful means of achieving this goal.

The problem of extending the bacterial expression system to many eukaryotic membrane proteins is more challenging. Our work with the bovine OGCP shows that it is possible to overexpress a eukaryotic membrane protein with six trans-membrane spans at a high level in the form of inclusion bodies. This is already a big step forward, particularly as the protein in the inclusion bodies has been refolded in vitro and reconstituted into phospholipid vesicles in an active form (12). Inclusion bodies of overexpressed proteins have provided a valuable source for obtaining active material for structural studies of many globular proteins, and a priori there is no reason why a similar approach could not be followed for many membrane proteins, provided that the C41(DE3) and related host strains can be adapted to provide abundant inclusion bodies as starting materials. Particularly encouraging in this respect is the in vitro assembly of the light-harvesting chlorophyll a/b membrane complex from prosthetic groups and proteins denatured in lithium dodecyl sulfate, and formation of two-dimensional crystals from the reassembled complex (18). More recently, other members of the family of mitochondrial transport proteins have been reconstituted in an active form by the procedure described for OGCP (12), using material from inclusion bodies obtained by overexpression of the protein in *E. coli* (23, 34). Likewise, an olfactory receptor, a member of the family of seven-helix receptors, has been reconstituted from inclusion bodies by a similar method (25).

The processes whereby intrinsic membrane proteins of *E. coli* are folded and inserted into its inner membrane are poorly understood, but they are likely to involve chaperones and other ancillary proteins. The experiments described above with *E coli* F-ATPase subunit b show that transcription may determine whether or not these subsequent events in protein biogenesis are effective. It remains to be shown whether these processes of folding and membrane insertion can be used efficiently by eukaryotic proteins, or whether the eukaryotic proteins must be adapted so that they are processed satisfactorily by the bacterium. Alternatively, it may be possible to adapt the processes themselves appropriately by mutation and selection.

REFERENCES

1. **Arechaga, I., B. Miroux, S. Karrasch, and J. E. Walker.** Unpublished results.

1a. **Blackwell, J. R., and R. Horgan.** 1991. A novel strategy for production of a highly expressed recombinant protein in an active form. *FEBS Lett.* **295:**10–12.

2. **Brosius, J.** 1984. Toxicity of an overproduced foreign gene product in *Escherichia coli* and its use in plasmid vectors for the selection of transcription terminators. *Gene* **27:**161–172.

3. **Buchanan, S. K. and J. E. Walker.** 1996. Large scale chromatographic purification of F_1F_0-ATPase and complex I from bovine heart mitochondria. *Biochem. J.* **318:**343–349.

4. **Chalfie, M., Y. Tu, G. Euskirchen, W. W. Ward, and D. C. Prasher.** 1994. Green fluorescent protein as a marker for gene expression. *Science* **263:**802–805.

5. **Collinson, I. R., M. J. Runswick, S. K. Buchanan, I. M. Fearnley, J. M. Skehel, M. J. van Raaij, D. E. Griffiths, and J. E. Walker.** 1994. The F_0 membrane domain of ATP synthase from bovine heart mitochondria: purification, subunit composition and reconstitution with F_1-ATPase. *Biochemistry* **33:**7971–7978.

6. **Collinson, I. R., J. M. Skehel, I. M. Fearnley, M. J. Runswick, and J. E. Walker.** 1996. The F_1F_0-ATPase complex from bovine heart mitochondria: the molar ratio of the subunits in the stalk linking the F_1 and F_0 domains. *Biochemistry* **35:**12640–12646.

7. **Collinson, I. R., M. J. van Raaij, M. J. Runswick, I. M. Fearnley, J. M. Skehel, G. Orriss, B. Miroux, and J. E. Walker.** 1994. ATP synthase from bovine heart mitochondria: *in vitro* assembly of a stalk complex in the presence of F_1-ATPase and in its absence. *J. Mol. Biol.* **242:**408–421.

8. **de Boer, P. A. J., R. E. Crossley, and L. I. Rothfield.** 1988. Isolation and properties of *min B*, a complex genetic locus involved in correct placement of the division site in *Escherichia coli*. *J. Bacteriol.* **170:**2106–2112.

9. **Doherty, A. J., B. A. Connolly, and A. F. Worrall.** 1993. Overproduction of the toxic protein bovine pancreatic DNAse I in *Escherichia coli* using a tightly controlled T7 promoter based vector. *Gene* **136:**337–340.

10. **Dong, H., L. Nilsson, and C. G. Kurland.** 1995. Gratuitous overexpression of genes in *Escherichia coli* leads to growth inhibition and ribosome destruction. *J. Bacteriol.* **177:**1497–1504.

11. **Dujon, B.** 1996. The yeast genome project: what did we learn? *Trends Genet.* **12:**263–270.

12. **Fiermonte, G., J. E. Walker, and F. Palmieri.** 1993. Abundant bacterial expression and reconstitution of an intrinsic membrane transport protein from bovine mitochondria. *Biochem. J.* **294:**293–299.

13. **Fillingame, R. H.** 1992. H^+ transport and coupling by the F_0 sector of the ATP synthase: insights into the molecular mechanism of function. *J. Bioenerg. Biomembr.* **24:**493–497.

14. **Friedberg, E. C., G. C. Walker, and W. Siede.** 1995. Mutagenesis in prokaryotes, p. 465–522. *In* E. C. Friedberg, G. C. Walker, and W. Siede (ed.), *DNA Repair and Mutagenesis*, American Society for Microbiology, Washington, D.C.

15. **George, J. W., R. M. Brosh, Jr., and S. W. Matson.** 1994. A dominant negative allele of the *Escherichia coli* uvrD gene encoding DNA helicase II. *J. Mol. Biol.* **235:**424–435.

16. **Grisshammer, R., and C. G. Tate.** 1995. Overexpression of integral membrane proteins for structural studies. *Q. Rev. Biophys.* **28:**315–422.

17. **Gunn, F. J., C. G. Tate, and P. J. F. Henderson.** 1994. Identification of a novel sugar-H^+ symport protein, FucP, for transport of L-fucose into *Escherichia coli*. *Mol. Microbiol.* **12:**799–809.

18. **Hobe, S., S. Prytulla, W. Kühlbrandt, and H. Paulsen.** 1994. Trimerization and crystallization of reconstituted light-harvesting chlorophyll a/b complex. *EMBO J.* **13:** 3423–3429.

19. **Hockney, R. C.** 1994. Recent developments in heterologous protein production in *Escherichia coli*. *Trends Biotechnol.* **12:**456–463.

20. **Indiveri, C., V. Iacobazzi, N. Giangregorio, and F. Palmieri.** 1997. The mitochondrial carnitine carrier protein: cDNA cloning, primary structure and comparison with other mitochondrial transport proteins. *Biochem J.* **321:** 713–719.

21. **Iost, I., and M. Dreyfus.** 1995. The stability of *Escherichia coli* lacZ mRNA depends upon the simultaneity of its synthesis and translation. *EMBO J.* **14:**3252–3261.

22. **Kamata, H., S. Akiyama, H. Morosawa, T. Ohta, T. Hamamoto, T. Kambe, Y. Kagawa, and H. Hirata.** 1992. Primary structure of the alanine carrier protein of thermophilic bacterium PS3. *J. Biol. Chem.* **267:**21650–21655.

23. **Kaplan, R. S.** 1996. High-level bacterial expression of mitochondrial transport proteins. *J. Bioenerg. Biomembr.* **28:** 41–47.

24. **Kaplan, R. S., J. A. Mayor, and D. O. Wood.** 1993. The mitochondrial tricarboxylate transport protein. cDNA cloning, primary structure, and comparison with other mitochondrial transport proteins. *J. Biol. Chem.* **268:**13682–13690.

25. **Kiefer, H., J. Krieger, J. D. Olszewski, G. Von Heijne, G. D. Prestwich, and H. Breer.** 1996. Expression of an olfactory receptor in *Escherichia coli*: purification, reconstitution, and ligand binding. *Biochemistry* **35:**16077–16084.

26. **Krämer, R., and F. Palmieri.** 1992. Metabolite carriers in mitochondria, p. 359–384. *In* L. Ernster (ed.), *Molecular Mechanisms in Bioenergetics*. Elsevier Science Publishers, Amsterdam.

27. **Kurland, C. G., and H. Dong.** 1996. Bacterial growth inhibition by overproduction of protein. *Mol. Microbiol.* **21:**1–4.

28. **Makarova, O. V., E. M. Makarov, R. Sousa, and M. Dreyfus.** 1995. Transcribing of *Escherichia coli* genes with mutant T7 RNA polymerases: stability of *lacZ* mRNA inversely correlates with polymerase speed. *Proc. Natl. Acad. Sci. USA* **92:**12250–12254.

29. **Miroux, B., V. Frossard, S. Raimbault, D. Ricquier, and F. Bouillaud.** 1993. The topology of the brown adipose tissue mitochondrial uncoupling protein determined with antibodies against its antigenic sites revealed by a library of fusion proteins. *EMBO J.* **12:**3739–3745.

30. **Miroux, B., and J. E. Walker.** 1996. Overexpression of proteins in *Escherichia coli*: mutant hosts that allow synthesis of some membrane proteins and globular proteins at high levels. *J. Mol. Biol.* **260:**289–298.

30a. **Miroux, B., and J. E. Walker.** Unpublished results.

31. **Murli, S., and G. C. Walker.** 1993. SOS mutagenesis. *Curr. Opin. Genet. Dev.* **3:**719–725.

32. **Orriss, G. L., M. J. Runswick, I. R. Collinson, B. Miroux, I. M. Fearnley, J. M. Skehel, and J. E. Walker.** 1996. The δ- and ε-subunits of bovine F_1-ATPase interact to form a heterodimeric subcomplex. *Biochem. J.* **314:** 695–700.

33. **Palmieri, F., F. Bisaccia, L. Capobianco, V. Dolce, G. Fiermonte, V. Iacobazzi, C. Indiveri, and L. Palmieri.** 1996. Mitochondrial metabolite transporters. *Biochim. Biophys. Acta* **1275:**127–132.

34. **Palmieri, L., F. Palmieri, M. J. Runswick, and J. E. Walker.** 1996. Identification by bacterial expression and functional reconstitution of the yeast genomic sequence encoding the mitochondrial dicarboxylate carrier protein. *FEBS Lett.* **399:**299–302.

35. **Prasher, D. C.** 1995. Using GFP to see the light. *Trends Genet.* **11:**320–323.

36. **Ricquier, D., L. Casteilla, and F. Bouillaud.** 1991. Molecular studies of the uncoupling protein. *FASEB J.* **5:** 2237–2242.

37. **Saraste, M., and J. E. Walker.** 1982. Internal sequence repeats and the path of polypeptide in mitochondrial ADP/ATP translocase. *FEBS Lett.* **144:**250–254.

38. **Simmons, L. C., and D. G. Yansura.** 1996. Translational level is a critical factor for the secretion of heterologous proteins in *Escherichia coli*. *Nat. Biotechnol.* **14:**629–634.

39. **St. Johnston, D., D. Beuchle, and C. Nüsslein-Volhard.** 1991. *Staufen*, a gene required to localize maternal RNAs in the *Drosophila* egg. *Cell* **66:**51–63.

40. **Studier, F. W., A. H. Rosenberg, J. J. Dunn, and J. W. Dubendorff.** 1990. Use of T7 RNA polymerase to direct expression of cloned genes. *Methods Enzymol.* **185:**60–89.

41. **Tucker, J., and R. Grisshammer.** 1996. Purification of a rat neurotensin receptor expressed in *Escherichia coli*. *Biochem. J.* **317:**891–899.

42. **Valenzuela, M. S., E. V. Ikpeazu, and K. A. Siddiqui.** 1996. *E. coli* growth inhibition by a high copy number derivative of plasmid pBR322. *Biochem. Biophys. Res. Commun.* **219:**876–883.

43. **van Raaij, M. J., G. L. Orriss, M. G. Montgomery, M. J. Runswick, I. M. Fearnley, J. M. Skehel, and J. E. Walker.** 1996. The ATPase inhibitor protein from bovine heart mitochondria: the minimal inhibitory sequence. *Biochemistry* **35:**15618–15625.

44. **Walker, J. E.** 1994. The regulation of catalysis in ATP synthase. *Curr. Opin. Struct. Biol.* **4:**912–918.

45. **Walker, J. E., and M. J. Runswick.** 1993. The mitochondrial transport protein super-family. *J. Bioenerg. Biomembr.* **25:**435–446.

46. **Walker, J. E., R. Lutter, A. Dupuis, and M. J. Runswick.** 1991. Identification of the subunits of F_1F_0-ATPase from bovine heart mitochondria. *Biochemistry* **30:**5369–5378.

47. **Walker, J. E., and M. Saraste.** 1996. Membrane protein structure: editorial overview. *Curr. Opin. Struct. Biol.* **6:** 457–459.

48. **Way, M., B. Pope, M. Hawkins, and A. G. Weeds.** 1990. Identification of a region in segment 1 of gelsolin critical for actin binding. *EMBO J.* **9:**4103–4109.

The Mechanisms of mRNA Degradation in Bacteria and Their Implication for Stabilization of Heterologous Transcripts

URBAN LUNDBERG, VLADIMIR KABERDIN, AND ALEXANDER VON GABAIN

48

The level of gene expression in bacteria depends primarily on the efficiencies of transcription, translation, and degradation of mRNA. Compared to transcription and translation, relatively little is known about the process that controls mRNA degradation. However, recent studies, mostly employing *Escherichia coli*, have shed some light on *cis*- and *trans*-acting factors determining the stability of mRNA in bacteria and have unraveled the rate-limiting and follow-up steps leading to the disintegration of intact transcripts into nucleotides. *cis*-acting determinants are segments of mRNA molecules defined by their sequence and/or their structure, while *trans*-acting factors are entities that target mRNA, either to catalyze degradation or to modulate its susceptibility toward degradation (Fig. 1A).

In exponentially growing *E. coli* cells, mRNA half-lives of different transcripts range from less than 30 s to more than 15 min (which is almost as long as the generation time of a fast-growing *E. coli* cell), disclosing an enormous potential to adjust the intracellular levels of transcripts by means of differences in stability (106). Relatively short-lived transcripts will be submitted to only a few rounds of translation before decay ensues and further translation is prevented. Expression of such genes ceases, therefore, soon after their transcription has been shut off, thereby allowing their tight regulation. In contrast, cells may afford long half-lives of transcripts derived from housekeeping genes whose expression does not rely on rapid adjustments. A relatively long half-life of mRNA seems also to be an economical means to achieve a high level of expression, since mRNA turnover is a major metabolic expenditure of the cells' total energy consumption. A relatively stable transcript (*hoxS* mRNA) with a half-life of 5 to 7 h has been reported to exist in *Alcaligenes eutrophus* growing at extremely slow growth rates (109). Long-lived mRNAs have also been found in bacteria that have gone through differentiation stages (126). One may speculate that dormant stable transcripts may be stored in these cells for later needs in the life cycle, as it has also been observed in metazoa (for example, see references 38 and 123). Additionally, the half-lives of single transcripts and entire pools of mRNA are the subject of regulation, meaning that stability can alter in response to environmental changes (106, 122) or owing to autoregulatory circuits (68, 83).

mRNA half-lives can be defined by two principal values, chemical half-life and functional half-life. Chemical half-life measures either the loss of radiolabeled RNA after pulse-labeling cells with a radiolabeled RNA precursor or the delay of transcripts after transcription has been blocked, e.g., inhibition of RNA polymerase by rifampicin. The combination of gel-blotting and recombinant DNA techniques has made it possible to follow the fate of single intact transcripts and their decaying products during degradation. Functional half-life measures the inactivation of a transcript by virtue of monitoring the residual synthesis of the corresponding protein product after the transcription has been blocked, thereby following the rate of decay of the functional transcript.

In the following we aim to introduce the most important concepts which we believe have to be accounted for when mRNA degradation is analyzed and which, therefore, may be important for optimization of biotechnological processes dealing with the production of heterologous proteins in microorganisms.

48.1. THE MODE OF mRNA DECAY AND *cis*-ACTING DETERMINANTS OF STABILITY

The search for determinants that confer (in)stability (Fig. 1A) to mRNA or that are targets for initiation of decay has been preceded by studies aiming to follow the direction of the degradation processes, i.e., addressing the question of whether degradation ensues from the 5′ to the 3′ end or vice versa. The earliest effort to determine a direction of mRNA decay (98, 99) showed that 5′ proximal segments of the *trp* mRNA are less stable than distal ones, suggesting that *E. coli* mRNA degradation may follow a 5′ to 3′ directionality. Studies of the *E. coli pap* pilus transcript (9) and the *Bacillus subtilis sdh* mRNA (92) provided additional evidence supporting this idea. Moreover, Cannistraro and Kennell (20) directly demonstrated that the degradation of the relatively long polycistronic *lac* transcript starts from the 5′ region even before its transcription has been completed. In contrast, the 3′ segment of the *Rhodobacter capsulatus* mRNA coding for the light harvesting proteins has been reported to decay prior to the 5′ terminal part (15), whereas the monocistronic *E. coli* β-lactamase (*bla*) mRNA or the polycistronic *trmD* transcript were found to decay with the same rate for all segments, including the 5′ and 3′ untranslated regions (UTRs) (19, 137). Taken together, these data indicate that there is no general scheme

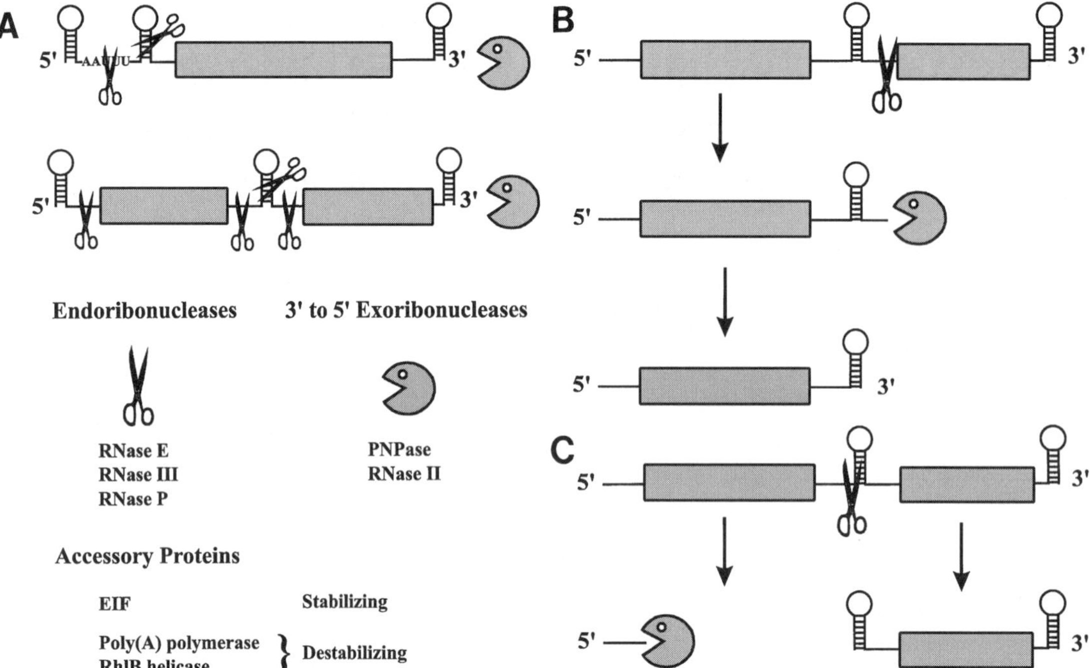

FIGURE 1 *cis-* and *trans-*acting factors involved in mRNA degradation in *E. coli.* (A) Most common *cis-*acting factors are structural elements located in the 5′ UTR, intercistronic regions, or at the 3′ end of mRNA. These structures serve as targets for nucleases digesting them, RNA binding proteins protecting them (e.g., EIF), or enzymes making them more susceptible for degradation [e.g., helicase or poly(A) polymerase]. (B and C) Examples where the 5′ part of a polycistronic transcript is more stable than the 3′ part (B) or vice versa (C). Disjunction of the two segments is performed by an initial endoribonucleolytic cleavage, e.g., by RNase E. In the case of a stable 5′ part (B), exoribonucleases degrade the segment upstream of the initial endoribonucleolytic cleavage up to the first encountered stable stem-loop structure to generate the stable 5′ part. The unstable downstream segment is possibly degraded by the combination of endo- and exoribonucleases. In the case of an unstable 5′ part (C), 3′–5′ exoribonucleases probably do not encounter any stable stem-loop structures upon the initial cleavage. The 5′ part is rapidly degraded while the remaining 3′ part is often stabilized by a resulting 5′ terminal stem-loop structure.

describing RNA decay pathways of *E. coli* mRNAs. Furthermore, stability of individual mRNAs is also likely to be modulated by the action of specific (in)stability determinants distributed within the transcript.

48.1.1. Influence of the Transcript's Length

The length of mRNAs in *E. coli* can vary from a few hundred nucleotides for monocistronic transcripts, to many thousand nucleotides for operons such as *gal* and *lac*. It can be envisaged that longer transcripts are more susceptible to degradation than shorter ones owing to an expected increase in the number of sites cleaved by endoribonucleases. However, a search for a correlation between half-life and length of mRNA in *E. coli* does not support such an idea, e.g., the stabilities of the monocistronic *lpp* transcript (325 nucleotides long; half-life 15 min) versus the *trxA* transcript (493 nucleotides; half-life 2 to 3 min) (5, 61). Furthermore, large polycistronic transcripts do not necessarily decay more rapidly than some extremely short monocistronic mRNAs, e.g., the *trmD* mRNA (2,100 nucleotides long; half-life 3 min) versus the *trxA* mRNA (493 nucleotides; half-life 2 to 3 min) (5, 19). The insignificance of length has also been demonstrated by an internal deletion of the *bla* mRNA that restored the reading frame and did not affect the stability of the RNA (16).

48.1.2. 5′ Determinants

Genetic approaches and recombinant DNA techniques have made it possible to analyze the stability of mutant and chimeric bacterial transcripts and to recognize that stability is mainly controlled by determinants located at the 5′ or 3′ UTR. The significance of the 5′ UTR for mRNA stability control was acknowledged by early studies on the *trp* transcript that also led to the conclusion that the degradation proceeds in a 5′ to 3′ direction (143). 5′ regions of relatively stable mRNAs have been identified by their ability to confer stability to otherwise less stable transcripts upon swapping of the corresponding 5′ UTRs (13, 16, 56, 92, 128, 143). The stability of the 5′ UTRs of the *E. coli ompA* mRNA, the *Staphylococcus aureus ermA* and *ermC* gene transcripts, and the *B. subtilis sdh* mRNA is regulated in response to environmental changes. This phenomenon will be addressed in a later section. In regard to their regulated stability, it is remarkable that their 5′ determinants seem to be necessary and sufficient to confer the pattern of stability regulation to report gene transcripts where they replace the natural 5′ UTRs (13, 45, 92). The 5′ UTRs of the *E. coli ompA* mRNA and the *S. aureus ermC* transcript are among the best-characterized 5′ determinants of mRNA stability (for review, see reference 12).

The secondary structure of the *ompA* 5′ UTR and its impact on mRNA stability has been elucidated by a number of in vivo and in vitro studies, including phylogenetic comparisons (for review, see reference 12). It is composed of two stem-loop structures located between the 5′ terminus and the ribosome binding site (30). The removal of the internal (second from the 5′ end) stem-loop structure has no effect on the ability to confer stability to the mRNA (46). In contrast, the 5′ terminal hairpin structure is essential for the relatively long half-life of the *ompA* mRNA (46). Extending the 5′ end with as few as five non-base-paired nucleotides abrogates the stabilizing effect of the terminal hairpin structure. It has been proposed that the base-pairing of the 5′ terminal nucleotide is an essential trait for the function of 5′ stability determinants (46). In agreement with this proposal, Bouvet and Belasco (18) have been able to provide data suggesting that stable stem-loop structures located 5′ to the cleavage site can act in a sequence-independent fashion as 5′ stability determinants. Although it has been suggested that they interfere with the rate-limiting endoribonucleolytic cleavage, the mechanism by which 5′ stability determinants protect their mRNA needs to be further investigated.

48.1.3. 3′ Determinant

While stem-loop structures at the 5′ end of bacterial transcripts seem not to occur frequently, most bacterial transcripts have at least one terminal stem-loop structure at the 3′ end (for review, see reference 60). One function of these structures is to protect the transcript's 3′ end against degradation by 3′–5′ exoribonucleases. Transcripts without such structures are extremely vulnerable to rapid degradation, so that even the presence of a 5′ stabilizer is able to only partially suppress a destabilization (93). The initial observation that an unprotected 3′ end of an mRNA is rapidly degraded up to the next stem-loop structure came from the study of the 3′ end formation of the *trp* transcript (101). Subsequently, numerous studies have confirmed the notion that stem-loop structures at the 3′ end are capable of stabilizing upstream RNA (71, 105, 115, 140). Interestingly, a single stem-loop structure normally seems to exert the same stability to a transcript as a series of such structures at the 3′ end (115, 137). Finally, 3′ protective hairpin structures in general seem not to influence the rate-limiting endoribonucleolytic cleavage in the upstream RNA segment that they protect against exoribonucleases. This last statement is best illustrated by the finding that stem-loop structures, inserted downstream of the principal site of endoribonucleolytic initiation of mRNA decay, do not prolong the lifetime of the upstream RNA segment (29). Thus, 3′ terminal stem-loop structures are necessary but not sufficient to determine the half-life of a transcript.

48.1.4. Internal Stem-Loop Structures

Internal hairpin structures often exist within polycistronic transcripts. It has been shown that endoribonucleolytic initiation of mRNA degradation occurring downstream of an intercistronic stem-loop structure leads to 3′–5′ exoribonucleolytic degradation up to this hairpin (15, 29, 71, 72) (Fig. 1B). Similarly, the initial cleavage taking place upstream of the internal stem-loop structure (9) promotes rapid degradation of the upstream RNA segment (Fig. 1C). In both cases internal stem-loop structures are able to separate polycistronic transcripts into cistron segments with different stabilities. Moreover, these examples demonstrate that intercistronic stem-loop structures, depending on the nature of the transcript, can function as 5′ or 3′ stabilizers.

On the other hand, there are also circumstances in which an internal stem-loop structure itself serves as a site for endoribonucleolytic cleavage altering the stability of flanking regions. For example, the internal stem-loop structure of the primary transcript of the *metY-nusA-infB* operon (120) is processed by RNase III. This causes a decrease in the stability of the RNA segment positioned downstream of the RNase III cleavage site.

Collectively, these data demonstrate an important regulatory function of internal stem-loop structures that can direct differential degradation of the adjacent (downstream and upstream) segments of mRNA and, hence, make a crucial contribution to differential expression of genes within an operon.

48.2. ENZYMES INVOLVED IN mRNA DEGRADATION IN *E. coli*

It is now accepted that endoribonucleolytic cleavage(s) controls the rate-limiting step of mRNA degradation of most mRNAs in *E. coli*. Although there are a large number of endoribonucleases identified in *E. coli* (39, 40), only three RNases (RNase E, RNase III, and RNase P) have been implicated in endoribonucleolytic cleavage of mRNA. Despite quite different sizes, structures, and subunit compositions (Table 1), their common denominator is the Mg^{2+}-dependent catalysis and the production of a 5′ phosphoryl group upon cleavage. The cleavage sites of all three enzymes are specified by structural requirements that severely restrict their frequency of occurrence in the cellular pool of mRNA sequences. Interestingly, all three enzymes seem to have homologs in eukaryotic cells (37, 62, 89, 139), and counterparts of RNase E and RNase P have even been found in Archaea (37, 50), underlining the primordial origin and general importance of these enzymes in RNA processing.

48.2.1. RNase P

RNase P, an enzyme composed of a 14-kDa protein and a catalytic RNA subunit of 377 nucleotides, plays an essential role in the 5′ maturation of pre-tRNA (for review, see reference 2) and has only recently been shown to be involved in mRNA metabolism. It has been found to cut an RNase E cleavage product derived from the *his* transcript of *Salmonella typhimurium*. In this case a preceding RNase E cleavage is a prerequisite for the subsequent processing by RNase P (1). In the absence of the RNase E cleavage, the *his* transcript fails to form the proper structure needed for the RNase P cleavage. Cleavage by RNase P yields a downstream cleavage product that is stabilized as compared to the unprocessed precursor. Thus, the role of the RNase P in this case seems to support the formation of a 5′ stability determinant, as described above.

48.2.2. RNase III

RNase III, a dimeric enzyme of 25 kDa monomers, was first recognized as a nuclease that participates in the maturation of rRNA (125). Its substrate specificity has been exten-

TABLE 1 Enzymes involved in mRNA degradation in *E. coli*

Enzyme	Subunits and composition	Size (kDa)	Substrate preferences and function	References
Endoribonuclease				
RNase E	Part of degradosome	118	Cleaves in AU-rich single-stranded regions	32, 53, 66, 75, 80, 82, 84, 85, 94, 95, 117
RNase III	Homodimer	25	Sequence-dependent cleavage of double-stranded regions	35, 73, 124
RNase P	Catalytic RNA in complex with protein	377 nucleotides; 14 kDa	Processing of pre-tRNA-like structures	1, 2
Exoribonuclease				
Polynucleotide phosphorylase	Homotrimer or part of degradosome	77	3′ to 5′ processive degradation irrespective of the sequence	26, 88, 93, 116
RNase II	ND[a]	64	3′ to 5′ processive degradation irrespective of the sequence	88
Polymerase				
Poly(A) polymerase I	ND	50	Addition of poly(A) to free 3′ OH ends	6, 22, 117, 140
Poly(A) polymerase II	ND	36	Addition of poly(A) to free 3′ OH ends	21

[a]ND, not determined.

sively studied; it prefers double-stranded RNA, but the exact sequence and structure requirements that specify cleavage are still unknown. The cleavage can occur in either one or both strands of the substrate duplex (74). RNase III is not essential for cell growth (7, 131); a mutant with a deficient *rnc* gene shows no change in the rate of bulk mRNA degradation (3). The major phenotype is the accumulation of unprocessed 30S ribosomal transcripts and a change in expression of about 10% of the cellular proteins (54, 130). Moreover, RNase III autoregulates its intracellular concentration by participating in rate-limiting degradation of its mRNA (10). This indicates that processing and degradation of a subset of mRNAs are controlled by RNase III. Several examples of transcripts are known where RNase III cleavage is involved in the rate-limiting step that leads to degradation of the mRNA; e.g., the transcripts of the *rpsO-pnp*, the *metY-nusA-infB*, the *rnc-era-recO*, and the *dicB* operons (10, 48, 73, 83, 116, 120). In most of these examples RNase III cleavage removes a stem-loop structure from the polycistronic transcript, which is located upstream of one of the coding regions. Cleavage results in a dramatic increase in the degradation of the downstream segment. In the case of the gene encoding the exoribonuclease polynucleotide phosphorylase (PNPase), RNase III cleaves the transcript into a product that is needed for the reported autoregulatory mechanism of this operon (124). RNase III has also been reported to be involved in activation of translation; examples are the transcripts derived from the *N* gene of phage λ and the *0.3* gene of phage T7 (43, 57). RNase III processing does not affect the stability of these transcripts but leads to formations of the 5′ UTRs, which then become accessible initiation sites of translation (35).

48.2.3. RNase E

RNase E was initially identified as an activity that processes the ribosomal 9S RNA into p5S RNA; in a temperature-sensitive mutant, 9S RNA was found to accumulate at nonpermissive temperature (4, 53). The activity was attributed to an enzyme named RNase E, encoded by the *rne* locus (53, 97). Kuwano et al. (75) isolated a temperature-sensitive mutant that at nonpermissive temperature has impaired bulk mRNA degradation but no significant effect on translation (111). The mutation has been located at a locus coined *ams* for altered mRNA stability. Initial studies seemed to suggest that the *rne* gene does not affect mRNA degradation (3). Later it was shown that both *rne* (*rne-3071*) and *ams* (*ams-1*) mutations, which inhibit cell growth at nonpermissive temperature, specify the same gene and that they both affect the stability of bulk mRNA and impair RNase E cleavage (8, 91, 103, 133).

Attempts to clone a DNA fragment that could complement the mutations led to the isolation of the *rne/ams* locus but with truncations of the region coding for the carboxy-terminal part of the protein (27, 31, 36). The complete gene was isolated by an independent approach aiming to identify a possible *E. coli* myosin homolog, using antibodies directed against a heavy-chain myosin from *Saccharomyces cerevisiae*. In this way a high-molecular-weight protein (Hmp) was identified, and its gene sequence revealed that its 5′ proximal part is identical to the *rne/ams* gene but that it codes for a much longer protein than the previously identified *rne/ams* gene fragments (24, 25). Thus, the same gene has been specified by three different terms: *rne*, *ams*, and *hmp*. Today, one prefers to refer to it as the *rne* gene, coding for RNase E. The complete open reading frame codes for a protein of 1,061 amino acids with the predicted molecular mass of 118 kDa (24, 25), which in its isolated form seems to be sufficient to catalyze RNase E-specific cleavages (33). The apparent molecular mass of 180 kDa seen upon sodium dodecyl sulfate-polyacrylamide gel electrophoresis has been shown to be a result of three proline-rich regions (84). The earlier-mentioned *rne-3071* and *ams-1* (today referred to as *rne-1*) mutations have been attributed to amino acid substitution at the codon positions 68 and 66, respectively (85). Furthermore, analysis of *rne* gene variants and their corresponding protein products have disclosed that the N-terminal half of the enzyme is sufficient for the cleavage activity and that the protein, in addition to its catalytic site, carries a second (arginine-rich) RNA binding domain (84, 132). The enzyme's ability to function in its truncated form can explain why incomplete gene segments have been found to complement the *rne* mutations and why the enzyme's activity has been attributed to a falsely designated, smaller endoribonuclease (27, 31, 36, 78, 80).

The *rne* gene encodes a transcript of about 3,500 nucleotides in length, which occurs at a relatively low level in wild-type strains but accumulates to relatively high levels in an *rne* mutant strain (68). The synthesis of RNase E is autoregulated by cleavage of its mRNA (68, 102). Analysis of diauxic behaviors in *E. coli* cultures revealed a transient down-regulation of the *rne* transcript and a subsequent reduction of RNase E activity, suggesting that the expression of RNase E is also under environmental control (11). This observation and another study support the notion that media composition affects the degradation pattern of the *rne* transcript (141).

The *rne* gene is essential for cell growth: the phenotype of the *rne* mutant at nonpermissive temperature is the reduced RNase E activity, which is concomitant with a prolongation of bulk mRNA degradation. RNase E cleavages in mRNA were identified as early as 1988 and connected with mRNA processing and degradation (104). Lundberg et al. (78) have been able to show that the cleavages implicated in the regulated stability of the *ompA* mRNA (90) are controlled by the *rne* gene. In the same study it was shown that cleavage leads to rapid and immediate degradation of the cleavage products (78). RNase E cleavage in the antisense RNA, RNA I, controlling the copy number of the plasmid pBR322, has been demonstrated to be coupled to the further degradation following the initial cleavage (76). Interestingly, recent work has demonstrated the ability of RNase E to shorten RNA poly(A) tails in vitro. However, it is not clear whether this activity of RNaseE is able to assist PNPase and RNase II in degrading the 3′ ends of mRNAs in vivo (61a). Currently, there is a long catalog of transcripts processed and degraded by RNase E. Considering the data and the fact that RNase E-like proteins and genes have been identified or predicted in a large number of bacteria and organelles, it is safe to implicate a key role for RNase E in bacterial mRNA processing and degradation (32, 49, 58, 69, 121, 134).

The requirements that define an RNase E cleavage site are not well understood. Tomcsányi and Apirion (135) proposed for the first time a consensus motif for an RNase E cleavage site that was based on the sequence similarities found between the cleavages in 9S RNA and RNA I. Ehretsmann et al. (44) compared more RNase E cleavage sites and suggested a pentanucleotide consensus sequence [A/G]AUU[A/U]. Later studies showed that RNase E cleavages, particularly the one in RNA I, are not simply determined by a consensus sequence; they rather seem to cleave within AU-rich regions of single-stranded RNA (77, 87).

Other investigations have suggested that the stem-loop structures found in the vicinity of the cleavage sites facilitate the accessibility for RNase E to its target sites (34, 44, 81). However, the rate of cleavage has been shown to be 20 times faster in an oligonucleotide encompassing the core cleavage site of RNA I than in the complete RNA I molecule containing, in addition, multiple secondary structures (86). This finding does not support the idea that RNase E needs a secondary structure to recognize its cleavage sites in RNA (34, 44), and suggesting that stem-loop structures are not needed for RNase E cleavage. However, stem-loop structures may regulate efficiency of RNase E-mediated degradation by means of modulation of the accessibility of the cleavage site to nuclease attack (18, 82, 86).

48.2.4. Exoribonucleases

In *E. coli* only enzymes that degrade RNA in a 3′ to 5′ direction, so-called 3′–5′ exoribonucleases, have been identified (40). Two enzymes of this type, RNase II and PNPase (55, 108), have been shown to be important for the exoribonucleolytic degradation of mRNA and its endoribonucleolytic cleavage products (41, 42). Inactivation of either of them is tolerated by the cell without any obvious phenotypic changes; thus, it seems that the two exoribonucleases can functionally compensate each other (41). If both RNase II and PNPase are conditionally inactivated, cells begin to accumulate small fragments derived from mRNA and are not viable at nonpermissive temperature (42). RNase II is a highly processive and hydrolytic enzyme converting RNA into nucleotide monophosphates and is strongly impeded by secondary structures (108). It is a monomer with a molecular mass of 64 kDa (144). PNPase was initially isolated as a homotrimer of a 77-kDa subunit; later, a more complex pentameric form of 320 to 360 kDa was also isolated (55). In contrast to RNase II, PNPase is a phosphorolytic enzyme that converts RNA polymers into nucleotide diphosphates, and the reaction is reversible (55).

As discussed in the previous section, mRNA in bacteria normally have stem-loop structures at their 3′ ends, which severely impair the action of the 3′–5′ exoribonucleases (28, 30, 88, 101, 105, 115). Thus, the degrading action of PNPase and RNase II relies on endoribonucleolytic cleavage providing the free 3′ OH extremity needed as entry site. However, it is not clear how an endoribonucleolytic cleavage product of a transcript "sealed" at its 3′ end by an intact stem-loop structure can be degraded by the two exoribonucleases. In this connection it is worth mentioning that exoribonuclease impeding factor (EIF) (Fig. 2), a protein interacting with stem-loop structures, increases in vitro the resistance of such structures toward the action of PNPase (26), suggesting that in vivo additional factors may modulate the function of 3′ protective structures.

48.2.5. Poly(A) Polymerases

Identification of a poly(A) polymerase activity that utilizes ATP to extend the 3′ end of RNA with a poly(A) tail (6, 119) was first reported in 1962 (6). However, a role for

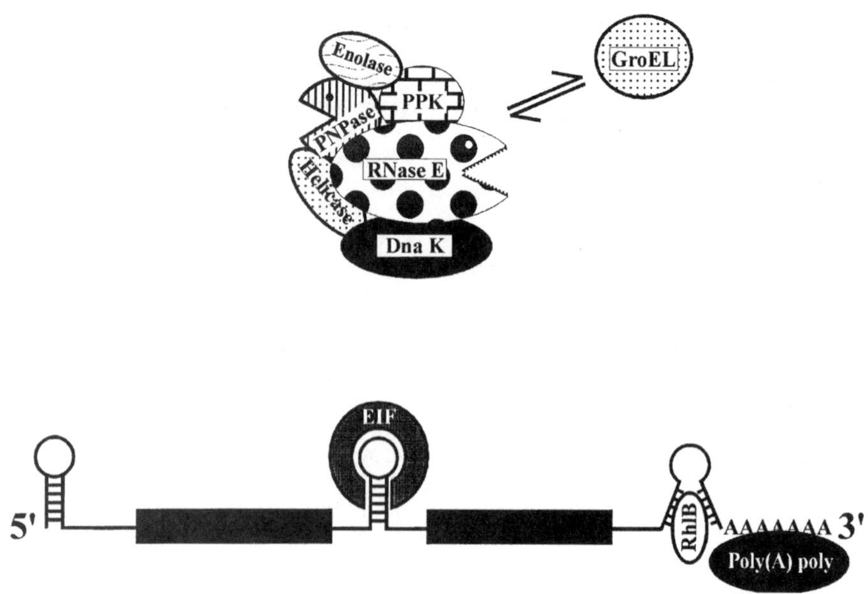

FIGURE 2 Putative structure of the *E. coli* RNA degradosome. The components RNase E, PNPase, RhlB helicase, enolase, DnaK, and PPK, as well as GroEL, that are strictly or facultatively associated with the degradosome are shown (top). The possible targets of an mRNA for the accessory proteins (see Fig. 1A) are shown below.

bacterial poly(A) polymerases in the metabolism of mRNA was recognized only recently. These studies were significantly accelerated after discovery of the *pcnB* gene (plasmid copy number) encoding *E. coli* poly(A) polymerase I, a 50-kDa protein (22).

Initial investigations concerning the occurrence of poly(A) tails in bacteria disclosed that approximately 1% of pulse-labeled RNA is polyadenylated in a wild-type *E. coli* strain. In a double mutant with conditionally deficient PNPase and RNase II (see previous section), this value was found to increase to 6% at nonpermissive temperature (22). The average length of the poly(A) tails in wild-type *E. coli* is between 10 and 40 nucleotides, and in the double mutant the length increases to more than 100 nucleotides (110). The inactivation of poly(A) polymerase I in *E. coli* results in a reduction of more than 90% of the number of poly(A) tailed transcripts and in a retarded degradation of many RNA species (59, 110, 142). Recent findings demonstrate that polyadenylation is involved in mRNA degradation by modulating accessibility of transcript to exoribonucleases (59, 64, 110, 142) whose action is usually inhibited by the 3′ stem-loop structures but can be primed by addition of single-stranded tails.

Although these data decipher the function of poly(A) polymerase I, we must mention that most recent work has provided evidence for the existence of a second poly(A) polymerase in *E. coli* (21), thus indicating that the function of polyadenylation in mRNA can only be fully understood after defining the relative role of these two enzymes in RNA metabolism and processing.

48.2.6. Degradosomes

Miczak and Apirion (94) presented the first data suggesting that RNase E might form a "processosome" with other RNases. They isolated a complex containing RNase E, RNase III, and RNase P from *E. coli*. More recent experiments have shown that RNase E is part of a different complex coined a degradosome (23, 95, 117, 118). In the degradosome RNase E is associated with PNPase in a high-molecular-weight complex (23, 117). In addition to PNPase, several other proteins have been identified in the degradosome: the RNA helicase RhlB, a member of the DEAD-box protein family (95, 118); the heat shock proteins GroEL and DNaK (95); and an enolase, a glycolytic protein of unknown role in RNA metabolism, which previously was isolated as the beta subunit of PNPase (95, 118). More recently, polyphosphate kinase (PPK) has been identified as a minor component of the degradosome (17). The DEAD-box protein found in the degradosome has been demonstrated to be able to assist PNPase in degrading 3′ stem-loop structures (118). Polyphosphate kinase in the degradosome has been suggested to maintain an appropriate microenvironment by eliminating inhibitory polyphosphates and nucleotide diphosphates and regenerating ATP (17).

rne mutant cells that express an RNase E that lacks the carboxy-terminal part, which seems to be responsible for the interaction with PNPase, are still viable (70). Furthermore, shorter RNase E versions, lacking the *E. coli* homologous carboxy-terminal part, seem to occur in other bacteria, such as the cyanobacteria *Synechocystis* sp. (69). Thus, these data, together with the above-mentioned notion that PNPase can be complemented by RNase II, suggest that the presence of PNPase in the degradosome may not be conserved in the bacterial world.

48.3 REGULATION OF mRNA DEGRADATION

mRNA stability in bacteria has been found to take part in numerous regulatory processes that are able to adjust a transcript's intracellular level in response to environmental changes. Bacterial growth rate and growth stage were identified as parameters that can affect the stability of single transcripts in *E. coli* and *B. subtilis*, respectively (106, 122). Regulation of mRNA stability has also been observed in connection with cold shock in *E. coli* (47). Nutritional stress in *E. coli*, such as the combination of slow growth rate and anaerobiosis, leads to a general reduction of RNA turnover. This could be attributed to a lowered rate of RNA synthesis that is offset by an increased stability of bulk mRNA (51).

The mechanisms behind these regulated changes in mRNA stability are far from being understood. The growth-rate-dependent stability of the *E. coli ompA* mRNA seems to be ruled by the rate of RNase E cleavage in the 5′ UTR (78, 80, 90). However, in this case the cascade of events that sense the growth rate and set the rate of RNase E cleavage remains obscure. Two different RNA binding activities, RBA1 and RBA2, able to bind to *ompA* mRNA have been identified (52). RBA1 increases in *E. coli* upon nutritional stress; this increase has been correlated to a reduction in mRNA degradation (52).

As mentioned earlier, 5′ UTRs of the *S. aureus ermA* and *ermC* transcripts are sufficient to mediate their stability regulation. They encode erythromycin resistance, and their stabilities are induced by the presence of erythromycin in the medium (13, 128). In both instances erythromycin leads to stalling of ribosomes in the 5′ regions preceding the coding regions of the *ermC* and *ermA* gene transcripts. The mechanisms of the erythromycin-induced mRNA stability have been further investigated, and it has been suggested that the stalled ribosomes act as a barrier to endoribonucleolytic degradation in the 5′ to 3′ direction (14, 129).

48.4. THE ROLE OF RIBOSOMES

Since mRNA in bacteria is usually assembled in polysomes, a number of studies have aimed to elucidate how the efficiency of translation of mRNA can interfere with mRNA decay (for review, see reference 114). Very early observations indicated that inhibition of translation by antibiotics affects mRNA stability (for example, see references 63, 113, and 136). The effect of chloramphenicol on the stability of two monocistronic transcripts, *bla* and *ompA*, differing by a factor of five in stability has been investigated. It was found that chloramphenicol, although it significantly retards the degradation of both transcripts, does not seem to alter their relative difference in stability in fast-growing cells (79). It was suggested that the rate of the decay initiating cleavage in both mRNAs is proportionally reduced by the ribosomes stalled in the coding region (79). Introduction of premature stop codons into mRNA revealed a destabilizing effect, e.g., in the case of the *trp* transcript (100). However, introduction of a series of stop codons at various positions in the coding region of the moncistronic *bla* mRNA disclosed that the impact on mRNA stability of the premature translational stop is position dependent (107). In that study it was found that translation of as few as 56 N-terminal codons (of about 300) is sufficient to confer a normal half-life to the mutant transcript (107). It was concluded that most of the segments, even in the absence of ribosomes, are inert toward too-early nuclease attacks. Since the stability of the *bla* mRNA is controlled by the 5′ UTR (16), it is interesting to note that insufficient or no loading of ribosomes onto the *bla* transcript facilitates its degradation (107). One could interpret this finding to mean that an entire lack of ribosomes favors a configuration of the 5′ UTR that dramatically increases its vulnerability toward rate-limiting cleavages. Similarly, it has been observed for the transcript of the *puf* operon of *R. capsulatus* that extending the coding region such that a downstream hairpin seems to be "ironed out" by translating ribosomes causes destabilization, which is attributed to the deprivation of the upstream segment of its protective 3′ structure (29, 72).

Bacterial mRNAs are usually synthesized and translated simultaneously so that the speed of ribosome matches that of RNA polymerase (67, 96). However, expression of heterologous proteins in *E. coli* is frequently based on vector systems using T7 RNA polymerase, an enzyme that transcribes eight times faster than its *E. coli* counterpart. The resulting desynchronization of transcription and translation seems to unmask nuclease-susceptible sites and to cause unusually rapid degradation of such heterologous transcripts. In the case of a hybrid *lac* mRNA transcribed by T7 RNA polymerase, overproduction of DEAD-box helicases (65) or improvement of ribosome binding (66) has been shown to restore a normal half-life. In addition, it has been shown that RNase E cleavage sites in the coding region are protected by translating ribosomes (66). In contrast, fusion transcripts containing the 5′ determinant of the *ermC* gene transcript were able to be stabilized by ribosomes stalling in the short open reading frame in the presence of erythromycin, although the downstream main coding region was entirely deprived of its translating ribosomes (12). Thus, the stabilization effected by the stalled ribosomes in the 5′ region seems to be independent of the protection that translating ribosomes might confer to the transcript downstream of the stability determinant. Many studies have tried to score the influence of translation efficiency on mRNA stability, i.e., the frequency of loading ribosomes onto the transcript. However, most of these investigations are difficult to compare with each other since experimental data are frequently obtained from studies of transcripts whose decay mechanisms have not yet been elucidated or are poorly understood. Although all these studies underline the importance of ribosomes for mRNA stability, their exact influence must be seen in connection with the *cis-* and *trans-*acting factors that control the degradation independent of translating ribosomes.

48.5. BIOTECHNOLOGICAL IMPLICATIONS

Stability of mRNA in *E. coli* and in closely related gram-negative organisms, as well as in gram-positive organisms such as *B. subtilis*, relies on stability determinants mostly residing in the 5′ and 3′ regions. Such determinants can be employed to design relatively stable fusion transcripts that should help to improve the yield of heterologous gene expression and to study the impact of mRNA degradation on gene expression in heterologous systems. Thus, such structures may be a natural strategy to elevate the half-life of bacterial messengers above the average level. An example of a gene expression system employing RNA stabilizing determinants is the pET vector system, where a stem-loop structure is located at the 5′ end of the transcript produced by T7 RNA polymerase (112, 127). In case such a strategy fails to yield an improvement in gene expression, *E. coli* mutants with conditional deficiencies of mRNA degradative enzymes are helpful tools to elucidate the rate-limiting step. The application of these *E. coli* mutants may even be helpful for analyzing the degradation pathway of transcripts expressed in far distantly related gram-negative organisms, since expression of mRNA species from *R. capsulatus* have been found to follow the same path of degradation in *E. coli* as in *R. capsulatus* (16). For example, a conditional knockout of RNase E, RNase III, or RNase P in mutant strains may answer the question of whether the stability of a given transcript is regulated by endoribonucleolytic cleavages. This information can then be extended to remove cleavage sites, which in turn could stabilize the transcript in question. Another approach is to probe heterologous transcripts in *E. coli* strains, allowing the conditional double knockout of PNPase and RNase II, which should answer the question of whether the 3′ stem-loop structure has the desired protective function against these exoribonucleases.

Changes in mRNA stability in response to growth stage and growth rate should be seriously considered when gene expression is critical in large-scale bacterial batch or continuous cultures. In this respect, again 5′ and 3′ UTRs are frequently necessary and sufficient to confer the regulated stability to heterologous transcripts. The reduction of mRNA turnover in cells shifted from fast- to slow-growing conditions (e.g., shift from aerobiosis to anaerobiosis) may provide an effective strategy to achieve a high level of gene expression with a minimal consumption of energy (51).

The identification and the apparent evolutionary conservation of the key players controlling mRNA stability in *E. coli* are the reason for the application of our present knowledge about mRNA degradation to less well studied microorganisms of biotechnological importance. Antibodies and gene probes are therefore helpful tools for the identification of homologs of the mRNA delay machinery, even in remotely related bacteria. Even complementation assays using conditional lethal *E. coli* mutants with deficiencies in mRNA degradation are a promising strategy to identify genes involved in mRNA degradation from other organisms. Using this concept, a human cDNA clone has been identified that complements the RNase E-deficient phenotype in *E. coli* (138).

We thank S. Wallace, J. Jacobsen, and Dr. U. Bläsi for critically reading the manuscript. This work has been supported by grants of the Austrian Research Fund, FWF, and by grants of the Austrian Federal Ministry of Science, Transport and the Arts (to A.v.G). We are aware that the list of references used in this chapter is far from being complete, and we apologize to our esteemed colleagues for the publications which we have not been able to cite.

REFERENCES

1. **Alifano, P., F. Rivellini, C. Piscitelli, C. M. Arraiano, C. B. Bruni, and M. S. Carlomagno.** 1994. Ribonuclease E provides substrate for ribonuclease P-dependent processing of a polycistronic mRNA. *Genes Dev.* **8:**3021–3031.
2. **Altman, S., L. Kirsebom, and S. Talbot.** 1993. Recent studies of ribonuclease P. *FASEB J.* **7:**7–14.
3. **Apirion, D., and D. Gitelman.** 1980. Decay of RNA in RNA processing mutants of *Escherichia coli*. *Mol. Gen. Genet.* **177:**139–154.
4. **Apirion, D., and A. B. Lassar.** 1978. A conditional lethal mutant of *Escherichia coli* which affects the processing of ribosomal RNA. *J. Biol. Chem.* **253:**1738–1742.
5. **Arraiano, C. M., S. D. Yancey, and S. R. Kushner.** 1988. Stabilization of discrete mRNA breakdown products in *ams, pnp, rnb* multiple mutants of *Escherichia coli*. *J. Bacteriol.* **170:**4625–4633.
6. **August, J. T., J. Ortiz, and J. Hurwitz.** 1962. Ribonucleic acid-dependent ribonucleotide incorporation. *J. Biol. Chem.* **237:**3786–3793.
7. **Babitzke, P., L. Granger, J. Olszewski, and S. R. Kushner.** 1993. Analysis of mRNA decay and rRNA processing in *Escherichia coli* multiple mutants carrying a deletion in RNase III. *J. Bacteriol.* **175:**229–239.
8. **Babitzke, P., and S. Kushner.** 1991. The Ams (altered messenger stability) protein and ribonuclease E are encoded by the same structural gene of *Escherichia coli*. *Proc. Natl. Acad. Sci USA* **88:**1–5.
9. **Båga, M., M. Göransson, S. Normark, and B. E. Uhlin.** 1988. Processed mRNA with differential stability in the regulation of *E. coli* pilin gene expression. *Cell* **52:**197–206.
10. **Bardwell, J. C. A., P. Regnier, S. M. Chen, Y. Nakamura, M. Grunberg-Manago, and D. Court.** 1989. Autoregulation of RNase III operon by mRNA processing. *EMBO J.* **8:**3401–3407.
11. **Barlow, T. M., M. Berkmen, D. Georgellis, L. Bayer, S. Arvidson, and A. von Gabain.** 1998. RNase E, the major player in mRNA degradation, is down-regulated in *Escherichia coli* during a transient growth retardation (diauxic lag). *Biol. Chem.* **379:**33–38.
12. **Bechhofer, D.** 1993. 5′ mRNA stabilizers, p. 31–52. *In* J. G. Belasco, and G. Brawerman (ed.), *Control of Messenger RNA Stability*. Academic Press, New York.
13. **Bechhofer, D. H., and D. Dubnau.** 1987. Induced mRNA stability in *Bacillus subtilis*. *Proc. Natl. Acad. Sci. USA* **84:**498–502.
14. **Bechhofer, D. H., and K. Zen.** 1989. Mechanism of erythromycin-induced *ermC* mRNA stability in *Bacillus subtilis*. *J. Bacteriol.* **171:**5803–5811.
15. **Belasco, J. G., J. T. Beatty, C. W. Adams, A. von Gabain, and S. N. Cohen.** 1985. Differential expression of photosynthesis genes in *R. capsulatus* results from segmental differences in stability within the polycistronic *rxcA* transcript. *Cell* **40:**171–181.
16. **Belasco, J. G., G. Nilsson, A. von Gabain, and S. N. Cohen.** 1986. The stability of *E. coli* gene transcripts is dependent on determinants localized to specific mRNA segments. *Cell* **46:**245–251.
17. **Blum, E., B. Py, A. J. Carposis, and C. F. Higgins.** 1997. Polyphosphate kinase is a component of the *Escherichia coli* RNA degradosome. *Mol. Microbiol.* **26:**387–398.
18. **Bouvet, P., and J. G. Belasco.** 1992. Control of RNase E-mediated RNA degradation by 5′-terminal base pairing in *E. coli*. *Nature* **360:**488–491.
19. **Byström, A. S., A. von Gabain, and G. R. Björk.** 1989. Differentially expressed *trmD* ribosomal protein operon of *Escherichia coli* is transcribed as a single polycistronic mRNA species. *J. Mol. Biol.* **208:**575–586.
20. **Cannistraro, V. J., and D. Kennell.** 1985. Evidence that the 5′ end of *lac* mRNA starts to decay as soon as it is synthesized. *J. Bacteriol.* **16:**820–822.
21. **Cao, G.-J., J. Pogliano, and N. Sarkar.** 1996. Identification of the coding region for a second poly(A) polymerase in *Escherichia coli*. *Proc. Natl. Acad. Sci. USA* **93:**11580–11585.

22. **Cao, G. J., and N. Sarkar.** 1992. Identification of the gene for an *Escherichia coli* poly(A) polymerase. *Proc. Natl. Acad. Sci. USA* **89:**10380–10384.

23. **Carpousis, A. J., G. Van Houwe, C. Ehretsmann, and H. M. Krisch.** 1994. Copurification of *E. coli* RNase E and PNPase: evidence for a specific association between two enzymes important in RNA processing and degradation. *Cell* **76:**889–900.

24. **Casaregola, S., A. Jacq, D. Laoudj, G. McGurk, S. Margarson, M. Tempete, V. Norris, and I. B. Holland.** 1992. Cloning and analysis of the entire *Escherichia coli ams* gene: *ams* is identical to *hmp1* and encodes a 114 kDa protein that migrates as a 180 kDa protein. *J. Mol. Biol.* **228:**30–40.

25. **Casaregola, S., A. Jacq, D. Laoudj, G. McGurk, S. Margarson, M. Tempete, V. Norris, and I. B. Holland.** 1994. Corrigendum: cloning and analysis of the entire *Escherichia coli ams* gene. *J. Mol. Biol.* **238:**867.

26. **Causton, H., B. P. Py, R. S. McLaren, and C. F. Higgins.** 1994. mRNA degradation in *Escherichia coli*: a novel factor which impedes the exoribonucleolytic activity of PNPase at stem-loop structures. *Mol. Microbiol.* **14:**731–741.

27. **Chauhan, A. K., A. Miczak, L. Taraseviciene, and D. Apirion.** 1991. Sequencing and expression of the *rne* gene of *Escherichia coli*. *Nucleic Acids Res.* **19:**125–129.

28. **Chen, C.-Y. A., J. T. Beatty, S. N. Cohen, and J. G. Belasco.** 1988. An intercistronic stem-loop structure functions as mRNA decay terminator necessary but insufficient for *puf* mRNA stability. *Cell* **52:**609–619.

29. **Chen, C.-Y. A., and J. G. Belasco.** 1990. Degradation of *pufLMX* mRNA in *Rhodobacter capsulatus* is initiated by nonrandom endonucleolytic cleavage. *J. Bacteriol.* **172:** 4578–4586.

30. **Chen, L.-H., S. A. Emory, A. L. Bricker, P. Bouvet, and J. G. Belasco.** 1991. Structure and function of a bacterial mRNA stabilizer: analysis of the 5′ untranslated region of *ompA* mRNA. *J. Bacteriol.* **173:**4578–4586.

31. **Claverie-Martin, F., M. R. Diaz-Torres, S. D. Yancey, and S. R. Kushner.** 1989. Cloning the altered mRNA stability (*ams*) gene of *Escherichia coli* K-12. *J. Bacteriol.* **171:**5479–5486.

32. **Condon, C., H. Putzer, D. Luo, and M. Grunberg Manago.** 1997. Processing of *Basillus subtilis thrS* leader mRNA is RNase E dependent in *Escherichia coli*. *J. Mol. Biol.* **268:** 235–242.

33. **Cormack, R. S., J. Genereaux, and G. A. Mackie.** 1993. RNase E activity is conferred by a single polypeptide: overexpression, purification and properties of the *ams/rne/hmp1* gene product. *Proc. Natl. Acad. Sci. USA* **90:** 9006–9010.

34. **Cormack, R. S., and G. Mackie.** 1992. Structural requirements for the processing of *Escherichia coli* 5S ribosomal RNA by RNase E *in vitro*. *J. Mol. Biol.* **228:**1078–1090.

35. **Court, D.** 1993. RNA processing and degradation by RNase III, p. 71–116. *In* J. Belasco and G. Brawerman (ed.) *Control of Messenger RNA Stability*. Academic Press, New York.

36. **Dallmann, G., K. Dallmann, A. Sonin, A. Miczak, and D. Apirion.** 1987. Expression of the gene for the RNA processing ribonuclease E in plasmids. *Mol. Gen.* (*Life Sci. Adv.*) **6:**99–107.

37. **Darr, S. C., J. W. Brown, and N. R. Pace.** 1992. The varieties of ribonuclease P. *Trends Biochem. Sci.* **17:**178–182.

38. **De Leon, V., A. Johnson, and R. Bachvarova.** 1983. Half-lives and relative amount of stored and polysomal ribosomes and poly(A)$^+$ mRNA in mouse oocytes. *Dev. Biol.* **98:**400–408.

39. **Deutscher, M. P.** 1985. *E. coli* RNases: making sense of alphabet soup. *Cell* **40:**731–732.

40. **Deutscher, M. P.** 1993. Ribonuclease multiplicity, diversity, and complexity. *J. Biol. Chem.* **268:**13011–13014.

41. **Donovan, W. P., and S. R. Kushner.** 1983. Amplification of ribonuclease II (rnb) activity in *Escherichia coli* K-12. *Nucleic Acids Res.* **11:**265–275.

42. **Donovan, W. P., and S. R. Kushner.** 1986. Polynucleotide phosphorylase and ribonuclease II are required for cell viability and mRNA turnover in *Escherichia coli*. *Proc. Natl. Acad. Sci. USA* **83:**120–124.

43. **Dunn, J. J., and F. W. Studier.** 1975. Effect of RNase III cleavage on translation of bacteriophage T7 messenger RNAs. *J. Mol. Biol.* **99:**487–499.

44. **Ehretsmann, C. P., A. J. Carpousis, and H. M. Krisch.** 1992. Speci[illegible]y of *E. coli* endoribonuclease RNase E: *in vivo* and *in vitro* analysis of mutants in bacteriophage T4 mRNA processing site. *Genes Dev.* **6:**149–159.

45. **Emory, S. A., and J. G. Belasco.** 1990. The *ompA* 5′ untranslated RNA segment functions in *E. coli* as a growth-rate-regulated mRNA stabilizer whose activity is unrelated to translational efficiency. *J. Bacteriol.* **172:** 4472–4481.

46. **Emory, S. A., P. Bouvet, and J. G. Belasco.** 1992. A 5′-terminal stem-loop structure can stabilize mRNA in *E. coli*. *Genes Dev.* **6:**135–148.

47. **Fang, L., W. Jiang, W. Bae, and M. Inouye.** 1997. Promoter-independent cold-shock induction of *cspA* and its derepression at 37°C by mRNA stabilization. *Mol. Microbiol.* **23:**355–364.

48. **Faubladier, M., K. Cam, and J. P. Bouche.** 1990. *Escherichia coli* cell division inhibitor *DicF* RNA of the *dicB* operon: evidence for its generation *in vivo* by transcription termination and by RNase III and RNase E dependent processing. *J. Mol. Biol.* **212:**461–471.

49. **Fleischmann, R. D., M. D. Adams, O. White, R. A. Clayton, E. F. Kirkness, A. R. Kerlavage, C. J. Bult, J.-F. Tomb, B. A. Dougherty, J. M. Merrick, K. McKenney, G. Sutton, W. FitzHugh, C. Fields, J. D. Gocayne, J. Scott, R. Shirley, L.-I. Liu, A. Glodek, J. M. Kelley, J. F. Weidmann, C. A. Phillips, T. Spriggs, E. Hedblom, M. D. Cotton, T. R. Utterback, M. C. Hanna, D. T. Nguyen, D. M. Saudek, R. C. Brandon, L. D. Fine, J. L. Fritchman, J. L. Fuhrmann, N. S. M. Geoghagen, C. L. Gnehm, L. A. McDonald, K. V. Small, C. M. Fraser, H. O. Smith, and J. C. Venter.** 1995. Whole genome random sequencing and assembly of *Haemophilus influenzae Rd*. *Science* **269:**496–512.

50. **Franzetti, B., B. Sohlberg, G. Zaccai, and A. von Gabain.** 1997. Biochemical and serological evidence for an RNase E-like activity in halophilic *Archae*. *J. Bacteriol.* **179:**1180–1185.

51. **Georgellis, D., T. Barlow, S. Arvidson, and A. von Gabain.** 1993. Retarded RNA turnover in *Escherichia coli*. *Mol. Microbiol.* **9:**375–381.

52. **Georgellis, D., B. Sohlberg, F. V. Hartl, and A. von Gabain.** 1995. Identification of GroEL as a constituent of an mRNA protection complex in *Escherichia coli*. *Mol. Microbiol.* **16:**1259–1268.

53. **Ghora, B. K., and D. Apirion.** 1978. Structural analysis and *in vitro* processing to p5 rRNA of a 9S RNA molecule isolated from an *rne* mutant of *E. coli*. *Cell* **15:**1055–1066.

54. **Gitelman, D. R., and D. Apirion.** 1980. The synthesis of some proteins is affected in RNA processing mutants of *Escherichia coli*. *Biochem. Biophys. Res. Commun.* **96:**1063–1070.

55. **Gogefroy-Colburn, T., and M. Grunberg-Manago.** 1972. Polynucleotide phosphorylase, p. 533–574. *In* P. D. Boyer (ed.), *The Enzymes*. Academic Press, New York.

56. **Gorski, K., J.-M. Roch, P. Prentki, and H. M. Krisch.** 1985. The stability of bacteriophage T4 gene 32 mRNA: A 5′ leader sequence that can stabilize mRNA transcript. *Cell* **43:**461–469.

57. **Guarneros, G., L. Kameyama, L. Orozco, and F. Velasquez.** 1988. Retroregulation of an *int-lacZ* gene fusion in a plasmid system. *Gene* **72:**129–130.

58. **Hagege, J. M., and S. N. Cohen.** 1997. A developmentally regulated Streptomyces endoribonuclease resembles ribonuclease E of *Escherichia coli*. *Mol. Microbiol.* **25:** 1077–1090.

59. **Hajnsdorf, E., F. Braun, J. Haugel-Nilsen, and P. Regnier.** 1995. Polyadenylylation destabilizes the *rpsO* mRNA of *Escherichia coli*. *Proc. Natl. Acad. Sci USA*. **92:**3973–3977.

60. **Higgins, C. F.** 1993. The role of the 3′ end in mRNA stability and decay, p. 13–30. *In* J. Belasco and G. Brawerman (ed.) *Control of Messenger RNA Stability*. Academic Press, New York.

61. **Hirashima, A., G. Ghilds, and M. Inouye.** 1973. Differential inhibitory effects of antibiotics on the biosynthesis of envelope proteins of *Escherichia coli*. *J. Mol. Biol.* **79:** 373–389.

61a. **Huang, H., J. Liao, and S. N. Cohen.** 1998. Poly(A)- and poly(U)-specific RNA 3′ tail shortening by *E. coli* ribonuclease E. *Nature* **391:**99–102.

62. **Iino, Y., A. Sugimoto, and M. Yamamoto.** 1991. *S. pombe pac1*, whose overexpression inhibits sexual development, encodes a ribonuclease III-like RNase. *EMBO J.* **10:**221–226.

63. **Imamoto, F., and Y. Kano.** 1971. Inhibition of transcription of the tryptophan operon in Escherichia coli by block in initiation of translation. *Nat. New Biol.* **232:**169–173.

64. **Ingle, C. A., and S. R. Kushner.** 1996. Development of *in vitro* mRNA decay system for *Escherichia coli*: poly(A) polymerase I is necessary to trigger degradation. *Proc. Natl. Acad. Sci. USA* **93:**12926–12931.

65. **Iost, I., and M. Dreyfus.** 1994. mRNA can be stabilized by DEAD-box proteins. *Nature* **372:**193–196.

66. **Iost, I., and M. Dreyfus.** 1995. The stability of *Escherichia coli lacZ* mRNA depends upon the simultaneity of its synthesis and translation. *EMBO J.* **14:**3252–3261.

67. **Jacquet, M., and A. Kepes.** 1971. Initiation, elongation and inactivation of lac messenger RNA in *Escherichia coli* studied by measurement of its beta-galactosidase synthesizing capacity *in vivo*. *J. Mol. Biol.* **60:**453–472.

68. **Jain, C., and J. G. Belasco.** 1995. RNase E autoregulates its synthesis by controlling the degradation rate of its own mRNA in *Escherichia coli*: unusual sensitivity of the rne transcript to RNase E activity. *Genes Dev.* **9:**84–96.

69. **Kaneko, T., S. Sato, H. Kotani, A. Tanaka, E. Asamizu, Y. Nakamura, N. Miyajima, M. Hirosava, M. Sugiura, S. Sasamoto, T. Kimura, T. Hosouchi, A. Matsuno, A. Muraki, N. Nakazaki, K. Naruo, S. Okumura, S. Shimpo, C. Takeuchi, T. Wada, A. Watanabe, M. Yamada, M. Yasuda, and S. Tabata,** 1996. Sequence analysis of the genome of the unicellular cyanobacterium *Synechocystis* sp. Strain PCC6803. II. Sequence determination of the entire genome and assignment of potential-coding regions. *DNA Res.* **3:**109–136.

70. **Kodo, M., K. Yamanaka, T. Mitani, H. Niki, T. Ogura, and S. Hiraga.** 1996. RNase E polypeptides lacking a carboxyl-terminal half suppress a *mukB* mutation in *Escherichia coli*. *J. Bacteriol.* **178:**3917–3925.

71. **Klug, G., C. W. Adams, J. Belasco, B. Doerge, and S. N. Cohen.** 1987. Biological consequences of segmental alterations in mRNA stability: effect of deletion of the intercistronic hairpin loop region of the *R. capsulatus puf* operon. *EMBO J.* **6:**3515–3520.

72. **Klug, G., and S. N. Cohen.** 1990. Combined actions of multiple hairpin loop structures and sites of rate-limiting endonucleolytic cleavage determine differential degradation rates of individual segments within polycistronic *puf* operon. *J. Bacteriol.* **172:**5140–5146.

73. **Koraimann, G., C. Schroller, H. Graus, D. Angerer, K. Teferle, and G. Högenauer.** 1989. Expression of gene 19 of the conjugative plasmid R1 is controlled by RNase III. *Mol. Microbiol.* **9:**717–727.

74. **Krinke, L., and D. L. Wulff.** 1990. The cleavage specificity of RNaseIII. *Nucleic Acids Res.* **18:**4809–4815.

75. **Kuwano, M., H. Ono, H. Endo, K. Hori, K. Nakamura, Y. Hirota, and Y. Ohnishi.** 1977. Gene affecting longevity of messenger RNA: a mutant of *Escherichia coli* with altered mRNA stability. *Mol. Gen. Genet.* **154:**279–285.

76. **Lin-Chao, S., and S. N. Cohen.** 1991. The rate of processing and degradation of antisense RNAI regulates the replication of Co1E1-type plasmids *in vivo*. *Cell* **65:**1233–1242.

77. **Lin-Chao, S., T.-T. Wong, K. J. McDowall, and S. N. Cohen.** 1994. Effects of nucleotide sequence on the specificity of *rne*-dependent and RNase E-mediated cleavages of RNA I encoded by the pBR322 plasmid. *J. Biol. Chem.* **269:**10797–10803.

78. **Lundberg, U., Ö Melefors, B. Sohlberg, D. Georgellis, and A. von Gabain.** 1995. RNase K: one letter less in the alphabet soup. *Mol. Microbiol.* **17:**595–596.

79. **Lundberg, U., G. Nilsson, and A. von Gabain.** 1988. The differential stability of the Escherichia coli ompA and bla mRNA at various growth rates is not correlated to the efficiency of translation. *Gene* **72:**141–149.

80. **Lundberg, U., A. von Gabain, and Ö. Melefors.** 1990. Cleavages in the 5′ region of the *ompA* and *bla* mRNA control stability: studies with an *E. coli* mutant altering mRNA stability and a novel endoribonuclease. *EMBO J.* **9:**2731–2741.

81. **Mackie, G., and J. L. Genereaux.** 1993. The role of RNA structure in determining RNase E-dependent cleavage sites in the mRNA for ribosomal protein S20 *in vitro*. *J. Mol. Biol.* **234:**998–1012.

82. **Mackie, G. A., J. L. Geneareaux, and S. K. Masterman.** 1997. Modulation of the activity of RNase E *in vitro* by RNA sequences and secondary structures 5′ to cleavage sites. *J. Biol. Chem.* **272:**609–616.

83. **Matsunaga, J., E. L. Simons, and R. W. Simons.** 1996. RNase III autoregulation: structure and function of *rncO*, the posttranscriptional "operator." *RNA* **2:**1228–1240.

84. **McDowall, K. J., and S. N. Cohen.** 1996. N-terminal domain of the *rne* gene product has RNase E activity and is non-overlapping with the arginine-rich RNA-binding site. *J. Mol. Biol.* **255:**349–355.

85. **McDowall, K. J., R. G. Hernandez, S. Lin-Chao, and S. N. Cohen.** 1993. The *ams-1* and *rne-3071* temperature-sensitive mutations in the *ams* gene are in close proximity to each other and cause substitutions within a domain that resembles a product of the *Escherichia coli mre* locus. *J. Bacteriol.* **175:**4245–4249.

86. **McDowall, K. J., V. R. Kaberdin, S.-W. Wu, S. N. Cohen, and S. Lin-Chao.** 1995. Site-specific RNase E cleavage of oligonucleotides and inhibition by stem-loops. *Nature* **374:**287–290.

87. **McDowall, K. J., S. Lin-Chao, and S. N. Cohen.** 1994. A+U content rather than particular nucleotide sequence determines the specificity of RNase E cleavage. *J. Biol. Chem.* **269:**10790–10796.

88. **McLaren, R. S., S. F. Newbury, G. S. C. Dance, H. C. Causton, and C. F. Higgins.** 1991. mRNA degradation by processive 3′–5′ exoribonucleasis *in vitro* and the implications for prokaryotic mRNA decay *in vivo*. *J. Mol. Biol.* **221:**81–95.

89. **Mead, D. J., and S. G. Oliver.** 1983. Purification and properties of a double-stranded ribonuclease from the yeast *Saccharomyces cerevisiae*. *Eur. J. Biochem.* **137:**501–507.

90. **Melefors, Ö., and A. von Gabain.** 1988. Site-specific endonucleolytic cleavages and the regulation of stability of *E. coli ompA* mRNA. *Cell* **52:**893–901.

91. **Melefors, Ö., and A. von Gabain.** 1991. Analysis of *ompA* mRNA decay and rRNA processing indicates that *ams* and *rne* denote the same *E. coli* gene locus. *Mol. Microbiol.* **5:**857–864.

92. **Melin, L., L. Rutberg, and A. von Gabain.** 1989. Transcriptional and posttranscriptional control of the *Bacillus subtilis* succinate dehydrogenase operon. *J. Bacteriol.* **171:** 2110–2115.

93. **Mertens, N., E. Remaut, and W. Fiers.** 1996. Increased stability of phage T7g10 mRNA is mediated by either a 5′- or 3′-terminal stem-loop structure. *Biol. Chem.* **377:** 811–817.

94. **Miczak, A., and D. Apirion.** 1991. Location of the RNA-processing enzymes RNase III, RNase E and RNase P in the Escherichia coli cell. *Mol Microbiol.* **6:**1801–1810.

95. **Miczak, A., V. R. Kaberdin, C.-L. Wei, and S. Lin-Chao.** 1996. Proteins associated with RNase E in a multicomponent ribonucleolytic complex. *Proc. Natl. Acad. Sci. USA* **93:**3865–3869.

96. **Miller, O. L., Jr., B. A. Hamkalo, and C. A. Thomas, Jr.** 1970. Visualization of bacterial genes in action. *Science* **169:**392–395.

97. **Misra, T. K., and D. Apirion.** 1979. RNase E, an RNA processing enzyme from *Escherichia coli*. *J. Biol. Chem.* **254:**11154–11159.

98. **Morikawa, N., and F. Imamoto.** 1969. Degradation of tryptophan messenger. *Nature* **223:**37–40.

99. **Morse, D. E., R. Mosteller, C. Baker, and C. Yanofsky.** 1969. Direction of in vivo degradation of tryptophan messenger RNA-A correction. *Nature* **223:**40–43.

100. **Morse, D. E., and C. Yanofsky.** 1969. Polarity and the degradation of mRNA. *Nature* **224:**329–321.

101. **Mott, J. E., J. L. Galloway, and T. Platt.** 1985. Maturation of *Escherichia coli* tryptophan operon mRNA: evidence for 3′ exonucleolytic processing after *rho*-dependent termination. *EMBO J.* **4:**1887–1891.

102. **Mudd, E. A., and C. F. Higgins.** 1993. *Escherichia coli* endoribonuclease RNase E: autoregulation of expression and site-specific cleavage of mRNA. *Mol. Microbiol.* **9:** 557–568.

103. **Mudd, E. A., H. M. Krisch, and C. F. Higgins.** 1990. RNase E, an endoribonuclease, has a general role in the chemical decay of *Escherichia coli* mRNA: evidence that *rne* and *ams* are the same genetic locus. *Mol. Microbiol.* **4:**2127–2135.

104. **Mudd, E. A., P. Prentki, D. Belin, and H. M. Krisch.** 1988. Processing of unstable bacteriophage T4 gene 32 mRNA into a stable species requires *Escherichia coli* ribonuclease E. *EMBO J.* **7:**3601–3607.

105. **Newbury, S. E., N. H. Smith, E. C. Robinson, I. D. Hiles, and C. F. Higgins.** 1987. Stabilization of translationally active mRNA by prokaryotic REP sequences. *Cell* **48:**297–310.

106. **Nilsson, G., J. G. Belasco, S. N. Cohen, and A. von Gabain.** 1984. Growth-rate dependent regulation of mRNA stability in *Escherichia coli*. *Nature* **312:**75–77.

107. **Nilsson, G., J. G. Belasco, S. N. Cohen, S. N., and A. von Gabain.** 1987. Effect of premature termination of translation on mRNA stability depends on the site of ribosome release. *Proc. Natl. Acad. Sci. USA* **84:**4890–4894.

108. **Nossal, N. G., and M. F. Singer.** 1968. The processive degradation of individual polynucleotide chains. *J. Biol. Chem.* **243:**913–922.

109. **Oelmüller, U., H. G. Schlegel, and C. C. Friedrich.** 1990. Differential stability of mRNA species of *Alcaligenes eutrophus* soluble and particulate hydrogenases. *J. Bacteriol.* **172:**7057–7064.

110. **O'Hara, E. B., J. A. Chekanova, C. A. Ingle, Z. R. Kushner, E. Peters, and S. R. Kushner.** 1995. Polyadenylation helps regulate mRNA decay in *Escherichia coli*. *Proc. Natl. Acad. Sci. USA* **92:**1807–1811.

111. **Ono, M., and M. Kuwano.** 1979. A conditional lethal mutation in an *Escherichia coli* strain with a longer chemical lifetime of mRNA. *J. Mol. Biol.* **129:**343–357.

112. **Panayotatos, N., and K. Truong.** 1985. Cleavage within an RNase III site can control mRNA stability and protein synthesis *in vivo*. *Nucleic Acids Res.* **7:**2227–2240.

113. **Pato, M. L., P. M. Bennett, and K. von Meyenburg.** 1973. Messenger ribonucleic acid synthesis and degradation in *Escherichia coli* during inhibition of translation. *J. Bacteriol.* **116:**710–718.

114. **Petersen, C.** 1993. Translation and mRNA stability in bacteria: a complex relationship, p. 117–145. *In* J. Belasco and G. Brawerman (ed.) *Control of Messenger RNA Stability*. Academic Press, New York.

115. **Plamann, M. D., and G. V. Stauffer.** 1990. *Escherichia coli glyA* mRNA decay: the role of secondary structure and the effects of the *pnp* and *rnb* mutations. *Mol. Gen. Genet.* **220:**301–306.

116. **Portier, C., L. Dondon, M. Grunberg-Manago, and P. Regnier.** 1987. The first step in the functional inacti-

vation of the *Escherichia coli* polynucleotide phosphorylase messenger is a ribonuclease III processing at the 5′ end. *EMBO J.* **6:**2165–2170.

117. **Py, B., H. Causton, E. A. Mudd, and C. F. Higgins.** 1994. A protein complex mediating mRNA degradation in *Escherichia coli*. *Mol. Microbiol.* **14:**717–729.

118. **Py, B., C. F. Higgins, H. M. Krisch, and A. J. Carpousis.** 1996. A DEAD-box RNA helicase in the *Escherichia coli* RNA degradosome. *Nature* **381:**169–172.

119. **Ramanarayanan, M., and P. R. Srinivasan.** 1976. Further studies on the isolation and properties of polyriboadenylate polymerase from *Escherichia coli* PR7 (RNase I-pnp) *J. Biol. Chem.* **251:**6274–6286.

120. **Regnier, P., and M. Grunberg-Manago.** 1989. Cleavage by RNase III in the transcripts of the *metY-nusA-infB* operon of *Escherichia coli* releases the tRNA and the decay of the downstream mRNA. *J. Mol. Biol.* **210:**293–302.

121. **Reith, M., and J. Munholland.** 1995. Complete nucleotide sequence of the *Porphyra purpurea* chloroplast genome. *Plant Mol. Biol. Rep.* **13:**333–335.

122. **Resnekov, O., L. Rutberg, and A. von Gabain.** 1990. Changes in the stability of specific mRNA species in response to growth stage in *Bacillus subtilis*. *Proc. Natl. Acad. Sci. USA* **87:**8355–8359.

123. **Richter, J. D.** 1991. Translational control during early development. *Bioessays* **13:**179–183.

124. **Robert-LeMeyer, M., and C. Portier.** 1992. *E. coli* polynucleotide phosphorylase expression in autoregulated through an RNase III-dependent mechanism. *EMBO J.* **11:**2633–2641.

125. **Robertson, H. D., R. F. Webster, and N. D. Zinder.** 1968. Purification and properties of ribonuclease III from *Escherichia coli*. *J. Biol. Chem.* **243:**82–91.

126. **Romeo, J. M., and D. R. Zusman.** 1992. Determinants of an unusually stable mRNA in the bacterium *Myxococcus xanthus*. *Mol. Microbiol.* **6:**2975–2988.

127. **Rosenberg, A. H., B. N. Lade, D.-S. Chui, S.-W. Lin, J. J. Dunn, and F. W. Studier.** 1987. Vectors for selective expression of cloned DNAs by T7 RNA polymerase. *Gene* **56:**125–135.

128. **Sandler, P., and B. Weisblum.** 1988. Erythromycin-induced stabilization or *ermA* messenger RNA in *Staphylococcus aureus* and *Bacillus subtilis*. *J. Mol. Bol.* **203:** 905–915.

129. **Sandler, P., and B. Weisblum.** 1989. Erythromycin-induced ribosome stall in the ermA leader: a barricade to 5′-to-3′ nucleolytic cleavage of the ermA transcript. *J. Bacteriol.* **171:**6680–6688.

130. **Takata, R., T. Mukai, and K. Hori.** 1987. RNA processing by RNase III is involved in the synthesis of *Escherichia coli* polynucleotide phosphorylase. *Mol. Gen. Genet.* **209:**28–32.

131. **Takiff, H. E., T. Baker, T. Copeland, S. M. Chen, and D. L. Court.** 1992. Locating essential *Escherichia coli* genes by using mini-Tn10 transposones: The *pdxJ* operon. *J. Bacteriol.* **174:**1544–1553.

132. **Taraseviciene, L., G. Björk, and B. E. Uhlin.** 1995. Evidence for an RNA binding region in the *Escherichia coli* processing endoribonuclease RNase E. *J. Biol. Chem.* **270:**26391–26398.

133. **Taraseviciene, L., A. Miczak, and D. Apirion.** 1991. The gene specifying RNase E (*rne*) and a gene affecting mRNA stability (*ams*) are the same gene. *Mol. Microbiol.* **5:**851–856.

134. **Taraseviciene, L., S. Naureckiene, and B. E. Uhlin.** 1994. Immunoaffinity purification of the *Escherichia coli rne* gene product. *J. Biol. Chem.* **269:**12167–12172.

135. **Tomcsányi, T., and D. Apirion.** 1985. Processing enzyme ribonuclease E specifically cleaves RNA I, an inhibitor of primer formation in plasmid DNA synthesis. *J. Mol. Biol.* **185:**713–720.

136. **Varmus, H. E., R. L. Perlman, and I. Pastan.** 1971. Regulation of *lac* transcription in antibiotic-treated *E. coli*. *Nat. New Biol.* **230:**41–44.

137. **von Gabain, A., J. G. Belasco, J. L. Schottel, A. C. Y. Chang, and S. N. Cohen.** 1983. Decay of mRNA in *Escherichia coli*: investigation of the fate of specific segments of transcripts. *Proc. Natl. Acad. Sci USA* **80:**653–657.

138. **Wang, N., and S. N. Cohen.** 1994. *ard-1*: a human gene that reverses the effects of temperature-sensitive and deletion mutants in the *Escherichia coli rne* gene and encodes an activity producing RNase E-like cleavages. *Proc. Natl. Acad. Sci. USA* **91:**10591–10595.

139. **Wennborg, A., B. Sohlberg, D. Angerer, G. Klein, and A. von Gabain.** 1995. A human RNase E-like activity that cleaves RNA sequences involved in mRNA stability control. *Proc. Natl. Acad. Sci. USA* **92:**7322–7326.

140. **Wong, H. C., and S. Chang.** 1986. Identification of a positive retroregulator that stabilizes mRNAs in bacteria. *Proc. Natl. Acad. Sci USA* **83:**3233–3237.

141. **Woo, W.-M., and S. Lin-Chao.** 1997. Processing of the *rne* transcript by an RNase E-independent amino acid-dependent mechanism. *J. Biol. Chem.* **272:**15515–15520.

142. **Xu, F., and S. N. Cohen.** 1995. RNA degradation in *Escherichia coli* regulated by 3′ adenylation and 5′ phosphorylation. *Nature* **374:**180–183.

143. **Yamamoto, T., and F. Imamoto.** 1975. Differential stability of *trp* messenger RNA synthesized originating at the *trp* promoter and p_L promoter of lambda *trp* phage. *J. Mol. Biol.* **92:**289–309.

144. **Zilhao, R., L. Camelo, and C. M. Arraiano.** 1993. DNA sequencing and expression of the gene *rnb* encoding *Escherichia coli* ribonuclease III. *Mol. Microbiol.* **8:**43–51.

Methods for Optimizing Industrial Enzymes by Directed Evolution

HUIMIN ZHAO, JEFFREY C. MOORE, ALEX A. VOLKOV, AND FRANCES H. ARNOLD

49

Enzymes exhibit exquisite catalytic power unmatched by conventional catalysts. Their many applications range from serving as catalysts for chemical synthesis to use in diagnostic testing, foods, and pharmaceuticals. However, naturally occurring enzymes are often not well suited for industrial applications. Problems include enzyme instability and low catalytic activity on nonnatural substrates and in nonnatural environments. Because the extensive structural and mechanistic information required to guide rational approaches to engineering improved enzymes is available for only a tiny fraction of known sequences, alternative approaches are needed. Directed evolution has proven very effective for modifying enzymes in the absence of such knowledge. By directed evolution we have been able to tailor enzyme functions never required in the natural environment. Properties that can be improved include stability, catalytic activity, activity towards new substrates, expression level in a heterologous host, and others (7).

This laboratory has focused on developing both the methods and strategies for design by directed evolution and demonstrating them by engineering novel, industrially useful enzymes (1, 2). As outlined in Fig. 1, the first step in directed evolution is to create molecular diversity starting from a target gene or a family of related genes. The diversity can be created by introducing mutations and/or by recombination. The gene products are sorted by screening or selection, and those genes encoding improved products can be returned for further generations of evolution. This evolutionary process can be repeated until the goal is achieved, or until there is no further improvement.

We will illustrate the process of directed evolution and the methods presented here with two enzymes: the serine protease subtilisin E and *p*-nitrobenzyl (pNB) esterase. pNB esterase can be used during the synthesis of certain β-lactam antibiotics to deprotect a pNB ester intermediate (19). We will focus on the three major components of directed evolution efforts: (i) creating diversity by random point mutagenesis and/or in vitro recombination, (ii) sorting the resulting enzyme libraries, and (iii) data analysis, required for the accurate selection of positives and to improve further experiments.

49.1. CREATING ENZYME DIVERSITY

In contrast to natural evolution, directed evolution has a defined goal, and the key processes—mutation, recombination, and selection or screening—must be carefully controlled by the experimenter. Because of the strict limitations on library sorting imposed by screening and selection, the mutation rate must be tuned to the power of the sorting method (1, 7). Useful, reasonably sized libraries can be created by introducing multiple mutations in a particular region or across a limited number of positions (e.g., by combinatorial mutagenesis with oligonucleotide cassettes [12]). However, such an approach will exclude the many useful solutions found in unexpected places (2). Protein engineers are becoming increasingly aware that many protein functions are not confined to a small number of amino acids but are affected by residues far from active sites. We therefore usually try to evolve the entire gene, rather than target particular positions. Because we can search only 10^4 to 10^5 enzyme variants, even with a good screening method, single-amino-acid-substitution libraries are preferred. Larger numbers of variants can, of course, be searched by good selections (10^8–10^9), allowing one to examine double- or even triple-amino-acid-substitution libraries, depending on the sequence length (7).

Most random mutagenesis methods create mutations in single bases. Because of degeneracy of the genetic code, this provides access to only about six amino acid substitutions instead of 19, thus significantly reducing the potential diversity. In addition, many mutagenesis methods are not really random, further limiting the number of amino acid substitutions actually accessible in a given experiment. For example, the mutagenesis method we use, error-prone PCR (8), shows a strong bias for transitions over transversions. In vitro recombination methods include DNA shuffling (15), random-priming recombination (14), and the staggered extension process (18). All these methods have a (controllable) level of associated point mutagenesis. With recombination, directed evolution can begin from multiple, closely related starting points rather than from a single sequence (5). Recombination of existing functional sequences (i.e., homologous enzymes) will create another level of diversity that point mutagenesis cannot generate. A library of recombined genes may provide an excellent starting point for creating novel functions.

49.1.1. Random Point Mutagenesis

In random point mutagenesis, the two most important factors to consider are mutation frequency and mutation bias.

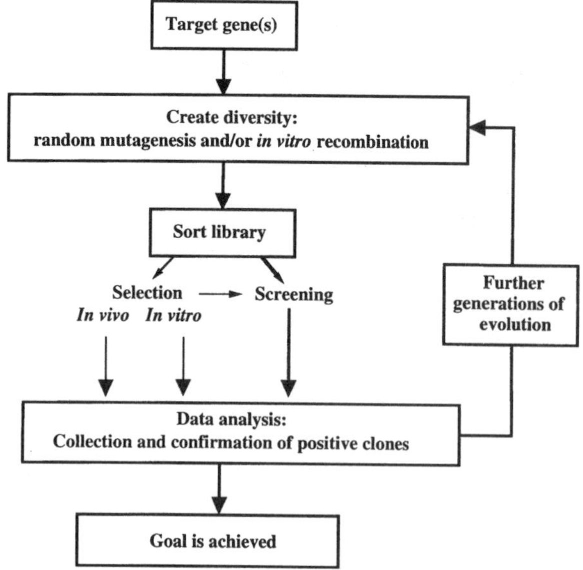

FIGURE 1 Flow chart for directed enzyme evolution.

Mutation frequency is the average number of mutations per gene and is usually reported as a percentage. The target mutation frequency can be calculated from the length of the DNA coding sequence and an estimate of the number of mutations per sequence desired. For instance, a desirable mutation level for directed evolution is ~2 to 5 base substitutions per gene (1, 10). Thus, for a 1-kb sequence, the mutation frequency should be ~0.2 to 0.5%. Although numerous methods for making random DNA mutations exist, we choose to use error-prone PCR because the procedure is simple, rapid, and robust and, most importantly, the mutation frequency can be precisely controlled.

Error-prone PCR does not create truly random DNA substitutions (e.g., a common bias of error-prone PCR is a high occurrence of A-to-G substitutions). Some bias can be tolerated in directed evolution experiments, but large variations in error frequency usually cannot. Bias affects the *location* of mutations (e.g., in error-prone PCR, mutations at AT base pairs occur much more frequently than mutations at GC base pairs) as well as their *type* (A is more frequently substituted with G). To a first approximation, however, the A's are well distributed throughout the gene, and mutations are still occurring throughout the protein. Additionally, error-prone PCR mutagenizes the A's on both DNA strands during synthesis, effectively doubling the location of mutations. PCR modifications reduce bias but do not eliminate it (3, 13).

The error-prone PCR method we routinely use was originally outlined by Leung and coworkers (8) and further examined by Cadwell and Joyce (3) and Shafikhani et al. (13). A series of mutation frequencies ranging from 0.11 to 2% have been obtained under different reaction conditions. In particular, the mutation frequency can be controlled (from 0.11 to 0.49%) simply by adjusting the concentration of manganese in the reaction mixture (13). The following protocol is used in the directed evolution of subtilisin E. The overall mutation frequency is ~0.2%, or 2 base changes per gene, on average.

PROTOCOL

Error-Prone PCR

1. Prepare purified plasmid DNA.
2. Prepare a 10× mutagenic buffer containing 70 mM $MgCl_2$, 500 mM KCl, 100 mM Tris (pH 8.3 at 25°C), and 0.1% (wt/vol) gelatin.
3. Prepare a 10× deoxynucleoside triphosphate (dNTP) mix containing 2 mM dGTP, 2 mM dATP, 10 mM dCTP, and 10 mM TTP.
4. Prepare a solution of 5 mM $MnCl_2$. (Do not combine with the 10× PCR buffer, which would result in precipitation.)
5. Combine 10 μl of 10× mutagenic buffer, 10 μl of 10× dNTP mix, 30–50 pmol of each primer, 2 fmol of template DNA (~10 ng for an 8-kb plasmid), and an amount of distilled H_2O that brings the volume up to 96 μl.
6. Add 3 μl of 5 mM $MnCl_2$. Mix well.
7. Add 1 μl of *Taq* polymerase (5 U/μl, Promega). Mix gently.
8. Run a PCR program: 14 cycles of 30 s at 94°C, 30 s at 50°C, and 30 s at 72°C. (For a gene of more than 1 kb, increase the extension time at 72°C accordingly.)
9. Purify the reaction products by using a Promega PCR DNA purification kit.
10. Run a small portion of the purified products on an agarose gel to estimate the yield of full-length gene (typically, a yield of 0.5–1.0 μg per reaction is obtained).
11. Digest with appropriate restriction enzymes and clone into expression vector.

The overall mutation frequency in error-prone PCR is the product of three parameters: the error rate (fidelity) of the polymerase in the reaction conditions, the length of the mutagenized gene, and the number of effective doubling cycles. Varying the concentration of $MnCl_2$ will affect only the error rate of the polymerase. In the above protocol, the overall mutation frequency is ~0.1% at 0 mM $MnCl_2$ and ~0.5% at 0.5 mM $MnCl_2$. The number of effective doublings is another easily adjustable parameter. This can be altered by adjusting either the PCR cycle numbers or the ratio of template to primers.

Because polymerase fidelity also depends on the nature of target sequences, different mutation rates will be observed on different sequences, even when exactly the same reaction conditions are used. A straightforward way to assess the overall mutation frequency and the nature of mutations is to sequence a few random clones from the amplified population. However, sequencing is time-consuming and expensive. A simple and perhaps more meaningful alternative is to estimate the mutation frequency from the fraction of active clones from the amplified population (13, 17). The activity profiles of small samplings from libraries of subtilisin E variants produced by error-prone PCR with various $MnCl_2$ concentrations are plotted in Fig. 2. Clones exhibiting at least 10% of wild-type subtilisin activity have been scored as active. The frac-

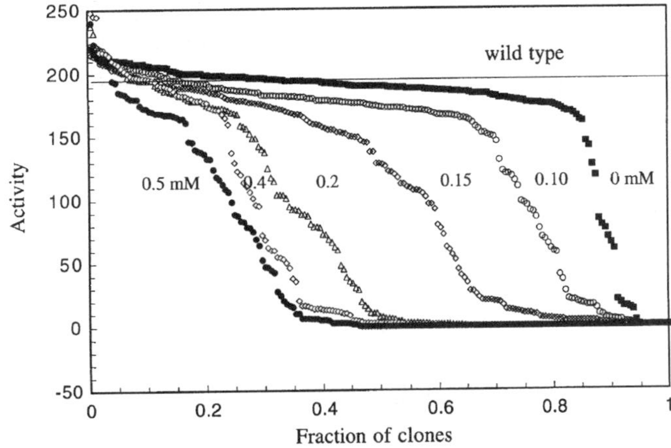

FIGURE 2 Effects of $MnCl_2$ concentration on the activities of enzymes from a random mutant library prepared by error-prone PCR.

tion of active clones varied from 90% at 0 mM $MnCl_2$ to 30% at 0.5 mM $MnCl_2$. Even a concentration difference as small as 0.05 mM makes a clear difference in the activity profile. For subtilisin E, 65% active clones corresponds to a mutation frequency of 2 base changes per gene, or 0.2%. Even though different enzymes will show different activity profiles for different mutation rates, the fraction of active clones is a convenient index of mutation frequency and can be used as a diagnostic check for the successful creation of the desired randomly mutated library.

49.1.2. In Vitro Recombination

A convenient method for performing recombination of DNA sequences has been described by Stemmer (15). This method, known as "DNA shuffling," involves enzymatic digestion of the parental DNA into short fragments followed by reassembly of the fragments into full-length genes. Since the fragments are free to associate with complementary fragments from other, similar genes, mutations from one parent can be combined with mutations from another parent(s) to generate novel combinations. The technique can also remove mutations which are deleterious or do not contribute to the desired property (neutral). The original method has been modified to simplify it and to yield better control over the associated mutagenic rate (9, 17). We have recently developed two alternative random recombination methods that do not involve digestion of the parent genes (14, 18). The recombination effected by these methods is similar to that obtained by DNA shuffling.

When recombining genes from multiple improved variants, it is important to first consider the recombined library size (11). The various in vitro recombination methods can recombine any number of parent genes. However, the resulting libraries may be impossibly large, too large to find further improvements in function. If each mutation sorts independently from the others, the probability of generating a sequence containing all mutations M present in N separate sequences is $1/N^M$ (11). When four improved pNB esterases are recombined (each parent contained one mutation responsible for increased activity), the probability of producing the sequence containing all four mutations is 1 in 256, assuming all mutations are sorted randomly. This probability, and therefore the screening requirements, decreases rapidly as the number of mutations and sequences increases. For example, consider the two cases of shuffling the four most improved pNB esterases versus that of shuffling a pool containing all 15 sequences deemed more active than the parent for that generation. The latter pool of 15 improved sequences contained the 4 most active variants as well as 11 more variants that are less active, but still better than the parent sequence. While recombining all 15 sequences should eventually lead to a more "fit" enzyme, the screening requirements may be beyond the screening capability. The four-variant pool will, in fact, provide the most fit recombined variants if screening is limited to a few hundred or a few thousand clones.

The in vitro recombination protocol included here is a modified high-fidelity DNA shuffling protocol (17) with a mutagenic rate of only 0.05%.

PROTOCOL

DNA SHUFFLING PROTOCOL

1. Preparation of genes to be shuffled by restriction enzyme digestion of plasmid DNA. Purify DNA templates from an agarose gel after electrophoresis and dissolve in 10 mM Tris-HCl (pH 7.4). DNA concentrations can be estimated by electrophoresis or spectrophotometry. Mix the templates 1:1 for a total of ~2 μg. (More DNA is preferred, but it should not exceed 5 μg. Otherwise, the following digestion reaction will not be complete.)
2. Combine 2.5 μl of 1 M Tris-HCl (pH 7.5), 2.5 μl of 200 mM $MnCl_2$, mixed DNA fragments, and distilled H_2O to bring the volume to 48.5 μl. Equilibrate this mixture at 15°C for 5 min on a thermocycler and then add 1.5 μl of 1:200 diluted DNase I (10 U/μl, Boehringer Mannheim). Perform digestion at 15°C and terminate after 2 min by adding 10 μl of 200 mM EDTA (incubation time may be shorter or longer depending on the amount of DNA). Confirm the fragment size on 2% agarose gel. Fragments can be purified on a Centri-Sep column (Princeton Separations, Inc., Adelphia, N.J.) or eluted directly from the gel, if specific fragment sizes are required.
3. Fragment reassembly. Combine 10 μl of purified fragments, 5 μl of 10× *Pfu* buffer, 5 μl of 10× dNTP mix

(2 mM of each dNTP), and 0.3 μl of *Pfu* polymerase (Stratagene, La Jolla, Calif.). Overlay the reaction mixture with 30 μl of mineral oil (no oil is needed if lids are also heated). PCR program: 3 min at 96°C, followed by 40 cycles of 1 min at 94°C, 1 min at 55°C, and 1 min + 5 s/cycle at 72°C; followed by 7 min at 72°C.

4. PCR amplification of reassembled products. Use 1 μl of this reaction and 1:10, 1:20, and 1:50 dilutions of it as templates in a 25-cycle PCR reaction. PCR conditions (100 ml final volume): 30 pmol of each primer, 1× *Taq* buffer, 1.5 mM $MgCl_2$, 0.2 mM each dNTP, and 2.5 U of *Taq*/*Pfu* (1:1) mixture. PCR program: 2 min at 96°C; 10 cycles of 30 s at 94°C, 30 s at 55°C, 45 s at 72°C; followed by another 14 cycles of 30 s at 94°C, 30 s at 55°C, 45 s + 20 s/cycle at 72°C; and, finally, 7 min at 72°C. This program gives a single band at the correct size. If initial DNA templates were prepared by PCR amplification, reassembled products may require amplification with nested primers separated by 50 to 100 bp from the original primers.
5. Purify the reaction products by using a Promega PCR DNA purification kit.
6. Run a small portion of the purified products on an agarose gel to estimate the yield of full-length gene (typically, a yield of 0.5 to 1.0 μg per reaction is obtained).
7. Digest with appropriate restriction enzymes and clone into expression vector.

StEP RECOMBINATION PROTOCOL

1. Purify plasmid DNAs from a dam-positive strain (DH5a, XL1-Blue) or amplify target genes in a PCR reaction.
2. Combine 5 μl of 10× PCR buffer, 3 μl of 25 mM $MgCl_2$, 5 μl of dNTP mix (2 mM of each dNTP), 1 to 20 ng of each template DNA, 30 to 50 pmol of each primer, and distilled H_2O up to 49.5 μl. Mix well. Run several reactions with different template:primer ratios.
3. Add 0.5 μl of *Taq* polymerase. Mix gently.
4. Run 80 to 100 extention cycles: 30 s at 94°C and 5 to 15 s at 55°C.
5. Run 5 to 10 μl of the StEP reaction on an agarose gel. Expected reaction products are full-size amplified gene(s), smear, or a combination of both.
6. If PCR products were used in step 2, go to step 7. Combine 2 μl of the StEP reaction, 1 μl of *Dpn*I reaction buffer, 6 μl of H_2O, and 1 μl of *Dpn*I restriction endonuclease. Incubate at 37°C for 1 h to remove parent plasmid DNA.
7. Amplify target gene in a standard PCR reaction using 1 μl from the previous reaction and 1:10, 1:20, and 1:50 dilutions of it as templates. If initial DNA templates were prepared by PCR amplification, reassembled products may require amplification with nested primers separated by 50 to 100 bp from the original primers.
8. Run a small portion of the amplified products on an agarose gel to determine the yield and quality of amplification. Select a reaction with high yield and low amount of nonspecific products.
9. Purify the selected reaction products using a Promega PCR DNA purification kit.
10. Digest with appropriate restriction enzymes and clone into expression vector.

As demonstrated elsewhere (17), the mutagenic frequency can be controlled over a wide range, from 0.05 to 0.7%, by the inclusion of Mn^{2+} or Mg^{2+}, by the choice of DNA polymerase, and/or by the use of restriction enzyme digestion to prepare the starting DNA. If high fidelity is not required, it may be more convenient to prepare sufficient starting DNA by conventional PCR amplification (step 1). *Taq* or other polymerases can be used in the reassembly and final PCR amplification steps.

A protocol for recombination by the staggered extension process (StEP) (18) is also provided.

The finite error frequency associated with DNA shuffling has been used by Stemmer (15) to supply point mutations for recombination and evolution. When starting with a single sequence, we prefer to use error-prone PCR under controlled conditions to generate libraries of variants. When more than one improved sequence is identified during screening, the sequences can be recombined.

49.2. LIBRARY SORTING

Given a thoughtful approach to generating the mutant library, the development of an efficient method to search for the desired properties is probably the single most important element determining the success of a directed evolution experiment. Libraries can be sorted by selection or screening. Various in vivo and in vitro selection methods have been established (6). The prerequisite to biological selection is the generation of a function which confers a growth or survival advantage to the host organism. Selection can be a very efficient search mechanism, allowing an exhaustive search of libraries of 10^6 and more variants. The disadvantage of biological selection is that the property or protein of interest cannot be decoupled from the biological function. Thus, it can be difficult or even impossible to explore novel functions such as activity in a highly nonnatural environment (16).

The sorting method chosen should reflect the desired features as much as possible. Solutions can arrive through unanticipated mechanisms—an all too common experience when selections are used! It is also often the case that improperly designed screens will allow uninteresting mutants to pass the sorting criteria. We have found, for example, that bacteria grown in a controlled laboratory medium exhibit substantial increases in activity as a result of directed evolution efforts. When the bacteria are transferred to the industrial growth medium and conditions, however, the increased activity may no longer be apparent. Thus, it is important to include in the sorting as many of the tasks the enzyme (and the organism) is expected to accomplish as possible. This can be accomplished in a single, comprehensive screen, or, better, by using a tiered screening protocol, in which positives identified during a rapid assay are verified in a more rigorous series of tests.

49.2.1. Selection

In a biological selection, the enzyme is linked to the host organism survival such that only those organisms possessing the desired trait can grow. When the environmental conditions are set properly, the wild type (or parent) and mutants performing more poorly than wild type do not survive; all surviving colonies are positives. Large libraries (whose sizes are limited only by transformation efficiency) can be sorted in this way. Selections can be useful for evolving drug resistance enzymes, enzymes responsible for providing nutrients to supplement an auxotrophy, and thermostability when an essential enzyme in a thermophile can be replaced with the enzyme of interest, to name a few common examples.

Unfortunately, designing a biological selection around many enzymes' activities is difficult, if not impossible. Selections often take a large amount of time to implement, with no guarantee of success. Additionally, many desired features are nonnatural and by definition cannot be coupled to the growth and survival of the host organism. Where a selection is possible, care must be taken to ensure that positives are a result of mutations in the targeted enzyme, rather than changes somewhere else in the organism. Other in vitro selection approaches for novel catalysts which include binding to a transition state or substrate analog attached to a column, or catalysis of a single reaction followed by trapping, are also problematic because they do not probe catalytic efficiency directly.

49.2.2. Screening

Screening is the most flexible sorting method for directed evolution (16). Experimental conditions can be tailored to meet the desired criteria. Additionally, the screens are often done in much the same way that the enzyme is traditionally assayed (e.g., spectrophotometrically), so a screen can be implemented relatively quickly. The screening conditions should ensure that the expected small enhancements brought about by single-amino-acid substitutions can be measured. This is usually accomplished by adjusting the assay conditions to the point where the wild-type (or parent) enzyme is near (but not below) the lower detection limit. Improvements over the wild-type enzyme will then be clearly discernible. (If wild type is below the detection limit, small improvements may not be measurable.) If a replica plate of the clones is prepared for storage, the clones can be subjected to an array of lethal procedures, and the original clones can be recovered. For example, many intracellular enzymes require cell lysis before they can be assayed. This procedure could be performed in a screen, and when positives are found, they can be regrown from the replica plate. Furthermore, multiple measurements can be made on each sample to account for variability in protein expression levels or to check other key enzyme properties. The biggest drawback of screening is that the size of the library that can be screened is limited, usually to $\sim 10^4$, although the use of robotic hardware can increase this number by 10- to 100-fold.

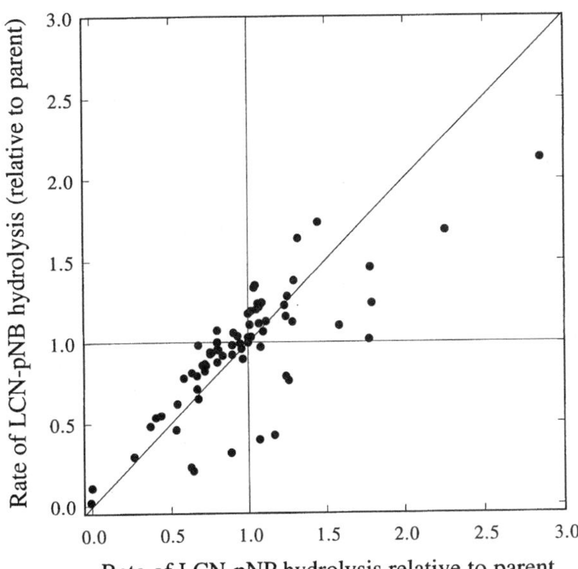

FIGURE 3 Comparison of activities of pNB esterase variants towards desired LCN-pNB substrate and LCN-pNP substrate used in screening.

Although the importance of screening under conditions close to what the enzyme would encounter in an actual process cannot be overstated, dealing with large numbers of variants requires approximations. (Screening 10^4 variants in high-density, 10,000-liter fermentations is clearly not feasible.) When approximations are made, the corre-

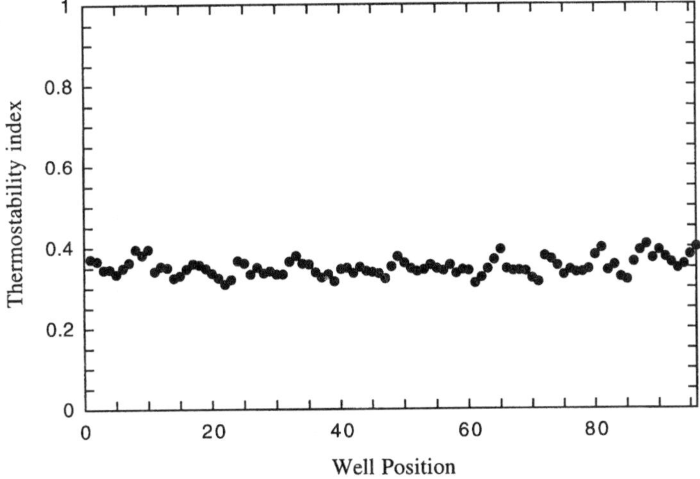

FIGURE 4 Variation in the thermostability index of wild-type subtilisin E clones in a 96-well plate assay.

lation between the approximate and actual conditions should be checked, if at all possible. In the directed evolution of the pNB esterase (10), two important approximations were made in the screening. First, the desired loracarbef pNB ester hydrolysis reaction is usually assayed by high-pressure liquid chromatography (HPLC), which is unsuitable for rapid screening. We therefore devised a screening assay based on the loracarbef *p*-nitrophenyl (pNP) ester, which provides a colorimetric signal. To validate the screening method, we compared the activities of a set of mutants towards the pNB and pNP substrates, as shown in Fig. 3. If the screening reaction perfectly mimicked the desired reaction, all the points would lie on the 45° line. Although there will be some false positives and false negatives in this screening reaction, the rapid screen provides sufficient information to make a rough cut of positive clones. Thus, we needed to retest only a small number of clones by HPLC to verify improved pNB esterase activity and to select which of the clones should parent the next generation. The second approximation concerned the concentration of organic solvent in the assay. The dimethylformamide (DMF) concentration desired for the ultimate application was 25 to 30%, which was too high to generate any measurable activity at the beginning of the evolution experiments. Initial screening experiments were therefore performed in 10 to 15% DMF. The activities towards LCN-nNP in 5 and 25% DMF were found to be strongly correlated. Thus, it was appropriate to screen in lower DMF concentrations, at least at the beginning.

Primary screening can often be performed in petri dishes. This involves either placing the substrate directly in the petri dishes, applying substrate to the already grown colonies, or transferring colonies to a second petri dish containing substrate. The assay conditions are then set so that wild-type enzyme generates a weak positive signal, allowing observation of small (and large) improvements. When environmental conditions of interest preclude growth, and the enzyme is secreted (e.g., when assaying proteases in organic solvents), the organisms can be grown on one or more filters. A "master" filter containing the colonies can be maintained, and an assay filter(s) capturing secreted enzymes can be placed on plates in any environment. One problem with this approach is the inability to keep a known wild-type colony on each plate as a control for plate-to-plate variability. Because transformation mixes are usually plated directly onto the plates, the colony density depends on the ligation and transformation efficiency. Wild type is then placed on separate plates and under different colony densities. This difference in colony density on a plate can dramatically alter the intensity of a wild-type signal and makes a control of this type less useful than expected. Positives found in this way should be regrown

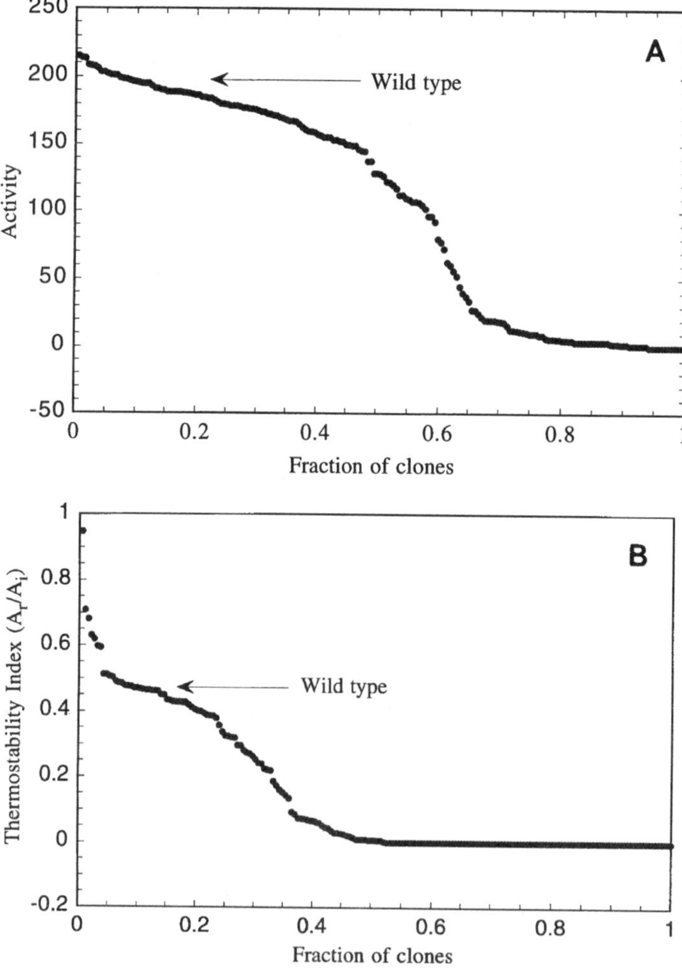

FIGURE 5 (A) Typical profile of the activity of enzyme variants in a random mutant library. (B) Thermostability of enzyme variants from the same mutant library.

on the same plate with wild type and reassayed. (For an example of this approach, see reference 4.)

Multiwell plates are very useful for screening large libraries, especially now that robotic equipment designed to handle these plates is becoming more widely available. Although one can pick, resuspend, and assay colonies in 96-well plates, growing liquid cultures in the plates is generally less tedious. Liquid cultures allow the inclusion of wild-type controls under conditions identical to those of clones from the library to be screened and also tend to yield more uniform and reliable data. Growth for reasonably long durations (36 h) can be achieved in regular incubators, provided there is some means to humidify the shaker. We routinely use beakers filled with water and paper towels in the shaker for this purpose.

When we set up an assay in 96-well plates, we always evaluate several plates that contain only wild-type clones. This enables us to determine the reproducibility of the assay and whether there is any positional variability in the plate. We can thus identify systematic errors due to pipetting, heat transfer, etc., and determine the values of the screening parameter that should be associated with true positives. For example, in screening thermostable subtilisin E variants, each well of a 96-well plate contained an identical subtilisin E clone growing in 200 μl of Schaeffer's (SG) medium. This plate was incubated for 20 h, and cells were spun down. Five microliters of supernatant from each well was transferred into two plates, into which we added 15 μl of SG medium (to avoid a problem with evaporation during incubation at high temperature). One plate was used to measure initial activity (after 5 min of incubation at 65°C), and the other was used to measure residual activity (after 20 min of incubation at 65°C). The ratio of residual activity to initial activity was taken as an index of thermostability (18). The 65°C incubation was performed in an oven on an aluminum block machined to closely contact standard multiwell plates for uniform heating. Data from a control plate containing only wild-type clones are shown in Fig. 4. A small amount of systematic variability is observed, possibly owing to uneven heating of the wells. In general, however, the activity values do not vary widely outside the range of 30 to 40% residual activity. This provides confidence that variants with more than 40% residual activity are likely to be more thermostable variants of subtilisin E.

49.3. DATA ANALYSIS

Screening hundreds of 96-well plates can generate a tremendous amount of data. Often the screen involves multiple measurements which have to be manipulated mathematically in order to determine whether a particular clone is more fit than the parent enzyme. For example, thermostability assays such as the subtilisin E assay described above use two measurements of activity, initial and residual. Screening pNB esterase for activity in DMF required three measurements: an absorbance reading of the cell suspension for the purpose of estimating cell density, an activity assay in a low concentration of DMF, and an activity assay in a high concentration of DMF. The cell-density measurement allowed us to determine which clones had increased total activity (increased specific enzyme activity and/or increased expression), while the ratio of activity measurements gave values which were independent of enzyme concentration and which reflected specific changes in the enzyme's ability to function in higher DMF concentrations. (Similar sorts of estimates can be made visually with petri dish screens that have less precision.)

These types of measurements can be easily performed using a 96-well or 384-well plate reader interfaced to a computer. By downloading the data directly into a spreadsheet program, the calculations described above and the relationships between different properties can be quantified (as in Fig. 3 and 4). Additionally, as already shown in the above activity profile, spreadsheets permit sorting the data, so that the screening results can be better evaluated. For example, data from one or more 96-well plates can be sorted from best to worst and then plotted, as in Fig. 5, to give "fitness profiles" from which additional interesting characteristics of the library and the property can be deduced. Apart from the application as a diagnostic check of mutation frequency, these profiles also allow us to assess the evolvability of a particular property of the enzyme. Figure 5 shows two different fitness profiles for activity and thermostability. Wild type is located near the top of the activity profile in Fig. 5A, while quite a few clones have thermostability indices larger than wild type in Fig. 5B. It thus appears that there are more ways to improve thermostability than activity.

We thank Pim Stemmer for his advice and assistance with the DNA shuffling method. This work was supported by the U.S. Office of Naval Research and the U.S. Department of Energy's program in Biological and Chemical Technologies Research within the Office of Industrial Technologies, Energy Efficiency and Renewables.

REFERENCES

1. **Arnold, F. H.** 1996. Directed evolution: creating biocatalysts for the future. *Chem. Eng. Sci.* **51:**5091–5102.
2. **Arnold, F. H.** 1998. Designed by directed evolution. *Accounts Chem. Res.* **31:**125–131.
3. **Caldwell, R. C., and G. F. Joyce.** 1992. Randomization of genes by PCR mutagenesis. *PCR Methods Appl.* **2:**28–33.
4. **Chen, K. Q., and F. H. Arnold.** 1993. Tuning the activity of an enzyme for unusual environments—sequential random mutagenesis of subtilisin E for catalysis in dimethylformamide. *Proc. Natl. Acad. Sci. USA* **90:**5618–5622.
5. **Crameri, A., S.-A. Raillard, and W. P. C. Stemmer.** 1998. DNA shuffling of a family of genes from diverse species accelerates directed evolution. *Nature* **391:** 288–291.
6. **Kast, P., and D. Hilvert.** 1997. Three dimensional structural information as a guide to protein engineering by genetic selection. *Curr. Opin. Struct. Biol.* **7:**470–479.
7. **Kuchner, O., and F. H. Arnold.** 1997. Directed evolution of enzyme catalysts. *Trends Biotechnol.* **15:**523–530.
8. **Leung, D. W., E. Chen, and D. V. Goeddel.** 1989. A method for random mutagenesis of a defined DNA segment using a modified polymerase chain reaction. *Technique* **1:**11–15.
9. **Lorimer, I. A. J., and I. Pastan.** 1995. Random recombination of antibody single-chain fv sequences after fragmentation with DNase I in the presence of Mn^{2+}. *Nucleic Acids Res.* **23:**3067–3068.
10. **Moore, J. C., and F. H. Arnold.** 1996. Directed evolution of a para-nitrobenzyl esterase for aqueous-organic solvents. *Nat. Biotechnol.* **14:**458–467.
11. **Moore, J. C., H. M. Jin, O. Kuchner, and F. H. Arnold.** 1997. Strategies for the in vitro evolution of protein func-

tion: enzyme evolution by random recombination of improved sequences. *J. Mol. Biol.* **272:**336–347.

12. **Reidhaar-Olson, J. F., and R. T. Sauer.** 1988. Combinatorial cassette mutagenesis as a probe of the information content of proteins. *Science* **241:**53–57.
13. **Shafikhani, S., R. A. Siegel, E. Ferrari, and V. Schellenberger.** 1997. Generation of large libraries of random mutants in *Bacillus subtilis* by PCR-based plasmid multimerization. *Biotechniques* **23:**301–310.
14. **Shao, Z., H. Zhao, L. Giver, and F. H. Arnold.** 1998. Random priming in vitro recombination: an effective tool for directed evolution. *Nucleic Acids Res.* **26:**681–683.
15. **Stemmer, W. P. C.** 1994. DNA shuffling by random fragmentation and reassembly: in vitro recombination for molecular evolution. *Proc. Natl. Acad. Sci. USA* **91:**10747–10751.
16. **Zhao, H., and F. H. Arnold.** 1997. Combinatorial protein design: strategies for screening protein libraries. *Curr. Opin. Struct. Biol.* **7:**480–485.
17. **Zhao, H., and F. H. Arnold.** 1997. Optimization of DNA shuffling for high-fidelity recombination. *Nucleic Acids Res.* **25:**1307–1308.
18. **Zhao, H., L. Giver, Z. Shao, J. A. Affholter, and F. H. Arnold.** 1998. Molecular evolution by staggered extension process (StEP) in vitro recombination. *Nat. Biotechnol.* **16:**258–261.
19. **Zock, J., C. Cantwell, J. Swartling, R. Hodges, T. Pohl, K. Sutton, P. Rosteck, Jr., D. McGilvray, and S. Queener.** 1994. The *Bacillus subtilis* pnbA gene encoding p-nitrobenzyl esterase—cloning, sequence and high-level expression in *Escherichia coli*. *Gene* **151:**37–43.

Metabolic Pathway Engineering of Aromatic Compounds

RICHARD J. LADUCA, ALAN BERRY, GOPAL CHOTANI, TIMOTHY C. DODGE, GUILLERMO GOSSET, FERNANDO VALLE, JAMES C. LIAO, JIMMY YONG-XIAO, AND SCOTT D. POWER

50

The study of aromatic molecules has its roots in the emergence of structural organic chemistry in the 1830s. The concept that molecules can be built from a series of chemical (or biochemical) transformations where certain substituents remain fixed in their position and chirality was first enunciated by Wohler and Liebig in their definitive work on benzoic acid (80). These studies were pursued in earnest by a large number of natural products chemists, eager to understand the structure and synthesis of a large number of plant aromatic compounds. Our knowledge of the major biosynthetic route to aromatic compounds is a product of these investigations, specifically the pioneering work of Eijkman, who in 1885 discovered and characterized the key aromatic intermediate, shikimic acid, from the Japanese shikimi tree *Illicium religiosum* (29). In the years following, Eijkman (30, 31) and Fischer and Dangschat (33–36) determined the structure of this new molecule and, in so doing, determined by 1937 the pivot point for aromatic biosynthesis. The importance of this work and the role of shikimic acid in the common, or shikimic acid, pathway was finally demonstrated by Davis in 1950 (20, 21) and became the focus for work over the next 50 years into the production of aromatic metabolites via this common pathway.

Today, each of the intermediates and their corresponding genetic loci are known for a wide variety of microorganisms, especially for the gram-negative bacteria *Escherichia coli*, *Klebsiella* spp., and *Salmonella typhimurium* (61, 79). The availability of these genes and the rapid modifications possible with genetic engineering have led to a variety of products from this pathway through the overexpression, deregulation, and reengineering of whole pathway segments. This offers the opportunity to devise methods for the fermentation production of low-cost aromatic amino acids and important related metabolites.

50.1. LIMITATIONS OF BIOCATALYTIC SYSTEMS

Current fermentation-based, biocatalytic whole-cell processes for the production of chemicals require the in situ generation of enzyme catalysts, the development of cellular biomass, and the maintenance of other, sometimes unnecessary cellular functions, in addition to the production of the desired end product. As a consequence, a significant fraction of the input carbon substrate, typically glucose, that is used to generate the desired product is used to facilitate cell growth and maintenance. In addition, complex carbon and nitrogen sources are generally utilized as raw materials to facilitate cell growth in these processes, thereby complicating and adding costs to downstream processing of chemical products. In a typical batch fermentation process, each of these biomass-building functions must be repeated with every new fermentation, thus wasting valuable raw materials and increasing processing cycle times.

From a kinetic standpoint, carbon flow or movement of substrates within a desired metabolic pathway or network of enzymes is constantly compromised by diversion into other metabolic processes. These include substrate transport into the cell, feedback regulation of inhibitors within enzyme systems, internal diffusion coefficients of reactants, antagonistic biocatalytic processes, multiple pathway branch points, turnover of intermediates and cofactors, and product efflux from the cell. The yield of end product is thus a complex function of both biocatalysis and cell maintenance.

A further complication associated with the biocatalytic production of chemicals is that appropriate host organisms rarely possess the complete pathway for production of the desired intermediate or final product. This dictates that methods be devised for identifying optimal enzymes from other sources whose genetic information can be included in the host organism to complete the metabolic pathway. Traditional tools of DNA mutagenesis, enzyme selection, and process development can go only part of the way to optimizing and controlling these processes. Numerous techniques have been developed to realize the metabolic goal of converting these inefficient processes into highly productive biocatalytic processes for the manufacture of chemical products.

50.2. BUILDING THE PRIMARY METABOLIC PATHWAY: ENGINEERING AROMATIC AMINO ACID BIOSYNTHESIS

50.2.1. Background

In a later section, we provide an overview of the aromatic compounds that have been produced via the aromatic amino acid pathway. These technologies were built upon the fundamental knowledge gained during the creation of a recombinant *E. coli* strain for the overproduction of tryptophan (see reference 8 for a review). Thus, tryptophan production in *E. coli* is used here as a model to illustrate the general strategies of pathway engineering for the production of aromatic compounds. Related work on the development of phenylalanine-producing strains of *E. coli* will also be discussed where appropriate.

Figure 1 shows a schematic of the aromatic amino acid pathway. For the sake of illustration, the circuits of transcriptional and allosteric control of aromatic biosynthesis shown in Fig. 1 are those that exist in *E. coli*. However, the same general approaches described below for removing aromatic pathway regulation would apply to other organisms that have different patterns of regulation of aromatic biosynthesis. This has been demonstrated by the extensive pathway engineering of *Corynebacterium glutamicum* for tryptophan production (49).

50.2.2. Engineering the Aromatic Amino Acid Pathway

50.2.2.1. General Strategies

Although there are no hard-and-fast rules with respect to the steps of engineering a biosynthetic pathway for metabolite overproduction, certain strategic steps appear repeatedly among the examples of successful pathway engineering efforts. These steps are (i) enhancement of the first enzymatic reaction that commits carbon flow from central metabolism to the pathway of interest, (ii) removal of transcriptional and allosteric regulation within the pathway, (iii) identifying and relieving rate-limiting pathway steps, and (iv) preventing loss of carbon flow to competing pathways. Other considerations are development of the host strain, plasmid stabilization without antibiotics, metabolic burden (genetic load), and product toxicity.

50.2.2.2. Deregulation of Pathway Control Points

The first enzymatic step of aromatic biosynthesis, a condensation of the two pathway precursors erythrose 4-phosphate (E4P) and phosphoenolpyruvate (PEP), is catalyzed by 3-deoxy-D-arabinoheptulosonate 7-phosphate (DAHP) synthase. In *E. coli*, three isozymes of DAHP synthase exist (61). These isozymes, encoded by the *aroG*, *aroF*, and *aroH* genes, are subject to transcriptional and allosteric control by phenylalanine, tyrosine, and tryptophan, respectively (see Fig. 1).

Amplification (and deregulation) of DAHP synthase is fundamental to the overproduction of aromatic compounds, as this step commits carbon from central metabolism to aromatic biosynthesis. Prior to the application of molecular biology to pathway engineering, DAHP synthase activity was amplified by selecting mutants resistant to toxic analogs of the aromatic amino acids. In *E. coli*, these regulatory mutants were generally found to contain mutations in the *tyrR* gene, which controls expression of *aroF* (45), although mutations in the TYR R boxes upstream of *aroF* can also lead to constitutive expression (19). While the approach of using toxic analogs can result in enough

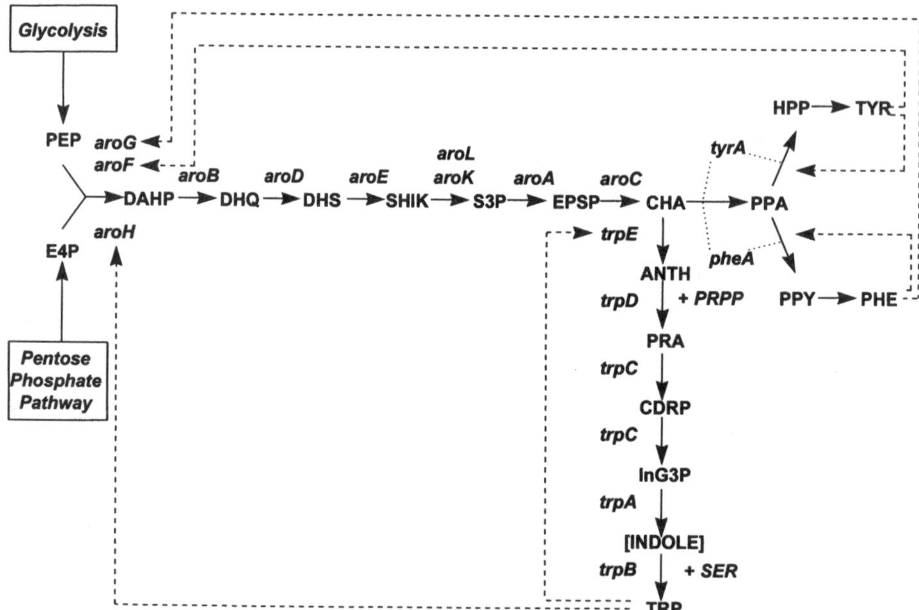

FIGURE 1 Pathway of aromatic amino acid biosynthesis and the genes (in italics) encoding each pathway enzyme. The dashed lines indicate both transcriptional and allosteric control exerted by the aromatic amino acid end products at the indicated pathway step. The shikimate kinase (*aroL*) step is the only exception, being regulated (by tyrosine) only at the transcriptional level. ANTH, anthranilate; CDRP, 1-(*o*-carboxyphenylamino)-1-deoxyribulose 5-phosphate; CHA, chorismate; DHQ, 3-dehydroquinate; DHS, 3-dehydroshikimate; EPSP, 5-enolpyruvylshikimate 3-phosphate; HPP, 4-hydroxyphenylpyruvate; InG3P, indole 3-glycerolphosphate; PHE, phenylalanine; PPA, prephenate; PPY, phenylpyruvate; PRA, phosphoribosyl anthranilate; PRPP, 5-phosphoribosyl-α-pyrophosphate; S3P, shikimate 3-phosphate; SER, L-serine; SHIK, shikimate; TRP, tryptophan; TYR, tyrosine.

derepression of DAHP synthase to allow for end-product overproduction, it is limited by the inability to increase the gene copy number. Nevertheless, selection for analog resistant mutants remains a useful tool for isolating allosterically insensitive mutant forms of DAHP synthase and other pathway enzymes (see below).

With the advent of molecular biology came rapid progress in our ability to overproduce metabolites. Amplification of desired activities was greatly enhanced by the ability to clone genes on multicopy plasmids. Many different cloning vectors are now available that range in copy number from 1 to 2,000 (67, 75), although the true copy number can vary depending on the size of the cloned insert as well as the genetic background of the host. For production of aromatic compounds in *E. coli*, the following vectors (with their approximate copy numbers in brackets) have been used: RP4 [1–2], pSC101 [4–6], RSF1010 [10–50], p15 [20], and ColEI [20–50] (1, 8, 56).

The promoter system used for gene overexpression can be a critical factor in developing strains for metabolite overproduction. Promoter activity studies in *E. coli* have shown that the range of promoter strengths can span almost two orders of magnitude (23). For overproduction of aromatic compounds, the necessary genes have been placed under control of promoters such as P_{lacUV5} (8), P_{tac} (60), λ P_L (72), P_{trp} (56), P_{tna} (74), and P_{luxI} (60). The rationale for using such promoters is to achieve high-level, controlled expression. While high-level expression is indeed achieved with such constructs, controlled expression is generally not attained, at least in a practical sense. For example, control of promoters such as *lac* and *tac* can be affected by glucose concentration. In a fed-batch fermentation, where glucose supply is limiting, *lac* and *tac* promoters can be very "leaky," i.e., essentially constitutive. Control of these promoters by overexpression of the $lacI^q$ gene can be achieved, but induction then requires addition of β-D-isopropylthiogalactopyranoside, which is expensive and affects cell metabolism. Control of the temperature-regulated λ P_L promoter system can be achieved in a fed-batch fermentation, but the rapid temperature fluctuations (heat shock) required for induction are not practical on a large scale and induce physiological responses that may have negative consequences. We have generally found that constitutive expression of cloned genes is best, provided that the gene is cloned under control of a suitable promoter and is present in a fixed (but sufficiently high) copy number.

While increasing the copy number and transcription level of DAHP synthase greatly increases carbon commitment to aromatic biosynthesis, ensuring unimpeded flow of carbon from central metabolism to the aromatic pathway also requires that DAHP synthase be made resistant to feedback inhibition by the normal allosteric effector. Many feedback-resistant mutants of DAHP synthase have been described, and the genes encoding these mutant enzymes have now been sequenced (62, 78). These feedback-resistant DAHP synthases were initially identified by the resistance that they confer to toxic analogs of aromatic amino acids. Once the specific mutations conferring allosteric insensitivity were known, additional beneficial mutations were introduced by site-directed mutagenesis. In addition to their importance in overproduction of aromatic compounds, the mutant DAHP synthases have provided considerable insight into the structure and mechanism of the enzymes (62).

In addition to the transcriptional and allosteric regulation exerted at the level of DAHP synthase, the initial enzymatic step of each of the terminal pathway branches leading to tryptophan, phenylalanine, and tyrosine is also tightly regulated. As such, overproduction of aromatic amino acids (and products derived via extension of the pathway) requires deregulation of the appropriate pathway step (*pheA*, *trpE*, *tyrA*; see Fig. 1). The same combination of molecular biology and classical selection methods used to deregulate DAHP synthase have been successfully used to remove regulation at the *trpE* and *pheA* steps of the aromatic pathway (9, 11, 57).

50.2.2.3. Identifying and Relieving Rate-Limiting Pathway Steps

In general, not every step of a biosynthetic pathway is regulated. Carbon flux is usually controlled by regulating strategic pathway steps. For example, in *E. coli*, carbon flow to aromatic biosynthesis is achieved by regulation of DAHP synthase. The steps of the common aromatic pathway (between DAHP and chorismate; see Fig. 1), with the exception of shikimate kinase (encoded by *aroL*), are not regulated by aromatic end products (61).

We have found that in *E. coli*, overproduction of tryptophan requires amplification (and deregulation) only of DAHP synthase and the enzymes of the tryptophan branch of the pathway (8). Since none of the intermediates of the common aromatic pathway accumulate in the culture medium, the level of the enzymes of the common aromatic pathway must be sufficient to handle the increased carbon flow afforded by overexpression of DAHP synthase. However, when carbon entry into the aromatic pathway is increased by increasing precursor availability (described in a later section), several steps in the common pathway become rate limiting (9, 22). In order to identify and relieve each rate-limiting step, a painstaking process using mutants blocked in each step of the common aromatic pathway was used. Plasmids containing the cloned DAHP synthase were introduced into each mutant, and the levels of the common pathway intermediates present in culture supernatants of each strain were measured. Accumulation of intermediates upstream of the mutational blocks was taken as evidence of a rate limitation at an earlier pathway step. The gene encoding the limiting pathway enzyme was then amplified and the analysis repeated. By systematically "walking" down the pathway, all of the rate-limiting pathway steps were identified. Ultimately, by combining several of the common pathway genes onto a single, low-copy-number plasmid, the increased carbon flow into aromatic biosynthesis was translated into increased end product (in this case, tryptophan) (9).

Another valuable tool for identifying rate-limiting pathway steps is feeding pathway intermediates to a production strain. We have successfully used this approach both in resting-cell experiments and in actual fed-batch fermentations. Essentially, this is an in vivo assay for all of the enzyme steps between the fed intermediate and the end product. If the intermediate is transported into the cell and converted efficiently to end product, the possibility of rate-limiting steps between the intermediate and the end product is eliminated.

50.2.2.4. Avoiding Loss of Carbon Flow to Competing Pathways

Chorismate is the branch point intermediate that leads not only to the aromatic amino acids, but to other aromatic compounds as well. A priori one might expect that in order to overproduce a single aromatic compound, the pathway branches leading to the other aromatic products would need to be eliminated. However, blockage of these competing pathways can result in the need to provide nutritional supplements to the culture medium, which adds to the production cost. We have found it unnecessary to block the *pheA* and *tyrA* steps of the aromatic pathway in order to overproduce tryptophan. The normal regulation governing expression of the *pheA* and *tyrA* genes in *E. coli* is evidently adequate to prevent any significant loss of carbon flow to phenylalanine and tyrosine (provided that the *trpE* gene is overexpressed; see Fig. 1). On the other hand, loss of carbon flow to tyrosine was significant in phenylalanine-producing strains of *E. coli* and led to the development of an elegant excision technology whereby the cloned *tyrA* gene was integrated into the host chromosome and then precisely excised (by temperature upshift) during the phenylalanine production phase of the fermentation (6).

50.2.3. Other Considerations

50.2.3.1. Host Strain Development

During the creation of a strain for overproduction of metabolites, much of the emphasis is placed on cloning and deregulating the requisite genes. However, equally important is the development of the host strain. Examples of factors to be considered are resistance to toxic levels of the product, elimination of undesirable enzymatic activities, and plasmid stabilization measures. Classical approaches to isolating mutants that have the desired properties are effective, especially when selecting for mutants resistant to high levels of the product. However, when elimination of certain activities is necessary, classical approaches can be tedious, involving random mutagenesis followed by several enrichment steps. Molecular biology methods have greatly enhanced our ability to specifically inactivate undesirable chromosomally encoded activities as well as to integrate useful genetic constructions into the host chromosome.

Integration of gene cassettes into the *E. coli* chromosome generally can be accomplished in two steps, provided that the target gene has been cloned. First, the desired changes are made to the cloned segment on a multicopy plasmid. Second, the altered construct is exchanged with its chromosomal allele by in vivo recombination. The second step requires selection for the integration event followed by elimination of the free replicating plasmid. The selection process requires that an antibiotic resistance gene or other suitable selectable marker be cloned within the region that is to be exchanged with the chromosomal allele. We have found that not all antibiotic resistance markers provide a high enough level of resistance (when integrated in single copy) of the *E. coli* host for growth on selective medium. Only chloramphenicol (39), kanamycin (2), and gentamycin (66) have worked well in our hands.

There are several methods for chromosomal integration via allele replacement in *E. coli*. We have used *E. coli* JC7623 (ATCC 47002), which is defective in its ability to replicate ColEI-based plasmids, extensively for this purpose. JC7623 contains the *recB21*, *recC22*, *sbcB15*, and *sbcC201* mutations (82) and depends on *recF* for recombination (53). Under the appropriate selective conditions, there is strong pressure for homologous recombination between the plasmid-borne and chromosomal alleles, as well as for loss of replicating plasmid. Once the desired integration has been obtained in the JC7623 host, the region of the chromosome containing the integrated cassette can be easily moved to other strains of *E. coli* by transduction with phage $P1_{vir}$.

50.2.3.2. Plasmid Stabilization

Plasmid stability is an important component of the development of recombinant strains for metabolite overproduction and can require modification of both the plasmid and the host. For economic and environmental reasons, large-scale use of antibiotics to maintain plasmid presence is not possible. Furthermore, depending on the mechanism of antibiotic resistance, supplementation of fermentation medium with antibiotics does not guarantee plasmid retention.

Several methods have been developed for stabilizing plasmids without antibiotic selection. For production of aromatic compounds, addition of the partition regions (the *par* region of pSC101 or the $parA^+$ and $parB^+$ regions from plasmid R1) or the mini-F region (from the *E. coli* F factor) has been used to stabilize plasmids containing the *E. coli trp* operon (58, 68, 69, 83). Nutritional selection schemes based on the *valS* and *trp* genes have also been used in *E. coli* strains for overproduction of aromatic compounds (58, 65).

50.2.3.3. Overcoming Product Toxicity

Product toxicity is a major problem for overproduction of many metabolites. Isolation of mutants resistant to high concentrations of the product is the simplest solution to this problem. However, other alternatives have been described. For example, Azuma et al. (5) reported that addition of nonionic detergents to *E. coli* tryptophan fermentations eliminated the product toxicity problem by decreasing the solubility of the tryptophan, causing it to precipitate before reaching a toxic level. This approach could also prove to be beneficial to downstream processing.

50.2.3.4. Metabolic Burden

Metabolic burden is a term used to describe the negative consequences of excessive gene expression on the production of metabolites or recombinant proteins (4, 7, 10, 44, 81). Since overproduction of metabolites can require overexpression of several genes (both native and heterologous), metabolic burden can be a major obstacle to efficient production.

Overproduction of aromatic compounds in *E. coli* requires overexpression of at least one gene (encoding DAHP synthase; see Fig. 1) to increase carbon commitment to the aromatic pathway. Thus, overexpression of the DAHP synthase gene represents the minimal metabolic burden condition for production of any aromatic compound. With each extension of the pathway to produce a new compound (described in a later section), additional genes need to be overexpressed. Consequently, the production potential may decrease as additional gene expression is required. Production of indigo via the tryptophan pathway represents the extreme case with respect to production of aromatic compounds, requiring overexpression of at least 10 genes from both *E. coli* and *Pseudomonas putida* (9). Figure 2 shows the difference in potential carbon flow to aromatic biosynthesis in a strain with minimal metabolic burden versus an isogenic strain carrying multiple plasmids expressing all the necessary genes for indigo production.

The results shown in Fig. 2 illustrate the importance of minimizing expression of the genes required for overproduction of a metabolite. One way that this can be achieved

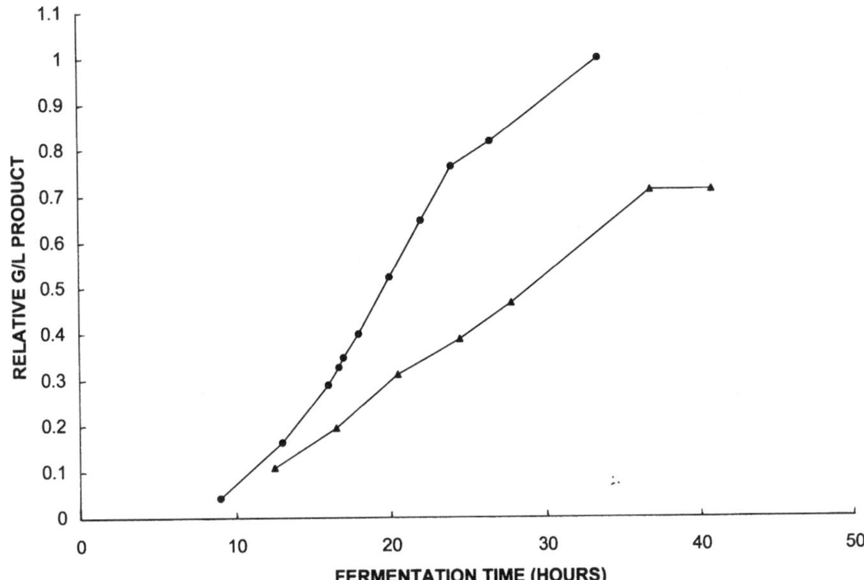

FIGURE 2 Effect of metabolic burden on potential carbon flow to aromatic biosynthesis. Comparison of 3-dehydroshikimate production, normalized as relative carbon flow betwen a strain without plasmids (●, upper curve) and an isogenic strain carrying multiple plasmids required for indigo production (▲, lower curve).

is by the methods described above for integrating gene cassettes into the chromosome. However, only by carrying out a careful analysis of minimum required gene expression levels can the effects of metabolic burden be overcome.

50.3. MAXIMIZING RAW MATERIAL COMMITMENT TO PRIMARY METABOLIC PATHWAYS

50.3.1. Manipulation of Central Metabolism

For production of metabolites, terminal pathways are often the first targets for engineering. However, after all the bottlenecks in the terminal pathways have been removed, central metabolism becomes the ultimate limiting path that controls the carbon flux to the terminal pathways. The central metabolism supplies monomers, energy, and reducing power for biosynthesis of proteins, nucleic acids, and other cellular components. The metabolic rates and flux distribution in these pathways are well controlled, and attempts to alter these pathways are likely to face resistance from many regulatory mechanisms in the cell. To redirect the metabolic fluxes in the central metabolism, it is imperative to understand how these pathways are regulated.

E. coli uses the phosphotransferase system (PTS) for transport of glucose and some other sugars. In this process, phosphoenolpyruvate (PEP) is converted to pyruvate, which drives the transport process and phosphorylates glucose at the same time. PEP and pyruvate act as important metabolic nodes in the pathway, and at least four enzymes compete to determine the direction and magnitude of the flux to biosynthesis and catabolic pathways: PEP carboxykinase (Pck), PEP carboxylase (Ppc), PEP synthase (Pps), and pyruvate kinase (PykI and PykII), as shown in Fig. 3. These enzymes are responsible for switching between glycolysis and gluconeogenesis.

When *E. coli* grows on glucose, PEP is converted to pyruvate via both Pyk and the PTS, and from PEP to oxaloacetate (OAA) via Ppc. When the cell grows on pyruvate, pyruvate is converted to PEP via Pps and from PEP to OAA via Ppc to supply precursors for cell growth. When the cell grows on succinate, OAA is converted to PEP via Pck, but the flux from PEP to pyruvate can be either positive or negative, depending on the activity of the malic enzymes. Apparently, the cell has evolved effective mechanisms to modulate these enzymes to control the direction and magnitudes of these metabolic fluxes in response to nutritional conditions. The control of these fluxes is exerted at both the transcriptional and activity levels. However, mechanisms for controlling gene expression of these enzymes are incompletely known. A study with *pck-lacZ* operon fusion (42) showed that the transcription of *pck* is under catabolite repression and is induced at the onset of stationary phase via a mechanism involving cyclic AMP and the Crp protein. The regulation of *pps* is linked to the PTS. The negative regulator of the *fru* operon, *fruR* (recently renamed *cra*) (64), has been shown to be required for *pps* transcription (16, 17, 40). An operon fusion of pps-lacZ showed that the *pps* gene in *S. typhimurium* is under catabolite repression, but the addition of cyclic AMP in the presence of glucose does not relieve this repression (70). In addition to the genetic control, these enzymes are regulated at the activity level. Ppc is inhibited by aspartate and malate and activated by acetyl coenzyme A, fructose 1,6-diphosphate, GTP, guanosine-5′-diphosphate-3′-diphosphate, and fatty acids (47, 48, 54, 73). Pck is allosterically activated by calcium (41) and inhibited by ATP and PEP (50). Pps is activated by the adenylate energy charge and inhibited by ADP, AMP, α-ketoglutarate, malate, OAA, and PEP (18). PykI is activated by fructose 1,6-diphosphate and inhibited by GTP and succinyl coenzyme A (77). PykII is inhibited by phosphate ions and

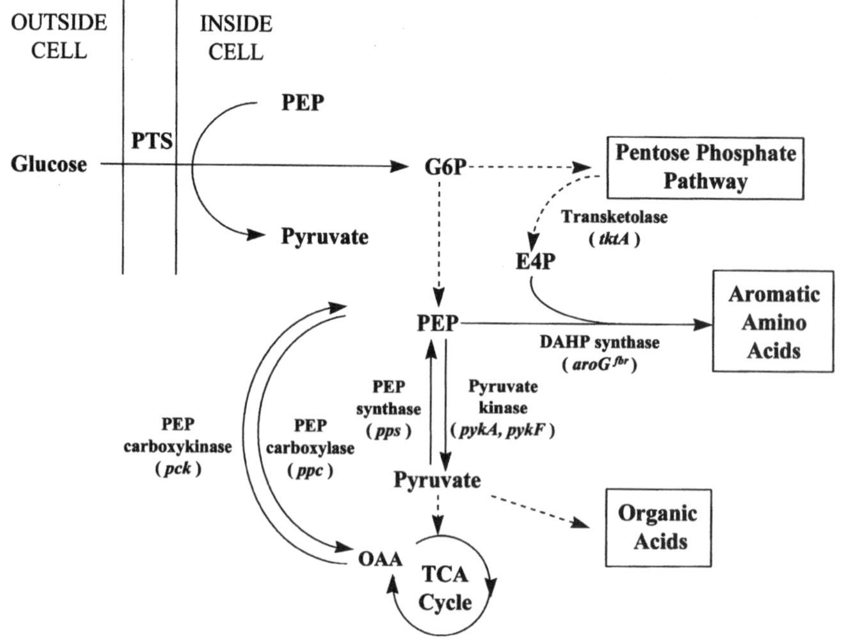

FIGURE 3 Central metabolic pathways related to the formation and consumption of PEP. Dashed lines indicate a sereis of reactions; genes encoding enzymes are in italics. G6P, glucose 6-phosphate; TCA, tricarboxylic acid.

activated by nucleoside monophosphate and ribose 5-phosphate (76).

Several publications (52, 59, 71) provide useful strategies for alteration of central metabolism. Mutations have been introduced at many key steps, and their effects have been measured in terms of changes in concentrations and fluxes of the product(s) and by-product(s). These measurements were made from the fermentation data obtained through comparison of the test (mutated) and control (wild-type) strains. Overexpression of Pps, Pck, Pyk, and Ppc resulted in various changes that are not always predictable (13, 14). Overexpression of Pps and Pck increased the glucose consumption rate and the acid formation rate (12, 59). Overexpression of Pyk or Ppc achieved the reverse effects (12). The underlying mechanisms for these effects are still under investigation.

50.3.2. Engineering Central Metabolism for Improved Production of Aromatic Compounds in *E. coli*

Manipulation of central metabolism to improve the production of aromatic compounds in *E. coli* has been studied intensively (see reference 8 for review). In principle, the objective has been to increase the commitment of the two aromatic pathway precursors erythrose 4-phosphate (E4P) and PEP (see Fig. 1) to aromatic biosynthesis by increasing their intracellular levels. The pathways of central metabolism related to E4P and PEP are also depicted in Fig. 3. Increased availability of E4P has been achieved by amplifying the *E. coli tktA* gene, which encodes transketolase, one of the two pentose phosphate pathway enzymes responsible for E4P synthesis (27). Increased PEP availability has been achieved by inactivation of the PEP:glucose PTS (37), inactivation of both isozymes of PyK (43), and amplification of Pps (59) (see Fig. 3). Gosset et al. (43) showed that the above approaches, when combined, have a synergistic effect on carbon commitment to aromatic biosynthesis. In the best case, a strain where the PTS and both isozymes of Pyk were inactivated and transketolase was amplified, carbon flow to aromatic compounds was increased 20-fold (43).

50.4. PRODUCTION OF AROMATIC COMPOUNDS

50.4.1. Overview of Ways of Extending the Aromatic Pathway for New Products

The aromatic amino acid pathway of *E. coli* is being used for the production of its industrially important end products tryptophan and phenylalanine. Tryptophan is used as an animal feed supplement, whereas phenylalanine is used in the manufacture of the artificial sweetener aspartame. However, beyond the natural end products, many other compounds of commercial interest could potentially be produced by this pathway (38) (Fig. 4). These include other compounds naturally produced by *E. coli* as well as products available through recruitment of enzymes from heterologous sources.

50.4.2. Examples of Production of Aromatic Compounds by Fermentation

50.4.2.1. *E. coli* Pathway and Products

Tryptophan

Tryptophan is an essential amino acid in the diets of higher animals. Typical feed for swine and poultry consists of grains such as corn and soybeans. Tryptophan is one of the first amino acids to become limiting in such a diet. As

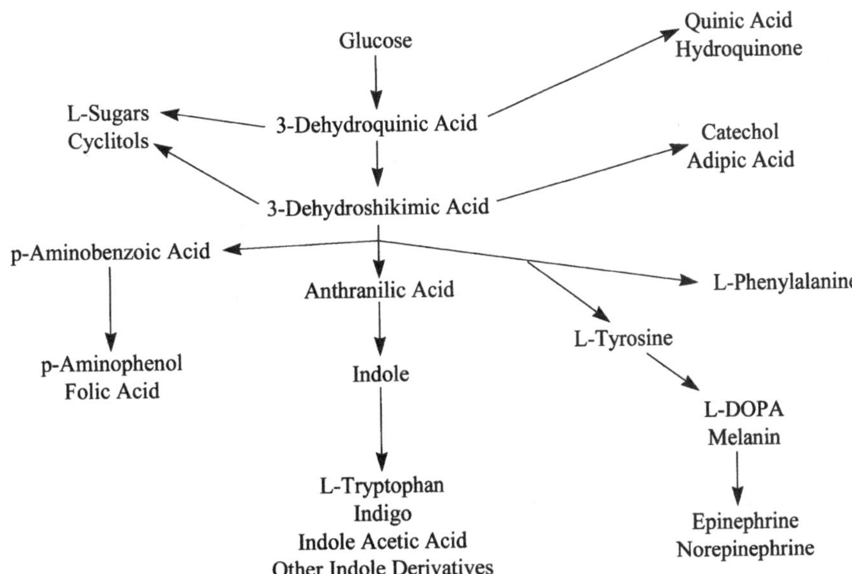

FIGURE 4 Representation of the multiple routes from the aromatic amino acid biosynthetic pathway to commercially important products. L-Dopa (3,4-dihydroxyphenylalanine) is the hydroxylation product of tyrosine. In many instances, the conversion of each intermediate into a product involves several enzymatic steps. Detailed descriptions of the metabolic pathways can be found in references 8, 9, and 24–28.

such, it can be more cost effective to supplement the feed with tryptophan that has been produced by fermentation. Development of this process required the amplification of one of the genes for the committed step into the aromatic pathway, such as *aroG*, and the gene for the tryptophan-specific branch of the pathway, *trpEDCBA* (8). Also, the gene products of *aroG* and *trpE* were rendered insensitive to the natural allosteric inhibitors phenylalanine and tryptophan. The resulting *E. coli* strain was found to produce tryptophan from glucose at a high rate and yield (8).

Phenylalanine

Phenylalanine is one of the principal ingredients in the artificial sweetener aspartame. In order to produce large volumes of this important food ingredient, an effective phenylalanine fermentation was developed. The strain employed was constructed in a way similar to that for the tryptophan fermentation (6). Once again, the committed step to the aromatic pathway was amplified and rendered feedback insensitive. The phenylalanine-specific portion of the pathway was also overexpressed.

50.4.2.2 Side Pathways from Other Organisms Placed into *E. coli*

Quinic Acid

Quinic acid is currently isolated from plants and used as a starting material in many chiral chemical syntheses. It was also the original starting material for production of hydroquinone and benzoquinone, materials with usage rates of tens of thousands of metric tons per year. However, strains have now been constructed which can convert glucose directly to quinic acid (28). The key to this development was the cloning of a gene from *Klebsiella pneumoniae* that is capable of converting the pathway intermediate dehydroquinic acid into quinic acid. This gene was placed into a strain of *E. coli* lacking *aroD*, the step responsible for consumption of dehydroquinic acid. By utilizing this block, material is diverted from the aromatic pathway into quinic acid.

Catechol and Adipic Acid

Catechol is used as a substrate for the synthesis of a large number of compounds used as flavors, fragrances, pharmaceuticals, and pesticides. Adipic acid is used in the production of nylon, which consumes over 100,000 metric tons per year. Both catechol and adipic and acid are currently derived from petrochemical feedstocks. Methods similar to those used for the production of quinic acid were employed to create strains capable of producing catechol and adipic acid from glucose (26). Once again, a block in the common portion of the aromatic amino acid pathway was employed, this time at the *aroE* locus, halting production of dehydroshikimic acid. Two genes from *K. pneumoniae* (*aroZ* and *aroY*, encoding 3-dehydroshikimate dehydratase and protocatechuate decarboxylase, respectively) are needed to convert dehydroshikimic acid to catechol. An additional gene from *Acinetobacter calcoaceticus* (*catA*, encoding catechol 1, 2-dioxygenase) is required to convert the catechol further to *cis, cis*-muconic acid. This compound can be catalytically hydrogenated to adipic acid.

Indigo

Indigo is one of the largest-volume textile dyes, with over 13,000 metric tons produced per year. Since the late 1800s it has been derived almost exclusively from aniline. The key enzyme required for the microbial production of indigo is a dioxygenase from *P. putida* (32). Naphthalene dioxygenase converts indole to indoxyl, most likely through a diol intermediate. Indoxyl is also the final step in the chemical synthesis of indigo, when indoxyl is air oxidized to form indigo. Indole, toxic to *E. coli* at levels as low as 200 mg/liter, is not normally found in cells growing on glucose. Tryptophanase is capable of converting tryptophan to indole and pyruvate but is repressed during growth on glucose. Tryptophan synthetase, the enzyme responsible for the production of tryptophan, traps indole within the enzyme complex. However, mutations have been made in the *trpB* protein subunit that allow for the release of free indole (55). This free indole can then be acted upon by naphthalene dioxygenase to produce indigo directly from glucose.

50.5. NEW DIRECTIONS IN THE ENGINEERING OF METABOLIC PATHWAYS

The intent of evolution in biological systems has been to establish species which are more fit for survival, as opposed to being fit for the commercial synthesis of chemicals. How then has it been possible to establish efficient primary enzyme networks (i.e., the sequential biocatalytic pathway to a targeted product) as well as sustaining, or secondary, enzyme networks (i.e., those enzymatic elements within a microbial system which enhance and sustain the productive life span of the process), which are required for the biocatalytic production of chemicals? In large part, much of the success that has been realized in metabolic pathway engineering efforts can be attributed to advances in recombinant DNA technologies. Enzyme engineering, mutagenesis, and gene recruitment methodologies have been primarily responsible for the adaptation of biocatalytic machinery for the synthesis of chemicals in microbial systems.

A major thrust in biotechnology has evolved around the use of recombinant DNA techniques which allow useful genes, isolated from wild organisms, to be expressed at commercially relevant levels in well-characterized, domesticated bacterial hosts. The focus of these efforts has been to identify and express genes for improved biocatalytic activities required in metabolic pathways for targeted compounds. Enzyme recruitment efforts have also been utilized to facilitate the isolation of gene activities from nature to fill gaps or extend existing enzymatic pathways to new chemical products. To date, these technologies have depended upon the use of culturable microorganisms which are used as a source for recruitment of industrially interesting characteristics and genes, e.g., ability to metabolize useful substrates such as cellulose or glucose, unique metabolic enzymes that can be used to modify or augment industrial processes, or novel pathways that may degrade or detoxify toxic chemicals. Genes from homologous as well as heterologous microbial systems have been recruited to establish primary enzyme networks for the synthesis of chemical products. These tools have been utilized to establish metabolic pathways for the synthesis of numerous compounds, such as ethanol (46); quinoid compounds, including quinic acid and hydroquinone (28); catechol and derivatives of catechol (25, 26); polyhydroxybutyrate (69); aromatic compounds, such as tryptophan (9), phenylalanine (6, 45), and dehydroshikimate (24); indigo dye (9); ascorbic acid (3); and many other examples too numerous to list.

A fundamental limitation in the development of novel metabolic pathways stems from our incomplete knowledge of the enzymes participating in biocatalytic pathways as well as our inability to tap biodiversity for the potential variety of pathways that may exist in nature. Since most of our knowledge is based upon enzymes produced by organisms that are readily cultured in the laboratory, and since we now know that culturable organisms represent an infinitesimal fraction of the naturally occurring species, it is safe to conclude that the natural environment harbors an enormous diversity of uncharacterized enzymes participating in a myriad of pathways.

A significant new direction in the engineering of metabolic pathways is the development of technologies for the direct recruitment of genetic information from unculturable microorganisms. Gene isolation and high-throughput robotic screening methodologies have recently been developed which allow for the direct isolation of interesting gene activities from environmental samples (63). Direct recruitment of gene activities from nature offers the potential for significant enhancements in the speed and efficiency of discovery of unique and useful biocatalysts.

Protein engineering technologies have also played a key role in the development of biocatalysts that have improved functionality (i.e., commercial fit) for use in metabolic pathways. These efforts rely upon structural information to create enzymes with altered catalytic activities. The method requires knowledge of the catalytic active site of the biocatalyst to permit targeted mutagenesis of the gene sequence encoding this region. Despite this limitation, significant improvements in catalytic activity can be realized. Protein engineering efforts that focused on a *Corynebacterium* 2,5-diketo-D-gluconic acid reductase activity utilized in the biocatalytic production of 2-keto-L-gulonic acid (a precursor to ascorbic acid) have resulted in the generation of enzyme variants with significantly enhanced catalytic activities (51).

A new direction for enzyme engineering has focused on methodologies for the rapid directed evolution of targeted genes for enzymes having specific industrial applications. Structural information on the targeted gene is not required in these directed evolution approaches to enzyme engineering. At the heart of this technology are selection procedures which ensure the enrichment, from randomly mutagenized populations of the targeted gene, of gene activities with "improved" biocatalytic performance. Enriched populations are remutagenized, challenged, and enriched again under selective conditions. In this way, it has been possible to speed up and direct evolution in a microorganism to establish a unique combination of mutations in a targeted gene. The approach has been successfully utilized to rapidly evolve subtilisin to be hundreds of times more active than a wild-type ancestor under reaction conditions that are not experienced in nature, i.e., in high concentrations of dimethylformamide (15). Directed evolution methodologies require a rapid and sensitive screening or selection methodology to enhance populations of variants expressing the properties of choice. The approach represents a promising alternative to more conventional protein engineering methodologies for the development of biocatalysts with improved properties and functionalities.

50.6 AFTERWORD

The study of the anabolic and catabolic reactions has been at the forefront of biology and chemistry for over 150 years. This collection of research has led to the construction of microbial systems challenging classical chemical synthesis in yield and productivity and remaining unchallenged for producing chirality. The microbial synthesis of products of the aromatic pathway has led the way in low-cost, large-scale fermentation production systems for compounds such as tryptophan, phenylalanine, and, through pathway extension, the organic dye indigo. The original work, which focused on simple overexpression of enzymes at branch points of pathways, now gives way to reengineering central metabolism, improving yield, and leading to even lower-cost fermentation systems. Finally, it is our belief that the extension of the normal pathways of cellular metabolism with enzymes from novel sources will allow the development of a next generation of low-cost fermentation products.

REFERENCES

1. **Aiba, S., H. Tsunekawa, and T. Imanaka.** 1982. New approach to tryptophan production by *Escherichia coli*: genetic manipulation of composite plasmids *in vitro*. *Appl. Microbiol. Biotechnol.* **43:**289–297.
2. **Alexeyev, M. F., I. N. Shokolenko, and T. P. Chroughan.** 1995. Improved antibiotic-resistance gene cassettes and omega elements for *Escherichia coli* vector construction and *in vitro* deletion/insertion mutagenesis. *Gene* **160:**63–67.
3. **Anderson, S., C. B. Marks, R. Lazarus, J. Miller, K. Stafford, J. Seymour, D. Light, W. Rastetter, and D. Stell.** 1985. Production of 2-keto-L-gulonate, an intermediate in L-ascorbate synthesis, by genetically modified *Erwinia herbicola*. *Science* **230:**144–149.
4. **Andersson, L., S. Yang, P. Neubauer, and S.-O. Enfors.** 1996. Impact of plasmid presence and induction on cellular responses in fed batch cultures of *Escherichia coli*. *J. Biotechnol.* **46:**255–263.
5. **Azuma, S., H. Tsunekawa, M. Okabe, R. Okamoto, and S. Aiba.** 1993. Hyperproduction of L-tryptophan via fermentation with crystallization. *Appl. Microbiol. Biotechnol.* **39:**471–476.
6. **Backman, K., M. J. O'Conner, A. Maruya, E. Rudd, D. McKay, R. Balakkrishnan, M. Radjai, V. DiPasquantonio, D. Shoda, R. Hatch, and K. Venkatsubramanian.** 1990. Genetic engineering of metabolic pathways applied to the production of phenylalanine. *Ann. N. Y. Acad. Sci.* **589:**16–24.
7. **Bentley, W. E., N. Jirjalili, D. C. Andersen, R. H. Davis, and D. S. Kompala.** 1990. Plasmid-encoded protein: the principal factor in the "metabolic burden" associated with recombinant bacteria. *Biotechnol. Bioeng.* **35:**668–681.
8. **Berry, A.** 1996. Improving production of aromatic compounds in *Escherichia coli* by metabolic engineering. *Trends Biotechnol.* **14:**250–256.
9. **Berry, A., S. Battist, G. Chotani, T. Dodge, S. Peck, S. Power, and W. Weyler.** 1995. Biosynthesis of indigo using recombinant *E. coli*: development of a biological system for the cost-effective production of a large volume chemical, p. 1121–1129. *In Proceedings of the Second Biomass conference of the Americas: Energy, Environment, Agriculture, and Industry*. National Renewable Energy Laboratory, Golden, Colo.
10. **Bhattacharya, S. K., and A. K. Dubey.** 1995. Metabolic burden as reflected by maintenance coefficient of recombinant *Escherichia coli* overexpressing target gene. *Biotechnol. Lett.* **17:**1155–1160.
11. **Caligiuri, M. G., and R. Bauerle.** 1991. Identification of amino acid residues involved in feedback regulation of the anthranilate synthase complex from *Salmonella typhimurium*. Evidence for an amino-terminal regulatory site. *J. Biol. Chem.* **266:**8328–8335.
12. **Chao, Y. P., and J. C. Liao.** 1993. Alteration of growth yield by overexpression of *ppc* and *pck* in *Escherichia coli*. *Appl. Environ. Microbiol.* **59:**4261–4265.
13. **Chao, Y. P., and J. C. Liao.** 1994. Metabolic responses to substrate futile cycling in *Escherichia coli*. *J. Biol. Chem.* **269:**5122–5126.
14. **Chao, Y. P., R. Patnaik, W. D. Roof, R. F. Young, and J. C. Liao.** 1993. Control of gluconeogenic growth by *pps* and *pck* in *Escherichia coli*. *J. Bacteriol.* **175:**6939–6944.
15. **Chen, K., and F. Arnold.** 1993. Tuning the activity of an enzyme for unusual environments: sequential random mutagenesis of subtilisin for catalysis in dimethylformamide. *Proc. Natl. Acad. Sci. USA* **90:**5618–5622.
16. **Chin, A. M., D. A. Feldheim, and M. H. Saier, Jr.** 1989. Altered transcriptional patterns affecting several metabolic pathways in strains of *Salmonella typhimurium* which overexpress the fructose regulon. *J. Bacteriol.* **171:**2424–2434.
17. **Chin, A. M., B. U. Feucht, Jr., and M. H. Saier.** 1987. Evidence for the regulation of gluconeogenesis by the fructose phosphotransferase system in *Salmonella typhimurium*. *J. Bacteriol.* **169:**897–899.
18. **Chulavatnatol, M., and D. E. Atkinson.** 1973. Phosphoenolpyruvate synthetase from *Escherichia coli*. *J. Biol. Chem.* **248:**2712–2715.
19. **Cobbett, C. S., and M. L. Delbridge.** 1987. Regulatory mutants of the *aroF-tyrA* operon of *Escherichia coli* K-12. *J. Bacteriol.* **169:**2500–2506.
20. **Davis, B. D.** 1950. Nutritionally deficient bacterial mutants isolated by means of penicillin. *Experientia* **6:**41–50.
21. **Davis, B. D.** 1951. The role of shikimic acid. *J. Biol. Chem.* **191:**315–325.
22. **Dell, K. A., and J. W. Frost.** 1993. Identification and removal of impediments to biocatalytic synthesis of aromatics from D-glucose: rate-limiting enzymes in the common pathway of aromatic amino acid biosynthesis. *J. Am. Chem. Soc.* **115:**11581–11589.
23. **Deuschle, U., W. Kammerer, R. Gentz, and H. Bujaurd.** 1986. Promoters of *Escherichia coli*: a hierarchy of *in vitro* strength indicates alternate structures. *EMBO J.* **5:**2987–2994.
24. **Draths, K. M., and J. W. Frost.** 1990. Genomic direction of synthesis during plasmid-based biocatalysis. *J. Am. Chem. Soc.* **112:**9630–9632.
25. **Draths, K. M., and J. W. Forst.** 1991. Conversion of D-glucose into catechol: the not so common pathway of aromatic biosynthesis. *J. Am. Chem. Soc.* **113:**9361–9363.
26. **Draths, K. M., and J. W. Forst.** 1994. Environmentally compatible synthesis of adipic acid from D-glucose. *J. Am. Chem. Soc.* **116:**399–400.
27. **Draths K. M., D. L. Pompliano, D. L. Conley, J. W. Forst, A. Berry, G. L. Disbrow, R. J. Staversky, and J. C. Lievense.** 1992. Biocatalytic synthesis of aromatics from D-glucose: the role of transketolase. *J. Am. Chem. Soc.* **114:**3956–3962.
28. **Draths, K. M., T. L. Ward, and J. W. Forst.** 1992. Biocatalysis and nineteenth century organic chemistry: con-

version of D-glucose into quinoid organics. *J. Am. Chem. Soc.* **114:**9725–9726.

29. **Eijkman, J. F.** 1885. Uber die wesentlichen Bestandteile von Illicium religiosum Sieb. (shikimi-no-kik): Eugenol; Shikimen und Shikimol (Safrol), Protokatechusaure, Shikiminsaure, Shkimipikrin. *Rec. Trav. Chim.* **4:**32–54.
30. **Eijkman, J. F.** 1886. Sur l'Illicium religiosum. *Rec. Trav. Chim.* **5:**299–304.
31. **Eijkman, J. F.** 1891. Ueber die Shikimisaure. *Berichte* **24:** 1278–1303.
32. **Ensley, B. D., B. J. Ratzkin, T. D. Osslund, M. J. Simon, L. P. Wackett, and D. T. Gibson.** 1983. Expression of naphthalene oxidation genes in *Escherichia coli* results in the biosynthesis of indigo. *Science* **222:**167–169.
33. **Fischer, H. O. L., and G. Dangschat.** 1934. Konstitution der Shikimisaure. *Helv. Chim. Acta* **17:**1200–1207.
34. **Fischer, H. O. L., and G. Dangschat.** 1935. Abbau der Shikimisaure zur Aconitsaure. *Helv. Chim. Acta* **18:**1204–1206.
35. **Fischer, H. O. L., and G. Dangschat.** 2935. Zur Konfiguration der Shikimisaure. *Helv. Chim. Acta* **18:**1206–1213.
36. **Fischer, H. O. L., and G. Dangschat.** 1937. Uber die Konfiguration der Shikimisaure und ihren Abbau zur Glucodesonsaure. *Helv. Chim. Acta* **20:**705–716.
37. **Flores, N., J. Xiao, A. Berry, F. Bolivar, and F. Valle.** 1996. Pathway engineering for the production of aromatic compounds in *Escherichia coli*. *Nat. Biotechnol.* **14:**620–623.
38. **Frost, J. W., and K. M. Draths.** 1995. Biocatalytic synthesis of aromatics from D-glucose: renewable microbial sources of aromatic compounds. *Annu. Rev. Microbiol.* **49:** 557–579.
39. **Fuqua, W. C.** 1992. An improved chloramphenicol resistance gene cassette for site directed marker replacement mutagenesis. *BioTechniques* **12:**223–225.
40. **Geerse, R. H., J. van der Pluijm, and P. W. Postma.** 1989. The repressor of the PEP: fructose phosphotransferase system is required for the transcription of the pps gene of *Escherichia coli*. *Mol. Gen. Genet.* **218:**348–352.
41. **Goldie, A. H., and B. D. Sanwal.** 1980. Allosteric control by calcium and mechanism of desensitization of phosphoenolpyruvate carboxykinase of *Escherichia coli*. *J. Biol. Chem.* **255:**1399–1405.
42. **Goldie, H.** 1984. Regulation of transcription of the *Escherichia coli* phosphoenolpyruvate carboxykinase locus: studies with *pck-lacZ* operon fusions. *J. Bacteriol.* **159:** 832–836.
43. **Gosset, G., J. Y. Xiao, and A. Berry.** 1996. A direct comparison of approaches for increasing carbon flow to aromatic biosynthesis in *Escherichia coli*. *J. Ind. Microbiol.* **17:**47–52.
44. **Hong, H., L. Nilsson, and C. G. Kurland.** 1995. Gratuitous overexpression of genes in *Escherichia coli* leads to growth inhibition and ribosome destruction. *J. Bacteriol.* **177:**1497–1504.
45. **Im, S. W. K., H. Davidson, and J. Pittard.** 1971. Phenylalanine and tyrosine biosynthesis in *Escherichia coli* K-12: mutants derepressed for 3-deoxy-D-arabinoheptulosonic acid 7-phosphate synthetase (*phe*), 3-deoxy-D-arabinoheptulosonic acid 7-phosphate synthetase (*tyr*), chorismate mutase T-prephenate dehydrogenase, and transaminase A. *J. Bacteriol.* **108:**400–409.
46. **Ingram, L. O., F. Alterthum, K. Ohta, and D. S. Beall.** 1990. Genetic engineering of *Escherichia coli* and other enterobacteria for ethanol production. *Dev. Ind. Microbiol.* **31:**21–30.
47. **Izui, K.** 1970. Kinetic studies on the allosteric nature of phosphoenolpyruvate carboxylase from *Escherichia coli*. *J. Biochem.* **68:**227–238.
48. **Izui, K., T. Nishikido, K. Ishihara, and H. Katsuki.** 1970. Studies on the allosteric effectors and some properties of phosphoenolpyruvate carboxylase. *J. Biochem.* **68:**215–226.
49. **Katsumata, R., and M. Ikeda.** 1993. Hyperproduction of tryptophan in *Corynebacterium glutamicum* by pathway engineering. *Bio/Technology* **11:**921–925.
50. **Krebs, A., and W. A. Bridger.** 1980. The kinetic properties of phosphoenolpyruvate carboxykinase of *Escherichia coli*. *Can. J. Biochem.* **58:**309–318.
51. **Lazarus, R. A., M. Hurle, S. Anderson, and D. B. Powers.** December 1994. U. S. Patent 5,376,544.
52. **Liao, J. C., Y. P. Chao, and R. Patnaik.** 1994. Alteration of the biochemical valves in central metabolism of *Escherichia coli*. *Ann. N Y. Acad. Sci.* **745:**21–34.
53. **Mahajan, S. K.** 1988. Pathways of homologous recombination in *Escherichia coli*, p. 87–140. *In* R. Kuchlerlapati and G. R. Smith (ed.), *Genetic Recombination*. American Society for Microbiology, Washington, D.C.
54. **Morikawa, M., K. Izui, M. Taguchi, and H. Katsuki.** 1980. Regulation of *Escherichia coli* phosphoenolpyruvate carboxylase by multiple effectors in vivo. I. Estimation of the activities in the cells grown on various compounds. *J. Biochem.* **87:**441–449.
55. **Murdock, D., B. D. Ensley, C. Serdar, and M. Thalen.** 1993. Construction of metabolic operons catalyzing the *de novo* biosynthesis of indigo in *Escherichia coli*. *Bio/Technology* **11:**381–386.
56. **Nagahari, K., T. Tanaka, F. Hishinumma, M. Kuroda, and K. Sakaguchi.** 1977. Control of tryptophan synthetase amplified by varying the numbers of composite plasmids in *Escherichia coli* cells. *Gene* **1:**141–152.
57. **Nelms, J., R. M. Edwards, J. Warwick, and I. Fotheringham.** 1992. Novel mutations in the *pheA* gene of *Escherichia coli* K-12 which result in highly feedback inhibition-resistant variants of chorismate mutase-prephenate dehydratase. *Appl. Environ. Microbiol.* **58:** 2592–2598.
58. **Nilsson, J., and G. Skogman.** 1986. Stabilization of *Escherichia coli* tryptophan-production vectors in continuous culture: a comparison of three different systems. *Bio/Technology* **4:**901–903.
59. **Patnaik, R., W. D. Roof, R. F. Young, and J. C. Liao.** 1992. Stimulation of glucose catabolism in *Escherichia coli* by a potential futile cycle. *J. Bacteriol.* **174:**7527–7532.
60. **Patnaik, R., R. G. Spitzer, and J. C. Liao.** 1995. Pathway engineering for production of aromatics in *Escherichia coli*: confirmation of stoichiometric analysis by independent modulation of *AroG*, *TktA* and *Pps* activities. *Biotechnol. Bioeng.* **46:**361–370.
61. **Pittard, A. J.** 1987. Biosynthesis of aromatic amino acids, p. 368–394. *In* F. C. Neidhardt, J. L. Ingraham, K. B. Low, B. Magasanik, M. Schaechter, and H. E. Umbarger (ed.),

Escherichia coli and Salmonella typhimurium. Cellular and Molecular Biology. American Society for Microbiology, Washington, D. C.

62. **Ray, J. M., C. Yanofsky, and R. Bauerle.** 1988. Mutational analysis of the catalytic and feedback sites of the tryptophan-sensitive 3-deoxy-D-*arabino*-heptulosonate-7-phosphate synthase of *Escherichia coli*. *J. Bacteriol.* **170:** 5500–5506.
63. **Robertson, D. E., E. J. Mathur, R. V. Swanson, B. L. Marrs, and J. M. Short.** 1995. The discovery of new biocatalysts from microbial diversity. *SIM News* **46(1):**3–8.
64. **Saier M. H., Jr., and T. M. Ramseier.** 1996. The catabolite repressor/activator (*cra*) protein of enteric bacteria. *J. Bacteriol.* **178:**3411–3417.
65. **Sakoda, H., and T. Imanaka.** 1990. A new way of stabilizing recombinant plasmids. *J. Ferm. Bioeng.* **69:**75–78.
66. **Schweizer, H. P.** 1993. Small broad-host-range gentamycin resistance gene cassettes for site-specific insertion and deletion mutagenesis. *BioTechniques* **15:**831–833.
67. **Shi, J., and P. Biek.** 1995. A versatile low-copy-number cloning vector derived from plasmid F. *Gene* **164:**55–58.
68. **Skogman, G., J. Nilsson, and P. Gustafsson.** 1983. The use of a partition locus to increase stability of tryptophan-operon-bearing plasmids in *Escherichia coli*. *Gene* **23:**105–115.
69. **Slater, S., T. Gallaher, and D. Dennis.** 1992. Production of poly-(3-hydroxybutyrate-co-3-hydroxyvalerate) in a recombinant *Escherichia coli* strain. *Appl. Environ. Microbiol.* **58:**1089–1094.
70. **Smyer, J. R., and R. M. Jeter.** 1989. Characterization of phosphoenolpyruvate synthase mutants in *Salmonella typhimurium*. *Arch. Microbiol.* **153:**26–32.
71. **Stock, J. B., A. J. Ninfa, and A. M. Stock.** 1989. Protein phosphorylation and regulation of adaptive response in bacteria. *Microbiol. Rev.* **53:**450–490.
72. **Sugimoto, S., M. Yabuta, N. Kato, T. Seki, T. Yoshida, and H. Taguchi** 1987. Hyperproduction of phenylalanine by *Escherichia coli*: application of a temperature-controllable expression vector carrying the repressor-promotor system of bacteriophage lambda. *J. Biotechnol.* **5:**237–253.
73. **Taguchi, M., K. Izui, and H. Katsuki.** 1977. Activation of *Escherichia coli* phosphoenolpyruvate carboxylase by guanosine-5′diphosphate-3′-diphosphate. *FEBS Lett.* **77:** 270–272.
74. **Terasawa, M., M. Inui, Y. Uchida, M. Kobayashi, Y. Kurusu, and H. Yukawa.** 1991. Application of tryptophanase promoter to high expression of the tryptophan synthase gene in *Escherichia coli*. *Appl. Microbiol. Biotechnol.* **34:**623–627.
75. **Uhlin, B. E., S. Molin, P. Gustafsson, and K. Nordstrom.** 1979. Plasmids with temperature-dependent copy number for amplification of cloned genes and their products. *Gene* **6:**91–106.
76. **Waygood, E. B., M. K. Rayman, and B. D. Sanwal.** 1975. The control of pyruvate kinases of *Escherichia coli*. II. Effectors and regulatory properties of the enzyme associated by ribose-5-phosphate. *Can. J. Biochem.* **53:**444–454.
77. **Waygood, E. B., and B. D. Sanwal.** 1974. The control of pyruvate kinases of *Escherichia coli*. I. Physicochemical and regulatory properties of the enzyme activated by fructose 1, 6-diphosphate. *J. Biol. Chem.* **249:**265–274.
78. **Weaver, L. M., and K. M. Herrmann.** 1990. cloning of an *aroF* allele encoding a tyrosine-insensitive 3-deoxy-D-*arabino*-heptulosonate 7-phosphate synthase. *J. Bacteriol.* **172:**6581–6584.
79. **Weiss, U., and J. M. Edwards.** 1980. *The Biosynthesis of Aromatic Compounds*. Wiley, New York.
80. **Wohler, F., and J. Liebig.** 1832. Untersuchungen uber das Radikal der Benzoesaure. *Ann. Pharm.* **3:**249–282.
81. **Wood, T. K., and S. W. Peretti.** 1990. Depression of protein synthetic capacity due to cloned-gene expression in *E. coli*. *Biotechnol. Bioeng.* **36:**865–878.
82. **Wyman, A. R., L. B. Wolfe, and D. Botstein.** 1985. Propagation of some human DNA sequences in bacteriophage λ vectors requires mutant *E. coli* hosts. *Proc. Natl. Acad. Sci. USA* **82:**2880–2884.
83. **Yukawa, H., Y. Kurusu, M. Shimazu, M. Terasawa, A. Ohta, and I. Shibuya.** 1985. Stabilization by the mini-F fragment of a pBR322 derivative bearing the tryptophan operon in *Escherichia coli*. *Agric. Biol. Chem.* **49:**3619–3622.

Polyhydroxyalkanoates: Biodegradable Polymer

SANG YUP LEE AND JONG-IL CHOI

51

Polyhydroxyalkanoates (PHAs) are the polymers of hydroxyalkanoates (Fig. 1), which are accumulated as a carbon and energy storage material in various microorganisms, usually under the condition of limiting nutritional elements such as N, P, S, O, or Mg in the presence of excess carbon source (1, 6, 9, 13, 17, 28, 48, 84). After the discovery of poly(3-hydroxybutyrate) [P(3HB)] from the bacterium *Bacillus megaterium* in 1926 (54), a large variety of PHAs possessing different numbers of main chain carbon atoms and different types of R-pendent groups have been reported (91). Figure 2 shows a transmission electron micrograph of the P(3HB) granules appearing as electrolucent bodies in the cells of recombinant *Escherichia coli* harboring the *Alcaligenes eutrophus* PHA biosynthesis genes.

Recently, the problems concerning the global environment and solid waste management have created much interest in the development of biodegradable plastics (45, 95). Among the several biodegradable polymers under development, PHAs have been drawing much attention because of their material properties, which are similar to those of conventional plastics (such as polypropylene) or elastomers, and their complete biodegradability (5, 8, 49, 86). The copolymer poly(3-hydroxybutyrate-co-3-hydroxyvalerate) [P(3HB-co-3HV)], developed by Zeneca Bio Product (formerly ICI, Cleveland, U.K.), has been produced on a fairly large scale and sold under the trade name of BIOPOL (8, 10). These activities have recently been sold to Monsanto (St. Louis, Mo.). However, because of the high production cost of PHAs compared with that of petrochemical-based polymers, much effort has been devoted to reduce the price of PHAs by developing better bacterial strains, more efficient fermentation, and more economical recovery processes (48, 49). A number of articles are available on the production of PHAs (8, 11, 30, 43, 49, 51, 69), the biochemistry and molecular biology of PHA synthesis (84, 85, 87, 88), polymer recovery (23, 29, 73, 74), and polymer characteristics (4, 5, 12, 18, 28, 60).

In this chapter, methods for the analysis of PHAs, biochemical and molecular biological methods for studying PHA biosynthesis, fermentation strategies for the production of PHAs, polymer recovery methods, and applications are described.

51.1. ISOLATION OF PHA PRODUCERS AND ANALYSIS OF PHA

51.1.1. Staining of PHA

Accumulation of PHA granules in microorganisms can be detected by electron or phase-contrast microscopic analysis. Bacteria accumulating PHAs can also be easily identified on solid medium (see below), since they appear as more turbid colonies than the cells not producing PHAs. However, for the definite visualization of PHA producers, staining methods are useful (41, 62, 78). Sudan black B or Nile blue sulfate has been used for staining PHAs as follows.

PROTOCOL

Nutrient Agar Plate for Rapid Screening of PHA Producers

The composition of the nutrient agar plate is (per liter): Glucose, 20 g; $(NH_4)_2SO_4$, 2 g; KH_2PO_4, 13.3 g; $MgSO_47H_2O$, 1.2 g; citric acid, 1.7 g; trace element solution, 10 ml [per liter: $FeSO_47H_2O$, 10 g; $ZnSO_47H_2O$, 2.25 g; $CuSO_45H_2O$, 1 g; $MnSO_44\text{-}5H_2O$, 0.5 g; $CaCl_22H_2O$, 2 g; $Na_2B_4O_710H_2O$, 0.23 g; $(NH_4)_6Mo_7O_{24}$, 0.1 g; 35% HCl, 10 ml]; Bacto agar, 15 g.

Glucose can be replaced by other carbon sources for isolating microorganisms that produce PHAs from different carbon sources.

Staining with Sudan black B

1. Prepare a heat-fixed film of the specimen grown under PHA-accumulating conditions on a slide and immerse it in a filtered solution of 0.3% (wt/vol) Sudan black B (in ethylene glycol). Stain for 5 to 15 min (determine the time experimentally).
2. Drain and air dry the slide.
3. Immerse and withdraw the slide several times in xylene, and blot dry with absorbent paper.
4. Counterstain for 5 to 10 s with 0.5% (wt/vol) aqueous safranin.

$$\left[-O-\overset{\displaystyle H}{\underset{\displaystyle R}{C}}-(CH_2)_n-\overset{\displaystyle O}{\overset{\|}{C}}- \right]_{100\text{-}30{,}000}$$

n = 1	R = hydrogen	Poly(3-hydroxypropionate)
	R = methyl	Poly(3-hydroxybutyrate)
	R = ethyl	Poly(3-hydroxyvalerate)
	R = propyl	Poly(3-hydroxyhexanoate)
	R = pentyl	Poly(3-hydroxyoctanoate)
	R = nonyl	Poly(3-hydroxydodecanoate)
n = 2	R = hydrogen	Poly(4-hydroxybutyrate)
	R = methyl	Poly(4-hydroxyvalerate)
n = 3	R = hydrogen	Poly(5-hydroxyvalerate)
	R = methyl	Poly(5-hydroxyhexanoate)
n = 4	R = hexyl	Poly(6-hydroxydodecanoate)

FIGURE 1 General structure of PHA (17, 49, 84).

5. Rinse the slide with tap water and blot dry.
6. Examine under oil immersion. PHA inclusions appear as blue-black droplets, while cytoplasmic parts of the organism appear pink.

Staining with Nile sulfate

1. Prepare and filter a 1% aqueous solution of Nile blue A just before use. Stain a heat-fixed smear of the bacteria containing PHAs with this solution for 10 min at 55°C in a coplin staining jar.
2. Wash the slide briefly with tap water to remove excess stain, and place it in 8% aqueous acetic acid solution for 1 min.
3. Wash with tap water and blot dry.
4. Remoisten with tap water, and place a coverslip over the smear.
5. Place immersion oil on the coverslip, and examine the preparation under an epifluorescence microscope, using an exciter filter that provides an excitation wavelength of approximately 460 nm. If the test is positive, PHA granules exhibit a bright orange fluorescence. Other cell inclusion bodies such as a polyphosphate do not fluoresce in this manner.

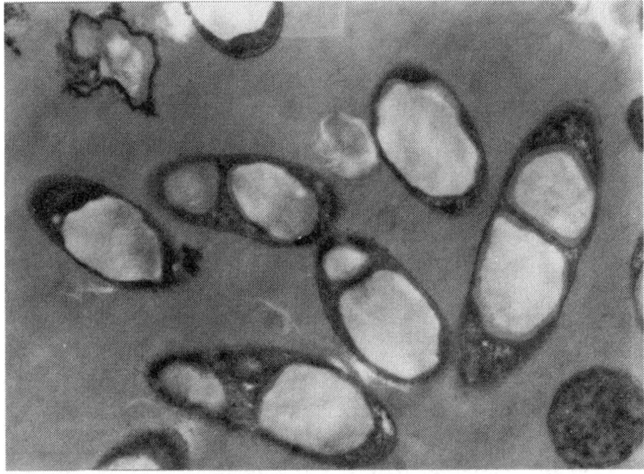

FIGURE 2 Transmission electron micrograph of recombinant *E. coli* accumulating a large amount of P(3HB).

51.1.2. Analysis of PHA

These staining methods are useful for the detection and rough quantification of PHAs. However, more accurate measurement is required for the quantitative analysis of PHAs. To quantify and distinguish constituents of PHAs, several methods have been suggested, including gas chromatography (7, 32), spectrophotometric assay of crotonic acid (44), capillary isotachophoresis (93), fluorometric analysis (14), and thermogravimetric analysis (22). Of these methods, gas chromatography provides the most convenient, reliable, and accurate way to quantitatively analyze PHAs.

PROTOCOL

Analysis of PHA by Gas Chromatography

1. Centrifuge culture broth containing 10 to 20 mg of cells (dry cell weight) for 15 min at 3,000 × *g* and 4°C.
2. Remove supernatant and add 2 ml of chloroform plus 2 ml of solution A (made of 97 ml of methanol, 3 ml of sulfuric acid, and 0.3 g of benzoic acid per 100 ml of solution). Make the standard containing 0.003 to 0.01 g of PHA in the same condition.
3. Incubate the solution in boiling water bath for more than 5 h.
4. Cool the solution at room temperature.
5. Add 1 ml of distilled water and mix by vortexing.
6. Centrifuge the solution for 5 min at 3,000 × *g*.
7. Take out 1 ml of bottom (chloroform layer) for gas chromatographic analysis. The concentration of PHA is determined by comparing the PHA peak with that of benzoic acid, which is used as an internal standard.

In this method, intracellular PHA granules are methanolyzed to the methyl esters of 3-hydroxyalkanoates. Different hydroxyalkanoates can be distinguished by their different retention times in gas chromatography. Under the reported conditions (32), the gas chromatographic analysis of P(3HB-co-3HV) results in two peaks: the first peak is methyl-3-hydroxybutyrate, and the second peak methyl-3-hydroxyvalerate. For the analysis of medium-chain-length PHAs, the peaks appearing represent (in order) 3-hydroxyhexanoate, 3-hydroxyoctanoate, 3-hydroxydecanoate, 3-hydroxydecanoate, 3-hydroxydodecanoate, and 3-hydroxy-*cis*-5-tetradecenoate.

The PHA content is defined as the ratio of PHA concentration to cell concentration (dry cell weight per unit volume). Residual cell concentration is defined as cell concentration minus PHA concentration.

51.2. BIOCHEMISTRY AND MOLECULAR BIOLOGY OF PHA BIOSYNTHESIS

51.2.1. PHA Biosynthetic Pathway

Several different metabolic pathways for the synthesis of PHAs are known to exist in bacteria. As reviewed by

Steinbuchel (84), four different pathways have been identified for the synthesis and incorporation of almost all the known constituents of bacterial PHAs.

First, in *A. eutrophus*, acetyl coenzyme A (acetyl-CoA) is converted to P(3HB) in three enzymatic steps. A biosynthetic β-ketothiolase catalyzes the formation of a carbon-carbon bond of two acetyl-CoA moieties. An NADPH-dependent acetoacetyl-CoA reductase catalyzes the reduction of acetoacetyl-CoA formed in the first reaction to D-(−)-3-hydroxybutyryl-CoA. The third reaction of this pathway is catalyzed by the PHA synthase, which links the D-(−)-3-hydroxybutyryl-moiety to an existing polymer molecule by an ester bond.

The second type is the five-step PHA biosynthetic pathway in a photosynthetic bacterium, *Rhodospirillim rubrum*. In this pathway, acetoacetyl-CoA formed by the β-ketothiolase is reduced by an NADH-dependent reductase to L-(+)-3-hydroxybutyryl-CoA. The latter is then converted to the D-(−)-3-hydroxybutyryl-CoA by two enoyl-CoA hydratases. The final step of this pathway is again the polymerization catalyzed by the PHA synthase.

The third type of PHA-biosynthetic pathway seems to be active in most pseudomonads belonging to the ribosomal RNA homology group I. In this pathway, the intermediates from the β-oxidation of the activated fatty acids derived from alkanes, alkanols, or alkanoates are directed to PHA biosynthesis. For example, *Pseudomonas oleovorans* accumulates a polyester that contains 3-hydroxyoctanoate and 3-hydroxyhexanoate as the main constituents when cultivated on octane, octanol, or octanoate.

The fourth type of pathway is found in almost all pseudomonads belonging to the rRNA homology group I, except *P. oleovorans*. Copolymers consisting of medium-chain-length 3-hydroxyalkanoates are synthesized from acetyl-CoA. The precursors for PHA synthesis are derived from de novo fatty acid synthetic pathways. The biosynthesis of PHA mainly consisting of 3-hydroxydecanoate in *Pseudomonas aeruginosa* follows this pathway.

51.2.2. Assay of PHA Biosynthetic Enzymes

The activities of the three PHA biosynthetic enzymes, PHA synthase (21), β-ketothiolase (61), and NADPH-dependent acetoacetyl-CoA reductase (68), can be assayed as follows.

PROTOCOL

Resuspend centrifuged cells (4 to 5 g of wet cell weight) in 300 ml of M9-salts medium, and divide into three aliquots for the assay of PHA synthase, β-ketothiolase, and acetoacetyl-CoA reductase activities. M9-salts medium contains (per liter): Na_2HPO_4, 6 g; KH_2PO_4, 3 g; NaCl, 0.5 g; NH_4Cl, 1 g. Collect cells by centrifugation, resuspend them in an appropriate 2-ml lysis buffer, and disrupt by sonication. The composition of lysis buffer and reaction mixture for the assay of each enzyme is shown below.

PHA Synthase

Lysis buffer
50 mM KH_2PO_4 (pH 7.0) containing 5% glycerol

Reaction mixture (total volume: 300 μl)

300 mM KH_2PO_4, pH 7.0	150 μl
9.72 mM 3-hydroxybutyryl-CoA	30.9 μl
Cell lysate	x μl
H_2O	to 300 μl

Assay procedure

1. Preincubate the assay mixture for 10 min at 25°C before adding cell lysate.
2. After adding cell lysate, incubate the mixture for 14 min. Remove aliquots (40 μl) from the reaction mixture at regular time intervals (2 min). Stop the reaction by adding 100 μl of 5% trichloroacetic acid.
3. Pellet the precipitated protein by centrifugation for 10 min.
4. Add an aliquot (125 μl) of supernatant to 675 μl of 500 mM KH_2PO_4 (pH 7.5). Add DTNB [(5,5′-dithiobis(2-nitrobenzoic acid), 10 μl of a 10 mM stock in 500 mM KH_2PO_4 (pH 7.5)] to this mixture and incubate for 2 min at room temperature.
5. Follow the absorbance at 410 nm at regular time intervals. Assay both the soluble protein fraction and the crude sonicated lysate to determine both soluble and insoluble PHA synthase activity.

β-Ketothiolase

Lysis buffer
10 mM Tris-HCl (pH 7.3) containing 5 mM β-mercaptoethanol

Reaction mixture (total volume: 750 μl)

25 mM acetyl-CoA	75 μl
5 mM dithiothreitol	75 μl
4.0 mM NADH	75 μl
666.7 mM Tris-HCl, pH 7.4	75 μl
L-(+)-β-hydroxy-acyl-CoA dehydrogenase	10 μl
Cell lysate	x μl
H_2O	to 750 μl

Assay procedure
Place the reaction mixture in a cuvette. Follow the optical reading at 340 nm. β-Ketothiolase activity is measured by the rate of NAD^+ formation.

Acetoacetyl-CoA reductase

Lysis buffer
10 mM Tris-HCl, pH 8.0
1 mM EDTA
10 mM β-mercaptoethanol
20% glycerol
0.2 mM phenylmethylsulfonyl fluoride

Reaction mixture (total volume: 1,000 μl)

1.8 mM acetoacetyl-CoA	50 μl
1 mM NADPH	100 μl
1 M Tris-HCl, pH 8.0	100 μl
Cell lysate	x μl
H_2O	to 1,000 μl

Assay procedure
Place the reaction mixture in a cuvette. Follow the optical reading at 340 nm. Reductase activity is measured by the rate of NADPH oxidation.

Several different methods for the assay of these enzymes are also available (25–27).

51.2.3. Cloning of PHA Biosynthetic Genes

In 1988, the PHA biosynthetic genes of A. *eutrophus* were cloned in *E. coli* by three independent groups (66, 79, 82). These genes were sequenced and characterized in detail and were found to form an operon as *phbC-A-B*, coding for PHA synthase, β-ketothiolase, and NADPH-dependent acetoacetyl-CoA reductase, respectively. Thereafter, PHA synthases, the key enzyme of the PHA biosynthetic pathway, were cloned from a number of bacteria.

Several strategies for cloning the PHA synthase genes are summarized below.

1. Screening of genomic libraries for phenotypic complementation of PHA negative mutants, such as *A. eutrophus* $PHB^{-}4$ (DSM 541)
2. Screening of genomic libraries for conferring the ability to accumulate PHA to a non-PHA producer, such as *E. coli*
3. Screening of a genomic library using homologous gene probe designed after transposon mutanenesis
4. Screening of a genomic library using heterologous gene probe. This is possible owing to the high degree of homology among the cloned PHA synthase genes
5. Screening of a genomic library using the oligonucleotide probe, which consists of the highly conserved sequences of various PHA synthase genes
6. Screening of a genomic library for the existence of a clone that expresses the activities of the enzymes in the PHA synthetic pathway

Using one of these methods, the PHA synthase genes were cloned from *Acinetobacter* sp. (77), *Alcaligenes* sp. SH-69 (46), *A. eutrophus* H16 (66,79, 82), *Chromatium vinosum* D (56), *Ectothiorhodospira shaposhnikoviii* N1 (55), *Lamprocystis roseopersicina* 3112 (55), *Methylobacterium extorquens* IBT6 (100), *Nocardia corralina* (87), *Paracoccus denitrificans* (99), *P. aeruginosa* (96), *Pseudomonas citronellolis* (97), *Pseudomonas fluorescens* (87), *Pseudomonas mendocina* (97), *P. oleovorans* (34), *Pseudomonas putida* KT2442 (33), *Pseudomonas* sp. DSM 1650 (97), *Pseudomonas* sp. GP4BH1 (97), *Rhizobium meliloti* (98), *Rhodobacter sphaeroides* (35, 36), *Rhodococcus ruber* PP2 (67), *R. rubrum* Ha (36), *Syntrophomonas wolfei* (59), *Thiocapsa pfennigii* 9111 (55), and *Thiocystis violacea* 2311 (55, 57).

51.3. PHA PRODUCTION BY BACTERIAL FERMENTATION

Even though many different PHAs have been described in the literature, only a few of them have been produced in large quantity, characterized, and used to develop applications (48, 51). These are P(3HB), P(3HB-co-3HV), P(3HV) (89), poly(4-hydroxybutyrate) [P(4HB)] (92), poly(3-hydroxyhexanoate-co-3-hydroxyoctanoate) [P(3HHx-co-3HO)]. Among many different microorganisms that are known to synthesize PHAs, only a few bacteria have been employed for the production of PHAs. These include *A. eutrophus* (37–39, 64), *Alcaligenes latus* (24, 103), *Azotobacter vinelandii* (63), several strains of methylotrophs (40, 94), *P. oleovorans* (70, 72), and recombinant *E. coli* (20, 52, 53). Each bacterium requires different conditions for growth and PHA production. PHA producers can be divided into two groups depending on the culture condition required for PHA synthesis (49). The first group of bacteria, such as *A. eutrophus*, *P. oleovorans*, and several methylotrophs, require the limitation of an essential nutritional element such as N, P, Mg, K, O, or S for the efficient synthesis of PHA from an excess carbon source. The second group of bacteria, such as *A. latus*, a mutant strain of *A. vinelandii*, and recombinant *E. coli* harboring the *A. eutrophus* PHA biosynthetic genes, do not require nutrient limitation for PHA synthesis and can accumulate polymer during growth.

51.3.1. Production of P(3HB)

P(3HB) is the best-known PHA and is accumulated by many microorganisms. *A. eutrophus*, *A. latus*, *A. vinelandii*, methylotrophs, and recombinant *E. coli* have been successfully employed for the production of P(3HB).

51.3.1.1. *A. eutrophus*

A. eutrophus, the bacterium employed for the commercial production of PHA by Zeneca, accumulates a large amount of P(3HB) (up to 80% of dry cell mass) when nitrogen or phosphate is completely depleted. The medium composition and fermentation conditions successfully used for the fed-batch culture of *A. eutrophus* are shown below (38).

The initial medium (pH 6.8) for fed-batch culture contains (per liter): glucose, 20 g; $(NH_4)_2SO_4$, 4 g; KH_2PO_4, 13.3 g; $MgSO_4 7H_2O$, 1.2 g; citric acid, 1.7 g; trace element solution, 10 ml (per liter: $FeSO_4 7H_2O$, 10 g; $ZnSO_4 7H_2O$, 2.25 g; $CuSO_4 5H_2O$, 1 g; $MnSO_4$-$5H_2O$, 0.5 g; $CaCl_2 2H_2O$, 2 g; $Na_2B_4O_7 10H_2O$, 0.23 g; $(NH_4)_6Mo_7O_{24}$, 0.1 g; 35% [vol/vol] HCl, 10 ml). The seed culture (100 ml) is prepared by growing cells in a 250-ml flask on a reciprocal shaker at 30°C. The initial volume of fed-batch culture is 800 ml. Temperature and pH are controlled at 30°C and 6.8, respectively. The pH is controlled with 2 N HCl solution and 28% (vol/vol) NH_4OH solution, which is replaced by 5 N NaOH/KOH solution during the period of nitrogen limitation. If phosphate limitation instead of nitrogen limitation is to be applied, the phosphate concentration in the initial medium is adjusted just enough to support cell growth to a desired concentration. Dissolved oxygen (DO) concentration is maintained above 20% of air saturation by manipulating the agitation speed and the aeration rate up to 1,000 rpm and 1.5 l/min, respectively. The glucose concentration in the feeding solution is 700 g/liter. Various feeding strategies, including the pH-stat, DO-stat, and glucose concentration control, have been tested for the fed-batch culture of *A. eutrophus*. Glucose concentration control (at 15 g/liter) by using the on-line glucose analyzer was found to be the best method for high-cell-density culture of *A. eutrophus* and efficient P(3HB) production. The time of applying nutrient limitation is important to obtain high productivity of P(3HB). Kim et al. (38) reported that P(3HB) concentrations of 71 and 92 g/liter could be obtained by limiting nitrogen at the cell concentrations of 30 and 55 g/liter, respectively. By further delaying the nitrogen limitation, until the cell concentration reached 70 g/liter, a final P(3HB) concentration of 121 g/liter could be achieved.

51.3.1.2. *A. latus*

A. latus can accumulate PHA during growth and utilize sucrose and, therefore, beet and cane molasses as a carbon source. Because of these advantages, there has been much interest in developing an efficient process for the production of P(3HB) by *A. latus* (24). *A. latus* can be cultivated in the following medium to produce a high concentration of P(3HB) with high productivity by employing the pH-stat feeding strategy (103). The initial medium (pH 7.0) contains (per liter): sucrose, 20 g; $(NH_4)_2SO_4$, 2 g; KH_2PO_4, 1.5 g; $Na_2HPO_412H_2O$, 9 g; $MgSO_37H_2O$, 0.2 g; $FeCl_2H_2O$, 60 mg; $CaCl_22H_2O$, 10 mg; trace element solution, 1 ml (per liter: H_3BO_3, 0.3 g; $CoCl_26H_2O$, 0.2 g; $ZnSO_47H_2O$, 0.1 g; $MnCl_24H_2O$, 30 mg; $Na_2MoO_42H_2O$, 30 mg; $NiSO_47H_2O$, 28 mg; $CuSO_45H_2O$, 10 mg). Three different feeding solutions used were sucrose solution, 900 g/liter; ammonia solution, 28% (vol/vol); mineral medium formulated from elemental analysis data of A. *latus* cell mass. Three feeding solutions are supplied simultaneously but with different flow rates so that sucrose and mineral medium supplied can be coupled to ammonia feeding, which is used for pH control.

Recently, it was demonstrated that nitrogen limitation could significantly enhance PHA production for A. *latus* as well (101). Cells were first cultured by the DO-stat feeding strategy without applying nitrogen limitation. Nitrogen limitation was applied at a cell concentration of 76 g/liter, and the sucrose concentration was maintained within 5 to 20 g/liter. After 8 h of nitrogen limitation, the cell concentration, P(3HB) concentration, and P(3HB) content reached 111.7 g/liter, 98.7 g/liter, and 88%, respectively, resulting in a productivity of 4.94 g of P(3HB) per liter per h (101).

51.3.1.3. Methylotrophs

Methylotrophs have attracted much research interest owing to their ability to use an inexpensive carbon source, methanol. Mass production of P(3HB), 149 g/liter of P(3HB), by nitrogen-limited fed-batch cultivation of *Protomonas extorquens* has been reported (94). However, a long cultivation time of 170 h resulted in the low productivity of 0.88 g of P(3HB) per liter per h. In another study, by controlling the concentration of methanol within the range of 2 to 3 g/liter and potassium concentration below 25 mg/liter, 250 g/liter of cell mass and 130 g/liter of P(3HB) could be obtained by the fed-batch culture of *Methylobacterium organophilum* (40). However, low P(3HB) content (ca. 50%) seems to be the major problem of employing methylotrophs. The following medium can be used for the efficient production of P(3HB) by fed-batch culture of M. *organophilum*. Potassium limitation can be applied when the cell concentration reaches a relatively high value (ca. 80.9 g/liter) (40). Initial medium (pH 7.0) contains (per liter): methanol, 5 g; $(NH_4)_2SO_4$, 0.8 g; KH_2PO_4, 0.8 g; $Na_2HPO_412H_2O$, 3.0 g; trace element. Feeding solutions used were methanol; acid mineral solution (per liter: H_3PO_4, 270 g; $MgSO_47H_2O$, 169 g; $FeSO_47H_2O$, 1.932 g; $MnSO_44H_2O$, 0.244 g; $ZnSO_47H_2O$, 0.529 g; $CoCl_26H_2O$, 0.403 g; H_3BO_3, 0.122 g; $CuCl_22H_2O$, 0.344 g; $CaCl_22H_2O$, 0.554 g); and alkaline solution (per liter: NaOH, 113.7 g; KOH, 100 g). The mineral solution was supplied along with methanol in a ratio of 0.075 (milliliter of mineral feed/milliliter of methanol feed).

51.3.1.4. Recombinant *E. coli* harboring the *A. eutrophus* PHA biosynthetic genes

Use of recombinant E. *coli* harboring the A. *eutrophus* PHA biosynthetic genes is attractive for the production of PHA since PHA accumulation up to 90% of dry cell mass has been reported. Recombinant E. *coli* harboring the A. *eutrophus* PHA biosynthetic genes can accumulate polymer during growth without nutrient limitation (52). Therefore, the development of a nutrient feeding strategy that provides optimal cell growth and polymer accumulation at the same time is crucial for the efficient production of P(3HB) (49, 52).

A stable high-copy-number plasmid containing the A. *eutrophus* PHA biosynthetic genes under the control of the native promoter has been used for the production of PHA to a high concentration. P(3HB) could be efficiently produced (ca. 80 g/liter) by the pH-stat fed-batch culture of recombinant E. *coli* in a complex medium (53). However, P(3HB) concentration of only 16 g/liter could be produced in a defined medium. P(3HB) production was considerably enhanced in a defined medium supplemented with various complex nitrogen sources, such as tryptone, yeast extract, peptone, Casamino Acids, beef extract, cotton seed hydrolysate, casein hydrolysate, collagen hydrolysate, corn steep liquor, or soy bean hydrolysate (50). Supplementation of amino acids or oleic acid also enhanced P(3HB) synthesis (52). The reasons that P(3HB) synthesis was enhanced by the addition of complex nitrogen sources, amino acids, or oleic acid can be explained by considering the primary metabolism of E. *coli*. Since P(3HB) is synthesized from acetyl-CoA by three enzymatic conversion steps, the synthesis of P(3HB) is dependent on the amount of acetyl-CoA available. Among the most directly competing pathways are the formation or synthesis of acetate, citrate, and fatty acids. In a defined medium, the amount of acetyl-CoA available for P(3HB) synthesis would be less than that in a semidefined or complex medium, because acetyl-CoA must be used to synthesize the biosynthetic intermediates and to generate energy. Availability of NADPH, which is a cofactor of the second enzyme (acetoacetyl-CoA reductase) in PHB synthesis, is also important. Inefficient regeneration of NADPH will lead to inefficient P(3HB) synthesis. Therefore, addition of complex nitrogen sources, amino acids, or oleic acid to the defined medium reduces the amounts of NADPH and acetyl-CoA required for the synthesis of the metabolic intermediates (52). A typical fermentation condition that allows efficient production of P(3HB) is shown below (50). The initial medium (pH 7.0) for fed-batch culture contains (per liter); glucose, 10 g; tryptone, 2 g; thiamine, 0.01 g; $(NH_4)_2SO_4$, 4 g; KH_2PO_4, 13.3 g; $MgSO_47H_2O$, 1.2 g; citric acid, 1.7 g; trace element solution, 10 ml (per liter: $FeSO_47H_2O$, 10 g; $ZnSO_47H_2O$, 2.25 g; $CuSO_45H_2O$, 1 g; $MnSO_44$-$5H_2O$, 0.5 g; $CaCl_22H_2O$, 2 g; $Na_2B_4O_710H_2O$, 0.23 g; $(NH_4)_6Mo_7O_{24}$, 0.1 g; 35% (vol/vol) HCl, 10 ml). The feeding solution contains (per liter): glucose, 700 g; $MgSO_47H_2O$, 20 g; tryptone, 17.5 g; thiamine, 0.1 g. The substrate feeding strategy is the pH-stat using setpoint of a high limit. When the pH becomes higher than 6.9, a feeding nutrient solution corresponding to 20 g of glucose (28.6 ml of feeding solution) is added for a definite on-time (as a pulse). Ammonia water (28% [vol/vol]) used as a nitrogen source during the fed-batch culture, which is automatically supplied when the pH drops below 6.8.

Recently, it was also reported that a high concentration of P(3HB) (149 g/liter) could be produced with a high productivity of 3.4 g of P(3HB) per liter per h in a chemically defined medium by fed-batch culture of a filamentation-suppressed recombinant E. *coli* strain (102).

51.3.2. Production of P(3HB-co-3HV)

Copolymer P(3HB-co-3HV) has better mechanical properties than homopolymer P(3HB) and has been industrially produced by Zeneca (by Monsanto since 1996). The procedure for the production of P(3HB-co-3HV) is similar to that of P(3HB) except that propionate or valerate is fed together with a major carbon source in the polymer accumulation phase. In general, final coplymer composition, dry cell mass, and productivity of PHA are dependent on the ratio of propionate (or valerate) to the major carbon source in the feed. The 3HV fraction in the copolymer increases with the increasing ratio of propionate (or valerate) to the major carbon source. However, the optimal ratio should be determined, since elevated levels of these acids inhibit cell growth and, consequently, PHA productivity. In *A. eutrophus*, P(3HB-co-3HV) is synthesized as follows if propionic acid is used alone or in combination with another substrate (84). Intracellular propionate is converted to propionyl-CoA by propionyl-CoA synthetase, which is usually converted to the Krebs-cycle intermediate succinyl-CoA by subsequent reactions catalyzed by propionyl-CoA carboxylase, methylmalonyl-CoA racemase, and methylmalonyl-CoA mutase. The biosynthetic β-ketothiolase catalyzes the condensation of propionyl-CoA with acetyl-CoA to 3-ketovaleryl-CoA, which is then reduced to D-(−)-3-hydroxyvaleryl-CoA by the acetoacetyl-CoA reductase. The hydroxyvaleryl moiety is then covalently linked to the growing chain of polyester by the PHA synthase.

On the other hand, recombinant *E. coli* harboring the *A. eutrophus* PHA biosynthetic genes does not possess an efficient system for the uptake and conversion of propionate to propionyl-CoA. Two approaches can be taken to produce P(3HB-co-3HV) in recombinant *E. coli*. When genetic regulation of fatty acid uptake and utilization is eliminated by *atoC* (Con) and *fadR* mutations, the enzymes involved in acetate and propionate utilization are constitutively expressed for the efficient utilization of propionate (81). In another approach, P(3HB-co-3HV) can be produced by any *E. coli* strain from glucose and propionate by first growing cells on acetate or oleic acid, which allows efficient propionate uptake and utilization (104).

51.3.3. Production of Medium-Chain-Length PHAs (MCL-PHAs)

Medium-chain-length PHAs (MCL-PHAs) consisting of 6 to 14 carbon atoms were first detected in the cells of *P. oleovorans* grown on *n*-octane (16). MCL-PHAs are semicrystalline elastomers exhibiting a low melting point and high elongation to break and can be potentially used as a biodegradable rubber (15). Accumulation of MCL-PHAs occurs under nitrogen limitation when cells are grown on MCL-alkanes, alkanols, or alkanoic acids. The composition of the accumulated PHA has a close relation with the carbon source employed (71); 3-hydroxyoctanoate was the main constituent of PHA accumulated from octane, octanol or octanoic acid.

Fed-batch and continuous cultivation of *P. oleovorans* have been carried out for the production of MCL-PHAs (70, 72). When *n*-octane was used as carbon substrate in a two-liquid-phase cultivation, the DO concentration in the culture broth became too low owing to low oxygen solubility in *n*-octane. In these cultures, it was necessary to design a reactor allowing very efficient oxygen transfer. Preusting et al. (72) developed a reactor system that has a high volumetric oxygen transfer coefficient of 0.49 s^{-1} at the air flow rate and stirred speed of 2 liter/min and 2,500 rpm, respectively. Fed-batch culture of *P. oleovorans* in this reactor resulted in cell and PHA concentrations of 37.1 and 12.1 g/liter, respectively.

A typical condition for the fed-batch culture of *P. oleovorans* is shown below (72). The initial medium contains (per liter): *n*-octane, 100 ml; KH_2PO_4, 8.0 g; $K_2HPO_43H_2O$, 8.3 g; $Na_2HPO_412H_2O$, 4.2 g; $(NH_4)_2SO_4$, 4.6 g; NH_4Cl, 0.1 g; polypropyleneglycol 2000, 1 ml; 1 M $MgSO_47H_2O$, 3.0 ml; 0.1 M $CaCl_22H_2O$, 1.5 ml; microelement stock solution, 1.5 ml (per liter: $FeSO_47H_2O$, 40.0 g; $MnSO_4H_2O$, 10.0 g; $CoCl_26H_2O$, 4.0 g; $ZnSO_47H_2O$, 2.0 g; $MoO_4Na_22H_2O$, 2.0 g; $CuCl_22H_2O$, 1.0 g; $Al_2(SO_4)_316H_2O$, 1.3 g). The feeding solution contains (per liter): $(NH_4)_2SO_4$, 215 g; $MgSO_47H_2O$, 28.7 g. The DO is maintained at 30% of air saturation, and 4 N NaOH is used to maintain the pH at 7.0. Cells are first grown batchwise until ammonium is completely consumed. When ammonium becomes limiting, indicated by a sudden increase of the DO, nutrient feeding is started at the constant rate of 4 ml/h.

Two-phase continuous culture of *P. oleovorans* has also been carried out (70). With the optimization of culture conditions, including the composition of the feeding solution, the steady-state cell concentration and PHA productivity of 11.6 g/liter and 0.58 g of PHA per liter per h, respectively, could be reached. Since *P. oleovorans* was stable in a chemostat for at least 1 month, continuous cultivation may also be a useful strategy for the production of MCL-PHAs by *P. oleovorans*.

51.4. METABOLIC ENGINEERING

Metabolic engineering principles can be used to broaden the utilizable substrate range, to enhance PHA production, and to produce novel PHAs (49). Cloning of the PHA synthase genes from various bacteria allowed construction of several recombinant strains that produce PHAs more efficiently. When a broad-host-range plasmid containing the *A. eutrophus* PHA biosynthetic genes was introduced into the same *A. eutrophus* strain, the activities of PHA biosynthetic genes could be considerably increased. In flask cultures, the PHA production rate of recombinant *A. eutrophus* was 1.24 times higher than that of the parent strain (65). Heterologous expression of PHA biosynthetic genes in various bacteria has also been reported. Expression of the *A. eutrophus* PHA biosynthetic genes in pseudomonads, which are unable to synthesize P(3HB), allowed the production of P(3HB) (90). Production of P(4HB) by a genetically engineered mutant *A. eutrophus* is a good example of producing a novel polyester by metabolic engineering (92). Unusual PHAs can also be produced by heterologous expression of PHA biosynthetic genes. Recombinant *P. putida* harboring the PHA biosynthetic genes of *T. pfennigii* was able to accumulate a ter-polyester consisting of 3-hydroxybutyrate, 3-hydroxyhexanoate, and 3-hydroxyoctanoate when cultivated on octanoate (55). As mentioned earlier, recombinant *E. coli* strains harboring the *A. eutrophus* PHA biosynthesis genes have been developed and were used for the production of a high concentration of P(3HB) with a high PHA content and high productivity (52). Since *E. coli* can utilize sucrose, lactose, and xylose in addition to glucose, the use of cheap carbon substrates such as molasses, whey, and hemicellulose substrates is possible. In another example, a recombinant *Klebsiella aerogenes* harboring the *A. eutrophus* PHA biosynthetic genes was used for the production of P(3HB) from molasses (105). Recently, transgenic plants *Arabidopsis thaliana* harboring the *A. eutrophus* PHA biosynthetic genes have been developed with the aim of ultimately reducing the price of PHA to close to that of starch (69). However, owing to the low accumulation of PHA, much progress is needed for commercialization.

These are some of the good examples of applying metabolic engineering principles to the PHA producers. There is little doubt that this approach will generate a super-strain for the cost-effective production of PHAs.

51.5. RECOVERY OF PHA

A number of methods have been developed for the recovery of PHA. Most of these methods have been developed with *A. eutrophus* producing P(3HB). Biomass containing PHA can be divided into two parts, PHA and non-PHA cellular materials (NPCM). PHA can be recovered by direct extraction of PHA or digestion of NPCM.

51.5.1. Extraction of PHA

The method that is most often used for extraction of PHA is direct extraction from biomass with various solvents (74, 83): chlorinated hydrocarbons (chloroform and 1,2-dichloroethane), azeotrophic mixtures (1,1,2-trichloroethane with water; chloroform with either methanol, ethanol, acetone, or hexane), or cyclic carbonates (ethylene carbonate and 1,2-propylene carbonate).

51.5.1.1. Extraction of P(3HB) with Chloroform

1. Collect bacterial cells by centrifugation and measure the dry cell weight.
2. Remove lipids by adding methanol (ca. 40 times the volume of the collected cells) and incubate at 95°C for 1 h.
3. Recover cells by vacuum filtration. To remove methanol completely, incubate cells in a dry oven for several minutes.
4. Add chloroform (50 times the dry cell weight) and incubate at 95°C for 10 min.
5. Cool to room temperature, and mix solution overnight with stirring.
6. Filter the solution to remove cell debris.
7. Precipitate P(3HB) by adding a mixture of methanol and water (7:3 [vol/vol]) (5 volumes of chloroform) to the filtrate.
8. Filter the precipitated P(3HB), wash with acetone, and dry.

When cyclic carbonates are used as solvents, P(3HB) can be precipitated by simply cooling the solution. However, cyclic carbonates are relatively expensive and have high boiling points, which makes the reuse of these solvents infeasible. Because PHAs consisting of different monomer units have different properties, the specific pairs of extraction solvent and precipitation nonsolvent should be used for efficient recovery. One of the major problems of solvent extraction is that large amounts of solvent and nonsolvent are required for polymer recovery owing to the high viscosity of even dilute PHA solution (5% [wt/vol]). It makes the extraction method economically unattractive, even after the recycling of the solvents. Therefore, the extraction method is useful only for small-scale analytical purposes.

51.5.2. Digestion of NPCM

Owing to the difficulties of solvent extraction in a large-scale recovery process, several methods of employing sodium hypochlorite (3), enzymes (29), and alkaline or acidic solutions (47) for the digestion of NPCM have been developed.

51.5.2.1. P(3HB) Recovery by Hypochlorite Digestion of NPCM

1. Dilute sodium hypochlorite solution with distilled water. The stock solution of sodium hypochlorite contains 5.68 g of Cl, 7.8 g of NaOH, and 32 g of Na_2CO_3 in 100 ml. The optimal hypochlorite concentrations should be determined for each bacterium by experiments.
2. By mixing the P(3HB)-containing biomass with hypochlorite solution, NPCM are efficiently digested. Then separate P(3HB) granules from the aqueous fraction containing cell debris by centrifugation.
3. Rinse the recovered P(3HB) with distilled water, centrifuge again, and finally rinse with acetone.

The hypochlorite digestion method is a simple and rapid method for the recovery of P(3HB). However, severe degradation of P(3HB) was observed during the digestion: P(3HB) having the molecular weight (MW) of 600,000 was recovered after hypochlorite digestion of P(3HB) having the original MW of 1,200,000 (3). To reduce the extent of polymer degradation, hypochlorite digestion combined with surfactant pretreatment was considered (73). In this method, 1% (wt/vol) biomass was mixed with 1% (wt/vol) surfactant solution at 25°C for 15 min. The aqueous portion was then removed by centrifugation at 4,000 × *g* for 15 min. The collected cells were washed twice with distilled water and were treated with hypochlorite solution for 1 min. A high purity of greater than 97% with less degradation of P(3HB) could be achieved. In another study, treatment with the dispersion of chloroform and sodium hypochlorite solution resulted in less degradation of P(3HB) and high polymer purity (23).

The enzymatic digestion method has been used commercially by Zeneca to produce P(3HB-co-3HV). In this method, cells are collected by centrifugation, heat-treated at 80°C for 1 h to denature cellular materials, digested with proteolytic enzymes at 55°C, and finally washed with anionic surfactant to solubilize NPCM (29).

In general, the purity and yield of recovery can be affected by PHA content and the characteristics of the bacteria employed. When the PHA content is high, the recovery process becomes simple and results in high purity and yield. Therefore, it is important to obtain high PHA content by developing a suitable fermentation strategy. It should be mentioned that PHA can be recovered more efficiently from some bacteria than others. Recombinant *E. coli* harboring the *A. eutrophus* PHA biosynthetic genes and *A. vinelandii* can be easily lysed by simple treatment (48, 52, 63). These strains may be useful as PHA producers, since recovery cost can be considerably reduced.

51.6. ANALYSIS OF POLYMER PROPERTIES

The properties of PHA accumulated in bacteria are affected by several factors, such as producing strains, culture conditions, and recovery method. In this section, several analytical procedures for determining polymer properties are presented.

51.6.1. Molecular Weight and Molecular Weight Distribution

P(3HB) is soluble in a number of common solvents, such as chloroform, trifluoroethanol, 1,2-dichloroethane, or dichloroacetic acid. Therefore, the molecular weight can be determined by light scattering, osmometry, sedimentation analysis, intrinsic viscosity measurement, and gel permeation chromatography (GPC) (28, 58).

Mark-Houwink-Sakurada parameters for P(3HB) in various solvents relating intrinsic viscosity to molecular weight can be found in the literature.

$$[\eta] = K(\mathrm{MW})^{\alpha}(\mathrm{dl\ g^{-1}})$$

For P(3HB) in chloroform at 30°C, K and α were found to be 1.66×10^{-4} and 0.76, respectively (28).

Using the universal calibration method, the molecular weight and molecular weight distribution of P(3HB) can be analyzed by GPC. As an example, a specific GPC system equipped with Shodex 80M and K-802 columns and RID-6A refractive index detector is operated at 40°C for the analysis. Chloroform is used as the eluent at a flow rate of 0.8 ml/min, and a sample concentration of 0.25 mg/ml is used. Polystyrene standards with low polydispersity having the number-average molecular weights of 1,200 to 2,900,000 are used to construct a calibration curve. The number-average molecular weight and weight-average molecular weight can be calculated by using a Shimadzu chromatopac C-R7A with a GPC program (42). However, determination of MW by GPC seems to be troublesome owing to the poor correlation between the P(3HB) and polystyrene standard. Therefore, the light scattering method or intrinsic viscosity measurement seems to be a better choice for the determination of the MW of PHA.

51.6.2. Crystallinity and Melting Temperature

The crystal structure of PHA has been studied by X-ray diffraction or thermal analysis by differential scanning calorimetry. X-ray diffraction studies have been carried out for P(3HB-co-3HV) with a Philips PW 1050/81 powder diffractometer controlled by a PW 1710 unit, using nickel-filtered Cu Kα radiation (λ = 0.1542 nm; 40 KV; 30 mA) (76). The lattice constants are calculated from the well-determined positions of 8 to 10 of the most intense reflections, by least-square refinements. The crystal sizes are determined by the Sherner equation from 020 reflection. Crystallinity is determined from diffracted intensity data by comparing the relative area under the crystalline peak with that from amorphous scatter (76).

To investigate the morphological state of p(3HB) granules in vivo, the melting temperature and the enthalpy of fusion of the lyophilized cell powder containing P(3HB) can be measured by differential scanning calorimetry (76). The samples are initially heated from room temperature to 200°C. The melting temperature is determined from the melting endothermy. After 1 min. the samples are rapidly quenched in liquid nitrogen and then heated once again to 200°C. Glass transition temperature can be determined from this second scanning. The crystallinity of P(3HB) is estimated from the enthalpy of fusion.

51.6.3. Nuclear Magnetic Resonance (NMR) Study on PHA

High-resolution ^{13}C NMR spectroscopy of live cells has been used to show the physical state of P(3HB) in vivo (2). Cells are pelleted from media by centrifugation for 15 min at 4,000 × g, 4°C, resuspended in isotonic saline (20% D_2O) for NMR, and placed in a 10-mm-diameter sample tube. NMR spectra are obtained at 100.6 MHz on a Bruker 400-MHz spectrometer. Waltz decoupling is used while temperatures are maintained by the flow of evaporated liquid nitrogen over the sample.

For the characterization of MCL-PHA produced by *P. oleovorans*, ^{1}H NMR spectra can be recorded on a Varian VXR 300 NMR spectrometer operating at 300 MHz. A sample is prepared in chloroform-*d* (20 to 30 mg/ml). The spectra are recorded at ambient temperature (20°C) with 24.0-μs pulse width and 2.0-s relaxation delay. The 75-MHz ^{13}C-NMR spectra are recorded at 20°C on a 20- to 30-mg/ml $CDCl_3$ solution (71).

^{13}C NMR studies can also be used for the investigation of metabolic pathways involved in PHA biosynthesis (31).

51.7. BIODEGRADATION OF PHA

A number of parameters can influence the rate of biodegradation of PHA. These include the type of environment, microbial population, the availability of water, temperature, the shape and thickness of the plastic material, surface texture, porosity and crystallinity, and the presence of other components in the plastic, such as fillers or coloring agents.

Several laboratory methods that determine the degradability of plastic materials are based on direct visual observation of plastic material, quantitative determination of microbial growth on plastic material (measured by turbidity or protein and phospholipid content), polymer utilization by microorganisms (by gravimetric or visual observation), determination of changes in polymer characteristics during incubation (by tensile strength or molecular weight), and determination of microbial activities (titrimetric, electrochemical, or manometric). However, these laboratory test methods are difficult to employ in the natural environment. Therefore, field or in situ experiments in which the samples are placed in the desired environments are required for the determination of biodegradability (5).

PHA is finally degraded into carbon dioxide and water in aerobic conditions and into methane, water, and carbon dioxide in anaerobic conditions. Degradation of P(3HB) seems to occur in two steps. First, P(3HB) is degraded into 3HB oligomers, preferentially 3HB dimers, by the extracellular P(3HB) depolymerase. Second, an oligomer hydrolase cleaves the dimers into 3HB monomers. A number of aerobic and anaerobic PHA-degrading bacteria and fungi have been isolated, and the extracellular PHA depolymerases from several bacteria have been cloned and characterized (48).

51.8. APPLICATION

PHAs have been drawing considerable industrial interest as candidates for biodegradable and/or biocompatible plastics and elastomers for a wide range of applications, including packaging, controlled release, and biomedical materials (43, 48). Early investigations of PHA granules by electron microscopy after freeze-etching showed that the polymer in the granules underwent a cold drawing process indicating the plastic nature of the polyester and suggesting that it could be processed as a conventional thermoplastic (19). In addition to its potential as plastic material, PHA is also a useful source for stereoregular compounds that can serve as chiral precursors for the chemical synthesis of optically active substances, particularly in the synthesis of certain drugs or insect pheromones (80).

It is particularly significant to find that the degradation product of P(3HB), D-(−)-3-hydroxybutyrate, is a common intermediate metabolite present in all higher animals. A low-molecular-weight P(3HB), consisting of 100 to 200 monomer units, has been found in prokaryotic and eukaryotic organisms and seems to function as a component of an ion channel through cell membranes. Recently, it has been detected in relatively large amounts in human blood plasma (75). Therefore, it is highly plausible that implanting P(3HB) in mammalian tissues would not be toxic. Several sample products made of BIOPOL are on the market. These include shampoo bottles, disposable razors, disposable utensils, BIOPOL-coated paper, golf tees, and fishing nets. When the price of PHA is further lowered by the development of an efficient fermentation/recovery process, PHA will replace many petrochemical-derived plastics in the applications that require biodegradability.

REFERENCES

1. **Anderson, A. J., and E. A. Dawes.** 1990. Occurrence, metabolism, metabolic role, and industrial uses of bacterial polyhydroxyalkanoates. *Microbiol. Rev.* **54:**450–472.
2. **Barnard, G. N., and J. K. M. Sanders.** 1989. The poly-β-hydroxybutyrate granule in vivo. *J. Biol. Chem.* **264:** 3286–3291.
3. **Berger, E., B. A. Ramsay, J. A. Ramsay, C. Chavarie, and G. Braunegg.** 1989. PHB recovery by hypochlorite digestion of non-PHB biomass. *Biotechnol. Techniques* **3:** 227–232.
4. **Bonthrone, K. M., J. Clauss, D. M. Horowitz, B. K. Hunter, and J. K. M. Sanders.** 1992. The biological and physical chemistry of polyhydroxyalkanoates as seen by NMR spectroscopy. *FEMS Microbiol. Rev.* **103:**269–278.
5. **Brandl, H., R. Bachofen, J. Mayer, and E. Wintermantel.** 1995. Degradation and applications of polyhydroxyalkanoates. *Can. J. Microbiol.* **41**(Suppl. 1)**:**143–153.
6. **Brandl, H., R. A. Gross, R. W. Lenz, and R. C. Fuller.** 1990. Plastics from bacteria and for bacteria: poly(β-hydroxyalkanoates) as natural, biocompatible, and biodegradable polyesters. *Adv. Biochem. Eng. Biotechnol.* **41:**77–93.
7. **Braunegg, G., B. Sonnleitner, and R. M. Lafferty.** 1978. A rapid gas chromatographic method for the determination of poly-β-hydroxybutyric acid in microbial biomass. *Eur. J. Appl. Microbiol. Biotechnol.* **6:**29–37.
8. **Byrom, D.** 1987. Polymer synthesis by microorganisms: technology and economics. *Trends Biotechnol.* **5:**246–250.
9. **Byrom, D.** 1991. *Biomaterials: Novel Materials from Biological Sources.* Stockton, New York.
10. **Byrom, D.** 1992. Production of poly-β-hydroxybutyrate: poly-β-hydroxyvalerate copolymers. *FEMS Microbiol. Rev.* **103:**247–250.
11. **Byrom, D.** 1994. Polyhydroxyalkanoates, p. 5–33. *In* D. P. Mobley (ed.), *Plastics from Microbes: Microbial Synthesis of Polymers and Polymer Precursors.* Hanser, Munich.
12. **Cox, M. K.** 1994. Properties and applications of polyhydroxyalkanoates, p. 120–135. *In* Y. Doi and K. Fukuda (ed.), *Biodegradable Plastics and Polymers.* Elsevier, Amsterdam.
13. **Dawes, E. A.** 1990. *Novel Biodegradable Microbial Polymers.* Kluwer Academic, Dordrecht, The Netherlands.
14. **Degelau, A., T. Scheper, J. E. Bailey, and C. Guske.** 1995. Fluorometric measurement of poly-β-hydroxybutyrate in *Alcaligenes eutrophus* by flow cytometry and spectrofluorometry. *Appl. Microbiol. Biotechnol.* **42:**653–657.
15. **De Koning, G. J. M., H. M. M. van Bilsen, P. J. Lemstra, W. Hazenberg, B. Witholt, H. Preusting, J. G. van der Galien, A. Schirmer, and D. Jendrossek.** 1994. A biodegradable rubber by crosslinking poly(hydroxyalkanoate) from *Pseudomonas oleovorans. Polymer* **35:**2090–2097.
16. **De Smet, M. J., G. Eggink, B. Witholt, J. Kingma, and H. Wynberg.** 1983. Characterization of intracellular inclusions formed by *Pseudomonas oleovorans* during growth on octane. *J. Bacteriol.* **154:**870–878.
17. **Doi, Y.** 1990. *Microbial Polyesters.* VCH, New York.
18. **Doi, Y., K. Mukai, K. Kasuya, and K. Yamada.** 1994. Biodegradation of biosynthetic and chemosynthetic polyhydroxyalkanoates, p. 39–51. *In* Y. Doi and K. Fukuda (ed.), *Biodegradable Plastics and Polymers.* Elsevier, Amsterdam.
19. **Dunlop, W. F., and A. W. Robards.** 1973. Ultrastructural study of poly-β-hydroxybutyrate granules from *Bacillus cereus. J. Bacteriol.* **114:**1271–1280.
20. **Fidler, S., and D. Dennis.** 1992. Polyhydroxyalkanoate production in recombinant *Escherichia coli. FEMS Microbiol. Rev.* **103:**231–236.
21. **Gerngross, T. U., K. D. Snell, O. P. Peoples, A. J. Sinskey, E. Csuhai, S. Masamune, and J. Stubbe.** 1994. Overexpression and purification of the soluble polyhydroxyalkanoate synthase from *Alcaligenes eutrophus*: evidence for a required posttranslational modification for catalytic activity. *Biochemistry* **33:**9311–9320.
22. **Hahn, S. K., and Y. K. Chang.** 1995. A themogravimetric analysis for poly(3-hydroxybutyrate) quantification. *Biotechnol. Techniques* **9:**873–878.
23. **Hahn, S. K., Y. K. Chang, B. S. Kim, and H. N. Chang.** 1994. Optimization of microbial poly(3-hydroxybutyrate) recovery using dispersions of sodium hypochlorite solution and chloroform. *Biotechnol. Bioeng.* **44:**256–261.
24. **Hangii, U. J.** 1990. Pilot scale production of PHB with *Alcaligenes lattus*, p. 65–70. *In* E. A. Dawes (ed.), *Novel Biodegradable Microbial Polymers.* Kluwer Academic, Dordrecht, The Netherlands.
25. **Haywood, G. W., A. J. Anderson, L. Chu, and E. A. Dawes.** 1988. Characterization of two 3-ketothiolases in the polyhydroxyalkanoate synthesizing organism *Alcaligenes eutrophus. FEMS Microbiol. Lett.* **52:**91–96.
26. **Haywood, G. W., A. J. Anderson, L. Chu, and E. A. Dawes.** 1988. The role of NADH-and NADPH-linked acetoacetyl-CoA reductases in the poly-3-hydroxyalkanoate synthesizing organism *Alcaligenes eutrophus. FEMS Microbiol. Lett.* **52:**259–264.
27. **Haywood, G. W., A. J. Anderson, and E. A. Dawes.** 1989. The importance of PHA-synthase substrate specificity in polyhydroxyalkanoate synthesis by *Alcaligenes eutrophus. FEMS Microbiol. Lett.* **57:**1–6.
28. **Holmes, P. A.** 1988. Biologically produced PHA polymers and copolymers, p. 1–65. *In* D. C. Bassett (ed.), *Developments in Crystalline Polymers,* vol. 2. Elsevier, London.
29. **Holmes, P. A., and G. B. Lim.** March 1990. Separation process. U.S. patent 4,910,145.
30. **Hrabak, O.** 1992. Industrial production of poly-β-hyroxybutyrate. *FEMS Microbiol. Rev.* **103:**251–256.

31. **Huijberts, G. N. M., T. C. De Rijk, P. De Waard, and G. Eggink.** 1994. ^{13}C nuclear magnetic resonance studies of *Pseudomonas putida* fatty acid metabolic routes involved in poly(3-hydroxyalkanoate) synthesis. *J. Bacteriol.* **176:** 1661–1666.

32. **Huijberts, G. N. M., H. van der Wal, C. Wilkinson, and G. Eggink.** 1994. Gas-chromatographic analysis of poly(3-hydroxyalkanoates) in bacteria. *Biotechnol. Techniques* **8:** 187–192.

33. **Huisman, G. W.** 1991. Poly(3-hydroxyalkanoates) from *Pseudomonas putida*: from DNA to plastic. Ph.D. dissertation. Rijksuniversiteit Groningen, Groningen, The Netherlands.

34. **Huisman, G. W., E. Wonink, R. Meima, B. Kazemier, P. Terpstra, and B. Witholt.** 1991. Metabolism of poly(3-hydroxyalkanoates) by *Pseudomonas oleovorans*: identification and sequences of genes and function of the encoded proteins in the synthesis and degradation of PHA. *J. Biol. Chem.* **266:**2191–2198.

35. **Hustede, E., and A. Steinbuchel,** 1993. Characterization of the polyhydroxyalkanoate gene locus of *Rhodobacter sphaeroides*. *Biotechnol. Lett.* **15:**709–714.

36. **Hustede, E., A. Steinbuchel, and H. G. Schlegel.** 1992. Cloning of poly(3-hydroxybutyric acid) synthase genes of *Rhodobacter sphaeroides* and *Rhodospirillum rubrum* and heterologous expression in *Alcaligenes eutrophus*. *FEMS Microbiol. Lett.* **93:**73–80.

37. **Ishizaki, A., and K. Tanaka.** 1991. Production of poly-β-hydroxybutyric acid from carbon dioxide by *Alcaligenes eutrophus* ATCC 17697^{T}. *J. Ferment. Bioeng.* **71:**254–257.

38. **Kim, B. S., S. C. Lee, S. Y. Lee, H. N. Chang, Y. K. Chang, and S. I. Woo.** 1994. Production of poly(3-hydroxybutyric acid) by fed-batch culture of *Alcaligenes eutrophus* with glucose concentration control. *Biotechnol. Bioeng.* **43:**892–898.

39. **Kim, B. S., S. C. Lee, S. Y. Lee, H. N. Chang, Y. K. Chang, and S. I. Woo.** 1994. Production of poly-(3-hydroxybutyric-co-3-hydroxyvaleric acid) by fed-batch culture of *Alcaligenes eutrophus* with substrate feeding using on-line glucose analyzer. *Enzyme Microb. Technol.* **16:** 556–561.

40. **Kim, S. W., P. Kim, H. S. Lee, and J. H. Kim.** 1996. High production of poly-β-hydroxybutyrate (PHB) from *Methylobacterium organophilum* under potassium limitation. *Biotechnol. Lett.* **18:**25–30.

41. **Kitamura, S., and Y. Doi.** 1994. Staining method of poly(3-hydroxyalkanoic acids) producing bacteria by nile blue. *Biotechnol. Techniques* **8:**345–350.

42. **Koizumi, F., H. Abe, and Y. Doi.** 1995. Molecular weight of poly(3-hydroxybutyrate) during biological polymerization in *Alcaligenes eutrophus*. *Pure Appl. Chem.* **A32:**759–774.

43. **Lafferty, R. M., B. Korsatko, and W. Korsatko.** 1988. Microbial production of poly-β-hydroxybutyric acid, p. 135–176. *In* H. J. Rehm and G. Reed (ed.), *Biotechnology*, vol. 6b. Verlagsgesellschaft, Weinheim, Germany.

44. **Law, J. H., and R. A. Slepecky.** 1961. Assay of poly-β-hydroxybutyric acid. *J. Bacteriol.* **82:**33–36.

45. **Leaversuch, R.** 1987. Industry weighs the need to make polymer degradable. *Mod. Plastics* **64:**52–55.

46. **Lee, I., S. Nam, Y. Rhee, and J. Kim.** 1996. Cloning and functional expression in *Escherichia coli* of polyhydroxyalkanoate synthase (*phaC*) gene from *Alcaligenes* sp. SH-69. *J. Microbiol. Biotechnol.* **6:**309–314.

47. **Lee, I. Y., H. N. Chang, and Y. H. Park.** 1995. A simple method for recovery of microbial poly-β-hydroxybutyrate by alkaline solution treatment. *J. Microbiol. Biotechnol.* **5:** 238–240.

48. **Lee, S. Y.** 1996. Bacterial polyhydroxyalkanoates. *Biotechnol. Bioeng.* **49:**1–14.

49. **Lee, S. Y.** 1996. Plastic bacteria? Progress and prospects for polyhydroxyalkanoate production in bacteria. *Trends Biotechnol.* **14:**431–438.

50. **Lee, S. Y., and H. N. Chang.** 1994. Effect of complex nitrogen source on the synthesis and accumulation of poly(3-hydroxybutyric acid) by recombinant *Escherichia coli* in flask and fed-batch cultures. *J. Environ. Polymer Degrad.* **2:**169–176.

51. **Lee, S. Y., and H. N. Chang.** 1995. Production of poly-(hydroxyalkanoic acid). *Adv. Biochem. Eng. Biotechnol.* **52:**27–58.

52. **Lee, S. Y., and H. N. Chang.** 1995. Production of poly(3-hydroxybutyric acid) by recombinant *Escherichia coli* strains: genetic and fermentation studies. *Can. J. Microbiol.* **41**(Suppl. 1)**:**207–215.

53. **Lee, S. Y., K. S. Yim, H. N. Chang, and Y. K. Chang.** 1994. Construction of plasmids, estimation of plasmid stability, and use of stable plasmids for the production of poly(3-hydroxybutyric acid) in *Escherichia coli*. *J. Biotechnol.* **32:**203–211.

54. **Lemoigne, M.** 1926. Products of dehydration and of polymerization of β-hydroxybutyric acid. *Bull. Soc. Chem. Biol.* **8:**770–782.

55. **Liebergesell, M., F. Mayer, and A. Steinbuchel.** 1993. Analysis of polyhydroxyalkanoic acid-biosynthesis genes of anoxygenic phototrophic bacteria reveals synthesis of a polyester exhibiting an unusual composition. *Appl. Microbiol. Biotechnol.* **40:**292–300.

56. **Liebergesell, M., and A. Steinbuchel.** 1992. Cloning and nucleotide sequences of genes relevant for biosynthesis of poly(3-hydroxybutyric acid) in *Chromatium vinosum* strain D. *Eur. J. Biochem.* **209:**135–150.

57. **Liebergesell, M., and A. Steinbuchel.** 1993. Cloning and molecular analysis of the poly(3-hydroxybutyric acid) biosynthesis genes of *Thiocystis violacea*. *Appl. Microbiol. Biotechnol.* **38:**493–501.

58. **Marchessault, R. H., K. Okamura, and C. J. Su.** 1970. Physical properties of poly(β-hydroxybutyrate). II. Conformational aspects in solution. *Macromolecules* **3:**735–740.

59. **McInerney, M. J., D. A. Amos, K. S. Kealy, and J. Palmer.** 1992. Synthesis and function of polyhydroxyalkanoates in anaerobic syntropic bacteria. *FEMS Microbiol. Rev.* **103:**195–206.

60. **Mergaert, J., C. Anderson, A. Wouters, J. Swings, and K. Kersters.** 1992. Biodegradation of polyhydroxyalkanoates. *FEMS Microbiol. Rev.* **103:**317–322.

61. **Nishimura, T., T. Saito, and K. Tomita.** 1978. Purification and properties of β-ketothiolase from *Zoogloea ramigera*. *Arch. Microbiol.* **116:**21–27.

62. **Ostle, A. G., and J. G. Holt.** 1982. Nile Blue A as a fluorescent stain for poly-β-hydroxybutyrate. *Appl. Environ. Microbiol.* **44:**238–241.

63. **Page, W. J., and A. Comish.** 1993. Growth of *Azotobacter vinelandii* UWD in fish peptone medium and simplified extraction of poly-β-hydroxybutyrate. *Appl. Environ. Microbiol.* **59:**4236–4244.

64. **Park, C. H., and V. K. Damodaran.** 1994. Biosynthesis of poly(3-hydroxybutyrate-co-3-hydroxyvalerate) from ethanol and pentanol by *Alcaligenes eutrophus*. *Biotechnol. Prog.* **10:**615–620.

65. **Park, J., H. Park, T. Huh, and Y. Lee.** 1995. Production of poly-β-hydroxybutyrate by *Alcaligenes eutrophus* transformants harboring cloned *phbCAB* genes. *Biotechnol. Lett.* **17:**735–740.

66. **Peoples, O. P., and A. J. Sinskey.** 1989. Poly-β-hydroxybutyrate biosynthesis in *Alcaligenes eutrophus* H16. Identification and characterization of the PHB polymerase gene (*phbC*). *J. Biol. Chem.* **264:**15298–15303.

67. **Pieper, U., and A. Steinbuchel.** 1992. Identification, cloning and molecular characterization of the poly(3-hydroxyaikanoic acid) synthase structural gene of the Gram-positive *Rhodococcus ruber*. *FEMS Microbiol. Lett.* **96:**73–80.

68. **Ploux, O., S. Masamune, and C. T. Walsh.** 1988. The NADPH-linked acetoacetyl-CoA reductase from *Zoogloea ramigera*. *Eur. J. Biochem.* **174:**177–182.

69. **Poirier, Y., C. Nawrath, and C. Somerville.** 1995. Production of polyhydroxyalkanoates, a family of biodegradable plastics and elastomers, in bacteria and plants. *Bio/Technology* **13:**142–150.

70. **Preusting, H., W. Hazenberg, and B. Witholt.** 1993. Continuous production of poly(3-hydroxyalkanoates) by *Pseudomonas oleovorans* in a high-cell-density, two-liquid-phase chemostat. *Enzyme Microb. Technol.* **15:**311–316.

71. **Preusting, H., A. Nijenhuis, and B. Witholt.** 1990. Physical characteristics of poly(3-hydroxyalkanoates) and poly(3-hydroxyalkenoates) produced by *Pseudomonas oleovorans* grown on aliphatic hydrocarbons. *Macromolecules* **23:**4220–4224.

72. **Preusting, H., R. van Houten, A. Hoefs, E. K. van Langenberghe, O. Favre-Bulle, and B. Witholt.** 1993. High cell density cultivation of *Pseudomonas oleovorans*: growth and production of poly(3-hydroxyalkanoates) in two-liquid phase batch and fed-batch systems. *Biotechnol. Bioeng.* **41:**550–556.

73. **Ramsay, J. A., E. Berger, B. A. Ramsay, and C. Chavarie.** 1990. Recovery of poly-β-hydroxybutyric acid granules by a surfactant-hypochlorite treatment. *Biotechnol. Techniques* **4:**221–226.

74. **Ramsay, J. A., E. Berger, R. Voyer, C. Chavarie, and B. A. Ramsay.** 1994. Extraction of poly-3-hyroxybutyrate using chlorinated solvents. *Biotechnol. Techniques* **8:**589–594.

75. **Reusch, R. N., A. W. Sparow, and J. Gardiner.** 1992. Transport of poly-β-hydroxybutyrate in human plasma. *Biochim. Biophys. Acta* **1123:**33–40.

76. **Scandola, M., G. Ceccorulli, M. Pizzoli, and M. Gazzano.** 1992. Study of the crystal phase and crystallization rate of bacterial poly (3-hydroxybutyrate-co-3-hydroxyvalerate). *Macromolecules* **25:**1405–1410.

77. **Schembri, M. A., R. C. Bayly, and J. K. Davies.** 1994. Cloning and analysis of the polyhyroxyalkanoic acid synthase gene from an *Acinetobacter* sp.: evidence that the gene is both plasmid and chromosomally located. *FEMS Microbiol. Lett.* **118:**145–152.

78. **Schlegel, H. G., R. Lafferty, and I. Krauss.** 1970. The isolation of mutants not accumulating poly-β-hydroxybutyric acid. *Arch. Microbiol.* **71:**283–294.

79. **Schubert, P., A. Steinbuchel, and H. G. Schlegel.** 1988. Cloning of the *Alcaligenes eutrophus* genes for synthesis of poly-β-hydroxybutyric acid (PHB) and synthesis of PHB in *Escherichia coli*. *J. Bacteriol.* **170:**5837–5847.

80. **Seebach, von D., and M. Zuger.** 1982. Uber die depolymerisierung von poly-(R)-3-hydroxy-buttersaureester (PHB). *Helv. Chim. Acta* **65:**495–503.

81. **Slater, S., T. Gallaher, and D. Dennis.** 1992. Production of poly-(3-hydroxybutyrate-co-3-hydroxyvalerate) in a recombinant *Escherichia coli* strain. *Appl. Environ. Microbiol.* **58:**1089–1094.

82. **Slater, S. C., W. H. Voige, and D. E. Dennis.** 1988. Cloning and expression in *Escherichia coli* of the *Alcaligenes eutrophus* H16 poly-β-hydroxybutyrate biosynthetic pathway. *J. Bacteriol.* **170:**4431–4436.

83. **Stageman, J. F.** 1984. Eur. patent 0,124,309.

84. **Steinbuchel, A.** 1991. Polyhydroxyalkanoic acids, p. 124–213. *In* D. Byrom (ed.), *Biomaterials: Novel Materials from Biological Sources*. Stockton, New York.

85. **Steinbuchel, A.** 1991. Recent advances in the knowledge of the metabolism of bacterial polyhydroxyalkanoic acids and potential impacts on the production of biodegradable thermoplastics. *Acta Biotechnol.* **11:**419–427.

86. **Steinbuchel, A.** 1992. Biodegradable plastics. *Curr. Opin. Biotechnol.* **3:**291–297.

87. **Steinbuchel, A., K. Aerts, W. Babel, C. Follner, M. Liebergesell, M. H. Madkour, F. Mayer, U. P. Furst, A. Pries, H. E. Valentin, and R. Wieczorek.** 1995. Considerations on the structure and biochemistry of bacterial polyhydroxyalkanoic acid inclusion. *Can. J. Microbiol.* **41**(Suppl. 1)**:**94–105.

88. **Steinbuchel, A., E. Hustede, M. Liebergesell, U. Pieper, A. Timm, and H. Valentin.** 1992. Molecular basis for biosynthesis and accumulation of polyhydroxyalkanoic acids in bacteria. *FEMS Microbiol. Rev.* **103:**217–230.

89. **Steinbuchel, A., and G. Schmack.** 1995. Large-scale production of poly(3-hydroxyvaleric acid) by fermentation of *Chromobacterium violaceum*, processing, and characterization of the homopolyester. *J. Environ. Polymer Degrad.* **3:** 243–258.

90. **Steinbuchel, A., and P. Schubert.** 1989. Expression of the *Alcaligenes eutrophus* poly(β-hydroxybutyric acid)-synthetic pathway in *Pseudomonas* sp. *Arch. Microbiol.* **153:**101–104.

91. **Steinbuchel, A., and H. E. Valentin.** 1995. Diversity of bacterial polyhydroxyalkanoic acid. *FEMS Microbiol. Lett.* **128:**219–228.

92. **Steinbuchel, A., H. E. Valentin, and A. Schonebaum.** 1994. Application of recombinant gene technology for production of polyhyroxyalkanoic acids: biosynthesis of poly(4-hydroxybutyric acid) homopolyester. *J. Environ. Polymer Degrad.* **2:**67–74.

93. **Sulo, P., D. Hudecova, A. Propperova, and I. Basnak.** 1996. Rapid and simple analysis of poly-β-hydroxybutyrate content by capillary isotachophoresis. *Biotechnol. Techniques* **10:**413–418.

94. **Suzuki, T., T. Yamane, and S. Shimizu.** 1986. Mass production of poly-β-hydroxybutyric acid by fed-batch cul-

ture with controlled carbon/nitrogen feeding. *Appl. Microbiol. Biotechnol.* **24:**370–374.

95. **Swift, G.** 1993. Directions for environmentally biodegradable polymer research. *Acc. Chem. Res.* **26:**105–110.

96. **Timm, A., and A. Steinbuchel.** 1992. Cloning and molecular analysis of the polyhydroxyalkanoic acid gene locus of *Pseudomonas aeruginosa* PAO1. *Eur. J. Biochem.* **209:**15–30.

97. **Timm, A., S. Wiese, and A. Steinbuchel,** 1994. A general identification of polyhydroxyalkanoic acid synthase genes from pseudomonads belonging to the rRNA homology group I. *Appl. Microbiol. Biotechnol.* **40:**669–675.

98. **Tombolini, R., S. Povolo, A. Buson, G. Laterza, A. Morea, A. Squartini, S. Casella, and M. P. Nuti.** 1994. Insertional inactivation of the PHA-synthase gene in *Rhizobium meliloti* leads to strains unable to synthesize poly-β-hydroxybutyrate (PHB). *In* Abstracts of the 4th International Symposium on Bacterial Polydroxyalkanoates. Montreal, Quebec, August 14–18.

99. **Ueda, S., T. Yabutani, A. Maehara, and T. Yamane.** 1996. Molecular analysis of poly(3-hydroxyalkanoate) synthesis gene from a methylotrophic bacterium, *Paracoccus denitrificans. J. Bacteriol.* **178:**774–779.

100. **Valentin, H. E., and A. Steinbuchel.** 1993. Cloning of the *Methylobacterium extorquens* polyhydroxyalkanoic acid synthase structural gene. *Appl. Microbiol. Biotechnol.* **39:**309–317.

101. **Wang, F., and S. Y. Lee.** 1997. Poly(3-hydroxybutyrate) production with high productivity and high polymer content by a fed-batch culture of *Alcaligenes latus* under nitrogen limitation. *Appl. Environ. Microbiol.* **63:**3703–3706.

102. **Wang, F., and S. Y. Lee.** 1997. Production of poly(3-hydroxybutyrate) by fed-batch culture of filamentation-suppressed recombinant *Escherichia coli. Appl. Environ. Microbiol.* **63:**4765–4769.

103. **Yamane, T., M. Fukunage, and Y. W. Lee.** 1996. Increased PHB productivity by high-cell-density fed-batch culture of *Alcaligenes latus*, a growth-associated PHB producer. *Biotechnol. Bioeng.* **50:**197–202.

104. **Yim, K. S., S. Y. Lee, and H. N. Chang.** 1996. Synthesis of poly(3-hydroxybutyrate-co-3-hydroxyvalerate) by recombinant *Escherichia coli. Biotechnol. Bioeng.* **49:**495–503.

105. **Zhang, H., V. Obias, K. Gonyer, and D. Dennis.** 1994. Production of polyhydroxyalkanoates in sucrose-utilizing recombinant *Escherichia coli* and *Klebsiella* strains. *Appl. Environ. Microbiol.* **60:**1198–1205.

ENVIRONMENTAL BIOTECHNOLOGY

VI

ENVIRONMENTAL BIOTECHNOLOGY EMPLOYS A DIVERSE SET of methodological approaches to explore and exploit the natural biodiversity of microorganisms and their enormous metabolic capacities. This field includes the application of microorganisms for improvement of environmental quality, the discovery of microorganisms with metabolic potentials that can be employed for industrial applications, and the use of molecular methods for assessing the natural distributions of microorganisms in the environment and the ecological functions they perform. The characteristics that distinguish environmental biotechnology from other fields of biotechnology are the necessity of achieving microbial functions in complex environments that are not subject to the precise experimental control that can be achieved in bioreactors and the examination of individual microorganisms and their functions in complex diverse microbial communities.

Bioremediation, which exploits the metabolic capacities of microorganisms to remove pollutants from the environment, is among the most heralded applications of environmental biotechnology. It has successfully been applied to major oil spills and numerous contaminated sites, thereby reducing the environmental damage caused by those pollutants and decreasing the risks they pose to human health. Most applications of bioremediation have been based on biostimulation, relying on the natural activities of indigenous microorganisms and using environmental modifications, such as addition of mineral nutrients, to stimulate the rates at which microorganisms metabolize pollutants. Such approaches are particularly effective in the remediation of petroleum hydrocarbon-polluted environments. Additionally, some applications of bioremediation of contaminated soils and waters involve bioaugmentation, in which cultures of microorganisms with specific pollutant-degrading capabilities are added to a polluted site.

Compared to the bioremediation of petroleum hydrocarbons, greater complexities occur with regard to metal and chlorinated xenobiotic-contaminated environments. Many chlorinated xenobiotic compounds are resistant to microbial attack and persist in the environment. In some cases cometabolism can be employed for the remediation of environments polluted with such compounds, as this approach enables microorganisms growing on one growth substrate to gratuitously biotransform a recalcitrant pollutant. Recombinant DNA methodologies can also be used to genetically engineer microorganisms with the capacity to degrade many of these compounds, and many research efforts have been examining this methodological approach.

Concern about the environmental impact of genetically engineered microorganisms has greatly constrained the possibility of deliberately releasing recombinant microorganisms for environmental remediation. Using such recombinant microorganisms may be possible within contained bioreactors, but their broader environmental applications will depend upon new understanding of ecological functions and risk assessments related to populations of introduced organisms.

As greater emphasis is being placed on pollution prevention than on remediation, microorganisms are also being considered for their potential uses as biocatalysts. Biotechnology can play an important role because of its environmental advantages and its economic competitiveness in a number of industrial sectors. At an equivalent level of production, biocatalysts can reduce materials and energy consumption, as well as pollution and waste. A wide range of "environmentally friendly chemicals" can be made from biomass, including biodegradable plastics and other novel biopolymers that do not accumulate and cause environmental damage. Biotechnology may also be used to produce cleaner fuels (for example, by selectively removing sulfur from diesel fuel and gasoline) that reduce the release of atmospheric pollutants. Alternative biomass-based fuels, such as bioethanol, also can reduce the buildup of atmospheric carbon dioxide, thereby reducing global warming.

Biomining and biological control are additional important commercial applications of environmental biotechnology. Biomining is used for the recovery of copper and uranium; it is based upon the oxidation of metal sulfides so as to increase the solubility of the metal and its recovery by leaching. In biological control, microorganisms are used as pesticides and herbicides. The effectiveness of microbial pesticides and herbicides is based on the natural antagonism or pathogenicity of specific microorganisms toward particular plants and animals. Biological control is considered an environmentally more friendly method for controlling weeds, insects, and other pests populations than are applications of chemicals for controlling these unwanted nuisance populations.

Besides the applications of microorganisms to reduce pollution and to support recovery of materials from the environment, environmental biotechnology is playing an important role in exploring natural ecological function. In many instances molecular approaches are providing new avenues for understanding the roles of microorganisms within soils and waters that contribute to the maintenance of environmental quality and ecological productivity. Modern molecular techniques are facilitating studies on complex natural communities. Analyses of microbial communities often depend upon nucleic acid amplification techniques such as the PCR to make rare populations detectable, gene probes and hybridization to detect specific genes and diagnostic nucleic acid sequences, and additional methods such as restriction digestion and electrophoretic separation to detect polymorphisms and specific strain characteristics.

Molecular methods are especially useful in assaying the diversity of complex microbial communities in their natural habitats, such as within biofilms. By revealing the natural functions of microorganisms in the environment, it is possible to control some of those microbial activities, for example, to limit biocorrosion. It is also possible to carry out biosprospecting for microorganisms with specific metabolic functions that are useful for industrial processes or that produce natural products that have commercial value, for example, as medicinals. In this manner the exploration of the natural microbial world can be exploited by environmental biotechnology.

Bioprospecting

JOY E. M. WATTS, ANNALIESA S. HUDDLESTON-ANDERSON, AND ELIZABETH M. H. WELLINGTON

52

The application of molecular techniques to the study of microorganisms in situ within natural environments has greatly improved our understanding of the distribution and diversity of microbial populations. However, we are still far from achieving complete integration of data derived from functional analysis with surveys of taxonomic diversity. Some examples can be found of studies attempting to unravel the interrelationship between functional diversity, sample activity, and taxonomic diversity: ammonia-oxidizers (97), 2,4-dichlorophenoxyacetic acid degraders (49), and streptomycin producers (39, 40). It is still generally the case that the commercial exploitation of microbial activities is dependent on the isolation of an individual strain with the desired function from an environmental sample.

There is now a considerable body of evidence that indicates that less than 1% of bacteria in the environment have been cultured or are represented in culture collections (11, 21, 75, 103, 104, 112). This discrepancy between diversity analyzed in situ and in vitro with isolates may be due to very inadequate selective isolation techniques coupled with a lack of understanding of the physiological status of the cells in situ. Whatever the reasons for the majority of bacteria remaining uncultivated, they represent a rich source of diversity that has yet to be screened for useful activities (11). Even less is known about the proportion of eukaryotic microorganisms that are uncultivated. The molecular phylogenetic approach to studying diversity using small-subunit (SSU) rRNA sequences has radically changed our ability to define microbial diversity (41). Phylogenetic analysis of sequence data has indicated that for the amitochondriate organism line of descent, represented by pathogens such as *Giardia*, the lack of nonpathogenic representatives may indicate an important group in the environment as yet undetected (76). It is now possible to characterize microbial communities without any cultivation steps using SSU rRNA (41), but relating this data to functional diversity is still a major problem.

The sequence-based methods have provided an ideal opportunity to bioprospect for microorganisms in diverse environmental samples. Consideration must be given to the Convention on Biological Diversity (27), which is an international treaty signed by member states of the United Nations. It recognizes the sovereign rights of states over their natural resources, and many countries have now put in place legislation to protect these rights and limit access for biodiversity prospecting (43). The objective is to allow developing countries in particular to exploit biological resources such that mutually beneficial partnerships can be developed with industrialized countries. Ideally, exploitation should lead to protection of these resources.

Bioprospecting for microorganisms has been practiced since microbes were first exploited over a thousand years ago. However, we now have the potential to exploit directly the gene pool in a sample without the need to culture. The first fully functional gene to be isolated from soil was a polyketide synthase involved in the production of polyketide antibiotic biosynthesis (91). Owing to the clustering of antibiotic biosynthesis genes and conservation of sequences across different genetic backgrounds, it was possible to amplify an entire gene using degenerate PCR primers targeting sequences in adjacent genes. The potential for exploitation of the microbial gene pool is enormous.

In this review of methods for accessing microbial diversity, we have included approaches for the extraction of both DNA and cells from environmental samples. The recovery of biomass from samples without any cultivation steps provides an opportunity for extraction of larger DNA fragments of greater purity. In addition, the physiological properties of the cells can be studied. Selective cell capture techniques are also reviewed, as they provide more powerful selectivity for cell isolation. Finally, methods for the analysis of the sample activity are included, as these will be useful for screening samples prior to DNA extraction and construction of gene banks.

52.1. SAMPLING

52.1.1. Sample Sites

As new molecular techniques allow a greater examination of the distribution and diversity of microorganisms in soil, there is increasing evidence that a directed approach to sample choice may be more successful in detecting microorganisms with the activity of interest. This directed approach has been taken for the discovery of thermostable enzymes; thermophilic environments ranging from hydrothermal vents, hot springs, and compost have successfully yielded useful organisms (120). These enzymes have improved catalytic activity and stability at very high temperatures, which has proved useful for certain industrial applications. Microorganisms isolated for their antibiotic-producing potential have traditionally been isolated from terrestrial environments. However, the selection of diverse habitats, including marine environments, may yield more novel metabolites (47).

52.1.2. Sampling Procedure

Owing to the heterogeneous nature of many natural ecosystems, the sample size and number of replicates taken must be considered to ensure that the maximum diversity is obtained. The statistics of sampling are reviewed by McSpadden-Gardener and Lilley (64). It is generally accepted that the four replicates of between 50 and 100 g should be taken at each site; these can either be investigated independently or pooled and mixed (71). Once samples are taken, subsamples are then used for isolations. For investigations in which DNA is extracted directly from the sample, sample sizes have been used that range from 100 mg (81) to 100 g (95). Similar diversity was detected from soil samples that ranged from 1 g to 10 g (40), although the same concentration of the DNA was used in each method of analysis (e.g., PCR, blotting) so in theory the actual sample size is not relevant. It is important to maximize recovery of DNA from the sample (see section 52.4.2).

52.1.2.1. Terrestrial Samples

Agricultural soils usually exhibit vertical gradients in texture, structure, and biota owing to the addition of nutrients and oxygen at the surface (107a). These gradients are important when sampling, as usually the first 10 to 25 cm of the soil profile is removed for samples, but if anaerobic microorganisms are required, then the samples must be taken much lower in the soil profile, often by using specialized equipment and techniques.

Sampling of bulk soil is achieved by selecting sites that are at least 20 cm away from plant tubers and plant stems. Rhizosphere sampling is done by removing as much root material as possible from the sampling site. Loose soil is removed by shaking; the rhizosphere soil that adheres to the root surface is removed by washing the roots with sterile dispersing solution.

Once samples have been collected, they can either be air dried prior to analysis and stored at room temperature (71) or stored at −20°C. We recommend the latter method as it ensures that maximum diversity within the sample is maintained, although some cells will be lost. If the sample is frozen, any freeze-thaw cycles should be avoided until the sample is processed, as this may cause cell lysis; 10% glycerol can be added to the soil prior to freezing to protect non-spore-forming bacteria (107). If the sample cannot be frozen, then processing should be carried out as quickly as possible, as storage over 3 months has been shown to affect the proportions and numbers of microorganisms present in soil samples (98).

52.1.2.2. Aquatic Samples

The sample size used from aqueous samples is also quite variable and is dependent on the cell capture technique used. For both cell capture and DNA extraction, the biomass is concentrated by filtration (79) prior to cultivation or cell lysis. Different filtration techniques can be applied, the simplest of which is membrane filtration, in which samples of between 10 and 20 liters of water can be processed (22). This method has approximately 90% efficiency for cell capture, which can be reduced by particulate matter blocking the filter. For large volumes, tangential flow filtration should be used, enabling up to 8,000 liters of water to be processed (26).

52.2. SAMPLE PREPARATION

Microorganisms present in soils or sediments may be strongly attached to soil particles or trapped inside aggregate structures and pores. Extensive pretreatment of the samples is required before either cell/DNA extraction or immunomagnetic capture can be achieved. There are two main types of soil dispersion—physical and chemical—and it will depend on the environmental matrix and the microorganism of interest which method is appropriate.

52.2.1. Physical Dispersion

The physical treatments fall into three categories: shaking, homogenization (blending) and ultrasonication. Homogenization is most often used for pretreatment of soil samples, either by use of a Waring blender or stomacher. Shaking the sample material is probably the least effective for dispersion but is suitable for highly sensitive cells and bacteriophage; it is often used with dilution plate isolations, in which the sample is diluted to allow enumeration as a viable count. Ultrasonic treatment is the most disruptive physical force for soil samples and may be a useful pretreatment for heavy clay soil (74). Various studies have shown that ultrasonication is likely to cause more extensive lysis than other dispersion methods; this can be avoided by the use of mild ultrasonic treatment (38).

52.2.2. Chemical Dispersion

If microbial cells can be desorbed from soil particles by chemical means, then a gentle shaking regimen will suffice for efficient extraction. Chemical dispersion can be achieved by the use of chelating agents, as this reduces the electrostatic attraction between soil particles and cells by exchanging polyvalent cations surrounding clay particles for monovalent cations. The ion-exchange resin Chelex 100 has been widely used for the efficient extraction of biomass, but other dispersants used include Tris buffer and sodium hexametaphosphate (116).

Soil treated with an ion-exchange resin may still leave microorganisms attached to soil particles by means of polymeric gums, thus making inorganic extractants redundant. These extracellular polymers, which bind cells to soil surfaces and to one another, are a complex mixture of polysaccharides of plant and microbial origin. Detergents can be added whose action will be to dissolve extracellular hydrophobic material such as gums (63). Herron and Wellington (33) combined chelex extraction with detergents and filtration to concentrate and extract streptomycete spores from soil.

52.3. CELL EXTRACTION

After the environmental matrix has been dispersed, a number of techniques can be used to remove the biomass fraction. A major advantage of biomass extraction is that the microbial population is concentrated; this can lower detection limits for specific cells in the environment (60, 61). A common method for the removal of environmental contaminants and concentration of cells is the combination of low- and high-speed centrifugation steps. This method can be manipulated to yield different components of the microbial biomass (33, 107). Further purification of the biomass fraction can be achieved by density gradient centrifugation using Percoll (2) and Nycodenz (3), yielding a very pure bacterial fraction.

Elutriation may also be used to separate cells from a dispersed environmental matrix; suspended particles sediment at different velocities determined by their size and density. This technique can be applied to soils after chemical dispersion (63); this allows the elimination centrifugation steps and the accompanying loss of cells. When dealing with large sample sizes, some form of filtration would be required to harvest cells to overcome practical problems, but this may result in cell damage (37).

To avoid sedimentation of the bacterial fraction, flow cytometry may be used (84). This can separate mixtures of particles by size and light scattering as the specimen is passed through an aperture and presented to a focused light beam. This method has not yet been applied to a wide range of terrestrial samples, as it requires highly dispersed extracts; however, it can sort and collect cells on the basis of viability and immunofluorescence of selected target groups using fluorescent dyes.

52.3.1. Immunomagnetic Capture

Recovery of a specific group of bacteria from environmental samples can be achieved by using immunomagnetic capture (IMC) techniques. Microscopic magnetic beads coated with monoclonal or polyclonal antibodies specific for the taxon of interest can be used to capture and purify cells. Specific antibodies can be coated directly onto the beads or added to the sample. It is preferable and more convenient to use commercially prepared antibody-coated beads. Indirect IMC is accomplished in two ways for environmental samples: first, the antibacterial antibodies are added directly to the sample to react with the species of interest. In this case, blocking agents must first be added to the sample to avoid nonspecific binding; gelatin has proved suitable for this (69). Then the beads coated with the secondary antibody (sheep anti-rabbit, for example) are added. After the formation of a primary/secondary antibody complex, the target cells are removed from the sample by using magnetic or electromagnetic fields (88). The second indirect method involves beads coated with primary and secondary antibody complex. They are added directly to the sample, and once the primary antibody has reacted with the microorganism of interest, it can be captured by a magnet. Indirect methods allow greater sensitivity, as steric interference from the close proximity of the magnetic bead to the primary antibody is reduced. The choice of method is dependent on the type of environmental sample and the microorganism of interest.

52.3.1.1. Selective Capture

Traditionally, IMC techniques have been used for the detection of medically important bacterial species in food (5, 93) and water (42, 66). Selective capture using polyclonal or monoclonal antibodies can also be applied to the problem of recovering cells from environmental samples. Morgan et al. (67) used IMC techniques to isolate *Pseudomonas putida* directly from lake water. Bacterial flagella were used as target antigens for cell capture. The antibody was added directly to the lake water sample, and the cells that were attached could be removed by magnetic particle concentrator. The recovery rate was determined as 20% at concentrations ranging from 10^2 to 10^5 CFU ml^{-1}.

IMC methods allow the detection of specific bacteria when traditional isolation techniques have not been successful, because colony formation on selective media is not required. *Thermodesulfobactetium mobile* (15), a thermophilic sulfate-reducing bacterium that was thought to be present in oil field waters, was only detected by capture techniques. Antibodies were raised against the cell wall lipopolysaccharides; these were then attached to magnetic particles and added to the sample. The main type of cells were serologically and morphologically similar to *T. mobile*, indicating the presence of this microorganism when culturing methods failed to detect it.

Indirect IMC can also be used for environmental samples and may be more successful in soil systems than the direct method. The indirect method was used to recover *Streptosporangium fragile* from soil (69). Antibodies were raised to *S. fragile* spores in rabbits and then used for the indirect capture method by addition to a soil sample following chemical dispersion, using the method of Herron and Wellington (33). After the antibodies had bound to the target spores, this complex was recovered using sheep anti-rabbit antibody-coated Dynabeads, removed by using a magnetic particle concentrator. This allowed the capture of up to 83% of the spore population in the soil when used in conjunction with soil dispersal steps.

The direct method was used for recovery of streptomycete spores from soil, using beads coated with a specific monoclonal antibody (116). The inclusion of blocking agents was necessary to avoid nonspecific binding of the beads to soil.

52.3.1.2. Nonselective Cell Recovery

IMC methods can be used to isolate microbial groups from the environment. Polyclonal antibodies that have reduced specificity can be used (31). Polyclonal antibodies were used in conjunction with indirect IMC to recover other actinomycetes groups in addition to the target group, *S. fragile*, from soil (69).

Uncultured bacteria cannot be used to raise antibodies, so other capture techniques with greatly reduced specificity are required. There are nonspecific binding agents that will recover a bacterial fraction from complex samples such as soil. An example of such a binding agent is the plant lectin group of carbohydrate-binding proteins (77). There are now many types of commercially available lectins, which are able to bind to specific sugar residues on cell walls. Porter et al. (84) have used lectin-coated (concanavalin A) magnetic beads to recover uncultured viable bacterial cells from soil. Other sources of less specific binding molecules include agglutinins from the haemolymph of invertebrates such as earthworms (80). Also, protein A and protein G are cell wall proteins from staphylococci and streptococci, respectively, and can bind to a broad range of antigens (114). The less specific binding agents provide a tool to concentrate cells from different environments to allow examination of a clean microbial fraction without cultivation.

52.3.2. Recovery of Cells

Efficient cell recovery is dependent on the sample being fully dispersed (see section 52.2), but other factors are important in environmental sample analysis. The antibody must be strongly bound to the magnetic beads to withstand the shear forces produced by washing the environmental sample (67). The amount of antibody present on the magnetic beads is a critical factor, as good levels of binding are dependent on the antibodies being spread evenly across the bead. When excessive levels of antibody are attached to the bead, binding is inhibited by interference from other antibody molecules. Another important factor when using IMC methods in environmental samples is the number of beads added to the sample (69).

After use of IMC techniques for microorganism capture, the cell fraction is relatively free from environmental contamination, facilitating further molecular analysis, such as DNA extraction from cells and recovery of large fragments of DNA. In addition, the method is ideal for isolation of slow-growing fastidious bacteria, which are often outcompeted on isolation plates.

There are problems associated with using IMC methods as bioprospecting tools; certain groups of uncultured bacteria may have completely different cell wall structures that even the most broad specificity molecule may not recognize. It may be possible to raise antibodies without cultivation if cells can be extracted and concentrated from a sample (see section 52.3.1.2).

52.4. CAPTURE OF GENES FROM ENVIRONMENTAL SAMPLES

Molecular methods can be used to investigate taxonomic and functional diversity of specific natural communities and to screen for novel genes that can be used in pathway engineering (91). These methods can also be applied to an environmental sample as a primary screen to assess the potential diversity of the sample. The PCR products can be subjected to diversity profiles using techniques such as DGGE (denaturing gradient gel electrophoresis) (70), RFLP (restriction fragment length polymorphism) (55), ARDRA (amplified ribosomal DNA restriction analysis) (68), and TRFLP (terminal restriction fragment length polymorphism) (62).

New methods are being developed to survey the diversity of environmental DNA that do not involve PCR. Guschin et al. (30) used oligonucleotide microchips as "geosensors" for the detection of specific microorganisms in environmental samples. This technique enables multiple sequences to be placed on a single chip so that the sample can be simultaneously analyzed for the presence of many diverse target groups. This method is expensive owing to the cost of manufacturing the chips, which can only be used once.

52.4.1. Screening for Specific Genes versus Environmental Library Construction

Two approaches can be taken for the direct capture of genes. Either DNA or cell concentrate is extracted from the soil and screened for the specific gene of interest, or an environmental library is constructed from the total DNA prior to screening. The choice of method is dependent on what is required from the environmental sample. If analyzing the soil for a specific gene of interest, DNA extraction followed by PCR screening allows a large number of samples to be screened. Construction of an environmental library allows a more detailed examination of one sample and is perhaps better suited to the discovery of novel genes via expression of a target activity. Another advantage of constructing a library is that a permanent record of the soil can be obtained and larger fragments can be captured rather than amplified for a survey of genetic diversity.

There are several disadvantages associated with environmental library construction owing the complexity of soil DNA (103). First, a very large number of clones are required if the library is going to be truly representative of the environment from which it has been constructed. For example, if the average streptomycete chromosome is 8 Mb, and there is a population of 10^6 present in 1 g of soil, at least 2×10^8 clones would be required to construct a library if fragments of 40 kb were ligated into a vector. Second, the logistics of screening so many clones presents considerable problems. The packaging of DNA extracted directly from soil is likely to give greater diversity than libraries obtained from extracted cells. Impurities are more likely to be present with cell lysis in situ, so steps must be taken to ensure that all the impurities are removed to improve the likelihood of packaging (see section 52.4.2). Screening of unknown genes is entirely dependent on expression of an activity that can be compromised by the choice of host, vector, and construction of clone in addition to choice of growth conditions for screening. The cloning of entire pathways for the production of metabolites such as antibiotics will be highly problematic owing to the above constraints and the need to clone fragments over 50 kb. The choice of host can be assisted by analysis of the taxonomic diversity in the community DNA, its GC content, and the diversity within the environmental sample. For further details of vectors, hosts, and expression, see Sambrook et al. (89).

Environmental samples can be rapidly screened for the presence of specific genes (36), allowing the analysis of many samples, although the level of sensitivity is not readily deduced owing to deficiencies in DNA extraction with different samples and variations in purity of the resultant DNA (54). Total community DNA has been analyzed for taxonomic diversity via 16S rRNA gene targets and for functional diversity using a wide range of target genes. Walia et al. (111) probed the soil with *cbpABCD* to detect polychlorinated biphenyl degraders; the *nahA* gene for naphthalene dioxygenase was used to estimate the napthalene-degrading population (90); a number of structural gene probes (*tfdA-e*) were used in soil to examine 2,4-dichlorophenoxyacetic acid degraders (48), and the *dhlA* gene was used to detect the presence of dichloropropene degraders (108). Uncultured groups can be screened by this approach (10), although there is no guarantee that all the cells in the sample have been lysed and DNA extracted.

52.4.2. DNA Extraction from Environmental Samples

Numerous methods for the extraction of DNA from environmental samples have been reported (16, 46, 73, 94, 96, 106, 121) and compared (59, 87). Because of the heterogeneous nature of soil and the different lysing efficiency for vegetative cells, spores, and resistant propagules, there is no one method suitable for all soils. There are three steps that can be altered according to the cell type, soil type, and the quality of DNA required (e.g., fragment size, purity, and yield). These steps are extraction, lysis, and purification and are summarized in Table 1 with one example of each procedure. For the lysis and purification steps, a combination of the different methods can be employed.

Soil DNA extraction is either done indirectly by extracting the cell biomass from the soil with subsequent lysis (44, 85) or by direct lysis of the cells in situ (16, 106) (Table 1). Direct lysis increases the probability of extract-

TABLE 1 Methods for the extraction and purification of DNA from soil

Step	Method	Example	Comment
Extraction procedure	Direct lysis	Bead beating (16)	Accesses total DNA, high yield, increased fragmentation (approx. 23 kb)
	Indirect lysis (lysis of extracted biomass)	Nycodenz density centrifugation (85)	Large fragments (over 48 kb), yield can be low, does not access the total microbial population
Lysis	Mechanical	Bead beating (16)	Accesses total DNA, high yield, increased fragmentation (approx. 23 kb)
	Enzymatic	Proteinase K (35)	Useful lysis agent, used in combination with other lysis methods, can be expensive
	Chemical	SDS (19)	Used hot or cold, basis of most indirect extraction procedures, useful for Gm -ves, will not lyse spores
	Differential temperatures	Liquid nitrogen (110)	Lyses most cell types
Purification	Chemical precipitation	Potassium acetate (16)	Precipitates proteins
	Chromatography	Sephadex-50 spin column (86)	Removes humic acids
	Cesium chloride centrifugation	See reference 94	Ensures extremely pure DNA, time-consuming and expensive
	Magnetic capture	See reference 45	Enables the purification of specific DNA sequences
	Electrophoresis	Electroelute from agarose gel and purify using standard chromosomal techniques (83)	Separates DNA from humic acids, reduces DNA yield

ing the total microbial population (9), whereas cell extraction techniques may be selective and exclude mycelial propagules (35). Direct lysis is either done by physical disruption using Bead-Beaters (Braun) or Ribolysers (Hybaid) or chemically treating soil suspensions with a combination of chemicals and enzymes. Cresswell et al. (16) compared a mechanical method (bead beating) with a chemical method (sodium dodecyl sulfate, SDS) using soil inoculated with streptomyctes, which are spore-forming mycelial bacteria. The bead-beating method extracted DNA from both mycelia and spores; DNA was extracted only from the mycelia when SDS lysis was used.

The extraction and lysis methods chosen will greatly affect the genetic diversity of the DNA obtained. Vigorous cell disruption techniques will result in more sheared DNA. Indirect lysis methods can give DNA fragments of over 48 kb, whereas lysis in situ may result in DNA fragments of between 0.1 and 25 kb (87). The shearing effect can be reduced by optimization of the buffers used. Those with a high salt concentration will reduce shearing (67), and addition of blocking reagents (such as skimmed milk powder) to the soil can also have the same effect (110). Other methods that disperse the sample and cause lysis include blending and sonication; sonication causes shearing of DNA (81). Most methods require the presence of nuclease inhibitors such as EDTA (29) and low temperature to prevent DNA loss. A wide range of chemical agents for cell lysis have been described (Table 1).

The yield of DNA obtained varies from soil to soil and depends on the method of extraction. The average yield is 20 μg from 1 g of soil (94). This accounts for approximately half the theoretical yield: if 1 g of soil contains 10^{10} bacterial cells, with 3 to 5 fg of DNA per cell, then 30 to 50 μg of DNA would be expected from each g of soil. However, some methods may obtain higher yields. Porteous et al. (83) reported a yield of 100 μg of DNA from 1 g of soil (wet weight) using a hot enzyme/chemical lysis protocol that included guanidine isothiocyanate.

DNA extracted from soil and sediments may contain impurities, and when precipitated it can have a brown appearance. This is caused by humic acids present in the sample, and their quantity and composition will vary depending on the sample type. Soils with a high organic content, such as forest soils and compost, have the highest levels of humic acids. Contaminants range from 0.7 to 3.3 μg/ml of extracted DNA and inhibit restriction enzymes at concentrations between 0.8 and 51.7 μg/ml and *Taq* polymerases at 0.24 to 0.48 μg/ml (101). Humic acids also contaminate DNA samples from sediments. Rochelle et al. (86) observed that the humic content of samples varied according to the depth of the sample. Humic contaminants can be reduced by adding absorption agents such as polyvinylpolypyrrolidone (PVPP) to the extraction buffer (29) or can be removed after extraction. DNA can be prepared for PCR by diluting out the humic acids (52). Straub et al. (99) used Sephadex G-50 columns followed by Chelex-100 columns to produce DNA that was amplifiable by PCR. PVPP columns can also be used (4). These columns can all be made in the lab and can reduce the DNA processing time from soil to PCR product to approximately 8 h.

Following purification steps, the total community DNA, although clean enough for PCR, may require further purification for restriction or packaging. Precipitation steps can be included, such as potassium acetate and spermine HCl (16) or ammonium acetate (19) (Table 1). Clean DNA can be obtained by cesium chloride density gradient purification (94). Humic acids are electrophoresed faster than nucleic acids, so they can be effectively separated, either by electroelution (35, 82, 83), agarose-polyvinylpyrrolidone gel extraction (102) or agarose extraction. Chandler et al. (14) demonstrated that PCR sensitivity could be increased up to a factor of 10^4 by electroelution. If only specific gene sequences are required, it is possible to selectively purify them using magnetic capture (45) by hybridizing the genes of interest to tagged magnetic beads, which are removed and analyzed using PCR.

The purity of extracted DNA can be checked by measuring the ratio of absorbance from A_{230}/A_{260}, which is reduced if phenolic contaminants are present. Universal eubacterial PCR primers can be used as a positive control with the addition of 1 μl of DNA to the sample.

52.5. DETECTION OF FUNCTIONAL ACTIVITY IN ENVIRONMENTAL SAMPLES

The relationship between metabolic diversity and genotypic diversity is not fully understood (119). Microorganisms do play a key role in the environment, being responsible for the decomposition of polymers and degradation of environmental contaminants (1, 56).

A total sample approach may be used when bioprospecting for a specific activity, such as assaying for the ability to degrade a specific polymer or a toxic compound. Environmental samples can be analyzed for specific enzyme activities as a preliminary screen, and with positive samples subsequent steps can be taken to isolate the specific microorganisms.

The analysis of the functional activity of the total sample requires the metabolic activity to act as an indicator of the microbial community present in the sample. There are a number of limitations; microorganisms with novel metabolic activities, is present in low numbers in the sample, may not be detected because the analysis will not be sensitive enough. In addition, sample activities do not always closely correlate with the microbial community present.

52.5.1. Metabolic Analysis of Environmental Samples (Biolog)

The Biolog system (Biolog, Inc., Hayward, Calif.) was originally developed to allow identification of bacteria by monitoring sole-carbon-source utilization patterns (51, 65). The Biolog GN plates consist of 96 wells—95 carbon sources and a control well. The wells also contain a redox dye (tetrazolium violet), which changes color when electrons are donated from NADH to the electron transport chain, indicating that the carbon source is being oxidized (6). This color change can be monitored by measuring changes in absorbance in the well.

The system was adapted for community analysis by Garland and Mills (24). The bacterial fraction from an environmental sample is added directly to the wells, and the carbon-source utilization patterns are analyzed. These patterns of respiration are thought to give a metabolic fingerprint of the sample, representing the functional attributes of the bacterial community with respect to a suite of substrates (7). The Biolog plate represents 95 separate enrichments; therefore, it is a rapid method allowing a preliminary screen of the sample for metabolic activity.

Metabolic fingerprints have allowed the discrimination of different microbial communities in environments, including the rhizosphere, inoculated rhizosphere, and bioreactor monitoring (25). It has been used to analyze differences in metabolic activity in different sizes of particles in the same soil (115) and for comparative analysis of plant phyllospheres (34). Biolog analysis has also allowed the definition of shifts in microbial communities when placed under heavy metal stress (53) and hydrocarbon pollution (118).

The Biolog analysis could be used as a primary screening step to obtain samples that contain diverse microbial communities. Altering the incubation conditions can allow metabolic profiling of specific groups such as thermophiles. The contents of the well can then be removed and the DNA extracted and sequenced to identify the microorganisms that are growing on each substrate (113).

The limitations of the Biolog system include the selectivity of the assay, which is based on growth and respiration in an aerobic system. This will cause the strictly anaerobic bacteria lacking enzymes from the electron transport chain to be excluded (115). The microtitre wells in the Biolog plate require growth to indicate degradation and will therefore have most of the limitations of a cultivation technique (17).

52.5.1.1. Biolog MT Plates

Biolog plates can be used in a more defined way as a bioprospecting tool for the detection of specific abilities by using Biolog MT plates. The MT plate contains the same nutrient base and color chemistry as the GN plate, but without the carbon sources. These plates allow the addition of specific carbon sources to the plate, to screen for their degradation. This approach can be useful in the detection of bacterial phenotypes for degradation of xenobiotics. The Biolog MT plates can also be used for direct assessment of the biodegradative abilities of the microbial community present in a sample, avoiding a primary selection step.

Lee et al. (58) used Biolog to screen for the ability to biodegrade five classes of anionic surfactant; they also found that the method was reproducible and rapid. The MT plates have been used with different environmental samples as an approach to phenotypic screening. Substrates include samples from kraft-pulp effluent treatment plants (109) and terrestrial subsurface sites (57), for biodegradation of substrates such as chlorinated aromatics (109), isocyanate-based polymers (72), and a number of nonvolatile toxic organic compounds (28, 100).

Identification of active degraders in the wells can be achieved by removing the well contents and subculturing on suitable media. If the microorganisms are difficult to culture, a DNA extraction from the well solution can be performed. It is also possible to clone the DNA from the well and then screen the clones for target metabolic pathways.

52.5.2. Sample Enzyme Activity

Functional activities of environmental communities can be deduced from enzyme assays of samples (50). Extracellular enzymes are known to play an important role in decomposition processes in environments such as soil (92). Enzyme activity can provide useful information on the functional diversity and total enzyme activity (12, 32). Various enzymes such as oxidoreductases, hydrolases, and transferases (78) have been used as indicators of the metabolic activities of soil microfloras. Enzyme activities can be sensitive indicators of environmental damage (18, 23). When determining microbial activities in samples that may be contaminated with heavy metals, enzyme activity may not be a suitable measure, as the assay can be affected by low levels of copper (13).

Analysis of total sample enzyme activity has been developed to measure potential activity and not active in situ activity. In soil samples, enzymes can also be absorbed onto soil clay constituents and retain a proportion of their original activity. Residual activity can be long lived and unrelated to the levels of microbial biomass (12). Also, great care must be taken to try to maintain in situ conditions such as temperature and moisture content, and the enzyme assay should be performed as quickly as possible after removing the sample from the environment.

52.5.3. Bioprospecting for Specific Enzymes

Enzymes are extremely useful in industry, and their ability to degrade large molecules and catalyze reactions has many commercial applications. To detect enzymes, an enrichment approach may be used; for example, to detect chitinase enzymes, the sample is baited with insect exoskeletons, which act as an enrichment for the chitinolytic microorganisms. Following incubation, enzymes can be extracted directly from the soil, and selective isolation of chitinolytic groups can also be achieved. The enrichment technique has been successful in identifying new enzymes. One isolate that was found in a cemetery in Copenhagen is now used in laundry detergents (105).

Wirth and Wolf (117) used a microplate colorimetric assay to measure the activities of five different enzymes extracted directly from soil samples. This method was successful in detecting chitinase activity and allowed a rapid analysis of many different environmental samples, although the sensitivity of the method may not be sufficient to detect some populations.

52.6. CONCLUSIONS

The extraction of total community DNA has been achieved from soil, sediment, peat, clay, manure, sewage sludge, and vegetation and from samples representing diverse habitats in terrestrial and aquatic environments. The complexity of DNA extracted from soil has been estimated at approximately 10^{10} bp (103). Perhaps the limiting factor in the exploitation of this gene pool is the need for very high throughput screens to process the large numbers of clones generated from such a gene bank. The microbial diversity in natural environments has yet to be fully characterized and represents an astonishing biological resource.

REFERENCES

1. **Aelion, C. M., and P. M. Bradley.** 1991. Aerobic biodegradation potential of subsurface microorganisms from a jet fuel-contaminated aquifer. *Appl. Environ. Microbiol.* **57:**57–63.
2. **Bakken, L. R.** 1985. Separation and purification of bacteria from soil. *Appl. Environ. Microbiol.* **49:**1482–1487.
3. **Bakken, L. R., and V. Lindahl.** 1995. Recovery of bacterial cells from soil, p. 9–27. *In* J. D. van Elsas and J. T. Trevors (ed.), *Nucleic Acids in the Environment.* Springer-Verlag, Berlin.
4. **Berthelet, M., L. G. Whyte, and C. W. Greer.** 1996. Rapid, direct extraction of DNA from soils for PCR analysis using polyvinylpolypyrrolidone spin columns. *FEMS Microbiol. Lett.* **138:**17–22.
5. **Blake, M. R., and B. C. Weimer.** 1997. Immunomagnetic detection of *Bacillus stearothermophilus* spores in food and environmental samples. *Appl. Environ. Microbiol.* **63:** 1643–1646.
6. **Bochner, B.** 1989. Sleuthing out bacterial identities. *Nature* **339:**157–158.
7. **Bochner, B.** 1989. Breathprints at the microbial level. *ASM News* **55:**536–539.
8. **Bogosian, G., L. E. Sammons, P. J. L. Morris, J. P. O'Neill, M. A. Heitkamp, and D. B. Weber.** 1996. Death of *Escherichia coli* K-12 strain W3110 in soil and water. *Appl. Environ. Microbial* **62:**4114–4120.
9. **Boivin-Jahns, V., R. Ruimy, A. Bianchi, S. Daumas, and R. Christen.** 1996. Bacterial diversity in a deep-subsurface clay environment. *Appl. Environ. Microbiol.* **62:**3405–3412.
10. **Brauns, L. A., M. C. Hudson, and J. D. Oliver.** 1991. Use of the polymerase chain reaction in detection of culturable and nonculturable *Vibrio vulnificus* cells. *Appl. Environ. Microbiol.* **57:**2651–2655.
11. **Bull, A. T., M. Goodfellow, and J. H. Slater.** 1992. Biodiversity as a source of innovation in biotechnology. *Annu. Rev. Microbiol.* **46:**219–252.
12. **Burns, R. G.** 1982. Enzyme activities in soil: location and possible role in microbial ecology. *Soil. Biol. Biochem.* **14:** 423–427.
13. **Chander, K., and C. Brookes.** 1991. Is the dehydrogenase assay invalid to estimate microbial activity in copper contaminated soils? *Soil. Biol. Biochem.* **23:**909–915.
14. **Chandler, D. P., R. W. Schreckhise, J. L. Smith, and H. Bolton.** 1997. Electroelution to remove humic compounds from soil DNA and RNA extracts. *J. Microbiol. Methods.* **28:**11–19.
15. **Christensen, B., T. Torsvik, and T. Lein.** 1992. Immunomagnetic capture of Thermophilic Sulphate Reducing bacteria from North-sea-oil field waters. *Appl. Environ. Microbiol.* **58:**1244–1248.
16. **Cresswell, N., V. A. Saunders, and E. M. H. Wellington.** 1991. Detection and quantification of Streptomyces violaceolatus plasmid DNA in soil. *Lett. Appl. Microbiol.* **13:** 193–197.
17. **Degens, B. P., and J. A. Harris.** 1997. Development of a physiological approach to measuring the catabolic diversity of soil microbial communities. *Soil. Biol. Biochem.* **29:**1309–1320.
18. **Dick, R. P., P. E. Rasmussen, and Kerle, E. A.** 1988. Influence of long term residue management on soil enzyme activities in relation to soil chemical properties of a wheat fallow system. *Biol. Fertil. Soils* **6:**159–164.
19. **Dijkmans, R., A. Jagers, S. Kreps, J. M. Collard, and M. Mergeay.** 1993. Rapid method for purification of soil DNA for hybridization and PCR analysis. *Microb. Releases* **2:**29–34.
21. **Felske, A., H. Rheims, A. Wolterink, E. Stackebrandt, and A. D. L. Akkermans.** 1997. Ribosome analysis reveals prominent activity of an uncultured member of the class Actinobacteria in grassland soils. *Microbiology* **143:** 2983–2989.
22. **Fuhrman, J. A., D. E. Comeau, A. Hagstrom, and A. M. Chan.** 1988. Extraction from natural planktonic microorganisms of DNA suitable for molecular biological studies. *Appl. Environ. Microbiol.* **54:**1426–1429.

23. **Garcia, C., and T. Hernandez.** 1997. Biological and biochemical indicators in derelict soils subject to erosion. *Soil. Biol. Biochem.* **29:**171–177.

24. **Garland, J. L., and A. L. Mills.** 1991. Classification and characterization of heterotrophic microbial communities on the basis of patterns of community-level sole-carbon-source utilization. *Appl. Environ. Microbiol.* **57:**2351–2359.

25. **Garland, J. L., and A. L. Mills.** 1994. A community-level physiological approach for studying microbial communities, p. 77–83. *In* K. Ritz, J. Dighton, and K. E. Giller (ed.), *Beyond the Biomass*. Wiley Sayce Publication, British Society of Soil Science, United Kingdom.

26. **Giovannoni, S. J., E. F. Delong, T. M. Schmidt, and N. R. Pace.** 1990. Tangential flow filtration and preliminary phylogenetic analysis of marine picoplankton. *Appl. Environ. Microbiol.* **56:**2572–2575.

27. **Glowka, L., F. Burhenne-Guilmin, H. Synge, J. A. McNeely, and L. Gündling. 1994.** *A Guide to the Convention on Biological Diversity*. Environmental policy and law paper No. 30. IUCN Gland, Cambridge.

28. **Gordon, R. W., T. C. Hazen, and C. B. Fliermans.** 1993 Rapid screening of bacteria capable of degrading toxic organic compounds. *J. Microbiol. Methods* **18:**339–347.

29. **Gray, J. P., and R. P. Herwig.** 1996. Phylogenetic analysis of bacterial communities in marine sediments. *Appl. Environ. Microbiol.* **62:**4049–4059.

30. **Guschin, D. Y., B. K. Mobarry, D. Proudnikov, D. A. Stahl, B. E. Rittmann, and A. D. Mirabekov.** 1997. Oligonucleotide microchips as genosensors for determinative and environmental studies in microbiology. *Appl. Environ. Microbiol.* **63:**2397–2402.

31. **Harboe, N. M. G., and A. Ingild.** 1983. Immunisation, isolation of immunoglobulins and antibody titre determination. *Scand. J. Immunol.*. **17:**345–351.

32. **Hayano, K.** 1986. Cellulase complex in a tomato field soil: induction, localisation and some properties. *Soil. Biol. Biochem.* **18:**215–219.

33. **Herron, P. R., and E. M. H. Wellington.** 1990. New method for extraction of streptomycete spores from soil and application to the study of lysogeny in sterile amended and nonsterile soil. *Appl. Environ. Microbiol.* **56:** 1406–1412.

34. **Heuer, H., and K. Smalla.** 1997. Evaluation of community-level catabolic profiling using BIOLOG GN microplates to study microbial community changes in potato phyllospher. *J. Microbiol. Methods* **30:**49–61.

35. **Hilger, A. B., and D. D. Myrold.** 1991. Method for extraction of frankia DNA from soil. *Agric. Ecosystems Environ.* **34:**107–113.

36. **Holben, W. E., J. K. Jansson, B. K. Chelm, and J. M. Tiedje.** 1988. DNA probe method for the detection of specific microorganisms in the soil bacterial community. *Appl. Environ. Microbiol.* **54:**703–711.

37. **Hopkins, D. W., S. J. MacNaughton, and A. G. O'Donnell.** 1991. A dispersion and differential centrifugation technique for representively sampling microorganisms from soil. *Soil Biol. Biochem.* **23:**217–225.

38. **Hopkins, D. W., and A. G. O'Donnell.** 1992. Methods for extracting bacterial cells from soil, p. 104. *In* E. M. H. Wellington and J. D. van Elsas (ed.), *Genetic Interactions between Microorganisms in the Environment*. Manchester University Press, Manchester, United Kingdom.

39. **Huddleston, A. S., and E. M. H. Wellington.** 1997. Detection of molecular diversity of actinomycete soil communities, abstr. S135. *In* Annual Meeting Program and Abstracts, Society for Industrial Microbiology, Reno, Nev.

40. **Huddleston, A. S., N. Cresswell, M. C. P. Neves, J. E. Beringer, S. Baumberg, D. I. Thomas, and E. M. H. Wellington.** 1997. Molecular detection of steptomycin-producing Streptomycetes in Brazilian soils. *Appl. Environ. Microbiol.* **63:**1288–1297.

41. **Hugenholtz, P., and N. R. Pace.** 1996. Identifying microbial diversity in the natural environment: a molecular phylogenetic approach. *Trends Biotechnol.* **14:**190–197.

42. **Huq, A., R. R. Colwell, R. Rahman, A. Ali, M. A. R. Chowdhury, S. Parveen, D. A. Sack, and E. Russek-Cohen.** 1990. Detection of *Vibrio cholerae* O1 in the aquatic environment by fluorescent-monoclonal antibody and culture methods. *Appl. Environ. Microbiol.* **56:**2370–2373.

43. **Iwu, M. M.** 1996. Implementing the biodiversity treaty: how to make international cooperative agreements work. *Trends Biotechnol.* **14:**78–83.

44. **Jackman, S. C., H. Lee, and J. T. Trevors.** 1992. Survival, detection and containment of bacteria. *Microb. Releases* **1:**125–154.

45. **Jacobsen, C. S.** 1995. Microscale detection of specific bacterial DNA in soil with magnetic capture-hybridization and PCR amplification assay. *Appl. Environ. Microbiol.* **61:** 3347–3352.

46. **Jacobsen, C. S., and O. F. Rasmussen.** 1992. Development and application of a new method to extract bacterial DNA from soil based on separation of bacteria from soil with cation-exchange resin. *Appl. Environ. Microbiol.* **58:** 2458–2462.

47. **Jensen, P. R., and W. Fenical.** 1994. Strategies for the discovery of secondary metabolites from marine bacteria. Ecological perspectives. *Annu. Rev. Microbiol.* **48:**559–584.

48. **Ka, J. O., W. E. Holben, and J. M. Tiedje.** 1994. Use of gene probes to aid in the recovery and identification of functionally dominant 2, 4-dichlorophenoxyacetic acid-degrading populations in soil. *Appl. Environ. Microbiol.* **60:**1116–1120.

49. **Kamagata, Y., R. R. Fulthorpe, K. Tamura, H. Takami, L. J. Forney, and J. M. Tiedje.** 1997. Pristine environments harbor a new group of oligotrophic 2,4-dichlorophenoxyacetic acid-degrading bacteria. *Appl. Environ. Microbiol.* **63:**2266–2272.

50. **Kanazawa, S., and Z. Filip.** 1986. Distribution of microorganisms, total biomass, and enzyme activities in different particles of brown soil. *Microb. Ecol.* **12:**205–215.

51. **Klingler, J. M., R. P. Stowe, D. C. Obenhuber, T. O. Groves, S. K. Mishra, and D. L. Pierson.** 1992. Evaluation of the Biolog automated microbial identification system. *Appl. Environ. Microbiol.* **58:**2089–2092.

52. **Knaebel, D. B., and R. L. Crawford.** 1995. Extraction and purification of microbial DNA from petroleum-contaminated soils and detection of low numbers of toluence, octane and pesticide degraders by multiplex polymerase chain reaction and Southern analysis. *Mol. Ecol.* **4:**579–591.

53. **Knight, B. P., S. P. McGrath, and A. R. Chaudri.** 1997. Biomass carbon measurements and substrate utilization patterns of microbial populations from soils amended with

cadmium, copper, or zinc. *Appl. Environ. Microbiol.* **63:** 39–43.

54. **Krsek, M., and E. M. H. Wellington.** 1997. Do we ever see the real picture?! Selectivity of DNA extraction techniques, abstr. S139. *In* Annual Meeting Program and Abstracts, Society for Industrial Microbiology, Reno, Nev.
55. **Laguerre, G., M. Allard, F. Revoy, and N. Amarger.** 1994. Rapid identification of rhizobia by restriction fragment length polymorphism analysis of PCR-amplified 16S rRNA genes. *Appl. Environ. Microbiol.* **60:**56–63.
56. **Lamar, R. T., and D. M. Dietrich.** 1990. *In-situ* depletion of pentachlorophenol from contaminated soil by *Phanerochaete* spp. *Appl. Environ. Microbiol.* **56:**3093–3100.
57. **Lee, B. D., R. M. Lehman, and F. S. Colwell.** 1992. Screening of microbes from deep-subsurface environments for the ability to degrade non-volatile organic contaminants, p. 384. Abstr. 92nd Annu. Meet. Am. Soc. Microbiol. 1992. American Society for Microbiology, Washington, D.C.
58. **Lee, C., N. J. Russell, and G. F. White.** 1995. Rapid screening for bacterial phenotypes capable of biodegrading anionic surfactants: development and validation of microtitre plate method. *Microbiology* **141:**2801–2810.
59. **Leff, L. G., R. M. Kernan, J. V. McArthur, and L. J. Shimkets.** 1995. Identification of aquatic *Burkholderia* (Pseudomonas) *cepacia* by hybridization with species specific gene probes. *Appl. Environ. Microbiol.* **61:**1634–1636.
60. **Lindahl, V.** 1996. Improved soil dispersion procedures for total bacterial counts, extraction of indigenous bacteria and cell survival. *J. Microbiol. Methods* **25:**279–286.
61. **Lindahl, V., and L. R. Bakken.** 1995. Evaluation of methods for extraction of bacteria from soil. *FEMS Microbiol. Ecol.* **16:**135–142.
62. **Liu, W., T. L. Marsh, H. Cheng, and L. J. Forney.** 1997. Characterization of microbial diversity by determining terminal restriction fragment length polymorphisms of genes encoding 16S rRNA. *Appl. Environ. Microbiol.* **63:**4516–4522.
63. **MacDonald, R. M.** 1986. Sampling soil microfloras—optimization of density gradient centrifugation in Percoll to separate microorganisms from soil suspension. *Soil. Biol. Biochem.* **18:**407–410.
64. **McSpadden-Gardener, B. M., and A. Lilley.** 1997. Application of common statistical tools, p. 501–522. *In* J. D. van Elsas, E. M. H. Wellington, and J. T. Trevors (ed.), *Modern Soil Microbiology*. Marcel Dekker, Inc., New York.
65. **Miller, J. M., and D. L. Rhoden.** 1991. Preliminary evaluation of Biolog, a carbon source utilization method for bacterial identification. *J. Clin. Microbiol.* **29:**1143–1147.
66. **Mitchell, B. A., J. A. Milbury, A. M. Brookins, and B. J. Jackson.** 1994. Use of immunomagnetic capture on beads to recover Listeria from environmental samples. *J. Food Protection* **57:**743–745.
67. **Morgan, J. A. W., C. Winstanley, R. W. Pickup, and J. R. Saunders.** 1991. Rapid immunocapture of *Pseudomonas putida* cells from lake water by using bacterial flagella. **57:**503–509.
68. **Moyer, C. L., F. C. Dobbs, and D. M. Karl.** 1994. Estimation of diversity and community structure through restriction fragment polymorphism distribution analysis of bacterial 16S rRNA genes from a microbial mat at an active, hydrothermal vent system, Loihi Seamount, Hawaii. *Appl. Environ. Microbiol.* **60:**871–879.
69. **Mullins, P. H., H. Gürtler, and E. M. H. Wellington.** 1995. Selective recovery of *Streptosporangium fragile* from soil by indirect immunomagnetic capture. *Microbiology* **141:**2149–2156.
70. **Muyzer, G., E. C. de Waal, and A. G. Uitterlinden.** 1993. Profiling complex microbial populations by denaturing gradient gel electrophoresis analysis of polymerase chain reaction-amplified genes coding for 16S rRNA. *Appl. Environ. Microbiol.* **59:**695–700.
71. **Nesme, X., M. Vaneechoute, S. Orso, B. Haste, and J. Swings.** 1995. Diversity and genetic relatedness within genera Xanthomonas and Sterotrophomonas using restriction endonuclease site differences of PCR-amplified 16S ribosomal RNA gene. *Syst. Appl. Microbiol.* **18:**127–135.
72. **Odocha, I., S. Wang, I. Horacek, S. Wong, I. Kresta, and J. Graves.** 1994. Biodegradability assessment of isocyanate based polymers, p. 463. Abstr. 94th Annu. Meet. Am. Soc. Microbiol. 1994. American Society for Microbiology, Washington, D.C.
73. **Ogram, A., G. S. Sayler, and T. Barkay.** 1987. The extraction and purification of microbial DNA from sediments. *J. Microbiol. Methods* **7:**57–66.
74. **Ozawa, T., and M. Yamaguchi.** 1986. Fractionation and estimation of particle-attached and unattached Bradyrhizobium-japonicum strains in soils. *Appl. Environ. Microbiol.* **52:**911–914.
75. **Pace, N. R.** 1996. New perspective on the natural microbial world: molecular microbial ecology. *ASM News* **62:** 463–470.
76. **Pace, N. R.** 1997. A molecular view of microbial diversity and the biosphere. *Science* **276:**734–740.
77. **Patchett, R. A., A. F. Kelly, and R. G. Kroll.** 1991. The adsorption of bacteria to immobilised lectins. *J. Appl. Bacteriol.* **71:**277–284.
78. **Paul, E. A., and F. E. Clark.** 1996. *Soil Microbiology and Biochemistry*. Academic Press, San Diego, Calif.
79. **Paul, J. H., and S. L. Pichard.** 1989. Specificity of cellular DNA-binding sites of microbial populations in a Florida reservoir. *Appl. Environ. Microbiol.* **55:**2798–2801.
80. **Payne, M. J., S. Campbell, and R. G. Kroll.** 1993. Separation of bacteria using agglutinins isolated from invertebrates. *J. Appl. Bacteriol.* **74:**276–283.
81. **Picard, C., C. Ponsonnet, E. Paget, X. Nesme, and P. Simonet.** 1992. Detection and enumeration of bacteria in soil by direct DNA extraction and polymerase chain reaction. *Appl. Environ. Microbiol.* **58:**2717–2722.
82. **Porteous, L. A., and J. L. Armstrong.** 1993. A simple mini-method to extract DNA directly from soil for use with polymerase chain-reaction amplification. *Curr. Microbiol.* **27:**115–118.
83. **Porteous, L. A., J. L. Armstrong, R. J. Seidler, and L. S. Watrud.** 1994. An effective method to extract DNA from environmental samples for polymerase chain reaction amplification and DNA fingerprint analysis. *Curr. Microbiol.* **29:**301–307.
84. **Porter, J., R. Pickup, and C. Edwards.** 1997. Evaluation of flow cytometric methods for the detection and viability assessment of bacteria from soil. *Soil. Biol. Biochem.* **29:** 91–100.
85. **Prieme, A., J. I. B. Sitaula, A. K. Klemedtsson, and L. R. Bakken.** 1996. Extraction of methane-oxidizing

bacteria from soil particles. *FEMS Microbiol. Ecol.* **21:** 59–68.

86. **Rochelle, P. A., J. C. Fry, R. J. Parkes, and A. J. Weightman.** 1992. DNA extraction for 16S rRNA gene analysis to determine genetic diversity in deep sediment communities. *FEMS Microbiol. Lett.* **100:**59–66.

87. **Saano, A., K. Lindström, and J. D. van Elsas.** 1995. Eubacterial diversity in Finnish forest soil, abstr. P1-3.9. *In* 7th International Symposium on Microbial Ecology, Brazilian Society for Microbiology, Santos, Brazil.

88. **Safarik, I., M. Safariková, and S. J. Forsythe.** 1995. The application of magnetic separations in applied microbiology. *J. Appl. Bacteriol.* **78:**575–585.

89. **Sambrook, J., E. F. Fritsch, and T. Maniatis.** 1989. *Molecular Cloning. A Laboratory Manual.* Cold Spring Harbor Laboratory, Cold Spring Harbor, N.Y.

90. **Sanseverino, J., C. Werner, J. Fleming, B. Applegate, J. M. King, and G. S. Sayler.** 1993. Molecular diagnostics of polycyclic aromatic hydrocarbon biodegradation in manufactured gas plant soils. *Biodegradation* **4:**303–321.

91. **Seow, K. T., G. Meurer, M. Gerlitz, E. Wendt Pienkowski, C. R. Hutchinson, and J. Davies.** 1997. A study of iterative type II polyketide synthases, using bacterial genes cloned from soil DNA: a means to access and use genes from uncultured microorganisms. *J. Bacteriol.* **179:**7360–7368.

92. **Sinsabaugh, R. L., R. K. Antibus, and A. E. Linkins.** 1991. An enzymatic approach to the analysis of microbial activity during plant litter decomposition. *Agri. Ecosystems Environ.* **34:**43–54.

93. **Skjerve, E., L. M. Rørvik, and Ø. Olsvik.** 1990. Detection of *Listeria monocytogenes* in foods by immunomagnetic separation. *Appl. Environ. Microbiol.* **56:**3478–3481.

94. **Smalla, K., N. Cresswell, L. C. Mendoncahagler, A. Wolters, and J. D. van Elsas.** 1993. Rapid DNA extraction protocol from soil for polymerase chain reaction-mediated amplification. *J. Appl. Bacteriol.* **74:**78–85.

95. **Steffan, R. J., and R. M. Atlas.** 1988. DNA amplification to enhance detection of genetically engineered bacteria in environmental samples. *Appl. Environ. Microbiol.* **54:**2185–2191.

96. **Steffan, R. J., J. Goksoyr, A. K. Bej, and R. M. Atlas.** 1988. Recovery of DNA from soils and sediments. *Appl. Environ. Microbiol.* **54:**2908–2915.

97. **Stephen, J. R., A. E. McCaig, Z. Smith, J. I. Prosser, and T. M. Embley.** 1996. Molecular diversity of soil and marine 16S rRNA gene sequences related to β-subgroup ammonia-oxidizing bacteria. *Appl. Environ. Microbiol.* **62:**4147–5154.

98. **Stotzky, G., R. D. Goos, and M. I. Timonin.** 1962. Microbial changes in soil as a result of storage. *Plant Soil* **16:**1–18.

99. **Straub, T. M., I. L. Pepper, M. Abbaszadegan, and C. P. Gerba.** 1994. A method to detect enteroviruses in sewage sludge-amended soil using the PCR. *Appl. Environ. Microbiol.* **60:**1014–1017.

100. **Strong-Gunderson, J. M., A. V. Palumbo, and A. O. Scarborough.** 1992. New method for rapidly determining microbial utilization of volatile contaminants, p. 371. Abstr. 92nd Annu. Meet. Am. Soc. Microbiol. 1992. American Society for Microbiology, Washington, D. C.

101. **Tebbe, C. C., and W. Vahjen.** 1993. Interference of humic acids and DNA extracted directly from soil in detection and transformation of recombinant-DNA from bacteria and yeast. *Appl. Environ. Microbiol.* **59:**2657–2665.

102. **Thornhill, R. H., J. G. Burgess, and T. Matsunaga.** 1995. PCR for direct detection of indigenous uncultured magnetic cocci in sediment and phylogenetic analysis of amplified 16S ribosomal DNA. *Appl. Environ. Microbiol.* **61:**495–500.

103. **Torsvik, V., J. Goksoyr, and F. L. Daae.** 1990b. High diversity in DNA of soil bacteria. *Appl. Environ. Microbiol.* **56:**782–787.

104. **Torsvik, V., K. Salte, R. Sorheim, and J. Goksoyr.** 1990a. Comparison of phenotypic diversity and DNA heterogeneity in a population of soil bacteria. *Appl. Environ. Microbiol.* **56:**776–781.

105. **Trombly, J.** 1995. Engineering enzymes for better bioremediation. *Environ. Sci. Technol.* **29:**560–564.

106. **Tsai, Y. L., and B. H. Olson.** 1991. Rapid method for direct extraction of DNA from soil and sediments. *Appl. Environ. Microbiol.* **57:**1070–1074.

107. **Turpin, P. E., K. A. Maycroft, C. L. Rowlands, and E. M. H. Wellington.** 1993. An ion-exchange based extraction method for the detection of salmonellas in soil. *J. Appl. Bacteriol.* **74:**181–190.

107a. **van Elsas, J. D., and K. Smalla.** 1997. Methods for sampling soil microbes, p. 383–391. *In* C. J. Hurst (ed.) *Manual of Environmental Microbiology.* American Society for Microbiology, Washington, D.C.

108. **Verhagen, C., E. Smit, D. B. Janssen, and J. D. van Elsas.** 1995. Bacterial dichloropropene degradation in soil-screening soils and involvement of plasmids carrying the *dhlA* gene. *Soil Biol. Biochem.* **12:**1547–1557.

109. **Victorio, L., D. G. Allen, K. A. Gilbride, and S. N. Liss.** 1993. Biodegradation profiles of microbiol communities in wastewater treatment systems, p. 70. Abstr. Joint Annu. Meet. Society for Industrial Microbiology and Canadian Society of Microbiology.

110. **Volossiouk, T., E. J. Robb, and R. N. Nazar.** 1995. Direct DNA extraction for PCR-mediated assays of soil organisms. *Appl. Environ. Microbiol.* **61:**3972–3976.

111. **Walia, S., A. Kahn, and N. Rosenthal.** 1990. Construction and applications of DNA probes for detection of polychlorinated biphenyl-degrading genotypes in toxic organic-contaminated soil environments. *Appl. Environ. Microbiol.* **56:**254–259.

112. **Ward, D. M., R. Weller, and M. M. Bateson.** 1990. 16S ribosomal-RNA sequences reveal numerous uncultured microorganisms in a natural community. *Nature* **345:**63–65.

113. **Watts, J. E. M., and E. M. H. Wellington.** 1997. Analysis of bacterial communities in heavy metal contaminated soil, p. 83. *In* Annual Meeting Program and Abstracts, Society for Industrial Microbiology. Reno, Nev.

114. **Widjojoatmodjo, M. N., A. C. Fluit, R. Torensma, and J. Verhoef.** 1993. Comparison of immunomagnetic beads coated with protein-A, protein-G, or goat anti-mouse immunoglobulins—applications in enzyme immunoassays and immunomagnetic separations. *J. Immunol. Methods.* **165:**11–19.

115. **Winding, A.** 1994. Fingerpainting bacterial soil communities using Biolog microtitre plates, p. 85–94. *In* K.

Ritz, J. Dighton, and K. E. Giller (ed.), *Beyond the Biomass*. Wiley Sayce Publication, British Society of Soil Science, United Kingdom.

116. **Wipat, A., E. M. H. Wellington, and V. A. Saunders.** 1994. Monoclonal antibodies for *Streptomyces lividans* and their use for immunomagnetic capture of spores from soil. *Microbiology* **140:**2067–2076.

117. **Wirth, S. J., and G. A. Wolf.** 1992. Micro-plate colourimetric assay for *endo*-acting cellulase, xylanase, chitinase, 1,3-β-glucanase and amylase extracted from forest soil horizons. *Soil Biol. Biochem.* **24:**511–519.

118. **Wünsche, L., L. Brüggemann, and W. Babel.** 1995. Determination of substrate utilization patterns of soil microbial communities: an approach to assess population changes after hydrocarbon pollution. *FEMS Microbiol. Ecol.* **17:**295–306.

119. **Zak, J. C., M. R. Willig, D. L. Moorhead, and H. G. Wildman.** 1994. Functional diversity of microbial communities: a quantitative approach. *Soil. Biol. Biochem.* **26:**1101–1108.

120. **Zamost, B. L., H. K. Nielsen, and R. L. Starnes.** 1991. Thermostable enzymes for industrial applications. *J. Ind. Microbiol.* **8:**71–81.

121. **Zhou, J., M. A. Bruns, and J. M. Tiedje.** 1996. DNA recovery from soils of diverse composition. *Appl. Environ. Microbiol.* **62:**316–322.

Biological Control of Foliar Pathogens and Pests with Bacterial Biocontrol Agents

STEVEN E. LINDOW AND MARK WILSON

53

The use of microorganisms for the control of plant diseases and other microbially mediated problems of plants that limit their productivity has received considerable attention. Currently, most diseases of plants are managed by the use of chemical pesticides, which are applied to the foliage, seeds, or soil. The environmental consequences of such pesticide applications, the development of resistance to pesticides, and the high cost of development of replacement pesticides has stimulated efforts to find biological agents that can be used to control these microbial plant pests. Considerable work has been done to understand the ecology and epidemiology of both foliar and soilborne plant pathogens and pests as well as the microorganisms that can be applied to mitigate those microbial pests or the disorders that they cause. There are numerous parallels between the biological control strategies and concepts applied to foliar pathogens and pests and soilborne pathogens. Several reviews on the topic of biological control of diseases on the phylloplane have appeared (3, 4, 9, 18, 19, 66, 80, 81). In this chapter, we will consider the important issues that must be considered when a biological control strategy is being developed for a new system.

53.1. ECOLOGY OF THE PLANT PATHOGEN OR PEST

An understanding of the autecology of the deleterious microbes that are the focus of a biological control effort is needed to best determine how the disease process can be interrupted by manipulating the microbial community on leaves. Most often, the goal of biological control is to establish sufficiently large population sizes of one or more bacterial biocontrol agents on plants. The composition of the phyllosphere microbial community is thus altered in such a way that the population size of a bacterial plant pathogen or pest, which may exist on leaves before initiating infection of the plant, is reduced. Since there is a strong relationship between the epiphytic population size of a bacterial plant pathogen and the likelihood of subsequent plant infection (49, 79), antagonists that reduce population sizes of the pathogen can reduce the likelihood of disease. In a similar fashion, the probability that a frost-sensitive plant will freeze and hence be damaged at a given temperature in the range of about 0 to −5°C is proportional to the logarithm of the population size of ice nucleation-active bacteria (25, 51, 57). Because of this, frost control can be achieved by establishing antagonistic bacteria on plants that reduce the population size of ice nucleation-active bacteria (50, 52–55, 57, 60, 61).

It has recently been recognized that certain bacteria can induce changes in host plant susceptibility to infection by plant pathogenic microbes (1, 36, 89, 90). It is clear that the population dynamics of the pathogen and its interactions with plants have a substantial impact on the approach to biological control that can be attempted. For example, most bacterial plant pathogens are thought to attain large population sizes on plants by a process of multiplication, using nutrients that leak onto leaves as the primary carbon and energy source (5, 19, 28, 30). Under favorable weather conditions, population sizes can increase from negligible levels on newly expanded leaves or recently opened flowers to more than 10^7 cells per gram of leaf tissue by growth over several days on a leaf (21, 27, 29, 30, 50, 59). Biological control of such plant pathogens thus might be achieved either by preventing the multiplication of cells on leaves, or by eradication of established populations before infection of plants can occur.

Clearly, if the approach of biological control is to prevent population sizes from increasing on leaves with time, manipulations of the microbial community, such as by applying a bacterial antagonist, would have to occur early in the development of the plant. Typically, the largest changes in the microbial community have been observed soon after application of nutrients or bacterial antagonists (19, 55, 57, 60, 61). Because of this, it might be expected that if infection occurred early in plant development, shortly after phytopathogens developed large population sizes on leaves, then manipulations of the microbial community would be large, and biological control would be more effective. Conversely, if microbial communities cannot be easily altered after establishment, as has been often observed (61), biological control of diseases or frost occurring later in plant development may be difficult to achieve.

It has recently been demonstrated that the population size of bacteria on some plants is determined more by the rate of immigration from nearby plants that harbor large epiphytic bacterial population sizes (32, 58). On plants such as navel orange, growth apparently contributes little to the populations of epiphytic bacteria (58). Immigration, rather than epiphytic growth, also largely determines the population sizes of many plant pathogenic fungi (41). Because of this, strategies of biological control that rely on prevention of increases in epiphytic populations would not be expected to be effective.

53.2. SOURCE OF ANTAGONISTS

Biological control of diseases will usually be most effective if the antagonistic microbes used to manipulate microbial communities have a high degree of fitness in the habitat in which they need to be established. That is, a large and stable population size of a bacterial antagonist often maximizes the degree of biological control (42, 47, 52–55). Unfortunately, little is known of the factors that determine fitness of bacteria on leaf surfaces (6). Thus, strains with desirable characteristics, such as antibiotic production, that have been isolated in another habitat such as soil often are found to be poor biological control agents when introduced onto leaves, because they either grow or survive poorly in this habitat (84). There have been no reported successes in increasing the fitness of such strains on leaves by genetic manipulation, largely owing to the fact that multiple phenotypes additively contribute to epiphytic fitness (2, 6), and most of the genetic determinants for such fitness determinants have not yet been characterized.

For this reason, it is usually assumed that bacteria that represent the predominant microflora on leaves have a high degree of fitness and that strains recovered from leaves will be good colonists if inoculated onto other leaves of the same plant species. As noted above, such an assumption is probably false for those plant species in which immigration, and not growth, accounts for much of the epiphytic bacterial population size. Some bacterial strains, such as *Pseudomonas fluorescens* strain A506, with the ability to establish and maintain large population sizes under a variety of environmental conditions and with substantial biological control ability have been isolated from the predominant epiphytic microflora (62, 94). Too little is known of the habitat specificity of such bacteria to know if strains optimally useful for biological control must be isolated from the specific plant on which biological control is being implemented.

Recent studies of the genetic diversity of bacterial strains colonizing the leaf surface of sugar beets suggest that a given bacterial strain is maximally fit only on a particular plant species and even for a limited period of time during that plant's development (82, 83). Specifically, over a period of several years a variety of bacterial genotypes were found on sugar beet leaves at various times of the year, and particular genotypes of *P. fluorescens* that were abundant in the middle of the growing season were not common on plants earlier or later in the growing season. Other studies had also suggested that considerable succession of bacterial strains occurs on leaf surfaces over the period of plant development, presumably owing to changes in the chemical and/or physical environment on the leaf surface (6). This suggests that bacterial colonists isolated from plants of early phenological development might not be optimally fit at later stages of development and vice versa. Clearly, the process of strain selection is not trivial and may have a major impact on the extent to which microbial communities can be manipulated to achieve biological disease control.

53.3. EMPIRICAL APPROACHES TO SELECTION OF BIOCONTROL AGENTS

As a consequence of our limited knowledge of the processes leading to changes in microbial communities upon introduction of microbial strains, the most common approach to selecting bacterial antagonists of a disease is to use a nonbiased approach in which randomly isolated strains that might be assumed to be fit on plants are screened for biological control activity. This approach has the advantage that strains that might have a great ability to inhibit disease would not be overlooked simply because they did not express a phenotype that had been hypothesized to contribute to antagonism in culture. This approach is most expedient when an assay for the effectiveness of the potential biocontrol agent in either reducing the population size of a pathogen or in reducing disease is relatively easy, in that large numbers of potential biocontrol agents can be evaluated. Typically, bacterial strains vary greatly in their ability to inhibit pathogen growth on leaves; while a small percentage can cause great reductions in the colonization of pathogens, others have negligible effect (54). Thus, it may be necessary to evaluate the biological control efficacy of hundreds of strains if no prior selection of phenotypes that might contribute to biological control has been made.

The most commonly used selection strategy involves the establishment of individual bacterial strains or mixtures of strains on plants prior to challenge inoculation of plants with a pathogen. In this way, the potential biocontrol agent is usually present as a relatively high proportion of the total microbes on a leaf prior to the arrival of a pathogen, since greenhouse-grown plants lacking a large leaf-surface microbial community (71) are usually used in the assay. In this approach, potential biocontrol agents are assessed for their ability to prevent "invasion" of leaves by pathogenic strains (42, 43). Because the reduction in population size of an invading pathogenic strain on leaves is usually strongly related to the population size of the previously established potential biocontrol agent, sufficient inoculum of the potential biocontrol agent, and/or sufficient time to allow substantial growth of the applied potential biocontrol agent after inoculation, must be provided to observe appreciable biological control of the pathogen or the disease that it incites. While coinoculation of similar numbers of cells of a potential biocontrol agent and pathogen can result in reductions in pathogen numbers relative to application of the pathogen alone, much greater inhibition of pathogen growth is usually seen when antagonist population sizes are much greater than the pathogen (43, 60). It can be expected that prior colonization of leaves with bacterial strains that inhibit the growth of pathogens by preemptive utilization of limited nutrients will be required to maximize biological control of disease by strains operating by this mechanism (42, 43, 95, 96).

Since the goal of most strategies of biological disease control is to reduce the population size of pathogenic strains on asymptomatic leaves, thereby reducing the probability that disease will occur, the amount of inoculum of pathogens used in experiments to test the efficacy of biological control can greatly influence the apparent efficacy of biological control. For example, if large amounts of inoculum of a bacterial plant pathogen are applied to plants harboring an effective biocontrol agent, then disease would be likely even if no further growth of the pathogen occurred on leaves. Since most biocontrol agents prevent disease by preventing growth of the pathogen on plants, application of large amounts of inoculum would circumvent the interactions among the microbes on leaves that otherwise would have led to a reduction in pathogen numbers and hence disease compared to plants lacking the biocontrol agent. Thus, many previous efforts to identify effective biocontrol agents of disease, such as fire blight of pear and apple caused by *Erwinia amylovora*, most likely underestimated the ability of bacterial strains to inhibit the growth of this pathogen, since high cell numbers of *E. amylovora* were applied to treated plants (62, 67, 84). It is likely the inoculum levels of *E. amylovora* encountered by blossoms of pear and apple under field conditions are far lower than those used to initiate the high frequency of disease in flowers inoculated in many laboratory experiments (38, 62). Therefore, as few cells of the pathogen that are required to initiate measurable levels of disease should be used in biological control assays so that potential biocontrol agents that may be effective in field conditions are not overlooked.

53.4. STRATEGIC APPROACHES TO SELECTION OF BIOCONTROL AGENTS

Empirical selection of potential biocontrol agents from large random collections of leaf-associated bacteria can be very time-consuming and expensive; hence, in those systems in which a specific mechanism or phenotype has been conclusively demonstrated to be involved in disease suppression, it may be more efficient to prescreen the random collection for the presence of the desired biocontrol phenotype. Such pre-screened potential biocontrol agents can then be subjected to a more exhaustive empirical selection procedure involving both greenhouse and field assays. This second-level selection takes into account other traits that are also likely to be involved, such as fitness under fluctuating environmental conditions or competitive ability in complex microbial communities, which will be important in a successful biocontrol agent but which are difficult or impossible to quantify in vitro or in simple in planta assays.

53.4.1. Resource Competition

Competition for nutrients on the leaf surface is possibly one of the most important mechanisms involved in the biological control of foliar bacterial pathogens, which must colonize the leaf surface to high population sizes prior to infection, and of necrotrophic fungal pathogens, in which the spores or mycelium require exogenous nutrients for germination and/or growth prior to penetration. In fact, nutrient competition between microorganisms on leaf surfaces is probably one of the most fundamental interactions occurring in this environment. However, the relative importance of nutrient competition compared to other mechanisms, such as antibiosis or induced host responses, in reducing the growth and or pathogenic potential of a pathogen is undoubtedly pathosystem specific. The following examples illustrate situations in which preemptive resource utilization is probably the predominant mechanism involved in the biological control of a particular pathogen and in which knowledge of the nature of resource competition can allow strategic selection of potential biocontrol agents.

The fire blight pathogen *E. amylovora* colonizes the stigmatic surfaces of blossoms and multiplies to high population sizes (91). Following the occurrence of rain or dew, the bacteria are washed or migrate in a film of moisture down the style and onto the nectary, from which they are able to invade the nectarthodes (97). Prior colonization of the stigmatic surfaces, or indeed the nectary, by saprophytic bacteria such as *Pantoea agglomerans* (*Erwinia herbicola*) (92) or *P. fluorescens* (94) reduces subsequent colonization by the pathogen *E. amylovora*. This preemptive exclusion probably involves to a large extent prior utilization of nutrients and/or space between the columnar papillae cells of the stigma. This knowledge can be applied to the strategic selection of biocontrol agents that effectively colonize the stigmas of blossoms in a "cut-flower assay" (75). Effective colonists of the nectary could also be selected in this fashion, or one could select strains that are able to grow to high cell densities in concentrated sugar solutions in vitro (76), such as would be encountered during colonization of the nectarial surface.

The role of nutrient competition has been demonstrated more clearly in the biological control of ice nucleation-active *Pseudomonas syringae* and in *P. syringae* pv. tomato. The incidence of frost injury to frost-sensitive plant tissues is proportional to the logarithm of the population size of ice-nucleating strains of *P. syringae* and other bacterial species (51); hence, factors that reduce the population size of ice-nucleating bacteria also reduce the incidence of frost injury (51, 52). Reductions in the population size of an ice nucleation-active *P. syringae* strain can be achieved by prior colonization by an engineered non-ice nucleation-active *P. syringae* strain (55) or by naturally occurring non-ice nucleation-active saprophytes (62). It was demonstrated that on nitrogen-sufficient plants, the leaf-associated population of ice nucleation-active *P. syringae* is primarily carbon limited and secondarily nitrogen limited (95), suggesting that prior utilization of substitutable carbon sources on the leaf surface was responsible for preemptive exclusion of the ice-nuclearing *P. syringae* strain. Furthermore, using replacement series experiments, Wilson and Lindow (96) showed that there was an inverse relationship between nutritional similarity, determined from in vitro carbon source utilization profiles, and coexistence between the ice nucleation-active *P. syringae* strains and other leaf-associated saprophytes. This relationship would predict that the effectiveness of a leaf-associated bacterium in the preemptive exclusion of an ice nucleation-active or other *P. syringae* strain would be proportional to the nutritional similarity between the two species. Hence, strategic selection of potential biocontrol agents of multiple, nutritionally diverse ice nucleation-active bacteria might be achieved by determination of the composite in vitro nutrient utilization profile of the ice nulceation-active bacteria and screening for leaf-associated bacteria with the maximum degree of overlap with this composite nutritional profile.

The predicted relationship between preemptive exclusion effectiveness and nutritional similarity, estimated from in vitro carbon source utilization profiles, has been examined using the pathogen *P. syringae* pv. tomato on tomato leaves. Using a collection of naturally occurring bacteria isolated from tomato leaves, Ji et al. (33, 37) determined that there was a significant positive correlation between preemptive exclusion efficiency (as measured by the biological control of bacterial speck) and the in vitro nutritional similarity between each leaf-associated bacterium and *P. syringae* pv. tomato. A similar relationship was also demonstrated using Tn5-induced catabolic mutants of the efficient preemptive biocontrol agent *P. syringae* TLP2, each of which was altered in catabolism of one or more carbon sources in vitro (34, 35). These data strongly suggest that a primary screening strategy for biocontrol agents of bacterial speck of tomato could be based on the ability to utilize the maximum number of carbon sources from the in vitro carbon source utilization profile of selected *P. syringae* pv. tomato strains. However, one word of caution; although this appears to be a viable strategy for selection of biocontrol agents of ice nucleation-active *P. syringae* and of *P. syringae* pv. tomato, this approach does not appear to be appropriate for all foliar bacterial pathogens and was not effective for selection of biocontrol agents of *Xanthomonas axonopodis* pv. vesicatoria (15).

Competitive utilization of carbon sources in the phyllosphere has also been demonstrated to occur between leaf-associated bacteria and the grey mold pathogen *Botrytis cinerea*. The spores of necrotrophic fungal pathogens such as *B. cinerea* may leak and reabsorb some nutrients during the germination process or they may require exogenous nutrients for spore germination and growth. Competitive utilization of these nutrients by leaf-associated bacteria reduces the germination frequency of *B. cinerea* spores (7, 8, 10, 11) and may also suppress mycelial growth of the grey mold pathogen. In the case of a necrotrophic foliar pathogen such as *B. cinerea*, strategic selection of a biocontrol agent might be achieved by determination of which exogenous nutrients are required for spore germination or hyphal elongation and screening of leaf-associated bacteria in vitro for rapid growth on these compounds.

53.4.2. Siderophore-Mediated Iron Competition

Siderophore-mediated competition for iron represents a specific case of nutrient competition. Studies on the role of siderophore-mediated iron competition in the biological control of fungal pathogens have primarily been conducted in rhizospere systems. In the rhizosphere, there have been more or less equal numbers of studies indicated that siderophores are involved in biological control (13, 17, 45, 64) and that siderophores are not involved in biological control (23, 24, 39, 73). Even among those studies indicating a role for siderophores, relatively few have convincingly demonstrated involvement of siderophores in iron competition (77). In the other cases, there appears to be some evidence that siderophores are involved in microbially mediated induced systemic resistance (14) (see below). The possibility of iron limitation and competition in the phyllosphere has been elegantly investigated using an iron-responsive pyoverdine siderophore promoter fused to a promoterless ice nucleation protein gene (*pvd-inaZ*) (65).These studies showed that while some limitation for iron existed among the population of cells of leaves, iron availability was greater than that expected based on the chemistry of iron in this habitat (65). Furthermore, iron limitation sensed by *P. syringae* cells containing the reporter construct was increased by prior inoculation of leaves with the biological control agent *P. fluorescens* A506 (57a).

Having demonstrated the involvement of a siderophore in the biological control of a particular foliar pathogen or pest, either in siderophore-mediated iron competition or in the induction of host physiological responses, additional potential biocontrol agents could be selected that express high concentrations of siderophores in vitro, particularly in low iron media, such as King's medium B, or media amended with iron chelators such as EDTA or HBED [*N*,*N*′-di-(2-hydroxybenzoyl)-ethylenediamine-*N*,*N*′-diacetic acid]. However, one must realize that since siderophore production is iron-regulated, in planta siderophore production and in vitro siderophore production are not necessarily correlated. Alternatively, a biocontrol agent could be created by engineering constitutive siderophore production, introducing heterologous receptors for the pathogen's siderophore, or heterologous siderophore biosynthetic genes and receptors (16, 68, 77).

53.4.3. Antibiosis

The contribution of bacteriocins and antibiotics to biological control of bacterial pathogens of plants remains somewhat contradictory. For example, both narrow-spectrum bacteriocins and broad-spectrum antibiotics have both been proposed to be involved in the biological control of *E. amylovora* by *E. herbicola* (*Pantoea* spp.) (31, 94). However, even under controlled conditions, antibiotics apparently account for only part of the efficacy of inhibitory strains of *E. herbicola* (94). In most cases, the use of in vitro antibiosis to select biocontrol agents of bacterial pests and pathogens has been discredited owing to lack of quantitative correlation between in vitro antibiosis and in planta activity. The immature pear fruit assay, an assay consisting of an immature pear slice into which a well is cut and which is inoculated first with the potential biocontrol agent and then with *E. amylovora*, although perhaps preferable to selection of biocontrol agents based upon in vitro inhibition zones, also tends to select for antibiotic producers and is, therefore, subject to the same limitations. Hence, for bacterial pathogens, at least, strategic selection based upon in vitro or even in planta antibiotic production is probably not wise (2).

The situation may be somewhat different with regard to foliar fungal pathogens. Fluorescent *Pseudomonas* spp. produce numerous secondary metabolites that exhibit antifungal activity in vitro, including acetylphloroglucinols (e.g., 2,4-diacetylphloroglucinol); oomycin A; phenazines (e.g., phenazine-1-carboxylic acid); pyocyanine; and the pyrroles, pyoluteorin and pyrrolnitrin. Biological control of *Septoria tritici* and *Puccinia recondita* on wheat by *P. fluorescens* strain PFM2 was attributed to antibiotic production (47, 48). *E. herbicola* (*Pantoea* spp.) produces both antibacterial bacteriocins and antibiotics and antifungal antibiotics. The antifungal antibiotic was detected in plant tissues in the crown region of wheat plants treated with *E. herbicola* B247 (40); hence, it could contribute to the biological control of foliar fungal pathogens. In such situations, there may be some justification for a preliminary in vitro screen for the potential to produce antifungal antibiotics. Such a screening should be performed on more than one type of medium, preferably including nutrient-poor media or even a medium based upon leaf washates; it should only be qualitative (i.e., the quantity produced in vitro almost certainly will not correlate with the quantity produced in planta); and it should not be too exclusive (although the time, space, and funds available for the second-level empirical screen also factor into the rigor of the preliminary screening).

53.4.4. Rhizobacteria-Mediated Induced Systemic Resistance

In recent years, it has been discovered that certain rhizobacteria have the capacity to induce physiological changes within the plant that provide systemic protection against a broad range of foliar pathogens, including bacteria (*P.*

syringae pv. phaseolicola [1]; *P. syringae* pv. tomato [36]; *P. syringae* pv. tabaci [74]; *P. syringae* pv. lachrymans [90]), fungi (*Colletotrichum orbiculare* [89, 90]; *Peronospora tabacina* [85]), and viruses (cytomegalovirus [78]). This systemic protection is superficially similar to the systemic acquired resistance that can be induced by treatment of leaves with a necrotizing pathogen or certain chemicals such as 2,6-dichloroisonicotinic acid. Pathogen-mediated systemic acquired resistance is a salicylic acid-dependent process, which is eliminated in plants expressing the *nahG* tansgene which encodes production of salicylate hydroxylase, and involves production of pathogenesis-related proteins in the plant, believed to be involved in plant defense. The signaling pathways involved in rhizobacteria-induced systemic resistance (ISR) do not appear to be salicylic acid-dependent or involve pathogenesis-related protein accumulation, and it has been proposed that signaling may involve jasmonate or ethylene (86). Unfortunately, little is known about the microbial component(s) or product(s) that may be involved in the induction of ISR; however, both surface lipopolysaccharides (46) and siderophores, such as salicylic acid and pseudobactin (14, 45), have been proposed as candidates.

Although one might prescreen potential biocontrol agents for in vitro siderophore production (see above) or salicylate production (a novel bioassay for salicylic acid based on a *nah-lux* fusion was developed by Press et al. [74]), not all siderophores appear to have the ability to induce ISR, and microbial salicylic acid is apparently not involved in all systems (74). If in the future pathogenesis-related gene expression is demonstrated to be involved in rhizobacterial ISR, a prescreen might be developed using seedlings of transgenic PR1a-Gus plants (72). Alternatively, the possible involvement of ethylene in the signal transduction pathway in microbially mediated ISR could facilitate the development of an in vitro prescreen. For example, Glick et al. (20) found that the ability to utilize 1-aminocylcopropane-1-carboxylate (ACC) as a nitrogen source in vitro, indicating the presence of the enzyme ACC deaminase, is correlated with microbial ability to promote growth, presumably owing to interference with tissue ethylene levels, since ACC is an ethylene precursor. Since there is a lot of interest in the possible commercial use of plant growth-promoting rhizobacteria for the induction of systemic resistance against both foliar and soilborne pathogens, the development of a rapid in vitro or in planta screen for the ISR phenotype would be extremely useful.

53.4.5. Induction of Resistance by Foliar Bacteria

Although preemptive exclusion/suppression of leaf surface populations of *P. syringae* pathogens provides effective reductions in the severity and incidence of disease caused by pathovars of this species, the same approach does not appear to be so effective against diseases caused by *Xanthomonas* spp. Burkett (12) showed that treatment of cabbage leaves with the incompatible pathovar *Xanthomonas campestris* pv. malvacearum, in suspension with a small amount of organosilicone surfactant, which facilitated stomatal penetration, resulted in "immunization" of the leaves against the black rot pathogen *X. campestris* pv. campestris. While no selection of strains is involved here, except the selection of an incompatible pathovar, one might develop an in planta seedling screen for maximal induction of protection against a compatible pathogen by leaf-associated bacteria.

53.5. EVALUATION OF BIOLOGICAL CONTROL EFFICACY

There are two general methods for evaluating the efficacy of bacteria in biological control. Clearly, the most direct method is to simply measure the incidence or severity of disease on plants treated with a potential biocontrol agent compared with that on a control plant not treated with such a strain. Initiating disease on plants is often not a trivial undertaking, however, since often very specific environmental conditions must be provided to enable the infection process to be initiated. Because of the strong environmental influence on disease of most plants, the incidence of disease also usually varies substantially from one experiment to the next unless great care is taken to carefully control experimental conditions. Such experiment-to-experiment variability can complicate the interpretation of assays in which bacterial strains are being compared for their ability to control disease. In addition, large plants such as trees, plants that are not easily propagated in a greenhouse, and seasonal plant parts such as flowers or fruit limit when and where biological control studies can be undertaken.

Since biocontrol agents most often simply reduce the likelihood of disease rather than completely preventing it, quantitative assays of disease are required to properly assess biological control efficacy. For those diseases that are characterized by discrete lesions or a systemic symptom in an entire plant, disease can be readily quantified by counting the number of lesions on a given plant or plant part or by determining the fraction of plants that have become systematically infected. If such evaluations are not possible, then the severity of disease must be estimated. Unfortunately, visual estimates of disease are often poor and are influenced greatly by the observer, and by the size, shape, and number of lesions on a leaf (69). For this reason, visual estimates of disease are much more accurate and quantitative if visualized disease can be compared to pictorial reference guides, showing different levels of disease (70); such disease assessment keys have been constructed for a variety of diseases. Recent advances in computer technology have allowed digital images of diseased plants to be analyzed to provide rapid and accurate estimates of disease (63). Such approaches are recommended for future studies because of the great improvement in disease assessment that they provide.

Since disease severity may be difficult to reproduce under laboratory or greenhouse conditions, measurements of pathogen abundance are often useful as indicators of the biological control efficacy of potential biocontrol agents. As noted above, biological control of most bacterial diseases and many fungal diseases is to reduce the ability of the pathogen to increase after immigration to a plant. Thus, the efficacy of an antagonist to reduce disease can be predicted from its ability to prevent the growth of pathogens on leaves. Owing to the relationship between the logarithm of bacterial populations and either subsequent disease incidence (79) or frost injury (51, 57), accurate estimates of biological control potential can be made from quantitative estimates of the reduction of population sizes of the target microbe in the presence of a potential biocontrol agent.

There are many factors that must be considered to obtain a reliable and unbiased estimate of pathogen population sizes on leaves. The bacterial population size varies greatly from leaf to leaf (26, 27, 31) and even from one portion of a single leaf to another (44). Population sizes of particular bacterial pathogens such as *P. syringae* have been shown to vary by over 100,000-fold on leaves of the same age and appearance in a plant canopy (26, 27). Bacterial population sizes among leaves are usually best described by a log-normal distribution (26, 27). This great variation of bacterial population sizes places great constraints on sampling strategies. For example, bulking of multiple leaves in a few samples greatly expedites processing of samples of leaves but will result in a great overestimation of the mean bacterial population size on those leaves (26, 44). Bulking also has been shown to obscure differences in mean bacterial population sizes that occur in different treatments (44). Thus, while it takes more effort, it is possible to obtain more information and a better differentiation of bacterial population sizes in different treatments from an analysis of bacterial population sizes on a relatively large number (ca. 20) of individual leaves.

The normal procedure for quantifying bacterial population sizes is to perform dilution plating analysis of suspensions of bacteria recovered from either washings or macerates of colonized leaves (28, 30, 71). This procedure works well for those plant-associated bacteria that are readily culturable (93). Unfortunately, this method requires considerable labor and materials and so is prohibitively expensive to employ to quantify bacteria on the thousands of leaves that might need to be assayed in the course of evaluating many potential biocontrol agents.

An effective method has recently been developed to facilitate estimations of bacterial population sizes on a large number of leaves. The population size of Ice$^+$ bacteria on leaves can be estimated from the nucleation temperature of leaves harboring such species (25, 56). The threshold freezing temperature of leaves harboring different numbers of cells of an Ice$^+$ *P. syringae* strain differed substantially (25, 56). The nucleation temperature of leaves increased linearly from about -5 to -2°C as the population size of an Ice$^+$ *P. syringae* strain increased logarithmically from about 10^2 to 10^7 cells per leaf (56). While few leaves containing less than about 10^6 cells per gram (fresh weight) froze at assay temperatures of -2.75°C or higher, nearly all leaves froze at those temperatures when population sizes of this strain increased to about 10^7 cells per gram (fresh weight) (56). The leaf-freezing assay was capable of differentiating samples that varied by approximately three- to fivefold in mean bacterial population size (56).

Ice nucleation genes have been cloned and characterized from several Ice$^+$ bacterial strains (22, 87, 88, 98). In all cases, a single gene confers ice nucleation activity. All gram-negative bacterial species into which ice nucleation genes have been transferred have expressed ice nucleation activity (22, 87). Since nearly all bacterial plant pathogens are gram-negative, they can be converted to an Ice$^+$ phenotype by introduction of a bacterial ice nucleation gene, and their population size can be estimated by a plant-freezing assay as noted above.

The utility of rapidly estimating changes in bacterial population sizes mediated by potential biocontrol agents by a plant-freezing assay was recently demonstrated (67). *E. amylovora*, which is not Ice$^+$, expressed high ice nucleation activity when the *iceC* gene from *P. syringae* was introduced on a stable plasmid (67). The population size of *E. amylovora* on individual pear flowers was easily estimated from the nucleation temperature of flowers inoculated with this strain (67). The relative ability of several hundred randomly isolated bacterial strain from pear flowers to inhibit the growth of *E. amylovora* in flowers when preinoculated onto flowers was easily determined from the relative freezing temperature of flowers challenge-inoculated with the Ice$^+$ *E. amylovora* strain. Such an approach should be readily applicable to assaying the ability of potential biocontrol agents to prevent the growth of other pathogens on other plant species. Clearly, the likelihood of identifying bacterial strains that are highly effective in biological control will be improved by employing such powerful methods to quantify the results of interactions of microbial strains on plants.

53.6. SUMMARY

As discussed in this review, it is clear that there are many considerations for developing strategies for selecting effective biological control agents. Clearly, if sufficient information is known about the processes involved in the interactions of microbes in a given habitat, then powerful methods can be employed to identify microbes that will greatly alter the normal pattern of colonization of a plant and/or the process of infection. It should be the goal of microbial ecologists to better define both the nature of plant surfaces that are the site for these interactions and microbial traits that mediate antagonistic interactions. Despite our lack of such detailed information for most systems, it is still possible to employ effective empirical screening methods that can identify effective biological control agents. It is hoped that such empirical methods will identify microbial strains with such great efficacy in biological disease control that such successes will further spur the research in microbial ecology of plant-associated microorganisms that will be needed to identify even more effective strains for future use.

REFERENCES

1. **Alstrom, S.** 1991. Induction of disease resistance in common bean susceptible to halo blight bacterial pathogen after seed bacterization with rhizosphere pseudomonads. *J. Gen. Appl. Microbiol.* **37:**495–501.
2. **Andrews, J. H.** 1985. Strategies for selecting antagonistic microorganisms from the phylloplane, p. 31–44. *In* C. E. Windels and S. E. Lindow (ed.), *Biological Control on the Phylloplane*. American Phytopathological Society, St. Paul, Minn.
3. **Andrews, J. H.** 1990. Biological control in the phyllosphere: realistic goal or false hope? *Can. J. Plant Pathol.* **12:**300–307.
4. **Andrews, J. H.** 1992. Biological control in the phyllosphere. *Annu. Rev. Phytopathol.* **30:**603–635.
5. **Beattie, G. A., and S. E. Lindow.** 1995. The secret life of bacterial colonists of leaf surfaces. *Annu. Rev. Phytopathol.* **33:**145–172.
6. **Beattie, G. A., and S. E. Lindow.** 1995. Epiphytic fitness of phytopathogenic bacteria: physiological adaptations for growth and survival, p. 1–28. *In* J. L. Dangl (ed.), *Bacterial Pathogenesis of Plants and Animals: Molecular and Cellular Mechanisms*. Springer-Verlag, Berlin.
7. **Blakeman, J. P.** 1975. Germination of *Botrytis cinerea* conidia *in vitro* in relation to nutrient conditions on leaf surfaces. *Trans. Br. Mycol. Soc.* **65:**239–247.

8. **Blakeman, J. P., and I. D. S. Brodie.** 1977. Competition for nutrients between epiphytic micro-organisms and germination of spores of plant pathogens on beetroot leaves. *Physiol. Plant Pathol.* **10:**29–42.
9. **Blakeman, J. P., and N. J. Fokkema.** 1982. Potential for biological control of plant disease on the phylloplane. *Annu. Rev. Phytopathol.* **20:**167–192.
10. **Brodie, I. D. S., and J. P. Blakeman.** 1975. Competition for carbon compounds by a leaf surface bacterium and conidia of *Botrytis cinera*. *Physiol Plant Pathol.* **6:**125–135.
11. **Brodie, I. D. S., and J. P. Blakeman** 1976. Competition for exogenous substrates *in vitro* by leaf surface microorganisms and germination of conidia of *Botrytis cinerea*. *Physiol. Plant Pathol.* **9:**227–239.
12. **Burkett, J. E.** 1977. Field evaluation and physiological characterization of plant immunization as a disease control strategy. M. S. thesis. Auburn University, Auburn, Ala.
13. **Buysens, S., K. Heungens, J. Poppe, and M. Hofte.** 1996. Involvement of pyochelin and pyoverdin in suppression of Pythium-induced damping-off of tomato by *Pseudomonas aeruginosa* 7NSK2. *Appl. Environ. Microbiol.* **62:**865–871.
14. **De Meyer, G., and M. Hofte.** 1997. Salicylic acid produced by the rhizobacterium *Pseudomonas aeruginosa* 7NSK2 induces resistance to leaf infection by *Botrytis cinerea* on bean. *Phytopathology* **87:**588–593.
15. **Dianese, A. C., and M. Wilson.** 1996. Evaluation of the importance of nutritional similarity between biocontrol agents and the target pathogen in biocontrol of *Xanthomonas campestris* pv. *vesicatoria* on tomato. *Phytopathology* **86:**S50.
16. **Dowling, D. N., Boesten, D. J. O'Sullivan, P. Stephens, J. Morris, and F. O'Gara.** 1996. Genetically modified plant-microbe interacting strains for potential release into the rhizosphere, p. 408–414. *In* E. Galli, S. Silver, and B. Witholt (ed.), *Pseudomonas: Molecular Biology and Biotechnology*. American Society for Microbiology, Washington, D.C.
17. **Duijff, B. J., P. A. H. M. Bakker, and B. Schippers.** 1994. Suppression of Fusarium wilt of carnation by *Pseudomonas putida* WCS358 at different levels of disease incidence and iron availability. *Biocontrol Sci. Technol.* **4:** 279–288.
18. **Elad, Y.** 1993. Microbial suppression of infection by foliar plant pathogens. *IOBC Bull.* **16:**3–7.
19. **Fokkema, N. J.** 1993. Opportunities and problems of control of foliar pathogens with micro-organisms. *Pesticide Sci.* **37:**411–416.
20. **Glick, B. R., D. M. Karaturovic, and P. C. Newel.** 1995. A novel procedure for rapid isolation of plant growth promoting pseudomonads. *Can. J. Microbiol.* **41:**533–536.
21. **Gross, D. C., Y. S. Cody, E. L. Proebsting, Jr., G. K. Radamaker, and R. A. Spotts.** 1983. Distribution, population dynamics, and characteristics of ice nucleation active bacteria in deciduous fruit tree orchards. *Appl. Environ. Microbiol.* **46:**1370–1379.
22. **Gurian-Shermann, D., and S. E. Lindow.** 1993. Bacterial ice nucleation: significance and molecular basis. *FASEB J.* **9:**1338–1343.
23. **Hamdan, H., D. M. Weller, and L. Tomashow.** 1991. Relative importance of fluorescent siderophores and other factors in biological control of *Gaeumannomyces graminis* var. *tritici* by *Pseudomonas fluorescens* 2-79 and M4-80R. *Appl. Environ. Microbiol.* **57:**3270–3277.
24. **Henry, M. B., J. M. Lynch, and T. R. Fermor.** 1991. Role of fluorescent siderophores in the biocontrol of *Pseudomonas tolaasii* by fluorescent pseudomonad antagonists. *J. Appl. Bacteriol.* **70:**104–108.
25. **Hirano, S. S., L. S. Baker, and C. D. Upper.** 1985. Ice nucleation temperature of individual leaves in relation to population sizes of ice nucleation active bacteria and frost injury. *Plant Physiol.* **77:**259–265.
26. **Hirano, S. S., E. V. Nordheim, D. C. Arny, and C. D. Upper.** 1982. Lognormal distribution of epiphytic bacterial populations on leaf surfaces. *Appl. Environ. Microbiol.* **44:**695–700.
27. **Hirano, S. S., D. I. Rouse, M. K. Clayton, and C. D. Upper.** 1995. *Pseudomonas syringae* pv. *syringae* and bacterial brown spot of bean: a study of epiphytic phytopathogenic bacteria and associated disease. *Plant. Dis.* **79:** 1085–1093.
28. **Hirano, S. S., and C. D. Upper.** 1983. Ecology and epidemiology of foliar plant pathogens. *Annu. Rev. Phytopathol.* **21:**243–269.
29. **Hirano, S. S., and C. D. Upper.** 1989. Diel variation in population size and ice nucleation activity of *Pseudomonas syringae* on snap bean leaflets. *Appl. Environ. Microbiol.* **55:**623–630.
30. **Hirano, S. S., and C. D. Upper.** 1990. Population biology and epidemiology of *Pseudomonas syringae*. *Annu. Rev. Phytopathol.* **28:**155–177.
31. **Ishimaru, C., K. M. Eskridge, and A. K. Vidaver.** 1991. Distribution analysis of naturally occurring epiphytic populations of *Xanthomonas campestris* pv. *phaseoli* on dry beans. *Phytopathology* **82:**262–268.
32. **Jacques, M.-A., L. L. Kinkel, and C. E. Morris.** 1995. Population sizes, immigration, and growth of epiphytic bacteria on leaves of different ages and positions of field-grown endive (*Cichorium endivia* var. *latifolia*). *Appl. Environ. Microbiol.* **61:**899–906.
33. **Ji, P., H. L. Campbell, S. E. Lindow, and M. Wilson.** 1996. Determination of the importance of nutritional similarity between biocontrol agents and the target pathogen in biocontrol of *Pseudomonas syringae* pv. *tomato* on tomato. *Phytopathology* **86:**S50.
34. **Ji, P., S. E. Lindow, and M. Wilson.** 1998. Role of nutritional similarity and population size of *Pseudomonas syringae* TLP2 mutants in pre-emptive exclusion of *P. s.* pv. tomato. Abstr. 98th Gen. Meet. Am. Soc. Microbiol. 1998. American Society for Microbiology, Washington, D.C.
35. **Ji, P., M. Wilson, and H. L. Campbell.** 1997. Molecular approaches to determine the role of pre-emptive carbon source use in the biocontrol of bacterial speck. *In* Proc. IOBC Workshop, Molecular Approaches in Biological Control, Delemont, Switzerland, Sept., 15–18, 1997.
36. **Ji, P., M. Wilson, H. L. Campbell, and J. W. Kloepper.** 1977. Rhizobacterial-mediated induced systemic resistance for the control of bacterial speck of fresh-market tomato, p. 273–276. *In* A. Ogoshi, K. Kobayashi, Y. Homma, F. Kodama, N. Kondo, and S. Akino (ed.), *Plant Growth-Promoting Rhizobacteria: Present Status and Future Prospects*. OECD Press, New York.
37. **Ji, P., M. Wilson, H. L. Campbell, and S. E. Lindow.** 1997. Determination of the importance of pre-emptive

carbon source use by nonpathogenic bacteria in the biocontrol of bacterial speck of tomato. *Phytopathology* **87:** S48.

38. **Johnson, K. B., V. O. Stockwell, R. J. McLaughlin, D. Sugar, J. E. Loper, and R. G. Roberts.** 1993. Effect of bacterial antagonists on establishment of honey bee-dispersed *Erwinia amylovora* in pear blossoms and on fire blight control. *Phytopathology* **83:**995–1002.
39. **Keel, C., and G. Defago.** 1991. The fluorescent siderophore of Pseudomonas fluorescens strain CHAO has no effect on the suppression of root diseases of wheat, p. 136–142. *In* N. Keel, W. Koller, and J. Defago (ed.), *PGPR — Progress and Prospects*.
40. **Kempf, H.-J., P. H. Bauer, and M. N. Schroth.** 1993. Herbicolin A associated with crown and roots of wheat after seed treatment with *Erwinia herbicola* B247. *Phytopathology* **83:**213–216.
41. **Kinkel, L. L., J. H. Andrews, and S. V. Nordheim.** 1989. Fungal immigration dynamics and community development on apple leaves. *Microb. Ecol.* **18:**45–58.
42. **Kinkel, L. L., and S. E. Lindow.** 1993. Invasion, exclusion, and coexistence among intraspecific bacterial epiphytes. *Appl. Environ. Microbiol.* **59:**3447–3454.
43. **Kinkel, L. L., and S. E. Lindow.** 1997. Microbial competition and plant disease biocontrol, p. 128–138. *In* D. Andow, D. Ragsdale, and R. Nyvall (ed.), *Ecological Interactions and Biological Control*. Westview Press, New York.
44. **Kinkel, L., M. Wilson, and S. E. Lindow.** 1995. Effects of scale on estimates of epiphytic bacterial populations. *Microb. Ecol.* **29:**283–297.
45. **Leeman, M., F. M. den Ouden, J. A. van Pelt, F. P. M. Dirkx, H. Steijl, P. A. H. M. Bakker, and B. Schippers.** 1996. Iron availability affects induction of systemic resistance to Fusarium wilt of radish by *Pseudomonas fluorescens*. *Phytopathology* **86:**149–155.
46. **Leeman, M., J. A. van Pelt, M. den Ouden, M. Heinsbroek, P. A. H. M. Bakker, and B. Schippers.** 1995. Induction of systemic resistance against Fusarium wilt of radish by lipopolysaccharides of *Pseudomonas fluorescens*. *Phytopathology* **85:**1021–1027.
47. **Levy, E., Z. Eyal, S. Carmely, Y. Kashman, and I. Chet.** 1989. Suppression of *Septoria tritici* and *Puccinia recondita* by an antibiotic-producing fluorescent pseudomonad. *Plant Pathol.* **38:**564–570.
48. **Levy, E., F. J. Gough, K. D. Berlin, P. W. Guiana, and J. T. Smith.** 1992. Inhibition of *Septoria tritici* and other phytopathogenic fungi and bacteria by *Pseudomonas fluorescens* and its antibiotics. *Plant Pathol.* **41:**335–341.
49. **Lindemann, J., D. C. Arny, and C. D. Upper.** 1984. Use of an apparent infection threshold population of *Pseudomonas syringae* to predict incidence and severity of brown spot of bean. *Phytopathology* **74:**1334–1339.
50. **Lindow, S. E.** 1982. Population dynamics of epiphytic ice nucleation active bacteria on frost sensitive plants and frost control by means of antagonistic bacteria, p. 395–416. *In* P. H. Li and A. Sakai (ed.), *Plant Cold Hardiness*. Academic Press, New York.
51. **Lindow, S. E.** 1983. The role of bacterial ice nucleation in frost injury to plants. *Annu. Rev. Phytopathol.* **21:**363–384.
52. **Lindow, S. E.** 1983. Methods of preventing frost injury caused by epiphytic ice nucleation active bacteria. *Plant Dis.* **67:**327–333.
53. **Lindow, S. E.** 1985. Strategies and practice of biological control ice nucleation active bacteria on plants, p. 293–311. *In* N. Fokkema (ed.), *Microbiology of the Phyllosphere*. Cambridge University Press, Cambridge.
54. **Lindow, S. E.** 1985. Integrated control and role of antibiosis in biological control of fireblight and frost injury, p. 83–115. *In* C. Windels and S. E. Lindow (ed.), *Biological Control on the Phylloplane*. American Phytopathological Society, St. Paul, Minn.
55. **Lindow, S. E.** 1987. Competitive exclusion of epiphytic bacteria by Ice mutants of *Pseudomonas syringae*. *Appl. Environ. Microbiol.* **53:**2520–2527.
56. **Lindow, S. E.** 1993. Novel method for identifying bacterial mutants with reduced epiphytic fitness. *Appl. Environ. Microbiol.* **59:**1586–1592.
57. **Lindow, S. E.** 1995. Control of epiphytic ice nucleation-active bacteria for management of plant frost injury, p. 239–256. *In* R. E. Lee, G. J. Warren, and L. V. Gusta (ed.), *Biological Ice Nucleation and Its Applications*. American Phytopathological Society, St. Paul, Minn.

57a. **Lindow, S. E.,** Unpublished data.

58. **Lindow, S. E., and G. A. Andersen.** 1996. Influence of immigration in establishing epiphytic bacterial populations on navel orange. *Appl. Environ. Microbiol.* **62:**2978–2987.
59. **Lindow, S. E., D. C. Arny, and C. D. Upper.** 1978. Distribution of ice nucleation active bacteria on plants in nature. *Appl. Environ. Microbiol.* **36:**831–838.
60. **Lindow, S. E., D. C. Arny, and C. D. Upper.** 1983. Biological control of frost injury. I: an isolate of *Erwinia herbicola* antagonistic to ice nucleation-active bacteria. *Phytopathology* **73:**1097–1102.
61. **Lindow, S. E., D. C. Arny, and C. D. Upper.** 1983. Biological control of frost injury. II: establishment and effects of an antagonistic *Erwinia herbicola* isolate on corn in the field. *Phytopathology* **73:**1102–1106.
62. **Lindow, S. E., G. McGourty, and R. Elkins.** 1996. Interactions of antibiotics with *Pseudomonas fluorescens* strain A506 in the control of fire blight and frost injury to pear. *Phytopathology* **86:**841–848.
63. **Lindow, S. E., and R. R. Webb.** 1983. Quantification of foliar plant disease symptoms by microcomputer-digitized video image analysis. *Phytopathology* **73:**520–524.
64. **Loper, J. E.** 1988. Role of fluorescent siderophore production in biological control of *Pythium ultimum* by a *Pseudomonas fluorescens* strain. *Phytopathology* **78:**166–172.
65. **Loper, J. E., and S. E. Lindow.** 1994. A biological sensor for iron available to bacteria in their habitats on plant surfaces. *Appl. Environ. Microbiol.* **60:**1934–1941.
66. **Lukezic, F. L., K. T. Leath, M. Jones, and F. G. Levine.** 1990. Efficiency and potential use in crop protection of the naturally occurring resident antagonists on the phylloplane, p. 793–812. *In* R. R. Baker and P. E. Dunn (ed.), *New Directions in Biological Control: Alternatives for Suppressing Agricultural Pests and Diseases*. Alan R. Liss, Inc., New York.
67. **Mercier, J., and S. E. Lindow,** 1996. A method involving ice nucleation for the identification of microorganisms antagonistic to *Erwinia amylovora* on pear flowers. *Phytopathology* **86:**940–945.
68. **Moenne-Loccoz, Y., B. McHugh, P. M. Stephens, F. I. McConnell, J. D. Glennon, D. N. Dowling, and F.**

O'Gara. 1996. Rhizosphere competence of fluorescent Pseudomonas sp. B24 genetically modified to utilize additional ferric siderophores. *FEMS Microbiol. Ecol.* **19:** 215–225.

69. **Nutter, F. W., M. L. Gleason, J. H. Jenco, and N. C. Christians.** 1993. Assessing the accuracy, intra-rater repeatability, and inter-rater reliability of disease assessment systems. *Phytopathology* **83:**806–812.

70. **Nutter, F. W., and P. M. Schultz.** 1995. Improving the accuracy and precision of disease assessments: selection of methods and use of computer-aided training programs. *Can. J. Plant Pathol.* **17:**174–184.

71. **O'Brien, R. D., and S. E. Lindow.** 1989. Effect of plant species and environmental conditions on epiphytic population sizes of *Pseudomonas syringae* and other bacteria. *Phytopathology* **79:**619–627.

72. **Park, S. K., A.-L. Moyne, S. Tuzun, C. H. Kim, and J. W. Kloepper.** 1997. Induction of PR1a promoter in a transgenic tobacco reporter system by selected PGPR strains which induce resistance, p. 251. *In* A. Ogoshi, K. Kobayashi, Y. Homma, F. Kodama, N. Kondo, and S. Akino (ed.), *Plant Growth-Promoting Rhizobacteria: Present Status and Future Prospects*. OECD Press, New York.

73. **Paulitz, T. C., and J. E. Loper.** 1991. Lack of a role for fluorescent siderophore production in the biological control of Pythium damping-off of cucumber by a strain of *Pseudomonas putida*. *Phytopathology* **81:**930–935.

74. **Press, C. M., M. Wilson, S. Tuzun, and J. W. Kloepper.** 1997. Salicylic acid produced by *Serratia marcescens* is not the primary determinant of induced systemic resistance in cucumber or tobacco. *Mol. Plant Microbe Interact.* **10:**761–768.

75. **Pusey, P. L.** 1996. Crab apple blossoms as a model system for fire blight biocontrol research. *Acta Horticulturae* **411:** 289–293.

76. **Pusey, P. L.** 1997. Effect of synthetic nectar on bacteria evaluated as biological control agents for fire blight of pome fruits. *Phytopathology* **87:**580.

77. **Raaijmakers, J. M., L. van der Sluis, M. Koster, P. A. H. M. Bakker, P. J. Weisbeek, and B. Schippers.** 1995. Utilization of heterologous siderophores and rhizosphere competence of fluorescent *Pseudomonas* spp. *Can. J. Microbiol.* **41:**126–135.

78. **Raupach, G. S., L. Liu, J. F. Murphy, S. Tuzun, and J. W. Kloepper.** 1996. Induced resistance in cucumber and tomato against cucumber mosaic cucumovirus using plant growth-promoting rhizobacteria (PGPR). *Plant Dis.* **80:** 891–894.

79. **Rouse, D. I., E. V. Nordheim, S. S. Hirano, and C. D. Upper.** 1985. A model relating the probability of foliar disease incidence to the population frequencies of bacterial plant pathogens. *Phytopathology* **75:**505–509.

80. **Sutton, J. C.** 1994. Biocontrol of aerial plant diseases: perspectives and application of epidemiology and microbial ecology, p. 140–149. *In* Proc. 4th Symposium on Biological Control. EMBRAPA CPACT, Brazil.

81. **Sutton, J. C., and G. Peng.** 1993. Manipulation and vectoring of biocontrol organisms to manage foliage and fruit diseases in cropping systems. *Annu. Rev. Phytopathol.* **31:** 473–493.

82. **Thompson, I. P., M. J. Bailey, R. J. Ellis, A. K. Lilley, P. J. McCormack, K. J. Purdy, and P. B. Rainey.** 1995. Short-term community dynamics in the phyllosphere microbiology of field-grown sugar beet. *FEMS Microbiol. Ecol.* **16:**205–211.

83. **Thompson, I. P., R. J. Ellis, and M. J. Bailey.** 1995. Autecology of a genetically modified fluorescent pseudomonad on sugar beet. *FEMS Microb. Ecol.* **17:**1–14.

84. **Thomson, S. V., M. N. Schroth, W. J. Moller, and W. O. Reil.** 1976. Efficacy of bactericides and saprophytic bacteria in reducing colonization and infection of pear flowers by *Erwinia amylovora*. *Phytopathology* **66:**1457–1459.

85. **Tuzun, S., J. Juarez, W. C. Nesmith, and J. Kuc.** 1992. Induction of systemic resistance in tobacco against metalaxyl-tolerant strains of *Peronospora tabacina* and the natural occurrence of the phenomenon in Mexico. *Phytopathology* **82:**425–429.

86. **Van Wees, S. C. M., C. M. J. Pieterse, A Trijssenarr, Y. A. M. Van't Westende, and L. C. Van Loon.** 1997. Differential induction of systemic resistance in Arabidopsis by biocontrol bacteria. *Mol. Plant Microbe Interact.* **10:** 716–724.

87. **Warren, G. J.** 1995. Identification and analysis of *ina* genes and proteins, p. 85–100. *In* R. E. Lee, G. J. Warren, and L. V. Gusta (ed.), *Biological Ice Nucleation and Its Applications*. American Phytopathological Society, St. Paul, Minn.

88. **Warren, G., and P. Wolber.** 1991. Molecular aspects of microbial ice nucleation. *Mol. Microbiol.* **5:**239–243.

89. **Wei, G., J. W. Kloepper, and S. Tuzun.** 1991. Induction of systemic resistance of cucumber to *Colletotrichum orbiculare* by select strains of plant growth-promoting rhizobacteria. *Phytopathology* **81:**1508–1512.

90. **Wei, G., J. W., Kloepper, and S. Tuzun.** 1996. Induced systemic resistance to cucumber diseases and increased plant growth by plant growth-promoting rhizobacteria under field conditions. *Phytopathology* **86:**221–224.

91. **Wilson, M., H. A. S. Epton, and D. C. Sigee.** 1989. *Erwinia amylovora* infection of hawthorn blossom. II. The stigma. *J. Phytopathol.* **127:**15–28.

92. **Wilson, M., H. A. S. Epton, and D. C. Sigee.** 1993. Interactions between *Erwinia herbicola* and *Erwinia amylovora* on the stigma of hawthorn. *Phytopathology* **82:**914–918.

93. **Wilson, M., and S. E. Lindow.** 1992. Relationship of total viable and culturable cells in epiphytic populations of *Pseudomonas syringae*. *Appl. Environ. Microbiol.* **58:**3980–3913.

94. **Wilson, M., and S. E. Lindow.** 1993. Interactions between the biological control agent *Pseudomonas fluorescens* A506 and *Erwinia amylovora* in pear blossoms. *Phytopathology* **83:**117–123.

95. **Wilson, M., and S. E. Lindow.** 1994. Ecological similarity and coexistence of epiphytic ice-nucleating (Ice^+) *Pseudomonas syringae* strains and a non-ice-nucleating (Ice^-) biological control agent. *Appl. Environ. Microbiol.* **60:** 3128–3137.

96. **Wilson, M., and S. E. Lindow.** 1994. Coexistence among epiphytic bacterial populations mediated through nutritional resources partitioning. *Appl. Environ. Microbiol.* **60:** 4468–4477.

97. **Wilson, M., D. C. Sigee, and H. A. S. Epton.** 1990. *Erwinia amylovora* infection of hawthorn blossom. III. The nectary. *J. Phytopathol.* **128:**62–74.

98. **Wolber, P. K.** 1993. Bacterial ice nucleation. *Adv. Microb. Physiol.* **34:**203–237.

Biomarkers and Bioreporters To Track Microbes and Monitor Their Gene Expression

JANET K. JANSSON AND FRANS J. DE BRUIJN

54

To understand the behavior, mode of action, and efficacy of genetically engineered microorganisms (GEMs) and other microbes of interest, it is important to have specific and sensitive methods for monitoring the cells and for assaying their activity. Sometimes a specific enzyme activity is of interest, such as during bioremediation of toxic compounds, and methods are also required to assess these specific activities in environmental samples. Such methods permit risk assessment analyses which can entail a description of the parent organism, the biochemical nature of the engineered trait, the fate and survival of the GEM in the environment, as well as possible effects of introduced GEMs on indigenous microbial populations (40).

However, the study of specific microorganisms in environmental samples is complicated by the tremendous complexity of natural ecosystems. For example, the soil environment has been approximated to contain between 10^9 and 10^{10} microbial cells of more than 10^3 distinct genotypes per gram (108). To monitor a specific genotype in these complex environments, molecular microbial ecology methods have been developed (4, 47, 49, 59, 82). Many of these methods are based on tracking cells tagged with biomarkers or bioreporters.

A biomarker is defined as an introduced gene that encodes a scorable phenotype. Biomarker tagged cells can also be monitored in environmental samples on the basis of their unique DNA sequence by DNA probes or by PCR amplification, although these methods, including genomic fingerprinting approaches, will not be described in this chapter (see reviews in references 4, 48, and 117).

A bioreporter is defined as an introduced gene with a readily assayable phenotype that is under the control of a specific promoter. Bioreporters can be used to monitor gene expression under specific environmental conditions. Bioreporter genes can also be used as industrial or environmental biosensors to detect the presence of particular chemicals or environmental conditions (see chapter 55 of this volume).

A number of diverse biomarkers/bioreporters are currently available. Their scorable phenotypes and the relevant methods for their detection vary considerably; therefore, the use of a given biomarker/bioreporter will depend on the particular environmental or industrial samples to be analyzed.

54.1. BIOMARKER GENES

54.1.1. Antibiotic Resistance Genes

The first genes to be used as biomarkers were those encoding resistance to particular antibiotics, such as the *nptII* (or Km^r) gene, encoding resistance to kanamycin and neomycin (e.g., see references 45 and 118). Other antibiotic resistance genes used as markers encode resistance to spectinomycin/streptomycin, chloramphenicol, tetracycline, ampicillin, and others (see reviews in references 55 and 99). The advantage of using antibiotic resistance genes as biomarkers lies in the ease with which tagged bacteria can be identified and selected for on antibiotic-containing media. Environmental samples can be plated on selective medium containing an antimicrobic so that only cells with the specific antibiotic resistance biomarker genes will grow. Some antibiotics, such as rifampin, are more selective than others and are more useful in complex environments containing large numbers of indigenous microbes (99). One major disadvantage of the use of antibiotic resistance genes as biomarkers for bacteria released into nature is the potential for transfer of the antibiotic resistance genes to pathogenic bacteria, thereby exacerbating the problem of the spread of drug resistance (55).

One of the first GEMs to be considered for widespread dissemination and agricultural application in the United States was a *Sinorhizobium meliloti* strain marked with a biomarker gene conferring resistance to streptomycin and spectinomycin, in addition to genes for the enhancement of biological nitrogen fixation (7). The U.S. Environmental Protection Agency (EPA) conducted a review of the potential hazards associated with the use of these antibiotic resistance biomarker genes (http://www.epa.gof/opptintr/biotech/factdft6.htm). Their conclusion was that the most likely hazardous scenario would be the transfer of streptomycin resistance to plant pathogens phylogentically related to *S. meliloti*, such as *Agrobacterium tumefaciens*. However, owing to the expected low probability of transfer and the fact that in the United States streptomycin has not been used in the treatment of crown gall disease, caused by *A. tumefaciens*, it was concluded that the genetically modified *Sinorhizobium* strain was not likely to represent a significant risk.

Although antibiotic resistance is a commonly used biomarker for detection of GEMs, concern about the introduction of large numbers of microbes carrying antibiotic resistance genes into nature has led to the development and use of alternative biomarker genes, with a reduced potential for undesired environmental effects.

54.1.2. Heavy Metal Resistance Genes

An alternative to antibiotic resistance as a selective biomarker is heavy metal resistance (for a review, see reference 68). The use of heavy metal resistance still enables the selection of biomarked populations on growth medium containing heavy metals, while the threat to the environment upon the release of organisms tagged with these metal resistance genes is considered to be minimal. However, one drawback of the use of heavy metal resistance biomarkers is that laboratory hazards are increased, since many of the metals used for resistance selection are toxic. For example, mercury resistance genes and arsenic resistance genes are excellent selective markers (44). However, mercury and arsenic acid are toxic, and considerable care is required during the preparation and use of growth media containing these compounds. Other heavy metal resistance biomarkers include resistances to zinc, copper, nickel, cadmium, chromate, cobalt, thallium, lead, antimony, and silver (68). One example of an environmental application of heavy metal resistance genes as biomarkers was the use of a gene cassette carrying the *czc* genes, encoding resistance to cadmium, zinc, and cobalt, to follow gene transfer by GEMs in soil (107).

One attractive recent option is the use of tellurite resistance (90). Unlike many of the other heavy metals, tellurite is nontoxic at concentrations used for selection, while remaining highly selective against the indigenous microflora (90). Moreover, colonies expressing the tellurite resistance gene are easily visualized as black colonies on medium containing tellurite (90).

54.1.3. Ice Nucleation Genes

The first GEM to be intentionally released in the United States was a mutant derivative of *Pseudomonas syringae* that reduced frost damage to crops (60). The mutant strain carries a deletion in the gene encoding an ice-nucleating protein that otherwise serves as a catalyst for ice formation on plants. The Ice$^-$ strain was found to prevent frost damage via preemptive colonization of the young leaves and prevention of colonization of the Ice$^+$ *P. syringae* strain (60). This field release study also paved the way for future risk assessment trials of other GEMs, since relevant governmental regulations were developed during this trial.

Recently, the ice nucleation "freezing assay" has been used to assess the potential for different biocontrol microorganisms to control the plant pathogen *Erwinia amylovora* tagged with the *iceC* gene. Ice$^+$ *E. amylovora* cells were coinoculated with different bacterial and yeast strains onto pear flowers, and the threshold freezing temperature of the flowers was determined. This assay was a good example of a quantitative assessment of reduction of freezing damage by different strains (67). Although there are few additional examples of the use of ice nucleation as a biomarker, the main application of this system to date has been for use as a bioreporter (see section 54.2.2).

54.1.4. Genes Encoding Chromogenic Substrate Cleavage Enzymes

A number of genes encoding metabolic enzymes that are capable of cleaving chromogenic substrates have been described and used successfully as biomarkers (Table 1). The β-galactosidase (*lacZ*) gene of *Escherichia coli* was one of the first genes of this class to be used to tag bacteria (28). Usually, it is used in combination with the *lacY* gene, encoding the enzyme lactose permease, which transports lactose across the cell membrane. The *lacZY* system is very simple to use and allows clear distinction of *lacZY*-marked colonies grown on medium containing X-Gal or other chromogenic, fluorescent, or chemiluminescent substrates (Table 1). One disadvantage with *lacZY* as a biomarker is that there is often a background of β-galactosidase activity in the environment, produced by indigenous microorganisms, such as enteric bacteria. Therefore, when *lacZY* is used as a biomarker for environmental samples, it is often used in combination with another biomarker to enhance selectivity.

The *celB* gene encodes the thermostable enzyme β-glucosidase, which also has a high β-galactosidase activity (94). The high degree of stability of this enzyme (half-life of 85 h at 100°C), allows the inactivation of background β-galactosidase activity, and the same inexpensive chromogenic substrates used to detect β-galactosidase activity encoded by the *lacZ* biomarker gene can be employed to identify *celB*-tagged microbes (94) (Table 1).

The *gusA* (or *uidA*) gene encodes the enzyme β-glucuronidase, which can cleave a variety of substrates, including chromogenic, fluorescent, and chemiluminescent compounds (Table 1) (94, 124). The GUS marker is particularly useful for monitoring plant-associated bacteria, owing to the absence of GUS enzyme activity in plant tissue or common microbes associated with plants. For example, GUS activity has been used to monitor *Azospirillum* spp. (15, 114), *Rhizobium* spp. (101, 126), and *Azoarcus* spp. (46) on plant surfaces. However, in some other environments, such as soil, the background of indigenous GUS$^+$ bacteria is similar to that of LAC$^+$ bacteria, and therefore, in these situations the problems with marker specificity are similar (126).

The catechol 2,3-dioxygenase (*xylE*) gene of *Pseudomonas putida* has also been used as a biomarker. The catechol 2,3-dioxygenase enzyme (C23O) will cleave catechol to generate a cleavage product (β-hydroxymuconic semialdehyde) with a bright yellow color, facilitating easy screening of *xylE*-tagged colonies (91) (Table 1). One advantage of *xylE* is that no special medium is required for growth of colonies since the substrate is sprayed onto the plate after growth. Although the background populations of bacteria containing *xylE* genes are usually very low, it is nevertheless often useful to use an additional marker for added selection (see below). The *xylE* gene has been used for several applications, including as a biomarker for monitoring *P. putida* in freshwater or soil microcosms (75, 128, 129) and for monitoring streptomycetes in soil (127).

54.1.5. Bioluminescence Genes

Another class of biomarkers that have been extensively used to track microbes is based on bioluminescence as the scorable phenotype. Bioluminescence markers are either of bacterial (*lux* genes) or eukaryotic (*luc* genes) origin. Bacteria tagged with bioluminescent markers can be easily identified on the basis of light production by a variety of methods (see section 54.3.4) (Fig. 1). Also, there is no background of *luc* activity in the microbial population in general and virtually no background of *lux* in most environments, with the exception of marine environments containing naturally bioluminescent bacteria. Therefore, the

TABLE 1. Examples of metabolic markers that cleave a substrate to produce a colored, chemiluminescent, or fluorescent product

Biomarker/bioreporter	Chromogenic, fluorescent, or luminescent substrate	Indicator change
lacZ (β-galactosidase) and *celB* (β-glucosidase)	MacConkey agar	Colorless to red
	5-Bromo-4-chloro-3-indolyl-β-D-galactopyranoside (X-Gal)	Coloress to blue
	o-Nitrophenyl-β-D-galactopyranoside (ONPG)	Colorless to yellow
	Fluorescein-di-β-D-galactopyranoside (FDG)	Fluorescent product
	Chlorinated derivative of 1,2-dioxetane phenyl-β-D-galactopyranoside (Cl-AMPGD)	Luminescent product
	Methylumbelliferyl-β-D-galactopyranoside (MUG)	Fluorescent product
xylE (catechol 2,3-dioxygenase)	Catechol	Colorless to yellow
phoA (alkaline phosphatase)	5-Bromo-4-chloro-3-indolyl phosphate	Colorless to blue
gusA or *uidA* (β-glucuronidase)	5-Bromo-4-chloro-3-indolyl-β-D-glucuronide (X-GlucA)	Colorless to blue
	Indoxyl-β-D-glucuroide	Colorless to indigo
	Fluorescein glucuronide	Fluorescent product
	5-Bromo-6-chloro-3-indolyl-β-D-glucuronide (Magenta-GlucA)	Colorless to magenta
	4-Methylumbelliferyl glucuronide (4-MUG)	Fluorescent product
	1,2-Dioxetane arylglucuronide (glucuron)	Luminescent product

bioluminescence phenotype is a very specific biomarker, with negligible background problems. An additional advantage of bioluminescence is that the light yield is quantitative and is directly proportional to the number of luminescent cells.

Although light is produced as a result of the luciferase enzymatic activities in both prokaryotic and eukaryotic systems, the corresponding enzymes and their reaction mechanisms differ substantially. The bacterial luciferase enzyme is encoded by two genes (*luxA* and *luxB*) and carries out the following reaction (R is a long-chain fatty aldehyde):

$$RCHO + FMNH_2 + O_2 \rightarrow RCOOH + FMN + H_2O + light$$

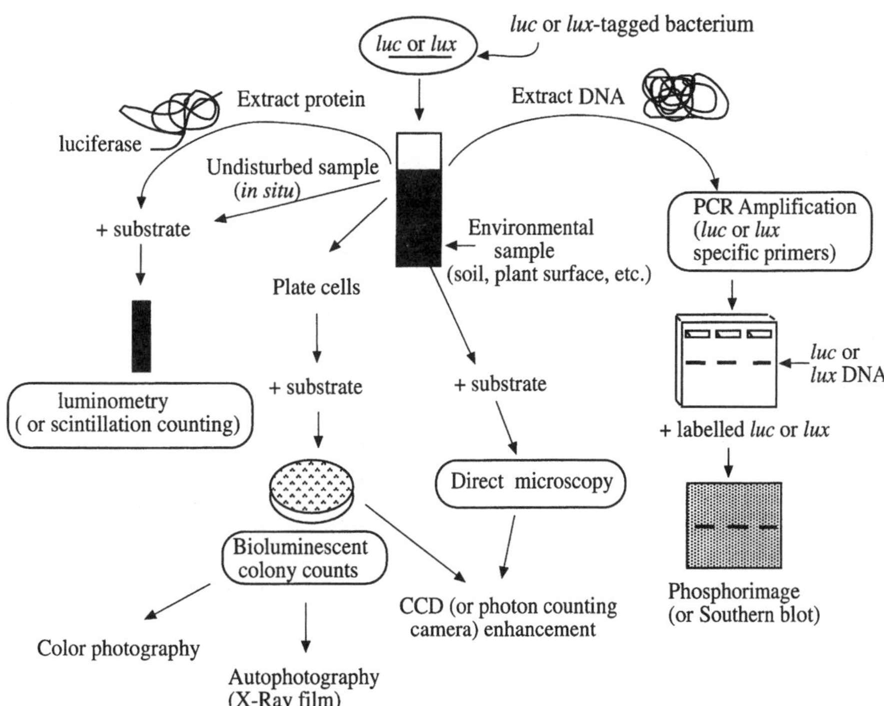

FIGURE 1 Methods for monitoring bioluminescent cells in environmental samples.

The *luxAB* genes can be introduced into other bacteria, which then produce light upon the addition of the aldehyde substrate (e.g., decanal) for the luciferase-catalyzed reaction. The requirement for adding the cofactor in environmental samples can be circumvented by introducing other genes of the *lux* operon, responsible for the synthesis of the aldehyde (*luxCD* and *E*). However, synthesis of the substrate places an energy burden on the cells, and it has been observed that cells tagged with the entire *lux* operon exhibit reduced viability compared to cells tagged with *luxAB* alone (1, 26).

The firefly luciferase enzyme, encoded by the *luc* gene, catalyzes the following reaction:

$$\text{luciferin} + \text{ATP} + O_2 \rightarrow \text{oxyluciferin} + PP_i + CO_2 + \text{light}$$

A number of highly homologous but very distinct *luc* genes have been isolated from firefly and click beetle species. The luciferase enzymes encoded by the genes isolated from these sources emit light of distinct wavelengths or colors (131). This property has been shown to be useful for distinguishing more than one *luc* gene-tagged GEM in a single sample.

Bioluminescence has been used for monitoring of several different bacterial species in a diversity of environmental samples. Luminometry is a useful technique for nondisruptive sampling of light output in soil. The majority of applications to date have used the *lux* biomarker, although the use of *luc* as a biomarker is increasing (for a review, see reference 82). For example, the *lux* biomarker has been used to monitor *Pseudomonas fluorescens* (65) and *E. coli* in soil (86). Shaw et al. (96) have used the *luxAB* genes to monitor a biomarker-tagged strain of *Xanthomonas campestris*, a phytopathogenic bacteria that causes black rot (96). The cells could be visualized in the plant phyllosphere using a charge-coupled device (CCD) camera and image analysis software. de Weger et al. (26) used the entire *luxABCDE* operon, and Beauchamp et al. (5) used Tn*5*-*luxAB* to introduce *luxAB* genes into the chromosome of different pseudomonads. In both cases, luminescent cells could be detected by luminometry and in the rhizosphere using a CCD camera (5, 26). The *lux*-based procedure was recently refined for rapid determination of the pattern of colonization of pseudomonads in the wheat rhizosphere (27). Luminometry, dilution plate counts, and CCD imaging were also used to monitor rhizosphere colonization by *luxAB*-tagged *Enterobacter cloacae* (85).

Examples of use of the *luc* gene as a biomarker include monitoring of *Rhizobia* spp. in soil (11, 92), *E. coli* in fresh water, and marine sediment microcosms (71) and cyanobacteria in Baltic Sea sediment microcosms (73). In these experiments, luminescent cells were monitored by luminometry or by counting luminescent colonies on selective medium.

54.1.6. Green Fluorescent Protein Genes

A relatively new biomarker gene is the green fluorescent protein (GFP) gene (*gfp*) originally isolated and cloned from the jellyfish *Aequorea victoria* (13). The wild-type GFP protein autofluoresces by emitting green light with a wavelength of 509 nm, upon excitation with UV light in the 395-nm range. A unique aspect of GFP-mediated fluorescence is that no cofactors or exogenous substrates are required, with the sole exception of oxygen which is briefly necessary for autoxidation of the GFP protein to facilitate autofluorescence (for a review, see reference 19). Various mutant derivatives of GFP have been created with shifted excitation or emission spectra for enhanced fluorescence properties (17, 18, 23, 41, 42). For example, one of the mutants (P4) is a GFP variant with the same excitation maximum as wild-type GFP, but with a blue-shifted emission maximum, resulting in a blue fluorescent protein (BFP) (41). In particular, many of the "red-shifted" GFP variants, having the excitation wavelength shifted toward a higher wavelength, have been found to be useful owing to the increased solubility of the GFP protein and enhanced fluorescence intensity in bacteria (17, 41). In addition, the longer wavelength of fluorescence is more compatible with common filters used for fluorescence microscopy and flow cytometry (see section 54.3.5).

The *gfp* marker gene system has proved to be ideal in environmental studies for the detection of single bacterial cells tagged with this biomarker gene. The GFP fluorescence intensity has been found to be sufficient to detect single cells with a single chromosomal insertion of the *gfp* gene by epifluorescence microscopy (31, 106), flow cytometry (105, 106, 109), or confocal microscopy (31, 106).

Other applications of GFP as a biomarker in environmental studies have included study of horizontal transfer of plasmids between two populations of *P. putida* at the level of single cells using GFP as a tag on the conjugal TOL plasmid (14). Single GFP-tagged cells have also been distinguished in activated sludge communities (31, 77). In addition, *R. meliloti* (37) and pseudomonads (6, 106) have been directly visualized on plant roots by confocal microscopy. Although we are primarily discussing bacteria in this chapter, it is interesting to note that the filamentous fungus *Aureobasidium pullulans*, when tagged with the GFP gene, could be easily detected by epifluorescence microscopy or confocal microscopy and distinguished on leaf surfaces (116).

These examples demonstrate that the use of GFP as a biomarker enables direct visualization of individual marked bacteria in environmental samples, without the need for sample disruption or stain addition. In addition, owing to the extreme stability of the GFP protein and its independence from cellular energy reserves or substrate addition, GFP is an ideal biomarker for enumeration of the total number of specific tagged cells, regardless of their energy status. For example, GFP-tagged pseudomonads remained fluorescent even after long-term starvation (106). Therefore, the GFP biomarker holds tremendous promise for elucidation of specific bacterial numbers and their behavior, colonization, distribution, interaction, and movement in situ in a diversity of environmental sample types.

54.1.7. Combinations of Biomarker Genes

Biomarker combinations are being increasingly utilized as a means for selective quantification of specific bacteria in nature. Combinations have included *luxAB*/*lacZY* (34), *lacZY*/*xylE*/Km^r (22, 104), and *gfp*/*luxAB* (110). Multiple biomarkers are used for a number of reasons: (i) to increase the specificity of detection, (ii) to increase the sensitivity of detection, (iii) to allow more versatile tracking approaches, and (iv) to enable the tracking of more than one tagged strain simultaneously. In addition, monitoring of two biomarkers integrated at distinct chromosomal locations can be used as a method for assessing genome stability (104).

Sometimes, the use of one biomarker gene is not sufficiently specific owing to the presence and expression of activity in the indigenous microbial population. This is particularly problematic in complex ecosystems, such as soil, often containing indigenous microbes that have β-galactosidase activity (e.g., reference 34) or GUS activity (125). Therefore, it is useful to be able to track a second biomarker, such as *luxAB* (34), *xylE*, or *Km*^{r} (22, 104, 125). For example, Wilson et al. (125) found a significant proportion of the soil population to be capable of growth on kanamycin-containing selective medium, and a significant proportion of CFU (approximately 20%) were GUS-positive. However, no colonies were found to be both kanamycin-resistant and GUS-positive (125), suggesting that the use of the combined biomarkers was sufficiently selective against the background microbial population.

The combination of *gfp* and *luxAB* biomarkers has also been demonstrated to be useful for monitoring both bacterial numbers and their physiological activity in soil (110). As previously mentioned, the GFP biomarker is useful for enumeration of *gfp*-tagged bacterial cell populations, regardless of their metabolic status (106). Therefore, the comparison of the numbers of GFP-fluorescing cells with the luciferase activity of the population can be useful for simultaneous determination of total cell counts and the energy status of the cells (110).

A combination of biomarker genes is useful for tagging different microbial strains that can be distinguished in the same environmental sample. For example, the use of both chromogenic substrates 5-bromo-4-chloro-3-indolyl-β-D-galactopyranoside (X-Gal) and 5-bromo-6-chloro-3-indolyl-β-D-glucuronide (Magenta-GlucA) were employed to track two populations of bacteria in nitrogen-fixing nodules on legume plants tagged either with the *gusA* or *celB* genes in parallel (94). Similarly, two *Rhizobia* spp. were tagged with different eukaryotic luciferase genes, one with the firefly luciferase (*luc*) gene and another with the click beetle luciferase gene (*lucOR*). Rhizobial colonies tagged with the firefly luciferase enzyme emitted light at 560 nm (yellow), whereas colonies tagged with the click beetle enzyme emitted light at 593 nm (orange), in the presence of the luciferin substrate, allowing an easy differential identification of the two marked strains on the same agar plates (12).

54.2. BIOREPORTER GENES

Bioreporter genes have been extensively used in microbial genetics and microbial ecology to monitor the regulation of expression of genes (promoters) to which they have been fused. They have proved to be extremely useful for study of environmental control of gene expression and to isolate promoters that are induced under particular physiological conditions. The latter property is important for the design of GEMs carrying beneficial genes whose expression needs to be carefully regulated. Expression-probe vectors have been generated that create both transcriptional and translational fusions to genes/promoters of interest. Many of the genes described above that constitute biomarker genes are also very useful bioreporter genes. Therefore, in this section, we will primarily review their relative utility to monitor gene expression, both qualitatively and quantitatively, under different physiological/environmental conditions.

54.2.1. Antibiotic Resistance Genes as Bioreporters

A number of antibiotic resistance genes have been used as bioreporters in a variety of gram-negative and gram-positive bacteria. The genes conferring kanamycin/neomycin (*nptII* or Km^r) or chloramphenicol (Cm^r or *cat*) resistance upon their host have been particularly useful, owing to the fact that the expression of these bioreporter genes can be not only "selected for" in growth medium but also quantified via enzyme assays for neomycin phosphotransferase (NptII) and chloramphenicol acetyltransferase (CAT) activity, respectively. These genes have been employed to monitor qualitative and quantitative gene expression in a wide variety of bacteria. For example, promoterless Cm^r genes were used to produce random reporter gene fusions in *X. campestris* to identify plant-inducible genes in this plant pathogen (78). Cells that were resistant to chloramphenicol on plants were identified as a result of induction of CAT activity from a plant-inducible promoter (78). Other bioreporters based on antibiotic genes include the tetracycline resistance (Tc^r) gene. However, these bioreporters can be used only for qualitative assessment of gene expression.

54.2.2. Ice Nucleation Gene as Bioreporter

The ice nucleation gene *inaZ* (or *iceC*) is useful as a sensitive and quantitative bioreporter (for a review, see reference 59), since ice nucleation activity increases with the second power of the concentration of the ice protein and can be quantitatively expressed as the logarithm of ice nuclei formed per CFU (29, 58). The ice nucleation reporter gene has been used to study iron-responsive genes of *P. syringae* (61), expression of different host promoters of *Zymomonas mobilis* (29), expression of plant-induced *P. syringae* phytopathogenicity genes (84), expression of phenazine biosynthesis genes of *Pseudomonas aureofaciens* (38), and as a sensor of copper on plants (88).

While the ice nucleation phenotype is extremely useful for both qualitative and quantitative reporter assays in the laboratory and on selected surfaces (e.g., plant leaves), its utility as a bioreporter in the environment (in situ) is otherwise relatively limited, and the environmental impact of the release of large numbers of Ice$^+$ bacteria remains unknown.

54.2.3. Genes Encoding Chromogenic Substrate Cleavage Enzymes as Bioreporters

All of the biomarker genes listed in section 54.1.4 that encode chromogenic substrate cleavage enzymes are also highly efficient bioreporters, owing to the fact that their activity can be quantitatively determined (Table 1).

In terms of free-living bacteria, the *lacZ* gene has been the first and one of the most frequently used bioreporter genes, especially in studies on enteric bacteria under laboratory conditions (for a review, see reference 98). However, the utility of this bioreporter gene is not restricted to enteric bacteria. For example, the *lacZ* gene has been used as a bioreporter to monitor carbon dioxide-induced genes of the marine bacterium *Pseudomonas* sp. strain S91 (102).

There are several examples of the use of the *lacZ* bioreporter to monitor gene expression of plant-associated bacteria. For example, *lacZ* activity was used to monitor expression of *nif* genes of *Azospirillum brasilense* or *Bradyrhizobium japonicum* on plant roots or soybean nodules, respectively (3, 122); hormone-controlled gene expression in the plant pathogen *P. syringae subsp. savastanoi* (36); *Erwinia stewartii* gene promoters involved in pathogenicity gene expression on corn plants (35); *P. fluorescens* genes responding to the phytopathogenic fungus *Pythium ultimum* on sugar beet plants (33); symbiotically important exopolysaccharide biosynthesis genes in the *R. meliloti*-alfalfa symbiosis (52); and a wheat root exudate-inducible promoter in *P. fluorescens* (119).

Thus, *lacZ* is a versatile bioreporter for examining the effect of both biotic and abiotic factors on a variety of bacteria. However, its utility in environmental settings, such as in soil and in association with plants, is hindered by the presence of endogenous microbial or plant β-galactosidases (8). Although protocols are available to reduce background β-galactosidase activities in these natural settings, and the use of the *celB*-encoded thermostable β-galactosidase enzyme allows for a more general application of this bioreporter in situ (94), nevertheless other bioreporter systems, including *gusA*, *lux*/*luc*, and *gfp*, may be more amenable to environmental studies.

The *gusA* gene bioreporter system was developed by Jefferson and colleagues (50, 126) and has proved to be particularly useful in genetic and ecological studies of plant-associated bacteria (e.g., rhizobia) (126). As previously mentioned (section 54.1.4), plants and rhizobia lack endogenous β-glucuronidase activity, and therefore there are no background problems when assaying this bioreporter for plant-microbe interactions in the absence of soil. For example, the *gusA* bioreporter gene has been used to study expression of nodulation (*nod*) genes of *R. loti* (80) and to study the temporal and spatial expression pattern of symbiotic genes of *R. meliloti* in planta (95).

The *xylE* gene has also been used as a bioreporter to study gene expression in a variety of bacteria with environmental applications, including carbon catabolite repression-regulated chitinase promoters of *streptomyces plicatus* (24), agglutination genes in *P. putida* species interacting with plants (9), and gene expression in mycobacteria (20). While the *xylE* gene is a versatile bioreporter for a highly diverse set of bacteria, C230 activity has been reported to be dependent on the physiological status of bacterial cells and the availability of oxygen (75). Therefore, the interpretation of quantitative *xylE* bioreporter activities must be subjected to careful consideration of these environmental parameters.

The *phoA* bioreporter gene, encoding alkaline phosphatase (see above), has been used for several applications including study of the secretion or periplasmic transfer of (chimeric) proteins in various bacterial systems, including *Clavibacter xyli* (112) and expression of the exopolysaccharide biosynthesis genes of *R. meliloti* (87). A mini-transposon carrying a promoterless *phoA* gene for the generation of bioreporter gene fusions, which should be useful for the analysis of gram-negative bacteria in general, has been described by De Lorenzo et al. (25). However, the use of the *phoA* bioreporter gene in environmental samples is also limited by the presence of endogenous alkaline phosphatase activity in resident microbial communities.

Therefore, the main problem with most of the bioreporters in this category, as mentioned above for their use as biomarkers, lies in the background enzymatic activity of indigenous bacteria, giving them limited utility in complex environmental samples, such as soil.

54.2.4. Luciferase Genes as Bioreporters

Transposons with promotorless *lux* genes have been constructed for identification of environmentally responsive genes in a number of microorganisms, including cyanobacteria (130), pseudomonads (53, 120), *Streptomyces* spp. (100), and rhizobia (21, 69).

Fusions of *lux* genes to specific promoters have been used to analyze the *pelE* promoter of *Erwinia chrysanthemi* (39), a plant-inducible gene of *X. campestris* (51), arsenic and cadmium resistance operons of *Staphylococcus aureus* (16), a regulatory gene for naphthalene catabolism (*nahG*) of *P. fluorescens* (43), and a regulatory gene involved in isopropylbenzene catabolism in *P. putida* (93).

The eukaryotic *luc* genes have also been used as bioreporters in a variety of applications. For example, Cebolla et al. (12) have compared the expression of *luc* and *lucOR* from both constitutive and regulated promoters in different bacterial strains. In another study, the *luc* gene was fused to the λPr promoter, and the luciferase activity was used to assay gene expression in a variety of gram-negative bacteria, including rhizobia and pseudomonads (79). Luciferase activity from *R. meliloti* cells tagged with the *luc* gene could even be demonstrated in intact nodules (79), demonstrating the utility of this bioreporter for in situ activity measurements. Lampinen et al. (57) have used *lucGR* as a bioreporter to follow gene induction upon a temperature shift to 42°C, to assay the efficiency of protein synthesis induction or inhibition under various conditions, such as after the addition of toxic compounds (57).

The luciferase bioreporters have several advantages over most other reporter systems owing to the sensitivity and quantitative potential of the assays. A major advantage of luciferase as a bioreporter is that its activity can be detected without disruption of the cells. Also, the light production is subject only to instrumental background and therefore has sufficient sensitivity for spatial determinations of cellular activity in situ. In addition, owing to the relatively short duration of expression of luminescence activity, luciferase is ideal for the study of temporal changes in gene expression with changing environmental conditions. Disadvantages of the luciferase bioreporters are their dependence on substrate addition (unless the entire *lux* operon is used) and cellular energy reserves for expression, sometimes limiting their utility in environmental settings.

54.2.5. Green Fluorescent Protein (*gfp*) Genes as Bioreporters

The *gfp* gene has several advantages as a bioreporter. Its small size (approximately 700 bp) facilitates DNA manipulations and fusions. Also, since GFP does not require any exogenous substrates or cofactors, it can function as a reporter under conditions that are not growth dependent. For example, *gfp* expression can be detected in both rapid and slow-growing mycobacteria, which is a significant advantage over other bioreporters, such as *phoA* (54). Additionally, expression of *gfp* can be monitored in real time by fluorescence or confocal microscopy, since the GFP protein is relatively resistant to photobleaching, and GFP-expressing single cells can be identified by flow cytometry (for recent reviews, see references 105 and 123).

For transcriptional fusions (promoter probes), it is useful for fluorescence to be expressed from a variety of host promoters. Tn*5*- and Tn*3*-derived promoter probe transposons using the *gfp* reporter gene have been constructed by several groups (10, 64, 103) with varying degrees of sensitivity of detection. Mini-Tn*5* GFP transposons with a cleverly modified *gfp* gene have been reported to detect monocopy *gfp* bioreporter gene fusions in *Pseudomonas* spp., and in *Alcaligenes eutrophus*. The fluorescence of the reporter GFP protein could be observed by eye in colonies and by fluorescence microscopy, on a single-cell level (103). In addition, Miller and Lindow (70) constructed an optimized *gfp* cloning cassette with sufficient fluorescence to allow expression from fairly weak promoters on low-copy-number plasmids in pseudomonads and *E. coli* bacteria, which should be applicable for other gram-negative bacteria as well.

The different *gfp* mini-transposons, cloning vectors, or cassettes based on the *gfp* reporter gene have begun to be used to monitor gene expression in different microbial cell types, although the majority of published examples today are with eukaryotic cells. For example, *gfp* has been used as a specific reporter of expression of the *cotE* gene that encodes a protein involved in the formation of the spore coat in *Bacillus subtilis* (121). To study the pattern of expression of genes involved in toluene degradation in an artificial community, consisting of seven microbes, Møller et al. (74) fused *gfp* to the *Pu* (upper pathway) and *Pm* (meta pathway) promoters from the TOL plasmid, which were induced by addition of benzyl alcohol or 3-methylbenzoate, respectively. The spatial distribution of strains in the biofilms was examined by a combination of GFP expression analysis and the use of fluorescent 16S rRNA targeting probes, using a confocal microscope (74). This approach could prove to be a useful method for investigating structure-function relationships in microbial communities.

In elegant experiments combining flow cytometry with the *gfp* reporter gene, acid-inducible promoters of *Salmonella typhimurium* were isolated (113). This approach could be applied to isolate environmentally controlled promoters in any bacterial species in which the GFP protein can be expressed. In addition, *gfp* has been shown to be a good bioreporter for the detection of nitrogen fixation (*nif*) gene expression in grass-associated diazotrophs, such as *Azoarcus* sp. (32).

One potential disadvantage of the use of the *gfp* gene as a reporter of gene expression is the extreme stability of the GFP protein. Whereas stability of the protein is advantageous for the tracking of cells harboring biomarker genes, it can be problematic in studies of temporal changes in gene expression using bioreporter genes, since, once the reporter protein is synthesized, it will persist. Recently, an unstable variant of GFP was constructed that is more susceptible to degradation by cellular proteases and therefore has a shorter half-life in cells (2). This should alleviate the potential problem.

54.2.6. Combinations of Reporter Genes

Combinations of reporter genes are useful for provision of an internal transformation and expression control (reviewed in reference 8). The combination of chemiluminescent β-galactosidase and luciferase assays are ideal, since cell extracts can be prepared using the same lysis buffers, and light is the reaction product in both cases (63). By comparison of these combined reporters, it is possible to distinguish between negative results generated by low-level expression of one promoter and negative results generated by low assay sensitivity. Reporter combinations can also be used for analysis of gene expression from multiple promoters. For example, reporter vectors have been constructed based on a combination of *gusA* and *lacZ* bioreporters for analysis of bidirectional promoter regions (80, 115).

54.3. PRACTICAL CONSIDERATIONS

In general, most of the methods for detection of microorganisms tagged with biomarkers or expression of bioreporters are similar. One important difference is that in the case of many bioreporters, an enzyme activity assay is more commonly used, whereas the biomarker phenotype is more commonly assayed by scoring of colonies. However, the principal mechanisms behind these different monitoring approaches are the same. We will briefly describe some of the more common assays used for monitoring the various biomarkers and bioreporters with an emphasis of application in environmental samples such as soil, plant surfaces, etc.

For all marker or reporter genes the following steps are necessary:

1. Selection of the appropriate construct
2. Introduction of the marker/reporter into the target organism
3. Comparison of the tagged strain to the wild-type strain
4. Assessment of phenotype stability and/or expression (for reporter genes) under relevant conditions (e.g., environmental samples)

54.3.1. Antibiotic or Heavy Metal Resistance

Bacteria tagged with antibiotic or heavy metal resistance genes as biomarkers are routinely identified as colonies grown on agar medium containing the antibiotic or heavy metal at a concentration that is inhibitory to growth of the non-tagged wild-type strain. This selectivity is advantageous for prevention of growth or interference from members of the indigenous microbial community in nature.

It is important to be careful in choosing a growth medium when using heavy metal resistance, since the metals may react with some medium ingredients and become inactivated or precipitate (68). Most resistance markers rely primarily on growth as a selective indicator. One exception is tellurite resistance. In this case, colonies of bacteria tagged with the tellurite resistance gene turn black on minimal medium containing tellurite. Therefore, visual identification of bacteria tagged with tellurite resistance is facilitated (90).

Selective plate counting remains one of the most sensitive techniques for enumeration of specific bacteria in the environment, if they are culturable. However, one problem with reliance on plate counting for detection of biomarked cells is that it is known that microorganisms in nature can be viable but nonculturable. This viable but nonculturable state is currently thought to be a response of the cells to stress, such as starvation or temperature extremes (for review, see reference 76). Therefore, if the goal of monitoring is to accurately count the total viable cell population, it is advisable to use a method that does not rely on culturability.

54.3.2. Cleavage of Chromogenic, Fluorescent, or Chemiluminescent Substrates

Usually, the metabolic markers *lacZY*, *gus*, *xylE*, *phoA*, or *celB* are used to identify colonies growing on agar medium. Therefore, these biomarkers also suffer from the same limitations as the resistance markers described above, since only the culturable cells will be detected. The primary advantage of these markers is that rapid visual screening is possible.

For all of the metabolic bioreporters, the enzyme activity assay can be performed in cuvettes or in microtiter plates. Cell extract is added to the appropriate assay buffer containing the substrate of choice (Table 1) (see reference

124 for experimental details). After incubation the samples can be read on an enzyme-linked immunosorbent (ELISA) plate reader, spectrophotometer, luminometer, or fluorimeter, depending on the product of the reaction. The optical density at the appropriate wavelength is compared with a standard curve for quantitative determination of enzyme activity. Chemiluminometry-mediated detection of enzyme activity is the most sensitive technique, allowing the detection of a smaller number of cells; however, the substrates are more expensive than more conventional substrates.

The GUS marker is particularly useful for plant-associated bacteria, since GUS-tagged cells can be localized on plant tissue by using substrates that are hydrolyzed to yield a colored precipitate, such as 5-bromo-4-chloro-3-indolyl-β-D-glucuronide (X-GlucA) (Table 1).

xylE-tagged cells can be detected by spraying colonies with catechol. Those that turn yellow owing to accumulation of β-hydroxymuconic semilaldehyde are positive. Care must be taken to distinguish $C230^+$ colonies from bacteria that naturally produce yellow colonies (91). An enzymatic assay is also possible for C230 activity (81). However, C230 is inactivated by oxygen. This can be used advantageously to distinguish living cells from lysed cells, since the C230 activity is rapidly inactivated when freed from the protective cell environment (91).

54.3.3. Ice Nucleation

Simple assays have been developed for rapid scoring of the ice nucleation phenotype of tagged strains. In the droplet freezing assay (61), drops of bacterial cultures are placed on a sheet of aluminum foil coated with paraffin and exposed to a temperature of −9°C on a refrigerated alcohol bath. Only the drops of bacteria expressing the ice nucleation biomarker gene will freeze at this temperature, and the frozen drops can be easily visualized. Since ice nucleation is a fairly unique phenotype, formation of ice from tagged bacteria can be assayed directly in environmental samples, such as soil slurries (59). The droplet freezing assays are extremely sensitive, since only a single ice nucleation event is necessary to cause a droplet to freeze (59). Ice nucleation activity can be calculated using appropriate computer software and subsequently expressed as the logarithm of ice nuclei per CFU. The assay is both qualitative and quantitative, since the number of ice nuclei in the sample is proportional to the threshold freezing temperature (59).

54.3.4. Bioluminescence

A number of techniques have been employed to detect luciferase activity of microbes tagged with either *lux* or *luc* biomarker genes, including (i) detection of light-emitting colonies, (ii) luminometry (or scintillation counting), and (iii) CCD camera detection (or photon camera counting) with associated image analysis software (Fig. 1) (see reviews with detailed experimental approaches in references 72 and 83). Another light detection option is to use fiber optics or light pipes to collect light from the sample (26, 43). In addition, it is possible to detect and quantify the specific *luc* or *lux* DNA by PCR amplification (Fig. 1) (for example, see references 71 and 73).

The amount of light produced by bacteria tagged with the *luc* or *lux* genes varies according to the gene used, the gene copy number, and the strength of the promoter driving transcription (1, 12, 120). Lampinen et al. (56) found that the *luc* bioreporter was approximately 10 times more sensitive as a bioreporter in *Bacillus* spp. than were similar constructs made with the bacterial luciferase genes, presumably owing to the higher quantum yield of the eukaryotic luciferase enzyme reaction. To optimize sensitivity of detection, *E. coli* cells were tagged with *luc* cloned behind a strong promoter on a multicopy-number plasmid, enabling fewer than 10 cells to be detected by luminometry (71). In another study, single *luxAB*-tagged *P. syringae* cells were detected by CCD-enhanced microscopy when the *lux* genes were expressed from a strong promoter on a multicopy-number plasmid (97).

If there is sufficient light emission from, for example, colonies of cells with *luxAB* or *luc* cloned behind a strong promoter on a multicopy-number plasmic, it is usually possible to visualize light-emitting colonies by eye in a dark room after addition of the appropriate substrate, *n*-decanal (for *luxAB*) or luciferin (for *luc*). Signals from weakly luminescent colonies may be enhanced by exposure to X-ray film, by imaging with a CCD camera, or by a photon-counting camera (Fig. 1). For imaging of low light levels (which could take exposure times of 10 min or longer), the plate with colonies should be kept in a light-tight box to prevent stray light from interfering with the signal.

Luminometry is a rapid and simple technique for accurate quantification of light output from broth cultures or bacterial suspensions. Scintillation counting can also be used to quantify light output, but luminometry is more commonly used. Luminometers exist in a variety of formats, from simple single-chamber models to microtiter plate formats. Bioluminescence can be monitored by luminometry in complex environmental samples, such as soil, although there is some masking of the light output owing to particulate matter and quenching of the light by humic acids and other light-absorbing materials in the sample. The detection limits of luminometry for *lux*- or *luc*-tagged cells are usually 10- to 100-fold higher in soil, compared to those in solution (1, 71, 86). The limit of luminometric detection is also dependent on the gene copy number and strength of the promoter driving transcription, as described above. For example, Möller et al. (73) found that the detection limit for the cyanobacterium *Synechocystis* sp. strain 6803, with a single chromosomal *luc* insertion, was on the order of 4×10^3 cells per g of sediment. The detection limit of bioluminescence is similar to that obtained by PCR amplification in soil or sediment samples (e.g., see reference 71). Using optimal conditions to enhance the sensitivity of detection, Rattray et al. (85) reported a minimum detection limit of 90 and 500 cells per ml of liquid culture on per g of soil, respectively, for *Enterobacter cloacae* tagged with *luxAB* on a multicopy number plasmid under control of a strong constitutive promoter (85).

Numbers of luminescent cells may also be underestimated in environmental samples owing to the energy dependence of the luciferase reaction. For example, Mahaffee et al. (62) found that the number of bioluminescent cells was underestimated under field conditions when compared to nontagged populations, although there was no observed loss in fitness of the engineered strain compared to the wild type using immunological methods. Therefore, the underestimation was presumably due to the substrate and energy limitations for the *luxAB*-encoded luciferase protein (62).

One way to overcome these problems is to partially purify, or extract, the bacterial cells from the environmental sample before measurement in the luminometer (71). Alternatively, luciferase activity may be quantified in cell lysates from soil or sediment (73). The main advantage of luciferase activity determinations in cell lysates is the independence of the assay on cellular ATP (or $FMNH_2$) levels, since ATP is directly added as a reagent. However, activity of the luciferase protein can also be inhibited by a range of substances found in environmental samples, such as chloride ions and salts. Therefore, depending on the sample type, some protein purification from the sample might be necessary. A higher light output has been obtained from environmental samples by concentration of extracted protein onto membrane filters before light measurement (73). In addition, it is possible to account for the effect of light quenching by addition of an internal standard of pure luciferase to the sample (72, 73).

Therefore, for most applications luminometry is used to directly assay light activity in soil as a qualitative detection method or to quantify relative metabolic activity changes in the population. If *lux* or *luc* genes are transcribed at a constant rate, light emission will be an indirect indicator of the metabolic activity of the cells. Population activity determinations in liquid or soil samples may be made without addition of nutrients, to assess in situ activity (59, 66, 82). Alternatively, nutrients may be added, and the increase in luminescence can be used to assess potential activity of a specific population (65, 96). Luciferase activity can also be used to determine metabolic activity of viable but nonculturable cells (30).

Bioluminescence biomarkers and bioreporters can also be used to assay the pattern of specific cell colonization or specific enzyme activities, respectively, directly on plant surfaces, as described in the examples given in the previous sections. The volatile substrate (*n*-decanal) for the *lux* bioreporter is often preferred for detection of enzyme activity on plant surfaces, since the vapor is more easily accessible than the liquid luciferin substrate used for the *luc* bioreporter. Luminescent cells on plant surfaces can be detected either by X-ray film exposure (autophotography; e.g., see reference 5), by CCD imaging (85, 96), or by using fiber optic light guides (26). Detection of the light emission with CCD cameras or photon-counting cameras allow nondestructive localization of the distribution of microcolonies. However, because single luminescent cells are rarely visible, cell aggregates or microcolonies with sufficient light output are required for detection on the plant surface by either method. Even so, the detection limit for luminescence-based detection is approximately 10^3 to 10^4 CFU/cm^2 of the root surface, which is about 1,000 times more sensitive than β-galactosidase assays (26). However, these techniques cannot discriminate between rhizosphere, rhizoplane, and internal root colonization (5), unlike GFP detection techniques (see below).

54.3.5. GFP Fluorescence

Methods that have previously been developed for detection of fluorescent cells, such as flow cytometry, confocal microscopy, or epifluorescence microscopy, are also ideal for detection of GFP-tagged cells (Fig. 2). In addition, GFP-fluorescent colonies can be visualized on agar plates, or fluorescence intensity of *gfp*-tagged cultures can be quantified by spectrofluorimetry (Fig. 2). Methods for detection of *gfp*-marked cells have recently been reviewed (105, 109), and the reader is directed to these reviews for more experimental details.

One particular advantage of visualization of *gfp*-tagged cells over that of other fluorescent staining approaches, such as fluorescent antibodies or fluorescent oligonucleotide probes targeted to 16S rRNA, is that *gfp*-tagged cells can be visualized in situ with minimal sample disturbance and without fixation or staining. The pattern of colonization or distribution of individual *gfp*-tagged cells can be directly visualized in environmental samples, such as sludge communities (31, 77), or on plant surfaces (6, 37, 106) by scanning laser confocal microscopy; colonization at a grosser level can be visualized by fluorescent stereomicroscopy (104a).

As mentioned above (section 54.1.6), cells tagged with optimized mutant derivatives of *gfp* expressed from a strong promoter are often intensely fluorescent, allowing single-cell detection at a level not previously possible with other biomarkers currently available. GFP-tagged cells can normally be easily visualized without the necessity for image enhancement by a CCD camera. Two copies of *gfp* inserted into the chromosome in tandem result in even brighter fluorescence, which can be advantageous for some applications, such as detection by flow cytometry (109, 111).

Pure cultures of GFP-tagged bacteria can be analyzed by flow cytometry to monitor the fluorescence intensity of single cells, to enumerate the proportion of fluorescent cells in the total cell population (89, 106), and to determine microbial association with eukaryotic cells (54, 113). Flow cytometers are very convenient and allow rapid screening of thousands of fluorescent cells in a few minutes, but the cost of a flow cytometer can be prohibitive. A flow cytometer measures and analyzes optical properties of single cells passing through a focused laser beam. Since the flow cytometer detects fluorescence from single cells, the proportion of fluorescent cells in a population can be accurately determined. An internal standard consisting of a known concentration of fluorescent microspheres can be incorporated into the sample for accurate quantitative measurements of the number of GFP-fluorescent cells in the sample (105, 106, 109). For more complex environmental samples, such as soil, it is necessary to first extract the bacterial cell fraction from the soil before injection into the flow cytometer (110).

For environmental samples, it is important that the GFP fluorescence of tagged cells is more intense than the intrinsic fluorescence from other bacteria, organic material, or other components of the sample. For example, background fluorescence can present a problem when analyzing *gfp*-tagged bacteria in the rhizosphere of plants or within plant tissues. However, the use of the confocal laser microscope to visualize *gfp* expression allows the distinction of GFP-mediated versus plant tissue-derived fluorescence. In addition, special filters are now available that are optimized for GFP fluorescence, thus reducing the background fluorescence interference for detection by epifluorescence microscopy (105).

Finally, visualization of GFP-fluorescing colonies cultured on solid media is a useful method for enumeration of culturable *gfp*-tagged bacteria in a sample. GFP-fluorescing colonies can readily by visualized under light illumination that excites GFP fluorescence. Special lamps are commercially available that emit the appropriate wavelength for excitation of GFP fluorescence (such as blue lamps, for red-shifted mutants of GFP) (105).

54.4. CONCLUDING REMARKS

There now exist a variety of biomarkers and bioreporters that can be applied as tools for monitoring bacterial numbers or gene expression/activity, respectively, in the environment. The choice of the appropriate tool depends on

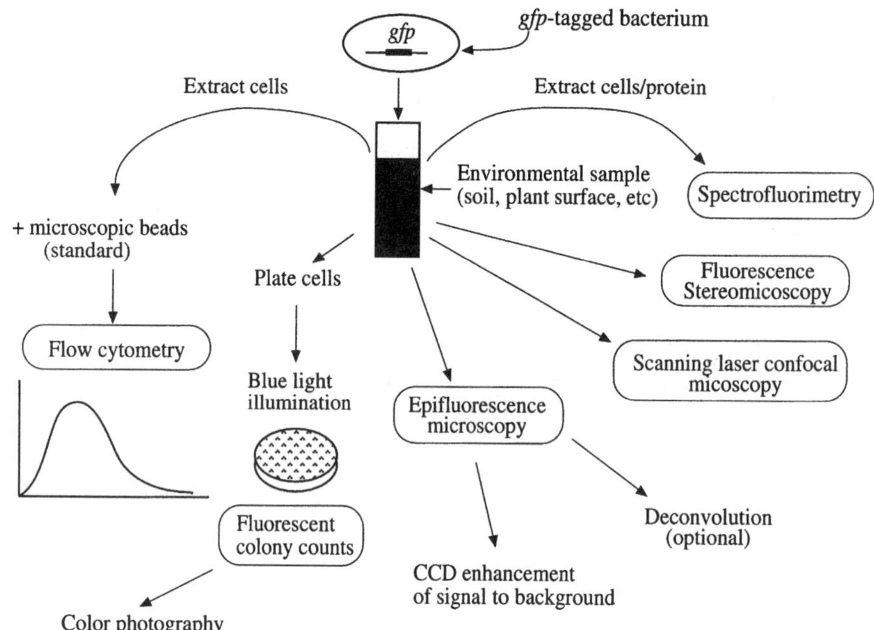

FIGURE 2 Methods for monitoring fluorescent cells in environmental samples.

the strain, the environmental sample type, and the specific application or questions to be addressed. In some cases a single biomarker or bioreporter is sufficient for addressing a given application, whereas in other cases it may be advantageous to combine markers or reporters to enhance specificity or sensitivity or to assay different parameters simultaneously. As these molecular tools become more refined, we will be able to explore with greater precision and accuracy the behavior of individual microbial cells in nature. These tools are essential for risk assessment of GEMs but also are increasingly being applied to gaining an understanding of "natural" microbes in environmental settings.

We gratefully acknowledge support from the Swedish Foundation for Strategic Research, the Carl Tryggers Foundation, the Swedish Council for Engineering Sciences, the Swedish Environmental Research Foundation (JKJ), and the NSF Center for Microbial Ecology and the Department of Energy (FJdB). We also thank Riccardo Tombolini (Department of Biochemistry, Stockholm University) for his useful comments during the writing of this chapter.

References

1. **Amin-Hanjani, S., A. Meikle, L. A. Glover, J. I. Prosser, and K. Killham.** 1993. Plasmid and chromosomally encoded luminescence marker systems for detection of *Pseudomonas fluorescens* in soil. *Mol. Ecol.* **2:**47–54.
2. **Anderson, J. B., C. Sternberg, L. K. Poulsen, S. P. Bjørn, M. Givskov, and S. Molin.** 1998. New unstable variants of green fluorescent protein for studies of transient gene expression in bacteria. *Appl. Environ. Microbiol.* **64:**2240–2246.
3. **Arséne, F., S. Katupitlya, I. R. Kennedy, and C. Elmerich.** 1994. Use of *lacZ* fusions to study the expression of *nif* genes of *Azospirillum brasilense* in association with plants. *Mol. Plant Microbe Interact.* **7:**748–757.
4. **Atlas, R. M.** 1992. Molecular methods for environmental monitoring and containment of genetically engineered microorganisms. *Biodegradation* **3:**137–146.
5. **Beauchamp, C. J., J. W. Kloepper, and P. A. Lemke.** 1993. Luminometric analyses of plant root colonization by bioluminescent pseudomonads. *Can. J. Microbiol.* **39:**434–441.
6. **Bloemberg, G. V., G. A. O'Toole, B. J. J. Lugtenberg, and R. Kolter.** 1997. Green fluorescent protein as a marker for *Pseudomonas* spp. *Appl. Environ. Microbiol.* **63:**4543–4551.
7. **Bosworth, A. H. M. K. Williams, A. Albrecht, R. Kwiatkowski, J. Beynon, T. R. Hankinson, C. W. Ronson, F. Cannon, T. J. Wacek, and E. W. Triplett.** 1994. Alfalfa yield response to inoculation with recombinant strains of *Rhizobium meliloti* with an extra copy of *dctABD* and/or modified *nifA* expression. *Appl. Environ. Microbiol.* **60:**3815–3832.
8. **Bronstein, E., J. Fortin, P. E. Stanley, G. S. A. B. Stewart, and L. J. Kricka.** 1994. Chemiluminescent and bioluminescent reporter gene assays. *Anal. Biochem.* **219:**169–181.
9. **Buell, C. R., and A. J. Anderson.** 1993. Expression of the *aggA* locus of *Pseudomonas putida in vitro* and *in planta* as detected by the reporter gene, *xylE*. *Mol. Plant Microbe Interact.* **3:**331–340.
10. **Burlage, R. S., Z. K. Yang, and T. Mehlhorn.** 1996. A transposon for green fluorescent protein transcriptional fusions: application for bacterial transport experiments. *Gene* **173:**53–58.
11. **Cebolla, A., F. Ruiz-Berraquero, and A. J. Palomares.** 1993. Stable tagging of *Rhizobium meliloti* with the firefly luciferase gene for environmental monitoring. *Appl. Environ. Microbiol.* **59:**2511–2519.
12. **Cebolla, A., M. E. Vazquez, and A. J. Palomares.** 1995. Expression vectors for the use of eukaryotic luciferases as

bacterial markers with different colors of luminescence. *Appl. Environ. Microbiol.* **61:**660–668.

13. **Chalfie M., Y. Tu, G. Euskirchen, W. W. Ward, and D. C. Prasher.** 1994. Green fluorescent protein as a marker for gene expression. *Science* **263:**802–805.
14. **Christensen, B. B., C. Sternberg, and S. Molin.** 1996. Bacterial plasmid conjugation on semi-solid surfaces monitored with the green fluorescent protein (GFP) from *Aequorea victoria* as a marker. *Gene* **173:**59–65.
15. **Christiansen-Weniger, C., and J. Venderleyden.** 1993. Ammonium-excreting *Azospirillum* sp. become intracellularly established in maize (*Zea mays*) para-nodules. *Biol. Fertil. Soils* **17:**1–8.
16. **Corbisier, P., G. Ji, G. Nuyts, M. Mergeay, and S. Silver.** 1993. *luxAB* gene fusions with the arsenic and cadmium resistance operons of *Staphylococcus aureus* plasmid pI258. *FEMS Microbiol. Lett.* **110:**231–238.
17. **Cormack, B. P., R. H. Valdivia, and S. Falkow.** 1996. FACS-optimized mutants of the green fluorescent protein (GFP). *Gene* **173:**33–38.
18. **Crameri, A., E. A. Whitehorn, E. Tate, and W. P. C. Stemmer.** 1996. Improved green fluorescent protein by molecular evolution using DNA shuffling. *Nat. Biotechnol.* **14:**315–319.
19. **Cubitt, A. B., R. Heim, S. R. Adams, A. E. Boyd, L. A. Gross, and R. Y. Tsien.** 1995. Understanding, improving and using green fluorescent proteins. *Trends Biochem. Sci.* **20:**448–455.
20. **Curcic, R., S. Dhandayuthapani, and V. Deretic.** 1994. Gene expression in mycobacteria: transcriptional fusions based on *xylE* and analysis of the promoter region of the response regulator from *Mycobacterium tuberculosis*. *Mol. Microbiol.* **13:**1057–1064.
21. **De Bruijn, F. J., M. E. Davey, B. McSpadden-Gardener, A. Millcamps, J. L. W. Rademaker, D. Ragatz, M. L. Schultz, P. Struffi, and J. Stoltzfus.** 1997. Molecular approaches in microbial ecology to assess genomic diversity and stress-induced gene expression in plant-associated diazotrophs. *In* C. Elmerich, E. Kondorosi, and W. E. Newton (ed.), *Proc. 11th International Congress of Nitroigen Fixation*. Kluwer Academic Publishers, Dordrecht, The Netherlands.
22. **De Leij, F. A. A. M., E. J. Sutton, F. M. Whipps, J. S. Fenlon, and J. M. Lynch.** 1995. Field release of a genetically modified *Pseudomonas fluorescens* on wheat: establishment, survival and dissemination. *Bio/Technology* **13:**1488–1492.
23. **Delgrave, S., R. E. Hawtin, C. M. Silva, M. M. Yang, and D. C. Youvan.** 1995. Red-shifted excitation mutants of the green fluorescent protein. *Bio/Technology* **13:**151–154.
24. **Delic, I., P. Robbins, and J. Westpheling.** 1992. Direct repeat sequences are implicated in the regulation of two *Streptomyces* chitinase promoters that are subject to carbon catabolite control. *Proc. Natl. Acad. Sci. USA* **89:**1885–1889.
25. **De Lorenzo, V., M. Merrero, U. Jakubzik, and K. N. Timmis.** 1990. Mini-Tn5 transposon derivatives for insertion mutagenesis, promoter probing and chromosomal insertion of cloned DNA in gram-negative eubacteria. *J. Bacteriol.* **172:**6568–6572.
26. **De Weger, L. A., P. Dunbar, W. F. Mahaffee, B. J. J. Lugtenberg, and G. S. Sayler.** 1991. Use of bioluminescence markers to detect *Pseudomonas* spp. in the rhizosphere. *Appl. Environ. Microbiol.* **57:**3641–3644.
27. **de Weger, L. A., I. Kuiper, A. J. Vanderbij, and B. J. J. Lugtenberg.** 1997. Use of a *lux*-based procedure to rapidly visualize root colonisation by *Pseudomonas fluorescens* in the wheat rhizosphere. *Antonie van Leeuwenhoek* **72:**365–372.
28. **Drahos, D., B. C. Hemming, and S. McPherson.** 1986. Tracking recombinant organisms in the environment: β-galactosidase as a selectable non-antibiotic marker for fluorescent pseudomonads. *Bio/Technology* **4:**439–444.
29. **Drainas, C., G. Vartholomatos, and N. J. Panopoulos.** 1995. The ice nucleation gene from *Pseudomonas syringae* as a sensitive gene reporter for promoter analysis in *Zymomonas mobilis*. *Appl. Environ. Microbiol.* **61:**273–277.
30. **Duncan, S., L. A. Glover, K. Killham, and J. I. Prosser.** 1994. Luminescence-based detection of activity of starved and viable but nonculturable bacteria. *Appl. Environ. Microbiol.* **60:**1308–1316.
31. **Eberl, L., R. Schulze, A. Ammendola, O. Geisenberger, R. Erhart, C. Sternberg, S. Molin, and R. Amann.** 1997. Use of green fluorescent protein as a marker for ecological studies of activated sludge communities. *FEMS Microbiol. Lett.* **149:**77–83.
32. **Egener, T., T. Hurek, and B. Reinholdhurek.** 1998. Use of green fluorescent protein to detect expression of *nif* genes of *Azoarcus* sp. BH72, a grass-associated diazotroph, on rice roots. *Mol. Plant Microbe Interact.* **11:**71–75.
33. **Fedi, S., E. Tola, Y. Moënne-Loccoz, D. N. Dowling, L. M. Smith, and F. O'Gara.** 1997. Evidence for signaling between the phytopathogenic fungus *Pythium ultimum* and *Pseudomonas fluorescens* F113: *P. ultimum* represses the expression of genes in *P. fluorescens* F113, resulting in altered ecological fitness. *Appl. Environ. Microbiol.* **63:**4261–4266.
34. **Flemming, C. A., K. T. Leung, H. Lee, J. T. Trevors, and C. W. Greer.** 1994. Survival of *lux-lac*-marked biosurfactant-producing *Pseudomonas aeruginosa* UG2L in soil monitored by nonselective plating and PCR. *Appl. Environ. Microbiol.* **60:**1606–1616.
35. **Frederick, R. D., D. R. Majerczak, and D. L. Coplin.** 1993. *Erwinia stewartii* WtsA, a positive regulator of pathogenicity gene expression, is similar to *Pseudomonas syringae* pv. *phaseolicola* HrpS. *Mol. Microbiol.* **9:**477–485.
36. **Gaffney, T. D., O. Da Costa e Silva, T. Yamada, and T. Kosuge.** 1990. Indoleacetic acid operon of *Pseudomonas syringae* subsp. *savastanoi*: transcription analysis and promoter identification. *J. Bacteriol.* **172:**5593–5601.
37. **Gage, D. J., T. Bobo, and S. R. Long.** 1996. Use of green fluorescent protein to visualize early events of symbiosis between *Rhizobium meliloti* and alfalfa (*Medicago sativa*). *J. Bacteriol.* **178:**7159–7166.
38. **Georgakopoulos, D. G., M. Hendson, M. Panopoulos, and M. N. Schroth.** 1994. Analysis of expression of a phenazine biosynthesis locus of *Pseudomonas aureofaciens* pgs12 on seeds with a mutant carrying a phenazine biosynthesis locus ice nucleation reporter gene fusion. *Appl. Environ. Microbiol.* **60:**4573–4579.
39. **Gold, S., S. Nishio, S. Tsuyumu, and N. T. Keen.** 1992. Analysis of the *pelE* promoter in *Erwinia chrysantemi* EC16. *Mol. Plant Microbe Interact.* **5:**170–178.
40. **Gustafsson, K., and J. K. Jansson.** 1993. Ecological risk assessment of the deliberate release of genetically modified microorganisms. *Ambio* **22:**236–242.

41. **Heim, R., D. C. Prasher, and R. Y. Tsien.** 1994. Wavelength mutations and posttranslational autoxidation of green fluorescent protein. *Proc. Natl. Acad. Sci. USA* **91:** 12501–12504.

42. **Heim, R., and R. Y. Tsien.** 1996. Engineering green fluorescent protein for improved brightness, longer wavelengths and fluorescence resonance energy transfer. *Curr. Biol.* **6:**178–182.

43. **Heitzer, A., K. Malachowsky, J. E. Thonnard, P. R. Bienkowski, D. C. White and G. S. Sayler.** 1994. Optical biosensor for environmental on-line monitoring of naphthalene and salicylate bioavailability with an immobilized bioluminescent catabolic reporter bacterium. *Appl. Environ. Microbiol.* **60:**1487–1494.

44. **Herrero, M., V. de Lorenzo, and K. H. Timmis.** 1990. Transposon vectors containing non-antibiotic resistance selection markers for cloning and stable chromosomal insertion of foreign genes in gram-negative bacteria. *J. Bacteriol.* **172:**6557–6567.

45. **Holben, W. E., J. K. Jansson, B. K. Chelm, and J. M. Tiedje.** 1988. DNA probe method for the detection of specific microorganisms in the soil bacterial community. *Appl. Environ. Microbiol.* **54:**703–711.

46. **Hurek, T., B. Reinhold-Hurek, and M. Van Montagu.** 1994. Root colonization and systemic spreading of *Azoarcus* sp. strain BH72 in grasses. *J. Bacteriol.* **176:**1913–1923.

47. **Jansson, J. K.** 1995. Tracking genetically engineered microorganisms in nature. *Curr. Opin. Biotechnol.* **6:**275–283.

48. **Jansson, J. K., and T. Leser.** 1996. Quantitative PCR of environmental samples, 2.7.4, p. 1–19. *In* A. D. L. Akkermans, J. D. van Elsas, and F. J. de Bruijn (ed.), *Molecular Microbial Ecology Manual.* Kluwer Academic Publishers, Dordrecht, The Netherlands.

49. **Jansson, J. K., and J. Prosser.** 1997. Quantification of the presence and activity of specific microorganisms in nature. *Mol. Biotechnol.* **7:**103–120.

50. **Jefferson, R. A., and K. J. Wilson.** 1991. The GUS gene fusion system. *Plant Mol. Biol. Man.* **B14:**1–33.

51. **Kamoun, S., and C. I. Kado.** 1990. A plant-inducible gene of *Xyanthomonas campestris* pv. *campestris* encodes an exocellular component required for growth in the host and hypersensitivity on non-hosts. *J. Bacteriol.* **172:**5165–5172.

52. **Keller, M., A. Roxiau, W. M. Weng, M. Schmidt, J. Quandt, K. Niehaus, D. Jording, W. Arnold, and A. Pühler.** 1995. Molecular analysis of the *Rhizobium meliloti mucR* gene regulating the biosynthesis of the expolysaccharides succinoglycan and galactoglucan. *Mol. Plant Microbe Interact.* **8:**267–277.

53. **Kragelund, L., C. Hosbond, and O. Nybroe.** 1997. Distribution of metabolic activity and phosphate starvation response of *lux*-tagged *Pseudomonas fluorescens* reporter bacteria in the barley rhizosphere. *Appl. Environ. Microbiol.* **63:**4920–4928.

54. **Kremer, L., A. Baulard, J. Estaquier, O. Paulain-Godefroy, and C. Locht.** 1995. Green fluorescent protein as a new expression marker in mycobacteria. *Mol. Microbiol.* **17:**913–922.

55. **Kruse, H., and J. Jansson.** 1997. *The Use of Antibiotic Resistance Genes as Marker Genes in Genetically Modified Organisms*. Norwegian Pollution Control Authority, SFT. Report 97:03. Oslo, Norway.

56. **Lampinen, J., L. Koivisto, M. Wahlsten, P. Mänsälä, and M. Karp.** 1992. Expression of luciferase genes from different origins in *Bacillus subtilis*. *Mol. Gen. Genet.* **232:**498–504.

57. **Lampinen, J., M. Virta, and M. Karp.** 1995. Use of controlled luciferase expression to monitor chemicals affecting protein synthesis. *Appl. Environ. Microbiol.* **61:**2981–2989.

58. **Lindgren, P. B., R. Frederick, A. G. Govindarajan, M. J. Panopoulos, B. J. Staskawicz, and S. E. Lindow.** 1989. An ice nucleation reporter system: identification of inducible pathogenicity genes in *Pseudomonas syringae* pv. *phaseolicola*. *EMBO J.* **8:**1291–1301.

59. **Lindow, S. E.** 1995. The use of reporter genes in the study of microbial ecology. *Mol. Ecol.* **4:**555–566.

60. **Lindow, S. E., and N. J. Panopoulos.** 1988. Field tests of recombinant Ice-*Pseudomonas syringae* for biological frost control in potato, p. 121–138. *In* M. Sussman, C. H. Collins, F. A. Skinner, and D. E. Stewart-Tull (ed.), *The Release of Genetically-Engineered Micro-Organisms*. Academic Press, London.

61. **Loper, J. E., and S. E. Lindow.** 1994. A biological sensor for iron available to bacteria in their habitats on plant surfaces. *Appl. Environ. Microbiol.* **60:**1934–1941.

62. **Mahaffee, W. F., E. M. Bauske, J. W. L. Van Vuurde, J. M. Van Der Wolf, M. Van Den Brink, and J. W. Kloepper.** 1997. Comparative analysis of antibiotic resistance, immunofluorescent colony staining and a transgenic marker (bioluminescence) for monitoring the environmental fate of a rhizobacterium. *Appl. Environ. Microbiol.* **63:**1617–1622.

63. **Martin, C. S., P. A. Wight, A. Dobretsova, and I. Bronstein.** 1996. Dual luminescent-based reporter gene assay for luciferase and β-galactosidase. *BioTechniques* **21:**520–524.

64. **Matthysse, A. G., S. Stretton, C. Dandie, N. C. McClure, and A. E. Goodman.** 1996. Construction of GFP vectors for use in gram-negative bacteria other than *Escherichia coli*. *FEMS Microbiol. Lett.* **145:**87–94.

65. **Meikle, A., L. A. Glover, K. Killham, and J. I. Prosser.** 1994. Potential luminescence as an indicator of activation of genetically modified *Pseudomonas fluorescens* in liquid culture and in soil. *Soil Biol. Biochem.* **24:**881–892.

66. **Meikle, A., K. Killham, J. I. Prosser, and L. A. Glover.** 1992. Luminometric measurement of population activity of genetically modified *Pseudomonas fluorescens* in the soil. *FEMS Microbiol. Lett.* **99:**217–220.

67. **Mercier, J., and S. E. Lindow.** 1996. A method involving ice nucleation for the identification of microorganisms antagonistic to *Erwinia amylovora* on pear flowers. *Phytopathology* **86:**940–945.

68. **Mergeay, M.** 1995. Heavy metal resistances in microbial ecosystems, p. 1–17. *In* A. D. L. Akkermans, J. D. van Elsas, and F. J. deBruijn (ed.), *Molecular Microbial Ecology Manual.* Kluwer Academic Publishers, Dordrecht, The Netherlands.

69. **Milcamps, A., D. M. Ragatz, P. O. Lim, K. A. Berger and F. J. de Bruijn.** Isolation of carbon and nitrogen starvation-induced loci of *Rhizobium meliloti* by Tn*5*-*luxAB* mutagenesis. *Microbiology*, in press.

70. **Miller, W. G., and S. E. Lindow.** 1997. An improved GFP cloning cassette designed for prokaryotic transcriptional fusions. *Gene* **191:**149–153.

71. **Möller, A., K. Gustafsson, and J. K. Jansson.** 1994. Specific monitoring by PCR amplification and bioluminescence of firefly luciferase gene-tagged bacteria added to environmental samples. *FEMS Microbiol. Ecol.* **15:**193–206.

72. **Möller A., and J. K. Jansson.** 1998. Detection of firefly luciferase-tagged bacteria in environmental samples. *Methods Mol. Biol.* **102:**269–283.

73. **Möller, A., A. M. Norrby, K. Gustafsson, and J. K. Jansson.** 1995. Luminometry and PCR-based monitoring of genetically modified cyanobacteria in Baltic Sea microcosms. *FEMS Lett.* **129:**43–50.

74. **Moller, S., C. Sternberg, J. B. Andersen, B. B. Christensen, J. L. Ramos, M. Givskov, and S. Molin.** 1998. In situ gene expression in mixed-culture biofilms—evidence of metabolic interactions between community members. *Appl. Environ. Microbiol.* **64:**721–732.

75. **Morgan, J. A. W., C. Winstanley, R. W. Pickup, J. G. Jones, and J. R. Saunders.** 1989. Direct phenotypic and genotypic detection of a recombinant pseudomonad population released into lake water. *Appl. Environ. Microbiol.* **55:**2537–2544.

76. **Oliver, J. D.** 1993. Formation of viable but nonculturable cells, p. 239–272. *In* S. Kjelleberg (ed.), *Starvation in Bacteria.* Plenum Press, New York.

77. **Olofsson, A.-C., A. Zita, and M. Hermansson.** 1998. Floc stability and adhesion of green-fluorescent-protein-marked bacteria to flocs in activated sludge. *Microbiology* **144:**519–528.

78. **Osbourn, A. E., C. E. Barver, and M. J. Daniels.** 1987. Identification of plant induced genes of the bacterial pathogen *Xanthomonas campestris* pv. *campestris* using a promoter-probe plasmid. *EMBO J.* **6:**23–28.

79. **Palomares, A. J., M. A. DeLuca, and D. R. Helinski.** 1989. Firefly luciferase as a reporter enzyme for measuring gene expression in vegetative and symbiotic *Rhizobium meliloti* and other gram-negative bacteria. *Gene* **81:**55–64.

80. **Parry, S. K., S. B. Sharma, and E. A. Terzaghi.** 1994. Construction of a bidirectional promoter probe vector and its use in analysing *nod* gene expression in *Rhizobium loti.* *Gene* **81:**55–64.

81. **Pickup, R. W., and J. R. Saunders.** 1990. Detection of genetically engineered traits among bacteria in the environment. *Trends Biotechnol.* **8:**329–335.

82. **Prosser, J. I.** 1994. Molecular marker systems for the detection of genetically modified microorganisms in the environment. *Microbiology* **140:**5–17.

83. **Prosser, J. I., E. A. S. Rattray, K. Killham, and L. A. Glover.** 1996. *Lux* as a marker gene to track microbes, 6.1.1., p. 1–17. *In* A. D. L. Akkermans, J. D. van Elsas, and F. J. de Bruijn (ed.), *Molecular Microbial Ecology Manual.* Kluwer Academic Publishers, Dordrecht, The Netherlands.

84. **Rahme, L. G., M. N. Mindrinos, and N. J. Panopoulos.** 1992. Plant and environmental sensory signals control the expression of *hrp* genes in *Pseudomonas syringae* pv. *phaeolicola. J. Bacteriol.* **174:**3499–3507.

85. **Rattray, E. A. S., J. I. Prosser, L. A. Glover, and K. Killham.** 1995. Characterization of rhizosphere colonization by luminescent *Enterobacter cloacae* at the population and single-cell levels. *Appl. Environ. Microbiol.* **61:**2950–2957.

86. **Rattray, E. A. S., J. I. Prosser, K. Killham, and L. A. Glover.** 1990. Luminescence-based nonextractive technique for *in situ* detection of *Escherichia coli* in soil. *Appl. Environ. Microbiol.* **56:**3368–3374.

87. **Reuber, T. L., S. L. Long, and G. C. Walker.** 1991. Regulation of *Rhizobium meliloti* exo genes in free-living cells and in planta examined using Tn*phoA* fusions. *J. Bacteriol.* **173:**426–434.

88. **Rogers, J. S., E. Clark, G. Cirvilleri, and S. E. Lindow.** 1994. Cloning and characterization of genes conferring copper resistance in epiphytic ice nucleation active *Pseudomonas syringae* strains. *Phytopathology* **84:**891–897.

89. **Ropp, J. D., C. J. Donahue, D. Wolfgang-Kimball, J. Hooley, J. Y. W. Chin, R. A. Hoffman, R. A. Cuthbertson, and K. D. Bauer.** 1995. *Aequorea* green fluorescent protein analysis by flow cytometry. *Mol. Microbiol.* **17:**901–912.

90. **Sanchez-Romero, J. M., R. Diaz-Orejas, and V. de Lorenzo.** 1998. Resistance to tellurite as a selection marker for genetic manipulations of *Pseudomonas* strains. *Appl. Environ. Microbiol.* **64:**4040–4046.

91. **Saunders, J. R., R. W. Pickup, J. A. Morgan, C. Winstanley, and V. A. Saunders.** 1996. *XylE* as a marker gene for microorganisms 6.1.3., p. 1–12. *In* A. D. L. Akkermans, J. D. van Elsas, and F. J. de Bruijn (ed.), *Molecular Microbial Ecology Manual.* Kluwer Academic Publishers, Dordrecht, The Netherlands.

92. **Selbitschka, W., M. Hagen, S. Maier, and A. Pühler.** 1992. The construction of bioluminescent and GUS-positive strains of Rhizobium for use in risk assessment studies, p. 267–272. *In* R. Casper and J. Landsmann (ed.), *The Biosafety Results of Field Tests of Genetically Modified Plants and Microorganisms.* Biologische Bundesanstalt für Land und Forstwirtschaft, Braunschweig, Germany.

93. **Selifonova, O. V., and R. W. Eaton.** 1996. Use of an *ipb-lux* fusion to study regulation of the isopropylbenzene catabolism operon of *Pseudomonas putida* RE204 and to detect hydrophobic pollutants in the environment. *Appl. Environ. Microbiol.* **62:**778–783.

94. **Sessitsch, A., K. J. Wilson, A. D. L. Akkermans, and W. M. De Vos.** 1996. Simultaneous detection of different *Rhizobium* strains marked with either the *Escherichia coli gusA* gene or the *Pyrococcus furiosus celB* gene. *Appl. Environ. Microbiol.* **62:**4191–4194.

95. **Sharma, S. B., and E. R. Signer.** 1990. Temporal and spatial regulation of the symbiotic genes of *Rhizobium meliloti* in planta revealed by transposon Tn*5-gusA*. *Genes Dev.* **4:**344–356.

96. **Shaw, J. J., F. Dane, D. Geiger, and J. W. Kloepper.** 1992. Use of bioluminescence for detection of genetically engineered microorganisms released into the environment. *Appl. Environ. Microbiol.* **58:**267–273.

97. **Silcock, D. J., R. N. Waterhouse, L. A. Glover, J. I. Prosser, and K. Killham.** 1992. Detection of a single genetically modified bacterial cell in soil by using charge coupled device-enhanced microscopy. *Appl. Environ. Microbiol.* **58:**2444–2448.

98. **Slauch, J. M., and T. J. Silhavy.** 1991. Genetic fusions as experimental tools. *Methods Enzymol.* **204:**213–248.

99. **Smit, E., K. Werners, and J. D. Van Elsas.** 1995. Antibiotic resistance as a marker for tracking bacteria in the

soil ecosystem, 6.1.6, p. 1–15. *In* A. D. L. Akkermans, J. D. van Elsas, and F. J. de Bruijn (ed.), *Molecular Microbial Ecology Manual*. Kluwer Academic Publishers. Dordrecht, The Netherlands.

100. **Sohaskey, C. D., H. Im, and A. T. Schauer.** 1992. Construction and application of plasmid- and transposon-based promoter-probe vectors for *Steptomyces* spp. that employ a *Vibrio harveyi* luciferase reporter cassette. *J. Bacteriol.* **174:**367–376.

101. **Streit, W., K. Kosch, and D. Werner.** 1992. Nodulation competitiveness of *Rhizobium leguminosarum* bv. *Phaseoli* and *Rhizobium tropici* strains measured by glucuronidase (*gus*) gene fusion. *Biol. Fertil. Soils* **14:**140–144.

102. **Stretton, S., K. C. Marshall, I. W. Dawes, and A. E. Goodman.** 1996. Characterisation of carbon dioxide-inducible genes of the marine bacterium, *Pseudomonas* sp. S91. *FEMS Microbiol. Lett.* **140:**37–42.

103. **Suarez, A., A. Guttler, M. Stratz, L. H. Staendner, K. N. Timmis, and C. A. Guzman.** 1997. Green fluorescent protein-based reporter systems for genetic analysis of bacteria including monocopy applications. *Gene* **196:** 69–74.

104. **Thompson, I. P., A. K. Lilley, R. J. Ellis, P. A. Bramwell, and M. J. Bailey.** 1995. Survival, colonization and dispersal of genetically modified *Pseudomonas fluorescens* SBW25 in the phytosphere of field grown sugar beet. *Bio/Technology* **13:**1493–1497.

104a. **Tombolini, R.** Unpublished observations.

105. **Tombolini, R., and J. K. Jansson.** 1998. Monitoring of GFP tagged bacterial cells. *Methods Mol. Biol.* **102:**285–298.

106. **Tombolini, R., A. Unge, M. E. Davey, F. J. de Bruijn, and J. K. Jansson.** 1997. Flow cytometric and microscopic analyses of GFP-tagged *Pseudomonas fluorescens*. *FEMS Microbiol. Ecol.* **22:**17–28.

107. **Top, E., M. Mergeay, D. Springael, and W. Verstraete.** 1990. Gene escape model: transfer of heavy metal resistance genes from *Escherichia coli* to *Alcaligenes eutrophus* on agar plates and in soil samples. *Appl. Environ. Microbiol.* **56:**2471–2479.

108. **Torsvik, V., J. Goksøyr, and F. L. Daae.** 1990. High diversity in DNA of soil bacteria. *Appl. Environ. Microbiol.* **56:**782–787.

109. **Unge, A., R. Tombolini, M. E. Davey, F. J. de Bruijn, and J. K. Jansson.** 1997. GFP as a marker gene, 6.1.13, p. 1–16. *In* A. D. L. Akkermans, J. D. van Elsas, and F. J. de Bruijn (ed.) *Molecular Microbial Ecology Manual*. Kluwer Academic Publishers. Dordrecht, The Netherlands.

110. **Unge, A., R. Tombolini, L. Mølbak, and J. K. Jansson.** Simultaneous monitoring of cell number and metabolic activity of specific bacterial populations using a dual *gfp/luxAB* marker system. Submitted for publication.

111. **Unge, A., R. Tombolini, A. Möller, and J. K. Jansson.** 1997. Optimization of GFP as a marker for detection of bacteria in environmental samples, p. 391–394. *In* J. W. Hastings, L. J. Kricka, and P. E. Stanley (ed.), *Bioluminescence and Chemiluminescence: Molecular Reporting with Photons*. John Wiley & Sons, Sussex, United Kingdom.

112. **Uratani, B. B., S. C. Alcorn, B. H. Tsang, and J. L. Kelly.** 1995. Construction of secretion vectors and use of heterologous signal sequences for protein secretion *in Clavibacter xyli* subsp. *cynodontis*. *Mol. Plant Microbe Interact.* **8:**892–898.

113. **Valdivia, R. H., A. E. Hromockyj, D. Monack, L. Ramakrishnan, and S. Falkow.** 1996. Applications for green fluorescent protein (GFP) in the study of host-pathogen interactions. *Gene* **173:**47–52.

114. **Van de Broek, A., J. Michiels, and A. Van Gool.** 1993. Spatial-temporal colonization patterns of *Azospirillum brasilense* on the wheat root surface and expression of the bacterial *nifH* gene during association. *Mol. Plant Microbe Interact.* **2:**261–266.

115. **Van den Eede, G., R. Deblaere, and K. Goethals.** 1992. Broad host range and promoter selection vectors for bacteria that interact with plants. *Mol. Plant Microbe Interact.* **5:**228–234.

116. **Vanden Wymelenberg, A. J., D. Cullen, R. N. Spear, B. Schoenike, and J. H. Andrews.** 1997. Expression of green fluorescent protein in *Aureobasidium pullulans* and quantification of the fungus on leaf surfaces. *BioTechniques* **23:**686–690.

117. **van Elsas, J. D., and A. Wolters.** 1995. Polymerase chain reaction (PCR) analysis of soil microbial DNA, 2.7.2, p. 1–10. *In* A. D. L. Akkermans, J. D. van Elsas, and F. J. de Bruijn (ed.), *Molecular Microbial Ecology Manual*. Kluwer Academic Publishers. Dordrecht, The Netherlands.

118. **van Elsas, J. D., A. C. Wolters, C. D. Clegg, H. M. Lappin-Scott, and J. M. Anderson.** 1994. Fitness of genetically modified *Pseudomonas fluorescens* in competition for soil and root colonization. *FEMS Microbiol. Ecol.* **13:**259–272.

119. **Van Overbeek, L. S., and J. D. van Elsas.** 1995. Root exudate-induced promoter activity in *Pseudomonas fluorescens* mutants in the wheat rhizosphere. *Appl. Environ. Microbiol.* **61:**890–898.

120. **Waterhouse R. N., H. White, D. J. Silcock, and L. A. Glover.** 1993. The cloning and characterisation of phage promoters directing high expression of luciferase in *Pseudomonas syringae* pv. *phaseolicola*, allowing single cell and microcolony detection in planta. *Mol. Ecol.* **2:**285–294.

121. **Webb, C. D., A. Decatur, A. Teleman, and R. Losick.** 1995. Use of green fluorescent protein for visualization of cell-specific expression and subcellular protein localization during sporulation in *Bacillus subtilis*. *J. Bacteriol.* **177:**5906–5911.

122. **Weidenhaupt, M., H.-M. Fischer, G. Acuna, J. Sanjuan, and H. Hennecke.** 1993. Use of promoter-probe vector system in the cloning of a new NifA-dependent promoter (*ndp*) from *Bradyrhizobium japonicum*. *Gene* **129:**33–40.

123. **Welsh, S., and S. A. Kay,** 1997. Reporter gene expression for monitoring gene transfer. *Curr. Opin. Biotechnol.* **8:**617–622.

124. **Wilson, K. J.** 1996. GUS as a marker to track microbes, 6.1.5, p. 1–25, *In* A. D. L. Akkermans, J. D. van Elsas, and F. J. de Bruijn (ed.). *Molecular Microbial Ecology Manual*. Kluwer Academic Publishers, Dordrecht, The Netherlands. 6.1.5:1–25.

125. **Wilson, K. J., A. Sessitsch, and A. Akkermans.** 1994. Molecular markers as tools to study the ecology of microorganisms, p. 149–156. *In* K. Ritz, J. Dighton, and K. E. Giller (ed.) *Beyond the Biomass*. British Society of Soil Science, Wiley-Sayce, United Kingdom.

126. **Wilson, K. J., A. Sessitsch, J. Corbo, K. E. Giller, A. D. L. Akkermans, and R. A. Jefferson.** 1995. β-glucuronidase (GUS) transposons for ecological and

genetic studies of rhizobia and other gram-negative bacteria. *Microbiology* **141:**1691–1705.

127. **Wipat, A., E. M. H. Wellington, and V. A. Saunders.** 1991. *Steptomyces* marker plasmids for monitoring survival and spread of streptomycetes in soil. *Appl. Environ. Microbiol.* **57:**3322–3330.

128. **Wistanley, C., J. P. Carter, M. Seasman, J. A. W. Morgan, R. W. Pickup, and J. R. Saunders.** 1993. A comparison of the survival of stable and unstable chromosomally-located *xylE* marker cassettes as an indicator of cell division within populations of *Pseudomonas putida* released into lake water and soil. *Microb. Rel.* **2:**97–107.

129. **Wistanley, C., J. A. W. Morgan, R. W. Pickup, and R. Saunders.** 1991. Use of a *xylE* marker gene to monitor the survival of recombinant *Pseudomonas putida* populations in lake water by culture on non-selective media. *Appl. Environ. Microbiol.* **57:**1905–1913.

130. **Wolk, C. P., Y. Cai, and J. M. Panoff.** 1991. Use of transposon with luciferase as a reporter to identify environmentally responsive genes in a cyanobacterium. *Proc. Natl. Acad. Sci. USA* **88:**5355–5359.

131. **Wood, K. V., Y. A. Lam, H. H. Seliger, and W. D. McElroy.** 1989. Complementary DNA coding click beetle luciferases can elicit bioluminescence of different colors. *Science* **244:**700–702.

Bioremediation

RONALD M. ATLAS AND RONALD UNTERMAN

55

Bioremediation is a pollution treatment technology that uses biological systems to catalyze the destruction or transformation of various chemicals to less harmful forms. Extensive descriptions of actual cases in which bioremediation has been employed, the utility and limitations of this technology, and its scientific and engineering underpinnings have been extensively reviewed (7, 13, 19, 26, 28, 42, 43, 50, 58, 71–74, 90, 97, 105, 110, 117, 118, 129). The general approaches to bioremediation are to (i) monitor the natural biodegradation process (intrinsic bioremediation), (ii) carry out environmental modification, such as through nutrient application and aeration (biostimulation), and (iii) add microorganisms (bioaugmentation). The end products of effective bioremediation, such as water and carbon dioxide, are nontoxic and can be accommodated without harm to the environment and living organisms.

While conventional technologies call for moving large quantities of toxic waste-contaminated soil to incinerators, bioremediation typically can be performed on-site and requires only simple equipment. Bioremediation, though, is not the solution for all environmental pollution problems. Like other technologies, bioremediation is limited in the materials that it can treat, by conditions at the treatment site, and by the time available for treatment (66).

Where applicable, bioremediation is a cost-effective means of restoring environmental quality. Its cost-effectiveness as compared to chemical and physical treatment technologies, especially for dilute contaminants, is the main driving force for the use of bioremediation (Table 1). For example, pump-and-treat technologies often require years to decades to achieve the desorption and mobilization of sorbed organics from aquifer and vadose zone materials. These methods can contain the migration of a plume of toxic chemicals; however, the length of time for cleanup creates long-term liability, high operating costs over the total period of time, and above-ground contaminated wastewater that must be destroyed or taken off-site for disposal. These are significant problems in using these methods. Another problem in the remediation of subsurface contamination though pump-and-treat methods is the dependence upon the flow of either air or water through the contaminated zones, which often contain impermeable "lenses" that preclude good transfer of the contaminants to the fluid media. Full remediation of the subsurface regions and determination of no nonaqueous-phase liquids (NAPLs), i.e., free product as a concentrated zone of contamination, are continuing problems. Many companies have therefore tried to develop and implement direct subsurface bioremediation technologies. Within these subsurface approaches, there is a hierarchy based on ease of cleanup, time frame, and cost.

55.1. BIODEGRADABILITY

The underlying basis of bioremediation is the natural process of biodegradation, which can reduce the concentrations of pollutants and completely oxidize some organic pollutants to CO_2, H_2O, NO_3^-, and other inorganic components that can be accommodated in the environment. Microorganisms are especially useful for bioremediation because of their great metabolic diversity, which often includes the ability to metabolize pollutants such as petroleum and chlorinated hydrocarbons. It is the harnessing of these microbiological activities and their implementation in advanced engineering applications that form the basis for the increasing use of biotreatment systems for biodegradation of even the most difficult hazardous chemicals.

The chemical structures of organic pollutants have a profound influence on the abilities of microorganisms to metabolize them, especially with respect to the rates and extent of their biodegradation (6). Biodegradability is essential for the bioremediation of organic pollutants. Some organic compounds are readily biodegraded, whereas others are recalcitrant (nonbiodegradable) (5). Low- to mid-molecular-weight hydrocarbons and alcohols are representative of readily biodegradable chemicals. Xenobiotic compounds, especially halocarbons, tend to be resistant to biodegradation, and some, such as some dioxins, may be recalcitrant compounds that are more difficult to biodegrade and present the greatest challenges for successful bioremediation. As a rule, branched and polynuclear compounds are more difficult to degrade than straight-chain and simple monoaromatic molecules, and increasing the degree of halogenation generally decreases the rates of biodegradation of a series of related compounds. Furthermore, trichloroethane (TCE), a widely distributed solvent, and higher-molecular-weight polychlorinated biphenyls (PCBs), which are widely used in transformers and various

TABLE 1 Economic benefits of bioremediation

Application[a]	Physical/chemical treatment	Bioremediation	Benefit
Petroleum-hydrocarbon-contaminated soil—urban brownfield site	Excavation and off-site disposal; projected cost, $3 million	Bioventing on site; projected cost, $0.2 million	Project cost savings, $2.8 million
Petroleum-hydrocarbon-contaminated soil—natural gas processing plant	Excavation and off-site disposal; projected cost, $3 million	Bioventing on site; projected cost, $0.2 million	Project cost savings, $2.8 million
Petroleum-hydrocarbon-contaminated groundwater—gasoline from underground storage tank	Pumping, treating with air stripping and skimming; projected cost, $2 million and time period greater than 12 years	Soil vapor extraction bioventing; projected cost, $0.25 million and time period less than 1 year	Project cost savings, $1.75 million and over a decade in time saving
TCE-contaiminated soil and groundwater—industrial site	Pump and treat; projected cost, $20 million	Bioventing; projected cost, $2 million	Project cost savings, $18 million
Multiple contaminant—Superfund site	Physical capping; projected cost estimated at $25 million	In situ biotreatment; estimated cost, $5 million	Project cost savings, $20 million
Multiple contaminant—Superfund site containing BTEX and arsenic	Pump and treat and capping; cost estimated at $50 million	In situ biostimulation, oxygen sparging, and bioventing plus biological immobilization of metals; estimated cost, $2 million	Project cost savings, $48 million
Chlorinated hydrocarbon—industrial site	Excavation; cost estimated at $15 million	Bioventing; estimated cost, $2 million	Project cost savings, $13 million
Marine oil spill	Physical washing; cost estimated at $1.1 million per kilometer of oiled shoreline	Biostimulation through fertilizer addition; cost estimated at $0.005 million per kilometer of oiled shoreline	Projected cost savings of over $1 million per kilometer of oiled shoreline

[a]BTEX, benzene, toluene, ethylbenzene, xylenes; TCE, trichloroethene.

chlorinated pesticide preparations (69, 95, 137), are more difficult to bioremediate because of the requirement to degrade these chemicals cometabolically. Dichloromethane and monochlorobiphenyls are much easier to bioremediate because microorganisms can utilize these compounds as sole carbon and energy sources.

55.1.1. Petroleum Hydrocarbons

Susceptibility of petroleum hydrocarbons to biodegradation varies with the chemical structure and molecular weight of the hydrocarbon molecule (12, 104). Short-chain alkanes are toxic to many microorganisms and relatively difficult to biodegrade. *n*-Alkanes of intermediate chain length (C_{10}-C_{24}) are degraded most rapidly. Very long chain alkanes become increasingly resistant to biodegradation. As the chain length increases and the alkanes exceed a molecular weight of 500, the alkanes cease to serve as carbon sources. The initial microbial attack on alkanes occurs by enzymes that have a strict requirement for molecular oxygen, that is, monooxygenases (mixed-function oxidases) or dioxygenases (30, 124). Most frequently, the initial attack is directed at the terminal methyl group, forming a primary alcohol that, in turn, is further oxidized to an aldehyde and fatty acid. The fatty acids subsequently are converted to CO_2 and H_2O.

Aromatic compounds, especially of the condensed polynuclear type, are degraded more slowly than alkanes. Many condensed polynuclear aromatic compounds are degraded only with difficulty or not at all (38, 59, 123). One reason for resistance to biodegradation is that induction of the enzymes responsible for polynuclear aromatic hydrocarbon degradation in some cases depends upon the presence of lower-molecular-weight aromatics (68).

Both mononuclear and polynuclear aromatic hydrocarbons are oxidized by dioxygenases to form catechols. The dihydroxylated aromatic ring is cleaved and converted to formic acid, pyruvic acid, and acetaldehyde. Condensed aromatic ring structures, if degradable, are also attacked by dihydroxylation and the opening of one of the rings. The opened ring is degraded to pyruvic acid and CO_2, and a second ring is attacked in the same fashion.

Some aromatic compounds, such as benzene and toluene, can be metabolized under anaerobic conditions (40, 41, 60, 64, 92–94), with the oxygen used in the ring hydroxylation coming from H_2O (142). Although the biodegradation of hydrocarbons is definitely much slower under anaerobic conditions than under aerobic ones, the rates of anaerobic toluene and *m*-xylene biodegradation are significant enough for use in aquifer bioremediation (151, 152).

Alicyclic compounds are frequently unable to serve as the sole carbon source for microbial growth unless they have a sufficiently long aliphatic side chain, but they can be degraded via cometabolism by two or more cooperating microbial strains with complementary metabolic capabilities (11).

55.1.2. Halocarbons

Various microorganisms produce dehalogenases and carry out reductive dehalogenation (130). For example, some sulfate-reducing bacteria transform tetrachloroethane to TCE and *cis*-1,2-dichloroethene by anaerobic dehalogenation of halocarbons (18). Similarly, tetrachloroethene (perchloroethene) is subject to stepwise dechlorination to vinyl chloride and subsequently to ethane (98). Cometabolic degradation of perchloroethene with stepwise dechlorination has been demonstrated by a methanogenic bacterial consortium growing on acetate (56, 143).

Various halogenated solvents are subject to biodegradation (145). Extensive aerobic degradation of TCE, a widely distributed halocarbon pollutant, has been demonstrated by methane-utilizing microbial consortia (52, 91, 96). The low specificity of methane monooxygenase allows the conversion of TCE to TCE epoxide, which subsequently spontaneously hydrolyzes to polar products (formic, glyoxylic, and dichloroacetic acids) utilizable by microorganisms. TCE is also attacked by toluene and phenol dioxygenases. In test sediment columns incubated under aerobic conditions, Enzien et al. (48) found approximately 90% biodegradation of TCE and tetrachloroethylene, suggesting the potential for simultaneous aerobic and anaerobic biotransformation processes under bulk aerobic conditions.

For chlorobenzenes and other haloaromatics, aerobic biodegradability typically decreases with the number of halosubstituents, but the same extensively halogenated aromatics are dechlorinated under anaerobic conditions with relative ease. As the halosubstituents are sequentially removed, the products are less and less likely to be dehalogenated further, and monochlorobenzene is not dechlorinated anaerobically (98). Hexachlorobenzene is dechlorinated relatively rapidly to 1,3,5-trichlorobenzene, but further dechlorination to dichloro- and monochlorobenzenes is slow and incomplete. In contrast, mono- and dichlorobenzenes are degraded aerobically with relative ease by various *Pseudomonas* and *Alcaligenes* strains that use dioxygenases to produce chlorocatechols. The aerobic biodegradation of trichloro- and tetrachlorobenzenes is more difficult, but some *Pseudomonas* strains with such capabilities have been isolated (120).

Although relatively resistant to biodegradation, a number of microorganisms transform PCBs in a similar manner (39, 51, 137). PCB biodegradation is carried out aerobically by the white rot fungus *Phanerochaete* (35), *Acinetobacter* (3), and *Alcaligenes* (22), and anaerobically by reductive dehalogenation (21, 113). Degradation of PCBs typically is by cometabolism and is enhanced by the addition of less chlorinated analogs such as dichlorobiphenyl (4, 109). Extensive degradation of some PCB congeners has been found in soils and aquatic waters and sediments (21, 22, 31, 109). The specific congeners exhibit different degrees of susceptibility to biodegradative transformations depending upon the degree of chlorine substitution and the positions of the chlorine atoms. The more extensively chlorinated PCBs are more likely to be dechlorinated by anaerobic dehalogenation than by the less chlorinated ones. Chlorine atoms in *meta* and *para* positions are dechlorinated in preference to those in *ortho* positions.

From laboratory and field studies, 50 and 85% biodegradation of PCB mixtures (Aroclors 1242 and 1248) has been demonstrated. Depending on the target cleanup level, this can translate into the ability to bioremediate PCB-contaminated soils aerobically at starting concentrations of up to 500 to 1,000 ppm. In terms of the commercial Aroclor mixtures, it is clear that only Aroclors 1221, 1242, and 1248 are currently amenable to direct aerobic biodegradation. Aroclors 1254 and 1260 are too highly chlorinated to be reasonably degraded by existing aerobic bacterial strains. Because the products of anerobic dechlorination are readily biodegradable by aerobic strains, a two-stage anaerobic/aerobic treatment can destroy a substantial level of PCBs. Thus, by using both anaerobic and aerobic systems in conjunction, one can now anticipate the ability to degrade even higher concentrations of the lower-chlorinated Aroclors (1242, 1248) as well as the higher-chlorinated Aroclors (1254, 1260).

55.1.3. Chlorophenols

Various microorganisms are capable of degrading chlorophenols under both aerobic and anaerobic conditions (39). Pentachlorophenol is converted by a monooxygenase to tetrachlorohydroquinone through the oxidative elimination of the chlorine *para* to the phenolic hydroxyl. Stepwise dechlorination to 2,5-dichlorohydroquinone is followed by ring opening. Anaerobically, pentachlorophenol is reductively dechlorinated stepwise to phenol. Phenol can be metabolized further anaerobically to methane and CO_2.

55.1.4. Nitroaromatics

Nitro-substitution of aromatic rings has a strong negative effect on biodegradability, and compounds with several nitro-substituents tend to be recalcitrant (62). The biodegradation of nitroaromatic compounds tends to be slow and often leads to bound or polymerized residues in soils and sediments. Extensively nitro-substitued aromatics are more easily transformed under reductive than under oxidative conditions. Under anaerobic and microaerophilic conditions, the nitro groups of TNT can be reduced one by one to amino groups, but each subsequent reduction is slower and less complete (108, 127). If conditions are shifted from anaerobic to aerobic, instead of ring cleavage, the partially reduced intermediates form very complex and mutagenic azo condensation products.

55.2. APPLICABILITY OF BIOREMEDIATION

From an applications perspective, the chemical target(s) for bioremediation must be viewed not only in terms of structure (biodegradability), but also in terms of the matrix containing the target. Thus, the applicability of bioremediation can be considered for each of the environmental states of matter (i) solids (soil, sediments, sludges), (ii) liquids (groundwater, industrial wastewater), and (iii) gases (industrial air emissions, soil vent gas). Additional consideration must be given when the pollutants occur in subsurface environments (saturated and vadose zones). Continued developments in both the microbiological and engineering applications will result in an increased use of advanced biodegradation systems. This will bring safe, natural, and cost-effective solutions for remediation and pollution control problems for many organics pollutants.

55.2.1. Intrinsic Bioremediation

Intrinsic bioremediation is a process whereby the natural microflora and environmental conditions exist for natural attenuation of a pollutant to safe levels within an acceptable time frame. This is generally the first choice for biotreatment because it requires no intervention, just monitoring of the natural process of biodegradation.

The lack of specific guidance on appropriate sampling and analytical procedures to ensure that intrinsic bioremediation measurements generate quality data is of concern. The extent to which intrinsic bioremediation is ultimately embraced will depend, to a large degree, on the valid characterization of site conditions. The American Petroleum Institute has sponsored studies to examine sampling and analytical methods, both in the field and in the laboratory, to determine their effects on the measurement of geochemical indicators of intrinsic bioremediation. These studies became the basis for a report on the implications of the sampling and analytical procedures, and for a guidance manual on using the procedures. This guidance manual is intended to be a resource for practitioners of intrinsic bioremediation in the following areas:

- Scoping field investigations: selection of sampling and analytical methods that meet project-specific and site-specific needs.
- Performing field investigations: allowing those implementing field investigations to understand how sampling and field analytical techniques can affect the data collected. Provides procedures that will improve the representative quality of the collected data.
- Evaluation of field investigation data: allowing those responsible for evaluation of geochemical indicators of intrinsic bioremediation to consider potential biases introduced into data through the sampling and analytical techniques employed in the site investigation.

55.2.2. Biostimulation

When the natural rates of biodegradation are inadequate, biostimulation of indigenous microbial populations to remediate the target chemicals is often employed. For these sites, it is demonstrated that a natural degradative population exists within the contaminated zone but that proper environmental conditions are insufficient for microbial activity. Measurements of the physical and chemical properties of field samples can reveal the physicochemical limitations to biodegradative activities, which can then be modeled to indicate the critical limiting factors (114).

Some common environmental limitations to biodegradation of pollutants are excessively high waste concentrations, lack of oxygen, unfavorable pH, lack of mineral nutrients, lack of moisture, and unfavorable temperature. A variety of methods that modify environmental conditions can be employed to enhance the rates of biodegradative activities by the indigenous microbial populations. Once the limiting environmental conditions are corrected, the ubiquitous distribution of microorganisms allows, in most cases, for a spontaneous enrichment of the appropriate microorganisms.

55.2.2.1. Oxygenation

The availability of molecular oxygen has a profound effect on the biodegradation of various compounds (Table 2). Oxygen limitation often is the most troublesome problem facing in situ bioremediation for hydrocarbons and other pollutants that are biodegraded aerobically. This problem is especially profound when the soil is waterlogged (27, 76, 78, 88, 144). For in situ bioremediation of surface soil, oxygen availability is best ensured by providing adequate drainage. Air-filled pore space in the soil facilitates diffusion of oxygen to hydrocarbon-utilizing microorganisms, while in waterlogged soil, oxygen diffusion is extremely slow and cannot keep up with the demand of heterotrophic decomposition processes. For sites where substantial concentrations of decomposable hydrocarbons create a high oxygen demand in soil, the rate of oxygen diffusion is inadequate even in well-drained and light-textured soils. Cultivation (ploughing and rototilling, for example) has been used to turn the soil and ensure its maximal access to atmospheric oxygen. Adding dilute solutions of hydrogen peroxide in appropriate and stabilized formulations can also be used to supply oxygen for hydrocarbon biodegradation (8, 25, 32, 33, 134, 150). The decomposition of hydrogen peroxide releases oxygen, which can support aerobic microbial metabolism. To avoid formation of gas pockets and microbial toxicity, the practical concentration of hydrogen peroxide in injected water is kept around 100 ppm (33, 150).

In many cases, hydrocarbons have contaminated soil and groundwater from leaking underground storage tanks (44). If excavation and surface spreading of the contaminated soil (on-site bioremediation) is not feasible, in situ treatment of the undisturbed soil is performed. Intrinsic or in situ bioremediation is increasingly becoming a popular alternative to other cleanup methods because it is cheaper; however, it requires careful monitoring to ensure safe and effective performance (66). In situ bioremediation based largely on the injection of oxygen is being extensively used in groundwater treatment (34, 70). The lead technologies for biostimulation involving the introduction of oxygen to the subsurface are bioventing and biosparging. Other methods that may be used in conjunction with bioventing and biosparging include (i) the addition of vapor-phase carbon (cosubstrate), nitrogen, or phosphorus, (ii) heating of the subsurface via radio frequency methods, and (iii) hydrofracturing to increase permeability and zones of influence.

While oxygen availability can severely limit the biodegradation of hydrocarbons, other compounds are more rapidly degraded under anaerobic conditions than they are when molecular oxygen is available. Thermodynamic con-

TABLE 2 Metabolic processes for bioremediation of various compounds[a]

Aerobic	Aerobic cometabolic	Anaerobic
Aromatic compounds	Chlorinated solvents	Chlorinated solvents
BTEX	TCE	Perchloroethylene
Phenol	DCE	TCE
Styrene	Vinyl chloride	DCE
Chlorobenzene(s)		
Aniline	Bromoform	Nitroaromatics
Nitrobenzene	MTBE	Munitions (TNT)
Naphthalene	HCFCs	High-MW PCBs
PAHs	High-MW PAHs	
	PCBs	
Solvents		
Ethanol		
Methanol		
Acetone		
Chlorinated solvents		
Methylene chloride		
Methylchloride		
Sulfur compounds		
H_2S		
CS_2		
Fertilizers		
Ammonia		

[a]BTEX, benzene, toluene, ethylbenzene, xylenes; DCE, dichloroethene; HCFC, hydrochlorofluorocarbon; MTBE, methyl tertiary butylether; PAH, polycylic aromatic hydrocarbon; PCB, polychlorinated biphenyl; TCE, trichloroethene.

siderations, as well as practical experience, indicate that dechlorination of halocarbon compounds is favored under anaerobic conditions (130). For the bioremediation of highly chlorinated compounds, sequential anaerobic and aerobic treatment may well be the most practical solution (1, 137). During an initial anaerobic phase (either intrinsic or stimulated), the higher-molecular-weight PCB congeners are dehalogenated to form lower-molecular-weight compounds with fewer chlorines. Subsequently, under aerobic conditions the lower-molecular-weight congeners are degraded to carbon dioxide, water, and chloride ions. This approach has been demonstrated for the bioremediation of PCB-contaminated sediments in the Hudson River in New York and in the Sheboygan River in Wisconsin. In these studies, large steel casings were driven into contaminated sediments, and nutrients were added following anaerobic dehalogenation of the PCBs. Then agitation of the sediments and forced aeration were used to create aerobic conditions. The lower-molecular-weight PCB congeners were biodegraded under these conditions. Thus, biodegradation of both the higher- and lower-molecular-weight PCB congeners was achieved by sequential anaerobic and aerobic bioremediation.

55.2.2.2. Nutrients

Biodegradation rates can be limited by the available concentrations of various nutrients. Since microorganisms require nitrogen and phosphorus for incorporation into biomass, the availability of these nutrients within the same area as the hydrocarbons is critical. Under conditions in which nutrient deficiencies limit the rate of petroleum biodegradation, the beneficial effect of fertilization with nitrogen and phosphorus has been conclusively demonstrated and offers great promise as a countermeasure for combating oil spills (112) and other hydrocarbon spills. Atlas and Bartha (16) developed an oleophilic nitrogen and phosphorus fertilizer that places the nitrogen and phosphorus at the oil-water interface, the site of active oil biodegradation. Because the fertilizer is oleophilic, it remains with the oil and is not rapidly diluted from the site. In the aftermath of the *Amoco Cadiz* oil spill of 1978 in France, a commercial oleophilic fertilizer was developed by Elf Aquitaine (84, 85, 125, 132, 136). The product, called Inipol EAP 22, is a microemulsion that contains urea as a nitrogen source, lauryl phosphate as a phosphate source, and oleic acid as a carbon source to boost the populations of hydrocarbon-degrading microorganisms. Inipol EAP 22 was also used extensively for the treatment of the 1989 *Exxon Valdez* spill in Prince William Sound, Alaska.

For the bioremediation of pollutants in surface soils, it is generally easy to add nutrients as agricultural fertilizers. However, getting nutrients to subsurface soil and groundwater populations is more complex. One method that has been explored is the use of electrical currents to help move the nutrients to the pollutants. This approach to in situ remediation utilizing low-level direct-current electric potential differences to transport ionic nutrients has been used for enhancing the biodegradation process to clean radioactively contaminated soils (2).

A remediation technology using hydraulic soil fracturing to form channels, and electric currents to drive nutrients through the soil, called the lasagna process, has been developed to improve the treatment of contaminated soils and groundwater (75). The approach, which integrates electrokinetics with desorption and degradation in treatment zones of the contaminated soil, could treat contamination due to organic and inorganic compounds and mixed wastes. Pilot testing results have shown that the process is an effective alternative for treating contamination in low-permeability soils.

55.2.2.3. Bioavailability

A general problem related to the bioremediation of certain xenobiotic pollutants in soils and sediments is bioavailability (7, 114). It is a common experience that a pollutant is degraded rapidly when it freshly enters the soil, but the biodegradation rate then levels off and stops, with some percentage of the pollutant still present in the soil, and conditions still favorable for degradation. Extraction and chemical analysis can detect these pollutants, i.e., they have not become bound residues, yet microorganisms no longer appear to have access to them. In the case of "old spills," remediated years after the pollution event, much of the pollutant may be in nonbioavailable form, rendering bioremediation efforts futile or at best incomplete. The nature of the pollutant, the nature of the soil, other pollutants, and the time elapsed since the spill all have an influence on bioavailability.

55.2.3. Bioaugmentation

Seeding with pollutant-degrading bacteria (bioaugmentation) can be used to augment the biodegradative capabilities of the indigenous microbial populations (147). Bioaugmentation involves the introduction of microorganisms into the natural environment for the purpose of increasing the rate and/or extent of pollutant biodegradation. The rationale for this approach is that indigenous microbial populations may not be capable of degrading xenobiotics or the wide range of potential substrates present in complex pollutant mixtures.

Bioaugmentation is used in cases where intrinsic bioremediation or biostimulation does not work because of insufficient or unacclimated bacterial populations. This is generally the case for only very recalcitrant chemicals. Selected strains of bacteria with the desired catalytic capabilities and other characteristics are injected directly into the contaminated zone, along with nutrients if necessary. This technology can be complemented through the use of radio frequency heating and hydrofracturing. Bioaug entation can also be used in cases where—although the natural population may be sufficient to achieve a biostimulation remediation—it may be desirable to increase the rate of degradation and shorten the time frame for full-scale remediation. Several commercial mixtures of naturally occurring microorganisms are being marketed for use in degrading oil pollution in contained treatment bioreactors as well as for in situ applications (9, 10). Often, however, the added costs for this approach are not justified.

The white rot fungus *Phanerochaete*, which has extensive biodegradative capabilities, has been proposed for bioremedition of many more complex pollutant compounds (23, 37). *Phanerochaete* can biodegrade pesticides, such as DDT (36); munitions, such as TNT (49); high-molecular-weight polynuclear aromatics, such as benzo(a)pyrene (35); and plastics, such as polyethylene (87).

Pseudomonas species have also been widely touted for their use as seed cultures based on their extensive biodegradative capacities. *Pseudomonas* and related species are often selected because bioaugmentation with certain specific microbial strains has been shown to increase the rate of removal of the pollutants such as TCE from groundwater. These bacterial cultures can promote transformation of TCE in the absence of inducing compounds such as phenol or toluene. A bacterial strain that has been shown to work effectively in bioaugmentation is *Burkholderia (Pseudomonas) cepacia* G4; in comparison to conventional bioremediation techniques for groundwater that employ indigenous microorganisms and addition of cosubstrates, the use of *B. cepacia* cultures removes twice the amount of TCE from groundwater (103).

55.2.3.1. Molecular Breeding

The absence of a catabolic pathway for a xenobiotic compound appears to be the ultimate obstacle to its biodegradative cleanup. But this obstacle is no longer an absolute one. Biochemical pathways are under constant evolution, and plasmid-mediated genetic information exchange between microbial strains can greatly accelerate this process. In each case, selective pressure is applied to evolve the desired characteristic, such as utilization of a recalcitrant xenobiotic compound. Typically, an enrichment culture is started in a chemostat that is fed small concentrations of the xenobiotic compound and higher concentrations of a related but utilizable substrate. During the enrichment procedure, which may take weeks or even months, the concentration of the utilizable substrate is gradually decreased, while the concentration of the xenobiotic compound is increased. Spontaneous mutants with increased ability to utilize the xenobiotic compound are selected for under these circumstances. Spontaneous mutation may be supplemented by treating a part or all of the chemostat culture with UV light or chemical mutagens.

Advances in the directed evolution of microorganisms have been aided by the recognition that many genes coding for the biodegradation of xenobiotics are located on transposable chromosomal elements (transposons) or on plasmids (46). Evidence indicates that the transfer and recombination of such movable genetic elements plays an important role in the evolution of antibiotic and heavy-metal resistance and utilization of novel substrates in natural environments (65). It is possible to augment the evolution of new degradative pathways by feeding into a chemostat enrichment microorganisms known to harbor plasmids, which encode for portions of the desired biodegradative pathway. Exchange, recombination, and amplification of genetic information under selective pressure, along with spontaneous and induced mutation, can greatly accelerate an evolutionary process. Combinations of the described techniques have yielded specially constructed *Pseudomonas* strains capable of degrading a wide array of chlorobenzoates and chlorophenols (67, 116). After the introduction of these strains into a waste stream of a chemical manufacturing plant, the removal of haloaromatic xenobiotics was greatly improved (81, 82).

Plasmid-assisted molecular breeding also produced a *B. cepacia* strain with the capacity to grow on the herbicide 2,4,5-T, although no 2,4,5-T-degrading microorganism could be isolated previously from the environment (79). Subsequently, it was demonstrated that the treatment of a heavily contaminated soil sample with this engineered *B. cepacia* effectively eliminated the 2,4,5-T contaminant while growing on it (57). The soil detoxified in this manner was subsequently able to support the growth of herbicide-sensitive plants.

TABLE 3 Genetic engineering solutions and benefits

Limitation	Genetic engineering solution	Benefit
Incomplete degradation	1. Uncoupling metabolism from degradation	1. Support activity with inexpensive, nontoxic substrates
	2. Deregulate genetic controls	2. Eliminate the need for toxic-inducing substrates 3. Achieve difficult cleanup
Low rates of degradation	1. Select high-performance host organisms	1. Use smaller, less expensive bioreactors
	2. Remove degradative "bottlenecks"	2. Decrease fermentation/treatments costs
Recalcitrant target compounds	1. Add substitution-specific functions (e.g., dehalogenation activity)	1. Increase range of treatable compounds
	2. Alter enzyme specificity	2. Increase substrate range of single organisms
Formation of toxic intermediates	1. Reroute metabolities	1. Extend treatment life
	2. Add complementary activities/pathways	2. Extend range of treatable compounds
Chemical mixtures (e.g., PCBs, mixed organic wastes)	1. Combine metabolic activities	1. Decrease fermentation costs (single organism)
	2. Broaden substrate specificity	2. Select environmentally robust host

55.2.3.2. Genetically Engineered Microorganisms

One approach for providing the enzymatic capability to degrade diverse pollutants is to use genetic engineering to create microorganisms with the capacity to degrade a wide range of compounds. This approach has a number of potential advantages (Table 3). Recombinant DNA technology holds great promise for developing microbial strains that can aid in the environmental removal of toxic chemicals (77, 80, 99, 135). A hydrocarbon-degrading pseudomonad engineered by Chakrabarty was the first organism that the Supreme Court of the United States ruled, in a landmark decision, could be patented. However, considerable controversy surrounds the release of such genetically engineered microorganisms into the environment, and field testing of these organisms has been delayed by the issues of safety, containment, and potential for ecological damage (63, 131). Given the current regulatory framework for the deliberate release of genetically engineered microorganisms, it is unlikely that any genetically engineered microorganisms will be utilized for full-scale bioremediation in the near future.

55.2.3.3. Adhesion-Deficient Microorganisms

An important factor limiting the applicability of bioaugmentation is the natural adhesive properties of native bacteria. Adhesion limits the microorganisms' penetration through soil and rock matrices. In an in situ field study of *B. cepacia* G4 by Nelson et al. (107), the movement of injected microorganisms was severely retarded. The hydraulic flow of the aquifer was approximately 48 feet per day, but none of the injected microorganisms were observed in a monitoring well 10 feet away from the injection point until 6 days after injection. Clearly, more aquifer material could be treated if the organisms or the aquifer conditions were altered to reduce adsorption and retardation of bacterial movement.

A specialized adhesion-deficient TCE-degrading strain of *B. cepacia* has been developed for in situ bioremediation of TCE-contaminated aquifers. In a field demonstration, this strain was injected into a bedrock aquifer during pneumatic fracturing. The strain was rapidly dispersed throughout the aquifer to a radius of approximately 25 feet. Chlorinated solvent concentrations in the aquifer decreased from approximately 20 mg/l to 5 mg/l during the first 21 days after bacterial injection (146). In a second field demonstration, the organisms were injected into a semiconfined sandy aquifer, and their migration through the aquifer and degradation of TCE was measured at a series of monitoring wells over a distance of approximately 40 feet. Hydraulic control within the test plot was maintained by recirculating groundwater from a down-gradient recovery well into the six up-gradient injection wells. A "control" plot was operated without added bacteria. After a signal injection of organisms, chlorinated ethene concentrations within the test plot were significantly decreased. Furthermore, viable cells of this bacterium were recovered at monitoring wells throughout the test plots. No loss of chlorinated ethenes was observed in the control plot, and no significant decreases in perchloroethene or TCE concentrations were observed in either the control or test plots (128).

55.3. APPLICATIONS OF BIOREMEDIATION TO VARIOUS CONTAMINANTS AND SITES

55.3.1. Hydrocarbon-Contaminated Soils and Aquifers

Hydrocarbons in soils and waters are the most often treated pollutants using bioremediation. Petroleum-contaminated soils and aquifers constitute about 60% of the sites where bioremediation is being used in field demonstrations or full-scale operations (73, 74). Several hundred hydrocarbon-contaminated sites have been identified where bioremediation is being tested as a possible cleanup technology, and bioremediation has great potential for treating these soils (101, 102, 126). In most cases the treatment of hydrocarbon-contaminated environments involves biostimulation (20, 86, 100) (Fig. 1).

A typical approach for bioremediation of hydrocarbon-contaminated soil begins with microbiological and chemical characterization of the site. Chemical analysis includes the measurement of nutrients (e.g., forms and concentrations of nitrogen and phosphorus) and evaluation of any other chemical factors that may be toxic to microbial growth. Microbiological testing includes the enumeration of both the total number of aerobic bacteria in the samples and the number of petroleum-degrading bacteria present. Low numbers of bacteria can indicate that the addition of bacteria may be warranted. In addition, the direct measurement of intrinsic air permeabilities of the soil can be accomplished in the field or in the laboratory. Taken in combination, these parameters can be used to select the most cost-effective strategy to meet the goals of a specific site.

In situ bioremediation techniques for hydrocarbon-contaminated subsurface systems have been reviewed and evaluated in several reports (32, 89, 115, 134, 139, 148, 149). Aquifer bioremediation depends strongly on local geological and hydrological conditions, and the engineering aspects of the process are beyond the scope of this chapter. In broad outline, an attempt is usually made to isolate the contamination plume from wells and other sensitive areas. This may be accomplished physically by using barriers such as cement, bentonite, or grout injection, or dynamically by pumping the polluted portion of the aquifer. In the latter

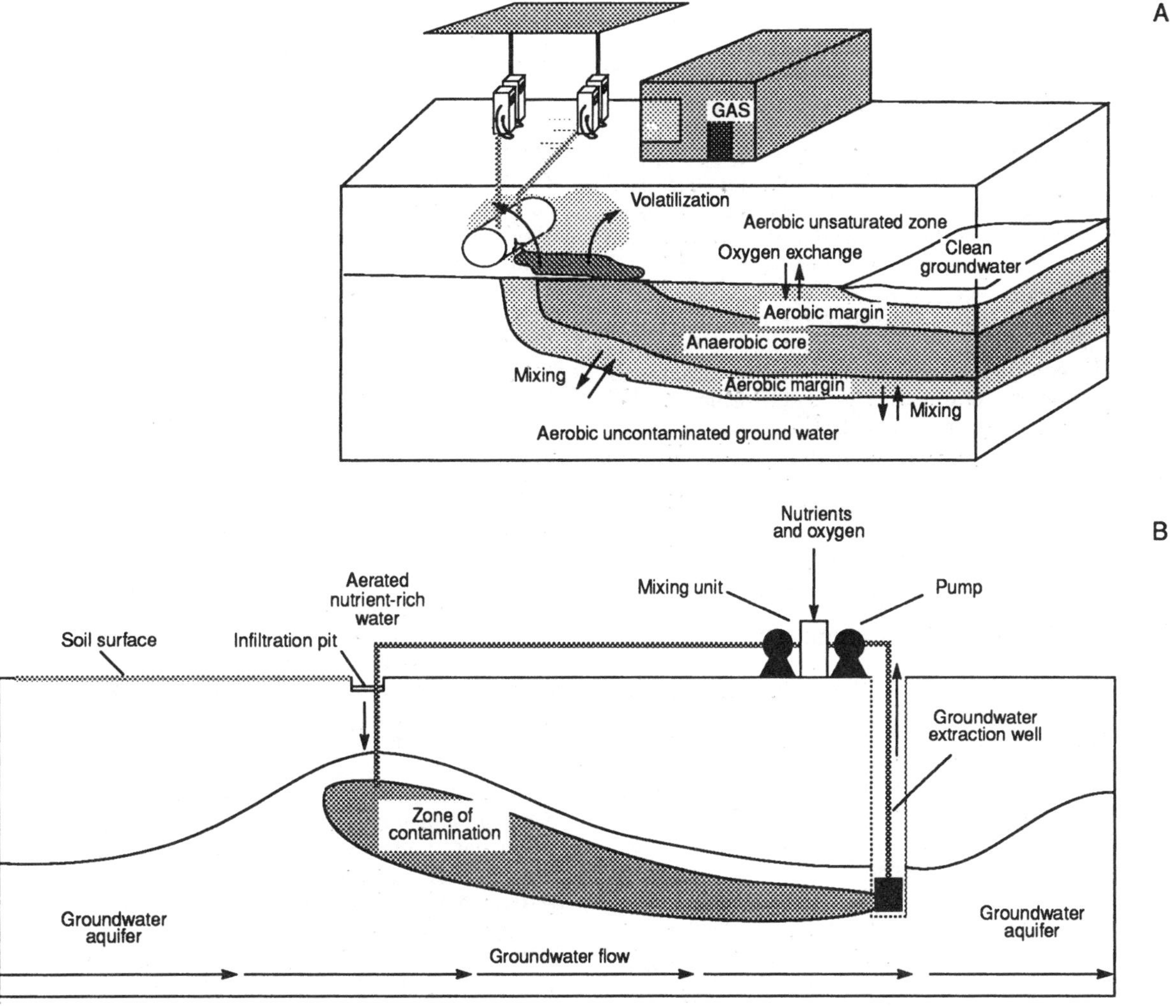

FIGURE 1 (A) Diagram showing leakage from an underground storage tank. The contaminated soil and groundwater can be bioremediated in situ or ex situ. (B) Diagram of in situ bioremediation of subsurface hydrocarbon-contaminated soil and groundwater.

case, groundwater flow is redirected toward rather than away from the contaminated plume, thus preventing the spreading of contamination.

Besides containment, the pumping process recovers free-flowing hydrocarbon and contaminants dissolved in water. In theory, prolonged pumping could eventually flush out all the contaminant, but the solubility characteristics of hydrocarbons make reliance on physical flushing alone prohibitively slow and expensive. Hydrocarbons have high affinities to soil particles (141). To clean an aquifer by simple water flushing may take tens to hundreds of years and several thousand pore volumes of water in the contaminated portion of the aquifer. Physical flushing may be facilitated by the addition of dispersants and emulsifiers through injection wells around the periphery of the contamination plume, while water is continuously withdrawn from the center of the plume. However, the use of combined biodegradation and emulsification by hydrocarbon-utilizing microorganisms is usually more efficient and economical (24, 32, 134). In a common modification, the pumped contaminated groundwater may be aerobically treated above ground for several hours prior to returning it to the aquifer replenished with oxygen and, if necessary, with mineral nutrients.

Volatile pollutants that are nonmiscible with water may be removed from vadose (water-unsaturated) soils by air stripping or venting. Air is pumped underground, volatilizing and removing the pollutant. In its simplest form, the treatment transfers the pollutant to the atmosphere where it might be eventually photodegraded. In more controlled treatments, the pollutant is removed from the effluent air by activated carbon and/or biofilters. In bioventing, the air injection is managed in such a way that the contaminated air stream passes through layers of uncontaminated soil. During this passage, the soil acts as a biofilter and removes, by biodegradation, some or most of the pollutant. Some filtration after treatment may still be necessary. As an example, in 1993, the U.S. Air Force completed a full-scale soil bioventing project to clean up a 27,000-gallon jet-fuel spill at Hill Air Force Base in Utah. During the 18-month project, jet-fuel residues in soil were reduced from an average total petroleum hydrocarbon concentration of about 900 mg/kg to less than 10 mg/kg. In this bioventing operation, 60% of the hydrocarbons removed from the soil volatilized directly into the atmosphere, and the remaining 40% were biodegraded to carbon dioxide and water.

55.3.2. Halocarbon-Contaminated Soils and Aquifers

Bioremediation of aquifers contaminated with halogenated aromatics, haloethanes, and halomethanes presents additional complex problems (25, 83, 148). While some of these materials are dehalogenated anaerobically, others cannot serve as substrates under either aerobic or anaerobic conditions and are attacked only cometabolically in the presence of methane or toluene, and only under oxidative conditions.

Since methanotrophs, through their production of methane monooxygenase, are able to degrade TCE, dichloroethene, and vinyl chloride by cometabolism (52), several investigators have used methanotrophs for the bioremediation of sites contaminated with these halogenated compounds (Fig. 2). A field test to determine the validity of in situ bioremediation has shown a 98% reduction of TCE concentrations in groundwater (45). McCarty et al. (96) found that they could stimulate indigenous methanotrophic populations. They developed a model in situ bioremediation treatment that would require 5,200 kg of methane and 19,200 kg of oxygen in order to convert 1,375 kg of chlorinated hydrocarbons from an aquifer of 480,000 m^3 containing a total contaminant load of 1,617 kg of halogenated compounds.

A variety of bioreactor configurations have been studied to overcome limitations associated with ex situ, cometabolic degradation of TCE and other volatile organic compounds. These designs have attempted to minimize competitive interactions between the primary substrate and the cometabolite (TCE) by temporal or spatial separation of the growth substrate and the TCE. Both gas-phase (for vapor treatment) and fluidized-bed (for direct groundwater treatment) bioreactors have been designed for the bioremediation of TCE (47, 53, 54). The gas-phase bioreactor was designed to balance TCE mass transfer with biodegradation capacity. Under typical operation, greater than 90% of the TCE is destroyed from a contaminated groundwater or air stream. This bioreactor has been demonstrated to perform equally well with *B. cepacia* G4 and strains of *Pseudomonas mendocina* as biocatalysts in the system (47). This bioreactor system has demonstrated stable performance over 10 months of continuous operation. Maximal TCE degradation rates were observed at concentrations as high as 300 μM (40 ppm). Competitive inhibition was observed for degradation of phenol and TCE in mixtures. This is consistent with the observation that K_s values for both TCE and phenol are similar and that TCE degradation is a cometabolic process in which an enzyme whose normal substrate is phenol also attacks TCE (55, 106, 122). The economic analysis generated as part of these projects indicates typical savings of 70 to 80% using the biological treatment system as compared to carbon adsorption.

Although an extensive body of research clearly demonstrates that PCB biodegradation occurs under both aerobic and anaerobic conditions, and that this degradation activity is widespread, the challenge that technologists face is transitioning this microbiology, biochemistry, and genetics into an efficient and cost-effective bioremediation process. There have been many claims of commercial PCB bioremediation technologies; however, none of these has yet survived rigorous technical and analytical scrutiny. Hydrophobic substrates such as PCBs tend to elude analysts, and many claims by commercial PCB-bioremediation vendors have been based on PCB disappearance, which can most often be attributed to repartitioning and redistribution within the biotreatment system, or to other abiotic losses and artifacts. The characteristic mark of these systems is no congener specificity (121, 138).

In situ bench-scale microcosms as well as field pilot studies have been conducted to optimize in situ PCB soil biotreatment processes. This work has focused on dilute PCB contamination (less than 100 mg/kg) because of the current rate limitations on in situ treatments. Parameters being addressed include bacterial dosing regimen, moisture content, nutrient addition, and additives. Results to date have shown 20 to 65% PCB biodegradation in soils contaminated with 10 to 100 ppm in the laboratory.

Current PCB bioremediation development programs continue to focus on (i) optimization of conditions for growth and activity of PCB-degrading bacteria, both aerobic and anaerobic, (ii) isolation and characterization of superior and novel PCB degraders, (iii) genetic engineering of PCB pathways, (iv) development of various physical and

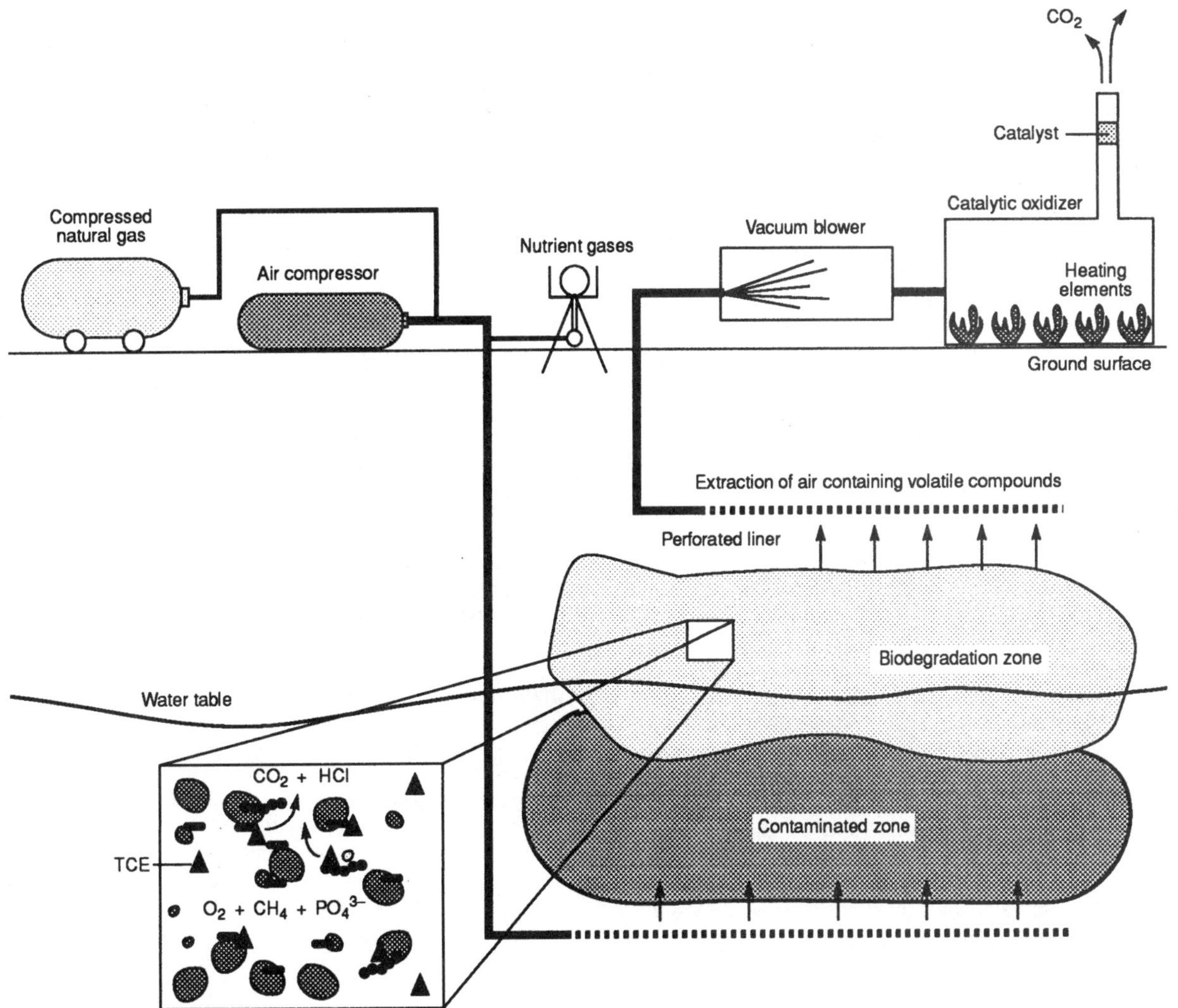

FIGURE 2 Bioventing system for treating TCE-contaminated groundwater based upon injection of air and the cosubstrate methane to stimulate a bacterial consortium that can utilize methane and cometabolize TCE. Much of the TCE is biodegraded in the treatment zone. Residues in the extracted air are destroyed in a catalytic oxidizer.

chemical pretreatments and cotreatments for improving the effectiveness of biotreatment steps, and (v) design and testing of field pilot systems.

55.3.3. Marine Oil Spills

Marine oil spills are major pollution events that can have a dramatic impact on coastal ecosystems. A number of technologies, including bioremediation, have been developed to mitigate the effects of marine oil spills (17). Few studies have been conducted in open waters, and the effectiveness of bioremediation in such cases remains unproven (133). Nutrient supplementation in coastal zones has been shown to be an effective means of stimulating microbial hydrocarbon biodegradation, thereby reducing the impact of marine oil pollutants (14, 15, 29, 111, 133). In some cases, such as the *Amoco Cadiz* spill, there were sufficient inputs of inorganic nitrogen and phosphates to support extensive biodegradation, and it was possible to rely upon intrinsic bioremediation (61). This approach to bioremediation simply employed monitoring to follow the natural microbial removal of the petroleum pollutants.

In other cases the rate of petroleum biodegradation in coastal marine environments is severely limited by the available concentrations of fixed forms of nitrogen and phosphate. In these cases, fertilizer addition is effective for stimulating the hydrocarbon biodegradative activities of the indigenous microbial populations (15, 111, 133). The oleophilic fertilizer Inipol EAP 22 was one of the fertilizers that was extensively used in the cleanup of the 1989 *Exxon Valdez* Alaskan oil spill. This fertilizer formulation is effective on rocky shorelines but has not worked well on some oil-polluted fine-grain sand beaches (133).

The problem with many of the studies on bioremediation of marine oil spills is that they take place following real tanker accidents when the emphasis is on emergency response and not scientific study. Planning and proper controls are often limited so that interpretation of data typically is ambiguous. Especially controversial is whether seeding with hydrocarbon-degrading microbial cultures is a useful adjunct to nutrient addition. To answer some of the lingering questions about the effectiveness of bioremediation for treating marine oil spills, an experimental oil spill study was conducted along the shoreline of Delaware Bay to determine if bioremediation with inorganic mineral nutrients and/or microbial inoculation enhanced the removal of crude oil contaminating a sandy beach and to calculate intrinsic and enhanced biodegradation rates (140). This study shows that significant hydrocarbon biodegradation also occurred in untreated plots. However, significant differences were observed in the biodegradation rates of total alkane and total aromatic hydrocarbons between treated and untreated plots. There were no significant differences between plots treated with nutrients alone and plots treated with nutrients and the inoculum microorganisms. First-order rate constants for the disappearance of alkanes and polycyclic aromatic hydrocarbons showed loss patterns typical of biodegradation. Thus, nutrient addition was the critical factor in stimulating hydrocarbon biodegradation, and seeding with nonindigenous bacterial cultures was not necessary for bioremediation of the oil pollutants. The general consensus is that indigenous populations of hydrocarbon-degrading microbial populations are adequate and that the critical variable for bioremediation is overcoming the environmental factors limiting the growth and activity of these populations.

A novel approach to supplying fertilizers that does require seeding with nonindigenous cultures has been developed in Israel (119). In this approach the fertilizer, which is a proprietary polymeric formulation, cannot be utilized as a source of carbon and energy by most microorganisms. Thus, the nitrogen and phosphate nutrients in the fertilizer are not available to the indigenous microbial populations and cannot be used alone for bioremediation. A hydrocarbon-utilizing bacterium that can also attack the polymeric fertilizer and utilize the nitrogen and phosphate it contains has been isolated. This bacterium is added along with the polymeric fertilizer to oil-contaminated shorelines. This approach gives a selective advantage to the seed bacterial because they are the only ones that can gain direct access to the nutrients. By avoiding competition for the nutrients in the fertilizer, the hydrocarbon biodegradative activities of the seed bacteria are greatly favored and more effective bioremediation can be achieved. This bioremediation treatment has been shown to be effective in the removal of oil pollutants from sand beaches along the Mediterranean Sea. It is, however, uncertain that this approach is worth the cost.

55.3.4. Metal-Contaminated Soils

In most cases the applicability of bioremediation depends upon the abilities of microorganisms to biodegrade the pollutants to safe levels and to produce nontoxic end products. However, immobilization of heavy metals, for example, by sulfide-producing microorganisms, is an effective means of treating some metal-contaminated sites. Bioconcentration, i.e., uptake without transformation, also can be used to treat heavy-metal-contaminated soils and waters. This forms the basis for phytoremediation in which plants are used to extract metals from contaminated soils and waters.

Phytoremediation is emerging as a promising method for decontaminating metal-contaminated soils and waters. Various plants concentrate heavy metals, including radionuclides and lead, in the plant biomass. The concentrations of hazardous metals in soils can be reduced, and leaching into groundwaters can be prevented by planting Indian mustard (*Brassica juncea*) or ragweed (various *Ambrosia* species) among other plants. Indian mustard can accumulate up to 40% of their biomass as heavy metals, including lead. Demonstration projects have been conducted at various sites, including Chernoble in Russia, which was heavily contaminated with radionuclides from a nuclear reactor accident. Once the plants accumulate the heavy metals, they can be harvested and the plant biomass incinerated, leaving the metals as a residue; metals recovered in this manner can be reused.

55.4. CONCLUSIONS

Bioremediation is a cost-effective technology for the treatment of a variety of pollutants and contaminated sites. Its applicability and potential for success depend upon three factors: the properties of the pollutant (biodegradability of the specific chemical pollutants); the microbial community (enzymatic capacity to metabolize the pollutant chemicals); and the environment (properties permitting or limiting microbial growth and metabolism of the polluting chemicals). The applicability and approaches to bioremediation depend upon these three factors. In many cases bioremediation relies upon the intrinsic degradative capacity of the indigenous microorganisms to remove the pollutant without further treatment (intrinsic bioremediation); in other cases the rates of metabolism of the indigenous microbial community are increased through environmental engineering (biostimulation); and in yet other cases the microbial community is altered through seeding with specialized cultures (bioaugmentation). To date, most commercial applications of bioremediation depend upon naturally occurring microorganisms and most have targeted hydrocarbon-contaminated sites. Approaches to the bioremediation of chlorinated compounds and metals are more complex but have been proceeding, primarily through demonstration projects. The use of genetically engineered microorganisms and ex situ treatments for remediation of additional industrial pollutants, including TCE, are likely to gain increased importance in the near future.

REFERENCES

1. **Abramowicz, D. A.** 1990. Aerobic and anaerobic biodegradation of PCBs: a review. *Crit. Rev. Biotechnol.* **10:** 241–251.
2. **Acar, Y. B., E. E., Ozsu, A. N. Alshawabkeh, M. F. Rabbi, and R. J. Gale.** 1996. Enhance soil bioremediation with electric fields. *Chemtech* **26:**40–45.
3. **Adriaens, P., and D. D. Focht.** 1990. Continuous coculture degradation of selected polychlorinated biphenyl congeners by *Acinetobacter* spp. in an aerobic reactor system. *Environ. Sci. Technol.* **24:**1042–1049.
4. **Adriaens, P. H., P. E. Kohler, and D. Kohler-Staub.** 1989. Bacterial dehalogenation of chlorobenzoates and coculture biodegradation of 4,4′-dichlorobiphenyl. *Appl. Environ. Microbiol.* **55:**887–892.
5. **Alexander, M.** 1965. Biodegradation: problems of molecular recalcitrance and microbial falibility. *Adv. Appl. Microbiol.* **7:**35–80.
6. **Alexander, M.** 1981. Biodegradation of chemicals of environmental concern. *Science* **211:**132–138.
7. **Alexander, M.** 1994. *Biodegradation and Bioremediation*. Academic Press, San Diego, Calif.
8. **American Petroleum Institute.** 1987. *Field Study of Enhanced Subsurface Biodegradation of Hydrocarbons Using Hydrogen Peroxide as an Oxygen Source*. American Petroleum

Institute Pub. 4448. American Petroleum Institute, Washington, D.C.

9. **Applied Biotreatment Association.** 1989. *Case History Compendium*. Applied Biotreatment Association, Washington, D.C.
10. **Applied Biotreatment Association.** 1990. *The Role of Biotreatment of Oil Spills*. Applied Biotreatment Association, Washington, D.C.
11. **Atlas, R. M.** 1981. Microbial degradation of petroleum hydrocarbons: an environmental perspective. *Microbiol. Rev.* **45:**180–209.
12. **Atlas, R. M. (ed.).** 1984. *Petroleum Microbiology*. Macmillan, New York.
13. **Atlas, R. M.** 1995. Bioremediation. *Chem. Eng. News* (April) **3:**32–42.
14. **Atlas, R. M.** 1995. Bioremediation of petroleum pollutants. *Int. Biodeterior. Biodegrad.*, p. 317–327.
15. **Atlas R. M.** 1996. Slick solutions. *Chemistry in Britain* **32:** 42–45.
16. **Atlas, R. M., and R. Bartha.** 1973. Stimulated biodegradation of oil slicks using oleophilic fertilizers. *Environ. Sci. Technol.* **7:**538–541.
17. **Atlas, R. M., and R. Bartha.** 1992. Hydrocarbon biodegradation and oil spill bioremediation. *Adv. Microb. Ecol.* **12:**287–338.
18. **Bagley, D. M., and J. M. Gossett.** 1990. Tetrachloroethene transformation to trichloroethene and *cis*-1, 2-dichloroethene by sulfate-reducing enrichment cultures. *Appl. Environ. Microbiol.* **56:**2511–2516.
19. **Baker, K. H., and D. S. Herson.** 1994. *Bioremediation*. McGraw-Hill, New York.
20. **Bartha, R.** 1986. Biotechnology of petroleum pollutant biodegradation. *Microb. Ecol.* **12:**155–172.
21. **Bedard, D. L., and J. F. Quensen III.** 1995. Microbial reductive dechlorination of polychlorinated byphenyls, p. 127–216. *In* L. Y. Young and C. E. Cerniglia (ed.), *Microbial Transformation and Degradation of Toxic Organic Chemicals*. Wiley-Liss, New York.
22. **Bedard, D. L., R. E. Wagner, and M. J. Brennan.** 1987. Extensive degradation of Arochlors and environmentally transformed polychlorinated biphenyls by *Alcaligenes eutrophus* H850. *Appl. Environ. Microbiol.* **53:**1094–1102.
23. **Bennett, J. W., and B. D. Faison.** 1997. Use of fungi in biodegradation, p. 758–765. *In* C. J. Hurst, G. R. Knudsen, M. J. McInerney, L. D. Stetzenbach, and M. V. Walter (ed.), *Manual of Environmental Microbiology*. ASM Press, Washington, D.C.
24. **Beraud, J. F., J. D. Ducreux, and C. Gatellier.** 1989. Use of soil-aquifer treatment in oil pollution control of underground waters, p. 53–59. *In* Proceedings of the 1989 Oil Spill Conference. American Petroleum Institute, Washington, D.C.
25. **Berwanger, D. J., and J. F. Barker.** 1988. Aerobic biodegradation of aromatic and chlorinated hydrocarbons commonly detected in landfill leachate. *Water Pollut. Res. J. Can.* **23:**460–475.
26. **Borden R. C., T. M. Vogel, J. M. Thomas, and C. H. Ward.** 1994. *Handbook of Bioremediation*. Lewis Publishers, Boca Raton, Fla.
27. **Bossert, I., and R. Bartha.** 1984. The fate of petroleum in soil ecosystems, p. 435–473. *In* R. M. Atlas (ed.), *Petroleum Microbiology*. Macmillan, New York.
28. **Boulding, J. R.** 1995. *Practical Handbook of Soil, Vadose Zone, and Ground-water Contamination: Assessment, Preventation, and Remediation*. Lewis Publishers, Boca Raton, Fla.
29. **Bragg, J. R., R. C. Prince, E. J. Harner, and R. M. Atlas.** 1994. Effectiveness of bioremediation for the *Exxon Valdez* oil spill. *Nature* **368:**413–418.
30. **Britton, L. N.** 1984. Microbial degradation of aliphatic hydrocarbons, p. 89–129. *In* D. T. Gibson (ed.), *Microbial Degradation of Organic Compounds*. Marcel Dekker, New York.
31. **Brown, J. F., D. L. Bedard, and M. J. Brennan.** 1987. Polychlorinated biphenyl dechlorination in aquatic sediments (river sediments). *Science* **236:**709–712.
32. **Brown, R. A., R. D. Norris, and G. R. Brubaker.** 1985. Aquifer restoration with enhanced bioreclamation. *Pollut. Eng.* **17:**25–28.
33. **Brown, R. A., R. D. Norris, and R. L. Raymond.** 1984. Oxygen transport in contaminated aquifers, p. 441–450. *In Proceedings of the Conference on Petroleum Hydrocarbons and Organic Chemicals in Ground Water—Prevention, Detection, and Restoration*. National Water Well Association, Worthington, Ohio.
34. **Brubaker, G. R.** 1995. The boom is *in situ* bioremediation. *Civil Eng.* **65:**38–42.
35. **Bumpus, J. A.** 1989. Biodegradation of polycyclic aromatic hydrocarbons by *Phanerochaete chrysosporium*. *Appl. Environ. Microbiol.* **55:**154–158.
36. **Bumpus, J. A., and S. D. Aust.** 1987. Biodegradation of DDT [1,1,1-trichloro-2,2-bis(4-chlorophenyl) ethane] by the white rot fungus *Phanerochaete chrysosporium*. *Appl. Environ. Microbiol.* **53:**2001–2028.
37. **Bumpus, J. A., M. Tien, D. Wright, and S. D. Aust.** 1985. Oxidation of persistent environmental pollutants by a white rot fungus. *Science* **228:**1434–1436.
38. **Cerniglia, C. E.** 1984. Microbial transformation of aromatic hydrocarbons, p. 99–128. *In* R. M. Atlas (ed.), *Petroleum Microbiology*. Macmillan, New York.
39. **Chaudhry, G. R., and S. Chapalamadugu.** 1991. Biodegradation of halogenated organic compounds. *Microbiol. Rev.* **55:**59–79.
40. **Coates, J. D, R. T. Anderson, and D. R. Lovley.** 1996. Oxidation of polycyclic aromatic hydrocarbons under sulfate-reducing conditions. *Appl. Environ. Microbiol.* **62:** 1099–1101.
41. **Coates, J. D., R. T. Anderson, J. C. Woodward, E. J. P. Phillips, and D. R. Lovley.** 1996. Anaerobic hydrocarbon degradation in petroleum-contaminated harbor sediments under sulfate-reducing and artificially imposed iron-reducing conditions. *Environ. Sci. Technol.* **30:**2784–2789.
42. **Cookson, J. T., Jr.** 1995. *Bioremediation Engineering: Design and Application*. McGraw-Hill, New York.
43. **Crawford, R. L., and D. L. Crawford (ed.).** 1996. *Bioremediation: Principles and Applications*. University Press, Cambridge.
44. **Dowd, R. M.** 1984. Leaking underground storage tanks. *Environ. Sci. Technol.* **18:**309–312.
45. **Duba, A. G., K. J. Jackson, M. C. Jovanovich, R. B. Knapp, and R. T. Taylor.** 1996. TCE remediation using *in situ*, resting-state bioaugmentation. *Environ. Sci. Technol.* **30:**1982–1990.

46. **Eaton, R. W., and K. N. Timmis.** 1984. Genetics of xenobiotic degradation, p. 694–703. *In* M. J. Klug and C. A. Reddy (ed.), *Current Prespectives in Microbial Ecology*. American Society for Microbiology, Washington, D.C.
47. **Ensley, B. D., and P. R. Kurisko.** 1994. A gas lift bioreactor for removal of contaminants from the vapor phase. *Appl. Environ. Microbiol.* **60:**285–290.
48. **Enzien, M. V., F. Picardal, T. C. Hazen, R. G. Arnold, and C. B. Fliermans.** 1994. Reductive dechlorination of trichloroethylene and tetrachloroethylene under aerobic conditions in a sediment column. *Appl. Environ. Microbiol.* **60:**2200–2204.
49. **Fernando, T., J. A. Bumpus, and S. D. Aust.** 1990. Biodegradation of TNT (2,4,6-trinitrotoluene) by *Phanerochaete chrysosporum*. *Appl. Environ. Microbiol.* **56:**1666–1671.
50. **Flathman, P. E., D. E. Jerger, and J. H. Exner.** 1994. *Bioremediation-field Experience*. Lewis Publishers, Boca Raton, Fla.
51. **Focht, D. D.** 1997. Aerobic biotransformation of polychlorinated biphenyls, p. 811–814. *In* C. J. Hurst, G. R. Knudsen, M. J. McInerney, L. D. Stetzenbach, and M. V. Walter (ed.), *Manual of Environmental Microbiology*. ASM Press, Washington, D.C.
52. **Fogel, M. M., A. R. Taddeo, and S. Fogel.** 1986. Biodegradation of chlorinated ethanes by a methane-utilizing mixed culture. *Appl. Environ. Microbiol.* **51:**720–724.
53. **Folsom, B. R., A. K. Bohner, T. Burick, and W. J. Guarini.** 1995. Two-stage bioreactor to destroy chlorinated and nonchlorinated organic groundwater contaminants. *In Proceedings of the Third International In Situ and On-Site Bioreclamation Symposium*. San Diego, Calif., April 24–28.
54. **Folsom, B. R., and P. J. Chapman.** 1991. Performance characterization of a model bioreactor for the biodegradation of trichloroethylene by *Pseudomonas cepacia* strain G4. *Appl. Environ. Microbiol.* **57:**1602–1608.
55. **Folsom, B. R., P. J. Chapman, and P. H. Pritchard.** 1990. Phenol and trichloroethylene degradation by *Pseudonomas cepacia* G4: kinetics and interactions between substrates. *Appl. Environ. Microbiol.* **56:**1279–1285.
56. **Galli, R., and P. L. McCarty.** 1989. Biotransformation of 1,1,1-trichloroethane, trichloromethane, and tetrachloromethane by a *Clostridium* sp. *Appl. Environ. Microbiol.* **55:** 837–844.
57. **Ghosal, D., I. S. You, D. K. Chatterjee, and A. M. Chakrabarty.** 1985. Microbial degradation of halogenated compounds. *Science* **228:**135–142.
58. **Gibson, D. T., and G. S. Saylor.** 1992. *Scientific Foundations of Bioremediation: Current Status and Future Needs*. American Academy of Microbiology, Washington, D.C.
59. **Gibson, D. T, and V. Subramanian.** 1984. Microbial degradation of aromatic hydrocarbons, p. 181–252. *In* D. T. Gibson (ed.), *Microbial Degradation of Organic Compounds*. Plenum Press, New York.
60. **Grbic-Galic, D., and T. M. Vogel.** 1987. Transformation of toluene and benezene by mixed methanogenic cultures. *Appl. Environ. Microbiol.* **53:**254–260.
61. **Gundlach, E. R., P. D. Boehm, M. Marchand, R. M. Atlas, D. M. Ward, and D. A. Wolfe.** 1983. The fate of *Amoco Cadiz* oil. *Science* **221:**122–129.
62. **Hallas, L. E., and M. Alexander.** 1983. Microbial transformation of nitroaromatic compounds in sewage effluent. *Appl. Environ. Microbiol.* **45:**1234–1241.
63. **Halvorson, H. O., D. Pramer, and M. Rogul (ed.).** 1985. *Engineered Organisms in the Environment: Scientific Issues*. American Society for Microbiology, Washington, D.C.
64. **Harding, G. L.** 1997. Bioremediation and the dissimilatory reduction of metals, p. 806–810. *In* C. J. Hurst, G. R. Knudsen, M. J. McInerney, L. D. Stetzenbach, and M. V. Walter (ed.), *Manual of Environmental Microbiology*. ASM Press, Washington, D.C.
65. **Hardy, K.** 1981. *Bacterial Plasmids*. Aspects of Microbiology Series, no. 4. American Society for Microbiology, Washington, D.C.
66. **Hart, S.** 1996. *In situ* bioremediation: defining the limits. *Environ. Sci. Technol.* **30:**398–401.
67. **Hartmann, J., W. Reineke, and H. J. Knackmuss.** 1979. Metabolism of 3-chloro, 4-chloro-, and 3,5-dichlorobenzoate by a pseudomonad. *Appl. Environ. Microbiol.* **37:**421–428.
68. **Heitkamp, M. A., and C. E. Cerniglia.** 1988. Mineralization of polycyclic aromatic hydrocarbons by a bacterium isolated from sediment below an oil field. *Appl. Environ. Microbiol.* **54:**1612–1614.
69. **Hill, I. R., and S. J. L. Wright (ed.).** 1978. *Pesticide Microbiology*. Academic Press, London.
70. **Hinchee, R. E. (ed.).** 1994. *Air Sparging for Site Remediation*. Lewis Publishers, Boca Raton, Fla.
71. **Hinchee, R. E., B. C. Alleman, R. E. Hoeppel, and R. N. Miller.** 1994. *Hydrocarbon Bioremediation*. Lewis Publishers, Boca Raton, Fla.
72. **Hinchee, R. E., D. B. Anderson, F. B. Metting, Jr., and G. D. Sayles (ed.).** 1994. *Applied Biotechnology for Site Remediation*. Lewis Publishers, Boca Raton, Fla.
73. **Hinchee, R. E., and R. F. Olfenbuttel (ed.).** 1991. *In Situ Bioreclamation: Applications and Investigations for Hydrocarbon and Contaminated Site Remediation*. Butterworth-Heinemann, Boston.
74. **Hinchee, R. E., and R. F. Olfenbuttel (ed.).** 1991. *On-Site Bioreclamation: Processes for Xenobiotic and Hydrocarbon Treatment*. Butterworth-Heinemann, Boston.
75. **Ho, S. V., P. W. Sheridan, C. J. Athmer, M. A. Heitkamp, J. M. Brackin, D. Weber, and P. H. Brodsky.** 1995. Integrated *in situ* soil remediation technology: the lasagna process. *Environ. Sci. Technol.* **29:**2528–2534.
76. **Huddleston, R. L., and L. W. Cresswell.** 1976. Environmental and nutritional constraints of microbial hydrocarbon utilization in the soil, p. 71–72. *In* Proceedings of the 1975 Engineering Foundation Conference: The Role of Microorganisms in the Recovery of Oil. NSF/RANN, Washington, D.C.
77. **Jain, R. K., and G. S. Sayler.** 1987. Problems and potential for *in situ* treatment of environmental pollutants by engineered microorganisms. *Microbiol. Sci.* **4:**59–63.
78. **Jamison, V. M., R. L. Raymond, and J. O. Hudson, Jr.** 1975. Biodegradation of high-octane gasoline in groundwater. *Dev. Ind. Microbiol.* **16:**305–312.
79. **Kellogg, S. T., D. K. Chatterjee, and A. M. Chakrabarty.** 1981. Plasmid assisted molecular breeding—new technique for enhanced biodegradation of persistent toxic chemicals. *Science* **214:**1133–1135.
80. **Kilbane, J. J.** 1986. Genetic aspects of toxic chemical degradation. *Microb. Ecol.* **12:**135–146.

81. **Knackmuss, H. J.** 1983. Xenobiotic degradation in industrial sewage: haloaromatics as target substances, p. 173–190. *In* C. F. Phelps and P. H. Clarke (ed.), *Biotechnology*. The Biochemical Society, London.

82. **Knackmuss, H. J.** 1984. Biochemistry and practical implications of organohalide degradation, p. 687–693. *In* M. J. Klug and C. A. Reddy (ed.), *Current Perspectives in Microbial Ecology*. American Society for Microbiology, Washington, D.C.

83. **Kuhn, E. P., P. J. Colberg, J. L. Schnoor, O. Wanner, A. J. B. Zehnder, and R. P. Schwartzenbach.** 1985. Microbial transformations of substituted benzenes during infiltration of river water to groundwater: laboratory column studies. *Environ. Sci. Technol.* **19:**961–968.

84. **LaDousse, A., C. Tallec, and B. Tramier.** 1987. Progress is enhanced oil degradation, abstr. 142. *In* Proceedings of the 1987 Oil Spill Conference. American Petroleum Institute, Washington, D.C.

85. **LaDousse, A., and B. Tramier.** 1991. Results of 12 years of research in spilled oil bioremediation: Inipol EAP 22, p. 577–581. *In* Proceedings of the 1991 International Oil Spill Conference. American Petroleum Institute, Washington, D.C.

86. **Leahy, J. G., and R. R. Colwell.** 1990. Microbial degradation of hydrocarbons in the environment. *Microbiol. Rev.* **54:**305–315.

87. **Lee, B., A. L. Pometto, and A. Fratzke.** 1991. Biodegradation of degradable plastic polyethylene by *Phanerochaete* and *Streptomyces* species. *Appl. Environ. Microbiol.* **57:** 678–685.

88. **Lee, K., and E. M. Levy.** 1991. Bioremediation: waxy crude oils stranded on low-energy shorelines, p. 541–547. *In* Proceedings of the 1991 International Oil Spill Conference. American Petroleum Institute, Washington, D.C.

89. **Lee, M. D., J. T. Wilson, and C. H. Ward.** 1987. *In situ* restoration techniques for aquifers contaminated with hazardous wastes. *J. Hazardous Material* **14:**71–82.

90. **Levin, M. A., and M. A. Gealt.** 1993. *Biotreatment of Industrial and Hazardous Waste*. McGraw Hill, New York.

91. **Little, C. D., A. V. Palumbo, and S. E. Herbes.** 1988. Trichloroethylene biodegradation by a methane-oxidizing bacterium. *Appl. Environ. Microbiol.* **54:**951–956.

92. **Lovley, D. R., J. D. Coates, J. C. Woodward, and E. J. P. Phillips.** 1995. Benzene oxidation coupled to sulfate reduction. *Appl. Environ. Microbiol.* **61:**953–958.

93. **Lovley, D. R., J. C. Woodward, and F. H. Chapelle.** 1994. Stimulated anoxic biodegradation of aromatic hydrocarbons using Fe(III) ligands. *Nature* **370:**128–131.

94. **Lovley, D. R., J. C. Woodward, and F. H. Chapell.** 1996. Rapid anaerobic benzene oxidation with a variety of chelated Fe(III) forms. *Appl. Environ. Microbiol.* **62:**288–291.

95. **Maugh, T. H.** 1973. DDT: an unrecognized source of polychlorinated biphenyls. *Science* **180:**578–579.

96. **McCarty, P. L., L. Semprini, M. E. Dolan, T. C. Harmon, C. Tiedeman, and S. M. Gorelick.** 1991. *In situ* methanotrophic bioremediation for contaminated groundwater at St. Joseph, MI, p. 16–40. *In* R. E. Hinchee and R. F. Olfenbuttel (ed.), *On-Site Bioreclamation: Processes for Xenobiotic and Hydrocarbon Treatment*. Butterworth-Heinemann, Boston.

97. **Means, J., and R. E. Hinchee.** 1994. *Emerging Technology for Bioremediation of Metals*. Lewis Publishers, Boca Raton, Fla.

98. **Mohn, W. M., and J. M. Tiedje.** 1992. Microbial reductive dehalogenation. *Microbiol. Rev.* **56:**482–507.

99. **Mongkolsuk, S., P. S. Lovett, and J. E. Trempy.** 1992. *Biotechnology and Environmental Science: Molecular Approaches*. Plenum Press, New York.

100. **Morgan, P., and R. J. Watkinson.** 1989. Hydrocarbon biodegradation in soils and methods for soil biotreatment. *Crit. Rev. Biotechnol.* **8:**305–333.

101. **Mueller, J. G., P. J. Chapman, and P. H. Pritchard.** 1989. Creosote-contaminated sites: their potential for bioremediation. *Environ. Sci. Technol.* **23:**1197–1201.

102. **Mueller, J. G., D. P. Middaugh, and S. E. Lantz.** 1991. Biodegradation of creosote and pentachlorophenol in contaminated groundwater: chemical and biological assessment. *Appl. Environ. Microbiol.* **57:**1277–1285.

103. **Munakata-Marr, J., P. L. McCarty, M. S. Shields, M. Reagin, and S. C. Francesconi.** 1996. Enhancement of trichloroethylene degradation in aquifer microcosms bioaugmented with wild type and genetically altered *Burkholderia* (*Pseudomonas*) *cepacia* G4 and PR1. *Environ. Sci. Technol.* **30:**2045–2053.

104. **National Research Council.** 1985. *Oil in the Sea: Inputs, Fates and Effects*. National Academy Press, Washington, D.C.

105. **National Research Council.** 1993. *In Situ Bioremediation*. National Academy Press, Washington, D.C.

106. **Nelson, H. J. K., S. O. Montgomery, W. R. Mahaffrey, and P. H. Pritchard.** 1987. Biodegradation of trichloroethylene and involvement of an aromatic biodegradation pathway. *Appl. Environ. Microbiol.* **53:**949–954.

107. **Nelson, M. J., J. V. Kinsella, and T. Montoya.** 1990. *In situ* biodegradation of TCE contaminated groundwater. *Environ. Prog.* **9:**190–196.

108. **Nishino, S. F., and J. C. Spain.** 1997. Biodegradation and transformation of nitroaromatic compounds, p. 776–783. *In* C. J. Hurst, G. R. Knudsen, M. J. McInerney, L. D. Stetzenbach, and M. V. Walter (ed.), *Manual of Environmental Microbiology*. ASM Press, Washington, D.C.

109. **Novick, N. J., and M. Alexander.** 1985. Cometabolism of low concentrations of propachlor, alachlor, and cycloate in sewage and lake water. *Appl. Environ. Microbiol.* **49:**737–743.

110. **OECD.** 1994. *Biotechnology for a Clean Environment*. Organization for Economic Cooperation and Development, Paris.

111. **Prince, R. C.** 1993. Petroleum spill bioremediation in marine environments. *Crit. Rev. Microbiol.* **19:**217–242.

112. **Pritchard, P. H., and C. F. Costa.** 1991. EPA's Alaska oil spill bioremediation project. *Environ. Sci. Technol.* **25:**372–379.

113. **Quensen, J. F., J. M. Tiedje, and S. A. Boyd.** 1988. Reductive dechlorination of polychlorinated biphenyls by anaerobic microorganisms from sediments. *Science* **242:**752–754.

114. **Ramaswami, A., and R. Luthy.** 1997. Measuring and modeling physiochemical limitations to bioavailability and biodegradation, p. 721–729. *In* C. J. Hurst, G. R.

Knudsen, M. J. McInerney, L. D. Stetzenbach, and M. V. Walter (ed.), *Manual of Environmental Microbiology*. ASM Press, Washington, D.C.

115. **Raymond, R. L., V. W. Jamison, and J. O. Hudson.** 1976. Beneficial stimulation of bacterial activity in ground waters containing petroleum products, p. 319–327. *In Water—1976*. American Institute of Chemical Engineers, New York.
116. **Reineke, W., and H. J. Knackmuss.** 1979. Construction of haloaromatic utilizing bacteria. *Nature* **277:**385–386.
117. **Riser-Roberts, E.** 1992. *Bioremediation of Petroleum Contaminated Sites*. CRC Press, Boca Raton, Fla.
118. **Rosenberg, E. (ed.).** 1993. *Microorganisms to Combat Pollution*. Kluwer Academic Publishers, Dordrecht, The Netherlands.
119. **Rosenberg, E., R. Legmann, A. Kushmaro, R. Taube, E. Adler, and E. Z. Ron.** 1992. Petroleum bioremediation—a multiphase problem. *Biodegradation* **3:**337–350.
120. **Sander, P., R. M. Wittich, P. Fortnagel, H. Wilkes, and W. Franke.** 1991. Degradation of 1,2,4-trichloro- and 1,2,4,5-tetrachlorobenzene by *Pseudomonas* strains. *Appl. Environ. Microbiol.* **57:**1430–1440.
121. **Shannon, M. J. R., and R. Unterman.** 1993. Evaluating bioremediation: distinguishing fact from fiction. *Annu. Rev. Microbiol.* **47:**715–38.
122. **Shields, M. S., S. O. Montgomery, P. J. Chapman, S. M. Cuskey, and P. H. Pritchard.** 1989. Novel pathway of toluene catabolism in the trichloroethylene-degrading bacterium, G4. *Appl. Environ. Microbiol.* **55:** 1624–1629.
123. **Shuttleworth, K. L., and C. E. Cerniglia.** 1997. Practical methods for the isolation of polycyclic aromatic hydrocarbon (PAH)-degrading microorganisms and the determination of PAH mineralization and biodegradation intermediates, p. 766–775. *In* C. J. Hurst, G. R. Knudsen, M. J. McInerney, L. D. Stetzenbach, and M. V. Walter (ed.), *Manual of Environmental Microbiology*. ASM Press, Washington, D.C.
124. **Singer, M. E., and W. R. Finnerty.** 1984. Microbial metabolism of straight-chain and branched alkanes, p. 1–59. *In* R. M. Atlas (ed.), *Petroleum Microbiology*. Macmillan, New York.
125. **Sirvins, A., and M. Angles.** 1986. Development and effects on marine environment of a nutrient formula to control pollution by petroleum hydrocarbons. *NATO ASI Series* **G9:**357–404.
126. **Song, H. G., X. Wang, and R. Bartha.** 1990. Bioremediation potential of terrestrial fuel spills. *Appl. Environ. Microbiol.* **56:**652–656.
127. **Spain, J. C. (ed.).** 1995. *Biodegradation of Nitroaromatic Compounds*. Plenum Press, New York.
128. **Steffan, R. J., K. Sperry, C. W. Condee, M. Walsh, W. Guarini, and A. Thomas.** *In situ* remediation of trichloroethelene (TCE)-contaminated groundwater by bioaugmentation. *In Proceedings of Tenth International IGT Symposium on Gas, Oil and Environmental Biotechnology and Site Remediation Technologies*. December 8–10, 1997, Orlando, Fla. In press.
129. **Stoner, D. L.** 1994. *Biotechnology for the Treatment of Hazardous Waste*. Lewis Publishers, Boca Raton, Fla.
130. **Suflita, J. M., A. Horowitz, D. R. Shelton, and J. M. Tiedje.** 1982. Dehalogenation: a novel pathway for the anaerobic biodegradation of haloaromatic compounds. *Science* **214:**1115–1117.
131. **Sussman, M., C. H. Collins, F. A. Skinner, and D. E. Stewart-Tull (ed.).** 1988. *Release of Genetically-Engineered Microorganisms*. Academic Press, London.
132. **Sveum, P., and A. LaDousse.** 1989. Biodegradation of oil in the Arctic: enhancement by oil-soluble fertilizer application, p. 439–446. *In* Proceedings of the 1989 Oil Spill Conference. American Petroleum Institute, Washington, D.C.
133. **Swannell, R. P. J., K. Lee, and M. McDonagh.** 1996. Field evaluations of marine oil spill bioremediation. *Microbiol. Rev.* **60:**342–365.
134. **Thomas, J. M., M. D. Lee, P. B. Bedient, R. C. Borden, L. W. Carter, and C. H. Ward.** 1987. *Leaking Underground Storage Tanks: Remediation with Emphasis on in situ Bioreclamation*. EPA/600/S2-87/008. U.S. Environmental Protection Agency, Ada, Okla.
135. **Timmis, K., R. J. Steffan, and R. Unterman.** 1994. Designing microorganisms for the treatment of toxic wastes. *Annu. Rev. Microbiol.* **48:**525–557.
136. **Tramier, B., and A. Sirvins.** 1983. Enhanced oil biodegradation: a new operational tool to control oil spills, p. 115–119. *In* Proceedings of the 1983 Oil Spill Conference. American Petroleum Institute, Washington, D.C.
137. **Unterman, R.** 1996. A history of PCB biodegradation, p. 209–253. *In* R. L. Crawford and D. L. Crawford (ed.), *Bioremediation: Principles and Applications*. University Press, Cambridge.
138. **Unterman, R., C. D. Chunn, and M. J. R. Shannon.** 1991. Isolation and characterization of a PCB-degrading bacterial strain exhibiting novel aerobic congener specificity, abstr. Q-49, p. 284. *In* Abstr. 91st Gen. Meet. Am. Soc. Microbiol. 1991. American Society for Microbiology, Washington, D.C.
139. **Vanloocke, R., R. DeBorger, J. P. Voets, and W. Verstraete.** 1975. Soil and groundwater contamination by oil spills: problems and remedies. *Int. J. Environ. Studies* **8:**99–111.
140. **Venosa, A. D., M. T. Suidan, B. A. Wrenn, K. L. Strohmeier, J. R. Haines, B. L. Eberhart, D. King, and E. Holder.** 1996. Bioremediation of an experimental oil spill on the shoreline of Delaware Bay. *Environ. Sci. Technol.* **30:**1764–1776.
141. **Verstraete, W., R. Vanlooke, R. deBorger, and A. Verlinde.** 1976. Modeling of the breakdown and the mobilization of hydrocarbons in unsaturated soil layers, p. 98–112. *In* J. M. Sharpley and A. M. Kaplan (ed.), Proceedings of the Third International Biodegradation Symposium. Applied Science Publishers, London.
142. **Vogel, T. M., and D. Grbic-Galic.** 1986. Incorporation of oxygen from water into toluene and benzene during anaerobic fermentative transformation. *Appl. Environ. Microbiol.* **52:**200–202.
143. **Vogel, T. M., and P. L. McCarty.** 1985. Biotransformation of tetrachloroethylene to trichloroethylene, dichloroethylene, vinyl chloride and carbon dioxide under methanogenic conditions. *Appl. Environ. Microbiol.* **49:** 1080–1083.
144. **von Wedel, R. J., J. F. Mosquera, C. D. Goldsmith, G. R. Hater, A. Wong, T. A. Fox, W. T. Hunt, M. S. Paules, J. M. Quiros, and J. W. Wiegand.** 1988. Bac-

terial biodegradation and bioreclamation with enrichment isolates in California. *Water Sci. Technol.* **20:**501–503.

145. **Wackett, L. P.** 1997. Biodegradation of halogenated solvents, p. 784–789. *In* C. J. Hurst, G. R. Knudsen, M. J. McInerney, L. D. Stetzenbach, and M. V. Walter (ed.), *Manual of Environmental Microbiology*. ASM Press, Washington, D.C.

146. **Walsh, M., T. Boland, J. Liskowitz, M. DeFlaun, and R. Steffan.** Remediation of a low permeability TCE contaminated bedrock, part 2. Pneumatic injection of constitutive TCE degrading organisms. *In Bioremediation in Rock Masses*. American Society for Chemical Engineers, Washington, D.C., in press.

147. **Walter, M. V.** 1997. Bioaugmentation, p. 753–757. *In* C. J. Hurst, G. R. Knudsen, M. J. McInerney, L. D. Stetzenbach, and M. V. Walter (ed.), *Manual of Environmental Microbiology*. ASM Press, Washington, D.C.

148. **Wilson, B. H., G. B. Smith, and J. F. Rees.** 1986. Biotransformations of selected alkylgbenzenes and halogenated aliphatic hydrocarbons in methanogenic aquifer material: a microcosm study. *Environ. Sci. Technol.* **20:**997–1002.

149. **Wilson, J. T., and C. H. Ward.** 1987. Opportunities for bioreclamation of aquifers contaminated with petroleum hydrocarbons. *Dev. Ind. Microbiol.* **27:**109–116.

150. **Yaniga, P. M., and W. Smith.** 1984. Aquifer restoration via accelerated *in situ* biodegradation of organic contaminants, p. 451–470. *In* Proceedings of the Conference on Petroleum Hydrocarbons and Organic Chemicals in Ground Water—Prevention, Detection, and Restoration. National Water Well Association, Worthington, Ohio.

151. **Zeyer, J., P. Eicher, J. Dolfing, and P. R. Schwarzenbach.** 1990. Anaerobic degradation of aromatic hydrocarbons, p. 33–40. *In* D. Kamely, A. Chakrabarty, and G. S. Omenn (ed.), *Biotechnology and Biodegradation*. Gulf Pub. Co., Houston, Tex.

152. **Zeyer, J., E. P. Kuhn, and R. P. Schwarzenbach.** 1986. Rapid microbial mineralization of toluene and 1,3-dimethylbenzene in the absence of molecular oxygen. *Appl. Environ. Microbiol.* **52:**944–947.

Microbial Biodiversity: Strategies for Its Recovery

JAMES M. TIEDJE AND JEFFREY L. STEIN

56

56.1. MICROBIAL DIVERSITY ON EARTH

56.1.1. What is the Extent of Microbial Diversity on Earth?

The microbial world traditionally consists of all organism groups that can be seen only with a microscope, i.e., the fungi, bacteria, archaea, algae, protozoa, and a number of newer and less known lower eucaryotes. If one considers the biodiversity of all these groups, the task is enormous and the subject of many specialized volumes. Several volumes have been devoted to microbial diversity (1, 11). Hence, this chapter will focus on the bacteria with some comment on the fungi, the two groups of greatest interest in industrial microbiology, and on recovering diversity through retrieving DNA from nature.

The extent of bacterial diversity is unknown and without a rationally based extrapolation. This is not the case for fungi. The number of known species is about 72,000 (26). Hawksworth (24) has conservatively estimated the number of fungal species to be about 1.6 million based on an experimentally determined ratio of six unique fungal species per plant species times the 270,000 plant species in the world, a reasonably well accepted number (26). This does not account for species associated with insects and perhaps different ratios of fungal species per plant species in the tropical and polar regions. The case for bacteria is far more primitive. Only about 4,200 species have been described. It is widely recognized that this represents, at best, only 0.1 to 1% of the organisms in nature. No rational attempt has been made to extrapolate the global extent of bacterial species as has been done for fungi, because too many coefficients for such an extrapolation are unknown. Several lines of evidence, however, do suggest that the number of bacterial species is much higher than the currently known number of species. First, DNA reannealing studies done by Torsvik et. al. (53) on Norwegian forest soil DNA showed that there are 4,000 nonhomologous bacterial-sized genomes in a 30-g sample of soil. If one then factors in the 70% DNA-DNA hybridization criterion for a species, the unknown relationship of this number to soil sample size as well as a typical species abundance profile, it would not be difficult to reason that a gram of fertile soil could contain several to many thousand species. Second, small subunit ribosomal DNA genes (SSU rDNA) obtained from soil in many studies show extremely high diversity with virtually no resampling of the same clone, even in very large clone libraries (8, 31, 34, 61). Furthermore, most of the sequences from these clones do not match sequences in the database within 1 to 2% similarity, a rough estimate of species level resolution. This high level of SSU rDNA diversity is consistent with the evidence of high diversity from the DNA reannealing studies. Third, new isolate collections from nature often show that one-third to two-thirds of the strains in the new collection do not match known species with an acceptable level of identity in existing fatty acid methyl ester (FAME), BIOLOG, API, and classical phenotypic databases or descriptions. This indicates that even among the culturable isolates, diversity is much higher than represented by the described species. Fourth, there is no evidence that there is a decline in the reporting of new bacterial taxa, if the annual weight increase in the *International Journal of Systematic Bacteriology* is any indication. Furthermore, the new taxa are not simply new species, but often new genera or even families. Some of these taxa would be expected to include species-rich groups. More problematic is the number of very unique organisms that are isolated but die on the shelf because there are few funds to support bacterial systematics, and the effort needed to describe the especially unusual organism can be daunting. We will never know whether these orphans harbor unique physiology, produce valuable pharmaceuticals, or play major roles in biogeochemical cycles.

While the above evidence documents that the extent of bacterial diversity is very high, the absence of information on how different organisms are at different spatial scales and to what degree there are geographic species severely limits our ability to estimate global bacterial diversity. May (32) described a relationship that showed that biodiversity increases as the body length of the organism decreases, at least down to organisms of several millimeters in length. For organisms smaller than this, however, there may not be separate species in different climatic and geographic regions as there are for larger organisms. This is an important issue that remains to be resolved. Furthermore, the above discussion does not consider the variety of different niches known to harbor procaryotes, e.g., special symbioses, extreme environments (temperature, pH, salt, pressure, and their combinations), and novel energy-generating biochemistry. Procaryotes have been on Earth perhaps 3.8 billion years, much longer than higher life forms. Evolution should have created enormous diversity during this extensive period of time. Some extinction would of course have occurred, but the planetary changes during this time are not ones thought to have been particularly lethal to most procaryotes. Hence, this long evolutionary period would also argue for high procaryotic diversity. Given all the above arguments, a global bacterial diversity of 10^5 to 10^6 organisms would not be unreasonable. The Global Biodiversity Assessment uses the estimate of 10^6 for bacteria (26).

56.1.2. Why is Microbial Diversity Important?

Microbial diversity has several important values to society and to the Earth's ecosystem.

- Microorganisms are of critical importance to the sustainability of life on our planet, including recycling elements on which primary productivity depends, producing and consuming gases important to maintaining our climate, and destroying the wastes of human civilization.
- Discoveries of microbial biodiversity expand the frontiers of knowledge about the strategies and limits of life, including microbes that live at the extreme conditions known for life and ones that have evolved novel redox couples for capturing energy.
- Microbial diversity represents the largest untapped reservoir of biodiversity for potential discovery of new biotechnology products, including new pharmaceuticals, new enzymes, new specialty chemicals, or new organisms that carry out novel processes.
- Microbes often play key roles in conservation of higher organisms and in restoration of degraded ecosystems. Hence, microbial diversity goes hand in hand with goals for maintenance of higher organism diversity.

When many think of industrial interest in microbiology, it is often assumed that biotechnology products are the only focus, but all of the above points are important to industry. Microbial treatment of waste streams is critical to acceptable and economic industrial practice in the modern world; microbes and their enzymes that withstand extreme conditions have obvious potential value for new processes; and restoration of mining, production, and harvesting areas are critical to some industries' existence. These topics are also important to the increasing focus by industry on "green" chemistries and to life cycle analysis of new products.

Beyond the more global values of biodiversity, there are variants in particular organism traits, i.e., particular phenotypes, that are of interest. Classes of traits for which variation (diversity) may be of value include particular kinetic properties, especially K_m and V_{max} values; tolerances to high or low temperatures, high or low pH, and high solute concentrations, e.g., Na^+, high or low pressure; the ability to attach or to be mobile; the use of particular electron donors or acceptors; and the avoidance of predators, i.e., grazing. Other phenotypic traits of biodiversity that can be of interest are the rates, yields, and efficiencies of production of desired products; enzymes with unique properties such as high temperature or alkaline stability, or high turnover at low temperatures; particular regulatory properties; and genetic stability of strains.

The above phenotypic properties are often why particular strains are sought and why particular strains become a focus in research and production. The range of diversity in these traits in nature, or whether there are evolutionary trade-offs, e.g., between K_m and V_{max}, is usually not known. New genetic combinatorial approaches are now used in an attempt to create diversity in the laboratory since it may be easier to recover diversity there than from nature. But, in principle, given the 3-billion-year time span of evolutionary history of procaryotes, a high degree of natural diversity in many of these traits can be expected to be extant.

56.1.3. The Problem

Given the probable existence of a huge variety of interesting microbial phenotypes, the challenge is the recovery of those traits for study or use. This entails several challenges, including strategies for site selection and sampling, the major problem of nonculturable microbes, DNA recovery and its expression, and efficient screening. Two major strategies are now used for recovery of microbial biodiversity from nature, and they are the subjects of two sections of this chapter—recovery of culturable organisms and recovery of the genetic blueprint, DNA.

56.1.4. Where is New Diversity To Be Found?

Two factors determine where particular organisms reside: their selection, i.e., growth and colonization in a particular habitat, and the degree of dispersal. Before evaluating these two forces, however, microbiologists have the problem of establishing whether an organism found in a habitat really successfully competes and lives in that habitat versus being a visitor. The latter is problematic because of the great ability of some microbes to survive long periods in environments outside of where they successfully compete. Such transients can be found in many places and do not factor into using ecological principles to evaluate distribution of diversity. As a practical matter, however, any population that is present in high numbers in a particular habitat must be considered a living resident of that habitat, because only with growth could it achieve such dominance.

A key ecological principle is that the most fit organism is the most successful in its niche. Hence, different niches would be expected to harbor different organisms, i.e., biodiversity. Major selective forces for the resulting populations are the energy source, especially the type and amount of carbon; the electron acceptors, especially oxygen or types of alternative electron acceptors used under anaerobic conditions; and stress factors that might demand tolerance features, e.g., resistance to pH or temperature extremes. Most of these factors are chemical; hence, site chemistry would be expected to be a major determinant of diversity. A sampling strategy that is based on different site chemistries should yield higher diversity. Site chemistry could mean a difference in some populations under oak trees versus maple trees versus a particular grass, or a calcareous soil versus a weathered clay soil. Few studies have been done (indeed few have been feasible) to actually define how the level of microbial diversity changes with a gradient of, for example, soil chemistry. As one example, different serogroups of *Bradyrhizobium japonicum* are known to occupy soils of slightly different pH (1 to 1.5 units), an example of a fine level of phenotypic difference dictated by a small change in soil chemistry.

Since the heterotrophic community in soil and water is usually energy starved, the particular carbon resources should be a dominant driving force of selection. An example of the profound influence of vegetation type on microbial community structure is shown in Fig. 1. In this case,

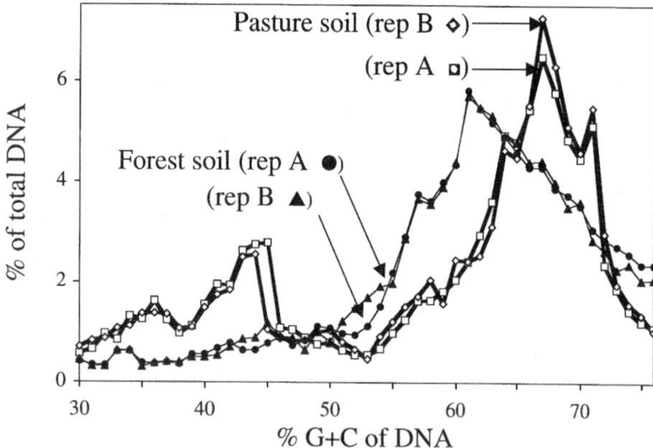

FIGURE 1 Percent guanine + cytosine (G+C) profile of DNA extracted from Hawaiian soil of identical parent material, climate, and vegetative cover until one portion was converted to pasture for cattle grazing approximately 80 years ago (35). Data for two replicate soil samples are shown.

soil DNA was extracted from two adjacent soils of identical history and chemistry until the native forest was replaced by pasture on a portion of the site approximately 80 years ago. Approximately one-quarter of the soil DNA had a different guanine + cytosine (G+C) composition as a result of the shift to pasture vegetation (35). Furthermore, some of the DNA of the same G+C composition under the two vegetation types is likely to be from very different organisms with different traits. This is supported by finding different SSU rRNA sequences in soils of the same G+C content under the two different vegetations (35). Hence, vegetation is a strong driving force in bacterial community selection and should be a primary consideration in schemes to recover new diversity.

The G+C fractionation method shown in Fig. 1 is also a methodology that allows one to recover DNA from more minor members of the community or to target DNA from particular taxonomic groups defined by G+C content. This method is described by Holben and Harris (27). Briefly, DNA is mixed with *bis*-benzimidazole (Hoechst 33258), which binds to the adenine and thymidine and changes the buoyant density of the DNA in proportion to the G+C content. A gradient of G+C concentrations can then be established by equilibrium density gradient ultracentrifugation. Fractions can be collected with a fraction collector, and the DNA in particular fractions can be amplified by PCR (34), cloned, hybridized to probes, or analyzed as appropriate. The G+C content of each fraction is established by using a standard curve relating G+C content to density measured with a refractometer.

Another type of selective habitat is that provided by a host organism such as a particular part of a plant, insect, animal, or other high organism. Bacterial colonizers of these habitats are often specialists and are not free living. As a result, they may be difficult to cultivate. Nonetheless, such symbionts are thought to be a rich source of microbial diversity. In a few cases where bacterial colonists of invertebrates have been extensively examined, the recovery of novel genera and species has been high, suggesting that the invertebrate hosts are a rich source of novel bacterial diversity (28). Since there are an estimated 10^8 insect species (26), this could make insects a very rich source of microbial biodiversity.

The second major factor that determines where new microbial diversity is located is the degree of dispersal of microbial species relative to the local rate of accumulated genetic variation. Another way to evaluate this question is to consider what is the degree of bacterial endemism, i.e., the extent of localized genetic and phenotypic uniqueness (9, 19). As discussed previously, the answer to the question has a major impact on the extent of microbial diversity on Earth.

The Center for Microbial Ecology has begun to evaluate this question by examining the degree of endemism in organisms that grow on 3-chlorobenzoic acid, a fairly rare trait in nature (19). Approximately 150 isolates were examined from six continental regions, which included four undisturbed Mediterranean ecosystems (western Australia, southwest South Africa, central Chile, and central California), and two undisturbed boreal forest ecosystems in north central Canada and in northwest Russia. Using repetitive extragenic palindromic PCR (rep-PCR) to indicate genotype, no globally dispersed genotypes were found but several genotypes were regionally endemic. This study suggested a high degree of endemism in this population. The degree of endemism may, however, vary with the type of species. Some species may disperse, survive, colonize, and be more genetically stable, while others may not.

For host-associated organisms, whose hosts are endemic, one would expect some level of endemism. For the non-host-associated microorganisms, the degree of bacterial endemism is far from clear. However, cosmopolitan subspecies or ecovars are likely not the rule. Hence, a sampling strategy that considers distinct geographic regions is probably wise in attempts to recover new diversity. Production of the same antibiotic is a trait that is not often conserved in the same species or subspecies, and hence sampling that uses a geographic as well as site chemistry strategy should be beneficial.

56.2. BIODIVERSITY OF CULTURABLE BACTERIA

56.2.1. What Level of Bacterial Diversity Matters?

In this sense, "level" means what level of taxon resolution, e.g., genus, species, subspecies, ecovar (analogous to pathovar), and in a slightly different context, what degree of genetic difference among the organisms. The pragmatic answer is the degree of resolution that is important for the particular problem (52). If it is a virulent versus a weakly virulent pathogen of the same subspecies, then a fine degree of resolution is critical. If it is a comparison of pine tree and cereal grain rhizosphere populations, a more moderate level of resolution is reasonable, or if it is a comparison of ocean and hydrothermal vent communities, then a coarse level of resolution may be adequate. The pragmatic approach places little emphasis on the species or any other level in a classification hierarchy.

Bacteriologists, however, cannot avoid the need for defining a distinct kind of bacteria, that is, a species, because it is needed to communicate among microbiologists, with other scientists, and in the world of practitioners, e.g., for patents, diagnostics, quality control, quarantine, or international material transport. Hence, we need to continue to evolve the species concept for bacteria. The current recognized bacterial species definition derives from an ad hoc committee on reconciliation of approaches to bacterial systematics (58). The committee proposed a number of recommendations to combine genotypic and phenotypic data leading to a polyphasic approach, which is the current standard for bacterial taxonomy. This committee proposed "a bacterial species as a group of strains, including the type strain, sharing 70% or more DNA-DNA relatedness." Furthermore, phenotypic characteristics should agree with the phylogenetic data, and it was recommended that a genospecies, although distinct, that cannot be differentiated on any known phenotypic grounds cannot be renamed. However, the 70% criterion presents its own technical limitations, including the need for pure cultures, imprecision of the method, lack of suitability for reference databasing, and the virtual impossibility of making all pairwise hybridization comparisons in a collection. This criterion also lacks a theoretical basis that would explain why two-thirds of the genome should be more highly conserved than some other portion, and it cannot be related to the role of natural selection in determining differential reproductive success that provides the theoretical basis for the species criteria among eucaryotic organisms. It does have the advantage, however, over other species definitions of providing a quantifiable criterion.

Until the relative phylogenetic relevance of any kind of information can be dependably evaluated for bacteria, distinction of one kind of bacterium from all others should be based on every kind of information that can be obtained by the methods available, rather than on one or a few kinds of traits. This has led to some new attempts to define bacterial species. At a biodiversity workshop in 1992, the following definition was proposed (10). A bacterial species can be defined as "a group of related organisms that is distinguished from similar groups by a constellation of significant genotypic, phenotypic and ecologic characteristics." A bacterial species defined in this way is at once a naturally occurring group of like organisms and a taxon sufficiently defined by the collected properties of the group to distinguish the species from closely related groups. With this meaning, a bacterial species should be comparable to eucaryotic species in its adequacy as a unit of diversity in both systematics and ecology. An addition to this definition not present in others is the recognition of ecological characteristics as a determining criterion, which helps bring the microbial criteria in line with reality for higher organisms. Obviously, the particular species definition used is a major factor in any estimate of the global extent of the bacterial species, but it seems reasonable that the definition should have a meaning as similar as possible to that for higher organisms. More recently, DeVoss et. al. (14) offered a possible definition of the polyphasic species as "a group of strains which originated from a common ancestor population in which the steady generation of genetic diversity and recombination after the introduction of foreign DNA, resulted in clones with different degrees of variation but still sharing a significant degree of DNA relatedness and with a common phenotype." Currently there are a large number of techniques to determine a common phenotype. These include standardized phenotypic tests, such as BIOLOG and API galleries; chemotaxonomic traits such as fatty acid profiles, polyamines, and sodium dodecyl sulfate-polyacrylamide gel electrophoresis patterns of cellular proteins; immunologic data; antibiotic resistance; and morphological characteristics. For genotypic characterization there are also a variety of techniques including amplified rDNA restriction analysis; sequencing of the 16S and 23S rRNA gene; ribotyping; rDNA intergeneric spacer region restriction analysis; amplified fragment length polymorphism; rep-PCR using REP, BOX, or ERIC primers; randomly amplified polymorphic DNA analysis; and DNA-DNA hybridization.

In recent years, SSU rRNA gene sequencing has become an extremely popular method to help identify new isolates as well as nonculturable microbes when their DNA is cloned from microbial communities. The rRNA gene, however, is highly conserved, which makes this methodology alone inadequate for providing insight into the physiology and ecology of the organism. It cannot be relied on to routinely provide a species-level identification according to any of the above definitions. Figure 2, reprinted from Stackebrandt and Goebel (48), illustrates this point. Some strains with greater than 99% 16S rRNA sequence homology do not meet the greater than 70% DNA-DNA reassociation criterion. The authors point out that "the strength of sequence analysis is to recognize the level at which DNA paring studies need to be performed, which certainly applies to (rRNA) similarities of 97% and higher" (48). Because of the great current interest in and ease of rRNA analysis, there is a danger that the species-level identification and new description will not be adequately done, leading to misrepresentation of microbial diversity information.

56.2.2. Isolation Strategies

56.2.2.1. Requirements

Skilled microbiologists believe that many nonculturable bacteria will become "formerly nonculturable" with careful investigation into the physiology and nutrition of the mixed culture, patience, and cleverness. There are many success stories, but all have one thing in common, that it takes considerable time to finally isolate a fastidious nonculturable microbe. Nothing, however, substitutes for the personal skill and insight that some microbiologists have for nurturing difficult-to-grow bacteria, the "green thumb" of microbiology. The skill of the scientist is the most important strategy.

The ability to track the target population during enrichment is extremely important to gaining insight into what conditions favor or inhibit growth of this organism over the others in the community. Isolation conditions can be constructed from this information. Identification of a means to track the target organism's growth can be challenging without having a pure culture to identify the unique target. Microscopy remains a primary method if there are any morphological clues to distinguish the target organism. Antibodies, oligonucleotide probes, and stains can also aid the recognition in the microscopic field. Nonmicroscopic molecular methods and chemotaxonomic methods can also be useful if there is a signature to be tracked. Effort spent on identifying a means to track the target population has a good probability of leading to a successful isolation.

One difficulty in isolating nonculturable microbes is that the cell separation step and the cultivation step are usually commingled so that one does not know which is the bottleneck. An ability to resolve those two components to determine which remains the problem helps focus the effort on the real problem preventing isolation.

56.2.2.2. Mimicking the Organism's Habitat

The classical enrichment technique was based on mimicking the organism's habitat and using natural selection to

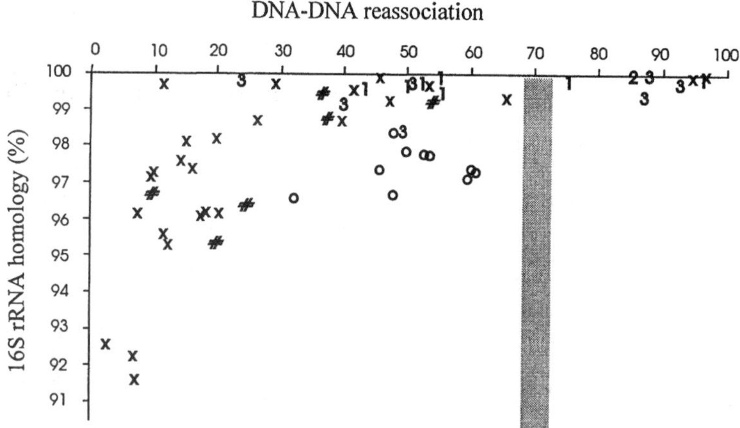

FIGURE 2 Comparison of 16S rRNA homology and DNA-DNA reassociation values. ×, membrane filter method; #, renaturation rate method; 1, renaturation rate method; 2, renaturation rate method; 3, renaturation rate method; O, S1 nuclease method. The bar indicates the DNA threshold value for species delineation. Reprinted from reference 48.

enrich the desired population over other members of the community. The technique has been powerful in microbiology, but it can do even more if carried to another level of sophistication. All aspects of mimicking the environment need to be considered. For example, might phosphorus (used as a buffer) be inhibitory, could the atmospheric oxygen concentration be inhibitory, or are the inorganic ions in the medium supplied in inappropriate concentrations or ratios compared to those in the natural habitat? As an example, a very low ionic strength medium was recently found to be necessary to cultivate the formerly nonculturable acidophilic methanotroph (12); their natural peat bog habitat has an ionic strength much lower than that of conventional mineral media. The carbon content of standard media, such as tryptic soy broth and nutrient broth, has been shown to completely inhibit growth of many important groups of soil bacteria (59). The R2A medium (41), originally designed as a relatively low-nutrient medium to recover oligotrophs from waters, has proved much more satisfactory for cultivating a variety of new and diverse types of bacteria from soil and water. A major recommendation in using the mimicry approach is to thoroughly consider all aspects of the targeted population's natural and laboratory environment and to make sure they are as similar as possible.

Rather than trying to construct optimum concentrations of the strain's resources in a defined medium, an alternative is to use gel-stabilized gradients or combinations of gradients (16). In principle, each of the three spatial dimensions can be used for a gradient to supply different concentrations of resources. The organism is then allowed to find its own optimum for growth within this combination of gradients. Once that combination is identified, those concentrations can be used in a defined medium. Steady-state gradients also more closely mimic the natural condition where the resource pool size is maintained by its natural turnover.

Some bacteria cannot be cultivated because they are members of a tightly interdependent food web. Cocultures, helper bacteria, and extracts from these bacteria can be used to help cultivate the desired organisms. As an example, isolation of the anaerobic fatty acid-oxidizing bacteria was initially successful only because of the use of cocultures of hydrogen consumers (60).

56.2.2.3. Patience

Many of the difficult-to-cultivate organisms from nature seem to be naturally very slow-growing organisms. This is especially true of a number of the obligate anaerobes and some of the aerobic oligotrophs. In the former case, it is not unusual for a skilled microbiologist to take 2 years to isolate a novel anaerobe. Many microbiologists, having grown up with *Escherichia coli* or *Pseudomonas* spp. in laboratory exercises, mistake slow growth for no growth. Hence, patience is needed to determine that growth actually has occurred, and turbidity many never be seen.

Hattori's group has effectively shown how both low-nutrient media and long incubation times (2 months) can be used to recover an order of magnitude higher numbers of bacteria from soil (59). Furthermore, the physiological and taxonomic features of the bacteria appearing on plates after a 1-month incubation period were very different from those that appeared in the first week of incubation (59). This cultivation approach shows that the number of CFU can come closer to approximating the number of organisms seen by direct count. The oligotrophic media they use is 1/100-strength nutrient broth. The oligotrophic isolates, which constitute 82% of those colonies found in the second month (Fig. 3), do not grow on full-strength complex media. Hence, this method shows that patience for the length of incubation and recognizing the appearance of tiny colonies yields a large number of newly culturable strains (59).

56.2.2.4. Suppressing "Weedy" Bacteria

A major limitation in isolation of new bacterial diversity is fast-growing, nonfastidious bacteria that out-compete slow growers in enrichments or on agar plates. These bacteria, which have the characteristics of weeds, are particularly problematic for isolation of slow-growing bacteria that are present in low numbers. The first strategy to reduce growth of weedy bacteria is to reduce the carbon concentration in the medium, as illustrated in Fig. 3. Fast-growing bacteria (high μ_{max}) typically have high K_s values, while

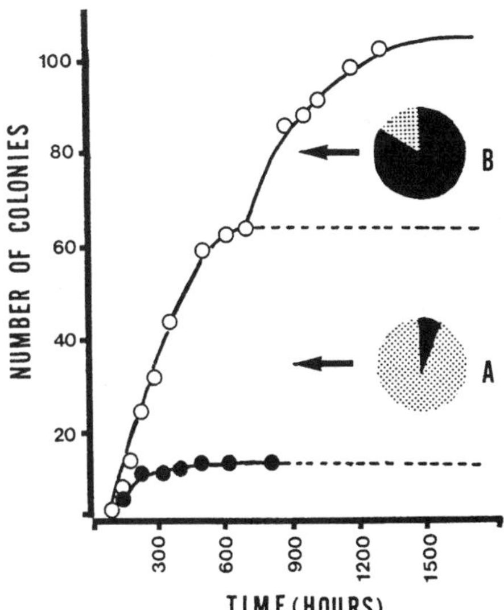

FIGURE 3 Comparison of the number of colonies on nutrient agar (●) and dilute (1/100) nutrient agar (○). The proportion of oligotrophic isolates at incubation time is shown by the solid portion of the pie chart. (A) 5% of the isolates were oligotrophs; (B) 82% were oligotrophs. Reprinted from reference 59 by permission.

slow-growing bacteria have a low K_s. Hence, low substrate concentrations often allow the slow growers to outcompete the fast growers (30). Since carbon is the most limiting resource to heterotrophs in nature, it is to be expected that the predominant bacteria in nature are specialized slow growers. A more sophisticated approach to recover low-μ_{max}, low-K_s microorganisms is by use of a chemostat fed very low concentrations of substrate. This is an effective but labor-intensive isolation strategy.

A second approach to reduce weedy bacteria as well as to impose some selection is to use antibiotics in the enrichment. If one has a means to track the target organism or its activity, appropriate antibiotics can often be identified.

A third approach to reduce weedy bacteria is to omit combined nitrogen from the medium and determine if the desired population fixes N_2. Nitrogen fixation is widespread in bacterial groups, and some slow-growing bacteria can fix nitrogen.

56.2.2.5. Resuscitation and Avoiding the Stress of Cultivation

Some bacteria are injured in nature, and hence care must be taken for their recovery. Also, some bacteria adapted to natural conditions cannot withstand the conditions of cultivation. Low carbon concentrations in the media, microaerophilic conditions, a supply of critical growth factors, and use of moderate temperatures are means to aid the transition to laboratory cultivation.

56.2.2.6. Physical Separation

Physical separation of target cells from the rest of the community is an alternative to enrichment for the separation phase of isolation. Methods of physical separation include motility to an attractant, micromanipulation more recently by laser tweezers, antibody-linked magnetic beads, and separation by density gradient centrifugation in a nontoxic matrix.

56.2.3. Has the Isolated Strain Been Seen Before?

Once a new isolate has been obtained, the first question is often whether it is novel and, if so, how different it is from what has been seen before. This question can be directed at two levels: is the strain in the collection and hence already counted or screened, or, at a taxonomic level, has it been identified as belonging to an established taxon? To answer these questions, rapid methods are needed so that resources can be focused on discovery of the novel types. This is often termed dereplication by microbiologists in industry. Methods such as FAME and Biolog are automated rapid systems that can be a first-level screen for certain organisms. Recently, the use of genomic fingerprints obtained by PCR amplification using primers binding to randomly interspersed repetitive DNA sequences (rep-PCR) has gained favor. This method is rapid and reproducible, is suitable for a database, and provides the highest level of taxonomic resolution of any current PCR-based method (56). There are three primer sets that have been found to work in most bacteria; these are known as REP, BOX, and ERIC. Of these, BOX is the most generally useful because it works reliably on new strains and gives the most amplicons and hence a higher degree of resolution among strains. These fingerprints reflect chromosome structure and hence provide resolution at the species-subspecies-ecovar level. Rep-PCR of most bacteria can be done from the cells by using the PCR protocol to lyse enough cells to provide target DNA for the primers. In some cases the DNA must first be extracted, which makes the method more time consuming. The gel images can be digitized, stored in TIFF files, and analyzed by software clustering programs. A complete protocol for use of rep-PCR and GelCompar software for pattern analysis and database construction has been described (40). An example of BOX-, ERIC-, and REP-PCR used to differentiate among *Bradyrhizobium* and reference strains is shown in Fig. 4.

56.2.4. Fungal Biodiversity: Isolation and Identification

Certain fractions of the mycota of temperate soils are reasonably well characterized, and methods for the isolation of these fungi (5, 7, 20, 29, 37) and their identification (4, 15, 21, 36, 57) are available. Other general references that are useful in the identification of soil fungi include von Arx (56a), Farr et al. (17), and Hawksworth et al. (25). Barron (6) provides a synthesis of the biology, identification, and isolation of a unique subset of soil fungi, the nematode-destroying fungi. The saprotrophic basidiomycetes, in all probability the most important group of fungi in the initial delignification of plant debris entering soil, are poorly represented by methods aimed at isolating soil fungi and are greatly underrepresented in discussions of the soil mycota (51). New methods for the selective isolation of these and other underrepresented groups of soil fungi are presented by Thorn et al. (50) and Bills et al. (7). Since the saprotrophic basidiomycetes in culture lack sufficient morphological characters or literature for their identification, identification is best attempted using DNA sequencing (e.g., nuclear 18S or 25S rDNA) and placement in a phylogenetic framework of known sequences.

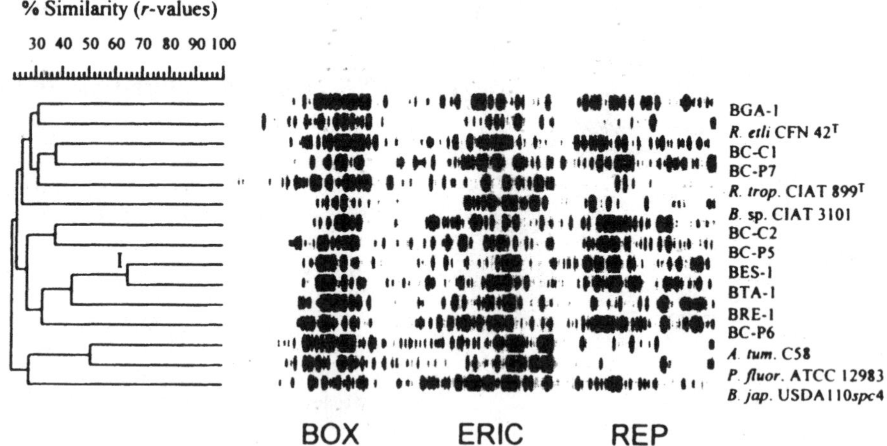

FIGURE 4 Product-moment/UPGMA cluster analysis of linearly combined BOX-, ERIC-, and REP-PCR genomic fingerprints of nodule isolates of *Bradyrhizobium* sp. from endemic woody legumes of the Canary Islands. On the scale, *r* values are expressed as percentages. *R. etli*, *Rhizobium etli*; *R. trop.*, *Rhizobium tropici*; *A. tum.*, *Agrobacterium tumefaciens*; *P. fluor.*, *Pseudomonas fluorescens*; *B. jap.*, *Bradyrhizobium japonicum*. Reprinted from reference 56 by permission.

56.3. RECOVERING BIODIVERSITY USING ENVIRONMENT DNA

56.3.1. Accessing Uncultivated Microbes

The art and science of cultivating microbial isolates has provided the basis for virtually all of our fundamental knowledge of microbial physiology and diversity. Within the past decade, however, there has been a growing appreciation that the microbes amenable to growth in the dense, monocultural state referred to as "pure culture" may not be representative of most microbes in the environment. This appreciation has been driven largely by results from the application of molecular techniques on nucleic acids extracted directly from environmental samples. With few exceptions, these studies have shown that the organisms readily cultivated from environmental samples are frequently minor components of the resident microbial population. The percentage of (culturable) microbes varies among environments, with a general theme being that nutrient-poor environments tend to harbor microbes that are more recalcitrant to growth in laboratory culture. There have been many explanations given for this recalcitrance, with most centering on the scrupulous nutritional requirements of the microbes, the lack of appropriate surface attachment materials, or on the difficulty in determining the precise conditions required to revive cells from dormant stages. Perhaps the most encompassing explanation for unculturability is that microbes in natural populations have evolved within the context of a microbial community where the cross-feeding and signaling relationships that have developed among individuals are extremely difficult to duplicate in a laboratory setting.

To circumvent the difficulties of cultivation, several general methods are available to directly describe microbial communities, including microscopic examination (2), flow cytometric methods (23), immunological approaches (54), and the analysis of lipids (18). Perhaps the most widely used approach, however, is to exploit the information content stored in nucleic acids extracted directly from environmental samples to examine the ecology, phylogeny, and physiology of microbes from natural populations. The great advantage in analyzing nucleic acids lies in the fact that a nucleotide sequence represents a high-resolution map to an organism's evolutionary history. Areas on this map can be read and compared among different organisms to derive information on phylogenetic affiliations and physiological potential. In addition, physiological activity, in the form of gene expression, can be determined by examining mRNA extracted from environmental samples (38).

By far the most extensively applied nucleic acid analysis method for biodiversity studies is the coupling of the PCR with 16S rRNA gene analysis. A cursory key-phrase search on Medline using the term "16S PCR" revealed 1,129 citations, a testimony to the broad application and ease of use of this method. The power of this approach comes from the ability to amplify, clone, and analyze homologous regions of 16S rDNA from vanishingly small amounts of sample DNA. This is particularly important when the sample is from environments where DNA is difficult to extract or separate from contaminants, such as in certain soil environments (42), or when the resident microbial population is present at very low abundance, such as in oligotrophic waters (22) or in deep subsurface environments (62). Variations of the 16S approach such as denaturing gradient gel electrophoresis (44) or in situ hybridization (2) have been used to examine the complexity and structure of microbial communities, further extending the utility of this approach. Results from these studies have revolutionized our view of microbial diversity and have provided the information required to organize micro- and macroorganisms into the universal tree of life.

Despite the power of this approach, a limitation to using PCR on the 16S or other gene sequence is that prior knowledge of the sequence is required to construct the oligomers used to prime the reaction. Even the so-called 16S "universal primers," which contain degenerate sites to accommodate base variations in conserved regions, do not match all 16S sequences (55) and will likely vary considerably from a subset of those yet to be discovered. Another drawback is that 16S sequences allow only limited inference of physiological potential. For example, the 16S-based phylogeny of the recently discovered and apparently ubiquitous Group I archaeoplankton suggests that they are hyperthermophiles when in actuality they are most abundant in cold marine environments and possess enzymes adapted to low temperature (13, 45). Conversely, protein-coding gene sequences amplified from environmental samples can rarely be definitively tied to the 16S sequence of the source organism, making the sequence difficult to place within a broader phylogenetic context. It would thus clearly be advantageous to couple the power of the 16S probing approach with one that could access other portions of an uncultivated organism's genome. This would allow a coupling of phylotype to phenotype that could rapidly expand our knowledge of the phylogeny, physiology, and ecology of uncultivated microbes.

56.3.2. Environmental Genomics

One means to connect phylotype to phenotype in uncultivated organisms is to directly clone DNA from environmental samples, probe the resultant environmental library with labeled 16S oligomers or genes, then sequence the hybridizing clones to identify the resident 16S gene along with any protein-coding regions that may provide clues to the organism's physiology. Even though the hybridization of 16S probes to a library is subject to similar constraints in uncovering novelty as in using PCR, annealing a single probe to a target gene is a less stringent event than PCR priming and may allow the discovery of more 16S variants. The initial report of this approach was by Schmidt et al. (46), who discovered numerous unique eubacterial phylotypes in an environmental lambda library constructed from marine picoplankton. This study, however, focused on the 16S sequences and did not describe protein-coding regions on the clones carrying the 16S genes. The recent advent of more efficient sequencing methods and robust bioinformatics tools and databases have made practical the rapid analysis of protein-coding regions that are contiguous with the 16S sequence on larger clones. The ends of such large clones can also serve as probes against the library to isolate overlapping clones to identify additional informative sequences. Carrying this approach to its natural end, one can envision that the entire genome of an uncultivated organism could be described given a sufficiently large library. This environmental genomics approach has recently been applied to describe the physiological potential of the group I archaeoplankton that had previously been known only from partial 16S sequences. For this study, an F-factor-based vector, the fosmid, was used to construct a stable 100-million-bp library with an average insert size of approximately 40 kbp per clone (49). An archaeal-specific probe was used to isolate a clone from this library. Sequence analysis of this clone revealed that it contained the entire 16-23S operon as well as several protein-coding genes, hypothetical proteins and unidentified open reading frames. These sequences were sufficient to provide clues to the physiology of this uncultivated organism. For example, sequence from the end of the rRNA operon showed that the archaeoplankton lack the target sequences for streptomycin and erythromycin, the peptidyl transferase domain of the 23S rRNA, suggesting that they are resistant to these antibiotics. Conversely, a conserved His residue in the elongation factor 2 gene indicated a susceptibility to diphtheria toxin protein. This clone also carried a gene encoding a form of glutamate semialdehyde aminotransferase involved in the synthesis of a precursor of chlorophyll, suggesting that the organism may have the ability to harvest energy from light. In a subsequent study (45) the DNA polymerase gene isolated from another group I archaeal clone (*Cenarchaeum symbiosum*) was subcloned and expressed. Analysis of the gene product showed that the polymerase was thermolabile, confirming that the archaeoplankton are not inactive refugees from hydrothermal vent environments but rather are likely to be significant members of many cold marine ecosystems. It is clear from these studies that environmental genomics can offer new insights into the physiological potential of uncultivated microbes and their subsequent role in ecological communities. In addition, this approach can help define the metabolic plasticity of microbes in ways that can help predict responses to environmental stresses or the conditions needed to eventually cultivate these organisms in the laboratory.

The constraints involved in cloning environmental DNA differ considerably from those involved in using PCR. Primary among these is the amount of DNA required. Leaving aside for a moment the inherent biases involved in PCR and cloning, consider that a comprehensive library of 16S molecules can be amplified from as little as 1 ng of DNA from a mixed microbial community. This amount of starting DNA would represent the combined genomes of approximately 10^5 bacteria, the amount in approximately 100 μl of coastal seawater or 1 mg of soil. To construct a comprehensive genomic library, by contrast, requires 10 to 1,000 times more DNA depending upon the type of vector/host system used and the quality of the DNA preparation.

56.3.3. Screening Environmental Libraries

By overcoming cloning constraints, the complete microbial diversity of an environmental sample can be made accessible in the form of a recombinant library. In effect, one could propagate uncultivated organisms in the form of genome fragments in surrogate hosts. Given the sheer numbers of microbes in natural samples, screening an environmental library for a particular gene fragment or activity can be challenging. For example, Torsvik et al. (53) estimated that a single 30-g sample of forest soil contained some 4,000 different genomes. Converting this sample to a plasmid library of 5-kbp average insert size would require 20×10^6 clones to have the desired fivefold genome coverage needed to ensure complete representation. Screening a library of this size by standard filter lift hybridization would require 4,000 agar plates containing 5,000 clones per plate, a daunting task by any measure. Alternative screening approaches are clearly needed to thoroughly screen environmental libraries. One such approach is to multiplex, i.e., create clone pools that can be screened by hybridization, expression, or PCR for an initial signal that can be broken out by subsequent screening. The size of the initial pools is determined by the sensitivity of the assay. For example, an initial PCR screen can be used to reduce the clones from the 4,000 plates above to 40 pools. A positive signal from one or more of these pools can be followed by screening successively smaller pools. Another approach, which can be used in concert with multiplexing, is to apply high-throughput robotic screening (43). Such systems are now in place that can screen tens of thousands of clones per day for multiple enzyme activities (43, 47).

56.3.4. Barriers and Challenges

Even though the concept of diverse, uncharacterized pools of unculturable microbes populating natural environments appears to have been largely accepted, there are key tasks that remain to be addressed in characterizing these organisms. First, we have yet to learn sufficient information about an uncultivated organism from sequence data to grow it under laboratory conditions. Laboratory culture still remains an important goal and is still the only way to accurately measure many physiological traits. Second, the complete genome sequence of an uncultivated organism remains only an intriguing concept. Funding, refinement of bioinformatic tools, and a consensus on which organism to tackle have yet to be resolved. Likely candidates would be a symbiont that can be readily separated from its host or a microbe that represents a large fraction of a natural population. A candidate group that meets both of these criteria is the Group I archaea, which are a major fraction of the Antarctic picoplankton (13) and also occur as a specific symbiont of a temperate-water marine sponge (39). A final goal that would benefit the field of environmental genomics would be to make a prediction of a physiological trait from sequence data and then later confirm expression of this trait with field or laboratory measurements. Several groups are now poised to accomplish this task and in doing so will further increase the interest in examining members of this largely uncharacterized, yet undoubtedly important group of microbes.

We thank Greg Thorn, Univ. of Wyoming, for sharing references and insight into fungal isolation, identification, and diversity. J.M.T. thanks colleagues at the Center for Microbial Ecology for discussions and collaborative research that have provided new insight on microbial diversity, and the National Science Foundation for research support (NSF DEB9120006).

REFERENCES

1. **Allsopp, D., R. R. Colwell, and D. L. Hawksworth.** 1995. *Microbial Diversity and Ecosystem Function.* CAB International, Wallingford, United Kingdom.
2. **Amann, R., J. Snaidr, M. Wagner, W. Ludwig, and K. H. Schleifer.** 1996. In situ visualization of high genetic diversity in a natural microbial community. *J. Bacteriol.* **178:**3496–3500.
4. **Barron, G. L.** 1968. *The Genera of Hyphomycetes from Soil.* Williams & Wilkins Co., Baltimore.
5. **Barron, G. L.** 1971. Soil fungi. *Methods Microbiol.* **4:**405–427.
6. **Barron, G. L.** 1977. *The Nematode-Destroying Fungi.* Canadian Biological Publications Ltd., Guelph, Ontario, Canada.
7. **Bills, G. F., M. Christensen, M. Powell, and R. G. Thorn.** Saprobic soil fungi. *In* G. Mueller, A. Y. Rossman, and G. F. Bills (ed.), *Measuring and Monitoring Fungal Diversity.* Smithsonian Institution Press, Washington, D.C., in press.
8. **Borneman, J., P. W. Skroch, K. M. O'Sullivan, J. A. Palus, N. G. Rumjanek, J. L. Jansen, J. Nienhuis, and E. W. Triplett.** 1996. Molecular microbial diversity of an agricultural soil in Wisconsin. *Appl. Environ. Microbiol.* **62:**1935–1943.
9. **Castenholz, R. W.** 1996. Endemism and biodiversity of thermophilic cyanobacteria. *Nova Hedwigia* **112:**33–47.
10. **The Center for Microbial Ecology.** http://www.cme.msu.edu/cme.
11. **Colwell, R. R., U. Simidu, and K. Ohwada.** 1996. *Microbial Diversity in Time and Space.* Plenum Press, New York.
12. **Dedysh, S. N., N. S. Panikov, and J. M. Tiedje.** 1988. Acidophilic methanotrophic communities from *Sphagnum* peat bogs. *Appl. Environ. Microbiol.* **64:**922–929.
13. **DeLong, E. F., K. Y. Wu, B. B. Prezelin, and R. V. Jovine.** 1994. High abundance of Archaea in Antarctic marine picoplankton. *Nature* **371:**695–697.
14. **DeVos, P., J. Swings, M. Gillis, P. Vandamme, B. Pot, and K. Kersters.** 1996. Bacterial species in polyphasic taxonomy, p. 122–126. *In* R. A. Samson, J. A. Stalpers, O. van der Mei, and A. H. Stouthamer (ed.), *Culture Collections to Improve the Quality of Life.* Centralbureau voor Schimmelcultures, Baarn, The Netherlands.
15. **Domsch, K. H., W. Gams, and T.-H. Anderson.** 1980. *Compendium of Soil Fungi,* vol. 1 and 2. Academic Press, London.
16. **Emerson, D., R. M. Worden, and J. A. Breznak.** 1994. A diffusion gradient chamber for studying microbial behavior and separating microorganisms. *Appl. Environ. Microbiol.* **60:**1269–1278.
17. **Farr, D. F., G. F. Bills, G. P. Chamuris, and A. Y. Rossman.** 1989. *Fungi on Plants and Plant Products in the United States.* APS Press, St. Paul, Minn.
18. **Frostegard, A., S. O. Petersen, E. Baath, and T. H. Nielsen.** 1997. Dynamics of a microbial community associated with manure hot spots as revealed by phospholipid fatty acid analyses. *Appl. Environ. Microbiol.* **63:**2224–2231.
19. **Fulthorpe, R. R., A. N. Rhodes, and J. M. Tiedje.** 1998. High levels of endemicity of 3-chlorobenzoate-degrading soil bacteria. *Appl. Environ. Microbiol.* **64:**1620–1627.
20. **Gams, W.** 1992. The analysis of communities of saprophytic microfungi with special reference to soil fungi, p. 183–223. *In* W. Winterhoff (ed.), *Fungi in Vegetation Science.* Kluwer Academic Publishers, Dordrecht, The Netherlands.
21. **Gams, W.** 1993. *Supplement and Corrigendum to the Compendium of Soil Fungi.* IHW-Verlag, Eching, Germany.
22. **Giovanonni, S. J., T. B. Britschgi, C. L. Moyer, and K. G. Field.** 1990. Genetic diversity in Sargasso Sea bacterioplankton. *Nature* **345:**3387–3394.
23. **Guindulain, T., J. Comas, and J. Vives-Rego.** 1997. Use of nucleic acid dyes SYTO-13, TOTO-1, and YOYO-1 in the study of *Escherichia coli* and marine prokaryotic populations by flow cytometry. *Appl. Environ. Microbiol.* **63:**4608–4611.
24. **Hawksworth, D. L.** 1990. The fungal dimension of biodiversity: magnitude, significance, and conservation. *Mycol. Res.* **95:**641–655.
25. **Hawksworth, D. W., P. M. Kirk, B. C. Sutton, and D. N. Pegler.** 1995. *Ainsworth & Bisby's Dictionary of the Fungi,* 8th ed. CAB International, Wallingford, United Kingdom.
26. **Heywood, V. H. (ed.)** 1995. *Global Biodiversity Assessment,* p. 1140. Cambridge University Press, Cambridge.
27. **Holben, W. E., and D. Harris.** 1995. DNA-based monitoring of total bacterial community structure in environmental samples. *Mol. Ecol.* **4:**627–631.
28. **Kane, M. D., and N. E. Pierce.** 1994. Diversity within diversity: molecular approaches to studying microbial in-

teractions with insects, p. 509–524. *In* E. Schierwater, B. Streit, G. P. Wagner, and R. DeSalle (ed.), *Molecular Ecology and Evolution: Approaches and Applications*. Berkhauser Verlag, Basel.

29. **Kendrick, W. B., and D. Parkinson.** 1990. Soil fungi, p. 49–68. *In* D. L. Dindal (ed.), *Soil Biology Guide*. John Wiley and Sons, New York.

30. **Konings, W. N., and H. Veldkamp.** 1980. Phenotypic responses to environmental change, p. 161–191. *In* D. C. Ellwood, J. N. Hedger, M. J. Latham, J. M. Lynch, and J. H. Slater (ed.), *Contemporary Microbial Ecology*. Academic Press, New York.

31. **Liesack, W., and E. Stackebrandt.** 1992. Occurrence of novel groups of the domain *Bacteria* as revealed by analysis of genetic material isolated from an Australian terrestrial environment. *J. Bacteriol.* **174:**5072–5078.

32. **May, R. M.** 1986. The search for patterns in the balance of nature's advances and retreats. *Ecology* **67:**1115–1126.

33. **Mueller, G., A. Y. Rossman, and G. F. Bills (ed.).** *Measuring and Monitoring Fungal Diversity*. Smithsonian Institution Press, Washington, D.C., in press.

34. **Nüsslein, K., and J. M. Tiedje.** 1988. Characterization of a young Hawaiian soil bacterial community for its dominant and rare members using small subunit rDNA amplified from DNA fractionated by its guanine and cytosine composition. *Appl. Environ. Microbiol.* **64:**1283–1289.

35. **Nüsslein, K.** 1999. Influence of soil age and vegetative cover on microbial community composition: a ribosomal DNA analysis of Hawaiian soils. Ph.D. thesis. Michigan State University, East Lansing.

36. **O'Donnell, K. L.** 1979. *Zygomycetes in Culture*. Dept. of Botany, University of Georgia, Athens, Georgia.

37. **Parkinson, D.** 1994. Filamentous fungi, p. 329–350. *In Methods of Soil Analysis*, Part 2. *Microbiological and Biochemical Properties*. SSSA Book Series no. 5, Soil Science Society of America, Madison, Wis.

38. **Paul, J. H.** 1996. Carbon cycling: molecular regulation of photosynthetic carbon fixation. *Microb. Ecol.* **32:**231–245.

39. **Preston, C. M., K. Y., Wu, T. F. Molinski, and E. F. DeLong.** 1996. A psychrophilic crenarchaeon inhabits a marine sponge: *Cenarchaeum symbiosum* gen. nov., sp. nov. *Proc. Natl. Acad. Sci. USA* **93:**6241–6246.

40. **Rademaker, J. L. W., F. J. Louws, U. Rossbach, P. Vinuesa, and F. J. de Bruijn.** Computer-assisted pattern analysis of molecular fingerprints and database construction, 7.1.3. *In* A. D. L. Akkermans, J. D. van Elsas, and F. J. de Bruijn (ed.), *Molecular Microbial Ecology Manual*. Kluwer Academic Publishers, Dordrecht, The Netherlands, in press.

41. **Reasoner, D. J., and E. E. Geldreich.** 1985. A new medium for the enumeration and subculture of bacteria from potable water. *Appl. Environ. Microbiol.* **49:**1–7.

42. **Rheims, H., F. A. Rainey, and E. Stackebrandt.** 1996. A molecular approach to search for diversity among bacteria in the environment. *J. Indust. Microbiol.* **17:**159–169.

43. **Robertson, D. E., E. J. Mathur, R. V. Swanson, B. L. Marrs, and J. M. Short.** 1996. The discovery of new biocatalysts from microbial diversity. *SIM News* **46:**3–8.

44. **Santegoeds, C. M., S. C. Nold, and D. M. Ward.** 1996. Denaturing gradient gel electrophoresis used to monitor the enrichment culture of aerobic chemoorganotrophic bacteria from a hot spring cyanobacterial mat. *Appl. Environ. Microbiol.* **62:**3922–3928.

45. **Schleper, C., R. V. Swanson, E. J. Mathur, and E. F. DeLong.** 1997. Characterization of a DNA polymerase from the uncultivated psychrophilic archaeon *Cenarchaeum symbiosum*. *J. Bacteriol.* **179:**7803–7811.

46. **Schmidt, T. M., E. F. DeLong, and N. R. Pace.** 1991. Analysis of a marine picoplankton community by 16S rRNA gene cloning and sequencing. *J. Bacteriol.* **173:**4371–4378.

47. **Short, J.** 1997. Recombinant approaches for accessing biodiversity. *Nat. Biotechnol.* **15:**1322–1323.

48. **Stackebrandt, E., and B. M. Goebel.** 1994. Taxonomic note: a place for DNA-DNA reassociation and 16S rRNA sequence analysis in the present species definition in bacteriology. *Intl. J. Syst. Bacteriol.* **44:**846–849.

49. **Stein, J. L., T. L. Marsh, K. Y. Wu, H. Shizuya, and E. F. DeLong.** 1996. Characterization of uncultivated prokaryotes: isolation and analysis of a 40-kilobase-pair genome fragment from a planktonic marine archaeon. *J. Bacteriol.* **178:**591–599.

50. **Thorn, G., C. A. Reddy, D. Harris, and E. A. Paul.** 1996. Isolation of saprophytic basidiomycetes from soil. *Appl. Environ. Microbiol.* **62:**4288–4292.

51. **Thorn, R. G.** 1997. The fungi in soil, p. 63–127. *In* J. D. van Elsas, J. T. Trevors, and E. M. H. Wellington (ed.), *Modern Soil Microbiology*. Marcel Dekker, New York.

52. **Tiedje, J. M.** 1995. Approaches to the comprehensive evaluation of prokaryote diversity of a habitat, p. 73–87. *In* D. Allsopp, D. L. Hawksworth, and R. R. Colwell (ed.), CAB International, Wallingford, United Kingdom.

53. **Torsvik, V., J. Goksoyr, and F. L. Daae.** 1990. High diversity in DNA of soil bacteria. *Appl. Environ. Microbiol.* **56:**782–787.

54. **Tuomi, P., T. Torsvik, M. Heldal, and G. Bratbak.** 1997. Bacterial population dynamics in a meromictic lake. *Appl. Environ. Microbiol.* **63:**2181–2188.

55. **Vergin, K. L., E. Urbach, J. L. Stein, E. F. DeLong, and S. J. Giovanonni.** 1998. Screening of a fosmid library of marine environmental genomic DNA fragments reveals four clones related to the Planctomycetales. *Appl. Environ. Microbiol.* **64:**3075–3078.

56. **Vinuesa, P., J. L. W. Rademaker, F. J. de Bruijn, and D. Werner.** 1998. Genotypic characterization of *Bradyrhizobium* strains nodulating endemic woody legumes of the Canary Islands by PCR-restriction fragment length polymorphism analysis of genes encoding 16S rRNA (16S rDNA) and 16S-23S rDNA intergenic spacers, repetitive extragenic palindromic PCR genomic fingerprinting, and partial 16S rDNA sequencing. *Appl. Environ. Microbiol* **64:**2096–2104.

56a. **von Arx, J. A.** 1981. *The Genera of Fungi Sporulating in Pure Culture*. J. Cramer, Vaduz, Germany.

57. **Watanabe, T.** 1994. *Pictorial Atlas of Soil and Seed Fungi*. Lewis Publishers, Boca Raton, Fla.

58. **Wayne, L. G., D. J. Brenner, R. R. Colwell, P. A. D. Grimont, O. Kandler, M. I. Drichevsky, L. H. Moore, W. E. C. Moore, R. G. E. Murray, E. Stackebrandt, M. P. Starr, and H. G. Trüper.** 1987. Report of the ad hoc committee on reconciliation of approaches to bacterial systematics. *Int. J. Syst. Bacteriol.* **37:**463–464.

59. **Whang, K., and T. Hattori.** 1988. Oligotrophic bacteria from rendzina forest soil. *Antonie van Leeuwenhoek* **54:**19–36.

60. **Wolin, M. J., and T. L. Miller.** 1982. Interspecies hydrogen transfer: 15 years later. *ASM News* **48:**561–565.

61. **Zhou, J., B. Xia, R. V. O'Neill, L.-Y. Wu, A. V. Palumbo, and J. M. Tiedje.** Unpublished data.

62. **Zlatkin, I. V., M. Schneider, F. J. de Bruijn, and L. J. Forney.** 1996. Diversity among bacteria isolated from the deep subsurface. *J. Indust. Microbiol.* **17:**219–227.

Deliberate Release of Recombinant Microorganisms

M. J. BAILEY, A. K. LILLEY, I. P. THOMPSON, F. A. A. M. DE LEIJ, J. M. WHIPPS, AND J. M. LYNCH

57

The purpose of this overview is to consider a variety of approaches to the study of the performance and activity of microbial inocula in soil and plant-associated habitats. Particular emphasis will be placed on studies that underpin the development of genetically modified bacteria with commercial potential. Using examples drawn from our own experience with phytosphere pseudomonads in different crops, we will consider the selection, modification, and release of active bacteria. Consideration of the underlying principles of microbial ecology will be made with reference to the delivery of inocula for predictable colonization densities, the persistence and potential impact of sustained populations, and how environmental impact may be assessed. Impact evaluations are particularly difficult and require understanding of microbial population function and interaction. To illustrate these points, we contrast assessments of alterations in community activity and diversity and discuss the consequences of natural gene flow mediated by mobile genetic elements indigenous to the microflora of the target habitat.

Genetically modified microorganisms (GMMs) have potential in biological control, plant growth promotion, degradation of toxic compounds in certain polluted environments, and mineral leaching (Table 1) (59, 65, 88). Despite at least a decade of study, and the obvious benefits that could be realized from the release of GMMs, concern remains about their potential impact on the environment. In general, releases (particularly in Europe) have been restricted to small controlled studies in which information on the fate and impact of GMMs has been gathered for the provision of risk assessment data (5, 16, 85, 88). These restrictions will remain until a better understanding of the wider ecological issues is obtained. Microorganisms not only have the capacity to multiply, but they are also subject to evolutionary selection pressures, including the exchange of genetic information, which increases the uncertainty of the fate and effect that novel genetic traits might have. It is therefore necessary to differentiate the real from perceived risks. These considerations precede the commercialization of engineered microbial inocula but may equally apply to the release of any strain. To date, environmental releases are undertaken with prototype strains that have been purposely constructed to optimize detection but that avoid the use of traits that represent any perceived risk. Understandably, the ecological activity of vector bacteria on release provides a baseline against which the effect of novel expressed traits might be determined. Consents for release are generally granted on a case-by-case basis (37). This position has not changed in the last decade despite the lack of direct evidence, with the analytical methods used, of unintentional harm resulting from deliberate releases. This may be the result of cautious approaches used in the construction of GMMs and the fact that fully competent inocula expressing introduced and/or overexpressed biologically active molecules have not been widely released in the field situation. It remains incumbent upon those who wish to release GMMs to provide a risk assessment, which in general requires the following information:

- the characteristics of donor and recipient organisms, vector DNA, and GMMs
- the conditions of the release and of the receiving environment
- the interaction with the environment, namely survival, multiplication, dissemination, and environmental impact of the GMMs
- the sensitivity, specificity, and practicality of monitoring techniques, waste treatment, and emergency responses

When considering impact, perturbation, and risk associated with the release of GMMs, it is not possible to provide all encompassing guidelines, but a large number of publications have been produced that cover different aspects of the subject (8, 25, 88). One significant obstacle to providing a generalized method for assessing the risk (potential) is the diversity of microbes themselves. Consider the diversity recognized within a taxon, with the large number of traits that can be modified or introduced and the multivariant habitats that are targets for release, and the problem is all too clear.

To be able to identify risks associated with the release of a GMM, information on the survival, establishment, and dissemination of GMMs in the environment is required. Such studies consider the colonization, spread, and persistence of the GMM (autecology) at the localized, microhabitat level as well as considering dispersal on a gross level, e.g., movement through a field crop or adjacent habitats. Furthermore, information on the frequency of transfer of the introduced genetic material to or from indigenous microorganisms in natural environments, and within and be-

TABLE 1 Selected examples of published GMM field releases[a]

Bacterium	Trait	Year	Reference
Pseudomonas aureofaciens	*lacZy* on Tn*7*	1987	35
Pseudomonas syringae	Ice (*ina*) minus	1987	49
Agrobacterium radiobacter	Deletion *ina*	1988	31
Clavibacter xyli	*B. thuringiensis* endotoxin	1988	36
Pseudomonas fluorescens	*lacZY* on Tn*5*	1988	36
Bradyrhizobium japonicum	Insertion of *nifA* and *nifO* in *ino* region	1989	13
Bacillus thuringiensis	*B. thuringiensis* endotoxin and Tet^r	1990	88
Pseudomonas corrugata	Tn*7*::*lacZY*	1990	61
Pseudomonas fluorescens	*lacZY*	1990	22
Rhizobium leguminosarum	Trifolitoxin genes	1990	80
Xanthomonas campestris	*lux*	1990	64
Pseudomonas fluorescens	*lacZY*	1991	55
Pseudomonas putida	*lux*	1991	64
Pseudomonas syringae	*lemA*	1991	28
Rhizobium meliloti	*dctABD* and *nifA*	1992	10
Pseudomonas fluorescens	*lacZy*, *aph*, *xylE*	1993	17, 75
Pseudomonas syringae	Tn*5*::Kan^r	1993	9
Pseudomonas fluorescens	*lacZy*, *aph*, *xylE*	1994	43, 44, 75
Pseudomonas fluorescens (pQBR103) (*tra*$^+$, Hg^r)	*lacZy*, *aph*, *xylE*	1994	42
Rhizobium leguminosarum biovar *viciae*	Tn*5*:: Kan^r	1994	29
Pseudomonas fluorescens	*lacZ*	1995	84
Pseudomonas fluorescens	Tn*5*::*cryIVB*	1995	2
Pseudomonas fluorescens	Bioluminescence Tn*4431*	1997	51

[a]To date, over 50 notified environmental releases have taken place worldwide. For further information contact OECD biotechnology information service (http://www.oecd.org/ehs/service.htm), biotechnology information for the USDA (http://www.inform.umd.edu: 8080/Ed/Res/Topic/AgrEnv/Biotecnology) and the UNIDO biosafety server (http//www.binas.unido.org/binas).

tween species and genera, is needed. Information is also needed in relation to the interaction of GMMs and the introduced gene products with other species and/or biological processes in the environment where releases are intended (synecology). To increase familiarity with the safe use of GMMs, impact has to be assessed with respect to the resident microbiota. A number of steps for determining the ecological impact of GMMs have been proposed (62, 63, 70, 76). In developing effective inocula, particularly those carrying novel traits, data on the level, duration, and effectiveness of expression at the local and global level are needed to verify the effectiveness of the introduction but also to evaluate impact on coresidents and distant communities.

In contemplating the release of GMMs, stepwise assessments are adopted that are conducted through laboratory, microcosm (simulated field conditions), controlled field releases (small scale), and finally open-field releases to test efficacy and evaluate risk. Irrespective of the origin of the inoculum, the introduction of a large number of organisms to any habitat is likely to have some effect, even if only short term. This "inoculum effect" has been demonstrated with the introduction both unmodified bacteria and GMMs. Unmodified microbes have been released into the environment for many decades, particularly in the agricultural practices of biological control and inoculation of crop plants with nitrogen-fixing microorganisms. Only a few studies have attempted to determine the effect of these introductions on indigenous microbial communities (24, 26). Weller (86) observed a net increase in total fluorescent pseudomonads but not in total bacterial populations after the introduction of a *Pseudomonas fluorescens* 2-79 onto wheat. Yuen and Schroth (89) found no effect on the population density of gram-negative bacteria, fluorescent pseudomonads, or actinomycetes on zinnia roots after the introduction of *P. fluorescens* E6. In contrast, Kloepper and Schroth (34) inoculated potato seed pieces with *P. fluorescens* strains A1, E6, or B10, which resulted in reduced populations of gram-positive bacteria. Assessments of the ecological consequences beyond the target species are fundamental when evaluating the impact associated with the release of genetically modified biological agents.

An understanding of the function and activity of the microbial inocula in microcosms, as a prerequisite to the granting of consent to release a GMM, will permit an accurate assessment of potential impact or harm to known target species or taxa. Such instances include the use of a bacterium that is pathogenic or toxigenic, or, by the introduction of additional genetic material, acquires antagonistic or antibiotic functions. When evaluating the potential of any GMM, the consideration must be whether the modification per se affects activity, host range, or toxicity in respect to the wild type. These studies, undertaken in both experimental microcosms and following field releases in a number of countries (Table 1), have proved invaluable to our understanding of the ecology and population dynamics of bacteria. These investigations provide basic information needed for the generic assessment of GMM biosafety.

Essentially two types of release have been undertaken with modified bacteria. First, many experiments have investigated the autecology of the release strain and its wider ecological interactions in the environment. These experiments have provided insights into the microbial ecology and behavior of GMMs in the environment. Second, releases have been used to evaluate GMM with intended beneficial activities and to study efficacy, scale-up, and important elements of safety testing.

57.1. ISOLATION OF BACTERIA FOR MODIFICATION AND FIELD RELEASE

57.1.1. Rationale

It is axiomatic that an understanding of the underlying principles of microbial ecology (how communities or individuals survive, persist, and colonize particular habitats) is required for the selection, design, and introduction of inocula with potential for environmental management. These are extremely complex issues interdependent on a series of (possibly) unpredictable multifactorial interactions. Therefore, the choice of candidate strain(s) is essential to the design of appropriate experiments to obtain a more intimate understanding of how cells perceive and interact with the environment and other microbes. Typically, studies have focused on the selection of candidates and their genetic modification to express novel phenotypes to facilitate detection. These investigations have provided considerable insight to the behavior of inocula and how inocula interact with the target environment. But because of practical issues, and the appropriate caution related to the use of GMMs, most investigations have been undertaken in constructed microcosms (including physically contained field studies) rather than under open-field conditions.

The validation of microcosm-derived data against actual field studies is fundamental, as most laboratory or microcosm investigations are presumed to reflect cellular activity and population ecology in the field. Advances in the theory of microbial ecology provide accessible parameters against which community succession can be judged. The development of models that predict how inocula might behave are invaluable for testing assumptions of risk assessment related to the use of GMMs. Furthermore, testable models assist the development of effective inocula whose activity can be targeted at appropriate habitats.

57.1.2. Selection of an Indigenous Pseudomonad

Our own investigations were undertaken as part of a wider program to study the microbial ecology of the phytosphere and to develop inocula suitable as vectors of novel traits, for example, in the biocontrol of phytopathogenic fungi. Our criteria were based on the belief that to sustain function, inocula must be relatively easy to apply, must establish a viable community, and must present little or no detrimental impact on targeted or aligned habitats. To be effective, it was determined that suitable organisms should be selected from the target habitat. Following assessments of the microbiology (community succession, population dynamics) of sugar beet (73), supported by studies in wheat (39), grown at the field sites proposed for later release, candidate bacteria were selected and modified (6). Investigations were undertaken using monocultures in simple constructed microcosms. However, for studies of the relevance of microbial function, more realistic investigations were required that simulated the natural world, or which were actually undertaken in the field. A candidate was selected that was able to colonize the root and leaf surface at ecological densities following simple seed inoculation of sugar beet. Contrasts were drawn against wheat, in which seed dressing was effective for rhizosphere establishment, but spray application at tillering was required to provide inocula to the phyllosphere. These comparisons between different crop plants grown at different sites provided reliable contrasts for the inocula under test (19, 20, 74).

57.2. GENETIC MODIFICATION OF CANDIDATE BACTERIA SUITABLE FOR FIELD RELEASE

According to the nature of the experiment, genes expressing a wide variety of traits may be inserted into the bacterial genome. In most experiments, genes will be inserted to facilitate the tracking and reliable recognition of the GMM in environmental samples. In studies of the ecology of the bacterial strain, these marker genes will be the full extent of the genetic manipulation. A variety of marker genes are available (30) with their own merits and limitations. These approaches range from the random selection of spontaneous mutants resistant to antibiotics to the construction of genetically modified isolates to express novel traits useful for monitoring cell activity (e.g., calorimetric or bioluminescent markers) (56) or for selection against the indigenous background (novel substrate utilization, antibiotic resistance). The choice of marker must be in context and meet the needs of the investigation. Certain reporters require more sophisticated apparatus, whereas others can be detected as a phenotype by direct plating methods. All are suitable for nucleic acid-based detection.

57.2.1. Autecology of *P. fluorescens* SBW25

Investigations were undertaken to define the phytosphere microbiology of sugar beet and wheat grown at the field sites used for field releases. These studies provided baseline knowledge of community diversity, which was needed for the assessment of impact on the bacterial community. Isolates were identified by phenotypic analysis of total fatty acid methyl ester content using gas chromatography. Isolates of the most abundant component, the fluorescent pseudomonads, were further characterized by genetic fingerprinting by comparing macro restriction patterns following the enzyme digest of total extracted DNA and electrophoresis under conditions of field inversion and hybridization with the rRNA-DNA operon to contrast restriction fragment length polymorphism patterns. Sensitive (accurate, reliable, and rapid) identification methods are essential for the description of diversity. Defining the community in both space and time provides a clearer understanding of the microbial ecology in terms of the succession and rate of turnover of particular populations. It also allows the identification of presumed keystone or abundant types and the possibility of testing the assumption that the proliferation of particular types is indicative of its chance selection under conditions favorable to its phenotype. Sugar beet appears to be colonized by a diverse community of fluorescent pseudomonads comprising particular ecotypes, root, soil, and leaf colonizers whose dynamics suggest periodic adaptation to prevailing climactic and biotic conditions (58, 71, 72). Strains isolated from plants collected throughout each of three growing seasons predominantly clustered together. However, within a given sample, approximately 12% were strains repeatedly isolated within and between years. Such persistence and periodicity indicates specific selection and proliferation. This dynamic process maintains the overall density but drives community succession. If these observations hold for other habitats, then considerable care must be taken in the selection and development of inocula for environmental use. Knowledge of the extent of the genotypic and phenotypic plasticity of an isolate, and how that plasticity is regulated under conditions of selection and competition, affects the ability of an individual to adapt to changing conditions and survive.

57.2.2. Genetic Modification of an Indigenous Pseudomonad Isolate

The candidate organism, a plasmid-free, nonpathogenic, ribosomal RNA group 1 fluorescent pseudomonad, *P. fluorescens* SBW25, was selected for marking from the natural microflora of the sugar beet plant surface because it occurred at high densities and was also able to colonize other plants. Two constitutively expressed marker gene cassettes, *aph-xylE* (kanamycin resistance and catechol 2,3-dioxygenase activity) and *lacZY* (β-galactosidase and lactose permease), allowing modified pseudomonads uniquely to utilize lactose as a sole carbon source (7), were inserted as single copies into separate sites on the bacterial chromosome (6) and mapped approximately 1 Mb distant on the 6.6-Mb bacterial chromosome (57) (Fig. 1). The marker genes were chosen because they had been shown to be well expressed in pseudomonads and because they provide phenotypes that facilitate highly sensitive detection (ca. 1 CFU g^{-1}) by simple substrate utilization in bacteriological media (6) or calorimetric assays (15) or by PCR (12). In addition, *lacZY* genes had already been widely used for monitoring pseudomonads in the environment (23). The marker cassettes were introduced by site-directed homologous recombination following the transformation of recipient SBW25 bacteria with modified chromosomal DNA. This approach avoided the retention of vector sequence or other genetic motifs, i.e., transposons (29) or disarmed transposons (7, 27). The chromosomal location of markers was preferred over plasmids to minimize the possibility of gene transfer, growth disadvantage owing to inocula carrying plasmids, or the loss of marked plasmid resulting in an inability to detect the introduced bacteria. The modified derivative we developed, *P. fluorescens* SBW25EeZY6KX (*lacZ, lacY, aph, xylE*) survived and naturally colonizes plants following seed inoculation (17, 19, 74, 75), can be easily and accurately monitored in the natural environment, and permits the direct assessments of genetic stability and gene transfer to be undertaken (6, 43).

57.2.3. Evaluation of Marker Gene Transfer

Many release experiments have been designed to assess in vitro and in situ horizontal spread of marker genes. In our own study, separate chromosomal sites were selected to specifically provide a rapid method for the direct assessment of marker gene transfer. One cassette was located close to the chromosomal *oriC* to assess transduction events (Fig. 1), and as the marker genes were constitutively expressed, all modified bacteria could be isolated as blue colonies on pseudomonad selection agar supplemented with 5-bromo-4-chloro-3-indolyl-β-D-galactopyranoside (X-Gal) and kanamycin. White colonies would represent indigenous kanamycin-resistant pseudomonads, GMM mutants that had lost the ability to express *lacZ*, or an indigenous population of pseudomonads to which the *aph-xylE* cassette had been transferred. Investigation confirmed that none of the combined phenotypes could be detected in bacteria isolated from the soil or plants sampled at the release sites. The design of the GMM was also biased toward detecting natural gene flux. As the most probable recipients of any transferred chromosomal gene are likely to be closely related bacteria, chromosomal sites were selected for marking so that disruption did not affect the fitness of the marked bacteria, but also on the basis of the relative genetic conservation of the sites in the indigenous pseudomonad community. A collection of pseudomonad isolates, representative of the phytosphere microbial community, were screened for shared homology to the two chromosomal fragments of SBW25 into which the markers were to be introduced. Fragment (Ee), the site to be marked with the *lacZY* cassette, hybridized to only 6% of the population, whereas the other site, fragment 6, the site to be marked with the *aph-xylE* cassette, hybridized with 100% of the isolates. The extent of the shared homology and the potential of fragment 6, marked with *aph-xylE*, to be mobilized, recombine, and be expressed in representatives of the indigenous microbial community was confirmed in vitro (6). Therefore, careful selection, design, and release strategies for the GMM allowed direct assessments of autecology and synecology using inocula indigenous to the target habitat and provided a direct assay for estimating chromosomal mobilization activity in natural populations based primarily on the detection of a characteristic, novel phenotype (aph^+, $xylE^+$, $lacZY^-$).

57.3. MICROCOSM EVALUATION AND FIELD STUDIES

Prior to obtaining permission from the regulatory authorities, for any field release it is obligatory to undertake investigations in microcosms or simulated environments. Therefore, a number of factors need to be considered, not least the appropriate design of systems for prerelease evaluation. Van Overbeek and van Elsas (84) have observed that the introduction of bacteria into soil not only results in specific localizations but also provides bacteria in an altered physiological state probably dissimilar to that of the indigenous microbial community. In related studies the source and preparation of inocula have proved critical. The fitness and persistence of inocula prepared under different in vitro growth conditions were compared with those of inocula washed directly from leaves, and it was confirmed that specific factors are essential (48, 87).

In microcosms and greenhouses using natural soil and plants, studies may not accurately reflect the normal physiological condition of bacterial strains in the open environment, because the methods of preparation, density of cells effectively introduced, and experimental conditions all limit the validity of the data. However, microcosm and greenhouse studies play a valuable role in evaluating the different methods of releasing GMMs and in characterizing their behavior. These studies will aim to deliver the inocula to the target site of the experiment and may involve leaf inoculation by spraying, soil drenches, or seed dressing. Through careful experimental design, employing approaches that simulate the natural environment and the metabolic status of the inocula, realistic insight into natural population dynamics can be obtained. In our comparative

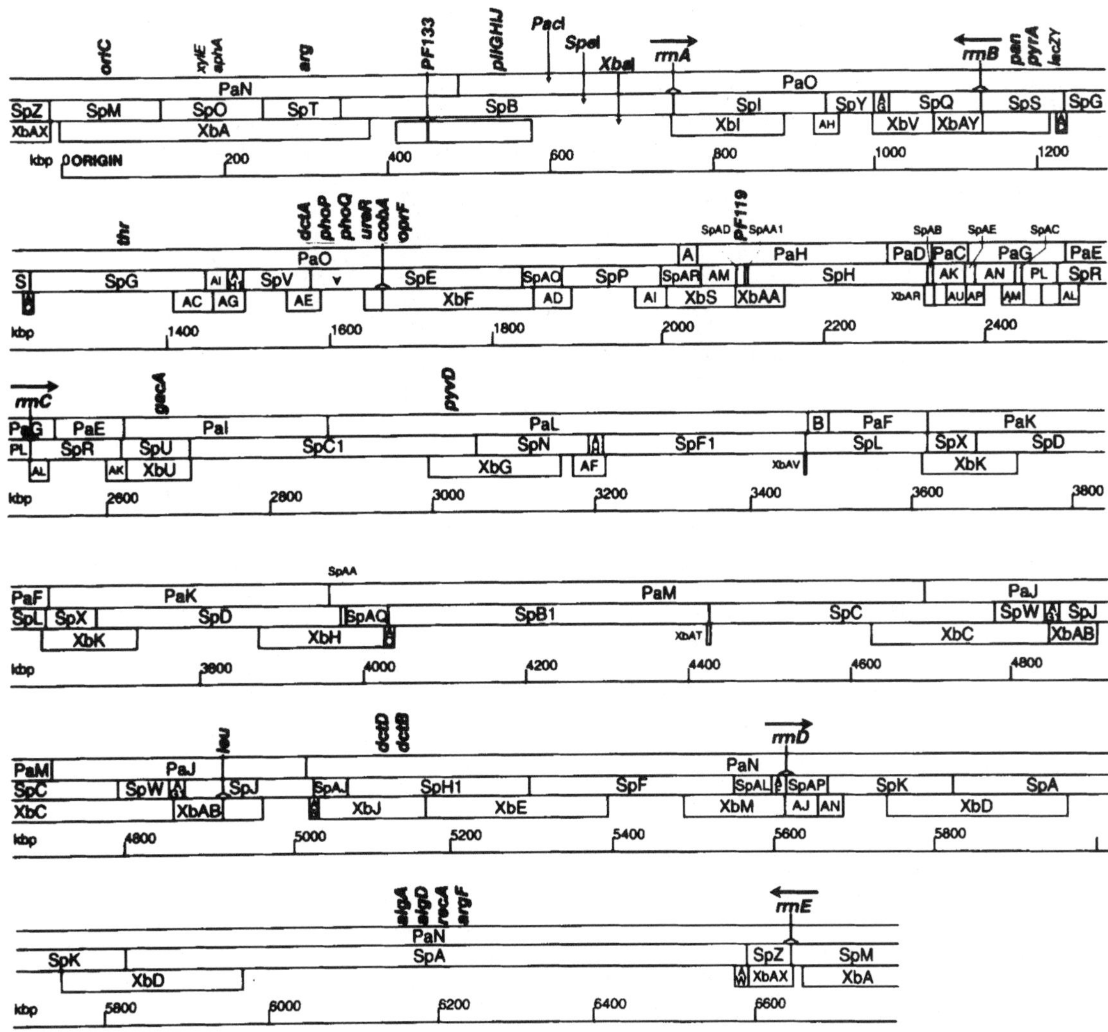

FIGURE 1 Physical map of SBW25 (after reference 57). The locations of the two marker gene cassettes are included, *aph-xylE* (close to *oriC*) and *lacZY* (adjacent to *rrnB*). Both markers were inserted into presumed noncoding sites by site-directed homologous recombination (6).

studies in sugar beet, inocula were introduced as a seed dressing that naturally colonized the germinating seeds and developing plants. Through this route of inoculation it was assumed that the physiological condition and distribution in the phytosphere of the naturalized population of the introduced bacteria were similar to those of entirely indigenous populations (45, 58, 73, 74).

Careful consideration should also be made of the practical limitations imposed by the number of samples that can be collected, processed, and accurately analyzed with the resources available. For example, it is important to resolve whether it is more valid to increase the number of replicates (plots) at the expense of the number of individual samples (plants) or vice versa. To this end, and to better understand seasonal variation, we undertook two releases over consecutive years at the Wytham site, Oxford. These data were compared with data from the prerelease greenhouse trials (Fig. 2). In all cases the inocula were introduced as seed dressings (<10^7 CFU/seed). As with all field experiments, the design has to be robust and provide data suitable for statistical analysis. At this stage, consultation with a statistician was considered essential to optimize the sampling strategy and ensure factors such as sample independence. Problems of variation can be reduced by randomized sampling of plants and replication of "treatment blocks" (Fig. 2). For meaningful comparison, all samples were collected at the same time of day (early morning) on every sampling occasion to avoid diurnal influences (71). In the first field investigation in 1993, six phytosphere habitats were compared (immature, mature, and senescing leaves and rhizosphere soil, rhizoplane, and root cortex), but because of practical limitations during the second release in 1994, only three tissue types (mature leaf, immature leaf, and rhizoplane) were sampled from three individual plants taken from each of the three replicate treatment plots (n = 9 for each treatment). This regimen increased the power of the tests and allowed plant-to-plant variation to be determined.

57.4. PRACTICAL CONSIDERATIONS IN THE ASSESSMENT OF HORIZONTAL GENE FLOW

Because of their complexity, different components of risk evaluation take on paramount importance for various re-

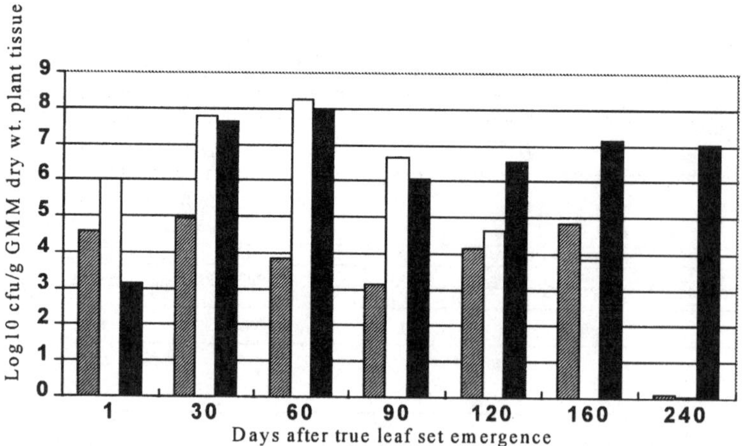

FIGURE 2 Phyllosphere population dynamics of *P. fluorescens* SBW25EeZY6KX; field and greenhouse sugar beet. Washed overnight cultures were introduced to pelleted sugar beet seeds and planted at 3 to 5 cm in field soil. Data for true emerging leaves (12 to 16 days after sowing) contrast colonization density and persistence of the inocula. Shaded bars, 1993; open bars, 1994; solid bars, greenhouse growth. Note that there were different population densities between 1993 and 1994, where local conditions at the time of seed germination affected preemptive colonization by inocula in the field (75). These data underline the need for comparative field studies to normalize natural variation and show that under appropriate conditions field conditions can be simulated in microcosms. Corrections for frost or rain were not made in the greenhouse. Sustained population densities observed in mature greenhouse-grown plants (>120 days) were not recorded in the field. In the greenhouse, inocula were still colonizing the phyllosphere up to the end of the experiment, 531 days after planting (74).

combinant microbes, while some risk components are common to nearly all GMMs. The risk of horizontal gene transfer should be considered in all cases. The role of mobile genetic elements and the mechanisms of environmental gene transfer (conjugation, transformation, and transduction) that contribute to gene flux between plant-associated microbial communities has been reviewed (4, 50). Genetic recombination, and more specifically the horizontal transfer of genes within and between bacterial populations, is commonly regarded as an important mechanism in the selective adaptation of bacteria to changes in the local environment. In soil and plant environments, conjugative transfer of autonomous plasmid replicons is probably the most important and best-studied means by which such potentially advantageous genetic determinants are mobilized (4, 41, 79). Studies of plant-associated microbial communities have demonstrated that bacteriophages may also contribute to gene flow between introduced populations (33).

A variety of laboratory-based soil-microcosm experiments have evaluated conjugal transfer among bacteria. Although plasmid transfer is generally a rare event in the oligotrophic environment of bulk soil, gene mobilization between introduced strains can be promoted by conditions that improve either their survival or metabolic activity, such as nutrient amendment or sterilization (50). Compared to the bulk soil, the root habitat is known to support higher densities and metabolically more active populations of bacteria and has been shown to promote plasmid transfer in microcosm and field studies (42, 43, 53, 60, 66, 82, 83).

Detailed investigations of bacterial population genetics and the role of mobile genetic elements in the movement of genetic information have revealed that the exchange between individuals is probably a frequent event. Our own findings demonstrate the apparent abundance and activity of mechanisms in natural habitats for the mobilization of DNA between populations. For example, large conjugative plasmids have been shown to be ubiquitous to, and move within, the pseudomonad community associated with the plant surface (44). The transfer of plasmids to recipient populations of bacteria can be stimulated by environmental factors (46). Troxler and colleagues (81) constructed fluorescent pseudomonad recipients and donors carrying Tn*5* and IncP plasmids functional in chromosomal mobilization activity and demonstrated that at high densities, transfer was possible in the rhizosphere of gnotobiotic plants. As discussed above, many releases have considered chromosomal gene transfer and have been monitored for the transfer of introduced markers. There have, however, so far been no reports of the in situ transfer of chromosomal DNA between bacteria during a field release. Therefore, it appears that the frequency of natural chromosomal gene transfer is low.

57.4.1. Inocula as Recipients of Naturally Transferred Genetic Elements

In any release experiment, a major potential source of genetic and phenotypic plasticity will be the acquisition of horizontally transferred DNA. We have reported studies focused on monitoring GMM acquisition of mobile genetic elements known to persist in the indigenous microflora (44). In these studies, the marked strain (*P. fluorescens* SBW25EeZY6KX) was introduced as a sugar beet seed dressing in a field experiment to study its ecology and genetic stability. This GMM colonized the phytosphere throughout the growing season and was used as a recipient "sink" to demonstrate the transfer of plasmids conferring mercuric resistance (Hg^r) from the indigenous community (43) in the absence of any deliberately imposed selection. Mercuric resistance was targeted as the selectable environmental phenotype because of its genetic diversity and common association with conjugative plasmids. Such plasmids are ubiquitously distributed, including environments such as our field site where the levels of mercury in soil are too low to impose positive selection for the carriage of mercuric resistance genes.

Transconjugant colonies were isolated from root and leaf samples collected within a 5-week midseason period of crop development. This result was replicated in two separate release experiments conducted over consecutive years. This study was limited to the isolation of transconjugants acquiring a single transferable phenotype (mercuric resistance) and probably therefore underestimated the extent of plasmid acquisition by *P. fluorescens* SBW25EeZY6KX in the phytosphere.

57.4.2. Potential Role of Acquired Plasmids in Adaptation and Survival of Inocula in the Phytosphere

While plasmid acquisition may impose burdens and reduce the fitness of GMMs in the environment, there is the possibility that GMM behavior may be modified. Following the experiments above in which the GMM acquired plasmids conferring mercuric resistance, a field release was undertaken to monitor the effects of carriage of one of these indigenous plasmids (pQBR103) on the ecology of the marked strain. While no difference in the tissues colonized or the patterns of colonization could be detected between the plasmid-bearing and plasmid-free strains, phytosphere population densities of *P. fluorescens* SBW25EeZY6KX declined significantly when carrying pQBR103. At about 100 days after planting, a simultaneous selection for plasmid-carrying hosts was observed in the phytosphere of field-grown plants. The recovery of these populations, to densities indistinguishable from those of plasmid-free inocula (4×10^5 CFU/g in the rhizosphere), demonstrated that plasmid pQBR103 confers a specific fitness advantage to host bacteria (42).

The key observations made demonstrate the capacity of plasmids to generate novel genotypes in GMM hosts that not only modify the behavior of transconjugants but also affect their ability to persist and proliferate. As with natural bacterial populations, the genetic and phenotypic stability of inocula released into the natural environment therefore cannot be predicted. This is highly relevant to the considerations and concerns voiced regarding the open release of GMMs, as considerable interest has been shown in the potential of manipulated genes to transfer from the donor and create novel phenotypes in the indigenous populations.

57.5. ASSESSING EFFICACY, AUTECOLOGY, AND IMPACT

Assessing the efficacy, autecology, and impact of a bacterial release requires the careful monitoring of GMM persistence, spread, interaction with other organisms, and effect on target and surrounding ecosystems. Where bacteria are released with specific traits such as biocontrol, evaluation of activity will be specific to the experiment. These data will need to be integrated with assessments of the level of persistence, spread, and activity of the released bacteria. The monitoring of persistence and spread will be resolved by the markers carried by the GMM and may need to continue at the release site for several years. It is necessary to consider the possibility of GMM relocation or changing habitat, including community succession, with their effect on the sustained activity of deliberately released inocula. These are important issues. The ability of iocula to persist and colonize requires that they survive and be able to exploit resources in competition with other residents or immigrants. Preemptive colonization leading to clonal expansion by inocula may establish a high population density which is rarely sustained. Nevertheless, it is probable that most habitats have evolved a specialized microbial community in which the temporal selection of dominant strains reflects local environmental changes that drive the succession of strains. In the case of SBW25EeZY6KX, we have shown that this strain can be efficiently relocated within the sugar beet crop by caterpillars (47) and that, as described above, the acquisition of indigenous plasmids can alter patterns of behavior (42).

A carefully chosen and marked release strain is unlikely to have major effects on ecosystem function. The monitoring of impact is more likely to be concerned with detecting the displacement of indigenous strains or detecting the effects of biotic interactions on microbial diversity. In the macro ecology of plants and animals, there is increasing evidence that the productivity, functioning, and sustainability of ecosystems may depend on their biological diversity (52, 77, 78). Although microbial activity is essential to ecosystem functioning and sustainability, it is less clear what effect perturbations (such as the release of bacteria) may have on community diversity and what effect in turn this may have on ecosystem productivity and stability. Within habitats, however, functional redundancy is commonly observed, and displacement of bacterial strains or altered diversity is not usually considered alarming.

There are a substantial number of useful non-GMM field studies considering the impact of perturbations on bacterial diversity. Diversity measurements have often been used in microbial ecology to study environmental stress (3, 32, 40, 55a). In many studies the control and perturbed habitats have been compared using diversity indices. Although various diversity indices may be used to characterize community structures, each makes assessments according to either relative species richness, evenness of species distribution, or a combination of the two. These diversity indices may also be applied to other taxonomic levels, or to data from sole-carbon-source utilization assays (24, 69).

A perennial method for describing diversity has been the characterization of bacterial isolates on agar media. One difficulty is the need for large sample sizes to provide ecologically significant data. Taxonomic characterization of substantial numbers of isolates has been reported by using sodium dodecyl sulfate-polyacrylamide gel electrophoresis (38), fatty acid methyl ester (FAME) analysis (45, 73), and physiological attributes. Particularly interesting is the work of Gilbert and colleagues (26), who detected perturbations in the rhizosphere bacterial community when soybean seeds were coated with the biological control (of damping off) agent *Bacillus cereus* UW85 before planting. From two field sites they isolated and characterized over 2,600 bacteria using 40 to 50 physiological attributes. By means of multivariate analysis of variance of the scores for each attribute test and discriminant analysis, they were able to detect perturbations in the communities of root-colonizing bacteria. Notably, they recorded the inoculum effect, even when the inoculated strain was no longer isolated, on the basis of physiological attributes. The bacteria were clustered, and the relative frequencies of these clusters were used to identify which clusters were most significant in discriminating among the communities (26).

Others have sought to assay changes in diversity from community fatty acid profiles (90) or to assay changes in community potential from patterns of collective carbon utilization and other physiological traits. For example, the impact of three forest-clearing regimens on forest soil microbial community structure were distinguished when the patterns of whole community sole-carbon-source utilization were recorded by inoculating total bacteria isolated into Biolog GN plates (see reference 31a).

During the field releases of *P. fluorescens* SBW25EeZY6KX, comparison were made between bacteria sampled from GMM-inoculated, wild-type inoculated, and uninoculated plants (5, 16). In the sugar beet studies (75) they were characterized to species and cluster level by FAME analysis, and the effects of the GMM on bacterial diversity were analyzed. In the early period of crop development, perturbations in community diversity were detected (75a). As the crop developed, these perturbations decreased until differences in the patterns of taxa colonizing GMM-treated and untreated plants were not detected. Using another approach, bacterial communities were characterized by the distribution of r (fast) and K (slow) growth strategists (1) indicated by seven classes defined by their time of appearance of colonies on agar media (21). These classes have been linked to a modified diversity (Shannon) index to produce an ecophysiological index to evaluate the perturbation following release of *P. fluorescens* SBW25EeZY6KX onto a wheat crop (18).

57.6. CONCLUDING REMARKS

In the last 5 years, considerable advances have been made in the understanding of the molecular basis of bacterial gene regulation and expression. These investigations have begun to unravel the complex interactions between microorganisms and their environment (11) and to demonstrate the presence of highly developed sensing systems that enable individuals and populations to adapt rapidly to changes in the local environment. An understanding of the genetic traits and ecology of microbes is fundamental for the successful development of functional microbial inoculants. In some instances an objective of the release may be to actively alter the ecological balance of the target habitat to favor the metabolic activity and proliferation of inocula over the indigenous microflora. This is especially relevant in the use of GMMs for bioremediation, biofertilization, or the control of plant disease. Identifying essential functions and related genes has led to the characterization of certain molecules necessary for biocontrol and rhizosphere competence. The description of the gene products, the related genes, and the regulatory factors that control expression in situ will lead to the development of more effective inocula. Such inocula will, by definition, be the products of genetic modification, either by the introduction of novel traits or by alterations that target expression to increase yield or direct the timing of expression to coincide with the appropriate environmental signal. All these factors improve efficacy.

Increases in efficacy are needed, because the field performance of microbial inocula has been inconsistent. The reasons for variable success include loss of ecological competence; spontaneous mutation either of the gene itself or of global regulatory genes that control the expression of secondary metabolites (e.g., many antibiotics and catabolic gene products); variable expression of the trait (introduced or nascent) due to local environmental conditions that impact on cellular metabolism; and poor survival, lack of fitness, and inability to compete (e.g., loss of rhizosphere competence). The application of molecular genetics in combination with detailed ecological studies may provide additional solutions, either through the prescribed control of the environment to enhance efficacy of inocula, or by the genetic modification of the inocula itself. The concern, therefore, lies with the impact of the functional organisms, the trait they express, and their ability to persist and proliferate in the environment. Even under extremes where targeted population densities are seen to fall, perturbations have to be large for an effect to be observed. An alternative is the direct evaluation of ecosystem function. One approach is that of measuring the relative activity of soil-microflora-associated enzymes (e.g., chitobiosidases, *N*-acetyl glucosaminidase, phosphatase, urease [54]), and another is the direct assessment of the rate of nutrient uptake by plants. We have successfully applied this approach by monitoring ^{15}N as a tracer (14).

However, it remains difficult to assess the effect that an introduced GMM has on the target and surrounding habitat. The accurate description of the diversity of a community, except at a very generalized functional level, remains beyond our current capabilities. But with the continued development of molecular methodologies, new and sensitive assays are being developed that register changes in community diversity and record perturbations to function. These techniques will not only enhance our understanding of microbial ecology but will also provide the tools to make more accurate assessments of risk. At present, to ensure reasonable survival of the applied inocula, most effective investigations have concentrated on the modification of indigenous bacteria isolated from habitats that are the same as or similar to those intended for the releases. With the advance in our ability to identify novel traits necessary for bacterial survival and activity in the natural environment, and with the ever-improving status of the practical applications of the principles of microbial ecology for improving inocula preparation, there will be an increasing demand for the use of GMMs. In the near future this demand will probably concentrate on their application in bioremediation and as replacements or adjuncts in plant protection and crop yield improvement. If the current progress in the development and cautious use of GMMs continues to demonstrate that no deleterious perturbations occur to a given habitat, then microbial inocula can contribute to superior strategies to solve field-based problems. In these advances the relevance of isolated natural isolates should not be forgotten. By applying molecular methodologies, more effective (preferred) "wild types" may be detected from the vast natural gene pool that will prove to be the imagined "super bugs" for prescribed function.

REFERENCES

1. **Andrews, J. H., and R. F. Harris.** 1985. r- and K-selection and microbial ecology. *Adv. Microb. Ecol.* **9:**99–147.
2. **Araujo, M. A. N., L. C. Mendonca-Hagler, A. N. Hagler, and J. D. van Elsas.** 1995. Competition between a

genetically modified *Pseudomonas fluorescens* and its parent in sub-tropical mirocosms. *Rev. Microbiol.* **26:**1–15.

3. **Atlas, R. M.** 1984. Use of microbial diversity measurements to assess environmental stress, p. 540–545. *In* M. J. Klug and C. A Reddy (ed.), *Current Perspectives in Microbial Ecology. American Society for Microbiology*, Washington, D.C.
4. **Bailey, M. J., A. K. Lilley, and J. D. Diaper.** 1996. Gene transfer in the phyllosphere, p. 103–123. *In* C. E. Morris, P. Nicot, and C. Nguyen (ed.), *Microbiology of Aerial Plant Surfaces*. Plenum Publishing Corp., New York.
5. **Bailey, M. J., A. K. Lilley, R. J. Ellis, P. A. Bramwell, and I. P. Thompson.** 1997. Microbial ecology, inocula distribution and gene flux within populations of bacteria colonising the surface of plants: case study of a GMM field release in the UK, p. 479–500. *In* J. D. Van Elsas, J. T. Trevors, and E. M. Wellington (ed.), *Modern Soil Microbiology*. Marcel Dekker, New York.
6. **Bailey, M. J., A. K. Lilley, I. P. Thompson, P. B. Rainey, and R. J. Ellis.** 1995. Site directed chromosomal marking of a fluorescent pseudomonad isolated from the phytosphere of sugar beet; stability and potential for marker gene transfer. *Mol. Ecol.* **4:**755–764.
7. **Barry, G. F.** 1988. A broad host range shuttle system for gene insertion into the chromosome of Gram-negative bacteria. *Gene* **71:**75–84.
8. **Bazin, M. J., and J. M. Lynch (ed.).** 1994. *Environmental Gene Releases: Models, Experiments and Risk Assessment*, p. 166. Chapman & Hall, London.
9. **Beattie, G. A., and S. E. Lindow.** 1994. Comparison of the behaviour of epiphytic fitness mutants of *Pseudomonas syringae* under controlled and field conditions. *Appl. Environ. Microbiol.* **60:**3799–3808.
10. **Bosworth, A. H., M. K. Williams, K. A. Albrecht, R. Kwiatkowski, J. Beynon, T. R. Hankinson, C. W. Ronson, F. Cannon, T. J. Wacek, and E. W. Triplett.** 1994. Alfalfa yield response to inoculation with recombinant *Rhizobium meliloti* with an extra copy of *dctABD* and/or modified *nifA* expression. *Appl. Environ. Microbiol.* **60:** 3815–3832.
11. **Bowen, G. D., and A. D. Rovira.** 1976. Microbial colonization of plant roots. *Annu. Rev. Phytopathol.* **14:**121–144.
12. **Bramwell, P. A., R. V. Barallon, H. J. Rogers, and M. J. Bailey.** 1995. Extraction of DNA from the phylloplane, p. 56–77. *In* A. D. L. Akkermans, J. D. Van Elsas, and F. J. DeBruijn (ed.), *Molecular Microbial Ecology Manual*. Kluwer Academic Publishers, Dordrecht, The Netherlands.
13. **Breitenbeck, G. A., and T. Hankinson.** 1992. Dispersal and persistance of recombinant *Bradyrhizobium japonicum*. *Agron. J.* **84:**259.
14. **Brimecombe, M. J., F. A. A. M. De Leij, and J. M. Lynch.** 1998. Effect of genetically modified *Pseudomonas fluorescens* on the uptake of nitrogen of pea plants from ^{15}N enriched organic residues. *Lett. Appl. Microbiol.* **26:** 155–160.
15. **De Leij, F. A. A. M., M. J. Bailey, J. M. Lynch, and J. M. Whipps.** 1993. A simple most probable number technique for the sensitive recovery of a genetically engineered *Pseudomonas aureofaciens* from soil. *Lett. Appl. Microbiol.* **16:**307–310.
16. **De Leij, F. A. A. M, M. J. Bailey, J. M. Whipps, I. P. Thompson, P. A. Bramwell, and J. M. Lynch.** 1998. Release of genetically modified micro-organisms and biomonitoring, p. 70–100. *In* J. M. Lynch and A. Wiseman (ed.), *Environmental Biomonitoring. The Biotechnology Ecotoxicology Interface*. Cambridge University Press, Cambridge.
17. **De Leij, F. A. A. M., E. J. Sutton, J. M. Whipps, J. S. Fenlon, and J. M. Lynch.** 1995. Field release of a genetically modified *Pseudomonas fluorescens* on wheat: establishment, survival and dissemination. *Bio/Technology* **13:** 1488–1992.
18. **De Leij, F. A. A. M., E. J. Sutton, J. M. Whipps, J. S. Fenlon, and J. M. Lynch.** 1995. Impact of field release of genetically modified *Pseudomonas fluorescens* on indigenous microbial populations of wheat. *Appl. Environ. Microbiol.* **61:**3443–3453.
19. **De Leij, F. A. A. M., E. J. Sutton, J. M. Whipps, and J. M. Lynch.** 1994. Spread and survival of a genetically modified *Pseudomonas aureofaciens* in the photosphere of wheat and soil. *Appl. Soil Ecol.* **1:**207–218.
20. **De Leij, F. A. A. M., C. E. Thomas, M. J. Bailey, J. M. Whipps, and J. M. Lynch.** 1998. Effect of insertion site and metabolic load on the environmental fitness of genetically modified *Pseudomonas fluorescens*. *Appl. Environ. Microbiol.* **64:**2634–2638.
21. **De Leij, F. A. A. M., J. M. Whipps, and J. M. Lynch.** 1994. The use of colony development for the characterisation of bacterial communities in soil and on roots. *Microb. Ecol.* **27:**81–97.
22. **Drahos, D. J.** 1991. Field testing of genetically engineered micro-organisms. *Biotechnol. Adv.* **9:**157–171.
23. **Drahos, D. J., G. F. Barry, B. C. Hemmin, E. J. Brandt, E. L. Kline, H. D. Skipper, D. A. Kluepfel, D. T. Gooden, and T. A. Hughes.** 1992. Spread and survival of genetically marked bacteria in situ, p. 147–159. *In* J. C. Fry and M. J. Day (ed.), *Release of Genetically Engineered and Other Micro-Organisms*. Cambridge University Press, Cambridge.
24. **Ellis, R. J., I. P. Thompson, and M. J. Bailey.** 1995. Metabolic profiling as a means of characterising plant-associated microbial communities. *FEMS Microbiol. Ecol.* **16:**9–18.
25. **Fry, J. C., and M. J. Day (ed.).** 1992. *Release of Genetically Engineered and Other Micro-organisms*, p. 178. Cambridge University Press, Cambridge.
26. **Gilbert, G. S., M. K. Clayton, J. Handelsman, and J. L. Parke.** 1996. Use of cluster and discriminant analysis to compare rhizosphere bacterial communities following biological perturbation. *Microb. Ecol.* **32:**123–147.
27. **Herrero, M., V. DeLorenzo, and K. N. Timmis** 1990. Transposon vectors containing non-antibiotic resistance selection markers for cloning and stable chromosomal insertion in gram-negative bacteria. *J. Bacteriol.* **172:**6557–6567.
28. **Hirano, S. S., D. K. Willis, and C. D. Upper.** 1992. Population dynamics of Tn-5 induced non-lesion forming mutant of *Pseudomonas syringae* pv. *syringae* on bean plants in the field. *Phytopathology* **82:**1067.
29. **Hirsch, P. R., and J. D. Spokes.** 1994. Survival and dispersion of genetically modified rhizobia in the field and genetic interactions with native strains. *FEMS Microbiol. Ecol.* **15:**147–160.

30. **Hurst, C. J., G. Knudsen, M. J. McInerney, L. D. Stetzenbach, and M. V. Walter (ed.).** 1997. *Manual of Environmental Microbiology*. American Society for Microbiology, Washington D.C.

31. **Jones, D. A., and A. Kerr.** 1989. *Agrobacterium radiobacter* K1026, a genetically engineered derivative of a strain K84, for biological control of crown gall. *Plant Disease* **73:**15–18.

31a. *Journal of Microbiological Methods*. Special issue. **30.**

32. **Kennedy, A. C., and K. L. Smith.** 1995. Soil microbial diversity and the sustainability of agricultural soils. *Plant and Soil* **170:**75–86.

33. **Kidambi, S. P., S. Ripp, and R. V. Miller.** 1994. Evidence for phage-mediated gene transfer among *Pseudomonas aeruginosa* strains on the phylloplane. *Appl. Environ. Microbiol.* **60:**496–500.

34. **Kloepper J. W., and M. N. Schroth.** 1981. Relationship of in vitro antibiosis of plant growth-promoting rhizobacteria to plant growth and the displacement of root microflora. *Phytopathology* **71:**1020–1024.

35. **Kluepfel, D. A., E. J. Kline, H. Skipper, T. A. Hughes, and D. T. Gooden.** 1991. The release and tracking of genetically engineered bacteria in the environment. *Phytopathology* **81:**348–152.

36. **Kostka, S. J.** 1990. The design and execution of successive field releases of genetically engineered microorganisms, p. 167–176. *In* D. R. Mackenzie and S. C. Henry (ed.), *Biological Monitoring of Genetically Engineered Plants and Microbes*. Agricultural Research Institute, Bethesda, Md.

37. **Krimsky, S., R. G. Wrubel, I. G. Naess, S. B. Levy, R. E. Wetzler, and B. Marshall.** 1995. Standardised microcosms in risk assessment. *BioScience* **45:**590–599.

38. **Lambert, B., P. Meire, H. Joos, P. Lens, and J. Swings.** 1990. Fast-growing, aerobic, heterotrophic bacteria from the rhizosphere of young sugar-beet plants. *Appl. Environ. Microbiol.* **56:**3375–3381.

39. **Legard, D. E., M. P. McQuilken, J. M. Whipps, J. S. Fenlon, T. R. Fermor, I. P. Thompson, M. J. Bailey, and J. M. Lynch.** 1994. Studies of seasonal changes in the microbial populations in the phyllosphere of spring wheat; a prelude to the release of a genetically modified microorganism. *Agric. Ecosystems Environ.* **50:**87–101.

40. **Leung, K., L. S. England, M. B. Cassiday, J. T. Trevors, and S. Weir.** 1994. Microbial diversity in soil: effect of releasing genetically engineered micro-organisms. *Mol. Ecol.* **3:**413–422.

41. **Levy, S. B., and R. V. Miller (ed.).** 1989. *Gene Transfer in the Environment*. McGraw-Hill Book Company, New York.

42. **Lilley, A. K., and M. J. Bailey.** 1997. Impact of pQBR103 acquisition and carriage on the phytosphere fitness of *Pseudomonas fluorescens* SBW25: burden and benefit. *Appl. Environ. Microbiol* **63:**1584–1587.

43. **Lilley, A. K., and M. J. Bailey.** 1997. The acquisition of indigenous plasmids by a genetically marked pseudomonad population colonising the phytosphere of sugar beet is related to local environmental conditions. *Appl. Environ. Microbiol.* **63:**1577–1583.

44. **Lilley, A. K., M. J. Bailey, M. J. Day, and J. C. Fry.** 1996. Diversity of mercury resistance plasmids obtained by exogenous isolation from the bacteria of sugar beet in three successive seasons. *FEMS Microbiol. Ecol.* **20:**211–228.

45. **Lilley, A. K., J. C. Fry, M. J. Bailey, and M. J. Day.** 1996. Comparison of aerobic heterotrophic taxa isolated from four root domains of mature sugar beet (*Beta vulgaris*). *FEMS Microbiol. Ecol.* **21:**231–242.

46. **Lilley, A. K., J. C. Fry, M. J. Day, and M. J. Bailey.** 1994. In situ transfer of an exogenously isolated plasmid between *Pseudomonas* spp in sugar beet rhizosphere. *Microbiology* **140:**27–33.

47. **Lilley, A. K., R. S. Hails, J. S. Cory, and M. J. Bailey.** 1997. The dispersal and establishment of pseudomonad populations in the phyllosphere of sugar beet by phytophagous caterpillars. *FEMS Microbiol. Ecol.* **24:**151–158.

48. **Lindow, S. E.** 1991. Determinants of epiphytic fitness in bacteria, p. 295–314. *In* J. H. Andrews and S. S. Hirano (ed.), *Microbial Ecology of Leaves*. Springer-Verlag, New York.

49. **Lindow, S. E., and N. J. Panpoloulos.** 1988. Field tests of recombinant ice$^-$ *Pseudomonas syringae* for biological frost control in potato, p. 121–138. *In* M. Sussman, C. Collins, F. Skinner, and D. Stuart-Tull. (ed.), *Release of Genetically Engineered Micro-organisms*. Academic Press, San Diego, Calif.

50. **Lorenz, M. G., and W. Wackernagel.** 1995. Bacterial gene transfer by natural genetic transformation in the environment. *Microbiol. Rev.* **58:**563–602.

51. **Mahaffee, W. F., and J. W. Kloepper.** 1997. Bacterial communities of the rhizosphere and endorhiza associated with field grown cucumber plants inoculated with a plant growth promoting rhizobacterium or its genetically modified derivative. *Can. J. Microbiol.* **43:**344–353.

52. **Naeem, S., L. J. Thompson, S. P. Lawler, J. H. Lawton, and R. M. Woodfin.** 1994. Declining biodiversity can alter the performance of ecosystems. *Nature* **368:**734–737.

53. **Naik, G. A., L. N. Bhat, B. A. Clopacle and J. M. Lynch.** 1994. Transfer of broad host range antibiotic resistance plasmids in soil microcosms. *Curr. Microbiol.* **28:** 209–215.

54. **Naseby, D. C., and J. M. Lynch.** 1997. Rhizosphere and soil enzymes as indicators of perturbations caused by enzyme substrate addition and inoculation of a genetically modified strain of *Pseudomonas fluorescens* on wheat soil. *Soil Biol. Biochem.* **29:**1353–1362.

55. **Parke, J. L., R. M. Zablotowicz, and R. E. Rand.** 1992. Tracking a *lacZY* marked strain of *Pseudomonas fluorescens* in a Wisconsin field trial. *Phytopathology* **82:**1177–1178.

55a. *Plant and Soil*. Special issue. **170.**

56. **Prosser, J. I.** 1994. Molecular marker systems for detection of genetically engineered micro-organisms in the environment. *Microbiology* **140:**5–17.

57. **Rainey, P. B., and M. J. Bailey.** 1996. Physical and genetic map of the *Pseudomonas fluorescens* SBW25 chromosome. *Mol. Microbiol.* **19:**521–533.

58. **Rainey, P. B., M. J. Bailey, and I. P. Thompson.** 1994. Phenotypic and genotypic diversity of fluorescent pseudomonads isolated from field grown sugar beet. *Microbiology* **140:**2315–2331.

59. **Ramos, R. L., E. Duque, and M. I. Ramos-Gonzalez.** 1991. Survival in soils of an herbicide-resistant *Pseudomonas putida* bearing a recombinant TOL plasmid. *Appl. Environ. Microbiol.* **57:**260–266.

60. **Richaume, A., E. Smit, G. Faurie, and J. D. van Elsas.** 1992. Influence of soil type on the transfer of plasmid Rp4(p) from *Pseudomonas fluorescens* to introduced recipient and to indigenous bacteria. *FEMS Microbiol. Ecol.* **101:**281–292.

61. **Ryder, M.** 1994. Key issues in the deliberate release of genetically manipulated bacteria. *FEMS Microbiol. Ecol.* **15:**139–146.

62. **Seidler, R. J.** 1994. Evaluation of methods for detecting ecological effects from genetically modified microorganisms and microbial pest control agents in terrestrial systems, p. 99–122. *In* M. J. Bazin and J. M. Lynch (ed.), *Environmental Gene Releases: Models, Experiments and Risk Assessment.* Chapman & Hall, London.

63. **Seidler, R. J., and J. Settle (ed.).** 1991. *The Use and Development of Environmentally Controlled Chambers (Mesocosms) for Evaluating Biotechnological Products.* US EPA/600/9-91/013. U.S. Environmental Protection Agency, Washington, D.C.

64. **Shaw, J. J., F. Dane, D. Geiger, and J. W. Kloepper.** 1992. Use of bioluminescence for detection of genetically engineered micro-organisms released into the environment. *Appl. Environ. Microbiol.* **58:**267–273.

65. **Short, K. A., R. J. Seidler, and R. H. Olsen.** 1990. Competitive fitness of *Pseudomonas putida* induced or constitutively expressing plasmid-mediated degradation of 2,4-dichlorophenoxyacetic acid (TED) in soil. *Can. J. Microbiol.* **36:**821–826.

66. **Smit, E., J. D. van Elsas, J. A. van Veen, and W. M. Devos.** 1991. Detection of plasmid transfer from *Pseudomonas fluorescens* to indigenous bacteria in soil by using bacteriophage Phi-R2F for donor counter selection. *Appl. Environ. Microbiol.* **57:**3482–3488.

69. **Staddon, W. J., L. C. Duchesne, and J. T. Trevors.** 1997. Microbial diversity and community structure of post-disturbance forest soils as determined by sole-carbon-source utilisation patterns. *Microb. Ecol.* **34:**125–130.

70. **Stotzky, G.** 1990. *Methods to Measure the Influence of Genetically Engineered Bacteria on Ecological Processes in Soil.* US EPA/600/3-90/011. U.S. Environmental Protection Agency. Washington, D.C.

71. **Thompson, I. P., M. J. Bailey, R. J. Ellis, A. K. Lilley, P. J. McCormack, K. J. Purdy, and P. B. Rainey.** 1995. Short term community dynamics in the phyllosphere microbiology of field grown sugar beet. *FEMS Microbiol. Ecol.* **16:**205–211.

72. **Thompson I. P., M. J. Bailey, R. J. Ellis, and K. J. Purdy.** 1993. Subgrouping of bacterial populations by cellular fatty acid composition. *FEMS Microbiol. Ecol.* **102:**75–84.

73. **Thompson, I. P., M. J. Bailey, J. S. Fenlon, T. R. Fermor, A. K. Lilley, J. M. Lynch, P. J. McCormack, M. P. McQuilken, K. J. Purdy, P. B. Rainey, and J. M. Whipps.** 1993. Quantitative and qualitative seasonal-changes in the microbial community from the phyllosphere of sugar-beet (*Beta-vulgaris*). *Plant and Soil* **150:**177–191.

74. **Thompson, I. P., R. J. Ellis, and M. J. Bailey.** 1995. Autecology of a genetically modified fluorescent pseudomonad on sugar beet. *FEMS Microbiol. Ecol.* **17:**1–14.

75. **Thompson, I. P., A. K. Lilley, R. J. Ellis, P. A. Bramwell, and M. J. Bailey.** 1995. Survival, colonisation and dispersal of genetically modified *Pseudomonas fluorescens* SBW25 in the phytosphere of field grown sugar beet. *Bio/Technology* **13:**1493–1497.

75a. **Thompson, I. P., A. K. Lilley, R. J. Ellis, and M. J. Bailey.** Unpublished results.

76. **Tiedje J. M., R. K. Colwell, Y. L. Grossman, R. E. Hodgson, R. E. Lenski, R. N. Mack, and P. J. Regal.** 1989. The planned introduction of genetically modified organisms: ecological considerations and recommendations. *Ecology* **70:**298–315.

77. **Tilman, D.** 1996. Biodiversity: population versus ecosystem stability. *Ecology* **72:**350–363.

78. **Tilman, D., D. Wedin, and J. Knops.** 1996. Productivity and sustainability influenced by biodiversity in grassland ecosystems. *Nature* **379:**718–720.

79. **Trevors, J. T., T. Barkay, and A. W. Bourquin.** 1987. Gene-transfer among bacteria in soil and aquatic environments: a review. *Can. J. Microbiol.* **33:**191–198.

80. **Triplett, E. W., and M. J. Sadowsky.** 1992. Genetics of competition for nodulation of legumes. *Annu. Rev. Microbiol.* **46:**399–428.

81. **Troxler, J., P. Azelvandre, M. Zala, G. Defago, and D. Haas.** 1997. Conjugative transfer of chromosomal genes between fluorescent pseudomonads in the rhizosphere of wheat. *Appl. Environ. Microbiol.* **63:**213–219.

82. **van Elsas, J. D., M. E. Starodub, and J. T. Trevors.** 1988. Bacterial conjugation between Pseudomonads in the rhizosphere of wheat. *FEMS Microbiol. Ecol.* **53:**299–306.

83. **van Elsas, J. D., L. S. van Overbeek, and M. Nikkel.** 1989. Detection of plasmid RP4 transfer in soil and rhizosphere, and the occurrence of homology to RP4 in soil bacteria. *Curr. Microbiol.* **19:**375–381.

84. **van Overbeek, L. S., and J. D. van Elsas.** 1995. Root exudate promoter activity in *Pseudomonas fluorescens* mutants in the wheat rhizosphere. *Appl. Environ. Microbiol.* **61:**890–898.

85. **van Veen, J. A., and C. E. Heijnen.** 1994. The fate and activity of micro-organisms introduced into soil, p. 63–71. *In* B. M. Pankhurst (ed.), *Soil Biota.* CSIRO, East Melbourne, Australia.

86. **Weller, D. M.** 1983. Colonization of wheat roots by a fluorescent pseudomonad suppressive to take-all. *Phytopathology* **73:**1548–1553.

87. **Wilson M., and S. E. Lindow.** 1993. Effect of phenotypic plasticity on epiphytic survival and colonisation by *Pseudomonas syringae.* *Appl. Environ. Microbiol.* **59:**410–416.

88. **Wilson, M., and S. E. Lindow.** 1993. Release of recombinant micro-organisms. *Annu. Rev. Microbiol.* **47:**913–944.

89. **Yuen, G. Y., and M. N. Schroth.** 1986. Interactions of *Pseudomonas fluorescens* strain E6 with ornamental plants and its effects on the composition of root-colonizing microflora. *Phytopathology* **76:**176–180.

90. **Zelles, L., R. Rackwitz, Q. Y. Bai, T. Beck, and F. Beese.** 1995. Discrimination of microbial diversity by fatty acid profiles of phospholipids and lipopolysaccharides in differently cultivated soils. *Plant and Soil* **170:**115–122.

Biofilms and Biocorrosion

MADILYN FLETCHER

58

Whenever solid surfaces occur in environments containing microorganisms, some of those organisms will attach and become immobilized. If conditions favor growth, cells replicate, frequently extracellular polysaccharides are produced, and such colonization of the surface results in the formation of a multicellular layer of organisms, termed a biofilm. In nature, biofilms contain a diverse assemblage of both micro- and macroorganisms, which are spatially and temporally distributed according to functional relationships, such as mutualism, competition, antagonism, and predation. Biofilms also occur on surfaces in laboratory cultures, although they may be less noticeable and are often ignored. However, those organisms colonizing the culture vessel surfaces are likely to differ physiologically from their freely suspended counterparts (57).

There is an increasing interest in the formation, composition, and function of microbial biofilms because of their significance in a wide range of industrial, clinical, and commercial situations. Biofilms interfere with the function and efficiency of pipelines, heat exchangers, and the movement of ships through water. They provide reservoirs and protective habitats for pathogens on implants and prosthetic devices, on food processing surfaces, and in drinking water systems. Biofilms appear to contribute to corrosion processes on metals such as iron and steel (42), manganese (44), and copper (36). Attached bacteria or microbial biofilms have also been associated with the degradation of rocks, such as marble (156), or the oxidation of minerals, such as arsenopyrite by *Thiobacillus ferrooxidans* (50). Biofilms can also be advantageous, such as in trickling filter systems or possibly through exclusion of other, deleterious types of organisms (30).

Recent years have seen new approaches for analyzing biofilm formation and composition, particularly for real-time characterization and molecular analysis. These have resulted in significant advances in our understanding of functional and spatial relationships among organisms in biofilms, and have considerably increased the potential for developing strategies for their control. Advances in the development of molecular probes are allowing identification of microorganisms and their distributions, and are beginning to provide information on physiological activities. Image and computer-enhanced analysis has allowed measurements of the kinetics of colonization processes, while scanning confocal laser microscopy and atomic force microscopy are enabling three-dimensional characterization of biofilms in real time. This chapter focuses on two principal types of biofilm studies: (i) those focusing on the initial stages of attachment and the factors that control adhesion to solid surfaces, and (ii) studies on the composition, successional development, and spatial and functional relationships of constituent organisms.

58.1. COLONIZATION OF SURFACES

For over 25 years, researchers have investigated the attachment of microorganisms to test surfaces to determine the adhesive properties of different species or types of microorganisms, to identify specific substratum chemistries that resist colonization, and to determine environmental conditions, e.g., temperature, pH, and electrolyte concentration, that control the attachment process. Although there have been numerous attempts to identify fundamental processes that control attachment, a wealth of information has illustrated the high degree of diversity among bacterial adhesion strategies and mechanisms. Thus, the design of any adhesion study is extremely important, and the conditions must be tightly defined and controlled to suit the specific question being addressed.

58.1.1. Adhesion Properties of Microorganisms

The number of bacteria that attach to surfaces and the rates of their attachment depend not only on the species, but also on the strain and on its physiological phenotype. Adhesion properties can be influenced by a range of environmental factors, either through their effect on the physiology of the organism or on the physiochemistry of the adhesion interaction. Thus, not only must test organisms be selected on the basis of their being representative of the environment under study, but attention must be paid to whether adhesion phenotypes have been altered through laboratory selection or growth conditions. Attachment properties can be affected by nutrient sources, concentration, and flux (49, 88, 109, 117). In batch culture, adhesion is sometimes affected by the growth phase (132, 144); with some organisms, attachment is greater in log phase, whereas with others, attachment properties increase into stationary phase. Mutations or phase variations in cell surface components, such as the fibrillar layer on *Streptococcus salivarius* (163) or in lipopolysaccharide in *Pseudomonas* species (103, 135, 170), have been shown to alter adhesion characteristics. Altered adhesion phenotypes can be selected by serial passaging in liquid medium (34), by repeated transfer through porous media columns (41), or by collection on surfaces in a low-dilution-rate chemostat (136). Factors that influence extracellular polymer production can also influence adhesion. Carbon source, carbon-nitrogen ratio, carbon flux, or nutrient concentration may influence polymer production or adhesive ability in various ways, depending on the organism (105, 109, 117).

58.1.2. Selection of Solid Surfaces for Experimentation

Numerous studies have demonstrated that numbers of cells and strength of binding are influenced by surface properties, particularly electrostatic charge (80, 140), surface free energy (80, 154, 160, 168, 169) and a related parameter, hydrophobicity (133, 154, 168, 169), and texture (10). Cleaning procedures are extremely important for experiments with chemically defined surfaces, as high-energy surfaces such as glass are easily contaminated by adsorbents in solution or the atmosphere. Such surfaces should be cleaned thoroughly with organic solvents and/or concentrated inorganic acid (1), and rinsed thoroughly in double-distilled water. Autoclaving should not be used for sterilization to avoid deposition of volatile compounds. Commercially available surfaces, e.g., glass, organic polymers, and metals, have been commonly used as test substrata. Attempts have also been made to alter adhesion by chemical modification of polymeric materials, such as by glow discharge treatment of plastics to make them more water-wettable (80).

More recently, highly defined, chemically homogeneous test substrata have been produced from alkanethiol self-assembled monolayers (SAMs) (78, 168, 169). SAMs comprise monolayers of alkanethiols, which self-assemble along the length of the alkane chains and bind in an orderly array to a metal (usually gold)-coated substratum by the thiol group at one end of the alkane (9). The chemistry of the surface exposed to the surrounding medium containing the bacteria is determined by the functional groups at the terminal ends of the alkanes. So far, SAMs with terminal hydroxyl and methyl (168, 169) and oligo (ethylene glycol) (78) have demonstrated that adhesion is considerably reduced on highly hydrophilic surfaces containing hydroxyl or oligo (ethylene glycol) groups, but the numbers of cells attached depend to some extent on the test organisms. There is considerable potential for the use of SAMs as highly homogeneous, chemically defined test substrata, since a range of terminal functional groups are commercially available or can be synthesized.

58.1.3. Static Systems versus Flow Systems

The simplest approach for exposing test substrata to bacteria is to place the test surface horizontally or vertically in a static system, or one with controlled agitation or shaking (53). Devices are needed to hold test surfaces in position (134, 160) and can be designed to suit the test surface size and volume of cell suspension. Microtiter plates have also been used for larger scale, multiple-surface adhesion assays (33). However, we have found a large degree of variability in microtiter plate wells from the standpoint of numbers of attached bacteria, and replication of results may be hard to achieve. Other devices have been designed to maintain different rates of laminar flow conditions, such as the rotating disc apparatus (2), radial flow reactor (58), or rotating annular reactor (31). After the adhesion test period, care must also be given to fixation, staining, and processing of attached bacteria for determination of attached biomass. If binding is weak, cells can be arbitrarily removed or redistributed on surfaces by erratic rinsing procedures, so these must be carefully controlled and measured (56).

58.1.4. Measurement of Attached Numbers or Biomass

Numbers or biomass of attached bacteria can be measured by destructively sampling test surfaces and evaluating them by microscopy, biochemical analysis of biomass components, radiolabeling cells, or applying molecular probes. Alternatively, attachment numbers and rates can be evaluated in real time by on-line computer analysis (see below). Most often, attached cell numbers have been determined by brightfield or fluorescent microscopy (161) combined with staining of cells (53) or polymers (3). Alternatively, phase-contrast, differential interference contrast (162), and interference reflection microscopy (55) have been used to visualize attached cells. Cells can be stained with standard stains, such as crystal violet, or with fluorochromes, e.g., acridine orange (54, 111, 128), 4′6-diamidino-2-phenylindole (DAPI) (77, 131), Hoechst dyes 33258 or 33342 (128), or 5-cyano-2, 3-ditolyl tetrazolium chloride (CTC) (77, 142, 149). By combinations of stains, estimates can also be made of relative activities of cells. For example, information on activity of attached *Klebsiella pneumoniae* was provided by intracellular accumulation of CTC, a soluble redox indicator reduced by respiring bacteria to fluorescent CTC-formazan crystals, and staining by rhodamine 123 (Ph 123), which can be incorporated into bacteria in relation to proton motive force (171).

Numerous cell constituents have been used as biochemical markers of attached cell biomass. These methods vary in their specificity and sensitivity, and the amount of biomass present is often a major criterion when selecting an appropriate analysis. Valuable information on viable biomass can be obtained by quantitative analysis of total phospholipid ester-linked fatty acids (PLFA), which occur in the intact membrane of viable cells (167). After death, the phosphate group is hydrolyzed, leaving diglyceride with the same signature fatty acids as the phospholipids. Thus, comparison of ratios of phospholipid fatty acids to diglyceride fatty acids gives an indication of relative values of viable and nonviable biomass. Indicators of physiological status include poly-β-hydroxyalkanoate (PHA) in bacteria (52, 122) or triglyceride in microeukaryotes (64), when quantified in relationship to PLFA. An increase in specific *trans* monoenoic PLFA, compared to *cis* isomers, is believed to be an indication of physiological stress (69, 75). These approaches require appropriate instrumentation and expertise for interpreting results but provide sensitive evaluations of total community biomass and structure.

Biomass of attached cells has been indirectly measured by extracting ATP and quantifying it by the luciferin/luciferase reaction (72). DNA can be quantified by fluorometry with Hoechst staining (129). Gram-negative bacteria can be measured as lipopolysaccharide using the *Limulus* amoebocyte lysate (43). Other, less commonly used markers are lipid A of lipopolysaccharide (127), teichoic acids in gram-positive bacteria (63), and muramic acid (51). Immunochemical techniques may be used for specific organisms, such as enzyme-linked lectinsorbent assay for measurement of *Staphylococcus epidermidis* biofilms. This method is based on a phosphatase-labeled wheat germ agglutinin, which bound with GlcNAc β-1, 4n, a component of the extracellular biofilm polymer (155).

Sensitive and accurate determination of attached cell biomass can be achieved by prelabeling cells with a radioisotope and precalibrating radioactivity with some other measure of biomass, such as direct or viable counts. Effective substrates for incorporation of radioisotopes have been ^{14}C- or ^{3}H-leucine (110) or ^{14}C-acetate (113).

58.1.5. Bacterial Surface Properties and Adhesives

Attempts have been made to identify the surface characteristics of bacteria that determine adhesiveness. Because of the roles that electrostatic charge and surface energy interactions, including hydrophobicity (143, 158), are thought to play in attachment mechanisms, methods have been developed to assess overall cell surface charge, surface free energy, or hydrophobicity of microorganisms. Hydrophobic interaction chromatography (HIC) (102) and electrostatic interaction chromatography (130) provide relative values of cell hydrophobicity and surface charge, respectively. Cell surface charge has also been assessed by electrophoretic potential with dedicated instrumentation (16, 102, 158). For example, negative mobility of *Desulfovibrio desulfuricans* was found to approach 0 with growth in batch culture and correlated with cell aggregation and biofilm formation on mild steel (16). Bacterial adherence to hydrocarbons (BATH), in which partitioning of a suspension of microorganisms between an aqueous phase and a hydrocarbon (e.g., hexane, hexadecane) phase is determined by optical density changes, is easy to apply and has been commonly used (11, 143, 154). Approximation of thermodynamic parameters, e.g., surface free energy or critical surface tension, has been obtained by measuring the contact angles of water or other liquids on lawns of cells (20, 140, 159), although the degree to which drying of the cell lawn influences the interpretation of the measured angle is not yet clear. Attempts have been made to identify correlations between the different approaches for measuring hydrophobicity. For example, a comparison of BATH, HIC, and the salt aggregation test under different growth conditions demonstrated good agreement between the salt aggregation test and BATH, whereas a weak correlation was obtained between HIC and other test methods (45). In a separate comparison of BATH, HIC, salt aggregation, and contact angle measurements, there was little correlation among methods when applied to 29 streptococci strains (157). Thus, the significance of hydrophobicity measurements is not clear, and at best such measurements provide only relative values for different organisms or physiological states of a given organism.

Attempts have also been made to identify the specific polymers involved in adhesion. Molecular approaches have enabled enormous progress in the identification of bacterial adhesins involved in specific receptor-ligand interactions in bacterial-host systems. Progress has been slower with the adhesives enabling adhesion to nonbiological surfaces, presumably because of the lack of specificity and the diversity of the polymers involved. Three genes involved in the production of the holdfast proteins of *Caulobacter crescentus* have been identified (92). Transposon (Tn*917*) mutagenesis of *S. epidermidis* demonstrated that a 60-kDa protein located at the cell surface was required for initial attachment to polystyrene surfaces and that this protein was probably an autolysin (74). Similarly, transposon mutagenesis has been used with freshwater and soil isolates of *Pseudomonas fluorescens* to demonstrate the involvement of flagella in adhesion to soil particles (40, 41) and of lipopolysaccharide in adhesion to polystyrene and sand particles (170).

Some information on polymer characteristics was provided by interference reflection microscopy, a technique that utilizes destructive or constructive interference between light reflected from the bacterial and substratum surfaces to visualize separation distance between the two surfaces (65, 66, 106). Exposure of attached pseudomonads to various electrolyte solutions and observation of corresponding changes in separation distance indicated that the adhesive polymers, presumably polysaccharides, condensed when exposed to electrolytes, probably because of screening or cross-linking of constituent anionic groups (55, 106).

58.1.6. Studies of Bacterial Detachment

Some investigations have focused on the release, detachment, or desorption of cells from surfaces. The simplest approach has been to determine numbers of released cells after a surface with attached microorganisms has been transferred to a cell-free solution (110). One approach was to measure resulting numbers of viable counts in the medium containing hydroxylapatite rods with attached *Streptococcus mutans*, following addition of an endogenously produced enzyme that induces release of the organisms' own surface proteins (96). More detailed, real-time data on desorption rates and residence times of cells have also been determined by applying image analysis to flow cell studies of an estuarine isolate (169).

58.2. BIOFILM CHARACTERIZATION

Until recently, it has been extremely difficult to probe the three-dimensional structure of biofilms without destructive sampling, so limited information on spatial relationships and dynamic responses to environmental conditions was available. With the introduction of scanning confocal laser microscopy, new insights have been realized about the morphology of biofilms. When combined with molecular and physiological probes, new imaging techniques are providing information on functional and spatial relationships never before possible.

58.2.1. Real-Time Measurements

58.2.1.1. Flow Chambers

For real-time computer analysis, a variety of flow chambers have been employed (19, 47, 87, 148, 152, 168). These range from chambers machined for laminar flow (113, 115, 168), to capillary tubes (145), to chambers constructed from coverslips and microscope slides (27). Some flow chambers, such as the radial flow reactor, have been designed to control, measure and modify shear stresses on developing biofilms (114) (see also above). Such controlled studies have demonstrated that biofilm characteristics can be altered by shear stresses; for example, higher C:N ratios and total fatty acids in *Pseudomonas atlantica* biofilms were associated with higher shear stresses (114).

By interfacing microscopic imaging with computer analysis, numerical data on attached cell numbers, area coverage, and biovolume of attached cells can be accurately measured over time (19, 23, 27, 94, 95). By capturing successive images, such as at 2-min intervals (168), data can be analyzed to determine rates of deposition, rates of desorption, and residence time of cells (168, 169). Patterns of growth of organisms on surfaces have also been tracked. For example, growth of an *Acinetobacter* sp. in different nutrient conditions was monitored (79), and growth of a marine strain on hydrophobic or hydrophilic surfaces resulted in tightly packed cells and microcolonies or in sparsely dispersed cells with $>$100-μm-long, pole-anchored chains, respectively (35).

58.2.1.2. Microscopic Monitoring

The scanning confocal laser microscope uses a scanning laser beam to image a given optical plane, or optical section, eliminating scattered light from above and below that section. Optical sections can be collected through the specimen and then used to produce a three-dimensional reconstructed image (26, 93). Early observations of *Vibrio* and *Pseudomonas* species biofilms demonstrated that they could be quite open structures, with distinct, heterogeneous arrangements of cells, associated polymers, and space (93). Similar "channeling" has been observed in aerobic biofilms in fixed film reactors containing a groundwater enrichment and being fed petroleum-contaminated groundwater (107) and in *Pseudomonas putida* biofilms degrading toluene (119).

The scanning confocal laser microscope depends on specimen fluorescence for imaging, and bacteria have been visualized by negative fluorescent staining (fluorescence exclusion) or direct staining with fluorochromes, such as acridine orange or fluorescein isothiocyanate (25, 165). Fluorescent probes can be used to locate specific chemical moieties in the biofilm. Species- or strain-specific antibodies, conjugated with fluorescein or rhodamine (22), can be used to detect and identify given organisms (60, 120, 139). Polysaccharides can be visualized by using fluorescein isothiocyanate and other conjugated lectins (24, 25). The potential for use of genetic probes is enormous. 16S rRNA targeting probes have been used to detect and determine the three-dimensional distribution of toluene-degrading *P. putida* in a waste gas treatment biofilter (119), sulfate-reducing bacteria (SRB) in photosynthetic biofilms from a sewage treatment plant biofilter (137), and *Nitrosomonas* and *Nitrobacter* species in an aquaculture system trickling filter (146).

Physiological properties of biofilms have also been probed. The biocidal effect of trisodium phosphate was found to be influenced by biofilm age, flow rate of the surrounding medium, and topography of the surface (89). The influence of the fluoroquinolone fleroxacin on the physical structure of mature *P. fluorescens* biofilms was also assessed (90). Shifts in biofilm architecture and heterogeneity have been induced by alterations in nutrient composition (118), illustrating the dynamic nature of biofilm structure.

Fluorescent microscopy and quantitative image analysis of cryoembedded specimens have been used to evaluate distribution of active cells in biofilms. CTC combined with DAPI staining showed decreasing respiratory activity of *K. pneumoniae* and *Pseudomonas aeruginosa* in response to treatment with monochloramine (77).

There have been numerous studies of the ultrastructure of biofilms by applying transmission (29), scanning (32, 67), or environmental scanning (100) electron microscopy. Environmental scanning electron microscopy allows visualization of samples without coating and in their hydrated state, and the resultant advantages have been illustrated in comparisons of scanning electron and environmental scanning electron microscopic observations of biofilms (100, 151). Energy-dispersive X-ray analysis can be combined with electron microscopy to determine elemental distributions, such as the localization of iron, manganese, and aluminum in biofilms formed on glass in river water (101).

Also promising is the application of scanning tunneling microscopy and atomic force microscopy, which can be used to image specimens in aqueous solutions and allow visualization of biomolecules at the nanometer scale without dehydration (21). Atomic force microscopy has the advantage that it can be applied to nonconducting materials and is therefore better suited to biological materials (14, 141). It has been used to image biofilms associated with pitting corrosion of copper (17), carbon steel (14), and stainless steel (12, 14, 150). Observations include the patchy distribution of the SRB biofilm on stainless steel, capsule, and flagellum of SRB, and nanometer resolution of a mosaic-like pattern on an SRB cell (14).

58.2.1.3. On-Line Monitoring of Microbial Biomass

In flow cells, microbial biomass can be monitored by measuring fluorescence of NADH, tryptophan, or chlorophyll (4), bioluminescence (e.g., as with *Vibrio harveyi*) (115), or luminescence of genetically engineered bacteria with *lux* operon insertions (113). For example, correlations ($p < 0.05$) were observed between bioluminescent, viable, and direct bacterial counts in an evaluation of antifouling coatings in a laminar flow cell (115). Quantitative measurements of luminescence or fluorescence are possible, such as with an ammeter-photomultiplier-fiber optic system (112).

Chemical shifts in biofilm composition can be monitored by attenuated total reflection/Fourier transform infrared spectrometry (123). Amide bands can be used to assess increase in bacterial biomass, production of PHA by *Pseudomonas cepacia* (125), differences in alginate synthesis by mutants of *P. aeruginosa* (166), or accumulations of exopolymers associated with copper corrosion (62).

58.2.1.4. Measurement and Monitoring with Microelectrodes

A major advance in characterizing biofilm chemistry at the microscale has been made possible by the development of various types of microelectrodes, which evaluate biofilm chemistry at the microscale. For example, a 10-μm-tip probe developed for chlorine detection measured micromolar concentrations of chlorine in *P. aeruginosa* and *K. pneumoniae* biofilms (38) and indicated heterogeneity in biomass and chlorine distribution. Gradients of O_2, H_2S, and pH were measured in biofilms containing SRB and phototrophs (137). Similar measurements in aerobic trickling filter biofilms demonstrated that oxygen was respired in the 0.2- to 0.4-μm surface layers, whereas sulfate reduction took place in deeper, anoxic portions of the biofilm (91). Gradients of O_2 and NO_3^- were also determined in biofilms from a trickling filter in an aquaculture recirculating system (146). Microsensors have provided important information on pH and O_2 profiles in biofilms associated with microbially induced corrosion (98, 99). Measurements showed O_2 being consumed in tubercles (anodic areas) on corroding mild steel and being reduced electrochemically at cathodic sites, indicating that these two O_2 concentration cells were the driving force for corrosion (98) (see also below).

58.2.2. Genetic Characterization of Biofilms

Rapid advances in the characterization of microorganisms through analysis of nucleic acids have created new approaches for evaluating the structure of biofilms. The two principal approaches are (i) to detect and identify constituent organisms using oligonucleotide probes targeting rRNA or DNA coding for rRNA and (ii) to profile total community structure by extraction and analysis of nucleic acids, usually in combination with PCR amplification of sequences. Microorganisms on organic particulates in lake water (164) or on surfaces exposed to drinking water (104) were analyzed with rRNA-directed fluorescent oligonucleotide probes targeting the domains *Bacteria*, *Archaea*, and *Eukarya* and/or the α-, β-, and γ-subclasses of the class *Proteobacteria*. Phylogenetic probes have been used to detect SRB (18, 137, 138), *Nitrosomonas* and *Nitrobacter* spp. (146), and sequenced isolates belonging to the β-subclass of *Proteobacteria* (84).

Total community diversity can be profiled by denaturing gradient gel electrophoresis based on PCR-amplified genes coding for 16S rRNA (121); when applied to an aerobic biofilm and combined with hybridization analysis using an oligonucleotide probe specific for the V3 region of SRB, results suggested that these organisms were present. Analysis of stable low-molecular-weight RNA (5S rRNA and transfer RNA) has been used to characterize total communities associated with particles in the Chesapeake Bay (15). Electrophoretic gels of extracted RNA were subjected to neural net analysis, a computer approach that "trains" a computer to recognize patterns in data that are impossible to analyze by standard statistical methods (108). Neural nets are particularly useful for analyzing noisy, chaotic, and unpredictably nonlinear data. Numerical values obtained from neural net analysis were then subjected to cluster analysis to determine relationships among communities and environmental parameters (126). In this study, particle-associated communities were shown to differ from free-living organisms in the winter, whereas they demonstrated little difference in the summer, when organic nutrient levels were higher.

58.2.3. Measuring Activity in Biofilms

Numerous methods to analyze species composition or chemistry of biofilms may also be indirect assessments of activity. Examples include measurements of PLFA, PHA, microsensor measurements of O_2, and molecular probes targeting phylogenetic groups with specific physiological functions, e.g., nitrifiers (116, 146) or SRB (137). Other measurements address activity (61), such as the determination of incorporation of super ^{32}P into phospholipid or thymidine incorporation into nucleic acids in river biofilms (59).

Enzyme activity of attached organisms has been assessed by measuring hydrolysis of a dipeptide analog and radiolabeled bovine serum albumin (BSA) by a marine *Pseudomonas* species attached to particles (68). Methylcoumarinyl-amide-leucine (MCA-leucine) was used to evaluate hydrolysis of peptides by measuring increase in fluorescence as MCA-leucine was hydrolyzed to leucine and the fluorochrome methylcoumarine (76). BSA was labeled by ^{14}C-methylation of lysyl residues and amino termini (83). Only ~25% of MCA-leucine became adsorbed to particles, and most of the substrate was hydrolyzed by free-living bacteria. In contrast, almost 100% of BSA adsorbed to particles and was hydrolyzed almost exclusively by attached cells (68).The relationship between substratum hydrophobicity and hydrolysis of adsorbed protein by an attached *Pseudomonas* species was measured using ribulose-1,5-bisphosphate carboxylase (RuBPCase) labeled with [^3H]borohydride (86, 153). Percent degradation of RuBPCase decreased with increasing surface hydrophobicity. These results were probably the result of a complex interaction among levels of bacterial attachment, protein adsorption, and the availability of adsorbed protein.

Activity has also been assessed by determining distribution of adenine nucleotide pools and adenylate energy charge across frozen sections of *P. aeruginosa* biofilms (85). Adenylate energy charge was generally low (up to 0.6 units), with highest values of energy charge and ATP at the surface. Measuring expression of specific genes is possible through the use of molecular reporter systems, such as *lacZ* fused to an alginate *algC* promoter in *P. aeruginosa* (37). Expression of *algC* was found to increase in biofilm cells, compared to freely suspended cells. This correlated with an increase in alginate production, measured by Fourier transform infrared spectrometry, uronic acid analysis, and an alginate-specific enzyme-linked immunosorbent assay.

58.2.4. Characterization of Biofilms Related to Corrosion

Studies addressing the influence of microorganisms on corrosion processes have ranged from measurement of the dissolution of metal in the presence of microorganisms, such as copper solvency in the presence of *Arthrobacter*, *Acidovorax*, and *Bacillus* species (36), to determination of changes in open circuit corrosion potential in the presence of microorganisms (42). With metals, microorganisms can contribute to corrosion processes by changing the chemical environment at a metal surface and thereby influencing electrochemical events. Environmental factors that can influence such corrosion include pH, temperature, and concentration of inorganic solutes (e.g., sulfate and ferric ions) (73) and organic solutes in the surrounding medium.

Bacterial biofilms on metal surfaces can create localized areas of reduced oxygen by microbial assimilation of oxygen, preventing its diffusion to deeper layers of the biofilm. Values of pH may be much reduced in deeper layers of biofilms owing to accumulation of fermentation products (70). When such biofilms are discretely distributed over the surface so that some areas are not colonized, localized concentration cells of oxygen and other chemical species are created, contributing to foci of microbially induced corrosion.

SRB have been implicated in localized corrosion of ferrous metals (97). Typically, the anodic dissolution of iron is coupled to the production of neutral hydrogen atoms at the cathode, and if the hydrogen is not removed, the reaction should cease when a complete layer of hydrogen is formed over the cathode (42). SRB sustain electrochemical corrosion by providing a sink for H used as an electron donor during sulfate reduction. Also, hydrogen sulfide produced by SRB can contribute to the corrosion reactions (48). A range of media have been evaluated for their effectiveness in isolation of SRB, and a modified iron sulfite medium was found to yield the highest numbers (39).

The corrosion of stainless metals, e.g., steel and copper alloys, becomes possible with the deposition of a microbial film and apparently requires metabolic activity (147). This transition of a stainless metal to a corroding one is termed ennoblement, which entails an increase in corrosion potential (E_{corr}) and cathodic current density. Manganese- and iron-oxidizing bacteria have been implicated in localized corrosion on stainless steel, primarily because of their presence along with increased levels of iron and manganese in tubercles at corrosion sites. Presumably, they contribute to corrosion by localized cells of reduced oxygen that allow growth of SRB, and possibly by mineralization of manganese, which results in ennoblement (44).

A number of approaches measuring the influence of microorganisms on corrosion processes measure corrosion current between the system anode and cathode. Devices have been developed to measure corrosion in flow cell systems containing pure cultures of test organisms or mixed communities (166). Examples include a four-sided electrode (124) measuring open circuit potential and the reciprocal of the polarization resistance obtained from electrochemical impedance spectroscopy (46, 166). The multi-electrode probe provided multiple measurements, enabling statistical treatment of results (124). To evaluate localized corrosion at a specific site, a concentric electrode system was developed that contained a small anode (0.031 cm^2) and a large cathode (4.87 cm^2) separated by a Teflon ring and electrochemically driven by a current density of 11 $\mu A\ cm^{-2}$ (8). Microorganisms were then added, the potential between the anode and cathode shut off, and the current between the electrodes monitored with a zero resistance ammeter (71). Corrosion current could be produced in the presence of SRB (28) or a mixture of SRB and *Vibrio* species (7), but microbial activity was required only for initiation of corrosion, which then proceeded independently of microbial activity (6).

Numerous studies have attempted to identify relationships between the presence of microorganisms and corrosion by isolating and characterizing organisms from corroding sites (60). Examples are polymer-producing *Pseudomonas* species isolated from copper water pipes with pitting corrosion (5) and enrichments from an oilfield seawater injection system, identified as *Desulfobacter* sp. by oligonucleotide 16S rRNA probes (18). Similarly, numerous studies have tested the ability of specific organisms to foster corrosion by measuring the progress of corrosion in laboratory systems (36). The corrosion rates of mild steel were determined by weight loss and related to the density of attached SRB, while also monitoring exopolymer and dissolved H_2S (13). Different species of SRB were found to vary in their ability to accelerate corrosion (13). In contrast, some studies have demonstrated that corrosion is inhibited by the presence of aerobic microorganisms, such as *Pseudomonas fragi* (82), which deplete oxygen at the surface (81).

58.3. CONCLUSIONS

An enormous range of approaches and techniques are available for the evaluation of biofilm formation, composition, and architecture. The technique that is most appropriate for a given study depends on the specific question being addressed, the level of sensitivity and resolution required (which often depends on the mass and complexity of the biofilm), and the instrumentation and expertise available. Recent developments in molecular tools and microscopic imaging have led to new understanding of biofilm structure and function. Creative applications of these newer technologies to complex biofilm questions will continue to provide new understanding and solution to formerly intractable problems.

REFERENCES

1. **Abbott, A., P. R. Rutter, and R. C. W. Berkeley.** 1983. The influence of ionic strength, pH, and a protein layer on the interaction between *Streptococcus mutans* and glass surfaces. *J. Gen. Microbiol.* **129:**439–445.
2. **Abbott, A., R. C. W. Berkeley, and P. R. Rutter.** 1980. Sucrose and the deposition of *Streptococcus mutans* at solid liquid interfaces, p. 116–142. *In* R. C. W. Berkeley, J. M. Lynch, J. Melling, P. R. Rutter, and B. Vincent (ed.), *Microbial Adhesion to Surfaces*. Ellis Horwood, Chichester, United Kingdom.
3. **Allison, D. G., and I. W. Sutherland.** 1984. A staining technique for attached bacteria and its correlation to extracellular carbohydrate production. *J. Microbiol. Methods* **2:**93–99.
4. **Angell, P., A. Arrage, M. W. Mittelman, and D. C. White.** 1993. On-line, non-destructive biomass determination of bacteria by fluorometry. *J. Microbiol. Methods* **18:** 317–327.
5. **Angell, P., and A. H. L. Chamberlain.** 1991. The role of extracellular products in copper colonisation. *Int. Biodeterior.* **27:**135–143.
6. **Angell, P., J.-S. Luo, and D. C. White.** 1994. Mechanisms of reproducible microbial pitting of 304 stainless steel by a mixed consortium containing sulfate-reducing bacteria, p. 157–168. *In* T. Naguy (ed.) *Proceedings of Tri-service Conference on Corrosion*. Army Materials Laboratory, Washington, D.C.
7. **Angell, P., J.-S. Luo, and D. C. White.** 1994. Microbially sustained pitting corrosion of 304 stainless steel in anaerobic seawater. *Corrosion Sci.* **37:**1058–1096.
8. **Angell, P., and D. C. White.** 1995. Is metabolic activity by biofilms with sulfate-reducing bacterial consortia essential for long-term propagation of pitting corrosion of stainless steel? *J. Ind. Microbiol.* **15:**329–332.
9. **Bain, C. D., J. Evall, and G. M. Whitesides.** 1989. Formation of monolayers by the coadsorption of thiols on gold: variation in the head group, tail group, and solvent. *J. Am. Chem. Soc.* **111:**7155–7164.
10. **Baker, J. H.** 1984. Factors affecting the bacterial colonization of various surfaces in a river. *Can. J. Microbiol.* **30:** 511–515.
11. **Bar-Ness, R., N. Avrahamy, T. Matsuyama, and M. Rosenberg.** 1988. Increased cell surface hydrophobicity of a *Serratia marcescens* NS 38 mutant lacking wetting activity. *J. Bacteriol.* **170:**4361–4364.
12. **Beech, I. B.** 1994. The use of atomic force microscopy for studying biodeterioration of metal surfaces due to the formation of biofilms, p. 4–14. *In* J. Weber and M. Kedro (ed.), *COST 511 Interaction of Microbial Systems with Industrial Materials*, Proc. 2nd COST 511 Workshop, Brest, France. European Commision Directorate-General, Telecommunications, Information Industries and Innovation, Luxembourg.
13. **Beech, I. B., C. W. S. Cheung, C. S. P. Chan, M. A. Hill, R. Franco, and A. R. Lino.** 1994. Study of parameters implicated in thebiodeterioration of mild steel in the

presence of different species of sulphate-reducing bacteria. *Mar. Biofouling Corr.* **34:**3–4.

14. **Beech, I. B., S. W. S. Cheung, D. B. Johnson, and J. R. Smith.** 1996. Comparative studies of bacterial biofilms on steel surfaces using atomic force microscopy and environmental scanning electron microscopy. *Biofouling* **10:**65–77.
15. **Bidle, K. D., and M. Fletcher.** 1995. Comparison of free-living and particle-associated bacterial communities in the Chesapeake Bay by stable low-molecular-weight RNA analysis. *Appl. Environ. Microbiol.* **61:**944–952.
16. **Bradley, G., and D. T. Pritchard.** 1990. Surface charge characteristics of sulphate-reducing bacteria and the initiation of a biofilm on mild steel surfaces. *Biofouling* **2:** 299–310.
17. **Bremer, P. J., G. G. Geesey, and B. Drake.** 1992. Atomic force microscopy examination of the topography of a hydrated bacterial biofilm on a copper surface. *Curr. Microbiol.* **24:**223–230.
18. **Brink, D. E., I. Vance, and D. C. White.** 1994. Detection of Desulfobacter in oil field environments by non-radioactive DNA probes. *Appl. Miocrobiol. Biotechnol.* **42:** 2–3.
19. **Busscher, H. J., and H. C. van der Mei.** 1995. Use of flow chamber devices and image analysis methods to study microbial adhesion. *Methods Enzymol.* **253:**455–477.
20. **Busscher, H. J., A. H. Weerkamp, H. van der Mei, A. W. J. van Pelt, H. P. de Jong, and J. Arends.** 1984. Measurement of the surface free energy of bacterial cell surfaces and its relevance for adhesion. *Appl. Environ. Microbiol.* **48:**980–983.
21. **Bustamante, C., A. E. Erie, and D. Keller.** 1994. Biochemical and structural applications of scanning force microscopy. *Curr. Opin. Struct. Biol.* **4:**750–760.
22. **Caldwell, D. E.** 1995. Cultivation and study of biofilm communities, p. 64–79. *In* H. M. Lappin-Scott and J. W. Costerton (ed.), *Microbial Biofilms*. Cambridge University Press, Cambridge.
23. **Caldwell, D. E., and J. J. Germida.** 1985. Evaluation of difference imagery for visualizing and quantitating microbial growth. *Can. J. Microbiol.* **31:**35–44.
24. **Caldwell, D. E., D. R. Korber, and J. R. Lawrence.** 1992. Confocal laser microscopy and digital image-analysis in microbial ecology. *Adv. Microb. Ecol.* **12:**1–67.
25. **Caldwell, D. E., D. R. Korber, and J. R. Lawrence.** 1992. Imaging of bacterial cells by fluorescence exclusion using scanning confocal laser microscopy. *J. Microbiol. Methods* **15:**249–261.
26. **Caldwell, D. E., D. R. Korber, and J. R. Lawrence.** 1993. Analysis of biofilm formation using 2D *vs* 3D digital imaging. *J. Appl. Bacteriol.* **74:**S52–S66.
27. **Caldwell, D. E., and J. R. Lawrence.** 1986. Growth kinetics of *Pseudomonas fluorescens* microcolonies within the hydrodynamic boundary layers of surface microenvironments. *Microb. Ecol.* **12:**299–312.
28. **Campaignolle, S., J.-S. Lou, D. C. White, J. Guezennec, and J. L. Crolet.** 1993. Stabilization of localized corrosion on carbon steel by sufate-reducing bacteria, p. paper 302, Corrosion/93. National Association of Corrosion Engineers, Houston, Tex.
29. **Chan. R., S. D. Acres, and J. W. Costerton.** 1982. The use of specific antibody to demonstrate the glycocalyx, K99 pili, and the spatial relationship of K99$^+$ enterotoxigenic *E. coli* in the ileum of colostrum-fed calves. *Infect. Immun.* **37:**1170–1180.
30. **Chan, R. C. Y., G. Reid, R. T. Irvin, A. W. Bruce, and J. W. Costerton.** 1985. Competitive exclusion of uropathogens from human uroepithelial cells by *Lactobacillus* whole cells and cell wall fragments. *Infect. Immun.* **47:** 84–89.
31. **Characklis, W. G.** 1990. Laboratory biofilm reactors, p. 55–89. *In* W. G. Characklis and K. C. Marshall (ed.), *Biofilms*. John Wiley & Sons, Inc., New York.
32. **Coutinho, C. M. L. M., F. C. M. Magalhaes, and T. C. Araujo-Jorge.** 1993. Scanning electron microscope study of biofilm formation at different flow rates over metal surfaces using sulphate reducing bacteria. *Biofouling* **7:**19–27.
33. **Cowan, M. M., and M. Fletcher.** 1987. Rapid screening method for detection of bacterial mutants with altered adhesion abilities. *J. Microbiol. Methods* **7:**241–249.
34. **Cuperus, P. L., H. C. van der Mei, G. Reid, A. W. Bruce, A. E. Khoury, P. G. Rouxhet, and H. J. Busscher.** 1992. The effects of serial passaging of lactobacilli in liquid medium on their physico-chemical and structural surface characteristics. *Cells Mat.* **2:**271–280.
35. **Dalton, H., L. Poulsen, P. Halasz, M. Angles, A. Goodman, and K. Marshall.** 1994. Substratum-induced morphological changes in a marine bacterium and their relevance to biofilm structure. *J. Bacteriol.* **176:**6900–6906.
36. **Davidson, D., B. Beheshti, and M. W. Mittelman.** 1996. Effects of *Arthrobacter* sp., *Acidovorax delafieldii*, and *Bacillus megaterium* colonisation on copper solvency in a laboratory reactor. *Biofouling* **9:**279–292.
37. **Davies, D., A. Chakrabarty, and G. Geesey.** 1993. Exopolysaccharide production in biofilms: substratum activation of alginate gene expression by *Pseudomonas aeruginosa*. *Appl. Environ. Microbiol* **59:**1181–1186.
38. **De Beer, D., R. Srinivasan, and P. Stewart.** 1994. Direct measurement of chlorine penetration into biofilms during disinfection. *Appl. Environ. Microbiol.* **60:**4339–4344.
39. **De Bruyn, E. E., and T. E. Cloete.** 1993. Media for the detection of sulphide-producing bacteria in industrial water systems. *J. Microbiol. Methods.* **17:**261–271.
40. **DeFlaun, M. F., B. M. Marshall, E.-P. Kulle, and S. B. Levy.** 1994. Tn5 insertion mutants of *Pseudomonas fluorescens* defective in adhesion to soil and seeds. *Appl. Environ. Microbiol.* **60:**2637–2642.
41. **DeFlaun, M. F., A. S. Tanzer, A. L. McAteer, B. Marshall, and S. B. Levy.** 1990. Development of an adhesion assay and characterization of an adhesion-deficient mutant of *Pseudomonas fluorescens*. *Appl. Environ. Microbiol.* **56:** 112–119.
42. **Dexter, S. C.** 1993. Role of microfouling organisms in marine corrosion. *Biofouling* **7:**97–127.
43. **Dexter, S. C., J. D. J. Sullivan, J. I. Williams, and S. W. Watson.** 1975. Influence of substrate wettability on the attachment of marine bacteria to various surfaces. *Appl. Microbiol.* **30:**298–308.
44. **Dickinson, W. H., and Z. Lewandowski.** 1996. Manganese biofouling and the corrosion behavior of stainless steel. *Biofouling* **10:**79–93.
45. **Donlon, B., and E. Colleran.** 1993. A comparison of different methods to determine the hydrophobicity of acetogenic bacteria. *J. Microbiol. Methods* **17:**27–37.

46. **Dowling, N. J. E., E. E. Stansbury, D. C. White, S. W. Borenstein, and J. C. Danko.** 1989. On-line electrochemical monitoring of microbially induced corrosion, p. 5-1–5-17. *In* G. J. Licina (ed.), *Microbial Corrosion: 1988 Workshop Proceedings EPRI R-6345*. Research Project 8000-26. Electric Power Research Institute, Palo Alto, Calif.

47. **Duxbury, T., B. A. Humphrey, and K. C. Marshall.** 1980. Continuous observations of bacterial gliding motility in a dialysis microchamber: the effects of inhibitors. *Arch. Microbiol.* **124:**169–175.

48. **Edyvean, R. G. J.** 1991. Hydrogen sulphide—a corrosive metabolite. *Int. Biodeterior.* **27:**109–120.

49. **Ellwood, D. C., J. R. Hunter, and V. M. C. Longyear.** 1974. Growth of *Streptococcus mutans* in a chemostat. *Arch. Oral Biol.* **19:**659–664.

50. **Fernandez, M. G. M., C. Mustin, P. De Donato, O. Barres, P. Marion, and J. Berthelin.** 1995. Occurrences at mineral-bacteria interface during oxidation of arsenopyrite by *Thiobacillus ferrooxidans*. *Biotechnol. Bioeng.* **46:** 13–21.

51. **Findlay, R. H., D. J. W. Moriarty, and D. C. White.** 1983. Improved method of determining muramic acid from environmental samples. *Geomicrob. J.* **3:**135–150.

52. **Findlay, R. H., and D. C. White.** 1983. Polymeric beta-hydroxyalkanoates from environmental samples and *Bacillus megaterium*. *Appl. Environ. Microbiol.* **45:**71–78.

53. **Fletcher, M.** 1976. The effects of proteins on bacterial attachment to polystyrene. *J. Gen. Microbiol.* **94:**400–404.

54. **Fletcher, M.** 1979. A microautoradiographic study of the activity of attached and free-living bacteria. *Arch. Microbiol.* **122:**271–274.

55. **Fletcher, M.** 1988. Attachment of *Pseudomonas fluorescens* to glass and influence of electrolytes on bacterium-substratum separation distance. *J. Bacteriol.* **170:**2027–2030.

56. **Fletcher, M.** 1990. Methods for studying adhesion and attachment to surfaces. *Methods Microbiol.* **20:**251–284.

57. **Fletcher, M.** 1991. The physiological activity of bacteria attached to solid surfaces. *Adv. Microb. Physiol* **32:**53–85.

58. **Fowler, H. W., and A. J. McKay.** 1980. The measurement of microbial adhesion, p. 143–161. *In* R. C. W. Berkeley, J. M. Lynch, J. Melling, P. R. Rutter, and B. Vincent (ed.), *Microbial Adhesion to Surfaces*. Ellis Horwood, Chichester, United Kingdom.

59. **Freeman, C., and M. A. Lock.** 1995. Isotope dilution analysis and rates of super(32)P incorporation into phospholipid as a measure of microbial growth rates in biofilms. *Water Res.* **29:**789–792.

60. **Gaylarde, C. C.** 1990. Advances in detection of microbiologically induced corrosion. *Int. Biodeterior.* **26:**11–22.

61. **Gaylarde, C. C.** 1992. Sulphate-reducing bacteria which do not induce accelerated corrosion. *Int. Biodeterior. Biodegrad.* **30:**331–338.

62. **Geesey, G. G., and P. J. Bremer.** 1990. Applications of Fourier transform infrared spectrometry to studies of copper corrosion under bacterial biofilms. *Mar. Technol. Soc. J.* **24:**36–43.

63. **Gehron, M. J., J. D. Davis, G. A. Smith, and D. C. White.** 1984. Determination of the gram-positive bacterial content of soils and sediments by analysis of teichoic acid components. *J. Microbiol Methods* **2:**165–176.

64. **Gehron, M. J., and D. C. White** 1982. Quantitative determination of the nutritional status of detrital microbiota and the grazing fauna by triglyceride glycerol analysis. *J. Exp. Mar. Biol. Ecol.* **64:**145–158.

65. **Gingell, D., and I. Todd.** 1979. Interference reflection to microscopy: a quantitative theory for image interpretation and its application to cell-substratum separation measurement. *Biophys. J.* **26:**507–526.

66. **Godwin, S., M. Fletcher, and R. Burchard.** 1989. Interference reflection microscopic study of sites of association between gliding bacteria and glass substrata. *J. Bacteriol.* **171:**4589–4594.

67. **Gomez de Saravia, S. G., M. F. L. de Mele, and H. A. Videla.** 1989. An assessment of the early stages of microfouling and corrosion of 70:30 copper nickel alloy in the presence of two marine bacteria. *Biofouling* **1:**213–222.

68. **Griffith, P. C., and M. Fletcher.** 1991. Hydrolysis of protein and model dipeptide substrates by attached and nonattached marine *Pseudomonas* sp. strain NCIMB 2021. *Appl. Environ. Microbiol.* **57:**2186–2191.

69. **Guckert, J. B., M. A. Hood, and D. C. White.** 1986. Phospholipid, ester-linked fatty acid profile changes during nutrient deprivation of *Vibrio cholerae*: increases in the *trans/cis* ratio and proportions of cyclopropyl fatty acids. *Appl. Environ. Microbiol.* **52:**794–801.

70. **Guezennec, J., N. J. E. Dowling, J. Bullen, and D. C. White.** 1994. Relationship between bacterial colonization and cathodic current density associated with mild steel surfaces. *Biofouling.* **8:**133–146.

71. **Guezennec, J., D. C. White, and J. L. Crolet.** 1992. Stabilization of localized corrosion on carbon steel by SRB, p. 1–10, *UK Corrosion/92*. The Institute of Corrosion, London.

72. **Harber, M. J., R. Mackenzie, and A. W. Asscher.** 1983. A rapid bioluminescence method for quantifying bacterial adhesion to polystyrene. *J. Gen. Microbiol.* **129:**621–632.

73. **Hassan, R. S., R. M. Hassan, and L. C. P. Oh.** 1990. Effect of sulphate and ferric ions on metal corrosion in in-vitro fermentation by sulphate-reducing bacteria. *Biofouling* **2:**101–111.

74. **Heilmann, C., M. Hussain, G. Peters, and F. Gotz.** 1997. Evidence for autolysin-mediated primary attachment of *Staphylococcus epidermidis* to a polystyrene surface. *Mol. Microbiol.* **24:**1013–1024.

75. **Heipieper, H.-J., R. Diffenbach, and H. Keweloh.** 1992. Conversion of *cis* unsaturated fatty acids to *trans*, a possible mechanism for the protection of phenol degrading *Pseudomonas putida* P8 from substrate toxicity. *Appl. Environ. Microbiol.* **58:**1827–1852.

76. **Hoppe, H.-G.** 1983. Significance of exoenzymatic activities in the ecology of brackish water: measurements by means of methylumbelliferyl-substrates. *Mar. Ecol. Prog. Ser.* **11:**299–308.

77. **Huang, C., F. Yu, G. McFeters, and P. Stewart.** 1995. Nonuniform spatial patterns of respiratory activity within biofilms during disinfection. *Appl. Environ. Microbiol.* **61:** 2252–2256.

78. **Ista, L., H. Fan, O. Baca. and G. Lópex.** 1996. Attachment of bacteria to model solid surfaces: oligo(ethylene glycol) surfaces inhibit bacterial attachment. *FEMS Microbiol Lett.* **142:**59–63.

79. **James, G. A., D. R. Korber, D. E. Caldwell, and J. W. Costerton.** 1995. Digital image analysis of growth and

starvation responses of a surface-colonizing *Acinetobacter* sp. *J. Bacteriol.* **177:**907–915.

80. **Jansen, B., and W. Kohnen.** 1995. Prevention of biofilm formation by polymer modification. *J. Ind. Microbiol.* **15:** 391–396.
81. **Jayaraman, A., E. Cheng, J. Earthman, and T. Wood.** 1997. Axenic aerobic biofilms inhibit corrosion of SAE 1018 steel through oxygen depletion. *Appl. Microbiol. Biotechnol.* **48:**11–17.
82. **Jayaraman, A., J. C. Earthman, and T. K. Wood.** 1997. Corrosion inhibition by aerobic biofilms on SAE 1018 steel. *Appl. Microbiol. Biotechnol.* **47:**62–68.
83. **Jentoft, N., and D. G. Dearborn.** 1983. Protein labeling by reductive alkylation. *Methods Enzymol.* **91:**570–579.
84. **Kalmbach, S., W. Manz, and U. Szewzyk.** 1997. Isolation of new bacterial species from drinking water biofilms and proof of their in situ dominance with highly specific 16S rRNA probes. *Appl. Environ. Microbiol.* **63:**4164–4170.
85. **Kinniment, S. L., and J. W. T. Wimpenny.** 1992. Measurements of the distribution of adenylate concentrations and adenylate energy charge across *Pseudomonas aeruginosa* biofilms. *Appl. Environ. Microbiol.* **58:**1629–1635.
86. **Kirchman, D. L., D. L. Henry, and S. C. Dexter.** 1989. Adsorption of proteins to surfaces in seawater. *Mar. Chem.* **27:**201–217.
87. **Kjelleberg, S., B. A. Humphrey, and K. C. Marshall.** 1982. Effect of interfaces on small, starved marine bacteria. *Appl. Environ. Microbiol.* **43:**1166–1172.
88. **Knox, K. W., L. N. Hardy, L. J. Markevics, J. D. Evans, and A. J. Wicken.** 1985. Comparative studies on the effect of growth conditions on adhesion, hydrophobicity, and extracellular protein profile of *Streptococcus sanguis* G9B. *Infect. Immun.* **50:**545–554.
89. **Korber, D., A. Choi, G. Wolfaardt, S. Ingham, and D. Caldwell.** 1997. Substratum topography influences susceptibility of Salmonella enteritidis biofilms to trisodium phosphate. *Appl. Environ. Microbiol.* **63:**3352–3358.
90. **Korber, D. R., G. A. James, and J. W. Costerton.** 1994. Evaluation of fleroxacin activity against established *Pseudomonas fluorescens* biofilms. *Appl. Environ. Microbiol.* **60:** 1663–1669.
91. **Kuhl, M., and B. B. Jorgensen.** 1992. Microsensor measurements of sulfate reduction and sulfide oxidation in compact microbial communities of aerobic biofilms. *Appl. Environ. Microbiol.* **58:**1164–1174.
92. **Kurtz, H. D., Jr., and J. Smit.** 1994. The *Caulobacter crescentus* holdfast: identification of holdfast attachment complex genes. *FEMS Microbiol. Lett.* **116:**175–182.
93. **Lawrence, J., D. Korber, B. Hoyle, J. Costerton, and D. Caldwell.** 1991. Optical sectioning of microbial biofilms. *J. Bacteriol.* **173:**6558–6567.
94. **Lawrence, J. R., and D. E. Caldwell.** 1987. Behavior of bacterial stream populations within the hydrodynamic boundary layers of surface microenvironments. *Microb. Ecol.* **14:**15–27.
95. **Lawrence, J. R., P. J. Delaquis, D. R. Korber, and D. E. Caldwell.** 1987. Behavior of *Pseudomonas fluorescens* within the hydrodynamic boundary layers of surface microenvironments. *Microb. Ecol.* **14:**1–14.
96. **Lee, S., Y. Li, and G. Bowden.** 1996. Detachment of *Streptococcus mutans* biofilm cells by an endogenous enzymatic activity. *Infect. Immun.* **64:**1035–1038.
97. **Lee, W., Z. L. Andowski, P. H. Nielsen, and W. A. Hamilton.** 1995. Role of sulfate-reducing bacteria in corrosion of mild steel: a review. *Biofouling* **8:**165–194.
98. **Lee, W., and D. de Beer.** 1995. Oxygen and pH microprofiles above corroding mild steel covered with a biofilm. *Biofouling* **8:**273–280.
99. **Lee, W., Z. Lewandowski, S. Okabe, W. G. Characklis, and R. Avci.** 1993. Corrosion of mild steel underneath aerobic biofilms containing sulfate-reducing bacteria. Part 1: At low dissolved oxygen concentration. *Biofouling* **7:**197–216.
100. **Little, B. J., P. A. Wagner, R. I. Ray, R. Pope, and R. Scheetz.** 1991. Biofilms: an ESEM evaluation of artefacts introduced during SEM preparation. *J. Ind. Microbiol.* **8:** 213–222.
101. **Lunsdorf, H., I. Brummer, K. Timmis, and I. Wagner-Dobler.** 1997. Metal selectivity of in situ microcolonies in biofilms of the Elbe river. *J. Bacteriol.* **179:**31–40.
102. **Mafu, A., D. Roy, J. Goulet, and L. Savoie.** 1991. Characterization of physicochemical forces involved in adhesion of *Listeria monocytogenes* to surfaces. *Appl. Environ. Microbiol.* **57:**1969–1973.
103. **Makin, S. A., and T. J. Beveridge.** 1996. The influence of A-band and B-band lipopolysaccharide on the surface characteristics and adhesion of *Pseudomonas aeruginosa* to surfaces. *Microbiology* **142:**299–307.
104. **Manz, W., U. Szewzyk, P. Ericsson, R. Amann, K. Schleifer, and T. Stenstrom.** 1993. In situ identification of bacteria in drinking water and adjoining biofilms by hybridization with 16S and 23S rRNA-directed fluorescent oligonucleotide probes. *Appl. Environ. Microbiol.* **59:**2293–2298.
105. **Marshall, K. C., R. Stout, and R. Mitchell.** 1971. Mechanism of the initial events in the sorption of marine bacteria to surfaces. *J. Gen. Microbiol.* **68:**337–348.
106. **Marshall, P. A., G. I. Loeb, M. M. Cowan, and M. Fletcher.** 1989. Response of microbial adhesives and biofilm matrix polymers to chemical treatments as determined by interference reflection microscopy and light section microscopy. *Appl. Environ. Microbiol.* **55:**2827–2831.
107. **Massol-Deya, A., J. Whallon, R. Hickey, and J. Tiedje.** 1995. Channel structures in aerobic biofilms of fixed-film reactors treating contaminated groundwater. *Appl. Environ. Microbiol.* **61:**769–777.
108. **Masters, T.** 1993. *Practical Neural Network Recipes in C++*. Academic Press, New York.
109. **McEldowney, S., and M. Fletcher.** 1986. Effect of growth conditions and surface characteristics of aquatic bacteria on their attachment to solid surfaces. *J. Gen. Microbiol.* **132:**513–523.
110. **McEldowney, S., and M. Fletcher.** 1987. Adhesion of bacteria from mixed cell suspension to solid surfaces. *Arch. Microbiol.* **148:**57–62.
111. **Meyer-Reil, L.-A.** 1978. Autoradiography and epifluorescence microscopy combined for the determination of number and spectrum of actively metabolizing bacteria in natural waters. *Appl. Environ. Microbiol.* **36:**506–512.

112. **Mittelman, M. W., J. M. H. King, G. S. Sayler, and D. C. White.** 1992. On-line detection of bacterial adhesion in a shear gradient with bioluminescence by a *Pseudomonas fluorescens* (lux) strain. *J. Microbiol. Methods* **15:** 53–60.

113. **Mittelman, M. W., L. L. Kohring, and D. C. White.** 1992. Multipurpose laminar-flow adhesion cells for the study of bacterial colonization and biofilm formation. *Biofouling* **6:**39–51.

114. **Mittelman, M. W., D. E. Nivens, C. Low, and D. C. White.** 1990. Differential adhesion, activity, and carbohydrate: protein ratios of *Pseudomonas atlantica* monocultures attaching to stainless steel in a linear shear gradient. *Microb. Ecol.* **19:**269–278.

115. **Mittelman, M. W., J. Packard, A. A. Arrage, S. L. Bean, P. Angell, and D. C. White.** 1993. Test systems for determining antifouling coating efficacy using on-line detection of bioluminescence and fluorescence in a laminar-flow environment. *J. Microbiol. Methods* **18:**51–60.

116. **Mobarry, B., M. Wagner, V. Urbain, B. Rittmann, and D. Stahl.** 1996. Phylogenetic probes for analyzing abundance and spatial organization of nitrifying bacteria. *Appl. Environ. Microbiol.* **62:**2156–2162.

117. **Molin, G., I. Nilsson, and L. Stenson-Holst.** 1982. Biofilm build-up of *Pseudomonas putida* in a chemostat at different dilution rates. *Eur. J. Appl. Microbiol. Biotechnol.* **15:**218–222.

118. **Møller, S., D. R. Korber, G. M. Wolfaardt, S. Molin, and D. E. Caldwell.** 1997. Impact of nutrient composition on a degradative biofilm community. *Appl. Environ. Microbiol.* **63:**2432–2438.

119. **Møller, S., A. Pedersen, L. Poulsen, E. Arvin, and S. Molin.** 1996. Activity and three-dimensional distribution of toluene-degrading *Pseudomonas putida* in a multispecies biofilm assessed by quantitative in situ hybridization and scanning confocal laser microscopy. *Appl. Environ. Microbiol.* **62:**4632–4640.

120. **Morin, P., A. Camper, W. Jones, D. Gatel, and J. C. Goldman.** 1996. Colonization and disinfection of biofilms hosting coliform-colonized carbon fines. *Appl. Environ. Microbiol.* **62:**4428–4432.

121. **Muyzer, G., E. C. De Waal, and A. G. Uitterlinden.** 1993. Profiling of complex microbial populations by denaturing gradient gel electrophoresis analysis of polymerase chain reaction-amplified genes coding for 16S rRNA. *Appl. Environ. Microbiol.* **59:**695–700.

122. **Nickels, J. S., J. D. King, and D. C. White.** 1979. Poly-beta-hydroxybutyrate accumulation as a measure of unbalanced growth of the estuarine detritial mcirobiota. *Appl. Environ. Microbiol.* **37:**459–465.

123. **Nivens, D. E., J. Q. Chambers, T. R. Anderson, A. Tunlid, J. Smit, and D. C. White.** 1993. Monitoring microbial adhesion and biofilm formation by attenuated total reflection/Fourier transform infrared spectroscopy. *J. Microbiol. Methods* **17:**199–213.

124. **Nivens, D. E., R. Jack, A. Vass, J. B. Guckert, J. Q. Chambers, and D. C. White.** 1992. Multi-electrode probe for statistical evaluation of microbiologically influenced corrosion. *J. Microbiol. Methods* **16:**47–58.

125. **Nivens, D. E., J. Schmitt, J. Sniateki, T. Anderson, J. Q. Chambers, and D. C. White.** 1993. Multi-channel AFT/FT-IR spectrometer for on-line examination of microbial biofilms. *Appl. Spectrosc.* **47:**668–671.

126. **Noble, P. A., K. D. Bidle, and M. Fletcher.** 1997. Natural microbial community compositions compared by a back-propagating neural network and cluster analysis of 5S rRNA. *Appl. Environ. Microbiol.* **63:**1762–1770.

127. **Parker, J. H., G. A. Smith, H. L. Fredrickson, J. R. Vestal, and D. C. White.** 1982. Sensitive assay, based on hydroxy-fatty acids from lipopolysaccharide lipid A for gram-negative bacteria in sediments. *Appl. Environ. Microbiol.* **44:**1170–1177.

128. **Paul, J. H.** 1982. Use of Hoechst dyes 33258 and 33342 for enumeration of attached and planktonic bacteria. *Appl. Environ. Microbiol.* **43:**939–944.

129. **Paul, J. H., and G. I. Loeb.** 1983. Improved microfouling assay employing a DNA-specific fluorochrome and polystyrene as substratum. *Appl. Environ. Microbiol.* **46:** 338–343.

130. **Pedersen, K.** 1980. Electrostatic interaction chromatography: a method for assaying the relative surface charge of bacteria. *FEMS Microbiol. Lett.* **12:**365–367.

131. **Porter, K. G., and Y. S. Feig.** 1980. The use of DAPI for identifying and counting aquatic microflora. *Limnol. Oceangr.* **25:**943–948.

132. **Powell, M. S., and N. K. H. Slater.** 1982. Removal rates of bacterial cells from glass surfaces by fluid shear. *Biotechnol. Bioeng.* **24:**2527–2537.

133. **Pringle, J., and M. Fletcher.** 1986. Influence of substratum hydration and adsorbed macromolecules on bacterial attachment to surfaces. *Appl. Environ. Microbiol.* **51:** 1321–1325.

134. **Pringle, J. H., and M. Fletcher.** 1983. Influence of substratum wettability on the attachment of freshwater bacteria to solid surfaces. *Appl. Environ. Microbiol.* **45:**811–817.

135. **Pringle, J. H., M. Fletcher, and D. C. Ellwood.** 1983. Selection of attachment mutants during the continuous culture of *Pseudomonas fluorescens* and relationship between attachment ability and surface composition. *J. Gen. Microbiol.* **129:**2557–2569.

136. **Ragout, A., F. Sineriz, R. Kaul, D. Guoqiang, and B. Mattiasson.** 1996. Selection of an adhesive phenotype of *Streptococcus salivarius* subsp. *thermophilus* for use in fixed-bed reactors. *Appl. Microbiol. Biotechnol.* **46:**126–131.

137. **Ramsing, N., M. Kuhl, and B. Jorgensen.** 1993. Distribution of sulfate-reducing bacteria, O_2, and H_2S in photosynthetic biofilms determined by oligonucleotide probes and microelectrodes. *Appl. Environ. Microbiol.* **59:** 3840–3849.

138. **Raskin, L., B. E. Rittmann, and D. A. Stahl.** 1996. Competition and coexistence of sulfate-reducing and methanogenic populations in anaerobic biofilms. *Appl. Environ. Microbiol.* **62:**3847–3857.

139. **Reid, G., and H. J. L. Brooks.** 1985. A fluorescent antibody staining technique to detect bacterial adherence to urinary tract epithelial cells. *Stain Technol.* **60:**211–217.

140. **Rijnaarts, H. H. M., W. Norde, E. J. Bouwer, J. Lyklema, and A. J. B. Zehnder.** 1993. Bacterial adhesion under static and dynamic conditions. *Appl. Environ. Microbiol.* **59:**3255–3265.

141. **Roberts, C. J., P. M. Williams, M. C. Davies, D. E. Jackson, and S. J. B. Tendler.** 1994. Atomic force microscopy and scanning tunneling microscopy: refining techniques for studying biomolecules. *Trends Biotechnol.* **12:**127–132.

142. **Rodriguez, G. G., D. Phipps, K. Ishiguro, and H. F. Ridgway.** 1992. Use of a fluorescent redox probe for direct visualization of actively respiring bacteria. *Appl. Environ. Microbiol.* **58:**1801–1808.

143. **Rosenberg, M., and S. Kjelleberg.** 1986. Hydrophobic interactions: role in bacterial adhesion. *Adv. Microb. Ecol.* **9:**353–393.

144. **Rosenberg, M., and E. Rosenberg.** 1985. Bacterial adherence at the hydrocarbon-water interface. *Oil Petrochem. Poll.* **2:**155–162.

145. **Rutter, P., and R. Leech.** 1980. The deposition of *Streptococcus sanguis* NCTC 7868 from a flowing suspension. *J. Gen. Microbiol.* **120:**301–307.

146. **Schramm, A., L. Larsen, N. Revsbech, N. Ramsing, R. Amann, and K. Schleifer.** 1996. Structure and function of a nitrifying biofilm as determined by in situ hybridization and the use of microelectrodes. *Appl. Environ. Microbiol.* **62:**4641–4647.

147. **Scotto, V., R. DiCintio, and G. Marcenaro.** 1985. The influence of marine aerobic microbial film on stainless steel corrosion behaviour. *Corr. Sci.* **25:**185–194.

148. **Sjollema, J., H. J. Busscher, and A. H. Weerkamp.** 1988. Deposition of oral streptococci and polystyrene latices onto glass in a parallel plate flow cell. *Biofouling* **1:**101–112.

149. **Smith, J., and G. McFeters.** 1996. Effects of substrates and phosphate on INT (2-(4-iodophenyl)-3-(4-nitrophenyl)-5-phenyl tetrazolium chloride) and CTC (5-cyano-2,3-ditolyl tetrazolium chloride) reduction in *Escherichia coli*. *J. Appl. Bacteriol.* **80:**209–215.

150. **Steele, A., D. Goddard, and I. B. Beech.** 1994. The use of atomic force microscopy in the study of biodeterioration of stainless steel in the presence of bacterial biofilms. *Int. Biodeterior. Biodegrad.* **34:**35–46.

151. **Sutton, N., N. Hughes, and P. Handley.** 1994. A comparison of conventional SEM techniques, low temperature SEM and the electroscan wet scanning electron microscope to study the structure of a biofilm of Streptococcus crista CR3. *J. Appl. Bacteriol.* **76:**448–454.

152. **Szewzyk, U., and B. Schink.** 1988. Surface colonization by and life cycle of *Pelobacter acidigllici* studied in a continous-flow microchamber. *J. Gen. Microbiol.* **134:**183–190.

153. **Tack, B. F., J. Dean, D. Eilat, P. E. Lorentz, and A. Schecter.** 1980. Tritium labelling of proteins to high specific activity by reductive methylation. *J. Biol. Chem.* **255:**8842–8847.

154. **Taylor, G. T., D. Zheng, M. Lee, P. J. Troy, G. Gyananath, and S. K. Sharma.** 1997. Influence of surface properties on accumulation of conditioning films and marine bacteria on substrata exposed to oligotrophic waters. *Biofouling* **11:**31–57.

155. **Thomas, V., B. Sanford, R. Moreno, and M. Ramsay.** 1997. Enzyme-linked lectinsorbent assay measures N-acetyl-D-glucosamine in matrix of biofilm produced by *Staphylococcus epidermidis*. *Curr. Microbiol.* **35:**249–254.

156. **Urzi, C., S. Lisi, G. Criseo, and A. Pernice.** 1991. Adhesion to and degradation of marble by a *Micrococcus* strain isolated from it. *Geomicrobiol. J.* **9:**2–3.

157. **van der Mei, H. C., A. H. Weerkamp, and H. J. Busscher.** 1987. A comparison of various methods to determine hydrophobic properties of streptococcal cell surfaces. *J. Microbiol. Methods* **6:**277–287.

158. **van Loosdrecht, M. C. M., J. Lyklema, W. Norde, G. Schraa, and A. J. B. Zehnder.** 1987. Electrophoretic mobility and hydrophobicity as a measure to predict the initial steps of bacterial adhesion. *Appl. Environ. Microbiol.* **53:**1898–1901.

159. **van Pelt, A. W. J., H. C. van der Mei, H. J. Busscher, J. Arends, and A. H. Weerkamp.** 1984. Surface free energies of oral streptococci. *FEMS Microbiol. Lett.* **25:**2–3.

160. **van Pelt, A. W. J., A. H. Weerkamp, M. H. W. J. Uyen, H. J. Busscher, H. P. de Jong, and J. Arends.** 1985. Adhesion of *Streptococcus sanguis* CH3 to polymers with different surface free energies. *Appl. Environ. Microbiol.* **49:**1270–1275.

161. **Walker, J. T., and C. W. Keevil.** 1994. Study of microbial biofilms using light microscope techniques. *Mar. Biofouling Corr.* **34:**3–4.

162. **Walker, J. T., D. Wagner, W. Fischer, and C. W. Keevil.** 1994. Rapid detection of biofilm on corroded copper pipes. *Biofouling* **8:**55–63.

163. **Weerkamp, A., P. Handley, A. Baars, and J. Slot.** 1986. Negative staining and immunoelectron microscopy of adhesion-deficient mutants of *Streptococcus salivarius* reveal that the adhesive protein antigens are separate classes of cell surface fibril. *J. Bacteriol.* **165:**746–755.

164. **Weiss, P., B. Schweitzer, R. Amann, and M. Simon.** 1996. Identification in situ and dynamics of bacteria on limnetic organic aggregates (lake snow). *Appl. Environ. Microbiol.* **62:**1998–2005.

165. **Wentland, E., P. Stewart, C. Huang, and G. McFeters.** 1996. Spatial variations in growth rate within *Klebsiella pneumoniae* colonies and biofilm. *Biotechnol. Progr.* **12:**316–321.

166. **White, D. C., A. A. Arrage, D. E. Nivens, R. J. Palmer, J. F. Rice, and G. S. Sayler.** 1996. Biofilm ecology: on-line methods bring new insights into MIC and microbial biofouling. *Biofouling* **10:**3–16.

167. **White, D. C., W. M. Davis, J. S. Nickels, J. D. King, and R. J. Bobbie.** 1979. Determination of the sedimentary microbial biomass by extractable lipid phosphate. *Oecologia* **40:**51–62.

168. **Wiencek, K. M., and M. Fletcher.** 1995. Bacterial adhesion to hydroxyl- and methyl-terminated alkanethiol self-assembled monolayers. *J. Bacteriol.* **177:**1959–1966.

169. **Wiencek, K. M., and M. Fletcher.** 1997. Effects of substratum wettability and molecular topography on the initial adhesion of bacteria to chemically defined substrata. *Biofouling* **11:**293–311.

170. **Williams, V., and M. Fletcher.** 1996. *Pseudomonas fluorescens* adhesion and transport through porous media are affected by lipopolysaccharide composition. *Appl. Environ. Microbiol.* **62:**100–104.

171. **Yu, F. P., and G. A. McFeters.** 1994. Rapid in situ assessment of physiological activities in bacterial biofilms using fluorescent probes. *J. Microbiol. Methods* **20:**1–10.

SECONDARY METABOLITES

THIS SECTION DISCUSSES THE BROAD TOPIC of secondary metabolites. Over the past decade there has been enormous growth in basic information relating to this exciting field, and it has been captured in great depth in the following chapters. Khetan and Hu provide a number of examples of metabolic engineering approaches to secondary metabolite biosynthesis. The Champness article includes an overview of genes involved in regulation of secondary metabolite biosynthesis, whereas Meurer and Hutchinson provide a broad view of information on the structural genes for synthesis of various classes of natural products. Sutcliffe et al. provide a detailed review of antibiotic resistance genes and approaches for cloning and analysis. Finally, the chapter by Allison and Klaenhammer provides an updated overview of bacteriocin genetics and industrial applications.

Metabolic Engineering of Antibiotic Biosynthetic Pathways

ANURAG KHETAN AND WEI-SHOU HU

59

Most producing microorganisms of industrial biochemicals and pharmaceuticals are subject to extensive genetic manipulations and modifications before they are employed for manufacturing purposes. Even after they are used in industrial production, they still undergo a series of strain improvement programs to give them the desired productivity and growth characteristics. Antibiotic producers are no exception. Most wild-type antibiotic producers synthesize only a minute concentration of product, or produce many undesirable side products. Strain improvement is critical for the development of a successful product.

With recombinant DNA technology, these antibiotic-producing organisms can be altered in a more directed manner. Metabolic engineering of secondary metabolism can possibly lead to the synthesis of a new or known antibiotic(s) in an organism predisposed to producing another antibiotic or in one not producing any at all. One may apply metabolic engineering to eliminate the biosynthesis of undesirable by-products. It is also possible to change flux distributions of compounds involved in antibiotic biosynthesis to increase the yields or productivity.

In a very broad sense, conventional strain improvement is the oldest form of metabolic engineering. Using a narrower definition, the term metabolic engineering is often applied only to the cases in which modern molecular genetic tools are applied to engineer cells. Since a large number of secondary metabolites produced by a variety of microorganisms are being explored for potential applications or for industrial production, it is virtually impossible to prescribe a path along which the modifications can be introduced to achieve desired results.

In this chapter, we address the general principles of metabolic engineering of antibiotic biosynthesis for improving antibiotic yield and productivity. Different strategies can be adopted, depending on our knowledge of the microorganism and the biosynthetic pathway, whether the genes coding for the biosynthesis enzymes are known, and whether the molecular genetic tools are available. If such knowledge and tools are yet to be developed, one may opt for conventional methods of strain improvement instead of metabolic engineering. In the case that the structural and regulatory genes for the biosynthesis are known, directed strain improvement may be desirable. There are rare cases in which the pathway is completely characterized, and the genes coding for each of the enzymes have been cloned. Under such conditions, it is possible to predict rate-limiting steps and introduce specific perturbations in the system. In other cases, it may be desirable to empower a new host microorganism to produce the product by cloning in the entire pathway. The strategies in making those decisions will be discussed in this chapter.

59.1. OVERVIEW OF STRAIN IMPROVEMENT

For most antibiotics, from the time of their discovery to reaching field application, the titer of the producing strain has increased by orders of magnitude. While medium development and process engineering have had major roles, strain improvement has been the key to increasing the final titer of antibiotics in fermentation. Through such strain improvement the producing strains are also gradually adapted to economically more desirable substrates, and the undesirable by-products are eliminated.

When the molecular genetic tools are underdeveloped for the producing microorganism, conventional strain improvement is the obvious method of choice. These conventional methods, by treating the secondary metabolite-producing system as a "black box," are also useful in improving the production of the metabolites for which the enzymes and the genes for the biosynthetic pathway are little known. These methods, although being perceived as old-fashioned and less elegant, are nevertheless effective. A major drawback is that they are labor intensive and often require extensive screening. In dealing with the newly discovered secondary metabolites for which the metabolic pathway is not known, the conventional strain improvement methodology is still among the most effective tools. A typical strain improvement program involves first generating genotype variants in the population. This is achieved by inducing mutations using physical or chemical means and/or by recombination among different strains. Those with improved phenotype properties are then isolated by selection or screening.

Many possible mechanisms can contribute to an increased antibiotic production in a producing microorganism. An increased flux of a precursor in primary metabolism may lead to an increased productivity. Basically, any metabolic engineering or conventional method for increasing the flux of the precursor in primary metabolism can be applied to increase its flux into the secondary metabolism.

Some antibiotics are toxic even to the producing microorganisms. An increased resistance of the producer to the antibiotics can lead to an enhanced productivity. A number of reviews have focused on the applications of such conventional methodology (35, 36, 43).

59.2. METABOLIC ENGINEERING OF SECONDARY METABOLISM FOR HIGHER PRODUCTIVITY

Since metabolic engineering aims to manipulate the metabolic network involved in the antibiotic biosynthesis, detailed knowledge of the biochemical pathways, the genes encoding for the enzymes, and the regulation of the expression of these genes is beneficial. In addition to such biochemical knowledge and the availability of genetic tools, the employment of various analytical tools for determining the bottlenecks of the biosynthesis is also important. However, in a general sense, even when the knowledge of the bottleneck steps is unknown, it is still possible to attempt a "rational" increase in pathway fluxes. The general strategies, as shown in Fig. 1, are outlined below.

59.2.1. Manipulation of Structural Genes

59.2.1.1. Amplifying the Entire Pathway

An approach of metabolic engineering is to amplify the entire pathway, or a large portion of the pathway, in a gene cluster responsible for the biosynthesis of the secondary metabolite. In general, the "pathway" being amplified entails the reaction steps beginning from putting the building blocks—the precursors—together, but it does not involve the biosynthesis of the precursors. As an example, the entire gene cluster coding for all the enzymes in the cephamycin pathway from *Streptomyces cattleya* was cloned as a 29.3-kb DNA segment (10). This was accomplished by screening for cephamycin-producing transformants derived from a cephamycin-nonproducing *Streptomyces lividans*, after "shotgun" cloning. Subsequently, an extra copy of this whole biosynthetic gene cluster was introduced into *Streptomyces lactamgens*, a native producer of cephamycin C, and led to a two- to threefold increase in yields. However, the cause of the increase was not characterized; it could be due to extra copies of either the structural genes or the regulatory genes.

59.2.1.2. Amplifying Segment of the Pathway

Although it has been possible to amplify the genes (possibly involving both regulatory genes and structural genes) encoding for the entire biosynthetic pathway for secondary metabolites, such a practice is still rare and often may not be necessary or most efficient. It is a more common practice to amplify only a single gene or a portion of the biosynthetic pathway to enhance the product formation. This approach is most effective if the segment of genes being amplified contains the gene encoding for the rate-limiting enzyme. However, there are also cases in which no single enzyme in the biosynthetic pathway is a dominant rate-limiting enzyme, but rather multiple enzymes restrict the flux of the product formation. In the cases where the rate-

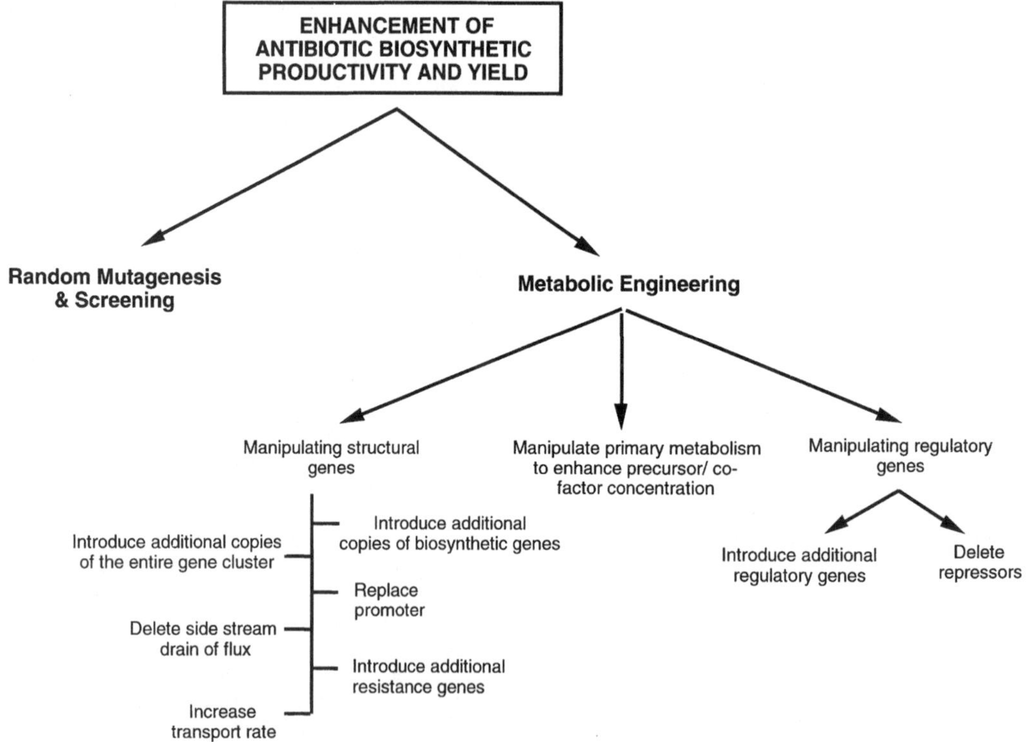

FIGURE 1 Strain improvement. Conventional strain improvement engineers the metabolism in a random manner. The mutated strains have to be screened or selected for the desired endpoint property. Metabolic engineering using molecular genetic tools uses a preexisting knowledge database to produce strains that have directed and precise changes.

limiting enzyme is known, the amplification of the gene encoding for that particular enzyme will certainly lead to an enhanced production.

On the other hand, even if the rate-limiting enzymes are unknown, the introduction of different segments of the gene cluster encoding the biosynthetic pathway will allow for the identification of the rate-limiting enzymes and also lead to increased productivity. For the latter, the identification of rate-limiting enzymes by introducing extra copies of the gene, the segment of the gene can be introduced by plasmid and remain in plasmid. Conversely, once the rate-limiting enzyme is known, it is best to integrate the segment of the genes into the chromosome for stable expression of the enzyme, as a deleterious "plasmid effect" (41) on antibiotic titers has been reported in transformants with plasmids.

In a study involving tetracenomycin, a polyketide synthesized by a type II polyketide multienzyme complex in *Streptomyces glaucescens*, enhancement of intermediates and the final product was observed with extra copies of the biosynthetic genes (14). Polyketide synthase (β-ketoacyl: acyl carrier protein synthase) and the acyl carrier protein have been postulated to play a role in the early steps of polyketide biosynthesis. By enhancing the gene dosage of these enzymes on multicopy plasmids with their expression under the control of a strong promoter, the accumulation of an intermediate TcmD3 increased by almost 30-fold, and tetracenomycin C production increased by 30%.

59.2.1.3. Enhanced Antibiotic Resistance

Antibiotic producers have resistance mechanisms to protect themselves from the adverse effects of the antibiotics they produce. Mechanisms of cellular self-protection include drug inactivation by chemical modification, target site modification, drug binding, and reduction of the intracellular concentration using an efflux pump system and/or a low permeability to the antibiotic (13). Genes coding for antibiotic biosynthesis and resistance are typically clustered, and their expression is interdependent physiologically.

In *Streptomyces kanamyceticus* and *Streptomyces fradiae*, producers of kanamycin and neomycin, respectively, aminoglycoside 6′-*N*-acetyltransferase provides the host with resistance to these antibiotics by *N*-acetylation of the compound. Crameri and Davies (11) subcloned the gene for aminoglycoside 6′-*N*-acetyltranferase and introduced it into the producers on a high-copy vector, pIJ702. The transformed *S. kanamyceticus* showed three- to fourfold-enhanced final titers of kanamycin. Transformed *S. fradiae* had up to sevenfold-enhanced titers of neomycin. The transformed strains showed an increased level of resistance against the antibiotics as well.

In many cases, cellular resistance to the produced antibiotic is inducible. Constitutive expression of the resistance gene in such producers may lead to enhanced productivity. Constitutively resistant strains of *S. fradiae* and *Streptomyces lincolnensis* had higher yields of tylosin and lincomycin, respectively (45).

59.2.1.4. Membrane Transport

Many industrial microorganisms produce antibiotics at levels exceeding tens of grams per liter at a similar level of biomass. In a highly productive antibiotic manufacturing process, the titer of antibiotics in the fermentation broth can reach as high as 100 mM. Without an active transport mechanism for the product, the intracellular concentration of the antibiotics must be higher than that in the fermentation broth, in order for the excretion to proceed. For such highly productive microorganisms, there must be either an active transport mechanism for the excretion of the product, or a very high intracellular concentration of the product in a facilitated membrane transport system. The latter is unlikely, since such a high concentration of product would have caused physiological changes (such as pH) in cytoplasm. Alternatively, there must be a transporter or membrane binding protein to allow for a localized high product concentration in the cytosic side of the membrane.

Improving the transport system for the secondary metabolite production can potentially have a very high payoff. Antibiotic transport proteins (and putative ones) have been found in many antibiotic-producing actinoimycetes. A class of these transporters relies on proton-dependent transmembrane gradients. Another class, termed ABC (ATP binding cassette) transporters, has also been found in many antibiotic-producing actinomycetes (32).

With an increasing awareness of the importance of antibiotic transport system in the last few years, it is likely that we are going to see a successful application of this strategy in the near future. Gene amplification has been held responsible for overexpression of membrane translocases in drug-resistant tumors (25). A similar mechanism might be employed to increase the rate of antibiotic transport from the producer, if there is a specific transporter. On the other hand, if the transport is via a passive mechanism, one of the ways to enhance transport may be by introducing additional copies of a nonspecific transporter.

59.2.2. Manipulation of Regulatory Genes

Regulation of antibiotic biosynthesis occurs at both genetic and biochemical levels. In *Streptomyces* spp. that produce multiple secondary metabolites, pleiotropic regulators control some or all of those biosynthetic pathways, while the pathway specific activators typically control the production of only a single antibiotic. In addition to pathway specific activators, pathway specific repressors (9) also regulate antibiotic biosynthesis in some cases.

Although most regulatory molecules that control production are intracellular, there are a few extracellular effector molecules with hormone-like properties. Notable examples are A-factor in *Streptomyces griseus* (19) and virginiae butanolides in *Streptomyces virginiae* (46). A-factor and virginiae butanolides are autoregulators whose synthesis turns on their own production and triggers both sporulation and antibiotic production. They are excreted extracellularly to induce antibiotic production and sporulation in neighboring cells (1).

The regulation at the biochemical level involves the control of enzyme activity, including feedback inhibition, activation, and inactivation. Microorganisms also control the flux of precursors and cofactors into secondary metabolism by a variety of means to ensure that these precursors and cofactors are not diverted when they are needed for cell growth. This type of control differs from conventional feedback inhibition, repression, and induction, because it is affected by the balance of fluxes in a metabolic network rather than involving only a single biosynthetic pathway.

From the regulatory point of view, there are many approaches of metabolic engineering that may lead to increased productivity. It is possible to increase the copy number of the positive regulatory gene to enhance antibiotic production. It is also possible to delete the negative regulator (repressor) to allow for constitutive expression of the pathway. Furthermore, the regulatory gene can be made inducible so that the time profile of antibiotic production can be manipulated. However, the current level of understanding on the regulatory mechanism of secondary metabolism and the interactions between the metabolic networks of the primary and secondary metabolism does not provide us with a way of predicting the consequences of altering the regulatory structure. In most cases, the consequences have to be observed experimentally, and no a priori quantitative analysis is feasible.

59.2.2.1. Amplifying Positive Regulatory Genes

Pathway-Specific Regulators

On the genetic level, regulation involves the induction or repression of the expression of the enzymes in the biosynthetic pathways. Positive regulator genes encode activator proteins that bind to the promoter of the structural genes for the biosynthesis of secondary metabolites. These activators, or the transcription of these positive regulatory genes, are themselves under another upper layer of transcriptional control. Increasing the gene product concentration of these positive regulatory genes often leads to an increased transcription of the biosynthetic genes, which in turn results in a higher production of secondary metabolites. Although it has not been reported for secondary metabolism, it is also possible to alter the regulation of the expression of these positive regulatory genes. For example, by changing the promoter for the positive regulatory genes to a constitutive promoter or to an inducible promoter, one may be able to alter the temporal profile of secondary metabolite production.

A regulatory gene, *srm*R, encoding a transcriptional activator is required for the initiation of transcription of the biosynthetic genes. It has been identified to have a role in the biosynthesis of spiramycin, a macrolide produced by *Streptomyces ambofaciens* (16). Introduction of the gene on a multicopy vector increased the final spiramycin titer from 100 to 500 μg/ml. A pathway-specific regulator, *cca*R, has also been identified in the *Streptomyces clavuligerus* cephamycin C gene cluster (34, 44). Additional copies of this regulatory gene on a multicopy plasmid led to an almost twofold-enhanced production of both cephamycin and clavulanic acid (34).

Global Regulators

Besides the pathway-specific regulators, global and pleiotropic regulators in *Streptomyces* spp. are also potential targets of genetic manipulation for enhancing secondary metabolite production. The *abs*A gene in *Streptomyces coelicolor* (3) appears to code for one of the components of a eubacterial two-component sensor kinase-response regulator involved in the negative regulation of antibiotic production. Disruption of the *abs*A gene led to the overproduction of the antibiotics actinorhodin and undecylprodigiosin. A similar two-component signal tranduction system, *cutRS*, has been found in *S. lividans* (7). Insertion disruption of either of the two genes resulted in about a threefold increase in actinorhodin synthesis. In *S. lividans* another such pleiotropic regulator, *afsR2*, which encodes a 63-amino-acid protein, has been identified. When expressed on a high-copy-number plasmid, it stimulates both actinorhodin and undecylprodigiosin production (42).

59.2.2.2. Deletion of Repressor

Genes encoding for transcriptional repressors have been identified in some of the antibiotic biosynthetic gene clusters. These include *mmyR* in the methylenomycin and *actII-orf I* (4) in the actinorhodin gene clusters in *S. coelicolor*; *tcmR* (17) in the tetracenomycin C gene cluster in *S. glaucescen*; *dnrO* (33) in the daunorubicin gene cluster in *Streptomyces peucetius*; and *jadR2* (47) in the jadomycin B gene cluster in *Streptomyces venezuelae*. MmyR mediates pathway-specific negative regulation, and its disruption led to the overproduction of methylenomycin in *S. coelicolor* (9). Some of the other repressors have been shown to regulate adjacent resistance genes. The deletion or inactivation of such repressors may have a positive effect on antibiotic production. However, it is possible that such a genetic alteration will affect the temporal profiles of antibiotic production but not necessarily the overall production.

59.3. METABOLIC ENGINEERING OF WELL-CHARACTERIZED PATHWAYS

A thorough understanding of the biochemical pathway, genetic makeup, and regulation of biosynthesis is critical for metabolic engineering of secondary metabolites. When such information is available, it is possible to measure the fluxes of the precursors and the intermediates involved and to identify the rate-controlling steps. It also provides a rational basis for metabolic engineering (Fig. 2).

59.3.1. Kinetic Analysis Is Combined with Genetic Manipulations

Antibiotic biosynthetic pathways typically consist of a large number of enzyme-mediated reactions. A kinetic model for the biosynthetic system can help one understand the dynamics of the pathway and identify the rate-limiting steps. Sensitivity analysis, or metabolic control analysis (18, 22,

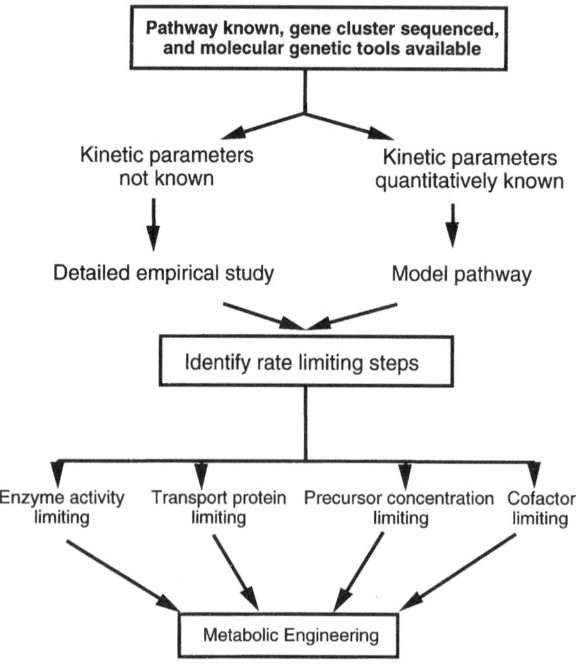

FIGURE 2 Rational metabolic engineering. A well-characterized biosynthetic pathway allows prediction of the rate-limiting steps.

26), can be carried out to identify steps that are exerting the strongest influence on the overall rate of biosynthesis.

A kinetic model typically involves setting up material balance equations for the intermediates in the reaction series making up the biosynthetic pathway. The intracellular concentration of each intermediate is a balance of its synthesis, its consumption by subsequent reactions, and the dilution due to biomass expansion (cell growth). A set of simultaneous differential equations (one for each intermediate) describes the changes of concentrations of the intermediates as a function of time.

$$\frac{dc_i}{dt} = \sum_{j=1}^{k} \alpha_{ij} r_j - \mu c_i \qquad i = 1,\ldots n$$

where c_i denotes the intracellular concentration of the reaction intermediate i, n is the number of intermediates in the system, r_j is the activity of the enzymes catalyzing the reaction j, k is the number of reactions in the system, α_{ij} is the stoichiometric coefficient, and μ is the specific growth rate. μ is defined as follows:

$$\mu = \frac{1}{x}\frac{dx}{dt}$$

where x is the cell concentration. The characteristics of each enzyme reaction are usually assumed to be Michaelis-Menten type. As a first approximation, the kinetic parameters (K_m and v_{max}) determined in vitro can be used for calculations. Experimental values of precursor concentrations and time profiles of enzyme activities provide a convenient starting point for simulation. Furthermore, as a first approximation, one can assume all the cofactors are at saturation concentrations.

Having the system of differential equations to describe the kinetics of antibiotic production and the associated parameters for enzyme kinetics, a pseudo-steady-state assumption can be made. The specific productivity of antibiotics can then be simulated. In general, the time profile of enzyme activities changes over the course of fermentation; therefore, a pseudo-steady-state calculation needs to be repeated for different time points. Ideally, the calculated time profile of antibiotic productivity should resemble the experimental observation, at least qualitatively. Such exercise of simulation of antibiotic production can be repeated to examine various assumptions, for example, the effect of cofactor or cosubstrate concentrations (24).

With the kinetic model, metabolic control analysis can be performed to locate the possible rate-controlling steps. A control coefficient is used for such evaluation. The control coefficient is basically the fractional change of the product formation rate caused by a fractional change of the concentration of a particular enzyme or precursor:

$$\xi_i^P = \frac{E_i}{P}\frac{\partial P}{\partial E_i}$$

where ξ_i is flux control coefficient, E_i is the enzyme concentration being evaluated, and P is the steady-state product flux. A control coefficient with a large value implies that the particular enzyme or precursor has a larger influence on the flux than that with a lower value. The sum of the control coefficients of all enzyme reactions equals 1. The step with the highest value of control coefficient is considered rate-limiting. The identification of the rate-limiting step allows one to pinpoint the enzyme that should be considered the first candidate for gene amplification or deregulation in metabolic engineering.

Malmberg and Hu (27) applied this approach to analyze cephalosporin biosynthesis, for both *S. clavuligerus* and the fungus, *Cephalosporium acremonium* (28). They identified the condensation reaction forming δ-(L-α-aminoadipyl)-L-cysteinyl-D-valine (ACV) tripeptide as the rate-limiting step for both organisms. The simulation of the in vivo kinetics of the cephamycin biosynthesis in wild-type *S. clavuligerus* and subsequent sensitivity analysis also identified the biosynthesis of an intermediate, α-aminoadipic acid (α-AAA), as having the most significant rate-controlling strength among the three precursors of tripeptides. To enhance the flux of α-AAA, an additional copy of the *lat* gene was introduced into the *S. clavuligerus* chromosome by homologous integration (29). *lat* codes for lysine ε-aminotransferase (LAT), an enzyme that mediates the first step in the conversion from lysine to α-AAA. The recombinant strain showed two- fivefold-enhanced productivity (30) and an enhanced intracellular level of the α-AAA.

59.3.2. Increasing Promoter Strength of a Rate-Limiting Enzyme

Once the rate-limiting enzymes involved in the biosynthetic pathway are identified, one can increase the gene dosage of the rate-limiting enzyme in order to increase the product formation rate. Alternatively, one can perform a promoter replacement for the rate-limiting enzyme to increase its transcription level and enzyme activity. Either approach will be effective as long as the enzyme concerned is not strongly feedback regulated.

In the metabolic engineering of penicillin-producing *Aspergillus nidulans*, ACV synthetase was overexpressed by replacing its native promoter with an inducible ethanol dehydrogenase promoter, *alcAp* (23). *A. nidulans* grown with 10 mM cyclopentanone inducer was estimated to have a 100-fold-enhanced ACV synthetase level compared to the control that was grown without the inducer. Penicillin final titers were increased 30-fold as compared to wild type when fermentations were carried out under induction conditions. When the same approach was used to overexpress two downstream enzymes, isopenicillin N synthetase (IPNS) and acyl-CoA:6-aminopenicillanic acid acyltransferase (AAT), no significant enhancement in penicillin titers was observed, even though the transcript level and the enzyme activity were considerably increased (15).

59.3.3. Redirecting the Metabolic Flux to the Branch Leading to the Product

In a divergent branched pathway, one branch usually leads to the final product, while the other results in wasted resources. The reaction intermediates can be channeled to the final product either by blocking the undesirable branch or by increasing the reaction rate toward the final product. In cephalosporin C production with a commercial strain of *C. acremonium*, an intermediate in the cephalosporin biosynthesis, penicillin N, accumulated in the fermentation broth significantly. The average molar ratio of extracellular penicillin N to cephalosporin C was found to be 0.3 (38). From this observation it was reasoned that the conversion of penicillin N to the six-membered ring containing deacetoxycephalosporin was slower as compared to the upstream pathway. As a result, penicillin N accumulated intracellularly, resulting in high-level secretion to the culture broth. To reduce the diversion of intracellular penicillin N to excretion, enzyme deacetoxycephalosporin C (DAOC) was overexpressed by introducing plasmids with the gene *cefEF*. In a transformant that had the plasmid integrated on the chromosome, the production of the antibiotic increased by 15% in pilot plant-scale fermentations. The recombinant strain had an enhanced level of DAOC synthase and did not accumulate intracellular penicillin N.

Another successful example of redirecting the metabolic flux led to the development of a strain of *Streptomyces pristinaespiralis*, in which the accumulation of an undesirable product, pristinamycin PII_B, was eliminated (37). PII_B is an intermediate in the pathway leading to PII_A. The gene coding for PII_A synthase, which causes the conversion of PII_B to PII_A, was overexpressed, resulting in a strain that produced only PII_A.

Both the above examples amplify an enzyme to redirect metabolic flux to a desired product. It is interesting to note that the branched pathway involved in these cases is atypical in the sense that the side reaction is excretion. This strategy of redirecting metabolic flux was successful, partly owing to the absence of strong feedback regulation along the pathway. If the branch point had been a "rigid node" (40), merely increasing the enzyme level of downstream reaction may not have resulted in a favorable redistribution of fluxes.

59.3.4. Disrupting Genes Leading to a Side Product

Many antibiotic producers synthesize a number of antibiotics. Some of them may be the intermediates in the biosynthetic pathway, some by-products in branched pathways, others entirely different molecules. In many cases only one or a small number are desired. The presence of the undesired by-products burdens downstream separation and constrains the overall yield. It is desirable to increase the selectivity of production by eliminating the side products. This can be achieved by deleting either the entire pathway leading to the side product or rendering inactive the first enzyme of that branched pathway.

A successful effort of enhancing product selectivity involved the production of antiparasitic avermectins. *Streptomyces avermitilis* produces eight different but structurally very similar avermectins, A1a, A1b, A2a, A2b, B1a, B2b, B2a, and B2b. Of these, only B1a and B1b possess useful antiparasitic activity. In addition, B2a is useful as the substrate for the manufacture of semisynthetic avermectin, Invermectin B1a, which is the most potent anthelmintic compound among the products. The downstream process challenge is to separate the useful components from the rest and to remove an additional antibiotic, oligomycin, which is also produced by the strain. Since the major components of the biosynthetic pathway have been elucidated and the gene cluster coding for the enzymes has been located, it is possible to devise an alternative process using a strain derived by metabolic engineering. Ikeda and Omura (20) used random mutagenesis to isolate a strain, K2021, that produces only A1a, A2a, B1a, and B2a and another strain, K2034, that produces only B1a, B1b, B2a, and B2b. Strain K2034 appears to have a mutation in a structural gene, *aveD*, coding for a biosynthetic enzyme, while K2021 is disrupted in the incorporation of a branched-chain fatty acid precursor derived from L-valine. Recombinants that had both the mutations were derived by protoplast fusion from these strains. They were shown to produce only two components, B1a and B2a. Subsequently, the gene *aveC* responsible for a dehydration step converting B2a to B1a was disrupted. The resulting clone, K2099, produced only a single avermectin B2a. However, this clone still produced the toxic polyketide oligomycin. To disrupt oligomycin production, a transposon Tn*4560* was used to produce disruptant clones in the wild-type *S. avermitilis*. Chromosomal DNA fragments were subsequently subcloned from these disruptants into a temperature-sensitive plasmid and used to disrupt the oligomycin biosynthesis in K2099, yielding a strain producing only a single desired avermectin.

59.3.5. Biosynthesis of a Product Previously Made Semi-synthetically

Biotransformation and chemical conversion of secondary metabolites are frequently practiced to give antibiotics more desireable properties. Molecules produced by microorganisms are typically subjected to further processing steps. Instead of multistep processing, a single-step synthesis can potentially be achieved by introducing enzymes from another organism into the producing organism. Essentially, this involves combining two complementary pathways to channel the metabolic flux in secondary metabolism into a new product. Usually, this is done by introducing new enzymes into the already highly productive industrial producers to take advantage of the already existing high productivity in the industrial strain.

7-aminodeacetoxycephalosporanic acid (7-ADCA) or 7-aminocephalosporanic acid (7-ACA) are the two main starting materials for semisynthetic cephalosporins. These starting materials are prepared from cephalosporins by removing D-α-aminoadipyl side chain chemically or enzymatically. Isogai et al. (21) introduced two heterologous bacterial genes coding for a D-amino oxidase and a cephalosporin acylase into a producer strain of *C. acremonium* to produce 7-ACA from cephalosporin C. To ensure expression of the genes in the fungal strain, the genes were modified by the addition of expression signals from the cloned *C. acremonium* alkaline protease gene. The developed strain synthesized and secreted 7-ACA, demonstrating the potential for using metabolic engineering to introduce additional antibiotic biosynthetic pathway segments. Another biosynthetic process has been developed recently for 7-ADCA and 7-ACA (12). Genes coding for enzymes from *S. clavuligerus* and *C. acremonium* were introduced in a penicillin-production strain of *P. chrysogenum*. The transformants, when grown in an adipic acid-containing medium, produced adipyl side chain containing cephalosporins. The adipyl side chain was removed through a final amidase-mediated transformation, to yield the desired cephalosporin intermediates.

59.3.6. Synthesizing Antibiotics in Heterologous Strains

Heterologous expression of entire gene clusters for a complete pathway or segment of a pathway has been used for the production of nonnative antibiotics in a number of *Streptomyces* spp. and fungi (31, 39). Production of an antibiotic in a heterologous strain is desirable for many reasons. Among these are higher precursor flux, better resistance to the end product, and fewer by-products in the recipient organism. Also, it might be attractive to carry out all metabolic engineering in a heterologous host, for which better genetic tools are available or the physiological characteristics are better known.

DAOC was produced in an industrial strain of *P. chrysogenum* by transforming it with two hybrid genes, *cef*D_h and *cef*E_h, coding for an isopenicillin N epimerase and DAOC synthase, respectively (5). A new branch was introduced in the pathway to siphon off the intermediate isopenicillin N, resulting in the production of DAOC. However, the resulting strain retained the capability of producing penicillin V because the pathway leading to penicillin V was not disrupted. The disruption was difficult because the production strain had multiple copies of the genes coding for acyltransferase, responsible for converting the isopenicillin N to penicillin V.

59.4. CONCLUSIONS

Metabolic engineering is increasingly being used to exploit the immense potential of secondary metabolism in microbes. Advancement in molecular genetic tools has contributed greatly to furthering our understanding of the genetic structure, as well as physiological regulation, of secondary metabolite synthesis. This has in turn greatly enhanced our ability to manipulate secondary metabolism. However, thus far the metabolic engineering of secondary metabolite production has been largely restricted to engineering "local" pathways, namely, the pathways directly related to biosynthesis. The physiological interactions with other pathways and cellular functions, such as energy production and consumption reactions and the regulation of ribosomal activities, are largely unheeded for lack of fundamental understanding. As genomic information becomes more readily available in the next decade, we will see the basic tools of metabolic engineering evolve. Despite the large number of secondary metabolites successfully developed and the tremendous enhancement in production achieved by generations of industrial microbiologists, the full potential of the microbe's ability to synthesize secondary metabolites has yet to be exploited.

This work was supported in part by NIH grant GM55850.

REFERENCES

1. **Beppu, T.** 1995. Signal transduction and secondary metabolism: prospects for controlling productivity. *Trends Biotechnol.* **132:**264–269.
2. **Bibb, M.** 1996. The regulation of antibiotic production in *Streptomyces coelicolor* A3(2). *Microbiology* **142:**1335–1344.
3. **Brian, P., P. J. Riggle, R. A. Santos, and W. C. Champness.** 1996. Global negative regulation of *Streptomyces coelicolor* antibiotic synthesis mediated by an absA-encoded putative signal transduction system. *J. Bacteriol.* **178:** 3221–3231.
4. **Caballero, J. L., F. Malpartida, and D. A. Hopwood.** 1991. Transcriptional organization and regulation of an antibiotic export complex in the producing *Streptomyces* culture. *Mol. Gen. Genet.* **228:**372–380.
5. **Cantwell, C., R. Beckmann, P. Whiteman, S. W. Queener, and E. P. Abraham.** 1992. Isolation of deacetoxycephalosporin C from fermentation broths of *Penicillium chrysogenum* transformants: construction of a new fungal biosynthetic pathway. *Proc. R. Soc. Lond.* (*Biol.*) **248:** 283–289.
6. **Champness, W. C., and K. F. Chater.** 1994. Regulation and integration of antibiotic production and morphological differentiation in *Streptomyces* spp., p. 61–93. *In* P. Piggot (ed.), *Regulation of Bacterial Differentiation*. American Society for Microbiology, Washington, D.C.
7. **Chang, H. M., M. Y. Chen, Y. T. Shieh, M. J. Bibb, and C. W. Chen.** 1996. The cutRS signal transduction system of *Streptomyces lividans* represses the biosynthesis of the polyketide antibiotic actinorhodin. *Mol. Microbiol.* **21:** 1075–1085.
8. **Chater, K. F., and M. J. Bibb.** 1997. Regulation of bacterial antibiotic production, p. 57–105. *In* H. Kleinkauf and H. von Dohren (ed.), *Products of Secondary Metabolism*, 2nd ed., vol. 7. VCH, Weinheim, Germany.
9. **Chater, K. F., and C. J. Bruton.** 1985. Resistance, regulatory and production genes for the antibiotic methylenomycin are clustered. *EMBO J.* **4:**1893–1897.
10. **Chen, C. W., H.-F. Lin, C. L. Kuo, H.-L. Tsai, and J. F.-Y. Tsai.** 1988. Cloning and expression of a DNA sequence conferring cephamycin C production. *Bio/Technology* **6:**1222–1224.
11. **Crameri, R., and J. E. Davies.** 1986. Increased production of aminoglycosides associated with amplified antibiotic resistance genes. *J. Antibiot.* **39:**128–135.
12. **Crawford, L., A. M. Stepan, P. C. Mcada, J. A. Rambosek, M. J. Conder, V. A. Vinci, and C. D. Reeves.** 1995. Production of cephalosporin intermediates by feeding adipic acid to recombinant *Penicillium chrysogenum* strains expressing ring expansion activity. *Bio/Technology* **13:**58–62.
13. **Cundliffe, E.** 1989. How antibiotic-producing organisms avoid suicide. *Annu. Rev. Microbiol.* **43:**207–233.
14. **Decker, H., R. G. Summers, and C. R. Hutchinson.** 1994. Overproduction of the acyl carrier protein component of a type II polyketide synthase stimulates production of tetracenomycin biosynthetic intermediates in *Streptomyces glaucescens*. *J. Antibiot.* **47:**54–63.
15. **Fernandez-Canon, J. M., and M. A. Penalva.** 1995. Overexpression of two penicillin structural genes in *Aspergillus nidulans*. *Mol. Gen. Genet.* **246:**110–118.
16. **Geistlich, M., R. Losick, J. R. Turner, and R. N. Rao.** 1992. Characterization of a novel regulatory gene governing the expression of a polyketide synthase gene in *Streptomyces ambofaciens*. *Mol. Microbiol.* **6:**2019–2029.
17. **Guilfoile, P. G., and C. R. Hutchinson.** 1992. Sequence and transcriptional analysis of the *Streptomyces glaucescens* tcmAR tetracenomycin C resistance and repressor gene loci. *J. Bacteriol.* **174:**3651–3658.
18. **Heinrich, R., and T. A. Rapoport.** 1974. A linear steady-state treatment of enzymatic chains. General properties, control and effector strength. *Eur. J. Biochem.* **42:**89–95.
19. **Horinouchi, S., and T. Beppu.** 1994. A-factor as a microbial hormone that controls cellular differentiation and secondary metabolism in *Streptomyces griseus*. *Mol. Microbiol.* **12:**859–864.
20. **Ikeda, H., and S. Omura.** 1995. Control of avermectin biosynthesis in *Streptomyces avermitilis* for the selective production of a useful component. *J. Antibiot.* **48:**549–562.
21. **Isogai, T., M. Fukagawa, I. Aramori, M. Iwami, H. Kojo, T. Ono, Y. Ueda, M. Kohsaka, and H. Imanaka.** 1991. Construction of a 7-aminocephalosporanic acid (7ACA) biosynthetic operon and direct production of 7ACA in *Acremonium chrysogenum*. *Bio/Technology* **9:** 188-191.

22. **Kacser, H., and J. A. Burns.** 1973. The control of flux, p. 65–104, *In* D. D. Davies (ed.), *Rate Control of Biological Processes*. Cambridge University Press, Cambridge.

23. **Kennedy, J., and G. Turner.** 1996. Delta-(L-alpha-aminoadipyl)-L-cysteinyl-D-valine synthetase is a rate limiting enzyme for penicillin production in *Aspergillus nidulans*. *Mol. Gen. Genet.* **253:**189–197.

24. **Khetan, A., L. H. Malmberg, D. H. Sherman, and W. S. Hu.** 1996. Metabolic engineering of cephalosporin biosynthesis in *Streptomyces clavuligerus*. *Ann. N.Y. Acad. Sci.* **782:**17–24.

25. **Lewis, K.** 1994. Multidrug resistance pumps in bacteria: variations on a theme. *Trends Biochem. Sci.* **19:**119–123.

26. **Liao, J. C., and J. Delgado** 1993. Advances in metabolic control analysis. *Biotechnol. Prog.* **9:**221–233.

27. **Malmberg, L. H., and W. S. Hu.** 1991. Kinetic analysis of cephalosporin biosynthesis in *Streptomyces clavuligerus*. Biotechnol. Bioeng. **38:**941–947.

28. **Malmberg, L. H., and W. S. Hu.** 1992. Identification of rate-limiting steps in cephalosporin C biosynthesis in *Cephalosporium acremonium*: a theorectival analysis. *Appl. Microbiol. Biotechnol.* **38:**122–128.

29. **Malmberg, L. H., W. S. Hu, and D. H. Sherman.** 1993. Precursor flux control through targeted chromosomal insertion of the lysine epsilon-aminotransferase (*lat*) gene in cephamycin C biosynthesis. *J. Bacteriol.* **175:**6916–6924.

30. **Malmberg, L. H., W. S. Hu, and D. H. Sherman.** 1995. Effects of enhanced lysine epsilon-aminotransferase activity on cephamycin biosynthesis in *Streptomyces clavuligerus*. *Appl. Microbiol. Biotechnol.* **44:**198–205.

31. **Malpartida, F., and D. A. Hopwood.** 1984. Molecular cloning of the whole biosynthetic pathway of a *Streptomyces* antibiotic and its expression in a heterologous host. *Nature* **309:**462–464.

32. **Mendez, C., and J. A. Salas.** 1998. ABC transporters in antibiotic-producing actinomycetes. *FEMS Microbiol. Lett.* **158:**1–8.

33. **Otten, S. L., J. Ferguson, and C. R. Hutchinson.** 1995. Regulation of daunorubicin production in *Streptomyces peucetius* by the dnrR2 locus. *J. Bacteriol.* **177:**1216–1224.

34. **Perez-Llarena, F. J., P. Liras, A. Rodriguez-Garcia, and J. F. Martin.** 1997. A regulatory gene (ccaR) required for cephamycin and clavulanic acid production in *Streptomyces clavuligerus*: amplification results in overproduction of both beta-lactam compounds. *J. Bacteriol.* **179:**2053–2059.

35. **Queener, S. W., and D. H. Lively.** 1986. Screening and selection for strain improvement, p. 155–169. *In* A. L. Demain and N. A. Solomon (ed.), *Manual of Industrial Microbiology and Biotechnology*. American Society for Microbiology, Washington, D.C.

36. **Rowlands, R. T.** 1984. Industrial strain improvement: mutagenesis and random screening procedures. *Enzyme Micro. Technol.* **6:**3–10.

37. **Sezonov, G., V. Blanc, N. Bamasjacques, A. Friedmann, J. L. Pernodet, and M. Guerineau.** 1997. Complete conversion of antibiotic precursor to pristinamycin IIa by overexpression of *Streptomyces pristinaespiralis* biosynthetic genes. *Nat. Biotechnol.* **15:**349–353.

38. **Skatrud, P. L.** 1992. Genetic engineering of a beta-lactam antibiotic biosynthetic pathways in filamentous fungi. *Trends Biotechnol.* **10:**324–329.

39. **Smith, D. J., M. K. Burnham, J. Edwards, A. J. Earl, and G. Turner.** 1990. Cloning and heterologous expression of the penicillin biosynthetic gene cluster for *Penicillum chrysogenum*. *Bio/Technology* **8:**39–41.

40. **Stephanopoulos, G., and J. J. Vallino.** 1991. Network rigidity and metabolic engineering in metabolite overproduction. *Science* **252:**1675–1681.

41. **Thomas, D. I., J. H. Cove, S. Baumberg, C. A. Jones, and B. A. Rudd.** 1991. Plasmid effects on secondary metabolite production by a streptomycete synthesizing an anthelmintic macrolide. *J. Gen. Microbiol.* **137:**2331–2337.

42. **Vogtli, M., P. C. Chang, and S. N. Cohen.** 1994. afsR2: a previously undetected gene encoding a 63-amino-acid protein that stimulates antibiotic production in *Streptomyces lividans*. *Mol. Microbiol.* **14:**643–653.

43. **Vournakis, J. N., and R. P. Elander.** 1983. Genetic manipulation of antibiotic-producing microorganisms. *Science* **219:**703–709.

44. **Waters, N. J., B. Barton, and A. J. Earl.** August 1994. International patent WO 9418326.

45. **Weisblum, B.** March 1983. U.S. patent 4376823.

46. **Yamada, Y., T. Nihira, and S. Sakuda.** 1997. Butyrolactone autoregulators, inducers of virginiamycin in *Streptomyces virginiae*: their structures, biosynthesis, receptor proteins, and induction of virginiamycin biosynthesis, p. 63–79. *In* W. R. Strohl (ed.), *Biotechnology of Antibiotics*, 2nd ed. Marcel Dekker, New York.

47. **Yang, K., L. Han, and L. C. Vining.** 1995. Regulation of jadomycin B production in *Streptomyces venezuelae* ISP5230: involvement of a repressor gene, *jadR2*. *J. Baceriol.* **177:**6111–6117.

Cloning and Analysis of Regulatory Genes Involved in *Streptomyces* Secondary Metabolite Biosynthesis

WENDY CHAMPNESS

60

The vast array of medically and agriculturally useful streptomycete secondary metabolites are generally produced in a growth-phase-dependent profile (33, 34). In this chapter I review the genetic elements known to be involved in regulation of the streptomycete antibiotics. Additional reviews on antibiotic regulation have been published (12, 26, 30, 51, 57). I also discuss experimental approaches for further exploration of the molecular genetic regulatory mechanisms.

60.1. PATHWAY-SPECIFIC REGULATORS

It is a general paradigm that the biosynthetic genes for an antibiotic are clustered. The *Streptomyces coelicolor act* genes for biosynthesis of actinorhodin are the prototypical examples; these genes occupy a region of approximately 25 kb. Transcription of the *act* genes, which occurs in several polycistronic mRNAs, depends on the product of a cluster-linked gene named *actII-ORF4*. ActII-ORF4 has been termed a "pathway-specific regulator." Discovery of *actII-ORF4* stemmed from cosynthesis tests of a collection of *act* (actinorhodin-nonproducing) mutants (119). *actII* mutants failed to cosynthesize with any other *act* mutant (119); these were later found to be complemented by a cloned DNA sequence (82). An activator function has been ascribed to ActII-ORF4 because *act* biosynthetic genes are not expressed in the *actII-ORF4* mutants (39, 49), and cloned extra copies of *actII-ORF4* cause actinorhodin overproduction (29).

60.1.1. Importance of Pathway-Specific Activators as a Limiting Factor in Antibiotic Synthesis

Two lines of evidence suggest that temporal regulation of the pathway-specific activators is largely responsible for growth-phase-dependent antibiotic production in defined media (48, 137a). First, accumulation of the *actII-ORF4* and *redD* (see below) transcripts, as assessed by nuclease protection assays, is limited to the postexponential growth period. Second, introduction of plasmid-borne copies of either *actII-ORF4* or *redD* causes exponential-phase antibiotic production.

60.1.2. Examples of Pathway-Specific Regulators

A pathway-specific activator for the *S. coelicolor* antibiotic undecylprodigiosin, *redD*, has been identified by criteria similar to those for *actII-ORF4* (83, 95, 120). Additional pathway-specific activators are listed in Table 1.

Among these regulators, the StrR, DnrN, and DnrI proteins have been demonstrated to be DNA binding proteins. The DnrI protein binds to DNA in the region of several *dnr* promoters. DnrI lacks an identifiable helix-turn-helix DNA binding motif (e.g., 94) and may define a new class of DNA binding protein (135, 140). The *S. coelicolor* ActII-ORF4 and RedD proteins are related to DnrI; their amino acid sequences share 33 to 37% amino acid similarity (39, 95, 135). DNA binding activity has not yet been shown for either ActII-ORF4 or RedD, but an *actII-ORF4* clone stimulates daunorubicin production, and cloned *dnrI* partially complements mutations in *actII-ORF4*, suggesting that ActII-ORF4's function is similar to that of DnrI (135, 140).

Several of the antibiotic pathway-specific activators are members of a novel family of regulatory proteins named SARPs (for *Streptomyces* antiobiotic regulatory proteins). The SARP family includes ActII-ORF4 and RedD, which regulate actinorhodin and undecylprodigiosin, respectively, in *S. coelicolor*; DnrI, which regulates daunorubicin in *Streptomyces peucetius*; SnoA, which regulates nogalamycin in *Streptomyces nogalater*; CcaR, which regulates cephamycin and clavulanic acid in *Streptomyces clavuligerus*; and the pleiotropic AfsR of *S. coelicolor*, which regulates multiple antibiotics (reviewed in reference 151). One of these proteins, DnrI, has been shown to bind promoter regions of the daunorubicin gene cluster (140). The SARPs are predicted (151) to contain N-terminal OmpR-like DNA binding domains (93), which bind promoter regions at heptameric direct repeats.

The *dnrN* gene, also associated with the *dnr* cluster, is required for transcription of the *dnrI* regulatory gene (46, 81). DnrN is related to the two-component response regulator class of proteins, but binding of purified DnrN to the *dnrI* promoter region does not require phosphorylation, nor does mutation of the aspartate residue commonly phosphorylated in response regulators prevent DNA binding (46). Moreover, the *dnrN* gene lacks a companion sensor kinase

TABLE 1 Streptomycete antibiotic pathway-specific regulators

Antibiotic	Gene	Criteria in discovery	Function
Actinorhodin (Act)	*actII-ORF4*	1. Mutant failed to cosynethesize with *act* mutants (119) 2. Mutant failed to express *actIII* (49) 3. Cloned extra copies elicit actinorhodin overproduction (29)	1. Activation of *act* genes 2. Member of SARP family of DNA-binding regulators (151)
Undecylprodigiosin (Red)	*redD*	1. Mutant failed to cosynthesize with *red* mutants (120) 2. Mutant failed to express *redE,F* (95) 3. Cloned extra copies elicited undecylprodigiosin overproduction (29, 95)	1. Activation of *red* genes 2. Member of SARP family of DNA-binding regulators (151)
	redZ	1. Site of *pwb* (for pigmented while bld) suppressor mutation (48a) 2. Disruption caused undecylprodigiosin-minus phenotype (150) 3. Disruption caused reduction in *redD* transcript (150)	1. Activation of *redD* expression (150) 2. Related to response regulators but lacks phosphorylation pocket (48a) 3. Related to DnrN (48a)
Daunorubicin	*dnrI*	1. Disruption caused daunorubicin-minus phenotype (135) 2. Disruption caused loss of *dnr* biosynthetic gene expression (81)	1. Activation of *dnr* genes 2. DNA binding activity in promoter region of *dnr* transcripts (140) 3. Member of SARP family of DNA-binding regulators (151)
	dnrN	1. Disruption caused daunorubicin-minus phenotype (81) 2. Disruption caused loss of *dnrI* expression (46, 81)	1. Activation of *dnrI* expression (46) 2. Related to response regulators but no evidence for phosphorylation (68, 107, 135)
Streptomycin	*strR*	1. Cloned *strR* activated a *str* biosynthetic gene promoter (35)	1. Activation of *str* genes (35) 2. DNA binding protein (35)
Spiramycin	*srmR*	1. Mutant failed to cosynthesize with *srm* mutants (115) 2. Mutant failed to express *srm* biosynthetic genes (47) 3. Cloned extra copies elicited spiramycin overproduction (47)	1. Activation of *srm* genes 2. Lacks similarity to database proteins (47)
Bialaphos	*brpA*	1. Mutant lacked seven bialaphos transcripts and 27 proteins (6, 52)	1. Activation of *brp* genes 2. Putative helix-turn-helix DNA binding motif (114)
Methylenomycin	*mmyR*	1. Insertion mutation or deletion resulted in methylenomycin overproduction (31)	1. Putative repressor related to *tetR* family (31)
Cephamycin, clavulanic acid	*ccaR*	1. Sequence homology (111)	1. Member of SARP family of DNA binding regulators (151)
Clavulanic acid	*claR*	1. Sequence homology (108) 2. Disruption caused loss of "late" gene transcription (108)	1. Member of LysR family of transcription regulators (108)

protein-encoding gene. These observations suggest that DnrN's regulatory activity is modulated by a mechanism other than phosphorylation (46).

The *S. coelicolor red* cluster contains a homolog of *dnrN* named *redZ*. Disruption of *redZ* results in a defect in undecylprodigiosin production, as well as *redD* transcription, suggesting that *redZ*'s role in regulation is mediated through activation of *redD* (11, 150).

One of the few examples of negatively acting pathway-specific regulators is the *mmyR* gene of the *S. coelicolor* methylenomycin gene cluster. Defined by disruption and deletion mutations that result in methylenomycin overproduction, the *mmyR* gene is predicted to encode a product that resembles repressor proteins of the *tetR* family (reviewed in references 30 and 31).

Pathway-specific regulators have not been reported for some of the large antibiotic gene clusters, such as erythromycin (36). In another example, the rapamycin biosynthetic gene cluster of *Streptomyces hygroscopicus* occupies more than 100 kb. Sequencing DNA adjacent to the cluster has revealed a sensor kinase-response regulator gene pair (126), but whether or not these genes function in rapamycin regulation is not known.

60.2. REVIEW OF GENETIC MECHANISMS RELATED TO GROWTH-PHASE DEPENDENCE OF SECONDARY METABOLISM

Genetic regulation of secondary metabolism involves mechanisms for detecting and integrating information about population density and nutritional starvation and, perhaps, growth rate, cell cycle, and other physiological and environmental parameters. In this section I discuss the genetically best characterized of these systems.

60.2.1. Population Density and Intercellular Signaling

A large family of γ-butyrolactones function at very low effective concentrations as chemical cellular signaling molecules in streptomycetes, regulating both morphogenesis and secondary metabolism (61). The *Streptomyces griseus* γ-butyrolactone A-factor is the best-studied example. Required for streptomycin production (reviewed in reference 62), A-factor is detectable in culture medium just before streptomycin production commences. Genetic studies on A-factor's role in *S. griseus* (50, 65, 91, 92) have identified a specific A-factor receptor protein (ArpA) that functions as a DNA binding (106) repressor during early stages of growth. A-factor is produced in a growth-phase-dependent manner and binds to ArpA to release it from its DNA targets, a critical one of which is postulated to be gene X, which in turn regulates expression of the so-called *adp* genes (for A-factor-dependent proteins). One *adp* gene, *adpB*, has been implicated in morphological regulation (141), whereas another is hypothesized to regulate streptomycin synthesis by binding to and activating the *strR* pathway-specific regulator.

Compounds related to the *S. griseus* A-factor function in antibiotic production in diverse streptomycetes (61). Another example involves the *Streptomyces virginiae* virginiamycin inducers, and genes for autoregulator receptors predicted to function as DNA binding repressors have also been cloned from *S. virginiae* (104) and *Streptomyces* sp. strain FRI-5 (146).

In early work, *S. griseus* A-factor seemed not to play a role in antibiotic production or morphologenesis in *S. coelicolor* (reviewed in reference 26), despite its importance in *S. griseus*. However, recent work has established an important role for γ-butyrolactone-mediated regulation in *S. coelicolor*. *S. coelicolor* produces at least four low-molecular-weight compounds (70a) that are capable of causing precocious antibiotic production in the wild-type strain. The structure of one of these, provisionally named ScbI, has been determined and is similar to that of A-factor. Moreover, ScbI antagonizes specific DNA binding by an *S. coelicolor* homolog of ArpA (137). Two additional ArpA homologs have been cloned and partially characterized genetically: *cprA* encodes a positive regulator of both morphogenesis and antibiotic production, and *cprB* encodes a negative regulator, also affecting both morphogenesis and antibiotic production (105).

An additional system of intercellular signaling is exemplified by an aspect of *bld* (for "bald," i.e., nonsporulating; see section 60.3 below) mutant behavior. Many such mutants participate in a phenomenon of "extracellular complementation" of their sporulation defects. This effect occurs between pairs of mutants growing in close proximity and has been hypothesized to reflect the operation of a cascade of signaling steps that control synthesis of a morphogenetic protein, SapB (152, 153). Whether or not this putative signaling cascade also regulates secondary metabolism is not yet known (97).

60.2.2. Nutritional Depletion

Among the numerous complex nutritional effects (e.g., 7, 121) on antibiotic synthesis that have been documented (reviewed in references 33 and 34), the following *S. coelicolor* phenomena have been genetically characterized.

60.2.2.1. Carbon metabolism

A phenotype displayed by many *bld* mutants of *s. coelicolor* is a defect in regulation of carbon-source utilization (112, 113). *bldB* mutants are the most pleiotropically affected, abnormally expressing a variety of metabolic operons in the absence of their normal inducers (e.g., galactose, glycerol). These observations suggest a genetic connection between carbon utilization and differentiation (112).

Signal molecules of the homoserine lactone (HSL) class (e.g., 45) are important to antibiotic regulation in various gram-negative bacteria, including *Escherichia coli* and members of the *Erwinia* and *Pseudomonas* genera (8, 9, 66, 136). Regulation by an HSL in *E.coli* may involve a starvation-response pathway in which the RspA protein affects HSL-dependent induction of the stationary-phase-specific sigma factor σ^S by degrading intracellular HSL (67). An *S. coelicolor* homolog of RspA, SpaA, has been identified and partially characterized (124). *spaA* was cloned from the *S. coelicolor* cosmid library by probing with an *rspA* homolog from *Streptomyces ambofacines*, which was identified adventitiously by sequence relatedness. Disruption of the *S. coelicolor spaA* locus reduced and delayed actinorhodin and undecylprodigiosin production on nutritionally poor (mannitol minimal) medium at low colony density, whereas at high colony density Act was overproduced, suggesting a role for *spaA* in a signaling pathway. The reported failure to detect extracellular interactions between *spaA*$^+$ and *spaA* mutant cultures (124) would be consistent with *spaA* involvement in an intracellular starvation-sensing pathway, such as that proposed for HSL in *E. coli* (156).

60.2.2.2. ppGpp

Antibiotic biosynthesis has been correlated with ppGpp accumulation in a variety of studies in *Streptomyces* species (10, 71, 85, 100–103, 130, 134). These observations prompted isolation and characterization of mutants defective in ppGpp metabolism. Two classes of *S. coelicolor* mutants have been studied: *relC* mutants, defective in the L11 protein that activates ribosome-associated ppGpp synthetase in amino acid starvation conditions; and *relA* mutants, defective in the ribosome-dependent ppGpp synthetase RelA. Mutants of both classes are deficient in antibiotic production (22, 23, 103).

The *S. coelicolor relA* gene was cloned through a PCR-based approach (23), using degenerate oligonucleotide primers corresponding to amino acid blocks conserved in the *relA* genes of *E. coli*, *Mycobacterium leprae*, *Streptococcus equisimilis*, and *Vibrio* sp. (21, 41, 90). A null allele of the *relA* gene affects *S. coelicolor* actinorhodin and undecylprodigiosin production on some media but not others. Specifically, this contrast (22) has been observed in conditions of nitrogen limitation, in which the Δ*relA* strain produces no actinorhodin or undecylprodigiosin; otherwise, the Δ*relA* strain produces both actinorhodin and undecylprodigiosin normally.

60.2.3. Phosphoprotein Signal Transduction

60.2.3.1. Two-Component Signal Transduction

In prokaryotes, the two-component signal transduction system is employed in numerous situations in which a cell expresses specific genes in response to a specific environmental factor (109, 110, 132, 133). The general paradigm is that a sensor protein senses an environmental signal and then autophosphorylates. A companion protein, which is often a transcriptional regulator, then receives the phosphoryl group, with the result being modulation of its activity as a regulator.

At present, three *S. coelicolor* two-component systems have been shown to be important to antibiotic regulation: *absA1*/*absA2* (14), *cutR*/*cutS* (28), and *afsQ1*/*afsQ2* (69). The *afsQ1*/*afsQ2* system exerts a positive effect on antibiotic production when cloned in a low-copy plasmid, but neither *afsQ1* nor *afsQ2* is required for antibiotic synthesis, at least in the growth conditions under which disruption mutants were studied (69). *absA1*/*absA2* and *cutR*/*cutS* function as negative regulators, and disruption mutations result in antibiotic overproduction (14, 28). The signals sensed by the sensor kinases, AbsA1, CutS, and AfsQ2, have not been defined, nor is it known whether these genes function is related or in independent signaling pathways.

60.2.3.2. Ser-Thr-Tyr Phosphoprotein Signal Transduction

Several Ser-Thr-Tyr phosphoproteins have been identified in streptomycetes (53, 54, 86, 142, 149). In eukaryotes, such proteins are critical in signaling pathways. In prokaryotes, the two-component-type signaling proteins are widely used, but Ser-Thr-Tyr phosphorylations are less common (60, 157).

The AfsR regulatory protein is the best-studied example of this class of protein. The *afsR* gene was discovered because multiple copies of *afsR* elicit overproduction of the antibiotic pigments actinorhodin and undecylprodigiosin (66, 131). AfsR is phosphorylated on Ser and Thr residues by the adjacently encoded AfsK kinase, which autophosphorylates on Ser and Tyr residues (86). The signal to which the AfsK-AfsR system responds is not known.

A phosphotyrosine phosphatase, encoded by the *ptpA* gene (78), was also discovered as a multicopy antibiotic-enhancing clone (59). Thus far, the target of *ptpA* action has not been identified, but it is not AfsK (59).

Additional *S. coelicolor* protein kinases are encoded by the *pkaA* and *pkaB* genes (142). Their roles in antibiotic regulation have not been reported.

60.2.4. Sigma Factors

Many bacteria, including pseudomonads, *E. coli*, and *Bacillus* sp., use alternative sigma factors to regulate growth-phase-dependent antibiotic gene expression (66, 84, 125). In *S. coelicolor*, the actinorhodin and undecylprodigiosin pathway-specific activators are presumed to be transcribed in vivo by holoenzyme containing the major (18, 139) vegetative sigma factor HrdB (12, 44). A second sigma factor, HrdD, can direct transcription from these genes' promoters in vitro but has no obligate in vivo transcriptional role, since disruption of *hrdD* causes no defect in actinorhodin or undecylprodigiosin production (19, 20, 44).

Involvement of an alternative sigma factor in streptomycete antibiotic synthesis has been observed in the case of sigma E, a member of the ECF (extracytoplasmic function) sigma factor subfamily (80). In *Streptomyces antibioticus*, a *sigE* null mutant fails to make actinomycin (70). The in vivo sigma E targets responsible for this phenotype are not yet known.

60.3. THE *bld* GENES

An additional genetic aspect of streptomycete antibiotic regulation involves genes identified by *bld* mutations, so named because they block aerial mycelium formation and therefore cause a bald phenotype. Most mutants that have been isolated for their Bld$^-$ phenotype are also blocked for antibiotic production (25, 89; reviewed in references 26 and 30); for example, in *S. coelicolor*, mutations in many *bld* genes affect all four of the known antibiotics, whereas some *bld* mutations affect only one or two of the antibiotics. The mechanisms by which *bld* genes regulate either morphogenesis and antibiotic production are not understood in detail, bur recent molecular characterizations are beginning to provide clues.

The most extensively studied *bld* gene is *bldA*. The *S. coelicolor bldA* gene plays a pleiotropic role in antibiotic synthesis, as well as in morphological differentiation. First defined by mutations that caused a bald phenotype (aerial mycelium-minus), the *bldA* gene encodes the only tRNA that translates a leucine codon (UUA) that occurs rarely in *Streptomyces* genes (75, 76). The *actII-ORF4* (actinorhodin-specific activator) gene includes a UUA codon; thus, actinorhodin production requires *bldA* function (39). Similarly, the undecylprodigiosin-specific activator *redZ* contains a UUA codon (150). In addition, production of methylenomycin and calcium-dependent antibiotic required *bldA*, although much less is known about the role of *bldA* in these cases. The *bldA* gene is also important in *S. griseus* morphological and physiological development; *S. griseus bldA* mutants exhibit a pleiotropic sporulation-minus, antibiotic-minus phenotype (74, 88).

The growth-phase profile of *bldA* transcript abundance has varied in two studies. In one (48), *bldA* transcripts were present throughout the culture's growth in liquid; in another (77), *bldA* showed growth-phase regulation on plate-grown cultures as well as in liquid cultures.

The *bldB* (113) and *bldD* (37) genes have been cloned by complementation, sequenced, and partially characterized. Neither of their sequences resemble any known proteins, but both are predicted to have DNA binding activity.

The *bldG* mutations (25) affect a complex locus that shows a high degree of similarity to *Bacillus subtilis* anti-sigma/anti-anti-sigma genes (reviewed in reference 37a). The *bldG103*) mutation maps to an open reading frame (ORF) encoding a product with significant similarity to the anti-anti-sigmas, and the ORF downstream, which is likely to be cotranscribed, shows significant similarity to anti-sigmas (147).

The *bldK* locus includes five ORFs that encode subunits of the oligopeptide permease family of ATP-binding (ABC) membrane-spanning transporters (99). A *bldK*-imported signal is proposed to function at the first step of the putative signaling cascade involved in morphological differentiation. A partially characterized factor that can be purified from conditioned media is a strong candidate for this signal (98).

The *brgA* gene mutates to a Bld^- phenotype but was discovered (129) as the site of a mutation conferring resistance to an inhibitor of ADP-ribosyltransferase (3-aminobenzamide). The *brgA* mutant, as well as *bldB*, *C*, and *H* mutants, show defects in protein ADP-ribosylation; the connection of this phenotype to differentiation has not been established.

60.4. GLOBAL REGULATION OF SECONDARY METABOLITES

The known antibiotics of *S. coelicolor* are subject to global regulation that is, in part, separate from sporulation regulation (4). The genetic elements responsible for global regulation (e.g., 1) are now being discovered through a variety of experimental approaches undertaken in several laboratories (Table 2). This work has primarily utilized *S. coelicolor* and the closely related *Streptomyces lividans*. These species, especially *S. coelicolor*, offer substantial advantages over other streptomycetes for the study of global regulation. *S. coelicolor* strains produce three characterized antibiotics—actinorhodin, undecylprodigiosin, and calcium-dependent antibiotic—and those strains that carry the SCP1 plasmid produce a fourth antibiotic, methylenomycin. Not only are these antibiotics' gene clusters relatively well characterized (25, 30–32, 38, 39, 57, 58, 73, 82, 83, 119, 120, 155), but the actinorhodin and undecylprodigiosin antibiotics are dramatic pigments. These attributes make genetic study of coordinate regulation relatively easy. Moreover, the absence of proprietary interests in *S. coelicolor* antibiotics greatly enhances information exchange.

Coordinate regulation of the two pigments actinorhodin and undecylprodigiosin was first observed (50, 64) in the course of a study of A-factor's possible role in *S. griseus* and *S. coelicolor* antibiotic production. One outcome of this work was discovery of the *afsR* gene as a plasmid-cloned sequence that caused Act and Red overproduction (64).

Next, the *abs* regulatory loci were discovered in a mutant hunt based on the following rationale (2, 4). Many *S. coelicolor bld* mutants are blocked for all four antibiotics as well as sporulation, suggesting the existence of a genetic pathway coregulating both morphological and physiological differentiation. If this pathway branched, with one branch regulating morphological differentiation and the other branch regulating physiological differentiation (antibiotic synthesis), identification of genes functioning in the latter branch might follow from isolation of mutants blocked for all four antibiotics without being blocked in sporulation. Isolation of mutants with such a phenotype (Abs^-) led to definition of the *absA* (4) and *absB* (2) loci.

60.5. GENERALITY OF REGULATORS THROUGHOUT THE GENUS

Essentially all of the knowledge regarding streptomycete global regulation has been obtained through study of *S. coelicolor* and *S. lividans*. This is due in part to the experimental advantages of these strains and in part to the product-oriented emphasis of much streptomycete research. Future research will profitably continue to focus on *S. coelicolor* and *S. lividans*, unless another species supplants these as a model system for genetics. The scientific precedent for working out the nature of regulatory mechanisms in model systems and then generalizing these mechanisms to the broader biological world is clear. Thus, the general approach to understanding global regulation would be, first, to fully exploit the opportunities to discover and characterize genes and their functions in *S. coelicolor*/*S. lividans*; second, to clone homologs from other actinomycetes; and third to create mutations in actinomycete "x" by gene replacement.

An example of the power of such an approach is the study of *spo0A* function in a diverse array of endospore-forming bacteria (15). The *spo0A* gene is a two-component-type response regulator required for sporulation in *B. subtilis*. Presumptive homolog of *spo0A* were cloned, using PCR, from eight *Bacillus* and six *Clostridium* species. Gene disruptions attempted in two of these bacteria, *Bacillus anthracis* and *Clostridium acetobutylicum*, demonstrated that the *spo0A* requirement for sporulation was conserved in these bacteria.

PCR has been used to clone various streptomycete regulatory genes. The general approach has been to design oligonucleotide primers from highly conserved amino acid sequences. Among numerous examples in the literature are the cloning of the *S. coelicolor relA* (23) gene, which used primers corresponding to conserved amino acid blocks from the *E. coli relA* gene, and cloning of the *S. coelicolor* protein kinase genes *pkaA* and *pkaB* (142), which used primers corresponding to consensus sequences in Ser/Thr kinases. A general approach for using degenerate PCR primers to amplify homologs is shown in Table 3.

60.6. APPROACHES TO IDENTIFICATION OF REGULATORY GENES

60.6.1. Mutant Hunts

60.6.1.1. The Screen

The first step in planning a mutant hunt is to devise a screen for the desired type of mutant. A critical aspect of the planning process is to predict the mutant phenotypes that are likely to result from mutations in regulatory genes. For example, the Abs^- phenotype was hypothesized following the rationale that if a gene existed that globally regulated *S. coelicolor's* four antibiotics but not sporulation, mutation in such a gene might create a mutant that spor-

TABLE 2 Global regulatory genes[a]

Global regulator			Effect on *S. coelicolor* antibiotics		
Locus	Gene	Function	Mutant alleles	Gene-specific knockout	Multicopy
absA1/A2	*absA1*	Two-component sensor kinase (14)	UV alleles: Abs⁻	Δ; Act, Red ↑ (5)	Act, Red ↑ (5)
	absA2	Two-component response regulator (14)		Δ; Act, Red ↑ (5)	ND
cutR/S	*cutR*	Two-component response regulator (28)	Disruption: Act, Red ↑	ND	Act ↓
	cutS	Two-component sensor kinase (28)	Disruption: Act, Red ↑	ND	Not reported
afsQ1/Q2	*afsQ1*	Two-component response regulator (69)	Disruption: no phenotype	ND	Act, Red ↑
	afsQ2	Two-component sensor kinase (69)	Disruption: no phenotype	ND	No phenotype
afsR/K/R2	*afsR*	Ser-Thr phosphoprotein (86); member DnrI family (135, 151)	Disruptions: Act$^{\pm}$ Red$^{\pm}$ (63)	media-dependent Act⁻ Red⁻ CDA$^{\pm}$ (42)	Act, Red ↑
	afsK	Serine-threonine protein kinase (86)	Act$^{\pm}$ Red$^{\pm}$	ND	Act, Red ↑
	afsR2 (*afsS*)	Unknown (87, 143)		ND	Act, Red ↑
ptpA	*ptpA*	Phosphotyrosine phosphatase (78), target not known (59)	59	ND	Act, Red ↑
aba	*abaA*	Unknown (40)	Decreased Act, Red, CDA	ND	Act, Red ↑
micX	ND	Unknown, possibly antisense (117, 118)	ND	ND	Act, Red ↑
mia	ND	90-nt sequence (16, 27, 116), unknown function	ND	ND	Act ↓ Red ↓ CDA ↓

[a]ND, not determined; nt, nucleotide; Act, actinorhodin; Red, undecylprodigiosin; CDA, calcium-dependent antibiotic.

ulated but failed to produce any of the four antibiotics (see section 60.4 above). In the hunt (2, 4) for the hypothesized Abs⁻ mutants, the screen involved visual observation of the actinorhodin (Act) and undecylprodigiosin (Red) pigments, as well as sporulation, on plate-grown colonies arising from mutagenized spores. Mutations in genes coordinately regulating *act* and *red* genes would be predicted to affect both pigments, whereas mutation to an *act* or *red* pathway gene would create an Act⁻ Red⁺ or Act⁻ Red⁻ phenotype, respectively. Unpigmented, sporulating colonies could then be screened by bacterial inhibition plate assay for loss of the other two antibiotics, methylenomycin and calcium-dependent antibiotic.

In addition to the Abs⁻ phenotype, a second phenotype that might result from mutation to a global regulator would be overproduction of multiple antibiotics. Indeed, some mutations in both the *absA1/A2* (14) and *cutR/S* (28) two-component gene systems result in precocious hyperproduction of antibiotics.

TABLE 3 Use of degenerate PCR primers

1. Choose a 7- to 10-amino-acid conserved region.
2. Use preferred codons from streptomycete genes (154).
3. Selection of amino acids with the least degeneracy is desirable because it provides the greatest specificity.
4. Avoid a degenerate base at the 3′ terminus.
5. Omit the last base of the terminal codon unless the amino acid is Met or Trp.
6. Six to nine extensions that contain restriction enzyme sites can be added to the 5′ terminus to facilitate cloning. In this case, the primer could be as short as 6 amino acid codons in length.
7. Deoxyinosine may be used for highly degenerate codons.

60.6.1.2. Choice of Mutagenesis Protocol

Mutagenesis with a transposon offers the clear advantage that subsequent cloning of the corresponding wild-type allele is generally much easier than if a mutant allele is generated through chemical or UV mutagenesis. But a disadvantage of transposon mutagenesis is that the range of possible mutants is limited. In general, only mutant phenotypes that result from gene disruption will be obtained, and potentially interesting and informative mutant phenotypes that could result from point mutations will be missed. For example, the Abs$^-$ mutant phenotype of the original *absA* mutants does not result from knockout mutation (14). Thus, if the screen for Abs$^-$ mutants had used transposon mutagenesis, the *absA* locus would have gone undiscovered.

60.6.1.3. Chemical or Physical Mutagens

UV is an effective mutagen for *S. coelicolor*. It is the mutagen of choice, affording ease and safety in carrying out mutagenesis. *S. coelicolor* lacks a photoreactivation system, so mutagenesis protocols can be completed in normal room light (56). The subsequent repair of UV-induced lesions is an error-prone process that results in mutation. The spectrum of UV-induced mutations is broad, with all base-pair changes represented and G-C to A-T transitions predominating (43). A convenient UV source is a hand-held UV lamp such as that used for visualizing ethidium bromide-strained DNA. A simple protocol for UV mutagenesis appears in Table 4. Chemical mutagens such as *N*-methyl-*N'*-nitro-*N*-nitrosoguanidine (MNNG) are also effective (56).

60.6.1.4. Transposon Mutagenesis in Streptomycetes

At present, an efficient, practical system for transposon mutagenesis of *S. coelicolor* or *S. lividans* does not exist. Although transposons can be used to obtain mutants in these strains, the insertion patterns are often complex. Candidate mutants often carry multiple insertions, and careful genetic crosses must be done to establish that a specific Tn insertion is responsible for a mutant phenotype. Since a system for generalized transduction does not exist for these strains, these crosses require the plasmid-mediated mating approach. An additional difficulty is that the insertion site often contains the plasmid vector as well as the transposon.

The lone example of a gene cloned via transposon mutagenesis is *bldK*, and the investigators reported that the above-mentioned problems slowed the cloning process (99). Other mutants obtained in transposon mutagenesis protocols have been described in the literature (112, 122), but the putatively mutated genes have not yet been reported to be cloned.

60.6.1.5. Frequency of Mutants

Following UV or MNNG mutagenesis, loss-of-function mutations have occurred at a frequency of about 10^{-3} to 10^{-4} in various published mutant hunts, e.g., *absB* and *bldG* mutants (2, 25). Hence, if a desired phenotype can result from gene inactivation, approximately 10^3 to 10^4 survivors of mutagenesis could be expected to carry a mutation in a given gene. If several genes participate in a pathway, and any can mutate to the phenotype sought, the mutant frequency will be several times higher.

A regulator's role as a negative or positive regulator is critical to the issue of mutant frequency. For example, if a gene product functions as a negative regulator of antibiotic production, a loss-of-function mutation will not cause an antibiotic-minus phenotype, but a mutation that alters the gene so as to enhance the repressive effect of its product might do so. Such a mutant allele could be expected to result from only a limited number of mutational changes and might occur as rarely as 1 in 10^6 survivors of mutagenesis.

TABLE 4 Protocol for UV mutagenesis of *S. coelicolor*

1. Transfer 1-ml aliquots of a spore suspension, at 1×10^8 spores/ml, to each of a series of sterile plastic petri dishes, each containing 4 ml of water.
2. Expose the spore suspensions, uncovered, to a UV lamp (254 nm) for varying times and at varying distances. The petri dish should be gently agitated during exposure so spores are not shaded. A suggested starting point would be 20 cm for 20-, 40-, 60-, 80-, and 100-s exposures.
3. Determine, by plating on the media chosen for the mutant screen, the surviving titer for each sample.
4. A dose that results in 1% survival should be chosen for mutant screening and further mutagenesis.

60.6.1.6. Genetic Instabilities in Streptomycetes and Choice of Parental Strain

Phenotypic instabilities in antibiotic production have been documented in many reports (e.g., 128; reviewed in reference 144). The genetic basis for high-frequency phenotypic instability is often deletion, usually but not always occurring at the ends of the linear chromosome. The deletions may be as large as 2 Mb and are often accompanied by DNA amplification and deletion of regions termed AUD, for amplifiable unit of DNA (e.g., references 13 and 145). The relationships between the DNA sequences involved in the amplifications and deletions and the loci involved in antibiotic regulation have not been established for any of the phenomena reported. Hence, these high-frequency phenotypic instabilities, which can occur at a frequency of 0.1 to 1% of spores, can seriously complicate a genetic analysis of antibiotic regulation.

Fortunately, *S. coelicolor* is relatively stable in its antibiotic production phenotype, and the strain J1501 is especially phenotypically stable (24). Thus, mutant hunts can be expected to identify mutants altered in antibiotic regulation genes that arise as rarely as 1 in 10^5 survivors of mutagenesis (4). Strain J1501 also carries several genetic markers that facilitate genetic mapping experiments, as well as the Pgl characteristic that facilitates use of the phage ϕC31-derived cloning vectors (56). Strain J1501 produces lower quantities of actinorhodin and undecylprodigiosin than does M145 but is more phenotypically stable than this strain. It may be relevant to J1501's stability that the strain has already undergone deletion of a portion of its genome (72). Some differences in the regulatory characteristics of J1501- and M145-derived strains have been observed (see below), but it is not known whether the deletion of J1501 is related to any phenotypic differences.

J1501 and all other *S. coelicolor* and *S. lividans* strains do exhibit a high-frequency phenotype named "Scarlet" (127), in which the red antibiotic undecylprodigiosin is overproduced (24). The frequency of Scarlet colonies is as high as 0.1%, increasing to 1 to 5% after UV treatment, freezing, or protoplasting. The genetic basis for undecylprodigiosin overproduction is not known, but the Scarlet phenotype correlates with deletion of a chloramphenicol resistance (Cam^r) gene. Thus, any antibiotic-overproducing mutant isolated should be tested for Cam^r (at 20 μg/ml), and all Cam^s strains should be discarded.

60.6.1.7. Has the Abs$^-$ Phenotype Been Saturated?

It is likely that additional antibiotic regulatory genes could be discovered through mutant hunts. In the mutant hunt that yielded *absA* and *absB* mutants, additional Abs⁻ isolates were discovered but only partially characterized (3). It is not known whether any of these lie in *relA* or *abaA*. Moreover, mutants of the *afsR* locus were not discovered.

It is important to note that J1501 served as the parent strain in the above-mentioned mutant hunt. This may have limited the range of mutants found. For example, in the case of *relA*, a disruption mutation caused a less severe antibiotic-minus phenotype in strain J1501 than in another *S. coelicolor* strain more closely related to M145 (23).

A second noteworthy parameter of future mutant hunts would be choice of media. Media dependence of the antibiotic-minus phenotype has been evaluated in several cases. Both Δ*afsR* and Δ*relA* mutant strains exhibit a more severe antibiotic-minus phenotype on SMMS media than on R2YE (23, 42). This has been partly attributed, in the case of Δ*afsR*, to phosphate concentration and, for Δ*relA*, to nitrogen limitation. The *absA* and *absB* mutants also exhibit varying degrees of antibiotic deficiencies on common media (Table 5).

60.6.1.8. Has the Bld⁻ Phenotype Been Saturated?

A recent mutant hunt, using MNNG as a mutagen, yielded new *bld* loci (97). Clearly, numerous important genes remain to be discovered. Screens at low colony density can avoid extracellular rescue of mutants by neighboring wild-type colonies (97). Also, as in the case of the Abs⁻ phenotype, media conditions affect the Bld⁻ mutants (Table 5), which, with the exceptions of *bldB* and *brgA*, sporulate (and some produce antibiotics) when grown on poor carbon sources such as mannitol or maltose. Hence, varying the growth conditions in mutant hunts would likely yield new mutants.

60.6.2. Cloning-Based Approaches

60.6.2.1. Multicopy Enhancement of *S. lividans* Antibiotics

One cloning-based approach to identification of pleiotropic antibiotic regulatory genes, which has been a rich source of regulatory genes, exploits the *S. lividans* failure to produce actinorhodin (Act) in most laboratory conditions. This Act⁻ phenotype results from an as yet undetermined regulatory defect, not a biosynthetic defect. Shotgun cloning of chromosomal DNAs from various streptomycetes, in plasmid libraries introduced into *S. lividans*, has resulted in discovery of the genes listed in Table 6. The cloned genes were isolated from Act⁺ (Blue) transformants. In all cases, these cloned genes enhance production of *S. lividans* undecylprodigiosin, and when introduced into *S. coelicolor*, they also enhance production of both actinorhodin and undecylprodigiosin.

The *S. lividans* strain used for isolation of several clones was HH21, a strain defective in production of A-factor (see section 60.2.1.). Stimulation of A-factor accompanied stimulation of actinorhodin and undecylprodigiosin by these clones, although the relationship between actinorhodin and undecylprodigiosin and A-factor regulation in *S. lividans* and *S. coelicolor* is not known.

In addition to its enhancement of actinorhodin and undecylprodigiosin in both *S. lividans* and *S. coelicolor*, the cloned *afsR* locus restores these two antibiotics to the *S. coelicolor absA* and *absB* strains (27). The cloned *afsQ1* gene likewise restores actinorhodin and undecylprodigiosin to an *absA* (Abs⁻ phenotype) strain, but not to an *absB* mutant (69).

Can more regulators be found by screening for enhancement of antibiotic production in *S. lividans*? The answer to this question is almost certainly "yes." Since the reported screens have not led to rediscovery of the same sequences, it is unlikely that this approach has yet identified all possible genes. Furthermore, introduction of heterologous DNA will likely identify additional new genes.

60.6.2.2. Complementation versus Suppression

A cautionary note regarding analysis of cloned regulatory sequences relates to the abundance of genes with antibiotic-enhancing abilities. The observations that several cloned genes can suppress antibiotic-nonproducing mutants have come not only from direct tests of cloned genes (69) but also from experiments intended to recover truly complementing genes. Thus, the *afsR* (64), *afsQ* (69), *ptpA* (59), and *hrdB* (11) genes were cloned in *afsB* (64) mutants, and the *afsR* locus and other loci were cloned in an *absB* mutant (27). Hence, analysis of novel cloned sequences obtained through attempts to complement a mutant defect should include a careful distinction as to whether the cloned sequence encodes a complementing or suppressive gene.

60.6.2.3. Multicopy Inhibition

A fruitful approach to cloning regulatory sequences is through multicopy inhibition. A medium- to high-copy plasmid library (e.g., pIJ702 in *S. coelicolor*) is introduced into "wild-type" protoplasts, and transformants with an antibiotic-minus phenotype are evaluated. In one such screen, the *mia* sequence was cloned (27, 116) because of its ability to inhibit synthesis of antibiotics, but not to block sporulation.

The multicopy inhibition approach can produce clones of various types: a binding site for an activator (e.g., reference 138), a gene encoding a repressor, a sequence encoding an antisense RNA, or a gene fragment encoding a truncated protein with a dominant-negative effect.

TABLE 5 Various antibiotic-minus (Ab⁻) phenotypes on common media

Mutant strain	SMMS (42, 134)	R2YE (56)	Glucose minimal (56)	Low phosphate minimal	References
Δ*afsR*	Ab⁻	Ab⁺	NR[a]	NR	42
Δ*relA*	Ab⁻	Ab⁺	NR	Ab⁺	22
absA542	Ab⁻	Ab⁻	Ab⁻	Act±	3, 4, 24
absB120	Ab⁻	Ab±	Ab⁻	Ab⁻	2, 3, 24
bldA	NR	Ab⁻	Ab⁻	Red±	30, 150

[a]NR, not reported.

TABLE 6 Pleiotropic regulatory genes identified by stimulation of actinorhodin in *S. lividans*

Gene (ref.)	Gene source	*S. lividans* host	Vector (copy number)
afsR (64)	*S. coelicolor*	HH21	pIJ41 (5)
afsQ1/Q2 (69)	*S. coelicolor*	HH21	pIJ41 (5)
ptpA (59)	*S. coelicolor*	HH21	pIJ41 (5)
micX (117)	*Streptomyces fradiae*	TK21	pIJ486 (30-100)
abaA (40)	*S. fradiae*	TK21	pIJ486 (30-100)
abaB (123)	*S. antibioticus*	TK21	pIJ486 (30-100)

60.7. LOGIC OF GENETIC CIRCUITS: POSITIVE AND NEGATIVE REGULATORS

In one theoretical consideration (96) of the predicted correlation of demand for gene expression in an organism's natural environment with the molecular mode for gene control, the prediction was made that if the demand for gene expression is high, a system will be under positive control. Alternatively, if the demand for gene expression is low, a system will be under negative control. However, in the relatively well understood example of secondary metabolite regulation in *Bacillus* spp., the stepwise flow of information involves both positive and negative regulatory steps (reviewed in reference 84).

In a simple regulatory system, such as that depicted in Fig. 1 (pathway 1), the predicted phenotypes created by loss-of-function mutations in positive and negative regulators will differ: loss of a positive regulator (B) will cause loss of antibiotic production, but loss of a negative regulator (A) may cause hyperproduction, such as that observed for the *absA1/A2* and *cutR/S* genes (14, 28).

In a more complex regulatory system, the elements of a dependent pathway could be arranged as in the hypothetical example of pathway 2 (Fig. 1). In this example, loss-of-function mutation in either A or D (positive regulators) will cause loss of antibiotics. As in the simple system above, hyperproduction of antibiotics can result from loss of a negative regulator such as C. But loss of the negative regulator B creates the loss of antibiotic phenotype. Hence, an even number of negative regulatory steps can appear to behave as positive regulatory action.

The molecular mechanisms of gene function need not be known to supply such logical schemes to an experimental system. Even in the absence of any knowledge regarding genes' products, such formal models can be used to predict the genetic structure and regulatory hierarchies in pathways.

Pathway 1

A	⊣ B	→ Antibiotics
OFF	ON	ON
ON	OFF	OFF (Ab$^-$ phenotype)

Pathway 2

A	→ B	⊣ C	⊣ D	→ Antibiotics
ON	ON	OFF	ON	ON (Ab$^+$)
OFF	OFF	ON	OFF	OFF (Ab$^-$)

FIGURE 1 Simple and complex regulatory systems. The arrows indicate positive regulation and the bars indicate negative regulation.

One important application of such a formal model lies in suppressor analysis. Consideration of a pathway's logical structure can be used to predict phenotypes that might result from suppressor mutations. As an example, consider pathway 2. Loss-of-function mutation in either gene A or B results in the Ab$^-$ phenotype. But either A$^-$ or B$^-$ (or A$^-$B$^-$) could be suppressed by a null C$^-$ mutation, and antibiotic production would be restored. Thus, this simple form of suppressor analysis can yield interesting and important mutants.

An example from *Bacillus* sp. illustrates these genetic relationships. In *Bacillus* species, *abrB* is a negative regulator of some stationary-phase gene expression (reviewed in reference 84). *abrB* was identified by mutations that partially suppressed the pleiotropic developmental phenotype of *spo0A* mutations. Spo0A negatively regulates *abrB*; hence, the *spo0A* and *abrB* activities are represented by gene B and C, respectively.

60.7.1. Genetic Approaches to Establishing Relationships of Elements within a Network

60.7.1.1. Null Phenotype

Determination of the null phenotype of a gene requires that a chromosomal gene-specific deletion be created. If a gene is clearly the sole gene in a transcription unit, a deletion/replacement mutation constructed with a resistance marker is a convenient null allele. Such a deletion should remove almost all of the gene's coding region since a truncated protein may retain activity that obscures the true null phenotype. Likewise, disruption mutations created through Campbell-type recombination of a nonreplicating vector containing a cloned fragment that is internal to a transcription unit should usually not be considered null mutations. In these experiments the cloned fragment must be at least 500 bp to allow homologous recombination. Thus, the amino-terminal truncated protein will be at least 200 or more amino acids in length. Two examples of gene fragments expressing significant activity are the RelA′ (22, 23) and AfsR fragments (42, 86).

If the gene of interest is known to be in an operon, or if downstream sequencing has not excluded that possibility, the possibility of a polar effect on downstream gene expression must be considered. A gene-specific deletion in an operon is best constructed as an in-frame deletion to avoid polar effects on expression of any downstream genes. Moreover, evaluation of a mutant strain's phenotype should include demonstration that a strain's phenotype can be complemented by the corresponding wild-type allele.

60.7.1.2. Dependence Pathways

The genetic interactions of a dependent pathway can be modeled as discussed above. Such models can be developed through exploring the epistatic relationships of a series of mutant strains. For example, in pathway 2, the mutants A^-, B^-, and D^- each have an Ab^- phenotype, and C^- has a hyperproducing phenotype. The double mutants B^-C^- and C^-D^- would have different phenotypes: C^- would be epistatic in the first case (creating the hyperproducing phenotype), but D^- would be epistatic in the second case (creating the Ab^- phenotype).

If the mutant phenotypes of two strains are identical, then double-mutant tests can be useful in determining whether the two genes in question function in the same pathway or in different pathways. For example, in pathway 2, the double mutants A^-B^-, B^-D^-, and A^-D^- would all be predicted to exhibit a phenotype like that of any of the three single mutants. In contrast, if a double-mutant phenotype is more severe than the phenotype of either single mutant, the classical interpretation is that the two genes function in different, partially redundant, pathways.

60.7.1.3. Networks

It is already clear that control of secondary metabolite synthesis involves a complex network of regulatory genes, not a single dependent pathway. Nevertheless, the approaches discussed here should help to clarify the interactions between the many regulatory genes because the overall networks will consist of dependence pathways.

60.8. PERSPECTIVES

It is probable that numerous antibiotic regulatory genes remain to be discovered. The lack of an efficient transposon mutagenesis system for *S. coelicolor* has limited the number of genes that have been identified and subsequently cloned. Continuing efforts are directed at developing a useful system, as well as a generalized transduction system (149a). Genome sequencing will reveal many candidate regulatory genes. Mutational analyses of such genes will likely often reveal phenotypes too subtle to have been observed in most mutant hunts (e.g., *spaA*; see section 60.2.2.1).

Identifying the *S. coelicolor* homologs of regulators found in other species will further expand the gene collection. The recent work on γ-butyrolactone signaling in *S. coelicolor* (see section 60.2.1) exemplifies the dramatic progress that can result from combining two systems: the extensive knowledge on A-factor biology regulation that has been obtained in *S. griseus* and the genetic resources available in *S. coelicolor*.

The pathway relationships among regulatory genes have only begun to be explored. The use of gene fusions to facilitate systematic analyses of expression dependency relationships has been powerfully applied in some bacterial systems, e.g., *B. subtilis* and *M. xanthus*. In streptomycetes (e.g., 17), however, such studies have been thwarted by the anomalous and unreliable behavior of many common reporter genes (e.g., 79). The now widely used green fluorescent protein (GFP) may prove to be useful for streptomycetes. Moreover, GFP fusions may allow analyses of gene expression compartmentalization. Indeed, current attempts to develop GFP as a reporter for temporal and spatial gene expression in streptomycetes are yielding promising results. Predicted patterns of spatial gene expression were observed using both fluorescence and confocal microscopy when a variant high-GC GFP gene was coupled to either an inducible or sporulation-specific promoter (135a). Dissection of the cell-specific regulatory mechanisms that operate during streptomycete differentiation—in one such example the polyketide synthase for Act is evidently expressed only in vegetative hyphae, whereas the spore pigment polyketide synthase is expressed only in developing spores (reviewed in reference 55)—will greatly further our understanding of the interrelated phenomena of morphogenesis and secondary metabolism.

I thank M. Bibb, M. Buttner, K. Chater, D. Hodgson, S. Horinouchi, D. Hutchinson, B. Leskiw, J. Nodwell, M. Paget, K. Ueda, and J. Westpheling for communicating unpublished results. I apologize to those colleagues whose work may have been inappropriately cited or omitted. Work in my laboratory is supported by the National Science Foundation (grant MCB9604055).

REFERENCES

1. **Aceti, D., and W. Champness.** Global transcriptional regulation of *Streptomyces coelicolor* antibiotic genes by the *absA* and *absB* loci. *J. Bacteriol.* **180:**3100–3106.
2. **Adamidis, T., and W. Champness.** 1992. Genetic analysis of *absB*: a *Streptomyces coelicolor* locus involved in global antibiotic regulation. *J. Bacteriol.* **174:**4622–4628.
3. **Adamidis, T., and W. Champness.** Unpublished results.
4. **Adamidis, T., P. Riggle, and W. Champness.** 1990. Mutations in a new *Streptomyces coelicolor* locus which globally block antibiotic biosynthesis but not sporulation. *J. Bacteriol.* **172:**2962–2969.
5. **Anderson, T., and W. Champness.** Unpublished results.
6. **Anzai, H., T. Murakami, S. Imai, A. Satoh, K. Nagaoka, and C. J. Thompson.** 1987. Transcriptional regulation of bialaphos biosynthesis in *Streptomyces hygroscopicus*. *J. Bacteriol.* **169:**3482–3488.
7. **Astrurias, J. A., P. Liras, and J. F. Martin.** 1990. Phosphate control of *pabS* gene transcription during candicidin biosynthesis. *Gene* **93:**79–84.
8. **Bainton, N. J., B. W. Bycroft, S. R. Chhabra, P. Stead, L. Gledhill, P. J. Hall, C. E. D. Rees, M. K. Winson, G. P. C. Salmond, G. S. A. B. Stewart, and P. Williams.** 1992. A general role for the *lux* autoinducer in bacterial cell signaling: control of antibiotic biosynthesis in *Erwinia*. *Gene* **116:**87–91.
9. **Bainton, N. J., P. Stead, S. R. Chhabra, B. W. Bycroft, G. O. C. Salmond, G. S. A. B. Stewart, and P. Williams.** 1992. *N*-(3-Oxohexanoyl)-L-homoserine lactone regulates carbapenem antibiotic production in *Erwinia carotovora*. *Biochem. J.* **288:**997–1004.
10. **Bascaran, V., L. Sanchez, C. Hardisson, and A. F. Braña.** 1991. Stringent response and initiation of secondary metabolism in *Streptomyces clavuligerus*. *J. Gen. Microbiol.* **137:**1625–1634.
11. **Bibb, M. J.** Personal communication.
12. **Bibb, M.** 1996. The regulation of antibiotic production in *Streptomyces coelicolor* A3(2). *Microbiology* **142:**1335–1344.
13. **Birch, A., A. Haüsler, C. Rüttener, and R. Hütter.** 1991. Chromosomal deletion and rearrangement in *Streptomyces glaucescens*. *J. Bacteriol.* **173:**3531–3538.
14. **Brian, P., P. J. Riggle, R. A. Santos, and W. C. Champness.** 1996. Global negative regulation of *Streptomyces coelicolor* antibiotic synthesis mediated by an *absA*-encoded putative signal transduction system. *J. Bacteriol.* **178:** 3221–3231.

15. **Brown, D. P., L. Canova-Raeva, B. D. Green, S. R. Wilkinson, M. Young, and P. Youngman.** 1994. Characterization of the *spo0A* homologues in diverse *Bacillus* and *Clostridium* species identifies as probable DNA-binding domain. *Mol. Microbiol.* **14:**411–426.
16. **Brown, G., and W. Champness.** Unpublished results.
17. **Bruton, C. J., E. P. Gutherie, and K. F. Chater.** 1991. Phage vectors that allow monitoring of secondary metabolism genes in *Streptomyces*. *Bio/Technology* **9:**652–656.
18. **Buttner, M. J.** 1989. RNA polymerase heterogeneity in *Streptomyces coelicolor* A3(2). *Mol. Microbiol.* **3:**1653–1659.
19. **Buttner, M. J., K. F. Chater, and M. J. Bibb.** 1990. Cloning, disruption, and transcriptional analysis of three RNA polymerase sigma factor genes of *Streptomyces coelicolor* A3(2). *J. Bacteriol.* **172:**3367–3378.
20. **Buttner, M. J., and C. G. Lewis.** 1992. Construction and characterization of *Streptomyces coelicolor* A3(2) mutants that are multiply deficient in the nonessential *hrd*-encoded RNA polymerase sigma factors. *J. Bacteriol.* **174:**5165–5167.
21. **Cashel, M. D., R. Gentry, V. J. Hernandez, and D. Vinella.** 1996. The stringent response, p. 1458–1496. *In* F. C. Neidhardt, R. Curtiss III, J. L. Ingraham, E. C. C. Lin, K. B. Low, Jr., B. Magasanik, W. S. Reznikoff, M. Riley, M. Schaechter, and H. E. Umbarger (ed.), *Escherichia coli and Salmonella: Cellular and Molecular Biology*, 2nd ed. ASM Press, Washington, D.C.
22. **Chakraburtty, R., and M. Bibb.** 1997. The ppGpp synthetase gene (relA) of *Streptomyces* coelicolor A3(2) plays a conditional role in antibiotic production and morphological differentiation. *J. Bacteriol.* **179:**5854–5851.
23. **Chakraburtty, R., J. White, E. Takano, and M. Bibb.** 1996. Cloning, characterization and disruption of a (p)ppGpp synthetase gene (*relA*) of *Streptomyces coelicolor* A3(2). *Mol. Microbiol.* **19:**357–368.
24. **Champness, W.** Unpublished results.
25. **Champness, W. C.** 1988. New loci required for *Streptomyces coelicolor* morphological and physiological differentiation. *J. Bacteriol.* **170:**1168–1174.
26. **Champness, W. C., and K. F. Chater.** 1994. Regulation and integration of antibiotic production and morphological differentiation in *Streptomyces* spp, p. 61–94. *In* P. Piggot, C. Moran, and P. Youngman (ed.), *Regulation of Bacterial Differentiation*. American Society for Microbiology, Washington, D.C.
27. **Champness, W., P. Riggle, T. Adamidis, and P. Vandevere.** 1992. Identification of genes involved in regulation of *S. coelicolor* antibiotic synthesis. *Gene* **115:**55–60.
28. **Chang, H.-M., J. Y. Chen, Y. T. Shieh, M. J. Bibb, and C. W. Chen.** 1996. The *cutRS* signal transduction system of *Streptomyces lividans* represses the biosynthesis of the polyketide antibiotic actinorhodin. *Mol. Microbiol.* **21:**1075–1085.
29. **Chater, K. F.** 1990. The improving prospects for yield increase by genetic engineering in antibiotic-producing streptomycetes. *Bio/Technology* **8:**115–121.
30. **Chater, K. F., and M. J. Bibb.** 1997. Regulation of bacterial antibiotic production, p. 57–105. *In* H. Kleinkauf, and H. von Döhren (ed.), *Products of Secondary Metabolism*, vol. 6, *Biotechnology*. VCH, Weinheim, Germany.
31. **Chater, K. F., and C. J. Bruton.** 1985. Resistance, regulatory and production genes for the antibiotic methylenomycin are clustered. *EMBO J.* **4:**1893–1897.
32. **Chong, P. P., S. M. Podmore, H. M. Kieser, M. Redenbach, K. Turgay, M. Marahiel, D. A. Hopwood, and C. P. Smith.** 1998. Physical identification of a chromosome locus encoding biosynthetic genes for the lipopeptide calcium-dependent antibiotic (CDA) of *Streptomyces coelicolor* A3(2). *Microbiology* **144:**193–199.
33. **Demain, A. L.** 1992. Microbial secondary metabolism: a new theoretical frontier for academia, a new opportunity for industry. *Ciba Found. Symp.* **171:**3–23.
34. **Demain, A. L., and A. Fang.** 1995. Emerging concepts of secondary metabolism in actinomycetes. *Actinomycetologica* **9:**98–117.
35. **Distler, J., K. Mansouri, G. Mayer, M. Stockmann, and W. Pipersberg.** 1992. Streptomycin biosynthesis and its regulation in streptomycetes. *Gene* **115:**105–111.
36. **Donadio, S., D. Stassi, J. B. McAlpine, M. J. Staver, P. J. Sheldon, M. Jackson, S. J. Swanson, E. Wendt-Pienkowski, W. Yi-Guang, B. Jarvis, C. R. Hutchinson, and L. Katz.** 1993. Recent developments in the genetics of erythromycin formation, p. 257–265. *In* R. H. Baltz, G. D. Hegeman, and P. L. Skatrud (ed.), *Industrial Microorganisms: Basic and Applied Molecular Genetics*. American Society for Microbiology, Washington, D.C.
37. **Elliot, M., F. Damji, R. Passantino, K. Chater, and B. Leskiw.** 1998. The *bldD* gene of *Streptomyces coelicolor* A3(2): a regulatory gene involved in morphogenesis and antibiotic production. *J. Bacteriol.* **180:**1549–1555.

37a. **Errington, J.** 1996. Determination of cell fate in *Bacillus subtilis*. *Trends Genet.* **12:**31–34.

38. **Feitelson, J., F. Malpartida, and D. A. Hopwood.** 1986. Genetic and biochemical characterization of the *red* cluster of *Streptomyces coelicolor* A3(2). *J. Gen. Microbiol.* **131:**2431–2441.
39. **Fernandez-Moreno, M. A., J. L. Caballero, D. A. Hopwood, and F. Malpartida.** 1991. The *act* gene cluster contains regulatory and antibiotic export genes, direct targets for translational control by the *bldA* tRNA gene of *Streptomyces coelicolor*. *Cell* **66:**769–780.
40. **Fernandez-Moreno, M. A., A. J. Martin-Triana, E. Martinez, J. Niemi, H. M. Kieser, D. A. Hopwood, and F. Malpartida.** 1992. *abaA*, a new pleiotropic regulatory locus for antibiotic production in *Streptomyces coelicolor*. *J. Bacteriol.* **174:**2958–2967.
41. **Flärdh, K., T. Axberg, N. H. Albertson, and S. Kjelleberg.** 1994. Stringent control during carbon starvation of marine *Vibrio* sp. strain 14: molecular cloning, nucleotide sequence and deletion of the *relA* gene. *J. Bacteriol.* **176:**5949–5957.
42. **Floriano, B., and M. J. Bibb.** 1996. *afsR* is a pleitropic but conditionally required regulatory gene for antibiotic production in *Streptomyces coelicolor* A3(2). *Mol. Microbiol.* **21:** 385–396.
43. **Friedberg, E. C., G. C. Walker, and W. Siede.** 1995. *DNA Repair and Mutagenesis*, p. 465–522. ASM Press, Washington, D.C.
44. **Fujii, T., H. C. Gramajo, E. Takano, and M. J. Bibb.** 1996. *redD* and *actII*-ORF4, pathway-specific regulatory genes for antibiotic production in *Streptomyces coelicolor* A3(2), are transcribed in vitro by an RNA polymerase holoenzyme containing σ^{hrdD}. *J. Bacteriol.* **178:**3402–3405.
45. **Fuqua, W. C., S. C. Winans, and E. P. Greenberg.** 1994. Quorum sensing in bacteria: the LuxR-LuxI family of cell

density-responsive transcriptional regulators. *J. Bacteriol.* **176:**269–275.

46. **Furuya, K., and C. R. Hutchinson.** 1996. The DnrN protein of *Streptomyces peucetius*, a pseudo-response regulator, is a DNA-binding protein involved in the regulation of daunorubicin biosynthesis. *J. Bacteriol.* **178:**6310–6318.

47. **Geistlich, M., R. Losick, J. Turner, and R. Nagaraja Rao.** 1992. Characterization of a novel regulatory gene governing the expression of a polyketide synthase gene in *Streptomyces ambofaciens*. *Mol. Microbiol.* **6:**2019–2029.

48. **Gramajo, H. E., E. Takano, and M. J. Bibb.** 1993. Stationary phase production of the antibiotic actinorhodin in *Streptomyces coelicolor* A3(2) is transcriptionally regulated. *Mol. Microbiol.* **7:**837–845.

48a. **Guthrie, E., C. Flaxman, J. White, D. A. Hodgson, M. J. Bibb, and K. F. Chater.** 1998. A response-regulator-like activator of antibiotic synthesis from *Streptomyces coelicolor* A3(2) with an amino-terminal domain that lacks a phosphorylation pocket. *Microbiology* **144:**727–738.

49. **Hallam, S. E., F. Malpartida, and D. A. Hopwood.** 1988. Nucleotide sequence, transcription and deduced function of a gene involved in polyketide antibiotic synthesis in *Streptomyces coelicolor*. *Gene* **74:**305–320.

50. **Hara, O., S. Horinouchi, S. Uozumi, and T. Beppu.** 1983. Genetic analysis of A-factor synthesis in *Streptomyces coelicolor* A3(2) and *Streptomyces griseus*. *J. Gen. Microbial.* **129:**2939–2944.

51. **Hodgson, D. A.** 1992. Differentiation in actinomycetes, p. 407–440. *In* S. Mohan, C. Dow, and J. A. Cole (ed.), *Prokaryotic Structure and Function: A New Perspective*. Cambridge University Press, Cambridge.

52. **Holt, T. G., C. Chang, C. Laurentwinter, T. Murakami, J. I. Garrels, J. E. Davies, and C. J. Thompson.** 1992. Global changes in gene expression related to antibiotic synthesis in *Streptomyces hygroscopicus*. *Mol. Microbiol.* **6:** 969–980.

53. **Hong, S.-K., M. Kito, T. Beppu, and S. Horinouchi.** 1991. Phosphorylation of the *afsR* product, a global regulatory protein for secondary metabolite formation in *Streptomyces coelicolor* A3(2). *J. Bacteriol.* **173:**2311–2318.

54. **Hong, S. K., A. Matsumoto, S. Horinouchi, and T. Beppu.** 1993. Effects of protein kinase inhibitors on *in vitro* protein phosphorylation and cellular differentiation of *Streptomyces griseus*. *Mol. Gen. Genet.* **263:**347–354.

55. **Hopwood, D. A.** 1997. Genetic contributions to understanding polyketide synthases. *Chem. Rev.* **97:**2465–2495.

56. **Hopwood, D. A., M. J. Bibb, K. F. Chater, T. Kieser, C. J. Bruton, H. M. Kieser, D. J. Lydiate, C. P. Smith, J. M. Ward, and H. Schrempf.** 1985. *Genetic Manipulation of Streptomyces. A Laboratory Manual.* The John Innes Foundation. Norwich, United Kingdom.

57. **Hopwood, D., K. Chater, and M. Bibb.** 1995. Genetics of antibiotic production in *Streptomyces coelicolor* A3(2), p. 71–108. *In* L. C. Vining and C. Stuttand (ed.), *Regulation and Biochemistry of Antibiotic Production*, Butterworth-Heinemann, Newton, Mass.

58. **Hopwood, D. A., and H. M. Wright.** 1983. CDA is a new chromosomally-determined antibiotic from *Streptomyces coelicolor* A3(2). *J. Gen. Microbiol.* **129:**3576–3579.

59. **Horinouchi, S.** Personal communication.

60. **Horinouchi, S.** 1993. "Eucaryotic" signal transduction systems in the bacterial genus *Streptomyces*. *Actinomycetologica* **7:**68–87.

61. **Horinouchi, S., and T. Beppu.** 1992. Autoregulatory factors and communication in actinomycetes. *Annu. Rev. Microbiol.* **46:**377–398.

62. **Horinouchi, S., and T. Beppu.** 1994. A-factor as a microbial hormone that controls cellular differentiation and secondary metabolism in *Streptomyces griseus*. *Mol. Microbiol.* **12:**859–864.

63. **Horinouchi, S., M. Kito, M. Nishiyama, K. Furuya, S.-K. Hong, K. Miyake, and T. Beppu.** 1990. Primary structure of AfsR, a global regulatory protein for secondary metabolite formation in *Streptomyces coelicolor* A3(2). *Gene* **95:**49–56.

64. **Horinouchi, S., O. Hara, and T. Beppu.** 1983. Cloning of a pleiotropic gene that positively controls biosynthesis of A-factor actinorhodin, and prodigiosin in *Streptomyces coelicolor* A3(2) and *Streptomyces lividans*. *J. Bacteriol.* **155:** 1238–1248.

65. **Horinouchi, S., H. Suzuki, M. Nishiyama, and T. Beppu.** 1989. Nucleotide sequence and transcriptional analysis of the *Streptomyces griseus* gene (*afsA*) responsible for A-factor biosynthesis. *J. Bacteriol.* **171:**1206–1210.

66. **Huisman, G. W., and R. Kolter.** 1994. Regulation of gene expression at the onset of stationary phase in *Escherichia coli*, p. 21–40. *In* P. Piggot, C. P. Moran, and P. Youngman (ed.), *Regulation of Bacterial Differentiation*. American Society for Microbiology, Washington, D.C.

67. **Huisman, G. W., and R. Kolter.** 1994. Sensing starvation: a homoserine lactone-dependent signaling pathway in *Escherichia coli*. *Science* **265:**537–539.

68. **Hutchinson, C. R.** Personal communication.

69. **Ishizuka, H., S. Horinouchi, H. M. Kieser, D. A. Hopwood, and T. Beppu.** 1992. A putative two-component regulatory system involved in secondary metabolism in *Streptomyces* spp. *J. Bacteriol.* **174:**7585–7594.

70. **Jones, G. H., M. S. B. Paget, L. Chamberlin, and M. J. Buttner.** 1997. Sigma-E is required for the production of the antibiotic actinomycin in *Streptomyces coelicolor*. *Mol. Microbiol.* **23:**169–178.

70a. **Kawabuchi, M. Y. Hara, T. Nirhira, and Y. Yamada.** 1997. Production of butyrolactone autoregulators by *Streptomyces coelicolor* A3(2). *FEMS Microbiol. Lett.* **157:**81–85.

71. **Kelly, K. S., K. Ochi, and G. H. Jones.** 1991. Pleiotropic effects of a *relC* mutation in *Streptomyces antibioticus*. *J. Bacteriol.* **173:**2297–2300.

72. **Kieser, H. M., T. Kieser, and D. Hopwood.** 1992. A combined genetic and physical map of the chromosome of *Streptomyces coelicolor* A3(2). *J. Bacteriol.* **174:**5496–5507.

73. **Kirby, R., and D. Hopwood.** 1977. Genetic determination of methylenomycin synthesis by the SCP1 plasmid *Streptomyces coelicolor* A3(2). *J. Gen. Microbiol.* **98:**239–252.

74. **Kwak, J., L. A. McCue, and K. E. Kendrick.** 1996. Identification of *bldA* mutants of *Streptomyces griseus*. *Gene* **171:**75–78.

75. **Lawlor, B., H. Baylis, and K. Chater.** 1987. Pleiotropic morphological and antibiotic deficiencies result from mutations in a gene encoding a tRNA-like product in *Streptomyces coelicolor* A3(2). *Genes Dev.* **1:**1305–1310.

76. **Leskiw, B., E. Lawlor, J. Fernandez-Abalos, and K. Chater.** 1991. The use of a rare codon specifically during development. *Mol. Microbiol.* **5:**2861–2867.

77. **Leskiw, B., R. Mah, E. Lawlor, and K. Chater.** 1993. Accumulation of *bldA*-specified tRNA is temporally regulated in *Streptomyces coelicolor* A3(2). *J. Bacteriol.* **175:** 1995–2005.
78. **Li, Y., and W. R. Strohl.** 1996. Cloning, purification, and properties of a phosphotyrosine protein phosphatase from *Streptomyces coelicolor* A3(2). *J. Bacteriol.* **178:**136–142.
79. **Lindley, H. K., V. J. Deeble, U. Peschke, M. O'Neill, S. Baumberg, and J. Cove.** 1995. Dependence on reporter gene of apparent activity in gene fusions of a *Streptomyces griseus* streptomycin biosynthesis promoter. *Can. J. Microbiol.* **41:**407–417.
80. **Lonetto, M. A., K. L. Brown, K. E. Rudd, and M. J. Buttner.** 1994. Analysis of the *Streptomyces coelicolor sigE* gene reveals the existence of a subfamily of eubacterial RNA polymerase σ factors involved in the regulation of extracytoplasmic functions. *Proc. Natl. Acad. Sci. USA* **91:**7573–7577.
81. **Madduri, K., and C. R. Hutchinson.** 1995. Functional characterization and transcriptional analysis of the dnR_1 locus, which controls daunorubicin biosynthesis in *Streptomyces peuceteus*. *J. Bacteriol.* **177:**1208–1215.
82. **Malpartida, F., and D. A. Hopwood.** 1984. Molecular cloning of the whole biosynthetic pathway of a *Streptomyces* antibiotic and its expression in a heterologous host. *Nature* **309:**462–464.
83. **Malpartida, F., J. Niemi, R. Navarrete, and D. A. Hopwood.** 1990. Cloning and expression in a heterologous host of the complete set of genes for biosynthesis of the *Streptomyces coelicolor* antibiotic undecylprodigiosin. *Gene* **93:**91–950.
84. **Marahiel, M. A., M. M. Nakano, and P. Zuber.** 1993. Regulation of peptide antibiotic production in *Bacillus*. *Mol. Microbiol.* **7:**631–636.
85. **Martinez-Costa, O. H. P. Arias, N. M. Romero, V. Parro, R. P. Mellado, and F. Malpartida.** 1996. A *relA*/*spoT* homologous gene from *Streptomyces coelicolor* A3(2) controls antibiotic biosynthetic genes. *J. Biol. Chem.* **271:** 10627–10634.
86. **Matsumoto, A., S.-K. Hong, H. Ishizuka, S. Horinouchi, and T. Beppu.** 1994. Phosphorylation of the AfsR protein involved in secondary metabolism in *Streptomyces* species by a eukaryotic-type protein kinase. *Gene* **146:**47–56.
87. **Matsumoto, A., H. Ishizuka, T. Beppu, and S. Horinouchi.** 1995. Involvement of a small ORF downstream of the *afsR* gene in the regulation of secondary metabolism in *Streptomyces coelicolor* A3(2). *Actinomycetologica.* **9:**37–43.
88. **McCue, L., J. Kwak, J. Wang, and K. Kendrick.** 1996. Analysis of a gene that suppresses the morphological defect of bald mutants of *Streptomyces griseus*. *J. Bacteriol.* **178:**2867–2875.
89. **Merrick, M. J.** 1976. A morphological and genetic mapping study of bald colony mutants of *Streptomyces coelicolor* A3(2). *J. Gen. Microbiol.* **96:** 299–315.
90. **Metzger, S., I. B. Dror, E. Aizenman, G. Schreiber, M. Toone, J. D. Friesen, M. Cashel, and G. Glaser.** 1988. The nucleotide sequence and characterization of the *relA* gene of *Escherichia coli*. *J. Biol. Chem.* **263:**15699–15704.
91. **Miyake, K., S. Horinouchi, M. Yoshida, N. Chiba, K. Mori, N. Nogawa, N. Morikawa, and T. Beppu.** 1989. Detection and properties of A-factor-binding protein from *Streptomyces griseus*. *J. Bacteriol.* **171:**4298–4302.
92. **Miyake, K., T. Kuzuyama, S. Horinouchi, and T. Beppu.** 1990. The A-factor-binding protein of *Streptomyces griseus* negatively controls streptomycin production and sporulation. *J. Bacteriol.* **172:**3003–3008.
93. **Mizuno, T., and I. Tanaka.** 1997. Structure of the DNA-binding domain of the OmpR family of response regulators. *Mol. Microbiol.* **24:**665–667.
94. **Molnár, I., and Y. Murooka.** 1993. Helix-turn-helix DNA-binding motifs of *Streptomyces*: a cautionary note. *Mol. Microbiol.* **8:**783.
95. **Narva, K. E., and J. S. Feitelson.** 1990. Nucleotide sequence and transcriptional analysis of the *redD* locus of *Streptomyces coelicolor* A3(2). *J. Bacteriol.* **172:**326–333.
96. **Neidhardt, F., and M. Savageau.** 1996. Regulation beyond the operon, p. 1310–1324. *In* F. C. Neidhardt, R. Curtiss III, J. L. Ingraham, E. C. C. Lin, K. B. Low, Jr., B. Magasanik, W. S. Reznikoff, M. Riley, M. Schaechter, and H. E. Umbarger (ed.), *Escherichia coli and Salmonella*. ASM Press, Washington, D.C.
97. **Nodwell, J.** Personal communication.
98. **Nodwell, J. R., and R. Losick.** 1998. Purification of an extracellular signaling molecule involved in production of aerial mycelium by *Streptomyces coelicolor*. *J. Bacteriol.* **180:**1334–1337.
99. **Nodwell, J. R., K. McGovern, and R. Losick.** 1996. An oligopeptide permease responsible for the import of an extracellular signal governing aerial mycelium formation in *Streptomyces coelicolor*. *Mol. Microbiol.* **22:**881–893.
100. **Ochi, K.** 1986. Occurrence of the stringent response in *Streptomyces* sp. and its significance for the inhibition of morphological and physiological differentiation. *J. Gen. Microbiol.* **132:**2621–2631.
101. **Ochi, K.** 1987. A *rel* mutation abolishes the enzyme induction needed for actinomycin synthesis by *Streptomyces antibioticus*. *Agric. Biol. Chem.* **51:**829–835.
102. **Ochi, K.** 1990. *Streptomyces relC* mutants with an altered ribosomal protein ST-L11 and genetic analysis of a *Streptomyces griseus relC* mutant. *J. Bacteriol.* **172:**4008–4016.
103. **Ochi, K.** 1990. A relaxed (*rel*) mutant of *Streptomyces coelicolor* A3(2) with a missing ribosomal protein lacks the ability to accumulate ppGpp, A-factor and prodigiosin. *J. Gen. Microbiol.* **136:**2405–2412.
104. **Okamoto, S., K. Najamura, T. Nihira, and Y. Yamada.** 1995. Virginiae butanolide binding protein from *Streptomyces virginiae*: evidence that *vbrA* is not the virginiae butanolide binding protein and reidentification of the true binding protein. *J. Biol. Chem.* **270:**12319–12326.
105. **Onaka, H., T. Hanagawa, and S. Horinouchi.** 1998. Involvement of two A-factor receptor homologs in *Streptomyces coelicolor* A3(2) in the regulation of secondary metabolism and morphogenesis. *Mol. Microbiol.* **28:**743–754.
106. **Onaka, H., and S. Horinouchi.** 1997. DNA-binding activity of the A-factor receptor protein and its recognition DNA sequences. *Mol. Microbiol.* **24:**991–1000.
107. **Otten, S. L., J. Ferguson, and C. R. Hutchinson.** 1995. Regulation of daunorubicin production in *Streptomyces peucetius* by the $dnrR_2$ locus. *J. Bacteriol.* **177:**1216–1224.
108. **Paradkar, A. S., K. A. Aidoo, and S. E. Jensen.** 1998. A pathway-specific transcriptional activator regulates

late steps of clavulanic acid biosynthesis in *Streptomyces clavuligerus*. *Mol. Microbiol.* **27:**831–844.

109. **Parkinson, J.** 1995. Genetic approaches for signaling pathways and proteins. *In* J. A. Hoch and T. J. Silhavy (ed.), *Two-Component Signal Transduction*. American Society for Microbiology, Washington, D.C.

110. **Parkinson, J. S., and E. C. Kofoid.** 1992. Communication modules in bacterial signaling proteins. *Annu. Rev. Genet.* **26:**71–112.

111. **Pérez-Llarena, F. J., P. Linas, A. Rodríguez-García, and J. F. Martín.** 1997. A regulatory gene (*ccaR*) required for cephamycin and clavulanic acid production in *Streptomyces clavuligerus*: amplification results in overproduction of both β-lactam compounds. *J. Bacteriol.* **179:**2053–2059.

112. **Pope, M., B. Green, and J. Westpheling.** 1996. The *bld* mutants of *Streptomyces coelicolor* are defective in the regulation of carbon utilization, morphogenesis and cell-cell signalling. *Mol. Microbiol.* **19:**747–756.

113. **Pope, M. K., B. Green, and J. Westpheling.** 1998. The *bldB* gene encodes a small protein required for morphogenesis, antibiotic production, and catabolite control in *Streptomyces coelicolor*. *J Bacteriol.* **180:**1556–1562.

114. **Rabaud, A., M. Zalacain, T. G. Holt, R. Tizard, and C. J. Thompson.** 1991. Nucleotide sequence analysis reveals linked *N*-acetyl hydrolase, thioesterase, transport, and regulatory genes encoded by the bialaphos biosynthetic gene cluster of *Streptomyces hygroscopicus*. *J. Bacteriol.* **173:**4454–4463.

115. **Richardson, M. A., S. Kuhstoss, M. L. B. Huber, L. Ford, O. Godfrey, J. R. Turner, and R. N. Rao.** 1990. Cloning of spiramycin biosynthetic genes and their use in constructing *Streptomyces ambofaciens* mutants defective in spiramycin biosynthesis. *J. Bacteriol.* **172:**3790–3798.

116. **Riggle, P., and W. Champness.** Unpublished data.

117. **Romero, N., V. Parro, F. Malpartida, and R. Mellado.** 1992. Heterologous activation of the actinorhodin biosynthesis pathway in *Streptomyces lividans*. *Nucleic Acids Res.* **20:**2767–2772.

118. **Romero, N. M., and R. P. Mellado.** 1995. Activation of the actinorhodin biosynthetic pathway in *Streptomyces lividans*. *FEMS Microbiol. Lett.* **127:**79–84.

119. **Rudd, B. A. M., and D. A. Hopwood.** 1979. Genetics of actinorhodin biosynthesis by *Streptomyces coelicolor* A3(2). *J. Gen. Microbiol.* **114:**35–43.

120. **Rudd, B. A. M., and D. A. Hopwood.** 1980. A pigmented mycelial antibiotic in *Streptomyces coelicolor*: control by a chromosomal gene cluster. *J. Gen. Microbiol.* **119:**333–340.

121. **Sánchez, L., and A. F. Braña.** 1996. Cell density influences antibiotic biosynthesis in *Streptomyces clavuligerus*. *Microbiology* **142:**1209–1220.

122. **Schauer, A., A. Nelson, and J. Daniel.** 1991. Tn 4563 transposition in *Streptomyces coelicolor* and its application to isolation of new developmental mutants. *J. Bacteriol.* **173:**5060–5067.

123. **Schen, A.-K., E. Martínez, J. Soliveri, and F. Malpartida.** 1997. *abaB*, a putative regulator for secondary metabolism in *Streptomyces*. *FEMS Microbiol. Lett.* **177:**29–36.

124. **Schneider, D., C. J. Bruton, and K. F. Chater.** 1996. Characterization of *spaA*, a *Streptomyces coelicolor* gene homologous to a gene involved in sensing starvation in *Escherichia coli*. *Gene* **177:**243–251.

125. **Schnider, U., C. Keel, C. Blumer, J. Troxler, G. Défago, and D. Haas.** 1995. Amplification of the housekeeping sigma factor in *Pseudomonas fluorescens* CHAO enhances antibiotic production and improves biocontrol abilities. *J. Bacteriol.* **177:**5387–5392.

126. **Schwecke, T., J. F. Aparicio, I. Molnár, A. König, L. E. Khaw, S. F. Haydock, M. Oliynyk, P. Caffrey, J. Cortés, J. B. Lester, G. A. Böhm, J. Staunton, and P. F. Leadlay.** 1995. The biosynthetic gene cluster for the polyketide immunospooressant rapamycin. *Proc. Natl. Acad. Sci. USA* **92:**7839–7843.

127. **Sermonti, G., A. Petris, M. R. Micheli, and L. Lanfaloni.** 1977. A factor involved in chloramphenicol resistance in *Streptomyces coelicolor* A3(2): its transfer in the absence of the fertility factor. *J. Gen. Microbiol.* **100:** 347–353.

128. **Shaw, P. D., and J. Piwonarski.** 1977. Effects of ethidium bromide and acriflavin on streptomycin production by *Streptomyces bikiniesis*. *J. Antibiot.* **30:**404–408.

129. **Shima, J., A. Penyige, and K. Ochi.** 1996. Changes in patterns of ADP-ribosylated proteins during differentiation of *Streptomyces coelicolor* A3(2) and its developmental mutants. *J. Bacteriol.* **178:**3785–3790.

130. **Simuth, J., J. Hudec, H. T. Chan, O. Danyi, and J. Zelinka.** 1979. The synthesis of highly phosphorylated nucleotides, RNA and protein by *Streptomyces aureofaciens*. *J. Antibiot.* **32:**53–58.

131. **Stein, D., and S. N. Cohen.** 1989. A cloned regulatory gene of *Streptomyces lividans* can suppress the pigment deficiency phenotype of different developmental mutants. *J. Bacteriol.* **171:**2258–2261.

132. **Stock, J. B., A. J. Ninfa, and A. J. Stock.** 1989. Protein phosphorylation and regulation of adaptive responses in bacteria. *Microbiol. Rev.* **53:**450–490.

133. **Stock, J., M. Surette, M. Levit, and P. Park.** 1995. Two-component signal transduction systems: structure-function relationships and mechanisms of catalysis. *In* J. Hoch and T. Shilhavy (ed.), *Two-Component Signal Transduction*. American Society for Microbiology, Washington, D.C.

134. **Strauch, E., E. Takano, H. A. Baylis, and M. J. Bibb.** 1991. The stringent response in *Streptomyces coelicolor* A3(2). *Mol. Microbiol.* **5:**289–298.

135. **Stutzman-Engwall, K. J., S. L. Otten, and C. R. Hutchinson.** 1992. Regulation of secondary metabolism in *Streptomyces* spp. and the overproduction of daunorubicin in *Streptomyces peucetius*. *J. Bacteriol.* **174:**144–154.

135a. **Sun, J., G. Kelemen, M. Buttner, and M. Bibb.** Personal communication.

136. **Swift, S., M. K. Winson, P. F. Chan, N. J. Bainton, M. Birdsall, P. J. Reeves, C. E. D. Rees, S. R. Chhabra, P. J. Hill, J. P. Throup, B. W. Bycroft, G. P. C. Salmond, P. Williams, and G. S. A. B. Stewart.** 1993. A novel strategy for the isolation of *luxI* homologues: evidence for the widespread distribution of a LuxR:LuxI superfamily in enteric bacteria. *Mol. Microbiol.* **10:**511–520.

137. **Takano, E., R. Chakraburtty, M. Bibb, T. Nihira, and Y. Yamada.** Personal communication.

137a. **Takano, E., H. C. Gramajo, E. Strauch, N. Andres, J. White, and M. J. Bibb.** 1992. Transcriptional regulation of the *redD* transcriptional activator gene accounts for growth-phase-dependent production of the antibiotic undecylprodigiosin in *Streptomyces coelicolor* A3(2). *Mol. Microbiol.* **6:**2797–2804.

138. **Tan, H., and K. Chater.** 1993. Two developmentally-controlled promoters of *Streptomyces coelicolor* A3(2) that resemble the major class of motility-related promoters in other bacteria. *J. Bacteriol.* **175:**933–940.

139. **Tanaka, K., T. Shiina, and H. Takahashi.** 1991. Nucleotide sequence of genes *hrdA*, *hrdC*, and *hrdD* from *Streptomyces coelicolor* A3(2) having similarity to *rpoD* genes. *Mol. Gen. Genet.* **229:**334–340.

140. **Tang, L., A. Grimm, Y.-X. Zhang, and C. R. Hutchinson.** 1996. Purification and characterization of the DNA-binding protein DnrI, a transcriptional factor of daunorubicin biosynthesis in *Streptomyces peucetius*. *Mol. Microbiol.* **22:**801–813.

141. **Ueda, K.** Personal communication.

142. **Urabe, H., and H. Ogawara.** 1995. Cloning, sequencing and expression of serine/threonine kinase-encoding genes from *Streptomyces coelicolor* A3(2). *Gene* **153:**99–104.

143. **Vogtli, M., P.-C. Chang, and S. N. Cohen.** 1994. *afsR2*: a previously undetected gene encoding a 63-amino-acid protein that stimulates antibiotic production in *Streptomyces lividans*. *Mol. Microbiol.* **14:**643–654.

144. **Volff, J.-N., and J. Altenbuchner.** 1998. Genetic instability of the *Streptomyces* chromosome. *Mol. Microbiol.* **27:**239–246.

145. **Volff, J.-N., C. Eichenseer, P. Viell, W. Piendl, and J. Altenbuchner.** 1996. Nucleotide sequence and role in DNA amplification of the direct repeats composing the amplifiable element AUD1 of *Streptomyces lividans* 66. *Mol. Microbiol.* **21:**1037–1047.

146. **Waki, M., T. Nihira, and Y. Yamada.** 1997. Cloning and characterization of the gene (*farA*) encoding the receptor for an extracellular regulatory factor (IM-2) from *Streptomyces* sp. strain FRI-5. *J. Bacteriol.* **179:**5131–5137.

147. **Warawa, J., K. Bilida, and B. Leskiw.** Personal communication.

149. **Waters, B., D. Vujaklija, M. R. Gold, and J. Davies.** 1994. Protein tyrosine phosphorylation in streptomycetes. *FEMS Microbiol. Lett.* **120:**187–190.

149a. **Westpheling, J.** Personal communication.

150. **White, J., and M. Bibb.** 1997. *bldA* dependence of undecylprodigiosin production in *Streptomyces coelicolor* A3(2) involves a pathway-specific regulatory cascade. *J. Bacteriol.* **179:**627–633.

151. **Wietzorrek, A., and M. Bibb.** 1997. A novel family of proteins that regulates antibiotic production in *Streptomyces* appears to contain an OmpR-like DNA-binding fold. *Mol. Microbiol.* **25:**1183–1184.

152. **Willey, J., R. Santamaria, J. Guijarro, M. Geistlich, and R. Losick.** 1991. Extracellular complementation of a developmental mutation implicates a small sporulation protein in aerial mycelium formation. *Cell* **65:**641–650.

153. **Willey, J., J. Schwedock, and R. Losick.** 1993. Multiple extracellular signals govern the production of a morphogenetic protein involved in aerial mycelium formation by *Streptomyces coelicolor*. *Genes Dev.* **7:**895–903.

154. **Wright, F., and M. J. Bibb.** 1992. Codon usage in the G + C-rich *Streptomyces* genome. *Gene* **113:**55–65.

155. **Wright, H. M., and D. A. Hopwood.** 1976. Identification of the antibiotic determined by the SCP1 plasmid of *Streptomyces coelicolor* A3(2). *J. Gen. Microbiol.* **95:** 96–106.

156. **Zambrano, M. M., and R. Kolter.** 1996. Gasping for life in stationary phase. *Cell* **86:**181–184.

157. **Zhang, C.-C.** 1996. Bacterial signalling involving eukaryotic-type protein kinases. *Mol. Microbiol.* **20:**9–15.

Genes for the Biosynthesis of Microbial Secondary Metabolites

GUIDO MEURER AND C. RICHARD HUTCHINSON

61

Investigations of the molecular genetics of antibiotic production in microorganisms began in the early 1980s, shortly after the recombinant DNA tools essential for research on microbial genetics had been invented. In 1975 Hopwood and coworkers showed that the *Streptomyces coelicolor* genes for the production of methylenomycin (*mmy*) resided on the SCP1 plasmid (94, 207). For a short time thereafter, it was believed that the involvement of plasmids in antibiotic production was widespread and might be responsible for the often-noted instability of antibiotic production. Although we now know that this is not true, the observation made by Hopwood and his colleagues (94, 207) certainly stimulated further study of the molecular genetics of antibiotic production by *Streptomyces* spp., the ubiquitous soil bacteria from which a large number of antibiotics have been isolated. A facile loss of antibiotic productivity, when seen, is believed to stem largely from the fact that streptomycetes have linear chromosomes with unstable ends prone to deletion (197) where the clusters of antibiotic biosynthesis genes often reside. Fortunately, this feature has seldom complicated genetic investigations of antibiotic production, and many sets of antibiotic biosynthesis genes have been cloned and characterized during the past 15 years.

The antibiotic production genes normally are located in the microbial chromosome as a cluster of the structural, regulatory, and self-resistance genes. This principle, first demonstrated in 1984 by Malpartida and Hopwood (118) through the isolation of the actinorhodin gene cluster from *S. coelicolor*, was made possible by the following developments: isolation of plasmids and phages from streptomycetes (14, 108); development of transformation and transfection methods with these entities (15, 17, 181), which depended in part on the cloning and analysis of antibiotic resistance genes to use as selectable markers (187); and isolation of mutations specifically blocking antibiotic production (154). From this auspicious beginning came the wealth of information available today about the genes governing secondary metabolism and antibiotic production in bacteria. Knowledge about secondary metabolism genes in fungi stems from work on penicillin (157) and 6-methylsalicylic acid (11) between 1985 and 1990. We summarize this information as a guide to those who wish to capture yet another set of secondary metabolism genes from bacteria or fungi; as a consequence, the coverage is illustrative rather than comprehensive.

61.1. TOOLS FOR GENE CLONING

61.1.1. Plasmid and Phage Vectors

Indigenous plasmids are common in actinomycetes (79), and the principal gene cloning vectors are derivatives of the high-copy-number pIJ101 (92) and low-copy-number SLP1 (18) and SCP2 (14, 110) plasmids first isolated from different streptomycetes (reviewed in references 77–79). The vectors typically carry antibiotic resistance genes enabling selection of transformants with apramycin (*aacIV*), neomycin/kanamycin (*aphII*), thiostrepton (*tsr*), or viomycin (*vph*). Special features that can facilitate cloning experiments are found in pIJ702 (90), pOJ446 (19, 123), pSET152 (100), pHM11 (109, 130), pDH5 (72), pKC1139 (19, 132), pKC505 (152), pRM5 (124), and pITSM107 (89) (Table 1). Many of these vectors can be used in both *E. coli* and *Streptomyces* spp., although such vectors are often less stable than those that contain only DNA from streptomycetes (e.g., pIJ702 [90], pIJ922 [110], pIJ301 [92], and pIJ61 [18]).

Transformants of *Streptomyces* spp. and other actinomycetes are customarily prepared by introducing the plasmid DNA with the help of polyethylene glycol and calcium ions into protoplasts, made from young mycelia by treatment with lysozyme under hypertonic conditions, followed by regeneration of the cell wall on a hypertonic solid growth medium (17, 78). A recent report indicates that, when the incoming DNA has a region enabling homologous recombination with the host's genome, protoplasts can be transformed more efficiently with denatured circular DNA than with the double-stranded form and also with denatured linear DNA (136). Introduction of the plasmid DNA by electroporation of young mycelia has been successful in only a few cases (54, 147). Conjugal transfer of plasmids from *E. coli* to the *Streptomyces* sp. is a way to avoid the use of protoplasts and restriction barriers, although this method has not been evaluated for a wide range of species (19, 123). If the *Streptomyces* donor and recipient strains can be properly marked to enable counterselection of the donor, a suitable plasmid vector in some cases may be transferred by mating the two *Streptomyces* strains (79). The use of either conjugal transfer or electroporation is desirable, when possible, because the formation and regeneration of protoplasts are time consuming and not possible for all genera or species of actinomycetes. Dairi et

TABLE 1 Examples of gene cloning vectors

Vector name	Characteristics	Reference
pIJ702	DNA inserts disrupt the tyrosinase gene to give colorless clones in a dark-colored background of melanin-producing colonies	90
pIJ699	A positive selection vector, itself lethal to the host without DNA inserts	93
pOJ446	Cosmid vector that can undergo conjugal transfer from *E. coli* to *Streptomyces* spp. owing to the presence of *oriT*	19, 123
pSET152	The presence of the *attP* site and *int* genes of the ϕC31 phage permits site-specific integration into the *Streptomyces* genome	100
pHM11	Minicircle insertion element permits random insertion into the *Streptomyces* genome	109, 130
pDH5	The single-strand form of this vector transforms the host at higher frequency than the double-stranded form	72
pKC1139	The temperature sensitivity facilitates loss of the vector at >37°C, which promotes integration of the plasmid into the chromosome as the first step in gene replacement	19, 132
pKC505	Cosmid vector useful for cloning 18 to 35-kb DNA segments	152
pRM5	Gives high-level expression of secondary metabolism genes	124
pITSM107	Temperature-inducible gene expression	89
pWHM3	An *E. coli–Streptomyces* shuttle vector derived from pIJ486 and commonly used in the authors' laboratories	196
pIJ922	Low-copy-number, broad-host-range *Streptomyces* plasmid made from SCP2*	110
pIJ486	High copy number, broad-host-range *Streptomyces* plasmid made from pIJ101	199

al. (37) describe an example of the difficulty one can encounter with a new genus.

Transformants can also be obtained by transduction or transfection of the host using phage vectors, most of which are derivatives of ϕC31, a broad-host-range phage initially isolated from *S. coelicolor* (108). KC301, KC401, and KC515 typify such vectors (27). Since phage vectors, compared with plasmids, require more effort to use and accept considerably smaller DNA segments, they are not as popular but should be considered in cases where the other methods of DNA transfer fail (105).

61.1.2. Transposons

Mutagenesis and gene cloning with transposons are still at a rudimentary stage for actinomycetes. Although numerous insertion elements have been found, only IS493 found at Eli Lilly and Company (174) has been developed into a widely used transposon (172, 173). However, new transposon-based tools are continually being developed (201). In contrast, transposon tagging is the preferred method for identifying antibiotic biosynthesis genes in *Pseudomonas* (209) and *Bacillus* spp. (30).

61.2. STRATEGIES FOR CLONING SECONDARY METABOLISM GENES

61.2.1. Complementation of Mutations Blocking Production

Mutations that specifically block the production of secondary metabolites, normally evidenced by the accumulation of one of the metabolites' precursors, help identify the steps in the pathway and provide the means for making firm gene-enzyme connections. Cross-feeding relationships among the members of a set of independently isolated mutations, when demonstrable, cement the connections and can establish the sequence of biosynthetic steps. Consequently, such mutations are invaluable aids to cloning the structural genes, i.e., the ones encoding the biosynthetic enzymes, through experiments that aim to restore metabolite production by introducing DNA clone from the wild organism into a representative of each type of mutant isolated. The desired genes can be obtained by the shotgun cloning method or by first choosing DNA that is closely linked to a self-resistance gene for the metabolite or some other useful marker, such as a homolog of one of the genes sought. Subcloning experiments then are used to identify the smallest piece of DNA that restores production to a particular mutant, followed by sequence analysis to characterize the physical limits of the gene and, by comparison with well-understood proteins, provide an inkling of the role that its protein product plays, if this has not already been revealed by the effect of the mutation.

This approach was used to clone all the genes for actinorhodin production from *S. coelicolor* (118), thereby demonstrating the clustering of secondary metabolism structural, resistance, and regulatory genes, a paradigm that has held true for both bacteria and fungi. Shortly thereafter, clusters of genes for the production of tetracenomycin (129), streptomycin (137), and bialaphos (131) were cloned in the same way (Table 2). Chater and Bruton (28) used an interesting variation of this method to identify and clone the *mmy* genes. Instead of first isolating blocked mutants, they devised a way to couple the insertional inactivation of the *mmy* genes with small segments of DNA cloned in the phage vector KC515; thiostrepton resistance transductants of an *S. coelicolor* strain with the *mmy* genes on SCP1 were then screened for loss of methylenomycin production due to the recombination between the homologous DNA carried by the vector and the *mmy* locus, and the phages released upon induction were retested for their ability to recreate *mmy* strains. The entire *mmy* gene cluster was then cloned by chromosome walking (29).

61.2.2. Search for Homologous Genes

Now that a large number of antibiotic production genes and gene clusters have been cloned from actinomycetes and fungi (Table 2), it is often possible to use a known gene to clone the genes for a newly discovered metabolite. Because there are only a few different types of secondary metabolic pathways, and considerable homology exists among genes encoding functionally related enzymes, if the genomic DNAs have a similar G+C content the known gene often will hybridize to the homologous DNA, which then can be cloned and characterized. The PCR method can also be used to clone the homologous gene, especially when the homology involves a highly conserved region amenable to PCR with degenerate primers. Once the desired gene is obtained, which can be established via the loss of metabolite formation as a result of targeted gene disruption, then the remaining genes for the metabolite's biosynthesis will be found in the surrounding DNA. Although cloning by DNA homology provides less initial information about the biosynthetic pathway than the isolation of a set of blocked mutants, it often is the faster way to identify and characterize the production genes.

This approach has been most successful for the polyketide metabolites, oligopeptide antibiotics, and bacterial metabolites that contain deoxysugar residues made from nucleotide diphosphate derivatives of glucose 1-phosphate. Since the sequence of the enzymes catalyzing the condensation reaction between the acylthioester intermediates of polyketide biosynthesis are very similar, especially around their active site regions, it is common to observe hybridization between known type I or type II β-ketoacyl:acyl carrier protein synthase genes and the homologous DNA from the organism that produces a polyketide. The genes most often used are the *Saccharopolyspora erythraea eryA* genes for type I polyketide synthases (PKSs) (34, 46) and the *S. coelicolor actI* genes (52, 117) or their *Streptomyces glaucescens tcmKL* homologs (12) for the type II PKSs. If this method is not successful, highly conserved regions of PKSs can be targeted by PCR to see if DNA can be amplified from the polyketide producer, as demonstrated by work on bacterial (10) and fungal (91) type I PKSs. With either approach, the involvement of the cloned DNA in the production of a particular polyketide must be established by gene disruption or enzymatic assay of the gene product, because microorganisms often contain more than one set of PKS genes. For instance, *Streptomyces peucetius* contains four regions hybridizing to the *actI* and *tcmKL* genes, but only one region is responsible for doxorubicin production (180). In view of the structural differences between the type I and type II PKS enzymes (see below), and *eryA* probe should logically be used to clone a new type I PKS gene and an *actI* probe should be used to clone a type II PKS gene, as illustrated by the cloning of the rapamycin (162) and daunorubicin (180) biosynthesis genes. However, Schupp et al. (160) were able to clone a type I PKS gene for soraphen A biosynthesis from themyxobacterium *Sorangium cellulosum* by using the *graI* gene, and *actI* homolog from the granaticin-producing *Streptomyces violaceoruber* (167).

61.2.3. Protein Isolation Followed by Gene Cloning

Knowledge about the sequence of a purified enzyme from a secondary metabolic pathway, or the availability of antibodies to that enzyme, provides a secure way to identify the corresponding gene as well as the rest of the gene cluster. This approach has been used less often than the two methods described above, because such enzymes typically have low titers in wild-type organisms, even though they can be purified by the use of specific and sensitive assays. The need for the chemical synthesis of unusual substrates (unless the enzyme catalyzes a step that uses a common primary metabolite) has also hindered the application of this method. On the other hand, for organisms whose DNA has a high G+C content, like actinomycetes, the gene probes synthesized in accord with the biased codon usage (13) have much less degeneracy than those made from *E. coli* or human proteins and therefore often given clear-cut results in DNA hybridization reactions. The first genes for the biosynthesis of the penicillins and cephalosporins (157), macrolide antibiotics (tylosin) (53), bacterial sesquiterpenes (pentalenolactone) (25), and oligopeptide antibiotics (gramicidin S) (99) were cloned in this manner.

61.2.4. Expression of Secondary Metabolism Genes and Gene Clusters

Expression of a set of several genes in a suitable host followed by detection of the metabolite formed is another secure way, in theory, to clone a particular cluster of secondary metabolism genes. For this approach to be successful, the following requirements have to be met: (i) the cloning vector must accept large DNA segments and replicate or integrate stably in the host; (ii) the host must be able to express all of the genes; (iii) all of the enzymes have to be postranslationally modified, when necessary, and any needed cofactors have to be supplied by the host; and (iv) the product formed must not be toxic to the host, or a resistance gene has to be cloned together with the structural genes. The first requirement can be met by cosmid vectors like KC505 (152), but the second one often restricts the choice of host organism to the same family or

TABLE 2 Examples of cloned secondary metabolism genes

Strategy for primary cloning	Antibiotic	Compound class	Producing organism	Reference(s)
Complementation of blocked mutants	Actinorhodin	Polyketide	*Streptomyces coelicolor*	118
	Tetracenomycin C	Polyketide	*Streptomyces glaucescens*	129
	Streptomycin	Aminoglycoside	*Streptomyces griseus*	137
	Bialaphos	Oligopeptide	*Streptomyces hydroscopicus*	131
	Methylenomycin	Cyclopentenoid	*Streptomyces coelicolor*	28, 29
	Clavulanic acid	β-Lactam	*Streptomyes clavuligerus*	73
	Carbapenem	β-Lactam	*Streptomyces fulvoviridis*	125, 133
	Avermectin	Macrolide	*Streptomyces avermitilis*	178
	Fortimicin	Aminoglycoside	*Micromonospora olivasterospora*	38, 138
	Undecylprodigiosin	Pyrrole	*Streptomyces coelicolor*	50, 119
	Oxytetracycline	Polyketide	*Streptomyces rimosus*	20, 24
	Penicillin	β-Lactam	*Penicillium chrysogenum*	45
	Sterigmatocystin	Polyketide	*Asperigillus nidulans*	23, 212
Genetic screening methods	Daunorubicin	Polyketide	*Streptomyces peucetius*	180
	Fengycin	Peptide	*Bacillus subtilis*	30
	Rapamycin	Macrolide	*Streptomyces hydroscopicus*	162
	Soraphen A	Macrolide	*Sorangium cellulosum*	160
	Granaticin	Polyketide	*Streptomyces violaceoruber*	167
	Milbemycin	Macrolide	*Streptomyces griseochromogenes*	148
	Mithramycin	Polyketide	*Streptomyces argillaceus*	103
	FK506	Macrolide	*Streptomyces* sp. strain MA6548	128
	Frenolicin	Polyketide	*Streptomyces roseofulvus*	16
	Griseusin B	Polyketide	*Streptomyces griseus*	213
	Oleandomycin	Macrolide	*Streptomyces antibioticus*	71, 186
	Jadomycin B	Polyketide	*Streptomyces venezuelae*	69
	Urdamycin A	Polyketide	*Streptomyces fradiae*	40
	Bleomycin	Peptide/polyketide	*Streptoverticillum* sp. strain ATCC 15003	48
	Cephalosporin	β-Lactam	*Cephalosporium acremonium*	66, 158
	Penicillin	β-Lactam	*Aspergillus nidulans*	111, 205
	Macrocyclictrichothecenes	Sesquiterpene	*Myrothecium roridum*	194
	Aflatoxin	Polyketide	*Aspergillus parasiticus*	51, 193, 210
	Rubradirin	Ansamycin	*Streptomyces achromogenes*	171
	Rifamycin	Ansamycin	*Amycolatopsis mediterranei*	3, 161
	Balhimycin	Peptide	*Amycolatopsis mediterranei*	145
	whiE spore pigment	Polyketide	*Streptomyces coelicolor*	39
Biochemical approaches	Tylosin	Macrolide	*Streptomyces fradiae*	53
	Pentalenolactone	Sesquiterpene	*Streptomyces* strain UC5319	25
	Gramicidin S	Oligopeptide	*Bacillus brevis*	99
	6-Methylsalicylic acid	Polyketide	*Pencillium patulum*	11
	Trichothecenes	Sesquiterpene	*Fusarium sporotrichioides*	74
	Cyclosporin A	Oligopeptide	*Tolypocladium inflatum*	202
	Chloroeremomycin	Glycopeptide	*Amyclatopsis orientalis*	195a
Expression of genes and gene clusters	Cephamycin C	β-Lactam	*Streptomyces cattleya*	31
	PD 116740	Polyketide	*Streptomyces* strain WP 4669	76
	Tetrangulol	Polyketide	*Streptomyces rimosus*	76
	Tetrangomycin	Polyketide	*Streptomyces rimosus*	76
	Daunorubicin	Polyketide	*Streptomyces peucetius*	180
	Penicillin	β-Lactam	*Penicillium chrysogenum*	170
	Puromycin	Aminonucleoside	*Streptomyces alboniger*	101
	Nikkomycin	Nucleoside	*Streptomyces tendae*	21
	Carbomycin	Macrolide	*Streptomyces thermotolerans*	49
	Candicidin	Macrolide	*Streptomyces griseus*	58
	Sisomycin	Aminoglycoside	*Micromonospora inyoensis*	60, 61
	Nosiheptide	Thiopeptide	*Streptomyces actuosus*	32, 47
	Actinomycin D	Peptide	*Streptomyces antibioticus*	86

genera that produces the metabolite (e.g., *Streptomyces* genes are seldom expressed in *E. coli* or *Bacillus* sp.) nor are fungal genes expressed in bacteria except as the cDNA form. Even if all the genes are expressed, specialized cofactors or modifying enzymes may be absent. Nonetheless, this approach was used to shotgun-clone the cephamycin C production genes from *Streptomyces cattleya* into *Streptomyces lividans* (31). After prescreening cosmids for hybridization to suitable gene probes, it was also used to clone angucycline biosynthesis genes (76) and most of the daunorubicin production genes (180) into this host and the *Penicillium chrysogenum* penicillin biosynthesis genes into *Neurospora crassa* and *Aspergillus nidulans* (170). Primary clones obtained by one of the methods discussed above have also been used to facilitate cloning of an entire gene cluster and expression in a heterologous host, for example, actinorhodin (117, 118), tetracenomycin (41, 129), puromycin (101), and nikkomycin (21).

61.2.5. Genome Sequencing

Now that it is possible to sequence the entire genomes of microorganisms in less than a year, the resulting data can be analyzed for the presence of putative antibiotic production genes by searching for homologs of the types of genes listed in Table 2. The "red genes" for the biosynthesis of undecylprodigiosin and related red pigments of *S. coelicolor* have been identified in this way (70), and a putative polyketide/peptide synthase of *Bacillus subtilis* has been revealed (1).

61.3. EXAMPLES OF SECONDARY METABOLISM GENES AND THEIR FUNCTIONS

61.3.1. Polyketides

Polyketides are the largest class of microbial products and have a carbon skeleton made from simple carboxylic acids in a manner resembling the biosynthesis of fatty acids. They are produced by two types of PKS enzymes: multifunctional type I, in which each protein has numerous catalytic activities, and type II, in which each protein usually is monofunctional. As a consequence, the corresponding genes fall into two classes, defined by the *eryAI-III* (34, 46, 203) and 6-methylsalicylic acid synthase (11) genes for the respective type I bacterial and fungal PKSs, and the *actI* (52) and *tcmK*, *tcmL*, and *tcmM* (12) genes for the type II bacterial PKS (Fig. 1). Specific bacterial type I PKSs have been described that use acetyl-, 3-amino-5-hydroxybenzoyl, oxygenated cyclohexyl-, or propionyl-coenzyme A (CoA) as the starter unit; malonyl-CoA and its 2-methyl or 2-ethyl homologs commonly serve as the chain extender units that undergo decarboxylative condensation with the starter unit or the growing β-ketoacylthioester carbon chain. The known fungal PKSs use acetyl- or hexanoyl-CoA as the starter unit and malonyl-CoA exclusively for chain extension. The β-ketothioester produced in the first condensation reaction can be left unaltered or, as illustrated in Fig. 2, can undergo (i) reduction to a β-hydroxythioester, (ii) reduction and dehydration to an α,β-unsaturated thioester, (iii) reduction, dehydration, and further reduction to a saturated thioester, as in the biosynthesis of long-chain fatty acids, or (iv) C methylation in fungal PKSs (135, 191).

Thousands of polyketide metabolites have been found in nature because, in theory, a huge number of different structures can be made by combining these four individual reactions in different ways along with different starter and extender units. The structural possibilities are further magnified by having two possible stereochemistries at each chiral center in the PKS-derived carbon skeleton and by decorating the latter compound through additional biochemical transformations involving oxygenation, reduction, methylation, acylation, glycosylation, etc. As an illustration of these facts, we have summarized the biosynthesis of erythromycin A and sterigmatocystin, made by bacterial and fungal type I PKSs, respectively, and tetracenomycin C and doxorubicin, made by bacterial type II PKSs (fungal type II PKSs are unknown) in Fig. 3 through 6.

In the initial stage of the erythromycin pathway in *S. erythraea*, deoxyerythronolide B synthase (DEBS) is produced by the *eryAI*, *eryAII*, and *eryAIII* genes (Fig. 3), which were identified initially by their close linkage to several other structural genes near the *ermE* gene for resistance to erythromycin A (188, 203). DEBS consists of homodimers of the DEBS1, DEBS2, and DEBS3 subunits (177), each of which contains two modules (Fig. 1A), defined as the set of domains having the unique collection of active sites required for assembly of a particular three-carbon unit derived from propionyl- or 2-methylmalomyl-CoA (46). In addition, the first module has two unique sites for loading the propionyl-CoA starter unit; once this happens, the linear intermediates are assembled by chain extension with (2*S*)-2-methylmalonyl-CoA (120) in a processive manner during which each active site is used only once and the individual modules precisely recognize the structure of the growing carbon chain so that the macrolide precursor is assembled in the correct order (195). Then the acyclic 6-deoxyerythronolide B precursor is cyclized intramolecularly upon release from the enzyme by the thioesterase domain at the C-terminal end of DEBS3 (35, 87) to give the macrolide product 6-deoxyerythronolide B (Fig. 3). The two bimodular proteins of each subunit are arranged in a parallel, possibly helical fashion (177) such that the ketosynthase active site at the "beginning" of one module in the protein is closest to the acyl carrier protein active site at the "end" of the identical module in the other protein (head-to-tail arrangement) (88). This model enables the two proteins in any one module to share active sites (59, 88) and allows a module to function independently of the others.

Erythromycin A biosynthesis proceeds from 6-deoxyerythronolide B by C6 hydroxylation to produce erythronolide B, O3 glycosylation with L-mycarose to form erythromycin C, glycosylation at O5 with D-desosamine to make erythromycin D, C12 hydroxylation to produce erythromycin C, and finally O methylation of mycarose to make the cladinose found in erythromycin A (Fig. 3). The genes for these steps (shown in Fig. 3) were identified largely by gene sequencing and disruption or replacement and are clustered about the *eryA* genes, with the *eryB* and *eryC* deoxysugar biosynthesis genes (55, 56, 155, 183) interdispersed among those encoding the EryF (204) and EryK (175) CYP450 hydroxylases and the EryG O methyltransferase (144) (Fig. 3). Deoxysugar biosynthesis is still poorly understood at the biochemical level, even though it occurs in many bacteria (95, 189). Consequently, the functions of several of the enzymes encoded by the *eryB* and *eryC* genes have not been determined (55, 56, 155,

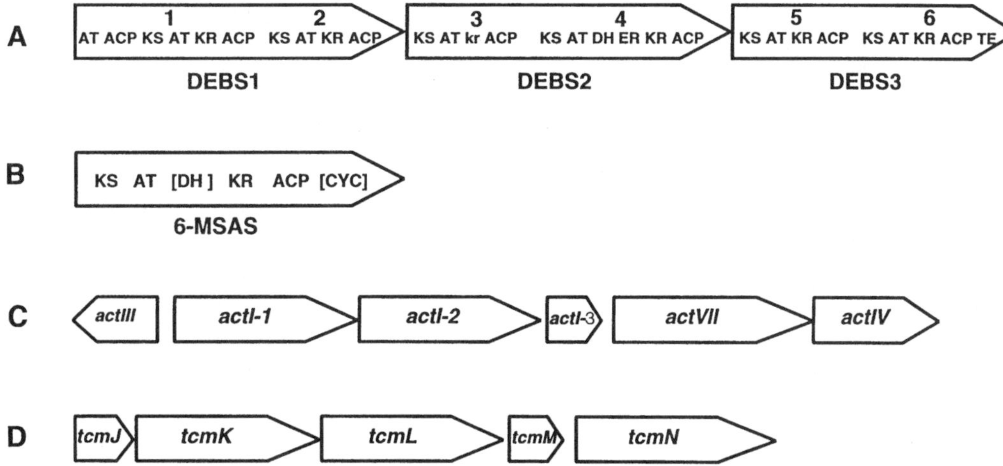

FIGURE 1 Microbial polyketide synthase genes and gene products. (A) The *S. erythraea eryA* genes encoding 6-deoxyerythronolide B synthase B synthase (DEBS). The active sites within the two modules of each gene are indicated inside a wedge, symbolizing the gene and the direction of gene transcription, its size being proportionate to the size of the gene product. The abbreviations for the enzymatic functions of the active sites in each module are ACP, acyl carrier protein; AT, acyltransferase; DH, dehydratase; ER, enoyl reductase; KR, keto reductase (kr, inactive allele); KS, β-ketoacylsynthase; TE, thioesterase. (B) The *Penicillium patulum* 6-methylsalicylic acid synthase (6-MSAS) gene. See panel A for abbreviations for the enzymatic functions of the active sites; [DH] and [CYC] designate putative dehydratase and cyclase domains, respectively, that have not been identified by sequence analysis. (C) The *S. coelicolor act* genes. The name of each gene is indicated inside a wedge, symbolizing the gene and the direction of gene transcription, its size being proportionate to the size of the gene product. Gene functions are explained in the text. (D) The *S. glaucescens tcm* genes. The name of each gene is indicated inside a wedge, symbolizing the gene and the direction of gene transcription, its size being proportionate to the size of the gene product. Gene functions are explained in the text.

183). It also is not known if any of the *ery* genes regulate erythromycin A production, because none of the deduced gene products resemble known regulatory proteins from other metabolic pathways, and no mutations blocking expression of all the *ery* production genes have been isolated.

Sterigmatocystin biosynthesis in *A. nidulans* (212) and aflatoxin biosynthesis in *Aspergillus flavus* (210) and *Aspergillus parasiticus* (26, 51) both involve type I PKSs (sterigmatocystin is an intermediate in the aflatoxin pathway). These enzymes differ from DEBS in two important ways: (i) they require a dedicated two-subunit fatty acid synthase, different from the one used by the *A. nidulans* fatty acid synthase (22) and presumed to associate directly with the PKS (200), to make the hexanoyl-CoA primer (22, 115, 193, 200); and (ii) they use each active site interactively to make the 20-carbon framework of 10-deoxynorsolorinic acid, the aromatic product of both PKSs (Fig. 4). Four active sites—for β-ketoacylsynthase, acyltransferase, acyl carrier protein, and thioesterase domains—can be recognized in the sequences of the StcA (212) and PksA (26, 51) fungal PKSs, similar to the organization of 6-methylsalicylic acid synthase (Fig. 1B). By analogy to the bacterial type II PKSs, these sites would be sufficient to form 10-deoxynorsolorinic acid from hexanoyl- and malonyl-CoA. The remaining 25 or more genes for sterigmatocystin (23) and aflatoxin (193) biosynthesis are clustered about the PKS and fatty acid synthase genes and include the *aflR* global regulator (206, 211) required for transcription of the structural genes (23). Hence, the organization and regulation of these genes resembles the bacterial paradigm for secondary metabolism, but their G+C content does not support a bacterial origin (23), as was suggested for the fungal penicillin biosynthesis genes (33, 150, 157). Many of the sterigmatocystin and aflatoxin genes have been characterized by sequence analysis and disruption (23, 192), and the mechanism of some of the purified enzymes has been elucidated (as in reference 126).

Genetic and biochemical studies of the biosynthesis of actinorhodin in *S. coelicolor* and tetracenomycin C in *S. glaucescens* led to the paradigm for bacterial polyketides made by a type II PKS. The polyketide encoded by the *tcmJ*, *tcmK*, *tcmL*, *tcmM*, and *tcmN* genes (Fig. 1D) (12, 184, 185) assembles the 20-carbon skeleton of tetracenomycin C, with the assistance of the *fabD* gene (8, 182) to load malonyl-CoA onto the TcmM acyl carrier protein (Fig. 5). As in *S. coelicolor*, where the *actI*, *actIV* and *actVI* (52), *actIII* (68) (Fig. 1C), and the homologous *fabD* (151) genes govern formation of the bicyclic octaketide precursor of actinorhodin, a multienzyme complex consisting of the TcmK and TcmL β-ketoacyl:acyl carrier protein synthase subunits (an $\alpha\beta$ heterodimer), the TcmM acyl carrier protein, and the FabD malonyl-CoA:acyl carrier protein acyltransferase is believed to create a TcmM-bound decaketide. This decaketide is cyclized by TcmN to a dissociable but tightly bound tricyclic aromatic product, Tcm F2 (8, 165). TcmJ facilitates Tcm F2 formation but nevertheless is dispensable (8, 185). Tcm F2 synthesis has been achieved in vitro by purification and reconstitution of the Tcm PKS (8). Cyclization of Tcm F2 to the tetracyclic Tcm F1, oxidation, stepwise O methylation, and a further oxidation of

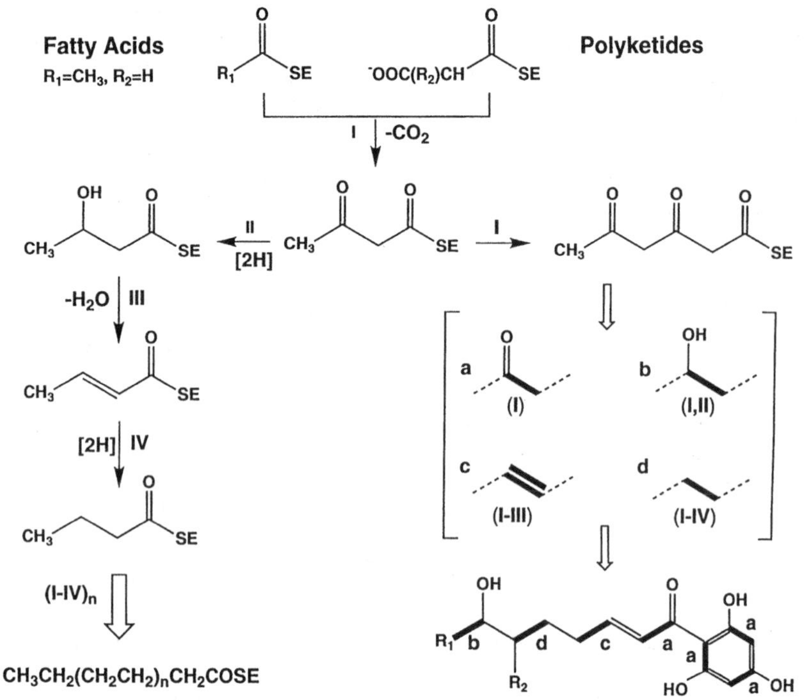

FIGURE 2 Comparison of the biochemical steps featured in fatty acid and polyketide biosynthesis. In step I of the fatty acid pathway, decarboxylative condensation takes place between acetyl-CoA (the starter unit; $R_1 = CH_3$) and malonyl-CoA (the carbon chain extender unit; $R_2 = H$) to produce a β-ketobutyrylthioester. In step II, the β-ketogroup is reduced to a hydroxyl group. In step III, dehydration occurs to produce a double bond conjugated to the thioester carbonyl, and in step IV, the double bond is reduced to form the butyrylthioester. The product of this four-step process undergoes the same reactions "n" times, as indicated by $(I–IV)_n$, to form a long-chain fatty acid. E, enzyme; [2H], reduction. In polyketide biosynthesis, each of the four steps can be used in any combination, as designated by (I), (I, II), (I–III) and (I–IV) in the bracketed section, to make the functional groups indicated by a, b, c, and d typically found in a polyketide. The letters a, b, c, and d indicate the origin of the different parts of the hypothetical polyketide structure drawn at the bottom right side, in which R_1 and R_2 could be any one of a variety of alkyl or aromatic (R_1 only) groups.

three positions results in Tcm C (Fig. 5). The function of all of the *tcm* genes was established through sequence analysis, isolation of metabolites accumulated by *tcm* mutants, expression of *tcmP* (42), purification and mechanistic analysis of the TcmM (166), TcmN (165), TcmI (185), TcmH (185), and TcmG (42) enzymes, and characterization of the *tcmA* and *tcmR* resistance and repressor genes (64, 65). As in the case of the *ery* genes, expression of the *tcm* genes from the four promoters identified (42, 65) does not appear to be controlled by a cluster-associated regulatory gene.

In the gene cluster for doxorubicin biosynthesis by *S. peucetius* and *Streptomyces* sp. strain C5 (reviewed in reference 179), the *dpsG* acyl carrier protein gene is located about 10 kb upstream of the *dpsA* and *dpsB* genes for the β-ketoacyl:acyl carrier protein synthase subunits (62, 208) instead of immediately downstream of them as in the other clusters of type II PKS genes so far studied. The three polyketide cyclase genes, *dpsF* (62, 208), *dpsH* (57), and *dpsY* (104), also are situated at different places in the cluster instead of in one region as in most other cases. Moreover, the *dpsC* and *dpsD* genes, although dispensable (149), nonetheless ensure that the PKS uses propionyl-CoA instead of acetyl-CoA as the starter unit to build the 21-carbon decaketide (179), which then is reduced at C11 (*dpsE*) and cyclized (*dpsF*, *dpsH*, *dpsY*) to 12-deoxyaklanonic acid (Fig. 6), the counterpart of Tcm F2. ε-Rhodomycinone, the aglycone precursor of doxorubicin, is made from 12-deoxyaklanonic acid by oxidation at C12 (*dnrG*), methylation of the carboxyl (*dnrC*), and cyclization of the fourth ring to make aklaviketone (*dnrD*), followed by C7 reduction (*dnrE*) and C11 oxidation (*dnrF*). TDP-L-daunosamine, a 3-amino-2,3,6-trideoxyhexose, is synthesized from TDP-glucose by the products of the *dnmL*, *dnmJ*, *dnmQ*, *dnmU*, *dnmV*, *dnmT*, and *dmmZ* genes, then attached to ε-rhodomycinone to form rhodomycin D (44) by the glycosyltransferase encoded by *dnmS*. Only the *dnrP*, *dnrK*, and *doxA* genes are required to convert rhodomycin D to doxorubicin (44). The *dnrH* (163) and *dnrX* (104) genes seem to govern further metabolism of dauno- and doxorubicin to the acid-sensitive baumycin glycosides (104, 163), wherease *dnrU* specifies C13 reduction of daunorubicin to its dihydroform (107). The role of the *dnrV* gene remains unsettled (107). Positively acting transcription factors encoded by the *dnrI* (112), *dnrN* (139), and *dnrO* (140) genes control the expression of the structural and self-resistance genes (*drrA* and *drrB* [63], *drrC* [106], and *drrD* [2]) required for doxorubicin production.

61.3.2. β-Lactams

β-Lactam and oligopeptide antibiotics are made from amino acids, and, when more than one is involved, the

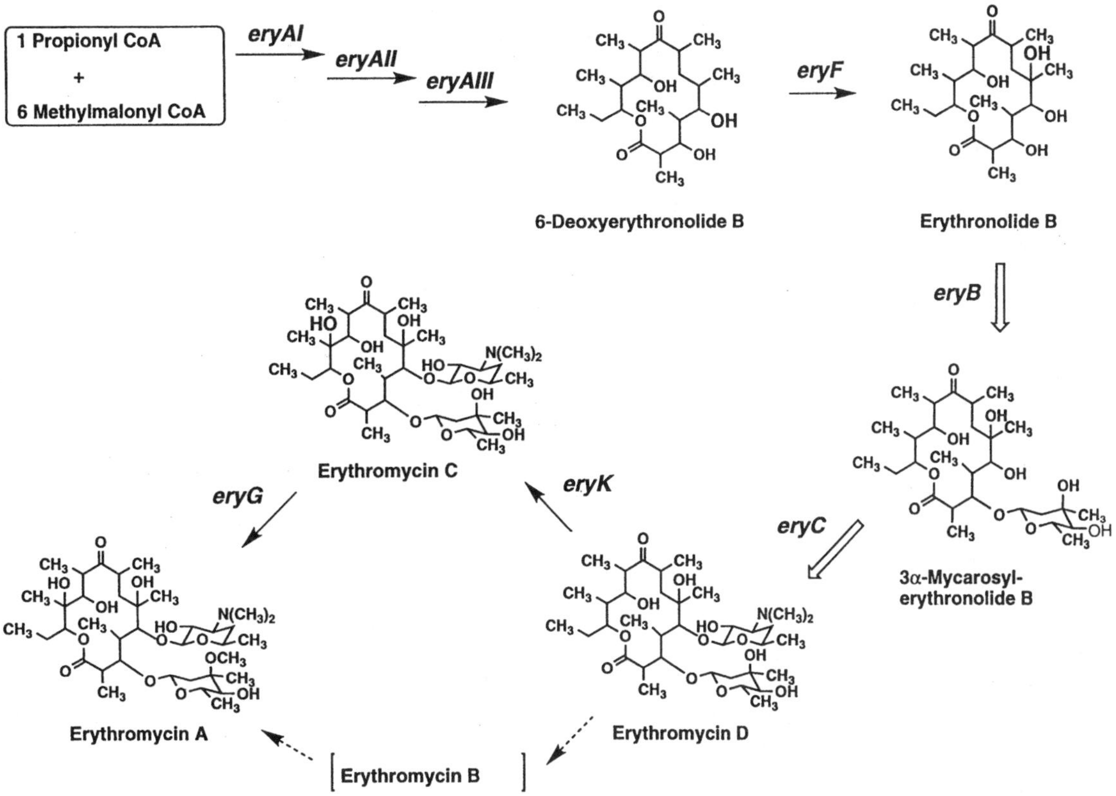

FIGURE 3 Biosynthetic pathway of erythromycin A. The functions of the genes shown above or beside the arrows are explained in the text. The dotted arrows indicate a minor pathway for the conversion of erythromycin D to erythromycin A.

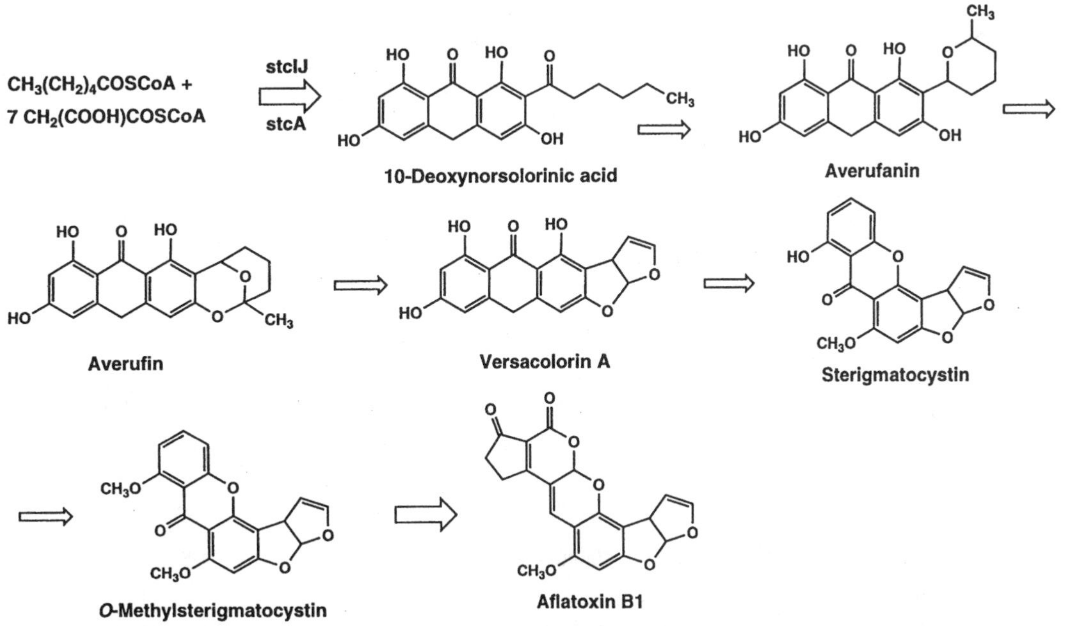

FIGURE 4 The key intermediates in the biosynthetic pathway of sterigmatocystin and aflatoxin B1. Open arrows represent several enzymatic steps. The functions of the genes shown above the first arrow are explained in the text.

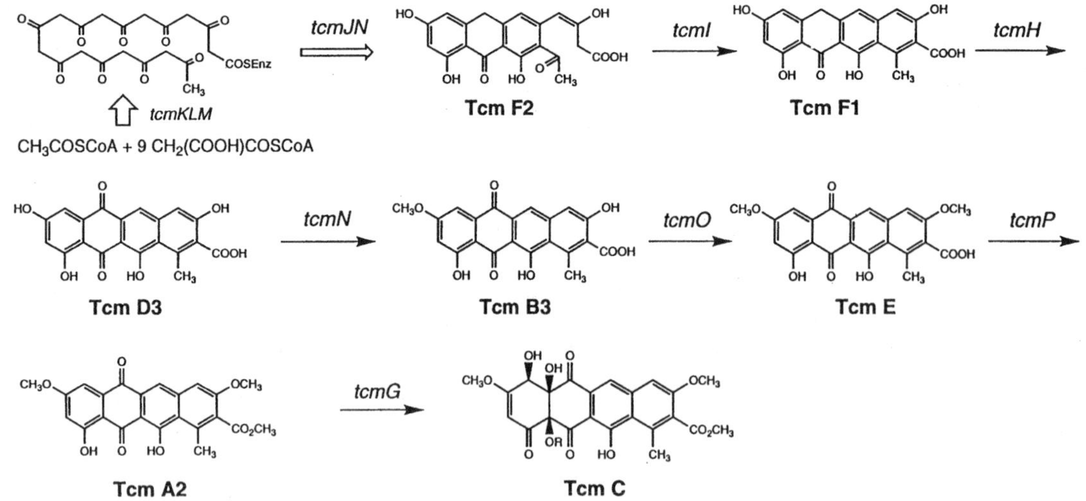

FIGURE 5 Biosynthetic pathway of tetracenomycin C. Open arrows represent several enzymatic steps; solid arrows represent one enzymatic step. The functions of the genes shown above the arrows are explained in the text.

process of carbon framework assembly resembles the bacterial type I mechanism for polyketide biosynthesis because the choice and sequence of amino acids are determined by a modular, thiotemplate mechanism. Penicillins and cephalosporins, the best-known β-lactam antibiotics, are fungal metabolites whose biosynthesis and genetics have been studied in *P. chrysogenum*, *A. nidulans*, and *Acremonium chrysogenum* (85, 143). Bacteria also make β-lactams; the biosynthesis of cephamycin C in *Steptomyces clavuligerus* and *Nocardia lactamdurans* and clavulanic acid in *S. clavuligerus* has been especially well studied (85, 143). Although there is much less molecular diversity among β-lactam antibiotics than among polyketides, their production is widespread among bacteria, suggesting that the β-lactam structural motif provides an effective means of self-defense.

Penicillin and cephalosporin biosynthesis (Fig. 7) begins with the formation of the tripeptide δ-(L-α-aminoadipyl)-L-cysteinyl-D-valine (ACV) from the three L-amino acids by ACV synthetase, encoded by *pcbAB* and a member of the thiotemplate enzyme family that make oligopeptides by a nonribosomal process (see next section). In bacteria, the

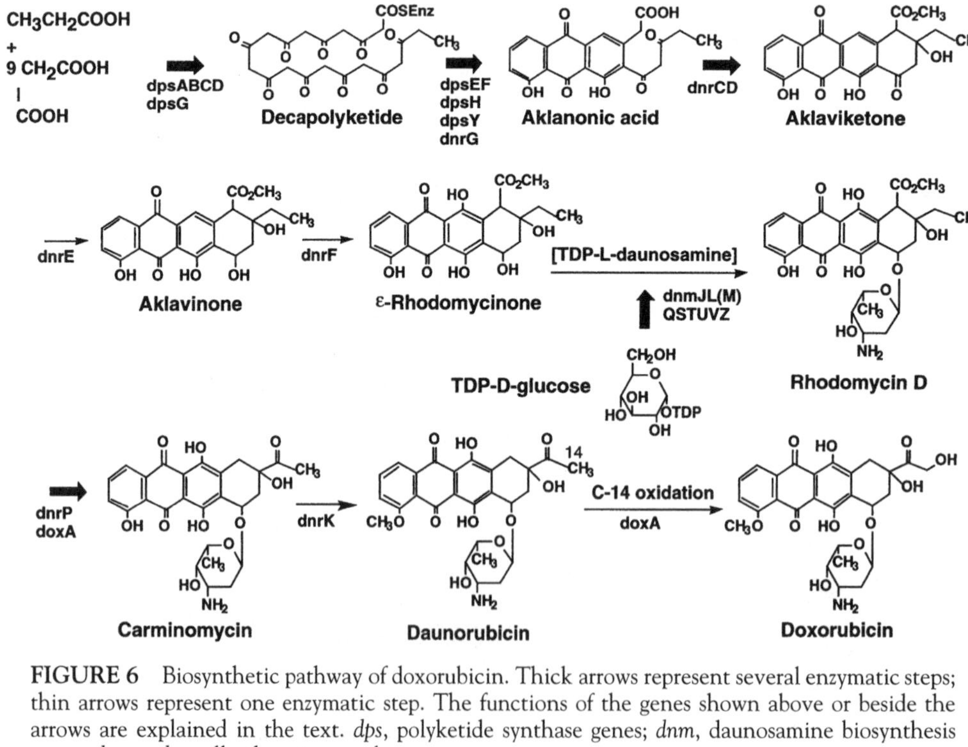

FIGURE 6 Biosynthetic pathway of doxorubicin. Thick arrows represent several enzymatic steps; thin arrows represent one enzymatic step. The functions of the genes shown above or beside the arrows are explained in the text. *dps*, polyketide synthase genes; *dnm*, daunosamine biosynthesis genes; *dnr* or *dox*, all other structural genes.

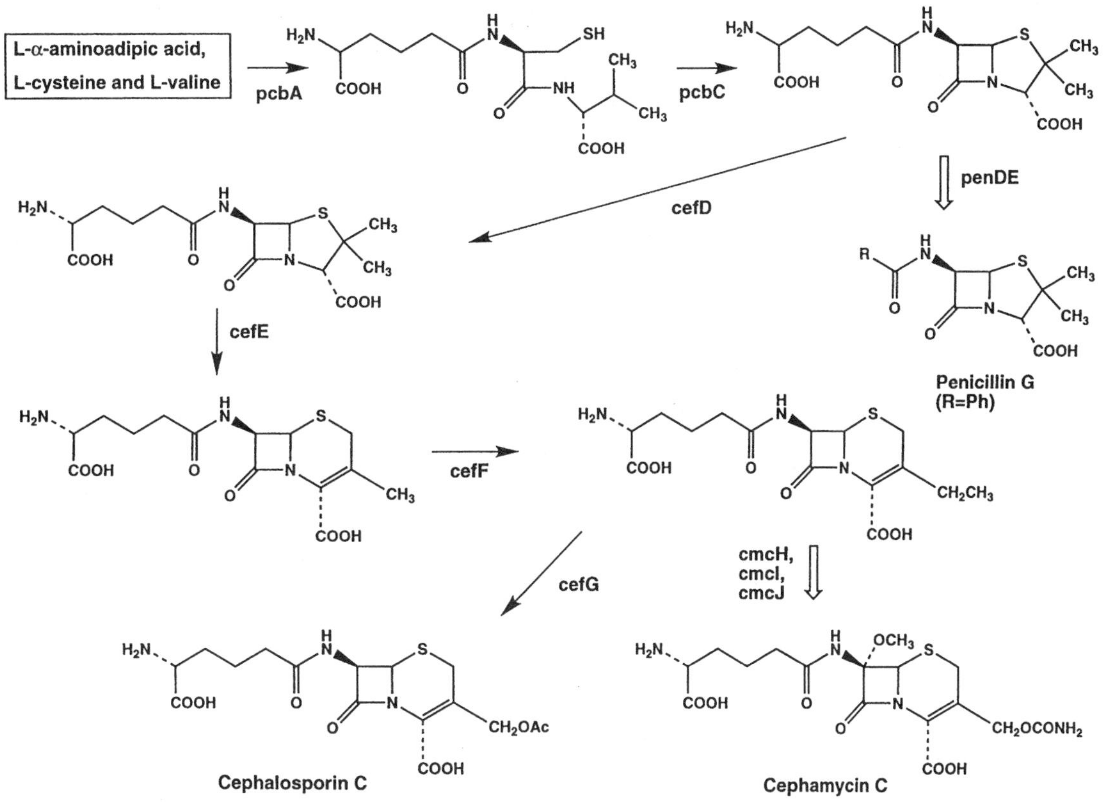

FIGURE 7 Biosynthetic pathways of bacterial and fungal β-lactam antibiotics. Open arrows represent several enzymatic steps; solid arrows represent one enzymatic step. The functions of the genes shown above or beside the arrows are explained in the text.

α-aminoadipic acid comes from the catabolism of lysine by lysine ε-aminotransferase, an enzyme encoded by one of the penicillin biosynthesis genes; this same amino acid is an offshoot of lysine biosynthesis in fungi (190). Purification of the next enzyme in the pathway, isopenicillin N synthase (IPNS) (75), followed by cloning and analysis of the *pcbC* gene that encodes IPNS (157), actually opened the way to studies of the genetics of penicillin and cephalosporin biosynthesis. IPNS has also been useful for the synthesis of β-lactam analogs (80). Fungi typically produce hydrophobic penicillins by replacing the L-α-aminoadipyl side chain of isopenicillin N with a substituted acetyl group commonly obtained from an acid present in or added to the growth medium, using acyl-CoA:6-aminopenicillanic acid acyltransferase encoded by the *penDE* gene (9). Interestingly, the organization of the gene clusters containing the *pcbAB* gene for tripeptide formation, *pcbC* and *penDE* in fungi and *pcbAB* and *pcbC* in bacteria, is highly conserved (45, 169). Organisms that make cephalosporins and cephamycins do not have *penDE* genes; instead, the L-α-aminoadipyl side chain of isopenicillin N is epimerized to the D-α-aminoadipyl isomer (*cefD* encodes the epimerase [97]); then the thiazole ring is expanded to a dihydrothiazine ring by deacetoxycephalosporin C synthase. The latter activity is associated with deacetylcephalosporin C synthase as part of a bifunctional enzyme encoded by *cefEF* in fungi (158), whereas two separate enzymes are used in bacteria (96, 98). Deacetylcephalosporin C undergoes acetylation of the C3 hydroxyl in fungi (*cefG* [G]) to give cephalosporins, but it undergoes carbamoylation followed by C7 hydroxylation and O methylation in *S. clavuligerus* to give cephamycin C (85, 143) (Fig. 7).

Cephamycin and clavulanic acid biosynthesis are governed by adjacent regions of the *S. clavuligerus* genome, which may facilitate coordinate production of the β-lactam antibiotic and β-lactamase inhibitor. Clavulanic acid is made from D-glycerate and L-arginine via the intermediates shown in Fig. 8; several of the enzymes and genes for this process have been identified (4, 73, 141, 143). The mechanism of the β-lactam synthetase enzyme (encoded by the *orf3* gene [4]) that forms the initial β-lactam ring system (Fig. 8) is quite different from that of the IPNS involved in β-lactam biosynthesis in bacteria and fungi. *S. clavuligerus* contains two clavaminate synthase isoenzymes (156), only one of which is required for clavulanic acid biosynthesis (the role of the other enzyme is unknown) (122). Self-resistance genes have been identified in the cluster of cephamycin C and clavulanic acid biosynthesis genes (142) but are not necessary in the fungi that produce β-lactams. The biosynthesis of carbapenem antibiotics such as thienamycin (Fig. 8) is still poorly understood, although the deduced products of some of the genes for 1-carbapen-2-em-3-carboxylic acid production (Fig. 8) resemble enzymes of the clavulanic acid pathway (125).

61.3.3. Oligopeptide Antibiotics

The nonribosomal process for the assembly of the small pepetides known as oligopeptide antibiotics is another example of the modular, thiotemplate mechanism for which large, multifunctional enzymes specify the choice of sub-

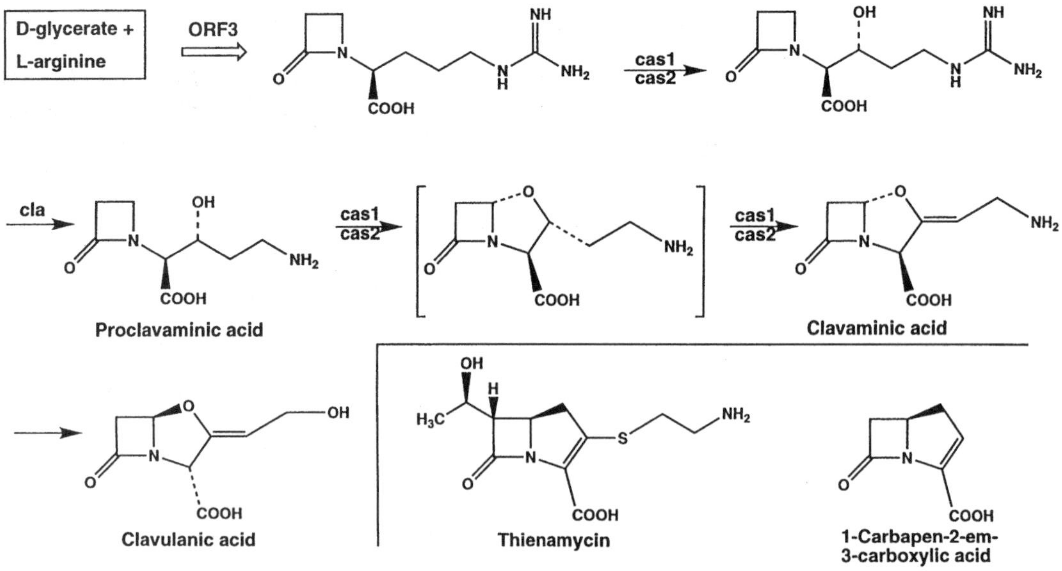

FIGURE 8 Key intermediates in the biosynthesis of clavulanic acid. The structures of two carbapenem antibiotics are shown in the inset. The functions of the genes shown above the arrows are explained in the text.

strate and sequence of the precursor assembly reactions (198, 214). These antibiotics are made by bacteria and fungi and embody considerable structural diversity, since the amino acid at each position in the peptide may be used unchanged or undergo C2 epimerization, C or N methylation, halogenation, hydroxylation, or desaturation, as well as other types of oxidation during or after assembly of the carbon backbone. Furthermore, oligopeptide antibiotics may contain nonprotein amino acids of unusual structure (formed before as well as after incorporation into the peptide) and may be cyclic or linear, in addition to the occasional presence of fatty acid-derived acyl groups or sugars attached to an amino acid side chain. Interestingly, substrate recognition and attachment via a thioester bond and formation of the peptide bond, as well as epimerization or N methylation, are all specified by the domains of the modules that dictate the reactions that take place each time an amino acid is added to the growing peptide. The enzyme subunits may contain as few as three or as many as 48 modules and have a very high molecular weight (cyclosporin synthetase is a 1.7-million-dalton protein that catalyzes 40 reactions [102, 191]).

Surfactin (Fig. 9), an acyl peptidolactone containing seven amino acids and a fatty acid-derived side chain esterified with two of them, is a well-studied oligopeptide antibiotic. The heptapeptide portion is made by an oligopeptide synthetase composed of at least four subunits (214); the *srfAA* and *srfAB* genes each encode a trimodular synthetase, whereas *srfAC* encodes a monomodular synthetase with a thioesterase domain at its C terminus. The *srfAD* gene encodes a separate thioesterase, which, along with the thioesterase domain of SrfAC, is required for surfactin production (159). Although the enzyme that begins assembly in most oligopeptide synthetase has an adenylation module with a domain that recognizes the first amino acid to be used and activates it as an acyl-adenylate, the SrfAA enzyme also contains a separate condensation domain in front of the first module that may be involved in attaching the fatty acid to the glutamate residue. Each of the other modules of the SrfA protein has domains that recognize the proper amino acid, catalyze its reaction with ATP to form an acyl-adenylate, and form a thioester between the resulting activated amino acid and the thiol group of the 4′-phosphopantheine residue attached to the thioester domain. These reactions all take place to load each module of SrfAA, SrfAB, and SrfAC with the correct amino acid thioester; then an elongation domain in each module catalyzes the reaction between the thioester of the first amino acid with the amino group of the next amino acid to form the peptide bond and so forth until the heptapeptide is assembled. Since surfactin contains D-leucine at two sites, one module in each of the SrfAA and SrfAB proteins also contains an epimerization domain that converts the enzyme-bound L-leucine to its D-enantiomer. Once assembly of the heptapeptide thioester is completed, the thioesterase domain of SrfAC together with the SrfAD

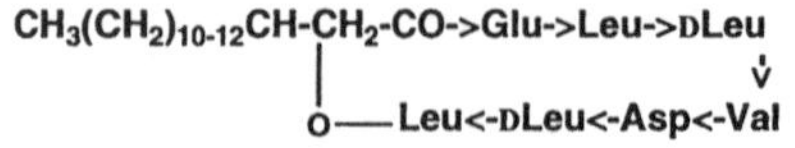

$CH_3(CH_2)_{10-12}CH-CH_2-CO$->Glu->Leu->DLeu

O—Leu<-DLeu<-Asp<-Val

Surfactin

FIGURE 9 The structure of surfactin. The customary three-letter abbreviations for the amino acids are used; “->” indicates the directionality of peptide bond formation, from the N to C terminus.

enzyme releases the oligopeptide product. The timing of acyl peptidolactone formation has not been elucidated.

61.4. INDUSTRIAL APPLICATIONS OF CLONED SECONDARY METABOLISM GENES

If a secondary metabolite found to have some value in human or animal medicine, agriculture, aquaculture, or related areas has enough commercial potential to justify the development of higher-producing strains, a strain improvement program often will be initiated to improve yields and fermentation characteristics. Both goals can be targeted simply at lowering the cost of metabolite production, although the production of novel metabolites by genetic engineering has recently become an ancillary aim. Secondary metabolism genes increasingly are being used in strain improvement programs to supplement or even supplant the traditional approach, which involves random mutation followed by screening for the desired mutants. Since a comprehensive review of this type of work is beyond the scope of this chapter, interested readers may consult specialized reviews (5–7, 81–83, 134, 146) and papers about the following antibiotics for highlights: pristinamycins (164), β-lactams (36, 43, 116, 168), actinorhodin (114), doxorubicin (107), epirubicin (113), and erythromycin (84, 121, 127, 153, 176).

REFERENCES

1. **Albertini, A. M., T. Caramori, F. Scoffone, C. Scotti, and A. Galizzi.** 1995. Sequence around the 159° region of the *Bacillus* genome: the *pksX* locus spans 33.6 kb. *Microbiology* **141:**299–309.
2. **Ali, A., and C. R. Hutchinson.** Unpublished results.
3. **August, P. R., L. Tang, Y. J. Yoon, S. Ning, R. Muller, T. W. Yu, M. Taylor, D. Hoffmann, C. G. Kim, X. Zhang, C. R. Hutchinson, and H. G. Floss.** 1998. Biosynthesis of the ansamycin antibiotic rifamycin: deductions from the molecular analysis of the *rif* biosynthetic gene cluster of *Amycolatopsis mediterranei* S699. *Chem. Biol.* **5:**69–79.
4. **Bachmann, B. O., R. Li, and C. A. Townsend.** 1998. β-Lactam synthetase: a new biosynthetic enzyme. *Proc. Natl. Acad. Sci. USA* **95:**9082–9086.
5. **Baltz, R. H.** 1997. Molecular genetic approaches to yield improvement in actinomycetes, p. 49–62. *In* W. R. Strohl (ed.), *Biotechnology of Antibiotics*, 2nd ed. Marcel Dekker, New York.
6. **Baltz, R. H.** 1998. Genetic manipulation of antibiotic-producing *Streptomyces*. *Trends Microbiol.* **6:**76–83.
7. **Baltz, R. H., and T. J. Hosted.** 1996. Molecular genetic methods for improving secondary-metabolite production in actinomycetes. *Trends Biotechnol.* **14:**245–250.
8. **Bao, W., E. Wendt-Pienkowski, and C. R. Hutchinson.** 1998. Reconstitution of the iterative type II polyketide synthase for tetracenomycin F2 biosynthesis. *Biochemistry* **37:**8132–8138.
9. **Barredo, J. L., P. van Solingen, B. Díez, E. Alvarez, J. M. Cantoral, A. Kattevilder, E. B. Smaal, M. A. Groenen, A. E. Veenstra, and J. F. Martín.** 1989. Cloning and characterization of the acyl-coenzyme A: 6-aminopenicillanic-acid-acyltransferase gene of *Penicillium chrysogenum*. *Gene* **83:**291–300.
10. **Basinki, M., M. Bierman, and B. E. Schoner.** 1993. Cloning of polyketide biosynthetic genes using colony hybridization and PCR. *Dev. Ind. Microbiol.* **33:**237–246.
11. **Beck, J., S. Ripka, A. Siegner, E. Schlitz, and E. Schweizer.** 1990. The multifunctional 6-methylsalicylic acid synthase gene of *Penicillium patulum*. *Eur. J. Biochem.* **192:**487–498.
12. **Bibb, M. J., S. Biro, H. Motamedia, J. F. Collins, and C. R. Hutchinson.** 1989. Analysis of the nucleotide sequence of the *Streptomyces glaucescens tcmI* genes provides key information about the enzymology of polyketide antibiotic biosynthesis. *EMBO J.* **8:**2727–2736.
13. **Bibb, M. J., P. R. Findlay, and M. W. Johnson.** 1984. The relationship between base composition and codon usage in bacterial genes and its use for the simple and reliable identification of protein-coding sequences. *Gene* **30:**157–166.
14. **Bibb, M. J., R. F. Freeman, and D. A. Hopwood.** 1977. Physical and genetical characterization of a second sex factor, SCP2, for *Streptomyces coelicolor* A3(2). *Mol. Gen. Genet.* **154:**155–166.
15. **Bibb, M. J., J. L. Schottel, and S. N. Cohen.** 1980. A DNA cloning system for interspecies gene transfer in antibiotic-producing *Streptomyces*. *Nature* **284:**526–531.
16. **Bibb, M. J., D. H. Sherman, S. Omura, and D. A. Hopwood.** 1994. Cloning, sequencing and deduced functions of a cluster of *Streptomyces* genes probably encoding biosynthesis of the polyketide antibiotic frenolicin. *Gene* **142:**31–39.
17. **Bibb, M. J., J. M. Ward, and D. A. Hopwood.** 1978. Transformation of plasmid DNA into *Streptomyces* at high frequency. *Nature* **274:**398–400.
18. **Bibb, M. J., J. M. Ward, T. Kieser, S. N. Cohen, and D. A. Hopwood.** 1981. Excision of chromosomal DNA sequences from *Streptomyces coelicolor* forms a novel family of plasmids detectable in *Streptomyces lividans*. *Mol. Gen. Genet.* **184:**230–240.
19. **Bierman, M., R. Logan, K. O'Brien, E. T. Seno, R. Nagaraja Rao, and B. E. Schoner.** 1992. Plasmid cloning vectors for the conjugal transfer of DNA from *Escherichia coli* to *Streptomyces* spp. *Gene* **116:**43–49.
20. **Binnie, C., M. Warren, and M. J. Butler.** 1989. Cloning and heterologous expression in *Streptomyces lividans* of *Streptomyces rimosus* genes involved in oxytetracycline biosynthesis. *J. Bacteriol.* **171:**887–895.
21. **Bormann, C., V. Mohrle, and C. Bruntner.** 1996. Cloning and heterologous expression of the entire set of structural genes for nikkomycin synthesis from *Streptomyces tendae* Tü901 in *Streptomyces lividans*. *J. Bacteriol.* **178:**1216–1218.
22. **Brown, D. W., T. H. Adams, and N. P. Keller.** 1996. *Aspergillus* has distinct fatty acid synthases for primary and secondary metabolism. *Proc. Natl. Acad. Sci. USA* **93:**14873–14877.
23. **Brown, D. W., J.-H. Yu, H. S. Kelkar, M. Fernandes, T. C. Nesbitt, N. P. Keller, T. H. Adams, and T. J. Leonard.** 1996. Twenty-five co-regulated transcripts define a sterigmatocystin gene cluster in *Aspergillus nidulans*. *Proc. Natl. Acad. Sci. USA* **93:**1418–1422.
24. **Butler, M. J., E. J. Friend, I. S. Hunter, F. S. Kaczmarek, D. A. Sugden, and M. Warren.** 1989. Molecular cloning of resistance genes and architecture of a linked gene clus-

ter involved in biosynthesis of oxytetracycline by *Streptomyces rimosus*. *Mol. Gen. Genet.* **215:**231–238.

25. **Cane, D. E., J.-K. Sohng, C. R. Lamberson, S. M. Rudnicki, Z. Wu, M. D. Lloyd, J. S. Oliver, and B. R. Hubbard.** 1994. Pentalenene synthase. Purification, molecular cloning, sequencing and high-level expression in *Escherichia coli* of a terpenoid cyclase from *Streptomyces* UC5319. *Biochemistry* **33:**5846–5857.

26. **Chang, P.-K., J. W. Cary, J. Yu, D. Bhatnagar, and T. E. Cleveland.** 1995. The *Aspergillus parasiticus* polyketide synthase gene *pksA*, a homolog of *Aspergillus nidulans wA*, is required for aflatoxin B1 biosynthesis. *Mol. Gen. Genet.* **248:**270–277.

27. **Chater, K. F.** 1986. *Streptomyces* phages and their applications to *Streptomyces* genetics, p. 119–158. *In* S. W. Queener and L. E. Day (ed.), *The Bacteria*, vol. 9. *Antibiotic-Producing Streptomyces*. Academic Press, New York.

28. **Chater, K. F., and C. J. Bruton.** 1983. Mutational cloning in *Streptomyces* and the isolation of antibiotic production genes. *Gene* **26:**67–78.

29. **Chater, K. F., and C. J. Bruton.** 1985. Resistance, regulatory and production genes for the antibiotic methylenomycin are clustered. *EMBO J.* **4:**1893–1897.

30. **Chen, C.-L, L.-K. Chang, Y.-S. Chang, S.-T. Liu, and J. S.-M. Tschen.** 1995. Transposon mutagenesis and cloning of the genes encoding the enzymes of fengycin biosynthesis in *Bacillus subtilis*. *Mol. Gen. Genet.* **248:**121–125.

31. **Chen, C. W., H.-F. Lin, C. L. Kuo, H.-L. Tsai, and J. F. Y. Tsai.** 1988. Cloning and expression of a DNA sequence conferring cephamycin C production. *Bio/Technology* **6:**1222–1224.

32. **Cho, H., T. Horisaka, H. Ohkishi, M. M. Nakano, and H. Ogawara.** 1988. Cloning of biosynthesis and resistance gene of nosiheptide, p. 129. Abstr. International Symposium on the Biology of Actinomycetes 1988, Tokyo.

33. **Cohen, G., D. Shiffman, M. Mevarech, and Y. Aharonowitz.** 1990. Microbial isopenicillin N synthase genes: structure, function, diversity and evolution. *Trends Biotechnol.* **8:**105–111.

34. **Cortes, J., S. F. Haydock, G. A. Roberts, D. J. Bevitt, and P. F. Leadlay.** 1990. An unusually large multifunctional polypeptide in the erythromycin-polyketide synthase of *Saccharopolyspora erythraea*. *Nature* **346:**176–178.

35. **Cortes, J., K. E. Wiesmann, G. A. Roberts, M. J. Brown, J. Staunton, and P. F. Leadlay.** 1995. Repositioning of a domain in a modular polyketide synthase to promote specific chain cleavage. *Science* **268:**1487–1489.

36. **Crawford, L., A. M. Stepan, P. C. McAda, J. A. Rambosek, M. J. Conder, V. A. Vinci, and C. D. Reeves.** 1995. Production of cephalosporin intermediates by feeding adipic acid to recombinant *Penicillium chrysogenum* strains expressing ring expansion activity. *Biotechnology* **13:**58–62.

37. **Dairi, T., Y. Hamano, Y. Igarashi, T. Furumai, and T. Oki.** 1997. Protoplasting and regeneration of strains belonging to the genus *Actinomadura*. *Actinomycetologia* **11:**1–5.

38. **Dairi, T., T. Ohta, E. Hashimoto, and M. Hasegawa.** 1992. Organization and nature of fortimicin A (astromicin) biosynthetic genes studied using a cosmid library of *Micromonospora olivasterospora* DNA. *Mol. Gen. Genet.* **236:**39–48.

39. **Davis, N. K., and K. F. Chater.** 1990. Spore colour in *Streptomyces coelicolor* A3(2) involves the developmentally regulated synthesis of a compound biosynthetically related to polyketide antibiotics. *Mol. Microbiol.* **4:**1679–1691.

40. **Decker, H., and S. Haag.** 1995. Cloning and characterization of a polyketide synthase gene from *Streptomyces fradiae* Tu2717, which carries the genes for biosynthesis of the angucycline antibiotic urdamycin A and a gene probably involved in its oxygenation. *J. Bacteriol.* **177:**6126–6136.

41. **Decker, H., and C. R. Hutchinson.** 1993. Transcriptional analysis of the *Streptomyces glaucescens* tetracenomycin C biosynthesis gene cluster. *J. Bacteriol.* **175:**3887–3892.

42. **Decker, H., H. Motamedi, and C. R. Hutchinson.** 1993. The nucleotide sequence and heterologous expression of *tcmG* and *tcmP*, biosynthetic genes for tetracenomycin C in *Streptomyces glaucescens*. *J. Bacteriol.* **175:**3876–3886.

43. **DeModena, J. A., S. Gutierrez, J. Velasco, F. J. Fernandez, R. A. Fachini, J. L. Galazzo, D. E. Hughes, and J. F. Martin.** 1993. The production of cephalosporin C by *Acremonium chrysogenum* is improved by the intracellular expression of a bacterial hemoglobin gene. *Bio/Technology* **11:**926–929.

44. **Dickens, M. L., N. D. Priestley, and W. R. Strohl.** 1997. In vivo and in vitro bioconversion of ε-rhodomycinone glycoside to doxorubicin: functions of DauP, DauK and DoxA. *J. Bacteriol.* **179:**2641–2650.

45. **Diez, B., S. Gutierrez, J. L. Barredo, P. van Solinger, L. H. M. van der Voort, and J. F. Martin.** 1990. The cluster of penicillin biosynthetic genes. Identification and characterization of the *pcbAB* gene encoding the α-aminoadipyl-cysteinyl-valine synthetase and linkage to the *pcbC* and *pcbDE* genes. *J. Biol. Chem.* **265:**16358–16365.

46. **Donadio, S., M. J. Staver, J. B. McAlpine, S. J. Swanson, and L. Katz.** 1991. Modular organization of genes required for complex polyketide biosynthesis. *Science* **252:**675–679.

47. **Dosch, D. C., W. R. Strohl, and H. G. Floss.** 1988. Molecular cloning of the nosiheptide resistance gene from *Streptomyces actuosus* ATCC 25421. *Biochem. Biophys. Res. Commun.* **156:**517–523.

48. **Du, L., R. Yang, C. Sanchez, M. Chen, D. J. Edwards, and B. Shen.** 1998. Bleomycin biosynthesis in *Streptoverticillum*: the search for a polyketide and peptide hybrid biosynthetic system, poster P.12. *In* Book of Abstracts, International Interdisciplinary Conference Polyketides II. The Royal Society of Chemistry, University of Bristol, United Kingdom.

49. **Epp, J. K., S. G. Burgett, and B. E. Schoner.** 1987. Cloning and nucleotide sequence of a carbomycin-resistance gene from *Streptomyces thermotolerans*. *Gene* **53:**73–83.

50. **Feitelson, J. S., and D. A. Hopwood.** 1983. Cloning of a *Streptomyces* gene for an O-methyltransferase involved in antibiotic biosythesis. *Mol. Gen. Genet.* **190:**394–398.

51. **Feng, G. H., and T. J. Leonard.** 1995. Characterization of the polyketide synthase gene (*pksL1*) required for aflatoxin biosynthesis in *Aspergillus parasiticus*. *J. Bacteriol.* **177:**6246–6254.

52. **Fernandez-Moreno, M. A., E. Martinez, L. Boto, D. A. Hopwood, and F. Malpartida.** 1992. Nucleotide sequence and deduced functions of a set of cotranscribed genes of *Streptomyces coelicolor* A3(2) including the polyketide syn-

thase for the antibiotic actinorhodin. *J. Biol. Chem.* **267:** 19278–19290.

53. **Fishman, S. E., K. Cox, J. L. Larson, P. A. Reynolds, E. T. Seno, W.-K. Yeh, R. Van Frank, and C. L. Hershberger.** 1987. Cloning genes for the biosynthesis of a macrolide antibiotic. *Proc. Natl. Acad. Sci. USA* **84:**8248–8252.
54. **Fitzgerald, N. B., R. S. English, J. S. Lampel, and T. J. Vanden Boom.** 1998. Sonication-dependent electroporation of the erythromycin-producing bacterium *Saccharopolyspora erythraea*. *Appl. Environ. Microbiol.* **64:**1580–1583.
55. **Gaisser, S., G. A. Böhm, J. Cortes, and P. F. Leadlay.** 1997. Analysis of seven genes from the *eryAI-eryK* region of the erythromycin biosynthetic gene cluster in *Saccharopolyspora erythraea*. *Mol. Gen. Genet.* **256:**239–251.
56. **Gaisser, S., G. A. Böhm, N. Dhillon, M. C. Raynal, J. Cortes, and P. F. Leadlay.** 1998. Analysis of *eryBI*, *eryBIII* and *eryBVII* from the erythromycin biosynthetic gene cluster in *Saccharopolyspora erythraea*. *Mol. Gen. Genet.* **258:**78–88.
57. **Gerlitz, M., G. Meurer, E. Wendt-Pienkowski, K. Madduri, and C. R. Hutchinson.** 1997. The effect of the daunorubicin *dpsH* gene on the choice of starter unit and cyclization pattern reveals that type II polyketide synthases can be unfaithful yet intriguing. *J. Am. Chem. Soc.* **119:**7392–7393.
58. **Gil, J. A., and D. A. Hopwood.** 1983. Cloning and expression of a p-aminobenzoic acid synthetase gene of the candicidin-producing *Streptomyces griseus*. *Gene* **25:**119–32.
59. **Gokhale, R. S., J. Lau, D. E. Cane, and C. Khosla.** 1998. Functional orientation of the acyltransferase domain in a module of the erythromycin polyketide synthase. *Biochemistry* **37:**2524–2528.
60. **Goldberg, S., and J. Romero.** 1987. Molecular cloning of the sisomycin resistance gene and biosynthetic complex of *Micromonospora inyoensis*. Poster presentation, UCLA Symposium on *Streptomyces* Genetics. 10 January 1986.
61. **Goldberg, S. L., J. G. Romero, and Y. M. Deo.** 1990. Cloning and characterization of the sisomicin-resistance gene from *Micromonospora inyoensis*. *J. Antibiot.* **43:**992–999.
62. **Grimm, A., K. Madduri, A. Ali, and C. R. Hutchinson.** 1994. Characterization of the *Streptomyces peucetius* ATCC 29050 genes encoding doxorubicin polyketide synthase. *Gene* **151:**1–10.
63. **Guilfoile, P. G., and C. R. Hutchinson.** 1991. A bacterial analog of the *mdr* gene of mammalian tumor cells in present in *Streptomyces peucetius*, the producer of daunorubicin and doxorubicin. *Proc. Natl. Acad. Sci. USA* **88:**8553–8557.
64. **Guilfoile, P. G., and C. R. Hutchinson.** 1992. Sequence and transcriptional analysis of the *Streptomyces glaucescens tcmAR* tetracenomycin C resistance and repressor gene loci. *J. Bacteriol.* **174:**3651–3658.
65. **Guilfoile, P. G., and C. R. Hutchinson.** 1992. The *Streptomyces glaucescens* TcmR protein represses transcription of the divergently oriented *tcmR* and *tcmA* genes by binding to an intergenic operator region. *J. Bacteriol.* **174:** 3659–3666.
66. **Gutierrez, S., B. Diez, E. Montenegro, and J. F. Martin.** 1991. Characterization of the *Cephalosporium acremonium pcbAB* gene encoding alpha-aminoadipyl-cysteinyl-valine synthetase, a large multidomain peptide synthetase: linkage to the *pcbC* gene as a cluster of early cephalosporin biosynthetic genes and evidence of multiple functional domains. *J. Bacteriol.* **173:**2354–2365.
67. **Gutierrez, S., J. Velasco, F. J. Fernandez, and J. F. Martin.** 1992. The *cefG* gene of *Cephalosporium acremonium* is linked to the *cefEF* gene and encodes a deacetylcephalosporin C acetyltransferase closely related to homoserine O-acetyltransferase. *J. Bacteriol.* **174:**3056–3064.
68. **Hallam, S. E., F. Malpartida, and D. A. Hopwood.** 1988. Nucleotide sequence, transcription and deduced function of a gene involved in polyketide antibiotic synthesis in *Streptomyces coelicolor*. *Gene* **74:**305–320.
69. **Han, L., K. Yang, E. Ramalingam, R. H. Mosher, and L. C. Vining.** 1994. Cloning and characterization of polyketide synthase genes for jadomycin B biosynthesis in *Streptomyces venezuelae* ISP5230. *Microbiology* **140:**3379–3389.
70. **Harris, D.** Sequence data available at www.sanger.ac.uk/projects/S_coelicolor/.
71. **Hernández, C., C. Olano, C. Méndez, and J. A. Salas.** 1993. Characterization of a *Streptomyces antibioticus* gene cluster encoding a glycosyltransferase involved in oleandomycin inactivation. *Gene* **134:**139–140.
72. **Hillemann, D., A. Puhler, and W. Wohlleben.** 1991. Gene disruption and gene replacement in *Streptomyces* via single stranded DNA transformation of integration vectors. *Nucleic Acids Res.* **19:**727–731.
73. **Hodgson, J. E., A. P. Fosberry, N. S. Rawlinson, H. N. M. Ross, R. J. Neal, J. C. Arnell, A. J. Earl, and E. J. Lawlor.** 1995. Clavulanic acid biosynthesis in *Streptomyces clavuligerus*: gene cloning and characterization. *Gene* **166:**49–55.
74. **Hohn, T. M., and P. D. Beremand.** 1989. Isolation and nucleotide sequence of a sesquiterpene cyclase gene from the trichothecene-producing fungus *Fusarium sporotrichioides*. *Gene* **79:**131–138.
75. **Hollander, I. J., Y.-Q. Shen, J. Heim, S. Wolfe, and A. L. Demain.** 1984. A pure enzyme catalyzing penicillin biosynthesis. *Science* **224:**610–612.
76. **Hong, S.-T., J. R. Carney, and S. J. Gould.** 1997. Cloning and heterologous expression of the entire gene clusters for PD 116740 from *Streptomyces* strain WP 4669 and tetrangulol and tetrangomycin from *Streptomyces rimosus* NRRL 3016. *J. Bacteriol.* **179:**470–476.
77. **Hopwood, D. A., M. J. Bibb, K. F. Chater, and T. Kieser.** 1987. Plasmid and phage vectors for gene cloning and analysis in *Streptomyces*. *Methods Enzymol.* **153(D):**116–165.
78. **Hopwood, D. A., M. J. Bibb, K. F. Chater, T. Kieser, C. J. Bruton, H. M. Kieser, D. J. Lydiate, C. P. Smith, J. M. Ward, and H. Schrempf.** 1985. *Genetic Manipulation of Streptomyces: a Laboratory Manual*. The John Innes Foundation, Norwich, United Kingdom.
79. **Hopwood, D. A., T. Kieser, D. J. Lydiate, and M. J. Bibb.** 1986. *Streptomyces* plasmids: their biology and use as cloning vectors, p. 159–230. *In* S. W. Queener and L. E. Day (ed.), *The Bacteria*, vol. 9. *Antibiotic-Producing Streptomyces*. Academic Press, New York.
80. **Huffman, G. W., P. D. Gesellchen, J. R. Turner, R. B. Rothenberger, H. E. Osborne, F. D. Miller, J. L. Chap-**

man, and S. W. Queener. 1992. Substrate specificity of isopenicillin N synthase. *J. Med. Chem.* **35:**1897–1914.

81. **Hutchinson, C. R.** 1994. Drug synthesis by genetically engineered microorganisms. *Bio/Technology* **12:**375–380.

82. **Hutchinson, C. R.** 1997. Antibiotics from genetically engineered microorganisms, p. 683–702. *In* W. R. Strohl (ed.), *Biotechnology of Antibiotics*, 2nd ed. Marcel Dekker, New York.

83. **Hutchinson, C. R.** 1998. Combinatorial biosynthesis for new drug discovery. *Curr. Opin. Microbiol.* **1:**319–329.

84. **Jacobsen, J. R., C. R. Hutchinson, D. E. Cane, and C. Khosla.** 1997. Precursor-directed biosynthesis of erythromycin analogs by an engineered polyketide synthase. *Science* **277:**367–369.

85. **Jensen, S. E., and A. L. Demain.** 1995. Beta-lactams, p. 239–268. *In* L. C. Vining and C. Stuttard (ed.), *Genetics and Biochemistry of Antibiotic Production*. Butterworth-Heineman, Boston.

86. **Jones, G. H., and D. A. Hopwood.** 1984. Molecular cloning and expression of the phenoxazinone synthase gene from *Streptomyces antibioticus*. *J. Biol. Chem.* **259:** 14151–14157.

87. **Kao, C. M., G. Luo, L. Katz, D. E. Cane, and C. Khosla.** 1994. Engineered biosynthesis of a triketide lactone from an incomplete modular polyketide synthase. *J. Am. Chem. Soc.* **116:**11612–11613.

88. **Kao, C. M., R. Pieper, D. E. Cane, and C. Khosla.** 1996. Evidence for two catalytically independent clusters of active sites in a functional modular polyketide synthase. *Biochemistry* **35:**12363–12368.

89. **Kataoka, M. R. Tatsuta, I. Suzuki, S. Kosona, T. Seki, and T. Yoshida.** 1996. Development of a temperature-inducible expression system for *Streptomyces* spp. *J. Bacteriol.* **178:**5540–5542.

90. **Katz, E., C. J. Thompson, and D. A. Hopwood.** 1983. Cloning and expression of the tyrosinase gene from *Streptomyces antibioticus* in *Streptomyces lividans*. *J. Gen. Microbiol.* **129:**2703–2714.

91. **Keller, N. P., D. Brown, R. A. E. Butchko, M. Fernandes, H. Kelkar, C. Nesbitt, S. Segner, D. Bhatnager, T. E. Cleveland, and T. H. Adams.** 1995. A conserved polyketide mycotoxin gene cluster in *A. nidulans*, p. 263–277. *In* J. L. Richard (ed.), *Molecular Approaches to Food Safety Issues Involving Toxic Microorganisms*. Alalan, Ft. Collins, Colo.

92. **Kieser, T., D. A. Hopwood, H. M. Wright, and C. J. Thompson.** 1982. pIJ101, a multi-copy broad host-range *Streptomyces* plasmid: functional analysis and development of DNA cloning vectors. *Mol. Gen. Genet.* **185:**223–238.

93. **Kieser, T., and R. E. Mellon.** 1988. Plasmid pIJ699, a multi-copy positive selection vector for *Streptomyces*. *Gene* **65:**83–91.

94. **Kirby, R., L. F. Wright, and D. A. Hopwood.** 1975. Plasmid-determined antibiotic synthesis and resistance in *Streptomyces coelicolor*. *Nature* **254:**265–267.

95. **Kirschning, A., A. F. W. Bechthold, and J. Rohr.** 1997. Chemical and biochemical aspects of deoxysugars and deoxysugar oligosaccharides. *Top. Curr. Chem.* **188:**1–84.

96. **Kovacevic, S., and J. R. Miller.** 1991. Cloning and sequencing of the beta-lactam hydroxylase gene (*cefF*) from *Streptomyces clavuligerus*: gene duplication may have led to separate hydroxylase and expandase activities in the actinomycetes. *J. Bacteriol.* **173:**398–400.

97. **Kovacevic, S., M. B. Tobin, and J. R. Miller.** 1990. The beta-lactam biosynthesis genes for isopenicillin N epimerase and deacetoxycephalosporin C synthetase are expressed from a single transcript in *Streptomyces clavuligerus*. *J. Bacteriol.* **172:**3952–3958.

98. **Kovacevic, S., B. J. Weigel, M. B. Tobin, T. D. Ingolia, and J. R. Miller.** 1989. Cloning, characterization, and expression in *Escherichia coli* of the *Streptomyces clavuligerus* gene encoding deacetoxycephalosporin C synthetase. *J. Bacteriol.* **171:**754–760.

99. **Krause, M., M. A. Marahiel, H. von Dohren, and H. Kleinkauf.** 1985. Molecular cloning of an ornithine-activating fragment of the gramicidin S synthetase 2 gene from *Bacillus brevis* and its expression in *Escherichia coli*. *J. Bacteriol.* **162:**1120–1125.

100. **Kuhstoss, S., M. A. Richardson, and R. N. Rao.** 1991. Plasmid cloning vectors that integrate site-specifically in streptomycetes. *Gene* **97:**143–146.

101. **Lacalle, R. A., J. A Tercero, and A. Jimenez.** 1992. Cloning of the complete biosynthetic gene cluster for an aminonucleoside antibiotic, puromycin, and its regulated expression in heterologous hosts. *EMBO J.* **11:**785–792.

102. **Lawen, R., and R. Zocher.** 1990. Cyclosporin synthetase. The most complex peptide synthesizing multienzyme polypeptide so far described. *J. Biol. Chem.* **265:** 11355–11360.

103. **Lombo, F., G. Blanco, E. Fernandez, C. Mendez, and J. A. Salas.** 1996. Characterization of *Streptomyces argillaceus* genes encoding a polyketide synthase involved in the biosynthesis of the antitumor mithramycin. *Gene* **172:**87–91.

104. **Lomovskaya, N., Y. Doi-Katayama, S. Filippini, C. Nastro, L. Fonstein, M. Gallo, A. L. Colombo, and C. R. Hutchinson.** 1998. The *Streptomyces peucetius dpsY* and *dnrX* genes govern early and late steps of daunorubicin and doxorubicin biosynthesis. *J. Bacteriol.* **180:** 2379–2386.

105. **Lomovskaya, N., L. Fonstein, X. Ruan, D. Stassi, L. Katz, and C. R. Hutchinson.** 1997. Gene disruption and replacement in the rapamycin-producing *Streptomyces hygroscopicus* strain ATCC 29253. *Microbiology* **143:**875–883.

106. **Lomovskaya, N., S.-K. Hong, S.-U. Kim, K. Furuya, L. Fonstein, and C. R. Hutchinson.** 1996. The *Streptomyces peucetius drrC* gene encodes a UvrA-like protein essential for daunorubicin and doxorubicin resistance and production. *J. Bacteriol.* **178:**3238–3245.

107. **Lomovskaya, N., S. L. Otten, Y. Doi-Katayama, L. Fonstein, X.-C. Liu, T. Takatsu, A. Inventi-Solari, S. Filippini, F. Torti, A. L. Colombo, and C. R. Hutchinson.** Doxorubicin overproduction in *Streptomyces peucetius* through the cloning and characterization of the *dnrU* ketoreductase and *dnrV* genes, and the *doxA* CYP450 hydroxylase gene. *J. Bacteriol.*, in press.

108. **Lomovskaya, N. D., K. F. Chater, and N. M. Mkrtumian.** 1980. Genetics and molecular biology of *Streptomyces* bacteriophages. *Microbiol. Rev.* **44:**206–229.

109. **Lydiate, D. J., H. Ikeda, and D. A. Hopwood.** 1986. A 2.6 kb DNA sequence of *Streptomyces coelicolor* A3(2) which functions as a transposable element. *Mol. Gen. Genet.* **203:**79–88.

110. **Lydiate, D. J., F. Malpartida, and D. A. Hopwood.** 1985. The *Streptomyces* plasmid SCP2*: its functional analysis and development into useful cloning vectors. *Gene* **35:**223–235.

111. **MacCabe, A. P., M. B. R. Riach, S. E. Unkles, and J. R. Kinghorn.** 1990. The *Aspergillus nidulans npeA* locus consists of three contiguous genes required for penicillin biosynthesis. *EMBO J.* **9:**279–287.

112. **Madduri, K., and C. R. Hutchinson.** 1995. Functional characterization and transcription analysis of the *dnrR$_1$* locus that controls daunorubicin biosynthesis in *Streptomyces peucetius*. *J. Bacteriol.* **177:**1208–1215.

113. **Madduri, K., J. Kennedy, G. Rivola, A. Inventi-Solari, S. Filippini, A. L. Colombo, and C. R. Hutchinson.** 1998. Production of the antitumor drug epirubicin (4′-epidoxorubicin) and its precursor, 4′-epidaunorubicin, by a genetically engineered strain of *Streptomyces peucetius*. *Nat. Bitechnol.* **16:**69–74.

114. **Magnolo, S. K., D. L. Leenutaphong, J. A. DeModena, J. E. Curtis, J. E. Bailey, J. L. Galazzo, and D. E. Hughes.** 1991. Actinorhodin production by *Streptomyces coelicolor* and growth of *Streptomyces lividans* are improved by the expression of a bacterial hemoglobin. *Bio/Technology* **9:**473–476.

115. **Mahanti, N., D. Bhatnagar, J. W. Carey, J. Joubrain, and J. E. Linz.** 1996. Structure and function of a *fas-1A*, a gene encoding a putative fatty acid synthetase directly involved in aflatoxin biosynthesis in *Aspergillus parasiticus*. *Appl. Environ. Microbiol.* **62:**191–195.

116. **Malmberg, L. H., W. S. Hu, and D. H. Sherman.** 1993. Precursor flux control through targeted chromosomal insertion of the lysine ε-aminotransferase (*lat*) gene in cephamycin C biosynthesis. *J. Bacteriol.* **175:**6916–6924.

117. **Malpartida, F., S. E. Hallam, H. M. Kieser, H. Motamedi, C. R. Hutchinson, M. J. Butler, D. A. Sugden, M. Warren, C. R. Bailey, C. McKillop, G. O. Humphreys, and D. A. Hopwood.** 1987. Sequence homology between *Streptomyces* genes coding for synthesis of different polyketides and its use to clone antibiotic biosynthesis genes. *Nature* **325:**818–821.

118. **Malpartida, F., and D. A. Hopwood.** 1984. Molecular cloning of the whole biosynthetic pathway of a *Streptomyces* antibiotic and its expression in a heterologous host. *Nature* **309:**462–464.

119. **Malpartida, F., J. Niemi, R. Navarrete, and D. A. Hopwood.** 1990. Cloning and expression in a heterologous host of the complete set of genes for biosynthesis of the *Streptomyces coelicolor* antibiotic undecylprodigiosin. *Gene* **93:**91–99.

120. **Marsden, A. F. A., P. Caffrey, J. F. Aparicio, M. S. Loughran, J. Staunton, and P. F. Leadlay.** 1994. Stereospecific acyl transfers on the erthromycin-producing polyketide synthase. *Science* **263:**378–380.

121. **Marsden, A. F. A., B. Wilkinson, J. Cortes, N. J. Dunster, J. Staunton, and P. F. Leadlay.** 1998. Engineering broader specificity into an antibiotic-producing polyketide synthase. *Science* **279:**199–201.

122. **Marsh, E. N., M. D.-T. Chang, and C. A. Townsend.** 1992. Two isozymes of clavaminate synthase central to clavulanic acid formation: cloning and sequencing of both genes from *Streptomyces clavuligerus*. *Biochemistry* **31:**12648–12657.

123. **Mazodier, P., R. Petter, and C. Thompson.** 1989. Intergeneric conjugation between *Escherichia coli* and *Streptomyces* species. *J. Bacteriol.* **171:**3583–3585.

124. **McDaniel, R., S. Ebert-Khosla, D. A. Hopwood, and C. Khosla.** 1993. Engineered biosynthesis of novel polyketides. *Science* **262:**1546–1550.

125. **McGowan, S. J., M. Sebalhia, S. O'Leary, K. R. Hardie, P. Williams, G. S. A. B. Stewart, B. W. Bycroft, and G. P. C. Salmond.** 1997. Analysis of the carbapenem gene cluster of *Erwinia carotovora*: definition of the antibiotic biosynthetic genes and evidence for a novel β-lactam resistance mechanism. *Mol. Microbiol.* **26:**545–556.

126. **McGuire, S. M., J. C. Silva, E. G. Casillas, and C. A. Townsend.** 1996. Purification and characterization of versicolorin B synthase from *Aspergillus parasiticus*. Catalysis of the stereodifferentiating cyclization in aflatoxin biosynthesis essential to DNA interaction. *Biochemistry* **35:**11470–86.

127. **Minas, W., P. Brunker, P. T. Kallio, and J. E. Bailey.** 1998. Improved erythromycin production in a genetically engineered industrial strain of *Saccharopolyspora erythraea*. *Biotechnol. Prog.* **14:**561–566.

128. **Motamedi, H., S.-J. Cai, A. Shafiee, and K. O. Elliton.** 1997. Structural organization of a multifunctional; polyketide synthase involved in the biosynthesis of the macrolide immunosuppressant FK506. *Eur. J. Biochem.* **244:**74–80.

129. **Motamedi, H., and C. R. Hutchinson.** 1987. Cloning and heterologous expression of a gene cluster for the biosynthesis of tetracenomycin C, the anthracycline antitumor antibiotic of *Streptomyces glaucescens*. *Proc. Natl. Acad. Sci. USA* **84:**4445–4449.

130. **Motamedi, H., A. Shafiee, and S.-J. Cai.** 1995. Integrative vectors for heterologous gene expression in *Streptomyces* spp. *Gene* **160:**25–31.

131. **Murakami, T., H. Anzai, S. Imai, A. Satoh, K. Nagaoka, and C. J. Thompson.** 1986. The bialaphos biosynthetic genes of *Streptomyces hygroscopicus*: molecular cloning and characterization of the gene cluster. *Mol. Gen. Genet.* **205:**42–50.

132. **Muth, G., B. Nussbaumer, W. Wohlleben, and A. Puehler.** 1989. A vector system with temperature-sensitive replication for gene disruption and mutational cloning in streptomycetes. *Mol. Gen. Genet.* **219:**341–348.

133. **Nakata, K., S. Horinouchi, and T. Beppu.** 1989. Cloning and characterisation of the carbapenem and biosynthetic genes from *Streptomyces fulvoviridis*. *FEMS Microbiol. Lett.* **57:**51–56.

134. **Nielsen, J.** 1998. The role of metabolic engineering in the production of secondary metabolites. *Curr. Opin. Microbiol.* **1:**319–329.

135. **Offenzeller, M., G. Santer, K. Totschnig, Z. Su, H. Moser, R. Traber, and E. Schneider-Scherzer.** 1996. Biosynthesis of the unusual amino aci (4*R*)-4-[(*E*)-2-butenyl]-4-methyl-L-threoine of cyclosporin A: enzymatic analysis of the reaction sequence including identification of the methylation precursor in a polyketide pathway. *Biochemistry* **35:**8401–8412.

136. **Oh, S. H., and K. F. Chater.** 1997. Denaturation of circular or linear DNA facilitates targeted integrative trans-

formation of *Streptomyces coelicolor* A3(2): possible relevance to other organisms. *J. Bacteriol.* **179:**122–127.

137. **Ohnuki, T., T. Imanaka, and S. Aiba.** 1985. Self-cloning in *Streptomyces griseus* of an *str* gene cluster for streptomycin biosynthesis and streptomycin resistance. *J. Bacteriol.* **164:**85–94.
138. **Ohta, T., E. Nagano, T. Dairi, and M. Hasegawa.** 1988. DNA homology between the genes for the biosynthesis of fortimicin group antibiotics, p. 131. Abstr. International Symposium on the Biology of Actinomycetes 1988, Tokyo.
139. **Otten, S. L., J. Ferguson, and C. R. Hutchinson.** 1995. Regulation of daunorubicin production in *Streptomyces peucetius* by the *dnrR*$_2$ locus. *J. Bacteriol.* **177:**1216–1224.
140. **Otten, S. L., and C. R. Hutchinson.** Unpublished results.
141. **Paradkar, A. S., K. A. Aidoo, and S. E. Jensen.** 1998. A pathway-specific transcriptional activator regulates late steps of clavulanic acid biosynthesis in *Streptomyces clavuligerus*. *Mol. Microbiol.* **27:**831–843.
142. **Paradkar, A. S., K. A. Aidoo, A. Wong, and S. E. Jensen.** 1996. Molecular analysis of a β-lactam resistance gene encoded within the cephamycin gene cluster of *Streptomyces clavuligerus*. *J. Bacteriol.* **178:**6266–6274.
143. **Paradkar, A. S., and S. E. Jensen.** 1997. Comparative genetics and molecular biology of β-lactam biosynthesis, p. 241–277. *In* W. R. Strohl (ed.), *Biotechnology of Antibiotics*, 2nd ed. Marcel Dekker, New York.
144. **Paulus, T. J., J. Tuan, V. E. Luebke, G. T. Maine, J. P. DeWitt, and L. Katz.** 1990. Mutation and cloning of *eryG*, the structural gene for erythromycin C O-methyltransferase from *Saccharopolyspora erythraea* and expression of *eryG* in *Escherichia coli*. *J. Bacteriol.* **172:** 2541–2546.
145. **Pelzer, S., W. Reichert, M. Huppert, D. Heckmann, and W. Wohlleben.** 1997. Cloning and analysis of a peptide synthetase gene of the balhimycin producer *Amycolatopsis mediterranei* DSM5908 and development of a gene disruption/replacement system. *J. Biotechnol.* **56:** 115–128.
146. **Piepersberg, W.** 1994. Pathway engineering in secondary metabolite-producing actinomycetes. *Crit. Rev. Biotechnol.* **14:**251–285.
147. **Pigac, J., and H. Schrempf.** 1995. A simple and rapid method of transformation of *Streptomyces rimosus* R6 and other streptomycetes by electroporation. *Appl. Environ. Microbiol.* **61:**352–356.
148. **Powell, N., and P. Dyson.** 1998. Cloning of genes encoding the milbemycin polyketide synthase of *Streptomyces griseochromogenes*, poster P. 17. *In* Book of Abstracts, International Interdisciplinary Conference Polyketides II. The Royal Society of Chemistry, University of Bristol, United Kingdom.
149. **Rajgharia, V., and W. R. Strohl.** 1997. Minimal *Streptomyces* sp. strain C5 daunorubicin polyketide biosynthesis genes required for aklanonic acid formation. *J. Bacteriol.* **179:**2690–2696.
150. **Ramon, D., I. Carramolino, C. Patino, F. Sanchez, and M. A. Penalva.** 1987. Cloning and characterization of the isopenicillin N synthetase gene mediating the formation of the beta-lactam ring in *Aspergillus nidulans*. *Gene* **57:**171–181.
151. **Revill, W. P., M. J. Bibb, D. A. Hopwood.** 1995. Purification of a malonyltransferase from *Streptomyces coelicolor* A3(2) and analysis of its genetic determinant. *J. Bacteriol.* **177:**3946–3952.
152. **Richardson, M. A., S. Kuhstoss, P. Solenberg, N. A. Schaus, and R. Nagaraja Rao.** 1987. A new shuttle vector, pKC505, for streptomycetes: its use in the cloning of three different spiramycin-resistance genes from a *Streptomyces ambofaciens* library. *Gene* **61:**231–241.
153. **Ruan, X., A. Perada, D. L. Stassi, D. Zeidner, R. G. Summers, M. Jackson, A. Shivakumar, S. Kakavas, M. J. Stover, S. Donadio, and L. Katz.** 1997. Acyltransferase domain substitutions in erythromycin polyketide synthase yield novel erythromycin derivatives. *J. Bacteriol.* **179:**6416–6425.
154. **Rudd, B. A., and D. A. Hopwood.** 1979. Genetics of actinorhodin biosynthesis by *Streptomyces coelicolor* A3(2). *J. Gen. Microbiol.* **114:**35–43.
155. **Salah-Bey, K., M. Doumith, J.-M. Michel, S. Haydock, J. Cortes, P. F. Leadlay, and M.-C. Raynal.** 1998. Targeted gene inactivation for the elucidation of deoxysugar biosynthesis in the erythromycin producer *Saccharopolyspora erythraea*. *Mol. Gen. Genet.* **257:**542–553.
156. **Salowe, S. P., E. N. Marsh, and C. A. Townsend.** 1990. Purification and characterization of clavaminate synthase from *Streptomyces clavuligerus*: an unusual oxidative enzyme in natural product biosynthesis. *Biochemistry* **29:** 6499–6507.
157. **Sampson, S. M., R. Belagaje, D. T. Blankenshipo, J. L. Chapman, D. Perry, P. L. Skatrud, R. M. VanFrank, E. P. Abraham, J. E. Baldwin, S. W. Queener, and T. D. Ingolia.** 1985. Isolation, sequence determination and expression in *Escherichia coli* of the isopenicillin N synthetase gene from *Cephalosporium acremonium*. *Nature* **318:**191–194.
158. **Sampson, S. M., J. E. Dotzlaf, M. L. Slisz, G. W. Becker, R. M. van Frank, L. E. Veal, W. Yeh, J. R. Miller, S. W. Queener, and T. D. Ingolia.** 1987. Cloning and expression of the fungal expandase/hydroxylase gene involved in cephalosporin biosynthesis. *Bio/Technology* **5:**1207–1216.
159. **Schneider, A., and M. A. Marahiel.** 1998. Genetic evidence for a role of thioesterase domains, integrated in or associated with peptide synthetases, in non-ribosomal peptide biosynthesis in *Bacillus subtilis*. *Arch. Microbiol.* **169:**404–410.
160. **Schupp, T., C. Toupet, B. Cluzel, S. Neff, S. Hill, J. J. Beck, and J. M. Ligon.** 1995. A *Sorangium cellulosum* (Myxobacterium gene cluster for the biosynthesis of the-macrolide antibiotic soraphen A: cloning, characterization, and homology to polyketide synthase genes from Actinomycetes. *J. Bacteriol.* **177:**3673–3679.
161. **Schupp, T., C. Toupet, N. Engel, and S. Goff.** 1998. Cloning and sequence analysis of the putative rifamycin polyketide synthase gene cluster from *Amycolatopsis mediterranei*. *FEMS Microbiol. Lett.* **159:**201–207.
162. **Schwecke, T., J. F. Aparicio, I. Molnar, A. Konig, L. E. Khaw, S. F. Haydock, M. Oliynyk, P. Caffrey, J. Cortes, J. B. Lester, G. A. Bohm, J. Staunton, and P. F. Leadlay.** 1995. The biosynthesis gene cluster for the polyketide immunosuppressant rapamycin. *Proc. Natl. Acad. Sci. USA* **93:**7839–7843.

163. **Scotti, C., and C. R. Hutchinson.** 1996. Enhanced antibiotic production by manipulation of the *Streptomyces peucetius dnrH* and *dnmT* genes involved in doxorubicin (adriamycin) biosynthesis. *J. Bacteriol.* **178:**7316–7321.

164. **Sezonov, G., V. Blanc, N. Bamas-Jacques, A. Friedmann, J.-L. Pernodet, and M. Guerineau.** 1997. Complete conversion of antibiotic precursor to pristanamycin IIA by overexpression of *Streptomyces pristinaespiralis* biosynthetic genes. *Nat. Biotechnol.* **15:**349–353.

165. **Shen, B., and C. R. Hutchinson.** 1996. Deciphering the mechanism for the assembly of aromatic polyketides by a bacterial polyketide synthase. *Proc. Natl. Acad. Sci. USA* **93:**6600–6604.

166. **Shen, B., R. G. Summers, H. Gramajo, M. J. Bibb, and C. R. Hutchinson.** 1992. Purification and characterization of the acyl carrier protein of the *Streptomyces glaucescens* tetracenomycin C polyketide synthase. *J. Bacteriol.* **174:**3818–3821.

167. **Sherman, D. H., F. Malpartida, M. J. Bibb, H. M. Kieser, M. J. Bibb, and D. A. Hopwood.** 1989. Structure and deduced function of the granaticin-producing polyketide synthase gene cluster of *Streptomyces violaceoruber* TU22. *EMBO J.* **8:**2717–25.

168. **Skatrud, P. L., A. J. Teitz, T. D. Ingolia, C. A. Cantwell, D. L. Fisher, J. L. Chapman, and S. W. Queener.** 1989. Use of recombinant DNA to improve production of cephalosporin C by *Cephalosporium acremonium. Bio/Technology* **7:**477–485.

169. **Smith, D. J., M. K. R. Burnham, J. H. Bull, J. E. Hodgson, J. M. Ward, P. Browne, J. Brown, B. Barton, A. J. Earl, and G. Turner.** 1990. β-Lactam antibiotic biosynthetic genes have been conserved in clusters in prokaryotes and eukaryotes. *EMBO J.* **9:**741–747.

170. **Smith, D. J., M. K. R. Burnham, J. Edwards, A. J. Earl, and G. Turner.** 1990. Cloning and heterologous expression of the penicillin biosynthesis gene cluster from *Penicillium chrysogenum. Bio/Technology* **8:**39–41.

171. **Sohng, J. K., T. J. Oh, J. J. Lee, and C. G. Kim.** 1997. Identification of a gene cluster of biosynthetic genes of rubradirin substructures in *S. achromogenes* var. *rubradiris* NRRL3061. *Mol. Cells* **7:**674–681.

172. **Solenberg, P. J., and R. H. Baltz.** 1991. Transposition of Tn5096 and other IS493 derivatives in *Streptomyces griseofuscus. J. Bacteriol.* **173:**1096–1104.

173. **Solenberg, P. J., and R. H. Baltz.** 1994. Hypertransposing derivatives of the *Streptomyces* insertion sequence IS493. *Gene* **147:**47–54.

174. **Solenberg, P. J., and S. G. Burgett.** 1989. Method for selection of transposable DNA and characterization of a new insertion sequence, IS493, from *Streptomyces lividans. J. Bacteriol.* **171:**4807–4813.

175. **Stassi, D., S. Donadio, M. J. Staver, and L. Katz.** 1993. Identification of a *Saccharopolyspora erythraea* gene required for the final hydroxylation step in erythromycin biosynthesis. *J. Bacteriol.* **175:**182–189.

176. **Stassi, D. L., S. J. Kakavas, K. A. Reynolds, G. Gunawardana, S. Swanson, D. Zeidner, M. Jackson, H. Liu, A Buko, and L. Katz.** 1998. Ethyl-substituted erythromycin derivatives produced by directed metabolic engineering. *Proc. Natl. Acad. Sci. USA* **95:**7305–7309.

177. **Staunton, J., P. Caffrey, J. F. Aparicio, G. A. Roberts, S. S. Bethell, and P. F. Leadlay.** 1996. Evidence for a double-helical structure for modular polyketide synthases. *Nat. Struct. Biol.* **3:**188–192.

178. **Steicher, S. L., C. L. Ruby, P. S. Paress, and J. B. Sweasy.** 1989. Cloning genes for avermectin biosynthesis in *Streptomyces avermitilis*, p. 44–52. *In* C. L. Hershberger, S. W. Queener, and G. Hegeman (ed.), *Genetics and molecular biology of industrial microorganisms*. Society for Industrial Microbiology, Fairfax, Va.

179. **Strohl, W. R., M. L. Dickens, V. B. Rajgarhia, A. J. Woo, and N. D. Priestley.** 1997. Anthracyclines, p. 577–657. *In* W. R. Strohl (ed.), *Biotechnology of Antibiotics*, 2nd ed. Marcel Dekker, New York.

180. **Stutzman-Engwell, K. J., and C. R. Hutchinson.** 1989. Multigene families for anthracycline antibiotic production from *Streptomyces peucetius. Proc. Natl. Acad. Sci. USA* **86:**3135–3139.

181. **Suarez, J. E., and K. F. Chater.** 1977. DNA cloning in *Streptomyces*: a bifunctional replicon comprising pBR322 inserted into a *Streptomyces* phage. *Nature* **286:**527–529.

182. **Summers, R. G., A. Ali, B. Shen, W. A. Wessel, and C. R. Hutchinson.** 1995. The malonyl-coenzyme A:acyl carrier protein acyltransferase of *Streptomyces glaucescens*: a possible link between fatty acid and polyketide biosynthesis. *Biochemistry* **34:**9389–9402.

183. **Summers, R. G., S. Donadio, M. J. Staver, E. Wendt-Pienkowski, C. R. Hutchinson, and L. Katz.** 1997. Sequencing and mutagenesis of genes from the erythromycin biosynthetic gene cluster of *Saccharopolyspora erythraea* that are involved in L-mycarose and D-desosamine production. *Microbiology* **143:**3251–3262.

184. **Summers, R. G., E. Wendt-Pienkowski, and C. R. Hutchinson.** 1992. Nucleotide sequence of the *tcmNO* region of the tetracenomycin C biosynthetic gene cluster of *Streptomyces glaucescens* and evidence that the *tcmN* gene encodes as multifunctional cyclase/dehydratase/O-melthyltransferase. *J. Bacteriol.* **174:**1810–1820.

185. **Summers, R. G., E. Wendt-Pienkowski, H. Motamedi, and C. R. Hutchinson.** 1993. The *tcmVI* region of the tetracenomycin C biosynthesis gene cluster of *Streptomyces glaucescens* encodes three enzymes: the tetracenomycin F1 monooxygenase, tetracenomycin F2 cyclase, and, most likely, a second cyclase. *J. Bacteriol.* **175:** 7571–7580.

186. **Swan, D. G., A. M. Rodriguez, C. Vilches, C. Mendez, and J. A. Salas.** 1994. Characterisation of a *Streptomyces antibioticus* gene encoding a type I polyketide synthase which has a unusual coding sequence. *Mol. Gen. Genet.* **242:**358–362.

187. **Thompson, C. J., J. M. Ward, and D. A. Hopwood.** 1980. DNA cloning in *Streptomyces*: resistance genes from antibiotic-producing species. *Nature* **286:**525–527.

188. **Thompson, C. J., J. M. Ward, and D. A. Hopwood.** 1982. Cloning of antibiotic resistance and nutritional genes in streptomycetes. *J. Bacteriol.* **151:**668–677.

189. **Thorson, J. S., and H. W. Liu.** 1994. Pathways and mechanisms in the biogenesis of novel deoxysugars by bacteria. *Annu. Rev. Microbiol.* **48:**223–256.

190. **Tobin, M. B., S. Kovacevic, K. Maadduri, J. A. Hoskins, P. L. Skatrud, L. C. Vining, C. Stuttard, and J. R. Miller.** 1991. Localization of lysine ε-aminotransferase (lat) and δ-(L-α-aminoadipyl)-L-cysteinyl-D-valine (ACV) synthetase (*pcbAB*) genes from *Streptomyces clavuligerus* and production of lysine ε-

aminotransferase activity in *Escherichia coli*. *J. Bacteriol.* **173**:6258–6264.

191. **Traber, R.** 1997. Biosynthesis of cyclosporins, p. 279–314. *In* W. R. Strohl (ed.), *Biotechnology of Antibiotics*, 2nd ed. Marcel Dekker, New York.

192. **Trail, F., N. Mahanti, and J. E. Linz.** 1995. Molecular biology of aflatoxin biosynthesis. *Microbiology* **141**:755–765.

193. **Trail, F., N. Mahanti, M. Rarick, R. Hehigh, S.-H. Liang, R. Zhou, and J. E. Linz.** 1995. Physical and transcriptional map of an aflatoxin gene cluster in *Aspergillus parasiticus* and functional disruption of a gene involved early in the aflatoxin pathway. *Appl. Environ. Microbiol.* **61**:2665–2673.

194. **Trapp, S. C., T. M. Hohn, S. McCormick, and B. B. Jarvis.** 1998. Characterization of the gene cluster for biosynthesis of macrocyclic trichothecenes in *Myrothecium roridum*. *Mol. Gen. Genet.* **257**:421–432.

195. **Tsukamoto, N., J.-A. Chuck, G. Luo, C. M. Kao, C. Khosla, and D. E. Cane.** 1996. 6-Deoxyerythronolide B synthase 1 is specifically acylated by a diketide intermediate at the β-ketoacyl-acyl carrier protein synthase domain of module 2. *Biochemistry* **35**:15244–15248.

195a. **Vangeningen, A. M. A., P. N. Kirpatrick, D. H. Williams, B. R. Harris, J. K. Kershaw, N. J. Lennard, M. Jones, S. J. M. Jones, and P. J. Solenberg.** 1998. Sequencing and analysis of genes involved in the biosynthesis of a vancomycin group antibiotic. *Chem. Biol.* **5**: 155–162.

196. **Vara, J. M. Lewandowska-Sharbek, Y.-G. Wang, S. Donadio, and C. R. Hutchinson.** 1989. Cloning of genes governing the deoxysugar portion of the erythromycin biosynthesis pathway in *Saccharopolyspora erythraea* (*Streptomyces erythraeus*). *J. Bacteriol.* **171**:5872–5881.

197. **Volff, J.-N., and J. Altenbuchner.** 1998. Genetic instability of the *Streptomyces* chromosome. *Mol. Microbiol.* **27**:239–246.

198. **von Döhren, H., and H. Kleinkauf.** 1997. Enzymology of peptide synthetases, p. 217–240. *In* W. R. Strohl (ed.), *Biotechnology of Antibiotics*, 2nd ed. Marcel Dekker, New York.

199. **Ward, J. M., G. R. Janssen, T. Kieser, M. J. Bibb, M. J. Buttner, and M. J. Bibb.** 1986. Construction and characterisation of a series of multi-copy promotor-probe vectors for *Streptomyces* using the aminoglycoside phosphotransferase gene from Tn*5* as indicator. *Mol. Gen. Genet.* **203**:468–475.

200. **Watanabe, C. M. H., D. Wilson, J. E. Linz, and C. A. Townsend.** 1996. Demonstration of the catalytic roles and evidence for the physical association of type I fatty acid synthases and a polyketide synthase in the biosynthesis of aflatoxin B1. *Chem. Biol.* **3**:463–469.

201. **Weaden, J., and P. Dyson.** 1998. Transposon mutagenesis with IS*6100* in the avermectin-producer *Streptomyces avermitilis*. *Microbiology* **144**:1963–1970.

202. **Weber, G., K. Schörgendorfer, E. Schneider-Scherzer, and E. Leitner.** 1994. The peptide synthetase catalyzing cyclosporine production in *Tolypocladium niveum* is encoded by a giant 45.8-kilobase open reading frame. *Curr. Genet.* **26**:120–125.

203. **Weber, J. M., J. O. Leung, G. T. Maine, R. H. B. Potenz, T. J. Paulus, and J. P. DeWitt.** 1990. Organization of a cluster of erythromycin genes in *Saccharopolyspora erythraea*. *J. Bacteriol.* **172**:2372–2383.

204. **Weber, J. M., J. O. Leung, S. J. Swanson, K. B. Idler, and J. B. McAlpine.** 1991. An erythromycin derivative produced by targeted gene disruption in *Saccharopolyspora erythraea*. *Science* **252**:114–117.

205. **Weigel, B. J., S. G. Burgett, V. J. Chen, P. L. Skatrud, C. A. Frolik, S. W. Queener, and T. D. Ingolia.** 1988. Cloning and expression in *Escherichia coli* of isopenicillin N synthetase genes from *Streptomyces lipmanii* and *Aspergillus nidulans*. *J. Bacteriol.* **170**:3817–3826.

206. **Woloshuk, C. P., K. R. Foutz, J. F. Brewer, D. Bhatnagar, T. E. Cleveland, and G. A. Payne.** 1994. Molecular characterization of *aflR*, a regulatory locus for aflatoxin biosynthesis. *Appl. Environ. Microbiol.* **60**: 2408–2414.

207. **Wright, L. F., and D. A. Hopwood.** 1976. Identification of the antibiotic determined by the SCP1 plasmid of *Streptomyces coelicolor* A3(2). *J. Gen. Microbiol.* **95**:96–106.

208. **Ye, J., M. L. Dickens, R. Plater, Y. Li, J. Lawrence, and W. R. Strohl.** 1994. Isolation and sequence analysis of polyketide synthase genes from the daunomycin-producing *Streptomyces* sp. strain C5. *J. Bacteriol.* **176**: 6270–6280.

209. **Young, S. A., S. K. Park, C. Rodgers, R. E. Mitchell, and C. L. Bender.** 1992. Physical and functional characterizaton of the gene cluster encoding the polyketide phytotoxin coronatine in *Pseudomonas syringae* pv. glycinea. *J. Bacteriol.* **174**:1837–1843.

210. **Yu, J., P.-K. Chang, J. W. Cary, M. Wright, D. Bhatnagar, T. E. Cleveland, G. A. Payne, and J. E. Linz.** 1995. Comparative mapping of aflatoxin pathway gene clusters in *Aspergillus parasiticus* and *Aspergillus flavus*. *Appl. Environ. Microbiol.* **61**:2365–2371.

211. **Yu, J.-H., R. A. E. Butchko, M. Fernandes, N. P. Keller, T. J. Leonard, and T. H. Adams.** 1995. Conservation of structure and function of the aflatoxin regulatory gene *aflR* from *Aspergillus nidulans* and *A. flavus*. *Curr. Genet.* **29**:549–555.

212. **Yu, J.-H., and T. J. Leonard.** 1995. Sterigmatocystin biosynthesis in *Aspergillus nidulans* requires a novel type I polyketide synthase. *J. Bacteriol.* **177**: 4792–4800.

213. **Yu, T. W., M. J. Bibb, W. P. Revill, and D. A. Hopwood.** 1994. Cloning, sequencing, and analysis of the griseusin polyketide synthase gene cluster from *Streptomyces griseus*. *J. Bacteriol.* **176**:2627–2634.

214. **Zuber, P., and M. A. Marahiel.** 1997. Structure, function and regulation of genes encoding multidomain peptide synthetases, p. 187–216. *In* W. R. Strohl (ed.), *Biotechnology of Antibiotics*, 2nd ed. Marcel Dekker, New York.

Antibiotic Resistance Mechanisms of Bacterial Pathogens

JOYCE A. SUTCLIFFE, JOHN P. MUELLER, AND ERIC A. UTT

62

Three conditions must be met for an antibiotic to be effective against bacteria. First, a susceptible target must exist in the cell. Second, the antibiotic must penetrate the bacteria and reach the target in sufficient quantity. Third, the antibiotic must not be inactivated or modified before interacting with the specific target. Thus, there are three main mechanisms by which bacteria circumvent the bactericidal or bacteriostatic action of an antibacterial agent. These include target modification, antibiotic inactivation or modification, and drug efflux and/or permeability changes at the cell surface. It has become increasingly apparent over the last decade that these resistance mechanisms do not exist in isolation. The level of resistance to one antibiotic may be determined by the cumulative effect of two or more distinct resistance mechanisms.

Several comprehensive reviews on antibiotic resistance mechanisms have been published recently, and readers should consult these references for resistance mechanisms not covered in this chapter (64, 71, 82, 110, 149, 237, 285, 299, 316, 323, 340, 361, 363, 367). Our focus is to summarize recent developments in resistance mechanisms in pathogenic bacteria with specific emphasis on clinically important antibiotic families, including β-lactams, macrolides, lincosamides, streptogramins, tetracyclines, quinolones, and aminoglycosides (see Tables 1, 2, 3). Since the mechanisms of antibacterial resistance in aerobic and anaerobic bacteria are similar, we emphasize primarily aerobic pathogens. However, we highlight differences in anaerobic resistance determinants with a specific emphasis on *Bacteroides* spp. We apologize for not citing all pertinent references, but space limitations have forced us to be eclectic and to emphasize recent publications.

62.1. TARGET MODIFICATIONS

62.1.1. β-Lactams

Multiple enzymes that bind penicillin (PBPs) are involved in the terminal stages of peptidoglycan synthesis (63, 84, 111, 132, 189, 323, 333). The high-molecular-weight PBPs are often multimodular, with an N-terminal transglycosylase domain and a C-terminal transpeptidase domain. Transpeptidases catalyze the reaction, which results in a covalent cross-link between peptidoglycan disaccharide units. An active site serine within the transpeptidase region mimics the structure of the natural substrate and becomes irreversibly acylated by β-lactams.

Resistance to β-lactams can arise by target modification if normal PBPs are altered by either point mutations or recombinational events resulting in low-affinity mosaic genes or by acquisition of supplementary low affinity (63, 84, 132, 323, 346). These changes result in reduced binding of penicillin by PBPs and other derivatives quantified by competitive binding experiments using radiolabeled drug and gel fluorography. Usually, single amino acid substitutions in one PBP result in low-level resistance to penicillin or cephalosporins, with high levels of resistance requiring multiple amino acid substitutions in a single PBP (i.e., PBP3) as seen in *Escherichia coli* (132) or *Pseudomonas aeruginosa* (116) or alterations in multiple PBPs (63, 84, 132, 323, 346). Because penicillin is a substrate analog, restructuring of the active site must be subtle, requiring changes that discriminate between drug binding and the true substrate in the transpeptidase reaction (323). In transformation-competent species such as *Streptococcus pneumoniae*, *Neisseria meningitis*, *Neisseria gonorrhoeae*, and *Haemophilis influenzae*, hybrid low-affinity PBPs have arisen through interspecies exchange of portions of PBPs (63, 84, 189, 323, 333, 346). In *S. pneumoniae*, high-level penicillin resistance requires the remodeling and retention of genes of at least three of the five PBPs (PBP1a, PBP2x, and PBP2b) (63, 132, 172, 346). However, resistance to cephalosporins requires changes only in PBP1a and PBP2x, largely because of the inherent low affinity that cephalosporins have for PBP2b (129). The PBP remodeling in pneumococci also requires concomitant alterations in peptidoglycan (104). The extent of recombination in nature is highlighted by the fact that multiple, distinct clones of penicillin-resistant meningococci and pneumococci can be discerned in clinical isolates (63, 84, 323, 346).

Low-level resistance to methicillin has been described in *Staphylococcus aureus* strains that possess PBPs with reduced affinity for penicillin (42, 127, 134, 345). However, the majority of methicillin-resistant staphylococci have *mecA* (26, 42, 56), a foreign gene that was acquired from an evolutionarily distinct species, *Staphylococcus sciuri* (67, 309). *mecA* resides chromosomally and encodes an additional peptidoglycan transpeptidase, PBP2a (PBP2′), that functions when the normal three requisite PBPs (PBP1, PBP2, and PBP3) are acylated by a β-lactam (26, 42, 56,

290). However, the presence of *mecA* is not sufficient for expression of methicillin resistance (27); there are multiple chromosomal *fem* (factors essential for methicillin resistance) genes that are required for expression of high-level methicillin resistance (26, 42, 56, 79).

Enterococci are intrinsically resistant to penicillin owing to the presence of a low-affinity PBP (98, 100). In different enterococcal species, the low-affinity PBP is PBP4 (*Enterococcus mundtii* 582, *Enterococcus hirae* 1258), PBP5 (some strains of *Enterococcus faecalis, Enterococcus faecium, E. hirae* 9790, and *Enterococcus casseliflavus*), PBP6 (some strains of *E. faecalis, E. faecium, Enterococcus durans, Enterococcus avium*, and *Enterococcus gallinarum*), or PBP5fm (*E. faecium*) (98, 100, 165, 186, 276, 315, 386). Low-affinity PBPs from multiple species of enterococci are immunologically related to one another (100) and more homologous to PBP2a than their own resident PBPs (89). In addition to the low-affinity PBP5 protein, *E. hirae* can express another low-affinity PBP, PBP3r, that is plasmid-mediated and immunologically related to PBP5 (276). The low-affinity PBP appears to be the bactericidal target of penicillin, as saturation of PBP3r correlates with the killing kinetics (100). Overproduction of PBP5 often accounts for moderate resistance, while changes in specific amino acids in a region between conserved SDN and KTG motifs provide high-level ampicillin resistance (99, 100, 386). High-level resistance can also be due to the presence of other normal PBPs with reduced affinity (7, 165) and is rarely due to the acquisition of a plasmid-encoded β-lactamase (52, 154, 207, 229). Anaerobes like *Bacteroides* spp. typically evade β-lactams by possession of a β-lactamase (285); however, there have been reports in *Bacteroides fragilis* of PBP3 (382), PBP2 (275), or PBP1 complex (365) with reduced affinity for cephalosporins.

PBP-independent penicillin resistance determinants have been described in *E. coli* (355, 364), pneumococci (121, 124), and staphylococci (27, 79, 169, 205). In the laboratory, a deletion that encompasses *mreBCD* (121), a region involved in determining cell shape, results in mecillinam resistance in *E. coli* in the absence of alteration to PBP2, the only mecillinam-sensitive target. Mecillinam resistance can also be conferred by increased levels of ppGpp, the effector of the stringent response, and is probably mediated at the level of transcription of an unknown gene whose product is involved in mecillinam sensitivity (353). Mutations that map to *aroK*, one of the two *E. coli* shikimate kinases, confer mecillinam resistance (354); it is possibly a second activity of AroK, one involved in cell division regulation, that alters mecillinam sensitivity. In laboratory-derived strains of pneumococci, non-PBP-dependent determinants have surfaced (121, 124). Single-step mutants selected with either piperacillin, a typical lytic penicillin, or cefotaxime, a cephalosporin that does not induce cellular lysis, are unexpectedly competence defective as well as resistant to the selecting β-lactam (121, 124). Some strains have mutations in *cpoA* or *ciaH* that encode domains homologous to the glycosyltransferase superfamily (121) or domains related to the superfamily of bacterial histidine protein kinases (124), respectively. Additionally, more non-PBP genes will surface as other classes of mutations are delineated (121). Although analysis of these mutations is expected to shed light on how penicillin resistance develops, neither a mutation in *cpoA* or a mutation in *ciaH* has been shown to reside in penicillin-resistant clinical strains of pneumococci (121, 124). In *S. aureus*, the multiplicity of *fem* genes is a prime example of non-PBP genes involved in modulating methicillin susceptibility (27, 79). A number of *fem* gene products are involved in peptidoglycan biosynthesis (170). More recently, *llm*, a gene encoding a hydrophobic, presumably membrane-associated protein (205), and *fmt*, a gene encoding a protein that has a hydropathy profile similar to that of *S. aureus* PBPs and contains two of the three conserved motifs shared by PBPs and β-lactamases (169), have been identified as non-PBPs contributing to β-lactam resistance.

62.1.2. Fluoroquinolones

The fluoroquinolones have two closely related intracellular targets, DNA gyrase and topoisomerase IV, enzymes involved in DNA replication and in maintaining the superhelical density of DNA. Both are type II topoisomerases that mediate ATP-dependent passage of one DNA duplex through a transient enzyme-bridged double-strand break in another DNA segment (160, 223, 288, 368). Both enzymes function as a heterodimeric complex, A_2B_2 referring to the GyrA and GyrB subunits of DNA gyrase and C_2E_2 referring to the ParC and ParE subunits of topoisomerase IV. The GyrA and ParC subunits are homologs and participate in the breakage and reunion reaction; the GyrB and ParE homologs each contain an ATP-binding domain. A ternary complex of enzyme, DNA, and drug prevents the religation of, and/or enhances, the double-stranded DNA breaks, leading to downstream bactericidal events.

The primary mechanism of resistance to fluoroquinolones in *E. coli* results from an alteration in a defined region of the GyrA protein (residues 67 to 106) termed the quinolone resistance-determining region (QRDR) (379). Mutations in the highly conserved residues Ser-83 and Asp-87 are seen with notable frequency; these residues are close to the catalytic Tyr-122, which is involved in the transient DNA breakage and reunion (97, 288). In *E. coli*, inhibition of topoisomerase IV by quinolones becomes apparent only after gyrase has become refractory to quinolones owing to the presence of at least one mutation within the GyrA QRDR (43, 136, 162, 173). Although a single base change in GyrA can mediate high-level resistance to unfluorinated quinolones (257, 298, 305), high-level fluoroquinolone-resistant mutants generally have two mutations in GyrA and one to two mutations in the corresponding QRDR (residues 64 to 103) of ParC (136). Low-level quinolone resistance can also be mediated by *gyrB* or *parE* mutations (43, 234, 373, 375), and a QRDR for the GyrB subunit has been defined (380). In general, *gyrA* appears to be the initial target for fluoroquinolones in other gram-negative species, including *H. influenzae, Helicobacter pylori, N. gonorrhoeae, Salmonella typhimurium, Klebsiella pneumoniae, Citrobacter freundii, Enterobacter cloacae*, and *Serratia marcescens* (73–75, 106, 109, 164, 225, 245, 324, 381).

Interestingly, topoisomerase IV is the primary target of fluoroquinolones, as discerned in the stepwise acquisition of ciprofloxacin resistance or in the evaluation of the fluoroquinolone-resistant clinical gram-positive pathogens *S. aureus* and *S. pneumoniae* (78, 95, 115, 151, 155, 240, 259, 272, 358, 374). It also appears that topoisomerase IV may be the primary target in enterococcal strains a well (158, 171, 339). The majority of mutations map in the QRDR of *parC* (*grlA* in *S. aureus*), but mutations in *parE* (*grlB* in *S. aureus*) have been described (101, 272). It appears likely that topoisomerase IV may be inherently more susceptible in gram-positive pathogens. This notion, however, is inconsistent with the observation that the primary target can be modulated by the quinolone challenge, as recently seen with sparfloxacin-selected mutants in *S. pneumoniae* (260). In these stepwise-selected mutants, mutations in the GyrA QRDR appeared prior to mutations in the ParC QRDR.

Both recombinational DNA repair (error-free) and SOS-mediated DNA repair (error-prone), as observed in *recA* and/or *lexA* mutants, respectively, have been implicated in attenuating the susceptibility of *E. coli* cells to downstream effects of ternary complex formation (143) (Table 1). Recently, mutations in *recG*, a helicase that catalyzes branch migration of Holliday junctions, were found to affect quinolone susceptibility in *S. aureus* (242). It appears that the normal dosage of this enzyme is sufficient to repair the quinolone-induced DNA damage, because overproduction of RecG did not confer additional resistance to susceptible *S. aureus* strain. Further, *E. coli hipA* and *hipQ* mutant strains exhibit reduced killing by quinolones (36, 367). In addition, the *mar* locus appears to provide a *hip*-independent bactericidal resistance mechanism to fluoroquinolones (114) (Table 1).

Jacoby and colleagues have identified a multiresistance, broad-host-range conjugative plasmid from clinical isolates of *K. pneumoniae* and *E. coli* that confers resistance to fluoroquinolones when transferred to *Enterobacteriaceae* strains and *P. aeroginosa* (211). Although quinolone resistance

TABLE 1 Antibiotic resistance determinants of pathogenic bacteria

Resistance gene	Antibiotic(s)	Mechanism
Target-mediated		
PBPs	β-Lactams	Hyperproduction of PBPs or proteins with reduced affinity
mecA	β-Lactams	Low-affinity PBP2a
gyrA, *gyrB*	Quinolones	Alteration of DNA gyrase
parC, *parE*	Quinolones	Alteration of topoisomerase IV
tetM, *tetO*, *tetP*, *tetQ*, *tetS*, *tetT*, *tetU*, *otrA*	Tetracyclines	GTP-dependent tetracycline dissociation
rpsL	Tetrcyclines, streptomycin	Alteration of ribosomal protein S12
miaA	Tetracyclines	Modification of adenosine-37 in tRNA
rteA/B	Tetracyclines	Two-component regulatory systems for *tetQ*
erm	Macrolides, lincosamides, Streptogramin B	rRNA methylase
rrn	Macrolides	rRNA point mutations
rrs	Streptomycin	Alteration of 16S RNA
vanA, *vanB*, *vanC1*, *vanC2*, *vanC3*, *vanD*	Vancomycin, teicoplanin	Ligase with alterated substrate specificity
vanS/R	Vancomycin, teicoplanin	Two-component regulatory system
vanH	Vancomycin, teicoplanin	Pyruvate dehydrogenase
vanX	Vancomycin, teicoplanin	D-Ala-D-Ala dipeptidase
vanZ	Teicoplanin	Unknown
Antibiotic inactivation[a]		
β-lactamases	β-Lactams	Esterases
AACs	Aminoglycosides	Acetyltransferases
APHs	Aminoglycosides	Phosphotransferases
ANTs	Aminoglycosides	Nucleotidyltransferases
ereA	Macrolides	Esterase
ereB	Macrolides	Esterase
mphA	Macrolides	Phosphorylase
mphB	Macrolides	Phosphorylase

(*Continued on next page*)

TABLE 1 Antibiotic resistance determinants of pathogenic bacteria (*Continued*)

Resistance gene	Antibiotic(s)	Mechanism
mphC	Macrolides	Phosphorylase
mphX	Macrolides	Phosphorylase and/or esterase
satA	Streptogramin A	Acetyltransferase
vat	Streptogramin A	Acetyltransferase
vatB	Streptogramin A	Acetyltransferase
vgb	Streptogramin B	Hydrolyase
linA	Lincomycin, clindamycin	Nucleotidyltransferase
linA′	Lincomycin	Nucleotidyltransferase
linB	Lincomycin	Nucleotidyltransferase
tetX	Tetracyclines	NADP-requiring oxidoreductase
Novel (unclassified)		
ciaH	Cefotaxine	Sensor kinase for competence development?
cpoA	Pipericillin	Glycosyl-transferase? competence-defective
ppGpp	Mecillinam	Increase in ppGpp can result in resistance
mreBCD	Mecillinam	Gene products determine cell shape
aroK	Mecillinam	Gene product is involved in cell division regulation
fem	Methicillin	Where known, gene products are involved in peptidoglycan biosynthesis
llm	β-Lactams	Unknown; membrane-associated protein
fmt	β-Lactams	Unknown; gene product has homology to PBPs
recG	Quinolones	Repair of DNA damage due to quinolone treatment
recA	Quinolones	Repair of DNA damage due to quinolone treatment
lexA	Quinolones	Repair of DNA damage due to quinolone treatment
hipA/Q	Quinolones	Unknown; protects cells against bactericidal effects
mar	Quinolones	Unknown; protects cells against bactericidal effects
qnr	Quinolones	Transferable plasmid-mediated quinolone resistance
otrC	Tetracyclines	Unknown mechanism in *Streptomyces* sp.

[a]AAC, acetyl coenzyme A-dependent *N*-acetylation by acetyltransferases; ANT, ATP-dependent *O*-adenylation by nucleotidyltransferases; APH, ATP-dependent *O*-phosphorylation by phosphotransferases.

provided by the plasmid was low in wild-type strains, the plasmid facilitated the selection of chromosomal mutations that conferred an elevated resistance phenotype. Futhermore, in certain clinical isolates deficient in outer membrane porins, the plasmid enhanced quinolone resistance with an MIC greater than 256 μg/μl. Analysis of quinolone accumulation, drug inactivation, protection of quinolone targets, and repair of quinolone-induced damage remains to be explored.

Many of the commercially available fluoroquinolones are poorly active against most anaerobic bacteria (285). Studies to determine the mechanism(s) of resistance have not been reported. A single report consistent with resistance being due to decreased permeability has been noted for clinical isolates of *Bacteroides* sp. (161). It will be interesting to follow resistance emergence as new fluoroquinolones like trovafloxacin and sparfloxacin with anaerobic activity are used (44, 203).

62.1.3. Tetracyclines

Tetracyclines were used clinically during the 1950s and 1960s, largely because of their broad spectrum of activity. Clinically useful analogs are bacteriostatic, inhibiting growth by binding to a high-affinity site on the 30S subunit, thereby weakening the ribosome-tRNA interaction (59, 68, 91, 306). Atypical tetracyclines like chelocardin are bactericidal, presumably through their interactions with the cytoplasmic membrane (255, 256); these agents proved too toxic to be used as antibiotics. The prevalent resistance mechanisms, efflux and ribosomal protection, limit the clinical utility of the order tetracyclines and do not confer resistance to the atypical tetracyclines (59). More recently, a new class of tetracyclines has been introduced, the glycylcyclines (328). These compounds were designed to obviate the clinically relevant resistance mechanisms (28, 286, 293, 306, 320).

Ribosomal protection as a mechanism of resistance was discovered in streptococci (46, 209). Bacteria possessing ribosomal protection (RP) genes are moderately resistant to tetracycline, minocycline, and doxycycline. Even though the level of resistance is intermediate, RP genes have the widest host range of any tetracycline determinant, including gram-negative genera (but excluding enterics), gram-positive genera, and anaerobes (293). Currently, eight classes of RP genes are recognized in pathogenic gram-positive and gram-negative bacteria (62, 262, 318) (Table 1), including *otrA*, a determinant originally found in the oxytetracycline-producing strain *Streptomyces rimosus* and, more recently, in clinical strains of *Streptomyces* and *Mycobacterium* spp. (85, 262). With one exception, this family of proteins shares ~40% amino acid sequence similarity to TetM, the prototypical determinant first described (46, 209, 318). These proteins can be divided into four groups based on their sequences: TetM, TetO, and TetS (class I); OtrA and TetB(P) (class II); the widespread TetQ as well as the newly described TetT (class III); and TetU, a distinct protein found in *E. faecium* that shares <20% amino acid identity to other members of the RP family (class IV) (62, 292, 293). The sequence similarity within this family of genes has facilitated the development of class-specific PCR primers (62, 293, 294).

All of the ribosomal protection proteins (RPP) have G domains in their N termini that are homologous to those of elongation factors EF-G and EF-Tu (47, 340). The RPPs have more sequence homology to EF-G than to EF-Tu, maintaining the G′ subdomain responsible for modulating the binding of guanine nucleotides (1, 69). Thus, it is not surprising that TetM, like EF-G, has GTPase activity that is highly stimulated in the presence of ribosomes (49). However, the reaction catalyzed by EF-G-GTP, a conformational transition of the ribosome that permits translocation of peptidyl-tRNA and deacylated tRNA from the A and P sites to the P and E sites of the ribosomes, respectively, is not tetracycline sensitive. The step that is tetracycline sensitive is the EF-Tu promoted delivery of aminoacyl-tRNA to the A site. The protection afforded by TetM is not the result of a simple substitution for either EF-G or EF-Tu, as shown in either a defined protein synthesis system (49) or in complementation experiments using temperature-sensitive mutants of either elongation factor (47). TetM also fails to compete with tetracycline for binding to the A site (49, 122); instead, TetM promotes displacement of bound tetracycline, reducing the half-life for dissociation of tetracycline from 70 s to 5 s at 37°C. GTP hydrolysis is required to fully release the antibiotic. It is not clear if TetM acts alone in promoting dissociation or in concert with EF-G, perhaps at a step just prior to A-site binding of aminoacyl-tRNA (49).

Several investigators noted that the MICs to tetracycline increased in cells containing TetM or TetO when they were preexposed to subinhibitory concentrations of tetracycline. The expression of TetM and TetO proteins appears to be regulated at the transcriptional level, unlike the tetracycline efflux proteins that appear to be regulated at the translational level (293, 306). Consistent with transcriptional regulation has been the finding of both short and long mRNA transcripts for *tetM* in strains carrying Tn*916* (326). Transcriptional analysis of other RP genes has not been reported. Although an upstream region of TetO is required for full expression (359), transcriptional attenuation does not appear to be involved in the regulation of this homologous RPP. An interesting twist is found in the regulation of *tetQ*, the universal tetracycline resistance determinant in *Bacteroides* spp. (285). This determinant is inducible and under the control of a two-component regulatory system, the products of *rteA* (sensor) and *rteB* (regulator) (325). RteA and RteB, along with RteC (undefined role), are also involved in the transferability of *tetQ* (325). Nearly all (80 to 90%) strains of *Bacteroides* are resistant to tetracycline, and the majority contain *tetQ*; some tetracycline-resistant strains of *Bacteroides ureolyticus* have been characterized to harbor a *tetM*-related determinant (285). Another anaerobic genus, *Mobiluncus*, can harbor either TetO or TetQ (293). *Clostridium perfringens* uses both ribosomal protection [*tetB*(P) or *tetM*] and efflux [*tetA*(P), *tetK*, or *tetL*] to resist tetracycline.

Although RP resistance determinants are typically found on mobile genetic elements and may have originated from tetracycline-producing strains (85, 293), the similarity of RPPs to endogenous enzymes involved in protein synthesis suggests modification of a common evolutionary ancestral elongation factor (340). Further evolution has occurred via homologous recombination, resulting in mosaic *tetM* genes. These genetic exchanges are made possible through the availability of genes on mobile elements (250). Further, the G+C contents in the TetM determinants from *Ureaplasma urealyticum*, *S. aureus*, *N. gonorrhoeae*, and *N. meningitis* are ~40%, consistent with the notion that the *tetM* genes have spread from gram-positive bacteria into gram-negative species (293).

Mutations in two genes, *miaA* and *rpsL*, in gram-negative bacteria have been described that reduce the level of tetracycline resistance mediated by TetM or TetO (48, 341). MiaA modifies the adenosine at position 37 in tRNA, next to the anticodon. RpsL encodes the S12 protein in the 30S subunit. Both mutations affect the anticodon-codon interaction: the modified tRNA stacks better, enhancing binding to the codon, while S12 acts indirectly to stabilize the amino acid-tRNA · Ef-Tu · GTP at the A site via its contact with 16S RNA. These results are consistent with tetracycline's mechanism of action and suggest that RPP interacts near or at the elongation factor binding site on the ribosome (48).

62.1.4. Macrolides/Lincosamides/Streptogramins

Macrolides, lincosamides, and streptogramin B (MLS_B) antibiotics bind to the 50S ribosomal subunit and interfere with the elongation of nascent polypeptide chains (65, 68). Other reports suggest that macrolides interfere with 50S ribosomal subunit assembly and that this reaction is equally susceptible to inhibition by macrolides (57). Cross-resistance to MLS_B antibiotics is conferred when a specific adenine residue (A2058 in the *E. coli* numbering scheme) in 23S rRNA is dimethylated (54, 80, 152, 384) or replaced by another nucleotide (216, 297, 352); this resistance is referred to as MLS_B resistance because these alterations prevent the binding of drugs from the structurally distinct classes of antibiotics (68, 177, 361, 363). Both mutational and postranscriptional alterations cluster in the peptidyltransferase region in domain V of 23S rRNA, and footprinting experiments with individual drugs confirm that the three classes of antibiotics have overlapping binding sites (68, 224). However, A2058 was not protected by streptogram in B in footprinting experiments (224); thus, it is not as clear how modification of this residue confers resistance to this antibiotic class.

Although there have been laboratory reports of mutations in ribosomal proteins that provide erythromycin resistance, none have been isolated in clinical pathogens (68, 361). Presumably, these mutations are not compatible with the infectious state. Recently, dominant point mutations in 23S rRNA have been described in pathogenic strains. Investigations into the resistance development associated with use of the second-generation macrolides in treatment of atypical pathogens, *Mycobacterium intracellulare* (216) and *H. pylori* (352), have led to the finding that mutations in the 23S rRNA domain V confer resistance to MLS_B antibiotics. Nucleotide changes at the cognate positions A2058 and G2059 were found in *M. intracellulare* and *H. pylori*, while substitutions at G2057 were found to confer low-level macrolide resistance in propionibacteria (297). Mutations at this latter site generally confer resistance to 14- and 15-membered macrolides but not to 16-membered macrolides or lincosamides (361).

Mutations in domain II have also been implicated in erythromycin resistance and clindamycin hypersusceptibility (361, 363), but none of these mutations have been described in clinically resistant pathogens. These mutations include changes that alter the translational attenuation of the "E-peptide" (MSLKV) located in domain II of 23S rRNA (343). Tenson et al. (344) have determined that translation of the E-peptide in *trans* or in *cis*, with preferences of Ile or Leu at position 3 and a hydrophobic amino acid at position 5, is required for erythromycin resistance. However, exogenously added E-peptide (≤1 mM) does not confer erythromycin resistance (344), suggesting that E-peptide must act only when part of a nascent ribosome (363).

By far, the most widespread mechanism of resistance to MLS_B antibiotics is mediated by an Erm methylase (177, 179, 361, 363). Erm methylases are a family of highly related proteins that use *S*-adenosylmethionine as a methyl donor to modify a single adenine residue (A2058) in nascent 23S rRNA (179, 361, 363). In aerobic pathogenic bacteria, the most predominant *erm* genes are *ermA* (staphylococci), *ermB* (= *ermAM* = *ermBC* = *ermB*-like; streptococci, enterococci, and enteric bacteria), *ermC* (= *ermM*; staphylococci), and the newly described *ermTR* (*Streptococcus pyogenes*, *Streptococcus milleri* (153, 179, 308) (Table 1). Additionally, there appears to be an *ermTR*-like gene in *S. pneumoniae* (330), but whether it represents a distinct *erm* class is unknown. In anaerobic species, closely homologous (>95%) *erm* genes, *ermF* (Tn*4351*), *ermFS* (Tn*4551*), and *ermFU* (conjugal element) are found in *B. fragilis*, *Bacteroides ovatus*, and *Bacteroides vulgaris* (179, 285). *Corynebacterium diphtheriae* has *ermCD*, a gene more homologous to *lmrB*, the methylase found in *Streptomyces lincolnensis*, the producer of lincomycin (141, 385), than to the *erm* genes found in other pathogenic species. *ermP* in *C. perfringens* and *ermZ* in *Clostridium difficile* are nearly identical to *ermB* and probably represent the results of intergenic transmission between staphylococci, streptococci, or enterococci and *Clostridium* spp. (29, 126). *ermQ* is the predominant MLS_B-resistant determinant in *C. perfringens* and represents a distinct *erm* class by hybridization (30). Although it is suspected that the *erm* genes seen in pathogenic species encode a dimethylase, direct proof of this is lacking for all species. Recently, the solution structure of ErmB from a clinical strain of *S. pneumoniae* was published (383); features of the catalytic domain of ErmB are conserved even when compared to other non-Erm methyltransferases, but the substrate binding domain is unique.

Regulation of *erm* is usually mediated by a translational attenuation mechanism (179, 361–363). Secondary structure of regions 5′ to the coding sequence sequesters the Shine-Dalgarno sequence and initiator codon (AUG) for the methylase in a stem-loop structure. Further upstream, there are one (*ermB* or *ermC*) or two (*ermA*, *ermTR*) short, leader open reading frames that potentiate ribosome stalling and permit alternative mRNA conformations that fail to sequester the Shine-Dalgarno Sequence and AUG. Thus erythromycin can induce the synthesis of its own resistance determinant, the *erm* methylase. The leader peptide sequence for *ermC* has been analyzed by Weisblum and colleagues, and four amino acids, IFVI, are critical for induction (213). However, not only the sequence and placement of the leader peptide(s) are critical; so also are the structural features of the antibiotic (157, 362). It is important to note that use of noninducers like 16-membered macrolides in the clinical environment has resulted in the selection of constitutively resistant mutants (181, 363). The *Bacillus licheniformis ermK* gene is regulated differently by transcriptional attenuation owing to the presence of a rho factor-independent termination site in the leader sequence of the nascent *ermK* mRNA (174). Expression of the closely related *ermJ* from *Bacillus anthracis* (163) is also likely to be regulated in a similar manner.

Most enterococci are intrinsically resistant to clindamycin. In fact, this characteristic can be useful in bacterial identification of *E. faecalis* (176). Clindamycin resistance rates range from 0 to >20% for *Bacteroides* spp., largely owing to the presence of one of the *erm* genes (285). *E. faecalis* is also intrinsically resistant to streptogramin A antibiotics, including Synercid (31, 238). Synercid is a mixture of dalfopristin (streptogramin A component) and quinupristin (streptogramin B component) that has recently been used in compassionate trials in the United States for multidrug-resistant *E. faecium* (238). The clinical efficacy reported has been moderate (50 to 60%), consistent with the report that *E. faecium* strains containing methylated ribosomes are susceptible in in vitro assays but not in rabbit endocarditis models (93). This is likely due to reduced diffusion of dalfopristin at the infection site.

62.1.5. Aminoglycosides

Aminoglycosides interact with 16S rRNA and inhibit protein synthesis (68). They can be classified according to their pattern of substitution as 4,5-disubstituted deoxystreptamines (neomycin B) or 4,6-disubstituted deoxystreptamines (kanamycin A, tobramycin, gentamicin, amikacin) (299). Both of these classes are bactericidal, and their interaction with the 30S ribosomal subunit impedes EF-G binding, resulting in a block in translocation. Streptomycin and spectinomycin constitute another structural class; steptomycin inhibits initiation of protein synthesis and is bactericidal, whereas spectinomycin is bacteriostatic as it binds reversibly to the 30S ribosomal subunit (68, 299).

Streptomycin resistance in *Mycobacterium tuberculosis* is due to one of two target modifications: *rrs* (16S RNA) or *rpsL* (S12) (230). *M. tuberculosis* strains resistant to streptomycin that do not have changes in one or both of these genes have been described, but the mechanism for resistance has not been investigated (137). Interestingly, streptomycin-resistant *Mycobacterium avium* complex bacilli do not manifest either resistance strategy (279). Aminoglycoside-modifying enzymes, the predominant form of resistance in other species, have been described in *M. tuberculosis* (137). *S. aureus* or *E. faecalis* strains with high-level streptomycin resistance can also harbor *rpsL* mutations (90, 176).

Several pathogenic species of bacteria have chromosomally encoded housekeeping enzymes that fortuitously modify aminoglycosides. *Providencia stuartii* contains a 2′-*N*-acetyltransferase [AAC(2′)-Ia] that is expressed at low levels in wild-type strains. Normally, this enzyme catalyzes the addition of an acetate group to the C-6 position of peptidoglycan-bound *N*-acetylmuramyl residues (269). However, when overexpressed as a result of a mutation in one of the many genes involved in regulation of its expression (204), the acetate group can be transferred from peptidoglycan to the 2′ amino group of aminoglycosides like gentamicin and tobramycin, resulting in clinically significant resistance (219, 269, 287). Likewise, *S. marcescens* has a chromosomal AAC(6′)-Ic acetyltransferase that confers high-level aminoglycoside resistance upon transcriptional activation (311). The fast-growing mycobacterial species (*Mycobacterium fortuitum* complex, *Mycobacterium smegmatis*, *Mycobacteruim phlei*, and *Mycobacterium vaccae*) have been credited with a 3-*N*-acetyltransferase-III resident in their chromosomes (348). Although this enzyme has a broad substrate specificity, it does not correlate with clinical aminoglycoside resistance but rather appears to be responsible for conversion of citrate from oxaloacetate and acetyl coenzyme A. Among the gram-positive bacteria, a 6′-*N*-acetyltransferase [AAC(6′)-Ii] appears to be responsible, in part, for the intrinsic resistance of *E. faecium* (66, 370). The presence of this gene therefore results in failure of therapeutic penicillin-aminoglycoside synergism when aminoglycosides with free 6′-amino groups are used.

Aminoglycoside resistance is universal in anaerobic bacteria (285). This is not due to inherent resistance at the level of the target, but rather to an inability to import the aminoglycoside in the absence of an electron transport system. Transport of aminoglycosides across the cytoplasmic membrane is an energy-dependent process fueled by the electron transport system (45, 335). It is interesting to note that *aarF* mutants in *P. stuartii* are ubiquinone deficient; hence, these strains may be aminoglycoside resistant, partly owing to the absence of a functional electron transport system (204).

62.1.6. Glycopeptides

Glycopeptide antibiotics, vancomycin and teicoplanin, interfere with the transglycosylation reaction in peptidoglycan (16, 17, 92, 113, 218, 228, 356). The glycopeptides bind with high affinity to the D-Ala-D-Ala termini of the target molecule, lipid-PP-disaccharide pentapeptide; dimerization of vancomycin facilitates its binding. The inhibitory activity of teicoplanin is enhanced by membrane anchoring via its C-10 acyl side chain, thereby increasing the avidity of ligand binding at the target site (25).

Glycopeptide resistance in enterococci is mediated by a cluster of proteins that are able to sense physiological changes imparted by the action of vancomycin. The system includes a two-component regulatory system comprising a sensor histidine kinase (VanS) and a transcriptional activator (VanR) and the products of three genes, *vanA* (or *vanB*), *vanH*, and *vanX*, that alter the lipid-PP-disaccharide to terminate in D-Ala-D-lactate. Vancomycin binds with a 1,000-fold reduced affinity to the depsipeptide, allowing cells to be insensitive to the presence of the glycopeptide. A dehydrogenase encoded by the VanH gene generates D-lactate from pyruvate and, working in concert with a VanA- or VanB-encoded depsipeptide ligase, produces an ester-linked dipeptide D-Ala-D-lactate that is inserted into the stem peptide precursor. VanA and VanB have weak D-Ala-D-Ala ligase activity but have the added, novel activity of activating D-hydroxy acids (356). VanX, a D-Ala-D-Ala dipeptidase, reduces the pool of normal dipeptide precursor.

Two clusters of *van* genes are found in clinical strains, *vanA* and *vanB* (16, 17, 92, 113, 218, 228, 356). The *vanA* cluster is encoded on a plasmid or a chromosomal copy of a transposable element, Tn*1546*, and cells harboring Tn*1546* can usually be distinguished phenotypically by having high-level resistance to both vancomycin and teicoplanin. This cluster often includes two accessory genes, *vanY* and *vanZ*. VanY is a carboxypeptidase that helps ensure that no normal pentapeptides reside in the mature peptidoglycan. The function of *vanZ* is unknown; however, overproduction of VanZ confers low-level teicoplanin resistance in the absence of detectable pentapeptide modification (14). In addition to the requisite genes (*vanSRHBX*), the *vanB* cluster has two accessory genes, *vanY* and *vanW*, the latter of unknown function. This cluster is found on the composite transposon Tn*1547*, which is often part of a larger, conjugative element (90 to 250 kb) located on the chromosome (92). Cells containing *vanB* are resistant to vancomycin but remain susceptible to teicoplanin, as the latter is not an inducer of the *vanSRHBX* operon. Constitutive mutations of the *vanB* cluster can arise, resulting in a vancomycin-and teicoplanin-resistant phenotype. VanB has also been found in other species, including *Streptococcus bovis* (280), heightening the concern that penicillin-resistant pneumococci may become vancomycin resistant. Recently, a strain of *E. faecium* constitutively resistant to vancomycin and to low levels of teicoplanin was found to produce a lipid-PP-disaccharide depsipeptide by virtue of a novel VanD ligase that has 69% identity to VanA and VanB (271).

Intrinsic, constitutive resistance to glycopeptides is found in several pathogenic enterococcal species, including *E. gallinarum* (*vanC1*), *E. casseliflavus* (*vanC2*), and *E. flavescens* (*vanC3*). In these organisms, both a D-Ala-D-Ala ligase (*ddl*) and a D-Ala-D-Ser ligase (*vanCx*) reside (236, 264), resulting in a mix of lipid-PP-disaccharide-pentapeptides and low-level vancomycin resistance. High-level resistance can be found in *E. gallinarum* and *E. casseliflavus* when the *vanA* gene cluster is present (86). High-level vancomycin resistance is intrinsically mediated by Ddl ligases that catalyze the synthesis of the depsipeptide D-Ala-D-Lac in *Leuconostoc mesenteroides*, *Pediococcus pentosaceus*, and *Lactobacillus* sp. (33, 35, 131). VanA, VanB, and presumably Van D are not closely related to their depsipeptide Ddl counterparts in the lactic acid bacterial species (88). Interestingly, their origin appears to be from the Ddl ligases residing in glycopeptide-producing organisms, *Amycolatopsis orientalis* and *Streptomyces toyocaensis* (208). The sequence of the different Van ligases can vary in clinical enterococcal strains. Multiplex PCR-restriction fragment length polymorphism assays can distinguish *vanA*, *vanB*, *vanC1*, and *vanC2/3* genes in enterococci. Sequence analysis has revealed sequence polymorphisms within *vanB* and *vanC2*, with little variation noted in the *vanA* and *vanC1* genes (265).

Alteration of lipid-PP-disaccharide pentapeptide precursors in vancomycin-resistant strains is well documented (33–35, 76, 77, 130, 131, 208, 291). However, very little of the low-affinity target resides in mature peptidoglycan, largely owing to the action of D,D-carboxypeptidases and transpeptidases (34, 35, 76, 77). Although there is not a big change in the murein chemical composition of the pentapeptide, an increase in the amount of tetrapeptide or tripeptide/tetrapeptide stem peptides is seen in peptidoglycan from vancomycin-resistant *E. faecalis* (*vanA*) or *E. faecium* (*vanB*) (76, 77, 130), respectively. Other alterations of note in *E. faecium* D366 (*vanB*) are increases in alanine ester content in lipoteichoic acid and penicillin tolerance. The strain is hypothesized to be tolerant because the modified lipoteichoic acid has reduced autolysin binding capacity (125).

The infectious disease community has been increasingly concerned with the possible spread of the vancomycin-resistant determinants from enterococci to the more aggressive pathogens, methicillin-resistant *S. aureus* and coagulase-negative staphylococci (16, 17, 92, 113, 218, 228, 334, 356). These concerns are legitimate given the history of genetic exchange between these two genera. However, the methicillin-resistant *S. aureus* strains reported in early 1997 with reduced susceptibility to vancomycin did not carry any of the known vancomycin resistance operons (139, 253, 313, 314, 342). Rather, the mechanism appears to be novel. Phenotype changes in septa placement have been characterized in the methicillin-resistant, vancomycin-intermediate *S. aureus* strain from Japan (139). Peptidoglycan from this strain is poorly cross-linked, with a higher percentage of monomers than normal (342).

Initially, a membrane protein of 35 to 39 kDa was found in clinical isolates of *Staphylococcus epidermidis*, *Staphylococcus haemolyticus*, and *S. aureus* with low-level resistance to vancomycin (253, 313). This protein has more recently been characterized as Ddh, a NAD+-dependent D-lacate dehydrogenase with homology to the D-hydroxy acid dehydrogenase (VanH in vancomycin-resistant enterococci (38). However, overproduction of Ddh does not result in D-lactate-containing precursors, and inactivation of *ddh* does not alter the low-level glycopeptide resistance in *S. aureus*. In teicoplanin- and vancomycin-resistant *S. haemolyticus* strains, 1.7% of the total cytoplasmic peptiodoglycan precursors were shown to contain UDP-muramyl-tetrapeptide-D-lactate (33). This small amount does not appear to be sufficient to confer resistance. However, the composition of the cross bridge, either Ala-Gly-Ser-Gly_2 or Gly_2-Ser-Gly_2, was found to be different, with an additional serine replacing a Gly residue in 13.6% of the muropeptides. The changes in the cross bridge, along with the small amount of stem peptides ending in D-lactate, may account for the resistance. Clinical isolates of coagulase-negative staphylococci appear to be heterogeneous, with cells in the population capable of growing at concentrations up to 50 μg of teicoplanin per ml (314). These clinical strains phenotypically resemble a laboratory-derived, highly vancomycin-resistant *S. aureus* strain in that sub-MIC concentrations of teicoplanin inhibited autolysis and caused cells to aggregate until the antibiotic level was reduced by sequestration. Thus, it is still uncertain what the mechanism(s) of vancomycin resistance are in stapylococci, but they appear to be multifactorial.

62.2. ANTIBIOTIC INACTIVATION

62.2.1. β-Lactams

In clinical isolates, the principal resistance mechanism to β-lactam antibiotics is mediated by a β-lactamase enzyme, which inactivates the drug by hydrolyzing the amide bond of the β-lactam. Over the last 40 years, β-lactamases have become increasingly responsible for β-lactam antibiotic resistance in gram-negative bacteria. Virtually every β-lactam currently used can be inactivated by at least one β-lactamase, as evidenced by the more than 200 β-lactamases that have been cataloged in the last 50 years (51). There are so many that a Web site has been created on the Internet to catalog and update the amino acid substitutions by which individual enzymes are defined (http://www.lahey.hitchcook.org/pages/lhc/studies/webt.htm).

Among gram-positive cocci, the only β-lactamase of clinical significance is the inducible, narrow-spectrum staphylococcal β-lactamase, which rapidly hydrolyzes penicillin G, ampicillin, carbenicillin, and related compounds. It is much less active against cephalosporins. Approximately 90% of *S. aureus* strains causing infections in communities and hospitals produce β-lactamase. Resistance due to β-lactamase is rare in *E. faecium* but can be found in *E. faecalis*, where it is constitutively expressed and often associated with high-level resistance to gentamicin (176).

Rarely, staphylococcal strains are resistant to methicillin in the absence of *mecA*. Some of these strains hyperproduce β-lactamase and, when coupled with a decrease in permeability, will express borderline resistance (42). Recently, strains have been detected that produce methicillinase and appear borderline resistant phenotypically. Furthermore, strains that have modified PBPs (PBPs 1, 2, or 4) with lower affinity for methicillin can be detected among the *mecA* mutant strains. Small-colony variants that tend to persist in tissues after antibiotic treatment may also show increased resistance to methicillin owing to slow growth and reduced uptake (42).

Among the gram-negative bacilli the situation is more complex, and numerous studies have demonstrated that these organisms produce different β-lactamases with distinct and/or overlapping spectra of activity. Several schemes have been proposed for the classification of β-lactamases. The most recent and complete scheme developed by Bush and colleagues (51) combines elements of previous schemes and correlates classification with the molecular structure of the enzyme. β-Lactamases are classified into four groups on the basis of amino acid sequence and functional characteristics, including preferred antibiotic substrate and inhibition by clavulanic acid (51). Briefly, group 1 includes cephalosporinases encoded on transmissible plasmids that are poorly inhibited by the β-lactamase inhibitor clavulanate. Group 2 enzymes are generally inhibited by clavulanate and subdivided into six classes based on preferential hydrolysis of various β-lactam antibiotics. Group 3 differs from the other groups in that the enzymes have an active-site zinc rather than serine. The group 3 metallo-β-lactamases hydrolyze a broad spectrum of substrates, including the carbapenem imipenem. Group 4 consists of a small group of primarily chromosomal, inhibitor-resistant penicillinases. Often clinical isolates produce multiple chromosomal β-lactamases belonging to different functional groups and molecular classes, or both a species-specific chromosomally encoded enzyme and two to three plasmid-determined β-lactamases (215).

Resistance in gram-negative bacilli to third-generation cephalosporins has emerged rapidly and extensively over the last 10 years (188, 189). Two widespread mechanisms of resistance, hyperproduction of AmpC enzymes and extended-spectrum β-lactamases (ESBLs), have compromised the efficacy of cephalosporins but not that of carbapenems. Virtually all gram-negative bacteria harbor chromosomally mediated AmpC cephalosporinases. When chromosomally encoded, these enzymes are inducible in *Enterobacter* sp., *Citrobacter* sp., *Serratia* sp., and *P. aeroginosa* and are constitutive in *E. coli* (22–24, 51, 105, 215). Selection of constitutive (derepressed) mutants from strains normally expressing this enzyme inducibly is responsible for the emergence of multiple β-lactam resistance in these species (304). Strains that hyperproduce AmpC cephalosporinases are responsible for a number of clinical failures of β-lactam antibiotics in the *Enterobacteriaceae* family. Inducible β-lactamases are clinically relevant when the inducing molecule is a substrate, such as ampicillin. Many good inducers, such as cefoxitin and imipenem, are not good substrates for these enzymes, but use of these agents has been associated with hyperproduction of chromosomally encoded gram-negative β-lactamases as a result of promoter-up mutations or the alteration or loss or regulatory products. Hyperproduction of group 1 AmpC enzymes arises readily and spontaneously via mutational loss of a peptidoglycan recycling enzyme, AmpD, and confers resistance to all β-lactams except carbapenems (148). Carbapenems are probably spared because their influx into the periplasmic space outpaces their slow hydrolysis (282). Hyperproduction of AmpC β-lactamases in combination with loss of permeability appears to mediate resistance to imipenem among some strains of *Enterobacter* and *Serratia*. In *P. aeruginosa*, loss of the "carbapenem-specific" porin OprD (D2 porin) occurs in up to 17% of cases in which imipenem is used for *P. aeroginosa* infection (53, 241, 281). Loss of outer membrane porins may also cause imipenem resistance in *Enterobacter* and *Klebsiella* spp. when combined with hyperproduction of an AmpC enzyme (41, 282).

The AmpC family of related cephalosporinases have migrated from a chromosomal location to plasmids and have been seen in the "transmission mode" with increasing frequency in gram-negative bacteria. Clinically relevant plasmid-encoded *ampC* variants include MIR-1, BIL-1, CMY-1/2, MOX-1, LAT-1/2, ACT-1, and FOX-1/2/3 (23, 24, 41, 51, 105, 206, 215). Strains that harbor plasmid-encoded AmpC enzymes are resistant to β-lactamase inhibitor/β-lactam drug combinations, including cephamycins, cephalosporins, and monobactams. Susceptibility only to cefepime and imipenem is retained.

The second blow to third-generation cephalosporins was the emergence of ESBLs, typically given three-letter, one-number codes (150, 215). Most ESBLs result from mutations of established plasmid forms, including TEM-1 (from Temoniara, a patient's name) and SHV-1 (signifying sulfhydryl-variable). Intraspecies and interspecies gene transfer coupled with the widespread use of ampicillin and related β-lactams explains the successful dissemination of these plasmid genes among gram-negative bacilli (135, 317). The most common β-lactamase in gram-negative organisms, accounting for >80% of all plasmid-encoded β-lactamases, is TEM-1, responsible for transferable ampicillin resistance among *Enterobacteriaceae* and *Pseudomonas* strains worldwide (226). TEM-1 hydrolyzes all β-lactam antibiotics with the exception of some broader-spectrum cephalosporins, but in general possesses little or no activity against cephamycins or carbapenems. SHV-1 predominates in *K. pneumoniae*. In the past 10 years, selective pressure has led to the development of more than 73 unique TEM-1- and SHV-1-β-lactamase variants (51). The rate of resistance to third-generation cephalosporins is increasing in nosocomial *Enterobacteriaceae* strains (249).

Since the 1980s, the TEM and SHV ESBLs have been identified in virtually every genus of *Enterobacteriaceae* (215). These enzymes confer moderate- to high-level resistance to all third-generation cephalosporins; however, they are readily antagonized in vitro by the various β-lactamase inhibitor agents (clavulanic acid, sulbactam, or tazobactam) (215). Compared to their parental types, ESBLs have one to four amino acid replacements. Substitutions that alter the active site of TEM to broaden its spectrum also lower its catalytic efficiency. To compensate for this loss of activity, two to four amino acid substitutions are often required before the modified TEM enzyme can efficiently hydrolyze enough cephalosporin to impart clinical resistance. Alternatively, the bacterium can increase production of the weakened enzyme as a result of up-promoter mutations or insertion of insertion-like elements carrying strong promoters (201, 202). Mutations at other sites can correct stability defects and thereby add to the enzyme's plasticity, further enhancing the cell's ability to withstand treatment with β-lactam antibiotics and inhibitors (144). *K. pneumoniae*, in which these enzymes were originally discovered, remains the most prevalent ESBL-producing species (41, 50). Although most strains remain susceptible to imipenem and other carbapenems, there is concern that porin loss could render these strains resistant to carbapenems.

Recently, a novel class of TEM-1-derived β-lactamases resistant to β-lactamase inhibitors (group 2br) has been identified (37, 51, 133, 188–190). These new inhibitor-resistant TEMs have a lower affinity for these inhibitors and have retained efficient hydrolytic properties (51, 150, 166). Groups 2br and 3 β-lactamases are increasingly found in *E. coli*, *K. pneumoniae*, *Proteus mirabilis*, and *C. freundii*. In addition to ESBLs, non-SHV, non-TEM derivatives have been characterized, including the MEN-1, CTX-MI, CTX-M2, PER-1, and PER-2 enzymes (22, 188–190, 215). They have been isolated worldwide, and all are refractory to β-lactamase inhibitors.

The fact that imipenem remains active against AmpC-derepressed strains and those with ESBLs is a major clinical advantage. Nevertheless, this carbapenem is not refractory to all β-lactamases. An increasing number of carbapenem-hydrolyzing β-lactamases have been described in the last few years, especially in Japan (215, 270, 283, 284). At the molecular level, these enzymes belong to either the group 2f β-lactamases with serine at the active site or the group 3 metallo-β-lactamase group (51). Group 2f enzymes include IMI-1 and NmcA from *E. cloacae* (248, 249, 284) and Sme-1 from *S. marcescens* (377). Both organisms are broadly resistant due to the production of multiple β-lactamases. IMI-1 from *E. cloacae* is produced in strains harboring both AmpC- and TEM-type enzymes, whereas the Sme-1 enzyme in *S. marcescens* is produced in addition to the chromosomal AmpC enzyme.

Metallo-β-lactamases are readily distinguished because they are refractory to inhibition by β-lactamase inhibitors and can hydrolyze nearly all β-lactams, including imipenem. Enzymes from emerging pathogens such as *Aeromonas* spp., *Stenotrophomonas maltophilia*, and *Burkholderia cepacia*, have also been characterized and are the subject of recent reviews (270, 283, 357). Of these, *S. maltophilia* presents the greatest challenge, as imipenem resistance is chromosomally mediated and virtually universal in all strains. In the past 5 years, plasmid-mediated variants from *B. fragilis*, *S. marcescens*, *K. pneumoniae*, and *P. aeruginosa* have been reported in Japan (18, 147, 215, 360, 382). The metallo-β-lactamase from *S. marcescens*, IMP-1, is mobile on an integron-like element, *intI3*, and has spread to *Pseudomonas putida*, *K. pneumoniae*, and *Alcaligenes* spp. (307). IMP-1 hydrolyzes both carbapenems and the most recently developed fourth-generation cephalosporins. Thus, metallo-β-lactamases have the potential for widespread dissemination and are perhaps the most formidable β-lactamases currently known.

Many anaerobic bacteria produce β-lactamases (285); virtually all (>90%) of the *Bacteroides* isolates in the United States produce β-lactamases, and 25% produce high levels of β-lactamases. Most β-lactamases from the *B. fragilis* group are constitutive, chromosomally encoded cephalosporinases that are species specific and have been placed in group 2e (51, 150). As noted above, an increasing number of *Bacteroides* isolates produce metallo-β-lactamases (18, 376). Like other resistance genes in this species, the transcriptional initiation signals for the metallo-β-lactamase gene *ccrA* are provided by an insertion element integrated within the promoter rather than from a natural promoter (285).

62.2.2. Aminoglycosides

Of the five mechanisms of bacterial resistance to the aminoglycosides, that mediated by drug-modifying enzymes is clinically the most prevalent. Unlike β-lactamases, in which the antibiotic is hydrolyzed, resistance to aminoglycosides involves enzymes that catalyze cofactor-dependent drug modification of hydroxy or amino groups on the aminocyclitol residues (70). These enzymes include many different types of acetyltransferases, phosphotransferases, and nucleotidyl transferases that vary greatly in their substrate specificity and in the degrees to which they inactivate different aminoglycosides (70, 72). As with β-lactamases, single amino acid changes can have significant effects on substrate profiles; however, the introduction of a new aminoglycoside antibiotic often results in multiple inactivating genes coresiding in the same strain. Enzymatically modified aminoglycosides do not interact with the ribosomal target and, thus, do not inhibit protein synthesis (372).

Over 50 different enzymes have been identified as aminoglycoside modifiers, most of which are unrelated at the nucleotide sequence level and are derived from a variety of microbial origins (310). The three classes of aminoglycoside-modifying enzymes differ in the nature of the sites modified: acetyl coenzyme A-dependent *N*-acetylation by acetyltransferases (AAC), ATP-dependent *O*-phosphorylation by phosphotransferases (APH), and ATP-dependent *O*-adenylation by nucleotidyltransferases (ANT) (72). The confusing nomenclature describing these enzymes has recently been simplified. Shaw et al. (310) have proposed a uniform terminology in which the predicted specificity of group transfer is delineated by an arabic numeral in parentheses, and the subfamily, based on the aminoglycoside resistance profile, is designated by a roman numeral sometimes followed by a letter indicating a specific gene. *Enterobacteriaceae* tend to produce AAC(3)I, II, and IV, AAC(6′)I, ANT(2″), and APH(3′); *P. aeruginosa* produces AAC(3)I, AAC(3)III, AAC(6′)I or II, ANT(2″), and APH(3′); staphylococci and *E. faecalis* often produce ANT(4′)(4″) or a bifunctional AAC(6′)/APH(2″) activity.

Gentamicin, kanamycin, and tobramycin resistance in staphylococci and high-level gentamicin resistance in enterococci are usually mediated by the bifunctional enzyme (96). This combination also effectively protects strains against all available aminoglycosides except streptomycin and spectinomycin. The enzymes are usually, though not always, encoded on a transposon (Tn*924*) that resides on a multiresistance plasmid; however, others are chromosomally located (112, 370), as seen in staphylococci harboring Tn*5405* encoding APH(3′)-III (provides resistance to kanamycin, neomycin, and amikacin) (81).

The most common aminoglycoside resistance enzymes in the gram-negative pathogens are the AAC(6′)s, which acetylate the 1, 3, 2′, or 6′ positions of aminoglycosides. Within this class, 16 genes that encode AAC(6′) enzymes have been identified. The aminoglycoside isepamicin has a free amino group at the 6′ position but is poorly acetylated. Nevertheless, resistance to this drug has been detected in several bacteria strains isolated recently (55, 370, 371).

Aminoglycoside resistance determinants are, with a few exceptions, encoded on plasmids and transposons in bacteria, which explains the rapid dissemination of resistance. Of particular interest is the finding that aminoglycoside resistance genes have been found on conjugative and nonconjugative plasmids of a variety of incompatibility types and associated with different transposons. Therefore, it is not surprising that an increasing number of strains appear to produce two or more enzymes that are active against a broad range of aminoglycosides (118, 214). A frequent combination that has occurred over the last few years is gentamicin-modifying enzymes [ANT(2″) and AAC(3)-I] combined with AAC(6′)-I, resulting in broad-spectrum resistance to gentamicin, tobramycin, netilmicin, kanamycin, and amikacin.

Miller and colleagues have recently discovered that aminoglycoside resistance is more complex than inactivation (219, 220). The inherent complexity includes a combination of permeability changes in addition to the plasmid-mediated aminoglycoside resistance mechanisms and preexisting chromosomally encoded modifying enzymes.

Several unique aminoglycoside-modifying enzymes have been identified. A strain of *Streptococcus agalactiae* was described that carried the chimeric gene *aac6′-aph2″* on a chromosomal copy of Tn*3706* (142). The amino acid sequence was nearly identical to that of the bifunctional enzyme encoded in Tn*4001* in *S. aureus* and Tn*5281* in *E. faecalis* (140). Also of note is the novel gentamicin resistance gene *aph*(2″)-Ic, which is present in strains of *E. gallinarum*, *E. faecium*, and *E. faecalis* (60). Strains carrying this gene exhibit only moderate aminoglycoside resistance (256 to 512 μg/ml); however, the resistance is clinically significant, as these strains resist the bactericidal activity of the synergistic combination of ampicillin plus gentamicin. A crytptic streptomycin resistance gene, *aadS*, has been described in the *Bacteroides* transposon Tn*4551*, positioned downstream from *ermFS* (319). The *aadS* gene is expressed only when activated by a chromosomal mutation in *trans*. Recently, a novel spectinomycin phosphotransferase was identified in strains of *Legionella pneumophila* (332). Although the authors did not demonstrate the position that was modified, the *Legionella aph* has the most extensive similarity with APH(3′) enzymes (332). This is the first example of a clinical strain that uses phosphorylation rather than nucleotidyl modification to inactivate spectinomycin.

62.2.3. Macrolides, Lincosamides, and Streptogramins

Unlike target modification, inactivation of MLS_B antibiotics confers resistance only to structurally related antibiotics. Esterases, phosphotransferases, acetyltransferases, hydrolases, and nucleotidyltransferases have been identified in strains resistant to members of the MLS_B antibiotics. High-level erythromycin resistance in gram-negative enterics can be attributed to the synthesis of macrolide 2′-phosphotransferases (MPH) or esterases that cleave the macrocyclic lacton (246, 247, 251, 252). *mphA*, *mphB*, and *mphC* encode MPH(2′)I, MPH(2′)II, and MPH(2′)III, respectively (246, 247, 251, 252). MPH(2′)I and II enzymes inactivate only 14- and 15-membered macrolides, whereas MPH(2′)III has extended its phosphorylating capacity to 16-membered macrolides. The plasmid-borne genes *ereA* and *ereB* encode erythromycin esterases that alter the 2′-hydroxy group of 14- and 15-membered macrolides and have been found in clinical strains of *E. coli* (11–13, 21, 258). Inactivation of macrolides in gram-positive pathogens is confined to staphylococci (119, 369). The enzyme encoded by the *S. aureus mphX* gene appears to be an esterase by analysis of the inactivated substrate. However, the amino acid sequence of MphX appears homologous to the superfamily of phosphotransferases (119), retaining both conserved domains seen in enzymes that phosphorylate aminoglycosides, streptogramin A, or viomycin.

Gram-negative bacteria are intrinsically resistant to lincosamides and streptogramin A or B antibiotics, largely owing to endogenous efflux pumps (330). Lincosamide inactivation in enterococci and staphylococci has been described (20, 39, 40, 178, 179). Staphylococci harbor a 3-lincomycin 4-clindamycin O-nucleotidyltransferase encoded by one of two closely related genes, *linA* and *linA′*. *linB*, a gene distinct from either of these, but with similarity to the *B. subtilis* kanamycin nucleotidyltransferase, has been reported in clinical isolates of *E. faecium* (39, 40).

Resistance to streptogramin type A antibiotics in staphylococci is conferred by one or two acetyltransferases, *vat* or *vatB* (3, 6). Both *vat* genes exhibit significant similarity to the *E. faecium satA*, which also encodes a streptogramin A acetyltransferase (289). Other uncharacterized mechanisms mediating streptogramin A resistance are exemplified by *S. aureus* strains that do not contain any of the known resistance determinants (2). Mechanisms specific for streptogramin A may quickly limit the utility of Synercid (176). The *vgb* gene encodes a hydrolase that splits the lactone ring of streptogramin B (180, 181). The *vat* and *vgb* genes often reside together in *S. aureus* on plasmids flanked by two copies of the insertion sequence *IS257* (2, 193).

62.2.4. Tetracyclines

Bacteroides strains can harbor *tetX*, a 44-kDa NADPH-dependent enzyme that inactivates tetracycline by oxidation (321, 322). The *tetX* gene has been found on two closely related *Bacteroides* transposons that also carry an *erm* (321). The TetX protein is active only under aerobic conditions and is not functional in *Bacteroides*. However, its presence and potential for mobility represent an untapped reservoir for newer agents that obviate the more common efflux and target modification mechanisms.

62.3. DRUG EFFLUX/PERMEABILITY CHANGES

Efflux pumps are often responsible for intrinsic resistance to antibiotics. These pumps can range in their substrate specificity from the generalized, such as the *acrB* multidrug resistance efflux pump of *E. coli* (233), to the more specific, tetracycline efflux proteins (182). In either case, the pump protein resides in the cytoplasmic membrane, acting alone or in concert with other proteins to effect efflux. Based on amino acid homology and membrane topology, four classes of efflux pumps are found in bacteria (267) (Table 2); two classes are specific to bacteria. The small multidrug resistance (SMR) family is composed of proteins that have a subunit size of ~100 amino acids and transverse the membrane four times. All members of the SMR family contain a highly conserved glutamate residue within the transmembrane domain (268, 300). The resistance/nodulation/division (RND) family members have a larger subunit size of ~1,000 amino acids encoding 12 membrane-spanning regions. The other two families in which bacterial efflux proteins have been delineated are the ATP-binding cassette (ABC) and the major facilitator (MFS) superfamilies. Members of the ABC superfamily usually have 12 transmembrane segments and are part of a multicomponent apparatus that facilitates efflux of natural and synthetic antibacterial agents. Transporters (≥400 amino acids) that are part of the MFS span the membrane 12 or 14 times. Members of the two superfamilies may require accessory proteins, often encoded in the same operon as the efflux protein, to facilitate transport across the dual membrane topology of gram-negative bacteria. Some efflux systems use a dedicated outer membrane channel, while other systems make use of a common membrane channel (351). A linker protein that is periplasmically located facilitates export between the inner and outer membrane. Energy for three of the four families is derived from the proton motive force. Only the ABC superfamily utilizes ATP hydrolysis to drive efflux. For *E. coli*, *H. influenzae*, and *Mycosplasma genitalium*, 29, 6, and 2 putative pumps, respectively, have been deduced (301). In *E. coli* and *H. influenzae*, MFS-type pumps predominate, whereas *M. genitalium* encodes only ABC-type efflux pumps, as expected, since it lacks an electron transport system requisite for proton motive force generation.

The *E. coli acr* drug efflux system, a member of the RND family, is one of the first bacterial efflux systems studied (231, 232). The operon is composed of two genes, *acrA* and *acrB*, that encode periplasmic and integral membrane proteins, respectively. They act in concert with TolC, an outer membrane porin, to outwardly direct the transport of a broad array of antibiotics to the extracellular environment (199, 200) (Table 2). Although this system probably evolved to protect *E. coli* against bile salts and fatty acids, the *acrAB* · *tolC* pump appears to be responsible for the intrinsic resistance of this organism to commercially available antibacterials. For example, in the absence of a functional pump, cells are four- to eightfold more susceptible to potent quinolones and β-lactams. An *acrAB*-like pump that contributes to erythromycin resistance has been described in *H. influenzae* (303).

TABLE 2 Efflux/impermeability

Resistance gene[a]	Antibiotic(s)	Host species
RND family		
acrAB	Oxazolidinones, streptogramin B, novobiocin, macrolides, fusidic acid, rifampin, penicillins, tetracyclines, puromycin, hygromycin A	*E. coli*
acr-like	Macrolides	*H. influenzae*
acrEF	Same as *acrAB*	*E. coli*
mexAB-oprM	β-Lactams, quinolones, tetracylines, chloramphenicol, cerulenin, thiolactomycin, trimethoprim, β-lactamase inhibitors	*P. aeruginosa*
mexCD-oprJ	Cerulenin, quinolones, tetracyclines, macrolides, novobiocin, chloramphenicol, fourth-generation cephalosporins (cefepime, cefpirome)	*P. aeruginosa*
mexEF-oprN	β-Lactamase inhibitors, quinolones, chloramphenicol, trimethoprim, carbapenems (imipenem)	*P. aeruginosa*
mtrCDE	Erythromycin, rifampin, penicillins	*N. gonorrhoeae*
SMR family		
smr (=*qacC*, *qacD*, *ebv*)	Lipophilic cations	*S. aureus*
qacE	Like Smr	*K. aerogenes*
*qacE*Δ*1*	Like Smr	Gram-negative bacteria
emrE (=*mvrC*, *ebr*)	Tetracylines, erythromycin, sulfadiazine	*E. coli*
tehAB	Crystal violet, proflavin, ethidium bromide, tetraphenylarsonium chloride	*E. coli*
HI0511	Unknown	Homolog of TehA in *H. influenzae*
MFS superfamily		
norA	Hydrophilic quinolones, macrolides, novobiocin, chloramphenicol, fourth-generation cephalosporins (cefepime, cefpirome)	*S. aureus*
qacA	Monovalent and divalent organic cations	*S. aureus*
qacB	Monovalent organic cations	*S. aureus*
mdfA	Rifampin, tetracycline, puromycin, erythromycin, chloramphenicol, aminoglycosides, fluoroquinolones	*E. coli*
emrB	Hydrophobic quinolones, thiolactomycin	*E. coli*
emrD	Uncouplers	*E. coli*
mcbEFG	Sparfloxacin, sulfloxacin, levofloxacin	*E. coli*
lfrA	Hydrophobic quinolones	*M. smegmatis*
efpA	Efflux substrates unknown	MFS-type efflux pump like *norA*; found in mycobacteria
bcr (=*sur*)	Bicylomycin, sulfathiazole	*E. coli*
tetA-E	Tetracyclines	Gram-negative bacteria
tetG	Tetracyclines	*Vibrio anguillarum*
tetH	Tetracyclines	*Pasteurella multocida*
tetI	Tetracyclines	*Escherichia*, *Pseudomonas* spp.
tetK	Tetracyclines	Gram-positive bacteria, *Haemophilus* sp.
tetL	Tetracyclines	*Veillonella* sp.
otrB	Tetracyclines	*Streptomyces* sp.
mefA	Macrolides	*S. pyogenes*, *S. agalactiae*
mefE	Macrolides	*S. pneumoniae*, *E. faecium*
ptr	Pristinamycin I (streptogramin B), pristinamycin II (streptogramin A), rifampin	*Streptomyces pristinaespiralis*

(*continued*)

TABLE 2 (*Continued*)

Resistance gene[a]	Antibiotic(s)	Host species
ABC superfamily		
msrA	Macrolides, streptogramin B	Coagulase-negative staphylococci, *S. aureus*
msrB	Macrolides, streptogramin B	*Staphylococcus xylosus*
vga	Streptogramin A	*S. aureus*, coagulase-negative staphylococci
vgaB	Streptogramin A	*S. aureus*, coagulase-negative staphylococci
Permeability		
ompF, *ompC*	β-Lactams	Outer membrane porin in *E. coli*
oprD	Carbapenems (imipenem)	Outer membrane porin in *P. aeruginosa*
impA	Erythromycin, chloramphenicol, norfloxacin, chenodeoxycholate	*M. smegmatis*, *M. tuberculosis*
Undefined		
Spne	Norfloxacin, ciprofloxacin	*S. pneumoniae*
Efae	Norfloxacin, chloramphenicol	*E. faecalis*, *E. faecium*
Smal	Quinolones, tetracyclines, chloramphenicol	*S. maltophilia*

[a]RND, resistance/nodulation/division; SMR, small multidrug resistance; MFS, major facilitator; ABC, ATP-binding cassette.

AcrAB is regulated by *acrR*, *marA*, *soxS*, and *robA* (Table 3) (107, 175). AcrR, homologous to TetR, encodes a repressor that is found upstream of the *acrAB* operon. MarA and SoxS are transcriptional activators whose synthesis is primarily increased in response to antibiotics (and other hydrophobic agents) or environmental stress, respectively. Along with MarR and SoxR, they constitute examples of two-component signaling systems (221, 254). The interplay between the two systems is highlighted by the partial dependence of a SoxR constitutive mutant on a functional *mar* locus to promote antibiotic resistance and the failure of MarA, even when overexpressed, to comple-

TABLE 3 Global regulatory elements and/or transcriptional regulators

Resistance gene	Function	Host species
marA	Global regulator of *mar* and *acrAB*, among others	*E. coli*
soxS	Global regulator of *soxRS*	*E. coli*
ramA	Transcriptional activator of RomA (?) or AcrAB	*K. pneumoniae*, *E. cloacae*, *Serratia* sp.
pqrA	Transcriptional activator, resulting in Mdr phenotype	*P. vulgaris*
robA	Transcriptional activator, resulting in Mdr phenotype	*E. coli*, *E. cloacae*; others
Cje	Global regulator? Efflux?	*C. jejuni*
acrR	Transcriptional repressor of *acrAB*	*E. coli*; others
mexR (=*nalB*?)	Transcriptional regulator of *mexAB* · *oprM*	*P. aeruginosa*
nfxB	Transcriptional repressor of *mexCD* · *oprJ*	*P. aeruginosa*
nfxC	Transcriptional activator of *mexEF* · *oprN*	*P. aeruginosa*
emrR (=*mprA*)	Transcriptional repressor of *emrAB* and *mar*; activator of microcin cluster (*mcb*)	*E. coli* *S. typhimurium*
mtrR	Transcriptional repressor of *mtrCDE*	*N. gonorrhoeae*
norR	Transcriptional repressor of *norA*	*S. aureus*
qacR	Transcriptional repressor of *qacA*	*S. aureus*
tetR	Transcriptional; repressor of *tet* efflux genes	Multiple species

ment a *soxRS* deletion in response to certain superoxide-enhancing (and stressful) agents. Both SoxS and MarA activate the expression of *acrAB* and *micF*, an antisense mRNA that negatively regulates the synthesis of OmpF, a major porin in *E. coli* (221, 254). The cell attempts to balance the down-regulation of OmpF by increasing OmpC synthesis; however, the pore size of OmpC is inherently smaller. Thus, the slower rate of drug entry, in combination with an up-regulated intrinsic efflux system that has a broad substrate specificity, allows the cell to maintain resistance to multiple antibiotic classes (243). The RobA protein in *E. coli* is homologous to SoxS and MarA and uses the AcrAB efflux system and activation of *micF* to effect a multidrug-resistant phenotype (107, 338). All three regulons can confer resistance to many antibiotics; however, the phenotype is largely mediated through regulation of the two common target genes, *acrAB* and *micF* (200, 221, 338). *E. cloacae* also encodes a *RobA*-like protein (83% identical) that has been shown to activate *micF* and, presumably, the *acrAB* equivalent in this species. Other homologs of *acrAB* exist in *E. coli* (244). AcrEF encodes a periplasmic linker and RND pump, respectively. Although this pump is not expressed to a significant degree under laboratory conditions, overexpression of this locus can rescue the hypersusceptible phenotype of *acrAB* mutants.

A superfamily of genes with homology to MarR and SoxS is present in bacteria, with many of its members involved in the bacterial adaptive response to host signals (221). A second *mar* locus was identified in *E. coli*; mutations in *mprA* (recently renamed *emrR*) result in a multidrug-resistant phenotype (192, 327). EmrR is located upstream of *emrAB*, a multidrug-efflux pump (see below). It is a transcriptional repressor of *emrAB* and the microcin biosynthetic (*mcb*) operon. Interestingly, EmrR is a positive activator of an internal promoter, P2, of the *mcb* operon, leading to enhanced synthesis of *mcbEFG*, components of the microcin pump (191). RamA has been described in *K. pneumoniae* and *E. cloacae* as a transcriptional activator of RomA, an outer membrane protein, and, possibly other AcrAB-type efflux systems (108). In *Proteus vulgaris*, mutations in *pqrA*, a transcriptional activator, result in multidrug resistance (107, 146). Recently, a mutation characterized by drug efflux of ciprofloxacin, minocycline, and pefloxacin and the appearance of two novel outer membrane proteins has been described in mutants of *Campylobacter jejuni* selected in vitro by either pefloxacin or cefotaxime (58, 107). These mutations are cross resistant to penicillins, cephalosporins, chloramphenicol, novobiocin, carbapenems, and macrolides as well. The complexity of efflux pump regulation exemplifies the capacity of bacteria to respond efficiently to noxious environmental agents, including antibiotics.

P. aeruginosa demonstrates intrinsic resistance to a large number of antibiotics that are typically effective against other gram-negative bacteria. It was originally thought that the outer cell membrane of this bacterium, which exhibits low permeability to low-molecular-weight hydrophobic molecules, was solely responsible for the intrinsic resistance (9, 243). However, existence of multidrug efflux systems in *P. aeruginosa* is now recognized as a major contributor to intrinsic resistance in this species (184, 244). The first multidrug efflux operon identified in *P. aeruginosa* was the *mexAB* · *oprM* system (117), a member of the RND family. MexA is a lipoprotein that facilitates efflux by linking the functions of MexB, the cytoplasmic membrane pump, with OprM, the outer membrane component. This pump was found to be partially dependent upon TonB, an energy-coupling protein that has been implicated in the opening (gating) of outer membrane channels responsible for iron-siderophore uptake (185). Upstream from the *mexAB* operon is *mexR*, a repressor with homology to MarR but not AcrR, which directly represses expression of *mexAB* · *oprM* and of itself. However, a single amino acid substitution (R69T) endows *mexR* with activator activity, as observed in *nalB* mutants (278). Whether another mutation outside of *mexR* contributes to the NalB phenotype is not entirely clear (278). The MexAB-OprM system has also recently been shown to export β-lactamase inhibitors as well as β-lactam antibiotics (185).

Two other RND-type efflux systems have been uncovered in *P. aeruginosa*, *mexDC* · *oprJ* and *mexEF* · *oprN* (167, 217). *mexCd* · *oprJ* is under negative transcriptional control by *nfxB*, a gene that encodes a repressor and maps in a divergent operon upstream from the efflux components (277, 312). Thus, the multidrug resistance phenotype of a *nfxB* mutant is a result of overexpression of the *mexCD* · *oprJ* operon and includes increased resistance to tetracyclines, chloramphenicol, macrolides, quinolones, and fourth-generation cephalosporins (cefepime, cefpirome) but hypersusceptibility to most other β-lactam antibiotics. How the system extrudes cephalosporins, which do not cross the cytoplasmic membrane, remains unclear. The *mexEF* · *oprN* system was uncovered by defining mutations that conferred increased resistance to the antibiotic sparfloxacin (167). Other substrates for the *mexEF* · *oprN* efflux system include chloramphenicol, trimethoprim, quinolones, and carbapenems, including imipenem (Table 2). However, the resistance to imipenem results from the loss of OprD, a porin protein that is down-regulated in *mexEF* · *oprN* mutants (281). There is evidence that the true function of this system is involved in the excretion of pyocyanin, a potentially toxic secondary metabolite generated during the biosynthesis of aromatic compounds. Unlike the *mexCD* · *oprJ* system, this system is under positive regulation by a protein belonging to the LysR family of transcriptional activators, NfxC. Further, *mexCD* · *oprJ* and *mexEF* · *oprN* are not expressed constitutively like *mexAB* · *oprM*. However, in vitro selection with quinolones yields mutations resulting in hyperexpression of any one of the three efflux systems, with *mexCD* · *oprJ* preferentially affected when the selecting agent is a fluoroquinolone with a piperazine ring (168).

The MtrR repressor of *N. gonorrhoeae* regulates the transcription of the *mtrCDE* operon (128). The *mtrR* gene is located downstream of *mtrCDE* and is transcribed in a divergent manner (195). Clinical strains resistant to erythromycin were most often found to have a single-base-pair deletion in the 13-bp inverted repeat located within the promoter regions between the two operons (195). This pump protects cells against erythromycin and other naturally occurring hydrophobic agents, such as bile salts and fatty acids. Higher-level resistance to hydrophobic agents was also found to depend on the production of the full-length lipooligosaccharide (196).

Several SMR family members reside in gram-negative bacteria. An efflux system identified in *E. coli*, EmrE, belongs to the SMR family (183). This protein, also known as MvrC or Ebr, is chromosomally encoded. There is no known transmembrane linker or outer membrane channel associated with this transporter. Moreover, reconstituted EmrE catalyzes efflux of erythromycin and tetracycline, in addition to lipophilic monovalent cations (378). EmrE may be the ethidium and phosphonium efflux system described by Midgley (267). A member of this family found in *S. aureus*, Smr, has also been purified, reconstituted into liposomes, and found to be an independent transporter of lipophilic cations (123). While it is clear that the electrochemical gradient provides the energy to fuel these pumps, the oligomeric structure that constitutes a functional EmrE or Smr pump is not known (300). The *Klebsiella aerogenes qacE* gene was originally found as part of an integron on plasmid R751. A disrupted derivative of this gene, *qacEΔ1*, is widespread among gram-negative species, largely owing to its residency at the 3′ end of the integron (266). An additional member of the SMR family, TehA, has been described in *E. coli* (347). The 36-kDa pump encodes a protein with 10 (rather than 4) transmembrane segments (TMS) that works most efficiently in conjunction with TehB. The TehA protein contains a conserved glutamic acid in the first TMS and is functionally similar to other SMR members that export organic cations. Although resistance to known antibiotics has not been mapped to this locus, this locus may become an obstacle for new antimicrobial classes. *H. influenzae* also appears to encode a close homolog to TehA, HI0511 (267).

The MFS family of transporters includes multidrug-resistant pumps like NorA, QacA, and QacB in *S. aureus* and MdfA, EmrAB, EmrD, McbEFG, and Bcr in *E. coli*, as well as efflux pumps that appear to have a more restricted substrate specificity, i.e., the tetracycline-specific pumps, the macrolide-specific pumps, chloramphenicol pumps, and some quinolone-specific pumps (175, 243, 300, 301) (Table 2). NorA encodes an efflux determinant for fluoroquinolone resistance, with specificity for hydrophilic quinolones (156). The *norA* gene resides on the *S. aureus* chromosome, and the resistance phenotype results from a promoter-up mutation. Single-base-pair changes within the coding sequence can alter the specificity of the pump for norfloxacin, i.e., a change from alanine to aspartic acid at position 362 reduces norfloxacin resistance (300). Just upstream from *norA* is *norR*, a potential repressor of *norA*, which has amino acid homology to TetR and QacR (see below) (267). The regulation of a related pump in *B. subtilis*, *bmr*, has been elucidated and serves as a useful paradigm for gram-positive MFS transporters (239). The *qacA* and *qacB* multidrug efflux genes from *S. aureus* differ by seven nucleotides and confer resistance to organic cations and disinfectants such as cetrimide, benzalkonium chloride, and chlorohexidine (268). A transcriptional repressor, QacR, with homology to TetR, has been identified upstream of *qacA* (266). QacB, unlike QacA, extrudes monovalent cations rather than divalent cations and is harbored on the heavy metal resistance plasmid pSK23 (198). An alanine at position 323 in QacB versus the corresponding aspartate in QacA is responsible for monovalent cation specificity (266). Both QacA and QacB are members of a subgroup of MFS proteins that contain 14 TMS domains rather than the 12 in NorA. Like NorA, the substrate specificity of QacA/B can also be altered by coding sequence changes, which often lie in the C-terminal regions (300).

In *E. coli*, the MdfA pump confers resistance to cationic or zwitterionic lipophilic antibiotics such as rifampin, tetracycline, and puromycin as well as to the uncharged antibiotic chloramphenicol and to nonaromatic compounds like erythromycin and, to a lesser extent, the aminoglycosides kanamycin and neomycin (87). Although many models depict an interaction of multidrug-resistant pumps with their substrates from within the bilayer, MdfA may interact with some of its more hydrophilic substrates in the aqueous environment. MdfA is highly hydrophobic but contains the acidic glutamate residue in TMS 1 as is found in other transporters, most notably, members of the SMR family (268, 300).

EmrB confers resistance to nalidixic acid and thiolactomycin in *E. coli* (183). Although EmrA, the periplasmically located linker protein, is not required for efflux, EmrA can, along with presumably TolC, facilitate export of substrates outside the cell (192). EmrD is an MFS transporter that was identified as a protein involved in cellular adaptation to low energy shock induced by uncouplers (235). Like EmrB, it likely uses an accessory protein, EmrC, and an outer membrane porin to optimally extrude noxious substrates (183, 244). The McbEFG pump, previously regarded as an export and immunity system specific for microcin B17, has been shown to confer resistance to a restricted set of quinolones (191) (Table 2). The *M. smegmatis* LfrA protein, with related homologs in *M. tuberculosis* and *M. avium*, extrudes hydrophilic quinolones (187, 337). EfpA, a putative efflux pump with similarity to LfrA and QacA, is found in both drug-susceptible and drug-resistant strains of *M. tuberculosis* and in some strains of *M. avium*, *Mycobacterium bovis*, and *M. intracellulare* (83). The substrate(s) for this pump remains elusive. Bcr is a chromosomally encoded pump in *E. coli* that confers resistance to bicyclomycin and sulfathiazole (183).

Efflux is a major mechanism of tetracycline resistance in clinical strains, and tetracycline-specific pumps are found in many bacterial species (182). Aerobic and anaerobic gram-negative bacteria most commonly contain one of five classes of tetracycline pumps, encoded by *tetA-E*. In addition to these five efflux determinants, TetG has been identified in *Vibrio anguillarum*, TetH in *Pasteurella multocida*, and TetI in *E. coli* and *P. aeruginosa* (267, 293). Although different classes of tetracyline pumps can be defined by hybridization, these MFS pumps apparently function as electroneutral antiporters, effluxing tetracycline complexed with a metal ion (like magnesium) in exchange for a proton (175, 293, 306). The genes are generally inducible, with the tetracycline molecule interacting directly with the dimeric form of a repressor protein to effect induction (32). TetK and TetL are found primarily in gram-positive species, including clinical strains of *Staphylococcus*, *Enterococcus*, *Streptococcus*, *Streptomyces*, *Nocardia*, *Clostridium*, *Listeria*, *Peptostreptococcus*, and *Mycobacterium* spp. (293). Examples of genera with gram-negative strains harboring pumps more typically found in gram-positive strains include *Haemophilus* (TetK) and *Veillonella* (TetL) (293). The OtrB protein, originally described in *Streptomyces* spp. producing oxytetracycline, is also found in pathogenic species of *Streptomyces* and *Mycobacterium* (262). All of the tetracycline-specific pumps confer resistance to tetracycline but not to minocycline, with the exception of TetB (293, 306). The glycylcyclines obviate the efflux systems (28, 320).

An efflux pump, designated *mef* was recently identified in clinical isolates of *S. pneumoniae* and *S. pyogenes* (61, 331, 336). Sequencing of the genes from pneumococci or group A streptococci revealed two genes, *mefE* and *mefA*, respectively, that are 88% identical. The MefE protein was demonstrated to efflux only 14- and 15-membered macrolides, and resistance to 16-membered macrolides was not seen even in the presence of inducing concentrations of erythromycin (61, 336). Strains harboring the *mef* gene remain susceptible to streptogramin B and clindamycin. Interestingly, the Mef protein conferred macrolide resistance when cloned into *E. coli*. More recently, *mefA* has been described in *S. agalactiae* (10) and in group C and group G streptococci (159), and *mefE* has been found in moderately erythromycin-resistant strains of *E. faecium* (102).

S. epidermidis harbors a plasmid with an *msrA* determinant, an ABC-type transporter specific for 14- and 15-membered macrolides and streptogramin B (295, 296). A nearly identical gene resides in *S. aureus*, although it appears to be uncommon (119, 212, 329). Although *msrA* contains the typical "6 by 6" (six TMS separated by a Q-linker from a second set of nearly identical six TMS), MsrB in *S. xylosus* contains only six TMS and likely functions as a dimer (222). The Vga and VgaB pumps in *S. aureus* and coagulase-negative staphylococci efflux streptogramin A and represent a threat to the newly launched Synercid (4,5). Interestingly, the *S. pristinaespiralis ptr* gene extrudes not only pristinamycin I (streptogramin B-type) and pristinamycin II (streptogramin A-type) but also rifampin (267). However, Ptr is a member of the MFS rather than the ABC transporters. Homologs that encode macrolide-specific pumps are found in macrolide producer strains; they belong to the superfamily of ABC transporters (243, 301).

Other species that have efflux pumps thought to be responsible for intrinsic resistance to certain antibiotics include *S. pneumoniae*, *E. faecalis*, and *E. faecium* (19, 197). The susceptibility of *S. pneumoniae* to the fluoroquinolones norfloxacin and ciprofloxacin is enhanced in the presence of reserpine, a plant alkaloid that interferes with multidrug transporters (19). Further, selection of cells resistant to ethidium results in coresistance to fluoroquinolones. Strains of enterococci were shown to have energy-driven efflux of norfloxacin, tetracycline, and chloramphenicol (197). The determinants responsible for the intrinsic resistance in pneumococci or enterococci have not been characterized (197). An endogenous pump that confers multidrug resistance to quinolones, tetracycline, and chloramphenicol was also detected in *S. maltophilia* (8).

Permeability in *M. tuberculosis* and *M. smegmatis* can be altered by changing the synthesis of inositol, an important component in lipoarabinomannan in the cell envelope. Mutations in *impA*, the gene encoding inositol monophosphate phosphatase, render *M. smegmatis* more resistant to chloramphenicol, erythromycin, norfloxacin, and chenodeoxycholate (263).

The literature has recently been saturated with a number of excellent reviews on microbial efflux pumps. Readers should consult these articles for a more detailed discussion of specific efflux paradigms (107, 138, 175, 182, 183, 267, 300–302, 351).

62.4. IDENTIFICATION OF RESISTANCE DETERMINANTS

As highlighted above, studies of a wide variety of bacterial pathogens have identified numerous genetic loci associated with antibiotic resistance. Investigators have used DNA-based technologies ranging from traditional cloning approaches to the application of cutting-edge genetic techniques to identify antibiotic resistance determinants in bacterial pathogens. Historically, the primary approach used to elucidate the molecular mechanism of resistance determinants has involved direct molecular cloning of bacterial resistance genes. Experimentally, genomic libraries of the resistant pathogen are constructed, and vectors containing random fragments from the library are introduced into a susceptible recipient bacterium. The identification of the resistance determinant is usually straightforward as resistance is often a dominant selection. Standard backcross experiments confirm that the recombinant plasmid can transfer antibiotic resistance to susceptible strains and define the limits of the resistance cassette. The initial identification can be further simplified if the bacterial resistance is plasmid mediated and can be transferred by natural genetic exchange (i.e., conjugation).

In addition to identifying novel antibiotic resistance-associated genetic loci by direct selection, rapid and definitive molecular genetic techniques have been adapted for detection of specific genetic determinants associated with antibiotic resistance. For example, investigators have used hybridization or PCR techniques to detect genes that encode antibiotic resistance (273, 274). Persing and colleagues have compiled a series of excellent and comprehensive articles that provide detailed information on the approach and discuss the practical applications, strengths, and weaknesses of each PCR application (273, 274). Such target genes include *S. aureus mecA* (227), *S. pneumoniae* PBPs 2b and 2x (120, 129), *Enterococcus vanABC* (103), genes coding for aminoglycoside-modifying enzymes (350), *Enterobacteriaceae tem/shv* (201), *erm* genes (15, 329), and the *rpoB* and *katG* genes of *M. tuberculosis* (145, 349, 366). El Solh and colleagues have published PCR primers for *vga*, *vgb*, *vat*, *vatB*, and *satA* (2, 194). PCR primers have been identified that detect tetracycline efflux (261) and RP genes (62, 294) from a wide variety of microbes. Bergeron and coworkers have also developed PCR assays specific for 25 clinically relevant antibiotic resistance genes associated with various bacterial pathogens (21). Positive PCR identification of known resistance determinants requires confirmation by additional methods, including DNA sequence analysis. The molecular detection of mutations that confer antibiotic resistance can be confirmed by single-strand conformational polymorphism analysis (94) or dideoxy fingerprinting (94).

Advanced DNA-based methods such as heteroduplex analysis, solid-phase hybridization, single-stranded conformation polymorphism, and direct genome sequencing are becoming available and can be included in one's arsenal of approaches to define antibiotic resistance determinants in emerging and reemerging pathogens. These genetic methods are clearly limited in that not all resistance mechanisms are known and each requires knowledge of the genes encoding the drug target and mutations associated with specific resistance phenotypes. For some types of antibacterial resistance, the diversity of responsible genetic mechanisms exceeds the capabilities of current detection technology. Nevertheless, current methodologies facilitate rapid determination of known resistance determinants in emerging bacterial pathogens. Moreover, sensitive and rapid genotypic detection of specific pathogens and associated drug resistance may ultimately lead to more judicious use of antimicrobials and the development of narrow-spectrum antibacterial agents.

62.5. CONCLUSION

Bacteria have developed numerous and often elegant mechanisms to resist the action of antibiotics. Study of the molecular mechanisms and epidemiology of antibiotic-resistant bacteria has revealed the extraordinary versatility of human pathogenic bacteria in their adaptations to survive and even thrive in an antibacterial-rich environment. Clearly, the elucidation of the biochemical and molecular mechanisms of resistance is critical to the design of effective new antibiotics. Multidrug resistance is increasing, with strains armed with resistance determinants to virtually all antibacterial agents. Fortunately, a substantial effort is being invested in the discovery of new classes of novel antibacterial agents that may be able to address and overcome the problem of multidrug-resistant bacteria. Not only will early detection of resistance genes aid in the discovery of new members of current drug classes, but use of strains with well-characterized resistance determinants will help in the discovery of novel classes that obviate the known resistance mechanisms. One approach to avoid encountering preexisting antibiotic resistance is to seek new targets against which antibacterial agents have not been previously developed. New technologies like combinatorial chemistry, genomics for identifying and evaluating new antibiotic targets, and advances in structural biology will accelerate the process of antibacterial drug discovery. In addition to the challenge facing the biopharmaceutical industry, the medical community can help by using new agents judiciously to minimize the selection of preexisting and novel resistant determinants.

We apologize that space constraints sometimes obligated us to cite reviews rather than primary research papers. We thank Paul F. Miller for a critical review of the manuscript.

REFERENCES

1. **Æevarsson, A., E. Brazhnikov, M. Garber, J. Zheltonosova, Y. Chirgadze, S. Al-Karadaghi, L. A. Svensson, and A. Liljas.** 1994. Three-dimensional structure of the ribosomal translocase: elongation factor G from *Thermus thermophilus*. *EMBO J.* **13:**3669–3677.
2. **Allignet, J., S. Aubert, A. Morvan, and N. el Solh.** 1996. Distribution of genes encoding resistance to streptogramin A and related compounds among staphylococci resistant to these antibiotics. *Antimicrob. Agents Chemother.* **40:** 2523–2528.
3. **Allignet, J., and N. el Solh.** 1995. Diversity among gram-positive acetyltransferases inactivating streptogramin A and structurally related compounds and characterization of a new staphylococcal determinant, *vatB*. *Antimicrob. Agents Chemother.* **39:**2027–2036.
4. **Allignet, J., and N. el Solh.** 1997. Characterization of a new staphylococcal gene, *vgaB*, encoding a putative ABC transporter conferring resistance to streptogramin A and related compounds. *Gene* **202:**133–138.
5. **Allignet, J., V. Loncle, and N. el Solh.** 1992. Sequence of a staphylococcal plasmid gene, *vga*, encoding a putative ATP-binding protein involved in resistance to virginiamycin A-like antibiotics. *Gene* **117:**45–51.
6. **Allignet, J., V. Loncle, C. Simenel, M. Delepierre, and N. el Solh.** 1993. Sequence of a staphylococcal gene, *vat*, encoding an acetyltransferase inactivating the A-type compounds of virginiamycin-like antibiotics. *Gene* **130:** 91–98.
7. **Al-Obeid, S., L. Gutmann, and R. Williamson.** 1990. Modification of penicillin-binding proteins of penicillin-resistant mutants of different species of enterococci. *J. Antimicrob. Chemother.* **26:**613–618.
8. **Alonso, A., and J. L. Martínez.** 1997. Multiple antibiotic resistance in *Stenotrophomonas maltophilia*. *Antimicrob. Agents Chemother.* **41:**1140–1142.
9. **Angus, B. L., A. M. Carey, D. A. Caron, A. M. B. Kropinski, and R. E. W. Hancock.** 1982. Outer membrane permeability in *Pseudomonas aeruginosa*: comparison of a wild type and antibiotic supersusceptibile mutant. *Antimicrob. Agents Chemother.* **21:**299–309.
10. **Arpin, C., H. Daube, and C. Quentin.** 1997. *Streptococcus agalactiae* resistant to macrolides but sensitive to lincosamides harbor the novel resistance efflux mechanism, *mefA*, abstr. C-73, p. 58. *In* Program and Abstracts of the 37th Interscience Conference on Antimicrobial Agents and Chemotherapy. American Society for Microbiology, Washington, D.C.
11. **Arthur, M., A. Andermont, and P. Courvalin.** 1987. Distribution of erythromycin esterase and rRNA methylase genes in members of the family *Enterobacteriaceae* highly resistant to erythromycin. *Antimicrob. Agents Chemother.* **31:**404–409.
12. **Arthur, M., D. Autissier, and P. Courvalin.** 1986. Analysis of the nucleotide sequence of the *ereB* gene encoding the erythromycin esterase type II. *Nucleic Acids Res.* **14:** 4987–4999.
13. **Arthur, M., and P. Courvalin.** 1986. Contribution of two different mechanisms to erythromycin resistance in *Escherichia coli*. *Antimicrob. Agents Chemother.* **30:**694–700.
14. **Arthur, M., F. Depardieu, C. Molinas, P. Reynolds, and P. Courvalin.** 1995. The *vanZ* gene from *Enterococcus faecium* BM4147 confers resistance to teicoplanin. *Gene* **154:** 87–92.
15. **Arthur, M., C. Molinas, C. Mabilat, and P. Courvalin.** 1990. Detection of erythromycin resistance by the polymerase chain reaction using primers in conserved regions of *erm* rRNA methylase genes. *Antimicrob. Agents Chemother.* **34:**2024–2026.
16. **Arthur, M., P. Reynolds, and P. Courvalin.** 1996. Glycopeptide resistance in enterococci. *Trends Microbiol.* **4:** 401–407.
17. **Arthur, M., P. E. Reynolds, F. Depardieu, S. Evers, S. Dutka-Malen, R. Quintiliani, Jr., and P. Courvalin.** 1996. Mechanisms of glycopeptide resistance in enterococci. *J. Infect.* **32:**11–16.
18. **Bandoh, K., K. Wantanabe, Y. Muto, Y. Tanaka, N. Kato, and K. Ueno.** 1992. Conjugal transfer of imipenem resistance in *Bacteroides fragilis*. *J. Antiobiot.* **45:**542–547.
19. **Baranova, N. N., and A. A. Neyfakh.** 1997. Apparent involvement of a multidrug transporter in the fluoroquinolone resistance of *Streptococcus pneumoniae*. *Antimicrob. Agents Chemother.* **41:**1396–1398.
20. **Barcs, I.** 1993. Different kinetics of enzymatic inactivation of lincomycin and clindamycin in *Staphylococcus aureus*. *J Chemother.* **5:**215–222.
21. **Barthelemy, P., D. Autissier, G. Gerbaud, and P. Courvalin.** 1984. Enzymatic hydrolysis of erthromycin by a strain of *Escherichia coli*: a new mechanism of resistance. *J. Antibiot.* **37:**1692–1696.
22. **Bauernfeind, A., I. Stemplinger, R. Jungwirth, P. Mangold, S. Amann, E. Akalin, O. Ang, C. Bal, and J. M.**

Casellas. 1996. Characterization of β-lactamase gene *bla* PER-2, which encodes extended-spectrum class A β-lactamase. *Antimicrob. Agents Chemother.* **40:**616–620.

23. **Bauernfeind, A., I. Stemplinger, R. Jungwirth, R. Wilhelm, and Y. Chong.** 1996. Comparative characterization of the cephamycinase bla_{CMY-1} gene and its relationship with other β-lactamase genes. *Antimicrob. Agents Chemother.* **40:**1926–1930.
24. **Bauernfeind, A., S. Wagner, R. Jungwirth, R. Wilhelm, and Y. Chong.** 1997. A novel class C β-lactamase (FOX-2) in *Escherichia coli* conferring resistance to cephamycins. *Antimicrob. Agents Chemother.* **41:**2041–2046.
25. **Beauregard, D. A., D. H. Williams, M. N. Gwynn, and D. J. Knowles.** 1995. Dimerization and membrane anchors in extracellular targeting of vancomycin group antibiotics. *Antimicrob. Agents Chemother.* **39:**781–785.
26. **Berger-Bächi, B.** 1997. Strategies of methicillin resistance in *Staphylococcus aureus*. *Biospektrum* **1997:**17–20.
27. **Berger-Bächi, B., A. Strässle, J. E. Gustafson, and F. H. Kayser.** 1992. Mapping and characterization of multiple chromosomal factors involved in methicillin resistance in *Staphylococcus aureus*. *Antimicrob. Agents Chemother.* **36:** 1367–1373.
28. **Bergeron, J., M. Ammirati, D. Danley, L. James, M. Norcia, J. Retsema, C. A. Strick, W.-G. Su, J. Sutcliffe, and L. Wondrack.** 1996. Glycylcyclines bind to the high-affinity tetracycline ribosomal binding site and evade Tet(M)- and Tet(O)- mediated ribosomal protection. *Antimicrob. Agents Chemother.* **40:**2226–2228.
29. **Berryman, D., and J. I. Rood.** 1995. The closely related *ermB-ermAM* genes from *Clostridium perfringens*, *Enterococcus faecalis* (pAMβ1), and *Streptococcus agalactiae* (pIP501) are flanked by variants of a directly repeated sequence. *Antimicrob. Agents Chemother.* **39:**1830–1834.
30. **Berryman, D. I., M. Lyristis, and J. I. Rood.** 1994. Cloning and sequence analysis of *ermQ*, the predominant macrolide-lincosamide-streptogramin B resistance gene in *Clostridium perfringens*. *Antimicrob. Agents Chemother.* **38:** 1041–1046.
31. **Beyer, D., and K. Pepper.** 1998. The streptogramin antibiotics: update on their mechanism of action. *Exp. Opin. Invest. Drugs* **7:**591–599.
32. **Biburger, M., C. Berens, T. Lederer, T. Krec, and W. Hillen.** 1998. Intragenic suppressors of induction-deficient TetR mutants: localization and potential mechanism of action. *J. Bacteriol.* **180:**737–741.
33. **Billot-Klein, D., L. Gutmann, D. Bryant, D. Bell, J. van Heijenoort, J. Grewal, and D. M. Shlaes.** 1996. Peptidoglycan synthesis and structure in *Staphylococcus haemolyticus* expressing increasing levels of resistance to glycopeptide antibiotics. *J. Bacteriol.* **178:**4696–4703.
34. **Billot-Klein, D., L. Gutmann, S. Sable, E. Guittet, and J. van Heijenoort.** 1994. Modification of peptidoglycan precursors is a common feature of the low-level vancomycin-resistant VanB-type *Enterococcus* D366 and of the naturally glycopeptide-resistant species *Lactobacillus casei*, *Pediococcus pentosaceus*, *Leuconostoc mesenteroides*, and *Enterococcus gallinarum*. *J. Bacteriol.* **176:**2398–2405.
35. **Billot-Klein, D., R. Legrand, B. Schoot, J. van Heijenoort, and L. Gutmann.** 1997. Peptidoglycan structure of *Lactobacillus casei*, a species highly resistant to glycopeptide antibiotics. *J. Bacteriol.* **179:**6208–6212.
36. **Black, D. S., B. Irwin, and H. S. Moyed.** 1994. Autoregulation of *hip*, an operon that affects lethality due to inhibition of peptidoglycan of DNA synthesis. *J. Bacteriol.* **176:**4081–4091.
37. **Blazquez, J. M. R. Baquero, R. Canton, I. Alos, and F. Baquero.** 1993. Characterization of a new TEM-type beta-lactamase resistant to calvulanate, sulbactam, and tazobactam in a clinical isolate of *Escherichia coli*. *Antimicrob. Agents Chemother.* **37:**2059–2063.
38. **Boyle-Vavra, S., B. L. M. de Jonge, C. C. Ebert, and R. S. Daum.** 1997. Cloning of the *Staphylococcus aureus ddh* gene encoding NAD^+-dependent D-lactate dehydrogenase and insertional inactivation in a glycopeptide-resistant isolate. *J. Bacteriol.* **179:**6756–6763.
39. **Bozdogan, B., L. Berrezouga, and R. LeClercq.** 1997. A new resistance gene, *linB*, conferring resistance to lincomycin be inactivation in *Enterococcus faecium* HM1025, abst. C-66, p. 57. *In* Program and Abstracts of the 37th Interscience Conference on Antimicrobial Agents and Chemotherapy. American Society for Microbiology, Washington, D.C.
40. **Bozdogan, B., L. Berrezouga, and R. Leclercq.** 1997. Resistance to lincosamides by nucleotidylation associated with conjugative transfer of a large chromosomal element in *Enterococcus faecium*. *Adv. Exp. Med. Biol.* **418:**491–493.
41. **Bradford, P. A., C. Urban, N. Mariano, S. J. Projan, J. J. Rahal, and K. Bush.** 1997. Imipenem resistance in *Klebsiella pneumoniae* is associated with the combination of ACT-1, a plasmid-mediated AmpC β-lactamase, and the loss of an outer membrane protein. *Antimicrob. Agents Chemother.* **41:**563–569.
42. **Brakstad, O. B., and J. A. Mæland.** Mechanisms of methicillin resistance in staphylococci. *APMIS* **105:**264–276.
43. **Breines, D. M., S. Ouabdesselam, E. Y. Eg, J. Tankovic, S. Shah, C. J. Sousy, and D. C. Hooper.** 1997. Quinolone resistance locus *nfxD* of *Escherichia coli* is a mutant allele of the *parE* gene encoding a subunit of topoisomerase IV. *Antimicrob. Agents Chemother.* **41:**175–179.
44. **Brighty, K. E., and T. D. Gootz.** 1997. The chemistry and biological profile of trovafloxacin. *J. Antimicrob. Chemother.* **39:**1–14.
45. **Bryan, L. E., S. K. Kowand, and H. M. Van Den Elzen.** 1979. Mechanism of aminoglycoside antibiotic resistance in anaerobic bacteria: *Clostridium perfringens* and *Bacteroides fragilis*. *Antimicrob. Agents Chemother.* **15:**7–13.
46. **Burdett, V.** 1986. Streptococcal tetracycline resistance mediated at the level of protein synthesis. *J. Bacteriol.* **165:**564–569.
47. **Burdett, V.** 1991. Purification and characterization of Tet(M), a protein that renders ribosomes resistant to tetracycline. *J. Biol. Chem.* **266:**2872–2877.
48. **Burdett, V.** 1993. tRNA modification activity is necessary for Tet(M)-mediated tetracycline resistance. *J. Bacteriol.* **175:**7209–7215.
49. **Burdett, V.** 1996. Tet(M)-promoted release of tetracycline from ribosomes is GTP dependent. *J. Bacteriol.* **178:**3246–3251.
50. **Bush, K.** 1996. Is it important to identify extended-spectrum beta-lactamase-producing isolates? *Eur. J. Clin. Microbiol. Infect. Dis.* **15:**361–364.

51. **Bush, K., G. A. Jacoby, and A. A. Medeiros.** 1995. A functional classification scheme for β-lactamases and its correlation with molecular structures. *Antimicrob. Agents Chemother.* **39:**1211–1233.

52. **Bush, L. M., J. Calmon, C. L. Cherney, M. Wendeler, P. Pitsakis, J. Poupard, M. E. Levison, and C. C. Johnson.** 1989. High-level penicillin resistance among isolates of enterococci. *Ann. Intern. Med.* **110:**515–520.

53. **Calandra, G., F. Ricci, C. Wang, and K. Brown.** 1986. Cross-resistance and imipenem. *Lancet* **ii:**340–341.

54. **Calcutt, M. J., and E. Cundliffe.** 1990. Cloning of a lincosamide resistance determinant from *Streptomyces caelestis*, the producer of celesticetin, and characterization of the resistance mechanism. *J. Bacteriol.* **172:**4710–4714.

55. **Casin, I. F. Bordon, P. Bertin, A. Coutrot, I. Podglajen, R. Brasseur, and E. Collatz.** 1998. Aminoglycoside 6′-*N*-acetlytransferase variants of the Ib Type with altered substrate profile in clinical isolates of *Enterobacter cloacae* and *Citrobacter freundii*. *Antimicrob. Agents Chemother.* **42:** 209–215.

56. **Chambers, H. F.** 1997. Methicillin resistance in staphylococci: molecular and biochemical basis and clinical implications. *Clin. Microbiol. Rev.* **10:**781–791.

57. **Champney, W. S., and R. Burdine.** 1996. 50S ribosomal subunit synthesis and translation are equivalent targets for erythromycin inhibition in *Staphylococcus aureus*. *Antimicrob. Agents Chemother.* **40:**1301–1303.

58. **Charvalos, E., Y. Tselentis, M. M. Hamzehpour, T. Köhler, and J.-C. Pechere.** 1995. Evidence for an efflux pump in multidrug-resistant *Campylobacter jejuni*. *Antimicrob. Agents Chemother.* **39:**2019–2022.

59. **Chopra, I., P. M. Hawkey, and M. Hinton.** 1992. Tetracyclines, molecular and clinical aspects. *J. Antimicrob. Chemother.* **29:**245–277.

60. **Chow, J. W., M. J. Zervos, S. A. Lerner, L. A. Thal, S. M. Donabedian, D. D. Jaworski, S. Tsai, K. J. Shaw, and D. B. Clewell.** 1997. A novel resistance gene in *Enterococcus*. *Antimicrob. Agents Chemother.* **41:**511–514.

61. **Clancy, J., J. Petitpas, F. Dib-Hajj, W. Yuan, M. Cronan, A. V. Kamath, J. Bergeron, and J. A. Retsema.** 1996. Molecular cloning and functional analysis of a novel macrolide-resistance determinant, *mefA* from *Streptococcus pyogenes*. *Mol. Microbiol.* **22:**867–879.

62. **Clermont, D., O. Chesneau, G. de Cespédes, and T. Horaud.** 1997. New tetracycline resistance determinants coding for ribosomal protection in streptococci and nucleotide sequence of *tet*(T) isolated from *Streptococcus pyogenes* A498. *Antimicrob. Agents Chemother.* **41:**112–116.

63. **Coffey, T. J., C. G. Dowson, M. Daniels, and B. G. Spratt.** 1995. Genetics and molecular biology of β-lactam-resistant pneumococci. *Microb. Drug Resist.* **1:**29–34.

64. **Coleman, K., M. Athalye, A. Clancey, M. Davison, D. J. Payne, C. R. Perry, and I. Chopra.** 1994. Bacterial resistance mechanisms as therapeutic targets. *J. Antimicrob. Chemother.* **33:**1091–1116.

65. **Corcoran, J. W.** 1984. Mode of action and resistance mechanisms of macrolides, p. 231–259. *In* S. Omura (ed.), *Macrolide Antibiotics*. Academic Press, Orlando, Fla.

66. **Costa, Y., M. Galimand, R. Leclercq, J. Duval, and P. Courvalin.** 1993. Characterization of the chromosomal *acc(6′)-Ii* gene specific for *Enterococcus faecium*. *Antimicrob. Agents Chemother.* **37:**1896–1903.

67. **Couto, I., H. de Lencastre, E. Severina, W. Kloos, J. Webster, R. Hubner, I. Santos Sanches, and A. Tomasz.** 1996. Ubiquitous presence of a *mecA* homologue in natural isolates of *Staphylococcus sciuri*. *Microb. Drug Resist.* **2:**377–391.

68. **Cundliffe E.** 1990. Recognition sites for antibiotics, p. 479–490. *In* W. E. Hill, A. Dahlberg, R. A. Garrett, P. B. Moore, D. Schlessinger, and J. R. Warner (ed.), *The Ribosome: Structure, Function, and Evolution*. American Society for Microbiology, Washington, D.C.

69. **Czworkowski, J., J. Wang, T. A. Steitz, and P. B. Moore.** 1994. The crystal structure of elongation factor G complexed with GDP, at 2.7 A resolution. *EMBO J.* **13:**3661–3668.

70. **Davies, J.** 1991. Aminoglycoside-aminocyclitol antibiotics and their modifying enzymes, p. 790–809. *In* V. Lorean (ed.), *Antibiotics in Laboratory Medicine*. Williams and Wilkins, Baltimore, Md.

71. **Davies, J.** 1994. Inactivation of antibiotics and the dissemination of resistance genes. *Science* **264:**375–382.

72. **Davies, J., and G. D. Wright.** 1997. Bacterial resistance to aminoglycoside antibiotics. *Trends Microbiol.* **5:**234–240.

73. **Deguchi, T., A. Fukuoka, Y. Yasuda, M. Nakano, S. Ozeki, E. Kanematsu, Y. Nishino, S. Ishihara, Y. Ban, and Y. Kawada.** 1997. Alterations in the GyrA subunit of DNA gyrase and ParC subunit of topoisomerase IV in quinolone-resistant clinical isolates of *Klebsiella pneumoniae*. *Antimicrob. Agents Chemother.* **41:**699–701.

74. **Deguchi, T., M. Yasuda, M. Nakano, S. Ozeki, E. Kanematsu, Y. Kawada, T. Ezaki, and I. Saito.** 1996. Uncommon occurrence of mutations in the *gyrB* gene associated with quinolone resistance in clinical isolates of *Neisseria gonorrhoeae*. *Antimicrob. Agents Chemother.* **40:** 2437–2438.

75. **Deguchi, T., M. Yasuda, M. Nakano, S. Ozeki, E. Kanematsu, Y. Nishino, S. Ishihara, and Y. Kawada.** 1997. Detection of mutations in the *gyrA* and *parC* genes in quinolone-resistant clinical isolates of *Enterobacter cloacae*. *J. Antimicrob. Chemother.* **40:**543–549.

76. **de Jonge, B. L., D. Gage, and S. Handwerger.** 1996. Peptidoglycan composition of vancomycin-resistant *Enterococcus faecium*. *Microb. Drug Resist.* **2:**225–229.

77. **de Jonge, B. L., S. Handwerger, and D. Gage.** 1996. Altered peptidoglycan composition in vancomycin-resistant *Enterococcus faecalis*. *Antimicrob. Agents Chemother.* **40:** 863–869.

78. **de la Campa, A. G., E. García, A. Fenoll, and R. Muñoz.** 1997. Molecular basis of three characteristic phenotypes of pneumococcus: optochin-sensitivity, coumarin-sensitivity, and quinolone-resistance. *Microb. Drug Resist.* **3:**177–193.

79. **de Lencastre, H., and A. Tomasz.** 1994. Reassessment of the number of auxiliary genes essential for the expression of high-level methicillin resistance in *Staphylococcus aureus*. *Antimicrob. Agents Chemother.* **38:**2590–2598.

80. **Denoya, C., and D. Dubnau.** 1989. Mono- and dimethylating activities and kinetic studies of the *ermC* 23S rRNA methyltransferase. *J. Biol. Chem.* **264:**2615–2624.

81. **Derbise, A., S. Aubert, and N. el Solh.** 1997. Mapping the regions carrying the three contiguous antibiotic resistance genes *aadE*, *sat4*, and *aphA-3* in the genomes of

staphylococci. *Antimicrob. Agents Chemother.* **41:**1024–1032.

82. **Dever, L. A., and T. S. Dermody.** 1991. Mechanisms of bacterial resistance to antibiotics. *Arch. Intern. Med.* **151:** 886–895.

83. **Doran, J. L., Y. Pang, K. E. Mdluli, A. J. Moran, T. C. Victor, R. W. Stokes, E. Mahenthiralingam, B. N. Kreiswirth, J. L. Butt, G. S. Baron, J. D. Treit, V. J. Kerr, P. D. Vanhelden, M. C. Roberts, and F. E. Nano.** 1997. *Mycobacterium tuberculosis efpA* encodes an efflux protein of the QacA transporter family. *Clin. Diagn. Lab. Immunol.* **4:**23–32.

84. **Dowson, C. G., T. J. Coffey, and B. G. Spratt.** 1994. Origin and molecular epidemiology of penicillin-binding protein-mediated resistance to β-lactam antibiotics. *Trends Microbiol.* **2:**361–366.

85. **Doyle, D., K. J. McDowall, M. J. Butler, and I. S. Hunter.** 1991. Characterization of an oxytetracycline-resistance gene, *otrA*, of *Streptomyces rimosus*. *Mol. Microbiol.* **5:**2923–2933.

86. **Dutka-Malen, S., B. Blaimont, G. Wauters, and P. Courvalin.** 1994. Emergence of high-level resistance to glycopeptides in *Enterococcus gallinarum* and *Enterococcus casseliflavis*. *Antimicrob. Agents Chemother.* **38:**1675–1677.

87. **Edgar, R., and E. Bib.** 1997. MdrA, and *Escherichia coli* multidrug resistance protein with an extraordinarily broad spectrum of drug recognition. *J. Bacteriol.* **179:**2274–2280.

88. **Elisha, B. G., and P. Courvalin.** 1995. Analysis of genes encoding D-alanine:D-alanine ligase-related enzymes in *Leuconostoc mesenteroides* and *Lactobacillus* spp. *Gene* **152:** 79–83.

89. **el Kharroubi, A., P. Jacques, G. Piras, J. van Beeumen, J. Coyette, and M. M. Ghuysen.** 1991. The *Enterococcus hirae* R40 penicillin-binding protein 5 and the methicillin-resistant *Staphylococcus aureus* penicillin-binding protein 2′ are similar. *Biochem. J.* **280:**463–469.

90. **el Solh, N., N. Moreau, and S. D. Ehrlich.** 1986. Molecular cloning and analysis of *Staphylococcus aureus* chromosomal aminoglycoside resistance genes. *Plasmid* **15:** 104–118.

91. **Epe, B., P. Woolley, and H. Hornig.** 1987. Competition between tetracycline and tRNA at both P and A sites of the ribosome of *Escherichia coli*. *FEBS Lett.* **213:**443–447.

92. **Evers, S., R. Quintiliani, Jr., and P. Courvalin.** 1996. Genetics of glycopeptide resistance in enterococci. *Microb. Drug Resist.* **2:**219–223.

93. **Fantin, B., R. Leclercq, L. Garry, and C. Carbon.** 1997. Influence of inducible cross-resistance to macrolides, lincosamides, and streptogramin B-type antibiotics in *Enterococcus faecium* on activity of quinupristin-dalfopristin in vitro and in rabbits with experimental endocarditis. *Antimicrob. Agents Chemother.* **41:**931–935.

94. **Femlee, T. A., Q. Liu, A. C. Whelen, D. Williams, S. S. Sommer, and D. H. Persing.** 1995. Genotypic detection of *Mycobacterium tuberculosis* rifampin resistance: comparison of single-strand conformation polymorphism and dideoxy fingerprinting. *J. Clin. Microbiol.* **33:**1617–1623.

95. **Ferrero, L., B. Cameron, and J. Crouzet.** 1995. Analysis of *gyrA* and *grlA* mutations in stepwise-selected ciprofloxacin-resistant mutants of *Staphylococcus aureus*. *Antimicrob. Agents Chemother.* **39:**1554–1558.

96. **Ferretti, J. J., K. S. Gilmore, and P. Courvalin.** 1986. Nucleotide sequence analysis of the gene specifying the bifunctional 6′-aminoglycoside acetyltransferase 2″-aminoglycoside phosphotransferase enzyme in *Streptococcus faecalis* and identification and cloning of gene regions specifying the two activities *J. Bacteriol.* **167:**631–638.

97. **Fisher, L. M., M. Oram, and S. Sreedharan.** 1992. DNA gyrase: mechanism and resistance to 4-quinolone antibacterial agents, p. 145–155. *In* T. Andoh, H. Ikeda, and Oguro (ed.), *Molecular Biology of DNA Topoisomerases and Its Application to Chemotherapy*. Proceedings of the International Symposium on DNA Topoisomerase in Chemotherapy, Nagoya, Japan, November 18–20, 1991. CRC Press, Boca Raton, Fla.

98. **Fontana, R., R. Cerini, P. Longoni, A. Grossato, and P. Canepari.** 1983. Identification of a streptococcal penicillin-binding protein that reacts very slowly with penicillin. *J. Bacteriol.* **155:**1343–1350.

99. **Fontana, R., A. Grossato, L. Rossi, Y. R. Cheng, and G. Satta.** 1985. Transition from resistance to hypersusceptibility to β-lactam antibiotics associated with loss of a low-affinity penicillin-binding protein in a *Streptococcus faecium* mutant highly resistant to penicillin. *Antimicrob. Agents Chemother.* **28:**678–683.

100. **Fontana, R., M. Ligozzi, F. Pittaluga, and G. Satta.** 1996. Intrinsic penicillin resistance in enterococci. *Microb. Drug Resist.* **2:**209–213.

101. **Fournier, B., and D. C. Hooper.** 1998. Mutations in topoisomerase IV and DNA gyrase of *Staphylococcus aureus*: pleiotropic effects on quinolone and coumarin activity. *Antimicrob. Agents Chemother.* **42:**121–128.

102. **Fraimow, H., and C. Knob.** 1997. Amplification of macrolide efflux pumps *msr* and *mef* from *Enterococcus faecium* by polymerase chain reaction, abst. A-125, p. 22. *In* Program and Abstracts of the 97th American Society for Microbiology General Meeting. American Society for Microbiology, Washington, D.C.

103. **Free, L., and D. F. Sahm.** 1996. Detection of enterococcal vancomycin resistance by multiplex PCR, p. 150–155. *In* D. H. Persing (ed.), *PCR Protocols for Emerging Infectious Diseases*. American Society for Microbiology, Washington, D.C.

104. **Garcia-Bustos, J., and A. Tomasz.** 1990. A biological price of antibiotic resistance: major changes in the peptidoglycan structure of penicillin-resistant pneumococci. *Proc. Natl. Acad. Sci. USA* **87:**5415–5419.

105. **Gazouli, M., L. S. Tzouvelekis, E. Prinarakis, V. Miriagou, and E. Tzelepi.** 1996. Transferable cefoxitin resistance in enterobacteria from Greek hospital and characterization of a plasmid-mediated group 1 β-lactamase (LAT-2). *Antimicrob. Agents Chemother.* **40:** 1736–1740.

106. **Gensberg, K., Y. F. Jin, and L. J. Piddock.** 1995. A novel *gyrB* mutation in a fluoroquinolone-resistant clinical isolate of *Salmonella typhimurium*. *FEMS Microbiol. Lett.* **132:**57–60.

107. **George, A. M.** 1996. Multidrug resistance in enteric and other gram-negative bacteria. *FEMS Microb. Lett.* **139:** 1–10.

108. **George, A. M., R. M. Hal, and H. W. Stokes.** 1995. Multidrug resistance in *Klebsiella pneumoniae*: a novel gene, *ramA*, confers a multidrug resistance phenotype in *Escherichia coli*. *Microbiology* **141:**1909–1920.

109. **Georgiou, M., R., R. Muñoz, F. Román, R. Cantón, R. Gómez-Lus, J. Campos, and A. G. de la Campa.** Ciprofloxacin-resistant *Haemophilus influenzae* strains possess mutations in analogous positions of GyrA and ParC. *Antimicrob. Agents Chemother.* **40:**1741–1744.

110. **Georgopapadakou, N. H.** 1996. Mechanisms of antibiotic resistance, p. 72–85. *In* S. S. Long, C. G. Probe, and L. K. Pickering (ed.), *Principles and Practice of Pediatric Infectious Diseases*, Churchill Livingstone, Ltd., Edinburgh.

111. **Ghuysen, J. M.** 1994. Molecular structures of penicillin-binding proteins and β-lactamases. *Trends Microb.* **2:** 372–380.

112. **Gillespie, M. T., B. R. Lyon, L. J. Messerotti, and R. A. Skurray.** 1987. Chromosome- and plasmid-mediated gentamicin resistance in *Strephylococcus aurerus* encoded by Tn*4001*. *J. Med. Microbiol.* **24:**139–144.

113. **Gin, A. S., and G. G. Zhanel.** 1996. Vancomycin-resistant enterococci. *Ann. Pharmacother.* **30:**615–624.

114. **Goldman, J. D., D. G. White, and S. B. Levy.** 1996. Multiple antibiotic resistance (*mar*) locus protects *Escherichia coli* from rapid cell killing by fluoroquinolones. *Antimicrob. Agents Chemother.* **40:**1266–1269.

115. **Gootz, T. D., R. Zaniewski, S. Haskell, B. Schmieder, J. Tankovic, D. Girard, P. Courvalin, and R. J. Polzer.** 1996. Activity of the new fluoroquinolone trovafloxacin (CP-99219) against DNA gyrase and topoisomerase IV mutants of *Streptococcus pneumoniae* selected in vitro. *Antimicrob. Agents Chemother.* **40:**2691–2697.

116. **Gotoh, N., K. Nunomura, and T. Nishino.** 1990. Resistance of *Pseudomonas aeruginosa* to cefsulodin: modification of penicillin-binding protein 3 and mapping of its chromosomal gene. *J. Antimicrob. Chemother.* **25:**513–523.

117. **Gotoh, N., H. Tsujimoto, K. Poole, J. Yamagishi, and T. Nishino.** 1995. The outer membrane protein OprM of *Pseudomonas aeruginosa* is encoded by *oprK* of the *mexA-mexB-oprK* multidrug resistance operon. *Antimicrob. Agents Chemother.* **39:**2567–2569.

118. **Gray, G. S., R. T.-S. Huang, and J. Davies.** 1983. Aminocyclitol resistance in *Staphylococcus aureus*: presence of plasmids and aminocyclitol-modifying enzymes. *Plasmid* **9:**147–152.

119. **Grebe, T., J. Cheng, and J. Sutcliffe.** 1998. Unpublished data.

120. **Grebe, T., and R. Hakenbeck.** 1996. Penicillin-binding proteins 2b and 2x of *Streptococcus pneumoniae* are primary resistance determinants for different classes of β-lactam antibiotics. *Antimicrob. Agents Chemother.* **40:** 829–834.

121. **Grebe, T., J. Paik, and R. Hakenbeck.** 1997. A novel resistance mechanism against β-lactams in *Streptococcus pneumoniae* involves CpoA, a putative glycosyltransferase. *J. Bacteriol.* **179:**3342–3349.

122. **Grewal, J., E. K. Manavathu, and D. E. Taylor.** 1993. Effect of mutational alteration of Asn-128 in the putative GTP-binding domain of tetracycline resistance determinant Tet(O) from *Campylobacter jejuni*. *Antimicrob. Agents Chemother.* **37:**2645–2649.

123. **Grinius, L. L., and E. B. Goldberg.** 1994. Bacterial multidrug resistance is due to a single membrane protein which functions as a drug pump. *J. Biol. Chem.* **269:** 29998–30004.

124. **Guenzi, E., A. M. Gase, M. A. Sicard, and R. Hakenbeck.** 1994. A two-component signal-transducing system is involved in competence and penicillin susceptibility in laboratory mutants of *Streptococcus pneumoniae*. *Mol. Microbiol.* **12:**505–515.

125. **Gutmann, L., S. Al-Obeid, D. Billot-Klein, E. Ebnet, and W. Fischer.** 1996. Penicillin tolerance and modification of lipoteichoic acid associated with expression of vancomycin resistance in VanB-type *Enterococcus faecium* D366. *Antimicrob. Agents Chemother.* **40:**257–259.

126. **Hächler, H., B. Berger-Bächi, and F. H. Kayser.** 1987. Genetic characterization of a *Clostridium difficile* erythromycin-clindamycin resistance determinant that is transferable to *Staphylococcus aureus*. *Antimicrob. Agents Chemother.* **31:**1039–1045.

127. **Hackbarth, C. J., T. Kocagoz, S. Kocagoz, and H. F. Chambers.** 1995. Point mutations in *Staphylococcus aureus* PBP2 gene affect penicillin-binding kinetics and are associated with resistance. *Antimicrob. Agents Chemother.* **39:**103–106.

128. **Hagman, K. E., C. E. Lucas, J. T. Balthazar, L. Snyder, M. Nilles, R. C. Judd, and W. M. Shafer.** 1997. The MtrD protein of *Neisseria gonorrhoeae* is a member of the resistance/nodulation/division protein family constituting part of an efflux system. *Microbiology* **143:**2117–2125.

129. **Hakenbeck, R., S. Tornette, and N. F. Adkinson.** 1987. Interaction of non-lytic β-lactams with penicillin-binding proteins in *Streptococcus pneumoniae*. *J. Gen. Microbiol.* **133:**755–760.

130. **Handwerger, S., M. J. Pucci, K. J. Volk, J. Liu, and M. S. Lee.** 1992. The cytoplasmic peptidoglycan precursor of vancomycin-resistant *Enterococcus faecalis* terminates in lactate. *J. Bacteriol.* **174:**5982–5984.

131. **Handwerger, S., M. J. Pucci, K. J. Volk, J. Liu, and M. S. Lee.** 1994. Vancomycin-resistant *Leuconostoc mesenteroides* and *Lactobacillus casei* synthesize cytoplasmic peptidoglycan precursors that terminate in lactate. *J. Bacteriol.* **176:**260–264.

132. **Hedge, P. J., and B. G. Spratt.** 1985. Resistance to β-lactam antibiotics by remodelling the active site of an *E. coli* penicillin-binding protein. *Nature* **318:**478–480.

133. **Henquell, C., C. Chanal, D. Sirot, R. Labia, and J. Sirot.** 1995. Molecular characterization of nine different types of mutants among 107 inhibitor-resistant TEM β-lactamases from clinical isolates of *Escherichia coli*. *Antimicrob. Agents Chemother.* **39:**427–430.

134. **Henze, U. U., and B. Berger-Bächi.** 1995. *Staphylococcus aureus* penicillin-binding protein 4 and intrinsic beta-lactam resistance. *Antimicrob. Agents Chemother.* **39:** 2415–2422.

135. **Heritage, J., P. M. Hawkey, N. Todd, and J. J. Lewis.** 1992. Transposition of the gene encoding TEM-12 extended-spectrum β-lactamase. *Antimicrob. Agents Chemother.* **36:**1981–1986.

136. **Hesig, P.** 1996. Genetic evidence for a role of *parC* mutations in development of high-level fluoroquinolone resistance in *Escherichia coli*. *Antimicrob. Agents Chemother.* **40:**879–885.

137. **Heym, B., N. Honore, C. Truffot-Pernot, A. Banerjee, C. Schurra, W. R. Jacobs, Jr., J. D. A. van Embden, J. H. Grosset, and S. T. Cole.** 1994. Implications of multidrug resistance for the future of short-course chemo-

therapy of tuberculosis: a molecular study. *Lancet* **344:** 293–298, 1049–1053.

138. **Higgins, C. F.** 1992. ABC transporters: from microorganisms to man. *Annu. Rev. Cell Biol.* **8:**67–113.

139. **Hiramatsu, K., H. Hanaki, T., Ino, K. Yabuta, T. Oguri, and F. C. Tenover.** 1997. Methicillin-resistant *Staphylococcus aureus* clinical strain with reduced vancomycin susceptibility. *J. Antimicrob. Chemother.* **40:** 135–136.

140. **Hodel-Christian, S. L., and B. E. Murray.** 1991. Characterization of the gentamicin resistance transposon Tn*5281* from *Enterococcus faecalis* and comparison to staphylococcal transposons Tn*4001* and Tn*4031*. *Antimicrob. Agents Chemother.* **35:**1147–1152.

141. **Hodgson, A. L. M., J. Krywult, and A. J. Radford.** 1990. Nucleotide sequence of the erythromycin resistance gene from *Corynebacterium* plasmid pNG2. *Nucleic Acids Res.* **18:**1891.

142. **Horaud, T., G. deCéspédes, and P. Trieu-Cuot.** 1996. Chromosomal gentamicin resistance transposon Tn*3706* in *Streptococcus agalactiae* B128. *Antimicrob. Agents Chemother.* **40:**1085–1090.

143. **Howard, B. M. A., R. J. Pinney, and J. T. Smith.** 1993. Function of the SOS process in repair of DNA damage induced by modern 4-quinolones. *J. Pharm. Pharmacol.* **45:**658–662.

144. **Huang, W., J. Petrosino, M. Hirsch, P. S. Shenkin, and T. Palzkill.** 1996. Amino acid sequence determinants of beta-lactamase structure and activity. *J. Mol. Biol.* **258:** 688–703.

145. **Imboden, P., S. Cole, T. Bodmer, and A. Telenti.** 1993. Detection of rifampin resistance mutations in *Mycobacterium tuberculosis* and *M. leprae*, p. 519–526. *In* D. H. Persing, T. F. Smith, F. C. Tenover, and T. J. White (ed.), *Diagnostic Molecular Microbiology: Principles and Applications*. American Society for Microbiology, Washington, D.C.

146. **Ishida, H., H. Fuziwara, Y. Kaibori, T. Horiuchi, K. Sato, and Y. Osada.** 1995. Cloning of multidrug resistance gene *pqrA* from *Proteus vulgaris*. *Antimicrob. Agents Chemother.* **39:**453–457.

147. **Ito, H., Y. Arakawa, S. Ohuska, R. Wacharotayankun, N. Kato, and M. Ohta.** 1995. Plasmid-mediated dissemination of the metallo-β-lactamase gene bla_{IMP} among clinical isolated strains of *Serretia marcesens*. *Antimicrob. Agents Chemother.* **39:**824–829.

148. **Jacobs, C., B. Joris, M. Jamin, K. Klarsov, J. Van Beeumen, D. Mengin-Lecreulx, J. van Heijenoort, T. J. Park, S. Normark, and J. M. Frere.** 1995. AmpD, essential for both β-lactamase regulation and cell wall recycling, is a novel cytosolic *N*-acetylmuramyl-L-alanine amidase. *Mol. Microbiol.* **15:**553–559.

149. **Jacoby, G. A.** 1996. Antimicrobial-resistant pathogens in the 1990s. *Annu. Rev. Med.* **47:**169–179.

150. **Jacoby, G. A., and A. A. Medeiros.** 1991. More extended-spectrum β-lactamases. *Antimicrob. Agents Chemother.* **35:**1697–1704.

151. **Janoir, C., V. Zeller, M. D. Kitzis, N. J. Moreau, and L. Gutmann.** 1996. High-level fluoroquinolone resistance in *Streptococcus pneumoniae* requires mutations in *parC* and *gyrA*. *Antimicrob. Agents Chemother.* **40:**2760–2764.

152. **Jenkins, G., and E. Cundliffe.** 1991. Cloning and characterization of two genes from *Streptomyces lividans* that confer inducible resistance to lincomycin and macrolide antibiotics. *Gene* **108:**55–62.

153. **Johnston, N. J., J. C. S. de Azavedo, P. Chang, S. Tyler, M. Coulthart, W. Johnson, and D. E. Low.** 1998. Identification of the novel erythromycin resistance methylase gene *ermTR* in *Streptococcus milleri*, abstr. 1D-20, p. 61. *In* ASM Conference on Streptococcal Genetics, 26–29 April 1998, France. American Society for Microbiology, Washington, D.C.

154. **Jones, R. N., H. S. Sader, M. E. Erwin, and S. C. Anderson.** 1995. Emerging multiply resistant enterococci among clinical isolates. I. Prevalence data from 97 medical center surveillance study in the United States. Enterococcus Study Group. *Diagn. Microbiol. Infect. Dis.* **21:**85–93.

155. **Kaatz, G. W., and S. M. Seo.** 1998. Topoisomerase mutations in fluoroquinolone-resistant and methicillin-susceptible and -resistant clinical isolates of *Staphylococcus aureus*. *Antimicrob. Agents Chemother.* **42:**197–198.

156. **Kaatz, G. W., S. M. Seo, and C. A. Ruble.** 1991. Mechanisms of fluoroquinolone resistance in *Staphylococcus aureus*. *J. Infect. Dis.* **163:**1080–1086.

157. **Kamimiya, S., and B. Weisblum.** 1997. Induction of *ermSV* by 16-membered-ring macrolide antibiotics. *Antimicrob. Agents Chemother.* **41:**530–534.

158. **Kanematsu, E., T. Deguchi, M. Yasuda, T. Kawamura, Y. Nishino, and Y. Kawada.** 1998. Alterations in the GyrA subunit of DNA gyrase and the ParC subunit of DNA topoisomerase IV associated with quinolone resistance in *Enterococcus faecalis*. *Antimicrob. Agents Chemother.* **42:**433–435.

159. **Kataja, J., H. Sepp, H. Sarkkinen, and P. Huovinen.** 1997. Different erythromycin resistance mechanisms in Group C and G streptococci in Finland, abstr. C-74, p. 59. *In* Program and Abstracts of the 37th Interscience Conference on Antimicrobial Agents and Chemotherapy. American Society for Microbiology, Washington, D.C.

160. **Kato, J., Y. Nishimura, R. Imamura, H. Niki, S. Higara, and H. Suzuki.** 1990. New topoisomerase essential for chromosome segregation in *E. coli*. *Cell* **63:**393–404.

161. **Kato, N., M. Miyauchi, Y. Muto, K. Watanabe, and K. Ueno.** 1988. Emergence of fluoroquinolone resistance in *Bacteroides fragilis* accompanied by resistance to β-lactam antibiotics. *Antimicrob. Agents Chemother.* **32:**1437–1438.

162. **Khodursky, A. B., E. L. Zechiedrich, and N. R. Cozzarelli.** 1995. Topoisomerase IV is a target of quinolones in *Escherichia coli*. *Proc. Natl. Acad. Sci. USA* **92:**11801–11805.

163. **Kim, H.-S., E.-C. Choi, and B.-K. Kim** 1993. A macrolide-lincosamide-streptogramin B resistance determinant from *Bacillus anthracis* 590: cloning and expression of *ermJ*. *J. Gen. Microbiol.* **139:**601–607.

164. **Kim, J. H., E. H. Cho, K. S. Kim, H. Y. Kim, and Y. M. Kim.** 1998. Cloning and nucleotide sequence of the DNA gyrase *gyrA* gene from *Serratia marcescens* and characterization of mutations in *gyrA* of quinolone-resistant clinical isolates. *Antimicrob. Agents Chemother.* **42:**190–193.

165. **Klare, I., A. C. Rodloff, J. Wagner, W. Witte, and R. Hakenbeck.** 1992. Overproduction of a penicillin-binding protein is not the only mechanism of penicillin resistance in *Enterococcus faecium*. *Antimicrob. Agents Chemother.* **36:**783–787.

166. **Knox, J. R.** 1995. Extended spectrum and inhibitor resistant TEM-type β-lactamases. *Antimicrob. Agents Chemother.* **39:**2593–2601.

167. **Köhler, T., M. Michea-Hamzehpour, U. Henze, N. Gotoh, L. K. Curty, and J. C. Pechere.** 1997. Characterization of MexE-MexF-OprN, a positively regulated multidrug efflux system of *Pseudomonas aeruginosa*. *Mol. Microbiol.* **23:**345–354.

168. **Köhler, T., M. Michea-Hamzehpour, P. Plesiat, A.-L. Kahr, and J.-C. Pechere.** 1997. Differential selection of multidrug efflux systems by quinolones in *Pseudomonas aeruginosa*. *Antimicrob. Agents Chemother.* **41:**2540–2543.

169. **Komatsuzawa, H., M. Sugai, K. Ohta, T. Fujiwara, S. Nakashima, J. Suzuki, C. Y. Lee, and H. Suginaka.** 1997. Cloning and characterization of the *fmt* gene which affects the methicillin resistance level and autolysis in the presence of Triton X-100 in methicillin-resistant *Staphylococcus aureus*. *Antimicrob. Agents Chemother.* **41:**2355–2361.

170. **Koop, R., M. Roos, J. Wecke, and H. Labischinski.** 1996. Staphylococcal peptidoglycan interpeptide bridge biosynthesis: a novel antistaphylococcal target? *Microb. Drug Resist.* **2:**29–41.

171. **Korten, V., W. M. Huang, and B. E. Murray.** 1994. Analysis by PCR and direct DNA sequencing of *gyrA* mutations associated with fluoroquinolone resistance in *Enterococcus faecalis*. *Antimicrob. Agents Chemother.* **38:** 2091–2094.

172. **Krauss, J., M. van der Linden, T. Grebe, and R. Hakenbeck.** 1996. Penicillin-binding proteins 2x and 2b as primary PBP targets in *Streptococcus pneumoniae*. *Microb. Drug Resist.* **2:**183–186.

173. **Kumagai, Y., J.-I. Kato, K. Hoshino, T. Akasaka, K. Sato, and H. Ikeda.** 1996. Quinolone-resistant mutants of *Escherichia coli* DNA topoisomerase IV *parC* gene. *Antimicrob. Agents Chemother.* **40:**710–714.

174. **Kwak, J. H., E. C. Choi, and B. Weisblum.** 1991. Transcriptional attenuation control of *ermK*, a macrolide-lincosamide-streptogramin B resistance determinant from *Bacillus licheniformis*. *J. Bacteriol.* **173:**4725–4735.

175. **Lawrence, L. E., and J. F. Barrett.** 1998. Efflux pumps in bacteria: overview, clinical relevance, and potential pharmaceutical target. *Exp. Opin. Invest. Drugs* **7:**199–217.

176. **Leclercq, R.** 1997. Enterococci acquire new kinds of resistance. *Clin. Infect. Dis.* **24:**S80–S84.

177. **Leclercq, R., and P. Courvalin.** 1991. Bacterial resistance to macrolide, lincosamide, and streptogramin antibiotics by target modification. *Antimicrob. Agents Chemother.* **35:**1267–1272.

178. **Leclercq, R., and P. Courvalin.** 1991. Intrinsic and unusual resistance to macrolide, lincosamide, and streptogramin antibiotics in bacteria. *Antimicrob. Agents Chemother.* **35:**1273–1276.

179. **Leclercq, R., and P. Courvalin.** 1993. Mechanisms of resistance to macrolides and functionally related antibiotics, p. 125–141. *In* A. J. Bruskier, J. P. Butzler, H. C. Neu, and P. M. Tulkens (ed.), *Macrolides—Chemistry, Pharmacology, and Clinical Uses*. Arnette Blackwell, Paris.

180. **Le Goffic, F., M. L. Capmau, J. Abbe, C. Cerceau, A. Dublanchet, and J. Duval.** 1977. Plasmid-mediated pristinamycin resistance: PH1A, a pristinamycin 1A hydrolase. *Ann. Microbiol. Inst. Pasteur* **128:**471–474.

181. **Le Goffic, F., M. L. Capmau, M. L. Bonnet, C. Cerceau, C. J. Soussy, A. Dublanchet, and J. Duval.** 1977. Plasmid-mediated pristinamycin resistance: PACIIA, a new enzyme which modifies pristinamycin IIA. *J. Antibiot.* **30:**665–669.

182. **Levy, S. B.** 1992. 1992. Active efflux mechanisms for antimicrobial resistance. *Antimicrob. Agents Chemother.* **36:**695–703.

183. **Lewis, K.** 1994. Multidrug resistance pumps in bacteria: variations on a theme. *Trends Biochem. Sci.* **19:**119–123.

184. **Li, X. Z., D. M. Livermore, and H. Nikaido.** 1994. Role of efflux pump(s) in intrinsic resistance of *Pseudomonas aeruginosa*: resistance to tetracycline, chloramphenicol, and norfloxacin. *Antimicrob. Agents Chemother.* **38:** 1732–1741.

185. **Li, X.-Z., L. Zhang, R. Srikumar, and K. Poole.** 1998. β-lactamase inhibitors are substrates for the multidrug efflux pumps of *Pseudomonas aeruginosa*. *Antimicrob. Agents Chemother.* **42:**399–403.

186. **Ligozzi, M., F. Pittaluga, and R. Fontana.** 1993. Identification of a genetic element (*psr*) which negatively controls expression of *Enterococcus hirae* penicillin-binding protein 5. *J. Bacteriol.* **175:**2046–2051.

187. **Liu, J., H. E. Takiff, and H. Nikaido.** 1996. Active efflux of fluoroquinolones in *Mycobacterium smegmatis* mediated by LfrA, a multi-drug efflux pump. *J. Bacteriol.* **178:**3791–3795.

188. **Livermore, D.** 1995. β-lactamases in laboratory and clinical resistance. *Clin. Microbiol. Rev.* **8:**557–584.

189. **Livermore, D. M.** 1996. Are all β-lactams created equal? *Scand. J. Infect. Dis. Suppl.* **101:**33–43.

190. **Livermore, D. M., and J. D. Williams.** 1996. β-Lactams: mode of action and mechanisms of bacterial resistance. p. 502–578. *In* M. Lorian (ed.), *Antibiotics in Laboratory Medicine*. Williams and Wilkins, Baltimore, Md.

191. **Lomovskaya, O., F. Kawai, and A. Martin.** 1996. Differential regulation of the *mcb* and *emr* operons of *Escherichia coli*: role of *mcb* in multidrug resistance. *Antimicrob. Agents Chemother.* **40:**1050–1052.

192. **Lomovskaya, O., and K. Lewis.** 1992. *emr*, and *Escherichia coli* locus for multidrug resistance. *Proc. Natl. Acad. Sci. USA* **89:**8938–8942.

193. **Loncle, V., J. Allignet, and N. el Solh.** 1991. Genetic analysis of RP59500 resistance in 85 *Staphylococcus* spp. hospital isolates, abstr. 1366, p. 326. *In* Program and Abstracts of the 31st Interscience Conference on Antimicrobial Agents and Chemotherapy. American Society for Microbiology, Washington, D.C.

194. **Loncle, V., A. Casetta, A. Buu-hoi, and N. el Solh.** 1993. Analysis of pristinamycin-resistant *Staphylococcus epidermidis* isolates responsible for an outbreak in a Parisian hospital. *Antimicrob. Agents Chemother.* **37:**2159–2165.

195. **Lucas, C. E., J. T. Blathazar, K. E. Hagman, and W. M. Shafer.** 1997. The MtrR repressor binds the DNA

sequence between the *mtrR* and *mtrC* genes of *Neisseria gonorrhoeae*. *J. Bacteriol.* **179:**4123–4128.

196. **Lucas, C. E., K. E. Hagman, J. C. Levin, D. C. Stein, and W. M. Shafer.** 1995. Importance of the lipooligosaccharide structure in determining gonococcal resistance to hydrophobic antimicrobial agents resulting from the *mtr* efflux system. *Mol. Microbiol.* **16:**1001–1009.

197. **Lynch, C., P. Courvalin, and H. Nikaido.** 1997. Active efflux of antimicrobial agents in wild-type strains of enterococci. *Antimicrob. Agents Chemother.* **41:**869–871.

198. **Lyon, B. R., and R. A. Skurray.** 1987. Antimicrobial resistance of *Staphylococcus aureus*: genetic basis. *Microbiol Rev.* **51:**88–134.

199. **Ma, D., D. N. Cook, M. Alberti, N. G. Pon, H. Nikaido, and J. E. Hearst.** 1993. Molecular cloning and characterization of *acrA* and *acrE* genes of *Escherichia coli*. *J. Bacteriol.* **175:**6299–6313.

200. **Ma, D., D. N. Cook, M. Alberti, N. G. Pon, H. Nikaido, and J. E. Hearst.** 1995. Genes *acrA* and *acrB* encode a stress-induced efflux system of *Escherichia coli*. *Mol. Microbiol.* **16:**45–55.

201. **Mabilat, C., and S. Goussard.** 1993. PCR detection and identification of genes for extended spectrum β-lactamases, p. 553–562. *In* D. H. Persing, T. F. Smith, F. C. Tenover, and T. J. White (ed.), *Diagnostic Molecular Microbiology: Principles and Applications*. American Society for Microbiology, Washington, D.C.

202. **Mabilat, C., S. Goussard, W. Sougakoff, R. C. Spencer, and P. Courvalin.** 1990. Direct sequencing of the amplified structural gene and promoter of the extended-spectrum β-lactamase TEM-9 (RHH-1) of *Klebsiella pneumoniae*. *Plasmid* **23:**27–34.

203. **MacGowan, A. P., K. E. Bowker, H. A. Holt, M. Wootton, and D. S. Reeves.** 1997. Bay 12-8039, a new 8-methoxy-quinolone: comparative in vitro activity with none other antimicrobials against anaerobic bacteria. *J. Antimicrob. Chemother.* **40:**503–509.

204. **Macinga, D. R., G. M. Cook, R. K. Poole, and P. N. Rather.** 1998. Identification and characterization of *aarF*, a locus required for production of ubiquinone in *Providencia stuartii* and *Escherichia coli* and for expression of 2′-*N*-acetyltransferase in *P. stuartii*. *J. Bacteriol.* **180:**128–135.

205. **Maki, H., T. Yamaguchi, and K. Murakami.** 1994. Cloning and characterization of a gene affecting the methicillin resistance level and the autolysis rate in *Staphylococcus aureus*. *J. Bacteriol.* **176:**4993–5000.

206. **Marchese, A., G. Arlet, G. C. Schito, P. H. Lagrange, and A. Philippon.** 1998. Characterization of FOX-3, an *ampC*-type plasmid-mediated β-lactamase from an Italian isolate of *Klebsiella oxytoca*. *Antimicrob. Agents Chemother.* **42:**464–467.

207. **Markowitz, S. M., V. D. Wells, D. S. Williams, C. G. Stuart, P. E. Coudorn, and E. S. Wong.** 1991. Antimicrobial susceptibility and molecular epidemiology of β-lactamase-producing aminoglycoside-resistant isolates of *Enterococcus faecalis*. *Antimicrob. Agents Chemother.* **35:**1075–1080.

208. **Marshall, C. G., G. Broadhead, B. K. Leskiw, and G. D. Wright.** 1997. D-Ala-D-Ala ligase from glycopeptide antibiotic-producing organisms are highly homologous to the enterococcal vancomycin-resistance ligases VanA and VanB. *Proc. Natl. Acad. Sci. USA* **94:**6480–6483.

209. **Martin, P., P. Trieu-Cuot, and P. Courvalin.** 1986. Nucleotide sequence of the *tetM* tetracycline resistance determinant of the streptococcal conjugative shuttle transposon Tn*1545*. *Nucleic Acids Res.* **14:**7047–7058.

210. **Martineau, F., F. J. Picard, P. H. Roy, M. Ouellette, and M. G. Bergeron.** 1998. Species-specific and ubiquitous DNA-based assay for identification of *Staphylococcus aureus*. *J. Clin. Microbiol.* **36:**618–623.

211. **Martinez-Martinez, L., A. Pascual, and G. A. Jacoby.** 1998. Quinolone resistance from a transferable plasmid. *Lancet* **351:**797–799.

212. **Matsuoka, M., K. Endou, S. Saitoh, M. Katoh, and Y. Nakajima.** 1995. A mechanism of resistance to partial macrolide and streptogramin B antibiotics in *Staphylococcus aureus* clinically isolated in Hungary. *Biol. Pharm. Bull.* **18:**1482–1486.

213. **Mayford, M., and B. Weisblum.** 1990. The *ermC* leader peptide: amino acid alterations leading to differential efficiency of induction by macrolide-lincosamide-streptogramin B antibiotics. *J. Bacteriol.* **172:**3772–3779.

214. **McGowan, J. E., P. M. Terry, T. S. R. Huang, C. L. Houk, and J. Davies.** 1979. Nosocomial infections with gentamicin-resistant *Staphylococcus aurerus*: plasmid analysis as an epidemiologic tool. *J. Infect. Dis.* **140:**864–872.

215. **Medeiros, A. A.** 1997. Evolution and dissemination of β-lactamases accelerated by generations of β-lactam antibiotics. *Clin. Infect. Dis.* **24:**S19–45.

216. **Meier, A., P. Kirschner, B. Springer, V. A. Steingrub, B. A. Brown, R. J. Wallace, and E. C. Böttger.** 1994. Identification of mutations in the 23S ribosomal RNA gene of clarithromycin resistant *Mycobacterium intracellulare*. *Antimicrob. Agents Chemother.* **38:**381–384.

217. **Michea-Hamzephour, M., J. C. Pechere, P. Plesiat, and T. Kohler.** 1995. OprK and OprM define two genetically distinct multidrug efflux systems in *Pseudomonas aeruginosa*. *Antimicrob. Agents Chemother.* **39:**2392–2396.

218. **Michel, M., and L. Gutmann.** 1997. Methicillin-resistant *Staphylococcus aureus* and vancomycin-resistant enterococci: therapeutic realities and possibilities. *Lancet* **349:**1901–1906.

219. **Miller, G., F. Sabatelli, R. Hare, and J. Waitz.** 1980. Survey of aminoglycoside resistance patterns. *Rev. Ind. Microbiol.* **21:**91–104.

220. **Miller, G. H., F. J. Sabatelli, R. S. Hare, Y. Glupszynski, P. Mackey, D. Shlaes, K. Shimizu, K. J. Shaw, and the Aminoglycoside Resistance Study Group.** 1997. The most frequent aminoglycoside resistance mechanisms—changes with time and geographic area: a reflection of aminoglycoside usage patterns? *Clin. Infect. Dis.* **24:**S46–S62.

221. **Miller, P. F., and M. C. Sulavik.** 1996. Overlaps and parallels in the regulation of intrinsic multiple-antibiotic resistance in *Escherichia coli*. *Mol. Microbiol.* **21:**441–448.

222. **Milton, E. D., C. L. Hewitt, and C. R. Harwood.** 1992. Cloning and sequencing of a plasmid-mediated erythromycin resistance determinant from *Staphylococcus xylosus*. *FEMS Microbiol. Lett.* **97:**141–147.

223. **Mizuuchi, K., L. M. Fisher, M. H. O'Dea, and M. Gellert.** 1980. DNA gyrase action involves the introduction of transient double strand breaks into DNA. *Proc. Natl. Acad. Sci. USA* **77:**1847–1851.

224. **Moazed, D., and H. F. Noller.** 1987. Chloramphenicol, erythromycin, carbomycin and vernamycin B protect overlapping sites in the peptidyl transferase region of 23S ribosomal RNA. *Biochimie* **69:**879–884.

225. **Moore, R. A., B. Beckthold, S. Wong, A. Kureishi, and L. E. Bryan.** 1995. Nucleotide sequence of the *gyrA* gene and characterization of ciprofloxacin-resistant mutants of *Helicobacter pylori*. *Antimicrob. Agents Chemother.* **39:** 107–111.

226. **Mugnier, P., P. Dubrous, I. Casin, G. Ariet, and E. Collatz.** 1996. TEM-derived extended spectrum *β*-lactamases in *Pseudomonas aeruginosa*. *Antimicrob. Agents Chemother.* **40:**2488–2493.

227. **Murakami, K., and W. Minamide.** 1993. PCR identification of methicillin-resistant *Staphylococcus aureus*, p. 539–542. *In* D. H. Persing, T. F. Smith, F. C. Tenover, and T. J. White (ed.), *Diagnostic Molecular Microbiology: Principles and Applications*. American Society for Microbiology, Washington, D.C.

228. **Murray, B.** 1997. Vancomycin-resistant enterococci. *Am. J. Med.* **102:**284–293.

229. **Murray, B. E., B. Mederski-Samoraj, S. K. Foster, J. L. Brunton, and P. Hardford.** 1986. In vitro studies of plasmid-mediated penicillinase from *Streptococcus faecalis* suggest a staphylococcal origin. *J. Clin. Invest.* **77:**289–293.

230. **Musser, J. M.** 1995. Antimicrobial agent resistance in mycobacteria: molecular genetic insights. *Clin. Microbiol. Rev.* **8:**496–514.

231. **Nakamura, H.** 1965. Gene controlling resistance to acriflavin and other basic dyes in *Escherichia coli*. *J. Bacteriol.* **90:**8–14.

232. **Nakamura, H.** 1966. Acriflavin-binding capacity of *Escherichia coli* in relation to acriflavin sensitivity and metabolic activity. *J. Bacteriol.* **92:**1447–1452.

233. **Nakamura, H.** 1968. Genetic determination of resistance to acriflavin, phenyl alcohol, and sodium dodecyl sulfate in *Escherichia coli*. *J. Bacteriol.* **96:**987–996.

234. **Nakamura, S., M. Nakamura, T. Kojima, and H. Yoshida.** 1989. *gyrA* and *gyrB* mutations in guinolone-resistant strains of *Escherichia coli*. *Antimicrob. Agents Chemother.* **33:**254–255.

235. **Naroditskaya, V., M. J. Schlosser, N. Y. Fang, and K. Lewis.** 1993. An *E. coli* gene *emrD* is involved in adaptation to low energy shock. *Biochem. Biophys. Res. Commun.* **196:**803–809.

236. **Navarro, F., and P. Courvalin.** 1994. Analysis of genes encoding D-Ala-D-Ala ligase-related enzymes in *Enterococcus casseliflavus* and *Enterococcus flavescens*. *Antimicrob. Agents Chemother.* **38:**1788–1793.

237. **Neu, H. C.** 1992. The crisis in antibiotic resistance. *Science* **257:**1064–1073.

238. **Neu, H. C., N. Chin, and J. Gu.** 1992. The in-vitro activity of new streptogramins, RP 59599, RP 57669 and RP 54476, alone and in combination. *J. Antimicrob. Chemother.* **30:**83–94.

239. **Neyfakh, A. A.** 1992. The multidrug efflux transporter of *Bacillus subtilis* is a structural and functional homolog of the *Staphylococcus* NorA protein. *Antimicrob. Agents Chemother.* **36:**484–485.

240. **Ng, E. Y., M. Trucksis, and D. C. Hooper.** 1996. Quinolone resistance mutations in topoisomerase IV: relationship to the *flqA* locus and genetic evidence that topoisomerase IV is the primary target and DNA gyrase is the secondary target of fluoroquinolones in *Staphylococcus aureus*. *Antimicrob. Agents Chemother.* **40:**1881–1888.

241. **Ngugen Van, J. C., and L. Gutmann.** 1994. Antibiotic resistance due to reduced permeability in gram-negative bacteria. *Presse Med.* **23:**522–531.

242. **Niga, T., H. Yoshida, H. Hattori, S. Nakamura, and H. Ito.** 1997. Cloning and sequencing of a novel gene (*recG*) that affects the quinolone susceptibility of *Staphylococcus aureus*. *Antimicrob. Agents Chemother.* **41:** 1770–1774.

243. **Nikaido, H.** 1994. Prevention of drug access to bacterial targets: permeability barriers and active efflux. *Science* **264:**382–388.

244. **Nikaido, H.** 1996. Multidrug efflux pumps of gram-negative bacteria. *J. Bacteriol.* **178:**5853–5859.

245. **Nishino, Y., T. Deguchi, M. Yasuda, T. Kawamura, M. Nakano, E. Kanematsu, S. Ozeki, and Y. Kawada.** 1997. Mutations in the *gyrA* and *parC* genes associated with fluoroquinolone resistance in clinical isolates of *Citrobacter freundii*. *FEMS Microbiol. Lett.* **154:**409–414.

246. **Noguchi, N., A. Emura, H. Matsuyama, K. O'Hara, M. Sasatsu, and M. Kono.** 1995. Nucleotide sequence and characterization of erythromycin resistant determinant that encodes macrolide 2′-phosphotransferase I in *Escherichia coli*. *Antimicrob. Agents Chemother.* **39:**2359–2363.

247. **Noguchi, N., J. Katayama, and M. Kono.** 1996. Cloning and nucleotide sequence of the *mphB* gene for macrolide 2′-phosphotransferase II in *Escherichia coli*. *FEMS Microbiol. Lett.* **144:**197–202.

248. **Nordmann, P., S. Mariotte, T. Naas, R. Labia, and M. H. Nicolas.** 1993. Biochemical properties of a carbapenem-hydrolyzing *β*-lactamase from *Enterobacter cloacae* and cloning of the gene into *Escherichia coli*. *Antimicrob. Agents Chemother.* **37:**939–946.

249. **Nordmann, P., and T. Naas.** 1997. The increasing problem of resistance to cephalosporins. *Curr. Opin. Infect. Dis.* **10:**435–439.

250. **Oggioni, M. R., C. G. Dowson, J. M. Smith, R. Provvedi, and G. Pozzi.** 1996. The tetracycline resistance gene *tet*(M) exhibits mosaic structure. *Plasmid* **35:**156–163.

251. **O'Hara, K., T. Kanda, and M. Kono.** 1988. Structure of a phosphorylated derivative of oleanodomycin, obtained by reaction of oleandomycin with an extract of an erthromycin resistant strain of *Escherichia coli*. *J. Antibiot.* **41:**823–827.

252. **O'Hara, K., T. Kawabe, K. Taniguchi, A. Nakamura, and T. Sawai.** 1997. A new macrolide 2′-phosphotransferase in *E. coli*, MPH(2′)-III, abstr. C-67, p. 57. *In* Program and Abstracts of the 37th Interscience Conference on Antimicrobial Agents and Chemotherapy. American Society for Microbiology, Washington, D.C.

253. **O'Hare M. D., and P. E. Reynolds.** 1992. Novel membrane proteins present in teicoplanin-resistant, vancomycin-sensitive, coagulase-negative *Staphylococcus* spp. *J. Antimicrob. Chemother.* **30:**753–768.

254. **Okusu, H., D. Ma, and H. Nikaido.** 1996. AcrAB efflux pump plays a major role in the antibiotic resistance phe-

notype of *Escherichia coli* multiple-antibiotic-resistance phenotype (Mar) mutants. *J. Bacteriol.* **178:**306–308.

255. **Oliva, B., and I. Chopra.** 1992. Tet determinants provide poor protection against some tetracyclines: further evidence for division of tetracycline into two classes. *Antimicrob. Agents Chemother.* **36:**876–878.

256. **Oliva, B., G. G. Gordon, P. McNicholas, G. Ellestad, and I. Chopra.** 1992. Evidence that tetracycline analogs whose primary target is not the bacterial ribosome cause lysis of *Escherichia coli*. *Antimicrob. Agents Chemother.* **36:** 913–919.

257. **Oram, M., and L. M. Fisher.** 1991. 4-Quinolone resistance mutations in the DNA gyrase of *Escherichia coli* clinical isolates identified by using the polymerase chain reaction. *Antimicrob. Agents Chemother.* **35:**387–389.

258. **Ounissi, H., and P. Courvalin.** 1985. Nucleotide sequence of the gene *ereA* encoding the erythromycin esterase in *Escherichia coli*. *Gene* **35:**271–278.

259. **Pan, X.-S., J. Ambler, S. Mehtar, and L. M. Fisher.** 1996. Involvement of topoisomerase IV and DNA gyrase as ciprofloxacin targets in *Streptococcus pneumoniae*. *Antimicrob. Agents Chemother.* **40:**2321–2326.

260. **Pan, X.-S., and L. M. Fisher.** 1997. Targeting of DNA gyrase in *Streptococcus penumoniae* by sparfloxacin: selective targeting of gyrase or topoisomerase IV by guinolones. *Antimicrob. Agents Chemother.* **41:**471–474.

261. **Pang, Y., T. Bosch, and M. C. Roberts.** 1994. Single polymerase chain reaction for the detection of tetracycline-resistant determinants TetK and TetL. *Mol. Cell. Probes* **8:**417–422.

262. **Pang, Y., B. A. Brown, V. A. Steingrube, R. J. Wallace, Jr., and M. C. Roberts.** 1994. Tetracycline resistance determinants in *Mycobacterium* and *Streptomyces* species. *Antimicrob. Agents Chemother.* **38:**1408–1412.

263. **Parish, T., J. Liu, H. Nikaido, and N. G. Stoker.** 1997. A *Mycobacterium smegmatis* mutant with a defective inositol monophosphate phosphatase gene homolog has altered cell envelope permeability. *J. Bacteriol.* **179:**7827–7833.

264. **Park, I. S. C. H. Lin, and C. T. Walsh.** 1997. Bacterial resistance to vancomycin: overproduction, purification, and characterization of VanC2 from *Enterococcus casseliflavus* as a D-Ala-D-Ser ligase. *Proc. Natl. Acad. Sci. USA* **94:**10040–10044.

265. **Patel, R., J. R. Uhl, P. Kohner, M. K. Hopkins, J. M. Steckelberg, B. Kline, and F. R. Cockerill III.** 1998. DNA sequence variation within *vanA*, *vancB*, *vanC-1*, and *vanC2/3* genes of clinical *Enterococcus* isolates. *Antimicrob. Agents Chemother.* **42:**202–205.

266. **Paulsen, I. T., M. H. Brown, T. G. Littlejohn, B. A. Mitchell, and R. A. Skurray.** 1996. Multidrug resistance proteins QacA and QacB from *Staphylococcus aureus*: membrane topology and identification of residues involved in substrate specificity. *Proc. Natl. Acad. Sci. USA* **93:**3630–3635.

267. **Paulsen, I. T., M. H. Brown, and R. A. Skurray.** 1996. Proton-dependent multidrug efflux systems. *Microbiol. Rev.* **60:**575–608.

268. **Paulsen, I. T., R. A. Skurray, R. Tam, M. H. Saie, Jr., R. J. Turner, J. H. Weiner, E. B. Goldberg, and L. I. Grinius.** 1996. The SMR family: a novel family of multidrug efflux proteins involved with the efflux of lipophilic drugs. *Mol. Microbiol.* **19:**1167–1175.

269. **Payie, K. G., and A. J. Clarke.** 1997. Characterization of gentamicin 2′-*N*-acetyltransferase from *Providencia stuartii*: its use of peptidoglycan metabolites for acetylation of both aminoglycosides and peptidoglycan. *J. Bacteriol.* **179:**4106–4114.

270. **Payne, D. J.** 1993. Metallo-β-lactamases—a new therapeutic challenge. *J. Med. Microbiol.* **39:**93–99.

271. **Périchon, B. P. Reynolds, and P. Courvalin.** 1997. VanD-type glycopeptide-resistant *Enterococcus faecium* BM4339. *Antimicrob. Agents Chemother.* **41:**2016–2018.

272. **Périchon, B., J. Tankovic, and P. Courvalin.** 1997. Characterization of a mutation in the *parE* gene that confers fluoroquinolone resistance in *Streptococcus pneumoniae*. *Antimicrob. Agents Chemother.* **41:**1166–1167.

273. **Persing, D., D. A. Relman, and F. C. Tenover.** 1996. Genotypic detection of antimicrobial resistance, p. 33–58. *In* D. H. Persing (ed.), *PCR Protocols for Emerging Infectious Diseasees*. American Society for Microbiology, Washington, D.C.

274. **Persing, D. H., T. F. Smith, F. C. Tenover, and T. J. White (ed.).** 1993. *Diagnostic Molecular Microbiology: Principles and Applications*. American Society for Microbiology, Washington, D.C.

275. **Piddock, L. J. V., and R. Wise.** 1987. Cefoxitin resistance in *Bacteroides* species: evidence indicating two mechanisms causing decreased susceptibility. *J. Antimicrob. Chemother.* **19:**161–170.

276. **Piras, G., A. el Kharroubi, J. van Beumen, E. Coeme, J. Çoyette, and J. M. Ghuysen.** 1990. Characterization of an *Enterococcus hirae* penicillin-binding protein 3 with low penicillin affinity. *J. Bacteriol.* **172:**6856–6862.

277. **Poole, K. N. Gotoh, H. Tsujimoto, Q. Zhao, A. Wada, t. Yamasaki, S. Neshat, J. Yamagishi, X. Z. Li, and T. Nishino.** 1996. Overexpression of the *mexC-mexD-oprJ* efflux operon in *nfxB*-type multidrug-resistant strains of *Pseudomonas aeruginosa*. *Mol. Microbiol.* **21:**713–724.

278. **Poole, K., K. Tetro, Q. Zhao, S. Neshat, D. E. Heinrichs, and N. Bianco.** 1996. Expression of the multidrug resistance operon *mexA-mexB-oprM* in *Pseudomonas aeruginosa*: *mexR* encodes a regulator of operon expression. *Antimicrob. Agents Chemother.* **40:**2021–2028.

279. **Portillo-Gomez, L., J. Nair, D. A. Rouse, and S. L. Morris.** 1995. The absence of genetic markers for streptomycin and rifampicin resistance in *Mycobacterium avium* complex strains. *J. Antimicrob. Chemother.* **36:**1049–1053.

280. **Poyart, C., C. Pierre, G. Quesne, B. Pron, P. Berche, and P. Trieu-Cuot.** 1997. Emergence of vancomycin resistance in the genus *Streptococcus*: characterization of a *vanB* transferable determinant in *Streptococcus bovis*. *Antimicrob. Agents Chemother.* **41:**24–29.

281. **Quinn, J. P., E. J. Dudek, C. A. DiVencenzo, D. A. Lucks, and S. A. Lerner.** 1986. Emergence of resistance to imipenem during therapy for *Pseudomonas aeroginosa* infections. *J. Infect. Dis.* **154:**289–294.

282. **Raimondi, A., A. Traverso, and H. Nikaido.** 1991. Imipenem- and meropenem-resistance mutants of *Enterobacter cloacae* and *Proteus rettgeri* lack porins. *Antimicrob. Agents Chemother.* **35:**1174–1180.

283. **Rasmussen, B. A., and K. Bush.** 1997. Carbapenem-hydrolyzing β-lactamases. *Antimicrob. Agents Chemother.* **41:**223–232.

284. **Rasmussen, B. A., K. Bush, D. Keeney, Y. Yang, R. Hare, C. O'Gara, and A. A. Medeiros.** 1996. Characterization of IMI-1 β-lactamase, a novel class A carbepenem-hydrolyzing enzyme from *Enterobacter cloacae*. *Antimicrob. Agents Chemother.* **40:**2080–2086.

285. **Rasmussen, B. A., K. Bush, and F. P. Tally.** 1997. Antimicrobial resistance in anaerobes. *Clin. Infect. Dis.* **24:** S110–S120.

286. **Rasmussen, B. A., Y. Gluzman, and F. P. Tally.** 1994. Inhibition of protein synthesis occurring on tetracycline-resistant, TetM-protected ribosomes by a novel class of tetracyclines, the glycylcyclines. *Antimicrob. Agents Chemother.* **38:**1658–1660.

287. **Rather, P. N., E. Orosz, K. J. Shaw, R. Hare, and G. Miller.** 1993. Characterization and transcriptional regulation of the 2′-*N*-acetyltransferase gene from *Providencia stuartii*. *J. Bacteriol* **175:**6492–6498.

288. **Reece, R. J., and A. Maxwell.** 1991. DNA gyrase: structure and function. *Crit. Rev. Biochem. Mol. Biol.* **26:**335–375.

289. **Rende-Fournier, R., R. Leclercq, M. Galimand, J. Duval, and P. Courvalin.** 1993. Identification of the *satA* gene encoding a streptogramin A acetyltransferase in *Enterococcus faecium* BM4145. *Antimicrob. Agents Chemother.* **37:**2119–2125.

290. **Reynolds, P. E., and C. Fuller.** 1986. Methicillin-resistant strains of *Staphylococcus aureus*: presence of an additional penicillin-binding protein in all strains examined. *FEMS Microbiol. Lett.* **33:**251–254.

291. **Reynolds, P. E., H. A. Snaith, A. J. Maguire, S. Dutka-Malen, and P. Courvalin.** 1994. Analysis of peptidoglycan precursors in vancomycin-resistant *Enterococcus gallinarum* BM4174. *Biochem. J.* **301:**5–8.

292. **Ridenhour, M. B., H. M. Fletcher, J. E. Mortensen, and L. Daneo-Moore.** 1996. A novel tetracycline-resistant determinant, *tet*(U), is encoded on the plasmid pKq10 in *Enterococcus faecium*. *Plasmid* **35:**71–80.

293. **Roberts, M. C.** 1996. Tetracycline resistance determinants: mechanisms of action, regulation of expression, genetic mobility, and distribution. *FEMS Microbiol. Rev.* **19:**1–24.

294. **Roberts, M. C., Y. Pang, D. E. Riley, S. L. Hillier, R. C. Berger, and J. N. Krieger.** 1993. Detection of TetM and TetO tetracycline resistance genes by polymerase chain reaction. *Mol. Cell. Probes* **7:**387–393.

295. **Ross, J. I., E. A. Eady, J. H. Cove, and S. Baumberg.** 1996. Minimal functional system required for expression of erythromycin resistance by *msrA* in *Staphylococcus aureus* RN4220. *Gene* **183:**143–148.

296. **Ross, J. I., E. A. Eady, J. H. Cove, W. J. Cunliffe, S. Baumburg, and J. C. Wootten.** 1990. Inducible erythromyin resistance in staphylococci is encoded by a member of the ATP-binding transport super-gene family. *Mol. Microbiol.* **4:**1207–1214.

297. **Ross, J. I. E. A. Eady, J. H. Cove, C. E. Jones, A. H. Ratyal, Y. W. Miller, S. Vyakrnam, and W. J. Cunliffe.** 1997. Clinical resistance to erythromycin and clindamycin in cutaneous propionibacteria isolated from acne patients is associated with mutations in 23S rRNA. *Antimicrob. Agents Chemother.* **41:**1162–1165.

298. **Ruiz, J., F. Marco, P. Goñi, F. Gallardo, J. Menas, A. Trilla, T. Jimenex de Anta, and J. Vila.** 1995. High frequency of mutations at codon 83 of the *gyrA* gene of quinolone-resistant clinical isolates. *J. Antimicrob. Chemother.* **36:**737–738.

299. **Russell, A. D., and I. Chopra.** 1996. *Understanding Antibacterial Action and Resistance*, p. 46–49. Ellis Horwood, New York.

300. **Saier, M. H., Jr., I. T. Paulsen, and A. Matin.** 1997. A bacterial model system for understanding multidrug resistance. *Microb. Drug Resist.* **3:**289–295.

301. **Saier, M. H., Jr., I. T. Paulsen, M. K. Sliwinski, S. Pao, R. A. Skurray, and H. Nikaido.** 1998. Evolutionary origins of multidrug and drug-specific efflux pumps in bacteria. *FASEB* **J12:**265–274.

302. **Saier, M. H., Jr., R. Tam, and J. Reizer.** 1994. Two novel families of bacterial membrane proteins concerned with nodulation, cell division, and transport. *Mol. Microbiol.* **11:**841–847.

303. **Sanchez, L., W. Pan, M. Vinas, and H. Nikaido.** 1997. The *acrAB* homolog of *Haemophilus influenzae* codes for a functional multidrug efflux pump. *J. Bacteriol.* **179:** 6855–6857.

304. **Sanders, C. C., and W. E. Sanders.** 1992. β-lactam resistance in gram-negative bacteria: global trends and clinical impact. *Clin. Infect. Dis.* **15:**824–839.

305. **Schmid, M. B.** 1990. A locus affecting nucleoid segregation in *Salmonella typhimurium*. *J. Bacteriol.* **172:**5416–5424.

306. **Schnappinger, D., and W. Hillen.** 1996. Tetracyclines: antibiotic action, uptake, and resistance mechanisms. *Arch. Microbiol.* **165:**359–369.

307. **Senda, K. Y. Arakawa, K. Nakashima, H. Ito, S. Ichiyama, K. Shimokata, N. Kato, and M. Ohta.** 1996. Multifocal outbreaks of metallo-β-lactamase-producing *Pseudomonas aeruginosa* resistant to broad-spectrum β-lactams including carbapenems. *Antimicrob. Agents Chemother.* **40:**349–353.

308. **Seppälä, H., M. Skurnik, H. Soini, M. C. Roberts, and P. Huovinen.** 1998. A novel erythromycin resistance methylase gene (*ermTR*) in *Streptococcus pyogenes*. *Antimicrob. Agents Chemother.* **42:**257–262.

309. **Shangwei, W., C. Piscitelli, H. de Lencastre, and A. Tomasz.** 1996. Tracking the evolutionary origin of the methicillin resistance gene: cloning and sequencing of a homologue of *mecA* from a methicillin susceptible strain of *Staphylococcus sciuri*. *Microb. Drug Resist.* **2:**435–441.

310. **Shaw, K. J., P. N. Rather, R. S. Hare, and G. H. Miller.** 1993. Molecular genetics of aminoglycoside resistance genes and familiar relationships of the aminoglycoside-modifying enzymes. *Microbiol. Rev.* **57:**138–163.

311. **Shaw, K. J., P. Rather, F. J. Sabetelli, P. Mann, H. Manayyer, R. Mierzwa, G. Petrikkos, R. S. Hare, G. H. Miller, P. Bennett, and P. Downey.** 1992. Characterization of the chromosomal *aac(6′)-Ia* gene from *Serratia marcescens*. *Antimicrob. Agents Chemother.* **36:**1447–1455.

312. **Shiba, T., K. Ishiguro, N. Takemoto, H. Koibichi, and K. Sugimoto.** 1995. Purification and characterization of the *Pseudomonas aeruginosa* NfxB protein, the negative regulator of the *nfxB* gene. *J. Bacteriol.* **177:**5872–5877.

313. **Shlaes, D. M., J. H. Shlaes, S. Vincent, L. Etter, P. D. Fey, and R. V. Goering.** 1993. Teicoplanin-resistant *Staphylococcus aureus* expresses a novel membrane protein and increases expression of penicillin-binding pro-

tein 2 complex. *Antimicrob. Agents Chemother.* **37:**2432–2437.

314. **Sieradzki, K., P. Villari, and A. Tomasz.** 1998. Decreased susceptibilities to teicoplanin and vancomycin among coagulase-negative methicillin-resistant clinical isolates of staphylococci. *Antimicrob. Agents Chemother.* **42:**100–107.
315. **Signoretto, C., M. Boaretti, and P. Canepari.** 1994. Cloning, sequencing, and expression in *Escherichia coli* of the low-affinity penicillin binding protein of *Enterococcus faecalis*. *FEMS Microbiol. Lett.* **123:**99–106.
316. **Silver, L. L., and K. A. Bostian.** 1993. Discovery and development of new antibiotics: the problem of antibiotic resistance. *Antimicrob. Agents Chemother.* **37:**377–383.
317. **Sirot, D., C. Chanal, R. Labia, M. Meyran, J. Sirot, and R. Cluzel.** 1989. Comparative study of five plasmid-mediated ceftazidimases isolated in *Klebsiella pneumoniae*. *J. Antimicrob. Chemother.* **25:**343–351.
318. **Sloan, J., L. M. McMurry, D. Lyras, S. B. Levy, and J. I. Rood.** 1994. The *Clostridium perfringens* TetP determinant comprises two overlapping genes: *tetA*(P), which mediates active tetracycline efflux, and *tetB*(P), which is related to the ribosomal protection family of tetracycline-resistance determinants. *Mol. Microbiol.* **11:**403–415.
319. **Smith, C. J., C. Owen, and L. Kirby.** 1992. Activation of a cryptic streptomycin-resistance gene in the *Bacteroides erm* transposon, Tn*4551*. *Mol. Microbiol.* **6:**2287–2297.
320. **Someya, Y., A. Yamaguchi, and T. Sawai.** 1995. A novel glycylcycline, 9-(*N*,*N*-dimethylglycylamido)-6-demethyl-6-deoxytetracycline, is neither transported nor recognized by the transposon Tn*10*-encoded methal-tetracycline/H^+ antiporter. *Antimicrob. Agents Chemother.* **39:**247–249.
321. **Speer, B. S., L. Bedzyk, and A. A. Salyers.** 1991. Evidence that a novel tetracycline resistance gene found on two *Bacteroides* transposons encodes an NADP-requiring oxidoreductase. *J. Bacteriol.* **173:**176–183.
322. **Speer, B. S., and A. A. Salyers.** 1989. Novel aerobic tetracycline resistance gene that chemically modifies tetracycline. *J. Bacteriol.* **171:**148–153.
323. **Spratt, B.** 1994. Resistance to antibiotics mediated by target alterations. *Science* **264:**388–393.
324. **Stein, D. C., R. J. Danaher, and T. M. Cook.** 1991. Characterization of a *gyrB* mutation responsible for low-level nalidixic acid resistance in *Neisseria gonorrhoeae*. *Antimicrob. Agents Chemother.* **35:**622–626.
325. **Stevens, A. M., N. B. Shoemaker, L.-Y, Li, and A. A. Salyers.** 1993. Tetracycline regulation of genes on *Bacteroides* conjugative transposons. *J. Bacteriol.* **175:**6134–6141.
326. **Su, Y. A., P. He, and D. B. Clewell.** 1992. Characterization of the *tet*(M) determinant of Tn*916*: evidence for regulation by transcription attenuation. *Antimicrob. Agents Chemother.* **36:**769–778.
327. **Sulavik, M. C., L. F. Gambino, and P. F. Miller.** 1995. The MarR repressor of the multiple antibiotic resistance (*mar*) operon in *Escherichia coli*: prototypical member of a family of bacterial regulatory proteins involved in sensing phenolic compounds. *Mol. Med.* **1:**436–446.
328. **Sum, P.-E., V. Lee, and R. T. Testa, J. J. Hlavka, G. A. Ellestad, J. D. Bloom, Y. Gluzman, and F. P. Tally.** 1994. Glycylcyclines. 1. A new generation of potent antibacterial agents through modification of 9-aminotetracyclines. *J. Med. Chem.* **37:**184–188.
329. **Sutcliffe, J., T. Grebe, A. Tait-Kamradt, and L. Wondrack.** 1996. Detection of erythromycin resistant determinants by PCR. *Antimicrob. Agents Chemother.* **40:**2562–2566.
330. **Sutcliffe, J., and A. Tait-Kamradt.** Unpublished data.
331. **Sutcliffe, J., A. Tait-Kamradt, and L. Wondrack.** 1996. *Streptococcus pneumoniae* and *Streptococcus pyogenes* resistant to macrolides but sensitive to clindamycin: a common resistance pattern mediated by an efflux system. *Antimicrob. Agents Chemother.* **40:**1817–1824.
332. **Suter, T. M., V. K. Viswanathan, and N. P. Cianciotto.** 1997. Isolation of a gene encoding a novel spectinomycin phosphotransferase from *Legionella pneumophila*. *Antimicrob. Agents Chemother.* **41:**1385–1388.
333. **Sutherland, R.** 1990. Bacterial resistance to β-lactam antibiotics: problems and solutions. *In* E. Jucker (ed.), *Progress in Drug Research*, vol. 41, p. 95–149. Birkhäuser Verlag, Basel, Switzerland.
334. **Tabaqchali, S.** 1997. Vancomycin-resistant *Staphylococcus aureus*: apocalypse now? *Lancet* **350:**1644–1655.
335. **Taber, H. W., J. P. Mueller, P. F. Miller, and A. S. Arrow.** 1987. Bacterial uptake of aminoglycoside antibiotics. *Microbiol. Rev.* **51:**439–457.
336. **Tait-Kamradt, A., J. Clancy, M. Cronan, F. Dib-Hajj, L. Wondrack, W. Yuan, and J. Sutcliffe.** 1997. MefE is necessary for the erythromycin-resistant M phenotype in *Streptococcus pneumoniae*. *Antimicrob. Agents Chemother.* **41:**2251–2255.
337. **Takiff, H. E., M. Cimino, M. C. Musso, T. Weisbrod, R. Martinez, M. B. Delgado, L. Salazar, B. R. Bloom, and W. R. Jacobs, Jr.** 1996. Efflux pump of the proton antiporter family confers low-level fluoroquinolone resistance in *Mycobacterium smegmatis*. *Proc. Natl. Acad. Sci. USA* **93:**362–366.
338. **Tanaka, T., T. Horii, K. Shibayama, K. Sato, S. Ohsuka, Y. Arakawa, K. Yamaki, K. Takagi, and M. Ohta.** 1997. RobA-induced multiple antibiotic resistance largely depends on the activation of the AcrAB efflux. *Microbiol. Immunol.* **41:**697–702.
339. **Tankovic, J., F. Mahjoubi, P. Courvalin, J. Duval, and R. Leclercq.** 1996. Development of fluoroquinolone resistance in *Enterococcus faecalis* and role of mutations in the DNA gyrase *gyrA* gene. *Antimicrob. Agents Chemother.* **40:**2558–2561.
340. **Taylor, D. E., and A. Chau.** 1996. Tetracycline resistance mediated by ribosomal protection. *Antimicrob. Agents Chemother.* **40:**1–5.
341. **Taylor, D. E., C. A. Trieber, G. Trescher, and M. Bekkering.** 1998. Host mutations (*miaA* and *rpsL*) reduce tetracycline resistance mediated by Tet(O) and Tet(M). *Antimicrob. Agents Chemother.* **42:**59–64.
342. **Tenover, F. C., M. V. Lancaster, B. C. Hill, C. D. Steward, S. A. Stocker, G. A. Hancock, C. M. O'Hara, N. C. Clark, and K. Hiramatsu.** 1998. Characterization of staphylococci with reduced susceptibilities of vancomycin and other glycopeptides. *Antimicrob. Agents Chemother.* **36:**1020–1027.

343. **Tenson, T., A. Deblasio, and A. Mankin.** 1996. A functional peptide encoded in the *Escherichia coli* 23S rRNA. *Proc. Natl. Acad. Sci. USA* **93:**5641–5646.

344. **Tenson, T., L. Q. Xiong, P. Kloss, and A. S. Mankin.** 1997. Erythromycin resistance peptides selected from random peptide libraries. *J. Biol. Chem.* **272:**17425–17430.

345. **Tomasz, A., H. B. Drugeon, H. M. de Lencastre, D. Jabes, and L. McDougall.** 1989. New mechanism of methicillin resistance in *Staphylococcus aureus*: clinical isolates that lack the PBP 2a gene and contrain normal-penicillin-binding proteins with modified penicillin-binding capacity. *Antimicrob. Agents Chemother.* **33:** 1869–1874.

346. **Tomasz, A., and R. Munoz.** 1995. β-lactam antibiotic resistance in gram-positive bacterial pathogens of the upper respiratory tract: a brief overview of mechanisms. *Microb. Drug Resist.* **1:**103–109.

347. **Turner, R. J., D. E. Taylor, and J. H. Weiner.** 1997. Expression of *Escherichia coli* TehA gives resistance to antiseptics and disinfectants similar to that conferred by multidrug resistance efflux pumps. *Antimicrob. Agents Chemother.* **41:**440–444.

348. **Udou, T., Y. Mizuguchi, and R. J. Wallace, Jr.** 1989. Does aminoglycoside-acetyltransferase in rapidly growing mycobacteria have a metabolic function in addition to aminoglycoside inactivation? *FEMS Microbiol. Lett.* **48:** 227–230.

349. **Uhl, J. R., G. S. Sandhu, B. C. Kline, and F. R. Cockerill III.** 1996. PCR-RFLP detection of point mutations in the catalase-peroxidase gene (*katG*) of *Mycobacterium tuberculosis* associated with isoniazid resistance, p. 144–149. *In* D. H. Persing (ed.), *PCR Protocols for Emerging Infectious Diseases*. American Society for Microbiology, Washington, D.C.

350. **Van de Klundert, J. A. M., and J. S. Vliegenthart.** 1993. PCR detection of genes coding for aminoglycoside-modifying enzymes, p. 547–552. *In* D. H. Persing, T. F. Smith, F. C. Tenover, and T. J. White (ed.), *Diagnostic Molecular Microbiology: Principles and Applications*. American Society for Microbiology, Washington, D.C.

351. **Van Veen, H. W., and W. N. Konings.** 1997. Drug efflux proteins in multidrug resistant bacteria. *Biol. Chem.* **378:** 769–777.

352. **Versalovic, J., D. Shortridge, K. Kibler, M. V. Griffy, J. Beyer, R. K. Flamm, S. K. Tanaka, D. Y. Graham, and M. F. Go.** 1996. Mutations in 23S rRNA are associated with clarithromycin resistance in *Helicobacter pylori*. *Antimicrob. Agents Chemother.* **40:**477–480.

353. **Vinella, D., R. D'Ari, and A. Jaffe.** 1992. Penicillin binding protein 2 is dispensable in *Escherichia coli* when ppGpp synthesis is induced. *EMBO J.* **11:**1493–1501.

354. **Vinella, D., B. Gagny, D. Joseleau-Petit, R. D'Ari, and M. Cashel.** 1996. Mecillinam resistance in *Escherichia coli* is conferred by loss of a second activity of the AroK protein. *J. Bacteriol.* **178:**3818–3828.

355. **Wachi, M., M. Doi, S. Tamaki, W. Park, S. Nakahima-Iijima, and M. Matsuhashi.** 1987. Mutant isolation and molecular cloning of the *mre* genes, which determine cell shape, sensitivity to mecillinam, and amount of penicillin-binding proteins in *Escherichia coli*. *J. Bacteriol.* **169:**4935–4940.

356. **Walsh, C. T., S. L. Fisher, I.-S. Park, M. Prahalad, and Z. Wu.** 1996. Bacterial resistance to vancomycin: five genes and one missing hydrogen bond tell the story. *Chem. Biol.* **3:**21–28.

357. **Walsh, T. R., W. A. Neville, M. H. Haran, D. Tolson, D. J. Payne, J. H. Bateson, A. P. Macgowan, and P. M. Bennett.** 1998. Nucleotide and amino acid sequences of the metallo-β-lactamase, ImiS, from *Aeromonas veronii* bv. *sobria*. *Antimicrob. Agents Chemother.* **42:**436–439.

358. **Wang, T., M. Tanaka, and K. Sato.** 1998. Detection of *grlA* and *gyrA* mutations in 344 *Staphylococcus aureus* strains. *Antimicrob. Agents Chemother.* **42:**236–240.

359. **Wang, Y., and D. E. Taylor.** 1991. A DNA sequence upstream of the *tet*(O) gene is required for full expression of tetracycline resistance. *Antimicrob. Agents Chemother.* **35:**2020–2025.

360. **Watanabe, M., S. Iyobe, M. Inoue, and S. Mitsuhashi.** 1991. Transferable imipenem resistance in *Pseudomonas aeruginosa*. *Antimicrob. Agents Chemother.* **35:**147–151.

361. **Weisblum, B.** 1995. Erythromycin resistance by ribosome modification. *Antimicrob. Agents Chemother.* **39:** 577–585.

362. **Weisblum, B.** 1995. Insights into erythromycin action from studies of its activity as inducer of resistance. *Antimicrob. Agents Chemother.* **39:**797–805.

363. **Weisblum, B.** 1998. Macrolide resistance. *Drug Resist. Updates* **1:**29–41.

364. **Westling-Häggström, B., and S. Normark.** 1975. Genetic and physiological analysis of an *envB* spherelike mutant of *Escherichia coli* K-12 and characterization of its transductants. *J. Bacteriol.* **123:**75–83.

365. **Wexler, H. M., and S. Halebian.** 1990. Alterations to the penicillin-binding proteins in the *Bacteroides fragilis* group: a mechanism for non-β-lactamase mediated cefoxitin resistance. *J. Antimicrob. Chemother.* **26:**7–20.

366. **Williams, D. L., C. W. Limbers, L. Spring, S. Jayachandra, and T. P. Gillis.** 1996. PCR-heteroduplex detection of rifampin-resistant *Mycobacterium tuberculosis*, p. 122–129. *In* D. H. Persing (ed.), *PCR Protocols for Emerging Infectious Diseases*. American Society for Microbiology, Washington, D.C.

367. **Wolfson, J., D. Hooper, G. McHugh, M. Bozza, and M. Swartz.** 1990. Mutants of *Escherichia coli* K-12 exhibiting reduced killing by both quinolone and β-lactam antibiotics. *Antimicrob. Agents Chemother.* **34:**1938–1943.

368. **Wolfson, J. S., and D. C. Hooper.** 1989. Fluoroquinolone antimicrobial agents. *Clin. Microbiol. Rev.* **2:**378–424,

369. **Wondrack, L., M. Massa, B. V. Yang, and J. Sutcliffe.** 1996. A clinical strain of *Staphylococcus aureus* inactivates and effluxes macrolides. *Antimicrob. Agents Chemother.* **40:**992–998.

370. **Wright, G. D., and P. Ladak.** 1997. Overexpression and characterization of the chromosomal aminoglycoside 6′-*N*-acetyltransferase from *Enterococcus faecium*. *Antimicrob. Agents Chemother.* **41:**956–960.

371. **Wu, H. Y., G. H. Miller, M. G. Blanco, R. S. Hare, and K. J. Shaw.** 1997. Cloning and characterization of an aminoglycoside 6′-*N*-acetyltransferase gene from *Citrobacter freundii* which confers an altered resistance profile. *Antimicrob. Agents Chemother.* **41:**2439–2447.

372. **Yamada, T., D. Tipper, and J. Davies.** 1968. Enzymatic inactivation of streptomycin by R factor-resistant *Escherichia coli*. *Nature* **219:**288–292.

373. **Yamagishi, J.-I, Y. Furutani, S. Inoue, T. Ohue, S. Nakamura, and M. Shimizu.** 1981. New nalidixic acid resistance mutations related to DNA gyrase activity. *J. Bacteriol.* **148:**450–458.

374. **Yamagishi, J.-I., T. Kojima, Y. Oyamada, K. Fujimoto, H. Hattori, S. Nakamura, and M. Inoue.** 1996. Alterations in the DNA topoisomerase IV *grlA* gene responsible for quinolone resistance in *Staphylococcus aureus*. *Antimicrob. Agents Chemother.* **40:**1157–1163.

375. **Yamagishi, J.-I., H. Yoshida, M. Yamayoshi, and S. Nakamura.** 1986. Nalidixic acid-resistant mutations of the *gyrB* gene of *Escherichia coli*. *Mol. Gen. Genet.* **204:** 367–373.

376. **Yang, Y., B. A. Rasmussen, and K. Bush.** 1992. Biochemical characterization of the metallo-β-lactamase CcrA from *Bacteroides fragilis* TAL3636. *Antimicrob. Agents Chemother.* **36:**1155–1157.

377. **Yang, Y., P. C. Wu, and D. M. Livermore.** 1990. Biochemical characterization of a β-lactamase that hydrolyzes penems and carbapenems from two *Serratia marcescens* isolates. *Antimicrob. Agents Chemother.* **34:** 755–758.

378. **Yerushalmi, H., M. Lebendiker, and S. Schuldiner.** 1995. EmrE, and *Escherichia coli* 12-kDa multidrug transporter, exchanges toxic cations and H^+ and is soluble in organic solvents. *J. Biol. Chem.* **270:**6856–6863.

379. **Yoshida, H., M. Bogaki, M. Nakamura, and S. Nakamura.** 1990. Quinolone resistance-determining region in the DNA gyrase *gyrA* gene of *Escherichia coli*. *Antimicrob. Agents Chemother.* **34:**1271–1272.

380. **Yoshida, H., M. Bogaki, M. Nakamura, L. M. Yamanaka, and S. Nakamura.** 1991. Quinolone resistance-determining region in the DNA gyrase *gyrB* gene of *Escherichia coli*. *Antimicrob. Agents Chemother.* **35:**1647–1650.

381. **Yoshida, H., M. Nakamura, M. Bogaki, and S. Nakamura.** 1990. Proportion of DNA gyrase mutants among quinolone-resistant strains of *Pseudomonas aeruginosa*. *Antimicrob. Agents Chemother.* **34:**1273–1275.

382. **Yotsuji, A., J. Mitsuyama, R. Hori, T. Yasuda, I. Saikawa, M. Inoue, and S. Mitsuhashi.** 1988. Mechanism of action of cephalosporins and resistance caused by decreased affinity for penicillin-binding proteins in *Bacteroides fragilis*. *Antimicrob. Agents Chemother.* **32:**1848–1853.

383. **Yu, L., A. M. Petros, A. Schnuchel, P. Zhong, J. M. Severin, K. Walter, T. F. Holzman, and S. W. Fesik.** 1997. Solution structure of an rRNA methyltransferase (ErmAM) that confers macrolide-lincosamide-streptogramin antibiotic resistance. *Nat. Struct. Biol.* **4:** 483–489.

384. **Zalacain, M., and E. Cundliffe.** 1991. Cloning of *tlrD*, a fourth resistance gene, from the tylosin producer, *Streptomyces fradiae*. *Gene* **97:**137–142.

385. **Zhang, H.-Z., H. Schmidt, and W. Piepersberg.** 1992. Molecular cloning and characterization of 2 lincomycin-resistant genes, *lmrA* and *lmrB*, from *Streptomyces lincolnensis* 78-11. *Mol. Microbiol.* **6:**2147–2157.

386. **Zorzi, W., X. Y. Zhou, O. Dardenne, J. Lamotte, D. Raze, J. Pierre, L. Gutmann, and J. Coyette.** 1996. Structure of the low-affinity penicillin-binding protein 5 PBP5fm in wild-type and highly penicillin-resistant strains of *Enterococcus faecium*. *J. Bacteriol.* **178:**4948–4957.

Genetics of Bacteriocins Produced by Lactic Acid Bacteria and Their Use in Novel Industrial Applications

G. E. ALLISON AND T. R. KLAENHAMMER

63

Bacteriocins are antimicrobial peptides, proteins, or protein complexes produced by several bacterial species. Among lactic acid bacteria (LAB), bacteriocin producers have been identified in the *Lactococcus*, *Pediococcus*, *Lactobacillus*, *Leuconostoc*, *Carnobacterium*, *Enterococcus*, and *Streptococcus* genera. The diversity of the bacteriocins and their wide spectra of activity reflect the broad habitats in which LAB compete, ranging from the intestinal tracts of animals to fermenting food substrates. Bacteriocins produced by LAB have been of particular interest because of their existing and potential ability to act as natural preservatives or biopreservatives in foods, and several excellent reviews have been written on this topic (21, 61, 126). The biochemical and genetic characterization of bacteriocins in LAB has also been investigated from both applied and fundamental perspectives.

Bacteriocins have been categorized by their biochemical properties, and the following four groups are currently recognized (72): class I, lantibiotics; class II, small, heat-stable peptides; class III, large, heat-labile proteins; and class IV, complex bacteriocins composed of protein and carbohydrate or lipid moieties. The biochemistry, mode of action, and genetics of the class I and class II bacteriocins have been the most fully characterized to date. The lantibiotics are small peptides that contain the unusual amino acids lanthionine, methyllanthionine, dehydroalanine, and dehydrobutyrine, which result from complex posttranslational modifications. The class II bacteriocins do not contain modified amino acids and are usually small, cationic peptides. Although the biological characteristics of these peptides may vary, many of the class I and II peptides act through the formation of pores in the membranes of sensitive cells (reviewed in references 1, 32, and 143). Each bacteriocin system is also characterized by a unique immunity factor that protects the producer from its own bacteriocin. Several immunity factors have been identified and studied. The data suggest that there may be three different mechanisms of immunity (reviewed in references 1 and 118): the immunity factor may interact with the bacteriocin receptor; the immunity factor may inhibit and/or block formation of a pore or channel; or the bacteriocin may be imported and inactivated in the cell.

The study of the genetics of bacteriocin production has provided a much-needed "back door" into the study of transport and regulatory processes of LAB. In addition to the genes encoding bacteriocin(s) and immunity factor (s), several other gene products are also required for bacteriocin production. There are two mechanisms through which bacteriocins are transported to the extracellular environment. Most bacteriocins are "exported" via ATP binding cassette transporters (ABC transporters) (reviewed in references 56, 68, and 99). In addition to the ABC transporter, an accessory factor is also required for transport of the class II peptides. Bacteriocins exported via ABC transporters are synthesized as prebacteriocins characterized by a rather conserved N-terminal extension of 14 to 30 amino acids which is cleaved at an invariant Gly^{-2} Gly^{-1} $*X^{+1}$ (GlyGly) processing site during transport. The gene encoding the ABC transporter is usually located near the bacteriocin gene and may occur in the same or a different operon. More recently, bacteriocins that are secreted via the general secretory pathway have been identified (81, 86, 147). Like other secreted proteins, these bacteriocins are characterized by the presence of a typical signal peptide that is required for recognition by the Sec machinery. The signal peptide is removed at an Ala-X-Ala cleavage site during secretion. The regulatory mechanisms of several bacteriocins have been characterized and discovered to involve two-component regulatory systems, consisting of a membrane-bound sensor and a cytoplasmic response regulator (reviewed in references 75 and 99). The environmental signal that is recognized by the sensor is usually a peptide that may or may not be the same as the bacteriocin itself. Upon interacting with the extracellular bacteriocin or inducing factor, the sensor interacts with the response regulator, which then acts as a transcriptional activator to turn on the genes involved in bacteriocin production. The genes encoding the two-component regulators are usually located adjacent to the bacteriocin gene and/or the transporters but are frequently in a separate operon.

There is increasing interest in the molecular biology of bacteriocin systems and their application in the genetic engineering of LAB. For example, the transport machinery can be utilized for the export or secretion of heterologous enzymes or proteins. The two-component regulatory systems have already been used for controlled production of homologous or heterologous proteins by LAB. In both cases, applications of these tools have the potential to improve and expand the present uses of LAB. Several applications and potential uses of these systems will be discussed. Furthermore, there is increasing interest in using bacteriocins and their immunity factors as selective agents and markers, respectively. This type of selection could be classified as "food-grade." This term has been used to designate systems that are ultimately targeted for use in foods and should, therefore, be safe, well-defined, acceptable in foods, and preferably derived from an organism that is generally recognized as safe (GRAS) (26, 107). GRAS organisms are those that have a history of safe use in foods and include many LAB. These and other uses of bacteriocins and their genetics in food-grade cloning systems will be addressed. Interest in using bacteriocins and their producers in roles other than biopresevation has been generated, and several examples will be reviewed.

63.1. THE GENETICS OF BACTERIOCIN PRODUCTION

63.1.1. Bacteriocins and Immunity Factors

The biochemistry and genetics of the class II bacteriocins and lantibiotics, especially nisin, are the most characterized to date and will, therefore, be the focus of this review.

63.1.1.1. Class II Bacteriocins

The class II bacteriocins are a heterogeneous group of small, heat-stable peptides (reviewed in reference 99). In general, these peptides are cationic and are characterized by hydrophobic and/or amphipathic regions. The class II peptides have been grouped into subclasses (72). Class IIa peptides are referred to as the "pediocin-like" or "listeria-active" peptides. These peptides are characterized by the presence of a conserved motif, the YGNVG box, which is usually located within the N terminus of the mature peptide. The pediocin-like peptides are the only bacteriocins characterized thus far that are homologous to one another, showing 40 to 60% identity; however, they differ in their antimicrobial activities (9). Examples of the class IIa bacteriocins include pediocin PA-1, leucocin A, mesentaricin Y105, carnobacteriocins A, BM1, and B2, sakacin A, sakacin P, and enterocin A (reviewed in reference 99). The class IIb peptides are two-component bacteriocins that require both peptides for optimum inhibitory activity. There appear to be two types of these bacteriocins. The first type or subclass involves one peptide that is active on its own but is "enhanced" in the presence of the second peptide. Lactacin F is an example of this type of two-component bacteriocin (5). A second type is characterized by both peptides' being required for activity, neither one having any activity on its own. Lactococcin G is an example (93, 103). Other two-component bacteriocins that have been reported are thermophilin 13 (85), lactococcin M/N (130), plantaricin S (69), plantaricins PlnEF and PlnJK (30), and brochocin C (126). The class IIc peptides are referred to as the thiol-activated peptides, which require a reduced cysteine for activity. Lactococcin B is the only example of this subclass (138). Other heat-stable class II bacteriocins are lactococcin A (132), acidocin B (81, 137), and divergicin A (147). Colicin V, produced by *Escherichia coli*, has been classified as a class II bacteriocin based on its biochemical and genetic characteristics (41, 58).

Class II bacteriocins may be chromosomal or plasmid-encoded (reviewed in references 32, 72, and 99). The bacteriocin gene encodes a prebacteriocin characterized by additional amino acids on the N terminus. There are two types of additions that are determined by the transport mechanism (see below). An N-terminal extension of bacteriocins exported via ABC transporters is characterized by the presence of an invariant GlyGly processing site or a signal peptide characteristic of proteins secreted by the general secretory pathway. The bacteriocin gene(s) is usually located in an operon that also contains the gene encoding the immunity factor. The immunity factors for the class II bacteriocins are usually small and cationic. The primary sequences of the immunity proteins are as heterogeneous as the bacteriocins themselves, with only immunity factors of the pediocin-like bacteriocins showing homology among one another (9, 49, 111). Cross immunity among these bacteriocins has been reported, suggesting a conserved mode of action and immunity in this class (35, 125). The immunity factors for the class IIb bacteriocins seem to share some similar motifs. The immunity factors for lactococcin M/N (LciM) (130), lactococcin G (100), and lactacin F (LafI) (48) are each predicted to have four transmembrane regions (TMRs). Between the second and third TMR of LciM and LafI are several charged amino acids. The C termini of both proteins are also charged. This may indicate that the two-component bacteriocins also have a common mode of action and/or immunity (6). The immunity factors for lactococcin A (LciA) (102, 141) and carnobacteriocin B2 (CbiB2) (112) have been characterized. The data indicate that LciA interacts with the lactococcin A receptor, thus preventing a productive interaction between the bacteriocin and cell (141). Quadri et al. (112) have proposed that CbiB2 may interfere with or inhibit pore formation by the bacteriocin from the cytoplasmic side. Similar to CbiB2, the putative immunity factor for the mesentaricin Y105 system has been located in the cytoplasm (20).

63.1.1.2. Lantibiotics

Lantibiotics are produced by a variety of LAB and many other organisms. There are two classes of lantibiotics. Type A lantibiotics are elongated, amphiphilic peptides whose primary mode of action is disruption of the cytoplasmic membrane (reviewed in references 49 and 68). Examples of type A lantibiotics include nisin, lactacin 481, and several other lantibiotics mainly produced by LAB. Type B lantibiotics are globular peptides that inhibit specific enzyme functions (reviewed in references 53 and 68). Both types of lantibiotics are characterized by the presence of unusual amino acids that result from complex posttranslational modifications. The lantibiotics, their biochemistry, mode of action, and genetics have been extensively reviewed (68, 101). Nisin is, by far, the most characterized of the type A lantibiotics and will be the only lantibiotic discussed in this review.

The nisin gene, *nisA*, encodes a peptide of 57 amino acids. The N-terminal 23 amino acids are removed during transport. The nature of the nisin N-terminal extension is very different from that of the class II bacteriocins (Fig. 1), the reasons for which may be twofold: (i) the ABC transport system of nisin is slightly different from that of the class II bacteriocins (see below), and (ii) it has been proposed that the N-terminal extension may be required for proper modification of the peptide, possibly mediating recognition of the prepeptide by the modification enzymes or maintaining the proper conformation for modification (123, 136). The modifying enzymes, NisB and NisC, are encoded directly downstream of *nisA*. The mechanism of action of these enzymes is unknown, but the hydrophobic-

```
Class II Bacteriocins

Lactococcin A            MK---------NQLNFNIVSDEELSEANGG
Lactococcin M            MK---------NQLNFEILSDEELQGINGG
Lactococcin N            MKKDEANTFKEYSSSFAIVTDEELENINGS

Pediocin PA1             MK------------KIEKLTEKEMANIIGG
     Consensus                          F S/TD-EL--I-GG

Mesentaricin Y105        M------TNMKSVEAYQQLDNQNLKKVVGG
Leucocin A               MK---------PTESYEQLDNSALEQVVGG

Carnobacteriocin A       M------------NNVKELSIKEMQQVTGG
Carnobacteriocin B2      M------------NSVKELNVKEMKQLHGG
Carnobacteriocin BM1     MK------------SVKELNKKEMQQINGG

Lactacin F LafA          MK------------QFNYLSHKDLAVVVGG
Lactacin F LafX          MK----------------LNDKELSKIVGG

Sakacin P                M------------EKFIELSLKEVTAITGG

Plantaricin A            MK--------IQIKGMKQLSNKEMQKIVGG
Plantaricin J            MTVNKMIKDLDVVDAFAPISNNKLNGVVGG
Plantaricin F            MK------------KFLVLRDRELNAISGG

     Consensus                          F--LS-KE---I-GG
                                      Y/V   N D   V

Colicin V                MR---------------TLTLNELDSVSGG


Class II-type lantibiotic

Lactococcin DR           MK------EQNSFNLLQEVTESELDLILGA


Lantibiotics

Nisin                               MSTK-DFNLDLVSVSKKDSGASPR
Subtilin                            MSKFDDFDLDVVKVSKQDSKITPQ

Sec-dependent

Divergicin A                      MKKQILKGLVIVVCLSGATFFSTPQASA
Acidocin B                      MVTKYGRNLGLSKKVELFAIWAVLVVALLLATA
```

FIGURE 1 Amino acid sequence of the N-terminal extensions of class II (30, 48, 56, 99) and lantibiotic bacteriocins (101). The signal sequences of Sec-dependent divergicin A (147) and acidocin B (81) are also shown. Regions of homology are underlined, and consensus sequences are indicated.

ity plots of these proteins suggest that they may be membrane associated (37, 123). These enzymes act on the prepeptide, modifying only the mature portion, which is then transported. The mature, modified form of nisin is characterized by five ring structures and seven modified amino acids.

Immunity to nisin requires expression of the *nisI* gene and the *nisFEG* operon (reviewed in reference 118). NisI is a secreted lipoprotein that is anchored to the cytoplasmic membrane. Increased expression of NisI increases immunity against nisin but does not confer total protection (38, 78). The NisEF proteins show homology to ABC transporters (importers, see below) and are somewhat similar to the transporters involved in microcin B17 immunity (122). The NisG protein is a very hydrophobic protein. It is proposed that the NisEFG proteins are involved in pumping nisin in so that it can be inactivated. The genes encoding the nisin structural gene, modification enzymes, transporter, and immunity factors are chromosomally located on a conjugative transposon that also encodes sucrose utilization in *Lactococcus* (*Lc.*) *lactis* (113).

The mode of action of class II bacteriocins and lantibiotics has been extensively reviewed (1, 17, 32, 61, 143). The main target of bacteriocins is the cytoplasmic membrane of sensitive cells. The hydrophobic and amphiphilic nature of these bacteriocins assists in the interaction with and formation of pores in the cytoplasmic membrane. The bacteriocin peptides, particularly the class II peptides, are very similar to many of the eukaryotic antimicrobial peptides, which include the magainins, cecropins, defensins, and many others (reviewed in reference 84).

63.1.2. Bacteriocin Transport Systems

63.1.2.1. Bacteriocins Secreted by the General Secretory Pathway

The general secretory (Sec) pathway is the major route through which proteins are delivered to the extracellular medium. Proteins secreted by the Sec pathway are synthesized in the cell as preproteins characterized by the presence of a signal peptide on the amino terminus. While not conserved at the sequence level, the signal peptides among bacteria have conserved domains consisting of a charged N terminus followed by a stretch of hydrophobic residues and a polar C terminus with a consensus cleavage site, Ala-X-Ala (reviewed in references 109 and 124). The signal peptide directs the preprotein to the Sec machinery and is cleaved by a signal peptidase during secretion. The components of the Sec machinery have been extensively reviewed (109, 124). The *secY* gene of *Lactococcus* spp. has been sequenced and shows similarity to its *E. coli* and *Bacillus subtilis* counterparts (74), indicating that the Sec system among LAB is homologous to that in other bacteria.

Sec-dependent bacteriocins are an anomaly in the bacteriocin world. Based on the presence of a typical signal peptide, there have only been three bacteriocins identified that utilize this system: divergicin A produced by *Carnobacterium divergens* LV13 (147), acidocin B produced by *Lactobacillus* (*Lb.*) *acidophilus* (81), and lactococcin 972 produced by *Lc. lactis* (86). Preliminary characterization of lactococcin 972 has only recently been reported. The divergicin A signal peptide has been translationally fused to alkaline phosphatase and introduced into *E. coli*. Alkaline phosphatase activity was detected in the periplasm, indicating that the divergicin A signal peptide is recognized by the *E. coli* Sec system (147). Both divergicin A and acidocin B are class II bacteriocins, whereas lactococcin 972 is a heat-sensitive, homodimeric bacteriocin that does not act on the cytoplasmic membrane (87). Other proteins secreted by the Sec pathway in LAB include the following (reviewed in references 26, 73, 80, and 106): S-layer proteins, proteinases, amylase, and insulinase. Signal peptides of unknown origin have been isolated from *Lactococcus* sp. (reviewed in reference 26) and *Lactobacillus plantarum* (64) using secretion signal selection vectors. Many LAB are not prodigious protein secretors (106). Furthermore, little is known about the Sec pathway in LAB relative to other bacteria, yet utilization of this system has great potential in strain improvement (see following sections). Identification and characterization of secreted proteins, therefore, are very important.

63.1.2.2. ABC Transporters

In sequencing the region encoding the bacteriocin, genes encoding proteins homologous to ABC transporters have been identified and shown to be necessary for bacteriocin production (reviewed in reference 99). ABC transporters are so named because of the presence of an ATP binding cassette in the transporter protein. They are found in eukaryotes and prokaryotes, where they are involved in the import or export of a variety of substrates. Regardless of origin or function, the transporter consists of four domains (60): two domains are highly hydrophobic, each consisting of six transmembrane helices; the other two domains are involved in ATP binding and hydrolysis, thus providing the energy and/or conformational change necessary for transport. In eukaryotes, all four domains are usually found in one polypeptide (60). In prokaryotes, there are two configurations depending on function (reviewed in reference 39): importers or bacterial periplasmic permeases involve two separate proteins, one of which contains the ATP binding domain and the other, the transmembrane regions; exporters have one polypeptide characterized by six transmembrane regions and an ATP binding site and are proposed to act as homo- or heterodimers.

The hemolysin A system of *E. coli* is the prototype ABC exporter among bacteria. Three proteins are involved in the transporter complex (reviewed in reference 39): HlyB, the 707-amino-acid ABC transporter, characterized by the transmembrane and ATP binding motifs; HlyD, the 477-amino-acid accessory factor, characterized by a cytoplasmic domain (ca. 1 to 60 amino acids), a transmembrane domain (ca. 61 to 84 amino acids), and a periplasmic domain (ca. 84 to 477 amino acids), and TolC. It is proposed that a dimer of HlyD forms a channel through the periplasmic space by interacting with a dimer of HlyB, anchored in the cytoplasmic membrane, and TolC, located in the outer membrane (reviewed in reference 39). The 1,023-amino-acid hemolysin A protein, HlyA, is directed to the transport machinery by an export signal located in the ca. 60 C-terminal amino acids. Data have suggested that certain "contact residues" in the export signal (39) and/or certain regions of the transmembrane domains of HlyB (39) may be important in the recognition process. The export signal is not removed during transport.

The genes encoding the ABC transporters and accessory factors of the class II bacteriocins are usually found adjacent to the bacteriocin gene and may or may not be located in the same operon (99). The ABC transporters of the class II bacteriocins are typically ca. 720 amino acids in length and are highly conserved (56). The ATP-binding domains are located in the C terminus, and there are six putative transmembrane regions in the N terminus (reviewed in reference 56). A cysteine protease domain has been identified in the N terminus of the class II ABC transporters, and its role in cleaving the N-terminal extension of the bacteriocin prepeptide has been demonstrated (56, 142). A recent report on the processing of colicin V suggests that a soluble cytoplasmic factor(s), in addition to the ABC transporter, is required for fast, efficient processing of the prebacteriocin (148). The N-terminal extensions of various bacteriocins show a great degree of homology among one another (Fig. 1). Recently, inducing factors involved in regulation of bacteriocin production have been identified (see section below). These inducing factors are small peptides that are synthesized as prepeptides characterized by an N-terminal extension with the same conserved motifs as found in the bacteriocin. It is assumed that they are also processed and exported by bacteriocin ABC transporters. The importance of the GlyGly motif has been shown in site-directed mutagenesis studies. Mutation of either glycine residue of the LafA or ColV N-terminal extension results in loss of detectable bacteriocin activity in the extracellular medium, although one of the colicin V mutations was reported as being slightly "leaky" (48, 52). In the colicin V mutants, intracellular bacteriocin activity was detected (52). These data indicate that the GlyGly motif is not only involved in cleavage but also functions as an export signal (56). Other reports of the presence of intracellular, active bacteriocin suggest that processing and transport may be uncoupled (3a, 142). The accessory factor of the class II systems is usually ca. 470 amino acids, with the accessory factor of pediocin only 174 amino acids (88). The primary sequence of the accessory factors predicts the presence of an N-terminal transmembrane helix. The topology of LcnD has recently been investigated, and the cytoplasmic (amino acids 1 to 21), transmembrane (amino acids 22 to 43), and extracellular (amino acids 43 to 474) domains have been identified (47). Although the role of the accessory factor is unknown, it is essential for bacteriocin export (139, 142). There have been two reports, however, of cases in which the accessory factor has not been required for transport. *Lactobacillus johnsonii* NCK65 can export both lactacin F peptides; however, this strain is an isogeneic derivative of the wild-type lactacin F producer. It has a ca. 10-kb deletion in its chromosome that eliminates the bacteriocin operon and ORF1, an incomplete open reading frame that encodes the putative accessory factor. It is unknown if the ABC transporter is present in the genome of NCK 65. Huhne et al. (66) have reported that *Lactobacillus sake* Lb790, a bacteriocin-negative strain, can produce sakacin P when the genetic determinants are introduced but does not require the presence of the sakacin P ABC accessory factor SppT. Hybridization experiments did not detect an *sppT* homolog in the *L. sake* Lb790 genome. It is possible that either of these backgrounds contains alternative ABC-type export systems that export the bacteriocin or complement the missing accessory factor. The family of ABC transporters to which the class II systems belong also include the *Streptococcus pneumoniae* ComAB transporters involved in the development of competence (56, 67) and lantibiotic transporters LcnDR3 and CylB (see below). A schematic representation of the class II ABC transporters is shown in Fig. 2.

Similar to the genetic organization of the class II systems, the gene encoding the ABC transporter of the lantibiotic systems may or may not be located in the same operon as the structural lantibiotic gene (reviewed in reference 68). The gene encoding the nisin ABC transporter, *nisT*, is located between the genes encoding the modification enzymes. NisT and other lantibiotic ABC transporters contain the conserved transmembrane and ATP-binding domains. Unlike the prototype and class II systems, however, an accessory factor is not required. It has been proposed that NisB, NisC, and NisT form a membrane-associated complex, and it is possible that the association with the modification enzymes serves the same function as the accessory factor (123). A separate signal peptidase, which is homologous to the subtilisin-like serine peptidases, is encoded by many of the lantibiotic operons. In the nisin system, the signal peptidase NisP is encoded downstream of the *nisBCTI* genes and is within the same operon. The NisP protein is secreted and anchored in the cytoplasmic membrane by its hydrophobic C terminus (135). It is proposed that the prepeptide becomes modified and transported by the NisBCT complex; then the N-terminal extension is cleaved as it is released from the cell. The N-terminal extensions of the lantibiotics, therefore, are quite different from those of the class II prebacteriocins shown in Fig. 1. The *Enterococcus faecalis* hemolysin/bacteriocin and *Lc. lactis* lactococcin DR are both lantibiotics; however, their ABC transporters seem to be a combination of the lantibiotic and class II systems. The ABC transporters of these lantibiotics contain the N-terminal proteolytic domain of class II ABC transporters, and the N-terminal extensions are similar to the N-terminal extensions of class II bacteriocins (Fig. 1). Other lantibiotics with N-terminal extensions characterized by a GlyAla or

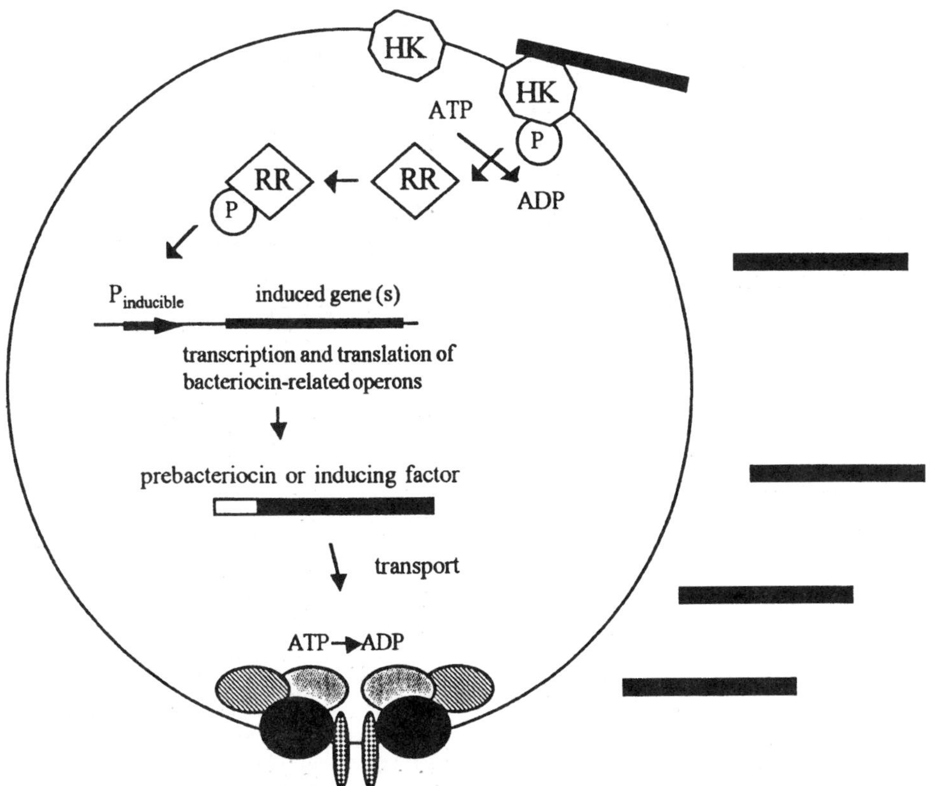

FIGURE 2 Schematic representation of the regulatory and transport processes involved in production of class II bacteriocins, as adapted from Nes et al. (99) and Kuipers et al. (76). HK and RR, histidine kinase and response regulator of the two-component regulatory systems. Ⓟ, phosphate. Open bar, N-terminal extension of the prebacteriocin or preinducing factor. Solid bar, mature bacteriocin or inducing factor. Thatched, shaded, and solid ovals, ABC transporter N-terminal proteolytic domain, C-terminal ATP-binding domain, and transmembrane domain, respectively. Dotted ovals, ABC transporter accessory factor.

GlyGly processing site are streptococci A-FF22, streptococci A-M49, and salvaricin A (reviewed in reference 101). Whether or not class II-type ABC transporters are also involved in the production of these lantibiotics remains to be determined. These lantibiotics may represent a subclass that will be referred to here as the class II-type lantibiotics.

63.1.3. Two-Component Regulatory Systems of Bacteriocins

In sequencing the bacteriocin-encoding loci of many different organisms, genes encoding proteins homologous to those involved in two-component regulatory systems have been found. These genes are usually located adjacent to the genes encoding the bacteriocin(s) or transporters and are frequently located in a separate operon (reviewed in reference 99). Two-component regulatory systems are found in a variety of prokaryotes and a few eukaryotes, where they act as "mediators" between the extracellular and intracellular environments (104, 127). Two-component regulatory systems consist of a membrane-bound histidine kinase (HK), characterized by an N-terminal transmembrane domain that interacts with the stimulus and a conserved C-terminal domain involved in kinase activity, and a response regulator (RR), characterized by a conserved N-terminal domain involved in phosphorylation and a variable C-terminal domain that frequently includes a DNA binding motif implicated in transcriptional activation (104, 127). When "stimulated," the HK becomes phosphorylated. The phosphate is then passed on to the RR, thus activating it. Numerous two-component regulatory systems have been identified in bacteria. Recently, two-component regulatory systems have been characterized in the nisin, sakacin P, plantaricin A, and carnobacteriocin systems.

Of the two-component regulatory systems associated with bacteriocin production, that of nisin is the best characterized. Although it has been reviewed elsewhere (75), it will be briefly discussed here. The genes encoding the HK and RR are in a separate operon downstream of the major nisin operon. The environmental signal recognized by the HK is nisin itself. Within the nisin-encoding locus, two nisin-inducible promoters have been characterized P_{nisA} and P_{nisF}. The former is located immediately upstream of *nisA*, and P_{nisF} directs transcription of an operon that encodes the microcin B17-type immunity factors (see previous section). When comparing the nucleotide sequences of P_{nisA} and P_{nisF}, the region between the −35 and −10 sequences is very similar, suggesting that these sequences may be important in interacting with the RR (22). Regulation of the nisin system has been shown to be at the transcriptional level (76). Less than 0.1 ng of nisin per ml is sufficient to induce transcription (22). The amount of transcription initiated from P_{nisA} and P_{nisF} is dependent upon the amount of nisin added (76). At a given amount of nisin, the transcription from P_{nisA} is greater (approximately fourfold) than that from P_{nisF} (22). Induction is specifically mediated by nisin and nisin analogs (76). Expression of P_{nisR}, which regulates the operon containing the RR and HK, is independent of nisin, but it is a relatively strong promoter (22).

There have been several reports on the preliminary characterization of the two-component regulatory systems of the following class II bacteriocins: sakacin P of *Lb. sake* (35), plantaricins produced by *Lb. plantarum* C-11 (30), and the carnobacteriocins of *Carnobacterium piscicola* LV17 (110). The sakacin P and plantaricin regulatory systems are very similar. Bacteriocin production in both systems is triggered by the presence of extracellular inducing factors, the ORFY gene product of *L. sake* and plantaricin A of *Lb. plantarum*. The primary structure of the inducing factors is very similar to the actual bacteriocins in that they are synthesized as prepeptides with a characteristic N-terminal extension common among ABC transporter-dependent class II bacteriocins, but they tend to be smaller than their respective bacteriocins. It is assumed that they are exported by the corresponding bacteriocin ABC transporters. The ORFY gene product exhibits antibacterial activity at concentrations approximately 100- to 1,000-fold higher than sakacin P; plantaricin A does not exhibit bacteriocin activity. Amounts of inducing factors required are very low: only 0.2 ng of the ORFY gene product per ml (35) and 1 ng of plantaricin A per ml (29) are required to induce sakacin A and plantaricin production, respectively. Upon induction, transcripts are evident within 30 min, and corresponding extracellular bacteriocin activity occurs within 2 to 2.5 h. Bacteriocin activity and extracellular inducing factors peak in the early stationary phase, at which time the inducible transcripts disappear. Unlike nisin, production of sakacin P is independent of the amount of inducing factor added. In the sakacin P system, P_{sakP} is the only inducible promoter characterized thus far. In the plantaricin system, P_{plnA}, P_{plnJ}, P_{plnM}, P_{plnE}, and P_{plnG} are all responsive to induction. While differences in the response to induction among the promoters have been noted, these have yet to be quantified (30).

Regulation of carnobacteriocin production by *C. piscicola* LV17 appears to be more complex and involves characteristics of both the lantibiotic and class II systems (110, 120). This strain produces three bacteriocins: CbnA is encoded by a 49-MDa plasmid; CbnB2 is encoded by a 40-MDa plasmid; and CbnBM1 is chromosomally encoded. Plasmid curing experiments produced a derivative containing the 49-MDa plasmid (LV17A), which produces CbnA, and a derivative containing the 40-Mda plasmid (LV17B), which produces both CbnB2 and CbnBM1 (3). It has been determined that carnobacteriocin production is autoinduced: purified CbnA2 (sulfoxynated form of CbnA), CbnBM1, CbnB2, and CbnB1 (sulfoxynated form of BM1) induce CbnA production by LV17A; CbnA2 and CbnB2 induce bacteriocin production by LV17B (120). Therefore, carnobacteriocin production is not only autoregulated but also cross-regulated (120). The sequence of the locus encoding CbnB2 has recently been extended. Genes encoding two-component regulatory proteins, CbnK (histidine kinase) and CbnR (response regulator), have been identified upstream of *cbnB2* (110). The *cbnS* gene, which encodes a small peptide containing a typical N-terminal extension characterized by a GlyGly processing site, is located upstream of *cbnK*. It has been determined that CbnS is one of the inducing factors for the carnobacteriocin regulatory system (110). When compared to the inducing properties of CbnB2, 20-fold less CbnS was required to elicit a response (both in the nanomolar range). Four inducible promoters have been identified, and by using β-glucuronidase as a reporter, the degree of induction in response to the presence of CbnB2/CbnS has been determined to be $P_{cbnBM1} >> P_{cbnX} > P_{cbnB2} > P_{cbnS}$ (110). As in the nisin system, regulation of carnobacteriocin production occurs at the transcriptional level (119). A schematic representation of the class II-type two-component regulatory system is shown in Fig. 2.

When the nucleotide sequences of the inducible promoters in the class II systems are compared, the general organization appears to be conserved (30). Upstream of the −35 region of these promoters are two direct repeats of 9 to 10 nucleotides. The absolute sequence of the repeats is conserved within each system but not among them. A 12- to 13-nucleotide spacer region is located between the direct repeats. Additional promoters in the sakacin P system, and in the sakacin A system, have this motif. Assuming that this motif is involved in the response generated by the two-component system, it appears that several inducible promoters have yet to be characterized (30).

Lactacin B, a class II bacteriocin produced by *Lb. acidophilus* NCFM/N2, is also inducible (10). A protein of ca. 58 kDa that is produced by a lactacin B-sensitive strain, *Lactobacillus delbrueckii* subsp. *lactis*, induces significant bacteriocin production (32- to 64-fold more bacteriocin activity) (10). It remains to be determined if a two-component regulatory system mediates this induction effect.

The two-component regulatory systems of bacteriocins are the first regulatory systems identified in LAB that are induced by peptides. A review on inducible and regulated systems in LAB has been published (75). Other two-component regulatory systems in gram-positive organisms that are homologous to the class II bacteriocin regulatory systems and also "sense" peptide inducers or pheromones are the following (reviewed in references 30, 57, 105): toxin production in *Staphylococcus aureus* and *Clostridium perfringens* and the development of competence in *S. pneumoniae* and *B. subtilis*. Bacteriocin production, toxin production, and development of competence are growth phase dependent. As the culture grows, the peptide inducers gradually accumulate to a certain critical concentration that triggers induction of the inducible genes. This is a common mechanism of communication among bacteria and is referred to as quorum sensing.

63.2. NOVEL APPLICATIONS OF THE BACTERIOCIN GENETICS, PRODUCERS, AND PEPTIDES

63.2.1. Export and Secretion of Heterologous Peptides and Proteins Using the Bacteriocin Transport Systems

Recently, there has been interest in improving the performance, competitiveness, and potential applications of starter cultures. One way of achieving this is through expression of extracellular enzymes that enable LAB to utilize nutritional sources more efficiently or to utilize additional nutritional sources. There have been numerous reports of such cloning and expression experiments that have resulted in LAB that produce and secrete amylases, proteases, levanase, lysostaphin, cellulase, glucanases, lipases, and several other enzymes (reviewed in references 26, 73, 92, 106). Many of these experiments have been successful even when using the promoter and signal peptides of the heterologous enzyme, however, in many cases, the levels of expression have been nonoptimal (reviewed in references 26 and 73). Another area of interest is the application of LAB as vaccine or protein delivery vehicles to the gastrointestinal and urogenital tracts (reviewed in references 73, 108, and 145). The antigen or protein may be expressed intracellularly, extracellularly, or attached to the cell surface. For vaccine delivery, the production of antigen must be very high to elicit an immune response. Since most LAB are not prodigious protein secretors, utilization of the native secretion and export signals will help to optimize secretion of antigens, proteins, and enzymes.

The secretion and export systems of bacteriocins are attractive mechanisms through which these objectives can be achieved. To date, most heterologous secretion/export experiments with bacteriocin systems have involved heterologous bacteriocins only (Tables 1 and 2). Several investigators (89, 126, 147) are interested in using this approach to construct industrial LAB that produce multiple bacteriocins, thus increasing the inhibitory activity and overall effectiveness of the cultures used in food biopreservation. The following is an overview of the studies conducted in this area.

63.2.1.1. Sec-Dependent Heterologous Expression

Divergicin A is a class II bacteriocin characterized by the presence of a typical Sec-dependent signal peptide with the conserved Ala-X-Ala cleavage site (147). It is produced by *C. divergens* LV13 and is inhibitory against other carnobacteria strains only. Extensive studies on divergicin A and utilization of its signal peptide have been conducted. When only the divergicin A structural gene and immunity gene (where necessary) are introduced into various *Carnobacterium* strains, *Lc. lactis*, and *E. coli*, divergicin A activity is detected, indicating that it is secreted by heterologous hosts (147). The divergicin A signal peptide has also been used to direct secretion of the following bacteriocins by several LAB (Table 1): leucocin A, colicin V, mesentaricin Y105, lactococcin A, carnobacteriocin B2, and the brochocin A peptides. Colicin V, a class II bacteriocin produced by *E. coli*, is active against a limited range of *E. coli* strains, including the 0157:H7 serotype (98). Brochocin C is a class II two-component bacteriocin, isolated from *Brochothrix campestris*, and displays an inhibitory spectrum equivalent to that of nisin A (126). Heterologous expression of these bacteriocins by LAB significantly increases the antibacterial activity. Production of colicin V is particularly important since none of the bacteriocins produced by LAB are active against gram-negative bacteria, and these experiments may set the example for heterologous expression of other colicins by food-grade LAB (90, 98, 134). Utilization of the general secretion pathway offers the advantage that only a small amount of DNA needs to be cloned into the heterologous host to achieve production and secretion, an important consideration in the construction of recombinant food-grade organisms. The divergicin A signal peptide has also been used to direct secretion of alkaline phosphatase in *E. coli*, and this is the only example of its use in the secretion of a nonbacteriocin.

Acidocin B is a Sec-dependent class II bacteriocin secreted by *Lb. acidophilus* M46 (81). The genetic determinants for acidocin B have been sequenced, and the bacteriocin has been heterologusly expressed in *Lb. plantarum* and *Lactobacillus fermentum* (81, 137). Within the *Lb. acidophilus* homology group, acidocin B is only the fourth Sec-dependent protein to be identified, the others are α-amylase of *Lactobacillus amylovorus* (44), and the S-layer A and B proteins of *Lb. acidophilus* ATCC4356 (16). The *Lb. acidophilus* homology group includes *Lb. acidophilus*, *Lb. amylovorus*, *Lb. crispatus*, *Lb. gasseri*, *Lb. johnsonii*, and *Lb. gallinarum*. This group of organisms, along with other species of lactobacilli, including *Lactobacillus reuteri*,

TABLE 1 Heterologous secretion of bacteriocins via the Sec pathway

Bacteriocin (origin)	Signal peptide	Heterologously expressed by	Reference(s)
Divergicin A (*C. divergens* LV13)	Divergicin A	*C. piscicola* LV17A, LV17B, LV17C, and UAL26; *Lc. lactis* MG1363, IL1403; *E. coli*	147
Carnobacteriocin B2 (*C. piscicola* LV17B)	Divergicin A	*C. divergens* LV13, UAL278; *C. piscicola* LV17C, UAL26; *Lc. lactis* IL 1403	89
Mesentaricin Y105 (*Leu. mesenteroides*)		Various industrial LAB	12
Colicin V (*E. coli*)		*C. piscicola* LV17C, UAL26; *C. divergens* LV13	90
Brochocin C peptides (*B. campestris*)		*C. divergens* LV13	90
Leucocin A		*Leu. gelidum*	
Acidocin B (*Lb. acidophilus* M46)	Acidocin B	*Lb. plantarum, Lb. fermentum*	81, 137
Lactococcin B (*Lc. lactis*) Pediocin PA-1 (*P. acidilactici*)	Proteinase P	*Lc. lactis*	42
Sakacin P (*Lb. sake*)	PelP	*E. coli*	54

are the major LAB found in the gastrointestinal tract of humans and animals. Identification and characterization of secretion signals in these organisms will be important in their successful application in heterologous protein delivery systems.

There have been two other reports of Sec-mediated bacteriocin production. A translational fusion was constructed to make chimeric proteins consisting of the proteinase P signal peptide of *Lc. lactis* and mature lactococcin B or pediocin PA-1 (42) (Table 1). Active lactococcin B and pediocin PA-1 were secreted by *Lc. lactis*. Furthermore, the PelB signal peptide of *E. coli* was fused to the mature peptide of sakacin P (54). Active sakacin P was produced by *E. coli*. This is the first report of large-scale production of LAB bacteriocins by *E. coli*, which is an excellent system for biochemical and mutagenesis studies (54).

63.2.1.2. Export of Heterologous Bacteriocins Using the ABC Transporters

The ABC transporter-dependent bacteriocins show a great deal of utility in heterologous expression experiments. These experiments have been conducted in three ways: (i) export of one bacteriocin by another bacteriocin's ABC transporter, (ii) fusion of the N-terminal extension of one bacteriocin to the mature form of another bacteriocin, followed by export by either ABC transporter, and (iii) expression of the complete ABC system and corresponding bacteriocin in a bacteriocin-negative background. A summary of these experiments is outlined in Table 2. Many bacteriocins and N-terminal/bacteriocin chimeric peptides can be properly processed and exported by heterologous ABC transporters, and the range of organisms and bacteriocins for which heterologous export can be achieved is quite extensive (Table 2). Export via the heterologous exporter may be as efficient as that in the native host (129); however, there have also been many reports of decreased activity upon heterologous export (4, 7, 36, 133) and even examples of unsuccessful heterologous export (36, 133). In most cases, the reason for decreased expression or lack of expression has not been determined, but there may be many explanations, including the following: poor transcription or translation when signals nonnative to the expression host are used; inefficient recognition of the N-terminal extension by a heterologous ABC transporter; alteration of the conformation of the processing site in a chimeric prebacteriocin; interaction with other factors secreted by the expression host, thus reducing activity of the bacteriocin; and size of the heterologous bacteriocin (see below).

Of particular interest, when the genes encoding lactococcin DR and its modification enzyme are introduced into *Lc. lactis* IL1403, active lactococcin DR is produced and exported (114). Although IL1403 does not produce a bacteriocin, the genes encoding the lactococcin transporters, *lcnC* and *lcnD*, are present in the IL1403 genome (139). The native lactococcin DR transporter, LcnDR3, is in the same family as the class II transporters (56) (see previous section), and it is possible that LcnC complements LcnDR3. It would be interesting to determine if the accessory factor LcnD is involved in heterologous transport of lactococcin DR or if the modification enzyme interacts directly with LcnC. It is proposed that the modification enzyme(s) of lantibiotics fulfills the function of the accessory factor (see previous section). This is the only report of a class II transport system exporting a lantibiotic. It will be interesting to determine if heterologous export of class II-type lantibiotics (see previous section) by class II ABC transporters is a general phenomenon.

Just as the Sec pathway can be utilized to secrete bacteriocins characteristically exported by ABC systems, divergicin A can also be transported by several ABC transporters (Table 2). Therefore, it has been shown that the class II ABC transporters can export a variety of bac-

TABLE 2 Heterologous export of class II bacteriocins in LAB

Bacteriocin (origin)	N-terminal extension (origin)	Heterologously expressed by organism (transporters and native bacteriocin)	Reference(s)
Divergicin A (*C. divergens* LV13)	Leucocin A (*Leu. gelidum* UAL187)	*Leu. gelidum* UAL187-22 (LcaCD, leucocin A); *E. coli* MC4100(pHK22) (CvaAB, colicin V); *Lc. lactis* IL1403(pMB500) (LcnCD, lactococcins)	134
	Colicin V (*E. coli*)	*Leu. gelidum* UAL187-22; *E. coli* MC1400(pHK22)	
	Lactococcin A (*Lc. lactis*)	*Leu. gelidum* UAL187-22; *Lc. lactis* IL1403(pMB500); *Lc. lactis* IL1403 (LcnCD, lactococcins)	
Colicin V	Leucocin A	*Lc. lactis* IL1403(pMB500)	134
Plantaricin A (*L. plantarum* C11)	Plantaricin A	*Lb. sake* Lb706B (SapTE, sakacin A)	29
Lactacin F peptides, LafA and LafX (*Lb. johnsonii* VPI11088)	LafA and LafX	*C. piscicola* LV17 (carnobacteriocins producer); *Leu. gelidum* UAL187-22	4, 7
Mesentaricin Y105 (*Leu. mesenteroides*)	Mesentaricin Y105	*Lb. johnsonii* NCK64 and NCK65 (lactacin F producers)	49
Pediocin PA-1 (*P. acidilactici*)	Pediocin PA-1; lactococcin A	*Lc. lactis* (LcnCD, lactococcins)	36
Lactococcin A	Pediocin PA-1	*Lc. lactis* (LcnCD, lactococcins)	36
Piscicolin 61 (*C. piscicola*)	Piscicolin 61	*Lb. sake* (SapTE, sakacin A)	62
Lactococcin A	Lactococcin A	*Leu. gelidum* UAL187-22	133
Helveticin J (*Lb. helveticus*)	Helveticin J	*Lb. johnsonii* NCK64 and NCK65	72
Lactococcin DR (*Lc. lactis*)	Lactococcin DR	*Lc. lactis* IL1403 (LcnCD, lactococcins)	114
Various bacteriocins	Sakacin A (*Lb. sake*)	*Lb. sake* (SapTE, sakacin A)	8

teriocins, including a lantibiotic and a Sec-dependent bacteriocin. All that is required to direct the heterologous export is the presence of an appropriate N-terminal extension with a GlyGly processing site. If a bacteriocin producer is used as the expression host, the amount of DNA required for cloning is tremendously reduced, as is the case for the Sec systems. In the case of lantibiotics, slightly more DNA would be required since the modification and possibly the immunity genes would be required.

Most reports of heterologous export of lantibiotics have involved nisin and subtilin, a lantibiotic closely related to nisin that is produced by *B. subtilis*. The N-terminal extension or leader region of nisin is 57% identical to subtilin (Fig. 1), and the mature part of nisin is approximately 60% identical to subtilin (79). When the gene encoding the complete nisin peptide is introduced into *B. subtilis*., nisin is not produced (59, 115). When a subtilin N-terminal extension/nisin chimera is introduced into *B. subtilis*, an inactive form of modified, processed nisin is produced (59). When this same construct is introduced into nisin-producing *Lc. lactis*, the mature part of nisin is correctly modified but the leader is not cleaved (79). When the mature part of nisin is fused to a subtilin/nisin chimeric leader, active nisin is produced by *B. subtilis* (115). Finally, when the subtilin N-terminal extension is fused to a nisin/subtilin chimera, the chimeric lantibiotic is processed, modified and has the same inhibitory activity as nisin (18). Therefore, heterologous expression of lantibiotics is not as straightforward as that for the class II peptides, presumably because of the specific interactions between the N-terminal extension and modification enzymes, and between the N-terminal extension and the signal peptidase. Although heterologous expression of the modification enzymes in conjunction with the lantibiotic has not been reported, it may be possible to minimize or eliminate these problems. This type of experiment, however, would be limited to lantibiotic producers only.

The ABC transporters have also been shown to be functional in heterologous genera, species, and strains. The results from these types of experiments are listed in Table 3. Reports of heterologous expression in related strains were not included, since most of these experiments have been successful and this is a common strategy used in the analysis of the genetic determinants of bacteriocin production. It should also be noted that increasing the copy number of the bacteriocin or transporter has resulted in dramatically increased bacteriocin production (36, 139, 142). For example, when a plasmid encoding LcnC and LcnD is introduced into *Lc. lactis* IL1403 (see above), lactococcin production increases 10-fold (139). This may be an impor-

TABLE 3 Expression of ABC transporters and corresponding bacteriocins in heterologous hosts

Bacteriocin/ immunity	Transporter/ accessory	Origin	Heterologously expressed by	Reference(s)
Lactococcin A/LciA	LcnC/LcnD	*Lc. lactis*	*P. acidilactici* PA-1	19
Pediocin PA-1/PedB	PedD/PedC	*P. acidilactici* PA-1	*Lc. lactis* IL1403 *Pediococcus pentosaceous* PPE1.2 *E. coli*	19, 88, 142
Leucocin A/LcaB	LcaC/LcaD	*Leu. gelidum* UAL187	*Lc. lactis* IL1403	133

tant consideration when improving bacteriocin production in industrial strains.

The use of the LAB bacteriocin ABC transporters in heterologous export of nonbacteriocin peptides and proteins has not been reported. The only class II transporter that has been used to export a nonbacteriocin is the colicin V system. Several colicin V (CvaC) and alkaline phosphatase (PhoA) chimeras were constructed and introduced into $CvaA^{+/-}$ (accessory factor) and $CvaB^{+/-}$ (ABC transporter) backgrounds (52). All chimeric proteins demonstrated alkaline phosphatase activity that was dependent on the presence of both CvaA and CvaB, with the CvaC57/PhoA and CvaC39/PhoA chimeras having the highest activities (38 times and 32 times the activity of the $CvaA^-B^-$ backgrounds, respectively). Analysis of total cell extracts showed that the chimeric proteins were processed in the presence of CvaAB; however, cellular localization studies showed that the chimeras were associated with the inner membrane, presumably on the periplasmic side. Therefore, CvaAB proteins were able to translocate the chimeras across the cytoplasmic membrane but were unable to completely release them from the cell (52). The hemolysin A-ABC export system of *E. coli*, however, has been shown to direct the export of a variety of proteins large and small (reviewed in reference 39), including antigens, Sec-dependent proteins, cytoplasmic proteins, and other proteins that have not been able to be transported via the Sec pathway.

Heterologous export studies with other gram-negative ABC transporters indicate that the maximum size of a protein to be heterologously exported may be limited by the size of the original substrate (40). Hemolysin A of *E. coli* is the largest substrate transported by ABC exporters, and therefore this system shows the greatest utility for heterologous export. Size may also play a factor in the inability of the colicin V transporters to release the PhoA fusions from the cell. While the heterologous export ability of the gram-positive ABC transporters has not been fully explored, preliminary data suggest that specificity and/or size may impose limitations. The pediocin PA-1 transporter, PedCD, is incapable of exporting lactococcin A or a chimeric peptide consisting of mature lactococcin A fused to the pediocin PA-1 N-terminal extension (36). The authors propose that the pediocin PA-1 machinery is specific for its substrate (36). (The mature forms of lactococcin A and pediocin PA-1 are 54 and 44 amino acids in size, respectively.) Mature colicin V is 88 amino acids and is the largest class II bacteriocin characterized thus far. When a chimeric peptide consisting of colicin V fused to the leucocin A N-terminal extension is introduced into Lc. lactis IL1403(pMB500) and *Leuconostoc* (*Leu.*) *gelidum* UAL187-22, it is exported by *Lc. lactis* but not by *Leu. gelidum* even though the native N-terminal extension is used (134). The lactococcins exported by LcnCD in IL1403 range in size from 77 amino acids (LcnN) to 47 amino acids (LcnB), whereas leucocin A produced by *Leu. gelidum* is only 37 amino acids (colicin V is over twice the size). The leucocin A transporters, however, can export divergicin A (46 amino acids) and lactococcin A (54 amino acids), which are approximately 1.2 and 1.5 times larger than leucocin A, respectively.

Whether or not an upper size limit will be a factor in the export of heterologous proteins by the bacteriocin ABC transporters remains to be determined; however, it is likely that there is no lower limit. The inducing factors that have been identified (see previous section) range in size from 19 to 26 amino acids and are presumably exported by the respective bacteriocin transporters (the corresponding bacteriocins are between 43 and 34 amino acids, respectively) (30, 35). It is likely that the bacteriocin transporters would be capable of exporting eukaryotic antimicrobial peptides (AMPs), which range in size from ca. 15 to 50 amino acids. Many of these AMPs, like most bacteriocins, act on the cytoplasmic membrane through the formation of pores. AMPs such as the defensins, magainins, and cecropins exhibit very broad ranges of inhibitory activity, including gram-negative bacteria, gram-positive bacteria, enveloped viruses, and parasites (84). There is considerable interest in using AMPs in biopreservation of food (15).

Eukaryotic AMPs are also secreted by the gut epithelium, and it is proposed that they may influence gut microflora (121). It may be possible, therefore, to reduce and/or control the presence of certain pathogens in the gastrointestinal or urogenital tracts using genetically engineered LAB to deliver and produce these peptides in vivo. To date, these AMPs have not been directly produced by bacteria, presumably because of their inhibitory affect on the producer. The success of such heterologous expression experiments in LAB will depend upon the sensitivity of these organisms to the desired AMP. Preliminary characterization of the sensitivity of *Lb. johnsonii* VPI11088 to magainin has shown that very high concentrations of the peptide (80 μg/ml) only marginally affect culture growth (3b).

The transport systems of bacteriocins have the potential to be used in secretion or export of heterologous proteins. Undoubtedly, utilization of these native signals will help to optimize production of heterologous proteins, particularly eukaryotic proteins such as antigens or AMPs.

63.2.2. Regulated Expression Systems in LAB: Utilization of the Bacteriocin Regulatory Mechanisms

In recent years, there has been increased interest in the characterization and applications of regulated gene expression systems in LAB. Such systems would enable these organisms to be engineered to produce specific proteins under certain conditions in a cultured food product or in a large-scale, food-grade fermentation system. These systems could also be used to study the effects of overproducing certain enzymes in a fermentation. For example, the overproduction of proteinase, peptidase, and lipase by lactococcal cheese cultures could be used to study the effects that these different enzymes have on flavor development. Tightly regulated systems would also enable toxic genes to be cloned into and then "turned on" under certain circumstances. Furthermore, there is interest in metabolic engineering of LAB, and the ability to control expression or overexpression of a certain enzyme(s) in a pathway would enable the study of the effects on the end products. Finally, there is increasing interest in using LAB to deliver antigens, proteins, and enzymes to the gastrointestinal or urogenital tracts of humans and domestic animals. One successful strategy, used in the oral immunization of mice against tetanus toxin, has involved "loading" the bacteria, intracellularly, with the antigen (tetanus toxin fragment C) prior to administration (145, 146). Therefore, the ability to direct the production of huge amounts of protein in LAB may become increasingly important. Since all of these applications are directed at food-grade fermentations, the regulatory system should be derived from GRAS organisms, and the "inducer" must be either food-grade or characteristic of the food or environment in which the recombinant organism is to be introduced. The two-component regulatory systems and inducing factors of bacteriocins satisfy these requirements and show great potential in regulated expression of heterologous proteins by LAB.

The nisin system is the first bacteriocin regulatory system to be used for expression of heterologous proteins. Overproduction of the following proteins in *Lc. lactis* has been directed by using nisin induction of P_{nisA}: β-glucuronidase (22, 24), peptidase N (24), holin and lysin of bacteriophage ϕUS3 (23), tributyrin esterase (63), lipase A (77), and NADH:H_2O oxidase (25). The ability to produce toxic genes such as the bacteriophage holin and lysin indicates that the system is very tightly regulated. It has been shown that in the absence of nisin, transcripts from P_{nisA} and P_{nisF} are undetected (22, 76). Additional factors have been found to affect expression and induction from P_{nisA} and P_{nisF} (24). For example, when the β-glucuronidase gene is translationally fused to the start codon of *nisA*, activity is increased sixfold compared to that from a transcriptional fusion to P_{nisA}. When the peptidase N gene was translationally fused to *nisA* in the same way, approximately half of the total intracellular protein consisted of PepN after induction with nisin (24). In the same publication, it was also reported that the nisin system has been used to produce homologous or heterologous proteins at levels ranging from 2 to 60% of the total intracellular protein (24). At present, this is the highest level of heterologous protein production reported in *Lc. lactis*. The inducible T7 RNA polymerase expression system is the only other system that has reported high levels of protein production, up to 22% of soluble intracellular protein (146).

Another "layer" of control can be introduced by using genetically engineered nisin analogs. T2S nisin Z is over 10-fold more effective at inducing P_{nisA} than is wild-type nisin A or Z (76). It was calculated that only five molecules of the T2S inducer per cell were required to activate transcription (76). The M17W analog of nisin Z has twice the inducing effect (76). From a practical perspective, compared to a given amount of the wild-type lantibiotics, one-tenth the amount of T2S nisin Z and one-half the amount of M17W nisin Z have the same inducing capacity on P_{nisA}. This may have important implications in an in vivo situation where either the nature of the fermentate or the initial numbers of starter culture could impose limitations on the effectiveness or availability of the inducing factor. Furthermore, when C-terminal amino acids of nisin A are removed, the resulting truncated versions encompassing only one-third to one-half of the peptide are still able to induce P_{nisA}, albeit at reduced efficiencies (33, 76). This has important implications in vivo where nisin may be subjected to proteolytic activity.

The nisin regulatory system is also functional in heterologous hosts. The *nisRK* genes were integrated into the chromosome of a non-nisin-producing *Lc. lactis* strain (24). When this integrant was transformed with a recombinant plasmid containing *gusA* under the control of P_{nisA}, β-glucuronidase activity was observed in a nisin-dependent, dose-response manner. When induced with equivalent amounts of nisin, the recombinant integrant showed a 30-fold higher response than the nisin producer (24). The authors propose that nisin may be bound by the immunity factors in the nisin-producing strain; therefore, more nisin is required to produce the same inducing effect (24). The regulatory system of nisin has also been functionally expressed in heterologous strains of the *Lactobacillus* and *Leuconostoc* genera (77).

Of the class II two-component regulatory systems that have been characterized, heterologous gene expression has been demonstrated in the carnobacteriocin B2 system only. The CbnB2 and CbnS inducible promoters were fused to β-glucuronidase to study relative responses to induction (110). Like the nisin system, the two-component regulatory systems of the class II bacteriocins have been heterologously expressed in other LAB. For example, when *cbnRK* was cloned into *Lc. lactis*, induction of P_{cbnB2} was observed when using either inducer. Furthermore, the *plnA* operon, which encodes the inducing factor, HK, and RR, has been functionally expressed in heterologous host *Lb. sake* Lb706B (29). PlnA was detected in the supernatant, presumably heterologously transported by the sakacin A ABC transporters, and transcription from P_{plnA} was observed (29).

Two-component regulation of bacteriocins is becoming a powerful system for use in regulated gene expression in LAB. The better-characterized nisin system illustrates the many advantages that these systems have to offer. By manipulating the following variables, a 1,000-fold variation in expression has been reported (24): P_{nisA} versus P_{nisF}; transcriptional versus translational fusions; and nature of the strain used (immune versus nonimmune). This can result in production of heterologous proteins at levels of 2 to 60% of total intracellular protein (24). While not as thoroughly characterized as nisin, the class II two-component regulatory systems also have potential to be used as inducible expression systems. Each of the class II systems differs in the number of inducible promoters and the extent to which each promoter is induced, thus increasing the options available to be exploited. With so many variables to manipulate, it is possible to fine-tune a system for a specific application. Furthermore, the ability to functionally express the regulatory proteins and achieve inducible expression in heterologous hosts potentially opens this technology for use in any LAB strain that is amenable to manipulation. How these systems function in vivo, where conditions can be unpredictable, remains to be determined. When nisin-producing strains are used in food fermentations, nisin is produced and detected in samples taken at different times during the process (reviewed in reference 21). Also, nisin or nisaplin can be added to a variety of foods as a biopreservative and has been reisolated from these foods after storage (21). In the case of nisin, therefore, under circumstances reported in the literature, the "inducer" has been shown to be present and should be available to trigger the expression system. When the carnobacteriocin producer *C. piscicola* LV17 is applied to raw meat under anaerobic conditions at low temperatures, however, growth is unpredictable and the presence of the bacteriocin has not been detected (91). The authors suggest that the bacteriocins may be unstable or present at a level below the detection limit (91). The applied experiments remain to be performed, but in the meantime, these expression systems offer wonderful tools to use under controlled conditions in the laboratory.

Since the concentrations of the inducers required to trip induction are so low (nanomolar range), it may be possible to develop these systems as biosensors for in vivo detection of bacteriocins. A "reporter bacterium" could be constructed such that the genes encoding the HK and RR regulatory proteins are introduced. The respective inducible promoter could be fused to a reporter gene such as the gene encoding green fluorescent protein. The inducible promoter/reporter gene construct would also be introduced into the recombinant reporter bacterium, which would not contain the gene for the bacteriocin or inducing factor. Experiments could then be conducted in which the reporter bacterium and bacteriocin-producing strains are introduced into the fermentation. Bacteriocin production could then be visualized by the reaction induced in the reporter. This could be very useful in determining where, when, and how much bacteriocin is produced, or as a tool to monitor bacteriocin production over time. This method has the added advantage that an immediate reading can be made, whereas many of the traditional methods that rely on direct inhibition assays can take up to 24 h to yield results.

63.2.3. Applications of Bacteriocins and Their Genetics in Food-Grade Cloning Systems

A great deal of effort has been devoted to improving LAB through genetic engineering. It may take considerable time, however, before recombinant LAB are approved for use in probiotics and the food industry. It is anticipated that the approval process may be accelerated by applying the following (26, 107): utilization of self-cloning, development of food-grade vectors, and integration of recombinant DNA into the chromosome. Self-cloning is defined as reintroducing native genes back into the desired host, which means nothing new will be introduced into the food system (26). Consequently, it is predicted that this type of recombinant organism may be among the first to be approved for use in the marketplace (26). There is considerable interest in using bacteriocins and their genetics in the development of food-grade vectors and markers, and as loci for integration events.

Food-grade vectors must be well defined, thoroughly characterized, sequenced, and derived from GRAS organisms (26, 107). Many plasmids in LAB have been identified, and several have been developed into cloning vectors (reviewed in references 73 and 106). The lactococcal pWV01 and enterococcal pAMβ1 plasmids are commonly used as a base for designing cloning vectors (26). The origins of both of these plasmids also replicate in a wide variety of hosts, both gram-positive and gram-negative, a trait making them unsuitable as a food-grade vector. Several plasmids native to different genera and species have also been characterized, modified with antibiotic resistance markers, and used in transformation experiments (reviewed in references 73 and 106), while other native plasmids have been used in the development of food-grade vectors (see below). Many bacteriocins are plasmid encoded, with the majority of the plasmids exceeding 20 kb. Bacteriocins encoded by smaller plasmids include the following (reviewed in references 32 and 72): acidocin B of *Lb. acidophilus* M46, acidocin 8912 of *Lb acidophilus* TK8912; pediocin PA-1/AcH and SJ-1 of *Pediococcus acidilactici*, and divergicin A of *C. divergens* LV13. Of these, several have shown potential to be used as cloning vectors. Acidocin B is encoded by a 14-kb plasmid, pCV461. The restriction map of pCV461 has been determined, and a 1.1-kb region encoding the bacteriocin has been sequenced (81, 137). An erythromycin resistance gene has been introduced into pCV461 through homologous recombination, and the resulting plasmid was able to confer antibiotic resistance and bacteriocin production in the heterologous host *Lb. plantarum* (137). Another *Lactobacillus* bacteriocin, acidocin 8912, is encoded by a plasmid of 14 kb, pLA103 (71). Using an optimized electroporation protocol, pLA103 was introduced, unmarked, into an isogeneic variant. After transformation, the cells were diluted, plated, and overlaid with an acidocin 8912-sensitive strain. Transformants were selected for their ability to inhibit the indicator, and a transformation frequency of 2.8×10^5 per μg of DNA was reported (71). Given the size of the plasmid, this is an excellent transformation frequency for *Lb. acidophilus*. The 8.9-kb pSM74 plasmid, which encodes pediocin PA-1/AcH from *P. acidilactici* AcH, has been completely sequenced; however, its potential as a food-grade cloning vector has not been reported (96). The plasmid encoding divergicin A, pCD3.4, is a small plasmid of 3.4 kb, and the complete sequence has been determined (147). A chloramphenicol resistance gene has been cloned into a unique site, and the recombinant plasmid was transformed into a variety of heterologous carnobacteria strains, where it conferred chloramphenicol resistance and divergicin production/immunity. Plasmid pCD3.4 is a high-copy-number plasmid in this *Carnobacterium* sp, and there is a great deal of interest in developing it as a food-grade vector (126, 147). pSM74 is the second native plasmid to be characterized from *Pediococcus* sp., and pCD3.4 is the first native plasmid to be characterized from *Carnobacterium* sp. Thorough characterization of these plasmids will be required before they can be used as food-grade cloning vectors.

Food-grade cloning vectors will not be allowed to contain antibiotic resistance markers. Food-grade markers that have been used to create food-grade vectors are as follows. A mutation in the *lacF* gene in the *lac* operon can be complemented by cloning the wild-type gene on a plasmid and selecting on lactose indicator plates (27). Nonsense suppressor genes have been cloned and used to isolate a mutant of *Lc. lactis* that has a nonsense mutation in the purine biosynthetic pathway (28). The nonsense suppressor genes are then used as a selectable marker in the auxotrophic background when selected for on purine-free medium (28). The selective agents in both cases, lactose and purine-free medium, are characteristic of milk and dairy products. The nisin resistance determinant, Nsr, has also been cloned into a food-grade vector. The *nsr* gene encodes a hydrophobic protein that has no relation to the other nisin immunity determinants, and it has been found in strains that do not produce nisin or utilize sucrose (51). The *nsr* gene has been cloned into two different plasmids of lactococcal origin, pVS40 (7.8 kb) and pMF011 (7.6 kb) (50, 144). Transformants are selected on agar plates containing nisin (7 IU of nisin per ml or 20 to 40 IU of nisin per ml, respectively). The lactacin F immunity determinant (LafI) has been used to select for lactobacilli transformed with a recombinant plasmid encoding LafI (6). The lactacin F producer *Lb. johnsonii* was used to create the selective medium on which the transformants were plated and selected (6). In experiments in which the lactacin F determinants were conjugally transferred to a bacteriocin-negative background, lactacin F was also used to select for lactacin F-producing transconjugants (97). Spontaneous lactacin F-resistant mutants of the lactacin F-negative recipient arose at frequencies of approximately 10^{-6} to 10^{-7} (97). Another bacteriocin that has been used as a selective agent is lactacin 3147 (117). In an experiment in which the lactacin 3147 determinants were conjugally transferred to an industrial cheese-making strain, lactacin 3147 (400 activity units [AU]/ml) and streptomycin were used to select for the bacteriocin-producing transconjugants (117). Therefore, bacteriocins and their respective resistance and immunity determinants may be used as food-grade selective agents and markers. Addition of bacteriocin (e.g., nisin), or inclusion of both the bacteriocin and immunity genes as markers will ensure the presence of the selective agent during growth or in a food fermentation. As with antibiotics, there is always the possibility of background problems due to spontaneous resistance as reported above. This problem may be circumvented by using more than one bacteriocin marker or by using a bacteriocin in combination with another food-grade selective agent. LAB that produce multiple bacteriocins have been constructed (Tables 1 and 2). By including two or three bacteriocins in the selective medium, the probability of spontaneous resistance is greatly reduced. Other food-grade markers that have been suggested include the genes that encode for sucrose, xylose, starch, or lactose (β-galactosidase) utilization, complementation of essential genes, and bacteriophage insensitivity (reviewed in reference 26).

Finally, integration of recombinant DNA into the chromosome offers several advantages, including stability and a significantly decreased risk of transfer of the recombinant DNA. Two types of integration experiments have been conducted in LAB (reviewed in references 26 and 73). Chromosomal sequences are cloned onto a plasmid that is marked but unable to replicate in LAB. Upon introduction of the "suicide vector" and plating on selective medium, only those organisms that have the plasmid integrated into the chromosome through homologous recombination can grow. Suicide plasmids used in integration include a variety of *E. coli* and *B. subtilis* replicons (reviewed in reference 26). A second type of integration experiment involves the use of a temperature-sensitive plasmid. Chromosomal sequences are cloned into the temperature-sensitive plasmid, transformed into LAB, and propagated at the permissive temperature. When the temperature is increased and selection maintained, the plasmid must integrate through homologous recombination into the host chromosome. The resulting recombinant can then be maintained at the elevated temperature, or a second crossover can be attempted by successive propagation at the permissive temperature in the absence of selection. The plasmid can then "pop out" of the chromosome in such a way that the desired mutation is created. Plasmids used for temperature-sensitive integration include the pG$^+$host series (temperature-sensitive derivative of pWV01) (13) and pSA3 (11). Which method is chosen depends on the transformation frequency of the target host and the amount of DNA to be used for the homologous recombination (13). The bacteriocin-encoding loci of LAB offer an attractive location for the directed integration of DNA into a well-defined region of the chromosome. The nisin locus has been used in homologous recombination experiments for two purposes: functional analysis of the genes encoding the nisin determinants, and as a method to produce genetically engineered nisin variants. In the latter case, the nisin gene is mutated, cloned into pG$^+$host6 (erythromycin resistant), and introduced into a background in which the nisin gene is deleted but the rest of the nisin determinants are intact (33, 34). Single-crossover and double-crossover events are selected for using erythromycin and nisin, respectively. The background containing the deleted nisin gene has a lower level of immunity than a strain that produces nisin or the variants; therefore, the integrants are selected for over the background (33). Furthermore, the integrated gene, mutated *nisA* in this case, is in single copy under the control of the nisin-inducible P_{nisA}. Therefore, genes integrated in this manner can be induced by the addition of nisin or, if *nisA* is included in the integration event along with the desired gene, natural nisin production could trigger the system. The resulting integrant would be food grade from the perspective that there are no vector sequences in its chromosome, but since nisin is encoded on a conjugative transposon it may be transferred to other lactococcal strains.

Homologous recombination has also been conducted in the *laf* operon of the lactacin F producer *Lb. johnsonii* (6). The *lafI* gene was replaced by a cassette in which the gene was disrupted. The double-crossover event in this case could be selected for by loss of immunity to lactacin F (6). Depending on the regulatory mechanism involved, it may also be possible to integrate heterologous genes into the bacteriocin-encoding genes and select for the double crossover with a bacteriocin-negative phenotype. Therefore, food-grade homologous recombination within the bacteriocin locus may be detected by alteration in bacteriocin production or immunity. Many bacteriocins that have been characterized are naturally present and produced by industrial strains. Examples include natural nisin producers isolated from a commercial sauerkraut fermentation (55) and a lactococcin A, B, M/N producer isolated from an Irish cheese factory (95). Therefore, directed, detectable, and well-defined integrations can be made into industrial cultures using already characterized genetic sequences.

In the future, it may be possible to develop bacteriocins and their genetics into other cloning and food-grade "tools." For example, there has only been one positive-selection cloning vector developed for gram-positive organisms (31). In a bacteriocin-sensitive background, production of a bacteriocin, in most cases, is lethal. A positive-selection cloning vector could be constructed such that it contains the mature part of a bacteriocin fused to a Sec-dependent signal sequence (to minimize the amount of DNA required). A strong promoter could be cloned upstream of the chimeric bacteriocin gene. A multiple-cloning site could be introduced either within the bacteriocin gene or between the promoter and bacteriocin gene. Only when production of the bacteriocin is disrupted by the insertion of a fragment would recombinant organisms be able to grow. Utilization of this system would be limited to those strains that are sensitive to the bacteriocin, and the effectiveness would rely on the "killing power" of the bacteriocin. It will be very important that the development of resistance to the bacteriocin be very low. Development of resistance has been reported for the lactococcins (130, 131, 140), lactacin F (97), and pediocin PA-1 (142).

63.2.4. Novel Applications of Bacteriocin Producers and Peptides

Many bacteriocins and their producers have been used in the biopreservation of food, usually in addition to another form of preservation such as modified atmosphere packaging or storage at low temperatures (reviewed in references 61, 91, 126). Recent reports, however, have begun to show how LAB bacteriocin producers and nisin can be used in applications other than biopreservation.

Bacteriocin producers may be used to accelerate cheese ripening and control the levels of nonstarter LAB. Characterization of an industrial cheese starter, *Lc. lactis* subsp. *lactis* biovar diacetylactis DP3286, indicated that it produces a bacteriocin that causes sensitive cells to lyse and release intracellular enzymes (95). Further characterization of this strain indicates that it produces lactococcins A, B, and M/N (95). Lactococcins A, B, and M/N are active against lactococci only and individually do not cause lysis (143). Utilization of this strain as an adjunct culture in the manufacture of cheddar cheese resulted in increased lysis of the starter cultures, measured by the presence of lactate dehydrogenase and peptidase N in the cheese juice (46). Elevated levels of free amino acids were also reported to result in a product with less bitterness and higher grading scores on flavor and texture than the control cheese made with a bacteriocin-negative, isogeneic derivative of 3286 (reviewed in references 46 and 94). Applications of bacteriocins in accelerated cheese ripening offer many advantages, including no special approval and no extra costs (46). Bacteriocin producers can also be used to control nonstarter LAB in cheese (NSLAB). The contribution of NSLAB in cheese flavor is unknown, but it has been determined that in some cases, off flavors can be attributed to NSLAB. In particular, certain mesophilic lactobacilli can catalyze the formation of histamine, which can become toxic at high levels (reviewed in reference 70). Utilization of LAB strains producing bacteriocins has been shown to reduce the levels of NSLAB. Application of enterocin 4, enterococcin EFS2, and nisin producers resulted in decreased numbers of histamine-forming lactobacilli and consequent reduction in the amount of histamine (70). In another study, utilization of lactococci producing lactacin 3147 in the manufacture of cheddar cheese resulted in significantly lower levels of NSLAB (117). By controlling the numbers of NSLAB in cheddar cheese, Ryan et al. (117) proposed that lactacin 3147-producing strains could be used to study the role of NSLAB in cheese flavor. In general, these studies indicate that it may be possible to control flavor development and other characteristics of fermented foods by using LAB that produce bacteriocins.

Pharmaceutical, veterinary, and microbiological applications of nisin have been reported. A study involving the use of nisin in mouthwash showed that it is effective against plaque buildup and gingivitis in beagle dogs (65). Nisin also had a significantly lower stain index than the positive control (65). Interest in using nisin in antibacterial ointments for skin infections has been reported (129). Atopic dermatitis is a skin infection caused by *S. aureus* and is characterized by inflammation and drying (129). Nisin was added to a variety of topical formulations, and its activity against *S. aureus* was assayed. While very effective against *S. aureus* in aqueous formulations, activity was significantly decreased in the hydrophobic bases that are preferred in the treatment of atopic dermatitis (129). Since nisin is very active against *S. aureus* and other gram-positives bacteria and is nontoxic and nonallergenic, it has great potential for use as a novel treatment for skin infections (129). Nisin is active against *Helicobacter pylori*, which causes peptic ulcers. Nisin has been patented as a prophylactic agent in the prevention of *H. pylori* colonization (14). Nisin has also been demonstrated to be effective in the treatment of bovine mastitis (21). Mastitis is caused by *Staphylococcus* and *Streptococcus* species and is usually treated with antibiotics by intramammary administration. This results in milk containing antibiotic residues that cannot be marketed. Intramammary administration of nisin cures mastitis, with the added advantage that it poses no threat to the safety of the milk supply during treatment. Furthermore, nisin is being used as a preventative agent against mastitis in a pre- and post-milking teat dip (21). Nisin has been used in the development of a selective medium for *Mycoplasma* spp. (2). Detection of *Mycoplasma* spp. in clinical samples is frequently complicated by the presence of *Acholeplasma* spp., inhabitants of the respiratory, urogenital, and intestinal tracts of animals. The addition of nisin to agar or broth medium inhibited *Acholeplasma* spp. and provided clearer results than digitonin, which is the selective agent currently used (2). It is possible that nisin and other bacteriocins could be used in the development of selective media for gram-negative organisms. Therefore, the existing and potential uses of nisin are varied. Nisin serves as a good model for potential applications of other bacteriocins that are not as well characterized.

Utilization of bacteriocins in pharmaceutical, veterinary, and microbiological applications will require large quantities of purified bacteriocins for testing. Purification of bacteriocins from culture supernatants is frequently a laborious and expensive process. However, chemical synthesis of class II bacteriocins has been reported (43, 45) and is an excellent way to generate gram quantities of peptides for testing. Alternatively, production of bacteriocins and lantibiotics by industrial strains such as *B. subtilis* or *E. coli* may also be an economical method for producing bacteriocins on an industrial scale (53). Efforts to express nisin by *B. subtilis* (discussed in a previous section) have yielded limited success. Pediocin PA-1 and sakacin P are the only LAB class II bacteriocins that have been produced by *E. coli* (54, 88). Production of large amounts of sakacin P by *E. coli* was achieved by translationally fusing the gene encoding mature sakacin P to the PelB signal sequence, indicating that secretion and overproduction of class II bacteriocins by *E. coli* are possible (54).

Utilization of bacteriocins in pharmaceutical, veterinary, and microbiological applications may also require alteration of certain characteristics of bacteriocins such as solubility, stability, or activity. For example, nisin exhibits optimal activity and solubility at very low pH, thus limiting the formulations in which it can be used. Through genetic engineering, alteration of single amino acids in nisin has resulted in significantly improved solubility at neutral pH with minimal effects on activity (116). The inhibitory activity of bacteriocins has also been improved through genetic engineering. Examples include the following: mutation of Asn-42 to Lys in sakacin P increases its activity against *Listeria monocytogenes* (54); mutation of Cys-24 in lactococcin B to various other amino acids increases activity against lactococci (140); and many modifications of nisin have been made that exhibit elevated activity against certain indicators (reviewed in reference 32). Novel inhibitory activity may be obtained by combining different domains of bacteriocins. Fimland et al. (43) reported the creation of chimeric bacteriocins consisting of combinations of the N-terminal and C-terminal domains of the class IIa listeria-active peptides. The inhibitory activity of the chimeric bacteriocins was quite different from that of the peptide from which the N-terminal domain was derived but similar to that from which the C-terminal domain was derived, suggesting that the C terminus determines the antibacterial activity of the class IIa bacteriocins (43). Increased and expanded antimicrobial activity has also been observed when bacteriocins are combined with chelators, lysozyme, and detergents (reviewed in reference 21). Therefore, many tools are available to "design" or "customize" a bacteriocin for various applications.

63.3. FUTURE PERSPECTIVES

The study of bacteriocins produced by LAB has provided a much-needed back door into the genetics of these organisms. The regulatory and transport systems of bacteriocins offer numerous possibilities for genetic engineering of LAB. Continued characterization of bacteriocin systems should lend further insight into these processes in LAB. For example, helveticin J produced by *Lactobacillus helveticus* and caseicin 80 produced by *Lactobacillus casei* are regulated by the SOS system (reviewed in reference 72). Little is known about the SOS response in LAB.

Utilization of bacteriocins in pharmaceutical and veterinary applications may offer one alternative to antibiotic therapy. Antibiotic-resistant bacteria are becoming a real threat to society, and the development of new antibiotics is lagging behind (128). Bacteriocins and AMPs target the cytoplasmic membrane, unlike present antibiotics, which are aimed at inhibition of protein synthesis, cell wall construction, and DNA replication. Effective administration of AMPs, however, may be challenging.

The role of bacteriocins and their producers in food systems, either for the purpose of biopreservation or for other reasons (see above), is an active area of research. The role of bacteriocin production in other natural systems, however, remains to be determined. For example, many intestinal lactobacilli produce bacteriocins, but it remains to be determined if they are involved in the ability of these organisms to compete in the gastrointestinal tract (reviewed in reference 73). It may be possible that other inhibitory factors produced by LAB play a major role in the competitiveness of these organisms in this environment. It will be interesting to determine if LAB genetically engineered to deliver and produce AMPs and proteins targeted against enteric pathogens can play a role in favorable modification of the gastrointestinal microflora.

Another area of interest is utilization of the lantibiotic modification enzymes for creating novel modified peptides (53). Recent data suggest that the nisin prepeptide interacts with the modification enzymes (123). Understanding of the mechanism of modification and determination of the recognition signals may allow engineering of novel modified AMPs with improved therapeutic uses. While a great deal of work remains to be done in this area of lantibiotics, the posttranslational modifications of microcin B17 are beginning to be unraveled (82). Microcin B17 is a modified peptide produced by *E. coli* that inhibits DNA gyrase. It has also been determined that the 26-amino-acid leader peptide of microcin B17 is essential for the posttranslational modifications (82, 83).

In the past decade there has been an explosion of research on bacteriocins of LAB. Originally, the study of bacteriocins was fueled by their potential use in biopreservation. However, it has become clear that bacteriocins serve as a valuable tool for studying many other processes in LAB, many of which can be used for improving and expanding the applications of LAB probiotic and starter culture strains. Further characterization of currently known bacteriocins and newly identified ones will continue to lend insight into these industrially important bacteria.

REFERENCES

1. **Abee, T.** 1995. Pore-forming bacteriocins of Gram-positive bacteria and self-protection mechanisms of producer organisms. *FEMS Microbial. Lett.* **129:**1–9.

2. **Abu-Amero, K. K., M. A. Halablab, and R. J. Miles.** 1996. Nisin resistance distinguishes *Mycoplasma* spp. from *Acholeplasma* spp. and provides a basis for selective growth media. *Appl. Environ. Microbiol.* **62:**3107–3111.

3. **Ahn C., and M. E. Stiles.** 1992. Mobilization and expression of bacteriocin plasmids from *Carnobacterium piscicola* isolated from meat. *J. Appl. Bacteriol.* **73:**217–228.

3a. **Allison, G. E., et al.** Unpublished data.

3b. **Allison, G. E., and T. R. Klaenhammer.** Unpublished data.

4. **Allison, G. E., C. Ahn, M. E. Stiles, and T. R. Klaenhammer.** 1995. Utilization of the leucocin A export system in *Leuconostoc gelidum* for production of a *Lactobacillus* bacteriocin. *FEMS Microbiol. Lett.* **131:**87–93.

5. **Allison, G. E., C. Fremaux, and T. R. Klaenhammer.** 1994. Expansion of bacteriocin activity and host range upon complementation of two peptides encoded within the lactacin F operon. *J. Bacteriol.* **176:**2235–2241.

6. **Allison, G. E., and T. R. Klaenhammer.** 1996. Functional analysis of the gene encoding immunity to lactacin F, *lafI*, and its use as a *Lactobacillus*-specific, food-grade genetic marker *Appl. Environ. Microbiol.*, **62:**4450–4460.

7. **Allison, G. E., R. W. Worobo, M. E. Stiles, and T. R. Klaenhammer.** 1995. Heterologous expression of the lactacin F peptides by *Carnobacterium piscicola* LV17. *Appl. Environ. Microbiol.* **61:**1371–1377.

8. **Axelsson, L., M. Bjornslett, and A. Holck.** 1996. Expression of bacteriocins using the transport and regulatory system of sakacin A, abstr. C9. Presented at the Fifth Symposium on Lactic Acid Bacteria: Genetics, Metabolism and Applications. Veldhoven, The Netherlands.

9. **Aymerich, T., H. Holo, L. S. Havarstein, M. Hugas, M. Garriga, and I. F. Nes.** 1996. Biochemical and genetic characterization of enterocin A from *Enterococcus faecium*,

a new antilisterial bacteriocin in the pediocin family of bacteriocins. *Appl. Environ. Microbiol.* **62:**1676–1682.

10. **Barefoot, S. F., Y. R. Chen, T. A. Hughes, A. B. Bodine, M. Y. Shearer, and M. D. Hughes.** 1994. Identification and purification of a protein that induces production of the *Lactobacillus acidophilus* bacteriocin lactacin B. *Appl. Environ. Microbiol.* **60:**3522–3528.
11. **Bhowmik, T., L. Fernandez, and J. L. Steele.** 1993. Gene replacement in *Lactobacillus helveticus*. *J. Bacteriol.* **175:** 6341–6344.
12. **Biet, F., J. M. Berjeaud, R. W. Worobo, Y. Cenatiempo, and C. Fremaux.** 1996. Secretion of mesentaricin Y105 by industrial lactic acid bacteria, abstr. C55. Presented at the Fifth Symposium on Lactic Acid Bacteria: Genetics, Metabolism and Applications. Veldhoven, The Netherlands.
13. **Biswas, I., A. Gruss, S. D. Ehrlich, and E. Maguin.** 1993. High-efficiency gene inactivation and replacement system for gram-positive bacteria. *J. Bacteriol.* **175:**3628–3635.
14. **Blackburn, P., and S. J. Projan.** April 1994. Pharmaceutical bacteriocin combinations and methods for using the same. U.S. patent 5,304,540.
15. **Board, R. G.** 1995. Natural antimicrobials from animals, p. 40–57. *In* G. W. Gould (ed.), *New Methods of Food Preservation*. Blackie Academic and Professional, Glasgow, Scotland.
16. **Boot, H. J., C. P. Kolen, and P. H. Pouwels.** 1995. Identification, cloning, and nucleotide sequence of a silent S-layer protein gene of *Lactobacillus acidophilus* ATCC 4356 which has extensive similarity with the S-layer protein gene of this species. *J. Bacteriol.* **177:**7222–7230.
17. **Bruno, M. E. C., and T. J. Montville.** 1993. Common mechanistic action of bacteriocins from lactic acid bacteria. *Appl. Environ. Microbiol.* **59:**3003–3010.
18. **Chakicherla, A., and J. N. Hansen.** 1995. Role of the leader and structural regions of prelantibiotic peptides as assessed by expressing nisin-subtilin chimeras in *Bacillus subtilis* 168, and characterization of their physical, chemical, and antimicrobial properties. *J. Biol. Chem.* **270:** 23533–23539.
19. **Chikindas, M. L., K. Venema, A. M. Ledeboer, G. Venema, and J. Kok.** 1995. Expression of lactococcin A and pediocin PA-1 in heterologous hosts. *Lett. Appl. Microbiol.* **21:**183–189.
20. **Dayem, M. A., Y. Fleury, G. Devilliers, E. Chaboisseau, R. Girard, P. Nicolas, and A. Delfour.** 1996. The putative immunity protein of the Gram-positive bacteria *Leuconostoc mesenteroides* is preferentially located in the cytoplasmic compartment. *FEMS Microbiol. Lett.* **138:** 251–259.
21. **Delves-Broughton, J., P. Blackburn, R. J. Evans, and J. Hugenholtz.** 1996. Applications of the bacteriocin, nisin. *Antonie Leeuwenhoek* **69:**193–202.
22. **de Ruyter, P. G. G. A., O. P. Kuipers, M. M. Beerthuyzen, I. van Alen-Boerrigter, and W. M. de Vos.** 1996. Functional analysis of promoters n the nisin gene cluster of *Lactococcus lactis*. *J. Bacteriol.* **178:**3434–3439.
23. **de Ruyter, P. G. G. A., O. P. Kuipers, L. C. Bijl, and W. M. de Vos.** 1996. Controlled gene expression in *Lactococcus lactis*, abstr. H46. Presented at the Fifth Symposium on Lactic Acid Bacteria: Genetics, Metabolism and Applications. Veldhoven, The Netherlands.
24. **de Ruyter, P. G. G. A., O. P. Kuipers, and W. M. de Vos.** 1996. Controlled gene expression systems for *Lactococcus lactis* with the food-grade inducer nisin. *Appl. Environ. Microbiol.* **62:**3662–3667.
25. **de Vos, W. M.** 1996. Metabolic engineering of sugar catabolism in lactic acid bacteria. *Antonie Leeuwenhoek* **70:** 223–242.
26. **de Vos, W. M., and G. F. M. Simons.** 1994. Gene cloning and expression systems in *Lactococci*, p. 52–105. *In* M. J. Gasson and W. M. de Vos (ed.), *Genetics and Biotechnology of Lactic Acid Bacteria*. Blackie Academic and Professional, Glasgow, Scotland.
27. **de Vos, W. M., I. van Alen-Boerrigter, R. J. van Rooyen, B. Reiche, and W. Hengstenberg.** 1990. Characterization of the lactose-specific enzymes of the phosphotransferase system in *Lactococcus lactis*. *J. Biol. Chem.* **265:**22554–22560.
28. **Dickely, F., D. Nilsson, E. B. Hansen, and E. Johansen.** 1995. Isolation of *Lactococcus lactis* nonsense suppressors and construction of a food-grade cloning vector. *Mol. Microbil.* **15:**839–847.
29. **Diep, D. B., L. S. Håvarstein, and I. F. Nes.** 1995. A bacteriocin-like peptide induces bacteriocin synthesis in *Lactobacillus plantarum* C11. *Mol. Microbiol.* **18:**631–639.
30. **Diep, D. B., L. S. Håvarstein, and I. F. Nes.** 1996. Characterization of the locus responsible for the bacteriocin production in *Lactobacillus plantarum* C11. *J. Bacteriol.* **178:**4472–4483.
31. **Djordjevic, G. M., and T. R. Klaenhammer.** 1996. Positive selection, cloning vectors for Gram-positive bacteria based on a restriction endonuclease cassette. *Plasmid* **35:** 37–45.
32. **Dodd, H. M., and M. J. Gasson.** 1994. Bacteriocins of lactic acid bacteria, p. 211–251. *In* M. J. Gasson and W. M. de Vos (ed.), *Genetics and Biotechnology of Lactic Acid Bacteria*. Blackie Academic and Professional, Glasgow, Scotland.
33. **Dodd, H. M., N. Horn, W. C. Chan, C. J. Giffard, B. W. Bycroft, G. C. K. Roberts, and M. J. Gasson.** 1996. Molecular analysis of the regulation of nisin immunity. *Microbiology* **142:**2385–2392.
34. **Dodd, H. M., N. Horn, C. J. Giffard, and M. J. Gasson.** 1996. A gene replacement strategy for engineering nisin. *Microbiology* **142:**47–55.
35. **Eijsink, V. G., M. B. Brurberg, P. H. Middelhoven, and I. F. Nes.** 1996. Induction of bacteriocin production in *Lactobacilus sake* by a secreted peptide. *J. Bacteriol.* **178:** 2232–2237.
36. **Emond, E.** 1996. Characterisation moléculaire de la pediocine PA-1 et des mécanismes de la maturation et de l'exportation des bacteriocines de la classe II. Ph.D. thesis. Université Laval, Quebec City, Quebec, Canada.
37. **Engelke, G., E. Z. Gutowski, M. Hammelmann, and K. D. Entian.** 1992. Biosynthesis of the lantibiotic nisin: genomic organization and membrane localization of the NisB protein. *Appl. Environ. Microbiol.* **58:**3730–3743.
38. **Engelke, G., Z. Gutowskieckel, P. Kiesau, K. Siegers, M. Hammelmann, and K. D. Entian.** 1994. Regulation of nisin biosynthesis and immunity in *Lactococcus lactis* 6F3. *Appl. Environ. Microbiol.* **60:**814–825.
39. **Fath, M. J., and R. Kolter,** 1993. ABC transporters: bacterial exporters. *Microbiol. Rev.* **57:**995–1017.

40. **Fath, M. J., R. C. Skvirsky, and R. Kolter.** 1991. Functional complementation between bacterial MDR-like export systems: colicin V, alpha-hemolysin, and *Erwinia* protease. *J. Bacteriol.* **173:**7549–7556.

41. **Fath, M. J., L. H. Zhang, J. Rush, and R. Kolter.** 1994. Purification and characterization of colicin V from *Escherichia coli* culture supernatants. *Biochemistry* **33:**6911–6917.

42. **Fayard, B., F. Prevots, J. Kok, G. Venema, and K. Venema.** 1996. Secretion of lactococcin B and pediocin PA-1 by the *sec*-dependent pathway, abstr. C65. Presented at the Fifth Symposium on Lactic Acid Bacteria: Genetics, Metabolism and Applications. Veldhoven, The Netherlands.

43. **Fimland, G., O. R. Blingsmo, K. Sletten, G. Jung, I. F. Nes, and J. Nissen-Meyer.** 1996. New biologically active hybrid bacteriocins constructed by combining regions from various pediocin-like bacteriocins: the C-terminal region is important for determining specificity. *Appl. Environ. Microbiol.* **62:**3313–3318.

44. **Fitzsimons, A., P. Hols, J. Jore, R. J. Leer, M. O'Connell, and J. Delcour.** 1994. Development of an amylolytic *Lactobacillus plantarum* silage strain expressing the *Lactobacillus amylovorus* alpha-amylase gene. *Appl. Environ. Microbiol.* **60:**3529–3535.

45. **Fleury, Y., M. A. Dayem, J. J. Montagne, E. Chaboisseau, J. P. LeCaer, P. Nicolas, and A. Delfour.** 1996. Covalent structure, synthesis, and structure-function studies of mesentericin Y 105(37), a defensive peptide from Gram-positive bacteria *Leuconostoc mesenteroides*. *J. Biol. Chem.* **271:**14421–14429.

46. **Fox, P. F., J. M. Wallace, S. Morgan, C. M. Lynch, E. J. Niland, and J. Tobin.** 1996. Acceleration of cheese ripening. *Antonie Leeuwenhoek* **70:**271–297.

47. **Franke, C. M., K. J. Leenhouts, A. J. Haandrikman, J. Kok, G. Venema, and K. Venema.** 1996. Topology of LcnD, a protein implicated in the transport of bacteriocins from *Lactococcus lactis*. *J. Bacteriol.* **178:**1766–1769.

48. **Fremaux, C., C. Ahn, and T. R. Klaenhammer.** 1993. Molecular analysis of the lactacin-F operon. *Appl. Environ. Microbiol.* **59:**3906–3915.

49. **Fremaux, C., Y. Hechard, and Y. Cenatiempo.** 1995. Mesentericin Y105 gene clusters in *Leuconostoc mesenteroides* Y105. *Microbiology* **141:**1637–1645.

50. **Froseth, B. R., and L. L. McKay.** 1991. Development and application of pFM011 as a possible food-grade cloning vector. *J. Dairy Sci.* **74:1445–1453.**

51. **Froseth, B. R., and L. L. McKay.** 1991. Molecular characterization of the nisin resistance region of *Lactococcus lactis* subsp. *lactis* biovar diacetylactis DRC3. *Appl. Environ. Microbiol.* **57:**804–811.

52. **Gilson, L., H. K. Mahanty, and R. Kolter.** 1990. Genetic analysis of an MDR-like export system: the secretion of colicin V. *EMBO J.* **9:**3875–3884.

53. **Hansen, J. N.** 1993. Antibiotics synthesized by posttranslational modification. *Annu. Rev. Microbiol.* **47:**535–564.

54. **Harmark, K., M. A. Lim, G. E. Schepers, R. E. H. Wettenhall, A. J. Hillier, and B. E. Davidson.** 1996. Heterologous expression of a cystibiotic in *E. coli* and probing for structure/function relationships by random mutagenesis, abstr. C54. Presented at the Fifth Symposium on Lactic Acid Bacteria: Genetics, Metabolism and Applications. Veldhoven, The Netherlands.

55. **Harris, L. J., H. P. Fleming, and T. R. Klaenhammer.** 1992. Characterization of two nisin-producing *Lactococcus lactis* subsp. *lactis* strains isolated from a commercial sauerkraut fermentation. *Appl. Environ. Microbiol.* **58:**1477–1483.

56. **Håvarstein, L. S., D. b. Diep, and I. F. Nes.** 1995. A family of bacteriocin ABC transporters carry out proteolytic processing of their substrates concomitant with export. *Mol. Microbiol.* **16:**229–240.

57. **Håvarstein, L. S., P. Gaustad, I. F. Nes, and D. A. Morrison.** 1996. Identification of the streptococcal competence-phermone receptor. *Mol. Microbiol.* **21:**863–869.

58. **Håvarstein, L. S., H. Holo, and I. Fe. Nes.** 1994. The leader peptide of colicin V shares consensus sequences with leader peptides that are common among peptide bacteriocins produced by Gram-positive bacteria. *Microbiology* **140:**2383–2389.

59. **Hawkins, G.** 1990. Investigation of the site and mode of action of the small protein antibiotic subtilin and development and characterization of an expression system for the small protein antibiotic nisin in *Bacillus subtilis*. PhD thesis. University of Maryland, College Park.

60. **Higgins, C. F.** 1992. ABC transporters: from microorganisms to man. *Annu. Rev. Cell Biol.* **8:**67–113.

61. **Hill, C.** 1995. Bacteriocins: natural antimicrobials form microorganisms, p. 22–39. *In* G. W. Gould (ed.), *New Methods of Food Preservation*. Blackie Academic and Professional, Glasgow, Scotland.

62. **Holck, A., U. Schillinger, I. Saeterdal, and L. Axelsson.** 1995. Heterologous expression of piscicolin 61, a bacteriocin from *Carnobacterium piscicola* 61, in *Lactobacillus sake* LB706-X, abstr. pG-26. Presented at the Lactic Acid Bacteria Conference. Cork, Ireland.

63. **Holland, R., O. Kuipers, J. C. Brown, V. L. Crow, and T. Coolbear.** 1996. Lactococcal tributyrin esterase: substrate selectivity and impact during cheese maturation, abstr. K1. Presented at the Fifth Symposium on Lactic Acid Bacteria: Genetics, Metabolism and Applications. Veldhoven, The Netherlands.

64. **Hols, P., T. Ferain, d. Garmyn, N. Bernard, and J. Delcour.** 1994. Use of homologous expression-secretion signals and vector-free stable chromosomal integration in engineering of *Lactobacillus plantarum* for alpha-amylase and levanase expression. *Appl. Environ. Microbiol.* **60:**1401–1413.

65. **Howell, T. H., J. P. Fiorellini, P. Blackburn, S. J. Projan, J. de la Harpe, and R. C. Williams.** 1993. The effect of a mouthrinse based on nisin, a bacteriocin, on developing plaque and gingivitis in beagle dogs. *J. Clin. Periodontol.* **20:**335–339.

66. **Huhne, K., L. Axelsson, A. Holck, and L. Krockel.** 1996. Analysis of the sakacin P gene cluster from *Lactobacillus sake* LB674 and its expression in sakacin-negative *Lb. sake* strains. *Microbiology* **142:**1437–1448.

67. **Hui, F. M., L. X. Zhou, and D. A. Morrison.** 1995. Competence for genetic transformation in *Streptococcus pneumoniae*: organization of a regulatory locus with homology to two lactococcin A secretion genes. *Gene* **153:**25–31.

68. **Jack, R., G. Bierbaum, C. Heidrich, and H. G. Sahl.** 1995. The genetics of lantibiotic biosynthesis. *Bioessays* **17:**793–802.

69. **Jimenez-Diaz, R., J. L. Ruiz-Barba, D. P. Cathcart, H. Holo, I. F. Nes, K. H. Sletten, and P. J. Warner.** 1995. Purification and partial amino acid sequence of plantaricin S, a bacteriocin produced by *Lactobacillus plantarum* LPCO10, the activity of which depends on the complementary action of two peptides. *Appl. Environ. Microbiol.* **61:**4459–4463.

70. **Joosten, H. M. L. J., and M. Nunez.** 1996. Prevention of histamine formation in cheese by bacteriocin-producing lactic acid bacteria. *Appl. Environ. Microbiol.* **62:**1178–1181.

71. **Kanatani, K., K. Yoshida, T. Tahara, K. Yamada, H. Miura, M. Sakamoto, and M. Oshimura.** 1992. Transformation of *Lactobacillus acidophilus* TK8912 by electroporation with pULA105E plasmid. *J. Ferment. Bioeng.* **74:** 358–362.

72. **Klaenhammer, T. R.** 1993. Genetics of bacteriocins produced by lactic acid bacteria. *FEMS Microbiol. Rev.* **12:** 39–86.

73. **Klaenhammer, T. R.** 1995. Genetics of intestinal lactobacilli. *Int. Dairy J.* **5:**1019–1058.

74. **Koivula, T., I. Palva, and H. Hemila.** 1991. Nucleotide sequence of the *secY* gene from *Lactococcus lactis* and identification of conserved regions by comparison of four SecY proteins. *FEBS Lett.* **288:**114–118.

75. **Kok, J.** 1996. Inducible gene expression and environmentally regulated genes in lactic acid bacteria. *Antonie Leeuwenhoek* **70:**129–145.

76. **Kuipers, O., P., M. M. Beerthuyzen, P. G. de Ruyter, E. J. Luesink, and W. M. de Vos.** 1995. Autoregulation of nisin biosynthesis in *Lactococcus lactis* by signal transduction. *J. Biol. Chem.* **270:**27299–27304.

77. **Kuipers, O. P., M. M. Beerthuyzen, R. J. Seizen, T. Coolbear, and R. Holland.** 1996. Cloning and controlled overexpression of a gene of *Lactococcus lactis* E8, encoding an intracellular esterase that has sequence similarities to extracellular lipases, abstr. K2. Presented at the Fifth Symposium on Lactic Acid Bacteria: Genetics, Metabolism and Applications. Veldhoven, The Netherlands.

78. **Kuipers, O. P., M. M. Beerthuyzen, R. J. Siezen, and W. M. Devos.** 1993. Characterization of the nisin gene cluster *nisABTCIPR* of *Lactococcus lactis*—requirement of expression of the *nisA* and *nisI* genes for development of immunity. *Eur. J. Biochem.* **216:**281–291.

79. **Kuipers, O. P., H. S. Rollema, W. M. de Vos, and R. J. Siezen.** 1993. Biosynthesis and secretion of a precursor of nisin Z by *Lactococcus lactis*, directed by the leader peptide of the homologous lantibiotic subtilin from *Bacillus subtilis*. *FEBS Lett.* **330:**23–27.

80. **Kunji, E. R. S., I. Mierau, A. Hagting, B. Poolman, and W. N. Konings.** 1996. The proteolytic systems of lactic acid bacteria. *Antonie Leeuwenhoek* **70:**187–221.

81. **Leer, R. J., J. M. van der Vossen, M. van Giezen, J. M. van Noort, and P. H. Pouwels.** 1995. Genetic analysis of acidocin B, a novel bacteriocin produced by Lactobacillus acidophilus. *Microbiology* **141:**1629–1635.

82. **Li, Y.-M., J. C. Milne, L. L. Madison, R. Kolter, and C. T. Walsh.** 1996. From peptide precursors to oxazole and thiazole-containing peptide antibiotics: microcin B17 synthase. *Science* **274:**1188–1194.

83. **Madison, L. L., E. L. Vivas, Y. M. Li, C. T. Walsh, and R. Kolter.** 1997. The leader peptide is essential for the post-translational modification of the DNA-gyrase inhibitory microcin B17. *Mol Microbiol.* **23:**161–168.

84. **Maloy, W. L., and U. P. Kari.** 1995. Structure-activity studies on magainins and other host defense peptides. *Biopolymers* **37:**105–122.

85. **Marciset, O., M. C. Jeronimusstratingh, B. Mollet, and B. Poolman.** 1997. Thermophilin 13, a nontypical antilisterial poration complex bacteriocin, that functions without a receptor. *J. Biol. Chem.* **272:**14277–14284.

86. **Martinez, B., A. Rodriguez, and J. E. Suarez.** 1996. Lactococcin 972, a plasmid- encoded bacteriocin whose active form is a homodimer that does not induce membrane damage, abstr. C13. Presented at the Fifth Symposium on Lactic Acid Bacteria: Genetics, Metabolism and Applications. Veldhoven, The Netherlands.

87. **Martinez, B., J. E. Suarez, and A. Rodriguez.** 1996. Lactococcin 972: a homodimeric lactococcal bacteriocin whose primary target is not the plasma membrane. *Microbiology* **142:**2393–2398.

88. **Marugg, J. D., C. F. Gonzalez, B. S. Kunka, A. M. Ledeboer, M. J. Pucci, M. Y. Toonen, S. A. Walker, L. C. Zoetmulder, and P. A. Vandenbergh.** 1992. Cloning, expression, and nucleotide sequence of genes involved in production of pediocin PA-1, a bacteriocin from *Pediococcus acidilactici* PAC1.0. *Appl. Environ. Microbiol.* **58:**2360–2367.

89. **McCormick, J. K., R. W. Worobo, and M. E. Stiles.** 1996. Expression of the antimicrobial peptide carnobacteriocin B2 by a signal peptide-dependent general secretory pathway. *Appl. Environ. Microbiol.* **62:**4095–4099.

90. **McCormick, J. K., R. W. Worobo, and M. E. Stiles.** 1996. *sec*-dependent expression of class II bacteriocins, abstr. C31. Presented at the Fifth Symposium on Lactic Acid Bacteria: Genetics, Metabolism and Applications. Veldhoven, The Netherlands.

91. **McMullen, L. M., and M. E. Stiles.** 1996. Potential for use of bacteriocin-producing lactic acid bacteria in the preservation of meats. *J. Food Prot.* Suppl.:64–71.

92. **Mercenier, A., P. H. Pouwels, and B. M. Chassy.** 1994. Genetic engineering of lactobacilli, leuconostocs and *Streptococcus thermophilus*, p. 253–293. *In* M. J. Gasson and W. M. de Vos (ed.), *Genetics and Biotechnology of Lactic Acid Bacteria*. Blackie Academic and Professional, Glasgow, Scotland.

93. **Moll, G., K. T. Ubbink, H. H. Hildeng, M. J. Nissen, I. F. Nes, W. N. Konings, and A. J. Driessen.** 1996. Lactococcin G is a potassium ion-conducting, two-component bacteriocin. *J. Bacteriol.* **178:**600–605.

94. **Morgan, S., D. Murray, R P. Ross, and C. Hill.** 1996. Application of a bacteriocin-producing lactococcal strain in the acceleration of cheddar cheese ripening, abstr. C21. Presented at the Fifth Symposium on Lactic Acid Bacteria: Genetics, Metabolism and Applications. Veldhoven, The Netherlands.

95. **Morgan, S., R. P. Ross, and C. Hill.** 1995. Bacteriolytic activity caused by the presence of a novel lactococcal plasmid encoding lactococcins A, B, and M. *Appl. Environ. Microbiol.* **61:**2995–3001.

96. **Motlagh, A., M. Bukhtiyarova, and B. Ray.** 1994. Complete nucleotide sequence of pSMB 74, a plasmid encoding the production of pediocin AcH in *Pediococcus acidilactici*. *Lett. Appl. Microbiol.* **18:**305–312.

97. **Muriana, P. M., and T. R. Klaenhammer.** 1987. Conjugal transfer of plasmid-encoded determinants for bacteriocin

production and immunity in *Lactobacillus acidophilus* 88. *Appl. Environ. Microbiol.* **53:**553–560.

98. **Murinda, S. E., R. F. Roberts, and R. A. Wilson.** 1996. Evaluation of colicins for inhibitory activity against diarrheagenic *Escherichia coli* strains, including serotype O157:H7. *Appl. Environ. Microbiol.* **62:**3196–3202.
99. **Nes, I. F., D. B. Diep, L. S. Håvarstein, M. B. Brurberg, V. Eijsink, and H. Holo.** 1996. Biosynthesis of bacteriocins in lactic acid bacteria. *Antonie Leeuwenhoek* **70:**113–128.
100. **Nes, I. F., L. S. Håvarstein, and H. Holo.** 1995. Genetics of non-lantibiotic bacteriocins, p. 645–651. *In* J. J. Ferretti, M. S. Glimore, and T. R. Klaenhammer (ed), *Genetics of Streptococci, Enterococci, and Lactococci*, vol. 85. S. Kargr, Basel.
101. **Nes, I. F., and J. R. Tagg.** 1996. Novel lantibiotics and their pre-peptides. *Antonie Leeuwenhoek* **69:**89–97.
102. **Nissen-Meyer, J., L. S. Håvarstein, H. Holo, K. Sletten, and I. F. Nes.** 1993. Association of the lactococcin-A immunity factor with the cell membrane—purification and characterization of the immunity factor. *J. Gen. Microbiol.* **139:**1503–1509.
103. **Nissen-Meyer, J., H. Holo, L. S. Havarstein, K. Sletten, and I. F. Nes.** 1992. A novel lactococcal bacteriocin whose activity depends on the complementary action of two peptides. *J. Bacteriol.* **174:**5686–5692.
104. **Parkinson, J. S., and E. C. Kofoid.** 1992. Communication modules in bacterial signaling proteins. *Annu. Rev. Genet.* **26:**71–112.
105. **Pestova, E. V., L. S. Håvarstein, and D. A. Morrison.** 1996. Regulation of competence for genetic transformation in *Streptococcus pneumoniae* by an auto-induced peptide pheromone and a two-component regulatory system. *Mol. Microbiol.* **21:**853–862.
106. **Pouwels, P. H., and R. J. Leer.** 1993. Genetics of lactobacilli: plasmids and gene expression. *Antonie Leeuwenhoek* **64:**85–107.
107. **Pouwels, P. H., and R. J. Leer.** 1995. Food-grade vectors for lactic acid bacteria, abstr. 4.1. Presented at the Conference on Bacteriocins of Lactic Acid Bacteria: Applications and Fundamentals. Banff, Alberta, Canada.
108. **Pouwels, P. H., R. J. Leer, and W. J. A. Boersma.** 1996. The potential of *Lactobacillus* as a carrier for oral immunization: development and preliminary characterization of vector systems for targeted delivery of antigens. *J. Biotechnol.* **44:**183–192.
109. **Pugsley, A. P.** 1993. The complete general secretory pathway in Gram-negative bacteria. *Microbiol. Rev.* **57:** 50–108.
110. **Quadri, L. E. N., M. Kleerebezem, O. P. Kuipers, W. M. de Vos, K. L. Roy, J. C. Vederas, and M. E. Stiles.** 1997. Characterization of a locus from *Carnobacterium piscicola* LV17B involved in bacteriocin production and immunity: evidence for global inducer-mediated transcriptional regulation. *J. Bacteriol.* **179:**6163–6171.
111. **Quadri, L. E. N., M. Sailer, K. L. Roy, J. C. Vederas, and M. E. Stiles.** 1994. Chemical and genetic characterization of bacteriocins produced by *Carnobacterium piscicola* LV17B. *J. Biol. Chem.* **269:**12204–12211.
112. **Quadri, L. E. N., M. Sailer, M. R. Terebiznik, K. L. Roy, J. C. Vederas, and M. E. Stiles.** 1995. Characterization of the protein conferring immunity to the antimicrobial peptide carnobacteriocin B2 and expression of carnobacteriocins B2 and BM1. *J. Bacteriol.* **177:**1144–1151.
113. **Rauch, P., J., and W. M. de Vos.** 1992. Characterization of the novel nisin-sucrose conjugative transposon Tn*5276* and its insertion in *Lactococcus lactis*. *J. Bacteriol.* **174:**1280–1287.
114. **Rince, A., A. Dufour, S. Le Pogam, D. Thuault, C. M. Bourgeois, and J. P. Le Pennec.** 1994. Cloning, expression, and nucleotide sequence of genes involved in production of lactococcin DR, a bacteriocin from *Lactococcus lactis* subsp. *lactis*. *Appl. Environ. Microbiol.* **60:** 1652–1657.
115. **Rintala, H., T. Graeffe, L. Paulin, N. Kalkkinen, and P. E. J. Saris.** 1993. Biosynthesis of nisin in the subtilin producer *Bacillus subtilis* ATCC6633. *Biotechnol. Lett.* **15:** 991–996.
116. **Rollema, H. S., O. P. Kuipers, P. Both, W. M. de Vos, and R. J. Siezen.** 1995. Improvement of solubility and stability of the antimicrobial peptide nisin by protein engineering. *Appl. Environ. Microbiol.* **61:**2873–2878.
117. **Ryan, M. P., M. C. Rea, C. Hill, and R. P. Ross.** 1996. An application in cheddar cheese manufacture for a strain of *Lactococcus lactis* producing a novel broad-spectrum bacteriocin, lacticin 3147. *Appl. Environ. Microbiol.* **62:**612–619.
118. **Saris, P. E. J., T. Immonen, M. Reis, and H. S. Sahl.** 1996. Immunity to lantibiotics. *Antonie Leeuwenhoek* **69:** 151–159.
119. **Saucier, L., A. S. Paradkar, L. S. Frost, S. E. Jensen, and M. E. Stiles.** 1996. Bacteriocin production and transcriptional analysis in *Carnobacterium piscicola* (*maltaromicus*) LV17, abstr. C33. Presented at the Fifth Symposium on Lactic Acid Bacteria: Genetics, Metabolism and Applications. Veldhoven, The Netherlands.
120. **Saucier, L., A. Poon, and M. E. Stiles.** 1995. Induction of bacteriocin in *Carnobacterium piscicola* LV17. *J. Appl. Bacteriol.* **78:**684–690.
121. **Selsted, M. E., S. I. Miller, A. H. Henschen, and A. J. Ouellette.** 1992. Enteric defensins: antibiotic peptide components of intestinal host defense. *J. Cell Biol.* **118:** 929–936.
122. **Siegers, K., and K. D. Entian.** 1995. Genes involved in immunity to the lantibiotic nisin produced by *Lactococcus lactis* 6F3. *Appl. Environ. Microbiol.* **61:**1082–1089.
123. **Siegers, K., S. Heinzmann, and K. D. Entian.** 1996. Biosynthesis of lantibiotic nisin—posttranslational modification of its prepeptide occurs at a multimeric membrane-associated lanthionine synthetase complex. *J. Biol. Chem.* **271:**12294–12301.
124. **Simonen, M., and I. Palva.** 1993. Protein secretion in *Bacillus* species. *Microbiol. Rev.* **57:**109–137.
125. **Skeie, M., M. B. Brurger, P. H. Middelhoven, G. Fimland, J. Nissen-Mayer, I. F. Nes, and V. G. H. L. Eijsink.** 1996. Structure-function studies of pediocin-like bacteriocins (PLBs), abstr. C18. Presented at the Fifth Symposium on Lactic Acid Bacteria: Genetics, Metabolism and Applications. Veldhoven, The Netherlands.
126. **Stiles, M. E.** 1996. Biopreservation by lactic acid bacteria. *Antonie Leeuwenhoek* **70:**331–345.
127. **Stock, J. B., A. J. Ninfa, and A. M. Stock.** 1989. Protein phosphorylation and regulation of adaptive responses in bacteria. *Microbiol. Rev.* **53:**450–490.

128. **Travis, J.** 1994. Reviving the antibiotic miracle. *Science* **264:**360–362.

129. **Valenta, C., A. Bernkop-Schnurch, and H. P. Rigler.** 1996. The antistaphylococcal effect of nisin in a suitable vehicle: a potential therapy for atopic dermatitis in man. *J. Pharm. Pharmacol.* **48:**988–991.

130. **van Belkum, M. J., B. J. Hayema, R. E. Jeeninga, J. Kok, and G. Venema.** 1991. Organization and nucleotide sequences of two lactococcal bacteriocin operons. *Appl. Environ. Microbiol.* **57:**492–498.

131. **van Belkum, M. J., J. Kok, and G. Venema.** 1992. Cloning, sequencing, and expression in *Escherichia coli* of *lcnB*, a third bacteriocin determinant from the lactococcal bacteriocin plasmid p9B4-6. *Appl. Environ. Microbiol.* **58:**572–577.

132. **van Belkum, M. J., J. Kok, G. Venema, H. Holo, I. F. Nes, W. N. Konings, and T. Abee.** 1991. The bacteriocin lactococcin A specifically increases permeability of lactococcal cytoplasmic membranes in a voltage-independent, protein-mediated manner. *J. Bacteriol.* **173:**7934–7941.

133. **van Belkum, M., and M. E. Stiles.** 1995. Molecular characterization of genes involved in the production of the bacteriocin leucocin A from *Leuconostoc gelidum*. *Appl. Environ. Microbiol.* **61:**3573–3579.

134. **van Belkum, M. J., R. W. Worobo, and M. E. Stiles.** 1996. The double-glycine-type leader peptides direct the secretion of peptides by bacteriocin ABC transporters, abstr. C26. Presented at the Fifth Symposium on Lactic Acid Bacteria: Genetics, Metabolism and Applications. Veldhoven, The Netherlands.

135. **van der Meer, J., J. Polman M. M. Beerthuyzen, R. J. Siezen, O. P. Kuipers, and W. M. de Vos.** 1993. Characterization of the *Lactococcus lactis* nisin A operon genes *nisP*, encoding a subtilisin-like serine protease involved in precursor processing, and *nisR*, encoding a regulatory protein involved in nisin biosynthesis. *J. Bacteriol.* **175:**2578–2588.

136. **van der Meer, J., H. S. Rollema, R. J. Siezen, M. M. Beerthuyzen, O. P. Kuipers, and W. M. de Vos.** 1994. Influence of amino acid substitutions in the nisin leader peptide on biosynthesis and secretion of nisin by *Lactococcus lactis*. *J. Biol. Chem.* **269:**3555–3562.

137. **van der Vossen, J. M. B. M., M. H. M. van Herwijnen, R. J. Leer, B. Ten Brink, P. H. Pouwels, and J. H. J. Huis in't Veld.** 1994. Production of acidocin B, a bacteriocin of *Lactobacillus acidophilus* M46 is a plasmid-encoded trait—plasmid curing, genetic marking by *in vivo* plasmid integration, and gene transfer. *FEMS Microbiol. Lett.* **116:**333–340.

138. **Venema, K., T. Abee, A. J. Haandrikman, K. J. Leenhouts, J. Kok, and G. Venema.** 1993. Mode of action of lactococcin B, a thiol-activated bacteriocin from *Lactococcus lactis*. *Appl. Environ. Microbiol.* **59:**1041–1048.

139. **Venema, K., M. H. Dost, P. A. Beun, A. J. Haandrikman, G. Venema, and J. Kok.** 1996. The genes for secretion and maturation of lactococcins are located on the chromosome of *Lactococcus lactis* IL1403. *Appl. Environ. Microbiol.* **62:**1689–1692.

140. **Venema, K., M. H. R. Dost, G. Venema, and J. Kok.** 1996. Mutational analysis and chemical modification of Cys24 of lactococcin B, a bacteriocin produced by *Lactococcus lactis*. *Microbiology* **142:**2825–2830.

141. **Venema, K., R. E. Haverkort, T. Abee, A. J. Haandrikman, K. J. Leenhouts, L. de Leij, G. Venema, and J. Kok.** 1994. Mode of action of LciA, the lactococcin A immunity protein. *Mol. Microbiol.* **14:**521–532.

142. **Venema, K., J. Kok, J. D. Marugg, M. Y. Toonen, A. M. Ledeboer, G. Venema, and M. L. Chikindas.** 1995. Functional analysis of the pediocin operon of *Pediococcus acidilactici* PAC1.0: PedB is the immunity protein and PedD is the precursor processing enzyme. *Mol. Microbiol.* **17:**515–522.

143. **Venema, K., G. Venema, and J. Kok.** 1995. Lactococcal bacteriocins: mode of action and immunity. *Trends Microbiol.* **3:**299–304.

144. **von Wright, A., S. Wessels, S. Tynkkynen, and M. Saarela.** 1990. Isolation of a replication region of a large lactococcal plasmid and use in cloning of a nisin resistance determinant. *Appl. Environ. Microbiol.* **56:**2029–2035.

145. **Wells, J. M., K. Robinson, K. M. Chamberlain, K. M. Schofield, and R. W. F. LePage.** 1996. Lactic acid bacteria as vaccine delivery vehicles. *Antonie Leeuwenhoek* **70:**317–330.

146. **Wells, J. M., P. W. Wilson, P. M. Nortonk, M. J. Gasson, and R. W. Le Page.** 1993. *Lactococcus lactis*: high-level expression of tetanus toxin fragment C and protection against lethal challenge. *Mol. Microbiol.* **8:**1155–1162.

147. **Worobo, R. W., M. J. Van Belkum, M. Sailer, K. L. Roy, J. C. Vederas, and M. E. Stiles.** 1995. A signal peptide secretion-dependent bacteriocin from *Carnobacterium divergens*. *J. Bacteriol.* **177:**3143–3149.

148. **Zhong, X., R. Kolter, and P. C. Tai.** 1996. Processing of colicin V-1, a secretable marker protein of a bacterial ATP binding cassette export system, requires membrane integrity, energy, and cytosolic factors. *J. Biol. Chem.* **271:**28057–28063.

Author Index

Subject Index